Proceedings of the *Exxon Valdez* Oil Spill Symposium

Funding support for the publication of this proceedings was provided by

***Exxon Valdez* Oil Spill Trustee Council**

PROCEEDINGS OF THE *EXXON VALDEZ* OIL SPILL SYMPOSIUM

EDITED BY

STANLEY D. RICE
ROBERT B. SPIES
DOUGLAS A. WOLFE
BRUCE A. WRIGHT

American Fisheries Society Symposium 18

Proceedings of the Exxon Valdez Oil Spill Symposium

Held at Anchorage, Alaska, USA
2–5 February 1993

American Fisheries Society
Bethesda, Maryland
1996

The American Fisheries Society Symposium series is a registered serial. Suggested citation formats follow.

Entire book

Rice, S. D., R. B. Spies, D. A. Wolfe, and B. A. Wright, editors. 1996. Proceedings of the *Exxon Valdez* oil spill symposium. American Fisheries Society Symposium 18, Bethesda, Maryland.

Article within the book

Carlson, P. R., and K. A. Kvenvolden. 1996. Tracking *Exxon Valdez* oil from beach to deepwater sediments of Prince William Sound, Alaska. Pages 109–120 *in* S. D. Rice, R. B. Spies, D. A. Wolfe, and B. A. Wright, editors. Proceedings of the *Exxon Valdez* oil spill symposium. American Fisheries Society Symposium 18.

Cover photograph: Prince William Sound, April 1989
by John Hyde, Alaska Department of Fish and Game

Library of Congress Catalog Card Number: 96-85358

ISBN 0-913235-95-4

ISSN 0892-2284

Printed in the United States of America on recycled, acid-free paper.

American Fisheries Society
5410 Grosvenor Lane, Suite 110
Bethesda, Maryland 20814-2199, USA

CONTENTS

Symbols and Abbreviations

The following symbols and abbreviations may be found in this book without definition. Also undefined are standard mathematical, statistical, and chemical symbols given in most dictionaries.

α	probability of type I error (false rejection of null hypothesis)
β	probability of type II error (false acceptance of null hypothesis)
°C	degrees Celsius
cm	centimeter
Co.	Company
Corp.	Corporation
cov	covariance
d	day
dL	deciliter
DC	District of Columbia
df	degrees of freedom
E	east
E	expected value
e	base of natural logarithm
e.g.	(exempli gratia) for example
et al.	(et alii) and others
etc.	et cetera
°F	degrees Fahrenheit
F	filial generation
g	gram
h	hour
ha	hectare
Hz	hertz
i.e.	(id est) that is
Inc.	Incorporated
k	kilo (10^3, as a prefix)
L	liter
log	logarithm
Ltd.	Limited
m	meter (as a suffix or by itself); milli (10^{-3}, as a prefix)
min	minute
mol	mole
n	nano (10^{-9})
N	north
N	sample size
NS	not significant
P	probability
p	para (as a chemical prefix)
Pa	pascal
pH	negative log of hydrogen ion activity
ppb	parts per billion
ppm	parts per million (per 10^6) milligrams per liter in the metric system
R	correlation or regression coefficient (multiple)
r	correlation or regression coefficient (simple)
S	south
s	second
SD	standard deviation
SE	standard error
sp	species (singular)
spp	species (plural)
U.S.	United States (as adjective)
V, var	variance
W	west
μ	micro (10^{-6}, as a prefix)
′	minute (angular)
″	second (angular)
%	percent (per hundred)
‰	parts per mille (per thousand, 10^3)

PREFACE

The *Exxon Valdez* oil spill on 24 March 1989 was one of those singular events in history that changes peoples lives. Like other rare events of such momentous nature, almost all of those people who were affected by this tragedy—the largest tanker oil spill in United States history—can tell you where they were and what they were doing when the news of the oil spill in Prince William Sound, Alaska, hit them. For many of these people, the changes wreaked by the oil spill still affect their lives and livelihoods over 6 years later.

The authors of this volume of papers are among those whose lives were jolted and changed by the oil spill. Largely government workers for state or federal natural resource agencies, or university scientists, they put their regular jobs and preoccupations in abeyance as the enormous effort was mounted—on hours or days notice—to determine, document, and attempt to understand the impacts of this oil spill on the abundant and valuable natural resources of Prince William Sound and the Gulf of Alaska. The wide variety of information the authors report in this volume was obtained through long and arduous efforts that began urgently but were painstakingly extended, in most cases, for several years.

When the federal and Alaska governments began their natural resources damage assessment of the *Exxon Valdez* oil spill, information was gathered primarily to support litigation to recover monetary compensation (damages) for the injuries caused to natural resources. From 1989 through 1991, the results of the government's damage assessment research were thought to be litigation sensitive and were not disclosed to the public. As the imposed silence lengthened, frustration rose, not only among an increasingly impatient and vocal public but also among the researchers themselves, who were learning much of interest and value.

In October 1991, the state and federal governments reached an out-of-court settlement with Exxon Corp. for over US$1 billion in damages. With this settlement achieved, the cloak of secrecy was finally lifted on the government's research results. A major symposium—The *Exxon Valdez* Oil Spill Symposium—was organized and held in Anchorage, Alaska, in February 1993. For 4 days, and with several concurrent sessions, researchers presented their results to large audiences of other scientists, resource managers, lawyers, oil company representatives, and many members of the public ranging from commercial fishers to Alaska Natives to "the person on the street." Almost everyone who participated considered this symposium a major watershed event of lasting importance.

This proceedings contains a majority of the papers presented at the *Exxon Valdez* Oil Spill Symposium. Because of the rush to get verbal and summary information to the public at the symposium, most of the formal writing occurred after the conference. Since the symposium in early 1993, much of the information contained in the proceedings has been refined or extended, and in some cases it is new. All of these papers received thorough peer review before they were accepted for publication; more than 100 reviewers contributed their time to this effort.

The overriding purpose of the proceedings is to provide a lasting record of the unprecedented effort on behalf of the governments to determine the extent and nature of injuries caused by the *Exxon Valdez* oil spill. No other oil spill has been as intensively investigated.

The key participants in the *Exxon Valdez* oil spill damage assessment were members of three federal and three Alaska resource management agencies. Representatives of these agencies made up the *Exxon Valdez* Oil Spill Trustee Council, which wholeheartedly supported the symposium and its proceedings both conceptually and financially. The members of the Trustee Council were: Charles Cole, Attorney General for the State of Alaska; Steve Pennoyer, Director, Alaska Region of the National Marine Fisheries Service; John Sandor, Commissioner, Alaska Department of Environmental Conservation; Mike Barton, Regional Forester, U.S. Forest Service; Carl Rosier, Commissioner, Alaska Department of Fish and Game; and Curt McVee, Special Assistant to the Secretary, U.S. Department of the Interior.

In addition to the Trustee Council, the Alaska Chapter of the American Fisheries Society and the Alaska Sea Grant Program cosponsored the symposium and proceedings. The Alaska Sea Grant Program largely organized the symposium, and the American Fisheries Society has undertaken the publication and distribution of the proceedings. Numerous others, in addition to the editors and authors of the proceedings, contributed valuable time and services to the success of this effort. Too numerous to be named here, they know who they are, and they should be justifiably proud of their participation.

As of this writing, tremendous efforts are being made to restore the natural resources injured by the spill, these efforts having been made possible by the oil spill settlement monies. Research into the nature of some lasting injuries continues, and new injuries are being documented that were not apparent earlier. The restoration program is now entirely in the public eye, as it has been since the settlement and the symposium. Public interest has not faltered with the years, and no program to my knowledge has devoted any greater effort to public involvement than the restoration program of the Trustee Council.

For many, putting the oil spill behind is not yet possible, but sharing the information and knowledge obtained from study of this oil spill is. Not all the answers are known; some will never be. Yet, many questions have been answered by the authors in this volume. This knowledge is one of the valuable legacies that the oil spill has left us.

Byron Morris
Chief, Office of Oil Spill Damage
Assessment and Restoration
National Marine Fisheries Service
Juneau, Alaska
6 October 1995

American Fisheries Society Symposium 18:1–16, 1996

The Effects of the *Exxon Valdez* Oil Spill on the Alaskan Coastal Environment

R. B. Spies

Applied Marine Sciences, 2155 Las Positas Court, Suite S, Livermore, California 94550, USA

S. D. Rice

National Marine Fisheries Service, Auke Bay Laboratory, 11305 Glacier Highway Juneau, Alaska 99801, USA

D. A. Wolfe

National Oceanic and Atmospheric Administration, Office of Ocean Resources Conservation and Assessment, 6001 Executive Boulevard, Room 323, Rockville, Maryland 20877, USA

B. A. Wright

National Marine Fisheries Service, Office of Oil Spill Damage Assessment and Restoration Post Office Box 210029, Auke Bay, Alaska 99821, USA

Abstract.—The 1989 *Exxon Valdez* oil spill had an immediate acute effect on seabirds, bald eagles *Haliaeetus leucocephalus*, marine mammals, and intertidal communities in Prince William Sound and portions of the northern Gulf of Alaska. Longer-term impacts, some lasting several years or more, were found on Pacific herring *Clupea pallasi*, pink salmon *Oncorhynchus gorbuscha*, Dolly Varden *Salvelinus malma*, cutthroat trout *O. clarki*, and the intertidal and subtidal environments in Prince William Sound. The spill and the subsequent cleanup activity also had impacts on important coastal archaeological sites, the use of marine resources by subsistence communities, and the social structure of spill-area communities. The oil was distributed over hundreds of kilometers from Bligh Reef in Prince William Sound well into the Gulf of Alaska. Beached oil penetrated deeply into cobbled beaches and still persists in some areas beneath the surface layer of rocks and under mussel beds 5 years after the surface cleanup. Dispersion, recovery, transformation, evaporation, photodegradation, and especially bacterial degradation have been the main processes removing the oil from the ecosystem. The determination of injury and recovery and the restoration program were undertaken by the United States and Alaska governments through the *Exxon Valdez* Oil Spill Trustee Council. Several sources of uncertainty, particularly the lack of prespill data on marine resources, have hampered a full determination of the spill's effects, leading to a variety of interpretations. However, numerous species sustained significant damage, especially sea surface dwellers such as seabirds and marine mammals. The intertidal communities were severely impacted, and early life history stages of intertidally spawning fish were affected. Recovery is occurring at different rates for different species. Constraints on recovery and the underlying causes of declines apparent in some species before the spill are the subject of several ecosystem-based studies. The practical legacy from the scientific studies of resource damage assessment and restoration are a greater understanding of how a rich subarctic marine ecosystem can be affected by a large petroleum spill, provision of new and improved tools for management of natural resources, and a better understanding of the marine ecosystem of Prince William Sound and the northern Gulf of Alaska.

The supertanker *Exxon Valdez*, carrying a full cargo of more than 75 million liters of North Slope crude oil, went off course before midnight 24 March 1989 after leaving the southern terminus of the trans-Alaska pipeline at Valdez, Alaska. Shortly thereafter the ship ran aground on Bligh Reef in northeastern Prince William Sound. As nearly 42 million liters of crude oil poured from the hull and meandered southwesterly under calm conditions, many marine scientists knew that the big spill some had long feared had started. Within 3 d a northerly gale descended on Prince William Sound to blow the oil slick beyond any hope of containment. As the oil was being driven onto hundreds of kilometers of Prince William Sound beaches and then through the southwest passages of the sound towards the productive seabird colonies of the Gulf of Alaska and the distant beaches of Shelikoff Strait, it was apparent that taking account of this spill along the remote and jagged coastline of the northern Gulf of Alaska would be a Herculean effort. Nonetheless, the resource agencies had already started a comprehensive assessment of the spill's effects.

This was the largest oil spill ever to have occurred in U.S. waters and the largest anywhere this far north. The northern Gulf of Alaska is known for its rich marine life: millions of seabirds; abundant marine mammals, including sea otters *Enhydra lutris*, killer whales *Orcinus orca*, humpback whales *Megaptera novaeangliae*, northern sea lions *Eumetopias jubatus*, harbor seals *Phoca vitulina*, and porpoises; five species of Pacific salmon *Oncorhynchus* spp., Pacific herring *Clupea pallasi*, and groundfish that support multimillion-dollar fisheries; and the intertidal and subtidal communities along its coasts that provide subsistence for coastal villages of Alaskan natives. The nearly pristine conditions and the abundant wildlife in the Gulf of Alaska are a magnet for tourism. Tourism and fisheries are key components of the Alaskan cash economy, especially in coastal areas.

The spill occurred in early spring just before the annual spawning by Pacific herring in Prince William Sound. Millions of salmon fry were also about to emerge from their gravel spawning beds to be washed into the nearshore waters and sustained by the spring plankton bloom. Young seals and sea otter pups were still experiencing an unusually cold winter; sea ice had not melted in the backs of many bays. Thousands of wintering sea ducks remained along the crenulated coasts of Prince William Sound and the outer Kenai Peninsula. Other seabirds were converging on their breeding colonies in the Gulf of Alaska.

Marine scientists were suddenly thrust into a marathon of planning meetings and logistical arrangements, followed by long days and weeks of field work under very difficult and sometimes dangerous conditions. The pace continued for many months, and for some it slowed little for several years. Vacations and leave were canceled; many people pushed themselves up to and beyond the point of exhaustion.

Among the difficulties that scientists faced repeatedly was the lack of up-to-date data on the abundance of organisms and data on the effects of oil on subarctic wildlife. There were no answers to such questions as "How much oil on the plumage of a murre produces hypothermia?" and "What concentration of petroleum hydrocarbons in the bile of a harbor seal is harmful?" They were fortunate, however, to have a body of data gathered in the late 1970s and early 1980s under an interagency agreement between the Department of Commerce (National Oceanic and Atmospheric Administration, NOAA) and the Department of the Interior as part of the Outer Continental Shelf Environmental Assessment Program (OCSEAP). This database contained inventories of many resources, including counts of sea otters and seabirds, that were valuable in assessing prespill conditions. Even these prespill data, however, were several years old. For example, the last quantitative survey of sea otters in Prince William Sound had been in 1984–1985. Also, the National Marine Fisheries Service's (NMFS) Auke Bay Laboratory had been actively investigating the effects of oil on commercially important species of marine fishes and invertebrates, mainly under OCSEAP (Rice et al. 1984). This work provided a seminal body of oil toxicology literature on some commercially important Alaskan fishes and invertebrates.

The government agencies participating in the oil spill assessment were organized into a Trustee Council, following the general form of Natural Resource Damage Assessments under the federal Comprehensive Environmental Resources Conservation and Liability Act (CERCLA). The three state trustees were the Commissioner of the Alaska Department of Fish and Game, the Commissioner of the Alaska Department of Environmental Conservation, and the Attorney General. The three federal trustees were the Administrator of NOAA, the Secretary of the Interior, and the Chief of the U.S. Forest Service on behalf of the Secretary of Agriculture. The trustees or their designates met frequently to guide the scientific studies. The Trustee Council process incorporated scientific peer review from a group of independent and mainly academic scientists from the United States and Canada, and a Chief Scientist position was created in 1990 to further guide the overall scientific program.

In October 1991 a settlement of the civil and criminal claims of the government against the Exxon Corp. was reached. The civil settlement established a trust, to be overseen by the Trustee Council, that included total payments of US$900 million through the year 2001. A memorandum of agreement established the trust "for the purposes of restoring, replacing, enhancing, or acquiring the equivalent of natural resources injured as a result of the oil spill and the reduced and lost services provided by such resources." Thus the restoration phase of the activities began, though damage continued to be assessed through field studies and data analysis.

In February 1993 the Trustee Council sponsored a 3-d symposium in Anchorage, Alaska, to report the initial results of oil spill studies. One hundred eighteen papers and posters were presented, but although the symposium was open to all, Exxon Corp. contractors did not offer papers. This proceedings volume contains 61 papers, representing the major studies presented at the Anchorage symposium, which have been peer reviewed and revised since the conference.

Two other collections of scientific papers related to the oil spill have been published: *Marine Mammals and the Exxon Valdez* (Loughlin 1994) and *Exxon Valdez Oil Spill: Fate and Effects in Alaskan Waters* (Wells et al. 1995). The second book contains mainly studies supported by Exxon Corp.; it is the proceedings of a 1993 symposium in Atlanta, Georgia, sponsored by the American Society for Testing and Materials.

Fate of the Spilled Oil

Like all crude petroleum, the Prudhoe Bay crude oil that spilled from the *Exxon Valdez* was an extremely complex mixture of organic compounds formed over geological time from ancient animal and plant materials. Crude oils from different geological formations vary considerably in chemical composition, and thus also in their ultimate fates when introduced in the environment. As soon as the *Exxon Valdez* oil spilled into the sea, many physical, chemical, and biological processes began to fractionate the oil into various physicochemical components and to transport and transform those components at rates dependent upon both their chemistry and the prevailing environmental conditions.

During its initial spreading on the water surface, the oil slick was moved by water currents and wind, and the more volatile constituents began to evaporate. After the wind speed rose dramatically on the afternoon of 26 March 1989, the slick moved rapidly toward the southwest, while the oil was further fractionated and transformed by processes of evaporation, dissolution, dispersion, emulsification, adsorption onto suspended particulate materials, and grounding on shorelines.

Analyses of petroleum aromatic hydrocarbons and total petroleum hydrocarbons in seawater over extensive areas indicated that dispersion of colloidal oil was a dominant mechanism of transport in the water column during the early periods after the spill (Neff et al. 1990; Neff 1991; Short and Harris 1996a, this volume). Analyses of petroleum hydrocarbons in caged mussels indicated that dispersed oil in the water column remained biologically available in some areas through the summer of 1990 (Short and Harris 1996b, this volume), and biologically available hydrocarbons had greatly increased over prespill conditions (Short and Babcock 1996, this volume).

During transport and fractionation, some oil constituents were oxidized by photochemical catalysis and bacterial metabolism (Braddock et al. 1995 and 1996, this volume). Augmented by shoreline cleanup activities and natural wave action, oil-contaminated sediments eroded from the more exposed intertidal zones into nearshore subtidal areas (Carlson and Kvenvolden 1996; O'Clair et al. 1996; both this volume), where they were resuspended by waves (Short et al. 1996, this volume) and relocated by coastal circulation. *Exxon Valdez* oil was generally not detected in deeper (>20 m) offshore sediments except in a few places close to heavily oiled shorelines (Carlson and Kvenvolden 1996; O'Clair et al. 1996).

Collectively, these processes accounted for the observed decomposition, or weathering, of the *Exxon Valdez* oil in the environment and its widespread dispersion and dilution in Prince William Sound and the nearby Gulf of Alaska. Oil persisted longer along more protected shoreline, such as the deeply buried (70–100 cm) sediments in the central platforms of well-armored cobble–boulder beaches (Michel and Hayes 1991, 1992) and the sediments within byssal mats underlying some densely populated mussel beds (Babcock et al. 1996; Harris et al. 1996; both this volume). Based on best estimates for the various processes (including recovery by cleanup activities) that affected oil distribution and transformation in the environment, a mass balance has been constructed for the overall fate of the *Exxon Valdez* oil (Wolfe et al. 1994; Wolfe 1995). Figure 1 summarizes the distribution of oil in major compartments from the time of the spill through September 1992.

About 20% of the spilled oil mass was estimated to have evaporated. About 40% beached within Prince William Sound, and an additional 7–11% probably beached along the shorelines of the Kenai and Alaska peninsulas and Kodiak Island. Cleanup operations recovered approximately 14% of the spilled oil by skimming (8–9%) and solid waste recovery (5–6%). By fall 1992 only an estimated 2% of the original oil mass remained on the beaches (Table 1). Except in certain mussel beds where relatively fresh oil remained (Babcock et al. 1996; Harris et al. 1996), this residual beached oil was highly weathered. Winter storms and biodegradation contributed substantially to the removal of oil stranded on the shorelines, but shoreline cleanup activities also promoted dispersion of oil from the beaches into the adjacent waters. The contributions of these various dispersive and weathering processes were discussed in much greater detail elsewhere (Wolfe et al. 1994; Wolfe 1995).

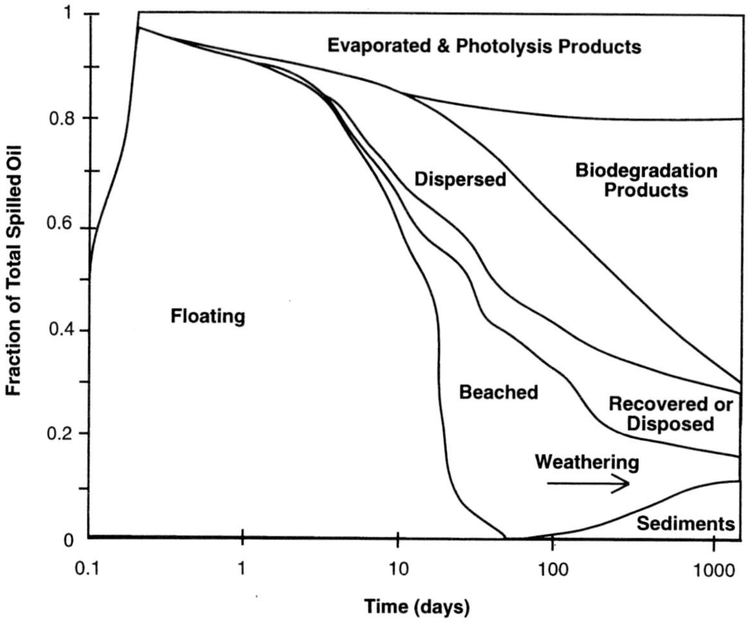

FIGURE 1.—Fate of *Exxon Valdez* oil, from the time of the spill to 1 October 1992 (Wolfe et al. 1994; Wolfe 1995). The uncertainties associated with the final estimates are shown in Table 1.

Effects on Mammals and Birds

Because they rely on the insulating properties of their pelage or plumage, some mammals and nearly all seabirds are vulnerable to the direct effects of crude oil. When fur or feathers are matted by oil, heat loss to the cold water quickly results in hypothermia. The initial effects of *Exxon Valdez* oil on mammals and birds were documented by the thousands of carcasses collected in the first few months after the spill. The long-term implications for mammal and bird populations are not so clear, however. Most birds and mammals have limited or indirect economic importance and are not subjected to the same level of monitoring as are commercially important species of fish. Data on prespill populations were relatively scarce, both spatially and temporally, especially for birds.

Sea otters, which maintain their body temperature with an enormous intake of invertebrate prey, rely almost solely on their extremely dense fur for protection from the cold. In contrast, pinnipeds and cetaceans have layers of subdermal blubber to provide insulation and rely less on their pelage. It is not surprising, then, that the proportional losses of sea otters seemed to be greater than those of harbor seals. Of perhaps 2,200 harbor seals in Prince Wil-

TABLE 1.—Summary mass balance estimates for various compartments of *Exxon Valdez* oil as of 1 October 1992, as fractional masses of the total oil originally spilled.

Compartment	Estimate	Minimum	Maximum
Subtidal sediments[a]	0.13	0.11	0.16
Floating	0.00	0.00	0.00
Beached[a]	0.02	0.001	0.04
Recovered and disposed	0.14	0.11	0.18
Dispersed in water column[a]	0.01	0.00	0.03
Aqueous biodegradation and photolysis products	0.50	0.40	0.60
Atmospheric photolysis products	0.20	0.18	0.25
Total	1.00	0.80	1.26

[a]Primarily as highly weathered refractory residuals (>90% in offshore sediments and dispersed fractions; >60% in nearshore subtidal and beached fractions).

liam Sound in 1989, estimated losses were about 300 (13%) (Frost et al. 1994b); of 10,000 sea otters in the sound, about 2,800 were lost in the immediate aftermath of the spill (28%) (Garrott et al. 1993). Most of these acute losses probably resulted from hypothermia, but inhalation of fumes, grooming of oiled pelts, and absorption of oil through the skin may have contributed to some deaths. Some studies examined possible chronic injuries to sea otters as a result of the spill (Loughlin et al. 1996, this volume). The poor survival of rehabilitated sea otters may indicate that the effects of capture and toxic stress interacted to suppress immunity and repair processes, resulting in permanent debilitation.

The persistence of detectable hydrocarbons in the bile of harbor seals 1 year after the spill attests to a surprising, lasting exposure of this species to petroleum, whether it resulted from dietary sources or from remobilization of fat-stored hydrocarbons from the initial exposure (Frost et al. 1994a). Whether this longer-term exposure had any effect beyond the initial mortalities within an already declining population is not known.

Harbor seals and killer whales were two of the few species for which up-to-date population data were available. Molting surveys of harbor seals at index colonies within Prince William Sound had been done in 1983, 1984, and 1988 (Frost et al. 1994b). Survey work had also been carried out for killer whales in Prince William Sound every year from 1984 through 1988 (Matkin et al. 1994). Was the disappearance of the 13 killer whales from AB pod between autumn 1988 and summer 1990 caused by the spill? The coincidence with the spill strongly suggests an effect, yet a credible mechanism of exposure and acute toxicity of oil has not been identified for this species (Dahlheim and Matkin 1994). The picture is further confounded because the AB pod had been taking fish from longlines and photographs gave evidence of bullet holes in some individuals. In 1994, however, researchers discovered that five more members were missing from the AB pod. Further, pod integrity had deteriorated; one AB subgroup was swimming with another pod, an unprecedented switch for resident killer whales (C. O. Matkin, North Gulf Oceanic Society, personal communication).

The *Exxon Valdez* oil spill killed probably between 100,000 and 645,000 seabirds, based on carcass recovery and modeling of mortality patterns (Ford et al. 1996; Piatt and Ford 1996; both this volume). Major uncertainties are associated with any estimate, so the potential range of mortalities is quite large and, of course, controversial. For example, M. Fry (University of California-Davis, unpublished data) preferred an estimate of about 500,000; Weins (1995) dismissed any estimate much greater than the 36,000 carcasses actually recovered. The best estimate may be about 250,000 birds killed by the spill (Piatt and Ford 1996). Despite the uncertainties associated with the estimates, this was clearly the greatest demonstrated mortality of birds from any oil spill (Piatt and Lensik 1989).

Recent seabird census data were sparse in 1989. Prespill declines in several seabird populations had been indicated by surveys in the early 1970s and middle 1980s, so long-term changes in populations of the northern Gulf of Alaska confound the interpretation of injury from the spill (Piatt and Anderson 1996, this volume). The effects of the *Exxon Valdez* oil spill may have added cumulatively to these declines (see the experience with common murres *Uria aalge* in California: Ainley and Lewis 1974; Ainley and Boekelheide 1990).

In the case of murres (*Uria* spp.), the direct spill-related mortalities may have led to population effects at some breeding locations. Although prespill data are poor, fewer breeders appeared to be present at oiled colonies. The effects on productivity during 1989–1991 were uncertain, but by 1993 reproductive success and timing were within normal bounds at the Barren Island colonies. Murres do not begin to breed until they are 5–7 years old (Harris et al. 1994). To the extent that reproductive success was reduced after the spill, the effects of that low productivity might not be realized until 1994 or later.

Marbled murrelets *Brachyramphus marmoratus*, small alcids that forage in open water, were greatly affected by the spill (Kuletz 1996, this volume). Based on recovered carcasses and estimates of total mortality, more than 8,000 individuals were killed in Prince William Sound, perhaps 5–10% of the species' population. The stronghold of this species is Prince William Sound, the Kodiak Archipelago, and southeast Alaska. At the southern end of its range, in Washington, Oregon, and California, its population is perilously low. This species had declined in Prince William Sound prior to the spill for unknown reasons.

Black-legged kittiwakes *Rissa tridactyla* are surface-foraging seabirds common in the spill area and also vulnerable to spilled oil. Two hundred forty-one carcasses of this species were recovered from 26 March to 1 August 1989, representing perhaps a total mortality of 1,200–2,400 of an estimated spill-wide area population of 268,000 nesting individuals. Kittiwake chick productivity was lower in oiled than in non-oiled areas (Irons 1996, this volume). Compared with 1984–1988 productivity (chicks/nest), both oiled and non-oiled areas had lower productivity after the spill (1989–1994). Chick production in 1989 was marginally less than expected (P = 0.06) based on prespill productivity. In postspill years (1990–1994) all of the colonies have had relatively poor production. The reason for this change is unclear, but may relate to poor food availability.

Pigeon guillemots *Cepphus columba* are another common alcid in Prince William Sound. Six hundred fourteen pigeon guillemot carcasses were recovered from beaches in the oil-spill area during summer 1989, which may have represented about 10% of the guillemot spill mortality. This species had also apparently declined during the 1970s and 1980s. After the spill, however, there were greater declines of pigeon guillemots on oiled shorelines of the Naked Island group in Prince William Sound than on non-oiled shorelines in the same area, suggesting a measurable effect of the spill on the population (Oakley and Kuletz 1996, this volume). Nesting chronology, clutch size, and the mean number of chicks hatched per nest did not change after the spill (1989–1991) relative to 1978–1980. Fledging success, however, was lower in oiled areas, which may be attributable to greater postspill rates of predation. There were also indications of an effect of oil on reproductive success of black oystercatchers *Haematopus bachmani*; the comparison was based on one oiled and one non-oiled area (Sharp et al. 1996, this volume).

About 150 dead bald eagles *Haliaeetus leucocephalus* were recovered, and it is not certain how many total deaths they represent; 300 may be a conservative estimate. Surveys throughout the spill area in 1989 revealed a decline of nesting success in oiled western Prince William Sound relative to the eastern sound, but this difference did not occur in 1990 (Bernatowicz et al. 1996, this volume). The Prince William Sound population probably recovered by 1992 (Bowman et al. 1995).

Effects on Fish and Fisheries

The impacts of spilled oil on fish were of the utmost concern in 1989, because fisheries are critical to the economy and subsistence of residents of the spill area. The impacts were immediate; fisheries for black cod (sablefish *Anoplopoma fimbria*) and shrimp were closed within the first few days of the spill, mainly because retrieving longlines and pots through slicks of oil would have contaminated the surfaces of fish and shrimp. Salmon and herring fishery closures continued through the rest of 1989 because shoreline cleanup activities resuspended oil in the nearshore areas where these fisheries are conducted, and fouling of both gear and products was likely.

For the sensitive stages of fish (eggs, larvae, and juveniles) the March spill could not have occurred at a worse time. In April Pacific herring spawn inshore and fry of pink salmon *Oncorhynchus gorbuscha* emerge from the gravel spawning redds and enter the nearshore marine environment. Larvae from other species are also distributed in the upper water column and, therefore, were exposed to the effects of the surface slicks (Norcross and Frandsen 1996, this volume). The very upper surface of the sea—the surface microlayer, which concentrates hydrophobic contaminants many times over their water column concentrations—is often rich in marine life and may contain large numbers of contaminant-sensitive fish eggs (Hardy et al. 1987). Slicks from the oil spill, which occurred regularly in many nearshore areas for months, probably had a major effect in this poorly studied environment, but there are no data to confirm this.

Previous spills have caused fishery closures because of the possibility of contamination during harvest, but damage to fish themselves—especially at the population level—has rarely been documented. The *Exxon Valdez* spill was different, however. It occurred in a relatively pristine environment where the regional economy depends greatly on fisheries, and two management agencies, the Alaska Department of Fish and Game and the National Marine Fisheries Service, had programs that could readily adapt to measuring the effects of the spill on fishes and fisheries. Both agencies mobilized considerable internal resources to protect sensitive areas and assess damage in the first few days after the spill. The best information on the biology and population dynamics of a fishery resource was for pink salmon and sockeye salmon *O. nerka*.

Juvenile salmon were directly affected by the spill in 1989 and their eggs may have been affected through 1993. Several lines of evidence indicated immediate hydrocarbon uptake and exposure and subsequent effects of oil on early life history stages of pink salmon. Exposure to oil was documented by visible oil in the stomachs of salmon fry (Sturdevant et al. 1996, this volume), by measurements of polynuclear aromatic hydrocarbons (PAHs) in salmon fry (Carls et al. 1996b, this volume), by increases of the enzyme cytochrome P450 in salmon fry tissues (Carls et al. 1996b; Willette 1996, this volume) and in alevins from contaminated intertidal streams (Wiedmer 1996, this volume), and by increases in P450 and bile hydrocarbon metabolites in Dolly Varden *Salvelinus malma* and several benthic species (Collier et al. 1996, this volume). Impacts on growth were demonstrated for tagged pink salmon juveniles from hatcheries (Willette 1996), for wild juvenile pink salmon (Wertheimer and Celewycz 1996, this volume) and for tagged Dolly Varden and cutthroat trout *O. clarki* (Hepler et al. 1996, this volume). The impacts on growth were all in 1989. No changes in food availability were detected (Celewycz and Wertheimer 1996, this volume; Wertheimer and Celewycz 1996), so these decreases in growth were probably attributable to oil exposure and the physiological costs of detoxification. The ability of oil-contaminated food to elevate P450 activity and reduce growth was confirmed for pink salmon in laboratory exposures (Carls et al. 1996a, this volume). Collier et al. (1996) also measured elevated metabolites of aromatic hydrocarbons in several species of fishes in 1990 and 1991, including walleye pollock *Theragra chalcogramma* up to 640 km from the wreck site. There is concern about the effects long-term, low-level exposures might have (Collier et al. 1996).

The effects of oil exposure on juvenile pink salmon were estimated in the southwestern sound during spring and summer 1989. These fish were of the 1988 brood year that emerged as fry in spring 1989 and returned as adults in 1990. The total adult returns of both hatchery and wild pink salmon to Prince William Sound in 1990 were at record levels, but probably fewer than would have appeared had the spill not occurred. The return of adult pink salmon to the southwestern sound in 1990 was estimated to have been 1.9 million fish fewer than would have occurred without the effects of the spill on the growth of juvenile salmon in 1989 (Geiger et al. 1996; Templin et al. 1996; both this volume). Fry from other areas of the sound also migrated through the oiled area in 1989 and may have experienced slower growth and higher mortality.

The evidence for long-term damages grew from management studies by the Alaska Department of Fish and Game to better predict run strength for pink salmon. In autumn, agency personnel dig eggs from index streams draining to Prince William Sound to measure egg densities and survival. These studies were expanded after the oil spill to include about a dozen each of oiled and non-oiled streams. Egg mortalities were higher in oiled than in non-oiled streams from 1989 through 1993, but not after 1993 (Bue et al. 1996, this volume). The egg mortality study provides the most compelling evidence of long-term damage to fish from the oil spill, and it has stimulated a series of studies, still in progress, to examine the possible inheritance of damage by progeny from oil-impacted adults.

There was no evidence that sockeye salmon were directly affected by the spill, though smolts and adults returning to spawning areas had to swim through oil-tainted areas. However, the fishery in upper Cook Inlet was closed to forestall of fouling of gear and public perception of tainting. Without a commercial harvest, many more adults than desirable returned to some freshwater spawning and rearing lakes bordering the inlet, and subsequent production of fry and smolts was poor there. The ecological basis for these declines is under continuing study (Schmidt et al. 1996, this volume).

Pink salmon was not the only commercially important species directly affected by the spill; the early life history stages of Pacific herring were also affected. Contrasts of oiled and non-oiled areas indicated that oil caused increased larval abnormalities in Pacific herring (Brown et al. 1996, this volume).

Other important postspill events in the fisheries not dealt with at any length in this proceedings are the poor returns of hatchery and wild stock pink salmon in 1992 and 1993 and the crash of the Pacific herring population in 1993. These population reductions were limited to Prince William Sound and raised questions about the planktonic community during the juvenile rearing seasons in the early 1990s (for pink salmon) and the potential lingering effects of adult oil exposure in 1989–1990 (on herring). A Sound Ecosystem Assessment was launched in 1994 to investigate physical and biological factors that may control survival of juveniles of both species. Other researchers are investigating two pathogens isolated from herring at the time of the crash and the possibility that oil exposure suppressed herring immune systems. Results of these current studies will be reported elsewhere.

Effects on Intertidal and Subtidal Communities

Floating oil was deposited on intertidal areas under the influence of wind and water currents. Some of the stranded oil was lifted off the intertidal substrates on subsequent tides, and much of this oil was redeposited higher on the shore. The cumulative effect of this process was to establish a zone of greatest deposition high in the intertidal zone, a so-called bathtub ring. Oil not only coated the upper intertidal zone, it penetrated between cobbles and into soft sediments. Cleaning these substrates was difficult, requiring intensive effort in spring and summer 1989 and follow-up in 1990 and 1991. The cleanup involved thousands of workers and massive logistical support, and it ultimately cost Exxon Corp. about $2 billion. The cleaning removed large amounts of oil from intertidal areas, but it also damaged habitats (Mearns 1996, this volume). The hot-water washes were particularly destructive.

Several papers in this volume address the effects of shoreline treatment. Lees et al. (1996) outline the general effects of cleanup on the intertidal zone. Ebert and Lees (1996) report a field experiment indicating that snails *Nucella lamellosa* on oiled beaches and on oiled and cleaned beaches had poorer growth and survival in 1991 than snails on non-oiled beaches. Driskell et al. (1996) show that the lower intertidal fauna on mixed soft-substrate beaches cleaned with hot water recovered less by 1992 than the fauna on similar beaches that had been cleaned by other means or not at all. Houghton et al. (1996) confirm this outcome for sheltered rocky shores.

In addition to these area-specific studies of treatment effects, a regional survey of intertidal effects was carried out by personnel at the University of Alaska (Highsmith et al. 1996; Stekoll et al. 1996; Sundberg et al. 1996; van Tamelen and Stekoll 1996; all this volume). The study covered all types of intertidal environments (sheltered rocky, estuarine, open coast, etc.) and found fewer invertebrates, mussels, barnacles, limpets and brown algae *Fucus gardneri* in oiled than in non-oiled areas; differences were greatest in the middle and upper intertidal zones. Marine oligochaetes, normally not abundant in most rocky intertidal areas, were enhanced on oiled shorelines, as has been shown around natural oil seepages (Spies and Davis 1979). In addition to the immediate mortality caused by the spill, recolonization of affected shorelines by algae and invertebrates was inhibited by oil coatings on intertidal rocks (Duncan and Hooten 1996, this volume).

To clarify the role of various intertidal processes in manifesting damage and in recovery, several studies were undertaken in Herring Bay on Knight Island (Hooten and Highsmith 1996, this volume). In Herring Bay, as was true elsewhere in the spill area, the greatest and longest-lasting effects were in the upper intertidal zone. By 1991, the limpets *Tectura persona* and *Lottia sitkana* had not recovered in the upper intertidal zone of Herring Bay. *Fucus gardneri* had recovered in the middle but not in the upper intertidal zone by 1992 (van Tamelen and Stekoll 1996), apparently because of poor survival of spores after settlement, desiccation of sporlings, and poorer spore production by plants in the upper zone. Residual oil may also have inhibited settlement of sporlings.

In contrast, small invertebrates constituting the meiofauna seemed tolerant of oil. Experimentally oiled sediments placed near the low-tide line in Herring Bay were abundantly colonized by meiofauna within a few days (Fleeger et al. 1996, this volume). A year after the spill, epibenthic harpacticoid copepods, which are important food for juvenile salmon, were more abundant on heavily oiled than on lightly oiled beaches (Wertheimer et al. 1996, this volume).

A conscious decision was made not to clean mussel beds because of potential adverse effects on predators. Consequently, many mussel beds on soft sediments remained contaminated with relatively high oil concentrations years after the spill. Highly contaminated mussel beds within Prince William Sound are reported by Babcock et al. (1996). High concentrations of oil were found in more than 30 mussel beds more than 3 years after the spill. Sediment concentrations of oil were quite variable in underlying sediments, but ranged to 60,000 μg total petroleum hydrocarbons per gram of sediment. The mussels in the overlying beds had oil concentrations that averaged about 1% of the values in the sediments (Harris et al. 1996), but up to 8 μg PAH/g were reported within tissues. The highest concentrations of oil in sediments and tissues of mussels were found in a few mussel beds on soft sediments, and some beds were the target of a limited experimental clean-up effort in 1994.

The apparent impacts on the subtidal communities were not as striking as those in the intertidal zone. This is not surprising because neither visible oil nor high concentrations of oil were found in the subtidal sediments. The studies of subtidal algae and large epibenthic invertebrates (Dean et al. 1996a, 1996b; both this volume) did not detect differences in total density, biomass, or percentage cover of macroalgae due to oil. In contrast, the epibenthic studies, which included five species of sea stars and a crab, done in 1990 and then again in 1993, revealed that one sea star, *Dermasterias imbricata*, and the helmet crab *Telmessus cheiragonus* were fewer at the oiled sites than at the non-oiled sites. Recovery of these species was apparent in shallow water but not yet complete in the deeper portions within eelgrass beds in 1993. The spill did not have a discernable negative effect on the abundance of subtidal fishes (Laur and Haldorson 1996, this volume).

The subtidal communities of a fjordic embayment on the northwestern side of Herring Bay were studied in October 1989 (after the spill had contaminated this environment) and again in 1990, 1991, and 1993 (Jewett et al. 1996, this volume). A large die-off of invertebrates occurred in fall 1989, but another one occurred in 1993 after oil had decreased approximately 95%. The study site has a 4-m-deep sill on its seaward side, and this environment may be subject to periodic hypoxia. Oil toxicity may have contributed directly to the initial die-off or it might have exacerbated hypoxic stress in fall 1989.

Effects on Archaeological Sites

Ancient peoples lived around Prince William Sound and the wider spill area. Many important archaeological sites subsided into the intertidal zone during the 1964 earthquake, and some of these became oiled in 1989. Laboratory studies showed that crude oil contamination could render carbon radioisotope dating unusable for these sites (Bittner 1996, this volume). Physical abuse was even more serious; assessment of 35 archaeological sites showed that 71% had been damaged by oil spill response and cleanup activities or by vandalism (Bittner 1996). Vandalism had the worst effect on archaeological sites, and the increase in vandalism was a direct result of people learning about the sites during the cleanup.

Effects on Subsistence

More than a dozen Alaska native communities exist in the spill zone, and all of them depend largely on food collected from the sea. In the year after the spill, the subsistence harvest declined by as much as 77% (Fall and Field 1996, this volume). Villagers were uncertain about the safety of subsistence foods, which could have been contaminated by oil. In response to this concern, the Alaska Oil Spill Health Task Force coordinated a risk assessment of subsistence foods from the spill area. Bolger et al. (1996, this volume) report that the risk of getting cancer from eating finfish was near zero. Fish can transform and quickly excrete many aromatic hydrocarbons, greatly limiting the deposition of these compounds in edible flesh (Hom et al. 1996, this volume). The risk of getting cancer from eating mollusks, which can accumulate some hydrocarbons, was somewhat higher in 1989. The task force advised that subsistence foods showing visible oil not be consumed. In 1991, samples of shellfish from some locations around Chenega, Windy Bay, Kodiak Village, and Old Harbor consistently had total aromatic hydrocarbons exceeding 100 ng/g wet weight (moderately or highly contaminated); the highest concentration (18,000 ng/g) was in a mussel sample from Windy Bay (Brown et al. 1996, this volume).

Psychological and Social Impacts on Humans

Most people can remember where they were and what they were doing when major dramas unfolded: the assassination of President Kennedy, the 1964 Alaskan earthquake, the explosion of the space shuttle *Challenger*. When the *Exxon Valdez* spilled oil into Prince William Sound, people throughout the world felt distressed, dismayed, angry, and unable to help. They wanted to clean up the oil, save the birds and beaches, and punish those responsible. What about the people who lived or recreated in the spill area and made their living or otherwise found sustenance from the sea? Survey and ethnographic data collected by a study sponsored by the Oiled Mayors (mayors of communities in the spill area) show a significant relationship between measures of exposure to the spill and cleanup and measures of general anxiety disorder, symptoms of depression, substance abuse, and domestic violence (Russell et al. 1996, this volume). Picou and Gill (1996, this volume) found that the *Exxon Valdez* spill produced patterns of chronic psychological stress that differentially affected communities and individuals. The people who lived and worked in or visited Prince William Sound and the adjacent spill areas were forever affected by the spill. Their lives will never be the same and their level of trust in industry and government is permanently diminished.

Discussion

Sources of Uncertainty in the Assessment of Damages

As mentioned at the outset, with a few exceptions such as harbor seals, little up-to-date information existed on the status of affected populations at the time of the spill. Many of the studies reported in this volume are therefore based on comparisons of oiled and non-oiled environments. It is likely that oiled and non-oiled environments would systematically differ in ways other than oiling. For example, the same currents that greatly influenced the spread, direction of travel, and eventual deposition of oil might also bring different amounts of food, suspended sediment, and water motion to these environments. These natural factors probably gave rise to faunal differences that would be intrinsically difficult to unravel after the toll taken by the spill. Other natural geographic differences could be related to factors such as physiography, freshwater runoff, local climate, and nutrients. Larger organisms might differ in their degree of parasitism between oiled and non-oiled study areas. Choosing multiple, widely scattered oiled and non-oiled study sites, balanced between mainland and island habitats, might help randomize such effects, but there are no guarantees that they are not confounded with the effects of oiling.

In some instances (e.g., birds and otters), injury to the population can be estimated directly from the numbers of carcasses recovered during the spill. Even in these cases, however, it may be difficult to reconcile estimates of animals killed with population estimates made before and after the spill. Inherent problems of experimental design (Hilborn 1996, this volume), and attendant uncertainties in results can lead to considerable controversy over both data interpretation and conclusions about the impact of the spill.

Initial Mortality versus Chronic and Sublethal Effects

The *Exxon Valdez* oil spill took a large immediate toll. But were there significant chronic and sublethal effects from the lingering oil? Many of the damage assessments pursued this question in one way or another. In many cases there was not a clear, compelling answer to this question, which led some people to cynicism over the scope of the scientific effort (e.g., Wheelwright 1994). Accordingly, it is worth reviewing this overarching question here.

The acute toxicity of fresh oil is due largely to low-molecular-weight fractions that rapidly weather in the environment, and it is well known from other oil spills that the most serious effects occur during the early stages of a spill (Neff 1979). The *Exxon Valdez* spill had many obvious and undisputed initial effects. Over the first several weeks and months following the spill, most of the acutely toxic compounds were lost to evaporation and dissolution. Over the course of the next 3 years, the beached petroleum was largely removed from the intertidal shorelines as a combined result of natural wave action (mainly during winter storms), cleanup activities, and weathering processes, including natural biodegradation and bioremediation. Without doubt, however, the residual shoreline oil exerted acutely toxic effects in the first few months after the spill and sublethal effects thereafter. The sublethal effects were most likely due to toxicity of high-molecular-weight PAHs, which play a relatively minor role in immediate effects but persist in the environment to become more important in the long term.

Oil weathers at different rates in different environments; some protected sites still contained substantial concentrations of relatively undegraded *Exxon Valdez* oil in 1994–1995, for example. Furthermore, cleanup activities extended over 3 years on many shorelines, and while they promoted removal of oil from the beaches, they also caused continuing mortality of intertidal flora and fauna. In many cases, these continuing (chronic) effects may have been largely masked by the overwhelming initial injury and by the dynamics of early biological recovery.

Sublethal and chronic effects were investigated with respect to egg mortality and larval growth of pink salmon; reproductive success, larval deformations, genetic aberrations, and disease of Pacific herring; histopathological alterations of pink salmon, herring, Dolly Varden, cutthroat trout, and sea otters; juvenile mortality and blood chemistry of sea otters; dietary change and acute-phase proteins in the blood of river otters *Lutra canadensis*; reproductive success and blood chemistry of bald eagles; reproductive success of pigeon guillemots and black-legged kittiwakes; and growth and byssal thread production by mussels. In some cases differences apparently due to oil exposure were found; in other cases no effects were found or differences could not be unequivocally ascribed to the spill.

Reduced populations and lower survival of sea otters in oiled versus non-oiled areas of Prince William Sound through 1991 were accompanied by subtle sublethal differences in several blood measures (Rebar et al. 1994). Histopathological lesions in a small sample of the sea otter carcasses recovered from the sound (Lipscomb et al. 1994) have been used as evidence for sublethal effects on sea otters that were exposed to oil but not killed immediately (Ballachey et al. 1994). In 1990 and 1991, mortalities of prime-age sea otters and recently weaned juveniles in oiled areas were higher than normal (Rotterman and Monnett 1991), suggesting a chronic effect of the spill through that period.

Reproductive success of bald eagles was depressed adjacent to moderately or heavily oiled beaches of Prince William Sound and the Alaska Peninsula during 1989, but it rebounded in 1990. The extent to which this sublethal effect was due to disturbance from the oil spill cleanup or to oil exposure is unknown (Bernatowicz et al. 1996). Eagle blood chemistry did not indicate an effect of the spill.

Exxon Valdez oil remained detectable in many intertidal and shallow subtidal sediments through 1992, but it had undergone extensive weathering and concentrations in most areas were no longer acutely toxic after 1990–1991. Standardized tests with the amphipods *Rhepoxynius abronius* or *Ampeliscus abdita* indicated residual sediment toxicity in the intertidal zones of several heavily oiled sites during 1990, but the numbers of toxic sites and the levels of toxicity declined markedly in 1991 (Boehm et al. 1995; Wolfe et al. 1996, this volume). No significant toxicity in subtidal sediments was attributable to *Exxon Valdez* oil in either year. These results do not preclude ongoing toxic effects in highly sensitive species in some areas, but they do support a conclusion that direct effects of the oil in the intertidal zone were largely over by 1991, when major cleanup activities ceased.

Although detectable sublethal effects on organisms were generally confined to the first year or two following the spill, one surprising exception was the systematically greater mortality of pink salmon eggs in oiled than in non-oiled streams from 1989 through 1992 (Sharr et al. 1994). This differential mortality continued into 1993 but was not statistically discernible in 1994 and 1995. Lingering toxicity from oil-contaminated gravel, inherited oil-induced genetic damage, and natural preexisting differences in egg mortality have been advanced to explain the 4-year mortality difference. None of these hypotheses have been substantiated, although recent spawning experiments with returning pink salmon strongly suggest that heritable differences in egg mortality rates exist between populations in different streams (Sharr et al. 1994). Whether these differences existed before the spill is not known.

Some have speculated that the crash of the Pacific herring population in Prince William Sound during 1992 and 1993 was due to weakened immunity to viral hemorrhagic septicemia and lowered resistance to fungal infections (*Icthyophonus*); increases of both diseases were associated with the decline. Work on the relationship between herring disease and oil exposure continued into 1995, but the cause of the crash—whether natural, oil-induced, or both—may not be resolvable.

Distinguishing initial acute effects oil from chronic sublethal effects is further obscured by the occurrence of secondary, density-dependent aftereffects of the initial acute mortality, and by substantial delays, or lags, between the initial exposure and its ultimate expression at the population level. Nevertheless, it can be seen from the above examples that chronic sublethal effects most likely attributable to residual oil occurred for several years after the *Exxon Valdez* spill. The number of cases in which effects were postulated and investigated was larger than those in which effects were found. This is to be expected. A major spill in a complex, biologically rich environment raises numerous concerns and we have little power to predict precisely the subtler effects that might occur.

Use of Water Concentrations of Petroleum to Infer Effects

The reported water column concentrations of less than 10 μg PAH/L (Short and Harris 1996a) were less than those previously reported to kill fish and invertebrates in short-term or chronic exposures (e.g., Capuzzo 1987). There are, however, other routes of exposure to consider. For example, it appears that juvenile pink salmon ingested small particles of oil suspended in the water (Carls et al. 1996b) and such particles also would have been available to copepods, ctenophores, mussels, tunicates, sponges, and other filter-feeding marine organisms. Exposure to dissolved and suspended oil in the water column was augmented by simultaneous exposure from direct contact with slicks and sheens, toxic components of oil in interstitial pore water (where it can be more concentrated than in the overlying water), oil-contaminated sediment, and oil-contaminated prey.

Variable Rates of Recovery

During the period of recovery, the persistence of damage to any particular biological resource probably depends upon (1) the magnitude of the initial injury, (2) continuing injury (e.g., toxicity), (3) secondary density-dependent effects, (4) delayed expression of effects, and (5) intrinsic rates of natural recovery. These factors differ for every category of injured resource, and recovery must be examined case by case.

Measuring recovery can be problematic if the magnitude of the damages is uncertain, the tools available are imprecise, or the population has continued a prespill decline. For example, if only several hundred bald eagles in Prince William Sound were killed by the spill and the population was several thousand birds (as indicated by the last good census in 1984), the impact of the spill on the population was about 10% and probably within the error of the survey technique. The postspill counts of eagles were all higher than 1984 counts, which suggests that the population is resilient enough to recover its losses in a few years (chick production returned to normal in 1990). The bald eagle was therefore listed as "recovering" in the Restoration Plan (EVOS Trustee Council 1994). The system as a whole cannot be declared fully recovered, even though many populations appear to have returned to normal over much of the spill area. Population estimates are inherently imprecise, and lack of significant differences between pre- and postspill population numbers is not by itself an adequate basis for judging either recovery or lack of effect.

Markers of exposure and sublethal impacts following an oil spill may help corroborate continuing chronic effects from oil exposure. Absence of sublethal effects does not necessarily mean that full recovery of the population has occurred, rather that the direct effects of oil have probably ceased. In addition, the decline of stress markers with uncertain causality in organisms from oiled areas tends to support the hypothesis of spill-related effects in those organisms.

The Evolution of the Spill Scientific Program

The spill had serious effects on many species, and many species have not yet fully recovered. Continued study of these resources is required if we are to better understand the full extent of the impacts, how they occurred, and the nature of recovery. Large year-to-year and longer changes in the climate and other environmental factors in Prince William Sound and the Gulf of Alaska may complicate or preclude simple models of population disruption and recovery (Bailey et al. 1995). The Trustee Council's restoration program has paid increasing attention to such processes and changes and less to obtaining simple counts of organisms, which were given greater weight in the immediate wake of the spill. These ecological studies should yield a more lasting legacy from the spill scientific program.

In 1994 and 1995, after the symposium, the emphasis of the Trustee Council's restoration program shifted toward obtaining a better understanding of ecological processes affecting spill-impacted species and their recovery. Multidisciplinary programs have been created to investigate the physical and biological oceanography of Prince William Sound as it relates to production, growth, and predation mortality of juvenile pink salmon and Pacific herring (the Sound Ecosystem Assessment); the factors affecting population recovery of sea otters, river otters, pigeon guillemots, and harlequin ducks *Histrionicus histrionicus* (the Nearshore Vertebrate Predators Program); and the relationship between forage fish quality and abundance and seabird reproduction (Alaskan Predator Ecosystem Experiment). These programs represent a logical evolution of the scientific program from damage assessment and population census to process-oriented research.

We maintain that it would have been very difficult, if not impossible, to assess the spill's effects based solely on ecological processes; our knowledge of oil's effects on these processes is too uncertain. Without an appreciation of injury to each affected species in terms of mortality, growth, and reproduction, it would have been difficult to move into restoration with any confidence. However, the studies of dynamic intertidal processes begun during 1990 in Herring Bay and described in this volume demonstrate the early commitment of the Trustee Council to process-oriented ecological work when it was feasible.

We are often asked by the public "Is the oil gone?" and "Is the environment back to normal?" In answering these questions we find it useful to distinguish two kinds of general effects: the continuing recovery of populations from the initial effects of the spill in 1989 and 1990, and the possible continuing effects of the oil remaining in the environment since that time. Despite the above-mentioned uncertainties, we have a much better understanding of recovery than of chronic effects. For instance, oil concentrations in the heavily contaminated mussel beds are probably still high enough to affect at least the fauna in the associated sediments and possibly the predators feeding continually on oiled mussels. At present, however, we have found no adverse effects that can be attributed unequivocally to oil in mussel beds.

Conclusions

- The *Exxon Valdez* oil spill released 42 million liters of oil, which contaminated hundreds of kilometers of shoreline. The oil was cleaned up by humans for 2 years, and it has been cleaned by nature for several more.
- The *Exxon Valdez* oil spill, at a minimum, resulted in:

(1) the direct mortality of an estimated 3,500–5,500 sea otters, 300 harbor seals, and 250,000 or more birds of 90 species, including common murres, and thick-billed murres *U. lomvia*, bald eagles, marbled murrelets, Kittlitz's murrelets *Brachyramphus brevirostris*, ancient murrelets

Synthliboramphus antiquus, pigeon guillemots, two species of puffins, four species of loons, and three species of cormorants;
(2) significant reductions in the populations of many intertidal organisms, including algae, barnacles, limpets, amphipods, isopods, worms, and fishes, over a very extensive area from Prince William Sound to the Kodiak Archipelago and the Alaska Peninsula;
(3) significant reductions in populations of subtidal organisms in Prince William Sound;
(4) increased mortality of Pacific herring eggs in 1989 and increased incidences of aberrations in herring larvae during 1989 and 1990;
(5) increased mortality of wild-stock pink salmon eggs in oiled stream beds in 1989–1992 and abnormal fry and reduced juvenile growth in 1989, which may have resulted in loss of nearly 2 million pink salmon from the 1990 harvest;
(6) poorer growth of Dolly Varden and cutthroat trout for 2 years following the spill;
(7) increased vandalism of archaeological resources during the cleanup; and
(8) reduced use of subsistence resources.

• Many estimates of injury are very uncertain, mainly because up-to-date prespill data on the affected populations was lacking.

• Injuries from the *Exxon Valdez* oil spill accrued from a variety of direct, indirect, and delayed effects from exposures that ranged from acute to chronic and differed widely among resources.

• Recovery rates have been resource-specific, varying with the magnitude and persistence of direct and indirect impacts and with intrinsic characteristics of population dynamics.

• As of September 1992, approximately 20% of the oil had evaporated or photolysed in the atmosphere, 50% had been biologically degraded, 2% remained in beaches, and 13% had settled into marine sediments of Prince William Sound and the Gulf of Alaska. Most of the oil remaining in sediments consisted of highly weathered, refractory components not readily identifiable as *Exxon Valdez* oil. Relatively unweathered oil still remained, however, in protected sediments under rock armor and in some mussel beds in Prince William Sound.

Acknowledgments

We thank Lori Ewing of the National Marine Fisheries Service's Auke Bay Laboratory and Stan Senner, Science Liason for the Trustee Council, for their editorial assistance.

References

Ainley, D. J., and R. J. Boekelheide. 1990. Seabirds of the Farallon Islands: ecology, structure and dynamics of an upwelling system community. Stanford University Press, Stanford, California.

Ainley, D. J., and T. J. Lewis. 1974. The history of Farallon Island bird populations, 1854–1972. Condor 76:432–446.

Babcock, M. M., G. V. Irvine, P. M. Harris, J. A. Cusick, and S. D. Rice. 1996. Persistence of oiling in mussel beds three and four years after the *Exxon Valdez* oil spill. Pages 286–297 *in* Rice et al. (1996).

Bailey, K. M., and eleven coauthors. 1995. ENSO events in the northern Gulf of Alaska, and effects on selected marine fisheries. California Cooperative Oceanic Fisheries Investigations Reports 36:78–96.

Ballachey, B. E., J. L. Bodkin, and A. R. DeGange. 1994. An overview of sea otter studies. Pages 47–59 *in* Loughlin (1994).

Bernatowicz, J. A., P. F. Schempf, and T. D. Bowman. 1996. Bald eagle productivity in south-central Alaska in 1989 and 1990 after the *Exxon Valdez* oil spill. Pages 785–797 *in* Rice et al. (1996).

Bittner, J. E. 1996. Cultural resources and the *Exxon Valdez* oil spill: an overview. Pages 814–818 *in* Rice et al. (1996).

Boehm, P. D., D. S. Page, E. S. Gilfillan, W. A. Stubblefield, and E.J. Harner. 1995. Shoreline ecology program for Prince William Sound, Alaska, following the *Exxon Valdez* oil spill: part 2—chemistry and toxicology. Pages 347–397 *in* P. G. Wells, J. N. Butler, and J. S. Hughes, editors. *Exxon Valdez* oil spill: fates and effects in Alaskan waters. American Society for Testing and Materials Special Technical Publication 1219, Philadelphia.

Bolger, M., S. H. Henry, and C. D. Carrington. 1996. Hazard and risk assessment of crude oil contaminants in subsistence seafood samples from Prince William Sound. Pages 837–843 *in* Rice et al. (1996).

Bowman, T. D., P. F. Schempf, and J. A. Bernatowicz. 1995. Bald eagle survival and population dynamics in Alaska after the *Exxon Valdez* oil spill. Journal of Wildlife Management 59:317–324.

Braddock, J. F., J. E. Lindstrom, and E. J. Brown. 1995. Distribution of hydrocarbon-degrading microorganisms in sediments from Prince William Sound, Alaska, following the *Exxon Valdez* oil spill. Marine Pollution Bulletin 30(2):125–132.

Braddock, J. F., J. E. Lindstrom, T. R. Yeager, B. T. Rasley, and E. J. Brown. 1996. Patterns of microbial activity in oiled and unoiled sediments in Prince William Sound. Pages 94–108 *in* Rice et al. (1996).

Brown, E. D., and seven coauthors. 1996. Injury to the early life history stages of Pacific herring in Prince William Sound after the *Exxon Valdez* oil spill. Pages 448–462 *in* Rice et al. (1996).

Brown, D. W., and nine coauthors. 1996. Survey of Alaskan subsistence invertebrate seafoods collected in 1989–1991 to determine exposure to oil spilled from the *Exxon Valdez*. Pages 844–855 *in* Rice et al. (1996).

Bue, B. G., S. Sharr, S. D. Moffitt, and A. K. Craig. 1996.

Effects of the *Exxon Valdez* oil spill on pink salmon embryos and preemergent fry. Pages 619–627 *in* Rice et al. (1996).

Capuzzo, J. M. 1987. Biological effects of petroleum hydrocarbons in the sea: assessments from experimental results. Pages 343–410 *in* D. F. Boesch and N. N. Rabalais, editors. Long-term environmental effects of offshore oil and gas development. Elsevier Applied Science, London.

Carls, M. G., and six coauthors. 1996a. Growth, feeding, and survival of pink salmon fry exposed to food contaminated with crude oil. Pages 608–618 *in* Rice et al. (1996).

Carls, M. G., A. C. Wertheimer, J. W. Short, R. M. Smolowitz, and J. J. Stegeman. 1996b. Contamination of juvenile pink and chum salmon by hydrocarbons in Prince William Sound after the *Exxon Valdez* oil spill. Pages 593–607 *in* Rice et al. (1996).

Carlson, P. R., and K. A. Kvenvolden. 1996. Tracking *Exxon Valdez* oil from beach to deepwater sediments of Prince William Sound, Alaska. Pages 109–120 *in* Rice et al. (1996).

Celewycz, A. G., and A. C. Wertheimer. 1996. Prey availability to juvenile salmon after the *Exxon Valdez* oil spill. Pages 564–577 *in* Rice et al. (1996).

Collier, T. K., C. A. Krone, M. M. Krahn, J. E. Stein, S.-L. Chan, and U. Varanasi. 1996. Petroleum exposure and associated biochemical effects in subtidal fish after the *Exxon Valdez* oil spill. Pages 671–683 *in* Rice et al. (1996).

Dahlheim, M. E., and C. O. Matkin. 1994. Assessment of injuries to Prince William Sound killer whales. Pages 163–171 *in* Loughlin (1994).

Dean, T. A., S. C. Jewett, D. R. Laur, and R. O. Smith. 1996a. Injury to epibenthic invertebrates resulting from the *Exxon Valdez* oil spill. Pages 424–439 *in* Rice et al. (1996).

Dean, T. A., M. S. Stekoll, and R. O. Smith. 1996b. Kelps and oil: the effects of the *Exxon Valdez* oil spill on subtidal algae. Pages 412–423 *in* Rice et al. (1996).

Driskell, W. B., A. K. Fukuyama, J. P. Houghton, D. C. Lees, A. J. Mearns, and G. Shigenaka. 1996. Recovery of Prince William Sound intertidal infauna from *Exxon Valdez* oiling and shoreline treatments, 1989 through 1992. Pages 362–378 *in* Rice et al. (1996).

Duncan, P. B., and A. J. Hooten. 1996. Influence of residual and applied oil on intertidal algal recruitment. Pages 238–248 *in* Rice et al. (1996).

Ebert, T. A., and D. C. Lees. 1996. Growth and loss of tagged individuals of the predatory snail *Nucella lamellosa* in areas within the influence of the *Exxon Valdez* oil spill in Prince William Sound. Pages 349–361 *in* Rice et al. (1996).

EVOS (*Exxon Valdez* Oil Spill) Trustee Council. 1994. *Exxon Valdez* oil spill restoration plan. EVOS Trustee Council, Anchorage, Alaska.

Fall, J. A., and L. J. Field. 1996. Subsistence uses of fish and wildlife before and after the *Exxon Valdez* oil spill. Pages 819–836 *in* Rice et al. (1996).

Fleeger, J. W., T. C. Shirley, M. G. Carls, and M. A. Todaro. 1996. Meiofaunal recolonization experiment with oiled sediments. Pages 271–285 *in* Rice et al. (1996).

Ford, R. G., and seven coauthors. 1996. Total direct mortality of seabirds from the *Exxon Valdez* oil spill. Pages 684–711 *in* Rice et al. (1996).

Frost, K. J., L. F. Lowry, E. H. Sinclair, J. Ver Hoef, and D. C. McAllister. 1994b. Impacts on distribution, abundance and productivity of harbor seals. Pages 97–118 *in* Loughlin (1994).

Frost, K. J., C. A. Manen, and T. L. Wade. 1994a. Petroleum hydrocarbons in tissues of harbor seals from Prince William Sound and the Gulf of Alaska. Pages 331–358 *in* Loughlin (1994).

Garrott, R. A., L. L. Eberhardt, and D. M. Burn. 1993. Mortality of sea otters in Prince William Sound following the *Exxon Valdez* oil spill. Marine Mammal Science 9:343–359.

Geiger, H. J., B. G. Bue, S. Sharr, A. C. Wertheimer, and T. M. Willette. 1996. A life history approach to estimating damage to Prince William Sound pink salmon caused by the *Exxon Valdez* oil spill. Pages 487–498 *in* Rice et al. (1996).

Hardy, J., S. Kiesser, L. Antrim, A. Stubin, R. Kocan, and J. Strand. 1987. The sea-surface microlayer of Puget Sound: part I. Toxic effect on fish eggs and larvae. Marine Environmental Research 23:227–249.

Harris, M. P., D. J. Halley, and R. L. Swann. 1994. Age of first breeding in common murres. Auk 111:207–209.

Harris, P. M., S. D. Rice, M. M. Babcock, and C. C. Brodersen. 1996. Within-bed distribution of *Exxon Valdez* crude oil in Prince William Sound blue mussels and underlying sediments. Pages 298–308 *in* Rice et al. (1996).

Hepler, K. R., P. A. Hansen, and D. R. Bernard. 1996. Impact of oil spilled from the *Exxon Valdez* on survival and growth of Dolly Varden and cutthroat trout in Prince William Sound. Pages 645–658 *in* Rice et al. (1996).

Highsmith, R. C., and six coauthors. 1996. Impact of the *Exxon Valdez* oil spill on intertidal biota. Pages 212–237 *in* Rice et al. (1996).

Hilborn, R. 1996. Detecting population impacts from oil spills: a comparison of methodologies. Pages 639–644 *in* Rice et al. (1996).

Hom, T., U. Varanasi, J. E. Stein, C. A. Sloan, K. L. Tilbury, and S.-L. Chan. 1996. Assessment of the exposure of subsistence fish to aromatic compounds after the *Exxon Valdez* oil spill. Pages 856–866 *in* Rice et al. (1996).

Hooten, A. J., and R. C. Highsmith. 1996. Impacts on selected intertidal invertebrates in Herring Bay, Prince William Sound, after the *Exxon Valdez* oil spill. Pages 249–270 *in* Rice et al. (1996).

Houghton, J. P., D. C. Lees, W. B. Driskell, S. C. Lindstrom, and A. J. Mearns. 1996. Recovery of Prince William Sound intertidal epibiota from *Exxon Valdez* oiling and shoreline treatments, 1989 through 1992. Pages 379–411 *in* Rice et al. (1996).

Irons, D. B. 1996. Size and productivity of black-legged kittiwake colonies in Prince William Sound before and after the *Exxon Valdez* oil spill. Pages 738–747 *in* Rice et al. (1996).

Jewett, S. C., T. A. Dean, and D. R. Laur. 1996. Effects of the *Exxon Valdez* oil spill on benthic invertebrates in an oxygen-deficient embayment in Prince William Sound, Alaska. Pages 440–447 *in* Rice et al. (1996).

Kuletz, K. J. 1996. Marbled murrelet abundance and breeding activity at Naked Island, Prince William Sound, and Kachemak Bay, Alaska, before and after the *Exxon Valdez* oil spill. Pages 770–784 *in* Rice et al. (1996).

Laur, D., and L. Haldorson. 1996. Coastal habitat studies: the effect of the *Exxon Valdez* oil spill on shallow subtidal fishes in Prince William Sound. Pages 659–670 *in* Rice et al. (1996).

Lees, D. C., J. P. Houghton, and W. B. Driskell. 1996. Short-term effects of several types of shoreline treatment on rocky intertidal biota in Prince William Sound. Pages 329–348 *in* Rice et al. (1996).

Lipscomb, T. P., R. K. Harris, A. H. Rebar, B. E. Ballachey, and R. J. Haebler. 1994. Pathology of sea otters. Pages 265–279 *in* Loughlin (1994).

Loughlin, T. R., editor. 1994. Marine mammals and the *Exxon Valdez*. Academic Press, New York.

Loughlin, T. R., B. E. Ballachey, and B. A. Wright. 1996. Overview of studies to determine injury caused by the *Exxon Valdez* oil spill to marine mammals. Pages 798–808 *in* Rice et al. (1996).

Matkin, C. O., G. M. Ellis, M. E. Dahlheim, and J. Zeh. 1994. Status of killer whales in Prince William Sound, 1985–1992. Pages 141–162 *in* Loughlin (1994).

Mearns, A. J. 1996. *Exxon Valdez* shoreline treatment and operations: implications for response, assessment, monitoring, and research. Pages 309–328 *in* Rice et al. (1996).

Michel, J., and M. O. Hayes. 1991. Geomorphological controls on the persistence of shoreline contamination from the *Exxon Valdez* oil spill. Research Planning, Inc. report (HMRB 91-2) to NOAA (National Oceanic and Atmospheric Administration) Hazardous Materials Response Branch, Seattle.

Michel, J., and M. O. Hayes. 1992. Evaluation of the condition of Prince William Sound shorelines following the *Exxon Valdez* oil spill and subsequent shoreline treatment. Volume 1. 1991 geomorphological shoreline monitoring survey. NOAA (National Oceanic and Atmospheric Administration) Technical Memorandum NOS (National Ocean Service) ORCA (Office of Ocean Resources Conservation and Assessment) 67, Seattle.

Neff, J. M. 1979. Polycyclic aromatic hydrocarbons in the marine environment. Applied Science Publishers, London.

Neff, J. M. 1991. Water quality in Prince William Sound and the Gulf of Alaska. Arthur D. Little, Inc., Cambridge, Massachusetts.

Neff, J. M., P. D. Boehm, L. Haas, and P. J. Kinney. 1990. Petroleum hydrocarbons in the water column of Prince William Sound, Alaska. Pages 426–443 *in* M. L. Spaulding and M. Reed, editors. Oil spills: management and legislative implications. American Society of Civil Engineers, New York.

Norcross, B. L., and M. Frandsen. 1996. Distribution and abundance of larval fishes in Prince William Sound, Alaska, during 1989 after the *Exxon Valdez* oil spill. Pages 463–486 *in* Rice et al. (1996).

Oakley, K. L., and K. J. Kuletz. 1996. Population, reproduction, and foraging of pigeon guillemots at Naked Island, Alaska, before and after the *Exxon Valdez* oil spill. Pages 759–769 *in* Rice et al. (1996).

O'Clair, C. E., J. W. Short, and S. D. Rice. 1996. Contamination of intertidal and subtidal sediments by oil from the *Exxon Valdez* in Prince William Sound. Pages 61–93 *in* Rice et al. (1996).

Piatt, J. F., and P. Anderson. 1996. Response of common murres to the *Exxon Valdez* oil spill and long-term changes in the Gulf of Alaska marine ecosystem. Pages 720–737 *in* Rice et al. (1996).

Piatt, J. F., and R. G. Ford. 1996. How many seabirds were killed by the *Exxon Valdez* oil spill? Pages 712–719 *in* Rice et al. (1996).

Piatt, J. F., and C. J. Lensik. 1989. *Exxon Valdez* bird toll. Nature 342:865–866.

Picou, J. S., and D. A. Gill. 1996. The *Exxon Valdez* oil spill and chronic psychological stress. Pages 879–893 *in* Rice et al. (1996).

Rebar, A. H., B. E. Ballachey, D. L. Bruden, and K. Kloecker. 1994. Hematology and clinical chemistry of sea otters captured in Prince William Sound, Alaska, following the *Exxon Valdez* oil spill. U.S. Fish and Wildlife Service, Natural Resources Damage Assessment report, Marine Mammal Study 6, Anchorage, Alaska.

Rice, S .D., and seven coauthors. 1984. Effects of petroleum hydrocarbons on Alaskan aquatic organisms: a comprehensive review of all oil-effects research on Alaskan fish and invertebrates conducted by the Auke Bay Laboratory, 1970–81. NOAA (National Oceanic and Atmospheric Administration) Technical Memorandum NMFS (National Marine Fisheries Service) F/NWC-67, Auke Bay, Alaska.

Rice, S. D., R. B. Spies, D. A. Wolfe, and B. A. Wright, editors. 1996. Proceedings of the *Exxon Valdez* oil spill symposium. American Fisheries Society Symposium 18, Bethesda, Maryland.

Rotterman, L. M., and C. Monnett. 1991. Mortality of sea otter weanlings in eastern and western Prince William Sound, Alaska during the winter of 1990–91. U.S. Fish and Wildlife Service, Natural Resources Damage Assessment report, Marine Mammal Study 6, Anchorage, Alaska.

Russell, J. C., M. A. Downs, J. S. Petterson, and L. A. Palinkas. 1996. Psychological and social impacts of the *Exxon Valdez* oil spill and cleanup. Pages 867–878 *in* Rice et al. (1996).

Schmidt, D. C., K. E. Tarbox, B. E. King, L. K. Brannian, G. B. Kyle, and S. R. Carlson. 1996. Kenai River sockeye salmon: an assessment of overescapements as a cause of the decline. Pages 628–638 *in* Rice et al. (1996).

Sharp, B. E., M. Cody, and R. Turner. 1996. Effects of the *Exxon Valdez* oil spill on the black oystercatcher. Pages 748–758 *in* Rice et al. (1996).

Sharr, S., J. E. Seeb, B. G. Bue, A. Craig, and G. D. Miller. 1994. Injury to salmon eggs and preemergent fry in Prince William Sound. Oil Spill Restoration Office,

Exxon Valdez Oil Spill Restoration Project final report (Project 93003). Anchorage, Alaska.

Short, J. W., and M. M. Babcock. 1996. Prespill and postspill concentrations of hydrocarbons in mussels and sediments in Prince William Sound. Pages 149–166 *in* Rice et al. (1996).

Short, J. W., and P. M. Harris. 1996a. Chemical sampling and analysis of petroleum hydrocarbons in near-surface seawater of Prince William Sound after the *Exxon Valdez* oil spill. Pages 17–28 *in* Rice et al. (1996).

Short, J. W., and P. M. Harris. 1996b. Petroleum hydrocarbons in caged mussels deployed in Prince William Sound after the *Exxon Valdez* oil spill. Pages 29–39 *in* Rice et al. (1996).

Short, J. W., D. M. Sale, and J. C. Gibeaut. 1996. Nearshore transport of hydrocarbons and sediments after the *Exxon Valdez* oil spill. Pages 40–60 *in* Rice et al. (1996).

Spies, R. B., and P. H. Davis. 1979. The infaunal benthos of a natural oil seep in the Santa Barbara Channel. Marine Biology 50:227–237.

Stekoll, M. S., and seven coauthors. 1996. Coastal habitat injury assessment: intertidal communities and the *Exxon Valdez* oil spill. Pages 177–192 *in* Rice et al. (1996).

Sturdevant, M. V., A. C. Wertheimer, and J. L. Lum. 1996. Diets of juvenile pink and chum salmon in oiled and non-oiled nearshore habitats in Prince William Sound, 1989 and 1990. Pages 578–592 *in* Rice et al. (1996).

Sundberg, K., L. Deysher, and L. McDonald. 1996. Intertidal and supratidal site selection using a geographical information system. Pages 167–176 *in* Rice et al. (1996).

Templin, W. D., J. S. Collie, and T. J. Quinn, II. 1996. Run reconstruction of the wild pink salmon fishery in Prince William Sound, 1990–1991. Pages 499–508 *in* Rice et al. (1996).

van Tamelen, P. G., and M. S. Stekoll. 1996. Population response of the brown alga *Fucus gardneri* and other algae in Herring Bay, Prince William Sound, to the *Exxon Valdez* oil spill. Pages 193–211 *in* Rice et al. (1996).

Weins, J. A. 1995. Recovery of sea birds following the *Exxon Valdez* oil spill: an overview. Pages 854–893 *in* P. G. Wells, J. N. Butler, and J. S. Hughes, editors. *Exxon Valdez* oil spill: fates and effects in Alaskan waters. American Society for Testing and Materials Special Technical Publication 1219, Philadelphia.

Wells, P. G., J. N. Butler, and J. S. Hughes, editors. 1995. *Exxon Valdez*: fates and effects in Alaskan waters. American Society for Testing and Materials Special Technical Publication 1219, Philadelphia.

Wertheimer, A. C., N. J. Bax, A. G. Celewycz, M. G. Carls, and J. H. Landingham. 1996. Harpacticoid copepod abundance and population structure in Prince William Sound, one year after the *Exxon Valdez* oil spill. Pages 551–563 *in* Rice et al. (1996).

Wertheimer, A. C., and A. G. Celewycz. 1996. Abundance and growth of juvenile pink salmon in oiled and non-oiled locations of western Prince William Sound after the *Exxon Valdez* oil spill. Pages 518–532 *in* Rice et al. (1996).

Wheelwright, J. 1994. Degrees of disaster. Simon & Schuster, New York.

Wiedmer, M., M. J. Fink, J. J. Stegeman, R. Smolowitz, G. D. Marty, and D. E. Hinton. 1996. Cytochrome P-450 induction and histopathology in preemergent pink salmon from oiled spawning sites in Prince William Sound. Pages 509–517 *in* Rice et al. (1996).

Willette, M. 1996. Impacts of the *Exxon Valdez* oil spill on the migration, growth, and survival of juvenile pink salmon in Prince William Sound. Pages 533–550 *in* Rice et al. (1996).

Wolfe, D. A. 1995. Fate and toxicity of spilled oil from the *Exxon Valdez*. Final report to the State of Alaska–Federal Trustees for Natural Resources Damage Assessment. NOAA (National Oceanic and Atmospheric Administration) Coastal Monitoring and Bioeffects Assessment Division, Silver Spring, Maryland.

Wolfe, D. A., and eleven coauthors. 1994. The fate of the oil spilled from the *Exxon Valdez*. Environmental Science and Technology 28:561A–568A.

Wolfe, D. A., and six coauthors. 1996. Toxicity of intertidal and subtidal sediments contaminated by the *Exxon Valdez* oil spill. Pages 121–139 *in* Rice et al. (1996).

American Fisheries Society Symposium 18:17–28, 1996

Chemical Sampling and Analysis of Petroleum Hydrocarbons in Near-Surface Seawater of Prince William Sound after the *Exxon Valdez* Oil Spill

JEFFREY W. SHORT AND PATRICIA M. HARRIS

National Marine Fisheries Service
Alaska Fisheries Science Center, Auke Bay Laboratory
11305 Glacier Highway, Juneau, Alaska 99801–8626, USA

Abstract.—To assess contamination from spilled oil after the grounding of the *Exxon Valdez* on 24 March 1989, we sampled 32 locations in Prince William Sound, Alaska, for petroleum hydrocarbons in seawater at depths of 1 and 5 m. Most locations were sampled in each of three periods between 31 March and 8 May 1989. Triplicate seawater samples were taken at each location and depth and were extracted into dichloromethane. Of 501 samples taken, 243 were analyzed for 22 alkane and 43 polynuclear aromatic hydrocarbon (PAH) analytes. Results show that PAHs from *Exxon Valdez* crude oil contaminated seawater to a depth of at least 5 m and at concentrations up to 6.24 $\pm$ 0.63 μg/L at nine contiguously located stations in the region encompassing northwest Montague Strait–northern Knight Island–Smith Island, 1–2 weeks after the grounding. Highest PAH concentrations were near heavily oiled beaches, which apparently were a reservoir of spilled oil that continued to contaminate adjacent seawater at PAH concentrations up to 1.59 μg/L after 5 weeks. Polynuclear aromatic hydrocarbons from *Exxon Valdez* crude oil were available to marine fauna during the first few weeks after the grounding, especially in nearshore waters. Measured PAH concentrations, however, were well below levels that are acutely toxic to marine fauna.

Weather conditions after the *Exxon Valdez* oil spill provided a unique opportunity to investigate hydrocarbon contamination of seawater from a large, coldwater oil spill. In comparison with what occurs after oil spills in more temperate conditions, ambient seawater and air temperatures of about 4°C after the *Exxon Valdez* oil spill resulted in decreased evaporation and increased solubility of petroleum hydrocarbons. After 3 d of calm weather, a storm occurred; thus, spilled oil initially formed a continuous slick on the sea surface and then was thoroughly mixed with subsurface seawater by wave action. These conditions promoted solution of sparingly soluble petroleum hydrocarbons into the seawater. Thus, concentrations of these hydrocarbons below the oil slick should have been among the highest observed directly beneath an oil spill.

The effects of this spill are scientifically significant because the surface waters of Prince William Sound before the spill were nearly pristine (Karinen et al. 1993). Confounding petroleum hydrocarbons from sources other than this oil spill were generally absent. Consequently, the chemical fate and biological effects of oil from the *Exxon Valdez* oil spill should be unambiguous.

We sampled seawater along and adjacent to the path of the oil spill in Prince William Sound for 5 weeks, beginning 1 week after the grounding of the *Exxon Valdez*, to determine the concentrations of petroleum hydrocarbons dissolved and dispersed down into the water column. Our objective was to document petroleum hydrocarbon concentrations at depths of 1 m and 5 m. We collected 501 samples of seawater at 32 locations; 243 of these samples were analyzed for 22 alkane and 43 polynuclear aromatic hydrocarbon (PAH) analytes.

Methods

Methods of sample collection, preparation, and analysis are presented in detail by Short and Harris (1993). A brief description of these protocols follows.

Sample Collection and Extraction

Thirty-two stations were sampled along or adjacent to the oil path inside Prince William Sound at least once in three sampling periods between 31 March and 8 May 1989. Each period lasted 4–7 d. Station locations, path of the spilled *Exxon Valdez* oil, and prevailing ocean currents (Galt et al. 1991) are shown in Figure 1.

At each sampling station and time, triplicate seawater samples were collected at both 1-m and 5-m depths. Samples were collected by pumping seawater through a stainless steel tube into a 2-L vacuum flask with a peristaltic suction pump. The seawater sample contacted only stainless steel or glass during collection and was not exposed to hydrocarbon vapors. The sampling tube was protected from surface

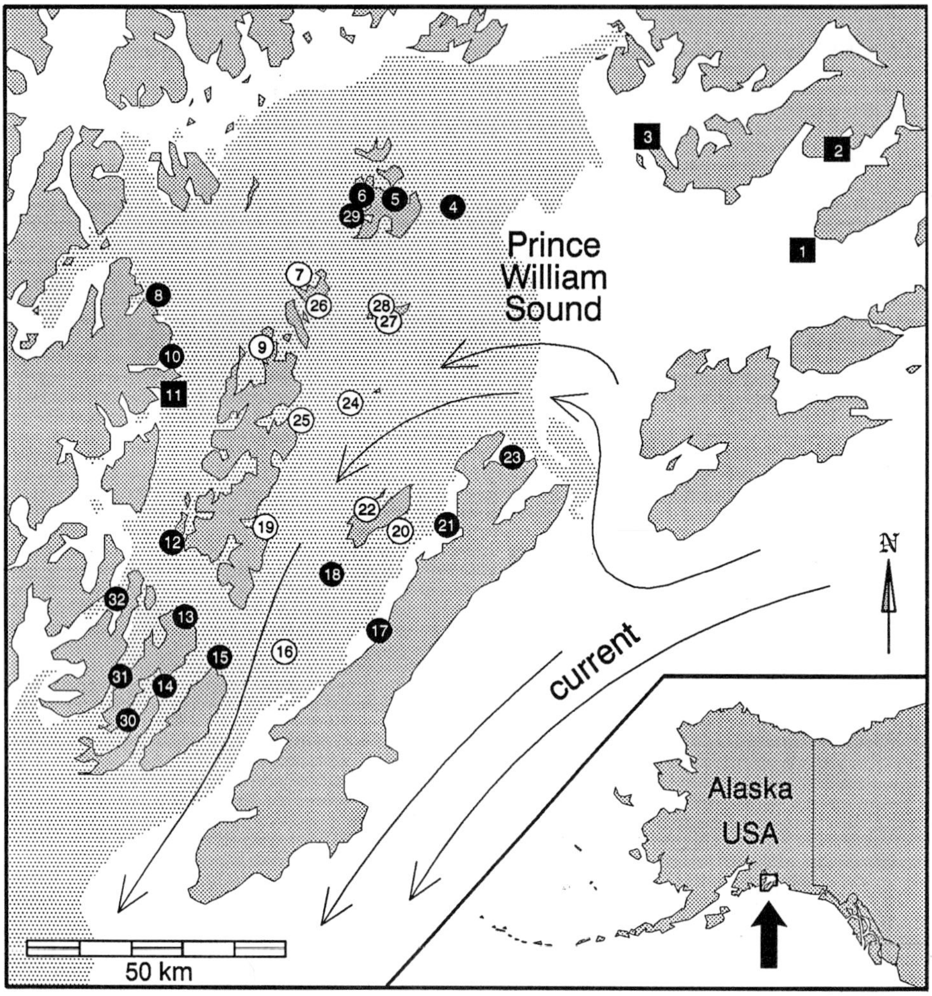

FIGURE 1.—Station locations in Prince William Sound where subsurface seawater hydrocarbons were measured after the *Exxon Valdez* oil spill of 24 March 1989. Station numbers within symbols correspond to the station numbers of Tables 2, 3, and 4. Arrows indicate the direction of prevailing oceanic currents through the sound (Galt et al. 1991). Shaded area indicates the path of surface oil through the sound. Open circles are stations where polynuclear aromatic hydrocarbons from the spilled oil were definitely detected. Solid squares are control stations. Solid circles are other stations sampled.

oil-slick contamination by covering the penetrating end of the tube with a plastic cover that was subsequently forced off with air pressure after the tube was at the sampling depth.

Each 900-mL seawater aliquot collected was extracted twice with a total of 75 mL dichloromethane within 1 min of collection. The dichloromethane used for extractions during the first sampling period was less pure than that used in subsequent periods, but it was used because of its availability. The combined extracts of each sample were stored at ambient temperature (approximately 5°C) for up to 5 d and then cooled to −20°C. The sampling apparatus was rinsed three times with dichloromethane before sampling the next depth and location.

Seawater sample extracts were prepared and analyzed by procedures described in detail by Short et al. (1996, this volume).

Polynuclear Aromatic Hydrocarbon Quality Assurance

Method detection limits (MDLs) were experimentally estimated for each calibrated PAH analyte (listed in Table 1) by procedures described by Glaser et al. (1981) and generally ranged from 0.020 to 0.040 μg/L. Detection limits for uncalibrated PAH analytes were arbitrarily assumed as the MDL of the most closely related calibrated PAH analyte.

Accuracy and precision of the analysis were evaluated based on results from analyzed reference samples. Mean accuracy, determined as the ratio of the mean amount of calibrated PAH in the reference samples of the 21 batches analyzed for this study and the amount added, ranged from 88.8% to 107.3%. Corresponding coefficients of variation (CVs, ratio of the SD and the mean of the calibrated PAHs in reference samples) ranged from 2.7% to 17.2% (median, 7.4%). Analytes that were associated with deuterated, surrogate standard recoveries of <40% or >150% were treated as missing data, affecting 2.1% of the PAH data.

Calibrated PAHs above MDLs were detected six times in analysis of 25 calibrated PAHs in each of the 21 method blanks. Four of these were attributable to naphthalene. Mean concentration of naphthalene in the dichloromethane used during the first sampling period was equivalent to 0.137 ± 0.019 μg/L seawater (95% confidence interval). Both 1- and 2-methylnaphthalene were also consistently measured in this dichloromethane at 0.018 ± 0.004 and 0.022 ± 0.003 μg/L, respectively, which were just below the MDLs of these PAHs. Uncalibrated PAHs above estimated MDLs were detected 31 times in the analysis of 18 uncalibrated PAHs in each of the 21 method blanks.

Alkane Hydrocarbon Quality Assurance

Method detection limits were not estimated for alkanes. Precision of the analysis was evaluated based on results from the two reference samples analyzed with each batch. Mean accuracy, determined as the ratio of the mean amount of alkane found in one milliliter samples of NIST-supplied standard labeled QA-CH-2 analyzed for this study and the amount present, ranged from 86.2% to 103.8%. Corresponding CVs ranged from 5.0% to 16.0% (median, 6.7%).

Alkanes were detected 23 times at equivalent seawater concentrations greater than 100 ng/L in the analysis of 21 alkanes in each of the 21 method blanks. Eleven of these were attributable to decane, undecane, and dodecane inconsistently detected in the method blanks prepared from remaining dichloromethane that was used to extract samples in the first sampling period.

TABLE 1.—Polynuclear aromatic hydrocarbons (PAHs) and alkane hydrocarbons measured for this study. Asterisk indicates PAHs used to calculate product-moment correlation coefficient. Abbreviations are C = calibrated PAH and U = uncalibrated PAH.

PAHs	U or C	Alkanes
Naphthalene	C	*n*-Decane (C10)
*2-Methylnaphthalene	C	*n*-Undecane (C11)
*1-Methylnaphthalene	C	*n*-Dodecane (C12)
*Biphenyl	C	*n*-Tridecane (C13)
2,6-Dimethylnaphthalene	C	*n*-Tetradecane (C14)
*C-2 Naphthalenes	U	*n*-Pentadecane (C15)
Acenaphthylene	C	*n*-Hexadecane (C16)
Acenaphthene	C	*n*-Heptadecane (C17)
2,3,5-Trimethylnaphthalene	C	Pristane
*C-3 Naphthalenes	U	*n*-Octadecane (C18)
*C-4 Naphthalenes	U	Phytane
*Fluorene	C	*n*-Nonadecane (C19)
*C-1 Fluorenes	U	*n*-Eicosane (C20)
*C-2 Fluorenes	U	*n*-Heneicosane (C21)
*C-3 Fluorenes	U	*n*-Docosane (C22)
*Dibenzothiophene	C	*n*-Tricosane (C23)
*C-1 Dibenzothiophenes	U	*n*-Tetracosane (C24)
*C-2 Dibenzothiophenes	U	*n*-Pentacosane (C25)
*C-3 Dibenzothiophenes	U	*n*-Hexacosane (C26)
*Phenanthrene	C	*n*-Heptacosane (C27)
Anthracene	C	*n*-Octacosane (C28)
1-Methylphenanthrene	C	*n*-Nonacosane (C29)
*C-1 Phenanthrene/Anthracenes	U	*n*-Triacontane (C30)
*C-2 Phenanthrene/Anthracenes	U	*n*-Dotriacontane (C32)
*C-3 Phenanthrene/Anthracenes	U	*n*-Tetratriacontane (C34)
*C-4 Phenanthrene/Anthracenes	U	
Fluoranthene	C	
Pyrene	C	
C-1 Fluoranthene/Pyrenes	U	
Benz[*a*]anthracene	C	
Chrysene	C	
C-1 Chrysenes	U	
C-2 Chrysenes	U	
C-3 Chrysenes	U	
C-4 Chrysenes	U	
Benzo[*b*]fluoranthene	C	
Benzo[*k*]fluoranthene	C	
Benzo[*e*]pyrene	C	
Benzo[*a*]pyrene	C	
Perylene	C	
Indeno[1,2,3-*c,d*]pyrene	C	
Dibenzo[*a,h*]anthracene	C	
Benzo[*g,h,i*]perylene	C	

Data Analysis

The similarity of relative PAH concentrations measured in seawater and measured in *Exxon Valdez* oil (EVO) was determined with the product-moment correlation coefficient, based on concentrations of 19 selected PAHs (Table 1). The 19 PAH concentrations in a sample of seawater or EVO were considered components of a vector for that seawater sample or for EVO. The correlation coefficient was calculated as the inner product of these two normalized vectors. Mean sample PAH concentrations were used when available. Vector components were reduced by the number of these 19 selected PAHs that were below the MDL in the sample considered.

The significance of these correlation coefficients was evaluated with Student's *t*-test, where *t* is calculated as

$$t = r\sqrt{(n - 2)/(1 - r^2)}$$

and compared with critical values of *t* with $n - 2$ degrees of freedom; *r* is the calculated correlation coefficient. Distributions of PAH in a sample were considered significantly correlated with distributions in EVO if the calculated *t* exceeded the critical value of *t* at the type I probability level of 0.05 or less, with at least 9 of the 19 selected PAHs detected above MDLs in the sample.

The 19 PAHs were selected based on three criteria: (1) presence in EVO at readily detectable relative concentration, (2) absence of extraneous contamination sources (this excludes naphthalene), and (3) linear independence. This latter criterion precluded using both 1-methylphenanthrene and C-1 phenanthrenes, for instance, because these were not linearly independent. When there was a choice between calibrated and uncalibrated PAHs, we selected uncalibrated PAHs.

Total PAHs attributable to EVO (EVO-PAHs) were estimated in samples in which PAHs were highly significantly ($P < 0.01$) correlated with corresponding PAHs in EVO. This estimate was the sum of measured PAHs used to calculate *r* plus an amount of naphthalene proportional to measured amounts of 1- and 2-methylnaphthalene, consistent with this same proportion in EVO. In every case, the naphthalene calculated this way as attributable to EVO was less than the measured amount of naphthalene in the sample.

We calculated an aromatic hydrocarbon enrichment factor (AHEF) as an indicator of dissolved EVO-PAHs. This AHEF was calculated for samples containing EVO-PAHs as the ratio of measured EVO-PAH concentration and a mean predicted EVO-PAH concentration based on the alkane measurements for the sample. The mean predicted EVO-PAH concentration for a sample was calculated in three steps. First, the ratio of EVO-PAHs and *n*-nonadecane (C-19) in crude oil was multiplied by the measured concentration of *n*-nonadecane in the sample, yielding a predicted EVO-PAH concentration for the sample based on *n*-nonadecane. Second, this calculation was repeated for each larger *n*-alkane through *n*-heptacosane (C-27), yielding a total of nine predicted EVO-PAH concentrations for the sample, based on the nine *n*-alkanes from *n*-nonadecane through *n*-heptacosane. Finally, the mean predicted EVO-PAH concentration was calculated as the mean of the three lowest predicted EVO-PAH concentrations calculated for the sample. Alkanes in the range C-19 through C-27 were used because they were probably the least abundant alkanes coming from natural sources in Prince William Sound seawater (Karinen et al. 1993), they were abundant in EVO, and they are both nonvolatile and insoluble in seawater so they may serve as marker compounds for particulate EVO. A mean of the three lowest predicted EVO-PAH concentrations was calculated to minimize bias caused by the possible presence of these *n*-alkanes from natural sources.

Results

The concentrations of calibrated PAHs, uncalibrated PAHs, and alkanes determined for this study, as well as results of analysis of crude oil from the hold of the *Exxon Valdez*, are listed in Short and Harris (1993). The following text provides a summary and interpretation of these results.

Polynuclear Aromatic Hydrocarbons

Polynuclear aromatic hydrocarbons consistent with EVO were clearly present in seawater in the first sampling period at 11 contiguously located stations (Table 2). These stations included northern and eastern Knight Island, Eleanor Island, Smith Island, Green Island, and northern and western Montague Strait. Beaches of these islands were among the first to be contaminated by the spilled oil.

These 11 stations all satisfied two criteria: (1) at least 9 of the 19 selected PAHs were detected at one of the depths sampled during the first sampling period, and (2) these PAHs were significantly correlated ($P < 0.05$) with corresponding PAHs in EVO. Polynuclear aromatic hydrocarbon analytes were detected in these samples according to their relative concentration in EVO. In samples in which only nine selected PAHs were detected, these generally corresponded to the nine highest relative PAH concentrations in EVO (Short and Harris 1993). Every station where more than 10 selected PAHs were detected during the first sampling period is included in Table 2, indicating that whenever numerous PAHs were detected, they were significantly correlated with corresponding PAHs in EVO.

The 11 stations listed in Table 2 constituted less

TABLE 2.—Seawater sampling stations and depths where measured polynuclear aromatic hydrocarbons (PAHs) were significantly correlated with *Exxon Valdez* oil PAHs after the grounding of the *Exxon Valdez* in Prince William Sound. Listed for each of three sampling periods in 1989 is the correlation coefficient, *r*, of (*N*) selected PAHs in *Exxon Valdez* oil with corresponding PAHs detected above method detection limits at the station and depth, and the type I error probability, P ($N - 2$ df). The abbreviation N/A indicates samples were not analyzed. A blank space indicates fewer than nine selected PAH analytes were detected (see *Methods*). The time periods for the three sampling periods were 31 March–4 April, 12–15 April, and 2–8 May, respectively.

Sampling station (station number)	Depth (m)	First sampling period			Second sampling period			Third sampling period		
		r	(*N*)	*P*	*r*	(*N*)	*P*	*r*	(*N*)	*P*
Adjacent to heavily oiled beaches										
Northwest Bay (7)	1	0.872	(18)	<0.001	0.860	(16)	<0.001	0.917	(15)[a]	<0.001
	5	0.907	(18)	<0.001	0.661	(10)	<0.05		(8)	
Herring Bay (9)	1	0.906	(14)	<0.001	0.913	(18)	<0.001	0.952	(13)	<0.001
	5	0.930	(16)	<0.001	0.782	(9)	<0.05		(3)	
Snug Harbor (19)	1	0.888	(14)	<0.001	0.941	(17)	<0.001	−0.122	(12)	NS
	5		(6)		0.982	(13)	<0.001		(1)	
Bay of Isles (25)	1	0.864	(12)	<0.001		N/A			(3)	
	5	0.708	(15)	<0.01		N/A			(1)	
Southeast Eleanor Island (26)	1	0.952	(18)	<0.001	0.951	(16)	<0.001	0.542	(15)	<0.05
	5	0.964	(17)	<0.001	0.823	(14)	<0.001	0.307	(10)	NS
N Smith Island (28)	1	0.915	(17)	<0.001	0.888	(13)	<0.001		(7)	
	5	0.926	(12)	<0.001	0.604	(11)[a]	<0.05		(7)	
Other stations										
S Montague Strait (16)	1		(7)			N/A			N/A	
	5	0.949	(9)	<0.001		N/A			N/A	
S Green Island (20)	1		(2)			N/A			(4)	
	5	0.667	(10)	<0.05		N/A			(1)	
N Green Island (22)	1		(2)			N/A			(1)	
	5	0.543	(15)	<0.05		N/A			(4)	
N Montague Strait (24)	1	0.945	(13)	<0.001		N/A			(5)	
	5	0.905	(15)	<0.001		N/A			(1)	
S Smith Island (27)	1	0.960	(18)	<0.001		(7)			(7)	
	5	0.931	(13)	<0.001		(8)		0.609	(11)	<0.05

[a]C-1 fluorenes not included in calculation of *r*; concentration of C-1 fluorenes in this sample was anomalously high.

than 40% of the stations sampled, yet they accounted for more than 85% of the calibrated PAHs detected in the first sampling period (naphthalene excluded). Also, PAHs that were not present at readily detectable relative concentrations in EVO were not detected in samples from any of these 11 stations. In particular, unsubstituted PAHs with four or five aromatic rings, which are characteristic of combustion products, were not detected in any samples in this study.

Analytes consistent with EVO were identified at fewer of the sampling stations after the first sampling period. In the second sampling period, PAHs in samples were strongly correlated ($P < 0.001$) with PAHs in EVO at 5 of the 11 initially identified stations; however, PAHs were not determined at another 5 of these stations (Table 2). In the third sampling period, PAHs in samples were strongly correlated at only 2 of the 11 initially identified stations, with PAHs not determined at only 1 of the remaining stations (Table 2). Analytes consistent with EVO were never identified at any station in the second or third sampling periods that were not also identified in the first period. In addition, more of the selected PAHs were usually detected in the first period than in later periods at all stations and depths.

In the first sampling period, the highest summed concentrations of EVO-PAHs were adjacent to beaches that were heavily oiled. Concentrations of EVO-PAHs at these stations ranged from 1.26 ± 0.40 μg/L seawater (mean ± 95% confidence interval, $N = 3$ here and in the cases that follow, except where noted) to 6.24 ± 0.63 μg/L seawater (Table 3). Concentrations of EVO-PAHs were somewhat lower at stations listed in Table 2 that were not adjacent to heavily contaminated beaches, ranging from 0.91 ± 1.81 μg/L ($N = 2$) to 2.23 ± 0.62 μg/L. Mean concentrations of EVO-PAHs were usually somewhat higher at the 1-m depth than at the 5-m depth, but often not significantly higher.

Concentrations of EVO-PAHs generally declined after the first sampling period. These concentrations were usually lower by a factor of about 2 from the first to the second sampling periods (Table 3). By the third sampling period, PAHs could be attrib-

TABLE 3.—Total polynuclear aromatic hydrocarbons (PAHs) attributed to *Exxon Valdez* oil (EVO) at 1-m and 5-m depths in Prince William Sound after the grounding of the *Exxon Valdez*. Polynuclear aromatic hydrocarbons were attributed to EVO if they were highly correlated ($P < 0.01$) with PAHs in EVO (see Table 2). Listed below are the sums of the seawater PAH concentrations (μg/L) attributed to EVO, calculated as described in *Methods*, together with sampling station, depth, and sampling period. Concentrations for the first sampling period are means of triplicate samples ±95% confidence intervals; others are based on single samples. The abbreviation N/A indicates samples were not analyzed. A blank space indicates PAHs were not sufficiently correlated with EVO.

Sampling station (station number)	Depth (m)	First sampling period	Second sampling period	Third sampling period
Adjacent to heavily oiled beaches				
Northwest Bay (7)	1	4.72 ± 1.18	2.18	1.59[a]
	5	3.06 ± 0.59	1.29	
Herring Bay (9)	1	2.86 ± 2.77[b]	4.00	0.92
	5	2.46 ± 0.50	0.83	
Snug Harbor (19)	1	6.24 ± 0.63	2.41	
	5		1.03	
Bay of Isles (25)	1	1.31 ± 0.34	N/A	
	5	1.58 ± 2.65[b]	N/A	
Southeast Eleanor Island (26)	1	3.08 ± 0.66	1.60	
	5	2.80 ± 0.67	1.64	
N Smith Island (28)	1	2.51 ± 0.56	1.48	
	5	1.26 ± 0.40		
Other stations				
S Montague Strait (16)	1	0.96 ± 0.22	N/A	N/A
	5	0.91 ± 1.81[b]	N/A	N/A
N Montague Strait (24)	1	1.11 ± 0.34	N/A	
	5	1.10 ± 0.49	N/A	
S Smith Island (27)	1	2.23 ± 0.62		
	5	1.12 ± 0.32		

[a]C-1 fluorenes not included in calculation of total; concentration of C-1 fluorenes in this sample was anomalously high.
[b]Based on duplicate samples; third sample not analyzed.

uted to EVO only at the 1-m depth at two stations that were adjacent to heavily contaminated beaches (Table 3).

Four stations, Olsen Bay (station 2), Gravina Point (1), Snug Corner Cove (3), and Granite Bay (11), were outside the path of the spilled oil (Figure 1) and were considered control stations to evaluate hydrocarbon concentrations from sources not related to the oil spill. Except for naphthalene and C-2 naphthalenes, PAHs, especially calibrated PAHs, were rarely detected at control stations in the first sampling period. Naphthalene and C-2 naphthalenes were consistently detected at all four stations and at both sampling depths. Naphthalene concentrations were elevated, ranging from 0.323 ± 0.077 μg/L (Granite Bay, 1 m) to 0.778 ± 0.637 μg/L (Gravina Point, 5 m), whereas C-2 naphthalenes were lower, ranging from 0.034 ± 0.034 μg/L (Granite Bay, 1 m) to 0.215 ± 0.814 μg/L (Gravina Point, 1 m). Only one other calibrated PAH (dimethylnaphthalene) was detected at one of these stations at a low concentration of 0.050 ± 0.054 μg/L at Gravina Point, 5 m. Other uncalibrated PAHs were detected sporadically at low concentrations at these control stations; the detection frequency ranged from 0 (Snug Corner Cove, 5 m) to 9 (Gravina Point, 5 m). At these stations, PAHs were never significantly correlated with corresponding PAHs of EVO ($P > 0.3$).

Calibrated PAHs were detected more frequently at control stations in the second and third periods. Both 1- and 2-methylnaphthalene were detected at least at one depth at all the control stations except Gravina Point, which was not sampled in the second or third periods. Concentrations of these PAHs ranged up to 0.076 μg/L. Also, biphenyl was detected at Olsen Bay and at Snug Corner Cove at concentrations ranging up to 0.055 μg/L. Otherwise, the distribution, detection frequency, and concentrations of detected PAHs at these stations were similar to those of the first sampling period, with one notable difference: nine of the selected PAHs were detected at Olsen Bay (5 m) in the third period, and these were significantly correlated with PAHs in EVO ($P < 0.05$).

The distribution, detection frequency, and concentrations of detected PAHs at the other stations of this study not listed in Table 2 were usually not different from those at control stations. At these stations, too few PAHs were detected to reliably identify sources, and detected PAHs were generally at low concentrations. Naphthalene, however, was consistently detected at all stations and at both depths at concentrations usually well above the MDL. The C-2 naphthalenes were almost always detected, except in the third sampling period, although concentrations were often low.

The AHEF calculated for 9 of the 11 stations listed in Table 2 indicated the presence of both particulate oil and dissolved PAHs at these stations. The AHEF values for these stations in the first sampling period ranged from 0.70 to 6.07 (Table 4). Values near 1 indicate that PAHs were associated with particulate oil; values substantially greater than 1 indicate the presence of predominantly dissolved PAHs. The enrichment factor was substantially greater than 1 at Northwest Bay, Herring Bay, and Snug Harbor, ranging from 3.90 to 6.07 in the first sampling period; it was generally closer to 1 at the other stations (Table 4). Values tended to re-

TABLE 4.—Aromatic hydrocarbon enrichment factors (AHEFs) of *Exxon Valdez* oil PAHs (EVO-PAHs) at 1-m and 5-m depths in Prince William Sound after the grounding of the *Exxon Valdez*. The AHEF is the ratio of measured EVO-PAHs and expected EVO-PAHs, based on measurements of *n*-alkanes from nonadecane through heptacosane in samples of seawater and EVO (see *Methods*). The AHEF is listed below for each station and sampling period where EVO-PAHs were determined (Table 3). The abbreviation N/A indicates samples were not analyzed. A blank space indicates EVO-PAHs were not determined.

Sampling station (station number)	Depth (m)	First sampling period	Second sampling period	Third sampling period
Northwest Bay (7)	1	4.42	3.91	1.84
	5	3.90	2.53	
Herring Bay (9)	1	4.19	3.74	0.42
	5	4.84	1.59	
Snug Harbor (19)	1	6.07	4.90	
	5		2.08	
Bay of Isles (25)	1	1.32	N/A	
	5	3.37		
Southeast Eleanor Island (26)	1	0.86	3.75	
	5	1.14	3.74	
N Smith Island (28)	1	1.25	4.78	
	5	2.48		
S Montague Strait (16)	1	1.41	N/A	N/A
	5	1.33	N/A	N/A
N Montague Strait (24)	1	0.70	N/A	
	5	1.12	N/A	
S Smith Island (27)	1	0.76		
	5	0.87		

main well above 1 at the stations listed in Table 2 with sufficient data in the second sampling period, ranging from 1.59 to 4.90.

Alkanes

Relative alkane concentrations changed substantially from the first to the third sampling periods at stations where oil contamination was negligible. We calculated mean alkane concentrations for six of these stations sampled in the first and third sampling periods (Figures 2 and 3, respectively): Eshamy Bay, Granite Bay, Main Bay, Olsen Bay, Port Chalmers, and Snug Corner Cove. These six stations were near the margin or outside the spill path and were sampled in both periods. In the first period, the normal alkanes C-19, C-21, and C-25 through C-34 were prominent at these stations (Figure 2), whereas the other normal alkanes and pristane were detected only sporadically at low concentrations. By the third period, however, concentrations of many alkanes (especially C-10, C-11, and pristane) increased substantially (Figure 3).

Alkane concentrations were generally elevated in the first sampling period at stations identified as contaminated by EVO in Tables 3 and 4, especially at the stations where the AHEF was relatively low. At stations and depths listed in Table 4 where the AHEF was greater than 2, mean concentrations of all alkanes except C-19 were consistently higher than corresponding alkanes at stations where EVO contamination was negligible (Figures 2 and 4). At stations and depths where the AHEF was less than 2, mean alkane concentrations—especially the normal alkanes from C-19 through C-26—were usually higher still (Figures 4 and 5). At both contaminated and uncontaminated stations in the first sampling period, a local, relative minimum normal alkane concentration occurred at C-22 or C-23.

By the third sampling period, alkane concentrations at uncontaminated stations were generally similar to concentrations at stations where contamination by EVO had been detected (Short and Harris 1993).

Discussion

Polynuclear Aromatic Hydrocarbons

Our results indicate that PAH concentrations in seawater at the stations listed in Table 2 are attributable to EVO (although additional sources of napthalene are indicated). We reached this conclusion because (1) the identity and concentrations of PAHs in water samples were significantly correlated with PAHs in EVO, (2) the geographic distribution of PAHs in water in Prince William Sound corresponded to the path of the spilled oil, (3) PAHs were generally absent from areas marginal or distant from the path of the spilled oil, (4) PAHs persisted in water adjacent to heavily oiled beaches in the second and third sampling periods, and (5) there was no evidence for a plausible alternative explanation for the observed results.

The contiguous location of the stations listed in Table 2 and the similarity of 1-m and 5-m PAH concentrations in the first sampling period indicate a large mass of contaminated seawater in the region of northwest Montague Strait–northern Knight Island–Smith Island. The storm that occurred 3–5 d after the tanker grounding may have created this contaminated water mass by generating persistent, small droplets of the spilled crude oil in the seawater column that were being flushed out of the sound during the first sampling period. This phenomenon would account for several results: (1) the geographic contiguity of the stations with elevated PAHs, including the open-water stations in Montague Strait; (2) the AHEFs of the open-water sta-

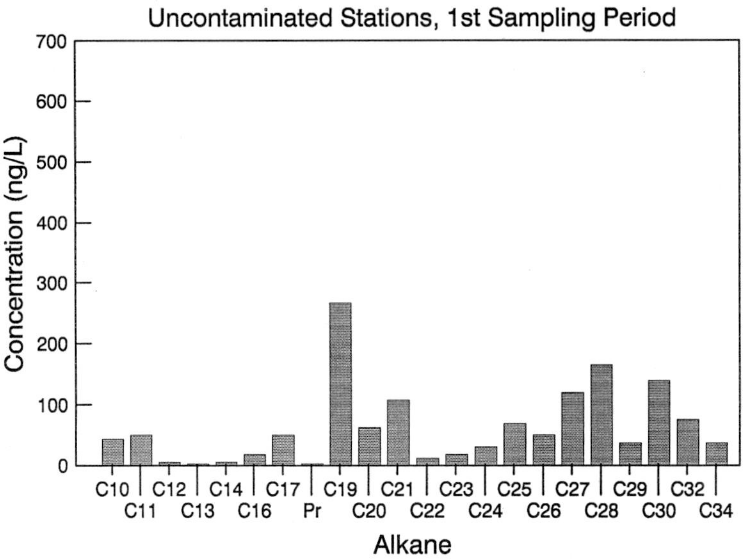

FIGURE 2.—Mean alkane concentrations (ng/L seawater) for six stations where oil contamination was negligible during the first sampling period (31 March–4 April 1989). Each mean alkane concentration is based on samples from Eshamy Bay, Granite Bay, Main Bay, Olsen Bay, Port Chalmers, and Snug Corner Cove ($N = 36$) taken during the first sampling period. The abbreviation Pr stands for pristane. Data are from Table III-3 in Appendix III of Short and Harris (1993).

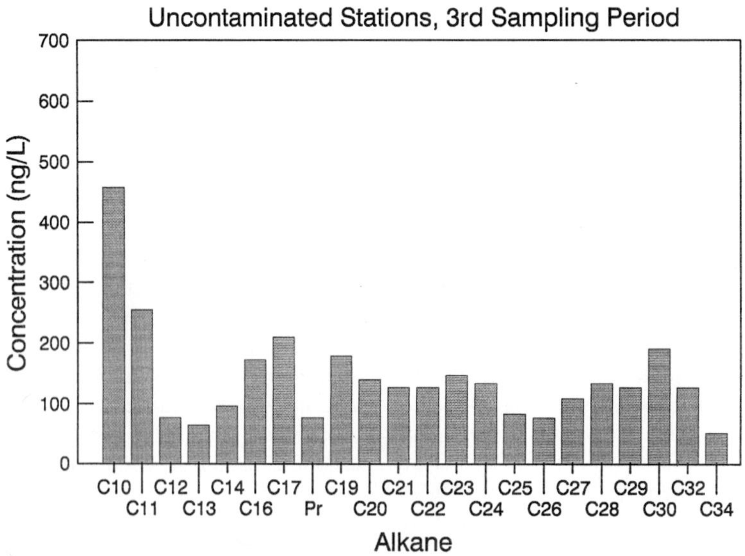

FIGURE 3.—Mean alkane concentrations (ng/L seawater) for six stations where oil contamination was negligible during the third sampling period (2–8 May 1989). Each mean alkane concentration is based on samples from Eshamy Bay, Granite Bay, Main Bay, Olsen Bay, Port Chalmers, and Snug Corner Cove ($N = 12$) taken during the third sampling period. The abbreviation Pr stands for pristane. Data are from Table III-9 in Appendix III of Short and Harris (1993).

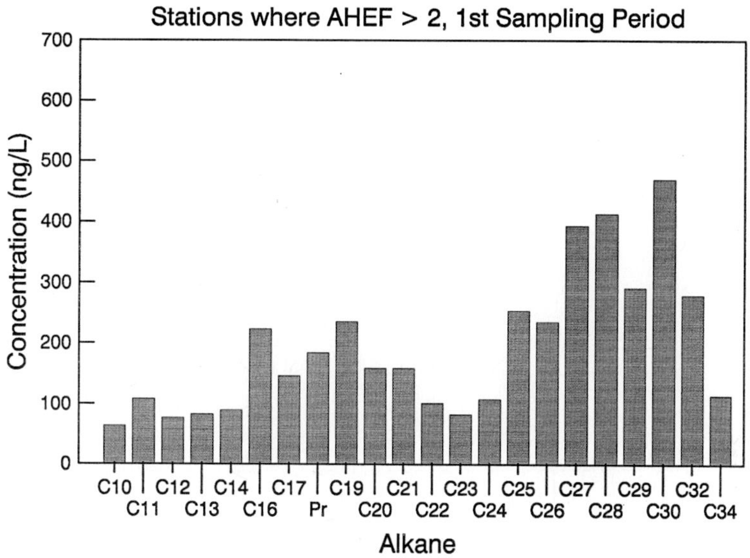

FIGURE 4.—Mean alkane concentrations (ng/L seawater) for first sampling period stations and depths listed in Table 4 where the aromatic hydrocarbon enrichment factor (AHEF) is greater than 2. Each mean alkane concentration is based on samples from Bay of Isles (5 m), Herring Bay, N Smith Island (5 m), Northwest Bay, and Snug Harbor (1 m) ($N = 19$) taken during the first sampling period (31 March–4 April 1989). The abbreviation Pr stands for pristane. Data are from Table III-3 in Appendix III of Short and Harris (1993).

tions (near 1), which were consistent with PAHs associated with particulate oil; (3) the similarity of PAH concentrations at 1-m and 5-m depths, indicating nearly homogeneous mixing of dispersed oil to a depth of at least 5 m; and (4) the general absence of EVO-PAHs at the open-water sites by the second sampling period.

Heavily oiled beaches apparently were reservoirs of spilled EVO that continued to contaminate adjacent seawater after the first sampling period. Stations where elevated seawater EVO-PAHs were detected in the second and third sampling periods were stations that had been contaminated in the first sampling period, were among the first beaches to be heavily oiled by the spill, and were enclosed within bays. These conditions probably resulted in the stable association of relatively unweathered oil with these beaches. The PAHs in this oil could reenter adjacent seawater as oil-coated particles mixed with seawater by wave action during subsequent storms or as PAHs that dissolve in seawater from beached oil or refloated surface slicks. The high AHEFs associated with the elevated PAH concentrations at these stations indicate that dissolution of PAHs was important at these stations.

Extraneous contamination of these samples with oil (e.g., by initial contact of the sampling apparatus with the oil slick) was unlikely for several reasons. Stations with low concentrations of seawater PAHs were interspersed in the sampling sequence with stations with high PAH concentrations, indicating that PAHs at contaminated stations did not contaminate samples from uncontaminated stations. Analysis of method blanks and the extraction solvent used in the first sampling period obviate laboratory contamination or the solvent as significant PAH sources. We observed almost no other vessel traffic in most of the areas contaminated by the spilled oil in the first sampling period when we measured the highest PAH concentrations, making refined petroleum products associated with marine transportation an unlikely source of extraneous contamination. Finally, combustion sources can be eliminated as significant because unsubstituted 4-to-5-ring aromatic hydrocarbons characteristic of combustion sources were not detected. The quality assurance results indicated that these compounds would have been detected if they were present at concentrations indicative of significant combustion sources. Therefore, all our results were consistent with the identification of EVO as the source of the elevated PAHs we detected.

The concentrations of PAHs measured in this study were consistent with similar measurements of

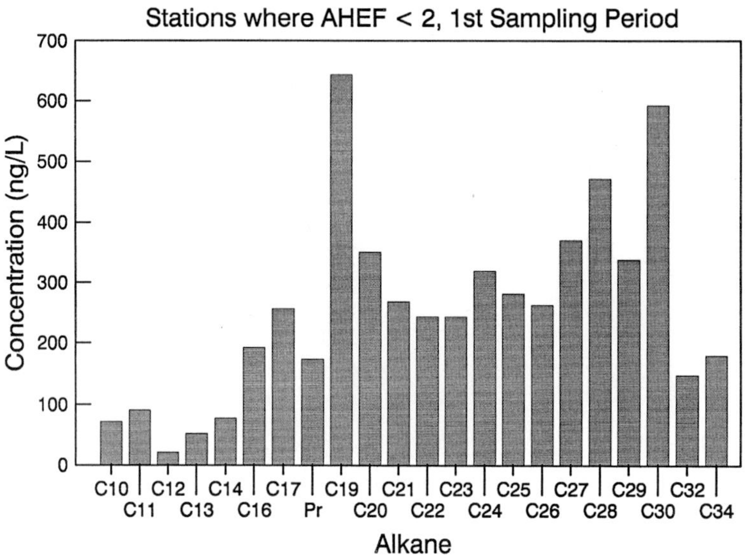

FIGURE 5.—Mean alkane concentrations (ng/L seawater) for first sampling period stations and depths listed in Table 4 where the aromatic hydrocarbon enrichment factor (AHEF) is less than 2. Each mean alkane concentration is based on samples from Bay of Isles (1 m), N Montague Strait, N Smith Island (1 m), Southeast Eleanor Island, S Montague Strait, and S Smith Island (N = 29) taken during the first sampling period (31 March–4 April 1989). The abbreviation Pr stands for pristane. Data are from Table III-3 in Appendix III of Short and Harris (1993).

seawater PAH concentrations after other coldwater oil spills. For example, Law (1978) found that a suite of PAHs similar to the suite we measured ranged up to 1.7 μg/L and 1.1 μg/L at 1-m and 5-m depths, respectively, below the oil slick from the *Ekofisk* blowout in April 1977 in the North Sea; and work quoted therein indicates PAH concentrations ranged up to 8.1 μg/L near the source of the oil.

The generally decreasing detection frequency and concentrations of EVO-PAHs with time indicate the operation of well-known mechanisms of PAH dilution and decomposition. The prevailing ocean current that continuously flushes Prince William Sound may account for the dilution and removal of the high initial PAH concentrations, as may microbial decomposition (Ballerini et al. 1981). The generally decreasing EVO-PAH concentrations we observed suggest that these concentrations may have been higher before our first sampling period.

Additional sources of naphthalene apart from EVO were evident in all samples, suggesting sampling contamination sources for naphthalene only. This conclusion comes from the ubiquity of naphthalene at a minimum apparent concentration of about 0.4 μg/L in the first sampling period. A similar pattern of naphthalene detection persisted at lower minimum concentrations in the second and third sampling periods. Although some of this naphthalene resulted from contamination levels equivalent to about 0.13 μg/L in the dichloromethane used in the first sampling period, the remainder is indicative of additional unknown contamination sources.

Alkanes

Determination of alkanes derived from EVO was obscured by variable amounts of alkanes from diverse natural sources. The relatively high concentrations of C-19 and C-21 alkanes at stations where contamination was negligible probably were derived from marine phytoplankton, whereas the C-25 through C-34 alkanes are characteristic of terrestrial plant waxes. Alkanes C-19, C-21, and C-25–C-34 were frequently detected in mussels in Prince William Sound before this oil spill (Karinen et al. 1993). The general increase in alkane concentrations, including pristane, from the first to the third sampling periods, coincided with and probably was caused by the spring phytoplankton bloom that is characteristic of Alaska coastal marine waters. In particular, Karinen et al. (1993) noted a dramatic spring increase in pristane concentration in mussels in the sound in 1977–1980, which they speculated was associated with detrital material derived from marine copepods. Pristane associated in this manner would have been extracted and detected by our methods, and our observation of increased pristane concentration in seawater in the third sampling period compared with the first is consistent with a planktonic or detrital source.

The relatively high concentrations and temporal variability of the alkanes interfered with determination of the form of EVO in seawater. The presence of normal alkanes from C-19 through C-27 from biological sources may bias interpretation of the AHEF because predicted EVO-PAHs were calculated on the assumption that all of the alkanes at the three lowest concentrations within this interval came from EVO. Although we constructed our AHEF index to take advantage of the general relative minimum of alkanes between C-19 and C-27, the presence of additional alkanes from sources other than EVO would tend to bias interpretation of these AHEFs in favor of particulate oil. As a result, dissolved EVO-PAHs are probably more widespread than indicated in Table 4.

Mechanism of Dissolution

The relative concentrations of dissolved EVO-PAHs in the water samples suggest that dissolution kinetics and not solubility of individual PAHs determined the composition of dissolved PAHs in the water column. Relative dissolved PAH concentrations correlated strongly with those of EVO, indicating that the composition of dissolved PAHs matched that of the oil. The initial relative rates of dissolution of individual PAHs are determined by differences in their molecular attractive forces to the seawater phase and the remaining oil phase, as well as by the relative concentrations of the PAHs in the oil. Differences in these attractive forces among PAHs are approximately proportional to molecular surface area, which varies by less than a factor of 2 among the EVO-PAHs (naphthalene through C-4 phenanthrene). Because the molecular surface areas of the PAHs are virtually equivalent, the dissolution rate of these compounds in water can be attributed to the relative concentrations of the PAHs in oil, which vary by more than 30-fold.

Solubility, in contrast, is not an important factor because the volume of affected seawater in Prince William Sound was much greater than the volume of the spilled oil. Consequently, solubility limits on EVO-PAH concentrations were never approached. These kinetic considerations may be derived from a dynamic equilibrium interpretation of the equilibrium concentrations of sparingly soluble petroleum hydrocarbons in water at high ratios of water and oil phases (Shiu et al. 1988).

Relation to Biological Effects

Although readily detectable, concentrations of EVO-PAHs were well below levels that are acutely toxic to marine fauna (Brodersen et al. 1977; Rice et al. 1979). Our data, however, demonstrate that PAHs from *Exxon Valdez* crude oil were available to subsurface marine fauna during the first few weeks after the oil spill, especially in nearshore, near-surface waters that are particularly productive biologically.

Acknowledgments

The authors gratefully acknowledge the care devoted to the chemical analysis of these samples given by Marie Larsen, Larry Holland, and Josefina Lunasin at the Auke Bay Laboratory, and the assistance of Robert Weise, master of the FV *Kanak* during the first sampling cruise.

References

Ballerini, D., J. Ducreax, and J. Riviere. 1981. Laboratory simulation of the microbiological degradation of crude oil in a marine environment. Pages 27–36 *in* Ecological study of the *Amoco Cadiz* oil spill. Report of the NOAA (National Oceanic and Atmospheric Administration)-CNEXO (Centre National pour l'Exploitation des Oceans) Joint Scientific Commission. U.S. Department of Commerce, NOAA, and CNEXO, Paris.

Brodersen, C. C., S. D. Rice, J. W. Short, T. A. Mecklenburg, and J. F. Karinen. 1977. Sensitivity of larval and adult Alaskan shrimp and crabs to acute exposures of the water-soluble fraction of Cook Inlet crude oil. Pages 575–578 *in* Proceedings, 1977 oil spill conference (prevention, behavior, control, cleanup). American Petroleum Institute Publication 4284, Washington, DC.

Galt, J. A., W. J. Lehr, and D. L. Payton. 1991. Fate and transport of the *Exxon Valdez* oil spill. Environmental Science and Technology 25:202–209.

Glaser, J. A., D. L. Foerst, G. D. McKee, S. A. Qave, and W. L. Budde. 1981. Trace analyses for wastewaters. Environmental Science and Technology 15:1426–1435.

Karinen, J. F., M. M. Babcock, D. W. Brown, W. D. MacLeod, Jr., L. S. Ramos, and J. W. Short. 1993. Hydrocarbons in intertidal sediments and mussels from Prince William Sound, Alaska, 1977–1980: characterization and probable sources. NOAA (National Oceanic and Atmospheric Administration) Technical Memorandum NMFS (National Marine Fisheries Service)-AFSC-9, Alaska Fisheries Science Center, Seattle.

Law, R. J. 1978. Determination of petroleum hydrocarbons in water, fish, and sediments following the *Ekofisk* blow-out. Marine Pollution Bulletin 9:321–324.

Rice, S. D., D. A. Moles, T. T. Taylor, and J. F. Karinen. 1979. Sensitivity of 39 Alaskan marine species to Cook Inlet crude oil and no. 2 fuel oil. Pages 549–554 *in* Proceedings, 1979 oil spill conference (prevention, behavior, control, cleanup). American Petroleum Institute Publication 4308, Washington, DC.

Shiu, W. Y., A. Maijanen, A. L. Y. Ng, and D. Mackay. 1988. Preparation of aqueous solutions of sparingly soluble organic substances: II. multicomponent systems—hydrocarbon mixtures and petroleum products. Environmental Toxicology and Chemistry 7:125–137.

Short, J. W., T. J. Jackson, M. L. Larsen, and T. L. Wade. 1996. Analytical methods used for the analysis of hydrocarbons in crude oil, tissues, sediments, and seawater collected for the natural resources damage assessment of the *Exxon Valdez* oil spill. American Fisheries Society Symposium 18:140–148.

Short, J. W., and P. Harris. 1993. Petroleum hydrocarbons in near-surface seawater of Prince William Sound, Alaska, following the *Exxon Valdez* oil spill. I: chemical sampling and analysis. Natural Resources Damage Assessment Air/Water 3 Final Report. NOAA (National Oceanic and Atmospheric Administration)/NMFS (National Marine Fisheries Service) Office of Oil Spill Damage Assessment and Restoration, Auke Bay, Alaska.

American Fisheries Society Symposium 18:29–39, 1996

Petroleum Hydrocarbons in Caged Mussels Deployed in Prince William Sound after the *Exxon Valdez* Oil Spill

JEFFREY W. SHORT AND PATRICIA M. HARRIS

National Marine Fisheries Service, Alaska Fisheries Science Center, Auke Bay Laboratory, 11305 Glacier Highway, Juneau, Alaska 99801, USA

Abstract.—Mussels *Mytilus trossulus*, initially free of hydrocarbons, were deployed in nearshore waters along the path of oil spilled from the *Exxon Valdez* to determine the biological availability and persistence of petroleum-derived hydrocarbons. Mussels were deployed at 22 locations inside Prince William Sound and 16 locations outside the sound at depths of 1, 5, and 25 m for 2–8 weeks. We conducted four successive deployments in 1989 after the spill and two in 1990 and 1991. Mussels were analyzed for 23 alkane and 43 polynuclear aromatic hydrocarbon (PAH) analytes. Concentrations of PAHs in mussels deployed along the oil's path decreased with depth, time, and distance from heavily oiled beaches. The highest concentration of total PAHs was 45,200 ng/g dry tissue weight at Herring Bay, l-m depth, 1–2 months after the spill. Concentrations were nearly that high at north Smith Island and Snug Harbor. Lower concentrations with PAH composition similar to *Exxon Valdez* crude oil were detected at other locations inside Prince William Sound, except at the control site, Olsen Bay. Concentrations of PAHs in mussels deployed inside Prince William Sound declined by late summer 1989. In 1990 and 1991, PAHs could only be detected near heavily oiled beaches at concentrations that were usually below 500 ng/g. Alternative sources of petroleum hydrocarbons in the mussels were negligible compared with *Exxon Valdez* crude oil. Petroleum hydrocarbons similar to those of *Exxon Valdez* crude oil were detected only sporadically outside the sound in 1989 and were generally below detection limits in 1990 and 1991. Accumulation of hydrocarbons derived from *Exxon Valdez* crude oil by deployed mussels indicates that these hydrocarbons were available to subsurface marine fauna, such as fish and invertebrates during the summer after the spill, especially in shallow waters near oiled beaches. Oil accumulated from seawater at these trophic levels would serve as a route of oil exposure through ingestion by fauna at higher trophic levels.

Subsurface seawater concentrations of petroleum hydrocarbons immediately after the *Exxon Valdez* oil spill of 24 March 1989 in Prince William Sound, Alaska, were relatively low and generally decreased to the detection limits within a few weeks (Short and Harris 1996, this volume). After this time, additional petroleum hydrocarbons may have been mobilized into subsurface seawater by disturbance of persistent environmental reservoirs of oil, such as by the action of storm waves or cleaning on oiled beaches. These events may release pulses of oil into seawater, possibly associated with fine sediments, and these pulses may escape detection by direct chemical sampling because of both temporal and spatial heterogeneity. Nonetheless, these pulses of oil may present a chronic, low-concentration source of oil that is available to marine organisms through ingestion, with impacts that are not well understood.

Caged bay mussels *Mytilus trossulus* that are initially free of petroleum hydrocarbons are particularly well suited for monitoring sporadically distributed hydrocarbons in seawater (NRC 1980; Phelps and Galloway 1980; Wolfe et al. 1981). *Mytilus edulis*, a closely related species, can filter about 340 L of seawater daily (Wildish and Miyares 1990); hence, the volume of seawater effectively sampled by mussels in 1 month can be 10,000 times greater than that sampled for direct chemical analysis. The biological relevance of detected hydrocarbons in mussels is immediate in that the hydrocarbons are obviously available biologically to the mussels and other filter feeders and to the diverse fauna at higher trophic levels.

Mussels can concentrate hydrocarbons in their lipids by factors as high as 10^6 (Murray et al. 1991), indicating that depuration is slower than accumulation, which results in persistence of some accumulated hydrocarbons for several weeks (Clark and Finley 1975; Ganning et al. 1983; Pruell et al. 1986). Such "bioconcentrated" hydrocarbons are proportionally easier to detect, and they indicate hydrocarbon concentrations in seawater integrated over the deployment period. Mussels do not metabolize petroleum hydrocarbons significantly (Vandermeulen and Penrose 1978); thus, accumulated hydrocarbons may be characteristic of the exposure source.

We deployed mussels in cages along and adjacent to the path of the spilled oil usually for 1 month in the summers of 1989, 1990, and 1991 to determine the distribution of persistent petroleum hydrocarbons in the subsurface seawater column. The mussels originated from a pristine site in southeast Alaska and were verified as free of petroleum hydrocarbons before deployment. We deployed cages of mussels at 22 stations inside Prince William Sound and 16 stations outside the sound near the Kenai and Alaska peninsulas and Kodiak Island. Mussels from 286 cages were analyzed for 23 alkane and 43 polynuclear aromatic hydrocarbon (PAH) analytes.

Methods

A complete description of the methods used for this study is presented by Short and Harris (1995). In this article, we briefly summarize these methods.

Mussel Collection, Deployment, and Retrieval

Bay mussels were collected a few days before each deployment from Admiralty Island, southeast Alaska. Mussels larger than 45-mm shell length were transported to deployment stations within 5 d for inside Prince William Sound, and within 10 d for outside the sound. Samples of mussels were taken to evaluate hydrocarbon burdens before deployment in mussel cages, which consisted of nylon mesh bags, each containing 20 mussels.

Caged mussels were deployed at 38 stations along or adjacent to the oil spill's path (Table 1 and Figures 1 and 2) at least once in the study period (1989–1991). Twenty-two of these stations were located inside Prince William Sound, and 16 were adjacent to the Kenai and Alaska peninsulas and Kodiak Island. Cages were deployed at depths of 1, 5, and 25 m.

Caged mussels usually were deployed at these stations for about 1 month, but deployment periods ranged from 2 to 8 weeks depending on vessel availability. Mussels were deployed for 2 months inadvertently at Herring Bay in 1989 because the cage could not be located after 1 month but was located after 2 months.

Survival and condition index were determined for retrieved mussels to evaluate mussel health during deployment. Surviving mussels were stored frozen at −18°C until analysis. The condition index was adapted from Higgens (1938) and Crosby and Gale (1990) and was calculated by dividing dry tissue weight by shell volume and multiplying by 100. We assumed that a stable or increased condition index indicated that mussels filtered enough water during deployment to at least maintain their condition.

Native mussels were collected in 1989 and 1990 from intertidal areas near caged mussel sites at Herring Bay, Olsen Bay, Sleepy Bay, Smith Island, and Snug Harbor to compare levels of biologically available hydrocarbons in the nearshore water column to levels in the adjacent intertidal area.

Hydrocarbon Analysis and Data Analysis

Mussel samples were analyzed for selected hydrocarbons at the Auke Bay Laboratory. The analytes, methods, detection limits, and the accuracy and precision of the methods are summarized elsewhere in this volume (Short et al. 1996a, this volume).

Exxon Valdez crude oil was identified as the source of PAHs in mussels if the relative concentrations of phytane and PAHs were consistent with those of *Exxon Valdez* crude oil, except for characteristic alterations of composition caused by weathering. These characteristic alterations include preferential loss of less-substituted PAH homologues and lower-molecular-weight PAHs. The PAHs prevalent in *Exxon Valdez* crude oil were summed and denoted as EVO-PAHs for mussels identified as contaminated by this oil. The EVO-PAHs included all the unsubstituted PAHs and the alkyl-PAH homologues in the unweathered crude oil (see Short et al. 1996a).

Hydrocarbon concentrations in mussel tissues are presented on a dry weight basis.

Results

Hydrocarbon Accumulation by Caged Mussels Inside Prince William Sound

Petroleum hydrocarbons derived from *Exxon Valdez* crude oil were detected at all stations and at all depths of the May 1989 deployment inside Prince William Sound, except at the control station at Olsen Bay. The pattern of relative concentrations of PAHs and of phytane were consistent with patterns in mussels on oiled beaches. The PAHs detected in the caged mussels included chrysenes, which obviates diesel oil derived from North Slope crude oil as a significant source of oil in the caged mussels.

Mussels were substantially free of petroleum hydrocarbons before deployment. Concentrations of the summed PAHs used to define the EVO-PAHs ranged up to 162 ng/g (median, 73 ng/g) in mussels before deployment. Total hydrocarbons ranged up to 47,300 ng/g, and phytane was consistently below detection limits.

Hydrocarbons characteristic of *Exxon Valdez* crude oil were highest in caged mussels deployed at 1-m depth near heavily oiled beaches in May 1989. The PAHs detected in these mussels typically con-

TABLE 1.—Deployment periods for caged mussels inside (stations numbered 1–22) and outside (stations numbered 23–38) Prince William Sound after the *Exxon Valdez* oil spill. Cages successfully retrieved from the three deployment depths of 1, 5, and 25 m are indicated by X; subscript numbers indicate depths of successful retrieval where some cages were lost.

	1989				1990				1991
Station name (station number)	May	Jun	Jul	Aug	May	Jun	Jul	Aug	May
Bainbridge Passage (1)	X	X	X	X		X_1		X_1	
Bay of Isles (2)						X_1		X_1	X
Block Island (3)				X_1		X_1			
Disk Island (4)						X		X	X
Elrington Passage (5)	X	X	X	X		X_1			
Eshamy Bay (6)									X
Green Island (7)						X_1			
Herring Bay (8)	X		X[a]	X	X	X_1	X_1	X	X
Herring Bay Berm (9)							X	X	X
Johnson Cove (10)	X	X	X	$X_{1,25}$		X_1		X_1	
MacLeod Harbor (11)						X_1		X_1	
Main Bay (12)	X	X	X	X	X_1			X_1	
Montague Strait (13)	$X_{1,5}$	X	X_1						
Northwest Bay (14)						X		X	X
Olsen Bay (15)	X	X	X	X	X	X		X	X
Outside Bay (16)	X	X	X			X_1		X_1	
Prince of Wales Passage (17)	X	X	X	$X_{5,25}$		X_1		X_1	
Sleepy Bay (18)						X	X_5		X
Smith Island (19)	$X_{1,5}$	X	X			X			
Snug Harbor (20)	X	$X_{1,5}$	X		X	X	$X_{5,25}$	X	X
Snug Harbor Berm (21)							$X_{1,5}$		X
Squire Island (22)	X	X	X_1	$X_{1,25}$					
Agnes Cove (23)							X_1		
Balboa Bay (24)				X_5					
Black Bay (25)			X	X			X_1		
Blue Fox Bay (26)			X	X			X_1		
Chignik Bay (27)				$X_{1,5}$					
Discoverer Bay (28)			X	$X_{5,25}$					
Hallo Bay (29)				X_1		X_5	X_1		
Katmai Bay (30)							X_1		
Kukak Bay (31)				$X_{1,5}$		X_1			
Port Chatham (32)			X	X					
Port Dick (33)			X	X					
Port Graham (34)				$X_{1,5}$					
Raspberry Bay (35)				X					
Sunny Cove (36)			X	X			X_1		
Tonsina Bay (37)			X	X			X_1		
Windy Bay (38)							X_1		

[a]Deployed in June and July, both collected in August.

sisted of the most persistent PAHs in *Exxon Valdez* crude oil. For example, in mussels deployed at 1-m depth in Herring Bay, a heavily oiled embayment, the most prominent PAHs detected included those homologues with the most alkyl-substitution and greatest molecular weight (Figure 3). Total hydrocarbon concentration, which included the unresolved complex mixture, alkane hydrocarbons, and PAHs, was about 808,000 ng/g dry tissue weight in these mussels, and mean concentrations of EVO-PAHs and of phytane were 45,200 and 6,430 ng/g. The amounts of EVO-PAH and of phytane relative to total hydrocarbons in these mussels is similar to respective proportions in *Exxon Valdez* crude oil: 800,000 ng total hydrocarbons in the crude oil includes 39,400 ng EVO-PAHs and 4,800 ng phytane (Figure 4).

Hydrocarbon concentrations in mussels deployed at 1 m at Smith Island and Snug Harbor were almost as high as those at Herring Bay and had similar composition. Mean EVO-PAHs in mussels at Smith Island and Snug Harbor were 15,200 and 6,770 ng/g, respectively, and concentrations of phytane and of total hydrocarbons varied proportionally (Figure 4). These three stations were adjacent to beaches that were heavily oiled by *Exxon Valdez* crude oil.

Hydrocarbons derived from *Exxon Valdez* crude oil in caged mussels generally decreased with deeper deployment depths. Mean EVO-PAH concentration in mussels at 5 m was 45 to 78% of the concentration at 1 m at the three stations adjacent to heavily oiled beaches in May 1989; at 25 m, it was 6 to 23% of the concentration at 1 m (Figure 5).

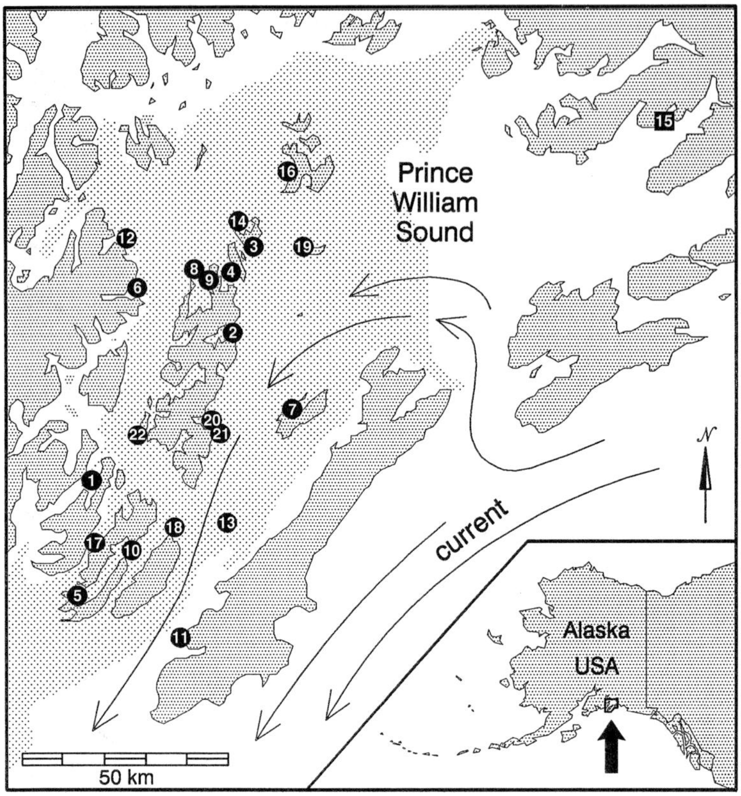

FIGURE 1.—Station locations in Prince William Sound where caged bay mussels *Mytilus trossulus* were deployed after the *Exxon Valdez* oil spill. Station numbers within symbols correspond to station numbers in Table 1. Arrows indicate direction of prevailing ocean currents (Galt et al. 1991). Shaded area indicates the path of surface oil. The control site is indicated by a solid square.

Concentrations of other hydrocarbons characteristic of *Exxon Valdez* crude oil decreased proportionally in these caged mussels.

At other stations along or adjacent to the path of the spilled oil in May 1989, distribution patterns of petroleum hydrocarbons in caged mussels were similar to those at stations adjacent to heavily oiled beaches, but concentrations were substantially lower. At the five stations along the path of the spilled oil (Bainbridge Passage, Elrington Passage, Johnson Cove, Prince of Wales Passage, and Squire Island), mean EVO-PAH concentrations in caged mussels deployed at 1-m depth ranged from 818 to 3,420 ng/g. Similarly, at the three stations near the margin of the oil's path (Main Bay, Montague Strait, and Outside Bay), mean concentrations ranged from 406 to 686 ng/g in caged mussels deployed at 1-m depth. Although petroleum hydrocarbon concentrations in caged mussels often decreased with deployment depth at these stations, the trend was not consistent, especially at the less-contaminated stations. At all these stations and deployment depths, however, relative concentrations of petroleum hydrocarbons in caged mussels were broadly similar to those in *Exxon Valdez* crude oil.

Hydrocarbons characteristic of petroleum were consistently lowest in caged mussels deployed at the control station at Olsen Bay. Of the 24 mussel cages deployed at Olsen Bay in this study from 1989 through 1991, the median summed concentrations of PAHs used to define EVO-PAHs was less than 80 ng/g (Figure 6). These summed PAH concentrations exceeded 300 ng/g in mussels of only one cage (25-m depth, May 1989), and naphthalene accounted for more than 70% of the PAHs in mussels of this cage. Total hydrocarbons in the mussels of the 24 cages at Olsen Bay ranged from 80 to 153,000 ng/g, and phytane was consistently below detection limits.

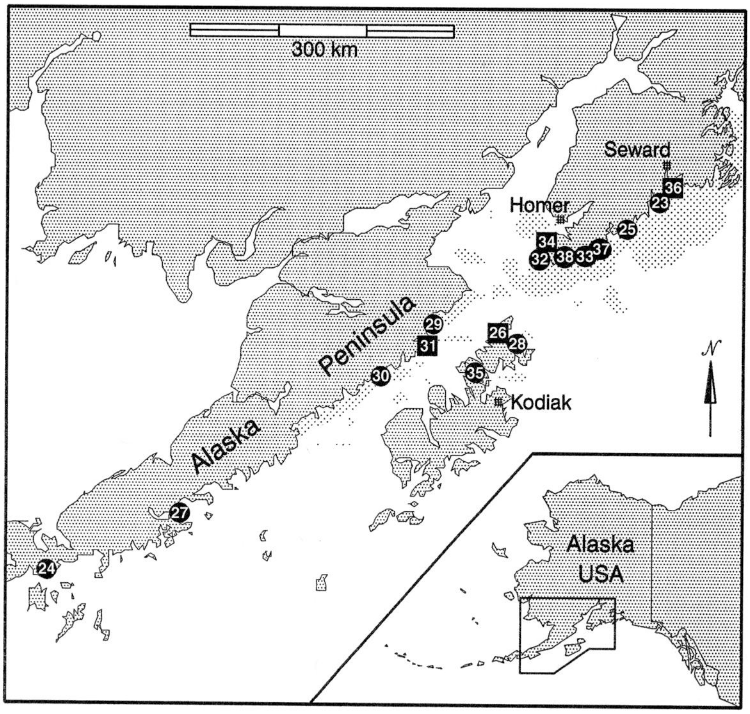

FIGURE 2.—Station locations in the northwestern Gulf of Alaska where caged bay mussels *Mytilus trossulus* were deployed after the *Exxon Valdez* oil spill. Station numbers within symbols correspond to station numbers in Table 1. Shaded area indicates the path of surface oil in the gulf. Control sites are indicated by a solid square.

Petroleum hydrocarbons consistently declined in successively deployed caged mussels after May 1989. At the three stations adjacent to heavily oiled beaches, EVO-PAH concentrations in caged mussels at 1-m depth after the June 1989 deployment were about 10–20% of that at corresponding stations and depth in the May 1989 deployment (Figure 7). Similar declines occurred at other depths and for other petroleum hydrocarbons.

Petroleum hydrocarbons were detected progressively less frequently and generally at lower concentrations in mussels deployed at other stations along or adjacent to the spill path after May 1989. We detected EVO-PAHs in 67% of mussel cages deployed during June 1989 inside Prince William Sound, and in 19 and 29% of mussel cages deployed during July and August, respectively. The highest concentrations of EVO-PAHs in these mussels were 800 ng/g, 533 ng/g, and 419 ng/g for the June, July, and August deployments, respectively.

Mussels deployed at Herring Bay in June 1989 and retrieved after 2 months accumulated substantially more petroleum hydrocarbons than did mussels deployed there in July 1989 and retrieved after 1 month. Concentrations of EVO-PAHs and of phytane were generally two to five times higher in the mussels deployed for 2 months compared with those deployed at corresponding depths for 1 month.

In 1990 and 1991, petroleum hydrocarbons were usually detected in caged mussels only at stations adjacent to heavily oiled beaches and only at low concentrations. The highest 1990 concentration of EVO-PAHs was 1,810 ng/g at Sleepy Bay in the July deployment at 5-m depth (the 1-m deployment was lost), followed by 558 and 878 ng/g at Northwest Bay 1-m depth in the June and August deployments, respectively. Lower concentrations of PAHs probably derived from *Exxon Valdez* crude oil were detected in caged mussels deployed at Bay of Isles, Herring Bay, Main Bay, Smith Island, and Snug Harbor, although the concentrations at these stations were often so low that source identification was less clear. Other petroleum hydrocarbons, such as phytane, often were not detected. Except for Main Bay, these stations were all adjacent to beaches that had been heavily oiled.

The highest EVO-PAH concentration in caged

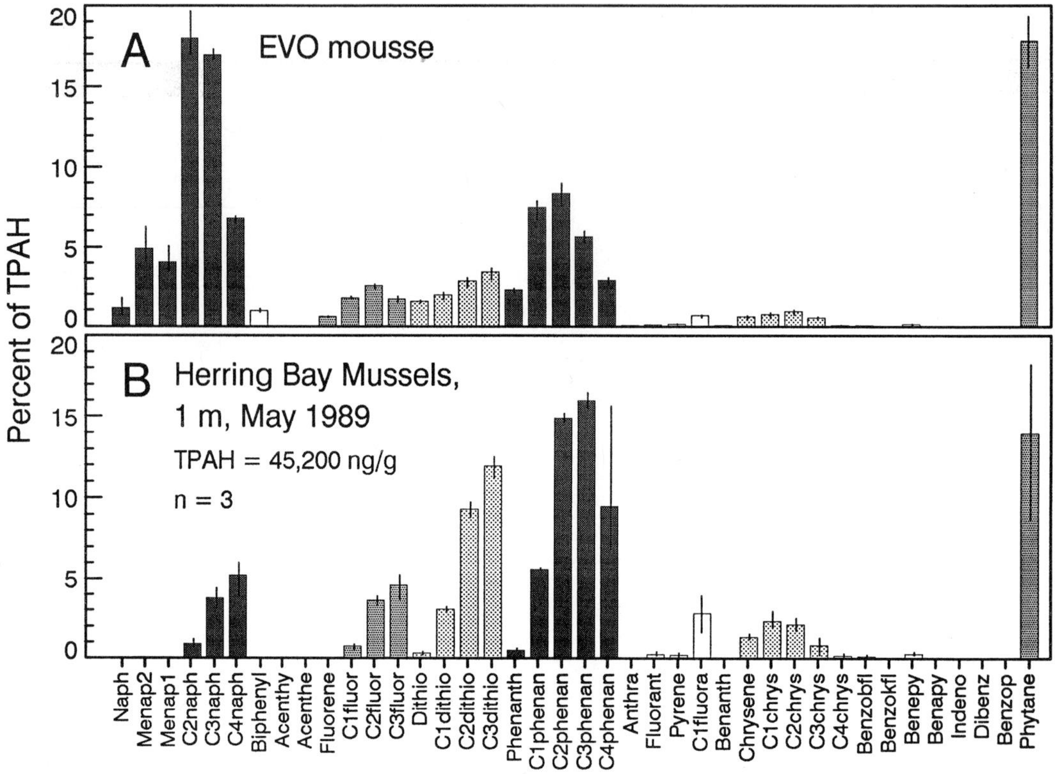

FIGURE 3.—Relative abundance of total polynuclear aromatic hydrocarbons (TPAH) and of phytane in (**A**) mousse oil collected from the sea surface at three locations off Knight Island 11 d after the spill (from Short et al. 1996b) and in (**B**) caged bay mussels *Mytilus trossulus* deployed at 1-m depth at Herring Bay in May 1989. Relative abundances were calculated as the ratio (percent) of the analyte to the sum of all the PAHs in the sample.

mussels deployed in 1991 was 511 ng/g at Northwest Bay, 5-m depth. Low concentrations of PAHs characteristic of petroleum also may have been present in caged mussels deployed at Bay of Isles, Herring Bay, and Sleepy Bay, where EVO-PAH concentrations ranged up to 198 ng/g and phytane was occasionally detected.

Pristane, a branched alkane that has both biogenic and petrogenic sources, showed large seasonal variation in caged mussels at all stations. Pristane concentration was highest in May, then declined thereafter to concentrations that were often below detection limits by fall. In May, pristane was often the most abundant alkane identified in caged mussels, with concentrations ranging to nearly 50,000 ng/g. At stations adjacent to heavily oiled beaches, pristane concentration in caged mussels was often much higher than expected on the basis of other petroleum hydrocarbons measured. For example, the ratio of pristane and phytane in unweathered *Exxon Valdez* crude oil is about 1.28 (see Figure 1b in Short et al. 1996a), but it was often greater than 5 in caged mussels deployed during May 1989, despite the high concentrations of other petroleum hydrocarbons accumulated by these mussels.

Native Mussels

Concentrations of petroleum hydrocarbons in caged mussels deployed at 1-m depth near heavily oiled beaches were consistently lower than in mussels collected from adjacent beaches. In 1989, the concentration of EVO-PAHs in caged mussels deployed at 1 m in Herring Bay during July was 8,780 ng/g, whereas the concentration in mussels collected from the adjacent beach was 95,900 ng/g. In the August deployment in Herring Bay, EVO-PAH concentrations were 4,220 and 27,300 ng/g in caged and beach mussels, respectively. At Smith Island, EVO-PAH concentrations were 1,360 and 48,300 ng/g in caged and beach mussels, respectively, in the

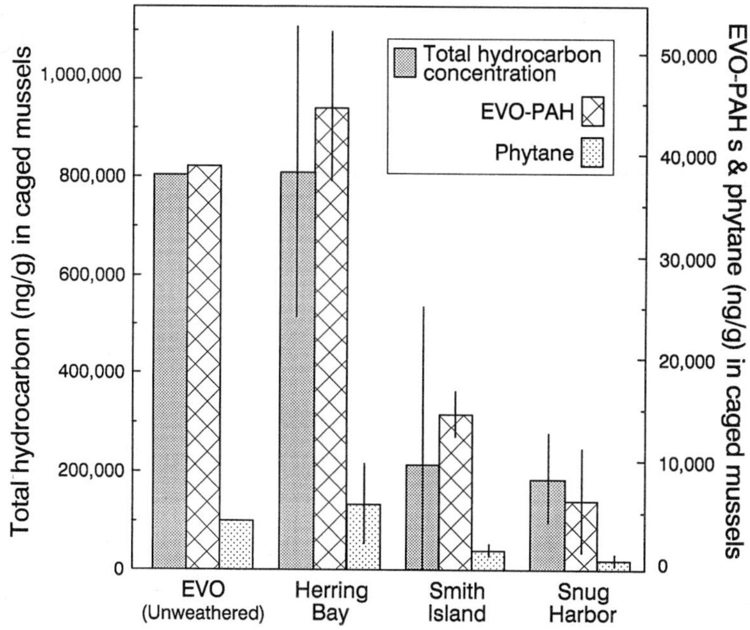

FIGURE 4.—Comparison of petroleum hydrocarbons characteristic of *Exxon Valdez* crude oil in caged bay mussels *Mytilus trossulus* deployed at 1-m depth at stations near heavily oiled beaches during May 1989 with the relative abundance of these hydrocarbons in unweathered crude oil from the hold of the *Exxon Valdez* (EVO). Compared are total measured hydrocarbons, the sum of the most abundant polynuclear aromatic hydrocarbons in *Exxon Valdez* crude oil (EVO-PAHs), and phytane, a branched alkane abundant in petroleum. In mussels, concentrations are nanograms per gram dry tissue weight with 95% confidence intervals; in EVO, it is assumed that there are 800,000 ng total hydrocarbons present from the oil in a 1-g sample.

July deployment. At Snug Harbor in August, caged and beach mussel EVO-PAH concentrations were 1,280 and 11,100 ng/g, respectively. In 1990, the highest EVO-PAH concentration in caged mussels was 1,810 ng/g at Sleepy Bay in the July deployment, but mussels collected simultaneously on the adjacent beach contained from 30,300 to 42,800 ng/g EVO-PAHs. In contrast, at Olsen Bay (control station), PAHs in caged and beach mussels were consistently near or below detection limits.

Outside Prince William Sound

In 1989, petroleum hydrocarbons from *Exxon Valdez* crude oil were evident at only two stations outside Prince William Sound, Tonsina Bay and Chignik Bay. At Tonsina Bay, EVO-PAH concentrations ranged up to 4,100 ng/g in caged mussels deployed during July 1989 but were barely detected (563 ng/g) in those deployed during August. At Chignik Bay, EVO-PAH concentrations ranged up to 2,690 ng/g in caged mussels deployed during August. At both stations, petroleum hydrocarbon

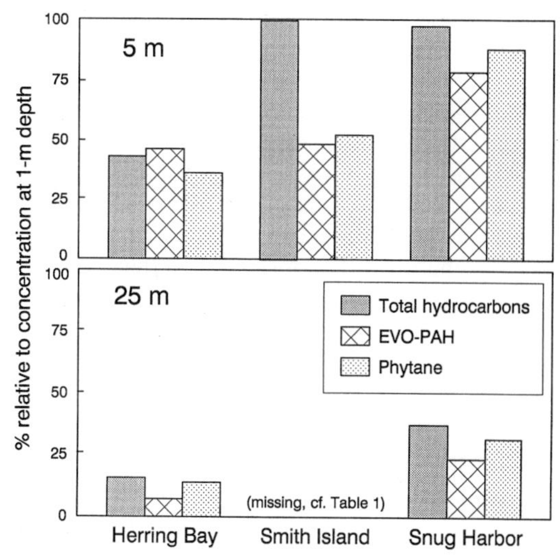

FIGURE 5.—Comparison of total measured hydrocarbons, the sum of the most abundant polynuclear aromatic hydrocarbons in *Exxon Valdez* crude oil (EVO-PAHs), and phytane in caged bay mussels *Mytilus trossulus* deployed at 5-m and 25-m depths at stations adjacent to heavily oiled beaches during May 1989, relative to 1-m deployment.

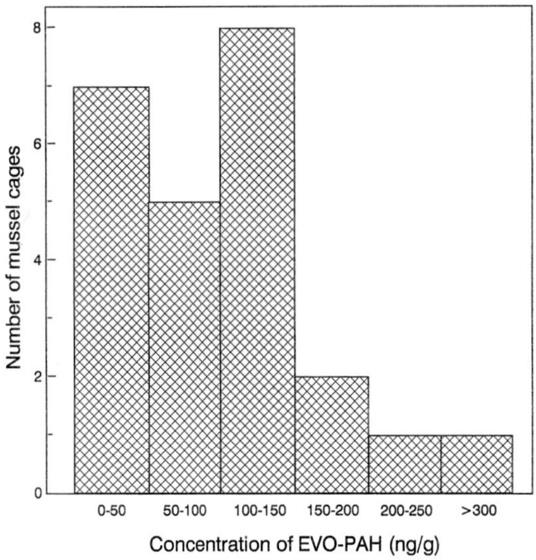

FIGURE 6.—Concentration distribution of EVO-PAHs (as defined in Figure 5) in all caged bay mussels *Mytilus trossulus* deployed at the Olsen Bay control station. The EVO-PAH concentration intervals are nanograms per gram dry tissue weight of mussels, and the *y*-axis indicates the number of observations in each interval. The mean EVO-PAH concentration of the sample in the interval exceeding 300 ng/g was 441 ng/g and was more than 70% naphthalene.

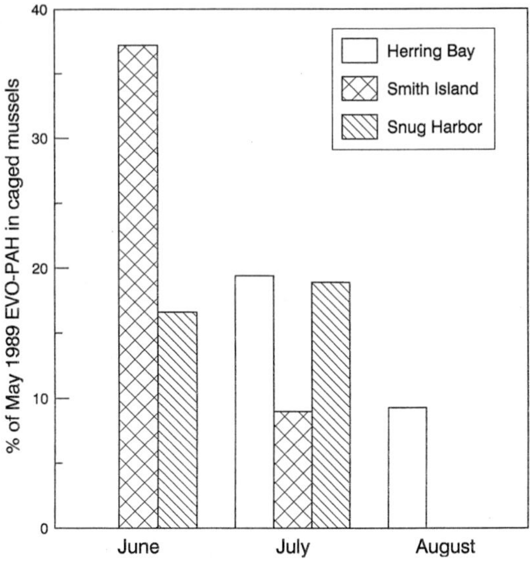

FIGURE 7.—Percent of May 1989 EVO-PAH (defined in Figure 5) concentration in caged bay mussels *Mytilus trossulus* deployed at 1-m depth at stations adjacent to heavily oiled beaches during summer 1989. Percentages are relative to EVO-PAH concentrations in caged mussels at the corresponding stations and depth in May 1989, when these concentrations were highest.

concentrations were highest in mussels deployed at 1-m depth. Phytane was consistently detected at both 1 m and 5 m at Chignik Bay (the 25-m depth mussel cage was lost) at concentrations ranging up to 3,060 ng/g but was detected only sporadically in mussels deployed at Tonsina Bay at concentrations ranging to 734 ng/g. Hydrocarbons characteristic of petroleum were generally near or below detection limits at the other stations outside Prince William Sound in 1989 and at all stations in 1990.

Mussel Mortality and Condition

Observed levels of mortality and condition indicate that caged mussels were able to maintain their preexposure condition and function effectively as water-sampling devices. Condition index of most exposed mussels increased during deployment; none had a condition index that decreased significantly. Mortality of caged mussels at sites inside Prince William Sound in 1989 averaged 6%. Mortality at Kenai, Alaska Peninsula, and Kodiak stations averaged 39% and was correlated ($r^2 = 0.748$) with the number of days mussels were held out of water during transport to the deployment station and the number of days they were deployed. Mortality at a station, however, was not correlated with the condition index of surviving mussels, indicating that survivors were able to function well in transplant locations.

Hydrocarbons in the water column or accumulated in mussel tissue did not affect short-term survival or condition. Significant differences in condition index were related to deployment depth and site but not to levels of tissue hydrocarbons. The condition index was generally higher in mussels at a depth of 1 m than at 5 and 25 m. Mussels at some oil-impacted sites (e.g., Outside Bay, Smith Island, and Snug Harbor; deployment 1, 1989, 1 m) had significantly higher condition indexes than mussels at all other sites, including Olsen Bay, and condition increased during deployment.

Discussion

Oil Dispersion from Oiled-Beach Reservoirs

Our results indicate that a substantial proportion of the oil spilled from the *Exxon Valdez* dispersed into the water column compartment and was subsequently transported into the Gulf of Alaska. The initially high concentrations of petroleum hydrocarbons that were ubiquitous at stations inside Prince William Sound in May 1989 (except at the control, Olsen Bay) uniformly declined in the months after the spill. The presence of oil in caged mussels at stations remote from oiled beaches, in particular at the station near the middle of Montague Strait in May 1989, indicates a very large volume of seawater contaminated at low aqueous concentration to depths of at least 25 m. Uncertainties associated with the aqueous oil concentrations, the volume of contaminated seawater, and the absolute flux of seawater through Prince William Sound during summer 1989 preclude a meaningful quantitative estimate of the volume of oil in this compartment, but the large seawater volumes and fluxes suggest a substantial fraction of the total spilled oil.

The association of the highest concentrations of petroleum hydrocarbons with stations adjacent to heavily oiled beaches suggests that heavily oiled beaches were the primary source of oil. These oiled beaches apparently were a reservoir of spilled oil that continued to contaminate adjacent seawater as the beached oil was subsequently dispersed into the water column by wave action or by cleanup activities.

The distribution of hydrocarbons we observed in the most contaminated mussels suggests that these mussels may have accumulated hydrocarbons by ingestion of small particles of whole oil. The relative hydrocarbon abundances in these mussels (e.g., Figures 3 and 4) are very similar to those characteristic of weathered *Exxon Valdez* oil in intertidal sediments (Short and Babcock 1996, this volume) and in subtidal sediment traps (Short et al. 1996b, this volume). These compositional similarities suggest that the routes through which oil hydrocarbons were incorporated by the caged mussels may include ingestion of whole, particulate oil.

The highest PAH concentrations measured in caged mussels in this study were comparable to the highest concentrations in a similar study using caged mussels for the *Amoco Cadiz* oil spill off Brittany, France, in March 1978. Wolfe et al. (1981) deployed mussels in cages in the Baie de Morlaix at 1-m depth for 2–4 weeks beginning about 5 weeks after the spill. A comparable station in our study was at Herring Bay in May 1989. Individual petroleum PAH analytes measured in these two studies fall well within the same order of magnitude, indicating that seawater dispersion of particulate oil from oiled beaches may be a common process, at least where sufficient wave energy is available to mobilize the oil.

Dispersion of oil into the water column continued on a much-reduced scale in 1990 and 1991. The close association of petroleum hydrocarbon concentrations in caged mussels with beaches that had been heavily oiled indicates that these beaches were persistent reservoirs of oil that locally contaminated adjacent seawater by the same mechanisms as in 1989. The consistently higher concentrations of petroleum hydrocarbons that we observed in mussels native to heavily oiled beaches compared with caged mussels deployed immediately adjacent to them corroborates these beaches as proximal sources and persistent reservoirs of oil. Together, these results imply that petroleum hydrocarbons characteristic of *Exxon Valdez* crude oil will remain detectable in mussels native to oiled beaches substantially longer than in caged mussels deployed adjacent to them because concentrations in caged mussels would decline below detection limits first.

The relative scarcity of petroleum hydrocarbons characteristic of *Exxon Valdez* crude oil in caged mussels at stations outside Prince William Sound is consistent with the proportions of spilled oil estimated to have been beached inside and outside the sound. Wolfe et al. (1994) estimated that approximately 40–45% of the original volume of spilled oil initially became beached within Prince William Sound, whereas only about 10% was carried outside the sound as floating oil, and much of this was deposited on beaches in the Kenai and Kodiak areas. The area traversed by the spilled oil outside the sound was much larger than inside the sound, and beaches were less heavily oiled and more widely and discontinuously distributed compared with those inside the sound.

Alternative Sources of Petroleum Hydrocarbons in Caged Mussels

Alternative sources of petroleum hydrocarbons evident in the caged mussels analyzed for this study were not supported by the relative distributions of individual hydrocarbons found in the mussels. Where elevated suites of petroleum PAHs were measured in the caged mussels, these suites consistently included chrysene and alkyl-substituted chrysenes in relative proportions consistent with those in weathered *Exxon Valdez* crude oil. In particular, the presence in caged mussels of suites of petroleum PAHs that included chrysenes precludes diesel oil as a significant source of PAHs in mussels. Suites of petroleum hydrocarbons that are similar to those of *Exxon Valdez* crude oil but lack chrysenes are characteristic of diesel oil distilled in Alaska from North Slope crude oil. This was the diesel oil that was primarily used commercially in Prince William Sound before and during the oil spill cleanup (Page et al. 1995).

Other sources of crude oil, such as submarine oil seeps, are excluded because of the uniformity of the relative PAH concentrations in caged mussels deployed near heavily oiled beaches (as well as at stations remote from these beaches) together with the consistent temporal decline in concentration and detection frequency of these hydrocarbons. Crude oil from submarine oil seeps along the southern coast of Alaska and east of Prince William Sound, which may be carried into the sound by prevailing oceanic currents (Page et al. 1995), characteristically contains low sulfur, resulting in relatively low suites of dibenzothiophene PAHs compared with other PAHs. Relative concentrations of dibenzothiophenes in the PAH-contaminated caged mussels, however, were uniformly consistent with the higher-sulfur crude oil spilled from the *Exxon Valdez*. Furthermore, it is improbable that the elevated concentrations and detection frequencies of petroleum PAHs in the caged mussels derived from seep oil would coincidentally be highest just after the *Exxon Valdez* oil spill and decline continuously thereafter.

Monitoring with Caged Mussels

The concentrations of petroleum hydrocarbons found in the caged mussels contrast with the low concentrations of hydrocarbons found by direct chemical analysis of seawater. The highest concentrations of petroleum hydrocarbons in caged mussels of this study were at the Herring Bay station in May 1989. Seawater hydrocarbon concentrations near this station were measured by direct chemical analysis and, by May 1989, approached the detection limits of the direct analysis (Short and Harris 1996). This contrast was noted by Vandermeulen (1981), who commented on the caged mussel deployment study of the *Amoco Cadiz* oil spill conducted by Wolfe et al. (1981). The greater sensitivity of mussels probably is attributable to the relative amounts of water sampled for direct chemical analysis (0.9 L) and filtered by a mussel in 1 month (around 10,000 L). Higher concentrations of petroleum hydrocarbons in the caged mussels are consistent with the higher volume of water they effectively "sample."

Our study corroborates the extreme sensitivity attributed to caged mussels as surrogate water-sampling devices. Petroleum hydrocarbons were detected in the caged mussels where the corresponding aqueous concentrations were either orders of magnitude below the detection limits practical for direct chemical analysis or else were so patchily distributed that the probability of sampling detectable concentrations was remote.

Stable or improved condition indicate mussels were able to adapt to transplant locations and function effectively for short-term (1-month) exposure periods. Hydrocarbon levels were not sufficiently high to cause cessation of feeding. If hydrocarbons in the water column or accumulated in tissue had any deleterious effect on short-term condition and survival of mussels, it was masked by other factors that favored condition and survival.

Hydrocarbon concentrations were more variable among replicates of caged mussels deployed outside Prince William Sound than among those deployed inside. The high variability may result from holding mussels for longer periods during transport before deployment. Mussels deployed outside Prince William Sound were generally very stressed, as evidenced by higher mortality.

The seasonally high ratio of pristane and phytane in caged mussels deployed in spring and early summer indicates that discriminating hydrocarbon sources based on this ratio must be used with caution. This ratio was high even in caged mussels that contained substantially elevated petroleum hydrocarbon concentrations, because of a biogenic pristane source in spring. Evidence for this biogenic pristane source was noted previously by Karinen et al. (1993).

Evidence for Bioavailability of Dispersed Oil

The presence of *Exxon Valdez* oil in the caged mussels of this study is prima facie evidence of the bioavailability of dispersed oil, at least to submerged mussels. Furthermore, it suggests that the dispersed oil may have provided a significant route of oil exposure to other marine fauna, such as larval invertebrates and fish. Some of the dispersed oil possibly was associated with plankton and thus could have been ingested by fauna feeding on oiled plankton. Plankton are often concentrated in the upper seawater column adjacent to beaches, where they may have incorporated dispersing oil. Although our study does not establish this exposure route, it does establish its plausibility.

In conclusion, our study clearly establishes that oil derived from the *Exxon Valdez* pervaded the seawater column to a depth of at least 25 m throughout the path of the spill in Prince William Sound during the first few months after the spill and became increasingly localized over the ensuing years. This submerged oil was clearly available to mussels and possibly to other fauna. Oil accumulated from seawater by mussels (and plausibly by other organisms) consequently would be available through ingestion to fauna at higher trophic levels.

Acknowledgments

We thank Marie Larsen, Larry Holland, and Josephina Lunasin for their careful hydrocarbon analyses of the samples reported herein.

References

Clark, R. C., Jr., and J. S. Finley. 1975. Uptake and loss of petroleum hydrocarbons by the mussel, *Mytilus edulis*, in laboratory experiments. U.S. National Marine Fisheries Service Fishery Bulletin 73(3):508–514.

Crosby, M. P., and L. D. Gale. 1990. A review and evaluation of bivalve condition index methodologies with a suggested standard method. Journal of Shellfish Research 9:233–237.

Galt, J. A., W. J. Lehr, and D. L. Payton. 1991. Fate and transport of the *Exxon Valdez* oil spill. Environmental Science and Technology 25:202–209.

Ganning, B., D. Broman, and C. Lindblad. 1983. Uptake of petroleum hydrocarbons by the blue mussel (*Mytilus edulis* L.) after experimental oiling and high pressure, hot water shore cleaning. Marine Environmental Research 10:245–254.

Higgens, E. 1938. Progress in biological inquiries, 1937. Bulletin of the U.S. Bureau of Fisheries, Administrative Report 30:1–70.

Karinen, J. F., M. M. Babcock, D. W. Brown, W. D. MacLeod, Jr., L. S. Ramos, and J. W. Short. 1993. Hydrocarbons in intertidal sediments and mussels from Prince William Sound, Alaska, 1977–1980: characterization and probable sources. NOAA (National Oceanic and Atmospheric Administration) Technical Memorandum NMFS (National Marine Fisheries Service)-AFSC-9, Seattle.

Murray, A. P., B. J. Richardson, and C. F. Gibbs. 1991. Bioconcentration factors for petroleum hydrocarbons, PAHs, LABs and biogenic hydrocarbons in the blue mussel. Marine Pollution Bulletin 22:595–603.

NRC (National Research Council). 1980. The international mussel watch. National Research Council, Washington, DC.

Page, D. S., P. D. Boehm, G. S. Douglas, and A. E. Bence. 1995. Identification of hydrocarbon sources on the benthic sediments of Prince William Sound and the Gulf of Alaska following the *Exxon Valdez* oil spill. Pages 41–83 *in* P. G. Wells, J. N. Butler, and J. S. Hughes, editors. *Exxon Valdez* oil spill: fate and effects in Alaskan waters. American Society for Testing and Materials Special Technical Publication 1219, Philadelphia.

Phelps, D. K., and W. B. Galloway. 1980. A report on the coastal environmental assessment stations (CEAS). Rapports et Process-Verbaux des Reunions Conseil International pur l'Exploration de la Mer 179:76–81.

Pruell, R. J., J. L. Lake, W. R. Davis, and J. G. Quinn. 1986. Uptake and depuration of organic contaminants by blue mussels (*Mytilus edulis*) exposed to environmentally contaminated sediment. Marine Biology 1:497–507.

Short, J. W., and M. M. Babcock. 1996. Prespill and postspill concentrations of hydrocarbons in mussels and sediments in Prince William Sound. American Fisheries Society Symposium 18:149–166.

Short, J. W., and P. M. Harris. 1995. Petroleum hydrocarbons in near-surface seawater of Prince William Sound, Alaska, following the *Exxon Valdez* oil spill II: analysis of caged mussels. Final report, *Exxon Valdez* natural resources damage assessment. NOAA (National Oceanic and Atmospheric Administration) NMFS (National Marine Fisheries Service) Office of Oil Spill Damage Assessment and Restoration, Auke Bay, Alaska.

Short, J. W., and P. M. Harris. 1996. Chemical sampling and analysis of petroleum hydrocarbons in near-surface seawater of Prince William Sound after the *Exxon Valdez* oil spill. American Fisheries Society Symposium 18:17–28.

Short, J. W., T. J. Jackson, M. L. Larsen, and T. L. Wade. 1996a. Analytical methods used for the analysis of hydrocarbons in crude oil, tissues, sediments, and seawater collected for the natural resources damage assessment of the *Exxon Valdez* oil spill. American Fisheries Society Symposium 18:140–148.

Short, J. W., D. M. Sale, and J. C. Gibeaut. 1996b. Nearshore transport of hydrocarbons and sediments after the *Exxon Valdez* oil spill. American Fisheries Society Symposium 18:40–60.

Vandermeulen, J. H. 1981. Contamination of marine populations by hydrocarbons. Pages 564–573 *in* Proceedings, 1979 international symposium centre oceanologique de Bretagne, Brest, France. Public Par Le Centre National Pour L'exploitation des Oceans, Paris.

Vandermeulen, J. H., and W. R. Penrose. 1978. Absence of aryl hydrocarbon hydroxylase (AHH) in three marine bivalves. Journal of the Fisheries Research Board of Canada 35:543–647.

Wildish, D. J., and M. P. Miyares. 1990. Filtration rate of blue mussels as a function of flow velocity: preliminary experiments. Journal of Experimental Marine Biology and Ecology 2:213–219.

Wolfe, D. A., R. C. Clark, Jr., C. A. Foster, J. W. Hawkes, and W. D. MacLeod, Jr. 1981. Hydrocarbon accumulation and histopathology in bivalve mollusks transplanted to the Baie de Morlaix and the Rade de Brest. Pages 599–616 *in* Proceedings, 1979 international symposium centre oceanologique de Bretagne, Brest, France. Publie Par Le Centre National Pour L'exploitation des oceans, Paris.

Wolfe, D. A., and eleven coauthors. 1994. The fate of the oil spilled from the *Exxon Valdez*. Environmental Science and Technology 28:561A–568A.

American Fisheries Society Symposium 18:40–60, 1996

Nearshore Transport of Hydrocarbons and Sediments after the *Exxon Valdez* Oil Spill

JEFFREY W. SHORT

National Marine Fisheries Service, Alaska Fisheries Science Center
Auke Bay Laboratory, 11305 Glacier Highway
Juneau, Alaska 99801-8626, USA

DAVID M. SALE

SnowOtter Environmental Consulting, 3859 Beach Drive SW, Seattle, Washington 98116, USA

JAMES C. GIBEAUT

University of Texas, Bureau of Economic Geology, University Station, Box X
Austin, Texas 78713, USA

Abstract.—Sediment traps were deployed in seawater depths of 10 m to 20 m to evaluate the transport of oil-laden sediments from oiled beaches to adjacent shallow subtidal sediments after the *Exxon Valdez* oil spill of 24 March 1989 in Prince William Sound, Alaska. Sediments collected by these traps between November 1989 and March 1991 were analyzed for petroleum hydrocarbons and for sedimentological and mineralogical characteristics. Benthic surface and core sediment samples were also collected and analyzed for hydrocarbons and for sedimentological characteristics, which are compared with similar results for trapped sediments.

Results of the hydrocarbon analyses of the trapped sediments indicate that *Exxon Valdez* oil was transported from oiled beaches to adjacent shallow subtidal sediments. Total polynuclear aromatic hydrocarbon (PAH) concentrations of trapped sediments were highest at the East Arm of Northwest Bay during the first trap deployment period, when concentrations exceeded 75,000 ng/g (dry weight basis). Total PAH concentrations of sediments trapped at the other three oiled locations ranged from 8,660 to 28,400 ng/g. Total PAH concentrations generally declined with time to 700–6,280 ng/g for sediments trapped during the March 1991–June 1991 deployment interval. Total PAH concentrations of sediments trapped at other stations were generally lower (<3,500 ng/g) and were usually compositionally distinct from *Exxon Valdez* oil.

These oil-laden suspended sediments were only slightly incorporated into adjacent benthic surface sediments. Total PAH concentrations derived from *Exxon Valdez* oil were less than 2,500 ng/g in shallow (10–20 m) benthic surface sediments at oiled locations and approached this value only in the East Arm of Northwest Bay. Elsewhere, total PAH concentrations derived from any source were generally well below 500 ng/g. Contributions to shallow subtidal benthic sediments of PAHs derived from other sources, including anthropogenic sources and PAH-laden sediments carried into Prince William Sound through Hinchinbrook Entrance, are generally well below 400 ng/g at locations within or near the path of the spill through the sound.

Sedimentological results indicate (1) mobilization of fine, oil-contaminated sediments from heavily oiled upper intertidal beaches by winter wave action and beach treatment activities and (2) residence times of oil in shallow epibenthic sediments on the order of months. The residence time increases with proximity to heavily oiled beaches where waves or treatment cause erosion but where adjacent subtidal areas are protected from wave and tidal energy. The conditions favorable for prolonged accumulations of elevated oil concentrations in shallow epibenthic sediments are restrictive, so that substantial and prolonged shallow subtidal sediment contamination is probably limited to relatively few locations. However, higher levels of contamination probably occur in deeper, lower energy, fine-grained sedimentation sites within oiled embayments.

The *Exxon Valdez* oil spill of 24 March 1989 in Prince William Sound, Alaska, resulted in extensive intertidal oiling of shorelines along the path of the oil. Of the approximately 41,400 m^3 of crude oil thought to have spilled initially, Wolfe et al. (1994) estimate that by mid-April 1989 about 40–45% had beached within Prince William Sound. These oiled shorelines may provide reservoirs of oil for persistent contamination of adjacent seawaters and subtidal sediments.

Some of the beached oil and oil-contaminated sediments may be transported to subtidal areas following reintroduction into nearshore waters by wave- and tide-induced processes and leaching by groundwater flowing through the intertidal zone from the backshore (Boehm et al. 1987; Owens et al. 1987). Shoreline cleanup activities contributed to subtidal deposition as oil and sediments escaped from containment booms (mainly occurring during the intensive mechanical operations in 1989). Subtidal deposition also occurred subsequent to berm-relocation efforts and excavations that exposed subsurface oil to winter storm activity. Bulk oil redispersed into the nearshore seawater column may be adsorbed onto inorganic particulate matter (Meyers and Quinn 1973) or be ingested by zooplankton and incorporated into feces that settle onto benthic surface sediments (Conover 1971; Bassin and Ichiye 1977; Clark and MacLeod 1977). These particles may have such low settling velocities that the oil is carried far from nearshore areas before settling.

The primary objective of this study is to examine the transport of *Exxon Valdez* crude oil from oiled shorelines through adjacent seawaters to subtidal sediments. Specifically, we sought to determine if sediments settling out of the water column in nearshore subtidal environments contained adsorbed petroleum hydrocarbons derived from *Exxon Valdez* crude oil and to determine the mechanism of transport. To accomplish these objectives, sediment traps were deployed to collect settling organic and mineral particulate matter offshore of oiled and unoiled shorelines. At some of these stations and during some deployment periods, we also collected benthic surface sediments and sediment core samples adjacent to the deployed sediment traps. The deployment stations included locations adjacent to heavily oiled shorelines and control locations remote from the path of the spilled oil through the sound. Collectively, results from chemical and geological analysis of these sediments verify transport of *Exxon Valdez* crude oil from oiled shorelines to adjacent subtidal sediments in the sound. Crude oil was associated with oiled particulate matter that had entered the seawater column primarily by winter storms and beach treatment activities.

Methods

A complete description of the methods used for this study is presented by Sale et al. (1994); a summary of these methods is presented below.

Study Area and Sampling Schedules

From November 1989 to August 1990, five suites of sediment traps were deployed at 10 m below mean lower low water (MLLW) at four oiled locations (Snug Harbor, Sleepy Bay, and both arms of Northwest Bay) and at a reference location at Port Fidalgo (Figure 1). Beginning in August 1990, additional traps were deployed to include more locations and depths (Table 1). Samples of benthic sediments were usually collected each time the sediment traps were retrieved, and benthic sediment core samples were collected at selected locations in June and September 1991 (Table 1).

The physical characteristics and oiling conditions at each study site are summarized in Table 2. The environmental sensitivity index (ESI) is a measure of the sensitivity of intertidal shorelines to perturbation and subsequent recovery, ranked according to substrate type (Michel and Hayes 1991). An ESI rank of 1 represents the least-sensitive type of shoreline, whereas a 10 is the most sensitive. The shoreline segments adjacent to sediment-trap sites are mixed sand and gravel beaches (ESI 6), gravel beaches (ESI 7), or sheltered rocky coasts (ESI 8). These three categories include 71% of the shoreline in Prince William Sound (Michel and Hayes 1991). Seven of the trap stations are in sheltered embayments, and the other stations are exposed to moderate- to high-energy waves.

Sample Collection

Trapped sediments.—Sediment traps were designed to capture sediments and particulates falling out of suspension. The sediment-trap design was based on guidelines from Woods Hole Oceanographic Institution (Woods Hole Oceanographic Institution 1989). Each trap consisted of three Schedule-40 polyvinyl chloride cylinders measuring 15 cm inside diameter and 1.2 m height. (Trap openings were about 1.3 m above the benthic surface.) Each cylinder was fitted with a 1 cm^2 grid baffle on the top and was filled with 230 g precombusted sodium chloride before deployment so as to suppress resuspension of collected sediments. The three cylinders were mounted on a 60 × 60 cm concrete base, filled with filtered seawater, and deployed and retrieved by divers assisted by shipboard winch and cable. After retrieval, trap contents were allowed to settle for at least 1 h, the top two-thirds of the water was decanted, and the remaining seawater was filtered through precombusted 9-mm glass-fiber filters after sieving removed macrofauna and macroflora. The filters and collected sediments were stored frozen for hydrocarbon and sedimentological analysis.

Epibenthic and benthic core sediments.—Epibenthic (0–2 cm) sediment samples were collected near the sediment traps by divers with precleaned stainless steel cups and spoons and prelabeled 250-ml glass

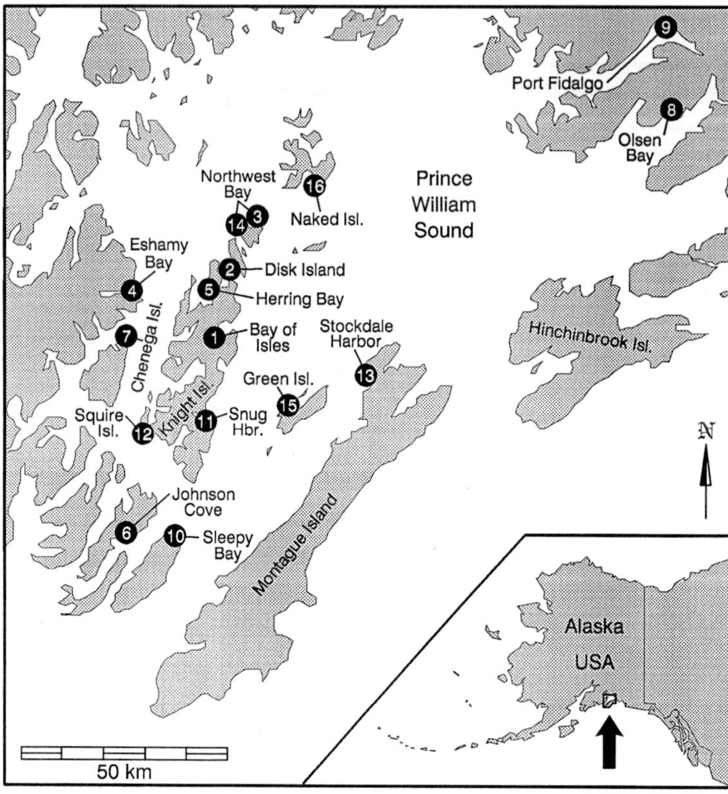

FIGURE 1.—Map of Prince William Sound, Alaska, showing the location of sampling stations of this study. Each station is identified by a number that corresponds to those in Tables 1 and 2.

jars, sealed under water immediately after sample collection. Benthic sediment core samples were collected by divers during trap retrieval with a precleaned stainless steel coring tube, 15 cm long with a diameter of 10 cm. Samples were frozen immediately after collection (benthic samples and cores) or filtration (trap samples) aboard ship.

Surface mousse oil.—Samples of mousse (emulsified oil) were collected from the sea surface at Snug Harbor, Bay of Isles, and Smith Island inside Prince William Sound, 11 d following the grounding of the TV *Exxon Valdez*. These collections were all within 100 m of very heavily oiled beaches. The mousse oil was collected off the sea surface by adhesion to a glass plate and then transferred to a sample collection jar.

Hydrocarbon Analysis

Sediments were prepared and analyzed by following procedures presented in detail by Short et al. (1996, this volume). Hydrocarbon concentrations are reported here on sediment dry weight basis. Ratios of sediment wet and dry weights were measured by dehydrating 1 g of wet sediment for 24 h at 60°C and weighing the remaining mass.

Sediment Grain Size and Clay Mineral Determination

Grain-size distributions of samples were analyzed using the general method of Folk (1974). Samples were rinsed with distilled water and allowed to settle for several days. After decanting the excess water, the samples were wet sieved through a 4-phi screen. The coarse fraction was dry sieved at 0.5- or 1-phi intervals, depending on the amount of sediment available. The mud fraction was separated by pipette analysis (Folk 1974) into simply clay and silt fractions for the sediment core fractions or into 1-phi intervals for trap and benthic sediment samples. The analysis of illite and chlorite clay minerals was performed according to Naidu and Mowatt (1983) and Biscaye (1965).

TABLE 1.—Sampling stations, locations, seawater depths, and sediment sample types for each deployment period in Prince William Sound. Station numbers correspond to those of Table 2 and Figure 1. Longitude and latitude are expressed as degrees = minutes = seconds. The first date indicates deployment of traps, the second indicates retrieval. Samples were taken at the end of each period.

					Samples taken at each deployment period[a]					
Station	Location	Depth (m)	Longitude	Latitude	Nov 1989– Jun 1990	Jun 1990– Aug 1990	Aug 1990– Mar 1991	Mar 1991– Jun 1991	Jun 1991– Sep 1991	Sep 1991– Mar 1992
1	Bay of Isles	20	147°42′21″	60°22′43″				T,B	T,TG,B,C	T,TGM,BG
1	Bay of Isles	10	147°42′24″	60°22′42″		B		B	T,TG,B,C	T,TGM,BG
2	Disk Island	20	147°39′37″	60°29′57″			B,BG	T,TGM,B,C	T,TG,B	T,TGM,BG
2	Disk Island	15	147°39′35″	60°29′57″			B,BG	T,TGM,B	T,TG,B,C	T,TGM,BG
2	Disk Island	10	147°39′37″	60°29′53″		B	T,B,BG	T,TGM,B,C	T,TG,B	T,TGM,BG
3	East Northwest Bay	20	147°34′43″	60°33′09″				T,TGM,B	T,TG,B,C	T,TGM,BG
3	East Northwest Bay	10	147°34′41″	60°33′06″	T,B	T,B	T,B,BG	T,TGM,B,C	T,TG,B	T,TGM
4	Eshamy Bay	10	147°58′07″	60°27′43″		B	T			
5	Herring Bay	10	147°42′48″	60°28′16″		B			T,TG,B,C	T,TGM
6	Johnson Cove	10	147°58′27″	60°03′58″		B	T			
7	N Chenega Island	10	148°00′14″	60°22′45″		B	T			
8	Olsen Bay	20	146°12′57″	60°44′31″					B,C	
8	Olsen Bay	10	146°12′57″	60°44′37″					B	T,TGM,BG
9	Port Fidalgo	10	146°17′11″	60°50′21″	T,B	T,B	T,B,BG			
10	Sleepy Bay	20	147°50′09″	60°04′04″			B,BG	T,TGM,B,C	T,TG,B	T,TGM,BG
10	Sleepy Bay	15	147°50′09″	60°04′02″			B,BG	T,TGM,B	T,TG,B,C	T,TGM,BG
10	Sleepy Bay	10	147°50′12″	60°03′58″	T,B	T,B	T,B,BG	T,TGM,B,C	T,TG,B	T,TGM,BG
11	Snug Harbor	20	147°43′34″	60°14′24″			B,BG	T,TGM,B,C	T,TG,B	T,TGM,BG
11	Snug Harbor	15	147°43′34″	60°14′20″			B,BG	T,TGM,B	T,TG,B,C	T,TGM,BG
11	Snug Harbor	10	147°43′43″	60°14′15″	T,B	T,B	T,B,BG	T,TGM,B,C	T,TG,B	T,TGM,BG
12	Squire Island	10	147°56′27″	60°14′34″		B	T			
13	Stockdale Harbor	10	147°11′40″	60°19′32″		B	T,B			
14	West Northwest Bay	12	147°36′08″	60°32′44″	T	T,B	T			
15	Green Island	10	147°29′04″	60°16′04″		B				
16	Naked Island	10	147°22′05″	60°37′32″		B				

[a]Sample types: T, sediment trap sample for hydrocarbon analysis; TG, sediment trap sample for grain-size analysis; TGM, sediment trap sample for grain-size, organic carbon–nitrogen, clay mineralogy; B, sample from upper 2 cm of benthic surface for hydrocarbon analysis; BG, sample from upper 2 cm of benthic surface for grain-size analysis; C, core sample for stratigraphic analysis, subsampled for hydrocarbon chemistry.

Benthic Sediment Core Stratigraphy

While still frozen, each core was split along its length with a coping saw and sampled at various stratigraphic intervals for hydrocarbon and grain-size analyses. Coping saw movements were limited to strokes of less than 2–3 cm to prevent smearing the sediments. After splitting, the core was scraped to reveal sedimentary structure and lithology. A 30× binocular microscope aided in core descriptions. A sketch was made of each core with detailed observations to the nearest millimeter; lithology was described using the classification scheme of Folk et al. (1970) and the Udden–Wentworth grain-size scale (Lewis 1984). Overall sediment color was determined by using a rock color chart following the Munsell color system. Initial qualitative classifications were modified as appropriate after consideration of quantitative grain-size analyses.

Data Analysis

Hydrocarbon data were accepted as comparable if the sample aliquot wet weight was at least 0.5 g and if the surrogate standard recoveries were more than 30% and less than 150% for all of the polynuclear aromatic hydrocarbon (PAH) surrogate standards except benzo[*a*]pyrene-d_{12} and perylene-d_{12}. Hydrocarbon analyte concentrations that were less than the experimentally determined method detection limits (adjusted for the sample aliquot weight) are treated as zero.

The relative concentrations of PAHs and of phytane in hydrocarbon samples are presented as proportions of the mean total PAH concentration of the sample. This permits comparison of hydrocarbon composition among samples from sediment traps, the epibenthos, and benthic core samples, and among samples of widely different total PAH concentrations.

TABLE 2.—Physical characteristics, oiling conditions, and treatment for intertidal shorelines adjacent to sediment-trap stations for 1989–1992 in Prince William Sound. Oiling conditions are initial for the year, reflecting previous years' treatment and winter storms. Naked and Green islands data are not presented because these study sites were eliminated in 1990. Station numbers correspond to those in Table 1 and Figure 1. Sources used for this table are Michel and Hayes (1991) for Environmental Sensitivity Index (ESI), oiling, and some treatment information; joint agency–Exxon shoreline assessment team (SCAT, SSAT, ASAP, MAYSAP, FINSAP) reports for oiling conditions; Pavia (1991) for oiling conditions and berm relocation; and Alaska Department of Environmental Conservation Daily Shoreline Assessment Reports for treatment information.

Station	Location	ESI Class[a]	Wave energy	Benthic sediments (Top 2 cm: 10–20 m depth)[b]	Initial 1989		
					Surf[c]	Sub[d]	Tx[e]
1	Bay of Isles	8	Low	Coarse silt–muddy sands	H	OP	Me
2	Disk Island	8, 6	Moderate–High	Coarse–muddy sand	H	OP	Me
3	East Northwest Bay	8, 6	Moderate	Coarse–muddy sand	H	OP	Me
4	Eshamy Bay	8, 6	Low		L–NO	NO	Me
5	Herring Bay	8	Low	Muddy, sandy gravel (10 m)	H	OP–OR	Me
6	Johnson Cove	8	Low		L	NO	None
7	North Chenega Island	8, 7	Moderate		H	OP–OR	Me
8	Olsen Bay	8, 6	Low	Coarse sandy mud (10 m)	NO	NO	None
9	Port Fidalgo	8, 6	Low	Very coarse–silty sand	NO	NO	None
10	Sleepy Bay	6	Moderate–High	Very coarse–muddy sand	H–M	OP	Me
11	Snug Harbor	6	Moderate–High	Fine–very fine sand	H–M	OP, OR	Me
12	Squire Island	8	Low		L–NO	NO	Me
13	Stockdale Harbor	8, 7	Moderate		NO	NO	None
14	West Northwest Bay	8, 6	Moderate–High		H	OP–OR	Me

Total PAH is calculated as the sum of all the PAHs measured, except for perylene, a naturally occurring PAH in Prince William Sound marine sediments. Variability of mean total PAH is given as ±1 SE.

The four moment measures for sediment grain-size data were calculated using the graphical method of Folk and Ward (1957) and the entire grain-size distribution for each sample. The graphical method was the most appropriate because of open tails in the size distributions.

Results

Hydrocarbons in Trapped Sediments

Stations adjacent to oiled beaches.—Sediment-trap concentrations of PAHs derived from *Exxon Valdez* crude oil were highest in traps deployed adjacent to four heavily oiled beaches during the first deployment period (November 1989–June 1990). Of these four stations, mean total PAH concentrations were highest at Northwest Bay–East Arm (75,500 ± 2,040 ng PAH/g sediment, N = 2) and Sleepy Bay (28,400 ± 1,980 ng/g, N = 3) and were lowest at Northwest Bay–West Arm (7,190 ± 1,390 ng/g, N = 3; Figure 2).

Total PAH concentrations of sediments trapped at the four stations adjacent to heavily oiled beaches generally declined in sediments trapped subsequent to the first trap-deployment interval. Total PAH concentrations in sediments trapped during the June–August 1990 deployment interval ranged from 1,820 ± 989 ng/g (N = 2) to 14,500 ± 178 ng/g (N = 2), or from 4% (at Northwest Bay–East Arm) to 88% (at Northwest Bay–West Arm) of those during the previous winter deployment at corresponding stations (Figure 3). Polynuclear aromatic hydrocarbon concentrations of sediments trapped during the August 1990–March 1991 deployment interval were comparable to those of corresponding stations collected during the June–August 1990 deployment interval, except at Northwest Bay–East Arm, where the total PAH concentration increased from 3,120 ng/g (N = 1) to 11,200 ng/g (N = 1; Figure 3). Total PAH concentrations of sediments trapped during the March–June 1991 deployment interval declined further at Sleepy Bay and Snug Harbor, to 5,780 ± 586 ng/g (N = 3) and 787 ± 147 ng/g (N = 3), respectively. However, at Northwest Bay–East Arm, the total PAH concentration of sediment trapped during the March–June 1991 deployment interval remained higher than that of the June–August 1990 deployment interval, at 6,280 ± 1,240 ng/g (N = 2; Figure 3).

The abundance of individual PAHs and of phytane relative to the total PAH in sediments collected during the initial deployment of the traps at the four stations adjacent to heavily oiled beaches is consistent with that of weathered *Exxon Valdez* crude oil. The most abundant PAHs measured in these sediments generally correspond to the most

TABLE 2.—Extended.

Station	Spring 1990			Spring 1991			Spring 1992		
	Surf	Sub	Tx	Surf	Sub	Tx	Surf	Sub	Tx
1	M–L	OP(berm)	Ma	M–L	OR	Ma	L	OR	Ma
2	M	OR	Ma	M–L	OR–OF	Ma	L	NO	Ma
3	M(berm)	OP–OR	Ma	L–NO	OP–OR	Ma	NO	OP–OR	None
4	L–NO	NO	None	L–NO	NO	None	NE	NE	None
5	H–M	OR	Ma	M–L	OR	Ma	NE	NE	None
6	L	NO	None	L	NO	None	NE	NE	None
7	L	OR–OF	Ma	L	OR	Ma	NE	NE	None
8									
9									
10	M	OP	BR, Me	M–L	OP–OR	BR–Me	L	OR	Ma
11	M–L	OP	BR	L–NO	NO	None	NE	NE	None
12	NO	NO	None	NE	NE	None			
13									
14	L	OP–OR	Me, Ma	NO	NO	None	NE	NE	None

[a]ESI: 1, exposed rocky coast; 2, exposed wave-cut platforms; 3, fine-grained sand beaches; 4, coarse-grained sand beaches; 5, exposed tidal flats; 6, mixed sand and gravel beaches; 7, gravel beaches; 8, sheltered rocky coasts; 9, sheltered tidal flats; 10, salt marshes. [b]Benthic sediment textural descriptions for sediments around traps are shown only for sites where grain-size analysis was available (BG samples in Table 1). [c]Surf = surface oiling condition, which here represents the average condition of shoreline segments adjacent to trap station. Original 1989 classification system is used for consistency: H, heavy; M, medium; L, light; NO, no oil; NE, not evaluated. [d]Sub = subsurface oiling condition, which here represents the condition of worst subsurface oiling lens in adjacent intertidal segments: OP, oil fills pore spaces; OR, oil residue on sediments; OF, oil film on sediments; NO, no oil. [e]Tx = treatment, which represents the most intrusive method used for that treatment season: Me, mechanical (high–low pressure, cold–hot water steam washing, heavy equipment removing beach sediments); BR, berm relocation using heavy equipment; Ma, manual (pick up of debris by hand and with shovels, raking and hand tilling, absorption pads and hand washing, bioremediation).

abundant PAHs in the mousse samples collected 11 d after the initial spill (compare Figure 2 and Figure 4A), except that the sediment trap PAHs show increasing preferential losses of lower-molecular weight and of less-substituted PAHs. The pattern of these preferential losses is very consistent among sediment trap replicates and among the four deployment stations (compare Figure 2A–D). In addition, note that (1) the high-molecular weight and unsubstituted PAHs that are not abundant in the mousse samples are also not abundant in the trapped sediments and (2) phytane, abundant in the weathered mousse, is also consistently abundant in these trapped sediments.

The abundance of individual PAHs and of phytane relative to the total PAHs in sediments trapped during the initial and subsequent deployment intervals are very similar (Figure 5). As noted above, this pattern of relative PAH abundances is broadly consistent with weathered *Exxon Valdez* crude oil. However, the pattern of relative PAH abundances for sediments trapped during the June–August 1990 deployment interval and the other deployment intervals differ in three respects: in the former, the relative abundance of the naphthalenes is relatively higher, the relative abundance of the chrysenes is relatively lower, and the most abundant homologues of dibenzothiophene and phenanthrene tend to be the dimethyl- rather than the trimethyl-homologues. These differences are consistent with *Exxon Valdez* crude oil that is less weathered (compare Figures 4A and 5).

Geographic distribution.—Sediment traps were deployed at nine additional stations during the August 1990–March 1991 deployment interval to determine the hydrocarbon content and composition of sediments trapped at the additional stations, including stations that were not adjacent to heavily oiled beaches (one of the nine traps was never retrieved). Mean total PAH concentrations consistent with weathered *Exxon Valdez* crude oil were evident at Disk Island and less clearly at Chenega Island, where total PAH concentrations were 3,500 ± 873 ng/g ($N = 2$) and 1,140 ng/g ($N = 1$), respectively. The Disk Island station is adjacent to a shoreline that was heavily oiled in 1989 and where oiling remained at least light through spring 1992, while initial heavy oiling at the Chenega Island station

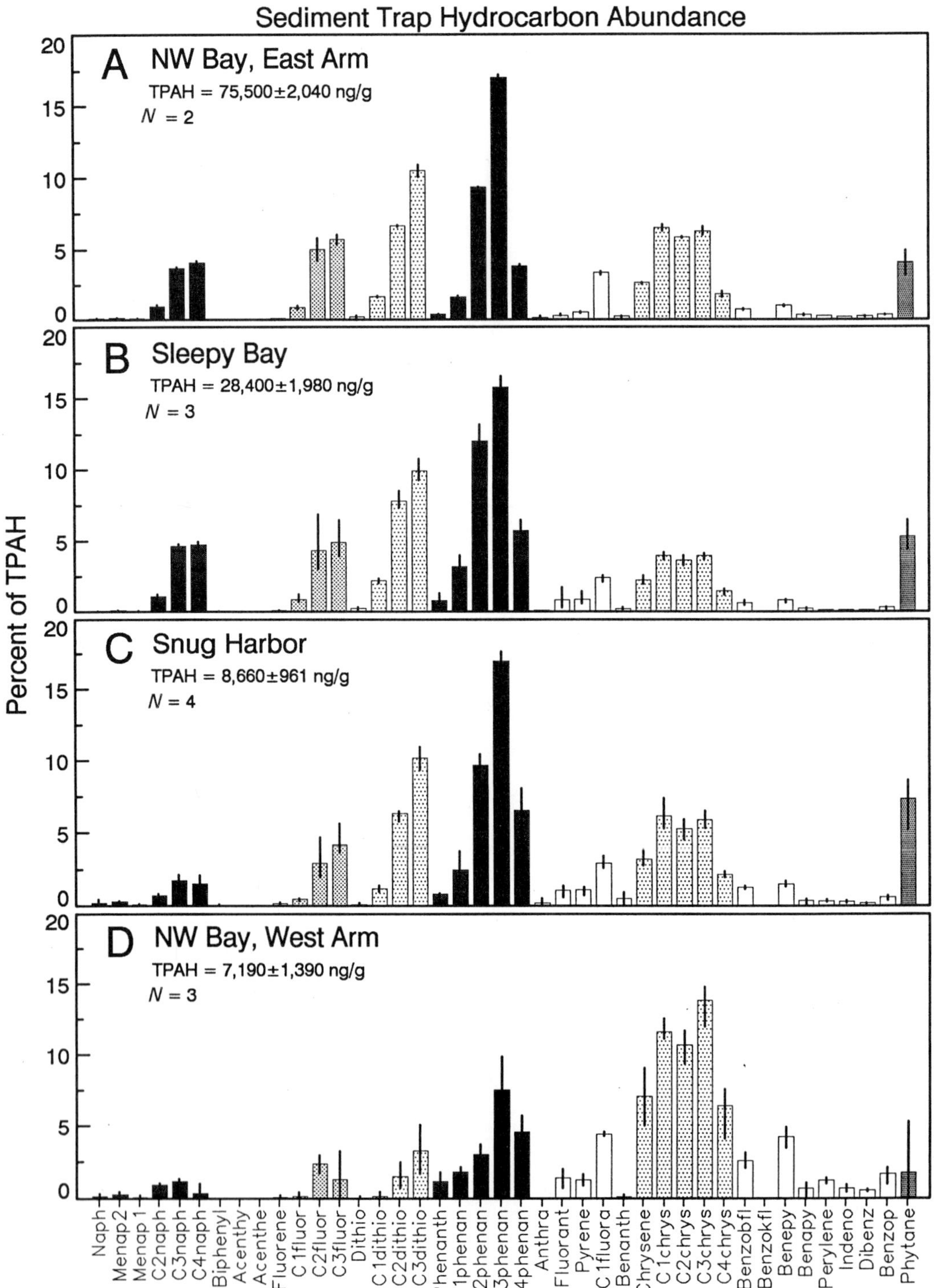

FIGURE 2.—Relative abundance of polynuclear aromatic hydrocarbons (PAH) and phytane in sediments collected in 10-m depth traps deployed from November 1989 to June 1990 at stations adjacent to heavily oiled beaches (**A–D**). Heavy bars indicate the proportion of the total PAH (TPAH) corresponding to the hydrocarbon indicated on the abcissa, and the associated vertical lines indicate the range of *N* replicated analyses. The mean TPAH concentration is also listed as nanograms PAH per gram of dry sediment weight.

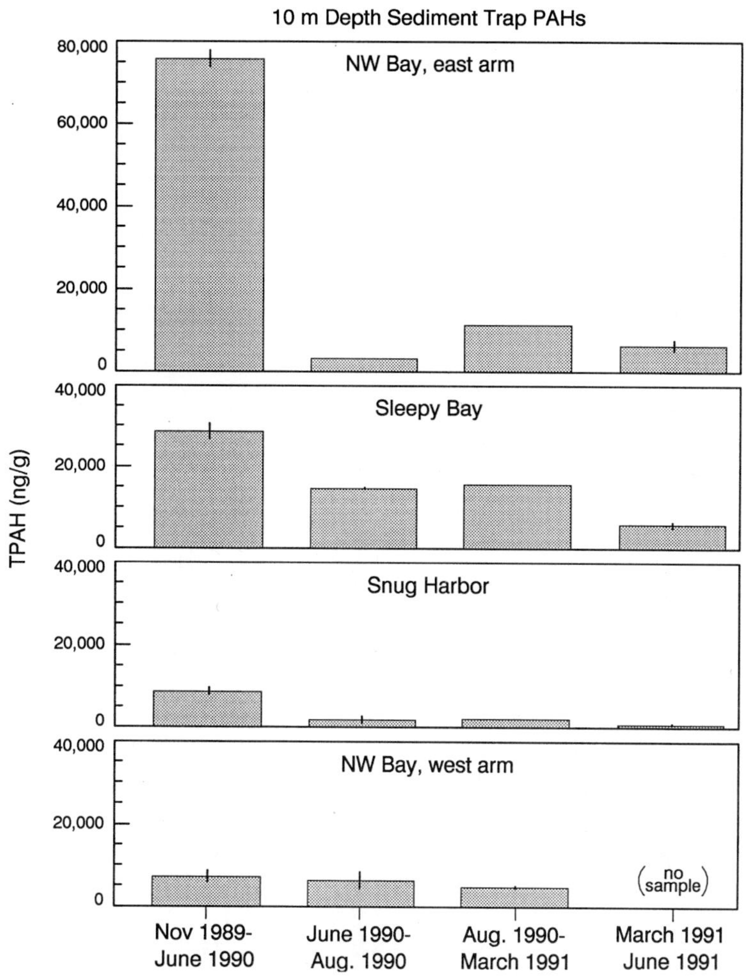

FIGURE 3.—Concentrations of total polynuclear aromatic hydrocarbons (TPAH) in sediment traps deployed during four sequential deployment intervals at stations adjacent to heavily oiled beaches after the *Exxon Valdez* oil spill of 24 March 1989 in Prince William Sound. The deployment intervals were (1) November 1989–June 1990, (2) June–August 1990, (3) August 1990–March 1991, and (4) March–June 1991. The TPAH units are nanograms of PAH per gram of dry sediment weight.

was reduced to sporadic patches of light oiling within the first year (1989–1990; see Table 2).

Polynuclear aromatic hydrocarbons in sediments trapped at Stockdale Harbor display an important alternative pattern of relative PAH abundance. The mean total PAH concentration in these sediments was at least 917 ± 415 ng/g ($N = 2$; the naphthalenes and methylnaphthalenes were not accurately measured because of unacceptable surrogate standard recoveries). The distribution of the relative PAH abundances indicate the presence of (1) dibenzothiophene homologue concentrations at about one tenth those of corresponding phenanthrene homologue concentrations; (2) substantial abundances of phenanthrene and chrysene, with monomethyl homologues of these the most abundant of each series, and abundances of more substituted homologues that decrease with degree of substitution; and (3) relatively low abundances of 4- and 5-ring unsubstituted PAHs (Figure 6). This distribution of relative PAH abundances is not consistent with weathered *Exxon Valdez* crude oil.

Sediments trapped at the Johnson Cove station during the August 1990–March 1991 deployment interval contained elevated PAH concentrations with almost the same distribution as at Stockdale

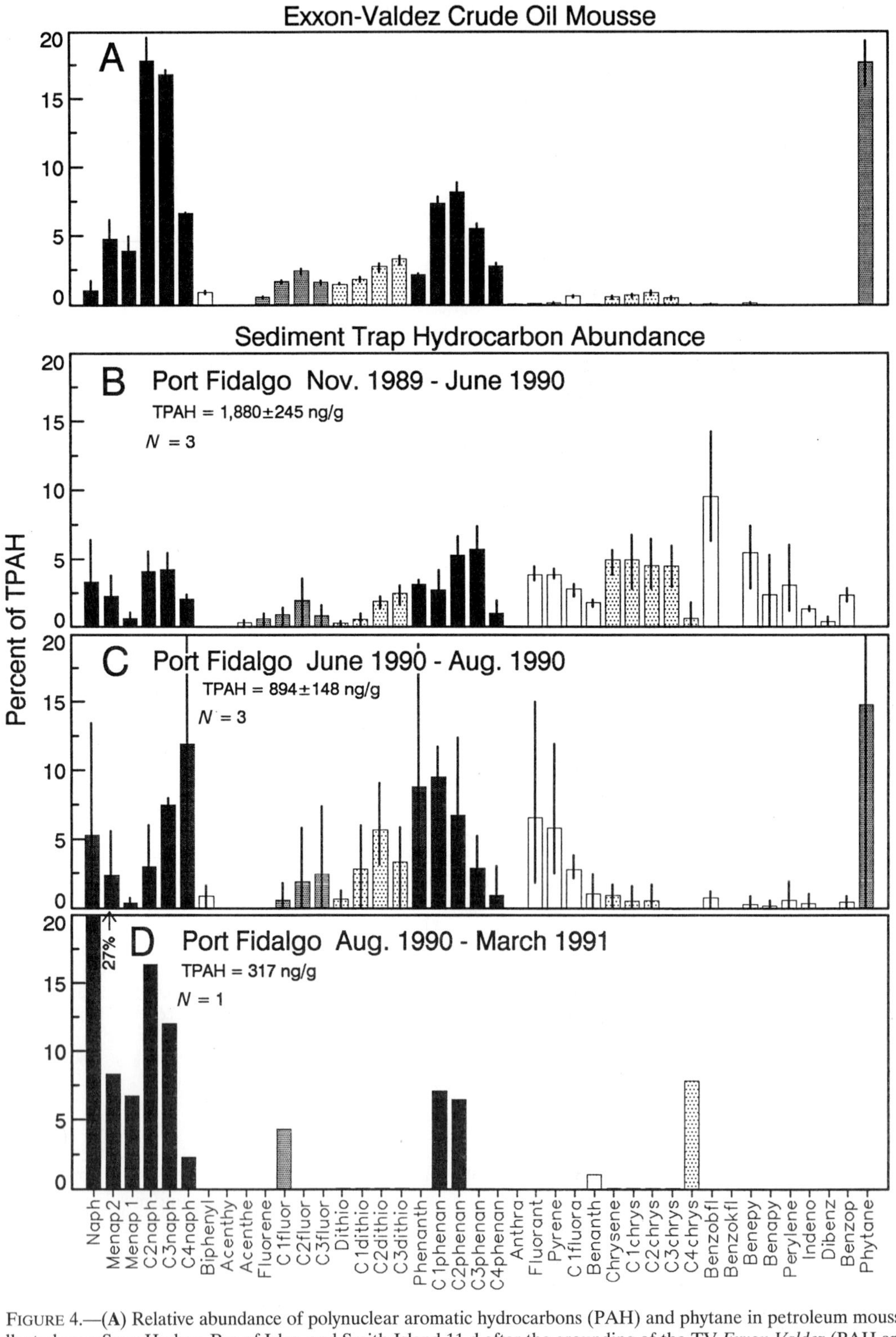

FIGURE 4.—(**A**) Relative abundance of polynuclear aromatic hydrocarbons (PAH) and phytane in petroleum mousse collected near Snug Harbor, Bay of Isles, and Smith Island 11 d after the grounding of the TV *Exxon Valdez* (PAH and phytane proportions from these three stations are composited for presentation in this figure); (**B–D**) relative abundance of PAH and phytane in sediments collected in 10-m depth traps deployed from November 1989 to June 1990, at the Port Fidalgo control station. Heavy bars indicate the proportion of the total PAH (TPAH) corresponding to the hydrocarbon indicated on the abcissa, and the associated vertical lines indicate the range of *N* replicated analyses. The mean TPAH concentration is also listed as nanograms of PAH per gram of dry sediment weight in **B–D**.

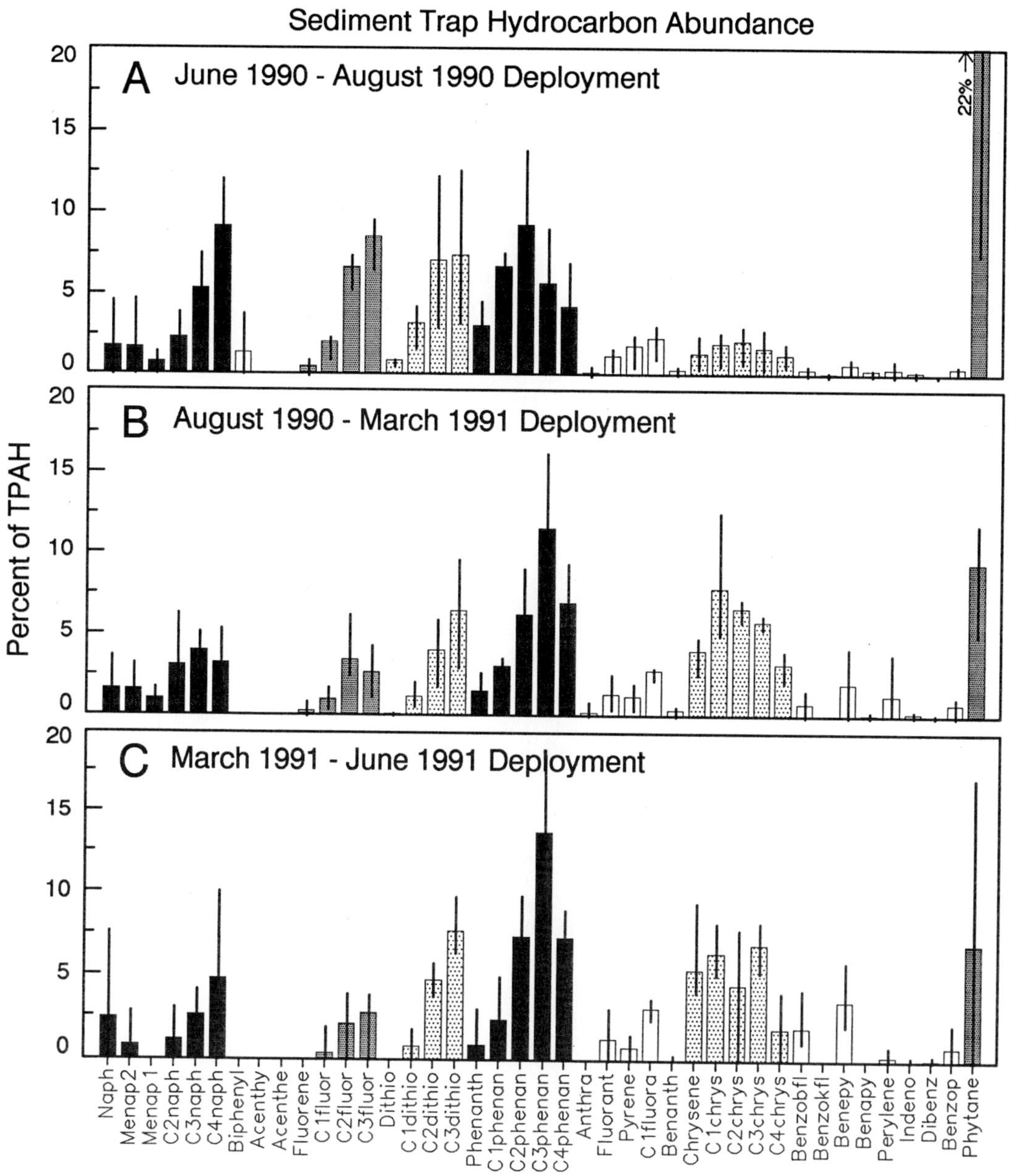

FIGURE 5.—Range of relative polynuclear aromatic hydrocarbons (PAH) and phytane abundances in sediments collected in 10-m depth traps deployed at four stations adjacent to heavily oiled beaches from (**A**) June 1990 to August 1990; (**B**) August 1990 to March 1991; and (**C**) March 1991 to June 1991. These ranges comprise all sediment-trap replicates for all four stations. TPAH = total PAH.

Harbor. The mean total PAH concentration was 2,610 ± 143 ng/g (N = 2), and although the unsubstituted 4- and 5-ring PAHs were relatively more abundant at Johnson Cove than at Stockdale Harbor, the PAHs were otherwise distributed similarly.

At two other stations, Eshamy Bay and Squire Island, mean total PAH concentrations were low (740 ng/g and 334 ± 23.7 ng/g, respectively), and although individual PAHs were detected less frequently, the overall distribution of PAHs was most similar to that at Stockdale Harbor.

During the March–June 1991 deployment inter-

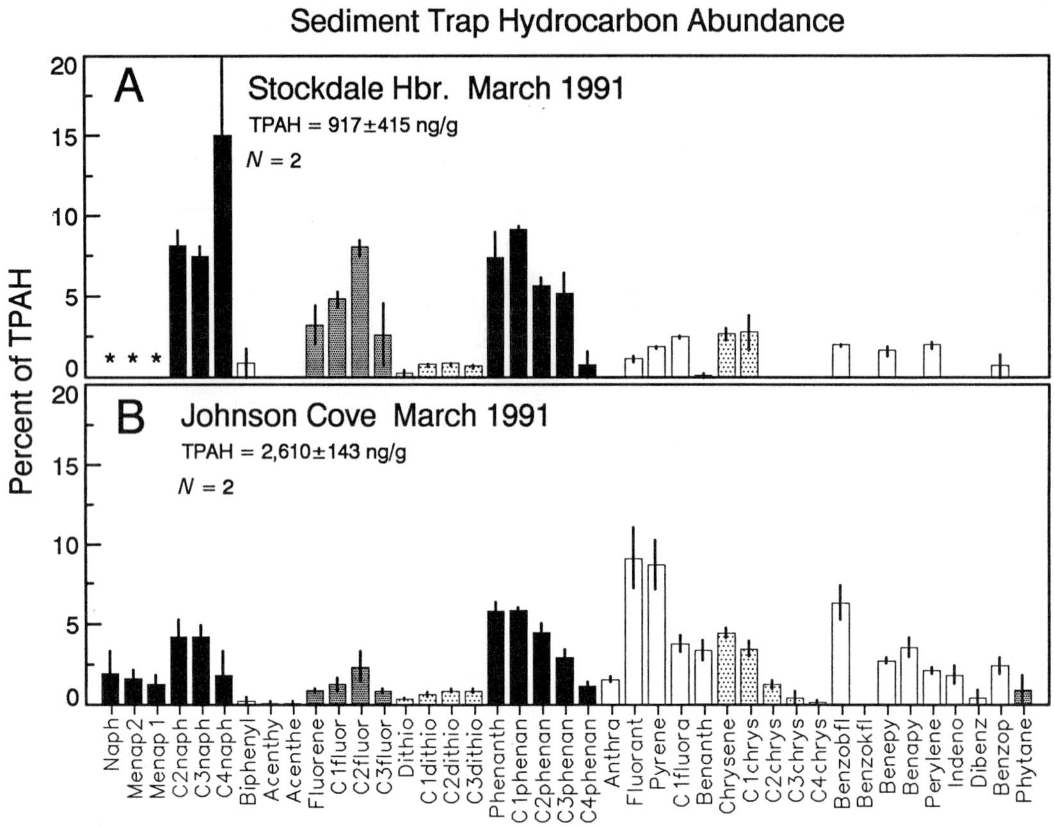

FIGURE 6.—Relative abundance of polynuclear aromatic hydrocarbons (PAH) in sediments collected in 10-m depth traps deployed from August 1990 to March 1991 at (**A**) Stockdale Harbor and at (**B**) Johnson Cove. Heavy bars indicate the proportion of the total PAH (TPAH) corresponding to the hydrocarbon indicated on the abcissa, and the associated vertical lines indicate the range of *N* replicated analyses. The TPAH units are nanograms of PAH per gram of dry sediment weight. Asterisk indicates data excluded because of unacceptable surrogate recoveries.

val, sediment traps were redeployed at the Disk Island station, and traps deployed at a station in the Bay of Isles in August 1990 were found. At Disk Island, mean total PAH concentrations of trapped sediments were 1,400 ± 192 ng/g ($N = 2$). These concentrations were substantially lower than those observed during the previous August 1990–March 1991 deployment interval and consisted primarily of naphthalenes and phenanthrenes; phytane was not detected in these sediments. At Bay of Isles, the total PAH concentration was 2,130 ± 750 ng/g ($N = 3$). The Bay of Isles station is adjacent to a beach that was heavily oiled, and the distribution of relative PAH abundances was consistent with weathered *Exxon Valdez* crude oil.

Port Fidalgo control station.—At the Port Fidalgo control station, the concentrations of trapped sediment PAHs were substantially lower than those at the four stations adjacent to oiled beaches during the first deployment period and exhibited a different distribution (compare Figures 2 and 4B). The mean total PAH concentration of sediments trapped at Port Fidalgo during the first deployment period was 1,880 ± 245 ng/g ($N = 3$), or less than 30% of the lowest station adjacent to an oiled beach during this period. In addition to the alkyl-substituted PAH homologues characteristic of petroleum sources, several unsubstituted homologues, including 4- and 5-ring PAHs, were relatively abundant in the trapped sediments at Port Fidalgo during the initial trap deployment (Figure 4B). Phytane was not detected in the sediments trapped during the initial deployment interval at Port Fidalgo. The total PAH concentration in sediments trapped during the deployment intervals June–August 1990 and August 1990–March 1991 were 894 ± 148 ng/g ($N =$

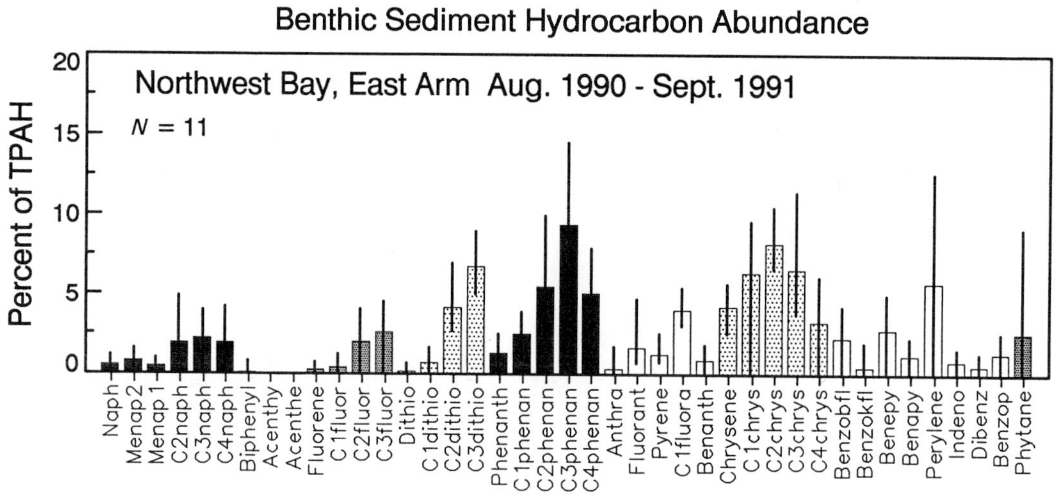

FIGURE 7.—Relative abundance of polynuclear aromatic hydrocarbons (PAH) and phytane in benthic surface sediments collected at Northwest Bay–East Arm at seawater depths ranging from 10 m to 20 m, from June 1990 through September 1991. Heavy bars indicate the proportion of the total PAH (TPAH) corresponding to the hydrocarbon indicated on the abcissa, and the associated vertical lines indicate the range of the 11 analyses included in this composite presentation.

3) and 317 ng/g (N = 1), respectively, and consisted primarily of naphthalenes, phenanthrenes, and fluoranthene-pyrenes, or just naphthalenes (Figure 4C and D). Phytane was detected sporadically in these sediments at concentrations near detection limits.

Hydrocarbons in Benthic Surface Sediments

Concentrations of individual PAHs in benthic surface sediments were often near or below detection limits, which precluded identification of relative PAH abundance patterns. Sediments that contained higher concentrations of total PAHs also contained greater numbers of PAH analytes measured above detection limits, and where comparisons were possible these samples usually had patterns of relative PAH abundances similar to those of corresponding trapped sediments.

Northwest Bay–East Arm.—Mean concentrations of total PAHs in benthic surface sediments were highest at Northwest Bay–East Arm, and the PAH composition was similar to that of PAHs in trapped sediments there. Mean total PAH concentrations at 10-m seawater depth increased from at least 426 ng/g (N = 1) in June 1990 to 2,080 ng/g (N = 1) in June 1991 and were above 1,200 ng/g in September 1991. Mean total PAH concentrations were generally similar on corresponding dates at 15-m and 20-m seawater depths, although at 20-m depth the concentration declined from more than 1,890 ng/g (N = 1) in March 1991 to 295–351 ng/g in June–September 1991. The pattern of relative PAH abundances in the benthic sediments at this station was consistently similar to the weathered *Exxon Valdez* crude oil pattern observed in the trapped sediments there (compare Figures 2, 4, and 7).

Bay of Isles, Northwest Bay–West Arm, and Disk Island.—Lower concentrations of total PAHs consistent with weathered *Exxon Valdez* crude oil are less clearly evident in benthic surface sediments at Bay of Isles, Northwest Bay–West Arm, and Disk Island. At Bay of Isles, mean total PAH concentrations for benthic sediments ranged from 39.9 ng/g (N = 1) to 269 ± 144 ng/g (N = 4) at 10-m seawater depths and ranged from 403 ng/g to 429 ng/g at 20-m seawater depths; the pattern of relative PAH abundances of the higher-concentration samples is similar to that of weathered *Exxon Valdez* crude oil. This was also the case at Northwest Bay–West Arm, although the mean total PAH concentration observed in benthic sediments there was lower, 223 ± 66.7 ng/g (N = 4). At Disk Island, mean total PAH concentrations at all depths were consistently less than 100 ng/g, except for one observation of 931 ng/g at 20-m seawater depth in September 1991. The pattern of relative PAH abundances in the anomalously high sample is similar to that of weathered *Exxon Valdez* crude oil, but in the remaining samples the concentrations of individual PAHs

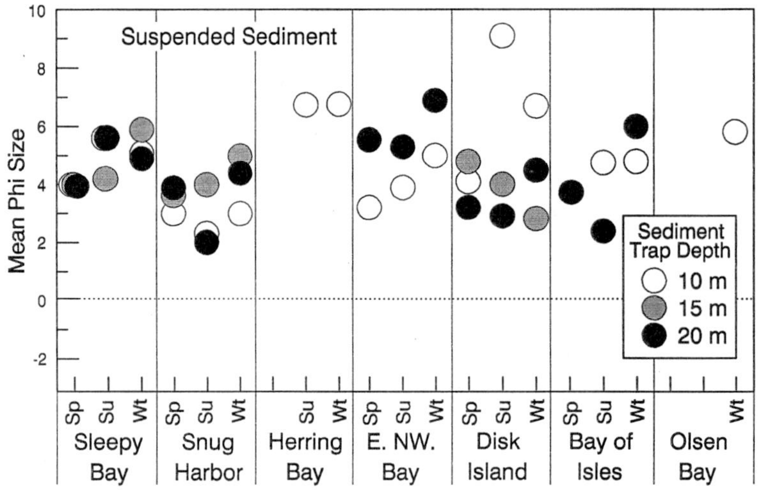

FIGURE 8.—Mean grain size of suspended sediment in phi units by location and season in Prince William Sound: Sp = spring deployment from March 1991 to June 1991; Su = summer deployment from June 1991 to September 1991; Wt = winter deployment from September 1991 to March 1992. Phi is the negative base 2 logarithm of the sediment size in millimeters ($-\log_2$ [size, in mm]). Larger phi values indicate finer-grained sediments.

were too frequently below detection limits to discern PAH distribution patterns.

Johnson Cove, Stockdale Harbor, and Sleepy Bay.—The mean total PAH concentration measured in Johnson Cove benthic sediments was 483 ± 102 ng/g (N = 3), and the pattern of relative PAH abundances was similar to that of the sediments trapped there. The similarity is also found at Stockdale Harbor, where benthic sediment PAH concentrations were measured at 245 ± 21.8 ng/g (N = 3) and 302 ng/g (N = 1). At Sleepy Bay, mean total PAH concentrations in benthic sediments ranged from 58.3 ± 26.4 ng/g (N = 4) to 357 ng/g (N = 1), and the higher-concentration samples had patterns of relative PAH abundances that were either similar to that at Stockdale Harbor, or at Northwest Bay.

Other stations.—Concentrations of total PAHs at the remaining stations sampled for benthic surface sediments were too low to discern PAH distribution patterns. At the Port Fidalgo control station, most PAHs were below detection limits, and mean total PAH concentrations were consistently less than 25 ng/g. Similarly, at Chenega Island, Eshamy Bay, Green Island, Herring Bay, Naked Island, and Snug Harbor, individual PAHs were frequently below detection limits, and mean total PAH concentrations were less than 125 ng/g. At Olsen Bay, a single measurement in September 1991 showed 267 ng/g total PAHs.

Suspended and Benthic Sediment Grain Sizes

All trapped, suspended sediments had a mean grain size of fine sand or finer, although there is some variability among sites and over time (Figure 8). The high-energy site at Snug Harbor had the coarsest sediments falling out of suspension, ranging from fine sand to medium silt. Traps set in Herring Bay in the summer and winter, eastern Northwest Bay in the winter at 20-m depth, and Disk Island in the summer at 10-m depth captured particularly fine sediments. Only the site in eastern Northwest Bay appeared to have a systematic trend in mean grain size with depth; here the sediments from 20 m were consistently finer-grained than sediments from 10 m.

At Sleepy Bay, Snug Harbor, eastern Northwest Bay, and the Bay of Isles, sediments collected during spring 1991 were generally coarser-grained than sediments collected during winter 1991–1992. At the Disk Island site, this pattern also occurred for sediments from 10- and 20-m depths but not from 15-m depth. A qualitative inspection of the trapped sediments indicated that this temporal pattern is

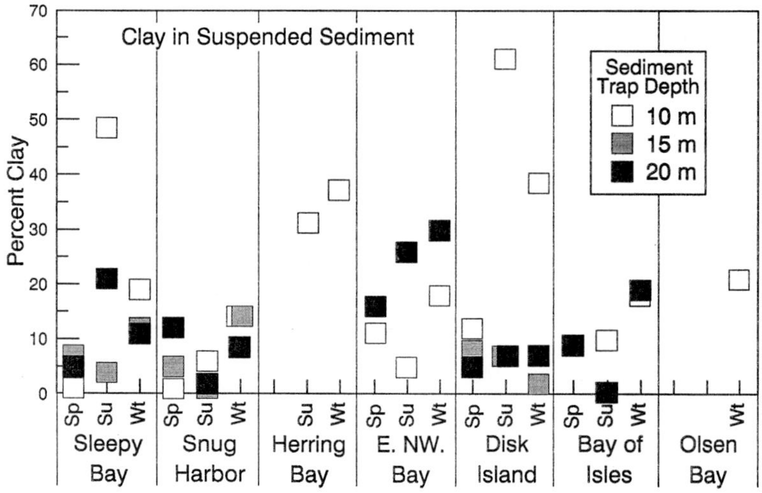

FIGURE 9.—Weight percent of clay-sized grains in suspended sediment by location and season in Prince William Sound: Sp = spring deployment from March 1991 to June 1991; Su = summer deployment from June 1991 to September 1991; Wt = winter deployment from September 1991 to March 1992.

caused by the relative abundance of diatoms in spring sediments compared to winter sediments. Diatom tests are discoidal in shape and are coarse silt to very fine sand sized.

Concentrations of clay-sized grains in suspended sediments ranged from 0% during summer in the Bay of Isles at 20-m depth to more than 60% collected during summer at Disk Island at 10-m depth (Figure 9). The only clear pattern among sites is that Snug Harbor, which is a high-energy site, consistently had low clay amounts of less than 15%. The deep, 20-m trap at Northwest Bay–East Arm consistently collected more clay-sized grains than the shallower 10-m trap, but at Disk Island the opposite was true. The other sites showed no depth-related patterns. Clay-sized sediment content is higher during winter than during spring and follows the pattern expected from the mean grain size parameter described above.

The presence of numerous benthic foraminifera and echinoderm spines in the core samples, but only a few benthic forams and no echinoderm spines in the trap samples, indicates that at least very fine-grained sand and coarser material was generally not resuspended and deposited in the sediment traps. Mud-sized grains, however, probably were resuspended from the bottom and deposited in the traps. The proportion of resuspended sediments to total sediments captured cannot be determined.

The mean grain size for bottom sediments taken adjacent to the traps in March and April of 1991 and 1992 ranged from pebble to silt sized (Table 3). There is no indication of a temporal shift in mean grain size (Figure 10) at Sleepy Bay, Snug Harbor, or Disk Island, but sediments in the East Arm of Northwest Bay became finer grained. The higher-energy sites at Sleepy Bay and Snug Harbor tend to have finer sediments than those at the relatively lower-energy sites of eastern Northwest Bay and the Bay of Isles. Sleepy Bay shows a trend of decreasing grain size from shallow to deeper water, but Northwest Bay, Disk Island, and the Bay of Isles show the opposite trend. Snug Harbor shows no trend either way.

Benthic sediments were coarser grained than suspended sediments collected during winter of 1991–1992, usually by more than 1 phi unit (Table 3). All areas except Snug Harbor had substantial amounts of gravel material on the bottom (10% or more). Benthic sediments at the Bay of Isles, Disk Island, the East Arm of Northwest Bay, and Olsen Bay were very poorly sorted to extremely poorly sorted, whereas sediments from the relatively high-energy locations of Sleepy Bay and Snug Harbor were sig-

TABLE 3.—Mean grain size and sorting comparisons of benthic and suspended sediments from study sites in Prince William Sound. Phi is the negative logarithm (base 2) of grain-size diameter in millimeters ($-\log_2$ diameter$_{mm}$). Larger phi values for mean grain size indicate finer-grained sediment, whereas smaller phi values indicate coarser-grained sediment. Sorting refers to the distribution of grain sizes in a sediment sample. High sorting values indicate a large range in grain size, and these sediments are considered poorly sorted. Depth refers to water depth (referenced to mean lower low water).

Location	Depth (m)	Mean grain size (phi)		Sorting	
		Benthic	Suspended	Benthic	Suspended
Bay of Isles, station 1	10	0.63	4.73	3.30	3.53
Disk Island, station 2	20	−0.48	4.56	2.90	2.22
Disk Island, station 2	15	0.47	2.67	3.81	2.65
Disk Island, station 2	10	0.97	6.60	0.89	4.35
East Northwest Bay, station 3	20	0.87	6.90	4.48	4.69
East Northwest Bay, station 3	10	4.47	5.04	3.37	2.86
Olsen Bay, station 8	10	5.15	5.70	3.97	2.91
Sleepy Bay, station 10	20	3.19	4.82	0.64	1.33
Sleepy Bay, station 10	15	3.27	5.87	0.96	1.93
Sleepy Bay, station 10	10	2.08	5.20	1.44	2.53
Snug Harbor, station 11	20	3.43	4.46	0.57	2.05
Snug Harbor, station 11	15	3.28	5.02	0.63	1.92

nificantly better sorted (moderately well sorted to poorly sorted). Gravel content was less at the better-sorted, high-energy sites. Suspended sediments were also substantially better sorted at the two high-energy sites than at the lower-energy sites; the suspended sediments were substantially less well sorted than the benthic sediments at Sleepy Bay and Snug Harbor. At the lower-energy sites, suspended sediments were very poorly to extremely poorly sorted, as was seen in the benthic sediments.

Clay Mineral Content of Suspended Sediments

The abundance ratio of chlorite to illite clay minerals ranged from 0.64, measured at the 10-m depth

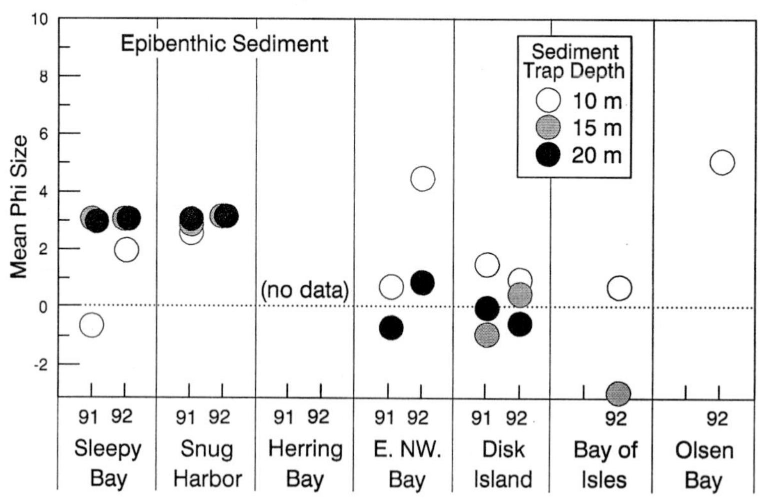

FIGURE 10.—Mean grain size of surficial bottom (benthic) sediment in phi units by location for retrievals in spring of 1991 and 1992. Sediments were taken from 0 to 2 cm of the sediment column. Phi is the negative base 2 logarithm of the sediment size in millimeters ($-\log_2$ [size, in mm]). Larger phi values indicate finer-grained sediments.

TABLE 4.—Chlorite and illite clay mineral content in suspended sediments by location and season in Prince William Sound: Sp = spring deployment from March 1991 to June 1991; Wt = winter deployment from September 1991 to March 1992. Depth refers to water depth (referenced to mean lower low water).

Location	Depth (m)	Illite (%)		Chlorite (%)		Chlorite–Illite ratio	
		Sp 1991	Wt 1991–1992	Sp 1991	Wt 1991–1992	Sp 1991	Wt 1991–1992
Bay of Isles, station 1	20	58	49	42	51	0.72	1.04
Bay of Isles, station 1	10		54		46		0.85
Disk Island, station 2	20	42	56	58	44	1.38	0.79
Disk Island, station 2	15	48	49	52	51	1.08	1.04
Disk Island, station 2	10	43	44	57	56	1.33	1.27
East Northwest Bay, station 3	20	53	45	47	55	0.89	1.22
East Northwest Bay, station 3	10	49	46	51	54	1.04	1.17
Herring Bay, station 5	10		44		56		1.27
Olsen Bay, station 8	10		61		39		0.64
Sleepy Bay, station 10	20	51	55	49	45	0.96	0.82
Sleepy Bay, station 10	15	49	46	51	54	1.04	1.17
Sleepy Bay, station 10	10	39	47	61	53	1.56	1.13
Snug Harbor, station 11	20	42	51	58	49	1.38	0.96
Snug Harbor, station 11	15	51	46	49	54	0.96	1.17
Snug Harbor, station 11	10	50	44	50	56	1.00	1.27

in Olsen Bay during winter 1991–1992, to 1.56, measured at the 10-m depth in Sleepy Bay during spring 1991 (Table 4). Most of the ratios fell between 0.8 and 1.3, with no spatial or temporal patterns. Most of the samples had ratios greater than 1.0, and every site except Olsen Bay had at least one sample greater than 1.0.

Benthic Sediment Cores

Consideration of the grain-size distribution of the upper 10–15 cm of the sediment column provides a more time-integrated view of the processes occurring at the sites, in contrast to the "snapshot" view obtained from the benthic sediment samples discussed above.

Benthic core sediments from the Disk Island and Snug Harbor stations are moderately sorted muddy sand. Cores at the remaining stations at Sleepy Bay, Northwest Bay–East Arm, the Bay of Isles, Herring Bay, and Olsen Bay have very poorly sorted mud, sand, and gravel; Olsen Bay is particularly muddy. The very poorly sorted core sediments are bimodal, with pebble and silt modes. Rock fragments are the dominant sediment type, but all stations have sand-sized to gravel-sized shell fragments and whole shells. All stations also have a few percent by weight of organic detritus that includes sea grass, macroalgae, wood fragments, spruce needles, and disseminated material.

Gastropod and bivalve shells are present at all the stations, and at Sleepy Bay there are barnacle fragments. Several bivalves and tube worms were found in life position. Remains of worm tubes were also present, and divers observed mounds of sediment created by worms at several stations. Benthic foraminifera were ubiquitous. Echinoid spines were abundant at the Disk Island, Snug Harbor, Northwest Bay–East Arm, and Bay of Isles stations. A few chitonous echinoderm pieces were also recovered in these cores. One core from the Bay of Isles had pellets. Some bioturbation was apparent in the cores, but physical sedimentary structures were preserved.

All cores except that from Olsen Bay were layered, indicating various levels of energy and various modes of sedimentation occurring over time. Layering typically consisted of gravel and shelly layers and layers with abundant organic debris or mud. Layers with normal grading (i.e., fining upward) and with erosional lower contacts are interpreted as storm deposits; these occurred at Sleepy Bay, Disk Island, Northwest Bay–East Arm, and the Bay of Isles. Cross-stratification was apparent in the sandy sediments of Disk Island and Snug Harbor and also in cores from Northwest Bay. The sediments at the relatively high-energy stations of Sleepy Bay, Disk Island, and Snug Harbor became finer-grained from 10-m to 20-m depth. These trends were caused by a decrease in wave energy with depth. The low-energy site in the Bay of Isles had an opposite trend, with finer-grained core sediment at 10-m than at 20-m depth. Northwest Bay–East Arm showed no difference in grain sizes of cores between 10-m and 20-m depths.

Discussion

Benthic Sediment Contamination by Exxon Valdez Oil

Our results indicate that benthic surface sediments at Northwest Bay–East Arm were contaminated to seawater depths as deep as 20 m by weathered crude oil from the *Exxon Valdez* oil spill and that the contaminating crude oil was associated with particulate matter mobilized off adjacent oiled beaches. Crude oil spilled from the TV *Exxon Valdez* is indicated as the proximate source of the PAHs that contaminated benthic surface sediments at Northwest Bay–East Arm for the following reasons: (1) the close correspondence of hydrocarbon analytes in the petroleum mousse collected from the sea surface 11 d after the spill, which was clearly from the spill, and the analytes in the sediments trapped during the initial deployment interval there; (2) the temporal association of the highest PAH concentrations in the trapped sediments during the first trap deployment interval succeeding the spill, followed by generally lower subsequent concentrations; (3) the similar patterns of relative PAH abundances in initially trapped sediments and in benthic surface sediments subsequently; and (4) the substantially lower concentrations and markedly different patterns of relative abundances of these PAHs in the deeper segments of sediment core samples, thereby obviating other background sources for the higher PAH concentrations of the benthic surface sediments.

The relatively high PAH concentrations found in the benthic sediments in the East Arm of Northwest Bay further corroborate the *Exxon Valdez* oil source. As discussed below, total PAH concentrations of shallow subtidal sediments along the spill path that are derived from PAH-laden sediments carried into the sound through Hinchinbrook Entrance usually do not exceed 400 ng/g. Total PAH concentrations of shallow subtidal sediments in the East Arm of Northwest Bay ranged up to 6 times higher, indicating that the contribution of PAHs derived from sediments imported into the sound is probably negligible at this site, at least to depths of 20 m.

Although the benthic surface sediments in the East Arm of Northwest Bay are contaminated with *Exxon Valdez* oil, the concentration of PAHs is much lower than that discovered in the sediment traps. There are two reasons for this discrepancy. The first involves the expression of total PAH concentration as percent by weight of sediment. Benthic sediments were much coarser grained than the suspended sediments caught in the traps (Figures 8 and 10). Thus, the level of contamination of the fine-grained fraction of the benthic surface sediments is diluted by the coarse-grained fraction, which may or may not carry PAHs. The second reason for the discrepancy is that wave or tide energy levels were high enough and fine-grained sediment supply low enough so that the highly contaminated mud-sized sediment mostly bypassed the site. This is the reason for the relatively coarse-grained and often gravely benthic sediments. Cores revealed a very poorly sorted, muddy, sandy, gravel in the upper 10 cm of the sediment column at this site. Crudely stratified shell hash interbedded with muddy, organic rich layers and organic debris that included spruce needles and wood fragments indicates that periods of relatively high energy, such as winter storms, were important in depositing the sediments recovered in the short cores. During the waning stages of storms and between storms, relatively fine-grained sediments may be deposited, but normal-energy levels from waves or tides are apparently high enough to remove fine-grained surface layers and form pebble lag deposits. Some mud is incorporated into storm deposits, particularly the upper portion that is deposited during waning storm periods, and all mud deposited between storms may not be removed before or during a subsequent high-energy event. Organisms, such as the whole clam recovered in life position in the 20-m core, may also work fine-grained sediment into the bottom. Much of the fine-grained sediment caught in the traps, however, ultimately bypassed the site and was either exported out of Northwest Bay or deposited in deeper, lower-energy sites within the bay.

The failure to consistently detect *Exxon Valdez* oil-derived PAHs in benthic surface sediments at the other locations where results from the sediment traps indicated an input of these PAHs is explained by higher-energy subtidal conditions. Along with variations in benthic topography, higher energy promoted the export of the finer-grained, oil-laden sediments either into open Prince William Sound or to deeper, lower-energy sites near these locations. Thus, the relatively high-energy subtidal conditions indicated at, for example, Disk Island and Sleepy Bay, led to relatively short residence times in shallow subtidal sediments for oil-laden sediments mobilized from the oiled upper intertidal zone of these locations. Even where less energetic subtidal conditions are indicated, such as at the Bay of Isles and Herring Bay, sediment movement was still sufficient to prevent substantial accumulations of oil-contaminated sediment in the shallow subtidal sites sampled at these locations.

Transport Mechanism of Exxon Valdez Oil-Contaminated Sediments

The nearly identical pattern of relative PAH abundances found in the sediment traps deployed initially at stations adjacent to oiled beaches, which persisted in these trapped sediments at least until June 1991, indicates *Exxon Valdez* crude oil was mobilized into the water column at these stations. The observation that concentrations of PAHs evidently derived from *Exxon Valdez* crude oil in trapped sediments at these stations generally declined with time after the spill, together with the proximity of persistent reservoirs of this oil on nearby beaches, suggests mobilization of oil from these reservoirs, either by fall and winter storms or by beach treatment activities. Suspended sediments collected during winter, however, generally have higher weight percents of clay-sized grains than spring and summer sediments (Figure 9). A qualitative inspection of the suspended sediments revealed that sand-sized, planktonic diatom tests were abundant in spring and summer sediments relative to winter sediments. Thus, the oil concentrations observed during the spring and summer may be depressed by the diluting effect of clean, newly produced biogenic sediment.

Jahns et al. (1991) concluded that the primary process for natural removal of shoreline oil by waves and tidal flushing is through the formation of an emulsion of micron-sized mineral particles, polar components of oil residue, and seawater that impeded the adhesion of oil to the larger rocks on the shorelines. Because these emulsion floc particles were composed mainly of seawater and fine-grained sediments, Stokes' law of settling velocities for individual grains suggests that the flocs could be transported great distances before settling and be widely dispersed. Other physical processes may however act to trap fine sediments in estuarine embayments (Drake 1976); physicochemical flocculation with other particles in the water column, for example, can produce flocculent settling rates that are an order of magnitude greater than those of the individual grains (Kranck 1975). The results of the sediment-trap hydrocarbon chemistry suggest that at least a portion of the shoreline oil was settling onto nearshore benthic sediments between 10 and 20 m below MLLW, depths that were generally within 75 to 400 m from the supratidal zone.

Beach treatment activities provided a secondary mechanism for mobilizing oil from beach reservoirs into adjacent seawater. These activities may account for the relatively less-weathered *Exxon Valdez* crude oil found in the sediment traps during summer 1990. During the 1990 cleanup season, beaches adjacent to several trap sites were tilled or raked, exposing subsurface oil. In some cases, oiled sediment was removed (although incompletely) or the beach was left "open" for oil removal by wave and tide energy. Sheens were commonly observed emanating from freshly tilled beaches. Only moderate weathering of subsurface oil occurring as concentrated layers of oiled sediment had occurred by 1991 (Michel and Hayes 1993). Such layers were present on beaches adjacent to a number of the trap sites.

Sources of Polynuclear Aromatic Hydrocarbons in Subtidal Sediments of Prince William Sound

There are two primary sources of PAHs in subtidal sediments of Prince William Sound. One source is natural PAH-laden sediments from outside the sound, and the other is from shoreline sediments contaminated with oil from the TV *Exxon Valdez*. The pattern of relative PAH abundance typified by Stockdale Harbor sediments closely matches the description given by Page et al. (1995) of sediment-borne PAHs that originate outside Prince William Sound. Page et al. (1995) speculate that this suite of PAHs originates as terrestrial or marine oil seeps that enter the Alaska coastal current east of the sound and are carried into the sound through Hinchinbrook Entrance, where current velocities drop, resulting in deposition of the naturally oiled sediments. The pattern of relative PAH abundances that characterizes this source has also been observed in deeper sediments of the sound by O'Clair et al. (1996, this volume) and in low intertidal sediments at Constantine Harbor (near Hinchinbrook Entrance) by Short and Babcock (1996, this volume). The PAHs measured in sediments trapped at Stockdale Harbor were almost certainly derived from this source, as were the other trapped and benthic sediments found with similar PAH distributions.

Molnia and Hein (1982) and Naidu and Mowatt (1983) concluded that the Copper River, via the central Gulf of Alaska to the east of Prince William Sound, is the dominant source of clay-sized sediment in the sound. Both teams showed that suspended sediments in the lower Copper River have high illite concentrations (average more than 50%) when compared to other coastal and shelf sediment sources in the northern Gulf of Alaska. Molnia and Hein (1982) interpreted the presence of illite-dominated sediments in Prince William Sound as indicating sediment influx from the Copper River. Naidu and Mowatt (1983) discovered small amounts of expandable clay-sized minerals in central Prince William Sound but an absence of these minerals in glacially derived coastal sediments of the sound. Expandable minerals are present in Copper River sediments and in sediments outside the sound, thereby also indicating an outside influx of sediment to the central sound. Images from *LANDSAT* (Burbank 1974; Molnia and Hein 1982) show that sediment plumes of the Copper River enter Prince William Sound through Hinchinbrook Entrance to the west. Furthermore, Royer (1981) showed that the freshwater-driven coastal current provides a mechanism for suspended sediment transport from the Copper River into Prince William Sound.

It is clear from the studies cited above and from this study that the deep central areas, and, to a lesser extent, the shallower areas throughout Prince William Sound receive illite-rich sediments from the Copper River that contain PAHs from natural sources outside the sound. The relative amounts of sediment from inside and outside Prince William Sound, however, are unknown. A study by Klein (1983) does indicate that illite is relatively more abundant in the deep (more than 300 m) parts of the sound just northwest of Hinchinbrook Entrance than in the shallower areas adjacent to islands to the southwest (such as Knight, Eleanor, Latouche, and Elrington islands). A remapping of Klein's data in terms of chlorite-to-illite ratios shows that this ratio is generally less than 0.6 in the central and northeastern sound northwest of Hinchinbrook Entrance but increases to the southwest along the east coast of Knight Island to more than 0.9. Higher ratios also occur around Eleanor and Ingot islands and are above 0.68. These trends indicate a decrease in the relative amounts of Copper River sediments away from the central sound northwest of Hinchinbrook Entrance and toward the sound's oiled islands.

Clay mineralogy of the suspended sediments collected during this study at 10- to 20-m depths strongly indicates a source from the adjacent littoral zone. This evidence is independent of the hydrocarbon chemistry discussed above. Chlorite-to-illite ratios of trapped sediments in this study are well above the values in the central sound and generally higher than any of the samples from Klein's study. The lowest ratio of the six oiled locations was 0.72, and most samples had values greater than 1.0 (Table 4). Klein's samples were taken from deeper than 40 m. Thus, the suspended sediments analyzed for our study were in shallower water and closer to shore than those used in Klein's study. The higher chlorite-to-illite ratios, therefore, indicate that the shorelines were the dominant source for clay-sized suspended sediments found in our traps.

The PAH concentrations measured in benthic surface samples at Stockdale Harbor, where PAHs were apparently derived primarily from natural sources external to the sound, together with concentrations measured in sediment core samples of similar PAH distribution, suggest that total PAH concentrations derived from these natural sources infrequently exceed 400 ng/g in benthic sediments at 10- to 20-m seawater depth, at least at locations along the spill path through the sound. Total PAH concentrations in benthic surface sediments at Stockdale Harbor remained below this value, where the input was demonstrated by the sediment-trap results. Also, total PAH concentrations in the deeper segments of sediment cores were usually much lower.

Conclusions

(1) Finer-grained intertidal sediments contaminated by *Exxon Valdez* oil were transported to adjacent shallow subtidal areas by waves associated with winter storms and by beach treatment activities. Oiled-sediment concentrations in the traps generally declined during the period November 1989–June 1991.

(2) Residence times of *Exxon Valdez* oil-contaminated sediments in the shallow subtidal areas adjacent to oiled beaches depended primarily on the energy available to transport these sediments to deeper, lower-energy sites within the embayments or to the open areas of Prince William Sound. At most sites where this sediment transport was observed, these residence times were probably less than 1 year. However, in the East Arm of Northwest Bay, the residence time may have been somewhat longer than 1 year.

(3) Conditions that promoted accumulation of oil-contaminated sediments in the shallow subtidal adjacent to oiled intertidal beaches include (a) beaches prone to onshore/offshore sediment transport caused by winter storms or treatment and (b) low-energy subtidal conditions caused by low tidal current velocities, low wave energy, or uneven bathymetry that forms pockets protected from tidal or wave action.

(4) Although benthic sediment contamination at 20 m and at shallower depths was low compared to suspended sediment concentrations, it is likely that higher levels of benthic sediment contamination are present in deeper, lower-energy fine-grained sedimentation sites within oiled embayments such as Northwest Bay and the Bay of Isles.

(5) Other sources of PAHs in shallow (to 20 m) subtidal sediments along the spill path through Prince William Sound, including local anthropogenic sources and natural sources external to the sound, rarely exceeded 400 ng/g total PAH. The relatively small contribution of PAH-laden sediments originating outside the sound to shallow subtidal sediments of the western sound is indicated by both hydrocarbon and mineralogical evidence.

References

Bassin, N. J., and T. Ichiye. 1977. Flocculation behavior of sediment and oil emulsions. Journal of Sedimentary Petrology 47:671–677.

Biscaye, P. E. 1965. Mineralogy and sedimentation of recent deep-sea clay in the Atlantic Ocean and adjacent seas and oceans. Geological Society of America Bulletin 76:803–832.

Boehm, P. D., and six coauthors. 1987. Comparative fate of chemically dispersed and beached crude oil in subtidal sediments of the arctic nearshore. Arctic 40 (Supplement 1):133–148.

Burbank, D. C. 1974. Suspended sediment transport and deposition in Alaskan coastal waters, with special emphasis on remote sensing by the ERTS-1 satellite. Master's thesis. University of Alaska, Fairbanks.

Clark, R. C., Jr., and W. D. MacLeod. 1977. Inputs, transport mechanisms, and observed concentrations of petroleum in the marine environment. Pages 91–224 *in* D. C. Malins, editor. Effects of petroleum on arctic and subarctic marine environments and organisms. Academic Press, New York.

Conover, R. J. 1971. Some relations between zooplankton and bunker C oil in Chedabucto Bay following the wreck of the tanker *Arrow*. Journal of the Fisheries Research Board of Canada 28:1327–1330.

Drake, D. 1976. Suspended sediment transport on shelves. Pages 127–158 *in* D. J. Stanley and D. J. P. Swift, editors. Marine sediment transport and environmental management. Wiley, New York.

Folk, R. L. 1974. Petrology of sedimentary rocks. Hemphill Publishing Company, Austin, Texas.

Folk, R. L., P. B. Andrews, and D. W. Lewis. 1970. Detrital sedimentary rock classification and nomenclature for use in New Zealand. New Zealand Journal of Geology and Geophysics 13:937–968.

Folk, R. L., and W. C. Ward. 1957. Brazos River bar: a study in the significance of grain-size parameters. Journal of Sedimentary Petrology 27:3–26.

Jahns, H. O., J. R. Bragg, L. C. Dash, and E. Owens. 1991. Natural cleaning of shorelines following the *Exxon Valdez* oil spill. Pages 167–176 *in* Proceedings, 1991 international oil spill conference. American Petroleum Institute Publication 4529, Washington, DC.

Klein, L. H. 1983. Provenances, depositional rates and heavy metal chemistry of sediments, Prince William Sound, south-central Alaska. Master's thesis. University of Alaska, Fairbanks.

Kranck, K. 1975. Sediment deposition from flocculated suspensions. Sedimentology 22:111–123.

Lewis, D. H. 1984. Practical sedimentology. Van Nostrand Reinhold Company Inc., New York.

Meyers, P. A., and J. G. Quinn. 1973. Association of hydrocarbons and mineral particles in saline solution. Nature 244:23–24.

Michel, J., and M. O. Hayes. 1991. Geomorphological controls on the persistence of shoreline contamination from the *Exxon Valdez* oil spill. Report of Research Planning, Inc. to NOAA (National Oceanic and Atmospheric Administration), Hazardous Materials Response Branch, Seattle.

Michel, J., and M. O. Hayes. 1993. Persistence and weathering of *Exxon Valdez* oil in the intertidal zone—3.5 years later. Pages 279–286 *in* Proceedings, 1993 international oil spill conference. American Petroleum Institute Publication 4580, Washington, DC.

Molnia, B. F., and J. R. Hein. 1982. Clay mineralogy of a glacially dominated, subarctic continental shelf: northeastern Gulf of Alaska. Journal of Sedimentary Petrology 52(2):515–527.

Naidu, A. S., and T. C. Mowatt. 1983. Sources and dispersal patterns of clay minerals in surface sediments from the continental-shelf areas off Alaska. Geological Society of America Bulletin 94:841–854.

O'Clair, C. E., J. W. Short, and S. D. Rice. 1996. Contamination of intertidal and subtidal sediments by oil from the *Exxon Valdez* in Prince William Sound. American Fisheries Society Symposium 18:61–93.

Owens, E. H., J. R. Harper, W. Robson, and P. D. Boehm. 1987. Fate and persistence of crude oil stranded on a sheltered beach. Arctic 40 (Supplement 1):109–123.

Page, D. S., P. D. Boehm, G. S. Douglas, and A. E. Bence. 1995. Identification of hydrocarbon sources in the benthic sediments of Prince William Sound and the Gulf of Alaska following the *Exxon Valdez* oil spill. Pages 41–83 *in* P. G. Wells, J. N. Butler, and J. S. Hughes, editors. *Exxon Valdez* oil spill: fate and effects in Alaskan waters. American Society for Testing and Materials Special Technical Publication 1219, Philadelphia.

Pavia, E. 1991. Shoreline evaluation group, final report, Prince William Sound. Report to the Alaska Department of Environmental Conservation, Juneau.

Royer, T. C. 1981. Baroclinic transport in the Gulf of Alaska—part II, a freshwater driven coastal current. Journal of Marine Research 39:251–266.

Sale, D. M., J. C. Gibeaut, and J. W. Short. 1994. Nearshore transport of hydrocarbons and sediments following the *Exxon Valdez* oil spill—subtidal study 3b. Final report, *Exxon Valdez* Natural Resources Damage Assessment. Report (Contract 18-6018-93) to the Alaska Department of Environmental Conservation, Juneau.

Short, J. W., and M. M. Babcock. 1996. Prespill and postspill concentrations of hydrocarbons in mussels

and sediments in Prince William Sound. American Fisheries Society Symposium 18:149–166.
Short, J. W., T. J. Jackson, M. L. Larsen, and T. L. Wade. 1996. Analytical methods used for the analysis of hydrocarbons in crude oil, tissues, sediments, and seawater collected for the natural resources damage assessment of the *Exxon Valdez* oil spill. American Fisheries Society Symposium 18:140–148.
Wolfe, D. A., and eleven coauthors. 1994. Fate of the oil spilled from the T/V *Exxon Valdez* in Prince William Sound, Alaska. Environmental Science and Technology 28:561A–568A.
Woods Hole Oceanographic Institution. 1989. Sediment trap technology and sampling. U.S. Global Ocean Flux Planning Report 10, Woods Hole, Massachusetts.

American Fisheries Society Symposium 18:61–93, 1996

Contamination of Intertidal and Subtidal Sediments by Oil from the *Exxon Valdez* in Prince William Sound

C. E. O'Clair, J. W. Short, and S. D. Rice

National Oceanic and Atmospheric Administration, National Marine Fisheries Service
Auke Bay Laboratory, 11305 Glacier Highway, Juneau, Alaska 99801, USA

Abstract.—In the 3 years after the *Exxon Valdez* oil spill, we sampled subtidal sediments at 45 sites in Prince William Sound to determine the geographical, bathymetric, and temporal distribution of oil in benthic sediments. We sampled sediments at mean lower low water (0 m) and at subtidal depths of 3, 6, 20, 40, and 100 m. Triplicate composite sediment samples were collected at each station in spring, summer, and fall each year (1989–1991). Our results are based on 1,278 sediment samples analyzed by gas chromatography–mass spectrometry.

At locations with oiled shorelines, sediments from 0 m, several meters below the zone where the oil initially stranded, had the greatest concentrations of total polynuclear aromatic hydrocarbons excluding perylene (TPAH) found in benthic sediments in 1989. The highest average TPAH concentration was 12,700 ng/g at 0 m at Disk Island in July 1989.

Shallow subtidal sediments (3–20 m) had a PAH composition pattern characteristic of weathered *Exxon Valdez* oil (EVO-PAH pattern) at 80% of the oiled locations in Prince William Sound in 1989. Contamination of subtidal sediments by *Exxon Valdez* oil reached a depth of at least 20 m at five of those locations sampled in 1989. The greatest concentration of oil was at shallow depths.

We found little evidence of *Exxon Valdez* oil in sediments from great depths (≥40 m); the TPAH concentration in these sediments was similar at oiled and reference locations. The EVO-PAH pattern was rarely present consistently in replicate samples from these depths. The TPAH concentration in sediments from ≥100 m usually exceeded that observed at 40 m.

At heavily oiled locations, the total *n*-alkane (TNA) concentration was greatest where the TPAH concentration was greatest and the EVO-PAH pattern was observed. The highest TNA concentration was 31,100 ng/g in intertidal sediments from Northwest Bay (July 1989). The TNA concentration varied widely among reference and assessment stations where the EVO-PAH pattern was absent.

After 1989, hydrocarbons matched the EVO-PAH pattern less often, *Exxon Valdez* oil hydrocarbons were more patchily distributed, and concentrations of these hydrocarbons decreased in benthic sediments. By 1990, the mean TPAH concentration in sediments from 0 m at many oiled sites had declined to 100–200 ng/g, comparable to that of the prespill background. In 1990 in shallow subtidal sediments, the EVO-PAH pattern was consistently present only at Block Island and Northwest Bay. No consistent trend in TPAH concentration was apparent with depth. In 1991 a further decrease in the TPAH concentration was observed at oiled sites; the EVO-PAH pattern was consistently present only at Northwest Bay.

In contrast to the intertidal region where *Exxon Valdez* oil caused major biological impacts, subtidal sediments probably did not have oil concentrations that were acutely toxic to benthic or demersal macrobiota; however, the oil in shallow subtidal sediments may have induced subtle changes in biotic communities in heavily oiled bays. At greater depths (≥40 m) contamination by *Exxon Valdez* oil was negligible. The changes, if any, that occurred in deep benthic communities as a result of this oil spill were probably obscured by natural variability.

After the grounding of the tanker *Exxon Valdez* in Prince William Sound, a substantial proportion of the 43.9 million liters of Prudhoe Bay crude oil released into the marine environment became stranded on the shoreline of the sound and the northeastern Gulf of Alaska. By 1 May 1989, about 41% of the spilled oil had been beached in the sound (Wolfe et al. 1994). Some of the beached oil and oil-contaminated fine sediments were reintroduced into nearshore waters by such processes as wave-induced redistribution, leaching by groundwater flowing through the intertidal region from the backshore, and shoreline cleanup activities (Boehm et al. 1987; Owens et al. 1987; Michel et al. 1991; Mearns 1996, this volume), and this beached oil and oiled fine sediments were subsequently transported to subtidal sediments at least as deep as 20 m in Prince William Sound (Short et al. 1996a, this volume). The mechanisms of transport to subtidal sediments may have included sorption to settling particulate matter and encapsulation into and subsequent settlement of zooplankton feces (Conover 1971; Meyers and Quinn 1973; Bassin and Ichiye 1977; Clark and MacLeod 1977). Wolfe et al. (1994) estimated that about 13% of the spilled oil was transported to the subtidal region and remained there as of October 1992, mostly in the Gulf of Alaska.

The purpose of this study was to determine the geographical and bathymetric distribution of petroleum hydrocarbons from the *Exxon Valdez* oil spill in subtidal sediments in Prince William Sound and to follow the temporal changes in the concentrations of these hydrocarbons. The chemical composition of the hydrocarbons was used to distinguish *Exxon Valdez* oil as the source of the hydrocarbons from other hydrocarbon sources, including vessel traffic, major forest fires, oil storage tanks ruptured during the Alaska earthquake of 1964, or natural polynuclear aromatic hydrocarbon (PAH) sources such as petroleum from coastal oil seeps (Kvenvolden et al. 1993; Page et al. 1995), or coal eroded from terrestrial deposits and introduced into the Alaska Coastal Current by riverine flow.

Information on the distribution and persistence of *Exxon Valdez* oil in subtidal sediments is useful in the reconstruction of a spatial–temporal mass balance of the oil in the spill area and can contribute to an assessment of the extent to which various shoreline treatments redistribute oil from the intertidal region to the subtidal region (Wolfe et al. 1994; Mearns 1996). More importantly, perhaps, knowledge of the concentration of *Exxon Valdez* oil hydrocarbons in subtidal sediments over space and time provides information on the level of exposure of subtidal biota to the toxic components of the oil.

Methods

Study Sites

We describe geographical position using three terms: location, site, and station. Location refers to a general area where one or more sampling sites were established (e.g., Northwest Bay). Site refers to a smaller geographical area containing the bathymetric transect used to sample benthic sediments (e.g., NWBay4 = site #4 in Northwest Bay). Station refers to a specific spot along a bathymetric transect where samples were collected (e.g., the 20-m station).

Sampling locations were selected to represent a variety of initial shoreline oiling conditions that range from non-oiled beaches outside the path of the spill, to heavily oiled beaches in north-facing embayments. Beach oiling classifications of heavily, moderately, or lightly oiled were based on shoreline oiling maps produced by the Alaska Department of Environmental Conservation; these classifications are collectively denoted as assessment locations. Reference locations are those where no oil came ashore.

TABLE 1.—Number of sites sampled and number of sediment samples analyzed from Prince William Sound during the first 3 years after the *Exxon Valdez* oil spill.

Year	Month	Number of sites	Number of samples analyzed
1989	May	7	42
	Jul	15	245
	Sep	13	116
	Nov–Dec	21	135
1990	May–Jun	26	164
	Jun–Jul	22	233
	Sep	12	94
1991	Apr–May	9	44
	Jun	12	166
	Sep	10	39

Sediments were sampled at 25 locations (45 sites) in Prince William Sound in 1989 (7 reference locations and 18 assessment locations; Table 1, Figure 1). Fourteen locations were studied intensively, typically involving sampling at six stations that extended from the intertidal region to 100 m in July. Two additional locations were sampled at three depths (intertidal, 3 m, and 6 m). Usually one site was established at each location, but up to four sites were sampled at some locations. In 1990, sediments were sampled at 25 locations (11 reference locations and 14 assessment locations). In 1991, the total number of locations sampled was 16 (7 reference locations and 9 assessment locations).

Sediment Collection

Intertidal collections were made at about mean lower low water (MLLW; 0 m; actual sampling elevation was within the range +0.5 to −1 m, depending on the distribution of fine sediments). At one site (Northwest Bay), additional intertidal sediment samples were collected at 2-m intervals along a 20-m transect run perpendicular to the shoreline beginning at MLLW and extending toward the upper intertidal zone. These samples were collected to assess the local variability of hydrocarbons in intertidal sediments with elevation on the shore.

Subtidal sediments were collected at depths of 3, 6, 20, 40, and 100 m below MLLW. Additional depths were sampled at some sites. Collections at 3, 6, and 20 m were made by divers on transects laid along the appropriate isobath. (In July 1989, the 20-m station was usually sampled with a Van Veen grab, except where remote sampling could not be conducted effectively, such as in boulder fields or dense algal beds.) Three samples, each a composite

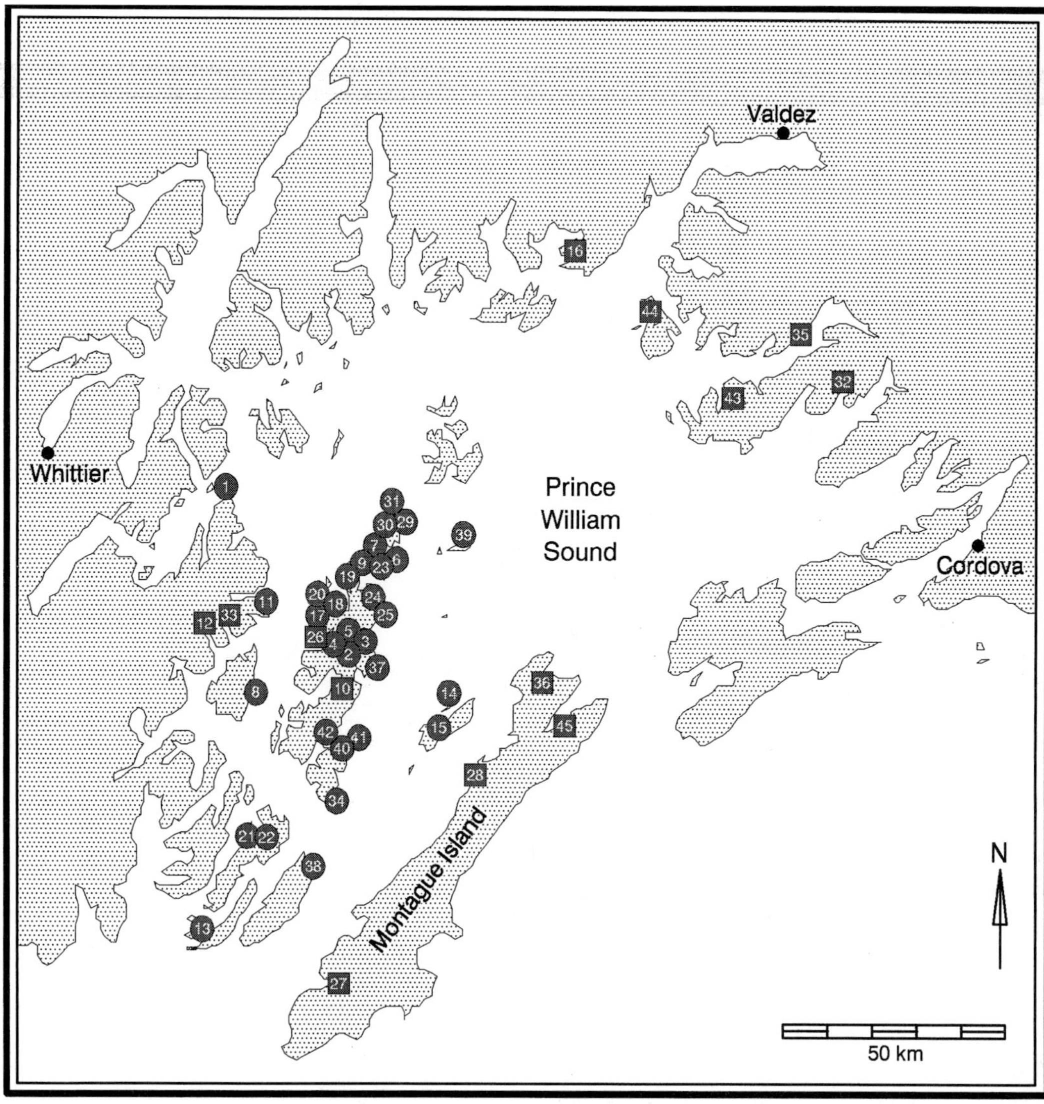

FIGURE 1.—Distribution of study sites in Prince William Sound. See Tables 2 and 3 for the geographical coordinates of each location sampled in 1989, 1990, and 1991. Numbered locations are (1) Applegate Island, (2) Bay of Isles 86, (3) Bay of Isles 90, (4) Bay of Isles bioremediation, (5) Bay of Isles, (6) Block Island 7, (7) Block Island 47, (8) Chenega Island, (9) Disk Island, (10) Drier Bay, (11) Eshamy Bay, (12) Ewan Bay, (13) Foxfarm, (14) Green Island 22, (15) Green Island, (16) Heather Bay, (17) Herring Bay 53, (18) Herring Bay 110, (19) Herring Bay 125, (20) Herring Bay, (21) Iktua Bay 2, (22) Iktua Bay, (23) Ingot Island, (24) northeast Knight Island bioremediation, (25) Knight Island, (26) lower Herring Bay, (27) Macleod Harbor, (28) Moose Lips Bay, (29) Northwest Bay 4, (30) Northwest Bay 5, (31) Northwest Bay, (32) Olsen Bay, (33) Paddy Bay, (34) Point Helen, (35) Port Fidalgo, (36) Rocky Bay, (37) Rua Cove, (38) Sleepy Bay, (39) Smith Island, (40) Snug Harbor 25, (41) Snug Harbor, (42) Snug Harbor meiofauna, (43) Two Moon Bay, (44) West Bay, (45) Zaikov Bay. Squares indicate reference locations; circles indicate assessment locations.

of eight subsamples collected randomly along a 30-m transect laid along the appropriate isobath, were taken at each of the shallow stations (0–20 m). Only one sample was analyzed from some triplicate sample sets to reduce analytical costs.

A Van Veen grab (1989) or a stainless steel Smith-McIntyre grab (1990–1991) was used to collect samples at 40- and 100-m depths. Three grab samples were taken at each depth. Four cores were removed from randomly selected points on the sur-

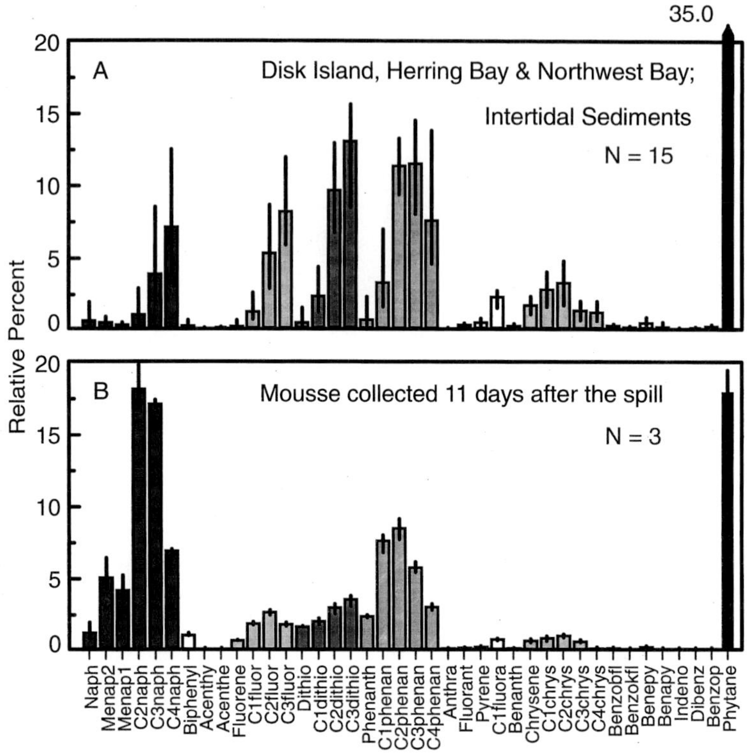

FIGURE 2.—Mean relative percents of polynuclear aromatic hydrocarbon compounds and phytane in intertidal sediments (A) at Disk Island, Herring Bay, and Northwest Bay combined and (B) in *Exxon Valdez* mousse collected 11 d after the oil spill of 24 March 1989. Error bars are ranges of *N* analyses.

face of the sediment contained in each grab. The subsamples were combined to form one sample per grab.

All samples were taken from the surface (top 0–2 cm) of the sediment with a 3.6-cm diameter stainless steel core tube and chrome-plated spatula or spoon. The core tube, spatula, and spoon were washed, dried, and rinsed with dichloromethane between sampling periods. Sample jars that were certified hydrocarbon-clean according to Environmental Protection Agency standards were used to store sediments. If these jars were unavailable, glass jars baked at 440°C or rinsed with dichloromethane were substituted. The jars were fitted with teflon-lined caps that were also rinsed with dichloromethane before use. Samples were kept cool after collection and frozen within a few hours.

In 1989, sampling in Prince William Sound was conducted during three periods by the National Oceanic and Atmospheric Administration (NOAA) (3–13 May, 1–18 July, and 5–11 September) and during one period (7 November to 8 December) by the Alaska Department of Environmental Conservation (Lindstrom and Weiner 1990). In 1990, sampling was conducted during three periods (31 May to 9 June, 27 June to 23 July, and 4–16 September). Dates of sampling in 1991 were 27 April to 2 May, 15–25 June, and 5–9 September.

Hydrocarbon and Data Analysis

Sediment samples were analyzed for selected hydrocarbons by the Geochemical and Environmental Research Group (GERG) at Texas A&M University. Short et al. (1996b, this volume) summarize the analytes, methods, detection limits, and the accuracy and precision of the methods used in the chemical analysis.

Results of the chemical analysis were screened on the basis of surrogate recoveries and minimum detection limits (MDLs). Individual analytes and the summary statistics affected by them (e.g., total polynuclear aromatic hydrocarbons including perylene [ΣPAH], total polynuclear aromatic hydrocarbons excluding perylene [TPAH], total normal alkanes [TNA; i.e., the sum of the normal alkanes], and total hydrocarbons) were excluded from the analysis if the recoveries of corresponding analyte surrogates fell outside the range of 30–150%. Concentrations of individual analytes reported by GERG below MDL were replaced by zeroes for our analyses. Hydrocarbon concentrations are reported on a dry weight basis to three significant figures when concentrations exceeded 10 ng/g, and to two significant figures for lower concentrations. A total of 1,443 sediment samples was analyzed for hydrocarbons from 45 sites over a period of 3 years. A group of 165 samples was excluded from the study because of extraneous contamination that occurred probably after sample collection (O'Clair et al. 1995).

The high sulfur content of *Exxon Valdez* oil helped to distinguish it from other PAH sources. In particular, concentrations of alkyl-dibenzothiophenes that reach at least 30% of the concentrations of alkyl-phenanthrenes are characteristic of higher-sulfur oils such as *Exxon Valdez* oil, and the presence of alkyl-chrysenes (at concentrations 10% or greater than those of alkyl-phenanthrenes) distinguishes this oil from products refined from it (Short et al. 1996b). Accordingly, we used the following criteria to compare hydrocarbon concentrations in sediments with those in *Exxon Valdez* oil. The pattern of PAH concentrations in the sediment samples was judged similar to the pattern in the oil if it consistently met each of three criteria in all replicated samples: (1) the ratio of alkyl-dibenzothiophenes (summed) to alkyl-phenanthrenes (summed) exceeded 0.30, (2) the ratio of alkyl-chrysenes (summed) to alkyl-phenanthrenes (summed) exceeded 0.10, and (3) the concentration of alkyl-phenanthrenes (summed) exceeded 5.0 ng/g. This latter criterion is necessary to insure that chrysenes would be detected if present.

The carbon preference index (CPI; Farrington and Tripp 1977) was also used to evaluate hydrocarbon sources. The index has the form

$$\text{CPI} = \frac{2(n\text{-C}_{27} + n\text{-C}_{29})}{n\text{-C}_{26} + 2n\text{-C}_{28} + n\text{-C}_{30}},$$

where n-C_i is the concentration (ng/g) of the n-alkane of carbon number i. The CPI is near 1 for oiled sediments; higher values indicate biogenic sources of these alkanes.

Composition patterns of PAHs are presented as mean relative percents in Figures 2, 3, and 5. Relative percent for each PAH is determined as the ratio of the PAH concentration and the TPAH concentration, multiplied by 100. Mean relative percent is the mean of these ratios for a specified set of samples. Variability of relative percents for each PAH is presented as the range in Figures 2 and 5 (where $N \leq 15$) and as 95% confidence intervals derived from arcsine-transformed relative percents in Figure 3 (where $N = 97$).

A t-test was used to test the differences between means of analyte concentrations when comparing pre- and postspill levels of PAHs in intertidal sediments at Olsen Bay and Rocky Bay.

Results

Heavily Oiled Locations, 1989

Oil became stranded extensively on shorelines in Prince William Sound, particularly in bays facing north, in the first weeks after the *Exxon Valdez* oil spill. Intensive cleaning activity took place at many of the most heavily oiled shorelines in summer 1989 and to a lesser extent in summer 1990. Some of the stranded oil was resuspended and deposited in the subtidal region over time by natural processes and by the shoreline treatment. The intensive treatment probably increased the variability among sampling periods of hydrocarbon concentrations in intertidal sediments at some locations in 1989.

In 1989, 10 assessment locations were sampled; 5 (Disk Island, Herring Bay, Northwest Bay, Sleepy Bay, and Snug Harbor) were heavily oiled in the upper intertidal zone. Of the remaining five locations, one was a protected bay that was sporadically oiled (Bay of Isles), two were in the southwestern passages of Prince William Sound along the exit path of the floating oil (Iktua Bay and Foxfarm), and two were near the periphery of the spill path (Green Island and Eshamy Bay).

Intertidal stations.—We found high TPAH concentrations in intertidal sediments (0-m station) at three of the heavily oiled locations: Disk Island, Herring Bay, and Northwest Bay (Table 2). Because we sampled at about MLLW, the concentrations of *Exxon Valdez* oil hydrocarbons in our intertidal sediments were probably lower than those at upper shore levels where the oil had become stranded. The highest mean TPAH concentration found during this study was 12,700 ± 2,760 ng TPAH/g dry sediment weight ($N = 3$; mean ± SE) in intertidal samples collected at Disk Island (Table 2). No intertidal sediment samples were collected subsequently from Disk Island in 1989, but samples collected from Northwest Bay in July 1989 contained 11,700 ± 3,280 ng/g ($N = 2$) and in September

TABLE 2.—Concentration of total polynuclear aromatic hydrocarbons excluding perylene (TPAH) in sediments from all stations at assessment locations where shores were moderately to heavily oiled in Prince William Sound by the *Exxon Valdez* oil spill, 1989–1991. Numbers in the body of the table are mean TPAH (ng/g), number of replicates analyzed (superscripts), and coefficient of variation (in parentheses). Numbers are preceded by an asterisk where the PAH composition pattern matched that of weathered *Exxon Valdez* oil for each replicate contributing to that data point. The abbreviation S denotes surrogate recoveries for one or more PAH analytes outside acceptable range. For "other" depths (last column), see footnotes. Bay of Isles, Foxfarm, and Rua Cove were moderately oiled. The rest were heavily oiled, although Sleepy Bay and Snug Harbor were heavily oiled only in parts.

Site			Depth		
Number	Name	Date	0 m	3 m	6 m
2	Bay of Isles 86	Nov 89		*485	
3	Bay of Isles 90	Nov 89		*510^2 (14.0)	* 625^3 (50.2)
		Jun 90		*241^2 (66.4)	* 473^3 (37.7)
4	Bay of Isles BR	Jul 90	S		* 909^2 (14.2)
5	Bay of Isles	Jul 89	408^3 (135.8)	383^3 (68.7)	26.9^3 (47.3)
		Sep 89	* 243^3 (11.2)	250^3 (120.8)	852^3 (138.1)
		Jun 90	* 1,480	84.0^3 (36.3)	106^3 (84.4)
		Jul 90	* 749^2 (72.0)	59.8^2 (38.9)	7.6^3 (44.6)
		Sep 90			88.3^3 (53.1)
		Jun 91	25.4^3 (65.0)	*124^3 (65.0)	64.4
6	Block Island 7	Nov 89		*539	
7	Block Island 47	Nov 89		*708^3 (12.7)	* 808^3 (27.2)
		Jun 90	* 945		* 762
		Jul 90	* 665^2 (61.0)	*700^2 (10.8)	* 389^2 (17.4)
9	Disk Island	Jul 89	*12,700^3 (37.6)	553^3 (54.1)	160
		Jun 90	* 131	* 41.5	38.4
		Jul 90	* 104	17.2^2 (129.9)	6.5^2 (19.9)
		Jul 91			
13	Foxfarm	May 89	* 290^3 (2.0)		98.2^3 (43.9)
		Jul 89	* 160^3 (103.2)	802^2 (3.8)	201^3 (18.7)
		Sep 89	68.5^2 (35.9)		97.8^3 (53.6)
		Jun 90	148^2 (11.4)	102^3 (67.8)	88.8^3 (60.0)
		Jul 90	65.2^2 (48.5)	5.5	47.1^2 (107.2)
		Sep 90	31.1^3 (99.3)	242^3 (95.3)	41.5^3 (73.7)
		Apr 91	57.6^2 (22.6)		534^2 (134.4)
		Sep 91	46.6^3 (72.6)		40.7^3 (72.9)
17	Herring Bay 53	Nov 89		*559^2 (14.5)	136^2 (19.6)
18	Herring Bay 110	Nov 89		70.0^2 (129.0)	156^2 (42.4)
18	Herring Bay 125	Nov 89		164^2 (33.9)	* 220^3 (50.5)
		Jun 90	116^3 (42.3)	* 62.3^2 (52.2)	88.7^3 (85.1)
20	Herring Bay	May 89	* 1,340^3 (37.0)		73.0^3 (35.1)
		Jul 89	* 1,840^3 (71.0)	*921^2 (27.9)	160^3 (89.4)
		Sep 89	* 769^3 (26.0)	*239^3 (26.6)	19.7^3 (102.0)
		Jun 90	* 215^3 (40.0)	*178^3 (37.0)	* 72.3^3 (158.5)
		Jul 90	* 949^3 (73.6)	S	14.2
		Sep 90	* 118^3 (41.9)	370^3 (97.3)	8.4^3 (15.4)
		Apr 91	* 101^2 (19.8)		519^3 (120.3)
		Jun 91	2,350^2 (138.2)	229^3 (123.8)	34.8^3 (22.6)
		Sep 91	41.7^3 (51.6)		* 105^3 (47.8)
29	Northwest Bay 4	Nov 89		*614	* 2,660
30	Northwest Bay 5	Dec 89		*194	
31	Northwest Bay	Jul 89	*11,700^2 (39.7)	*1,490^3 (88.6)	* 517^3 (43.1)
		Jul 89			
		Sep 89	* 2,860^3 (18.5)	* 684^2 (26.1)	*1,060^3 (16.3)
		Jun 90	* 166^3 (31.6)	*1,060	*1,000^3 (38.7)
		Jul 90	* 4,650	* 540^2 (0.6)	* 517^2 (26.6)
		Sep 90	* 360^2 (14.4)	* 422	911^3 (99.8)
		Jun 91	* 584^2 (71.9)	*1,710	457^2 (35.2)
37	Rua Cove	Nov 89		* 264	
38	Sleepy Bay	May 89	193^3 (13.0)		254^3 (35.5)
		Jul 89		157^3 (22.2)	40.1^3 (22.7)
		Sep 89	* 335^2 (9.9)	* 810^3 (107.2)	* 193^3 (6.3)
		Nov 89		125^3 (47.7)	* 358^3 (49.7)
		Jun 90	176^3 (72.2)	* 144	* 671^3 (6.4)
		Jul 90	80.9^2 (101.4)	106^2 (10.9)	163^2 (112.7)
		Sep 90	* 253^3 (45.3)	* 182^3 (11.2)	202^3 (92.4)
		May 91	18.2^3 (51.6)		940^3 (53.1)
		Jun 91	* 235^3 (26.9)	51.2^2 (22.6)	70.8^3 (5.2)
		Sep 91	29.3^2 (29.5)		37.9^2 (7.3)

TABLE 2.—Extended.

Site			Depth			
Number	Name	Date	20 m	40 m	100 m	Other
2	Bay of Isles 86	Nov 89	*580			
3	Bay of Isles 90	Nov 89	*416^{2} (35.7)			* 753^{2} (39.1)[a]
		Jun 90	*985^{3} (18.7)			
4	Bay of Isles BR	Jul 90				
5	Bay of Isles	Jul 89	420^{3} (24.8)	423^{3} (30.3)	811^{3} (9.8)	
		Sep 89				
		Jun 90	*331^{3} (31.3)			
		Jul 90	912^{2} (10.5)	62.0^{2} (25.2)	825^{2} (9.1)	
		Sep 90				
		Jun 91	965^{3} (33.1)	*3,670	1,340^{3} (46.9)	812^{3} (2.1)[b]
6	Block Island 7	Nov 89				
7	Block Island 47	Nov 89	*585^{3} (27.3)			* 903^{3} (15.7)[a]
		Jun 90	*227			
		Jul 90	*452^{3} (41.7)	142	977^{2} (36.5)	
9	Disk Island	Jul 89	406^{3} (142.3)	463^{2} (19.3)	1,360^{3} (9.6)	
		Jun 90	*485			
		Jul 90	186^{2} (2.0)	402^{2} (6.2)	812^{2} (1.8)	
		Jul 91			1,970	
13	Foxfarm	May 89				
		Jul 89	5.9^{3} (69.7)	13.8^{3} (12.2)		
		Sep 89				
		Jun 90	137			
		Jul 90	294^{2} (62.7)	174^{2} (34.2)	128^{2} (26.6)	
		Sep 90	697^{3} (108.6)			
		Apr 91				
		Sep 91				
17	Herring Bay 53	Nov 89	220^{2} (0.3)			2,300[a]
18	Herring Bay 110	Nov 89	344^{3} (32.8)			82.7^{2} (82.7)[a]
18	Herring Bay 125	Nov 89	273^{3} (36.3)			
		Jun 90	*107^{3} (24.7)			
20	Herring Bay	May 89				
		Jul 89	146^{3} (43.4)	93.5^{3} (82.3)	604^{3} (21.2)	
		Sep 89	172^{3} (112.8)			
		Jun 90	*134			
		Jul 90	S	S	382	
		Sep 90	647^{3} (78.7)			
		Apr 91				
		Jun 91	1,220	201^{3} (9.3)	481^{3} (2.3)	908[b]
		Sep 91				
29	Northwest Bay 4	Nov 89	* 588			*1,650[a]
30	Northwest Bay 5	Dec 89	193		193	
31	Northwest Bay	Jul 89	403^{3} (57.7)	261^{3} (41.6)	991^{3} (56.1)	* 474^{3} (57.8)[a]
		Jul 89				232^{2} (12.0)[c]
		Sep 89	366^{3} (44.0)			
		Jun 90	* 559			
		Jul 90	* 657^{2} (37.8)	245^{2} (13.4)	940^{2} (5.7)	
		Sep 90	*1,850^{3} (51.4)			
		Jun 91	* 375^{3} (26.5)	* 437^{2} (63.1)	1,250^{2} (68.8)	2,820[d]
37	Rua Cove	Nov 89				
38	Sleepy Bay	May 89				
		Jul 89	309^{3} (28.7)	163^{2} (2.8)	1,230^{2} (69.3)	
		Sep 89				
		Nov 89	* 506^{3} (27.4)			* 366^{2} (5.4)[a]
		Jun 90	* 784			
		Jul 90	391^{2} (26.3)	407^{2} (87.6)	355	
		Sep 90	*1,070^{2} (92.4)			
		May 91				
		Jun 91	238^{3} (47.8)	350^{3} (87.2)	596^{3} (33.5)	907^{2} (18.9)[b]
		Sep 91				

TABLE 2.—Continued.

Site		Date	Depth		
Number	Name		0 m	3 m	6 m
39	Smith Island	Dec 89		* 196	55.2
		Jun 90	386	379	346
40	Snug Harbor 25	Nov 89		132[2] (105.8)	133[3] (71.8)
		Jun 90	33.5[3] (53.1)	143[3] (67.8)	31.3[3] (103.4)
		Jul 90	* 2,550[2] (83.3)	S	94.4
		Jun 91	* 128[3] (2.8)	214[3] (91.0)	* 101[3] (8.0)
41	Snug Harbor	Jul 89	592[3] (60.6)	* 468[3] (32.9)	* 169[2] (52.0)
		Sep 89		2,090	
		Jul 90			
		Jun 91			
		Sep 91	3,610		2,180
42	Snug Harbor meiofauna	Jul 89	420[3] (65.4)	66.6[3] (132.6)	188[3] (78.2)
		Sep 89	1,030[3] (165.9)	8,780[2] (130.3)	*1,300[3] (37.0)
		Jun 90	177[3] (48.0)		527[3] (39.0)
		Sep 90	106[3] (20.3)		* 641[3] (35.8)
		May 91	* 127[3] (60.1)		332[3] (72.0)
		Sep 91	* 206[2] (39.5)		189

contained 2,860 ± 306 ng/g sediment ($N = 3$). The TPAH concentration ranged from 1,300 ng/g to 6,000 ng/g in sediments collected at 2-m intervals along a 20-m transect laid perpendicular to the shoreline at Northwest Bay in September 1989. At Herring Bay, intertidal sediment samples collected in May, July, and September 1989 had high TPAH concentrations (Table 2).

The PAH composition pattern in sediments at intertidal stations at the three assessment locations where the TPAH concentration was high was consistent with weathered *Exxon Valdez* oil. (Hereafter, this PAH composition pattern will be referred to as the EVO-PAH pattern.) The most abundant PAHs measured in these sediments generally corresponded to the most abundant PAHs reported by Short et al. (1996a) for mousse samples collected 11 d after the oil spill (Figure 2), except the sediment PAHs reflected greater losses of lower-molecular-weight and of less-substituted PAHs. The pattern of these losses was consistent among sediment replicates. In addition, it is noteworthy that (1) the high-molecular-weight and unsubstituted PAHs that were less abundant in the mousse samples were also less abundant in the sediments, and (2) phytane, which was abundant relative to other aromatic analytes in the weathered mousse, was consistently more abundant relative to those analytes in the sediments, mainly as a consequence of the differential PAH losses in the sediments. Finally, the normal alkanes with molecular weights greater than *n*-eicosane were present at concentrations that were approximately as high as the alkyl-substituted PAH homologues in the sediment samples, and the unresolved complex mixture (UCM) was large (i.e., present at concentrations greater than 10^6 ng/g in some samples).

The total *n*-alkane (TNA) concentration was highest when the TPAH concentration was highest and the EVO-PAH pattern was present (Tables 2 and 3). The highest TNA concentration of this study was 31,100 ng/g in intertidal sediments from Northwest Bay in July 1989; the same samples contained nearly the highest TPAH concentration of the study (11,700 ng/g; Table 2). (A TNA concentration was not available for the station at Disk Island where the TPAH concentration was 12,700 ng/g because of inadequate surrogate standard recoveries for those alkanes). The ratio of TNA to TPAH in these samples was 2.66, as compared with that for weathered *Exxon Valdez* oil (3.62–4.68 for mousse samples). The highest TNA concentration (31,100 ng/g) at an oil-contaminated assessment site was more than twice as high as the highest TNA concentration at any reference station or any assessment station where the EVO-PAH pattern was generally absent (Tables 2, 3, 4, and 5).

Shallow Subtidal Stations

We found substantially lower concentrations of *Exxon Valdez* oil hydrocarbons in shallow subtidal sediments than in intertidal sediments. In 1989, shallow subtidal (3–20 m) sediment samples were collected at all locations where intertidal samples had been collected, and at an additional six locations where intertidal samples had not been collected. At the three locations where we found the

TABLE 2.—Extended.

Site			Depth			
Number	Name	Date	20 m	40 m	100 m	Other
39	Smith Island	Dec 89	40.0			111[a]
		Jun 90	377			
40	Snug Harbor 25	Nov 89	* 265[3] (39.1)			* 141[3] (31.8)[a]
		Jun 90				
		Jul 90	125			
		Jun 91	* 219[3] (80.7)			
41	Snug Harbor	Jul 89	2,730[2] (2.3)	652[3] (20.6)	891[3] (11.4)	
		Sep 89	1,550			
		Jul 90		1,400[2] (60.0)	1,100[2] (14.7)	
		Jun 91		95.9	534[2] (19.0)	
		Sep 91				
42	Snug Harbor meiofauna	Jul 89				
		Sep 89	1,630[2] (30.5)			
		Jun 90				
		Sep 90				
		May 91				
		Sep 91				

[a]10 = m depth.
[b]140 = m depth.
[c]30 = m depth.
[d]55 = m depth.

highest mean TPAH concentrations similar in composition to weathered *Exxon Valdez* oil in intertidal sediments, we found a substantially lower mean TPAH concentration of similar composition in subtidal sediments. This finding occurred most often at Northwest Bay and less often at Herring Bay. The EVO-PAH pattern was not found subtidally at Disk Island. We also found the EVO-PAH pattern at relatively low TPAH concentrations in subtidal sediments at three locations where intertidal sediments were not sampled: Block Island, Rua Cove, and Smith Island.

In 1989 at Northwest Bay, the EVO-PAH pattern was consistently evident in subtidal sediments from 3 to 10 m. This pattern was less consistently observed in 20-m sediments. At the Northwest Bay 4 site, the EVO-PAH pattern appeared in all samples (unreplicated) collected from 3 to 20 m in November 1989; the TPAH concentration ranged from 588 to 2,660 ng/g (Table 2). At the NOAA Northwest Bay site in July 1989, mean TPAH concentration ranged from 474 ± 158 to 1,490 ± 762 ng/g (mean ± SE; $N = 3$) in subtidal sediments from depths of 3 to 10 m. Concentrations of unsubstituted PAHs were relatively high in many of these samples, indicating additional PAH sources, especially in two of the three replicate samples from the 3-m depth in July 1989 (Table 2). In the 20-m sediments collected from this site in 1989, mean TPAH concentration (± SE) ranged from 366 ± 93.0 to 403 ± 134 ng/g ($N = 3$). The PAH composition differed from that of weathered *Exxon Valdez* oil and was more variable among the replicates. At the Northwest Bay 5 site, one sample replicate each was analyzed from the 3-m and 20-m depths sampled in December 1989. The EVO-PAH pattern was present at 3 m (TPAH, 194 ng/g), but not at 20 m (TPAH, 193 ng/g).

As with TPAH, the TNA concentration decreased rapidly with increasing bathymetric depth at Northwest Bay in July 1989 (Table 3), whereas the carbon preference index (CPI) increased with increasing depth. At 3 m at the NOAA Northwest Bay site, the mean TNA concentration (± SE) was 4,130 ± 2,470 ng/g ($N = 3$), decreasing to 1,700 ± 232 ng/g ($N = 3$) at 20 m. The CPI increased from 1.27 at 3 m to 3.38 at 20 m.

At Herring Bay, the EVO-PAH pattern was occasionally found at 3-m and 6-m depths in 1989 (Table 2). At the NOAA Herring Bay site (sampled in July and September 1989) and Herring Bay site 53 (sampled in November 1989), this pattern was present in 3-m sediments where the mean TPAH concentration (± SE) ranged from 239 ± 36.7 ng/g ($N = 3$) to 921 ± 182 ng/g ($N = 2$). At Herring Bay site 125, the EVO-PAH pattern was present in 6-m sediments and the mean TPAH concentration was 220 ± 64.1 ng/g ($N = 3$), but at 3 m the PAH pattern differed from weathered *Exxon Valdez* oil and the mean TPAH concentration was lower (164 ± 39.3 ng/g, $N = 2$). The EVO-PAH pattern was not evident in sediments from the 3-m depth at Herring Bay site 110 nor in sediments from depths greater than 6 m at any Herring Bay site in 1989.

TABLE 3.—Concentration of total *n*-alkanes (TNA) in sediments from all stations at assessment locations where shores were moderately to heavily oiled in Prince William Sound by the *Exxon Valdez* oil spill, 1989–1991. Numbers in the body of the table are mean TNA (ng/g), number of replicates analyzed (superscripts), and coefficient of variation (in parentheses). The abbreviation S denotes surrogate recoveries of one or more alkanes outside acceptable range. For "other" depths (last column), see footnotes.

Site			Depth		
Number	Name	Date	0 m	3 m	6 m
2	Bay of Isles 86	Nov 89		1,250	
3	Bay of Isles 90	Nov 89		926^{2} (2.5)	5,780
		Jun 90		891^{2} (54.1)	1,430^{3} (32.2)
4	Bay of Isles BR	Jul 90	S		1,390
5	Bay of Isles	Jul 89	1,420^{2} (17.2)	8,300^{2} (11.3)	1,240^{3} (20.6)
		Sep 89	1,550^{2} (21.4)	429^{3} (18.3)	4,320^{3} (124.2)
		Jun 90	2,460	488^{3} (34.2)	477^{3} (29.6)
		Jul 90	1,990^{2} (8.7)	886^{2} (7.2)	375^{3} (11.1)
		Sep 90			232^{3} (52.7)
		Jun 91	76.3^{3} (47.7)	550^{3} (65.9)	217^{3} (54.7)
6	Block Island 7	Nov 89		S	
7	Block Island 47	Nov 89		S	S
		Jun 90	S		1,610
		Jul 90	3,530^{2} (0.8)	1,740^{2} (37.4)	1,350^{2} (32.2)
9	Disk Island	Jul 89	S	2,030^{3} (51.7)	1,210^{2} (40.3)
		Jun 90	4,170	736	479
		Jul 90	2,430^{2} (46.9)	1,130^{2} (46.9)	222^{2} (23.5)
		Jun 91			
13	Foxfarm	May 89	S		S
		Jul 89	620^{3} (35.5)	1,580^{3} (21.5)	599^{3} (24.0)
		Sep 89	S		203
		Jun 90	377^{2} (80.7)	266^{3} (64.6)	431
		Jul 90	397	S	144
		Sep 90	168	164^{3} (113.2)	65.9
		Apr 91	101		37.2
		Sep 91	142^{3} (46.9)		159^{3} (111.8)
17	Herring Bay 53	Nov 89		S	705
18	Herring Bay 110	Nov 89		88.2	S
19	Herring Bay 125	Nov 89		378	976
		Jun 90	928^{3} (13.3)	370^{2} (49.0)	398^{3} (19.2)
20	Herring Bay	May 89	4,080^{3} (47.7)		879^{3} (16.0)
		Jul 89	3,190^{2} (46.1)	4,870^{2} (44.9)	875^{3} (50.2)
		Sep 89	4,330^{2} (64.8)	1,690^{3} (39.4)	697^{3} (22.3)
		Jun 90	883^{3} (34.2)	1,030^{3} (22.1)	531^{2} (67.9)
		Jul 90	2,040	1,280	580^{3} (15.9)
		Sep 90	678^{3} (13.7)	179^{3} (46.6)	343^{3} (21.9)
		Apr 91	573^{2} (30.1)		1,480^{2} (66.4)
		Jun 91	419^{2} (17.9)	1,500^{2} (45.6)	237^{3} (21.6)
		Sep 91	296^{3} (43.0)		647^{3} (36.3)
29	Northwest Bay 4	Nov 89		1,700	S
30	Northwest Bay 5	Dec 89		940	
31	Northwest Bay	Jul 89	31,100	4,130^{3} (103.4)	2,540^{3} (26.7)
		Jul 89			
		Sep 89	10,500^{2} (8.4)	2,540^{3} (0.5)	3,600^{3} (14.4)
		Jun 90	821^{3} (41.7)	2,290	1,840^{3} (14.1)
		Jul 90	6,540^{2} (101.4)	1,080^{2} (1.0)	1,740^{2} (22.4)
		Sep 90	1,370^{3} (32.7)	797^{2} (7.9)	837^{3} (43.4)
		Jun 91	1,650^{3} (109.8)	719	781^{2} (47.8)
37	Rua Cove	Nov 89		848	
38	Sleepy Bay	May 89	647^{3} (22.4)		1,410^{3} (42.7)
		Jul 89		574^{3} (31.0)	259^{3} (15.5)
		Sep 89	S	1,350^{2} (57.0)	416^{2} (5.9)
		Nov 89		306^{3} (32.5)	778^{3} (43.8)
		Jun 90	805	458	2,640
		Jul 90	132	S	793^{2} (48.2)
		Sep 90	201^{2} (13.1)	369	401^{3} (10.5)
		May 91	131^{3} (17.7)		1,400^{3} (24.3)
		Jun 91	758^{3} (43.9)	78.4^{3} (15.1)	137^{3} (47.3)
		Sep 91	54.1^{2} (50.7)		119^{2} (4.6)
39	Smith Island	Dec 89		466	247
		Jun 90	963	978	896

TABLE 3.—Extended.

Site			Depth			
Number	Name	Date	20 m	40 m	100 m	Other
2	Bay of Isles 86	Nov 89	S			
3	Bay of Isles 90	Nov 89	1,220[2] (51.8)			1,740[a]
		Jun 90	7,180[3] (20.6)			
4	Bay of Isles BR	Jul 90				
5	Bay of Isles	Jul 89	8,290[2] (7.0)	2,580[3] (31.4)	4,160[3] (7.4)	
		Sep 89				
		Jun 90	1,580[3] (17.4)			
		Jul 90	8,930[2] (5.2)	996[2] (17.3)	3,220[2] (15.0)	
		Sep 90				
		Jun 91	8,220[3] (5.3)	6,730[3] (46.8)	4,250[3] (11.9)	1,760[3] (9.3)[b]
6	Block Island 7	Nov 89				
7	Block Island 47	Nov 89	S			S[a]
		Jun 90	1,120			
		Jul 90	S	372	2,210	
9	Disk Island	Jul 89	589	1,710	4,050[3] (5.4)	
		Jun 90	3,170			
		Jul 90	S		1,770	
		Jun 91				
13	Foxfarm	May 89				
		Jul 89	201[2] (71.5)	194[3] (56.0)		
		Sep 89				
		Jun 90	323			
		Jul 90	354	S	S	
		Sep 90	1,030[3] (69.1)			
		Apr 91				
		Sep 91				
17	Herring Bay 53	Nov 89	S			1,650[2] (47.8)[a]
18	Herring Bay 110	Nov 89	871			231[2] (19.7)[a]
19	Herring Bay 125	Nov 89	666[3] (12.4)			
		Jun 90	696[2] (31.2)			
20	Herring Bay	May 89				
		Jul 89	1,710[3] (40.8)	1,040[3] (64.7)	1,650[3] (33.2)	
		Sep 89	717[3] (26.4)			
		Jun 90	853			
		Jul 90	795	933[2] (7.4)	918	
		Sep 90	808[3] (43.9)			
		Apr 91				
		Jun 91	768	710[3] (59.8)	1,010[3] (18.5)	1,640[b]
		Sep 91				
29	Northwest Bay 4	Nov 89	S			S[a]
30	Northwest Bay 5	Dec 89	1,010			
31	Northwest Bay	Jul 89	1,700[3] (23.6)	927[3] (5.9)	1,720[2] (18.4)	1,700[3] (23.6)[a]
		Jul 89				1,170[2] (42.8)[c]
		Sep 89	1,660[3] (25.7)			
		Jun 90	S			
		Jul 90	1,940[2] (93.1)	1,210[2] (16.5)	2,650[2] (14.7)	
		Sep 90	3,060[3] (14.9)			
		Jun 91	1,340[3] (61.4)	1,120[2] (26.9)	735	S[d]
37	Rua Cove	Nov 89				
38	Sleepy Bay	May 89				
		Jul 89	704[3] (7.3)	490[3] (15.8)	1,320[3] (1.7)	
		Sep 89				
		Nov 89	S			566[2] (12.5)[a]
		Jun 90	S			
		Jul 90	1,700	2,390	1,500	
		Sep 90	1,530[2] (43.4)			
		May 91				
		Jun 91	889[3] (103.3)	352[3] (82.5)	713[3] (8.3)	1,280[2] (11.9)[b]
		Sep 91				
39	Smith Island	Dec 89	208			282[a]
		Jun 90	884			

TABLE 3.—Continued.

Site			Depth		
Number	Name	Date	0 m	3 m	6 m
40	Snug Harbor 25	Nov 89		214[2] (120.4)	233[2] (98.1)
		Jun 90	122[3] (82.0)	162[3] (33.1)	184[3] (32.3)
		Jul 90	4,640	1,560	607
		Jun 91	86.5[3] (7.7)	212[3] (51.2)	131[3] (31.4)
41	Snug Harbor	Jul 89	1,240[3] (31.1)	2,630[3] (10.7)	1,020[2] (23.0)
		Sep 89		2,470	
		Jul 90			
		Jun 91			
		Sep 91	S		S
42	Snug Harbor meiofauna	Jul 89	1,000[2] (12.6)	276[3] (18.5)	716[3] (24.1)
		Sep 89	744[2] (58.4)	S	3,240[2] (7.9)
		Jun 90	1,210		1,750[3] (28.2)
		Sep 90	974[3] (55.7)		3,090[2] (0.9)
		May 91	689[3] (53.8)		525[3] (20.5)
		Sep 91	665[2] (34.1)		1,270

Mean TPAH concentrations were consistently less than 350 ng/g in shallow subtidal sediments from depths greater than 3 m, except in one sample (Herring Bay 53, 10 m, TPAH = 2,300 ng/g; Table 2). The relative abundances of PAHs in this sample differed from that of weathered *Exxon Valdez* oil because of lower proportions of chrysenes, which indicates diesel oil as the most likely source. (Reference to a PAH in the plural—e.g., chrysenes—denotes the unsubstituted PAH together with the alkyl-substituted homologues collectively as a group.)

At Disk Island we did not observe the EVO-PAH pattern in subtidal sediments despite the presence of that pattern and high TPAH concentrations in intertidal sediments there. Mean TPAH concentrations in the shallow subtidal samples ranged from 160 ng/g (N = 1) to 553 ± 173 ng/g (N = 3; Table 2), with low concentrations of chrysenes and often low concentrations of dibenzothiophenes relative to other PAHs.

At Block Island the composition and concentration of hydrocarbons in subtidal sediments resembled those from Northwest Bay site 4. The EVO-PAH pattern was consistently present in all subtidal sediment samples from the 3-m to 20-m depths. The mean TPAH concentration ranged from 539 ng/g (N = 1) to 903 ± 81.9 ng/g (N = 3). The EVO-PAH pattern was present in the single subtidal sediment sample analyzed from Rua Cove (3-m depth) and in the single 3-m depth sample from Smith Island. The TPAH concentration in the Rua Cove and Smith Island samples was 264 and 196 ng/g, respectively.

Deep Subtidal Stations

The TPAH concentration and the PAH composition pattern in sediments at depths ≥40 m were similar at reference locations and locations where moderate to heavy beach oiling occurred in Prince William Sound. The EVO-PAH pattern was rarely present consistently among replicate samples at these depths regardless of sampling location (Tables 2 and 5). The PAH composition pattern differed from that of weathered *Exxon Valdez* oil in that concentrations of dibenzothiophenes were relatively low (Figure 3). At the 40-m depth, the median TPAH concentration was 277 and 308 ng/g at locations near moderately to heavily oiled beaches and reference locations, respectively, and the mode fell within the range 100 ng/g to 300 ng/g (Figure 4). At the 100-m depth and greater, the median concentration was 899 ng/g and 574 ng/g at moderately to heavily oiled locations and at reference locations, respectively, and the mode was above 400 ng/g (Figure 4). The TPAH concentration in sediments at depths ≥40 m was often higher than in sediments in the 3-m to 20-m depth range.

Moderately Oiled Locations, 1989

Intertidal stations.—The EVO-PAH pattern was occasionally found at relatively low concentrations at the intertidal stations at four moderately to heavily oiled locations sampled in 1989: Bay of Isles, Foxfarm, Sleepy Bay, and Snug Harbor (Table 2). The mean TPAH concentration was highest at Snug Harbor, where it ranged from 420 ± 159 to 1,030 ± 983 ng/g (N = 3), but the PAH pattern differed from the EVO-PAH pattern. The mean TPAH concentration at the other stations generally ranged from 160 to 408 ng/g, although by September 1989 the TPAH concentration at the Foxfarm intertidal station had declined to 68.5 ± 17.4 ng/g (N = 2).

Subtidal stations.—At Bay of Isles, Sleepy Bay, and Snug Harbor locations where a low TPAH

TABLE 3.—Extended.

Site		Date	Depth			
Number	Name		20 m	40 m	100 m	Other
40	Snug Harbor 25	Nov 89	742^3 (18.4)			270^3 (61.5)[a]
		Jun 90				
		Jul 90	403			
		Jun 91	353^2 (0.8)			
41	Snug Harbor	Jul 89	$4{,}410^2$ (44.0)	$3{,}570^3$ (15.1)	$2{,}270^3$ (19.2)	
		Sep 89	S			
		Jul 90		S	2,090	
		Jun 91		534^2 (36.5)	$1{,}390^2$ (12.6)	
		Sep 91				
42	Snug Harbor meiofauna	Jul 89				
		Sep 89	S			
		Jun 90				
		Sep 90				
		May 91				
		Sep 91				

[a] 10 = m depth.
[b] 140 = m depth.
[c] 30 = m depth.
[d] 55 = m depth.

concentration and EVO-PAH pattern were occasionally found in intertidal sediments, we found similar results for subtidal sediments. At the NOAA Bay of Isles site, the EVO-PAH pattern was generally not evident. The EVO-PAH pattern was found in two of three replicate samples from 6-m depth in September 1989 (Table 2). During other sampling periods and at other depths in the range 0–20 m, the PAH composition pattern differed from the EVO-PAH pattern because of low relative concentrations of chrysenes. At sites 86 and 90 in the Bay of Isles, where subtidal (but not intertidal) samples were collected, the EVO-PAH pattern was encountered more frequently. Results for these sites were similar to those for Block Island and Northwest Bay 4 (Table 2).

At Snug Harbor, the EVO-PAH pattern was observed at five stations: Snug Harbor site 25, 10-m and 20-m depths, November 1989; and the Snug Harbor meiofauna site, 6-m depth, September 1989 (Table 2). Other subtidal sediments often contained a relatively high TPAH concentration, but the composition patterns differed markedly from that of weathered *Exxon Valdez* oil. The PAH pattern char-

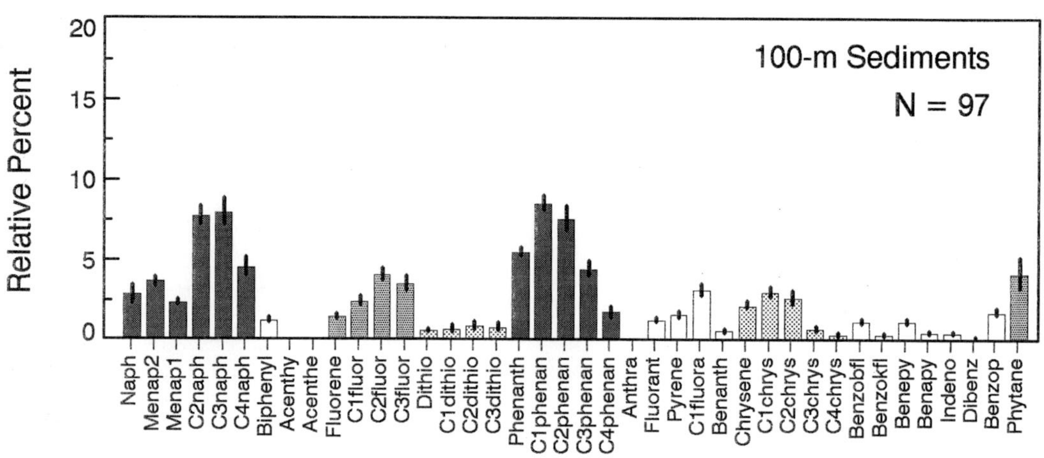

FIGURE 3.—Mean relative percents of polynuclear aromatic hydrocarbon compounds and phytane in sediments from ≥100 m, all sites in Prince William Sound combined, 1989–1991. Error bars are 95% confidence intervals derived from arcsine-transformed relative percentages.

TABLE 4.—Concentration of total *n*-alkanes (TNA) in sediments from all stations at reference locations (italics) and assessment locations where shores were lightly oiled in Prince William Sound by the *Exxon Valdez* oil spill, 1989–1991. Numbers in the body of the table are mean TNA (ng/g), number of replicates analyzed (superscripts), and coefficient of variation (in parentheses). Abbreviations are as follows: S denotes surrogate recoveries of one or more alkanes outside acceptable range; ND indicates that *n*-alkane concentrations were below detection limits; (T) indicates bioremediation treatment site; and (R) indicates bioremediation reference site. For "other" depth (last column), see footnotes.

Site			Depth		
Number	Name	Date	0 m	3 m	6 m
1	Applegate Island	Nov 89		S	
8	Chenega Island	Nov 89		S	S
		Jun 90	740	583	
		Jul 90	1,920	S	84.3
10	*Drier Bay*	Jul 90	7,400^{2} (37.9)	15,600^{2} (0.2)	8.890^{2} (10.4)
		Sep 90			5,310^{3} (8.9)
		Jun 91	244^{2} (69.1)	177^{3} (26.4)	909^{3} (55.4)
11	Eshamy Bay	May 89	S		S
		Jul 89	ND[c]	ND[c]	4,860^{2} (7.9)
		Sep 89	1,510	9,130^{2} (21.7)	3,430
		Jun 90	9,730^{3} (153.9)		1,780^{2} (52.1)
		Sep 90	689^{2} (30.4)		2,944
		Apr 91	2,470		1,589
		Sep 91	1,110^{2} (11.9)		2,970^{2} (33.7)
12	*Ewan Bay*	May 89	S		S
		Sep 89	760^{3} (46.2)	6,060^{2} (9.9)	823^{2} (90.1)
		Jun 90	1,840		752^{3} (26.4)
		Sep 90	683^{2} (39)		443^{2} (12.1)
		Apr 91	1,830		259
		Sep 91	1,040		342
14	Green Island 22	Nov 89		407	521^{3} (49.5)
		Jun 90	422^{3} (60.3)	297^{3} (112.2)	482^{3} (46.4)
15	Green Island	Jul 89	S	103	278
		Jun 90	1,220	743	
16	*Heather Bay*	Jul 89	3,690^{3} (31.1)	671^{3} (11.2)	934
21	Iktua Bay 2	Jul 89	608^{3} (74.1)	330	1010
22	Iktua Bay	May 89	S		S
		Jul 89	505^{3} (65.3)		
		Sep 89	S		S
		Jun 90	937^{2} (52.4)		710^{3} (67.6)
		Sep 90	124		977
		Apr 91	1,830^{3} (34.1)		865
		Sep 91	201^{2} (11.8)		755^{2} (13.3)
23	Ingot Island	Nov 89		S	S
24	Knight Island	Jul 90 (T)	146^{3} (29.8)		
	bioremediation	Jul 90 (R)	158^{2} (19.1)		339^{2} (3.6)
25	Knight Island	Nov 89		S	87.5^{3} (116.4)
		Jun 90	699	506	
		Jul 90	853^{2} (48.7)	163^{2} (54.4)	76.6^{2} (14.9)
26	*Lower Herring Bay*	Jul 90	1,120	292	414^{2} (58.5)
		Sep 90			193^{3} (37.7)
		Jun 91	94.7^{2} (78.8)	448^{2} (55.2)	471^{2} (48.6)
27	*MacLeod Harbor*	Jun 90	200	699	1,740
		Jul 90	206	586^{2} (16.3)	1,200
28	*Moose Lips Bay*	Jul 90	S	1,040	1,190^{2} (26.8)
		Jun 91	3.15^{2} (75.8)	55.5^{3} (66.5)	107^{3} (67.1)
32	*Olsen Bay*	Jul 89	13,200^{2} (7.9)	14,300^{3} (14.4)	2,760^{3} (7.4)
		Sep 89	517^{3} (3.7)	9,390^{3} (12.3)	3,900^{3} (2.7)
		May 90	274^{3} (10.8)	2,890^{2} (10.8)	1,300
		Jun 90	1,880	1,780	1,010^{2} (51.2)
		Sep 90	1,730^{3} (1.3)	2,370^{3} (20.9)	3,000^{3} (5.8)
		Apr 91	285^{3} (63.7)		1,400^{3} (12.6)
		Jun 91	581^{3} (18.4)	2,310^{3} (29.9)	1,800
		Sep 91	229^{2} (8.1)		1,030^{2} (60.9)
33	*Paddy Bay*	May 89	S		S
		Sep 89	288	19,500^{3} (13.3)	S
		Jun 90	6,180^{2} (4.6)		6,660^{2} (43.7)
		Sep 90	389^{2} (13.4)		3,140
		Apr 91			7,080
		Sep 91	198^{3} (10.4)		3,340
34	Point Helen	Nov 89		S	

TABLE 4.—Extended.

Site			Depth			
Number	Name	Date	20 m	40 m	100 m	Other
1	Applegate Island	Nov 89				
8	Chenega Island	Nov 89	252			108[a]
		Jun 90	785			
		Jul 90	S	S	S	
10	*Drier Bay*	Jul 90	$1,740^2$ (26.1)	$5,300^2$ (35.9)	S	
		Sep 90				
		Jun 91	1,710	$1,890^3$ (21.5)	$3,030^2$ (6.8)	$2,960^2$ (4.1)[b]
11	Eshamy Bay	May 89				
		Jul 89	$2,860^2$ (27.9)	$9,200^2$ (14.3)	$3,190^3$ (14.3)	
		Sep 89				
		Jun 90				
		Sep 90				
		Apr 91				
		Sep 91				
12	*Ewan Bay*	May 89				
		Sep 89				
		Jun 90				
		Sep 90				
		Apr 91				
		Sep 91				
14	Green Island 22	Nov 89	612^2 (2.2)			306^3 (46.7)[a]
		Jun 90	373^3 (43.4)			
15	Green Island	Jul 89	1,200	811	1,470	
		Jun 90				
16	*Heather Bay*	Jul 89	437^2 (33.2)	592^3 (13.4)	556^2 (0.6)	
21	Iktua Bay 2	Jul 89	S	692	1,610	
22	Iktua Bay	May 89				
		Jul 89				
		Sep 89				
		Jun 90				
		Sep 90				
		Apr 91				
		Sep 91				
23	Ingot Island	Nov 89	525			1,080[a]
24	Knight Island bioremediation	Jul 90 (T)				
		Jul 90 (R)				
25	Knight Island	Nov 89	448			25.0[a]
		Jun 90	693			
		Jul 90	642^2 (17.1)	670^2 (25.0)	$1,940^2$ (4.5)	
26	*Lower Herring Bay*	Jul 90	1,020	$2,890^2$ (2.1)	$5,930^2$ (5.1)	
		Sep 90				
		Jun 91	447^2 (43.1)	$1,420^2$ (30.0)	$1,680^2$ (38.5)	2,580[b]
27	*MacLeod Harbor*	Jun 90				
		Jul 90	S	1,760	$1,350^2$ (3.6)	
28	*Moose Lips Bay*	Jul 90	$4,990^2$ (118.4)	S		
		Jun 91	717^3 (11.1)	732^3 (8.0)	852^3 (14.2)	
32	*Olsen Bay*	Jul 89	$2,290^3$ (14.6)	720^3 (26.5)	1,600	
		Sep 89				
		May 90	$2,020^3$ (37.7)			
		Jun 90	715	$1,480^2$ (6.2)	$1,730^2$ (14.4)	
		Sep 90	$3,310^2$ (12.5)			
		Apr 91				
		Jun 91	S	$1,170^3$ (31.7)	$1,240^3$ (56.5)	
		Sep 91				
33	*Paddy Bay*	May 89				
		Sep 89				
		Jun 90				
		Sep 90				
		Apr 91				
		Sep 91				
34	Point Helen	Nov 89				

TABLE 4.—Continued.

Site		Date	Depth		
Number	Name		0 m	3 m	6 m
35	*Port Fidalgo*	Dec 89		943	478
		Jun 90	539	1,620	705
		Jul 90[d]	116[2] (31.1)	1,130	373[2] (53.8)
36	*Rocky Bay*	Jul 89	583[3] (35.5)	355[3] (1.3)	708[3] (11.7)
		Jun 90	545	730	436
		Jul 90	646	511[2] (6.3)	520[2] (6.9)
43	*Two Moon Bay*	Dec 89		ND[c]	
44	*West Bay*	Jun 90	317	1,700	
		Jul 90	194[2] (66.8)	1,020[2] (14)	5,100[2] (33.5)
45	*Zaikov Bay*	Jun 90	611	959	1,380
		Jul 90	150	605[2] (9.9)	S
		Jun 91	218[2] (109.9)	1,290[2] (2.1)	548[2] (77.5)

acteristic of combustion products (i.e., relatively high concentrations of higher-molecular-weight unsubstituted PAHs) was observed at two of the sites (NOAA site and Snug Harbor meiofauna site). The TPAH concentration exceeded 16,800 ng/g in one sample. Subtidal sediments from 3 and 6 m at Snug Harbor site 25 contained a low TPAH concentration, composed mainly of naphthalenes (Table 2).

At Sleepy Bay, the results for subtidal sediments were similar to those for intertidal sediments. The EVO-PAH pattern was found in samples from 3 to 20 m in September and November 1989 (Table 2). In these 2 months, the mean TPAH concentration ranged from 193 ± 7.0 ng/g to 810 ± 501 ng/g (N = 3) and usually showed a low variability in TPAH concentration among sample replicates. The highest mean TPAH concentration of samples collected earlier (May and July 1989) was 309 ± 51.2 ng/g (N = 3), and the PAH composition pattern of these samples differed from weathered *Exxon Valdez* oil.

The TNA concentration at moderately oiled stations where the EVO-PAH pattern was evident was relatively small and was dominated by n-alkanes from natural sources. The median TNA concentration at these stations was 774 ng/g; 80.6% of the samples contained a TNA concentration lower than 2,000 ng/g (N = 165). The median CPI was 2.71; 80.8% of the samples had a CPI greater than 1.5 (N = 193). The highest TNA concentration observed at these stations was greater than 8,000 ng/g in subtidal sediments at Bay of Isles in July 1989 (Table 3). The CPI for these samples exceeded 4.

Lightly Oiled Locations and Reference Locations, 1989

Intertidal stations.—The EVO-PAH pattern was not found at intertidal stations at the three lightly oiled locations sampled in 1989: Eshamy Bay, Green Island, and Iktua Bay (Table 5). Mean TPAH concentration at these stations was usually below 150 ng/g, with naphthalenes the most abundant PAHs present. However, mean concentrations of 219 ± 45 ng/g and 214 ± 38.2 ng/g (N = 3) were observed at Iktua Bay and Eshamy Bay, respectively, in September 1989.

At intertidal reference stations, PAHs were detected sporadically and at low concentration and did not exhibit the EVO-PAH pattern. The median of the mean TPAH concentrations of all the reference intertidal stations sampled during all 3 years of this study was 40 ng/g, with more than 83% of the observations below 300 ng/g (Figure 4). In contrast, TPAH concentrations exceeding 300 ng/g were found in sediments from 43% of the intertidal stations at moderately to heavily oiled locations over the 3 years of our study (Figure 4).

At the Rocky Bay intertidal reference station, a distinctive PAH composition pattern was observed with a mean TPAH concentration of 316 ± 0.2 ng/g (N = 2) in 1989 (Table 5). Distinctive characteristics of the pattern included (1) a general absence of dibenzothiophenes; (2) alkyl-phenanthrene and alkyl-chrysene concentrations that decreased with increasing alkyl substitution, with unsubstituted phenanthrene the most abundant PAH; and (3) the presence of unsubstituted four- and five-ring PAHs including fluoranthene, pyrene, benzofluoranthene, benzo[e]pyrene, and benzoperylene (Figure 5). In addition, phytane was present at low concentrations compared with the most abundant PAHs (Figure 5), normal alkanes larger than n-eicosane were generally near detection limits, and the unresolved complex mixture (UCM) was consistently low. These characteristics of both PAHs and alkanes were consistent among intertidal samples in both 1989 and

TABLE 4.—Extended.

Site			Depth			
Number	Name	Date	20 m	40 m	100 m	Other
35	*Port Fidalgo*	Dec 89	S			
		Jun 90				
		Jul 90[d]	299^{2} (40.0)	$1,310^{2}$ (26.9)	1,150	
36	*Rocky Bay*	Jul 89	622^{3} (15.8)	$1,180^{3}$ (22.4)	$5,440^{2}$ (20.8)	
		Jun 90	1,980			
		Jul 90	$1,020^{2}$ (27.5)	S	$1,640^{2}$ (28.8)	
43	*Two Moon Bay*	Dec 89	ND[c]			
44	*West Bay*	Jun 90	3,130			
		Jul 90	1,450	$2,140^{2}$ (16.6)	$1,570^{2}$ (19.9)	
45	*Zaikov Bay*	Jun 90	$1,200^{2}$ (17.5)			
		Jul 90	418^{2} (11.7)	$2,520^{2}$ (5.6)	1,150	
		Jun 91	615^{3} (24.8)	$2,530^{2}$ (7.0)	$1,090^{2}$ (11.1)	

[a] 10 = m depth.
[b] 140 = m depth.
[c] Wet weight not available.
[d] Sampled 28 June 1990.

1990, the only years that samples were collected at Rocky Bay.

The TPAH concentration was lowest at the Olsen Bay intertidal reference station (Table 5), where it was consistently below the median and usually included primarily naphthalenes. Naphthalenes were also the most prevalent PAHs in intertidal sediments of the other reference stations, where they accounted for more than half of the TPAH concentration in more than half of the samples.

Shallow subtidal stations.—In 1989, the TPAH concentrations in sediments from depths of 3 to 20 m at lightly oiled locations and at reference locations were generally low and did not show the EVO-PAH pattern (Table 5). Considering all reference locations together during the 3-year study period, 50% of the mean TPAH concentrations in subtidal sediments within the 3-m to 20-m depth range were below 175 ng/g and 85% were below 400 ng/g (N = 91; Figure 4). By comparison, 61% of the mean TPAH concentrations in subtidal sediments from moderately to heavily oiled locations were below 400 ng/g. At the remaining assessment locations where the EVO-PAH pattern was not observed, half of the mean TPAH concentrations within the 3-m to 20-m depth range were below 100 ng/g and 86% were below 400 ng/g (N = 49).

Various PAH composition patterns were observed at lightly oiled assessment and reference locations that did not show the EVO-PAH pattern. The most common patterns included (1) high (i.e., more than 50%) relative proportions of naphthalenes; (2) high proportions of unsubstituted, higher-molecular-weight PAHs; (3) patterns that differed from weathered *Exxon Valdez* oil because of low relative proportions of chrysenes; and (4) patterns that differed from weathered *Exxon Valdez* oil because of low proportions of dibenzothiophenes (e.g., Figure 5).

Deep subtidal stations.—The TPAH concentration often tended to increase with increasing bathymetric depth at particular sampling sites. This was especially evident at reference sites (Table 5). For example, at Olsen Bay in July 1989, the mean sediment TPAH concentration increased with depth from 25.7 ± 4.5 ng/g (N = 3) in the low intertidal zone to 574 ± 57.0 ng/g (N = 3) at a depth of 100 m (Table 5 and Figure 6). A similar pattern was seen at Olsen Bay in 1990 and 1991. At Rocky Bay, where the mean intertidal TPAH of 316 ng/g in July 1989 was substantially higher than at Olsen Bay, the increase with depth was less pronounced but still evident (Table 5 and Figure 6). At both sites, sediments from depths ≥40 m had PAH composition patterns typical of deeper sediments throughout Prince William Sound (Figure 3).

In contrast to the TPAH concentration, the total *n*-alkane (TNA) concentration did not show a consistent pattern of change with depth at reference and lightly oiled sites (Table 4). At some sites, the mean TNA concentration was usually highest in intertidal and shallow subtidal sediments (e.g., Olsen Bay); at other sites (e.g., Rocky Bay), the opposite was true. At sites such as Eshamy Bay, no consistent pattern was observed in TNA concentration with depth.

Heavily Oiled Locations, 1990 and 1991

Concentrations of TPAH in intertidal and subtidal sediments tended to decrease or become more variable in 1990 and 1991. The EVO-PAH pattern

TABLE 5.—Concentration of total polynuclear aromatic hydrocarbons excluding perylene (TPAH) in sediments from all stations at reference locations (italics) and assessment locations where shores were lightly oiled in Prince William Sound by the *Exxon Valdez* oil spill, 1989–1991. Numbers in the body of the table are mean TPAH (ng/g), number of replicates analyzed (superscripts), and coefficient of variation (in parentheses). Numbers are preceded by an asterisk where the PAH composition pattern matched that of weathered *Exxon Valdez* oil for each replicate contributing to that data point. Abbreviations are as follows: S denotes surrogate recoveries for one or more PAH analytes outside acceptable range; ND indicates that TPAH analytes were below detection limits; (T) indicates bioremediation treatment site and (R) indicates bioremediation reference site. For "other" depths (last column), see footnotes.

Site			Depth		
Number	Name	Date	0 m	3 m	6 m
1	Applegate Island	Nov 89		69.9	
8	Chenega Island	Nov 89		18.4	34.7
		Jun 90	180	253	
		Jul 90	S	5.8	18.4
10	*Drier Bay*	Jul 90	1,160^{2} (109.2)	1,200^{2} (39.9)	815^{2} (49.4)
		Sep 90			63.5^{3} (71.2)
		Jun 91	91.7	6.1^{2} (141.4)	307^{3} (94.7)
11	*Eshamy Bay*	May 89	95.8^{3} (27.9)		82.6^{3} (44.9)
		Jul 89	22.5^{2} (5.1)	12.9^{2} (38.2)	37.3^{3} (23.4)
		Sep 89	214^{3} (30.9)	840^{3} (86.0)	191^{3} (45.1)
		Jun 90	46.5^{3} (122.8)		34.3^{3} (123.1)
		Sep 90	12.2^{3} (98.0)		28.9^{3} (40.4)
		Apr 91	49.9		449
		Sep 91	13.1^{2} (1.7)		132^{2} (56.2)
12	*Ewan Bay*	May 89	375^{3} (43.5)		102^{3} (43.3)
		Sep 89	306^{3} (60.7)	3,823^{3} (113.3)	110^{3} (84.5)
		Jun 90	25.6^{3} (46.7)		13.4^{3} (47.0)
		Sep 90	17.0^{3} (111.8)		16.5^{3} (53.4)
		Apr 91	1,570^{2} (135.0)		10,300^{3} (62.6)
		Sep 91	17.4		20.0
14	Green Island 22	Nov 89		194^{3} (15.5)	113^{3} (106.4)
		Jun 90	95.8^{3} (45.0)	54.0^{3} (65.6)	116^{3} (40.3)
15	Green Island	Jul 89	118	22.2	30.3
		Jun 90	98.5	271	
16	*Heather Bay*	Jul 89	38.3^{2} (68.2)	69.7	181^{2} (119.1)
21	Iktua Bay 2	Jul 89	86.7^{3} (103.2)	55.0^{2} (119.0)	162^{3} (89.0)
22	Iktua Bay	May 89	130^{3} (30.7)		156^{3} (17.6)
		Jul 89	56.2^{3} (134.6)		
		Sep 89	219^{3} (35.6)		547^{3} (126.1)
		Jun 90	112^{3} (165.2)		116^{3} (158.1)
		Sep 90	5.9^{2} (40.2)		22.3^{3} (40.9)
		Apr 91	542^{3} (141.16)		535^{2} (95.1)
		Sep 91	20.3		123^{2} (29.2)
23	Ingot Island	Nov 89		34.6	238
24	Knight Island bioremediation	Jul 90 (T)	39.3		
		Jul 90 (R)	45.8^{2} (0.1)		23.8^{2} (70.9)
25	Knight Island	Nov 89		41.0^{2} (4.5)	22.4^{3} (166.5)
		Jun 90	229	193	
25	Knight Island	Jul 90	*1,340^{2} (96.6)	S	4.7
26	*Lower Herring Bay*	Jul 90	9.4^{2} (19.1)	11.7^{2} (106.4)	104^{2} (134.4)
		Sep 90			3.9
		Jun 91	15.1^{2} (93.9)	14.2^{2} (61.4)	74.6^{2} (99.8)
27	*MacLeod Harbor*	Jun 90	95.7	178	301
		Jul 90	163^{2} (7.3)	246^{2} (22.4)	231^{2} (68.8)
28	*Moose Lips Bay*	Jul 90	398^{2} (8.2)	423^{2} (45.6)	349^{2} (41.5)
		Jun 91	S	34.2	26.9^{3} (67.2)
32	*Olsen Bay*	Jul 89	25.7^{3} (30.5)	59.3^{3} (7.3)	24.2^{3} (9.7)
		Sep 89	8.4^{3} (121.8)	305^{3} (78.5)	57.7^{3} (46.3)
		May 90	6.5^{3} (54.4)	268^{2} (0.9)	42.3^{3} (40.2)
		Jun 90	42.7^{2} (86.6)	32.1^{2} (34.6)	13.9^{2} (52.1)
		Sep 90	27.6^{3} (2.2)	308^{3} (65.3)	28.6^{3} (17.0)
		Apr 91	36.2		109^{2} (19.1)
		Jun 91	21.8^{3} (158.0)	27.0^{3} (70.9)	* 211
		Sep 91	7.5		68.8^{2} (68.5)
33	*Paddy Bay*	May 89	52.9^{3} (91.3)		220^{3} (14.9)
		Sep 89	83.4^{3} (27.5)	350^{3} (46.2)	188^{3} (105.6)
		Jun 90	22.0^{3} (60.6)		26.5^{3} (40.6)
		Sep 90	11.0^{3} (112)		16.6^{3} (18.7)
		Apr 91			275^{3} (58.0)
		Sep 91	S		13.6
34	Point Helen	Nov 89		S	

TABLE 5.—Extended.

Site		Date	Depth			
Number	Name		20 m	40 m	100 m	Other
1	Applegate Island	Nov 89				
8	Chenega Island	Nov 89	73.0			S[a]
		Jun 90	300			
		Jul 90	22.1	72.6	227	
10	*Drier Bay*	Jul 90	360^{2} (22.8)	967^{2} (61.4)	$1{,}190^{2}$ (5.3)	
		Sep 90				
		Jun 91	546^{2} (96.6)	484^{3} (25.3)	$1{,}070^{3}$ (24.4)	$1{,}120^{2}$ (23.3)[b]
11	*Eshamy Bay*	May 89				
		Jul 89	577^{3} (54.8)	* 975	960^{3} (5.4)	
		Sep 89				
		Jun 90				
		Sep 90				
		Apr 91				
		Sep 91				
12	*Ewan Bay*	May 89				
		Sep 89				
		Jun 90				
		Sep 90				
		Apr 91				
		Sep 91				
14	Green Island 22	Nov 89	191^{3} (24.3)			25.5^{3} (42.0)[a]
		Jun 90	84.1^{3} (44.4)			
15	Green Island	Jul 89	283	253	955	
		Jun 90				
16	*Heather Bay*	Jul 89	340	S	57.7	
21	Iktua Bay 2	Jul 89	578	294	S	
22	Iktua Bay	May 89				
		Jul 89				
		Sep 89				
		Jun 90				
		Sep 90				
		Apr 91				
		Sep 91				
23	Ingot Island	Nov 89	S			S[a]
24	Knight Island bioremediation	Jul 90 (T)				
		Jul 90 (R)				
25	Knight Island	Nov 89	209^{3} (24.0)			S[a]
		Jun 90	240			
25	Knight Island	Jul 90	88.3^{2} (101.8)	144^{2} (69.8)	854^{2} (7.4)	
26	*Lower Herring Bay*	Jul 90	S	S	$*1{,}700^{2}$ (9.6)	
		Sep 90				
		Jun 91	34.9	186^{2} (72.8)	620^{2} (4.1)	782[b]
27	*MacLeod Harbor*	Jun 90				
		Jul 90	671^{2} (1.5)	884	462^{2} (61.8)	
28	*Moose Lips Bay*	Jul 90	533^{2} (46.6)	719		
		Jun 91	188^{3} (11.4)	244^{2} (1.8)	474^{3} (10.4)	
32	*Olsen Bay*	Jul 89	117^{3} (22.5)	135^{3} (54.2)	* 574^{3} (17.2)	
		Sep 89				
		May 90	148^{3} (50.8)			
		Jun 90	66.4^{2} (106.9)	230^{2} (6.6)	518^{2} (3.9)	
		Sep 90	594^{3} (87.8)			
		Apr 91				
		Jun 91	1,050	161^{3} (38.6)	575^{3} (10.8)	
		Sep 91				
33	*Paddy Bay*	May 89				
		Sep 89				
		Jun 90				
		Sep 90				
		Apr 91				
		Sep 91				
34	Point Helen	Nov 89				

TABLE 5.—Continued.

Site		Date	Depth		
Number	Name		0 m	3 m	6 m
35	*Port Fidalgo*	Dec 89		189^{3} (96.5)	54.9^{3} (74.7)
		Jun 90	181	194	155
		Jul 90[c]	6.3	9.7	5.5
36	*Rocky Bay*	Jul 89	316^{2} (0.1)	186^{2} (6.2)	306^{3} (28.3)
		Jun 90	291	* 439	193
		Jul 90	211^{2} (10.5)	126^{2} (1.0)	213^{2} (9.3)
42	*Two Moon Bay*	Dec 89		ND[d]	
43	*West Bay*	Jun 90	116	358	
		Jul 90	10.3^{2} (6.5)	29.6^{2} (7.5)	95.1^{2} (82.3)
44	*Zaikov Bay*	Jun 90	210	* 303	* 418
		Jul 90	8.5^{2} (49.6)	21.5^{2} (2.1)	41.4^{2} (12.8)
		Jun 91	216^{3} (97.3)	162^{2} (4.8)	366^{3} (106.7)

was frequently absent or inconsistently present in these sediments in 1990 and 1991 at stations where it had been observed in 1989. The EVO-PAH pattern was consistently found in shallow subtidal sediments into 1991 only at Northwest Bay (Table 2). Subtidal sediments showed the EVO-PAH pattern at 80% of the locations moderately to heavily oiled in 1989, 78% of oiled locations in 1990, and 57% of oiled locations in 1991. Contamination of subtidal sediments by *Exxon Valdez* oil at oiled locations reached a depth of at least 20 m at five locations in 1989 and six locations in 1990. Two locations showed contamination of sediments by *Exxon Valdez* oil at 20 m in 1991.

Intertidal stations.—In 1990, the EVO-PAH pattern was found at the same locations that it was found in 1989, including Disk Island, Herring Bay, and Northwest Bay (Table 2); however, because the concentrations of oil in the intertidal region were probably markedly altered by intensive shoreline treatment and by winter storms in 1989, some sites showed significant decreases in concentration or exhibited more variable concentrations in 1990 and 1991 than in 1989. At Disk Island, unreplicated samples from the 0-m station contained 131 and 104 ng/g TPAH in June and July 1990, respectively (Table 2). These concentrations were far lower than the TPAH concentration (12,700 ng/g) initially measured at Disk Island in July 1989. At two other heavily oiled locations the intertidal TPAH concentration varied greatly between sampling periods. Mean TPAH concentrations in intertidal sediments at Herring Bay and Northwest Bay in 1990 varied between sampling periods, ranging from 118 ± 28.6 ng/g to 949 ± 403 ng/g (N = 3) at Herring Bay and from 166 ± 30.3 ng/g (N = 3) to 4,650 ng/g (N = 1) at Northwest Bay (Table 2). The intertidal samples from Block Island in summer 1990 were the first collected at the 0-m station at that location. These samples contained mean TPAH concentrations of 945 ng/g (N = 1) and 665 ± 287 ng/g (N = 2) in June and July 1990, respectively (Table 2). The PAH composition pattern of these samples closely resembled that of sediments from other heavily oiled beaches, such as Disk Island and Northwest Bay, in 1989.

In 1991, the TPAH concentration generally decreased in intertidal sediments at Herring Bay and Northwest Bay, locations where the EVO-PAH pattern was found (Table 2). At Herring Bay, the pattern was present in sediments collected in April but not in sediments collected later in 1991. At Northwest Bay, the EVO-PAH pattern was found in intertidal sediments in June 1991, the only time in 1991 that Northwest Bay was sampled.

The TNA concentration generally decreased and the carbon preference index (CPI) generally increased with time at heavily oiled locations. At Northwest Bay the mean TNA concentration of intertidal sediments had declined to 1,650 ng/g (Table 3) and the CPI had risen to 1.44 by June 1991. At Herring Bay the mean TNA concentration of intertidal sediments decreased from 4,080 ± 1,120 ng/g (N = 3) in May 1989 to 419 ± 53.0 ng/g (N = 2) in June 1991; the CPI increased from 0.85 to 16.3.

Subtidal stations.—In 1990, the EVO-PAH pattern was consistently present in shallow subtidal sediments at Block Island and Northwest Bay and TPAH concentrations were relatively high (Table 2); with one exception (Northwest Bay, 6-m depth, September 1990), the EVO-PAH pattern was observed in all sample replicates and at each depth sampled from 3 to 20 m at these locations. The

TABLE 5.—Extended.

Site		Date	Depth			
Number	Name		20 m	40 m	100 m	Other
35	*Port Fidalgo*	Dec 89	172			
		Jun 90				
		Jul 90[c]	8.5	22.4^{2} (44.8)	223^{2} (1.8)	
36	*Rocky Bay*	Jul 89	287^{3} (16.6)	470^{3} (35.5)	$1{,}200^{3}$ (59.1)	
		Jun 90	* 800			
		Jul 90	290^{2} (7.6)	371^{2} (1.6)	673^{2} (17.3)	
42	*Two Moon Bay*	Dec 89	ND[d]			
43	*West Bay*	Jun 90	* 639			
		Jul 90	152^{2} (64.0)	234^{2} (2.1)	507^{2} (2.7)	
44	*Zaikov Bay*	Jun 90	543^{2} (12.4)			
		Jul 90	72.4^{2} (22.1)	749^{2} (10.6)	468^{2} (2.3)	
		Jun 91	240^{3} (78.2)	$1{,}020^{3}$ (15.4)	466^{2} (10.0)	

[a] 10 = m depth.
[b] 140 = m depth.
[c] Sampled 28 June 1990.
[d] Wet weight not available.

mean TPAH concentration in sediments from these depths at Block Island and Northwest Bay ranged from 227 ng/g ($N = 1$) to 1,850 ± 549 ng/g ($N = 3$) and equalled or exceeded 700 ng/g at least twice in each of these locations. The TPAH concentration usually varied with depth by a factor less than 3, and no consistent trends in TPAH concentration with depth were apparent.

The EVO-PAH pattern was more sporadically observed in shallow subtidal sediments at Disk Island and Herring Bay than at Block Island and Northwest Bay in 1990 (Table 2). The EVO-PAH pattern was present in the samples from 3-m and 20-m depths at Disk Island in June but was absent at the 6-m depth in June and at all shallow depths in July. At Herring Bay site #125, the pattern was observed at 3 and 20 m, and at the NOAA site it was present at 3–20 m in June 1990.

In 1991, the EVO-PAH pattern was found in shallow subtidal sediments mainly at Northwest Bay (at 3 and 20 m) and at Snug Harbor site 25 (at 6 and 20 m) (Table 2). Block and Disk islands were not sampled in 1991. The EVO-PAH pattern was observed at 6 m in September 1991 at Herring Bay, but the mean TPAH concentration was only 105 ± 29.0 ng/g ($N = 3$).

The concentration of *n*-alkanes in shallow subtidal sediments at heavily oiled locations followed a temporal trend similar to that observed in intertidal sediments at these locations. At Northwest Bay the TNA concentration in sediments at the 3-m depth declined from 4,130 ± 2,470 ng/g ($N = 3$) in July 1989 to 719 ± 44.5 ng/g in June 1991 (Table 3). The CPI increased from 1.27 to 2.89 at 3 m during the same period. At 6 m at Northwest Bay, the TNA concentration decreased from 2,540 ± 392 ng/g ($N = 3$) in July 1989 to 781 ± 264 ng/g ($N = 2$) in June 1991; the CPI increased from 2.37 to 3.56. The same trend was seen at the 3-m depth in Herring Bay (Table 3; the CPI increased from 1.55 to 6.94 between July 1989 and June 1991); however, at 6 m in Herring Bay, a declining trend in mean TNA concentration with time was less clearly indicated. The mean TNA concentration at 6 m in May 1989 was 879 ± 81.2 ng/g ($N = 3$), but in April 1991 it had increased to 1,480 ± 695 ng/g ($N = 2$). In June and September 1991, the mean TNA concentration was 237 ± 29.6 ng/g and 647 ± 136 ng/g ($N = 3$), respectively. The CPI at 6 m in Herring Bay increased from 2.30 in May 1989 to 6.34 in April 1991.

Moderately Oiled Locations, 1990 and 1991

Intertidal stations.—In 1990, the EVO-PAH pattern, though consistently present in intertidal sediments from Bay of Isles, was less consistently observed in intertidal sediments from Sleepy Bay and Snug Harbor. At Bay of Isles, the mean TPAH concentration in intertidal sediments ranged from 749 ± 381 ng/g ($N = 2$) to 1,480 ng/g ($N = 1$) in 1990 (Table 2). At Sleepy Bay, the EVO-PAH pattern was evident only in the September 1990 samples, where the mean TPAH concentration was 253 ± 66.2 ng/g ($N = 3$; Table 2). The mean TPAH concentration was lower and more variable in samples collected in June and July at Sleepy Bay, and these samples were either high in naphthalenes or low in chrysenes and dibenzothiophenes when compared with weathered *Exxon Valdez* oil. At Snug Harbor, the EVO-PAH pattern was evident only at the Snug Harbor 25 site in July 1990. The EVO-PAH pattern was not present in samples collected

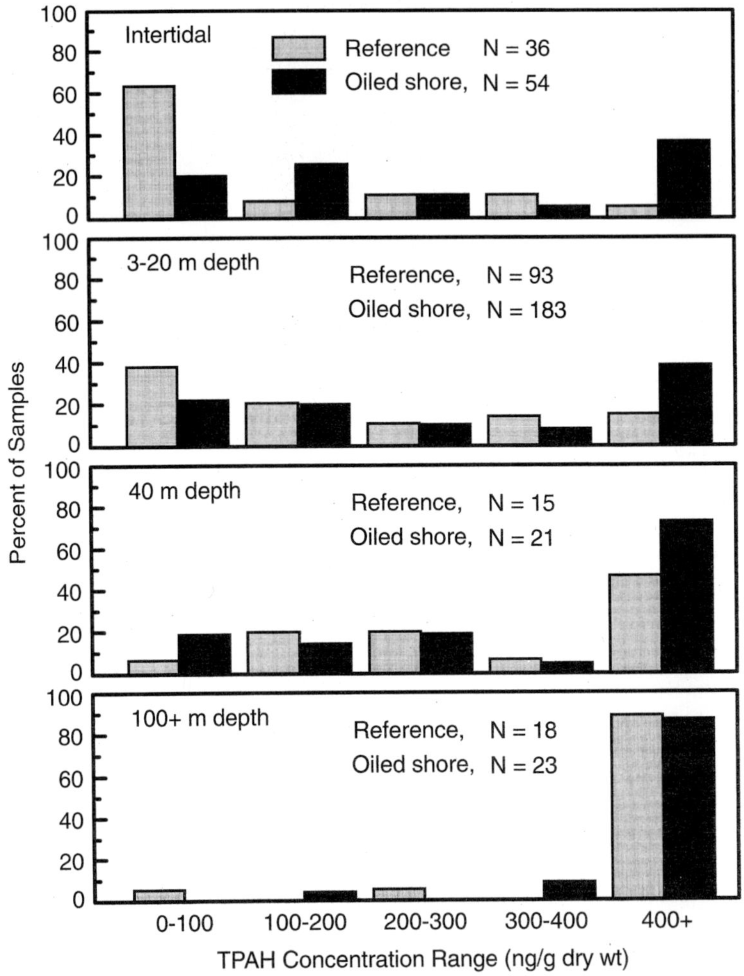

FIGURE 4.—Percentage of sediment samples in five concentration ranges of mean total polynuclear aromatic hydrocarbons (TPAH) in sediments in Prince William Sound at reference stations (italicized in Table 5) and at stations near moderately to heavily oiled beaches (listed in Table 2) from the intertidal region (0 m) and the subtidal region at bathymetric depths of 3–20 m, 40 m, and 100 m and greater during 1989–1991. The number of means included for each treatment × depth category is indicated by *N*.

in June 1990 at this site nor in samples collected from the Snug Harbor meiofauna site.

Concentrations of TPAH were low at the remaining assessment sites sampled in 1990; 62% of the sediments had less than 230 ng TPAH/g. The rest of the samples had either (1) high proportions of unsubstituted, high-molecular-weight PAHs or (2) a PAH composition analyte pattern that differed from weathered *Exxon Valdez* oil because of relatively low concentrations of chrysenes. Both of these analyte patterns often were observed in one replicate only.

In 1991, the TPAH concentration had generally decreased from 1990 in intertidal sediments at moderately oiled locations where the EVO-PAH pattern was found. The EVO-PAH pattern was observed in intertidal sediments at Sleepy Bay and Snug Harbor (Table 2). At Sleepy Bay, intertidal samples collected in June 1991 consistently showed the EVO-PAH pattern; however, replicate samples collected there in May and in September 1991 did not show this pattern. At Snug Harbor, the EVO-PAH pattern was found at site #25 and the meiofauna site. Elsewhere, the mean TPAH concentra-

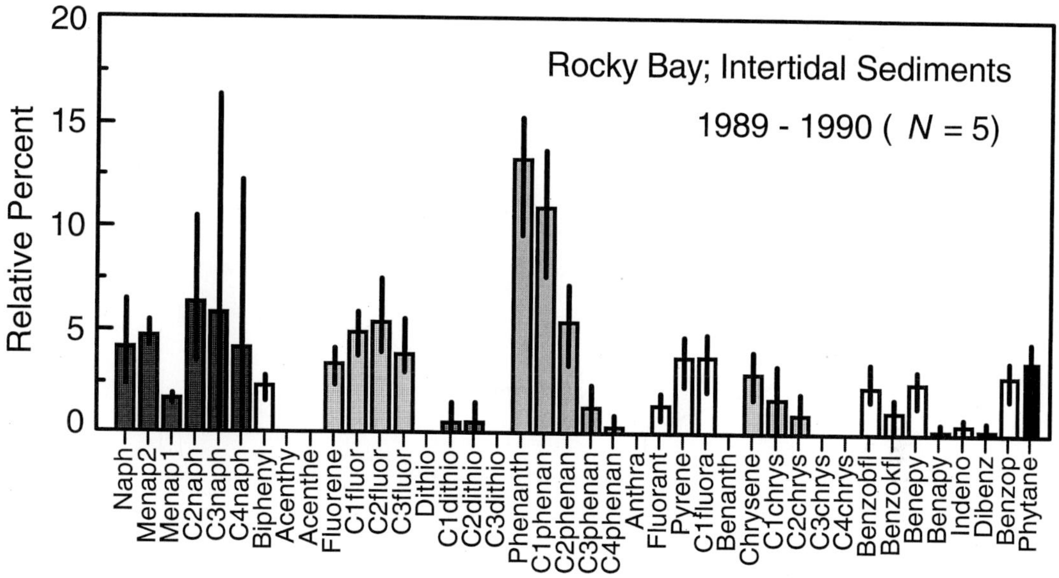

FIGURE 5.—Mean relative percents of polynuclear aromatic hydrocarbon compounds and phytane in intertidal sediments at Rocky Bay, 1989–1990. Error bars are ranges of $N = 5$ analyses.

tion in intertidal sediments was either (1) less than 100 ng/g (80% of remaining samples; Table 2) or (2) the sediments showed one or more of the alternative PAH composition patterns described above for intertidal sediments in 1990.

Shallow subtidal stations.—The EVO-PAH pattern was observed inconsistently in shallow subtidal sediments at Bay of Isles, Sleepy Bay, and Snug Harbor in 1990 (Table 2). At Bay of Isles, the EVO-PAH pattern was consistently present in sediments sampled from 3 through 20 m at two sites (Bay of Isles 90 and Bay of Isles Bioremediation [BR]; Table 2). At the NOAA Bay of Isles site, however, the pattern was evident only at 20 m in June. At Sleepy Bay, the EVO-PAH pattern was observed in shallow subtidal samples in June and September but not in July 1990. At Snug Harbor, the EVO-PAH pattern was observed only at 6 m at the meiofauna site in September 1990.

The number of stations at moderately oiled locations where the EVO-PAH pattern was observed in subtidal sediments further decreased in 1991. The EVO-PAH pattern was not observed in any subtidal sediments from Sleepy Bay in 1991 (Table 2). At the three stations where the pattern was observed in Bay of Isles and Snug Harbor in 1991 the highest mean TPAH concentration was 219 ± 102 ng/g ($N = 3$).

Discussion

Bathymetric Distribution of Exxon Valdez Oil: Contaminated Locations

Intertidal stations.—Oil was broadly distributed on beaches in Prince William Sound during the first few months after the *Exxon Valdez* oil spill. Most of the spilled oil that had not evaporated or dispersed naturally into the water column was deposited in the intertidal region and in the supralittoral fringe primarily in western Prince William Sound (Wolfe et al. 1994). Lower intertidal sediments, near mean lower low water (MLLW), at some locations were clearly contaminated by crude oil. The concentrations of individual aromatic analytes in samples from the Disk Island 0-m station usually averaged 1 to 3 orders of magnitude greater than the baseline concentrations of those same analytes reported by Karinen et al. (1993) for intertidal sites in the sound before the oil spill. *Exxon Valdez* oil was indicated as the source of the PAHs at Disk Island and Northwest Bay by (1) the close similarity between relative PAH abundances in the intertidal sediments and those in floating mousse collected 11 d after the spill; (2) high concentration of *n*-alkanes and CPI near 1, indicating a petroleum source of the alkanes; and (3) the proximity of the intertidal station to the heavily oiled upper intertidal shoreline where

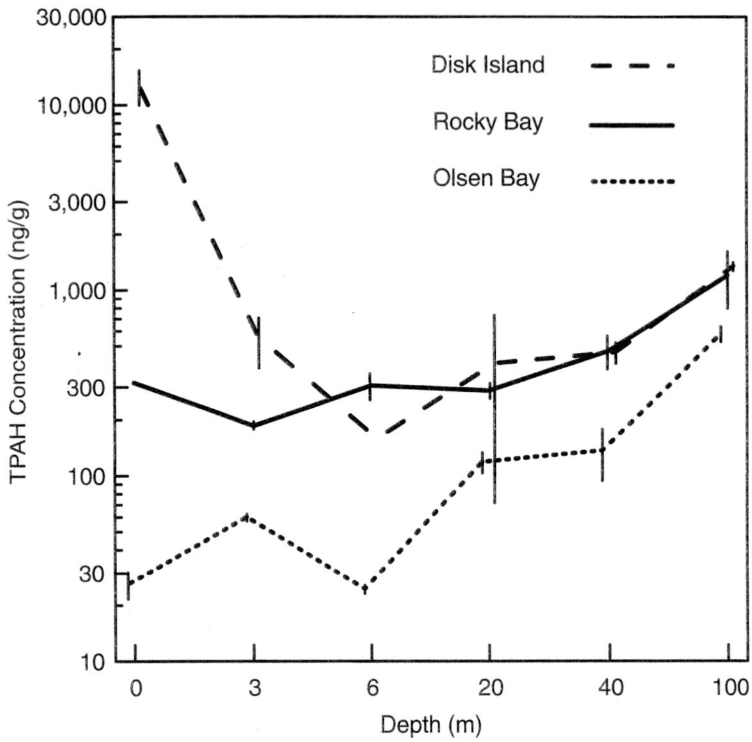

FIGURE 6.—Depth distribution of mean concentrations of total polynuclear aromatic hydrocarbons (TPAHs) at Disk Island, Olsen Bay, and Rocky Bay in July 1989. Axis scales are logarithmic. Error bars are ±1 SE.

oil was visually apparent in spring 1989 (indicating oil concentrations near percent levels).

Results from samples collected along a vertical intertidal transect in September 1989 at Northwest Bay corroborated the July 1989 results from that location. Every sample (collected at 2-m intervals) along the 20-m transect was qualitatively similar to samples collected at the 0-m depth in July 1989 in PAH distribution, *n*-alkane concentration, and CPI. Although the TPAH concentrations in the samples collected in September 1989 were lower by a factor of 2–10 compared with samples collected in July 1989, they were much higher than intertidal TPAH concentrations at reference stations. In addition, TPAH concentrations in samples from the 20-m vertical transect at Northwest Bay in September 1989 were between 1,300 and 6,000 ng/g, indicating that the PAHs were distributed homogeneously along the transect, rather than in isolated patches of high concentration. These findings corroborate the transport mechanism posited by Short et al. (1996a), which involves movements of oil from the heavily oiled upper intertidal zone to the less-heavily oiled lower intertidal zone and involves widespread dispersion of fine-grained, oil-contaminated sediments.

Lower intertidal sediments were also contaminated by *Exxon Valdez* oil at other locations, but at a lower TPAH concentration than initially found at Disk Island or Northwest Bay. The PAHs in the intertidal sediments of at least some of the stations at these locations (except Smith Island) were qualitatively similar to PAHs at Disk Island and at Northwest Bay in 1989. Moderate-to-heavy oiling was observed in the upper intertidal zone at all of the locations. At Smith Island, intertidal samples were not collected in December 1989, but oiling may be inferred from the PAH distribution found in the 3-m subtidal sample and from the heavy, upper intertidal oiling that occurred there. Contamination of many intertidal sediments by *Exxon Valdez* oil was confirmed by an associated high *n*-alkane concentration and a low CPI at stations where the TPAH concentration was also high. At stations where the TPAH concentration was lower, the magnitude and distribution of *n*-alkane concentrations were often confounded by alkanes from terrigenous sources, indicated by higher CPI values.

Total *n*-alkane concentrations ranging to more than 1,000 ng/g from terrigenous and marine sources were widespread in intertidal sediments at reference stations and at assessment stations where oiling in the upper intertidal zone was low or absent. The concentrations of *n*-alkanes from natural sources can dominate the TNA. The concentrations of *n*-alkanes at reference stations in the present study were similar to those in prespill intertidal sediments in Prince William Sound (Karinen et al. 1993). The alkanes were probably derived from terrestrial plant waxes, which are characterized by higher concentrations of odd carbon-numbered alkanes in the range C-20 to C-30 and above (Eglinton et al. 1962; Eglinton and Hamilton 1967; Kolattukudy 1976), and from marine sources including bacteria (Oro et al. 1967), blue-green algae (Winters et al. 1969), and planktonic and macrophytic algae (Clark and Blumer 1967; Blumer et al. 1971), which are characterized by higher concentrations of odd carbon-numbered alkanes smaller than C-20 (especially C-15, C-17, and C-19). The benthic sediments of boreal fjords (sensu Burrell 1988) commonly receive large amounts of terrigenous organic material from inflowing rivers and streams, chiefly in the form of plant detritus refractory to decomposition. Added to this material can be substantial amounts of phytodetritus derived from phytoplankton production and sedimented out of the euphotic zone (Burrell 1988). The deposition of particulate organic material from these sources can cause the accumulation of naturally occurring hydrocarbons in benthic sediments.

Subtidal stations.—Very little oil may have reached the bottom by direct transport from the water column immediately after the spill (see discussion of mechanisms of transport, below). Most of the contamination of subtidal sediments probably resulted from resuspension of oiled sediment particles from contaminated beaches, followed by sedimentation in the subtidal region. Consequently, contamination of subtidal sediments by *Exxon Valdez* oil was mainly confined to shallow bathymetric depths at locations where shorelines were (1) heavily oiled, (2) exposed to wave-action or shoreline treatment, and (3) where adjacent, shallow-subtidal sediment deposits occurred on low-gradient slopes. We found strong evidence of *Exxon Valdez* oil contamination in shallow subtidal sediments at Block Island, Herring Bay, and Northwest Bay, locations where the above conditions prevailed. The mean TPAH concentration in samples showing the EVO-PAH pattern was highest, consistent among replicate samples, and the EVO-PAH pattern was persistent at these locations. The EVO-PAH pattern was frequently found at depths to 20 m but rarely deeper. The EVO-PAHs were absent from greater depths at oiled locations probably because (1) insufficient energy was available from waves and currents to transport contaminated sediments far enough offshore to reach greater depths, and (2) as contaminated sediments spread to greater depths, they became dispersed and intermixed with uncontaminated sediments over a broader area, resulting in lower PAH concentrations that were more difficult to detect.

Environmental conditions were less favorable for transport of oil-contaminated intertidal sediments to subtidal depths at other heavily oiled locations. Evidence for *Exxon Valdez* oil contamination of subtidal sediments was less consistent at these locations, and the mean TPAH concentration attributable to *Exxon Valdez* oil was generally lower than at Northwest Bay, for example. The sampling stations at Bay of Isles and Snug Harbor were more sheltered from heavy wave action. Shorelines near some sampling stations at these locations were less heavily oiled initially than at Northwest Bay. Although oil was available for transport to the subtidal region, those factors promoting the transport of intertidal sediments to subtidal depths were less pronounced at these locations.

At Sleepy Bay, Smith Island, and perhaps Disk Island, shorelines were heavily oiled but were exposed to heavy wave-action, which may have precluded long sediment-residence times. Concentrations of terrigenous (i.e., high CPI) *n*-alkanes at the shallow subtidal depths at these locations were generally low. The sediments were probably less stable and did not act as sinks for terrigenous organic material. The general absence of *Exxon Valdez* oil-derived PAHs in the sediments is consistent with this conclusion. The Smith Island and Sleepy Bay sites are surrounded by or are adjacent to large passages where strong currents prevail. These conditions may have promoted greater export rates of suspended sediments, resulting in a shorter residence time of oil-contaminated subtidal sediments at these locations.

Bathymetric Distribution of Exxon Valdez Oil: Uncontaminated Locations

Intertidal stations.—Lower intertidal sediments at reference locations and at assessment locations that received little oil were not contaminated by *Exxon*

TABLE 6.—Comparison of polynuclear aromatic hydrocarbons in intertidal sediments at Olsen Bay and Rocky Bay during 1977–1980 (Karinen et al. 1993) and 1989–1990 (this study). Numbers in the body of the table are means and 95% confidence intervals. Significance levels (using *t*-test) are $P < 0.05^*$ and $P < 0.01^{**}$.

	Rocky Bay		Olsen Bay	
Polynuclear aromatic hydrocarbon	1989–1990 ($N = 6$)	1977–1980 ($N = 10$)	1989–1990 ($N = 22$)	1977–1980 ($N = 9$)
Naphthalene	12.5 ± 2.63**	4.50 ± 0.61[a]	3.61 ± 0.62**[b]	0.36 ± 0.17
1-Methyl naphthalene	4.39 ± 0.53*	2.29 ± 0.49	0.72 ± 0.21[b]	0.27 ± 0.18
2-Methyl naphthalene	13.1 ± 1.69*	7.74 ± 1.25	0.90 ± 0.16[b]	0.40 ± 0.17
2,6 Dimethyl naphthalene	8.36 ± 0.90**	4.68 ± 0.61	<2.40	<0.21
2,3,5 Trimethyl naphthalene	<2.40	<0.20	<2.40	<0.22
Biphenyl	6.47 ± 0.63**	3.62 ± 0.60	<2.40	<0.21[c]
Fluorene	9.43 ± 0.83	10.7 ± 1.47	<2.40	<0.22[d]
Dibenzothiophene	<2.00	1.78 ± 0.64	<2.00	<0.39
Phenanthrene	39.8 ± 4.52	37.9 ± 5.61	2.16 ± 0.42*	0.34 ± 0.14
1-Methyl phenanthrene	4.66 ± 0.67	2.41 ± 0.76[e]	1.22 ± 0.68	<0.28
Anthracene	<4.00	1.80 ± 0.45	<4.00	<0.16
Fluoranthene	4.09 ± 0.54**	9.74 ± 1.42	1.65 ± 0.61	<0.29
Pyrene	11.4 ± 1.68	12.6 ± 2.05	<3.00	<0.20
Benzanthracene	<1.40	1.31 ± 0.47	<1.40	<0.72
Chrysene	9.71 ± 1.87	13.0 ± 2.58	0.92 ± 0.31	<0.39
Benzo[*e*]pyrene	7.00 ± 1.13	8.90 ± 1.58	<2.40[f]	<0.89
Benzo[*a*]pyrene	<1.20	0.85 ± 0.69	<1.20[f]	<0.73
Perylene	7.52 ± 3.04**	33.3 ± 5.07	10.5 ± 1.49[f]	7.60 ± 1.71

[a] $N = 8$.
[b] $N = 21$.
[c] One sample 0.29 ng/g.
[d] One sample 0.58 ng/g.
[e] $N = 9$.
[f] $N = 15$.

Valdez oil. The TPAH concentrations at these locations were usually less than 100 ng/g and, where comparable, selected PAH concentrations were similar to those in intertidal sediments sampled during the period 1977–1980 in Prince William Sound (Karinen et al. 1993).

To establish baseline concentrations of petroleum hydrocarbons in Prince William Sound in case of an oil spill, a suite of PAHs was monitored in intertidal sediments from 1977 (when the Trans-Alaska Pipeline Terminal at Valdez began operations) to 1980 (Karinen et al. 1993). The accuracy and precision of the analytical methods used in that study were comparable to those of our study. Two of the locations monitored during that period (Olsen Bay and Rocky Bay) were also sampled during our study. Comparison of intertidal sediments from these locations showed general agreement in PAH concentrations between the two studies (Table 6). Several of the same PAH analytes were elevated in both studies, and the concentrations of the elevated analytes were often quite close. Most PAH analytes were near or below detection limits at Olsen Bay during both studies; the exception was perylene, which was present at the highest concentration of the PAH analytes in both studies at Olsen Bay and which did not differ significantly in concentration between studies ($P > 0.25$; Table 6). At Rocky Bay, concentrations of fluorene, phenanthrene, pyrene, chrysene, and benzo[*e*]pyrene were substantially greater than detection limits in both studies. The means of these concentrations did not differ significantly between studies ($P > 0.40$). Mean concentrations of the remaining three-ring PAHs (including dibenzothiophene) were near or below detection limits in both studies. In contrast, the two-ring PAHs in sediments from Rocky Bay were significantly greater in concentration in our study because of greater concentrations at the intertidal station in 1989. The concentration of fluoranthene was lower in our study.

The generally close agreement of the concentrations of PAH analytes at comparable stations between our study and that of Karinen et al. (1993) suggest that during the period 1977–1990 PAH concentrations changed little in the intertidal sediments of Prince William Sound that were not oiled by the *Exxon Valdez* oil spill. Karinen et al. (1993) found little evidence of PAH contamination at four stations (Bligh Island, Naked Island, Siwash Bay, and Olsen Bay) that they sampled in 1977–1980, based on the general absence of peaks in the representative chromatograms they presented for these stations. They found very low concentrations of PAH at Rocky Bay and at Constantine Harbor near the eastern entrance to the sound. Therefore, with the exception of two-ring PAHs, any increases in PAH concentrations in intertidal sediments of the sound before the oil spill probably were attributable to localized sources of PAH rather than to some large-scale, dispersed source that contaminated the sound generally. Nevertheless, a widespread source of two-ring PAHs cannot be ruled out because these PAHs were often found in intertidal and subtidal sediments in 1989 at reference and assessment sites. The source of these two-ring PAHs is not known.

Subtidal stations.—The general absence of *Exxon Valdez* oil-derived PAHs in subtidal sediments at reference locations and at assessment locations where shorelines were not heavily oiled indicates that subtidal sediment contamination by *Exxon Valdez* oil was localized, rather than widespread, along the spill trajectory through Prince William Sound. The few instances where the PAH composition pattern resembled that of weathered *Exxon Valdez* oil in subtidal sediments at these locations probably resulted from mixtures of PAHs from other sources. These few occurrences were distributed throughout the range of the subtidal depths sampled and sometimes appeared at locations away from the spill path in the sound (e.g., Olsen Bay). When the EVO-PAH pattern was observed in shallow subtidal sediments at these locations, it was present in one replicate only. We consider these isolated occurrences of the EVO-PAH pattern in sediments from reference locations and lightly oiled assessment locations to be largely spurious. We conclude that detectable contamination of subtidal sediments by *Exxon Valdez* oil was not widespread throughout the spill path, but rather was restricted to those relatively few locations where conditions favored subtidal accumulation. Most of the subtidal area within the spill path escaped detectable contamination.

Geographical Distribution of Exxon Valdez Oil

In general, the pattern of distribution of petroleum hydrocarbons in benthic sediments in Prince William Sound in 1989 was consistent with the pattern expected from maps of the trajectory of the *Exxon Valdez* oil spill compiled by NOAA and the Alaska Department of Environmental Conservation. We found *Exxon Valdez* oil in lower intertidal and subtidal sediments over a broad geographic range in the sound, from Northwest Bay at the north end of Eleanor Island to Foxfarm (intertidal sediments only) at the southern end of Elrington Island (Figure 1). Subtidal sediments contained *Exxon Valdez* oil at eight locations where oil had come ashore (oiled locations) in 1989. Those locations constituted 80% of oiled locations studied in 1989. Contamination of subtidal sediments by *Exxon Valdez* oil at oiled locations reached a depth of at least 20 m at five locations in 1989.

Temporal Changes in Exxon Valdez Oil Contamination

The general decline in 1990 and 1991 of mean TPAH concentrations in intertidal sediments at sites where *Exxon Valdez* oil was found in 1989 reflects the continued action of the mechanisms of dispersion that initially transported oil-contaminated sediments from heavily oiled, upper intertidal shorelines to the lower intertidal zone and the subtidal region. Shoreline treatment probably augmented the action of high-energy waves to suspend and disperse fine-grained, oil-contaminated sediments. By 1990 the mean TPAH concentration in lower intertidal sediments at many oiled sites had declined to levels comparable to the background TPAH concentration (i.e., to a concentration ≤200 ng/g). Additional decreases in TPAH concentration were observed at those sites sampled in 1991.

Concentrations of petroleum hydrocarbons attributable to *Exxon Valdez* oil decreased, and these hydrocarbons became more localized in subtidal sediments, as well, after 1989. By 1990 the EVO-PAH pattern was consistently present in shallow subtidal sediments only at Block Island and Northwest Bay. The EVO-PAH pattern was occasionally seen in subtidal sediments at five additional locations in 1990. No consistent trend in TPAH concentration with depth was apparent. Additional decreases in the TPAH concentration were observed at oiled sites in 1991. The EVO-PAH pattern was consistently present only at Northwest Bay in 1991, although the pattern was sporadically observed in subtidal sediments from Bay of Isles, Herring Bay, and Snug Harbor in 1991. In succeeding years the PAHs will probably continue to decline and become more patchily distributed.

Mechanisms of Transport of Exxon Valdez Oil to Subtidal Sediments

Wolfe et al. (1994) estimated that about 12% of the spilled oil was transported to the subtidal region by fall 1992. The accumulation of *Exxon Valdez* oil in subtidal sediments could have resulted from (1) sinking of oil, sedimentation of oil and clay colloids, or settling of biogenic, composite grains directly out of the water column shortly after the spill and (2) resuspension of oiled sediment particles from contaminated beaches by wave action and cleanup activities, followed by sedimentation in the subtidal region. Transport of oil to subtidal sediments directly from the water column can be mediated by three mechanisms that cause oil settling (Boehm et al. 1982a). The first involves an increase in the density of the oil by loss of soluble, volatile fractions to the extent that the oil density exceeds that of water, causing the oil to sink. This mode of transport can occur under some conditions when the initial oil density is close to that of water (Boehm et al. 1982a). The *Exxon Valdez* oil probably did not sink in this way (Galt et al. 1991).

The second mode of transport causing oil to settle involves the adsorption of oil films onto particulate matter, especially clay particles, which can result in the formation of spontaneous association colloids through the process of flocculation by electrolytic action (Bassin and Ichiye 1977). Physicochemical flocculation can increase settling rates up to about one order of magnitude (Drake 1976). There are apparently no definitive data on how much oil, associated with particulate matter, settled to the bottom of Prince William Sound shortly after the spill. Galt et al. (1991:206) presumed that oil did not settle to the bottom to "any significant extent" as the oil slick moved through the sound because "there was little or no sediment-laden water near the oil slick," but they presented no data on the amount of suspended sediment in the water near the slick. Payne et al. (1989) measured suspended particulate material (SPM) loads in Prince William Sound during the period 12–15 April 1989. They found SPM loads in the range <0.01 to 4.57 mg/L. From this, Payne et al. (1991) expected that very little oil-SPM interaction and sedimentation occurred in the first few weeks after the spill.

The third mechanism of transport of oil to the bottom involves encapsulation of oil into feces or pseudofeces after ingestion of oil droplets by copepods or suspension-feeding benthic invertebrates such as mussels (Conover 1971; Clark and MacLeod 1977). Biotic production of composite grains in this way can increase settling rates up to several orders of magnitude (Drake 1976). No data have been presented to date that would allow an assessment of the importance of this mechanism to oil settling during the *Exxon Valdez* oil spill. According to Koons and Jahns (1992), the U.S. Geological Survey concluded that the amount of oil that sank to the bottom during the first 2 months after this spill was insignificant, but they cite no reference.

The most likely pathway by which *Exxon Valdez* oil was transported to subtidal sediments, especially at certain heavily oiled locations, was resuspension of contaminated beach sediments followed by deposition nearshore. Sediments collected from sediment traps initially deployed during the period November 1989–June 1990 at depths of 10, 15, and 20 m near oiled and non-oiled sites showed hydrocarbon analyte patterns consistent with *Exxon Valdez* oil. Hydrocarbon concentrations in these sediments were highest at heavily oiled sites (Sleepy Bay, Northwest Bay, and Snug Harbor) and lowest in sediment traps placed at non-oiled sites (Port Fidalgo, Eshamy Bay, and Stockdale Harbor; Short et al. 1996a).

Other Hydrocarbon Sources in Prince William Sound

Intertidal and shallow subtidal sediments.—Our results indicate that the quantitative impact of PAHs from sources other than the *Exxon Valdez* oil spill on intertidal and shallow subtidal sediments in Prince William Sound was small. The TPAH concentrations of these sources were usually below 200 ng/g. Overall, intertidal and shallow subtidal sediments in the sound not affected by the oil spill remained substantially free of petroleum hydrocarbons. A mean TPAH concentration greater than 100 ng/g was occasionally found in intertidal sediments where *Exxon Valdez* oil was not indicated, but a concentration this high usually occurred sporadically among station replicates.

The probable sources of PAHs in intertidal and shallow subtidal sediments where *Exxon Valdez* oil was absent included diesel oil and products of pyrolysis associated with small (and often temporary) human settlements, asphalt from storage tanks at Valdez, and forest fires. Fish-processing plants or mining facilities have been located historically in Snug Harbor, Drier Bay, and near Sleepy Bay. Relatively small areas of upper intertidal sediments may have become contaminated by PAHs through human activity at these installations, and the contamination may have subsequently dispersed into lower intertidal sediments. Background PAHs could be transported to subtidal sediments by the same mechanism that transported terrigenous *n*-alkanes and *Exxon Valdez* oil to those sediments. Trace residues of California crude oil linked to storage tanks at Valdez and Whittier that may have ruptured during the Alaska earthquake of 1964 have been reported in intertidal sediments in Prince William Sound along the path of the *Exxon Valdez* oil spill (Kvenvolden et al. 1993). Finally, in certain of the last 30 years (1965–1995) large forest fires on the Kenai Peninsula have blanketed the entire sound with wood smoke for weeks, thereby providing an additional potential source of pyrogenic PAHs that were initially widely distributed in the sound and are now confined to localized areas.

Mixtures of PAHs from multiple sources may result in PAH composition patterns that resemble weathered *Exxon Valdez* oil at low TPAH concentration. In particular, diesel oil derived from North Slope crude oil together with PAHs characteristic of deep subtidal sediments (Figure 3) could result in alkyl-PAH ratios near those found in samples contaminated by *Exxon Valdez* oil (e.g., floating mousse oil collected from the oil slick 11 days after the spill). Compared with *Exxon Valdez* oil, the diesel oil lacked chrysenes and the deep-sediment PAHs were low in dibenzothiophenes. Mixtures of these two components may have contained all the PAHs at relative abundances sufficiently close to those of *Exxon Valdez* oil to satisfy the criteria we used for preliminary identification of that oil. This may account for the isolated cases where weathered EVO-PAH assemblages were identified in subtidal sediments at reference sites and at sites where *Exxon Valdez* oil was not found in intertidal sediments. These cases were usually single samples at a TPAH concentration less than 200 ng/g in shallow subtidal sediments, or they were samples at somewhat higher TPAH concentration in deep subtidal sediments. Some cases were near Hinchinbrook Entrance where the concentrations of PAHs were higher in subtidal sediments (see below).

Deep subtidal sediments.—Deep subtidal sediments were uniformly contaminated by PAHs derived from marine oil seeps or some other natural source of PAHs such as unburned coal (Tripp et al. 1981; Barrick et al. 1984). The PAH composition pattern characteristic of this source differs from weathered *Exxon Valdez* oil in that the coal contains lower relative abundances of dibenzothiophenes. Page et al. (1995) attributed this PAH composition pattern to coastal oil seeps in the eastern Gulf of Alaska near Katalla Island and Cape Yakataga. According to Page et al. (1995), the PAHs from the oil seeps are adsorbed to clay particles introduced into the Gulf of Alaska by glacier-fed streams and rivers, including the Copper River, and are subsequently carried into Prince William Sound through Hinchinbrook Entrance. Once inside the sound, the current velocity slows, resulting in increased deposition of oil-laden sediments. Another possible source may be microscopic coal particles eroded from terrestrial coal beds in the Copper River drainage and introduced into the Alaska Coastal Current together with other sediments transported by the Copper River, and which are subsequently carried into the sound through Hinchinbrook Entrance. Unburned coal and petroleum in sediments may be very difficult to distinguish at lower concentrations on the basis of PAH analyses (Tripp et al. 1981).

Our results are generally consistent with these explanations. The generally increasing mean TPAH concentration with depth below 20 m (Figure 6) associated with the PAH composition shown in Figure 3 was consistent with the deposition of PAH-laden sediments from natural sources. Concentrations were greatest near Hinchinbrook Entrance, where the current slows on entering the sound. Inside Prince William Sound, sediments containing these PAHs generally were restricted to deeper subtidal sediments.

As the distance from the source increases, the amount of suspended-sediment-bearing hydrocarbons presumably decreases as these sediments settle out of the water column. As a result, contributions of PAHs from natural sources to shallow subtidal sediments would be smaller within Prince William Sound than at Hinchinbrook Entrance. In fact, PAH concentrations were low in shallow subtidal sediments at locations within the sound that were not directly affected by the oil spill. The concentration of TPAH in shallow subtidal sediments at locations not directly affected by the spill was usually less than 200 ng/g; however, at some stations the TPAH concentration in shallow subtidal sediments contaminated by *Exxon Valdez* oil often exceeded 500 ng/g, which is clearly more than a small increase on a large natural PAH background.

Comparisons with Other Oil Spills

The proportion of spilled oil from the *Exxon Valdez* that entered subtidal sediments was similar to that entering subtidal sediments from other major spills. Wolfe et al. (1994) estimated that between 8 and 16% of the approximately 35,500 metric tons of oil spilled (9–18% of the unrecovered oil) from the *Exxon Valdez* oil spill was transported to subtidal sediments. Gundlach et al. (1983) estimated that 8% of the 223,000 metric tons of oil spilled from the *Amoco Cadiz* was deposited in subtidal sediments. A somewhat lower percentage was estimated by Johansson et al. (1980) for No. 5 fuel oil sedimented to the bottom after the *Tsesis* oil spill. Their estimate was 3–6% of the spilled oil (10–15% of the unrecovered oil). Lower still was the estimate of Boehm et al. (1982b) of the percentage of spilled oil that was eventually transported to subtidal sediments near the *Ixtoc* blowout site (0.5–3%). The proportion of spilled *Exxon Valdez* oil transported to subtidal sediments was similar to that transported to sediment (7–16%) in experimental releases of water-accommodated No. 2 fuel oil into marine microcosms (Gearing et al. 1979).

The concentrations of petroleum hydrocarbons measured in subtidal sediments after major oil spills have varied greatly among spills. Factors contributing to this variability include the magnitude of the spill, environmental conditions during the spill, configuration of the coastline where oil came ashore, and the analytical techniques used to estimate oil concentration in sediments. Nevertheless, maximum concentrations of oil in sediments after the *Exxon Valdez* oil spill were usually within an order of magnitude of the maximum concentrations in subtidal sediments after other major spills. We found the concentration of total hydrocarbons to range from 18 to 694 μg/g where the PAH composition pattern matched *Exxon Valdez* oil in the first year of the spill. Total polynuclear aromatic hydrocarbon concentration including perylene (ΣPAH) in these samples ranged from 0.1 to 2.7 μg/g. At embayments heavily oiled by the *Amoco Cadiz* spill, the concentration of petroleum hydrocarbons in subtidal sediments ranged from 3 to 28,457 μg/g dry weight during the first month after the spill (Marchand and Caprais 1981). Surficial subtidal sediments (0–5 cm) contaminated by oil from the Gulf War oil spill contained a total hydrocarbon concentration ranging from $<$5 to 900 μg/g dry weight. Total PAH concentration was in the range 1–7 μg/g (Michel et al. 1993). Concentration of total hydrocarbons in surficial sediments (0–15 cm) near the *Argo Merchant* wreck site ranged from 0.1 to 327 μg/g dry weight 2 months after the spill of No. 6 fuel oil from the vessel (Hoffman and Quinn 1979).

When compared with chronically polluted locations, the maximum concentration of total hydrocarbons that we found in shallow subtidal sediments (3–20 m) where the PAH composition pattern matched that of *Exxon Valdez* oil (694 μg/g) was similar in magnitude to the middle of the concentration range of hydrocarbons in sediments at such heavily polluted locations as New York Bight and Narragansett Bay. Farrington and Tripp (1977) found that the total hydrocarbon concentration ranged from 35 to 2,900 μg/g dry weight in sediments at a depth of 23–39 m in New York Bight. The total hydrocarbon concentration ranged from 4 to 1,650 μg/g dry weight in sediments at a depth of 5–55 m in Narragansett Bay (Hurtt and Quinn 1979).

Toxicological Effects of Oil in Subtidal Sediments

The toxicological effects of *Exxon Valdez* oil in shallow subtidal sediments were more difficult to identify compared with the pronounced effects observed in the intertidal region at heavily oiled sites because in the shallow subtidal region *Exxon Valdez* oil concentrations did not approach known acutely toxic levels. None of the ΣPAH concentrations in subtidal sediments where the *Exxon Valdez* oil composition pattern was present exceeded the "Effects Range-Low" (ERL) sediment toxicity threshold of 4,000 parts per billion proposed by Long and Morgan (1990; see also Long 1992) for total PAHs. Nevertheless, impacts on shallow (0–20 m) subtidal algae, eelgrass, and infaunal and epifaunal invertebrates were documented at oiled locations in Prince William Sound (Dean et al. 1996a, 1996b; Jewett et al. 1996; all this volume). Unequivocal attribution of the impacts to hydrocarbon toxicity was often precluded because of other factors, such as cleanup activities and bioremediation.

Oil toxicity may have been the primary cause of the apparent reduction in burrowing amphipod densities at oiled sites compared with non-oiled sites in 1990 (Jewett et al. 1996) despite relatively low PAH concentrations in most subtidal sediments in Prince William Sound. A comparable situation may have prevailed after the *Tsesis* oil spill, where the burrowing amphipod *Pontoporeia* sp. was eliminated from some sediments for at least 1 year (Linden et al. 1979; Elmgren et al. 1980) despite the absence of detectable quantities of hydrocarbons related to the spilled *Tsesis* oil in subtidal sediments (Boehm et al. 1982a). Measurements of hydrocarbons in surficial sediments may underestimate the concentration of hydrocarbons to which benthic organisms are exposed because much of the hydrocarbon material may be in the surface floc layer, which is difficult to sample (Boehm et al. 1982a); however, evidence to date for the accumulation of spilled oil in the surface floc layer is inconclusive (Gearing et al. 1979; Boehm et al. 1987).

Reliance on acute-toxicity data to estimate or predict damage in the environment has been problematical. Disturbance to natural communities may take place after an oil spill even though measured sediment hydrocarbon concentrations are less than those that are acutely toxic to individual species.

Processes (feeding, reproduction, growth, molting, etc.) and species sensitive to oil pollution are seldom evaluated in an ecological context, in which complex interactions with predators and competitors is taken into account. Consequently, the proposed ERL sediment toxicity threshold proposed by Long and Morgan (1990) may be appropriate for estimating effects on macroinvertebrates in the short term, but may be high when one considers subtle changes in natural communities over the long term.

Shallow subtidal communities in Prince William Sound apparently did not suffer major structural damage, unlike some heavily oiled and heavily cleaned intertidal areas (Houghton et al. 1993a, 1993b). Moreover, some subtidal communities have shown evidence of recovery (Dean et al. 1996a; Jewett et al. 1996).

Changes in the deep benthos were probably not complicated by *Exxon Valdez* oil contamination. This oil spill probably did not add significantly to the regional background levels of hydrocarbons at ≥40-m depths. The effect of temporal changes in factors such as larval settlement, food availability, and predation pressure on the structure of deep benthic communities may not have been independent of the *Exxon Valdez* oil spill. Nevertheless, considering that *Exxon Valdez* oil generally was not detected in sediments at greater depths, it is unlikely that this oil spill had a direct impact on the deep benthos.

Acknowledgments

A number of workers helped with the intertidal, dive, and grab sampling over the course of the study: We thank M. Baer, S. Benda, J. Braddock, T. Broullette, A. Chan, B. Choy, M. Cunningham, D. Dixon, R. Fletcher, K. Frank, L. Freese, P. Harris, M. Hilley, D. Hopkins, G. Hubner, J. Karinen, J. Lindstrom, D. Neander, C. Rebstock, D. Sale, D. Savage, B. Smith, T. Stinnett, R. Stone, C. Wilson, G. Winter, D. Wolfe, and T. Yeager. We are grateful to the captains and crews of the NOAA Ships *John N. Cobb*, *Davidson*, and *Fairweather* and the MV *Auklet* and MV *Big Valley*. M. Murphy and A. Bence read an early version of this article.

References

Barrick, R. C., E. T. Furlong, and R. Carpenter. 1984. Hydrocarbon and azaarene markers of coal transport to aquatic sediments. Environmental Science and Technology 11:846–854.

Bassin, N. J., and T. Ichiye. 1977. Flocculation behavior of sediment and oil emulsions. Journal of Sedimentary Petrology 47:671–677.

Blumer, M., R. R. L. Guillard, and T. Chase. 1971. Hydrocarbons of marine phytoplankton. Marine Biology 8:183–189.

Boehm, P. D., J. E. Barak, D. L. Fiest, and A. A. Elskus. 1982a. A chemical investigation of the transport and fate of petroleum hydrocarbons in littoral and benthic environments: the *Tsesis* oil spill. Marine Environmental Research 6:157–188.

Boehm, P. D., and six coauthors. 1987. Comparative fate of chemically dispersed and beached crude oil in subtidal sediments of the arctic nearshore. Arctic 40 (Supplement 1):133–148.

Boehm, P., A. D. Wait, D. L. Feist, and D. Pilson. 1982b. Chemical assessment-hydrocarbon analyses, section 2. *Ixtoc* oil spill assessment final report. U. S. Department of the Interior, Bureau of Land Management, (AA851-CTO-71), Washington, DC.

Burrell, D. C. 1988. Carbon flow in fjords. Oceanography and Marine Biology an Annual Review 26:143–226.

Clark, R. C., and M. Blumer. 1967. Distribution of N-paraffins in marine organisms and sediment. Limnology and Oceanography 12:79–87.

Clark, R. C., Jr., and W. D. MacLeod. 1977. Inputs, transport mechanisms, and observed concentrations of petroleum in the marine environment. Pages 91–224 *in* D. C. Malins, editor. Effects of petroleum on arctic and subarctic marine environments and organisms. Academic Press, New York.

Conover, R. J. 1971. Some relations between zooplankton and bunker C oil in Chedabucto Bay following the wreck of the tanker *Arrow*. Journal of the Fisheries Research Board of Canada 28:1327–1330.

Dean, T. A., S. C. Jewett, D. R. Laur, and R. O. Smith. 1996b. Injury to epibenthic invertebrates resulting from the *Exxon Valdez* oil spill. American Fisheries Society Symposium 18:424–439.

Dean, T. A., M. S. Stekoll, and R. O. Smith. 1996a. Kelps and oil: the effects of the *Exxon Valdez* oil spill on subtidal algae. American Fisheries Symposium 18: 412–423.

Drake, D. 1976. Suspended sediment transport on shelves. Pages 127–158 *in* D. J. Stanley and D. J. P. Swift, editors. Marine sediment transport and environmental management. Wiley, New York.

Eglinton, G., A. G. Gonzalez, R. J. Hamilton, and R. A. Raphael. 1962. Hydrocarbon constituents of the wax coating of plant leaves: a taxonomic survey. Phytochemistry 1:89–102.

Eglinton, G., and R. J. Hamilton. 1967. Leaf epicuticular waxes. Science 156:1322–1335.

Elmgren, R., S. Hansson, U. Larsson, and B. Sundelin. 1980. Impact of oil on deep soft bottoms. Pages 97–128 *in* J. J. Kineman, R. Elmgren, and S. Hansson, editors. The *Tsesis* oil spill. Report of the first year of scientific study. U.S. Department of Commerce, National Oceanic and Atmospheric Administration, Boulder, Colorado.

Farrington, J. W., and B. W. Tripp. 1977. Hydrocarbons in western north Atlantic surface sediments. Geochimica et Cosmochimica Acta 41:1627–1641.

Galt, J. A., W. J. Lehr, and D. L. Payton. 1991. Fate and transport of the *Exxon Valdez* oil spill. Environmental Science and Technology 25:202–209.

Gearing, J. N., and six coauthors. 1979. The rates of transport and fates of petroleum hydrocarbons in a controlled marine ecosystem, and a note on analytical variability. Pages 555–564 *in* Proceedings, 1979 oil spill conference (prevention, behavior, control, clean-up). American Petroleum Institute Publication 4308, Washington, DC.

Gundlach, E. R., P. D. Boehm, M. Marchand, R. M. Atlas, D. M. Ward, and D. A. Wolfe. 1983. The fate of *Amoco Cadiz* oil. Science 221:122–130.

Hoffman, E. J., and J. G. Quinn. 1979. Gas chromatographic analyses of *Argo Merchant* oil and sediment hydrocarbons at the wreck site. Marine Pollution Bulletin 10:20–24.

Houghton, J. P., and seven coauthors. 1993a. Evaluation of the condition of Prince William Sound shorelines following the *Exxon Valdez* oil spill and subsequent shoreline treatment: volume II 1991 biological monitoring survey. NOAA (National Oceanic and Atmospheric Administration) Technical Memorandum NOS (National Ocean Service) ORCA (Office of Ocean Resources and Conservation Assessment) 67, Seattle.

Houghton, J. P., and six coauthors. 1993b. Evaluation of the condition of Prince William Sound shorelines following the *Exxon Valdez* oil spill and subsequent shoreline treatment: volume II 1992 biological monitoring survey. NOAA (National Oceanic and Atmospheric Administration) Technical Memorandum NOS (National Ocean Service) ORCA (Office of Ocean Resources and Conservation Assessment) 73, Seattle.

Hurtt, A. C., and J. G. Quinn. 1979. Distribution of hydrocarbons in Narragansett Bay sediment cores. Environmental Science and Technology 13:829–836.

Jewett, S. C., T. A. Dean, and D. R. Laur. 1996. Effects of the *Exxon Valdez* oil spill on benthic invertebrates in an oxygen deficient embayment of Prince William Sound, Alaska. American Fisheries Society Symposium 18:440–447.

Johansson, S., U. Larsson, and P. Boehm. 1980. The *Tsesis* oil spill: impact on the pelagic ecosystem. Marine Pollution Bulletin 11:284–293.

Karinen, J. F., M. M. Babcock, D. W. Brown, W. D. MacLeod, Jr., L. S. Ramos, and J. W. Short. 1993. Hydrocarbons in intertidal sediments and mussels from Prince William Sound, Alaska, 1977–1980: characterization and probable sources. NOAA (National Oceanic and Atmospheric Administration) Technical Memorandum NMFS (National Marine Fisheries Service)-AFSC-9, Alaska Fisheries Science Center, Seattle.

Kolattukudy, P. E. 1976. Chemistry and biochemistry of natural waxes. Elsevier, New York.

Koons, C. B., and H. O. Jahns. 1992. The fate of oil from the *Exxon Valdez*—a perspective. Marine Technology Society Journal 26:61–69.

Kvenvolden, K. A., F. D. Hostettler, J. B. Rapp, and P. R. Carlson. 1993. Hydrocarbons in oil residues on beaches of islands of Prince William Sound, Alaska. Marine Pollution Bulletin 26:24–29.

Linden, O., R. Elmgren, and P. Boehm. 1979. The *Tsesis* oil spill: its impact on the coastal ecosystem of the Baltic Sea. Ambio 8:244–253.

Lindstrom, J. E., and A. Weiner. 1990. Petroleum hydrocarbon-induced injury to subtidal marine sediment resources. Report of Alaska Department of Environmental Conservation to the National Marine Fisheries Service, Auke Bay Laboratory, Juneau.

Long, E. R. 1992. Ranges in chemical concentrations in sediments associated with adverse biological effects. Marine Pollution Bulletin 24:38–45.

Long, E. R., and L. G. Morgan. 1990. The potential for biological effects of sediment-sorbed contaminants tested in the National Status and Trends Program. NOAA (National Oceanic and Atmospheric Administration) Technical Memorandum NOS (National Ocean Service) OMA (Office of Marine Assessment) 52, Seattle.

Marchand, M., and M.-P. Caprais. 1981. Suivi de la pollution de l'*Amoco Cadiz* dans l'eau de mer et les sediments marins. Pages 23–54 *in* G. Conan, L. Laubier, M. Marchand, and L. d'Ozouville, editors. *Amoco Cadiz*, fates and effects of the oil spill: proceedings of the international symposium. Le Centre National Pour L'Exploitation Des Oceans, Paris.

Mearns, A. J. 1996. *Exxon Valdez* shoreline treatment and operations: implications for response, assessment, monitoring, and research. American Fisheries Society Symposium 18:309–328.

Meyers, P. A., and J. G. Quinn. 1973. Association of hydrocarbons and mineral particles in saline solution. Nature 244:23–24.

Michel, J., M. O. Hayes, R. S. Keenan, T. C. Sauer, J. R. Jensen, and S. Narumalani. 1993. Contamination of nearshore subtidal sediments of Saudi Arabia from the Gulf War oil spill. Marine Pollution Bulletin 27: 109–116.

Michel, J., M. O. Hayes, W. J. Sexton, J. C. Gibeaut, and C. Henry. 1991. Trends in natural removal of the *Exxon Valdez* oil spill in Prince William Sound from September 1989 to May 1990. Pages 181–187 *in* Proceedings, 1991 oil spill conference (prevention, behavior, control, cleanup). American Petroleum Institute Publication 4529, Washington, DC.

O'Clair, C. E., J. W. Short, and S. D. Rice. 1995. Petroleum hydrocarbon-induced injury to subtidal marine sediment resources: subtidal study number 1. Final report, *Exxon Valdez* Natural Resources Damage Assessment. NMFS (National Marine Fisheries Service), Auke Bay Laboratory, Juneau, Alaska.

Oro, J., T. G. Tornabene, D. W. Nooner, and E. Gelpi. 1967. Aliphatic hydrocarbons and fatty acids of some marine and freshwater microorganisms. Journal of Bacteriology 93:1811–1818.

Owens, E. H., J. R. Harper, W. Robson, and P. D. Boehm. 1987. Fate and persistence of crude oil stranded on a sheltered beach. Arctic 40 (Supplement 1):109–123.

Page, D. S., P. D. Boehm, G. S. Douglas, and A. E. Bence. 1995. Identification of hydrocarbon sources in the benthic sediments of Prince William Sound and the

Gulf of Alaska following the *Exxon Valdez* oil spill. Pages 41–83 *in* P. G. Wells, J. N. Butler, and J. S. Hughes, editors. *Exxon Valdez* oil spill: fate and effects in Alaskan waters. American Society for Testing and Materials Special Technical Publication 1219, Philadelphia.

Payne, J. R., and eight coauthors. 1989. Oil-ice-sediment interactions during freezeup and breakup. NOAA (National Oceanic and Atmospheric Administration) OCSEAP (Outer Continental Shelf Environmental Assessment Program) Final Report 64, Anchorage, Alaska.

Payne, J. R., J. R. Clayton, Jr., G. D. McNabb, Jr., and B. E. Kirstein. 1991. *Exxon Valdez* oil weathering fate and behavior: model predictions and field observations. Pages 641–654 *in* Proceedings, 1991 oil spill conference (prevention, behavior, control, cleanup). American Petroleum Institute Publication 4529, Washington, DC.

Short, J. W., T. J. Jackson, M. L. Larsen, and T. L. Wade. 1996b. Analytical methods used for the analysis of hydrocarbons in crude oil, tissues, sediments, and seawater collected for the natural resources damage assessment of the *Exxon Valdez* oil spill. American Fisheries Society Symposium 18:140–148.

Short, J. W., D. M. Sale, and J. C. Gibeaut. 1996a. Nearshore transport of hydrocarbons and sediments after the *Exxon Valdez* oil spill. American Fisheries Society Symposium 18:40–60.

Tripp, B. W., J. W. Farrington, and J. M. Teal. 1981. Unburned coal as a source of hydrocarbons in surface sediments. Marine Pollution Bulletin 12:122–125.

Winters, K., P. L. Parker, and C. Van Baalen. 1969. Hydrocarbons of blue green algae: geochemical significance. Science 163:467–468.

Wolfe, D. A., and eleven coauthors. 1994. The fate of the oil spilled from the *Exxon Valdez*. Environmental Science and Technology 28:561A–568A.

American Fisheries Society Symposium 18:94–108, 1996

Patterns of Microbial Activity in Oiled and Unoiled Sediments in Prince William Sound

JOAN F. BRADDOCK, JON E. LINDSTROM, THOMAS R. YEAGER, AND BRIAN T. RASLEY

University of Alaska Fairbanks, Institute of Arctic Biology and Water Research Center Fairbanks, Alaska 99775, USA

EDWARD J. BROWN

University of Northern Iowa, College of Natural Science Cedar Falls, Iowa 50614, USA

Abstract.—Perturbations of microbial communities generally result in shifts in the composition of the microbial population and lead to differences in their activities. An increase in the biodegradation activity of naturally occurring populations of microorganisms can lead to substantial removal of petroleum from the environment. For these reasons, measurements of microbial populations are an important component of contaminated site assessment studies. Over a 3-year period following the *Exxon Valdez* oil spill in Prince William Sound, Alaska, we measured numbers of hydrocarbon-degrading microorganisms and hydrocarbon mineralization potentials of microorganisms in oiled and unoiled surface sediments from the shore through 100-m depth offshore. We used these techniques to assess the response of microbial populations to oiling. We present here the data from two oiled sites and two reference sites as an example of how these techniques can be used to monitor changes in sediment microbial populations with time after perturbation (oiling). We found both temporal and spatial variations in numbers and activity of hydrocarbon-degrading microorganisms, with higher values at the oiled sites than at reference sites.

Following the grounding of the TV *Exxon Valdez* on 24 March 1989, the National Oceanic and Atmospheric Administration (NOAA) organized a multi-investigator cruise to document the extent of oil contamination of coastal habitats in Prince William Sound and the Gulf of Alaska. This first survey cruise was followed by five cruises over the next 2 years organized as a joint effort by NOAA and the Alaska State Department of Environmental Conservation (ADEC). The purpose of these survey cruises was to document oil concentration distributions and to assess the relative ecological impacts of the spill on intertidal and subtidal areas. Our specific objective was to measure the effect of the spill on microbial numbers and activities in intertidal and subtidal sediments.

Assessment of microbial populations is an important component of damage surveys because a major fate of spilled petroleum depends on the ability of microorganisms to use hydrocarbons as a source of carbon and energy (Leahy and Colwell 1990). Additionally, patterns of hydrocarbon mineralization activity and distribution of hydrocarbon-degrading microorganisms can be used as an indication of in situ biodegradation of petroleum (Madsen et al. 1991). Measurements of total numbers of hydrocarbon-degrading microorganisms and assays for the mineralization potential of hydrocarbon fractions by these populations provide evidence of the presence of hydrocarbons that can be utilized by microorganisms. When sediments from a pristine environment are perturbed with oil, the distribution of hydrocarbon-degrading microorganisms reflects the extent, movement, and persistence of the contamination. In the course of these studies an enormous amount of data was collected on microbial populations in sediments (Braddock et al. 1990; Braddock et al. 1993). We compare the detailed results of microbial analyses of sediments for two sites (Northwest Bay and Snug Harbor) at which the shorelines were heavily oiled by the *Exxon Valdez* spill. These oiled sites are compared to reference sites (Rocky Bay and Zaikof Bay). In general, we found large differences in the numbers and activities of hydrocarbon-degrading microorganisms between the oiled sites and the reference sites. These results indicate that, shortly after oiling, similar microbial responses occurred after the *Exxon Valdez* spill as were observed after other major oil spills, such as that of the *Amoco Cadiz* (Ward et al. 1980). In addition, by following these responses with time we have found evidence of movement and depletion of microorganisms acclimated to hydrocarbon degradation.

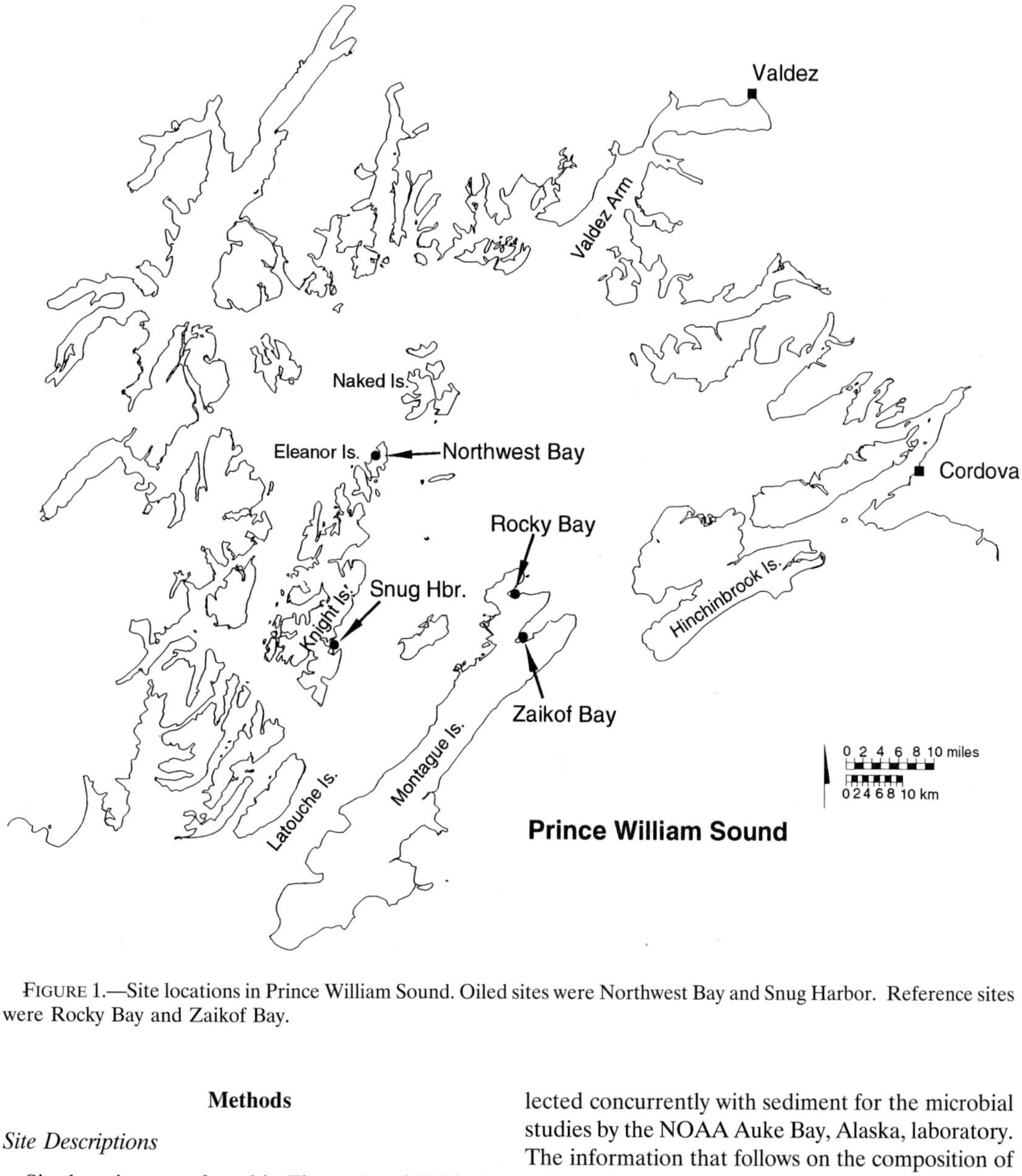

FIGURE 1.—Site locations in Prince William Sound. Oiled sites were Northwest Bay and Snug Harbor. Reference sites were Rocky Bay and Zaikof Bay.

Methods

Site Descriptions

Site locations are found in Figure 1 and Table 1. Northwest Bay and Snug Harbor are sites that were heavily oiled after the *Exxon Valdez* spill. Olsen Bay, Rocky Bay, and Zaikof Bay, were used as reference sites in this study. Although Rocky Bay is now thought to have been lightly oiled after the spill, we chose to use this site as a reference because it is in the same general geographical location as the heavily oiled sites and the degree of oiling was substantially less than those sites used as "oiled" sites.

Sediment samples for chemical analysis were collected concurrently with sediment for the microbial studies by the NOAA Auke Bay, Alaska, laboratory. The information that follows on the composition of collected sediments was obtained from analysis by the Auke Bay laboratory on samples from 1989. Northwest Bay, Eleanor Island, is characterized by cobble and gravel overlying a subsurface pea gravel and sand. Sediment collected was predominately sand with up to 15% silt onshore. The percentage silt increased in these samples with distance offshore to approximately 75% at 100-m depth. Snug Harbor, Knight Island, has a shoreline characterized by boulders (30–50 cm) covering coarse sand. Samples collected had approximately the same composition as those from Northwest Bay except

TABLE 1.—Site locations and sampling dates for studies of microbial activity in oiled and unoiled sediments in Prince William Sound.

Site	Latitude	Longitude	Sampling dates
Rocky Bay	60°20.28′N	147°08.15′W	17 Jul 1989, 8 Jun 1990, 4 Jul 1990, 24 Jun 1991
Zaikof Bay	60°16.58′N	147°02.01′W	8 Jun 1990, 3 Jul 1990, 25 Jun 1991
Olsen Bay	60°45.13′N	146°11.05′W	18 Jul 1989, 31 May 1990, 27 Jun 1990, 5 Sep 1990, 25 Jun 1991
Northwest Bay	60°33.05′N	147°34.62′W	12 Jul 1989, 26 Nov 1989, 5 Jun 1990, 9 Jul 1990, 6 Sep 1990, 15 Jun 1991
Snug Harbor	60°14.38′N	147°43.11′W	6 Jul 1989, 15 Nov 1989, 1 Jun 1990, 18 Jul 1990, 22 Jun 1991

that some of the samples collected offshore also had substantial amounts of clay. The two reference sites are located on Montague Island. Zaikof Bay is a gravel beach overlying sediment that is primarily sand. As in the oiled sites, the silt composition increased with distance offshore, with silt being the major constituent of samples from 100-m depth. Rocky Bay samples collected had a particle composition that was very similar to those of Northwest Bay and Zaikof Bay.

Both the Snug Harbor and Northwest Bay shoreline sites received various treatments over time after the spill. The Northwest Bay site was in a segment that was treated by manual pickup, tarball removal, washing, and bioremediation. The Snug Harbor site was located in a shoreline segment that was likewise treated with manual pickup, tarball removal and bioremediation (A. Mearns, NOAA, personal communication).

Sampling Protocol

Samples were collected on six cruises: RV *Fairweather*, 1 July to 22 August 1989; RV *Nautilus*, 7 November to 8 December 1989; RV *Cobb* (*Cobb* I), 31 May to 10 June 1990; RV *Davidson*, 27 June to 5 August 1990; RV *Cobb* (*Cobb* II), 5 September to 15 September 1990; and FV *Big Valley*, 15 June to 25 June 1991. At every site, samples were analyzed for the most probable numbers of hydrocarbon-oxidizing microorganisms and for mineralization potentials using radiolabeled hydrocarbon substrates. Total microscopic counts in sediments were also measured in samples collected in summer 1989 (*Fairweather*). During the three summer cruises (*Fairweather*, *Davidson*, and *Big Valley*), surface sediment samples were collected in the intertidal zone (referred to as beach or shoreline) and at 3-, 6-, 20-, 40-, and 100-m depths offshore at mean low tide. These surface sediments collected at depth will be referred to as depth samples. On the other cruises, samples were collected only in the intertidal zone and at the shallower depths (*Nautilus*, 0 and 3 m; and *Cobb* I/II, 0, 3, 6, and 20 m).

Sediment samples at the 40- or 100-m depths were collected using either a Van Veen grabber or Smith-McIntyre corer. The Van Veen grabber has the disadvantage of often disturbing the sediment sample but was consistent in being able to obtain surface sediment at depth in the often rocky bays of Prince William Sound. Composite samples were obtained by subsampling surface sediment (top 0–3 cm) collected by three grabs into sterile Whirl-Pak bags. Samples at the 3-, 6-, and 20-m depths were collected by scuba divers, whereas the shoreline samples were generally collected by a shore party in the low intertidal zone at as close to low tide as was feasible. When the intertidal zone sample was under water (when it could not be reached by launch at low tide), the samples were collected by divers. The intertidal (shoreline) and 3-, 6-, and 20-m samples were composites of eight subsamples collected at random intervals along a 30-m transect parallel to the shoreline. Only one bag of sediment was collected for each site at each depth on the *Fairweather* cruise (which limited statistical analysis of the data), whereas three replicate bags were collected at each depth for all subsequent cruises. During the *Fairweather* and *Nautilus* cruises, "pore water" samples were obtained for microbial analyses by collecting interstitial water from sediment-sampling holes dug at the intertidal stations. For every microbiology sample collected, there were associated samples collected for chemical analysis of hydrocarbons (collected and analyzed by National Marine Fisheries Service Laboratory, Auke Bay, Alaska). All sediment and water samples were placed into coolers at the time of collection for transport to the support vessel. Processing for microbiological analyses was performed on the support vessel within 3 h of collection.

Direct Counts of Bacteria

The direct count method was used to determine the total population of bacteria in all sediments and pore water samples. Ten grams (wet weight) of sediment were mixed with 90 mL of filtered (0.2-μm

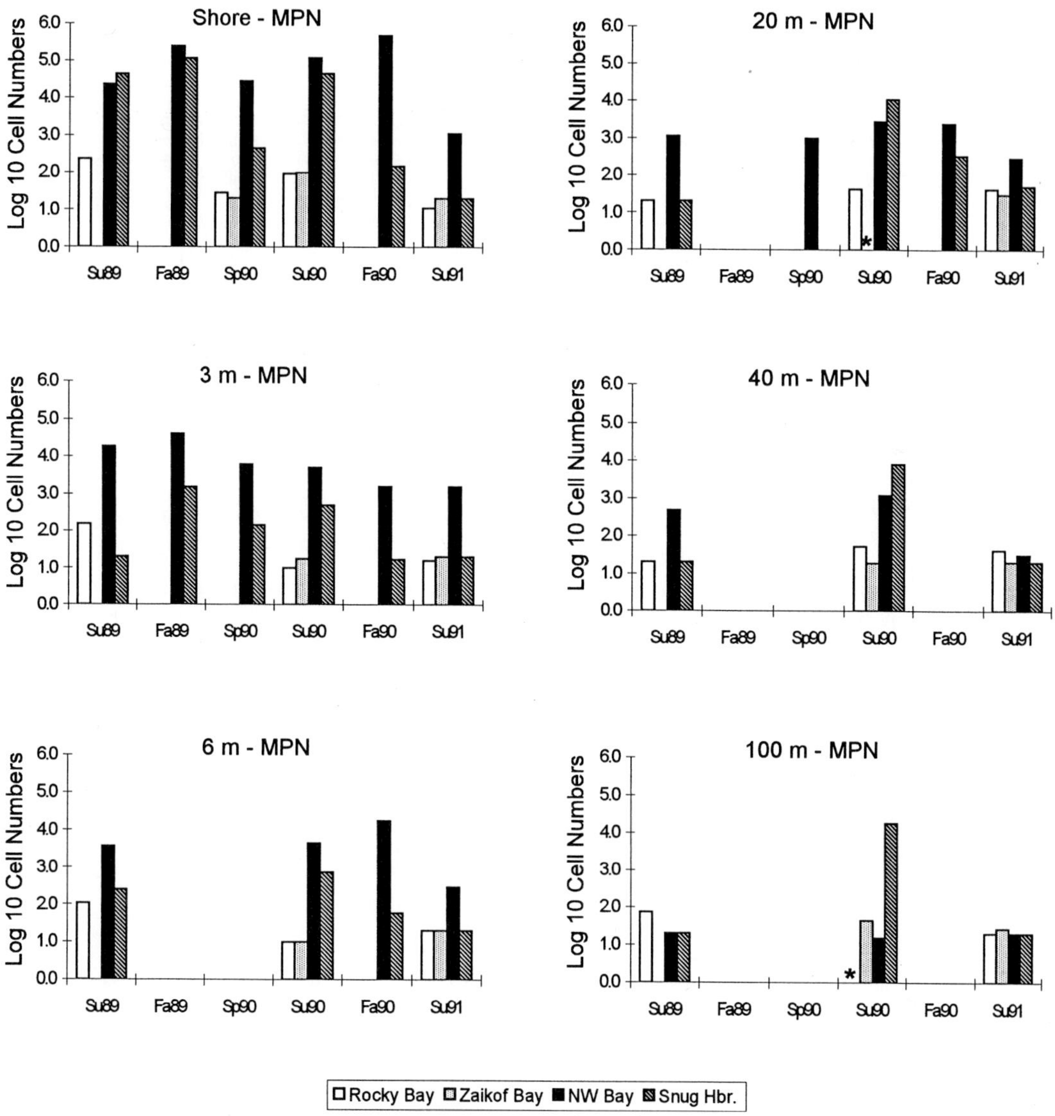

FIGURE 2.—Mean most probable number (MPN) of hydrocarbon-degrading microorganisms per gram dry weight of sediment in surface sediments with time for oiled (Northwest Bay and Snug Harbor) and reference sites (Rocky Bay and Zaikof Bay); Su = summer, Fa = fall, Sp = spring. Not all sites and depths were sampled on every cruise. An asterisk indicates that the sediment was sampled and a value of 0 cells/g dry weight of sediment was measured.

filter) saline solution. Duplicate aliquots were preserved for later counting in 3.7% formaldehyde for the sediment samples and 1.8% formaldehyde for the pore water samples, providing two subsamples for each sediment and each water sample (Montagna 1982). The formaldehyde-fixed samples were processed by vigorous shaking for 1 min followed by serial dilution with filtered water. The samples were then incubated for approximately 10 min with 5 μg/mL of the fluorochrome (DAPI, 4′-6-diamidino-2-phenylindole dihyrochloride) stain at 4°C in the dark after the method of Porter and Feig (1980). After incubation the samples were filtered onto prestained (Irgalan Black) polycarbonate filters (0.2 μm, 25-mm diameter; Poretics Corporation, Livermore, California) and counted at a magnification of 1250× with a Zeiss Standard microscope equipped with an epifluorescent light source. Only one dilution (yielding between 30 and 100 bacteria per field on a filter) (Kirchman et al. 1982) was counted per

TABLE 2.—Direct counts of microorganisms in sediments from the *Fairweather* cruise (summer 1989) in Prince William Sound. Rocky Bay and Olsen Bay are reference sites. Northwest Bay and Snug Harbor are oiled sites. The units for porewater samples are cells per milliliter ± 1 SD × 10^6.

	Mean number of cells/g dry weight sediment ± 1 SD × 10^6 at:			
Sample type	Rocky Bay	Olsen Bay	Northwest Bay	Snug Harbor
Shoreline	26 ± 23	690 ± 400	9 ± 4	48 ± 34
3 m	220 ± 130	740 ± 460	110 ± 31	47 ± 43
6 m	450 ± 250	440 ± 230	100 ± 29	59 ± 32
20 m	460 ± 330	400 ± 170	510 ± 140	170 ± 120
40 m	250 ± 170	570 ± 210	230 ± 89	98 ± 61
100 m	1,700 ± 780	1,700 ± 1,100	980 ± 870	480 ± 260
Porewater	2 ± 1	8 ± 4	2 ± 1	5 ± 2

sample. Ten fields per filter were counted (for a minimum of 300 cells), and the counts from duplicate samples were combined to provide 20 fields counted per sample. A mean and standard deviation were determined for the 20 fields and those values were corrected for background and multiplied by the appropriate dilution factors. Background bacterial numbers were determined by running 10-mL blanks of stained filtered water with the samples.

Most Probable Number of Hydrocarbon-Degrading Microorganisms

The number of hydrocarbon-degrading microorganisms in each sample was estimated using the

TABLE 3.—Most probable number of hydrocarbon-degrading microorganisms per gram (dry weight) of surface sediment for each cruise in Prince William Sound. Note that some sites and depths were not sampled on every cruise. An asterisk (*) signifies a site that was significantly different from the designated reference sites for each cruise at the 95% confidence level using the Mann–Whitney *U*-test. The units for pore water samples are most probable numbers per milliliter. Abbreviations are Su = summer, Fa = fall, Sp = spring, 89 = 1989, 90 = 1990, and 91 = 1991.

		Fairweather (Su89)	*Nautilus* (Fa89)		*Cobb* I (Sp90)	
Sample type	Site	Mean	Mean	SD	Mean	SD
Shore	Rocky Bay	2.3 × 10^2			2.8 × 10^1	1.3 × 10^1
	Zaikof Bay				<2.0 × 10^1	
	Northwest Bay	2.3 × 10^4	*2.4 × 10^5	7.0 × 10^3	*2.8 × 10^4	9.5 × 10^3
	Snug Harbor	4.5 × 10^4	*1.2 × 10^5	2.6 × 10^4	4.6 × 10^2	3.8 × 10^2
3 m	Rocky Bay	1.5 × 10^2				
	Zaikof Bay					
	Northwest Bay	1.9 × 10^4	*4.1 × 10^4	4.4 × 10^4	*6.2 × 10^3	5.3 × 10^3
	Snug Harbor	<2.0 × 10^1	*1.5 × 10^3	4.0 × 10^2	*1.4 × 10^2	7.5 × 10^1
6 m	Rocky Bay	1.1 × 10^2				
	Zaikof Bay					
	Northwest Bay	3.7 × 10^3				
	Snug Harbor	2.5 × 10^2				
20 m	Rocky Bay	<2.0 × 10^1				
	Zaikof Bay					
	Northwest Bay	1.1 × 10^3			*1.0 × 10^3	5.6 × 10^2
	Snug Harbor	<2.0 × 10^1				
40 m	Rocky Bay	<2.0 × 10^1				
	Zaikof Bay					
	Northwest Bay	4.8 × 10^2				
	Snug Harbor	<2.0 × 10^1				
100 m	Rocky Bay	7.2 × 10^1				
	Zaikof Bay					
	Northwest Bay	<2.0 × 10^1				
	Snug Harbor	<2.0 × 10^1				
Porewater	Rocky Bay	1.2 × 10^1				
	Zaikof Bay					
	Northwest Bay	1.4 × 10^4	2.3 × 10^5	2.7 × 10^5		
	Snug Harbor	1.4 × 10^2	2.5 × 10^4	2.1 × 10^4		

Sheen Screen most-probable-number technique (Brown and Braddock 1990). On the summer 1989 cruise (*Fairweather*), duplicate sets of cell well plates were prepared for each depth at each site. For all other cruises, one set of cell well plates was prepared from each replicate sediment sample at a given site and depth to yield triplicate values. The Sheen Screen plates were incubated at approximately 15°C for 3 weeks before being scored for disruption of the oil sheen. All values reported from the Sheen Screen technique have been corrected to dry weight sediment or milliliter of pore water.

Hydrocarbon Mineralization Potential

Radiorespirometry was used to assay the hydrocarbon mineralization potentials of microorganisms in sediment slurries (Brown et al. 1991). The radiolabeled hydrocarbons, [1-^{14}C]-hexadecane and [1,(4,5,8)-^{14}C]-naphthalene or [9-^{14}C]-phenanthrene were used as representatives of aliphatic and polycyclic aromatic hydrocarbons. The assay was designed to be independent of most of the complex factors regulating microbial hydrocarbon metabolism (including hydrocarbon availability) except microbial biomass and its potential to degrade hydrocarbons in each sample. Sediment was diluted (1:10) in sterile mineral salts medium (Bushnell-Haas Medium; Difco, Detroit, Michigan) amended with 2.5% NaCl. After shaking vigorously by hand for 1 min, the samples (10 mL) were pipetted into 40-mL pre-cleaned and sterilized glass incubation vials fitted with teflon-lined septa (I-Chem Research, Hayward, California).

Replicate vials of the 10-mL slurries from each of the sediment samples were injected with 50 μL of a 2-g/L (in acetone) solution of radiolabeled hydrocarbon. Acetone has been used as a hydrocarbon carrier in similar experiments (Bauer and Capone 1988; Guerin and Jones 1989; Shiaris 1989). The resulting initial concentration of added hydrocarbon was then 100 μg per vial (100 μg/g wet weight sediment; 10 μg/mL slurry; approximately 50,000 disintegrations per minute [dpm]/vial). By adding 100 μg of hydrocarbon substrate to each vial, the hydrocarbon mineralization potential was independent of the degree of oil contamination of the sediment tested (see Brown et al. 1991).

For the summer 1989 cruise (*Fairweather*), dupli-

TABLE 3.—Extended.

Davidson (Su90)		*Cobb* II (Fa90)		*Big Valley* (Su91)	
Mean	SD	Mean	SD	Mean	SD
1.0×10^{2}	6.2×10^{1}			1.1×10^{1}	1.0×10^{1}
9.1×10^{1}				$<2.0 \times 10^{1}$	
$*1.0 \times 10^{4}$	1.6×10^{4}	$*4.8 \times 10^{5}$	4.3×10^{5}	$*1.1 \times 10^{3}$	1.2×10^{3}
$*4.6 \times 10^{4}$	3.4×10^{4}	$*1.5 \times 10^{2}$	1.9×10^{2}	$<2.0 \times 10^{1}$	
1.8×10^{1}	1.6×10^{1}			1.6×10^{4}	1.6×10^{1}
$<2.0 \times 10^{1}$				$<2.0 \times 10^{1}$	
$*5.0 \times 10^{3}$	6.5×10^{3}	1.6×10^{3}		$*1.6 \times 10^{3}$	7.9×10^{1}
$*5.0 \times 10^{2}$	4.4×10^{2}	1.7×10^{1}	1.5×10^{1}	$<2.0 \times 10^{1}$	
$<2.0 \times 10^{1}$				2.0×10^{1}	2.3×10^{1}
$<2.0 \times 10^{1}$				$<2.0 \times 10^{1}$	
$*4.6 \times 10^{3}$	2.3×10^{3}	$*1.8 \times 10^{3}$	1.1×10^{4}	$*3.0 \times 10^{2}$	2.4×10^{2}
$*7.2 \times 10^{2}$	8.5×10^{2}	6.0×10^{1}	3.9×10^{1}	$<2.0 \times 10^{1}$	
0	0			4.2×10^{1}	1.7×10^{1}
4.3×10^{1}	3.7×10^{1}			3.1×10^{1}	5.3×10^{1}
2.8×10^{3}	1.4×10^{3}	$*2.4 \times 10^{3}$	9.7×10^{2}	$*2.9 \times 10^{2}$	1.2×10^{2}
1.1×10^{4}	1.1×10^{4}	3.3×10^{2}	3.3×10^{2}	4.6×10^{1}	3.1×10^{1}
1.9×10^{1}	1.7×10^{1}			4.3×10^{1}	2.7×10^{1}
5.4×10^{1}	0			$<2.0 \times 10^{1}$	
1.2×10^{3}	3.0×10^{2}			3.1×10^{1}	1.4×10^{1}
8.1×10^{3}	3.0×10^{3}			$<2.0 \times 10^{1}$	
4.5×10^{1}	4.9×10^{1}			$<2.0 \times 10^{1}$	
0	0			2.7×10^{1}	2.4×10^{1}
1.5×10^{2}	2.6×10^{2}			$<2.0 \times 10^{1}$	
1.8×10^{4}	1.9×10^{4}			$<2.0 \times 10^{1}$	

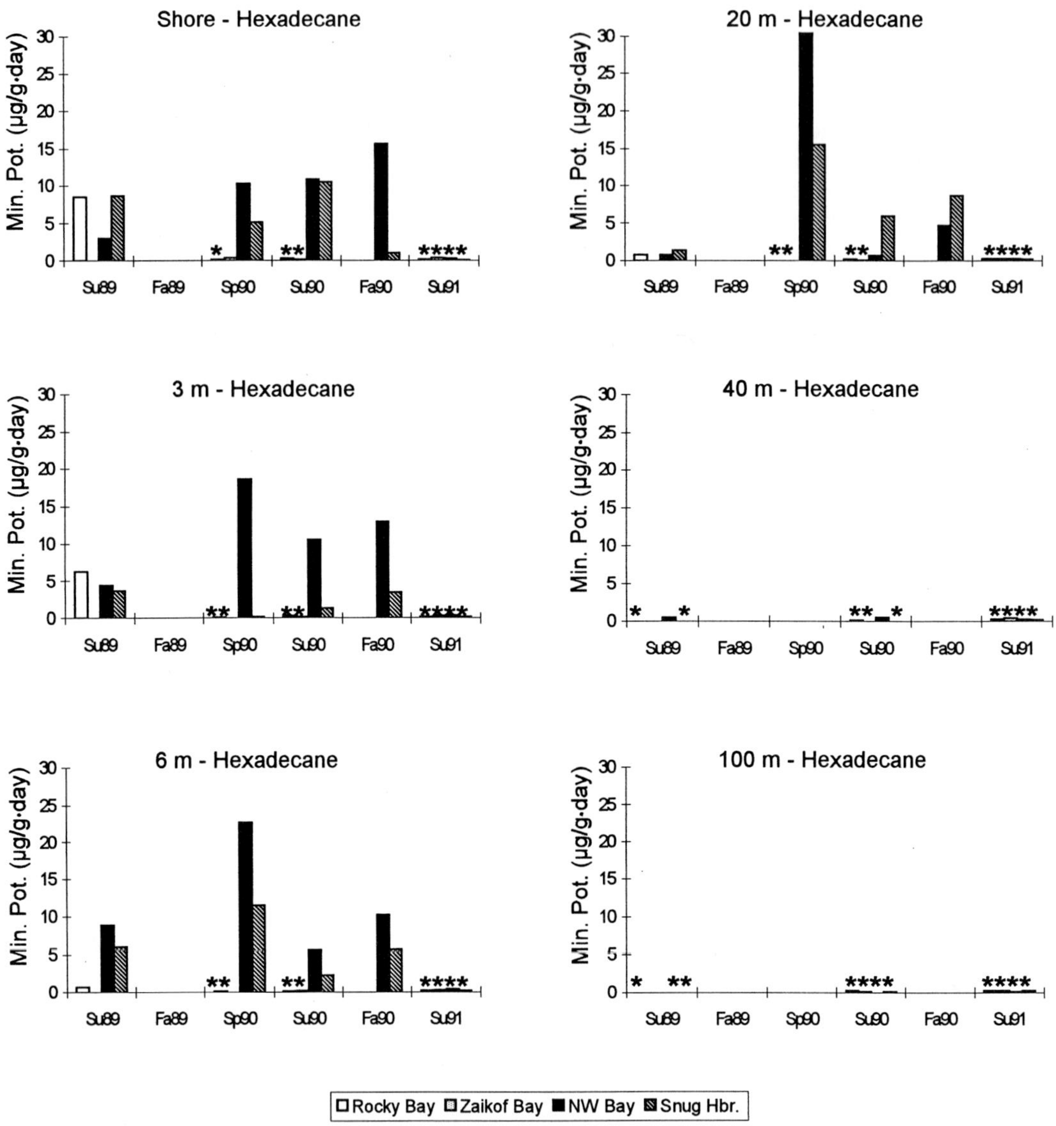

FIGURE 3.—Mean hexadecane mineralization potentials for samples incubated for 2 d for oiled (Northwest Bay and Snug Harbor) and reference sites (Rocky Bay and Zaikof Bay); Su = summer, Fa = fall, Sp = spring. Not all sites and depths were sampled on every cruise. An asterisk indicates that the sediment was sampled and the measured potential was ≤0.4 μg/g dry weight sediment.

cate vials were prepared from each sediment (one bag per depth per site) for three incubation times and two radiolabeled hydrocarbon substrates. In addition, "time-zero" killed controls were prepared at each site for each substrate. For all other cruises, three bags of sediment were collected for each depth at each site. From each bag, seven replicate vials of each concentration of each substrate were prepared: one time-zero killed control and three vials each at two incubation times. All vials were placed on a rotary shaker for the first 24 h and then stored off the shaker at approximately 15°C for the duration of the incubation period. At the end of the designated incubation period, samples were "killed" and the CO_2 fixed by adding 1 mL 10 N NaOH per vial. At the end of each cruise the vials were returned to the laboratory where they were acidified and the radiolabeled CO_2 stripped into

TABLE 4.—Mean hexadecane mineralization potentials in micrograms per (gram dry weight of sediment · day) after a 2-d incubation. The units for porewater samples are micrograms per (milliliter · day). Abbreviations are as in Table 3. No samples were taken during fall 1989 cruise on *Nautilus*.

		Fairweather (Su89)	*Cobb* I (Sp90)		*Davidson* (Su90)		*Cobb* II (Fa90)		*Big Valley* (Su91)	
Sample type	Site	Mean	Mean	SD	Mean	SD	Mean	SD	Mean	SD
Shore	Rocky Bay	8.5	0.2	0.7	0.3	0.4			0.2	0.0
	Zaikof Bay		0.4	0.7	0.2	0.4			0.4	0.1
	Northwest Bay	3.0	10.3	1.8	10.8	2.1	15.7	0.6	0.3	0.1
	Snug Harbor	8.6	5.1	1.4	10.5	0.7	1.1	0.7	0.1	0.0
3 m	Rocky Bay	6.3	0.1	0.8	0.3	0.4			0.2	0.0
	Zaikof Bay		0.0	0.0	0.2	0.4			0.3	0.1
	Northwest Bay	4.4	18.6	2.1	10.6	3.4	13.0	2.3	0.4	0.1
	Snug Harbor	3.7	0.2	0.9	1.3	0.4	3.5	1.0	0.2	0.0
6 m	Rocky Bay	0.7	0.1	0.7	0.1	0.4			0.2	0.0
	Zaikof Bay		0.0	0.0	0.2	0.5			0.3	0.1
	Northwest Bay	8.9	22.7	1.9	5.6	0.7	10.2	4.7	0.4	0.1
	Snug Harbor	6.0	11.5	2.2	2.2	0.9	5.7	1.6	0.2	0.1
20 m	Rocky Bay	0.8	0.0	0.0	0.2	0.4			0.3	0.0
	Zaikof Bay		0.0	0.0	0.1	0.4			0.3	0.1
	Northwest Bay	0.8	30.7	2.1	0.7	0.5	4.7	2.1	0.3	0.1
	Snug Harbor	1.4	15.5	1.3	6.0	2.0	8.7	2.4	0.2	0.1
40 m	Rocky Bay	0.0			0.1	0.4			0.3	0.1
	Zaikof Bay				0.0	0.0			0.4	0.1
	Northwest Bay	0.5			0.5	0.5			0.3	0.1
	Snug Harbor	0.0			0.0	0.0			0.2	0.1
100 m	Rocky Bay	0.0			0.3	0.6			0.3	0.1
	Zaikof Bay				0.1	0.5			0.3	0.1
	Northwest Bay	0.0			0.0	0.0			0.1	0.1
	Snug Harbor	0.0			0.1	0.7			0.3	0.1
Porewater	Rocky Bay	12.6								
	Zaikof Bay									
	Northwest Bay	0.2								
	Snug Harbor	3.8								

cocktail and counted on a model LSC 1800 liquid scintillation counter (Beckman Instruments, Irvine, California) by the procedure described in Brown et al. (1991).

The time-zero values for each isotope for each cruise served as negative controls and were averaged and subtracted from each mineralization potential sample to yield a corrected dpm value. The background dpm values were always less than 1% of added radiolabeled hydrocarbon. Because 100 μg hydrocarbon were added per gram of sediment, the assay has a sensitivity of approximately 0.6 μg/g sediment. In addition to the NaOH-killed time-zero controls, a series of sediment samples were inoculated, spiked with radioisotopes, and killed (autoclaved) at the beginning of the summer 1989 cruise. These samples were run approximately 4 months later to check for abiotic evolution of CO_2. The values from all of these samples were within the range for a scintillation cocktail blank. Positive controls showed that the purging system could recover greater than 99% of radiolabeled CO_2 from radiolabeled bicarbonate processed as if it were a sediment sample. The potential for carryover between samples was monitored by running blank controls through the purging line periodically. The values from blank controls run in this manner were always within the range for time-zero control samples. All reported values have been corrected to dry weight of sediment.

Results

The total numbers of bacteria as measured by epifluorescent microscopy direct counts varied over several orders of magnitude at different sites and depths. However, there did not appear to be a trend related to oiling (Table 2). Snug Harbor had the lowest total populations of bacteria at nearly every depth, but the populations at Northwest Bay essentially did not vary from the reference sites. Total bacterial counts ranged from 10^6 to 10^9 cells/g dry weight sediment, but most samples had populations of about 10^8 cells/g. The pore water samples had

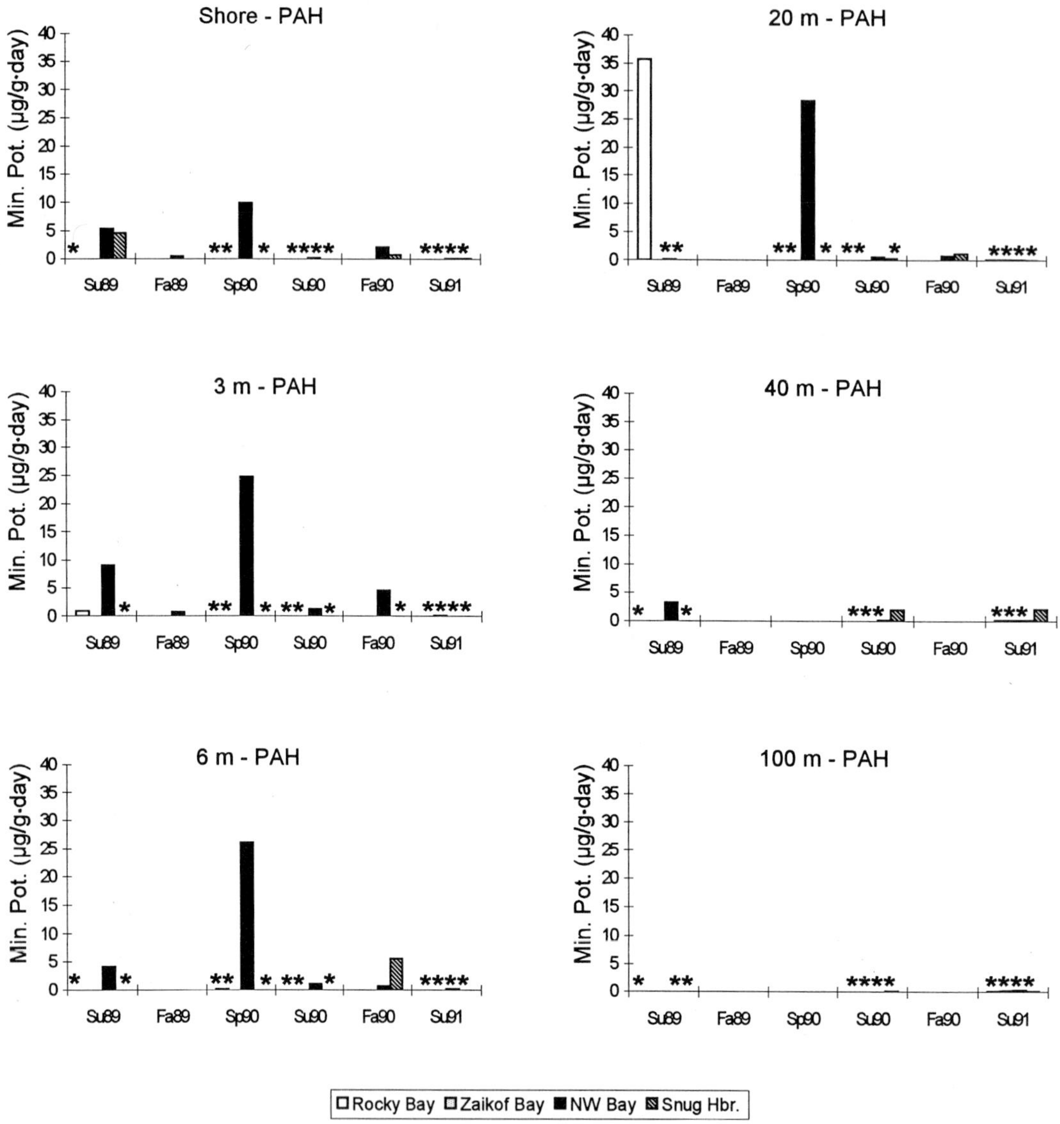

FIGURE 4.—Mean polycyclic aromatic hydrocarbon (PAH) mineralization potentials for samples incubated for 2 d for oiled (Northwest Bay and Snug Harbor) and reference sites (Rocky Bay and Zaikof Bay); Su = summer, Fa = fall, Sp = spring. Naphthalene was used in summer 1989 and phenanthrene was used on all subsequent cruises. Not all sites and depths were sampled on every cruise. An asterisk indicates that the sediment was sampled and the measured potential was ≤0.4 μg/g dry weight of sediment.

approximately 10^6 cells/mL. Although total counts of bacteria were apparently not changed by oiling in 1989, the most probable numbers of hydrocarbon-degrading microorganisms did vary widely between oiled sites and reference sites in 1989 and in subsequent years (Figure 2 and Table 3).

In the shoreline sediments, populations of hydrocarbon degraders were always higher in oiled sediment samples than in samples collected at reference sites. Although means and standard deviations are reported in Table 3, we found that, as in similar studies, (see Lindstrom et al. 1991) the heterogeneity of the environment sampled did not result in normally distributed data. Therefore, we used nonparametric statistical analysis to test for differences in distributed data between oiled and reference

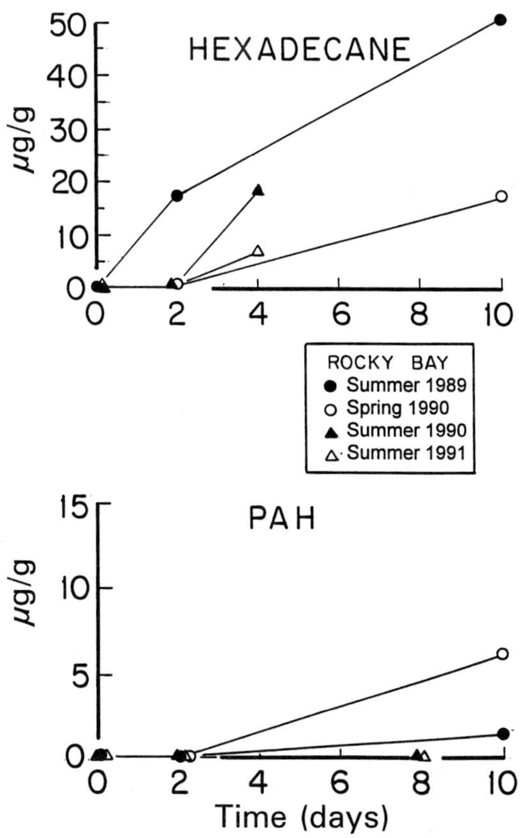

FIGURE 5.—Mean hydrocarbon mineralization potentials for hexadecane and polycyclic aromatic hydrocarbons (PAH) with time in shoreline sediments collected at the reference site Rocky Bay.

sites. Sediments in which the reported value was statistically greater than the values at reference sites at the 95% confidence level as determined by the Mann–Whitney *U*-test (Zar 1984) are indicated in Table 3. For this test, sites with values that were significantly greater than values at the designated reference sites (not always Zaikof Bay or Rocky Bay) for that cruise are indicated by an asterisk in Table 3 (see also Braddock et al. 1993). In general, oiled shoreline sediments initially showed high populations of hydrocarbon degraders, which diminished by 1991. Levels at Snug Harbor had dropped to near those of the reference sites by summer 1991; however, Northwest Bay still had populations that were significantly greater than those at the reference sites. A similar trend was observed at the 3-m and 6-m depths. There is an indication of a temporal increase in the numbers of hydrocarbon degraders at the 20-, 40-, and 100-m depths, particularly in the Snug Harbor sediments; these numbers reached a peak in summer 1990 in Northwest Bay and Snug Harbor. By 1991 mean probable numbers at the 40- and 100-m depth sediments had returned to values indistinguishable from those at the reference sites.

In addition to microbial numbers, we also measured microbial mineralization potentials for two hydrocarbon substrates. The potential for microbial populations to mineralize aliphatic hydrocarbons was measured using radiolabeled hexadecane (Figure 3 and the top part of Figures 5 through 8; Table 4). The hexadecane mineralization potentials were measured under partially optimized conditions after a 2-d incubation and again after either a 4- or 10-d incubation (see Figures 5A, 6A, 7A, and 8A). Mineralization potentials in samples incubated for 2 d (Figure 3) were relatively high in many samples collected from the shoreline through the 20-m depth until 1991. The maximum potentials after a 2-d incubation were at the 6- and 20-m depths in

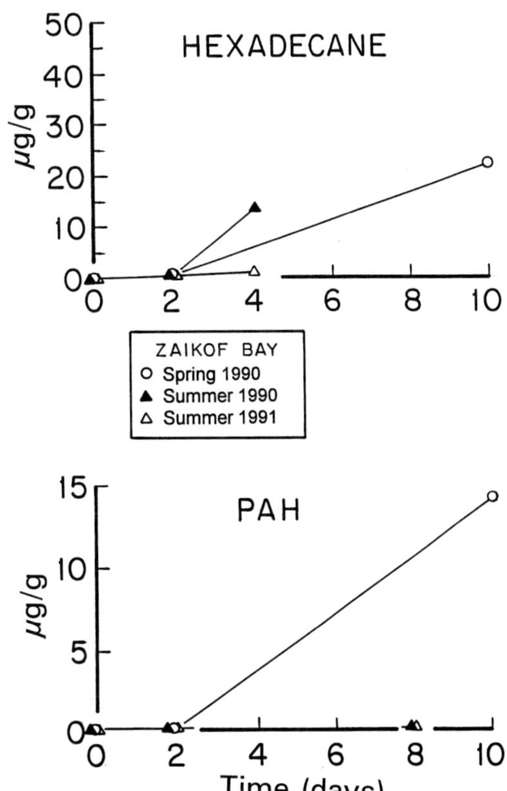

FIGURE 6.—Mean hydrocarbon mineralization potentials for hexadecane and polycyclic aromatic hydrocarbons (PAH) with time in shoreline sediments collected at the reference site Zaikof Bay.

summer 1990. This trend was seen in both the Snug Harbor and Northwest Bay sites. Virtually no microbial activity for hexadecane mineralization was measured in sediments collected from the 40- and 100-m depths for any cruise or in any sample collected in 1991. In the top parts of Figures 5 through 8 it can be seen that there were generally high potentials for hexadecane mineralization after a 2-d incubation. With the exception of the Rocky Bay site in 1989, there was never a positive potential after a 2-d incubation for samples collected from the reference sites (see also Brown et al. 1991).

Naphthalene (*Fairweather* cruise, Su89) and phenanthrene (all other cruises) were used as representative polycyclic aromatic hydrocarbons (PAH) (Figure 4 and the bottom part of Figures 5 through 8; Table 5). Overall PAH mineralization potentials were lower than the potentials observed for hexadecane in all sediments except at Northwest Bay in spring 1990, where substantial PAH potentials were measured in sediments from the shoreline through the 20-m depth. A high potential was also measured at Rocky Bay from the sediment from the 20-m depth in summer 1989. Essentially no activity for PAH was measured in any sediment from the 40- or 100-m depth. At the reference sites positive rates for PAH mineralization were never seen after a 2-d incubation (Figures 5 and 6). Positive potentials were measured after a 2-d incubation in about half of the sediments collected from the oiled sites (Figures 7 and 8).

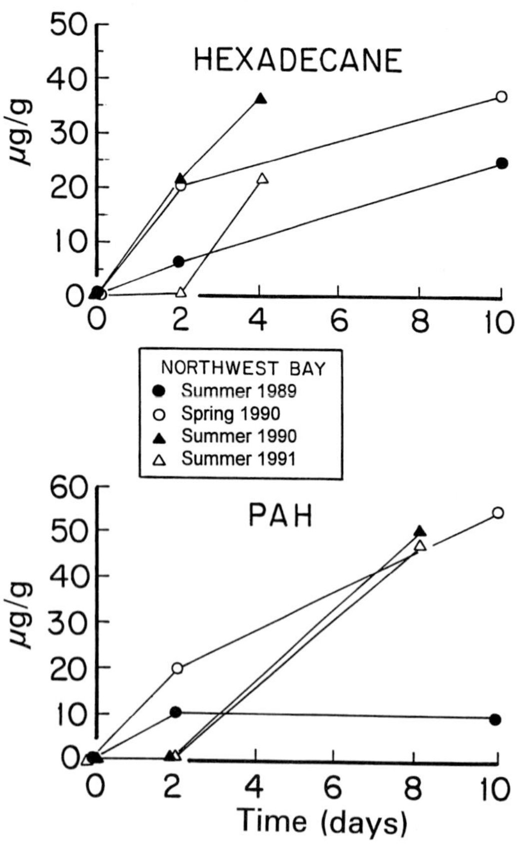

FIGURE 7.—Mean hydrocarbon mineralization potentials for hexadecane and polycyclic aromatic hydrocarbons (PAH) with time in shoreline sediments collected at the oiled site Northwest Bay. Hexadecane mineralization was not measured for fall 1989. Note the change in *y*-axis scale for PAH from that in Figures 5 and 6.

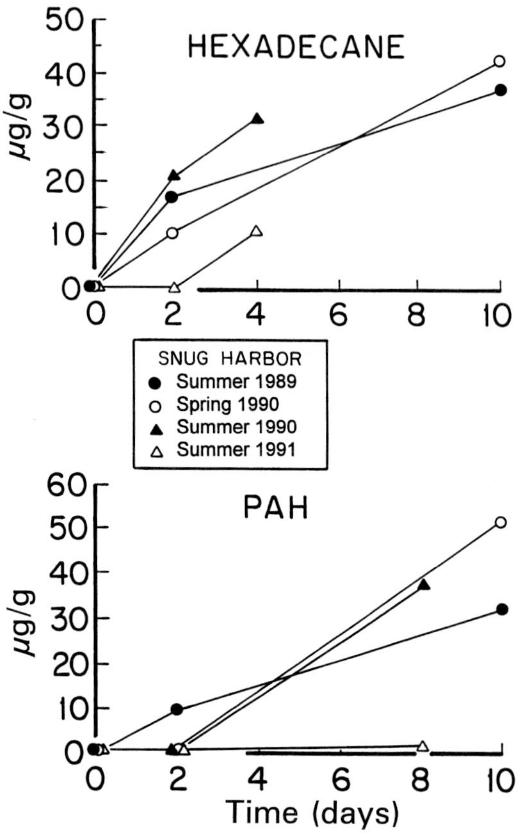

FIGURE 8.—Mean hydrocarbon mineralization potentials for hexadecane and polycyclic aromatic hydrocarbons (PAH) with time in shoreline sediments collected at the oiled site Snug Harbor. Hexadecane mineralization was not measured for fall 1989. Note the change in *y*-axis scale for PAH from that in Figures 5 and 6.

Discussion

Microorganisms capable of degrading a variety of petroleum hydrocarbons are widespread in marine

TABLE 5.—Mean polycyclic aromatic hydrocarbon mineralization potentials in milligrams per (gram dry weight of sediment · day) after a 2-d incubation. Naphthalene was used on the *Fairweather* cruise and phenanthrene was used on all subsequent cruises. The units for pore water samples are micrograms per (milliliter · day). Abbreviations are as in Table 3.

		Fairweather (Su89)	*Nautilus* (Fa89)		*Cobb* I (Sp90)		*Davidson* (Su90)		*Cobb* II (Fa90)		*Big Valley* (Su91)	
Sample type	Site	Mean	Mean	SD	Mean	SD	Mean	SD	Mean	S D	Mean	SD
Shore	Rocky Bay	0.0			0.0	0.0	0.0	0.0			0.0	0.0
	Zaikof Bay				0.0	0.0	0.0	0.0			0.0	0.0
	Northwest Bay	5.3	0.5	0.4	9.9	1.8	0.2	0.3	2.2	1.2	0.1	0.1
	Snug Harbor	4.6			0.0	0.4	0.0	0.4	0.8	1.2	0.1	0.1
3 m	Rocky Bay	0.9			0.0	0.6	0.0	0.0			0.0	0.0
	Zaikof Bay				0.0	0.0	0.0	0.0			0.1	0.1
	Northwest Bay	9.1	0.7	0.5	24.8	2.5	1.3	0.5	4.6	5.0	0.0	0.0
	Snug Harbor	0.0			0.0	0.0	0.0	0.0	0.0	0.0	0.0	0.0
6 m	Rocky Bay	0.0			0.1	0.5	0.0	0.0			0.0	0.0
	Zaikof Bay				0.0	0.0	0.0	0.0			0.0	0.0
	Northwest Bay	4.2			26.1	3.6	1.1	0.6	0.7	1.1	0.1	0.1
	Snug Harbor	0.0			0.0	0.0	0.0	0.0	5.7	8.2	0.0	0.0
20 m	Rocky Bay	35.7			0.0	0.0	0.0	0.4			0.1	0.1
	Zaikof Bay				0.0	0.0	0.0	0.0			0.1	0.1
	Northwest Bay	0.1			28.4	1.7	0.6	0.5	0.8	1.1	0.1	0.1
	Snug Harbor	0.0			0.0	0.0	0.3	0.4	1.2	1.3	0.1	0.1
40 m	Rocky Bay	0.0					0.0	0.0			0.1	0.1
	Zaikof Bay						0.0	0.0			0.2	0.2
	Northwest Bay	3.2					0.1	0.4			0.1	0.1
	Snug Harbor	0.0					2.0	1.5			0.1	0.1
100 m	Rocky Bay	0.0					0.0	0.0			0.1	0.1
	Zaikof Bay						0.0	0.0			0.2	0.1
	Northwest Bay	0.0					0.0	0.0			0.3	0.3
	Snug Harbor	0.0					0.1	0.7			0.1	0.1
Porewater	Rocky Bay	1.8										
	Zaikof Bay											
	Northwest Bay	0.6	0.4	0.4								
	Snug Harbor	2.1										

environments (Atlas et al. 1981; Leahy and Colwell 1990). Our results show that the numbers of hydrocarbon-degrading microorganisms vary by several orders of magnitude among sites sampled after the *Exxon Valdez* oil spill. Several studies have shown a positive correlation between the numbers of hydrocarbon-degrading microorganisms and oiling patterns in marine systems polluted by hydrocarbons (Colwell et al. 1978; Roubal and Atlas 1978; Ward et al. 1980; Lizarraga-Partida et al. 1991).

Enrichment of hydrocarbon degraders in the microbial community was apparent after the *Amoco Cadiz* oil spill. Ward et al. (1980) found that, of 10^7–10^8 bacteria/g sediment, there were 10^4–10^7 hydrocarbon-degrading bacteria per gram of sediment in oiled areas versus less than 5×10^2 cells/g sediment in unoiled areas. Similar numbers were found for the relative proportion of hydrocarbon degraders to total numbers of bacteria in shoreline sediments in 1989 in our study. Total bacterial populations in shoreline sediments during the *Fairweather* cruise (summer 1989) were approximately 10^7–10^8 cells/g sediment, whereas numbers of hydrocarbon-degrading microorganisms ranged from 10^4 to 10^5 cells/g in oiled sediments and less than 3×10^2 cells/g at the reference site (see Table 3). We have had the opportunity to follow these microbial populations over time since the *Exxon Valdez* spill. While overall numbers of hydrocarbon degraders have generally decreased since the spill, the numbers increased offshore at the 40- and 100-m depths, reaching a peak in 1990 and indicating that there may have been movement of oil to offshore sediments over time.

The only baseline data available for a geographical area near that affected by the *Exxon Valdez* oil spill are from studies performed about 10 years before the spill. Roubal and Atlas (1978) studied natural populations of hydrocarbon-degrading microorganisms in sediments along the coast of south-central Alaska and reported finding only 0.6–12.0 colony-forming units per gram of sediment. Cook

Inlet populations had values that were two or three orders of magnitude higher. The authors concluded that the relatively large number of hydrocarbon utilizers in some of the Cook Inlet sediments reflected a previous history of exposure to hydrocarbons. At the time of this 1978 study there were already several oil platforms producing oil in upper Cook Inlet where the higher numbers of hydrocarbon degraders were found.

In our study there was no statistically significant correlation between the total numbers of hydrocarbon-degrading microorganisms and the mineralization potentials for hexadecane. However, when high hexadecane mineralization potentials were measured, the numbers of hydrocarbon oxidizers were generally also high. Increased populations of Sheen Screen-positive organisms represent an increased potential for biodegradation of hydrocarbons in oil; but other factors such as salinity, temperature, mineral nutrient availability, oxygen availability, hydrocarbon concentration, and acclimation of the microbial population to a particular hydrocarbon component of the oil can affect the mineralization potential as well (Bartha and Atlas 1987; Leahy and Colwell 1990). Experiments in this study were designed to minimize as many of these factors as possible (including hydrocarbon availability) except the in situ microbial biomass and its potential to degrade the hydrocarbons added experimentally. Using this "optimal" procedure, rates of hexadecane, naphthalene, and phenanthrene mineralization among sites and sampling stations could be compared. For example, mineralization potential samples were run in a mineral salts medium with nutrients such as nitrogen and phosphorus, were well oxygenated, and were incubated at a constant temperature (15°C). In addition, a relatively large amount of hydrocarbon substrate was added to each sample to be assayed so that the final rate was dependent on added substrate rather than on the hydrocarbon substrate in the original sample. The concentrations of the specific hydrocarbon that we added exceeded ambient levels in heavily oiled samples collected in 1989 (see Brown et al. 1991). The reported potentials thus reflect the potential of the in situ microbial populations (Bartha and Atlas 1987; Aelion and Bradley 1991) to transform hydrocarbons when incubated in conditions that are standardized and essentially optimal.

Bartha and Atlas (1987) summarized the results of a number of published studies on biodegradation rates (potentials) of samples from marine systems. They found a range of 5–2,500 $\mu g/(g \cdot d)$ for seawater communities under partially optimized conditions. These rates for nutrient-enriched samples were found to be as much as 300-fold higher than for non-nutrient-enriched samples. Values from our study range from 0 to approximately 40 $\mu g/(g \cdot d)$. Therefore, our values fall at the low end of those reported by Bartha and Atlas for samples incubated under partially optimized conditions. A more recent study (Karl 1992) conducted about 6 weeks after the grounding of the *Bahia Paraiso* in Antarctica in 1989 found extremely low rates for hexadecane mineralization potential, 0.13–1.21 $pmol/(g \cdot d)$ in samples run without nutrient amendments and incubated at 1°C (suboptimal conditions).

We found apparent differences in microbial preference for hexadecane and naphthalene or phenanthrene in Prince William Sound sediments temporally and with depth. Hexadecane potentials nearly always exceeded those for PAH, except for sediments collected in the spring of 1990 where potentials for both hydrocarbon fractions were high in sediments from the shoreline through the 20-m depth (see Figures 3 and 4). However, there did not appear to be a significant preference for naphthalene over phenanthrene when potentials were measured for both substrates in samples collected at the same time. Both phenanthrene and naphthalene mineralization potentials were measured in shoreline and 3-m sediments in fall 1989. In these sediments the median degradation potentials for phenanthrene were approximately the same as those measured for naphthalene (data not shown). Hydrocarbon-polluted surficial sediments in Boston Harbor showed a similar relationship between phenanthrene and naphthalene utilization where turnover times measured for these hydrocarbons were similar at all sites (Shiaris 1989). We also measured potentials for benzo[*a*]pyrene in the spring of 1990 but found very low to no potentials at all sites and depths.

When incubated under optimal conditions, hexadecane and naphthalene and phenanthrene degradation potentials are useful indicators of previous exposure to hydrocarbons but are not good indicators of in situ mineralization rates. In concert with hydrocarbon concentration data, they can be used to estimate maximum biodegradation rates in situ. Similar data were used to estimate a biodegradation rate of 0.5 $\mu g/(g \cdot d)$ for sediments polluted by the *Amoco Cadiz* oil spill (Ward et al. 1980). Mineralization potential data using this method with non-nutrient-amended samples (not optimal) were obtained in a bioremediation monitoring study in Prince William Sound. These data, along with hydrocarbon chemistry data were used to estimate differences in biodegradation rates in situ for fertilized and unfertilized shoreline sediments. These rates were estimated to range from 1.9 to 10 μg/(g · d) on unfertilized beaches and up to several fold higher on fertilized beaches (Prince et al. 1990, 1994).

We believe that this study is unique with respect to the extent of information collected following a major oil spill. Sediments were sampled to measure microbial numbers and activity within a few months following the *Exxon Valdez* spill, and sampling continued for 2 years after the spill. We have chosen to describe here, in detail, our results from two oiled sites and two reference sites as an example of how these microbial techniques can be used to assess and monitor damage after an oil spill. The boundaries of the study were set by the purpose of accumulating evidence for litigation. Nevertheless, the results of this study provide valuable information on marine sediment microbial responses to hydrocarbon perturbations, processes that are extremely important in the removal of organic pollutants from the environment.

Acknowledgments

The authors acknowledge Chuck O'Clair and Alan Mearns for information on sediment particle size and shoreline treatments. We also thank Greg Winter and Mike Brockman for technical assistance and Doug Braddock for graphics. This work was supported by the National Oceanographic and Atmospheric Administration contract 50-DSNC-8-00141 to Science Applications International Corp. and contracts from the Alaska Department of Environmental Conservation through the Comprehensive Environmental Response, Compensation, and Liability Act. We also acknowledge the Water Research Center and Institute of Arctic Biology at the University of Alaska for their support in providing service to the state of Alaska.

References

Aelion, C. M., and P. M. Bradley. 1991. Aerobic biodegradation potential of subsurface microorganisms from a jet fuel-contaminated aquifer. Applied and Environmental Microbiology 57:57–63.

Atlas, R. M., P. D. Boehm, and J. A. Calder. 1981. Chemical and biological weathering of oil from the *Amoco Cadiz* spillage within the littoral zone. Estuarine Coastal and Shelf Science 12:589–608.

Bartha, R., and R. M. Atlas. 1987. Transport and transformation of petroleum: biological processes. Pages 287–341 *in* D. M. Boesch and N. N. Rabalais, editors. The long-term effects of offshore oil and gas development: an assessment and a research strategy. Elsevier Applied Science, New York.

Bauer, J. E., and D. G. Capone. 1988. Effects of cooccurring aromatic hydrocarbons on degradation of individual polycyclic aromatic hydrocarbons in marine sediment slurries. Applied and Environmental Microbiology 54:1649–1655.

Braddock, J. F., M. R. Brockman, J. E. Lindstrom, and E. J. Brown. 1990. Microbial hydrocarbon degradation in sediments impacted by the *Exxon Valdez* oil spill. Report to National Oceanic and Atmospheric Administration, contract 50-DSNC-8-00141, Washington, DC.

Braddock, J. F., B. T. Rasley, T. R. Yeager, J. E. Lindstrom, and E. J. Brown. 1993. Hydrocarbon mineralization potentials and microbial populations in marine sediments following the *Exxon Valdez* oil spill. Report to Alaska Department of Environmental Conservation, Juneau.

Brown, E. J., and J. F. Braddock. 1990. Sheen screen, a miniaturized most probable number method for enumeration of oil-degrading microorganisms. Applied and Environmental Microbiology 56:3895–3896.

Brown, E. J., S. M. Resnick, C. Rebstock, H. V. Luong, and J. Lindstrom. 1991. UAF radiorespirometric protocol for assessing hydrocarbon mineralization potential in environmental samples. Biodegradation 2:121–127.

Colwell, R. R., A. L. Mills, J. D. Walker, P. Garcia-Tello, and V. Campos-P. 1978. Microbial ecology studies of the *Metula* spill in the Straits of Magellan. Journal of the Fisheries Research Board of Canada 35:573–580.

Guerin, W. F., and G. E. Jones. 1989. Estuarine ecology of phenanthrene-degrading bacteria. Estuarine Coastal and Shelf Science 29:115–130.

Karl, D. M. 1992. The grounding of the *Bahia Paraiso*: microbial ecology of the 1989 Antarctic spill. Microbial Ecology 24:77–89.

Kirchman, D., J. Sigda, R. Kapuscinski, and R. Mitchell. 1982. Statistical analysis of the direct count method for enumerating bacteria. Applied and Environmental Microbiology 44:376–382.

Leahy, J. G., and R. R. Colwell. 1990. Microbial degradation of hydrocarbons in the environment. Microbiological Reviews 54:305–315.

Lindstrom, J. E., and six coauthors. 1991. Microbial populations and hydrocarbon biodegradation potentials in fertilized shoreline sediments affected by the T/V *Exxon Valdez* oil spill. Applied and Environmental Microbiology 57:2514–2522.

Lizarraga-Partida, M. L., F. B. Izquierdo-Vicuna, and I. Wong-Chang. 1991. Marine bacteria on the Campeche Bank oil field. Marine Pollution Bulletin 22:401–405.

Madsen, E. L., J. L. Sinclair, and W. C. Ghiorse. 1991. In situ biodegradation: microbiological patterns in a contaminated aquifer. Science 252:830–834.

Montagna, P. A. 1982. Sampling design and enumeration statistics for bacteria extracted from marine sediments. Applied and Environmental Microbiology 43: 1366–1372.

Porter, K. G., and Y. S. Feig. 1980. The use of DAPI for

identifying and counting aquatic microflora. Limnology and Oceanography 25:943–948.

Prince, R. C., J. R. Clark, and J. E. Lindstrom. 1990. Bioremediation monitoring program. Report to U.S. Coast Guard and Alaska Department of Environmental Conservation, Anchorage.

Prince, R. C., and sixteen coauthors. 1994. Bioremediation of the *Exxon Valdez* oil spill: monitoring safety and efficacy. Pages 107–124 *in* R. E. Hinchee, B. C. Alleman, R. E. Hoeppel, and R. N. Miller, editors. Hydrocarbon bioremediation. Lewis Publishers, Boca Raton, Florida.

Roubal, G., and R. M. Atlas. 1978. Distribution of hydrocarbon-utilizing microorganisms and hydrocarbon biodegradation potentials in Alaska continental shelf areas. Applied and Environmental Microbiology 35: 897–905.

Shiaris, M. P. 1989. Seasonal biotransformation of naphthalene, phenanthrene, and benzo[*a*]pyrene in surficial estuarine sediments. Applied and Environmental Microbiology 55:1391–1399.

Ward, D. M., R. M. Atlas, P. D. Boehm, and J. A. Calder. 1980. Microbial biodegradation and chemical evolution of oil from the *Amoco* spill. Ambio 9:277–283.

Zar, J. H. 1984. Biostatistical analysis, 2nd edition. Prentice-Hall, Englewood Cliffs, New Jersey.

American Fisheries Society Symposium 18:109–120, 1996

Tracking *Exxon Valdez* Oil from Beach to Deepwater Sediments of Prince William Sound, Alaska

PAUL R. CARLSON AND KEITH A. KVENVOLDEN
U.S. Geological Survey, 345 Middlefield Road, MS 999
Menlo Park, California, 94025, USA

Abstract.—Prince William Sound is a large fjord-type estuary system containing rugged bedrock islands dominated by coarse gravel beaches. The insular submarine slopes are covered with gravelly to muddy sediment. Bedrock and morainal ridges separate deep basins containing Holocene diatom-rich mud. Since the *Exxon Valdez* oil spill in the sound in March 1989, we have participated in four sediment-sampling cruises. Two months after the spill, oil had not reached the deepwater sediment at 19 sample sites. Reoccupation of nine sites, 14 months after the spill, produced no visible oil and no compelling evidence for oil contamination. Beaches and shallow-water sediments were sampled 17 months after the spill. Oil was present in various forms on all beaches visited. Geochemical characterizations indicate that tar found on two beaches was not from *Exxon Valdez* oil. At 11 shallow-water stations, geochemical markers suggest spilled oil in the sediment. In June 1992 (38 months after the spill), minor amounts of *Exxon Valdez* oil, as well as tar from earlier spills, were found on several beaches. In addition, sediment at two deepwater sites contained geochemical markers indicating trace amounts of the spilled *Exxon Valdez* oil.

Shortly after midnight on 24 March 1989, the tanker *Exxon Valdez* struck Bligh Reef in upper Prince William Sound, Alaska (Figures 1 and 2), spilling an estimated 43.9 million liters of North Slope crude oil into the water of this important fisheries and tourism region. Many teams of investigators studying the effects of the spill focused on the water column and the beaches. The efforts of the U.S. Geological Survey (USGS) in Prince William Sound were designed to track the geological fate of the oil in bottom sediment in the deeper parts of the sound. These studies began about 7 weeks after the grounding of the tanker with a brief (3 d) bottom-sediment-sampling cruise of the deepwater sediment sinks along the spill trajectory (Carlson and Reimnitz 1990). Altogether we have now undertaken four sediment-sampling cruises since the oil spill in March 1989: May 1989, MV *Farnella*; May 1990, National Oceanic and Atmospheric Administration (NOAA) ship *Davidson*; August 1990, RV *Karluk*; and June 1992, NOAA ship *McArthur*. This article provides a review and summary of the results of these cruises.

Geologic Setting

Prince William Sound is a large, complex fjord-type estuarine system adjacent to the northern Gulf of Alaska (Figures 1 and 2). The sound owes its configuration to colliding tectonic plates and multiple episodes of glaciation. Continuing active tectonism along this convergent margin results from the collision of the Pacific and the North American plates. Underlying Prince William Sound is the Prince William terrane, which was accreted to the North American plate in middle Eocene time (Jones et al. 1986) and consists of early Tertiary flysch-like sedimentary and tholeiitic volcanic rocks of the Orca Group (Winkler 1976). Extensive tectonism along this plate boundary resulted in mountain-building episodes and frequent earthquakes.

The Great Alaska Earthquake of 1964 had its epicenter at the northern edge of Prince William Sound. This earthquake (magnitude 8.4 on the Richter scale) resulted in about 8 m of uplift on Montague Island (Figure 2) in the southern part of the sound and from 1 to 2 m of uplift at Bligh Reef (Plafker 1969). The mountain building created regional cooling and coastal glaciation that resulted in numerous ice advances. During the Pleistocene, some of the glaciers achieved thicknesses of 2,000 m in Prince William Sound (von Huene et al. 1967). On 24 March 1989, ice that calved from the presently retreating Columbia Glacier (Figure 2) was reported to have drifted into the tanker lanes. To avoid hitting the ice, the *Exxon Valdez* tanker changed course toward Bligh Reef and ran aground (Lee 1990).

Gravel beaches dominate the exposed segments of bedrock islands in Prince William Sound, whose steep submarine slopes are covered with gravelly to muddy sediment. Submerged glacial morainal ridges consist of relict diamicts, indicating the presence of tidal currents with sufficient velocity to sweep away modern fine sediment. Between bedrock and morainal highs of the fjord floor, numer-

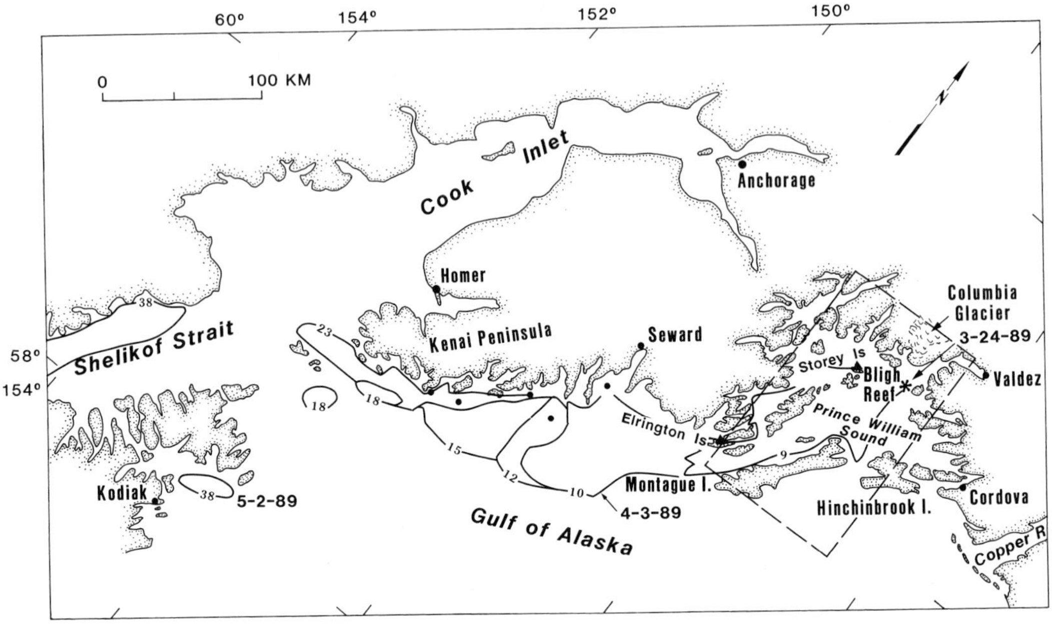

FIGURE 1.—Map of northern Gulf of Alaska showing Prince William Sound and Kenai Peninsula coast. Dashed rectangle shows location of Figure 2. Dots are sample sites outside of the area of Figure 2. Triangles mark locations on Elrington and Storey islands of first discovery of California tar. Numbered lines show locations of the leading front of the *Exxon Valdez* oil spill. The numbers indicate days after the spill on 24 March 1989. Positions of the spill front were compiled from a variety of maps made by the National Oceanic and Atmospheric Administration, Exxon Corp., the U.S. Coast Guard and the Alaska Department of Environmental Conservation.

ous deep basins (to water depths of 800 m) contain Holocene diatom-rich mud, as much as 200 m thick, that overlies irregular discontinuous deposits of probable glacial origin (von Huene et al. 1967). Minor amounts of coarse sediment are being introduced into the sound from a few fjord-head deltas and a bottom load from the Copper River delta through Hinchinbrook Entrance (Figures 1 and 2).

The dominant sedimentary material accumulating today in this fjord complex is fine, suspended glacial flour from the Copper River (Figure 1) that enters the sound through channels at each end of Hinchinbrook Island (Carlson et al. 1976; Molnia and Hein 1982; Klein 1983). In the northern part of the sound, the fine sediment has a secondary source, runoff from the adjacent land mass (Klein 1983). Marine and estuarine assemblages of diatoms are important components of the fine-grained sediment (E. Hemphill-Haley, USGS, personal communication). This diatom-rich glacial-marine sediment is deposited mainly in deep-sediment sinks throughout the sound (Figure 3A) at rates that vary from 0.3 to 0.6 cm/year (Klein 1983; Bothner et al. 1990). Much of the submarine-insular-slope and fjord wall environments are kept bare of fine sediment (Figure 3B) settling from the water column by complex current circulation within the sound.

Review of Sampling Program and Results

In May 1989 onboard the MV *Farnella*, we sampled bottom sediment by box coring at 20 stations (Figures 1 and 2) along the oil spill trajectory from Bligh Reef southwest through Prince William Sound (15 sites) and along the Kenai Peninsula (5 sites). Each site was chosen after a 3.5-kHz acoustic profile line was run across the prospective sample areas (Figure 3). Acoustic profile lines enabled us to identify sites of relatively thick accumulations of modern sediment. Box cores sampled postglacial, unconsolidated, diatom-rich mud that is accumulating in the deep basins of the sound today (Figure 3A). Some sites were selected on seafloor highs, principally glacial moraine substrate (Figure 3C). Cores from these sites contained pebbly sandy muds (Carlson and Reimnitz 1990). No visible oil was seen when the box cores were being subsampled,

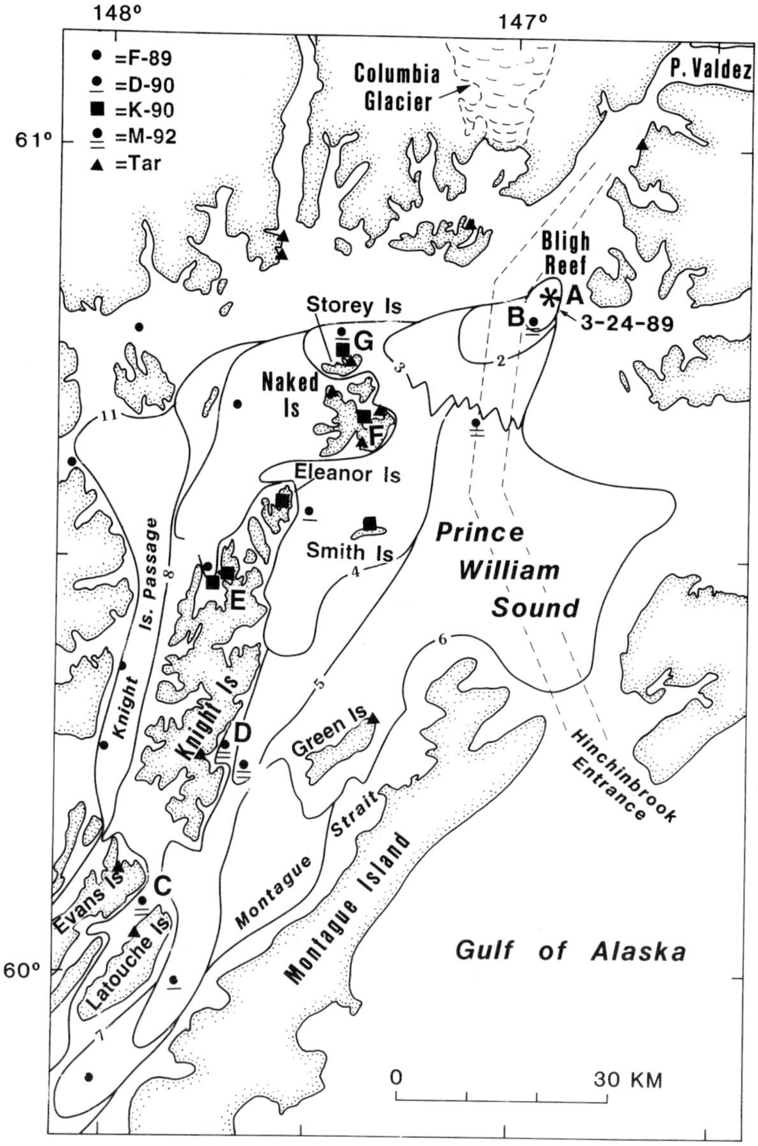

FIGURE 2.—Composite map of sample sites in Prince William Sound. Dots and squares indicate sample sites from four cruises (*Farnella*, F-89; *Davidson*, D-90; *Karluk*, K-90; *McArthur*, M-92). Underlined dots indicate resampling of F-89 sites: one line, resampling on D-90; two lines, resampling on both D-90 and M-92. Triangles show locations of tar samples. Locations A–G are discussed in the text. As in Figure 1, the numbered lines show locations of the leading front of the *Exxon Valdez* oil spill, which originated at Bligh Reef. The position of the front on consecutive days 2–8 and day 11 are indicated. Dashed lines delineate tanker lanes.

and, when viewed under ultraviolet light, no fluorescence was observed.

Sediment samples were analyzed by gas chromatography and mass spectrometry for aliphatic and aromatic hydrocarbons (see Rapp et al. [1990] and Kvenvolden et al. [1991] for description of geochemical methods). The geochemical results were compared with results obtained from fresh oil impounded from the *Exxon Valdez* tanker (location A, Figure 2) immediately after the spill. Oil contamination could not be positively identified in sediment at any of the 20 deepwater sites sampled 2 months

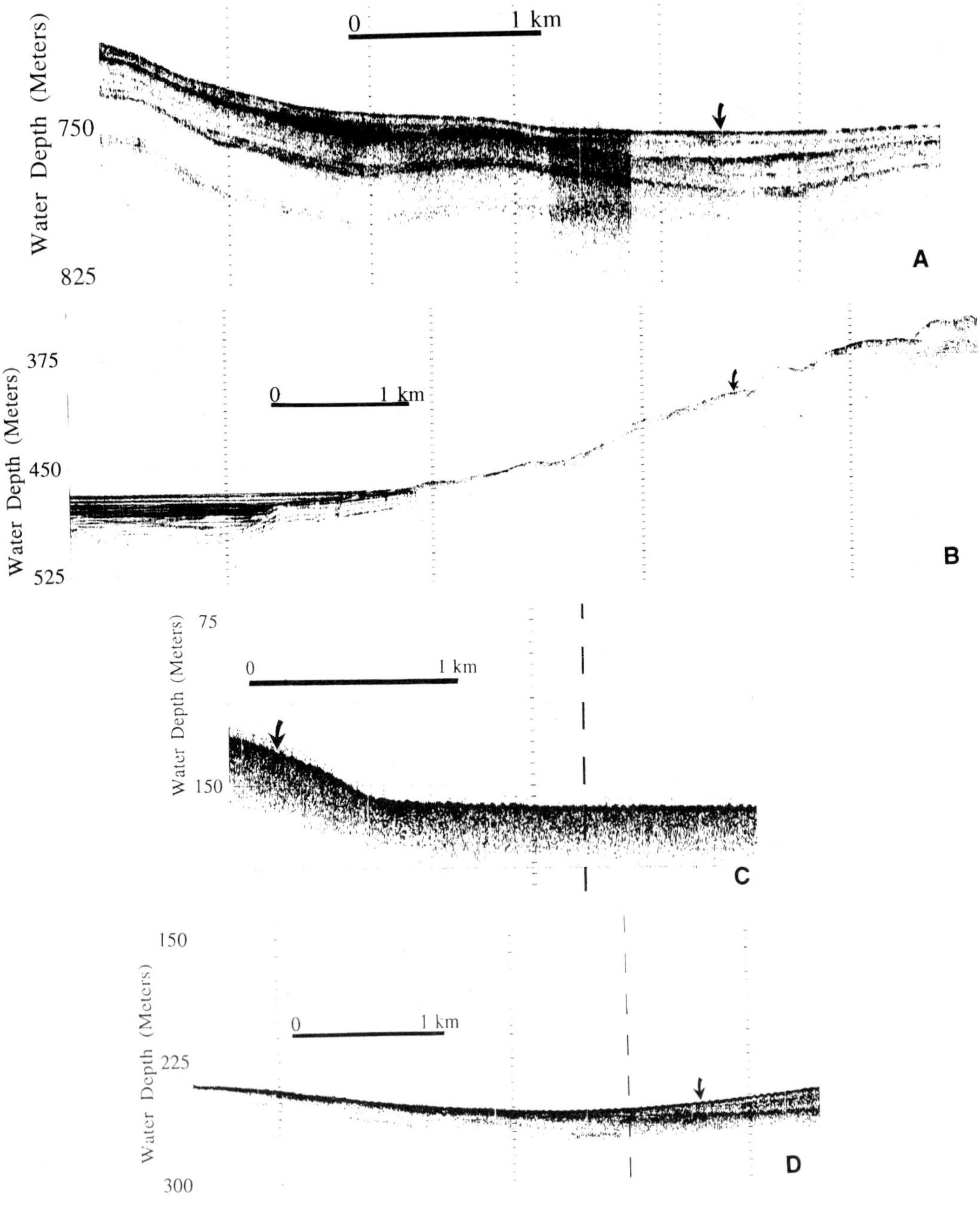

FIGURE 3.—Acoustic profiles (3.5-kHz system with a 1-s firing and recording rate and vertical exaggeration of ~10×) showing various features of the fjord floor and locations of sampling sites (arrows). **A**. Sample site on the west side of Naked Island (755-m water depth). The profile shows the nature of the sediment ponded in a deep part of the sound. **B**. Sample site on the southern flank of Bligh Reef, the obstacle struck by the *Exxon Valdez* tanker. The profile shows the change from a flat, sediment-covered fjord floor with numerous internal reflections to rugged bedrock that makes up the southern flank of the reef. **C**. Sample site on moraine at the mouth of Montague Strait. The profile shows the hard bottom typical of a glacial diamict. **D**. Sample site on the floor of Montague Strait. The profile shows the transition from an older diamict surface to a wedge-shaped layer of modern soft mud.

after the spill. Only at one site, near the south end of the sound, northeast of Latouche Island (location C, Figure 2), did the sediment extract have geochemical indicators of possible oil contamination (Rapp et al. 1990).

Eight deepwater sites in the sound (Figure 2) and one site southwest of Seward (Figure 1), originally sampled in 1989, were resampled in May 1990, 14 months after the oil spill, on a cooperative cruise with National Marine Fisheries Service (NMFS) on the NOAA RV *Davidson*. No visible signs of oil were present in the deepwater sediment. Subsequent geochemical analyses of 1990 sediment samples showed results that were similar to the results obtained with the 1989 samples at each site (Kvenvolden et al. 1991). Geochemical changes in deepwater samples from 1989 to 1990 were attributed to changes in the depositional environment, which may or may not have been caused by effects of the oil spill.

Seventeen months after the spill (August 1990), six islands and their insular slopes (Elrington, Knight, Eleanor, Smith, Naked, and Storey) within the trajectory of the spilled oil (Figures 1 and 2) were investigated using the USGS RV *Karluk* (Carlson et al. 1991). We found some oil on all of the beaches that we visited. The oil was in a variety of forms, including sheens of oil on water that percolated from the beach sediment, thin coatings of oil on sediment or rock, brown sticky mousse-like oil patches on sediment and driftwood, and tar or asphalt-like patches on rock.

In the case of the tar, two geochemically similar samples were found about 100 km apart on beaches on the north side of Storey Island and on the northwest side of Elrington Island (Figure 1). These tars were not from the *Exxon Valdez* oil spill because they are chemically and isotopically distinct from either the spill oil or the oiled sediment samples that we collected from the beaches. These tars exhibit the geochemical characteristics of oils from the Monterey Formation in southern California (Kvenvolden et al. 1993a). The other oil samples have geochemical characteristics similar to those of the spilled North Slope crude sample from the *Exxon Valdez* tanker; however, the oil samples from the beaches are at various stages of weathering degradation (Kvenvolden et al. 1993a; Hostettler and Kvenvolden 1994).

After sampling the six island beaches, we ran high-resolution acoustic profiles across their insular slopes (Figure 3). We then collected bottom sediment at sites selected from the profiles where modern fine sediment has been accumulating (Carlson et al. 1991). The insular slopes were mostly devoid of fine sediment, but small depressions with thin deposits were identified and sampled. The sediment ranged from mud through sandy mud to pebbly, sandy mud. Geochemical results suggest addition of spilled oil to sediment at 11 of the shallow-water sites occupied in 1990 (Kvenvolden et al. 1991). All of these samples were collected from beaches that were heavily impacted by oil spilled from the *Exxon Valdez* tanker, but none of the shallow-water samples contained visible amounts of the spilled oil.

Thirty-eight months after the oil spill (June 1992), on another cooperative cruise with NMFS aboard the NOAA RV *McArthur*, we found oil from the 1989 spill on beaches of Naked, Green, Knight, and Evans islands (Figure 2). Tar from sources other than the 1989 spill (e.g., from the California Monterey Formation) was also present on these island beaches and elsewhere (Kvenvolden et al. 1993b). On this cruise five deepwater sites, previously occupied in both 1989 and 1990 (Figure 2), were resampled. Initial analyses of the samples from two sites showed evidence of hydrocarbon contamination: location D (Figure 2) near the entrance to Snug Harbor of Knight Island and location C (Figure 2) off the northwest end of Latouche Island. Location C was the only deepwater site to show even a hint of oil contamination in 1989.

Summary of Geochemical Observations

In tracking oil residues from the *Exxon Valdez* oil spill into the sediment record of Prince William Sound, we used geochemical criteria based on both molecular and isotopic measurements (Kvenvolden et al. 1993a, 1993b). These kinds of measurements, made initially on a sample of North Slope crude oil impounded immediately after the oil spill, provided the basis for tracking the spilled oil. This oil contains a mixture of *n*-alkanes and isoprenoid hydrocarbons (pristane and phytane) that was rapidly degraded as the oil was weathered in the environments of the sound. Figure 4 illustrates, by means of gas chromatographic distributions of these aliphatic hydrocarbons, the changes that can take place. In the oil, the *n*-alkanes dominate and decrease in relative concentration with increasing molecular weight, ranging from about n-C_{11} to at least n-C_{40}. Residues from the beaches of various islands show losses in both *n*-alkanes and isoprenoid hydrocarbons and the development of an unresolved complex mixture (UCM) of hydrocarbons. These results demonstrate that distributions of *n*-alkanes and isoprenoid hydrocarbons are very limited in their ap-

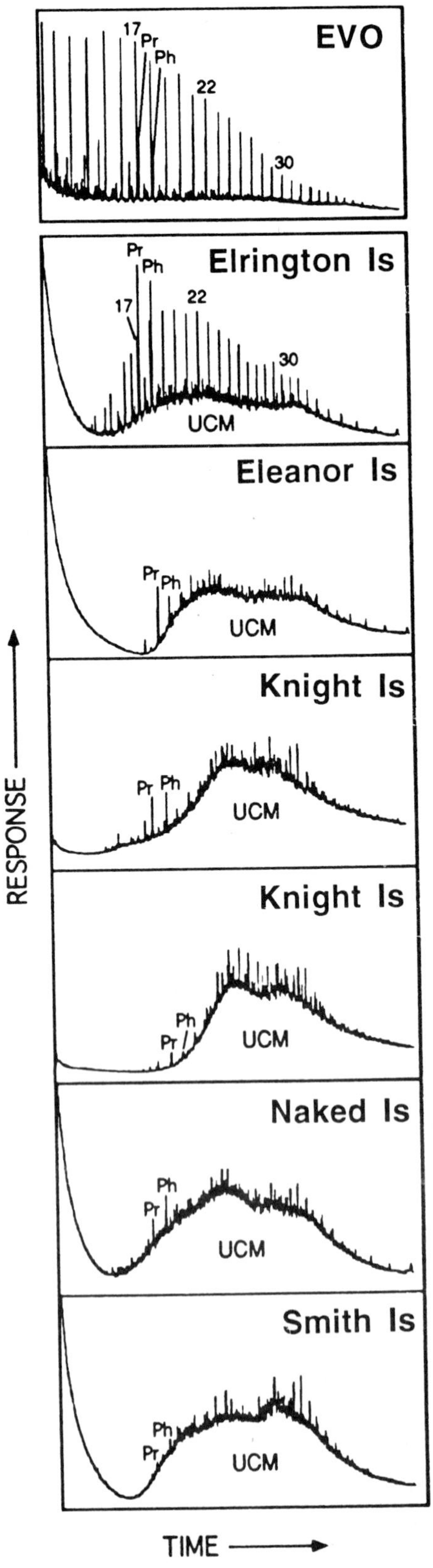

FIGURE 4.—Gas chromatograms of aliphatic hydrocarbons in a sample of North Slope crude oil impounded from the *Exxon Valdez* tanker (EVO) and in samples of oil residues from the *Exxon Valdez* oil spill found on beaches of islands (Elrington, Eleanor, Knight, Naked, and Smith) in Prince William Sound. Numbers on chromatograms denote carbon numbers of homologous series of *n*-alkanes. Pr, pristane; Ph, phytane; UCM, unresolved complex mixture.

plication to tracking the presence of the spilled oil. Nevertheless, pristane/phytane (Pr/Ph) ratios are shown in Table 1 to illustrate, as an example, the variability in the magnitude of this ratio for selected samples.

In contrast to the *n*-alkanes and isoprenoids, the polycyclic aliphatic hydrocarbons are less susceptible to rapid biodegradation (Volkman 1984) and were found to be more useful in tracking the spilled oil (Kvenvolden et al. 1993a; Hostettler and Kvenvolden 1994). The distributions of two classes of polycyclic aliphatic hydrocarbons were obtained from single-ion monitoring (SIM) mass chromatograms of mass/charge (m/z) 191 (terpanes) and m/z 217 (steranes). Mass chromatograms of the original spilled oil and of the extracts from a shallow-water sediment sample from Naked Island are shown in Figure 5; compound identifications are listed in Table 2. Four ratios calculated from m/z 191 mass chromatograms are given in Table 1:

(1) C_{31}-tricyclic terpane/$\alpha\beta$-hopane (C_{23}/C_{30}) is the ratio used to describe the relative proportions of the major member of the two major suites of terpanes (tricyclic and pentacyclic) in each sample.
(2) C_{30} $\alpha\beta$-hopane/$C_{29}\alpha\beta$-norhopane (C_{30}/C_{29}) is a ratio used for source-rock correlations (Palacas et al. 1984).
(3) C_{31} $\alpha\beta$-homohopane [$S/(S + R)$] is a hopane epimer ratio used as a maturity indicator (Ensminger et al. 1974); *S* is sinistral and *R* is recto configuration. With increasing maturity, the biologic configuration of 22*R* changes to a mixture of 22*R* and 22*S*, reaching an equilibrium ratio of 60:40 or an $S/(S + R)$ ratio of 0.6 at maturity (Mackenzie 1984).
(4) The triplet pattern commonly found in the m/z 191 mass chromatograms of samples from Prince William Sound consists of three peaks (in order of retention times): a C_{24} tetracyclic terpane and two probable epimers of a C_{26} tricyclic terpane; this triplet with the three peaks of approximately equal heights is very characteristic of North Slope crude oil in gen-

TABLE 1.—Summary of geochemical variables of samples from Prince William Sound, Alaska, in 1989, 1990, and 1992.

					m/z 191				*m/z* 2/7	
Station	Sample	Location and description of sample	Water depth (m)	Pr/Ph[a]	C_{23}/ C_{30}[b]	C_{30}/ C_{29}[c]	C_{31} $S/(S+R)$[d]	Triplet[e]	C_{29} $S/(S+R)$[f]	$\delta^{13}C$ (‰)
A	89-EVO	Bligh Reef, *Exxon Valdez* oil sample	0	1.4	0.71	1.4	0.60	2.0	0.45	−29.2
		Deepwater stations								
B	89-F-7	Near Bligh Reef, bottom sediment	394	4.8	0.12	1.6	0.50	0.67	0.18	
	92-M-9	Same	390	3.5	0.67	1.6	0.48	0.67	0.24	
C	89-F-15	N, Latouche Island, bottom sediment	240	4.6	0.59	1.3	0.53	3.7	0.42	
	92-M-33	Same		2.5	2.5	1.4	0.54	2.0	0.49	
D	89-F-4	E, Knight Island, bottom sediment	125	5.9	0.30	1.7	0.50	1.6	0.34	
	92-M-23	Same	114	3.9	0.83	1.6	0.56	2.0	0.45	
		Shallow-water stations								
E	90-K-15	N, Knight Island, bottom sediment	77		0.65	1.6	0.59	2.0	0.33	
F	90-K-40	N, Naked Island, bottom sediment	20		1.3	1.5	0.56	2.0	0.37	
G	90-K-35	N, Storey Island, bottom sediment	33		1.0	1.5	0.48	2.0	0.39	
		Beach stations								
E	90-K-18	N, Knight Island, oil on beach	0	1.0	0.59	1.3	0.59	2.0	0.49	−29.2
F	90-K-40	N, Naked Island, oil on beach	0	0.8	1.0	1.5	0.62	2.2	0.48	−29.4
G	90-K-37A	N, Storey Island, tar on beach	0	1.2	0.17	1.8	0.57	5.1	0.32	−23.9

[a] Pr/Ph, pristane (2,6,10,14-tetramethylpentadecane)/phytane (2,6,10,14-tetramethylhexadecane).
[b] C_{23}/C_{30}, C_{23}-tricyclic terpane/17α(H),21β(H)-hopane.
[c] C_{30}/C_{29}, 17α(H),21β(H)-hopane/17α(H),21β(H)-30-norhopane.
[d] C_{31} $S/(S+R)$, 17α(H),21β(H)-homohopane(22S)/17α(H),21β(H)-homohopane (22*S*) + (22*R*).
[e] Triplet, [C_{26}-tricyclic terpane (*S*?) + C_{26}-tricyclic terpane(*R*?)]/C_{24}-tetracyclic terpane.
[f] C_{29} $S/(S+R)$, 24-ethyl-5α(H),14α(H),17α(H)-cholestane (20*S*)/24-ethyl-5α(H),14α(H),17α(H)-cholestane (20*S*) + (20*R*); $\delta^{13}C$(‰), $[(^{13}C/^{12}C_{(sample)}/^{13}C/^{12}C_{(standard)}) - 1] \times 10^3$.

eral (Kvenvolden et al. 1985) and of *Exxon Valdez* spill oil in particular; a ratio for the triplet is formed by summing the peak heights of the two C_{26} tricyclic terpanes and dividing this sum by the peak height of the C_{24} tetracyclic terpane, in the *Exxon Valdez* spill oil this ratio is about 2.0 (Table 1).

One ratio calculated from the *m/z* 217 mass chromatogram is given in Table 1: C_{29} $\alpha\alpha\alpha$-ethylcholestane [$S/S(S+R)$]. This sterane epimer ratio has been used as a maturity indicator (Mackenzie et al. 1980). With increasing maturity, the biologic configuration of 20*R* changes to a mixture of 20*R* and 20*S* with an equilibrium distribution at maturity of about 1:1 or an $S/(S+R)$ ratio of about 0.5.

We used the terpane and sterane maturity parameters as source parameters because the spill oil is not expected to undergo significant maturation processes during the short time of the spill history. Thus we used conventional polycyclic aliphatic hydrocarbon source and maturity parameters to track the spill oil into the sediments of the sound.

Geochemical analyses of sediment in shallow-water samples (<125-m water depth), collected offshore from beaches that had been heavily impacted by the oil spill, suggested the presence of small quantities of the spilled oil. For example, the mass chromatograms comparing distributions of terpanes in the spilled oil and in an extract of sediment collected at 20-m water depth offshore from Naked Island (location F, Figure 2) are remarkably similar (Figure 5, A and B); however, sterane distributions are less similar (Figure 5, C and D). The triplet ratio in the extracts from sediments at locations E, F, and G measures 2.0, which correlates with the spilled oil ratio of 2.0 (Table 1). Although not conclusive, we believe that the correlation indicates the presence of *Exxon Valdez* oil in the shallow sediment.

Several deepwater locations were resampled dur-

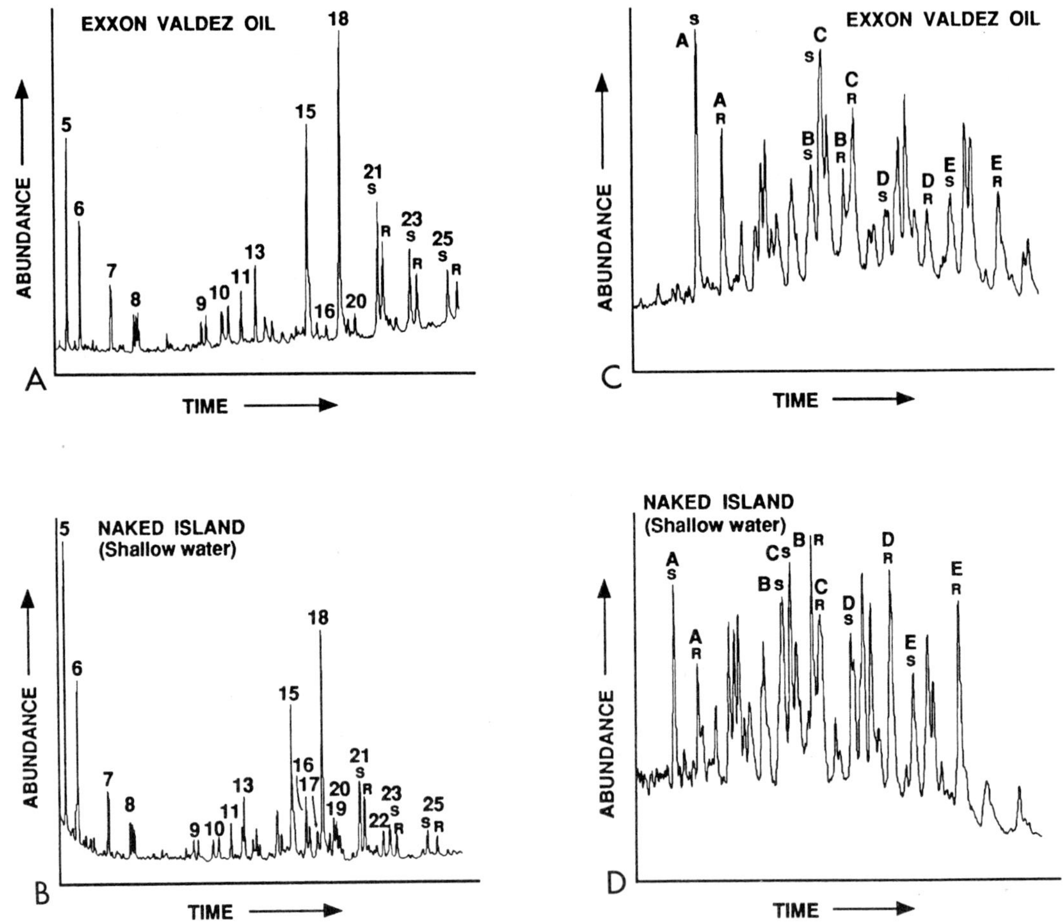

FIGURE 5.—Comparison of mass chromatograms of terpanes (*m/z* 191) and steranes (*m/z* 217) from the *Exxon Valdez* oil and from an extract of a sediment collected at a shallow-water depth of 20 m near Naked Island. Compound identifications are given in Table 2. **A**. Terpane distribution, *Exxon Valdez* oil (location A, Figure 2). **B**. Terpane distribution, sediment extract from Naked Island (location F, Figure 2). **C**. Sterane distribution, *Exxon Valdez* oil (location A, Figure 2). **D**. Sterane distribution, sediment extract from Naked Island (location F, Figure 2).

ing the course of 4 years, and our results suggest that spilled oil is now present at locations C and D (Figure 2). A comparison of mass chromatograms (*m/z* 191) from sediment extracts from these locations (Figure 6) show changes between 1989 and 1992 in a direction in which the distributions of terpanes in 1992 look more like the distribution in the spilled oil. Also, maturity and triplet ratios have changed during time to have values more like those observed in the spilled oil (Table 1). In contrast, at location B near Bligh Reef (Figure 2) the molecular ratios in sediment extracts have remained essentially unchanged since the oil spill (Table 1).

Eight oil residues from beaches of six islands (Storey, Naked, Eleanor, Smith, Knight, and Elrington; Figure 2) had distributions of polycyclic aliphatic hydrocarbons and carbon isotopic compositions of total extracts indicating two distinct sources (Kvenvolden et al. 1993a). Samples from beaches at locations E and F (Figure 2 and Table 1) are representative of the six oil residues that correlate with the *Exxon Valdez* oil. The sample from the beach at location G (Figure 2 and Table 1) is representative of the two oil residues, found in 1990, that are chemically and isotopically distinct from the *Exxon Valdez* oil. We have expanded this preliminary study of oil residues on beaches using carbon isotopic compositions of whole-oil residues as the principal method of identification and correlation (Kvenvolden et al. 1993b). Eight residues from beaches on

TABLE 2.—Identification of terpanes and steranes for the mass chromatograms in Figures 5 and 6.

Peak	Compound	
	Terpanes and triterpanes (m/z 191)	
1	C_{19} tricyclic terpane	C_{19}
2	C_{20} tricyclic terpane	C_{20}
3	C_{21} tricyclic terpane	C_{21}
4	C_{22} tricyclic terpane	C_{22}
5	C_{23} tricyclic terpane	C_{23}
6	C_{24} tricyclic terpane	C_{24}
7	C_{25} tricyclic terpane	C_{25}
8	Triplet: C_{24} tetracyclic terpane	C_{24}
	C_{26} tricyclic terpane (?*S*)	C_{26}
	C_{26} tricyclic terpane (?*R*)	C_{26}
9	C_{28} tricyclic terpanes (?*S* and *R*)	C_{28}
10	C_{29} tricyclic terpanes (?*S* and *R*)	C_{29}
11	18α(H)-22,29,30-trisnorneohopane (Ts)	C_{27}
12	22,29,30-trisnorhop-17(21)-ene	C_{27}
13	17α(H)-22,29,30-trisnorhopane (Tm)	C_{27}
14	17β(H)-22,29,30-trisnorhopane	C_{27}
15	17α(H),21β(H)-30-norhopane	C_{29}
16	17β(H),21α(H)-30-normoretane	C_{29}
17	18α and/or β(H)-oleanane	C_{30}
18	17α(H),21β(H)-hopane	C_{30}
19	17β(H),21β(H)-30-norhopane	C_{29}
20	17β(H),21α(H)-moretane	C_{30}
21	17α(H),21β(H)-homohopane(22 *S* and *R*)	C_{31}
22	17β(H),21β(H)-hop-22(29)-ene (diploptene)	C_{30}
23	17α(H),21β(H)-bishomohopane(22 *S* and *R*)	C_{32}
24	17β(H),21β(H)-homohopane	C_{31}
25	17α(H),21β(H)-trishomohopane(22 *S* and *R*)	C_{33}
	Steranes and diasteranes (m/z 217)	
A(*S*)	13β(H),17α(H)-diacholestane(20*S*)	C_{27}
A(*R*)	13β(H),17α(H)-diacholestane(20*R*)	C_{27}
B(*S*)	5α(H),14α(H),17α(H)-cholestane(20*S*)	C_{27}
C(*S*)	24-ethyl-13β(H),17α(H)-diacholestane(20*S*)	C_{29}
B(*R*)	5α(H),14α(H), 17α(H)-cholestane(20*R*)	C_{27}
C(*R*)	24-ethyl-13β(H),14α(H)-diacholestane(20 *R*)	C_{29}
D(*S*)	24-methyl-5α(H),14α(H),17α(H)cholestane(20*S*)	C_{28}
D(*R*)	24-methyl-5α(H),14α(H),17α(H)cholestane(20*R*)	C_{28}
E(*S*)	24-ethyl-5α(H),14α(H),17α(H)-cholestane(20*S*)	C_{29}
E(*R*)	24-ethyl-5α(H),14α(H),17α(H)-cholestane(20*R*)	C_{29}

six islands impacted by the *Exxon Valdez* oil spill have $\delta^{13}C$ values averaging $-29.3 \pm 0.1‰$ which correlates with the value of $-29.2‰$ of the spilled oil (Table 1). In contrast, the non-*Exxon Valdez* residues (15 samples from 12 locations throughout the sound) have an average $\delta^{13}C$ value of $-23.8 \pm 0.1‰$. We believe that these residues resulted from the Great Alaska Earthquake of 1964, when asphalt was spilled into the sound from destroyed storage-tank facilities in the oil town of Valdez. Thus, Prince William Sound was apparently contaminated with asphalt 25 years before the *Exxon Valdez* oil spill. Residues from both the 1964 earthquake and the 1989 tanker spills are still present.

Discussion

Circulation of Prince William Sound water is strongly influenced by the counterclockwise-flowing Alaska Coastal Current. This current is affected by freshwater discharge, which, according to Royer et al. (1990), was at a record low at the time of the *Exxon Valdez* spill in March 1989. They concluded that because of the lower freshwater discharge, the spilled oil advanced through the sound more slowly than it would have in a normal year. Under these conditions of lower discharge, the amount of suspended sediment carried by streams draining the large glaciers bordering the Gulf of Alaska was probably below normal. Floating oil, even after losing volatiles, is not dense enough to sink, unless bonded to particulate matter. If the amount of particulate matter is low, the probability of bonding decreases. This process might explain the absence of the spill oil in the deepwater samples collected on the 1989 MV *Farnella* cruise. On the other hand, if the lower freshwater discharge caused the water circulation in the sound to be slowed, the oil would have had more time to attach to sedimentary particles. However, the general absence of oil in the deepwater sediment-sinks 2 months after the spill suggests that the first scenario is more likely.

By the second summer after the spill, there was evidence that traces of oil were migrating from oil-impacted beaches down insular slopes to depths approaching 120 m. There was meager evidence that some deep-basin sediment had trace amounts of oil contamination. Preliminary study of samples collected in June 1992 suggests that hydrocarbon contamination from the March 1989 oil spill had reached the deeper-water sediment of the fjord in at least two places: Location D (Figure 2) near the mouth of Snug Harbor at the southeastern part of Knight Island and location C (Figure 2) at the northeastern end of the passage between Latouche and Evans islands.

On several of the beaches around the sound, tar not associated with the *Exxon Valdez* oil spill has been observed and sampled. This tar appears to have originated from the Monterey Formation in California. It is probably a legacy of the days when Alaska's oil needs were met by imported California oil products that were transported by barge to the Prince William Sound ports of Whittier and Valdez before the development of an Alaskan oil industry. The tar is widely dispersed throughout the sound and could be the result of the Great Alaska Earthquake of 1964, when asphalt storage tanks were destroyed in Valdez, spilling asphaltic tar into the sound.

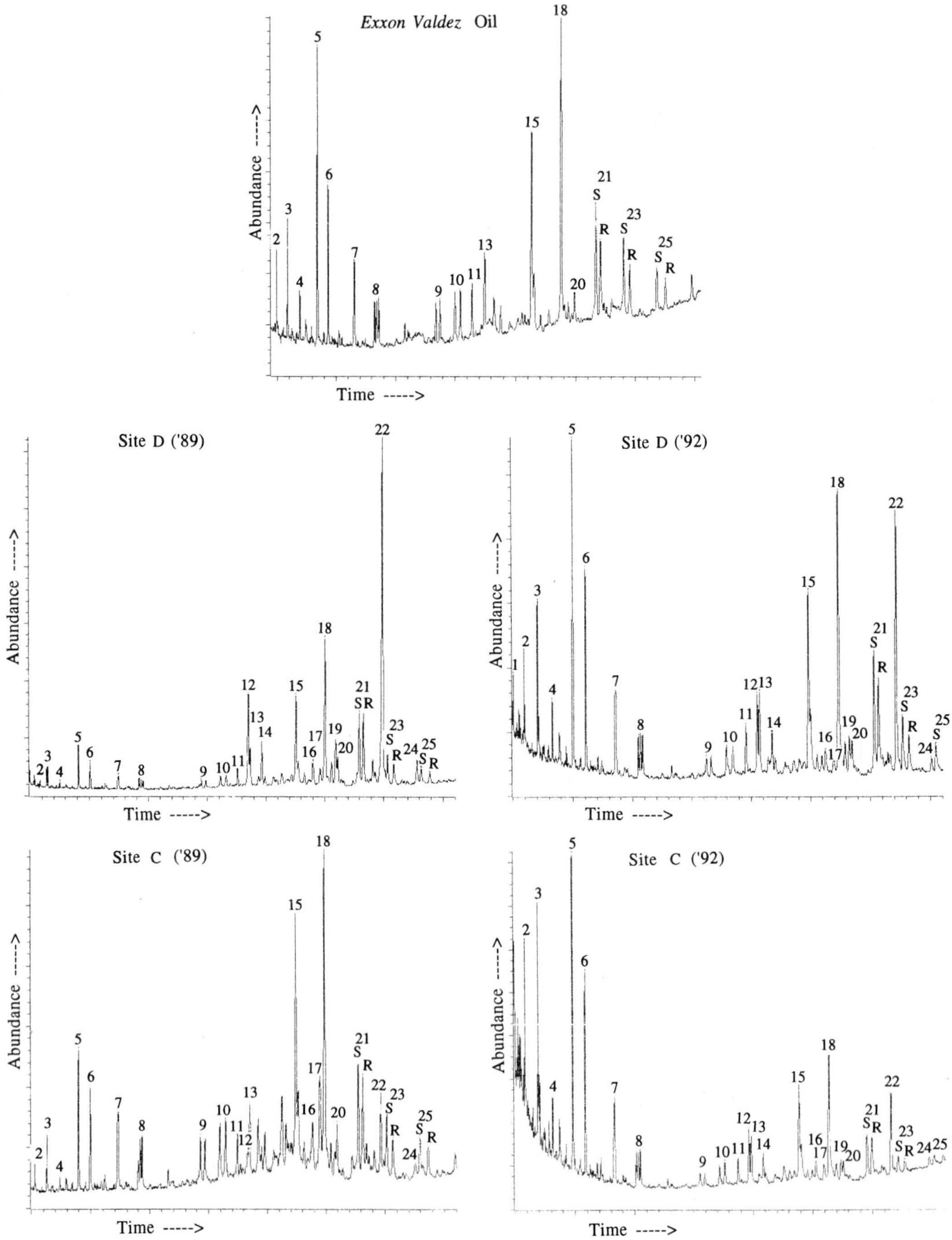

FIGURE 6.—Comparison of mass chromatograms (*m/z* 191) showing distributions of terpanes in *Exxon Valdez* oil and in extracts of sediment samples collected at locations C and D (Figure 2) in 1989 and 1992. Compound identifications are given in Table 2.

Acknowledgments

We are grateful to the officers, crew, and scientific parties aboard the MV *Farnella*, the National Oceanic and Atmospheric Administration (NOAA) ships *Davidson* and *McArthur*, and the U.S. Geological Survey (USGS) RV *Karluk* for their help on the cruises. Special appreciation is extended for cooperation and assistance provided on both NOAA cruises by chief scientist Bill Gronlund of the National Marine Fisheries Service. We thank Nancy and Jim Lethcoe of Alaskan Wilderness Sailing Safaris for providing location information vital to our finding tar on islands and shorelines of Prince William Sound. Thanks are also expressed to John Rapp, Fran Hostettler, and Brad Carkin (USGS, Menlo Park, California) and to Chuck Threlkeld and Augusta Warden (USGS, Denver, Colorado) for laboratory analyses. Constructive reviews were contributed by Ken Bird, Tom Lorenson, and Bernd Simoneit. Funding was provided by the Minerals Management Service and the USGS.

References

Bothner, M. H., C. M. Parmenter, and A. B. Brown. 1990. Plutonium and ^{210}Pb activities in two cores from Prince William Sound. Pages 1–12 *in* P. R. Carlson and E. Reimnitz, editors. Bottom sediment along oil spill trajectory in Prince William Sound and along Kenai Peninsula, Alaska. U.S. Geological Survey, Open-File Report 90-39D.

Carlson, P. R., P. W. Barnes, E. Hayden, and B. A. Carkin. 1991. Morphology and bottom sediment of Prince William Sound along the oil spill trajectory. Pages 1–30 *in* P. R. Carlson, editor. Sediment of Prince William Sound, beach to deep fjord floor, a year after the *Exxon Valdez* oil spill. U.S. Geological Survey, Open-File Report 91-631.

Carlson, P. R., B. F. Molnia, and E. Reimnitz. 1976. Dispersal, distribution, and thickness of holocene sediments on the continental shelf, northern Gulf of Alaska. Pages 63–64 *in* A. E. Fritsche et al., editors. The Neogene symposium. Society of Economic Paleontologists and Mineralogists, Pacific Section.

Carlson, P. R., and E. Reimnitz. 1990. Characterization of sample sites along the oil spill trajectory in Prince William Sound and the Gulf of Alaska. Pages 1–23 *in* P. R. Carlson and E. Reimnitz, editors. Bottom sediment along oil spill trajectory in Prince William Sound and along Kenai Peninsula, Alaska. U.S. Geological Survey, Open-File Report 90-39A.

Ensminger, A., A. van Dorsselaer, C. Spykerelle, P. Albrecht, and G. Ourisson. 1974. Pentacyclic triterpanes of the hopane type as ubiquitous geochemical markers—origin and significance. Pages 245–260 *in* B. Tissot and F. Brenner, editors. Advances in organic geochemistry 1973. Editions Technip, Paris.

Hostettler, F. D., and K. A. Kvenvolden. 1994. Geochemical changes in crude oil spilled from the *Exxon Valdez* supertanker into Prince William Sound, Alaska. Organic Geochemistry 21:927–936.

Jones, D. L., N. J. Silberling, P. J. Coney, and G. Plafker. 1986. Lithotectonic terrane map of Alaska (west of 141st meridian). U.S. Geological Survey Miscellaneous Field Study Map, MF-1874.

Klein, L. H. 1983. Provenances, depositional rates, and heavy metal chemistry of sediments, Prince William Sound, southcentral Alaska. Master's thesis. University of Alaska, Fairbanks.

Kvenvolden, K. A., J. B. Rapp, and J. H. Bourell. 1985. Comparison of molecular markers in crude oils and rocks from the North Slope of Alaska. Pages 593–617 *in* L. B. Magoon and G. E. Claypool, editors. Alaska north slope oil-rock correlation study. American Association of Petroleum Geologists Studies in Geology 20.

Kvenvolden, K. A., J. B. Rapp, and F. D. Hostettler. 1991. Tracking hydrocarbons from the *Exxon Valdez* oil spill in beach, shallow-water, and deepwater sediment of Prince William Sound, Alaska. Pages 69–98 *in* P. R. Carlson, editor. Sediment of Prince William Sound, beach to deep fjord floor, a year after the *Exxon Valdez* oil spill. U.S. Geological Survey Open-File Report 91-631.

Kvenvolden, K. A., F. D. Hostettler, J. B. Rapp, and P. R. Carlson. 1993a. Hydrocarbons in oil residues on beaches of islands of Prince William Sound, Alaska. Marine Pollution Bulletin 26:24–29.

Kvenvolden, K. A., P. R. Carlson, C. N. Threlkeld, and A. Warden. 1993b. Possible connection between two Alaskan catastrophes occurring 25 years apart (1964 and 1989). Geology 21:813–816.

Lee, M. R. 1990. The *Exxon Valdez* oil spill. CRS Issue Brief, Congressional Research Service, The Library of Congress, Washington, DC.

Mackenzie, A. S. 1984. Applications of biological markers in petroleum geochemistry. Pages 115–214 *in* J. Brooks, and D. Welte, editors. Advances in petroleum geochemistry, volume 1. Academic Press, London.

Mackenzie, A. S., R. L. Patience, J. R. Maxwell, M. Vandenbroucke, and B. Durand. 1980. Molecular parameters of maturation in the Toarcian shales, Paris Basin, France—1. Changes in the configuration of acyclic isoprenoid alkanes, steranes, and triterpanes. Geochimica et Cosmochimica Acta 44:1709–1721.

Molnia, B. F., and J. R. Hein. 1982. Clay mineralogy of a glacially dominated, subarctic continental shelf: northeastern Gulf of Alaska. Journal of Sedimentary Petrology 52:515–527.

Palacas, J. G., D. E. Anders, and J. D. King. 1984. South Florida Basin—a prime example of carbonate source rocks of petroleum. Pages 71–96 *in* J. G. Palacas, editor. Petroleum geochemistry and source rock potential of carbonate rocks. American Association of Petroleum Geologists Studies in Geology 18.

Plafker, G. 1969. Tectonics of the March 27, 1964 Alaska earthquake. U.S. Geological Survey Professional Paper 543-I.

Rapp, J. B., F. D. Hostettler, and K. A. Kvenvolden. 1990.

Comparison of *Exxon Valdez* oil with extractable material from deepwater bottom sediment in Prince William Sound and the Gulf of Alaska. Pages 1–26 *in* P. R. Carlson, and E. Reimnitz, editors. Bottom sediment along oil spill trajectory in Prince William Sound and along Kenai Peninsula, Alaska. U.S. Geological Survey Open-File Report 90-39B.

Royer, T. C., J. A. Vermersch, T. J. Weingartner, H. J. Niebauer, and R. D. Muench. 1990. Ocean circulation influencing the *Exxon Valdez* oil spill. Oceanography 3:3–10.

Volkman, J. K. 1984. Biodegradation of aromatic hydrocarbons in crude oils from the barrow sub-basin of western Australia. Organic Geochemistry 6:619–632.

von Huene, R., G. G. Shor, Jr., and E. Reimnitz. 1967. Geological interpretation of seismic profiles in Prince William Sound, Alaska. Geological Society of America Bulletin 78:259–268.

Winkler, G. R. 1976. Deep-sea fan deposition of the lower Tertiary Orca group, eastern Prince William Sound, Alaska. Pages R1–R20 *in* T. P. Miller, editor. Recent and ancient sedimentary environments in Alaska. Alaska Geological Society, Anchorage, Alaska.

American Fisheries Society Symposium 18:121–139, 1996

Toxicity of Intertidal and Subtidal Sediments Contaminated by the *Exxon Valdez* Oil Spill

DOUGLAS A. WOLFE[1]

National Oceanic and Atmospheric Administration, Ocean Resources Conservation and Assessment Coastal Monitoring and Bioeffects Assessment Division, Silver Spring, Maryland, 20910, USA

MARGARET M. KRAHN, ED CASILLAS, AND SEAN SOL

National Oceanic and Atmospheric Administration, National Marine Fisheries Service Northwest Fisheries Science Center, Seattle, Washington, 98112, USA

TIMOTHY A. THOMPSON AND JOHN LUNZ

Science Applications International Corp., Bothell, Washington, 98011, USA

K. JOHN SCOTT

Science Applications International Corp., Narragansett, Rhode Island, 02882, USA

Abstract.—To estimate the toxicity potential of sediments oiled by the *Exxon Valdez* spill, standardized toxicity tests were applied to intertidal (0 m) and subtidal (three to five depths ranging from 3 to 100 m) sediment samples from Prince William Sound and the nearby Gulf of Alaska during 1989–1991. In 1989, toxicity measured by Microtox® in shallow subtidal (0–6 m) sediments showed a generally decreasing trend with increasing distance from the spill center. At the Prince William Sound sites, however, where the ranges of total hydrocarbon concentrations were greatest (at 0 and 3 m) and where those concentrations were related most certainly to *Exxon Valdez* oil, the Microtox® test did not show clear dose-response relations to ultraviolet fluorescence signal. Laboratory tests confirmed the non-responsiveness of the Microtox® test to *Exxon Valdez* oil, and the test was not used further. In 1990 and 1991, sediment toxicities at oiled and non-oiled sites were compared using a sediment elutriate test with larval Pacific oysters *Crassostrea gigas* and a whole sediment test with amphipods *Ampelisca abdita*. The 1990 amphipod results indicated that (1) mortality was correlated with hydrocarbon concentrations in intertidal sediments, (2) intertidal toxicity was substantially higher than subtidal toxicity, and (3) mean toxicity of intertidal sediments at exposed sites was significantly higher than at reference sites. Mean mortalities in subtidal sediments (at 6, 20, and 40 m) were not significantly different between exposed and reference sites, although concentrations of total hydrocarbons were generally higher at the exposed sites down to 20 m. In 1991, toxicity was generally low at all depths at all sites. Although some 1991 sediment samples (from 0, 6, and 20 m) caused mortalities of test organisms that were significantly greater than in controls, no difference in mean toxicity was demonstrated between oiled and reference sites with amphipods or oyster larvae. Between 1990 and 1991, the relative toxicity of sediments from exposed sites (compared to reference sites) decreased for intertidal (0 m) sediments and may have increased slightly for shallow subtidal (6 m) sediments. These results suggest that residual toxicity was still present during summer 1990 in intertidal sediments from some sites that were heavily oiled in 1989. By 1991, the toxicity of these sediments had diminished to background levels. These studies demonstrated no significant toxicity to test amphipods or oyster larvae that was clearly associated with *Exxon Valdez* oil in subtidal sediments (6–40 m) during 1990–1991.

The 24 March 1989 grounding of the *Exxon Valdez* on Bligh Reef in Prince William Sound released approximately 43.9 million liters of Prudhoe Bay crude oil into the Alaskan marine environment (Kelso and Kendziorek 1991; Maki 1991). For several weeks after the spill much of this oil was distributed onto about 575 km of shoreline in Prince William Sound and about 1,170 km in the Gulf of Alaska (Maki 1991). As a combined result of cleanup activities and storm-related wave action, this beached oil was subjected to removal and rigorous physical dispersion throughout the subsequent summer and winter months, and a fraction of it migrated into adjacent shallow subtidal zones (Wolfe et al. 1994; O'Clair et al. 1996, this volume). As part of the State-Federal Natural Resources Damage Assessment, a coordinated effort was designed to assess the distribution, persistence, and potential effects of petroleum in the lower intertidal and associated subtidal areas affected by the *Exxon Valdez* oil spill. In this article we report the results of standard toxicity test protocols that were applied to assess the toxicity of sediments from the lower

[1]Present address: 109 Shore Drive, Beaufort, North Carolina, 28516, USA

intertidal and shallow (from 0 m to 100 m) subtidal zones during the first 3 years after the spill. Results from studies of replicate sediment samples taken at the same times and stations described here appear elsewhere for detailed hydrocarbon compositions and concentrations (O'Clair et al. 1996) and microbial hydrocarbon degradation potentials (SAIC 1990a; Braddock et al. 1996, this volume).

A considerable body of literature on the toxicity of Alaskan crude oil to arctic and subarctic marine organisms existed before the spill. Much of the early work in this area focused on the acute toxicities (generally 96-h exposures) of water-soluble fractions (WSF) of fresh Cook Inlet crude oil and Prudhoe Bay crude oil to a variety of species and life stages of commercially or recreationally important Alaskan marine organisms. Data on the acute toxicities of crude oil to marine organisms of interest have been summarized by Brodersen et al. (1977), Craddock (1977), Moles et al. (1979), Rice et al. (1976, 1977, 1979, 1984), and the National Research Council (NRC 1985). In general, LC50s for the WSF of Prudhoe Bay crude oil lie in the range of 0.1 to 2 mg/L for most organisms tested. Rice et al. (1981) demonstrated that the compositions of the WSF of Cook Inlet and Prudhoe Bay crude oils were very similar both to one another and to that of the discharge from the ballast treatment facility at Valdez.

Sublethal effects of oil exposure have also been studied extensively, through the use of long-term exposures (e.g., up to 40 d) to WSF of Alaskan crude oil or of prolonged exposure to oiled food or oiled sediments. Earlier work (which focused primarily on temperate organisms and crude oils from sources other than Alaska) was summarized by Anderson (1977), Johnson (1977), and Patten (1977). During the late 1970s and early 1980s, increased attention was given to arctic and subarctic organisms, especially relative to Alaskan and Canadian oils, and some of this work has been reviewed by Rice et al. (1984), Rice (1985), Wolfe (1985), the NRC (1985), and Karinen (1988). Although it has been well documented that spilled petroleum constituents may persist for years in muddy sediments and continue to affect resident organisms (Mayo et al. 1978; Teal et al. 1978; Dauvin and Gentil 1990), relatively little work has been performed to characterize the threshold conditions for biological responses and toxicity caused by petroleum associated with fine-grained sediments.

In an effort to quantify the potential effects of the spilled *Exxon Valdez* oil in the sediment environment, we applied recently developed standard sediment toxicity test protocols to intertidal and subtidal sediment samples from oiled and non-oiled sites in Prince William Sound and the adjoining Gulf of Alaska. In this work we were concerned primarily with (1) comparing sediment toxicity in relation to petroleum levels at oiled and non-oiled sites, (2) the potential transport of toxic petroleum constituents from the intertidal zone into adjacent subtidal sediments, and (3) the changes in toxicity with time.

Methods

Sampling Locations and Sediment Sampling

Sediment sampling was carried out from the National Oceanic and Atmospheric Administration (NOAA) ship *Fairweather* between 29 June and 22 August 1989 (Table 1), from the NOAA ship *Davidson* between 25 June and 5 August 1990 (Table 2), and from the charter vessel *The Big Valley* between 15 and 25 June 1991 (Table 2). All samples tested were composites of approximately equal amounts of surficial sediment, either from each of eight randomly located positions along a 30-m transect at 0-m, 3-m, 6-m, and 20-m depths or from each of three Van Veen grabs from 40-m and 100-m depths.

Surficial sediments (0–1 cm) were collected for hydrocarbon analysis by ultraviolet fluorescence (UVF) and for toxicity testing by Microtox® (1989 only) with a precleaned, noncontaminating spatula into precleaned 20-mL glass containers with teflon-lined caps. Two liters of surficial (0–2 cm) sediment were collected at each depth and site in 1990 and 1991 for toxicity testing with Pacific oyster *Crassostrea gigas* larvae and the amphipod *Ampelisca abdita*. These latter samples were kept cold (1–4°C) for up to 11 d before initiation of toxicity testing in Seattle, Washington. In all years, separate replicate composite samples (i.e., from the same randomly located positions or grabs, but composited and homogenized separately) were taken concurrently from the same stations for chemical analysis by gas chromatography–mass spectrometry (GC–MS) and sediment grain-size analyses, and results are discussed elsewhere for many of those samples (O'Clair et al. 1996). The 1989 and 1990 sites, selected on the basis of preliminary information on distribution of floating and/or beached oil, encompassed the full geographic range of significant spill exposure (i.e., from Bligh Island in Prince William Sound to Katmai Bay on the Alaska Peninsula), whereas the 1991 sampling for toxicity testing was restricted to 8 "exposed" sites and 7 "reference"

TABLE 1.—Locations and dates for sites within the *Exxon Valdez* oil spill zone that were sampled in 1989 for toxicity testing with the Microtox® protocol.

Site number	Site name	Sampling date	North latitude	West longitude
00	Kodiak Harbor	29 June	57°47.94	152°22.20
01	Fox Farm	1 July	59°58.43	148°10.50
02	Sawmill Bay area	2 July	60°00.16	147°58.90
03	Shelter Bay	3 July	60°06.52	147°58.90
04	Iktua Bay	4 July	60°06.00	147°59.70
05	Mummy Island	5 July	60°17.26	147°54.38
06	Snug Harbor	6 July	60°14.38	147°43.11
07	Green Island	7 July	60°16.30	147°26.30
08	Bay of Isles	8 July	60°23.00	147°44.90
09	Smith Island	9 July	60°31.79	147°20.75
10	Cabin Bay	10 July	60°39.35	147°26.30
11	Columbia Bay	11 July	60°39.00	147°01.40
12	Northwest Bay	12 July	60°33.05	147°34.70
13	Disk Island	13 July	60°29.90	147°39.50
14	Herring Bay	14 July	60°25.90	147°47.20
15	Eshamy Bay	15 July	60°26.82	147°58.50
16	Sleepy Bay	16 July	60°04.14	147°50.58
17	Rocky Bay	17 July	60°20.20	147°08.15
18	Olsen Bay	18 July	60°45.13	146°11.50
19	Knowles Bay	19 July	60°40.68	146°37.70
20	Fox Island	25 July	59°56.20	149°19.00
21	Agnes Cove	26 July	59°46.00	149°34.40
22	Taroka Arm	27 July	59°37.54	150°08.30
23	Black Bay	28 July	59°32.12	150°12.28
24	McArthur Cove	29 July	59°26.60	150°20.50
25	Tonsina Bay	30 July	59°18.70	150°55.00
26	Gore Point	31 July	59°14.23	150°58.79
27	Port Dick	1 August	59°17.25	151°08.75
28	Windy Bay	2 August	59°13.84	151°31.00
29	Chugach Bay	3 August	59°11.20	151°37.80
30	Seldovia Bay	4 August	59°25.85	151°44.30
31	Ursus Cove	5 August	59°30.80	153°45.40
32	Amakdedori Beach	6 August	59°16.50	154°07.80
33	Douglas Beach	7 August	59°00.00	153°29.50
34	Ushagat Island	8 August	58°56.97	152°17.61
35	Andreon Bay	9 August	58°30.20	152°25.10
36	King Cove	14 August	58°11.03	152°03.30
37	Cape Douglas	15 August	58°50.54	153°21.00
38	Hallo Bay	16 August	58°27.48	154°00.23
39	Katmai Bay	17 August	57°58.50	155°01.80
40	Halibut Bay	18 August	57°21.47	154°45.10
41	Wide Bay	19 August	57°26.42	156°15.12
42	Chignik Bay	20 August	56°19.68	158°25.40
43	Ivanof Bay	21 August	55°52.12	159°31.01
44	Zachary Bay	22 August	55°19.55	160°36.50

sites within Prince William Sound. The positions shown in Tables 1 and 2 are for the intertidal (mean low water, 0-m depth) sampling locations at each site; subtidal samples were taken at 3-m, 6-m, 20-m, 40-m, and 100-m depths located as close as possible to the intertidal locations. In 1989, samples from all depths were tested for toxicity, whereas in 1990 and 1991, only four depths (0 m, 6 m, 20 m, and 40 m) were tested.

Hydrocarbon Analyses

Ultraviolet fluorescence spectrophotometry.—Sediments were screened for total petroleum hydrocarbons, using different UVF analytical protocols in 1989 and 1990. In 1989, samples were extracted and analyzed onboard ship (SAIC 1989). A weighed portion (1–5 g) of sediment, mixed with an equal weight of anhydrous sodium sulfate, was sonicated for 30 s with two successive 5-mL volumes of high-performance liquid chromatography (HPLC)-grade methylene chloride. The extracts were decanted through a filter containing about 2 g anhydrous sodium sulfate and collected in a clean 25-mL graduated cylinder. Fluorescence was determined on diluted aliquots of these extracts, using excitation/emission wavelengths of 310/370 nm. Petroleum hydrocarbon concentration was estimated by comparison to a standard reference solution of *Exxon Valdez* oil in methylene chloride.

In 1990, sediment samples were frozen (−80°C) onboard ship and kept frozen till analyzed at the Northwest Fisheries Science Center in Seattle, Washington (Krahn et al. 1991, 1993). Sediment (1.0 g) was mixed with sodium sulfate (10 g), activated copper (1 mL), and methylene chloride (20 mL) in a centrifuge tube. The tubes were placed in a sonic bath for 15 min and then centrifuged at 1,500 revolutions per minute (rpm) for 5 min. The extracts were decanted into 50-mL concentrator tubes. Extraction was repeated twice more, using 10 mL of methylene chloride and 5-min sonication periods. The three extracts were combined, a polystyrene internal standard (100 μL; 50μg/μL) was added, and the solution was concentrated by evaporation to about 4 mL. An aliquot (15 μL) was injected onto an HPLC size-exclusion column and eluted with methylene chloride with a flow of 2.5 mL/min. Fluorescence was monitored at two different wavelength pairs, 260/380 nm and 290/335 nm (excitation/emission), optimized for phenanthrene and naphthalenes plus dibenzothiophenes, respectively. Oil concentrations were estimated by comparison of chromatographic areas with those obtained with a standard reference solution of *Exxon Valdez* oil.

Gas chromatography–mass spectrometry analyses.—Selected sediment samples from all 3 years were analyzed by GC–MS to provide confirmation and comparison for the UVF screening results. Methods used for sediment extraction, alumina–silica gel chromatographic cleanup of the extracts, and analysis by GC–MS have been described previously (Brooks et al. 1990; Krahn et al. 1993).

Toxicity Bioassays

Sediment samples were tested for toxicity using one or more of three standard bioassay protocols:

TABLE 2.—Names, locations, and oiling designations (R = reference; E = exposed) of sites sampled for toxicity testing and chemical analysis of sediments in 1990 and/or 1991. Sites lacking a 1991 site number were not sampled in that year.

1990 site number	1991 site number	Site name	Nominal oiling designation	Latitude	Longitude
01		Olsen Bay	R	60°44.80 N	146°13.10 W
02		Port Fidalgo	R	60°50.20 N	146°12.58 W
03		Smith Island	E	60°31.80 N	147°20.80 W
04	15	Zaikof Bay	R	60°16.85 N	147°02.10 W
05	14	Rocky Bay	R	60°20.30 N	147°08.20 W
06	01	West Bay	R	60°51.80 N	146°46.50 W
07	05	Herring Bay	E	60°26.54 N	147°47.13 W
08	03	Disk Island	E	60°29.80 N	147°39.70 W
09	04	Block Island	E	60°31.70 N	147°36.40 W
10	02	Northwest Bay	E	60°33.10 N	147°34.60 W
11		NE Knight Island	E	60°26.35 N	147°37.65 W
12	13	Bay of Isles	E	60°22.80 N	147°45.40 W
13		Green Island	E	60°16.20 N	147°26.10 W
14	10	MacLeod Harbor	R	59°53.21 N	147°45.80 W
15	11	Mooselips Bay	R	60°12.55 N	147°18.00 W
16	12	Snug Harbor	E	60°14.25 N	147°44.10 W
17	07	Chenega	E	60°19.85 N	148°00.45 W
18	06	Lower Herring Bay	R	60°24.40 N	147°47.80 W
19	08	Drier Bay	R	60°19.20 N	147°44.00 W
20	09	Sleepy Bay	E	60°03.95 N	147°50.35 W
21		Fox Farm	E	59°58.40 N	148°10.65 W
22		Sunny Cove	R	59°56.20 N	149°19.10 W
23		Agnes Cove	E	59°46.05 N	149°34.55 W
24		Black Bay	R	59°32.61 N	150°12.78 W
25		Chugach Bay	E	59°11.16 N	151°37.90 W
26		Tonsina Cove	E	59°19.75 N	150°54.90 W
27		Windy Bay	E	59°13.85 N	151°31.00 W
28		Hallo Bay	E	58°27.45 N	154°00.30 W
29		Katmai Bay	E	57°54.50 N	155°04.50 W

(1) a Microtox® assay with organic extracts of sediments, based on a 15-min inhibition of phosphorescence by *Photobacterium phosphoreum* (Schiewe et al. 1985; PSEP 1986); (2) a 10-d (static) exposure of adult amphipods *Ampelisca abdita* to whole sediments, with survival as the endpoint (ASTM 1990a); and (3) a 48-h (static) exposure of Pacific oyster larvae to sediment elutriates, with survival and development as endpoints (ASTM 1990b).

Microtox® assay.—At least one of the replicate composite sediment samples obtained at each depth from each sampling site in 1989 was analyzed for sediment toxicity based on the inhibition of bioluminescence in *P. phosphoreum* (15-min Microtox® assay). Organic extracts of the sediments were prepared and assayed for toxicity by the methods of Schiewe et al. (1985). The Microtox® assay is rapid, simple, and inexpensive; the bioassay results generally have correlated well in other studies with the results of other standard bioassays that use fish, amphipods, or bivalve larvae as test organisms (Chang et al. 1981; Williams et al. 1986; Giesy et al. 1988).

After the 1989 field survey, a laboratory study was performed to validate the ability of the Microtox® assay to measure oil toxicity. Naturally weathered *Exxon Valdez* oil was collected from Prince William Sound 11 d after the spill and stored at 4°C. A sediment sample from Zachary Bay, Alaska, with undetectable concentrations of aromatic hydrocarbons (Krahn et al. 1991), was spiked with the weathered *Exxon Valdez* oil and tested for toxicity with the Microtox® assay. Portions (100 g) of sediment were amended with 0, 0.5, 5, 25, 100, or 500 mg of the weathered oil. The samples were mixed for 24 h on a modified rock tumbler and allowed to equilibrate for 72 h (all at 4°C). Samples were then extracted and tested using the Microtox® procedure described above. The resultant concentrations of oil in sediment (0–5,000 parts per million [ppm]) simulated the range of oil concentrations in sediment observed in 1989.

Toxicity tests with amphipods.—The toxicity of test sediments to amphipods *Ampelisca abdita* was tested (SAIC 1990b, 1991) following the general procedures of Swartz et al. (1985) and Scott and Redmond (1990). Test organisms were collected in San Francisco Bay and shipped to Seattle by overnight courier. Five replicate 1-L beakers, each containing 175 mL of sediment and (approximately) 825 mL of filtered seawater from Alkai Point (Puget Sound, Washington) were set up for each sediment sample. A sixth replicate container was used for daily measurement of dissolved oxygen, pH, salinity, and temperature. Clean control sediments were collected with the test organisms in San Francisco Bay, and (in 1991) reference sediments also were collected from two sites in Puget Sound (West Beach on Whidbey Island, and Samish Bay). Replicate control sediments were tested with each group of test sediments. Twenty amphipods were added to each container, and the containers were maintained at 20°C for 10 d with gentle aeration. At the end of the test period, live amphipods were enumerated, and the mean mortality of test organisms (corrected for control mortality) was determined for each sample.

Toxicity tests with Pacific oyster larvae.—Survival of Pacific oyster larvae was tested in 48-h exposures to elutriates of test sediments (SAIC 1990b, 1991), following procedures described previously (Chapman and Morgan 1983; Chapman and Becker 1986). Sediment elutriates were prepared by vigorous mixing for 30 min of 200 g (wet weight) of test or control sediment into 1 L of filtered, sterilized seawater obtained in Puget Sound from Alkai Point (1990) or Duwamish Head (1991). After the suspensions settled for 1 h, the supernatant elutriate was decanted, and the settled solids were discarded. Five test replicates, each containing approximately 980 mL of elutriate, were inoculated with 20,000–40,000 developing oyster embryos within 2 h of fertilization for each sediment sample. After 48 h, three 10-mL aliquots from each test container were preserved in 5% buffered formalin. Normal and abnormal prodissoconch I larvae were counted, and the percent survival (normal survivors compared to initial density) and percent abnormality (abnormal survivors compared to total survivors) were calculated. Abnormality data were adjusted for the rates measured in control larvae that developed in control seawater.

TABLE 3.—Total petroleum hydrocarbon concentrations (mean ± SD, in parts per million) estimated by ultraviolet fluorescence in intertidal and subtidal Alaskan sediments from three regions within the *Exxon Valdez* oil spill zone in 1989.

Depth (m)	Mean	SD	Range	*N*
Prince William Sound (18 sites)				
0	841.30	2414.76	0.59–9,800	17
3	22.70	46.07	1.4–190	18
6	8.83	12.21	1.5–50	18
20	11.70	19.79	1.4–87	18
40	6.36	3.54	1.1–12	18
100	5.85	3.54	0.16–11	17
Kenai Peninsula (14 sites)				
0	7.66	14.78	0.44–55	14
3	3.85	4.63	0.55–17	14
6	4.28	6.09	0.7–23	14
20	11.30	12.64	0.37–46	14
40	12.53	23.16	0.33–77	13
100	6.41	5.62	0.76–21	13
Kodiak and Shelikof Strait area (11 sites)				
0	7.25	13.26	0.40–37	11
3	1.11	0.72	0.34–2.6	11
6	2.83	4.28	0.4–15	11
20	4.77	6.97	0.34–19	11
40	4.65	7.41	0.26–26	11
100	4.49	6.86	0.94–20	7

TABLE 4.—Microtox® response (EC50; mean ± SD, in mg sediment/mL) by depth in intertidal and subtidal Alaskan sediments from three regions within the *Exxon Valdez* oil spill zone in 1989.

Depth (m)	Mean	SD	Range	*N*
Prince William Sound (18 sites)				
0	12.48	11.39	1.4–38	13[a]
3	5.44	6.22	1.0–29	18
6	5.78	7.28	0.53–28	18
20	9.76	11.34	0.53–47	18
40	18.62	15.68	1.2–64	18
100	20.86	19.66	0.81–75	17
Kenai Peninsula (14 sites)				
0	24.64	11.39	13–49	11[a]
3	15.19	10.98	1.8–38	14
6	14.32	8.93	2.3–29	14
20	11.90	8.92	1.5–26	14
40	16.58	9.48	0.72–29	13
100	17.27	10.89	1.2–34	12
Kodiak and Shelikof Strait area (11 sites)				
0	18.41	10.23	5.0–35	11
3	19.45	12.84	2.8–45	11
6	16.23	8.67	4.7–31	11
20	18.68	10.33	3.9–38	11
40	20.79	13.96	2.9–50	10
100	16.30	11.41	3.2–31	7

[a]In four Prince William Sound samples, precipitation of oil precluded valid determination of EC50s. In four other samples, toxicity was too low to estimate EC50s at the dilutions used. See text.

Results

Microtox® Assays, 1989

Tables 3 and 4 present the means and standard deviations for petroleum hydrocarbon concentrations and Microtox® EC50s, respectively, from the 1989 surveys, by sampling region and depth. These means are depicted and compared graphically in Figure 1. At shallow depths (0–6 m), sediments from sites in Prince William Sound were more toxic on average and had higher hydrocarbon concentrations than sediments from outside the sound. Shallow sediments from sites along the Kenai Peninsula

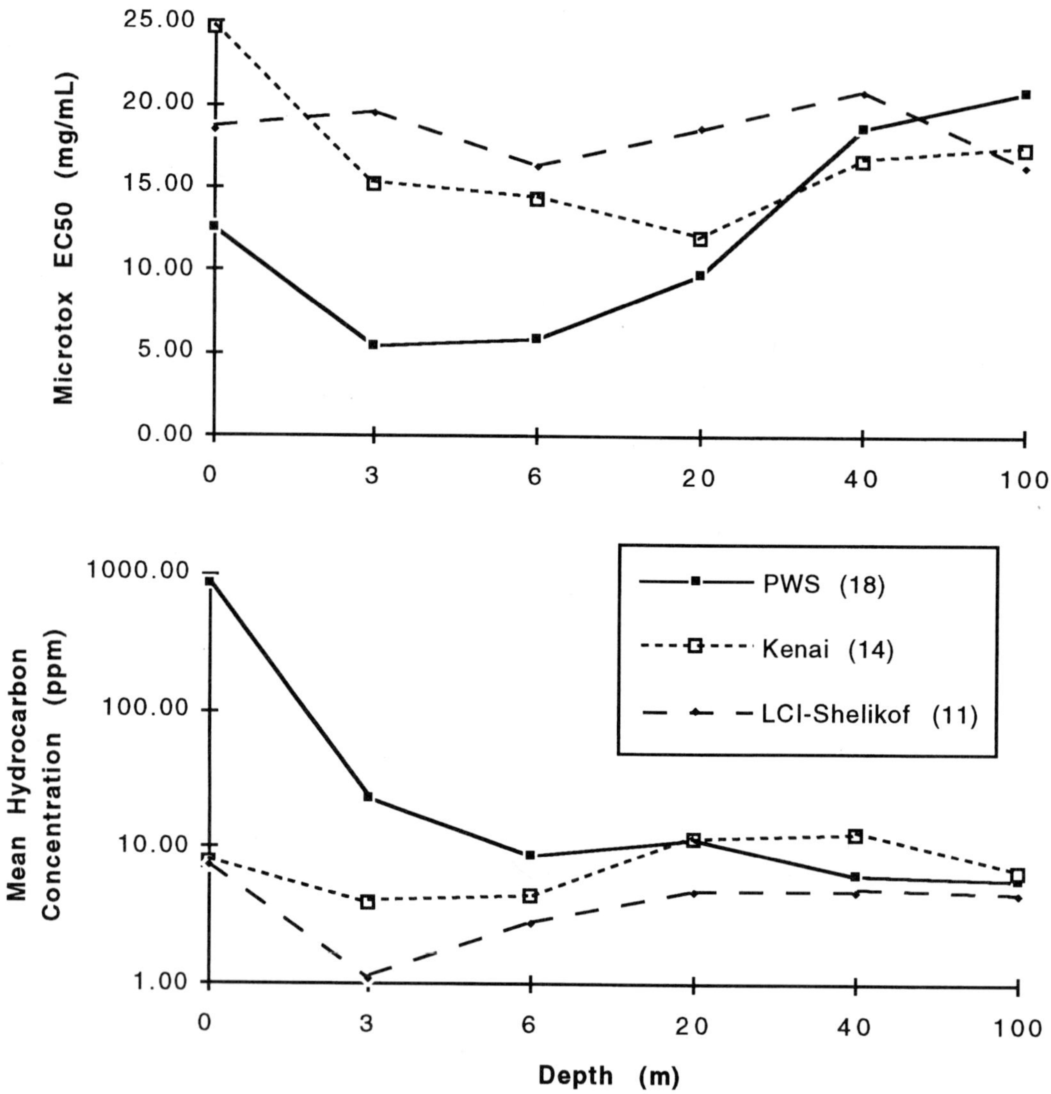

FIGURE 1.—Mean Microtox® EC50 (top) and mean hydrocarbon concentration (bottom) estimated by ultraviolet fluorescence for sediments from different regions and depths in 1989. The number of sites sampled in each region is in parentheses. See Tables 3 and 4 for estimates of variance. (PWS = Prince William Sound; LCI = lower Cook Inlet; ppm = parts per million)

exhibited toxicities and hydrocarbon concentrations very similar to those from Kodiak and the Alaska Peninsula (Figure 1). At greater depths (40–100 m), Prince William Sound sites were not notably different from sites outside the sound, with respect either to Microtox® response or to hydrocarbon concentration. At all depths except 40 m and 100 m, the UVF data showed a generally decreasing trend in hydrocarbon concentration with increasing distance from the spill site. This pattern was strongest in data from the 3- and 6-m depths. The Microtox® EC50s of sediment extracts from 3- to 20-m depths usually increased (i.e., were less toxic) as distance from the spill site increased.

At the Prince William Sound sites alone, where the total hydrocarbon concentrations exhibited the greatest range of values, hydrocarbon concentrations for the 0-m samples ranged over five orders of magnitude. It was not possible to calculate the EC50s for the four most heavily oiled samples, from

TABLE 5.—Results of Microtox® toxicity measurements on sediments amended with increasing amounts of Prudhoe Bay crude oil collected from Prince William Sound 11 d after the *Exxon Valdez* spill. Results are reported as the amount of sediment equivalents in the reaction mixture that would decrease the light output by 50% relative to a control. A lower EC50 value indicates greater toxicity to the photobacteria. Values in parentheses represent the lower and upper 95% confidence intervals for the EC50 value.

Oil concentration (μg oil/g sediment)	EC50 (mg sediment equivalents)
5,000	4.6 (3.9–5.5)
1,000	3.7 (3.2–4.4)
250	2.7 (2.2–3.4)
50	2.0 (1.5–2.6)
5	2.9 (2.5–3.3)
0	1.7 (1.5–2.4)

Disk Island, Smith Island, Northwest Bay, and Herring Bay; in the sediment sample from Rocky Bay toxicity was too low to estimate a valid EC50 at the standard test dilution. Without those samples, there was no significant rank correlation between toxicity and hydrocarbon concentration, and the most toxic intertidal samples from within Prince William Sound were from Olsen Bay and Columbia Bay, areas that were not affected by the spill.

The UVF response in the 3-m samples from Prince William Sound ranged over three orders of magnitude. At this depth, there was a significant inverse rank correlation (Spearman, $P = 0.046$) between UVF and EC50. However, the toxicity was not dose-dependent, and there were curious outliers: the most toxic samples were from Olsen Bay, with Rocky Bay (only slightly oiled by the spill) a close second. Heavily oiled Sleepy Bay, by contrast, was the least toxic site.

In the 6-m samples from Prince William Sound, UVF ranged over less than two orders of magnitude, and the most toxic samples were from two heavily oiled sites: Northwest Bay and Bay of Isles. Other heavily oiled sites (e.g., Snug Harbor, Smith Island, Herring Bay), however, exhibited toxicities similar to that at Olsen Bay. At 40 m, the UVF spanned just one order of magnitude, and there was no clear relationship between toxicity and hydrocarbon concentration.

The laboratory tests of the Microtox® procedure with oil-amended sediments showed no relationship between EC50 and the concentration of *Exxon Valdez* oil in the sediments over the four order-of-magnitude range in concentration (Table 5). This unexpected finding caused us to abandon the Microtox® assay in the 1990 and 1991 surveys of sediment toxicity.

Patterns of Hydrocarbon Distribution with Depth, 1989–1990

Both hydrocarbon concentrations and Microtox® EC50s varied considerably with depth and exhibited quite different patterns with depth at different stations. Three qualitative distributional patterns of UVF signal with depth were tentatively identified from analyses in both 1989 and 1990 (Figure 2): (1) moderate to very high intertidal values, usually associated with intertidal oiling, with decreasing values at greater depth; (2) a usually variable maximum value at 20–40 m, with lower values both intertidally and at greater depth; and (3) variable low values in the intertidal and shallow subtidal samples, usually increasing to a maximum at 40–100 m. These three qualitative patterns are illustrated (using 1990 data) in Figure 2, and stations exhibiting the different patterns are identified in Tables 6 and 7. Because the 1989 UVF and 1990 HPLC–UVF methods were different, the 1989 and 1990 values are not directly comparable, but maximum values for each site pattern are given in Tables 6 and 7 to permit comparison of relative concentrations among sites in either year.

Stations exhibiting pattern A consistently represented stations that showed strong evidence of significant intertidal oiling and that may have received either coincident or consequent subtidal exposure. Most of these sites were in Prince William Sound (Table 6), but a few were in the Kenai or Alaska Peninsula areas (Table 7). All of the sites exhibiting pattern A in 1990 also exhibited the HPLC elution pattern characteristic of Prudhoe Bay crude oil (Krahn et al. 1993). The Gulf of Alaska sites that exhibited pattern A in 1989 showed shifts either to pattern B or C in 1990 (Table 7). Sites exhibiting pattern B generally showed near minimal UVF values at 0 m (but samples from Green Island, Sleepy Bay, and Chugach Bay exhibited unusually high variability in intertidal replicates) and, with two exceptions (MacLeod Harbor in 1990 and Agnes Cove in 1989 and 1990), showed indications of Prudhoe Bay oil in subtidal samples. With the exception of MacLeod Harbor (1990, pattern B), all sites designated as non-oiled (or only slightly oiled) reference sites consistently exhibited pattern C and did not show the characteristic Prudhoe Bay crude oil pattern on HPLC elution (Table 6).

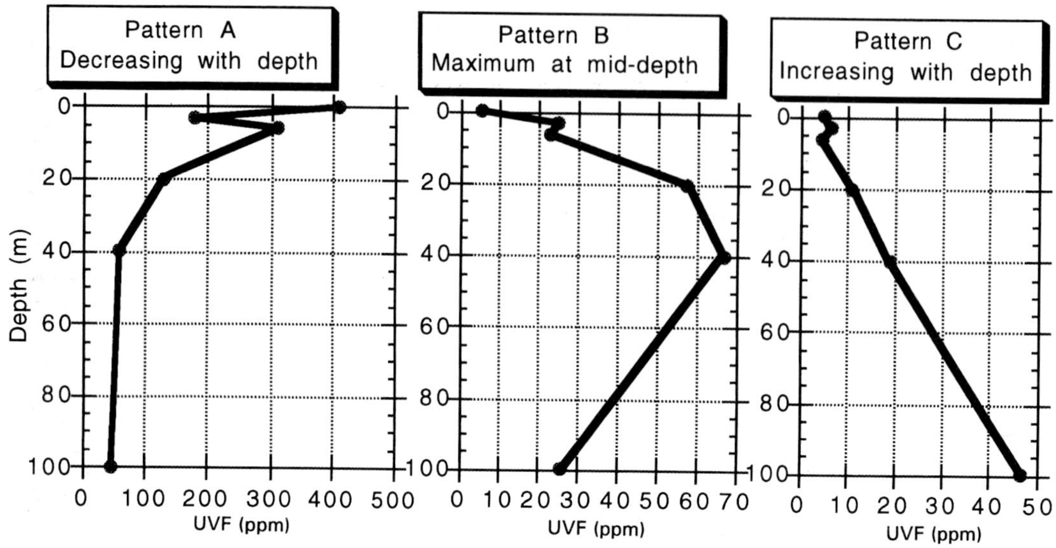

FIGURE 2.—Typical patterns of distribution with depth of hydrocarbons measured by ultraviolet fluorescence spectrometry (UVF) in 1990 at Northwest Bay (Pattern A), MacLeod Harbor (Pattern B), and Drier Bay (Pattern C). (ppm = parts per million)

TABLE 6.—Patterns of hydrocarbon distribution (see Figure 2) at sites in Prince William Sound, 1989–1990. Values are in parts per million for the depth with the greatest concentration at each site. Ultraviolet fluorescence analyses for 1989 were done by Science Applications International Corporation (1989) and for 1990 by Krahn et al. (1993).

	Pattern A		Pattern B		Pattern C	
Site name	1989	1990	1989	1990	1989	1990
Herring Bay[a,b]	290	970[c]				
Northwest Bay[a,b]	1,590	410[c]				
Snug Harbor[a,b]	27.4	100[c]				
Disk Island[a,b]	9,840	53[c]				
NE Knight Island[a]	NS	51[c]				
Chenega[a,b]	NS	41[c]				
Fox Farm[a]	51.0	29[c]				
Bay of Isles[a,b]	35.3	26[c]				
Smith Island[a]	2,420	23[c]				
Block Island[a,b]			NS	210[c]		
Sleepy Bay[a,b]			86.7	76.7[c]		
MacLeod Harbor[e,b]			NS	67		
Green Island[a]			21.1			79[c]
Olsen Bay[e]					4.99	54
Drier Bay[e,b]					NS	46.7
Zaikof Bay[e,b]					NS	33
Rocky Bay[e,b]					8.8	33
West Bay[e,b]					NS	30
Mooselips Bay[e,b]					NS	22
Lower Herring Bay[e,b]					NS	21
Port Fidalgo[e]					NS	20

[a]Exposed site.
[b]Site was also sampled in 1991.
[c]Weathered Prudhoe Bay crude oil pattern on high-performance liquid chromatography elution (Krahn et al. 1993).
[d]NS = no sample.
[e]Reference site.

TABLE 7.—Patterns of hydrocarbon distribution (see Figure 2) at sites in the Gulf of Alaska, 1989–1990. Values are in parts per million for the depth with the greatest concentration at each site. Ultraviolet fluorescence analyses for 1989 were done by Science Applications International Corporation (1989) and for 1990 by Krahn et al. (1993).

	Pattern A		Pattern B		Pattern C	
Site name	1989	1990	1989	1990	1989	1990
Tonsina Cove	55.1			37[a]		
Windy Bay	21.7			62[a]		
Agnes Cove			48.1	600		
Sunny Cove			77.4			9.7[a]
Chugach Bay			46.1			20.0[a]
Black Bay					21.1	18.0[a]
Hallo Bay	30.5					17.0[a]
Katmai Bay	37.0					12.0[a]

[a]Weathered Prudhoe Bay crude oil pattern on high-performance liquid chromatography elution (Krahn et al. 1993).

Amphipod and Pacific Oyster Bioassays, 1990–1991

Sampling in 1990 and 1991 was focused on the comparison of sets of selected exposed and reference sites in Prince William Sound (Table 8). Four reference sites were located on Montague Island, one on Bligh Island, and two on the west side of Knight Island. Some of these Prince William Sound reference sites may have been subjected to minor

TABLE 8.—Sites used for reference-versus-exposed comparisons of toxicity in Prince William Sound, 1990–1991.

Reference sites	Exposed sites
West Bay (1990 only)	Green Island (1990 only)
	Northeast Knight
Rocky Bay	Herring Bay
Mooselips Bay	Disk Island
MacLeod Harbor	Northwest Bay
Lower Herring Bay	Bay of Isles
Drier Bay	Snug Harbor
	Sleepy Bay
	Block Island
Zaikof Bay (1991 only)	Chenega Island

oiling from the spill—either by small amounts of floating oil soon after the spill or by very low levels of dispersed oil in the water column during the 18 months after the spill (Wolfe et al. 1994). West Bay and Green Island were sampled only in 1990, Zaikof only in 1991, and the other sites were sampled in both years. Sediment samples from 0 m, 6 m, 20 m, and 40 m were assayed by the amphipod and Pacific oyster larval tests, and the results are compared against UVF data from separate composite sediment samples taken concurrently.

Percent mortalities for amphipods and for Pacific oyster larvae exposed to sediments from the 0-m (intertidal) depths are shown in Figure 3, along with corresponding UVF hydrocarbon concentrations. The six reference sites are located on the left side of the figure; all had hydrocarbon concentrations of less than 5.1 ppm (Drier Bay), whereas the exposed sites ranged from 3.5 ppm (Green Island) to 26 ppm (Bay of Isles) to 970 ppm (Herring Bay). Mortality of control amphipods was high and variable in the 1990 tests (23 ± 12.1% [SD], 17 ± 17.9 %, and 28.8 ± 14.7% for the three test batches), and only the five most toxic exposed sites were significantly (P = 0.05) different from controls. Nonetheless the mean amphipod mortality at the 10 exposed sites was significantly greater (t-test, P = 0.02) than that at

TABLE 9.—Percent mortality (mean ± SD, corrected for control mortality) of test amphipods in 1990 at reference-versus-exposed sites in Prince William Sound (t-test: N = 6 versus 10).

	Reference		Exposed		
Depth (m)	Mean	SD	Mean	SD	Significance level
0	22.7	9.1	53.5	27.2	0.019
6	22.5	12.6	19.5	9.4	0.601
20	23.6	8.9	14.5	7.7	0.061
100	13.0	7.7	12.1	4.7	0.787

the 6 reference sites (Table 9). However, the toxicity of the Herring Bay sample with the highest UVF signal (970 ppm oil) was similar (Figure 3) to that of the West Bay sample, which had the lowest UVF (at <1.0 ppm).

Control survival of Pacific oyster larvae in reference seawater was only 70% in one test batch (compared to 88.3% and 98.0% in the other two) in 1990, but coefficients of relative variation (SD) were between ± 6.1% and ± 9.6% for all three control batches. The highest mortalities for Pacific oyster larvae exposed to sediment elutriates from the 0-m depth were observed at West Bay and Rocky Bay, two reference sites with very low hydrocarbon levels, and at Block Island (Figure 3). There were no significant differences between the exposed and reference site means for Pacific oyster either at 0 m or at any other depth (Table 10). Although there was a tendency toward greater mortality with increasing depth (both for the exposed group and for all sites taken together), these differences were not significant.

Figure 4 shows the mortalities for amphipods and for Pacific oyster larvae exposed to sediments from the 20-m depths. At this depth, mortality of amphipods was uniformly low, irrespective of hydrocarbon level, which ranged up to 120 ppm at Northwest Bay. Although a number of sites exhibited moderate 20-m sediment toxicity to Pacific oyster larvae, this toxicity was independent of hydrocarbon signal. Except for amphipods at 0-m, the exposed site means were not significantly different from the reference site means for either test organism at any depth (Tables 9 and 10).

The same sediment toxicity tests were performed again in 1991, but in this year the control mortality in the amphipod test was lower and less variable than in 1990 (14.2 ± 6.6%, 20.8 ± 13.2%, and 12.5 ± 4.2% mortality in three test batches), resulting in a more valid test of relative toxicity among samples. Control mortality of Pacific oyster larvae in reference seawater was also lower than in 1990; survival exceeded 85% in both test batches, with coefficients of relative variation (SD) between ± 9.1% and ± 9.5%. Statistically significant mortalities (test versus control) were observed with one or both test organisms for many of the sediment samples (Table 11).

Figure 5 shows the 1991 control-corrected mortality data for amphipods and Pacific oyster larvae at 0-m and 6-m depths at reference and exposed sites. For some 1991 samples (Lower Herring Bay and Snug Harbor at 0 m, and MacLeod Harbor at 6 m), the quantity of available sediment was inadequate to conduct both tests, and no oyster data were

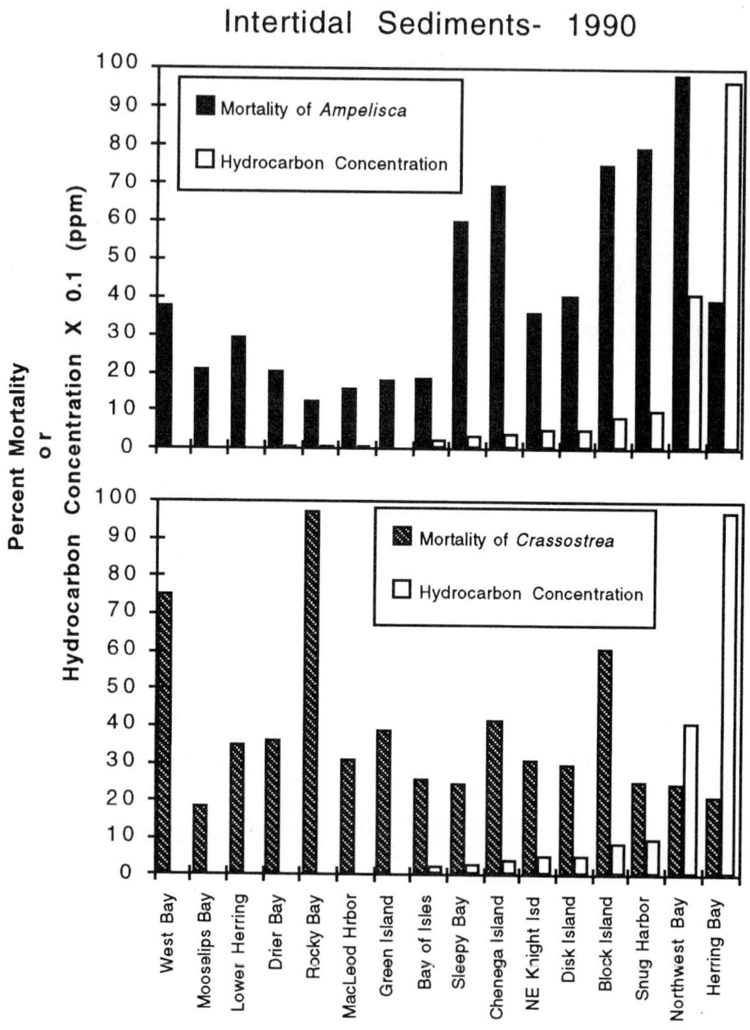

FIGURE 3.—Hydrocarbon concentrations (in parts per million [ppm] × 0.1 by ultraviolet fluorescence) and toxicity of intertidal (0 m) sediment samples collected in 1990. Toxicity is presented as percent mortality for the amphipod *Ampelisca abdita* (top) and for Pacific oyster *Crassostrea gigas* (bottom). Stations are arranged in ascending order of hydrocarbon concentration, with six reference sites (West Bay to MacLeod Harbor) at left and ten exposed sites (Green Island to Herring Bay) at right.

TABLE 10.—Percent mortality (mean ± SD, corrected for control mortality) of Pacific oyster *Crassostrea gigas* larvae in 1990 at reference-versus-exposed sites in Prince William Sound (t-test: N = 6 versus 10).

	Reference		Exposed		
Depth (m)	Mean	SD	Mean	SD	Significance level
0	48.7	30.5	32.4	12.0	0.147
6	43.3	19.7	46.2	18.5	0.770
20	36.7	8.7	48.9	17.6	0.140
100	40.2	21.4	58.0	22.8	0.171

generated. The other blank entries in Figure 5 represent 0% mortalities. For both the Pacific oyster and amphipod tests, the threshhold for statistical significance (relative to controls) in 1991 was at about 12–15 % mortality, depending the control batch. Experience with a very large number of toxicity tests (G. B. Thursby, SAIC, unpublished) has demonstrated, moreover, that mortalities in excess of 20% (survivals less than 80% of control) are significant in about 95% of all tests. This "80%

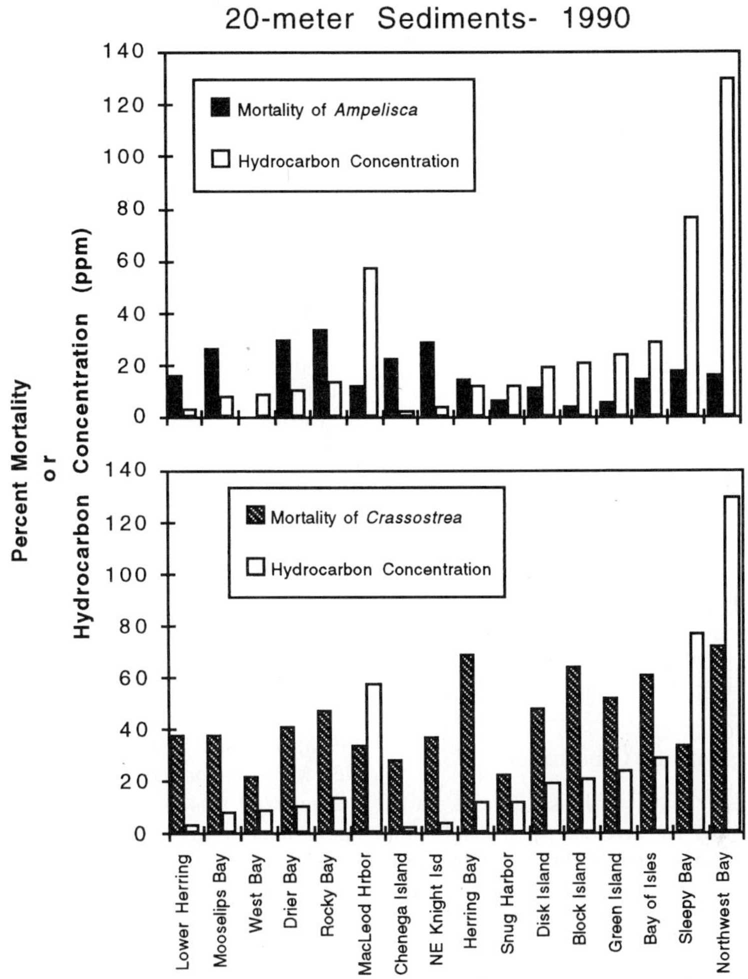

FIGURE 4.—Hydrocarbon concentrations (in parts per million [ppm] by ultraviolet fluorescence) and toxicity of 20-m sediment samples collected in 1990. Toxicity is presented as percent mortality for the amphipod *Ampelisca abdita* (top) and for Pacific oyster *Crassostrea gigas* (bottom). Stations are arranged in ascending order of hydrocarbon concentration, with six reference sites (Lower Herring to MacLeod Harbor) at left and ten exposed sites (Chenega Island to Northwest Bay) at right.

criterion" allows for the potential variability of control sediments across test batches. According to that criterion, sediments from 0 m exhibited toxicity to amphipods and/or to larval Pacific oysters at three exposed sites (Sleepy Bay, Northwest Bay and Bay of Isles—all sites that had exhibited significant toxicity to amphipods in the previous year) and at two reference sites (Mooselips Bay and more notably Drier Bay) (Figure 5). Significant toxicity to amphipods was also found with the 6-m sediment samples from Snug Harbor and Drier Bay (Figure 5). No significant toxicity in excess of 29% was seen for either test organism at either the 20-m or 100-m depths in 1991. There were no significant differences between mean toxicities at exposed and reference sites at any depth. The significant toxicity observed in sediments from shallow depths at Drier and Mooselips bays remains unexplained.

Because of the large differences in control survival in 1990, the data sets are not directly comparable between years. Qualitative insights are possible, however, from comparisons of ratios of the mean amphipod mortalities for exposed and reference sites across years (Table 12). This approach suggests a decrease in the relative toxicity of exposed-to-reference sites between 1990 and 1991 at 0

TABLE 11.—Sites at which sediment toxicity tests in 1991 revealed significant toxicity (*t*-test; $P = 0.05$) relative to controls, based on mortality of amphipods (A = *Ampelisca*) and/or Pacific oyster larvae (C = *Crassostrea*).

Station	Nominal oiling designation	Depth (m) 0	6	20	100
Northwest Bay[a]	E	A	A		
Snug Harbor[a]	E		A	A	
Block Island[a]	E		A, C	C	
Chenega Island[a]	E	C	C		
Bay of Isles	E	C	A		
Lower Herring Bay	R				
Mooselips Bay	R	A	C		
Drier Bay	R	A, C	A	A	
Rocky Bay	R		C		
Mooselips Bay	R	A	C		
Drier Bay	R	A, C	A	A	
Rocky Bay	R		C		
MacLeod Harbor	R	C			
Zaikof Bay	R				

[a]Sites with significant intertidal toxicity to amphipods in 1990.

m (from 3 to 1), and a possible increase (from 0.8 to 1.8) at 6 m. This inference is consistent with an overall decline in toxicity, and a concomitant shift of toxicity from the intertidal toward the shallow subtidal sediments.

Discussion and Conclusions

Similar UVF screening analysis methods have been used previously for investigations of oil spill impacts (e.g., at the *Argo Merchant* spill; Hoffman et al. 1979) and for monitoring programs involving discharges of petroleum hydrocarbon-containing constituents (e.g., Georges Bank Monitoring Program; Phillips et al. 1987). These methods offer potential benefits related to sensitivity, speed, and cost; however, the method is nonspecific and only semiquantitative, and the results may be difficult to interpret, especially when concentrations of the target oil are low relative to other sources of fluorescing materials.

Gas chromatographic analyses for sediment samples taken concurrently with this study indicate that oil from the *Exxon Valdez* generally was absent in samples from 40- or 100-m depths and rarely occurred in appreciable quantities below 3 m in 1989 (O'Clair et al. 1996). Other studies have shown that *Exxon Valdez* oil generally was absent in offshore deep sediments of Prince William Sound, and that, along with pyrogenic and biogenic sources, many of the hydrocarbons in those sediments originate from coastal petroleum seeps at Katalla (east of the Copper River) (Kvenvolden et al. 1991; Page et al. 1995; Carlson and Kvenvolden 1996, this volume).

The HPLC–UVF methodology employed here was able to distinguish *Exxon Valdez* crude oil from the sometimes substantial background signal from other sources in deep (40–100 m) sediments (but not to quantify degrees of mixture), and from an apparent diesel signal in some shallow bays outside the spill path, as verified by and correlated with GC–MS data (Krahn et al. 1993). The UVF screening approach appeared therefore to provide both a reliable indication of relative degree of contamination by fluorescent aromatic hydrocarbons of *Exxon Valdez* origin, particularly within the 0-, 3-, and 6-m depth strata and a valid basis for relating hydrocarbon concentrations to measured sediment toxicities.

Estimates of most probable numbers (MPN) of hydrocarbon-degrading bacteria provided an independent measure of oil exposure in these sediment samples (SAIC 1990a; Braddock et al. 1996). Hydrocarbon-degrading bacteria were highly elevated in oil-exposed intertidal sediments in 1989, showing high correlation with the estimates of oil concentration (UVF) reported here. Numbers of hydrocarbon-degrading bacteria in subtidal sediments, however, indicated only sporadic contamination with spilled oil in 1989, and nearly all of those occurrences were outside of Prince William Sound (SAIC 1990a; Braddock et al. 1992, 1996). In subtidal samples, MPN was not correlated with the UVF values.

Microtox® was used as the sole estimator of sediment toxicity in 1989. Microtox® EC50s did not show a clear dose-response relationship with hydrocarbon concentration, especially at high concentrations. Although toxicity was positively rank correlated with aromatic hydrocarbons (UVF) and was generally greatest at 0-, 3-, or 6-m depths at heavily oiled sites within Prince William Sound, the EC50s measured at some sites known to have been either non-oiled or only very lightly oiled by the spill (e.g., Olsen Bay, Rocky Bay, Chignik Bay) were about as low as at some of the most heavily oiled sites where the concentrations of total hydrocarbons were higher by 10- to 100-fold. The greatest differences in Microtox® response between Prince William Sound samples and Gulf of Alaska samples (Figure 1), as well as the strongest rank correlations of Microtox® response with UVF signal, occurred in shallow (3–6 m) subtidal sediments. At 0 m, where the hydrocarbon concentrations exhibited the greatest range, commensurate differences in toxicity were not measured with Microtox®. In deeper sediments (20–100 m), however, where *Exxon Valdez* oil was usually very low or absent, toxicity was still rank correlated with total hydrocarbons estimated by UVF. Taken together, these observations suggest that the Microtox® test (with organic extracts) is basically sen-

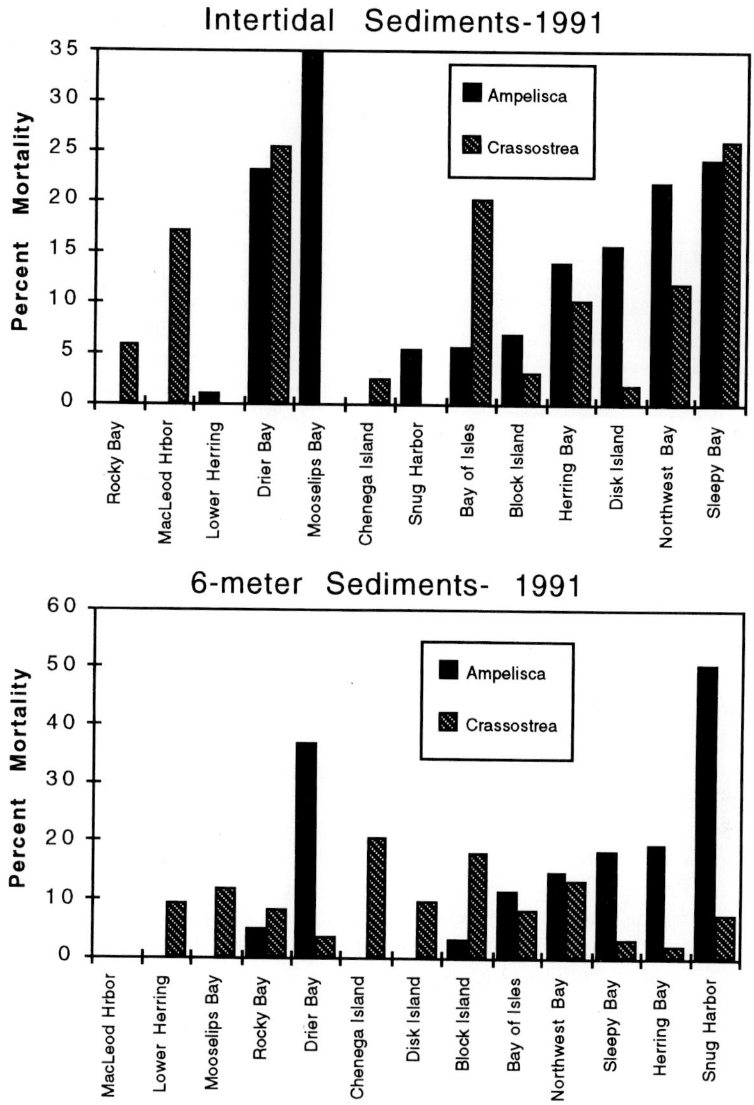

FIGURE 5.—Percent mortality of adult amphipod *Ampelisca* and larval Pacific oyster *Crassostrea* exposed to sediments collected in 1991 from 0-m (top) and 6-m (bottom) depths at five reference sites (left of slide) and eight exposed sites (at right of slide) in Prince William Sound. For Lower Herring Bay and Snug Harbor at 0 m and MacLeod Harbor at 6 m, the quantity of sediment was inadequate to conduct both tests, and no oyster data are available. The rest of the blanks represent actual 0% mortalities.

sitive to fluorescing aromatic constituents but that this sensitivity may be reduced in the presence of other oil constituents of fresh to moderately weathered oil, especially at high concentrations. Ho and Quinn (1993) presented Microtox® data that exhibited strong rank correlations of response with polycyclic aromatic hydrocarbon (PAH) levels in the aromatic fractions of organic extracts of sediments. As reported here, the Microtox® test was insensitive in direct laboratory tests to organic extracts of whole fresh or moderately weathered oil, even at very high hydrocarbon concentrations, although it gave a normal dose-response to a mixture of many of the PAH constituents of Prudhoe Bay crude oil (S. Sol et al., NOAA, unpublished). For this reason, the Microtox® test was not used in the subsequent surveys carried out in 1990 and 1991.

These observations lead us to speculate that rel-

TABLE 12.—Comparison of percent mortality (corrected for control mortality) of test amphipods in 1990 versus 1991 by depth.

Site type and year	Depth (m) 0	6	20	100
	Reference site (N = 5) means			
1990	19.8	25.9	23.6	10.2
1991	11.8	8.40	4.98	1.12
	Exposed site (N = 8) means			
1990	60.1	21.4	13.7	12.4
1991	11.7	14.8	5.58	0.96
	Ratio of exposed/reference means			
1990	3.0	0.83	0.58	1.1
1991	0.99	1.8	1.1	0.86

atively immobile, higher-molecular weight constituents of crude oil (perhaps the asphaltene–polar components), when present at high concentrations in a relatively fresh, unweathered form, may mask the bioavailability and toxicity of PAHs in the oil, at least as measured by Microtox® with organic extracts. The full toxicity potential of the PAHs may be manifested only after the oil has undergone some preliminary fractionation in the environment, such as through differential adsorption and partitioning processes. This hypothesis may warrant further examination in the future.

Most of the Microtox® EC50s reported here for oiled sediments from Prince William Sound are not remarkable, considering the extent of hydrocarbon contamination in some of the samples. The measured Microtox® EC50s in all samples at all depths ranged from 0.5 mg/mL (wet weight of sediments per unit volume of extract tested; Northwest Bay at 6 m) to 75 mg/mL. By contrast, a suite of coastal sediment samples taken recently along the length of Long Island Sound (Bricker et al. 1993) produced an EC50 range of about 0.006–1.3 mg/mL. Based on Microtox® results then, the sediments from representative coastal areas in Long Island Sound were about 50–100 times more toxic than those determined in 1989 for oiled subtidal sediments from Prince William Sound.

In 1990 and 1991, sediment toxicity was tested with amphipods and with Pacific oyster larvae. In 1990, sediment toxicity was demonstrated only for sediments collected from 0-m depth at some of the more heavily oiled sites (Northwest Bay, Snug Harbor, Block Island, Chenega Island, and Sleepy Bay), which were significantly toxic (60–95% mortalities) to test amphipods. At the 0-m depth, the mean sediment toxicity to amphipods for our 10 oiled sites in Prince William Sound was significantly higher than that for the 6 non-oiled (or very slightly oiled) reference sites. However, samples from some other heavily oiled sites, including Herring Bay, Disk Island, Northeast Knight Island, and Bay of Isles, exhibited relatively low toxicities to amphipods, comparable to those at sites that received no oil or were only very slightly oiled. No spill-related sediment toxicity to amphipods was demonstrated in 1990 samples from 6-, 20-, or 100-m depths, or to Pacific oyster larvae in samples from any depth.

In 1991, we observed generally lower and less variable control mortality in the toxicity tests (compared to 1990), which enabled detection of smaller statistically significant differences than in 1990. Although the mortality was lower in 1991, statistically significant toxicity (relative to controls) to amphipods and/or Pacific oysters was indicated in 1991 in sediments from 0 and/or 6 m at several oiled sites (most notably Sleepy Bay, Snug Harbor, and Bay of Isles). Most of these statistically significant differences, however, occurred at relatively high survival rates (i.e., greater than 80% of controls). Repetitive tests with amphipods have demonstrated that the protocol will reliably detect 20% mortality (relative to controls) in 90% of the tests, and it has been suggested that amphipod test results should be judged significant only when survival of test animals is both statistically different from the batch controls and less than 80% (Thursby and K. J. Scott, SAIC, unpublished). Amphipod survival less than 80% of controls was found at only three of eight oiled sites in 1991: Sleepy Bay and Northwest Bay at 0 m and Snug Harbor at 6 m. In addition, Pacific oyster survivals were less than 80% of controls for 0-m samples from Sleepy Bay and Bay of Isles. Although mean mortality was similar in 1991 for exposed and reference sites (Table 12), mortality of test amphipods was less than 5% at three (0 m) or four (6 m) of the five reference sites, compared to only one (0 m at Chenega Island) or three (6 m at Chenega, Disk, and Block Islands) of the eight oiled sites. However, sediments from two of the five reference sites exhibited significant toxicities to amphipods (Drier Bay at 0 and 6 m and Mooselips Bay at 0 m) and/or Pacific oysters (Drier Bay at 0 m), on the same order as the most toxic of the oiled sites. As a result, no statistically significant differences between oiled and non-oiled sites were demonstrated in 1991.

Our results with amphipods conclusively demonstrated spill-related sediment toxicity in 1990 in the lower intertidal zones at some of the most heavily oiled sites. Results with all three toxicity tests in all

3 years demonstrate the general absence of any spill-related sediment toxicity at depths of 20 m or greater. Although the remaining results obtained for shallow sediments (i.e., with Microtox® in 1989 and with amphipods and Pacific oyster in 1990–1991) are ambiguous, they are consistent with a general pattern of declining sediment toxicity in the lower intertidal zone from 1989 through 1991, accompanied by associated sediment toxicity down to at least 6 m depth in adjacent subtidal zones. Low residual sediment toxicity was still measurable in 1991 at 0 m and 6 m at many of the most heavily oiled sites, although the presence of significant toxicity at two of the non-oiled sites raises uncertainty about the spill-related causality. The decline in sediment toxicity at the 0 m depth between 1990 and 1991 may have been accompanied by a concomitant (small) increase in the toxicity of shallow (6–20 m) subtidal sediments in 1991. Overall, however, the measured chemical toxicity was low throughout the spill area in 1991, and it was not strongly correlated with oiling history at the sampling sites. The unexplained toxicity at non-oiled sites (most notably at Drier Bay) was similar in magnitude to that at oiled sites. Except for the positive tests with amphipods in 1990 for samples from 0-m depth at some of the most heavily oiled sites, the levels of toxicity indicated by these tests (e.g., the subtidal results with Microtox® in 1989 and with amphipods and Pacific oyster in 1990 and 1991) were much lower than is typically associated with contaminated sites elsewhere in the United States.

Boehm et al. (1995) reported similar results from sediment toxicity tests conducted with the amphipod *Rhepoxynius abronius* in 1990–1991. Sediment toxicity was indicated for intertidal sediment samples from several heavily oiled sites in 1990, compared to very few in 1991. Some of the observed toxicity was attributed to grain-size differences. Significant differences between oiled and non-oiled sites were found only for upper intertidal sediments from boulder–cobble beaches in 1990. Subtidal sediments (3–20 m) were not toxic to *R. abronius* in either 1990 or 1991.

Chemical toxicity of the spilled oil was probably greatest during the first month or two of the spill when the more toxic aromatic constituents were more abundant. Much of the acute toxicity of oil is attributed directly to the content of soluble aromatic compounds (Neff et al. 1976). In short-term exposures, molar toxicity appears to increase with number of aromatic rings (i.e., benzene < naphthalene < phenanthrene), at least up through three-ring compounds, and also with the extent of substitution (i.e., benzene < toluene < xylene < ethylbenzene, etc.) (Neff et al. 1976; Caldwell et al. 1977; Rice et al. 1977). The concentrations of monoaromatic hydrocarbons were largely depleted in the spilled *Exxon Valdez* oil by evaporative processes during the first 2–3 d, before the oil slick was widely dispersed and spread onto shorelines by winds. Similarly, the high winds that occurred during days 4–6 after the spill promoted the natural dispersion of oil into the water column (Wolfe et al. 1994), which probably also contributed to the early loss of some of the more toxic constituents from the floating oil that subsequently became beached. These processes undoubtedly reduced the effective toxicity substantially below the level that would have been present had the fresh oil been transported immediately into sheltered bays and shoreline sediments.

Other potential sources of spill-related toxicity are the oxidized derivatives formed either by photooxidation or by biodegradation of parent petroleum hydrocarbons. Such compounds have been found in harbor waters (Burns et al. 1990; Ehrhardt and Burns 1990) and in coastal waters contaminated by an oil spill (Ehrhardt and Burns 1993). Very limited information is available on the significance of either the polar constituents of crude oil or the intermediate oxidation products of petroleum hydrocarbons (whether from photooxidation or biodegradation) in terms of their potential for bioaccumulation and toxicity to resource organisms in the marine environment. Because these compounds have undergone preliminary oxidation and sometimes conjugation, they are more polar than their parent hydrocarbons and, as a result, generally would be more subject to excretion or depuration, less subject to bioaccumulation, more susceptible to further oxidation (or biodegradation if accumulated), and more susceptible to dilution and dispersion in the water column.

To determine the contribution of polar constituents to the toxicity of sediments contaminated by *Exxon Valdez* oil, Wolfe et al. (1995) compared the toxicities of polar and aromatic fractions of organic extracts from sediments from two sites in Prince William Sound. Large samples of sediment and pore water were collected in early September 1990 from the mid-intertidal zones at the Bay of Isles and Mooselips Bay sites. Methylene chloride extracts from these sediment and water samples were fractionated by liquid chromatography into aliphatic, aromatic, and polar fractions, and the aromatic and polar fractions were tested for relative toxicity by a number of tests. In addition to the Microtox® test and the bivalve larval mortality and development (with *Mytilus* sp. instead of Pacific oyster), the other tests were measures of genotoxicity. The samples from Bay of Isles (which showed obvious signs of heavy oil, both to the eye and to the GC) were only slightly more toxic (usually a factor of 2–5×) than the samples from Mooselips Bay, which was not oiled (as determined by GC) and which gave very low responses in all tests. However, for both sites, at the very high extract concentrations tested, the relative response to the polar and aromatic fractions were about the same in most tests. On the basis of these tests, it was concluded (Wolfe et al. 1995) that although the overall toxicity of the oil was low, at least a part of that toxicity was derived from polar constituents.

As with many marine oil spills, the initial shoreline oiling from the *Exxon Valdez* spill generally occurred primarily in the mid- to upper intertidal zones, where substantial mortalities occurred, through the cumulative effects of smothering, toxicity, and cleanup (especially high-pressure hot water) (Mearns 1996, this volume). While intertidal infauna are clearly highly vulnerable to exposure by oiling, observations at past oil spills have demonstrated that in some cases the subtidal infauna can also be exposed and can suffer significant mortality and changes in population density (Cross et al. 1978; Beslier et al. 1980; Cabioch et al. 1980; Elmgren et al. 1980, 1983). Furthermore, Addy et al. (1978) and Dauvin and Gentil (1990) showed that chronic oil pollution could also have deleterious effects on benthic invertebrate assemblages in relatively deep offshore waters. It is possible that some of these changes may have been secondary ecological effects of sublethal responses to oil exposure. For example, feeding behavior and survival of polychaetes *Abarenicola pacifica* and survival and condition index of deposit-feeding clams, stained macoma *Macoma inquinata*, were affected by prolonged exposure (11 d and 38–54 d, respectively) to Prudhoe Bay crude oil concentrations in the range of 800–1,200 ppm in sediments (Augenfeld 1980; Augenfeld et al. 1980). Observations of subtidal benthic effects after the oil spill in Prince William Sound are limited and the direct cause-effect mechanisms are not known (Dean et al. 1996a; Dean et al. 1996b; Jewett et al. 1996; all this volume). The information presented here suggests that acute chemical toxicity could have been a causal factor in 1989 and 1990 for observed biological effects in the lower intertidal and associated shallow subtidal zones in areas of Prince William Sound that were heavily impacted by the *Exxon Valdez* oil spill.

Acknowledgments

We thank Tom Ruszala, captain of the National Oceanic and Atmospheric Administration (NOAA) RV *Fairweather* in 1989 and NOAA RV *Davidson* in 1990, along with the officers and crew members of both ships for their support in the field during sample collection. Chuck O'Clair, Terri Stinnett, Link Freese, and Anthony Chan of the National Marine Fisheries Service (NMFS), Auke Bay Laboratory; Michael Hilley of the Alaska Department of Environmental Conservation; and Tom Yeager of the University of Alaska Institute of Water Resources provided invaluable assistance in collecting and processing sediment samples for analysis and testing. The 1989 UVF analyses were conducted by John Clayton and associates of Science Applications International Corp. Gina Ylitalo of the NMFS Northwest Fisheries Science Center provided technical assistance with the HPLC–UVF analyses in 1990. Terry Wade and associates at Texas A&M University performed the GC–MS analyses that supported the interpretation of the UVF analyses, and Carol-Ann Manen provided access through the Prince William Sound OIL database to these GC–MS data. Bernie Gottholm of NOAA, Office of Ocean Resources Conservation and Assessment, provided graphics support.

References

Addy, J. M., D. Levell, and J. P. Hartley. 1978. Biological monitoring of sediments in *Ekofisk* oil field. Pages 515–539 *in* Proceedings, conference on assessment of ecological impacts of oil spills. American Institute of Biological Sciences, Arlington, Virginia.

ASTM (American Society for Testing and Materials). 1990a. Standard guide for conducting solid phase 10-day static sediment toxicity tests with marine and estuarine infaunal amphipods, E 1367-90. Pages 1–24 *in* Annual book of ASTM standards. ASTM, Philadelphia.

ASTM (American Society for Testing and Materials). 1990b. Standard guide for conducting static acute toxicity tests starting with embryos of four species of saltwater bivalve molluscs, E 724–89 *in* Annual book of ASTM standards. ASTM, Philadelphia.

Anderson, J. W. 1977. Responses to sublethal levels of petroleum hydrocarbons: are they sensitive indicators and do they correlate with tissue contaminants? Pages 95–114 *in* D. A. Wolfe, editor. Fate and effects of petroleum hydrocarbons in marine organisms and ecosystems. Pergamon Press, New York.

Augenfeld, J. M. 1980. Effects of Prudhoe Bay crude oil contamination on sediment working rates of *Abarenicola pacifica*. Marine Environmental Research 3:307–313.

Augenfeld, J. M., J. W. Anderson, D. L. Woodruff, and J. L. Webster. 1980. Effects of Prudhoe Bay crude oil contaminated sediments on *Protothaca staminea* (Mollusca: Pelecypoda): hydrocarbon content, condition index, free amino acid levels. Marine Environmental Research 4:135–143.

Beslier, A., J. L. Birrien, L. Cabioch, C. Larsonneur, and L. LeBorgne. 1980. La pollution des Baies de Morlaix et de Lannion par les hydrocarbures de l'*Amoco Cadiz*: repartition sur les fonds et evolution. Helgolaender Meeresuntersuchungen 33:209–224.

Boehm, P. D., D. S. Page, E. S. Gilfillan, W. A. Stubblefield, and E. J. Harner. 1995. Shoreline ecology program for Prince William Sound, Alaska, following the *Exxon Valdez* oil spill: part 2-chemistry and toxicology. Pages 347–397 *in* P. G. Wells, J. N. Butler, and J. S. Hughes, editors. *Exxon Valdez* oil spill: fate and effects in Alaskan waters. American Society for Testing and Materials Special Technical Publication 1219, Philadelphia.

Braddock, J. F., B. T. Rasley, T. R. Yeager, J. E. Lindstrom, and E. J. Brown. 1992. Hydrocarbon mineralization potentials and microbial populations in marine sediments following the *Exxon Valdez* oil spill. Final report of the Institute of Arctic Biology and Water Research Center, University of Alaska to Alaska Department of Environmental Conservation Office of Restoration and Damage Assessment, Juneau.

Braddock, J. F., J. E. Lindstrom, T. R. Yeager, B. T. Rasley, and E. J. Brown. 1996. Patterns of microbial activity in oiled and unoiled sediments in Prince William Sound. American Fisheries Society Symposium 18:94–108.

Bricker, S. B., D. A. Wolfe, K. J. Scott, G. Thursby, E. R. Long, and A. Robertson. 1993. Sediment toxicity in Long Island Sound embayments. Pages 181–189 *in* M. S. Van Patten, editor. Long Island Sound research conference proceedings. Publication CT-SG-93-03, University of Connecticut, Groton.

Brodersen, C. C., S. D. Rice, J. W. Short, T. A. Mecklenburg, and J. F. Karinen. 1977. Sensitivity of larval and adult Alaskan shrimp and crabs to acute exposures of the water-soluble fraction of Cook Inlet crude oil. Pages 575–578 *in* Proceedings, 1977 oil spill conference (prevention, behavior, control, cleanup). American Petroleum Institute Publication 4284, Washington, DC.

Brooks, J. M., M. C. Kennicutt, II, T. L. Wade, A. D. Hart, G. J. Denoux, and T. J. McDonald. 1990. Hydrocarbon distributions around a shallow water multiwell platform. Environmental Science and Technology 24: 1079–1085.

Burns, K. A., M. G. Ehrhardt, J. MacPherson, J. A. Tierney, G. Kananen, and D. Connelly. 1990. Organic and trace metal contaminants in sediments, seawater and organisms from two Bermudan harbours. Journal of Experimental Marine Biology and Ecology 138:9–34.

Cabioch, L., J. C. Dauvin, J. Mora Bermudez, and C. Rodriguez Babio. 1980. Effets de la maree noire de l'*Amoco Cadiz* sur le benthos sublittoral du nord de la Bretagne. Helgolaender Meeresuntersuchungen 33: 92–208.

Caldwell, R. S., E. M. Caldarone, and M. H. Mallon. 1977. Effects of a seawater-soluble fraction of Cook Inlet crude oil and its major aromatic components on larval states of the dungeness crab, *Cancer magister* Dana. Pages 210–220 *in* D. A. Wolfe, editor. Fate and effects of petroleum hydrocarbons in marine organisms and ecosystems. Pergamon Press, New York.

Carlson, P. R., and K. A. Kvenvolden. 1996. Tracking *Exxon Valdez* oil from beach to deepwater sediments of Prince William Sound, Alaska. American Fisheries Symposium 18:109–120.

Chang, J. C., P. B. Taylor, and F. R. Leach. 1981. Use of the Microtox® assay system for environmental samples. Bulletin of Environmental Contamination and Toxicology 26:150–156.

Chapman, P. M., and S. Becker. 1986. Recommended protocols for conducting laboratory bioassays on Puget Sound sediments. Final report TC-3991-04, U.S. Environmental Protection Agency, Region 10, Seattle.

Chapman, P. M., and J. D. Morgan. 1983. Sediment bioassays with oyster larvae. Bulletin of Environmental Contamination and Toxicology 31:438–444.

Craddock, D. R. 1977. Acute toxic effects of petroleum on arctic and subarctic marine organisms. Pages 1–93 *in* D. C. Malins, editor. Effects of petroleum on arctic and subarctic marine environments and organisms, volume II, biological effects. Academic Press, New York.

Cross, F. A., W. P. Davis, D. E. Hoss, and D. A. Wolfe. 1978. Biological observations. Pages 197–215 *in* W. N. Hess, editor. The *Amoco Cadiz* oil spill. A preliminary scientific report. NOAA (National Oceanic and Atmospheric Administration)/EPA (Environmental Protection Agency) Special Report. NOAA Environmental Research Laboratories, Boulder, Colorado.

Dauvin, J.-C., and F. Gentil. 1990. Conditions of the Peracarid populations of subtidal communities in northern Brittany ten years after the *Amoco Cadiz* oil spill. Marine Pollution Bulletin 21:123–130.

Dean, T. A., S. C. Jewett, D. R. Laur, and R. O. Smith. 1996a. Injury to epibenthic invertebrates resulting from the *Exxon Valdez* oil spill. American Fisheries Society Symposium 18:424–439.

Dean, T. A., M. S. Stekoll, and R. O. Smith. 1996b. Kelps and oil: the effects of the *Exxon Valdez* oil spill on subtidal algae. American Fisheries Society Symposium 18:412–423.

Ehrhardt, M. G., and K. A. Burns. 1990. Petroleum-derived organic compounds concentrated from inshore waters in Bermuda. Journal of Experimental Marine Biology and Ecology 138:35–47.

Ehrhardt, M. G., and K. A. Burns. 1993. Hydrocarbons and related photooxidation products in Saudi Arabian coastal waters and hydrocarbons in underlying sediments and bioindicator bivalves. Marine Pollution Bulletin 27:187–197.

Elmgren, R., S. Hansson, U. Larsson, and B. Sundelin. 1980. Impact of oil on deep soft bottoms. Pages 97–126 *in* J. J. Kineman, R. Elmgren, and S. Hansson, editors. The *Tsesis* oil spill. NOAA (National Oceanic and Atmospheric Administration) Office of Marine Pollution Assessment, Boulder, Colorado.

Elmgren, R., S. Hansson, U. Larsson, B. Sundelin, and P. D. Boehm. 1983. The *Tsesis* oil spill: acute and long-term impact on the benthos. Marine Biology 73:51–65.

Giesy, J. P., and six coauthors. 1988. Comparison of three sediment bioassay methods using Detroit River sediments. Environmental Toxicology and Chemistry 7:483–498.

Ho, K. T. Y., and J. G. Quinn. 1993. Physical and chemical parameters of sediment extraction and fractionation that influence toxicity, as evaluated by Microtox®. Environmental Toxicology and Chemistry 12: 615–625.

Hoffman, E. J., J. G. Quinn, J. R. Jadamec, and S. H. Fortler. 1979. Comparison of UV fluorescence and gas chromatographic analyses of hydrocarbons in sediments from the vicinity of the *Argo Merchant* wreck site. Bulletin of Environmental Contamination and Toxicology 23:536–543.

Jewett, S. C., T. A. Dean, and D. R. Laur. 1996. Effects of the *Exxon Valdez* oil spill on benthic invertebrates in an oxygen-deficient embayment in Prince William Sound, Alaska. American Fisheries Society Symposium 18:440–447.

Johnson, F. G. 1977. Sublethal biological effects of petroleum hydrocarbon exposures: bacteria, algae, and invertebrates. Pages 271–318 *in* D. C. Malins, editor. Effects of petroleum on arctic and subarctic marine environments and organisms, volume II, biological effects. Academic Press, New York.

Karinen, J. F. 1988. Sublethal effects of petroleum on biota. Pages 294–328 *in* D. G. Shaw and M. J. Hameedi, editors. Environmental studies in Port Valdez, Alaska: a basis for management. Lecture notes on coastal and estuarine studies, volume 24. Springer-Verlag, Berlin.

Kelso, D. D., and M. Kendziorek. 1991. Alaska's response to the *Exxon Valdez* oil spill. Environmental Science and Technology 25:16–23.

Krahn, M. M., G. M. Ylitalo, J. Joss, and S.-L. Chan. 1991. Rapid semi-quantitative screening of sediments for aromatic compounds using sonic extraction and HPLC/fluorescence analysis. Marine Environmental Research 31:175–196.

Krahn, M. M., and seven coauthors. 1993. Comparison of high-performance liquid chromatography/fluorescence screening and gas chromatography/mass spectrometry analysis for aromatic compounds in sediments sampled after the *Exxon Valdez* oil spill. Environmental Science and Technology 27:699–708.

Kvenvolden, K. A., J. B. Rapp, and F. D. Hostettler. 1991. Tracking hydrocarbons from the *Exxon Valdez* oil spill in beach, shallow-water, and deep-water sediment of Prince William Sound, Alaska. Pages 69–98 *in* P. R. Carlson, editor. Sediment of Prince William Sound, beach to deep fjord floor, a year after the *Exxon Valdez* oil spill. U.S. Geological Survey, Open-File Report 91-631.

Maki, A. W. 1991. The *Exxon Valdez* oil spill: initial environmental impact assessment. Environmental Science and Technology 25:24–29.

Mayo, D. W., and six coauthors. 1978. Weathering characteristics of petroleum hydrocarbons deposited in fine clay marine sediments, Searsport, Maine. Journal of the Fisheries Research Board of Canada 35:552–562.

Mearns, A. J. 1996. *Exxon Valdez* shoreline treatment and operations: implications for response, assessment, monitoring, and research. American Fisheries Society Symposium 18:309–328.

Moles, A., S. D. Rice, and S. Korn. 1979. Sensitivity of Alaskan freshwater and anadromous fishes to Prudhoe Bay crude oil and benzene. Transactions of the American Fisheries Society 108:408–414.

NRC (National Research Council). 1985. Oil in the sea. National Academy Press, Washington, DC.

Neff, J. M., J. W. Anderson, B. A. Cox, R. B. Laughlin, Jr., S. S. Rossi, and H. E. Tatum. 1976. Effects of petroleum on survival, respiration and growth of marine animals. Pages 516–539 *in* Sources, effects and sinks of hydrocarbons in the aquatic environment. The American Institute of Biological Sciences, Washington, DC.

O'Clair, C. E., J. W. Short, and S. D. Rice. 1996. Contamination of intertidal and subtidal sediments by oil from the *Exxon Valdez* in Prince William Sound. American Fisheries Society Symposium 18:61–93.

Page, D. S., P. D. Boehm, G. S. Douglas, and A. E. Bence. 1995. Identification of hydrocarbon sources in the benthic sediments of Prince William Sound and the Gulf of Alaska following the *Exxon Valdez* oil spill. Pages 41–83 *in* P.G. Wells, J. N. Butler, and J. S. Hughes, editors. *Exxon Valdez* oil spill: fate and effects in Alaskan waters. American Society for Testing and Materials Special Technical Publication 1219, Philadelphia.

Patten, B. G. 1977. Sublethal biological effects of petroleum hydrocarbon exposures: fish. Pages 319–335 *in* D. C. Malins, editor. Effects of petroleum on arctic and subarctic marine environments and organisms, volume II, biological effects. Academic Press, New York.

Phillips, C. R., J. R. Payne, J. L. Lambach, G. H. Farmer, and R. R. Sims. 1987. Georges Bank monitoring program: hydrocarbons in bottom sediments and hydrocarbons and trace metals in tissues. Marine Environmental Research 22:33–74.

PSEP (Puget Sound Estuary Program). 1986. Recommended guidelines for conducting laboratory bioassays on Puget Sound sediments. Final report to U.S. Environmental Protection Agency, Seattle.

Rice, S. D. 1985. Effects of oil on fish. Pages 157–182 *in* F. R. Engelhardt, editor. Petroleum effects in the arctic environment. Elsevier Applied Science Publishers, New York.

Rice, S. D., S. Korn, C. C. Brodersen, S. A. Lindsay, and S. A. Andrews. 1981. Toxicity of ballast-water treatment effluent to marine organisms at Port Valdez, Alaska. Pages 55–61 *in* Proceedings, 1981 oil spill conference (prevention, behavior, control, cleanup).

American Petroleum Institute Publication 4334, Washington, DC.

Rice, S. D., A. Moles, T. L. Taylor, and J. F. Karinen. 1979. Sensitivity of 39 Alaskan marine species to Cook Inlet crude oil and No. 2 fuel oil. Pages 443–549 *in* Proceedings, 1979 oil spill conference (prevention, behavior, control, cleanup). American Petroleum Institute Publication 4308, Washington, DC.

Rice, S. D., and seven coauthors. 1984. Effects of petroleum hydrocarbons on Alaskan aquatic organisms: a comprehensive review of all oil-effects research on Alaskan fish and invertebrates conducted by the Auke Bay Laboratory, 1970–1981. NOAA (National Oceanic and Atmospheric Administration) Technical Memorandum NMFS (National Marine Fisheries Service)/NWC-67, Seattle.

Rice, S. D., J. W. Short, and J. F. Karinen. 1976. Toxicity of Cook Inlet crude oil and no. 2 fuel oil to several Alaskan marine fishes and invertebrates. Pages 394–406 *in* Sources, effects and sinks of hydrocarbons in the aquatic environment. The American Institute of Biological Sciences, Washington, DC.

Rice, S. D., J. W. Short, and J. F. Karinen. 1977. A review of comparative oil toxicity and comparative animal sensitivity. Pages 78–94 *in* D. A. Wolfe, editor. Fate and effects of petroleum hydrocarbons in marine organisms and ecosystems. Pergamon Press, New York.

SAIC (Science Applications International Corp.). 1989. Screening analysis for petroleum hydrocarbons in sediments and sediment pore waters by use of ultraviolet fluorescence spectrophotometry for *Exxon Valdez* damage assessment. Final report (Contract 50-DSNC-8-00141, Task No. 55-DSNC-9-00017). NOAA (National Oceanic and Atmospheric Administration) Office of Ocean Resources Conservation and Assessment, Rockville, Maryland.

SAIC (Science Applications International Corp.). 1990a. Microbial hydrocarbon degradation in sediments impacted by the *Exxon Valdez* oil spill. Final report (Contract 50-DSNC-8-00141, Task No. 55-DSNC-9-00016). NOAA (National Oceanic and Atmospheric Administration) Office of Ocean Resources Conservation and Assessment, Rockville, Maryland.

SAIC (Science Applications International Corp.). 1990b. Bioassay of toxicity associated with intertidal and subtidal sediments contaminated by the *Exxon Valdez* oil spill. Final report (Contract 50-DSNC-8-00141, Task No. 55-DSNC-0-00008). NOAA (National Oceanic and Atmospheric Administration) Office of Ocean Resources Conservation and Assessment, Rockville, Maryland.

SAIC (Science Applications International Corp.). 1991. Bioassay of toxicity associated with intertidal and subtidal sediments contaminated by the *Exxon Valdez* oil spill: second year. Final report (Contract 50-DSNC-1-00107). NOAA (National Oceanic and Atmospheric Administration) Office of Ocean Resources Conservation and Assessment, Rockville, Maryland.

Scott, J. K., and M. S. Redmond. 1990. The effects of a contaminated dredged material on laboratory populations of the tubiculous amphipod *Ampelisca abdita*. Pages 289–303 *in* U. M. Cowgill and L. R. Williams, editors. Aquatic toxicology and hazard assessment, volume 12. American Society for Testing and Materials Special Technical Publication 1027, Philadelphia.

Schiewe, M. H., E. G. Hawk, D. I. Actor, and M. M. Krahn. 1985. Use of a bacterial bioluminescence assay to assess toxicity of contaminated marine sediments. Canadian Journal of Fisheries and Aquatic Sciences 42:1244–1248.

Swartz, R. C., W. A. DeBen, J. K. P. Jones, J. O. Lamberson, and F. A. Cole. 1985. Phoxocephalid amphipod bioassay for marine sediment toxicity. Pages 284–307 *in* R. D. Cardwell, R. Purdy, and R. C. Bahner, editors. Aquatic toxicology and hazard assessment: seventh symposium. American Society for Testing and Materials Special Technical Publication 854, Philadelphia.

Teal, J. M., K. Burns, and J. Farrington. 1978. Analyses of aromatic hydrocarbons in intertidal sediments resulting from two spills of no. 2 fuel oil in Buzzards Bay, Massachusetts. Journal of the Fisheries Research Board of Canada 35:510–520.

Williams, L. G., P. M. Chapman, and T. C. Ginn. 1986. A comparative evaluation of marine sediment toxicity using bacterial luminescence, oyster embryo and amphipod sediment bioassays. Marine Environmental Research 19:225–249.

Wolfe, D. A. 1985. Fossil fuels: transportation and marine pollution. Pages 45–93 *in* I. W. Duedall, D. R. Kester, P. Kilho Park, and B. H. Ketchum, editors. Wastes in the ocean, volume 4. Energy wastes in the ocean. Wiley, New York.

Wolfe, D. A., and eleven coauthors. 1994. Fate of the oil spilled from the T/V *Exxon Valdez* in Prince William Sound, Alaska. Environmental Science and Technology 28(13):561A–568A.

Wolfe, D. A., K. J. Scott, J. R. Clayton, Jr., J. Lunz, J. R. Payne, and T. A. Thompson. 1995. Comparative toxicities of polar and non-polar organic fractions from sediments affected by the *Exxon Valdez* oil spill in Prince William Sound, Alaska. Chemistry and Ecology 10:137–156.

American Fisheries Society Symposium 18:140–148, 1996

Analytical Methods Used for the Analysis of Hydrocarbons in Crude Oil, Tissues, Sediments, and Seawater Collected for the Natural Resources Damage Assessment of the *Exxon Valdez* Oil Spill

JEFFREY W. SHORT
National Marine Fisheries Service, Auke Bay Laboratory, Alaska Fisheries Science Center 11305 Glacier Highway, Juneau, Alaska 99801, USA

THOMAS J. JACKSON
Texas A&M University, Geochemical and Environmental Research Group, 833 Graham Road College Station, Texas 77845, USA

MARIE L. LARSEN
Auke Bay Laboratory, Alaska Fisheries Science Center

TERRY L. WADE
Texas A&M University, Geochemical and Environmental Research Group

Abstract.—This paper describes the analytical methods used for quantitative determination of hydrocarbons in samples of crude oil, biological tissues, marine sediments, and seawater that were collected for the Natural Resources Damage Assessment of the *Exxon Valdez* oil spill.

The Natural Resources Damage Assessment of the *Exxon Valdez* oil spill involved the chemical analysis of more than 7,700 samples for aliphatic and aromatic hydrocarbons. The objective of these analyses was to accurately measure hydrocarbons derived from *Exxon Valdez* crude oil. The samples included crude oil, biological tissues, marine sediments, and seawater and were analyzed by nearly identical methods at two laboratories: Auke Bay Laboratory (ABL), National Marine Fisheries Service, in Juneau, Alaska (1,700 samples); and Geochemical and Environmental Research Group (GERG), Texas A&M University, in College Station, Texas (6,000 samples). The methods included rigorous controls to ensure that data produced by the two laboratories were comparable. The resulting database constitutes the largest set of environmental samples analyzed on a comparable basis for any oil spill. We describe the methods used at ABL and at GERG to determine concentrations of individual hydrocarbons for all of the studies included in this volume. Additional details are given elsewhere (ABL: Larsen et al. 1992; GERG: Krahn et al. 1993; Wade et al. 1993; Frost et al. 1994).

Analytes and Method Synopsis

The fundamental quantitative principle used to estimate the concentrations of hydrocarbons in the samples analyzed is the internal standard method, where isotopically similar surrogate hydrocarbons are used as internal standards. This approach involves comparison of known quantities of selected deuterated hydrocarbons (surrogate standards) that are added to the samples before preparation for analysis with amounts of chemically similar hydrocarbons detected by the analysis. Hydrocarbon concentrations in samples are determined using 5-point calibration by assuming deuterated surrogates mimic the behavior of the native hydrocarbons.

The 25 alkane hydrocarbons determined are listed in Table 1, and the surrogate standards used by each laboratory are also identified for each alkane. These alkanes include the normal alkanes containing 10–30 carbon atoms (inclusive), normal dotriacontane, normal tetratriacontane, pristane (2,6,10,14-tetramethylpentadecane), and phytane (2,6,10,14-tetramethylhexadecane). The laboratories were not required to report hentriacontane (C_{31}) and tritriacontane (C_{33}), although concentrations of these two alkanes were often determined (especially at GERG).

The 44 polynuclear aromatic hydrocarbons (PAH) determined are listed in Table 2, and the surrogate standards used by each laboratory are also identified for each PAH. These PAH include unsubstituted and alkyl-substituted homologues of PAH containing from two to five rings and dibenzothiophene homologues.

TABLE 1.—Alkane analytes, abbreviations, and associated internal[a] and surrogate[b] standards used at Geochemical and Environmental Research Group (GERG) and at Auke Bay Laboratory (ABL).

		GERG		ABL	
Alkane analyte	Abbreviation	Internal standard reference[a]	Surrogate reference[b]	Internal standard reference[a]	Surrogate reference[b]
Decane	n-C_{10}	A	1	B	1
Undecane	n-C_{11}	A	1	B	1
Dodecane	n-C_{12}	A	1	B	1
Tridecane	n-C_{13}	A	1	B	1
Tetradecane	n-C_{14}	A	1	B	2
Pentadecane	n-C_{15}	A	1	B	2
Hexadecane	n-C_{16}	A	1	B	2
Heptadecane	n-C_{17}	A	3	B	2
2,6,10,14-tetramethylpentadecane	Pristane	A	3	B	2
Octadecane	n-C_{18}	A	3	B	3
2,6,10,14-tetramethylhexadecane	Phytane	A	3	B	3
Nonadecane	n-C_{19}	A	3	B	3
Eicosane	n-C_{20}	A	3	B	3
Heneicosane	n-C_{21}	A	3	B	3
Docosane	n-C_{22}	A	3	B	4
Tricosane	n-C_{23}	A	4	B	4
Tetracosane	n-C_{24}	A	4	B	4
Pentacosane	n-C_{25}	A	4	B	4
Hexacosane	n-C_{26}	A	4	B	4
Heptacosane	n-C_{27}	A	4	B	4
Octacosane	n-C_{28}	A	4	B	4
Nonacosane	n-C_{29}	A	4	B	5
Triacontane	n-C_{30}	A	5	B	5
Dotriacontane	n-C_{32}	A	5	B	5
Tetratriacontane	n-C_{34}	A	5	B	5

[a]Internal standards: A = hexadecane–d_{34}; B = dodecylcyclohexane.
[b]Deuterated surrogate standards: 1 = dodecane–d_{26}; 2 = hexadecane–d_{34}; 3 = eicosane–d_{42}; 4 = tetracosane–d_{50}; 5 = triacontane–d_{62}.

Sample Storage and Chain of Custody

Each sample was assigned a unique sample identification number (SIN) at the time of collection. A custodial record of the samples using SINs documents the history of each sample from collection through analysis.

Sample Preparation and Extraction

Seawater.—Each 0.9-L seawater sample was extracted in the field successively with aliquots of 50 mL and 25 mL dichloromethane for 1 min, followed by 1 min for phase separation. The dichloromethane extracts were combined and stored at 5°C for up to 5 d and were then stored at −20°C until analysis. After storage, samples were spiked with 0.5 mL or less of hexane solutions of the deuterated alkane and PAH surrogate standards listed in Tables 1 and 2, filtered through 50 g anhydrous sodium sulfate, reduced in volume over a 60–70°C water bath, and exchanged with hexane to a final volume of 1 mL.

Tissues.—Tissue samples were dissected and mechanically macerated. The apparatus used for dissection and all glassware that contacted the sample subsequently were precleaned either by repeated rinsing with dichloromethane or by combustion at 400°C. Samples were dissected so that the external surfaces of the specimen did not contact the sampled tissues. An aliquot of the macerated sample was removed before surrogate standard-spiking for dry weight determination. Samples were macerated for 3 min with a Tekmar tissuemizer and then spiked with hexane solutions of the deuterated alkane and PAH surrogate standards listed in Tables 1 and 2. Seventy grams of anhydrous sodium sulfate were added for dehydration, followed by successive extractions with each of three 100-mL aliquots of dichloromethane. Each dichloromethane aliquot was mixed with the sample for 3 min with the tissuemizer, separated by filtration (ABL) or centrifugation (GERG), and combined with the successive extracts. The combined dichloromethane extracts were reduced in volume over a 60–70°C water bath and exchanged with hexane to a final volume of 1 mL. These concentrated hexane extracts were stored at −20°C for fractionation of alkane and PAH by column chromatography.

Sediments.—At ABL, hydrocarbons were extracted from sediments by prolonged room temper-

TABLE 2.—Polynuclear aromatic hydrocarbon (PAH) analytes, abbreviations, and associated internal[a] and surrogate[b] standards used at Geochemical and Environmental Research Group (GERG) and at Auke Bay Laboratory (ABL). Uncalibrated PAH are identified in this table as those that begin with C followed by a numeral from 1 to 4; all other PAH are calibrated. The numerals following the letter C of the uncalibrated PAH indicate the total number of carbon atoms borne by the alkyl substituents of the PAH.

		GERG		ABL	
PAH analyte	Abbreviation	Internal standard reference[a]	Surrogate reference[b]	Internal standard reference[a]	Surrogate reference[b]
Naphthalene	Naph	A	1	C	1
C1 naphthalene	C1naph	A	1	C	1
C2 naphthalene	C2naph	A	1	C	2
C3 naphthalene	C3naph	A	1	C	2
C4 naphthalene	C4naph	A	1	C	2
Biphenyl	Biphenyl	A	2	C	2
Acenaphthylene	Acenthy	A	2	C	2
Acenaphthene	Acenthe	A	2	C	2
Fluorene	Fluorene	A	2	C	2
C1 fluorene	C1fluor	A	2	C	2
C2 fluorene	C2fluor	A	2	C	2
C3 fluorene	C3fluor	A	2	C	2
Dibenzothiophene	Dithio	A	3	C	3
C1 dibenzothiophene	C1dithio	A	3	C	3
C2 dibenzothiophene	C2dithio	A	3	C	3
C3 dibenzothiophene	C3dithio	A	3	C	3
Anthracene	Anthra	A	3	C	3
Phenanthrene	Phenanth	A	3	C	3
C1 phenanthrene–anthracene	C1phenan	A	3	C	3
C2 phenanthrene–anthracene	C2phenan	A	3	C	3
C3 phenanthrene–anthracene	C3phenan	A	3	C	3
C4 phenanthrene–anthracene	C4phenan	A	3	C	3
Fluoranthene	Fluorant	B	3	C	3
Pyrene	Pyrene	B	3	C	3
C1 fluoranthene–pyrene	C1fluora	B	3	C	3
Benzo[*a*]anthracene	Benanth	B	4	C	4
Chrysene	Chrysene	B	4	C	4
C1 chrysene	C1chrys	B	4	C	4
C2 chrysene	C2chrys	B	4	C	4
C3 chrysene	C3chrys	B	4	C	4
C4 chrysene	C4chrys	B	4	C	4
Benzo[*b*]fluoranthene	Benzobfl	B	5	C	6
Benzo[*k*]fluoranthene	Benzokfl	B	5	C	6
Benzo[*e*]pyrene	Benepy	B	5	C	6
Benzo[*a*]pyrene	Benapy	B	5	C	6
Perylene	Perylene	B	5	C	5
Indeno[1,2,3-*cd*]pyrene	Indeno	B	5	C	6
Dibenzo[*ah*]anthracene	Dibenz	B	5	C	6
Benzo[*ghi*]perylene	Benzop	B	5	C	6
Specific isomers					
1-Methylnapththalene	Menap1	A	1	C	1
2-Methylnaphthalene	Menap2	A	1	C	1
2,6-Dimethylnaphthalene	Dimeth	A	2	C	2
2,3,5-Trimethylnaphthalene	Trimeth	A	2	C	2
1-Methylphenanthrene	Mephen1	A	3	C	3

[a]Internal standards: A = fluorene–d_{10}; B = benzo[*a*]pyrene–d_{12}; C = hexamethylbenzene.
[b]Deuterated surrogate standards: 1 = naphthalene–d_8; 2 = acenaphthene–d_{10}; 3 = phenanthrene–d_{10}; 4 = chrysene–d_{12}; 5 = perylene–d_{12}; 6 = benzo[*a*]pyrene–d_{12}.

ature agitation with dichloromethane using a procedure developed by Brown et al. (1980). About 10 g of wet sediment were removed from the sample for dry weight determination, and an additional 10–20 g of wet sediment were added to 80 g anhydrous sodium sulfate for dehydration and spiked with 0.5 mL of hexane solutions of the deuterated alkane and PAH surrogate standards listed in Tables 1 and 2. The mixture of sediment, sodium sulfate, and surrogate standards was agitated on a roller table for three successive periods of 16 h, 6 h, and 16 h with each of three successive 100-mL aliquots of dichloromethane, with the dichloromethane separated by filtration after each agitation period. The combined dichloromethane extracts were reduced in volume over a 60–70°C water bath and exchanged with hexane to a final volume of 1 mL. These concentrated hexane extracts were stored at −20°C until separation of alkane and PAH fractions by column chromatography.

At GERG, hydrocarbons were extracted from sediments by Soxhlet extraction with dichloromethane. An aliquot (1–5 g) of wet sediment was weighed from the sample for dry weight determination. A freeze-dried sediment sample (approximately 10 g dry weight) was combined with 150 mL dichloromethane, spiked with the deuterated alkane and PAH surrogate standards listed in Tables 1 and 2, and Soxhlet extracted for 4–8 h. The dichloromethane was filtered through anhydrous sodium sulfate, reduced in volume over a 60–70°C water bath, and exchanged with hexane to a final volume of 1 mL.

The hexane extracts were further purified by alumina chromatography (but were not resolved into alkane and PAH fractions) before instrumental analysis. Hexane extracts were eluted from columns that contained 10 g alumina, 10 g elemental copper, and 1 g anhydrous sodium sulfate with 100 mL dichloromethane, which was subsequently reduced in volume and exchanged with hexane to a final volume of 1 mL as above.

Dry Weight Determination

Aliquots of wet samples were weighed, dried at 50–60°C to constant weight, and reweighed to determine percentage moisture. These percentages were used to calculate dry weights from measured wet weights of sample aliquots used for hydrocarbon analyses whenever wet tissues or sediments were extracted.

Sample Extract Fractionation and Purification

Seawater.—Hydrocarbon extracts of seawater samples analyzed at ABL were not fractionated or purified by high-performance liquid chromatography (HPLC). These extracts were spiked, after final concentration, with internal standards (dodecylcyclohexane for alkanes; hexamethylbenzene for PAH) to determine recovery of the surrogate standards, and separate aliquots were analyzed for alkanes and PAH.

Tissues.—Alkane hydrocarbons and PAH extracted from tissue samples were separated by silica gel–alumina column chromatography. The chromatographic columns consisted of 5 g sand above 20 g silica gel (deactivated with 5% water) above 10 g alumina in a 20-mm internal diameter glass column filled with pentane. The alkane hydrocarbons eluted with approximately 50 mL pentane, followed by the PAH that eluted with approximately 200 mL 1:1 pentane:dichloromethane (by volume). The pentane solutions containing the alkane hydrocarbons were reduced in volume over a 60–70°C water bath, exchanged with hexane to a final volume of 1 mL, spiked with an internal standard for estimating recoveries of the deuterated alkane hydrocarbon surrogate standards added initially, and stored for analysis by gas chromatography with flame ionization detection. The pentane:dichloromethane solutions containing the PAH were reduced in volume over a 60–70°C water bath to a final volume of approximately 0.5 mL, and stored for further purification before analysis by gas chromatography–mass spectrometry.

Polynuclear aromatic hydrocarbons from the sample extracts were further purified by gel permeation HPLC. The injection volume was 0.5 mL into dichloromethane flowing at 7 mL/min through two size–exclusion gel columns (Phenomenex, phenogel, 22.5 mm × 250 mm, 100 Å pore size) connected sequentially. The initial 110-mL eluate was discarded, and the next 50 mL was concentrated over a 60–70°C water bath, exchanged with hexane to a final volume of approximately 1 mL, and then spiked with an internal standard for estimating recoveries of the deuterated aromatic hydrocarbon surrogate standards added initially.

Sediments.—At ABL, hydrocarbon extracts of sediment samples were fractionated using the same method as for tissue samples (described above), except that 20 g elemental copper and 8 g anhydrous sodium sulfate were added above the sand of the silica gel–alumina columns.

At GERG, hydrocarbon extracts of sediment samples were not fractionated or purified by HPLC. These extracts were spiked, after final concentration, with internal standards (hexadecane-d_{34} for alkanes; fluorene-d_{10} and benzo[*a*]pyrene-d_{12} for PAH) to determine recovery of the surrogate standards, and separate aliquots were analyzed for alkanes and PAH.

Instrumental Analysis

Polynuclear aromatic hydrocarbons.—In extracts PAH were separated and analyzed with a Hewlett-Packard 5890 gas chromatograph equipped with a 5970B mass selective detector (MSD). The injection volume was 1–4 μL into a splitless injection port at 300°C. The initial oven temperature was 40–60°C, increasing at 10°C per minute immediately after injection to a final temperature of 300°C. The chromatographic column was a 25-m or longer fused silica capillary (0.20- or 0.32-mm internal diameter) coated with 5% phenyl methyl silicone. The helium carrier gas was maintained at 70 kPa inlet pressure.

The chromatographic column eluted into the 70-eV electron impact MSD through a 300°C transfer line. The ionizer temperature and pressure were 240°C and 10^{-5} torr, respectively. The MSD was operated in the selected ion monitoring (SIM) mode, and the specific ions and retention time windows of the SIM Table were chosen to detect all the calibrated aromatic hydrocarbons, the surrogate standards, and the 200 most prominent peaks found on the total ion chromatogram of an aliquot of *Exxon Valdez* crude oil analyzed with the MSD operating in the scan mode. The MSD was tuned with mass 69, 219, and 502 fragments of perfluorotributylamine before each batch of samples was analyzed.

Calibrated PAH were identified based on retention time and ratio of two mass fragment ions characteristic of each hydrocarbon. Calibrated PAH are identified in Table 2 and include dibenzothiophene and the aromatic hydrocarbons in Standard Reference Material (SRM) 1491 supplied by the National Institute of Standards and Technology (NIST). Chromatographic peaks were identified as a calibrated aromatic hydrocarbon if both ions were codetected at retention times within ±0.15 min of the mean retention time of the hydrocarbon in the calibration standards and if the ratio of the confirmation ion to the quantification ion was within ±30% of the expected ratio.

Uncalibrated PAH include the alkyl-substituted isomers of naphthalene, fluorene, dibenzothiophene, phenanthrene, fluoranthene, and chrysene listed in Table 2. Uncalibrated PAH were identified by the presence, within a relatively wide retention time window, of a single mass fragment ion that is characteristic of the uncalibrated PAH sought. Wider retention time windows were necessary for the uncalibrated PAH because of the range of retention times of the various isomers that constitute an uncalibrated PAH homologue grouping (e.g., C3-phenanthrene).

Concentrations of calibrated PAH in extracts were estimated by a method employing multiple internal standards and a five-point calibration curve for each calibrated PAH. The deuterated surrogate standards that were initially spiked into each sample are treated as internal standards, where each surrogate compound is associated with one or more calibrated PAH (see Table 2). A calibration curve for each calibrated PAH and batch of samples analyzed was based on five different hexane dilutions of dibenzothiophene and NIST SRM 1491. Each calibration curve was derived from linear regression of (1) the ratio of MSD–SIM quantification ion response of the calibrated PAH to that of the associated deuterated surrogate standard and (2) the ratio of the amount of calibrated PAH to the amount of deuterated surrogate in the calibration standards.

Concentrations of uncalibrated PAH in extracts were determined with calibration curves and procedures for the most similar calibrated PAH (ABL) or the most similar unsubstituted PAH (GERG). The sum of the MSD–SIM responses to the quantification ion of each uncalibrated PAH homologue isomer was used in place of the calibrated PAH response in the procedure described above for calculating concentrations of calibrated PAH. For example, the fluorene calibration curve and procedure was used at ABL and at GERG for all the alkyl-substituted fluorenes identified, but the 1-methylphenanthrene calibration curve and procedure was used for all the alkyl-substituted phenanthrenes at ABL, whereas the phenanthrene calibration curve and procedure was used for all the alkyl-substituted phenanthrenes at GERG.

Alkanes.—Alkanes in extracts were separated and analyzed with a Hewlett-Packard 5890 gas chromatograph equipped with a flame ionization detector (FID). The injection volume was 1–4 μL into a splitless injection port at 300°C. The 60°C initial oven temperature was maintained for 1 min and then increased by 6°C/min to a final temperature of 300°C. The chromatographic column was the same as that used for PAH analysis (see above). The helium carrier gas flow rate was 0.80–2.0 mL/min, and the column effluent was combined with 34 mL/min helium or nitrogen make-up gas before entering the FID. The FID was operated with hydrogen and air flow rates of approximately 33 and 360–410 mL/min, respectively.

Alkane hydrocarbons were identified based on

TABLE 3.—Method detection limits (MDLs), in nanograms, of alkane analytes analyzed in sediment and tissue samples at Geochemical and Environmental Research Group (GERG) and at Auke Bay Laboratory (ABL). Method detection limits were estimated according to the procedure described in Appendix B, 40 CFR Part 136 and are presented here on an absolute mass basis; MDLs for specific alkanes in specific samples are calculated as the ratio of the absolute mass MDLs presented here to the sample aliquot weight.

	GERG		ABL	
Alkane analyte	Sediment	Tissue	Sediment	Tissue
Decane	124.6	124.6	61.8	134.0
Undecane	50.9	50.9	44.8	78.4
Dodecane	48.9	48.9	67.0	51.0
Tridecane	48.9	48.9	57.6	103.0
Tetradecane	101.0	101.0	45.0	196.0
Pentadecane	101.0	101.0	202.0	181.0
Hexadecane	54.8	54.8	88.4	284.0
Heptadecane	40.8	40.8	113.6	379.0
Pristane	61.7	61.7	210.0	204.0
Octadecane	35.9	35.9	242.0	147.0
Phytane	35.9	35.9	242.0	147.0
Nonadecane	16.6	16.6	55.0	173.0
Eicosane	31.9	31.9	108.4	197.0
Heneicosane	30.9	30.9	94.2	145.0
Docosane	23.3	23.3	62.6	54.1
Tricosane	12.8	12.8	222.0	107.0
Tetracosane	30.2	30.2	350.0	124.0
Pentacosane	35.9	35.9	75.4	73.0
Hexacosane	31.8	31.8	212.0	74.6
Heptacosane	26.4	26.4	170.2	141.0
Octacosane	25.0	25.0	140.8	129.0
Nonacosane	46.1	46.1	152.4	239.0
Triacontane	30.1	30.1	186.0	158.0
Dotriacontane	48.9	48.9	242.0	110.0
Tetratriacontane	80.2	80.2	144.6	64.4

TABLE 4.—Method detection limits (MDLs), in nanograms, of polynuclear aromatic hydrocarbon (PAH) analytes analyzed in sediment and tissue samples at Geochemical and Environmental Research Group (GERG) and at Auke Bay Laboratory (ABL). Method detection limits were estimated according to the procedure described in Appendix B, 40 CFR Part 136 and were estimated for calibrated PAH only; MDLs for uncalibrated PAH were assumed to be similar to those for homologous calibrated PAH. Method detection limits are presented here on an absolute mass basis; MDLs for specific PAH in specific samples are calculated as the ratio of the absolute mass MDLs presented here to the sample aliquot weight.

	GERG		ABL	
PAH analyte	Sediment	Tissue	Sediment	Tissue
Naphthalene	6	22.5	10.1	22.0
2-Methylnaphthalene	8	36.0	9.44	9.94
1-Methylnaphthalene	8	25.5	20.6	13.1
C1 naphthalene	16	61.5	30.0	23.0
2,6-Dimethylnaphthalene	24	25.5	19.5	7.03
C2 naphthalene	24	25.5	19.5	7.03
2,3,5-Trimethylnaphthalene	24	25.5	9.46	4.23
C3 naphthalene	24	25.5	9.46	4.23
C4 naphthalene	24	25.5	9.46	4.23
Biphenyl	24	15.0	43.0	7.52
Acenaphthylene	38	21.0	25.0	5.73
Acenaphthene	46	10.0	12.1	5.40
Fluorene	24	12.5	18.3	4.97
C1 fluorene	24	12.5	18.3	4.97
C2 fluorene	24	12.5	18.3	4.97
C3 fluorene	24	12.5	18.3	4.97
Dibenzothiophene	20	10.0	9.64	3.54
C1 dibenzothiophene	20	10.0	9.64	3.54
C2 dibenzothiophene	20	10.0	9.64	3.54
C3 dibenzothiophene	20	10.0	9.64	3.54
Phenanthrene	6	11.0	13.3	8.29
1-Methylphenanthrene	6	29.0	27.2	5.57
C1 phenanthrene–anthracene	6	29.0	27.2	5.57
C2 phenanthrene–anthracene	6	29.0	27.2	5.57
C3 phenanthrene–anthracene	6	29.0	27.2	5.57
C4 phenanthrene–anthracene	6	29.0	27.2	5.57
Anthracene	40	9.0	33.2	6.50
Fluoranthene	4	6.5	9.12	11.8
Pyrene	30	9.0	10.4	9.75
C1 fluoranthene	4	6.5	9.12	11.8
Benzo[*a*]anthracene	14	25.0	3.34	4.28
Chrysene	6	18.5	26.4	5.54
C1 chrysene	6	18.5	26.4	5.54
C2 chrysene	6	18.5	26.4	5.54
C3 chrysene	6	18.5	26.4	5.54
C4 chrysene	6	18.5	26.4	5.54
Benzo[*b*]fluoranthene	18	19.5	12.2	6.01
Benzo[*k*]fluoranthene	18	19.0	7.96	7.75
Benzo[*e*]pyrene	24	19.5	12.1	8.21
Benzo[*a*]pyrene	12	21.5	32.2	6.54
Perylene	34	10.0	11.7	9.41
Indeno[1,2,3-*cd*]pyrene	16	22.5	6.80	4.43
Dibenzo[*ah*]anthracene	26	20.0	4.94	3.76
Benzo[*ghi*]perylene	4	15.5	9.34	5.53

their retention times. Any peak detected above the integrator threshold within ±0.25% of the mean retention time of an alkane in the calibration standards was identified and quantified as that alkane. Concentrations of calibrated alkanes (listed in Table 1) were estimated by an internal-standard method employing a five-point calibration curve for each alkane. The deuterated surrogate standards that were initially spiked into each sample were treated as internal standards, where each surrogate compound was associated with a group of calibrated alkanes (see Table 1). A calibration curve for each calibrated alkane and batch of samples analyzed was based on five different hexane dilutions of the alkane standards. Each calibration curve was derived from linear regression of (1) the ratio of FID response of the alkane to that of the associated deuterated surrogate standard and (2) the ratio of the amount of calibrated alkane to the amount of deuterated surrogate in the calibration standards.

Amounts of uncalibrated alkane hydrocarbons and the cumulative amount of hydrocarbons in the unresolved complex mixture (UCM) were calculated with respective detector responses and the calibration curve for hexadecane. Flame ionization detector response attributable to the UCM was de-

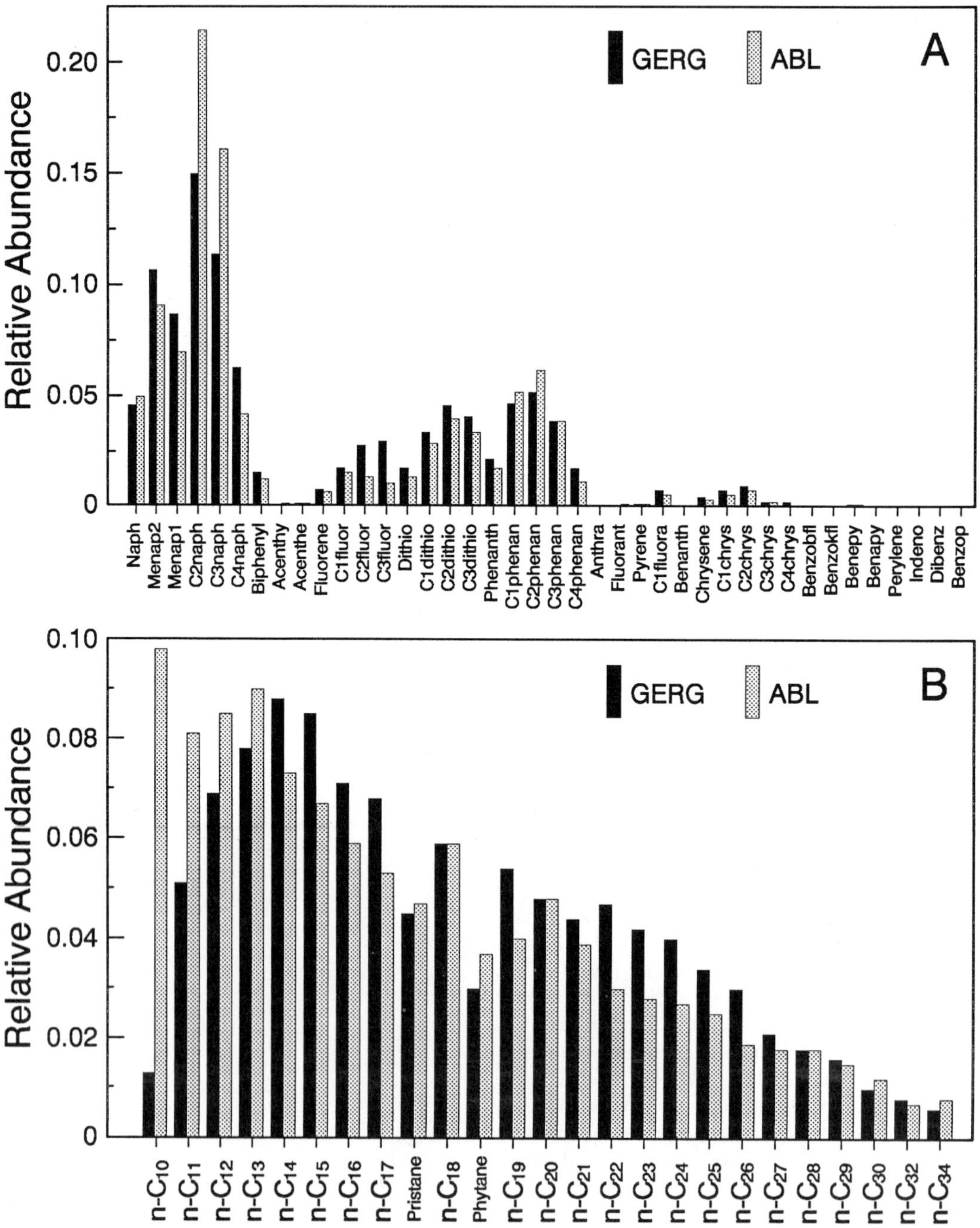

FIGURE 1.—Comparison of (**A**) polynuclear aromatic hydrocarbon (PAH) and (**B**) alkane results for crude oil spilled from the TV *Exxon Valdez* collected from the cargo (Auke Bay Laboratory, [ABL]) or from the sea surface within 24 h of the spill onset (Geochemical and Environmental Research Group [GERG]), and analyzed at ABL and at GERG using the methods described herein. Relative abundance is each analyte's proportion of the sum of the PAH or normal alkanes detected.

termined as the difference between the total FID response and the response caused by distinguishable peaks.

Quality Assurance

Quality control samples were analyzed with each batch of 12–20 samples to assess the accuracy and precision of the analysis and to verify the absence of laboratory contaminants introduced during analysis. Quality control samples for accuracy assessment were prepared from NIST hydrocarbon standards, and the results of the analysis of these samples with each batch was compared with hydrocarbon concentrations certified by NIST. More than 95% of the results from these quality control samples were within ±15% of the NIST-certified values. Precision was assessed by analysis of reference samples prepared at each laboratory and analyzed with each batch. Precision expressed as coefficient of variation was within ±20% for 95% of the reference samples analyzed. Absence of laboratory contaminants was verified by analysis of method blanks with each batch. The frequency with which contaminants were measured in amounts exceeding method detection limits (MDLs, see below) was less than 1%.

The MDLs were estimated for each calibrated alkane and PAH analyte using the procedure described in Appendix B, 40 CFR (Code of Federal Regulations) Part 136. These estimates of detection limit amounts are indicated for each calibrated alkane and PAH analyte in Tables 3 and 4. Detection limits expressed as concentrations in a sample may be found by dividing the amounts listed in Tables 3 or 4 by the weight of sample extracted. The MDLs for uncalibrated PAHs were not experimentally determined. Consequently, detection limits for these analytes were arbitrarily assumed to be the MDL of the most closely related calibrated PAH analyte (Table 4).

Comparative Analysis of *Exxon Valdez* Crude Oil

A sample of crude oil from the hold of the TV *Exxon Valdez* was analyzed at ABL using the methods described above, and results for PAH and for alkanes are compared in Figure 1 with results produced by GERG for a sample of this oil collected from the sea surface 24 h after the oil began to spill. The results in Figure 1 are presented as relative abundances, that is, as proportions of the PAH analyte with respect to the sum of the PAH analytes or as proportions of the alkane analyte with respect to the sum of the normal alkane analytes. The ABL results are from a single analysis, whereas the GERG results are means of 109 analyses. The relative abundances of the calibrated PAH and alkane analytes generally agree to within 1%, except for the lowest-molecular-weight alkanes, which probably evaporated from the GERG sample. The relative abundances of the uncalibrated PAH agree somewhat less well, especially the C2 and C3 alkylnaphthalenes. Reasons for this lack of agreement may include differences in the way uncalibrated PAH were integrated at the two laboratories and different response factors.

The spilled oil contains five prominent homologous series of PAH, in relative proportions that may be useful for identifying the oil in environmental samples. These PAH series include the naphthalenes, fluorenes, dibenzothiophenes, phenanthrene–anthracenes, and the chrysenes. As the oil weathered after the spill, preferential losses of the less alkyl-substituted and the lower-molecular-weight PAH may be expected (Boehm et al. 1981). The most persistent PAH should therefore include the alkyl-substituted dibenzothiophenes, phenanthrenes, and chrysenes. We therefore present two ratios of these PAH analytes for the unweathered oil for comparison with results for weathered oil presented elsewhere in this volume (e.g., O'Clair et al. 1996, this volume). These two ratios are (1) the sum of the alkyl-dibenzothiophenes divided by the sum of the alkyl-phenanthrenes (anthracenes) and (2) the sum of the alkyl-chrysenes divided by the sum of the alkyl-phenanthrenes (anthracenes). Based on results of the GERG analyses, these ratios (mean ± SD) are (1) 0.772 ± 0.114 and (2) 0.134 ± 0.056. Based on the single ABL analysis, these ratios are (1) 0.619 and (2) 0.094, respectively, and are not significantly different from the GERG results ($P > 0.18$).

References

Boehm, P. D., D. L. Fiest, and A. Elskus. 1981. Comparative weathering patterns of hydrocarbons from the *Amoco Cadiz* oil spill observed at a variety of coastal environments. Pages 159–173 *in Amoco Cadiz*, fates and effects of the oil spill. Centre National Pour L'Exploitation des Oceans, Paris.

Brown, D. C., L. S. Ramos, M. Y. Uyeda, A. J. Friedman, and W. S. MacLeod. 1980. Ambient temperature extraction of hydrocarbons from marine sediment-comparison with boiling-solvent extractions. Pages 313–326 *in* L. Petrakis and F. T. Weiss, editors. Petroleum in the marine environment. ACS (American Chemical Society) Symposium Series 185.

Frost, K. J., C. Manen, and T. L. Wade. 1994. Petroleum

hydrocarbons in tissue of harbor seals from Prince William Sound and the Gulf of Alaska. Pages 331–358 *in* T. R. Laughlin, editor. Impacts of the *Exxon Valdez* oil spill on marine mammals. Academic Press, New York.

Krahn, M. M., and nine coauthors. 1993. Comparison of HPLC/fluorescence screening and GC/MS analysis for aromatic compounds in sediments sampled after the *Exxon Valdez* oil spill. Environmental Science and Technology 27:699–708.

Larsen, M., L. Holland, D. Fremgen, J. Lunasin, M. Wells, and J. Short. 1992. Standard operating procedures for the analysis of petroleum hydrocarbons in seawater, marine sediments, and marine faunal tissue at the Auke Bay Laboratory. National Marine Fisheries Service, Auke Bay Laboratory, Juneau, Alaska.

O'Clair, C. E., J. W. Short, and S. D. Rice. 1996. Contamination of intertidal and subtidal sediments by oil from the *Exxon Valdez* in Prince William Sound. American Fisheries Society Symposium 18:61–93.

Wade, T. L., J. M. Brooks, M. C. Kennicutt, II, T. J. McDonald, J. L. Sericano, and T. J. Jackson. 1993. GERG trace organics contaminant analytical techniques. National Status and Trends Program for Marine Environmental Quality. NOAA (National Oceanic and Atmospheric Administration) Technical Memorandum NOS (National Ocean Service) ORCA (Office of Ocean Resources and Conservation Assessment) 71, Silver Spring, Maryland.

American Fisheries Society Symposium 18:149–166, 1996

Prespill and Postspill Concentrations of Hydrocarbons in Mussels and Sediments in Prince William Sound

JEFFREY W. SHORT AND MALIN M. BABCOCK

National Marine Fisheries Service, Alaska Fisheries Science Center, Auke Bay Fisheries Laboratory 11305 Glacier Highway, Juneau, Alaska 99801-8626, USA

Abstract.—Hydrocarbon concentrations in intertidal mussels and sediments were compared before and after crude oil from the TV *Exxon Valdez* contaminated beaches in Prince William Sound. Comparison of hydrocarbon concentrations in these samples at stations inside and outside the path of the oil and before and after landfall of the spilled oil were used to (1) evaluate the significance of petroleum hydrocarbon contaminants found after the spill, (2) establish restoration criteria for affected beaches, and (3) determine natural recovery rates for these beaches. Concentrations of polynuclear aromatic hydrocarbons (PAHs) near or below detection limits in mussels outside the path of the spilled oil, or collected before oil became beached, indicate that other sources of PAHs were negligible and that seawater in the spill area generally was free of anthropogenic hydrocarbon pollution. Contamination of mussels by PAHs derived from natural sources, such as marine oil seeps, was not detected. Contamination of mussels was highest during the first few weeks after the spill, followed by a consistent decline. Concentrations of hydrocarbons derived from *Exxon Valdez* oil were highest and most persistent in mussels near or on heavily oiled beaches. Contamination of lower intertidal sediments by PAHs derived from *Exxon Valdez* oil was less than 1,000 ng/g (dry weight basis), even when concentrations in adjacent upper intertidal sediments were higher by factors exceeding 10. Concentrations of spill-derived PAHs in lower intertidal sediments consistently declined during the years after the spill. At stations unaffected by the spilled oil, total PAH (TPAH) concentrations in lower intertidal sediments were usually less than 100 ng/g, except near Hinchinbrook Entrance. Analytical data on mussels and sediments for stations sampled before the spill and for unaffected stations can be used to establish restoration criteria for oiled beaches.

The TV *Exxon Valdez* oil spill of 24 March 1989 provided a unique opportunity to assess the impacts of spilled crude oil on a subarctic intertidal ecosystem, because a rigorous baseline study had been conducted that demonstrated the general absence of substantial hydrocarbon contamination before this spill (Karinen et al. 1993). As a result, it is possible to compare hydrocarbon concentrations in intertidal sediments and mussels before and after the spill. The impacts of the spilled oil on these sediments and mussels are therefore uniquely definitive: unique because no other major marine oil spill has occurred in a largely pristine environment where prior hydrocarbon contaminant concentrations had been established and definitive because the impacts of the spill on intertidal sediments and mussels are generally not obscured by other hydrocarbon contamination sources.

Shipment of Alaska North Slope crude oil through Prince William Sound began after the opening of the Trans-Alaska pipeline and the Valdez tanker terminal in 1977. At that time, our laboratory conducted a survey of selected hydrocarbons in sediments and mussels *Mytilus trossulus* for comparison with future evidence of hydrocarbon contamination associated with oil transshipment (Karinen et al. 1993). Samples of intertidal sediments and mussels were collected two to three times yearly during 1977–1979 and once in 1980 at eight sampling stations in Prince William Sound and were analyzed for alkane and (mainly) unsubstituted polynuclear aromatic hydrocarbons (PAHs).

Within 2 d after the *Exxon Valdez* oil spill, we resumed sampling at these stations, and all stations were sampled before landfall of any of the spilled crude oil. Sample collection after the oil spill involved use of collection methods identical to those used for the hydrocarbon-baseline-monitoring study of 1977–1980, and the hydrocarbon analytical methods that we used had comparable accuracy, precision, and detection limits (Brown et al. 1980; Short et al. 1996a, this volume). Additional alkyl-substituted PAHs were measured in samples collected after the oil spill, thereby providing better corroborating evidence for hydrocarbon source identification by significantly expanding the range of PAHs measured for the 1977–1980 hydrocarbon-baseline-monitoring study.

The primary objective of this study was to compare variations in temporal and geographic petroleum hydrocarbon concentrations in nearly 500 samples of lower intertidal sediments and mussels. Results of comparisons of petroleum hydrocarbon concentrations and compositions at stations before and after landfall of the spilled oil and inside and outside the path followed by the spilled oil through Prince William Sound, along with data collected during 1977–1980 (Karinen et al. 1993) can be used to evaluate the significance of petroleum hydrocarbon contaminants found at stations after the spill. These data can be used further to establish restoration criteria for affected beaches and to determine natural recovery rates for these beaches.

Study Area

Mussels and sediments were collected from 13 stations in Prince William Sound from 26 March 1989 through 2 May 1991 (Figure 1 and Table 1). Of the 13 stations, 6 were established in 1977 by Karinen et al. (1993), 7 were sampled in March 1989 before *Exxon Valdez* oil became beached in Prince William Sound, and 4 were outside the path of the spilled oil through the sound. These 13 stations were sampled up to five times in 1989 from March through August. Twelve of the stations were each sampled in April, June, and August of 1990, and in April or May of 1991 (except Constantine Harbor); the station at Green Island was only sampled in 1989.

Methods

Sample Collection

Sample collection methods were identical to those of Karinen et al. (1993). One mussel collection transect was located in the mussel band parallel to the waterline (circa +1-m tide level) at each station. Mussel samples were collected in triplicate by taking approximately 30 mussels (2–5 cm in length) each along the 30-m transect within 1 m above and 1 m below the line.

The sediment-collection transect line (30 m) was located parallel to the waterline at a tide level between −0.75 and +0.75 m. Sediment-sampling transects were usually directly seaward of the mussel-sampling transects, but at some stations they were offset along the shoreline in order to sample suitable fine sediment. Sediment cores were collected with a dichloromethane-rinsed, hydrocarbon-free stainless steel cookie cutter (diameter 3.2 cm × depth 1.25 cm). When necessary, a hydrocarbon-free spatula was thrust underneath the cutter to retain the sediment sample intact. Triplicate, pooled sediment samples were collected at each station, each consisting of 10 cores taken along the 30-m transect and within 1 m above and 1 m below the line. At the Elrington Island station, triplicate sediment samples were also collected from a visibly oiled area in the upper intertidal area in May 1989 for comparison with the lower intertidal sediments there. All samples from each site were placed in separate, dichloromethane-rinsed glass jars and were frozen within 2–3 h of collection.

Hydrocarbon and Data Analysis

Sediment and mussel samples were analyzed for selected hydrocarbons at the Auke Bay Laboratory. The analytes, methods, detection limits, and accuracy and precision of the methods are summarized elsewhere in this volume (Short et al. 1996a).

Results of the chemical analysis were screened on the basis of surrogate recoveries and method detection limits (MDLs). Individual analytes and the summary statistics affected by them (total PAH [TPAH] and total normal alkanes [TNA, see below]) were excluded from consideration if the recoveries of corresponding analyte surrogates fell outside the range 30–150%, which affected 16% of mussel samples and 3% of sediment samples. Concentrations of individual analytes reported below MDL were replaced by zeros for our analyses. Hydrocarbon concentrations are reported on a dry weight basis to three significant figures when concentrations exceeded 10 ng/g and to two significant figures for lower concentrations. A total of 256 mussel samples and 231 sediment samples were analyzed for hydrocarbons.

Results for PAHs and alkane hydrocarbons were summarized as mean TPAH and mean TNA. Total PAH is the sum of all PAHs detected excluding perylene, a naturally occurring PAH found in marine sediments of Prince William Sound; TNA is the sum of all normal alkanes detected. Mean TNA or TPAH are the means of the TNA or TPAH for replicated samples and are presented together with ±1 SE. The number of replicated samples is three, unless indicated otherwise.

Hydrocarbon composition patterns are presented as mean relative TPAH percents in Figures 3, 6, and 8. Relative percent for each PAH is determined as the ratio of the hydrocarbon concentration and the TPAH concentration, multiplied by 100. Mean relative percent is the mean of these ratios multiplied by 100 for a specified set of samples. Variability of relative percents for each PAH is presented as the range.

We relied on the persistence and relative abundance of alkyl-substituted homologues of dibenzothiophene and chrysene to distinguish between PAH suites derived from *Exxon Valdez* oil versus from alternative sources. In oil collected from the

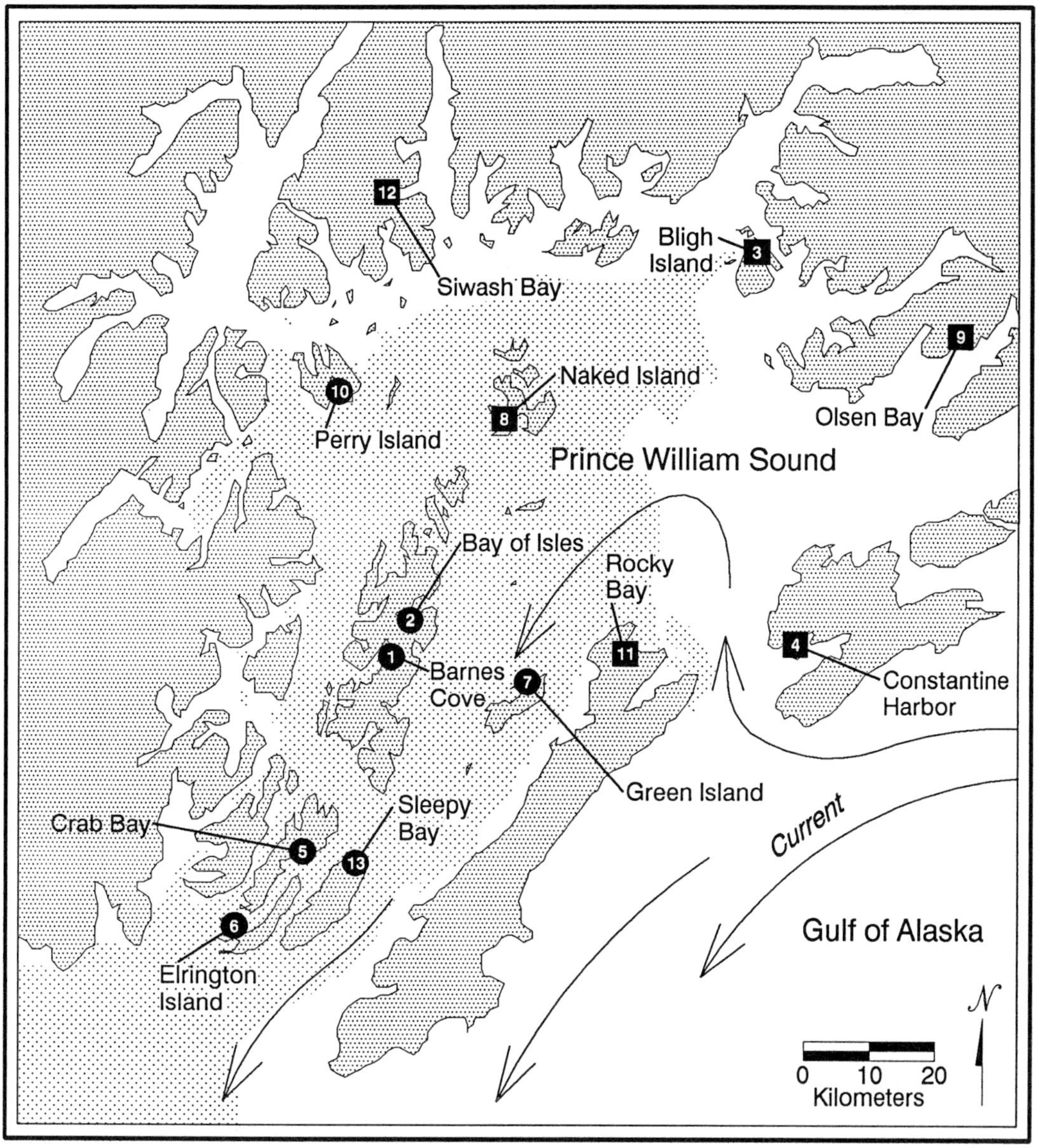

FIGURE 1.—Station locations in Prince William Sound, where mussels and sediments were sampled. Arrows indicate the direction of prevailing oceanic currents through the sound (Galt et al. 1991). Shaded area indicates the path of surface oil through the sound. Numbers in circles denote stations established in 1989 and numbers in squares indicate stations originally established in 1977 and described by Karinen et al. (1993).

cargo of the *Exxon Valdez* and analyzed at Auke Bay Laboratory, the ratio of alkyl-dibenzothiophenes (summed) to alkyl-phenanthrenes (summed) was 0.619, and the ratio of alkyl-chrysenes (summed) to alkyl-phenanthrenes (summed) was 0.094 (Short et al. 1996a). These same ratios in three samples of *Exxon Valdez* oil mousse collected from the sea surface 11 d after the spill (see Short et al. 1996b, this volume, for collection details) had respective ranges of 0.319–0.354 and 0.084–0.099 (see Figure 3A for the relative PAH composition of these mousse samples).

In sediment and mussel samples, the abundance of the alkyl-dibenzothiophenes relative to the alkyl-phenanthrenes was used to distinguish *Exxon Valdez* oil from other crude oils (or coals), and the abun-

TABLE 1.—Station locations and sample dates where mussel and sediment samples were collected from Prince William Sound. Italicized station names indicate stations established in 1977 described by Karinen et al. (1993). The symbol "x" indicates collection of sediment and mussel samples at the indicated station and date; the symbols "s" and "m," respectively, indicate collection of sediment or mussel samples only.

	1989					1990			1991
Station	Mar	Apr	May	Jun	Aug	Apr	Jun	Aug	Apr or May
Barnes Cove	x	x	x	m	x	x	x	x	x
Bay of Isles	x	x	x	m	x	x	x	x	x
Bligh Island	x		x	m	x	x	x	x	x
Constantine Harbor	s		s	m	s	x	x	s	
Crab Bay		x	x	m	x	x	x	x	x
Elrington Island			x	x	x	x	x	x	x
Green Island		m							
Naked Island	x	x	x	x	x	x	x	x	x
Olsen Bay		x	x	m	x	x	x	x	x
Perry Island			m		x	x	x	x	x
Rocky Bay	x		x	m	x	x	x	x	x
Siwash Bay	x		x		x	x	x	x	x
Sleepy Bay			x		x	x	x	x	x

dance of the alkyl-chrysenes relative to the alkyl-phenanthrenes was used to distinguish crude oils from products refined from crude oils, as follows: The pattern of PAH concentrations in the samples was judged similar to that in *Exxon Valdez* oil (and is denoted as EVO-PAH) if each of three criteria was consistently met in all replicated samples, such that (1) the ratio of alkyl-dibenzothiophenes (summed) to alkyl-phenanthrenes (summed) exceeded 0.29, (2) the ratio of alkyl-chrysenes (summed) to alkyl-phenanthrenes (summed) exceeded 0.05, and (3) the concentration of alkyl-phenanthrenes (summed) exceeded 50 ng/g in mussels or 20 ng/g in sediments (this latter criterion is necessary to ensure that chrysenes would be detected if present).

Polynuclear aromatic hydrocarbon patterns that differed from the EVO-PAH pattern because of relatively low alkyl-dibenzothiophene concentrations were judged to be from some alternative crude oil (or coal) source, whereas patterns that differed because of relatively low alkyl-chrysene concentrations were judged to be derived from diesel oil refined from North Slope crude oil. Also, PAH patterns that were not consistently evident in all members of a replicated sample set are denoted as sporadic.

The carbon preference index (CPI) (Farrington and Tripp 1977) was also used to evaluate hydrocarbon sources. The index has the form

$$\mathrm{CPI} = \frac{2(\mathrm{C27} + \mathrm{C29})}{\mathrm{C26} + 2 \cdot \mathrm{C28} + \mathrm{C30}}, \qquad (1)$$

where *Ci* is the concentration (ng/g) of the *n*-alkane of carbon number *i*. The CPI is near 1 for oiled sediments; higher values indicate biogenic sources of these alkanes.

Pentadecane, heptadecane, and nonadecane (i.e., C15, C17, and C19), were often the most prevalent alkanes found in uncontaminated mussels and sediments of Prince William Sound by Karinen et al. (1993), who ascribed them to marine algal sources. Accordingly, we denote these three alkanes collectively as marine-derived alkanes.

One-way analysis of variance was used to evaluate the significance of time variation of sediment TNA and TPAH means at Constantine Harbor and Rocky Bay.

Results

Mussels and intertidal sediments in Prince William Sound generally had low concentrations of PAHs and TNA before impact of *Exxon Valdez* oil. Mussels and sediments from several of the stations in this study were affected by the spill and showed increases in petroleum hydrocarbon concentrations that are directly attributable to *Exxon Valdez* oil.

Mussel Polynuclear Aromatic Hydrocarbons

Prespill concentrations.—Mussel samples taken from stations in Prince William Sound before landfall of *Exxon Valdez* oil had mean TPAH concentrations ranging from 143 ± 58.0 ng/g ($N = 2$) at Barnes Cove to 544 ± 316 ng/g at Rocky Bay (Figure 2). More than 50% of the PAHs of these samples were naphthalenes.

Postspill concentrations.—Mean TPAH concentrations in mussels increased to more than 100,000

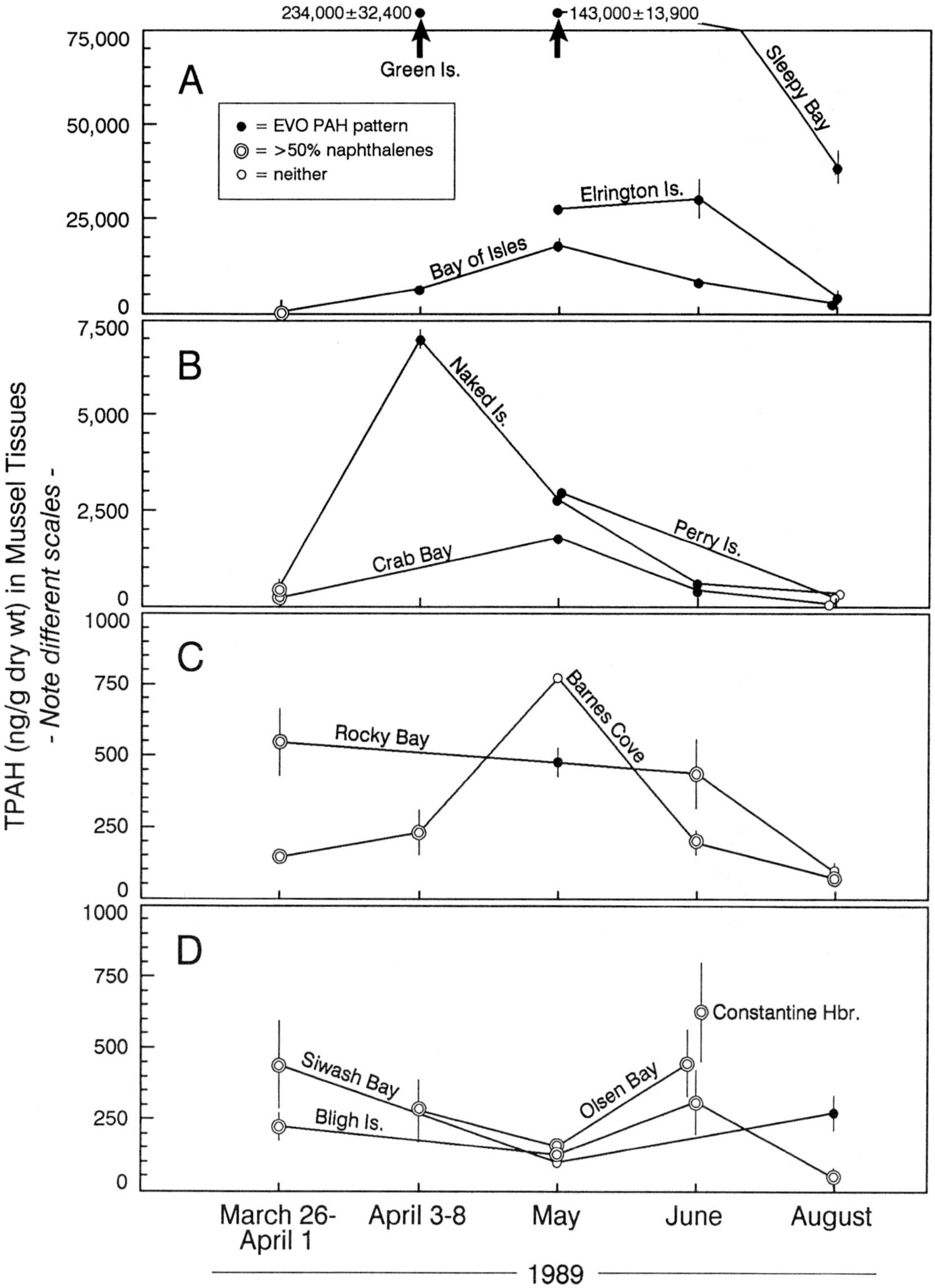

FIGURE 2.—Concentrations (ng/g dry tissue weight) of total polynuclear aromatic hydrocarbons (TPAH) in mussels collected from 13 Prince William Sound sites in spring and summer 1989. Filled symbols indicate PAH composition pattern characteristic of *Exxon Valdez* oil (EVO-PAH) as determined by criteria given in *Methods* section. Circled symbols indicate PAHs consisting mostly of naphthalenes. Vertical bars are ±SE. Note different scales used for **A–D**.

ng/g within 6 weeks at stations where beached oil was especially heavy (Figure 2). Concentrations of PAHs derived from *Exxon Valdez* oil were found in mussels at all stations within the path of the spilled oil through Prince William Sound and were highest during the first 3 months after the spill, followed by a steady decline. The highest TPAH concentrations found in mussels were in samples collected in April–June 1989 from Green Island (234,000 ± 32,400 ng/g, $N = 2$), Sleepy Bay (143,000 ± 13,900 ng/g), and Elrington Island (30,100 ± 5,060 ng/g), where heavy beach oiling was present on or near the mussel bed sampled; however, mean TPAH concentrations at Sleepy Bay and Elrington Island declined to 38,600 ± 4,330 ng/g and 4,480 ± 1,890 ng/g, respectively, by August 1989 (Figure 2A).

The PAH composition of mussels sampled from oiled beaches was similar to that of weathered *Exxon Valdez* oil. For example, the relative abundance of PAHs in mussels collected from Sleepy Bay in May 1989 was very similar to that of the three samples of floating mousse oil collected near eastern Knight Island 11 d after the spill (Figure 3, A and B).

The maximum TPAH concentrations were lower in mussels at stations where beach oiling was less apparent. At Bay of Isles, Naked Island, Perry Island, and Crab Bay, TPAH concentrations were consistently highest in April or May 1989 and were derived from EVO-PAHs. Mean TPAH concentrations ranged from 1,790 ± 13.4 ng/g ($N = 2$) at Crab Bay to 18,000 ± 1,540 ($N = 2$) at Bay of Isles (Figure 2, A and B).

Contributions of EVO-PAHs to mussels at remaining stations were relatively small. These hydrocarbons were evident only at Rocky Bay in May and at Siwash Bay in August, and mean TPAH concentrations were less than 500 ng/g in these samples (Figure 2, C and D).

Mean TPAH concentrations in 1989 were usually less than 500 ng/g in mussels from the stations also sampled in 1977–1980, and consisted mainly of naphthalenes. At Bligh Island, Constantine Harbor, and Olsen Bay, naphthalenes accounted for more than half of the TPAHs detected (Figure 2D). Similarly, naphthalenes consistently accounted for more than half of the TPAHs detected in all mussel samples collected in March before any oil spill impacts (compare A–D of Figure 2).

In 1990, mussel EVO-PAHs were consistently evident only in samples from Sleepy Bay. The PAH composition pattern differed from that depicted in Figure 3A for 11-d floating mousse oil because of preferential losses of lower-molecular-weight PAHs and lower-alkyl-substituted PAHs (Figure 3C). The mean TPAH concentration in Sleepy Bay samples was highest in August at 21,700 ± 1,510 ng/g, compared to 4,150 ± 205 ng/g in April and 1,830 ± 69.3 ng/g in June.

Mean TPAH concentrations in mussels from Elrington Island and Bay of Isles were also relatively higher in samples collected later in 1990, but the PAH composition indicated diesel oil contamination. Mean TPAH concentrations in mussels increased from 90.5 ± 41.8 ng/g in April 1990 to 3,860 ± 167 ng/g in August at Elrington Island, and from 205 ± 43.7 ng/g ($N = 2$) in April to 489 ng/g ($N = 1$) in June at Bay of Isles.

Mean TPAH concentrations in mussels were low at the remaining stations sampled in Prince William Sound in 1990. Of all the remaining stations and times sampled, the median of the mean TPAH concentrations found was 19 ng/g, and 80% of these means were less than 100 ng/g. The highest mean TPAH observed among these samples was at Olsen Bay in August at 250 ± 245 ng/g ($N = 2$), which occurred because of a single sample that was not corroborated by the replicate.

In April 1991, the EVO-PAH pattern persisted in mussels at Sleepy Bay, when the mean TPAH concentration was 2,030 ± 96.6 ng/g, and the PAH composition pattern was similar to the weathered *Exxon Valdez* oil pattern evident in these mussels in 1990 (compare C and D of Figure 3). At the remaining stations sampled in 1991, mean TPAH concentrations were all less than 165 ng/g.

In contrast to what was observed in 1989, mussel samples—including those that had low TPAH—generally contained naphthalenes as minor constituents in 1990 and 1991. Instead, the most abundant PAHs of these samples were usually alkyl-substituted PAHs sporadically measured near detection limits.

Mussel Alkane Hydrocarbons

Prespill concentrations.—Mussels collected from stations before landfall of *Exxon Valdez* oil showed mean TNA concentrations ranging from 296 ± 36.0 ng/g at Bligh Island to 2,910 ± 903 ng/g at Rocky Bay. At most of the stations, more than 50% of the alkanes were marine-derived (C15, C17, and C19) (Figure 4).

Postspill concentrations.—Mussels in Prince William Sound that contained the highest TPAH concentrations also contained the highest concentrations of TNA and of phytane. Mean TNA and phytane concentrations were highest in mussels

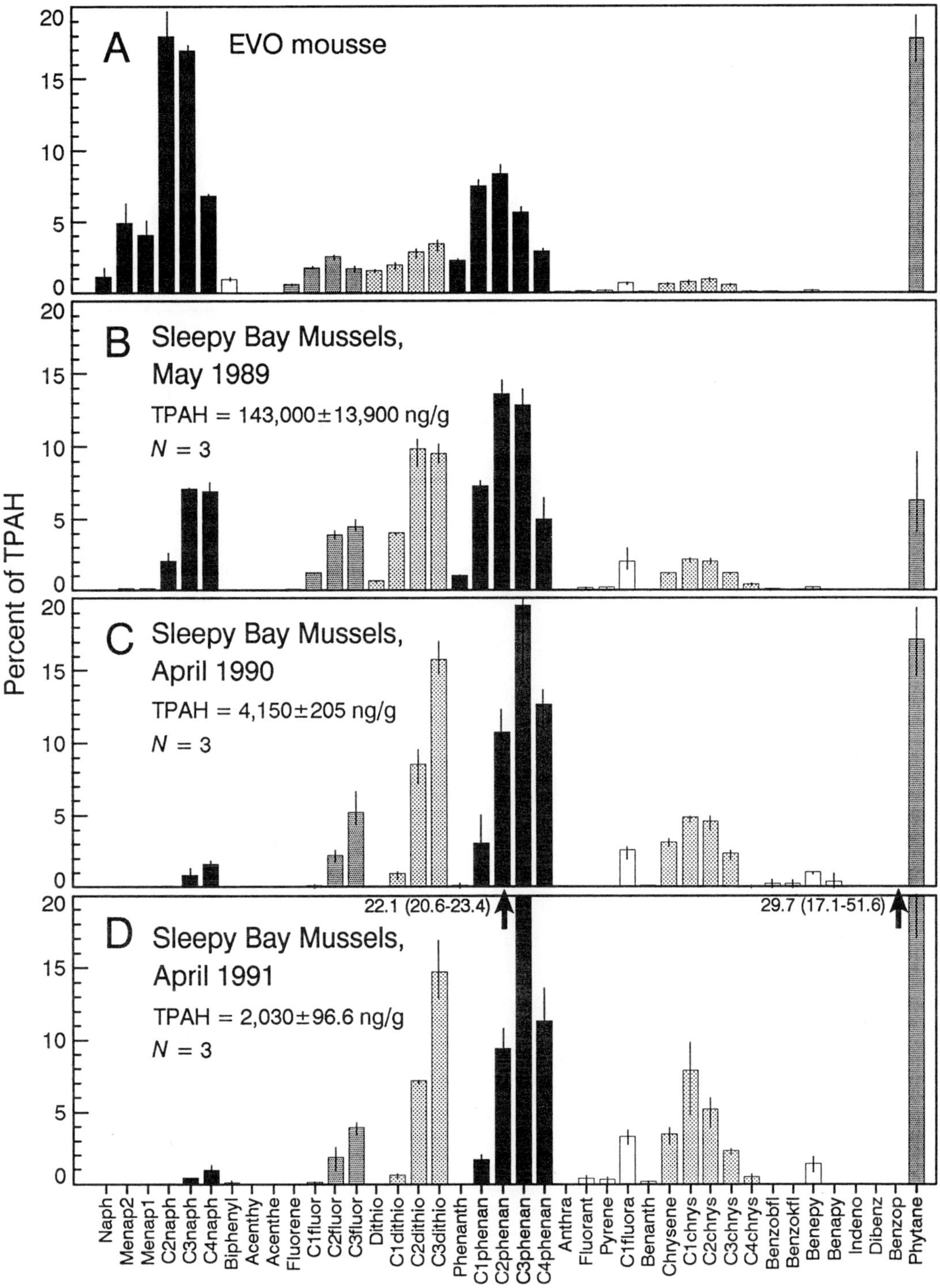

FIGURE 3.—Relative abundance of polynuclear aromatic hydrocarbons (PAHs) and phytane in mousse collected from the sea surface of western Montague Strait 11 d after the *Exxon Valdez* oil spill (**A**) and in mussels collected from Sleepy Bay in 1989 (**B**), 1990 (**C**), and 1991 (**D**). Relative abundances for each sample were calculated as the ratio (percent) of the analyte to the sum of all the PAHs (excluding perylene) in the sample; thick vertical bars indicate the mean of these calculations for replicate samples and thin vertical bars indicate the range. Also given are the mean total PAH (TPAH) concentrations (ng/g dry tissue weight) of the mussels ±SE.

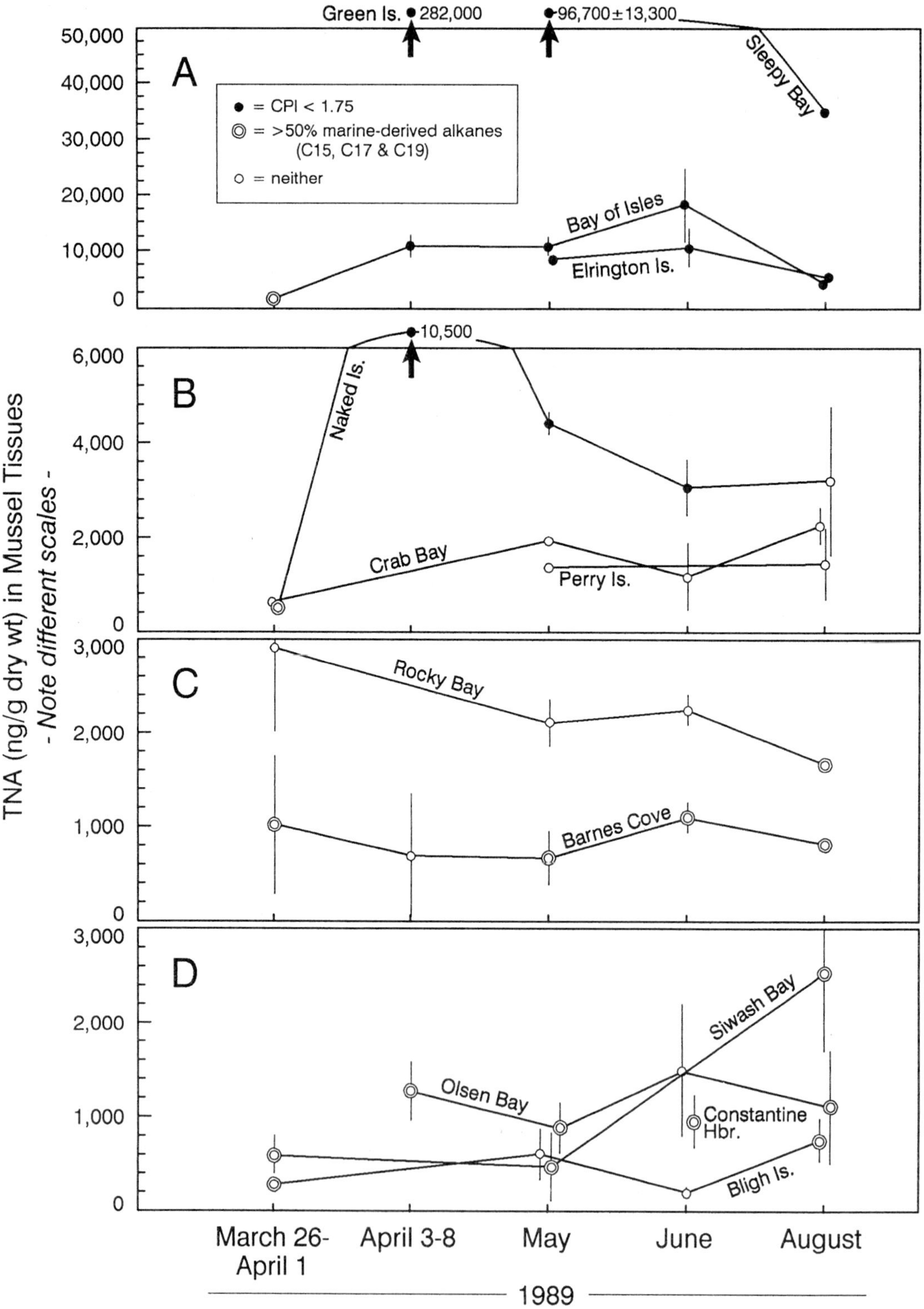

FIGURE 4.—Concentrations of total normal alkanes (TNAs; ng/g dry tissue weight) in mussels collected from Prince William Sound stations in spring and summer 1989. Filled symbols indicate carbon preference indexes (CPIs; see *Methods* section for definition) less than 1.75. Circled symbols indicate that more than 50% of the TNA consists of marine-derived alkanes C15, C17, and C19. Vertical bars are ±SE. Note different scales used for **A–D**.

sampled from April to June 1989 at Green Island, Sleepy Bay, Elrington Island, Bay of Isles, and Naked Island (Figure 4, A and B). Mean TNA concentrations were highest at Green Island (282,000 ng/g [$N = 1$]) and Sleepy Bay (96,700 ± 13,300 ng/g), and exceeded 9,000 ng/g at the other three named stations during this period. In contrast, TNA concentrations in mussels were consistently less than 3,000 ng/g at all of the remaining stations sampled in 1989, as well as in mussels sampled from Naked Island and Bay of Isles in March before oil impact there (Figure 4). Similarly, mean phytane concentrations were also highest at Green Island and Sleepy Bay at 15,900 ± 1,290 ng/g and 9,150 ± 2,430 ng/g, respectively, and ranged from 236 ng/g to 2,530 ± 380 ng/g at Elrington Island, Bay of Isles, Naked Island, Perry Island, and Crab Bay in mussels sampled from April to June 1989. Phytane concentrations in mussels at these stations generally varied with TNA concentrations; however, phytane concentrations were either near or below detection limits at the remaining stations sampled in 1989.

Alkane carbon preference indexes (CPIs) for mussels that contained higher concentrations of normal alkanes were consistently near 1 in 1989, indicating a petroleum source for these alkanes. Alkane CPIs of mussels sampled after March 1989 from Green Island, Sleepy Bay, Elrington Island, and Bay of Isles were consistently lower than 1.75 (Figure 4A), and also at Naked Island except in August. In contrast, normal alkanes of mussels sampled elsewhere, or sampled in March, consisted predominantly of the marine-derived alkanes C15, C17, and C19 and exhibited CPIs > 1.75 (Figure 4).

In 1990, concentrations of TNA and of phytane remained highest at the three stations where TPAH were highest. At Sleepy Bay, TNA concentrations in mussels ranged from 4,890 ± 1,090 ng/g (in June) to 15,500 ± 1,550 ng/g (in August), and phytane concentrations ranged from 414 ± 69.3 ng/g to 2,380 ± 701 ng/g. At Elrington Island and at Bay of Isles, TNA concentrations ranged from 1,950 ± 1,330 ng/g ($N = 2$) to 5,700 ± 1,270 ng/g ($N = 2$), and phytane concentrations ranged from near detection limits to 1,910 ± 265 ng/g. At the remaining stations, TNA concentrations ranged from below detection limits to 5,660 ng/g (except for one sample from Olsen Bay that contained 58,300 ng/g of C21 or a co-eluant), and phytane concentrations ranged from below detection limits to 394 ± 31.9 ng/g (at Crab Bay in August).

For 1990 samples, alkane CPIs could be meaningfully determined only for mussel samples from Sleepy Bay. At Sleepy Bay, alkane CPIs ranged from 0.77 to 1.25. Concentrations of normal alkanes in the range C25–C30 were too frequently below detection limits in most of the remaining mussel samples collected from Prince William Sound stations in 1990 for the CPI calculation to be meaningful. Normal alkanes of mussels sampled at these stations consisted predominantly of the marine-derived alkanes C15, C17, and C19.

In 1991, concentrations of TNA and of phytane were highest in mussels from Sleepy Bay. The TNA concentration in mussels at Sleepy Bay was 2,880 ± 733 ng/g, compared with levels that ranged from below detection limits to 936 ± 794 ng/g elsewhere. Similarly, the phytane concentration was 926 ± 251 ng/g at Sleepy Bay, compared with concentrations elsewhere that were usually below detection limits (exceptions included Crab Bay and Naked Island, where phytane concentrations were 411 ± 233 ng/g and 154 ± 154 ng/g, respectively). As in 1990, concentrations of normal alkanes in the range C25–C30 were too frequently below detection limits in most of the mussel samples collected from Prince William Sound stations in 1991 for the CPI calculation to be meaningful (except at Sleepy Bay, where the CPI was 1.55). Normal alkanes of mussels sampled at these stations consisted predominantly of the marine-derived alkanes C15, C17, and C19.

Sediment Polynuclear Aromatic Hydrocarbons

Prespill concentrations.—Mean TPAH concentrations in sediments collected before impact of *Exxon Valdez* oil ranged from 4.3 ± 4.3 ng/g at the Bay of Isles station to 601 ± 26.1 ng/g at Constantine Harbor.

Postspill concentrations.—In sediments, EVO-PAHs were most consistently evident at Sleepy Bay. Mean TPAH concentrations were highest during the first sampling period at this station in May 1989 at 924 ± 407 ng/g (Figure 5). Concentrations declined slightly by August, substantially by 1990 (<200 ng/g), and further by April 1991 to 33.1 ± 5.43 ng/g. The generally declining mean TPAH concentrations in sediments at Sleepy Bay corresponded to a similar pattern of decline in mussels there (Figure 5).

The PAH composition pattern in sediments at Sleepy Bay was similar to that of mussels that were sampled concurrently (compare Figures 3 and 6). In both sediments and mussels at Sleepy Bay, PAHs of lower molecular weight or with lower alkyl substitution were preferentially lost with time.

The EVO-PAHs were also detected at Elrington Island. In May 1989, the mean TPAH concentration

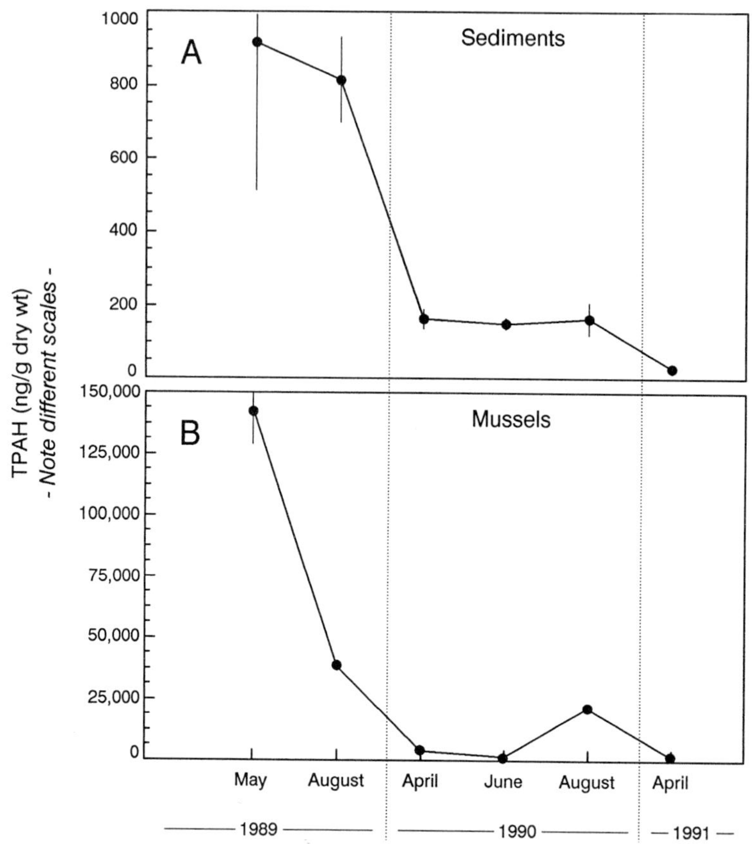

FIGURE 5.—Comparison of total polynuclear aromatic hydrocarbons (TPAH; ng/g dry mussel tissue or sediment weight) in sediments (**A**) and in mussels (**B**) at Sleepy Bay from spring 1989 to spring 1991. Samples were collected concurrently during the indicated periods, but from different places at each site. Vertical bars are ±SE. Note different scales used for **A** and **B**.

in sediment samples collected from the lower intertidal zone near the mussel sampling site was 117 ± 35.7 ng/g, compared with 11,600 ± 2,740 ng/g in adjacent sediments collected from the upper intertidal zone that were visibly oiled, and EVO-PAHs were evident at both sites. At the lower intertidal site, concentrations were consistently lower than 50 ng/g for the remainder of 1989 but increased to 163 ± 89.9 ng/g ($N = 2$) in April 1990, and EVO-PAHs were only sporadically evident in June 1989 and in April 1990. After April 1990, mean TPAH concentrations in sediments at this site were consistently less than 10 ng/g, and individual PAH concentrations were too low to indicate possible PAH sources.

The EVO-PAHs were also sporadically detected in sediments from Barnes Cove, Bay of Isles, Crab Bay, and Perry Island. In 1989, mean TPAH concentrations at these stations ranged to 427 ± 217 ng/g (at Barnes Cove). The highest mean TPAH concentrations in sediments from Bay of Isles, Crab Bay, and Perry Island were 155 ± 75.7 ng/g, 238 ± 86.8 ng/g, and 43.1 ± 31.0 ng/g, respectively. The PAH composition pattern of samples that contained the highest TPAH concentrations from each of these stations consistently matched that of weathered *Exxon Valdez* oil in 1989; however, the PAH composition pattern characteristic of weathered *Exxon Valdez* oil was not as evident in samples with a TPAH concentration less than 50 ng/g.

In 1990, EVO-PAHs were still consistently evident at Bay of Isles, where mean TPAH concentrations ranged from 76.4 ± 7.8 ng/g ($N = 2$) to 197 ± 14.9 ng/g ($N = 2$). At Barnes Cove and Crab Bay, TPAH levels ranged from 55.0 ± 13.6 ng/g to 111 ± 18.8 ng/g ($N = 2$) and often contained low propor-

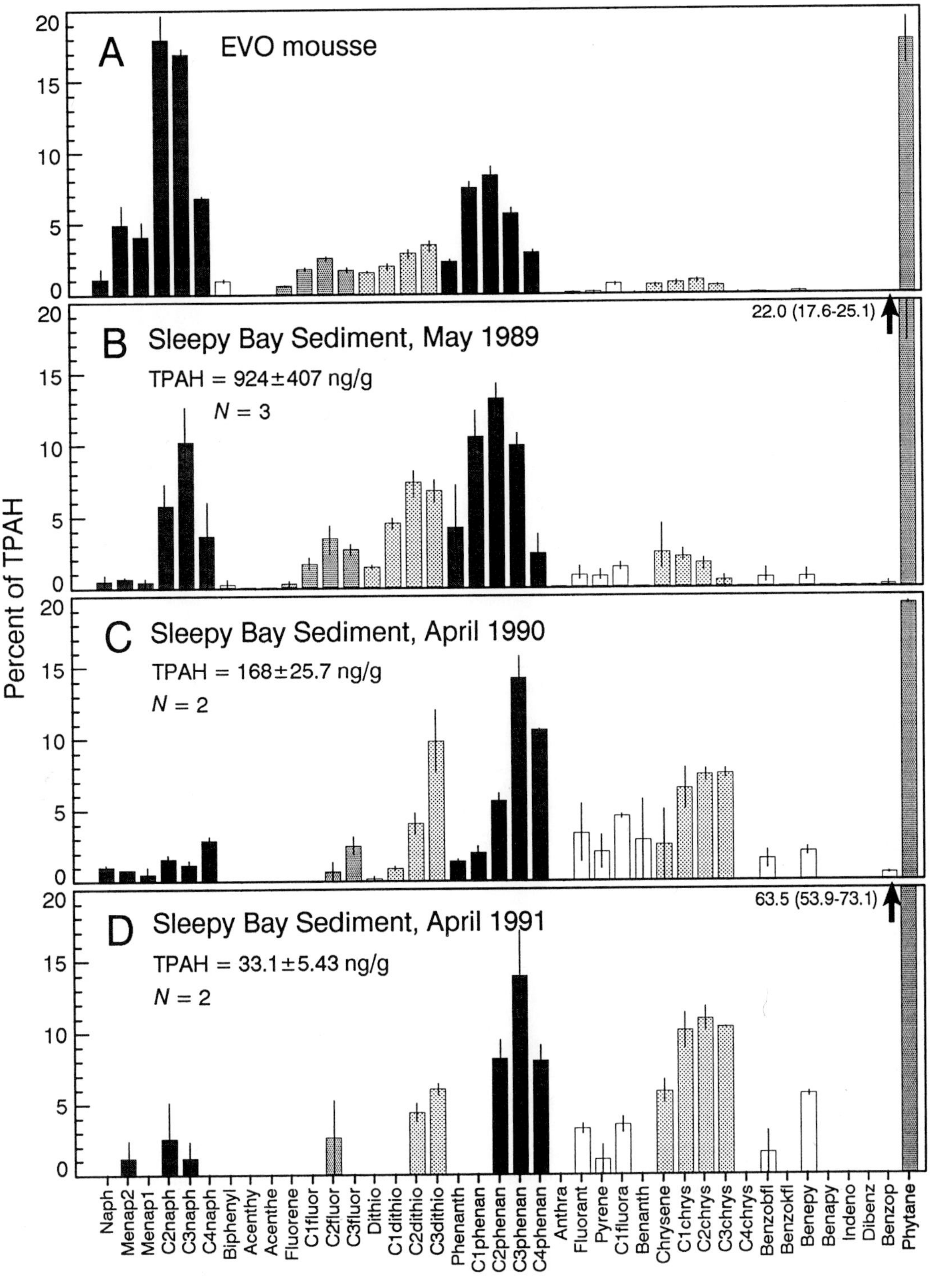

FIGURE 6.—Relative abundance of polynuclear aromatic hydrocarbons (PAHs) and phytane in mousse collected from the sea surface of western Montague Strait 11 d after the *Exxon Valdez* oil spill (**A**) and in sediments collected from Sleepy Bay in 1989 (**B**), 1990 (**C**), and 1991 (**D**). Relative abundances for each sample were calculated as the ratio (percent) of the analyte to the sum of all the PAHs (excluding perylene) in the sample; thick vertical bars indicate the mean of these calculations for replicate samples and thin vertical bars give the range. Also given are the mean total PAH (TPAH) concentrations (ng/g dry sediment weight) of the sediments ±SE.

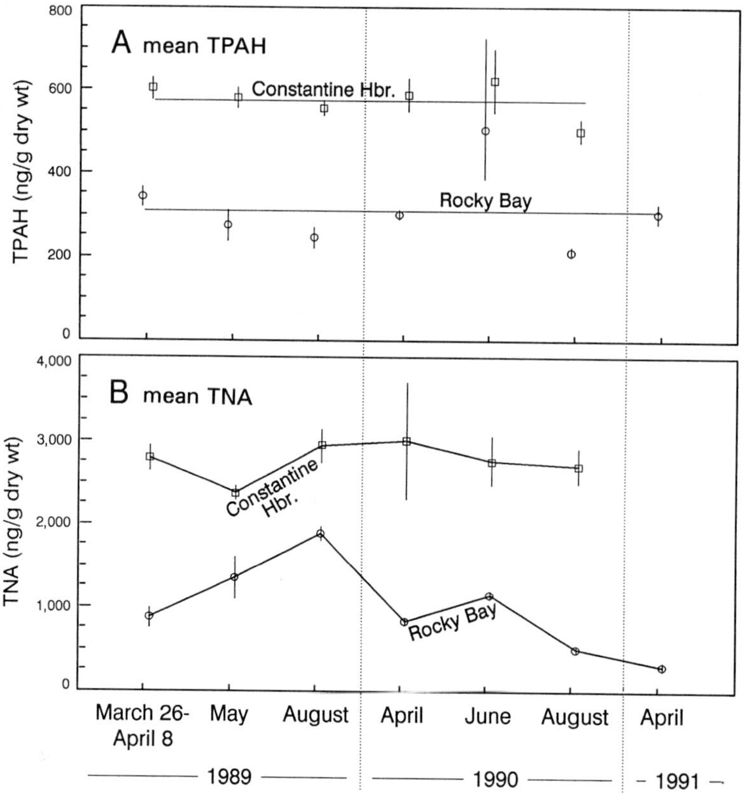

FIGURE 7.—Comparison of total polynuclear aromatic hydrocarbons (TPAH; **A**) and total normal alkanes (TNA; **B**) in sediments at Constantine Harbor and Rocky Bay from spring 1989 to spring 1991. Vertical bars are ±SE and horizontal lines indicate means over all sampling periods when these were not significantly different (analysis of variance: $P > 0.05$).

tions of dibenzothiophenes or chrysenes compared with phenanthrenes, which suggests sources other than *Exxon Valdez* oil. At Perry Island, mean TPAH concentrations ranged from 3.4 ± 0.10 ng/g ($N = 2$) to 19.0 ± 7.0 ng/g ($N = 2$), which are too low to indicate possible sources. Mean TPAH concentrations were similarly low at these stations in 1991, except at Barnes Cove, where the mean concentration was 236 ± 65.0 ng/g, and the PAH composition patterns were most similar to those of sediments at Constantine Harbor and Rocky Bay.

At Rocky Bay and Constantine Harbor, mean TPAH concentrations in sediments were very consistent and relatively high from 1989 through 1991, and the PAH composition pattern differed markedly from that of weathered *Exxon Valdez* oil. The mean TPAH concentration of all sediment samples from Rocky Bay was 300 ± 27.8 ng/g ($N = 18$) and from Constantine Harbor was 561 ± 15.2 ng/g ($N = 15$). Mean TPAH concentrations did not vary significantly with the date of sampling at either station (analysis of variance: Rocky Bay, $P > 0.14$ and Constantine Harbor, $P > 0.23$; Figure 7). At both stations, sediments contained relatively low concentrations of dibenzothiophene PAHs compared with other PAHs and substantial concentrations of alkyl-substituted PAHs compared with unsubstituted PAHs. Although similar, the relative PAH abundances at Rocky Bay and Constantine Harbor were not identical, in that the relative abundances of unsubstituted PAHs compared with homologous alkyl-substituted PAHs were somewhat higher at Rocky Bay than at Constantine Harbor (Figure 8).

Mean TPAH concentrations were usually well below 100 ng/g at the remaining stations in Prince William Sound where sediments were sampled and were lowest at Bligh Island (<35 ng/g). At Naked Island, Olsen Bay, and Siwash Bay, concentrations ranged to 113 ± 52.0 ng/g, but 68% of the means were below 50 ng/g. Consistent patterns of relative

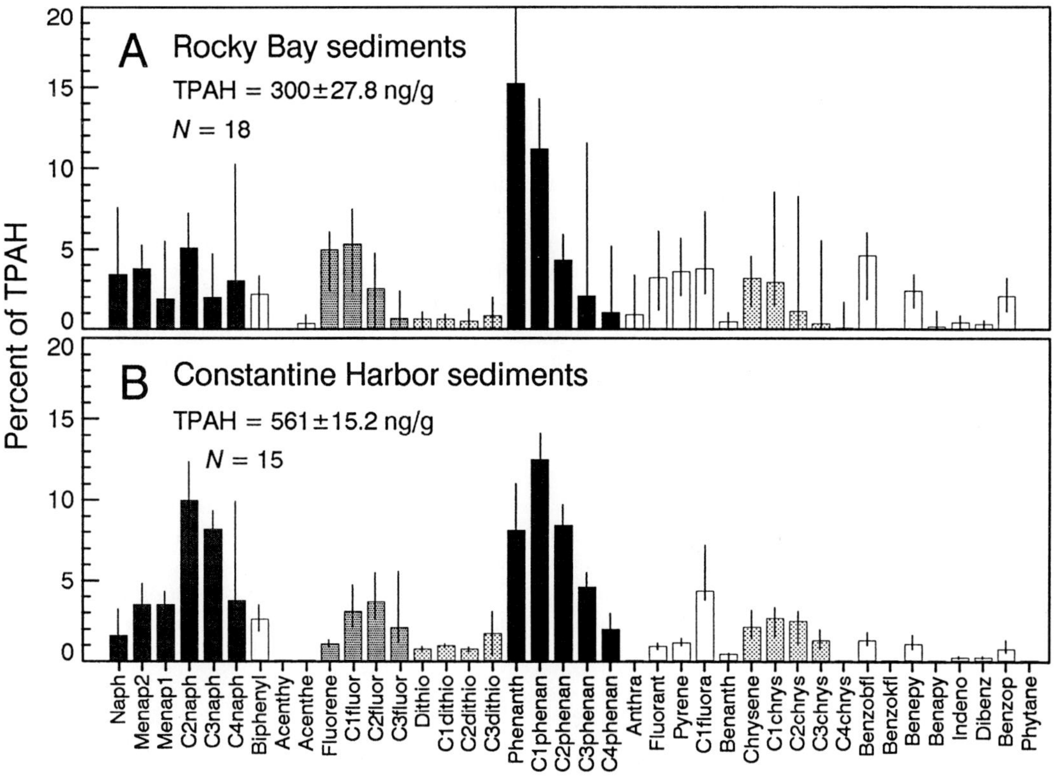

FIGURE 8.—Relative abundance of polynuclear aromatic hydrocarbons (PAHs) and phytane in sediments collected from Rocky Bay (**A**) and Constantine Harbor (**B**). Included are all sediment samples collected from these two stations during this study. Relative abundances for each sample were calculated as the ratio (percent) of the analyte to the sum of all the PAHs (excluding perylene) in the sample; thick vertical bars indicate the mean of these calculations for included samples and thin vertical bars indicate the range. Also given are the mean total PAH (TPAH) concentrations (ng/g dry sediment weight) ±SE.

PAH abundances were not evident in these samples, although naphthalenes were often the most abundant PAHs present.

Sediment Alkane Hydrocarbons

Prespill concentrations.—Mean TNA concentrations in sediments collected in Prince William Sound before landfall of *Exxon Valdez* oil ranged from 394 ± 35.7 ng/g at Naked Island to 3,050 ± 591 ng/g at Bay of Isles, and the individual alkanes present were derived mainly from indigenous sources.

Postspill concentrations.—Except at Elrington Island and Sleepy Bay, the alkanes found in sediments of Prince William Sound were derived primarily from indigenous sources. At the Elrington Island site that was visibly oiled in May 1989, the sediment TNA concentration was 10,700 ng/g (N = 1) and the CPI was 0.89; however, mean TNA concentrations in sediment samples collected from the lower intertidal zone near the mussel sampling site were consistently less than 300 ng/g. At Sleepy Bay, mean TNA concentrations of sediments declined from 3,280 ± 1,200 ng/g in May 1989 to 2,720 ± 330 ng/g in August and to less than 650 ng/g in 1990 and 1991. The CPIs of Sleepy Bay sediments were consistently less than 1.75, and phytane was consistently detected, which corroborated a petroleum source. In contrast, the normal alkanes in sediments of the remaining stations in Prince William Sound consisted mainly of odd-numbered-carbon chains with molecular weights equivalent to C15 or greater, had CPIs that were usually substantially greater than 2, and had phytane concentrations that were usually below detection limits.

Mean TNA concentrations were especially consistent over time in sediments at Constantine Harbor, similar to mean TPAH concentrations there. The mean TNA concentration was 2,780 ± 102 ng/g ($N = 16$) and did not vary significantly during the 2 years of this study (analysis of variance: $P > 0.5$; Figure 7B). The alkanes in Constantine Harbor sediments consisted primarily of odd-numbered-carbon normal alkanes with molecular weights equivalent to C21 or greater and had CPIs ranging from 3.82 to 6.63, which indicates a terrestrial origin for these alkanes. In contrast, mean TNA concentrations in samples from Rocky Bay were quite variable, despite the fact that mean TPAH concentrations did not vary significantly there. Mean TNA concentrations in sediments at Rocky Bay ranged from 323 ± 53.9 ng/g to 1,900 ± 80.4 ng/g and consisted primarily of marine-derived alkanes, with C17 the most abundant normal alkane present. Despite the higher TNA and TPAH concentrations in Constantine Harbor and Rocky Bay sediments, phytane was consistently below detection limits.

Perylene and Total Normal Alkanes in Sediments

Mean concentrations of TNA and of perylene in sediments of Prince William Sound generally varied similarly among stations where *Exxon Valdez* oil was not indicated. At stations where perylene concentrations in sediments were lowest (i.e., Elrington Island, Perry Island, and Sleepy Bay), mean TNA concentrations were less than about 600 ng/g, except for samples where *Exxon Valdez* oil was indicated. At stations where perylene concentrations in sediments were highest (i.e., Barnes Cove, Bay of Isles, Constantine Harbor, and Rocky Bay), mean TNA concentrations usually exceeded 1,000 ng/g, often ranged above 2,000 ng/g (except at Rocky Bay), and were highest at Barnes Cove at 5,650 ± 948 ng/g ($N = 2$). At the remaining stations where perylene concentrations were intermediate, mean TNA concentrations were intermediate also and ranged from 326 ± 73.1 ng/g ($N = 2$) to 2,100 ± 69.1 ng/g.

Discussion

Mussels

Crude oil that spilled from the TV *Exxon Valdez* clearly contaminated mussels at all of the sampling stations within the trajectory of the spilled oil through Prince William Sound in 1989. Contamination by this oil is indicated by (1) the consistent pattern of preferential PAH losses compared with *Exxon Valdez* oil (exemplified in Figure 3), found in these mussels in the spring and summer of 1989, (2) the simultaneous appearance of these PAH composition patterns in mussels after, but not before, the spilled oil had moved through the sound (Figure 2), (3) increases of phytane and of TNA with CPIs near 1 concurrent with the PAH increases (Figure 4), and (4) the general absence of these relative hydrocarbon abundance patterns and the general low TPAH concentrations in mussels at stations remote from the spill trajectory.

Mussels accumulated *Exxon Valdez* oil primarily as whole, particulate oil rather than by absorption of hydrocarbons dissolved in seawater. Accumulation of whole, particulate oil is indicated by the similarity of mussel PAH composition to EVO-PAH composition (Figure 2) concurrent with high TNA and phytane concentrations in mussels that contained high TPAH concentrations derived from *Exxon Valdez* oil. The solubility of alkanes in seawater is much lower than the solubilities of the PAHs, but the relative concentrations of these hydrocarbons in the mussels contaminated by *Exxon Valdez* oil was similar to the relative concentrations in weathered *Exxon Valdez* oil itself (e.g., Figure 3). This finding precludes intermediary partitioning of hydrocarbons into seawater before accumulation by the mussels and suggests direct accumulation of whole, particulate *Exxon Valdez* oil by the mussels.

The amount of *Exxon Valdez* oil accumulated by mussels in 1989 was strongly associated with proximity to oiled beaches. As depicted in Figures 2 and 5, mussels on heavily oiled beaches contained the highest concentrations of *Exxon Valdez* oil, which indicates that proximal oil sources contribute substantially to the oil burden in these mussels.

The dispersed, particulate *Exxon Valdez* oil accumulated by the mussels was obviously biologically available to them, which implies that this oil may also have been available to other fauna, such as larval fish, that feed on microplankton-sized prey. The oil ingested by the mussels may have been either discrete, small oil droplets dispersed in the water column or associated with planktonic prey. In either case, larval fish may have ingested oil because it was associated with prey items or by mistaking oil droplets for prey.

Mussels contaminated by *Exxon Valdez* oil are a vector for transmission of oil to mussel predators by ingestion. The widespread distribution of *Exxon Valdez* oil-contaminated mussels in 1989 suggests a similarly widespread availability of oil to mussel predators, such as marine snails, marine mammals, and birds. In general, the effects of ingested crude oil on these predators is not well known.

Widespread contamination of mussels at our stations by *Exxon Valdez* oil did not persist into 1990 (but see Babcock et al. 1994; Babcock et al. 1996, this volume). The absence of *Exxon Valdez* oil-contaminated mussels (except at Sleepy Bay) indicates that particulate *Exxon Valdez* oil in the seawater column was limited in 1990 to stations near heavily oiled beaches such as Sleepy Bay; however, the increasing *Exxon Valdez* oil concentrations in mussels at Sleepy Bay from April to August 1990 may have been the result of cleanup activities. Cleanup efforts mobilized this oil into adjacent seawaters and it may have accumulated to progressively higher concentrations in mussels through summer 1990 at Sleepy Bay, where beach-cleaning activities were substantial. The persistence of this oil in Sleepy Bay mussels up to April 1991 indicates that *Exxon Valdez* oil remaining after the cleanup continued to be a significant contamination source for mussels (and thus for mussel predators).

The increases in TPAH observed at Elrington Island and Bay of Isles during 1990 may have been indirect effects of cleanup efforts. The relative PAH abundances in mussels at these stations were consistent with diesel oil in August, when TPAH concentrations were highest. At Elrington Island, mean TPAH concentrations had a low coefficient of variation in August. This reproducibility suggests that the diesel oil was not the result of contamination during sampling. Alternatively, the high August TPAH concentrations at Elrington Island and at Bay of Isles may have been the result of small, local diesel oil spills that often attend high marine vessel traffic, such as was present near heavily oiled beaches during summer 1990; however, the general absence of diesel-derived PAHs at most stations in Prince William Sound in 1990 indicates that the magnitude of diesel oil contamination of marine waters there was small compared with the *Exxon Valdez* oil spill.

Comparison with Historical Data

Except for naphthalenes, concentrations of unsubstituted PAHs and selected methyl-substituted PAHs in Prince William Sound mussels collected from stations that were not affected by the spilled *Exxon Valdez* oil were generally below detection limits, which is consistent with results for comparable samples collected during the period 1977–1980 (Karinen et al. 1993). During this period, naphthalene and 1-methylnaphthalene were detected sporadically in mussels at concentrations near detection limits and most of the remaining PAHs considered in the Karinen et al. (1993) study were consistently below detection limits. Results were similar for the corresponding PAH analytes in this study, except that naphthalene and methylnaphthalene concentrations were substantially greater and more consistently detected in mussels collected during spring 1989. The results of Karinen et al. (1993) together with those of this study affirm that the seawater in Prince William Sound was generally free of petroleum hydrocarbons just before the *Exxon Valdez* oil spill, remained so at stations distant from the path of the spilled oil through the sound, and returned to a comparable state at the sampled stations by 1991. In particular, the results of these studies place strong constraints on the magnitude of other sources of anthropogenic PAHs in mussels (and hence in seawater) compared with *Exxon Valdez* oil, such as spilled diesel oil associated with beach cleanup efforts, commercial fishing, or recreational marine vessel traffic.

Sediments

Results of this study imply that contamination of lower intertidal sediments by *Exxon Valdez* oil was relatively slight compared with the heavy oiling that frequently affected upper intertidal sediments. Results from other Natural Resource Damage Assessment studies (O'Clair et al. 1996, this volume; Short et al. 1996b) indicate that transport of oil from the upper intertidal zone to the lower intertidal zone to the subtidal sediments is mediated primarily by wave (or cleanup) mobilization and dispersion of oiled sediments from the upper intertidal area to sediments at lower depths. The PAH concentrations decrease rapidly with distance from the local oil source. In this scenario, contamination of lower intertidal sediments would be transient, as *Exxon Valdez* oil available for mobilization is depleted. The generally decreasing PAH concentrations found in this study at Sleepy Bay (Figure 5) are consistent with this scenario, as are the low PAH concentrations found in lower intertidal sediments at Elrington Island, where sediment transport fluxes are high because of strong marine (Alaska Coastal) and tidal currents.

The relatively high PAH concentrations consistently found at Constantine Harbor and Rocky Bay clearly are not derived from *Exxon Valdez* oil. Karinen et al. (1993) attributed PAHs at these stations to local anthropogenic sources associated with marine vessels, based on their analysis of mainly unsubstituted PAHs; however, the additional analysis of the alkyl-substituted PAHs herein indicates that

TABLE 2.—Comparison of polynuclear aromatic hydrocarbons (PAHs) in sediments at Constantine Harbor and at Rocky Bay during the periods 1977–1980 (Karinen et al. 1993) and 1989–1991. Included are means of all PAHs measured in both studies (*N* given in parentheses). Units are nanograms PAH per gram dry sediment weight. The abbreviation ND = not detected.

	Constantine Harbor				Rocky Bay			
	1977–1980		1989–1991		1977–1980		1989–1991	
Analyte	Mean	SE	Mean	SE	Mean	SE	Mean	SE
Naphthalene	4.67 (11)	0.47	5.30 (16)	0.57	4.50 (8)	0.66	6.47 (19)	0.82
2-Methylnaphthalene	14.55 (13)	1.80	12.03 (16)	0.49	7.74 (10)	1.32	7.14 (19)	0.43
1-Methylnaphthalene	11.65 (13)	1.71	11.91 (16)	0.43	2.29 (10)	0.52	3.34 (19)	0.64
2,6-Dimethylnaphthalene	9.17 (13)	1.50	8.03 (16)	0.29	4.68 (10)	0.65	4.17 (19)	0.18
2,3,5-Trimethylnaphthalene	6.81 (13)	0.84	5.58 (16)	0.13	ND		0.55 (19)	0.10
Biphenyl	8.85 (13)	0.78	8.81 (16)	0.35	3.62 (10)	0.63	4.22 (19)	0.43
Fluorene	3.43 (13)	0.44	3.57 (16)	0.11	10.72 (10)	1.54	9.64 (19)	0.38
Dibenzothiophene	2.51 (13)	0.69	2.48 (16)	0.08	1.78 (10)	0.68	1.16 (19)	0.10
Phenanthrene	32.62 (13)	3.84	27.91 (16)	1.02	37.89 (10)	5.92	29.32 (19)	1.24
1-Methylphenanthrene	6.62 (12)	1.41	9.78 (16)	0.41	2.41 (9)	0.81	2.55 (19)	0.40
Anthracene	0.09 (13)	0.07	ND		1.80 (10)	0.47	1.60 (19)	0.37
Fluoranthene	4.03 (13)	1.48	3.10 (16)	0.10	9.74 (10)	1.50	5.97 (19)	0.54
Pyrene	7.27 (12)	2.58	3.76 (16)	0.09	12.61 (8)	2.19	6.90 (19)	0.51
Benzanthracene	2.07 (13)	0.85	1.32 (16)	0.10	1.31 (10)	0.49	1.08 (18)	0.32
Chrysene	8.16 (13)	1.89	7.17 (16)	0.34	13.03 (10)	2.72	6.48 (18)	0.79
Benzo[*e*]pyrene	5.29 (13)	1.47	3.57 (16)	0.17	8.90 (10)	1.67	4.70 (19)	0.43
Benzo[*a*]pyrene	0.40 (13)	0.33	ND		0.85 (10)	0.73	0.42 (19)	0.25
Perylene	80.46 (13)	15.41	17.97 (16)	0.74	33.30 (10)	5.34	13.79 (19)	0.87

these PAHs are probably derived primarily from geologic sources external to Prince William Sound. The relatively low proportions of dibenzothiophenes in these sediments clearly indicate a source other than *Exxon Valdez* oil. The pattern of relative PAH abundances we found in intertidal sediments at Constantine Harbor is consistent with the pattern that has been identified with petroleum seepage oil near Katalla, northeast Gulf of Alaska, that is swept into Prince William Sound through Hinchinbrook Entrance along with Copper River sediments by the Alaska Coastal current (Page et al. 1995) (see Figure 1). Although our data from these two stations generally confirm the source identification (a geologic source external to the sound) and transport mechanism described by Page et al. (1995), our data do not preclude alternative geologic PAH sources such as coal.

A geologic PAH source external to Prince William Sound is consistent with the invariance in PAH concentrations and with the relatively high perylene concentrations at Constantine Harbor and Rocky Bay. Mean TPAH concentrations were nearly constant during the 2 years of this study at both stations. Furthermore, intertidal sediment PAH concentrations measured both in this study and during 1977–1980 by Karinen et al. (1993) were nearly the same at these stations (Table 2), which indicates that these PAH concentrations have been almost constant for nearly 15 years. Both stations are very stable depositional environments, as indicated by their relatively high perylene concentrations (perylene is produced diagenetically in depositionally stable marine sediments that are anoxic and contain terrigenous or aquatic organic debris [Venkatesan 1988]). The invariance of PAH concentrations in these sediments suggests a general absence of anthropogenic PAHs during this 15-year period, at least in comparison with the magnitude of the geologic PAH source.

The temporal invariance suggests that the PAHs in the intertidal sediments at Constantine Harbor and Rocky Bay are not subject to weathering processes that characterize petroleum degradation in the marine environment (e.g., Figure 3). The high and persistent relative abundance of unsubstituted PAHs in these sediments further contraindicates weathering influences. In addition, the absence of phytane in these sediments and the absence of PAHs in mussels from these stations suggest that these PAHs are associated with a relatively refractory matrix. This matrix could be clay particles that have strongly adsorbed PAHs derived from petroleum seeps near Katalla, or alternatively, the matrix could be a form of coal in the Copper River drainage that has a pattern of relative PAH abundances that is low in dibenzothiophenes. Coal and petroleum are difficult to distinguish on the basis of PAH composition patterns (Tripp et al. 1981). In either case, the PAHs from this source were evidently not bioavailable, as indicated by the absence of these PAHs in the mussels at these two stations.

The relatively low concentrations of PAHs derived from the geologic PAH source external to Prince William Sound in sediments at stations along the spill path suggest substantial depletion of these PAHs from the upper water column soon after entering the sound through Hinchinbrook Entrance. The PAHs from this source were generally less than 100 ng/g in intertidal sediments at stations along the spill path, including stations with high depositional rates (as indicated by relatively high sediment perylene concentrations) at Barnes Cove and at Bay of Isles. The PAHs from this source are therefore probably not significant in intertidal sediments within the path of the spilled *Exxon Valdez* oil through the sound, although this source may contribute more substantially to subtidal sediments, especially to deeper subtidal sediments (O'Clair et al. 1996).

Weathering

Comparison of the relative PAH abundances of *Exxon Valdez* oil in sediments and in mussels at Sleepy Bay suggests that the composition of the spilled oil changed little after May 1989. Losses of less-alkyl-substituted PAH homologues and of lower-molecular-weight PAHs (especially naphthalenes) were marked in *Exxon Valdez* oil from Sleepy Bay sediments and mussels between early April and early May, but were progressively less so over the following 2 years. This pattern follows that depicted in Hostettler and Kvenvolden (1994) for phenanthrene and alkylated phenanthrenes. Consequently, the relative PAH abundance patterns depicted in Figures 3 and 6 should be reliable indicators of *Exxon Valdez* oil in sediments, mussels, and in other matrices where active biochemical catabolism or differential physical partitioning are not factors. Moreover, the relatively small changes in the PAH composition patterns between 1990 and 1991 suggest that the 1991 pattern may undergo subsequent alteration only very slowly, so that patterns substantially similar may be evident in contaminated sediments of Prince William Sound for several additional years. Similar patterns are shown by Michel and Hayes (1993) for two of their more protected stations in the intertidal areas of the sound.

Restoration Criteria

Results from this study for stations that were not affected by the oil spill may be used to establish quantitative restoration criteria for oiled beaches. In general, the hydrocarbon results of this study for stations outside the path of the spilled oil, and for stations sampled before spill impacts, confirm the conclusions of Karinen et al. (1993) that the intertidal zone of Prince William Sound was nearly pristine with respect to petroleum hydrocarbon contamination before the spill. This conclusion is based on the only sporadic occurrence of PAHs in intertidal sediments and mussels at concentrations near detection limits observed by Karinen et al. (1993) and herein at stations in western Prince William Sound that were not affected by the spill. The magnitude and composition of PAHs observed at these unimpacted stations could therefore be used as a standard for determining the restoration of oiled beaches. Specifically, oiled beaches might be considered as fully restored when the dry weight TPAH concentration of intertidal sediments is less than 100 ng/g and of mussels is less than 200–400 ng/g, with the higher mussel concentrations attributable mainly to naphthalenes.

Acknowledgments

The efforts of many people and agencies contributed to the success of this study. During the weeks immediately after the spill, the U.S. Coast Guard, the Alaska Department of Environmental Conservation, and the National Oceanic and Atmospheric Administration's Office of Aircraft Services provided transportation to the sampling stations. John F. Karinen, Auke Bay Laboratory, whose foresight more than a decade ago in establishing baseline hydrocarbon sampling stations along the tanker route in Prince William Sound made this project possible, worked tirelessly during the weeks after the spill to resample all original stations and establish new stations in the path of the floating oil. Other Auke Bay Laboratory staff who assisted in field sampling and/or helped with preparation of sampling gear and chain-of-custody procedures were Doris Alcorn, Nancy Barr, Christine C. Brodersen, Daniel Fremgen, Sidney Korn, Paula A. Johnson, Charles E. O'Clair, Julie Sharp-Dahl, Mona C. Wells, and Bruce L. Wing. Marie L. Larsen, Larry G. Holland, and Josefina G. Lunasin deserve recognition for producing quality analytical data. Finally, we thank K. A. Kvenvolden, P. R. Carlson, and one anonymous reviewer for several helpful comments.

References

Babcock, M. M., G. V. Irvine, P. M. Harris, J. A. Cusick, and S. D. Rice. 1996. Persistence of oiling in mussel beds three and four years after the *Exxon Valdez* oil spill. American Fisheries Society Symposium 18:286–297.

Babcock, M. M., P. M. Rounds, C. C. Brodersen, and S. D. Rice. 1994. 1991 and 1992 recovery monitoring and restoration of intertidal oiled mussel (*Mytilus trossulus*) beds in Prince William Sound impacted by the *Exxon Valdez* oil spill. NOAA (National Oceanic and Atmospheric Administration) Technical Memorandum NMFS (National Marine Fisheries Service)-AFSC-94-02, Seattle.

Brown, D. C., L. S. Ramos, M. Y. Uyeda, A. J. Friedman, and W. S. MacLeod. 1980. Ambient temperature extraction of hydrocarbons from marine sediment—comparison with boiling-solvent extractions. Pages 313–326 *in* L. Petrakis and F. T. Weiss, editors. Petroleum in the marine environment. American Chemical Society Symposium Series 185.

Farrington, J. W., and B. W. Tripp. 1977. Hydrocarbons in western north Atlantic surface sediments. Geochimica et Cosmochimica Acta 41:1627–1641.

Galt, J. A., W. J. Lehr, and D. L. Payton. 1991. Fate and transport of the *Exxon Valdez* oil spill. Environmental Science and Technology 25:202–209.

Hostettler, F. D., and K. A. Kvenvolden. 1994. Geochemical changes in crude oil spilled from the *Exxon Valdez* supertanker into Prince William Sound, Alaska. Organic Geochemistry 21(8/9):927–936.

Karinen, J. F., M. M. Babcock, D. W. Brown, W. D. MacLeod, Jr., L. S. Ramos, and J. W. Short. 1993. Hydrocarbons in intertidal sediments and mussels from Prince William Sound, Alaska, 1977–1980: characterization and probable sources. NOAA (National Oceanic and Atmospheric Administration) Technical Memorandum NMFS (National Marine Fisheries Service) AFSC-9, Seattle.

Michel, J., and M. O. Hayes. 1993. Persistence and weathering of *Exxon Valdez* oil in the intertidal zone—3.5 years later. Pages 279–286 *in* Proceedings, 1993 international oil spill conference. American Petroleum Institute Publication 4580, Washington, DC.

O'Clair, C. E., J. W. Short, and S. D. Rice. 1996. Contamination of intertidal and subtidal sediments by oil from the *Exxon Valdez* in Prince William Sound. American Fisheries Society Symposium 18:61–93.

Page, D. S., P. D. Boehm, G. S. Douglas, and A. E. Bence. 1995. Identification of hydrocarbon sources in the benthic sediments of Prince William Sound and the Gulf of Alaska following the *Exxon Valdez* oil spill. Pages 41–83 *in* P. G. Wells, J. N. Butler, and J. S. Hughes, editors. *Exxon Valdez* oil spill: fate and effects in Alaskan waters. American Society for Testing and Materials Special Technical Publication 1219, Philadelphia.

Short, J. W., T. J. Jackson, M. L. Larsen, and T. L. Wade. 1996a. Analytical methods used for the analysis of hydrocarbons in crude oil, tissues, sediments, and seawater collected for the natural resources damage assessment of the *Exxon Valdez* oil spill. American Fisheries Society Symposium 18:140–148.

Short, J. W., D. M. Sale, and J. C. Gibeaut. 1996b. Nearshore transport of hydrocarbons and sediments after the *Exxon Valdez* oil spill. American Fisheries Society Symposium 18:40–60.

Tripp, B. W., J. W. Farrington, and J. M. Teal. 1981. Unburned coal as a source of hydrocarbons in surface sediments. Marine Pollution Bulletin 12:122–125.

Venkatesan, M. I. 1988. Occurrence and possible sources of perylene in marine sediments—a review. Marine Chemistry 25:1–27.

American Fisheries Society Symposium 18:167–176, 1996

Intertidal and Supratidal Site Selection Using a Geographical Information System

KIMBAL SUNDBERG
Alaska Department of Fish and Game, Habitat and Restoration Division
333 Raspberry Road, Anchorage, Alaska 99518-1599, USA

LAWRENCE DEYSHER
Coastal Resources Associates, Inc.
1185 Park Center Drive, Suite A, Vista, California 92083, USA

LYMAN MCDONALD
Western Ecosystems Technology, Inc.
2003 Central Avenue, Cheyenne, Wyoming 82001, USA

Abstract.—A geographical information system (GIS) was employed in the selection of sampling sites for the intertidal and supratidal injury assessment of the *Exxon Valdez* oil spill. The use of GIS was suggested by (1) the large amounts of shoreline (potentially 2,003 kilometers) contaminated by this oil spill, (2) the extreme heterogeneity of shoreline types and degrees of oiling in the *Exxon Valdez* oil spill area, (3) the need to embody the various shoreline habitats and degrees of oiling in the assessment, and (4) the need to make statistical inferences regarding intertidal and supratidal injury to the universe of all oiled shorelines in this oil spill area. The shoreline affected by the *Exxon Valdez* oil spill was stratified into 15 classes based on habitat type and degree of oiling. Potential sampling sites were randomly drawn from the database with probability proportional to size. Field surveys conducted in 1989 to verify shoreline classifications, verify degree of oiling, and locate and mark shoreline segments for subsampling found many misclassified sites. Moderately to heavily oiled sites selected in the stratified random sample were retained and subjectively paired with reference (non-oiled or very lightly oiled) sites. The final quasi-experimental study design for intertidal and supratidal injury assessment consisted of 50 oiled sites paired with 47 reference sites. Randomly selected oiled sites maintained their probability-based selection status for inferences to the entire *Exxon Valdez* oil spill area. Recommendations are made concerning the future use of GIS for intertidal and supratidal injury assessments.

A primary goal of the Coastal Habitat Injury Assessment (CHIA) was to quantify injuries to intertidal and supratidal biota in the *Exxon Valdez* oil spill area. Information from this study was used in the Natural Resource Damage Assessment (NRDA) program conducted under the auspices of the *Exxon Valdez* oil spill trustees. The requirements of this study were dictated by the dual need to test the hypothesis of injury to intertidal and supratidal biota and to determine the nature and degree of injury over the full range of intertidal and supratidal habitats found in the oil spill area. Most oiled shoreline had received one or more types of cleanup treatment before or during this study; however, records were inadequate to determine the amount or type of cleanup activity on most specific study sites. The objective of the study was to assess the confounded effects of oil and cleanup on the shoreline biological community. We refer to the sites as treated, or oiled–cleaned, or oiled.

Geographical information systems (GIS) have the capability to collect, store, process, and display large amounts of spatial data (Burrough 1986). They have been used for a wide range of natural resources applications, including management, environmental planning, and assessment (Webb 1982; Dangermond 1991; Schaller 1992). Geographical information systems have also been used for cumulative environmental impact assessment where it was necessary to analyze empirical relationships between resource loss and overall environmental degradation (Johnston et al. 1988). However, the implementation of a full-scale GIS can be a long and expensive process involving multiple organizations (Antenucci et al. 1991). The need to initiate a rigorous intertidal and supratidal assessment covering a large geographic area required that a full-scale GIS be available within a short time. Fortunately, advanced GIS capabilities existed within state and federal resource agencies in Alaska at the time that the T/V *Exxon Valdez* spilled 42 million liters (11 million gallons) of North Slope crude oil into Prince William Sound on 24 March 1989.

The GIS capabilities used in this study were located in three federal agencies and one state agency. The U.S. Geological Survey (USGS), Earth Resources Observation System (EROS) field office provided primary support with assistance from the U.S. Forest Service, Chugach National Forest; the U.S. Fish and Wildlife Service, Alaska Region; and the Alaska Department of Natural Resources, Lands Records and Information Section. These agencies each developed GIS capabilities using similar hardware and software before 24 March 1989 and had a history of working cooperatively on applications and systems development.

The initial planning for this study was conducted during May and June 1989. During this time it was decided by the Trustee Council management team that intertidal, supratidal, and subtidal study sites would be selected using a probability-based approach that allowed for extrapolation of results to the universe of all treated sites in the *Exxon Valdez* oil spill area. During June 1989 the concept of using a GIS to select sampling sites was approved. A working group comprising the principal investigators and technical support staff undertook the review of available data and GIS capabilities and completed the task of defining user needs and determining the feasibility of a workable solution. By late June 1989 the system was designed, digitizing was underway, and the first test plots were available. In mid-July, a pilot field study was conducted in the Naked Island and northern Knight Island areas in Prince William Sound to test the site evaluation methodology. Following that, full-scale site selection commenced in the Prince William Sound (PWS), Cook Inlet–Kenai Peninsula (CIK), and Kodiak–Alaska Peninsula (KAP) regions. The first phase of site selection was completed in August 1989 concurrent with the first occupation of the sites and subsampling by the supratidal, intertidal, and subtidal study crews under the lead of the University of Alaska-Fairbanks (UAF).

Based on the results of the 1989 site-selection surveys and initial UAF subsampling, many sites were found to be misclassified on either degree of oiling, habitat type, or both. Additional treated and reference sites were selected in 1990 to address the intertidal and supratidal components of the CHIA. The subtidal component of the CHIA adopted a different site-selection strategy in 1990.

Methods

The study area was defined by the universe of all shorelines on USGS 1:63,360-scale topographic quadrangles where any shoreline had been documented as receiving grounded *Exxon Valdez* oil. This approach was intended to ensure a comprehensive sampling of all habitat types and degrees of oiling (ranging from heavy to none) in the spill area. The methodology for site-selection was designed to (1) test the hypothesis that oiling (treatment) had not caused injury to subtidal, intertidal, and supratidal biota; and (2) allow for extrapolation of injury determinations at study sites to the universe of all treated sites in the spill area. The GIS for site-selection "GEO" was constructed from three primary databases using ARC/INFO[1]:

1. The State Shoreline consisted of the approximate mean high water line digitized from the USGS 1:63,360 scale topographic quadrangle series in the spill affected area. The shoreline was modified by interpretation of 1:60,000 scale high-altitude color infrared photographs to add small islands and update post-1964 earthquake shoreline changes.
2. The Environmental Sensitivity Index (ESI) maps classified the shoreline into 19 geomorphologic types following the shoreline classification system described by Gundlach and Hayes (1978) and Hayes et al. (1980). The ESI rates the shoreline from 1 to 10 based on progressively higher sensitivity to oiling. The 1:60,000 scale maps were digitized from five map atlases covering the spill-affected area (RPI 1983a, 1983b, 1985, 1986; M. O. Hayes and C. H. Ruby, Research Planning Institute, unpublished data 1979). The ESI classification scheme was translated to five habitat types (Table 1) for this study.
3. The Oil Spill Impact (OSI) database was maintained by the Alaska Department of Environmental Conservation (ADEC) for the *Exxon Valdez* oil spill area (ADEC, unpublished data, 1989). The OSI classifies the shoreline into five oiling types: heavy, moderate, light, very light, and no observed oil at a nominal scale of 1:63,360. The OSI data were based on cumulative aerial and ground observations made by ADEC personnel using standardized evaluation criteria (Table 2). The OSI classification scheme was condensed to three oiling types for this study: heavy–moderate, light–very light, and no observed oil (reference).

Using GEO, the shoreline was stratified into 15 habitat–oiling classes: five habitat types multiplied

[1]ARC/INFO geographical information system software. Version 5.0. Environmental Systems Research Institute, Redland, California.

TABLE 1.—Translation of Environmental Sensitivity Index (ESI) shoreline types to Coastal Habitat Injury Assessment (CHIA) habitat types.

ESI number	CHIA habitat type	ESI shoreline type
1	Exposed rocky shores	Exposed rocky headlands
1		Exposed rocky shores
2		Wave-cut platforms
2		Exposed wave-cut platforms
3	Fine-textured beaches	Fine–medium-grained sand beaches
3		Fine-grained sand beaches
4		Coarse-grained sand beaches
5	Coarse-textured beaches	Exposed tidal flats
5		Mixed sand and gravel beaches
5		Exposed tidal flats (low biomass)
6		Mixed sand and gravel beaches
6		Gravel beaches
7		Gravel beaches
7		Exposed tidal flats
7A		Exposed tidal flats (moderate biomass)
7		Exposed tidal flats (moderate–high biomass)
8	Sheltered rocky shores	Sheltered rocky shores
9	Sheltered estuarine shores	Sheltered tidal flats
10		Marshes

by three oiling types. Shorelines within these strata were divided into segments ranging in length from 100 to 600 m (corresponding to arcs in the GIS). Arcs less than 100 m were removed from the target universe because they were too small to allow repetitive, destructive subsampling of multiple transects. Multiple visits with destructive subsampling within short shoreline segments could affect recruitment and introduce bias. Arcs were limited to 600 m to ensure that sites could be efficiently subsampled during one low-tide period. Following stratification, each arc was assigned a unique identification number.

Arcs were randomly selected without replacement from each of the 15 habitat–oiling classes with probability (approximately) proportional to size. Selection probabilities are approximate because the sites were randomly selected without replacement and there was no adjustment for the changing size of the stratum. The center points and arc numbers of the first 15 selected sites within each class were color plotted on 1:63,360 scale maps depicting shoreline, habitat–oiling type, upland ownership, and USGS topographic quadrangle boundaries.

Sites were visited in the field to verify mapped data and to locate and mark shoreline segments for subsampling by the CHIA crews. Surveys were performed during low to mid-tide levels. Tides in the study area are mixed semidiurnal. For purposes of this study, the intertidal zone was defined as the zone between mean lower low water and mean higher high water; the supratidal zone was defined by the zone between mean higher high water and extreme higher high water (generally the upper extent of visible oiling on oiled sites); and the subtidal zone was defined as the zone between mean lower low water and the 20-m isobath.

The center point of each arc was located on the shoreline using the GEO maps, USGS topographic maps, visual references, and compass bearings. The end points of each arc were marked and the length measured along the upper intertidal zone (generally above the line of *Fucus gardneri*). Each shoreline segment was surveyed along its entire length to classify and map oiling and habitat types. Sites that could not be safely reached by skiff or where beach

TABLE 2.—Alaska Department of Environmental Conservation criteria used for rating Oil Spill Impact (OSI).[a]

OSI type	Rating criteria
Heavy	>6-m-wide band or >50% coverage–penetration
Moderate	3- to 6-m-wide band or 10 to 50% coverage–penetration
Light	1- to 3-m-wide band or 1 to 10% coverage–penetration
Very light	<1-m-wide band or <1% coverage–penetration
No observed oil	No oil observed

[a]Oil spill impacts are evaluated solely on the amount of area covered or penetrated by oil between the mean high tide and mean low tide lines.

slopes were measured with a clinometer to exceed 35 degrees for more than 50% of the shoreline were rejected as inaccessible for intertidal sampling. Sites not matching the mapped habitat–oiling type were reclassified to the most appropriate type, but retained the original probability of selection (weight) for statistical inferences. Reclassification was accomplished by measuring the alongshore length of habitat–oiling units within each arc. The longest unit determined the habitat–oiling classification of the site. In 1989, the goal was to obtain four sampling sites in each of the heavy–moderate oiling strata and three sampling sites in each of the other strata for a total of 150 sites in the three regions.

For 1990, the CHIA was changed to a quasi-experimental design by retaining the randomly selected treated sites and matching each with a subjectively selected "reference" site. In true experiments, the treatment would be randomly assigned to one of the members of the pair, hence the term quasi-experiment is used to indicate that conclusions are limited by the protocol through which reference sites are selected. The objective was to pair each treated site with a reference site on the basis of physical characteristics known to influence intertidal community structure. We also subjectively selected additional paired heavy–moderate treated and reference sites in habitat categories where the sample size was below the target value.

Suitable pairs of reference sites and treated sites were first identified based on 1989 field survey data and site photographs. Treated sites that lacked suitable references from within the group of non-oiled sites sampled in 1989 were matched with new reference sites using the following procedure. Using GEO and other sources, five potential reference sites were identified in closest proximity to each treated site from shorelines classified as non-oiled or very lightly oiled. Reference was made to the 1989 Post-Treatment Shoreline Oiling Assessment (ADEC 1989a, 1990a, 1990b) to determine the oiling history of each potential reference site. Sites classified as having more than very light oiling by any shoreline survey were rejected from consideration as references. Potential reference sites were listed according to their distance from each treated site, with the closest site receiving the highest priority. All other sites classified as light–very light oiling were eliminated from the study in 1990. No damages to intertidal and supratidal biota were to be claimed for injuries to light–very light oiled sites and, as discussed above, some of the very lightly oiled sites were used as references.

Field surveys conducted at treated sites and at potential reference sites determined the closest match of physical and biological characteristics. The physical characteristics of greatest import were (1) substrate composition, (2) wave exposure, (3) beach slope, (4) proximity to sources of fresh water, and (5) nearshore bathymetry. General biological characteristics of the sites were also checked. These were limited to general indicators of community composition including the presence or absence of mussel beds, algal beds, gastropods, and barnacles. Potential reference sites were evaluated in order of priority until a suitable site was found. An attempt was made to measure reference-site lengths equal to the paired treated site, but always a minimum of 100 m of comparable habitat was marked.

Additional paired treated and reference sites were subjectively selected in habitat types that were below the target sample size of the 1989 studies. These sites were selected to test oiling–injury hypotheses and to incorporate some special interest study sites occupied by non-CHIA biologists in 1989. However, these sites were treated as censused subpopulations and were not used to extrapolate injury assessment to other oiled sites.

TABLE 3.—Results of 1989 site selection in the Prince William Sound (PWS), Cook Inlet–Kenai Peninsula (CIK), and Kodiak–Alaska Peninsula (KAP) regions.

Region	Universe	Potential sites	Sites surveyed	Sites reclassed
PWS	4,975	125	63	38
CIK	4,130	164	103	41
KAP	12,257	135	74	40
Total	21,362	424	240	119

Results

From a universe of 21,362 sites encompassing 9,173 km of shoreline, 424 potential sites were plotted on maps using GEO (Table 3). Site visits were conducted in rank order according to selection within strata during 15 July 1989 to 22 August 1989. A total of 240 sites were visited. Initially, it was intended that the three (or four in the case of moderate–heavy oiled strata) highest ranked accessible sites within each strata would be accepted if they were found to fit the GEO classification. However, it became apparent early in the survey that many sites did not fit the GEO classification for either habitat type, degree of oiling, or both. For example in PWS, the first region examined, 38 of 63 sites (60%) were found misclassified. Other regions had lower classification errors.

The selection of reference sites in the original, stratified random sample resulted in unforeseen sampling problems. The control strata were found to be "too large," containing habitat unrepresentative of the oiled strata. For example, in PWS all of the heavy–moderate treated sites were located on islands. Most island shorelines in the path of the *Exxon Valdez* oil spill exhibited some evidence of oiling during site examinations in 1989. Under the 1989 site-selection criteria, any evidence of oil, even one tar ball or tar spot, was cause for rejecting a reference site. With the exception of four sites, all PWS reference sites were located on the mainland. Most reference sites were in bays with more freshwater sources, lower salinity, and greater winter stress (lower temperatures, ice scour). Therefore, standard measurements of biotic health, including species richness, density, biomass, and recruitment, were not directly comparable between many GEO-identified treated sites and reference sites in the absence of oil.

Figures 1 and 2 compare the locations of 1989 and 1990 study sites in the Prince William Sound region. To preserve the statistical validity of the study design and allow statistical inferences to the most important subset of oiled shoreline, we retained all heavy–moderate oiled sites selected by the 1989 protocol, regardless of the initial classification, and subjectively paired them with the closest reference site having similar physical characteristics. We also allowed very lightly oiled shorelines as potential references in 1990. These very lightly oiled sites were interspersed with treated sites on the islands and improved the capability of the study to control extraneous factors and increase the precision and accuracy in measurement of oiling effects. This approach preserved the probability sample of heavy–moderate oiled sites based on the original stratification and database. Three important results were as follows:

1. Selected sites carried their original unequal weights from whatever original strata they were in.
2. The original frame (database and stratification) was retained to extrapolate results to the subset of accessible heavy–moderate treated sites.
3. The target number of sites in some strata was not attained (e.g., in the CIK region no accessible heavy–moderate oiled, exposed rocky shore sites were identified in 1989 among the highest ranked accessible sites within original strata).

In 1990, site-selection was focused on pairing suitable reference sites with treated sites and with purposively selecting treated sites to provide information in habitat categories where there were a deficit of randomly selected sites. The objective was to obtain a minimum of four replicate oiled-reference pairs for each habitat type in each region. Purposively selected sites represented only themselves in the extrapolation of results to the subset of accessible heavy–moderate treated sites within a given habitat type. Table 4 summarizes the results of the 1990 site-selection.

TABLE 4.—Results of 1990 site selection; regions are defined in Table 3. (T) = treated site; (C) = control site.

Region	Sites needed	Potential sites	Sites surveyed	Sites accepted
PWS	3 (T) 14 (C)	3 (T) 69 (C)	3 (T) 47 (C)	1 (T) 12 (C)
CIK	10 (T) 14 (C)	5 (T) 27 (C)	5 (T) 19 (C)	4 (T) 8 (C)
KAP	7 (T) 12 (C)	6 (T) 41 (C)	5 (T) 22 (C)	5 (T) 9 (C)
Total	20 (T) 40 (C)	14 (T) 137 (C)	13 (T) 88 (C)	10 (T) 29 (C)

Following the pairing of suitable 1989 reference sites with treated sites, an additional 20 treated and 40 reference sites were sought in 1990. Using GEO and other sources, 14 treated sites and 137 potential reference sites were identified for ground truthing. An insufficient number of heavy–moderate treated sites in some habitat types, most notably sheltered estuarine shores and exposed rocky shores in the CIK region, made it impossible to reach the desired goal of four treated sites per habitat type called for in the study plan.

Field surveys conducted during low tide series from 23 April 1990 to 11 June 1990 resulted in 101 sites being ground-truthed and 39 additional sites being accepted into the study. In three cases, it was necessary for a reference site to be "paired" (i.e., grouped) with more than one treated site. The final array of intertidal and supratidal study sites consisted of 50 treated sites paired with 47 reference sites. The numbers of sites for some habitat types including coarse-textured beaches, were larger than planned, whereas sample sizes for other habitat types, including sheltered estuarine shores, were smaller than planned. The randomly selected treated sites maintained their probability-based selection status for making statistical inferences within the resulting quasi-experimental design. Further details concerning the basis for statistical analyses are reported in McDonald et al. (1995). The results of subsampling of the sites are reported in Highsmith et al. (1996, this volume) and Stekoll et al. (1996, this volume).

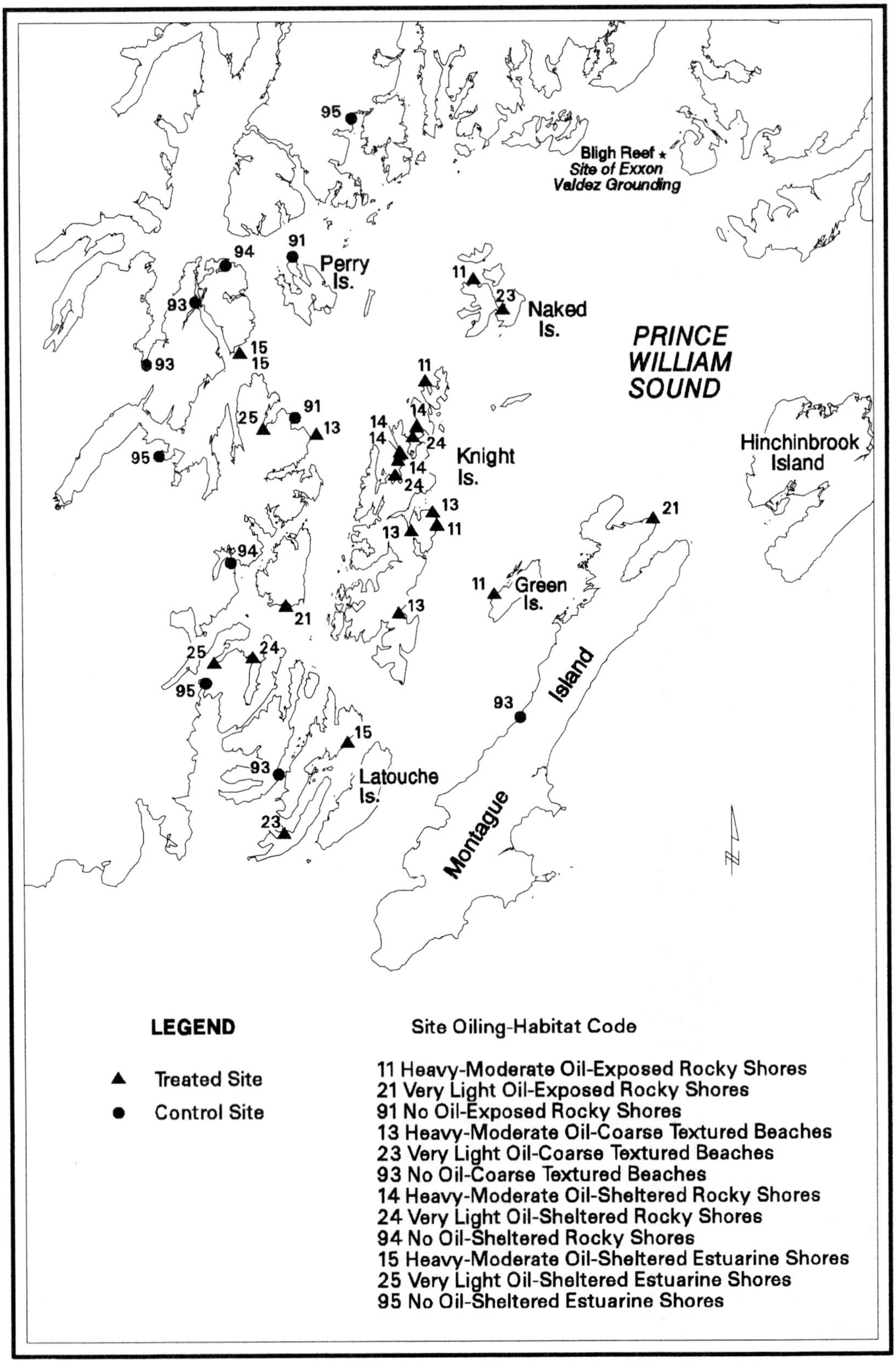

FIGURE 1.—1989 Coastal Habitat Injury Assessment (CHIA) study sites, Prince William Sound region. Map was produced by the Alaska Department of Natural Resources.

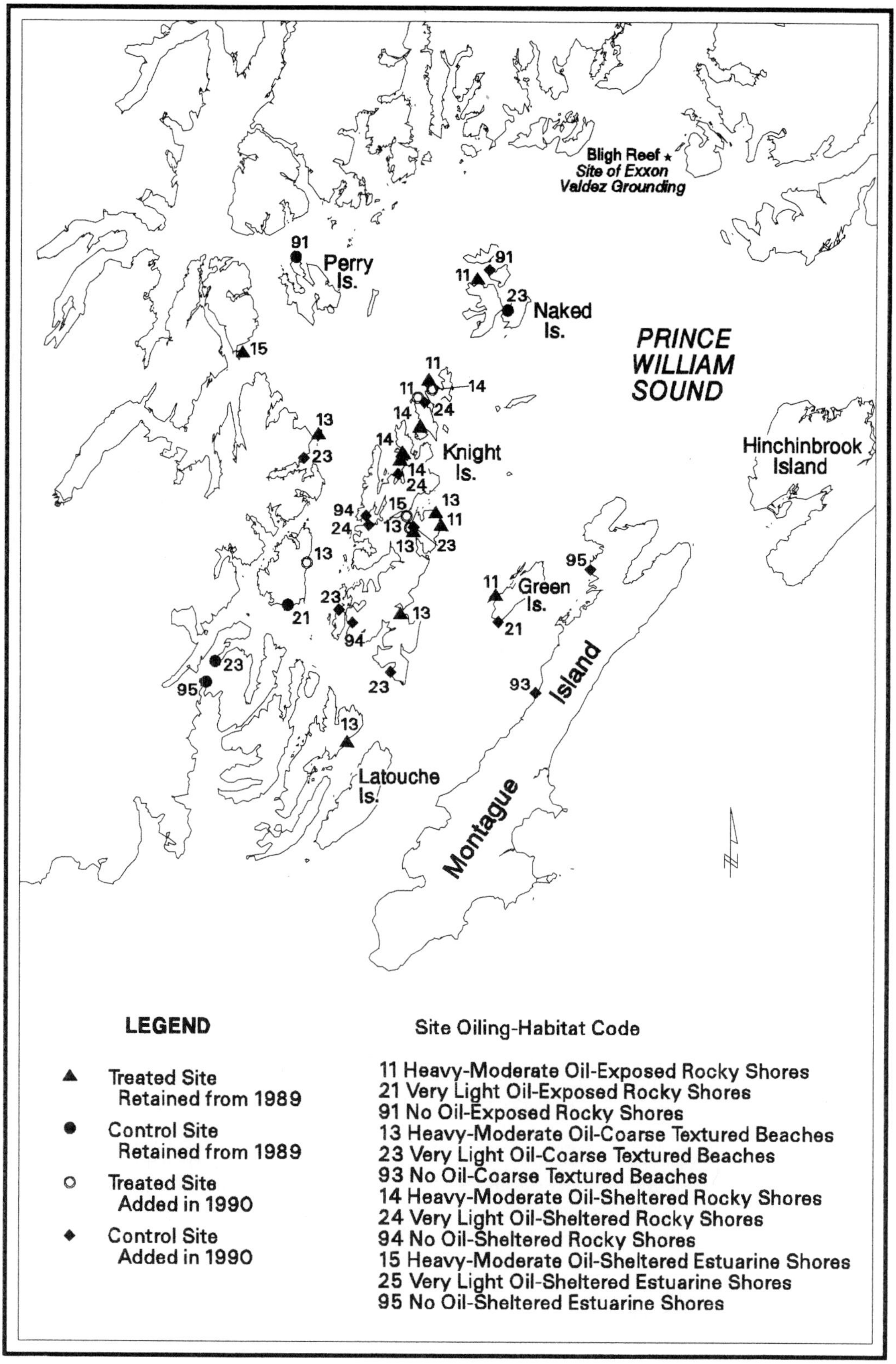

FIGURE 2.—1990 Coastal Habitat Injury Assessment (CHIA) study sites, Prince William Sound region. Map was produced by the Alaska Department of Natural Resources.

Discussion

The *Exxon Valdez* oil spill potentially contaminated 2,003 km of shoreline consisting of 501 km in PWS, 161 km in CIK, and 1,341 km in KAP (ADEC 1989b). The task of quantitating spill injuries to intertidal and supratidal biota presented a formidable challenge. In the spill area there were very little preexisting biological baseline data that could be used for assessment of intertidal and supratidal injury. Most of the existing baseline data were not in a form that could be used for Natural Resource Damage Assessment (NRDA) studies and occurred at sites that were not significantly oiled. These facts eliminated the possibility of conducting a classical "Before-After-Control-Impact" (BACI) design (Stewart-Oaten et al. 1986).

The use of a GIS to identify potential intertidal and supratidal study sites within a stratified random sample of treated and reference sites was suggested by (1) the extraordinary amount of shoreline and associated habitat potentially affected by the spill, (2) the extreme heterogeneity of shoreline types and oiling impacts in the *Exxon Valdez* oil spill area, and (3) the need to make inferences from intertidal and supratidal sampling to the universe of all spill-affected shorelines. The availability of shoreline sensitivity and oiling impact information for the spill area and the existence of in-place GIS capabilities and skilled GIS technicians enabled the rapid development of the programs and databases needed to construct a workable GIS.

Oil spills induce short-term and long-term effects on shoreline biota and there is great urgency to carry out damage assessment studies at the earliest time. During the *Exxon Valdez* oil spill there was neither sufficient time nor resources to independently validate existing complex data sets, including the State Shoreline, the ESI classification scheme, and the OSI database before implementing the Coastal Habitat Injury Assessment (CHIA) study plan. This study showed the need to validate data sets that will be used for damage assessment before implementing a study plan. Many misclassifications could be attributed to (1) the strong dependence on aerial overflight information in the ESI and OSI databases that was often inconsistent with characteristics that we observed in the intertidal zone during ground surveys, (2) the lag time of 2 to 4 months between shoreline oiling information in the OSI database and our site-selection surveys, (3) collapsing the 19 ESI types into five habitat types, and (4) digitizing errors, including one that misclassified 20 potential sites into the wrong habitat category.

The large number of sites misclassified in the GEO database initially appeared to threaten the validity of the study plan. This problem was averted by restricting the sample to heavy–moderate treated sites in the top 50 accessible sites in each region (40 in PWS) ranked according to order of selection within the original strata. However, reclassifying the sites and restricting the sample size resulted in more than the target number of sites in some categories and less than the target number of sites in others. This problem was rectified by randomly eliminating the surplus sites and filling in deficient categories with purposively selected, treated sites. Each selected heavy–moderate treated site was then subjectively matched with a reference site from either non-oiled shoreline or very lightly oiled shoreline in a quasi-experimental design. Reference sites were interspersed with the treated sites because of the spotty nature of contact of oil with the shoreline, resulting in improved capability of the study design to control extraneous factors and increase the precision and accuracy in measurement of oiling-cleanup effects. Subsampling results from the purposively selected sites represented only themselves and were not used to extrapolate results to the other treated shorelines. Final unequal weights of the accessible heavy–moderate treated sites for statistical analysis within the quasi-experimental design were obtained by Monte Carlo-simulated sampling from the original frame (database and stratification) using estimates of the misclassification rates (McDonald et al. 1995).

Using hindsight, what changes would the authors make in the design of the CHIA in response to the *Exxon Valdez* oil spill? A complete discussion of all alternatives would include most of modern field ecology, marine science, and statistical theory, but we will attempt to address several important points as they relate to our basic design and logistical constraints.

1. It is now known that a straight stratified random sample (same number of study sites with no problems of misclassification) would probably not have sufficient statistical power to detect important adverse effects in the universe of oiled shoreline. Scientists conducting independent studies for Exxon Corp. used within-site sampling variance (i.e., pseudoreplication) to artificially increase statistical power in tests of hypotheses (Gilfillan et al. 1995; Page et al. 1995). This practice increases the likelihood that ad-

verse effects will be declared statistically significant and is an excellent conservative approach from their point of view to help them meet their objective of 80% power to detect 20% effects. However, in NRDA cases, the trustees are charged with proving adverse effects. Use of pseudoreplication to artificially increase the power is unacceptable in presentation of results from the trustees' studies. The paired site quasi-experimental design with randomly selected treated sites and subjectively selected reference sites is one of the few designs that had sufficient power to "prove" effects by the usual scientific method. Of course, we did not know whether sufficient power existed when the study was being redesigned during 1989 to 1990. Also, in any experiment, the conclusions must be referenced to the protocol for selection of reference sites and measurement of variables.

2. We recommend equal size units with equal probability of selection within strata to simplify data recording, analyses, and future uses of the data.
3. Fewer misclassifications in the original database would be desirable. If we had known the high rate of misclassifications in the database, we probably would have selected a simple random sample from equal size units. Sites would then be visited in the rank order of selection until the desired number of sites were classified into the subpopulations (domains) of interest.
4. We would allow for very lightly oiled shorelines, including those with occasional tar balls or tar spots, to qualify as potential reference sites. The net result would have been better interspersion of reference sites and treated sites. Use of very lightly oiled shoreline as reference should yield conservative estimates of the effects of oil-cleanup on heavy–moderate treated sites.
5. At all sites, we would measure more quantified "covariates"; that is, uncontrolled extraneous physical factors (e.g., proximity to sources of fresh water, wave exposure) that influence the biological community.
6. We would devise unique sampling procedures for discrete biological communities of special interest, such as mussel beds.
7. Finally, with the growing use of GIS for oil spill response (Harper et al. 1991), database standards should be established to require ground truthing and estimates of classification errors. This would improve the application of GIS to future subtidal, intertidal, and supratidal injury assessments.

Acknowledgments

The authors acknowledge the following organizations and individuals who contributed to this study: Earl Becker, Susan Beyersdorfer, Jeff Bromaghin, Lana Shea, Lance Trasky, and Frances Inoue of the Alaska Department of Fish and Game; Richard McMahon, Dorothy Mortenson, and Jean Tamm of the Alaska Department of Natural Resources; Jim Gibeaut, Eric Gundlach, and Jerry Sexton of E-Tech Inc.; Malin Babcock of the National Marine Fisheries Service; Ray Highsmith, Peter Ribbons, Josh Schimel, and Mike Stekoll of the University of Alaska; Emily Binnian and Stan Moll of the U.S. Geological Survey Earth Resources Observation System Field Office; Bill Edwards and Dave Gibbons of the U.S. Department of Agriculture Forest Service; and Wally Erickson and Dale Strickland of WEST Inc.

References

ADEC (Alaska Department of Environmental Conservation). 1989a. Impact maps and summary reports of shoreline surveys of the *Exxon Valdez* spill site—Prince William Sound. Report to the *Exxon Valdez* Oil Spill Trustee Council, Anchorage, Alaska.

ADEC (Alaska Department of Environmental Conservation). 1989b. Fact sheet: *Exxon Valdez* oil spill. Oil Spill Public Information Office, Alaska Department of Environmental Conservation, Juneau.

ADEC (Alaska Department of Environmental Conservation). 1990a. Impact maps and summary reports of shoreline surveys of the *Exxon Valdez* spill site—Kodiak area. Report to the *Exxon Valdez* Oil Spill Trustee Council, Anchorage, Alaska.

ADEC (Alaska Department of Environmental Conservation). 1990b. Impact maps and summary reports of shoreline surveys of the *Exxon Valdez* spill site—Homer area. Report to the *Exxon Valdez* Oil Spill Trustee Council, Anchorage, Alaska.

Antenucci, J. C., K. Brown, P. L. Croswell, and M. J. Kevany. 1991. Implementation. Pages 211–236 *in* Geographic information systems: a guide to the technology. Van Nostrand Reinhold, New York.

Burrough, P. A. 1986. Principles of geographical information systems for land resources assessment. Monographs on soil and resources survey 12. Oxford University Press, New York.

Dangermond, J. 1991. Development and applications of GIS. Pages 101–108 *in* M. Heit and A. Shortreid, editors. GIS applications in natural resources. GIS World, Inc., Fort Collins, Colorado.

Gilfillan, E. S., D. S. Page, E. J. Harner, and P. D. Boehm. 1995. Shoreline ecology program for Prince William Sound, Alaska, following the *Exxon Valdez* oil spill, part 3: biology. Pages 398–443 *in* P. G. Wells, J. N. Butler, and J. S. Hughes, editors. *Exxon Valdez* oil spill: fate and effects in Alaskan waters. American

Society for Testing and Materials Special Technical Publication 1219, Philadelphia.

Gundlach, E. R., and M. O. Hayes. 1978. Vulnerability of coastal environments to oil pollution. Marine Technology Society Journal 12:18–27.

Harper, J. R., D. E. Howes, and P. D. Reimer. 1991. Shore-zone mapping system for use in sensitivity mapping and shoreline countermeasures. Pages 509–523 *in* Proceedings of the fourteenth arctic and marine oil spill program technical seminar. Environment Canada, Environmental Protection Conservation and Protection, Technology Development Branch, Ottawa.

Hayes, M. O., E. R. Gundlach, and C. D. Getter. 1980. Sensitivity ranking of energy port shorelines. Pages 697–709 *in* Proceedings of a specialty conference on ports. American Society of Civil Engineers, New York.

Highsmith, R. C., and six coauthors. 1996. Impact of the *Exxon Valdez* oil spill on intertidal biota. American Fisheries Society Symposium 18:212–237.

Johnston, C. A., N. E. Detenbeck, J. P. Bonde, and G. J. Niemi. 1988. Geographic information systems for cumulative impact assessment. Photogrammetric Engineering and Remote Sensing 54:1609–1615.

McDonald, L. L., W. P. Erickson, and M. D. Strickland. 1995. Survey design, statistical analysis, and basis for statistical inferences in coastal habitat injury assessment: *Exxon Valdez* oil spill. Pages 296–311 *in* P. G. Wells, J. N. Butler, and J. S. Hughes, editors. *Exxon Valdez* oil spill: fate and effects in Alaskan waters. American Society for Testing and Materials Special Technical Publication 1219, Philadelphia.

Page, D. S., E. S. Gilfillan, P. D. Boehm, and E. J. Harner. 1995. Shoreline ecology program for Prince William Sound, Alaska, following the *Exxon Valdez* oil spill, part 1: study design and methods. Pages 263–295 *in* P. G. Wells, J. N. Butler, and J. S. Hughes, editors. *Exxon Valdez* oil spill: fate and effects in Alaskan waters. American Society for Testing and Materials Special Technical Publication 1219, Philadelphia.

RPI (Research Planning Institute, Inc.). 1983a. Sensitivity of coastal environments and wildlife to spilled oil, Prince William Sound, Alaska: an atlas of coastal resources. Report to NOAA (National Oceanic and Atmospheric Administration) Office of Oceanography and Marine Services, Seattle.

RPI (Research Planning Institute, Inc.). 1983b. Sensitivity of coastal environments and wildlife to spilled oil, Shelikof Strait Region, Alaska: an atlas of coastal resources. Report to NOAA (National Oceanic and Atmospheric Administration) Office of Oceanography and Marine Services, Seattle.

RPI (Research Planning Institute, Inc.). 1985. Sensitivity of coastal environments and wildlife to spilled oil, Cook Inlet/Kenai Peninsula, Alaska: an atlas of coastal resources. Report to NOAA (National Oceanic and Atmospheric Administration) Office of Oceanography and Marine Services, Seattle.

RPI (Research Planning Institute, Inc.). 1986. Sensitivity of coastal environments and wildlife to spilled oil, Southern Alaska Peninsula, Alaska: an atlas of coastal resources. Report to NOAA (National Oceanic and Atmospheric Administration) Office of Oceanography and Marine Services, Seattle.

Schaller, J. 1992. GIS application in environmental planning and assessment. Computers, Environment and Urban Systems 16:337–353.

Stekoll, M. S., and seven coauthors. 1996. Coastal habitat injury assessment: intertidal communities and the *Exxon Valdez* oil spill. American Fisheries Society Symposium 18:177–192.

Stewart-Oaten, A., W. W. Murdoch, and K. R. Parker. 1986. Environmental impact assessment: pseudoreplication in time? Ecology 67:929–940.

Webb, R. P. 1982. A synopsis of natural resource management and environmental assessment techniques using geographic information system technology. Computers, Environment and Urban Systems 7:219–231.

American Fisheries Society Symposium 18:177–192, 1996

Coastal Habitat Injury Assessment: Intertidal Communities and the *Exxon Valdez* Oil Spill

MICHAEL S. STEKOLL
University of Alaska Fairbanks, Juneau Center, School of Fisheries and Ocean Sciences 11120 Glacier Highway, Juneau, Alaska 99801, USA

LAWRENCE DEYSHER
Coastal Resources Associates, Inc., 1185 Park Center Drive, Suite A, Vista, California 92083, USA

RAYMOND C. HIGHSMITH AND SUSAN M. SAUPE
University of Alaska Fairbanks, School of Fisheries and Ocean Sciences Fairbanks, Alaska 99775, USA

ZHANYANG GUO
University of Alaska Fairbanks, Juneau Center, School of Fisheries and Ocean Sciences

WALLACE P. ERICKSON, LYMAN MCDONALD, AND DALE STRICKLAND
Western EcoSystems Technology, 1402 South Greely Highway, Cheyenne, Wyoming 82007, USA

Abstract.—This paper summarizes results of intertidal community studies from the Natural Resource Damage Assessment, Coastal Habitat Injury Assessment (CHIA) Project carried out from spring 1990 through summer 1991. Injuries to the intertidal plant and invertebrate populations by the *Exxon Valdez* oil spill and subsequent cleanup treatment activities were documented over the entire area of the spill, from Prince William Sound to the Alaska Peninsula. A stratified random sample of oiled sites was selected and populations on matching reference (control) sites were compared. Statistically significant differences between oiled–treated and reference sites are interpreted as injury resulting from the oil spill. These differences are summarized for algal and invertebrate variables, such as percent cover, biomass, and abundance. Results indicate that the oil spill and subsequent cleanup treatment had serious effects that were evident throughout the entire region, at least until 1991, when the study ended. The organisms most often showing differences between oiled–treated and reference sites included *Fucus*, limpets, mussels, barnacles, littorine gastropods, and oligochaetes. However, the responses of the intertidal biota to the oil spill and treatment varied depending on the tidal elevation, habitat, and region. As of 1991 recovery of the intertidal biota showed varying patterns for the oiled–treated sites depending on the tide level, habitat, and region. Additional monitoring is needed to determine complete recovery of these affected areas.

The oil supertanker *Exxon Valdez* ran aground on 24 March 1989 on Bligh Reef in the northeastern part of Prince William Sound. About 40 million liters (11 million gallons) of Prudhoe Bay (also called North Slope) crude oil was discharged into Prince William Sound. The oil spread not only within the sound, but also polluted the Kenai Peninsula, Cook Inlet, the Kodiak Islands, and the Alaska Peninsula. About 1,300 km of the coastline of the Gulf of Alaska was contaminated.

Intertidal invertebrates and plants can be affected indirectly or directly by such an oil spill. Initial reduction in populations can occur as a result of mortality caused by direct contamination, smothering, or cleanup activities. Sublethal effects could include reduced growth rates or inhibited reproduction. Decreases in algal cover can affect invertebrate populations as a result of decreased food availability and reduced protection from desiccation or predation. All of these effects could contribute to decreased recolonization rates. In contrast, some invertebrate species may show increased populations after an oil spill and subsequent cleanup treatment; the increase may be caused by a reduction in the number of predators or, in the case of herbivores, by initial recolonization by rapidly growing ephemeral algal species (Southward 1982).

The *Exxon Valdez* oil spill, the largest oil spill in American history, has been well studied. Highsmith et al. (1996, this volume) and Stekoll et al. (1993 and unpublished data) reported the injuries and potential recovery processes of the intertidal invertebrates and macroalgae in the areas affected by this oil spill. Van Tamelen and Stekoll (1996, this volume) found shading by adult plants is critical for algal establishment in bare areas. Houghton et al. (1991, 1993), Lees et al. (1993), Driskell et al. (1993), and Gilfillan et al. (1995a, 1995b) also found

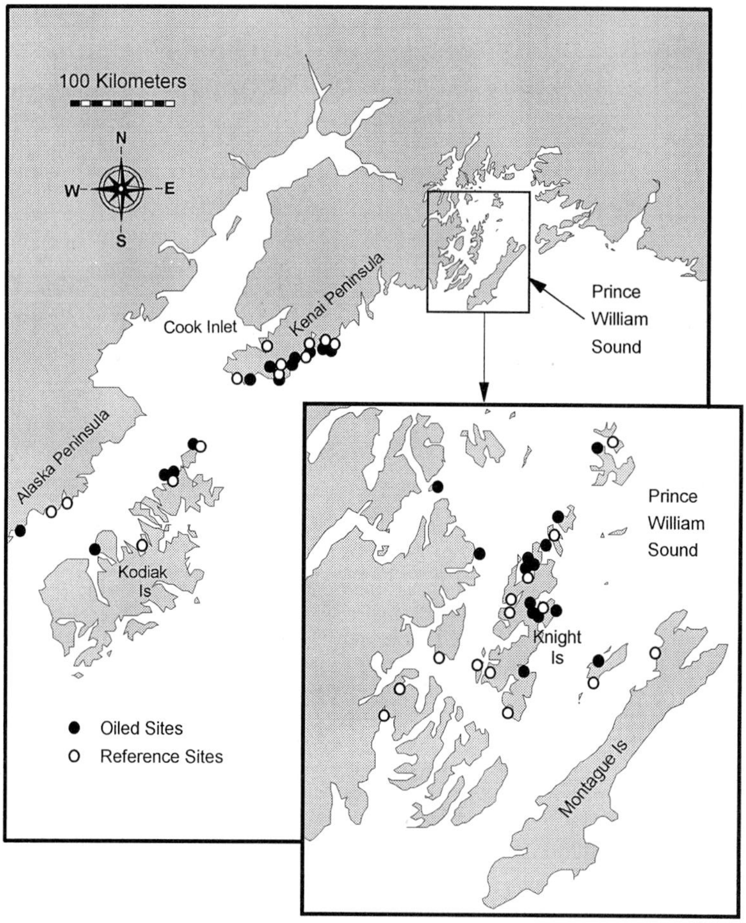

FIGURE 1.—Map of the study sites in the Gulf of Alaska. Is = Island.

that the oil spill and subsequent cleanup had significant impacts on intertidal communities. This paper was prepared to provide an overview of the results from the National Resource Damage Assessment, Coastal Habitat Injury Assessment (CHIA) project, which assessed injury to intertidal communities. Because of the large data sets collected, we present in this report only an overview of the effects of the oil spill over the entire affected area. More specific data are presented in the companion paper by Highsmith et al. (1996).

Methods

Site Selection

Detailed information for the site selection for this project is described by Sundberg et al. (1996, this volume). A brief overview is given here. The *Exxon Valdez* oil spill-impacted area within the Gulf of Alaska was divided into three regions: (1) Prince William Sound (PWS), (2) Cook Inlet and the Kenai Peninsula (CIK), and (3) Kodiak Island and the Alaska Peninsula (KAP)(Figure 1). Each region was stratified into four habitat types: sheltered rocky, coarse textured, exposed rocky, and estuarine. Oiled sites were selected randomly using a geographical information system made available from other agencies (Sundberg et al. 1996). Sites were designated as heavily, moderately, lightly, or not oiled on the basis of beach surveys conducted by the Alaska Department of Environmental Conservation in the summer of 1989 (originally as an unpublished digitized map database, but subsequently published: ADEC 1989, 1990a, 1990b). Ground truthing to verify both oiling level and habitat type was conducted by the Alaska Department of Fish and Game along with CHIA personnel (Sundberg et al. 1996). A control or reference site was selected to match each oiled site using physical criteria including habitat type, beach slope, beach texture, and wave exposure. Site locations are given in Table 1 and Figure 1. Comparisons were made for each matched site pair and across all site pairs of each habitat type within each region (McDonald et al. 1995). The experimental design precluded consideration of the type of cleanup treatment performed at oiled sites. Because treatment of sites varied considerably and was not well documented, we refer to the selected oiled sites as "oiled–treated" sites.

Site Layout and Quadrat Sampling

The site length was measured and marked along the mean higher high water (MHHW) line. Six equally spaced transects were randomly placed along the length of the beach, running from MHHW to mean lower low water (MLLW). Permanent quadrats (80 × 50 cm) were established along each transect at randomly located positions within each of the first 3 m of vertical drop (MVD) from MHHW. The tide level of MVD 3 corresponds to about 0.5 m above MLLW. During subsequent visits to a site, a new transect was placed 3 m to the left of each previous transect, and a new set of randomly placed quadrats was permanently established and sampled. Each quadrat was divided perpendicular to the water line into two 40- × 50-cm subquadrats. One subquadrat was left undisturbed to assess the effect of the oil spill on intertidal communities. All of the plant and animal biota were removed from one-half of the other subquadrat and preserved for later analyses. All visible surface organisms were collected, which included those that fell within the three-dimensional volume defined by the quadrat. Only plants whose holdfasts lay within the quadrat were collected. To collect infauna on nonbedrock substrates, the top 10 cm of sediment were collected and preserved. Samples were collected on four occasions: May–June (spring) 1990, July–August (summer) 1990, May (spring) 1991, and June–July (summer) 1991. For the last sampling trip, only percent cover data were collected. A listing of all variables tested and their overall means are given in Tables 2 and 3.

Percent Coverage Measurements

Percent cover data were collected using a point contact method. In spring 1991 percent cover observations were made on the subquadrat established during the spring of 1990 and a newly established quadrat during that visit. In summer 1991 observations were carried out on subquadrats established during late summer 1989, summer 1990, and a newly established quadrat during that visit. Percent cover was estimated for all macroalgae in the undisturbed quadrats (Table 2). All the data collected during each visit for each site were treated as one data set.

Biomass and Abundance

Blotted wet weights were determined on all macroalgae and invertebrates larger than 1 mm that were collected from the cleared subquadrats (Table 2). Abundance was determined for *Fucus* plants and for invertebrates collected in a subquadrat (Tables 2 and 3). Macroalgae and most mollusks, echinoderms, and crustaceans were identified to genus and species when possible. Polychaetes were classified to the family level and oligochaetes to the class level. Rare or difficult-to-classify organisms were identified only to the class or phylum level.

Fucus Attributes

Data were collected on several *Fucus* attributes, including plant length, reproductive index, blade inflation, relative amount and type of epiphytes, thallus damage, and plant regeneration (Table 3).

Statistical Procedures

For the majority of the data analyzed, the null hypothesis of no significant difference between a specific pair of matched oiled and reference sites was tested. The following procedure was used for invertebrate biomass and abundance, for macroalgal biomass and percent cover, and for *Fucus* density data. Levene's test (Milliken and Johnson 1984) was conducted to test for homogeneity of variances within sites of the region, habitat type, MVD, and time period. If variances were not significantly different at the $P = 0.01$ level, then ordinary one-way ANOVA were conducted to obtain the pooled estimate of variance. Pairs of oiled and reference sites were then compared by a two-tailed t-test using the pooled estimate of variance. If variances were significantly different, data were transformed to logarithms and the above tests were repeated. If the Levene's test remained significant with the transformed data, then regular t-tests with Satterwaite's correction for the degrees of freedom for unequal variances (Milliken and Johnson 1984) were conducted to compare matched pairs. Algal percent cover data were arcsin-square root transformed before Levene's test was applied. A two-sample ran-

TABLE 1.—Locations of sites surveyed in 1990 and 1991 Coastal Habitat Injury Assessment study. Site pairs are listed with the reference site first, followed by its paired oiled site. In two instances a reference site (Taylor Bay, Cook Inlet–Kenai; Foul Bay, Kodiak–Alaska Peninsula) was used for two oiled sites.

Site type and locality	Latitude	Longitude	Type
	Prince William Sound		
Exposed rocky			
Peak Island	60°42.50′N	147°24.70′W	Reference
Naked Island	60°42.00′N	147°26.90′W	Oiled
Chenega Island	60°16.70′N	148°04.00′W	Reference
East Knight Island	60°22.15′N	147°37.18′W	Oiled
Green Island	60°13.60′N	147°29.20′W	Reference
Green Island	60°15.90′N	147°29.00′W	Oiled
Coarse-textured			
Squire Island	60°15.85′N	147°55.60′W	Reference
Crafton Island	60°30.60′N	147°56.50′W	Oiled
Bay of Isles	60°22.05′N	147°41.67′W	Reference
Bay of Isles	60°21.80′N	147°42.00′W	Oiled
Whale Bay	60°12.60′N	148°17.90′W	Reference
Bay of Isles	60°21.80′N	147°41.90′W	Oiled
Little Bay	60°10.50′N	147°47.80′W	Reference
Snug Harbor	60°15.30′N	147°45.30′W	Oiled
Sheltered rocky			
Deer Cove	60°14.60′N	147°53.50′W	Reference
Northwest Bay	60°33.30′N	147°26.60′W	Oiled
Ingot Island	60°32.10′N	147°37.70′W	Reference
Disk Island	60°30.30′N	147°39.20′W	Oiled
Lower Herring Bay	60°23.35′N	147°49.55′W	Reference
Herring Bay	60°28.10′N	147°42.20′W	Oiled
Lower Herring Bay	60°22.58′N	147°49.21′W	Reference
Herring Bay	60°28.20′N	147°42.20′W	Oiled
Herring Bay	60°26.55′N	147°44.55′W	Reference
Herring Bay	60°27.60′N	147°42.90′W	Oiled
Estuarine			
Stockdale Harbor	60°17.20′N	147°12.30′W	Reference
Bay of Isles	60°23.00′N	147°42.60′N	Oiled
Puffin Cove	60°10.80′N	148°19.50′W	Reference
Applegate Island	60°37.70′N	148°08.35′W	Oiled
	Cook Inlet–Kenai		
Coarse-textured			
Southest Port Dick	59°15.45′N	151°07.18′W	Reference
One Haul Bay	59°13.10′N	151°13.60′W	Oiled
Chrome Bay	59°12.60′N	151°49.35′W	Reference
Chugach Bay	59°11.30′N	151°37.90′W	Oiled
Sadie Cove	59°27.90′N	151°20.30′W	Reference
West Arm, Port Dick	59°18.70′N	151°17.70′W	Oiled
Sheltered rocky			
Nuka Bay	59°28.45′N	150°21.65′W	Reference
McArthur Pass	59°27.70′N	150°22.60′W	Oiled
Chance Cove	59°29.09′N	150°18.21′W	Reference
Moring Cove	59°26.95′N	150°19.70′W	Oiled
Yalik Bay	59°27.30′N	150°36.60′W	Reference
Tonsina Bay	59°18.50′N	150°55.40′W	Oiled
Estuarine			
Taylor Bay	59°18.72′N	151°01.31′W	Reference
NE Nuka Island	59°23.50′N	150°37.45′W	Oiled
Tonsina Bay	59°18.60′N	150°57.20′W	Oiled
Berger Bay, Nuka Island	58°20.05′N	150°44.00′W	Reference
Tonsina Bay	59°18.50′N	150°56.70′W	Oiled

TABLE 1.—Continued.

Site type and locality	Latitude	Longitude	Type
	Kodiak–Alaska Peninsula		
Coarse-textured			
Dakavak Bay	58°02.85′N	154°38.85′W	Reference
Kashvik Bay	57°53.20′N	155°15.50′W	Oiled
Uganik Passage	57°49.60′N	153°11.20′W	Reference
Foul Bay	58°21.00′N	152°45.20′W	Oiled
Sheltered rocky			
East Perevalnie Island	58°38.10′N	152°20.51′W	Reference
Perevalnie Passage	58°37.90′N	152°22.00′W	Oiled
Takli Island	58°04.20′N	154°29.50′W	Reference
Chief Cove	57°42.50′N	153°54.00′W	Oiled
Foul Bay	58°20.52′N	152°45.70′W	Reference
Foul Bay	58°21.60′N	152°45.90′W	Oiled
Foul Bay	58°21.70′N	152°46.10′W	Oiled

domization test (Manly 1991) was used for the analysis of *Fucus* data involving ratios (e.g., proportion of plants with epiphytes).

Fisher's procedure for combining *P* values of independent analyses on matched pairs (Sokal and Rohlf 1981) was used to test for overall trends resulting from oiling within a given region, habitat, MVD, and time period. This procedure for drawing statistical inferences is designed in so far as possible after the philosophy of Fisher's least-significant-difference method, a procedure whose strategy is to minimize the total number of errors rather than only giving control over type I errors (Milliken and Johnson 1984).

Injury and recovery of the intertidal biological community was defined in terms of statistical variables, including the number of individuals of a particular species, biomass of a particular species, and number of species present. We considered injury to exist (and recovery to be incomplete) as long as any differences (positive or negative) can be detected between sets of matched pairs of oiled and reference sites. Many multivariate indices of community response have been proposed in the literature (e.g., canonical correspondence analysis by Page et al. 1995). Data from our study design with matched pairs is not easily analyzed by those multivariate indices because the inherent precision created by pairing is lost if that pair bond is broken. In this paper, we characterize the community in terms of biological variables mentioned above and use a direct and more understandable measure of community response, namely the percentage of biological variables that are significantly different (positive or negative) when tested on a set of matched pairs in a particular region, time period, habitat type, and tidal zone.

The data presented in Highsmith et al. (1996, and unpublished data) were used to prepare the graphs shown in this overview paper. Because the non-oiled (reference) sites determined the expected number of species present, the size of the data sets (i.e., the number of different variables) varies with the region and habitat sampled. Results are presented as the proportion (percentage) of traits measured that were significantly different between oiled and reference sites as determined by Fisher's combined test at both the $0.01 \leq P < 0.05$ and $P < 0.01$ levels of significance. A difference in the positive direction indicates that the values in the reference were greater than the values in the oiled sites. A difference in the negative direction indicates the converse.

Results

Time Sequences

In spring and summer 1990, about 12 to 16% of the variates measured from all habitats in all regions and tide levels showed statistically significant differences between oiled–treated and reference sites (Figure 2A). Values of variates from the reference sites were greater than those from oiled–treated sites in about 70% of the significant cases. In spring 1991 the number of statistically significant results remained at about 13%. In summer 1991 more than 20% of the variates measured (percent cover data only) were statistically significant. Many of the significant differences in these data can be attributed to reduced coverage of total algae and of the perennial alga *Fucus gardneri* in oiled–treated sites. The responses of the invertebrate species were mixed, depending on the region, habitat, and MVD, as discussed below.

TABLE 2.— Mean values of all variables tested averaged over all of the regions, habitat types, sampling times, and tide levels. A blank space indicates no tests were done for the variable in this category of data type. A value of 0.00 indicates trace amount present.

Variable	Cover (%)	Biomass (g/m²)	Abundance (number/m²)
Total uncolonized substrate	50.55		
Algae			
Total algae	30.29	803.77	
Perennials	24.07	700.61	
Annuals	11.04	103.16	
Ephemerals	7.03	44.80	
Total erect algae	33.98		
Total crustose algae	0.98		
Total coralline algae	0.16		
Total Chlorophyta	3.93	21.35	
Bladed greens	2.13	10.42	
Bladed greens (except *Prasiola–Rosenvingiella*)	1.91		
Kornmannia leptoderma		0.33	
Monostroma–Ulvaria		7.93	
Total *Ulva*		3.33	
Ulva spp.		2.29	
Ulva fenestrata		1.75	
Unidentified green blades		0.01	
Fine filamentous greens	1.77	9.72	
Unidentified green fine filaments		2.41	
Acrosiphonia	0.42	3.37	
Total *Chaetomorpha–Rhizoclonium*	0.11	0.32	
Chaetomorpha–Rhizoclonium spp.		0.35	
Chaetomorpha melagonium		0.01	
Cladophora spp.	1.27	6.80	
Gomontia polyrhiza		0.00	
Ulothrix spp.	0.11	0.03	
Urospora spp.	0.30	0.08	
Tubular and saccate greens	0.31	1.67	
Tubular greens (except *Derbesia*)	0.31		
Total *Blidingia–Enteromorpha*		1.67	
Blidingia–Enteromorpha spp.		1.08	
Enteromorpha clathrata		0.69	
Enteromorpha intestinalis		0.32	
Total Phaeophyta	24.72	651.26	
Bladed browns	21.04	620.52	
Unidentified bladed browns		0.08	
Alaria spp.	4.06	45.62	
Fucus cottonii	0.26	0.75	
Fucus gardneri	18.75	608.28	
Adult *Fucus gardneri* (>2 cm)	19.75		
Juvenile *Fucus gardneri* (≤2 cm)	0.57		
Fucus stipes	0.12		
Laminaria spp.		0.10	
Petalonia fascia	0.00	0.09	
Total *Punctaria*		0.00	
Punctaria spp.		0.00	
Punctaria lobata		0.00	
Coarse filamentous browns	0.42	3.40	
Analipus japonicus	0.13	0.06	
Chordaria flagelliformis		1.14	
Cystoseira geminata		0.13	
Desmarestia spp.		2.85	
Total *Dictyosiphon*	0.25	0.57	
Dictyosiphon foeniculaceus		0.49	
Dictyosiphon sinicola		0.41	
Eudesme virescens		0.09	
Myelophycus–Scytosiphon	0.36	2.77	
Filamentous browns (except *Sphacelaria*)	3.63		
Fine filamentous browns	3.63	27.34	

TABLE 2.—Continued.

Variable	Cover (%)	Biomass (g/m²)	Abundance (number/m²)
Ectocarpus sp.		0.06	
Elachista fucicola		0.04	
Pilayella sp.		26.59	
Sphacelaria sp.		1.87	
Saccate browns	0.22	1.10	
Leathesia–Colpomenia	0.22		
Soranthera ulvoidea	0.06	0.44	
Colpomenia peregrina		0.02	
Leathesia difformis		0.83	
Ralfsia fungiformis	0.08	0.65	
Total Rhodophyta	6.76	131.16	
Bladed reds	2.31	42.76	
Unidentified red blades		0.08	
Bangia spp.	0.13	0.00	
Constantinea simplex		0.07	
Neodilsea spp.		0.03	
Gigartinaceae group	0.98	20.84	
Total *Membranoptera*		0.00	
Membranoptera weeksiae		0.00	
Total *Palmaria*	1.47	24.20	
Palmaria spp.		0.68	
Palmaria callophylloides		14.44	
Palmaria hecatensis		14.29	
Palmaria mollis		6.06	
Phycodrys riggii	0.10	0.79	
Porphyra spp.	0.73	2.90	
Rhodymenia pertusa	0.01	0.03	
Tokidadendron kurilensis	0.32	3.95	
Coarse filamentous reds	3.47	78.52	
Ahnfeltia fastigata		0.20	
Cryptosiphonia woodii	0.42	5.34	
Dumontia contorta		0.05	
Endocladia muricata	0.89	3.70	
Gloiopeltis furcata	0.45	1.64	
Microcladia borealis	0.01	1.04	
Nemalion helminthoides		0.31	
Total Neorhodomela–Rhodomela	1.27	47.48	
Neorhodomela–Rhodomela spp.		38.65	
Neorhodomela larix		14.95	
Neoptilota asplenioides		0.50	
Ptilota spp.	0.69	20.53	
Odonthalia floccosa	1.39	22.28	
Fine filamentous reds	0.53	5.77	
Unidentified red fine filaments		0.01	
Antithamnion–Antithamnionella		0.05	
Audouinella purpurea		0.02	
Callithamnion pikeanum	0.01	0.46	
Ceramium sp.		0.07	
Erythrotrichia sp.		0.00	
Gloiosiphonia capillaris		0.00	
Polysiphonia–Pterosiphonia	0.52	5.67	
Scagelia–Hollenbergia		0.18	
Tiffaniella snyderae		0.06	
Saccate reds	0.51	9.65	
Halosaccion spp.	0.04	0.21	
Halosaccion glandiforme	0.50	9.59	
Calcified reds		8.48	
Articulate coralline algae	0.14	8.48	
Clathromorphum recliniatum		0.02	
Crustose reds	0.93		
Crustose coralline algae	0.08		
Hildenbrandia spp.	0.87		
Petrocelis spp.	0.13		
Invertebrates			
Lottidae		4.50	31.43
Lottia pelta		1.95	9.95
Tectura persona		1.22	8.36

TABLE 2.—Continued.

Variable	Cover (%)	Biomass (g/m^2)	Abundance ($number/m^2$)
Tectura scutum		0.66	3.13
Balanomorpha		280.14	3,119.98
Chthamalus dalli		6.28	838.90
Balanus glandula		50.96	369.97
Semibalanus balanoides		162.63	1,586.33
Lacuna spp.		0.30	106.45
Littorina spp.		27.88	852.71
Littorina sitkana		25.04	597.23
Littorina scutulata		2.78	255.24
Mytilus edulis		246.78	1,628.30
Turtonia minuta		0.25	236.42
Pagurus spp.		1.00	20.72
Nucella spp.		3.91	3.17
Nucella lima		1.83	1.62
Nucella lamellosa		2.05	1.25
Amphipoda		1.59	498.64
Isopoda		1.06	117.26
Errant polychaetes		0.85	140.78
Sedentary polychaetes		0.41	368.59
Oligochaetes		1.42	2,441.26
Total invertebrates		624.31	11,033.26

Regions

In all three regions about 12 to 15% of tests were significant when results are summarized over time, habitat type, and tide level (Figure 2B). Values were higher in reference sites in all three regions, especially in PWS and KAP. In PWS the taxa that were generally less abundant in oiled–treated sites included *Fucus*, the mussel *Mytilus*, the limpet *Tectura persona*, and some species of barnacles. Depending on the habitat type, abundance of littorines and infaunal annelids were sometimes significantly different between oiled and reference sites. In CIK the directions of the significant differences, either (+) or (−), were more evenly distributed compared with PWS and KAP. Many of the cases where oiled sites had higher values than reference sites occurred in coarse-textured habitats. *Fucus* cover was less in oiled–treated sites in the high intertidal, but greater in oiled–treated sites in the lower intertidal. The barnacle *Balanus glandula* and the gastropod *Littorina scutulata* showed higher values on CIK oiled–treated sites. In KAP, the algal group *Myelophycus–Scytosiphon* and members of the Gigartinaceae family were more abundant, while *Fucus* was generally less abundant at oiled sites. In KAP limpets and barnacles tended to be more abundant on reference sites, and oligochaetes were more abundant on oiled sites.

Habitat

Estuarine habitats were not sampled in KAP, and exposed rocky habitats were not sampled in CIK or KAP (Table 1). Thus, fewer variates were tested in these habitat types.

In both sheltered rocky and coarse-textured habitats, about 14% of the variates measured from all time periods, all regions, and all tide levels showed statistically significant results. About 8 to 11% of tests showed significantly higher values in the reference sites than in the oiled–treated sites, whereas about 5% of the tests showed significant differences in the reverse direction (Figure 2C). In the estuarine and exposed rocky habitats 8 to 13% of the tests were significantly higher in the reference sites than in the oiled–treated sites; about 3 to 4% were in the opposite direction.

At sheltered rocky sites *Fucus* generally had less coverage in oiled–treated sites. The alga *Gloiopeltis* and those in the Gigartinaceae family had greater coverage in oiled sites. *Fucus* plants in oiled sites had fewer fertile receptacles and had more attached epiphytes. Barnacles were generally less abundant at oiled sites. The upper intertidal barnacle *Chthamalus dalli* was usually more abundant and had greater biomass at oiled sites.

TABLE 3.—Mean values for *Fucus* attributes and for algal diversity indices averaged across all regions, habitat types, sampling times, and tide levels.

Attributes and indices	Units	Mean
Fucus attributes		
Plant density	$number/m^2$	603.30
Adult (>2 cm) density	$number/m^2$	169.50
Stipe density	$number/m^2$	17.56
Average plant length	cm	4.95
Size-class 1 (≤2 cm)	%	51.06
Size-class 2 (2.5–5 cm)	%	19.06
Size-class 3 (5.5–10 cm)	%	13.97
Size-class 4 (>10 cm)	%	15.90
Size-class 1 (≤2 cm) density	$number/m^2$	433.80
Size-class 2 (2.5–5 cm) density	$number/m^2$	93.89
Size-class 3 (5–10 cm) density	$number/m^2$	46.69
Size-class 4 (>10 cm) density	$number/m^2$	28.92
Damaged adults (>2 cm)	%	49.88
Damaged juveniles (≤2 cm)	%	15.35
Reproductive plant density	$number/m^2$	17.40
Receptacle density	$number/m^2$	278.28
Average reproductive index	per reproductive adult	3.65
Average number of receptacles	per reproductive plant	12.73
Reproductive adults (>2 cm)	%	16.43
Plants with regeneration	%	13.92
Inflated adults (>2 cm)	%	4.45
Average inflation index	per inflated adult	0.08
Adults (>2 cm) with epiphytes	%	18.88
Average epiphyte index	per adult with epiphytes	0.26
Algal indices		
Algal taxa richness	from biomass data	3.63
Algal taxa richness	from percent cover	1.76
Algal diversity index	from percent cover	0.15

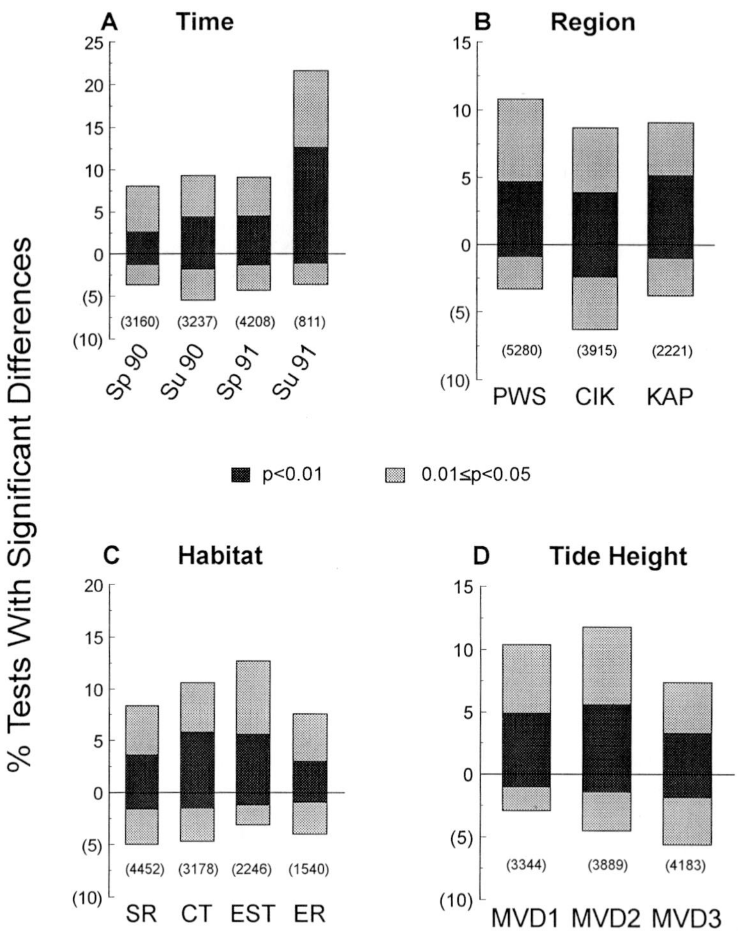

FIGURE 2.—Percentage of tests that showed significant ($P < 0.05$) overall differences (Fisher's combined test) between the means for variables measured in oiled–treated and reference sites. Positive values indicate comparisons that had the reference mean greater than the mean from the oiled sites; negative values indicate means from the oiled sites were greater than those from the reference sites. Numbers in parentheses are the total number of tests made for that data set. Sp = spring, Su = summer, SR = sheltered rocky, CT = coarse-textured, EST = estuarine, ER = exposed rocky, PWS = Prince William Sound, CIK = Cook Inlet and the Kenai Peninsula, KAP = Kodiak Island and the Alaska Peninsula, MVD = meter of vertical drop. (**A**) Tests grouped by the time at which samples were collected. The last sampling date includes only percent cover data. (**B**) Tests grouped by region (**C**) Tests grouped by habitat type. Estuarine habitats were not sampled in KAP. Exposed rocky habitats were sampled only in PWS. (**D**) Tests grouped by tidal zonation.

At oiled–treated, coarse-textured sites, coverage of *Fucus* was generally less than at reference sites. Oligochaetes and amphipods tended to be more abundant and polychaetes less abundant on coarse textured, oiled sites in all three regions. Barnacles, mussels, and littorines showed strong trends for differences in abundance and biomass throughout the intertidal zone.

Estuarine sites had lower coverage of total algae, ephemerals, perennials, *Fucus*, and filamentous brown algae at oiled–treated sites. Littorines, amphipods, and isopods were also generally less abundant and had less biomass at these oiled sites. Barnacles accounted for many of the statistically significant differences, but the direction of the differences varied depending on the region.

Data from exposed rocky sites (measured only in PWS) revealed lower coverage of *Fucus* and the red alga *Neorhodomela* spp. at oiled–treated sites, but greater coverage of the red, bladed alga, *Palmaria* spp. and species in the Gigartinaceae family. Barnacles were more abundant at oiled–treated sites, and littorines and annelids were more abundant at reference sites.

Zonation

When results are pooled across all time periods, habitat types, and regions, there were significant differences between oiled and reference sites at all three MVDs in about 12 to 17% of the variates measured (Figure 2D). The highest percentage of differences occurred at MVD 2 and the lowest at MVD 3. Values were generally higher at reference sites, especially for MVDs 1 and 2. For example, in 1991 *Fucus* percent cover was significantly greater in reference, sheltered rocky sites in all three regions in the upper intertidal. But *Fucus* percent cover was greater in oiled–treated sites at MVD 3. *Palmaria* spp. and species in the Gigartinaceae family are some of the other algae that had greater percent cover in oiled–treated sites in the lower intertidal zone.

Although there was not a large difference in the proportion of significant results among the three MVDs, in most cases different organisms were affected at each MVD. In many instances this difference may be caused by a shift in the normal distribution of an organism. That is, an organism occurred abundantly at an elevation lower or higher than would be expected. Some organisms, however, showed a consistent trend over all three tidal heights. For example, in most instances on coarse-textured sites in PWS and KAP, mussels and littorines showed higher values for both abundance and biomass on reference sites at all MVDs.

Patterns in the Intertidal Zone

The response of the intertidal communities to the oiling and treatment varied with region, habitat type, tidal zone, and kingdom (Figures 3–5). A more detailed discussion of the responses of specific taxa is presented in Highsmith et al. (1996). Some general trends are presented here.

One striking difference between the oiling–treatment effects on algal and invertebrate populations was that a greater proportion of the invertebrate taxa showed significant differences. The total percentage of significant differences detected for invertebrates was generally a few times higher than that for algae.

On sheltered rocky sites, values for invertebrate variables generally showed more differences between oiled–treated and reference sites in the lower two MVDs in both 1990 and 1991 (Figure 3). The general trend over time for invertebrates was fewer significant differences between oiled and reference sites in 1991 compared to 1990. However, for the algal data, the trend on sheltered rocky sites showed equal or increased proportion of significant differences between 1990 and 1991, especially in PWS. Most of the significant differences detected for algae on sheltered rocky sites in both 1990 and 1991 showed higher values on reference sites compared to oiled–treated sites, again, especially in PWS. In CIK the algal data did not show a strong direction effect; higher values occurred in approximately equal proportions at both oiled–treated and reference sites. In KAP the highest proportion of significant differences for algae was detected at MVD 2.

In coarse-textured sites the proportion of significant differences for invertebrates was much higher than for the algae in all three regions (Figure 4). The proportion of significant differences between oiled and reference sites was also higher for invertebrates in coarse-textured sites than in sheltered rocky sites. In CIK algae were sparse throughout the intertidal zone and, thus, very few significant differences were detected (Figure 4). However, invertebrates in CIK were relatively more abundant and had higher biomasses on oiled–treated sites, especially at MVD 3 (Figure 4). In contrast, most significant differences for both invertebrates and algae on coarse-textured habitats in PWS and KAP showed higher values on reference sites. The trend in PWS from 1990 to 1991 showed equal or increased proportions of significant differences, except for invertebrates at MVD 1. In that instance, there was a change from about 30% of the tests showing significant differences in 1990 to less than 3% in 1991. In CIK the general trend for invertebrates also showed an increased proportion of significant differences in 1991 compared with 1990. The KAP region showed a decreased proportion of significant differences between oiled and reference sites in 1991 compared to 1990, for the invertebrate data but an increase for the algal data.

In PWS and CIK estuarine habitats, both algal and invertebrate data showed increased proportions of significant differences in most cases in 1991 compared with 1990 (Figure 5). These increased proportions in 1991 indicate delayed injury in this habitat. Algal data from this habitat showed higher proportions of significant differences compared with the other habitat types. Almost all of the significant differences seen for algae in both PWS and CIK, for both visits, and at all MVDs, showed that values were higher on reference sites compared to oiled–treated sites. The invertebrate data at MVD 1 in PWS showed a similar trend for both 1990 and

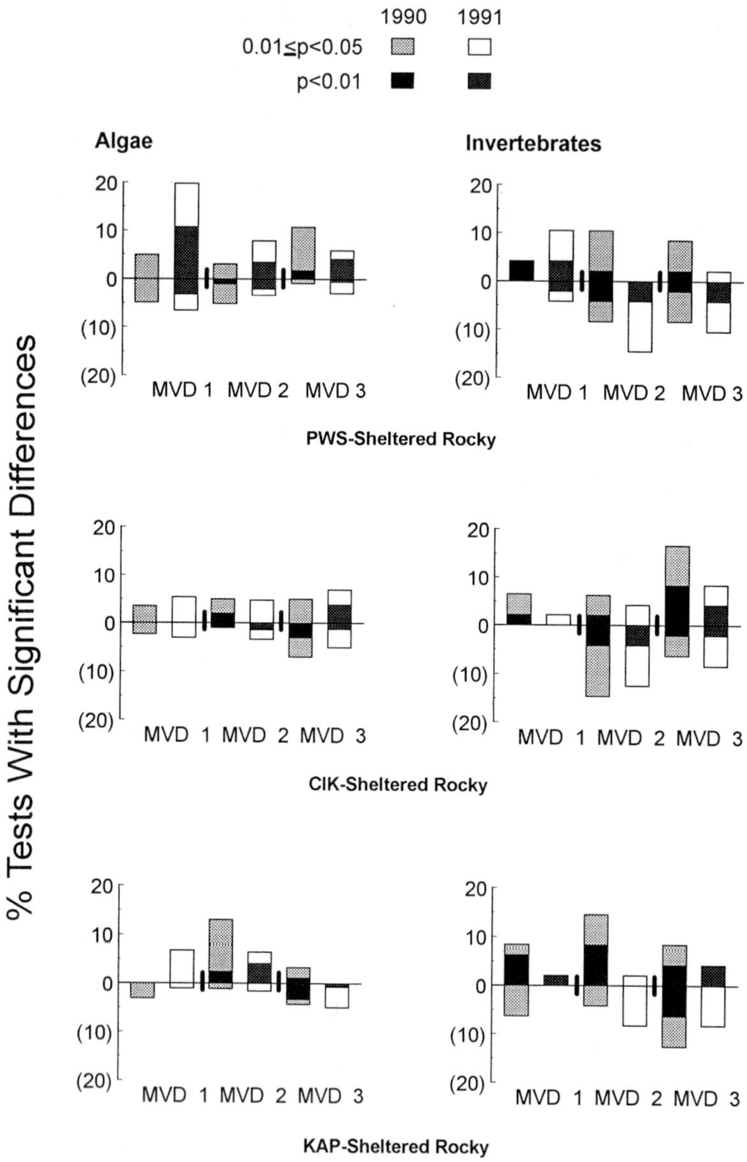

FIGURE 3.—Percentage of total overall tests (Fisher's combined test) that showed significant ($P < 0.05$) differences for biota in sheltered rocky habitats between reference and oiled–treated sites.

1991. At MVD 2 this trend did not persist into 1991. In CIK, invertebrate differences favored oiled–treated sites at MVDs 1 and 2 in 1990 but not in 1991.

At exposed rocky sites in PWS, algal and invertebrate data showed marked differences (Figure 5). Algal data showed low proportions of significant differences between oiled–treated and reference sites, and there was no noticeable tide level effect. Invertebrates showed a high proportion of significant differences, which varied with tide level. At MVD 1 more than 20% of the tests showed significant differences in 1990, with the majority of the values greater at the oiled–treated sites. By 1991 these differences had almost completely disappeared. In contrast, at MVD 2 all of the values for the variables with significant differences were greater in the reference sites. This proportion in-

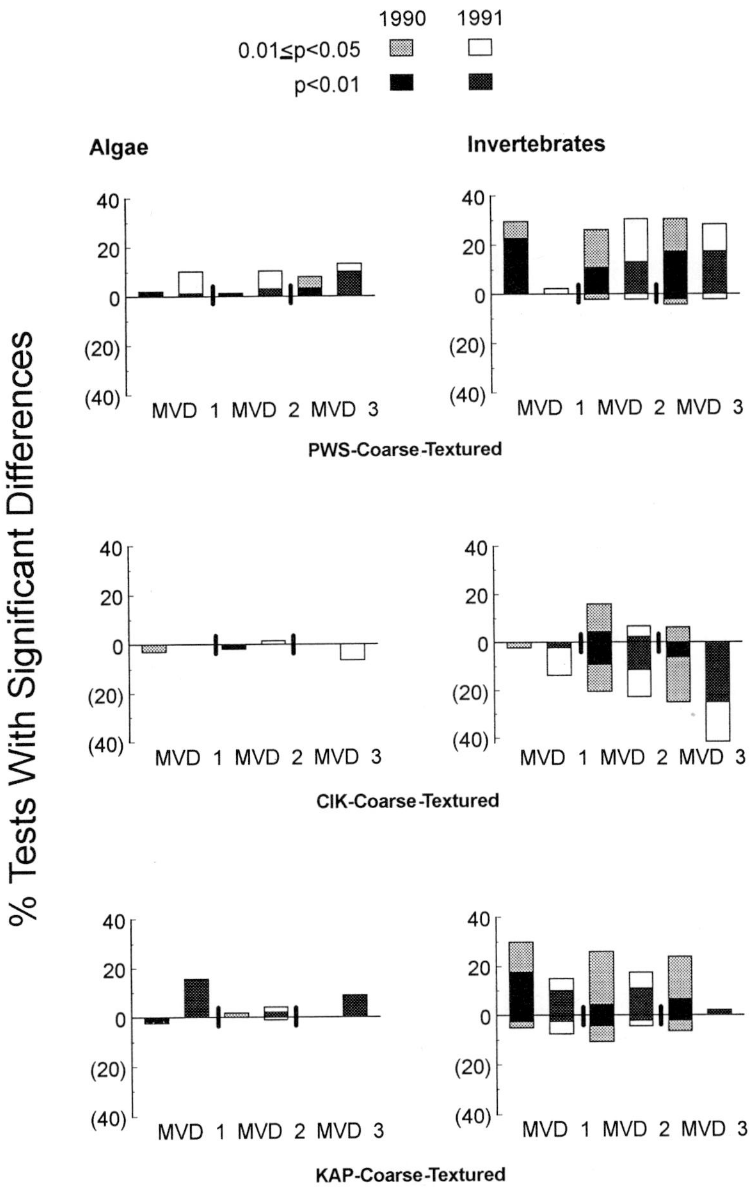

FIGURE 4.—Percentage of total overall tests (Fisher's combined test) that showed significant ($P < 0.05$) differences for biota in coarse-textured habitats between reference and oiled–treated sites.

creased to more than 40% in 1991, with most of the values remaining greater at the reference sites. At MVD 3 the direction of the significant differences switched between 1990 and 1991.

Discussion

The experimental design, an "After, Control-Impact Paired" design (ACIP; Dean et al. 1993) was selected to avoid problems similar to those encountered by Page et al. (1995) of high levels of variation that would lead to type II errors: falsely accepting the null hypothesis and concluding that there were no differences between oiled–treated and reference sites. By randomly selecting oiled sites and then matching these sites with unoiled reference sites using physical environmental criteria, such as wave

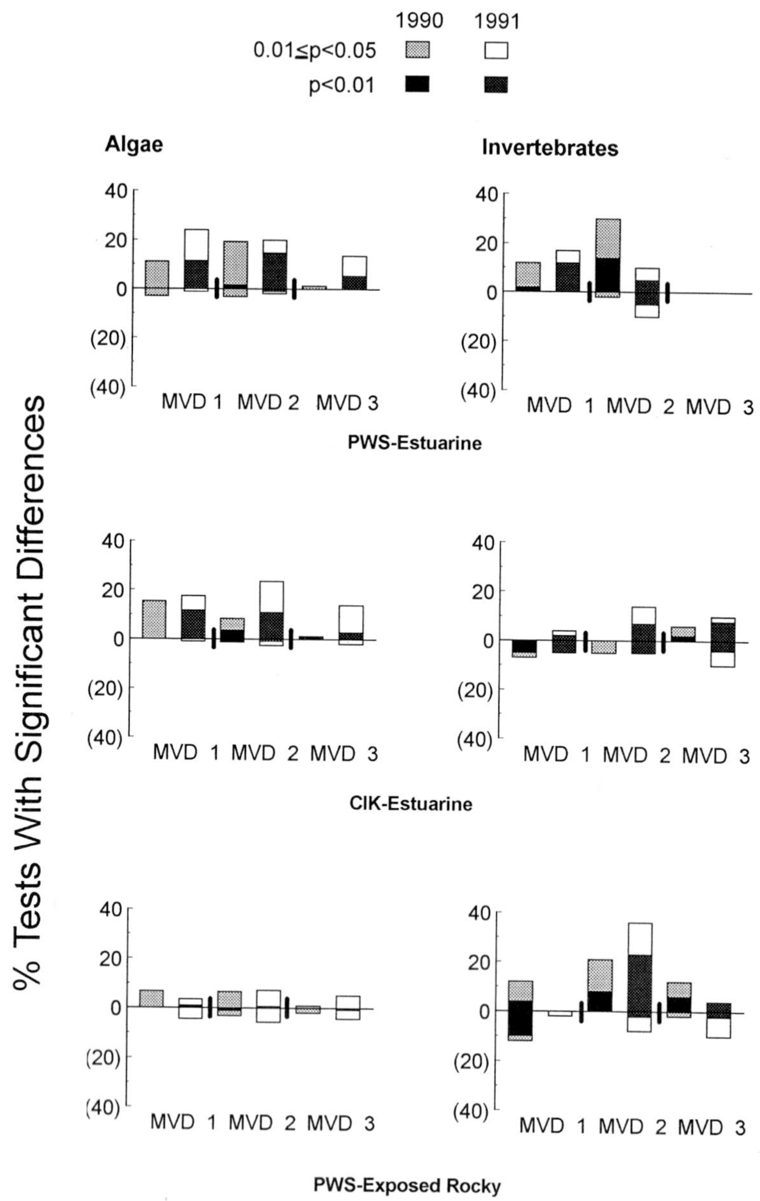

FIGURE 5.—Percentage of total overall tests (Fisher's combined test) that showed significant ($P < 0.05$) differences for biota in estuarine and exposed rocky habitats between reference and oiled–treated sites.

exposure, slope, aspect, fetch, beach texture, and proximity, the chances are improved that the paired sites would have similar biological communities in the absence of oiling.

For the results reported here there appears to be no consistent pattern into which all of the data fall. Some combinations of regions, habitats, and tide level showed many differences between oiled and reference sites, while others showed fewer differences. These inconsistencies are a reflection of the inherent differences in the communities themselves. For example, the upper intertidal zone in coarse-textured sites in PWS is a different community than in CIK, even though we have classified them both as upper intertidal, coarse-textured sites. In fact, the experimental design was stratified based on the premise that each region, habitat, and MVD would be different. Although the oil spill and subsequent treatment may have been responsible for the differences that were detected, each community responded differently depending on initial conditions and subsequent environmental factors. Additionally, the oil and cleanup treatment themselves were not consistently applied throughout the affected regions. The limitations of the data interpretation are not with respect to the question of whether the *Exxon Valdez* oil spill caused injury, but rather on being able to generalize as to what that injury was. Rather than being able to say species X or family Y was affected by the oil throughout the area of the spill, we are restricted to discussing injury patterns in the context of the region, the habitat type, and the tide level.

For discussion purposes it is assumed that the differences between values from reference and oil sites are a result of the oil spill and subsequent cleanup treatment activities. Any significant difference found by the Fisher's combined test is defined to be "injury," in the sense that the value is different than that expected if there had been no oil spill. An increase in biomass or abundance over the expected value in an oiled area is considered as much injury as a comparable decrease. Complete recovery is the state at which there is no statistical difference ($P \geq 0.05$) between the values being compared. With the above definitions in mind, it can be concluded from the presented data that the *Exxon Valdez* oil spill and subsequent cleanup activities caused substantial injury to the intertidal communities, with more than 10% of the tests showing significance (Figure 2A). Such injury was evident in 1990 and was still conspicuous in the summer of 1991 (Figure 2A).

We recognize that the data here contain some redundancies and autocorrelations. For example, the *Fucus* percent cover in some cases makes up almost all of the Total Algae category. It would be impossible to eliminate these correlations because of the presence of nonobvious indirect relationships. For example, does an increase in littorine biomass mean a decrease in *Fucus* percent cover? We have elected to include all of the data, with the assumption that such direct and indirect correlations will tend to act both to increase and to decrease the number of significant differences.

One of the most common aspects of the injury caused by this oil spill to the intertidal algae and invertebrates was a net loss in abundance and biomass. Generally, significantly higher values of abundance or biomass occurred more frequently at reference sites than at oiled sites throughout the area of the spill. The above phenomenon resulted mainly from reduced percent coverage, abundance, and biomass of both algae and invertebrates at the oiled–treated sites, with a corresponding increase in the amount of uncolonized substrate at oiled sites. An elevated proportion of unhealthy *Fucus* plants (lower ratio of productive plants, fewer receptacles per mature plant, and increased epiphytic coverage of plant surfaces) were components of the difference as well.

Decreases in abundance, biomass and cover of intertidal biota are common phenomena after exposure to an oil spill. Houghton et al. (1991) observed that some treated rocky beaches in Prince William Sound were totally stripped of flora and fauna at high and mid-intertidal zones, and the decline in populations in the lower intertidal zone was moderate. Several investigators have observed varying degrees of mortality or decline of intertidal organisms caused by oil spills and treatment worldwide. Among them are Smith (1968) for the *Torrey Canyon* (Kuwait crude oil and chemical treatment); Floc'h and Diouris (1980, 1981) for the *Amoco Cadiz* (crude oil and treatment); Thomas (1973, 1978) with the *Arrow* (Bunker C oil); and Rolan and Gallagher (1991) with the *Esso Bernicia* accident (Bunker C oil and treatment). Heavy losses of plants from physical damage and the elimination of *Fucus* and the salt-marsh plant *Spartina* occurred in some areas after the *Arrow* oil spill (Thomas 1973; Thomas and Harley 1973; Thomas 1978). Detachment of limpets occurred after a small spill of light fuel oil in Milford Haven, U.K. (Baker 1976). Following a major oil spill and treatment, increased variances in population variables are associated with the destruction or reduction of the intertidal biota. For instance, considerable fluctuations in densities and species composition at heavily oiled (Number 2 fuel oil) stations were reported after the *Florida* accident (Sanders 1978).

Alterations and fluctuations in the population structures of intertidal communities were effected by the *Exxon Valdez* oil spill. For example, algal coverage and biomass were reduced in most of the oiled sites, as discussed previously, but a number of taxa, such as *Gloiopeltis* spp. and species in the family of Gigartinaceae, showed increased cover and biomass within a year after the spill at oiled–treated, sheltered rocky habitats. High values from the oiled sites were observed at mid-intertidal zones at sheltered rocky habitats in all three regions, while reverse patterns occurred at other intertidal zones for most cases. Another example comes from the coarse-textured habitats. In both PWS and KAP coarse-textured habitats, invertebrates had lower values in the oiled sites, while in CIK, the opposite trend was found. Other studies have found a similar variety of responses as a result of oil spills and subsequent treatment (e.g., Neushul 1972; Sanders 1978).

Our data contradict some of the findings of others who have found that the high intertidal zones, where stranded oil accumulates, suffer the heaviest injuries and experience the slowest recovery after exposure to an oil spill (Neushul 1972; Thomas 1973, 1978; Houghton et al. 1991). Our data generally show that differences were greater at the mid-intertidal level (MVD 2), but each habitat and each region exhibited a different pattern. It does appear, however, that recovery from the effects of the *Exxon Valdez* oil spill, as measured by a decrease in the proportion of significant differences between oiled and reference sites, is slowest in the upper intertidal zone.

The time it takes for a spill-polluted intertidal community to return to its "normal" state depends upon the scale of the spill, the type of oil spilled, beach and water characteristics, and many other factors. Our data indicate that recovery of plants and animals in the *Exxon Valdez* oil spill-affected intertidal zones was proceeding slowly for certain variables. Injury to the intertidal communities could still be detected during the summer of 1991, two and a half years after the accident. Delay in recovery of the injured intertidal biota is probably the result of reduced numbers of mature organisms, decreased reproductive capacity of mature organisms, lower availability of food, or decreased canopy protection (P. G. van Tamelen and M. S. Stekoll, University of Alaska, unpublished data). After the *Tampico Maru* spill (diesel oil), considerable fluctuations and marked changes of the affected intertidal and subtidal communities were observed (North et al. 1965; Neushul 1972). It took at least 7 years for the area to return to a state resembling a normal ecosystem. It may take much longer for the *Exxon Valdez* oil spill-impacted habitats because recovery may be delayed in colder waters (Clark et al. 1978; Mann and Clark 1978).

The plants and animals of the intertidal community suffer a multitude of different effects during an oil spill and the subsequent cleanup activities. Reduced rates of photosynthesis of algae have been recorded after exposure to oil (Schramm 1972; Gordon and Prouse 1973). Intertidal invertebrates have been shown to suffer impaired reproductive capacity. They can also experience lower availability of food and destruction of protective algal habitat after a spill. All of these effects contribute to slow recolonization. However, enhanced growth of some salt-marsh plants exposed to weathered crude oil was observed several months following the *Torrey Canyon* spill (Stebbings 1970). Other investigators (Gilfillan et al. 1995a, 1995b) found that biota were enhanced by the *Exxon Valdez* oil spill, especially in the lower intertidal, coarse-textured habitats. Their explanation for these observations is that the residual oil enhanced the populations of oil-degrading bacteria, which in turn provided food for some intertidal invertebrates. Our own data support these observations only for certain tidal zones in selected habitats (CIK coarse textured, PWS sheltered rocky, MVD 3; see Figures 3 and 4). Opposite patterns occurred at other sites, such as at PWS and KAP coarse-textured beaches. The intertidal plants showed increases at oiled–treated sites mainly at MVD 3 in CIK and KAP sheltered rocky habitats. Any attempt to rationalize changes in biomass or abundance of intertidal biota in relation to oiling–treatment will be relevant only after examination of results from experimental studies, such as those carried out at Herring Bay (Hooten and Highsmith 1996, this volume; van Tamelen and Stekoll, unpublished data).

Conclusion

Among the three regions studied, the overall impacts to PWS and KAP intertidal communities were similar. Both regions experienced losses in percent cover, biomass, and abundance. The CIK region, on the other hand, experienced the greatest changes in community structure. For example, at MVD 3 in CIK sheltered rocky sites, more *Fucus* was observed at almost all of the oiled sites compared with the reference sites. Similarly, animal abundance and biomass were higher at the oiled sites in CIK coarse-textured habitats, even though algae were rarely observed at those sites. The causes for the differences between regions and habitats are unknown. Some mechanisms are being investigated by studies in Herring Bay (van Tamelen and Stekoll 1996 and unpublished data; Highsmith et al., unpublished data).

In concluding that there is injury to an oiled site or habitat using our approach, it is possible that we may be making type I errors, or, alternatively, that the differences we see are the result of factors other than injury from oiling or treatment. Type I errors are probably not attributable to the variances we encountered in these data. From a simple power analysis of the percent cover data, we concluded that differences would generally have to be at least 50% before we had an 80% probability of detecting them with a test of size $\alpha = 0.05$. Most of our statistically significant differences between oiled and reference sites were large and likely to be real. In many cases, the percentage of statistically significant results is in the neighborhood of 20 to 40%, which is much larger than 5%, the nominal probability of a type I error. On a univariate basis, a consistent finding of a difference over time, or over more than one tidal height, is also a check on a type I error. Differences resulting from factors other than the *Exxon Valdez* oil spill would require that reference and oiled sites are inherently dissimilar. For example, oiled sites may be located in areas of more current movement, which led to the initial oiling, and also was the mechanism causing the different community types. This problem of inherent differences is minimized by careful site selection, but ultimately it can only be tested by continued monitoring of the sites over a period of several years, until the community variates converge or stabilize. Our data do not show such a decrease in the percentage of significant differences from 1990 to 1991. An estimate of complete recovery is not possible after only 2 years of data collection. Data from Herring Bay indicate that sheltered rocky sites had not fully recovered by the end of the summer of 1993 (van Tamelen and Stekoll 1996).

Acknowledgments

This work was funded as part of the Natural Resources Damage Assessment (NRDA) Comprehensive Environmental Response Compensation, and Liability Act (CERCLA) process through the U.S. Forest Service by contract (#53-0109-9-00276) to the University of Alaska Fairbanks. We especially acknowledge the work of Mandy Lindeberg, Robin Jenne, Michelle Bourassa, Chirk Chu, John Noll, Tama Rucker, Vida Fruebis, and Stephanie Moreland for data acquisition, supervision of sorting, and data analyses. Tom Dean of Coastal Resources Associates contributed valuable advice on sampling design and data analysis. Gayle Hansen provided algal taxonomic expertise to the project.

References

ADEC (Alaska Department of Environmental Conservation). 1989. Impact maps and summary reports of shoreline surveys of the *Exxon Valdez* spill site—Prince William Sound. Report to the *Exxon Valdez* Oil Spill Trustee Council, Anchorage, Alaska.

ADEC (Alaska Department of Environmental Conservation). 1990a. Impact maps and summary reports of shoreline surveys of the *Exxon Valdez* spill site—Kodiak area. Report to the *Exxon Valdez* Oil Spill Trustee Council, Anchorage, Alaska.

ADEC (Alaska Department of Environmental Conservation). 1990b. Impact maps and summary reports of shoreline surveys of the *Exxon Valdez* spill site—Homer area. Report to the *Exxon Valdez* Oil Spill Trustee Council, Anchorage, Alaska.

Baker, J. M. 1976. Ecological changes in Milford Haven during its history as an oil port. Pages 55–66 *in* J. M. Baker, editor. Marine ecology and oil pollution: proceedings. Halsted Press, New York.

Clark, R. C., Jr., B. G. Patten, and E. E. DeNike. 1978. Observations of a cold-water intertidal community after 5 years of a low-level, persistent oil spill from *General M. C. Meigs*. Journal of the Fisheries Research Board of Canada 35:754–765.

Dean, T. A, L. McDonald, M. S. Stekoll, and R. R. Rosenthal. 1993. Damage assessment in coastal habitats: lessons learned from *Exxon Valdez*. Pages 695–697 *in* Proceedings, 1993 international oil spill conference: prevention, preparedness, response. American Petroleum Institute Publication 4580, Washington, DC.

Driskell, W. B., A. K. Fukuyama, J. P. Houghton, D. C. Lees, G. Shigenaka, and A. J. Mearns. 1993. Impacts on intertidal infauna: *Exxon Valdez* oil spill and cleanup. Pages 355–361 *in* Proceedings, 1993 international oil spill conference: prevention, preparedness, response. American Petroleum Institute Publication 4580, Washington, DC.

Floc'h, J.-Y., and M. Diouris. 1980. Initial effects of *Amoco Cadiz* oil on intertidal algae. Ambio 9(6):284–286.

Floc'h, J.-Y., and M. Diouris. 1981. Impact du petro de l'*Amoco Cadiz* sur les Algues de Portsall: suivi ecologique dans une anse tres polluee. Pages 381–391 *in Amoco Cadiz*: fates and effects of the oil spill. Proceedings of the international symposium. Centre National pour l'Exploitation des Oceans, Brest, France.

Gilfillan, E. S., D. S. Page, E. J. Harner, and P. D. Boehm. 1995a. Shoreline ecology program for Prince William Sound, Alaska, following the *Exxon Valdez* oil spill, part 3: biology. Pages 398–443 *in* P. G. Wells, J. N. Butler, and J. S. Hughes, editors. *Exxon Valdez* oil spill: fate and effects in Alaskan waters. American Society for Testing and Materials Special Technical Publication 1219, Philadelphia.

Gilfillan, E. S., T. H. Suchanek, P. D. Boehm, E. J. Harner, D. S. Page, and N. A. Sloan. 1995b. Shoreline impacts in the Gulf of Alaska following the *Exxon Valdez* oil spill. Pages 444–484 *in* P. G. Wells, J. N. Butler, J. S. Hughes, editors. *Exxon Valdez* oil spill: fate and effects in Alaskan waters. American Society for Testing and Materials Special Technical Publication 1219, Philadelphia.

Gordon, D. C., Jr., and N. J. Prouse. 1973. The effects of three oils on marine phytoplankton photosynthesis. Marine Biology 22:329–333.

Highsmith, R. C., and six coauthors. 1996. Impact of the

Exxon Valdez oil spill on intertidal biota. American Fisheries Society Symposium 18:212–237.

Hooten, A. J., and R. C. Highsmith. 1996. Impacts on selected intertidal invertebrates in Herring Bay, Prince William Sound, after the *Exxon Valdez* oil spill. American Fisheries Society Symposium 18:249–270.

Houghton, J. P., W. B. Driskell, D. C. Lees, and A. J. Mearns. 1991. Impacts of the *Exxon Valdez* spill and subsequent cleanup on intertidal biota—1 year later. Pages 467–475 *in* Proceedings, 1991 international oil spill conference: prevention. American Petroleum Institute Publication 4529, Washington, DC.

Houghton, J. P., A. K. Fukuyama, D. C. Lees, W. B. Driskell, G. Shigenaka, and A. J. Mearns. 1993. Impacts of intertidal epibiota: *Exxon Valdez* spill and subsequent cleanup. Pages 293–300 *in* Proceedings, 1993 international oil spill conference: prevention, preparedness, response. American Petroleum Institute Publication 4580, Washington, DC.

Lees, D. C., G. Shigenaka, and A. J. Mearns. 1993. Effects of shoreline treatment methods on intertidal biota in Prince William Sound. Pages 345–354 *in* Proceedings, 1993 international oil spill conference: prevention, preparedness, response. American Petroleum Institute Publication 4580, Washington, DC.

Manly, B. F. 1991. Randomization and Monte Carlo methods in biology. Chapman and Hall, London.

Mann, K. H., and R. B. Clark. 1978. Long-term effects of oil spills on marine intertidal communities. Journal of the Fisheries Research Board of Canada 35:745–795.

McDonald, L. L., W. P. Erickson, and M. D. Strickland. 1995. Design, statistical analysis, and basis for statistical inferences in coastal habitat injury assessment: *Exxon Valdez* oil spill. Pages 296–311 *in* P. G. Wells, J. N. Butler, and J. S. Hughes, editors. *Exxon Valdez* oil spill: fate and effects in Alaskan waters. American Society for Testing and Materials Special Technical Publication 1219, Philadelphia.

Milliken, G. A., and D. E. Johnson. 1984. Analysis of messy data, volume 1: designed experiments. Van Nostrand Reinhold Company, New York.

Neushul, M. 1972. The effects of pollution on populations of intertidal and subtidal organisms in southern California. Pages 165–172 *in* R. W. Holmes and F. A. DeWitt, Jr., editors. Santa Barbara oil symposium proceedings. U.S. Government Printing Office, Washington, DC.

North, W. J., M. Neushul, and K. A. Clendenning. 1965. Successive biological changes observed in marine cove exposed to a large spillage of mineral oil. Pages 335–354 *in* Pollutions marines par les microorganisms et les produits petrolies: Symposium de Monaco. Commission Internationale pour l'Exploration Scientifique de la Mer Mediterranee, Monaco.

Page, D. S., E. S. Gilfillan, P. D. Boehm, and E. J. Harner. 1995. Shoreline ecology program for Prince William Sound, Alaska, following the *Exxon Valdez* oil spill, part 1: study design and methods. Pages 263–295 *in* P. G. Wells, J. N. Butler, and J. S. Hughes, editors. *Exxon Valdez* oil spill: fate and effects in Alaskan waters. American Society for Testing and Materials Special Technical Publication 1219, Philadelphia.

Rolan, R. G., and R. Gallagher. 1991. Recovery of intertidal biotic communities at Sullom Vo following the *Esso Bernicia* oil spill of 1978. Pages 461–465 *in* Proceedings, 1991 international oil spill conference. American Petroleum Institute Publication 4529, Washington, DC.

Sanders, H. L. 1978. *Florida* oil spill impact on the Buzzards Bay benthic fauna: West Falmouth. Journal of the Fisheries Research Board of Canada 35:717–730.

Schramm, W. 1972. The effects of oil pollution on gas exchange in *Porphyra umbilicalis* when exposed to air. Proceedings of the International Seaweed Symposium 7:309–315.

Smith, E. J., editor. 1968. *Torrey Canyon* pollution and marine life. Cambridge University Press, Cambridge, UK.

Sokal, R. R., and F. J. Rohlf. 1981. Biometry. Freeman, New York.

Southward, A. J. 1982. An ecologist's view of the implications of the observed physiological and biochemical effects of petroleum compounds on marine organisms and ecosystems. Philosophical Transactions of the Royal Society of London Series B Biological Sciences 297:241–255.

Stebbings, R. E. 1970. Recovery of salt marsh in Brittany sixteen months after heavy pollution by oil. Environmental Pollution 1:163–167.

Stekoll, M. S., L. Deysher, and T. A. Dean. 1993. Seaweeds and the *Exxon Valdez* oil spill. Pages 135–140 *in* Proceedings, 1993 international oil spill conference: prevention, preparedness, response. American Petroleum Institute Publication 4580, Washington, DC.

Sundberg, K., L. Deysher, and L. McDonald. 1996. Intertidal and supratidal site selection using a geographical information system. American Fisheries Society Symposium 18:167–176.

Thomas, M. L. H. 1973. Effects of bunker C oil on intertidal and lagoonal biota in Chedabucto Bay, Nova Scotia. Journal of the Fisheries Research Board of Canada 30:83–90.

Thomas, M. L. H. 1978. Comparison of oiled and unoiled intertidal communities in Chedabucto Bay, Nova Scotia. Journal of the Fisheries Research Board of Canada 35:707–716.

Thomas, M. L. H., and M. L. Harley. 1973. Long term effects of bunker C oil from tanker *Arrow* on intertidal and lagoonal biota. Pages 84–85 *in* D. Mackay and W. Harrison, editors. Proceedings of the conference on oil and the Canadian environment.

van Tamelen, P. G., and M. S. Stekoll. 1996. Population response of the brown alga *Fucus gardneri* and other algae in Herring Bay, Prince William Sound, to the *Exxon Valdez* oil spill. American Fisheries Society Symposium 18:193–211.

American Fisheries Society Symposium 18:193–211, 1996

Population Response of the Brown Alga *Fucus gardneri* and Other Algae in Herring Bay, Prince William Sound, to the *Exxon Valdez* Oil Spill

PETER G. VAN TAMELEN AND MICHAEL S. STEKOLL

University of Alaska, School of Fisheries and Ocean Sciences
11120 Glacier Highway, Juneau, Alaska 99801, USA

Abstract.—Disturbance of populations of the brown alga *Fucus gardneri* and other intertidal algae by the *Exxon Valdez* oil spill and subsequent cleanup efforts was examined with an after-event, control–impact paired design study in Herring Bay, Prince William Sound. At oiled sites, both on rocky and coarse-textured shorelines, the larger size-classes of *Fucus* occurred in depressed numbers in 1990, but recovered to control levels by 1992. Populations in the upper intertidal zones recovered more slowly. In some areas a transient increase in ephemeral algae preceded *Fucus* repopulation. *Fucus* reproductive potential was significantly lower in plants from oiled areas in the upper intertidal zone. Reproductive capacity of the Herring Bay *Fucus* population had not recovered by the end of the field season in 1993. Populations in oiled areas that were cleaned with hot water and water under high pressure showed more damage and slower recovery.

Crude oil spilled from tankers can directly affect shore organisms in two ways. Physically, the oil can smother organisms, resulting in death or limiting acquisition of resources such as food, light, or nutrients (Foster et al. 1971). Biologically, the toxicity of oil may reduce growth rates of organisms, reproductive potentials, and survivorship (Wolfe 1977). There may be additional effects from shoreline cleaning methods that can negatively affect intertidal plants and animals, with the magnitude of the impact varying with the intensity of the effort and the method used (Foster et al. 1990; Houghton et al. 1996, this volume). Finally, because of reductions in the abundances of some organisms, other organisms not directly affected by an oil spill, but which interact strongly with the damaged populations, may also be influenced. Such indirect interactions are common in intertidal habitats (Dethier and Duggins 1984; van Tamelen 1987).

On 24 March 1989, the *Exxon Valdez* ran aground on Bligh Reef in northeastern Prince William Sound, spilling 40 million liters of North Slope crude oil. The spilled oil was transported by currents and prevailing winds to the south and west, impacting the shorelines of southwest Prince William Sound (Galt et al. 1991). Substantial amounts of oil entered Herring Bay on the north end of Knight Island in late March. During the summer of 1989, Herring Bay was a focal point for extensive cleanup operations to remove oil from impacted shorelines. A wide range of shoreline treatments was employed in Herring Bay, including hand cleaning, spraying (with water with pressures and temperatures ranging up to 140 psi and 160°F, respectively) repeated washings, and bioremediation (Nauman 1991). High-pressure, hot-water (HPHW) washing conducted from Omni and Maxi barges was applied to many sites and clearly contributed to the removal of organisms (Foster et al. 1990; Lees et al. 1996, this volume).

In Herring Bay, intertidal habitats are dominated by the alga *Fucus gardneri* (Silva) (hereafter *Fucus*). This perennial brown alga occurs at all tidal levels and forms a major structural component of the intertidal community. *Fucus* reproduces by producing eggs and sperm in small structures called conceptacles, which are located on the ends of inflated blades known as receptacles (Bold and Wynne 1978). After the eggs are fertilized and settle on the substrate, they grow into adult plants, forming dense beds. These *Fucus* beds provide habitat and shelter from terrestrial conditions and predators for a variety of intertidal invertebrates including limpets, periwinkles, sea stars, and various crustaceans (Menge 1976, 1978). Because of the scarcity of other algae in *Fucus* beds, *Fucus* appears to be the competitively dominant alga in this area, as it is on the west Atlantic coast (Menge 1976; Lubchenco 1978). Because *Fucus* is the major component of this intertidal community, accounting for up to 90% of the algal biomass (Stekoll et al. 1996, this volume), it is important to assess the impacts of major disturbances on this species. This study assessed the damage to populations of *Fucus* and other algae in Herring Bay caused by the *Exxon Valdez* oil spill.

TABLE 1.—Characteristics of study sites in Herring Bay, Prince William Sound.

Site[a]	Habitat[b]	Length (m)	Aspect[c]	Slope	Oiling level	Cleanup treatment[d]	Segment[e]
1C	SR	55	45	23	Light, very light	No treatment	KN5012
1X	SR	32	50	21	Heavy, moderate	Omni, HPHW, Bio	KN0133
2C	SR	38	260	27–51	Light, very light	No treatment	KN5002
2X	SR	42	290	28–37	Heavy, moderate	No treatment	KN0133
3C	VW	24	305	90	None	No treatment	KN5007
3X	VW	33	255	90	Moderate	Omni	KN0128
4C	CT	51	60	16	Very light	No treatment	KN5004
4X	CT	42	95	18	Moderate	Omni, HPHW	KN0131
5C	CT	37	270	14	Very light	No treatment	KN5004
5X	CT	30	290	14	Moderate	Omni, LPHW	KN0121

[a]C = control site; X = oiled site.
[b]SR = sheltered rocky; VW = vertical walls; CT = coarse-textured.
[c]Degrees from magnetic north.
[d]Probable cleanup treatment(s) is given. Omni = Omni boom; HPHW = high-pressure hot-water wash; Bio = bioremediation; LPHW = low-pressure hot-water wash.
[e]Exxon segment number.

Methods

Site Selection

Sites for these studies were selected by pairing sites from oiled and non-oiled areas in Herring Bay (Figure 1). The southeast corner of the bay retained ice until early April 1989, essentially excluding the oil slick. Therefore, control sites were restricted to the southeastern corner of the bay. The general procedure for selecting sites was to identify a workable area in the control section of the bay and then find an oiled area that resembled the control site as closely as possible. Site pairs were matched in as many physical characteristics as possible. The criteria used for matching sites included similarity in rock type, shoreline slope, directional and solar aspect, and wave exposure. Table 1 lists the five paired sites and gives some of their physical characteristics and probable shoreline treatments.

Salinity and temperature measurements at the water surface and at 1-m depth were recorded weekly at oiled and control sites in 1990 and 1991 and twice during 1992. In addition, the physical disturbance occurring at oiled and control coarse-textured sites was quantified by recording the number of missing or displaced rebar stakes used to mark the quadrat locations. This procedure was done at the beginning of the 1991 and 1992 field seasons at site pair 4C–4X.

Population Dynamics

The population structure of *Fucus* was monitored at five pairs of control and oiled sites: three sheltered rocky site pairs, consisting of bedrock of various slopes and including one vertical wall site; and two coarse-textured site pairs. Each site had 6 permanently marked, randomly placed quadrats (20 × 50 cm) in each of three tidal levels, for a total of 18 quadrats per site. At sheltered rocky sites, quadrats were marked at two corners with screws anchored in the substrate. At coarse-textured sites, quadrats were marked at two corners with rebar stakes driven into the gravel. At each site, six transect heads were located along the base of the *Verrucaria* zone, at mean higher high water 3.5 m above mean lower low water. The length of the site was measured and divided by six, giving a segment length. The segment length was then multiplied by a random number (0–1.0) after subtracting the quadrat width (20 cm) from the segment length. Adding the quadrat width to the new number gave the location of the first transect head. Each of the subsequent transect heads were located by adding the segment length to the location of the previous transect head. The upper right corner of each quadrat was located by measuring the length of the transect at 1 m of vertical drop (MVD), subtracting the length of the quadrat (50 cm), and multiplying by a random number. This was done for all three MVDs on each transect. The same random number was used for all MVDs on a given transect. A different number was generated for each transect.

The size-frequency distribution of *Fucus* was determined in each quadrat by measuring the maximum length of all visible *Fucus* plants to the nearest 0.5 cm without removing plants from the substratum. Only plants with visible fronds were counted, and each plant was recorded as being reproductive or nonreproductive. Also, the number of receptacles on each reproductive plant was recorded once during the first summer and on all sampling dates during subsequent field seasons. Percent cover of

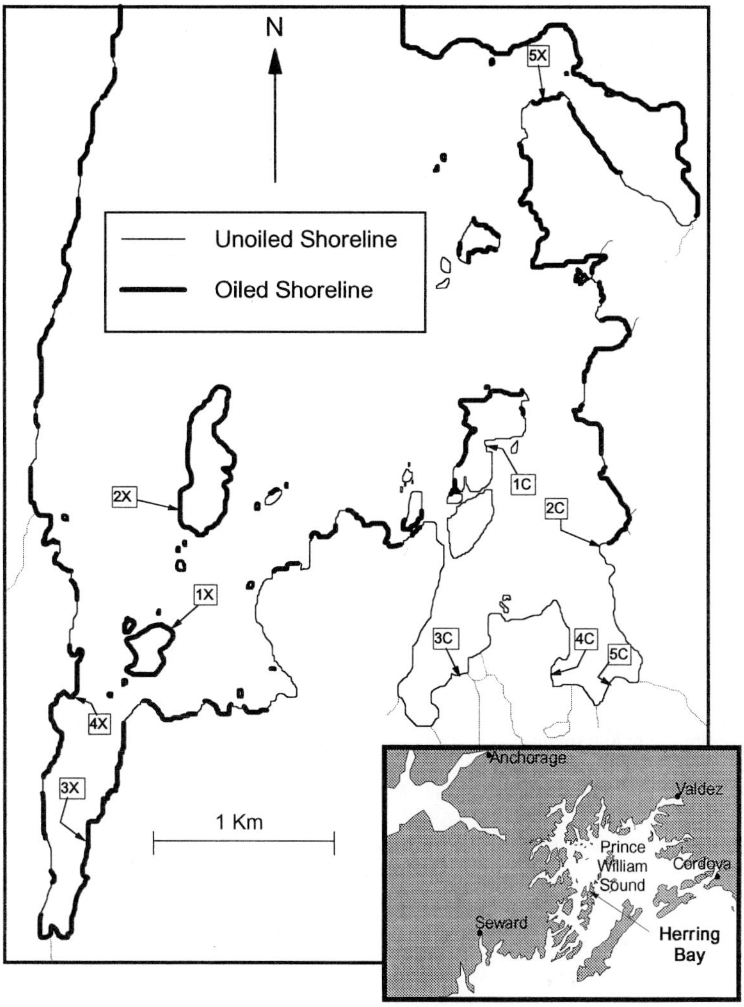

FIGURE 1.—Map of the study area showing the locations of the study sites in Herring Bay, Knight Island, Prince William Sound. Oiled sites are designated by an "X" suffix and reference sites by a "C" suffix.

Fucus and ephemeral algae—those species that are either annuals or that colonize denuded areas rapidly and are either outcompeted or grazed soon after colonization—was estimated by placing a systematic 50-point grid over the quadrat. Other perennial algae and sessile invertebrates were also recorded, but were too sparse to warrant analyses. The species directly beneath each point was recorded, and percent cover was calculated by multiplying the number of points for each species by two. No attempt was made to assess primary space coverage. All drift algae were removed before assessment of percent cover. The study plots were monitored every 2 weeks during a 3-month period in 1990. In 1991 the quadrats were visited on three occasions, once each in April, June, and August. During 1992 and 1993 plots were sampled twice, once each in May and August.

Statistical Methods

The same basic statistical procedure was followed for most observations. Statistical tests were applied to each sampling date separately because analyses of repeated measures would, in many cases, probably yield time-by-treatment interactions, and proper interpretation would be reduced to individual times. For any given type of data, comparisons were ultimately made for each pair of sites only, but where appropriate, the pooled estimate of variance for all

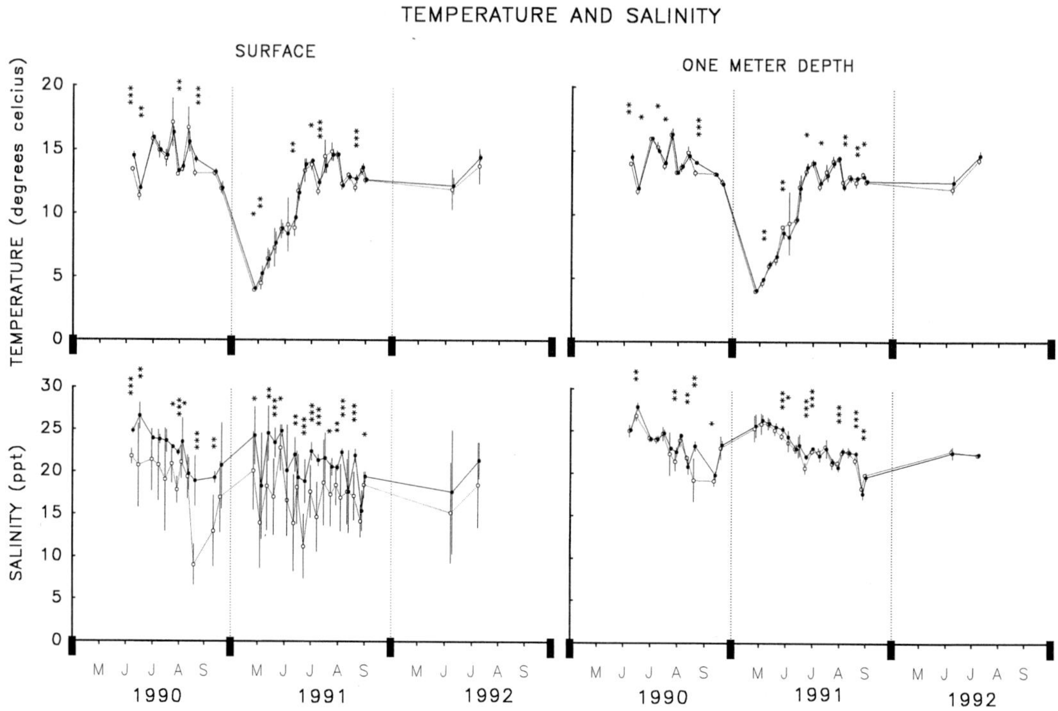

FIGURE 2.—The temperature and salinity at the water surface and at 1-m depth at oiled (filled circle) and control areas (open circles) of Herring Bay. Error bars represent 1 SD. Salinity is in parts per thousand, ppt. Asterisks indicate statistical differences detected using t-tests: * indicates $P < 0.05$, ** indicates $P < 0.01$, and *** indicates $P < 0.001$.

sites of a given habitat type was used. Raw data for all pairs of sites were checked for homogeneity of variances with Levene's test at the $P = 0.10$ level. If variances were unequal, then the data were subjected to either an arcsin (percent cover data) or a log transformation (all other data). If either the raw or transformed variances were equal, then a one-way analysis of variance (ANOVA) was performed on all sites, and contrasts between oiled and control sites within a pair were used to detect differences between oiled and control sites. If neither the raw nor the transformed variances were equal, then a regular t-test was used to compare each pair of sites. Before applying the t-test, however, raw data variances for each pair of sites were tested for homogeneity with the F-max test. If variances were not equal, then the raw data were transformed using either of the two transformations mentioned above. If the transformation failed to alleviate the heteroscedasticity, then a regular t-test with Satterthwaite's correction for nonhomogeneous variances was applied to the raw data. In some cases, simple t-tests were performed without attempting to use a pooled estimate of variance. When these simple t-tests were employed, the procedure above was applied as if neither the raw nor transformed data had homogeneous variances for all sites. All figures and tables represent raw means and 1 SE of the mean. No corrections for multiple comparisons were employed because there was such a large number of tests performed, but the level of statistical significance is indicated by one asterisk ($P < 0.05$), two asterisks ($P < 0.01$), or three asterisks ($P < 0.001$).

Results

Site Differences

Despite attempts to minimize physical differences between oiled and control sites, some differences remained. Differences in water temperature were occasionally detected between oiled and control parts of the bay (Figure 2). These differences, however, were small, within 1°C, compared to seasonal and weekly fluctuations of up to 10°C. On 57% of the sampling dates, the surface salinity was significantly higher in the oiled part of the bay than in the control side (Figure 2). The salinity at 1 m did not

TABLE 2.—The number of plots in the 4C–4X site pair in Herring Bay, Prince William Sound, with disturbed rebar stakes in the spring of 1991 and 1992 at three tidal levels. The total number of plots in each category was six. MVD, meter of vertical drop.

	Control		Oiled	
MVD	1991	1992	1991	1992
1	4	2	5	0
2	6	2	2	0
3	6	5	0	0
Total	16	9	7	0

show as many differences as the surface salinity, but differences were detected on 12 of the sampling dates (Figure 2).

Physical disturbance seems to have a greater impact on intertidal shorelines at control sites in Herring Bay. In spring of 1991, the control site 4C had a greater number of rebar stakes upturned and missing from all three tidal levels compared to its matched oiled site, 4X (Table 2; $\chi^2 = 3.52$, $0.10 > P > 0.05$). In 1992, more rebar stakes were disturbed at the control site while the rebar at the oiled site remained intact ($\chi^2 = 9.00$, $P < 0.05$). The source of this disturbance is unknown, but ice scouring is a strong possibility. Ice was observed in the southeastern section of the bay in spring 1991 and can be seen in aerial photographs taken in spring 1989 and 1990. Ice has not been observed at any time in the southwestern section of the bay.

Populations at Sheltered Rocky Sites

At site pair 1C–1X (pair 1), a gently sloping sheltered rocky pair, there were fewer large plants (>10 cm) at the oiled site in the first and second MVDs during 1990 and 1991 (Figure 3J, K). In the first MVD only one difference was statistically significant, because of low densities and high variability. On all dates in 1990 and 1991 at the second MVD, all differences in large plant abundance were statistically significant. In the third MVD on the last sampling date in 1992 and the first in 1993 there were more large plants at the oiled site (Figure 3L). Medium-sized (5.5–10.0cm) plants also showed lower abundances initially at the oiled sites during 1990 in the first and second MVDs, but only two of these differences were statistically significant (Figure 3G, H). There were more medium-sized plants at the oiled site at the end of summer 1991 (3 MVD) and in 1992 and 1993 (2 MVD), although only four of the differences were significant (Figure 3H, I). On most dates, the density of small (2.5–5 cm) plants was less in the first MVD at the oiled site in 1990, but there were no differences in subsequent years (Figure 3D). In the second MVD, however, there were more small plants at the oiled site in 1991 and 1992, but in 1993 there were no differences when compared with control sites (Figure 3E). The oiled site had significantly more germlings (0–2 cm plants) in the second MVD during 1990 and early in 1991 (Figure 3B). There were no other significant differences for this size-class.

A more steeply sloped sheltered rocky pair, 2C–2X (pair 2), had few significant differences in plant densities of plants in any size category (Figure 4). In the second and third MVDs in 1992 there were more germlings in the second and third MVDs at the control site (Figure 4B, C).

At the vertical wall sites, 3C–3X (pair 3), there were significantly more large plants (>10 cm) at the control site in the first MVD during 1990 and 1991 (Figure 5J). Differences were no longer significant in 1992, because of an increase in plant density at the oiled site. In the second MVD there were more large plants in 1992 and in May 1993 at the oiled site (Figure 5K). There were no differences in the number of large plants for any year in the third MVD (Figure 5L). Medium-sized plants (5.5–10 cm) showed significantly lower abundances in the second and third MVDs at the oiled site on a few dates in 1990 (Figure 5H, I). In May 1993 there were more medium-sized plants at the oiled site in the first MVD (Figure 5G). Small plants (2.5–5 cm) were also significantly less abundant at the oiled site in the first and third MVDs on half of the sampling dates in 1990 (Figure 5D, F). There were no other significant differences in the abundance of small plants except in the third MVD in 1992 (Figure 5, D–F). The trend, however, was toward higher abundances at the oiled site in 1991 and 1992 in the first and second MVDs. Germlings (0–2 cm) were significantly less abundant in the third MVD at the oiled site from 1990 into 1992 (Figure 5C), but there were more germlings in the second MVD at the oiled site at the end of summer 1990 (Figure 5B, C).

At pair 1, there were more reproductive plants at the control site in the first and second MVDs, but because of low numbers and high variability, the differences were significant on most sampling dates only in the second MVD (Figure 6A, D). There were no differences in the third MVD (Figure 6G). A similar pattern occurred for receptacles per quadrat, with significant differences on all in 1991–1992 dates in 2 MVD and in spring 1992 and 1993 in 1 MVD (Figure 7A, D, G). For pair 2, there were

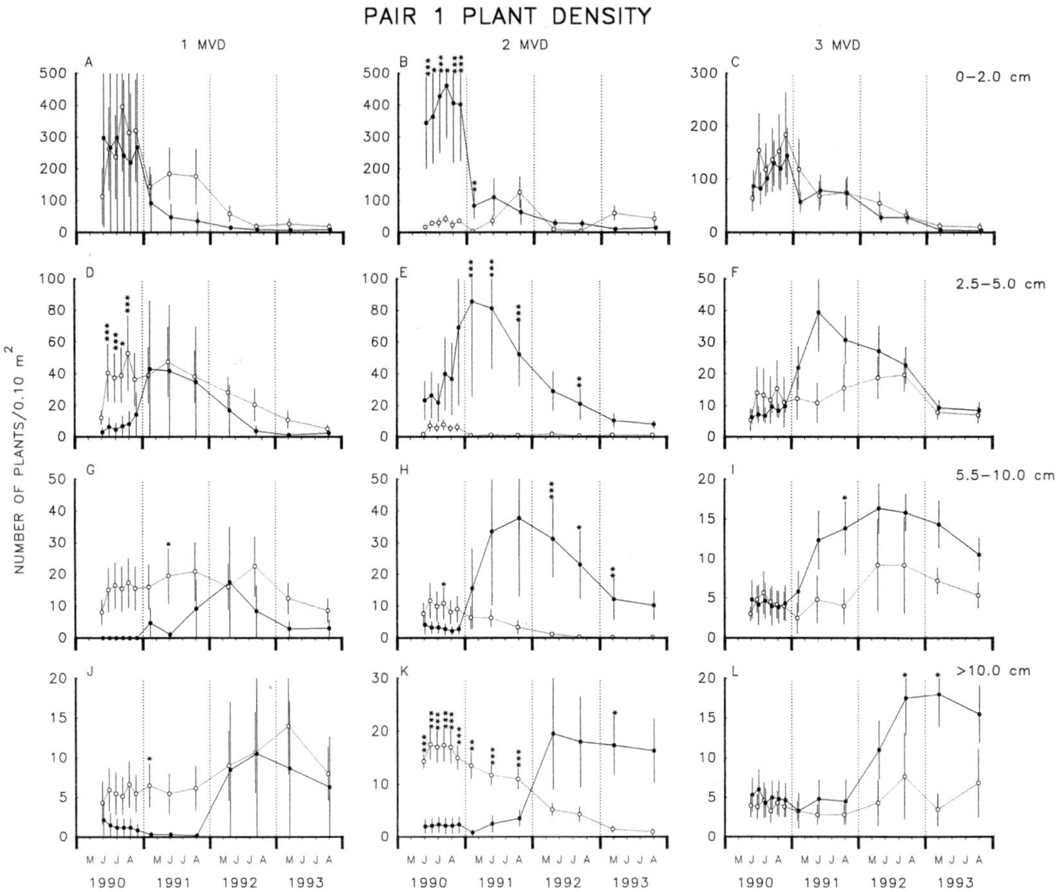

FIGURE 3.—The density of germlings (0–2.0 cm), small (2.5–5.0 cm), medium (5.5–10 cm) and large (>10 cm) *Fucus* plants at pair 1 at all three tidal levels, designated by 1, 2, and 3 MVD (meters of vertical drop). The oiled site is represented by solid circles and lines, and the control site is represented by open circles and dashed lines. The error bars represent 1 SEM. The vertical dashed lines represent divisions between field seasons, which ran from May (M on the *x*-axis) through August (A on the *x*-axis). Winter months are not represented on the graphs. Asterisks indicate statistically significant differences between oiled and control sites on the indicated sampling date as in Figure 2.

only two significant differences in the number of reproductive plants (Figure 6B, E, H). In the third MVD in 1991 there were more plants at the control site. In the second MVD in 1993 there were more plants at the oiled site. There were no significant differences in the number of receptacles per quadrat, although there were consistently more receptacles at the control site in the first MVD until the last sampling date in 1992 (Figure 7B). At pair 3, for 1991 and early 1992 there were significantly more reproductive plants and receptacles per quadrat on control sites in the first MVD (Figures 6C and 7C). The only other significant difference was that the number of reproductive plants was greater at the control site in the second MVD once in 1990 (Figure 6F).

The percent cover of *Fucus* and other algae was also different between control and oiled sites in some cases. At pair 1, *Fucus* cover was lower, but not significantly so, at the oiled site in the first MVD for all 4 years (Figure 8A). *Fucus* cover was significantly lower at the oiled site in the second MVD through June 1991 (Figure 8D). Although there was not any convergence of *Fucus* cover in the first MVD as of 1992, cover tripled from 1990 to 1992 in the second and third MVDs at the oiled site (Figure 8A, D, G). In May 1993 in the second MVD and at all times in 1993 in the third MVD, *Fucus* cover was significantly greater at the oiled site (Figure 8D, G). Weedy, ephemeral species, algae in the genera *Enteromorpha*, *Cladophora*, *Pilayella*, and other similar algal types had higher cover at the oiled site in

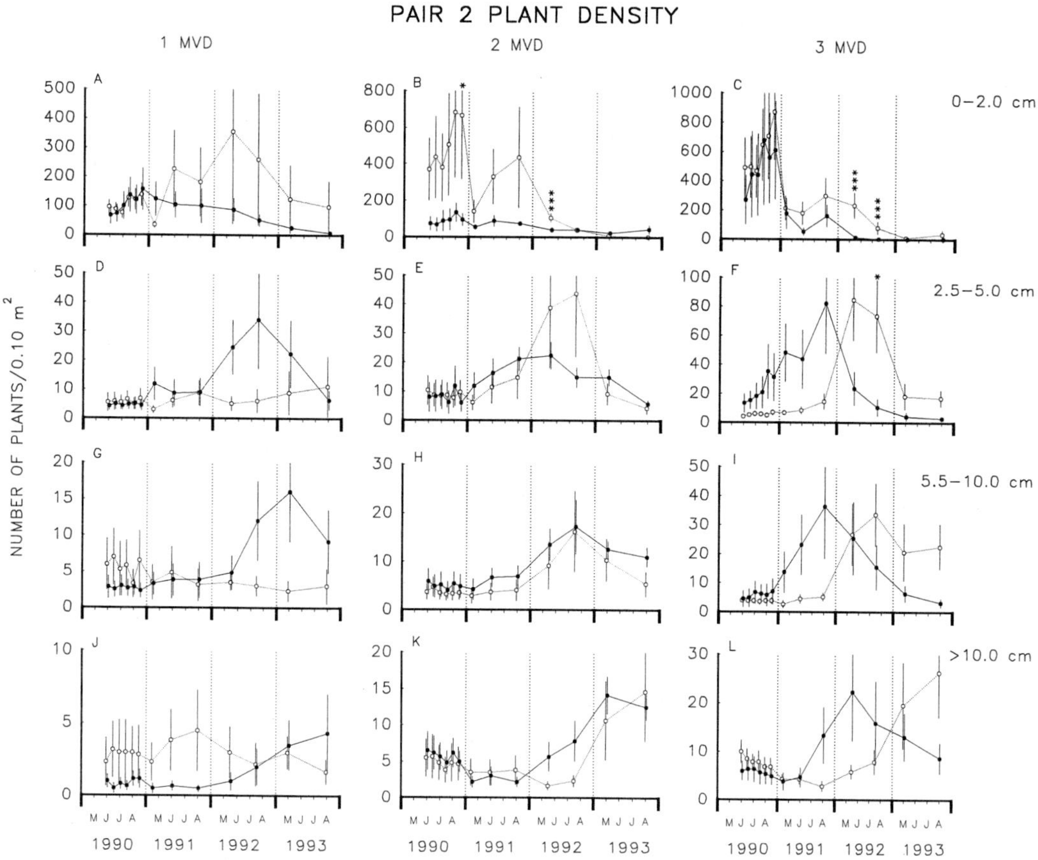

FIGURE 4.—The density of germlings, small, medium, and large *Fucus* plants at pair 2. Layout is the same as in Figure 3.

the second and third MVDs through the middle of 1991, but differences were only significant in the third MVD (Figure 9D, G).

There were no significant differences in the percent cover of *Fucus* at pair 2 (Figure 8B, E, H). There was, however, higher cover of ephemeral algae at the oiled site in the second and third MVDs on some dates in 1990 (Figure 9E, H). In 1991, however, the cover of ephemeral algae was greater at one time at the control site.

Pair 3 showed patterns in the percent cover data similar to those seen at pair 1 (Figures 8 and 9). In 1990 *Fucus* cover was lower at the oiled site at all levels with significant differences in the first and third MVDs (Figure 8C, F, I). In the third MVD, differences were only detectable until August 1990, but, in the first MVD, the differences extended into 1991. Percent cover of *Fucus* tended to increase at the oiled site in the second and third MVDs in 1991 and 1992, and in May 1993 *Fucus* cover was significantly greater at the oiled site in the second MVD (Figure 8F, I). Ephemeral species had significantly higher percent covers at all tidal levels at the oiled site in 1990 and early 1991 (Figure 9C, F, I). The percent cover of ephemerals at oiled sites tended to decline over time, especially in the first MVD, converging to control values.

The overall results reflect the patterns seen at pairs 1 and 2 (Table 3). In the first MVD there were more small and large plants at the control sites in 1990 and 1991. At the same level in 1991 there were also more medium and reproductive plants as well as more receptacles per quadrat at the control sites. By 1993 the number of medium plants was greater at the oiled sites. There was a higher percent cover of *Fucus* in 1990 at control sites. There was more ephemeral algae at oiled sites in spring 1991. In the second MVD there were more large plants, repro-

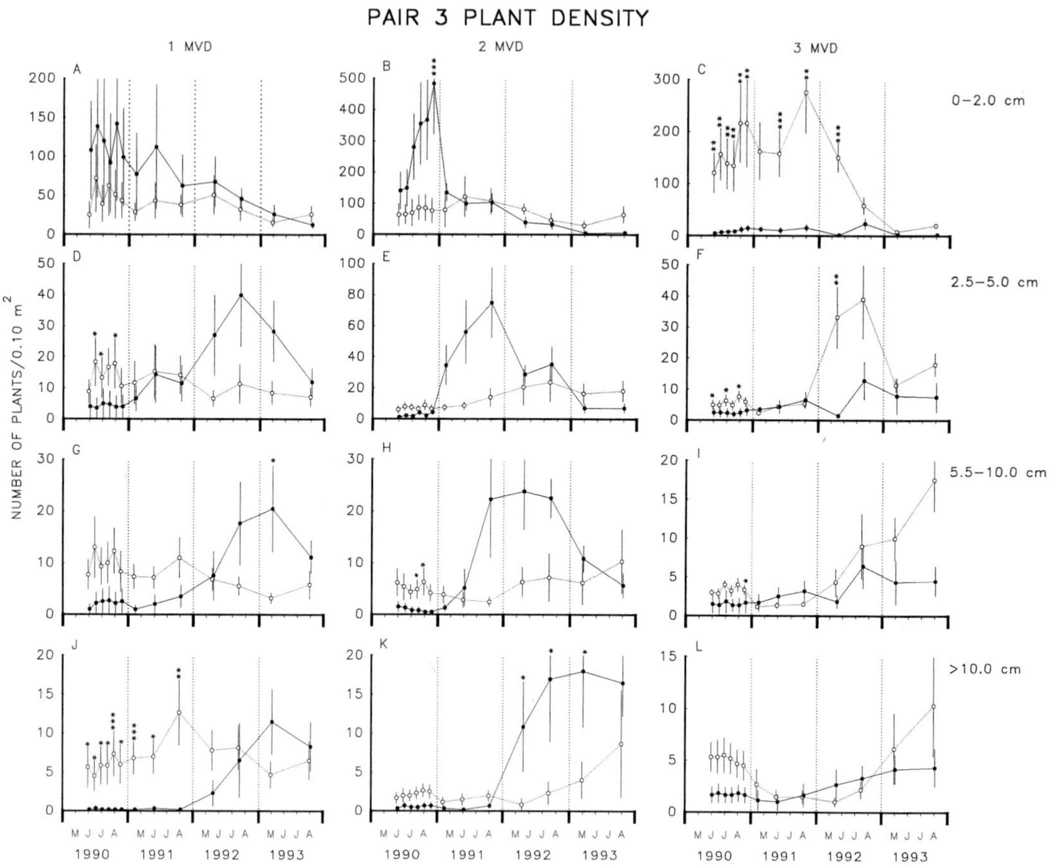

FIGURE 5.—The density of germlings, small, medium, and large *Fucus* plants at pair 3. Layout is the same as in Figure 3.

ductive plants, and receptacles in 1990 and 1991 at control sites. In 1990 at oiled sites there was a lower percent cover of *Fucus* and a higher percent cover of ephemeral algae. In 1990 there were more germlings at oiled sites and these plants grew into successively larger size-classes. At oiled sites there were more small plants in 1991, more medium plants in 1992, and more large plants in 1993. In 1993 the cover of *Fucus* was greater at oiled sites. In the third MVD there were more germlings in all years at control sites. There were also more small and medium plants in 1992 and 1993 at control sites. At control sites in 1990 the percent cover of *Fucus* was greater and in 1991 there were more reproductive plants. By 1993, however, there were more reproductive plants and receptacles per quadrat at oiled sites. In 1990 and 1991 there were more ephemeral algae at oiled sites.

Populations at Coarse-textured Sites

Algae were sparse in the first two MVDs at the two coarse-textured site pairs, so only third MVD results are reported here. The size distribution of *Fucus* plants at the coarse-textured sites (Figure 10) showed patterns somewhat similar to those given above for sheltered rocky sites. Germlings (0–2 cm) tended to be more abundant at the oiled sites, with significant differences occurring on two sampling dates at site pair 4C–4X (= pair 4, Figure 10A). At the 5C/5X site pair (= pair 5), there were no differences in the number of germlings (Figure 10B). At both site pairs there was no difference in the abundance of small plants (2.5–5.0 cm) in 1990, but, in 1991 and 1992, they became more abundant at oiled sites (Figs. 10C, D). There were significant differences at pair 4 on all 1991 and 1992 dates, but

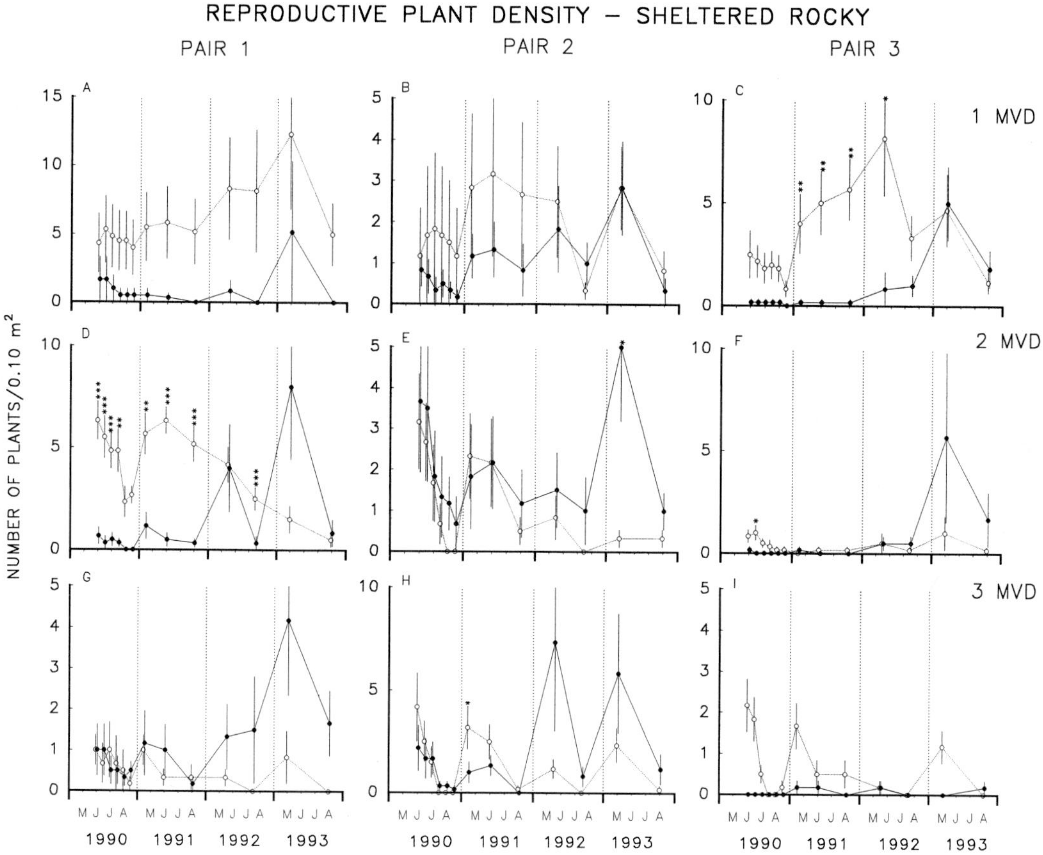

FIGURE 6.—The density of reproductive *Fucus* plants at all sheltered rocky site pairs at three tidal levels. Layout is the same as in Figure 3.

there was only one significant difference at pair 5 in 1991. The medium-sized plants (5.5–10 cm) at both site pairs became more abundant at the oiled sites in 1991 and significantly so in 1992 (Figure 10E, F). By the fall of 1993 the number of small and medium plants at the oiled sites had declined to control levels at both site pairs. Large plants (>10 cm) tended to be more abundant at control sites in 1990 and 1991, but this difference was only significant on the first sampling date at pair 4 (Figure 10G). In 1992 large plants became significantly more abundant at oiled sites at both site pairs, and this difference persisted through 1993, but was significant only at pair 4 (Figure 10G, H).

In the first 3 years, 1990–1992, there were no significant differences in the number of reproductive plants or the number of receptacles per quadrat at either coarse-textured site pair (Figure 11, A–D). In the first 2 years, however, there were always more reproductive plants and receptacles per quadrat at the control sites, but because of high variation and low numbers, none of these differences were statistically significant. In May 1993 there were significantly more reproductive plants at the oiled site 4X compared to 4C (Figure 11A). The numbers of reproductive plants and receptacles at pair 5 were also greater, but not significantly so, at the oiled site, as were the receptacles per quadrat at pair 4 (Figure 11, B–D).

The percent cover of *Fucus* was significantly lower at the oiled sites on a few sampling dates in 1990 at both coarse-textured site pairs (Figure 11E, F). In 1992 and 1993 at pair 4, *Fucus* cover was significantly higher at the oiled site. There were no differences in the percent cover of ephemeral algae at either site pair (Figure 11G, H).

Both coarse-textured site pairs showed similar temporal patterns (Figure 10). In 1990 at the

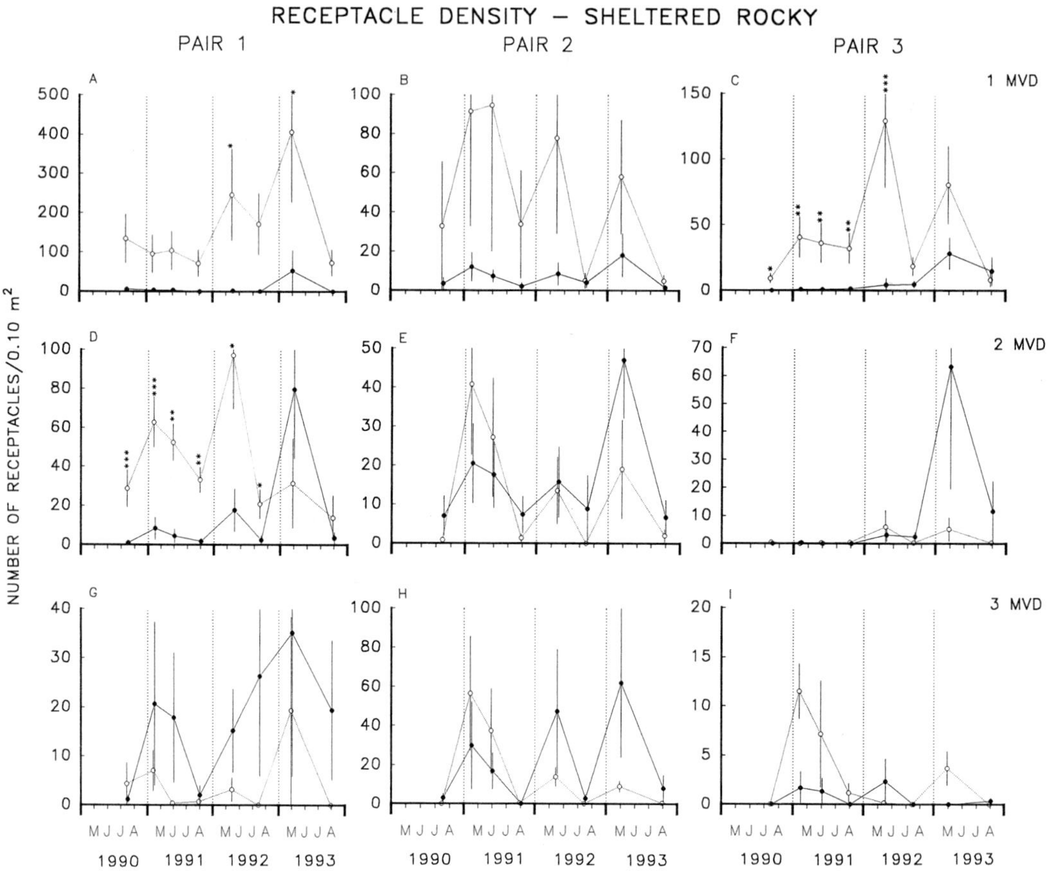

FIGURE 7.—The density of receptacles per quadrat at all sheltered rocky site pairs at three tidal levels (MVD). Layout is the same as in Figure 3.

oiled sites, there were fewer large plants and more germlings at each site pair. In 1991, the abundance of small plants had increased at both oiled sites, and in 1992 there were increases in the number of medium and large plants at both oiled sites. In 1993, the number of medium-sized plants decreased, but the number of large plants remained high at oiled sites. Thus, at both oiled sites, a cohort of plants can be seen to recruit in 1990 and grow to larger size-classes in subsequent years. This recolonization can also be seen in the gradual increase of the percent cover of *Fucus* at both oiled sites (Figure 11E, F).

The results of the overall tests indicate that in 1990 the percent cover of *Fucus* and perhaps the number of large plants was greater at the control sites (Table 4). In 1991 there were more germlings and small plants at oiled sites, which grew into medium and large plants in 1992 and into large plants in 1993. In 1992 and 1993 the percent cover of *Fucus* was greater at the oiled sites.

Reproductive Plant Quality

Examination of all reproductive plants at all sites (sheltered rocky and coarse-textured) revealed that at oiled sites reproductive plants were shorter than those at control sites in the second and third MVDs early in 1990, in the first and second MVDs at times in 1991, in all MVDs in 1992, and only in the first MVD in 1993 (Figure 12). There were not enough reproductive plants in the third MVD at the end of 1993 to perform the analysis. *Fucus* at control sites had more receptacles per plant in the first and second MVD in spring 1992 and in the first MVD in spring 1993 (Figure 12). There were no significant differences at any time in the third MVD.

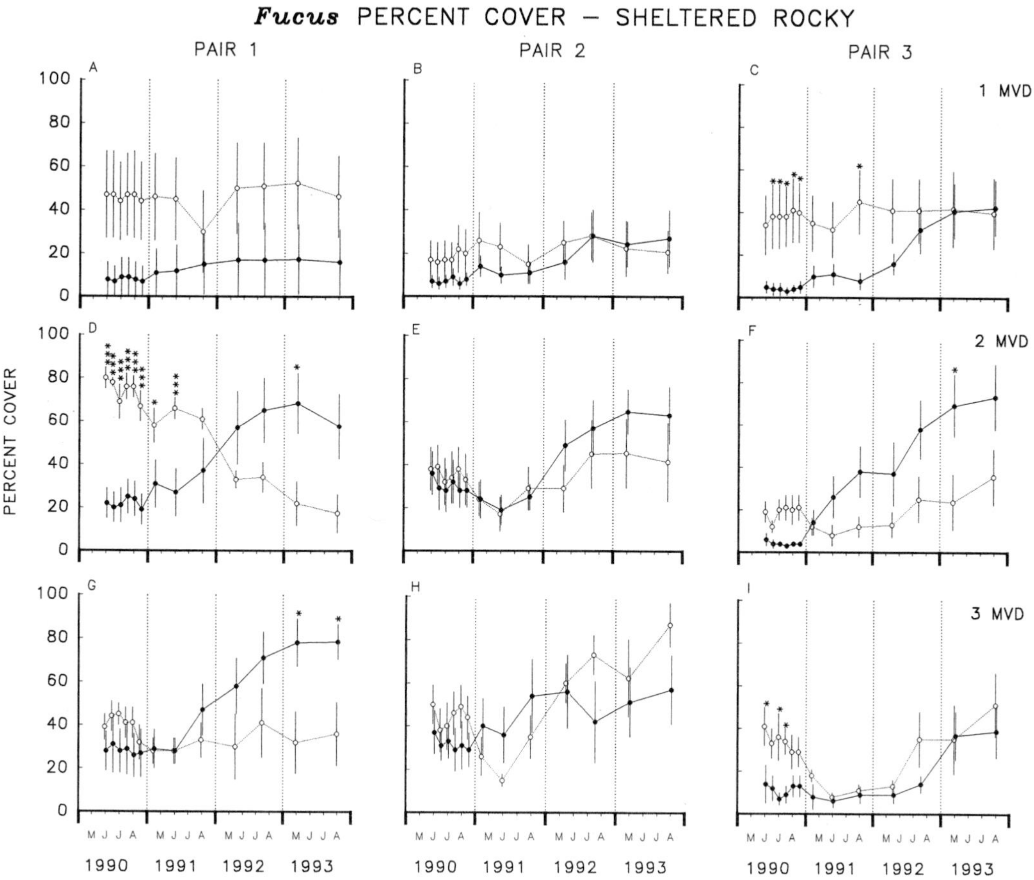

FIGURE 8.—The percent cover of *Fucus* at all sheltered rocky site pairs at three tidal levels (MVD). Layout is the same as in Figure 3.

Discussion

Physical Differences Among Sites

In most studies that compare impacted and unimpacted sites to assess the effects of natural or human-caused disasters, a major assumption is that the compared areas were environmentally and biologically similar before the event in question. There are many physical factors in Herring Bay that can affect intertidal algal populations. Rock temperatures in the intertidal can get as high as 43°C, and desiccation can be severe on sunny or windy days. Because of streams and freshets, surface salinities can be as low as 3.5‰. Waves produced by winter storms can move loose rocks and boulders and remove algae (Sousa 1979a, 1979b). Ice can form on the sea surface in isolated bays and coves, scouring the shoreline as the tide moves in and out (McCook and Chapman 1991). Given these factors, which potentially affect intertidal biota, observed differences between sites cannot be attributed solely to oil or shoreline treatment. It is possible, however, to evaluate the intensity of these physical factors at oiled and control sites in order to ascertain differences between sites, and, if differences exist, to estimate their effect on community structure and dynamics.

Because of lower salinities and greater physical disturbance at control sites in Herring Bay, algal populations would be expected to have lower densities than at oiled sites. The chronic presence of low salinities may reduce the densities of *Fucus* in the control areas. Most intertidal algae, however, can withstand salinities down to 10‰ (Lobban et al. 1985). The lowest salinities measured here (about 20‰) probably have little or no effect on *Fucus* populations. Physical disturbance, in the form of ice scouring, is more likely to lower *Fucus*

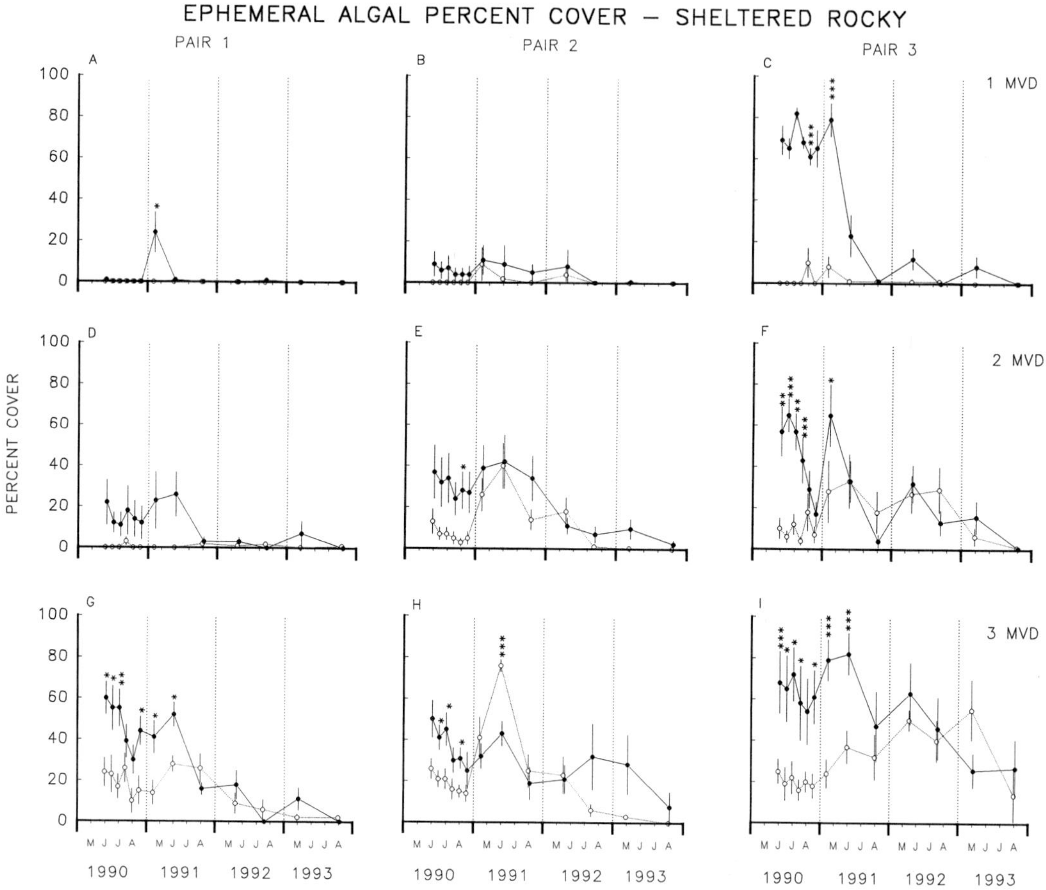

FIGURE 9.—The percent cover of ephemeral algae at all sheltered rocky site pairs at three tidal levels (MVD). Layout is the same as in Figure 3.

densities at the control sites where ice formation has been documented (McCook and Chapman 1991). These expectations suggest that our results underestimate actual differences between oiled and control sites. For example, if ice scouring is indeed more severe at control sites compared to oiled sites and removes *Fucus* from the intertidal, then observed lower densities of *Fucus* at oiled sites would be a conservative estimate of the actual decrease in *Fucus* density caused by the effects of the oil spill. Observed lower densities of plants or animals at oiled sites compared to control sites would indicate reductions first to a level equal to control levels in addition to the observed reductions to below control levels. In addition, some decreases in plant or animal populations may have occurred at oiled sites but were not detected in our studies because the reductions were to levels equal to those of control sites.

Effects on Fucus

The information gathered on *Fucus* size indicates that in some areas many larger plants were killed or removed by the oil spill and subsequent cleanup efforts. The removal of large plants at oiled sites resulted in fewer reproductive plants as well as fewer receptacles per quadrat in the first two MVDs. These results occurred mostly in the first two MVDs at sheltered rocky sites, showing that the effects of the spill on *Fucus* were concentrated in the upper portion of the intertidal zone in this habitat. Ephemeral algae, however, were more abundant in the third MVD, demonstrating that the effects of the spill were not limited to the upper intertidal. Also, at coarse-textured sites where algae were only abundant in the low intertidal, lower abundances of large plants were observed in the third MVD at oiled sites.

TABLE 3.—Results of overall tests for *Fucus* population dynamics studies at sheltered rocky sites in Herring Bay, Prince William Sound, 1990–1993. Each test combined all pairs of sites using Fisher's test for combined probabilities (Sokal and Rohlf 1981). Number of plants in four size classes (0–2, 2–5, 5–10, and >10 cm), number of reproductive plants (REP), number of receptacles per 0.10 m^2 (REC), and percent cover of *Fucus* (FUC) and ephemeral algae (EPH) were tested at 1, 2, and 3 meters of vertical drop (MVD). (○) indicates that the control site values are significantly greater ($P < 0.05$) than the values at oiled sites. A solid circle (●) indicates that the oiled values are significantly greater ($P < 0.05$) than the values at control sites. A blank space indicates no data.

MVD	Year	Date	0–2	2–5	5–10	>10	REP	REC	FUC	EPH
1	1990	25 Jun	NS	NS	NS	NS	NS		NS	NS
		7 Jul	NS	○	NS	○	NS		○	NS
		21 Jul	NS	○	NS	○	NS		○	NS
		5 Aug	NS	○	NS	○	NS	NS	○	NS
		16 Aug	NS	○	NS	○	NS		○	NS
		4 Sep	NS	NS	NS	○	NS		○	NS
	1991	1 May	NS	NS	NS	○	○	○	NS	●
		12 Jun	NS	○	○	○	○	○	NS	NS
		7 Aug	NS	○	○	○	○	○	NS	NS
	1992	29 May	NS	NS	NS	NS	○	○	NS	NS
		27 Aug	NS	NS	NS	NS	NS	NS	NS	NS
	1993	31 May	NS	NS	●	NS	NS	NS	NS	NS
		31 Aug	NS	NS	NS	NS	NS	NS	NS	NS
2	1990	25 Jun	●	NS	NS	○	○		○	●
		7 Jul	NS	●	NS	○	○		○	●
		21 Jul	●	NS	NS	○	○		○	●
		5 Aug	●	NS	○	○	○	○	○	●
		16 Aug	●	NS	NS	○	NS		○	●
		4 Sep	●	NS	NS	○	NS		○	NS
	1991	1 May	●	●	NS	○	○	○	NS	●
		12 Jun	NS	●	NS	○	NS	○	NS	NS
		7 Aug	NS	●	NS	○	○	○	NS	NS
	1992	29 May	○	NS	●	NS	NS	NS	NS	NS
		27 Aug	NS	●	NS	NS	NS	NS	NS	NS
	1993	31 May	NS	NS	●	●	NS	NS	●	NS
		31 Aug	NS	NS	NS	NS	NS	NS	●	NS
3	1990	25 Jun	○	NS	NS	NS	NS		NS	●
		7 Jul	○	NS	NS	NS	NS		NS	●
		21 Jul	○	NS	NS	NS	NS		○	●
		5 Aug	○	NS	NS	NS	NS	NS	NS	●
		16 Aug	○	NS	NS	NS	NS		NS	●
		4 Sep	○	NS	NS	NS	NS		NS	●
	1991	1 May	NS	NS	NS	NS	○	NS	NS	●
		12 Jun	○	NS	NS	NS	NS	NS	NS	●
		7 Aug	○	NS	NS	NS	NS	NS	NS	NS
	1992	29 May	○	○	NS	NS	NS	NS	NS	NS
		27 Aug	○	○	NS	NS	NS	NS	NS	NS
	1993	31 May	NS	NS	NS	NS	NS	NS	NS	NS
		31 Aug	○	○	○	NS	●	●	NS	NS

At some control sites, variations in the density of plants were observed over time. In particular, site 1C in the second MVD exhibited declines in the abundances of medium and large plants, reproductive plants, and percent cover of *Fucus* (Figures 3H, K, 6D, 8D). These declines may be due to natural variation in the abundance of *Fucus*, but the amount of foot traffic may have also contributed to these declines (Brosnan and Crumrine 1994). This site experienced a large amount of foot traffic because many experiments were performed there and it was relatively accessible. The other sheltered rocky sites did not have the same amount of foot traffic because of their steepness and did not show declines in *Fucus* densities.

In addition to differences in plant densities, the condition of reproductive plants at all sites showed differences between oiled and control sites. At oiled sites, plants were shorter and had fewer receptacles than at control sites. This result might be attributable to a larger proportion of relatively young reproductive plants at oiled sites. If this were the case,

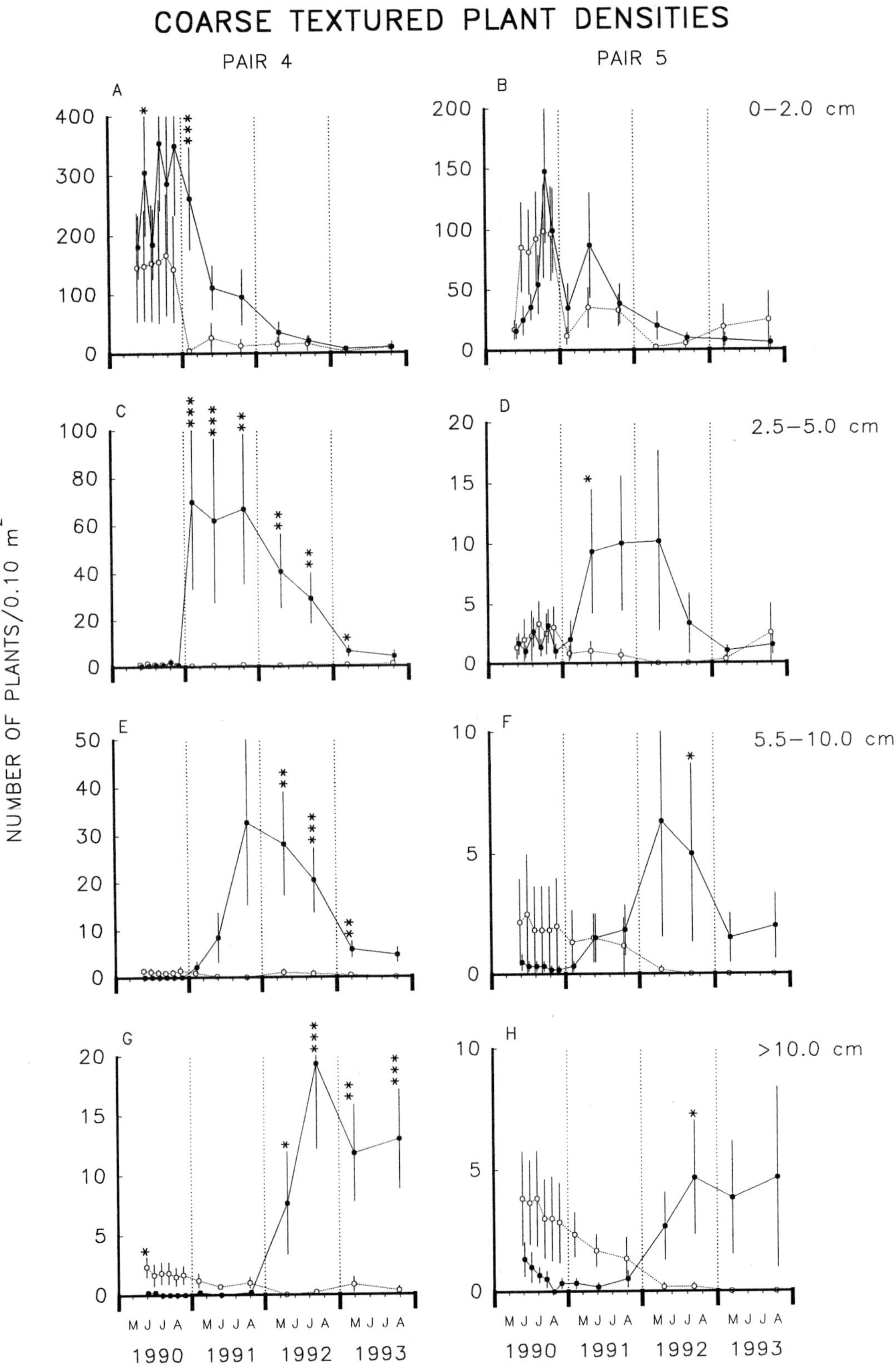

FIGURE 10.—The density of germlings, small, medium, and large *Fucus* plants in the third MVD at the coarse-textured site pairs. Layout is the same as in Figure 3.

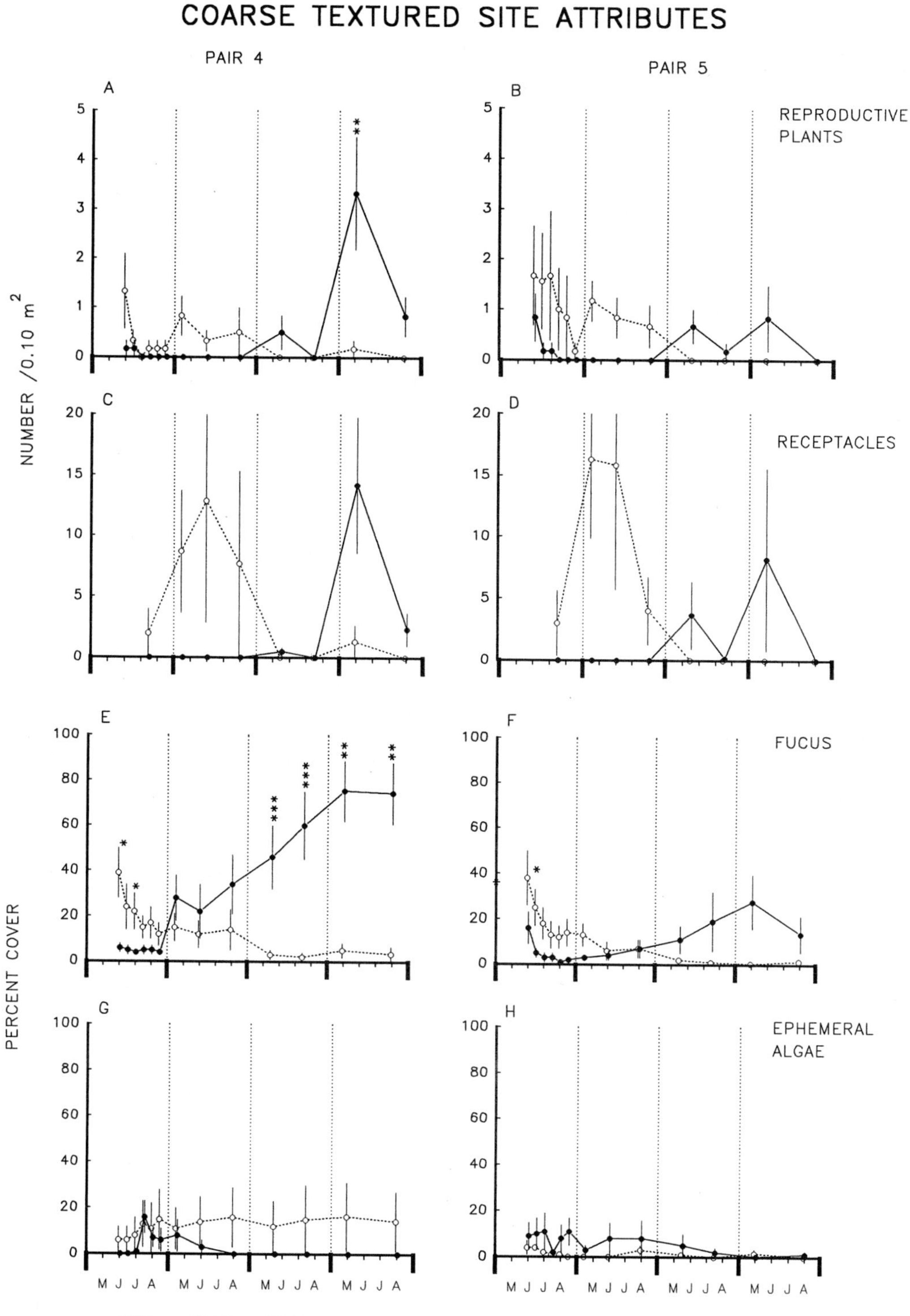

FIGURE 11.—The density of reproductive *Fucus* plants, receptacles per quadrat, and percent cover of *Fucus* and ephemeral algae at the coarse-textured site pairs in the third MVD. Layout is the same as in Figure 3.

TABLE 4.—Results of overall tests for *Fucus* population dynamics studies at coarse-textured sites in Herring Bay, Prince William Sound, 1990–1993. Each test combined all pairs of sites using Fisher's test for combined probabilities (Sokal and Rohlf 1981). Four size classes (0–2, 2–5, 5–10, and >10 cm), number of reproductive plants (REP), number of receptacles per 0.10 m^2 (REC), percent cover of *Fucus* (FUS) and ephemeral algae (EPH) were tested at the third MVD. Symbols and abbreviations used in body of table are defined in Table 3.

Year	Date	0–2	2–5	5–10	>10	REP	REC	FUC	EPH
1990	25 Jun	NS	NS	NS	○	NS		○	NS
	7 Jul	NS	NS	NS	NS	NS		○	NS
	21 Jul	NS	NS	NS	NS	NS		○	NS
	5 Aug	NS	NS	NS	NS	NS	NS	NS	NS
	16 Aug	NS	NS	NS	NS	NS		○	NS
	4 Sep	NS	NS	NS	NS	NS		NS	NS
1991	1 May	●	●	NS	NS	NS	NS	NS	NS
	12 Jun	NS	●	NS	NS	NS	NS	NS	NS
	7 Aug	NS	●	NS	NS	NS	NS	NS	NS
1992	29 May	NS	NS	●	NS	NS	NS	●	NS
	27 Aug	NS	NS	●	●	NS	NS	●	NS
1993	31 May	NS	●	NS	NS	NS	NS	●	NS
	31 Aug	○	NS	NS	●	NS	NS	●	NS

then the length and number of receptacles would be expected to increase over time at the oiled sites as plants grow and produce more receptacles. Little or no convergence was seen in either the length or number of receptacles per plant, indicating that these plants are recovering slowly because of low growth rates, rapid loss of plant fronds, limited production of receptacles, or a combination of these processes. The fact that the number of receptacles per plant was lower at oiled sites indicates a significant problem for recolonization. *Fucus* eggs tend to travel less than 0.5 m from their source plant (McConnaughey 1985; van Tamelen and Stekoll, unpublished data), therefore reductions in receptacle number could lead to reductions in egg settlement rate, leading to increased recovery times.

Lower percent coverage of *Fucus* at oiled sites in 1990 and 1991 was a result of the removal of large and reproductive plants. *Fucus* cover was lower at the same oiled sites and tidal levels where the density of large plants was reduced. The loss of the dominant alga also may have led to increases in the cover of weedy, ephemeral algal species such as *Cladophora*, *Scytosiphon*, and *Enteromorpha*. In many habitats, ephemeral species are indicative of recently disturbed areas where the competitive dominant has been removed (Lubchenco 1978; Sousa 1979b). There is, however, an alternative explanation to the observed increase in ephemeral algae. These types of algae are typically more susceptible to grazing by molluscan herbivores. Because postspill decreases in the abundance of these herbivores have been demonstrated in Herring Bay and other spill-affected regions (Hooten and Highsmith 1996, this volume; Houghton et al. 1996; Lees et al. 1996), the lack of herbivory may have allowed ephemeral algae to flourish.

The general results described above were not uniform at all sites. Only two of the sheltered rocky site pairs sampled—the gently sloping pair 1 and the vertical wall pair 3—showed patterns indicative of major disturbance. The remaining site pair—the intermediately sloped pair 2—showed little or no effect of the oil spill. The variability among sites may be partly the result of different cleanup treatments applied to the sites. Site 2X probably was not treated to remove oil, but sites 1X and 3X were treated (Table 1). Where *Fucus* canopy was removed, the rock was mostly free of oil, except for small amounts in cracks and crevices in the rock in 1990. There were often holdfasts still attached to the rock, indicating that *Fucus* was once abundant. These observations suggest that these sites were treated with HPHW in 1989.

It has been extremely difficult to ascertain the cleanup history of our particular sites because of incomplete records and the geographic scale of the record keeping. All of the oiled sites are in shoreline segments in which some part, but not necessarily all, of the segment received HPHW treatments, but we cannot be certain of the treatment applied to individual sites. Although the evidence is circumstantial, it is likely that cleanup efforts caused much of the decrease in *Fucus* canopy (Broman et al. 1983; Foster et al. 1990; Rolan and Gallagher 1991;

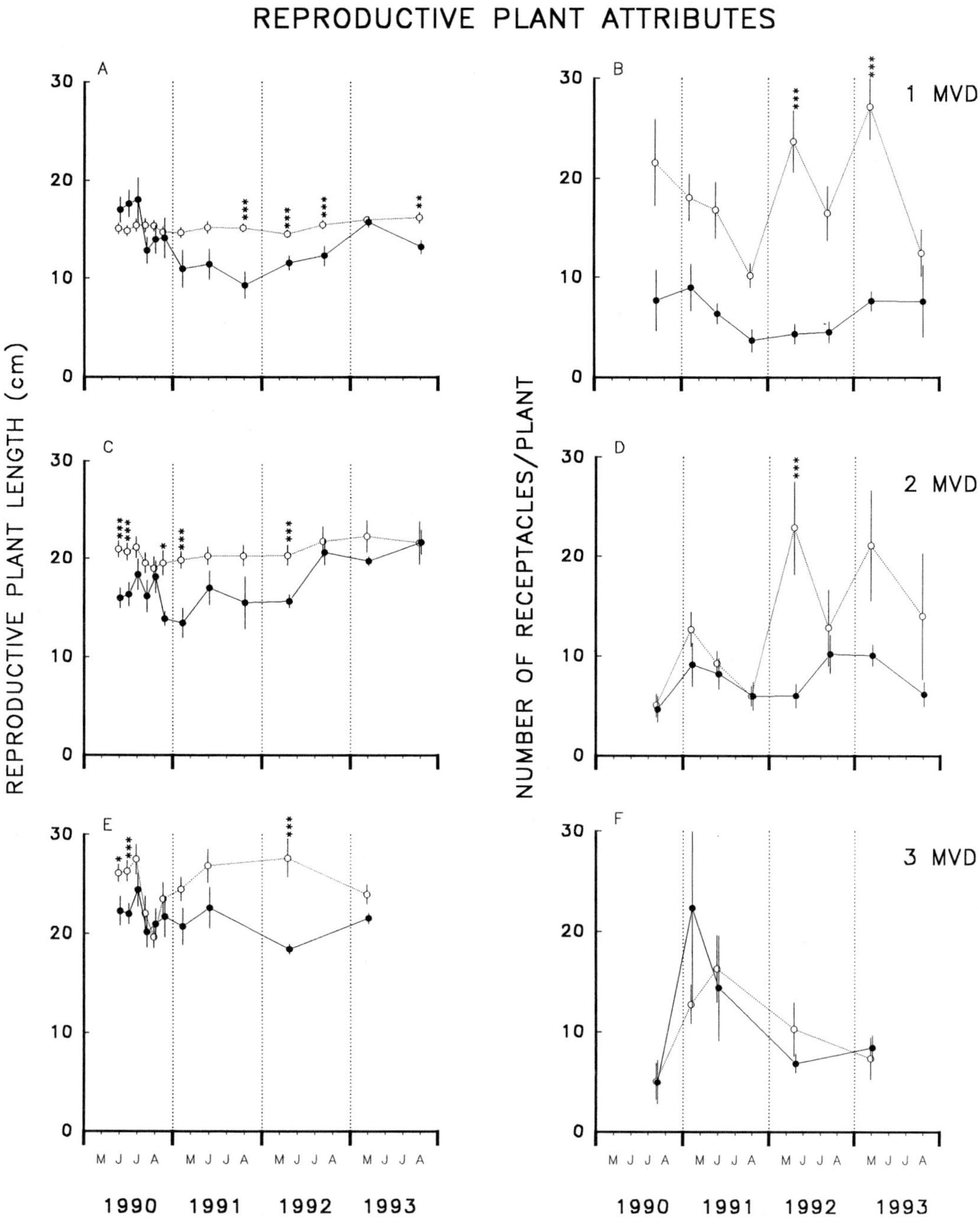

FIGURE 12.—The average length and number of receptacles per plant for all reproductive plants at all five site pairs sampled. The top graphs are for the first MVD, the middle graphs for the second MVD, and the bottom graphs for the third MVD. Layout is the same as in Figure 3.

Houghton et al. 1996; Lees et al. 1996). Photographs of an Omni boom in action show large amounts of floating *Fucus* within the containment boom during and after cleaning (Foster et al. 1990; Noerager and Goodman 1991). These plants probably came from the rock being cleaned, because the density of floating *Fucus* was visually much lower in surrounding areas.

Recovery Processes

Recovery of *Fucus* is evident at some oiled sites. For example, at site pair 1 in the second MVD there were more germlings (0–2.0 cm) in 1990 at the oiled site. Later in 1990 and early in 1991 there were more small plants (2.5–5.0 cm), in 1991 there were more medium plants (5.5–10 cm), and in 1992 and 1993 there was an increase in large plants at the oiled site (Figure 3B, E, H, K). Thus, over time the plants grew into successively larger size-classes.

Similar patterns were found at the vertical wall site pair 3C–3X and the coarse-textured site pair 4C–4X. The rate, magnitude, and variability of recovery, however, varies slightly among site pairs. Much of this variation could be explained by variations in environmental conditions and starting points. For example, recovery at the vertical site 3X was faster, had a greater magnitude (actually increasing above control densities), and had less variation compared to the gently sloping site 1X. The vertical walls faced northwest and rarely had direct sunlight at low tide, leading to lower desiccation stress. Lower stress would have the effect of enhancing recolonization, as recruitment may be limited by desiccation and heat stress (Brawley and Johnson 1991; van Tamelen and Stekoll, unpublished data).

As of 1993 recovery had begun in the upper intertidal zone at sheltered rocky sites, but was proceeding more slowly than in lower zones. The number of 5 to 10 cm and >10 cm plants increased in 1991 (5–10 cm) and in 1992 (>10 cm) at oiled sites in the first MVD to levels similar to those at control sites (Figures 3 and 5). These increases occurred earlier in lower tidal levels. A similar pattern of recovery was seen in Bristol Bay when larger numbers of small plants were observed in plots cleared of *Fucus* (Kendziorek and Stekoll 1984). Predictably, as the number of plants at oiled sites increased, the percent cover of *Fucus* also increased, especially as the plants grew to larger sizes.

At the end of the 1992 season in the mid-intertidal zone, there were dense beds of young *Fucus* plants just starting to become reproductive. Young *Fucus* plants were found in higher densities at oiled sites than in control areas because of their smaller size. In the future, it is expected that the density of *Fucus* plants will fall to values similar to those for the control sites. We estimate that *Fucus* may not fully recover in the mid-intertidal zone until the 1994 or 1995 season. In the upper intertidal zone, where desiccation stress is high, recovery of *Fucus* is proceeding slowly. The rate-limiting step for *Fucus* recolonization seems to be settlement of eggs and recruitment of new germlings. Some recruitment has occurred, but only in the deepest crevices in the rock surface. As the plants that have recruited into the upper intertidal zone grow, they will be able to ameliorate desiccation stress, further enhancing *Fucus* recruitment.

Impact of Cleanup Technologies

Because of the lack of adequate treatment records, it was not possible to experimentally assess the impact of various cleanup technologies on the intertidal community. There were few differences in the population structure of *Fucus* between the oiled site, 2X, and its matched control, 2C. Because 2X was probably not treated, this observation suggests that the oil per se had little effect on the dominant intertidal alga. Also, observations of adult *Fucus* plants at heavily oiled sites suggest that *Fucus* may be able to withstand a fairly high degree of oiling. Adult *Fucus* plants in other areas have been observed to be tolerant of oiling because of nonadherence of oil to the thallus and blades of the plants (Foster et al. 1971; Crothers 1983; but see Foster et al. 1990), making the plants resistant to the physical smothering by the oil. These observations suggest that the severe cleanup technologies, such as HPHW washings, applied in Herring Bay may have been more harmful to intertidal algal populations than leaving the oil in place. This hypothesis is consistent with the conclusions reached by other researchers (Broman et al. 1983; Foster et al. 1990; Rolan and Gallagher 1991; Houghton et al. 1996; Lees et al. 1996) in their analyses of the ecological effectiveness of oil spill cleanup technologies.

Acknowledgments

A. Bernhard, C. Carlson, A. Dreves, D. Knox, M. Lindeberg, and F. Roddy provided assistance in the field. D. Bausch, E. Berlow, R. Highsmith, A. Hooten, J. Houghton, and C. Peterson reviewed earlier drafts of this article. L. Deysher contributed to experimental design and methodology. W. Erikson and D. Strickland provided statistical advice.

References

Bold, H. C., and M. J. Wynne. 1978. Introduction to the algae: structure and reproduction. Prentice-Hall, Englewood Cliffs, New Jersey.

Brawley, S. H., and L. E. Johnson. 1991. Survival of fucoid embryos in the intertidal zone depends upon development stage and microhabitat. Journal of Phycology 27:179–186.

Broman, D., B. Ganning, and C. Lindblad. 1983. Effects of high pressure, hot water shore cleaning after oil spills on shore ecosystems in the northern Baltic proper. Marine Environmental Research 10:173–187.

Brosnan, D. M., and L. L. Crumrine. 1994. Effects of human trampling on marine rocky shore communities. Journal of Experimental Marine Biology and Ecology 177:79–97.

Crothers, J. H. 1983. Field experiments on the effects of crude oil and dispersant on the common animals and plants of rocky sea shores. Marine Environmental Research 8:215–239.

Dethier, M. N., and D. O. Duggins. 1984. An "indirect commensalism" between marine herbivores and the importance of competitive hierarchies. American Naturalist 124:205–219.

Foster, M., M. Neushul, and R. Zingmark. 1971. The Santa Barbara oil spill part 2: initial effects on intertidal and kelp bed organisms. Environmental Pollution 2:115–134.

Foster, M. S., J. A. Tarpley, and S. L. Dearn. 1990. To clean or not to clean: the rationale, methods, and consequences of removing oil from temperate shores. The Northwest Environmental Journal 6:105–120.

Galt, J. A., W. L. Lehr, and D. L. Payton. 1991. Fate and transport of the *Exxon Valdez* oil spill. Environmental Science and Technology 25:202–209.

Hooten, A. J., and R. C. Highsmith. 1996. Impacts on selected intertidal invertebrates in Herring Bay, Prince William Sound, after the *Exxon Valdez* oil spill. American Fisheries Society Symposium 18:249–270.

Houghton, J. P., D. C. Lees, W. B. Driskell, S. C. Lindstrom, and A. J. Mearns. 1996. Recovery of Prince William Sound intertidal epibiota from *Exxon Valdez* oiling and shoreline treatments, 1989 through 1992. American Fisheries Society Symposium 18:379–411.

Kendziorek, M., and M. S. Stekoll. 1984. Intraspecific competition and the management of the Bristol Bay herring-roe-on-kelp fishery. Hydrobiologia 116/117: 333–337.

Lees, D. C., J. P. Houghton, and W. B. Driskell. 1996. Short-term effects of several types of shoreline treatment on rocky intertidal biota in Prince William Sound. American Fisheries Society Symposium 18: 329–348.

Lobban, C. S., P. J. Harrison, and M. J. Duncan. 1985. The physiological ecology of seaweeds. Cambridge University Press, New York.

Lubchenco, J. 1978. Plant species diversity in a marine intertidal community: importance of herbivore food preference and algal competitive abilities. American Naturalist 112:23–39.

McConnaughey, J. 1985. The growth and regeneration of the rockweed, *Fucus distichus*, in Bristol Bay. Master's thesis. University of Alaska, Juneau.

McCook, L. J., and A. R. O. Chapman. 1991. Community succession following massive ice-scour on an exposed rocky shore: effects of *Fucus* canopy algae and of mussels during late succession. Journal of Experimental Marine Biology and Ecology 154:137–169.

Menge, B. A. 1976. Organization of the New England rocky intertidal community: role of predation, competition and environmental heterogeneity. Ecological Monographs 46:355–393.

Menge, B. A. 1978. Predation intensity in a rocky intertidal community: effect of an algal canopy, wave action and desiccation of predator feeding rates. Oecologia 34:17–35.

Nauman, S. A. 1991. Shoreline cleanup: equipment and operations. Pages 141–147 in Proceedings, 1991 international oil spill conference: prevention, behavior, control, cleanup. American Petroleum Institute Publication 4529, Washington, DC.

Noerager, J. A., and R. H. Goodman. 1991. Oil tracking, containment, and recovery during the *Exxon Valdez* response. Pages 193–203 *in* Proceedings, 1991 international oil spill conference: prevention, behavior, control, cleanup. American Petroleum Institute Publication 4529, Washington, DC.

Rolan, G. R., and R. Gallagher. 1991. Recovery of the intertidal biotic communities at Sullum Voe following the *Esso Bernicia* oil spill of 1978. Pages 461–465 *in* Proceedings, 1991 international oil spill conference: prevention, behavior, control, cleanup. American Petroleum Institute Publication 4529, Washington, DC.

Sokal, R. R., and F. J. Rohlf. 1981. Biometry, 2nd edition. Freeman, San Francisco.

Sousa, W. P. 1979a. Experimental investigations of disturbance and ecological succession in a rocky intertidal community. Ecological Monographs 49:227–254.

Sousa, W. P. 1979b. Disturbance in marine intertidal boulder fields: the non-equilibrium maintenance of species diversity. Ecology 60:1225–1239.

Stekoll, M. S., and seven coauthors. 1996. Coastal habitat injury assessment: intertidal communities and the *Exxon Valdez* oil spill. American Fisheries Society Symposium 18:177–192.

van Tamelen, P. G. 1987. Early successional mechanisms in the rocky intertidal: the role of direct and indirect interactions. Journal of Experimental Marine Biology and Ecology 112:39–48.

Wolfe, D. A., editor. 1977. Fate and effects of petroleum hydrocarbons in marine organisms and ecosystems. Pergamon Press, New York.

American Fisheries Society Symposium 18:212–237, 1996

Impact of the *Exxon Valdez* Oil Spill on Intertidal Biota

RAYMOND C. HIGHSMITH AND TAMA L. RUCKER
University of Alaska Fairbanks, School of Fisheries and Ocean Sciences
Institute of Marine Science, Fairbanks, Alaska 99775–7220, USA

MICHAEL S. STEKOLL
University of Alaska Fairbanks, Juneau Center, School of Fisheries and Ocean Sciences
11120 Glacier Highway, Juneau, Alaska 99801, USA

SUSAN M. SAUPE
University of Alaska Fairbanks, Institute of Marine Science

MANDY R. LINDEBERG AND ROBIN N. JENNE
University of Alaska Fairbanks, Juneau Center

WALLACE P. ERICKSON
Western EcoSystems Technology, 1402 South Greely Highway, Cheyenne, Wyoming 82007, USA

Abstract.—The Coastal Habitat Injury Assessment (CHIA) study was initiated to assess injury to biological resources in intertidal habitats affected by the *Exxon Valdez* oil spill. Intertidal communities were subjected to the most intense impacts of the spill and of subsequent cleanup operations. An understanding of the effects of the oil spill and cleanup on the intertidal zone is critical in determining the extent of injury and the recovery potential of the affected areas. The CHIA study encompassed the three major geographic areas affected by the spill: Prince William Sound, Cook Inlet–Kenai Peninsula, and Kodiak–Alaska Peninsula. Randomly selected oiled sites from various habitats were matched with non-oiled, reference sites. All epibenthic organisms and infauna were sampled and quantified. The results indicate that injury occurred to a wide range of taxa and community types. This article only reports the effects on the dominant taxa in the intertidal zone. The affected dominant taxa were the brown alga *Fucus gardneri*, the limpet *Tectura persona*, the barnacles *Chthamalus dalli*, *Semibalanus balanoides*, and *Balanus glandula*, the mussel *Mytilus trossulus*, two species of littorines—Sitka periwinkle *Littorina sitkana* and checkered periwinkle *L. scutulata*—and oligochaetes. Although abundances and biomasses of *C. dalli* and oligochaetes were enhanced after the spill, other major taxa were negatively affected by the spill either as a direct result of exposure or treatment or indirectly because of loss of habitat or food. Observed differences between oiled and reference populations varied across regions, habitats, and tidal heights. Differences for several taxa were still apparent as of our last observation in 1991, implying incomplete recovery.

The Coastal Habitat Injury Assessment (CHIA) study was initiated to assess injury to biological resources in intertidal habitats affected by the *Exxon Valdez* oil spill. The supertanker *Exxon Valdez* ran aground on Bligh Reef on 24 March 1989, resulting in the largest oil spill in U.S. history (40 million liters of Prudhoe Bay crude oil). The spill and subsequent cleanup severely affected the highly productive and diverse biota of the intertidal region, which supports many commercially and ecologically important species. The intertidal region is particularly vulnerable to oil spills because of stranding and subsequent persistence of oil in coarse- and fine-grained sediments. Many shorelines included in the study also were subjected to cleanup operations, such as high-pressure hot-water treatment, that effectively removed most organisms (Houghton et al. 1996, this volume).

The impact of oiling on intertidal regions is well documented for previous oil spills (Thomas 1973, Keizer et al. 1978, and Thomas 1978 for the *Arrow*; Southward and Southward 1978 for the *Torrey Canyon*; Bonsdorff 1981 for the *Antonio Gramsci*; Chasse 1978 for the *Amoco Cadiz*; and Clark et al. 1978 for the *General M. C. Meigs*). The degree of observable damage is dependent on a myriad of conditions, ranging from the extent of the spill and type of oil to habitat type, ambient weather conditions, season, and latitude. All prior investigations revealed some degree of intertidal damage and similar patterns of recovery based on successional events.

After the *Exxon Valdez* oil spill, several intertidal studies have reported various impacts to coastal resources. Houghton et al. (1996) and Lees et al. (1996, this volume) gave damage estimates for intertidal epifauna and infauna attributable to the spill and subsequent high-pressure hot-water treatment. Van Tamelen and Stekoll (1996, this volume and unpublished data) demonstrated severe and widespread damage to *Fucus* in Herring Bay, Prince William Sound. Hooten and Highsmith (1996, this volume) observed lowered densities of several key organisms on oiled sites. A reduction in prey availability for river otters *Lutra canadensis* (Duffy et al. 1994) and contamination of food items used by harlequin ducks *Histrionicus histrionicus* (S. M. Patten, Alaska Department of Fish and Game, unpublished data) were observed after the spill.

The objective of this study was to determine the effects of oiling and subsequent cleanup on key species in the intertidal community. The potential recovery of the various taxa was also examined.

Methods

Field sampling in all three major coastal regions affected by the *Exxon Valdez* oil spill—Prince William Sound (PWS), Cook Inlet–Kenai Peninsula (CIK), and Kodiak–Alaska Peninsula (KAP) (Figure 1)—was initiated in late summer 1989 and continued through summer 1991. Effects of the oil and subsequent cleanup were determined by comparison of intertidal communities on oiled and matched reference sites. Site selection, based on a stratified random sampling design (Sundberg et al. 1996, this volume), included beach segments in sheltered rocky, exposed rocky, coarse-textured, and estuarine habitats. These classifications were created as a result of the condensation of the shoreline types described in a Research Planning Institute (RPI) report to the National Oceanic and Atmospheric Administration (NOAA) (RPI 1983) and are explained in Sundberg et al. (1996). Sheltered rocky sites were composed of segments designated as "sheltered rocky" in the RPI report (RPI 1983). Coarse-textured sites were composed of "mixed sand and gravel," "gravel beaches," and "exposed tidal flats." Estuarine sites were a combination of "sheltered tidal flats" and "marshes." Exposed rocky sites were composed of "exposed rocky shores" and "wave cut platforms." Field procedures included the collection of epifaunal, infaunal, and algal samples from randomly placed quadrats at each of three tidal levels.

Site Selection

A detailed explanation of site selection methodology can be found in Sundberg et al. (1996). Essentially, a geographic information system (GIS), maintained by the Alaska Department of Natural Resources, was used to combine oiling and habitat data to select oiled study sites representative of the shoreline habitat affected by the *Exxon Valdez* oil spill. Moderately to heavily oiled sites were randomly selected from the four habitat types and matched with non-oiled or lightly oiled reference sites. Physical characteristics considered in matching reference sites to the oiled sites were geomorphology, wave exposure, slope and aspect, proximity to freshwater and nearshore bathymetry. Site selection was carried out in summer 1989 and spring 1990. Specific site locations and types are listed in Stekoll et al. (1996, this volume). In PWS all four habitat types were sampled. In CIK only sheltered rocky, coarse-textured, and estuarine sites were sampled. In KAP only sheltered rocky and coarse-textured habitats were sampled.

Site Layout and Quadrat Collection

Details of the experimental procedures are given in Stekoll et al. (1996). Using a random-numbers procedure, six transects were established at regular intervals along each beach from mean high water to mean lower low water (zero chart datum). Permanent 0.1-m^2 quadrats were randomly positioned along the transects within each of the first 3 m of vertical drop (MVD) from mean high water (Figure 2). Mean high water in PWS is about 3.4 m above mean lower low water.

All algae, invertebrates, and fish within the three-dimensional volume encompassed by the quadrats were removed, including infauna from the top 10 cm of sediment. A habitat type encountered in the quadrat that differed from the designated habitat type was eliminated from the sampling design. Samples were preserved in formalin for subsequent processing. Limited collections were made in fall 1989 (August–October) and in summer 1991 (June–July), and more extensive collections were done in spring 1990 (May–July), summer 1990 (July–September), and spring 1991 (May–June). Data presented here are for the 1990 and 1991 sampling periods.

Percent-cover data for *Fucus*, other algae, and uncolonized substrate were collected from photographs of the sampled areas in 1990 and directly from the field in 1991 using a point contact method. Percent-cover data came from undisturbed quadrats and from quadrats before they were cleared. Only the top algal layer was counted using this method. An 80-point grid was used to determine algal cover over a 0.4-m^2 area. Regressions were calculated using 1991 data to determine the corre-

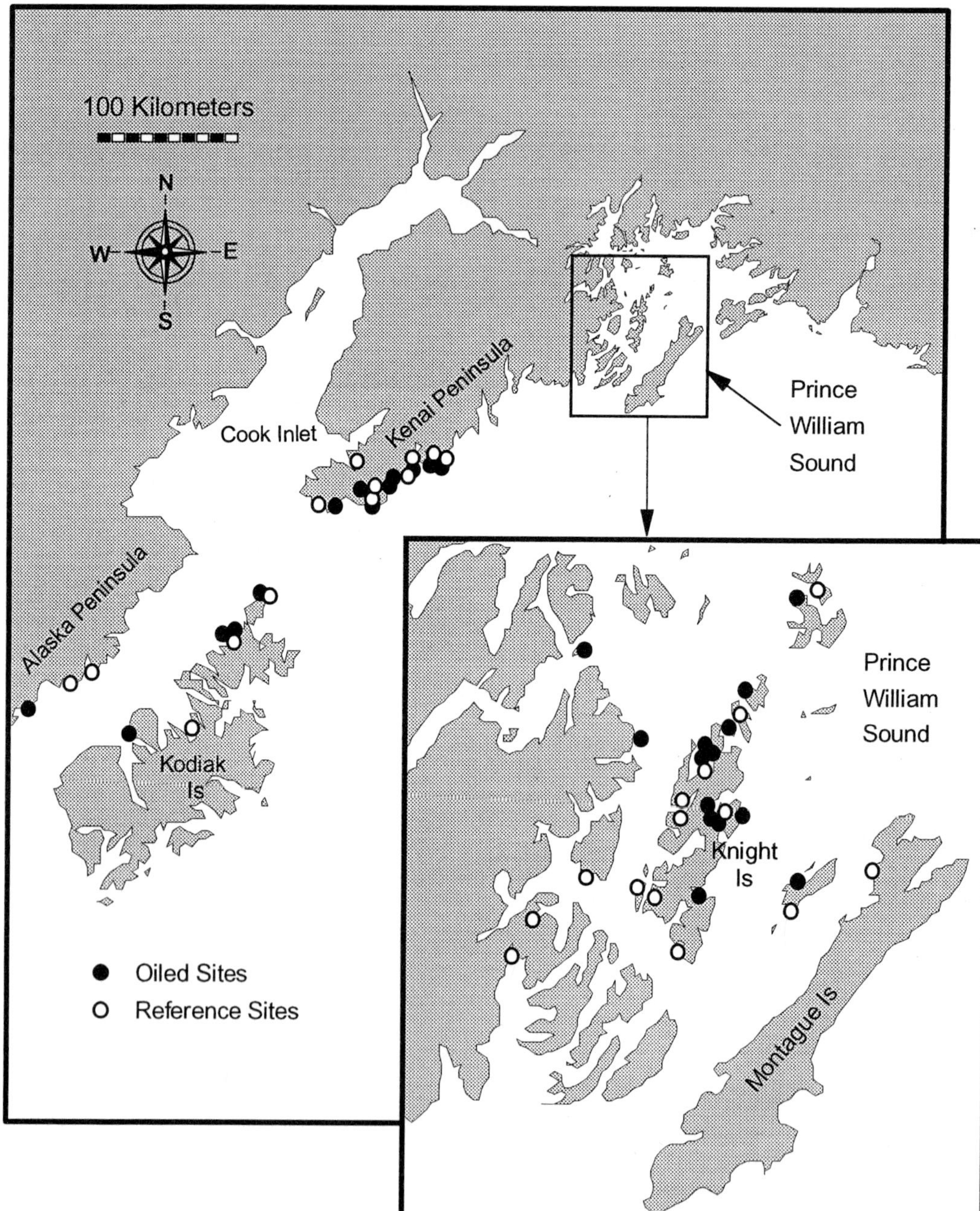

FIGURE 1.—Location map for the study sites sampled in the Coastal Habitat Injury Assessment study, 1989–1991.

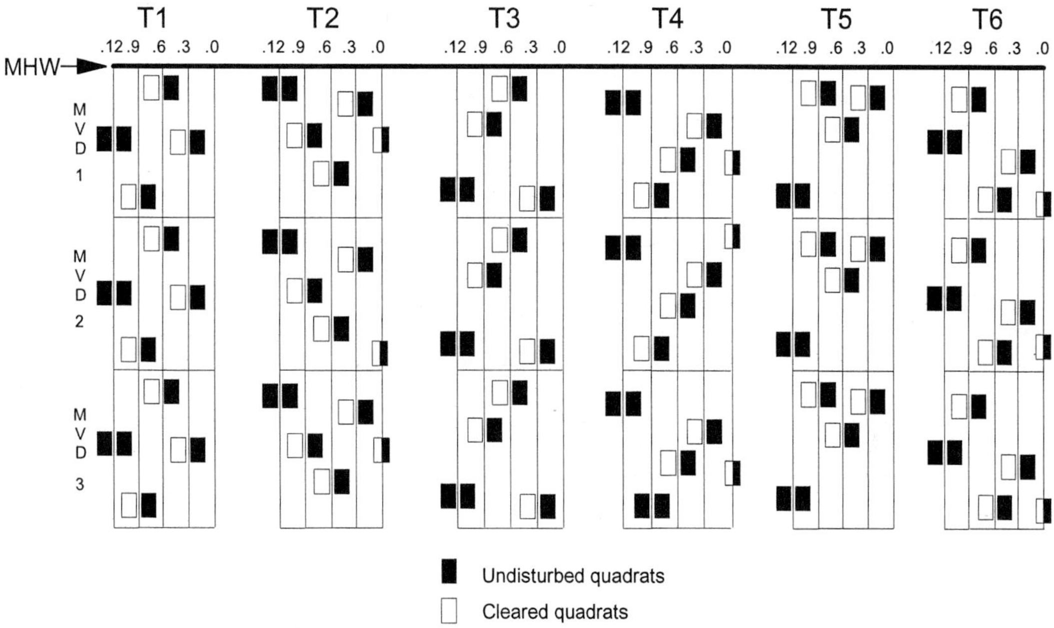

FIGURE 2.—Transect and quadrat layout on a typical site. Abbreviations are T = transect, MHW = mean high water, and MVD = meter of vertical drop below MHW. The numbers .0, .3, .6, .9, and .12 correspond to the locations of transects in 1989, spring 1990, summer 1990, spring 1991, and summer 1991, respectively. Only three transects were established in 1989, thus the presence of unsampled transects at T1.0, T3.0, and T5.0. During the summer 1991 sampling (.12 transects), only percent-cover algal data were recorded and no cleared quadrats were established.

lation between the photographic percent cover and percent cover estimated in the field; R^2 values for *Fucus* and substrate were 0.87 and 0.84, respectively.

Laboratory Procedures

All organisms were sorted and identified to the lowest possible taxonomic category. Blotted wet weights were determined for each taxon to 0.001 g. Invertebrate samples were rinsed through a 1-mm mesh sieve before sorting.

Statistical Analyses

The null hypothesis that no significant differences existed between oiled and matched reference sites was tested. Two-tailed *t*-tests were used to compare abundance, biomass, and percent cover for each taxonomic category at each MVD between paired sites. Statistical comparisons of abundance and biomass were expanded to all study sites within each region, habitat, and MVD using Fisher's procedure for combining *P* values of independent tests (Sokal and Rohlf 1981). Statistical inferences were further extended to all possible locations within the universe of a given region, habitat, and MVD using Stouffer's meta-analysis procedure (Folks 1984).

Impact trends are generally discussed here if either or both Fisher's or Stouffer's tests were significant at $P < 0.05$. A significant difference included either a significantly lower or higher value on the oiled sites in comparison with the reference sites. Significant results with both Fisher's and Stouffer's tests indicate a high probability that an overall oiling and cleanup effect existed for the particular taxa in a given habitat, region, MVD, and time period. Significant differences detected in 1991 and not in 1990 indicate possible delayed effects attributable to oiling and cleanup or may be a result of low power in 1990. Significant differences between oiled and reference sites detected in 1990 but not in 1991 were interpreted as evidence of recovery if statistical power was greater than 0.5, but because power diminishes with convergence of the data, such instances were rare. Recovery was also indicated if there were no significant differences and the means were converging; but with only 2 years of data, such estimations of convergence are tenuous. In the absence of obvious convergence, we can say only that the differences are not significant and make no inferences about recovery. Significant differences persisting through 1991 suggest incomplete recovery.

Statistical power analyses were conducted on all of the data (Highsmith et al. 1994). The power analysis procedure used 5,000 randomly generated data sets based on the original site means and variances. Fisher's and Stouffer's tests were performed on each of these data sets. The proportion of these tests that was significant at $P < 0.05$ is the power of the original test. In addition, an estimate of the power to detect a 50% difference in the means was also determined.

Results

An initial, major consequence of the spill and cleanup was the creation of unoccupied space in oiled habitats. In general, the occurrence of uncolonized substrate (i.e., bare rock) was greater in the oiled–treated areas (Figures 3–5). Differences were significant in several habitats and MVDs in 1990. Incomplete recovery was evident at sheltered rocky sites at MVD 1 in PWS and CIK and at MVD 2 in CIK and KAP (Figure 3). At coarse-textured sites recovery was incomplete at MVDs 2 and 3 in PWS (Figure 4). Convergence appeared to be occurring at CIK coarse-textured sites. In KAP more uncolonized substrate was observed at reference sites at MVDs 1 and 2 in coarse-textured habitats in summer 1990, a result of the fact that the oiled–treated sites had greater coverage of sand and mud (data not shown). Estuaries in both PWS and CIK showed significant differences in substrate cover through 1991 at all MVDs (Figure 5). The exposed rocky habitat in PWS had no significant differences at MVDs 2 and 3 (except at MVD 2 in June 1991) and convergence was seen at MVD 1 in 1991 (Figure 5).

The biomass and/or percent cover of the perennial brown alga *Fucus gardneri* was significantly less on the oiled–treated sites for most habitats in each area (Table 1; Figures 3–5). The *Fucus* biomass values showed fewer significant differences than did percent-cover values. This fact is attributable to the lower power of the biomass data because the percent-cover data used larger quadrats and often had a larger *N*.

There was significantly less *Fucus* at all oiled–treated habitats in PWS and CIK at MVD 1 in 1990 except at CIK coarse-textured sites, which had no plants at MVD 1 (Figures 3, 4) and at PWS estuarine sites (Figure 5). Significant differences persisted through 1991 at all habitats in PWS except at exposed rocky sites (Figure 5). In CIK differences were significant in 1991 at sheltered rocky and estuarine habitats (Figures 3, 5). In KAP there was less *Fucus* cover at oiled–treated sheltered rocky and coarse-textured sites as of the last sampling in 1991 (Figures 3, 4).

The pattern of less *Fucus* at oiled–treated sites in 1990 was repeated at MVD 2 at all sheltered rocky sites, at KAP coarse-textured sites, and in the PWS estuarine and exposed rocky sites (Figures 3–5). Values had not converged by the last sampling in 1991, except at CIK sheltered rocky sites. In addition, at PWS coarse-textured sites *Fucus* cover became significantly less at oiled–treated sites in 1991.

At MVD 3, *Fucus* populations did not exhibit a consistent trend across the spill area. In PWS at MVD 3, biomass and percent cover of *Fucus* were greater in 1990 at sheltered rocky and coarse-textured reference sites than at oiled–treated sites (Table 1; Figures 3, 4) but not at estuarine or exposed rocky sites. By 1991 the *Fucus* percent cover was greater in oiled–treated sheltered rocky sites (Figure 3), but the biomass values had converged (Table 1). In the coarse-textured habitat, significant differences remained (Table 1; Figure 4). *Fucus* biomass at MVD 3 at coarse-textured sites in the KAP region was greater at reference sites on all sampling dates but only significantly so in the summer of 1990; percent cover was significantly greater at reference sites through 1991.

In contrast to the above trend, *Fucus* populations in CIK were often enhanced on oiled–treated sites at MVD 3, especially in sheltered rocky habitats and to a lesser extent in coarse-textured habitats (Figures 3, 4). For example, the percent cover of *Fucus* in spring 1990 at CIK sheltered rocky sites was 53.4% at oiled–treated sites and 19.6% at reference sites; biomass was about five times greater at oiled–treated sites than at reference sites at this same time (Table 1). Values were not converging by 1991 at CIK sheltered rocky sites. The data indicate that *Fucus* (and to a lesser extent plants in the Gigartinaceae family) was occupying space that is normally populated by other algae, including *Acrosiphonia*, *Odonthalia*, *Polysiphonia–Pterosiphonia*, *Porphyra*, and especially *Alaria* (Table 2).

Abundance and biomass patterns varied for many invertebrate taxa. The following taxa, however, showed consistent patterns across time, regions, and/or habitats: the limpet *Tectura persona*, the barnacles *Chthamalus dalli*, *Balanus glandula*, and *Semibalanus balanoides*, the mussel *Mytilus trossulus* (formerly *M. edulis*; see McDonald and Koehn 1988; Geller et al. 1994), two species of littorines—Sitka periwinkle *Littorina sitkana* and checkered

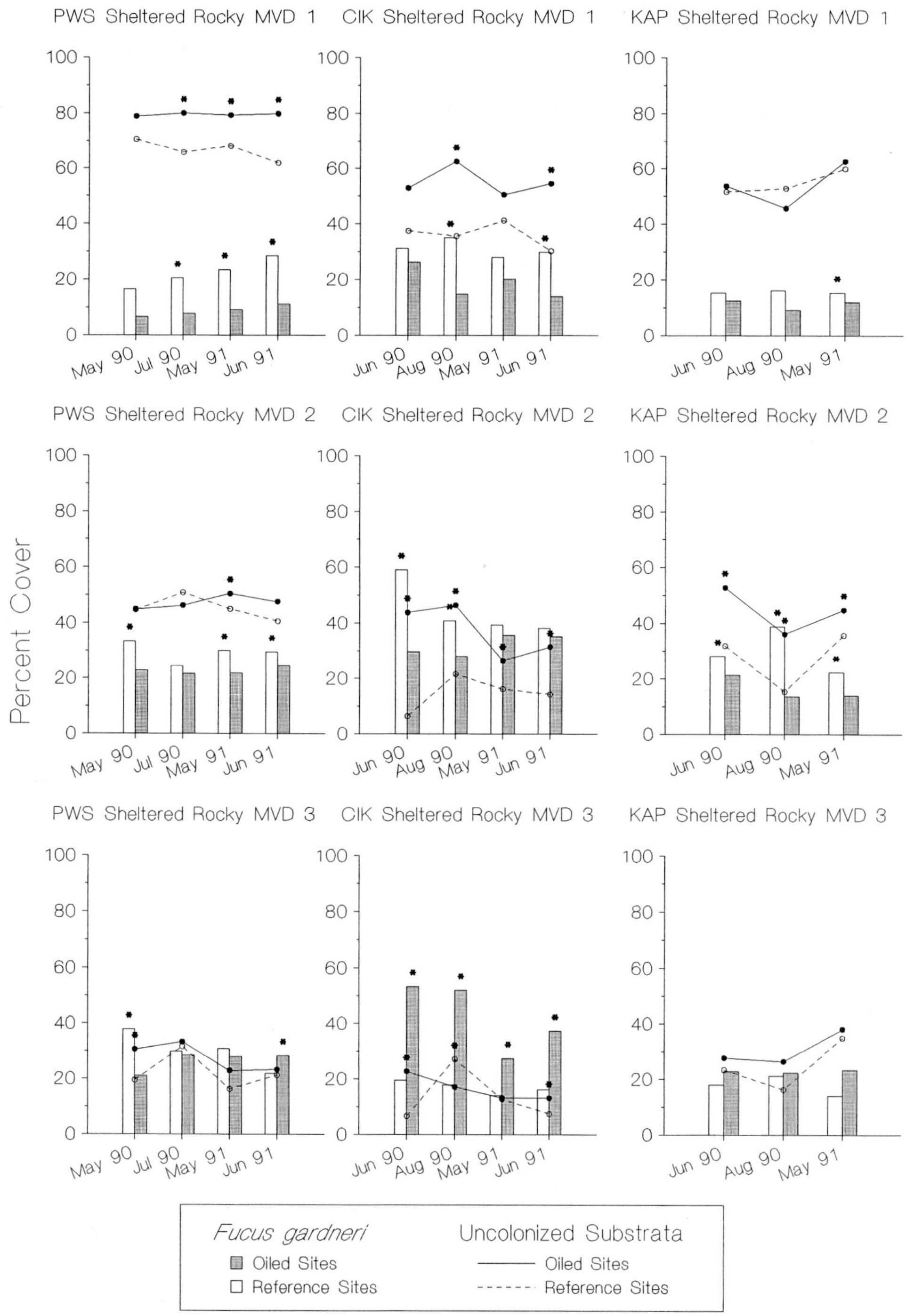

FIGURE 3.—Percent cover of both *Fucus gardneri* and uncolonized substrata at sheltered rocky sites for each meter vertical drop (MVD), 1990–1991. Locations are Prince William Sound (PWS), Cook Inlet–Kenai Peninsula (CIK), and Kodiak–Alaska Peninsula (KAP). Values are the grand mean of the quadrat means from each site. Asterisks indicate that the differences between the oiled and reference (non-oiled) sites were significant at $P < 0.05$ by either Fisher's or Stouffer's test.

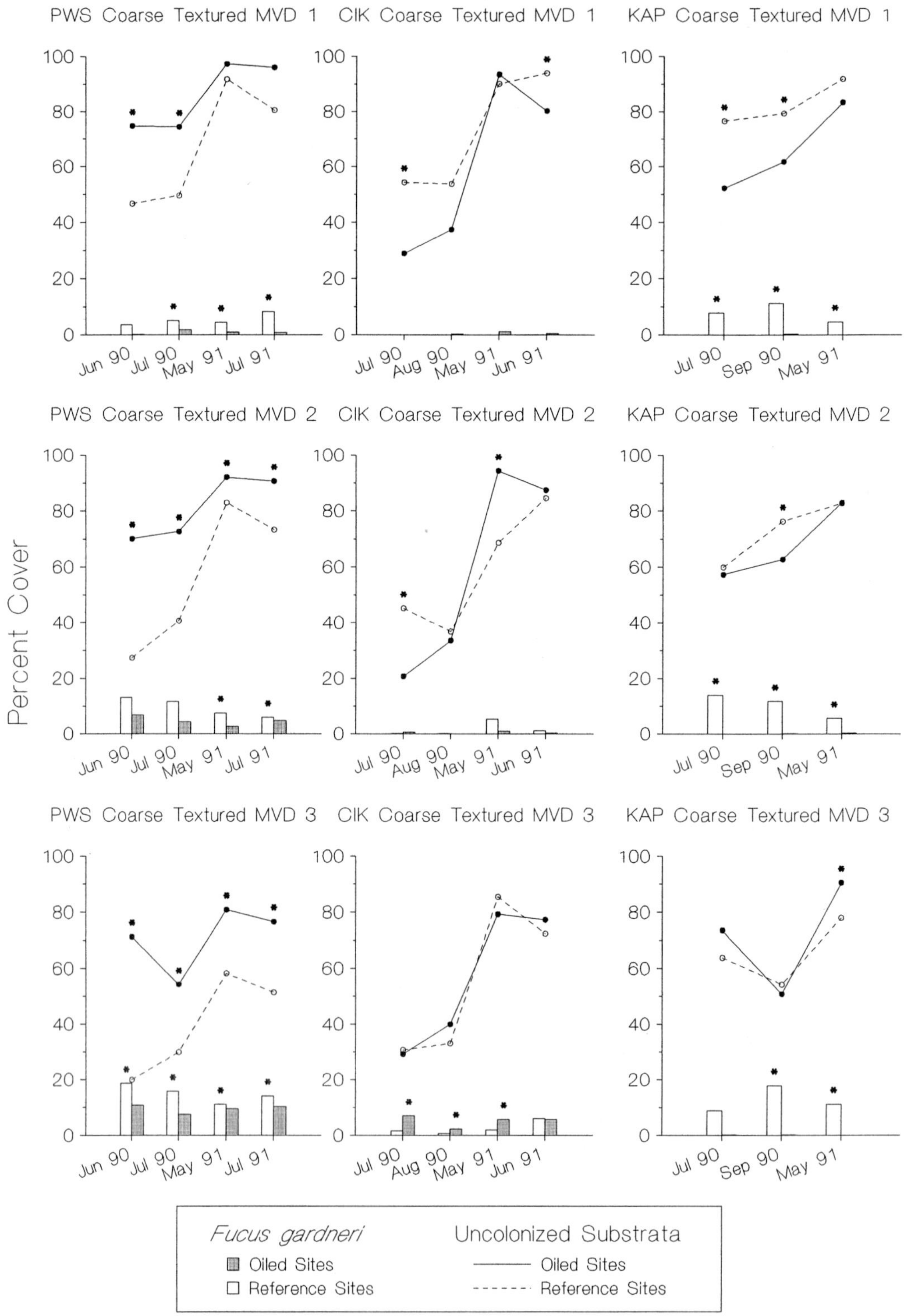

FIGURE 4.—Percent cover of both *Fucus gardneri* and uncolonized substrata at coarse-textured sites for each meter vertical drop (MVD), 1990–1991. Location abbreviations and statistics are as in Figure 3.

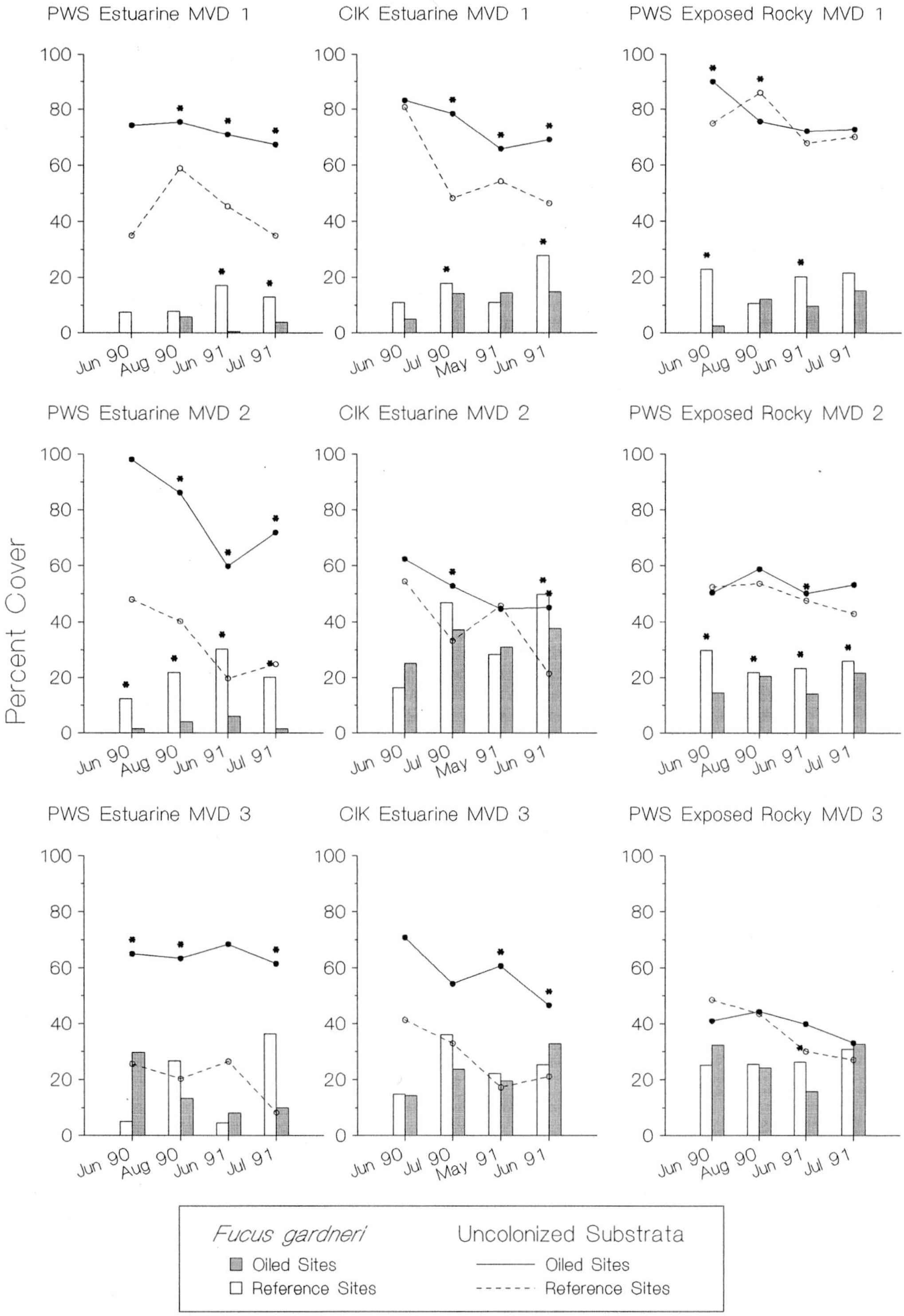

FIGURE 5.—Percent cover of both *Fucus gardneri* and uncolonized substrata at estuarine and exposed rocky sites for each meter vertical drop (MVD), 1990–1991. Conventions are as in Figure 3.

TABLE 1.—Summary of *Fucus gardneri* biomass data. Biomass means for each habitat type within a region are presented in grams per square meter for each meter vertical drop (MVD) below mean high water. Locations are Prince William Sound (PWS), Cook Inlet–Kenai Peninsula (CIK), and Kodiak–Alaska Peninsula (KAP). Statistically significant difference between reference (non-oiled) and oiled sites is indicated by an asterisk, $P < 0.05$. Statistical abbreviations are F = Fisher's combined test and S = Stouffer's consensus test. Spring samplings occurred May–July 1990 and May–June 1991. Summer sampling occurred July–September 1990.

	Spring 1990			Summer 1990			Spring 1991		
Location	Reference	Oil	F/S	Reference	Oil	F/S	Reference	Oil	F/S
				MVD 1					
Sheltered rocky									
PWS	913.8	248.4		1,002.4	309.8		1,173.2	358.5	*/*
CIK	2,066.9	902.2		1,460.5	655.3	*/*	1,316.3	194.7	
KAP	666.6	542.8		412.5	463.5		1,000.6	955.2	
Coarse-textured									
PWS	80.2	0.0		112.1	27.4		134.0	31.7	
CIK	0.0	0.1		0.0	1.5		0.3	34.7	
KAP	126.7	0.0		443.3	30.4		141.1	0.2	
Estuarine									
PWS	303.3	0.8	*/*	423.8	0.0		432.4	4.2	*/*
CIK	238.3	82.3	*/*	254.6	381.4		151.7	185.0	
Exposed rocky									
PWS	815.7	57.9		456.9	352.4		923.1	683.0	
				MVD 2					
Sheltered rocky									
PWS	1,742.3	1,300.3		1,149.6	1,354.4		1,538.3	1,488.6	
CIK	3,156.4	1,970.4		1,266.0	732.8	*/	1,332.6	1,997.5	/*
KAP	1,009.5	419.5	*/	2,016.4	191.2	*/*	506.0	737.1	
Coarse-textured									
PWS	664.2	3.8		391.2	69.6		478.7	24.2	
CIK	1.1	27.7		5.4	0.7		158.0	1.6	
KAP	778.1	0.0		131.2	0.8		141.0	18.4	
Estuarine									
PWS	468.8	24.6	*/*	545.9	0.0		1,545.1	110.3	*/
CIK	312.0	935.2		929.8	1,083.6		792.3	517.5	*/*
Exposed rocky									
PWS	1,493.9	958.1		590.9	268.2	*/*	656.0	510.5	
				MVD 3					
Sheltered rocky									
PWS	1,773.2	1,025.7		774.5	1,272.8		1,227.0	1,245.6	
CIK	575.1	2,865.9	*/*	715.1	2,568.0	*/*	449.2	1,064.7	
KAP	929.1	707.7		503.3	265.1		298.3	679.3	
Coarse-textured									
PWS	360.1	185.1		358.4	72.6	*/*	593.9	279.8	/*
CIK	56.8	208.4		114.8	80.1		1.1	157.8	
KAP	828.0	2.6		646.5	0.0		528.4	0.1	
Estuarine									
PWS	95.3	278.5		271.4	431.4		355.1	193.1	
CIK	254.1	246.9		342.8	382.8		319.6	420.7	
Exposed rocky									
PWS	1,069.4	1,196.5		396.2	878.0		1,751.8	1,132.2	

periwinkle *L. scutulata*—and oligochaetes. A summary of findings for each of these taxa follows. Abundance data are organized into tables and selected abundance and biomass trends are represented graphically. Because biomass trends generally reflect those of abundance, only the abundance data are included in tables.

Data for the limpet *T. persona*, primarily an upper intertidal inhabitant, showed more significant differences between oiled–treated and reference populations at MVD 1 than at other tidal heights (Table 3; Figure 6). In PWS and CIK sheltered rocky, KAP coarse-textured, and PWS exposed rocky habitats, abundances and biomasses were generally higher at reference sites at MVD 1. By 1991, significant differences were no longer detected at MVD 1 in coarse-textured and exposed rocky habitats and values were converging. For lower beach elevations in primarily coarse-textured habitat, abundances and biomasses were generally

TABLE 2.—Summary of Cook Inlet–Kenai Peninsula sheltered rocky habitat percent-cover data at 3 m vertical drop (MVD 3). Statistically significant difference between reference (non-oiled) and oiled sites is indicated by an asterisk, $P < 0.05$. Abbreviations are F = Fisher's combined test and S = Stouffer's consensus test.

	Spring 1991			Summer 1991		
Alga	Reference	Oil	F/S	Reference	Oil	F/S
	Perennials					
Fucus gardneri	14.0	27.4	*/*	16.1	37.3	*/*
Gigartinaceae	6.6	7.3		1.5	5.3	*/
Odonthalia floccosa	7.6	3.4	*/*	6.7	3.2	/*
Ptilota–Neoptilota	7.6	2.3		0.9	3.4	
	Annuals					
Acrosiphonia sp.	1.1	0.8		3.6	1.4	*/*
Alaria sp.	11.3	1.1	*/*	21.1	3.9	*/*
Palmaria sp.	11.4	8.2		4.9	5.8	
Polysiphonia–Pterosiphonia sp.	3.2	1.2	*/*	2.3	0.5	*/*
Porphyra sp.	0.6	1.0		13.8	3.5	*/
	Totals					
Total algae	74.9	78.3		87.3	79.5	
Perennials	37.9	47.4	*/*	28.9	53.2	*/*
Annuals	37.1	30.7		58.4	26.3	*/*

higher at oiled–treated sites when significant differences occurred. No differences were observed for *T. persona* from estuarine habitat. Other limpet species examined, *T. scutum* and *Lottia pelta*, showed inconsistent results, although a few significant differences were observed at lower tidal levels (data not shown).

Three barnacle species were examined in detail for abundance differences (Tables 4–6). Abundances and biomasses of *C. dalli* were higher for oiled–treated sheltered and exposed rocky sites in most cases when differences were significant (Table 4; Figure 7). The differences persisted through the last sampling period of 1991 in several cases. *Chthamalus dalli* generally was not found in high densities on coarse-textured sites and differences were not apparent. Significant differences were not observed for *C. dalli* in PWS estuaries. In CIK estuaries, although abundances and biomasses of *C. dalli* were higher on oiled–treated sites, (this barnacle was absent from reference sites), the differences were not significant.

Abundance and biomass data for *B. glandula* were not consistent among the three regions or even within a given habitat (Table 5). Most significant differences were for higher values at reference sites in PWS and KAP and higher values at oiled–treated sites in CIK, although the opposite pattern was observed in some instances. Several significant differences were still apparent in 1991, primarily in coarse-textured and estuarine habitats, signifying incomplete recovery.

A majority of the significant differences observed for *S. balanoides* revealed a higher abundance for the reference populations (Table 6). Significant differences remained in several cases in 1991, representing incomplete recovery.

Mytilus trossulus abundance and biomass differences, when significant, were higher on reference sites in most regions and habitats (Table 7; Figure 8); however, biomass was higher for mussels on oiled–treated sites in a few instances in which abundance was higher on reference sites (Figure 8). Apparently, the oiled population was composed of fewer, larger individuals compared with smaller more numerous mussels in reference populations. Significant differences were not found in sheltered rocky habitat at KAP and were observed only once in exposed rocky habitat (PWS). Statistical differences for sheltered rocky and PWS estuarine habitats were generally no longer apparent by 1991. Mussel populations from many coarse-textured and CIK estuarine habitats, however, still showed significant differences in 1991, suggesting incomplete recovery.

Significant differences in abundance of the Sitka periwinkle were observed between oiled–treated and reference populations (Table 8). In sheltered rocky, exposed rocky, and PWS estuarine habitats, values were higher on reference sites when significant. In coarse-textured habitat, abundances tended to be significantly higher at reference sites in PWS and KAP and higher at oiled–treated sites in CIK. A consistent trend was not observed for *L. sitkana* populations from CIK estuarine habitat.

The checkered periwinkle generally had higher abundances at oiled–treated sites in sheltered rocky habitat (Table 9). Values in exposed rocky habitat, when significant, were higher on reference sites at MVD 2 but higher on oiled–treated sites at MVD 1. The differences at MVD 1 were not significant by 1991. As observed for the Sitka periwinkle, abundances of the checkered periwinkle in coarse-textured and estuarine habitats generally were significantly higher at references sites in PWS and KAP and higher at oiled–treated sites in CIK. Biomass data, although not shown for the two littorine species, exhibited patterns similar to those for abundance.

Abundances and biomasses of oligochaetes were significantly higher on oiled–treated sites in more than half of the habitats and regions compared, particularly in coarse-textured habitat in CIK,

TABLE 3.—Mean densities (number/m^2) of *Tectura persona*. Abbreviations, statistics, and sampling dates are as in Table 1. The black diamond symbol indicates that the direction of the difference between habitat means is different from the direction of overall tests.

Location	Spring 1990			Summer 1990			Spring 1991		
	Reference	Oil	F/S	Reference	Oil	F/S	Reference	Oil	F/S
MVD 1									
Sheltered rocky									
PWS	43.3	15.7	*/	44.0	17.0	/*	34.3	7.3	*/*
CIK	5.6	6.4		6.1	6.1		3.9	4.6	/*◆
KAP	55.4	46.8		36.3	26.3		40.2	37.3	
Coarse-textured									
PWS	9.3	15.7		23.3	7.7		18.9	18.2	
CIK	0.6	5.0		1.7	2.2		0.0	10.6	
KAP	10.0	0.0	/*	8.0	0.0	*/*	2.5	0.0	
Estuary									
PWS	7.0	0.0		9.2	11.7		0.0	0.0	
CIK	6.7	2.2		0.0	5.0		3.9	7.9	
Exposed rocky									
PWS	22.2	1.7	*/*	6.1	5.0		6.1	18.1	
MVD 2									
Sheltered rocky									
PWS	11.9	12.2		15.4	14.7		6.0	3.3	
CIK	0.0	8.1		0.0	9.7		2.2	1.1	
KAP	35.8	3.8		16.3	1.3	*/	19.1	6.7	
Coarse-textured									
PWS	9.8	20.1		23.5	18.6		17.2	9.4	*/
CIK	0.6	11.1	/*	6.7	13.9		6.3	18.3	
KAP	4.2	1.7		1.7	5.0		7.5	2.5	
Estuary									
PWS	4.0	58.0		5.3	0.0		13.2	5.8	
CIK	6.1	0.6		0.0	0.6		1.9	1.3	
Exposed rocky									
PWS	11.7	6.7		3.3	1.7		0.6	1.7	
MVD 3									
Sheltered rocky									
PWS	0.0	0.0		0.0	0.8		0.4	0.0	
CIK	0.0	0.0		0.0	0.6		0.0	0.0	
KAP	0.8	0.0		0.0	0.0		0.0	0.0	
Coarse-textured									
PWS	1.4	8.8		0.0	9.8	*/*	0.8	6.8	
CIK	0.0	7.2		0.6	25.1	*/*	0.7	12.2	
KAP	17.5	13.3		2.5	20.8	*/	0.8	6.7	
Estuary									
PWS	0.0	50.0		0.0	13.0		0.0	0.0	
CIK	0.0	2.7		0.0	3.0		0.0	0.8	
Exposed rocky									
PWS	0.8	0.6		0.0	2.7		0.0	0.0	

where more than 90% of the total invertebrate community consisted of oligochaetes (Table 10; Figure 9). Exceptions were PWS exposed rocky and most sheltered rocky habitats.

Crustacea showing trends in specific regions or habitats were pagurids, amphipods, and isopods (data not shown). Pagurids were significantly more abundant in most PWS habitats on reference sites and in CIK habitats on oiled–treated sites. Amphipods were more abundant in PWS exposed rocky and KAP sheltered rocky habitats on reference sites and in CIK coarse-textured and estuarine habitats and KAP coarse-textured habitat on oiled–treated sites. Isopod abundance was higher at oiled–treated sites only in PWS sheltered rocky and CIK coarse-textured and estuarine habitats. All other significant differences for isopods indicated higher values on reference sites.

Other common taxa examined for abundance and biomass differences showed no consistent trends, primarily because of their low abundances. This group included two dog whelks—frilled dogwinkle *Nucella lamellosa* and file dogwinkle *N. lima*—the gastropod *Lacuna* spp., a bivalve—minute turton *Turtonia minuta*—and both errant and sedentary polychaetes.

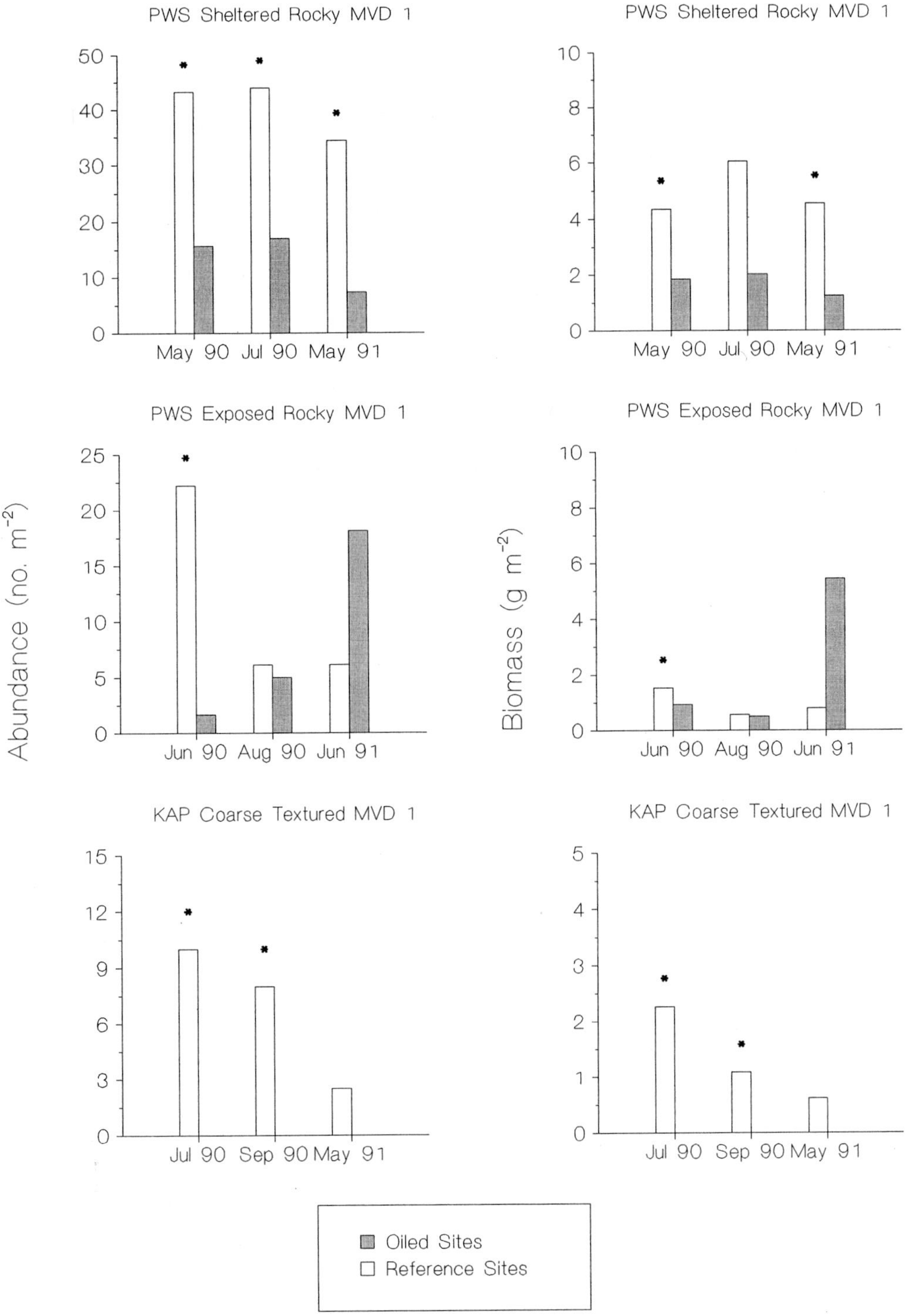

FIGURE 6.—Abundance and biomass trends for *Tectura persona* in selected habitat types, 1990–1991. Location abbreviations are as in Figure 3. Asterisks indicate that differences between the oiled and reference (non-oiled) sites were significant at $P < 0.05$ by either Fisher's or Stouffer's test.

TABLE 4.—Mean densities (number/m^2) of *Chthamalus dalli*. Abbreviations, statistics, and sampling dates are as in Table 1.

Location	Spring 1990			Summer 1990			Spring 1991		
	Reference	Oil	F/S	Reference	Oil	F/S	Reference	Oil	F/S
MVD 1									
Sheltered rocky									
PWS	294.4	330.2		341.7	668.7		998.0	2,977.0	*/
CIK	1,466.7	651.9		1,105.6	420.6	/*	1,757.8	929.9	
KAP	91.2	193.7		159.2	457.9	*/	60.6	181.7	
Coarse-textured									
PWS	95.1	3.1		10.4	66.3		17.5	276.3	
CIK	8.9	131.1		37.2	246.7		0.0	16.7	
KAP	8.3	0.0		1.0	0.0		80.8	0.0	
Estuary									
PWS	3.0	506.7		9.2	12.5		0.0	0.0	
CIK	0.0	153.3		0.0	206.7		0.0	235.6	
Exposed rocky									
PWS	17.2	1,241.1	*/*	86.7	891.4	*/*	1,245.0	4,019.5	
MVD 2									
Sheltered rocky									
PWS	237.5	627.0	*/*	554.7	347.3		3,731.5	5,698.1	*/*
CIK	2,302.4	5,979.2		5,852.2	4,131.0		4,854.2	5,932.2	*/*
KAP	34.2	285.8		204.2	461.1	*/	204.8	637.2	*/
Coarse-textured									
PWS	3.8	35.8		13.6	2.4		1.7	21.9	
CIK	0.0	0.0		2.8	27.8		0.0	1.7	
KAP	1.7	2.5		0.8	0.0		4.2	0.0	
Estuary									
PWS	0.0	0.0		0.0	0.0		0.0	0.8	
CIK	0.0	20.0		0.0	20.0		0.0	127.2	
Exposed rocky									
PWS	307.8	9,783.6		59.1	13,572.5	*/*	3,611.1	11,186.8	*/*
MVD 3									
Sheltered rocky									
PWS	177.5	52.3		29.9	379.2		50.1	1,321.0	*/
CIK	2,123.9	1,156.0		209.4	2,654.9	*/*	1,362.8	1,883.9	
KAP	69.2	70.3		32.1	52.0		15.8	21.3	
Coarse-textured									
PWS	16.3	8.2		15.1	11.3		49.3	12.3	
CIK	9.4	12.2		380.0	119.0		2.0	29.4	
KAP	4.2	54.2		36.7	2.5		25.8	100.8	
Estuary									
PWS	0.0	0.0		0.0	0.0		0.0	0.0	
CIK	0.0	2.2		0.0	128.0		0.0	78.3	
Exposed rocky									
PWS	673.4	11,621.1		146.4	186.4		2,191.3	4,724.3	

In summary, *Fucus*, limpets, mussels, and two of the three species of barnacles examined (*Balanus glandula* and *Semibalanus balanoides*) were negatively affected by the *Exxon Valdez* oil spill and subsequent cleanup activities. The other barnacle, *Chthamalus dalli*, and oligochaetes were positively affected in that abundance and biomass increased after the spill and cleanup. In most cases of significance, the Sitka periwinkle showed a negative oiling response in all habitats in PWS, in sheltered rocky habitat in CIK, and in coarse-textured habitat in KAP, whereas the checkered periwinkle had higher abundance and biomass on oiled–treated sites in sheltered rocky habitat in all regions, in coarse-textured habitat in CIK, and in the upper intertidal zone of exposed rocky habitat. Many of the observed differences diminished with time.

Discussion

Comprehensive examination of the community response to a perturbation is often the preferred method of assessing effects. It is also useful, however, to break the overall response into individual components. In this study a few dominant organisms were primarily responsible for the changes in

TABLE 5.—Mean densities (number/m^2) of *Balanus glandula*. Abbreviations, statistics, and sampling dates are as in Table 1. The black diamond symbol indicates that the direction of the difference between habitat means is different from the direction of overall tests.

	Spring 1990			Summer 1990			Spring 1991		
Location	Reference	Oil	F/S	Reference	Oil	F/S	Reference	Oil	F/S
				MVD 1					
Sheltered rocky									
PWS	108.5	90.5		165.0	81.7	*/	289.0	302.0	*/*◆
CIK	2,598.3	1,086.4		2,887.2	965.6	*/*	1,945.6	2,718.4	
KAP	866.7	841.9	/*	291.3	672.5		339.8	424.8	
Coarse-textured									
PWS	74.5	13.4		22.3	92.2		16.3	208.3	
CIK	195.6	87.8		16.7	262.8		23.3	276.7	
KAP	240.8	0.0	*/*	855.0	0.8	*/*	508.3	1.7	
Estuary									
PWS	49.5	65.0		16.7	25.0		85.8	0.0	*/*
CIK	5.6	228.5	*/*	2.8	153.1	*/*	14.4	198.2	*/*
Exposed rocky									
PWS	161.7	520.3		71.7	461.9	*/*	676.1	1,494.7	*/*
				MVD 2					
Sheltered rocky									
PWS	121.1	44.0	*/	120.7	61.3		101.0	122.0	
CIK	236.9	2,755.0		719.4	3,609.1		1,708.7	2,888.3	
KAP	133.8	316.7	*/*	211.7	313.4		230.8	422.5	
Coarse-textured									
PWS	86.2	66.5		119.2	31.9	*/	170.8	63.5	*/
CIK	20.0	31.7	/*	20.6	105.0	*/	144.4	74.1	*/*◆
KAP	382.5	22.5	*/*	466.1	2.5	*/*	583.3	0.8	/*
Estuary									
PWS	146.3	4.0		12.2	3.3		17.2	1.8	/*
CIK	37.2	159.9		60.8	213.2	*/*	34.2	253.2	*/*
Exposed rocky									
PWS	576.1	1,325.0	*/◆	392.9	305.3	*/*	682.8	959.6	
				MVD 3					
Sheltered rocky									
PWS	47.4	31.3		55.7	15.5		76.7	53.8	
CIK	18.3	742.9		30.6	790.0	/*	128.9	459.2	
KAP	76.3	19.1	*/*	35.4	31.5	*/*	72.1	14.5	*/*
Coarse-textured									
PWS	49.3	52.1		93.3	185.7		117.7	127.2	/*
CIK	25.0	201.4	/*	40.4	485.4	*/*	11.2	119.4	*/*
KAP	121.9	198.3	*/*◆	912.5	119.2	*/*	265.8	512.5	
Estuary									
PWS	6.7	6.7		16.4	14.0		30.0	16.0	
CIK	51.6	72.2		168.9	138.8		183.3	300.2	
Exposed rocky									
PWS	412.0	421.7		176.4	402.0		265.7	217.8	

community composition. Several organisms responded negatively to the oiling and/or cleanup because of sensitivity to the disturbance itself, loss of protective cover, or depletion of food. A few responded positively because of superior colonizing abilities, lowered predation rates, or increased availability of food or space.

The limpet *Tectura persona* and the mussel *Mytilus trossulus* were negatively affected by oiling and treatment. *Tectura persona*, a grazer in the upper intertidal zone, depends on a thin film of algae and diatoms for survival (Castenholz 1961; Nicotri 1977). The oiling and subsequent cleanup were especially damaging to upper intertidal areas, and resident limpets as well as their food were eliminated. Prior investigations into the sublethal effects of hydrocarbons on limpets revealed that limpets were unable to maintain pedal attachment to the substratum upon exposure to oil (Ehrsam et al. 1972; Dicks 1973). In the present study, the absence of *T. persona* on oiled beaches was well documented by abundance and biomass comparisons. A parallel study conducted in Herring Bay, Prince William Sound, resulted in similar findings for *T. persona*

TABLE 6.—Mean densities (number/m^2) of *Semibalanus balanoides*. Abbreviations, statistics, and sampling dates are as in Table 1. The black diamond symbol indicates that the direction of the difference between habitat means is different from the direction of overall tests.

Location	Spring 1990			Summer 1990			Spring 1991		
	Reference	Oil	F/S	Reference	Oil	F/S	Reference	Oil	F/S
MVD 1									
Sheltered rocky									
PWS	2,402.8	2,085.2		1,782.2	2,216.0		3,159.3	2,509.7	/*
CIK	5,682.2	2,333.2		1,628.9	1,845.6		856.7	1,297.9	
KAP	4,825.4	2,990.3	*/*	3,879.6	2,692.9		4,146.1	4,442.8	*/◆
Coarse-textured									
PWS	1,077.7	194.6	*/*	510.2	147.7		458.2	551.9	
CIK	30.0	223.9		13.3	238.3		48.0	194.4	
KAP	207.7	46.7		58.0	0.0	*/*	36.9	0.0	/*
Estuary									
PWS	2,960.2	1,245.0	*/*	546.3	124.2		490.0	1.7	
CIK	345.2	1,015.6		410.1	822.1		1,168.3	789.4	
Exposed rocky									
PWS	266.7	44.4		51.7	9.4		817.2	326.3	
MVD 2									
Sheltered rocky									
PWS	2,195.0	2,012.5		2,296.2	2,289.7		3,042.8	2,266.4	
CIK	2,188.6	1,212.2		3,690.0	3,642.0		3,530.1	4,080.0	
KAP	4,087.9	6,561.3	*/◆	11,008.8	7,960.9	*/	5,492.4	8,194.5	*/
Coarse-textured									
PWS	1,064.8	898.1		1,424.9	385.3	*/*	1,375.7	750.2	*/*
CIK	402.2	55.0	*/	411.7	541.7		377.0	228.5	*/◆
KAP	5.8	176.7		154.2	92.5		11.7	7.5	
Estuary									
PWS	1,605.0	34.0	*/*	1,414.7	249.2	*/*	746.9	134.9	*/
CIK	656.1	790.5		2,406.6	945.5	*/	1,514.7	1,295.1	
Exposed rocky									
PWS	1,484.3	195.3		735.5	29.7	*/*	697.8	946.4	
MVD 3									
Sheltered rocky									
PWS	770.3	1,546.0		1,448.7	1,130.2		1,050.1	602.1	
CIK	2,070.0	428.0		966.7	351.1	*/◆	293.3	243.3	
KAP	7,460.8	5,712.5		10,705.6	11,536.8		2,411.9	3,502.3	
Coarse-textured									
PWS	725.2	728.9	*/◆	1,740.2	2,954.6	*/◆	1,118.4	1,179.0	*/◆
CIK	709.4	406.3		1,448.3	1,065.2	*/◆	416.5	1,037.8	*/*
KAP	989.2	1,010.3		1,275.8	1,403.3		401.7	653.3	
Estuary									
PWS	6,455.0	1,260.0		1,539.7	408.5		1,338.4	550.0	
CIK	399.6	729.6		1,933.9	1,112.0		1,116.7	856.4	
Exposed rocky									
PWS	147.0	697.2		86.7	13.3	*/	683.1	37.2	

(Hooten and Highsmith 1996). Houghton et al. (1991), in a shoreline study of the sound after the *Exxon Valdez* oil spill, reported lowered abundances of limpets on oiled–treated sites in the upper intertidal zone.

The reduction in mussels on oiled sites was probably caused by a combination of events. Upon initial exposure, mussels covered with oil would no longer be able to feed after their gill filaments were fouled; if conditions did not improve, they could eventually die of starvation. In a laboratory study, shell-closure ability of blue mussels *M. edulis* was unaffected by crude oil; however, byssal thread activity was impaired (Swedmark et al. 1973), which may result in detachment from the substrate and subsequent death. Mussels generally congregate in bands or beds beneath which stranded oil can readily accumulate. Thus, oiled beds may provide a mechanism for chronic oiling. After the *Exxon Valdez* oil spill, many mussel beds in PWS and CIK remained contaminated with hydrocarbons (Babcock et al. 1996, this volume). Oil retention in substrate, therefore, may be contributing to the delayed recovery of mussel populations in some habitats.

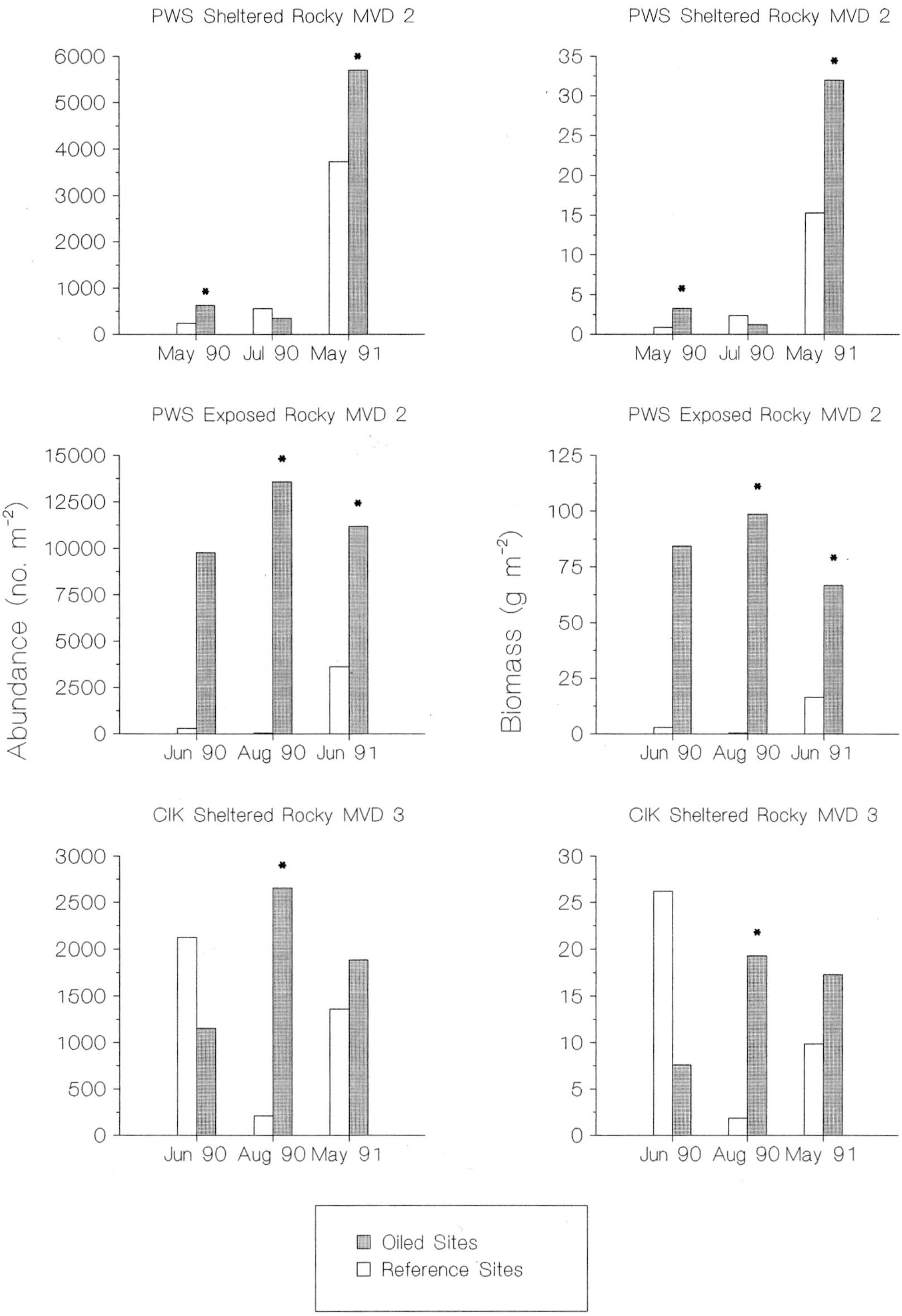

FIGURE 7.—Abundance and biomass trends for *Chthamalus dalli* in selected habitat types, 1990–1991. Location abbreviations are as in Figure 3. Statistics are as in Figure 6.

TABLE 7.—Mean densities (number/m^2) of *Mytilus trossulus*. Abbreviations, statistics, and sampling dates are as in Table 1. The black diamond symbol indicates that the direction of the difference between habitat means is different from the direction of overall tests.

Location	Spring 1990			Summer 1990			Spring 1991		
	Reference	Oil	F/S	Reference	Oil	F/S	Reference	Oil	F/S
MVD 1									
Sheltered rocky									
PWS	404.0	1,139.2		583.9	535.7		193.0	1,290.7	
CIK	6,635.6	489.3	/*	6,901.7	785.6	*/*	6,535.0	416.1	
KAP	303.7	114.1		67.1	183.5		14.3	95.8	
Coarse-textured									
PWS	408.6	42.2	*/*	444.2	307.6	*/*	343.0	484.0	/*◆
CIK	10.0	39.4		4.4	35.6	*/*	11.4	42.5	/*
KAP	4.2	3.3		130.0	0.0	*/*	29.2	0.0	
Estuary									
PWS	54.0	0.0	*/*	174.1	102.7		172.7	40.8	
CIK	45.6	80.0		414.4	129.9		245.6	293.6	
Exposed rocky									
PWS	208.3	254.2		70.6	1,081.4		1,177.8	7,846.4	
MVD 2									
Sheltered rocky									
PWS	4,794.9	2,725.7		4,290.1	5,125.7		4,138.9	2,125.9	
CIK	6,323.1	4,314.4		11,166.1	4,471.8		17,438.3	4,553.3	
KAP	511.3	664.4		551.3	787.0		490.7	1,080.1	
Coarse-textured									
PWS	839.0	484.1	*/*	1,119.6	1,283.5	*/*◆	794.6	929.6	*/*◆
CIK	135.0	158.9		206.7	197.2		188.2	260.0	
KAP	418.3	18.6	*/*	310.3	6.3	*/*	760.8	43.1	*/*
Estuary									
PWS	531.3	164.0	*/*	868.3	225.7	*/*	481.5	418.9	
CIK	327.5	312.0		735.6	333.0	*/*	494.3	287.6	*/
Exposed rocky									
PWS	2,748.2	4,735.8		3,454.9	1,722.8		3,357.2	4,123.1	*/*◆
MVD 3									
Sheltered rocky									
PWS	9,818.9	2,959.0	*/*	6,013.7	6,373.0		2,791.2	2,009.7	
CIK	9,475.6	2,075.1	*/*	4,211.1	1,711.6		12,343.9	2,855.8	*/*
KAP	1,284.3	2,203.5		1,293.1	2,295.2		794.8	1,145.2	
Coarse-textured									
PWS	2,241.0	1,010.6	*/*	1,591.8	2,287.6	*/◆	1,366.2	1,099.3	*/*
CIK	1,062.2	381.8		1,894.1	509.4		585.9	418.5	
KAP	771.7	111.8		1,513.3	148.7	*/*	362.5	191.6	/*
Estuary									
PWS	3,311.7	306.7		1,106.7	631.8		965.0	1,028.3	
CIK	478.2	42.3	*/*	709.4	372.5		1,557.8	285.3	*/*
Exposed rocky									
PWS	1,947.2	1,714.7		1,650.0	922.0		5,237.6	1,495.9	

The different response to the spill observed in some areas for the two species of littorines may be explained by their different reproductive strategies. Littorines initially died in oiled environments as a consequence of contamination and cleanup efforts, either directly or from loss of food or shelter. Both species inhabit the mid to upper intertidal zone, where effects of the spill and cleanup were most prevalent. The Sitka periwinkle, which is a direct developer with crawl-away young, may be slow to recolonize oiled areas, because the resident population was often eliminated. After the *Exxon Valdez* oil spill, the Sitka periwinkle was less abundant at oiled habitats in Herring Bay, Prince William Sound (Hooten and Highsmith 1996), and on various oiled habitats throughout the sound (Houghton et al. 1991). Conversely, the checkered periwinkle, with planktonic larvae, is capable of recolonizing a depopulated beach more rapidly than the direct-developing Sitka periwinkle (Gray 1979). The checkered periwinkle was found in higher densities on oiled than on reference shorelines in several regions and habitats in our study. It was the first littorine to reoccupy oiled shorelines in Cornwall following the *Torrey Canyon* spill (Southward and Southward 1978).

The barnacle *C. dalli* generally was more abun-

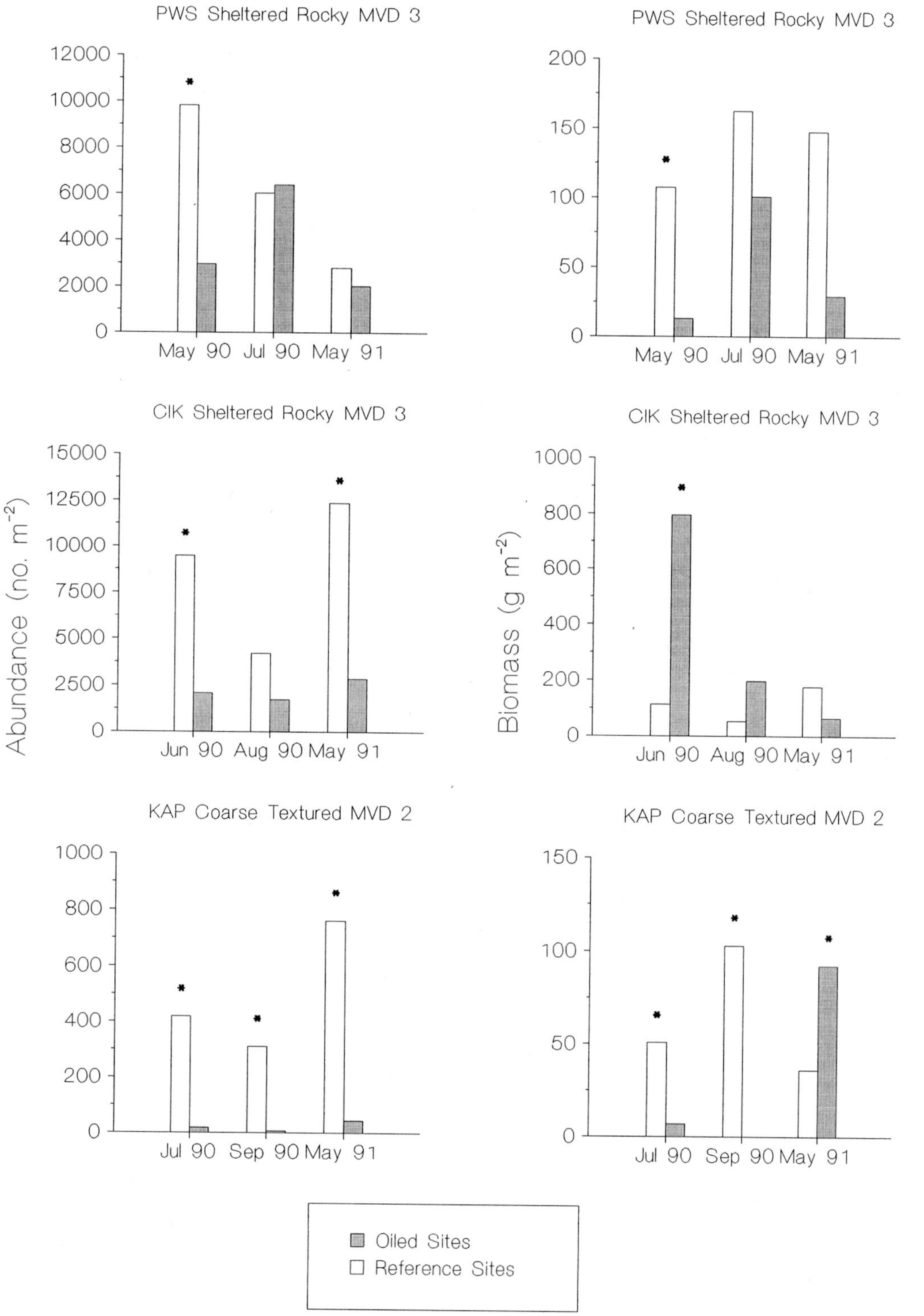

FIGURE 8.—Abundance and biomass trends for *Mytilus trossulus* in selected habitat types, 1990–1991. Location abbreviations are as in Figure 3. Statistics are as in Figure 6.

TABLE 8.—Mean densities (number/m^2) of *Littorina sitkana*. Abbreviations, statistics, and sampling dates are as in Table 1. The black diamond symbol indicates that the direction of the difference between habitat means is different from the direction of overall tests.

Location	Spring 1990			Summer 1990			Spring 1991		
	Reference	Oil	F/S	Reference	Oil	F/S	Reference	Oil	F/S
MVD 1									
Sheltered rocky									
PWS	223.9	195.2		257.9	215.3		213.3	307.3	
CIK	1,116.1	713.0	/*	1,857.2	413.9	*/*	1,024.4	632.7	
KAP	1,532.5	1,984.7		1,212.1	1,191.1		1,589.8	1,400.8	
Coarse-textured									
PWS	288.4	131.9	*/*	359.7	117.4	*/*	205.4	415.9	
CIK	128.3	378.3		86.7	421.1	*/◆	55.8	116.4	
KAP	328.3	15.8	*/*	1,485.0	4.2	*/*	309.7	4.4	*/*
Estuary									
PWS	98.2	8.3	*/*	77.7	15.8	*/*	398.0	28.3	*/*
CIK	101.7	151.2		244.2	353.1		137.2	398.2	*/◆
Exposed rocky									
PWS	995.0	373.9		745.0	245.0	/*	1,567.2	681.5	
MVD 2									
Sheltered rocky									
PWS	857.6	322.8		382.2	422.3		532.6	798.3	
CIK	921.9	507.2	*/	946.7	899.6		2,295.7	1,202.2	
KAP	1,975.0	2,402.4		2,478.8	2,348.0		3,070.4	2,717.9	
Coarse-textured									
PWS	470.5	147.6	*/*	431.5	453.0	*/*◆	587.1	1,111.2	
CIK	83.3	258.9	*/◆	208.9	452.6		416.3	366.4	
KAP	460.0	125.0	*/*	263.2	23.3	*/*	412.5	56.1	*/*
Estuary									
PWS	154.7	67.3	*/*	296.6	97.6		304.3	463.2	
CIK	333.5	285.0		355.7	374.2		581.2	637.6	*/◆
Exposed rocky									
PWS	1,033.0	588.1		3,083.2	165.8	*/*	2,462.8	404.2	*/*
MVD 3									
Sheltered rocky									
PWS	197.8	42.0		373.6	278.1		544.2	220.5	*/*
CIK	237.2	333.4		134.4	93.7		466.1	558.6	
KAP	1,724.0	1,837.4		1,877.9	2,387.1		985.2	2,588.0	
Coarse-textured									
PWS	218.4	242.2		174.4	676.6		483.0	455.2	
CIK	245.6	216.9		120.6	193.3	*/	214.4	418.3	
KAP	185.0	400.5		440.0	282.9		371.4	346.6	
Estuary									
PWS	143.3	198.9		240.0	231.7		141.7	136.3	
CIK	162.6	332.9		137.2	494.0		223.9	1,051.7	
Exposed rocky									
PWS	814.2	63.9		700.6	74.4	*/*	484.2	54.4	

dant and had higher biomass on oiled than on reference sites in most regions and habitats in which it occurred. In addition, at lower beach elevations, this typically splash-zone inhabitant was present in higher densities on oiled–treated than on reference sites in approximately 75% of observations. *Chthamalus dalli*, which is capable of producing several broods annually (Southward and Southward 1967; Barnes 1989), successfully settled in high numbers during late summer and fall of 1989. The space created by the spill and cleanup was quickly colonized by *C. dalli*, which probably encountered little spatial competition, few predators, and no whiplash effect because of the absence of large *Fucus* plants. Whiplash effect of *Fucus* on young barnacles can be fatal to the barnacles (Farrell 1989). In the absence of competitors and large *Fucus* plants, the small, fragile *C. dalli* may settle into areas in which it generally cannot persist. This barnacle also was an initial colonizer in a community succession study in Oregon (Farrell 1991). Another prominent barnacle in the intertidal zone, *Semibalanus balanoides*, produces only one brood each year in the spring (Crisp 1954; Barnes and Barnes 1968; Crisp 1968; Barnes 1989). When *S. balanoides* settled in spring 1990, *C. dalli* was already established, persisting as

TABLE 9.—Mean densities (number/m^2) of *Littorina scutulata*. Abbreviations, statistics, and sampling dates are as in Table 1. The black diamond symbol indicates that the direction of the difference between habitat means is different from the direction of overall tests.

	Spring 1990			Summer 1990			Spring 1991		
Location	Reference	Oil	F/S	Reference	Oil	F/S	Reference	Oil	F/S
MVD 1									
Sheltered rocky									
PWS	142.3	217.8		236.8	250.0		120.3	304.7	
CIK	115.0	380.9		110.0	235.6		477.2	424.7	/*◆
KAP	0.0	13.5	*/	0.8	11.3		0.0	1.3	
Coarse-textured									
PWS	260.8	22.7	*/*	333.3	79.3	*/*	162.6	106.4	*/*
CIK	0.0	0.6		0.0	1.7		0.0	0.6	
KAP	5.0	0.0	*/*	59.0	0.8	*/*	38.3	0.0	*/*
Estuary									
PWS	51.0	8.3	*/*	107.2	5.0		30.3	0.0	
CIK	0.6	2.8		9.8	3.7		8.9	5.7	
Exposed rocky									
PWS	21.1	153.9	*/*	91.1	652.8	*/*	275.6	506.5	
MVD 2									
Sheltered rocky									
PWS	548.3	447.7		264.2	432.0		314.1	743.6	*/
CIK	170.6	413.1	*/*	288.9	719.1	*/*	967.4	940.6	*/*◆
KAP	1.3	8.3		2.5	22.5	*/	29.3	48.8	
Coarse-textured									
PWS	550.2	172.1	*/*	805.0	236.9	*/*	845.5	898.7	*/*◆
CIK	3.9	3.3		1.1	10.6	*/*	5.3	5.0	
KAP	240.8	0.0	*/*	500.2	0.0	*/*	124.2	0.0	*/*
Estuary									
PWS	203.3	0.0	*/*	822.5	11.2	*/*	326.5	9.2	*/*
CIK	76.1	24.2		43.2	40.6	*/◆	91.5	25.4	
Exposed rocky									
PWS	426.1	2,728.1		1,064.9	360.0	*/*	1,542.2	1,075.2	/*
MVD 3									
Sheltered rocky									
PWS	185.1	108.0		94.5	95.1		472.2	291.9	
CIK	62.8	121.9		30.6	76.8		185.3	423.6	
KAP	7.2	11.4		5.8	7.5		130.8	8.4	
Coarse-textured									
PWS	1,117.3	404.6	*/*	1,549.7	711.6	*/	1,547.2	1,451.1	*/
CIK	4.4	33.4	*/*	11.8	39.7		2.0	23.9	*/*
KAP	103.3	0.0	*/*	221.7	2.5	*/*	172.2	0.0	/*
Estuary									
PWS	866.7	3.3	*/*	1,130.7	53.2		1,346.7	84.7	
CIK	30.0	17.6		64.4	31.8	*/◆	155.0	131.5	
Exposed rocky									
PWS	349.3	783.9		190.6	122.4		457.7	132.8	

the numerical dominant in many sheltered and exposed rocky areas into 1991. Eventually, the large barnacles *S. balanoides* and *Balanus glandula* should outcompete *C. dalli*, at which point the latter will again be limited primarily to the upper intertidal zone (Connell 1961). Farrell (1991) observed the eventual replacement of *C. dalli* by *B. glandula* in lower intertidal habitats.

Oligochaetes had higher abundances and biomasses on oiled sites in several habitats and regions. In coarse-textured habitats in CIK, for example, up to 95% of the total invertebrate community consisted of oligochaetes. High numbers may have resulted from high concentrations of hydrocarbon-degrading bacteria found in similar habitats (Braddock et al. 1996, this volume), which would provide a readily available source of food for the oligochaetes. Oligochaete abundance increased in a variety of habitats in prior pollution studies (Gray 1979; Davis and Spies 1980; Harrel 1985), and oligochaetes have been recognized as one of the earliest colonizers of environmentally disturbed areas (Gray 1979). Houghton et al. (1991) also reported a higher abundance of oligochaetes on oiled beaches in Prince William Sound after the *Exxon Valdez* oil spill.

Impact from the spill and cleanup varied by re-

TABLE 10.—Mean densities (number/m^2) of oligochaetes. Abbreviations, statistics, and sampling dates are as in Table 1. The black diamond symbol indicates that the direction of the difference between habitat means is different from the direction of overall tests.

	Spring 1990			Summer 1990			Spring 1991		
Location	Reference	Oil	F/S	Reference	Oil	F/S	Reference	Oil	F/S
MVD 1									
Sheltered rocky									
PWS	146.1	211.5		681.8	162.3	/*	179.0	929.3	
CIK	256.7	256.9		369.7	351.1		388.9	17.6	
KAP	362.1	627.9		1,413.8	522.1		510.4	316.7	/*◆
Coarse-textured									
PWS	1,180.8	1,156.8		1,577.1	788.3	*/*	1,069.4	1,955.5	
CIK	1,116.1	3,305.0		2,303.9	1,836.7		1,162.7	526.4	
KAP	900.5	4,964.8	*/*	132.0	16,190.9		132.2	6,882.7	*/
Estuary									
PWS	1,378.2	1,101.1		5,436.9	5,888.3		4,527.2	2,630.3	
CIK	4,813.0	4,106.5		1,431.1	2,825.5		4,289.4	4,510.9	
Exposed rocky									
PWS	1,162.8	8.3	*/*	1,113.3	49.4	/*	983.9	133.3	
MVD 2									
Sheltered rocky									
PWS	1,736.4	1,295.0	*/◆	1,671.6	2,972.7		151.7	708.5	
CIK	1,265.4	983.1		2,481.1	155.0	*/*	2,018.0	631.7	
KAP	4,637.5	1,058.5		1,202.9	1,816.9		789.0	456.9	
Coarse-textured									
PWS	1,962.8	1,984.4		1,765.8	1,527.1		2,142.1	3,159.5	*/
CIK	613.3	14,761.7	*/	364.4	4,885.0	*/*	1,679.4	10,831.9	
KAP	295.6	3,187.7	*/	162.4	1,427.3	*/	640.8	5,544.3	*/*
Estuary									
PWS	705.2	1,333.3	*/*	1,667.7	7,008.0	*/*	706.2	6,196.4	*/*
CIK	3,210.7	6,352.0	*/*	3,002.7	4,948.8		5,015.6	8,536.7	
Exposed rocky									
PWS	3,340.2	495.3	*/*	5,964.9	425.0	*/*	2,675.6	537.5	*/*
MVD 3									
Sheltered rocky									
PWS	2,665.0	841.0		1,076.0	1,665.7	*/	595.8	201.6	
CIK	1,486.7	85.4	/*	1,503.3	22.9	*/*	1,583.6	540.8	/*
KAP	348.7	2,260.6	*/*	5,961.9	3,662.2	/*◆	1,541.7	1,298.4	
Coarse-textured									
PWS	1,539.1	1,399.3		1,925.7	3,273.1	*/*	930.2	1,961.9	*/
CIK	706.7	16,342.3	*/*	879.8	4,777.4	*/*	726.5	3,258.0	*/*
KAP	939.4	379.4		230.8	395.9		895.0	1,879.4	
Estuary									
PWS	11.7	1,593.3	*/*	529.1	4,451.2	*/*	0.0	2,776.3	*/*
CIK	2,682.0	10,002.2		763.9	2,925.8	*/*	2,130.0	4,910.2	
Exposed rocky									
PWS	2,087.4	465.6	*/*	801.7	816.9		2,070.2	989.1	

gion. The CIK area generally had more taxa with a higher abundance or biomass on oiled–treated sites. In contrast, the PWS and KAP areas had the opposite trend, with generally higher abundance and biomass values at reference sites. For a more detailed discussion of regional and habitat differences, see Stekoll et al. (1996).

Comparisons of results for selected taxa in our study and in a similar *Exxon Valdez* oil spill study in Prince William Sound (Gilfillan et al. 1995) identified several differences that should be addressed. For taxa examined in both studies—limpets, littorines, and mussels—our study revealed significant differences in populations between oiled and reference sites in many cases. Differences generally reflected higher abundance and biomass for reference populations. Gilfillan et al. (1995), however, report no significant differences between oiled and reference sites for littorines or mussels at exposed or sheltered bedrock habitat in 1990. A possible explanation for these different results, at least for littorines, is that littorines were identified to species in our study and results for Sitka and checkered periwinkles were reported separately. Gilfillan et al.

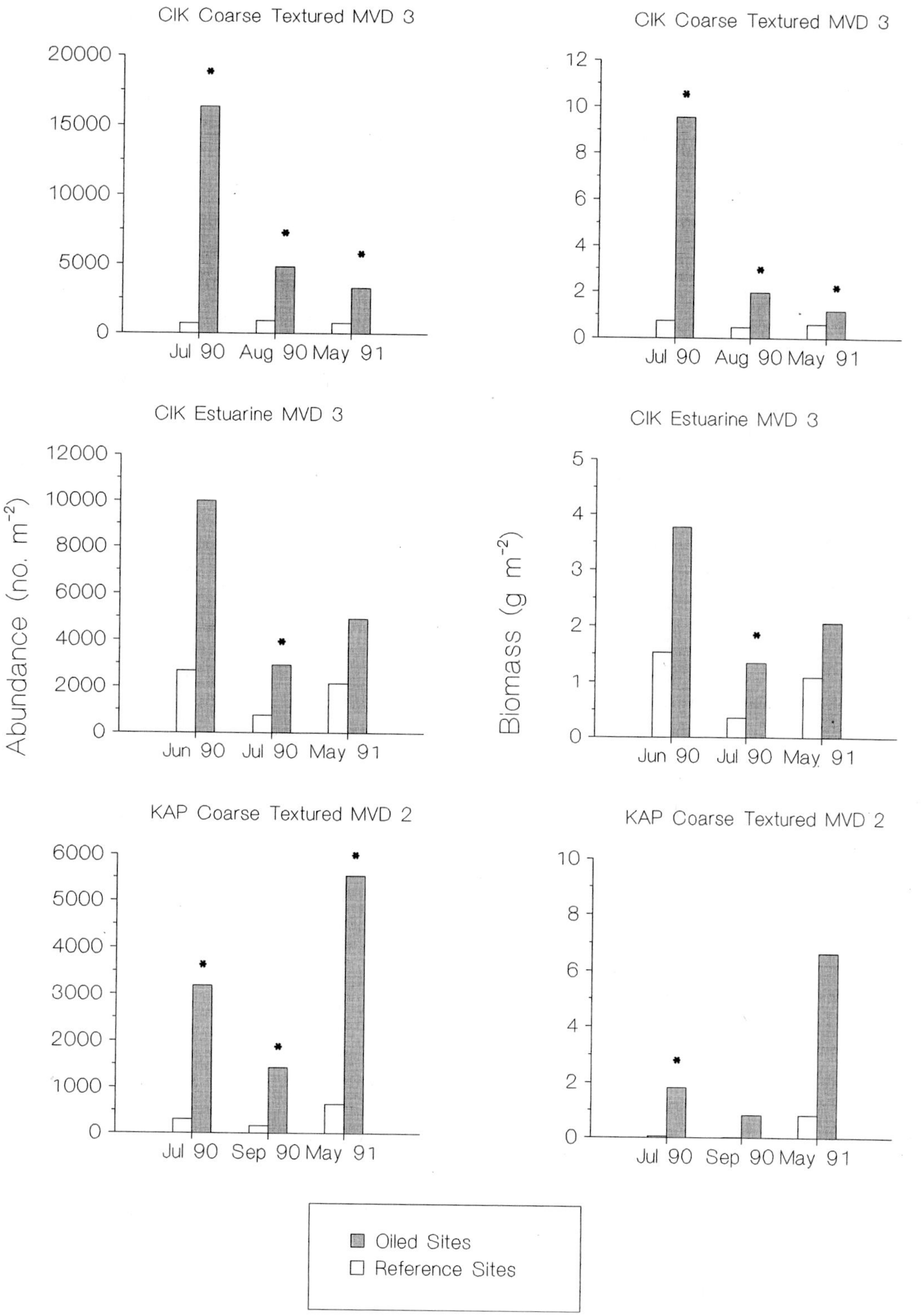

FIGURE 9.—Abundance and biomass trends for oligochaetes in selected habitat types, 1990–1991. Location abbreviations are as in Figure 3. Statistics are as in Figure 6.

(1995) reported abundance trends for *Littorina* spp., a compilation of data for the two species. Our study showed an opposite effect for the two species in some habitats (Sitka periwinkle had higher values on reference sites and checkered periwinkle had higher values on oiled–treated sites). By combining data for these two species with different dispersal capabilities, Gilfillan et al. (1995) may have masked the impact. We also pooled data for the two species of littorines (*Littorina* spp.): Significant differences between oiled and reference populations, although observed for the two species individually, were not always evident when the data were pooled. In fact, in 44% of all cases in which significant differences were observed for one or both species individually, significant differences were not found for the pooled data.

In the Gilfillan et al. (1995) study, limpets were significantly less abundant at oiled sites in one instance each for sheltered and exposed bedrock habitats in 1990 (upper intertidal regions of lightly oiled shorelines). The investigators consolidated limpets into the family Lottiidae, which they defined as the sum of Lottiidae and *Lottia pelta*. Our study showed a few cases of significance for the limpet *L. pelta* and for Lottiidae, but the majority of significant differences were observed for the high-zone limpet, *Tectura persona*, which was found in greater numbers on reference sites in several instances in 1990. Data were not reported for *T. persona* in the Gilfillan et al. (1995) study, which may explain their inability to detect impacts on limpets; however, Houghton et al. (1991) consolidated limpet data and reported significant reductions in mid tidal levels at oiled sites in sheltered rocky habitat in 1990.

Several studies have concluded that the damages incurred by intertidal communities were not totally the result of the *Exxon Valdez* oil spill but included a strong response to beach treatments, primarily high-pressure hot-water washing (Houghton et al. 1996; Lees et al. 1996; Mearns 1996, this volume). Prior investigations of oil-related perturbations also suggested a greater impact on marine communities as a result of treatment (Southward and Southward 1978). Houghton et al. (1996) report that intertidal epibiota survived the initial 3–4-month period after the *Exxon Valdez* oil spill and before treatment. Comparative studies of oiled–treated and oiled–untreated shorelines revealed lower abundance, diversity, and species richness on oiled–treated beaches and a slower rate of recovery than on oiled–untreated beaches. Hydraulic washing and berm relocation, treatments used on gravel and sand beach types, were thought to be largely responsible for infaunal damages observed in these habitats (Houghton et al. 1996). In our study, beach stratification was based primarily on habitat and oiling characteristics rather than on treatment history. Most documentation that exists for treatment histories is not detailed enough within a region to determine if our sites were subjected to specific treatments; thus, damages reported here represent response to a combination of oiling and cleanup events.

The sequence of events following the *Exxon Valdez* oil spill (initial impact, mortality, and subsequent recovery) paralleled that observed in prior oil spill studies. After the *Torrey Canyon* spill, there was an almost complete disappearance of gastropods, especially predatory whelks, littorines, and limpets (Southward and Southward 1978). Barnacle populations were unhealthy or had been destroyed. Damage was heaviest in the upper intertidal zone and recovery was first apparent on high-energy exposed rocky shores. The *Arrow* disaster in Chedabucto Bay, Nova Scotia (Thomas 1973, 1978) and the *Florida* grounding in West Falmouth, Massachusetts (Sanders 1978), resulted in similar losses of organisms and persistence of hydrocarbons in finer sediments. Although the environmental effects of a spill depend upon such factors as the extent of the spill, oil type, local weather conditions, season, sediment type, and latitude, similar patterns of injury have been identified for intertidal shorelines in past spill events.

The main driving force causing differences between oiled and reference sites in this study was the creation of bare substrata as a result of the oiling and cleanup efforts. The rate of recovery will depend on the rate of recolonization of denuded areas with possible complications caused by any residual oiling effects. Grazers such as limpets and littorines are generally most affected by an oiling event. Loss of these mollusks alters algal composition and successional stages, disrupting the entire community. In only a few cases in this study has recolonization followed a typical scenario of the "greening" of the intertidal region by opportunistic algae (Southward and Southward 1978), which persist because of the absence of the grazers. In some instances, barnacles were able to settle and may have competed with the algae for space (but see below).

Recolonization in some cases will be determined by a combination of biotic and abiotic factors. At the upper edge of the *Fucus* zone, successful recolonization will be determined by the availability of fertilized eggs and refugia from desiccation and from grazers (van Tamelen and Stekoll 1996). *Fu-*

cus, the dominant plant in the upper intertidal zone in all areas affected by the *Exxon Valdez* oil spill, was severely damaged in many habitats. Normally the adult *Fucus* plants form a canopy that protects germlings from desiccation (van Tamelen and Stekoll 1996). Barnacles may also play a role in *Fucus* recolonization by providing surface rugosity (Farrell 1991) and/or chemical cues (M. L. Carroll and R. C. Highsmith, unpublished data) for egg attachment. Because surface rugosity alone is insufficient for recolonization (van Tamelen and Stekoll 1996), recovery of the upper intertidal zone by *Fucus* will proceed slowly in areas where the *Fucus* overstory was damaged by the oil spill and cleanup.

Species interactions can affect recolonization patterns. For example, in some cases, the perennial alga *F. gardneri*, has actually taken advantage of the increase in uncolonized substratum in the lower zones, outcompeting annual and ephemeral algae. A specific example of this can be seen in CIK sheltered rocky habitats at MVD 3. The data indicate that the perennials *Fucus* and members of the algal family Gigartinaceae were replacing space normally occupied by the annuals *Acrosiphonia* sp., *Alaria* sp., *Palmaria* sp., *Polysiphonia–Pterosiphonia* spp., and *Porphyra* sp. *Alaria* is of particular interest because of its large biomass and its role as the major canopy species in the lower intertidal zone. Normally the fast-growing annual *Alaria* will outcompete perennials such as *Fucus*, but it is probable that the oil spill and cleanup removed the young *Alaria* sporophytes and cryptic gametophytes, creating bare substratum subsequently colonized by *Fucus*. Eventually, *Alaria* may be able to reestablish itself, but the *Fucus* canopy that is there now may preclude settlement and/or germination and growth of new *Alaria* plants.

Other specific interactions may affect recolonization. For example, in our study, the small barnacle *Chthamalus dalli* was prominent on oiled sites. Apparently, larger barnacle species such as *Balanus glandula* and *Semibalanus balanoides* had not settled in sufficient numbers by 1991 to competitively displace the smaller *C. dalli*. As young barnacles and mussels recruit into newly created space, populations of predatory snails may increase sharply in response to the sudden availability of prey. *Nucella* spp. were observed feeding on recently settled mussels in Prince William Sound in summer 1993 (S. Saupe, unpublished observation) and had cleared entire mussel bands on large stretches of beach. Such an impact may disrupt successional patterns and further delay recovery, as observed by Carroll (1994) for mussels in Kachemak Bay.

A final implication of the potential shoreline damage from the *Exxon Valdez* oil spill and cleanup is the impact on higher-order consumers that utilize food items present in the intertidal zone. Black oystercatchers feed in the intertidal, generally on mussels, limpets, chitons, and crabs (Hartwick 1978; Frank 1982; Marsh 1986a, 1986b). Other birds known to feed in the intertidal zone include harlequin ducks (on clams, chitons, limpets, hermit crabs, and mussels), Barrow's goldeneyes *Bucephala islandica*, and common goldeneyes *B. clangula*, and surf scoters *Melanitta perspicillata* (on large mussels and snails) (Patten, unpublished data). Patten reported reproductive failure in harlequins subsequent to the *Exxon Valdez* oil spill and concluded that failure was probably the result of ingestion of contaminated food. R. T. Bowyer and associates (University of Alaska Fairbanks, unpublished), in a river otter study in Prince William Sound, reported from fecal pellet analyses that the variety of prey taken declined by 33% between 1989 and 1990. River otters generally feed on intertidal and shallow subtidal invertebrates and fish. Other higher-order consumers known to utilize intertidal food resources include brown bears and black bears *Ursus americanus*, Sitka black-tailed deer, sea otters *Enhydra lutris*, turnstones, surfbirds, northwestern crows *Corvus caurinus*, pink salmon *Oncorhynchus gorbuscha*, chum salmon *O. keta*, Dolly Vardens *Salvelinus malma*, and cutthroat trout *O. clarki*. Thus, the oil spill and subsequent treatment affected not only resident plant and invertebrate species in the intertidal zone, but also mobile vertebrate predators that forage intertidally.

Acknowledgments

Funding for this project was provided by U.S. Forest Service (USFS) contract #53-0109-9-00276 to the University of Alaska Fairbanks as part of the Natural Resource Damage Assessment, Comprehensive Environmental Response, Compensation, and Liability Act (NRDA CERCLA) process. Several people provided invaluable assistance during the course of this study, including David Gibbons, USFS project manager; David Doudna, Coastal Habitat Injury Assessment project manager; Chirk Chu and Michelle Bourassa, data management; John Noll and Mark Vallarino, programming–data analyses; Tracy Asselin, Janet Ballek, Catherine Egan, Stephanie Moreland, Peter Shier, and Paul Will, sample processing–data analyses; Steve Sklavounos, data compilation; and Ken Coyle, editorial advice.

References

Babcock, M. M., G. V. Irvine, P. M. Harris, J. A. Cusick, and S. D. Rice. 1996. Persistence of oiling in mussel beds three and four years after the *Exxon Valdez* oil spill. American Fisheries Society Symposium 18:286–297.

Barnes, H., and M. Barnes. 1968. Egg numbers, metabolic efficiency of egg production and fecundity; local and regional variations in a number of common cirripedes. Journal of Experimental Marine Biology and Ecology 2:135–153.

Barnes, M. 1989. Egg production in cirripedes. Oceanography and Marine Biology Annual Review 27:91–166.

Bonsdorff, E. 1981. The *Antonio Gramsci* oil spill. Impact on the littoral and benthic ecosystems. Marine Pollution Bulletin 12:301–305.

Braddock, J. F., J. E. Lindstrom, T. R. Yeager, B. T. Rasley, and E. J. Brown. 1996. Patterns of microbial activity in oiled and unoiled sediments in Prince William Sound. American Fisheries Society Symposium 18:94–108.

Carroll, M. L. 1994. The ecology of an Alaskan rocky intertidal community: an experimental analysis of mechanisms regulating populations in Kachemak Bay, Alaska. Doctoral dissertation. University of Alaska, Fairbanks.

Castenholz, R. W. 1961. The effect of grazing on marine littoral diatom populations. Ecology 42:783–794.

Chasse, C. 1978. The ecological impact on and near shores by the *Amoco Cadiz* oil spill. Marine Pollution Bulletin 9:298–301.

Clark, R. C., Jr., B. G. Patten, and E. E. DeNike. 1978. Observations of a cold-water intertidal community after 5 years of a low-level, persistent oil spill from the *General M.C. Meigs*. Journal of the Fisheries Research Board of Canada 35:754–765.

Connell, J. H. 1961. The influence of interspecific competition and other factors on the distribution of the barnacle *Chthamalus stellatus*. Ecology 42:710–723.

Crisp, D. J. 1954. The breeding of *Balanus porcatus* (DaCosta) in the Irish Sea. Journal of the Marine Biological Association of the United Kingdom 33:473–496.

Crisp, D. J. 1968. Differences between North American and European populations of *Balanus balanoides* revealed by transplantation. Journal of the Fisheries Research Board of Canada 25:2633–2641.

Davis, P. H., and R. B. Spies. 1980. Infaunal benthos of a natural petroleum seep: study of community structure. Marine Biology 59:31–41.

Dicks, B. 1973. Some effects of Kuwait crude oil on the limpet *Patella vulgata*. Environmental Pollution 5:219–229.

Duffy, L. K., R. T. Bowyer, J. W. Testa, and J. B. Faro. 1994. Evidence for recovery of body mass and haptoglobin values of river otters following the *Exxon Valdez* oil spill. Journal of Wildlife Diseases 30:421–425.

Ehrsam, L. C., Jr., and seven coauthors. 1972. Biological assessment of diesel spill in the vicinity of Anacortes, Washington. Final report, Texas Instruments, Inc., Dallas.

Farrell, T. M. 1989. Succession in a rocky intertidal community: the importance of disturbance size and position within a disturbed patch. Journal of Experimental Marine Biology and Ecology 128:57–73.

Farrell, T. M. 1991. Models and mechanisms of succession: an example from a rocky intertidal community. Ecological Monographs 61:95–113.

Folks, J. L. 1984. Combination of independent tests. Pages 113–121 *in* P. R. Krishnaiah and P. K. Sen, editors. Handbook of statistics 4, nonparametric methods. North-Holland, New York.

Frank, P. W. 1982. Effects of winter feeding on limpets by black oystercatchers, *Haematopus bachmani*. Ecology 63:1352–1362.

Geller, J. B., J. T. Carlton, and D. A. Powers. 1994. PCR-based detection of mtDNA haplotypes of native and invading mussels on the northeastern Pacific coast: latitudinal pattern of invasion. Marine Biology 119:243–249.

Gilfillan, E. S., D. S. Page, E. J. Harner, and P. D. Boehm. 1995. Shoreline ecology program for Prince William Sound, Alaska following the *Exxon Valdez* oil spill, part 3: biology. Pages 398–443 *in* P. G. Wells, J. N. Butler, and J. S. Hughes, editors. *Exxon Valdez* oil spill: fate and effects in Alaskan waters. American Society for Testing and Materials Special Technical Publication 1219, Philadelphia.

Gray, J. S. 1979. Pollution-induced changes in populations. Philosophical Transactions of the Royal Society of London Series B Biological Sciences 286:545–561.

Harrel, R. C. 1985. Effects of a crude oil spill on water quality and macrobenthos of a southeast Texas stream. Hydrobiologia 124:223–228.

Hartwick, E. B. 1978. Some observations on foraging by black oystercatchers (*Haematopus bachmani* Audubon). Syesis 11:55–60.

Highsmith, R. C., M. S. Stekoll, W. E. Barber, L. McDonald, D. Strickland, and W. P. Erickson. 1994. Comprehensive assessment of coastal habitat. *Exxon Valdez* oil spill state/federal natural resource damage assessment. Final report of Coastal Habitat Study 1A. School of Fisheries and Ocean Sciences, University of Alaska, Fairbanks to Exxon Valdez Oil Spill Trustee Council, Oil Spill Information Office, Anchorage.

Hooten, A. J., and R. C. Highsmith. 1996. Impacts on selected intertidal invertebrates in Herring Bay, Prince William Sound, after the *Exxon Valdez* oil spill. American Fisheries Society Symposium 18:249–270.

Houghton, J. P., D. C. Lees, and T. A. Ebert. 1991. Evaluation of the condition of intertidal and shallow subtidal biota in Prince William Sound following the *Exxon Valdez* oil spill and subsequent shoreline treatment. NOAA (National Oceanic and Atmospheric Administration) Hazardous Materials Response Branch (HMRB 91-1), Seattle.

Houghton, J. P., D. C. Lees, W. B. Driskell, S. C. Lindstrom, and A. J. Mearns. 1996. Recovery of Prince William Sound intertidal epibiota from *Exxon Valdez* oiling and shoreline treatments, 1989 through 1992. American Fisheries Society Symposium 18:379–411.

Keizer, P. D., T. P. Ahern, J. Dale, and J. H. Vandermeulen. 1978. Residues of bunker C oil in Chedabucto Bay, Nova Scotia, six years after the *Arrow* spill. Journal of the Fisheries Research Board of Canada 35:528–535.

Lees, D. C., J. P. Houghton, and W. B. Driskell. 1996. Short-term effects of several types of shoreline treatment on rocky intertidal biota in Prince William Sound. American Fisheries Society Symposium 18: 329–348.

Marsh, C. P. 1986a. Impact of avian predators on high intertidal limpet populations. Journal of Experimental Marine Biology and Ecology 104:185–201.

Marsh, C. P. 1986b. Rocky intertidal community organization: the impact of avian predators on mussel recruitment. Ecology 67:771–786.

McDonald, J. H., and R. K. Koehn. 1988. The mussels *Mytilus galloprovincialis* and *M. trossulus* on the Pacific coast of North America. Marine Biology 99:111–118.

Mearns, A. J. 1996. *Exxon Valdez* shoreline treatment and operations: implications for response, assessment, monitoring, and research. American Fisheries Society Symposium 18:309–328.

Nicotri, M. E. 1977. Grazing effects of four marine intertidal herbivores on the microflora. Ecology 58:1020–1032.

RPI (Research Planning Institute, Inc). 1983. Sensitivity of coastal environments and wildlife to spilled oil, Prince William Sound, Alaska, an atlas of coastal resources. Report to NOAA (National Oceanic and Atmospheric Administration), Office of Oceanography and Marine Services, Seattle.

Sanders, H. L. 1978. *Florida* oil spill impact on the Buzzards Bay benthic fauna: West Falmouth. Journal of the Fisheries Research Board of Canada 35:717–730.

Sokal, R. R., and F. J. Rohlf. 1981. Biometry, 2nd edition. Freeman, New York.

Southward, A. J., and E. C. Southward. 1967. On the biology of an intertidal Chthamalid (Crustacea, Cirripedia) from the Chukchi Sea. Arctic 20:8–20.

Southward, A. J., and E. C. Southward. 1978. Recolonization of rocky shores in Cornwall after use of toxic dispersants to clean up the *Torrey Canyon* spill. Journal of the Fisheries Research Board of Canada 35: 682–706.

Stekoll, M. S., and seven coauthors. 1996. Coastal habitat injury assessment: intertidal communities and the *Exxon Valdez* oil spill. American Fisheries Society Symposium 18:177–192.

Sundberg, K., L. Deysher, and L. McDonald. 1996. Intertidal and supratidal site selection using a geographical information system. American Fisheries Society Symposium 18:167–176.

Swedmark, M., A. Granmo, and S. Kollberg. 1973. Effects of oil dispersants and oil emulsions on marine animals. Water Research 7:1649–1672.

Thomas, M. L. H. 1973. Effects of bunker C oil on intertidal and lagoonal biota in Chedabucto Bay, Nova Scotia. Journal of the Fisheries Research Board of Canada 30:83–90.

Thomas, M. L. H. 1978. Comparison of oiled and unoiled intertidal communities in Chedabucto Bay, Nova Scotia. Journal of the Fisheries Research Board of Canada 35:707–716.

van Tamelen, P. G., and M. S. Stekoll. 1996. Population response of the brown alga *Fucus gardneri* and other algae in Herring Bay, Prince William Sound, to the *Exxon Valdez* oil spill. American Fisheries Society Symposium 18:193–211.

American Fisheries Society Symposium 18:238–248, 1996

Influence of Residual and Applied Oil on Intertidal Algal Recruitment

P. BRUCE DUNCAN
U.S. Environmental Protection Agency, 1200-6th Avenue, OEA-095
Seattle, Washington 98101, USA

ANTHONY J. HOOTEN[1]
Coastal Resources Associates, Inc., 1185 Park Center Drive, Suite A
Vista, California 92083, USA

Abstract.—We studied the effect of stranded oil on algal recruitment in rocky intertidal communities in two locations in Prince William Sound, Alaska, between 1989 and 1992. Rocks collected from two oiled beaches and ceramic tiles were used as substrates for recruitment, and tarred and clean substrates were paired for each experiment. Two other treatments—caged tiles and tiles painted black—were used to examine how grazers and dark color affected recruitment. Recruitment of biota was consistently reduced on tarred substrates (by 50–100%), despite changes in location, experimental methods, and year. In all cases, tarred halves of rocks and tarred tiles had significantly lower algal cover than did clean surfaces. Similar reductions were observed for limpets and checkered periwinkles *Littorina scutulata*, indicating that the depressed algal recruitment was not caused by grazers feeding preferentially on tarred substrates. Cages did not successfully exclude grazers. Inside the cages, littorines, algae, and rockweed germlings *Fucus* sp. were more abundant and their recruitment was again significantly reduced on tarred tiles. Paint had no effect on recruitment, indicating that dark coloration could not account for the inhibition of recruitment by tar. The relationship between residual tar and recruitment success can be used in cleanup decisions for rocky shores if consideration is given to substrate composition, exposure to weathering, potential effects of cleaning on the established community, and availability of dispersing propagules.

After an oil spill, recovery of intertidal biota on oiled substrates can differ from natural succession because of residual toxicity of oil, large-scale mortality of recolonizing propagules, and loss of herbivores (Southward and Southward 1978). After the *Exxon Valdez* oil spill in March 1989, we were interested in the extent to which recovery of intertidal biota in Prince William Sound, Alaska, was affected by stranded oil. This information is useful to managers in deciding to what extent residual levels of oil should be removed.

Nelson (1982) suggested that the greatest effect of crude oil in rocky sites at northern latitudes is disruption of settlement and growth of juvenile organisms. We chose to test how the initial stages of succession of algae, namely recruitment and early growth, responded to residual oil. Here we define recruitment based on Connell (1985) as recently settled juveniles that have survived until censusing, which in our case was from 6 months to 2 years.

Many factors combine to affect the recruitment of algae. These factors were discussed extensively by Santelices (1990), who described how pools of suspended propagules are transported and settle onto substrates. Once settled, propagules depend on favorable conditions such as availability of nutrients and CO_2, microstructure of the substrate, temperature, and protection from desiccation. Mann and Clark (1978) argued that variability in the marine environment makes it difficult to distinguish effects attributable to oil spills. Effects of oil spills on intertidal biota generally have been evaluated by describing changes through time (e.g., Chassé 1978; Mann and Clark 1978; Floc'h and Diouris 1980), whereas few field studies have experimentally tested for the effects of oil in a controlled manner. For instance, Crapp (1971a) and Nelson (1981) applied crude oil to patches on intertidal rocks and found biota to be largely unaffected. Straughan (1971) evaluated settlement over time on asbestos tiles that had one-half soaked in oil. Even though the tiles were not replicated, there was consistently less algae on oiled halves of tiles than on non-oiled halves.

In this article, we report results from experiments conducted in two different locations in Prince William Sound in which we tested whether algal recruitment differed as a function of the presence or absence of oil (tar) on the settlement substrate. In one experiment we also evaluated the recruitment by *Fucus* germlings to examine whether the pattern

[1]Present address: AJH Environmental Services, 4005 Glenridge Street, Kensington, Maryland 20895, USA.

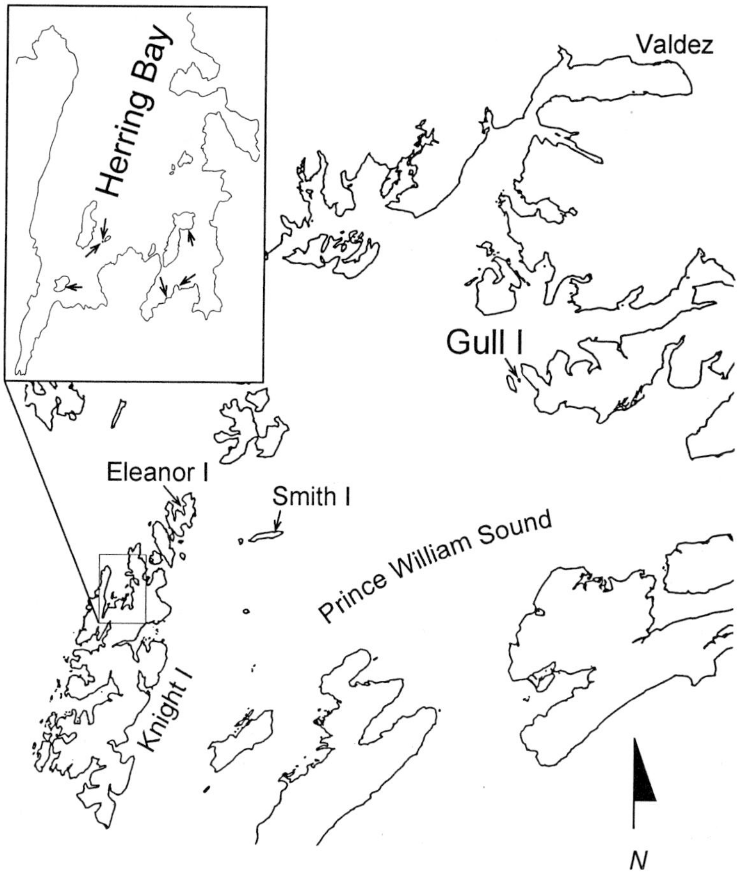

FIGURE 1.—Location of study sites in Prince William Sound: tarred rock collection sites at Eleanor and Smith islands; Gull Island, where the half-cleaned rocks were placed in 1989; and Herring Bay on Knight Island, where the tarred, painted, untreated, caged, and uncaged tile pairs were placed in 1990 and 1991 on six beaches.

we observed for the overall algal assemblage was reflected in one component. Furthermore, we measured the loss of tar from one subset of our experimental substrates and, in another subset, evaluated the effect of grazer exclusion cages on the algal response. Both of these factors influence algal recruitment (grazers, Santelices 1990; presence of tar, DeVogelaere and Foster 1994).

Methods

Gull Island (1989–1990)

Between 15 and 18 September 1989, tarred rocks were collected from two oiled beaches (Eleanor and Smith islands), cleaned, and placed onto a beach at Gull Island (Figure 1) that had a flourishing community of *Fucus* sp., filamentous algae, barnacles, limpets, littorine snails, mussels, and other organisms typical of this area (Feder and Bryson-Schwafel 1988). Gull Island lies to the east of Bligh Island and was outside of the path of floating oil, and no sign of stranded oil was noted during any visits.

Both of the contaminated sites (Smith and Eleanor islands) received heavy oiling. Galt et al. (1991) described how storms, wind, and currents combined to send high concentrations of oil throughout Prince William Sound. Most beaches were subjected to a variety of treatment efforts to remove the oil, including flooding, pressure washing, handwiping, and bioremediation (addition of fertilizer to enhance microbial degradation). The particular beach segment examined on Smith Island included an area where Corexit (a kerosene-based cleanser) had been tested. Despite the oil removal efforts, at both sites it was relatively easy to obtain rocks stained with residual tar by turning rocks over and using the oiled undersides.

Rocks with visually low levels of tarring (stained

rocks) were collected. The Alaska Department of Environmental Conservation (ADEC 1991) defined staining as "Oil ≤ 0.1 mm thick, cannot be easily scratched off with fingernail." At Eleanor Island, two levels of staining (high and low) were chosen. Stained rocks were cleaned by dipping one-half of each rock into methylene chloride or by removing tar with wipes soaked with methylene chloride. Rocks were also transplanted from the supratidal region on Gull Island into the experimental plot to test for any subsequent stranding of oil. These supratidal rocks were cleaned with wire brushes and sandpaper to remove soil, lichens, and mosses. The median length (longest axis) of all the rocks was approximately 16 cm, except for the Gull Island rocks collected from the supratidal which had a median length of 14.5 cm. Rocks were assigned randomly to one of two groups, for either initial or final analysis. The rocks were then placed randomly into the upper intertidal zone of Gull Island on a horizontal rock bench approximately 2 m × 20 m. Biota were cleared from the rocky substrate and the rocks were affixed with marine epoxy putty (Z-Spar©). The zone chosen was dominated by barnacles and *Fucus* (similar to the zone from +1.5 to +3.0 m above mean lower low water described by Juday and Foster 1990 for Green Island in Prince William Sound, and similar to the rockweed zone described by Houghton et al. 1991). Rocks for initial analysis were removed immediately; those for final analysis were retrieved 17 March 1990.

Measurements of algal colonization were made by bisecting each rock with a transect across the tarred and cleaned halves. Five microscope fields were examined on each half; each field was viewed with a grid containing 25 squares of 1 mm^2 each. Algal coverage was recorded as the number of squares covered and converted to percent cover. Species of algae were not determined, but some notes were made in the field and laboratory on types of algae and grazers present.

In addition to algal coverage, tar coverage was measured on rocks at the beginning and end of the exposure period. Wipes dipped in methylene chloride were used to clean a 3-cm^2 area on tarred and supratidal rocks. The wipes were placed in vials and sent to the laboratory. Material in the vials was extracted into methylene chloride and an aliquot was withdrawn, filtered, dried, and weighed. The laboratory detection limit of 1.0 mg/filter was based on the lower limit of the analytical balance of 0.1 mg and the extract dilution factor of 10. A blank sample (wipe saturated with methylene chloride) was below detection. Sample results of less than 1 mg/filter (non-detects) were assigned a value of 0.5 mg/filter. The results were then divided by 3 to convert from milligrams per filter to milligrams per square centimeter.

Herring Bay (1990–1992)

Herring Bay (Figure 1) is a north-facing embayment on Knight Island that received heavy oiling beginning 29 March 1989 (ADEC 1989). The data examined here are part of a larger study of the effects of stranded *Exxon Valdez* oil in Herring Bay (see Highsmith et al. 1993 for a more detailed description of the study sites and methods used). Six sheltered rocky beaches, three oiled and three non-oiled, were selected for study and were matched as closely as possible in substrate composition, slope, and orientation. The rough undersides of ceramic tiles were used as substrates to test the effect of tar on algal recruitment. The tiles were coated with crude oil taken from the *Exxon Valdez* in 1989 during lightering. The oil had been autoclaved and stored at room temperature in a dark glass jar 2 years before acquisition. Coated tiles were allowed to dry for about 1 month until tar-like consistency was achieved. In June 1990, six pairs of tarred and clean white clay tiles (11.4 cm × 11.4 cm) were placed randomly along the elevation contour, 2 m below mean higher high water at each of the six beaches. In June 1991, red clay tile pairs (7.3 cm × 7.3 cm) were added to each study site. At each of the six beaches, nine pairs were added as follows: three tarred and untarred pairs, three tarred and untarred pairs placed in 4-mm-mesh stainless steel cages, and three painted and unpainted pairs. Cages, attached with marine epoxy, were used in an attempt to exclude grazers. Painted tiles were used to test for the effect of dark color on recruitment. Tiles were given several coats of black spray paint and allowed to dry for several days. The resulting thickness of the paint was similar to the ADEC stain classification. Tile pairs were placed at least 50 cm apart. The length of beach at the six locations where tiles were placed ranged from 15 to 30 m.

At approximately 2-week intervals in June–August 1991 and 1992, algal cover was estimated visually and separate counts were made of *Fucus* germlings, limpets, and two snails, checkered periwinkle *Littorina scutulata* and Sitka periwinkle *L. sitkana*. Juvenile grazers were removed from the cages during each visit. Algal cover was recorded as a percentage; the counts of *Fucus* germlings and the grazers were converted to density (number/dm^2). Our analysis here is of the June 1992 results because this was the first census where algal recruitment was noticeable on either the 1991 or 1992 tiles and thus represents our best measurements of early recruitment.

Analyses

The experiments were designated G89 (Gull Island, rocks set out from 1989 to 1990), H90 (Herring Bay, tiles set out from 1990 to 1992), H91 (Herring Bay, tiles set out from 1991 to 1992), H91C (tiles in cages), and H91P (painted tiles).

A natural logarithm transformation that corrected for unequal variances (Bartlett's test, $P > 0.5$; Sokal and Rohlf 1981) was applied before analysis of the G89 tar coverage data. Data on algae were pooled across the three rock sources used on Gull Island (G89) and data on algae, *Fucus* germlings, and grazers were pooled across the six locations in Herring Bay (H90, H91). For G89 and H91 a high number of test substrates received no recruitment, and for H90 many tiles were lost by 1992. The percent cover, *Fucus* germling density, and grazer density data were analyzed with the Wilcoxon paired-sample test and the Mann–Whitney *U*-test (two-tailed; Zar 1984). Nonparametric tests were selected because the abundance data contained many zeros and a few very high values. Differences based on these data were highly skewed. For the Wilcoxon paired-sample test, a significant effect ($P < 0.05$, two-tailed) requires a minimum of six pairs with each difference of like sign (Zar 1984). Even with pooling, six pairs were not always achieved. In addition, following Zar (1984), differences of zero were ignored in the analysis and tied ranks were averaged. Although this procedure discarded data (especially pairs for which no recruitment occurred), it did test the null hypothesis that, when recruitment occurs, the abundance on the treated substrate is equal to the abundance on the untreated or cleaned substrate.

Results

Gull Island

The G89 rocks had no algae before being placed on Gull Island. Upon retrieval, the rocks were characterized in the field as having a green film. Littorine snails were observed on some of the rocks (on both tarred and untarred halves). Further inspection revealed the green film consisted of filamentous algae and diatom "scum." Some *Fucus* germlings were also noted. Littorine snails recovered from the rocks included checkered and Sitka periwinkles; limpets included *Lottia pelta* and *Tectura persona*; an occasional mussel (*Mytilus* sp.) was noted. Algal cover on the cleaned half of the rocks was always greater than or equal to the cover on the tarred half (Figure 2). The effect of tar on algal cover on the G89 rocks was highly significant and the magnitude was estimated as a 69% reduction in recruitment (Table 1). Field observations also revealed that much of the tar had been lost from the transplanted rocks. The loss was highly significant ($P = 0.0003$) with declines in methylene chloride-extractable material ranging from 51 to 81% over the 6-month period for the three types of rocks (Table 2). No tar or oil was ever observed on the supratidal rocks transplanted to the high intertidal zone. The wipes, however, did remove material (lichens, mosses, etc.) that remained on these rocks at the beginning of the experiment as well as diatoms and ephemeral algae present on the rocks at the end of the experiment. For both sampling dates, the mean amount of methylene chloride-extractable material recovered from the transplanted supratidal rocks was below the laboratory detection limit of 0.33 mg/cm^2 (Table 2).

Herring Bay

Algae that recruited to the tiles were mainly weedy, ephemeral species such as *Cladophora*, *Scytosiphon*, and *Enteromorpha* spp. and the rockweed *Fucus*. Grazers present on the tiles included the limpets *Lottia pelta*, *L. borealis*, *Tectura persona*, and limpets too small to identify to species in the field, as well as littorine snails. The checkered and Sitka periwinkles were present frequently enough to evaluate their response to the presence of tar and paint.

A large portion of the 1990 tiles were lost, presumably from ice scour during the two winters; only 15 of the 36 pairs set out were recovered in 1992. In contrast, all of the tiles (54 pairs) set out in 1991 were recovered in 1992. Of the tile pairs recovered (69 total), many had no recruitment on either tile. Overall, more than 40% of the pairs had no measured algae on either tile, and about 60% of the pairs had no *Fucus* recruitment. Similarly, grazers were absent from 33 to 70% of the pairs. Of all the tile pairs examined where recruitment occurred on at least one of the two tiles (excluding the painted tiles), there were only two cases (of 35 pairs) where algal cover was greater on the tarred tile and only one case (of 24 pairs) where *Fucus* germling density was greater on the tarred tile.

Similarly, limpets were always found in greater density on untarred tiles (of 21 pairs). Littorine snails (checkered and Sitka periwinkles) achieved greater densities on untarred tiles less frequently

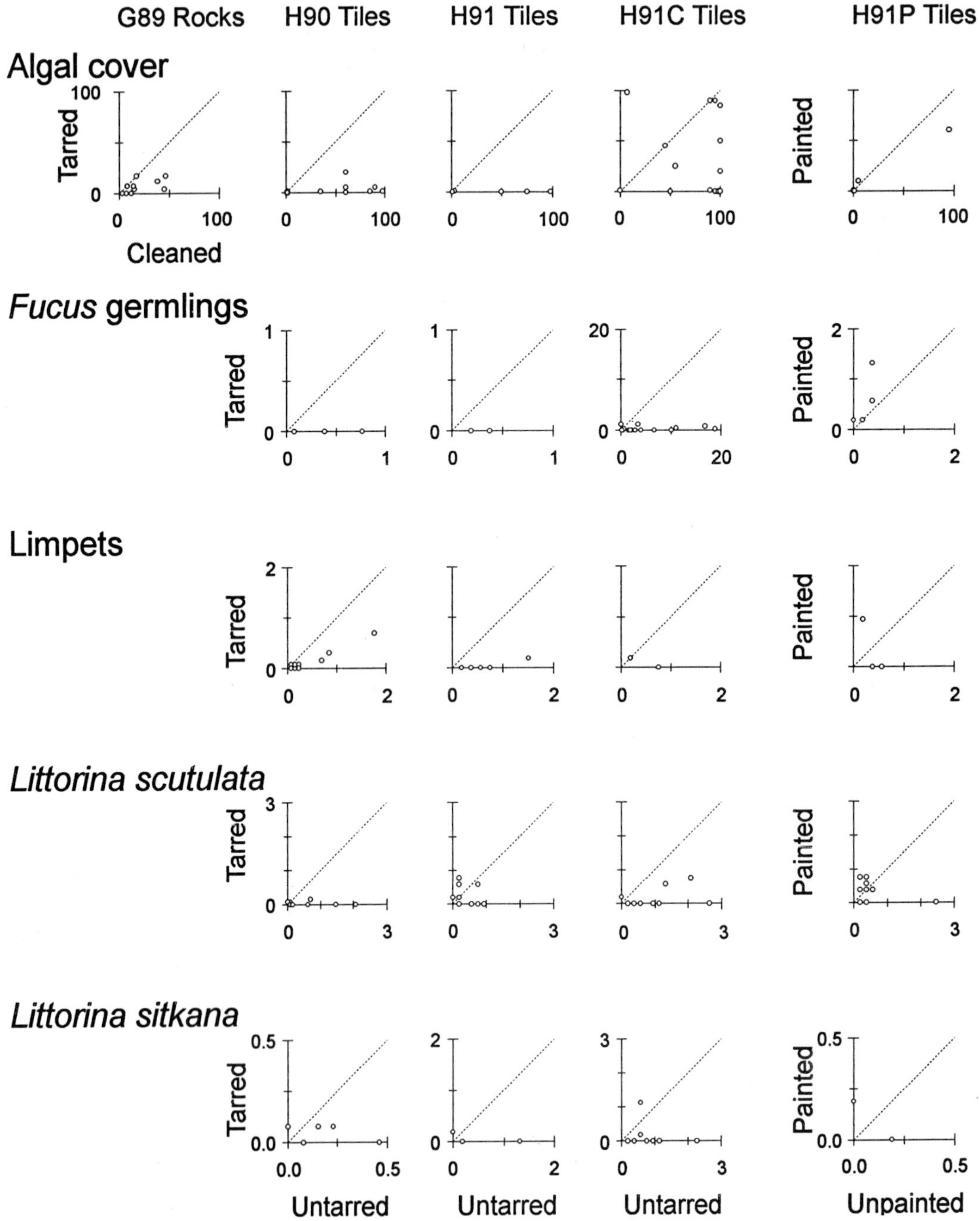

FIGURE 2.—Scatterplots of algal cover (%) and density (number/dm^2) of *Fucus* germlings and grazing gastropods (limpets, checkered periwinkles *Littorina scutulata*, and Sitka periwinkles *L. sitkana*) on untreated (or cleaned) and treated (tarred or painted) substrates in Prince William Sound. Experiment codes are G89 = Gull Island rocks placed there in 1989, H = Herring Bay, 90 = 1990 white tile pairs, 91 = 1991 red tile pairs, 91C = 1991 caged red tile pairs, and H91P = 1991 painted red tile pairs. Dotted lines represent equal cover or density on the two test substrates. Measurements of cover and density were made in March 1990 (Gull Island rocks) and June 1992 (Herring Bay tiles).

TABLE 1.—Mean algal cover and mean densities of *Fucus* germlings, limpets, checkered periwinkles *Littorina scutulata*, and Sitka periwinkles *L. sitkana* on clean and treated (tarred or painted) substrates at Gull Island (March 1990) and Herring Bay (June 1992). The magnitude of treatment effect was estimated as percent reduction from response on clean substrate: 100(1 − [treated mean/untreated mean]); *N* = pairs with recruitment on at least one of the two substrates and *N*w = pairs with nonzero difference. The *P* value is from the Wilcoxon paired-sample test of whether cover or density differed due to treatment (tar or paint); intervals for two-tailed test are 0.5, 0.2, 0.1, 0.05, 0.02, 0.01, 0.005, 0.001 (Zar 1984); when *N*w = 4, minimum *P* = 0.2. Absence of a *P* value means "not tested."

	Mean ± SE[b]				Significance	
Experiment[a]	Clean	Treated	Reduction (%)	*N*	*N*w	*P*
			Algae			
G89	20 ± 4.7	6.2 ± 2.0	69	11	10	<0.005
H90	41 ± 11	2.8 ± 1.6	93	12	11	0.001
H91	29 ± 14	0.1 ± 0.1	100	8	8	0.01
H91C	75 ± 9.0	39 ± 10	48	15	13	<0.02
H91P	14 ± 12	20 ± 19	−44	5	5	>0.5
			***Fucus* germlings**			
H90	0.50 ± 0.17	0 ± 0	100	4	4	0.2
H91	0.31 ± 0.06	0 ± 0	100	3	3	
H91C	5.2 ± 1.4	0.22 ± 0.10	96	17	17	<0.001
H91P	0.57 ± 0.27	0.24 ± 0.09	58	4	3	
			Limpets			
H90	0.38 ± 0.14	0.12 ± 0.06	68	12	11	0.001
H91	0.65 ± 0.16	0.03 ± 0.03	96	7	7	0.02
H91C	0.47 ± 0.28	0.09 ± 0.09	80	2	1	
H91P	0.42 ± 0.09	0.24 ± 0.24	44	4	4	>0.5
			Checkered periwinkles			
H90	0.59 ± 0.24	0.03 ± 0.02	94	9	8	<0.05
H91	0.38 ± 0.08	0.19 ± 0.07	50	13	11	<0.2
H91C	0.74 ± 0.22	0.13 ± 0.06	82	14	14	<0.005
H91P	0.53 ± 0.22	0.32 ± 0.10	39	10	9	>0.5
			Sitka periwinkles			
H90	0.16 ± 0.06	0.04 ± 0.02	73	7	7	<0.2
H91	0.50 ± 0.41	0.06 ± 0.06	88	3	3	
H91C	0.82 ± 0.20	0.17 ± 0.12	79	9	9	0.05
H91P	0.09 ± 0.09	0.09 ± 0.09	0	2	2	

[a] Experiment codes: G = Gull Island; H = Herring Bay; numbers denote years in which experiments began (1989–1991); C = tiles in cages; P = painted tiles.
[b] Percent coverage for algae; number/dm^2 for *Fucus*, limpets, and periwinkles.

than did the limpets (29 of 36 pairs and 16 of 19 pairs, respectively, for the two species).

Algal cover on the untarred tiles was greater than or equal to the cover on the tarred tiles (Figure 2). The effect of the tar was highly significant and the magnitude of the effect was estimated as a 48 to 100% reduction from equal cover (Table 1). Likewise, *Fucus* germling density was reduced on tarred tiles (Figure 2) by an estimated 96 to 100%, significantly so for the caged tiles (H91C; Table 1). In the other two experiments where recruitment of *Fucus* germlings was measured in response to the presence of tar (H90 and H91), the recruitment rate was too low to adequately test for a tar effect. In these experiments, germlings were only found on four or fewer tile pairs (Table 1); however, in both instances, the germlings were found only on the clean tiles.

The tar also affected grazers (limpets and snails; Figure 2); densities were reduced on tarred tiles by 50 to 96% (Table 1). Of the nine separate tests for the effect of tar on grazers (Table 1), two had low replication ($N \leq 3$) and could not be tested, five were significant ($P \leq 0.05$), and two were nonsignificant at $P < 0.2$. The danger of presenting results from nine separate tests is that one expects a proportion of false positives by chance. In this instance, the preponderance of significant results strongly indicates that the underlying effect (more grazers on untarred than tarred tiles) was not attributable to chance.

In the caged treatments (H91C; Table 1), about

TABLE 2.—Tar coverage, as methylene chloride-extractable material, at the beginning and end of winter 1989 on rocks transplanted to Gull Island in Prince William Sound. Loss of methylene chloride-extractable material from tarred rocks was significant ($P = 0.0003$; two-way analysis of variance with time and rock source as treatments; interaction was not significant, $P = 0.31$). The amount of methylene chloride-extractable material was below the detection limit of 0.33 mg/cm^2 on supratidal (untarred) rocks from Gull Island placed into the intertidal zone during the same period.

	1989		1990		
Source of rocks	Mean ± SE (mg/cm^2)	*N*	Mean ± SE (mg/cm^2)	*N*	Loss (%)
Eleanor Island, heavily stained	8.0 ± 1.9	5	1.5 ± 0.5	4	81
Eleanor Island, lightly stained	2.1 ± 0.7	5	1.0 ± 0.4	4	53
Smith Island	9.2 ± 1.3	5	4.5 ± 1.9	6	51
Gull Island supratidal	0.23 ± 0.04	5	0.30 ± 0.03	5	

twice as many tile pairs had algal recruitment and more than five times as many tile pairs had *Fucus* germling recruits, relative to the uncaged tiles (H91). Algal cover and *Fucus* germling density averaged over all 18 clean tiles increased from 13% and 0.05/dm^2 to 63% and 4.9/dm^2 on caged tiles. These differences were significant ($0.002 < P < 0.005$ for algal cover, $P < 0.001$ for *Fucus* germling density; two-tailed Mann–Whitney *U*-test; Zar 1984; means for the 18 tiles can be estimated by multiplying means in Table 1 by *N*/18). The cages were somewhat successful in excluding limpets, which were found in only 2 of 18 cages. Mean limpet density on the 18 clean, caged tiles was 0.05/dm^2 compared with 0.25/dm^2 on clean, uncaged tiles, but this difference was not significant ($0.1 < P < 0.2$; two-tailed Mann–Whitney *U*-test). In contrast, both littorine snails (checkered and Sitka periwinkles) recruited more frequently on caged than on uncaged tiles. Densities of these two species increased from 0.27/dm^2 and 0.08/dm^2 outside of cages to 0.58/dm^2 and 0.41/dm^2, respectively, inside cages (significantly so for Sitka periwinkle, $0.02 < P < 0.05$; but not for checkered periwinkle, $P > 0.2$; two-tailed Mann–Whitney *U*-test).

Unlike the tarred tiles, painted tiles did not affect recruitment. Estimates of an effect attributable to the paint ranged from a 58% reduction to a 44% enhancement, but none of these effects were significant (Table 1; $P > 0.5$ for algae, limpets, and checkered periwinkles; results for *Fucus* germlings and Sitka periwinkles could not be tested).

Discussion

Recruitment of algae was dramatically reduced on the tarred halves of rocks on Gull Island in 1990. This reduction was documented again in 1992 in Herring Bay, where our results showed that tar affected *Fucus* germlings and gastropod grazers as well as recruiting algae. Tests of both total algae and *Fucus* germling recruitment (Table 1) revealed a consistent and significant depression in recruitment from 48 to 100% because of the presence of weathered *Exxon Valdez* tar. DeVogelaere and Foster (1994) found that *Fucus* does not settle onto residual tar, and Straughan (1971) showed that algal recruitment is reduced on oiled asbestos tiles relative to untreated tiles. Although adult algae appear to be somewhat resistant to oil (e.g., Nelson 1982 and references therein), sporelings are known to be more sensitive (Thomas 1977; Notini 1978; Floc'h and Diouris 1980). Somewhat similarly, our results show that recruitment of *Fucus* germlings in Herring Bay was either equally or more sensitive to the presence of residual tar when compared with the overall algal recruitment (Table 1).

Like algal and *Fucus* germling recruitment, grazer recruitment on tarred tiles was reduced by a similar range in magnitude (50 to 96%; Table 1). Grazers can recruit via local migration as well as from the larval pool, and field experiments suggest the adult grazers may be less affected by applied crude oil than are juveniles (Nelson 1982). Therefore, we had expected less of an effect on grazers relative to effects on algae. Our results, however, are consistent with other work describing the sensitivity of intertidal grazers to oil spills (Chassé 1978; Mann and Clark 1978; Southward and Southward 1978; Highsmith et al. 1993; Hooten and Highsmith 1996, this volume) and are somewhat in contrast to cases documenting the resistance of grazers to oil spills (Peltier et al. 1976; Thomas 1977; Nelson 1982; Dethier 1991).

Tar may have affected recruiting biota in several ways. Algal propagules rely on complex cellular processes to secrete adhesive materials and develop fibrils for attachment (Santelices 1990). Compounds in the tar may have affected these processes or the tar may have changed the physical properties of the substrate and reduced the ability of algae to attach. Although fresh oil can be toxic to algae and grazers (Crapp 1971b; Thomas 1977; Juday and Foster 1990), we speculate, as have others (Nelson 1982, for example), that weathering reduced the toxicity of the tar on our substrates. Weathering of fresh oil can occur quickly. Lighter, more toxic fractions of the oil weather in hours to days and oil on exposed beaches can be removed by natural processes within several years (Foster et al. 1990). Weathering of tar from our rocks on Gull Island was substantial (Table 2). The weathering rates (50–80% loss over winter) were very similar to the rate of about 50% loss over winter and 95% loss after 11 months reported by DeVogelaere and Foster (1994) for rocky surfaces in Herring Bay. Despite the likelihood that weathering reduced the toxicity of the oil, whether and how residual tar may be toxic to biota recruiting to the upper rocky intertidal zone in Prince William Sound is unknown.

The presence of tar probably changed the physical properties of our rock and tile surfaces. The microstructure of substrate surfaces is known to affect settlement of seaweed spores (Harlin and Lindbergh 1977). Roughness, pits, and cracks can enhance settlement, reduce desiccation, and provide refuges from grazers (Santelices 1990). Harlin and Lindbergh (1977) have demonstrated that smooth surfaces receive the least amount of algal settlement. Recent work in Herring Bay has demonstrated that *Fucus* recruits preferentially to cracks and does not recruit onto tar (Highsmith et al. 1993; DeVogelaere and Foster 1994). After an oil spill, one might expect that a coating of oil could fill in cracks and pits on substrate surfaces and produce a smoother surface that would have a reduced level of recruitment.

Others have suggested that the dark color may cause the substrate to absorb more solar radiation at low tide and that elevated temperatures could affect the recolonization process (Straughan 1976). Our results showed no significant effect of painted tiles on recruitment by algae or *Fucus* germlings. This lack of effect of paint relative to tar suggests that at least the dark color of the tar cannot account for the inhibition of recruitment. Our painted tiles initially had a thinner physical coating than the tarred tiles; however, the tar quickly weathered to a stain level in most cases, and the painted tiles possessed a dark color much longer than the oiled tiles.

Factors other than substrate smoothness or the presence of a potentially toxic compound probably interacted to influence algal recruitment into the rocky intertidal zone in Prince William Sound. We acknowledge that the effect of tar on recruitment may have differed as a function of beach location, rock type, or amount of tar; however, we tested for and repeatedly found a significant effect of tar on recruitment despite the influence of other factors and the potential that pooling data might have obscured such an effect. We also acknowledge that substrate type, location within the intertidal zone, and distribution and abundance of grazers are extremely important (Santelices 1990). By using relatively uniform tiles in Herring Bay, we hoped to reduce the influence of substrate type on recruitment, yet the range of recruitment onto untarred substrates was even greater on tiles in Herring Bay (0–100% cover) than it was on the rocks on Gull Island (0–50% cover; Figure 2). This difference certainly could have been caused by location rather than substrate type. We placed the test substrates at random within a narrow vertical zone to minimize the influence of vertical changes. As Santelices (1990) has pointed out, recruitment is tremendously influenced by very localized conditions that determine water flow, turbulence, and the ability of propagules to settle out of a heterogeneous and patchy propagule pool. Thus perhaps it is not surprising that broad ranges in recruitment were encountered.

Within this variability in expected recruitment, however, we saw a clear increase in algal and *Fucus* germling recruitment when we employed cages with 4-mm-mesh opening in an attempt to exclude herbivores (Table 1). Although this response was anticipated (others such as Santelices [1990] have documented or reviewed the importance of grazers in organizing algal assemblages in the rocky intertidal zone), the increased recruitment to caged tiles was probably not caused by a reduction in grazers, because the cages were only partially successful. That partial success was limited to limpets, which were present in only two cages (compared with their presence on seven uncaged tile pairs; Table 1, *N* for H91C and H91) and at one-fifth the density found outside the cages. The difference in density, based on averages for 18 clean tiles, was not significant, however. In contrast, littorines were found more frequently within cages and, on clean tiles, had higher mean densities inside than outside cages (significantly so for Sitka periwinkles). This general, enhanced recruitment of algae, *Fucus*, and littorines is consistent with the well-documented effects that cages exert by providing shading, moisture retention, baffling, and even protection from adult plants whose fronds can sweep propagules off the substratum (Dayton 1971; Santelices 1990). These types of changes can also enhance survivorship of grazers such as littorines (Ricketts et al. 1985).

Could grazing by limpets and littorines have affected our algal and *Fucus* germling results? To produce the dramatic reductions we observed in algal and *Fucus* germling recruitment, these grazers would have had to preferentially remove the algae from the tarred halves. Such selective removal of algae and *Fucus* germlings from the tarred substrates is unlikely because grazers were significantly reduced on tarred tiles (Table 1). A similar reduction in grazers on tarred vertical rock faces has been observed in Herring Bay (Highsmith et al. 1993). Whether recruitment of grazers to our tiles was influenced more by the presence of algae or possible direct effects of the tar is currently unknown. Littorines were generally more abundant than limpets on our tarred tiles (Table 1), but this is not surprising given that they are known to be active on oiled substrates (Dethier 1991; DeVogelaere and Foster 1994) and that checkered periwinkles can achieve greater densities on tarred compared to untarred substrates (DeVogelaere and Foster 1994).

As opposed to studies that observe recovery of communities, our experiments introduced some artificial elements. We reduced the natural *Fucus* canopy during placement of the substrates, which may have reduced the availability of propagules. Our substrates contained no "seed bank" of microscopic forms of algae (Santelices 1990), which are known to inhabit natural substrates and are potentially available to repopulate a substrate if the existing algae are removed. We also did not evaluate whether the tar on our experimental substrates led to changes in algal species over time nor did we examine species interactions (physical or chemical), both of which are acknowledged as important in organizing intertidal algal assemblages (Santelices 1990). Certainly, the effect on early recruitment that we document here would change with time and it would be enlightening to know how such changes occur as a function of exposure, substrate type, position in the intertidal zone, and level of tar. Although we did not test for any effect of the methylene chloride on colonization, Highsmith et al. (1993) did not detect an effect of this solvent on invertebrate recruitment. The effects of the epoxy on colonization by algae are also unknown; however, the epoxy was used on all of the rocks, and anecdotal observations of algae and invertebrates using the cured epoxy for attachment sites suggest it was not toxic. Although we do not know the exact manner in which the residual tar in our experiments affected recruitment, our consistent results indicate that inhibition of recruitment was attributable to the presence of tar and not some other cause. The paired design helped reveal the response of algae to the tar despite variability caused by differences in surface texture of the substrates, heterogeneity of microenvironments, recruitment period, time of year, year, location within Prince William Sound, and loss of tar during the experiment. Finding similar, repeatable results despite these differences makes our conclusions about the effect of residual *Exxon Valdez* tar on algal recruitment in Prince William Sound particularly robust.

We have demonstrated that *Exxon Valdez* oil in Prince William Sound affected recruitment of biota, but can these findings be applied to policies and management of shoreline treatment for future oil spills? Previous authors who have examined policy (Foster et al. 1990; Lees et al. 1996, this volume) or previous spills (Thomas 1978) have concluded that the effects of mechanical shoreline treatment are more deleterious to intertidal biota than leaving oil in place. The evidence of rapid oil weathering presented here and elsewhere (DeVogelaere and Foster 1994) suggests that treatment of some shorelines may be unnecessary, especially at very exposed sites. Our results suggest effects of residual oil on algal recruitment and grazer behavior could occur for at least 2 years after a spill. Such effects are likely to persist for much longer in protected embayments and coves that contain fine or porous substrate and do not receive consistent annual recruitment (Highsmith et al. 1993). In these areas, it may be appropriate to employ treatment technologies that can effectively clean substrates for future recruitment (perhaps low-pressure, ambient-temperature washing), leaving as much epibiota intact as possible. Treated shorelines may then rely upon a clean substrate, local reproductive individuals, and infrequent recruitment pulses from elsewhere to assist in recovery of affected populations.

Future field studies, similar to ours, that focus on the relationship between the amount of residual tar, substrate type, and recruitment success, could help managers choose an appropriate cleanup level. Two important decisions are needed before this step, namely, (1) an acceptable definition of recovery and how it is best measured, and (2) an acceptable time to reach recovery. To accelerate recovery, consideration should be given to restoring organisms (Ganning et al. 1984), perhaps by enhancing local sources of reproductive individuals, especially for species that disperse only on the scale of meters to tens of meters.

Acknowledgments

The Gull Island study was supported in part by the Oil Spill Response Center (OSRC) of the Alaska Department of Environmental Conservation (ADEC) in Valdez (1989). Staff from the OSRC assisted with field work as well as conceptual development (A. H. Weiner) and algal counts (C. Fultz). Gravimetric analyses were conducted by the ADEC Douglas laboratory. R. Benda provided laboratory space at the Valdez High School. Work in Herring Bay was supported by funding from the *Exxon Valdez* Oil Spill Trustee Council to the University of Alaska, Fairbanks, Institute of Marine Science; R. C. Highsmith, coordinating principal investigator, contributed editorially to an early abstract. Staff of the Coastal Habitat Injury Assessment Project assisted in the field. We thank D. C. Lees, C. H. Peterson, anonymous reviewers, and, in particular, M. N. Dethier for constructive criticism. L. Fore provided critical review of statistical methods. The U.S. Environmental Protection Agency (EPA), through its Region 10 Office of Environmental Assessment, partially supported the research described here. This product has not been subjected to the Agency's peer review; therefore, it does not reflect the views of the EPA. The U.S. Government has a non-exclusive, royalty-free license in and to any copyright covering this article. Mention of trade names or commercial products does not constitute endorsement or recommendation for use by the EPA.

References

ADEC (Alaska Department of Environmental Conservation). 1989. Field notebooks, audio cassettes, and field notes, 1989 files (A-Jod). ADEC (Alaska Department of Environmental Conservation) Oil Spill Response Center, Anchorage.

ADEC (Alaska Department of Environmental Conservation). 1991. Glossary of terms for surface oil characters. ADEC (Alaska Department of Environmental Conservation) Oil Spill Response Center, Anchorage.

Chassé, C. 1978. The ecological impact on and near shores by the *Amoco Cadiz* oil spill. Marine Pollution Bulletin 9:298–301.

Connell, J. H. 1985. The consequences of variation in initial settlement vs. post settlement mortality in rocky intertidal communities. Journal of Experimental Marine Biology and Ecology 93:11–45.

Crapp, G. B. 1971a. Field experiments with oil and emulsifiers. Pages 114–128 *in* E. B. Cowell, editor. The ecological effects of oil pollution on littoral communities. Applied Science Publishers, New York.

Crapp, G. B. 1971b. The ecological effects of stranded oil. Pages 181–186 *in* E. B. Cowell, editor. The ecological effects of oil pollution on littoral communities. Applied Science Publishers, New York.

Dayton, P. K. 1971. Competition, disturbance, and community organization: the provision and subsequent utilization of space in a rocky intertidal community. Ecological Monographs 41:351–389.

Dethier, M. N. 1991. The effects of an oil spill and freeze event on intertidal community structure in Washington. Final Report of Friday Harbor Labs and Institute for Environmental Studies, University of Washington, Friday Harbor, to U.S. Department of the Interior, Minerals Management Service, Pacific OCS (Outer Continental Shelf) Region, Camarillo, California.

DeVogelaere, A. P., and M. S. Foster. 1994. Damage and recovery in intertidal *Fucus gardneri* assemblages following the *Exxon Valdez* oil spill. Marine Ecology Progress Series 106:263–271.

Feder, H. M., and B. Bryson-Schwafel. 1988. The intertidal zone. Pages 117–164 *in* D. G. Shaw and M. J. Hameedi, editors. Environmental studies in Port Valdez, Alaska. Lecture notes on coastal and estuarine studies, volume 24. Springer-Verlag, New York.

Floc'h, J.-Y., and M. Diouris. 1980. Initial effects of *Amoco Cadiz* oil on intertidal algae. Ambio 9:284–286.

Foster, M. S., J. A. Tarpley, and S. L. Dean. 1990. To clean or not to clean: the rationale, methods, and consequences of removing oil from temperate shores. Northwest Environmental Journal 6:105–120.

Galt, J. A., W. J. Lehr, and D. L. Payton. 1991. Fate and transport of the *Exxon Valdez* oil spill. Environmental Science and Technology 25:202–209.

Ganning, B., D. J. Reish, and D. Straughan. 1984. Recovery and restoration of rocky shores, sandy beaches, tidal flats, and shallow subtidal bottom impacted by oil spills. Pages 7–36 *in* J. Cairns, Jr. and A. L. Buikema, Jr., editors. Restoration of habitats impacted by oil spills. Butterworth Publishers, Boston.

Harlin, M. M., and J. M. Lindbergh. 1977. Selection of substrata by seaweeds: optimal surface relief. Marine Biology 40:33–40.

Highsmith, R. C., and seven coauthors. 1993. Herring Bay experimental and monitoring studies. Final status report. School of Fisheries and Ocean Sciences Restoration Project 102, University of Alaska, Fairbanks.

Hooten, A. J., and R. C. Highsmith. 1996. Impacts on selected intertidal invertebrates in Herring Bay, Prince William Sound, after the *Exxon Valdez* oil spill. American Fisheries Society Symposium 18:249–270.

Houghton, J. P., D. C. Lees, H. Teas, III, H. L. Cumberland, S. Landino, and T. A. Ebert. 1991. Evaluation of the intertidal and shallow subtidal biota in Prince William Sound following the *Exxon Valdez* oil spill and subsequent shoreline treatment. Report (HMRB 91-1) to NOAA (National Oceanic and Atmospheric Administration), Hazardous Materials Response Branch, Seattle.

Juday, G. P., and N. R. Foster. 1990. A preliminary look at effects of the *Exxon Valdez* oil spill on Green Island Research Natural Area. Agroborealis 22:10–17.

Lees, D. C., J. P. Houghton, and W. B. Driskell. 1996. Short-term effects of several types of shoreline treatment on rocky intertidal biota in Prince William Sound. American Fisheries Society Symposium 18: 329–348.

Mann, K. H., and R. B. Clark. 1978. Long-term effects of oil spills on marine intertidal communities. Journal of the Fisheries Research Board of Canada 35:791–795.

Nelson, W. G. 1981. Inhibition of barnacle settlement by *Ekofisk* crude oil. Marine Ecology Progress Series 5:41–43.

Nelson, W. G. 1982. Experimental studies of oil pollution on the rocky intertidal community of a Norwegian fjord. Journal of Experimental Marine Biology and Ecology 65:121–138.

Notini, M. 1978. Long-term effects of an oil spill on *Fucus* macrofauna in a small Baltic bay. Journal of the Fisheries Research Board of Canada 35:745–753.

Peltier, J. C., W. Dupont, and J. P. Dupont. 1976. Conséquences bioécologiques d'un déversement accidental d'hydrocarbures sur le platier du littoral haut-normand. Bilan un an après la marée noire du 26 novembre 1974. Société de Biologie de Rouen 2:442–448.

Ricketts, E. F., J. Calvin, J. W. Hedgepeth, and D. W. Phillips. 1985. Between Pacific tides, 5th edition. Stanford University Press, Stanford, California.

Santelices, B. 1990. Patterns of reproduction, dispersal and recruitment in seaweeds. Oceanography and Marine Biology Annual Review 28:177–276.

Sokal, R. R., and F. J. Rohlf. 1981. Biometry, 2nd edition. Freeman, San Francisco.

Southward, A. J., and E. C. Southward. 1978. Recolonization of rocky shores in Cornwall after use of toxic dispersants to clean up the *Torrey Canyon* spill. Journal of the Fisheries Research Board of Canada 35: 682–706.

Straughan, D. 1971. The influence of oil and detergents on recolonization in the upper intertidal zone. Pages 437–440 *in* Proceedings, 1971 joint conference on prevention and control of oil spills. American Petroleum Institute, Washington, DC.

Straughan, D. 1976. Temperature effects of crude oil in the upper intertidal zone. U.S. Environmental Protection Agency, EPA/600/2-76-127, Cincinnati, Ohio.

Thomas, M. L. H. 1977. Long term biological effects of bunker C oil in the intertidal zone. Pages 238–245 *in* D. A. Wolfe, editor. Fate and effects of petroleum hydrocarbons in marine organisms and ecosystems. Pergamon Press, New York.

Thomas, M. L. H. 1978. Comparison of oiled and unoiled intertidal communities in Chedabucto Bay, Nova Scotia. Journal of the Fisheries Research Board of Canada 35:707–716.

Zar, J. H. 1984. Biostatistical analysis, 2nd edition. Prentice-Hall, Englewood Cliffs, New Jersey.

American Fisheries Society Symposium 18:249–270, 1996

Impacts on Selected Intertidal Invertebrates in Herring Bay, Prince William Sound, after the *Exxon Valdez* Oil Spill

ANTHONY J. HOOTEN[1]
Coastal Resources Associates, Inc., 1185 Park Center Drive, Suite A
Vista, California 92083, USA

RAYMOND C. HIGHSMITH
University of Alaska, Institute of Marine Science
School of Fisheries and Ocean Sciences, Fairbanks, Alaska 99775-7220, USA

Abstract.—Population densities of several species of invertebrates were determined at seven pairs of matched oiled and control sites in Herring Bay, Knight Island, Prince William Sound, from 1990 to 1993. Significantly lower population densities were observed for two limpet species and the Sitka periwinkle *Littorina sitkana* in the middle and upper intertidal zones of oiled sites, especially during 1990 and 1991. The predatory snails *Nucella* spp. and the six-armed seastar *Leptasterias hexactis* were either not present or were found in very low densities, and no differences were observed between site pairs. Recovery of invertebrate density had occurred by the end of 1993 at oiled sites in the lower and middle intertidal zones but remained incomplete in the upper intertidal zone for the limpet, *Tectura persona* at sheltered rocky and coarse-textured sites and for the Sitka periwinkle at sheltered rocky sites.

Effects of oil pollution on marine intertidal biota have been well documented since the late 1960s (Teal and Howarth 1984). Previous studies reported reductions in epibenthic invertebrate populations resulting both from oil stranded on shore and from subsequent shoreline cleaning (Mann and Clark 1978; Southward and Southward 1978; Crothers 1983). Recovery times of affected populations have varied depending upon several factors, such as type of oil and degree of weathering, type of substrate, latitude, cleanup technologies used, and life histories of the species involved. Several studies have documented continued effects of oil 5 to 10 years after a spill (Clark et al. 1978; Southward and Southward 1978; Burns et al. 1994; Garrity et al. 1994; Levings et al. 1994).

On the third day after the *Exxon Valdez* ran aground on Bligh reef, a storm rapidly pushed the oil slick toward the western half of Prince William Sound and eventually through Knight Island Passage and into the Gulf of Alaska. Herring Bay (60°N, 147°W), located at the northern end of Knight Island and approximately 130 km SW of Bligh Reef, received a large volume of oil on or around 29 March 1989 (ADEC 1989a; Figure 1). With the exception of some southeastern and south-central shorelines, practically all of the bay experienced some degree of oiling, with many areas receiving heavy accumulations. Herring Bay became a center for many oil spill response and support activities, such as the oily waste incinerator, freshwater supply, and anchorage for the berthing vessels of staff and cleanup crews. At the height of spill response, Herring Bay had a human population approaching 1,000 (J. Bauer, Alaska Department of Environmental Conservation, personal communication).

From June through August 1989, shoreline treatment was intensive within Herring Bay. Shorelines were mapped and treatment was managed and monitored using a system of labeled shoreline segments adopted by the Coast Guard, Alaska Department of Environmental Conservation (ADEC), and Exxon Corp. Segments varied in length from hundreds of meters to kilometers and also varied in degree of oiling and treatment.

The treatment technology employed by Exxon Corp. evolved over several months, with early methods including hand-wiping of rocks and physical removal of oiled *Fucus* plants and drift algae. Shallow-draft vessels were outfitted with water pumps and hoses supplying low- to high-pressure cold water to shore-based crews, which forced oil from shore into nearshore waters for skimming. The use of water heated to various temperatures became a preferred treatment because of its expediency in freeing stranded oil. Steam treatment was also applied to some areas in Herring Bay by portable, high-pressure hand or "Lambda" wands, which emitted water up to 77°C (ISCC 1989; P. Endres, ADEC, personal communication). Omni Sweeps,

[1]Present address: AJH Environmental Services, 4005 Glenridge Street, Kensington, Maryland 20895, USA.

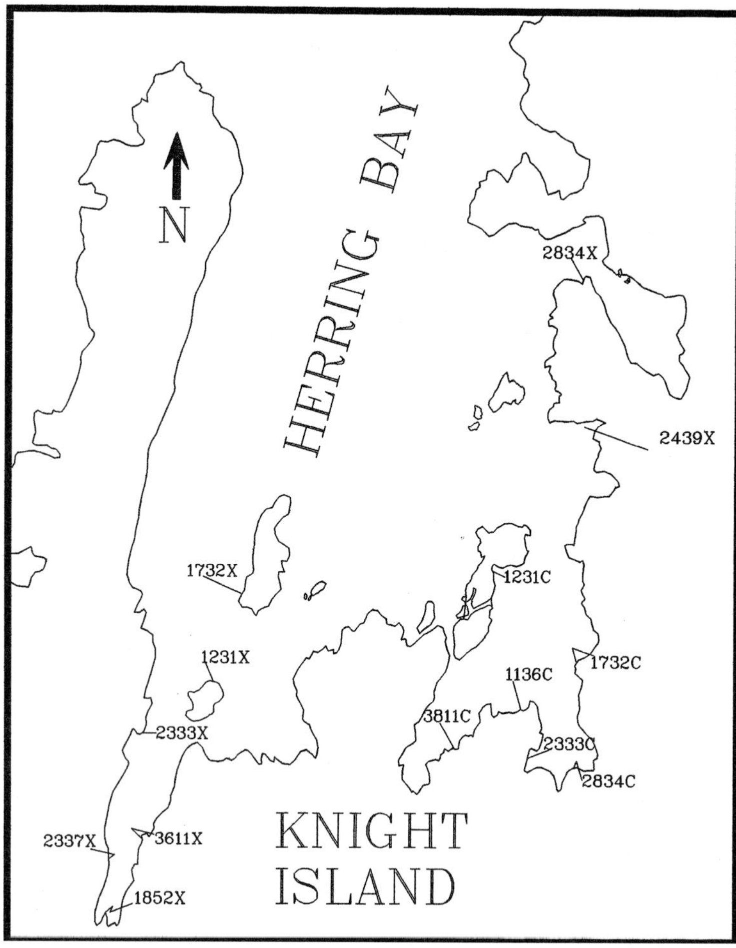

FIGURE 1.—Map of Herring Bay, Knight Island, Prince William Sound, showing the location of study sites, 1990–1993. Sites were assigned the following numbering convention: The first digit represents the habitat type (i.e., 1 = sheltered rocky, 2 = coarse-textured, and 3 = vertical wall); the second digit represents a general compass orientation in one of eight directions (i.e., 1 = N, 2 = NE, 3 = E and so on to 8 representing NW); the third digit represents a study or experiment number (3 for population dynamics studies); the fourth digit represents the replication number of site pairs. Letters represent control (C) or oiled (X) sites. The only site pairs not following this numbering convention were 1136C–1852X, and 3811C–3611X, as these sites were originally designated for other studies in 1990.

barge-mounted articulating arms that sprayed heated water (up to 60°C at 7 kg/cm^2), were perhaps the most effective treatment in removing oil from shorelines and were widely used in Herring Bay from June through August (ADEC 1989b; S. Jokiel, Chemtrak, personal communication). The extensive treatment of most oiled shorelines in Herring Bay was completed by 30 August 1989 (ADEC 1989b).

Herring Bay was also the location of 2 of 11 sites in Prince William Sound that were selected by the federal and state agencies and Exxon Corp. to be set aside from shoreline treatment (A. Maki and A. Teal, Exxon Corp., unpublished). These two sites were part of an early effort to isolate certain oiled shorelines so that studies might be designed to compare the effects of shoreline treatment versus no treatment and to monitor natural oil-weathering. Unfortunately, during the course of the response effort, several set-aside sites were compromised. The regulatory agencies monitoring treatment would not allow any heavily oiled sites to remain untreated, so only moderately to lightly oiled sites were available for study (J. Bauer, personal communication). At least two of the sites received manual or mechanical treatment or bioremediation in 1989 or 1990 (ADEC 1989b; J. Bauer and D. Gardner, personal communication), and one of the sites in Herring Bay was frequently used for landing by helicopters in 1989 and 1990 (A. J. H., personal observation).

In spring 1990, the Coastal Habitat Injury Assessment program established experimental and monitoring studies in Herring Bay to assess the effects of the oil spill and subsequent treatment on marine intertidal communities.

Here we compare density data for selected invertebrates from paired control and oiled sites from 1990 to 1993. Sitka periwinkles *Littorina sitkana*, dogwinkles *Nucella* spp., and six-armed seastar *Leptasterias hexactis*, were selected for study because they lack a free-swimming larval stage (which may limit species dispersal) and were hypothesized to recover slowly after reduction in population size over a large area. Limpets were included because they are important intertidal grazers, and in previous oil spills their reduction in density has been attributed to changes in algal community structure (Southward and Southward 1978).

Methods

Site Selection

Initially, our study was designed to include a range of sites with different oiling and treatment histories in Herring Bay, including non-oiled control sites and oiled sites that were washed with different treatments and bioremediated. Ideally, this combination of sites would have been replicated several times to achieve statistical rigor. After detailed field surveys in May 1990, review of the Exxon Corp., federal, and state shoreline assessment data, and detailed discussion with ADEC monitors working in Herring Bay in 1989, specific treatment histories could not be clearly determined for several reasons. Our study sites were much smaller than the minimum segment lengths used to monitor and record data on shoreline treatment. Therefore, although a segment may have had a certain type of treatment documented, it may not have been for the entire segment length, and exact location of treatment within a segment was usually not recorded (ADEC 1989b). Some heavily oiled segments required repeated treatments, which were often performed by treatment techniques different from the initial one. Furthermore, Omni Sweeps in Herring Bay were frequently removed before completion of assigned segments and relocated to other areas of higher treatment priority. Vacated segments or repeated treatments were often completed by other techniques, resulting in variable treatment histories (G. Lewis and J. Coxey, Chemtrak, personal communication). Herring Bay also received wide application of bioremediation on almost every oiled segment in 1989 or 1990 (ADEC 1989b, 1989c).

Therefore, a matched pair design of oiled and non-oiled study sites was adopted and shoreline treatment was not initially defined as a variable. Subsequent research and interviews with personnel familiar with the events in Herring Bay allowed us to learn where specific treatments occurred (Table 1). During site selection, control sites were restricted to the southeast corner of Herring Bay, where ice had prevented oil from entering in spring 1989 (AeroMap 1989). Most paired oiled sites were located in the lower-mid and western portion of the bay to closely match exposure and habitat type.

In 1990, permanent quadrats measuring 50 × 20 cm were established at five pairs of sites: three sheltered rocky sites (including one pair of vertical rock walls) and two coarse-textured beaches (Table 1; Figure 1). Sheltered rocky shores and coarse-textured beaches represent the most common habitat types found in Prince William Sound. The site pairs may have differed in substrate character and slope, but sites in each pair were matched as closely as possible for substrate composition, slope, direction and solar aspect, wave exposure, and common biological communities.

Each site selected for study was measured parallel to shore at the approximate mean high high water line (MHHW) zone, (i.e., at the base of the black lichen *Verrucaria*). Six transects perpendicular to shore were established at each site using a random numbers process. At each transect a tape was extended to the waterline, and tidal elevations of 1, 2, and 3 m below MHHW (or meters of vertical drop [MVD]) were determined using a meter stick and a site level. Within each MVD the position of one 50 × 20-cm quadrat was determined using a random procedure. Eighteen quadrats were established per site. The long dimension of the quadrats was perpendicular to the waterline. Within each quadrat, all Sitka periwinkles, *Nucella* spp., *Leptasterias hexactis*, and limpets were counted. Sites were sampled on seven dates in 1990 from June through September.

One sheltered rocky site pair (1136C and 1852X) and one coarse-textured oiled beach (2337X, matched to existing control site 2333C) were added to the study in 1991 to increase replication. Sites were sampled on six dates in 1991 from May until

TABLE 1.—List of study sites in Herring Bay and their physical characteristics. Abbreviations are SR = sheltered rocky; CT = coarse-textured; VW = vertical walls; Omni = Omni Sweeps; HWHP = hot water, high pressure; HWLP = hot water, low pressure. Site length is given in meters, orientation in degrees, and slope in percent. Oiling level, treatment, and segment numbers are from the Alaska Department of Environmental Conservation (ADEC 1989b, 1989c) maximum impact survey and treatment-monitoring data. The treatment represented for segment KN0133 (associated with site 1732X) is based on personal field observations and differs from that recorded by ADEC.

Site	Habitat type	Site length	Orientation	Slope	Oiling level	Treatment	Segment
1231C	SR	55	45	23	Light, very light	None	KN5011
1231X	SR	32	50	21	Heavy, moderate	Omni, HWHP	KN0133
1732C	SR	38	260	27–51	Light, very light	None	KN5002
1732X	SR	42	290	28–37	Heavy, moderate	None (Omni)	KN0133
1136C	SR	35	330	30	None	None	KN5006
1852X	SR	41.5	355	33	Heavy, moderate	Omni, HWHP	KN0129
2333C	CT	51	60	16	Very light	None	KN5004
2333X	CT	42	95	18	Moderate	Omni, HWHP	KN0131
2337X	CT	55	110	14	Moderate, light	Omni, HWLP	KN0130
2834C	CT	37	270	14	Very light	None	KN5004
2834X	CT	30	290	14	Moderate	Omni, HWLP	KN0121
3811C	VW	24	305	90	None	None	KN5007
3611X	VW	33	255	90	Moderate, heavy	Omni	KN0128

September. In 1992 and 1993, sites were sampled during two visits in May–June and July–August.

Data for each site pair and sample date were tested for variance homogeneity using Levene's test at the $P = 0.10$ level. Unequal variances were log transformed. Site pairs were segregated into sheltered rocky and coarse-textured groups. Data were pooled for oiled and control treatments, and site pairs for each habitat type served as blocks in the randomized block design (Bhattacharyya and Johnson 1977; Sokal and Rohlf 1981). A two-way analysis of variance without replication was used to test for differences in density for most species on each of the 18 sample dates from 1990 to 1993.

Site pair 3811C and 3611X was separated from the coarse-textured and sheltered rocky groups and analyzed as a single site pair using a *t*-test because the steep slope of the vertical walls was not comparable to the other habitat types. Species with low frequency (i.e., file dogwinkle *Nucella lima* and *Leptasterias hexactis*) were analyzed using randomization testing (Manly 1991). Null hypotheses were rejected when $P \leq 0.05$.

The two set-aside sites in Herring Bay were originally included in our study design; however, the grain size, orientation, and wave exposure made them poor matches to any control or oiled, treated site. Nonetheless, because the set-aside sites were the only areas known to have been oiled and not treated, we sampled one of the sites (2439X, Figure 1) on two dates in 1991 and on one date each in 1992 and 1993 to track changes in invertebrate population density. Descriptive data were generated for invertebrate densities at the set-aside site.

Hydrocarbon Sampling

Sediment hydrocarbon samples were not collected at the population dynamics sites, thus minimizing disturbance to the populations under study. Sediment hydrocarbon samples were collected in 1990 and 1991 at eight oiled and control site pairs located near the population dynamics study sites to document exposure in Herring Bay (Figure 2). Sediment samples were collected from within a 1-m radius to the left of a randomly placed study quadrat at 1, 2, and 3 MVD. Transects and quadrats were established in a manner identical to the population dynamics site pairs. Sediment samples were placed into two 40-ml sterile glass jars using hydrocarbon-free instruments and were stored and handled according to standard operating procedures adopted by the *Exxon Valdez* Oil Spill Trustees (C. Manen, National Oceanic and Atmospheric Administration, personal communication). Samples were not collected at some sites on some sample dates because sediment was scarce within the rocky substrate. The sediment samples were analyzed for petroleum hydrocarbons by a high-performance liquid chromatography (HPLC)–fluorescence screening method (Krahn et al. 1991).

Results

At the sheltered rocky pairs, a reduction of Sitka periwinkle is evident at 1 MVD at the oiled sites beginning in 1992 and densities were significantly less on both sample dates in 1993 (Figure 3A). At MVD 2 at sheltered rocky sites, Sitka periwinkle densities were generally lower at oiled sites, with

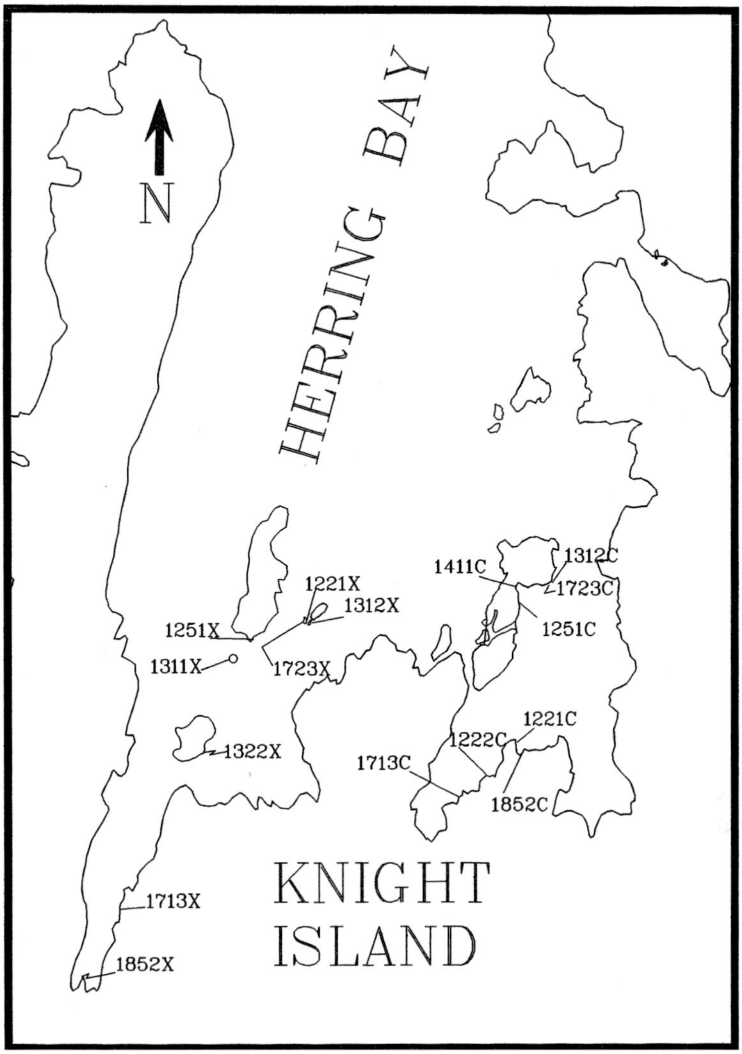

FIGURE 2.—Location of sites in Herring Bay sampled for sediment hydrocarbons, 1990–1991. See Figure 1 for explanation of numbering and lettering system.

significant differences in 1990 and on occasion in 1991–1992. Sitka periwinkle was less dense at oiled sites compared with controls at 1 and 2 MVD at the coarse-textured pairs during 1990 and early 1991 (Figure 3B). Lower Sitka periwinkle densities were also found at site 3611X compared with 3811C at MVD 1 and 2, with significant differences observed on several dates (Figure 3C).

Frilled dogwinkle *Nucella lamellosa* occurred in densities sufficient for statistical comparison at only one of the seven site pairs (1732), and there were no significant differences in densities between oiled and control sites (Figure 4). The other direct developers, file dogwinkle and *Leptasterias hexactis*, were either absent or found in very low densities and no significant differences between oiled and control sites were observed (Figure 5).

The upper intertidal limpet *Tectura persona* was generally less dense at the oiled sites in both sheltered rocky and coarse-textured habitats (Figure 6). At sheltered rocky sites, significant differences persisted through 1993 at MVD 1 and through 1992 at MVD 2 (Figure 6A). At coarse-textured sites, densities at oiled sites were significantly lower at MVD 1 on most sampling dates in 1990 and 1991, and again in June of 1993 (Figure 6B). For MVD 2 and

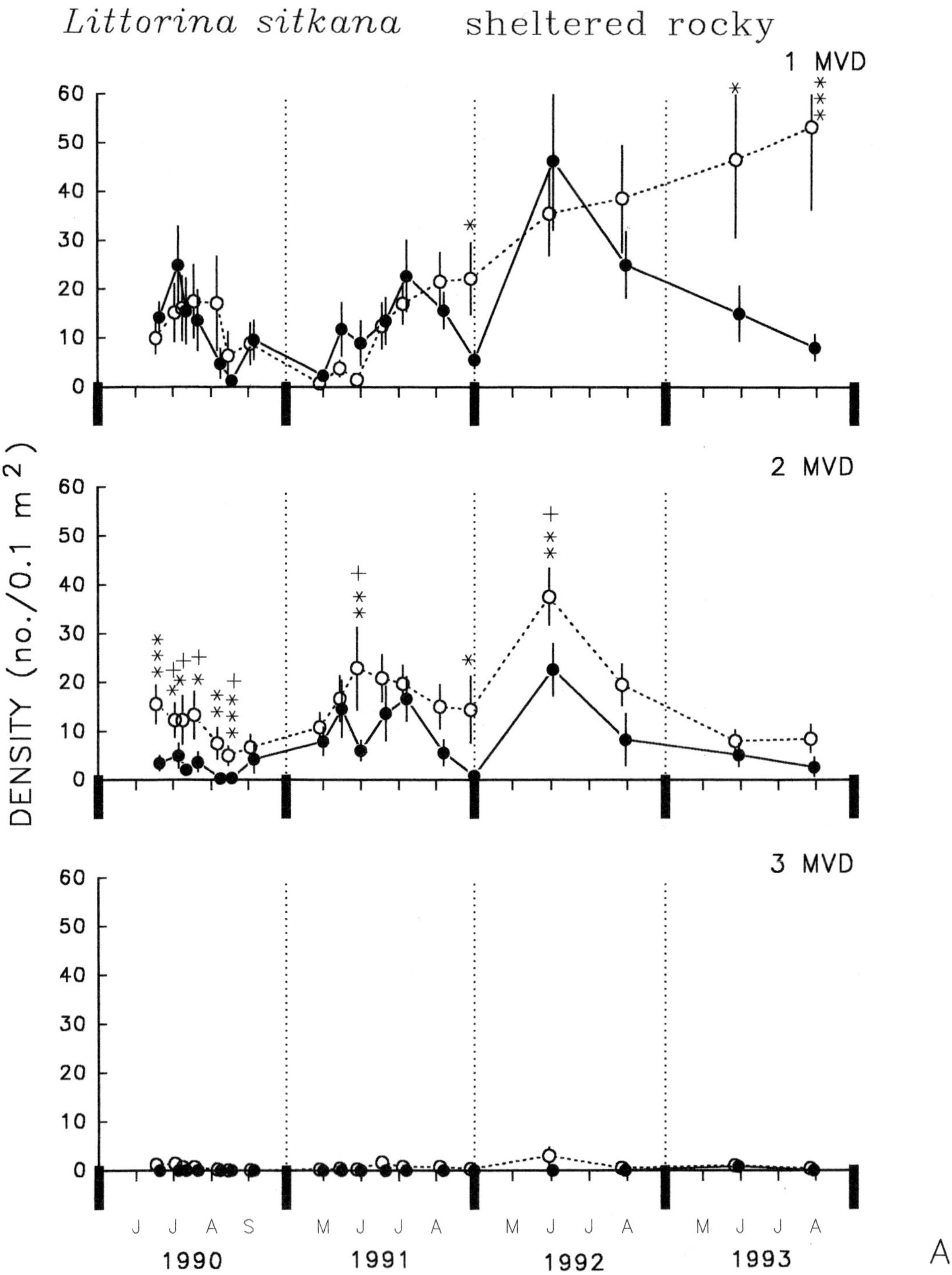

FIGURE 3.—Density of Sitka periwinkle *Littorina sitkana* at (**A**) sheltered rocky, (**B**) coarse-textured and (**C**) pair 3811C and 3611X at each tidal elevation (MVD = meter of vertical drop below mean high high water) from 1990 to 1993 sample seasons in Herring Bay. For **A** and **B**, $N = 12$ for dates 1–10 and $N = 18$ for dates 11–18; for **C**, $N = 6$. The oiled sites are represented by solid circles and lines, and the control sites are represented by open circles and dashed lines. The error bars represent 1 SE. Asterisks over a given sample date indicate statistically significant differences between oiled and control treatments from analysis of variance ($P \leq 0.05$*, $P \leq 0.025$**, and $P \leq 0.01$***). A plus sign indicates that $P \leq 0.05$ between site pairs within each habitat group. Vertical dotted lines separate sampling seasons (May–September). Months given are May (M), June (J), July (J), August (A), and September (S).

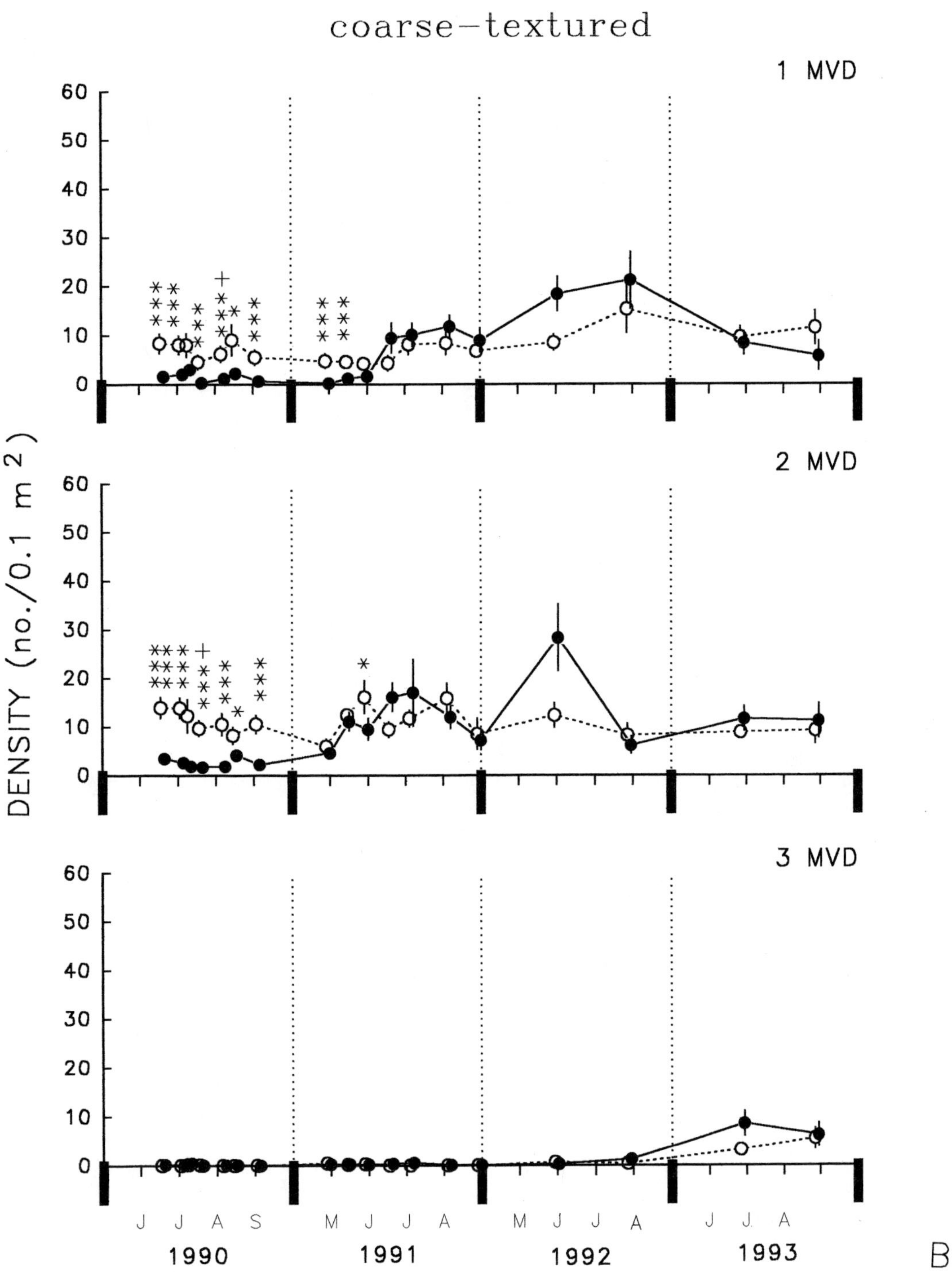

FIGURE 3.—Continued.

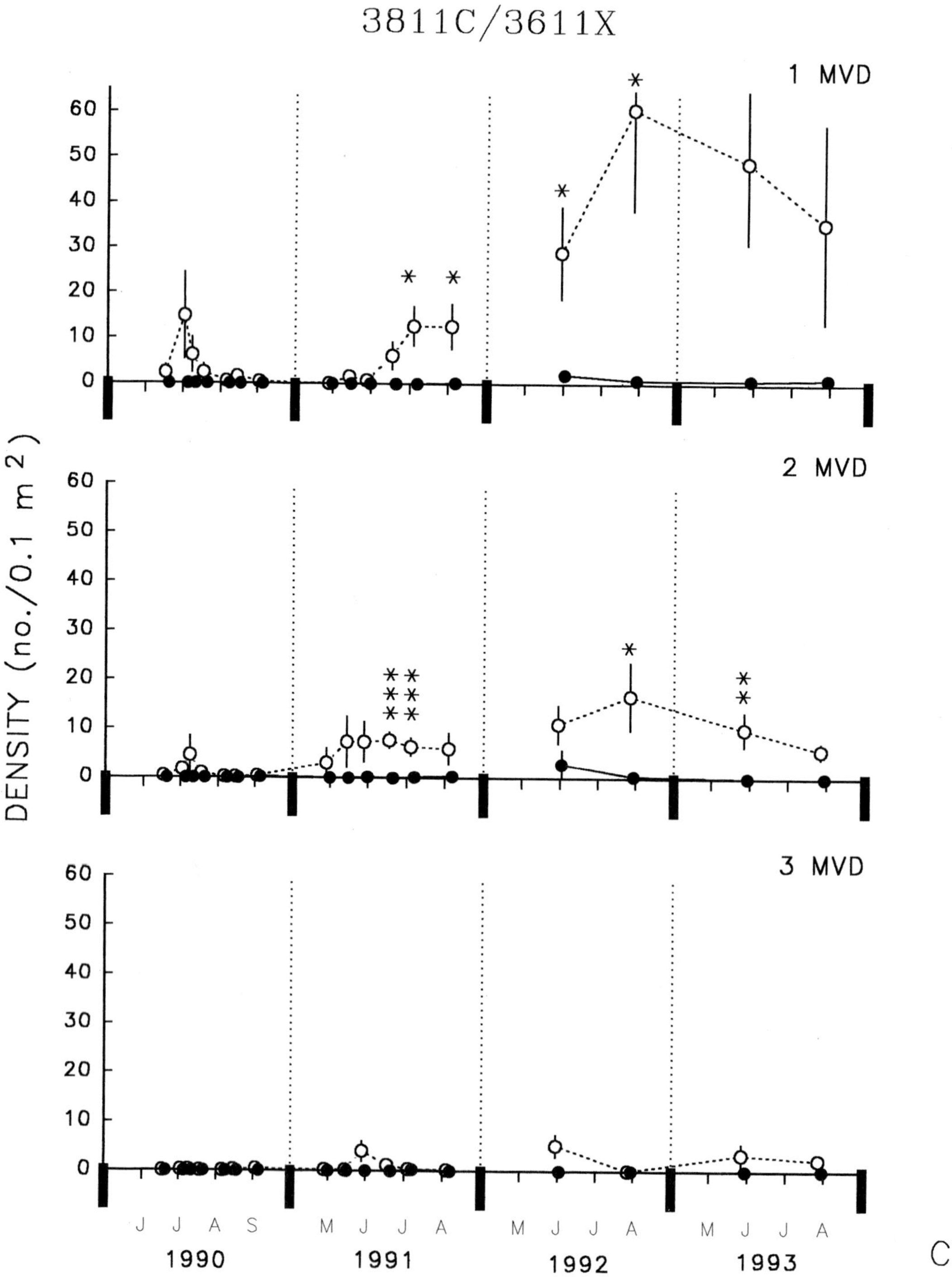

FIGURE 3.—Continued.

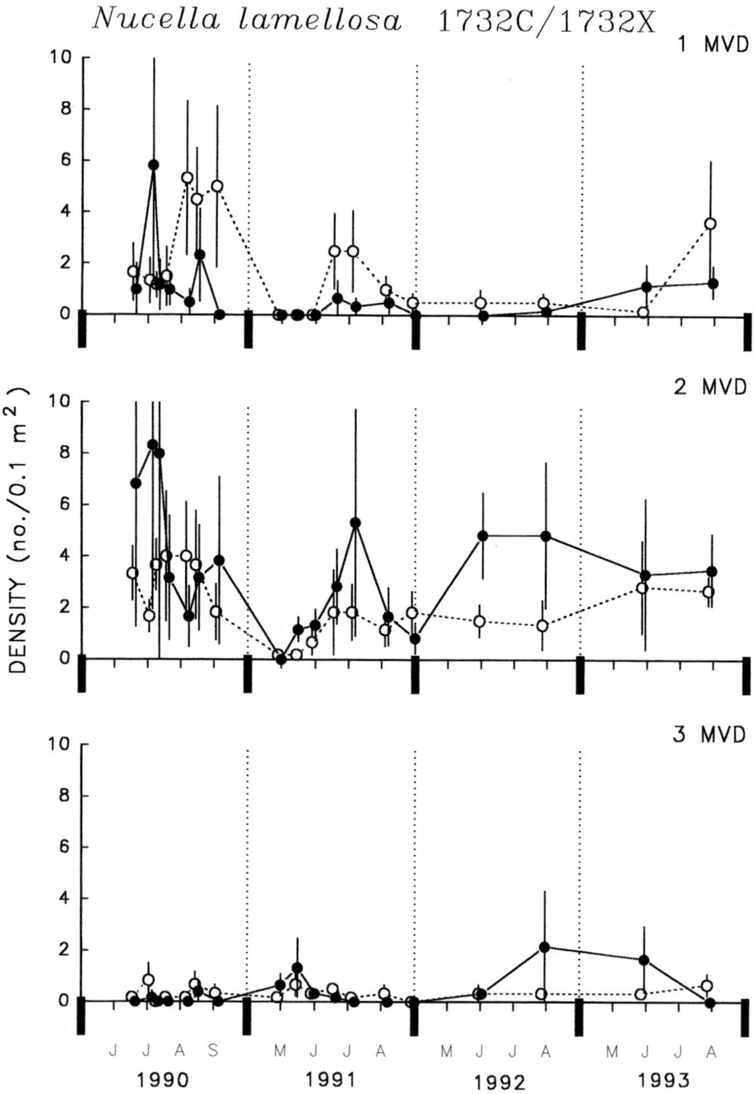

FIGURE 4.—Density of frilled dogwinkle *Nucella lamellosa* at site pair 1732C and 1732X at each tidal elevation during the 1990–1993 sample seasons in Herring Bay. See caption of Figure 3 for explanation of symbols and abbreviation.

3, no significant differences were found after June 1991. Densities at the vertical wall site pair (3811C and 3611X) were low, with one significant difference occurring at MVD 2 (Figure 6C).

Another limpet, *Lottia pelta*, occurs lower in the intertidal zone than *T. persona* and exhibited fewer differences between control and oiled sites, especially in sheltered rocky habitats (Figure 7A). Significantly greater densities of *L. pelta* occurred at control sites on one to three sample dates for each tidal level of sheltered rocky sites. At MVD 2 and 3 of coarse-textured sites, *L. pelta* densities increased more rapidly at oiled than at control sites in 1993 (Figure 7B). *L. pelta* densities on the vertical walls (3811C and 3611X) were significantly higher at control sites only at MVD 3 in June 1992 (Figure 7C).

At the set-aside site (2439X), Sitka periwinkle density was similar to that at the oiled sites of the sheltered rocky pairs at MVD 1 and 2 (Table 2; Figure 3A), including a reduction in density at MVD 1 from 1992 to 1993. File dogwinkle and *Leptasterias hexactis* occurred in very low densities. The mean densities of *T. persona* and *L. pelta* (Ta-

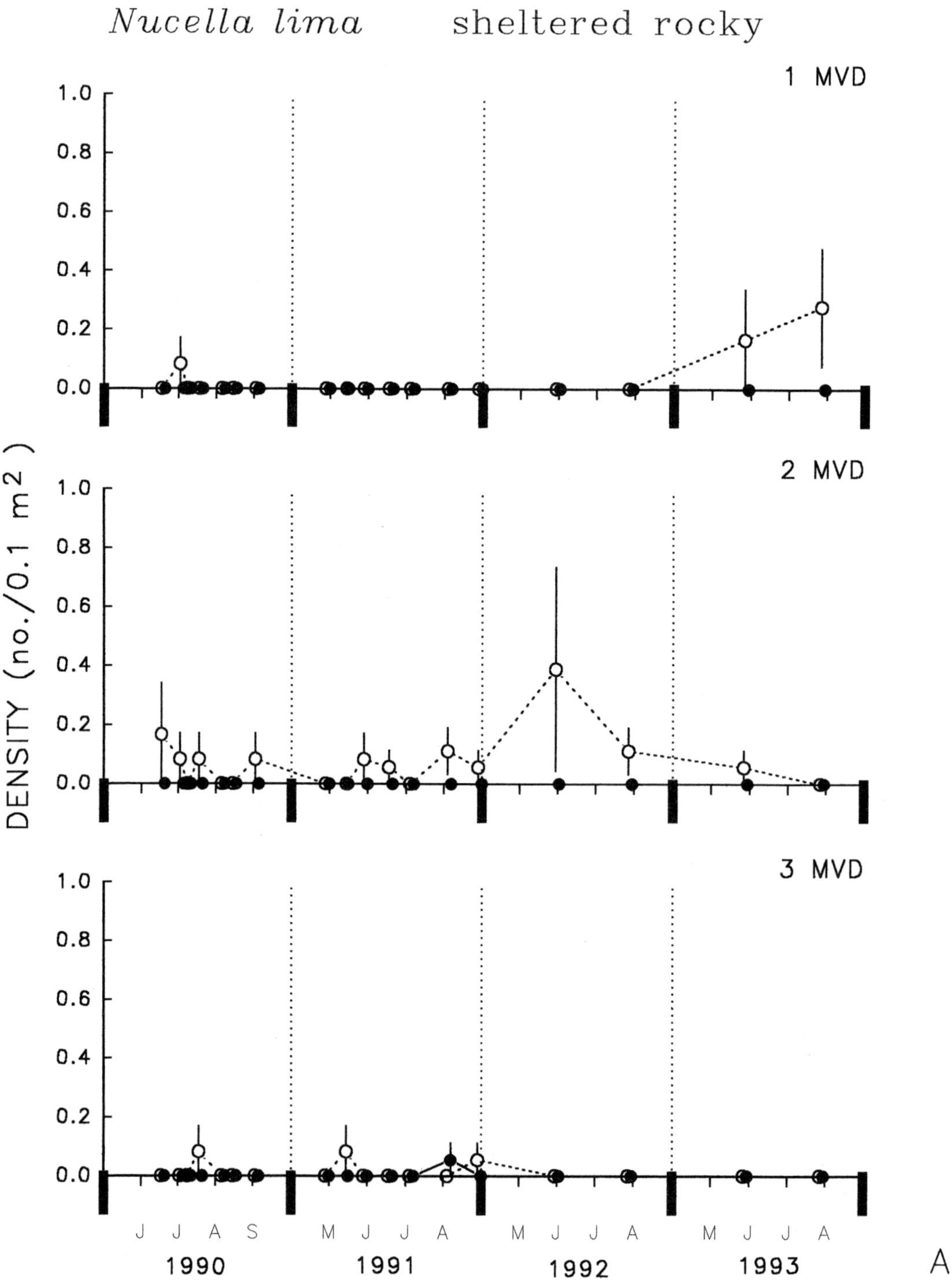

FIGURE 5.—Density of file dogwinkle *Nucella lima* at (**A**) sheltered rocky and (**B**) coarse-textured sites; (**C**) density of *Leptasterias hexactis* at sheltered rocky sites only. Data are for the 1990–1993 sample seasons in Herring Bay. See caption of Figure 3 for explanation of symbols and abbreviation.

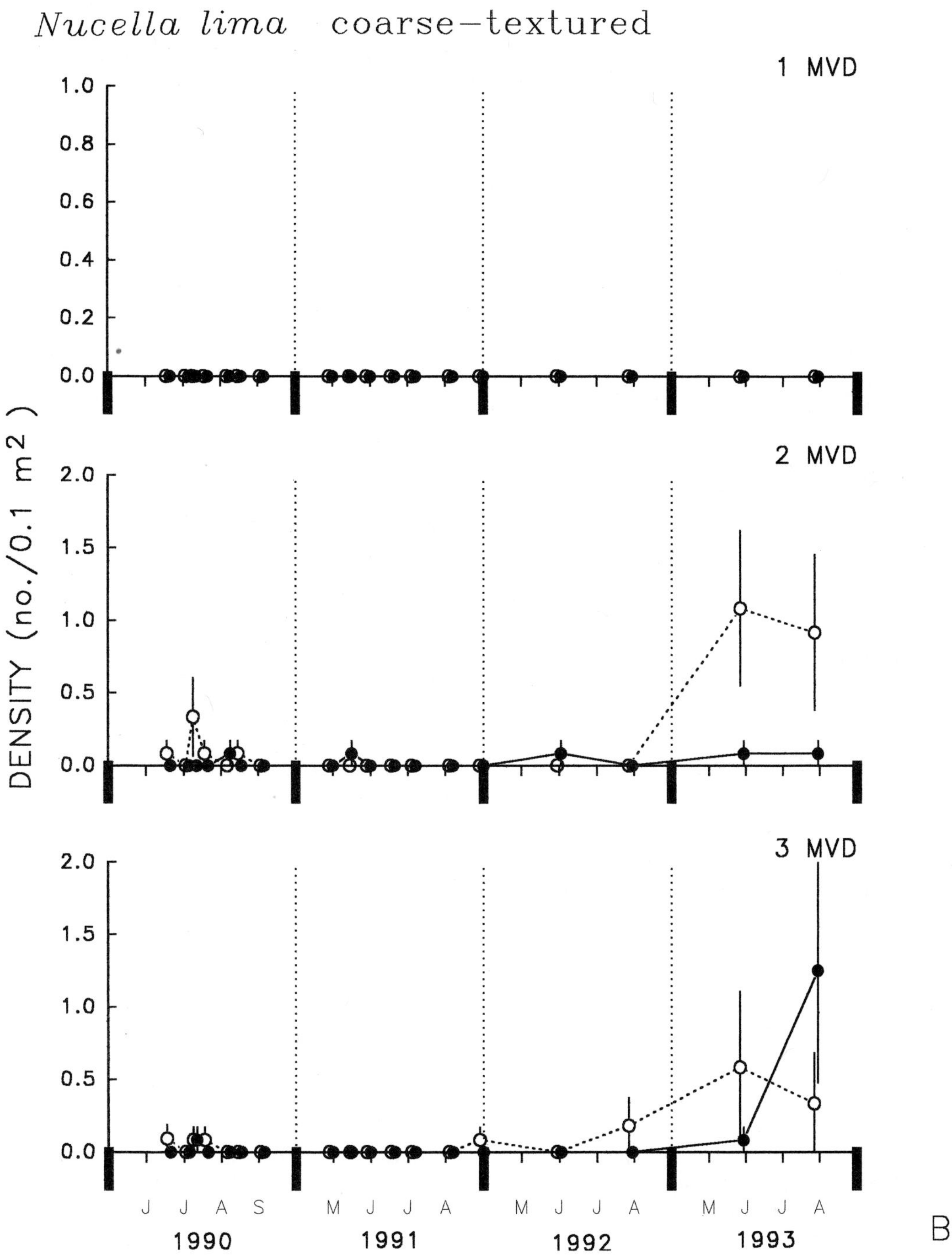

FIGURE 5.—Continued.

ble 2) appear consistent with those of the oiled sites for both habitat types (Figures 6 and 7).

Sediment hydrocarbon samples for 1990 and 1991 were pooled by MVD, and an index of hydrocarbon concentration was calculated by the HPLC–ultraviolet fluorescence method defined by Krahn et al. (1991; Table 3). Ultraviolet fluorescence indices for sediment hydrocarbons were significantly

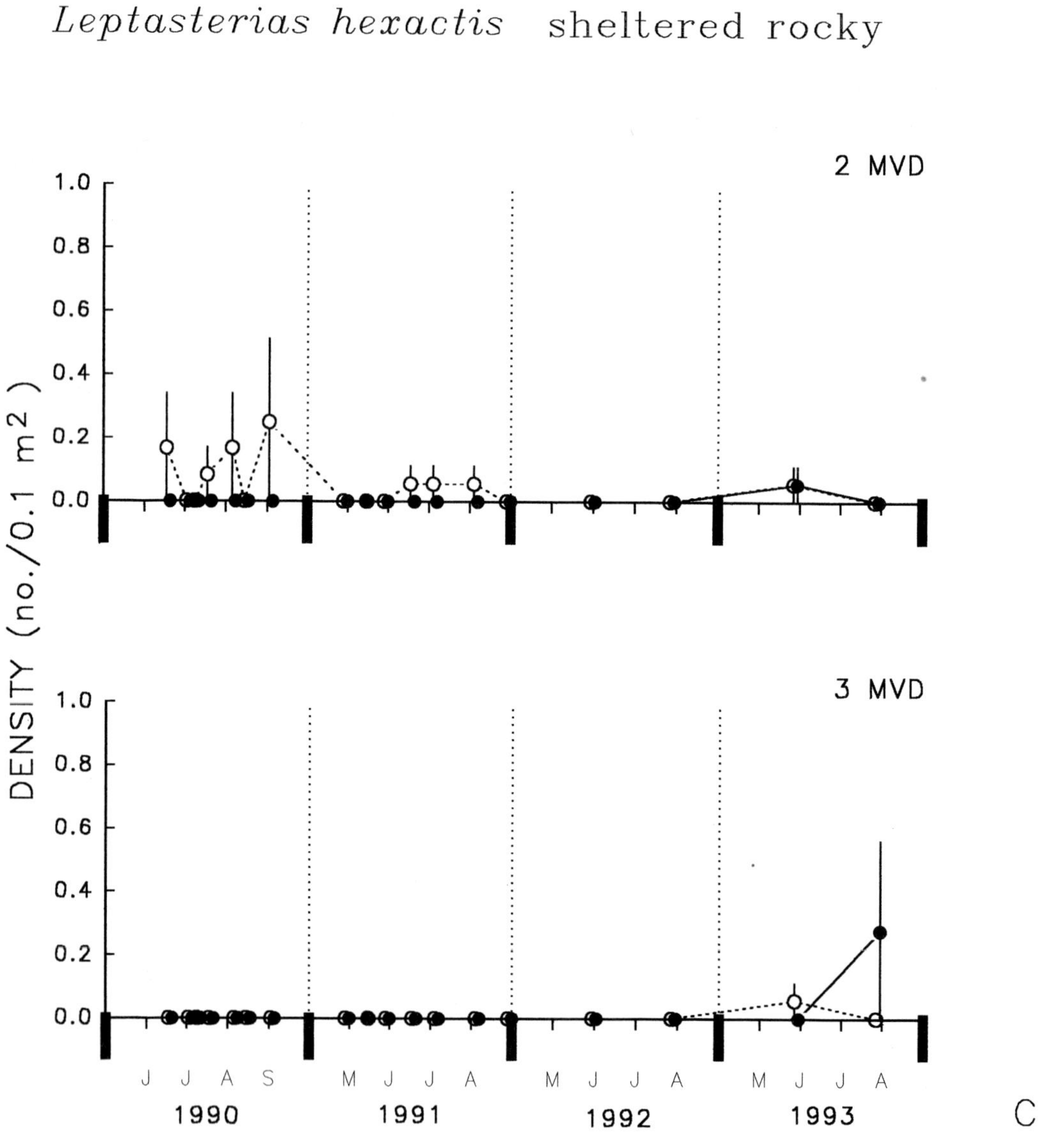

FIGURE 5.—Continued.

higher at oiled sites ($P < 0.002$, Mann–Whitney *U*-test for pooled control versus oiled sites; Table 4). The indices decreased by 40–60% at oiled sites from 1990 to 1991.

Discussion

Reductions in densities of invertebrates, particularly intertidal grazers such as limpets and periwinkles, have been reported for other oil spills (Nelson-Smith 1977; Mann and Clark 1978; Southward and Southward 1978), and our findings are consistent with these earlier studies. A major assumption of any study where sites are chosen after a perturbation is that control sites represent predisturbance conditions. In Herring Bay, the intertidal communities at the control sites were assumed to be similar to those at the oiled sites before the spill. Risks associated with this assumption were minimized by pairing sites that resembled each other in as many physical characteristics as possible. Segregating the site pairs by habitat type and similar slopes also minimized between-site differences during analysis.

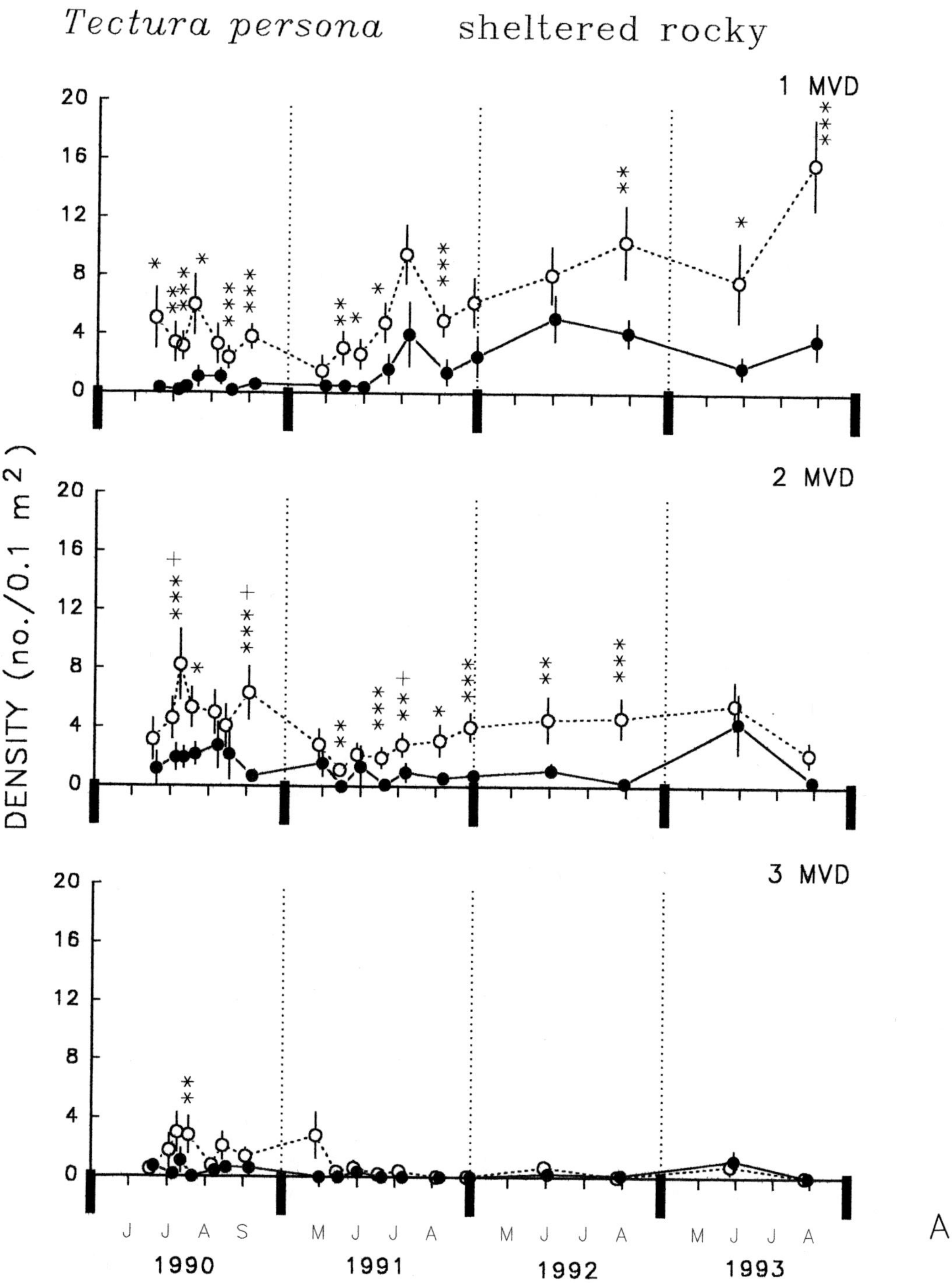

FIGURE 6.—Density of the limpet *Tectura persona* at (**A**) sheltered rocky, (**B**) coarse-textured, and (**C**) pair 3811C and 3611X at each tidal elevation during the 1990–1993 sample seasons in Herring Bay. See caption of Figure 3 for explanation of symbols and abbreviation.

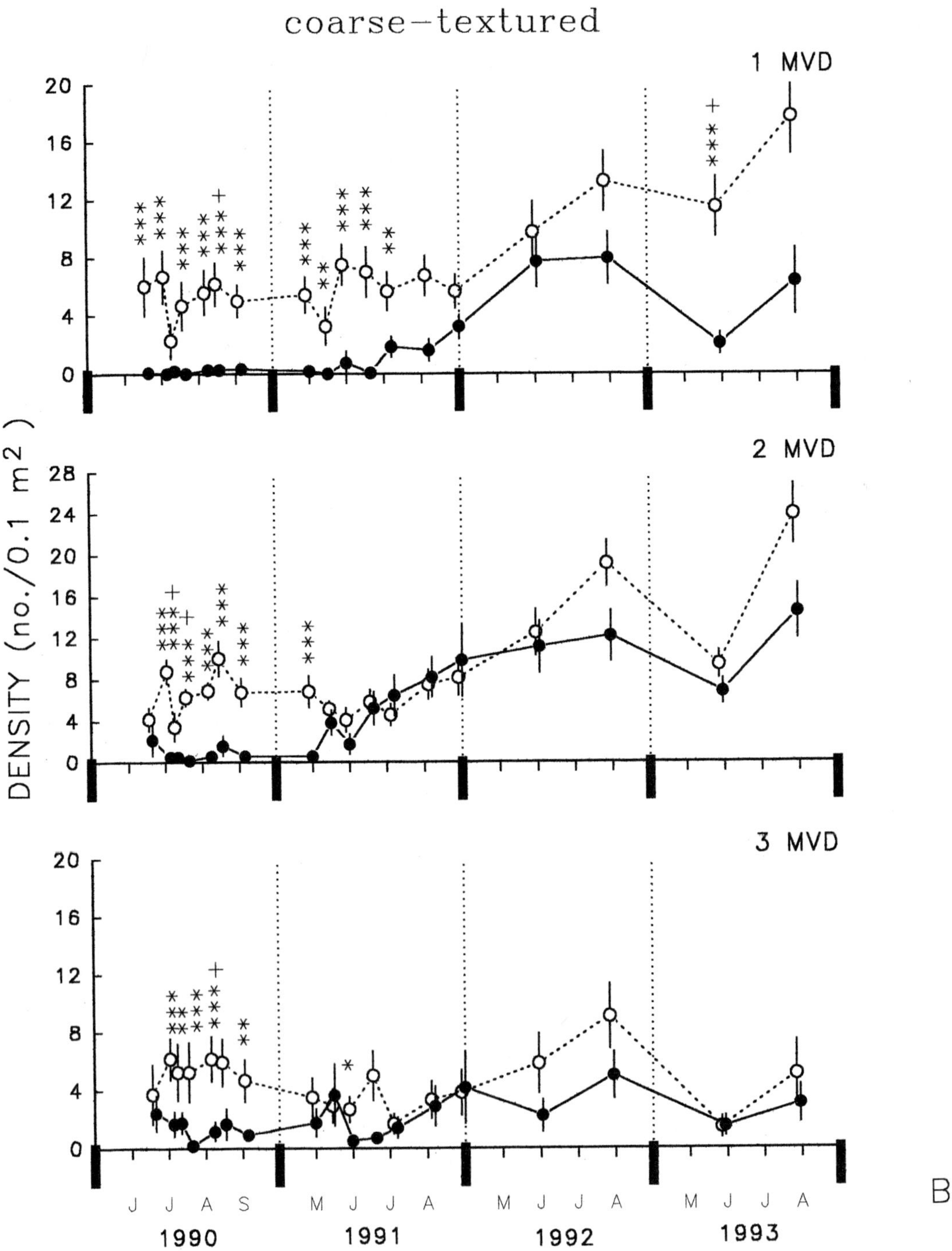

FIGURE 6.—Continued.

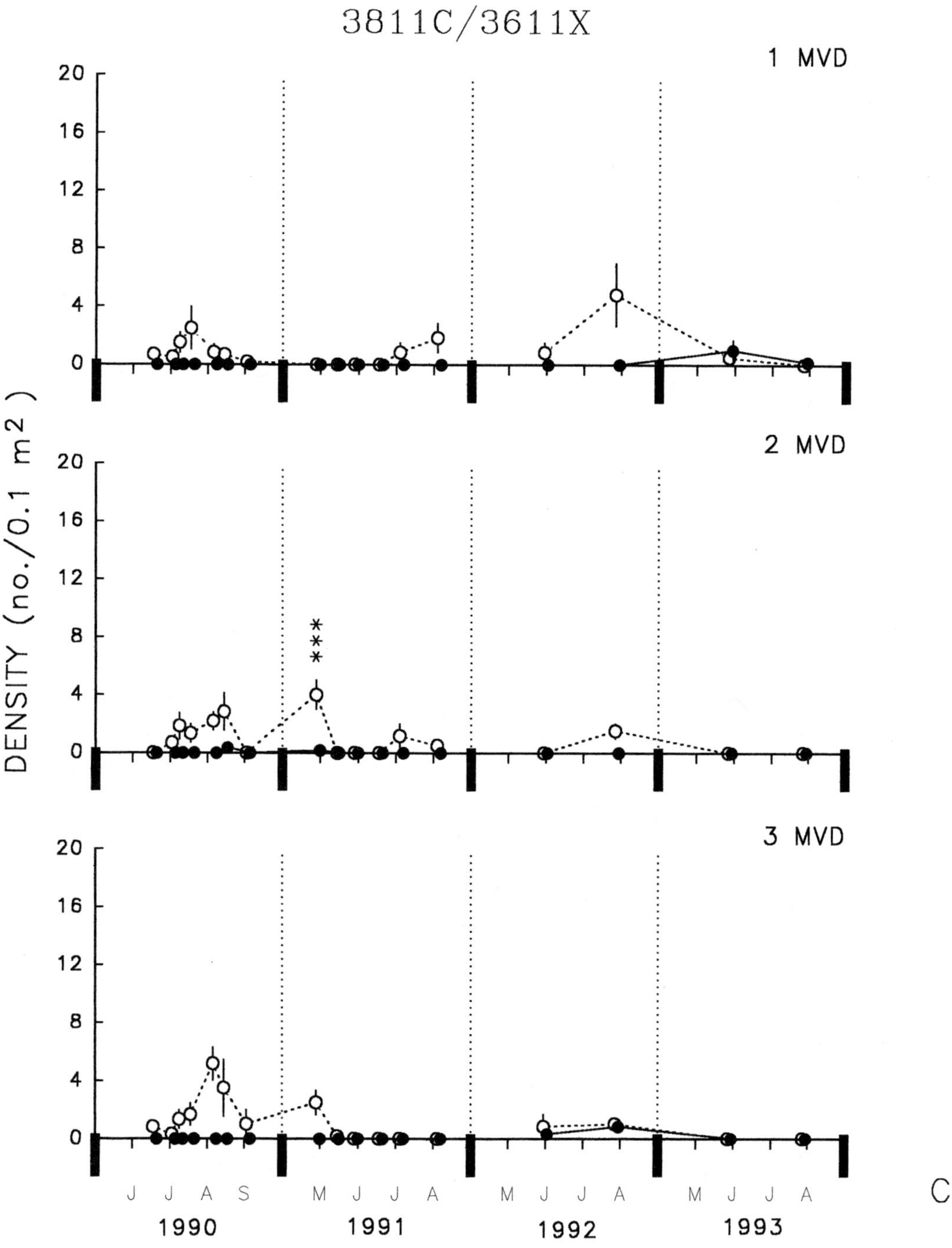

FIGURE 6.—Continued.

Other factors may potentially affect the populations in Herring Bay. Different sections of the bay have differences in salinity, ice scouring, and exposure to waves and storms, and these conditions probably contribute to variability between sites (Highsmith et al. 1993); however, when viewed over time, the data presented here show definite effects from the *Exxon Valdez* oil spill on several of the species at our study sites.

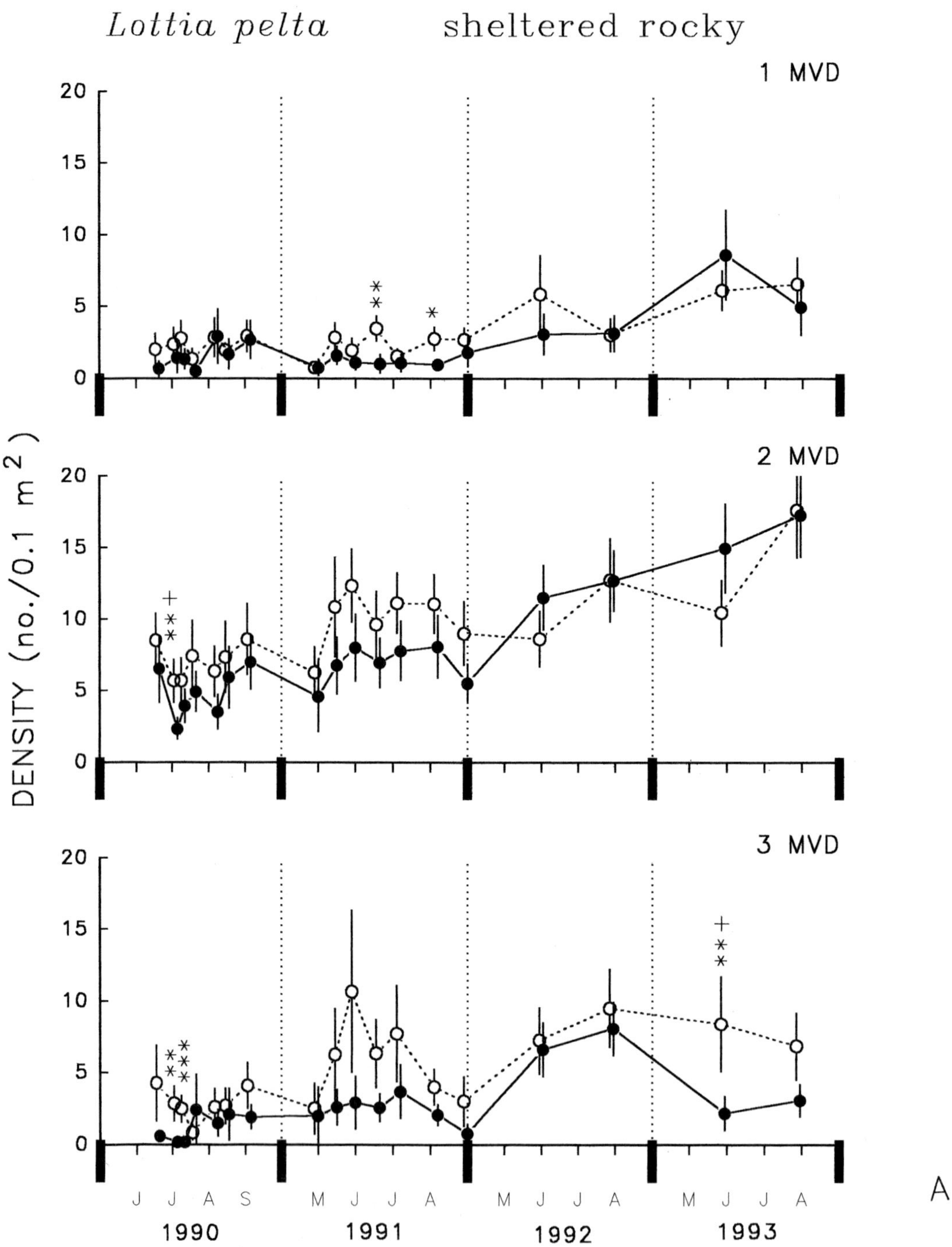

FIGURE 7.—Density of the limpet *Lottia pelta* at (**A**) sheltered rocky, (**B**) coarse-textured, and (**C**) pair 3811C and 3611X at each tidal elevation during the 1990–1993 sample seasons in Herring Bay. See caption of Figure 3 for explanation of symbols and abbreviation.

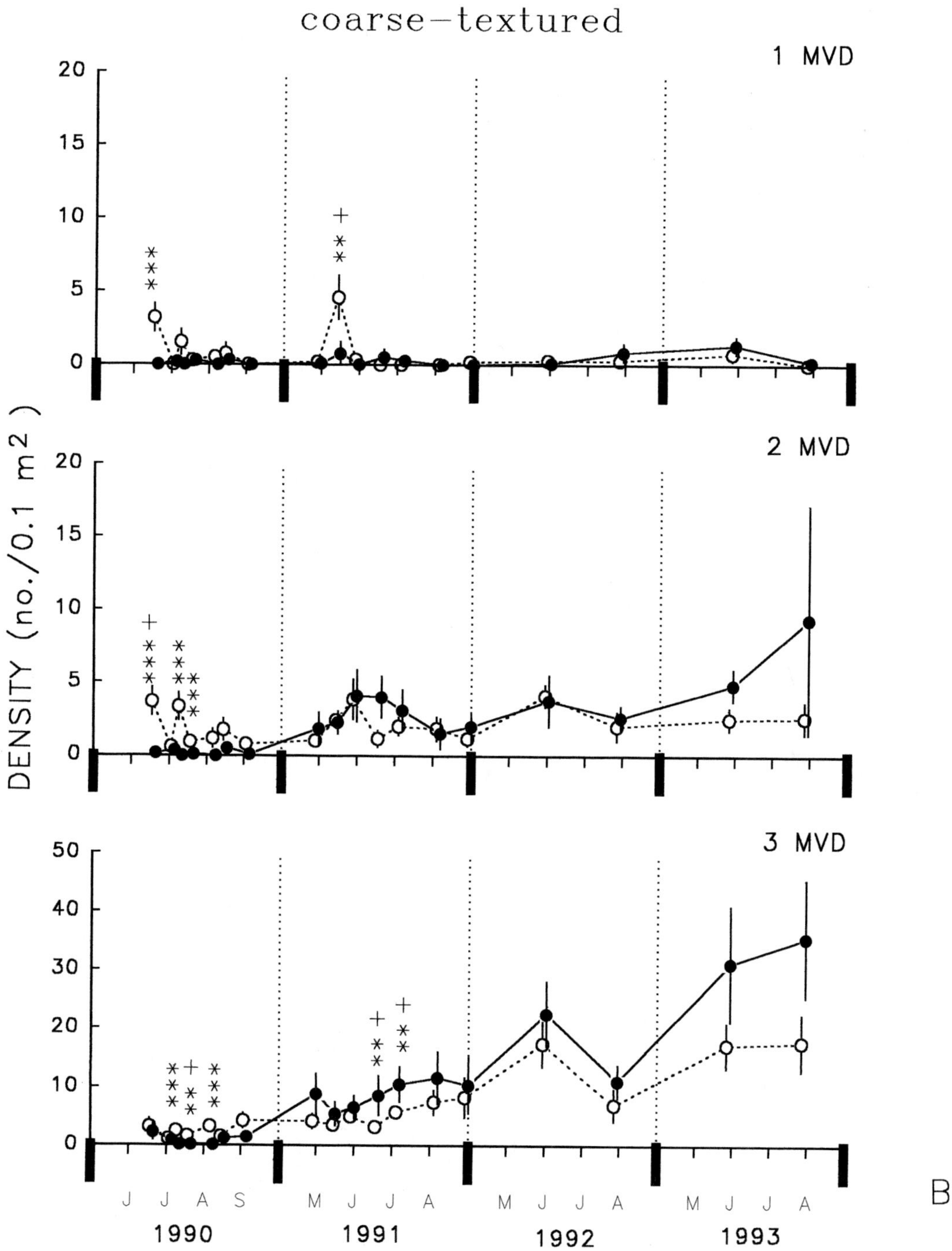

FIGURE 7.—Continued.

A main hypothesis of this population study was that invertebrates lacking dispersal phases in their life histories, such as Sitka periwinkle, *Nucella* spp., and *Leptasterias hexactis*, would take longer to recover on oiled or treated sites because of reduced adult densities and/or fecundities and the low probability of recruitment from elsewhere. Of the three direct-developers studied, only Sitka periwinkle ap-

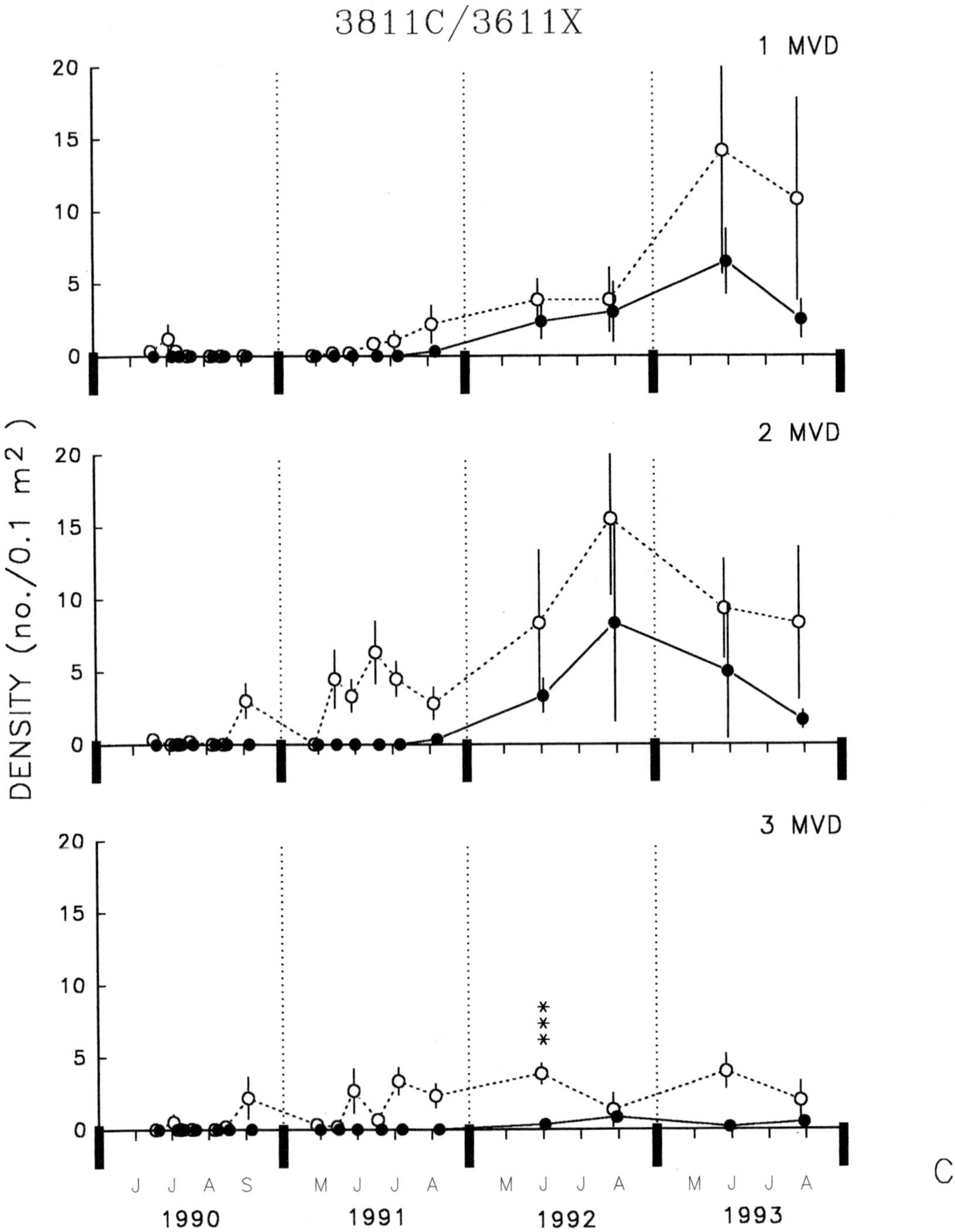

FIGURE 7.—Continued.

pears to have been somewhat affected by oil (Figures 3–5). In 1990 and the first half of 1991, densities of Sitka periwinkle were significantly lower in oiled areas: in the second MVD at sheltered rocky sites and in the first and second MVD at coarse-textured sites. In subsequent seasons, except for the first MVD at sheltered rocky sites, densities declined.

TABLE 2.—Mean density (number/0.1 m^2) and SE (in parentheses) for Sitka periwinkle *Littorina sitkana*, file dogwinkle *Nucella lima*, *Leptasterias hexactis*, *Tectura persona*, and *Lottia pelta* at set-aside site 2439X at three elevations (MVD = meter of vertical drop below mean high high water). The sample size (*N*) for 3 MVD on 5 September 1993 is 2; for all other MVDs and sample dates, *N* is 6.

Date	*Littorina sitkana*	*Nucella lima*	*Leptasterias hexactis*	*Tectura persona*	*Lottia pelta*
			1 MVD		
14 Jul 1991	9.6 (4.7)			4.8 (1.2)	0.67 (0.5)
14 Aug 1991	6.6 (1.6)			1.2 (0.6)	
5 Jun 1992	20.6 (5.7)			7.3 (2.0)	0.5 (0.34)
5 Sep 1993	5.16 (3.4)			6.3 (2.2)	0.67 (0.42)
			2 MVD		
14 Jul 1991	3.8 (2)			6.2 (1.8)	5.6 (3.1)
14 Aug 1991	5.6 (1.9)	0.16 (0.16)		5.5 (2.1)	3 (2.05)
5 Jun 1992	9 (3.3)			4.5 (1.9)	11 (5.2)
5 Sep 1993	7.6 (7.6)		0.5 (0.5)	2 (1.4)	18.3 (6.07)
			3 MVD		
14 Jul 1991	0.16 (0.16)	0.66 (0.66)		0.66 (0.66)	12.8 (3.2)
14 Aug 1991	0.16 (0.16)			2.2 (1.8)	16.8 (3.6)
5 Jun 1992	0.33 (0.33)			0.66 (0.33)	22 (8.7)
5 Sep 1993					4 (1)

Low densities of the predatory invertebrates *Leptasterias hexactis* and *Nucella* spp. occurred at both control and oiled sites. The few data generated for these species made their assessment difficult, a problem experienced also by other researchers (Houghton et al. 1991). *Leptasterias hexactis* was observed in greater frequency at the set-aside site than at any of the study plots (nine were counted at MVD 2 and 3 on 5 September 1993).

At the one site pair (1732) where large numbers of frilled dogwinkle were sampled, there were no differences in density between the oiled and control sites (Figure 4). These snails may have migrated to the lower intertidal or subtidal zone during the winter months, avoiding the initial shoreline oiling. Alternatively, site 1732C was lightly oiled, which could have affected the population of frilled dogwinkle at that site. Laboratory sensitivity studies using Cook Inlet crude and number 2 fuel oil, however, found *Nucella lima* to be one of the more oil-tolerant gastropods (Rice et al. 1979). Ebert and Lees (1996, this volume) found that survival and growth of frilled dogwinkle were lower at oiled sites (including one site in Herring Bay) than at control sites. A severe freeze occurred during a low-tide series in January 1989 (Carroll 1994). The impact of the freeze on the direct-developing species is unknown but may have reduced densities at both control and oiled sites.

All oiled sites were parts of segments that received Omni treatments (Table 1). Site 1732X is part of segment KN0133, which was heavily treated by Omni Sweeps (J. Coxey and G. Lewis, personal communication and Chemtrak boat logs). We suspect, however, that certain small sections were missed because of the number of mature, reproductive *Fucus* plants observed in the upper and mid intertidal zones, the lack of denuded *Fucus* stipes or basal discs common in areas known to have received treatment, and the presence of heavy tar layers still present in 1990 and 1991 at sites immediately adjacent to 1732X. The fact that the density of the frilled dogwinkle population at 1732X was not significantly lower at any time when compared with its density at the control site also suggests that treatment of the 1732X site did not occur.

The effects of shoreline treatment by Omni Sweeps were documented during controlled experiments conducted in 1989 (Houghton et al. 1991). Under high pressure, the hot water emitted by the sweeps clearly caused the death or removal of organisms; however, the densities of some organisms observed at the untreated set-aside site (2439X; Table 2) were often consistent with densities found at the oiled–treated sites (Table 2). Furthermore, in other studies in Herring Bay, grazer densities were found to be significantly lower at sites known to be oiled but not treated than at matched controls (Highsmith et al. 1993). Therefore, although shoreline treatment certainly confounded impacts of the *Exxon Valdez* oil spill, North Slope crude oil alone was probably responsible for significant reductions in populations of epibenthic invertebrates in Herring Bay.

For *Tectura persona*, densities tended to be higher at control sites than at oiled sites in the first

TABLE 3.—Ultraviolet (UV) fluorescence indices for sediment hydrocarbon samples taken in Herring Bay at matched oiled and control sites in 1990 and 1991: the larger the number, the greater the concentration of petroleum hydrocarbons. Sites followed by the letter C are controls; X denotes oiled sites. A dash means sediment was not available at a matched site or elevation. The abbreviation MVD refers to meter of vertical drop below mean high high water.

Control site	MVD	UV index	Oiled site	MVD	UV index
		1990			
1222C	1	−0.5	1322X	—	—
	2	−0.2		2	130
	3	1		3	160
1251C	2	−0.2	1251X	2	500
	3	0.7		3	23
1411C	1	0	1311X	1	1000
	2	−1		2	600
	3	−0.5		3	300
1312C	1	30	1312X	1	300
	2	1		2	80
	3	−1		3	190
1723C	1	1.6	1723X	1	100
	2	2		2	150
	3	3		3	130
1852C	2	−2	1852X	1	80
				2	0
				3	190
		1991			
1221C	2	0.20	1221X	—	—
	3	0.10			
1222C	1	3	1322X	—	—
	2	0			
	3	−1.60			
1312C	2	0	1312X	—	—
	3	6			
1231C	—	—	1231X	1	90
1251C	—	—	1251X	1	94
				2	100
				3	80
1411C	2	3	1311X	—	—
1713C	1	0.80	1713X	2	70
				3	50
1723C	1	1	1723X	1	450
	2	37			
	3	0			
1852C	1	14	1852X	—	—
	2	50			
	3	3			

and second MVD through both 1991 and 1992 and through 1993 in the first MVD (Figure 6), suggesting continued effects of the oiling or treatment. These observations correspond to the high hydrocarbon concentrations found in the upper intertidal zone at oiled sites (Table 4). For *Lottia pelta* (Figure 7), which occurs somewhat lower in the intertidal zone than *T. persona*, the trend is less clear, although densities were significantly greater at control sites in several comparisons in 1990 and 1991. The apparent increase in limpet populations at some of the control sites after 1990 may be attributed to recovery from the severe winter of 1989 (Carroll 1994).

Four years after the *Exxon Valdez* oil spill, recovery at the Herring Bay study sites appeared to be underway but remained incomplete in some cases, especially in the upper intertidal zone. Recovery of the upper intertidal zone may be tied largely to algal recolonization. Algae, particularly adult *Fucus* plants, were significantly reduced at the same oiled study sites when compared with matched controls (van Tamelen and Stekoll 1996, this volume).

Recruitment may also play a crucial role in invertebrate recovery at the study sites. Recruitment data from other Herring Bay studies (Highsmith et al. 1993) show consistently greater settlement at more exposed sites (i.e., the lower central and southwest sections of Herring Bay), suggesting that prespill populations were originally higher at the oiled sites than at the controls. The increase in the mean densities of *T. persona*, *L. pelta*, and Sitka

TABLE 4.—Mean ultraviolet fluorescence indices for sediment hydrocarbon samples taken in Herring Bay at matched control and oiled sites in 1990 and 1991. The abbreviation MVD refers to meter of vertical drop below mean high high water.

Year	MVD	N	Mean	SD
		Control sites		
1990	1	4	7.78	14.84
	2	6	−.07	1.42
	3	5	.64	1.56
1991	1	4	4.7	6.28
	2	5	18.04	23.72
	3	6	1.25	2.76
		Oiled sites		
1990	1	4	370	431.58
	2	6	243.33	245.16
	3	6	165.5	90.42
1991	1	3	211.33	206.70
	2	2	85	21.21
	3	2	65	21.21

periwinkle at the oiled sites of the coarse-textured habitats in mid-1991 arose largely from one exposed site (2834X), which received heavy recruitment not observed at the other sites (Figures 3B, 6B, and 7B). If continued monitoring demonstrates that species eventually reach greater population densities at the oiled sites than at the control sites, then the effects of the *Exxon Valdez* oil spill in Herring Bay may have been underestimated.

Acknowledgments

We thank the following persons for field assistance: G. Reedy, G. Hollowell, C. Sullivan, T. Lewis, M. Derenoff, P. van Tamelen, F. Roddy, K. Baldwin, R. Clune, B. Baxter, S. Saupe, T. Rucker, and T. Aselin. Assistance with logistics was provided by K. Ostercamp, M. Rental, C. Williams, S. Petit, S. Kennedy, D. Doudna, J. Smithhisler, M. McEwen, and P. Shier. Special thanks go to D. Doudna for project coordination and management over several seasons. L. McDonald, D. Strickland, and W. Erickson provided guidance with sample design and data analyses.

References

ADEC (Alaska Department of Environmental Conservation). 1989a. Field notebooks, audio cassettes, and field notes. 1989 files (A-Jod). ADEC, Oil Spill Response Center, Anchorage.

ADEC (Alaska Department of Environmental Conservation). 1989b. Daily shoreline assessment forms: shoreline segments within Herring Bay. ADEC, Oil Spill Response Center, Anchorage.

ADEC (Alaska Department of Environmental Conservation). 1989c. Five month maximum impact map for Herring Bay, Knight Island, 3/24/89 to 8/24/89, map projection UTM, zone 6. File HER5MNTH. ADEC, Oil Spill Response Center, Anchorage.

AeroMap. 1989. 1:2400 aerial color photographs of Herring Bay, 3 April 1989. AeroMap, Anchorage, Alaska.

Bhattacharyya, G. K., and R. A. Johnson. 1977. Statistical concepts and methods. Wiley, New York.

Burns, K. A., and six coauthors. 1994. The Galeta oil spill. II. Unexpected persistence of oil trapped in mangrove sediments. Estuarine Coastal and Shelf Science 38:349–364.

Carroll, M. L. 1994. The ecology of an Alaskan rocky intertidal community: an experimental analysis of mechanisms regulating populations in Kachemak Bay. Doctoral dissertation. University of Alaska, Fairbanks.

Clark, R. C., Jr., B. G. Patten, and E. E. DeNike. 1978. Observations of a cold-water intertidal community after 5 years of a low-level, persistent oil spill from the *General M. C. Meigs*. Journal of the Fisheries Research Board of Canada 35:754–765.

Crothers, J. H. 1983. Field experiments on the effects of crude oil and dispersant on common animals and plants of rocky sea shores. Marine Environmental Research 8:215–239.

Ebert T. A., and D. C. Lees. 1996. Growth and loss of tagged individuals of the predatory snail *Nucella lamellosa* in areas within the influence of the *Exxon Valdez* oil spill in Prince William Sound. American Fisheries Society Symposium 18:349–361.

Garrity, S. D., S. C. Levings, and K. Burns. 1994. The Galeta oil spill. I. Long-term effects on the physical structure of the mangrove fringe. Estuarine Coastal and Shelf Science 38:327–348.

Highsmith, R. C., and seven coauthors. 1993. Herring Bay experimental and monitoring studies. Final status report. Restoration Project 102. University of Alaska, School of Fisheries and Ocean Sciences, Fairbanks.

Houghton, J. P., D. C. Lees, H. Teas, III, H. L. Cumberland, S. Landino, and T. A. Ebert. 1991. Evaluation of the intertidal and shallow subtidal biota in Prince William Sound following the *Exxon Valdez* oil spill and subsequent shoreline treatment. Report to NOAA (National Oceanic and Atmospheric Administration), HMRB (Hazardous Materials Response Branch) 91-1, Seattle.

ISCC (Interagency Shoreline Cleanup Committee). 1989. Best management practices. Pages 3-1–3-29 *in Exxon Valdez* oil spill field shoreline treatment manual. ISCC, Valdez, Alaska.

Krahn, M. M., G. M. Ylitalo, J. Joss, and S.-L. Chan. 1991. Rapid, semi-quantitative screening of sediments for aromatic compounds using sonic extraction and HPLC/flourescence analysis. Marine Environmental Research 31:175–196.

Levings, S. C., S. D. Garrity, and K. A. Burns. 1994. The Galeta oil spill. III. Chronic reoiling, long-term toxicity of hydrocarbon residues and effects of epibiota in the mangrove fringe. Estuarine Coastal and Shelf Science 38:365–395.

Manly, B. F. J. 1991. RT: a program for randomization testing. Version 1.01. West, Inc., Cheyenne, Wyoming.

Mann, K. H., and R. B. Clark. 1978. Long-term effects of oil spills on marine intertidal communities. Journal of the Fisheries Research Board of Canada 35:791–795.

Nelson-Smith, A. 1977. Recovery of some British rocky seashores from oil spills and cleanup operations. Pages 190–207 *in* J. Cairns, Jr., K. L. Dickson, and E. E. Herricks, editors. Recovery and restoration of damaged ecosystems. University of Virginia Press, Blacksburg.

Rice, S. D., A. Moles, T. L. Taylor, and J. F. Karinen. 1979. Sensitivity of 39 Alaskan marine species to Cook Inlet crude oil and number 2 fuel oil. Pages 549–554 *in* Proceedings, 1979 oil spill conference (prevention, behavior, control, cleanup). American Petroleum Institute Publication 4308, Washington, DC.

Sokal, R. R., and F. J. Rohlf. 1981. Biometry: the principles and practice of statistics in biological research, 2nd edition. Freeman, San Francisco.

Southward, A. J., and E. C. Southward. 1978. Recolonization of rocky shores in Cornwall after use of toxic dispersants to clean up the *Torrey Canyon* spill. Journal of the Fisheries Research Board of Canada 35: 682–706.

Teal, J. M., and R. W. Howarth. 1984. Oil spill studies: a review of ecological effects. Environmental Management 8(1):27–44.

van Tamelen, P. G., and M. S. Stekoll. 1996. Population response of the brown alga *Fucus gardneri* and other algae in Herring Bay, Prince William Sound, to the *Exxon Valdez* oil spill. American Fisheries Society Symposium 18:193–211.

American Fisheries Society Symposium 18:271–285, 1996

Meiofaunal Recolonization Experiment with Oiled Sediments

J. W. FLEEGER
Louisiana State University, Department of Zoology and Physiology, Baton Rouge Louisiana 70803, USA

T. C. SHIRLEY
University of Alaska Fairbanks, Juneau Center for Fisheries and Ocean Science, Juneau Alaska 99801, USA

M. G. CARLS
National Marine Fisheries Service, Auke Bay Laboratory, Auke Bay, Alaska 99821, USA

M. A. TODARO
Universita Modena, Dipartimento di Biologie Animale, 41100 Modena, Italy

Abstract.—To gain insight into long-term studies of the effects of the *Exxon Valdez* oil spill on the meiobenthos, a colonization experiment was initiated in 1990. Unweathered *Exxon Valdez* crude oil was mixed with azoic sediment to prepare low (26 ± 3 $\mu g\ g^{-1}$ total aromatics) and high (210 ± 12 $\mu g\ g^{-1}$ total aromatics) treatments of oiled sediments. The resulting mixtures, and azoic sediment without oil as a control, were added to triplicate colonization trays and buried flush with the surface of two beaches near mean low water in Herring Bay, Prince William Sound, Alaska. Hydrocarbon concentrations tended to decline with time. Trays and ambient sediments were sampled by coring and harpacticoid copepod species enumerated on days 0, 1, 2, and 28. Harpacticoids (more than 40 species) were mostly phytal associates from surrounding eelgrass beds. Colonization was rapid (approaching ambient levels in 1–2 d), especially in control and low-oil treatments. High-oil treatments exhibited significantly reduced densities of total harpacticoids and of two rapidly colonizing species (*Halectinosoma* sp. and *Mesochra pygmaea*) especially on days 1 and 2. Two slower colonizers (*Paralaophonte perplexa* and *Amphiascus minutus*) were unaffected by oil additions. Detrended correspondence analysis identified oil effects on the harpacticoid assemblage among the treatment-date collections. Ambient samples from all dates segregated from experimentals and were tightly clustered. Day 1 and 2 low- and high-oil treatments clustered together as did control, low-, and high-oil sediments from day 28. Control day 1 and 2 collections were intermediate. Thus, results suggest that an oil effect on migration and colonization was detectable, but for fewer than 28 d.

The grounding of the *Exxon Valdez* on Bligh Reef in Prince William Sound, Alaska, on 24 March 1989 caused a 44 million-L spill of crude oil that oiled over 1,600 km of shoreline (Maki 1991). The exposed area was largely pristine before this spill; the only sources of hydrocarbons previously present in Prince William Sound were those from natural processes plus some localized anthropogenic hydrocarbons associated with boating (Karinen et al. 1993). Aromatic hydrocarbons, except possibly perylene and occasionally phenanthrene, did not occur naturally in Prince William Sound sediments (Karinen et al. 1993).

Adverse effects of hydrocarbons on the meiobenthos have been demonstrated (see Coull and Chandler 1992). Field studies following oil spills (Wormald 1976; Giere 1979; Elmgren et al. 1983; Bodin 1988, 1991), sample collection in contaminated sediments (Heip et al. 1988) and experimental additions of hydrocarbons to mesocosms (Frithsen et al. 1985; Stacey and Marcotte 1987; Warwick et al. 1988) have all been employed, and show that meiofaunal density declines may occur during exposure to the toxic fractions of polycyclic aromatic hydrocarbons (PAH). Laboratory studies (harpacticoid copepods have been investigated most frequently) demonstrate that the water-soluble fraction of PAH can prompt severe mortality of meiofauna (Kontogiannis and Barnett 1973; Barnett and Kontogiannis 1975) and that PAH can reduce reproductive output (Ustach 1979; Stacey and Marcotte 1987; Misitano and Schiewe 1990).

Nevertheless, many studies have failed to demonstrate a negative impact of PAH on meiofauna, especially when those studies use sediment-bound crude oil. Nematodes from natural hydrocarbon seeps show increases in densities (Montagna et al. 1987), and experimental oiling has been shown to increase or have no influence on meiofaunal densities in experimental additions of oil in the field (Fleeger and Chandler 1983; DeLaune et al. 1984; Smith et al. 1984; Spies et al. 1988; Feder et al.

1990). Density increases are usually assumed to be related to increased microbial food resources in oiled sediments (Montagna et al. 1987). Colonization by meiofauna, and especially by harpacticoid copepods, into oiled sediments has been shown to be rapid (on the order of days), especially at low concentrations of PAH or other pollutants (Alongi et al. 1983; Decker and Fleeger 1984; Palmer et al. 1988; Spies et al. 1988).

Changes in abundance following an oil spill may have several confounding causes; increases or decreases in mortality, reproduction, and migration/recruitment are all possible and may occur simultaneously. The colonizing and dispersing abilities of meiofauna have only recently been recognized, but appear to be surprisingly strong, especially in seagrass beds (Palmer 1988; Walters 1991; Kurdziel and Bell 1992). Our study was designed to determine the effect of unweathered *Exxon Valdez* crude oil on migration and colonization by meiofauna at the site of the spill. Faunal responses were examined in colonization experiments with and without the application of crude oil. Because many harpacticoids (typically the second most abundant meiofaunal group) are especially rapid colonizers with relatively good swimming abilities (Palmer 1988) and because they are important contributors to the diet of a number of Alaskan fishes (Cordell 1986; McCall 1992), harpacticoid species composition was examined in detail.

Methods

Approximately 333 L of sediment were collected in Auke Nu Cove (58°22′48″N, 134°41′39″W) in Auke Bay, Alaska, on 1 March 1990 from the surface of an eelgrass bed and placed in 19-L buckets. Buckets were subjected to three freeze-thaw cycles over 1 month. During the first thaw, sediment was washed through 2- and 0.36-mm sieves to remove large particles and macrofauna. Excess water was decanted throughout the freeze-thaw cycles, reducing sediment volume to 227 L. Examination revealed that the meiofauna had been killed, but not completely decomposed, by this process.

Sediment and unweathered *Exxon Valdez* crude oil were combined in a cement mixer. The interior of the mixer was washed with hot water and soap and rinsed thoroughly before use. Processing was sequential, starting with controls, so that lightly oiled sediments were not contaminated by preceding heavily oiled samples. The sediment was divided randomly into three equal treatment groups (control, low-, and high-oil). Forty liters of sediment were tumbled in the mixer for 15 min; then *Exxon Valdez* crude oil was added (except controls). The total mixture was tumbled an additional 1.5 h and then stored in 20-L buckets with snap-top lids. For the low-oil treatment, 380 ml of oil were added to yield 0.5% oil, and 1,250 ml of oil were added to yield 1.7% for the high-oil treatment. Processing was completed on 5 April 1990; buckets were stored at ambient air temperatures. The contents of each storage bucket were turned once (17 or 18 April) to retard the onset of anoxia.

Two 12-m transects were positioned intertidally along the upper margins of eelgrass beds in Herring Bay, Prince William Sound, in a lightly oiled (HBL) and in a heavily oiled cove (HBH). Oil ratings were based on Alaska Department of Environmental Conservation shoreline data. Nine 28 × 33 × 14 cm (width × length × height) trays were located at 1.5-m intervals along each transect. Tray walls were straight, and the bottoms and sides perforated with 3-mm holes drilled at 8-mm intervals to allow water circulation and drainage. Treatment positions along the transects were randomly assigned within three consecutive 4-m blocks. After addition of sediment, trays were buried flush with the natural substrate. Implements used to distribute sediment and collect samples were separated by treatment to avoid cross-contamination and were initially hydrocarbon free.

The transect at the high-oil site (HBH: 60°28′0″N, 147°41′12″W) was positioned in the soft substrate of a cobble-gravel beach along the upper margin of an eelgrass bed at approximately −0.6 m (with reference to mean low water) on 25 April 1990. Tray elevations were slightly lower on the 0-m end of the transect (−0.7 m) than at the 12-m end (−0.5 m) to accommodate variations in the substrate. Oil remaining from the *Exxon Valdez* spill was obviously present and may have entered trays.

Experimental sediment was added to each tray before burial. Approximately 1 cm of washed gravel, obtained from a stream near the transect, was added to the bottom of each tray to ensure that a sufficient quantity of sediment was available. Gravel was collected well above the maximum tide line (so it was not contaminated with oil), placed in experimental trays, and washed in a stream to eliminate fine sediment and organic debris. The HBH transect was exposed to air 20 times (1.8% of total time) during the entire 89-d study; emergence times ranged from 32 to 162 min.

Meiofauna were subsampled from each tray (at low tide) with hand-held piston corers (modified 60-ml plastic syringes, 2.66 cm in diameter); the upper 4 cm was retained. Corers were separated by treatment to avoid cross-contamination. Cores were collected in triplicate from each treatment tray on days 0, 1, 2, 28, and 89. Core positions were determined randomly on an *x–y* grid without replacement through day 2 and was completely random thereafter. Core placement was at least 2 cm from the tray margin to avoid possible edge effects. Additionally, eight ambient sediment samples were collected from haphazardly selected sites along the tidal level between treatment trays at each sampling period to determine the colonizing source pool of harpacticoids. All faunal samples were preserved in 10% formalin shortly after collection. Additional cores for faunal and hydrocarbon analysis were collected on day 89 (faunal cores were not analyzed).

Hydrocarbon samples were collected with a 3-cm diameter, chrome-plated brass tube; a spoon was slipped down beside the corer to cap it off at 4 to 6 cm. All equipment used for hydrocarbon sampling was prewashed with soap and hot water, rinsed, dried, and rinsed with acetone followed by dichloromethane. Corers were separated by treatment to avoid cross-contamination. Two samples were initially collected from each treatment tray before any mixing with ambient sediments. On days 2, 28, and 89, one sample was collected from each tray, plus three from randomly selected ambient sediment locations among trays. Core positions were chosen randomly along with meiofauna cores to avoid conflict between sample types. Hydrocarbon samples were placed directly into hydrocarbon-free glass jars with teflon lids and frozen as soon as possible after collection.

Hydrocarbon concentrations were determined by gas chromatography followed by mass spectrometry for aromatic fractions or by flame ionization detection for aliphatics. Samples were analyzed at the Geochemical and Environmental Research Group at Texas A & M University and reported via the PWSOIL database.[1] Hydrocarbon samples were filtered by deuterated recovery and method detection limits, and questionable samples (3% of total) were excluded using a procedure developed by Short and Heintz (Alaska Fisheries Center, unpublished data). Data filtration eliminated the initial low-and high-oil treatment observations at the HBH site (12 samples), but because these observations were replicates of those collected at the HBL transect, corresponding data were repeated as HBH values.

Polycyclic aromatic hydrocarbons are not normally present at measurable concentrations intertidally in Prince William Sound (except perylene and occasionally phenanthrene). Because there are natural sources of alkanes in the sound (Karinen et al. 1993), PAH were chosen as the primary measure of hydrocarbons in this study. Concentration changes in various subcomponents of PAH (e.g., naphthalenes, fluorenes, dibenzothiophenes, phenanthrenes, and chrysenes) were similar to changes in PAH, thus the more generalized PAH measurement was used to summarize the data. Some alkane data, including the unresolved complex mixture (UCM), are also presented.

Hydrocarbon concentrations were compared between sites for each treatment with analysis of variance (ANOVA) procedures. Because variance was proportional to the means, data were log transformed before analysis. To avoid pseudoreplication, multiple observations within single replicate pans were averaged before other calculations; this situation occurred on day 0 only. Tests were considered significant when $P \leq 0.05$.

Seawater temperature and salinity were measured with a self-contained sensing device located at an elevation of approximately −0.1 m near the HBL transect. Particle size analysis was performed on ambient and experimental sediments collected on all sampling dates. In the laboratory, sediment was dried and sieved through 2-mm and 63-μm screens. Sand and gravel were treated by the methods of the Geochemical and Environmental Research Group (GERG) (1989) and silt and clay were measured by pipette analysis (Folk 1974). Total organic carbon (TOC) content was measured by combustion in an induction furnace. Carbon dioxide production was quantified after removal of sulfur oxides and water vapor and conversion of CO to CO_2 (GERG 1989).

In the laboratory, meiofauna passing through a 0.5-mm sieve but retained on a 63-μm mesh sieve was examined. A sucrose flotation/centrifugation technique (Fleeger 1979) was used to extract harpacticoids. Samples were suspended in a dense sucrose solution (700 g sugar in 500 ml water) and centrifuged at 350 rpm. The supernatant was rinsed through a 63-μm sieve. The process was repeated at least twice; selected sediment pellets resulting from the centrifugation were examined for sorting efficiency. The technique was estimated to exceed 95% efficiency for nematodes and 98% for copepods in recently published studies of Alaskan meiofauna from similar habitats (Fleeger et al. 1989; Fleeger and Shirley 1990). The concentrated meiofauna was preserved in 5% buffered formalin and stained with

[1]Available from C. A. Manen, National Ocean Service, Office of Ocean Resource Conservation and Assessment, Damage Assessment Center, 1305 East-West Highway, Silver Spring, Maryland 20910.

rose bengal to aid sorting. All organisms were identified to major taxon under a stereo-dissection microscope and enumerated using ruled trays; harpacticoids were removed into separate vials. When high densities of meiofauna were present, predominant taxa (mainly nematodes) were subsampled using the technique of Sherman et al. (1984). Adult and larger copepodite harpacticoids were identified to sex (if mature) and species from collections on days 1, 2, and 28 from HBH.

Day 0 collections revealed that many harpacticoids were not completely decomposed through the freeze-thaw cycles. Subsequent collections contained harpacticoids similar in appearance to those from day 0: broken or missing setae and legs, and exoskeletons broken and covered with detritus. The presence or absence of the caudal setae was used to score all individuals as "live" or "dead" at the time of collection. Concurrent and unrelated work on harpacticoids using freeze-killed specimens in a laboratory flume showed that the caudal setae of dead harpacticoids is easily broken (J. W. Fleeger, personal observation).

Changes in abundance were analyzed using a randomized block design with Statistical Analysis System (SAS) generalized linear model (GLM) programming. A two-way ANOVA using oil treatments (control, low-, and high-oil) and dates (days 1, 2, and 28) was used to identify oil-related effects on colonization of total harpacticoids and abundant copepod species. Tukey's Studentized Range Test was used as an a posteriori test to compare means among treatments. Detrended correspondence analysis (as recommended by Gauch 1982) was used as a method of ordination to investigate patterns in harpacticoid community structure using the DECORANA computer program (Hill 1979). This ordination describes spatial or temporal patterns in communities by using shared species compositions. A multidimensional coordinate system is created that locates various entities (in this case collections for each treatment and date) and describes their relationship to each other. The mean density (across all replicates for a treatment-date combination) of the 40 most abundant species was used for this multivariate technique.

TABLE 1.—Particle-size composition of experimental and ambient sediments, measured initially and on day 89. Initial samples were collected on days 0, 1, or 2. HBL = lightly oiled cove in Herring Bay; HBH = heavily oiled cove in Herring Bay.

Particle	Initial Mean (%) (SE)	Day 89 Mean (%) (SE)
All experimental sediments		
Sand	36.8 (1.4)	40.7 (3.2)
Silt	59.6 (1.5)	57.3 (3.3)
Clay	3.6 (0.2)	2.0 (0.2)
Ambient sediments at HBL		
Sand	6.5 (1.3)	66.2 (21.7)
Silt	89.4 (3.5)	26.5 (22.5)
Clay	4.1 (2.2)	7.3 (2.2)
Ambient sediments at HBH		
Sand	2.6 (0.8)	19.0 (13.1)
Silt	94.1 (1.5)	38.2 (26.1)
Clay	3.4 (0.7)	42.9 (20.1)

Results

Water temperature and salinity fluctuated with tidal cycle, but means remained near 8.0°C and 29.0 ‰ over the first 28 days at HBL. Mean temperatures increased gradually from day 46 to 89 (8.9 to 16.2°C), while salinities declined (28.2 to 20.8‰). Air temperatures ranged from 2.7 to 14.0°C, but may have been influenced by evaporative cooling. Particle-size distributions for all experimental sediments were similar at both transect locations and across all treatments ($0.392 \leq P \leq 0.987$; Table 1). Silt was the dominant size range in the experimental sediments. There was evidence (significant slopes for sand and clay) that particle sizes changed slightly over the course of the experiment in ambient sediments: the percentages of sand and clay increased, while silt percentages decreased (Table 1). Sediment changes in oil-treated sediments were minimal.

Hydrocarbon Concentrations in Sediments

Hydrocarbon concentrations in experimental sediments (measured by TOC, PAH, and alkanes at HBL) correlated well with the percent oil added ($0.81 \leq r^2 \leq 0.94$, Figure 1). Initial PAH concentrations in the azoic sediments ranged from 4.6 $\mu g\ g^{-1}$ (control) to 210 $\mu g\ g^{-1}$ (high oil) (Table 2 and Figure 2). Initial PAH concentrations in ambient sediments, measured on days 2 and 3, were 0.15 ± 0.06 and 0.27 ± 0.03 $\mu g\ g^{-1}$ at HBL and HBH, respectively. The PAH composition in control sediments was unlike that of *Exxon Valdez* crude oil; the naphthalene component was relatively high and there was evidence of low-temperature combustion products. Benzo[*b*]fluoranthene and heavier compounds were present in control sediments, and concentrations of increasingly methyl-substituted compounds tended to decline with respect to their unsubstituted homologues. Concentrations in ambi-

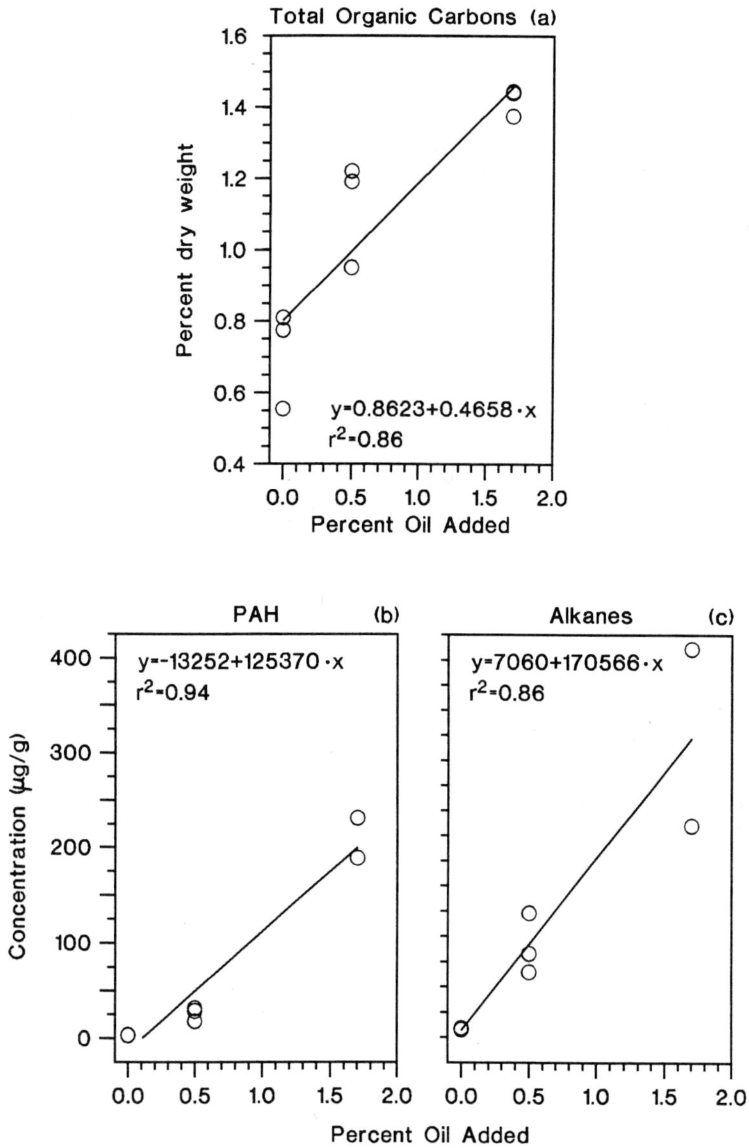

FIGURE 1.—Hydrocarbon concentration on the first day plotted against percentage of oil added. Fitted lines are linear: $y = a + b \cdot x$, where x = percent oil added. (**a**) Total organic carbons ($N = 18$). (**b**) Sum aromatics ($N = 17$). (**c**) Alkane concentrations, excluding the unresolved complex mixture ($N = 18$).

ent sediments were low and essentially constant; PAH composition was characteristic of weathered *Exxon Valdez* crude oil. In oil-treated sediments, PAH composition matched that of *Exxon Valdez* crude oil.

Results of ANOVAs suggest that concentrations of PAH and alkane hydrocarbons in experimental sediments depended primarily on oil treatment and time ($P < 0.001$). At the HBL site, concentrations in oil treatments tended to decline over the first 89 d (Figure 2). At the HBH site, concentrations tended to remain constant or rise slightly during the first 28 d, but declined by day 89 (Figure 2). Although hydrocarbon concentrations declined during the experiment, day 89 PAH and alkane concentrations remained significantly greater than control concentrations at HBL ($0.025 \leq P \leq 0.075$) and HBH ($P \leq 0.028$).

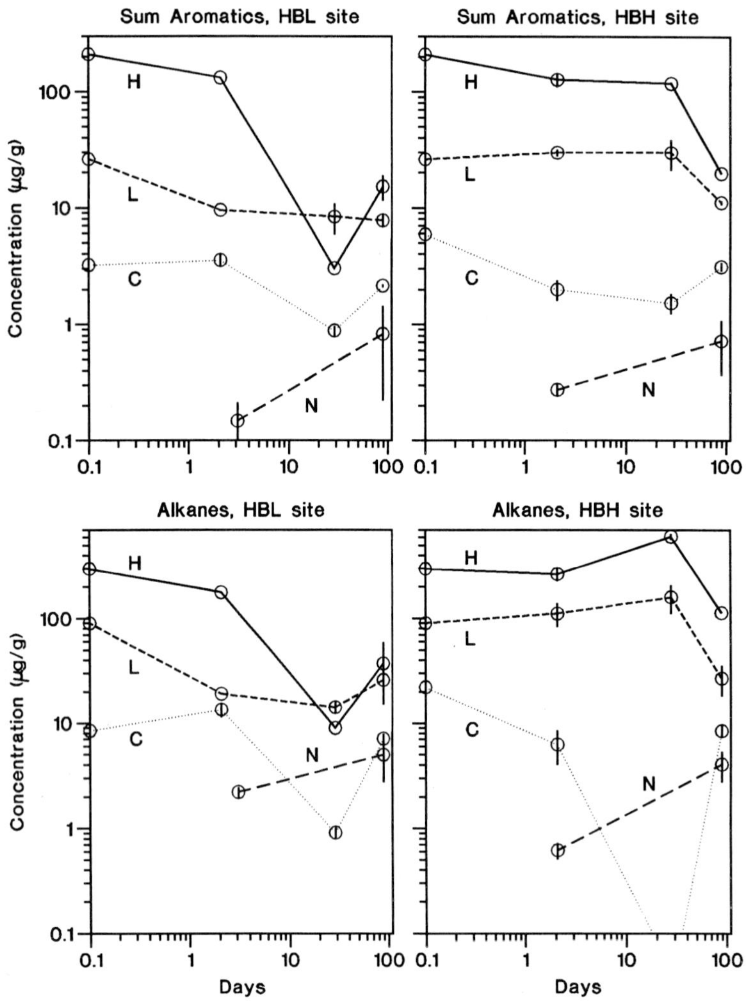

FIGURE 2.—Mean polycyclic aromatic hydrocarbon and alkane concentrations at both transect sites over time for each treatment: H = high oil; L = low oil; C = control; N = natural; HBL = lightly oiled cove at Herring Bay; HBH = heavily oiled cove at Herring Bay. Time data were adjusted upward 0.1 d to allow log transformation. Alkane concentrations in control sediments at HBH were below detection limits on day 28. Error bars are ± SE.

TABLE 2.—Initital hydrocarbon concentrations (µg/g) in treated and control sediments in Herring Bay. PAH = polycyclic aromatic hydrocarbons; UCM = unresolved complex mixture.

	Control	Low oil	High oil
Analyte	Mean (SE)	Mean (SE)	Mean (SE)
PAH	4.6 (0.6)	26 (3)	210 (12)
Alkanes	15.3 (4.6)	90 (11)	298 (51)
UCM	42.8 (5.7)	286 (23)	5,123 (15)
Sum	62.7	402	5,631

Harpacticoid Collections

All collections were dominated by harpacticoids with strongly prehensile first legs and a fusiform prehensile body shape, especially in the families Diosaccidae and Laophonitidae. Hicks and Coull (1983) and Bell et al. (1987) associate these characteristics with a phytal life style, suggesting that most of the more than 40 sediment-dwelling species cataloged in the present study were associated with

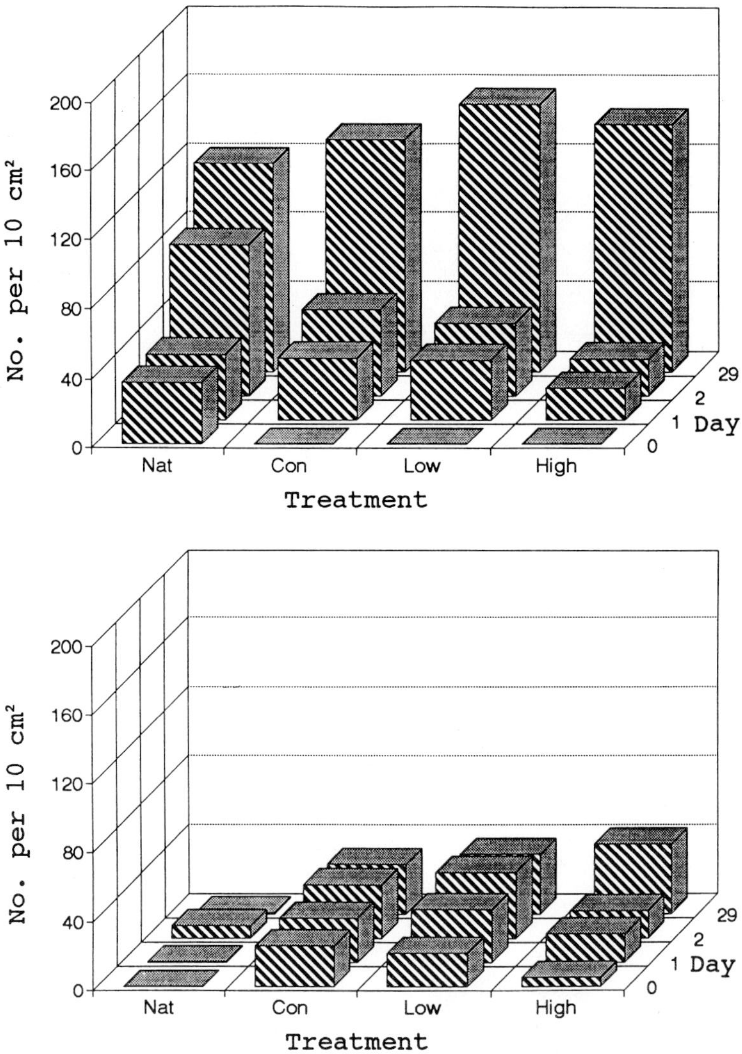

FIGURE 3.—Total copepod density (mean number individuals/10 cm^2) over the time course of the experiment. Upper figure represents the number of copepods scored as living; lower figure, the number of copepods scored as dead at the time of collection. Nat = ambient sediments; Con = control azoic sediments; Low = azoic sediment with low-oil additions; High = azoic sediments with high-oil additions.

surrounding eelgrass beds. Colonists were diverse, and all of the predominant species from the ambient sediments were collected in colonization trays (Table 3). Diversity, measured either as species richness (number of species) or as the Shannon-Wiener index, and relative abundance patterns did not change appreciably over time or differ greatly between ambient and experimental sediments (Table 3). Total harpacticoid densities in ambient sediments increased throughout the experimental period (Figure 3). Differential colonizing abilities were evident among the harpacticoid species (Figures 4 and 5). Among the more abundant species, *Halectinosoma* sp. and *Mesochra pygmaea* were good colonists (higher densities and relative abundances were observed after 1 or 2 d in the experimental trays, regardless of oil dose, compared to ambient sediments), while *Paralaophonte perplexa* and *Amphiascus minutus* were poor colonists (lower densities and relative abundances in trays compared to ambient sediments).

Two-way ANOVA was conducted in a random-

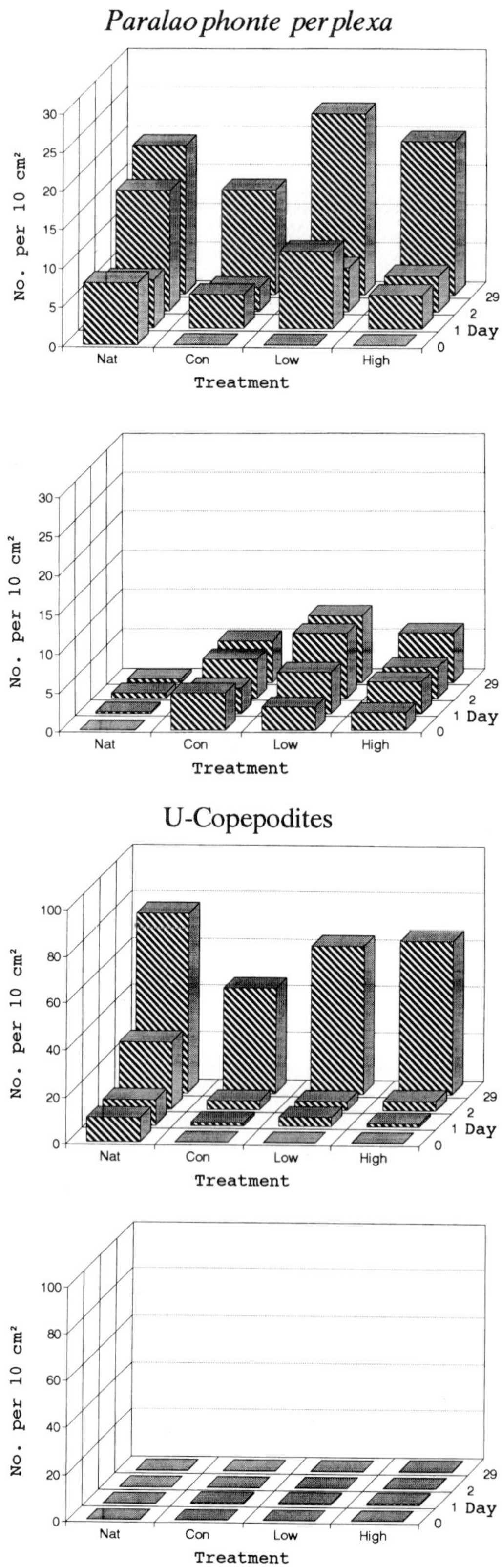

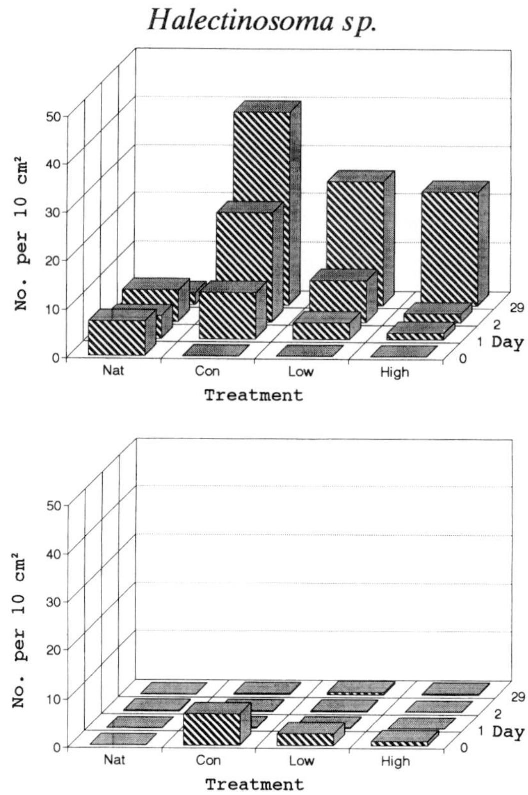

FIGURE 4.—Copepod density (mean number individuals/10 cm^2) for *Paralaophonte perplexa*, *Halectinosoma* sp., and unidentified (U) copepodites over the time course of the experiment. Upper figure represents the number of copepods scored as living; lower figure, the number of copepods scored as dead at the time of collection. Nat = ambient sediments; Con = control azoic sediments; Low = azoic sediment with low-oil additions; High = azoic sediments with high-oil additions.

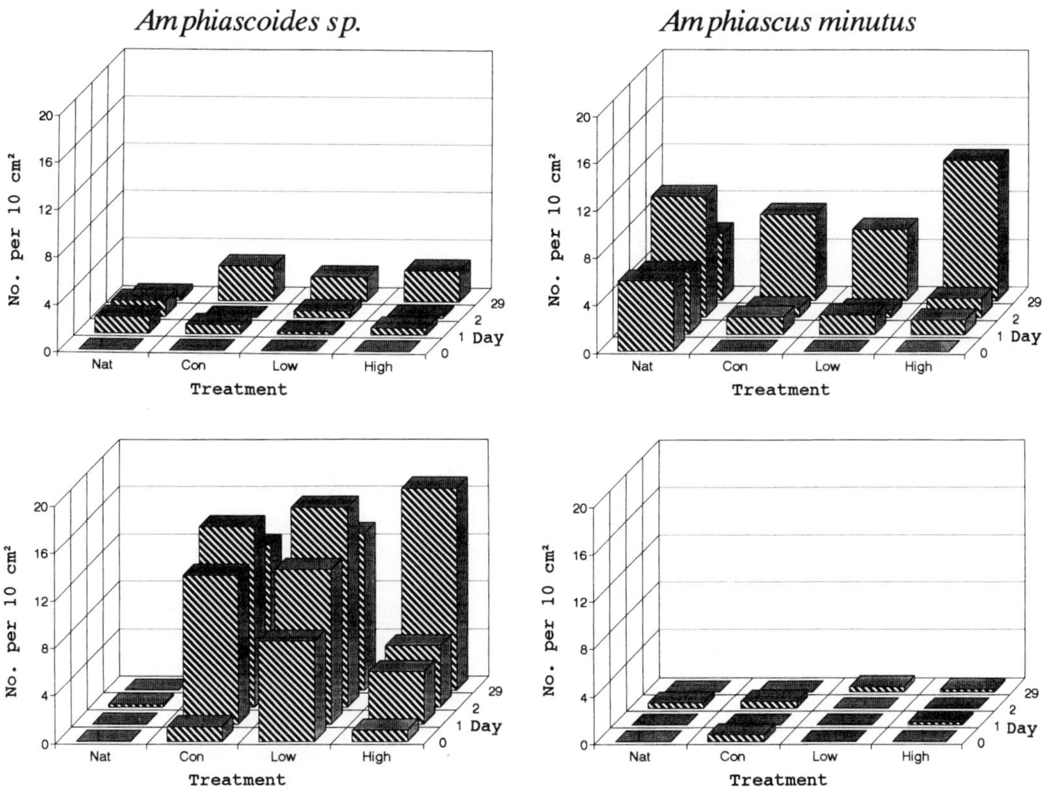

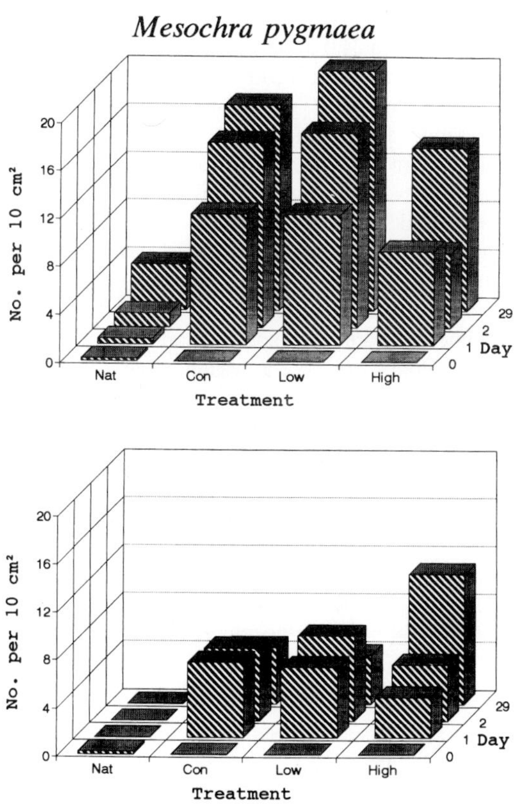

FIGURE 5.—Copepod density (mean number individuals/10 cm^2) for *Amphiascoides* sp., *Amphiascus minutus*, and *Mesochra pygmaea* over the time course of the experiment. Upper figure represents the number of copepods scored as living; lower figure, the number of copepods scored as dead at the time of collection. Nat = ambient sediments; Con = control azoic sediments; Low = azoic sediment with low-oil additions; High = azoic sediments with high-oil additions.

TABLE 3.—Relative abundances (%) and species diversity patterns for abundant harpacticoid species. Percent of the total represents the cumulative percentage represented by the abundant species and the unidentifiable (U) copepodites in a given date-treatment combination. Densities are found in Figures 3 through 8. Richness is the total number of species. *H'* is the Shannon-Wiener diversity index value.

Species or statistic	Ambient	Control	Low	High
	Day 1			
Mesochra pygmaea	1.2	29.8	31.0	41.9
Paralaophonte perplexa	16.0	11.6	28.1	22.6
Halectinosoma sp.	12.4	26.5	9.8	6.4
Amphiascus minutus	13.6	3.9	4.4	8.6
U copepodites	29.6	3.8	11.5	7.5
% of the total	72.8	75.6	84.8	87.0
Richness	16	20	19	10
H'	2.2	1.9	1.9	1.6
	Day 2			
Mesochra pygmaea	1.3	31.3	39.1	28.7
Paralaophonte perplexa	17.9	6.1	13.5	21.3
Halectinosoma sp.	7.5	45.5	20.8	8.3
Amphiascus minutus	11.7	1.6	1.4	7.4
U copepodites	32.7	9.2	9.7	19.4
% of the total	71.1	93.7	84.5	85.1
Richness	15	14	17	14
H'	2.4	1.7	1.9	2.1
	Day 28			
Mesochra pygmaea	3.2	12.7	12.9	9.5
Paralaophonte perplexa	16.0	10.1	15.1	13.8
Halectinosoma sp.	2.0	29.8	16.4	16.2
Amphiascus minutus	5.0	5.3	4.2	8.3
U copepodites	63.4	33.2	40.6	45.6
% of the total	89.6	91.1	89.2	93.4
Richness	15	14	20	13
H'	1.9	1.9	2.8	2.0

ized block design (tray replication was blocked) to test for date and treatment effects on harpacticoid abundance (Table 4). Analysis of variance treatments were collection day (1, 2, or 28) and oil treatment (control, low-, or high-oil). Tests were performed on the total live harpacticoids, unidentified copepodites, and the four most abundant species. Collection-day effects were significant in all taxa, probably reflecting a true increase in abundances of ambient harpacticoids during the springtime experimental period (Table 4). An oil effect (especially at the high-oil dosage) was observed in total copepods, and in *Halectinosoma* sp. and *M. pygmaea*. For total copepods and *M. pygmaea*, Tukey's Studentized Range Test indicated that densities in control and low-oiled sediments did not differ, but that densities in high-oiled sediments were significantly lower than in other treatments throughout the experiment. For *Halectinosoma* sp., means were significantly different in all experimental treatments; densities in control sediments were

TABLE 4.—Analysis of variance summary: probabilities of greater *F*-values for a randomized block design for total meiofauna and abundant harpacticoid species. Day refers to collection date 0, 2, or 28. Treatment refers to oil dose: control, low, or high.

Species or group	Block	Treatment	Day	Treatment × day
Total harpacticoids	0.1146	0.0040	0.0001	0.4440
Mesochra pygmaea	0.0011	0.0024	0.0074	0.7214
Paralaophonte perplexa	0.5433	0.1408	0.0001	0.8093
Halectinosoma sp.	0.3592	0.0001	0.0001	0.5222
Amphiascus minutus	0.0002	0.0681	0.0001	0.2087

significantly higher than in low-oiled treatments, and, in turn, densities in low-oiled were significantly higher than in high-oiled treatments. No treatment effects were observed in two other harpacticoids, *P. perplexa* and *A. minutus*.

Harpacticoids scored as "dead" at the time of collection were variable in density, but were found in all treatments and in all collections through day 28. Dead harpacticoids were always rare in ambient collections. Species-level identifications of day 0 collections reveal that the relict assemblage (resident in sediment rendered azoic) was generally similar to that found in HBH, and most species were shared in common. Most of the dead harpacticoids collected on subsequent dates also appeared to be relict individuals that decomposed slowly at the cool temperatures found in Prince William Sound. For example, "dead" *Amphiascoides* sp. was abundant in samples from azoic sediments on day 0, contributing about 25% of all dead specimens. Live *Amphiascoides* sp. was always rare in ambient collections and in experimental trays (Figure 5). Unidentifiable copepodites were virtually absent from all day 0 collections. Densities of unidentifiable copepodites increased with time in the ambient sediments and in the experimental trays (Figure 4). This colonization event was notable because no increase in the numbers of dead copepodites was observed in the experimental treatments, suggesting that there was little or no oil-induced mortality. *Halectinosoma* sp. was present in very low abundance in day 0 relict collections, but colonized azoic sediments rapidly. After day 0, the number of *Halectinosoma* sp. scored as dead were always low regardless of oil treatment, again suggesting low levels of oil-induced mortality (Figure 5).

Harpacticoid Community Ordinations

Detrended correspondence analysis was conducted to determine if an oiling effect influenced

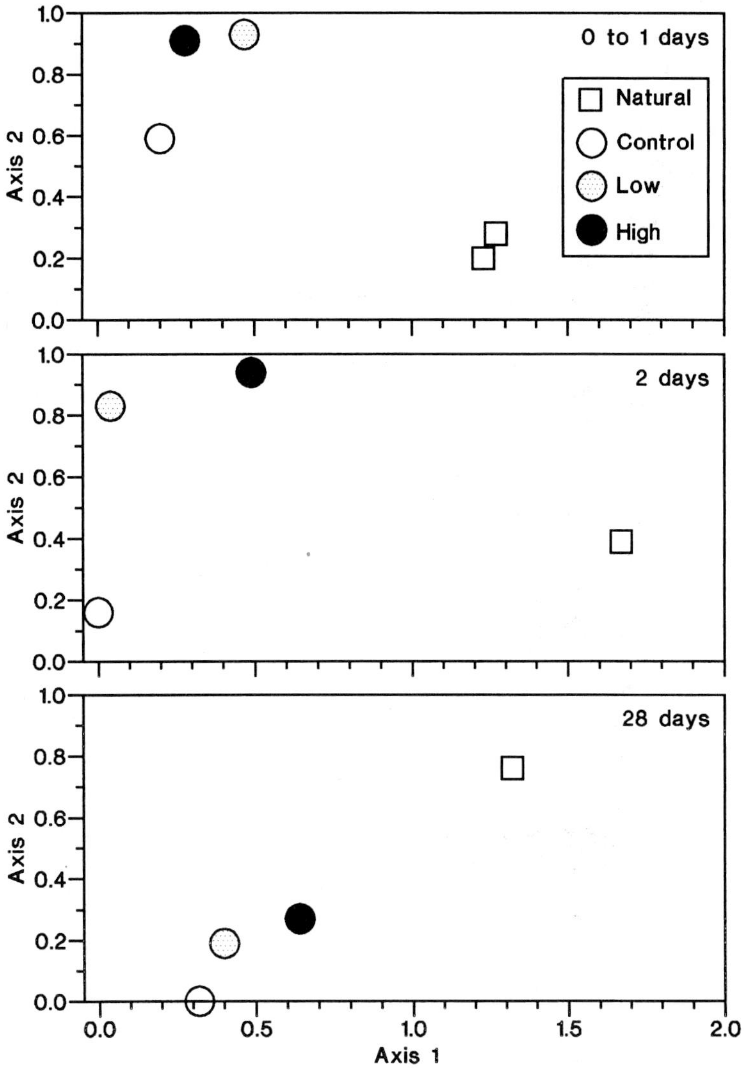

FIGURE 6.—Detrended correspondence analysis of the copepod assemblage in azoic and ambient sediments on various collection dates.

the copepod assemblage (Figure 6). Two axes comprised more than 70% of the variance. Axis 1 contained relatively high levels of variation, with values approaching 2 SDs overall (a variation of 4 SDs indicates complete faunal turnover). All ambient samples separated from experimentals on axis 1 and clustered near each other, indicating that experimental trays could be distinguished from the surrounding sediments in species composition on all collections. Other collections were separated on axis 2 with only 1 SD of faunal turnover. Days 1 and 2 low- and high-oiled treatments clustered together as did control, low-, and high-oiled sediments from day 28. Control day 1 collections were intermediate. Results therefore indicate that an oiling effect was present on days 1 and 2, but no difference between control and oiled sediments was apparent by day 28.

Discussion

Unweathered *Exxon Valdez* crude oil delayed, but did not preclude, colonization by meiofauna into azoic sediment in Prince William Sound. The effect on colonization was short lived, and copepod species densities and copepod species composition in oiled sediments and controls became indistinguishable from 2 to 28 days post-oiling. Two of eight abundant species showed reduced rates of recolonization. Aromatic hydrocarbon concentrations in oiled sediments tended to decline with time (Figure 2). Hydrocarbon-related effects on densities were found only in the most rapidly colonizing species (*Halectinosoma* sp. and *M. pygmaea*). Slower colonists (*P. perplexa* and *A. minutus*) may not have reached the trays in large numbers until oil concentrations diminished or until the effects of hydrocarbon additions were mitigated in other ways (see below).

The concentrations of *Exxon Valdez* crude oil used in this experiment were environmentally meaningful. Based on our observations, the maximum PAH concentration in the high-oil treatment (246 $\mu g\ g^{-1}$) was well within the range of PAH concentrations observed in Prince William Sound sediments after the *Exxon Valdez* oil spill (0–482 $\mu g\ g^{-1}$). Very low levels of PAH were observed in the sediments ambient to recolonization chambers. The initial levels of PAH were higher in control sediments (relict sediments from Auke Nu Cove rendered azoic) but were much less than levels in the low-oil treatment (Table 2). The most likely source of contamination to relict sediments was ship activity around the collection location.

Although hydrocarbon concentrations tended to decrease with time, effects on harpacticoid copepods may have been further mitigated by an even modest accumulation of uncontaminated ambient sediments into the colonization trays. Harpacticoids in muddy Alaskan sediments have a very shallow depth profile, with most living in the flocculent layer 2 mm in sediment depth (J. W. Fleeger and T. C. Shirley, personal observation). Hydrodynamic effects that may cause retention of flocculent sediments are minimized when trays are buried flush into the sediment (Snelgrove et al. 1992), as were ours, and grain size was unchanged over time (Table 1). Nevertheless, additions of uncontaminated flocculent sediments, which may be prone to adhere to oil-enriched sediments, could have provided a refuge from PAH for harpacticoids.

Four previous studies examined the effects of crude or seep oil on the colonization of meiofauna, and one, Alongi et al. (1983), used Prudhoe Bay crude oil. Three studies were unable to identify negative effects of oil on density, but found that abundances increased with low levels of oil addition (Alongi et al. 1983; Spies et al. 1988) or that oil increased colonization potential (Palmer et al. 1988). The fourth study, Decker and Fleeger (1984), found that oil retarded colonization rates (for fewer than 20 d in nematodes and polychaetes) but that oil had positive effects on density after 30 d for one species of harpacticoid copepod. Prudhoe Bay crude oil was also used in a controlled spill by Feder et al. (1990), who found either no effect or increases in abundance of harpacticoid copepods in Port Valdez, Alaska. Generally, studies that use crude oil in field additions (and thus as sediment-bound PAH) reveal weak evidence for mortality in meiofauna (Fleeger and Chandler 1983; DeLaune et al. 1984; Smith et al. 1984; Feder et al. 1990), whereas studies that use fuel oil in the water-soluble form identify stronger effects (Frithsen et al. 1985; Stacey and Marcotte 1987). Even so, Stacey and Marcotte (1987) found an increase in density of one harpacticoid species in a mesocosm experiment. Spies et al. (1988) compared seep oil with additions of organic detritus in a colonization study and concluded that both, in low concentrations, had similar effects on meiofaunal abundance. Thus it appears that the sediment-bound phase of many crude oils, including that of *Exxon Valdez* crude oil, is not particularly toxic and that changes in abundance may be related to increases in organic matter. A similar effect was noted by Stickle et al. (1990), who concluded that pink shrimp *Pandalus borealis* exposed to the water-soluble fraction of oil were far more vulnerable to oil effects on their survival and bioenergetics than shrimp exposed to oiled sediment.

Our harpacticoid species data further suggest that oil-induced mortality of colonists was minimal. Species-level comparisons between the colonizing source pool and the relict copepods (dead but not decomposed) retained in azoic sediments were made. Most of the obviously dead harpacticoids found in azoic sediments over the 28-d experimental period appeared to be relict individuals, rather than harpacticoids that may have colonized and then died from exposure to oil. *Amphiascoides* sp. was abundant in relict sediments on day 0 and contributed about 25% of all dead specimens, but it was always infrequent in ambient collections. Few live specimens of *Amphiascoides* sp. were recovered from colonization trays. Unidentifiable copepodites and *Halectinosoma* sp. were rare on day 0 in the azoic sediments but colonized rapidly without mortality.

Meiofauna have only recently been recognized as active colonizers (Palmer 1988). Sediment-dwelling meiofauna associated with seagrass beds, like those in our study, are especially mobile, displaying attributes of active emergence into the water column, rapid resettlement onto sediment and grass blades, and dispersal (Bell et al. 1989; Walters 1991; Kurdziel and Bell 1992). Copepods are especially active, moving back and forth from sediment to grass blades (Walters 1991; Webb and Parsons 1992). As such, seagrass meiofauna may serve as important model experimental subjects because their short-term migration rates are high and are more likely to be altered by an ephemeral environment perturbation.

In a few selected taxa, the presence of oil slowed, but certainly did not halt, meiofaunal migration and colonization. Although harpacticoids colonize sediments from the water column (Chandler and Fleeger 1983), hydrodynamic models suggest that they have only limited abilities to select locations at settlement (Palmer 1988). Nevertheless, harpacticoids occur disproportionately at sites that are favorable to them in the field and laboratory (Kern and Taghon 1986; Decho and Fleeger 1988; Fegley 1988; Kern 1990). Sun (1993) studied harpacticoid settlement and postsettlement movement in a flume and found that postsettlement behavior of harpacticoids is instrumental in determining local distributions. His work suggests that harpacticoids cannot select settlement locations but probably move through the benthic boundary layer to select nearby sites. Alternatively, harpacticoids may be stimulated to leave an unfavorable site either by swimming up into the flow or by allowing themselves to be carried away from the sediment surface. During colonization, *Halectinosoma* sp. and *M. pygmaea* probably entered all trays, regardless of oil treatment, but a disproportionate number probably emigrated from the high-oil treatment trays.

The influence of hydrocarbons on fish feeding is poorly known; however, PAH do not appear to interfere with chemoreception in most fishes, and fishes do not generally avoid PAH in the field (Klaprat et al. 1992). Therefore, fishes and invertebrates may feed on contaminated harpacticoids. Contact of meiofauna with oiled sediments and subsequent emigration from these sediments into the water column may provide a mechanism for introducing oil into the diets of other organisms, such as juvenile pink and chum salmon. In associated field studies, we found that juvenile pink salmon *Oncorhynchus gorbuscha* and chum salmon *O. keta* were contaminated with hydrocarbons from the *Exxon Valdez* oil spill and that ingestion was a likely route of contamination (Carls et al. 1996, this volume). Perhaps harpacticoids contaminated with *Exxon Valdez* hydrocarbons partially mediated transfer of hydrocarbons from sediments to fish.

Changes in abundance of meiofauna following oiling in the field can be rapid. Fleeger and Chandler (1983) found increases in density within 5 d, a time much shorter than that in which these metazoans can increase by reproductive activity (Hicks and Coull 1983). Our data suggest that because colonization readily occurred in oiled Prince William Sound sediments, any change in meiofaunal density associated with the *Exxon Valdez* spill is likely to be the result of changes in mortality or reproductive success rather than changes in migration rates. Furthermore, our experimental design allowed the identification of short-term effects, an opportunity not usually afforded investigators using relatively long intervals between sampling (Spies et al. 1988). Our results suggest that recolonization of an area that suffers high mortality after an oil spill around seagrass beds should occur quickly as long as a source pool of meiofauna remains available, especially given the long-distance dispersal capabilities of seagrass meiofauna (Kurdziel and Bell 1992).

References

Alongi, D. M., D. F. Boesch, and R. J. Diaz. 1983. Colonization of meiobenthos in oil-contaminated subtidal sands in the lower Chesapeake Bay. Marine Biology 72:325–335.

Barnett, C. J., and J. E. Kontogiannis. 1975. The effect of crude oil fractions on the survival of a tidepool copepod, *Tigriopus californicus*. Environmental Pollution 8:45–54.

Bell, S. S., G. R. F. Hicks, and K. Walters. 1989. Experimental investigations of benthic reentry by migrating meiobenthic copepods. Journal of Experimental Marine Biology and Ecology 130:291.

Bell, S. S., K. Walters, and M. O. Hall. 1987. Habitat utilization by harpacticoid copepods: a morphometric approach. Marine Ecology Progress Series 35:59–64.

Bodin, P. 1988. Results of ecological monitoring of three beaches polluted by the *Amoco Cadiz* oil spill: development of meiofauna from 1978 to 1984. Marine Ecology Progress Series 42:105–123.

Bodin, P. 1991. Perturbations in the reproduction cycle of some harpacticoid copepod species further to the *Amoco Cadiz* oil spill. Hydrobiologia 209:245–257.

Carls, M. G., A. C. Wertheimer, J. W. Short, R. M. Smolowitz, and J. J. Stegeman. 1996. Contamination of juvenile pink and chum salmon by hydrocarbons in Prince William Sound after the *Exxon Valdez* oil spill. American Fisheries Society Symposium 18:593–607.

Chandler, G. T., and J. W. Fleeger. 1983. Meiofaunal colonization of azoic estuarine sediment in Louisiana: mechanisms of dispersal. Journal of Experimental Marine Biology and Ecology 69:175–188.

Cordell, J. R. 1986. Structure and dynamics of an

epibenthic harpacticoid assemblage and the role of predation by juvenile salmon. Master's thesis. University of Washington, Seattle.

Coull, B. C., and G. T. Chandler. 1992. Pollution and meiofauna: field, laboratory and mesocosm studies. Oceanography and Marine Biology Annual Review 30:191–271.

Decho, A. W., and J. W. Fleeger. 1988. Microscale dispersion of meiobenthic copepods in response to food-resource patchiness. Journal of Experimental Marine Biology and Ecology 118:229–244.

Decker, C. J., and J. W. Fleeger. 1984. The effect of crude oil on the colonization of meiofauna into salt marsh sediments. Hydrobiologia 118:49–58.

DeLaune, R. D., C. J. Smith, W. H. Patrick, Jr., J. W. Fleeger, and M. D. Tolley. 1984. Effect of oil on salt marsh biota: methods for restoration. Environmental Pollution 36:207–227.

Elmgren, R., S. Hansson, U. Larsson, B. Sundelin, and P. D. Boehm. 1983. The *Tsesis* oil spill: acute and long-term impact on the benthos. Marine Biology 73:51–65.

Feder, H. M., A. S. Naidu, and A. J. Paul. 1990. Trace element and biotic changes following a simulated oil spill on a mudflat in Port Valdez, Alaska. Marine Pollution Bulletin 21:131–136.

Fegley, S. R. 1988. A comparison of meiofaunal settlement on the sediment surface and recolonization of defaunated sandy sediment. Journal of Experimental Marine Biology and Ecology 123:97–114.

Fleeger, J. W. 1979. Population dynamics of three estuarine meiobenthic harpacticoids (Copepoda) in South Carolina. Marine Biology 52:147–156.

Fleeger, J. W., and G. T. Chandler. 1983. Meiofauna responses to an experimental oil spill in a Louisiana salt marsh. Marine Ecology Progress Series 11:257–264.

Fleeger, J. W., and T. C. Shirley. 1990. Meiofaunal responses to sedimentation from an Alaskan spring bloom. II. Harpacticoid population dynamics. Marine Ecology Progress Series 59:239–247.

Fleeger, J. W., T. C. Shirley, and D. A. Ziemann. 1989. Meiofaunal responses to sedimentation from an Alaskan spring bloom. I. Major taxa. Marine Ecology Progress Series 57:137–145.

Folk, R. L. 1974. Petrology of sedimentary rocks. Hemphill Publishing Company, Austin, Texas.

Frithsen, J. B., R. Elmgren, and D. T. Rudnick. 1985. Responses of benthic meiofauna to long-term, low-level additions of number 2 fuel oil. Marine Ecology Progress Series 23:1–14.

Gauch, H. G. 1982. Multivariate analysis in community ecology. Cambridge University Press, New York.

GERG (Geochemical and Environmental Research Group). 1989. NOAA/*Exxon Valdez* oil spill assessment. Texas A&M University, GERG, Department of Oceanography, College Station.

Giere, O. 1979. The impact of oil pollution on intertidal meiofauna. Field studies after the *La Coruna*-spill, May 1976. Cahiers de Biologie Marine 20:231–251.

Heip, C., and six coauthors. 1988. Analysis of community attributes of the benthic meiofauna of Frierfjord/Langesundfjord. Marine Ecology Progress Series 46: 171–180.

Hicks, G. R. F., and B. C. Coull. 1983. The ecology of marine meiobenthic harpacticoid copepods. Oceanography and Marine Biology Annual Review 21:67–175.

Hill, M. O. 1979. DECORANA-A FORTRAN program for detrended correspondence analysis and reciprocal averaging. Cornell University Press, Ithaca, New York.

Karinen, J. F., M. M. Babcock, D. W. Brown, W. D. MacLeod, Jr., L. S. Ramos, and J. W. Short. 1993. Hydrocarbons in intertidal sediments and mussels from Prince William Sound, Alaska, 1977–1980: characterization and probable sources. NOAA (National Oceanic and Atmospheric Administration) Technical Memorandum NMFS (National Marine Fisheries Service)-AFSC-9, Seattle.

Kern, J. C. 1990. Active and passive aspects of meiobenthic copepod dispersal at two sites near Mustang Island, Texas. Marine Ecology Progress Series 60: 211–224.

Kern, J. C., and G. L. Taghon. 1986. Can passive recruitment explain harpacticoid copepod distributions in relation to epibenthic structure? Journal of Experimental Marine Biology and Ecology 101:1–23.

Klaprat, D. A., R. E. Evans, and T. J. Hara. 1992. Environmental contaminants and chemoreception in fishes. Pages 321–341 *in* T. J. Hara, editor. Fish chemoreception. Chapman and Hall, London.

Kontogiannis, J. E., and C. J. Barnett. 1973. The effect of oil pollution on survival of the tidal pool copepod, *Tigriopus californicus*. Environmental Pollution 4:69–79.

Kurdziel, J. P., and S. S. Bell. 1992. Emergence and dispersal of phytal-dwelling meiobenthic copepods. Journal of Experimental Marine Biology and Ecology 163:43–64.

Maki, A. W. 1991. The *Exxon Valdez* oil spill: initial environmental impact assessment. Environmental Science and Technology 25:24–29.

McCall, J. N. 1992. Source of harpacticoid copepods in the diet of juvenile starry flounder. Marine Ecology Progress Series 86:41–50.

Misitano, D. A., and M. H. Schiewe. 1990. Effect of chemically contaminated marine sediment on naupliar production of the marine harpacticoid copepod, *Tigriopus californicus*. Bulletin of Environmental Contamination and Toxicology 44:636–642.

Montagna, P. A., J. T. Bauer, D. Hardin, and R. B. Spies. 1987. Temporal variability and the relationship between benthic meiofaunal and microbial populations of a natural coastal petroleum seep. Journal of Marine Research 45:761–789.

Palmer, M. A. 1988. Dispersal of marine meiofauna: a review and conceptual model explaining passive transport and active emergence with implications for recruitment. Marine Ecology Progress Series 48:81–91.

Palmer, M. A., P. A. Montagna, R. B. Spies, and D. Hardin. 1988. Meiofauna dispersal near natural petroleum seeps in the Santa Barbara Channel: a recolonization experiment. Oil and Chemical Pollution 4:179–189.

Sherman, K. M., D. A. Meeter, and J. A. Reidenauer. 1984. A technique for subsampling an abundant taxon while completely sorting other taxa. Limnology and Oceanography 29:433–439.

Smith, C. J., R. D. DeLaune, W. H. Patrick, and J. W. Fleeger. 1984. Impact of dispersed and undispersed oil entering a Gulf Coast salt marsh. Environmental Toxicology and Chemistry 3:609–616.

Snelgrove, P. V. R., J. F. Grassle, and R. F. Petrecca. 1992. The role of food patches in maintaining high deep-sea diversity—field experiments with hydrodynamically unbiased colonization trays. Limnology and Oceanography 37:1543–1550.

Spies, R. B., D. D. Hardin, and J. P. Toal. 1988. Organic enrichment or toxicity? A comparison of the effects of kelp and crude oil in sediments of the colonization and growth of benthic infauna. Journal of Experimental Marine Biology and Ecology 124:261–282.

Stacey, B., and B. Marcotte. 1987. Chronic effect of number 2 fuel oil on population dynamics of harpacticoid copepods in experimental marine mesocosms. Marine Ecology Progress Series 40:61–68.

Stickle, W. B., M. A. Kapper, T. C. Shirley, M. G. Carls, and S. D. Rice. 1990. Bioenergetics and tolerance of the pink shrimp (*Pandalus borealis*) during long-term exposure to the water-soluble fraction and oiled sediment from Cook Inlet crude oil. Pages 87–106 *in* W. B. Vernberg, A. Calabrese, F. P. Thurberg, and F. J. Vernberg, editors. Pollution physiology of estuarine organisms. University of South Carolina Press, Columbia.

Sun, B. 1993. Interaction between hydrodynamics and sediment topography on meiofaunal abundance and distribution. Doctoral dissertation. Louisiana State University, Baton Rouge.

Ustach, J. F. 1979. Effects of sublethal oil concentrations on the copepod, *Nitocra affinis*. Estuaries 2:273–276.

Walters, K. 1991. Influences of abundance, behavior, species composition, and ontogenetic stage on active emergence of meiobenthic copepods in subtropical habitats. Marine Biology 108:207–215.

Warwick, R. M., M. R. Carr, K. R. Clarke, J. M. Gee, and R. H. Green. 1988. A mesocosm experiment of the effects of hydrocarbon and copper pollution of a sublittoral soft-sediment meiobenthic community. Marine Ecology Progress Series 46:181–191.

Webb, D. G., and T. R. Parsons. 1992. Winter-spring recruitment patterns of epiphytic harpacticoid copepods in a temperate-zone seagrass bed. Marine Ecology Progress Series 82:151–162.

Wormald, A. P. 1976. Effects of a spill of marine diesel oil on the meiofauna of a sandy beach at Picnic Bay, Hong Kong. Environmental Pollution 11:117–130.

American Fisheries Society Symposium 18:286–297, 1996

Persistence of Oiling in Mussel Beds Three and Four Years after the *Exxon Valdez* Oil Spill

MALIN M. BABCOCK
National Marine Fisheries Service, Alaska Fisheries Science Center Auke Bay Laboratory, 11305 Glacier Highway Juneau, Alaska 99801, USA

GAIL V. IRVINE
National Biological Service, Alaska Science Center 1011 East Tudor Road, Anchorage, Alaska 99503, USA

PATRICIA M. HARRIS
National Marine Fisheries Service, Alaska Fisheries Science Center

JOEL A. CUSICK
National Park Service, 2525 Gambell St., Room 107 Anchorage, Alaska 99503, USA

STANLEY D. RICE
National Marine Fisheries Service, Alaska Fisheries Science Center

Abstract.—Dense beds of the mussel *Mytilus trossulus* affected by *Exxon Valdez* crude oil in Prince William Sound and along the Kenai and Alaska peninsulas were intentionally left untreated during shoreline cleanup activities in 1989–1991. In 1992 and 1993, mussels and sediments from 70 mussel beds in Prince William Sound and 18 beds along the Kenai and Alaska peninsulas were sampled to establish the geographic extent and intensity of *Exxon Valdez* oil persisting in mussel beds. Sediments collected in 1992 and 1993 from 31 of the oiled mussel beds in the sound had total petroleum hydrocarbon (TPH) concentrations greater than 10,000 μg/g wet weight. The highest concentrations were in sediments collected from Foul Bay (62,258 $\pm$ 1,272 μg TPH/g, mean $\pm$ SE). Five of the 18 beds sampled along the Kenai Peninsula showed sediment TPH concentrations greater than 5,000 μg/g. The mean concentration of total polynuclear aromatic hydrocarbons (TPAH) in mussels from these same beds ranged up to 8.30 $\pm$ 0.26 μg/g (Squirrel Island) in Prince William Sound and 4.01 $\pm$ 1.54 μg/g along the Kenai Peninsula (Morning Cove, Pye Islands). Polynuclear aromatic hydrocarbon fingerprints of mussel tissue collected from surveyed sites indicated the contaminant source was *Exxon Valdez* oil. In 1993, mean TPH concentrations in sediments and mean TPAH concentrations in mussels were lower by more than 50% compared with these concentrations in 1992. Some beds showed little reduction in oil. Almost all the beds showing only small decreases in hydrocarbons were in protected, low-energy areas, where there probably was little remobilization of residual oil underlying the beds. This study has produced analytical evidence showing that substantial residual *Exxon Valdez* oil persists in sediments underlying mussel beds in the area affected by the spill. Residual crude oil is a source of chronic contamination of mussels and their predators. In the more-protected intertidal areas, natural flushing and remobilization of *Exxon Valdez* oil will be slow; some of these mussel beds potentially can be manually cleaned.

Based on the importance of dense beds of the mussel *Mytilus trossulus* (on finer, unconsolidated substrates) as food for higher consumers and as a community and physically stabilizing influence in the intertidal area, the *Exxon Valdez* oil spill Interagency Shoreline Cleanup Committee intentionally avoided cleanup and other treatment of these beds after the oil spill on 24 March 1989. It was hoped that natural processes would clean the beds in reasonable time.

In spring 1991, the persistence of crude oil in sediments underneath some dense mussel beds in Prince William Sound was apparent and caused concern among scientists from federal and state agencies. Because these beds were left intact and uncleaned, as an unanticipated consequence they could be a source of chronic hydrocarbon (HC) exposure to organisms inhabiting the near-surface and surface areas, thus providing a possible pathway for petroleum HCs to enter the food web for higher consumers. Persistent, high concentrations of HCs in mussels were identified as a possible source of impacts in several consumer species (Duffy et al. 1996, this volume; Sharp et al. 1996, this volume). These contaminated beds could also affect human subsistence users.

Pilot surveys and collections of sediments and mussels in 1991 confirmed the existence of substan-

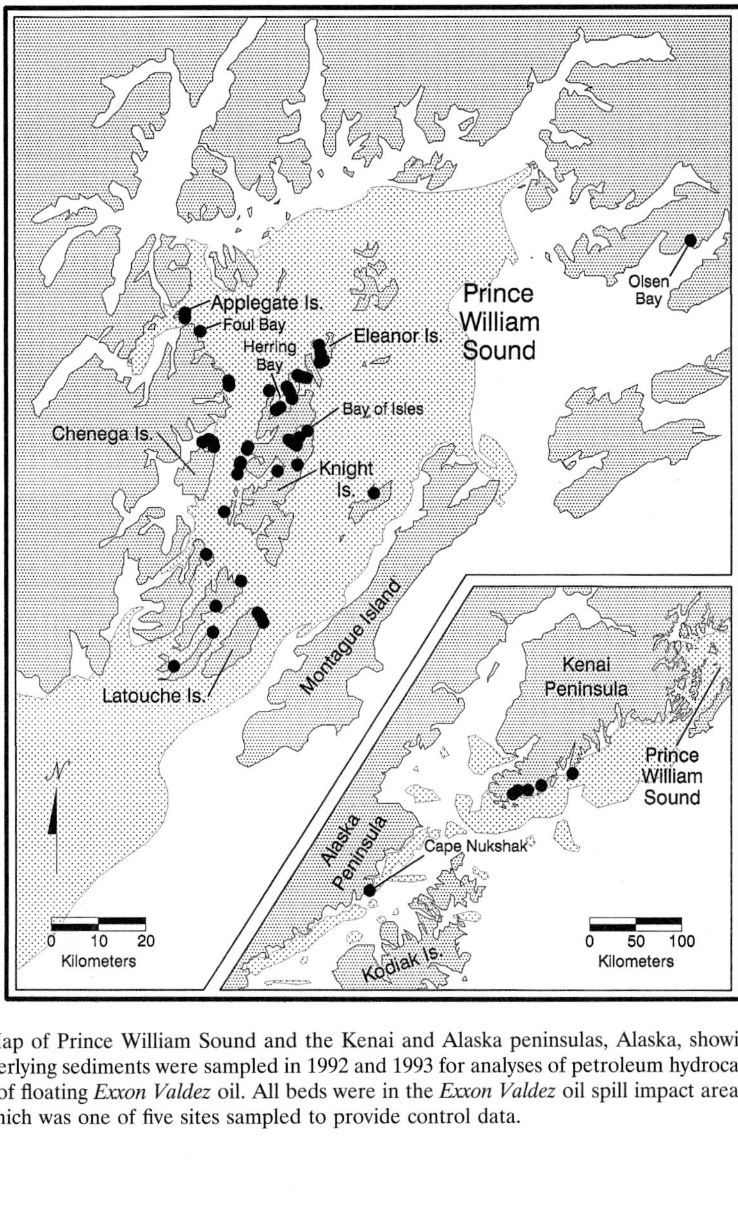

FIGURE 1.—Map of Prince William Sound and the Kenai and Alaska peninsulas, Alaska, showing locations where mussels and underlying sediments were sampled in 1992 and 1993 for analyses of petroleum hydrocarbons. Shaded area indicates extent of floating *Exxon Valdez* oil. All beds were in the *Exxon Valdez* oil spill impact area with the exception of Olsen Bay, which was one of five sites sampled to provide control data.

tial amounts of residual *Exxon Valdez* crude oil in sediments immediately underlying dense mussel beds and also in mussels (Babcock et al. 1994). This crude oil was mobile and odorous, and the concentrations within sediments and mussels were the highest found in 1991.

Extensive surveys in Prince William Sound and the Kenai and Alaska peninsulas were conducted to determine the extent of mussel bed contamination and to assess the magnitude of the problem. We document the geographic extent of oiled mussel beds in 1992 and 1993 and the intensity of oiling of mussels and underlying sediments within the sound and along the Kenai and Alaska peninsulas (Figure 1). Specifically, we (1) measured petroleum HCs in mussels and underlying sediments collected from 88 mussel beds in the oil spill impact area, (2) measured the loss of HCs between postspill years 3 and 4 at 37 sites, and (3) confirmed the match of oil in the mussel beds with spilled *Exxon Valdez* oil. These data on oiled mussel beds will allow monitoring of natural recovery and provide information for decisions on assisted recovery in areas with residual HCs in future years.

Study Area

Potentially oiled mussel beds within Prince William Sound and along the Kenai and Alaska peninsulas were tentatively identified with the help of several sources: (1) Alaska Department of Environmental Conservation's (ADEC's) extensive Shoreline Assessment records, (2) Alaska Department of Fish and Game (ADFG) researchers studying harlequin ducks *Histrionicus histrionicus*, and (3) U.S. Fish and Wildlife (USFWS) personnel working on black oystercatchers *Haematopus bachmani*. In addition, six previously documented sites showing little or no *Exxon Valdez* oil contamination (Short and Babcock 1996, this volume) were sampled for comparative purposes. Mussel beds were visited, and mussels and underlying sediments were sampled if oil was present. Surveys along the Kenai and Alaska peninsulas were conducted by the National Park Service (NPS), while similar surveys in Prince William Sound were conducted by the National Oceanic and Atmospheric Administration's Auke Bay Laboratory (ABL). Actual sampling of sediments and mussels was conducted by ABL, NPS, ADFG, and USFWS personnel during extensive fieldwork.

Sixty-five mussel beds in Prince William Sound and 13 along the Kenai and Alaska peninsulas were sampled in 1992. Repetitive sampling was conducted in 1993 to determine changes in HC concentrations; sites sampled were selected primarily on the basis of HC concentrations in sediments in 1992. In 1993, 31 beds were sampled in the sound, including five new sites not located in 1992. An additional nine sites sampled originally in 1992 were examined but not sampled because there was no visual or olfactory evidence of oil remaining. Along the Kenai and Alaska peninsulas, 15 sites were sampled in 1993, including five oiled mussel beds not located in 1992.

In Prince William Sound, the sampled mussel beds ranged in size from approximately 20 m^2 for a small bed on Disk Island to 700 m^2 for the large bed on the tombolo adjacent to Eleanor Island. Density of mussels ranged from thinly interspersed (288/m^2) at Aguliak Island to multiple layers (5,000/m^2) at Eleanor Island. Some beds sampled along the Kenai and Alaska peninsulas were smaller than the Disk Island bed. We estimated the total area of mussel beds sampled at 5,250 m^2 in Prince William Sound and 720 m^2 along the Kenai and Alaska peninsulas.

Most mussel beds were situated on mixed sand and gravel substrates, and mussels were usually fairly evenly dispersed throughout the sampling area; however, the presence of large cobbles and boulders created heterogeneity in many beds.

Methods

Sampling Procedures

The primary criteria for sampling mussels and sediments were the presence of moderately to densely packed mussels on sand- and gravel-sized sediments (<1 cm in diameter) and the detection of crude oil by visual or olfactory means. Mussel- and sediment-sampling methods were modified from those developed by ABL in previous years (Karinen et al. 1993; Babcock et al. 1994). A transect, usually 30 m long and parallel to the waterline (as topography allowed), was established through the middle of a mussel bed. The length of the transect varied according to size and topography and ranged from 10 m at a Disk Island site to 50 m at the Foul Bay site. At each of 8 to 10 places along the transect (within 1 m above or below) a small portion of mussels was overturned. Triplicate pooled subsamples of underlying sediment were collected (0–2 cm deep) by scooping sediment from each exposed location with a hydrocarbon (HC)-free stainless steel spoon into each of three 118-ml, HC-free glass jars. Similarly, triplicate pooled samples of 20–25 mussels (ranging from 25 to 40 mm in length) each were collected from the overturned portions and placed in three HC-free jars. Some variation of this basic sampling method occurred at the Kenai and Alaska peninsula sites, and at Prince William Sound sites used for the within-bed HC distribution study (Harris et al. 1996, this volume). At the Kenai and Alaska peninsula sites, three transects were laid: one through the center of the bed, another 1 m above and another 1 m below the original transect. Pooled samples of sediments and mussels were collected along each transect. Although these were not true replicate samples like most of the Prince William Sound samples were, collectively they represented an area of approximately the same breadth. All samples were cooled immediately, and frozen within 2–4 h.

Chemistry

All sediment samples were analyzed by ultraviolet fluorescence (UVF), as adapted from Krahn et al. (1991, 1993). For UVF screening, wet sediment samples were extracted twice with methylene chloride, then concentrated or diluted to match a calibration curve based on an *Exxon Valdez* oil standard. These extracts were analyzed using a high-performance liquid chromatograph equipped with a fluorescence detector. Excitation and emission spectra of the extracts were read at the phenanthrene wavelengths (260 and 380 nm), and values were calculated to estimate total petroleum HCs (TPH) based on the amount of phenanthrene in *Exxon Valdez* oil. Data are reported as micrograms per gram wet weight TPH. All sediment data have an N of 3 unless otherwise noted.

This procedure allowed economical screening of many samples and produced semiquantitative data, which were then used to select samples for further analyses by gas chromatography–mass spectroscopy (GC–MS) for quantitative determination of individual HC analytes. Selected sediments and all mussel samples were analyzed by GC–MS as described by Short et al. (1996, this volume). Whereas all mussel samples collected along the Kenai and Alaska peninsulas were analyzed, mussel samples from Prince William Sound were selected for GC–MS analyses based on UVF results of underlying sediments. Mussels from 36 of the 70 sampled beds in the sound were actually analyzed by GC–MS. Units presented in this paper are micrograms per gram dry weight total polynuclear aromatic HCs (TPAH), representing the sum of all measured aromatic HCs except perylene (produced by biogenic sources). Mussel data have an N of 3 unless otherwise noted.

Values produced by UVF can be used with reliability for estimating the oiling conditions of sediment. Figure 2 shows good agreement ($R^2 = 0.74$) between UVF and TPAH values and even better correlation ($R^2 = 0.83$) between UVF data and total HCs (the sum of all aromatic and alkane HC analytes, as determined by GC–MS).

Results

Geographic Extent and Intensity of Oiling

Sediments and mussels from visibly oiled intertidal areas in Prince William Sound and the Kenai and Gulf of Alaska areas had substantial amounts of oil in 1992 and 1993. Sediments were visibly oiled and odorous, and the oil was easily detected. Sheening was often visible without any manual disturbance. Intertidal oil in the sound was more mobile and less weathered than oil observed along the Kenai and Alaska peninsulas, where floating oil had usually progressed to a mousse stage before landfall. Sites examined and sampled for sediments and mussels were located within the entire *Exxon Valdez* oil spill impact area in Prince William Sound and along the Kenai Peninsula. In addition, one site at Cape Nukshak on the Alaska Peninsula and Olsen Bay in eastern Prince William Sound were sampled (Figure 1).

In Prince William Sound, sediments from 31 of the 70 mussel beds sampled showed HCs greater than 10,000 μg/g wet weight TPH (measured by UVF), and sediments from 13 beds were between 5,000 and 10,000 μg/g TPH (Figure 3). The five sites with the highest mean concentrations (±SE) of oil in sediments were Foul Bay in Prince William Sound (62,258 ± 1,558 μg/g, $N = 3$), a small islet in Herring Bay that was a site of experimental manipulation and intensive sampling (35,029 ± 8,793 μg/g, $N = 9$) (Harris et al. 1996), another experimental bed on northern Chenega Island (27,872 ± 4,359 μg/g, $N = 12$), a mussel bed on eastern Applegate Island (26,867 ± 2,107 μg/g, $N = 6$), and a bed on northern Knight Island (26,728 μg/g ± 43.6 μg/g, $N = 3$) (Figure 3).

Along the Kenai and Alaska peninsulas, 5 of 18 mussel beds sampled showed mean TPH concentrations in sediments greater than 5,000 μg/g; these were all on the Kenai Peninsula (Figure 3). The three sites with the highest mean TPH concentrations were Morning Cove in the Pye Islands (36,538 ± 24,381 μg/g), a second mussel bed in Morning Cove (22,307 ± 10,837 μg/g), and Windy Bay (9,394 ± 9,130 μg/g) ($N = 3$ unless otherwise noted; Figure 3). Moderately contaminated sediments (1,000–5,000 μg/g TPH) were documented at 13 mussel beds in Prince William Sound and 10 sites along the Kenai and Alaska peninsulas. These sites included the only oiled mussel bed located on the Alaska Peninsula, which was on Cape Nukshak (4,639 ± 2,298 μg/g TPH in 1993).

Sediment samples from the remaining mussel beds had mean oil concentrations greater than 1,000 μg/g TPH. In Prince William Sound, 5 of the remaining 13 sites represented stations established under the Natural Resource Damage Assessment Coastal Habitat 1B study (Short and Babcock 1996); these 5 sites were sampled intentionally to provide control data and were sites where data existed from previous years.

Substantial TPAH concentrations (>1.00 μg/g, dry weight, measured by GC–MS) were found in mussels from 16 beds in Prince William Sound and 5 beds along the Kenai Peninsula. In the sound, the highest mean TPAH concentrations (±SE) in mussels were found at Squirrel Island (8.30 ± 0.26 μg/g TPAH dry weight, $N = 2$), the Foul Bay site (8.09 ±

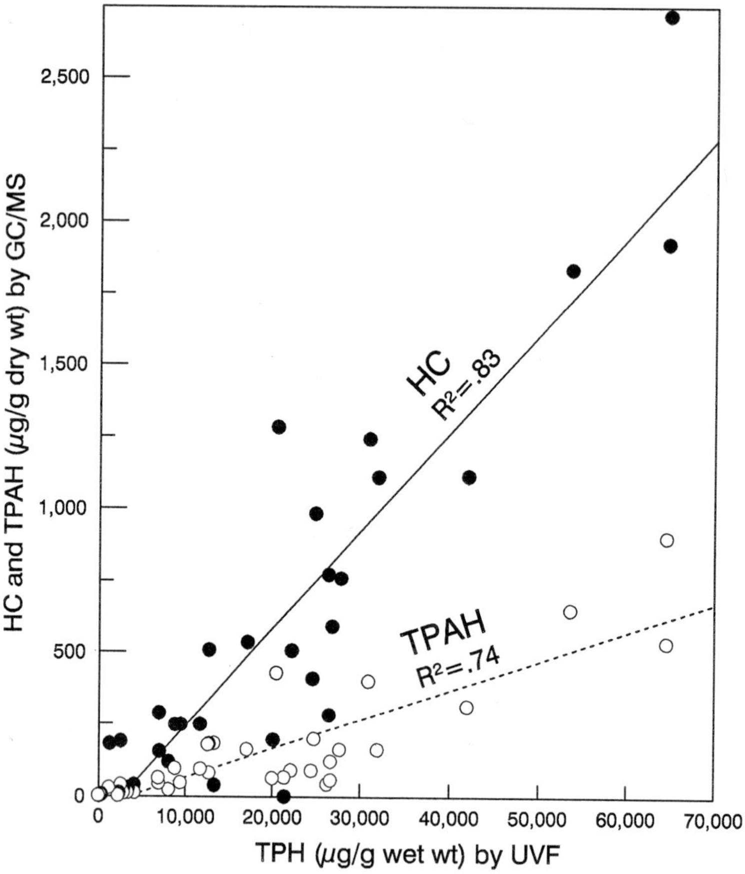

FIGURE 2.—Correlation of total petroleum hydrocarbons (TPH) in sediments collected from underneath dense mussel beds, measured by ultraviolet fluorescence (UVF), with the sum of polynuclear aromatic hydrocarbons (minus perylene, which is produced biogenically) (TPAH, open circles) and total hydrocarbons (HC, solid circles), measured by gas chromatography/mass spectroscopy (GC–MS), on the same sample. Samples were collected in Prince William Sound and the Kenai and Alaska peninsulas in 1992 and 1993 from the *Exxon Valdez* oil spill impact area.

1.13 µg/g), the large mussel bed on Eleanor Island (6.99 ± 4.83 µg/g), a Bay of Isles site (5.57 ± 0.94 µg/g), and a bed on Disk Island (4.90, N = 1) (Figure 4). Along the Kenai Peninsula, the highest TPAH concentrations in mussels from oiled beds were found at Morning Cove in the Pye Islands (4.01 ± 1.54 µg/g in 1993), another bed in Morning Cove (3.99 ± 0.51 µg/g in 1993), and Grim Beach in Tonsina Bay (2.91 ± 1.05 µg/g) (Figure 4).

Changes in Petroleum Hydrocarbons from 1992 to 1993

Total petroleum HCs in sediments showed some reductions from 1992 to 1993 in the 27 mussel beds in Prince William Sound and in the 10 beds along the Kenai and Alaska peninsulas that were examined in both years—3 and 4 years after the *Exxon Valdez* oil spill. The mean TPH concentration in sediments from oiled mussel beds in 1993 averaged 47.6 ± 37.4% (SD) of the concentrations found in 1992 (Figure 5A); 20 of the 37 beds showed less than a 50% reduction. Changes were variable, ranging from 0 to 100%. Sediments underlying five oiled mussel beds actually showed higher TPH values in 1993, but are shown at 100% in Figure 5A. Most of these higher values were within the SE, and did not differ between years.

A similar reduction was seen in mussels from 1992 to 1993. The mean TPAH concentrations in mussels in 1993 averaged 43.8 ± 30.0% (SD) of the 1992 values (Figure 5B). Mussels from three beds actually had higher concentrations but the differ-

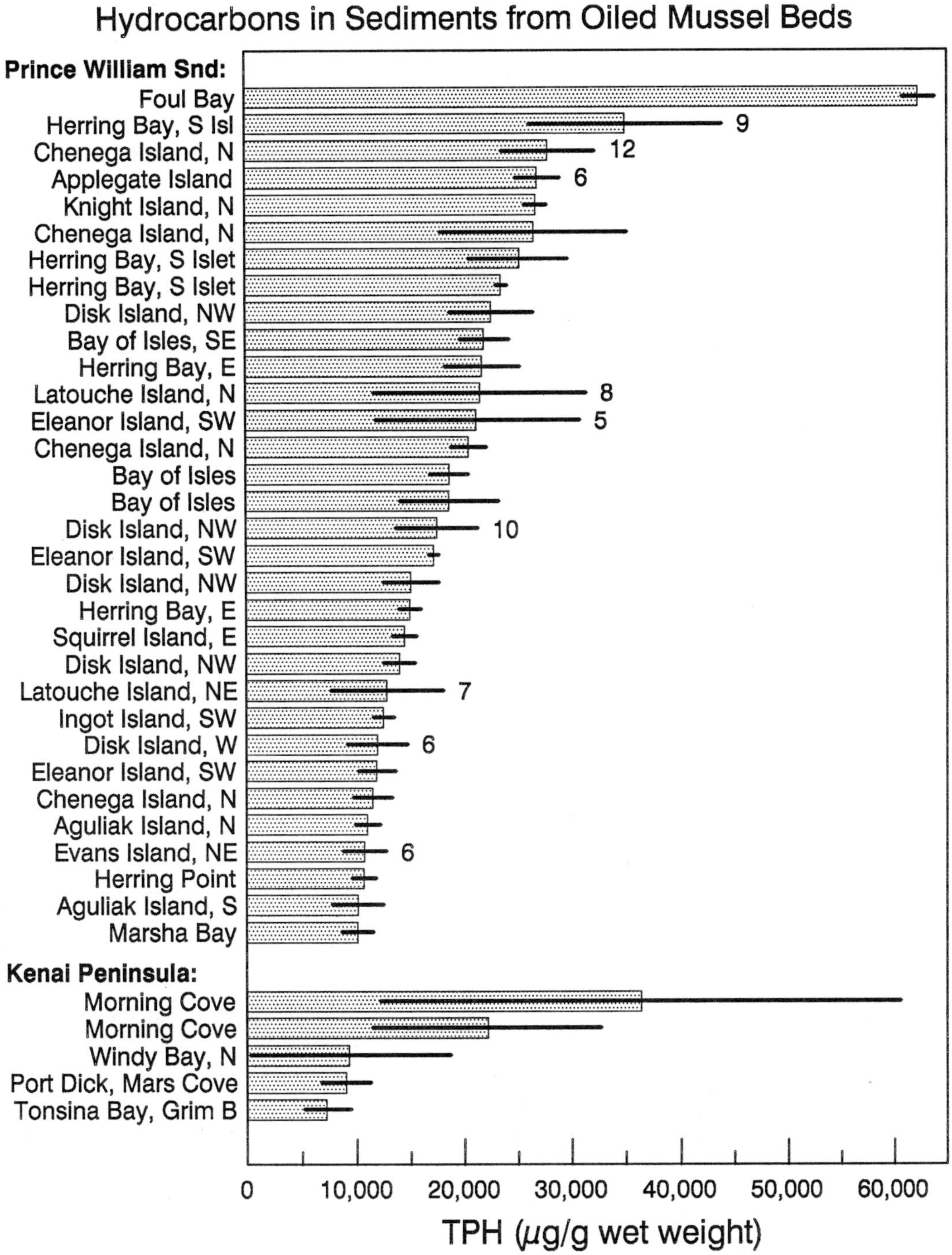

FIGURE 3.—Total petroleum hydrocarbons (TPH) in Prince William Sound, Alaska, mussel beds with underlying sediments exceeding 10,000 µg/g wet weight TPH and Kenai Peninsula oiled mussel beds exceeding 5,000 µg/g TPH as measured by ultraviolet fluorescence. Samples were collected in 1992 or 1993, 3 or 4 years after the *Exxon Valdez* oil spill. For some locations, there were multiple, distinctly separate, oiled mussel beds sampled. The compass directions S, W, E, N, NW, NE, and SW represent more-specific locations of the mussel beds within the geographic feature. Error bars are SEs; *N* is usually 3, but ranges to 12 (indicated to the right of the concentration bars) where there were multiple visits within a year or where spot samples were collected for another study (Harris et al. 1996). When sites were sampled in both 1992 and 1993, means from the higher TPH concentrations are presented.

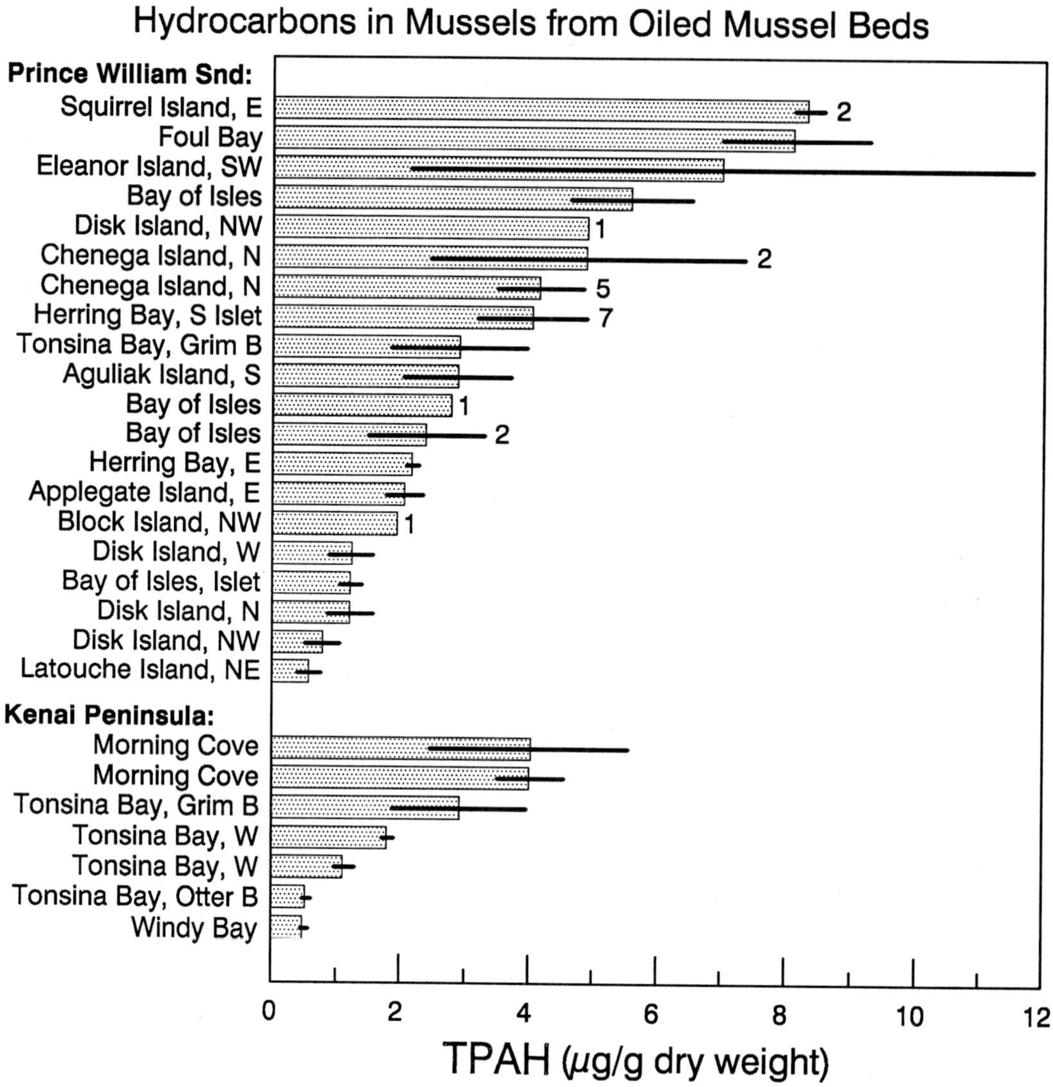

FIGURE 4.—Total polynuclear aromatic hydrocarbons (TPAH) in Prince William Sound and Kenai Peninsula mussels collected from oiled mussel beds with TPAH exceeding 0.50 μg/g dry weight as measured by gas chromatography/mass spectroscopy. Samples were collected in 1992 or 1993, 3 or 4 years after the *Exxon Valdez* oil spill. For some locations, there were multiple, distinctly separate, oiled mussel beds sampled. The compass directions S, W, E, N, NW, NE, and SW represent more specific locations of the mussel beds within the geographic feature. Error bars are SEs; *N* is usually 3, but ranges from 1 to 7 (as indicated to the right of the concentration bars) where there were multiple visits within a year or where spot samples were collected for another study (Harris et al. 1996). When sites were sampled in both 1992 and 1993, means from the higher TPAH concentrations are presented.

ences were not significant except for the Eleanor Island bed (discussed later). They are shown at 100%, or no change, for 1993 in Figure 5B.

Exxon Valdez Oil Fingerprints

The polynuclear aromatic HC patterns in mussels collected from the oiled beds in 1992 and 1993 were consistent with that of *Exxon Valdez* oil (Figure 6). Polynuclear aromatic HC fingerprints in Prince William Sound mussels (Figure 6, B through D) for both 1992 and 1993 and in mussels from the Kenai Peninsula from 1992 (Figure 6E) all show a pattern similar to that of *Exxon Valdez* oil (Figure 6A). The fingerprint of mussels from one bed along the Kenai

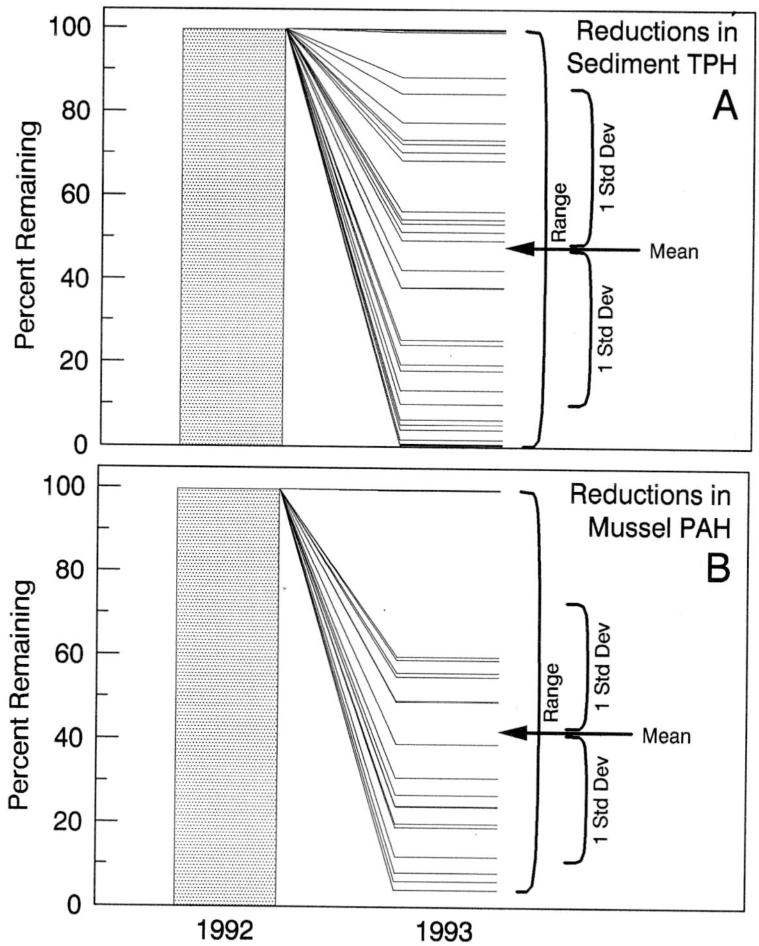

FIGURE 5.—Reduction in total petroleum hydrocarbons (TPH) in sediments (**A**) and total polynuclear aromatic hydrocarbons (TPAH) in mussels (**B**) from oiled mussel beds from 1992 to 1993. Sampled beds were within the *Exxon Valdez* oil spill impact area of Prince William Sound and the northeast Gulf of Alaska. All means of 1992 sediments are shown as 100%; paired sample data from 1993 represent the percent TPH remaining. Some means depicted at 100% in 1993 actually increased but are shown here at 100%. For sediments, *N* is 27; for mussels, *N* is 22.

Peninsula (Morning Cove, Pye Islands, collected in 1993) is characterized by an additional source of HCs, probably of pyrolytic origin, as shown in Figure 6F (compare the relative amounts of anthracene, fluoranthene, and pyrene with those in Figure 6, A–E).

Discussion

Mussels and sediments sampled during this study, in both 1992 and 1993, contained among the highest concentrations of *Exxon Valdez* oil found in mussels and sediments collected in any study sponsored by the *Exxon Valdez* Trustee Council since 1990, the year after the oil spill. In sediments, a mean TPH greater than 62,000 μg/g wet weight (measured by UVF) was found at one site, and a mean TPAH concentration greater than 8.00 μg/g dry weight (measured by GC–MS) was documented in mussels. In contrast, sediments from known control beds showed a mean TPH concentration less than 100 μg/g, and mussels from control beds had a mean TPAH concentration less than 0.30 μg/g, both of which approach the detection limits for the respective analytical methods. Historically (1977–1980), sediments and mussels collected from established stations along the shipping lane through Prince William Sound, as well as those collected in 1989 before landfall of the oil, exhibited little or no contaminat-

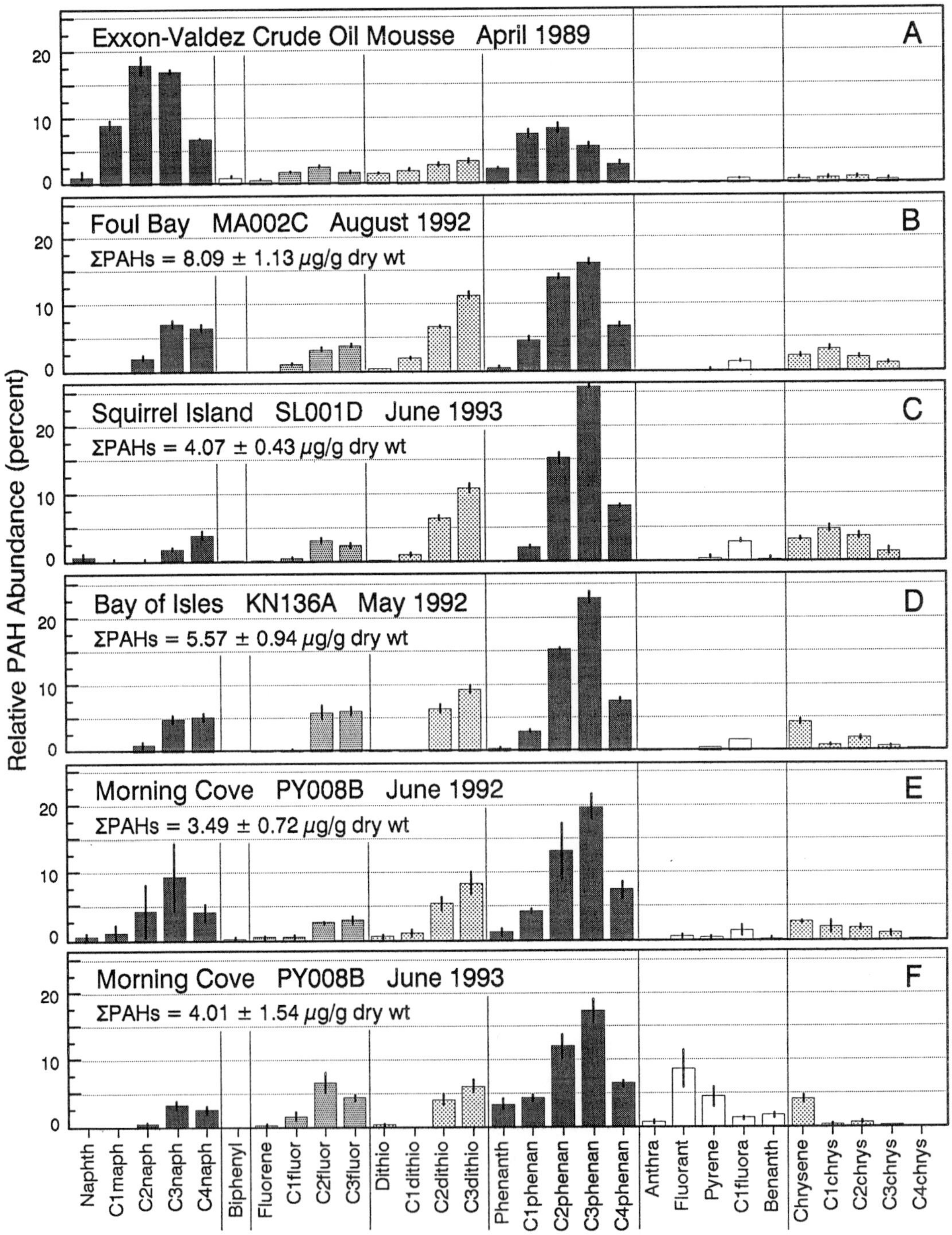

FIGURE 6.—Relative abundance of individual polynuclear aromatic hydrocarbons (PAH) in *Exxon Valdez* crude oil (**A**) and mussels from selected sites in Prince William Sound (**B**, **C**, **D**) and the Kenai Peninsula (**E**, **F**). Mean concentrations are proportional to total PAH to depict patterns independent of the actual quantities of PAH present in the mussels. Error bars are SEs; N is 3.

ing petroleum HCs (Karinen et al. 1993; Short and Babcock 1996).

The geographic distribution of these substantially contaminated mussel beds included almost the entire area of Prince William Sound and the Kenai Peninsula that was impacted by the spill (Figure 1). In addition, one oiled mussel bed was sampled along the Alaska Peninsula. Within the sound, sampled oiled mussel beds were bounded by Applegate Island and Foul Bay in the northwest, north Eleanor Island in the northeast, Bay of Isles on the east, and northern Elrington Island in the south. Most of the contaminated mussel beds were located within the Knight Island group, an area particularly affected by the *Exxon Valdez* oil spill. Extensive surveys were not conducted in the farthest southwestern portion of the sound. Most of the oiled beds sampled along the Kenai Peninsula were along the southwestern portion of the peninsula. The one oiled mussel bed located on the Alaska Peninsula (Cape Nukshak) was due west of northern Kodiak Island.

The most contaminated sites were within Prince William Sound rather than outside it, not only because of the volume of *Exxon Valdez* oil stranded on the beaches within the sound, but also because of the additional weathering of the oil before it landed on the intertidal areas along the Kenai and Alaska peninsulas. Most mussel beds within the sound were oiled in the first or second week after the spill when oil was not weathered into mousse. The floating oil coming ashore was thick in volume and very fluid. Intertidal exposure was sustained through several tidal cycles. As oil progressed into the northeast Gulf of Alaska, the total volume of the oil was much less (estimated at 7–11% of the total spilled) than that reaching landfall and becoming stranded in Prince William Sound (estimated at 41% of the total oil spilled) (Wolfe et al. 1994). Floating mousse in the Gulf of Alaska was patchy and more weathered, and repeated exposures and penetrations in intertidal areas were probably not sustained through as many tidal cycles as in Prince William Sound.

Many mussel beds throughout the study area showed natural reductions in oil—more than 50% for sites where there were data from both years; however, several showed little decrease in HC concentrations from 1992 to 1993. Although sediment samples from six beds showed increases in TPH and mussels from three beds showed increases in TPAH for 1993 over 1992, most of these values were within the margin of error.

One mussel bed that showed HC increases in both sediment and mussels from 1992 to 1993, the tombolo at Eleanor Island, incurred vigorous storm activity in winter 1992–1993. When we visited the site in June 1993, approximately 20% of the mussel layer had disappeared and the underlying sediments had been redistributed. In August even more of the mussel layer (total estimated at 30%) had eroded and disappeared, indicating that some disruption had continued during the summer. The mussel layer was probably more susceptible to erosion after the initial redistribution over the winter. This disruption of the bed and subsequent remobilization of *Exxon Valdez* oil entrained in the underlying sediments probably accounted for the marked increase in body burden of TPAH seen in the mussels (0.43 $\pm$ 0.14 [SE] μg/g in 1992 versus 6.99 $\pm$ 4.83 μg/g in 1993) and the simultaneous doubling of TPH in underlying sediments (10,070 $\pm$ 3,821 μg/g in 1992 versus 21,208 $\pm$ 9,436 μg/g in 1993).

Most of the beds that exhibited less than a 50% decrease in oil (TPH in sediments and TPAH in mussels) between 1992 and 1993 were either on sheltered beaches or had large cobbles–small boulders that armored the surface sediments and mussels. Vandermeulen and Gordon (1976) and Gundlach et al. (1982) found similar persistence of bunker C oil spilled from the *Arrow* and the *Metula* in intertidal areas of Chedabucto Bay, Nova Scotia, and the Strait of Magellan 5 and 6.5 years, respectively, after each accidental grounding. Residual crude oil remained on both the surface and in subsurface sediments but not specifically associated with mussel beds.

In addition to large differences in HC concentrations among sites, there was a high degree of variability of concentrations within the beds. The uneven distribution of residual crude oil was confirmed by our intensive sampling of selected beds (Harris et al. 1996) and by incidental samples collected at depths of 5–10 cm at three mussel beds during regular survey sampling. At two of these mussel beds, oil concentrations were significantly higher ($P < 0.01$, ANOVA [analysis of variance]) at the subsurface depths (5–7 cm) compared with surface sediment (0–2 cm) (Bay of Isles, 18,653 $\pm$ 3,791 versus 1,764 $\pm$ 827 μg/g TPH; Herring Bay, 35,029 $\pm$ 8,793 versus 5,473 $\pm$ 876 μg/g TPH), whereas at the mussel bed on Squirrel Island, surface sediments had a higher mean TPH than did subsurface sediments (14,467 $\pm$ 913 versus 3,499 μg/g, $N = 1$). The patchiness of distribution of *Exxon Valdez* oil was also shown by Michel and Hayes (1993a, 1993b). There were many references to subsurface "lenses" of oil throughout ADEC's

Shoreline Assessment Patrol reports, and the presence of subsurface lenses was also documented by Michel and Hayes (1993a, 1993b). These lenses were not necessarily associated with mussel beds. Oil at depth is probably less available as a source of chronic exposure to surface-dwelling organisms than oil that is only 0–2 cm below the surface.

We recognized the variability of oil distribution within these mussel beds (Harris et al. 1996) and designed pooled sampling strategies to minimize these effects and to characterize the overall HC concentration within each bed; however, variability was still high in many cases, as reflected by relatively high SEs for both sediments and mussels (Figures 3 and 4). Harris et al. (1996) showed high within-bed variability of *Exxon Valdez* oil distribution both horizontally and vertically.

The dense mussel layer probably provided a protective layer against natural environmental and weathering processes degrading oil entrained in underlying sediments. The lack of weathering was indicated by the fluidity and odor and by the similarity of profiles in the GC–MS fingerprints (Figure 6). Relative proportional distributions of polynuclear aromatic HC analytes in mussels shown in Figure 6 are consistent with those of *Exxon Valdez* oil mousse collected in April 1989 from floating material. The distribution patterns are also similar to those shown by Michel and Hayes (1993a) for a surface (1–5 cm) sediment sample taken in September 1991 (N13/2B—not associated with any mussel bed) from a sheltered, set-aside area in eastern Herring Bay. The main differences are relatively lower amounts of naphthalenes and lower amounts of C3- and C4-alkylated homologues, particularly C4-phenanthrene, for the mussels, as reflected in Figure 6. This finding indicates less degradation of the *Exxon Valdez* oil available for uptake by these mussels than in the sediments from their Herring Bay site. The patterns in our samples more closely resemble their profile from station N3/3B, a subsurface sample (40–45 cm deep) from Smith Island that they depicted as only intermediately degraded (Michel and Hayes 1993a).

Outside the sound, the polynuclear aromatic HC pattern for mussels collected in 1993 from Morning Cove, in the Pye Islands along the Kenai Peninsula, had HCs from a pyrolytic source in addition to *Exxon Valdez* oil. Mussels collected from other oiled beds along the Kenai Peninsula in 1993 showed either this pattern of multiple sources of contamination or had polynuclear aromatic HC concentrations too low for identification of particular contaminant sources. These HC distributions in mussels collected from most of these same beds in 1992 were similar to what would be expected if the only contamination source was *Exxon Valdez* oil.

As we have seen in this study, there are mussel beds that still contained substantial amounts of *Exxon Valdez* oil in 1993. There is evidence that for some beaches natural weathering and flushing are playing a role in reducing the amount of petroleum HCs that are available for uptake in mussels.

Although the geographic spread of oil-contaminated sites extends throughout the oil spill region and there is a high level of oil contamination in some places, the actual contaminated area is rather small. This begs the question: Should we be concerned about small physical areas? If there is disproportionate consumption by top predators, including subsistence harvesters, then there would be obvious concern.

In 3 or 4 years nature has not taken care of the oiling problem in many beds, and some of the more sheltered beds probably will continue to be sources of oil for a long time. Most of the 20 beds that showed less than a 50% decrease in TPH between 1992 and 1993 are in relatively protected areas that are not vulnerable to natural disturbances. Some of these beds are on fairly uniform mixed sand and gravel substrates and could be candidates for manual restoration. Certainly these beds need to be monitored for *Exxon Valdez* oil concentrations in mussels and underlying sediments until the concentrations approach prespill or background levels (Karinen et al. 1993; Short and Babcock 1996).

Acknowledgments

This study was partially funded by the *Exxon Valdez* Oil Spill Trustee Council under Restoration Studies 103 and 93036. We relied on and are grateful for the help of many in fulfilling the objectives of this study. Mark Brodersen and John Bauer of ADEC provided access to ADEC's shoreline assessment data. Sam Patten, Dan Gray, and Michael East of ADFG and Brad Andres of USFWS identified and sampled potential oiled mussel bed sites. Mark Carls, Scott Feldhausen, Dan Fremgen, Larry Holland, Marie Larsen, Debbie Shosteck, and Bruce Wright of ABL assisted in the field. Andy Gunther of Applied Marine Sciences, Livermore, California, also helped in our sampling program. Field surveys by NPS were conducted with the assistance of Laurel Lonergan and Nettie Kelly, with additional help from ADEC staff members Clara Crosby and Steve Ferguson. Special thanks to ABL's chemistry analytical staff members Jeffrey

Short, Marie Larsen, Larry Holland, and Josefina Lunasin for consistently producing quality analytical data.

References

Babcock, M. M., P. M. Rounds, C. C. Brodersen, and S. D. Rice. 1994. 1991 and 1992 recovery monitoring and restoration of intertidal oiled mussel (*Mytilus trossulus*) beds in Prince William Sound impacted by the *Exxon Valdez* oil spill. NOAA (National Oceanic and Atmospheric Administration) NMFS (National Marine Fisheries Service) AFSC (Alaska Fisheries Science Center) Processed Report 94-02, Juneau.

Duffy, L. K., R. T. Bowyer, J. W. Testa, and J. B. Faro. 1996. Acute phase proteins and cytokines in Alaskan mammals as markers of chronic exposure to environmental pollutants. American Fisheries Society Symposium 18:809–813.

Gundlach, E. R., D. D. Domeracki, and L. C. Thebeau. 1982. Persistence of *Metula* oil in the Strait of Magellan six and one-half years after the incident. Oil and Petrochemical Pollution 1(1):37–48.

Harris, P. M., S. D. Rice, M. M. Babcock, and C. C. Brodersen. 1996. Within-bed distribution of *Exxon Valdez* crude oil in Prince William Sound blue mussels and underlying sediments. American Fisheries Society Symposium 18:298–308.

Karinen, J. F., M. M. Babcock, D. W. Brown, W. D. Macleod, Jr., L. S. Ramos, and J. W. Short. 1993. Hydrocarbons in intertidal sediments and mussels from Prince William Sound, Alaska, 1977–1980: characterization and probable sources. NOAA (National Oceanic and Atmospheric Administration) Technical Memorandum NMFS (National Marine Fisheries Service) AFSC-9, Seattle.

Krahn, M. M., and nine coauthors. 1993. Comparison of high-performance liquid chromatography/fluorescence screening and gas chromatography/mass spectrometry analysis for aromatic compounds in sediments sampled after the *Exxon Valdez* oil spill. Environmental Science and Technology 27:699–708.

Krahn, M. M., G. M. Ylitalo, J. Joss, and S.-L. Chan. 1991. Rapid, semi-quantitative screening of sediments for aromatic compounds using sonic extraction and HPLC/fluorescence analysis. Marine Environmental Research 31:175–196.

Michel, J., and M. O. Hayes. 1993a. Persistence and weathering of *Exxon Valdez* oil in the intertidal zone–3.5 years later. Pages 279–286 *in* Proceedings, 1993 oil spill conference. American Petroleum Institute Publication 4580, Washington, DC.

Michel, J., and M. O. Hayes. 1993b. Evaluation of the condition of Prince William Sound shorelines following the *Exxon Valdez* oil spill and subsequent shoreline treatment: volume 1. 1991 geomorphological shoreline monitoring survey. NOAA (National Oceanic and Atmospheric Administration) Technical Memorandum NOS (National Ocean Service) ORCA 67, Seattle.

Sharp, B. E., M. Cody, and R. Turner. 1996. Effects of the *Exxon Valdez* oil spill on the black oystercatcher. American Fisheries Society Symposium 18:748–758.

Short, J. W., and M. M. Babcock. 1996. Prespill and postspill concentrations of hydrocarbons in mussels and sediments in Prince William Sound. American Fisheries Society Symposium 18:149–166.

Short, J. W., T. J. Jackson, M. L. Larsen, and T. L. Wade. 1996. Analytical methods used for the analysis of hydrocarbons in crude oil, tissues, sediments, and seawater collected for the natural resources damage assessment of the *Exxon Valdez* oil spill. American Fisheries Society Symposium 18:140–148.

Vandermeulen, J. H., and D. C. Gordon, Jr. 1976. Reentry of 5-year-old stranded bunker C fuel oil from a low-energy beach into the water, sediments, and biota of Chedabucto Bay, Nova Scotia. Journal of the Fisheries Research Board of Canada 33:2002–2010.

Wolfe, D. A., and eleven coauthors. 1994. The fate of the oil spilled from the *Exxon Valdez*. Environmental Science and Technology 28:561A–568A.

American Fisheries Society Symposium 18:298–308, 1996

Within-Bed Distribution of *Exxon Valdez* Crude Oil in Prince William Sound Blue Mussels and Underlying Sediments

PATRICIA M. HARRIS, STANLEY D. RICE, MALIN M. BABCOCK, AND CHRISTINE C. BRODERSEN

National Oceanic and Atmospheric Administration, National Marine Fisheries Service
Auke Bay Laboratory, Alaska Fisheries Science Center
11305 Glacier Highway, Juneau, Alaska 99801-8626, USA

Abstract.—The distribution of *Exxon Valdez* crude oil in sediments and its relationship to oiling in blue mussels *Mytilus trossulus* was examined in six Prince William Sound mussel beds in 1992. Distribution of oil in sediments within each bed was patchy and was related to several factors, including sample location on the beach, sample depth, and sediment grain size. Concentrations of total petroleum hydrocarbons as high as 30,000–40,000 μg/g wet weight were found within a few meters of concentrations two orders of magnitude less. The distribution of total polynuclear aromatic hydrocarbons (TPAH) in mussels overlying the sediments was also patchy. Concentration levels in mussels roughly followed concentrations in underlying sediments, but correlation at specific sampling points was low. Generally, TPAH in mussels averaged 1% or less of the TPAH in shallow sediments from the same area of the bed.

Similar proportions of individual analytes relative to TPAH in mussels, shallow sediments, and *Exxon Valdez* crude oil indicate that the latter was the source of contamination in sediment and mussels, that the oil in the sediments has only moderately weathered, and that oil taken in by mussels was particulate. Mussels depurate petroleum hydrocarbons rapidly if not chronically exposed; the contaminated sediments were the continuous source of *Exxon Valdez* crude oil in mussel tissues. The pattern of oil distribution in mussels and in sediments within a mussel bed implies that it was dispersed from sediments into surrounding water and was taken up by nearby mussels.

The chronic recontamination of mussels by *Exxon Valdez* crude oil in the sediments underlying them and the uneven distribution of this oil throughout mussel beds have important implications for studies monitoring concentrations and effects of oil remaining in mussel beds and for any mussel-bed restoration.

Concern was focused on beds of blue mussels *Mytilus trossulus*[1] affected by the *Exxon Valdez* oil spill when damage assessment studies suggested links between oiled mussels and negative impacts on several predator species (Duffy et al. 1996; Sharp et al. 1996; both this volume). Most oiled mussel beds were not intensively cleaned in 1989 and 1990 by the *Exxon Valdez* Oil Spill Interagency Shoreline Cleanup Committee because mussels are such an important food source in intertidal communities. Natural flushing has not cleaned the beds; oil has persisted longer and at much higher concentrations than expected especially in beds on finer sediments. In 1991, more than 2 years after the spill, the presence of *Exxon Valdez* oil in mussels and underlying sediments was confirmed in 13 mussel beds, and concentrations in these beds were among the highest measured in Prince William Sound that year. Total polynuclear aromatic hydrocarbons (TPAH) ranged as high as 10.3 μg/g dry weight in blue mussels and as high as 489 μg/g dry weight in sediments (Babcock et al. 1994). These persistent high concentrations and the importance of mussel beds as feeding areas for many species, gave oiled mussel beds an importance out of proportion to their relatively small size and numbers.

In 1991 and early 1992, we determined mean hydrocarbon levels in mussels and sediments in each mussel bed by pooling samples taken at several locations within a bed (Babcock et al. 1994). While sampling, we observed uneven oil distribution throughout each bed. In 1992 this study was begun to assess oil distribution in mussel bed sediments and its relationship to oiling in mussels. This understanding was needed for continued monitoring of oil remaining in mussel beds and is relevant to any mussel-bed restoration projects.

Site Selection

Six sampling sites were identified using Alaska Department of Environmental Conservation Shoreline Patrol Assessment records, information from other Natural Resource Damage Assessment principal investigators, and hydrocarbon data from mus-

[1]*Mytilus trossulus* Gould, 1850 (formerly *M. edulis* Linnaeus, 1758) McDonald and Koehn (1988).

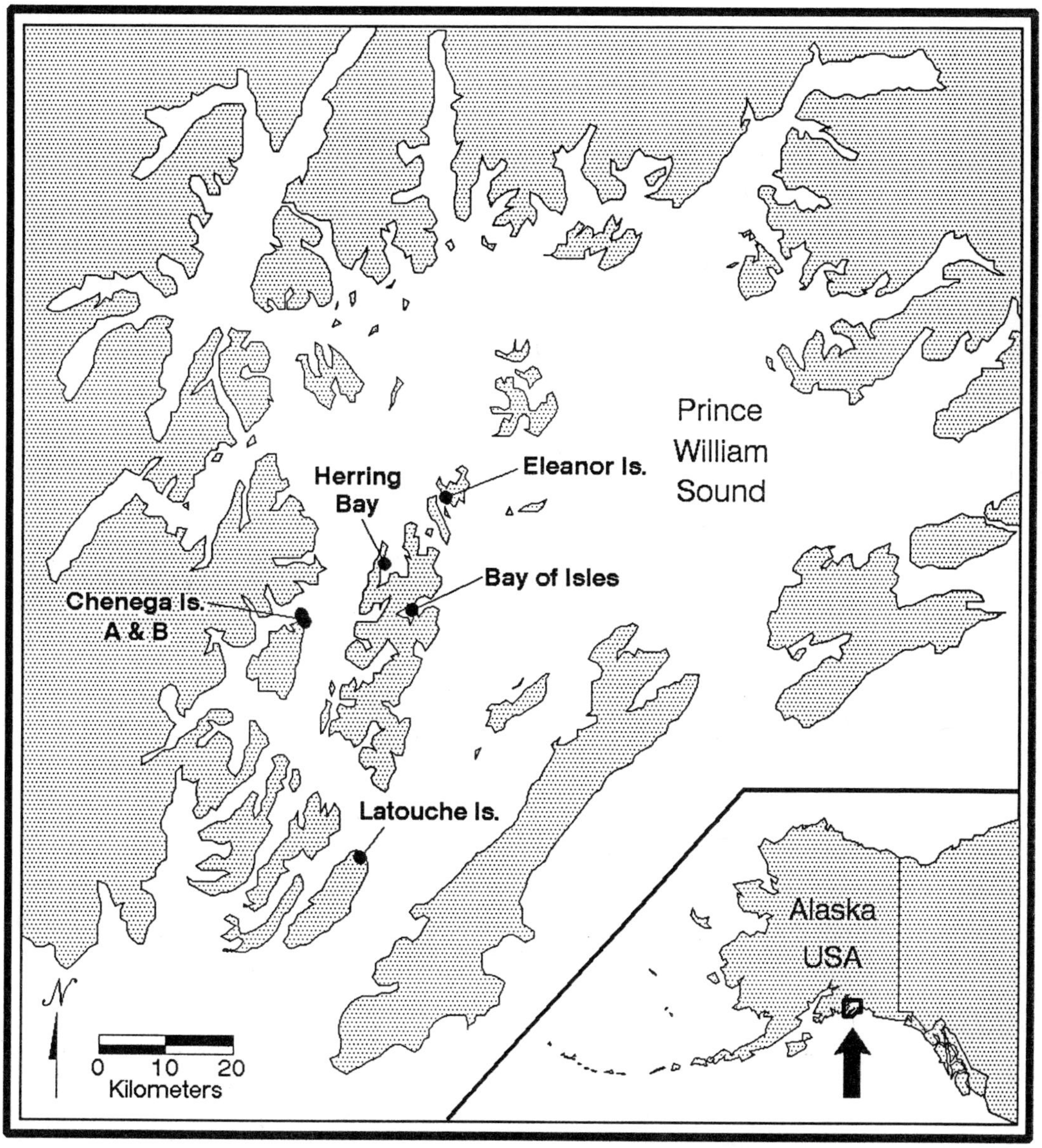

FIGURE 1.—Locations of sampled, oiled mussel beds in Prince William Sound.

sel beds sampled in 1991, and by making site visits early in 1992. Sites selected were Eleanor Island, Herring Bay, Chenega Island A and B, and Latouche Island (Figure 1). We refer to the larger Chenega Island bed as Chenega A and the smaller one as Chenega B.

Study beds were on gently (<5%) sloping beaches and ranged in area from approximately 10 to 700 m^2. The tidal range occupied by the beds was 0.1–1.7 m above mean lower low water, but the tidal range occupied by each bed was at most 0.9 m. In the lower portions of those beds, oil-contaminated water remained close to the surface of the bed even at the lowest tides; on the upper transects, subsurface water was below sampling depth. Sediments were gravel–granules over sand–silt, often with interspersed cobbles. Mussels within each bed were a uniform size (30–45 mm in shell length) and were generally evenly distributed at densities ranging from 800 to 5,000 mussels/m^2.

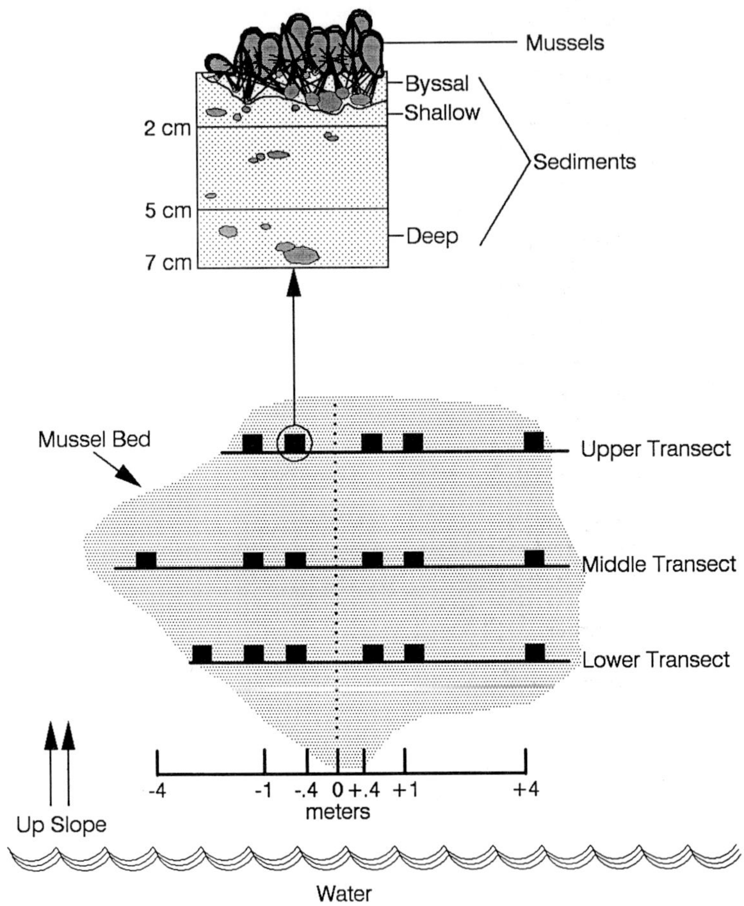

FIGURE 2.—Sediment depths sampled and typical sampling design at Chenega A, Herring Bay, and Eleanor Island mussel beds in Prince William Sound. Black squares represent sample quadrats. Bed shown is Chenega Island A.

Methods

Sampling

At three mussel beds, Chenega Island A, Herring Bay, and Eleanor Island, samples were collected in a grid laid out over the bed (Figure 2). At each bed, several transects were laid parallel to the water. Exact location of transects depended on bed size and topography; they were generally 2–5 m in length and 0.1 m vertical distance and 2–5 m horizontal distance from adjacent transects. Five specific 0.25 m × 0.25 m sampling quadrats were located along each transect, except in cases where bed shape limited the length of a transect. The irregular shapes of the three other beds, Latouche Island, Bay of Isles, and Chenega Island B, precluded sampling along transects, so these were sampled at three or four quadrats selected to best represent each bed. Additional mussels were collected from bedrock adjacent to the sampled mussel beds.

At each quadrat, separate samples of mussels, sediment attached to the byssal threads of the mussels, and shallow sediment underlying the byssal thread–sediment mat (0–2 cm deep) were collected in May 1992 for hydrocarbon analysis (Figure 2). At some Eleanor Island quadrats, coarseness of the sediment made collecting discrete byssal-mat samples difficult, so samples consisting of combined byssal mat and surface sediments were collected at these locations. Deeper sediments (5–7 cm) were later collected (August 1992) at some quadrats in the six beds.

Mussel density at each quadrat was estimated by counting mussels in one-half of the quadrat area. Each mussel sample consisted of 20 individuals, evenly distributed over the quadrat, and each sedi-

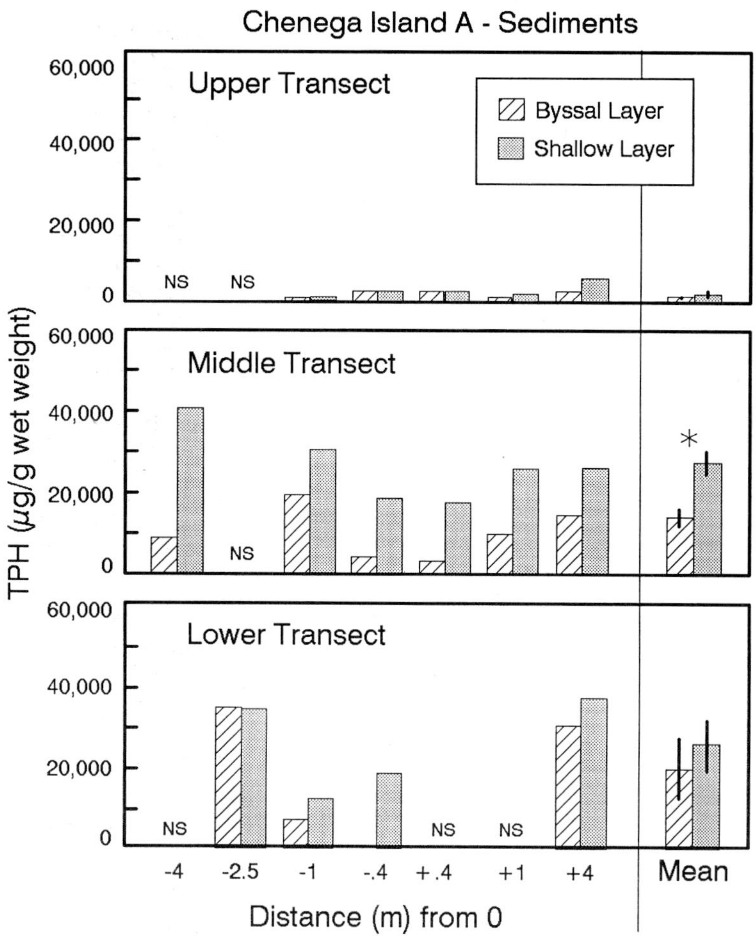

FIGURE 3.—Distribution of total petroleum hydrocarbons (TPH) determined by ultraviolet fluorescence in shallow and byssal layer sediments of the Chenega Island A mussel bed in Prince William Sound, May 1992. Distances of samples from the central axis of the mussel bed are on the *x*-axis. Mean TPH concentrations for each transect are at the right. Vertical lines = ±1 SE, asterisk = significant difference at $P \leq 0.05$, and NS = no sample collected.

ment sample consisted of 8–10 scoops (approximately 200 g total). Sampling instruments and sample jars were hydrocarbon free. Samples were frozen within 2–3 h after collection.

Observations of general sediment grain size were noted for each sampled mussel bed, and specific measurements were made of the percentage by dry weight of silt and clay (particles <62 µm) for Chenega A byssal and shallow sediment samples, following procedures of Krumbein and Pettijohn (1988).

Chemical Analyses

Two different types of analysis were used to measure petroleum hydrocarbons. Gas chromatography–mass spectroscopy (GC–MS) was used to measure precisely the concentrations of specific polynuclear aromatic hydrocarbons (listed in Table 2 of Short et al. 1996, this volume). Concentrations of these analytes (except perylene) were combined, giving TPAH as micrograms per gram dry weight of tissue or sediment. Proportions of individual polynuclear aromatic hydrocarbons (PAH) relative to TPAH were calculated and used to construct a PAH profile for each sample. Although GC–MS analysis is the most accurate measure of petroleum hydrocarbon content, it was more practical to use a faster screening technique for sediment samples. Ultraviolet fluorescence (UVF) fast screening (adapted from Krahn et al. 1993) was used to esti-

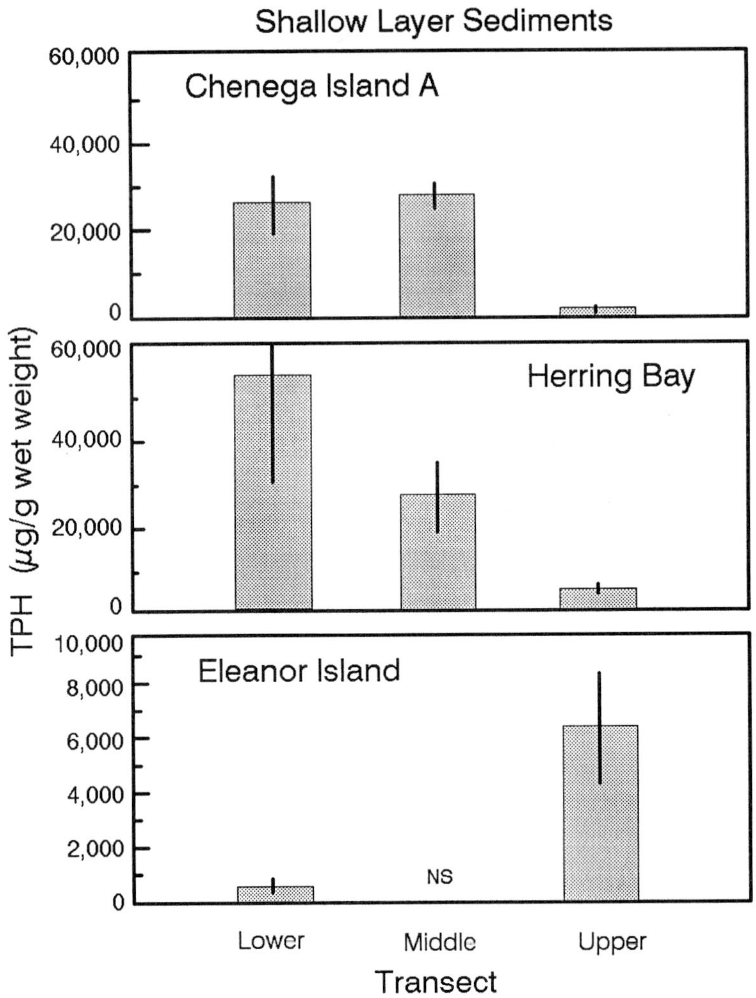

FIGURE 4.—Total petroleum hydrocarbons (TPH), measured by ultraviolet fluorescence, in shallow sediments in Chenega Island A, Herring Bay, and Eleanor Island mussel beds in Prince William Sound, May 1992 (N = 4–6). Note that the scale for Eleanor Island TPH differs from that of the other two beds. Vertical lines = ±1 SE and NS = no sample.

mate total petroleum hydrocarbons (TPH) in sediments as micrograms per gram wet weight. Concentrations of TPH in each sample were quantified by comparing the sample's fluorescence at the phenanthrene wavelength to the fluorescence of unweathered *Exxon Valdez* oil. Little error was anticipated comparing an unweathered standard with possibly weathered samples because phenanthrene is relatively less susceptible to weathering than most other aromatics. Note that TPAH and TPH concentrations are not directly comparable; TPAH is a quantitative measurement of a specific suite of polynuclear aromatic analytes, whereas TPH is a semiquantitative measure of total petroleum hydrocarbon concentrations (aromatic and aliphatic). All of the sediment samples in this study have been analyzed by UVF; all mussel samples and selected sediment samples have been analyzed by GC–MS. Some sediments were analyzed by GC–MS so TPAH and PAH profiles could be compared with TPAH and PAH profiles of *Exxon Valdez* oil and of mussel samples.

Data Analyses

Analysis of variance was used to compare sediment hydrocarbon concentrations in different transects, at different distances from the centers of the transects, and at different depths within quadrats. This test was also used to determine that there

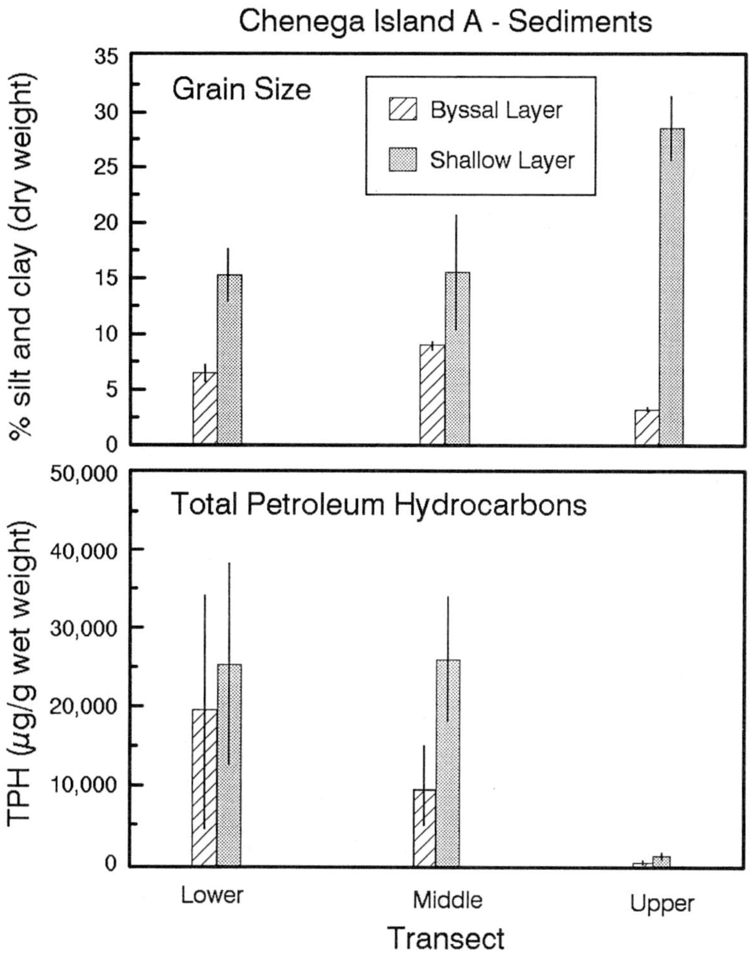

FIGURE 5.—Percentage (dry weight) of silt and clay particles (<62 μm) and total petroleum hydrocarbons (TPH) concentrations in the shallow and byssal layers along the three sampling transects (lower, middle and upper) in the Chenega Island A mussel bed, May 1992 (N = 2 or 3). Vertical lines = ±1 SE.

was no significant difference between shallow layer TPH measurements in May and in August, making it reasonable to assume that deep sediment samples taken in August could be compared with byssal and shallow sediment samples taken in May. Hydrocarbon concentrations in the byssal layer and in the shallow sediments of each sample quadrat at Chenega A and at Herring Bay were compared by an unbalanced two-way split plot analysis (Snedecor and Cochran 1956), where sample transects were considered blocks and sediment depth was the treatment. The fewer quadrats with separate byssal layer and shallow sediment samples at Eleanor Island were compared by paired t-test. The percentages of particles <62 μm in byssal sediments and in shallow sediments at each Chenega A quadrat were compared by paired t-test.

Analysis of variance was used to compare TPAH concentrations among Chenega A mussels and to examine the relationship between TPAH concentrations in mussels and in underlying sediments. Concentrations in sediments (TPH) and in mussel tissue (TPAH) were regressed against mussel density to determine the relationship between oiling and mussel density. All tests were considered significant at $P \leq 0.05$.

Results

Distribution of Total Petroleum Hydrocarbons in Sediments

The distribution of oil in sediments in each mussel bed was uneven (Figure 3). In the Chenega A bed, for example, TPH concentrations in shallow

TABLE 1.—Relationship between total petroleum hydrocarbon (TPH) concentrations (measured by ultraviolet fluorescence) in byssal layer, shallow, and deep sediments in six mussel beds in Prince William Sound. Samples were collected in May, June, and August 1992. Asterisks indicate significant difference in TPH concentrations between depths indicated (analysis of variance; $P \leq 0.05$).

Site	Sediment depth (cm) 0		0–2		5–7
Chenega Island A					
Upper	Byssal	<	Shallow	>	Deep
Middle	Byssal	<*	Shallow	>	Deep
Lower	Byssal	<	Shallow	>*	Deep
Herring Bay					
Upper	Byssal	<	Shallow	>	Deep
Middle	Byssal	<*	Shallow	>*	Deep
Lower	Byssal	<*	Shallow	>	Deep
Eleanor Island					
Upper	Byssal	<*	Shallow	<*	Deep
Lower	Byssal	<*	Shallow	<*	Deep
Chenega Island B	Byssal	<	Shallow	>	Deep
Bay of Isles	Byssal	<*	Shallow	>	Deep
Latouche Island	No data		Shallow	<*	Deep

sediments differed by as much as three orders of magnitude. Several variables, including location within a bed, sediment depth, and sediment grain size, showed some correlation with TPH, but relationships were not consistent at all beds.

In the Chenega A, Herring Bay, and Eleanor Island beds, TPH concentrations differed significantly between transects more than within transects, whereas horizontal distance from center of transect had no significant effect on concentration. Concentrations in the upper transects at Chenega A (Figure 4) and Herring Bay were far lower than concentrations in the middle and lower transects; however, at Eleanor Island the upper transect had the higher concentrations.

Depth of the sample was an important factor in sediment hydrocarbon loads. Total petroleum hydrocarbon concentrations in byssal layer sediments (sediments attached to the mussels) were lower than concentrations in the corresponding shallow (0–2 cm deep) sediments at all beds where byssal samples were taken (none were taken at Latouche). These differences were significant on the middle transect at Chenega A (Figure 3), the lower and middle transects at Herring Bay, and at Bay of Isles (Table 1). The relationship between TPH concentrations in shallow and deeper sediments (5–7 cm deep) was mixed. Concentrations of TPH were significantly higher in shallow sediments than in deeper sediments at Chenega A and Herring Bay but were lower at Latouche and Eleanor islands.

Sediment grain size appeared to be a factor in the retention of hydrocarbons in sediments. On all transects in the Chenega A bed, the percentage of silt and clay particles was lower in the byssal layer than in the shallow (0–2 cm deep) sediments, and TPH levels in the byssal sediments were lower relative to their paired shallow sediments (Figure 5). Across the whole bed, however, percentage of silt and clay in a sample did not correlate significantly with TPH. In the beds with coarser sediments, Eleanor and Latouche islands, the higher TPH levels were from the deeper (5–7 cm) samples, whereas in the finer textured beds, at Chenega A and B, Herring Bay and Bay of Isles, the higher TPH levels were in the shallow (0–2 cm) sediments.

Distribution of Total Polynuclear Aromatic Hydrocarbons in Mussels

The concentrations of TPAH in mussels were related to concentrations in underlying sediments (Figure 6) but were much lower. The mussels from the middle and upper transects at Chenega A contained about 1% of the TPAH in underlying sediments.

The distribution of petroleum hydrocarbons in mussels was related to the distribution in associated sediments, but the correspondence at each quadrat was not exact. Mussels with high oil concentrations came from transects with high oil concentrations in sediments. At Chenega A, for instance, highest concentrations were in mussels from the lower two transects; levels of TPAH in the mussels from the upper transect were similar to concentrations in mussels on nearby bedrock and at non-oiled Prince William Sound mussel beds (less than 0.15 μg/g) (Figure 6). Correlation of TPAH concentration in each mussel sample and its paired sediment sample, however, was relatively poor ($R^2 = 0.55$).

The overall distribution of oil in mussels was highly uneven within a bed and within a transect (Figure 7), just as it was in sediments. In the Chenega A bed, TPAH concentrations ranged over two orders of magnitude from 0.09 to 6.00 μg/g. Even within a single sampling transect, TPAH concentrations differed by one order of magnitude. Mean TPAH concentrations were significantly higher ($P \leq 0.05$) in mussels from the lower and middle transects than in mussels from the upper transect and bedrock.

There was no relationship between mussel density and levels of contamination in mussels ($R^2 = 0.003$) or in sediments ($R^2 = 0.021$) at specific quadrats.

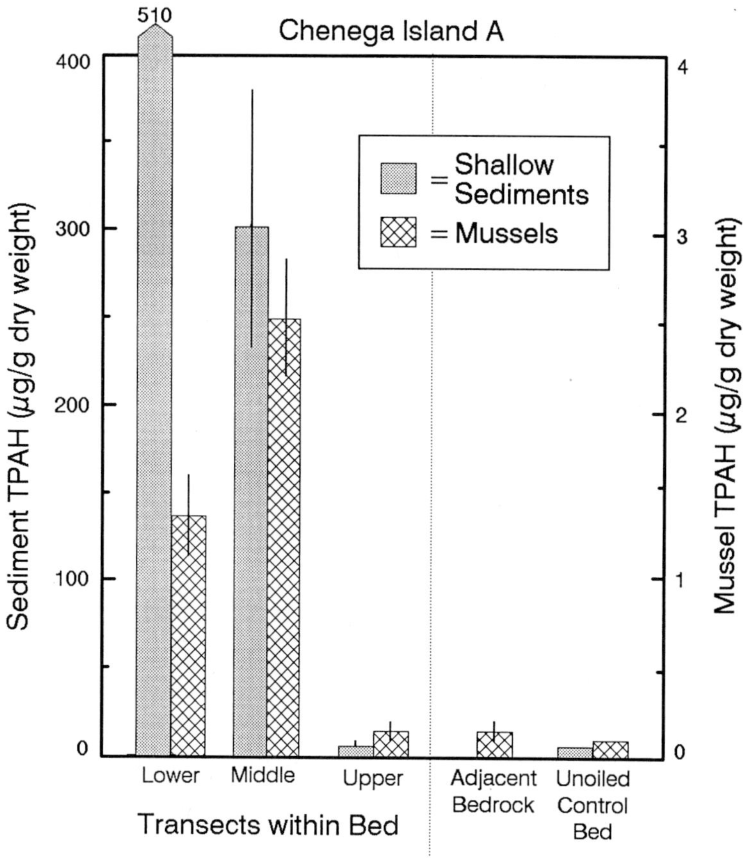

FIGURE 6.—Mean total polynuclear aromatic hydrocarbons (TPAH), measured by gas chromatography–mass spectroscopy, in shallow sediments (0–2 cm) and in mussels from the three sample transects in the Chenega Island A mussel bed (May 1992), adjacent bedrock (May 1992), and non-oiled control site at Barnes Cove (August 1992) (N = 2–6). Vertical lines = ±1 SE.

Total Polynuclear Aromatic Hydrocarbon Profiles

The PAH profiles in shallow sediment, byssal layer sediment, and mussel samples collected more than 3 years after the spill were strikingly similar to proportions in *Exxon Valdez* oil collected 1 week after the spill (Figure 8), indicating that the oil in these beds was *Exxon Valdez* oil and that it was only slightly weathered. The analyte patterns in sediments and mussels showed enrichment of the higher molecular weight compounds relative to *Exxon Valdez* oil (e.g., higher chrysene concentrations). Also in mussels and sediments, the more highly alkyl-substituted dibenzothiophenes and phenanthrenes were enriched relative to their parent compounds. The relative proportions of C3 dibenzothiophene to dibenzothiophene and C3 and C4 phenanthrene to phenanthrene increased from fresh *Exxon Valdez* oil to shallow sediments to byssal layer sediments to mussel tissues (Figure 8), indicating that the byssal sediments have undergone more weathering than the shallow sediments below them, and the oil in the mussels has weathered slightly more.

Discussion

Distribution of oil hydrocarbons in sediments within a mussel bed was uneven and was probably influenced by many factors, including degree and pattern of oiling, exposure or protection of a specific site, sediment grain size, and water flow through the bed. The importance of each of these factors and their interactions differed from bed to bed, making consistent relationships between a factor and oiling hard to discern. Exposure was an important factor at the Eleanor Island bed; patches

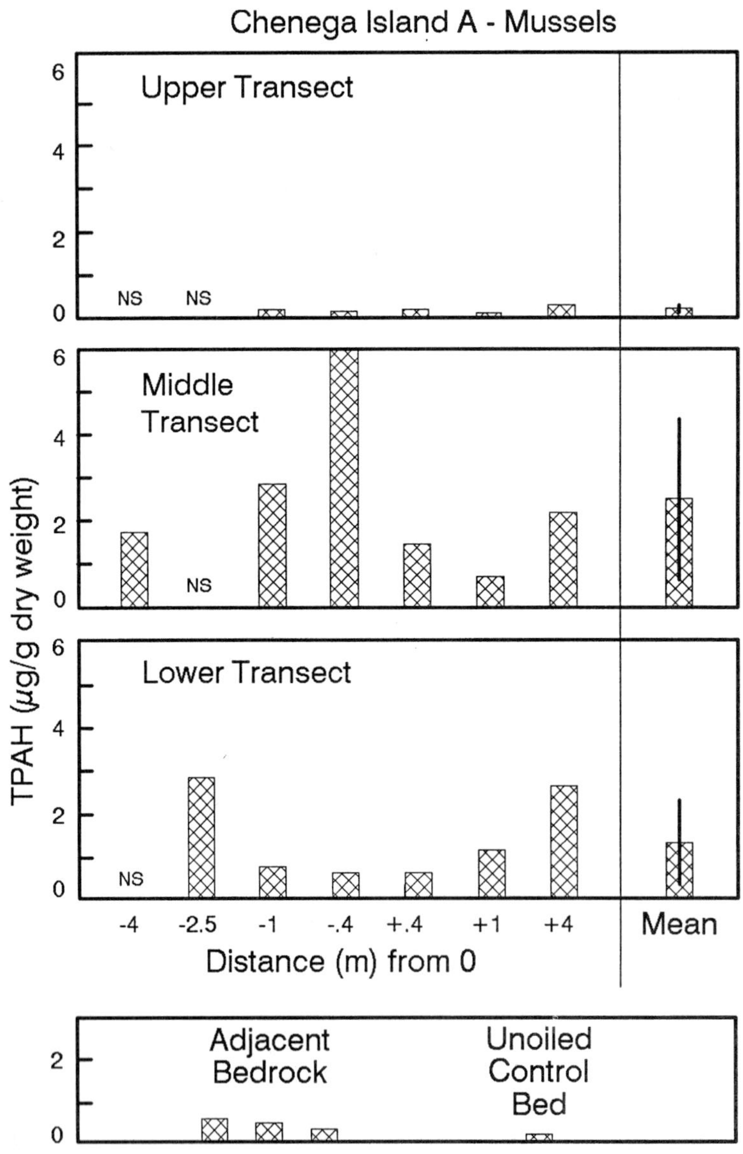

FIGURE 7.—Distribution of total polynuclear aromatic hydrocarbons (TPAH), measured by gas chromatography–mass spectroscopy, in mussels from the Chenega Island A bed (May 1992), grouped across sample transect and compared with mussels on adjacent bedrock (May 1992), and a non-oiled control site at Barnes Cove (August 1992). Abbreviation is NS = no sample and vertical lines = ±SE.

of mussels were removed from the lower transect by storm wave action over the course of this study. Flushing by previous storms or cleanup operations (Houghton et al. 1993) may have accounted for lower concentrations on that transect relative to the upper part of the bed. However, at the Chenega A and Herring Bay beds, differing wave exposure probably had minimal effect on within-bed concentration differences, because these beds were small and their sample transects were at virtually the same tidal elevation and subject to similar wave exposure. Surface and subsurface water flows within these beds may have accounted for the lower concentrations at the upper transects here. High percentages of silt and clay in soils are usually associated with increased oil retention by providing more

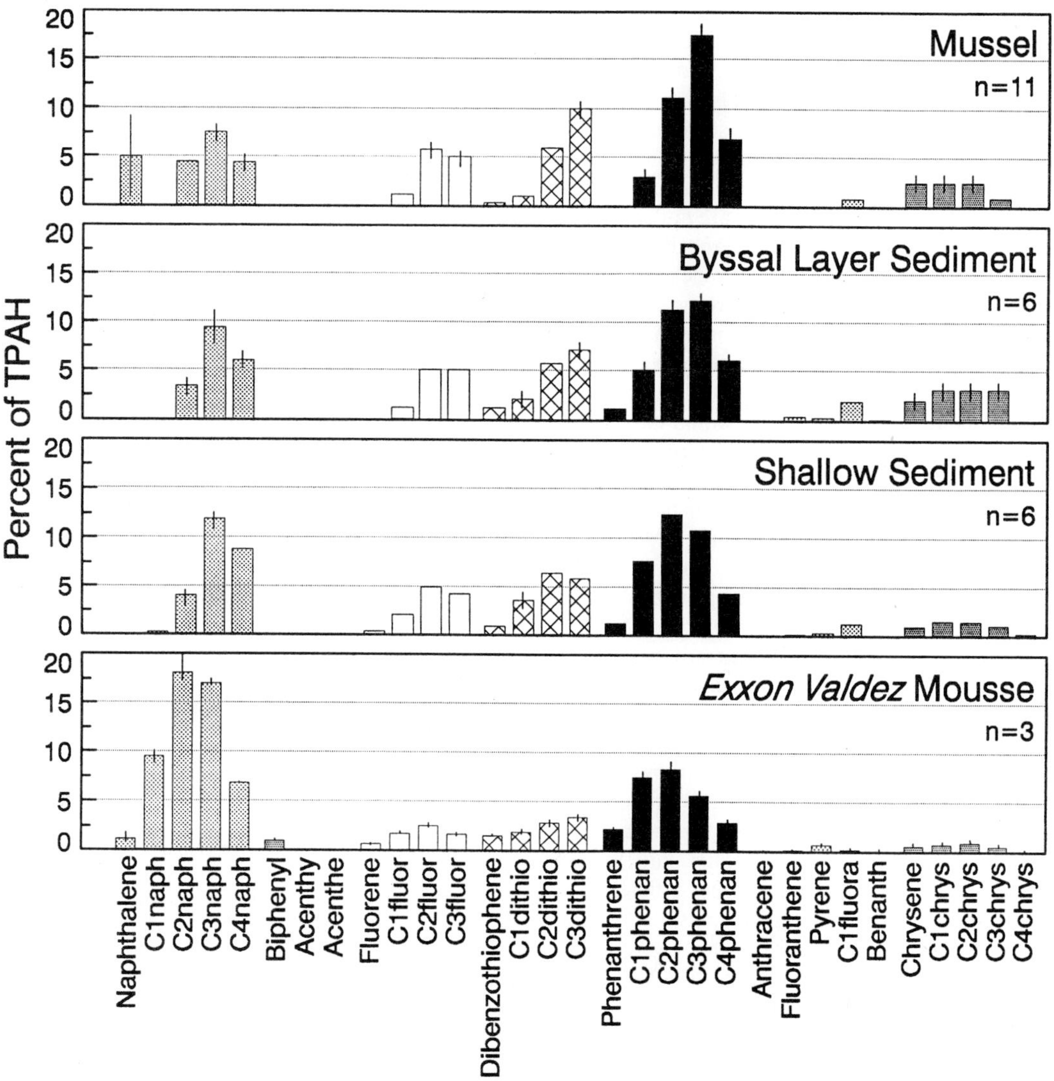

FIGURE 8.—Mean percent of individual polynuclear hydrocarbons (PAH) relative to total PAH (TPAH) in mussels, byssal sediment, shallow sediment (0–2 cm), and *Exxon Valdez* mousse as determined by gas chromatography–mass spectroscopy. Mussel, byssal sediment, and shallow sediment samples were collected from the lower and middle transects in the Chenega A mussel bed in May 1992. *Exxon Valdez* mousse was collected 4 April 1989. Vertical lines = ±1 SE. For many analytes the SE was 0.

surface area for oil adsorption than do coarser sediments (Wade and Quinn 1980). This relationship was characteristic of the byssal and shallow sediments at each sample quadrat at Chenega A. Fine sediments in the shallow layer here seemed to have retained the oil and prevented its movement into underlying sediments. At Latouche and Eleanor islands, where byssal and shallow sediments were visibly coarser than at Chenega A, TPH concentrations were higher in the deeper than in the shallow sediments at each quadrat; however, on a bedwide basis at Chenega A, this relationship was apparently overridden by other factors.

Localized density of mussels appeared not to affect the amount of oil in the mussels or in the sediments below them. In the beds studied, mussel densities were apparently sufficient to stabilize substrate sediments with byssal threads; higher densities in some quadrats were not related to higher concentrations of oil.

Hydrocarbon distribution in mussels was also uneven within a mussel bed and was related to sediment oiling. The matching PAH profiles indicate that the source of oil in mussels was particulate *Exxon Valdez* oil washing into the surrounding water from nearby contaminated sediments. Oil in mussels could not have been retained since the spill because mussels depurate if they are not chronically exposed (Lee et al. 1973; Clark and Finley 1975), and mussels on nearby bedrock, which would have received similar initial oiling, were no longer contaminated in 1992. Therefore, contamination measured in 1992 must have resulted from continued exposure. If mussels had taken in oil directly from the sediment under them, there would have been a more precise correlation between mussel and sediment contamination at specific quadrats. The more general distribution pattern in each bed of oil in mussels and in sediments implies that hydrocarbons were washed out of the sediment, dispersed locally into the water, and were taken up by the mussels from the water. The matching profiles also indicated that oil accumulated by mussels was particulate (rather than dissolved) because mussels were not enriched in water-soluble compounds.

In summary, oiled Prince William Sound mussel beds still contained high concentrations of oil 3 years after the *Exxon Valdez* oil spill, but the oil was distributed unevenly within these beds. *Exxon Valdez* crude oil, remaining relatively unweathered, was the source of the oil in mussel bed sediments and has continued to recontaminate mussels as it dispersed from sediments into the water surrounding the mussels. These findings have an impact on research related to the *Exxon Valdez* oil spill. Anyone studying the effects of oiled mussels on their predators or proposing to restore mussel beds should consider the persistent and unweathered nature of the oil in the sediments and the chronic exposure of nearby mussels. To devise a valid sampling design, studies monitoring *Exxon Valdez* oil concentrations in mussel beds must consider the uneven distribution of oil.

Acknowledgments

We thank the following people at Auke Bay Laboratory for their invaluable help in sample collection and analysis and in manuscript preparation and review: Mark Carls, Bruce Wright, Marie Larsen, Larry Holland, Josie Lunasin, John Grosenbeck, Dan Fremgen, Jeff Short, Michael Murphy, and Lori Ewing.

References

Babcock, M. M., P. M. Rounds, C. C. Brodersen, and S. D. Rice. 1994. 1991 and 1992 recovery monitoring and restoration of intertidal oiled mussel (*Mytilus trossulus*) beds in Prince William Sound impacted by the *Exxon Valdez* oil spill. NOAA (National Oceanic and Atmospheric Administration) Technical Memorandum NMFS (National Marine Fisheries Service)-AFSC 94-02, Alaska Fisheries Science Center, Seattle.

Clark, R. C., Jr., and J. S. Finley. 1975. Uptake and loss of petroleum hydrocarbons by the mussel, *Mytilus edulis*, in laboratory experiments. U.S. National Marine Fisheries Service Fishery Bulletin 73:508–514.

Duffy, L. K., R. T. Bowyer, J. W. Testa, and J. B. Faro. 1996. Acute phase proteins and cytokines in Alaskan mammals as markers of chronic exposure to environmental pollutants. American Fisheries Society Symposium 18:809–813.

Houghton, J. P., D. C. Lees, T. A. Ebert, and W. B. Driskell. 1993. Evaluation of the condition of Prince William Sound shorelines following the *Exxon Valdez* oil spill and subsequent shoreline treatment. Volume II, 1991 biological monitoring survey. NOAA (National Oceanic and Atmospheric Administration) Technical Memorandum NOS (National Ocean Service) ORCA (Office of Ocean Resources Conservation and Assessment) 67, Seattle.

Krahn, M. M., and nine coauthors. 1993. Comparison of high-performance liquid chromatography/fluorescence screening and gas chromatography/mass spectrometry analysis for aromatic compounds in sediments sampled after the *Exxon Valdez* oil spill. Environmental Science and Technology 27:699–708.

Krumbein, W. C., and F. J. Pettijohn. 1988. Manual of sedimentary petrography. Appleton-Centry-Crofts, Inc., New York. (Reprinted by Society of Economic Paleontologists and Mineralogists, Tulsa, Oklahoma.)

Lee, R. F., R. Sauerheber, and A. A. Benson. 1973. Petroleum hydrocarbons: uptake and discharge by the marine mussel *Mytilus edulis*. Science 177:344–346.

McDonald, J. H., and R. K. Koehn. 1988. The mussels *Mytilus galloprovincialis* and *M. trossulus* on the Pacific coast of North America. Marine Biology 99:111–118.

Sharp, B. E., M. Cody, and R. Turner. 1996. Effects of the *Exxon Valdez* oil spill on the black oystercatcher. American Fisheries Society Symposium 18:748–758.

Short, J. W., T. J. Jackson, M. L. Larsen, and T. L. Wade. 1996. Analytical methods used for the analysis of hydrocarbons in crude oil, tissues, sediments, and seawater collected for the natural resources damage assessment of the *Exxon Valdez* oil spill. American Fisheries Society Symposium 18:140–148.

Snedecor, G. W., and W. G. Cochran. 1956. Statistical methods applied to experiments in agriculture and biology, 5th ed. Iowa State University Press, Ames.

Wade, T. L., and J. G. Quinn. 1980. Incorporation, distribution and fate of saturated petroleum hydrocarbons in sediments from a controlled marine ecosystem. Marine Environmental Research 3:15–33.

American Fisheries Society Symposium 18:309–328, 1996

Exxon Valdez Shoreline Treatment and Operations: Implications for Response, Assessment, Monitoring, and Research

ALAN J. MEARNS

National Oceanic and Atmospheric Administration
Hazardous Materials Response and Assessment Division, Biological Assessment Team
Seattle, Washington 98115, USA

Abstract.—Existing operational and research information was examined to evaluate the effectiveness and effects of shoreline treatment and supporting operations during response to the March 1989 *Exxon Valdez* oil spill. The primary treatment methods were manual cleanup, aggressive shoreline washing with specially designed vessels, bioremediation, and, in 1990 and 1991, berm relocation. Other types of treatment (chemical cleaners, rock washing) were proposed, tested, and rejected. Treatment activity at some sites lasted for up to 2 months. Published data suggest that 1989 treatment removed between 4 and 9% of the stranded oil; additional data suggest that high-pressure hot-water washing was twice as effective as this average. About one-third of all shoreline segments in western Prince William Sound received high-pressure hot-water washing coupled in many areas with beach-face coldwater deluge. Calculations from shoreline biological-monitoring data suggest that although half of the midintertidal shoreline seaweed and other marine life survived oiling, these were nearly completely destroyed by the cleanup at the aggressively treated sites. The effectiveness of bioremediation and berm relocation was studied. Only anecdotal information exists for evaluating the effectiveness and effects of other technologies. Review of photos also suggests that aggressive treatment in midsummer 1989 produced sediment plumes that may have redistributed oil offshore of treatment sites. Operations supporting shoreline treatment were extensive: more than 10,000 people expended 20.5 million hours using more than 1,600 vessels and thousands of aircraft sorties. The literature suggests noise, low-flying aircraft, vessel avoidance (deflection), and anchor chain could have affected marine and terrestrial biota. Despite these concerns, shoreline treatment was undertaken under constraints that presumably did minimize additional impacts. Continuing studies of this spill should include treatment and supporting operations as potential sources of injury.

This article addresses one of the most controversial issues in oil spill cleanup: the effectiveness and effects of shoreline cleanup and support operations. The response to the 24 March 1989 *Exxon Valdez* oil spill—in which 43.9 million liters of oil was spilled—was an exceptional event for a remote wilderness. The marine, terrestrial, and aquatic life of Prince William Sound were exposed not only to oil but also to the heat, pressure, and chemicals from shoreline treatment and to noise, exhaust, and disturbances from equipment, vessels, aircraft, and the activities of thousands of people. Shoreline cleanup involved nearly every treatment tool available (and many new ones; Teal 1991). To support the cleanup, Exxon Corp. implemented and maintained a large program involving marine operations, logistics, safety and health, waste management, and demobilization (Exxon Production Research Corp. 1990; Nauman 1991). By 30 September 1989—half a year after the spill—more than 10,000 people had expended 20.5 million work hours in and around the sound and the Gulf of Alaska (Nauman 1991).

The emphasis of this article is on the ecological effects and benefits of shoreline treatment and support operations, collectively defined as "the response." There are several reasons why it is important to address the ecological effects of this response. First, if response effects can be identified and are significant, action can be taken to reduce them during responses to future spills. Second, knowledge of actual or potential impacts of response actions can guide research to help resolve the problems and perhaps come up with better response options. Finally, understanding response impacts will improve our knowledge about the effects of oil itself: It could help resolve conflicting information in the literature about what oil does and does not do to marine ecosystems and how response actions can enhance or impede ecological recovery. Accordingly, the aim of this article is to help improve future oil spill responses so that the maximum amount of oil is removed with the least amount of injury and maximum amount of ecological recovery.

Approach

This article is organized in three sections: shoreline treatment, operations, and implications. Two levels of the response are considered: (1) shoreline treatment itself, some of which injured marine life that had survived the oil, and (2) support operations, which may have contributed to injuries throughout the response area. Two kinds of effects are considered: direct effects on intertidal and subtidal marine life and effects on shoreline structure (geomorphology), which in turn may have indirect or long-term effects on intertidal marine life.

Five shoreline treatment questions are posed and discussed: (1) What shoreline treatment methods were used and where? (2) What criteria were used to select treatment methods? (3) How effective was shoreline treatment? (4) What were the effects of treatment on shoreline marine life? (5) How important were the effects compared to the effects of the untreated oil? The first three questions are addressed to acquaint the reader with the rationale, scope, extent, and effectiveness of the treatment activities. The remaining questions focus on the extent to which treatment may have contributed to the effects of the oil spill described elsewhere in this volume.

Sources of data and information for evaluating shoreline treatment technology and its effects are described in the appropriate sections of this article. All of this information comes from monitoring activities that were independent of the Natural Resource Damage Assessment (NRDA) studies that form the main body of this book. Some of the cited material comes from reports produced as part of the Longterm Shoreline Ecological Recovery Monitoring Program conducted by the Hazardous Materials Response and Assessment Division (HMRAD) of the National Oceanic and Atmospheric Administration (NOAA) in Seattle, Washington. This program, which will continue to the year 2000, has been funded jointly by various sources cited in the acknowledgments.

The section on support operations makes use of operational reports and publications to acquaint the reader with the wide range of activities that took place to support the shoreline treatment operations. It also attempts to give some measure of the extent to which these activities exceeded "natural" or background levels of activity in this remote area. Finally, because few measurements or observations were made on the effects of operations on marine life, relevant literature is cited to provide some idea of the potential for effects that need to be documented or considered in future spill response operations.

It is beyond the scope of this article to treat each of these subjects in the detail necessary to achieve the full intent of the objectives. Rather, enough information is presented to help identify implications for response, monitoring, assessment, and further research. The concluding section on implications provides some shoreline treatment recommendations, suggests opportunities for learning at future spills, identifies important research needs, and offers ideas on the significance of results.

Shoreline Treatment

By May 1989—2 to 3 months after the spill—most of the floating oil had either been removed by skimmers, left the coastal area, evaporated, degraded, or was stranded on the shoreline (Wolfe et al. 1994) Estimates of the amount of oiled shoreline vary. Exxon Corp. (1992) reported oil on at least 2,090 of the 5,470 km surveyed in 1989, or about 15% of the area's 14,480 km, whereas Michel and Hayes (1991) reported some 5,221 km were oiled to some degree, of which 912 km were moderately to heavily oiled. Regardless of the exact measurement, these moderately to heavily oiled shorelines were home to at least 400 species of marine plants and animals and became the focus of subsequent treatment and cleanup activities.

Treatment Methods

A wide range of methods—from manual pickup and hand-wiping to high-pressure washing—was used to remove oil from the shoreline. In 1989, six task forces, each composed of approximately 1,000 personnel, and more than 100 vessels were deployed in different areas to conduct the operations. Operational aspects of treatments are discussed in more detail in several reports (Exxon Production Research Corp. 1990; Teal 1991). Information most relevant to this report are reviewed below.

Manual treatment for picking up tar balls and pooled oil involved deployment and recovery of absorbent materials; tilling, raking, and shoveling with hand tools; removal of oiled debris and tarmats; and cutting and removal of seaweed (*Fucus* sp.).

Mechanical treatment was used in 1990 and 1991 to expose buried oil in boulder–cobble beaches. Mechanical treatment involved use of backhoes, tractors, front-end loaders, and other equipment to scoop, dig, and redistribute beach material. At some sites tractors pulled steel cable tines to agitate sediment, releasing oil ahead of rising tides (Piper 1993). In 1990 and 1991, berm relocation was used to open high, intertidal storm berms to expose buried oil and accelerate its weathering (Michel and Hayes 1991; Michel et al. 1991). Tractors were placed on shore and used to dig and flatten storm

FIGURE 1.—Close-up view of hot water discharging from spray head of an Omni-boom 1989. (Photograph from the collection of National Oceanic and Atmospheric Administration, Hazardous Materials Response and Assessment Division, Seattle.)

berms and redistribute the cobble to the midintertidal zone. Oil within the open berms was treated with slow-release nutrient capsules in an attempt to accelerate degradation (see bioremediation, below).

Washing to clean coated and remaining pooled oil off rocky surfaces involved use of cold, warm, or hot, low-, moderate-, or high-pressure seawater coupled with high-volume flushing (deluge or flooding) using header diffusers on upper beach facies (Exxon Production Research Corp. 1990; Teal 1991). To conduct the washing operations, Exxon Corp. provided an array of equipment and vessels including more than 600 commercial pressure washers (maximum nominal temperatures 90–121°C, 830–16,548 kPa) and 60 landing craft vessels that supplied either cold water for deluge and hand-held fire hoses (cold minis) or hot water (nominally, 60°C, 345–552 kPa, 946 L/m) for fire hoses. Generally, multiple hand-held 10-cm hoses, fitted with 3.8 or 5.1-cm nozzles were connected in series to delivery barge pumps. In addition, 13 maxi-barges supplied 511 L/m of 60°C water at 552–827 kPa (nominal) through fire hoses mounted on man lifts or cranes to wash vertical rock faces. Finally, several Omni-barges were built that supplied 1,893 L/m of 60°C water at 552–827 kPa (nominal) via spray heads mounted on an articulating crane (Figure 1).

In 1990, a proposal was developed to remove some boulder–cobble beaches, wash the material in a large, specially designed offshore rock washer machine, and then replace the material; the proposal was reviewed in detail and rejected (USCG 1993).

Bioremediation, to accelerate the chemical degradation of oil not otherwise removed by other treatments, involved application of slow-release pellets containing nitrogen and phosphorus fertilizer (Customblen™) and spraying a liquid oleophilic, urea-based fertilizer (Inipol™; Exxon Production Research Corp. 1990; Bragg et al. 1992). Despite considerable controversy about results of effectiveness tests and toxicity (Hoff 1993; Piper 1993), bioremediation in the form of nutrient applications was approved for use in 1989–1992. A microbial additive was tested at one site and rejected; however, several unauthorized microbial treatments were reported.

Chemicals (other than the bioremediation materials) were not approved for use during 1989–1991, but they were tested. During 1989 at least three sets of shoreline experiments were done to test cleaning agents and dispersants at sites on Disk, Knight, and Smith islands in Prince William Sound (Piper 1993).

Treatment Decision Process and Criteria

The strategy for treating shorelines in 1989 was an evolving process centered around a common theme, which was to remove as much of the oil as quickly as possible while minimizing impacts to surviving resources. For example, as stated by Exxon Production Research Corp. (1990:55) "The treatment approach was to remove, contain and collect the oil from the shoreline with the least environmental impact, particularly to keep oil off the lower intertidal zone and to avoid oiling other areas." To underscore the drive for an effective response, the U.S. Coast Guard commandant "made it clear that the administration expected an aggressive and highly visible shoreline cleanup effort" (USCG 1993).

After several weeks of manual labor activity and spot washing, it became apparent that some kind of expanded shoreline cleanup would be needed. An Interagency Shoreline Cleanup Committee (ISCC) chaired by NOAA provided the forum for presenting, evaluating, and recommending shoreline treatment methods (Knorr et al. 1991). The 14-member ISCC reviewed ecological, cultural, and social resource issues and considered and recommended resource protection and work priorities to the Federal On-Scene Coordinator (FOSC). In response to missing information identified by the ISCC, Exxon Corp. formed Shoreline Cleanup Assessment Teams that surveyed every meter of oiled shoreline in Prince William Sound. Shorelines in the spill area were divided into numbered segments that varied in size from a few tens of meters to many kilometers. Exxon Corp. then proposed cleanup methods for specific shoreline segments, which were reviewed by the ISCC, whose suggestions and recommendations were passed on to the FOSC for approval. By May 1989, the ISCC published and distributed a Prince William Sound shoreline treatment manual that contained numerous but simple guidelines for protecting resources and information and guidance about the effectiveness and possible effects of all viable treatment methods (ISCC 1989).

The extent to which ecological concerns were used as rationale for specific treatment technologies is clear in some cases but obscure in others. For example, bioremediation, if successful and safe (i.e., nontoxic) in field trials, offered the opportunity to minimize invading shorelines with people, equipment, vessels, and aircraft. In contrast, it is not clear that any ecological criteria were used to determine appropriate pressures and temperatures for the 1989 washing operations. Rather, pressures and temperatures were used that most effectively removed oil within the constraints of the technology itself. For example, the primary reason for using "hot" water and "high" pressure was strictly to remove the oil. During the course of the 1989 treatment operations, the U.S. Coast Guard first approved using temperatures up to 43°C and eventually up to 60°C to match the increasing pour-point temperature of the oil. And, as noted above, the pressures and temperatures reported were "nominal," based on outputs of the pumps and heat exchangers. Criteria for pressures and temperatures acceptable to the surviving marine life do not appear to have been considered. My preliminary calculations suggest that nozzle exit velocities could have resulted in forces that far exceeded thresholds for some intertidal organisms (Denny 1988).

The net result of these activities, interactions, and research (especially on bioremediation) was a 1989 shoreline cleanup program that emphasized invasive technology (washing) coupled with criteria (ecological windows, etc.) to avoid additional impacts on marine, aquatic, and terrestrial resources. In subsequent years, however, washing was eliminated as a primary tool, with a shift toward manual recovery, increased bioremediation, berm relocation, limits on the size of work crews, and additional constraints on shoreline activity and operations (Exxon Corp., unpublished data).

Extent of Treatment

Two databases developed during the response by NOAA's HMRAD were used by the U.S. Coast Guard to track treatment progress. These are the CAMEO™ (Computer Assisted Management of Emergency Operations) Valdez database (NOAA 1992a) and the Segment Summary database. Together these data sources list, by segment and sub division (subunits of segments), shoreline type, degree of oiling, type of material delivered (hot water, cold water, etc.), and the timing and duration of activity. The information was extracted from U.S. Coast Guard Strike Team daily shoreline-monitoring reports. Not included in these listings are data on shoreline chemical cleaner tests or on several unauthorized treatments (microbial applications). In addition, these data sources do not include information about the delivery systems employed (landing craft vessels, maxi-barges, pressure washers, etc.).

Overview.—These databases were used to examine the geographic and temporal extent of treatment. These sources list 1,149 shoreline segments and approximately 20 types of treatment actions (Table 1). In 1989, of the 1,149 segments surveyed, 519 occurred in Prince William Sound, 138 in the Kenai sector, and 426 in the Gulf of Alaska sector. Of those shoreline segments in Prince William Sound, 284 (55%) received bioremediation, 180 (35%) received hot water, 166 (32%) received warm water, 61 (12%) received high-pressure wash-

TABLE 1.—Numbers of Alaska shoreline segments treated by various shoreline cleanup methods after the *Exxon Valdez* oil spill, 1989–1991. (Treatments used in 1992 and beyond are not included.)

Treatment summary or method	1989	1990	1991
Segments surveyed	1,149	740	415
Not treated	89	366	344
Manual			
Manual pickup (hand tools)	51	368	94
Tar mat (removal)	0	221	0
Debris (removal)	42	0	0
Hand-wiping	130	0	0
Tilling, raking	0	75	38
Gross contaminant removal	170	0	0
Chopped SOR	0	0	7
Vegetation cutting	3	?	?
Snare boom	29	0	0
Hydraulic			
Steam clean	1	?	?
Water deluge	153	0	0
Flooding	44	0	0
Moderate-pressure	31	0	0
High-pressure	63	0	0
Hot-water wash	194	0	0
Spot wash	0	58	1
Mechanical			
Mechanical till	54	0	9
Sediment removal	?	?	?
Berm relocation	0	?	30
Chemical			
Dispersants	1	0	0
Shoreline cleaner	2	0	0
Bioremediation			
Nutrients	290	298	77
Enzymes	1	?	?
Microbial addition	1	?	?

ing, and 57 (11%) received no treatment in 1989. High-pressure washing was discontinued in 1990, when manual pickup and tar mat removal (in 368 and 221 segments, respectively, in all three sectors) became the primary nonchemical methods.

According to Exxon Production Research Corp. (1990), hot-water washing became the primary treatment method for nearly all treated shorelines in Prince William Sound in 1989, a statement that is not quite consistent with the data presented above; however, it was not used in 1990–1992. Bioremediation was the most frequently occurring treatment type in all sectors: 290 segments treated in 1989; 298 in 1990; 77 in 1991. The largest amount of treatment occurred in 1990 (Bragg et al. 1992). Prince William Sound shorelines received 2,244 individual treatments of Customblen™ (59,000 kg) and Inipol™ (395,500 L) in 1989–1991 (Bragg et al. 1992).

Variability in treatment effort.—Not all oiled shores were given equal levels of treatment. Treatment data for two areas of interest in the NOAA HMRAD research (Houghton et al. 1996, this volume)—the Eleanor Island area and Herring Bay on Knight Island—were examined to provide a more detailed idea of the distribution and cumulative use of these methods. Thirty-six of 37 segments within Herring Bay that were surveyed in spring 1989 were oiled to some degree (Figure 2A). Two of four segments in the southeast arm were lightly oiled. This area was apparently partially protected from oiling by ice flows and was used as a control area in some NRDA studies (Stekoll et al. 1993). Throughout the rest of the bay, 17 segments had moderate oiling and 12 had heavy oiling (Figure 2A). Up to 10 treatment methods were used to clean Herring Bay segments in 1989 (Figure 2B). The most frequently used methods were bioremediation (25 segments), hot-water wash (19 segments), water deluge (15 segments), gross contaminant pickup (11 segments), and warm-water wash (10 segments). Two-thirds (24) of the segments received three or more treatment methods and one received eight methods. Generally, the heavily oiled sites received more treatments than did the lightly oiled sites, but that was not always the case.

Figure 3 shows additional types of information concerning 1989 treatment at segments on Eleanor Island and the northern half of Ingot Island. All shoreline types typical of Prince William Sound are represented, including pebble and cobble in interior bays and protected channels, and rocky outcrops and boulder beaches along more exposed areas (Figure 3A). Heaviest oiling (Figure 3B) occurred along the northeastern coast of Eleanor Island, in Northwest Bay, and along the western shores of Eleanor and Ingot Islands. Moderate levels of oiling occurred in more protected areas.

The level of treatment effort varied among the segments. No segment along these shores was left untreated, and each segment received from three to ten treatment types (Figure 3C). The greatest number of treatment types (ten) were used on segment IN022 on northern Ingot Island. Moderate-to high-pressure washing, apparently using an Omni-boom, occurred at eight sites, mainly in the passage between Eleanor and Ingot islands (Figure 3D). Hot- and warm-water washing was used throughout the area with notable exceptions (Figure 3, E and F). Although fewer treatment types were used in some areas, these areas may have received substantial effort in terms of the duration of treatment activity,

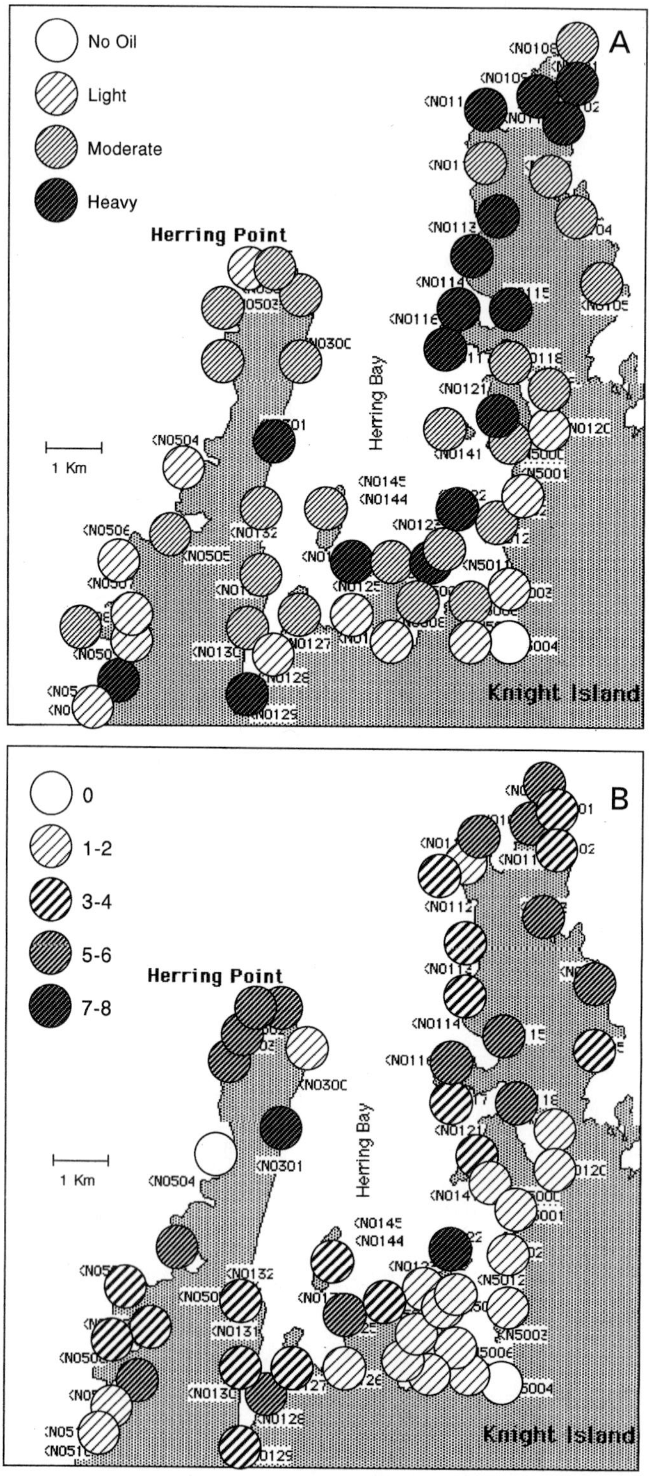

FIGURE 2.—Oiling and treatment conditions at 37 shoreline segments in Herring Bay, Knight Island, Prince William Sound, 1989: (**A**) degree of oiling and (**B**) number of treatment types. Based on data in the National Oceanic and Atmospheric Administration's Segment Summary database.

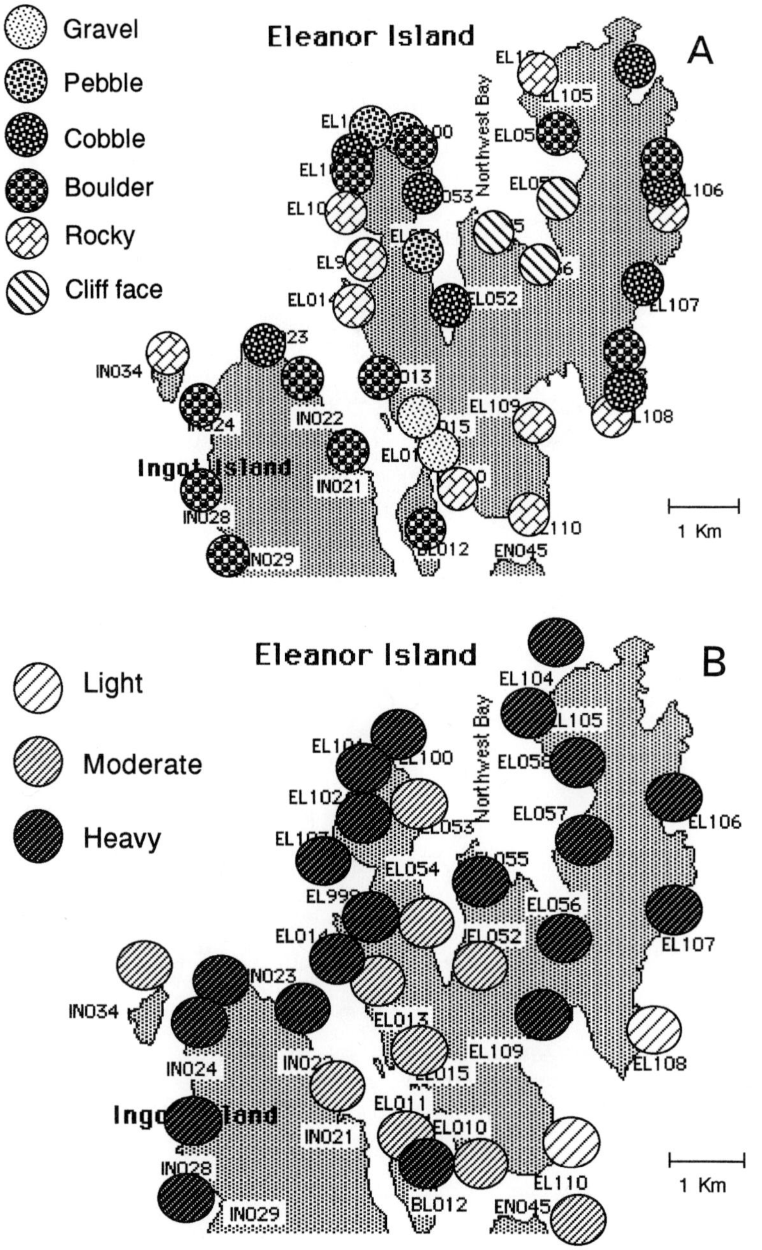

FIGURE 3.—Oiling and treatment conditions on 31 shoreline segments of Eleanor Island and vicinity, Prince William Sound, 1989: (**A**) predominant substrate types, (**B**) degree of oiling (**C**) cumulative number of treatment types used, (**D**) locations treated with moderate to high pressure, (**E**) locations treated with hot water, (**F**) locations treated with warm water, and (**G**) number of treatment days on site. Based on data in the National Oceanic and Atmospheric Administration's Segment Summary database.

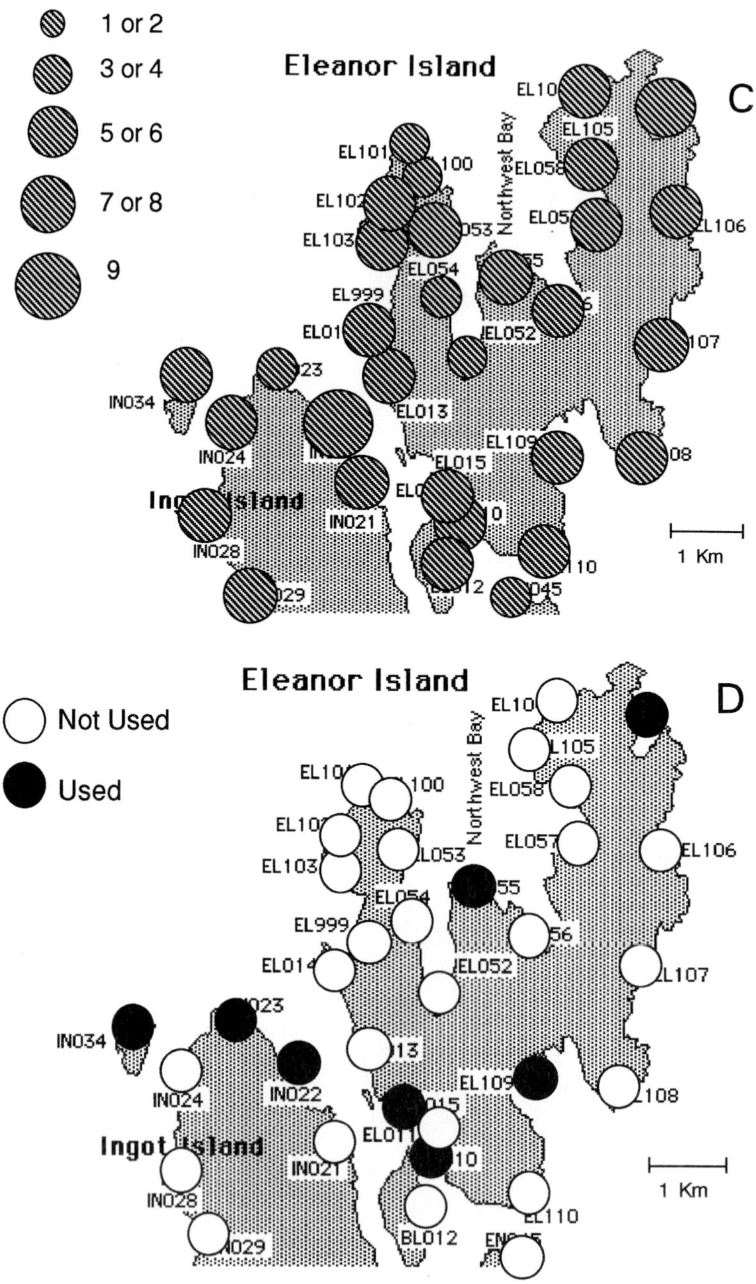

FIGURE 3.—Continued.

ranging from several to more than 70 days (Figure 3G).

There are other and possibly better ways to portray the levels of effort expended in treatment activity. The actual dates treatment took place (from late May to mid-August or from 60 to nearly 150 days postspill) may be important to those concerned about seasonal recruitment of marine organisms. A U.S. Coast Guard method for predicting estimates of work load (P. C. Olsen and W. R. Hamilton, U.S. Coast Guard, unpublished data) may relate better to the physical energy expended on each shoreline segment. Also, there are other information sources not fully utilized here, including independent shore-

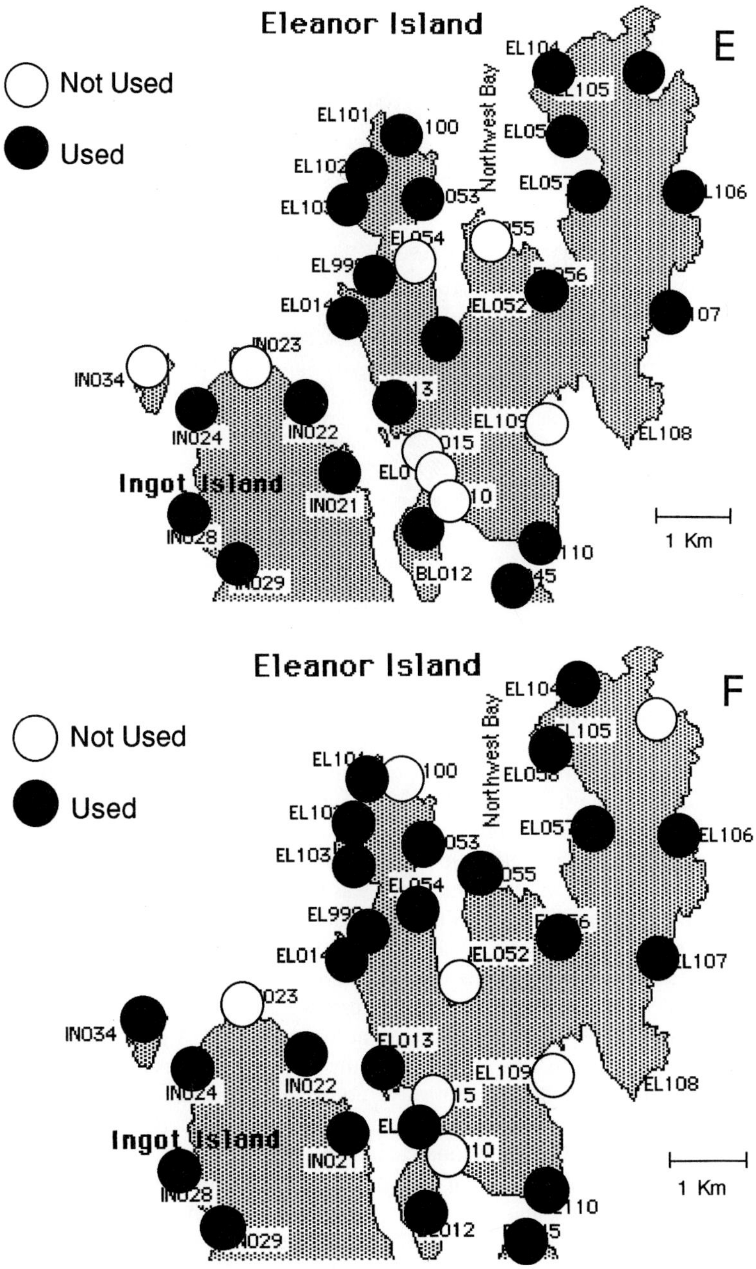

FIGURE 3.—Continued.

line treatment-monitoring notes collected by the Alaska Department of Environmental Conservation (ADEC) and, presumably, by Exxon Corp.

In summary, this discussion reveals that oiled shorelines of Prince William Sound received not only varying amounts of oil but also amounts of treatment activity that perhaps ranged over several orders of magnitude.

Treatment Effectiveness

How effective was treatment in removing oil? "Effectiveness" may be defined as the percent of oil

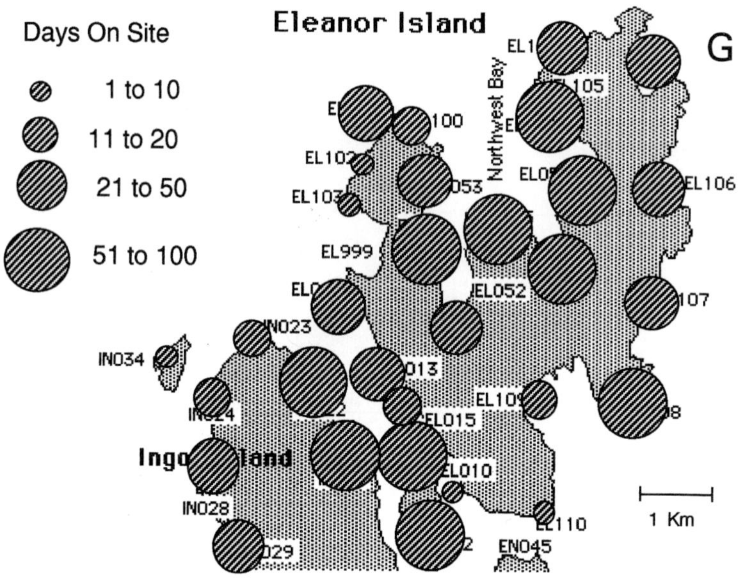

FIGURE 3.—Continued.

removed relative to the total amount stranded. Operationally, effectiveness of the cleanup operations was monitored daily in terms of amount of shoreline treated (in thousands of square yards or equivalents; Exxon Production Research Corp. 1990; Olsen and Hamilton, unpublished). Generally, treatment at a segment ceased when the rate of recovering oil dropped precipitously (R. Pavia, HMRAD, NOAA, personal communication).

Wolfe et al. (1994) estimated that treatment recovered between 4 and 9% of the total spilled oil (nominal best estimate, 5.5%) from beaches throughout the oiled area, with 50 to 75% (nominal best estimate, 70%) coming from Prince William Sound. Between 35 and 45% of the original oil stranded on shorelines (nominal best estimate, 41%; Wolfe et al. 1994). This yields a nominal best estimate recovery effectiveness for all methods used in Prince William Sound of 9.39% (range, 4.4–19.3%).

How effective were specific methods? Studies were done to document or compare the effectiveness of a few individual treatment actions. One was bioremediation, where studies suggested that at some sites this treatment may have increased the rate of natural degradation by up to fourfold above a natural shoreline degradation rate that was about 10-fold per year; however, enhanced biodegradation could not be demonstrated at other test sites (Hoff 1993). Berm relocation was also initially effective in decreasing the amount of oil buried in upper boulder–cobble shorelines, but after 3.5 years, buried oil remained at both treated and untreated sites (Michel and Hayes 1993).

Another group of studies was designed to compare the effectiveness of Omni-barge pressure washing with and without the use of the chemical shoreline cleaner, Corexit 9580 (Exxon Research and Engineering Co. 1990; NOAA, HMRAD, unpublished data). Two Omni-barge effectiveness tests were done on two adjacent cobble shorelines at Segment KN113 in northern Herring Bay on 22 July 1989. From these data I completed a calculation suggesting effectiveness was 14.9% at one site and 26.9% at the other (nominal best estimate, 20.9%).

Comparison of these estimates suggest that at a nominal effectiveness of 20.9% the Omni-barge was twice as effective at removing oil as the nominal best estimate for the entire recovery (i.e., 9.39%), supporting the conclusion that this was one of the more effective methods used to remove oil from Prince William Sound shorelines. No measurements exist to determine the effectiveness of the alternative washing procedures (cold water, low pressure, etc.) used in Prince William Sound.

What about the remaining oil? The nominal best estimate based on calculations from the numbers in Wolfe et al. (1994) is that 90.6% (range, 80.7–95.6%) of the originally stranded oil was not removed by the summer 1989 shoreline treatment operations; however, the remaining stranded oil obviously did not persist on the shoreline. Natural weathering processes such as biodegradation and photolysis removed oil during and after the course of the 1989 response. For example, Michel et al. (1991) reported that natural processes during winter 1989–1990 (October–March) removed 55% (on average) of the oil remaining at the end of the 1989 cleanup season (September). In addition, the shoreline treatment promoted physical removal of oil from the beaches without recovery.

Direct Effects of Treatment

Shoreline treatment had both direct and indirect effects on marine life and on the structure and composition of shorelines. The critical question is how much shoreline marine life in washed areas actually survived oiling? There are no published reports documenting conditions before and after oiling but before treatment. In 1990, more than 1 year after the spill, NOAA compared densities of marine organisms among several washed and unwashed (control or "set-aside") sites. At the time, the percent cover of the dominant seaweed, *Fucus* sp., was about half that at non-oiled sites, while it was less than 10% at the washed sites: thus, it is possible to conclude that as much as half of the *Fucus* sp. (and perhaps other biota) actually survived oiling (Mearns 1993.) Houghton et al. (1996) describe many species that survived oiling at several sites before treatment.

Manual treatment.—No information could be located that documented the effects of tilling, raking, shoveling, debris removal, or other forms of manual treatment on surviving marine life. On one occasion in Northwest Bay in 1990 I observed destruction of mussels and displacement of colonized boulders from upper to lower intertidal elevations. Manual treatment also involved harvest and removal of oiled seaweed. By 9 May 1989, for example, 76,500 m^3 of oiled seaweed had been removed from Knight Island (USCG 1993).

Mechanical treatment.—Storm berm relocation at three boulder–cobble study sites did not appear to affect recovery of intertidal biota (Houghton et al. 1992); however, recovery of beach profiles has been seriously delayed at several sites (Michel and Hayes 1993).

Cold- and hot-water, moderate- and high-pressure treatment.—Measured direct effects of high-pressure washing included displacement and killing of intertidal epibiota and infauna, including clams (Driskell et al. 1996, this volume; Houghton et al. 1996; Lees et al. 1996, this volume). Based on studies before and after treatment, it appears that as much as 90% of the midintertidal shoreline marine life that survived oiling could have been killed or displaced by high-pressure hot-water washing where it occurred (such as at locations shown in Figure 3, D–F).

Anecdotal data (i.e., notes in field logbooks of cleanup monitoring personnel) provide additional independent indications that washing injured surviving shoreline marine life at specific sites. The following are quotes from ADEC field personnel: "they plan to use 140F to wash the mussel bed . . . they shouldn't be treating mussel beds at all." "Visited sensitive lower tidal area that Omni- . . . had treated. Measure damage . . . due to treatment . . . only organisms that survived are some *Fucus* and some barnacles . . . dead or damaged entire *Fucus* bed. Sea stars, barnacle colonies damaged. Algae badly damaged . . . sea stars dead. " "Primary boom is full, no skimmer on sight, workers are blasting mussels off the rocks with steam, oil is escaping the secondary boom." "A lot of sediment is coming off this beach." Although caution is needed in interpreting such quotes (they may represent exceptions to accepted practices), they do provide independent confirmation that there were biological impacts of washing.

Bioremediation.—Direct effects of bioremediation on shoreline marine life were not studied; however, several kinds of toxicity and bioaccumulation studies were done on water, sediments, and biota from identifiable treated and untreated research sites. Water overlying nutrient-treated sites was apparently not toxic to mollusk larvae and other test organisms except at one site (Bragg et al. 1992). Also, there were no substantial differences between treated and untreated sites in terms of sediment mutagenicity; in both treated and untreated sites, mutagenicity declined by an order of magnitude during summer 1989, with treated sites possibly being slightly more mutagenic (Claxton et al. 1991).

On several shorelines visited in summer 1990 I observed very high concentrations of nutrient capsules (Customblen) in tide pools that otherwise contained no oil (Figure 4). The bottoms of some tide pools were completely covered with pellets. My impression was that, after application at an acceptable rate, the tides and waves redistributed and concentrated nutrient pellets into depressions. During low tides, this could have resulted in effective concentrations of nutrients far above nominal concentra-

FIGURE 4.—Customblen nutrient pellets (about 3–5 mm in diameter) and Sitka periwinkles *Littorina sitkana* congregated in a tide pool in Northwest Bay, July 1990. (Photograph by A. Mearns.)

tions computed for the larger treatment areas. Finally, from carbon and nitrogen stable isotope studies, it was clear that both oil carbon and nutrient nitrogen passed into the food web; however, it was difficult to determine if oil uptake in shoreline marine life differed between treated and untreated sites (Pritchard et al. 1991; Bragg et al. 1992).

Indirect Effects of Shoreline Treatment

Indirect effects of shoreline treatment are considerably more speculative than the direct effects; however, they should not be summarily ignored.

Subtidal release of oil.—Cleaning may have released oil in several forms. As intended, stranded oil was washed off the shoreline, refloated into boomed areas, and recovered by skimmers (Wolfe et al. 1994). Some sheen escaped booms (Figure 5), moved offshore, and either contributed to shoreline reoiling elsewhere or moved farther offshore. Washing also liberated contaminated sediment and organic (detritus) material (see quote, above.) In reviewing aerial photographs taken during shoreline operations in June and July 1989, NOAA Scientific Support Coordinators (SSCs) observed many shoreline sites with reddish-brown, turbid plumes moving along- and offshore of actively washed sites. The plumes could easily be observed moving under containment boom, and turbid water was observed several hundred meters offshore of active wash sites (Figures 6 and 7).

The oil content of these released sheen and sediments has not been measured. It can only be speculated that this material may have added dissolved and particulate-bound oil to subtidal areas. Such subtidal release could have contaminated non-oiled shoreline shellfish downstream of cleanup areas (thus raising the overall "background" concentration in animals such as mussels) and added to the exposure of subtidal bottomfish and shellfish through food chain accumulation (Collier et al. 1993).

Persistence of subsurface oil.—There is concern that hot-water washing may also have pushed oil deeper into intertidal substrate. At one site the FOSC and SSC concluded that washing did push oil further into the shoreline substrate; however, only one study was done to check this and no deepening of the oil was observed, presumably because high-pressure hot-water washing was used in conjunction with adequate shoreline (deluge) flushing (J. Michel, Research Planning Institute, Columbia, South Carolina, personal communication). Regardless, it is not clear that shoreline treatment in 1989 accelerated natural removal of subsurface oil (Houghton et al. 1996; Michel, personal communication). Both washed and unwashed sites contained persistent subsurface oil for several years (Michel et al. 1991; Houghton et al. 1996).

Effects on substrate.—There is some evidence to suggest that high-pressure washing removed fine-grain sediments, thus altering the size distribution and composition of sediments in heavily washed

FIGURE 5.—Aerial view of sheen oil moving off shoreline cleaning sites, June 1989.

mixed-soft sites. If true, this could affect the long-term recovery of infauna, including clams (Houghton et al. 1996).

Support Operations

Exxon Corp. organized and maintained a large program to support oil recovery and shoreline treatment activities. The program involved marine operations, logistics, safety and health, waste management and demobilization (Exxon Production Research Corp. 1990; Nauman 1991). It is beyond the scope of this article to describe all of these activities in detail or to conduct a detailed evaluation of actual or potential impacts of support operations. Below are some examples of concerns that may be worth further investigation.

FIGURE 6.—View of oil (black material inside primary containment boom) and suspended sediment plume (moving under boom and offshore to lower left) at a mixed-soft beach (segment EL-11) on Block Island, 17 June 1989. Thirteen hoses are in operation washing material into boom-enclosed water. This mixed-soft site is under continued study the National Oceanic and Atmospheric Administration, Hazardous Materials Response and Assessment Division (NOAA HMRAD) (see Houghton et al. 1996). (Photograph from the collection of NOAA, HMRAD, Seattle.)

FIGURE 7.—View of oil (black material inside primary containment boom) and suspended sediment plume (moving under boom and offshore to lower left and right) at a mixed-soft beach at or near Block Island, July 1989. (Photograph from the collection of the National Oceanic and Atmospheric Administration, Hazardous Materials Response and Assessment Division, Seattle.)

Noise

It was widely recognized that response operations were taking place in a remote wilderness where resources had not been exposed to industrial activity to the extent experienced in urban areas. Many operational guidelines and cautions promulgated by the ISCC and the U.S. Fish and Wildlife Service (USFWS) (Bergmann 1991) were offered to protect wildlife from noise and disturbances from low-flying and landing aircraft; however, to my knowledge, noise associated with operations was not measured either above or below the waterline. In evaluating Alaska offshore oil production operations on marine mammals, Malme et al. (1989) concluded that sound levels produced by vessels used for cargo hauling, fishing, and recreation are significant when several vessels operate in a relatively small area. Underwater noise can also alter the behavior and survival of both invertebrates (shrimp, crab) and fish (Manci et al. 1988). Many tens of operating vessels congregated in small bays of western Prince William Sound. Regardless of adherence to guidelines, noise from task forces supporting shoreline treatment operations could have affected marine life and wildlife by altering migrations and confusing communications of both terrestrial and marine animals.

Air Pollution

On several occasions while working aboard research vessels, I observed smoke and haze trapped by inversion layers in steep-sided bays such as Snug harbor and Windy Bay (Kenai Peninsula). Sources included vessel exhaust and garbage combustion.

Aircraft

Valdez aircraft sorties (take-offs and landings), which normally occurred at a rate of 8–10/d rose nearly 4,000% to 300–400/d in April 1989 and continued at a rate of 4,000–6,000/month through the summer (USCG 1993). Pilots were asked to stay at least 800 m offshore and 300 m above sensitive areas of wildlife concentrations, including eagle nests. I have not investigated the extent to which monitoring addressed these criteria nor the extent of violations; however, I did participate in a number of overflights that appeared to violate these guidelines. Coast Guard air operations filed a report of violation against a news helicopter in Lower Passage that landed at KN-103 with its tail rotor well over the boom and within 9 m of workers. The USFWS reported infractions, notably with media aircraft (USCG 1993). On several occasions I also observed helicopters blowing debris and intertidal marine organisms off sections of shoreline.

Anchors and Chain

Presumably, only anecdotal information is available to support the possibility that physical and other mechanical activities affected several habitats in or near Prince William Sound treatment areas. In particular, Stekoll et al. (1993:139), in their study of the effects of the spill on Prince William Sound subtidal habitats, noted that "increased boating activity and losses associated with the pulling of anchors may have been a primary cause of loss of (subtidal) algae."

Deflection

Moving vessels can deflect surface animals into oil. Several people at this symposium offered their own observations that sea otters were indeed oiled as a result of avoidance of fast-moving skiffs and other vessels. As a passenger aboard several fast vessels in Alaska, I observed many incidents of animals fleeing oncoming vessels. Migrating shoreline animals can also be deflected and trapped by skirted boom. For example, salmon fry were observed running into and congregating in boom skirts and "being eaten by sea terns who see them from the air" (16 May, Chronology of Events; USCG 1993:53).

Based on the foregoing, noise and disturbance from both monitoring and task force vessels, aircraft, and humans cannot be ruled out as factors affecting the distribution, feeding, and reproduction of plants and animals near the treatment areas.

Implications for Response, Assessment, Monitoring and Research

The aim of this article is not only to bring attention to treatment activities and operations that may have contributed to the effects as well as the successes of the *Exxon Valdez* spill response, but also to provoke discussion about what this information means to responders, researchers, and damage assessors. It is a difficult task to discuss impacts of cleanup and operations when so many people worked hard to minimize those impacts while at the same time trying to make progress in removing the oil; however, avoiding the issue could result in overlooking methods and strategies for protecting resources at future spills or in misinterpretation of sources of injury. Clearly, this treatment of the subject is far from complete and should be taken only as a sample of the kinds of questions and concerns that should be considered. With these limitations in mind, what is the possible ecological importance of impacts from the response and what are some implications for future responses, monitoring, assessment, and research?

Balancing the Scale: Weighing the Benefits

How important were the effects of treatment and operations? Was injury from the response worth the oil removed from the environment? With the information that now exists, it is worth speculating on the "balance sheet" for the treatment program.

No studies or monitoring activities were planned a priori to quantitate separately the impacts of shoreline treatment or operations over the entire operations area. The potential scale and magnitude of treatment and operational impacts must be inferred from a posteriori studies at a few sites and from anecdotal observations. For example, scaling up the NOAA observations on shoreline impacts observed at several test sites, it can be estimated that high-pressure hot-water washing in Prince William Sound destroyed more than 3,000 metric tons (mt) of seaweed. The total wet biomass of marine life killed directly by the oil was as follows: mammals, 125 mt; birds, 275 mt; seaweed (*Fucus* sp.) 5,800 mt (excluding cutting). Thus, of the total biomass killed by the oil and by the shoreline treatment (6,200 mt), more than 48% was attributable to shoreline treatment. Excluded from this summary is the biomass of several hundred species of other intertidal plants, mollusks, crustaceans, annelids, echinoderms, and other organisms.

Thus, from one—admittedly negative—point of view, it could be concluded that half of the 1989 shoreline operation was devoted to a shoreline treatment activity (Teal 1991) that left behind most of the stranded oil and killed half of all the biomass that was nominally injured by the entire event. Of course this potential impact must be balanced by the effectiveness and benefits of oil removal: a cleaner-looking oiled shoreline that may well have been more stable and less of a threat to wildlife and subsistence and commercial fisheries. It is also likely that the nominal recovery of 9.39% of the beached oil, along with additional oil removed (but not recovered) as a result of the treatment processes, may have accelerated the subsequent natural removal, particularly if it removed thick deposits. This viewpoint also does not accommodate site-specific benefits, such as protected rocky shorelines where effectiveness may have been much greater.

On the other hand, there are no estimates on what other resource injury was prevented by this potential sacrifice. Also, did high-pressure hot-water treatment speed recovery? For shoreline ecosystems the answer is no (Houghton et al. 1996). Washing delayed natural recovery of intertidal marine life. It also may have delayed natural oil removal, particularly if it increased penetration of the oil into the underlying cobble, pebble, and sand of boulder–cobble beaches.

Estimating the magnitude of potential impacts of other shoreline treatment activities and supporting operations is even more problematic. Moreover, these are largely hypothetical impacts. Vessel and aircraft noise could have ensonified the entire water and air masses of the various bays and inlets of western Prince William Sound to intensities disturbing to resident mammals and birds. Regardless of the intensity, it presumably did not occur at control or reference sites elsewhere, such as in eastern Prince William Sound, where non-oiled control sites were used by many researchers.

Implications for Future Responses

Response goal: how clean is clean enough?—It is a credit to all agencies involved in this spill that the need for and intensity of shoreline cleanup was hotly debated and evaluated (Knorr et al. 1991) and that many guidelines and criteria were established to prevent additional injury to resources (minimum aircraft elevations, limitations of treatment methods, etc). However, it is not clear that survival of marine life was an explicit part of the definition of "clean enough," only an adjunct condition. Perhaps the operational definition of "effectiveness" should be redefined to include maximum survival and recovery of marine life.

Shoreline treatment recommendations.—To protect surviving intertidal marine life, Houghton et al. (1996) provide specific recommendations for cleanup of protected rocky, boulder–cobble and mixed-soft shorelines: (1) do not attempt to remove all visible oil, (2) use simple mechanical methods to remove or vacuum pooled oil, (3) use low-pressure cool-water flushing on rocky and boulder–cobble shorelines, (4) limit pressures on mixed-soft beaches to avoid mobilizing sediment, (5) flush low-gradient mixed-soft beaches only when the tide is above the midtide level (*Fucus* zone); and (6) if necessary to clean the *Fucus* zone on high-gradient rocky beaches, do so only with ambient-temperature low-pressure washing and make sure oil is fully flushed onto the water surface for pickup (skimming). To these, I add that high-pressure hot-water washing can play a role in some situations, especially when restricted to hard substrates with little or no marine life.

Berm relocation is a viable option as a cleanup technique but has to be evaluated against site-specific conditions as described by Michel and Hayes (1993).

New treatment methods.—The ban on use of nontoxic shoreline cleaners should be reconsidered. Nondispersive shoreline cleaners can reduce the need for high-pressure hot-water washing and time spent on the beach. The limited use of bioremediation should be encouraged, provided it is accompanied by effectiveness and effects monitoring. These methods will not remove all visible oil nor will bioremediation do so quickly; however, they could deter invasive actions and thus protect much of the intertidal biota that survives oiling. They also will require less fuel, energy, human effort, and logistical support and thus reduce the potential impact of support activities.

Chemical applications (for bioremediation as well as cleaning) should be done in a manner and monitored so that high concentrations do not accumulate in sensitive habitats such as tide pools.

Set-asides and testing treatment technologies.—Tests of both traditional and promising new technologies should be accommodated at future spills. This requires appropriate test sites, including setting aside oiled shorelines of sufficient scale and protecting them from other forms of treatment. The few limited set-asides in the *Exxon Valdez* oil spill provided critical data for comparison with that from treated sites.

Operational monitoring.—A great deal of information associated with treatment and operations was collected during this spill. The unheralded day-to-day reporting of operational progress, oil recovery and variances, and the development and use of information management tools are among the success stories of this spill response that would be well to emulate elsewhere, keeping in mind that this information may take on great importance not only during the response, but long after. One concern is whether the scales and resolutions of operations and ecological assessment monitoring are comparable so that they can be compared in a meaningful way.

Planning.—Since the *Exxon Valdez* incident, state, regional, and port authorities have been developing detailed response plans using guidance from this and other recent spills (e.g., NOAA 1992b). Although not quantitated here, there is a connection between support operations and shoreline treatment: perhaps as much as half of the support operations was devoted to supporting major controversial shoreline treatment activities such as high-pressure hot-water washing. Response-planning authorities should consider possible larger-scale impacts of support operations (number of workers, fuel, food, traffic, waste management, etc.) when evaluating the efficacy of specific cleanup methods such as pressure washing. They should also develop a policy for set asides. In this regard, many lessons remain to be discovered from the *Exxon Valdez* shoreline cleanup.

Implications for Research

Many research questions emerge from the issues raised in this limited review. The oil spill and Natural Resource Damage Assessment research community can help considerably by supporting and conducting treatment effectiveness and effects studies. Monitoring must include documenting the amount and state of the oil, the effectiveness of removal, and the effects in terms of resource recovery. The concept of establishing untreated set asides is a critical factor. A few of many potential research issues are described below.

Shoreline washing.—A major need is for additional quantitative information on treatment techniques, including the long-term benefits as well as impacts. For the *Exxon Valdez* spill, additional evaluation of existing information is indicated. This includes a more thorough review of treatment-monitoring field notes for information on temperatures, hydraulic pressures, offshore turbidity plumes, and so forth. What is the effectiveness of various shoreline-washing scenarios on several shoreline types? What forces, velocities, and temperatures do surviving shoreline marine organisms actually experience during washing operations? Denny (1988) provides guidance for estimating these. Can temperature and velocity criteria be developed based on field or laboratory tolerance studies of marine plants and animals?

Alternative treatment technologies.—What is inherently wrong with using low-toxicity chemicals for treating shorelines and how should they be evaluated? As a result of research and monitoring at the *Exxon Valdez* spill, a considerable amount of quantitative information now exists for evaluating the effectiveness and effects of alternative treatment technologies and strategies (bioremediation, chemical cleaners). A more thorough review of this information may help resolve additional uncertainties about various treatment technologies.

Intentional oiling experiments.—Spill response research suffers from a lack of well-designed field experiments. There are only two ways to get needed information. One way is to conduct studies at spills of opportunity, not only during the response but long after it has ended. This is probably the worst approach because of the demands of the response process. The best way is to conduct treatment effectiveness and effects studies at intentional, controlled spills. With well-designed field experiments, it is possible to develop criteria for ecologically safe treatment methods and strategies. Unfortunately, federal and state regulatory agencies have been reluctant to permit such intentional activities, although that reluctance may be changing. The research community should work to assure authorities that such activities have considerable benefit and can be done safely.

Effects of support operations.—This article raises questions about the potential effects of noise, vessel and aircraft movement, and other activities associated with supporting spill responses. Research is needed to confirm whether such actions pose threats to surviving resources and, if so, how those threats can be minimized.

Planning.—Researchers can also get involved in the area-and port-planning process, providing guidance on set-aside and monitoring strategies and methods for evaluating the effectiveness and effects of proposed response operations.

Implications for Assessment of the Exxon Valdez Oil Spill

Most of the articles in this volume are directed at documenting injuries and damage attributed to the *Exxon Valdez* oil spill. By implication, the "responsible agent" is presumed to be the oil itself. In cases where exposure was explicitly documented and correlated with injuries, there can be little question that oil played a causative role; however, where exposure to oil was not explicitly documented, or where assessment studies were carried out after shoreline treatment, the treatment or supporting operations cannot be ruled out as factors causing or contributing to the injuries or damage. Thus, in assessing cause of injury, treatment activity should be considered an exposure variable. Therefore, readers of this and other reports on oil spills should use caution in ascribing biological effects to the oil.

The important question, then, is how can treatment information be used in the assessment of injuries? The basic answer is to consider shoreline treatments (along with oil concentrations) as variables to be correlated with effects observations. A primary constraint of this approach is whether the treatment information is at a level of resolution comparable to that of the effects and assessment studies. Treatment information based on segments of shoreline that are 1 km or more in length is too coarse for studies done on a few meters of that shoreline that may or may not have been given the nominal treatment. Perhaps higher-resolution treatment information can be derived from careful examination of treatment work-plan maps, where they are available, and from interviews with state personnel and contractors charged with cleanup. A summary of how NOAA HMRAD obtained this higher-resolution information from its study sites is given in Appendix A of Houghton et al. (1992).

Alternatively, treatment and operational information may be at an adequate scale to correlate with other larger-scale impacts such as prolonged effects of oil and operations on bird reproduction, bird distribution, fisheries landings, and marine mammal abundance and distribution.

Conclusion

The intensive response to the *Exxon Valdez* oil spill contributed to its impacts on marine life and marine ecosystems. The most effective shoreline treatment—high-pressure hot-water washing—was also the most damaging, killing marine life that otherwise survived oiling. There were also longer-term indirect impacts of pressure washing, such as changes in shoreline sediment structure, a change that could alter repopulation of shorelines by clams and other sediment-dwelling invertebrates. An alternative treatment strategy—bioremediation with chemical fertilizers—may have helped degrade the oil, but the supporting data are unclear on this point and questions remain unanswered about its effects. Shoreline ecosystems are still recovering from impacts of both the oil and treatment activities.

Treatment varied greatly among shorelines. Although treatment effort was generally proportional to the amount of oil present, effort and methods varied greatly because of differences in shoreline type and state of the oil.

Marine, aquatic, and terrestrial life may also have been affected by the operations that supported intensive shoreline treatment—especially vessel and aircraft operations and activities needed to support a human population of 10,000. Such impacts are less certain because they were not explicitly monitored; however, based on studies elsewhere, vessel and equipment noise was within ranges known to disturb shellfish, fish, birds, and mammals, and aircraft sorties and other human activities were frequent and extensive enough to impact birds.

Perhaps 10% (range, 4–19%) of the stranded oil was removed by treatment and cleanup operations. This removal was visually effective and probably prevented extensive reoiling; however, it is not clear to what extent marine, aquatic, and terrestrial life benefited in the long run from the intensive treatment and extensive support operations.

Future spill response operations would benefit by having a specific process for defining cleanup criteria (endpoints), setting aside untreated shorelines for monitoring and technology testing, better defining the operational ranges of cleanup equipment and its use, and explicitly testing the feasibility of new and alternative shoreline treatment technologies, including their support operations.

Indeed, additional research is needed to define the tolerances of marine life to both chemical and physical treatment parameters, notably pressure, temperature, and tolerances to chemicals such as fertilizers and shoreline cleaners. Much can be learned before the next major spill by conducting research on intentionally oiled shorelines.

Finally, to help distinguish effects of the oil from effects of the response operations, continued and ongoing studies of the *Exxon Valdez* oil spill should, to the extent possible, accommodate existing treatment information.

Acknowledgments

Ellen Shaw and Jackie MacDonald of Genwest, Edmonds, Washington, provided critical guidance in the use and evaluation of CAMEO™ Valdez and Segment Summary data. Several NOAA SSCs provided important checks on operational information and I especially thank Sharon Christopherson, John Whitney, and Gary Petrae. I also appreciate the guidance and support from Gary Shigenaka and Ruth Yender (BioAssessment Team, HMRAD, NOAA), Robert Pavia (Director, Scientific Support Coordination Branch, HMRAD), Jacqui Michel and Miles Hayes (Research Planning Institute, Columbia, South Carolina), Jonathan Houghton and Allan Fukuyama (PENTEC Inc., Edmonds, Washington), and Dennis Lees (Ogden Environmental, San Diego, California). On behalf of my colleagues, I also gratefully acknowledge information provided by John Bauer (ADEC and former staff member of the VECO Co.). Finally, I appreciate the patience and assistance of NOAA editor Lori Harris. Opinions expressed here do not necessarily reflect the policy of NOAA or HMRAD.

References

Bergmann, P. A. 1991. Implementation and revision of the wildlife protection guidelines for Alaska's federal on-scene coordinators. Pages 137–139 *in* Proceedings, 1991 oil spill conference. American Petroleum Institute Publication 4529, Washington, DC.

Bragg, J. R., R. C. Prince, J. B. Wilkinson, and R. M. Atlas. 1992. Bioremediation for shoreline cleanup following the 1989 Alaska oil spill. Exxon Research and Engineering Company, Florham Park, New Jersey.

Claxton, L. D., V. S. Houk, R. Williams, and F. Kremer. 1991. Effect of bioremediation on the mutagenicity of oil spilled in Prince William Sound, Alaska. Chemosphere 23:643–650.

Collier, T. K., and six coauthors. 1993. Oil exposure and effects in subtidal fish following the *Exxon Valdez* oil spill. Pages 301–305 *in* Proceedings, 1993 oil spill conference. American Petroleum Institute Publication 4580, Washington, DC.

Denny, M. W. 1988. Biology and the mechanics of the wave-swept environment. Princeton University Press, Princeton, New Jersey.

Driskell, W. B., A. K. Fukuyama, J. P. Houghton, D. C. Lees, A. J. Mearns, and G. Shigenaka. 1996. Recovery of Prince William Sound intertidal infauna from *Exxon Valdez* oiling and shoreline treatments, 1989 through 1992. American Fisheries Society Symposium 18:362–378.

Exxon Production Research Corp. 1990. Valdez oil spill technology. 1989 operations. Houston.

Exxon Research and Engineering Co. 1990. Large-scale field test of Corexit 9580, Smith Island, August 8–14, 1989. Exxon Research and Engineering Company Report EE.2DM.90 (1-22-90), Houston.

Exxon Corp. 1992. Three years after. Exxon Corp., Houston.

Hoff, R. 1993. Bioremediation: an overview of its development and use for oil spill cleanup. Marine Pollution Bulletin 26(9):476–481.

Houghton, J. P., D. C. Lees, W. B. Driskell, S. C. Lindstrom, and A. J. Mearns. 1996. Recovery of Prince William Sound intertidal epibiota from *Exxon Valdez* oiling and shoreline treatments, 1989 through 1992. American Fisheries Society Symposium 18:379–411.

Houghton, J. P., and seven coauthors. 1992b. Evaluation of the condition of Prince William Sound shorelines following the *Exxon Valdez* oil spill and subsequent treatment: volume 3 appendices to 1991 biological monitoring survey. NOAA (National Oceanic and Atmospheric Administration) Technical Memorandum NOS (National Ocean Service) ORCA 67, Seattle.

Houghton, J. P., and seven coauthors. 1992a. Evaluation of the 1991 condition of Prince William Sound shorelines following the *Exxon Valdez* oil spill and subsequent shoreline treatment: volume 2 biological monitoring survey. NOAA (National Oceanic and Atmospheric Administration) Technical Memorandum NOS (National Ocean Service) ORCA 67, Seattle.

ISCC (Interagency Shoreline Cleanup Committee). 1989. *Exxon Valdez* oil spill field shoreline treatment manual. NOAA (National Oceanic and Atmospheric Administration), Hazardous Materials Response and Assessment Division, Seattle.

Knorr, J., A. Teal, N. Lethcoe, S. Christopherson, and J. Whitney. 1991. The inter-agency shoreline cleanup committee: a cooperative approach to shoreline cleanup-the *Exxon Valdez* spill. Pages *in* Proceedings, 1991 international oil spill conference. American Petroleum Institute Publication 4529, Washington, DC.

Lees, D. C., J. P. Houghton, and W. B. Driskell. 1996. Short-term effects of several types of shoreline treatment on rocky intertidal biota in Prince William Sound. American Fisheries Society Symposium 18: 329–348.

Malme, C. I., and six coauthors. 1989. Analysis and ranking of the acoustic disturbance potential of petroleum industry activities and other sources of noise in the environment of marine mammals in Alaska. BBN Systems and Technology Corporation Report 6945, Cambridge, Massachusetts.

Manci, K. M., D. N. Gladwin, R. Villella, and M. G. Cavendish. 1988. Effects of aircraft noise and sonic booms on domestic animals and wildlife: a literature synthesis. U.S. Fish and Wildlife Service, National Ecology Research Center NERC-88/29, AFESTC TR 88-14, Fort Collins, Colorado.

Mearns, A. J. 1993. Recovery of shoreline ecosystems following the *Exxon Valdez* oil spill and subsequent treatment. Pages 466–479 *in* O. T. Magoon, editor. Proceedings, 8th symposium on coastal and ocean management. American Society of Civil Engineers, Washington, DC.

Michel, J., and M. O. Hayes. 1991. Geomorphological controls on the persistence of shoreline contamination from the *Exxon Valdez* oil spill. Report to NOAA (National Oceanic and Atmospheric Administration) Hazardous Materials Response Branch, Seattle.

Michel, J., and M. O. Hayes. 1993. Evaluation of the condition of Prince William Sound shorelines following the *Exxon Valdez* oil spill and subsequent shoreline treatment. Volume I: summary of results-geomorphological shoreline monitoring survey of the *Exxon Valdez* spill site, Prince William Sound, Alaska, September 1989–August 1992. NOAA (National Oceanic and Atmospheric Administration) Technical Memorandum NOS (National Ocean Service) ORCA 73, Seattle, WA.

Michel, J., M. O. Hayes, W. J. Sexton, J. C. Gibeaut, and C. Henry. 1991. Trends in natural removal of the *Exxon Valdez* oil spill in Prince William Sound from September 1989 to May 1990. Pages 181–187 *in* Proceedings, 1991 oil spill conference. American Petroleum Institute, Publication 4529, Washington, DC.

Nauman, S. A. 1991. Shoreline cleanup: equipment and operations. Pages 141–147 *in* Proceedings, 1991 oil spill conference. American Petroleum Institute, Publication 4529, Washington, DC.

NOAA (National Oceanic and Atmospheric Administration). 1992a. CAMEO™ Valdez: computer-aided management of emergency operations. NOAA Hazardous Materials Response and Assessment Division, Seattle.

NOAA (National Oceanic and Atmospheric Administration). 1992b. Template shoreline countermeasures manual. Temperate coastal environments. NOAA Hazardous Materials Response and Assessment Division, Seattle.

Piper, E. 1993. The *Exxon Valdez* oil spill. Final report, state of Alaska response. Alaska Department of Environmental Conservation, Anchorage.

Pritchard P. H., C. F. Costa, and L. Suit, editors. 1991. Alaska oil spill bioremediation project. Science advisory board draft report. U.S. EPA (Environmental Protection Agency) Report 600/9-91/046a, Environmental Research Laboratory, Gulf Breeze, Florida.

Stekoll, M., L. Deysher, and T. A. Dean. 1993. Seaweeds and the *Exxon Valdez* oil spill. Pages 135–140 *in* Proceedings, 1993 oil spill conference. American Petroleum Institute, Publication 4580, Washington, DC.

Teal, A. R. 1991. Shoreline cleanup-reconnaissance, evaluation, and planning following the *Exxon Valdez* oil spill. Pages 149–160 *in* Proceedings, 1991 oil spill conference. American Petroleum Institute Publication 4529, Washington DC.

USCG (United States Coast Guard). 1993. Federal on scene coordinator's report T/V *Exxon Valdez* oil spill. U.S. Coast Guard, Washington, DC.

Wolfe, D. A., and eleven coauthors. 1994. The fate of the spilled oil from the *Exxon Valdez*. Environmental Science and Technology 28(13):561A–568A.

American Fisheries Society Symposium 18:329–348, 1996

Short-Term Effects of Several Types of Shoreline Treatment on Rocky Intertidal Biota in Prince William Sound

DENNIS C. LEES

Ogden Environmental and Energy Services Co., Inc.
5510 Morehouse Drive, San Diego, California 92121, USA

JONATHAN P. HOUGHTON

Pentec Environmental, Inc., 120 West Dayton, Edmonds, Washington 98020, USA

WILLIAM B. DRISKELL

6536 20th Avenue NE, Seattle, Washington 98115, USA

Abstract.—Studies conducted in Prince William Sound during 1989 assessed short-term biological effects of treatment methods being considered or employed to treat oil-contaminated beaches. Treatment alternatives evaluated included (1) low-pressure warmwater wash (LP-WW), (2) high-pressure hot-water wash (HP-HW), (3) the dispersant Corexit 7664, and (4) the beach cleaner Corexit 9580 M2. Effects on the biota were assessed on the basis of changes in abundance of dominant taxa and in magnitude of selected assemblage attributes (e.g., percent cover by algae or animals and number of taxa) and, in some cases, observed mortality of organisms in the treatment areas. Significant reductions in one or more assemblage or population attributes and increases in the percent of dead mussels were observed in response to all types of treatment. The strongest and most consistent effects were observed following HP-HW treatment. These included heavy mortality in rockweed and mussels. Generally, the programs were not designed to allow discrimination among the potential causes of damage. However, temperature appeared to cause significant mortality in the dominant plants and in grazing and filter-feeding animals at HP-HW sites. In contrast, available data suggest that neither chemical nor LP-WW treatments caused significant thermal impacts in the intertidal biota. An intermediate level of changes, both increases and decreases, accompanied the chemical and LP-WW treatments, whereas the highest percentage of changes, nearly all decreases, accompanied HP-HWs. The severe and persistent effects observed in many areas subjected to HP-HW throughout the spill area were still conspicuous in our most recent survey in the sound (July 1995). The observed displacement and mortality of clams and mussels suggest that physical and possibly chemical effects may have been substantial in some cases. Although dispersant and beach cleaner treatments appeared to be accompanied by the smallest number of significant changes in abundance, this conclusion is weak because the LP-WW that accompanied chemical applications during the tests was less rigorous than when performed alone. Moreover, the surveys were not focused on enumerating abundance of the types of organisms that toxicity data indicate are more susceptible to the types of chemicals tested. In fact, probable dosages applied appear to substantially exceed published LC50s (median lethal concentrations) for Corexit 7664 and later formulations of Corexit for many types of crustaceans and molluscs relevant to intertidal assemblages in Prince William Sound.

A substantial amount of the 43.9 million liters of Alaska North Slope crude oil that spilled from the TV *Exxon Valdez* on 24 March 1989 was deposited on beaches in Prince William Sound (40 to 45%; Wolfe et al. 1996, this volume). Officials from Exxon Corp. and several state and federal agencies subsequently decided to carry out shoreline cleanup. Several hydraulic treatment methods were adapted or developed to remove oil from beaches of the sound (Nauman 1991), and some of these methods were applied in 1989. In addition, application of chemical dispersants and beach cleaners was considered for enhanced removal of oil.

To evaluate the efficacy of various treatment methods, Exxon Corp. and the agencies designed several quantitative field studies to document the efficiency in removal of oil and the short-term impacts of exposing the biota to treatment with warm or hot water, dispersants, or beach cleaners. We were contracted by Exxon Corp. to conduct the biological components of these studies from May through July 1989. Detailed reports for the dispersant Corexit 7664, the beach cleaner Corexit 9580 M2, and Omni-barge studies were prepared by Houghton et al. (1990a), Lees et al. (1990), and Houghton et al. (1990b), respectively.

The basic objective of the biological element of each study was to assess and compare the short-term biological impacts of two alternative treatment methods. The programs were independent, and no effort was made at that time to compare them. This paper presents data from three treatment effects studies and provides a general overview of the short-term biological effects of several types of treatment. As part of this presentation, we have also assessed the validity of the studies themselves.

Cleanup Methods Examined

Four treatment methods were evaluated: (1) low-pressure warmwater wash (LP-WW), (2) the dispersant Corexit 7664, (3) the beach cleaner Corexit 9580 M2, and (4) high-pressure hot-water wash (HP-HW). The first three methods were implemented with hand-held hose systems supplied from small landing craft or by backpack systems, whereas the fourth method was implemented by the Omni-barge. All of these methods were designed to remobilize oil that had weathered to a point where it was quite tacky and stuck firmly to the substrate. Both LP-WW and HP-HW methods were used extensively throughout Prince William Sound to clean large areas of shoreline, whereas the chemical cleaners were only used experimentally on small plots (Fiocco et al. 1991).

Dispersants and beach cleaners.—We studied the effects of the dispersant Corexit 7664 and the beach cleaner Corexit 9580 M2 in May and July 1989, respectively. Both treatments were two-phase operations involving a 1-h presoak with the chemical, but application procedures were different. In both cases, the chemicals were applied directly to the substrate while the rocks, sediment, and biota were exposed during low tide. For its presoak, Corexit 7664 (4%; Fiocco et al. 1991) was mixed with seawater and sprayed onto the beach at 32°C. The presoak was followed by 2 to 3 h of LP-WW with a Corexit–seawater solution at 32°C. In contrast, the presoak for the beach cleaner Corexit 9580 M2 involved application of full strength chemical at ambient temperature with a hand sprayer supplied from a backpack pressure tank. This beach cleaner treatment was also followed by 2 to 3 h of LP-WW with Corexit added to the seawater.

Low-pressure warm water.—To promote consistency with the dispersant and the beach cleaner tests, LP-WW, as practiced during the comparative tests, was also a two-phase operation. The first phase was a 1-h presoak with seawater at a nominal temperature of 32 or 60°C. In both cases, the presoak was followed by a 2-to 3-h wash phase. The rocks were sprayed with 60°C seawater applied with six to eight hand-operated 5-cm diameter fire hoses to remove the oil. To facilitate transport of the oil dislodged by the spraying into the containment boom systems and skimmers located just offshore, the beach was flooded at 3,785 L/min with ambient seawater sprayed uniformly along the beach from header hoses deployed along the high-tide line.

In tests comparing LP-WW with treatment by a dispersant or a beach cleaner, the individual treatments were conducted in adjacent areas about 30 m wide × 15 to 25 m from upper to lower tide levels and separated by 10-m-wide buffer zones. The different treatments were implemented within 2 d of each other.

High-pressure hot water.—In the HP-HW test, the effects of washing the substrate for 95 and 165 min with seawater heated to 60°C and sprayed onto the substrate at a pressure of about 690 kPa and a flow rate of about 1,890 L/min were compared. Because the flow was directed through six nozzles on a 1.83-m long spray head and the temperature was not moderated by ambient seawater flooding across the beach, this method was physically quite intense. Treatment areas were 15 m wide × about 20 m from upper to lower tide levels. The contrasted areas were adjacent and not separated by a buffer zone. The two treatments were performed consecutively on the same day.

Study Sites

All three studies were conducted at the north end of the Knight Island Archipelago (Figure 1). The Corexit 7664 study was conducted on the northwestern shore of Ingot Island. The Omni-barge study was conducted on the northeastern shore of Herring Bay on Knight Island. The Corexit 9580 M2 study was conducted near the southwestern corner of Disk Island. The Ingot Island and Herring Bay sites are generally west facing, whereas the Disk Island site is south facing.

In all studies, the major habitat was protected boulder–cobble. Considerable patchiness of substrate and biota was evident among and within the study sites. Differences in the degree of wave exposure, as manifest by the degree of angularity and smoothness in the rocks, probably influenced the development and nature of the biota as well as the size and heterogeneity in distribution of the rocks among the sites. Nevertheless, these areas were fairly typical of semi-exposed boulder–cobble shorelines in Prince William Sound (personal observation).

At Disk Island, slope, rock diameter, and percent cover were substantially different between the two test areas, with appreciably more mixed-soft habitat

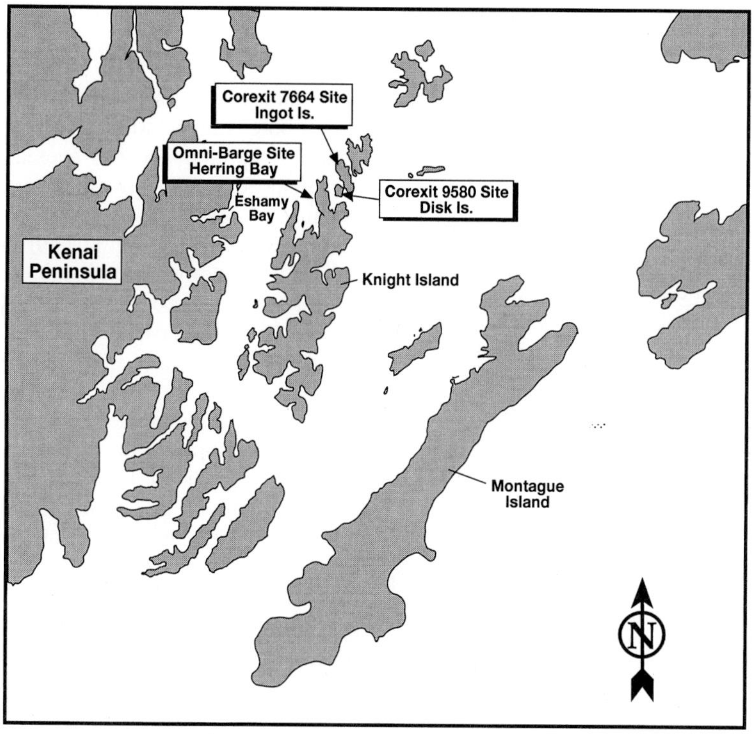

FIGURE 1.—Map showing short-term effects study sites in Prince William Sound.

in the lower intertidal zone at the LP-WW test area than in the Corexit test area. The problems caused by these differences will be discussed later.

Methods

The generic null hypothesis for these studies was that there would be no difference in the measurement for a site characteristic (e.g., number of species or density of a species) following a specific treatment. An operative assumption was that the variables examined would not change significantly over the 3 to 10 d between the pre- and posttreatment surveys except as a consequence of the treatment regime. We applied this assumption to comparisons of all variables examined, ranging from oil cover and oiling intensity to assemblage descriptors (e.g., number of species and abundance of individuals) and species abundance.

Sampling Design

In all test series, each plot was subjected to a different treatment. The contrasted tests were conducted sequentially rather than simultaneously. Test plots extended from upper to lower intertidal zones, but only the mid and upper levels, where most of the oil came to rest during the initial stranding, were supposed to be washed. Each treatment plot was surveyed at least at mid and lower elevations. No reference tests examining effects of a no-treatment alternative were included in any of the studies.

The approach was to compare pretreatment data on percent cover or density of the algae and invertebrates and species richness of the intertidal assemblage with posttreatment measurements from the same transects. We applied a stratified random design to reduce the influence of elevation effects. Mid and lower intertidal strata were designated at all sites on the basis of the biological assemblages. The lower stratum was established in the area of transition between the red algal zone and the lower *Fucus* (rockweed) zone (tidal elevation about 0.6 m above mean lower low water [MLLW]), and the middle stratum was placed at the middle of the upper *Fucus–Mytilus* zone (tidal elevation about 1.7 m above MLLW). The upper level was included in only the first of the studies (the Corexit 7664 study at Ingot Island) because we found that the low level of abundance and considerable patchiness at that level resulted in a level of variability that precluded adequate discrimination.

Sampling Methods

Within each stratum, we randomly selected the position of ten 0.25 m^2 quadrats along a transect line of a length appropriate to the study area. In the Corexit 7664 study, destructive quadrat sampling was employed so that different plots were examined in the pre- and posttreatment surveys. In the other studies, we attempted to reoccupy the same quadrats in pre- and posttreatment surveys. During the pretreatment survey, we marked the quadrats with stakes and took Polaroid photographs of the plots to facilitate detailed positioning of the quadrat frame during subsequent surveys.

The quantitative abundance data recorded for the biota in each quadrat included (1) individual counts for invertebrates other than barnacles and mussels and (2) visual estimates of percent cover for barnacles, mussels, and algae. Abundance was measured on exposed rock surfaces and overhangs, and on the underside of movable cobbles up to about 8 cm in diameter. In some areas, rockweed formed a substantial mat over an understory of smaller foliose and encrusting algae; because cover for each of these taxa was estimated separately, total algal cover sometimes sums to more than 100%.

Data Analysis

Summary statistics were prepared with the aid of common spreadsheet programs. Differences between pre- and posttreatment values were evaluated using randomization t-tests or analyses of variance (ANOVAs) (Edgington 1987). One-tailed tests were used to assess the generic hypotheses that abundance of a live taxon did not decrease significantly or abundance of a dead taxon did not increase significantly between pre- to posttest surveys. Two-tailed tests were used to assess the generic hypothesis that assemblage attributes did not change significantly between pre- to posttest surveys. Significance is noted at the 5 and 10% rejection levels, recognizing the risk that, because of the number of tests being run, a significant effect may be accepted by chance in some cases where one does not exist. Ultimate conclusions on the validity of observed effects were based on weight-of-evidence considerations rather than single tests. Thus, we based our conclusions on the strength of the summation of all comparisons made.

Results

Quantitative Assemblage Descriptors

Variables used as indicators for comparing the condition of the epibiota included algal and epifaunal cover, epifaunal density, average number of taxa per quadrat, species diversity, and density of dead epifauna. Additionally, we measured cover by dead rockweed *Fucus gardneri* at Herring Bay following the HP-HW by the Omni-barge.

Algal cover.—Before treatment, algal cover in the upper and mid-intertidal zones ranged from about 1 to 34% (Table 1). Most treatments appeared to reduce cover by live algae, but HP-HW at Herring Bay and LP-WW in the Corexit 9580 M2 test at Disk Island appeared to cause the most damage (Figure 2; $P < 0.01$). On average, the greatest decrease in algal cover occurred in the middle intertidal zone, where treatment and high algal cover overlapped (Table 1).

Pretreatment algal cover at the lower level ranged from about 36 to 117%. Both increases and decreases in cover were observed following treatment, but none of the changes was significant. Cover differed substantially between the LP-WW and Corexit sites at Disk Island. Although treatment was not supposed to extend into the lower intertidal zone, algal cover decreased at both durations of HP-HW at Herring Bay; the longer treatment (165-min) seemed to cause a more severe decline. The conflicting patterns at Ingot Island suggest a difference in treatment.

Epifaunal cover.—Epifaunal cover in the upper and middle intertidal, ranging from about 1 to 13%, was highest in the middle intertidal zone (Table 2). Although estimated cover was commonly 50% lower following treatment, low cover and high small-scale variability often combined to confound the statistical significance of the changes. Epifaunal cover in the algae-dominated lower intertidal zone was quite low (about 0.1 to 3%) and stable.

Epifaunal density.—Epifaunal density in the upper and mid-intertidal zones, ranging from about 7 to 80 individuals per quadrat before treatment, was generally higher in the midzone (Table 3). Large increases as well as decreases were observed following treatment, suggesting that impacts were not consistently strong. The strongest decreases (100-fold) occurred in the middle intertidal zone following the HP-HW tests ($P < 0.01$).

Epifaunal density in the lower tide zone, ranging from about 12 to 121 individuals per quadrat, did not change in a consistent manner. The significant decrease observed following the 165-min HP-HW

TABLE 1.—Algal cover (%) on substrate in study areas in Prince William Sound, by elevation, before and after treatment; LP-WW, low-pressure warmwater wash; HP-HW, high-pressure hot-water wash. Asterisks denote treatments for which pre- and posttreatment values differ significantly at $P \leq 0.1$* or $P \leq 0.05$**.

	Days relative to treatment								
	Pretreatment			Posttreatment					
Site, treatment; statistic	−8	−3	−1	0	1	2	5	6	8
Upper level									
Ingot, LP-WW		1.8		0.3					
Ingot, Corexit 7664*			1.3		7.8				
Mean			1.6						4.0
SD			0.3						3.8
Number of measurements			2						2
Mid level									
Ingot, LP-WW**		4.3		0.1					
Ingot, Corexit 7664			8.3		6.7				
Disk, LP-WW**			31.5						4.8
Disk, Corexit 9580**	7.9		5.3		1.5		1.6		
Herring, 165-min HP-HW*		34.2							0.5
Herring, 95-min HP-HW**		18.9							5.2
Mean			15.8						2.9
SD			11.7						2.4
Number of measurements			7						7
Lower level									
Ingot, LP-WW		95.5			60.8				
Ingot, Corexit 7664			62.5			71.6			
Disk, LP-WW			117.5					117.5	
Disk, Corexit 9580	41.7		36.3				34.6		
Herring, 165-min HP-HW		108.3							79.1
Herring, 95-min HP-HW		116.8							102.5
Mean			82.6						77.7
SD			32.6						27.0
Number of measurements			7						6

at Herring Bay reflects reduced density of the hermit crab *Pagurus hirsutiusculus* and the periwinkle *Littorina scutulata*. Moderate increases in density following both treatments in the Corexit 7664 study seem to represent an increase in abundance of scavengers such as hermit crabs, possibly in response to the increased availability of dead or injured organisms.

Species richness.—Clear patterns in species diversity were not apparent at any of the tidal elevations observed (Figure 3). Average number of taxa per quadrat on the upper and middle intertidal transects ranged from approximately four to eight before treatment (Table 4) and tended to exhibit declines following treatment. Strongest declines (57 to 74%) occurred at Herring Bay following HP-HW ($P < 0.01$). Lower numbers of species were also observed following all three LP-WWs.

The average number of taxa per quadrat in the lower tide zone ranged from approximately 8 to 18 before treatment. Strong declines were observed only at the HP-HW sites at Herring Bay ($P < 0.1$). The reason for the significant twofold increase observed following LP-WW during the Corexit 7664 study at Ingot Island is unclear.

Dead epifauna.—Density of dead epifauna, measured as an indication of mortality in animals such as crabs and mussels, averaged from 2.6 to 15.7 dead animals per quadrat at the upper and mid-intertidal levels before treatment and from 2.3 to 43.6 per quadrat after treatment. Dead specimens were most common at mid-intertidal levels, where mussels were most abundant (Table 5). Generally, the initial counts included mainly detached mussel shells that were gaping but still articulated. The large changes in abundance of dead animals observed following HP-HW reflected increases in numbers of attached but gaping mussel shells with intact tissues. Although these quantitative and qualitative observations suggest a strong impact from

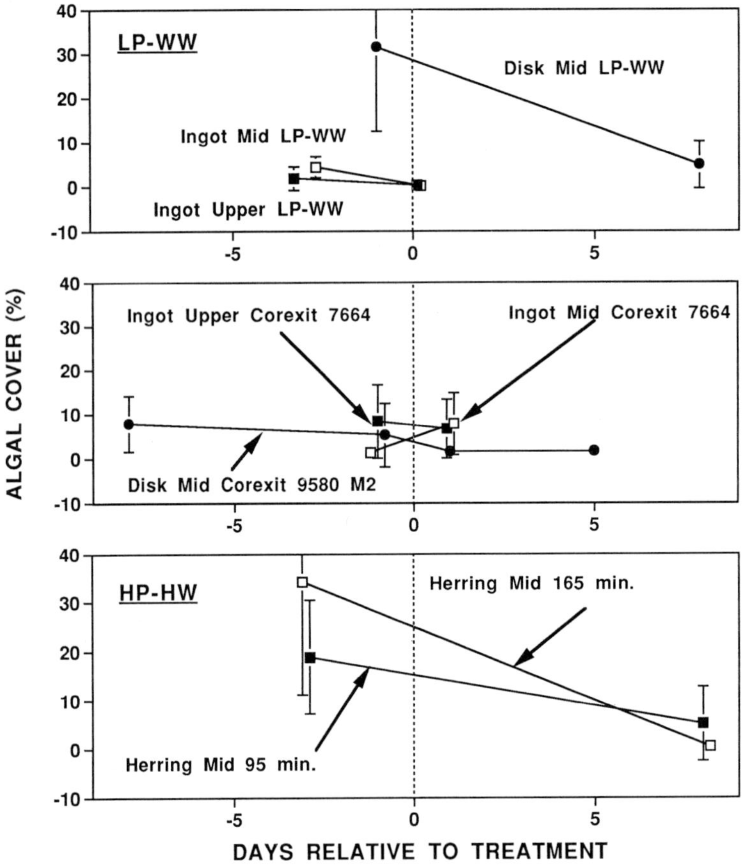

FIGURE 2.—Average cover by algae in quadrats on transects at upper and middle intertidal levels; LP-WW, low-pressure warmwater wash; HP-HW, high-pressure hot-water wash. Error bars represent 1SD.

HP-HW, only the change following the 95-min HP-HW was significant because of high variability related to the normal patchiness of mussels at this tide level.

Density of dead epifauna in the lower intertidal ranged from less than 0.1 to 13.4 animals per quadrat. Only the fourfold decline at the Corexit 9580 M2 site at Disk Island was significant. The decrease in density of dead animals following treatment in many tests probably reflects the tendency of the wash activities to flush loose materials from the areas.

Dead rockweed.—During the HP-HW study, the treatment process caused rockweed to change from a normal color to the blackish orange color assumed by accumulations of dead *Fucus* in wrack lines. We therefore assumed these plants had been killed by the treatment and recorded them as dead. The accuracy of this assumption has been borne out in subsequent studies where *Fucus* was absent in areas in which it was reported before exposure to HP-HW (Houghton et al. 1993a).

Dead *Fucus* at the mid-intertidal in HP-HW tests increased from 0% before treatment to 14 and 27% after the 95- and 165-min treatments, providing a strong indication of the rigorous effects of HP-HW (Figure 4). For both treatments, combined cover of live and dead *Fucus* following treatment was nearly equal to the cover recorded for live *Fucus* before treatment. On a population basis, the 95- and 165-min treatments caused 74 and 94% mortality, respectively.

Fucus with a similar appearance was not observed in any of the other tests. Subsequent observations of *Fucus* cover at numerous other sites throughout the sound suggest that HP-HW (60°C at about 690 kPa) caused heavy mortality in *Fucus* populations and led to widespread, long-term denudation of the rocks in this tide zone (Houghton et al. 1993a).

TABLE 2.—Epifaunal cover (%) in study areas in Prince William Sound, by elevation, before and after treatment; LP-WW, low-pressure warmwater wash; HP-HW, high-pressure hot-water wash. Asterisks denote treatments for which pre- and posttreatment values differ significantly at $P \leq 0.1^{*}$ or $P \leq 0.05^{**}$.

	Days relative to treatment								
	Pretreatment			Posttreatment					
Site, treatment; statistic	−8	−3	−1	0	1	2	5	6	8
Upper level									
Ingot, LP-WW		2.8		7.0					
Ingot, Corexit 7664**			5.5		2.3				
Mean			4.2						4.7
SD			1.4						2.4
Number of measurements			2						2
Mid level									
Ingot, LP-WW*		10.2		4.9					
Ingot, Corexit 7664**			10.8		4.7				
Disk, LP-WW**			12.1						6.4
Disk, Corexit 9580	2.5		0.9		1.7		1.2		
Herring, 165-min HP-HW		12.3							13.3
Herring, 95-min HP-HW		13.1							9.4
Mean			8.8						5.9
SD			4.6						4.0
Number of measurements			7						7
Lower level									
Ingot, LP-WW		1.3			0.5				
Ingot, Corexit 7664			0.1			1.1			
Disk, LP-WW			1.3					1.2	
Disk, Corexit 9580	2.1		1.9				2.8		
Herring, 165-min HP-HW		0.5							0.4
Herring, 95-min HP-HW		0.7							0.1
Mean			1.1						1.0
SD			0.7						0.9
Number of measurements			7						6

Damage to Hard-Shell Clams

The only area of mixed-soft substrate large enough to survey during these programs was the lower intertidal level in the Corexit study plot at Disk Island. In this area, density of burrowing hard-shell clams (*Protothaca staminea* and *Saxidomus giganteus*) declined from about 160 live clams/m^2 in the pretreatment survey to about 46 live clams/m^2 and about 13 dead clams/m^2 following treatment. Many other clams were found on the surface of the sediment following treatment. Half were dead or had fractured shells, whereas the other half were apparently alive but inactive. Unfortunately, similar habitat was not present in the paired LP-WW site, thus the relative importance of the physical and chemical effects cannot be evaluated.

Measurements of Water Temperature in Treatment Areas

Temperatures of ambient and wash water and thermal gradients across the test areas were measured in the Corexit and LP-WW sites during the Corexit 7664 study in May 1989. Ambient seawater temperatures during these tests were between 8 and 9°C. During the Corexit–seawater presoak phase of the Corexit test, the maximum temperature in pools where spray streams from the hoses impinged on the substrate was 13.5°C. In comparison, a maximum of 38°C was observed in similar pools during the LP-WW phase of this study. However, because the beach wash crews swept their spray streams back and forth over the substrate, it is unlikely that organisms were exposed to this high temperature for more than a few minutes at a time. Moreover, because of the high volume of water flooding across the beach from the headers above the tideline during the LP-WW phase, water temperatures declined to 14.5°C within 1 m of the spray stream and to between 9 and 12°C (within 3°C of ambient) in the flood water between the beach wash crew and the waterline.

TABLE 3.—Epifaunal density (number of individuals/0.25 m^2) in study areas in Prince William Sound, by elevation, before and after treatment; LP-WW, low-pressure warmwater wash; HP-HW, high-pressure hot-water wash. Asterisks denote treatments for which pre- and posttreatment values differ significantly at $P \leq 0.1$* or $P \leq 0.05$**.

	Days relative to treatment								
	Pretreatment			Posttreatment					
Site, treatment; statistic	−8	−3	−1	0	1	2	5	6	8
Upper level									
Ingot, LP-WW		46.4		6.8					
Ingot, Corexit 7664*			7.2		17.4				
Mean			26.8						12.1
SD			19.6						5.3
Number of measurements			2						2
Mid level									
Ingot, LP-WW		21.3		55.8					
Ingot, Corexit 7664			25.7		16.0				
Disk, LP-WW*			74.9						35.9
Disk, Corexit 9580	23.2		11.8		17.4		23.4		
Herring, 165-min HP-HW**		73.8							0.8
Herring, 95-min HP-HW**		80.2							0.2
Mean			44.4						21.4
SD			28.0						18.2
Number of measurements			7						7
Lower level									
Ingot, LP-WW		44.2			65.8				
Ingot, Corexit 7664			30.4			55.3			
Disk, LP-WW			12.2					10.9	
Disk, Corexit 9580	121.0		48.5				103.7		
Herring, 165-min HP-HW**		50.3							15.9
Herring, 95-min HP-HW		14.8							9.0
Mean			45.9						43.4
SD			33.8						34.8
Number of measurements			7						6

Patterns in Assemblage Attributes

Several taxa displayed substantial changes in abundance during the tests. Most of the significant changes were declines that probably indicate removal or relocation caused by vigorous washing.

Plants.—Significant declines in algae were not observed at upper levels (Table 6). However, significant declines were observed in *Fucus* at middle levels at Disk Island, at the LP-WW site on Ingot Island, and Herring Bay (Table 7). Significant declines were also observed at lower levels (Table 8) in *Fucus* at Ingot Island, filamentous green algae at Disk Island and Herring Bay, flagelliform brown algae at Herring Bay, *Mastocarpus* spp. at Disk Island, *Pylaiella littoralis* and *Ralfsia pacifica* at Disk Island and Herring Bay, and *Ulva–Monostroma* at Disk Island. In a few cases, significant increases in algal cover were observed. The increase in *Gloiopeltis furcata* at the upper level at Ingot Island may reflect improved observability of the turf-like alga as a consequence of washing (Table 6). Increases in *Ulva–Monostroma* at the lower level at Ingot Island may have resulted from accumulations of the algae following washing at higher levels (Table 8). However, because destructive sampling at Ingot Island prevented resampling in the same quadrats, both cases could merely reflect sampling variability.

Invertebrates.—Significant reductions in invertebrates were observed in *Balanus–Semibalanus* spp. following LP-WW at the middle level at Ingot Island (Table 7); periwinkles (*Littorina sitkana* following LP-WW at the upper level [Table 6] and Corexit treatment at the middle level [Table 7] at Ingot Island, LP-WW at the middle level at Disk Island [Table 7], and both durations of HP-HW at the middle level at Herring Bay [Table 7]; and *L. scutulata* following Corexit treatment at the middle level at Disk Island and both durations of HP-HW at the middle level at Herring Bay [Table 7]); limpets (Lottiidae, unidentified species) following

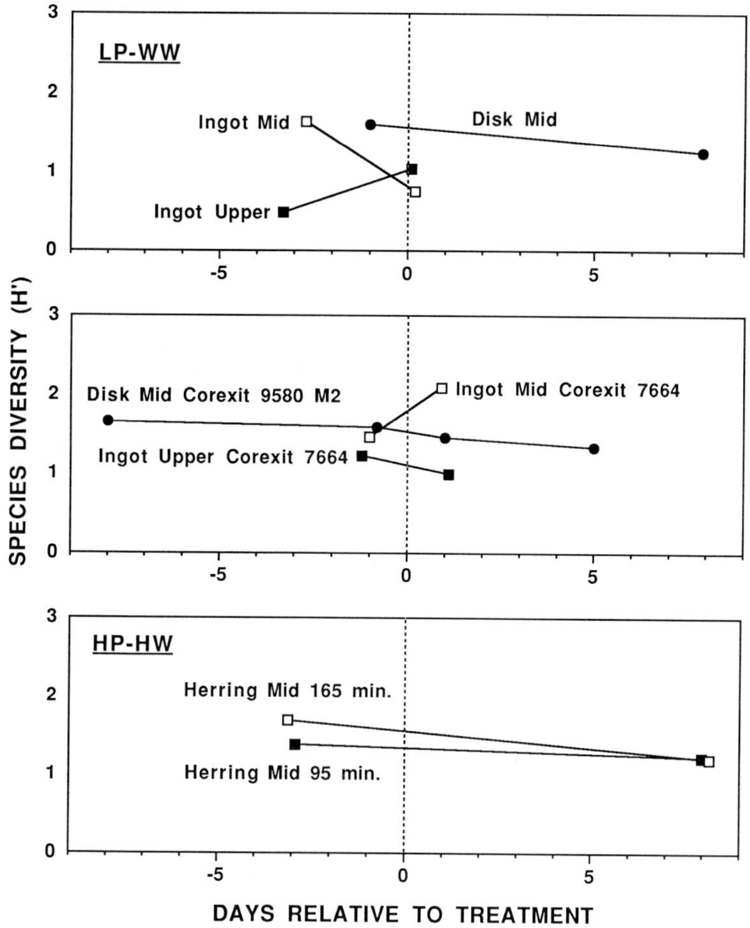

FIGURE 3.—Pooled species diversity in upper and middle intertidal zones; LP-WW, low-pressure warmwater wash; HP-HW, high-pressure hot-water wash.

LP-WW and Corexit treatments at upper levels at Ingot Island (Table 6), LP-WW and Corexit treatments at the middle level at Disk Island (Table 7), and both durations of HP-HW at the middle level at Herring Bay (Table 7); mussels (*Mytilus trossulus*) following Corexit treatment at the upper level (Table 6) and LP-WW and Corexit treatments at the middle level at Ingot Island (Table 7), LP-WW at the middle level at Disk Island (Table 7), and the 95-min HP-HW at the middle level at Herring Bay (Table 7); hermit crabs (*Pagurus hirsutiusculus*) following Corexit treatment at the lower level at Disk Island and the 165-min HP-HW at the lower level at Herring Bay (Table 8); and a whelk (*Searlesia dira*) following LP-WW at the lower level at Disk Island (Table 8). Significant increases that probably resulted from relocation by washing were observed in the periwinkle *L. scutulata* following Corexit treatment at the middle level (Table 7) and LP-WW and Corexit treatments at the lower level at Disk Island (Table 8). An increase that possibly reflects immigration in response to increased abundance of dislocated prey as a consequence of the washing process was observed at the lower level in a whelk (*S. dira*) following LP-WW at Ingot Island (Table 8). As noted above for the algae, the increases at Ingot Island could also reflect sampling variability.

Dead organisms.—Observed numbers of dead specimens of one species increased substantially at only one site. Cover by dead *Fucus* increased significantly at the middle level at Herring Bay following both HP-HWs (Table 7). Numbers of dead mussels also increased substantially but not significantly

TABLE 4.—Average number of taxa per quadrat in study areas in Prince William Sound, by elevation, before and after treatment; LP-WW, low-pressure warmwater wash; HP-HW, high-pressure hot-water wash. Asterisks denote treatments for which pre- and posttreatment values differ significantly at $P \leq 0.1$* or $P \leq 0.05$**.

	Days relative to treatment								
	Pretreatment			Posttreatment					
Site, treatment; statistic	−8	−3	−1	0	1	2	5	6	8
Upper level									
Ingot, LP-WW		4.4		2.8					
Ingot, Corexit 7664			4.4		3.8				
Mean			4.4						3.3
SD			0.0						0.5
Number of measurements			2						2
Mid level									
Ingot, LP-WW		7.7		4.3					
Ingot, Corexit 7664			5.3		7.3				
Disk, LP-WW			6.7						6.2
Disk, Corexit 9580	5.8		4.8		6.3		6.1		
Herring, 165-min HP-HW**		8.4							3.6
Herring, 95-min HP-HW**		7.6							2.0
Mean			6.6						5.1
SD			1.2						1.7
Number of measurements			7						7
Lower level									
Ingot, LP-WW**		8.2			16.7				
Ingot, Corexit 7664			12.6			10.2			
Disk, LP-WW			14.6					16.5	
Disk, Corexit 9580	17.8		14.6				15.0		
Herring, 165-min HP-HW**		10.4							7.1
Herring, 95-min HP-HW*		12.2							7.0
Mean			12.9						12.1
SD			2.9						4.1
Number of measurements			7						6

at this site following HP-HW. The appearance of large numbers of dead organisms at both HP-HW sites should be expected in view of the large body of literature showing that thermal limits of most intertidal organisms are less than 35°C (see below).

Patterns in Significant Changes—Weight of Evidence

In order to better compare the relative magnitude of treatment effects and interactions between treatment and elevation, increases and decreases in species abundance and assemblage attributes are summarized to explore the inference of the weight of evidence.

Species abundance.—The type and number of significant changes observed in species abundance varied considerably by type of treatment and tidal elevation. Generally, significant decreases were more common (78%) than increases (22%; Table 9). Decreases in abundance were relatively less frequent following the Corexit and LP-WW treatments than following HP-HW, where nearly all significant changes were decreases. Significant changes occurred in fewer than one-third of the total number of opportunities for change available for each treatment type.

The relative frequency of significant changes in abundance of taxa also varied somewhat by elevation (Table 9). Fewer significant changes were observed at the upper tide level, probably in part because that level is normally characterized by fewer species, low abundance of organisms, and high spatial variability. Nevertheless, in relative terms, significant changes were more frequent at the mid and upper tide levels than at the lower level. Significant changes in species abundance at upper and middle levels were predominantly declines (80 and 95%, respectively; Table 9), although periwinkles increased considerably at some locations. In contrast, changes at lower levels were more

TABLE 5.—Average number of dead epifauna per quadrat in study areas in Prince William Sound, by elevation, before and after treatment; LP-WW, low-pressure warmwater wash; HP-HW, high-pressure hot-water wash. Asterisks denote treatments for which pre- and posttreatment values differ significantly at $P \leq 0.1$* or $P \leq 0.05$**.

	Days relative to treatment								
	Pretreatment			Posttreatment					
Site, treatment; statistic	−8	−3	−1	0	1	2	5	6	8
Upper level									
Ingot, LP-WW		4.8		3.2					
Ingot, Corexit 7664			2.6		2.4				
Mean			3.7						2.8
SD			1.1						0.4
Number of measurements			2						2
Mid level									
Ingot, LP-WW		11.0		8.0					
Ingot, Corexit 7664			15.7		14.0				
Disk, LP-WW			9.6						9.4
Disk, Corexit 9580*	8.1		12.0		5.1		5.9		
Herring, 165-min HP-HW		8.9							24.0
Herring, 95-min HP-HW*		6.2							43.6
Mean			10.2						15.7
SD			2.8						12.9
Number of measurements			7						7
Lower level									
Ingot, LP-WW		0.0			0.2				
Ingot, Corexit 7664			0.0			0.0			
Disk, LP-WW			5.0					4.9	
Disk, Corexit 9580**	13.4		6.8				2.8		
Herring, 165-min HP-HW		0.7							0.2
Herring, 95-min HP-HW		0.4							0.0
Mean			3.8						1.3
SD			4.7						1.9
Number of measurements			7						6

evenly distributed between decreases and increases (58 versus 42%, respectively; Table 9). From the perspective of total number of opportunities for change at an elevation, significant changes occurred in only 10% of the possible cases at lower tide levels and more than four times as often at the middle level. These differences among tidal elevation probably reflect the position of the zone relative to washing activities (i.e., washing activities were not supposed to extend into the lower intertidal zone).

Assemblage attributes.—The type and number of significant changes observed in assemblage attributes also varied considerably by type of treatment and elevation. Significant decreases were relatively more common (80%) than increases (Table 10). The relative frequency of significant changes in assemblage attributes varied substantially by type of treatment. Significant changes were equally distributed between increases and decreases in the Corexit tests. In contrast, decreases far exceeded increases in the LP-WW and HP-HW tests. From the perspective of total opportunities for change, significant changes in assemblage attributes occurred at about the same rate in the Corexit and LP-WW tests (20 to 24%), but about twice as often in the HP-HW tests (45%; Table 10).

On the basis of tidal elevation, significant changes in assemblage attributes were numerically more common at the middle level than at the upper and lower levels. Significant changes in attributes at the mid and lower tide levels were dominated by decreases. In contrast, all of the significant changes at upper levels were increases (Table 10). From the perspective of total possible opportunities, changes were significant in only 13% of the cases at lower tide levels but in 43% of the cases at the middle levels (Table 10).

In summary, the weight of evidence from these studies implies that shoreline treatment by HP-HW causes considerable damage to intertidal assemblages and substantially more damage than the other types of treatments examined during these

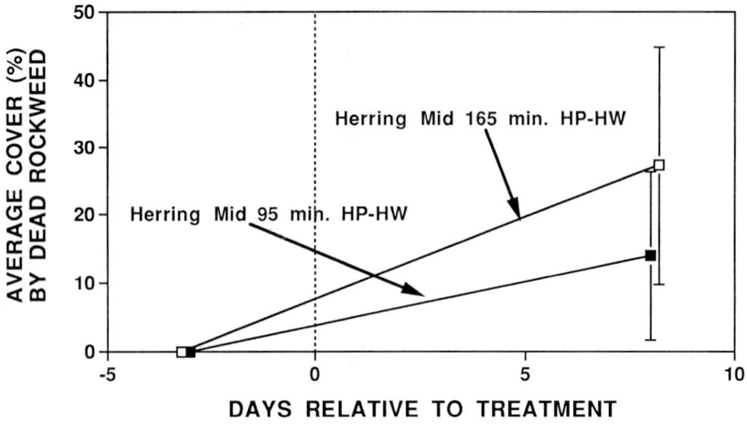

FIGURE 4.—Changes in average cover of dead rockweed in quadrats on transects at mid-intertidal levels at Herring Bay; HP-HW, high-pressure hot-water wash. Error bars represent 1SD.

studies. Moreover, the data imply that the damage is more severe in the assemblages living at mid-intertidal elevations. However, these implications must be tempered by consideration of weaknesses inherent in the study design.

Discussion

Study Weaknesses

Several consistent flaws in sampling design weakened the ability of the studies to evaluate treatment effects. Paired treatment sites in each study lacked strong comparability in terms of substrate and biota. For example, important factors varied considerably between the Corexit and LP-WW sites at Disk Island (algal cover, epifaunal cover and density, beach slope, and proportions of boulder, cobble, and gravel on the beach; Tables 1–3), Ingot Island (algal cover and epifaunal cover; Tables 1, 3), and Herring Bay (epifaunal density [Table 3]).

TABLE 6.—Change in cover (%) or density (N = mean number of individuals/0.25 m^2) of dominant taxa (only living organisms) on heavily oiled upper cobble beaches on Ingot Island during tests of selected treatment activities; LP-WW, low-pressure warmwater wash. Asterisks denote posttreatment means that differ significantly from the respective pretreatment means at $P \leq 0.1$* or $P \leq 0.05$**.

	LP-WW treatment area		Corexit 7664 treatment area	
Taxon	Pretest	Posttest	Pretest	Posttest
Gloiopeltis furcatus (%)	1.3	0.3	1.3	7.6*
Fucus gardneri (%)	0.5	0.0	0.0	0.2
Balanus sp. (%)	0.7	0.6	0.4	0.5
Chthamalus dalli (%)	0.1	0.3	0.1	0.0
Littorina scutulata (*N*)	0.6	0.0	0.2	0.2
Littorina sitkana (*N*)	72.0	7.0*	8.0	17.2
Lottiidae[a] (*N*)	2.6	0.2**	0.6	0.0
Mytilus trossulus (%)	2.0	6.2	5.0	1.7**
Mytilus (dead) (*N*)	2.0	2.5	2.6	2.4

[a]Unidentified species.

Moreover, it is unfortunate that, even though a large proportion of the protected shoreline in Prince William Sound is mixed-soft substrate, systematic evaluations of treatment effects were not conducted on these habitats. In the one instance where a study area included soft substrate (Corexit 9580 M2 treatment area), density of hard-shell clams declined by about 71%. Because similar habitat was not present in the paired LP-WW site, chemical toxicity cannot be ruled out. Of greater importance, however, is the failure of these studies to evaluate the effects of the various types of treatment on the clam assemblages, which comprise an important food resource for several important food webs in the sound.

Inconsistency in temperature regimes, duration of exposure, and intensity and method of washing between the LP-WW and chemical tests within a test set and between the two Corexit test sets is another weakness in the study designs. Substantial differences in treatment characteristics are described for the Corexit 7664 study in the *Methods* (above). Further differences in consistency, duration, and intensity of treatment were observed in the field (personal observation). The LP-WWs were less rigorous during the Corexit tests than when performed alone. These inconsistencies weaken the

TABLE 7.—Change in cover (%) or density (N = mean number of individuals/0.25 m^2) of dominant taxa (only living organisms) on heavily oiled middle cobble beaches in study areas in Prince William Sound during tests of selected treatment activities; LP-WW, low-pressure warmwater wash; HP-HW, high-pressure hot-water wash. Asterisks denote posttreatment means that differ significantly from the respective pretreatment means at $P \le 0.1$* or $P \le 0.05$**.

	Ingot Island				Disk Island						Herring Bay			
	LP-WW treatment area		Corexit 7664 treatment area		LP-WW treatment area		Corexit 9580 M2 treatment area				165-min HP-HW		95-min HP-HW	
Taxon	Pretest	Posttest	Pretest	Posttest	Pretest	Posttest	Pretest 1	Pretest 2	Posttest 1	Posttest 2	Pretest	Posttest	Pretest	Posttest
Gloiopeltis furcatus (%)	2.2	0.13*	1.3	5.5	0.0	0.0	0.0	0.0	0.0	0.0	0.2	0.1	0.5	0.2
Fucus gardneri (%)	2.2	0.0**	7.0	1.2	31.2	4.8**	7.7	5.2	1.5	1.6**	31.6	1.9**	17.8	5.0**
Balanus–Semibalanus sp. (%)	1.2	0.3*	0.0	0.3	5.7	4.2	0.8	0.2	0.8	0.6	8.8	11.0	9.2	8.2
Chthamalus dalli (%)	0.0	0.1	0.2	0.0	0.1	0.1	0.0	0.0	0.2	0.0				
Littorina scutulata (*N*)	3.0	2.0	4.3	0.0	44.2	28.1	6.9	4.9	11.6	14.7**	32.6	0.0**	64.2	0.0**
Littorina sitkana (*N*)	16.0	50.0	20.0	4.7*	21.3	7.1**	13.5	5.7	4.8	7.4	22.4	0.0**	10.6	0.0**
Lottiidae[a] (*N*)	1.3	2.5	1.0	2.0	7.5	0.7**	1.5	0.4	0.3	0.7**	17.8	0.0**	4.8	0.0*
Mytilus trossulus (%)	9.0	4.5*	10.7	4.3*	6.3	2.2**	1.7	0.7	0.8	0.6	3.5	2.3	3.9	1.2*
Mytilus (dead) (*N*)	7.3	6.5	15.7	14.0	44.5	9.4**	8.1	12.0	4.4	5.6	6.4	22.2	4.4	43.6
Fucus gardneri (dead) (%)					0.0	0.0	0.0	0.0	0.0	0.0	0.0	27.3**	0.0	14.0*

[a]Unidentified species.

conclusion that changes at both Corexit sites were less severe than at the LP-WW sites.

Moreover, temperature and other relevant physical and chemical variables were not systematically monitored over the course of the studies. Therefore, it is not possible to determine and compare exposure gradients for the various factors being tested.

Because of the proximity of the paired sites and timing differences in testing, the probability of cross-test interactions is high. Such interactions could explain observed pretest declines in abun-

TABLE 8.—Change in cover (%) or density (N = mean number of individuals/0.25 m^2) of dominant taxa (only living organisms) on heavily oiled lower cobble beaches in study areas in Prince William Sound during tests of selected treatment activities; LP-WW, low-pressure warmwater wash; HP-HW, high-pressure hot-water wash. Asterisks denote posttreatment means that differ significantly from the respective pretreatment means at $P \le 0.1$* or $P \le 0.05$**.

	Ingot Island				Disk Island					Herring Bay			
	LP-WW treatment area		Corexit 7664 treatment area		LP-WW treatment area		Corexit 9580 M2 treatment area			165-min HP-HW		95-min HP-HW	
Taxon	Pretest	Posttest	Pretest	Posttest	Pretest	Posttest	Pretest 1	Pretest 2	Posttest	Pretest	Posttest	Pretest	Posttest
Algae (%)													
Palmaria callophylloides					3.1	2.8	0.9	0.4	0.05**	0.1	0.0	0.0	0.1
Filamentous green algae	16.1	5.7	5.4	2.8	81.9	57.6**	16.7	10.3	11.0	48.2	30.3*	51.4	38.6
Flagelliform brown algae	0.2	0.3	0.5	0.4	0.7	0.3	0.1	0.1	0.0	0.3	0.0**	0.2	0.2
Filamentous red algae	0.2	0.0	1.0	0.4	0.2	0.2	0.3	0.4	0.4			0.1	0.0
Fucus gardneri	37.1	18.3**	27.0	34.1	20.4	23.9	8.4	6.1	5.8	26.6	20.9	30.6	35.0
Mastocarpus spp.	0.0	0.3	0.0	0.2	0.6	7.6**	0.3	0.5	0.9	0.1	0.0	0.3	0.2
Halosaccion glandiforme	0.0	0.2	0.0	0.2	0.3	0.3	0.8	0.7	0.4				
Pylaiella littoralis	4.0	2.8	5.4	16.7	1.5	12.6	2.2	0.5	0.0**	3.2	5.8*	1.7	2.4
Ralfsia pacifica	0.0	7.2	5.6	1.7	0.6	1.2	0.0	0.6	0.4**	9.0	1.7	1.5	0.0**
Ulva–Monostroma spp.	1.6	15.5**	6.0	9.7	1.9	1.2	0.7	0.3	0.2*				
Invertebrates													
Chthamalus dalli (%)	1.3	0.5	0.1	0.9	0.1	0.1	0.2	0.3	0.1	0.1	0.0	0.1	0.0
Littorina scutulata (*N*)			0.2	0.8	0.0	1.4**	0.1	0.0	0.7*	11.6	0.6		
Littorina sitkana (*N*)	0.2	0.0			0.1	0.2	0.3	0.0	0.3	0.4	0.0	0.2	0.0
Lottiidae[a] (*N*)	0.4	0.2	0.8	0.0	5.7	6.0	83.8	35.8	84.4	0.0	0.5	1.0	0.4
Onchidella borealis (*N*)	21.8	7.3	3.6	1.0			0.2	0.0	0.1				
Pagurus hirsutiusculus (*N*)	15.0	20.0	22.2	50.5	1.1	1.1	16.8	6.3	4.3**	36.9	14.2**	12.8	8.4
Searlesia dira (*N*)	0.0	1.2*	0.2	0.0	3.0	1.3*	11.6	5.2	12.0				

[a]Unidentified species.

TABLE 9.—Number of live taxa exhibiting significant changes in abundance by type of treatment and tide level in Prince William Sound; LP-WW, low-pressure warmwater wash; HP-HW, high-pressure hot-water wash.

		Number of taxa changing significantly					
		Intertidal level				Percent of signficant changes by treatment	Percent of possible changes by treatment
Treatment or statistic	Type of change	Upper	Mid	Lower	Total		
Corexit	Decrease	2	4	4	10	71	17
	Increase	1	1	2	4	29	7
LP-WW	Decrease	2	8	3	13	72	22
	Increase	0	0	5	5	28	9
HP-HW[a]	Decrease		9	4	13	93	26
	Increase		0	1	1	7	2
Total significant changes	Decrease	4	21	11	36	78	22
	Increase	1	1	8	10	22	6
Percent of significant changes	Decrease	80	95	58			
	Increase	20	5	42			
Percent of possible changes	Decrease	25	44	10			
	Increase	6	2	7			

[a]Upper tide levels were not surveyed in this treatment.

dance for several taxa at Disk Island where two surveys were conducted before treatment (e.g., *Littorina sitkana*, Lottiidae, live *Mytilus trossulus* in the midtide zone in the Corexit test [Table 7] and *Pylaiella littoralis* and *Pagurus hirsutiusculus* in the lower tide zone in the Corexit test [Table 8]).

The failure to include either oiled or non-oiled reference sites to evaluate the no-treatment alternative under both conditions of oiling constitutes a major design flaw because it results in the absence of data to demonstrate that the types and magnitude of changes observed during the various types of treatments did not occur universally throughout the sound at oiled and untreated sites. The approach in the Corexit tests was to determine the effects of the dispersant or beach cleaner by subtracting out the effects of the LP-WW, which was common to both treatments. This approach makes the unjustified assumption that the common treatment (LP-WW) is equally intense in both tests. It

TABLE 10.—Number of assemblage attributes exhibiting significant changes in abundance by type of treatment and tide level in Prince William Sound; LP-WW, low-pressure warmwater wash; HP-HW, high-pressure hot-water wash.

		Number of assemblage attributes changing significantly					
		Intertidal level				Percent of signficant changes by treatment	Percent of possible change by treatment
Treatment or statistic	Type of change	Upper	Mid	Lower	Total		
Corexit	Decrease	0	2	0	2	40	8
	Increase	3	0	0	3	60	12
LP-WW	Decrease	0	5	0	5	83	20
	Increase	0	0	1	1	17	4
HP-WW[a]	Decrease		6	3	9	100	45
	Increase		0	0	0	0	0
Total significant changes	Decrease	0	13	3	16	80	23
	Increase	3	0	1	4	20	6
Percent of significant changes by tide level	Decrease	0	100	75			
	Increase	100	0	25			
Percent of possible changes	Decrease	0	43	10			
	Increase	30	0	3			

[a]Upper tide levels were not surveyed in this treatment.

TABLE 11.—Comparison of density and percent cover between epibiota at untreated, unoiled and untreated, oiled rocky intertidal sites in Prince William Sound during 1989. All data are expressed as mean ± SD. Values in parentheses after months are numbers of 0.25-m² quadrats.

Taxon or statistic	Eshamy Bay—untreated and unoiled			Herring Bay—untreated and oiled		
	May (4)	Jun (10)	Sep (10)	May (10)	Jun–Jul (10)	Sep (10)
Algae (% cover)						
Filamentous green algae	0.1 ± 0.3	3.1 ± 6.7	0.9 ± 1.7	0.1 ± 0.3	0.9 ± 2.2	2.5 ± 4.1
Fucus gardneri	88.3 ± 7.7	62.0 ± 22.0	69.7 ± 18.6	40.3 ± 24.9	42.0 ± 26.4	38.6 ± 30.9
Gloiopeltis furcata	0.5 ± 1.0	0.6 ± 0.8	0.2 ± 0.3	0.0 ± 0.0	0.2 ± 0.3	0.0 ± 0.0
Neorhodomela larix	0.0 ± 0.0	1.5 ± 3.4	0.3 ± 0.9	3.3 ± 6.6	0.0 ± 0.0	0.3 ± 0.4
Neorhodomela oregona	0.4 ± 0.3	0.2 ± 0.3	3.4 ± 7.8	0.0 ± 0.0	0.4 ± 0.7	1.0 ± 1.6
Odonthalia floccosa	0.0 ± 0.0	0.0 ± 0.0	0.0 ± 0.0	0.1 ± 0.3	0.2 ± 0.6	0.0 ± 0.0
Pylaiella littoralis	0.0 ± 0.0	0.4 ± 0.9	0.1 ± 0.2	0.0 ± 0.0	0.0 ± 0.0	0.0 ± 0.0
Invertebrates (% cover)						
Balanus glandula	0.0 ± 0.0	0.0 ± 0.0	1.1 ± 1.7	0.0 ± 0.0	0.0 ± 0.0	0.0 ± 0.0
Balanus–Semibalanus spp.	8.2 ± 5.2	0.0 ± 0.0	0.0 ± 0.0	0.0 ± 0.0	0.0 ± 0.0	0.0 ± 0.0
Chthamalus dalli	0.5 ± 0.0	0.3 ± 0.3	0.2 ± 0.3	0.0 ± 0.0	19.8 ± 19.0	0.1 ± 0.2
Mytilus trossulus	14.8 ± 10.0	3.1 ± 3.1	6.3 ± 5.7	0.9 ± 0.6	2.1 ± 1.6	1.3 ± 0.8
Semibalanus balanoides	0.0 ± 0.0	6.7 ± 4.3	6.5 ± 5.2	12.2 ± 11.1	7.0 ± 7.8	26.0 ± 13.9
Invertebrates (number/0.25 m²)						
Anthopleura artemisia	1.8 ± 2.9	0.6 ± 1.3	0.0 ± 0.0	0.0 ± 0.0	0.0 ± 0.0	0.0 ± 0.0
Buccinum baeri	1.8 ± 3.5	0.2 ± 0.4	0.0 ± 0.0	0.0 ± 0.0	0.0 ± 0.0	0.0 ± 0.0
Littorina scutulata	0.0 ± 0.0	123.2 ± 120.3	88.0 ± 77.8	0.0 ± 0.0	0.0 ± 0.0	259.2 ± 260.3
Littorina sitkana	0.0 ± 0.0	137.3 ± 110.5	342.2 ± 352.1	0.0 ± 0.0	0.0 ± 0.0	57.9 ± 62.5
Lottiidae[a]	11.0 ± 3.9	12.4 ± 4.0	28.0 ± 19.2	11.6 ± 14.0	6.7 ± 7.8	33.5 ± 15.6
Nucella lamellosa	3.5 ± 4.7	2.7 ± 4.6	0.6 ± 1.6	0.0 ± 0.0	0.0 ± 0.0	0.0 ± 0.0
Nucella lima	0.0 ± 0.0	0.1 ± 0.3	0.2 ± 0.4	2.7 ± 2.9	3.2 ± 2.9	1.2 ± 1.0
Pagurus hirsutiusculus	2.0 ± 1.8	25.7 ± 58.9	27.6 ± 23.6	0.8 ± 1.0	0.8 ± 1.3	7.7 ± 10.5
Invertebrates (number dead/0.25 m²)						
Mytilus trossulus	21.5 ± 16.4	22.2 ± 23.7	13.4 ± 10.1	0.2 ± 0.6	0.0 ± 0.0	3.7 ± 2.5
Extent of oiling						
Oil cover (%)	0.0 ± 0.0	0.0 ± 0.0	0.0 ± 0.0	57.0 ± 27	0.0 ± 0.0	2.3 ± 4.6
Oil scale	0.0 ± 0.0	0.0 ± 0.0	0.0 ± 0.0	4.4 ± 0.5	1.7 ± 1.3	1.5 ± 2.0

[a]Unidentified species.

also assumes that only declines in population and assemblage attributes reflecting negative impacts will be observed and implies that increases reflect beneficial effects. In fact, in several instances, populations or assemblage attributes at the Corexit sites declined less than at the LP-WW sites or improved, suggesting that Corexit ameliorated the LP-WW effects. In reality, the LP-WW accompanying the Corexit 7664 test was considerably less rigorous than when applied by itself.

However, data collected for baseline and recovery studies conducted for Exxon Corp. (ERC et al. 1990a, 1990b) have been tabulated to simulate untreated non-oiled or untreated oiled reference sites (Table 11). All of these simulated reference sites had a biota strongly resembling that reported for the treated sites. However, all were truly rocky, whereas all of the treated sites were boulder–cobble. Eshamy Bay and Herring Bay, respectively, were selected to represent the four untreated non-oiled and two untreated oiled sites examined in the larger studies.

These data provide strong indications that the dramatic changes described herein for several factors at the treated sites did not occur at either type of simulated reference site during the period in which these tests were conducted (Table 11). At both Eshamy Bay and Herring Bay, approximately 60% of the taxa increased from May to September. In contrast, the average number of taxa increasing at the treated sites was only 17%. Moreover, all of the potentially significant changes in abundance at the simulated reference sites were increases whereas only 20% of the significant changes at the treated sites were increases (Tables 6–8).

Another flaw is that monitoring studies were not planned to assess chronic effects of chemical exposure or long-term effects. Failure to include such studies implies an assumption that chronic or long-term effects would be negligible. In the case of

HP-HW, the ongoing National Oceanic and Atmospheric Administration (NOAA)-sponsored monitoring program in the sound has shown that this assumption was definitely unfounded (e.g., Houghton et al. 1993a, 1993b); effects of HP-HW remained detectable in 1993.

Treatment Effects

The findings of this analysis of effects for shoreline treatment methods used or considered for use in the sound in 1989 appear to concur with conclusions regarding previous similar applications of these methods following several earlier oil spills. The American Petroleum Institute (API 1985) and Boesch et al. (1974) both reported a consensus among "shore ecologists" that neither steam nor dispersants should be used to remove oil from most beaches.

Temperature.—During these studies, it appears that the biota exposed to LP-WW and chemical cleaners generally was not subjected to lethally high temperatures. Although target temperatures (32 and 60°C) and pressure varied considerably between the Corexit and LP-WW components of the study, the temperatures to which the organisms were exposed were generally fairly similar (13.5 and 14.5°C) but somewhat above ambient temperature (8–9°C). These realized temperatures are well below general short-term thermal tolerances for most intertidal species (i.e., probably above 35°C; Gunter 1957; Naylor 1965).

In contrast, it is clear from a brief review of the literature that temperatures used in HP-HW will cause widespread mortality. Gunter (1957) , Naylor (1965), and Kinne (1970) indicate that most intertidal invertebrates subjected to temperatures exceeding 45°C for even short periods will suffer thermal shock and die. Littorine snails appear to have greater thermal tolerance than most other invertebrates; several species tolerate temperatures up to 47.9°C. The indicated thermal limit for a common mussel *Mytilus edulis* is 40.8°C. Graham et al. (1975) found that exposure to 40°C seawater for about 15 min killed up to 95% of the specimens of the mussel *M. "edulis"* (probably *M. galloprovincialis*). However, considering the fairly wide latitudinal range of those species, this limit may be high for the mussel species found in the sound (probably *M. trossulus*), which has a more northerly distribution (McDonald and Koehn 1988; Kepel and Ozolinsh 1992; Seed 1992). Tsuchiya (1983) reported mass mortality for intertidal populations of the mussel *M. "edulis"* (probably *M. trossulus*) in Japan following exposure to abnormally hot weather, during which body temperatures of the animals rose to at least 40°C. Fraenkel (1960) reported thermal limits of 40 to 41°C for the snail *Littorina littorea* and 36°C for the hermit crab *Pagurus longicarpus*. Later, he reported that the Mediterranean snail *Littorina neritoides* tolerated 1- to 2-h exposures to temperatures of 46 to 47°C when submerged, but 48 to 49°C when in air (Fraenkel 1961). Finally, he reported that thermal tolerance correlated positively with the elevation at which a species normally occurred, that is, high intertidal species had higher tolerance than low intertidal or subtidal species (Fraenkel 1966). For Japanese snails, thermal tolerance for high intertidal snails ranged up to 48.5°C, whereas low intertidal species only tolerated 42 to 43°C and those at the water's edge, only 39°C.

Dawson (1966) indicated that although some algae exhibit lethal limits in air similar to those reported above for invertebrates, they generally have lower tolerances when submerged in seawater (e.g., 42°C in air versus <35°C in seawater for *Bangia*, a notably tolerant alga). More recently, Lüning and Freshwater (1988) reported temperature limits for a number of algae and seagrasses common in the northeastern Pacific. Lethality was defined as the inability of a plant to maintain dissolved oxygen at or above 100% saturation in 10°C seawater after a 1-week exposure to a specified elevated temperature. They tested at 15, 18, 20, 23, 25, 28, and 30°C. The only species that survived a 1-week exposure to 30°C was eelgrass *Zostera marina*, which extends only into the lower intertidal zone in Prince William Sound. Among the dominant algae in middle intertidal rock in the sound, *Fucus gardneri* survived only up to 23°C; *Neorhodomela larix*, *Odonthalia floccosa*, *Mastocarpus papillatus* (gametophytic crust) and *Ulva fenestrata* survived only up to 25°C; and the sporophyte of *M. papillatus* survived only to 28°C. Algae examined in their study showed a strong positive correlation between normal intertidal elevation occupied by a species and its thermal tolerance, generally agreeing with the findings reported by Fraenkel (1966).

Since the *Torrey Canyon* oil spill, investigators have consistently reported that steam cleaning destroys the flora and fauna of the treated areas (Nelson-Smith 1972; Boesch et al. 1974; NRC 1985). Although it is not clear whether these discussions of steam cleaning include what has been termed, in Prince William Sound, high-pressure hot-water wash (i.e., 60°C seawater at about 690 kPa), it is clear from literature on thermal tolerance for intertidal organisms that the target temperature for

HP-HW is in a range that should cause severe damage. It is also clear that damage of a degree similar to that described for the *Torrey Canyon* cleanup occurred at the HP-HW sites in Prince William Sound. Moreover, Houghton et al. (1993a, 1993b) provide substantial evidence that this type of damage was widespread in parts of the sound that were oiled in 1989.

Physical and mechanical disturbance.—Observations of displacement and mortality in clams at Disk Island and mussels at Ingot Island suggest that damage from physical and mechanical effects and flushing may be substantial. Some of the declines observed in littorine snails may have been the result of flushing. Large numbers of mussels were dislodged in all of the tests; many also had cracked shells. However, most dislodged but undamaged mussels showed an unequivocal ability to reattach following the Corexit and LP-WW tests (Houghton et al. 1990a; Lees et al. 1990).

A large proportion of the clams in mixed-soft sediments at Disk Island were excavated by the treatment in the Corexit plot (Lees et al. 1990). Many of the excavated specimens were crushed or inactive and had not been able to rebury during several tidal cycles. However, this study was not designed in a manner that allowed discrimination among the thermal, chemical, and physical effects associated with the two treatment methods, so the causative factors could not be identified.

Rolan and Gallagher (1991) ascribed extensive damage to the intertidal biota from massive mechanical cleanup activities following the *Esso Bernicia* spill at Sullom Voe in the Shetland Islands. They reported that the damaged areas had not recovered 9 years later, whereas the untreated areas had recovered substantially within 1 year.

Chemical effects.—Of the types of treatment examined, dispersant and beach cleaner treatments appeared to be accompanied by the smallest number of significant changes in abundance or assemblage attributes. However, Fiocco et al. (1991) demonstrated that Corexit 9580 only marginally enhanced beach cleaning at the Disk Island site and concluded that Corexit 7664 was ineffective in removing oil at Ingot Island.

Moreover, the occurrence of a large number of dead and moribund clams on the surface of the sediment in the one dispersant test that included mixed-soft substrate (Corexit 9580) is reminiscent of dispersant effects reported by several authors and raises concerns. Smith (1968), in describing the effects of the cleanup following the *Torrey Canyon* spill, indicated that the intertidal and subtidal biota were damaged more at sites exposed to a variety of first-generation dispersants than at sites where these were not used; BP1002 was the most extensively applied dispersant. Heavy mortality was reported following surfacing by clams (*Mactra* and *Ensis*) and other burrowing and surface-dwelling invertebrates in soft subtidal habitats. It is now widely accepted that the mortality from these dispersants was caused by both the surfactants, usually poorly biodegradable, and the organic solvents (e.g., kerosene) that aid in mixing the surfactant into the oil to achieve dispersion, and that the solvents caused the greatest amount of toxicity. It is also recognized that the newer generations of chemicals have been formulated to be far less toxic than the first-generation products.

Consequently, although two studies have reported emergence of clams, polychaetes, or both following exposure to dispersants, most of the animals reportedly were able to rebury. For example, Boehm et al. (1985) reported that substantial numbers of clams and polychaetes moved to the surface of the sediment following exposure to about 50 μL/L crude oil dispersed with approximately 2% Corexit 9527 in the Baffin Island Oil Spill Project, but subsequently, Cross and Thomson (1982) reported that most of the polychaetes and bivalves ultimately reburied and that the loss was negligible. Additionally, Dekker and van Moorsel (1987) reported that exposing the clams *Macoma balthica* and *Cerastoderma edule* to controlled application of unspecified concentrations of Finasol OSR5 in artificial mud flats caused the clams to move temporarily to the surface of the sediment, but did not cause elevated mortality. They reported that most of the clams that surfaced in their mesocosm studies eventually reburied. They hypothesized, however, that under normal conditions on a natural mud flat, very few of the surfaced clams would have escaped predation by birds, crabs, or fish. We also believe, considering the abundance of sea stars and other predators and scavengers, that reburial is an unlikely scenario in Prince William Sound.

However, the mode of application used in the studies in Prince William Sound (spraying the substrate and biota directly with a full-strength or somewhat dilute solutions of the chemical) differs substantially from that used in the studies cited above (exposure of immersed animals to a highly dilute solution). After review of toxicity data for marine invertebrates and considering the mode of application, especially for Corexit 9580 M2, it is not surprising that clams were observed on the surface of the sediment. Generally, 96-h LC50s (median lethal concentration) for clams after exposure to later-generation dispersants such as Corexit 7664 and 9527 range from 0.1 to 1,500 μL/L (Tables 3–5 in NRC 1989). For the Pacific littleneck clam *Protothaca staminea*, a dominant clam in intertidal sediments in Prince William Sound and one that surfaced following the Corexit 9580 M2 test, published median lethal times for Corexit 9527 are more than 10 d at 10 μL/L and 3–5 d at 100 μL/L. It seems unlikely that full-strength beach cleaner sprayed directly on substrate exposed by low tide would be diluted to as low as 10,000 μL/L (a 100-fold dilution) during the 1-h presoak. In view of the toxicity data cited above, it seems likely that exposure to such high concentrations of Corexit 9580 M2 for even a brief period would cause substantial stress or mortality to clams and other infaunal organisms that are even less protected from dermal contact.

Viewed from this perspective, it appears that the application of dispersants directly onto exposed beaches at the dosages adopted in Prince William Sound (4,000 μL/L for Corexit 7664 at Ingot Island and 1,000,000 μL/L for Corexit 9580 M2 at Disk Island) is inappropriate and inconsistent with existing toxicity data. For Corexit 7664, a chemical for which the 96-h LC50 for mussels is 900 to 1,400 μL/L (NRC 1989), the chemical was applied at 4,000 μL/L for a 1-h presoak directly to an exposed beach supporting mussels. Available toxicity data for seagrasses, clams, a variety of crustaceans including amphipods and isopods, and sculpins (e.g., NRC 1989) indicate that these concentrations could have been damaging. However, the studies discussed herein were not designed in a manner that allowed rigorous enumeration of these taxa and even catastrophic damage would not have been detected.

Conclusions for this analysis can be summarized as answers to several questions.

(1) Were the treatments accompanied by biological damage?—Damage, as defined for these studies, was manifest as (1) a significant degree of reduction in one or more assemblage or population attributes or (2) increases in the percentage of dead mussels or *Fucus*. Although the type and degree of damage varied by type of treatment, cases of significant damage accompanied all types of treatment examined. In the case of the HP-HW tests, extensive and conspicuous mortality in dominant species of the intertidal assemblage was observed. In the case of the LP-WW and chemical tests, effects were generally more moderate and subtle.

(2) Where damage occurred, was it a response to thermal or chemical exposure or physical and mechanical disturbance?—All of the studies involved comparisons of combinations of two or three disturbance factors, namely, thermal and physical effects or chemical, thermal, and physical effects. However, the programs were not designed in a manner that allows the relative importance of potential causes of damage to be evaluated separately. Thus, the data set is somewhat inadequate for assessing this question. However, it includes some useful anecdotal information.

Although systematic monitoring of temperature gradients was not included in the design, limited data on temperature gradients downslope and across the test plots collected during the Corexit 7664 study suggest that the LP-WW components of the studies probably did not cause significant thermal impacts in the intertidal biota. In contrast, temperature was probably the principal cause of the significant mortality observed in the dominant plants and grazing and filter-feeding animals during the HP-HW study.

Observations of displacement and mortality in clams at Disk Island and mussels at Ingot Island suggest that physical effects may be substantial. Many specimens were dislodged or excavated and subsequently became crushed or inactive. Mussels demonstrated a greater ability than clams to become reestablished following these disturbances. Based on published toxicity literature for dispersants and beach cleaners, there is an appreciable potential that the observed displacement and mortality in clams at Disk Island is a result of chemical toxicity and that mortality in other organisms could have gone undetected.

(3) Which type of treatment caused the least amount of damage?—Of the types of treatment examined, dispersant and beach cleaner treatments were accompanied by the smallest number of significant changes in species abundance and assemblage attributes. However, the studies were not designed appropriately to detect effects in taxa most sensitive to these chemicals. LP-WW was accompanied by a moderate number of changes, whereas HP-HW was accompanied by the highest percentage of changes. Nearly all of the changes associated with HP-HW were decreases. Although it was not studied, the results of other studies (Houghton et al. 1991a, 1991b, and 1993a, 1993b) suggest that the no-treatment alternative would be accompanied by the smallest number of short- and long-term changes to the intertidal biota.

(4) What are the long-term consequences of the types of treatment? Was the observed damage long

lived?—Generally, the severity and persistence of the treatment effects appeared to vary by treatment type. HP-HW caused the most severe and persistent effects; denuded rock substrate was still a common sight at middle and upper tidal elevations in the sound in 1993 (Houghton et al. 1993a, 1993b). Effects of LP-WW were apparently less severe and persistent. The long-term consequences of the dispersant and beach cleaner applications cannot be evaluated because long-term monitoring has not been conducted at these sites.

Based on these conclusions, the following recommendations seem to be justified.

- High-pressure hot-water wash should not be used except locally on high-value habitats such as bird rookeries or marine mammal haulouts where and when the risk of contaminating a valuable resource is high.
- Dispersants and beach cleaners should not be applied directly to intertidal habitat until these chemicals have been demonstrated to be safe by realistic field studies conducted on appropriate substrates and biological assemblages in a manner that permits assessing the taxa most at risk from exposure to the specific chemical or the chemical–oil mixture.
- Proposed shoreline cleanup methodologies should be designed with due consideration to available information on the lethality of all potentially lethal elements of the approach (temperature, pressure, chemical toxicity, mechanical and physical damage). Before its actual implementation, the biological costs and benefits of cleanup and no-action alternatives should be compared with a rigorous ecological risk assessment.
- Methods evaluations should be carefully designed and include adequate control and reference sites. Sampling design should be prepared by a team of scientists and engineers with strong expertise in sampling design and statistics, appropriate types of biological assemblages, hydrocarbon sampling and analysis, fate and behavior of spilled crude oil, geomorphology, and cleanup methodologies.

Acknowledgments

We acknowledge the technical, logistical, and funding support of Exxon Corp., which funded and designed these studies. Field crews during the four studies performed admirably under the adverse conditions prevalent during the grueling field studies. These crews included personnel from Dames & Moore (David Erikson and Robert Morrow) and ERC (now Ogden) Environmental and Energy Services Co. (Steven Landino, Marion Patton, and Howard Teas). In addition, we are indebted to the skippers and crews of the various fishing vessels that acted as our transportation, dormitories, and laboratories while we were involved in these studies. Their efforts, enthusiasm, and fervor for collection of accurate and helpful information were quite helpful. We also thank Robert Smith of EcoAnalysis, Inc. for his suggestions and able statistical assistance. Finally, we appreciate the constructive criticism and suggestions of an anonymous reviewer and June Lindstedt-Siva.

References

API (American Petroleum Institute). 1985. Oil spill response options for minimizing adverse ecological effects. API, Washington, DC.

Boehm, P. D., W. Steinhauer, A. Requejo, D. Cobb, S. Duffy, and J. Brown. 1985. Comparative fate of dispersed and untreated oil in the arctic: Baffin Island oil spill studies 1980–83. Pages 561–569 *in* Proceedings, 1985 oil spill conference. American Petroleum Institute, Washington, DC.

Boesch, D. F., C. H. Hershner, and J. H. Milgram. 1974. Oil spills and the marine environment. Report to the Ford Foundation's energy policy project. Ballinger, Cambridge, Massachusetts.

Cross, W. E., and D. H. Thomson. 1982. Macrobenthos—1981 study results. Baffin Island oil spill (BIOS) working report. Report to Environment Canada, Ottawa.

Dawson, E. Y. 1966. Marine botany, an introduction. Holt, Rinehart and Winston, New York.

Dekker, R. , and G. W. N. M. van Moorsel. 1987. Effects of different oil doses, dispersant, and dispersed oil on macrofaunal in model tidal flat ecosystems. Pages 117–13 *in* J. Kuiper and W. J. Van Den Brink, editors. Fate and effects of oil in marine ecosystems. Martinus Nijhoff, Dordrect, Netherlands.

Edgington, E. S. 1987. Randomization tests. Marcel Dekker, New York.

ERC (Environmental and Energy Services Co.), Pentec Environmental, W. B. Driskell, and Dames & Moore. 1990a. Intertidal and shallow subtidal baseline studies in Prince William Sound, Alaska, study 5. Draft data report of Dames & Moore (Technical Report series—the *Exxon Valdez* oil spill studies) to Exxon Corp., New York.

ERC (Environmental and Energy Services Co.), Pentec Environmental, W. B. Driskell, and Dames & Moore. 1990b. Intertidal and shallow subtidal baseline studies in Prince William Sound, Alaska, study 4. Draft data report of Dames & Moore (Technical Report series—the *Exxon Valdez* oil spill studies), to Exxon Corp., New York.

Fiocco, R. J., and six coauthors. 1991. Development of Corexit 9580—a chemical beach cleaner. Pages 395–400 *in* Proceedings, 1991 international oil spill conference: prevention, behavior, control, cleanup.

American Petroleum Institute Publication 4529, Washington, DC.
Fraenkel, G. 1960. Lethal high temperatures for three marine invertebrates: *Limulus polyphemus*, *Littorina littorea* and *Pagurus longicarpus*. Oikos 11:171–182.
Fraenkel, G. 1961. Resistance to high temperatures in a Mediterranean snail, *Littorina neritoides*. Ecology 42: 604–606.
Fraenkel, G. 1966. The heat resistance of intertidal snails at Shirahama, Wakayama-ken, Japan. Publications of the Seto Marine Biological Laboratory 14:185–195.
Graham, J. W., R. W. Moncreiff, P. H. Benson, and J. N. Stock. 1975. Heat treatment for the control of marine fouling at coastal electric generating stations. Pages 926–930 *in* Proceedings of IEEE (Institute of Electrical and Electronics Engineers) Ocean '75. IEEE, New York.
Gunter, G. 1957. Temperature. Geological Society of America Memoir 67:159–184.
Houghton, J. P., D. E. Erikson, and W. B. Driskell. 1990b. Biological effects of beach treatment using an Omni barge on the nearshore zone of Herring Bay, July 1989. Report of Dames & Moore (Technical Report series—the *Exxon Valdez* oil spill studies) to Exxon Corp., New York.
Houghton, J. P., A. K. Fukuyama, D. C. Lees, and W. B. Driskell. 1993a. Evaluation of the 1992 condition of Prince William Sound littoral biota following the *Exxon Valdez* oil spill and subsequent shoreline treatment. Report of Pentec Environmental and Ogden Environmental and Energy Services to National Oceanic and Atmospheric Administration, Washington, DC.
Houghton, J. P., A. K. Fukuyama, D. C. Lees, H. J. Teas, III, T. A. Ebert, and W. B. Driskell. 1993b. Evaluation of the condition of Prince William Sound shorelines following the *Exxon Valdez* oil spill and subsequent shoreline treatment. NOAA (National Oceanic and Atmospheric Administration) Technical Memorandum NOS (National Ocean Service) ORCA 67, volume 2, Washington, DC.
Houghton, J. P., D. C. Lees, and W. B. Driskell. 1990a. Ingot Island Corexit 7664 beach treatment biological evaluations, May 17 to June 3, 1989. Dames & Moore (Technical Report series—the *Exxon Valdez* oil spill studies), Seattle.
Houghton, J. P., D. C. Lees, W. B. Driskell, and A. J. Mearns. 1991b. Impacts of the *Exxon Valdez* spill and subsequent cleanup on intertidal biota—1 year later. Pages 467–475 *in* Proceedings, 1991 international oil spill conference. American Petroleum Institute Publication 4529, Washington, DC.
Houghton, J. P., D. C. Lees, W. B. Driskell, H. J. Teas, III, H. Cumberland, and T. A. Ebert. 1991b. Evaluation of the condition of the intertidal and shallow subtidal biota in Prince William Sound following the *Exxon Valdez* oil spill and subsequent shoreline treatment. Report by ERC Environmental and Energy Services and Pentec Environmental, to National Oceanic and Atmospheric Administration, Washington, DC.
Kepel, A. A., and A. V. Ozolinsh. 1992. Morphometric analysis of species from the genus *Mytilus* (Mollusca, Bivalvia, Mytilidae) of Russian seas. Zoologichesky Zhurnal 71(9):33–40.
Kinne, O. 1970. Environmental factors. Marine ecology: a comprehensive, integrated treatise on life in oceans and coastal waters, volume 1, part 1. Wiley-Interscience, London.
Lees, D. C., J. P. Houghton, and W. B. Driskell. 1990. An evaluation of biological effects of exposure to Corexit 9580 M2 in the biota of the intertidal and shallow subtidal zones. Final data report of Dames & Moore (Technical Report series—the *Exxon Valdez* oil spill studies) to Exxon Corp., New York.
Lüning, K., and W. Freshwater. 1988. Temperature tolerance of northeastern Pacific marine algae. Journal of Phycology 24:310–315.
McDonald, J. H., and R. K. Koehn. 1988. The mussels *Mytilus galloprovincialis* and *M. trossulus* on the Pacific coast of North America. Marine Biology 99:111–118.
Nauman, S. A. 1991. Shoreline cleanup: equipment and operations. Pages 141–148 *in* Proceedings, 1991 international oil spill conference: prevention, behavior, control, cleanup. American Petroleum Institute Publication 4529, Washington, DC.
Naylor, E. 1965. Effects of heated effluents upon marine and estuarine organisms. Advances in Marine Biology 3:63–103.
Nelson-Smith, A. 1972. Oil pollution and marine ecology. Elek Science, London.
NRC (National Research Council). 1985. Oil in the sea: inputs, fates, and effects. National Academy Press, Washington, DC.
NRC (National Research Council). 1989. Using oil spill dispersants on the sea. National Academy Press, Washington, DC.
Rolan, G. R., and R. Gallagher. 1991. Recovery of the intertidal biotic communities at Sullom Voe following the *Esso Bernicia* oil spill of 1978. Pages 461–466 *in* Proceedings, 1991 international oil spill conference: prevention, behavior, control, cleanup. American Petroleum Institute Publication 4529, Washington, DC.
Seed, R. 1992. Systematics, evolution and distribution of mussels belonging to the genus *Mytilus*—an overview. American Malacological Bulletin 9(2):123–137.
Smith, J. E. 1968. *Torrey Canyon* pollution and marine life: a report by the Plymouth Laboratory of the Marine Biological Association of the United Kingdom. Cambridge University Press, Cambridge, UK.
Tsuchiya, M. 1983. Mass mortality in a population of the mussel *Mytilus edulis* L. caused by high temperature on rocky shores. Journal of Experimental Marine Biology and Ecology 66:101–111.
Wolfe, D. A., and six coauthors. 1996. Toxicity of intertidal and subtidal sediments contaminated by the *Exxon Valdez* oil spill. American Fisheries Society Symposium 18:121–139.

American Fisheries Society Symposium 18:349–361, 1996

Growth and Loss of Tagged Individuals of the Predatory Snail *Nucella lamellosa* in Areas within the Influence of the *Exxon Valdez* Oil Spill in Prince William Sound

THOMAS A. EBERT
San Diego State University, Department of Biology, San Diego, California 92182–4614, USA

DENNIS C. LEES
Ogden Environmental and Energy Services, Inc., 5510 Morehouse Drive, San Diego, California 92121–1709, USA

Abstract.—Growth and recapture rates of tagged individuals of the predatory frilled dogwinkle *Nucella lamellosa* were studied at areas in Prince William Sound, Alaska, that were within the influence of the *Exxon Valdez* oil spill. Snails were individually tagged with small plastic fish tags in spring and summer 1991, measured with calipers, and released at three types of sites: (1) unoiled, (2) oiled, and (3) oiled and washed with hot water. Sites were searched again in September 1991 and in July 1992 and all tagged individuals were measured. The Brody–Bertalanffy function was used as the basic growth model and was modified to accommodate seasonal variation in growth. Analysis showed a strong seasonal component to growth, with minimum growth during the winter. Growth was highest at sites that had not experienced oiling and was lowest at oiled locations. When growth of frilled dogwinkle was adjusted for degree of exposure to open sea conditions, growth was similar at unoiled and "oiled-and-washed" sites and higher than at oiled sites. Recapture rates of tagged snails were highest at unoiled sites and lowest at oiled locations; recapture rates at sites washed with hot water were intermediate. We conclude that oil from the *Exxon Valdez* depressed growth and recapture rates of tagged frilled dogwinkle in Prince William Sound and that treatment of oiled beaches with hot water improved growth conditions for frilled dogwinkle but did not return the sites to an unoiled state with respect to recapture rates.

Evaluation of the impact of a disaster, with a goal of predicting recovery of animal and plant species to particular distributions and abundances, can be viewed as a set of problems in population analysis that includes age- or size-specific growth, reproduction, and survival. Individual growth is a measure of general well-being in both vertebrates (e.g., Brisbin et al. 1986) and invertebrates (e.g., Black 1973; Brown and Howard 1985; de Nicola Giudici et al. 1987; Salazar and Salazar 1991). For many species, growth is directly related to reproductive output (e.g., Wootton 1979; Grahame and Branch 1985; Hughes 1986; MacDonald and Bourne 1987; Hiyama et al. 1988; Richards and Schnute 1990).

The following study provides preliminary information on growth and recapture rates of tagged individuals of the frilled dogwinkle or drill *Nucella lamellosa* in Prince William Sound, Alaska, for a 1-year period starting 2 years after the *Exxon Valdez* oil spill of 24 March 1989. This snail is the most abundant of several conspicuous intertidal predatory snails that are common in rocky intertidal habitats in the sound. The species prefers barnacles but also preys on mussels and other molluscs (Morris et al. 1980). Eggs are laid in capsules and a planktonic stage is lacking so that natural repopulation rates following a disaster are expected to be extremely low.

Study Sites

Study sites were located in western Prince William Sound within the area most heavily exposed to oil from the *Exxon Valdez* oil spill (Figure 1). The sites fall into three treatment categories: (1) sites that escaped direct oil contamination, which we call "no oil" or "unoiled"; (2) sites that were oiled and then left alone, which we call "oiled" or "oiled only"; and, (3) sites that were oiled and subsequently washed with high-pressure hot water, which we call "oiled and washed" or "washed with hot water." Because the hot-water process caused redistribution of oil into areas that had not been contaminated before treatment (see Driskell et al. 1996; Houghton et al. 1996; both this volume), the designation "washed with hot water" rather than "cleaned" is a more accurate description of the treatment. Washing with hot water was the most physically rigorous of shoreline treatments employed in Prince William Sound. It was used extensively throughout the sound to clean large areas of shoreline. This method involved washing the sub-

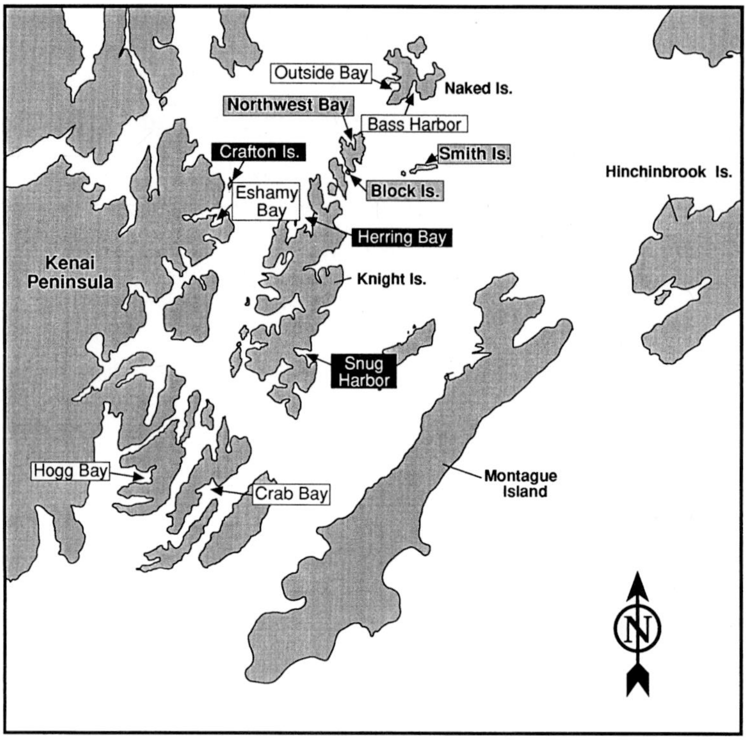

FIGURE 1.—Location of study sites in Prince William Sound, 1989–1991. "No oil" sites, with white backgrounds, are Bass Harbor, Crab Bay, Eshamy Bay, Hogg Bay, and Outside Bay; "oiled" sites, with black backgrounds, are Herring Bay, Snug Harbor, and Crafton Island; "oiled and hot-washed" sites, with gray backgrounds, are Smith Island, Block Island, and a rocky islet in Northwest Bay.

strate with seawater heated to 60°C and sprayed onto the substrate at a pressure of about 7 kg/cm^2 and a flow rate of about 1900 L/min through six nozzles on a 6-ft-long spray head. Washed sites also often were subjected to a variety of other treatments and documentation of these are, in some cases, in conflict. The detailed justification of our categories is presented elsewhere (Houghton et al. 1993b).

Work we present here was carried out at the mid-tidal elevation at most sites. However, at Bass Harbor, the tagged animals were found at the mid- and upper intertidal zone, whereas at Smith Island, the mid-intertidal habitat was mostly a boulder–cobble field and frilled dogwinkle occurred on bedrock outcrops in the lower mid-intertidal zone. Exposure to wave force varied from protected at several sites to more exposed at Smith Island (Table 1); most of the sites were protected. Exposure to insolation and desiccation (aspect) also varied, but more than half of the sites faced in a southerly direction. The remainder of the sites were north or west facing (Table 1).

In 1991, 2 years after the oil spill, frilled dogwinkles were dense at sites that had not been exposed to oil but tended to be uncommon or absent at sites that had been oiled and either left alone or washed with hot water. Small isolated populations of adult frilled dogwinkle were observed in restricted pockets at Herring Bay (oiled only) and on a rocky islet in Northwest Bay (oiled and washed) and appeared distributed throughout the lower intertidal zone and on bedrock outcrops that extended up to about +1.5 m above mean lower low water (MLLW). Food items for the snails were not equally abundant at all sites. Barnacles generally were more abundant at the unoiled sites than at oiled only and hot-washed sites. Mussels generally ranged from uncommon to absent at all of the sites that had been exposed to oil.

TABLE 1.—Physical and biological characteristics of sites in Prince William Sound used to study effects of oiling and shoreline treatment on growth and recapture of frilled dogwinkle. Exposure: 1 = very protected and 4 = moderately exposed; barnacle density in 1991: 1 = light and 3 = dense cover; mussel density: 0 = none and 2 = light cover.

Treatment	Site	Exposure	Aspect	Barnacles 1991	Mussels 1991
No oil	Bass Harbor	3	South	2	1
	Outside Bay	2	North	3	1
	Eshamy Bay	2	South	2	2
	Crab Bay	2	South	1	1
	Hogg Bay	3	South	3	2
Oiled	Crafton Island	1	West	1	0
	Herring Bay	1	South	2	0
	Snug Harbor	1	South	2	0
Oiled and hot washed	Northwest Bay islet	1	West	2	0
	Block Island	2	North	1	0
	Smith Island	4	North	1	0

Methods

Hydrocarbon Analysis

Samples of sediment and tissues of several common intertidal species were collected for analysis of polycyclic aromatic hydrocarbons (PAH) at all of the study sites with varying frequency. A subset of the sites was sampled in 1989 and all were sampled at least two times in subsequent years. Detailed presentation of the methods is given elsewhere (Houghton et al. 1993a). Sediment samples were composites from the upper 5 cm of substrate at five randomly chosen locations along the beach contour. Animal samples were composites of sufficient individuals to provide a minimum of 10 g of wet tissue. All samples were frozen and shipped for analysis to the Institute for Environmental Studies, Louisiana State University (Baton Rouge, Louisiana). Analytical methods were modified from procedures of Krahn et al. (1988).

Snail Tagging

In spring and summer 1991, individually numbered tags designed for fish fingerlings (Floy Tag and Manufacturing, Inc., Seattle, Washington 98105) were used to mark frilled dogwinkle in resident stocks where possible, or, at sites where populations had been decimated, to tag whelks imported from a reference population in Hogg Bay, an unoiled site in southwestern Prince William Sound (Figure 1). Tagging was done during April–May and July 1991 by permitting the shells to air dry, applying a small spot of marine epoxy (Seagoin' Poxy putty, Termalite Plastics, Newport Beach, California) to a shell, and then pressing a tag into the spot. Floy fingerling tags have a small hole at one end, and pressing the tag into the epoxy forced resin up through the hole to form a bead that riveted the tag to the shell. Following tagging, total length was measured to the nearest 0.01 cm with calipers. Samples of animals tagged in May were measured in July 1991; samples from most sites were again measured in September 1991 and July 1992. Time periods of up to 14 months were used to evaluate growth and recapture rates of frilled dogwinkle among the three treatments.

Calculation of Growth Parameters

The Brody–Bertalanffy growth model (von Bertalanffy 1938; Brody 1945; Ricker 1975) was used to describe growth of tagged frilled dogwinkle:

$$S_t = S_\infty(1 - be^{-Kt}). \quad (1)$$

As a difference equation, it can be written

$$S_{t+\Delta t} = S_\infty - (S_\infty - S_t)e^{-K(t+\Delta t)}. \quad (2)$$

One problem with the model is that it is necessary to assume that growth is independent of the time of year over which measurements are made so growth during 4 months of summer is assumed to be the same as growth during 4 months of winter. Clearly this is not a good assumption for animals living in Alaskan waters, therefore a growth equation was used that started with the Brody–Bertalanffy model (equation 1) and added two additional parameters, ϵ and t_A, to modify growth for seasonality during the year (Sager 1982; Sager and Gosselck 1986):

$$S_t = S_\infty(1 - be^{-Kt})\left[1 + \frac{(1-\epsilon)K}{2\pi}\sin 2\pi(t - t_A)\right]. \quad (3)$$

Sager's model is attractive because its second element operates to modify the first (the basic growth model, equation 1) with a component that varies with season. When the parameter $\epsilon = 1.0$, the model has no seasonal component, but seasonality becomes more pronounced as $\epsilon \rightarrow 0$.

Equation 3 was recast as a difference equation using size pairs (S_t and $S_{t+\Delta t}$), a time interval (Δt), and Julian day/365 (t):

$$S_{t+\Delta t} = S_\infty - (S_\infty - S_t)e^{-K(\Delta t)} \cdot \left[1 + \frac{(1-\epsilon)K}{2\pi} \sin 2\pi(t + \Delta t - t_A)\right]. \quad (4)$$

The parameters are

S_∞ = asymptotic size (mm),
K = the growth rate constant (year^{-1}),
ϵ = parameter measuring strength of the seasonal effect (equal to 1 with no effect), and
t_A = parameter that adjusts the time of minimum growth (year).

The parameters S_∞, K, ϵ, and t_A were estimated by nonlinear regression using the NONLIN module in SYSTAT (Evanston, Illinois). The SYSTAT command file for this procedure, which we call Model I, constrains ϵ between 0 and 1:

```
MODEL S2 = A − (A − S1) * EXP(−K * DT)
  * (1 + (1 − E) * K/6 .2831853 * ,
SIN(6.2831853 * (T2 − TA)))
LOSS = (E> = 0 AND E< = 1)
  * (S2 − ESTIMATE) ∧2+,
(E < 0 OR E > 1) * 1E25
```

Analyses also were performed by fixing $\epsilon = 0$, that is, fixing the seasonal effect at its maximum, and constraining t_A between 0 and 1. This is Model II:

```
MODEL S2 = A − (A − S1) * EXP(−K * DT)
  * (1 + K/6.2831853 * ,
SIN(6.2831853 * (T2 − TA)))
LOSS = (TA> = 0 AND TA< = 1)
  * (S2 − ESTIMATE)∧2 +,
(TA < 0 OR TA > 1) * 1E25
```

Finally, a SYSTAT model (Model III) was used that fixed $\epsilon = t_A = 0$, that is, maximum seasonal effect with a minimum at the beginning of the new year:

```
MODEL S2 = A − (A − S1) * EXP(−K * DT)
  * (1 + K/6.2831853 * ,
SIN(6.2831853 * T2))
```

A one-way analysis of variance (ANOVA) was used to compare growth among the three treatments of no oil, oiled, and hot washed. The product ($S_\infty K$), termed ω (Gallucci and Quinn 1979; Appeldoorn 1983; Duineveld and Jenness 1984), was used as the dependent variable:

$$\omega = S_\infty K \text{ (mm/year)}. \quad (5)$$

The parameter ω approximates the instantaneous growth rate of a newly hatched individual in millimeters per year. The larger the value of ω, the faster the growth. Analysis of covariance was used to examine whether additional variance of ω could be explained by covariates such as exposure, mussel density, and barnacle density (Table 1) using treatment as a factor.

Calculation of Recovery Rates of Tagged Individuals

Recapture rates of tagged frilled dogwinkle were estimated for sites from all three treatments. The two time periods considered appropriate were (1) April–May 1991 to June–July 1992 and (2) July 1991 to June–July 1992. The fraction recovered after roughly 12 or 14 months (S) is the number in 1992 divided by the original number released in 1991. Recovery (S) was adjusted to an annual rate (S_A) by

$$S_A = S^{1/\Delta t}, \quad (6)$$

where Δt is the difference in years from tagging in 1991 to July 1992. If all losses were caused by mortality this rate would be called the annual survival rate. However, even though substantial effort was invested in searching sites, the number of tagged snails recovered in 1992 obviously is the minimum number of what might have been present. Also, some tagged individuals may have migrated out of the study area and so were lost to the study, although still possibly alive in 1992. Movements of more than 20 m across rugged, algal-covered habitat were observed over a 2-month period. Accordingly, we call S "recapture rate" rather than "survival rate" and $1 - S$ is "loss rate" rather than "mortality rate."

Annual recapture rates (S_A) grouped by treatments were compared using a nonparametric Kruskal–Wallis ANOVA. Differences among groups represent differences in recapture rates, which result from some combination of mortality and behavioral differences. Analysis of covariance was used to examine whether additional variance of S could be explained by covariates such as exposure, mussel density, and barnacle density (Table 1) using treatment as a factor.

Results

Hydrocarbon Contamination

Patterns of PAH contamination in sediments, mussels *Mytilus trossulus*, and frilled dogwinkle at the study sites (Table 2) demonstrate trends based on the categories of no oil, oiled, and oiled and washed. Total PAH present in the sediments adjacent to the rocky sites where snails were tagged are significantly different in the three categories (df = 2, P = 0.0009). The trend is lowest PAH concentrations at the unoiled sites and highest concentrations in the sediments at sites that were oiled and washed (Figure 2A). Although this general trend is also shown for PAH concentrations in tissues of *M. trossulus* (Figure 2B) and frilled dogwinkle (Figure 2C) these relationships are not significant at α = 0.05. For *M. trossulus*, with df = 2, P = 0.10 and for frilled dogwinkle, P = 0.39. The general conclusion is that total PAH levels in sediments reflect the category descriptions of no oil, oiled, and oiled and washed. There is substantial variation in the *M. trossulus* and frilled dogwinkle measurements, and the signal may be obscured by the high degree of spatial and temporal patchiness in hydrocarbon distribution.

Generally, PAH concentrations in the tissues of mussels were substantially higher than in frilled dogwinkle (Table 2; Figure 2, B and C), which means that biomagnification of hydrocarbons did not appear to occur. Polycyclic aromatic hydrocarbon concentrations were erratic in both species and the correlation between average mussel and frilled dogwinkle tissue concentrations, log transformed, was 0.49 (P = 0.18).

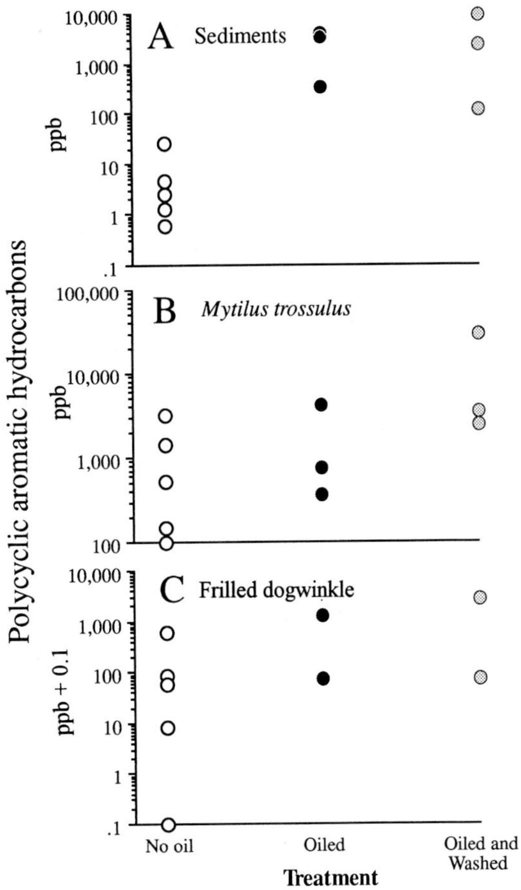

FIGURE 2.—Total polycyclic aromatic hydrocarbon (PAH) concentrations in parts per billion (ppb) in wet weight of sediments (**A**) and dry tissue weights of the mussel *Mytilus trossulus* (**B**) and the frilled dogwinkle (**C**). Values are means from 1990 to 1993 for sediments and mussels and from 1990 for drills. The categories "no oil" (white circles), "oiled" (black circles), and "oiled and hot-washed" (gray circles) are explained in the text.

Growth and Recovery of Tagged Frilled Dogwinkle

The data structure for growth analysis of frilled dogwinkle in Prince William Sound (Table 3) shows the numbers of size pairs and the time intervals separating measurements. A data set for a site is all possible size pairs together with the Julian years starting with January 1991 for both measurements. A single animal that was measured in May 1991 and recaptured in July and September 1991 and in July 1992 would be represented by six size pairs in the data set:

May 1991–July 1991,
May 1991–September 1991,
May 1991–July 1992,
July 1991–September 1991,
July 1991–July 1992,
September 1991–July 1992.

Nonlinear regression was used to estimate parameters. Equation (4) has four parameters, two from the basic Brody–Bertalanffy model, K and S_∞, and two for seasonal adjustment, ϵ and t_A. Our approach will be to determine whether all four parameters must be estimated or whether one or more can be fixed, which would simplify estimation as well as comparisons among treatments. Frilled dogwinkle at Hogg Bay will be used to illustrate our analysis.

There are 420 size pairs for animals at Hogg Bay,

TABLE 2.—Total polycyclic aromatic hydrocarbon (PAH) concentrations in sediments (parts per billion wet weight) and in tissues (parts per billion dry weight) of *Mytilus trossulus* and frilled dogwinkle in July of 1990 to 1993 in Prince William Sound. Further details of hydrocarbon sampling and analysis are given in Houghton et al. (1993a).

Treatment and site	July 1990	July 1991	July 1992	July 1993	Mean
Sediment					
No Oil					
Bass Harbor	0.5	6.0	0.90		2.5
Crab Bay rocky			27		27
Eshamy Bay	1.1	5.8	7.2		4.7
Hogg Bay	0.3		0.95		0.6
Outside Bay rocky	1.6		2.2	0.2	1.3
Oiled					
Herring Bay rocky	8,100	330	5,000	40	3,367
Snug Harbor rocky	12,000	3,900	2,000	1,400	4,825
Crafton Island	53	210	750	350	341
Oiled and hot washed					
Northwest Bay islet	22,000	11,000	2,300	2,600	9,475
Smith Island	35	290	36		120
Block Island	1,500	2,600	4,200	940	2,310
Mytilus trossulus					
No oil					
Bass Harbor	6,470	0	94		3,282
Crab Bay rocky	14	510		1,100	541
Eshamy Bay	1,500	3,700	130	350	1,420
Hogg Bay	200	16	84		100
Outside Bay rocky		85		210	148
Oiled					
Herring Bay rocky	4,400	8,100	600	4,200	4,325
Snug Harbor rocky	1,200	840	860	110	752
Crafton Island	540	170	310	450	368
Oiled and hot washed					
Northwest Bay islet	3,400	9,300	1,200	250	3,538
Smith Island	82,000	5,800		2,400	30,067
Block Island	2,700	2,100	3,800	1,400	2,500
Frilled dogwinkle					
No Oil					
Bass Harbor	8.2				
Crab Bay rocky	617				
Eshamy Bay	88				
Hogg Bay	59				
Outside Bay rocky	0.0				
Oiled					
Herring Bay rocky	74				
Snug Harbor rocky	1,300				
Crafton Island					
Oiled and hot washed					
Northwest Bay islet					
Smith Island	2,800				
Block Island	76				

distributed through times as shown in Table 3. Using Model I and attempting to estimate all four parameters, NONLIN (SYSTAT, Evanston, Illinois) was never able to achieve convergence and satisfy a tolerance even as low as 0.1. The parameter ϵ was constrained between 0 and 1 and, over several hundred iterations, ϵ continued to converge on 0 but at a very slow rate using the SIMPLEX algorithm (Table 4). NONLIN has two nonlinear algorithms, the SIMPLEX and the Quasi-Newton methods. Attempting to estimate four parameters could not be done using the Quasi-Newton method because of overflow errors, and the SIMPLEX method could not converge after several hundred iterations.

Because of nonconvergence when we attempted to estimate four parameters, we fixed $\epsilon = 0$ (Model II), which implies a maximum seasonal effect (equation 4): there is a time during the year when growth ceases. Such an assumption is reasonable for marine animals living at high latitudes (e.g., Sager and Gosselck 1986). For Hogg Bay snails, the residual sum of squares (SSE) increased 0.83 (0.06%) when the number of estimated parameters was reduced from four to three. What is gained by estimating three versus four parameters is sufficient stability to achieve convergence either with the SIMPLEX or Quasi-Newton algorithms.

The scatter of data points around the estimates from Model II with fixed $\epsilon = 0$ (Table 4) indicates a reasonable fit to the data (Figure 3). For all time periods, some individuals showed zero growth similar to results obtained by Spight (1981); these animals clearly have a substantial influence on how fitted lines pass through the data clouds. Variability is substantial and it is not possible with the data at hand to partition this variability into measurement variation, associated with the many persons involved in the study, and biological variability in growth among the snails.

A third analysis (Model III) was performed for the Hogg Bay data by fixing both ϵ and t_A at 0 (Table 4), which reduces the estimation problem to just two parameters, K and S_∞. The increase in SSE over four parameters was only 5.39 or 0.40%, which indicates that assuming that minimum growth occurred at $t_A = 0$ rather than at $t_A = 0.03$ was of little significance in goodness of fit of the model to the data.

A final analysis was done assuming no seasonal effect (equation 1); that is, $\epsilon = 1$. The increase in SSE over the SSE with four parameters was 132.5 or 9.80%, and the increase in SSE over the SSE with two estimated parameters with ϵ and t_A fixed at 0 was 127.11 or 9.36% . It is clear that a seasonal component to growth improves the fit to the data but also that it is reasonable to fix $\epsilon = 0$ and, possibly, also to fix $t_A = 0$.

Growth parameters for frilled dogwinkle at Hogg Bay can be used with equation (3) to draw a growth

TABLE 3.—Data structure for growth analysis of frilled dogwinkle in Prince William Sound, 1991–1992. Numbers are pairs of data values (N) with an initial and final shell length. Samples labeled "–Hogg" are animals that were imported from Hogg Bay; the Herring Bay sample is local plus Hogg imports combined.

Treatment	Site	5/91–7/91	5/91–9/91	5/91–7/92	7/91–9/91	7/91–7/92	9/91–7/92
No oil	Bass Harbor	106	79	59	20	18	13
	Crab Bay	88	142	57	52	25	37
	Eshamy Bay				183	111	66
	Hogg Bay	102	78	36	126	45	33
	Outside Bay	105	165	88	50	24	52
Oiled	Herring Bay	46	16	7	3	6	
	Snug Harbor–Hogg				55	15	2
	Crafton Island–Hogg				206	19	14
Oiled and hot washed	Northwest Bay islet				232	100	86
	Northwest Bay islet–Hogg	1	16	1	163	39	35
	Smith Island–Hogg				47		
	Block Island–Hogg	38	36	9	123	29	31

curve, once parameter b in equation (3) is estimated. Parameter b is $S_\infty - S_c$, where S_c is a correction needed in order to make a specific size match a specific age and date. We have selected the size at hatching, S_0, as 1.0 mm (Spight 1976) and time of hatching as 1 June. These values are approximately correct, but obviously both hatching size and time require further study. S_c was determined by numerical solution in the equation

$$S_0 = S_\infty - (S_\infty - S_c)e^{-Kt_0} \cdot \left[1 + \frac{(1-\epsilon)K}{2\pi} \sin 2\pi(t_0 - t_A)\right]. \quad (7)$$

S_0 is 1.0 mm and t_0 is 152/365 or 0.416438 (152 is June 1st):

$$S_c = -12.931. \quad (8)$$

S_c is similar to t_0 in formulations of the Brody–Bertalanffy equation (Ricker 1975) in the sense that its function is to anchor the growth curve on the time axis; changes in S_c do not change the estimates of K and S_∞.

The growth curve produced with parameters estimated for Hogg Bay (Figure 4) shows depression of growth during the winter and rapid growth during the summer. For comparison, when ϵ is fixed at 1 so there is no seasonal effect (equation 1), asymptotic shell length, S_∞, is about the same as with the more complex growth model (Table 3); however, K is much higher because of the assumption that growth was as good all year as it was during the summer months when most of the measurements were made. After several years, the two models provide very similar estimates of size versus age (Figure 4).

Results of nonlinear regression analysis for frilled dogwinkle at all sites (Table 5) show that, in all cases, Model III (two estimated parameters) had SSE nearly as small as Model II (three estimated parameters), and S_∞ and K were about the same for both models. Fixing both ϵ and t_A equal to zero does not change conclusions about growth, so we will compare treatments using the summary statistic of growth, ω (equation 5).

For unoiled sites ($N = 5$), ω ranged from 38.04 to

TABLE 4.—Growth analysis of frilled dogwinkle at Hogg Bay in Prince William Sound. $N = 420$ size pairs. Confidence limits are asymptotic SEs (SYSTAT, Evanston, Illinois); the SIMPLEX and Quasi-Newton are algorithms for estimating nonlinear parameters. SSE = the sum of squares of the deviations from the model; S_∞ = asymptotic length; K = the growth rate constant of the Brody–Bertalanffy equation; t_A = the time of year of minimum growth with January 1 = 0; $\epsilon = 0$ when annual fluctuation is maximum.

	Number of parameters			
Model or term	4	3	2	2
Algorithm	SIMPLEX	Newton	Newton	Newton
SSE	1352.22	1353.05	1357.61	1484.72
S_∞(mm)	29.49	29.49 ± 0.32	29.55 ± 0.23	29.059 ± 0.30
K (year^{-1})	1.389	1.389 ± 0.112	1.287 ± 0.078	1.841 ± 0.159
t_A	0.093	0.033 ± 0.016	0 (fixed)	0 (fixed)
ϵ	0.00001	0 (fixed)	0 (fixed)	1 (fixed)

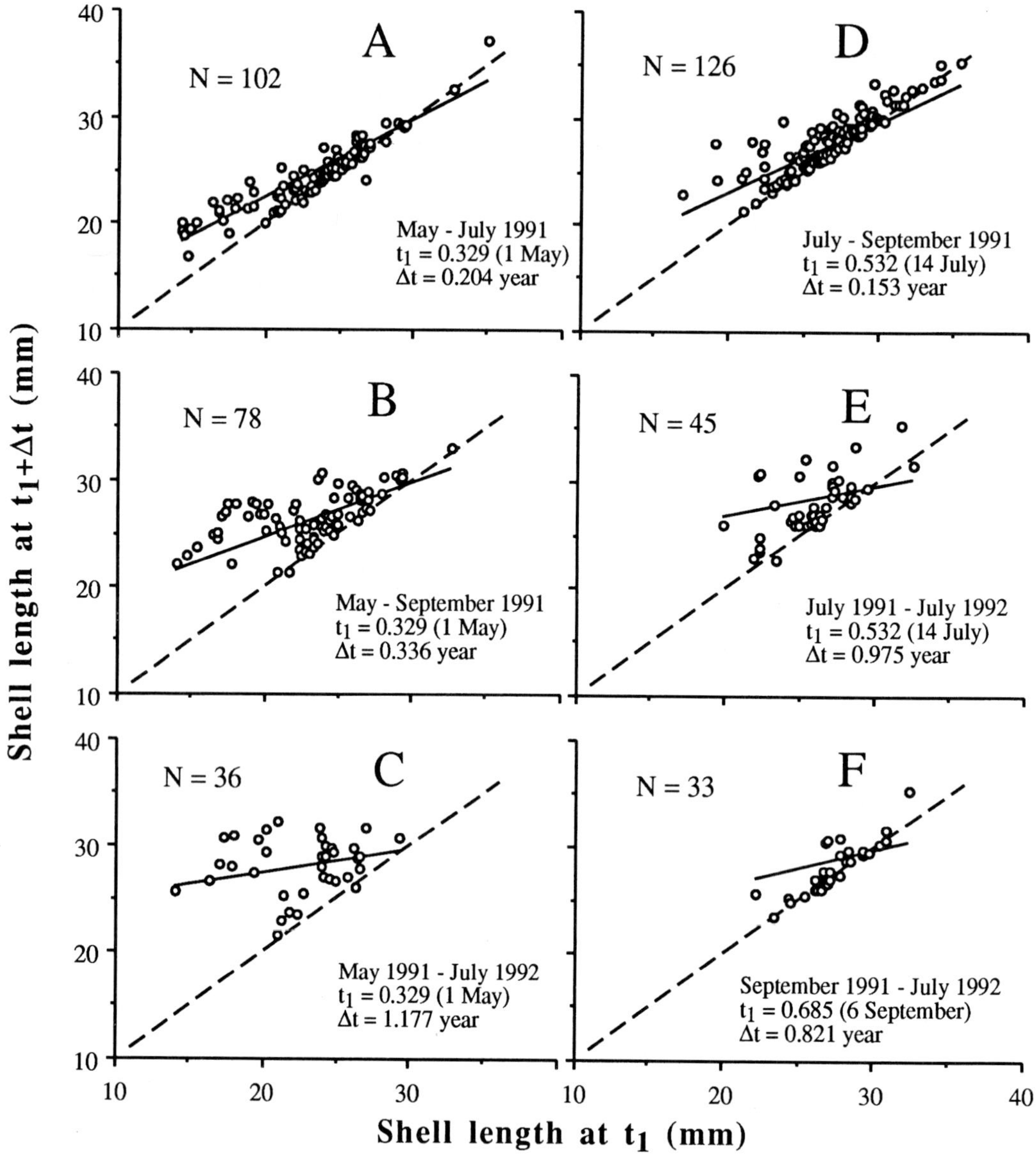

FIGURE 3.—Walford plots of growth of frilled dogwinkle at Hogg Bay for six time periods. Dashed line is the line of 0-growth; fitted line is for three parameters, S_∞ = 29.491 mm, K = 1.389 year^{-1}, and t_A = 0.033 with data values of original length (S at t_1) and length after Δt (S at $t_1 + \Delta t$); t_1 is the decimal fraction of a year starting with January 1 = 0 and Δt is time in years from t_1 to time of remeasurement.

52.08 with a mean of 44.09. For oiled only sites (N = 3), ω ranged from 16.60 to 32.56 with a mean of 24.04 and at hot washed beaches (N = 3), ω ranged from 21.08 to 48.93 with a mean of 36.85. Differences among treatments were tested by ANOVA (Table 6) with a resulting P = 0.049. There was no overlap of ω values for unoiled and oiled only sites (Figure 5A). Estimates of ω for snails at hot washed sites overlapped the ω values of those at unoiled sites and oiled only sites. Total PAH concentration in sediments was negatively correlated with ω (Figure 5B): $r = -0.562$ and P = 0.07 for ω versus $\log_e$(sediment PAH), and r = -0.593 and P = 0.05 for $\log_e(\omega)$ versus ln(sediment PAH). Northwest Bay islet is an outlier with a high value of ω as well as a high concentration of sedi-

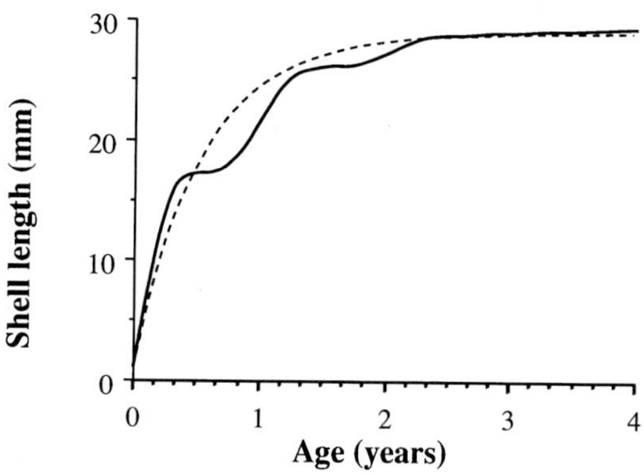

FIGURE 4.—Growth of frilled dogwinkle at Hogg Bay, Prince William Sound, 1991–1992. Solid line is with seasonal adjustments to growth (equation 3) and the dashed line is the simple Brody–Bertalanffy model (equation 1).

ment PAH. Average values of PAH in mussels at Northwest Bay islet (Table 2) were similar to such values at Block Island, another oiled and washed site. Northwest Bay islet had higher densities of barnacles in 1991 (Table 1) as well as being the least exposed of the oiled and washed sites. Whether any or some combination of these features is sufficient to explain the growth performance of frilled dogwinkle at Northwest Bay islet is unknown.

Other attributes of sites were examined to see whether densities of potential prey or relative exposure to open sea conditions might help to explain additional variance of ω. The analysis was cast as an analysis of covariance (ANCOVA) using $\log_e(\omega)$ as the dependent variable with treatment as a factor and environmental measurements as covariates. Initial analyses were tests for homogeneity of slopes and tests for exposure, mussel density, and barnacle density; all were not significant at $\alpha = 0.05$, which then moves the analysis on to tests for the significance of differences of treatment following adjustment by the covariate. Neither barnacle nor mussel densities were significant at $\alpha = 0.05$ with respect to explaining additional variation of ω; however, exposure was significant, with $P = 0.014$ (Table 7). The coefficient for exposure was negative (-0.087), which means that increasing exposure to surf conditions tends to decrease $\log_e(\omega)$. Adjusting for exposure made mean values of ω the same for no oil and oiled and washed treatments, with ω for oiled treatments much lower (Table 8).

For animals tagged in April–May, Δt ranged from 1.148 (Block–Hogg 1) to 1.175 years (Hogg 1). For animals tagged in June–July 1991, Δt ranged from 0.953 (Northwest Bay islet) to 0.993 (Hogg 2). Geometric means for estimates of annual recapture rates (Table 9) varied significantly by treatment (Table 10): Kruskal–Wallis ANOVA ($P = 0.02$). It is reasonable to conclude that recapture of tagged animals was different at sites with different treatments. The distribution of recapture rates (Figure 6) shows that the oiled only sites (Herring Bay, Snug Harbor, and Crafton Island) had the lowest rates and that the highest rates were at the unoiled locations of Bass Harbor, Outside Bay, Hogg Bay, and Crab Bay. Small sample sizes preclude a determination of whether recapture rates at unoiled sites

TABLE 5.—Summary of growth parameters for frilled dogwinkle in Prince William Sound, 1991–1992. Missing estimates for parameters (*) represents no convergence with any of the starting values that were tried. Missing estimates for SE means that the SIMPLEX method was used because the Quasi-Newton method could not converge because of round-off error. "–H" means that snails were imported from Hogg Bay; the sample from Herring Bay is for local and Hogg imports combined.

		Estimation with three parameters				Estimation with two parameters			
Treatment	Site	SSE	S_∞	K	t_A	SSE	S_∞	K	ω
No oil	Bass	1108.82	32.872 ± 0.342	1.179 ± 0.080	0.957 ± 0.033	1113.95	32.823 ± 0.245	1.270 ± 0.053	41.685
	Crab	1325.79	33.730 ± 0.273	1.514 ± 0.147	0.039 ± 0.031	1331.94	33.759 ± 0.259	1.417 ± 0.077	47.836
	Eshamy	*				1524.66	38.208 ± 0.246	1.068 ± 0.028	40.806
	Hogg	1353.06	29.491 ± 0.318	1.389 ± 0.112	0.033 ± 0.016	1357.61	29.554 ± 0.235	1.287 ± 0.078	38.036
	Outside	2008.44	28.329 ± 0.213	1.836 ± 0.127	0.0004 ± 0.0198	2008.44	28.379 ± 0.216	1.835 ± 0.108	52.075
Oiled	Herring	338.54	38.314 ± 1.655	0.718 ± 0.189	0.075 ± 0.059	344.85	38.724 ± 1.762	0.593 ± 0.123	22.966
	Snug–H	139.12	38.850 ± 3.909	0.557 ± 0.224	0.114 ± 0.050	141.88	40.296 ± 4.099	0.412 ± 0.137	16.602
	Crafton–H	645.45	29.439 ± 0.421	2.955 ± 0.672	0.276 ± 0.029	729.38	30.011 ± 0.487	1.085 ± 0.125	32.562
Oiled and hot washed	Northwest Bay	2766.65	47.056 ± 1.022	1.183 ± 0.130	0.085 ± 0.024	2797.53	48.139	0.965	46.454
	Northwest Bay–H	2121.87	36.318	1.773	0.090	2144.21	36.850 ± 0.850	1.395 ± 0.162	51.406
	Smith	118.74	34.602 ± 12.092	1.706 ± 3.493	0.760 ± 0.099	118.74	34.602 ± 1.176	1.172 ± 0.248	40.553
	Block–H	1138.27	37.965	0.608	0.048	1147.75	38.260	0.551	21.081

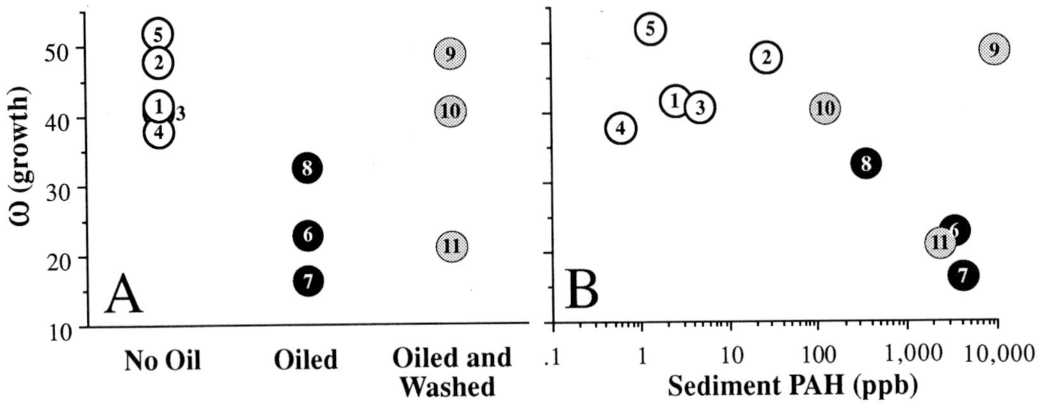

FIGURE 5.—Relationship between ω ($K \times S_\infty$ mm/year) and treatment for frilled dogwinkle populations in Prince William Sound (**A**); relationship between ω and sediment polycyclic aromatic hydrocarbon (PAH) concentrations in part per billion (ppb) (**B**). Numbers in data dots represent the following: 1–5 are sites with no oil (white), sites 6–8 were oiled (black), and 9–11 were oiled and hot washed (gray); 1 = Bass Harbor, 2 = Crab Bay rocky, 3 = Eshamy Bay, 4 = Hogg Bay, 5 = Outside Bay rocky, 6 = Herring Bay rocky, 7 = Snug Harbor rocky, 8 = Crafton Island, 9 = Northwest Bay islet, 10 = Smith Island, 11 = Block Island.

were different from those at the hot-washed sites; however, local animals at Northwest Bay islet, a hot-washed site, had the highest estimated recapture rate for tagged snails. The recapture rate of Hogg Bay animals transported to Northwest Bay islet was much lower and suggests that they fared less well than local snails.

An analyses of covariance were performed using recapture or presence rate (S) as the dependent variable, treatment as a factor, and exposure, mussel density, and barnacle density (Table 1) as covariates. None of these covariates explained significant amounts of variation, although exposure was closest with $P = 0.12$.

Discussion

Our work shows that tagged individuals of frilled dogwinkle had different responses of growth and recapture rates at unoiled, oiled only, and hot-washed beaches. Unoiled beaches were the best habitat, whereas oiled only beaches showed poorest growth and greatest losses of tagged snails. Hot-washed sites tended to be better habitats than oiled only locations, but the beaches treated with hot water were not in the same condition as unoiled ones even 2 to 3 years after the oil spill, at least with respect to probabilities of recapturing tagged snails. Because our work spanned just 1 year, it is not possible to suggest the rate at which the populations of frilled dogwinkle at the disturbed sites will reassume the traits of this species at unoiled sites. Also we are not able to identify the pathways that produced the effects. The oil itself is a possible cause of depressed rates of growth and recapture; however, the relationship between ω and PAH concentrations in frilled dogwinkle (Table 2), though nega-

TABLE 6.—Analysis of variance for the effect of different treatments (no oil, oiled only, and oiled and hot washed) on ω. Sites are replicates within treatments; $N = 11$, $r^2 = 0.53$.

Source	Sum of squares	df	Mean square	F-ratio	P
Treatment	753.5264	2	376.7632	4.5093	0.0488
Error	668.4146	8	83.5518		

TABLE 7.—Analysis of covariance of $\log_e(\omega)$ as the dependent variable, using treatments as a factor and exposure from Table 1 as a covariate; the interaction T × E was not significant ($P = 0.72$), and so slopes are homogeneous, which moves the analysis to a test for the significance of difference of adjusted means; $N = 11$; $r = 0.901$; $r^2 = 0.813$; the coefficient for exposure = −0.087: more exposed sites have smaller values of ω; ω approximates the instantaneous growth rate of a newly hatched frilled dogwinkle in millimeters per year.

Source	Sum of squares	df	Mean square	F-ratio	P
Treatment	1.067	2	0.533	13.795	0.004
Exposure	0.412	1	0.412	10.656	0.014
Error	0.271	7	0.039		

TABLE 8.—Mean values of $\log_e(\omega)$ for frilled dogwinkle in Prince William Sound, 1991–1992, adjusted by least squares (Table 7) to a common exposure (Table 1).

Source	$\log_e(\omega)$	SE	ω	N
No oil	3.743	0.089	42.2	5
Oiled	2.983	0.123	19.7	3
Oiled and hot washed	3.767	0.132	43.2	3

TABLE 10.—Kruskal–Wallis one-way analysis of variance for 10 cases. Dependent variable is recapture or presence rate (S) and the grouping variable is treatment.

Treatment	Count	Rank sum
No oil	5	40.0
Oiled	3	6.0
Oiled and hot washed	2	9.0
Kruskal–Wallis test statistic = 7.64; df = 2, P = 0.022		

tive, was not significant ($r = -0.06$, $P = 0.88$); the data set for PAH concentration in frilled dogwinkle is weak. Depression of ω may be the result of decimation of the food resources, namely, barnacles and mussels, although such a relationship could not be demonstrated with our data set. The pathway by which oil or related disturbances influence growth and recapture rates remains unknown.

There are several problems with our analysis that deserve comment. Sager's model (equation 3) is adequate as a beginning for analysis; however, the Brody–Bertalanffy model possibly should be scrutinized closely and, if needed, be replaced because in its difference-equation form (equation 2), it is always linear (Figure 2) whereas there are many cases where this is not true (e.g., Frank 1965; Ebert 1980; Tanaka 1982; Waters and Shay 1991; Kenner 1992). Some other growth model may be more suitable such as the Richards or Tanaka functions (Richards 1959; Ebert 1982; Tanaka 1982; Ebert and Russell 1993) but all have the disadvantage of requiring the estimation of an additional parameter.

The modeling of seasonality in the growth equations should be viewed as a first attempt. Given the rigorous winter conditions in Alaskan waters, it is quite likely that there is a prolonged period of no growth rather than just an instant as in equation (3) with $\epsilon = 0$. Sager and Gosselck (1986) present a model that permits a long period of zero growth:

$$S_t = S_\infty - be^{-Kt}\left[1 + \frac{(1-\epsilon)K}{2\pi(1-c)}\sin 2\pi \cdot \left[t - t_A - \frac{c}{2\pi}\sin 2\pi(t - t_A)\right]\right]; \quad (9)$$

we tried this model with the Hogg Bay data, which resulted in a second oscillation in the growth curve rather than an extension of the period of 0-growth. This problem in estimation probably was caused by not having samples during the wintertime. It would be very useful to have growth data from November through March.

TABLE 9.—Data structure for recapture of tagged frilled dogwinkle at sites with different treatments in Prince William Sound. Numbers are totals tagged at two times in 1991 and the number of these recovered in 1992. Recapture rate is adjusted to a time period of 1 year. Mean recapture rate per year is the geometric mean of all of the estimates at a site.

		Number tagged		Recaptured	
Treatment	Site and sample	April–May 1991	July 1991	Recapture rate per year	Mean rate per year
No oil	Bass Harbor	385		0.1974	0.1974
	Outside Bay	364		0.2920	0.1875
	Outside Bay–Hogg	46		0.1203	
	Hogg Bay #1	340		0.1480	0.1234
	Hogg Bay #2		265	0.1028	
	Crab Bay	277		0.2556	0.2556
	Eshamy Bay		333	0.3190	0.3190
Oiled	Herring Bay	62		0.0945	0.0600
	Herring Bay–Hogg	171		0.0308	
	Snug Harbor–Hogg		300	0.0440	0.0440
	Crafton Island–Hogg		301	0.0575	0.0575
Oiled and hot washed	Northwest Bay islet–Hogg 1	70		0.0249	0.1142
	Northwest Bay islet–Hogg 2		197	0.1780	
	Northwest Bay islet–Northwest Bay islet		283	0.3358	
	Block Island–Hogg 1	385		0.0921	0.1076
	Block Island–Hogg 2		170	0.1256	

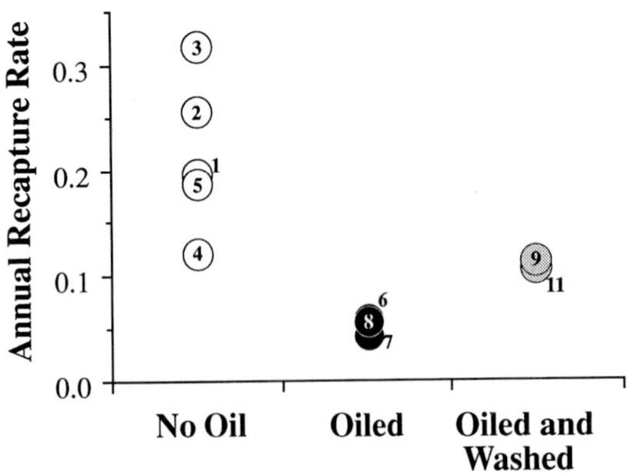

FIGURE 6.—Annual recapture rates of tagged frilled dogwinkle at sites with different treatments; points are geometric means of all samples at site (Table 8). Site numbers are the same as in Figure 5.

Our point in calling attention to these problems is to emphasize that our growth analysis is a first attempt to analyze data of the sort we present here. All other studies of seasonal components to growth (e.g., Cloern and Nichols 1978; Pauly 1981; Sager 1984) require age and size as data rather than size pairs (S_1 and S_2) and a time interval (Δt).

Our loss estimates for tagged animals probably are best interpreted as differences in survival, so mortality rates were highest at oiled only beaches with lingering effects at hot-washed beaches. However, we cannot rule out effects on behavior, so differences could be brought about by differential movement or differential cryptic behavior in habitats with different-sized boulders or algal cover. We know of no studies on the effects of oiling and washing on mollusc behavior.

We conclude that oiling or associated disturbances had a negative effect on the predatory frilled dogwinkle and that this effect was still evident 2 to 3 years after the *Exxon Valdez* oil spill. Treatment with hot water to remove oil improved both growth and recapture rates for frilled dogwinkle, at least for the period 1991 to 1992, but recapture rates on washed beaches were lower than recapture rates that were observed on unoiled beaches.

Acknowledgments

We acknowledge the technical, logistical, and funding support of several organizations: The study was contracted and funded by the National Oceanic and Atmospheric Administration (NOAA). Additional funding was provided by the U.S. Environmental Protection Agency (EPA), U.S. Coast Guard, Minerals Management Service (MMS), and the Marine Spill Response Corporation. Valuable staffing support in the form of field personnel was provided by NOAA, EPA, and MMS. Field crews performed admirably with enthusiasm and fervor during the grueling bouts of tagging and measuring the snails. The crews included R. Yender (EPA); S. S. Landino (MMS); H. Curl, R. Hoff, A. Mearns, and G. Shigenaka (NOAA); C. Vallee, H. Cumberland, and P. Hague (Ogden Environmental and Energy Services); A. Fukuyama and J. Houghton (Pentec Environmental, Inc.); and finally, R. Becker, skipper of the MV *Renown*, who vigorously and enthusiastically supported the tagging and recapture efforts and enhanced recapture efficiency by developing the innovative concept of "*Nucella* Lotto." We also thank D. Leslie (Department of Mathematics, San Diego State University), who checked our derivation of the difference equation (equation 4).

References

Appeldoorn, R. S. 1983. Variation in the growth rate of *Mya arenaria* and its relationship to the environment as analyzed through principal components analysis and the ω parameter of the von Bertalanffy equation. U.S. National Marine Fisheries Service Fishery Bulletin 81:75–84.

Black, R. 1973. Growth rates of intertidal molluscs as indicators of effects of unexpected incidents of pollution. Journal of the Fisheries Research Board of Canada 30:1385–1388.

Brisbin, I. L., Jr., G. C. White, P. B. Bush, and L. A. Mayack. 1986. Sigmoid growth analyses of wood ducks: the effects of sex, dietary protein and cadmium on parameters of the Richards model. Growth 50:41–50.

Brody, S. 1945. Bioenergetics and growth. Reinhold, New York.

Brown, B. E., and L. S. Howard. 1985. Assessing the effects of stress on reef corals—a review. Advances in Marine Biology 22:1–63.

Cloern, J. E., and F. H. Nichols. 1978. A von Bertalanffy growth model with a seasonally varying coefficient. Journal of the Fisheries Research Board of Canada 35:1479–1482.

de Nicola Giudici, M., L. Migliore, S. M. Guarino, and C. Gambardella. 1987. Acute and long-term toxicity of cadmium to *Idothea baltica* (Crustacea, Isopoda). Marine Pollution Bulletin 18:454–458.

Driskell, W. B., A. K. Fukuyama, J. P. Houghton, D. C. Lees, A. J. Mearns, and G. Shigenaka. 1996. Recovery of Prince William Sound intertidal infauna from *Exxon Valdez* oiling and shoreline treatments, 1989 through 1992. American Fisheries Society Symposium 18:362–378.

Duineveld, G. C. A., and M. I. Jenness. 1984. Differences

in growth rates of the sea urchin *Echinocardium cordatum* as estimated by the parameter ω of the von Bertalanffy equation applied to skeletal rings. Marine Ecology Progress Series 19:65–72.

Ebert, T. A. 1980. Estimating parameters in a flexible growth equation, the Richards function. Canadian Journal of Fisheries and Aquatic Sciences 37:687–692.

Ebert, T. A. 1982. Longevity, life history, and relative body wall size in sea urchins. Ecological Monographs 52:353–394.

Ebert, T. A., and M. P. Russell. 1993. Growth and mortality of subtidal red sea urchins (*Strongylocentrotus franciscanus*) at San Nicolas Island, California, USA: problems with models. Marine Biology 117:79–89.

Frank, P. W. 1965. Shell growth in a natural population of the turban snail, *Tegula funebralis*. Growth 29:395–403.

Gallucci, V. F., and T. J. Quinn, II. 1979. Reparameterizing, fitting and testing a simple growth model. Transactions of the American Fisheries Society 108: 14–25.

Grahame, J., and G. M. Branch. 1985. Reproductive patterns of marine invertebrates. Oceanography and Marine Biology an Annual Review 23:373–398.

Hiyama, Y., T. Kitahara, and T. Tokai. 1988. Numerical prediction of a relation among growth, reproduction and mortality in iteroparous fish populations. Researches on Population Ecology 30:267–278.

Houghton, J. P., D. C. Lees, W. B. Driskell, S. C. Lindstrom, and A. J. Mearns. 1996. Recovery of Prince William Sound intertidal epibiota from *Exxon Valdez* oiling and shoreline treatments, 1989 through 1992. American Fisheries Society Symposium 18:379–411.

Houghton, J. P., and seven coauthors. 1993a. Evaluation of the condition of Prince William Sound shorelines following the *Exxon Valdez* oil spill and subsequent shoreline treatment: volume 2: 1991 biological monitoring survey. NOAA (National Oceanic and Atmospheric Administration) Technical Memorandum NOS (National Ocean Service) ORCA (Office of Ocean Resources, Conservation and Assessment) 67, Seattle.

Houghton, J. P., and seven coauthors. 1993b. Evaluation of the condition of Prince William Sound shorelines following the *Exxon Valdez* oil spill and subsequent shoreline treatment: volume 3-appendices to 1991 biological monitoring survey. NOAA (National Oceanic and Atmospheric Administration) Technical Memorandum NOS (National Ocean Service) ORCA (Office of Ocean Resources, Conservation and Assessment) 67, Seattle.

Hughes, R. N. 1986. A functional biology of marine gastropods. Johns Hopkins University Press, Baltimore, Maryland.

Kenner, M. C. 1992. Population dynamics of the sea urchin *Strongylocentrotus purpuratus* in a central California kelp forest: recruitment, mortality, growth, and diet. Marine Biology 112:107–118.

Krahn, M. M., and seven coauthors. 1988. Standard analytical procedures of the NOAA national analytical facility, 1988: new HPLC cleanup and revised extraction procedures for organic contaminants. NOAA (National Oceanic and Atmospheric Administration) Technical Memorandum NMFS (National Marine Fisheries Service) F/NWC-153, Seattle.

MacDonald, B. A., and N. F. Bourne. 1987. Growth, reproductive output, and energy partitioning in weathervane scallops, *Patinopecten caurinus*, from British Columbia. Canadian Journal of Fisheries and Aquatic Sciences 44:152–160.

Morris, R. H., D. P. Abbott, and E. C. Haderlie. 1980. Intertidal invertebrates of California. Stanford University Press, Stanford, California.

Pauly, D. 1981. The relationships between gill surface area and growth performance in fish: a generalization of von Bertalanffy's theory of growth. Meeresforschung 28:251–282.

Richards, F. J. 1959. A flexible growth function for empirical use. Journal of Experimental Botany 10:290–300.

Richards, L. J., and J. T. Schnute. 1990. Use of a general dose-response model for rockfish fecundity-length relationships. Canadian Journal of Fisheries and Aquatic Sciences 47:1148–1156.

Ricker, W. E. 1975. Computation and interpretation of biological statistics of fish populations. Fisheries Research Board of Canada Bulletin 191.

Sager, G. 1982. Das Längenwashstum der Nordsee Seezunge (*Solea vulgaris* Quensel) und die Problematik der Jahresschwankungen. Anatomischer Anzeiger 151:160–178.

Sager, G. 1984. Saisonal-modifizierte Formen der abgewandelten Janoschek-Funktion. Gegenbaurs Morphologisches Jahrbuch 130:659–669.

Sager, G., and F. Gosselck. 1986. Investigation into seasonal growth of *Branchiostoma lanceolatum* off Heligoland, according to data by Courtney (1975). Internationale Revue der Gesamten Hydrobiologie 71: 701–707.

Salazar, M. H., and S. M. Salazar. 1991. Assessing site-specific effects of TBT contamination with mussel growth rates. Marine Environmental Research 32: 131–150.

Spight, T. M. 1976. Ecology of hatching size for marine snails. Oecologia 24:283–294.

Spight, T. M. 1981. How three rocky shore snails coexist on a limited food resource. Researches on Population Ecology 23:245–261.

Tanaka, M. 1982. A new growth curve which expresses infinite increase. Publications from the Amakusa Marine Biological Laboratory, Kyushu University 6:167–177.

von Bertalanffy, L. 1938. A quantitative theory of organic growth. Human Biology 10:181–213.

Waters, I., and J. M. Shay. 1991. A field study of the effects of water depth, order of emergence and flowering on the growth of *Typha glauca* shoots using the Richards model. Aquatic Botany 39:231–242.

Wootton, R. J. 1979. Energy costs of egg production and environmental determinants of fecundity in teleost fishes. Symposia of the Zoological Society of London 44:133–159.

American Fisheries Society Symposium 18:362–378, 1996

Recovery of Prince William Sound Intertidal Infauna From *Exxon Valdez* Oiling and Shoreline Treatments, 1989 through 1992

W. B. Driskell
6536 20th Avenue N.E., Seattle, Washington 98115, USA

A. K. Fukuyama[1] and J. P. Houghton
Pentec Environmental, Inc., Edmonds, Washington 98020, USA

D. C. Lees
Ogden Environmental and Energy Services Co., Inc., 5510 Morehouse Drive San Diego, California 92121, USA

A. J. Mearns and G. Shigenaka
National Oceanic and Atmospheric Administration/National Ocean Service/Office of Ocean Resources Conservation and Assessment, 7600 Sand Point Way N.E., Seattle, Washington 98115, USA

Abstract.—In 1989, following the massive crude oil spill from the T/V *Exxon Valdez*, a sampling program was established to assess effects on and track the recovery of low-intertidal infauna from mixed gravel–sand–silt (mixed-soft) beaches. At some sites, effects from oiling were compounded by impacts from high-pressure hot-water washing used in shoreline cleanup. Total abundance, species richness, species diversity, and abundance of several major taxa (polychaetes, bivalves, and gastropods) were significantly lower in hot-water-washed beaches than in unoiled beaches. Infauna at oiled sites that were not hot-water washed rebounded quickly following the disturbance. Although all study sites showed some degree of recovery by 1992, recovery of infauna at sites that were cleaned still lagged significantly behind the oiled sites.

There have been three major studies of impacts to and recovery of the Prince William Sound beaches following the 1989 *Exxon Valdez* oil spill. Two of the studies (Gilfillan et al. 1995; Highsmith et al. 1996, this volume) used stratified random designs to sample all habitats in the sound so that they might infer, extrapolate, and assess systemwide effects. Our study takes a different tack: if the effects and recovery processes can be well understood at a few sites, we may be able to suggest the type and extent of damages that occurred elsewhere. The first approach is a Natural Resource Damage Assessment (NRDA)-style injury-assessment approach most useful in litigation proceedings; the second is the type of directed inquiry most useful for future crisis-management decisions. In this context, our study has two objectives: (1) to assess the extent of impacts from oiling and hot-water washing at selected sites and (2) to monitor the recovery process. This article deals only with infauna on lower intertidal beaches. Epifauna is addressed elsewhere in this volume by Houghton et al. (1996, this volume).

The impacts of oil on infauna are numerous and varied depending on factors such as type of oil, severity of oiling, beach porosity, and sensitivity of species. But whatever the degree of causality, the negative effects to the infauna can be generalized as reduction in size or biomass, mortality, and reduced (or failed) reproductive success. Conversely, positive impacts may occur, for example, when oil-tolerant species bloom during the period of reduced competition–predation. Mortality, blooms, and changes in reproductive success can be measured by tracking species abundance over time, as we do in this study. Size and growth rates for a few selected mollusks have also been monitored to determine if there have been changes in size or biomass (see Houghton et al. 1993a, 1993b; Ebert and Lees 1996, this volume).

Most of the oiled beaches in Prince William Sound were "cleaned," most commonly using a high-pressure hot-water-washing technique. The technique involved various methods of dislodging the oil by spraying the intertidal zone with heated sea water (40–60°C) and then skimming up the oil as it flowed down the beach and refloated on the tide. Commonly, the hot water was directed at the beach using hose nozzles or using a large sprayer-head mounted on a mechanical arm. Several im-

[1]Present address: Fukuyama-Hironaka, Taxonomic and Environmental Services, 7019 157th St. SW, Edmonds, Washington 98026, USA.

pacts are expected from the hot-water-washing technique.

- Beach biota are killed from thermal shock.
- Fine sediment and infauna are lost by hydraulic flushing.
- Sediment and infauna are overturned with high-pressure spray.
- Motile fauna are dislodged from their rocks by high-pressure spray.
- Sessile fauna are buried with sluiced sediments.
- Surface oil is forced deeper by stirring the surface sediments.
- Oil is spread across the lower intertidal zones.

The resulting impacts to the infauna from hot-water washing are similar in nature to the oiling impacts and can be tracked using the same techniques. The physical changes, namely the weathering and loss of oil and the redistribution of sediments, are also monitored during the recovery process.

The infaunal study reported here is part of a larger intertidal and shallow-subtidal sampling program (see Houghton et al. 1993a, 1993b, 1996) begun initially by the authors for Exxon Corporation in 1989 but continued since 1990 as a National Oceanographic and Atmospheric Administration (NOAA) program jointly funded by NOAA, the Environmental Protection Agency (EPA), the Minerals Management Service (MMS), the United States Coast Guard (USCG), the American Petroleum Institute (API), and the Marine Spill Response Corporation (MSRC). The change in program sponsorship allowed public examination of the beach recovery status during the time when most other investigators' data sets remained sealed under the NRDA litigation process. The 1989 data set, proprietary to Exxon Corporation, was not released until April 1993 and is only partially analyzed in this paper. The 1989 data extend the biological record to early postspill and, in some cases, to prewashing conditions. Such information greatly increases the confidence of statements about pre-spill similarities of and impacts to stations within each site category.

Study Area

Prince William Sound is a protected fjord system located on the south-central coast of Alaska. Beaches in the protected embayments and small coves are primarily composed of a mixture of gravel, sand, and silt (mixed-soft), but may include cobbles and/or scattered boulders. The study area encompassed most of the central and southern sound with the mixed-soft beach study sites concentrated along the chain of islands stretching from Naked Island (in the central sound) to the ocean entrances. This portion of the sound lay in the path of the oil slick; many beaches on these islands were eventually oiled. Unoiled reference sites were located peripherally on the east, west, and northern regions in the sound and at one location amid the central island group (see Figure 1 in Houghton et al. 1996).

Two major natural events in the past 30 years have reshaped the composition of intertidal soft-bottom assemblages in the sound: the earthquake of 1964 and the population explosion of sea otters *Enhydra lutris*. The Alaskan earthquake of 1964 dramatically altered vast areas within the intertidal zone in Prince William Sound by vertically displacing areas up to 38 feet (Hanna 1971). Hard-shell clams were particularly affected; mortality was estimated to range from 11 to 100% (Haven 1971). During this study, the authors have found relict beds of mature butter clam *Saxidomus giganteus* shells at sites where butter clams are now absent or rare.

The sea otter population within Prince William Sound has rapidly expanded during the past 30 years (Lensink 1962; Garshelis et al. 1986). As a result, there have been dramatic changes in epifaunal and infaunal communities where the otter has become reestablished. Important prey species include clams, crabs, and sea urchins (Calkins 1978; Garshelis et al. 1986; VanBlaricom 1987; Kvitek et al. 1988). Predation by otters can be an important factor in accounting for differences in abundance and size structure of prey species. Larger-sized mussels are absent from areas within the sound where there is a heavy concentration of otters (VanBlaricom 1987). Smaller fauna are also affected; when otters dig for larger bivalves, they expose smaller-sized mollusks in the feeding pits to scavenging by other predators (Kvitek et al. 1992). Thus, otter feeding may account for some differences in shellfish abundance at some sites occupied by the study. The heavy mortality of otters during and after the spill should reduce their influence on intertidal assemblages; however, their eventual return will be yet another confounding factor in the recovery process.

Methods

Site Selection and Classification

The study sites were not selected by any formal randomized process. Responding during the first week of the spill (while the slick was still in mid-

TABLE 1.—Lower-intertidal stations sampled for infauna in Prince William Sound, July 1989–1992, grouped by treatment category. Category 3 sites were washed with hot water.

Sites	Degree of oiling	Years sampled				Latitude (N)	Longitude (W)
		1989	1990	1991	1992		
Category 1—unoiled							
Bainbridge Bight	None			x	x	60°7′2″	148°14′58″
Bass Harbor	None	x				60°39′1″	147°23′7″
Crab Bay site 1	None	x	x			60°4′19″	147°59′47″
Crab Bay site 2	None				x	60°4′19″	147°49′57″
Hanning Bay	None	x				60°57′19″	147°36′13″
Hogg Bay	None	x				60°4′55″	148°11′45″
Outside Bay	None	x	x	x	x	60°38′17″	147°27′1″
Port Chalmers	None	x				60°15′24″	146°40′48″
Sheep Bay	None	x	x	x	x	60°41′6″	145°56′11″
Category 2—oiled, not washed							
Bay of Isles	Light	x				60°23′5″	147°45′5″
Block Island	Heavy		x	x	x	60°31′44″	147°36′36″
Crafton Island	Light		x	x		60°30′13″	147°56′33″
Herring Bay	Light	x	x	x	x	60°27′25″	147°42′37″
Ingot Island	Moderate			x		60°31′38″	147°39′26″
Mussel Beach[a]	Light	x	x	x	x	60°32′6″	147°37′1″
Snug Harbor	Light	x	x	x	x	60°15′45″	147°45′53″
Category 3—oiled and washed							
Northwest Bay	Moderate	x	x	x	x	60°32′37″	147°36′13″
Shelter Bay	Moderate	x	x	x	x	60°7′6″	147°57′24″
Sleepy Bay	Moderate			x	x	60°3′56″	147°50′4″

[a]Reportedly washed with cool water.

sound), the goal in the initial design was to quickly select comparable sites that represent the major geomorphological habitat types (boulder–cobble, rocky, and mixed-soft) and to include oiled and unoiled sites within each habitat type. The sites selected had to be accessible by boat and/or helicopter. After the cleanup effort began in summer 1989, some of the oiled sites were washed with hot water, a rare few were washed with cool-water flushes or were officially designated as untreated "set asides" by governing agencies. Thus, although the sites were not originally selected in a strictly random fashion, the treatment categories were randomly assigned.

For this study, the sites were categorized according to their general degree of disturbance, based on the nature of their treatment. Bioremediation effects (applying Inipol or Customblen fertilizers) are confounded in these categories but are assumed to be nondeleterious. Site categories are

- category 1: unoiled reference sites;
- category 2: oiled sites, not treated or treated only with cool-water flushes and bioremediation fertilizers; and
- category 3: oiled sites, "cleaned" with hot-water flushes; most were also treated with bioremediation fertilizers.

In 1990, several sites were added to coordinate the biological monitoring program with the NOAA geomorphological monitoring program (Michel et al. 1991; Michel and Hayes 1993). Additional sites and stations were added in 1991 and 1992 to correct imbalances in the sampling design matrix. Sites sampled, degrees of oiling, and known treatment histories are listed in Table 1.

Field Methods

Biological sampling.—A random sampling design stratified by tidal elevation was initially used in 1989 and continued with some site changes from 1990 through 1992. At first, both mid and lower tidal elevations were sampled at various seasons. The mid-elevation stations turned out to be sparsely populated and were dropped in later years. Only the lower elevation infaunal data from June–July samplings are analyzed in this paper. Samples collected in 1993–1995 were analyzed separately and are not reported here.

Five 10.7-cm diameter (0.009 m^2) by 15-cm deep cores were collected in the vicinity of permanently marked locations (10 replicates in 1989) along a 30-m horizontal beach transect at each site. Each sample was field-sieved through a 1.0-mm screen and preserved in buffered 10% formalin. Samples were sorted and identified to lowest practical taxon, most recently at Pentec Environmental, Inc. (Edmonds, Washington).

A one-time study was conducted in April 1989 to

document the immediate impacts of the hydraulic flushing. Side-by-side areas were sampled in the heavily oiled West Arm of Northwest Bay, one side recently cleaned by hydraulic flushing of unknown temperature (presumably hot) and the other side oiled but not washed.

Sampling of physical factors.—Field-preserved whole sediment samples collected at sites in July 1991 and 1992 (N = 11, 12) were analyzed for grain size following the wet-volume-displacement procedures of McNeil and Ahnell (1964). Grain-size data are not available from 1989 or 1990.

Composite sediment samples were taken at each sampling elevation and frozen for analysis of polycyclic aromatic hydrocarbon (PAH) content. At most sites the chemistry sample was composited (using standard hydrocarbon collection techniques to avoid sample contamination) from surface sediments scooped approximately 3 cm deep at five locations immediately adjacent to the locations of the five infaunal cores. Duplicate samples were collected at a limited number of sites for quality assessment. Sediment hydrocarbon analyses were performed at the Institute for Environmental Studies, Louisiana State University (Baton Rouge, Louisiana).

In 1992, additional samples were collected for total organic carbon (TOC) and total Kjeldahl nitrogen (TKN) to complete the suite of pertinent sediment property covariates and for exploratory purposes as rough indices of available bulk sediment nutrients. The samples were purged of inorganic carbon, dried at 70°C, ground, and sieved through a 120-mesh screen. Total organic carbon was measured on a Dohrman DC-180 carbon analyzer (Rosemount Analytical, Inc., Santa Clara, California). Total Kjeldahl nitrogen was measured using EPA method 351.4. Analyses were performed by Analytical Resources Incorporated (Seattle, Washington).

All sediment samplings were nonreplicated samples (often composites) from each site. There is no measure of within-site variance for the sediment components.

Statistical Analyses

To assess infaunal impacts and monitor recovery, four lines of evidence were examined: (1) the typical suite of community biotic indices, namely, species richness (S), total abundance (N), and Shannon-Wiener diversity (H'); (2) the abundance of individual species and major taxonomic groups; (3) multivariate analyses of species abundance; and (4) sediment quality factors including grain-size composition, PAH, TOC, and TKN.

Two general null hypotheses were formulated for testing:

1. Diversity, total abundance, number of infaunal taxa, or density of selected invertebrate species in mixed-soft substrates do not differ significantly between site categories or years.
2. Physical sediment factors do not differ significantly between site categories.

Typical of constrained sampling plans, the infaunal sampling design has both strengths and weaknesses. The design was established primarily to allow monitoring of long-term recovery trends at specific sites, tracking the convergence of their differences over time. It is also well suited, by number of replicated samplings at each site, for comparisons between site categories at specific points in time (yearly snapshots). But because of the limited number of sites that could be sampled in each category and because no formal randomization was used in site selection, the design does not have the rigor for general statistical inference to all similar beaches in Prince William Sound. However, this study does suggest the nature and scope of impacts to expect from similarly oiled or washed beaches.

Data processing.—To produce tables for consistent analysis and comparison between years, the raw infaunal database was revised considerably. Irrelevant taxa that are typically epifauna or meiofauna were deleted, multiple taxonomic designations that were believed to represent a single species were consolidated, and polyspecific taxa were dropped. Calculations of total abundance, species richness, and species diversity were made using this refined data set. Another round of filtering, essentially dropping rare species, preceded the multivariate analyses.

Oversampling bias.—When the 1989 data set was finally incorporated into this study, a preliminary comparison of species richness pooled at each site in 1989 versus 1990 showed an impressive decrease in number of species at all sites, including reference sites. Species acquisition curves revealed that a large portion of the difference was the result of oversampling; 10 replicates were taken in July 1989 versus 5 in the succeeding years. To compensate, bootstrap estimates of the mean biotic indices (H', N, and S) for each site were calculated from individual replicates rather than from pooled replicates. In most cases, the bootstrapping made little adjustment relative to the pooled replicate means. For comparisons of individual taxon abundance (including multivariate analyses), pooled replicate means for each taxon were used.

Inferential statistics.—To assess impacts, randomization tests, namely one- and two-way analyses of variance (ANOVA) and both paired and unpaired *t*-tests, were used to evaluate the significance of differences between treatment categories. Randomization testing was chosen because it is insensitive to distributional non-normality and unequal variances (Potvin and Roff 1993). Only two-tailed test results were considered under the assumption that, depending on the taxon, either increases or decreases could occur with oiling or with treatment. The randomization routines were adapted from algorithms published by Edgington (1987).

In these tests, the mean of all replicate cores at a given site was used to represent each variable measured; thus, n equals the number of sites within each category. Some trends are noted without probability values where the results are considered biologically relevant (particularly if they are consistent across a time series of samplings), even though they are not statistically significant because of the limited replication of sites within site categories.

Before testing for differences in species compositions, infauna were lumped into major taxonomic groups (e.g., Polychaeta, Bivalvia). The groups were compared to determine significant category effects using a randomization ANOVA and then, when significant, for differences between treatment categories using post hoc randomization *t*-tests. Similar procedures were used (without lumping) to compare total abundance (N), species richness (S), and diversity (H').

In accordance with the objectives of this study, it was more important to reduce the type II error (failure to detect an effect) as opposed to committing a type I error (falsely identifying an effect). Given this importance and considering our limited replication of sites (roughly four per category), it was decided to adopt 0.10 as the alpha level of statistical significance for this study (see Houghton et al. 1996).

Multivariate analyses.—Principal components analysis (PCA), a multivariate ordination technique, was used to track site recovery trends. As in any other multivariate technique, the rare or sparsely distributed species were removed from the initial analysis in order to reduce the noise in the system. The remaining abundances then were log transformed to reduce population-scaling effects. Polyspecific groups were usually discarded to avoid distorting the discrimination between sites.

The resulting site scores were plotted on three-dimensional axes and examined for significant site clustering, which indicates site similarities, and for trends in movement between years, which indicate possible trends in successional recovery. The site scores were then correlated against various factors including individual species abundance, PAH concentrations, biotic indices, sediment grain distribution, and treatment categories to interpret the biological significance of the component axes. Species abundances were also correlated against site scores; the most highly correlated species were extracted from each component to help interpret the components and to identify key indicator species.

In the initial analyses, all sites from all years were incorporated into the PCA calculations, including sites that were dropped in later phases of the study. For this publication, only the sites sampled in 1992 are included to reduce the confusion of the graph and to emphasize the time-series data. The data set was also examined using detrended correspondence analyses to confirm the validity of the PCA analyses. The CANOCO software package (ter Braak and Prentice 1988; available from R. Fumas, Department of Mathematics, Cornell University) was used for all multivariate analyses.

Multivariate oversampling bias.—As discussed above, sampling 10 replicates in 1989 versus 5 in succeeding years induced an oversampling bias, which added rarer species and thereby increased 1989 species richness relative to the later years. An adjustment was made to correct for inferential testing, but no simple adjustment was possible for multivariates. This bias is thought to have no significant effect on the multivariate results because most rare species are deleted in the initial data screening process. As confirmation of nil effect, the ratio of the screened but unadjusted (biased) 1989 10-replicate species count to the screened 1990 5-replicate species counts used in the PCA approximated the same proportions obtained using unscreened adjusted 1989 counts to the unscreened 1990 counts (Table 2). For example, the Herring Bay PCA data set had 26 species in 1989 and 13 in 1990 using the screened species list, versus 55 and 25 species total from the adjusted species counts. The similarity of the ratios suggests that elimination of rare species before the multivariate analyses is adequate to compensate for the oversampling bias. No extraordinary adjustments are needed.

Results

Generally our program is designed to determine the effects of beach-cleaning activities based on an

TABLE 2.—Comparisons of relative species richness before and after 1989 sampling was adjusted from 10 to 5 replicates, and after rare taxa were removed for principal components analysis (PCA).

Site	Unadjusted number of taxa (pooled taxa) 10 rep:5 rep 1989:1990	Adjusted number of taxa (pooled taxa) 5 rep:5 rep 1989:1990	PCA data set number of taxa (screened) 10 rep:5 rep 1989:1990
Herring Bay	75:25	55:25	26:13
Mussel Beach	82:52	60:52	26:20
Northwest Bay West Arm	23:16	15:16	7:5
Outside Bay	85:57	62:57	35:27
Sheep Bay	77:50	53:50	29:24
Shelter Bay	54:26	39:26	12:7
Snug Harbor	74:30	53:30	21:16

inferential comparison of infauna at category 1, 2, and 3 sites. However, the program provided one case where we could directly evaluate immediate impacts of beach cleaning by comparison of pre- and postcleaning conditions.

Immediate Impacts of Hydraulically Cleaning Beaches

Dramatic visual impacts from washing mixed-soft sediment beaches were initially observed at the heavily oiled Northwest Bay West Arm site (Table 3). One part of the site appeared undisturbed. The other portion of the site was obviously washed, presumably using hot water (unconfirmed in the sketchy cleanup documentation) because several nearby sites within the embayment were observed being treated in this manner. The visible impacts in these side-by-side, washed and unwashed areas helped form the theory that similar damage might have been widespread at other washed sites.

Visually, the mixed-soft substrate in the washed section had a much lighter appearance than the adjacent heavily oiled section. Hydraulic disturbance of the sediment had greatly reduced the abundance of mussels *Mytilus trossulus* embedded in the surface gravels (0.2 per core versus 18.6 per core in the unwashed section; Table 3). Core samples contained a greatly reduced density of organisms on the washed section of beach (4.6 versus 76 per core counting all infaunal and epifaunal taxa; Table 3). The difference was primarily from a higher abundance of the small polychaetes *Orbiniella nuda* and *Pholoe minuta* and mussels in the unwashed area. Total number of taxa was 8 in the washed section and 20 in the unwashed section (Table 3; possibly underestimated because of the difficulty of field-processing the heavily oiled samples).

During a later visit to this same site in June 1989, the entire beach had been reworked hydraulically, causing substantial alterations in beach morphology. Pebbles had been washed downslope in large

TABLE 3.—Intertidal infauna abundance (number/0.009 m^2) in washed and unwashed areas, Northwest Bay West Arm lower mixed-soft station, Prince William Sound, April 1989. Min = minimum; Max = maximum.

Taxon or statistic	Washed area Mean	SD	Min	Max	Unwashed area Mean	SD	Min	Max
Decapoda, larvae	0.2	0.45	0	1	0.0	0.00	0	0
Calanoida	0.0	0.00	0	0	0.2	0.45	0	1
Eteone cf. *longa*	0.0	0.00	0	0	0.4	0.55	0	1
Harpacticus	0.0	0.00	0	0	0.2	0.45	0	1
Hemiptera, larvae	0.0	0.00	0	0	0.2	0.45	0	1
Hiatella arctica	0.0	0.00	0	0	0.2	0.45	0	1
Laophontidae	0.0	0.00	0	0	0.2	0.45	0	1
Littorina	0.0	0.00	0	0	0.2	0.45	0	1
Littorina sitkana	0.0	0.00	0	0	0.2	0.45	0	1
Lottiidae	0.0	0.00	0	0	0.2	0.45	0	1
Mollusca	0.0	0.00	0	0	0.2	0.45	0	1
Mysella tumida	0.0	0.00	0	0	0.2	0.45	0	1
Mytilus trossolus	0.2	0.45	0	1	18.6	20.24	0	52
Naineris quadricuspida	0.4	0.89	0	2	0.6	1.34	0	3
Nematoda	1.0	0.71	0	2	0.8	1.30	0	3
Nemertina	0.0	0.00	0	0	0.2	0.45	0	1
Nereis vexillosa	0.0	0.00	0	0	0.2	0.45	0	1
Odostomia–Alvania	0.2	0.45	0	1	0.0	0.00	0	0
Oligochaeta	0.6	0.89	0	2	6.0	5.15	1	14
Orbiniella nuda	0.6	0.89	0	2	37.6	25.04	7	62
Pholoe minuta	1.4	1.52	0	4	9.4	8.17	2	21
Polycladida	0.0	0.00	0	0	0.2	0.45	0	1
Total abundance	4.6				76			
Number of taxa	8				20			

windrows that eroded in steep scarps where cut by a local stream. Substantial sheening was evident as the tide flooded the area or where materials sloughed into the stream. Only seven taxa were found in 10 replicates. Gradually through the years the beach appearance has been restored, but the infauna shows only modest progress toward recovery. As of June 1992, the mean infaunal density of 22.4 individuals per core and mean species richness of 7.4 taxa per core remained among the lowest of any sites sampled in this study.

Species and Assemblage Indices

From the monitoring sites, the strongest summary evidence of the effects of high-pressure hot-water washing comes from the biotic indices. On the basis of treatment categories, category 3 sites were invariably lowest in species richness (S), abundance (N), and infaunal diversity (H') (Figure 1). In 1990, category 2 sites had reduced numbers of individuals and species but bloomed in 1992 to exceed the category 1 sites. The category differences were significantly different for each index in all 4 years (1989–1992, $P = 0.02$). By 1992, the differences between treatment categories had diminished, but abundance was still an average three to five times higher at category 1 and 2 sites compared to the category 3 sites; diversity and species richness were about twice as high.

Two-factor randomization ANOVAs were run on various data subsets to test the combined effects of the treatment categories and the year sampled on species richness (S) and number of individuals (N). For seven sites that were sampled for 4 years (1989–1992), both factors were significant: the treatment categories at $P = 0.001$ and 0.051, respectively, and the yearly variation at $P < 0.02$ for both factors. For 10 sites sampled in both 1991 and 1992, when presumably treatment differences would be partially mitigated by recovery, the number of individuals was still significantly different across treatment groups ($P = 0.049$), but species richness was not ($P = 0.24$). In the combined 1991–1992 data, yearly variation was also significant for both indices ($P =$ 0.004 and 0.046 for S and N, respectively). These results suggest that the ability to discern category differences (i.e., effects of oiling and treatment) from biotic indices is waning relative to the annual variation in the assemblages.

Major Infaunal Taxonomic Groups

The major infaunal taxa used in quantitative analyses following exclusion of meiofaunal and epifaunal taxa were Polychaeta, Sipunculida, Phoronida, Bivalvia, Gastropoda, Crustacea, and Echinodermata. Among these, only polychaetes, bivalves, gastropods, and crustaceans were common; the abundances of other taxonomic groups were orders of magnitude lower.

General distribution patterns for these major taxonomic groups are strongly related to type of treatment (Figure 2). Average density of all major taxa was highest in category 1 and 2 sites and lowest in category 3. If one assumes that category 1 represents the undisturbed condition and that categories 2 and 3 represent stages in recovery, these relationships suggest that the infaunal populations at category 2 sites suffered little impact relative to those at category 3 sites and had nearly recovered by 1992, while those at category 3 sites had been severely disturbed but are recovering. Crustaceans and polychaetes are rebounding more rapidly than gastropods or bivalves; of the latter taxa, the bivalves appear to be recovering at the slowest rate. The large spikes of gastropods in category 2 sites in 1989 and 1992 were mostly attributable to blooms of *Cingula* and Rissoidae at Herring Bay; the abundance of gastropods at the other category 2 sites without the blooms appear similar in numbers to the category 1 sites, with slightly fewer in 1989.

Community Dominants

The dominant taxa in 1992 varied considerably among the treatment categories but most markedly between both categories 1 and 2 versus category 3 (Table 4). Three major modes of distribution were observed for the dominant taxa: (1) taxa that appear only in the "healthy" assemblages at category 1 or 2 sites, (2) taxa that are "bloomers" appearing at category 2 but not category 3 sites, and (3) taxa that are omnipresent. Each of these modes suggests feeding-guild similarities or some functional tolerance for disturbed sites; however, this line of analysis will be deferred for further study. Only the trends are reported here.

Dominant and omnipresent in 1992 were the cumacean *Cumella vulgaris* and the polychaetes *Eteone longa* and *Pholoe minuta*. The latter bloomed at category 2 sites and at the recovering category 3 site, Northwest Bay. The amphipod *Paramoera* sp. 1 had a mixed response (discussed below): they appeared in very low numbers at category 1 and 2 sites but had a strong appearance in 1992 at category 3 sites.

The distribution patterns of the dominant mollusks was variable (Table 4). The pink bivalve Baltic

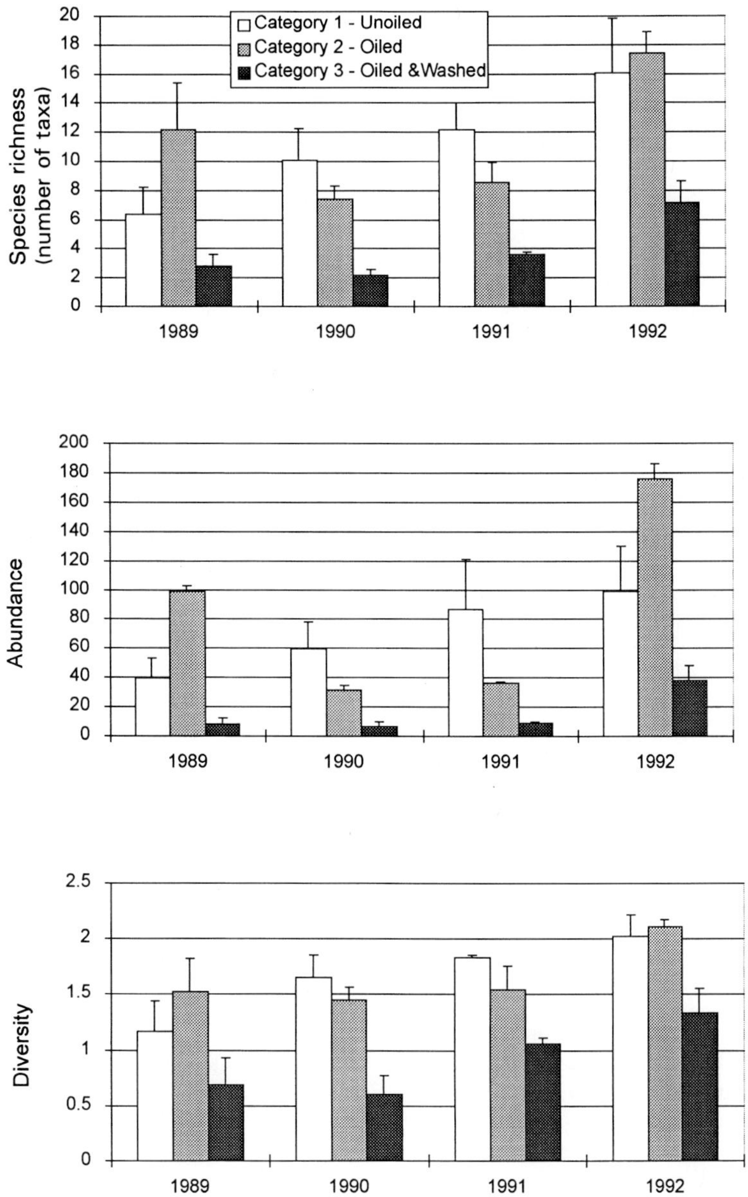

FIGURE 1.—Comparison of trends in species richness (number of taxa), infaunal abundance, and species diversity within each treatment category, Prince William Sound, 1989–1992. Mean category values were calculated from bootstrapped estimates of means at individual sites. Error bars represent standard error of mean. The legend in top histogram applies to all three histograms.

macoma *Macoma balthica* was abundant at category 1 sites, common in category 2 sites, but present only at one category 3 site, Northwest Bay. The small bivalve robust misella *Mysella tumida* was common at category 1 and 2 sites. At category 3 sites, it was rare in 1989, subsequently disappeared, and reappeared in 1992. The gastropod *Fartulum* sp. seems to be an opportunist; it occurred at several category 2 sites and at only one category 1 site, and was mostly rare at category 3 sites, it also bloomed in 1992. The tiny gastropod *Alvania compacta* was present in all categories and was increasingly abun-

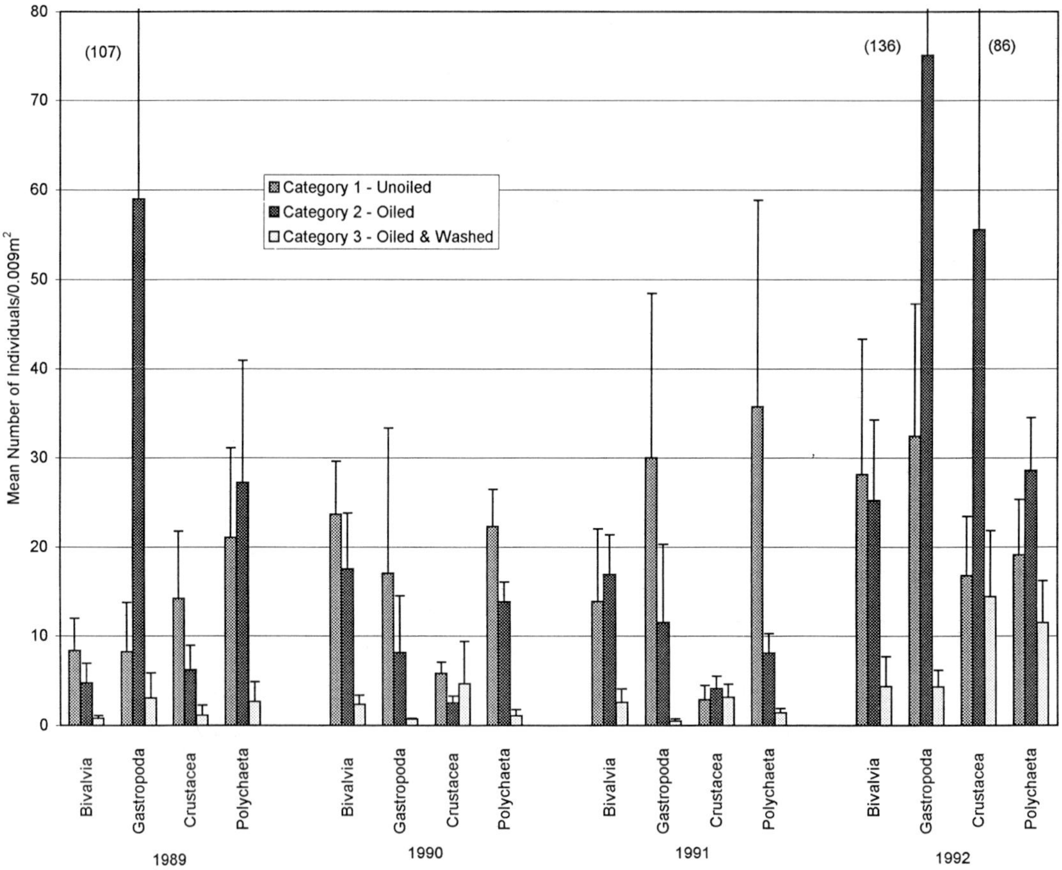

FIGURE 2.—Mean abundance per core (0.009 m^2) of major infaunal taxa within each treatment category, Prince William Sound, 1989–1992. Error bars represent standard error of mean.

dant in 1992. Pacific littleneck clams, *Protothaca staminea* also had good recruitment in 1991 and 1992, except at category 3 sites, where they remained in low abundance. The gastropod *Cingula* sp. was also very strongly represented in certain category 1 and 2 sites, present at others, but uncommon at category 3 sites.

The assemblage at category 3 sites represents a very limited subset of the taxa found at category 1 and 2 sites, and in general, the abundances have been much lower. An exception is the amphipod *Paramoera* spp., which bloomed in 1992 and was most abundant at category 3 sites. This genus is often associated with freshwater-influenced habitats, for example, in gravel or mixed substrates on beaches near the mouths of streams. It is uncertain whether these *Paramoera* species (currently unidentified) found in this study are associated with the disturbance from cleanup impacts or whether the category 3 sites are more strongly influenced by nearby beach streams. The latter is likely to be the case at the Sleepy Bay site but is less likely at Northwest Bay and Shelter Bay. As the recovery proceeds at these sites, the loss of *Paramoera* would suggest this species is a good indicator of disturbed sites. On the other hand, the persistence of *Paramoera* might suggest that category 3 sites have significant freshwater influence. *Paramoera* was also found at the Ingot Island and Bay of Isles category 2 sites but not at the Crab Bay category 1 site; each of these locations is freshwater influenced.

Chemical and Physical Measurements

Sediment hydrocarbons.—Based on field observations during the initial spill, it appeared that the stranded oil typically grounded in the upper and middle intertidal zones; it seeped down into the

TABLE 4.—Comparison of most abundant taxa per core (0.009 m^2) within treatment categories from lower mixed-soft stations, Prince William Sound, July 1992.

Category 1 sites (unoiled)			Category 2 sites (oiled)			Category 3 sites (oiled and washed)		
Species	Mean	Rank	Species	Mean	Rank	Species	Mean	Rank
Mysella tumida	17.7	1	*Cingula*	29.9	1	*Paramoera* sp. 1	6.7	1
Cingula	16.4	2	*Cumella vulgaris*	21.1	2	*Pholoe minuta*	3.8	2
Fartulum	7.1	3	*Mysella tumida*	14.7	3	*Alvania compacta*	2.8	3
Ampithoe	6.1	4	*Fartulum*	10.7	4	*Cumella vulgaris*	0.7	4
Cumella vulgaris	5.2	5	*Alvania compacta*	9.0	5	*Fartulum*	0.7	5
Protothaca staminea	4.6	6	*Pholoe minuta*	8.9	6	*Eteone spetsbergensis*	0.6	6
Alvania compacta	4.4	7	*Ampithoe*	8.1	7	*Mysella tumida*	0.5	7
Pholoe minuta	2.8	8	*Protothaca staminea*	7.5	8	*Eteone longa*	0.3	8
Macoma balthica	2.5	9	*Eteone longa*	5.8	9	*Protothaca staminea*	0.3	9
Pectinaria granulata	2.0	10	*Macoma inquinata*	5.2	10	*Syllis elongata*	0.2	10
Macoma inquinata	1.5	11				*Ampithoe*	0.2	10
Eteone longa	1.0	13	*Macoma balthica*	2.8	12	*Pectinaria granulata*	0.1	12
Syllis elongata	0.3	23	*Syllis elongata*	1.8	14	*Cingula*	0.07	26
Eteone spetsbergensis	0.1	33	*Pectinaria granulata*	1.0	18	*Macoma balthica*		
Paramoera sp. 1			*Eteone spetsbergensis*	0.3	27	*Macoma inquinata*		
			Paramoera sp. 1					

lower intertidal zone only at relatively few sites. However, at many sites during the subsequent hot-water washing, oil was remobilized and moved down across the relatively unaffected, lower intertidal zone. This scenario of oil distribution was reflected in the 1990 PAH concentrations (Table 5).

In 1990, the average total PAH concentrations at the category 2 sites were less at the intertidal and shallow subtidal elevations compared to similar elevations at category 3 sites. The highest PAH values of all the elevations and categories were found at the lower intertidal at category 3 sites which suggests that washing did transport oil down the beach. Polycyclic aromatic hydrocarbon concentrations were lower by at least an order of magnitude at category 1 sites relative to those at category 2 or category 3 sites. Polycyclic aromatic hydrocarbons at category 1 sites appeared to be primarily from pyrogenic or diesel sources.

TABLE 5.—Mean concentration (parts per million) of polycyclic aromatic hydrocarbons (PAHs) for mixed-soft sediments from Prince William Sound, 1990–1992.

Tidal elevation	Category 1 sites (unoiled)	Category 2 sites (oiled)	Category 3 sites (oiled and washed)
		1990	
Upper	0.0003 ± 0.0003	3.390 ± 6.470	1.32 ± 1.66
Middle	0.0009 ± 0.0006	4.350 ± 7.550	2.37 ± 5.47
Lower	0.0010 ± 0.0013	0.105 ± 0.105	12.88 ± 28.58
Subtidal	0.0012 ± 0.0060	0.448 ± 0.110	3.487 ± 3.01
		1991	
Upper		0.068	0.89
Middle		0.18 ± 0.22	2.40 ± 3.70
Lower	0.0006 ± 0.0003	0.89 ± 1.50	0.05 ± 0.049
Subtidal	0.0032 ± 0.0023	0.05 ± 0.054	0.20 ± 0.30
		1992	
Lower	0.0024 ± 0.0029	0.1722 ± 0.3401	0.0465 ± 0.0375

By 1991, PAH concentrations were lower at all category 2 elevations except at the lower elevation, suggesting weathering–degradation at the upper levels and perhaps erosional transport into the lower level. Polycyclic aromatic hydrocarbon concentrations at category 3 upper and middle elevations remained at about the same levels as in the previous year, while at the lower intertidal and shallow subtidal elevations, there was a dramatic decrease in PAH (presumably from higher rates of microbial degradation). Regression of H', N, and S versus total PAH concentrations showed significant negative correlations.

In 1992, the average concentrations of PAH in the lower intertidal zone were quite similar to those observed in 1990 with a significant decrease at category 3 sites. However, these averaged results disguise some confusing individual site trends. Looking at the individual values for sites at which sampling was repeated for 3 years (1990 through 1992, all categories), individual site concentrations actually increased at more than 70% of the sites. In contrast, PAH concentrations decreased at nearly 80% of the sites where sampling was repeated for only 2 years.

Given that the expected weathering trends of North Slope crude oil are fairly well known (and have been observed in this study and others) and that the analyses in this study were sufficiently sensitive to identify both new oiling (not from the *Exxon Valdez*) and pyrogenic products in various samples, the observed individual site results pose a dilemma. Michel et al. (1991) reported similar data anomalies relating to high spatial variability within sites and propose probable causal mechanisms such as variability in available surface area correlated with size of rocks on the beach. Nevertheless, the principal conclusion is that this data set from composited but nonreplicated samplings is inadequate to fully describe quantitative changes in hydrocarbon concentrations in the lower intertidal zone, even at sites that were sampled repetitively. However, noting that the differences between categories reflect at least an order of magnitude change and assuming that the measured magnitude of PAH concentrations is relatively accurate even if the absolute values are highly variable, this data set can be used for describing general trends. For example, the biotic indices N and S exhibited consistently strong and negative correlations with total sediment PAH in 1991 ($r = -0.52, -0.49$; $P = 0.007, 0.013$), yet by 1992 the indices exhibited nonsignificant correlations with total sediment PAH ($P = 0.13, 0.18$). This particular trend is important to note because one tentative conclusion of this study is that the residual oiling seems to have little continuing impact on infaunal community structure.

Sediment grain size, total organic carbon, and total Kjeldahl nitrogen.—Because infaunal distributions often correlate with grain-size fractions, the loss of fine-grain sediments (both organic and mineral) from the high-pressure flushing at category 3 sites is hypothesized to have significant effects on the assemblages. The loss of fines should reflect in the category 3 data as an initial abundance of coarser grained materials followed by the reacquisition of fines as time progresses. Grain-size data are not available from 1989 or 1990, but grain-size distributions obtained from 11 sites in 1991 and 12 sites in 1992 were used to evaluate the trends.

Suggestive trends were evident between categories in many of the sediment fractions, but no statistically significant temporal trends were found in the 1991 and 1992 data sets. There also were no statistically significant differences among the three categories when each year was analyzed separately for the fines, sand, or coarse fractions; however, the category 3 sites were consistently lower in sediment fines than the other sites.

Hypothetically, as lost fines are reacquired at category 3 sites, the bulk organic component of the fines should also increase and be reflected as shifts in TOC and TKN. In 1992, TOC and TKN were analyzed from four category 1, four category 2, and three category 3 sites. Total organic carbon was significantly different among the three categories ($P = 0.03$). Further comparisons showed significant differences between categories 1 and 3, and between categories 2 and 3 ($P = 0.05$ and 0.02, respectively) with lowest values from the category 3 sites. The category 2 sites had the highest mean TOC levels, possibly related to petroleum hydrocarbons in the sediments, but were not significantly different from category 1 sites. Although the expected differences in TOC were found, expected correlations with infaunal biotic indices are limited; TOC correlates with species diversity ($r = 0.53$) but with little else. Total organic carbon may be more important in future analyses when looking at feeding guilds or species succession.

From the TKN results, a spurious high reading of 2,190 parts per million (ppm) at Snug Harbor (category 2) was discarded as a probable sample preparation error (e.g., inclusion of the algal mat). Amongst the remaining data, a category effect was detected ($P = 0.07$). There was no difference between categories 1 and 2, but both were significantly different from category 3 ($P = 0.05$ for each). Like TOC, the TKN data correlates poorly with x and y PCA axes scores ($r = -0.12, 0.20$; see below) suggesting that in 1992, neither TKN nor TOC exerted much influence over mass species distributions.

Multivariate Analyses

Inferential statistics quantify the nature and degree of effects from oiling and cleanup, but recovery trends are best visualized from the graph of PCA site scores (Figure 3). An individual site's position is determined by both the abundance and composition of the infaunal assemblage. The annual progression for each site (reflecting shifts in the site's assemblage) is indicated by lines that connect the yearly data points. From the graphic relationships and the subsequent analytic comparisons with various physical factors and indices to assist interpretation, two major points are noted:

1. Annual shifts at impacted sites now mirror the shifts at nonimpacted sites, suggesting that recovery is occurring.
2. The sites are distributed along gradients of species richness, abundance, and treatment category; residual oiling is of minor concern.

PCA correlations.—Interpreting PCA results was limited to analyzing the eigenvectors for species importance and finding correlations between the

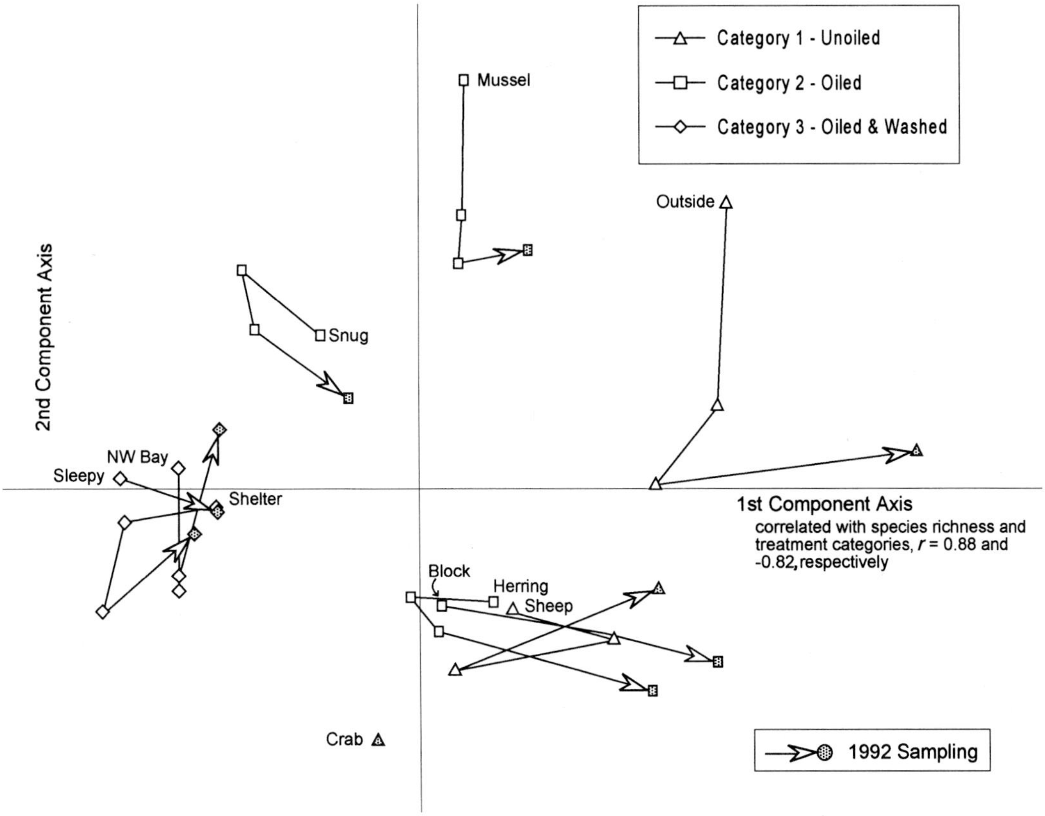

FIGURE 3.—Time-series plot of site scores from principal components analysis (PCA) of infaunal species abundance from Prince William Sound lower intertidal mixed-soft sites, 1989–1992. Site names are located at the beginning of each site's time series; arrows indicate shifts in assemblages between years. Interpretive correlations with various factors are noted marginally.

PCA scores and various factors. In Figure 3, the x-axis (the most important component) correlated -0.82 with treatment categories, 0.58 with total abundance, 0.76 with diversity, and 0.88 with species richness ($P < 0.001$ in all cases). These results integrate nicely with the results from the inferential analyses of biotic indices and physical factors. The extremely strong correlation with species richness makes the recovery trends particularly easy to spot. These correlations also confirm the general trend of impacted category 3 sites lying to the left and healthy category 1 sites lying to the right in Figure 3. Overall, total PAH from 1990 to 1992 ($N = 25$ of 34 samplings) correlated -0.43 ($P = 0.003$) with the x-axis but became nonsignificant in 1992 ($r = -0.37$, $P = 0.30$). These results suggest that residual oiling has influenced the site scatter but the effect is diminishing. Total Kjeldahl nitrogen and TOC measurements from a subset of 10 sites in 1992 correlated weakly and nonsignificantly with the x-axis site scores ($r = -0.12$ and 0.01).

In the initial PCA analyses (before dropping the miscellaneous sites), the y-axis correlated well with the coarsest grain sediment fractions ($r = 0.61$); however, in the final analysis with only the stations sampled in 1992, this relationship disappeared ($r = 0.18$). Among all factors, TKN correlated best ($r = 0.20$); however, the y-axis had no significant correlations with any of the examined physical factors.

Annual recovery trends.—The general trend among healthy or recovering sites is progression toward the right, toward sites with greater species richness (Figure 3). The richest site in the study is the category 1 site Outside Bay, which in 1992 showed a dramatic shift up-axis after a particularly good year of adding new species. Its movement up the *x*-axis is mirrored by most of the other sites sampled in 1992. For example, in the leftmost cluster the category 3 sites are generally shifting toward the right, as are Snug Harbor and Mussel Beach, both category 2 sites.

In the rightmost group, the pattern is more complex. The lower cluster of category 1 and 2 sites, Herring Bay, Sheep Bay, and Block Island, is moving parallel with the richest category 1 site, Outside Bay. The outlier, Crab Bay, a category 1 site sampled initially in 1992, is located in an embayment near the mouth of a stream, and although well populated, it is somewhat species impoverished.

The patterns of movement indicate considerable change at individual sites. Even the healthy, nonimpacted sites show sóme degree of movement, representing annual variation and regional community shifts. But the impacted sites display a characteristic pattern of movement. Sites such as Shelter Bay, Snug Harbor, Northwest Bay, and Sheep Bay moved initially down or to the left following impacts, and then, during recovery in subsequent years, crossed back near their earlier origins. During 1992, the category 3 site at Northwest Bay departed slightly from the pattern and made a dramatic leap away from the other category 3 sites. This shift was primarily caused by the appearance of several opportunistic polychaete species. In the lower right group, the unoiled Sheep Bay site has recovered from an unknown impact noted in 1991 (bent and dislodged rebar site markers suggest an ice-grounding incident).

Discussion

Impacts

Washing beaches.—From the battery of inferential testing and multivariate plots, the effects from high-pressure hot-water washing have been demonstrated to be both severe and long lasting. Previous studies of treatment effects following oil spills indicate that some cleanup methods often result in more short- and long-term ecological damage than that caused by the oil itself (Smith 1968; Lindstedt-Siva 1979; Baca et al. 1987; Houghton et al. 1996; Lees et al. 1996, this volume). Water pressures generated through fire-hose nozzles, pressure-wash nozzles, and the multiport nozzles of the Omnibarge equipment (to 7 kg/cm^2, 100 pounds/in^2) were sufficient to penetrate several centimeters deep into the surface sediments, sluicing the sediments and disrupting the infauna. Water temperatures commonly used for the *Exxon Valdez* cleanup (60°C on the vessels, probably slightly lower on the beach; Nauman 1991) are well above the acute tolerances of temperate intertidal species (e.g., Graham et al. 1975). The types of impacts seen in the side-by-side study at Northwest Bay in April 1989 (mortalities, displacement, loss of fines, burial, and oil dispersion) suggest that similar impacts could be expected to some degree at all mixed-soft sediment sites washed with hot water.

One problem with the category comparisons in this study lies in broadly defining high-pressure hot-water washing as a category of treatment, because this treatment encompasses a potentially wide variation of washing and cleaning techniques. For example, a given beach may receive a hotter or cooler wash either mechanically or by hand and for a longer or shorter duration than an adjacent section of beach. After washing, bioremediation fertilizer was commonly applied by hand, sometimes several times throughout the seasons (and years) of cleanup activities. Each of these factors is an unknown variable in this study.

The severity of washing (i.e., water temperature and duration of washing) is essentially undocumented and thus a confounding error; however, it may also be a trivial variable. From a few monitored studies (Lees et al. 1996), any contact with 60°C water was an acute shock for the coldwater Alaskan species. Minor changes in nozzle temperature would probably not make significant differences in impacts. Variation in duration of flushing would presumably cause significant differences in bulk sediment factors versus total wash time; that is, the longer the flush, the greater the residual proportion of coarse sediment fractions. But most of the surficial fines, those containing the most available and labile organics for the bulk of the shallow infauna, would likely disappear during the initial flushing disturbance, thus diminishing the effect of longer washing times. The mode of hot-water application (e.g., hand-held hoses versus enormous showerhead nozzles) may influence the patchiness of survivors, but its influence is impossible to document or quantify. And finally, with only three category 3 sites in the study, all of which were probably bioremediated, the effects of bioremediation will also remain confounded. All of these confounding errors are presumed to be a part of the variance of cleanup effects and are considered trivial in comparison to the magnitude of the category effect.

Speculating on the relative consequences of thermal versus hydraulic effects, it seems probable that thermal effects, though lethal, should be less disruptive to the recovery process. Recolonization would be rapid in an intertidal area that was depopulated thermally (a natural occurrence in some intertidal communities) but with sediment fines, TOC, and TKN still intact. In contrast, hydraulic flushing, regardless of the temperature, physically alters the habitat.

Oiling.—Although apparently less damaging to infauna than the high-pressure hot-water washes, oiling effects alone still had significant impact. The biotic indices at category 2 oiled sites dropped in 1990 and trailed the levels at category 1 sites (Figure 1) until 1992. In 1992, category 2 sites experienced blooms of several species, thus exceeding the category 1 sites in overall mean species richness and mean total abundance even though the dominant taxa were similar for both site categories. The recent bloom at the category 2 sites, the beginning recovery at category 3 washed sites, the modest overall correlation of PAH with the 1992 biotic indices and the poor correlation with 1992 PCA scores suggest that residual oiling currently has little impact (discussed below).

Several formal and informal reviewers were concerned that the category 3 sites tend as a group to be more heavily oiled than the category 2 sites to which they are often compared (Table 1); thus, when assessing differences in group averages, the cleaning effects are confounded with degree-of-oiling effects. Although the relative impacts of each effect cannot be fully assessed in this study, there is evidence from the Block Island site (category 2) and the Northwest Bay West Arm sites (categories 2 and 3) that oiling is less catastrophic than the effects from high-pressure hot-water cleaning. The adjacent washed and unwashed heavily oiled sites in the West Arm of Northwest Bay perhaps best demonstrate this point: the unwashed area sustained significantly less impact than the adjacent washed area (Table 3).

The lower tidal elevation at the Block Island site was also heavily oiled and undisturbed. Because of local geomorphology, the oil was still present in 1994 and seemingly unweathered in the shallow subsurface sediments. Despite the heavy residual oiling, this site has a rich infauna assemblage and one of the most abundant populations of Pacific littleneck clams compared to all the sites we have visited in Prince William Sound. In the multivariate analyses (Figure 3), the Block Island site scored in the lower-right cluster with Sheep Bay (unoiled category 1) and Herring Bay (lightly oiled category 2). Had heavy oiling caused a more significant impact than high-pressure hot-water cleaning, this heavily oiled site would be expected to score in the left-side cluster amidst the species-impoverished category 3 sites.

Describing the Block Island site as species rich and possessing abundant clam resources does not deny the impact of oiling. In a small-scale experiment transplanting Pacific littleneck clams across the gradient of oiling at the Block Island site, reduction in survival and growth correlated well with increased levels of oiling (Houghton et al. 1993a, 1993b). An expanded series of transplants were initiated in 1994 to further document this effect.

Infaunal indices.—The infaunal biotic indices for 1989 to 1992 continue to show a strong disparity between the category 1 or 2 sites versus the category 3 sites. Category 1 sites have shown fairly consistent annual increases in total abundance, while category 2 sites dipped in 1990 and then increased with a bloom in 1992 (Figure 1). Category 3 sites are still quite impoverished relative to the others, showing consistent but modest annual increases. In 1991 to 1992, two-factor ANOVAs accounting for the yearly variation still show significant differences among treatment categories in total abundance, although species richness is no longer significantly different.

Demonstrating the dynamic nature of beaches, species richness and diversity dipped slightly in 1990 at all site categories but returned in 1992. The progression of the category 1 Outside Bay site toward the right in the multivariate analysis (Figure 3) reveals the magnitude of shift or yearly variation that can occur at a site that was not impacted by oiling or cleaning. The shift implies the addition of species to the site's assemblage, a situation that suggests recovery is occurring from a natural impact (probably the extremely cold winter of 1988–1989 followed by El Niño warming). The accrual of species obviously cannot continue indefinitely; the next significant perturbation in environmental factors or population dynamics should cause a consolidation or negative shift in the assemblages. The significant differences in yearly variation of biotic indices and the annual shifts of category 1 sites in the multivariate analyses affirm that Alaskan beaches, like other beaches worldwide, are not static environments.

Dominant taxa.—Substantial dissimilarities in species composition were observed among the categories, especially contrasting category 1 or 2 with category 3. In 1992, the dominant species were fairly similar at category 1 and 2 sites and substantially different from the dominant taxa at category 3 sites (Table 4). Infauna at category 1 and 2 sites was generally dominated by polychaetes, bivalves, and gastropods, whereas category 3 had higher numbers of crustaceans and polychaetes (Figure 2). Typically, the dominant species at category 1 and 2 sites were either in different proportions or less abundant at category 3 sites. The sharp increase in relative abundance of polychaetes, gastropods, and crustaceans at category 3 sites between 1991 and 1992 indicates the progress of recovery (Figure 2).

Sediment quality.—The significant deficiencies in TOC and TKN at category 3 sites in 1992 confirm the sluicing effect from shoreline washing. Yet with this evidence, it is surprising that there were no significant differences in grain-size distributions. The increase in fines between 1991 and 1992 supports the 1989 field observations that fines were sluiced from the beaches, but apparently the differences are no longer statistically significant. Both TOC and TKN serve as rough indices of bulk sediment nutrients, the food resource of deposit feeders, and of the restoration of organic fines. However, TOC and TKN correlated poorly with PCA site scores, suggesting they exert little influence over gross species distributions in 1992. Total organic carbon correlates with species diversity but with little else. These sediment-quality factors are expected to be important covariates in further detailed analyses of individual species abundance and overall species succession.

The long-term trends in sediment PAH (concentrations decreasing at some sites yet increasing at others) suggest the lower intertidal PAH data set is probably inadequate to assess subtle effects, possibly because of the nonreplicated sampling design at each site (alternatively, seepage from subsurface residual deposits is also possible). Using the available data (presumed accurate for assessing gross magnitude effects), the 1992 biotic indices (H', N, and S) exhibited poor correlation with total sediment PAH in all cases. The PCA analyses did correlate with a PAH gradient along the x-axis. However, the effect disappeared when looking at only the 1992 data. This apparent lack of detectable community responses to a wide range of hydrocarbon concentrations (from below detection to 780 parts per billion) tentatively implies that residual hydrocarbon concentrations in the sediments are not appreciably influencing the condition or recovery of the infaunal community.

These suppositions from limited PAH data and subjective interpretation of PCA results are confirmed somewhat by analyses of Exxon Corporation's infaunal data from their area-wide, stratified random sampling of boulder–cobble and mixed-soft habitats in 1990 (Gilfillan et al. 1995). Rerunning their detrended canonical correspondence analyses of lower intertidal infauna from these two habitats demonstrated that the canonical variates, the degree of initial oiling, accounted for 9.9% and 15.7%, respectively, of the total variance. This estimate is probably low and confounded with cleanup effects but still verifies the magnitude of influence. By 1992, the effects should be even smaller as the oil continues to weather.

Species richness was negatively correlated (and marginally insignificant) with treatment categories ($r = -0.55$, $P = 0.13$) in 1992 data but was even more poorly correlated with sediment grain size, TOC, and TKN. Based on these relationships and the supposition that residual PAH are not exerting much influence, it is interesting to ask what factors related to the oil spill and the shoreline cleanup program are still acting to inhibit recovery at the category 3 sites. Possible factors include biological phenomena such as microbial or meiofaunal impacts, or successional events that depend on chance recruitment and the passage of time, or physical factors, primarily altered sediment quality as discussed above.

Recovery

As depicted in the multivariate plots (Figure 3), recovery appears to be underway at all impacted sites. In 1992, the category 3 hot-water-washed sites were just beginning to show modest increases in species richness and total abundance (Figure 1) but still remained clustered at the species-impoverished end of the PCA plot (Figure 3). At most category 2 sites the recovery was well along, with a few of the sites even clustered amid category 1 reference sites.

The category 1 sites have been fairly stable as a group from the context of biotic indices (Figure 1) but individually show substantial annual variation in the multivariate plots (Figure 3; e.g., from severity of storms, winter conditions, El Niño events, freshwater impacts, or population shifts). Intertidal infaunal assemblages are not static; they show considerable natural seasonal and long-term variability. In this regard, the beaches of Prince William Sound are not expected to achieve any quota levels of abundance or species richness during the recovery process. Instead, the sites should continue to converge in the multivariate plots toward the species-rich reference sites as differences diminish (see Houghton et al. 1996 for general discussion of recovery concepts).

In Prince William Sound, return to pre-spill conditions of populations is expected to occur more rapidly for smaller infaunal organisms than for longer-lived biota. Wormald (1976) showed return to normal for nematode and harpacticoid meiofaunal assemblages within 18 months following a diesel spill on a sandy beach in Hong Kong. Our data show an influx of surficial biota, amphipods, cumaceans, small bivalves, and gastropods during initial recovery.

The degree of impact may affect the "directness" of the recovery track. Sanders (1978) documented wide fluctuations in several infaunal factors at more heavily oiled sites over the first 3 years following the Buzzard's Bay spill of 1969. These fluctuations did not occur at the minimally oiled sites. Equating Sanders's degree of oiling with degree of disturbance in Prince William Sound, the category 3 sites may display dramatic shifts in the coming years while category 2 sites show a more direct trend.

The comparative rates of recovery may also prove interesting in the coming years (Figure 3). In one instance in which a category 1 site, Sheep Bay, was impacted by an unknown stress (probably ice) in 1991, full recovery was achieved by the following year. In contrast, recovery at two category 2 sites and the three category 3 sites has each required two or more years to reachieve their 1989 positions along the primary PCA axis. Are the oil-impacted sites less capable of recovering than the reference sites or was the nature of the disturbance more severe? And will Northwest Bay accelerate its recovery from the PCA cluster of category 3 sites?

In summary, infaunal assemblages from oiled mixed-soft beaches not subjected to high-pressure hot-water washing were generally indistinguishable from unoiled sites by midsummer 1991; high-pressure hot-water-washed sites remained in early stages of recovery in 1992. These results are confounded to an unknown degree by amount of initial oiling, severity and mode of treatment, and impacts from bioremediation.

Implications For Policymakers

This intertidal ecology study was initially designed to identify impacts from the oil spill and to monitor the progress of recovery. The results cannot be validly extrapolated to make statistical estimates of recovery for all of Prince William Sound. The study does, however, validate some commonsense biological principles that a regulating agency should consider in the aftermath of a spill.

At the time of the *Exxon Valdez* spill, few knew the extent of the impacts or the length of the recovery process incurred from high-pressure hot-water washes on beaches like those in Prince William Sound. As cleanup decisions were made, the environmental tradeoffs were unknown and concerns for intertidal life insupportable in the face of the other perceived environmental priorities and political pressures. This study puts information into the crisis managers' handbooks—we now know the costs to the intertidal habitat, and assemblages are substantial and the recovery period lengthy. This technique of shoreline cleanup should thus be used only with due circumspection.

References

Baca, B. J., T. E. Lankford, and E. R. Gundlach. 1987. Recovery of Brittany coastal marshes in the eight years following the *Amoco Cadiz* incident. Pages 459–464 *in* Proceedings, 1987 oil spill conference. American Petroleum Institute Publication 4452, Washington, DC.

Calkins, D. G. 1978. Feeding behavior and major prey species of the sea otter, *Enhydra lutris*, in Montague Strait, Prince William Sound, Alaska. U.S. National Marine Fisheries Service Fishery Bulletin 76:125–131.

Ebert, T. A., and D. C. Lees. 1996. Growth and loss of tagged individuals of the predatory snail *Nucella lamellosa* in areas within the influence of the *Exxon Valdez* oil spill in Prince William Sound. American Fisheries Symposium 18:349–361.

Edgington, E. S. 1987. Randomization tests. Statistics: textbooks and monographs, volume 77, 2nd edition. Marcel Dekker, New York.

Garshelis, D. L., J. A. Garshelis, and A. T. Kimken. 1986. Sea otter time budgets and prey relationships in Alaska. Journal of Wildlife Management 50:637–647.

Gilfillan, E. S., D. S. Page, E. J. Harner, and P. D. Boehm. 1995. Shoreline ecology program for Prince William Sound, Alaska, following the Exxon Valdez oil spill: part 3-biology. Pages 398–443 *in* P. G. Wells, J. N. Butler, and J. S. Hughes, editors. *Exxon Valdez* oil spill: fate and effects in Alaskan waters. American Society for Testing and Materials Special Technical Publication 1219, Philadelphia.

Graham, J. W., R. W. Moncreiff, and P. H. Benson. 1975. Heat treatment for the control of marine fouling at coastal electric generating stations. Pages 926–930 *in* Ocean '75, record of the combined meeting of 1975 IEEE (Institute of Electrical and Electronic Engineers) conference on engineering in the oceanic environment and 11th annual meeting of the Marine Technical Society. New York Institute of Electrical and Electronic Engineers.

Hanna, G. D. 1971. Introduction: biological effects of the earthquake as observed in 1965. Pages 15–34 *in* The great Alaska earthquake of 1964. National Academy of Sciences, Washington, DC.

Haven, S. B. 1971. Effects of land-level changes on intertidal invertebrates. Pages 83–136 *in* The great Alaska earthquake of 1964. National Academy of Sciences, Washington, DC.

Highsmith, R. C., and six coauthors. 1996. Impact of the *Exxon Valdez* oil spill on intertidal biota. American Fisheries Symposium 18:212–237.

Houghton, J. P., D. C. Lees, W. B. Driskell, S. C. Lindstrom, and A. J. Mearns. 1996. Recovery of Prince William Sound intertidal epibiota from *Exxon Valdez* oiling and shoreline treatments, 1989 through 1992. American Fisheries Symposium 18:379–411.

Houghton, J. P., and seven coauthors. 1993a. Evaluation of the condition of Prince William Sound shorelines following the *Exxon Valdez* oil spill and subsequent shoreline treatment: volume II 1991 biological monitoring survey. NOAA (National Oceanic and Atmospheric Administration) Technical Memorandum NOS (National Ocean Service) ORCA (Office of Ocean Resources, Conservation and Assessment) 67, Seattle.

Houghton, J. P., and six coauthors. 1993b. Evaluation of the condition of Prince William Sound shorelines following the *Exxon Valdez* oil spill and subsequent shoreline treatment: volume II 1992 biological monitoring survey. NOAA (National Oceanic and Atmospheric Administration) Technical Memorandum NOS (National Ocean Service ORCA (Office of Ocean Resources, Conservation and Assessment) 73, Seattle.

Kvitek, R. F., A. K. Fukuyama, B. S. Anderson, and B. Grimm. 1988. Sea otter foraging on deep-burrowing bivalves in a California coastal lagoon. Marine Biology 98:157–167.

Kvitek, R. F., J. S. Oliver, A. R. DeGange, and B. S. Anderson. 1992. Changes in Alaskan soft-bottom prey communities along a gradient in sea otter predation. Ecology 73:413–428.

Lees, D. C., J. P. Houghton, and W. B. Driskell. 1996. Short-term effects of several types of shoreline treatment on rocky intertidal biota in Prince William Sound. American Fisheries Society Symposium 18: 329–348.

Lensink, C. J. 1962. The history and status of sea otters in Alaska. Doctoral dissertation. Purdue University, Westville, Indiana.

Lindstedt-Siva, J. 1979. Ecological impacts of oil spill cleanup: are they significant? Pages 521–524 *in* Proceedings, 1979 oil spill conference. American Petroleum Institute Publication 4308, Washington, DC.

McNeil, W. J., and W. H. Ahnell. 1964. Success of pink salmon spawning relative to size of spawning bed materials. U.S. Fish and Wildlife Service, Special Scientific Report 469.

Michel, J., and M. O. Hayes. 1993. Persistence and weathering of *Exxon Valdez* oil in the intertidal zone—3.5 years later. Pages 279–286 *in* Proceedings, 1993 international oil spill conference: prevention, preparedness, and response. American Petroleum Institute Publication 4580, Washington, DC.

Michel, J., M. O. Hayes, W. J. Sexton, J. C. Gibeaut, and C. Henry. 1991. Trends in natural removal of the *Exxon Valdez* oil spill in Prince William Sound from September 1989 to May 1990. Pages 181–187 *in* Proceedings, 1991 international oil spill conference: prevention, behavior, control, cleanup. American Petroleum Institute Publication 4529, Washington DC.

Nauman, S. A. 1991. Shoreline cleanup: equipment and operations. Pages 141–147 *in* Proceedings, 1991 international oil spill conference: prevention, behavior, control, cleanup. American Petroleum Institute Publication 4529, Washington, DC.

Potvin, C., and D. A. Roff. 1993. Distribution-free and robust statistical methods: viable alternatives to parametric statistics? Ecology 74:1617–1628.

Sanders, H. L. 1978. *Florida* oil spill impact on the Buzzards Bay benthic fauna: West Falmouth. Journal of the Fisheries Research Board of Canada 10:717–730.

Smith, J. E. 1968. Torrey Canyon pollution and marine life. Cambridge University Press, New York.

ter Braak, C. J. F., and I. C. Prentice. 1988. A theory of gradient analysis. Advances in Ecological Research 18:271–317.

VanBlaricom, G. R. 1987. Regulation of mussel population structure in Prince William Sound, Alaska. National Geographic Research 3:501–510.

Wormald, A. P. 1976. Effects of a spill of marine diesel oil on the meiofauna of a sandy beach at Picnic Bay, Hong Kong. Environmental Pollution 11:117–130.

American Fisheries Society Symposium 18:379–411, 1996

Recovery of Prince William Sound Intertidal Epibiota from *Exxon Valdez* Oiling and Shoreline Treatments, 1989 through 1992

J. P. Houghton

Pentec Environmental, Inc., 120 Third Avenue South, Edmonds, Washington 98020, USA

D. C. Lees

Ogden Environmental and Energy Services Company, Inc.
5510 Morehouse Drive, San Diego, California 92121, USA

W. B. Driskell

6536 20th Avenue, Seattle, Washington 98115, USA

S. C. Lindstrom

University of British Columbia, Department of Botany
British Columbia, Vancouver, B. C. V6T 1Z4, Canada

A. J. Mearns

National Oceanic and Atmospheric Administration
National Ocean Service, Office of Ocean Resources, Conservation and Assessment
7600 Sand Point Way N.E., Seattle, Washington 98115, USA

Abstract.—Biological and chemical data from 4 years of quantitative monitoring have been analyzed for trends in recovery of biota inhabiting beaches in Prince William Sound from the effects of the 1989 *Exxon Valdez* oil spill and subsequent cleanup. Stratified random sampling was used from April 1989 through July 1992 to assess biota at sheltered rocky sites subjected to three degrees of disturbance (unoiled, oiled but not hot-water washed, and oiled plus hot-water washed) in 1989. Hot-water treatments used to remove crude oil from the beaches of the sound in 1989 have had severe short-term impacts on intertidal epibenthos. Some high-pressure hot-water treated rocky shores that were stripped of biota in 1989 showed little colonization by 1990. On other oiled rocky shores that were not high-pressure hot-water washed, the majority of the epibiotal community dominants, including rockweed, mussels, barnacles, limpets, drills, and littorines, survived. Several significant differences were noted between the epibiota on unoiled shores and on oiled shores that were not hot-water treated; an even greater number of significant differences in epibiota was noted between those oiled beaches that had not been treated and those that had been treated. In May 1991 epibiotal populations on oiled and untreated shores were still depressed below those on unoiled shores. By July 1991 most high-pressure hot-water washed rocky shores showed some recovery, having been colonized by opportunistic species with planktonic larvae; new growth from remnants of the original biota also contributed to recovery. No significant differences remained between unoiled stations and those that had been oiled and not treated; some significant differences remained between the biota of unoiled and hot-water washed shores. Effects of high-pressure hot-water treatments applied in 1989 remained visible in intertidal assemblages in some areas through 1992. By July 1992 recovery of most rocky shores had progressed considerably, and few differences remained between the unoiled and the high-pressure hot-water washed categories. Individual hot-water washed sites still showed altered community structure that is attributed to the treatment, however, and full recovery of intertidal epibiota is still several years away in these areas.

The TV *Exxon Valdez* ran aground on Bligh Reef in the northeastern part of Prince William Sound, Alaska, on 24 March 1989. Over the next several weeks, a substantial amount of the 40.9 million liters (10.8 million gallons) of spilled Alaska North Slope crude oil was deposited on beaches in the southern and western portions of the sound and on Gulf of Alaska beaches to the southwest. Beach cleanup activities began in Prince William Sound in May 1989 and continued with varying degrees of intensity throughout the summer, ultimately encompassing about 560 km (Harrison 1991) of the 780 km (Page et al. 1995) of oiled shoreline in the sound. At its peak, the cleanup force involved more than 11,000 persons and approximately 1,400 vessels (Carpenter et al. 1991). A primary method of shoreline treatment in 1989 was hydraulic flushing with heated water at moderate to high pressures. The Omni-barge and Maxi-barge were two of the several floating systems used to apply high-pressure heated (38–60°C) water to beaches in Prince William Sound in 1989 (Nauman 1991). Additional cleanup activities were carried out during the summers of 1989, 1990, and 1991, with manual and mechanical techniques often followed by bioremediation (treatment of oiled areas with fertilizer; Pritchard and Costa 1991).

Several studies conducted shortly after the spill provided support for concern about the effects of treatment on shoreline marine life. Studies sponsored by Exxon Corporation provide data that allow inference of the short-term effects of oiling and describe the immediate impacts of two types of hydraulic beach treatments. Both of these high-pressure hot-water washes clearly had significant, similar impacts on intertidal assemblages (Lees et al. 1996, this volume) and showed that

- the pretest intertidal assemblages included all of the dominant taxa that characterize unaffected beaches in Prince William Sound despite 4 months of heavy oiling, that is, the epibiota had largely survived the toxic and physical effects of oiling; and
- the hot water treatment resulted in significant ($P < 0.05$) mortalities of the dominant taxa in the middle intertidal zone and lesser impacts in the lower intertidal zone (below the area directly treated).

An implied working assumption of those directing spill cleanup activities appears to be that stranded oil in any amount is detrimental (e.g., to shoreline marine life, to wildlife, or to aesthetic values). A further assumption is that any biological damage resulting from cleanup activities is justified by the presumed accelerated recovery of the affected environments. Contradicting these assumptions is the consensus of most biological reviewers (Lindstedt-Siva 1979; Broman et al. 1983; Foster et al. 1990; Rolan and Gallagher 1991; Hawkins and Southward 1992) that shoreline treatments may do more immediate ecological harm than good, at least to intertidal resources. Moreover, in most habitats, certain methods of treatment may actually delay rather than accelerate environmental recovery. As noted by Lindstedt-Siva (1991), after protection of human health and safety, the overall goal of spill response should be to minimize the ecological impacts.

The National Oceanic and Atmospheric Administration (NOAA) Hazardous Materials Response and Assessments Division (HMRAD) provides technical advice to the Federal On-Scene Coordinator (FOSC) who directs oil spill response and cleanup activities. Following the *Exxon Valdez* spill, NOAA saw a need to learn more about the effects and effectiveness of shoreline treatment methods. Thus, the present study was initiated to address the following objectives:

- to assess and compare the impacts of oiling and shoreline treatment activities (specifically, effects of high-pressure hot-water washing) in important littoral habitats in the first through third years following the spill (1990–1992); and
- to evaluate rates of recovery over several years in areas receiving different levels of oiling and treatment.

This paper describes results of monitoring intertidal epibiota on high-pressure, hot-water-washed rocky shorelines in Prince William Sound. It is part of a broader study conducted under contract to NOAA. Limited reporting of data collected by the authors under contract to Exxon Corporation in 1989 is also included. Results of the infaunal component of the broader study are reported by Driskell et al. (1996, this volume), and detailed reports of hydrocarbon trends in sediments and mollusc tissues and of age and growth studies are provided by Houghton et al. (1993a, 1993b).

For purposes of this study, recovery is defined as the return to a state within the limits of predisturbance natural variability (Ganning et al. 1984; Hawkins and Southward 1992). We have used unoiled reference stations as one measure of this natural range of variability. In addition, monitoring of species abundances and assemblage variables over time is expected to reveal an initial period of directed change during early recovery, followed by achievement of some level of recurrent or oscillating equilibrium at which time recovery can be said to be complete.

To document progress toward recovery, our approach involved examination of a broad spectrum of variables representative of the status of and trends in a variety of intertidal assemblages and species. The intent was to produce a body of data covering potential responses of a wide range of biological indicators to hydrocarbon contamination and to various disturbances caused by shoreline treatment. The study was designed to separate the effects of hydrocarbon contamination and shoreline treatment and to compare rates and patterns of

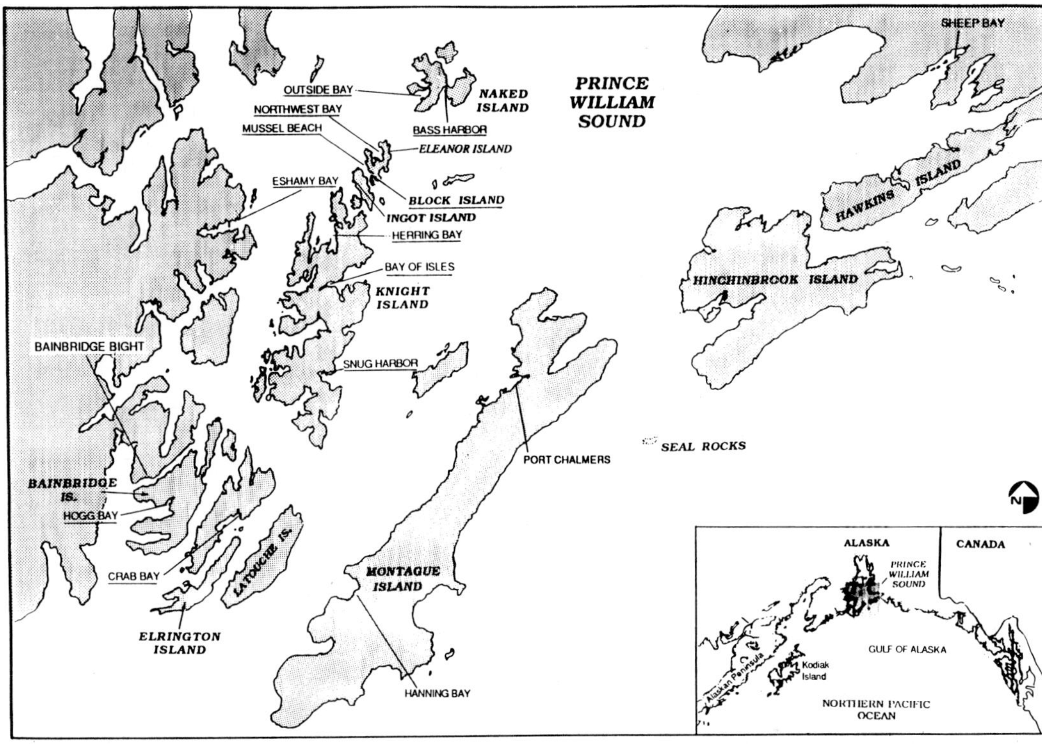

FIGURE 1.—Prince William Sound study area, with site locations underlined.

recovery in oiled but untreated areas with those in areas that were hot-water washed.

The Study Area

Prince William Sound is a protected fjord system located on the south-central coast of Alaska (Figure 1). Wave action from North Pacific storms is blocked by the outer line of islands. The winds, however, are only minimally abated by the low-lying peaks of those islands. This topography generates storm seas and chop that strike exposed shorelines with high-intensity wave action during storm events. Within embayments, wave energy may be minimal despite high wind forces because of limited fetch and frequent shifts in wind direction (Bascom 1964; Lethcoe and Lethcoe 1989). Tides are of the mixed semidiurnal type, with a mean tide level of about 1.8 m and an extreme range of more than 5 m.

The study area encompassed most of central and southern Prince William Sound (Figure 1). Rocky sites were concentrated along the chain of islands stretching from Naked Island (in the central sound), south-southwest through the Knight Island group, to the islands protecting the southwest entrances to the sound. This portion of the sound lay in the path of movement of oil from the *Exxon Valdez*, and many beaches in these islands were oiled.

Protected and semiprotected rocky intertidal habitats within Prince William Sound support an attached and motile epibiota similar in many respects to that seen on other protected beaches throughout the Northeast Pacific (Nybakken 1969; Kozloff 1973; Ricketts et al. 1985). These areas are visually dominated by rockweed *Fucus gardneri* in the central portion of the intertidal zone between about mean high water and mean lower low water. In Prince William Sound a suite of common rockweed-associated fauna grouped consistently in cluster analyses performed on epibiota sampled in 1990 (Houghton et al. 1991). These common taxa are mussels *Mytilus*, littorine snails (two species), limpets (several species), and barnacles (primarily *Semibalanus balanoides*). They are frequently joined by the hermit crab *Pagurus hirsutiusculus*, filamentous green algae (several taxa), and another barnacle *Chthamalus dalli*. This paper emphasizes monitoring results for these dominant taxa.

TABLE 1.—Rocky intertidal stations sampled in Prince William Sound by oiling–treatment category, 1989–1992. Hvy = heavy; Mod = moderate; Lt = light.

	Station category (degree of oiling)[a,b]		
Elevation	Category 1	Category 2	Category 3
Upper	Bass Harbor[d,e,g,h,i]	Herring Bay (Hvy)[d,e,g,h,i]	Mussel Beach S (Hvy)[d,e,g,h,i]
	Eshamy Bay[d,e,h,i]	Outside Bay (Lt)[d,e,g,h,i]	Northwest Bay Islet (Hvy)[d,e,g,h,i]
	Hogg Bay[d,e,g,h,i]	Snug (Hvy)[d,e,g,h,i]	Block Island (Hvy)[g,h,i]
			Elrington East (Hvy)[i]
			Mussel Beach N (Hvy)[i]
			Elrington Islet N (Hvy)[i]
			Elrington Islet W (Hvy)[i]
			Elrington Islet E (Hvy)[i]
Middle	Crab Bay[c,d,e,g,h,i]	Herring Bay (Hvy)[c,d,e,g,h,i]	Block Island (Hvy)[e,g,h,i]
	Eshamy Bay[c,d,e,h,i]	Outside Bay (Lt)[c,d,e,g,h,i]	Northwest Bay Islet (Hvy)[c,d,e,g,h,i]
	Hogg Bay[c,d,e,f,g,h,i]	Snug (Hvy)[c,d,e,g,h,i]	Northwest Bay West Arm (Mod)[d,h,i]
		Bay of Isles (Lt)[c,d,e,h]	Elrington East (Hvy)[d,i]
		Northwest Bay West Arm[j] (Mod)[d,h,i]	Elrington West (Hvy)[d,i]
			Mussel Beach N (Hvy)[i]
Lower	Crab Bay[c,d,e,h,i]	Snug (Lt)[c,d,e,h,i]	Northwest Bay Islet (Hvy)[c,d,e,g,h,i]
	Hogg Bay[c,d,e,h,i]	Outside Bay (Lt)[c,d,e,g,h,i]	Elrington East (Mod)[d,i]
	Eshamy Bay[c,d,f,h,i]		Elrington West (Mod)[d,i]
			Mussel Beach N (Mod)[i]

[a]Category 1 = unoiled; category 2 = oiled, untreated; category 3 = oiled, treated with hot water. Note: Stations categorized as oiled and treated are known to have been treated with some form of hydraulic flushing in 1989.

[b]Footnotes c–i denote sampling times: [c]April-May and July 1989. [d]September 1989. [e]July 1990. [f]September 1990. [g]May 1991. [h]July and September 1991. [i]July 1992.

[j]There is uncertainty regarding treatment history at this site; thus it was not included in any category analyses.

Methods

Sampling Design

A stratified random sampling design was used to assess important intertidal assemblage and population (individual taxon) characteristics for this study. Sampling was structured following Zeh et al. (1981) to obtain statistically reliable estimates of density or cover of macrobiota inhabiting rock surfaces (epibiota) within important life zones of typical habitats.

The intertidal sampling effort was initially stratified according to three habitat types important in Prince William Sound: rocky, boulder–cobble, and mixed. Multiple sites were selected in 1989 to represent each of these habitat types. During the early stages of the spill in spring 1989, sites were selected because they were in areas already impacted by the spill or in areas that appeared likely to be oiled; several of these sites were never oiled and thus became reference stations. Emphasis was placed on sampling in sheltered areas known to be more vulnerable to the long-term impact of oiling (National Research Council 1985) and in some cases in embayments with significant salmon streams. At the time of site selection in 1989, we had no knowledge of the ultimate degree of oiling, nor of the treatment to which a site would be subjected. Sites were added in subsequent years to provide biological sampling at sites already being sampled in the companion NOAA coastal geomorphology study (Michel and Hayes 1993) or to increase the replication within a habitat–treatment category. Only sheltered rocky habitats are covered in this paper (Table 1).

To represent important life zones (i.e., to further stratify the sampling), three elevations (stations) were typically sampled at each site: (1) near the upper limit of attached macrobiota, (2) in the upper portion of the broad rockweed-dominated zone, and (3) along the lower edge of this rockweed zone (Figure 2). Thus, in the terminology of this study, a location, such as Snug Harbor, can have both rocky and mixed-soft sites, and each site can have up to three stations or depth strata to represent different intertidal zones.

Site Classification, Oiling, and Treatment History

Information used to document the site designations was compiled from the authors' direct observations, Exxon Corporation and State of Alaska records of treatments applied to various beach segments, and conversations with others in the field during 1989 (see Appendix Table A-1 in Houghton et al. 1993a).

For statistical testing and for qualitative discus-

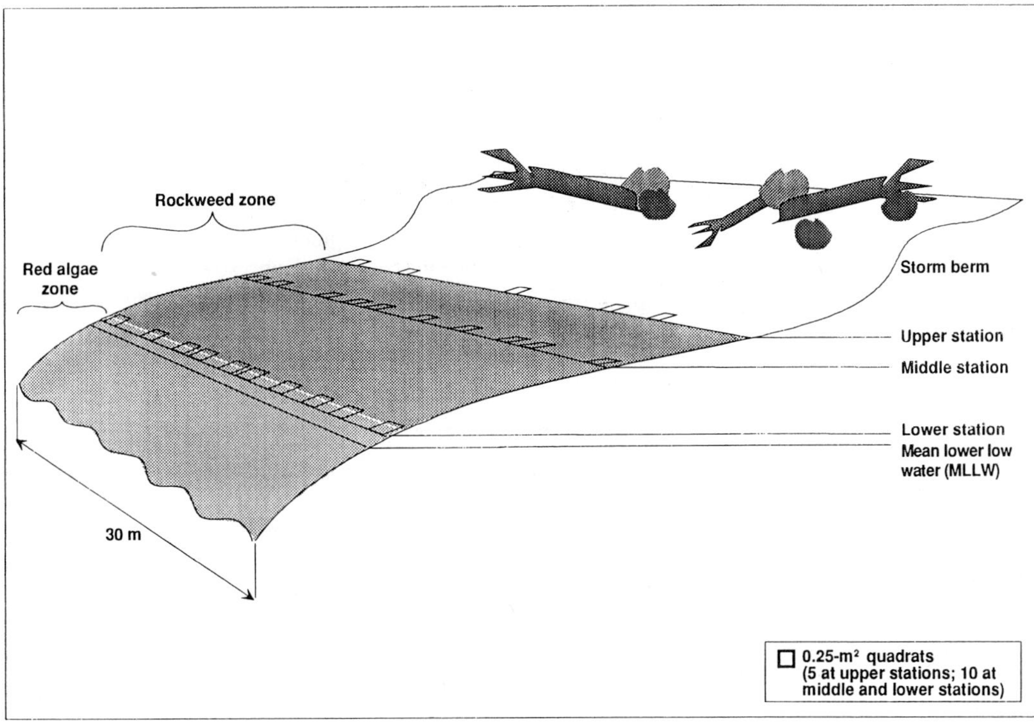

FIGURE 2.—Typical site layout.

sion purposes, stations were assigned to one of three categories to represent the range of possible stresses experienced in 1989:

- category 1: Unoiled in 1989—No significant oiling or treatment reported; considered reference stations.
- category 2: Oiled in 1989—Untreated (set aside) or treated with cool-water flushes in 1989 and/or bioremediation in 1989, 1990, or 1991.
- category 3: Oiled in 1989—Treated with high-pressure hot-water wash(es); most were also bioremediated in 1989, 1990, and/or 1991.

Stations at a given site may or may not be classified in the same category, depending on the site's known treatment history.

Foster et al. (1990), Broman et al. (1983), and Exxon-sponsored Omni-Barge and Corexit 9580 studies (Houghton et al. 1990a; Lees and Houghton 1990; Lees et al. 1993, 1996) provided data on the severity of short-term impacts on intertidal biota from high-pressure hot-water wash treatments. Other forms of treatment (manual pickup, low-temperature flushes, and bioremediation) had far fewer effects on intertidal habitats and biota (personal observation and Foster et al. 1990), and stations thus treated were classified as category 2. This distinction was consistent with the broader study goal of assessing the effect of the dominant treatment method used in 1989 (high-pressure hot-water washes) on long-term recovery.

Ten relatively small (<900 m) lengths of shoreline were set aside and protected from all forms of treatment except some bioremediation. Four of these (two each in Herring Bay and Snug Harbor) were included in the study. Because most oiled sites were treated with some form of bioremediation (nutrient enhancement) in one or more years, effects of bioremediation cannot be isolated within the sample design.

Each intertidal station was classified as to the degree of oiling experienced in 1989 based on first-hand observations of the authors or of NOAA personnel (Houghton et al. 1993a). Oiling was typically very uneven vertically over the intertidal zone and upper elevations were generally much more heavily oiled. As a result, stations at different elevations were independently classified at each site.

This sample design was established to monitor long-term recovery trends at sites with known oiling and treatment history. It is also well suited (by the number of replicate subsamples at each station) for comparison of pairs of stations with similar habitat but different oiling and/or treatment histories. Because only a limited number of stations could be sampled in each habitat–oiling–treatment category, and because sites were not selected by a strictly random process, the design is less well suited for statistical inference regarding the generalized impacts of oiling and treatment over all habitats in Prince William Sound with similar oiling and treatment histories.

Sampling Approach

Field methods.—The abundance of epibiota was measured at two or three intertidal elevations at each site. Five (upper elevation) and 5 to 10 (middle and lower elevation) 0.25-m^2 quadrats were located at positions randomly established during the first site visit along a 30-m sampling transect oriented along the beach contour (Figure 2). Quadrats were repositioned at the same orientation previously sampled with the aid of rebar stakes, spikes, or epoxy markers. Each quadrat was photographed and most taxa were identified in the field. Biological variables measured or estimated included algal cover (% by taxon) and numbers or percent cover (sessile species) of major epibenthic fauna. Relative cover estimates for biota, substratum types, and oiling were based on visual examination of the tops, sides, and overhangs within a quadrat, but rocks were not overturned.

Data analysis.—Detailed abundance data were used in calculations of total algal cover and total taxa present. Some qualitative observations of trends or patterns observed in the course of field surveys are reported without quantitative measurement or without demonstration of statistical significance. Nonparametric randomization analyses of variance (ANOVAs) and *t*-tests (Edgington 1987) were used to evaluate significance of differences overall and between treatment categories. In all tests, comparisons were only made within a given elevation stratum (i.e., middle stations were only compared with middle stations). The abundance data were then lumped into higher taxonomic groupings (e.g., all species of limpets into the family Lottiidae) to further examine treatment effects on selected species groups. In these tests the mean of all subsamples (quadrats) at a given station was used to represent each variable; thus, N = the number of stations within that category and elevation stratum where the variable in question was measured.

The statistical objectives and approach of this study have evolved year-by-year. Initially, the interest was merely to demonstrate impacts related to oiling or treatment—were there significant differences and what were they? In the initial 1990 snapshot view, it was important not to make a type II error, so the alpha critical value was set at 0.10. Following suggested unrestricted least-significant-difference procedures of Saville (1990), randomization ANOVAs were run for each habitat–tidal elevation category; if overall significance was found (alpha < 0.1), then one-way multicomparison *t*-tests were run (also using alpha of 0.1). It was accepted that the overall experiment-wise alpha may be increased (worse case when there are no differences among the treatments), but the alternative was to fix the multicomparison alpha at some minimal level and lose control of type II errors. From our field observations in 1989 (Lees et al. 1996), the nature and degree of immediate impacts of hot-water washing were known, so the risk of a type I error was minimized and certainly would be much less, for example, than the compounded alpha estimated by a Bonferroni calculation.

In succeeding years, the interest shifted to tracking recovery from effects previously observed. Although the analytical procedures are presented in the same manner (ANOVA of the current year's data followed by multiple comparisons), the primary units of experimental interest were the planned comparisons between treatments.

Aside from the hypothesis-testing procedures, the recovery process is also being tracked using multivariate ordinations (nonmetric multidimensional scaling). These analyses generally confirm the results of the inferential statistics (see Houghton et al. 1993b) and are not presented in this paper. However, these additional analyses support our "loosened-alpha" approach by independently confirming our general impression of recovery at individual sites.

Results

Direct Observations of Treatment Effects

At the time of this report, information is available on initial effects of oiling and/or treatment at two sites. Direct effects of both oiling and treatment in 1989 were documented at one site, a rocky islet in Northwest Bay, Eleanor Island. In addition, effects of treatment were quantified at a second rocky site in the west arm of Northwest Bay. These two sites were located in shoreline segments treated with high-pressure hot-water washing in June 1989. Discussion of these sites is important because there are no other comparable observations in 1989.

Northwest Bay Rocky Islet.—The Northwest Bay Islet middle rocky station (Figures 1 and 3 top) was sampled four times in 1989: during oiling (April), just before high-pressure hot-water washing (May), and twice after washing (June and September). On 7 April, the area was still being oiled with relatively fresh oil, the entire study site was covered with thick black oil, and tide pools were full of oil. Although it seemed unlikely that there would be many survivors, live seaweeds and animals were partially quantified (Table 2).

On 7 May, there was substantially less oil present on the rock and in the adjacent waters. Two tide pools that appeared full of oil on the April visit were filled with water covered by a heavy sheen of oil with some tar–mousse. The entire area, and especially tide-pool catch basins, had many dead invertebrates, including hermit crabs, limpets, chitons, and another grazing mollusk, *Siphonaria thersites*. However, despite the very heavy oiling that had persisted for a month, numerous living animals were counted (Table 2). Algae, except for the most heavily tarred rockweed at the upper elevations, remained alive and structurally sound. Under the mousse-covered rockweed at mid to upper elevations, there was a remarkable activity of littorine snails (*Littorina sitkana* and *L. scutulata*) and limpets, and much of the rock surface appeared free of oil. In many areas (here and elsewhere), it appeared that these grazers were actually cleaning oil from the rock surfaces. The majority of barnacles and mussels were alive.

In June 1989, following high-pressure hot-water washing, substantial quantities of oil remained on the surrounding waters, presumably mobilized, in part at least, by the beach treatment. The temperature of the water used to wash these areas was sufficient to kill small grasses and forbs (*Potentilla* sp.) and to kill leaves on alders (*Alnus* sp.) and spruce trees (*Picea* sp.) above the tide zone. Despite the treatment, virtually all portions of the study area (south side of the islet), from the uppermost high-tide swash line to the lowest exposed rocks, remained covered with oil to some degree, although quantities on the upper rock surfaces had been greatly reduced in most areas. Tide-pool sediments and sediments from the sand and gravel deposits in swales in the bedrock had been hydraulically flushed into the lower intertidal zone, where accumulations of >10 to 20 cm occurred. These sediments were very heavily oiled throughout. The crown of a heavily oiled cobble and pebble berm across the upper part of the island had been lowered by an estimated 0.5 m.

In the broad rockweed zone of the middle station, the biota had been decimated by the cleaning activity. Mean rockweed cover was reduced from 79.6% in May to 10.6% in June ($P < 0.01$, Table 2). Density of littorine snails and limpets declined from 128.8 and 5.0 per 0.25 m^2 to 0.5 and 0.2 per 0.25 m^2, respectively (all $P < 0.01$). Mussel cover dropped from 4.6 to 0.0% ($P < 0.01$), and large numbers of dead mussels were found (mean of 36.8/0.25 m^2 vs. 4.1/0.25 m^2 on 7 May). The total number of taxa present was reduced from 12 to 6. Only one tidepool sculpin (*Oligocottus maculosus*) was seen where numerous limpets, littorines, hermit crabs, and sculpins had resided before treatment. Little recovery was seen at the middle station by September 1989 (Table 2), except that the number of taxa reported had returned to 12.

The Northwest Bay Islet lower rocky station was located at the upper fringe of the red algal zone and, like the middle station discussed above, had been severely affected by cleaning when revisited in June 1989. The majority of algae present, especially the reds *Palmaria* sp. and *P. callophylloides*, were bleached and presumably dead. Cumulative algal coverage had declined from 142% to 71% (Table 3). The number of species increased between the two periods; two species of littorines were not present in May but were the dominant taxa in June (62/0.25 m^2 for the two species combined). This increase may have resulted from animals dislodged and carried downslope as the result of beach washing. They were gone by September 1989 (Table 3).

Northwest Bay West Arm.—In late 1989 and 1990 many areas were noted along the straight, steeply sloping rocky shorelines of Northwest Bay where biota had been mostly removed, exposing nearly bare rock; the condition was similar to that at the Northwest Bay Islet study site. Sharp vertical boundaries often separated these severely affected areas from adjacent rocky areas supporting a much richer biota and a near normal cover of rockweed, barnacles, and mussels (Figure 4 top). Repeated observations of the effects of hot-water washing during 1989 suggest that these differences were the result of differences in the intensity of washing.

One such area was sampled in 1989 and resampled in 1991 and 1992. Both a category 3, heavily treated, middle-elevation station (left side of photographs in Figure 4) and an adjacent reference station that had received little, if any, treat-

FIGURE 3.—Middle elevation rocky beaches at (top) category 3, Northwest Bay Islet, 11 July 1991 and (bottom) category 2, Herring Bay, 4 September 1990.

TABLE 2.—Mean abundance of important epibiota at the Northwest Bay Islet middle rocky station in Prince William Sound, 1989. # = number/0.25m^2; a = percent change not calculable; b = abundance not documented. Asterisks denote significant reduction (**$P < 0.05$) from May value or highly significant reduction (***$P < 0.01$) from May value.

Taxon or statistic	April 1989 (pretreatment) Mean	SD	May 1989 (pretreatment) Mean	SD	June 1989 (posttreatment) Mean	SD	September 1989 (posttreatment) Mean	SD	Percent change (May to Sept)
					Plants				
Endocladia muricata (%)	b	b	1.5	4.7	0	0	0	0	−100
Fucus gardneri (%)	72.1	28.0	79.6	23.4	10.6	9.4***	0	0	−100
Fucus gardneri, sporelings (%)	b	b	0	0	0	0	0.1	0.2	a
Neorhodomela larix (%)	1.5	2.4	2.0	6.3	0	0	0	0	−100
Neorhodomela oregona (%)	b	b	0.2	0.6	1.5	3.4	0	0	−100
Number of plant taxa			4		2		4		−75
Total plant cover (%)	73.6		83.3		12.1	***	0.1		−99
					Animals				
Balanus–Semibalanus spp. (%)	8.5	8.2	9.0	9.1	2.8	2.3***	0.2	0.3	−98
Littorina scutulata (#)	b	b	20.4	15.3	0.2	0.6***	7.0	7.9	−65
Littorina sitkana (#)	b	b	108.4	91.2	0.3	1.0***	0.2	0.4	−99
Lottiidae (#)	b	b	5.0	4.1	0.2	0.4***	0.2	0.6	−96
Mytilus cf. *trossulus* (%)	2.7	6.3	4.6	5.2	0	0***	0.2	0.4	−95
Pagurus hirsutiusculus (#)	b	b	0.1	0.3	0	0	0.2	0.4	100
Siphonaria thersites (#)	b	b	0	0	0	0	0.1	0.3	a
Number of animal taxa			8		4		8		15
					Dead animals				
Balanus–Semibalanus spp. (%)	b	b	0	0	0	0	3.9	4.5	a
Lottiidae (#)	b	b	0.2	0.6	0	0	0	0	−100
Mytilus cf. *trossulus* (#)	b	b	4.1	6.4	36.8	55.5**	6.0	7.4	46
Oil cover (%)	100.0	0	100.0	0	86.0	15.8**	21.2	19.2	−79

ment (right side of photographs) were sampled. Based on 1989 observations made by one of us (Houghton), we assume the two adjacent sites had similar prespill biological assemblages and a similar degree of oiling.

In September 1989 oil was present only at the category 3 site. The rocky substratum of the reference station was dominated by a thick cover of rockweed (87.5%) and two red algae (*Neorhodomela larix*, 6.3%; *N. oregona*, 8.3%; Table 4). Barnacles (*Chthamalus dalli*, 9.3% cover), predatory snails (*Nucella lamellosa*, 10.8/0.25 m^2), and limpets (22.3/0.25 m^2) were the dominant fauna. In the category 3 (treated) area live rockweed was very sparse (7% cover), and dead fronds were common. The biota was impoverished, but a small tide pool in one of the four upper intertidal quadrats in the treated area (see Figures F-3A, F-3B in Houghton et al. 1993a) had more representative animals and algae than the remainder of that area. Overall in the treated area, cover of *N. oregona*, *N. larix*, and *C. dalli* was lower than that in the untreated area. Numbers of limpets (0.9/0.25 m^2), and *N. lamellosa* (2.9/0.25 m^2) also were lower in the treated area, whereas littorine snails were more abundant in the treated area.

This site was revisited in 1991 and 1992. As described in Houghton et al. (1993a) and in the remaining data of Table 4, abundance and numbers of species at the category 3 site seemed to be converging with those of the reference site during this period.

In summary, both the effects of the oil and the early effects of high-pressure hot-water washing were readily apparent at these two sites in 1989.

Long-term Trends

General trends.—Three-year (1990–1992) trends in the abundances of various plant and animal taxa among the three broad treatment categories are depicted in Figures 5 through 16 and described below in detail. In general, there were more differences and more dramatic changes at the middle elevation stations than at the upper or lower stations.

Algal assemblages.—In the year following the spill, middle and lower elevations at category 3 sites had lower erect plant cover (Figure 5) and higher encrusting algal cover (Figure 6) than did other site categories. However, by July 1991, no apparent differences in erect plant cover existed among catego-

TABLE 3.—Mean abundance of important epibiota at the Northwest Bay Islet lower rocky station in Prince William Sound, 1989. # = number/0.25m^2; a = percent change not calculable; b = abundance not documented. Asterisks denote significant reduction (*$P < 0.1$; **$P < 0.05$) from May value or highly significant reduction (***$P < 0.01$) from May value.

Taxon or statistic	May 1989 (pretreatment)		June 1989 (posttreatment)		September 1989 (posttreatment)		Percent change (May to Sept)
	Mean	SD	Mean	SD	Mean	SD	
Plants							
Articulated coralline algae (%)	0.1	0.2	2.3	5.0	0	0	−100
Cryptosiphonia woodii (%)	0	0	0	0	3.7	7.8	a
Encrusting coralline algae (%)	0.1	0.2	8.9	16.9	0	0	−100
Encrusting algae (except corallines) (%)	0.8		11.1		8.7		870
Filamentous green algae (%)	20.4	16.4	7.1	16.2	3.0	3.1***	−86
Fucus gardneri (%)	15.4	11.3	22.4	16.5	16.8	18.4	9
Halosaccion glandiforme (%)	6.8	7.5	1.0	1.8	0.1	0.3***	−99
Neorhodomela larix (%)	12.6	12.4	0	0	3.5	9.4	−72
Neorhodomela oregona (%)	32.0	24.1	8.8	9.3	0.1	0.2***	−100
Palmaria callophylloides (%)	6.4	5.9	5.0	8.9	0	0***	−100
Palmaria spp. (%)	2.1	4.4	0	0	0	0*	−100
Rhodoglossum–Mastocarpus (%)	13.6	26.0	2.4	3.2	0.9	2.2*	−94
Ulva–Ulvaria spp. (%)	31.0	32.3	0.1	0.2	0	0***	−100
Number of plant taxa	13.0		18		13.0		0
Total plant cover (%)	141.7		70.7		40.6	***	−68
Animals							
Littorina scutulata (#)	0	0	50.2	109.3	0	0	a
Littorina sitkana (#)	0	0	12.0	33.0	0	0	a
Lottiidae (#)	3.4	3.9	0.4	0.7	2.3	3.5	−33
Mytilus cf. *trossulus* (%)	0	0	1.1	1.6	0.2	0.3	a
Pagurus hirsutiusculus (#)	0.2	0.5	2.0	3.7	2.7	4.3	1250
Searlesia dira (#)	3.6	3.6	0.6	1.0	0	0***	−100
Semibalanus balanoides (%)	0	0	0.4	0.7	0.3	0.4	a
Number of animal taxa	7		11		5		−29
Dead plants							
Articulated coralline algae (%)	0	0	b	b	0.5	0.9	a
Encrusting coralline algae (%)	0	0	b	b	12.7	21.9	a
Fucus gardneri (%)	0	0	b	b	2.3	3.4	a
Halosaccion glandiforme (%)	0	0	0.2	0.7	0	0	a
Neorhodomela spp. (%)	0	0	0.6	1.7	0	0	a
Dead animals							
Mytilus cf. *trossulus* (#)	0	0	5.1	8.2	0.3	1.0	a
Oil cover (%)	100.0	0	72.8	33.5	38.0	45.4	a

ries at middle stations, but the single lower category 3 site (Northwest Bay Islet) continued to have higher encrusting cover and lower erect plant cover compared to the other two site categories.

High variability precluded detection of statistically significant differences among treatment categories in upper elevation total algal cover in 1990 and 1991, but in 1992 there was a higher percent cover of encrusting phaeophyta (brown algae) at category 3 (treated) sites (Table 5). In early May 1991, heavily oiled upper stations had substantial coverage of opportunistic filamentous green algae (mainly *Ulothrix* spp.; Houghton et al. 1993a). No filamentous green algae were recorded at the two category 1 (unoiled) stations sampled. Also in May 1991, filamentous green algae were abundant at a single lightly oiled category 2 middle station and at category 3 middle stations but were almost totally absent at category 1 stations and at heavily oiled category 2 stations. Filamentous red algae (primarily *Bangia fuscopurpurea*, another opportunistic species usually more abundant at upper intertidal elevations) were also more abundant at category 3 middle stations in May 1991.

Qualitative comparisons in July 1990 showed greater average cover of encrusting brown, filamentous brown, and filamentous green algae—some of the more opportunistic members of the lower rocky assemblage—at the category 3 station than at reference stations. Important long-term components

FIGURE 4.—Northwest Bay West Arm rocky sites with category 3 site on left, reference site on right: (top) 10 July 1990 and (bottom) 11 July 1991.

TABLE 4.—Mean abundance of important epibiota at the Northwest Bay West Arm middle rocky station in Prince William Sound, September 1989 (N = 4,4), July 1991 (N = 5,5), and July 1992 (N = 5,5). a = percent change not calculable; b = abundance not documented. t-test: $^{*}P < 0.10$; $^{**}P < 0.05$; $^{***}P < 0.001$.

	1989			1991			1992		
	Category		difference	Category		difference	Category		difference
Lumped taxon	Reference	3	(%)	Reference	3	(%)	Reference	3	(%)
Plants (% cover)									
Elachista fucicola	b	b	b	3.8	0.1	−97**	0.4	0	−100
Encrusting coralline algae	1.4	0	−100	4.6	0.2	−96	2.7	0.2	−93
Encrusting noncoralline algae	9.5	15.2	60	18.0	9.4	−48	10.2	1.1	−89**
Filamentous Chlorophyta	0.1	2.1	1538	1.2	2.0	67	1.5	0.5	−67
Fucus gardneri	87.5	6.9	−92**	88.0	34.4	−61***	85.0	63.0	−26
Fucus gardneri (sporeling)	b	b	b	2.4	2.4	0	0.8	0.2	−75*
Gloiopeltis furcata	b	b	b	0.7	7.2	929**	2.5	4.6	84
Halosaccion glandiforme	0.4	0	−100	2.1	0	−100**	1.0	0	−100***
Mastocarpus papillatus	0.9	0.9	−2	1.4	0	−100**	1.0	0	−100
Neorhodomela larix	6.3	0	−100	5.6	0.1	−98	6.1	0	−100
Neorhodomela oregona	8.3	6.3	−23	11.4	3.4	−70*	5.2	2.4	−54
Pilayella littoralis	b	b	b	8.4	0.1	−99**	1.4	0.8	−43
Number of plant taxa	10.0	6.0	−40	10.6	7.0	−34	12.2	4.8	−61
Total plant cover (%)	115.6	31.6	−73	149.40	61.8	−59***	121.2	74.1	−39***
Animals (% cover or number/0.25m²)									
Chthamalus dalli (%)	9.3	4.6	−50	23.6	15.5	−34	9.2	12.6	37
Littorina scutulata (#)	0.3	9.9	3856**	10.2	312.8	2967**	12.2	433.6	3454***
Littorina sitkana (#)	1.8	2.2	26	62.6	11.6	−81*	6.2	83.8	1252
Lottiidae (#)	22.3	0.9	−96**	47.0	22.4	−52	45.0	42.2	−6
Mytilus cf. *trossulus* (%)	0	0.2	a	0.4	0.5	25	2.5	0.9	−64
Nucella lamellosa (#)	10.8	2.9	−73	7.0	0.6	−91***	7.2	14.6	103
Pagurus hirsutiusculus (#)	3.0	6.1	103	11.2	1.8	−84**	7.8	2.8	−64*
Semibalanus balanoides (%)	0	0.2	a	0.7	18.9	2600	0.9	11.4	1167**
Siphonaria thersites (#)	3.8	0	−100**	21.2	0.2	−99***	63.2	3.2	−***
Mean number of animal taxa	9.0	9.0	0	7.8	6.6	−15	12.6	9.8	−22**
Dead plants (% cover)									
Encrusting coralline algae	0.3	7.7	2960**	0	0	0	0	0	0
Fucus gardneri	1.5	9.0	497	0.2	0.3	50	0.2	0	−100
Oil cover (%)	0	18.4	a**	0	0	0	0	0	0

of the assemblage, such as several taxa of red algae (Rhodomelaceae, Gigartinaceae, *Halosaccion glandiforme*), were less abundant at the category 3 station than at stations from categories 1 and 2 (Figure 7). The lower abundance of rockweed at the lightly oiled category 2 stations is the result of dominance of red algae (Rhodomelaceae and Gigartinaceae) at those stations.

Endocladiaceae.—The family Endocladiaceae is represented in Prince William Sound by *Endocladia muricata*, primarily found at upper elevations, and *Gloiopeltis furcata*, an opportunistic species also normally most common at upper elevations. The Endocladiaceae showed a significant category effect ($P < 0.05$) at middle stations in July 1990, with the highest percent cover at the category 2 sites; no category effect was seen in 1991 (Table 6). In 1992 the Endocladiaceae again showed category differences ($P < 0.1$), with cover at category 3 middle stations significantly higher than at category 1 middle stations.

In 1990 through 1992, *Gloiopeltis* was found mainly in its normal distributional range at upper stations at categories 1 and 2 sites; in July 1991 this species was most abundant at middle stations at category 3 sites (Figure 8).

Other red algae.—Non-encrusting red algae are typical understory plants at middle elevations on rocky shores in Prince William Sound and may dominate the substratum at lower elevations. There were no significant category effects in distributions of red algae (all erect taxa other than Endocladiaceae) at upper or middle elevation stations (data not shown). The apparently greater, but nonsignificant, abundance of red algae at category 2 lower rocky stations (cf. category 1) throughout the study (Figure 7) is largely the result of site differences, with especially high abundance (>90%) at Outside

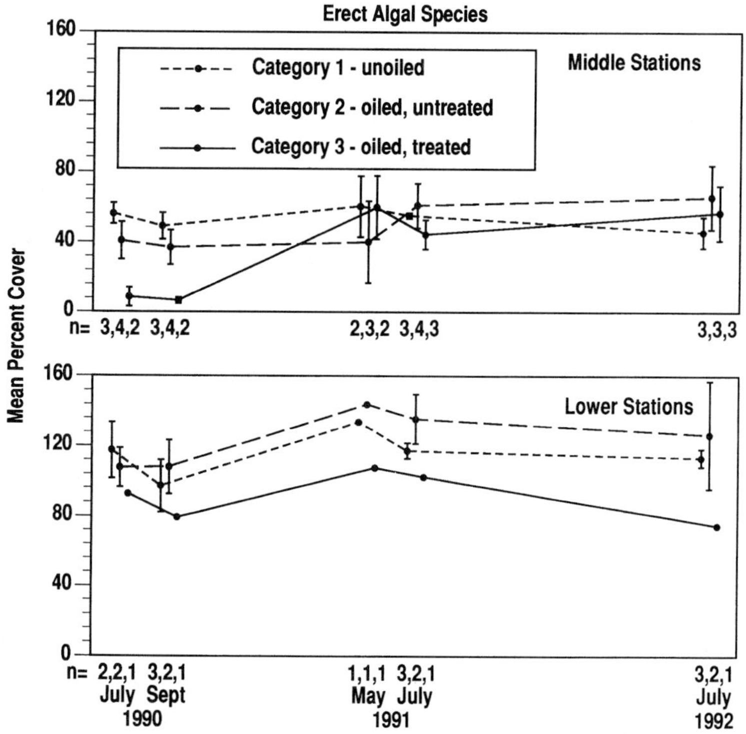

FIGURE 5.—Mean cumulative percent cover (±1 SE) of erect algal species from rocky sites, 1990–1992. The number of stations sampled (n) for each category (1, 2, 3) is shown below the x-axis.

Bay. Finally, as discussed above and below, the single category 3 lower elevation rocky transect at the Northwest Bay Islet suffered prolonged depression of non-encrusting red algae (Figure 7).

Although data on coralline algae are not presented here, there were impacts as cited in the discussion (below).

Fucus gardneri.—At middle-elevations, differences in *Fucus* cover among site categories were significant (ANOVA, $P < 0.05$) in July 1990 (Table 6; Figure 9). Differences between categories 1 and 2 were not significant, but persisted through May 1991, suggesting a continued effect of oiling alone. By July 1991 these small differences had disappeared, but cover at category 3 stations remained significantly below that at category 1 stations (P = 0.1). By the summer of 1992, *Fucus* cover at category 3 middle rocky stations had nearly doubled from that in 1991, and there were no longer any significant category differences.

However, at lower elevations, *Fucus* cover at the untreated category 2 lower stations remained low compared to the other site categories (Table 7; Figure 9). This probably resulted from the relatively minor impact of oil at lower stations or the high percent cover of erect red algae (Figure 7), which normally dominate algal cover in the lower intertidal.

Invertebrate assemblages.—The mean number of invertebrate taxa on rocky shores in Prince William Sound has been consistently higher at unoiled (category 1) sites than at oiled sites at all elevations (Tables 5 through 7). This relationship has diminished over time, however, and no significant category effects were seen in ANOVAs at any elevation in 1992. The overall increase in apparent faunal richness over time is at least partly the result of the initiation of identification of limpets (Lottiidae) to the species level where possible.

Littorina scutulata and L. sitkana.—There were no significant differences in the abundances of the two littorine species between categories 1 (unoiled) and 2 (oiled, untreated) stations (Tables 5 and 6; Figures 10 and 11). However, At upper rocky stations *L. scutulata* densities changed from near absence at category 3 sites in July 1990 to be most abundant at these sites in July 1992 (Figure 10). Abundances at middle and lower stations were

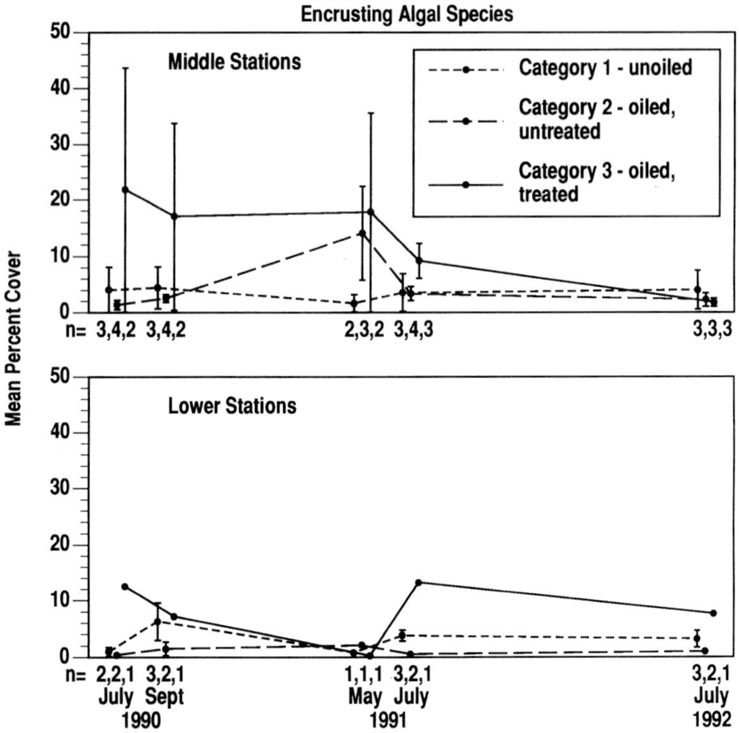

FIGURE 6.—Mean cumulative percent cover (±1 SE) of encrusting algal species from rocky sites, 1990–1992. The number of stations sampled (n) for each category (1, 2, 3) is shown below the x-axis.

much higher in 1991 and 1992 at category 3 sites than at the other site categories. However, despite the apparent strength of temporal trends seen in littorine abundance, differences among site categories were not significant (Tables 5 through 7).

The abundances of *L. sitkana* at the category 3 upper stations were relatively high in July 1990 (Figure 11), primarily because of concentrations of large individuals among boulders at one site. Abundance at category 3 upper stations declined through winter 1990–1991, but by July 1992 densities exceeded those at other site categories. At middle elevations at category 3 sites, *L. sitkana* densities reflected the impacts of high-pressure hot-water washes and were very low in 1990 and early 1991; numbers increased somewhat in July 1991 and matched the abundance at other site categories by July 1992 (Figure 11).

Lottiidae.—Limpets (all species grouped as Lottiidae) suffered nearly total mortality in the high-pressure hot-water wash tests reported by Lees et al. (1996). At category 2 and 3 upper elevation stations, few limpets were observed over all 3 years (Figure 12), suggesting strong effects of oiling. However, at middle elevation stations, there were virtually no differences in limpet abundance between categories 1 and 2 (Figure 12), suggesting that effects from oiling alone were relatively minor at this elevation. The depressed abundances throughout 1990 and early 1991 at category 3 middle stations ($P < 0.1$) indicate significant effects from high-pressure hot-water washing. Numbers at these category 3 middle stations gradually increased in the summer of 1991 (Figure 12), but remained smaller than at category 1 stations ($P = 0.1$; Table 6). However, by summer of 1992, limpet numbers had increased to the point where they exceeded those at the category 1 and 2 sites.

Limpets have pelagic larval stages, so recovery of populations was expected to occur quickly in the upper and middle elevations. However, juvenile limpets were not abundant at any elevation or category until 1992 (Houghton et al. 1993b). At this time, juveniles were abundant only at the middle and lower elevations: middle elevation category 2 sites had the most juveniles ($P < 0.10$; ANOVA; Houghton et al. 1993b). At the low elevation, how-

TABLE 5.—Mean abundance of important epibiota at upper rocky sites in Prince William Sound, July 1990–1992, by category. Values shown are means of all station means within the elevation–treatment category. Asterisks on treatment category 3 values indicate significant analysis of variance at $^{**}P < 0.05$. The category in parentheses was significantly different in a randomization t-test.

Lumped taxon/category	1990 1	1990 2	1990 3	1991 1	1991 2	1991 3	1992 1	1992 2	1992 3
Plants (% cover)									
Encrusting Phaeophyta	0.2	0.3	0.0	1.0	0.0	0.0	0.0 (3)	0.5 (3)	1.5**
Encrusting Rhodophyta	0.0	0.0	0.0	0.4	0.6	0.5	0.9	0.5	1.0
Endocladiaceae	0.0	0.0	0.0	3.4	2.0	0.0	1.1	2.5	0.1
Filamentous Chlorophyta	0.0	0.0	0.1	0.0	0.0	1.0	0.0	0.5	0.5
Fucus gardneri	0.4	4.2	0.6	0.3	2.2	0.2	1.5	6.5	2.2
Fucus gardneri (sporeling)	0.0	0.0	0.0	0.4	0.4	0.1	0.6	0.4	0.2
Cyanophyta	0.0	0.0	0.0	0.0 (2)	35.7	34.6	0.0	0.0	0.4
Verrucaria spp.	30.2	0.0	0.0	31.7	7.1	6.3	26.4	1.6	8.4
Mean number of plant taxa	2.3	1.7	1.5	4.3	5.0	4.7	5.3	4.7	7.0
Total plant cover (%)	32.1	6.1	0.9	37.3	48.4	46.5	30.8	13.2	14.6
Animals (% cover or number/0.25m²)									
Balanus glandula (%)	0.0 (2)	0.0	0.0	0.5	0.3	1.6	0.3 (2)	0.0	0.5
Balanus–Semibalanus spp. (%)	2.9	1.5	0.3	3.1	0.2	0.2	0.1	0.2	0.1
Chthamalus dalli (%)	0.0	0.0	0.0	1.3	0.2	0.2	1.8	0.4	0.4
Ligia sp. (#)	0.0	0.0	0.0	0.0	1.9	0.0	0.9	0.5	0.1
Littorina scutulata (#)	54.0	83.1	1.8	30.3	21.3	22.2	62.5	88.8	148.9
Littorina sitkana (#)	27.0	21.1	64.6	25.9	21.9	32.3	62.4	41.0	99.6
Lottiidae (#)	16.3	1.7	0.0	15.1	2.1	0.9	28.6	2.2	1.7
Musculus spp. (#)	0.0	0.0	0.0	0.0	0.0	0.0	1.2	0.0	0.0
Mytilus cf. *trossulus* (%)	0.8	0.6	0.0	0.7	0.5	0.1	0.3	0.2	0.1
Nucella lamellosa (#)	3.3	0.6	0.0	0.3	0.0	0.0	1.1	0.1	0.0
Semibalanus balanoides (%)	0.0	0.0	0.0	1.2	1.0	0.3	6.9	0.7	1.0
Mean number of animal taxa	7.0	6.7 (3)	4.0	8.0	7.0	5.7	10.0	7.3	8.3
Oil cover (%)	0.0	29.4	21.8	0.0	3.4	6.4	0.0	0.5	0.3
Number of stations	3	3	2	3	3	3	3	3	3

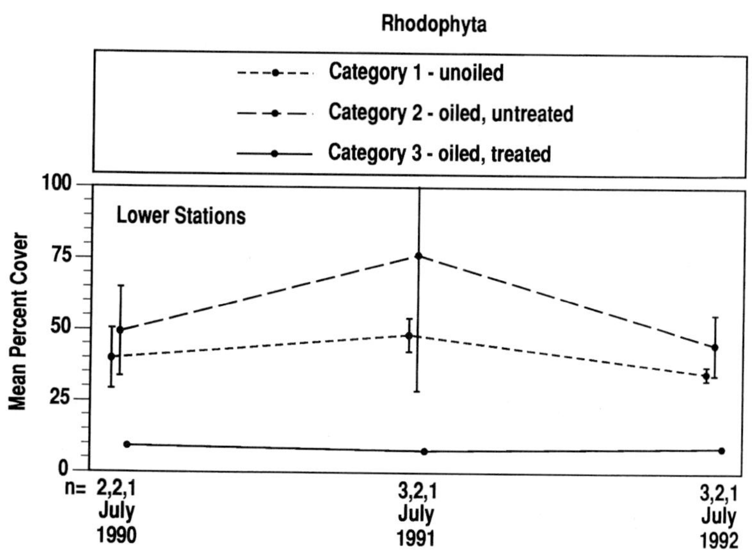

FIGURE 7.—Mean cumulative percent cover (±1 SE) of Rhodophyta from rocky sites, 1990–1992. Includes all erect taxa other than Endocladiaceae. The number of stations sampled (n) for each category (1, 2, 3) is shown below the x-axis.

TABLE 6.—Mean abundance of important epibiota at middle rocky sites in Prince William Sound, July 1990–1992, by category. Values shown are means of all station means within the elevation–treatment category. Asterisks on treatment category 3 values indicate significant analysis of variance at $^{*}P < 0.10$ or $^{**}P < 0.05$. The category in parentheses was significantly different in a randomization t-test.

Lumped taxon/category	1990			1991			1992		
	1	2	3	1	2	3	1	2	3
Plants (% cover)									
Elachista fucicola	0.3	2.1	0.3	0.3	1.4	0.2	1.2	0.3	0.0
Encrusting Phaeophyta	2.5	0.7	13.7	0.0	0.3	0.0	0.2 (3)	0.4 (3)	0.6*
Encrusting Rhodophyta	1.5	0.0	0.1	3.3	2.5	4.2	3.2	1.4	0.6
Endocladiaceae	0.2	1.1	0.4**	1.2	2.0	3.0	1.7 (3)	2.3	2.4*
Filamentous Chlorophyta	1.0	1.1	1.7	0.9	1.6	2.1	0.9	1.9	0.3
Filamentous Phaeophyta	0.3	2.2	0.7**	0.2	6.4	4.8	0.4	5.8	0.4
Fucus gardneri	52.0 (3)	34.8	3.0**	48.3 (3)	43.5	25.8	38.5	51.3	43.6
Fucus gardneri (sporeling)	0.7	1.9	1.2	1.3	1.8	1.3	0.7	0.6	0.1
Gigartinaceae	0.7	0.5	0.0	0.4	0.1	0.0	0.4	0.7	0.0
Halosaccion glandiforme	0.2	0.3	0.0	0.3	0.5	0.1	0.1	0.6	0.0
Rhodomelaceae–*Cryptosiphonia* spp.	1.2	0.7	2.0	1.1	0.8	2.4	0.4	1.5	1.5
Mean number of plant taxa	9.0	10.0	13.0	15.3	13.3	15.7	13.3	20.3	17.7
Total plant cover (%)	60.2	42.3	29.6	58.6	64.2	53.5	49.7	67.9	50.0
Animals (% cover or number/0.25 m²)									
Balanus glandula (%)	2.8	0.0	2.0	1.2	0.3	4.2	3.6	0.4 (3)	0.6
Chthamalus dalli (%)	1.8	3.3	0.1	2.4	7.7	5.8	3.4	8.9	6.7
Littorina scutulata	47.7 (3)	70.5	45.8	51.5	64.1	132.6	54.7	83.6	360.0
Littorina sitkana	66.1	43.2	4.2	123.9	45.4	40.1	91.6	79.2	68.0
Lottiidae	13.4 (3)	17.8	2.2	24.5 (3)	23.6	12.9	22.3	16.5	26.9
Lottiidae (juv.)	0.0	0.0	0.0	0.0	0.0	0.0	6.3	13.6	6.1
Mytilus cf. *trossulus* (%)	1.8	3.8	4.7	3.2	2.7	2.9	4.2	3.4	1.2
Mytilidae, (spat) (%)	0.4	0.8	0.2	2.8	0.3	0.4	3.1	0.4	0.5
Nucella lamellosa	4.1	0.8	0.0	2.9	0.6	0.2	3.7	1.0	7.3
Pagurus hirsutiusculus	4.0	3.4	0.5	3.7	2.3	1.6	11.9 (2,3)	5.1	2.5**
Semibalanus balanoides (%)	4.4	13.6	4.5	8.1	9.0	10.9	3.3	4.3	12.9
Semibalanus cariosus (%)	12.5	0.0	0.0	8.1	0.0	0.0*	5.5	0.0	0.0
Siphonaria thersites	2.6	1.6	0.0	3.8	2.2	0.1	1.4	4.4	1.7
Mean number of animal taxa	12.0	11.0	9.0	16.0	13.5	15.0	19.3	19.7	15.8
Oil cover (%)	0.0	0.4	4.3**	0.0	0.1	0.6	0.0	0.0	0.0
Number of stations	3	4	2	3	4	3	3	3	3

ever, juvenile limpet abundances were highest at the single category 3 site (Northwest Bay Islet).

Predatory snails.—Drills are represented in the assemblages sampled in this study primarily by *Nucella lamellosa*, although *N. lima*, *Searlesia dira*, and *Buccinum baerii* are common in some areas. *Nucella* spp. are usually more abundant at lower and middle elevations, but may make substantial foraging migrations to upper elevations.

Both oiling (category 2) and treatment (category 3) appear to have affected abundances of *Nucella* at lower elevations, with little recovery through July 1992 (Figure 13). At the middle category 3 station at Northwest Bay West Arm, numbers of *N. lamellosa* increased from 0.6/0.25 m^2 in July 1991 to 14.6/0.25 m^2 in July 1992 (Table 4). No Nucella were taken at the lower station at the Northwest Bay Islet before treatment, but *S. dira* was common (3.6/0.25 m^2; Table 3); few drills of any species have been taken at this station since 1989.

Qualitative observations in 1992 showed extensive recruitment of small *N. lamellosa* into other areas with high numbers of barnacles and newly settled mussels. It was expected that abundance of this species would be slow to recover at Category 2 and 3 sites because of the lack of a planktonic larval stage (Strathmann 1987), which limits its ability to recolonize areas with reduced populations. Results of tagging experiments (Houghton et al. 1993a) show *Nucella* to be highly motile as adults, however, and recolonization appears to be progressing through a combination of immigrations and local reproduction of surviving animals.

Hermit crabs.—Hermit crabs *Pagurus* spp. are important scavengers of the middle and lower elevations of rocky and mixed-soft intertidal habitats in Prince William Sound. Large aggregations are often found in association with any large detritus (e.g., salmon carcasses) or in refugia under rocks or in tide pools. This genus was less abundant (but not

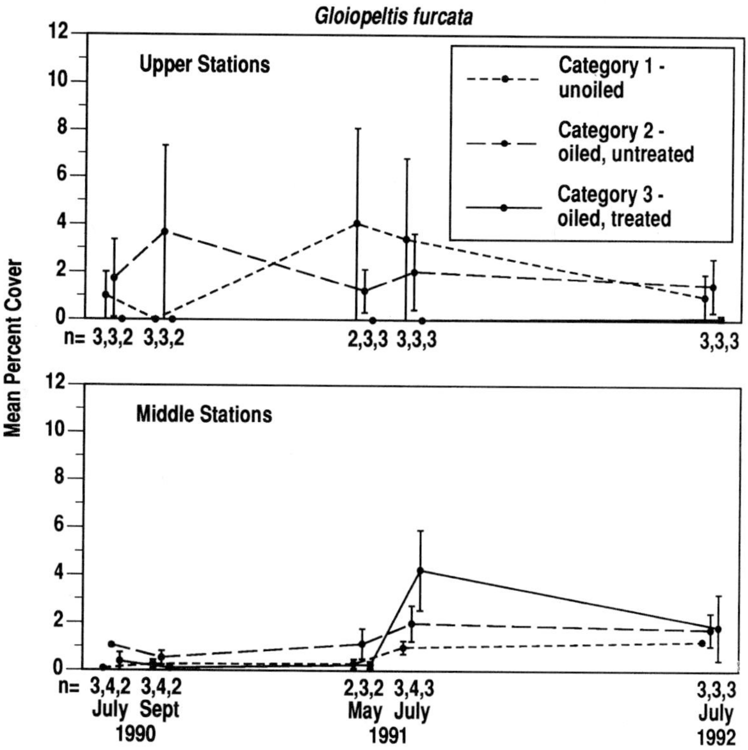

FIGURE 8.—Mean percent cover (±1 SE) of *Gloiopeltis furcata* from rocky sites, 1990–1992. The number of stations sampled (n) for each category (1, 2, 3) is shown below the x-axis.

significantly so) at category 2 and 3 middle rocky stations in 1990 (Figure 14), possibly because of mortalities from oiling and/or treatment. An increase in numbers of *P. hirsutiusculus* in 1992 at middle elevation category 1 sites resulted in a significant category effect in an ANOVA ($P < 0.05$; Table 6); there were also significant differences between category 1 and both category 2 and category 3 in t-tests ($P < 0.05$ and $P < 0.1$, respectively; Table 6). Because there were no significant differences in abundance among categories in either 1990 or 1991, it is unlikely that the significant differences in 1992 are explained by oiling or treatment impacts. Large increases of this species were also seen at the lower elevation category 3 site in 1991 and 1992 (Figure 14). Numbers of hermit crabs increased following treatment at the Northwest Bay Islet site (Tables 2 and 3); apparently, surviving animals congregated to exploit the large detrital resource of dead animals present following treatment.

Barnacles (Balanomorpha).—At upper rocky stations the opportunistic barnacle *Chthamalus dalli* was most abundant at the category 1 sites in 1991 and 1992 (Table 5). At middle stations a major set occurred at Category 2 and 3 sites between May and July 1991, and *Chthamalus* remained somewhat more abundant at these sites through 1992 (Table 6).

The percent cover for all species of barnacles combined at upper stations was consistently highest at Category 1 sites (Figure 15). This is explained in part by the set of another opportunistic barnacle, *Semibalanus balanoides*, in the spring of 1991. *S. balanoides* also successfully recruited to the lower Category 1 and 2 stations in 1990 and 1991 and to the middle Category 3 stations in 1991 (Figure 15). Because of this high abundance of *S. balanoides*, Category 3 middle stations had highest total cover of barnacles after July 1991. At the single lower elevation Category 3 station at Northwest Bay Islet, this species successfully colonized available space in 1990 and 1991 but disappeared by July 1992 (Table 7) probably as a result of predation as biological controls became reestablished.

Mussels (Mytilus cf. trossulus).—High-pressure hot-water wash treatment in 1989 resulted in high mortalities of blue mussels (Lees et al. 1996). At the

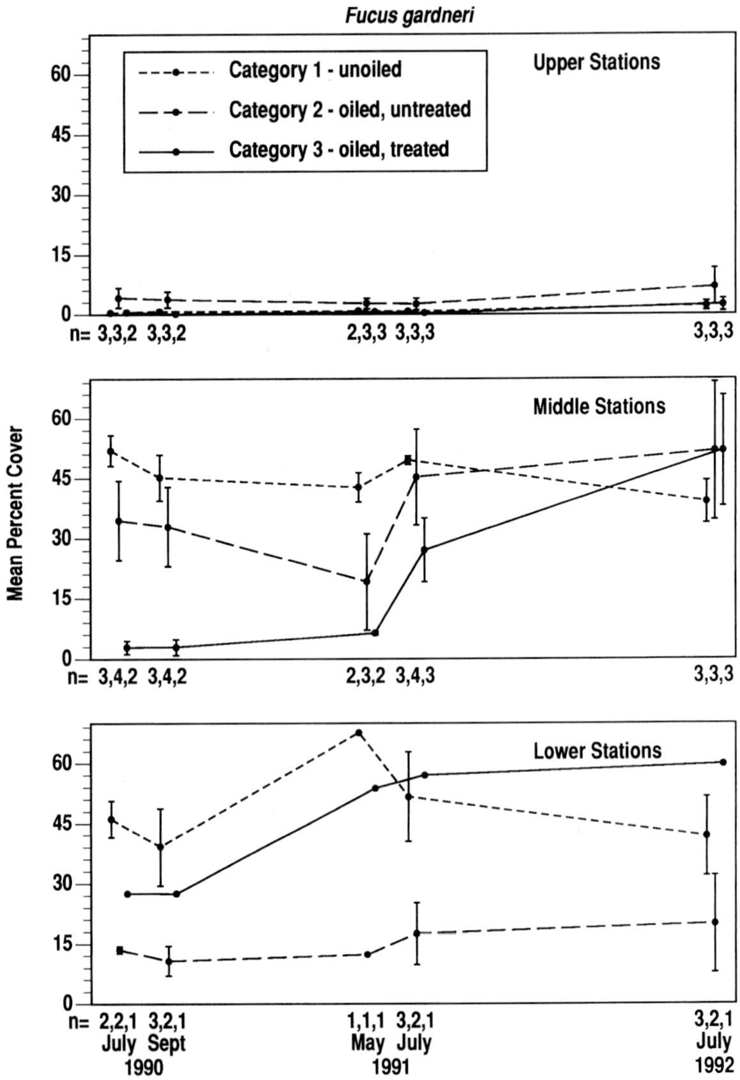

FIGURE 9.—Mean percent cover (±1 SE) of *Fucus gardneri* from rocky sites, 1990–1992. The number of stations sampled (*n*) for each category (1, 2, 3) is shown below the *x*-axis.

Northwest Bay Islet middle station, percent cover of mussels went from 4.6% before treatment (May 1989) to 0.2% following treatment (September 1989; Table 2) and remained below that level until July 1991. By July 1992 *Mytilus* cover (including spat) had increased to 2.6% at this station.

Overall patterns of mussel abundance are not striking except for generally, but not significantly, higher abundances at unoiled (category 1) sites in 1991 and 1992 (Figure 16). Large fluctuations in abundance at category 1 lower stations (Table 7) presumably reflect normal patterns of recruitment and predation at lower intertidal elevations.

Discussion

The primary objective of this study was to assess and compare the impacts of oiling and shoreline treatment activities (specifically, effects of high-pressure hot-water washing) in important littoral habitats in the first through third years following the spill (1990–1992). We conclude from these and

TABLE 7.—Mean abundance of important epibiota at lower rocky sites in Prince William Sound, July 1990–1992, by category. Values shown are means of all station means within the elevation–treatment category. Asterisks on treatment category 3 values indicate significant analysis of variance at $*P < 0.10$ or $***P < 0.01$. The category in parentheses was significantly different in a randomization *t*-test.

	1990			1991			1992		
Lumped taxon/category	1	2	3	1	2	3	1	2	3
Plants (% cover)									
Articulated coralline algae	0.0	0.0	0.0	0.4	0.2	0.0	0.2	2.1	0.1
Delesseriaceae	0.3	1.6	0.0	0.8	5.0	0.1***	1.5	4.6	0.0
Encrusting Phaeophyta	0.2	0.0	5.7	1.2	0.1	1.1	0.6	0.2	5.9
Encrusting Rhodophyta	0.2	0.1	0.6	1.1	0.2	5.1	1.5	0.2	0.8
Filamentous Chlorophyta	14.4	34.8	45.7	10.4	28.8	3.9	17.8	35.6	2.2
Filamentous Phaeophyta	3.2	6.5	6.7	1.8	15.9	25.7	5.5	14.4	0.3
Filamentous Rhodophyta	7.0	3.2	0.2	7.8	12.9	0.4	4.5	4.6	0.1
Foliose green algae	11.9	4.4	2.8	7.5	10.1	9.5	10.6	6.9	1.7
Fucus gardneri	45.4	13.7	27.6*	51.4	17.4	56.7	41.6	19.7	59.6
Gigartinaceae	6.7	4.4	1.9	3.7	4.4	0.7	6.6	2.9	1.2
Halosaccion glandiforme	7.2	0.2	0.4	10.0	1.5	0.2	6.9	0.3	0.5
Palmaria spp.	4.3	13.8	0.0	7.1	9.0	0.0	3.9	4.9	0.1
Rhodomelaceae–*Cryptosiphonia* spp.	14.6	25.4	6.7	12.1	27.7	3.1*	10.9	27.5	5.9*
Mean number of plant taxa	25.0	18.0	19.0	30.0	27.5	27.0	33.3	29.0	28.0
Total plant cover (%)	120.6	110.4	105.3	121.1	135.8	115.5	116.5	127.6	82.2
Animals (% cover or number/0.25 m²)									
Chthamalus dalli (%)	2.6	0.1	3.0	0.9	0.5	2.6	1.2	0.2	0.5
Encrusting bryozoan (%)	8.0	0.2	0.0	6.4	1.9	0.0	5.2	0.7	0.1
Lacuna spp.	0.2	0.0	0.1	154.1	2.1	7.1	4.8	0.8	1.0
Littorina scutulata	0.2	1.8	31.5	0.7	3.8	55.2	2.2	1.3	58.1
Littorina sitkana	0.0	0.3	0.0	0.0	0.1	7.8	0.9	0.5	1.4
Lottiidae	8.5	5.4	12.1	15.5	5.5	36.3	3.9	3.3	0.4
Lottiidae (juv.)	0.0	0.0	0.0	0.5	0.3	0.2	8.8	6.0	143.9*
Margarites spp.	0.2	0.0	0.0	3.3	0.0	0.0	3.4	0.0	0.0
Mytilus cf. *trossulus* (%)	2.7	0.2	0.5	0.0	0.2	3.2	0.0	0.2	0.0
Mytilidae, (spat) (%)	3.3	0.1	0.5	3.4	0.4	0.7	5.0	0.8	0.7*
Nucella lamellosa	31.3	0.6	0.1	30.6	1.9	0.0	16.7	1.8	0.0
Nucella spp.	0.0	0.4	0.0	10.3	0.0	0.0	7.2	0.2	0.0
Pagurus hirsutiusculus	2.3	0.5	2.1	3.3	1.8	11.2	3.6	6.2	23.6
Semibalanus balanoides (%)	0.0	1.3	5.7***	0.0	1.1	4.1	0.0	1.0	0.1
Semibalanus cariosus (%)	6.9	0.0	0.0	13.6	0.0	0.1	10.4	0.0	0.0
Mean number of animal taxa	19.0	14.0	12.0	25.0	18.5	14.0	23.3	17.0	21.0
Oil cover (%)	0.0	0.0	0.0	0.0	0.0	0.0	0.0	0.0	0.0
Number of stations	2	2	1	3	2	1	3	2	1

other data (Driskell et al. 1996; Lees et al. 1996) that although much of the rocky intertidal epibiota survived oiling, much of it was killed or displaced by high-pressure hot-water washing.

A second major objective was to evaluate rates of recovery over several years in areas receiving different levels of oiling and treatment. Overall algal cover at treated and oiled sites has recovered, that is, returned to the range of values found on unoiled sites by 1991. Beyond this, several populations of the intertidal community were still recovering in 1992, three summers after the spill and subsequent treatment. Low-growing turf algae reached overpopulation in mid-intertidal zones in treated areas during one or more of these years. Recovery of *Fucus* cover at middle elevation transects appeared to be delayed by a year in treated areas. Littorines rebounded with possibly enhanced populations, while limpets and predatory snails continued a slow recovery; there remained minor, but perhaps important, differences related to treatment in abundances of hermit crabs and barnacles. Therefore, overall recovery had not been completed by 1992, three summers after the spill and subsequent treatment.

Full assessment of the ecological effects of an event as large as the *Exxon Valdez* oil spill is a major undertaking involving many disciplines and approaches directed at improved understanding of induced changes in a variety of complex natural processes. The number and ranges of variables that must be considered to describe natural conditions are great in a system such as Prince William Sound. When additional factors associated with the spill

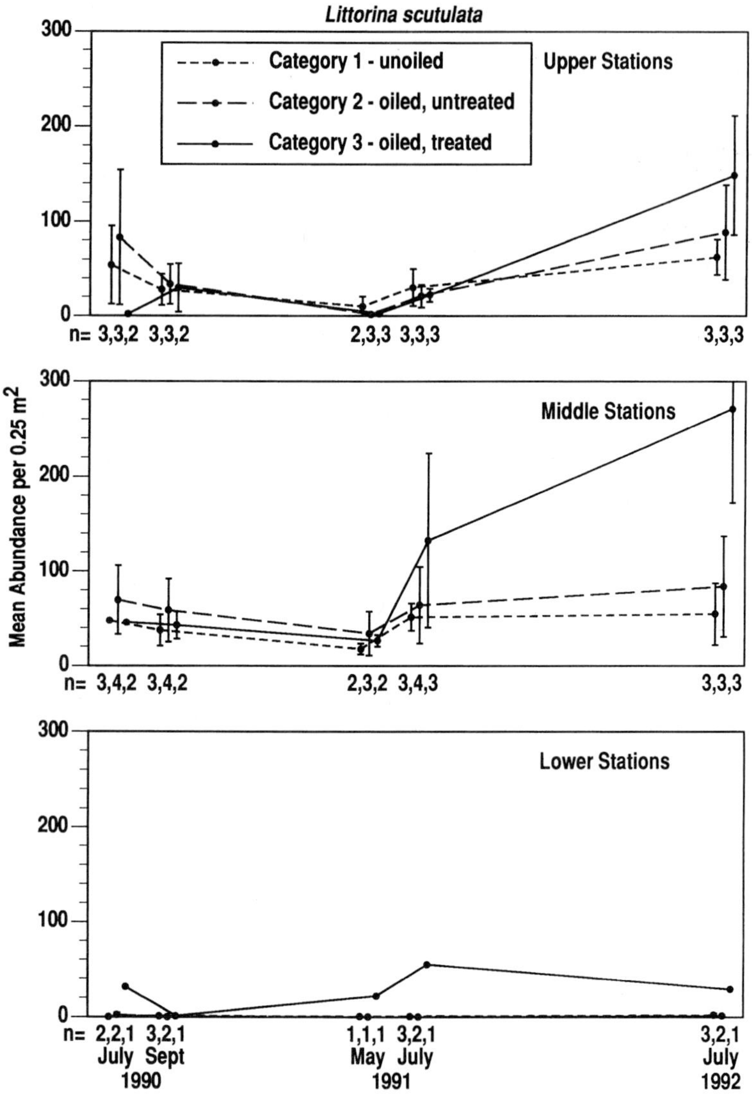

FIGURE 10.—Mean abundance (±1 SE) of *Littorina scutulata* adults from rocky sites, 1990–1992. The number of stations sampled (n) for each category (1, 2, 3) is shown below the x-axis.

and subsequent shoreline treatments must be considered, the task increases dramatically. Although the intent of this study was to explore only a narrow facet of the impact of the *Exxon Valdez* spill (i.e., effects of the spill and cleanup on intertidal assemblages), the task was nonetheless substantial.

Below we discuss some of the factors that must be taken into account in accepting these conclusions, offer suggestions about the future course of recovery, and present guidelines for shoreline washing at future spills in similar environments.

Initial Effects of Oiling and Treatment

Even though prespill data are generally unavailable, conclusions about the effects of oiling and treatment on intertidal epibenthic assemblages can be made by evaluating several types of information, including results from previous studies, and temporal trends among sites and within the three treatment categories. Data from 1989 (ERCE et al. 1990a, 1990b) showed that, before the oiling, the assemblages at several oiled and

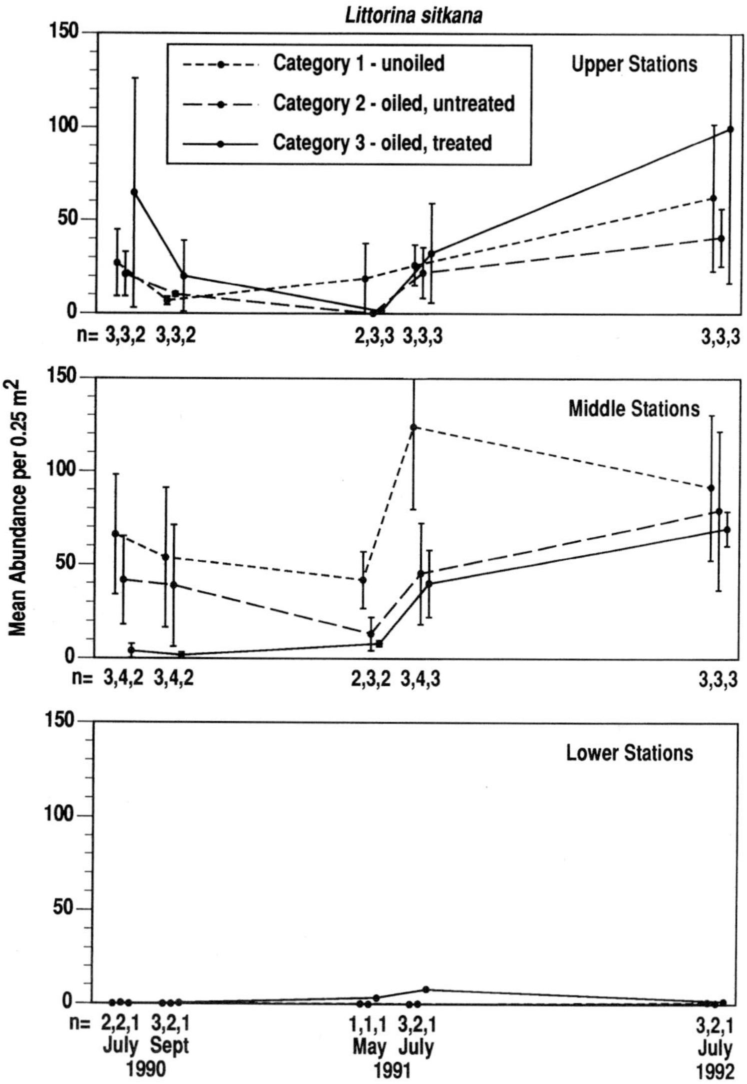

FIGURE 11.—Mean abundance (±1 SE) of *Littorina sitkana* adults from rocky sites, 1990–1992. The number of stations sampled (n) for each category (1, 2, 3) is shown below the x-axis.

treated (category 2 and 3) sites in this study had been typical of these assemblages at unoiled sites in the sound. Remnants of some of the dominant taxa (barnacle base plates, mussel byssus clumps, *Fucus* holdfasts) were evident at the two category 3 rocky middle stations sampled in July 1990 (Houghton et al. 1991). These remnants, coupled with the low occurrence of significant differences within this group of taxa among category 1 and 2 sites throughout the sound, confirm that biota on category 3 rocky beaches before treatment was similar to that on untreated beaches. The nature of the pretreatment community is also reflected by the evidence for survival of these dominant intertidal epibiota for several months on heavily oiled but as yet untreated beaches in 1989 (Lees et al. 1996).

Previous studies of treatment effects following oil spills indicate that some cleanup methods often result in more short- and long-term ecological damage than the oil itself (Smith 1968; Foster and Holmes 1977; Baca et al. 1987; Hawkins and Southward 1992; Lees et al. 1993, 1996). Hot-water washing, steam cleaning, or sand blasting of

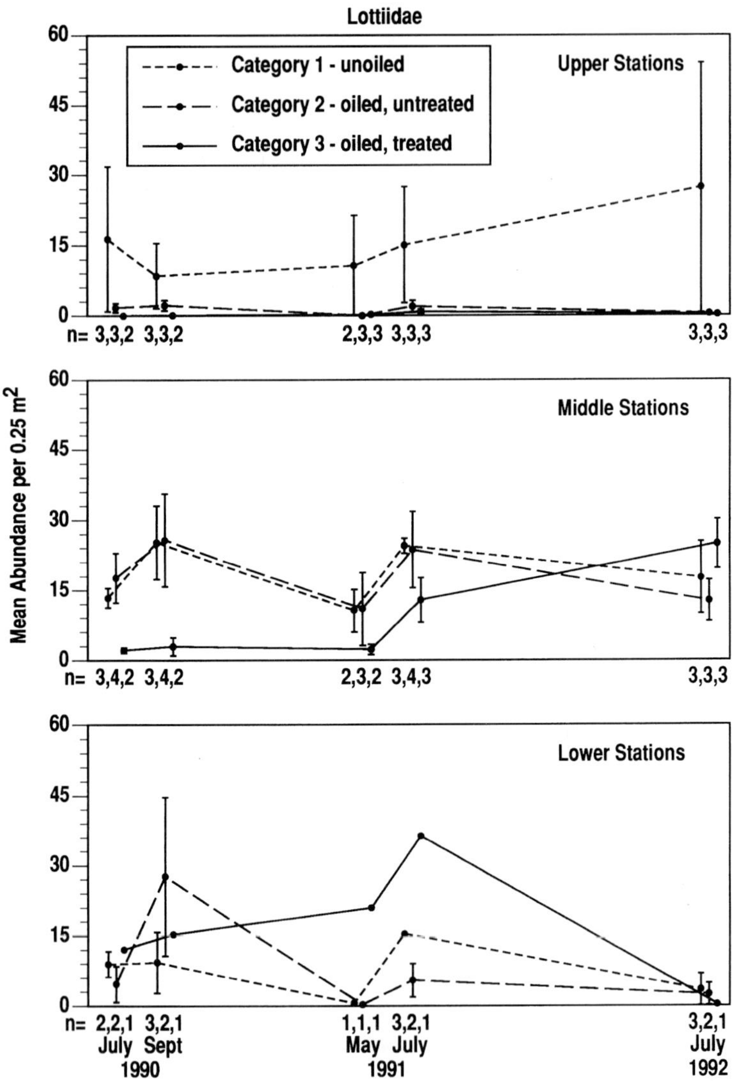

FIGURE 12.—Mean abundance (±1 SE) of Lottiidae adults from rocky sites, 1990–1992. The number of stations sampled (n) for each category (1, 2, 3) is shown below the x-axis.

oiled rocky shores can be expected to denude the rock surface of a great majority of biota (Lindstedt-Siva 1979). Water temperatures commonly used for the *Exxon Valdez* cleanup (60°C on the vessels, probably slightly lower on the beach; Nauman 1991; Mearns 1996, this volume) are well above the short-term lethal tolerances of temperate intertidal species (e.g., Graham et al. 1975; Mearns 1996). Water pressures generated through fire-hose nozzles, pressure-wash nozzles, and the multiport nozzles of the Omni-barge (to 7 kg/cm^2) were probably sufficient, added to thermal stress, to dislodge all but the most firmly attached barnacles and algae. The fact that numerous statistically significant ($P < 0.05$) pretreatment to posttreatment changes were documented in two 1989 hot-water studies (Lees et al. 1996) attests to the biological effects of high-pressure hot water. Even in hot-water washed beaches, patches of mussels and other biota can survive where protected from, or missed by, the direct force of the wash, as seen at our Block Island middle rocky station.

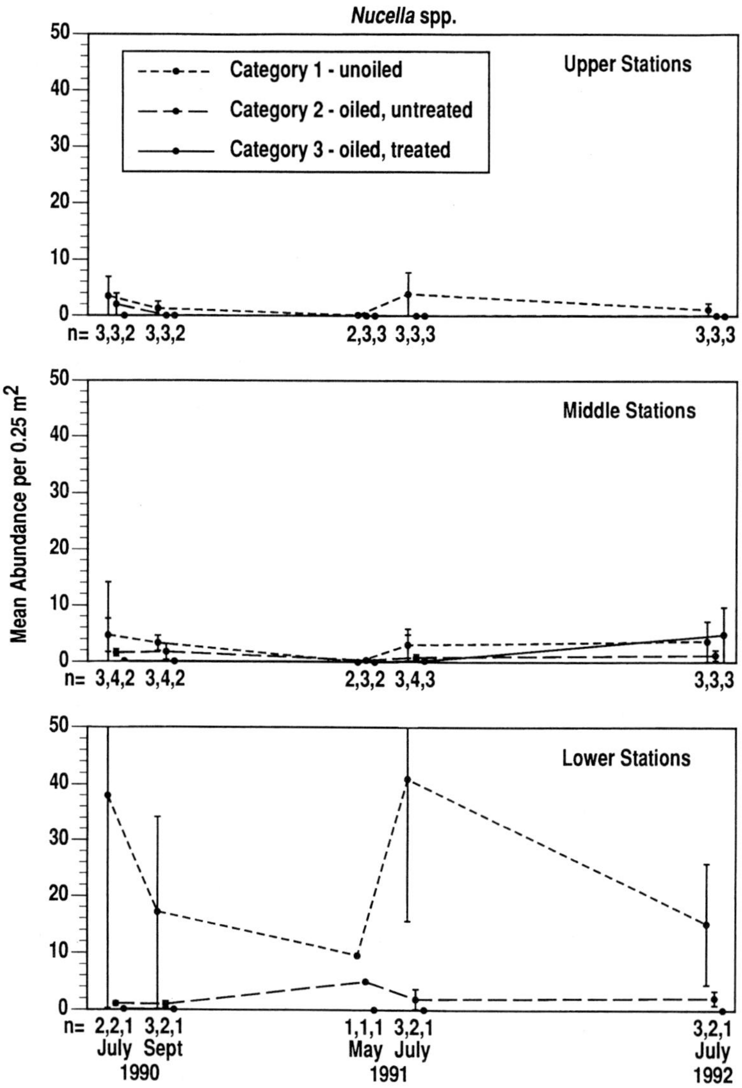

FIGURE 13.—Mean abundance (±1 SE) of *Nucella* spp. adults from rocky sites, 1990–1992. The number of stations sampled (n) for each category (1, 2, 3) is shown below the x-axis.

Definition of Recovery

In requiring that, for recovery to be complete, there must be a return to within the natural range of original functional and structural conditions, Ganning et al. (1984) and Hawkins and Southward (1992) imply a requirement for similarity between stressed assemblages and unstressed reference assemblages in biomass, species composition, age, and trophic structure. We interpreted this to mean oiled (category 2) and treated (category 3) stations are recovered when they are statistically indistinguishable from unoiled reference stations (category 1) for any measurement except by chance. This is the basis for our comparisons and conclusions. There are arguments both for why this definition may be too liberal and why it may be too conservative, however, that will not be discussed in this article.

Recovery Time

In Prince William Sound some rocky beaches that were substantially denuded by cleanup activities, as in the case of the Northwest Bay Islet site (Figure

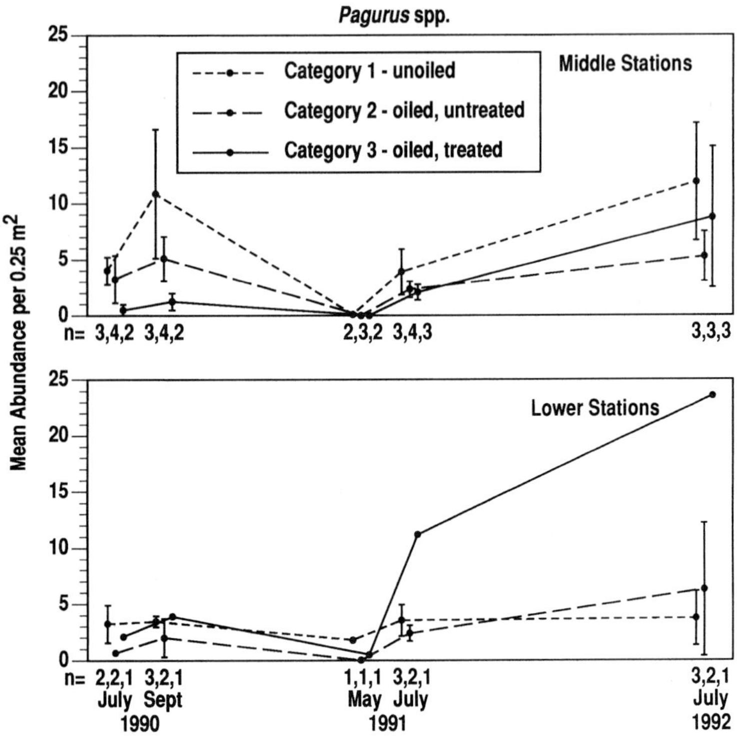

FIGURE 14.—Mean abundance (±1 SE) of *Pagurus* spp. adults from rocky sites, 1990–1992. The number of stations sampled (*n*) for each category (1, 2, 3) is shown below the *x*-axis.

3A), can be expected to take many more years to return to a prespill-like assemblage. Hawkins and Southward (1992) reported that most rocky intertidal assemblage dominants in Cornwall had generally recovered 15 years after the Torrey Canyon spill; Rolan and Gallagher (1991) reported that, 12 years after the Sullom Voe accident in the Shetland Islands, recovery was far from complete in rocky intertidal assemblages that had been vigorously cleaned with heavy equipment.

Sources of Variation—Environmental

Within a marine community, various species can recover at different rates depending on many natural forces and our ability to measure them, as well as on the status of the community itself.

Physical conditions.—Factors that influence the natural conditions of littoral resources—that is, conditions at the unoiled category 1 sites—include substrate, exposure, slope, tidal elevation, natural cycles of disturbance and recovery, and the nature of adjacent communities. The most important of these are discussed below.

On a local scale, exposure is critical in controlling recovery time. The southern exposure and smooth rock surface of much of our Northwest Bay Islet middle and upper elevation stations have greatly retarded recovery of this site relative to other hot-water washed rocky sites. Numerous fissures and pockets in the rock and an easterly or northern exposure at the Northwest Bay West Arm and Block Island rocky sites provide a more hospitable environment for colonization by rockweed zygotes. At these latter locations, zygotes settling in 1989 or 1990 grew well in 1991 and were largely responsible for the substantial recovery of rockweed seen at category 3 stations in July 1991 (Figure 8). Even at these sites, however, the upper limit of the recolonizing rockweed remains well below the prespill elevation (compare left and right sides of Figure 4 bottom). This upper band at these sites, and much of the Northwest Bay Islet bench (Figure 3 top), will be colonized only slowly, as established plants grow and provide the shade and moisture retention necessary for survival progressively upslope, until the physiological limits of the plant are reached. This

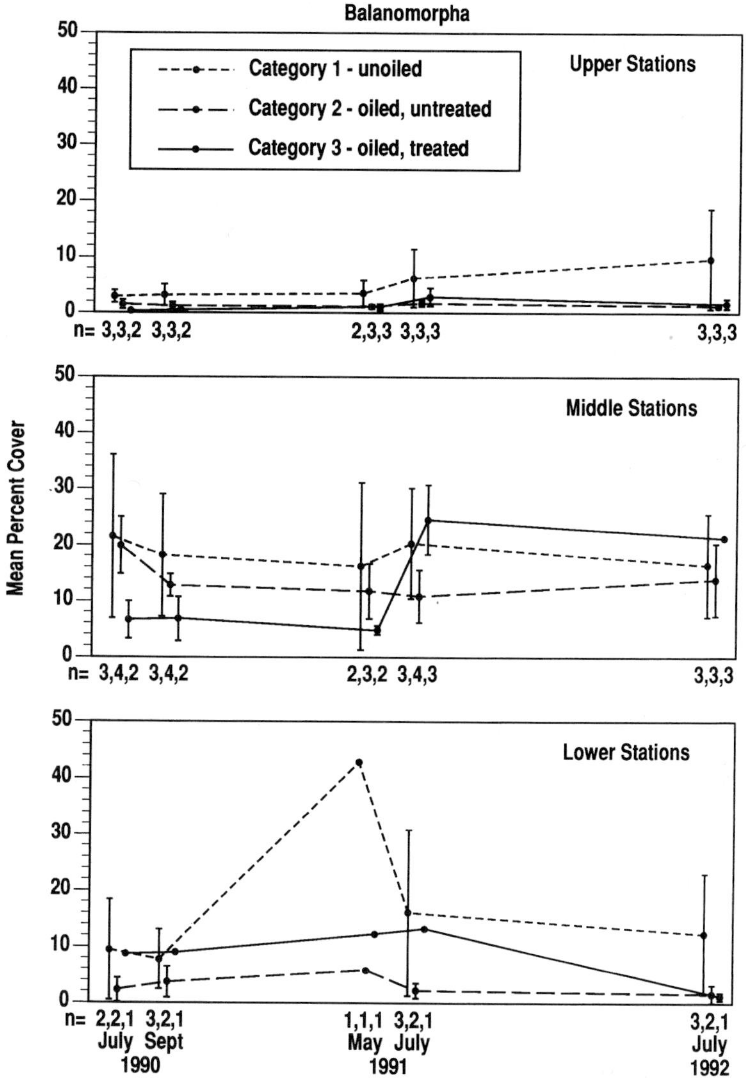

FIGURE 15.—Mean percent cover (±1 SE) of Balanomorpha from rocky sites, 1990–1992. The number of stations sampled (*n*) for each category (1, 2, 3) is shown below the *x*-axis.

slow colonization is an interesting contrast to conditions reported by Southward and Southward (1978), who found that the absence of grazers following the Torrey Canyon cleanup allowed a species of *Fucus* to recolonize rock faces well above its normal upper limit.

There are larger scale factors to consider as well. In 1989, Prince William Sound intertidal communities may still have been in a state of change following the 1964 Alaska earthquake, which uplifted our study sites from 1 to 2.5 m (Hanna 1971) and caused radical changes in intertidal biota (Haven 1971). Some components of these communities, especially mollusks, also were experiencing increasing predation pressure from the growing sea otter populations (Garshelis et al. 1986). The effects of the spill and subsequent treatment on conditions at our oiled category 2 and 3 sites will vary with all these factors. These effects depend further on the amount of oil stranded, degree of repetitive oiling, depth of penetration into the sediment, and the timing and nature of treatment activities (see review by Ganning et al. 1984).

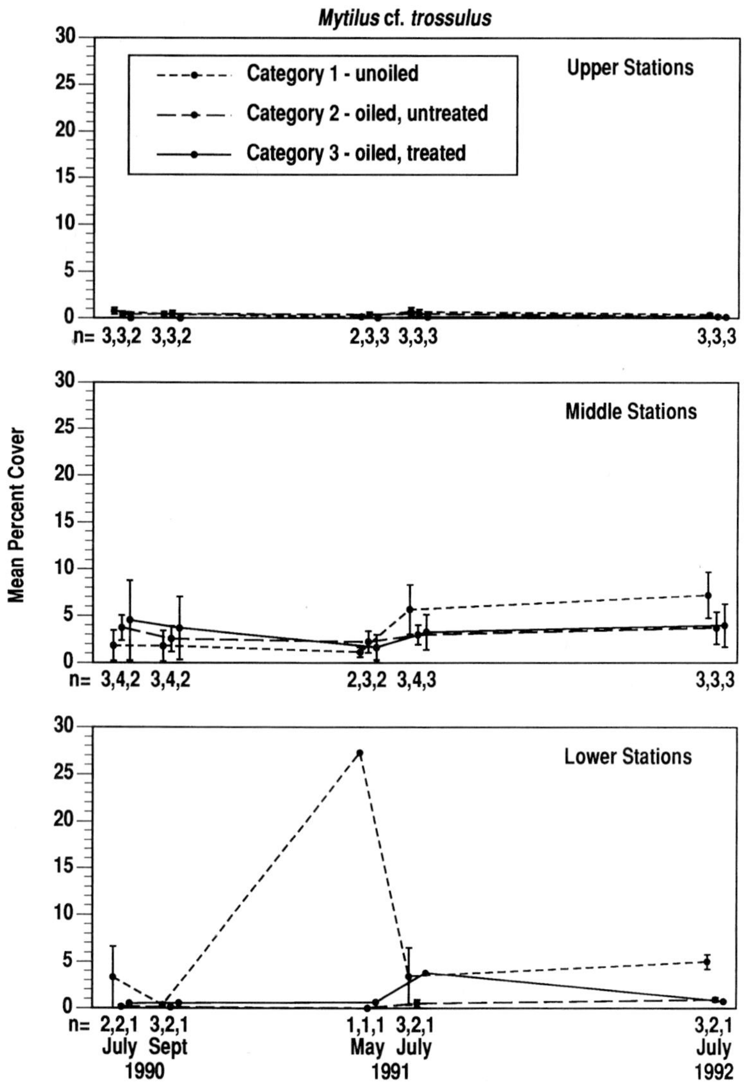

FIGURE 16.—Mean percent cover (±1 SE) of mussels *Mytilus* cf. trossulus from rocky sites, 1990–1992. The number of stations sampled (n) for each category (1, 2, 3) is shown below the x-axis.

Biological controls.—It is not surprising that recovery has progressed more slowly at hot-water washed sites with wider swings of dominance and die-off than at category 2 sites. This pattern has been observed in other oil-impacted rocky communities where important "keystone" species (sensu Paine 1966) have been largely eliminated (e.g., Southward and Southward 1978). Algal colonization was expected to proceed with greater initial success because grazers, such as limpets, chitons, and *Littorina sitkana*-lacking nearby adult populations and/or planktonic larvae, have been slow to control the early colonizers. Where early successional animals became established, as happened with *S. balanoides* at middle and upper intertidal elevations throughout the oiled areas in the sound, predators such as drills were slow to exploit them.

Southward and Southward (1978) reported that long-term recovery following the Torrey Canyon spill and cleanup with toxic dispersants followed classical successional stages. A major factor prolonging the return of denuded beaches (or experimental plots) to predisturbance conditions was the instability in the populations of initial colonizers.

Lack of biological controls (too few herbivores such as littorine snails and limpets to crop the algae and too few predators, such as *Nucella*, to crop the herbivores) caused wide fluctuations in numbers of the few species present (e.g., Nelson-Smith 1968; Southward and Southward 1978). Only a gradual return to the near-natural conditions of biological control allows reestablishment of some form of dynamic equilibrium around the central tendency that can be considered a normal assemblage (i.e., varying, but within the range of variation of the unaffected community).

Reductions in grazers, such as littorines and limpets, that often occur following spills (Southward and Southward 1978; National Research Council 1985) can be expected to result in rapid growth of opportunistic algae, as was observed at many areas in Prince William Sound in spring 1990. Limpets have pelagic larval stages, so recovery of populations was expected to occur quickly. Some species are known to settle gregariously (Dixon 1981), so rapid increases in numbers are possible once recruitment takes place. We found sharp increases in 1992 in limpet numbers at the category 3 lower station at the Northwest Bay Islet site, almost all of which were juveniles. This increase was probably the result of the limpets' reproductive capacity, an abundance of algae, and the reduced numbers of predators at this site.

Near-complete loss of mussels and rockweed at sites such as Northwest Bay Islet rocky stations has changed or eliminated several ecological niches. Absence of mature mussels and rockweed increases desiccation at the rock surface, removes areas of refuge from predators, and increases exposure to wave energy. Reduced numbers of limpets and littorines then allow rapid colonization by opportunistic algae, which may retard reestablishment of rockweed and red algae. The importance of these grazers in controlling algal community structure (Paine and Vadas 1969) and recolonization of rocky shores (Southward and Southward 1978) has long been recognized. Recent experimental manipulations in an Oregon rocky intertidal community (Farrell 1991) confirm that the presence or absence of limpets can also have a profound effect on recolonization by barnacles by physically dislodging newly settled animals.

Barnacle recruitment is known to vary greatly between years and is dependent on larval availability, settlement rates, and early mortality of recruits (e.g., Bertness et al. 1992). An expected rise in numbers of barnacles colonizing open space at the category 3 sites occurred in July 1991 at the middle elevation (primarily *S. balanoides*) but only to numbers already seen at the category 1 sites. This trend was also reported in the sound by Highsmith et al. (1996, this volume) and was seen at Sullom Voe by Rolan and Gallagher (1991). If succession occurs as expected on category 3 sites (i.e., to convergence with conditions at category 1 sites), we would expect *S. balanoides* to remain sparse at the lower stations and to be partially replaced at middle stations by *Balanus glandula*. In other temperate areas succession leads to gradual replacement of barnacles by *Fucus* and *Endocladia* (Farrell 1991). The delayed recovery of barnacle predators such as *Nucella* from the category 3 sites may temper the normal succession pattern, however, and prolong the dominance of *S. balanoides*.

Shoreline treatment.—Treatment itself is a highly variable activity influenced by another set of factors. In Prince William Sound, treatment ranged from simple manual pickup and removal of oiled debris, to multiple washings with heated high-pressure water, followed by application of fertilizers for bioremediation and in some cases excavations with heavy equipment (e.g., Nauman 1991). Even where cleaning was undertaken with only one type of equipment, results can vary with duration of treatment, temperature of wash water (Lees et al. 1996), weather conditions, tidal elevations treated, and tidal elevation during the treatment. This variation and incomplete documentation of exact treatment type and intensity applied to specific beaches leave some uncertainty as to the exact applications at our study sites and stations. The July 1989 Omni-Barge test (Lees et al. 1996) demonstrated that significant mortalities to dominant epibiota occurred following 95- and 165-min washings, with only marginally greater impacts associated with the extended washing time. Our design attempted to account for this large-scale (hundreds of meters) variability by sampling replicate sites and stations within each treatment category and habitat.

Individual members of the beach cleanup crews, working with various degrees of thoroughness, exerted a substantial influence over beach impacts on a smaller scale (meters or centimeters). Within the Omni-Barge test area, one of the quadrats sampled before and after treatment appeared to have been completely missed by the direct force of the water jets; most biota survived the treatment (Houghton et al. 1990). Replicated subsampling along a 30-m section of the beach contour using an adequately sized quadrat (5 to 10 replicates of a 0.25-m^2 quadrat cf. three replicates of a 0.031-m^2 quadrat used by

Gilfillan et al. 1995) assessed variability on this scale.

The hydraulic treatment applications most effective at oil removal (hot water at high pressure) are obviously detrimental to marine plants and animals. Different habitat characteristics, organisms, or groups of organisms respond in varying ways to different aspects of treatment. The principal mechanisms by which the hot-water high-pressure wash process could damage intertidal biota appear to be (1) hydraulic jetting or sand blasting of the epibiota with sand and gravel at all elevations, (2) translocation and deposition of fines and biota to lower elevations by erosion (placer mining), (3) transport of oil to lower elevations, (4) thermal (heat) shock, or (5) a combination of these. In addition, there may be substantial damage caused by movement of personnel and equipment on and across the beaches (e.g., Baker 1977), although this was probably of lesser concern on rocky habitats.

Sources of Variation—Assessment Strategy

The best way to get the information required by NOAA from this study would be to intentionally and uniformly oil and treat carefully selected and structurally uniform shorelines. This approach has been illegal in the United States and Canada for nearly two decades. The alternative—using a spill of opportunity such as that from the *Exxon Valdez*—was accomplished, but with many sacrifices in experimental design and control.

Site selection strategy.—The goal in designing this study was to control a number of variables through stratification within the rocky habitat type and within biological zones of importance (e.g., the rockweed zone) as suggested by Zeh et al. (1981). Pursuit of this goal was limited by several factors, including cost, logistics, and the desire to continue gathering data at sites where the authors collected comparable data in 1989. Several of these sites were first sampled within 2 weeks of the spill and either before or soon after initial oiling.

To allow statistically defensible extrapolation of results, all sites would be randomly selected from within the suite of all sites with similar substratum, exposure, and biological community and would differ only in the degree of oiling and/or treatment. This was attempted for the purposes of injury assessment for a Natural Resources Damage Assessment by Page et al. (1995) and Highsmith et al. (1996). In practice, because of the realities of natural variability, logistical constraints, and the urgency of the situation, this was not possible in the present study. Sites initially sampled in 1989 were established within habitat types considered the most sensitive to the effects of oiling and so that there were multiple stations in oiled and unoiled areas. Logistically, we had to be able to access the sites either by helicopter or by skiff, and there had to be a sufficient extent of the habitat in question to allow establishment of the site setup design. These sites were selected without knowledge of the degree of oiling relative to other sites and without knowledge of future cleanup actions that would be taken. Although this lack of knowledge does not constitute a random selection process, it does demonstrate that the sites were selected without bias with regard to ultimate degree of oiling or treatment.

Design limitations.—Study sites selected were monitored with two broad (and not entirely compatible) objectives in mind: (1) to provide postspill "snapshots" of conditions in key habitats subjected to oiling vis-à-vis those that were unoiled and those that were oiled and high-pressure hot-water washed; and (2) to establish differences between stations and site categories that could be monitored for convergence as recovery progresses.

Optimum design to allow generalizations at any one time regarding differences between site categories (i.e., to define effects of treatment vs. effects of oiling) would call for maximizing the number of sites. This increases the power of statistical tests using station means as replicates within each site category. Optimum design for tracking long-term recovery at individual stations (convergence with an unaffected station of similar habitat) would be met by having multiple random subsamples at each of the stations compared.

Our participation in, and analysis of, two hot-water test washings in 1989 (see Lees et al. 1996) and other research (e.g., Broman et al. 1983) led to a decision to isolate this form of treatment as a variable. No attempt was made to evaluate other forms of treatment or to distinguish among degrees of hot-water wash applications. The study plan was designed, in part, to document the persistence of the effects of 1989 hot-water washes, if they remained evident, over the broader area where hot-water treatments had been applied. Thus, the effects of the oil spill and subsequent shoreline treatment activities on rocky intertidal community structure were examined by building on 1989 data with repeated sampling in 1990, 1991, and 1992. Primary variables isolated in the sampling design were habitat type, tidal elevation, degree of oiling, and use of hot-water high-pressure treatments.

This design assumed that the site (habitat) groupings were similar enough to permit robust comparisons that identify major shifts in species dominance within the assemblages. We believe that the results, despite the limited replication of sites within each station category, accurately suggest the nature of continuing impacts and describe the direction of recovery. Validation of the 1990 to 1992 results will occur within the sequence of longer-term monitoring and trend analyses.

Review of the field design for this study reveals that a preponderance of oiled category 2 and treated category 3 upper and middle stations were heavily oiled; at the lower elevation, category 3 stations tended to be more heavily oiled than category 2 stations, leaving open the potential that some of the differences seen were the result of different oil impacts rather than treatment alone; the radical changes seen in the heavily oiled lower station at Northwest Bay Islet (Table 3) following treatment in 1989, however, provide strong evidence that this is not the case.

In summary, this field design was not established a priori using rigorous statistical procedures; field studies set up in response to emergencies seldom are. Rather, the study was designed to control as much variability as was reasonable (through habitat and elevational stratification) and to provide sufficient sampling area and replication to allow meaningful contrasts to be made. The data obtained provide both statistical and qualitative evidence of real differences in intertidal communities related to effects of oiling and high-pressure hot-water washing. The data also show that, as of 1992, many intertidal areas of the sound clearly had not recovered.

Statistical considerations.—Analysis of variance (randomization ANOVA) tests performed on epibiotal data collected in 1990 (middle stations) indicated that, for certain dominant taxa, there were significant category effects—that is, abundance varied significantly among treatment categories. Multiple comparison tests using the 1990 epibiota data (Houghton et al. 1991) identified significant ($P < 0.1$) differences in abundances of certain taxa between various permutation pairs of site categories that mirrored differences seen, in short-term impact studies, to result from hot-water high-pressure washes (e.g., Lees et al. 1996). The same approach, ANOVA for category effects followed by *t*-tests for significance of differences between pairs of site categories, was applied in 1991 and 1992. Relatively fewer taxa of epibiota continued to show category effects in the 1991 and 1992 ANOVAs.

Because a main purpose of this study was to assess the degree of recovery occurring over time, it was considered important to continue to test for differences between pairs of site categories, even for taxa for which no experiment-wise category effect remained. It is recognized that such multiple comparisons have a statistical penalty in the true experiment-wise alpha (type I error term): differences calculated to have an alpha of 0.1 in the multiple comparison randomization *t*-tests in fact represent differences with a greater than 1 in 10 chance of occurring randomly. The consistency of these trends for dominant organisms and the general patterns over time, however, provide confidence that they are not the result of chance alone.

Implications for Shoreline Treatment

This study poses two kinds of implications for cleaning rocky shores: (1) alternative shoreline cleanup strategies and methods and (2) spill response research.

Shoreline Treatment Alternatives

This study provides strong evidence that high-pressure hot-water washing significantly increased the mortality of intertidal epibiota and delayed recovery by one to several years. Clearly, removal (by high-pressure hot-water washing) of rockweed did not accelerate its recovery. As described in Mearns (1996) hot water (up to 60°C) was used to warm the oil above its pour point, so it would flow, while high pressure was effective in moving oil downslope into the water to be collected by booms and skimmers. Thus, this method was effective in removing visible oil, but at the cost of increasing mortality and delaying biological recovery.

In the future, responders should carefully balance the need to remove all visible oil from rocky shores against the need to protect remaining and surviving biota. Based on the recovery of non-hot-water-washed (category 2) sites, we suggest that oil remaining at these sites was considerably less of a threat to the structure and recovery of intertidal epibiota than that posed by high-pressure hot-water washing.

Therefore, we suggest that to enhance recovery of the rocky intertidal epibiota, the goal should not be to remove all visible oil, but rather to remove only the oil that is pooled and likely to refloat and re-oil other sites or oil that may form tar mats and asphalt pavements. Furthermore, the goal should be to remove this oil with minimal additional disturbance to the surviving biota, particularly in the mid- and lower intertidal elevations.

Methods for achieving these goals include manual pickup (with hand tools and absorbents), passive pickup with absorbent boom, direct vacuuming, and low-pressure ambient-temperature washing with high-volume flooding and, under special circumstances, high-pressure and high-pressure hot-water washing.

Results of this and other studies suggest some of the circumstances that may provide both limits and "windows of opportunity" for each of these methods. For example, long-term damage from hot-water washing of lower rocky intertidal areas was clearly evident at our Northwest Bay Islet rocky site. Most of the oil was initially stranded in the mid-, upper, and supralittoral elevations. Thus, to avoid oiling the lower zone, the uppermost elevations of this high-gradient rocky shore could have been washed when the tide was high and ebbing. Because very little biota was present in the upper and supralittoral elevations, high-pressure hot-water washing may indeed have been effective and relatively benign there. By contrast, the seaweed-rich mid-intertidal elevation could have been treated with low-pressure ambient-temperature washing.

There was a high degree of variability of epibiota obvious in late 1989 and in subsequent years on heavily oiled and treated beaches of the sound. This suggests that the operational guidelines that were invoked (i.e., don't wash the "green" zone) were variously adhered to (see Mearns 1996). With modifications to the operational guidelines and increased monitoring, damage may have been considerably decreased.

In summary, it is of paramount importance that biologists and ecologists knowledgeable in the diverse resources of the region in question be involved in establishing shoreline cleanup objectives. Obviously, these objectives should seek to balance the net ecological benefit (weigh the costs to one resource against the benefits to another) with economic considerations related to fisheries, subsistence use, and recreation. Removal of all of the oil from a beach just because it is there is a very poor cleanup objective from a biological perspective (e.g., if collateral damage is expected to be high or natural cleanup rapid). Although our data strongly indicate that the treatments applied in Prince William Sound were detrimental to intertidal life and delayed recovery of many areas, we recognize that there may be other overriding scientific or economic reasons for cleanup. The immediate need to wash Applegate Rocks and other areas where marine mammals were expected to pup in April or May 1989 is an example of a reasonable trade-off of intertidal productivity for the protection of critical habitat for other species.

Implications for Response Research

An objective in the next cleanup of a moderate-to large-sized spill should be to include controlled experiments to provide definitive information on the effectiveness and effects of cleanup methods applied. These experiments should include physical, chemical, and biological sampling of replicated pairs of cleaned and uncleaned sites stratified by habitat and cleanup method. Monitoring should be conducted before and after treatment and be continued to document impacts of treatment on long-term bioavailability and impacts of hydrocarbons and recovery of assemblages affected by the oiling and treatment.

There remain important and intriguing questions that should be addressed in continuing research on shoreline response methods done outside the venue of spill response. As described in more detail by Mearns (1996) and Lees et al. (1996), there is a clear need to better define the pressures and temperatures that can be tolerated by shoreline communities vis-à-vis the oil removal effectiveness of various combinations of temperature and pressure. We observed what appeared to be cleaning activity by snails and limpets in unwashed areas: do grazing animals that survive oiling actually aid cleanup?

If severe treatment is needed, how important is it to leave patches of untreated shoreline—refugia—to provide nearby sources of organisms for recolonizing barren (treated) areas? How big and how far apart should patches or refugia be located? Indeed, is there some level of incremental damage from which recovery is not greatly delayed? Although the high-pressure hot-water washes eliminated the majority of the flora and fauna at test-site rocky beaches in Prince William Sound (Lees et al. 1996), it does not appear that the extent of rocky beaches denuded by the treatments was as great as in the Torrey Canyon cleanup, where washes were supplemented with a toxic dispersant (Southward and Southward 1978). The dispersant reached into crevices and tide pools as it was swept along the beaches by longshore currents; the result was nearly complete mortality of fauna and severe impacts to flora over extended continuous reaches. By contrast, in Prince William Sound, variability in intensity of high-pressure hot-water treatment (horizontally and vertically) and refugia from direct washing in crevices and undersides of boulders prevented total mortality of the intertidal community.

It is well known that natural processes such as log battering (Dayton 1971) and ice scour (Lees et al. 1979; McCook and Chapman 1992) tend to clear patches of bare substrate on temperate rocky shores. These patches then undergo colonization by species with planktonic larvae, invasion from adjacent undamaged areas, and regrowth from remnants of holdfasts. The broader ecological implications of resetting the successional stage over larger areas of shoreline, as occurred in Prince William Sound, are not clear.

Recovery of *Fucus* would have been much more rapid had high-pressure hot-water washes not led to the complete removal of holdfasts. McCook and Chapman (1992) found that damaged holdfasts of two other species of *Fucus* sprouted adventitious shoots even when basal tissue had been cut to only 2-mm thickness. In most cases, *Fucus* at oiled category 2 sites in this study regenerated by vegetative growth of surviving plants, stems, or holdfasts, while at treated category 3 sites, recovery resulted primarily from settlement and growth of sporelings, a much slower process. Moreover, Stekoll et al. (1996, this volume) showed that loss of a majority of mature plants at affected sites greatly reduced the rate of settlement of *Fucus* zygotes; van Tamelen and Stekoll (1996, this volume) also showed that zygotes were more likely to settle in cracks and crevices than on smooth surfaces. Both of these observations are consistent with findings in this study and provide guidance for directed research on response methods.

Acknowledgments

This work was part of a broader study conducted under contract with NOAA, with funding provided NOAA, the U.S. Environmental Protection Agency (EPA), the U.S. Coast Guard, the Minerals Management Service, the American Petroleum Institute, the Marine Spill Response Corporation, and the Restitution Fund from the 1991 legal settlement between Exxon, the State of Alaska, and the United States. Limited logistical support was provided by Exxon Company, USA, in 1990 and May 1991.

We thank individuals who contributed their efforts and exceptionally good humor to the often arduous field work over the years: Allan Fukuyama of Fukuyama Hironaka Taxonomic and Environmental; Steve Landino, Peter Hague, and Charles Valle of Ogden; John Armstrong of EPA; Tim Reilly of the Marine Spill Response Corporation; Marguerite Holloway, writer for *Scientific American*; Ruth Yender of EPA (now with NOAA); Herb Curl and Rebecca Hoff of NOAA; Anne Holmquist of Chevron Oilfield Research Company; Garry Phillips, Stennie Medours, and Bud Stanley of the Texas Water Commission; Gerald Schimke of the California Office of Emergency Services; Jeff Wheelwright, author; Christina Rumbaitis Del Rio, student; and especially Sandra Lindstrom of the University of British Columbia for invaluable help in algal identification and field quantification.

Charley Henry and his co-workers at the Louisiana State University Institute for Environmental Studies analyzed the multitude of tissue and sediment hydrocarbon samples. We also thank the many peer reviewers who have donated their time to critical assessment of our study design and interpretations, including Michael Foster, Tom Dean, Charles Peterson, Chuck O'Clair, and Robert Paine. Finally, we extend our gratitude to Craig O'Connor, John Knauss, Sylvia Earle, Dave Kennedy, Jean Snider, and Robert Pavia of NOAA for their support of the project and their vigorous efforts to obtain the necessary funding and logistical support.

References

Baca, B. J., T. E. Lankford, and E. R. Gundlach. 1987. Recovery of Brittany coastal marshes in the eight years following the *Amoco Cadiz* incident. Pages 459 to 464 *in* Proceedings, 1987 oil spill conference. American Petroleum Institute Publication 4452, Washington, DC.

Baker, J. M. 1977. Responses of saltmarsh vegetation to oil spills and refinery effluents. Pages 529–542 *in* R. L. Jeffries and A. J. Davy, editors. Ecological processes in coastal environments. Blackwell: British Ecological Society, London.

Bascom, W. 1964. Waves and beaches, the dynamics of the ocean surface. Anchor Books, Doubleday, Garden City, New Jersey.

Bertness, M. D., S. D. Gaines, E. G. Stephens, and P. O. Yund. 1992. Components of recruitment in populations of the acorn barnacle Semibalanus balanoides (Linnaeus). Journal of Experimental Marine Biology and Ecology 156:199–215.

Broman, D., B. Ganning, and C. Lindblad. 1983. Effects of high pressure, hot water shore cleaning after oil spills on shore ecosystems in the Northern Baltic Proper. Marine Environmental Research 10:173–187.

Carpenter, A. D., R. G. Dragnich, and M. T. Smith. 1991. Marine operations and logistics during the *Exxon Valdez* spill cleanup. Pages 205–211 *in* Proceedings, 1991 international oil spill conference: prevention, behavior, control, cleanup. American Petroleum Institute Publication 4529, Washington, DC.

Dayton, P. K. 1971. Competition, disturbance, and community organization: the provision and subsequent utilization of space in a rocky intertidal community. Ecological Monographs 41:351–389.

Dixon, J. 1981. Evidence of gregarious settlement in the larvae of the marine snail *Collisella strigatella* (Carpenter). Veliger 24(2):181–184.

Driskell, W. B., A. K. Fukuyama, J. P. Houghton, D. C. Lees, A. J. Mearns, and G. Shigenaka. 1996. Recovery of Prince William Sound intertidal infauna from *Exxon Valdez* oiling and shoreline treatments, 1989 through 1992. American Fisheries Society Symposium 18:362–378.

Edgington, E. S. 1987. Randomization tests, 2nd edition. Marcel Dekker, New York.

ERCE (ERC Environmental and Energy Services Company, Inc.), Pentec Environmental, Inc., W. B. Driskell, and Dames & Moore. 1990a. Intertidal and shallow subtidal baseline studies in Prince William Sound, Alaska. Draft data report, Exxon Study 5. Dames & Moore, Seattle.

ERCE (ERC Environmental and Energy Services Company, Inc.), Pentec Environmental, Inc., W. B. Driskell, and Dames & Moore. 1990b. Intertidal and shallow subtidal baseline studies in Prince William Sound, Alaska. Draft data report, Exxon Study 4. Dames & Moore, Seattle.

Farrell, T. M. 1991. Models and mechanisms of succession: an example from a rocky intertidal community. Ecological Monographs 61:95–113.

Foster, M. S., and R. W. Holmes. 1977. The Santa Barbara oil spill: an ecological disaster? Pages 166–190 *in* J. Cairns, K. L. Dickson, and E. E. Herricks, editors. Recovery and restoration of damaged ecosystems. University Press of Virginia, Charlottesville.

Foster, M. S., J. A. Tarpley, and S. L. Dearn. 1990. To clean or not to clean: the rationale, methods, and consequences of removing oil from temperate shore. Northwest Environmental Journal 6:105–120.

Ganning, B., D. J. Reish, and D. Straughan. 1984. Recovery and restoration of rocky shores, sandy beaches, tidal flats, and shallow subtidal bottoms impacted by oil spill. Pages 7–35 *in* J. Cairns, Jr., and A. L. Buikema, Jr., editors. Restoration of habitats impacted by oil spills. Butterworth, Boston.

Garshelis, D. L., J. A. Garshelis, and A. T. Kimken. 1986. Sea otter time budgets and prey relationships in Alaska. Journal of Wildlife Management 50:637–647.

Gilfillan, E. S., D. S. Page, E. J. Harner, and P. D. Boehm. 1995. Shoreline ecology program for Prince William Sound, Alaska, following the *Exxon Valdez* oil spill: part 3: biology. Pages 398–443 *in* P. G. Wells, J. N. Butler, and J. S. Hughes, editors. *Exxon Valdez* oil spill: fate and effects in Alaskan waters. American Society for Testing and Materials, Special Technical Publication 1219, Philadelphia.

Graham, J. W., R. W. Moncreiff, and P. H. Benson. 1975. Heat treatment for the control of marine fouling at coastal electric generating stations. Pages 926–930 *in* Ocean '75, record of the combined meeting of 1975 IEEE (Institute of Electrical and Electronic Engineers) conference on engineering in the ocean environment and 11th annual meeting of the Marine Technical Society. New York Institute of Electrical and Electronic Engineers.

Hanna, G. D. 1971. Introduction: biological effects of the earthquake as observed in 1965. Pages 15–34 *in* The great Alaska earthquake of 1964. National Academy of Sciences, Washington, DC.

Harrison, O. R. 1991. An overview of the *Exxon Valdez* oil spill. Pages 313–319 *in* Proceedings, 1991 international oil spill conference: prevention, behavior, control, cleanup. American Petroleum Institute Publication 4529, Washington, DC.

Haven, S. B. 1971. Effects of land-level changes on intertidal invertebrates. Pages 83–136 *in* The great Alaska earthquake of 1964. National Academy of Sciences, Washington, DC.

Hawkins, S. J., and A. J. Southward. 1992. The Torrey Canyon oil spill: recovery of rocky shore communities. Pages 583–631 *in* G. W. Thayer, editor. Restoring the nation's marine environment. Maryland Sea Grant, College Park, Maryland.

Highsmith, R. C., and six coauthors. 1996. Impact of the *Exxon Valdez* oil spill on intertidal biota. American Fisheries Society Symposium 18:212–237.

Houghton, J. P., D. Erikson, and W. B. Driskell. 1990. Biological effects of beach treatment using an Omnibarge on the nearshore zone of Herring Bay, July 1989. Technical report to Exxon Company, USA. Dames & Moore Technical Report Series-The *Exxon Valdez* Oil Spill Studies. Dames & Moore, Seattle.

Houghton, J. P., D. C. Lees, H. Teas, H. Cumberland, S. Landino, and W. B. Driskell. 1991. Evaluation of the condition of intertidal and shallow subtidal biota in Prince William Sound following the *Exxon Valdez* oil spill and subsequent shoreline treatment, volume 1. Report (HMRB 91-1) to NOAA (National Oceanic and Atmospheric Administration) Hazardous Materials Response Branch, Seattle.

Houghton, J. P., and seven coauthors. 1993a. Evaluation of the condition of Prince William Sound shorelines following the *Exxon Valdez* oil spill and subsequent shoreline treatment, volume 2: 1991 biological monitoring survey. NOAA (National Oceanic and Atmospheric Administration) Technical Memorandum NOS (National Ocean Service) ORCA (Office of Ocean Resources, Conservation and Assessment) 67, Seattle.

Houghton, J. P., A. K. Fukuyama, D. C. Lees, P. J. Hague, P. M. Harper, and W. B. Driskell. 1993b. Evaluation of the 1992 condition of Prince William Sound littoral biota following the *Exxon Valdez* oil spill and subsequent shoreline treatment. Report to NOAA (National Oceanic and Atmospheric Administration), Hazardous Materials Response and Assessments Division, Seattle.

Kozloff, E. N. 1973. Seashore life of Puget Sound, the Strait of Georgia, and the San Juan Archipelago. University of Washington Press, Seattle.

Lees, D. C., J. P. Houghton, W. B. Driskell, D. E. Erikson, and D. E. Boettcher. 1979. Ecological studies of intertidal and shallow subtidal habitats in lower Cook Inlet, Alaska. Annual report to NOAA (National Oceanic and Atmospheric Administration)/OCSEAP by Dames & Moore, Seattle.

Lees, D. C., J. P. Houghton, and W. B. Driskell. 1993. Effects of shoreline treatment methods on intertidal biota in Prince William Sound. Pages 345–354 *in* Proceedings, 1993 international oil spill conference: prevention, preparedness, response. American Petroleum Institute Publication 4580, Washington, DC.

Lees, D. C., and J. P. Houghton. 1990. An evaluation of biological effects of exposure to Corexit 9580 M2 in the biota of the intertidal and shallow subtidal zones. Final data report to Exxon Company, USA. Dames & Moore Technical Report Series—The *Exxon Valdez* Oil Spill Studies. Dames & Moore, Seattle.

Lees, D. C., J. P. Houghton, and W. B. Driskell. 1996. Short-term effects of several types of shoreline treatment on rocky intertidal biota in Prince William Sound. American Fisheries Society Symposium 18: 329–348.

Lethcoe, J., and N. Lethcoe. 1989. Cruising guide to Prince William Sound. Prince William Sound Books, Valdez, Alaska.

Lindstedt-Siva, J. 1979. Ecological impacts of oil spill cleanup: are they significant? Pages 521–524 *in* Proceedings, 1979 oil spill conference. American Petroleum Institute Publication 4308, Washington, DC.

Lindstedt-Siva, J. 1991. U.S. oil spill policy hampers response and hurts science. Pages 349–352 *in* Proceedings, 1991 international oil spill conference. American Petroleum Institute Publication 4529, Washington, DC.

McCook, L. J., and A. R. O. Chapman. 1992. Vegetative regeneration of *Fucus* rockweed canopy as a mechanism of secondary succession on an exposed rocky shore. Botanica Marina 35:35–46.

Mearns, A. J. 1996. *Exxon Valdez* shoreline treatment and operations: implications for response, assessment, monitoring, and research. American Fisheries Society Symposium 18:309–328.

Michel, J., and M. O. Hayes. 1993. Persistence and weathering of *Exxon Valdez* oil in the intertidal zone −3.5 years later. Pages 279–286 *in* Proceedings, 1993 international oil spill conference: prevention, preparedness, response. American Petroleum Institute Publication 4580, Washington, DC.

National Research Council. 1985. Oil in the sea: inputs, fates, and effects. National Academy Press, Washington, DC.

Nauman, S. A. 1991. Shoreline cleanup: equipment and operations. Pages 141–147 *in* Proceedings, 1991 international oil spill conference: prevention, behavior, control, cleanup. American Petroleum Institute Publication 4529, Washington, DC.

Nelson-Smith, A. 1968. The effects of oil pollution and emulsifier cleansing on shore life in southwest Britain. Journal of Applied Ecology 5:97–107.

Nybakken, J. W. 1969. Pre-earthquake intertidal ecology of Three Saints Bay, Kodiak Island, Alaska. The Allen Press, Lawrence, Kansas.

Page, D. S., E. S. Gilfillan, P. D. Boehm, and E. J. Harner. 1995. Shoreline ecology program for Prince William Sound, Alaska, following the *Exxon Valdez* oil spill: part 1-study design and methods. Pages 263–295 *in* P. G. Wells, J. N. Butler, and J. S. Hughes, editors. *Exxon Valdez* oil spill: fate and effects in Alaskan waters. American Society for Testing and Materials, Special Technical Publication 1219, Philadelphia.

Paine, R. T. 1966. Food web complexity and species diversity. American Nature 100:65–75.

Paine, R. T., and R. L. Vadas. 1969. The effect of grazing by sea urchins, *Strongylocentrotus* spp., on benthic algal populations. Limnology and Oceanography 14: 710–714.

Pritchard, P. H., and C. F. Costa. 1991. EPA's Alaska oil spill bioremediation project. Environmental Science and Technology 25:372–379.

Ricketts, E. F., J. Calvin, and J. W. Hedgpeth. 1985. Between Pacific tides, 5th edition. Stanford University Press, Stanford, California.

Rolan, G. R., and R. Gallagher. 1991. Recovery of the intertidal biotic communities at Sullom Voe following the *Esso Bernicia* oil spill of 1978. Pages 461–465 *in* Proceedings, 1991 international oil spill conference: prevention, behavior, control, cleanup. American Petroleum Institute Publication 4529, Washington, DC.

Saville, D. J. 1990. Multiple comparison procedures: the practical solution. The American Statistician 44(2): 174–180.

Smith, J. E. 1968. Torrey Canyon pollution and marine life. Cambridge University Press, New York.

Southward, A. J., and E. C. Southward. 1978. Recolonization of rocky shores in Cornwall after use of toxic dispersants to clean up the Torrey Canyon spill. Journal of the Fisheries Research Board of Canada 10: 682–706.

Stekoll, M. S., and seven coauthors. 1996. Coastal habitat injury assessment: intertidal communities and the *Exxon Valdez* oil spill. American Fisheries Society Symposium 18:177–192.

Strathmann, M. F. 1987. Reproduction and development of marine invertebrates of northern Pacific coast: data and methods for the study of eggs, embryos, and larvae. University of Washington Press, Seattle.

van Tamelen, P. G., and M. S. Stekoll. 1996. Population response of the brown alga *Fucus gardneri* and other algae in Herring Bay, Prince William Sound, to the *Exxon Valdez* oil spill. American Fisheries Society Symposium 18:193–211.

Zeh, J. E., J. P. Houghton, and D. C. Lees. 1981. Evaluation of existing marine intertidal and shallow subtidal biological data. U.S. EPA (Environmental Protection Agency) Interagency Agreement D6-E693-EN, Seattle.

American Fisheries Society Symposium 18:412–423, 1996

Kelps and Oil: The Effects of the *Exxon Valdez* Oil Spill On Subtidal Algae

THOMAS A. DEAN
Coastal Resources Associates, Inc.
1185 Park Center Drive, Suite A, Vista, California 92083, USA

MICHAEL S. STEKOLL
Juneau Center for Fisheries and Ocean Sciences, 11120 Glacier Highway, Juneau, Alaska 99801, USA, and School of Fisheries and Ocean Sciences, University of Alaska Fairbanks, Fairbanks, Alaska 99775, USA

RICHARD O. SMITH
Coastal Resources Associates, Inc.
1185 Park Center Drive, Suite A, Vista, California 92083, USA

Abstract.—This study examined possible changes in the subtidal macroalgal populations as a result of the *Exxon Valdez* oil spill. The abundance and size distribution of dominant subtidal algae were measured in Prince William Sound 1 year after the spill. Population density, biomass, and cover were compared between oiled and control sites within each of three habitats: sheltered bays, moderately exposed points, and very exposed points with a surface canopy of *Nereocystis luetkeana*. Dominant macroalgae in these habitats were the kelps *Agarum cribrosum*, *Laminaria saccharina*, *L. groenlandica*, and *N. luetkeana*. There were no differences in the total density, biomass, or percentage cover of macroalgae between oiled and control sites. However, at least one of the dominant kelp species within each habitat was more abundant at the oiled sites. In addition, there were generally more small plants at oiled sites, suggesting recent recruitment or slower growth there. Recruitment at oiled sites may have been indicative of recovery from recent oil-related disturbance. Although these data suggest possible injury to kelps as a result of the spill, there were no apparent long-term impacts on subtidal populations of macroalgae.

On 24 March 1989, 43.9 million liters of Prudhoe Bay crude oil spilled into Prince William Sound, Alaska. The oil moved from Bligh Reef in a southwesterly direction, coating hundreds of miles of coastline from Prince William Sound to the Alaska Peninsula. Here we report the results of studies performed after the spill that examined its impact on macroalgal communities in Prince William Sound. Particular emphasis was placed on examining the impact of oil and associated cleanup activities on the subtidal kelps, *Agarum cribrosum*, *Laminaria* spp., and *Nereocystis luetkeana*.

The shallow subtidal habitat in Prince William Sound is characterized by dense macroalgal assemblages that extend from the intertidal to depths of about 20 m. Seaweeds are major primary producers in this system. They supply food and critical habitat for many commercially and ecologically important animals, including salmon, Pacific herring *Clupea pallasi*, crabs, sea otters *Enhydra lutris*, river otters *Lutra canadensis*, and marine birds (Rosenthal et al. 1977; Rosenthal 1980; Feder and Jewett 1987).

Results of studies performed after oil spills elsewhere (North et al. 1964; Foster et al. 1971; Guzman and Campodonico 1981; Peckol et al. 1990) indicate that oil had little impact on subtidal kelps. However, results of studies of previous spills also indicate that changes in species number, abundance, biomass, and diversity occurred within the subtidal benthic community when sizable quantities of oil reached the bottom (e.g., Cabioch et al. 1978; Kineman et al. 1980; Sanders et al. 1980). Furthermore, field studies have shown that intertidal populations of algae can be harmed by oiling (e.g., Smith 1968; Thomas 1973, 1978; Clark et al. 1978; Floc'h and Diouris 1980, 1981; Jackson et al. 1989; Rolan and Gallagher 1991). Laboratory studies have also suggested that oil may adversely affect reproduction in some brown algae (Steele 1977). Based on these later observations, and knowing that a sizable fraction of oil reached subtidal sediments after the *Exxon Valdez* spill (Wolfe et al. 1994; O'Clair et al. 1996; Short et al. 1996; both this volume), we had reason to suspect that there may have been adverse impacts on subtidal populations of macroalgae.

Methods

The effects of oiling and associated cleanup activities were examined by measuring the abundance, biomass, and size distribution of the dominant subtidal macroalgae at pairs of oiled and control sites using a stratified random sampling design. Sampling was done within three habitats: *Agarum–Laminaria* beds (areas where *A. cribrosum* and *L. saccharina* dominate) both in bays and on points within the Knight Island archipelago, and *Nereocystis* beds (Figure 1). Within each habitat, we selected oiled sites and then located a control site that matched each oiled site with respect to nonbiological factors other than oiling.

An initial selection of oiled sites was conducted in the laboratory in winter 1990. We identified habitats based on our experience in Prince William Sound and on anecdotal evidence from biologists familiar with the sound. Sites within habitats were initially chosen based on an overlay of oil information and habitat information on navigation charts. Oiled areas were identified based on maps produced from shoreline surveys of the extent of oiling in the summer and fall of 1989 (ADEC 1989; ADNR 1991). Areas that were classified as being moderately to heavily oiled in both surveys were identified as oiled areas. We selected 200-m sections of oiled shoreline in each habitat as sampling sites. The selection was based on the following hierarchy for order of preference: sites for which there were prespill biological data, sites previously sampled for hydrocarbons, sites sampled to examine the extent of injury in the intertidal zone (see Sundberg et al. 1996, this volume), and randomly selected sites within the habitat.

Controls were selected that were designated as non-oiled in both the summer and fall shoreline surveys. They were matched with selected oiled sites, as closely as possible, with regard to aspect, proximity to sources of freshwater input, slope, wave exposure, and water circulation. We randomly selected a matched control site if more than one existed within a specified habitat.

Within each habitat, the sampling was further stratified by depth. In the *Agarum–Laminaria* habitats, we sampled at two depth strata: 2–11 m and 11–20 m. Within the *Nereocystis* habitat, we sampled in a depth range of 2–8 m. All sampling was conducted between late May and early August 1990.

Sampling was done along randomly placed transects that ran parallel to shore within each depth stratum in each habitat. Three 30-m transects were sampled at each site in the bay and point habitats, whereas up to six transects were sampled in the *Nereocystis* habitat. The percentage cover, density, and biomass of understory algae were estimated in four 0.25-m^2 quadrats placed at random positions within four contiguous 7.5-m sections along each transect. All algae were identified and counted in the field, and those individuals greater than 10 cm in height were collected from the quadrats and returned to the boat. Each individual was identified, weighed, and measured. In the *Nereocystis* habitat, we also sampled *Nereocystis* density by counting all plants greater than 2 m in height within a 4-m swath along each transect.

The size distribution of *N. luetkeana* was determined by measuring the diameter of the stipe at a distance 1 m above the bottom. Stipes were measured for the first 20 plants observed along each transect, except where fewer than 20 individuals occurred. Biomass (wet weight) of *N. luetkeana* was estimated based on a regression of wet weight versus stipe diameter. Regressions of $\log_e$ wet weight versus stipe diameter were highly significant (r^2 = 0.60 to 0.89, $P < 0.01$ for all sites), and slopes of the regressions differed significantly among sites. As a result, separate regression equations were used to estimate wet weight within each site. The stipe diameter for 20–40 plants collected from near the transects were measured, and the plants were collected and weighed. Plants were selected in order to obtain as wide a range in sizes as possible.

Sediments were collected from each depth stratum at each study site and were analyzed to determine hydrocarbon concentrations. Scuba divers collected the sediment samples in precleaned, widemouthed 118-ml jars. The jars were used to scoop sediment to a depth of approximately 5 cm. A 10- to 100-g sample of sediment was obtained at each sampling location.

Samples were sent to the Technical Services Task Force, Analytical Chemistry Group, National Oceanic and Atmospheric Administration, Auke Bay, Alaska, for processing. The concentration of polycyclic aromatic hydrocarbons (PAHs) was measured using gas chromatography combined with a mass spectrometer detector. All samples for the bay habitats were analyzed, but only a few samples from the other habitats were analyzed.

For the percentage cover, abundance, and biomass estimates for each of the dominant taxa and for total PAH concentrations, we tested the null hypothesis of no significant difference between oiled and control sites using randomization analysis of variance (ANOVA) (Manly 1991). In most instances, separate analyses were performed for each depth stratum within each habitat and year. We tested for oil category (oil or control) as the main effect, with pair as a blocking factor.

The randomization procedure can be briefly sum-

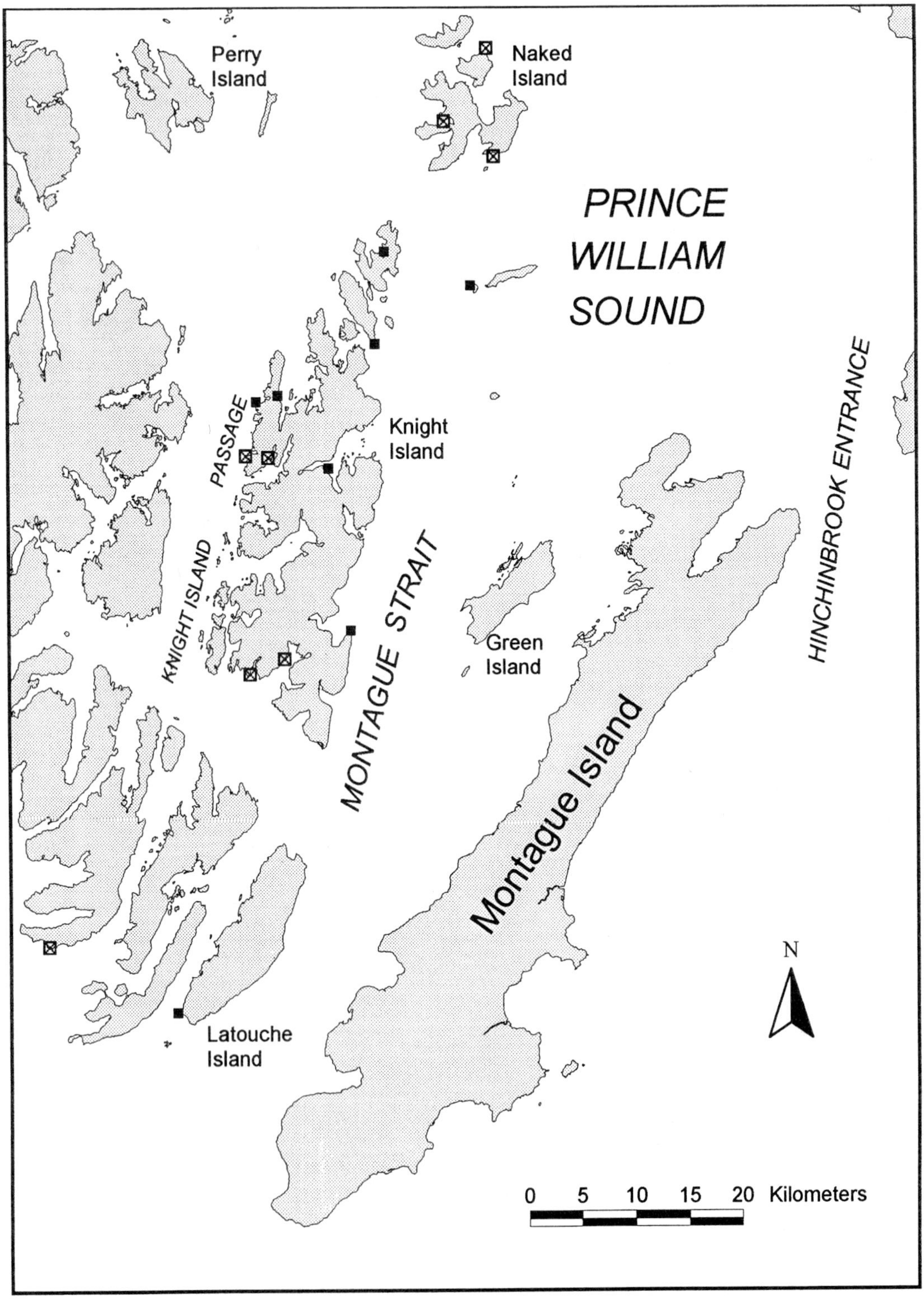

FIGURE 1.—Map of Prince William Sound showing the location of sampling sites in 1990. Filled symbols are oiled sites and open symbols are control sites.

TABLE 1.—Relative abundance (percentage) of understory kelps in the shallow subtidal zone in Prince William Sound, May–August 1990. Relative densities (of plants greater than 10 cm in height), relative biomass (wet weight), and relative cover are given for plants in sheltered bays, on moderately exposed points, and at highly exposed sites where *Nereocystis* was present. The "shallow" designation does not apply to the *Nereocystis* data.

Taxon	Shallow Density	Shallow Biomass	Shallow Cover	Deep Density	Deep Biomass	Deep Cover
			Bays			
Agarum cribrosum	69	68	63	67	88	76
Laminaria saccharina	27	32	34	21	12	16
Others	4	<1	3	12	<1	8
			Points			
Agarum cribrosum	60	51	55	67	79	75
L. saccharina	19	24	28	21	17	23
L. groenlandica	11	21	13	1	3	0
Others	10	4	4	11	1	2
			Nereocystis			
L. groenlandica	73	61	71			
Pleurophycus gardneri	8	12	11			
L. yezoensis	9	6	7			
A. cribrosum	6	4	7			
Others	4	17	4			

marized as follows: (1) A blocked ANOVA was performed, and a sum of squares (SS) produced. (2) Next, using the original data set, we randomly reassigned values for oil category to each station value. The ANOVA was then rerun on this new data set. (3) Step 2 was repeated 1,000 times. (4) The SS from the ANOVA of the original data set was compared with the SS of the 1,000 randomly drawn data sets to determine the proportion of instances in which the SS for the randomly drawn data exceeded the SS for the original data. This value is the significance level of the test as described by Fisher (1935). The significance level is interpreted in the same manner as for parametric procedures. If the randomly drawn SS exceeded the SS for the original data 10% of the time, then this is equivalent to $P = 0.10$ for an ANOVA.

Differences in size distribution (wet weight or stipe diameter) between oiled and control sites were tested using a randomization procedure similar to that described above, but using a Kolmogorov–Smirnov (KS) D value (Siegel 1956) as the test statistic. We performed a KS test on the original data set, randomly reassigned oil and control categories to plants found on a particular transect, and reran the KS test. Groups of plants from quadrats and transects, rather than individual plants, were randomized, this was done because the sizes of plants in close proximity to one another may have been correlated and were probably not indepen-

TABLE 2.—Mean densities (number/m^2), biomass (wet weight in g/m^2), and percentage cover for dominant macroalgae at oiled and control sites in bays within Prince William Sound, May–August 1990. Abbreviations are O = means from oiled sites, C = means from control sites, and P = probabilities associated with randomization analyses of variance comparing these means.

Taxon	Statistic	Shallow Density	Shallow Biomass	Shallow Cover	Deep Density	Deep Biomass	Deep Cover
Agarum cribrosum	O	6.9	703	30.9	4.6	293	22.1
	C	8.9	1,146	51.8	3.9	529	32.0
	P	0.48	0.16	0.25	0.69	0.26	0.31
Laminaria saccharina	O	3.7	429	34.5	1.0	93	9.5
	C	0.9	320	9.0	0.0	0	0.06
	P	0.03	0.71	0.04	<0.01	<0.01	<0.05
Total macroalgae	O	9.6	1,132	66.8	5.9	387	32.5
	C	11.3	1,766	63.0	4.1	529	38.9
	P	0.67	0.15	0.75	0.38	0.53	0.51

TABLE 3.—Mean densities (number/m²), biomass (wet weight in g/m²), and percentage cover for dominant kelps at oiled and control sites on points within Prince William Sound, May–August 1990. The abbreviation NA indicates no analysis because of insufficient data.

		Shallow			Deep		
Taxon	Statistic	Density	Biomass	Cover	Density	Biomass	Cover
Agarum cribrosum	O	25.8	983	43.1	14.7	667	30.9
	C	11.7	1,086	41.1	4.2	445	34.7
	P	0.06	0.80	0.85	0.02	0.45	0.72
Laminaria saccharina	O	15.7	575	28.1	1.9	105	11.0
	C	51.9	492	17.9	4.8	228	10.3
	P	0.49	0.75	0.22	0.26	0.80	0.90
L. groenlandica	O	3.3	318	6.6	0.3	58	0.1
	C	7.6	915	11.6	0.0	0.0	0.0
	P	0.24	0.28	0.59	NA	NA	NA
Total macroalgae	O	51.6	1,903	79.0	30.0	813	43.1
	C	32.8	2,690	75.3	17.4	678	45.3
	P	0.62	0.22	0.65	0.34	0.79	0.83

dent. The randomization process was repeated 1,000 times, and the *D* statistic from the randomized data sets was compared with the *D* statistic from the initial data set.

Slopes of regressions of stipe diameter versus wet weight for *Nereocystis luetkeana* at oiled versus control sites were compared using an analysis of covariance (Snedecor and Cochran 1967).

Results

The dominant subtidal macroalgae in sheltered bays within Prince William Sound were the stipate kelps *Agarum cribrosum* and *Laminaria saccharina* (Table 1). Together, these taxa generally constituted more than 90% of the total numbers of individuals, total biomass, and total cover of macroalgae in the bay habitat. On more exposed points, *A. cribrosum* dominated, along with a mixture of *L. saccharina* and *L. groenlandica*. *A. cribrosum* constituted 60% or more of the total number of individuals in both shallow and deep strata. Within the *Nereocystis* habitat, algal diversity was somewhat higher. At these exposed sites, there was a canopy of *N. luetkeana* with an understory of *L. groenlandica*, *L. yezoensis*, *Pleurophycus gardneri*, and *A. cribrosum*. Of the understory algae, *L. groenlandica* made up 73% of the individuals and 61% of the biomass. Each of the other species contributed less than 10% to the total number of individuals.

In 1990, there were no significant differences between oiled and control sites with respect to the density, biomass, or cover for all algal species combined (Tables 2, 3, and 4). However, there were differences for several individual taxa. Within the bay habitat, the density, cover, and biomass of *L. saccharina* in the deep stratum were significantly greater at oiled sites than at control sites; this was also true for the density and cover of *L. saccharina* in the shallow stratum (Table 2 and Figure 2). In

TABLE 4.—Mean densities (number/m²), biomass (wet weight in g/m²), and percentage cover for dominant macroalgae at oiled and control sites in *Nereocystis* beds within Prince William Sound, May–August 1990. Estimates of biomass for *Nereocystis* are the product of mean wet weight (from a regression of stipe diameter versus wet weight) and mean density. The abbreviation NA means no analysis was performed because variables were derived (for *Nereocystis* biomass) or because there were insufficient data. The abbreviation NM means not measured.

Taxon	Statistic	Density	Biomass	Cover
Nereocystis luetkeana	O	0.71	1,330	NM
	C	0.10	770	NM
	P	<0.01	NA	NA
Laminaria groenlandica	O	145.7	2,735	56.6
	C	88.9	3,154	49.7
	P	0.14	0.39	0.78
Pleurophycus gardneri	O	2.1	150	1.5
	C	6.7	624	11.1
	P	0.44	0.15	0.26
L. yezoensis	O	4.6	203	4.1
	C	7.1	777	6.7
	P	0.74	0.40	0.69
Agarum cribrosum	O	4.6	285	9.9
	C	2.9	32	0.0
	P	0.19	<0.01	NA
Total understory algae	O	158.0	3,699	76.7
	C	104.7	6,240	68.9
	P	0.15	0.16	0.83

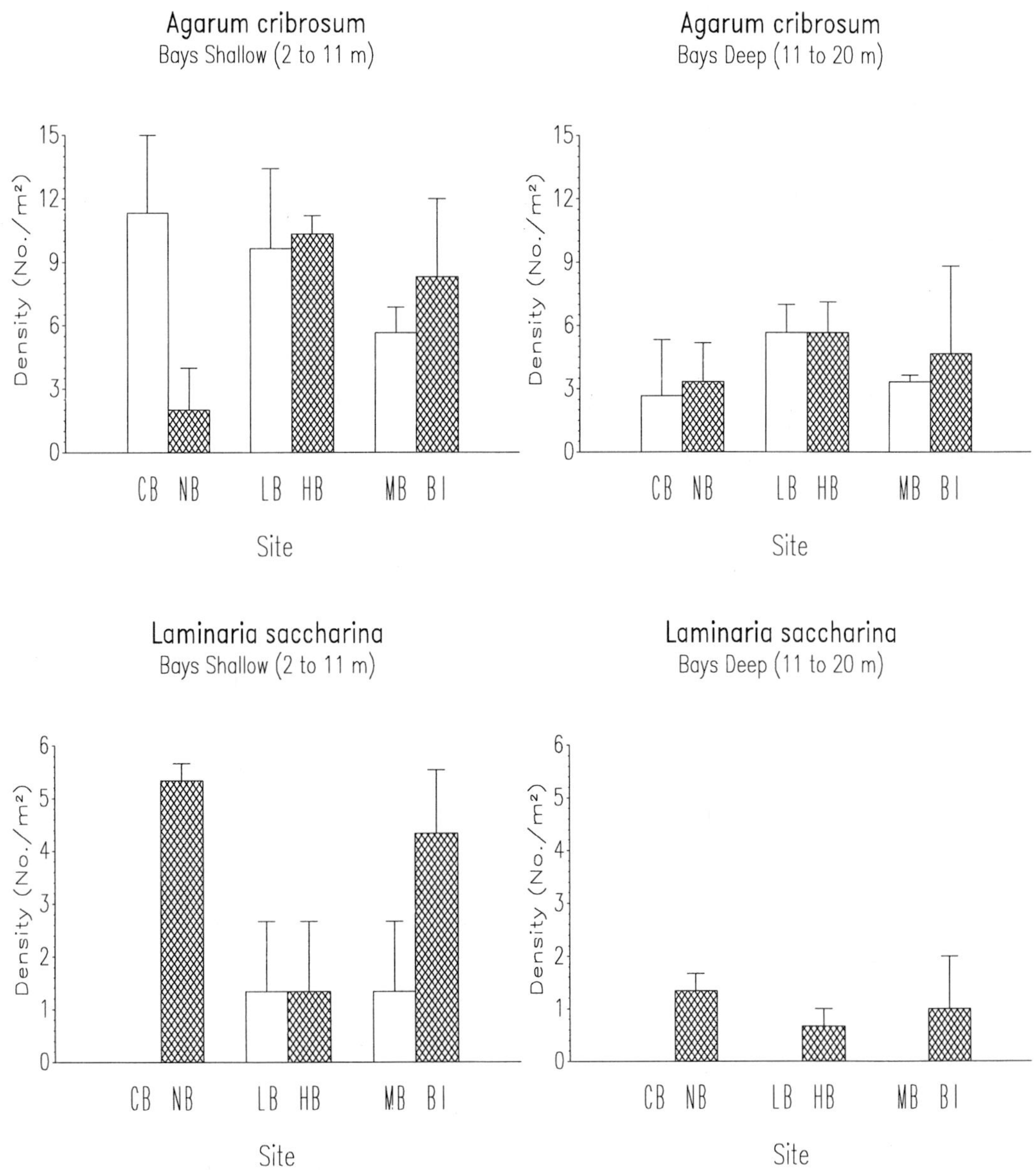

FIGURE 2.—Mean densities (number/$m^2 \pm 1$ SE) of dominant understory algae within oiled and control bays in 1990. Hatched bars are for oiled sites and open bars are for control sites. Site designations are as follows: CB = Cabin Bay, NB = Northwest Bay, LB = Lower Herring Bay, HB = Herring Bay, MB = Mummy Bay and BI = Bay of Isles.

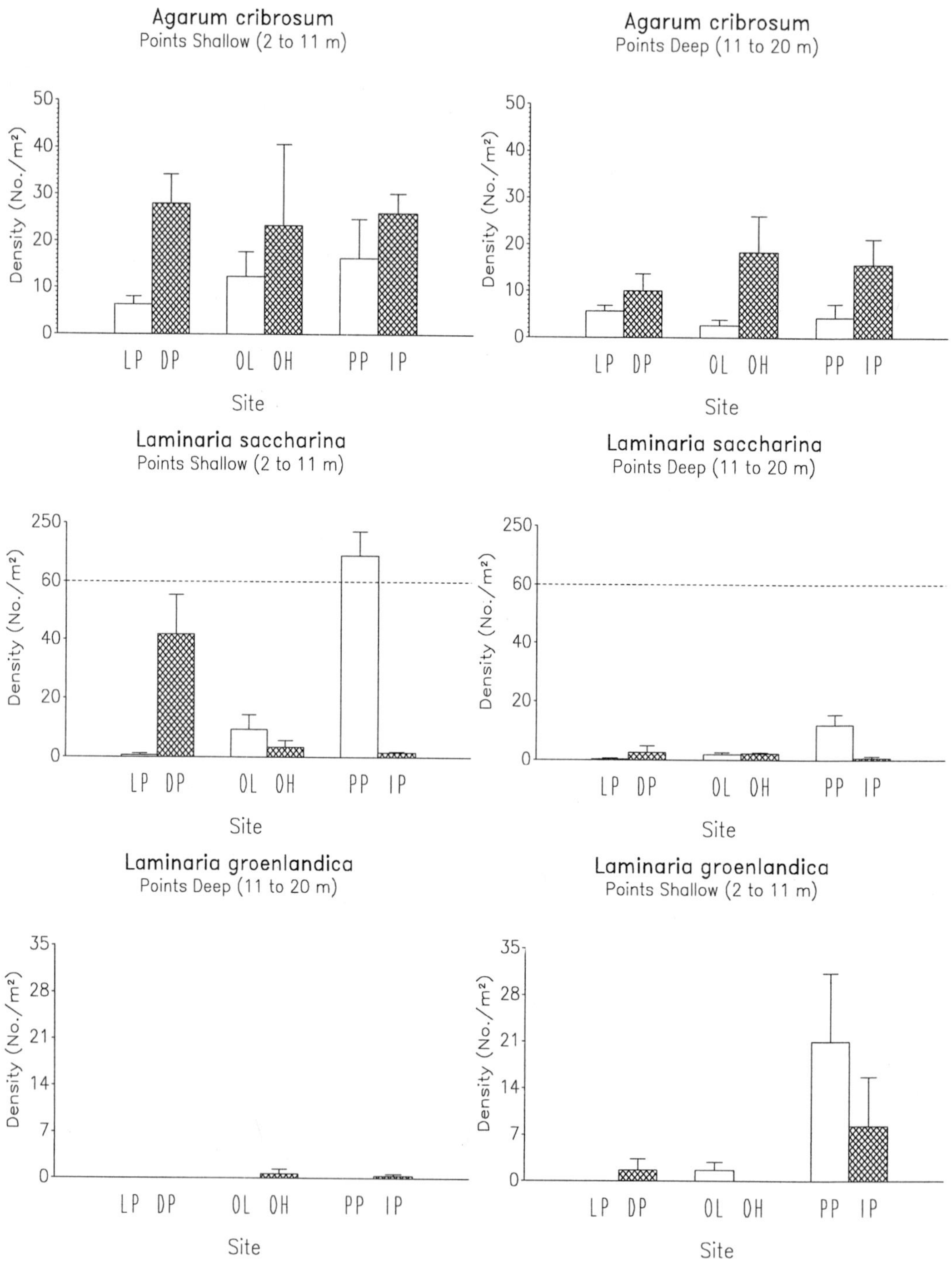

FIGURE 3.—Mean densities (number/m^2 $\pm$ 1 SE) of dominant understory algae at oiled and control points in 1990. Hatched bars are for oiled sites and open bars are for control sites. Site designations are as follows: LP = Lucky Point, DP = Discovery Point, OL = Outside Lower Herring Bay, OH = Outside Herring Bay, PP = Peak Point and IP = Ingot Point. The dashed horizontal line in the plot of density of *L. saccharina* on points indicates a change in scale.

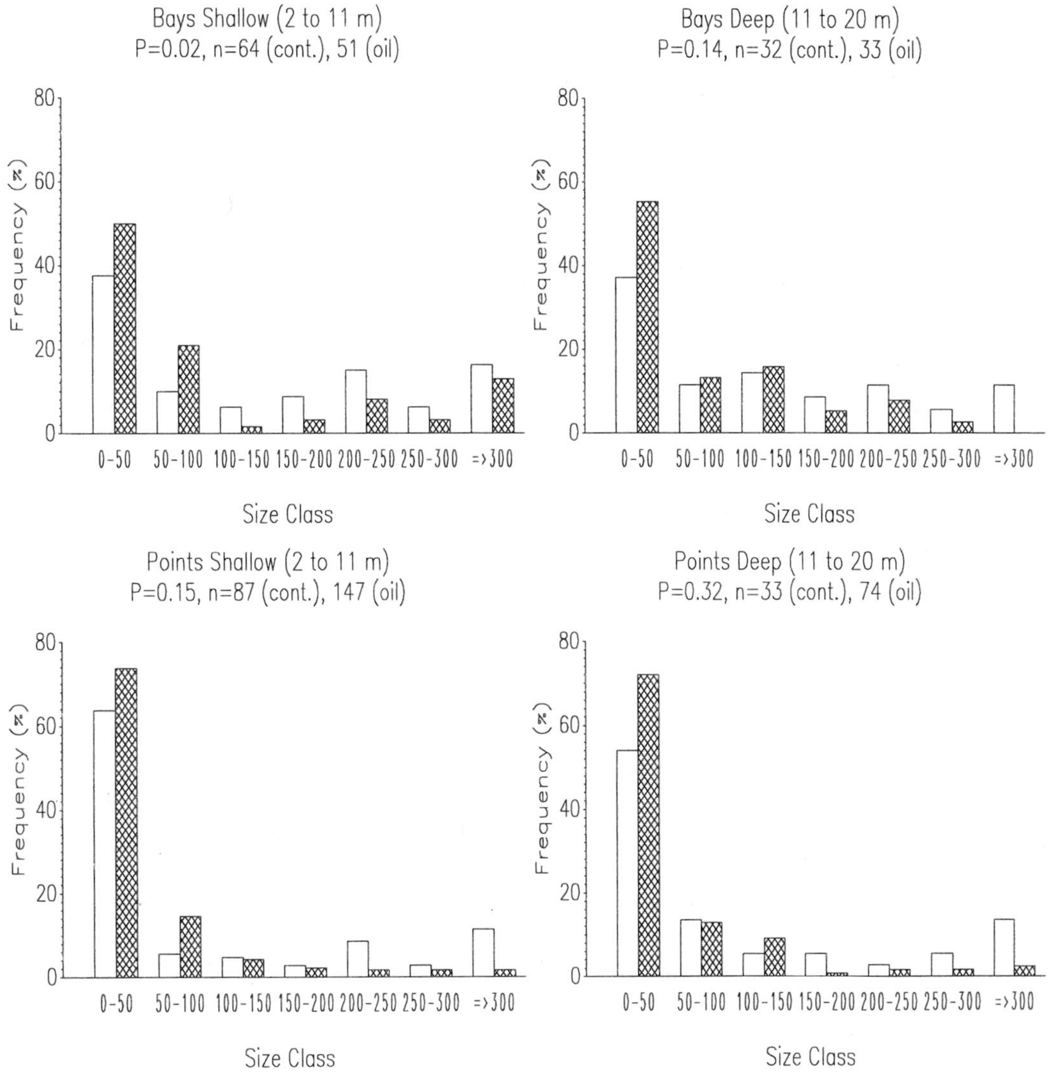

FIGURE 4.—Frequency distributions for wet weight (g) of *Agarum cribrosum* from bays and points within Prince William Sound in 1990. Hatched bars are for oiled sites and open bars are for control sites. The number of plants measured in each oiling category and the probability that distributions were similar are also shown.

shallow portions of the bay habitat, the density of *L. saccharina* was about four times as high at oiled sites than at control sites. In deeper parts of this habitat, *L. saccharina* averaged about $1/m^2$ at oiled sites, while there were none observed at the control sites. There were no differences in density of the other dominant alga, *A. cribrosum*, in the bay habitat.

At the somewhat more exposed points around the Knight Island archipelago, density of *A. cribrosum* was higher at oiled sites than at controls (Table 3 and Figure 3). Differences were significant in the deeper stratum ($P = 0.02$) and nearly so in the shallow stratum ($P = 0.06$). Differences were primarily the result of higher densities of small plants at the oiled sites (see below and Figure 4). There were no differences in the biomass or percentage cover of *A. cribrosum* (Table 3), or in density, biomass, or cover of *L. saccharina* or *L. groenlandica* (Table 3).

There appeared to be proportionally more small *A. cribrosum* at oiled sites than at control sites in

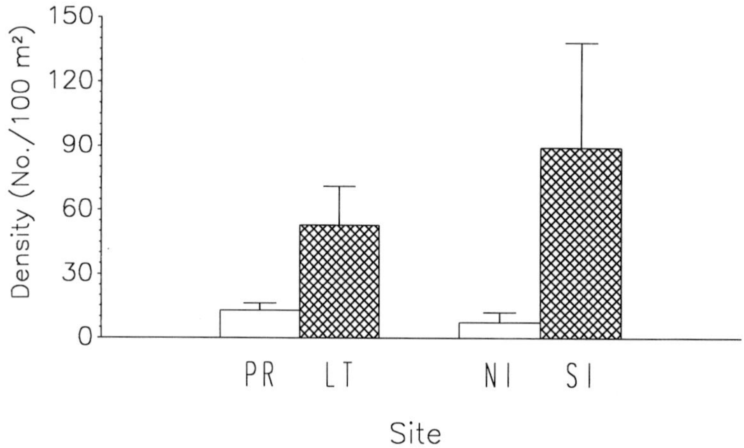

FIGURE 5.—Mean densities (number/100 m^2 ± 1 SE) of *Nereocystis luetkeana* at oiled and control sites in 1990. Hatched bars are for oiled sites and open bars are for control sites. Site designations are as follows: PR = Procession Rocks, LT = Latouche Point, NI = Naked Island, and SI = Little Smith Island.

both bay and point habitats (Figure 4). Within each depth stratum and habitat, there were proportionally more plants of less than 50 g at the oiled sites, although overall differences in size distribution were significant at only the shallow bay sites (P = 0.02).

Within the *Nereocystis* habitat, *N. luetkeana* density was significantly greater at the oiled sites than at controls (Table 4 and Figure 5). Densities averaged 0.7/m^2 at oiled sites and 0.1/m^2 at control sites. The differences in density were primarily the result of higher abundances of small plants at the oiled sites (Figure 6). As a result, the estimate of mean biomass was only about twice as great at oiled sites as at control sites (Table 4). Of the understory algae, the biomass of *A. cribrosum* was significantly greater at the oiled sites than at the control sites within the *Nereocystis* habitat (P < 0.01, Table 4). This was largely the result of relatively high biomass of *A. cribrosum* at one oiled site. There were no other significant differences between oiled and control sites with regard to the density, biomass, or percentage cover of any other understory algae.

As in the bay and point habitats, we also noted a

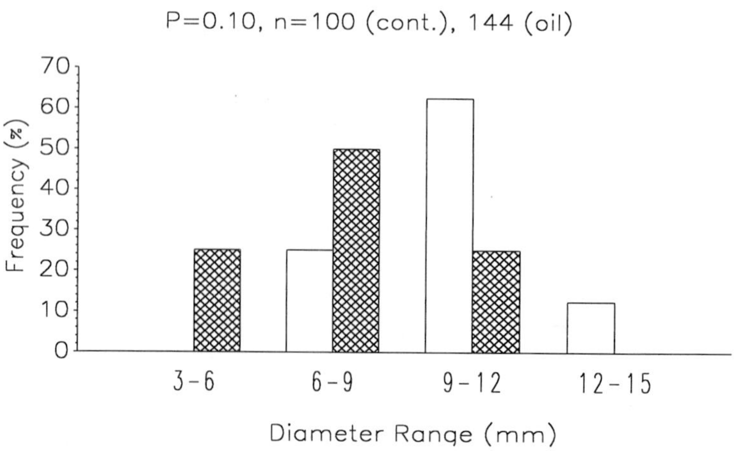

FIGURE 6.—Frequency distributions for stipe diameters of *Nereocystis luetkeana* at oiled and control sites in Prince William Sound in 1990. Hatched bars are for oiled sites and open bars are for control sites. The number of plants measured in each oiling category and the probability that the distributions were similar are also shown.

TABLE 5.—Regressions of stipe diameter versus $\log_e$ wet weight of *Nereocystis luetkaena* from each of four sites within Prince William Sound, May–August 1990. Also given are probabilities for tests of the hypothesis of equality of slopes for regresssions of oiled versus control sites within each pair of sites.

Site	Type	N	Slope	Intercept	r^2	P
Latouche Point	Oil	43	0.49	−2.99	0.71	<0.01
Procession Rocks	Control	21	0.70	−4.46	0.89	
Little Smith Island	Oil	38	0.33	−1.87	0.60	<0.01
Naked Island	Control	41	0.51	−3.16	0.81	

TABLE 6.—Mean concentrations (ng/g[1]) of polycyclic aromatic hydrocarbons and results of randomization analyses of variance testing for differences in mean concentrations between oiled and control sites in 1990. The abbreviation ND means no data available. The abbreviation ID means insufficient data for statistical test.

Habitat	Oil	Control	P
Bays–shallow	1,142	420	0.02
Bays–deep	406	98	<0.01
Points–shallow	147[a]	19[a]	ID
Points–deep	163	125[a]	0.46
Nereocystis	45[a]	ND	ID

[a]Samples from only one site were analyzed.
[b]Samples from only two sites were analyzed.

difference in the size distribution of the dominant alga between oiled and control sites in the *Nereocystis* habitat. There was a somewhat higher proportion of small *Nereocystis luetkeana* at oiled sites than at control sites (Figure 6). The median diameter of the stipe (1 m above the substratum) was between 6 and 9 mm at oiled sites and between 9 and 12 mm at control sites. These roughly correspond to median wet weights of 2 and 14 kg, respectively. Size distributions of 1-m stipe diameters did not differ significantly between oiled and control sites (P = 0.10, Figure 6). However, this was an underestimation of differences in biomass. For a given stipe diameter, plants at control sites were somewhat heavier, as indicated by the greater slope of the stipe diameter versus wet weight regression at control versus oiled sites (Table 5).

Within the bay habitat, concentrations of PAHs differed between oiled and control sites (Table 6). In the shallow depth stratum, PAH concentrations averaged 1,142 ng/g at the oiled sites and 420 ng/g at the control sites. In the deeper portions of the habitat, concentrations averaged 406 and 98 ng/g at oiled and control sites, respectively. Differences between oiled and control sites were statistically significant in the deep stratum (P = <0.01) and nearly so in the shallow stratum (P = 0.02). There were no differences noted in the deep-point habitat (P = 0.46), and there were too few hydrocarbon data analyzed from other habitats for statistical evaluation.

Discussion

A comparison between our survey and the prespill descriptions of the subtidal flora in Prince William Sound (Rosenthal et al. 1977; Rosenthal 1980) indicate that there were no drastic changes in the relative abundance of subtidal algae as a result of oiling. Quantitative data that allow for pre- versus postspill comparisons are available from only one site, Latouche Point (Table 7). In 1990, the mean density of *N. luetkeana* was lower and the density of *L. groenlandica* was higher than before the spill, in 1975. No attempt was made to test if these differences were statistically significant because different methodologies were used in the two surveys. However, these differences are not beyond the range of what one might expect from interannual variation in the absence of a spill (Rosenthal et al. 1977).

Although there were no radical changes in the relative abundance of algae, there was a suggestion of possible disturbance to the algal communities at oiled sites. There were higher densities of at least one of the dominant kelps at oiled sites, and there were generally more small plants there. There are several possible explanations for the above observations. First, larger plants at the oiled sites may have been removed in 1989, with recruitment to disturbed sites in the following winter or spring. Second, the growth rate of plants at the oiled sites may have been reduced by oiling, resulting in a population of smaller plants at oiled sites in 1990. Third, nutrient enrichment associated with oiling

TABLE 7.—Comparison of mean densities of dominant macroalgae at Latouche Point before (June 1976) and after (July 1990) the *Exxon Valdez* oil spill. Data from 1976 are from Rosenthal et al. (1977). The 1976 data are means from depths of 3.5–6 m. The 1990 data are from depths of 3–6 m.

Taxon	Mean density (number/m²) 1975	1990
Nereocystis luetkeana	1.2	0.3
Laminaria groenlandica[a]	6.8	38.0
L. yezoensis	1.5	0.3
Agarum cribrosum	0.3	0.3
Pleurophycus gardneri	3.1	3.3

[a]Rosenthal et al. (1977) identified this species as *L. dentigera*.

may have led to better recruitment of kelps at oiled sites. Finally, there may have been differences among sites that were unrelated to oiling.

We can not rule out that differences among sites were the result of factors other than oil, but we think that this is unlikely. Our suspicion is based on observations of benthic invertebrate populations at those same sites (Jewett et al. 1995; Dean et al. 1996, this volume). Densities of several benthic infaunal and epifaunal species differed between oiled and control sites in 1990, but not in subsequent years. The convergence of invertebrate population densities was concomitant with a decline in PAH concentrations, thereby providing evidence that, in the absence of oil, oiled and control sites were similar.

We also think it is unlikely that the oil in the water column or in the sediments directly impacted kelp plants. Polycyclic aromatic hydrocarbon concentrations were relatively low at all sites, averaging less than 1 μg/g sediment, and were probably too low to cause acute toxicity. In prior studies, oil has been shown to have little direct adverse impact on adult kelps (North et al. 1964; Foster et al. 1971; Guzman and Campodonico 1981; Peckol et al. 1990), except in cases of chronic exposure (Clark et al. 1978). Perhaps it is more likely that losses of kelp were caused by the dropping and dragging of anchors by vessels taking part in cleanup activities. In *Nereocystis* beds, additional damage to the floating fronds would be caused by the vessels' surface activities. There is no documentation of the amount and type of vessel activity that occurred just offshore of oiled coastlines. However, all shores adjacent to our selected oiled sites were subjected to intense cleaning activities that required extensive vessel support. The beach-cleaning activities also created plumes of suspended oiled sediment in the nearshore subtidal zone (A. Mearns, National Oceanic and Atmospheric Administration, personal communication), which could have caused damage by shading or abrading plants, thus leading to their poorer growth or survival.

Although there may have been possible impacts of the oil spill on subtidal algal populations, it is clear that these did not cause dramatic long-term changes in population structure. There were some differences in the abundance of kelp species between oiled and control sites, however, there tended to be greater abundances of these species at oiled sites, and total algal abundance and biomass did not differ among sites.

Acknowledgments

We thank the sponsor of this work, the U.S. Department of Agriculture Forest Service, for its support. We especially thank the more than 20 divers who helped us to collect the data, often under adverse conditions. We also thank Lyman McDonald, who assisted in sampling design, site selection, and statistical analysis. Richard Rosenthal willingly shared his extensive knowledge of Prince William Sound and its natural history. Dorothy Mortensen of the Alaska Department of Natural Resources helped to produce site maps.

References

ADEC (Alaska Department of Environmental Conservation). 1989. Impact maps and summary reports of shoreline surveys of the *Exxon Valdez* spill site—Prince William Sound. Report to the *Exxon Valdez* Oil Spill Trustee Council, Anchorage, Alaska.

ADNR (Alaska Department of Natural Resources). 1991. The *Exxon Valdez* oil spill natural resource damage assessment and restoration: report on oiling to environmentally sensitive shorelines. Report to the *Exxon Valdez* Oil Spill Trustee Council, Anchorage, Alaska.

Cabioch, L., J. C. Dauvin, and F. Gentil. 1978. Preliminary observations on pollution of the sea bed and disturbance of sublittoral communities in Northern Brittany by oil from the *Amoco Cadiz*. Marine Pollution Bulletin 9:303–307.

Clark, R. C., Jr., B. G. Patten, and E. E. DeNike. 1978. Observations of a cold-water intertidal community after 5 years of a low-level persistent oil spill from the *General M.C. Meigs*. Journal of the Fisheries Research Board of Canada 35:754–765.

Dean, T. A., S. C. Jewett, D. R. Laur, and R. O. Smith. 1996. Injury to epibenthic invertebrates resulting from the *Exxon Valdez* oil spill. American Fisheries Society Symposium 18:424–439.

Feder, H. M., and S. C. Jewett. 1987. The subtidal benthos. Pages 347–396 *in* D. W. Hood and S. T. Zimmerman, editors. The Gulf of Alaska: physical environment and biological resources. U.S. Minerals Management Service, Publication MMS 86-0095. U.S. Government Printing Office, Washington, DC.

Fisher, R. A. 1935. The design of experiments. Oliver and Boyd, Edinburgh, UK.

Floc'h, J. Y. and M. Diouris. 1980. Initial effects of *Amoco Cadiz* oil on intertidal algae. Ambio 9:284–286.

Floc'h, J. Y., and M. Diouris. 1981. Impact du petro de l'*Amoco Cadiz* sur les Algues de Portsall: suivi ecologique dans une anse tres polluee. Pages 381–391 *in Amoco Cadiz*: fates and effects of the oil spill. Centre National pour l'exploitation des oceans. Paris.

Foster, M. S., M. Neushul, and R. Zingmark. 1971. The Santa Barbara oil spill. Part 2: initial effects on intertidal and kelp bed organisms. Environmental Pollution 2:115–134.

Guzman, L., and I. Campodonico. 1981. Studies after the

Metula oil spill in the straits of Magellan, Chile. Pages 363–371 *in* Petroleum and the marine environment, Petromar 80. Graham and Trotman Limited, London.

Jackson, J. B. C, and seventeen coauthors. 1989. Ecological effects of a major oil spill on Panamanian coastal marine communities. Science 243:37–44.

Jewett, S. C., and six coauthors. 1995. The effects of the *Exxon Valdez* oil spill on shallow subtidal communities in Prince William Sound, Alaska 1989–1993. Report to the *Exxon Valdez* Oil Spill Trustee Council, Anchorage, Alaska.

Kineman, J. J., R. Elmgren, and S. Hansson. 1980. The *Tsesis* oil spill. NOAA (National Oceanic and Atmospheric Administration) Office of Marine Pollution Assessment, Bolder, Colorado.

Manly, B. 1991. Randomization and Monte Carlo methods in biology. Chapman and Hall, London.

North, W. J., M. Neushul, and K. A. Clendenning. 1964. Successive biological changes observed in a marine cove exposed to a large spillage of mineral oil. Pages 335–354 *in* Pollutions marines par les microrganismes et les produits petroliers. Monaco.

O'Clair, C. E., J. W. Short, and S. D. Rice. 1996. Contamination of intertidal and subtidal sediments by oil from the *Exxon Valdez* in Prince William Sound. American Fisheries Society Symposium 18:61–93.

Peckol, P., S. C. Levings, and S. D. Garrity. 1990. Kelp response following the *World Prodigy* oil spill. Marine Pollution Bulletin 10:473–476.

Rolan, R. G., and R. Gallagher. 1991. Recovery of intertidal biotic communities at Sullom VO following the *Esso Bernicia* oil spill of 1978. Cahiers de Biologie Marine 22:323–348.

Rosenthal, R. 1980. Shallow water fish assemblages in the northeastern Gulf of Alaska: habitat evaluation, species composition, abundance, spatial distribution and tropic interaction. Prepared for the National Oceanic and Atmospheric Administration, Outer Continental Shelf Environmental Assessment Project Office, Juneau, Alaska.

Rosenthal, R. J., D. C. Lees, and T. M. Rosenthal. 1977. Ecological assessment of sublittoral plant communities in northern Gulf of Alaska. Final report to NMFS (National Marine Fisheries Service), Auke Bay Laboratory, Juneau, Alaska.

Sanders, H. L., J. F. Grassle, G. R. Hampson, L. S. Morse, S. Garner-Price, and C. C. Jones. 1980. Anatomy of an oil spill: long-term effects from the grounding of the barge *Florida* off West Falmouth, Mass. Journal of Marine Research 38:265–380.

Short, J. W., D. M. Sale, and J. C. Gibeaut. 1996. Nearshore transport of hydrocarbons and sediments after the *Exxon Valdez* oil spill. American Fisheries Society Symposium 18:40–60.

Siegel, S. 1956. Nonparametric statistics for the behavioral sciences. McGraw-Hill, New York.

Smith, E. J., editor. 1968. Torrey Canyon pollution and marine life. Cambridge University Press, Cambridge, UK.

Snedecor, G. W., and W. G. Cochran. 1967. Statistical methods. Iowa State University Press, Ames.

Steele, R. L. 1977. Effects of certain petroleum products on reproduction and growth of zygotes and juvenile stages of the alga *Fucus edentatus* De La Pyl (Phaeophyceae: Fucales). Pages 138–142 *in* D. A. Wolfe, editor. Fate and effects of petroleum hydrocarbons in marine organisms and ecosystems. Pergamon Press, New York.

Sundberg, K., L. Deysher, and L. McDonald. 1996. Intertidal and supratidal site selection using a geographical information system. American Fisheries Society Symposium 18:167–176.

Thomas, M. L. H. 1973. Effects of bunker C oil on intertidal and lagoonal biota in Chedabucto Bay, Nova Scotia. Journal of the Fisheries Research Board of Canada 30:83–90.

Thomas, M. L. H. 1978. Comparison of oiled and unoiled intertidal communities in Chedabucto Bay, Nova Scotia. Journal of the Fisheries Research Board of Canada 35:707–716.

Wolfe, D. A., and eleven coauthors. 1994. The fate of the oil spilled from the *Exxon Valdez*. Environmental Science and Technology 28:560A–568A.

American Fisheries Society Symposium 18:424–439, 1996

Injury to Epibenthic Invertebrates Resulting from the *Exxon Valdez* Oil Spill

THOMAS A. DEAN
Coastal Resources Associates, Inc.
1185 Park Center Drive, Suite A, Vista, California 92083, USA

STEPHEN C. JEWETT
University of Alaska Fairbanks
School of Fisheries and Ocean Sciences, Fairbanks, Alaska 99775, USA

DAVID R. LAUR
Juneau Center for Fisheries and Ocean Sciences
11120 Glacier Highway, Juneau, Alaska 99801, USA

RICHARD O. SMITH
Coastal Resources Associates, Inc.

Abstract.—Injuries to subtidal populations of epibenthic invertebrates in Prince William Sound were assessed by comparing densities of dominant species at oiled versus non-oiled (control) sites between 1990 and 1993. Dominant taxa included five species of sea stars and one species of crab. Population densities of leather star *Dermasterias imbricata*, and helmet crab *Telmessus cheiragonus* were smaller at oiled sites than at controls in 1990. The same general pattern was observed in most habitats and depths sampled. There was possible injury to slender arm stars *Evasterias troschelii*, although the evidence for this was less convincing. The sunflower sea star *Pycnopodia helianthoides* was found in greater abundance at control sites within eelgrass beds, but in greater abundance at oiled sites in other habitats. Densities of two other species of sea stars did not differ between sites and were apparently unaffected by oiling.

In shallow portions of bays, populations of *Dermasterias* and *Telmessus* were fully recovered by 1993; however, recovery was less than complete within eelgrass beds. There were no significant differences in the density of either *Dermasterias* or *Telmessus* between oiled and control eelgrass sites in 1993, but average densities remained more than twice as high at the controls.

Polycyclic aromatic hydrocarbon (PAH) concentrations were higher at oiled sites, confirming our classification of sites with respect to oiling. There was a general correspondence between level of injury to epifaunal invertebrates and PAH concentrations. Injury was greatest and PAH levels were highest in shallow waters in eelgrass beds and bays. Levels of both injury and PAHs declined between 1990 and 1993. It was unclear whether the injuries were the result of acute toxicity of oil, sublethal effects of oiling, collateral injury from cleanup activities, or a combination of these factors.

Large epibenthic invertebrates are often injured by oiling. Crabs and echinoderms are particularly sensitive to oil, and their populations are often slow to recover (North et al. 1964; Cabioch et al. 1978; Chassé 1978; Krebs and Burns 1978; Cross et al. 1987; Jackson et al. 1989; Burger et al. 1991, 1992). For example, after the *Florida* oil spill off West Falmouth, Massachusetts, population densities of Atlantic marsh fiddler crabs *Uca pugnax* were greatly reduced and recovery required more than seven years (Krebs and Burns 1978). Similarly, North et al. (1964) found many dead sea urchins and sea stars after the *Tampico Maru* oil spill in Mexico. Follow-up studies indicated that urchin and sea star populations had not returned to prespill densities 6 years after the spill. Epibenthic invertebrates are exposed to oil through a number of pathways, including direct contact with oil in the water or on the sediments as well as by ingestion of oil through feeding. In addition, epifauna are susceptible to injury from cleanup activities, especially high-pressure hot-water washes that were extensively used after the *Exxon Valdez* oil spill (Houghton et al. 1996, this volume; Lees et al. 1996, this volume).

Epibenthic invertebrates are also extremely important as organizing forces in the nearshore community. They commonly serve as "keystone" species (sensu Paine 1966) that influence community structure because of their grazing or predatory activities. Both sea stars (e.g., Paine 1966, 1980; Menge 1982; Duggins 1983; O'Clair and Zimmerman 1986) and crabs (Peterson 1979), for example, can effectively prevent monopolization of the substratum by preying on the potential dominant occupier of space, thereby increasing community diversity.

We examine injury to and recovery of epibenthic invertebrates in the nearshore subtidal zone in Prince William Sound, Alaska, as a result of the *Exxon Valdez* oil spill. On 24 March 1989, 43.9 million liters of Prudhoe Bay crude oil were spilled after the *Exxon Valdez* accident, and oil coated about 560 km of shoreline within the sound (ADNR 1991). Much of the oil was removed by evaporation or cleanup activities; however, it is estimated that more than 50% of the oil either dispersed into the water column or remained deposited on the beaches (Wolfe et al. 1994). Some of this oil eventually found its way to subtidal sediments (Wolfe et al. 1994; O'Clair et al. 1996, this volume; Short et al. 1996, this volume). There are no quantitative data for the proportion of oil that was deposited subtidally, but it is estimated to be slightly less than 10% (Wolfe et al. 1994), which is in line with that observed in previous spills (Johansson et al. 1980; Gundlach et al. 1983). Although the impacts of oil are often less obvious in the subtidal than in the neighboring intertidal zone, these effects can be severe and long lasting (Michael 1977).

The shallow subtidal habitat in Prince William Sound is characterized by dense macroalgal and eelgrass beds that extend from the intertidal to depths of about 20 m. Common epibenthic invertebrates include a variety of sea stars and the helmet crab *Telmessus cheiragonus*. The most abundant stars are the sunflower sea star *Pycnopodia helianthoides*, the leather star *Dermasterias imbricata*, the slender arm star *Evasterias troschelii* the blood star *Henricia leviuscula*, and the rainbow sea star *Orthasterias koehleri*. Our study concentrated on these dominant taxa and examined the impacts of oil on their population densities. Sea urchins, common elsewhere in Alaska (e.g., Dayton 1975; Duggins 1980), are rare in Prince William Sound, presumably because of heavy predation by otters, which can severely limit the distribution of urchins (see for example Estes et al. 1978; Simenstad et al. 1978; Duggins 1980).

Methods

The effects of oiling and associated cleanup activities were examined by measuring the abundance of the dominant subtidal invertebrates at pairs of oiled and control sites using a stratified random sampling design. The population density of each taxon was measured in each of four habitats: Eelgrass *Zostera marina* beds in relatively shallow embayments characterized by soft substrata, *Agarum–Laminaria* beds (areas where *Agarum cribrosum* and *Laminaria saccharina* dominate) both in bays and on points within the Knight Island archipelago, and *Nereocystis* beds (Figure 1). Within each habitat, oiled sites were selected and then control sites were chosen that matched the oiled sites with respect to nonbiological factors other than oiling. Oiled sites were randomly chosen from each of several geographic regions within the western portion of Prince William Sound, where oiling of the shoreline was moderate to heavy.

Within each habitat, sampling was further stratified by depth. In the eelgrass habitat, sampling was conducted within eelgrass beds, generally at depths of from 1 to 3 m mean lower low water (MLLW). In the *Agarum–Laminaria* habitats, two depth strata were sampled: 2–11 m (shallow) and 11–20 m (deep). Within the *Nereocystis* habitat, sampling was conducted in a depth range of 2–8 m (the depth range for *Nereocystis*). All habitats and depths were sampled in summer 1990. Only two habitats, eelgrass and bays, were resampled in summer 1991 and again in summer 1993. Only the shallow depth stratum was sampled in the bay habitat after 1990.

Sampling was conducted within randomly placed transects that ran parallel to shore. Three 30-m long transects were sampled in each stratum at each site in the eelgrass, bay, and point habitats. Up to six transects per site were sampled in the *Nereocystis* habitat. Animals were counted within a 1-m wide band on either side of the transect. Only individuals readily visible upon moving aside the algae or eelgrass were counted, and no attempt was made to count animals that were under cobbles or boulders. The number of *P. helianthoides* was categorized within two size groups: animals that were less than approximately 10 cm in width (ray to ray) and animals greater than 10 cm. For all other species, only larger individuals were counted (sea stars greater than about 10 cm and crabs greater than 5 cm carapace width). Few newly recruited individuals were observed, except in the case of *Pycnopodia*.

Sediments were collected from each depth stratum at each study site and were analyzed to determine polycyclic aromatic hydrocarbon (PAH) concentrations. In 1990, all samples for the eelgrass and bay habitats were analyzed, but only a few samples from the other habitats were analyzed. In 1991 and 1993, only sediments from the eelgrass habitat were analyzed.

Results provided here were used to evaluate the effects of oiling and associated cleanup activities, and the rate of recovery of injured populations. We

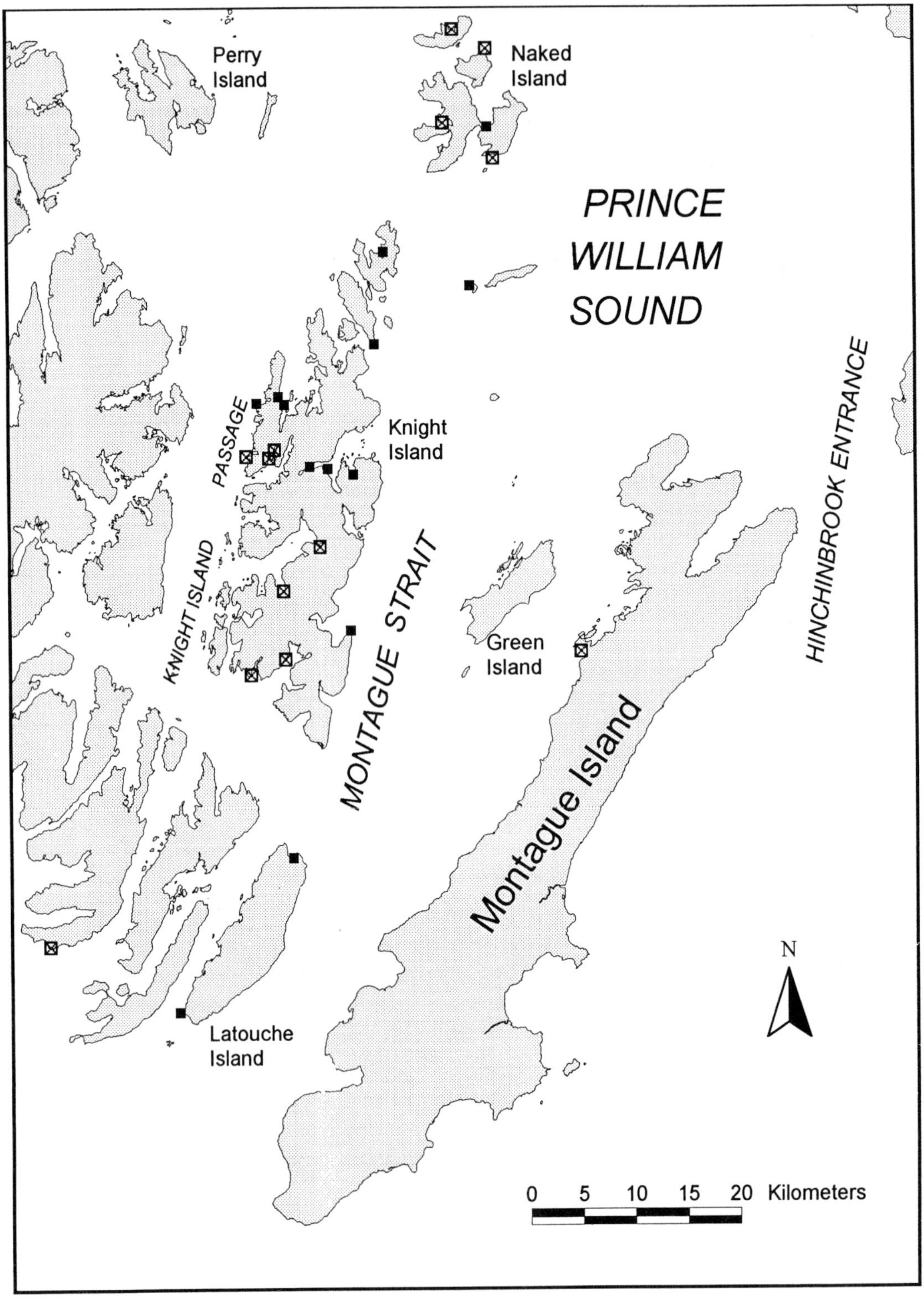

FIGURE 1.—Map of Prince William Sound showing the location of sampling sites. Filled symbols are oiled sites and open symbols are control sites.

relied primarily on differences observed between oiled and control sites in 1990 to infer injury. For the density of each of the dominant taxa, the null hypothesis of no significant difference between oiled and control sites was tested using a randomization analysis of variance (ANOVA) (Manly 1991). Randomization tests are robust with respect to assumptions of normality and homogeneity of variance and thereby eliminate the need for testing for these assumptions. Separate analyses were performed for each of the dominant taxa in each depth stratum within each habitat. Oil category (oil or control) was treated as the main effect, with pair as a blocking factor. The randomization procedure is described in Dean et al. (1996, this volume). For those taxa that occurred in several habitats and depths, differences between oiled and control sites were also tested in analyses that combined all habitats and depths.

Rates of recovery were assessed by comparing population densities over time and, more particularly, the relative differences between oiled and control sites over time. Several analyses were used to examine these effects. First, separate one-way randomization ANOVAs were conducted for each depth and habitat type in each year. These analyses utilized all available data (N = 3, 3, and 2 pairs of sites in 1990, 1991, and 1993, respectively, for bays; and N = 4, 5, and 3 sites, respectively, for the eelgrass habitat). Two-way randomization ANOVAs examined the effects of oiling and year, and their interaction. In the two-way analyses, only sites sampled in each of the 3 years were used (two pairs for bays and three pairs for eelgrass). Both oiling and year were treated as fixed effects.

For all statistical analyses, P values of less than 0.10 are reported as significant, however, we report relative significance ($P < 0.10$, $P < 0.05$, or $P < 0.01$) in all cases and make strong inferences based only on values of $P < 0.05$.

Chemical analyses of the collected sediments yielded values for several component hydrocarbon analytes. We present the total PAH concentrations for all analytes that were present. Randomization procedures, similar to those described above for biological variates, were used to determine statistical differences in total PAH concentrations between oiled and control sites.

Results

The dominant subtidal invertebrates in bays, on points, and in *Nereocystis* habitats were five species of sea stars (*Dermasterias imbricata*, *Evasterias troschelii*, *Pycnopodia helianthoides*, *Orthasterias koehleri*, and *Henricia leviuscula*) and one crab (*Telmessus cheiragonus*). In 1990, in the deeper stratum of the bays and in both shallow and deep strata on the points, *Telmessus* abundance was significantly greater at the control sites than at the oiled sites ($P < 0.05$, Table 1, Figure 2). Average abundances were 1.1/100 m^2 at control sites and less than 0.2/100 m^2 at the oiled sites. In addition, two sea star species were generally more abundant at control sites. *Dermasterias* density was significantly greater at control sites in the shallower bay ($P < 0.05$) and point ($P < 0.10$) habitats (Figure 3). *Evasterias* were more abundant at control sites in the shallow stratum at points and in *Nereocystis* habitats ($P < 0.10$, Figure 4). None of the other species showed significant reductions at any of the other oiled sites in 1990 (Table 1).

Within the eelgrass beds, only *Pycnopodia*, *Dermasterias*, and *Telmessus* were abundant. *Telmessus* were significantly more abundant at the control sites than at oiled sites in 1990 ($P < 0.01$, Table 1, Figure 2). Average densities were more than 3/100 m^2 at control sites but only 0.4/100 m^2 at oiled sites. In each of the four site pairs, average abundance was greater at the control sites. There were no significant differences for the sea star species within the eelgrass bed in 1990 (Table 1); however, average abundances of *Dermasterias* at control sites were about twice as high as those at oiled sites (Figure 3).

Telmessus and *Dermasterias* were common in most or all habitats. Therefore, we were able to examine the effects of oiling on these species over all habitats combined. The densities of *Telmessus* and *Dermasterias* were lower at heavily oiled sites than at control sites in 1990 (Table 1, Figures 2 and 3). Of the 12 pairs of sites where *Telmessus* occurred, 11 of these had higher abundance at the control site. The overall average abundance of helmet crabs was about four times greater at the control sites. *Dermasterias* abundance was greater at the control site in 10 pairs of sites, and greater at the oiled site in 3 pairs; average abundance was almost twice as great at control sites.

Within the bay habitat, populations of both *Telmessus* and *Dermasterias* appeared to have recovered fully by 1993 (Figure 5). In 1991, mean abundances of both species were still slightly greater at control sites, and neither differed significantly between sites (Table 1). By 1993, mean abundances of both species were slightly higher at the oiled sites. On the other hand, within the eelgrass bed, there were still indications of a possible

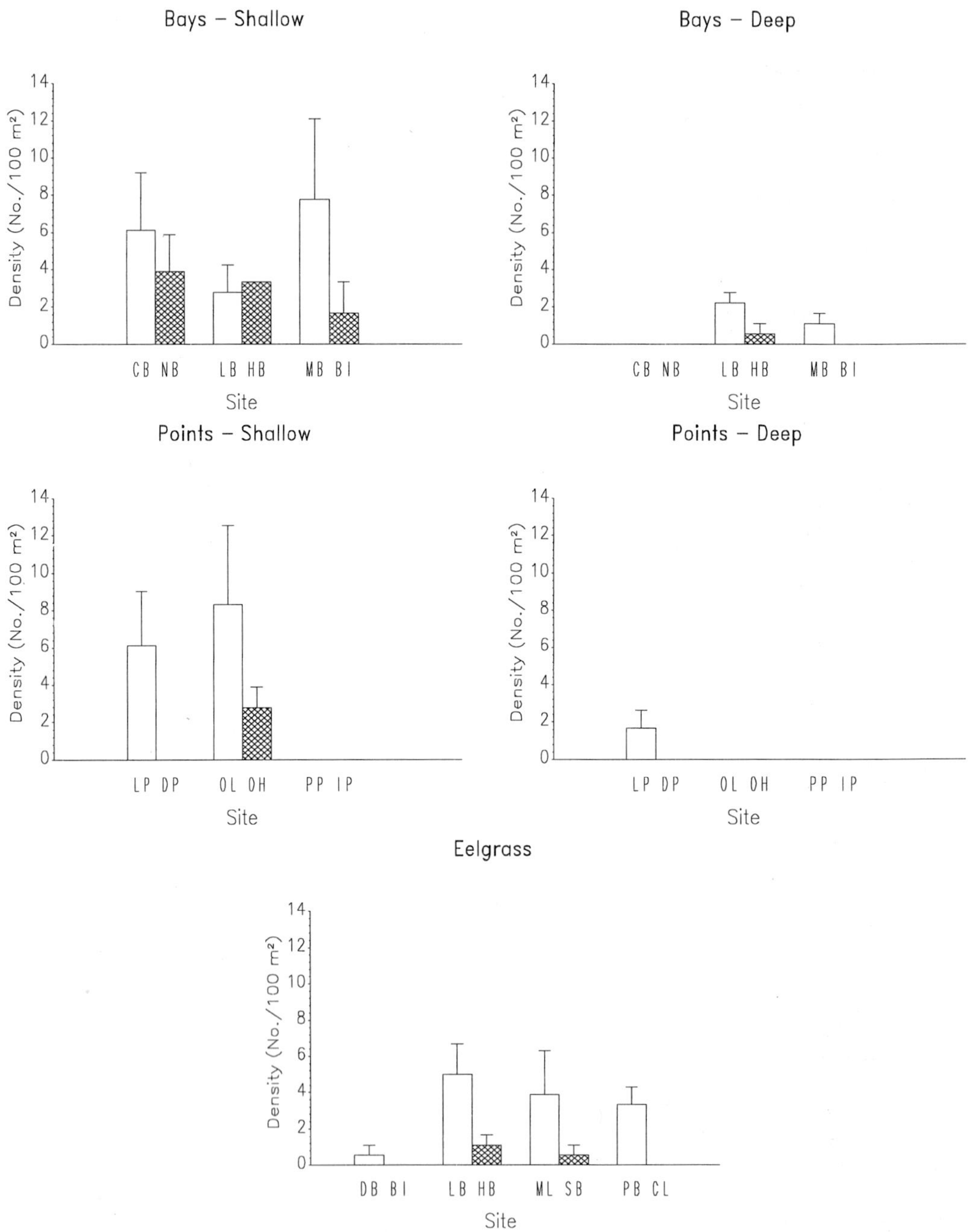

FIGURE 2.—Mean densities (number/100 $m^2 \pm 1$ SE) of *Telmessus cheiragonus* at oiled (shaded bars) and control sites (open bars) in 1990. Site names are as follows: CB = Cabin Bay, NB = Northwest Bay, LB = Lower Herring Bay, HB = Herring Bay, MB = Mummy Bay, BI = Bay of Isles, LP = Lucky Point, DP = Discovery Point, OL = Outside Lower Herring Bay, OH = Outside Herring Bay, PP = Peak Point, and IP = Ingot Point, DB = Drier Bay, ML = Mooselips Bay (Montague Island), SB = Sleepy Bay, PB = Puffin Bay (Peak Island), CL = Clammy Bay (Naked Island).

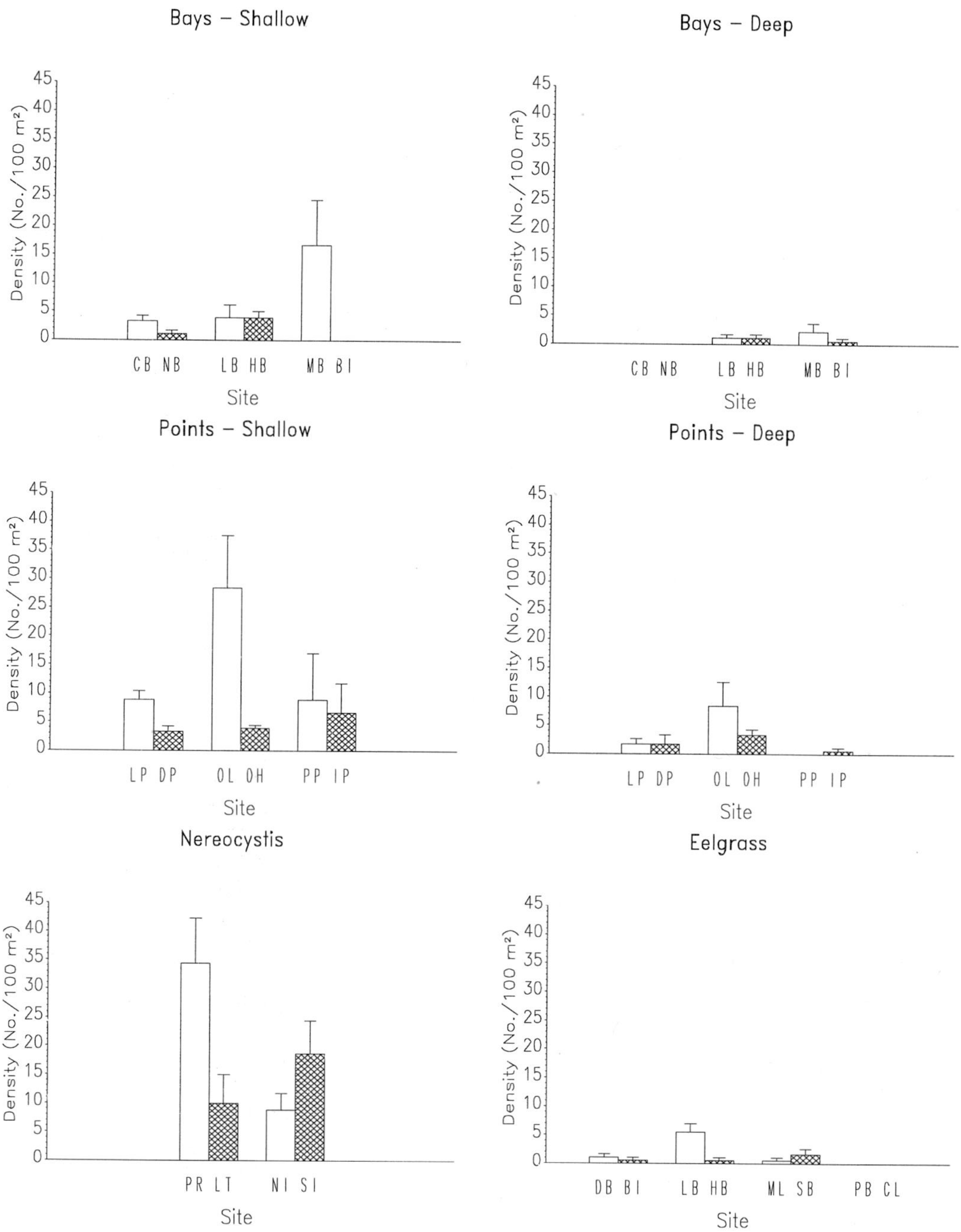

FIGURE 3.—Mean densities (number/100 $m^2 \pm 1$ SE) of *Dermasterias imbricata* at oiled (shaded bars) and control sites (open bars) in 1990. Site names are as indicated in Figure 2, with the following additions: PR = Procession Rocks, LT = Latouche Point, NI = Naked Island, and SI = Little Smith Island.

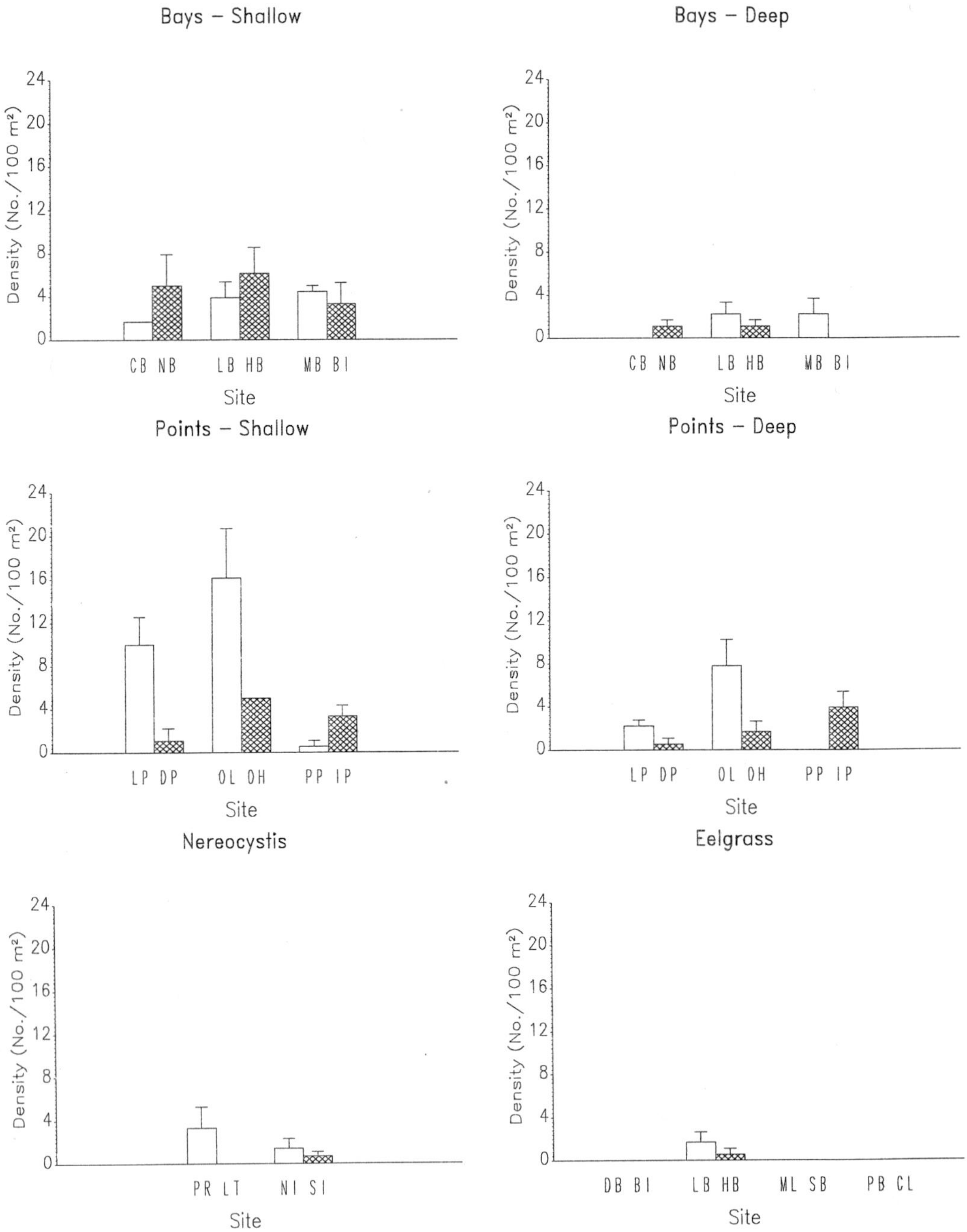

FIGURE 4.—Mean densities (number/100 $m^2 \pm 1$ SE) of *Evasterias troschelii* at oiled (shaded bars) and control sites (open bars) in 1990. Site names are as indicated in Figures 2 and 3.

TABLE 1.—Summary of results of randomization analyses of variance testing for differences in mean densities of dominant taxa between oiled and control sites in Prince William Sound within each depth stratum and year. Results are expressed as follows: minuses = significantly greater at the control sites, and pluses = significantly greater at the oiled sites; one symbol = $P < 0.10$, two symbols = $P < 0.05$, and three symbols = $P < 0.01$; open circles = no significant difference at $P < 0.10$; ID = insufficient data to test for differences.

Habitat	*Telmessus cheiragonus*	*Dermasterias imbricata*	*Evasterias troschelii*	*Pycnopodia helianthoides* (adult)	*Pycnopodia helianthoides* (juvenile)	*Orthasterias koehleri*	*Henricia leviuscula*
				1990			
Bays, shallow	O	– –	O	+	O	O	O
Bays, deep	– – –	O	O	O	O	O	ID
Eelgrass	– – –	O	ID	O	O	ID	ID
Points, shallow	– –	–	–	+	O	O	O
Points, deep	– – –	ID	O	O	O	O	O
Nereocystis	O	ID	–	O	O	O	O
All habitats	– – –	– –	–	O	O	–	O
				1991			
Bays, shallow	O	O	O	O	+ + +	O	ID
Eelgrass	O	– –	ID	– –	O	ID	ID
All habitats	O	– –	O	O	+	O	ID
				1993			
Bays, shallow	O	O	O	O	O	O	ID
Eelgrass	ID	O	ID	O	O	ID	ID
All habitats	O	O	O	+ +	O	O	ID

effect of oil as of July 1993. The pattern of higher mean abundance at the control sites persisted, with densities of both *Telmessus* and *Dermasterias* nearly three times greater at control sites than at oiled sites in 1993 (Figure 5). The differences in densities in 1993 were not significant (Table 1); however, when data from all years were considered together, there were significant differences observed between oiled and control sites ($P < 0.05$ for *Telmessus* and $P < 0.01$ for *Dermasterias*, Table 2 and Figure 5).

Also apparent was a significant decline in the population density of *Telmessus* within the eelgrass beds between 1990 and 1993 ($P < 0.05$), especially at control sites (Table 2 and Figure 5). We do not know if this decline was related to the effects of oil or to natural fluctuations in population density.

The patterns of abundance for *Pycnopodia* were quite different from those of other sea stars and varied between habitats (Figure 6). In 1990, densities of adults were generally greater at oiled sites within bay and point habitats, differences in density were marginally significant ($P < 0.10$) in the shallow portions of these habitats (Table 1 and Figure 6). The differences persisted to a large extent through 1993 (Figure 7). In 1990, there were 2.2 times as many adult *Pycnopodia* at oiled sites than at control sites in the shallow bay habitat. In 1993, there were 1.7 times more of these sea stars at oiled sites than at control sites. The differences in mean densities of adult *Pycnopodia* between sites remained significant ($P < 0.05$) when data from all 3 years were considered together (Table 2 and Figure 7).

In contrast, there was no significant difference in the densities of adult *Pycnopodia* at oiled versus control sites within the eelgrass habitat in 1990 (Table 1 and Figure 6). By 1991, densities were significantly greater at control sites ($P < 0.05$, Table 1 and Figure 7). There was also a significant interaction between oiling and year (Table 2), with mean densities being higher at control sites in 1990 and 1991, but higher at oiled sites in 1993 (Figure 7). These data suggest possible injury to populations of *Pycnopodia* in the eelgrass habitat, with recovery by 1993.

One striking result was the extremely large number of juvenile *Pycnopodia* present in 1991 and 1993 and the significant increase in juveniles over time (Figure 7). There were highly significant ($P < 0.01$) increases in the density of juveniles between 1990 and 1993 in both eelgrass and bay habitats (Table 2 and Figure 7). In 1991, newly settled *Pycnopodia* were observed in moderate abundance (more than $5/100\ m^2$) in the bay habitat and in high abundance (more than $20/100\ m^2$) in the eelgrass habitat. By 1993, the densities had more than doubled in these habitats.

In 1991, the density of these newly settled *Pycnopodia* at oiled sites was twice as high as that at control sites in the bay habitats and was slightly higher at control sites than at oiled sites in the eelgrass habitat. Differences between oiled and control sites were significant in the bay habitat ($P < 0.01$) but not in the eelgrass habitat ($P = 0.12$, Table 1 and Figure 7). Most of the individuals ob-

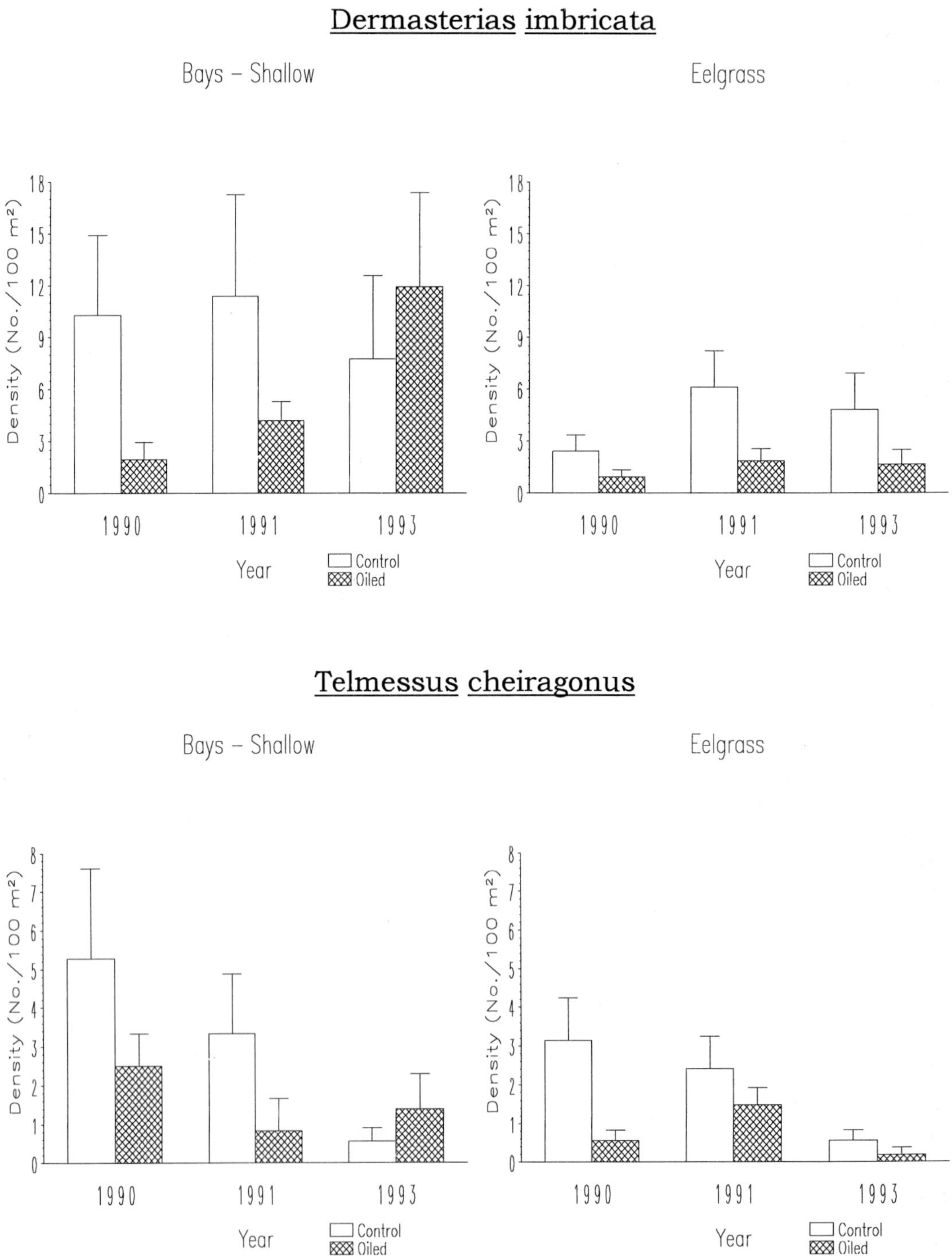

FIGURE 5.—Mean densities (number/100 m^2 ± 1 SE) of *Dermasterias imbricata* and *Telmessus cheiragonus* at oiled (shaded bars) and control sites (open bars) in 1990, 1991, and 1993.

TABLE 2.—Summary of probabilities from two-way randomization analyses of variance testing for differences in mean densities of dominant taxa at sites in Prince William Sound between oil category, year, and their interaction.

Variable	*Telmessus cheiragonus*	*Dermasterias imbricata*	*Pycnopodia helianthoides* (adult)	*Pycnopodia helianthoides* (juvenile)
		Bays, shallow		
Oil category	0.18	0.29	0.03	0.99
Year	0.12	0.73	<0.01	<0.01
Interaction	0.39	0.26	0.39	0.91
		Eelgrass		
Oil category	0.02	<0.01	0.96	0.98
Year	0.03	0.23	0.47	0.01
Interaction	0.26	0.60	<0.01	0.99
		Bays, shallow, plus eelgrass		
Oil category	0.01	0.04	0.05	0.93
Year	0.01	0.41	<0.01	<0.01
Interaction	0.12	0.34	0.28	0.99

served were attached to the blades of either algae or eelgrass and were apparently feeding on epifauna attached to the plants. There were no significant differences in densities of small *Pycnopodia* between oiled and control sites in 1993 (Table 1 and Figure 7).

Concentrations of PAHs were generally higher at oiled sites than at control sites in 1990 (Table 3). Within bay sites and eelgrass beds, average concentrations were more than twice as high at oiled sites than at controls. In the eelgrass beds and shallow portions of bays, PAH concentrations averaged approximately 1,100–4,000 ng/g at oiled sites and about 400–700 ng/g at controls.

Concentrations of PAH decreased markedly in 1991 and 1993. Within the eelgrass bed (the only habitat for which hydrocarbon data are available after 1990), concentrations of PAHs in 1993 were less than 11% of those observed in 1990, at both oiled and control sites. Concentrations remained somewhat higher at oiled sites than at controls in 1993 but did not differ significantly ($P = 0.38$).

Discussion

As with all assessments of the effects of a disturbance on ecological systems, the final decision as to whether injury has occurred often rests on the "weight of evidence." When one examines many biological variates as we have done, one expects to find differences between oiled and control sites even in the absence of an effect of an event such as an oil spill. However, the evidence for an effect becomes stronger if one finds repeated patterns that indicate the same effect over a variety of habitats, especially if these observations can be supported by knowledge of the effects of oil and of interactions between species. Therefore, we will focus our discussion of the effects of the *Exxon Valdez* oil spill on consistent patterns that emerge from statistical tests and on possible mechanisms of impact.

In our sampling design, statistical inferences can be made only for those sites sampled. The oiled sites, and transects within sites, were selected with a degree of randomness such that we could be relatively assured of eliminating bias with respect to oiling; however, we were unable to clearly define a universe of potentially oiled sites from the outset and therefore must rely on judgment rather than statistical inference to extrapolate from our results to the entire oiled area.

Making inferences from observed differences between oiled and control sites rests on the assumption that population densities would have been similar at those sites in the absence of a spill. It is possible that there were inherent differences between oiled and control sites, but the convergence of population densities at oiled and control sites over time and the concomitant decline in PAH levels suggest that differences were oil related.

There was strong evidence of an adverse effect of the *Exxon Valdez* oil spill on populations of two numerically dominant taxa: *Telmessus cheiragonus* and *Dermasterias imbricata*. *Telmessus* were found consistently in lower abundance at oiled sites in each of the three habitats in which they occurred (eelgrass, bays, and points). *Dermasterias* abundance was lower at oiled sites, especially in shallower portions of the habitats (Figure 3).

There was evidence of full recovery of these populations by 1993 within bays but only partial recov-

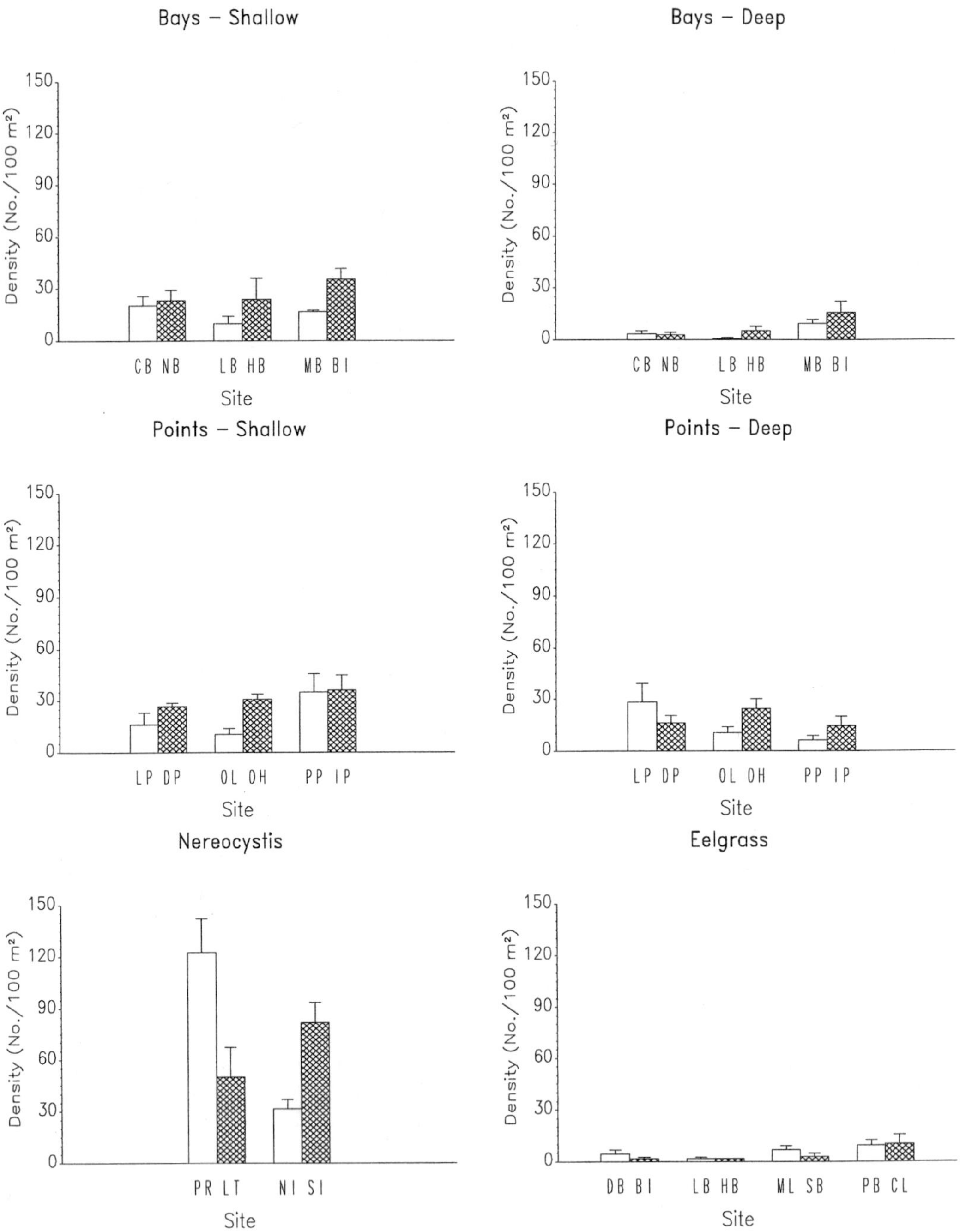

FIGURE 6.—Mean densities (number/100 m² ± 1 SE) of adult *Pycnopodia helianthoides* at oiled (shaded bars) and control sites (open bars) in 1990. Site names are as indicated in Figures 2 and 3.

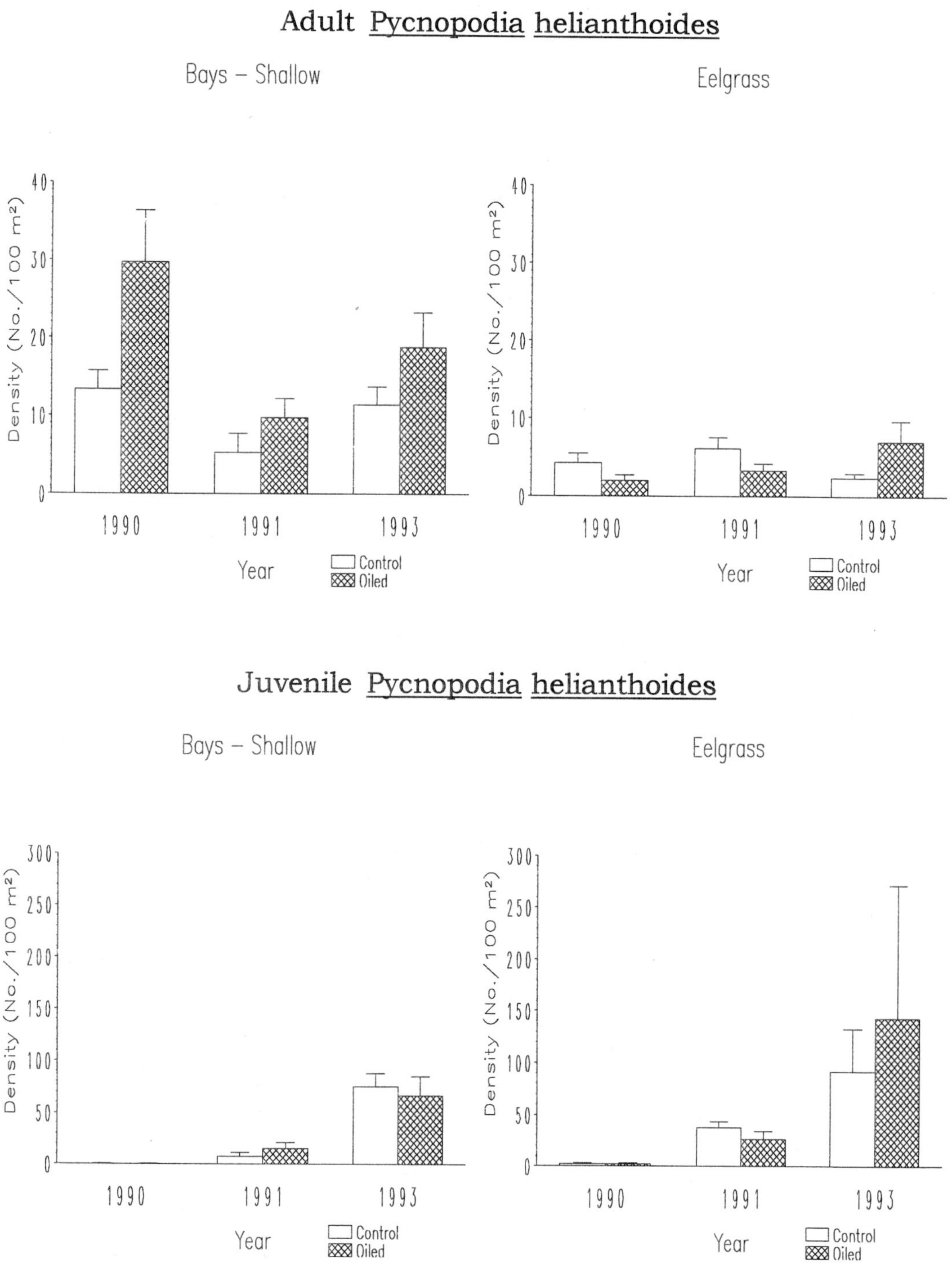

FIGURE 7.—Mean densities (number/100 $m^2 \pm 1$ SE) of adult and juvenile *Pycnopodia helianthoides* at oiled (shaded bars) and control sites (open bars) in 1990, 1991, and 1993.

TABLE 3.—Mean concentrations (ng/g) of polycyclic aromatic hydrocarbons and results of randomization analyses of variance testing for differences in mean concentrations between oiled and control sites in Prince William Sound. Abbreviations are ND = no data available, and ID = insufficient data to test for differences.

Habitat and year	Oil	Control	*P*
Bays, shallow			
1990	1,142	420	0.02
Bays, deep			
1990	406	98	<0.01
Eelgrass			
1990	3,990	691	0.08
1991	649	537	0.46
1993	196	74	0.38
Points, shallow			
1990	ND	19[a]	ID
Points, deep			
1990	163	125[b]	0.46
Nereocystis			
1990	45[a]	ND	ID

[a]Samples from only one site were analyzed
[b]Samples from only two sites were analyzed

ery within eelgrass beds. Mean densities of both *Telmessus* and *Dermasterias* were higher at oiled sites within bays than at controls by 1993, albeit not significantly so (Figure 5 and Table 1); on the other hand, densities of both species were still more than twice as high at control sites within the eelgrass habitat. The apparent recovery of *Telmessus* populations at oiled sites in shallow bays resulted largely from a decline in abundance at the control sites (Figure 5), which may have been the result of emigration from control sites to previously oiled areas. There was little evidence of crab recruitment, and a significant decline in the abundance of crabs was noted between 1990 and 1993.

The evidence for possible effects on sea stars other than *Dermasterias* was less convincing. In 1990, *Evasterias* were slightly more abundant at control sites than at oiled sites in shallow points and in the *Nereocystis* habitat, but this pattern was not evident in any of the other habitats or depths. Similarly, there were few significant differences between oiled and control sites with respect to densities of either *Orthasterias* or *Henricia*. Patterns of abundance suggested possible adverse impacts of oiling on adult *Pycnopodia* in eelgrass beds, however, these adults were generally more abundant in 1990 at oiled sites in shallow portions of bay and point habitats, and in 1991 young-of-year *Pycnopodia* were much more abundant at oiled sites in shallow bays.

Concentrations of PAHs were higher at sites adjacent to oiled beaches, thereby verifying our general classification of oiled and control sites. The PAH concentrations measured give a relative index of exposure between sites and between years but do not necessarily reflect the absolute levels of PAHs associated with *Exxon Valdez* oil. We did not "fingerprint" the hydrocarbons found in sediments, and it is likely that at least some of the PAHs were an indirect effect of the spill (e.g., fuel spills from vessels used in the cleanup of oil) or from natural sources; however, the higher levels of PAHs found at oiled sites and the decrease in PAH levels over time strongly suggest that most of the PAHs measured in 1990 were spill related.

There was a general correspondence between PAH concentrations and biological impacts, both on temporal and spatial scales. Differences in abundance of our study animals between oiled and control sites were greatest within eelgrass beds and in shallow portions of bays where PAH levels were highest, and differences between sites diminished over time as PAH levels diminished.

Possible sources of oil-related impacts include (1) toxicity of oil, (2) emigration from or avoidance of oiled sites, (3) mortality caused by cleanup activities, and (4) secondary trophic effects including reductions in prey availability or increases in predator abundance. We suspect that differences in population densities of leather stars between oiled and control sites were caused by the toxicity of oil. The PAH levels that we observed in sediments were well below levels shown to have acute toxic effects on most benthic organisms (Rice et al. 1977, 1979) or that alter benthic community structure (Long 1992); however, in some cases, levels of exposure to oil probably were much higher than indicated by our analysis of PAHs. Many of the organisms that we studied, especially the sea stars, often venture into the intertidal zone and probably were exposed to much higher levels of oil than were measured in subtidal sediments. Also, sea stars in the intertidal zone were subjected to the rigors of shoreline treatment, including hot-water high-pressure washing (Lees et al. 1996). *Dermasterias* often move into the lower intertidal areas as the tide rises, become exposed on steeper rock faces as the tide falls, and then release themselves from the rocks, falling into deeper water as the tide recedes further. Sea stars are sensitive to oil (O'Clair and Rice 1985) and narcosis of sea stars has been observed after oiling both in the laboratory (O'Clair and Rice 1985) and after an experimental spill in the field (Cross et al. 1987). In addition, sea stars (especially *Pycnopodia*) were observed overturned in the intertidal zone in heavily oiled areas of Prince William Sound several days after the *Exxon Valdez* spill (J. Bodkin, National Biological Service, personal communication), indicating narcosis or possible death as a result of oiling. Dead sea stars were also observed subtidally in a heavily oiled fjord in fall 1989 (Jewett et al. 1995), but it was unclear whether the deaths were attributable to oiling or to naturally occurring low oxygen levels at that site.

Impacts on *Telmessus* probably were the result of avoidance of oiled sites by adult crabs or toxicity of oil on larvae. Although oiling can be acutely toxic to adult crabs (Rice et al. 1977, 1979; Burger et al. 1992), *Telmessus* seldom venture into the intertidal zone, and adult crabs probably were not exposed to lethal concentrations of oil (or to the rigors of shoreline treatment). Furthermore, crabs can detect oil in relatively low concentrations (Pearson et al. 1980, 1981), are highly mobile, and may have effectively avoided heavily oiled sites. Oiling can also lead to behavioral modifications such as a poor righting response, which may in turn cause these crabs to be more susceptible to predation (Burger et al. 1991). The observed longer-term decline in the abundance of *Telmessus* was possibly the result of the effects of oil on recruitment of crabs. Larvae of crabs are particularly sensitive to oil, relative to adult crabs and to other invertebrate larvae (see review in Rice et al. 1977), and concentrations of hydrocarbons that were not lethal to adults may have impaired recruitment. Krebs and Burns (1978) noted similar long-term declines in population density and a general failure of recruitment of fiddler crabs for several years after the West Falmouth, Massachusetts, oil spill.

There is little direct evidence of the effects of observed changes in crabs or sea stars on the abundance of their predators or prey; however, the reduction in numbers of *Telmessus* probably affects its predators, especially sea otters *Enhydra lutris*. *Telmessus* are prey to a variety of birds and mammals (McConnaughey and McRoy 1979) and make up a sizable proportion of the diet of sea otters (Calkins 1978; Garshelis et al. 1986; Green and Brueggeman 1991). A reduction in the numbers of these crabs, along with reductions in other potential food sources for otters (Jewett et al. 1995; Driskell 1996, this volume), may lead to further stress on already impaired (Ballachey et al. 1994) otter populations. Sea stars, on the other hand, are only occasional prey for otters or other organisms, and losses of sea stars probably have little impact on higher trophic levels.

The reduction in the numbers of crabs and sea stars at oiled sites probably had some indirect effect on their prey. *Telmessus* consume large amounts of eelgrass and both *Telmessus* and *Dermasterias* prey on a variety of epifauna, many of which are attached to or closely associated with eelgrass (McConnaughey and McRoy 1979; Feder 1980). There were notably higher densities of epifauna within oiled eelgrass beds in 1990, especially of the small mussel, *Musculus* spp., and spirorbid worms (Jewett et al. 1995). We suspect that the increase in the numbers of epifauna may, in part, have resulted from a reduction in predation by both crabs and sea stars.

We conclude that populations of some epibenthic invertebrates, in particular helmet crabs and leather stars, were negatively affected by the *Exxon Valdez* oil spill. The causes of the negative impacts are not clear, but they appear to be related to toxicity of oil, collateral injury from cleanup activities, or a combination of these factors.

Acknowledgments

We thank the sponsors of this work, the USDA Forest Service and the Alaska Department of Fish and Game, for their support. We especially thank the more than 20 divers who helped us to collect the data, often under adverse conditions. We also thank Lyman McDonald, who assisted in sampling design and site selection. Richard Rosenthal willingly shared his extensive knowledge of Prince William Sound and its natural history. Charles Peterson, Robert Spies, and an anonymous reviewer provided valued reviews of the manuscript and of prior reports on which the manuscript was based.

References

ADNR (Alaska Department of Natural Resources and U.S. Fish and Wildlife Service). 1991. A report on oiling to environmentally sensitive shorelines. Report to the *Exxon Valdez* Oil Spill Trustee Council, Anchorage, Alaska.

Ballachey, B. E., J. L. Bodkin, and A. R. DeGange. 1994. An overview of sea otter studies. Pages 47–59 *in* T. R. Loughlin, editor. Marine mammals and the *Exxon Valdez*. Academic Press, New York.

Burger, J., J. Brzorad, and M. Gochfeld. 1991. Immediate effects of an oil spill on behavior of fiddler crabs (*Uca pugnax*). Archives of Environmental Contamination and Toxicology 20:404–409.

Burger, J., J. Brzorad, and M. Gochfeld. 1992. Effects of an oil spill on emergence and mortality in fiddler crabs (*Uca pugnax*). Environmental Monitoring and Assessment 22:107–115.

Cabioch, L., J. C. Dauvin, and F. Gentil. 1978. Preliminary observations on pollution of the sea bed and disturbance of sublittoral communities in Northern Brittany by oil from the *Amoco Cadiz*. Marine Pollution Bulletin 9:303–307.

Calkins, D. G. 1978. Feeding behavior and major prey species of the sea otter, *Enhydra lutris*, in Montague Strait, Prince William Sound, Alaska. U.S. National Marine Fisheries Service Fishery Bulletin 76:125–131.

Chassé, C. 1978. The ecological impact on and near shores by the *Amoco Cadiz* oil spill. Marine Pollution Bulletin 11:298–301.

Cross, W. E., C. M. Martin, and D. H. Thomson. 1987. Effects of experimental releases of oil and dispersed oil on arctic nearshore macrobenthos. 2. Epibenthos. Arctic 40:201–210.

Dayton, P. K. 1975. Experimental studies of algal canopy interactions in a sea otter-dominated kelp community at Amchitka Island, Alaska. U.S. National Marine Fisheries Service Fishery Bulletin 73:230–237.

Dean, T. A., M. S. Stekoll, and R. O. Smith. 1996. Kelps and oil: the effects of the *Exxon Valdez* oil spill on subtidal algae. American Fisheries Society Symposium 18:412–423.

Driskell W. B., A. K. Fukuyama, J. P. Houghton, D. C. Lees, A. J. Mearns, and G. Shigenaka. 1996. Recovery of Prince William Sound intertidal infauna from *Exxon Valdez* oiling and shoreline treatments, 1989 through 1992. American Fisheries Society Symposium 18:362–378.

Duggins, D. O. 1980. Kelp beds and sea otters: an experimental approach. Ecology 6:447–453.

Duggins, D. O. 1983. Starfish predation and the creation of mosaic patterns in a kelp-dominated community. Ecology 64:1610–1619.

Estes, J. A., N. S. Smith, and J. F. Palmisano. 1978. Sea otter predation and community organization in the western Aleutian Islands, Alaska. Ecology 59:822–833.

Feder, H. M. 1980. Asteroidea: the sea stars. Pages 117–135 *in* R. H. Morris, D. P. Abbott, and E. C. Haderlie, editors. Intertidal invertebrates of California. Stanford University Press, Stanford, California.

Garshelis, D. L., J. A. Garshelis, and A. T. Kimken. 1986. Sea otter time budgets and prey relationships in Alaska. Journal of Wildlife Management 50:637–647.

Green, G. A., and J. J. Brueggeman. 1991. Sea otter diets in a declining population in Alaska. Marine Mammal Science 7:395–401.

Gundlach, E. R., P. D. Boehm, M. Marchand, R. M. Atlas, D. M. Ward, and D. A. Wolfe. 1983. The fate of *Amoco Cadiz* oil. Science 221:122–129.

Houghton, J. P., D. C. Lees, W. B. Driskell, S. C. Lindstrom, and A. J. Mearns. 1996. Recovery of Prince William Sound intertidal epibiota from *Exxon Valdez* oiling and shoreline treatments, 1989 through 1992. American Fisheries Society Symposium 18:379–411.

Jackson, J. B. C., and seventeen coauthors. 1989. Ecological effects of a major oil spill on Panamanian coastal marine communities. Science 243:37–44.

Jewett, S. C., and six coauthors. 1995. The effects of the *Exxon Valdez* oil spill on shallow subtidal communities in Prince William Sound, Alaska 1989–1993. Report to the *Exxon Valdez* Trustee Council, Anchorage, Alaska.

Johansson, S., U. Larson, and P. Boehm. 1980. The *Tsesis* oil spill: impact on the pelagic ecosystem. Marine Pollution Bulletin 11:284–293.

Krebs, C. T., and K. A. Burns. 1978. Long-term effects of an oil spill on populations of the salt-marsh crab *Uca pugnax*. Journal of the Fisheries Research Board of Canada 35:648–649.

Lees, D. C., J. P. Houghton, and W. B. Driskell. 1996. Short-term effects of several types of shoreline treatment on rocky intertidal biota in Prince William Sound. American Fisheries Society Symposium 18: 329–348.

Long, E. R. 1992. Ranges in chemical concentrations in sediments associated with adverse biological effects. Marine Pollution Bulletin 24:38–45.

Manly, B. 1991. Randomization and Monte Carlo methods in biology. Chapman and Hall, London.

McConnaughey, T., and C. P. McRoy. 1979. ^{13}C label identifies eelgrass (*Zostera marina*) carbon in an Alaskan estuarine food web. Marine Biology 53:263–269.

Menge, B. A. 1982. Effects of feeding on the environment. Asteroidea. Pages 521–551 *in* M. A. Jangoux and J. M. Lawrence, editors. Echinoderm nutrition. A. A. Balkeman, Rotterdam.

Michael, A. D. 1977. The effect of petroleum hydrocarbons on marine populations and communities. Pages 129–137 *in* D. A. Wolfe, editor. Fate and effects of petroleum hydrocarbons in marine ecosystems and organisms. Pergamon Press, New York.

North, W. J., M. Neushul, and K. A. Clendenning. 1964. Successive biological changes observed in a marine cove exposed to a large spillage of mineral oil. Pages 335–354 *in* Proceedings of the symposium on pollution of marine organisms. Prod. Petrol., Monaco.

O'Clair, C. E., and S. D. Rice. 1985. Depression of feeding and growth rates of the seastar *Evasterias troschelii* during long-term exposure to the water-soluble fraction of crude oil. Marine Biology 84:331–340.

O'Clair, C. E., J. W. Short, and S. D. Rice. 1996. Contamination of intertidal and subtidal sediments by oil from the *Exxon Valdez* in Prince William Sound. American Fisheries Society Symposium 18:61–93.

O'Clair, C. E., and S. T. Zimmerman. 1986. Biogeography and ecology of intertidal and shallow subtidal communities. Pages 305–344 *in* D. W. Hood and S. T. Zimmerman, editors. The Gulf of Alaska: physical environment and biological resources. Minerals Management Service, Publication MMS 86-0095. U.S. Government Printing Office, Washington, DC.

Paine, R. T. 1966. Food web complexity and species diversity. American Naturalist 100:65–75.

Paine, R. T. 1980. Food webs: linkage, interaction strength and community infrastructure. Journal of Animal Ecology 49:667–685.

Pearson, W. H., S. E. Miller, J. W. Blaylock, and B. L. Olla. 1981. Detection of the water-soluble fraction of crude oil by the blue crab, *Callinectes sapidus*. Marine Environmental Research 5:3–11.

Pearson, W. H., P. C. Sugarman, D. L. Woodruff, J. W. Blaylock, and B. L. Olla. 1980. Detection of petroleum hydrocarbons by the Dungeness crab, *Cancer magister*. U.S. National Marine Fisheries Service Fishery Bulletin 78:821–826.

Peterson, C. H. 1979. The importance of predation and competition in organizing the intertidal epifaunal communities of Barnegat Inlet, New Jersey. Oecologia 39:1–24.

Rice, S. D., A. Moles, T. L. Taylor, and J. F. Karinen. 1979. Sensitivity of 39 Alaskan marine species to Cook inlet crude oil and no. 2 fuel oil. Pages 549–554 *in* Proceedings, 1979 oil spill conference (prevention, behavior, control, cleanup). American Petroleum Institute Publication 4308, Washington, DC.

Rice, S. D., J. W. Short, and J. F. Karinen. 1977. Comparative oil toxicity and comparative animal sensitivity. Pages 78–94 *in* D. A. Wolfe, editor. Fate and effects of petroleum hydrocarbons in marine ecosystems and organisms. Pergamon Press, New York.

Short, J. W., D. M. Sale, and J. C. Gibeaut. 1996. Nearshore transport of hydrocarbons and sediments after the *Exxon Valdez* oil spill. American Fisheries Society Symposium 18:40–60.

Simenstad, C. A., J. A. Estes, and K. W. Kenyon. 1978. Aleuts, sea otters, and alternate stable-state communities. Science 200:403–411.

Wolfe, D. A., and eleven coauthors. 1994. The fate of the oil spilled from the *Exxon Valdez*. Environmental Science and Technology 28:560A–568A.

American Fisheries Society Symposium 18:440–447, 1996

Effects of the *Exxon Valdez* Oil Spill on Benthic Invertebrates in an Oxygen-Deficient Embayment in Prince William Sound, Alaska

STEPHEN C. JEWETT

School of Fisheries and Ocean Sciences, University of Alaska Fairbanks, Fairbanks, Alaska 99775-1080, USA

THOMAS A. DEAN

Coastal Resources Associates, Inc., 1185 Park Center Dr., Suite A, Vista, California 92083, USA

DAVID R. LAUR

Marine Science Institute, University of California Santa Barbara Santa Barbara, California 93106, USA

Abstract.—The possible impacts of the *Exxon Valdez* oil spill on the infaunal and epibenthic community within a heavily oiled fjordic embayment in western Prince William Sound, Alaska, were examined. Initial observations within this embayment in fall 1989 revealed numerous dead and dying organisms. The infauna within the embayment was represented by a relatively rich assemblage of 24 taxa. By fall 1990, the infaunal community was reduced to only six taxa and was dominated by a single polychaete species. The observations of dead organisms in 1989 and the decline of the benthic community between 1989 and 1990 were coincident with high concentrations of hydrocarbons in the sediments. By 1991, hydrocarbon concentrations were greatly reduced, very few dead organisms were observed, and the benthic community had recovered to include 32 taxa. These data suggest a possible adverse impact of oiling in 1989 and 1990, followed by recovery. However, sampling in fall 1993 again revealed a greatly impoverished community (four taxa), concomitant with low hydrocarbon concentrations in the sediment and depleted dissolved oxygen in the bottom waters. These data suggest that although the *Exxon Valdez* oil spill may have been, in part, responsible for the dead organisms and the impoverished infaunal community in the years subsequent to the spill, such reductions of benthic infauna can occur as a result of natural hypoxia–anoxia.

Benthic invertebrates have been used successfully throughout the industrialized world as a basis for measuring the effects of manmade and natural disturbances. Numerous investigations have addressed the effects of anthropogenic sources, such as petroleum production activities (Boesch and Rabalais 1987), mining (Ellis and Hoover 1990), dredging (McCauley et al. 1977), pulp effluent (Pearson et al. 1986), and sewage (Anderlini and Wear 1992). Somewhat less attention has been given to the effects of natural sources, such as storms (Dobbs and Vozarik 1983), ice scouring (Carey and Ruff 1977), and feeding of marine organisms (Klaus et al. 1990). However, one source of natural perturbation, seasonal hypoxia–anoxia, has been widely recognized as exerting a strong influence on benthic communities worldwide (e.g., Tyson and Pearson 1991). Occasional anoxic conditions are caused by a combination of physically stratified water column, biological processes, and long residence times of water parcels and can result in drastic reductions in infaunal diversity, abundance, and biomass (e.g., Jørgensen 1980; Rosenberg 1980; Josefson and Rosenberg 1988).

During October 1989, when examining the effects of the *Exxon Valdez* oil spill (24 March 1989) on several shallow subtidal habitats, high mortality of invertebrates and fishes was observed in a heavily oiled fjordic embayment in western Prince William Sound, Alaska. We present the results from initial observations and subsequent sampling of this site and consider the effects on the benthos from the spill or the natural effects from hypoxia ($O_2 < 2$ mg/L) or anoxia ($O_2 = 0$ mg/L).

Study Sites

The main site examined is a small embayment (60°28′N, 147°42′W) in northeastern Herring Bay located along the northwestern side of Knight Island in Prince William Sound, Alaska (Figure 1). The 0.5-km^2 embayment has a gently sloping basin that is approximately 35 m at the greatest depth; a sill at the entrance is only 4 m deep. The macroalgae *Laminaria saccharina* and *Agarum cribosum* are dominant only inside the estuary entrance; lesser stands of eelgrass *Zostera marina* are near the entrance as well as adjacent to a small stream at the head of the embayment. At depths greater than 10

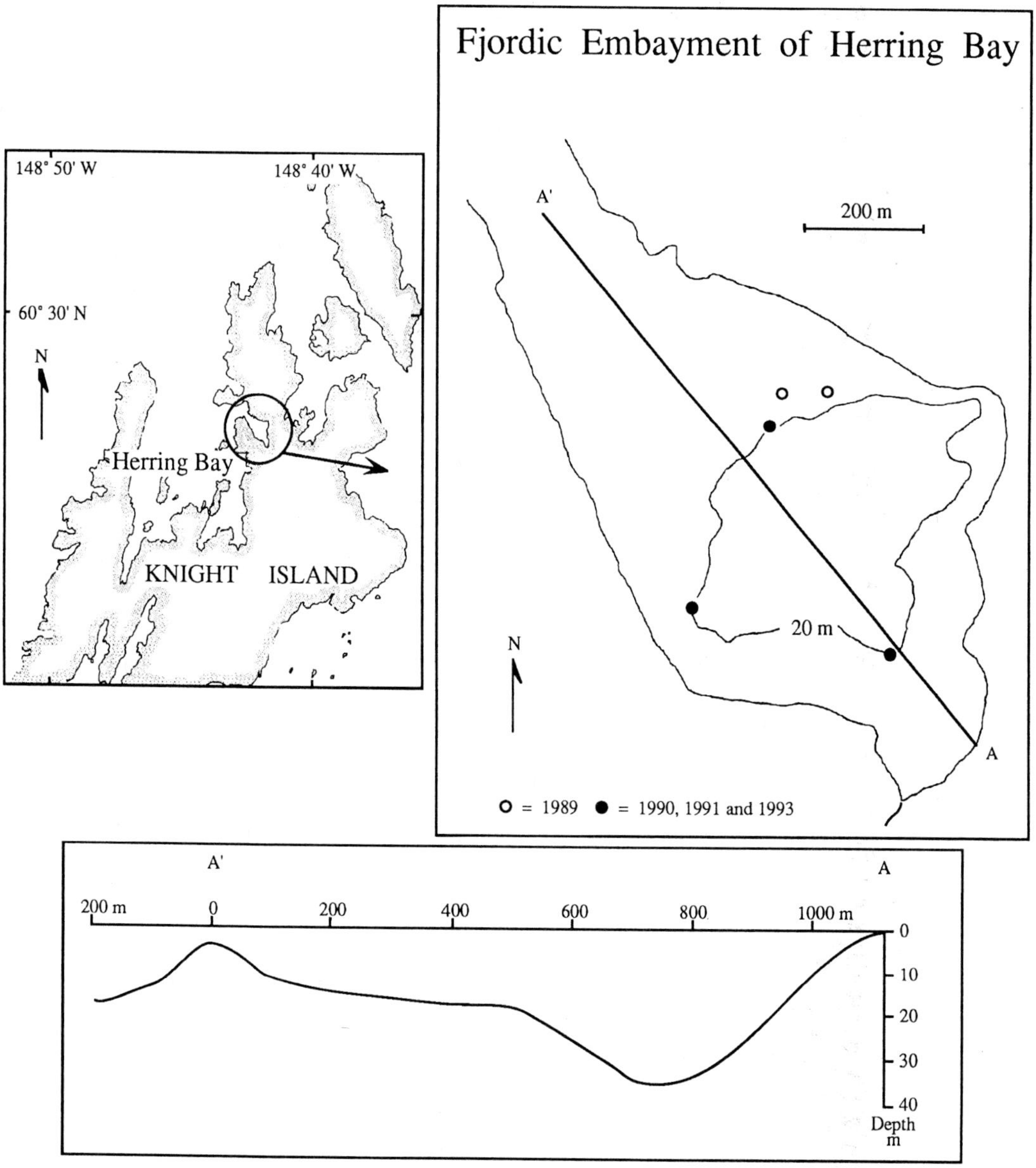

FIGURE 1.—Location of the subtidal study area within the fjordic embayment of Herring Bay, Prince William Sound, Alaska. Open (1989) and closed (1990, 1991, and 1993) circles represent approximate locations of sampling sites.

m, the substrate is composed mainly of flocculent detritus over mud and fine sand. This bay was considered moderate (10–50% coverage) to heavily (>50%) oiled after two shoreline surveys were conducted during the summer and fall of 1989 (ADEC 1989).

Methods

In October 1989 samples of infaunal invertebrates were collected from the embayment along two randomly placed 30-m transects within the 8 to 20-m depth stratum. A randomly placed 0.25-m^2 quadrat was sampled along each transect to a substrate depth of 10 cm by a diver-operated suction dredge. The collection bag on the suction dredge had a mesh size of 1 mm. A video of each transect was made by divers and a bathymetric chart of the embayment was made using a fathometer aboard a small boat. Additional videos were taken along unsampled transects through the deeper regions to document the extent of dead organisms observed.

Diver observations and density estimates of infauna were made from the Herring Bay site again in May and October 1990, August 1991, and September 1993. No sampling occurred in 1992. Because the initial zone of dead organisms in 1989 was mainly in deep waters, three stations were randomly established along the 20-m depth contour at each site. At each station, divers collected duplicate 0.1-m^2 suction-dredge samples of sediment for benthic infauna (see Figure 1 for approximate locations of sampling sites). Measurements of temperature, salinity, and dissolved oxygen were taken within <1 m of the bottom on all surveys except in October 1989. Oxygen measurements were made with a YSI Model 51B (accuracy 0.2 mg/L) meter. Temperature and salinity measurements were made with a YSI Model 33 S-C-T meter.

Two samples of sediment were collected at each sampling site: one was used to determine hydrocarbon concentrations and the other to determine grain size. Both sediment samples were collected by scooping a precleaned, widemouthed jar directly into the sediment to a 5-cm depth to obtain 10–100 g of sediment.

All hydrocarbon sediment samples were frozen and sent to the Technical Services Task Force, Analytical Chemistry Group (TSTF/ACG), National Oceanic and Atmospheric Administration, National Marine Fisheries Service, Auke Bay, Alaska, for processing. Because of the difficulty in identifying weathered *Exxon Valdez* crude oil, we report the dry weight concentrations of polycyclic aromatic hydrocarbons (summed PAH), which include compounds that resemble weathered *Exxon Valdez* crude oil, as well as compounds from sources other than the *Exxon Valdez* oil spill, and from biogenic sources (Table 1). The concentrations of those summed PAH analytes were used as indicators of the contribution of oil spilled by the *Exxon Valdez*. A one-way randomized analysis of variance (ANOVA) (Manly 1991) was performed to test the hypothesis of no significant differences among years with respect to the summed PAH concentrations.

Sediment samples for grain size were frozen and shipped to the University of Alaska Fairbanks for processing.

The benthic samples were returned to the laboratory at the University of Alaska Fairbanks for sorting, counting, weighing (blotted wet weight), and identification. The organisms included infaunal and small, slow-moving macrofaunal surface dwellers. Organisms excluded from the analyses were meiofauna (typically smaller than 1 mm) and highly mobile, nonbenthic forms such as calanoid copepods, mysids, euphausiids, chaetognaths, and fishes. Although most identifications were made to the family or (occasionally) higher taxonomic level, many common and dominant organisms were identified to the genus and species levels. Sediments were analyzed for grain size by the pipette-sieve method; the sediment types and grain-size distributions were defined following the conventional grain-size parameters stated in Folk (1980).

TABLE 1.—Polycyclic aromatic hydrocarbons (summed PAH) analyzed from sediments of the fjordic embayment in Herring Bay, Prince William Sound. These represent PAH from various sources, including *Exxon Valdez* crude oil, other petroleum products, and biogenic sources. The list of *Exxon Valdez* PAH represents PAH most characteristic of weathered *Exxon Valdez* crude oil (source: Technical Services Task Force, Analytical Chemistry Group, National Oceanic and Atmospheric Administration, National Marine Fisheries Service, Auke Bay, Alaska).

Summed PAH	*Exxon Valdez* PAH
Naphthalene	Naphthalene
1-methylnaphthalene	1-methylnaphthalene
2-methylnaphthalene	2-methylnaphthalene
C-2 naphthalenes	C-2 naphthalenes
C-3 naphthalenes	C-3 naphthalenes
C-4 naphthalenes	C-4 naphthalenes
Fluorene	Fluorene
C-1 fluorenes	C-1 fluorenes
C-2 fluorenes	C-2 fluorenes
C-3 fluorenes	C-3 fluorenes
Phenanthrene	Phenanthrene
C-1 phenanthrenes	C-1 phenanthrenes
C-2 phenanthrenes	C-2 phenanthrenes
C-3 phenanthrenes	C-3 phenanthrenes
C-4 phenanthrenes	C-4 phenanthrenes
Dibenzo[*b*,*d*]thiophene	Dibenzo[*b*,*d*]thiophene
C-1 dibenzo[*b*,*d*]thiophenes	C-1 dibenzo[*b*,*d*]thiophenes
C-2 dibenzo[*b*,*d*]thiophenes	C-2 dibenzo[*b*,*d*]thiophenes
C-3 dibenzo[*b*,*d*]thiophenes	C-3 dibenzo[*b*,*d*]thiophenes
Chrysene	Chrysene
C-1 chrysenes	C-1 chrysenes
C-2 chrysenes	C-2 chrysenes
C-3 chrysenes	
C-4 chrysenes	
C-1 fluoranthenes	C-1 fluoranthenes
Biphenyl	Biphenyl
Acenaphthylene	
Anthracene	
Acenaphthene	
Fluoranthene	
Pyrene	
Benzo[*a*]anthracene	
Benzo[*b*]fluoranthene	
Benzo[*k*]fluoranthene	
Benzo[*e*]pyrene	
Benzo[*a*]pyrene	
Perylene[a]	
Indeno[1,2,3-*c*,*d*]pyrene	
Dibenzo[*a*,*h*]anthracene	
Benzo[*g*,*h*,*i*]perylene	

[a]Perylene is derived from biogenic sources.

Results

In fall 1989 numerous dead and moribund organisms were observed within the fjordic embayment of Herring Bay at depths of approximately 11 m and deeper. In an extensive visual survey within this area (approximately 70 m^2), 44 dead animals were observed lying on the bottom, including 23 large polychaete worms, 11 sea stars *Pycnopodia helianthoides* and miscellaneous clams. Also observed were dead mobile organisms, such as pandalid shrimp, squid, and juvenile Pacific cod *Gadus macrocephalus*. In addition to the observed dead organisms, the substrate had a patchy, cobweb-like layer of the bacteria *Beggiatoa*. These colorless, sulfur-dependent, hemolithotrophic bacteria are typically associated with decaying organic material and low dissolved oxygen (Jørgensen 1980). A strong sulfur odor could be detected at depth by the divers. Although grain-size analysis was not accomplished, the sediments in these deeper waters were composed predominantly of mud.

Although the benthic–demersal community was obviously stressed as evidenced by the "dead zone," it still contained a relatively rich assemblage of 24 infauna taxa. However, signs of disturbance were evident in the moderately low Shannon—Wiener diversity index, H' (1.7), the moderately high Simpson dominance index, D (0.4), and the low abundance of sensitive burrowing amphipods (Ampithoidae; 16 individuals/m^2). The dominance index was mainly attributed to stress-resistant taxa, such as the bivalves *Lucina tenuisculpta* (Lucinidae; 2,544 individuals/m^2) and robust mysella *Mysella tumida* (Montacutidae; 256 individuals/m^2) and the polychaetes *Polydora socialis* (Spionidae; 368 individuals/m^2) and *Nephtys cornuta* (Nephtyidae; 336 individuals/m^2) (Table 2).

Analyses of oil from a limited suite of sediment samples yielded several hydrocarbon analytes, most presumably from *Exxon Valdez* oil spill, but some from nonspill sources. In October 1989 the average summed PAH concentration in the sediments from the embayment was 3,373 ng/g^1 (n = 2).

Similar surveys in the dead zone region of Herring Bay revealed fewer dead animals in 1990 than in 1989. In 90-m^2 surveys conducted in late May and early October, we saw scattered patches of *Beggiatoa*, one dead *Pycnopodia* (spring), one dead Pacific cod, and three dead worms (fall). More extensive qualitative visual searches in fall 1990 revealed some dead fishes (Pacific herring *Clupea pallasi*, Stichaeidae, and Pleuronectidae), but there were no concentrated dead zones as observed in 1989.

However, the infaunal community was obviously disturbed by the spring and fall of 1990. Diversity and number of taxa were extremely low in both surveys: $\leq$0.1 and <8 taxa, respectively. Furthermore, *L. tenuisculpta* was absent and the community was almost totally dominated by *N. cornuta* (98–99% of the infaunal abundance; 4,848 and 3,432 individuals/m^2 in May and October, respectively). Dissolved oxygen values at 0.5–1.0 m above the bottom averaged 3.6 mg/L in May, but were near 0 mg/L in October. The summed PAH concentration was 4,426 ng/g^1 (n = 2) by May 1990 (Table 2).

By mid-August 1991 the community in Herring Bay demonstrated dramatic signs of recovery. Diver surveys revealed no dead organisms, although *Beggiatoa* was still evident on the bottom. Almost all infaunal community parameters had recovered to or near levels observed in 1989. *N. cornuta* was still dominant (2,737 individuals/m^2), however, *L. tenuisculpta* was present again, but at low density (13 individuals/m^2). The dissolved oxygen during the August sampling, approximately 1 month earlier than the 1989 and 1990 fall samplings, averaged 9.7 mg/L. The summed PAHs declined to 542 ng/g^1 (n = 3).

One group of organisms that was present in 1991 in moderate density was burrowing amphipods (130 individuals/m^2). Representatives of this sensitive group included the families Ischyroceridae, Isaeidae, Dexaminidae, Phoxocephalidae, and Lysianassidae (Table 2).

The results from sampling in the fjordic embayment of Herring Bay in late September 1993 tended to resemble the sampling in October 1990 in that there were few taxa (4), low H' (0.02), high D (0.99), and no dissolved oxygen in the bottom waters. Nephtyid polychaetes, in particular *N. cornuta*, still dominated the fauna (2,352 individuals/m^2; 99.8% of all infaunal abundance; Table 2). Summed PAH concentration in the sediment was the lowest in 4 years, averaging only 478 ng/g^1 (n = 3). One-way randomized ANOVA revealed that there was a significant decline (P = 0.005) in summed PAHs over the 4-year period.

Discussion

The decline in abundance of benthic infauna and the occurrence of a large number of dead and dying organisms in the fjordic embayment of Herring Bay in 1989 and 1990, followed by recovery in 1991, apparently resulted from effects of the hydrocarbons present. A significant decline in polycyclic aromatic hydrocarbon concentrations was noted over

TABLE 2.—Variables for benthic invertebrates and characteristics of their environment in an oiled fjordic embayment in Herring Bay, Prince William Sound.

Statistic or taxon	18 Oct 1989	28 May 1990	1 Oct 1990	14 Aug 1991	25 Sept 1993
Area sampled (m^2)	0.50	0.60	0.60	0.60	0.60
Total abundance (individuals/m^2)	4,192.00	4,863.00	3,487.00	4,393.00	2,357.00
Total biomass (g/m^2)	79.44	19.02	16.20	17.80	12.67
Total taxa (family level and higher)	24.00	8.00	6.00	32.00	4.00
Shannon—Wiener diversity (H')	1.66	0.03	0.10	1.64	0.02
Simpson dominance (D)	0.39	0.99	0.97	0.41	0.99
Species richness (SR)	2.76	0.82	0.61	3.70	0.39
Bottom oxygen (mg/L)		2.2–5.0	0.0–0.1	8.7–11.2	0.00
Mean% mud		98.80		42.60	48.80
Mean summed PAH (ng/g^1) (dry weight)	3,373.32	4,426.28		542.21	478.29
Taxon abundance (individuals/m^2)					
Ampharetidae	64			145	
Amphipoda[a]				55	
Ampithoidae	16				
Balanidae	32				
Bivalvia[a]				82	
Caecidae	32				
Capitellidae				442	
Caprellidea				67	
Corambidae			22		
Dexaminidae				13	
Diastylidae				238	2
Dorvilleidae			7		
Gastropoda[a]	64				
Goniadidae	48				
Hesionidae			7	27	
Hiatellidae	32				
Isaeidae				13	
Ischyroceridae				23	
Lucinidae	2,544			13	
Lysianassidae				13	
Montacutidae	256				
Munnidae				80	
Myidae	48				
Nephtyidae	336	4,848	3,432	2,737	2,352
Opheliidae				133	
Phoxocephalidae				13	
Phyllodocidae	32				
Polyodontidae	48				
Pyrenidae	32				
Rissoidae	80				
Sigalionidae	32				
Spionidae	368			13	
Spirorbidae			7	133	
Syllidae				40	
Thyasiridae	32				

[a]Unidentified to family.

the 4-year period, with the greatest drop occurring between 1989 and 1990: 1,214 ng/g^1 in October 1989, 336 ng/g^1 in May 1990, 65 ng/g^1 in August 1991, and 56 ng/g^1 in September 1993. These findings are consistent with observations of the effects of hydrocarbons from prior spills. For example, a general decline in the abundance and diversity of benthic infauna was observed following the *Florida* oil spill in West Falmouth, Massachusetts (Sanders et al. 1980) and the *Amoco Cadiz* oil spill along the Brittany coast of France (Chassé 1978; Cabioch et al. 1980; Dauvin 1982). Total hydrocarbons were present in the shallow subtidal sediments in exceptionally high quantities (515,000 ng/g^1) 5 months after the *Amoco Cadiz* spill, but decreased (310,000 ng/g^1) within 2 years (Dauvin 1982). Among the possible causes of oil-related effects are chemical toxicity of aromatic derivatives; asphyxiation or entanglement resulting from direct physical coating; and a variety of reproductive, behavioral, and other sublethal disorders leading ultimately to long-term population changes. In the case of the *Exxon Valdez* oil spill, there also is the possibility of indirect effects of bioremediation and other cleanup efforts because such activities were observed at the oiled Herring Bay site (Alaska Department of Environmental Conservation Oil Spill Response Center, Anchorage, Alaska). Especially suggestive of the impact of oil was the mortality of mobile organisms such as fishes, squid, and, to a lesser extent, shrimp. Oil-related effects presumably altered the mobility of these organisms, thereby limiting escape.

The disappearance of amphipods within the embayment in 1990 could also be caused by oil because benthic amphipods are extremely sensitive to oil. Massive declines in ampeliscid amphipods were observed following the *Amoco Cadiz* spill (Chassé 1978; Cabioch et al. 1980). As noted in Herring Bay and elsewhere, amphipod recovery generally follows within 1 or 2 years after a spill, as hydrocarbon concentrations in sediments decline (Sanders et al. 1980; Dauvin 1982).

In spite of correlation between declines in the benthic community at the Herring Bay fjordic embayment and high concentrations of hydrocarbons, along with strong support from historical evidence, there is still doubt as to the relative contributions of oiling on observed changes in the benthic community. There are no pre-spill data on benthic communities within embayments similar to the one sampled in Herring Bay prior to the *Exxon Valdez* oil spill. As a result, an understanding of the effects of this oil spill on the benthic community comes only when one places observations of the conditions after the spill in the perspective of longer term trends within the community, especially with respect to natural perturbations that may cause declines in abundance similar to those observed in 1989–1990. One such natural perturbation that is widely observed in similar embayments with restricted water flow is the seasonal occurrence of low oxygen concentrations (Rosenberg 1980).

One might expect a silled embayment such as that in Herring Bay to have a depauperate benthic community because the sill restricts exchange between inside and outside waters, and seasonal density stratification limits exchange between bottom and surface waters. In many Scandinavian and Scottish fjords, a seasonal cycle of oxygen depletion and natural cycles of organic enrichment are brought about by these restrictions, resulting in a dominance of stress-tolerant benthic taxa and low diversity (Jørgensen 1980; Rosenberg 1980; Josefson and Rosenberg 1988; also see review by Pearson 1980) such as was observed in the fjordic portion of Herring Bay. These organisms (e.g., Lucinidae, Nephtyidae, and Spionidae) are periodically subjected to stress caused by, or confounded by, hypoxic ($O_2 <$ 2 mg/L) or anoxic ($O_2 = 0$ mg/L) conditions on the seafloor. Any hypoxic–anoxic conditions that may have been prevalent in 1989 were presumably prolonged or accentuated by the additional organic loading from organisms that died from oil contamination. This periodic phenomenon of low dissolved oxygen has also been documented in estuaries (Rosenberg 1977; Rhoads and Germano 1982) as well as along the open coast (e. g., Boesch and Rabalais 1991; Harper et al. 1991; Rabalais and Harper 1991). It has been called the "August Effect" in a New England estuary (Rhoads and Germano 1982).

Of the four periods when oxygen concentrations were measured, only the two earliest seasonal samplings, May 1990 and August 1991, had appreciable values (2.2–11.2 mg/L). Although our data are meager, we speculate that the oxygen concentrations rise in spring when the stratification breaks down and drop again in fall (September or October) when stratification reforms, as documented for other regions (e.g., Rosenberg 1977; Stachowitsch 1991). The conditions prevalent during summer are conducive to colonization from adjacent shallow depths as well as outside the embayment. Only the stress-tolerant taxa can apparently withstand the hypoxic–anoxic conditions that often arise in late fall. The community's recovery from disturbance tends to be rapid because the community is kept in an early successional state by the annual recurring hypoxia–anoxia (Boesch and Rosenberg 1981; Boesch and Rabalais 1991).

Many of the taxa that were dominant in the Herring Bay fjordic embayment, including Lucinidae, Nephtyidae, and Spionidae, have commonly been observed in stressed communities elsewhere and appear somewhat tolerant of low oxygen conditions. *Lucina tenuisculpta* and presumably other lucinid clams appear to be able to live where conditions are extreme and oxygen and food are limited (Yonge and Thompson 1976). *L. tenuisculpta* is in the same order as the stress-tolerant *Thyasira* genus, and several species of *Thyasira* (flexuose cleftclam *T. flexuosa*, *T. sarsi*, *T. mtokanagai*, and *T. miyadii*) have been reported from organically enriched and polluted substrates (see Table 1 in Pearson and Rosenberg 1978). *L. tenuisculpta* that dominated at Herring Bay in 1989 (61% of faunal abundance) were mainly older than 1 year; many were more than 3 years old. Therefore, the habitat in the Herring Bay site before the spill must have been conducive to support *L. tenuisculpta*, even though the region is periodically subjected to low oxygen concentrations.

Lizarraga-Partida (1974) reported *Nephtys cornuta* in semipolluted substrates in Ensenada Bay, Mexico, in areas enriched with organic material derived from sewage-fish waste. Pearson and Rosenberg (1978) gave several other examples of

Nephtys spp. (*N. incisa*, *N. hombergi*, *N. ciliata*, *N. longosetosa*, and *N. faneiscona*) appearing in organically enriched and polluted areas that were often low in dissolved oxygen. Busdosh (1978) found *Nephtys* spp. only in association with oiled substrates; they preferred oiled to clean sediment. Although *Nephtys* spp. are mainly predators, they also deposit-feed and thus can utilize the high organic loads associated with the decay of dead organisms.

Pearson and Rosenberg (1978) reported that several species of *Polydora* (Spionidae) occur in organically enriched and polluted substrates (i.e., *P. ciliata*, *P. ligni*, *P. kempi*, *P. paucibranchiata*, *P. quadrilobata*, and *P. nuchalis*. Members of this family are often observed as initial colonizers of disturbed substrates. *Polydora* spp. are able to colonize empty (defaunated) substrates rapidly and in high numbers, but they are also described as poor competitors (Grassle and Grassle 1974). Although spionid polychaetes were the second most abundant taxon in Herring Bay in 1989, they were absent in 1990, present in low density (13 individuals/m^2) in 1991, and absent again in 1993.

The disappearance of amphipods from the fjordic portion of Herring Bay is also consistent with observations made in other low-oxygen environments (Pavella et al. 1983; Harper et al. 1991). Benthic crustaceans, like amphipods, generally emigrate when dissolved oxygen is below 2.8 mg/L and are killed when values reach 0.4–0.7 mg/L (converted from milliliters per liter as cited in Tyson and Pearson 1991). Other mobile organisms, such as fishes and squid, are also prone to emigrate from areas stressed by low oxygen (Pavella et al. 1983).

In summary, the observations made at the fjordic portion of Herring Bay between 1989 and 1991 are consistent with the hypothesis that oil was a contributing cause of death to benthic organisms and also contributed to the decline in infaunal abundance. There were no observations of similar numbers of dead organisms after 1989, providing support to the hypothesis that our initial observations of death were related to the impacts of oil. However, in 1993, we noted declines in the benthic infauna similar to, or greater than, those observed in 1990, but in the absence of high concentrations of hydrocarbons. Therefore, although the declines in the benthic community in 1990 may have been, in part, related to oiling, these changes may also be attributed to natural occurrences of extremely low oxygen concentrations, as observed in 1993. With the data at hand, we have no way of partitioning the causes of declines in 1990 between oiling or natural oxygen depletion. We can only state that such declines are possible even in the absence of oiling.

This study points to two important generalizations concerning the evaluation of impacts of oiling or other perturbations. First, without appropriate temporal and spatial controls, it is often difficult to unequivocally assess the impacts of the perturbation. Unfortunately, such controls are often unavailable. For example, we had no data from embayments similar to the Herring Bay site before the spill, and there were few (if any) sites of comparable physiography in the area unaffected by the spill that could be used as spatial controls. Second, in the absence of these controls, longer-term observations of the community in question are imperative in order to make appropriate evaluations of impacts. Without our 1993 observations, implications regarding the potential impacts of oiling probably would have been overstated.

Acknowledgments

We thank the sponsors of this work, the U.S. Department of Interior Forest Service and the Alaska Department of Fish and Game, for their support. We especially thank the many divers who helped collect the data, often under adverse conditions; Kathy Omura and Max Hoberg who assisted in the sorting and taxonomy of benthic invertebrates; Chirk Chu, Christopher Kjellmark, and Richard Smith who assisted with data management and analyses; and the skippers and crews of our research vessels *Nip'N Tuck*, *Joanna Marie*, *Wm. A. McGaw*, *Kittiwake* II, and *Bering Explorer*.

References

ADEC (Alaska Department of Environmental Conservation). 1989. Impact maps and summary reports of shoreline surveys of the *Exxon Valdez* oil spill site-Prince William Sound. Report to the *Exxon Valdez* Oil Spill Trustee Council. Anchorage, Alaska.

Anderlini, V. C., and R. G. Wear. 1992. The effect of sewage and natural seasonal disturbances on benthic macrofaunal communities in Fitzroy Bay, Wellington, New Zealand. Marine Pollution Bulletin 24:21–26.

Boesch, D. F., and R. Rosenberg. 1981. Response to stress in marine benthic communities. Pages 215–225 *in* G. W. Barrett and R. Rosenberg, editors. Stress effects on natural ecosystems. Wiley, New York.

Boesch, D. F., and N. N. Rabalais. 1987. Long-term environmental effects of offshore oil and gas development. Elsevier Applied Sciences, New York.

Boesch, D. F., and N. N. Rabalais. 1991. Effects of hypoxia on continental shelf benthos: comparisons between the New York Bight and the northern Gulf of Mexico. Pages 27–34 *in* Tyson and Pearson (1991).

Busdosh, M. 1978. The effects of Prudhoe crude oil fractions on the arctic amphipods *Boeckosimus affinis* and *Gammarus zaddachi*. Doctoral dissertation. University of Louisville, Louisville, Kentucky.

Cabioch, L., J. C. Dauvin, J. Mora Bermudez, and C. Rodriguez Babio. 1980. Effects of the *Amoco Cadiz* oil spill on the sublittoral benthos, north of Brittany. Helgoländer Meeresuntersuchungen 33:192–208.

Carey A. G., and R. E. Ruff. 1977. Ecological studies of the benthos in the western Beaufort Sea with special reference to bivalve mollusks. Pages 505–530 *in* M. J. Dunbar, editor. Polar oceans. Arctic Institute of North America, Calgary, Alberta, Canada.

Chassé, C. 1978. The ecological impact on and near shores by the *Amoco Cadiz* oil spill. Marine Pollution Bulletin 11:298–301.

Dauvin, J.-C. 1982. Impact of *Amoco Cadiz* oil spill on the muddy fine sand *Abra alba* and *Melinna palmata* community from the Bay of Morlaix. Estuarine Coastal and Shelf Science 14:517–531.

Dobbs, F. C., and J. M. Vozarik. 1983. Immediate effects of a storm on coastal infauna. Marine Ecology Progress Series 11:273–279.

Ellis, D. V., and P. M. Hoover. 1990. Benthos of tailings beds from an abandoned coastal mine. Marine Pollution Bulletin 21:477–480.

Folk, R. L. 1980. Petrology of sedimentary rocks. Hemphills Publishing Co., Austin, Texas.

Grassle, J. F., and J. P. Grassle. 1974. Opportunistic life histories and genetic systems in marine benthic polychaetes. Journal of Marine Research 32:253–284.

Harper, D. E., Jr., L. D. Mckinney, J. M. Nance, and R. R. Salzer. 1991. Recovery responses of two benthic assemblages following an acute hypoxic event on the Texas continental shelf, northwestern Gulf of Mexico. Pages 49–64 *in* Tyson and Pearson (1991).

Jørgensen, B. B. 1980. Seasonal oxygen depletion in the bottom waters of a Danish fjord and its effect on the benthic community. Oikos 34:68–76.

Josefson, A. B., and R. Rosenberg. 1988. Long-term soft-bottom faunal changes in three shallow fjords, west Sweden. Netherlands Journal of Sea Research 22(2): 149–159.

Klaus, A. D., J. S. Oliver, and R. G. Kvitek. 1990. The effects of gray whale, walrus, and ice gouging disturbance on benthic communities in the Bering Sea and Chukchi Sea, Alaska. National Geographic Research 6:470–484.

Lizarraga-Partida, M. L. 1974. Organic pollution in Ensenada Bay, Mexico. Marine Pollution Bulletin 5:109–112.

Manly, B. 1991. Randomization and Monte Carlo methods in biology. Chapman and Hall, London.

McCauley, J. E., R. A. Parr, and D. R. Hancock. 1977. Benthic infauna and maintenance dredging: a case study. Water Research 11:233–242.

Pavella, J. S., J. L. Ross, and M. E. Chittenden. 1983. Sharp reductions in abundance of fishes and benthic macroinvertebrates in the Gulf of Mexico off Texas associated with hypoxia. Northeast Gulf Science 6:167–173.

Pearson, T. H. 1980. Macrobenthos of fjords. Pages 569–602 *in* H. J. Freeland, D. M. Farmer, and C. D. Levings, editors. Fjord oceanography. Plenum, New York.

Pearson, T. H., and R. Rosenberg. 1978. Macrobenthic succession in relation to organic enrichment and pollution of the marine environment. Oceanography and Marine Biology Annual Review 16:229–311.

Pearson, T. H., G. Duncan, and J. Nuttall. 1986. Long term changes in the benthic communities of Loch Linnhe and Loch Eil (Scotland). Hydrobiologia 142: 113–119.

Rabalais, N. N., and D. E. Harper, Jr. 1991. Studies of the benthic oxygen depletion phenomenon off Louisiana. Pages 57–63 *in* H.-J. Krock and D. E. Harper, Jr., editors. Proceedings of the American Academy of Underwater Sciences eleventh annual scientific diving symposium.

Rhoads, D. C., and J. D. Germano. 1982. Characterization of organism-sediment relations using sediment profile imaging: an efficient method of remote ecological monitoring of the seafloor (Remots TM System). Marine Ecology Progress Series 8:115–128.

Rosenberg, R. 1977. Benthic macrofaunal dynamics, production, and dispersion in an oxygen-deficient estuary of west Sweden. Journal of Experimental Marine Biology and Ecology 26:107–133.

Rosenberg, R. 1980. Effects of oxygen deficiency on benthic macrofauna in fjords. Pages 499–514 *in* H. J. Freeland, D. M. Farmer, and C. D. Levings, editors. Fjord oceanography. Plenum, New York.

Sanders, H. L., J. F. Grassle, G. R. Hampson, L. S. Morse, S. Garner-Price, and C. C. Jones. 1980. Anatomy of an oil spill: long-term effects from the grounding of the barge *Florida* off West Falmouth, Massachusetts. Journal of Marine Research 38:265–380.

Stachowitsch, M. 1991. Anoxia in the northern Adriatic Sea: rapid death, slow recovery. Pages 119–129 *in* Tyson and Pearson (1991).

Tyson, R. V., and T. H. Pearson, editors. 1991. Modern and ancient continental shelf anoxia. Geological Society of London Special Publications 58.

Tyson, R. V., and T. H. Pearson. 1991. Modern and ancient continental shelf anoxia: an overview. Pages 1–24 *in* Tyson and Pearson (1991).

Yonge, C. M., and T. E. Thompson. 1976. Living marine molluscs. William Collins Sons & Co. Ltd., Glasgow, UK.

American Fisheries Society Symposium 18:448–462, 1996

Injury to the Early Life History Stages of Pacific Herring in Prince William Sound after the *Exxon Valdez* Oil Spill

EVELYN D. BROWN
Alaska Department of Fish and Game, Division of Commercial Fisheries, Research, and Development Box 669, Cordova, Alaska 99574, USA

TIMOTHY T. BAKER
Alaska Department of Fish and Game, Division of Commercial Fisheries, Research, and Development 333 Raspberry Road, Anchorage, Alaska 99518, USA

JO ELLEN HOSE
Occidental College, Department of Biology, 1600 Campus Road, Los Angeles, California 90041, USA

RICHARD M. KOCAN
University of Washington, School of Fisheries, Fisheries Research Institute, Seattle, Washington 98195, USA

GARY D. MARTY[1]
University of California, Department of Medicine, School of Veterinary Science, Davis, California 95616, USA

MICHAEL D. MCGURK
Triton Environmental Consultants, Ltd.
120-13511 Commerce Parkway, Richmond, British Columbia, Canada V6V 2L1

BRENDA L. NORCROSS
University of Alaska, School of Fisheries and Ocean Science, Institute of Marine Science Fairbanks, Alaska 99775, USA

JEFF SHORT
National Marine Fisheries Service, Alaska Fisheries Science Center 11305 Glacier Highway, Juneau, Alaska 99801, USA

Abstract.—The *Exxon Valdez* oil spill occurred a few weeks before Pacific herring *Clupea pallasi* spawned in Prince William Sound. About half of the egg biomass was deposited within the oil trajectory, and an estimated 40 to 50% sustained oil exposure during early development. The resulting 1989 year-class displayed sublethal effects in newly hatched larvae, primarily premature hatch, low weights, reduced growth, and increased morphologic and genetic abnormalities. Genetic endpoints, especially anaphase aberration rates, were highly correlated with site-specific *Exxon Valdez* oil concentrations. Responses were specific and sensitive to this oil exposure. In newly hatched larvae, anaphase aberration rates were elevated at oiled sites, and in pelagic larvae genetic damage was greatest near oiled areas of southwestern Prince William Sound. Genetic damage in larvae from oiled areas progressively decreased during the 6-week study, but site-specific measures of instantaneous mortality suggest that a significant reduction (52.3%) in larval production occurred in 1989. Although approximately equal egg biomass was deposited in non-oiled and oiled areas, we estimated that oiled areas produced only 0.016×10^9 pelagic larvae compared with 11.82×10^9 non-oiled areas. Despite the estimated substantial decrease in larval production, reduced abundance in the 1989-year class recruiting as 4-year-old adults in 1993 could not be estimated because natural processes affecting recruitment are poorly understood; however, the 1989 year-class was a minority of the 1993 spawning population, one of the smallest cohorts observed in Prince William Sound, and it returned to spawn with an adult herring population reduced by approximately 75%, apparently because of a widespread epizootic.

The *Exxon Valdez* oil spill on 24 March 1989 occurred only a few days before the annual migration of Pacific herring *Clupea pallasi* into nearshore spawning areas of Prince William Sound. Because of temporal and spatial movements of the oil, the potential exposure of herring to oil was high. The herring migration follows a northeasterly pattern from the Montague–Green Island staging areas to-

[1]Present address: University of California, Department of Anatomy, Physiology, and Cell Biology, School of Veterinary Medicine, Davis, California 95616, USA.

ward Bligh Island (Alaska Department of Fish and Game, unpublished data); the oil trajectory moved in the opposite direction from Bligh Island southwestward through Montague Strait (Royer et al. 1990; Figure 1). Of the 43.9 million liters of *Exxon Valdez* oil spilled, only about 10% was carried out of Prince William Sound as floating oil, whereas an estimated 40–45% was beached within the sound (Wolfe et al., in press). An unknown proportion of *Exxon Valdez* oil was dispersed as particulates throughout the water column to a depth of at least 25 m. Particulate oil was bioavailable to organisms throughout the summer in 1989 after the spill and to a lesser degree in 1990, after beach-cleaning operations (Short and Harris 1996a, this volume). The spill occurred at the beginning of the spring–summer zooplankton bloom in Prince William Sound. Ingestion of oil-contaminated food can be a significant route for oil exposure, and copepods, one of the major prey items for juvenile herring (Cooney 1995), significantly bioaccumulate petroleum hydrocarbons (Conover 1971; Capuzzo 1987). Because of the potential for oil contamination, commercial herring fisheries were closed in 1989.

In 1989 Pacific herring deposited their eggs on Prince William Sound beaches in early to mid-April, and the eggs hatched in early May. Larval herring remained in nearshore areas until late spring; in early summer they were transported to the western areas of the sound and into Montague Strait (Norcross and Frandsen 1996, this volume). Because of the rapid dispersion of particulate *Exxon Valdez* oil throughout the subsurface waters within the oil trajectory and its persistence near oiled beaches into summer (Short and Harris 1996a), exposure potential existed for spawning adults, eggs, and larvae for at least the first few months after the spill. The early life stages, being the most sensitive to the toxic effects of oil (Capuzzo 1987), were therefore emphasized in Natural Resource Damage Assessment studies.

The objective of the herring damage assessment study was to determine the effects of the *Exxon Valdez* oil spill on Prince William Sound herring. Using both long-term (prespill) fisheries baseline data and damage assessment studies and also targeting critical processes such as growth and reproduction when possible, we attempted to link organismal-level effects with population effects. This article summarizes and synthesizes the results of damage assessment studies of the early life history stages of Pacific herring and describes damage to embryos and larvae. Studies on embryos, juveniles, and adults are continuing, and their results, along with more detailed information on oil spill injury to all of the life stages of herring, will be described in future publications (Hose et al., in press; Kocan et al., in press; McGurk and Brown, in press; B. L. Norcross, unpublished).

Overview of Damage Assessment Study Methods

Determination of the Oil Trajectory

Information on the geographic extent of the oil trajectory was combined from several sources, including overflight maps produced by the National Oceanic and Atmospheric Administration (Galt et al. 1991), Alaska Department of Environmental Conservation, and Alaska Department of Fish and Game aerial surveyors; shoreline oiling estimates from the Alaska Department of Natural Resources and visual surveys during sample collection; and satellite images of the oil trajectory (Dean et al. 1990). The comprehensive oil-trajectory map was used to determine potential exposure of herring to *Exxon Valdez* oil (Brown 1995; Figure 1). The presence of oil at specific adult, egg, and larval collection sites was verified using *Exxon Valdez* oil measurements in caged or native mussels (Short and Harris 1996a).

Exxon Valdez Oil Analysis

Concurrent with sampling of herring eggs, native bay mussels (*Mytilus trossulus*) were collected at −1 and −2 m mean lower low water (MLLW) adjacent to egg sampling sites in 1989 and at 0 and −1 m MLLW in 1990 (Figure 2). Mussel tissue was analyzed for 44 polycyclic aromatic hydrocarbons derived from *Exxon Valdez* oil (EVO-PAH) with gas chromatography (Short and Harris 1996a, 1996b, this volume). Method detection limits ranged from 0.020 to 0.040 μg/L for individual analytes and were based on 10-g samples. A limited number of herring egg and adult tissues were analyzed for EVO-PAH, but samples were considerably less than 10 g. These samples were not used to estimate oil exposure because of unacceptably high EVO-PAH detection limits; levels of oil known to cause sublethal effects would not be detected.

Embryo–Larval Studies

There are five herring spawning areas in Prince William Sound (North Shore, Northeast, Southeast, Naked Island, and Montague Island areas; Figure 1). From 1989 to 1991, egg mortality at a series of non-oiled and oiled sites throughout the sound's

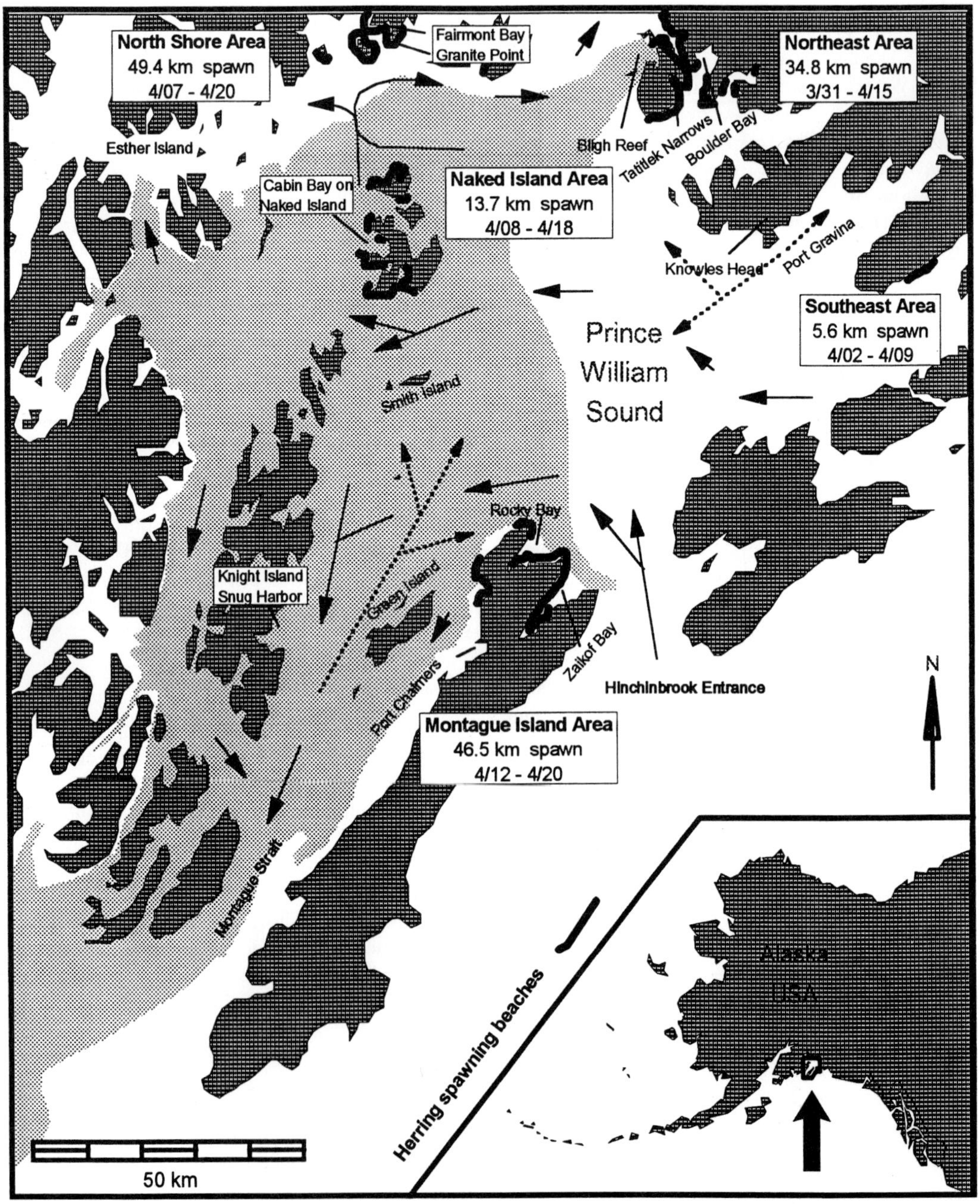

FIGURE 1.—The composite oil trajectory (light shaded area) of the *Exxon Valdez* oil spill (Galt et al. 1991), the five major Pacific herring spawning areas (circles), and the beaches receiving herring spawn in 1989 (heavy black lines), in Prince William Sound. Solid arrows indicate the general ocean circulation pattern in the sound. Dotted arrows indicate the migration of adult herring from over wintering grounds to spawning beaches.

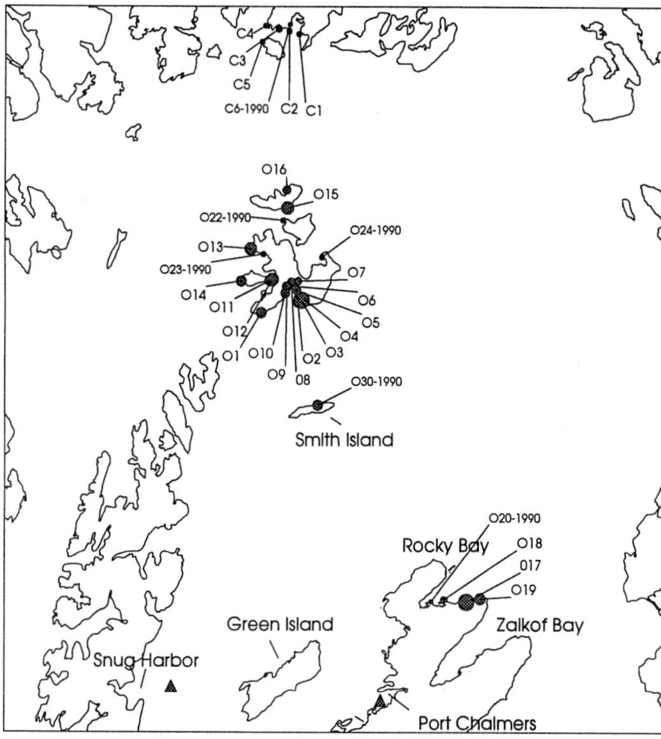

Mussel Tissue Hydrocarbon Samples (Circles)	Site No.	Mean Oil Concentration (ppb)	Standard Deviation
1989 Study Sites	C1	18.69	15.71
	C2	45.80	28.47
	C3	79.77	36.91
	C4	51.97	6.15
	C5	53.33	43.48
	O1	649.77	107.93
	O2	388.16	125.96
	O3	2,503.74	1,046.52
	O4	216.73	74.47
	O5	292.58	25.57
	O6	158.69	27.66
	O7	181.57	49.52
	O8	198.18	31.47
	O9	452.22	143.75
	O10	453.56	39.62
	O11	779.01	79.26
	O12	1,142.21	113.29
	O13	1,028.98	84.17
	O14	716.97	171.67
	O15	1,205.95	116.63
	O16	374.07	18.11
	O17	2,261.87	213.76
	O18	97.52	14.26
	O19	831.28	41.18
1990 Study Sites	C2	0.00	0.00
	C3	3.73	5.28
	C6	2.17	3.06
	O20	7.08	7.85
	O22	48.38	26.26
	O23	28.08	22.94
	O24	21.72	21.23
	O30	581.74	143.33
Bile Samples (Triangles)	**Location**	**Napthalene (ppb)**	**Phenanthrene (ppb)**
1989	Snug Harbor	140,000.00	28,000.00
1990	Port Chalmers	153,000.00	24,000.00
Reference	No Oil	22,000.00	2,000.00

FIGURE 2.—Sites where Pacific herring eggs were collected for density measurements and in situ and laboratory measurements of egg mortality, Prince William Sound, 1989 and 1990. Sites beginning with C were non-oiled (control) stations, those beginning with O were oiled stations. Shaded circles denote the concentrations in parts per billion (ppb, μg/L) of polycyclic aromatic hydrocarbons derived from *Exxon Valdez* oil in native bay mussels adjacent to egg collection sites; larger circles denote higher concentrations of *Exxon Valdez* oil. Most sites were sampled in 1989; 1990 sites are so labeled.

spawning areas was annually monitored in situ (Egg Deposition Study) (Biggs and Baker 1993; Brown 1995) and by returning subsamples to the laboratory for controlled incubations (Egg Incubation Study) (McGurk 1990b, 1991a, 1991b, 1993) (Figure 2). In 1990 a site in southeast Alaska, Sitka Sound, was also evaluated in the Egg Incubation Study (McGurk 1991a). The Egg Deposition Study yielded site-specific information on spawn timing and rates of egg density, biomass, and in situ mortality, which in 1990 and 1991 was corrected for egg loss (Biggs and Baker 1993). The Egg Incubation Study provided measurements of egg mortality, hatching success and dynamics, and length and weight of newly hatched larvae.

Subsamples of newly hatched larvae from the Egg Incubation Study were evaluated for sublethal morphologic, genetic, and histopathologic damage (Hose et al., in press). Morphologic damage was assessed using graduated severity indices (GSI); each skeletal, craniofacial, and/or finfold defect was scored and summarized for each individual assessed. Higher GSI scores indicated more severe malformations. Genetic damage was measured in pectoral fins by determining the percentage of anaphase–telophase mitotic configurations with chromosome or chromatid breaks (Kocan et al. 1982). Larvae were evaluated for histopathologic damage using standard processing procedures (Hose et al., in press). A laboratory experiment evaluated relationships between exposure to an oil–water dispersion of *Exxon Valdez* oil and hatching dynamics, embryonic stage sensitivity, embryo–larval survival, morphologic defects, and genetic damage (Kocan et al., in press).

In 1989 two studies measured the abundance and distribution of herring larvae nearshore (McGurk 1990a) and offshore (Norcross and Frandsen 1996; Norcross, unpublished). Planktonic larvae between hatch and the start of exogenous feeding were sampled in May and June near oiled and non-oiled spawning beds using bongo nets (McGurk 1990a). Site-specific abundances of newly hatched larvae were compared with egg abundances to estimate instantaneous mortality rates (McGurk and Brown, in press). Surviving pelagic larvae drifting offshore but within Prince William Sound were sampled using a Tucker trawl over a 3-month period (Norcross and Frandsen 1996; Norcross, unpublished). In both studies (nearshore and offshore trawl surveys), growth was measured using length–frequency analysis. Subsamples of larvae collected in May and June were evaluated for morphologic, genetic, and histopathologic damage as described above (Hose et al., in press).

Abundance Estimation

Fishery management assessments and estimates of the abundance of spawned eggs were augmented after the *Exxon Valdez* oil spill. In 1984 and 1988–1992, the total number of eggs deposited in Prince William Sound was estimated by egg deposition surveys; 1989–1991 estimates of egg deposition used a correction for eggs that were consumed or dislodged from the study area before assessment (egg loss). Methods for egg deposition surveys, estimates of the abundance of the spawning population, adult age composition sampling, and size at age sampling are described in detail by Biggs and Baker (1993).

Overview of Damage Assessment Study Results

Exposure

Hydrocarbon chemistry results from mussel tissue, rather than direct measurements in herring eggs, were a more reliable and useful index of site-specific oil exposure of herring embryos. Because of low sample volumes, *Exxon Valdez* oil analysis of herring eggs yielded unacceptably high detection limits, and *Exxon Valdez* oil was detected in eggs from only Cabin Bay on Naked Island. The analyte profile was identical to that of whole oil, which was observed adhering to the egg surface. Using the surrogate measurement of *Exxon Valdez* oil concentrations in adjacent native mussels, oil was detected in most spawning sites sampled within the oil trajectory (Hose et al., in press). In 1989 mean ($\pm$ SE) concentrations of EVO-PAH in mussels within the oil trajectory were significantly higher than at non-oiled areas (753 $\pm$ 22 μg/kg versus 49 $\pm$ 8 μg/kg, respectively; the majority of the PAH detected at non-oiled sites was naphthalene). Tarballs and sheens were visually observed at two of the Rocky Bay sites on Montague Island, two of the Bass Harbor sites, Naked Island sites located within oil booms, and sites near Cabin Bay on Naked Island. The variability of mussel *Exxon Valdez* oil concentrations at the 1989 herring sites was similar to the variability observed from other studies sampling within herring spawning areas (Short and Babcock 1996, this volume); therefore, the concentrations measured at our study sites appear representative of the entire 1989 spawning area.

Because egg deposition was not uniform, site-specific egg biomass measurements were used to calculate the proportion of eggs exposed to *Exxon Valdez* oil. Fifty-three percent of the total egg biomass was within the oil trajectory. Of the total eggs deposited, 35% were in the Montague Island area and 18% were in the Naked Island area. Measurements of EVO-PAH in native and caged mussels confirmed the "oiled" designation of all mussel study sites within the oil trajectory (Short and Babcock 1996; Short and Harris 1996a). Although oil was present along the outside of both Rocky and Zaikof bays on Montague Island (Figure 1), *Exxon Valdez* oil was documented in mussels only at Rocky Bay (Figure 2; most northeastern embayment); no mussels were collected in Zaikof Bay (immediately southeast of Rocky Bay), which contained approximately 13% of the total 1989 egg biomass (E. D. Brown, unpublished results). We therefore estimate that between 40 and 50% of the herring eggs were exposed to *Exxon Valdez* oil in 1989.

In 1990 EVO-PAH was not detected in herring spawn sites, except for Smith Island (Figure 2, site O30–1990). This site was sampled only in 1990 and its eggs represented less than 10% of the total deposited in Prince William Sound.

Effects on Embryos

Mortality was measured in situ at non-oiled and oiled sites until hatching. Contamination with EVO-PAH did not significantly affect mortality during embryonic incubation (Brown 1995). Eggs collected from oiled sites (particularly eggs from upper depths) and hatched in the laboratory exhibited lower survival than those from non-oiled areas, but the difference was not statistically significant (McGurk 1990b).

Effects on Newly Hatched Larvae

Site-specific differences in survival became more pronounced after hatching. Mean instantaneous rates of in situ egg-to-larva mortality were significantly lower at the non-oiled areas (0.187/d at

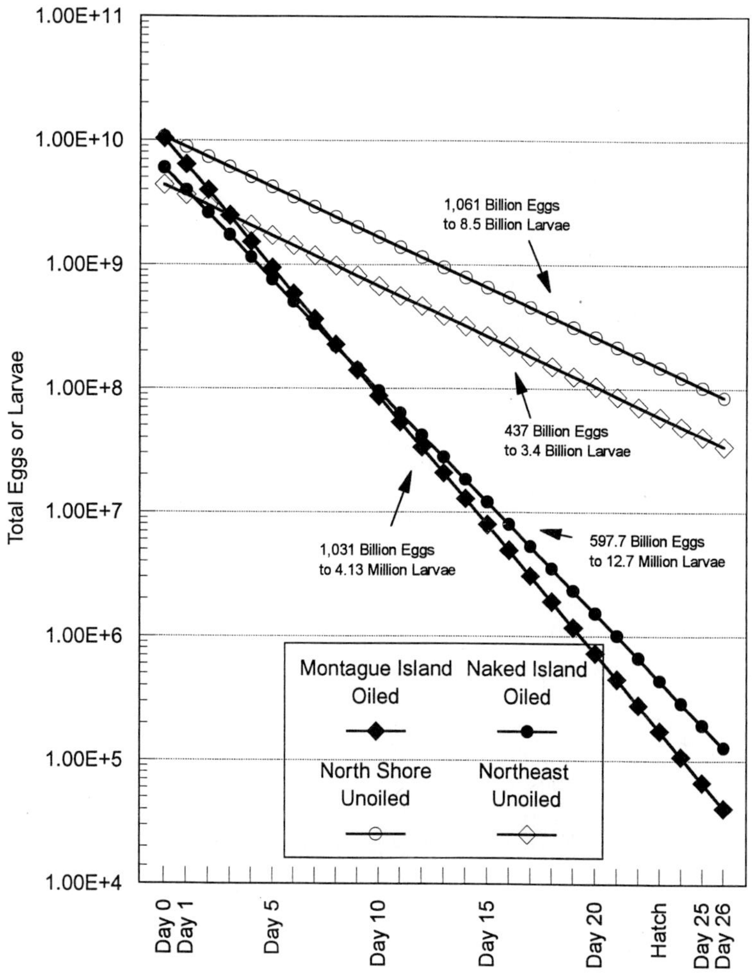

FIGURE 3.—Mortality of Pacific herring from embryo to posthatch (day 26 from fertilization), calculated as site-specific instantaneous mortality in the absence of any other estimates of larval survival that are not confounded with dispersal of larvae away from the beds, in two oiled and two non-oiled spawning sites, 1989 (McGurk and Brown, in press). The starting points for the curves are estimates of the actual numbers of eggs (logarithmic scale) deposited in each of the four major spawning areas.

Northeast and 0.186/d at North Shore) than at oiled areas (0.414/d at Naked Island and 0.477/d at Montague Island) (McGurk and Brown, in press); however, because of differences in beach characteristics, egg distributions, and larval dispersions between oiled and non-oiled areas, differences in egg–larval mortality estimates are not conclusive evidence of an oil effect. Site-specific instantaneous mortality rates, in the absence of any other estimates of larval survival that are not confounded with dispersal of larvae away from the egg beds, indicated that non-oiled areas produced more than 200 times more free-swimming larvae per billion eggs deposited 3 d after hatch than did the oiled areas (calculated by multiplying the initial abundance of eggs in each area by the mortality rate; Figure 3). The estimate of larval production surviving to the pelagic stage was 11.82 billion (11.82×10^9) from the Northeast and North Shore non-oiled areas compared to 16.82 million from the Montague Island and Naked Island areas. When mortality rates of each of the non-oiled areas were applied to each of the two oiled areas, larval production expected in the absence of oil was estimated. Expected larval production in the oiled areas was estimated to be 12.98 billion larvae, with a difference between expected

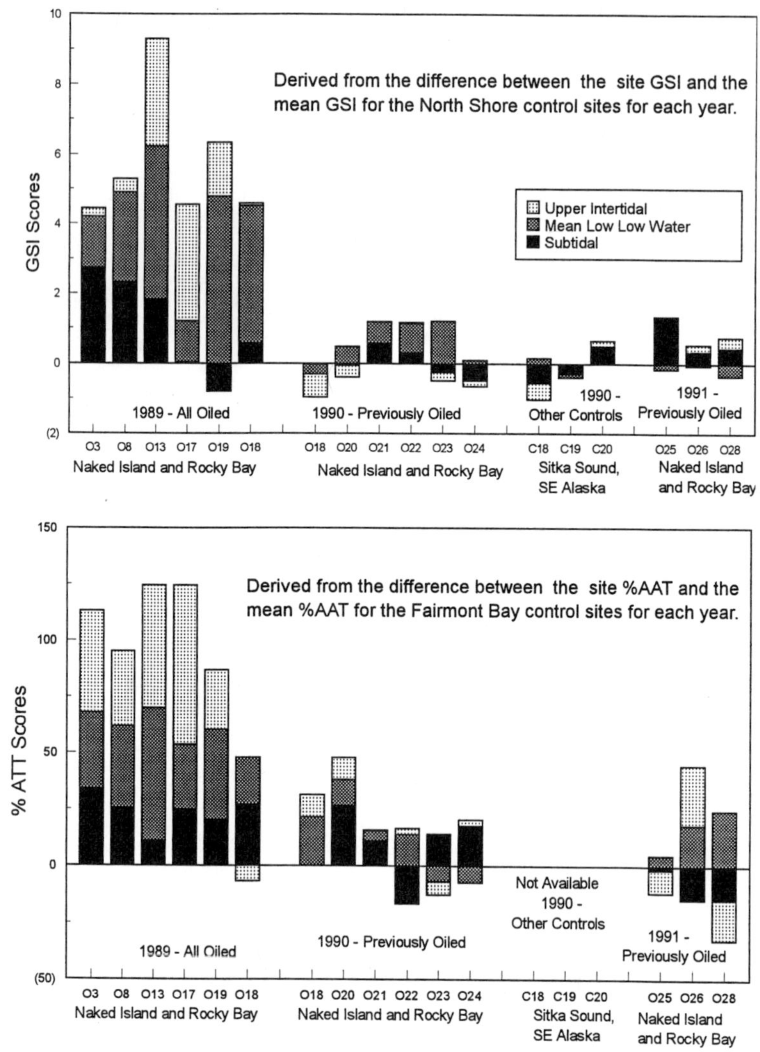

FIGURE 4.—Changes in graduated severity index (GSI) scores (**top**) and percentage anaphase aberrations (% AAT; **bottom**) for Pacific herring at the oiled sites and non-oiled sites in Galena Bay and Sitka Sound, southeast Alaska, relative to those indices in Fairmont Bay in the non-oiled North Shore area. The magnitude of the bars is calculated by subtracting mean scores by depth (upper intertidal, mean lower low water [0 m], and subtidal) at the Fairmont Bay sites from the mean scores for each site in the comparison.

and actual production of 12.96 billion larvae. This difference represents a 99.87% loss of potential larval production in oiled areas and a 52.3% loss overall. Posthatch larval weights and hatch timing were different between non-oiled and oiled areas in 1989 ($P < 0.01$). The strongest effect of oil was premature hatching. The presence of oil caused significant ($P < 0.01$) decreases in the 50% hatch date and in the fraction of larvae hatched in late stages of development. Hatching occurred sooner in the upper depths than in the lower depths and in oiled eggs than in non-oiled eggs. After corrections for age, posthatch larvae from oiled areas were 0.1 mm longer but weighed 4 μg less than larvae from non-oiled areas ($P < 0.01$) (McGurk 1990b). These measurements yield a condition index (weight/length^3) that is 7% less for oiled larvae.

Ratings of morphologic defects (graduated severity indices, GSI) were significantly higher at Naked Island and Rocky Bay on Montague Island than at Fairmont Bay in the non-oiled North Shore area, indicating more severe defects at the oiled sites

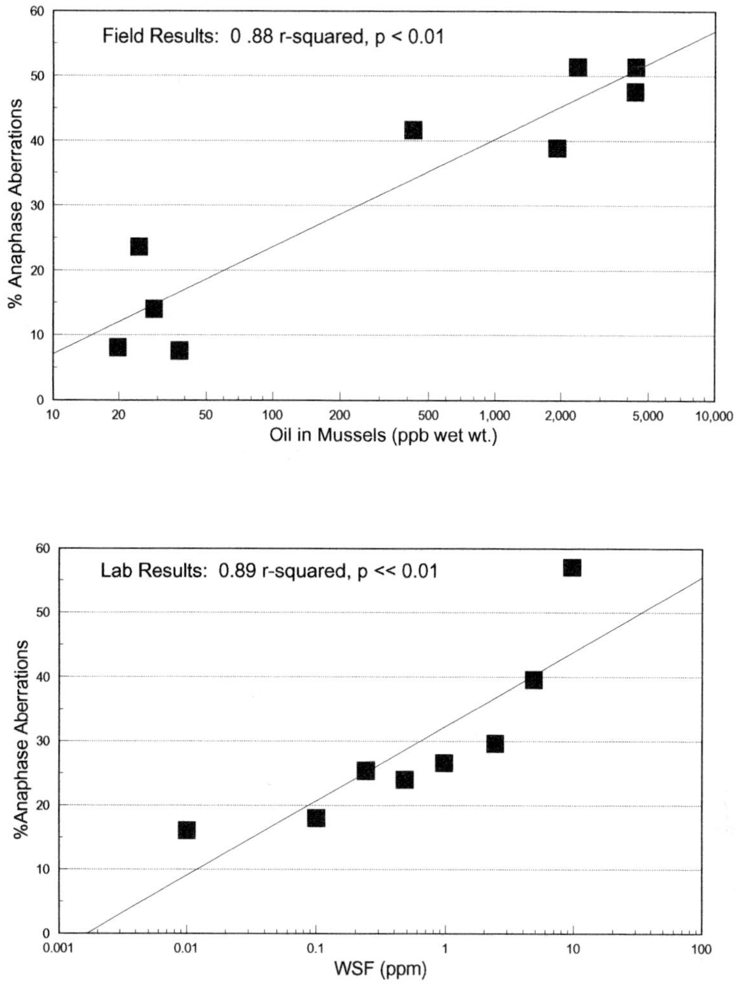

FIGURE 5.—Anaphase aberration rate in cells from posthatch larval Pacific herring as a measure of response to exposure level of *Exxon Valdez* oil (EVO) in the field (site-specific log-transformed EVO concentrations in mussels) and in the laboratory (log-transformed concentrations of high-molecular-weight, water-soluble fraction (WSF) of EVO in oil–water dispersions).

($P < 0.01$) (Figure 4, top). These abnormalities included skeletal, otic, or finfold defects and absent or abnormal lower jaws. Graduated severity index scores were significantly correlated ($P < 0.05$) with the log-transformed EVO-PAH concentrations present in adjacent mussels (Hose et al., in press). In 1989 GSI scores at all sites were higher than in successive years. Suboptimal laboratory incubation conditions present in 1989 were subsequently improved but would not have affected the magnitude of the difference observed between oiled and non-oiled treatments. In 1990 and 1991 GSI scores were generally similar between non-oiled and previously oiled areas. In 1990 a second non-oiled area, Sitka Sound in southeast Alaska, was evaluated as a control and yielded scores similar to those from North Shore. Relative differences in GSI scores at non-oiled areas in 1990 and 1991 were much less than differences between oiled and non-oiled stations in 1989. No histopathologic lesions attributable to *Exxon Valdez* oil exposure were observed in the 1990–1991 larvae.

Newly hatched larvae were also examined for genetic damage and analyzed with the same baseline approach as was used for assessing morphologic defects (Figure 4, bottom). At Fairmont Bay in the

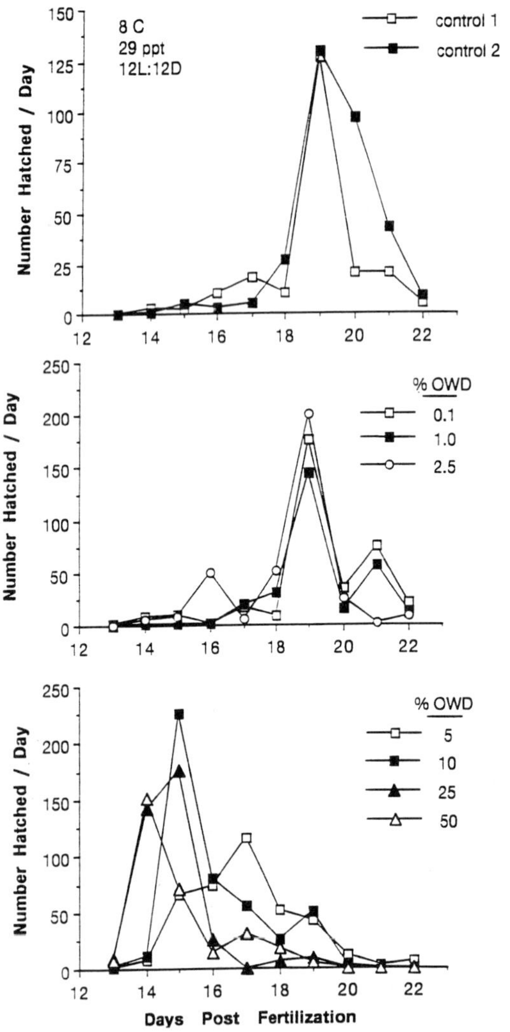

FIGURE 6.—Hatching dynamics of embryos of Prince William Sound Pacific herring exposed from fertilization to hatching to oil–water dispersions (OWD) measured by percent water-soluble fraction of Prudhoe Bay crude oil from the *Exxon Valdez* (Kocan et al., in press).

North Shore area, the anaphase aberration rate averaged 10% in 1989; this was not different from the 1990 mean of 18% but was significantly less than the 1991 value of 22% ($P < 0.05$). The 1989 to 1991 values suggest an upward trend in the baseline aberration rate; 1991 also had high in situ egg mortality rates (Brown 1995). In 1989, anaphase aberration rates were significantly elevated at all oiled sites ($P < 0.05$). Site-specific EVO-PAH concentrations were significantly correlated with the anaphase aberration rate ($r^2 = 0.88$, $P < 0.01$ and

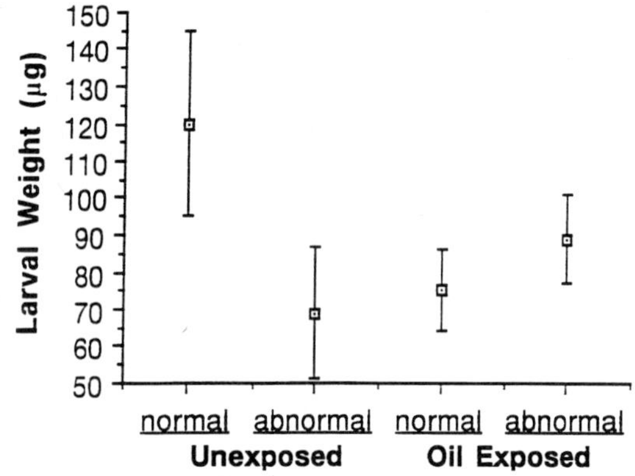

FIGURE 7.—Mean weight of newly hatched larvae of Pacific herring exposed continuously to oil–water dispersions as embryos (Kocan et al., in press).

Mantel's test, $P < 0.001$). After 1989 there were no genetic effects attributable to residual oil. In 1990 the anaphase aberration rate at Rocky Bay on Montague Island was higher than the North Shore rate, but the difference was much less than the 1989 difference.

Exposure of herring embryos in the laboratory to the water-soluble fraction of *Exxon Valdez* oil produced morphologic abnormalities, genetic damage, and other responses in newly hatched larvae identical to larval responses at oiled sites in 1989 (Kocan et al., in press) (Figure 5). The percentage of anaphase aberrations was a highly sensitive indicator of oil exposure (Figure 5). Premature hatch was well documented in exposures exceeding 5% oil–

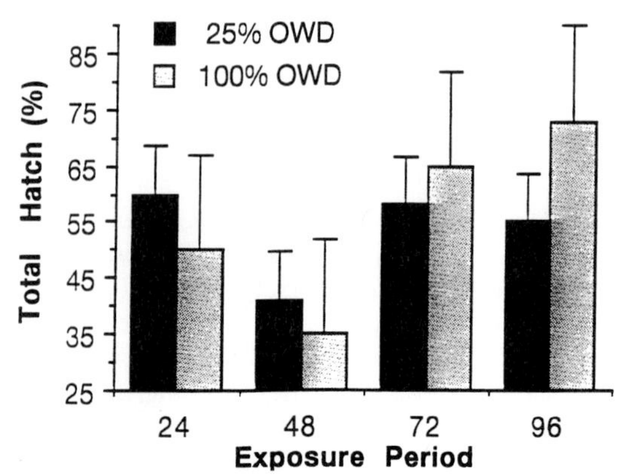

FIGURE 8.—Embryo development stage sensitivity for Pacific herring measured as percent of total hatch after exposure to the water-soluble fraction of *Exxon Valdez* oil–water dispersions (Kocan et al., in press).

water dispersion (OWD) or 0.48 mg/L of high-molecular-weight petroleum hydrocarbons (HMW; Figure 6). Larval weights were greatly reduced in abnormal larvae whether exposed to oil or not, but weights were also reduced in morphologically normal larvae exposed to oil (Figure 7). Exposure to 24-h pulses of OWD at different embryonic stages demonstrated that the earliest developmental stages (24–48 h postfertilization) were the most susceptible to toxicity (Figure 8). Differences in analytical techniques for *Exxon Valdez* oil preclude comparisons between effects levels in field and laboratory studies.

Effects on Pelagic Larvae

Two studies of growth in pelagic larvae were conducted in 1989, one evaluating herring larvae captured 1–5 km from spawning beds and the second assessing larvae farther offshore. Constant growth rates of nearshore larvae ranged from 0.10 to 0.17 mm/d among stations, and no difference was detected between oiled and non-oiled stations (McGurk and Brown, in press). Measurements of larvae captured at offshore stations yielded a growth rate of 0.15 mm/d (Norcross, unpublished), which is similar to the rate measured nearshore. More than 85% of larvae offshore were captured within the oil trajectory (Norcross and Frandsen 1996). In May, 61% of the larvae were collected just north of Naked Island (247 larvae/100 m^3) and 27% off Knight Island (108 larvae/100 m^3). By June, 87% of the larvae were captured in Montague Strait (69–733 larvae/100 m^3). Few larvae were collected in July.

Morphologic defects were evaluated in larvae collected near their spawning beaches and farther offshore in May and June 1989. Soon after hatching in May, larvae near non-oiled beaches (North Shore area and Tatitlek Narrows, Northeast area) had GSI scores significantly lower than those from oiled sites (equivalent of 0–16% abnormal). At Bass Harbor on Naked Island, 72% of the larvae were abnormal, and at Montague Island, where many premature larvae were collected, 80% were abnormal. In late May, larvae from the two non-oiled areas still had low GSI scores relative to those from the oiled stations (48–52% abnormal). In contrast, Norcross (unpublished) found that although 89% of the larvae captured offshore throughout Prince William Sound were severely deformed in May, only 24% were severely deformed in June. In both months, morphologic defects were significantly more severe among larvae from areas within the oil trajectory.

Pelagic larvae were also evaluated for genetic damage (J. E. Hose, unpublished data). In early May, anaphase aberration rates were low (6–8%) near non-oiled beaches of North Shore and Tatitlek Narrows. Anaphase aberration rates were significantly higher in nearshore larvae from Bass Harbor (33%) and Montague Island (28%) and in offshore larvae north of Naked Island (35%). In late May to early June, aberration rates of 2- to 4-week-old larvae decreased off oiled beaches. Although lower than earlier in May, the aberration rates of Bass Harbor larvae (21%) and Montague Island larvae (14%) were still significantly elevated over rates from the non-oiled areas near Fairmont Bay, Tatitlek Narrows, and Esther Island (8%). Values for larvae collected offshore in Hinchinbrook Entrance (16%) in late May to early June were similar to values for Montague Island (14%), where these larvae are thought to originate. During this same period, the aberration rate north of Naked Island declined from 35 to 22%. Three offshore sites within the western part of the oil trajectory also had elevated aberration rates (21 and 26%). No histopathological lesions attributable to *Exxon Valdez* oil exposure were observed in pelagic larvae (G. D. Marty, unpublished data).

Discussion

The *Exxon Valdez* oil spill occurred shortly before the most toxicologically sensitive period of embryo–larval development for Pacific herring (Capuzzo 1987; Kocan et al., in press), and a large proportion of the eggs deposited were exposed. In addition, 30–50% of the eggs were distributed near the surface (E. D. Brown, unpublished data), where exposure to the toxic microlayer (Kocan et al. 1987) of oil was likely during tidal cycles. Using site-specific egg biomass measurements, approximately 53% of total herring eggs deposited were within the oil trajectory. Although only a small portion of the spawn was visibly contaminated with crude oil, measurements of EVO-PAH in native mussels adjacent to our study sites demonstrated that most spawning areas within the oil trajectory were exposed to *Exxon Valdez* oil; only one area within the trajectory, Zaikof Bay, was not analyzed for this oil. The mean EVO-PAH concentration in mussels from oiled areas was 753 ppb wet weight (SE, 22 ppb); this estimate is similar to the 700 ppb (SE, 25 ppb) mean concentration measured at oiled sites in April in a more comprehensive study of Prince William

Sound (Short and Babcock 1996). Our estimate that between 40 and 50% of the total 1989 eggs deposited were exposed to *Exxon Valdez* oil is much greater than that estimated by Pearson et al. (1995), who found visible oil only on 7% of the shoreline with herring spawn. From recent studies (Short and Babcock 1996; Short and Harris 1996a; Wolfe et al., in press), it is apparent that much of the spilled oil was not visible because major portions dispersed into the water column or were subsequently dispersed from heavily oiled beaches to remote locations.

It would have been preferable to relate embryo–larval responses to actual oil concentrations in herring eggs rather than to the surrogate mussel measurement; however, the analytical sensitivity of the chemical data for the herring egg samples was severely compromised by the small quantity of tissue analyzed and the correspondingly high hydrocarbon detection limits. Thus, the presence of *Exxon Valdez* oil hydrocarbons in herring egg samples from Cabin Bay is strong evidence for contamination at this area, but the presence of *Exxon Valdez* oil cannot be ruled out in the remaining samples that did not show detectable EVO-PAH. In a corresponding 1989 study, Pearson et al. (1995) found *Exxon Valdez* oil concentrations in herring egg samples elevated above non-oiled areas only at Cabin Bay. Pearson et al. (1995) did not find any significant correlations between biological responses and oil exposure in 1989 or 1990 except for a positive association between PAH concentrations in eggs-on-kelp and developed eggs in the mid intertidal zone (0 to −2.4 m MLLW) for 1989. Their measurement of *Exxon Valdez* oil in herring eggs-on-kelp is biased toward underestimating the oil concentrations because kelp does not accumulate petroleum hydrocarbons (NRC 1985) and the proportion of kelp in their sample was not described.

Sublethal effects in posthatch and pelagic larval herring were substantial and measurable throughout the oil trajectory during their entire pelagic period. Morphologic defects and genetic damage were strongly associated with ambient *Exxon Valdez* oil concentrations in mussels ($P < 0.001$ to $P < 0.05$). The morphologic defects, premature hatch, and low larval weights observed in *Exxon Valdez* oil-exposed embryo–larval herring are not specific to oil exposure because they are generalized responses to stress during early development (von Westernhagen 1988; Purcell et al. 1990). Pearson et al. (1985) stated that oil toxicity in embryo–larval herring is manifest primarily as morphologic defects. This response has been consistently replicated in field studies of other oil spills using herring and other fish species (see Urho 1990) and in numerous laboratory exposures to crude oil (Lindén 1975; Lonning 1977; Lindén 1978), oil fractions (Smith and Cameron 1979; Carls and Rice 1989) and to individual PAH components of petroleum (Struhsaker et al. 1974; Hose et al. 1984; Stene and Lonning 1984). In the egg incubation experiment, the baseline scores for defect severity were higher in 1989 than in subsequent years; the cause has not been fully resolved. The magnitude of the difference between oiled and non-oiled areas was much greater than temporal or spatial variations among non-oiled areas (Figure 4). Statistical relationships between *Exxon Valdez* oil indices and morphologic defects were stronger with increasing molecular weight of the EVO-PAH, an association with substantial experimental evidence (Black et al. 1983).

Other *Exxon Valdez* oil-induced effects, such as premature hatch and reduced larval weights observed in both the laboratory and the field, would be expected to reduce posthatching survival. Smaller larvae with less-developed jaws would feed less efficiently and be more susceptible to predation (von Westernhagen 1988). McGurk and Brown (in press) found that instantaneous mortality rates between egg and early larval periods were two to three times higher at the oiled stations in Prince William Sound than at the non-oiled stations; however, physical differences between beaches at the two groups confounded interpretation of the data. These results are similar to those from studies with pink salmon that documented persistently lower survival of young from oiled streams until 1992 (Bue et al. 1996, this volume).

The response with the most specificity to crude oil contamination is genetic damage, as measured by the anaphase aberration rate. This endpoint is highly sensitive to mutagenic compounds (Kocan et al. 1982) (Figure 5), some of which are present in crude oils (Dipple 1982; Flesher and Meyers 1991), and to oil–water dispersion of *Exxon Valdez* oil in laboratory exposure (Kocan et al., in press). Significant responses in cells of marine organisms are typically observed in the low (ppb) range for organic mutagens (Kocan et al. 1982; Anderson et al. 1994). Anaphase aberration rates in newly hatched herring larvae were strongly correlated with EVO-PAH concentrations in adjacent mussels (Figure 5); aberration rates were also correlated with every measured PAH except naphthalene, which is nonmutagenic. In pelagic larvae, genetic damage was greatest near oiled areas of southwestern Prince William Sound and was undetectable at open water sites in Montague Strait and Hinchinbrook Entrance. This pattern is consistent with observations that oiled beaches served as persistent reservoirs of oil and that oil concentrations in open water were low (Short and Harris 1996a). Anaphase aberration rates progressively decreased at Naked and Montague islands during the 6-week period from hatching to metamorphosis. The decline in aberration rates could have resulted from mortality of severely damaged larvae, mixture of larvae from oiled and non-oiled areas, and reduced toxicity of the free-floating, weathered oil.

The 1989–1991 aberration rates in larvae from non-oiled areas suggest an upward trend in the baseline aberration rate. Rather than being caused by persistent oil, based on the analysis of in situ mussel tissue, the increase in 1991 may be related to the severe climatic conditions, unusually cold water temperatures, and enhanced mortality that year. In addition, the upward trend in the baseline aberration rate could have been effected by residual genetic damage in the returning adults exposed to oil. Whereas the adults may have spawned in non-oiled areas, they may have traversed oiled areas during the summer of 1989 and beyond in the normal course of seasonal movement; there are no data to support this theory of heritable genetic damage in the returning 1990 and 1991 adults.

That we can potentially track oil toxicity in a planktonic organism by monitoring genotoxic effects over space and time is noteworthy. In addition, the correlations suggest that genetic damage may be detected in the absence of observable deformities. Genetic measurements, although rarely performed in these situations, are a recommended component of oil spill assessments (NRC 1985). In the single oil spill in which genetic measurements of fish embryos were used (the *Argo Merchant* spill), genetic damage was observed near floating oil sheens and decreased with distance from the oil (Longwell 1977). Measurements of genetic damage would be expected to be specific for mutagenic events; however, the influences of natural environmental extremes (temperature, oxygen, etc.) and mortality have not been systematically investigated. Although in general both the anaphase aberration and malformation endpoints were generally correlated in larval herring and were of similar sensitivity in laboratory exposures to *Exxon Valdez* oil (Kocan et al., in press), genetic damage can be considered a more specific response to oil exposure.

The consequences of genetic damage during early development are not predictable with any certainty but probably include developmental malformations and decreased survival (Cross et al. 1987). The demonstration of genetic damage in somatic cells of herring larvae, however transitory, raises the possibility of germ-cell damage that could lead to decreased reproductive potential. Reduced growth and immune responsiveness result from generalized metabolic alterations caused by oil exposure (Capuzzo 1987).

We observed extremely low growth rates in nearshore herring larvae in May and in offshore larvae through June 1989. Other researchers have measured growth rates in larval Alaskan herring at much higher ranges: 0.30 mm/d (McGurk et al. 1993), 0.18–0.29 mm/d (McGurk 1984; Norcross, unpublished), and 0.31–1.48 mm/d for Alaskan herring rearing in Norway (Wespestad and Moksness 1990). Many of the lowest reported rates, 0.19–0.22 mm/d, resulted from starvation experiments (McGurk 1984). There was no evidence of reduced prey fields that would have affected growth in 1989 (Celewycz and Wertheimer 1996, this volume; McGurk and Brown, in press), and water temperatures were within normal ranges for Prince William Sound. The distribution of herring larvae in spring 1989 is consistent with potential exposure to dispersed oil from oiled beaches and by ingestion of oiled prey. Particulate oil is thought to have been bioavailable throughout the oil trajectory into June (Short and Harris 1996a).

Although estimates of the geographic extent and magnitude of damage are difficult to develop because of the complexities of the Prince William Sound ecosystem, we estimate that more than 40% of the 1989 year-class was affected by *Exxon Valdez* oil at toxic levels. The high correlation of increased chromosome aberrations at anaphase in larvae collected in oiled areas provides additional proof of exposure. Although we could not predict the exact proportion of herring exposed to *Exxon Valdez* oil concentrations sufficient to cause subsequent mortality, we could compare differences between expected and actual larval production using instantaneous mortality rates measured at control sites as baseline rates. Using the 1989 Prince William Sound estimates, the difference between expected (24.80 billion) versus actual (11.84 billion) larval production represents a 52.3% decrease in the sound; there was a 99.9% drop in larval production (12.98 billion expected versus 16.82 million actual) in the oiled areas alone. A decrease in abundance of this magnitude may be observable depending on the strength of the year-class. The 1989 year-class was a minority of the 1993 spawning population, one of the smallest cohorts observed in Prince William

Sound (Funk 1993), and returned to spawn with an adult herring population reduced by approximately 75% apparently because of the occurrence of a widespread epizootic (Meyers et al. 1994; Marty, unpublished data); however, natural variability and density-dependent effects cannot be ruled out as the cause of the small year-class and disease. There is a high degree of natural variability in herring recruitment; although environmental factors seem to cause a significant portion of that variability, the mechanisms involved remain elusive (Collie 1990). We currently do not predict recruitment because the processes driving it are poorly understood. Current studies on juvenile herring and their ecosystem (Cooney 1995) and on disease and reproductive impairment in adult herring in Prince William Sound may reveal enough information about herring recruitment to revisit herring damage assessment results and better estimate the effect of the *Exxon Valdez* oil spill on this Prince William Sound herring population.

Acknowledgments

We thank the many researchers who contributed to the implementation of this project or to preparation of the manuscript: National Marine Fisheries Service personnel (S. Rice, A. Moles, M. Carls, S. Korn, C. Broderson, J. Hamilton, and T. Rutecki), University of Alaska faculty (M. Stekoll, L. Haldorson, J. Collie, and T. Cooney), Canadian Department of Fisheries and Oceans biologists (M. Stocker, J. Schweigert, C. Haegele, D. Hay, and R. Tanasichuk), Alaska Department of Fish and Game (ADFG) scientists (J. Brady, B. Bechtol, D. Crawford, W. Donaldson, F. Funk, H. Geiger, D. Jones, S. Morstad, S. Sharr, K. Schultz, J. Wilcock, S. Fried, and L. Brannian), ADFG technicians (D. Coyer, S. Walker, T. McNeil, B. Haley, D. Norman, M. Miller, D. Nees, K. Becker, E. Wolfe, and S. Morgan). This article is contribution PP-115 of ADFG, Commercial Fisheries Management and Development Division, Juneau.

References

Anderson, S. L., J. Knezovich, and J. E. Hose. 1994. Genotoxic and developmental effects in sea urchins are sensitive indicators of effects of gentoxic chemicals. Environmental Toxicology and Chemistry 13: 1033–1041.

Biggs, E. D., and T. T. Baker. 1993. Studies on Pacific herring *Clupea pallasi* spawning in Prince William Sound following the 1989 *Exxon Valdez* oil spill, 1989–1992. Draft report for Natural Resource Damage Assessment Fish/Shellfish Study 11, *Exxon Valdez* Oil Spill Trustee Council, Anchorage, Alaska.

Black, J. A., W. J. Birge, A. G. Westerman, and P. C. Francis. 1983. Comparative toxicology of aromatic hydrocarbons. Fundamental and Applied Toxicology 3:353–358.

Brown, E. D. 1995. Injury to Prince William Sound following the *Exxon Valdez* oil spill: final report for Natural Resource Damage Assessment Fish/Shellfish Study 11. *Exxon Valdez* Oil Spill Trustee Council, Anchorage, Alaska.

Bue, B. G., S. Sharr, S. D. Moffitt, and A. K. Craig. 1996. Effects of the *Exxon Valdez* oil spill on pink salmon embryos and preemergent fry. American Fisheries Society Symposium 18:619–627.

Capuzzo, J. M. 1987. Biological effects of petroleum hydrocarbons: assessments from experimental results. Pages 343–410 *in* D. F. Boesch and N. N. Rabalais, editors. Long-term environmental effects of offshore oil and gas development. Elsevier Applied Science Publishers, London.

Carls, M. G., and S. D. Rice. 1989. Abnormal development and growth reductions of pollock *Theragra chalcogramma* embryos exposed to water-soluble fractions of oil. U.S. National Marine Fisheries Service Fishery Bulletin 88:29–37.

Celewycz, A. G., and A. C. Wertheimer. 1996. Prey availability to juvenile salmon after the *Exxon Valdez* oil spill. American Fisheries Society Symposium 18:564–577.

Collie, J. S. 1990. Herring population dynamics and management in Sitka Sound, Alaska. Pages 463–476 *in* Proceedings, international herring symposium. Alaska Sea Grant College Program AK-SG-91-01, Fairbanks.

Conover, R. J. 1971. Some relations between zooplankton and bunker C oil in Chedabucto Bay following the wreck of the tanker *Arrow*. Journal of the Fisheries Research Board of Canada 28:1327–1330.

Cooney, R. T. 1995. Sound ecosystem assessment (SEA)—an integrated science plan for the restoration of injured species in Prince William Sound. Draft final report to the *Exxon Valdez* Oil Spill Trustee Council, Anchorage, Alaska.

Cross, J. N., and six coauthors. 1987. Contaminant concentrations and toxicity of sea-surface microlayer near Los Angeles, California. Marine Environmental Research 23:307–323.

Dean, K. G., W. J. Stringer, J. E. Groves, K. Ahlnas, and T. C. Royer. 1990. The *Exxon Valdez* oil spill: satellite analyses. Pages 492–502 *in* Proceedings, conference on oil spills, management and legislative implications. American Society of Civil Engineers, New York.

Dipple, A. 1982. Polynuclear aromatic carcinogens. Pages 246–265 *in* M. Cooke, A. J. Dennis, and G. L. Fisher, editors. Polynuclear aromatic hydrocarbons: physical and biological chemistry. Springer-Verlag, New York

Flesher, J. W., and S. R. Meyers. 1991. Rules of molecular geometry for predicting carcinogenic activity of unsubstituted polynuclear aromatic hydrocarbons. Teratogenesis, Carcinogenesis and Mutagenesis 11: 41–54.

Funk, F. 1993. Forecast of the Pacific herring biomass in Prince William Sound, Alaska, 1993. Alaska Department of Fish and Game Regional Information Report 5J93, Juneau.

Galt, J. A. , W. J. Lehr, and D. L. Payton. 1991. Fate and transport of the *Exxon Valdez* oil spill. Environmental Science and Technology 25:202–209.

Hose, J. E., J. B. Hannah, H. W. Puffer, and M. L. Landolt. 1984. Histologic and skeletal abnormalities in benzo[*a*]pyrene-treated rainbow trout alevins. Archives of Environmental Contamination and Toxicology 13:675–684.

Hose, J. E., M. D. McGurk, G. D. Marty, D. E. Hinton, E. D. Biggs, and T. T. Baker. In press. Sublethal effects of the *Exxon Valdez* oil spill on herring embryos and larvae: morphologic, cytogenetic, and histopathological assessments, 1989–1991. Canadian Journal of Fisheries and Aquatic Sciences.

Kocan, R. M., J. E. Hose, E. D. Biggs, and T. T. Baker. In press. Laboratory evaluation of herring (*Clupea pallasi*) embryo sensitivity to Prudhoe Bay petroleum hydrocarbons, and in situ exposure of embryos at oiled and unoiled sites in Prince William Sound. Canadian Journal of Fisheries and Aquatic Sciences.

Kocan, R. M., M. L. Landolt, and K. M. Sabo. 1982. Anaphase aberrations: a measure of genotoxicity in mutagen-treated fish cells. Environmental Mutagenesis 4:181–189.

Kocan, R. M., H. V. Westernhagen, M. L. Landolt, and G. Furstenberg. 1987. Toxicity of sea-surface microlayer: effects of hexane extract on Baltic herring (*Clupea harengus*) and Atlantic cod (*Gadus morhua*) embryos. Marine Environmental Research 23:291–305.

Lindén, O. 1975. Acute effects of oil and oil/dispersant mixture on larvae of Baltic herring. Ambio 4(3):130–133.

Lindén, O. 1978. Biological effects of oil on early development of the Baltic herring *Clupea harengus membras*. Marine Biology 45:273–283.

Longwell, A. C. 1977. A genetic look at fish eggs and oil. Oceanus 20:46–48.

Lonning, S. 1977. The effects of crude ekofisk oil and oil products on marine fish larvae. Astarte 10:37–48.

McGurk, M. D. 1984. Ring deposition in the otoliths of larval Pacific herring, *Clupea harengus pallasi*. U.S. National Marine Fisheries Service Fishery Bulletin 82:113–120.

McGurk, M. D. 1990a. Early life history of Pacific herring: Prince William Sound herring larvae survey. Final report to NOAA (National Oceanic and Atmospheric Administration) NOS (National Ocean Service), 50ABNC-7-00141, Anchorage, Alaska.

McGurk, M. D. 1990b. Early life history of Pacific herring: 1989 Prince William Sound herring egg incubation experiment. Final report and appendices to NOAA (National Oceanic and Atmospheric Administration) NOS (National Ocean Service), 50ABNC-7-00141, Anchorage, Alaska.

McGurk, M. D. 1991a. Viable hatch of Pacific herring eggs from Prince William Sound and Sitka Sound, Alaska, in 1990. Final report to Alaska Department of Fish and Game, IHP-90-055, Cordova.

McGurk, M. D. 1991b. Viable hatch of Pacific herring eggs from Prince William Sound and Sitka Sound, Alaska, in 1991. Final report to Alaska Department of Fish and Game, IHP-91-042, Cordova.

McGurk, M. D. 1993. Revised appendices of Prince William Sound herring egg incubation experiments: 1989, 1990 and 1991. Final report to Alaska Department of Fish and Game, IHP-93-011, Cordova.

McGurk, M. D., and E. D. Brown. In press. Egg-larval mortality of Pacific herring in Prince William Sound, Alaska, after the *Exxon Valdez* oil spill. Canadian Journal of Fisheries and Aquatic Sciences.

McGurk, M. D., A. J. Paul, K. O. Coyle, D. A. Ziemann, and L. J. Haldorson. 1993. Relationship between prey concentration and growth, condition, and mortality of Pacific herring, *Clupea pallasi*, larvae in an Alaskan subarctic embayment. Canadian Journal of Fisheries and Aquatic Sciences 50:163–180.

Meyers, T. R., and six coauthors. 1994. Epizootic hemorrhages of the skin in Pacific herring *Clupea harengus pallasi* from Prince William Sound and Kodiak Island, Alaska, USA associated with the isolation of North American viral hemorrhagic septicemia virus (VHSV). Diseases of Aquatic Organisms 19:27–37.

Norcross, B. L., and M. Frandsen. 1996. Distribution and abundance of larval fishes in Prince William Sound, Alaska, during 1989 after the *Exxon Valdez* oil spill. American Fisheries Society Symposium 18:463–486.

NRC (National Research Council). 1985. Oil in the sea—inputs, fates and effects. National Academy Press, Washington, DC.

Pearson, W. H., E. Moksness, and J. R. Skalski. 1995. A field and laboratory assessment of oil spill effects on survival and reproduction of Pacific herring following the *Exxon Valdez* spill. Pages 626–661 *in* P. G. Wells, J. N. Butler, and J. S. Hughes, editors. *Exxon Valdez* oil spill: fate and effects in Alaskan waters. American Society for Testing and Materials Special Technical Publication 1219, Philadelphia.

Pearson, W. H., D. L. Woddruff, S. L. Kiesser, G. W. Fellingham, and R. A. Elston. 1985. Oil effects on spawning behavior and reproduction in Pacific herring (*Clupea harengus pallasi*). Final report (2311205196) to API (American Petroleum Institute), Environmental Affairs Department, Washington, DC.

Purcell, J. E., D. Grosse, and J. J. Glover. 1990. Mass abundances of abnormal Pacific herring larvae at a spawning ground in British Columbia. Transactions of the American Fisheries Society 119:463–469.

Royer, T. C., J. A. Vermersch, T. J. Weingartner, H. J. Niebauer, and R. D. Muench. 1990. Ocean circulation influencing the *Exxon Valdez* oil spill. Oceanography 3:3–10.

Short, J. W., and M. M. Babcock. 1996. Prespill and postspill concentrations of hydrocarbons in mussels and sediments in Prince William Sound. American Fisheries Society Symposium 18:149–166.

Short, J. W., and P. M. Harris. 1996a. Petroleum hydrocarbons in caged mussels deployed in Prince William

Sound after the *Exxon Valdez* oil spill. American Fisheries Society Symposium 18:29–39.

Short, J. W., and P. M. Harris. 1996b. Chemical sampling and analysis of petroleum hydrocarbons in the near-surface seawater of Prince William Sound after the *Exxon Valdez* oil spill. American Fisheries Society Symposium 18:17–28.

Smith, R. L., and J. A. Cameron. 1979. Effect of water soluble fraction of Prudhoe Bay crude oil on embryonic development of Pacific herring. Transactions of the American Fisheries Society 108:70–75.

Stene, A., and S. Lonning. 1984. Effects of 2-methylnaphthalene on eggs and larvae of six marine species. Sarsia 69:199–203.

Struhsaker, J. W., M. B. Eldridge, and T. Echeverria. 1974. Effects of benzene (a water soluble component of crude oil) on eggs and larvae of Pacific herring and northern anchovy. Pages 253–284 *in* F. J. Vernber and W. B. Vernberg, editors. Pollution and physiology of marine organisms. Academic Press, New York.

Urho, L. 1990. Impact of an oil spill on herring stock. Pages 569–582 *in* Proceedings, international herring symposium. Alaska Sea Grant College Program AK-SG-91-01, Fairbanks.

von Westernhagen, H. 1988. Sublethal effects of pollutants on fish eggs and larvae. Pages 253–346 *in* W. S. Hoar and D. J. Randall, editors. Fish physiology, volume XIA. Academic Press, New York.

Wespestad, V. G., and E. Moksness. 1990. Observations on growth and survival during the early life history of Pacific herring *Clupea pallasi* from Bristol Bay, Alaska, in a marine mesocosm. U.S. National Marine Fisheries Service Fishery Bulletin 88:191–200.

Wolfe, D. A., and eleven coauthors. In press. Fate of the oil spilled from the T/V *Exxon Valdez* in Prince William Sound, Alaska. Environmental Science and Technology.

American Fisheries Society Symposium 18:463–486, 1996

Distribution and Abundance of Larval Fishes in Prince William Sound, Alaska, during 1989 after the *Exxon Valdez* Oil Spill

Brenda L. Norcross and Michele Frandsen

University of Alaska Fairbanks, School of Fisheries and Ocean Sciences, Institute of Marine Science Fairbanks, Alaska 99775-1080, USA

Abstract.—Field studies were conducted in 1989 to determine spatial and temporal variation of finfish larvae in Prince William Sound in the months following the *Exxon Valdez* oil spill. This study contributes basic larval fish occurrence information for Prince William Sound. Unfortunately, it is biased as a baseline because it represents the ecosystem following an oil spill. The patterns of distribution are interpreted in the context of oceanographic patterns affecting transport of oil and larvae. Temporal and spatial distribution and abundance patterns are shown for the six most dominant taxa of larval fishes captured: walleye pollock *Theragra chalcogramma*, northern smoothtongue *Leuroglossus schmidti*, Pacific herring *Clupea pallasi*, capelin *Mallotus villosus*, rockfishes *Sebastes* spp., and flathead sole *Hippoglossoides elassodon*. Most larval fishes were in the upper 50 m of the water column and either concentrated in the oiled western portions of Prince William Sound or were transported through those areas by the same processes that moved the oil.

Several species of finfishes are sensitive to toxic hydrocarbons during the egg and larval stages (Kuhnhold 1974; Falk-Petersen and Kjorsvik 1987). Toxic effects of hydrocarbons on fish eggs can be both immediate and acute (Rice et al. 1976; Smith and Cameron 1979) or prolonged and sublethal (Kuhnhold 1977; Rice et al. 1978; Solbakken et al. 1984). The effects of oil pollutants on larvae depend on larval age, size, and species (L. Foyn and B. Serigstad, Institute of Marine Research, unpublished data). Environmental factors such as temperature and salinity also affect toxicity (Rice et al. 1978). The severity of these effects appears to be dependent on a complex interaction of factors, including spatial and temporal distribution of spawning, developmental stage of larvae, population dynamics of species, spill location and timing, and hydrodynamics of the area (Spaulding et al. 1985).

Not all fish larvae are affected equally by hydrocarbons. In Alaskan waters, Pacific herring *Clupea pallasi* are among the most sensitive to oil (Rice et al. 1979). Larvae exposed to oil display jaw deformities and cytogenetic defects (B. Norcross and colleagues, University of Alaska Fairbanks, unpublished). Clupeid larvae collected from the Gulf of Finland exhibit curvature of the posterior notochord (Urho 1991). Gadoid fishes in Norway show altered growth and development of eggs and larvae and morphogenic abnormalities (L. Foyn and B. Serigstad, unpublished data).

Before oil spill impacts on fish larvae can be assessed, there must be some baseline data about larval fish species composition, distribution, and abundance in the spill area. That information, together with field and laboratory studies on similar species, provides a basis for damage assessment. No such studies had been done before the TV *Exxon Valdez* oil spill. Therefore, the main objective of this study was to determine temporal and spatial presence of larval finfishes in Prince William Sound and to relate that to the general physical dispersal patterns affecting transport of oil and larvae in the region during 1989.

Methods

Field Collections

In response to the *Exxon Valdez* oil spill, six 1-week cruises were organized to determine the distribution of larval fishes around Prince William Sound in 1989 (Table 1). Four of the six cruises were conducted by the University of Alaska Fairbanks ship RV *Alpha Helix* (HX), one was by the National Oceanic and Atmospheric Administration (NOAA) ship *John Cobb* (CO), and one was by the chartered fishing vessel FV *Jennie Girl* (JG).

Strata representative of the characteristic hydrological, geological, and ecological areas within Prince William Sound were chosen to cover as much of the open area of the sound as possible. We used overflight maps of oil trajectories to geographically partition Prince William Sound into oiled (Naked Island and Montague) and non-oiled (southeast and north) areas (Figure 1). The specific station locations were randomly chosen within the representative strata. The same stations that were occupied on the first cruise were also occupied on following cruises (weather permitting), and additional stations were added as the season progressed.

TABLE 1.—Larval fish sampling effort in Prince William Sound after the 1989 Exxon Valdez Oil Spill.

Cruise or total	Dates	Number			Volume filtered (m^3)	
		Stations	Nets	Larvae	Mean	SD
HX121	5–11 April	8	13	796	397.4	193.7
HX123	5–11 May	15	42	13,243	42.1	16.5
HX125	1–6 June	15	34	4,037	59.0	21.8
CO8902	7–14 July	18	106	2,542	347.0	964.4
JG001	31 August–10 September	14	67	501	441.5	962.2
HX134	27 October–3 November	24	211	432	309.3	145.7

Early cruises were shared with other investigators and thus had limited time available for sampling larval fishes. The combination of number and location of stations was dependent upon what could be accomplished in 7 d, sampling 24 h per day. Conductivity-temperature-depth (CTD) profiles were taken at all stations at which fish were collected (Norcross et al. 1991).

The gear used to sample larvae during all cruises, except in October (cruise HX134), was a 1-m^2 Tucker Trawl rigged with two or three 505-μ mesh nets towed at 1 m/s. A TSK (Research Nets, Bothell, Washington) or General Oceanics (Miami, Florida) flowmeter was mounted in the mouth of the net to determine the volume of water filtered. A self-contained CTD (SeaBird SBE 19 Profiler, Bellevue, Washington) was attached to the net for each deployment and provided measurements of depth, temperature, and salinity. The net used on the final cruise (HX134) was a MOCNESS (Multiple Opening/Closing Net and Environmental Sensing System) (Wiebe et al. 1985) equipped with nine 1-m^2 505-μ mesh nets. During the first three cruises on the RV *Alpha Helix* (April, May, and June), sample depth was limited to 100 m. On later cruises, depths to 500 m were sampled. Oblique samples in increments of 0 to 25 m, 25 to 50 m, and 50 to 100 m were taken on all cruises, except in April when no 0 to 25 m sample was taken. Intervals of 100 to 200 m and 200 to 500 m were sampled in July (cruise CO8902), September (cruise JG001), and October (cruise HX134). Additionally, oblique tows of 0 to 50 m were sampled in June (cruise HX125), July, September, and October, and a 0 to 100-m interval was sampled in April (cruise HX121), July, September, and October.

After each tow, the contents of each net were immediately rinsed with seawater, sieved, and preserved in alcohol (100% ethanol or isopropyl alcohol) or 10% seawater-mixed formalin. Different preservatives were used in order to allow potential analysis of several factors (e.g., alcohol for aging of otoliths, formalin for identification of eggs). The initial preservative was changed after 24 h, 48 h, and again after 72 h.

Laboratory Analysis

Samples were analyzed at the Fisheries Oceanography Laboratory at the University of Alaska Fairbanks. We did not subsample or split samples. Using dissecting microscopes, the plankton were methodically sorted to isolate fish larvae and eggs.

Larval fishes were identified to lowest taxonomic level using criteria described by Matarese et al. (1989), Hart (1973), and R. Baxter (deceased, unpublished data). Additional larval identifications were made or confirmed by M. Busby and B. Vinter (National Marine Fisheries Service, personal communication). Standard length was measured to the nearest 0.01 mm using a Wild MZ3 dissecting microscope and BioQuant image-analysis system (Fowler 1988) interfaced to a computerized database. This process simultaneously enumerated the larvae because the length data were interfaced with the database containing the cruise information. Because samples had been preserved for more than 3 years before identification and measurement of larvae, lengths were not corrected for preservative shrinkage and condition factors could not be estimated. Fish larvae that were partially or wholly disintegrated or that could not be identified (<10% of total) were discarded and not used in the analysis. Abundance was standardized for volume of water filtered, and catch per unit effort (CPUE) was expressed as number of larvae per 100 m^3.

Statistical Analysis

Chi-square (χ^2) or analysis of variance (ANOVA); (F) followed by a Tukey honestly significant difference (HSD) multiple comparison test, as implemented in SYSTAT (Wilkinson 1988), were used to compare abundance of larvae among areas, sampling times and depths. Significance was established at values of $P < 0.05$, unless otherwise stated.

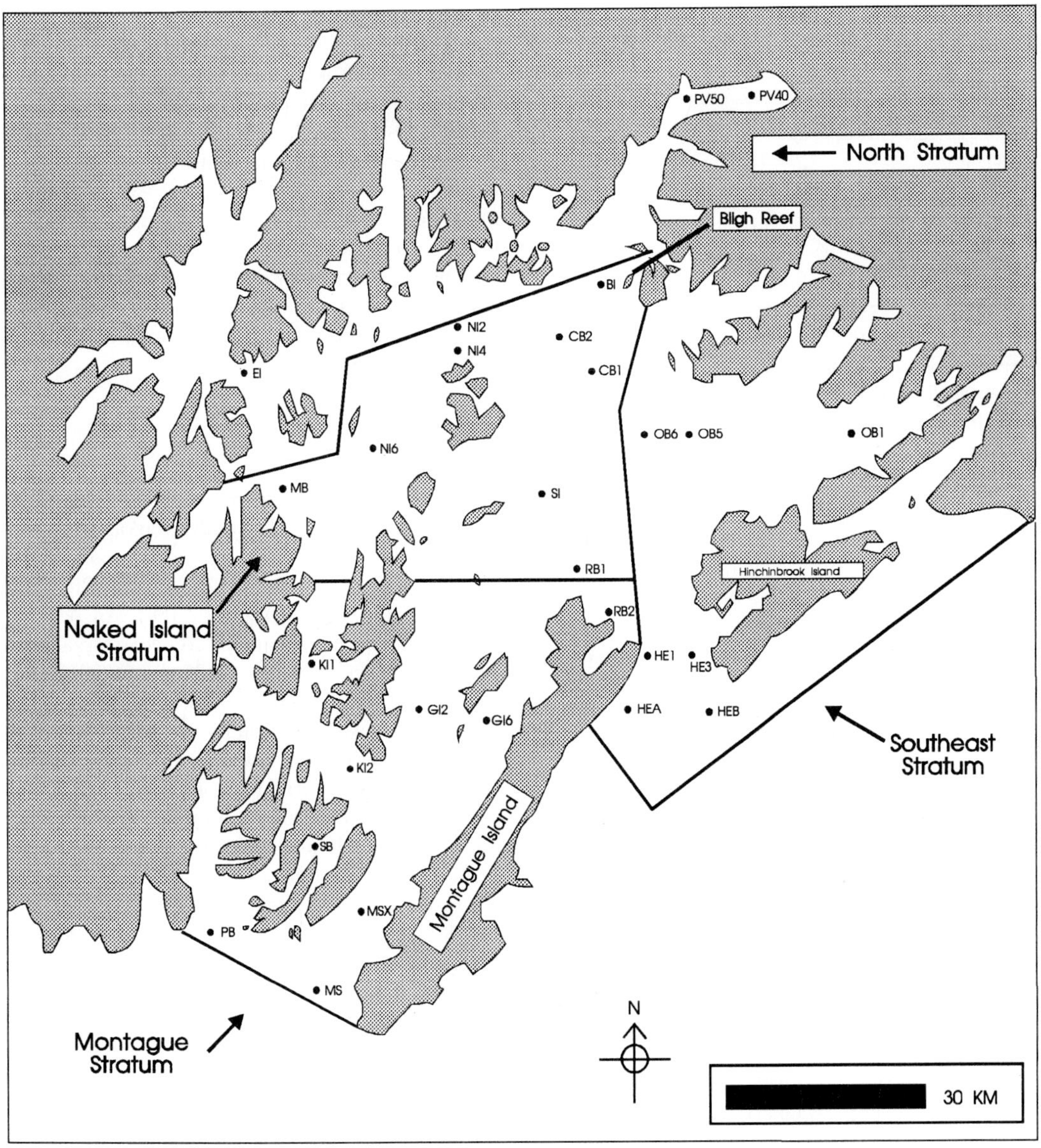

FIGURE 1.—Prince William Sound stations divided into oiled (Naked Island and Montague) and non-oiled (southeast and north) strata.

Results

During the 7 months (April–October) following the 1989 TV *Exxon Valdez* oil spill, six research cruises were conducted in Prince William Sound (Table 1). Altogether, 473 collections were made at various depth increments, with 21,551 fish larvae captured.

Water Characteristics

In April 1989 the open waters of Prince William Sound were nearly isothermal at 4°C (Norcross et al. 1991). Salinity gradually increased from 30 practical standard units (psu) at the surface to 32 psu near the bottom. By May the upper 10 to 20 m had warmed to 5 to 7°C with salinities of 29 to 31 psu,

TABLE 2.—Rank order of absolute abundance of dominant fish species captured in Prince William Sound in 1989.

Species	Total number of fish	Percent of total number
Walleye pollock	13,174	61.55
Northern smoothtongue	2,765	12.83
Pacific herring	1,991	9.24
Capelin	1,298	6.01
Rockfishes	933	4.33
Flathead sole	281	1.30
Other species	1,021	4.74

while the lower layers were 4 to 5°C and 32 psu. Warming progressed in June, with temperatures of 6 to 10°C and salinities of 28 to 30.5 psu; the warm surface layer had deepened to 20 to 50 m. The lower water layers were still 4 to 5°C with a salinity of 32 psu. In July the water temperature warmed to 6 to 15°C above the seasonal thermocline at approximately 50 m. The lower layers also increased in temperature to 5 to 6°C. The upper 5 to 10 m had salinities as low as 15 psu, but below that the salinity was uniformly near 34 psu. Stations near the ocean were cooler and better mixed than those well within the sound. By September surface waters were approximately 13°C at all locations, including those outside Hinchinbrook Entrance. The thermocline had deepened to approximately 50 m. The deep waters were 4 to 6°C. Salinities in September were similar to those in July, with a low-salinity shallow (<10 m) upper layer; however, there was a gradual increase in salinity to the seasonal halocline at approximately 50 m. Below that, salinities were 34 psu. By October the deep waters again were 4 to 6°C. The depth of the thermocline remained at about 50 m with temperatures reduced to 8 to 9°C. Surface waters within Prince William Sound cooled to 5°C, but remained as warm as 9°C outside of the sound. The very-low-salinity upper layer was gone, with surface layers again having a salinity of 28 to 30 psu. The halocline was still at about 50 m.

Presence and Abundance of Larval Fishes

Forty-three taxa of larval fishes were captured in Prince William Sound over 7 months. Gadidae, the single most abundant family of larval fishes, constituted 61.5% of all larvae captured. Other dominant families were at least an order of magnitude lower in abundance: Bathylagidae (13.4%), Clupeidae (9.2%), Osmeridae (6.1%), Scorpaenidae (5.3%), Pleuronectidae (2.1%), Cottidae (1.1%), Ammodytidae (0.4%), Stichaeidae (0.4%), and Bathymasteridae (0.3%). Six taxa made up 95.3% of the total catch for all cruises (Table 2): walleye pollock *Theragra chalcogramma*, northern smoothtongue *Leuroglossus schmidti*, Pacific herring, capelin *Mallotus villosus*, rockfishes *Sebastes* spp., and flathead sole *Hippoglossoides elassodon*.

Spatial and Temporal Distribution of Larval Fishes

The density of larvae peaked in May and decreased with time (Table 3). Most larval fishes were captured in May (82%). Larval abundance was lower in June (15%), July (17%), April (2%), September (<1%) and October (<1%). An ANOVA showed a highly significant difference in larval abundance among months (F = 8.25, $P < 0.001$) and a Tukey HSD indicated that abundance of larval fishes in May was significantly greater than that for the other 5 months sampled ($P < 0.001$).

Five of the six overall most abundant fishes, walleye pollock, northern smoothtongue, Pacific herring, flathead sole and rockfishes, were most abundant in May and June (Table 4). Walleye pollock made up 10% of the catch in April, 88% in May, and 43% in June. Though northern smoothtongue were collected in highest numbers in May, they constituted only 6% of the catch, which was dominated by walleye pollock. However, northern smoothtongues constituted 90% of April, 11% of June, 34% of July, 3% of September, and 35% of October catches. Pacific herring were 2% of May, 42% of June, and 3% of July collections. Capelins were found in greatest abundance in July (31%) and September (93%). Only one species, northern smoothtongue, was collected as larvae or very small juveniles during each of the 6 months.

Overall CPUE of larval fishes was not significantly different among strata (F = 0.457, P =

TABLE 3.—Number of fish larvae, standardized to catch per 100 m^3, by stratum and month in Prince William Sound in 1989.

Stratum	April	May	June	July	September	October
Southeast	1.35	1,320.27	35.29	1.19	3.28	1.02
North	58.25	511.16	101.09	12.37	0.00	0.43
Naked Island	7.44	942.82	180.38	9.11	2.59	0.40
Montague	1.62	408.59	284.26	14.39	0.54	0.33

TABLE 4.—Monthly densities (catches per 100 m^3) and sizes of principal fish larvae caught in Prince William Sound (all stations); N is the number of larvae caught and measured.

Species	Catch/100 m^3	Total length (mm)			N
		Mean ± SD	Minimum	Maximum	
April					
Walleye pollock	1.53	4.16 ± 0.46	3.13	5.28	79
Northern smoothtongue	13.61	9.04 ± 1.50	5.42	14.73	703
May					
Walleye pollock	650.98	6.61 ± 1.14	3.15	11.0	11,518
Northern smoothtongue	45.44	9.43 ± 2.23	6.40	37.20	804
Pacific herring	18.14	8.16 ± 1.20	5.98	15.84	321
Capelin	0.06		4.57		1
Rockfishes	10.34	4.26 ± 0.55	3.02	6.17	183
Flathead sole	11.47	5.12 ± 1.15	3.70	12.23	203
June					
Walleye pollock	82.58	13.88 ± 2.85	4.09	23.68	1,656
Northern smoothtongue	20.50	10.78 ± 2.72	4.84	21.76	411
Pacific herring	80.59	12.17 ± 1.87	7.93	19.91	1,616
Rockfishes	2.74	4.54 ± 0.59	3.33	5.80	55
Flathead sole	3.59	9.06 ± 2.80	4.65	17.22	72
July					
Walleye pollock	0.03	33.11 ± 5.59	26.38	44.00	12
Northern smoothtongue	1.96	17.47 ± 5.37	5.86	73.00	721
Pacific herring	0.15	12.88 ± 3.23	8.58	23.24	54
Capelin	1.80	10.30 ± 3.19	3.86	22.83	661
Rockfishes	1.84	4.44 ± 0.58	2.42	6.22	678
Flathead sole	0.02	21.96 ± 1.71	18.75	23.50	6
September					
Northern smoothtongue	0.05	40.64 ± 14.56	20.91	56.00	15
Capelin	1.42	13.14 ± 4.52	4.92	43.90	421
Rockfishes	0.06	4.07 ± 0.39	3.30	4.86	17
October					
Northern smoothtongue	0.17	26.36 ± 3.76	18.53	53.00	111
Capelin	0.32	28.00 ± 8.68	9.53	52.55	212

0.713). The sampling effort in the strata in Prince William Sound was equitable, except in the north station, where the effort was less than that at the other three. The total number of fishes caught shows this, with 17.53% of the larvae captured in north, 18.21% in Montague, 34.95% in southeast, and 29.31% in Naked Island. The catch per unit effort of larval fishes seemed to vary over time (month) and space (strata) ($\chi^2 = 770$, $P < 0.001$). The stratum where abundance was highest varied from north in April, to Montague in June and July, to southeast in May, September, and October (Table 3), but an ANOVA of the number of larval fishes captured in each stratum within a month did not show significant differences among strata.

Each species of fish larvae had a characteristic geographical distribution pattern (Table 5). Walleye pollock were the most abundant larvae in all four strata. Walleye pollock were low in abundance in April, but were found in great abundances in all areas of the sound in May (Figure 2). The highest abundance captured during May was in southeast, followed by Naked Island, north, and Montague strata. By June, most walleye pollock were in the western strata of Naked Island and Montague. For all months combined there was no significant dif-

TABLE 5.—Spatial distribution of individual fish taxa combined over all time periods, standardized to number of larvae per 100 m^3 among this study's strata in Prince William sound.

Species	Southeast	North	Naked Island	Montague
Walleye pollock	13.42	17.18	6.07	7.08
Northern smoothtongue	0.71	9.06	1.61	1.91
Pacific herring	0.14	0.64	0.79	3.31
Capelin	0.96	0.22	0.60	1.32
Rockfishes	0.09	0.62	1.19	0.73
Flathead sole	0.42	0.05	0.09	0.13

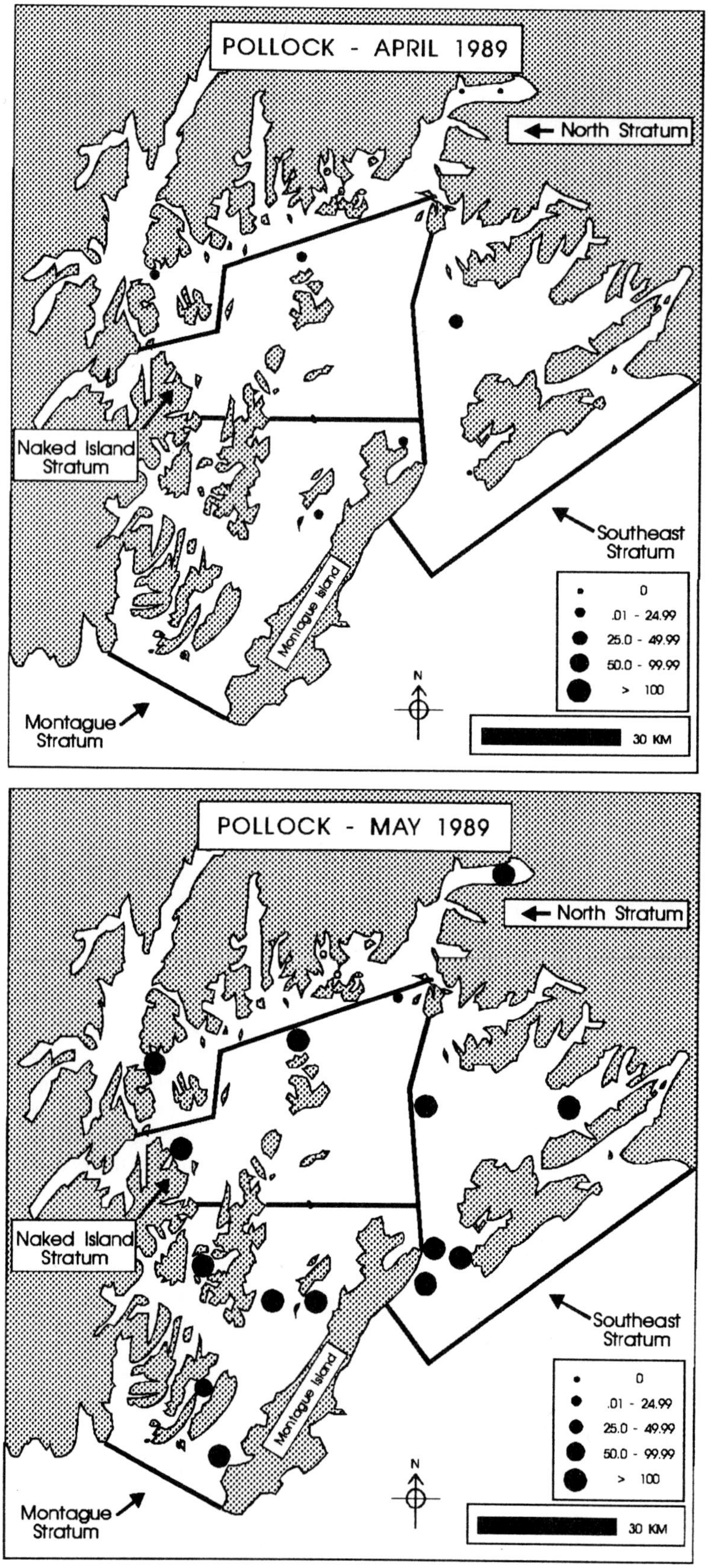

FIGURE 2.—Distribution of walleye pollock larvae in April, May, June, and July 1989, standardized to number per 100 m^3 water filtered.

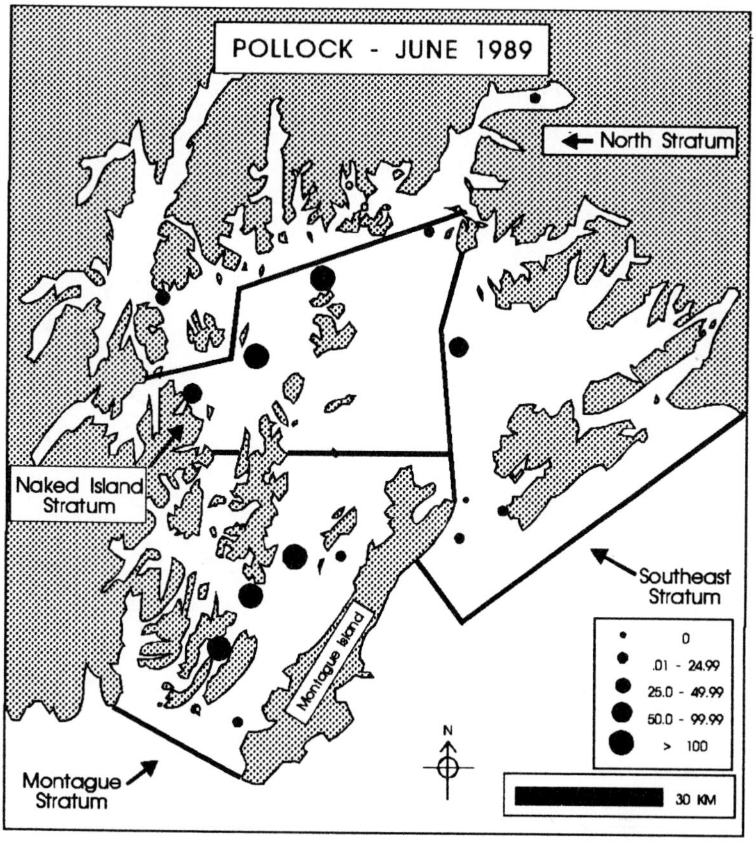

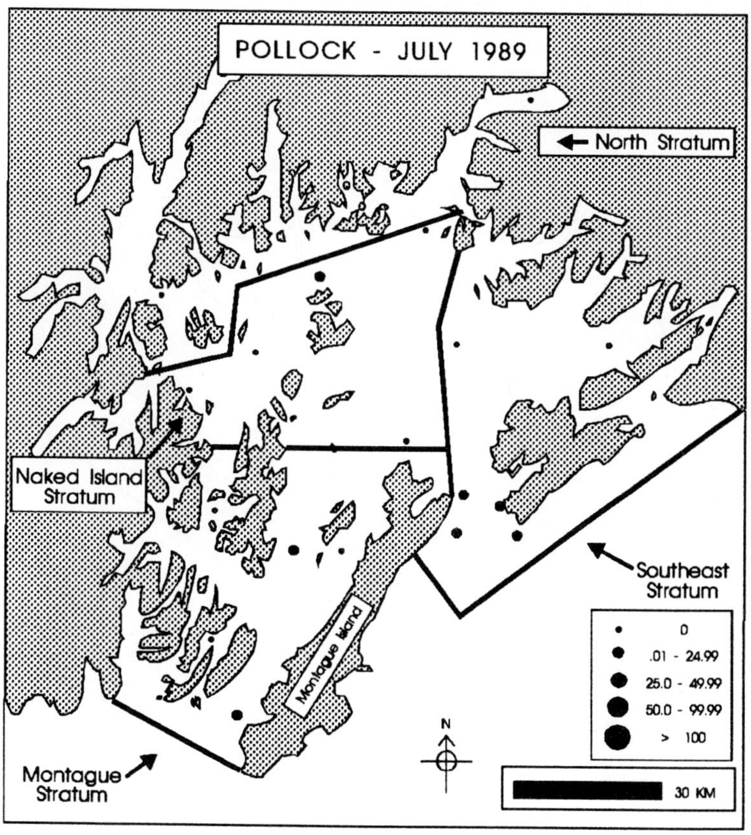

FIGURE 2.—Continued.

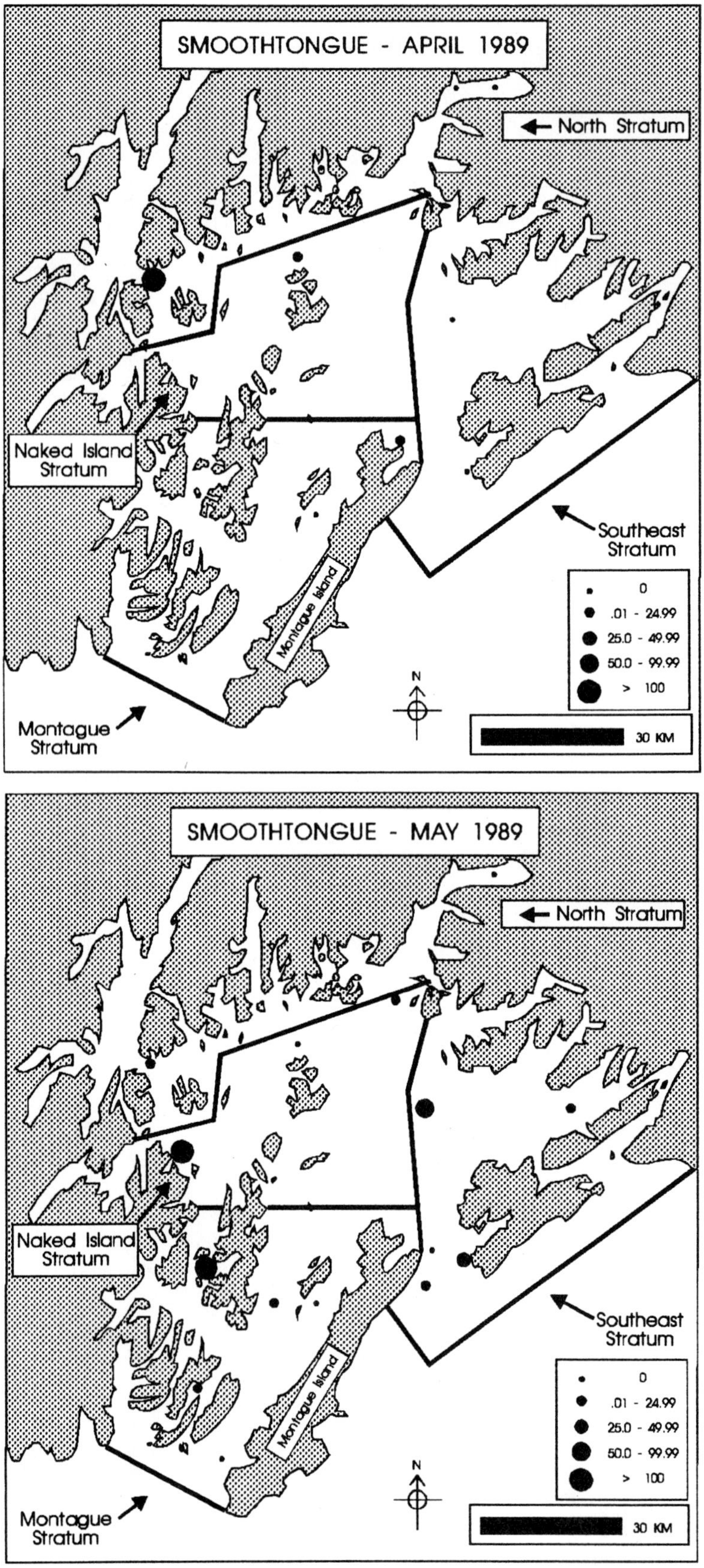

FIGURE 3.—Distribution of northern smoothtongue larvae in April, May, June, July, September, and October 1989, standardized to number per 100 m^3 water filtered.

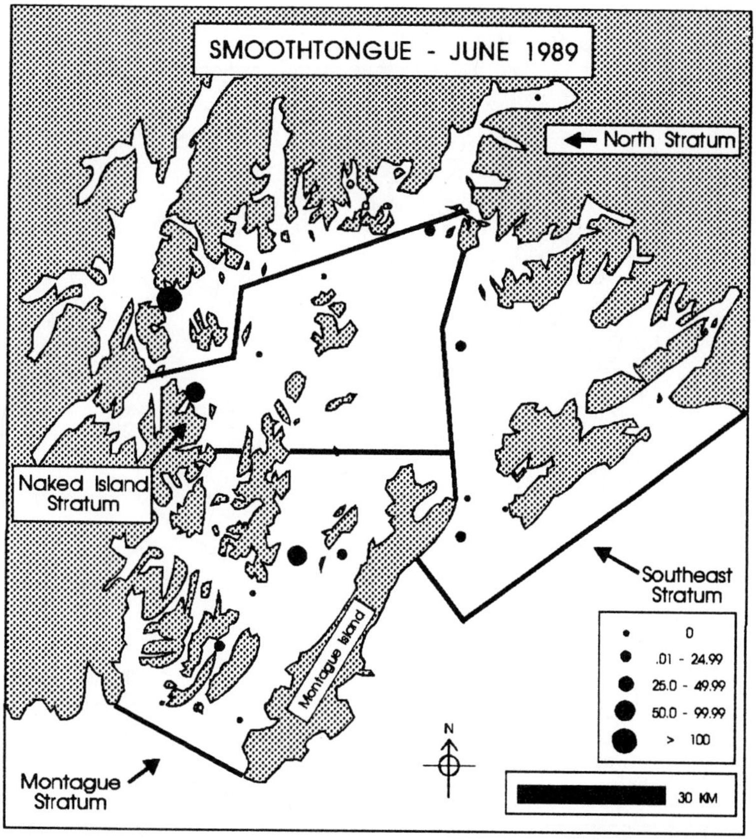

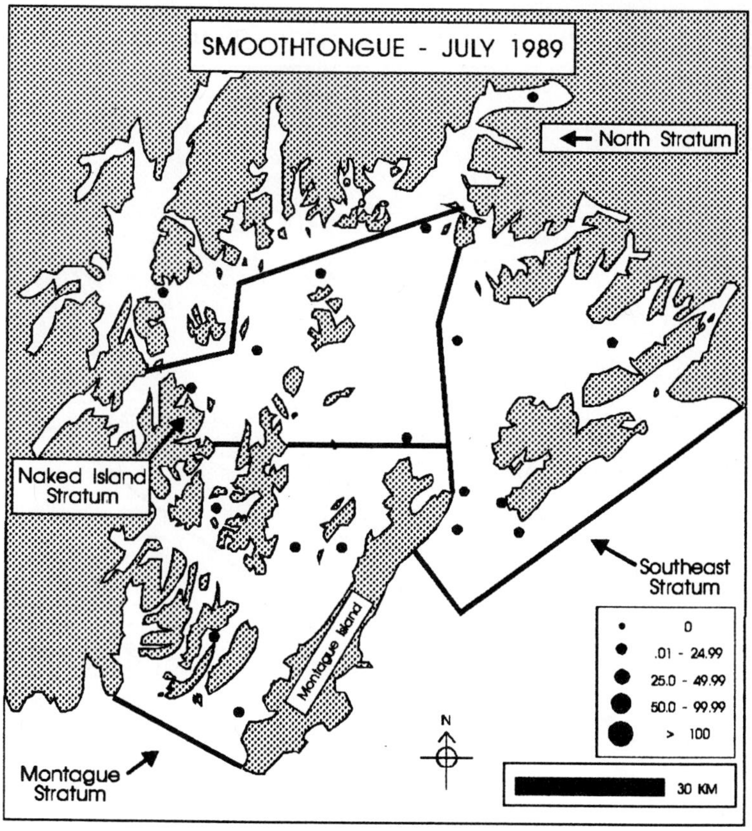

FIGURE 3.—Continued.

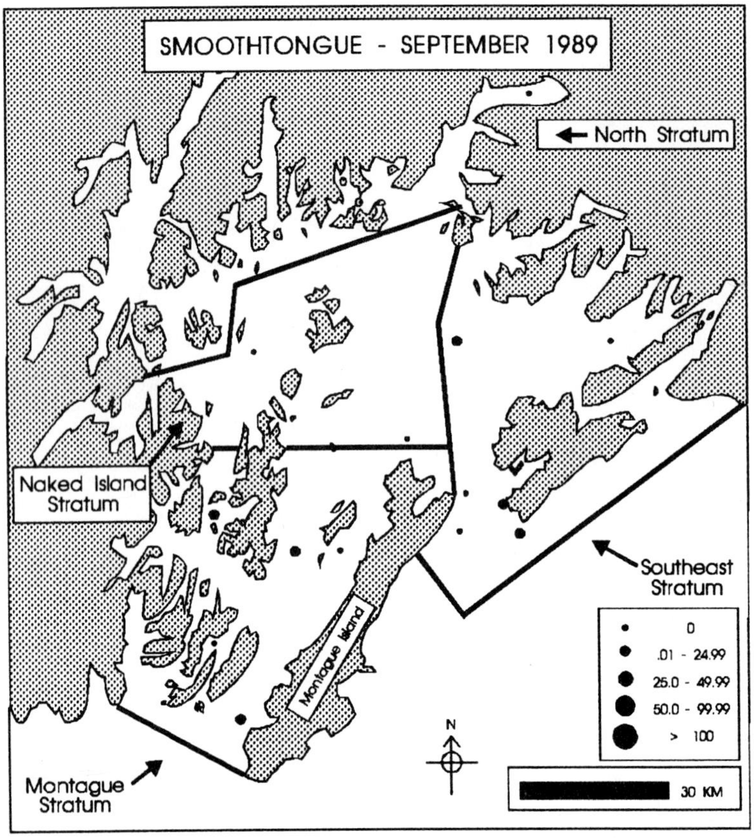

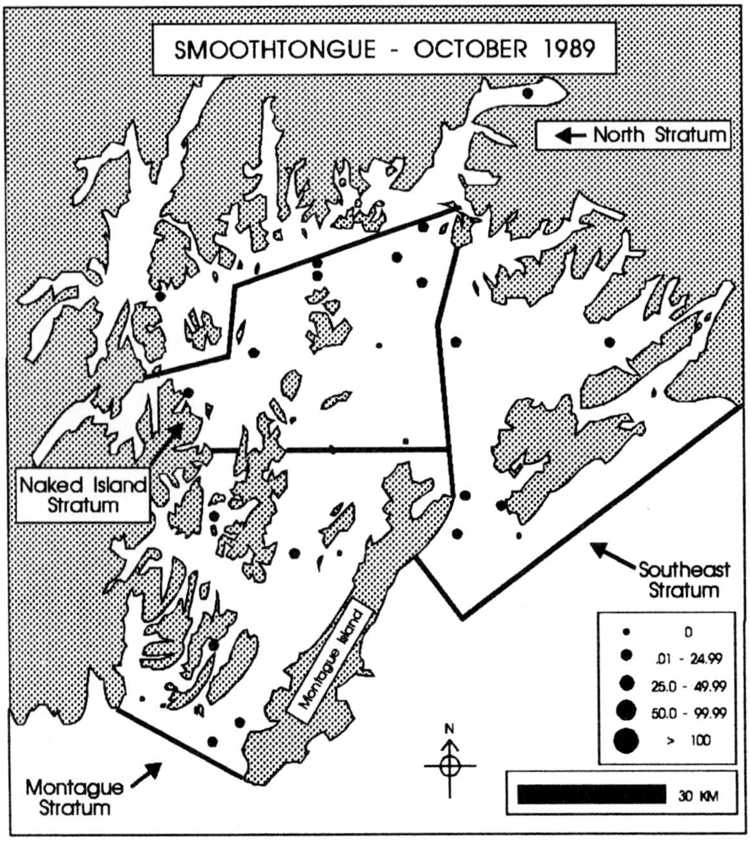

FIGURE 3.—Continued.

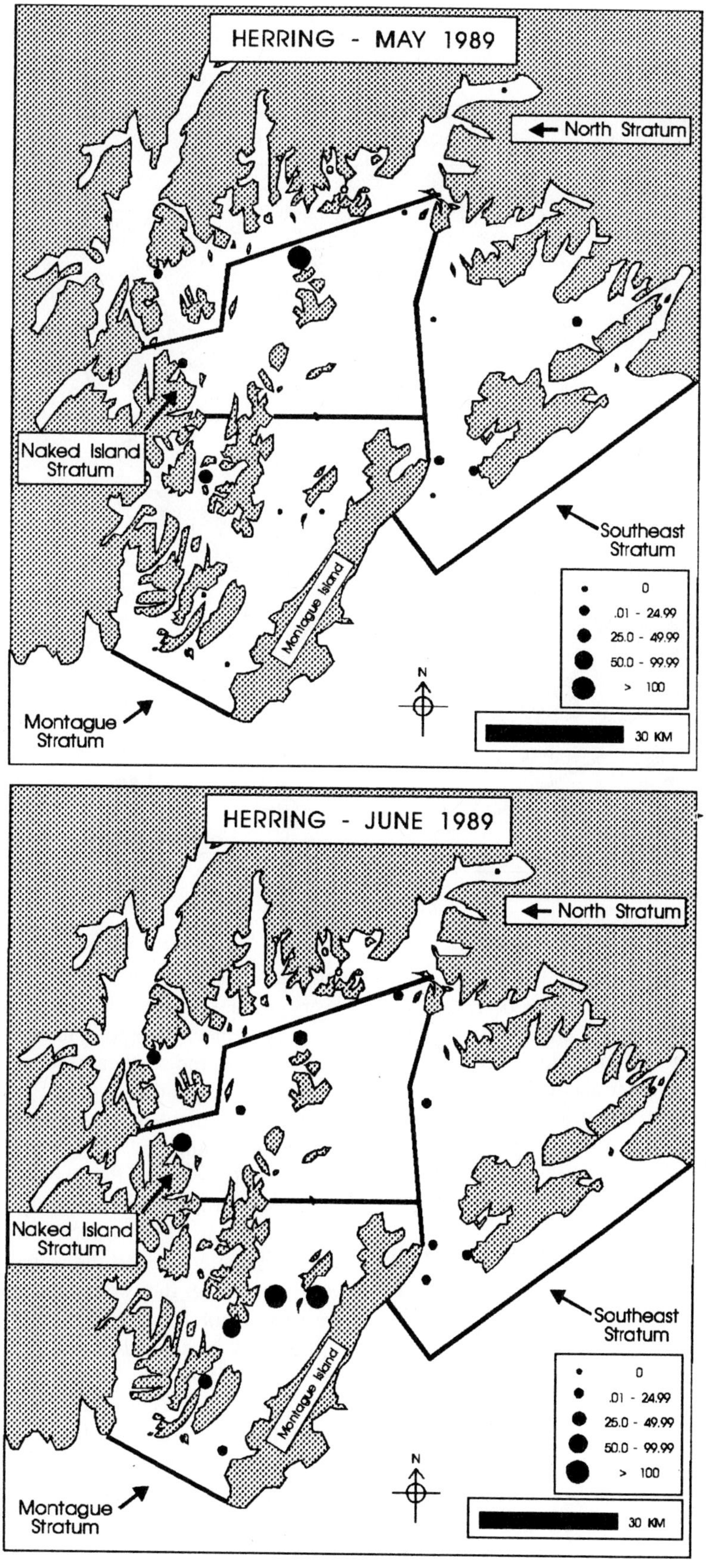

FIGURE 4.—Distribution of Pacific herring larvae in May, June, and July 1989, standardized to number per 100 m^3 water filtered.

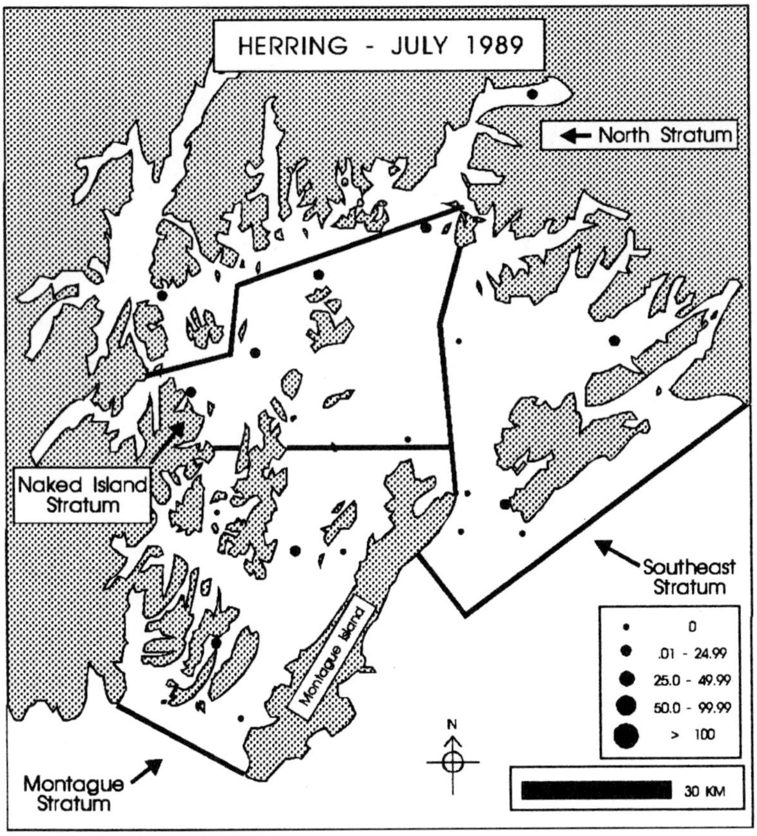

FIGURE 4.—Continued.

ference in the distribution of walleye pollock around the sound ($F = 0.771$, $P = 0.513$).

Northern smoothtongues concentrated on the western side of the sound and thus were less evenly distributed than the other major species (Figure 3), however, there was no significant difference in their distribution among strata over all months combined ($F = 0.820$, $P = 0.486$). They were usually abundant in north stratum, but were also important in Naked Island and Montague strata.

Pacific herring were concentrated near spawning areas (Norcross et al., unpublished) during May in Naked Island and Montague strata (Figure 4). In June larvae were mainly on the western side of the sound, especially in Montague stratum. By July there were only a few Pacific herring larvae small enough to be captured by our plankton net scattered evenly around the sound. For all months combined there was no significant difference in abundance of Pacific herring larvae among areas around the sound ($F = 1.115$, $P = 0.347$).

A few capelin were collected in Naked Island stratum in May, but by July capelin were found in every stratum, with the most in Montague (Figure 5). In September there were more capelin caught in Naked Island and southeast than in earlier months. By October capelin again were captured in every stratum. As with the other species, the differences in distribution within the sound were not statistically significant ($F = 0.596$, $P = 0.619$).

Larvae of rockfishes were found in the west and northwest areas of the sound in May and June and were distributed around the sound in July (Figure 6). In September they were missing from the northwest portion of the sound, but that did not affect the overall lack of significance in distribution among areas ($F = 1.009$, $P = 0.392$).

Flathead sole was the only species to concentrate in southeast stratum in May (Figure 7). In June flathead soles were mostly in Naked Island and Montague strata, whereas in July they were in Montague stratum only. Despite this apparent concentration in the western areas, it was not enough to

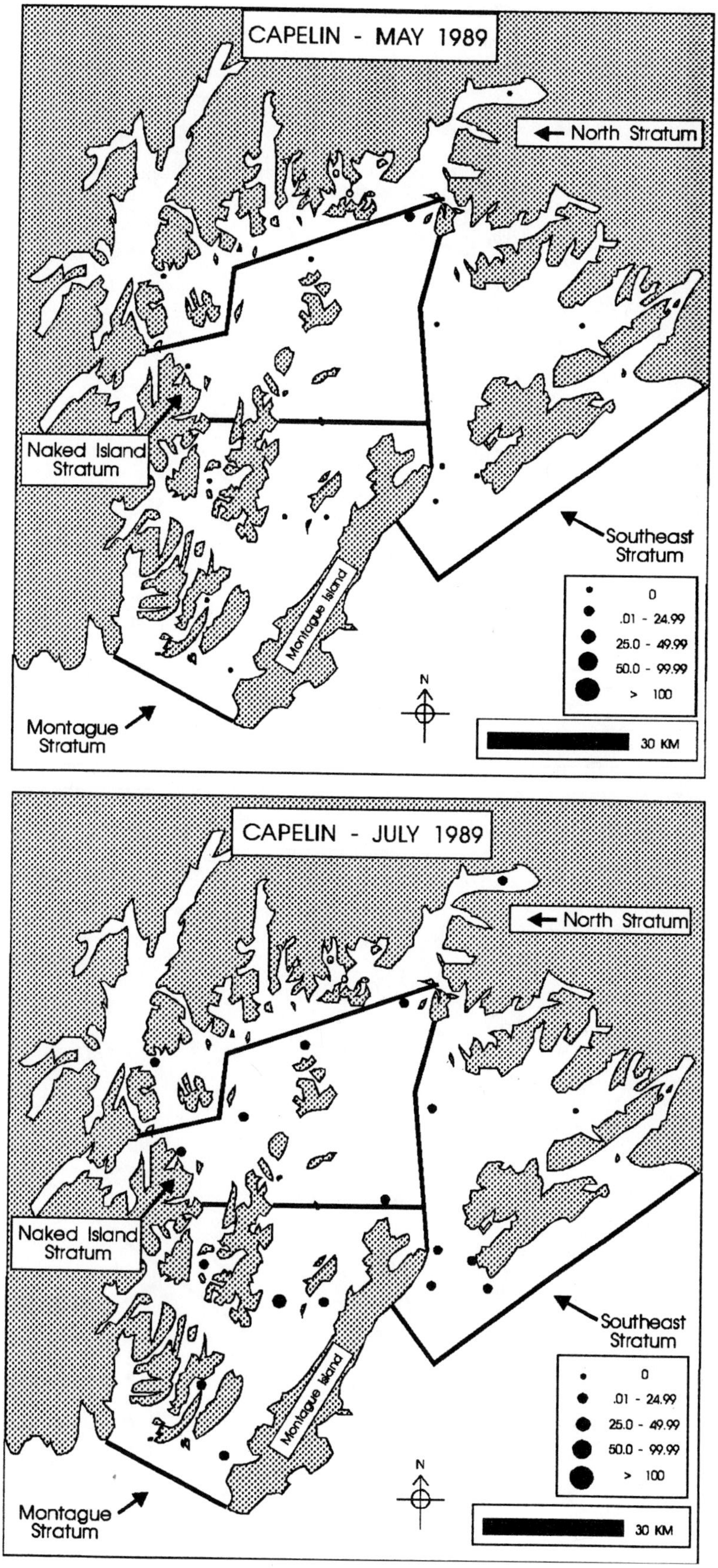

FIGURE 5.—Distribution of capelin larvae in May, July, September, and October 1989, standardized to number per 100 m^3 water filtered.

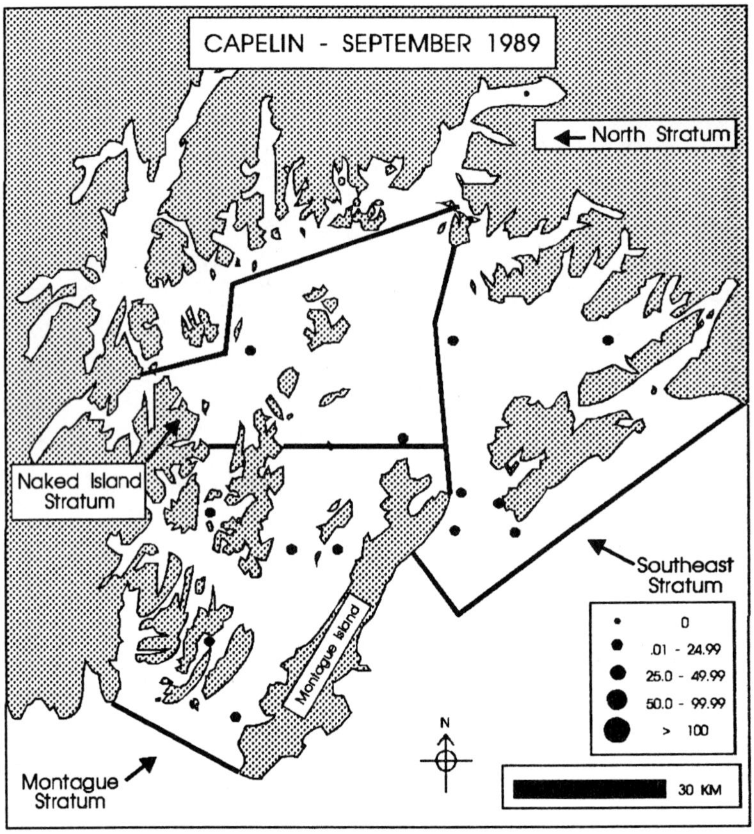

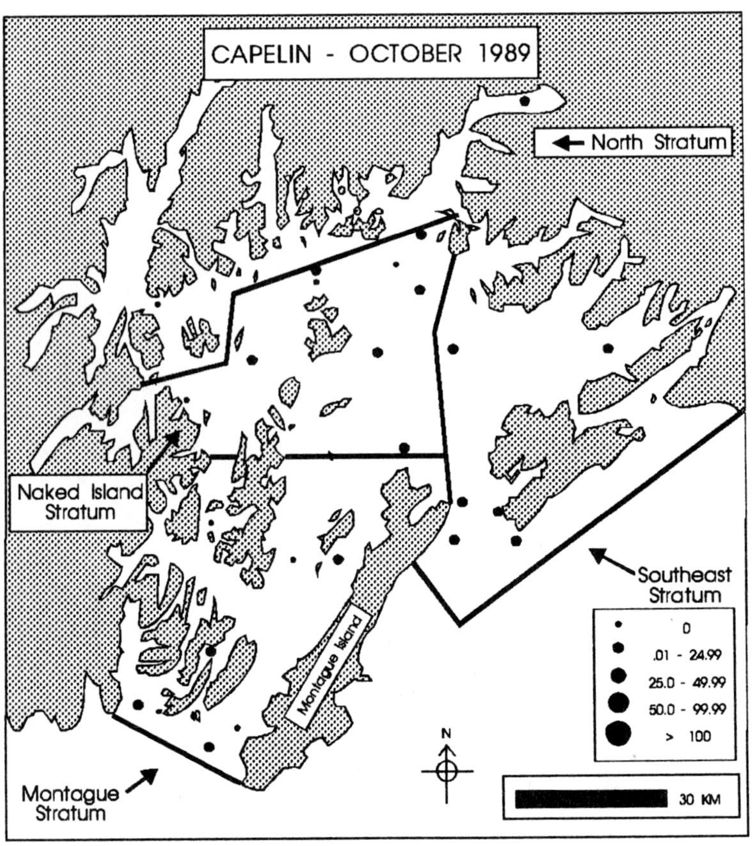

FIGURE 5.—Continued.

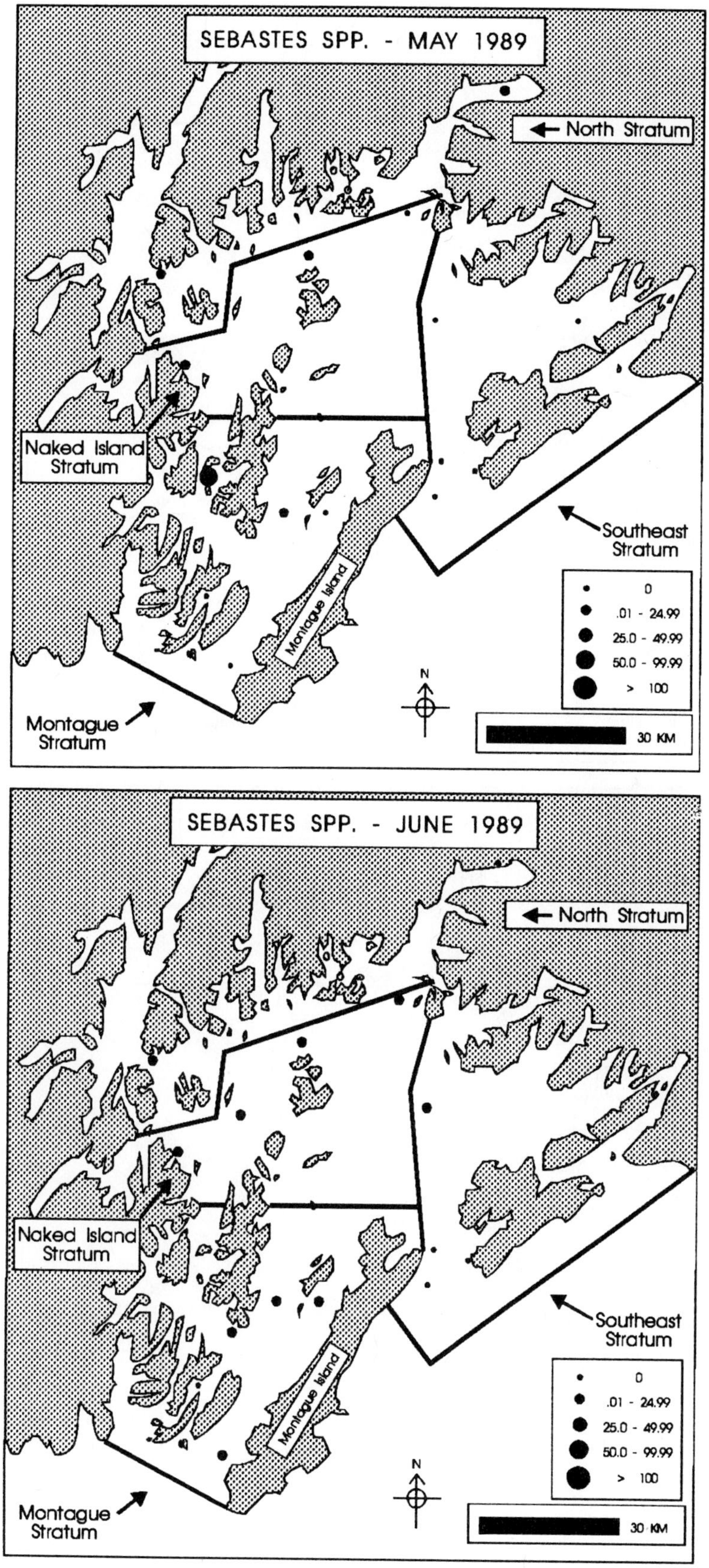

FIGURE 6.—Distribution of *Sebastes* spp. larvae in May, June, July, and September 1989, standardized to number per 100 m^3 water filtered.

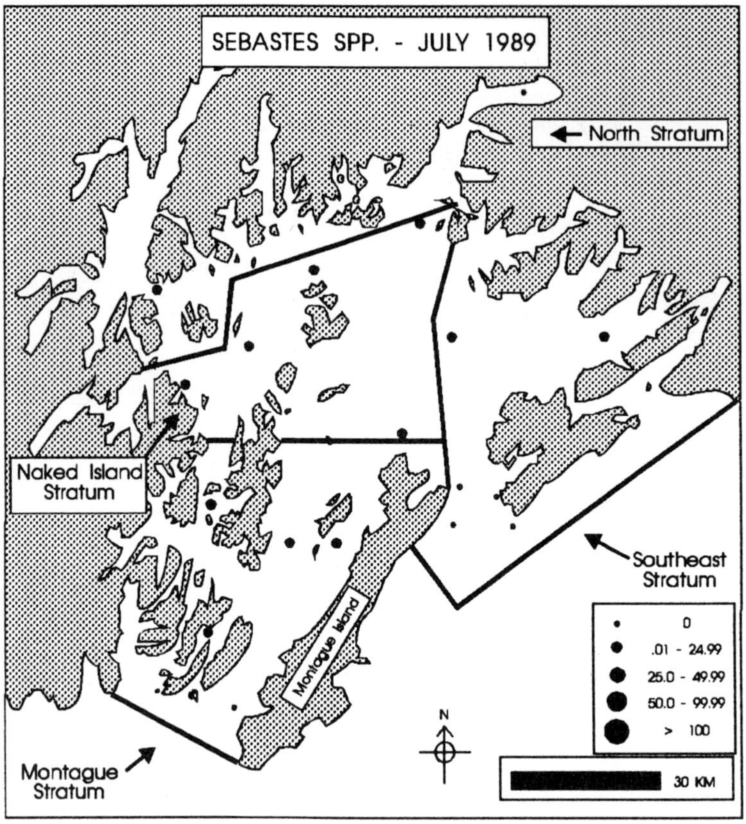

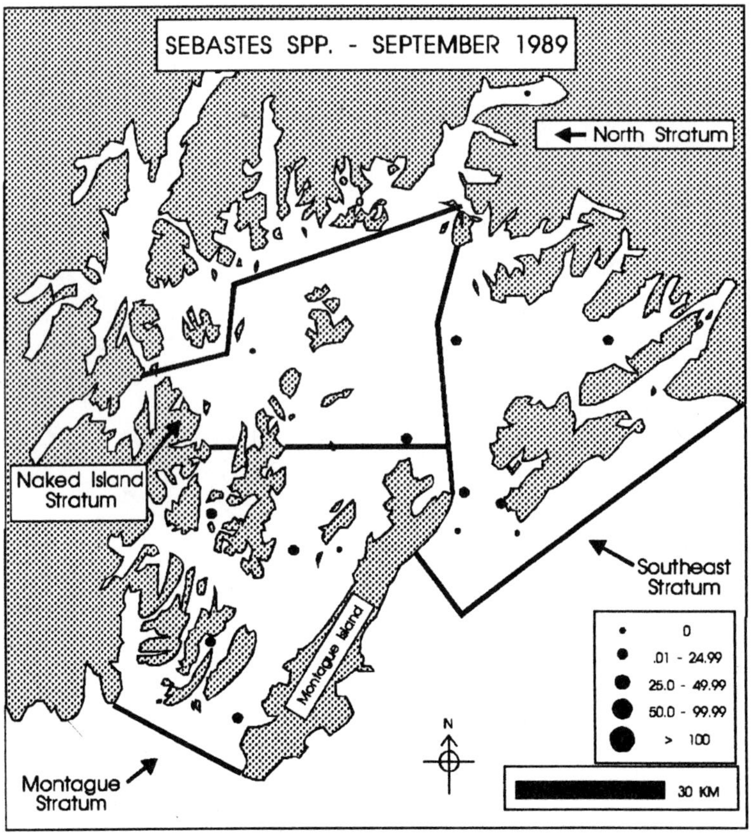

FIGURE 6.—Continued.

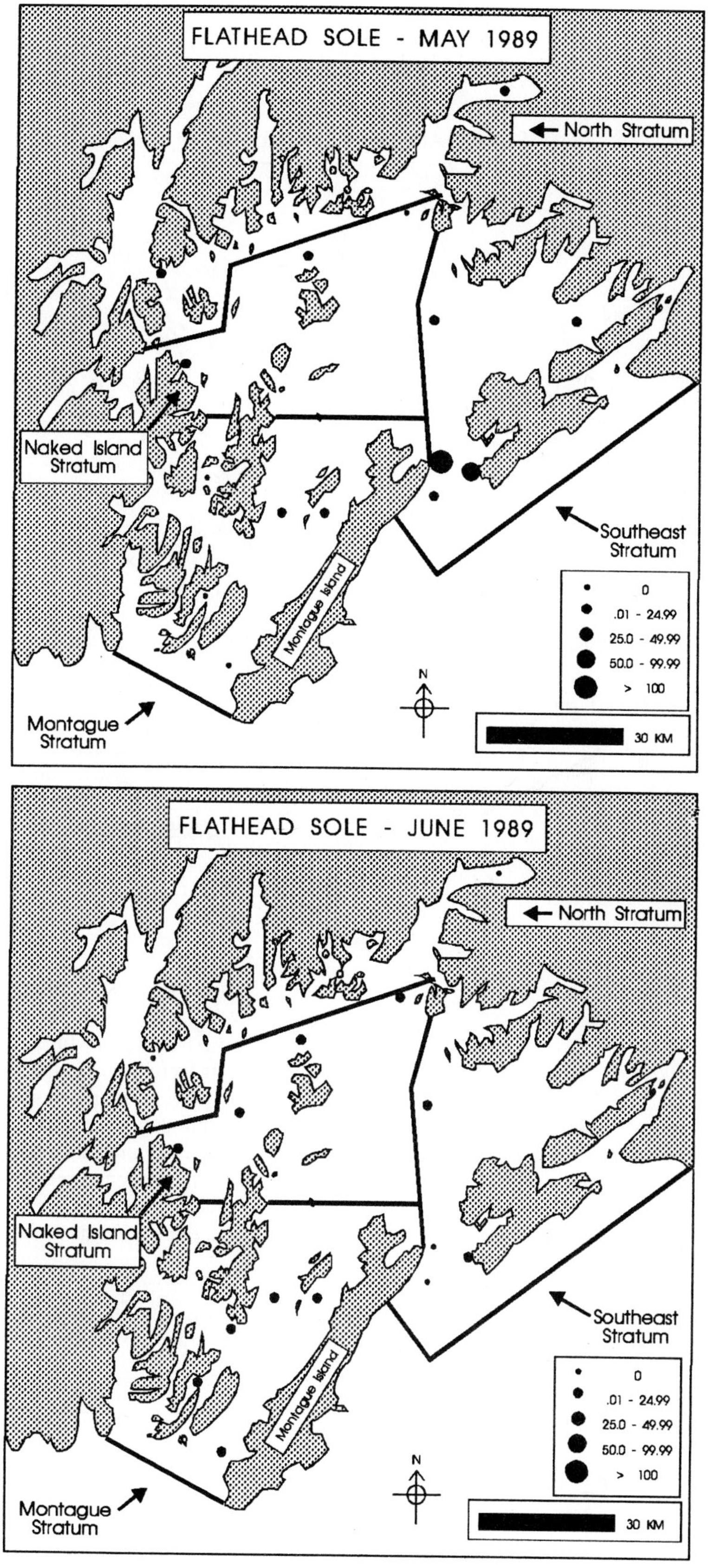

FIGURE 7.—Distribution of flathead sole larvae in May, June, and July 1989, standardized to number per 100 m^3 water filtered.

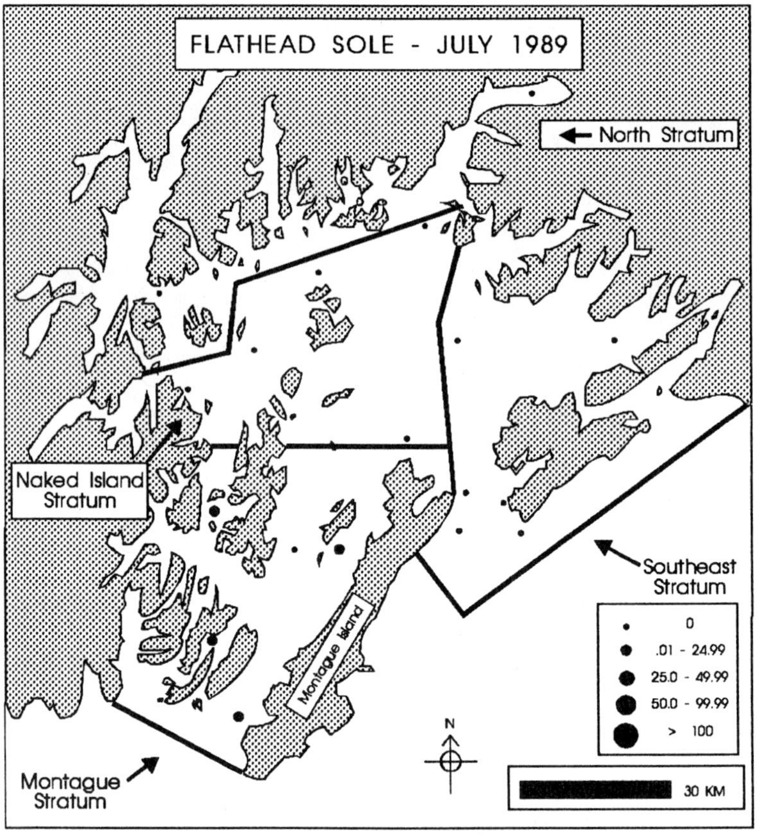

FIGURE 7.—Continued.

result in a significant difference (F = 1.115, P = 0.334).

Depth Distribution of Larval Fishes

All six species were captured principally in the upper 50 m (Table 6). In April walleye pollock and northern smoothtongues were captured between 0 and 100 m. In May most walleye pollock, Pacific herring, rockfishes, and flathead soles were in the upper 25 m, while most northern smoothtongues were between 25 and 50 m. The depth distribution of species was similar for June, except that most Pacific herring were found in the 25- to 50-m depth interval and many northern smoothtongues were in the 50- to 100-m depth range. The dominant larval fish in the upper 25 m in July was capelin. Northern smoothtongues were found mainly between 50 and 200 m. Each species, except flathead sole, was also found to a depth of 500 m in July. In September

TABLE 6.—Depth distribution of individual fish taxa combined over all time periods, standardized to number of larvae per 100 m^3 among this study's strata in Prince William Sound. Columns 0–50 m and 0–100 m, not sampled on all cruises, do not include fish from 0–25 m, 25–50 m, or 50–100 m.

Species	0–25 m	25–50 m	50–100 m	100–200 m	200 –500 m	0–50 m	0–100 m
Walleye pollock	110.02	7.64	0.75	0.01	0.01	0.22	0.18
Northern smoothtongue	1.39	18.26	1.51	0.85	0.29	0.26	3.27
Pacific herring	9.05	15.30	0.05	0.01	0.05	0.21	0.02
Capelin	1.92	2.56	0.72	0.17	0.09	1.84	0.60
Rockfishes	2.16	1.29	0.07	0.09	2.27	0.17	0.10
Flathead sole	2.13	0.53	0.02	0.00	0.00	<0.01	0.01

northern smoothtongues and capelins were found to a depth of 200 m, while rockfishes were found to 500 m. The two species found in October, capelin and northern smoothtongue, were collected at depths to 500 m.

Vertical distribution of fish larvae varied significantly. Abundance of walleye pollock at each depth interval sampled was statistically different from every other depth interval at $P < 0.001$. Abundance of northern smoothtongue larvae, a species not often found in the upper 25 m, was significantly different between all depth strata ($P < 0.001$). In contrast, Pacific herring, which were mainly found in the upper layers, only showed significant differences in depth distribution between 0 to 25 m and 100 to 200m, 25 to 50 m and 50 to 100 m, and 50 to 100 m and 100 to 200m ($P < 0.05$). Capelin, another species found mainly in the upper layers, showed significant differences in CPUE between shallow and deep strata: 0 to 25 m and 50 to 100 m; 50 to 100 m and 100 to 200 m; and 100 to 200 m and 200 to 500 m ($P < 0.002$). Rockfishes were found relatively evenly distributed vertically. The only significant difference was between 0 to 25 m and 200 to 500 m ($P < 0.05$). There were few flathead sole larvae, but there were significant differences between 0 to 25 m and 25 to 50 m ($P < 0.001$).

Length Analysis of Larval Fishes

We examined length distributions for the six most abundant species of fish. Modal lengths of walleye pollock larvae were 3.8 mm in April, 6.5 mm in May, and 13.0 mm in June; these values were quite similar to mean lengths (Table 4). The maximum size of walleye pollock larval lengths progressively increased from April through July, but the minimum size was 4.1 mm or less in April, May, and June.

Northern smoothtongues showed no change in size between the modal April length of 7.5 mm and the modal May length of 7.0 mm or between the somewhat larger mean lengths (Table 4), though the maximum size increased from April to May. The modal (10.0 mm) and mean lengths increased in June, but the range in sizes did not. The modal length in July was 15.0 mm, with the minimum size remaining small but the maximum size increasing. During the later months few northern smoothtongues were captured. Modal length was 26.5 mm for September and 27.5 mm for October. The minimum size increased, but the maximum size decreased.

Modal length of Pacific herring larvae was 7.5 mm in May. Yolk-sac larvae made up 32% of the May collection of Pacific herring, but no larvae in the later cruises had yolk sacs. In June the modal length was 11.5 mm. The average change in modal length over the 27 d between the mid-dates of the May and June collections was 4.0 mm or 0.15 mm/d. Using mean sizes of 8.2 mm in May and 12.2 mm in June (Table 4) results in an identical estimated growth of 0.15 mm/d. There was no change in modal size between June and July, though the mean size increased to 12.9 mm.

Capelin modal lengths (July:9.00 mm, September:12.0 mm, October:20 mm) progressively increased as did their means and ranges (Table 4). In contrast, rockfishes modal (May:4.0 mm, June:4.4 mm, July:4.6 mm, September:4.2 mm) and mean lengths and ranges did not increase (Table 4). Numbers of flathead sole larvae decreased as mean lengths and ranges increased (Table 4). Change in modal size from 4.5 mm in May to 9.0 mm in June was 0.17 mm/d.

Discussion

Abundance and Distribution of Larval Fishes

There are no previous records of fish larvae in the open waters of Prince William Sound. Therefore, it is not possible to say whether the community composition of larvae in 1989 is "normal" or affected by the *Exxon Valdez* oil spill. It is possible to report presence of species but not to document absence. Of the six most abundant species (Table 2), four are commercially important: walleye pollock, Pacific herring, rockfishes, and flathead sole. One species, capelin, is a primary forage fish. Little is known about the sixth species, northern smoothtongue.

Larval fish composition in Prince William Sound in 1989 is similar, but not identical, to that found in other areas of the Gulf of Alaska. In the western Gulf of Alaska from Cook Inlet to the Shumagin Islands, the 14 most abundant species included all but herring of the 6 most abundant species collected in this study (Rugen 1990). The collection of fish in Resurrection Bay, a glacial fjord, from April to July 1988 consisted of walleye pollock, flathead soles, Pacific herring and Pacific sand lances *Ammodytes hexapterus*, in decreasing order of abundance (Paul et al. 1991).

In northern southeast Alaska from Frederick Sound to Lynn Canal, Mattson and Wing (1978) sampled enclosed bays, the middle of deep channels, and the edge of deep channels from April to November 1972. Most fish were identified to family level. Gadids, followed by osmerids and bathylagids, were the three most abundant families captured. The most abundant gadid was walleye pollock and the predominant osmerid was probably capelin *M. villosus*, although identification was not made to the species level. Northern smoothtongue was the most abundant bathylagid. In decreasing order of importance, Scorpaenidae, Cottidae, Stichaeidae, Ammodytidae, Bathymasteridae, Cyclopteridae, and Pleuronectidae were captured, although not in high abundances. Clupeidae were expected but were not collected.

In a more restricted area of southeast Alaska, Auke Bay (Haldorson et al. 1993), the five dominant taxa in spring 1986 to 1989 were Osmeridae (not identified to species), Pacific sand lance, walleye pollock, flathead soles and rock sole *Pleuronectes bilineatus*. All of the dominant taxa in our study, except rockfishes, also were collected in Auke Bay, however northern smoothtongue and Pacific herring were not among the top five in the 4 years sampled. Both in the Gulf of Alaska and in southeast Alaska studies, walleye pollock larvae predominated as in Prince William Sound.

Circulation of Prince William Sound and the Exxon Valdez Oil Spill

The general circulation of Prince William Sound is thought of as flow inward through Hinchinbrook Entrance in the southeast and flow outward through Montague Strait in the southwest, with additional fresh water contributed from streams and glacier melt within the sound. Thus, oil spilled at Bligh Reef should only affect the western sound. According to overflight sightings, the oil spread throughout the western portion of the sound without affecting the eastern sound, which is consistent with a simulated drift model based on the strong winds experienced from 26 to 28 March (Venkatesh 1990). These strong northerly winds played a significant role in pushing the oil out of Prince William Sound and covering much of the Naked Island and Montague strata in its wake. After 10 d, most of the surface slick was out of the sound, yet oil was still leaking from the tanker (Royer et al. 1990) and continued to leach from beaches for many months. These strong winds only persisted for days (Royer et al. 1990). Under model simulations (Venkatesh 1990) of reduced winds, the oil is carried further west. With no wind but only residual circulation, the oil remains in the sound and affects the western islands and shore. Satellite-tracked drifters with drogues at 3-m depth were released in the sound from late 1989 to early 1990 to simulate the oil movement. Several of these drifters never left the sound (Royer et al. 1990), which supports Venkatesh's (1990) model with reduced wind.

Although the major flow is inward through Hinchinbrook Entrance, down-fjord winds that converge in the central sound can cause flow out through Hinchinbrook Entrance (Royer et al. 1990). Such flow was observed with an Acoustic Doppler Current Profiler (ADCP) in April 1989 (Norcross et al. 1991). The same effect, flow out through Hinchinbrook Entrance, can also occur when residual currents are removed from the simulated drift model (Venkatesh 1990). Under that scenario, the flow of the oil is almost straight south forced by the prevailing winds. Although the residual currents were never actually completely eliminated as in the model, they were reduced at the time of the spill. The *Exxon Valdez* oil spill occurred at the minimum in the seasonal circulation cycle. Additionally, the freshwater discharge that drives the residual currents was the lowest recorded discharge in 59 years in 1989 (Royer et al. 1990).

Distribution of Larval Fishes in Relation to Oil Distribution

The pattern of oil spilled from the *Exxon Valdez* in the days following the spill was easily traced by aircraft overflights (Wolfe et al. 1996, this volume). These sightings were used to produce maps of gross movement of the heaviest parts of the slick. Based on the sightings, the oil appeared to be present only in the western portion of the sound in our Naked Island and Montague strata.

In April, the time period covered by overflight maps, there were few larvae. Those present were northern smoothtongue that were concentrated in the non-oiled north stratum (Table 3; Figure 3). However, there are two complication factors. Because fish larvae are planktonic, those in the upper waters are expected to move with the same flow patterns that affect the trajectory of the oil. Also, even though oil was no longer visible from the air, toxins from the *Exxon Valdez* oil were still present to a depth of at least 25 m throughout the path of the spill in Prince William Sound for the first few months following the spill (Short and Harris 1996, this volume). Additionally, oil continued to be present on beaches (Wolfe et al. 1996) and in subtidal sediments (Braddock et al. 1996, this volume; O'Clair et al. 1996, this volume) for at least 2 years. Thus, after the initial dispersal of the visible slick and in places where there was no documented slick, oil was still available to the biota. The dispersed oil may have been a significant source of oil exposure to larval fishes (Short and Harris 1996).

In May when larval fishes were the most abundant, they were found in every stratum, but the highest densities, mainly of walleye pollock, were in southeast stratum (Table 3). Based on flow patterns of the sound, larvae from the non-oiled southeast stratum probably passed through the sound via the oiled waters of Montague stratum. The next most abundant fishes in May, northern smoothtongues and Pacific herring, were concentrated mainly in oiled Montague and Naked Island strata.

In May and June walleye pollock were found in the western sound in the strata through which the oil had passed. The numbers and sizes of walleye pollock larvae in Prince William Sound and Resurrection Bay in April and May 1989 (Müter and Norcross 1994) indicate spawning occurred during this time in both places. The size of larvae in May indicates they are less than 25 d old (Müter and Norcross 1994). Walleye pollock eggs are laid deep in the water column and rise to near the surface (Kendall and Kim 1989). In spring 1989 walleye pollock were concentrated above the pycnocline in the upper 25 m of the water column; therefore, walleye pollock could have been exposed first as embryos and then as larvae to oil from the *Exxon Valdez* spill.

Exposure of walleye pollock embryos to water-soluble fractions of oil can slow initial development, produce shorter larvae, and cause morphological abnormalities in larvae, including body curvatures, deformations of eye, brain, jaw, and intestine and fin erosion (Carls and Rice 1990). Average size of walleye pollock at hatch is 3 to 4 mm (Matarese et al. 1989). The smallest size of walleye pollock larvae that we collected was 3.13 mm. Oil and other stresses may produce premature hatch. Larval survival is complicated by both natural and anthropogenic effects, and there are no data from a non-oil-spill year to compare larval walleye pollock abundance in Prince William Sound to assess abundance changes. Thus, it is not possible to determine an effect of the *Exxon Valdez* oil spill on walleye pollock larvae from the available data.

The second most abundant fish, northern smoothtongue, occurred throughout Prince William Sound, with high abundances found mostly in the western portions of the sound. This species is a deep-sea smelt that coexists with walleye pollock and Pacific hake (*Merluccius productus*) in the Strait of Georgia, British Columbia (Mason and Phillips 1985). The eggs develop and hatch below 200 m, and following hatching, larvae ascend to approximately 40 to 90 m below the surface. Northern smoothtongue larvae metamorphose at approximately 22 mm at 90 d. By 120 d and 26 mm, the juveniles descend to depths greater than 150 m (Mason and Phillips 1985). Young larval northern smoothtongue were found in the upper 100 m, but were most abundant at 25 to 50 m of the water column in April, May, and June after the oil spill. No samples deeper than 100 m were taken during the first 3 months. By July northern smoothtongue were throughout the water column but were concentrated at 50- to 200-m depth and had increased in size.

Though northern smoothtongue is known to have very slow growth (0.16 mm/d) (Mason and Phillips 1985), there was no apparent change in modal size between April and May in Prince William Sound in 1989 and still little by June. This apparent lack of growth could be attributed to continued spawning, though a kurtotic curve would be expected in that instance. It may be that the larval cohort from April is missing and was replaced by a new cohort appearing in the May collections. Because there is only limited information about northern smoothtongue and no information about the effect of oil on this species, it is difficult to draw any conclusions about the effect of the *Exxon Valdez* oil spill.

The larval Pacific herring distributions found are in agreement with the general oceanographic circulation of Prince William Sound (Royer et al. 1990; Venkatesh 1990). Although Pacific herring spawned in April, Pacific herring larvae were found only in open waters of Prince William Sound in limited numbers at the beginning of May 1989. The highest concentrations of Pacific herring larvae were adjacent to or downstream from oiled spawning sites (Norcross et al., unpublished). By June Pacific herring were more abundant, well distributed in the sound, and found in greatest numbers in the waters flowing out through the Montague Strait region. Pacific herring larvae were predominantly in the upper layers above the pycnocline. As horizontal distribution of larvae is largely determined by upper layer flow (Müter and Norcross 1994), Pacific herring from all spawning locations probably passed through this region in a manner analogous to the flow of oil.

One cohort of Pacific herring larvae came from the major spawning that occurred between 31 March and 20 April and hatched between 30 April and 10 May (McGurk et al. 1990). The estimated change in size of 0.15 mm/d presented here is approximate because it is based only on increments in modal length over 27 d between the May and June cruises; however, it represents one of the lowest growth rates recorded for Pacific herring when compared to 0.31 to 1.48 mm/d (Wespestad and Moksness 1990), 0.30 mm/d (McGurk et al. 1993), and 0.18 to 0.29 mm/d (McGurk 1984a). The small sizes of larvae collected in May could be attributed to exposure of Pacific herring eggs to oil immediately after the oil spill (McGurk and Brown, in press). It is difficult to explain the apparent lack of size increase from June to July without invoking oil as an explanation (Norcross et al., unpublished).

When Pacific herring eggs are stressed by delayed spawning, physical conditions, or toxicity, the time to hatching decreases (Hay 1986). The emerging larvae may be premature and less viable. Exposure of herring eggs to petroleum is a stressor that has been noted to retard hatching and development (Struhsaker et al. 1974; Lindén 1978; von Westernhagen 1988) and reduce subsequent growth of herring larvae (Lindén 1978; Smith and Cameron 1979). Exposure of eggs to Prudhoe Bay crude oil for 12 h or longer reduced the size of newly hatched larvae (Smith and Cameron 1979). Larvae hatched in uncontaminated waters may be adversely affected when encountering an oil spill during larval drift (von Westernhagen 1988). Pacific herring exposed in the larval stage to oil are foreshortened with compressed myomeres and damaged notochords (Urho 1991), have inhibited feeding and swimming abilities (Struhsaker et al. 1974), and starvation exacerbates the negative effects of oil (von Westernhagen 1988). Some deformations of larvae persisted even after oil was no longer detectable (Urho 1991).

Jaw malformation seen in Pacific herring larvae in Prince William Sound in 1989 (Norcross et al., unpublished) may have restricted the feeding ability of Pacific herring larvae. Incidence of jaw malformation appeared to be related to smaller-sized herring larvae. Thus, the small lengths of larvae captured in May may be caused in part by poor growth because of physical inability to feed. The time from hatching to irreversible starvation is 28 to 37 d for Pacific herring larvae in water temperatures of 6 to 8°C (McGurk 1984b), such as those experienced in Prince William Sound in May of 1989. In the month between the May and June cruises, we assume these severely affected larvae died and were removed from the population. This is supported by the cytogenetic studies that revealed severe genetic defects in the Pacific herring larvae collected in May (Norcross et al., unpublished).

Capelins hatched after the heaviest concentrations of oil were out of the sound, and modal size continued to increase over time. Rockfishes did not appear to grow between April and September, but because these were analyzed as a genus complex this could be explained by sequential spawning of different species. Flathead soles showed continuous increase in size and decrease in numbers, indicating they were growing out of the planktonic stage. Because no more detailed data are available, an effect of oil on capelins, rockfishes, or flathead soles was not readily seen.

This study contributes basic larval fish occurrence information for Prince William Sound that can be used for future comparisons. Unfortunately, it is biased as a baseline because it represents the ecosystem following an oil spill. General physical oceanographic flow patterns are applied here to explain movements of larvae and oil within Prince William Sound during 1989. Still, there is not sufficient information to estimate population size nor to quatitate the effects of the *Exxon Valdez* oil spill (or a future oil spill) on the population structure of larval fishes in Prince William Sound.

Acknowledgments

We thank Doug Hay, Morgan Busby, Brenda Holladay, Franz Müter, and one anonymous reviewer for comments on this manuscript and Brenda Holladay for field collections, Milford Green for measurement of larval fish, and Mark Vallarino for technical support of the database. We also thank the crews and scientists aboard the RV *Alpha Helix*, the National Oceanic and Atmospheric Administration ship *John Cobb* and the FV *Jennie Girl*. Funding for this study was provided by Natural Resources Damage Assessment funds through the Alaska Department of Fish and Game RSA# EN1103435 and RSA# 11211167, the University of Alaska Natural Resources Fund, and the Institute of Marine Science at the University of Alaska Fairbanks. This chapter is contribution number 1540 of the Institute of Marine Science.

References

Braddock, J. F., J. E. Lindstrom, T. R. Yeager, B. T. Rasley, and E. J. Brown. 1996. Patterns of microbial activity in oiled and unoiled sediments in Prince William Sound. American Fisheries Society Symposium 18:94–108.

Carls, M. G., and S. D. Rice. 1990. Abnormal development and growth reductions of pollock *Theragra chalcogramma* embryos exposed to water-soluble fractions of oil. U.S. National Marine Fisheries Service Fishery Bulletin 88:29–37.

Falk-Petersen, I. B., and E. Kjorsvik. 1987. Acute toxicity tests of the effects of oils and dispersants on marine fish embryos and larvae: a review. Sarsia 72:411–413.

Fowler, D. A. 1988. BioQuant system IV. R & M Biometrics, Inc., Nashville, Tennessee.

Haldorson, L., M. Pritchett, D. Sterritt, and J. Watts. 1993. Abundance patterns of marine fish larvae during spring in a southeastern Alaskan bay. U.S. National Marine Fisheries Service Fishery Bulletin 91: 36–44.

Hart, J. L. 1973. Pacific fishes of Canada. Fisheries Research Board of Canada Bulletin 180.

Hay, D. E. 1986. Effects of delayed spawning on viability of eggs and larvae of Pacific herring. Transactions of the American Fisheries Society 115:155–161.

Kendall, A. W., and S. Kim. 1989. Buoyancy of walleye pollock (*Theragra chalcogramma*) eggs in relation to water properties and movement in Shelikof Strait, Gulf of Alaska. Canadian Special Publication of Fisheries and Aquatic Sciences 108:169–180.

Kuhnhold, W. W. 1974. Investigations on the toxicity of seawater-extracts of three crude oils on eggs of cod (*Gadus morhua* L.). Berichte der Deutschen Wissenschaftlichen Kommission fuer Meeresforschung 23: 165–180.

Kuhnhold, W. W. 1977. The effect of mineral oils on the development of eggs and larvae of marine species: a review and comparison of experimental data in regard to possible damage at sea. Rapports et Proces-Verbaux des Reunions Conseil International pour l'Exploration de la Mer 171:175–183.

Lindén, O. 1978. Biological effects of oil on early development of the Baltic herring, *Clupea harengus membras*. Marine Biology 45:273–283.

Mason, J. C., and A. C. Phillips. 1985. Biology of the bathylagid fish, *Leuroglossus schmidti*, in the Strait of Georgia, British Columbia, Canada. Canadian Journal of Fisheries and Aquatic Sciences 42:1144–1153.

Matarese, A. C., A. W. Kendall, Jr., D. M. Blood, and B. M. Vinter. 1989. Laboratory guide to early life history stages of northeast Pacific fishes. NOAA (National Oceanic and Atmospheric Administration) Technical Report NMFS (National Marine Fisheries Service) 80 Northwest and Alaska Fisheries Science Center, Seattle.

Mattson, C. R., and B. L. Wing. 1978. Ichthyoplankton composition and plankton volumes from inland coastal waters of Southeastern Alaska, April–November 1972. NOAA (National Oceanic and Atmospheric Administration) Technical Report NMFS (National Marine Fisheries Service) SSRF-723, Juneau, Alaska.

McGurk, M. D., and E. D. Brown. In press. Egg-larval mortality of Pacific herring in Prince William Sound, Alaska, after the *Exxon Valdez* oil spill. Canadian Journal of Fisheries and Aquatic Sciences.

McGurk, M. D., D. Warburton, and V. Komori. 1990. Early life history of Pacific herring: 1989 Prince William Sound herring larvae survey. National Ocean Service, Anchorage, Alaska.

McGurk, M. D., A. J. Paul, K. O. Coyle, D. A. Ziemann, and L. J. Haldorson. 1993. Relationships between prey concentration and growth, condition, and mortality of Pacific herring, *Clupea pallasi*, larvae in an Alaskan subarctic embayment. Canadian Journal of Fisheries and Aquatic Sciences 50:163–180.

McGurk, M. D. 1984a. Ring deposition in the otoliths of larval Pacific herring, *Clupea harengus pallasi*. U.S. National Marine Fisheries Service Fishery Bulletin 1:113–120.

McGurk, M. D. 1984b. Effects of delayed feeding and temperature on the age of irreversible starvation and on the rates of growth and mortality of Pacific herring larvae. Marine Biology 84:13–26.

Müter, F.-J., and B. L. Norcross. 1994. Distribution, abundance and growth of larval walleye pollock (*Theragra chalcogramma*) in an Alaskan fjord. U.S. National Marine Fisheries Service Fishery Bulletin 92:579–590.

Norcross, B. L., B. Holladay, M. Frandsen, and M. Vallarino. 1991. Injury to larval fish in Prince William Sound. Alaska Department of Fish and Game-OSIAR (Oil Spill Impact and Restoration) Division, Anchorage.

O'Clair, C. E., J. W. Short, and S. D. Rice. 1996. Contamination of intertidal and subtidal sediments by oil from the *Exxon Valdez* in Prince William Sound. American Fisheries Society Symposium 18:61–93.

Paul, A. J., J. M. Paul, K. O. Coyle, and R. L. Smith. 1991. Phytoplankton, zooplankton, and ichthyoplankton in Resurrection Bay, northern Gulf of Alaska in 1988. Alaska Sea Grant Report AK-SG-91-02, Fairbanks.

Rice, S. D., S. Korn, and J. F. Karinen. 1978. Lethal and sublethal effects on selected Alaskan marine species after acute and long-term exposure to oil and oil components. NOAA (National Oceanic and Atmospheric Administration) NMFS (National Marine Fisheries Service) OCSEAP Final Report 1, Boulder, Colorado.

Rice, S. D., A. Moles, T. L. Taylor, and J. F. Karinen. 1979. Sensitivity of 39 Alaskan marine species to Cook Inlet crude oil and number 2 fuel oil. Pages 549–554 *in* Proceedings, 1979 oil spill conference (prevention, behavior, control, cleanup). American Petroleum Institute Publication 4308, Washington, DC.

Rice, S. D., and seven coauthors. 1976. Acute toxicity and uptake-depuration studies with Cook Inlet crude oil, Prudhoe Bay crude oil, number 2 fuel oil and several subarctic marine organisms. National Marine Fisheries Service, Northwest Fisheries Center Processed Report, Seattle.

Royer, T. C., J. A. Vermersch, T. J. Weingartner, H. J. Niebauer, and R. D. Muench. 1990. Ocean circulation influencing the *Exxon Valdez* oil spill. Oceanography 3:3–10.

Rugen, W. C. 1990. Spatial and temporal distribution of larval fish in the western Gulf of Alaska, with emphasis on the period of peak abundance of walleye pollock (*Theragra chalcogramma*). Unpublished data report, Northwest Alaska Fisheries Center Processed Report 90-01, Seattle.

Smith, R. L., and J. A. Cameron. 1979. Effect of water-soluble fraction of Prudhoe Bay crude oil on embryonic development of Pacific herring. Transactions of the American Fisheries Society 108:70–75.

Short, J. W., and P. M. Harris. 1996. Petroleum hydrocarbons in caged mussels deployed in Prince William Sound after the *Exxon Valdez* oil spill. American Fisheries Society Symposium 18:29–39.

Solbakken, J. E., S. Tilseth, and K. H. Palmork. 1984. Uptake and elimination of aromatic hydrocarbons and a chlorinated biphenyl in eggs and larvae of cod, *Gadus morhua*. Marine Ecology Progress Series 16: 297–301.

Spaulding, M. L., and seven coauthors. 1985. Oil spill fishery impact assessment model: sensitivity to spill location and timing. Estuarine Coastal and Shelf Science 20:41–53.

Struhsaker, J. W., M. B. Eldridge, and T. Echeverria. 1974. Effects of benzene (a water-soluble component of crude oil) on eggs and larvae of Pacific herring and northern anchovy. Pages 253–284 *in* F. J. Vernberg and W. B. Vernberg, editors. Pollution and physiology of marine organisms. Academic Press, New York.

Urho, L. 1991. Impact of an oil spill on herring stock. Alaska Sea Grant Report AK-SG-91-01, Fairbanks.

Venkatesh, S. 1990. Model simulations of the drift and spread of the *Exxon Valdez* oil spill. Atmosphere-Ocean 28:90–105.

von Westernhagen, H. 1988. Sublethal effects of pollutants on fish eggs and larvae. Pages 253–346 in W. S. Hoar and D. J. Randall, editors. Fish physiology, volume 11, part A. Academic Press, New York.

Wespestad, V. G., and E. Moksness. 1990. Observations on growth and survival during the early life history of Pacific herring, *Clupea pallasi*, from Bristol Bay, Alaska, in a marine mesocosm. U.S. National Marine Fisheries Service Fishery Bulletin 88:191–200.

Wiebe, P. H., and seven coauthors. 1985. New developments in the MOCNESS, an apparatus for sampling zooplankton and micronekton. Marine Biology 87: 313–323.

Wilkinson, L. 1988. SYSTAT: the system for statistics. SYSTAT, Inc., Evanston, Illinois.

Wolfe, D. A., and six coauthors. 1996. Toxicity of intertidal and subtidal sediments contaminated by the *Exxon Valdez* oil spill. American Fisheries Society Symposium 18:121–139.

American Fisheries Society Symposium 18:487–498, 1996

A Life History Approach to Estimating Damage to Prince William Sound Pink Salmon Caused by the *Exxon Valdez* Oil Spill

HAROLD J. GEIGER

Alaska Department of Fish and Game, Division of Commercial Fisheries Management and Development, P. O. Box 25526, Juneau, Alaska 99802, USA; and the University of Alaska Fairbanks School of Fisheries and Ocean Sciences, Division of Fisheries, Fairbanks, Alaska 99775, USA

BRIAN G. BUE AND SAM SHARR

Alaska Department of Fish and Game, 333 Raspberry Road, Anchorage, Alaska 99518, USA

ALEX C. WERTHEIMER

National Marine Fisheries Service, Auke Bay Laboratory 11305 Glacier Highway Juneau, Alaska 99801, USA

T. MARK WILLETTE

Alaska Department of Fish and Game, 401 Railroad Avenue, Cordova, Alaska 99574, USA

Abstract.—The 1989 *Exxon Valdez* oil spill resulted in contamination of the habitat of juvenile pink salmon *Oncorhynchus gorbuscha*, including approximately 31% of the spawning streams and much of the nearshore marine rearing habitat in the southwest portion of Prince William Sound. In this study we combined the results of other studies to evaluate the loss of returning wild adult pink salmon attributable to the *Exxon Valdez* oil spill. We estimated that approximately 1.9 million wild adult pink salmon failed to return to Prince William Sound in 1990, primarily because of a lack of growth in the critical nearshore life stage when they entered seawater in spring 1989 during the height of the spill. This number of fish was approximately 28% of the potential wild-stock production from the southwestern part of the sound returning in 1990. We estimated that 60,000 wild adult pink salmon failed to return in 1991 and 70,000 in 1992 because of direct poisoning of salmon in the embryo stage. In both 1991 and 1992 this was 6% or less of the potential wild-stock production in the southwestern part of the sound and less than 2% of the potential wild production in the entire sound.

In the confusion after the *Exxon Valdez* oil spill, anything from no effect to devastation of populations of pink salmon *Oncorhynchus gorbuscha* in the southwestern part of Prince William Sound seemed plausible. In the weeks after the spill, the Alaska Department of Fish and Game and the National Marine Fisheries Service initiated several studies to (1) develop improved estimates of pink salmon escapement for pre- and postspill years, (2) measure egg (embryo) survival, (3) measure fry survival, (4) observe the nearshore condition and distribution of fry in Prince William Sound, (5) measure hatchery and wild salmon survival in the saltwater environment, and (6) survey the pink salmon habitat affected by the oil spill.

These studies have documented contamination of pink salmon embryos in oiled streams (Wiedmer et al. 1996, this volume) and juvenile salmon in the nearshore environment (Carls et al. 1996, this volume). The contamination has been linked to reduced survival of embryos in oiled streams (Bue et al. 1996, this volume) and to reduced growth of pink salmon in the marine environment (Wertheimer and Celewycz 1996; Willette 1996; both this volume). The reduced early growth in the marine environment has, in turn, been linked to reduced marine survival (Willette 1996). Here we synthesize the results of these studies of individual life history stages and estimate the loss of returning wild adult Prince William Sound pink salmon in the first three years after the oil spill.

Pink salmon populations in Prince William Sound are not static. Powerful forces that may have affected them in recent years include the Alaskan earthquake of 1964, an increasingly intense commercial fishery, tremendous increases in hatchery production, and the *Exxon Valdez* oil spill. The earthquake of 1964 severely damaged some stocks and eliminated some runs entirely (Roys 1971). In the mid-1980s, large runs of hatchery-produced pink salmon began to equal and then far outnumber the wild salmon (Eggers et al., in press). Hatchery stocks are an important input to the commercial fishing industry, but from an ecological perspective they are entirely different from the wild salmon. In this report we focus on the wild salmon, which play an important role in the Prince William Sound ecosystem and an

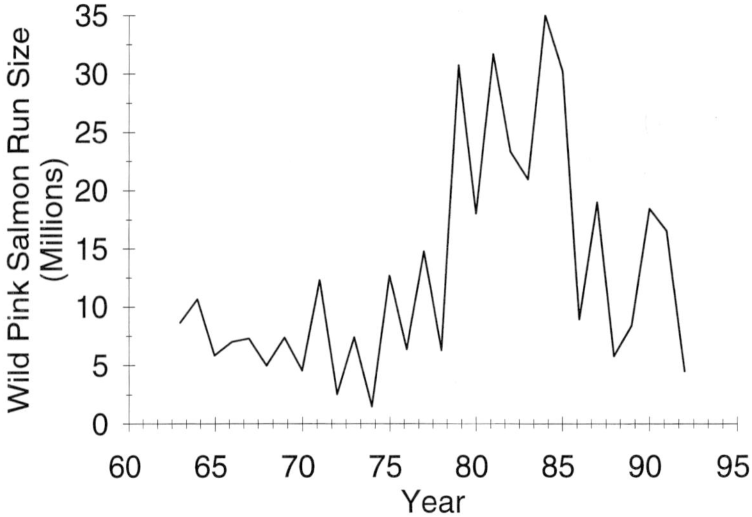

FIGURE 1.—The estimated size of Prince William Sound wild pink salmon runs from 1963 to 1992.

important economic role in the Prince William Sound fishing industry.

The size of pink salmon runs to Prince William Sound has fluctuated wildly (Figure 1) in response to unobserved phenomena in the freshwater and marine environments. With this much background variability, even extremely large damage would be entirely undetectable using before-and-after comparisons of any function of population size. To gain statistical power we focused on the affected life history stages, looking for differences in life history statistics between groups of fish classified as oiled and non-oiled in the same year. We refer to the summary of the life stage comparisons as the life history model.

We followed the 1988 brood year by estimating parental escapement, survival to emergence, and survival to emigration in 1989; by measuring growth and survival of coded-wire tag groups as they migrated through the oil in the nearshore environment in spring 1989; and by measuring their final return numbers in 1990. We then continued with the odd-year line, following the 1989 brood year as they incubated in the oiled and non-oiled streams in Prince William Sound, migrated in the relatively non-oiled nearshore areas in 1990, and returned in 1991. The 1990–1992 brood years were followed in a similar manner up to summer 1992. At each life stage for the 1988–1992 brood years we summarized the evidence that the presence of oil reduced survival and the evidence that a survival reduction was translated to reduced survival at the adult life stage as well. Figure 2 illustrates the life history pattern for pink salmon, showing the life stages and the points where oil detectably reduced survival in wild pink salmon.

The oil spill could have reduced survival at an early juvenile life stage, yet the final returns of adult salmon have remained unaffected because of compensatory mortality, namely, mortality that increases when density is high and decreases when density is low. In her review of the pink salmon literature, Alexandersdottir (1987) cited Heard's (1978) work in Sashin Creek in southeastern Alaska in 1967 and Donnelly's (1983) work in Kodiak to conclude compensatory mortality was linked to an overabundance of spawners and density-dependent survival could be a factor from spawning to outmigration.

If compensatory mortality effects are negligible at a particular life stage and all subsequent life stages, then estimating the effects of the oil spill are straightforward: the estimated proportional loss of production can simply be the measured decrease in estimated survival at that life stage. Alternatively, if compensatory mechanisms exist, the loss of production from the oil spill will be overestimated by assuming oil-induced mortality translates directly into reduced adult production. We examined the available evidence that compensatory mortality could be caused by oil-induced loss of production.

The effect of the *Exxon Valdez* oil spill on any important part of the biota will never be known exactly. Unfortunately, the failure to precisely mea-

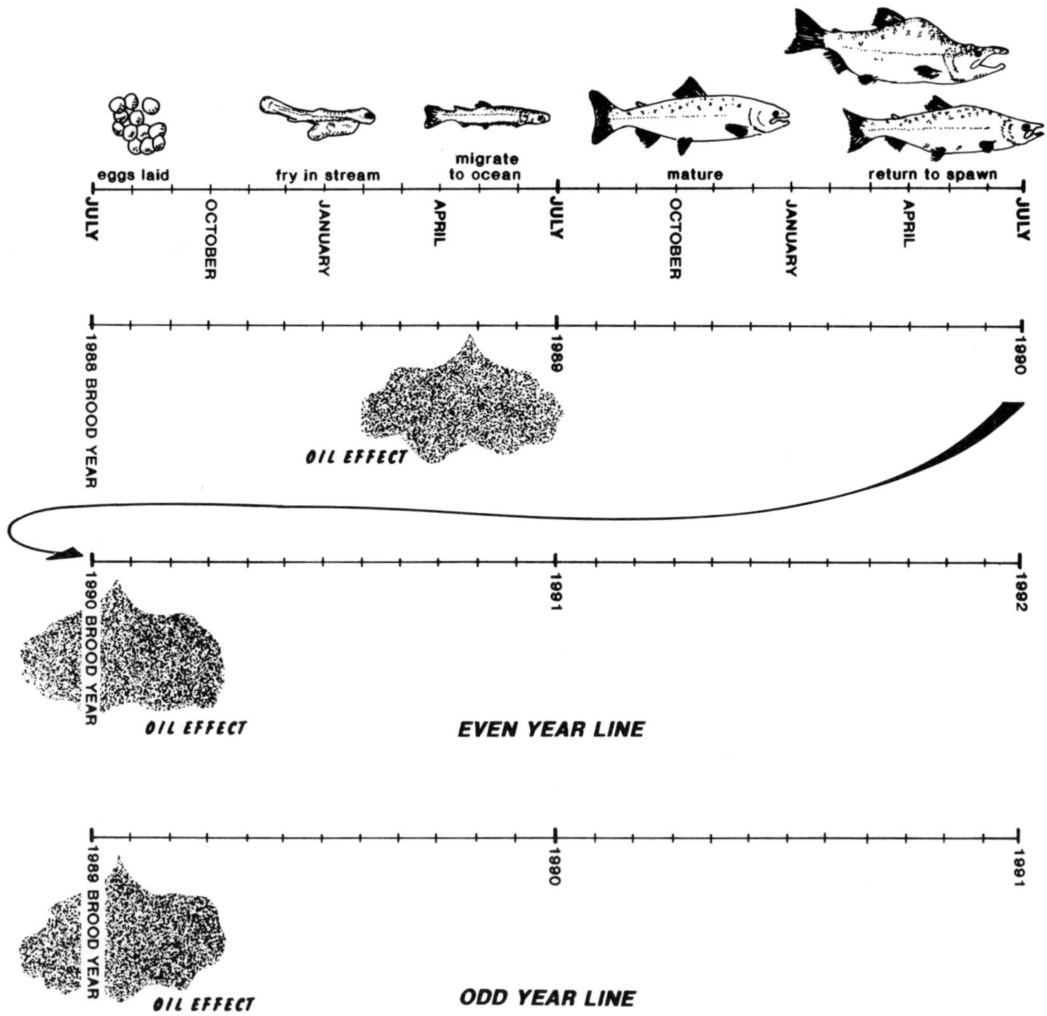

FIGURE 2.—Pink salmon life history stages. Stages affected by oil in Prince William Sound in brood years 1988, 1989, and 1990 are indicated.

sure injury from the oil spill is sometimes confused with justification for concluding the injury is nonexistent. Similarly, the ability to statistically detect some aspect of injury is sometimes confused with significance in a larger biological sense. Rather than perform a statistical test of the hypothesis that pink salmon populations were completely unchanged by the oil spill, we viewed our responsibility as to identify the biologically—not statistically—important sources of mortality and to provide the most reasonable estimates of the magnitude of loss, in units of adult salmon, consistent with the information available to us.

Methods

Prince William Sound Fishing Districts, Stocks, and Stock Size Estimates

Prince William Sound is divided into nine fishing districts, which correspond to stocks for the purposes of fisheries management (Figure 3). In much of our discussion we will be restricting our focus to the southwestern fishing district, where the greatest damage to freshwater spawning habitat occurred (Middleton et al. 1992). We used total production and revised estimates of district-specific escapement from a database maintained by the Alaska

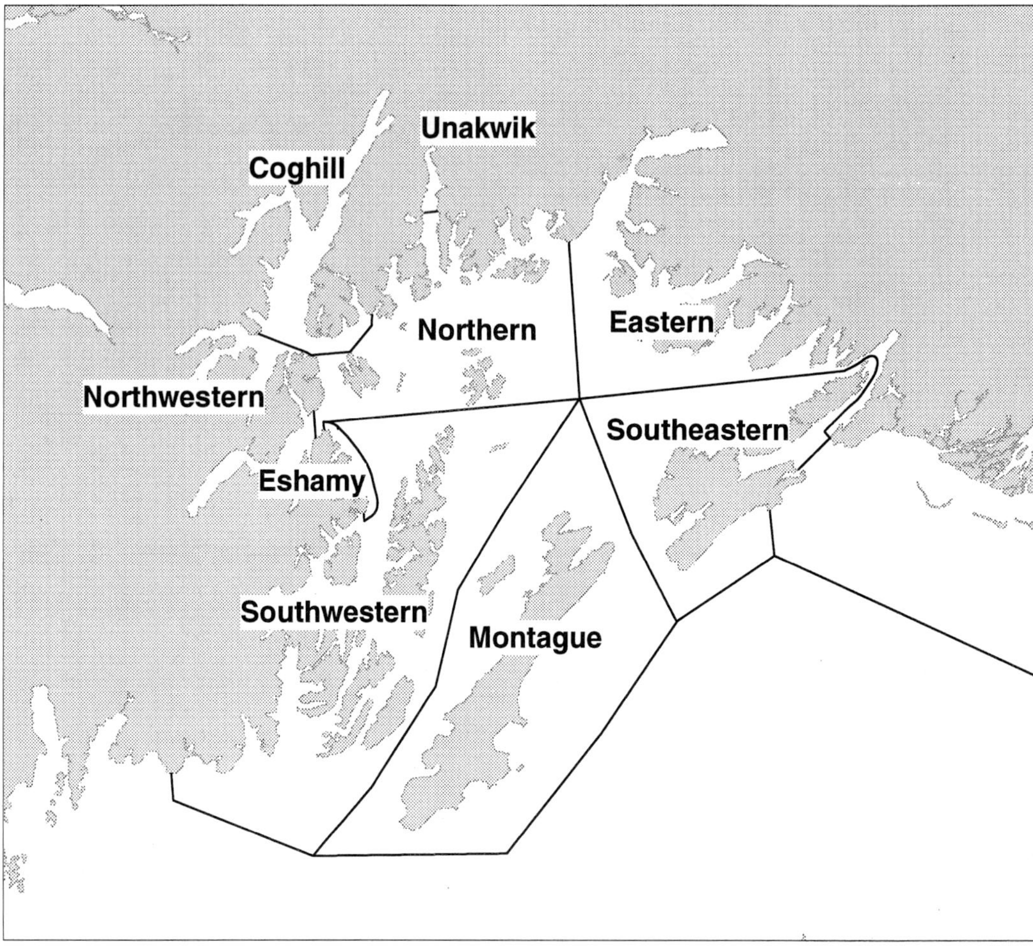

FIGURE 3.—Map of Prince William Sound showing the fishing districts.

Department of Fish and Game in Cordova, Alaska. These data were the most current as of June 1993.

Life History Parameters, Statistics, and Estimators

Let y_{t+2} represent the subsequent Prince William Sound wild pink salmon production in year $t + 2$ from an escapement of x_t spawners in brood year t. Let s_t represent the random variable that summarizes survival and fecundity. This is commonly called return per spawner and is usually expressed as some function of previous escapement (e.g., Ricker 1975). We are interested in the life history steps underlying the return per spawner when escapement has been restricted to approximately what managers considered optimal. Formally, the process is described by

$$y_{t+2} = s_t x_t. \quad (1)$$

In year $t + 2$, Prince William Sound salmon management attempts to harvest all of the return except a fixed escapement goal, x_{goal}. Because of management error, the actual escapement is x_{t+2}, and the process repeats.

For years when the oil spill reduced survival, let p_t be the fraction of the total number of eggs from brood year t that were potentially affected by oil sometime during that and subsequent life stages. Let s'_t be the return per spawner reflecting the reduction attributable to the oil spill, including the effects of compensatory mortality.

For years when a fraction of the potential production is reduced by an oil spill effect, we assume production is given by

$$y_{t+2} = [(1 - p_t)s_t + p_t s'_t]x_t. \tag{2}$$

Notice that the quantity $(1 - p_t)$ is the fraction of eggs unaffected by oil, and s_t is the return per spawner, so that $(1 - p_t)s_t x_t$ denotes the production from non-oiled streams. Similarly, $p_t s'_t x_t$ denotes the production from oiled streams. Production from both sources equals the total production.

In actual practice, managers observe $\hat{x}_t$, some function of x_t, where the function describes observational error. Similarly, $\hat{y}_{t+2}$ denotes the estimated total run, consisting of the catch, which is approximately known, and the estimated escapement.

To estimate parameters, we first assume no strong compensatory mortality mechanisms, or at least no compensatory mortality after the affected life stage. Without reference to stochastic models, we estimate parameters in equations (1) and (2) algebraically, assuming that s'_t is approximately equal to s_t multiplied by the estimated reduction in survival.

Letting r_t denote the product of reduction in survival at each affected life stage, estimates of s'_t and s_t are found as follows by replacing elements in equation (2) with their estimates and solving

$$\hat{s}'_t = \hat{s}_t \hat{r}_t \tag{3}$$

and then from equation (2)

$$\hat{s}_t = \frac{\hat{y}_{t+2}}{\hat{x}_t}[(1 - \hat{p}_t) + \hat{p}_t \hat{r}_t]^{-1}. \tag{4}$$

Production, expressed in units of adult salmon that would have returned had there been no oil spill, is then estimated by

$$\hat{y}_{\text{no oil}, t+2} = \hat{s}_t \hat{x}_t, \tag{5}$$

and the estimated loss in brood year t, at the adult population level, is given by

$$\hat{L}_t = \hat{y}_{\text{no oil}, t+2} - \hat{y}_{t+2}. \tag{6}$$

To estimate typical values of s_t, for the purpose of projecting future production from the 1991 and 1992 escapements, we considered the median recruits-per-spawner from 1963 to 1987. Information on the steps that lead to s_t from the pink salmon literature was used to provide additional justification for this estimate. Five stochastic steps lead to the return-per-spawner value, s_t: (1) the fraction of the run consisting of females, denoted Fem_t; (2) the average fecundity per female, denoted, Fec_t; (3) the fraction of the potential egg deposition that is actually deposited, denoted Dep_t; (4) the survival from eggs deposited to emergence, denoted Seg_t; and (5) the marine survival, denoted Smr_t. The overall return per spawner, s_t, is simply the product of each of these stochastic outcomes:

$$s_t = \text{Fem}_t \times \text{Fec}_t \times \text{Dep}_t \times \text{Seg}_t \times \text{Smr}_t. \tag{7}$$

Fraction of Potential Egg Deposition in Oiled Streams

Estimates of the fraction of eggs in oiled streams were obtained by estimating the fraction of streams affected by oil and adjusting for the relative egg density in oiled streams. We assumed that only streams in the southwestern fishing district were affected, ignoring some oiling on Montague Island and elsewhere.

To estimate the fraction of pink-salmon-spawning streams affected by oil, we needed information on oiling level and egg density for each stream in the study. In 1991 the aerial survey of spawning escapement was expanded to include a representative sample of oiled and control streams. To determine which streams in the 1991 aerial survey were affected by oil, we referred to Middleton et al. (1992), personal field observations, and conversations with other observers. In general, if a stream was classified as "heavy" through "very light" in Middleton et al. we considered it an oiled stream. In cases where Middleton et al. disagreed with our field observation, we relied on our observation. The peak spawner density from this survey was used as a proxy for the egg density in each stream in this survey. To estimate the fraction of the potential egg deposition in the southwestern fishing district that incubated in oiled streams, we calculated the weighted average of the proportion of streams classified as oiled in the 1991 survey, using the peak aerial escapement count in the 1991 aerial survey as the weights. Then, to estimate the fraction of the entire potential egg deposition in the whole sound affected by oil, we multiplied this weighted average by the fraction of the entire escapement that was in the southwestern fishing district.

Oil-Induced Mortality in the Freshwater Stage

Let m_v be the underlying mortality rate of juveniles, at some life stage, with oiling level v. If this is the only affected life stage, then the proportional reduction in survival at this life stage, defined as r above, is found by

$$r = \frac{(1 - m_{\text{oil}})}{(1 - m_{\text{unoil}})}. \tag{8}$$

To estimate r, we substituted the unweighted estimated average mortality in the embryos, using data described in Bue et al. (1996). Average mortality was not weighted by total eggs deposited at tide stage because estimates of total eggs deposited by tide stage were unavailable to us.

Fraction of Eggs Exposed in the Early Marine Stage

The fraction of eggs that could be affected by exposure to oil or whose fry could eventually be affected through early occupation of the marine environment was the most difficult parameter to estimate. In the end, we simply used the estimated fraction of the 1988 escapement that was in the southwestern fishing district, knowing that fry that emerged anywhere in the southwestern fishing district in 1989 were a short distance from exposure once they emigrated into the marine environment. We know from Willette (1996) that some fish from non-oiled areas emigrated into oiled marine areas and vice versa.

Results

Willette (1996) estimated that fry from non-oiled areas that were reared in oiled nearshore environments in 1989 experienced a marine survival of 72% of their non-oiled potential based on coded-wire tagging studies from the Wallace H. Noerenberg Hatchery.

We estimated that the survival from egg deposition to the fry stage was 92% of its potential in oiled areas for the 1989 brood year and 88% of its potential for the 1990 brood year (Table 1). Bue et al. (1996) found statistically detectable excess mortalities in the eyed-embryo stage, which they attributed to oil spill effects, but they failed to find statistically detectable effects in the preemergent fry stage of development. Our estimates of the proportional reduction in survival of oiled eggs relative to control eggs are reflected in the data in Table 2; also shown are estimates of overall embryo-stage mortality for oiled and non-oiled streams. Statistically detectable elevated mortalities were also noted by Bue et al. (1996) between oiled and non-oiled sites for the 1991 and 1992 brood years.

We estimated the fraction of pink salmon spawning streams contaminated with oil to be 31% of all spawning streams in the southwestern fishing district. Non-oiled streams had larger average peak escapement counts than oiled streams, meaning that 31% is expected to overstate the proportion of eggs affected in the southwestern fishing district. When we used peak spawner density from the 1991 areal escapement survey as a proxy for egg density,

TABLE 1.—Estimated overall embryo mortalities from oiled and control sites in Prince William Sound and estimated survival in oiled areas as percent of potential survival, by brood year. The estimates are based on unweighted average mortalities by tide stage from data described by Bue et al. (1996).

Brood year	Oiled sites	Control sites	Survival in oiled areas as percent of potential survival
1989	0.174	0.104	92%
1990	0.295	0.195	88%
1991	0.433	0.221	73%
1992	0.450	0.250	73%

as described above, an estimated 20% of the potential egg deposition in the southwestern fishing district was associated with streams classified as oiled. Using escapement information to estimate the fraction of the potential egg deposition in the entire sound potentially affected by oil in the freshwater stage, the estimate never exceeded 6% of the eggs (Table 2).

Estimated Adult Salmon that Failed to Return

In 1990, total production of wild pink salmon was estimated to be 18.40 million from an escapement of 4.99 million in the 1988 brood year. We assumed that 32% of the potential 1988 brood year was exposed to oil effects, using the proportional escapement in the southwestern fishing district, as described above. Based on the observed reduction of marine survival of coded-wire tagged fish from Willette (1996), we estimated the proportional reduction in survival to be 72%, as described above. Substituting these estimates for the appropriate quantities in equation (2) yields

$$18.40 = [(1 - 0.32)s_{1988} + 0.32(s_{1988} \times 0.72)] \times 4.99. \quad (9)$$

Solving for s_{1988} produced an estimated return per spawner in the absence of oil of 4.06. Applying this return per spawner to the 1988 escapement produced an estimate of 20.26 million adults that would have returned in the absence of oil. By subtraction, the estimated loss of adult pink salmon in 1990 was approximately 1.86 million. Key statistics and results for the 1988–1990 brood years are provided in Table 2. Using the same reasoning, we estimate the loss of adult pink salmon to be approximately 60,000 and 70,000 for the 1989 and 1990 brood years, respectively.

TABLE 2.—Estimates of oil-spill-induced losses of adult pink salmon production in Prince William Sound and other important statistics for brood years 1988, 1989, and 1990, and forecast future losses for brood years 1991 and 1992 based on the assumption that the genetic damage hypothesis is correct. Fish numbers are expressed in units of million adult salmon. Lost production is in the return 2 years after the brood year. Definitions are PED = potential egg deposition and SW district = southwestern fishing district.

	Estimated for brood year:			Projected for brood year:	
Statistic	1988	1989	1990	1991	1992
Soundwide escapement (millions)	4.99	6.89	6.18	9.66	2.71
Escapement to SW district (millions)	1.61	1.91	1.77	2.21	0.81
Subsequent soundwide production (millions)	18.4	16.54	4.50		
Estimated return per spawner in the absence of oil	4.06	2.41	0.74	1.79[a]	1.79[a]
Percent of PED potentially affected by oil	32%	6%	6%	5%	6%
Estimated survival of oil-affected fish relative to non-oiled fish	72%	92%	88%	73%	73%
Estimated production without oil spill effects (millions)	20.26	16.60	4.57	17.30	4.85
Estimated loss of adult fish (millions)	1.86	0.06	0.07	0.24	0.08
Loss as a percent of potential production in the SW district	28%	1%	6%	6%	5%

[a]Median value for brood years 1963–1987.

Compensatory Mortality

We found no evidence that compensatory mortality mechanisms operate on the scale of variation induced by oil spill effects. Figure 4 shows the new stock–recruitment history using the revised estimates of escapement. Although compensatory mortality mechanisms may control production in Prince William Sound at escapement extremes—such as when escapement reaches 20 million—for the important brood years of 1988, 1989, and 1990, the escapement remained very near what managers have traditionally thought of as optimum (the escapement goal in the old index units can be found in Geiger and Savikko 1991; an escapement goal has not yet been formulated in the units of the revised escapement estimates).

To see if compensatory mortality was operating in the embryo stage, we examined estimated egg survival, up to the time the survey was conducted, as a function of observed egg density for 1989, 1990, and 1991. Although some slight compensatory mortality might have occurred as densities exceeded 2,500 eggs per unit of surface area (0.186 m^2) in the lower tide stages in 1989, overall no strong compen-

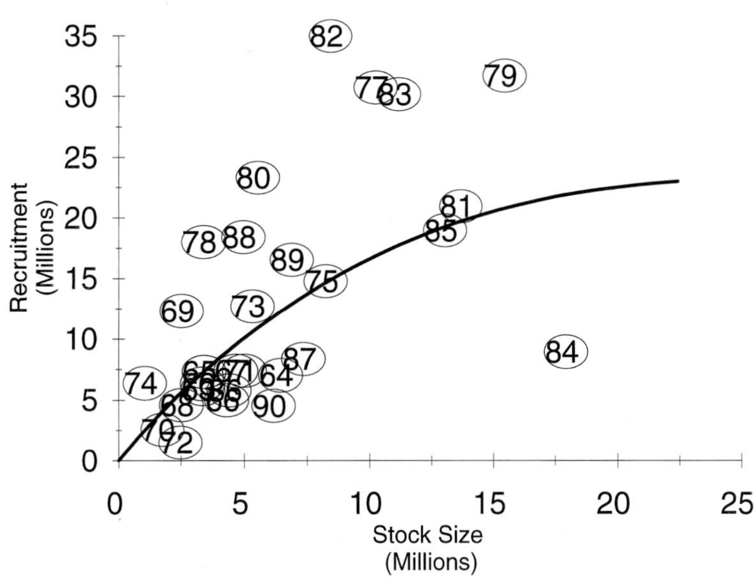

FIGURE 4.—Stock–recruitment history for Prince William Sound pink salmon. The numbers denote the brood year.

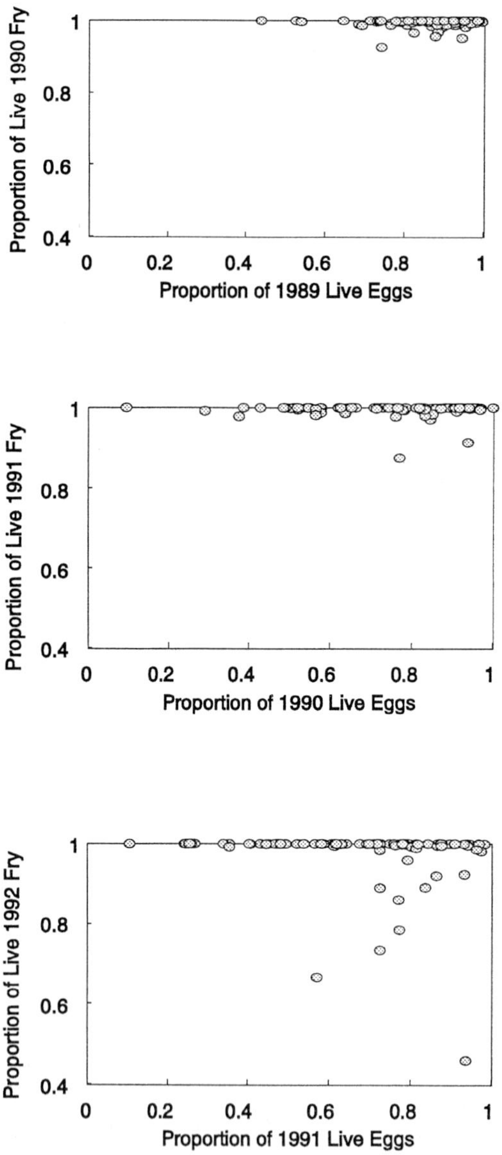

FIGURE 5.—Proportion of live fry paired with the previous proportion of live eggs for years 1989–1991 in Prince William Sound.

sation was observed. We were unable to detect a relationship between egg density and the empirical logit (Agresti 1990) of the egg survival (analysis of covariance, $P = 0.3135$) after adjusting for year, tide stage, and oiling level.

If oil-induced mortality is strongly compensated for in the preemergent fry stage, then the proportion of live fry would be strongly related, with negative slope, to the proportion of live eggs. Figure 5 shows this relationship for the 1989–1991 fry dig studies. These graphs show considerable mortality and variation in mortality from site to site, yet the measured mortality is low in the fry stage, irrespective of the mortality between egg deposition and the fry stage. Thus, no compensatory mortality was seen in this life stage.

To look for compensatory mortality in the marine life stages, we examined Prince William Sound pink salmon hatchery production as a function of fry released (from Eggers et al., in press, and recent estimates from Prince William Sound hatchery operators). This relationship is shown in Figure 6. The number of fry emigrating from freshwater systems shows a strong linear relationship to the number of returning adults, within the range of 0 to approximately 600 million hatchery fry. The slope of a smooth line through this data is approximately 5%—a reasonable average survival for pink salmon (Heard 1991). These data indicate no compensatory survival mechanisms in the marine life stages.

In summary, we found no evidence that oil-induced mortality in either the freshwater or the marine environment would be compensated for by density-dependent mechanisms in Prince William Sound pink salmon populations.

Typical Life History Values and the 1991 and 1992 Brood Years

The annual return per spawner is a random variable in equation (1). This random variable results from the realization of five previous stochastic steps, as described above. Table 3 shows our best estimate of approximate typical values for life history parameters at each of these steps from various sources in the pink salmon literature and from postspill unpublished coded-wire tagging studies of wild pink salmon (data available from the Alaska Department of Fish and Game, Cordova). These estimates led to the impression that typically about 1.65 recruits per spawner are generated from an escapement near the optimum level.

Using the revised estimates of escapement, the median return per spawner from the 1963 to 1987 brood years was 1.79. To ensure that this statistic represents years with moderate-to-low escapement, brood years with an escapement in excess of 10 million spawners were excluded, yet the median return per spawner remained unchanged.

The life history parameters from Table 3 lead to a similar picture of the typical return-per-spawner values that have been observed in the early 1990s in Prince William Sound. However, this image is quite

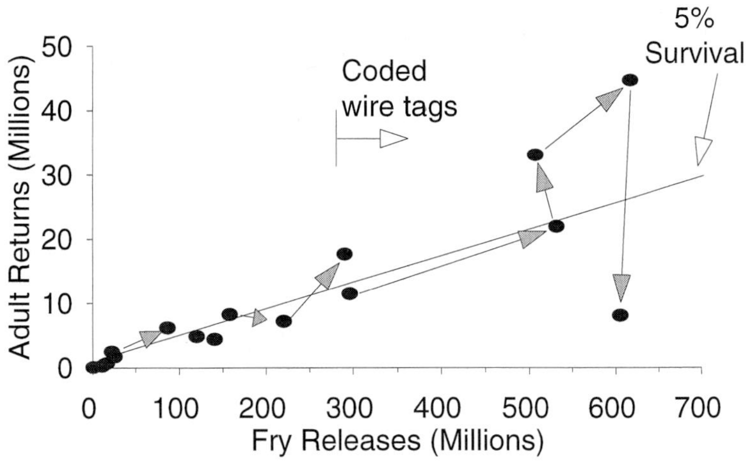

FIGURE 6.—Production of pink salmon from Prince William Sound hatcheries as a function of fry released. Coded-wire tags were first introduced into the estimation process in 1987.

different from the one based on previous escapement estimates (Heard 1991; Eggers et al. in press), where the typical return per spawner is thought to be closer to 4.0. We assumed 1.79 returns per spawner as a typical value, in the absence of oil spill effects, for projection of future production at the time this report was originally prepared in June 1993. The projected returns for the 1991 and 1992 brood years are 17.3 million and 4.9 million, in the absence of oil spill effects. If the estimated egg-stage survival ratio between oiled and control areas represents persistent oil spill effects, the projected return is reduced by approximately 240 thousand and 80 thousand for the 1991 and 1992 brood years, respectively.

Discussion

The studies we examined indicate that the *Exxon Valdez* oil spill resulted in a detectable loss of wild adult pink salmon. That conclusion by itself is not of much value without some measure of the magnitude of the loss and a context in which to view the magnitude. We estimated that more that 1 million adult pink salmon failed to return in 1990 because of oil-induced mortality. This loss may have been more than one-quarter of the potential wild production in the southwestern fishing district. Unless we have considerably underestimated the injuries in the 1989 and 1990 brood years, the loss of adult salmon was measurable, but small relative to other human-induced mortality, such as that caused by the commercial fishery. We concluded that the proportion of adult pink salmon missing from these brood years was less than 10% of the potential wild production in the southwestern fishing district and less than 2% of the potential wild production in the entire sound.

The 1990 pink salmon harvest was a record catch

TABLE 3.—Life history survival and fecundity values from the pink salmon literature. Let Fem be the fraction of the run consisting of females, Fec the average fecundity per female, Dep the fraction of the potential egg deposition (PED) that is actually deposited (AED = actual egg deposition), Seg the survival from eggs to emergence, and Smr the marine survival; $s_t = \text{Fem}_t \cdot \text{Fec}_t \cdot \text{Dep}_t \cdot \text{Seg}_t \cdot \text{Smr}_t$, s being return per spawner and t designating year.

Life stage	Conversion	Next stage	Typical value	Source
Escapement	× Fem	= Females	Fem ≈ 0.5	Assumed
Females	× Fec	= PED	Fec ≈ 1875	Heard (1991)
PED	× Dep	= AED	Dep ≈ 0.44	Heard (1991)
AED	× Seg	= Fry	Seg ≈ 0.20	Estimated by eye from data in Ignell (1988)
Fry	× Smr	= Adults	Smr ≈ 0.02	Unpublished coded-wire tag data from M. Willette, Alaska Department of Fish and Game, Cordova
Escapement	× s	= Adults	$s \approx 1.65$	By multiplication

at approximately 44.2 million pink salmon (Geiger and Savikko 1991). The large harvest in 1990 legitimately helps make the point that the *Exxon Valdez* oil spill did not devastate or cripple the commercial fishery after the year of the spill. However, the commercial harvest in 1990 is a poor indicator of the effect of the oil spill on the wild pink salmon resource in the affected areas of Prince William Sound. The survival of the 1988 brood year was strengthened by favorable nearshore conditions for pink salmon fry—a conclusion based on measured zooplankton abundance (Jeff Olsen, Prince William Sound Aquaculture Corporation, personal communication). The size of the harvest of wild salmon is masked by the harvest of hatchery salmon; the latter fish play a completely different ecological role in Prince William Sound. Long before the oil spill, the total catch of salmon in Prince William Sound was expected to increase throughout the 1980s because of increasing hatchery production. In 1989, fry from the 1988 brood year of wild salmon were joined by 507 million hatchery fry, the second largest release of pink salmon fry into the sound at that time.

Even though the salmon hatcheries largely masked the effect of oil on the wild salmon resource that returned in 1990, the oil had a detectable effect on the hatchery resource as well (Willette 1996). Of the total 507 million hatchery pink salmon released in 1989, 160 million fry were from the Armin F. Koerning Hatchery, which sits just off the path of the oil in the southwestern portion of the sound. Based on the observed growth reductions in Willette (1996), the potential loss of the hatchery resource in 1990 is in units of millions of adult salmon. The overall estimated survival for the 1988 brood year, which returned in 1990, was 4.24% for the Armin F. Koerning Hatchery. The survival at the Wallace H. Noerenberg Hatchery, a similar hatchery, but farther from the path of the oil, had an estimated survival of 8.49% (unpublished data available from Sam Sharr, Alaska Department of Fish and Game, Cordova)—approximately twice as large. The estimated survivals in years without an oil influence for the Armin F. Koerning and Wallace H. Noerenberg hatcheries were 6.8% and 8.8%, respectively, in 1987 (Peltz and Geiger 1990); 4.52% and 4.99%, respectively, in 1991; and 2.08% and 0.94%, respectively, in 1992 (Alaska Department of Fish and Game, unpublished data). Although not compelling in themselves, these data certainly are not in conflict with presumed large numbers of missing adult salmon that would have returned to the Armin F. Koerning Hatchery in 1990.

We also concluded that some wild adult salmon that would have returned in 1991 and 1992 from the 1989 and 1990 brood years were lost as a result of the oil spill. The primary mechanism of the loss was decreased embryo-stage survivals. The presence of oil is known to elevate juvenile salmon mortality (Moles et al. 1987). This appears to have happened in the contaminated streams in Prince William Sound, yet relatively few streams were oiled—approximately 31% of the streams, by our criteria, in that part of Prince William Sound most affected by oil. The oiled streams tended to face north, and were generally in the southwestern part of the sound. The oiled streams had a lower density of spawners when the density was observed in 1991.

We compared our estimates of the spatial distribution of the oil contamination to those presented in other studies, including those sponsored by the Exxon Corp. Maki et al. (1995) estimated that 14% of the streams in the affected area were oiled, but they considered a larger affected area than we did, including the Eshamy and Montague fishing districts. We found no classification disagreement in our sample with that of Bue et al. (1996). We did find two streams (16760 and 16610) of 10 that were in both our sample and that of Brannon et al. (1995), which they considered non-oiled yet we considered lightly or very lightly oiled. Moderate errors in the spawner density in these streams, moderate errors in number of streams oiled, or even moderate errors in the ratio of survival in oiled and non-oiled streams would not have affected the larger conclusions about the result of egg-stage damage in the 1989–1992 brood years: the lost wild adult salmon probably were far less than 10% of the potential wild adult production in the southwestern fishing district and less than a few percent of the wild potential production in the entire sound.

The most serious criticism of these estimates is that our assumed fraction of fish exposed in the nearshore environment is little more than a guess. Most authors believe the most critical period for pink salmon survival is when the fish first enter the marine environment (e.g., Alexandersdottir 1987; Heard 1991). Carls et al. (1996) showed that juvenile salmon were contaminated, and Willette (1996) observed the reduced growth and survival that might have been predicted from an understanding of pink salmon biology. Even without an exact count of the number of fish exposed, because of the extent of the oiling of the marine environment we surmised that the greatest potential for damage to wild pink salmon resulting from the *Exxon Valdez* oil spill came from exposure in the critical nearshore environment.

The second most serious criticism of this study is that we may have missed large, delayed mortality in the marine life stage in brood years after 1988 and may have missed mortality in the fry stages in all brood years. Direct examination of preemergent fry in oiled streams by Wiedmer et al. (1996) showed damage in individual preemergent fry from oiled streams. Unfortunately, Wiedmer et al. provided no quantitative assessment.

The most important unanswered question about pink salmon in this study is why the estimated egg-stage mortality did not converge to control levels, but instead became more dissimilar in 1991 and 1992. The egg-stage effects we attributed to the oil spill could simply reflect natural differences in sites. Bue et al. (1996) explained why we think this is unlikely and speculate that the difference may be the result of genetic damage. This hypothesis seems a reasonable if untested explanation, but this explanation is far from certain. *Exxon Valdez* oil was shown to cause chromosome aberrations in larval herring that incubated in oiled areas in Prince William Sound, and these aberrations were reproduced in laboratory studies (Brown et al. 1996, this volume; Hose et al., in press).

Taking all these uncertainties into consideration, we have attempted to gauge the order of magnitude of the wild pink salmon injury for each brood year, leaving the question of the larger significance of this loss for someone else to answer. This study points to the large potential for loss of adult salmon resulting from contamination in the early marine life stage. It is this point in particular that we hope is not lost when planning for and evaluating future oil spills in Alaska.

Acknowledgments

We gratefully acknowledge Ray Hilborn and Phil Mundy for their technical leadership and support for pink salmon studies after the *Exxon Valdez* oil spill. Carey D. McAllister provided a critical review of an earlier draft of the manuscript.

References

Agresti, A. 1990. Categorical data analysis. Wiley Interscience, New York.

Alexandersdottir, M. 1987. Life history of pink salmon (*Oncorhynchus gorbuscha*) in southeast Alaska and implications for management. Doctoral dissertation. University of Washington, Seattle.

Brannon, E. J., L. L. Moulton, L. G. Gilbertson, A. W. Maki, and J. R. Skalaki. 1995. An assessment of oil spill effects on pink salmon populations following the *Exxon Valdez* oil spill—part 1: early life history. Pages 548–584 *in* P. G. Wells, J. N. Butler, and J. S. Hughes, editors. *Exxon Valdez* oil spill: fate and effects in Alaskan waters. American Society for Testing and Materials Special Technical Publication 1219, Philadelphia.

Brown, E. D., and seven coauthors. 1996. Injury to the early life history stages of Pacific herring in Prince William Sound after the *Exxon Valdez* oil spill. American Fisheries Society Symposium 18:448–462.

Bue, B. G., S. Sharr, S. D. Moffitt, and A. K. Craig. 1996. Effects of the *Exxon Valdez* oil spill on pink salmon embryos and preemergent fry. American Fisheries Society Symposium 18:619–627.

Carls, M. G., A. C. Wertheimer, J. W. Short, R. M. Smolowitz, and J. J. Stegeman. 1996. Contamination of juvenile pink and chum salmon by hydrocarbons in Prince William Sound after the *Exxon Valdez* oil spill. American Fisheries Society Symposium 18:593–607.

Donnelly, R. F. 1983. Factors affecting the abundance of Kodiak Archipelago pink salmon (*Oncorhynchus gorbuscha*). Doctoral dissertation. University of Washington, Seattle.

Eggers, D. M., L. R. Peltz, G. B. Bue, and T. M. Willette. In press. Trends in abundance of hatchery and wild stocks of pink salmon in Cook Inlet, Prince William Sound, and Kodiak, Alaska. The international symposium on the biological interactions of enhanced salmonids.

Geiger, H. J., and H. Savikko. 1991. Preliminary forecasts and projections for the 1991 Alaska salmon fisheries and summary of the 1990 season. Alaska Department of Fish and Game, Division of Commercial Fisheries, Regional Information Report 5J91-01, Juneau.

Heard, W. R. 1978. Probable causes of streambed overcrowding — 1967 pink salmon (*Oncorhynchus gorbuscha*) spawners and survival of their progeny in Sashin Creek, Southeast Alaska. U.S. National Marine Fisheries Service Fishery Bulletin 76:569–582.

Heard, W. R. 1991. Life history of pink salmon (*Oncorhynchus gorbuscha*). Pages 119–230 *in* C. Groot and L. Margolis, editors. Salmon life histories. University of British Columbia Press, Vancouver.

Hose, J. E., M. D. McGurk, G. D. Marty, D. E. Hinton, E. D. Brown, and T. T. Baker. In press. Sublethal effects of the *Exxon Valdez* oil spill on herring embryos and larvae: morphologic, cytogenetic, and histopathologic assessments, 1989–1991. Canadian Journal of Fisheries and Aquatic Science.

Ignell, S. 1988. Effects of habitat variation on recruitment of pink salmon (*Oncorhynchus gorbuscha*) at Sashin Creek, Little Port Walter, Alaska. Master's thesis. University of Alaska, Juneau.

Maki, A. W., E. J. Brannon, L. G. Gilbertson, L. L. Moulton, and J. R. Skalaki. 1995. An assessment of oil spill effects on pink salmon populations following the *Exxon Valdez* oil spill, part 2: adults and escapement. Pages 585–625 *in* P. G. Wells, J. N. Butler, and J. S. Hughes, editors. *Exxon Valdez* oil spill: fate and effects in Alaskan waters. American Society for Testing and Materials Special Technical Publication 1219, Philadelphia.

Middleton, K., M. Fink, K. Sundet, and M. Kuwada. 1992.

Alaska Department of Fish and Game *Exxon Valdez* oil spill response operations report—habitat division. Alaska Department of Fish and Game, Habitat Division, Juneau.

Moles, A., M. M. Babcock, and S. D. Rice. 1987. Effects of oil exposure on pink salmon, *O. gorbuscha*, alevins in a simulated intertidal environment. Marine Environmental Research 21:49–58.

Peltz, L., and H. J. Geiger. 1990. A tagging study of the effects of hatcheries on the 1987 pink salmon fishery in Prince William Sound, Alaska. Pages 2–17 *in* H. J. Geiger, editor. Pilot studies in tagging Prince William Sound hatchery pink salmon with coded-wire tags. Alaska Department of Fish and Game, Division of Commercial Fisheries, Fishery Research Bulletin 90-02, Juneau.

Ricker, W. E. 1975. Computation and interpretation of biological statistics of fish populations. Bulletin 191 of the Fisheries Research Board of Canada.

Roys, R. S. 1971. Effect of tectonic deformation on pink salmon runs in Prince William Sound. National Academy of Sciences, Washington. DC.

Wertheimer, A. C., and A. G. Celewycz. 1996. Abundance and growth of juvenile pink salmon in oiled and non-oiled locations of western Prince William Sound after the *Exxon Valdez* oil spill. American Fisheries Society Symposium 18:518–532.

Wiedmer, M., M. J. Fink, J. J. Stegeman, R. Smolowitz, G. D. Marty, and D. E. Hinton. 1996. Cytochrome P-450 induction and histopathology in preemergent pink salmon from oiled spawning sites in Prince William Sound. American Fisheries Society Symposium 18:509–517.

Willette, M. 1996. Impacts of the *Exxon Valdez* oil spill on the migration, growth, and survival of juvenile pink salmon in Prince William Sound. American Fisheries Society Symposium 18:533–550.

American Fisheries Society Symposium 18:499–508, 1996

Run Reconstruction of the Wild Pink Salmon Fishery in Prince William Sound, 1990–1991

WILLIAM D. TEMPLIN[1]

University of Alaska Fairbanks, Juneau Center
School of Fisheries and Ocean Sciences, 11120 Glacier Highway, Juneau, Alaska 99801-8677, USA

JEREMY S. COLLIE

University of Rhode Island, Graduate School of Oceanography, Narragannsett Bay Campus
Narragansett, Rhode Island 02882, USA

TERRANCE J. QUINN, II

University of Alaska Fairbanks, Juneau Center, School of Fisheries and Ocean Sciences

Abstract.—We reconstruct the abundances of the 1990 and 1991 runs of pink salmon *Oncorhynchus gorbuscha* in Prince William Sound for use in damage assessment. The basic problem is to allocate the catch in mixed-stock fisheries to the component stocks. With no information to directly identify stocks, the run reconstruction uses available information on migration and the timing of escapement and harvest. A migration model moves the salmon backward in time from the spawning streams, through the districts in Prince William Sound to the Gulf of Alaska, adding the catch as the salmon pass through each fishery encountered. Two major assumptions in the model are (1) migration is governed by a transition matrix that is constant over time and (2) stocks are equally vulnerable to exploitation in a district. The run reconstruction suggests that the pink salmon stocks experienced different harvest rates and that the exploitation pattern differed in 1990 and 1991. The Coghill stock had high estimated harvest rates because it migrates through several districts where it encounters intense harvest pressure. The Montague and Southeastern stocks had low estimated harvest rates because very little fishing occurred in these districts, and they do not migrate through any other districts. Testing the sensitivity of the results to assumptions about migratory rates and timing revealed that errors in these assumptions have little effect on final estimates. The migration pathways are the largest source of uncertainty in the run reconstruction results. Because harvest rates differ among stocks, it is important to consider the total run of a stock, and not just its escapement, when assessing oil spill damage. However, uncertainty in the reconstructed run sizes may exceed the magnitude of the estimated damages.

Early in the process of choosing research projects to evaluate the effects of the TV *Exxon Valdez* oil spill on the wild pink salmon *Oncorhynchus gorbuscha* in Prince William Sound, outside reviewers recommended run reconstruction as a means to estimate stock-specific abundances. Although the total abundance of wild pink salmon may be computed from the total wild catch and the total escapement, determining the abundance of each stock requires stock-specific information. The necessary escapement information is available from the Alaska Department of Fish and Game, but currently we cannot directly measure the stock contribution to each catch from mixed-stock areas. This catch information is necessary because of the possibility of differential harvest rates between pink salmon stocks within the mixed-stock fishery. Only when harvest rates are the same between stocks will stock-specific escapement information be sufficient to index a stock's run size.

Methods to estimate the abundance of a population at a given location during a period of time have often been used in fisheries management. Depending on the available information and the desired output, these methods can provide valuable insight into the migratory behavior of the resource (Mundy and Mathisen 1981; Mundy 1982) or the interaction between the resource and the harvester (Schnute and Sibert 1983; Starr and Hilborn 1988). Models in the latter group are generally referred to as run reconstructions. In essence, the run reconstructions must estimate stock-specific catch contributions by constructing spatial and temporal abundance distributions of each stock using only catch, escapement, and migration data. When these distributions are considered in conjunction with the pattern of harvest, we can estimate the probable stock makeup of the catch. This method is well suited to pink salmon mixed-stock fisheries where empirical methods of stock identification fail and some other means of

[1]Present address: Alaska Department of Fish and Game, Genetics Laboratory, 333 Raspberry Road, Anchorage, Alaska 99518-1599, USA.

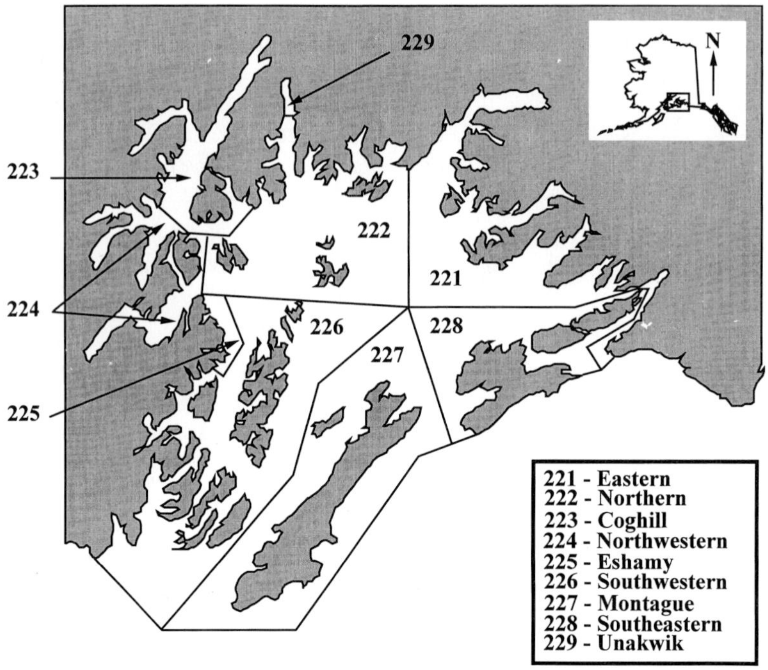

FIGURE 1.—Salmon management districts in Prince William Sound.

separating the stocks within the catch is necessary.

We present a reconstruction model for Prince William Sound and apply it to 1990 and 1991 data; then, we discuss the resulting estimate of abundance and harvest rates in relation to the limitations of the method. If the reconstruction is successful, then useful information for the restoration of salmon stocks harmed by the oil spill is obtained. Armed with improved knowledge of stock distributions and interactions within the mixed-stock fishery, the manager can act to mitigate the effects of fishing on the weakened stocks. In addition, other valuable stock-specific information can be derived from the reconstruction, such as daily and seasonal harvest rates, which are valuable to the fishery manager for analyzing the effects of management decisions on a given stock. By applying the reconstruction to historical data, one could derive stock-specific spawner–recruit data and estimate target harvest rates for each stock. To determine the robustness of the reconstruction, we study the sensitivity of the model output to changes in the input parameters, although we note that full validation of the model would require additional information on timing, which is not available.

Methods

To reconstruct the return migration of wild pink salmon through the fishery in Prince William Sound, we make the following assumptions: (1) Each salmon belongs to a stock. (2) Fish of each stock migrate purposefully to their stream of origin to spawn. (3) Once in the sound, salmon movement is described by a Markov process; that is, the movement of a fish in a given stock is solely determined by its location on a given day. The reconstruction begins on the last day a pink salmon remains alive in the streams in the sound and moves by daily steps backward in time to the day the first pink salmon enters the sound from the Gulf of Alaska.

The stocks are defined by escapement within nine management district boundaries that were set in the 1960s based on similarities between the spawning populations (Figure 1). The Unakwik (229) and the Eshamy (225) districts contain small spawning populations of pink salmon and are pooled with their geographically adjacent districts: Northern (222) and Southwestern (226), respectively. This consolidation leads to a model that considers seven stocks as they traverse seven fishery districts.

The model reconstructs the history of each stock

as it traverses the salmon fishery by accumulating fish in a series of pools as they pass through the sound: one pool for the Gulf of Alaska and seven pools of migrating fish, one in each district. Catch and escapement are removed from these pools, and migration is modeled by moving salmon between pools. By keeping track of the fish as they pass through these pools, the model develops stock-specific information on harvest rate, run size, and distributions in space and time.

The model is deterministic and assumes there is no observation error in the data, similar to the model presented by Starr and Hilborn (1988), except that we model movement differently. Their method models migration deterministically, using the "boxcar method" for migration timing, moving fish in "clumps" and requiring each group of fish to spend a fixed number of days within a district before moving on. We model migration stochastically, using a Markov chain (Taylor and Karlin 1984), which enables salmon to move independently of each other. We chose stochastic movement for the following reasons: (1) Computer programming is simpler. (2) Fish are allowed to move at varying rates. (3) Sensitivity analysis of the migration assumptions is simpler. (4) Using transition probabilities backward in time functionally creates an increase in the speed of migration as the season progresses, as shown below. Although the two methods are different, they both belong to the family of semi-Markov models.

Capital letters indicate values accumulated (pooled) across strata: time, stock, district. Underlined variables are vectors over the missing subscript. Bold letters indicate matrices and Greek letters indicate rates or probabilities. Each variable may be accompanied by the following subscripts:

t = day of the year (1, 2, 3, . . . , 365);
n = stock (1, 2, 3, . . . , 7);
d = district (1, 2, 3, . . . , 8).

The variables used in the model have the following definitions:

y = number of fish escaping the fishery
s = number of fish entering the stream pool
τ = number of days a fish spends in the mouth pool
c = stock-specific catch
C = observed catch
ρ = harvest rate
P = number of fish in harvestable pool before catch
P^* = number of fish in harvestable pool after catch
$\boldsymbol{\Theta}$ = matrix of migration probabilities
$\mathbf{I}$ = identity matrix
$\mathbf{U}$ = diagonal matrix with elements ρ.

The escapement data consisted of aerial counts of the number of fish alive in the streams observed, but the reconstruction required information on the daily arrival of fish to the streams. By assuming that all salmon of a given stock survive for a set number of days in freshwater (the stream life), and using the historical distributions of fish in the stream for odd and even years, provided by the Alaska Department of Fish and Game, we fitted a discrete normal timing curve to the total escapement of each stock as described by Templin (1995). The normal distribution was chosen as a first approximation to the actual escapement of each stock because it is a flexible, two-parameter distribution that approximates a wide range of possible distributions. More accurate descriptions of stock-specific escapement are possible but are beyond the scope of this project.

Salmon hold at the mouth of the stream for a number of days before ascending to the spawning gravel (Heard 1991). Because fishing is not allowed within 914 m of stream mouths, these fish are no longer part of the harvestable pool and are considered a part of the escapement pool. Assuming that all salmon hold for τ days before entering the stream, the number of salmon y_{tnd} of stock n daily escaping the fishery is

$$y_{tnd} = s_{t+\tau,nd}\,. \tag{1}$$

The escapement of a stock in all seven districts can be written as the 7 × 1 vector, $\underline{y}_{tn}$. Under the definition that a stock consists of all the fish that spawn within a district, the only nonzero element of $\underline{y}_{tn}$ is the nth element corresponding to the district stock n escapes in. We set τ equal to 5 d in all districts as a best estimate of the average time spent at the mouth (S. Sharr, Alaska Department of Fish and Game, personal communication).

The fishery in recent years has been concentrated in regions where pink salmon stocks mix as they migrate through the sound. Because currently there is no evidence of differential harvest of the stocks, all stocks within a pool of migrating salmon are assumed to be equally available to the fishing gear. Thus, the relative contribution of each stock to the observed catch was equal to its relative abundance in the migratory pool,

$$c_{tnd} = C_{td} \frac{P_{tnd}}{\sum_{n} P_{tnd}}\,. \tag{2}$$

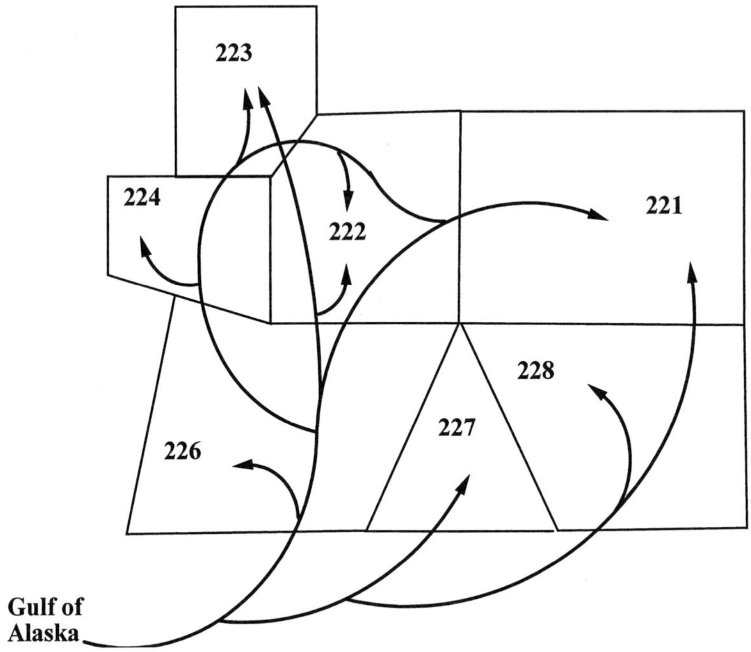

FIGURE 2.—Assumed pink salmon migration pathways between salmon management districts in Prince William Sound.

The daily harvest of wild salmon was estimated by removing the estimated daily catch of hatchery fish (Sharr et al. 1996) from the recorded daily harvest. The resulting daily harvest data are found in Templin (1995).

It follows from equation (2) that the daily fraction of available fish removed in the harvest, the harvest rate ρ_{td}, from that migratory pool can be calculated as

$$\rho_{td} = \frac{c_{tnd}}{P_{tnd}} = \frac{C_{td}}{\sum_{n} P_{tnd}}. \quad (3)$$

We assume that a salmon's return migration through Prince William Sound is unidirectional and follows the pathways in Figure 2. Computationally, this movement is controlled by the migration matrix $\mathbf{\Theta}_n$ (where θ_{ij} is the probability that a given fish migrated from district i to district j during the previous day, working in the backward direction). The diagonal elements of $\mathbf{\Theta}_n$ are the probabilities that fish stay in the same area and are calculated as $1 - \bar{t}^{-1}$, where $\bar{t}$ is the average number of days that a salmon spends in that district.

Alternatively, $\bar{t}^{-1}$ can be thought of as the probability of a successful transition. The θ_{ij} are estimated from the results of a pink salmon radiotagging experiment in 1992 (Templin 1995; H. Geiger, Alaska Department of Fish and Game, personal communication). Entry to the fishery from the Gulf of Alaska is subsumed into this matrix. Thus, all matrices and vectors have dimensions 8 × 8 and 8 × 1, respectively. Columns sum to 1, so that fish neither disappear nor appear after entering the migration.

The pools of migrating fish are central to the reconstruction method. It is in these pools that harvest, escapement, and migration interact to produce the spatial and temporal distributions of salmon. The daily size of the pool is a function of inputs and outputs. Because the model works backward in time, catch, escapement, and emigration are inputs, and immigration (from other districts or the Gulf of Alaska) is the output. On any given day, the order of events is arbitrarily assumed to be catch and escapement, then migration. Pool abundance is determined at the very beginning of the day before catch, escapement, and migration. We express this with an equation governing the size of the migratory pool backward in time:

$$\underline{P}_{tn} = \mathbf{\Theta}_n \underline{P}_{t+1,n} + \underline{c}_{tn} + \underline{y}_{tn}. \quad (4)$$

TABLE 1.—Stock-specific run reconstruction statistics of interest for the 1990 return year in seven salmon management districts in Prince William Sound. Escapement, harvest, run size, and parental stock size (1988 spawners) are in units of thousand fish. The 1988 spawners are the parents of the 1990 adult return.

Stock	Observed escapement	Estimated harvest	Estimated run size	Estimated harvest rate	1988 spawners	Return per spawner
221	1,558	5,749	7,306	0.79	1,407	5.19
222	549	2,276	2,826	0.81	615	4.60
223	154	1,313	1,467	0.89	138	10.63
224	451	467	918	0.51	341	2.69
226	1,913	3,400	5,313	0.64	1,710	3.11
227	475	11	486	0.02	315	1.54
228	1,082	10	1,092	0.01	461	2.37
Total	6,182	13,226	19,408	0.68	4,987	3.89

The lefthand side of equation (4) is the abundance in the migratory pools at the beginning of day t and the righthand side is the abundance in the migratory pool at the beginning of the following day $t + 1$ moved according to the migration matrix with the catch and the escapement on day t added.

Notice that estimating c_{tnd} requires knowledge of P_{tnd}, which in turn depends on the estimated value of c_{tnd}. We circumvented this problem by noting the harvest rate ρ_{td} was the same for all stocks. By rearranging equation (3), substituting into equation (4) and defining $\mathbf{U}_t$ to be a diagonal square matrix with elements ρ_{td}, we develop the recursive pool-size equation

$$(\mathbf{I} - \mathbf{U}_t)\,\underline{P}_{tn} = \mathbf{\Theta}_n\,\underline{P}_{t+1,n} + \underline{y}_{tn}, \qquad (5)$$

as an intermediate step to obtaining the final form of the recursive pool-size equation

$$\underline{P}_{tn} = (\mathbf{I} - \mathbf{U}_t)^{-1}[\mathbf{\Theta}_n\,\underline{P}_{t+1,n} + \underline{y}_{tn}]. \qquad (6)$$

Because we want information on the stock-specific catches c_{tnd}, which are not explicitly computed in equation (6), we add an extra computational step. By defining the 8×1 vector $\underline{P}^*_{tn}$ as the number of salmon remaining in each migratory pool after catch is removed, that is,

$$\underline{P}^*_{tn} = \mathbf{\Theta}_n\,\underline{P}_{t+1,n} + \underline{y}_{tn}, \qquad (7)$$

we then compute $\underline{c}_{tn}$ as

$$\underline{c}_{tn} = \underline{P}_{tn} - \underline{P}^*_{tn}. \qquad (8)$$

The total run size of each stock R_n is computed as

$$R_n = \sum_t 1'\underline{c}_{tn} + \sum_t 1'\underline{y}_{tn} = \sum_d \sum_t c_{tnd} + \sum_d \sum_t y_{tnd}, \qquad (9)$$

where $1'$ is the transpose of a 8×1 vector of ones. Also of interest is the seasonal harvest rate experienced by each stock ρ_n,

$$\rho_n = \frac{\sum_d \sum_t c_{tnd}}{R_n}. \qquad (10)$$

Results

In 1990, the early return of wild pink salmon to the sound was poor, but by late July escapements dramatically improved, and the 1990 harvest of 44.1 million pink salmon was the highest on record. Fishing was limited early in the season because of low escapements, but improving escapements in late July allowed fishing in every district of the sound. Much of the effort was concentrated in the mixed-stock areas in each district because wild stock returns were considered strong enough to sustain more intense harvest. Escapement goals were met in every district but 222 and 223. Stock 223 achieved less than 40% of the minimum desired escapement range (Geiger and Savikko 1991).

In 1990, the estimated total returns of each wild pink salmon stock from the run reconstruction range from 486,000 (stock 227) to 7.3 million (stock 221; Table 1). The estimated returns per spawner for each stock, calculated from the estimated stock-specific run size and the escapement in 1988 (S. Sharr, personal communication) range from 1.54 (stock 227) to 10.63 (stock 223), with a sound-wide average of 3.89. The estimate of 10.63 returns per spawner for stock 223, more than twice the value of the next most productive stock, is probably caused by too much of the harvest being attributed to stock 223 during the reconstruction.

The estimated seasonal harvest rates for each stock range from 1% (stock 228) to 89% (stock 223; Table 1). The harvest rate a stock experiences appears to be related to the number and intensity of the fisheries through which it migrates, and the harvest rates fall within three regional groups based

TABLE 2.—Stock-specific run reconstruction statistics of interest for the 1991 return year in seven salmon management districts in Prince William Sound. Escapement, harvest, run size, and parental stock size (1989 spawners) are in units of thousand fish. The 1989 spawners are the parents of the 1991 adult return.

Stock	Observed escapement	Estimated harvest	Estimated run size	Estimated harvest rate	1989 spawners	Return per spawner
221	2,319	1,306	3,624	0.36	1,849	1.96
222	853	642	1,495	0.43	472	3.17
223	378	666	1,044	0.64	248	4.21
224	525	161	686	0.23	374	1.83
226	2,406	4,151	6,557	0.63	2,103	3.12
227	1,384	0	1,384	0.00	870	1.59
228	1,794	0	1,794	0.00	974	1.84
Total	9,659	6,926	16,584	0.42	6,890	2.41

on similar experiences in the Prince William Sound salmon fishery. Stocks 227 and 228 have the lowest seasonal harvest rates, 2% and 1%, respectively, because they traverse only one fishery before escaping (Figure 2) and very little effort is expended fishing in either of these districts. Stocks 224 and 226 experience moderate harvest rates, 51% and 64%, respectively. Stock 224 must pass through two fisheries, whereas stock 226 only enters one, but district 226 is a very intense fishery: more than half of the total harvest in the sound is caught in this district. The northern stocks, 221, 222, and 223, experience the highest harvest rates, 79%, 81% and 89%, respectively, because of the gauntlet of fisheries through which they must pass.

Although the harvest of 37.3 million pink salmon in the sound in 1991 was the second highest on record, the run was late and compressed. For the first half of the season, low numbers of wild salmon in the index streams resulted in restricted fishing to avoid harvest of wild stocks, with much of the effort confined to hatchery terminal areas and special harvest areas. Concern about the buildup of hatchery fish in the terminal areas prompted the opening of parts of the migratory pathway to common property fishing in mid-August. At this time, wild stock escapement counts showed improvement and the entire sound was opened to seine harvest in late August. All stocks, except for stock 223, met or exceeded escapement goals (Geiger and Savikko 1992).

The estimated total stock-specific run sizes in 1991 from run reconstruction range from 686,000 (stock 224) to 6.6 million (stock 226; Table 2). Estimated harvest rates range from 0% (stocks 227 and 228) to 64% (stock 223), with a sound-wide average harvest rate of 42%. Stocks 221 and 222 experienced moderate harvest pressure in comparison with the previous year, while the harvest rates on stocks 223 and 226 remained high. This pattern of harvest rates is probably reflective of the pattern of fishing effort. With fishing effort limited until late in the season, stocks 221 and 222 were able to temporally bypass much of the harvest. When fishing was resumed, much of it was limited to special harvest areas where it was more likely that local salmon would be harvested, allowing stocks 221 and 222 to travel the migration corridors without encountering much harvest.

Estimated seasonal harvest rates and return-per-spawner values are not evenly distributed among the stocks (Tables 1, 2). Stocks that traverse more fisheries or constitute a large portion of the migratory pool during periods of large catches tend to have higher harvest rates. For example, the Coghill stock (223) is subject to harvest in four fishing districts (Figure 2): districts 222, 223, 224, and 226; a higher harvest rate for this stock occurs. In contrast, the Eastern stock (221) migrates through six districts (Figure 2): districts 221, 222, 223, 224, 226, and 228; but its earlier time of migration allows it to miss much of the harvest pressure in districts 222, 223, 224, and 226.

We explore the model's sensitivity to migration speed and run timing as a possible source of misallocation of salmon to stock 223. Because of the mechanics of the reconstruction, we expect changes in the relative composition of the pools of harvestable fish to affect the resulting distribution of harvest among stocks. To test this, the reconstruction program is run four more times. During the first two reconstructions the escapement entry timing of stock 223 is shifted by 10 d earlier and later to shift its temporal distribution within each of the fisheries it traverses. We also expect changes in the transition rates of stock 223 to affect an individual fish's vulnerability by changing its rate of travel, so the last two reconstructions are computed with the original

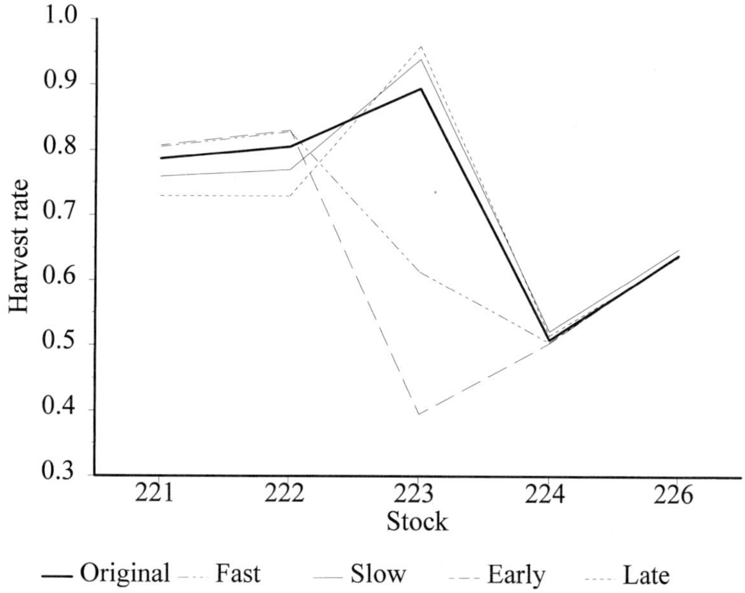

FIGURE 3.—Calculated harvest rates from sensitivity analysis of migration speed and escapement timing for stock 223 in 1990.

migration rates for stock 223 first doubled, then halved. This is accomplished by redetermining the $\boldsymbol{\Theta}$ matrix with the new migration rates.

Changes in the migration timing and migration speed of stock 223 do not have much effect on the estimated harvest rates on stocks other than stock 223 in 1990 (Figure 3). The effects of the resultant changes in the harvest rate on stock 223 are straightforward. Earlier migration moves many of the salmon through the fishery before the period of harvest, reducing their probability of capture and, consequently, the harvest rate on the stock. Later migration increases the estimated harvest rate, because the peak of the stock's abundance coincides with the peak of harvest effort. Faster migration decreases the availability of an individual fish, which reduces the harvest rate, whereas slower migration has the opposite effect. In all cases, changes in stock 223's harvest rate are inversely reflected in the harvest rate on stocks 221 and 222, indicating that the allocation of some of the salmon to stocks 221, 222, and 223 is dependent on assumptions about migration.

For the purpose of estimating lost pink salmon production, the estimated run size of each stock is sufficient, but the reconstruction also generates information on stock distributions. Such information helps the fishery manager to understand how decisions about fishing affect individual stocks. These distributions show the temporal distribution of a stock in each of the districts it traverses and the interaction of all the stocks within a single district. For example, consider the temporal abundance distributions of stock 223 in each of the districts it traverses (Figure 2) enroute to the spawning streams in district 223 in 1990 (Figure 4). As expected, district 226, the first district encountered upon leaving the Gulf of Alaska, has the earliest and greatest abundance of salmon from stock 223. Five stocks pass through district 226 during their migration: 221, 222, 223, 224, and 226 (Figure 2). Consequently, this district is the site of the fishery harvesting the greatest mixture of stocks, and information on the timing and abundance of each stock in this district is important (Figure 5).

Our model has features that capture biological realism not found in other run reconstruction models. Markovian movement backward in time, combined with a normal escapement distribution, allows the perceived mean residence time in the sound to change over the course of the season. The cumulative percent-of-total entry to the sound is compared with the cumulative percent-of-total entry to the streams for stock 227 in 1991 (Figure 6). This stock is selected for several reasons: (1) it experienced no harvest in 1991, (2) it traverses only one district (227), and (3) it does not interact with other stocks. Thus, the results for this stock are solely the result

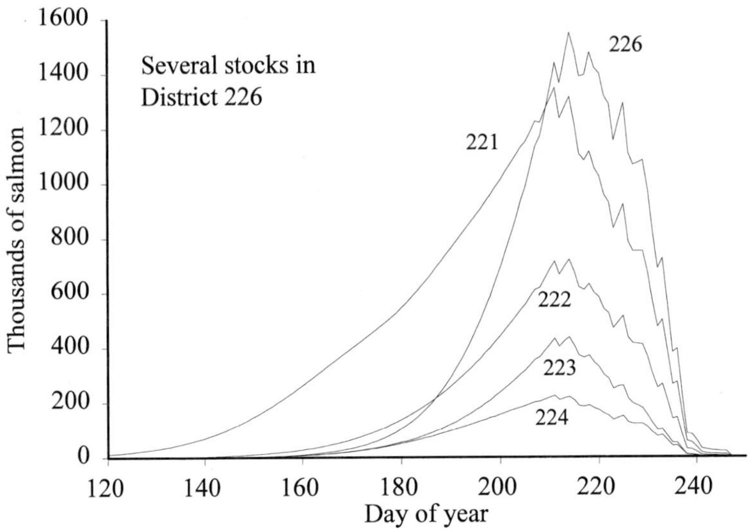

FIGURE 4.—Temporal abundance distributions of stock 223 in the districts it traversed in 1990.

of the stream-entry distribution as input and the probability of transition to the Gulf of Alaska. Because one curve denotes entry, and the other denotes exit, the difference between the two curves in the horizontal direction is an estimate of the mean residence time. At the point where 25% of this stock had entered the sound, the mean residence time was 17 d. The difference narrows to 12 d by the time 75% of the stock had entered the sound. So the model presupposes a change in perceived mean residence time of 5 d from the mid-early portion of the run, to the mid-late part of the run.

Discussion

The reconstructions of the 1990 and 1991 runs of pink salmon in Prince William Sound are plausible estimates of the run sizes for each stock. One of the major findings of the reconstruction is that not every stock is experiencing the same

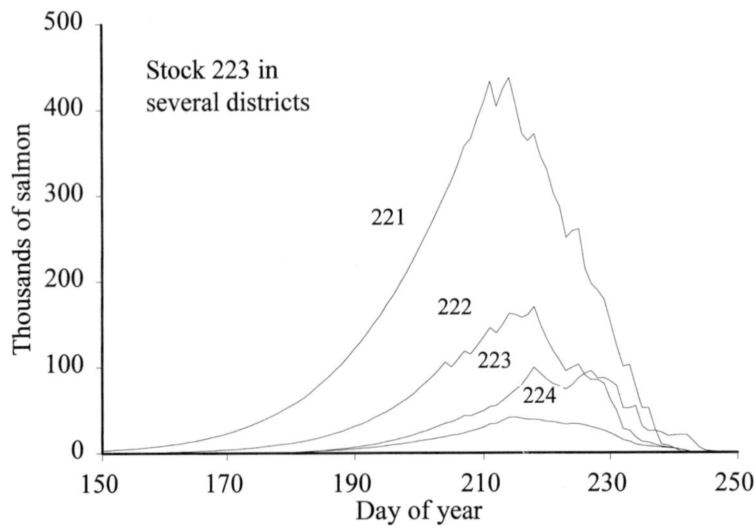

FIGURE 5.—Temporal abundance distributions of stocks 221, 222, 223, 224, and 226 in district 226 in 1990.

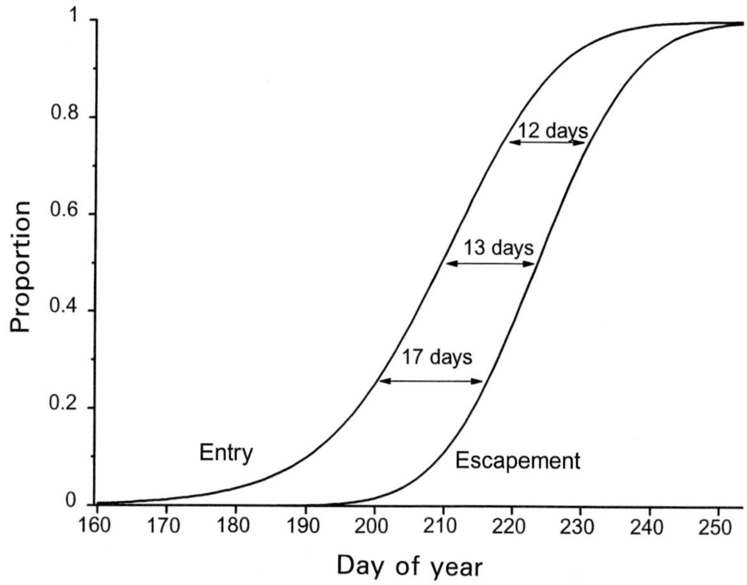

FIGURE 6.—Cumulative entry into the fishery and cumulative entry into the escapement for stock 227 in 1991.

harvest rate over the course of the season. Although this finding is intuitive, this is the first time a quantitative analysis of stock-specific harvest history has been possible in Prince William Sound. This comprehensive model of the pink salmon fishery, which combined several different sources of data, should be useful for making management decisions mitigating the effects of damage caused to the wild pink salmon stocks by the *Exxon Valdez* oil spill.

The weakest part of the method is our understanding of the migration behavior of the wild pink salmon in the sound. Our assumed migration pathways might not be flexible enough to account for the fish sometimes overshooting and reversing direction along migration pathways, which the 1992 tag study showed. Wandering and probing behaviors that pink salmon have been observed to exhibit before actual homing (Heard 1991) are not reflected in our model because we have little specific information on these actions. Starr and Hilborn (1988) discuss the necessity for good migration information when reconstructing a mixed-stock fishery, because of the possibility of misallocating harvest when migration behavior is misspecified.

Another serious obstacle to the use of reconstruction in Prince William Sound is the lack of auxiliary data to validate the model. The sensitivity analysis we present indicates that the method is internally insensitive to error, but this does not authenticate the usefulness of the model results. Empirical validation requires independent measurement of entry into the fishery (e.g., test fishery) or of stock-specific catch composition. Neither source of information is available for the Prince William Sound wild pink salmon fishery.

In neither 1990 nor 1991 was there an obvious reduction in the numbers of returning salmon per spawner (*R*/*S*) in the Southwestern stock (226), the district most heavily impacted by oil (Middleton et al. 1992). With a salmon life history model, Geiger et al. (1996, this volume) estimated an oil-related loss of 1.86 million adult salmon in 1990 that would have originated from the Southwestern stock. This loss translates into a reduction of 1.09 *R*/*S* for the Southwestern stock. From run reconstruction results alone, it is impossible to state what the returns to the Southwestern stock would have been without the oil spill. Sound-wide *R*/*S*'s in both 1990 and 1991 were higher than the 1963–1987 median of 1.79 (Geiger et al. 1996). In 1990, the *R*/*S* for the Southwestern stock was lower than the sound-wide *R*/*S* (Table 1); in 1991, the Southwestern stock *R*/*S* was higher than the sound-wide value (Table 2). The fact that no reduction in production is apparent from the run reconstruction results does not mean that no damage occurred. We conclude only that the extent of damage cannot be detected from empirical estimates of the numbers of returning adults.

Acknowledgments

We thank Hal Geiger and Sam Sharr for providing the data used in the reconstruction model and technical advice about the Prince William Sound pink salmon fishery. We also thank Richard Gates for programming assistance, Doug Eggers and Gordon Kruse for suggesting this study, and James Blick and Philip Mundy for reviewing the manuscript. This project was funded through the Alaska Department of Fish and Game as part of the State/Federal Natural Resource Damage Assessment, Study F/SF 28.

References

Geiger, H. J., B. G. Bue, S. Sharr, A. C. Wertheimer, and T. M. Willette. 1996. A life history approach to estimating damage to Prince William Sound pink salmon caused by the *Exxon Valdez* oil spill. American Fisheries Society Symposium 18:487–498.

Geiger, H. J., and H. Savikko. 1991. Preliminary forecasts and projections for 1991 Alaska salmon fisheries and summary of the 1990 season. Alaska Department of Fish and Game, Division of Commercial Fisheries, Regional Information Report 5J91-01, Juneau.

Geiger, H. J., and H. Savikko. 1992. Preliminary forecasts and projections for 1992 Alaska salmon fisheries and summary of the 1991 season. Alaska Department of Fish and Game, Division of Commercial Fisheries, Regional Information Report 5J92-05, Juneau.

Heard, W. R. 1991. Life history of pink salmon (*Oncorhynchus gorbuscha*). Pages 119–230 *in* C. Groot and L. Margolis, editors. Pacific salmon life histories. University of British Columbia Press, Vancouver.

Middleton, K., M. Fink, K. Sundet, and M. Kuwada. 1992. Alaska Department of Fish and Game *Exxon Valdez* oil spill response operations report—habitat division. Alaska Department of Fish and Game, Habitat Division, Juneau.

Mundy, P. R. 1982. Computation of migratory timing statistics for adult chinook salmon in the Yukon River, Alaska, and their relevance to fisheries management. North American Journal of Fisheries Management 4:359–370.

Mundy, P. R., and O. A. Mathisen. 1981. Abundance estimation in a feedback control system applied to the management of a commercial salmon fishery. Pages 81–98 *in* K. B. Haley, editor. Applied operations research in fishing. Plenum, New York.

Schnute, J., and J. Sibert. 1983. The salmon terminal fishery: a practical, comprehensive timing model. Canadian Journal of Fisheries and Aquatic Sciences 40: 835–853.

Sharr, S., and seven coauthors. 1996. Coded wire tag studies on Prince William Sound salmon, 1989–1991. Alaska Department of Fish and Game, Commercial Fisheries Management and Development Division, Regional Informational Report 2A96-15, Anchorage.

Starr, P., and R. Hilborn. 1988. Reconstruction of harvest rates and stock contribution in gauntlet salmon fisheries: application to British Columbia and Washington sockeye (*Oncorhynchus nerka*). Canadian Journal of Fisheries and Aquatic Sciences 45:2216–2229.

Taylor, H., and S. Karlin. 1984. An introduction to stochastic modeling. Academic Press, New York.

Templin, W. 1995. Reconstruction of wild pink salmon (*Oncorhynchus gorbuscha*) runs in Prince William Sound, Alaska. Master's thesis. University of Alaska, Fairbanks.

American Fisheries Society Symposium 18:509–517, 1996

Cytochrome P-450 Induction and Histopathology in Preemergent Pink Salmon from Oiled Spawning Sites in Prince William Sound

MICHAEL WIEDMER AND MARK J. FINK

Alaska Department of Fish and Game, Habitat and Restoration Division
333 Raspberry Road, Anchorage, Alaska 99518, USA

JOHN J. STEGEMAN AND ROXANNA SMOLOWITZ

Woods Hole Oceanographic Institution, Biology Department
Woods Hole, Massachusetts 02543, USA

GARY D. MARTY[1] AND DAVID E. HINTON

University of California, Department of Medicine, Davis, California 95616, USA

Abstract.—The March 1989 *Exxon Valdez* oil spill contaminated intertidal pink salmon *Oncorhynchus gorbuscha* spawning areas in Prince William Sound and the Gulf of Alaska. To determine if 8- to 26-month old oil remaining in some spawning areas produced physiological responses in developing pink salmon eggs and alevins, we conducted an initial assessment of cytochrome P-4501A induction and histopathologic lesion occurrence in preemergent pink salmon collected from oiled spawning substrates. Egg and alevin samples were collected from four oiled and five reference sites in Prince William Sound, Alaska, between December 1989 and May 1991. Immunohistochemical staining for cytochrome P-4501A was increased in alevins from 13 of 16 samples from oiled sites, but was not increased in any of the 7 samples from the reference sites. Cytochrome P-4501A induction was not detected in egg samples from either oiled or control sites. Persistent P-4501A staining through the end of the study was evidence for chronic exposure of two year-classes of pink salmon to hydrocarbon contamination. Histopathologic lesions were more frequent in alevins from oiled sites, but differences were not statistically significant, and lesion occurrence seemed dependent on developmental stage. These results provide evidence that pink salmon alevins developing in heavily oiled sites were exposed to hydrocarbons more than 2 years after the initial spill and that the hydrocarbons induced detectable physiological changes. Results of this study were used to develop appropriate treatments for oiled anadromous fish streams.

The 24 March 1989 grounding of the TV *Exxon Valdez* spilled 43.9 million liters of Prudhoe Bay crude oil into the waters of Prince William Sound. Currents and winds carried the oil through the islands of the southwestern Sound and into the Gulf of Alaska. The spill eventually oiled almost 2,000 km of shoreline, including a minimum of 213 streams supporting anadromous fish.

The state of Alaska Department of Fish and Game (ADFG) cooperated in the spill response program. In this effort, we focused on the need for, and the adequacy of, treatment of intertidal areas in and around anadromous fish streams. Because up to 75% of wild Prince William Sound pink salmon *Oncorhynchus gorbuscha* spawn intertidally, the ADFG was concerned that oiled beach sediments might affect developing pink salmon eggs and larvae.

In Prince William Sound, adult pink salmon spawn in inter- and supratidal stream channels from late July through early October. All pink salmon die within a few days of spawning. Eggs typically hatch in December and January, and alevins remain within the spawning substrate, absorbing their yolk sac, until emerging as free-swimming fry between April and June.

Fish are generally most sensitive to environmental pollutants as juveniles (McKim 1977; Moles et al. 1979). While oil remained in and near some spawning streams by fall of 1989, some individuals involved in the spill response believed that any remaining oil would be biologically inert. If remaining weathered oil was biologically inert, additional treatment would not be necessary. To determine if remaining oil was biologically inert, the ADFG began the initial assessment reported here. We analyzed wild preemergent pink salmon collected from oiled spawning substrates by (1) measuring cytochrome P-450 induction in eggs and alevins as an indicator of exposure to hydrocarbons and (2) documenting histopathological changes in alevins from oiled sites.

[1]Present address: University of California, Department of Anatomy, Physiology, and Cell Biology, School of Veterinary Medicine, Davis, California 95616, USA.

Cytochrome P-450 is a particular group of monooxygenase (MO) enzymes that mediates oxidation of petroleum hydrocarbons and other xenobiotics, thereby facilitating their excretion (Jimenez and Stegeman 1990). Studies of free-ranging fish exposed to petroleum contamination (Burns 1976; Payne 1976; Kurelec et al. 1977) and to general industrial contamination (Spies et al. 1982; Foureman et al. 1983; Spies et al. 1988) showed elevated MO activity associated with the cytochrome P-4501A, the hydrocarbon-inducible P-450.

Despite the potential of using histopathologic lesions as biological indicators of exposure to toxic environmental contaminants (Hinton and Lauren 1990), and the relatively common occurrence of large oil spills, comparatively few studies have been done on field-exposed fish. Histopathologic lesions have been demonstrated in free-ranging adult fish up to 2 years after an oil spill (Haensly et al. 1982). For larval fish, lesions have been reported after exposure to crude oil in the laboratory (Cameron and Smith 1980; Hawkes and Stehr 1982), but lesions in naturally exposed larvae have not previously been reported.

Study sites

Preemergent pink salmon were collected from intertidal stream spawning substrates at four oiled sites and five reference sites in southwestern Prince William Sound (Figure 1). Oiled sites were located in unnamed streams at Herring Point and Marsha Bay on Knight Island, Shelter Bay on Evans Island, and Sleepy Bay on Latouche Island. Reference sites were located in unnamed streams of Herring Bay on Knight Island, Kake Cove and the east side of Chenega Island, Shelter Bay, and north of Eshamy Bay. Fish samples designated "oiled" were collected in close proximity to visible oil while samples designated "reference" were collected from areas with no visible sources of potential contamination. Heavy oil deposition at the contaminated sites occurred primarily during late March and early April, 1989. Site observations indicated that potential exposure of developing pink salmon embryos to hydrocarbons was attributable primarily to chronic leaching from and direct exposure to contaminated sediments, rather than to reoiling from refloated oil.

Methods

Eggs and alevins were collected from spawning substrates in mid- to upper intertidal zones between December 1989 and May 1991 (two brood years). Eggs were collected in December 1989 and September 1990. Alevins were collected in December 1989, May and June 1990, and March and May 1991. Eggs averaged 6–7 mm in diameter; alevins averaged 30 mm long. Because of subfreezing temperatures, hand shovels, rather than egg pumps, were used to extract specimens from spawning gravels. Excavated substrates were placed and agitated immediately upstream of a 0.7-m × 1.0-m funnel net, where stream currents flushed the eggs and alevins into the net. Specimens were separated from debris and immediately placed in 10% phosphate-buffered formalin. Specimens collected in 1991 were anesthetized in tricaine methanesulfonate (MS-222) before fixation. Overall, 1–6 alevin samples from each oiled site (16 total samples) and 1 or 2 samples from each reference site (7 total samples) were collected. Two egg samples each from two oiled sites and one sample each from two reference sites were also collected. Each sample consisted of 4–6 alevins or 8–12 eggs. Four egg samples were incidently collected at three other sites. Two egg samples were collected from an oiled site near Point Countess, and one egg sample each was collected at non-oiled sites in Thumb Cove and at Point Countess (not identified in Figure 1).

Representative sediment samples (one per site) collected at the oiled sites in Marsha Bay, Shelter Bay, and Sleepy Bay in 1989, and at Herring Point in 1990, were analyzed for *n*-alkanes and isoprenoids with gas chromatography using flame ionization detection (GC–FID) and for aromatic hydrocarbons using gas chromatography with mass spectrometry (GC–MS) to verify that the oil source was North Slope crude. Hydrocarbon analysis of sediments was conducted by the Geotechnical and Environmental Research Group at Texas A&M University (College Station, Texas). No sediment samples from reference sites were analyzed. Lack of contamination at these sites was visually determined.

Alevin and egg samples were analyzed immunohistochemically at Woods Hole Oceanographic Institution (WHOI), Massachusetts, for induction of cytochrome P-4501A, the specific P-450 induced by polycyclic aromatic hydrocarbons (PAHs) (Stegeman and Lech 1991). One to six alevin samples from each of the four oiled sites (94 total alevins) and one to two alevin samples from each of the five reference sites (42 total alevins) were analyzed. One to three egg samples from each of the oiled sites and one egg sample from each of three reference sites were also analyzed. Tissues were processed routinely into paraffin. Eggs and alevins were em-

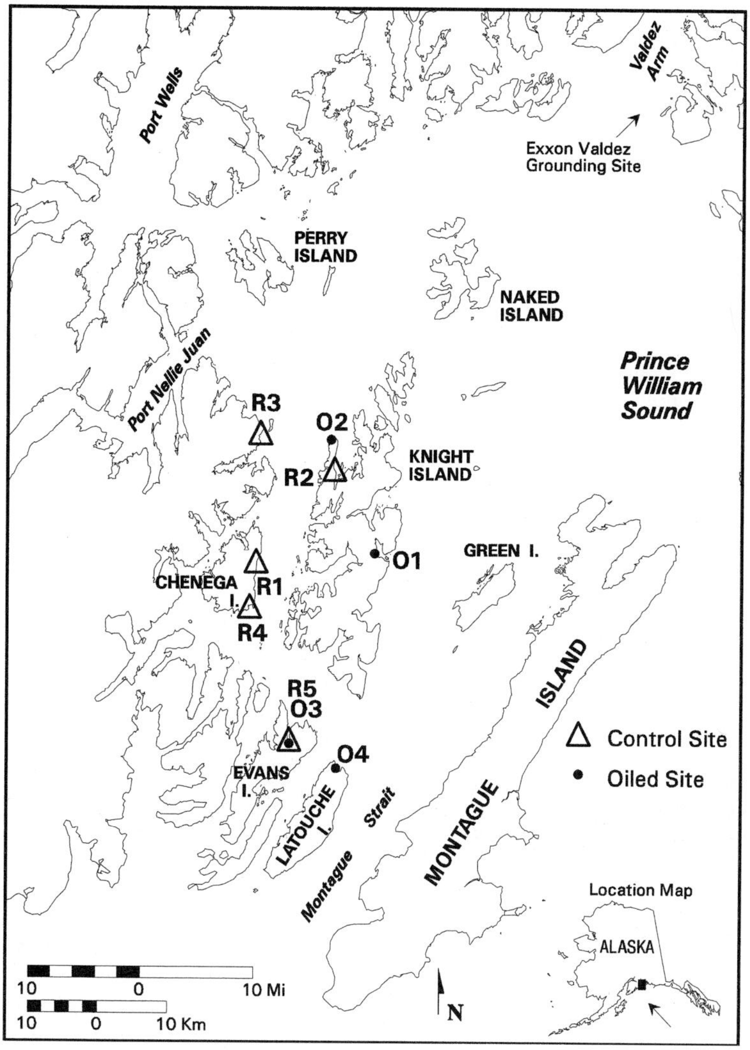

FIGURE 1.—Pink salmon egg and alevin collection sites in Prince William Sound, Alaska. Oiled sites are O1 = Marsha Bay, O2 = Herring point, O3 = Shelter Bay, O4 = Sleepy Bay. Reference sites are R1 = East side Chenega Island, R2 = Herring Bay, R3 = Eshamy Bay, R4 = Kake Cove, R5 = Shelter Bay. Sites O3 and R5 are located in two separate streams approximately 600 m apart.

bedded whole, in lateral-to-dorsolateral recumbency, with two fish or 8–12 eggs per paraffin block. Resultant blocks were trimmed and sections discarded until the approximate midsection was reached. Five-micron-thick sections were stained with monoclonal antibody (MAb) 1-12-3 and a peroxidase-labeled secondary antibody (Park et al. 1986; Smolowitz et al. 1991). Monoclonal antibody 1-12-3 is specific for P-4501A proteins and cross-reacts with P-4501A in all vertebrate species tested to date. For quality control, serial sections of at least one block per sample were stained identically except for substitution of the primary antibody with a nonspecific monoclonal antibody (of the same immunoglobulin G subclass). All sections were treated under identical conditions and duration. Staining intensity and occurrence of P-4501A was determined for tissues in each egg or alevin, and an overall, relative level of induction was determined for each sample: negative (0), very mild (1), mild (2), moderate (3), and strong or severe (4). After P-450 induction analysis, duplicate slides and embedded alevins were sent to the University of California at Davis (UC Davis) for histopathological analyses. Where liver was missing from duplicate slides, additional sections were cut at UC Davis. To avoid possible bias, background information on samples and collection sites was not provided to workers at WHOI or UC Davis until after analyses were completed.

Individual glass microscope slides containing two or three serial or near-serial longitudinal sections (two alevins per slide) were examined to determine the extent of autolysis, the presence of major organs, the relative abundance of liver glycogen and yolk stores (scored as none [0], minimal or moderate [1], or abundant [2]), and the presence of lesions. Frequently occurring lesions were ranked and scored relative to other similar lesions as follows: none (0), mild (1), moderate (2), or severe (3).

Egg and alevin samples were stratified by time and pooled for each site. The Wilcoxon–Mann–Whitney two-sample test (Conover 1980) was used to test for statistical differences between the median of cytochrome P-4501A induction and lesion scores from oiled and reference sites. Statistical tests were one-tailed and performed at the $\alpha = 0.10$ level. The null hypothesis for all tests was that oiled and reference sites were affected equally. Lesions used in scoring included epidermal atrophy (EA) and myofiber degeneration and/or necrosis (MDN). Scores for individual hepatocellular degeneration (IHD) and vacuolar degeneration of gastric glands (VDGG) were not used because of several missing values.

Results

Cytochrome P-4501A induction was observed in endothelial and epithelial cells of several organs from alevin samples from each of the oiled sites (13 of 16 samples) and appeared to be independent of posthatch developmental stage. All seven alevin samples from reference sites were negative for induction. In samples from oiled sites, organs exhibiting positive immunohistochemical staining included kidney, gill, liver, intestine, heart, brain, yolk sac, skin, peritoneal connective tissue, and pharyngeal epithelium (Table 1). Because all tissues were not visible in sections from every individual fish, some induced tissues may have gone undetected. Cytochrome P-4501A induction was not observed in any of the 10 egg samples, regardless of hydrocarbon exposure. In one instance where both eggs and alevins were simultaneously collected from the same microsite (Marsha Bay [O1], December 1989), alevins had mild-to-moderate induction while eggs had none.

Cytochrome P-4501A content was significantly elevated in alevin samples from oiled sites in May and June 1990 ($P < 0.067$) and in March 1991 ($P < 0.10$). Because only one oiled-site (O1) sample and one reference-site (R1) sample were collected in December 1989, no statistical tests were performed; however, the oiled site had mild staining for P-4501A expression, whereas the reference site had no staining (Figure 2). No statistical difference was detected in May 1991 ($P < 0.20$) (Figure 2). There was no significant change in induction intensity between 1990 and 1991.

Lesions were observed in 6 of 13 samples from oiled sites exhibiting elevated P-4501A (Table 1). In two samples collected from Marsha Bay (O1) in May 1991, 11 of 12 alevins had EA and VDGG, but P-4501A induction was negative. In three of seven samples from reference sites, alevins had lesions but no increase in P-4501A (Table 1). In 1990, EA was found in one of three reference samples (3 of 18 alevins); in 1991, VDGG or EA or both were found in two of four reference samples (4 of 24 alevins).

Contamination of oiled-site sediments by North Slope crude oil was confirmed.

In 1989–1990 and 1990–1991 brood years, histopathological lesions were observed in 8 of 16 samples (35 of 93 alevins, 38%) from oiled sites and 3 of 7 samples (7 of 42 alevins, 17%) from reference sites. Lesions occurring at more than one study site included EA (most common), MDN, and VDGG. Lesions were moderate or severe in 17 of 21 fish from oiled sites in 1990 compared with 7 of 14 fish from oiled sites in 1991. Lesions were absent in December 1989 and March 1991 samples. No statistically significant differences were detected in the occurrence of lesions between oiled and reference sites in May–June 1990 ($P < 0.27$) or May 1991 ($P < 0.40$). However, because of small sample size, there was insufficient power to provide for meaningful statistical analysis.

The occurrence of lesions appeared to correlate with absorption of the yolk sac. Lesions were absent in samples with abundant yolk stores from both oiled and reference sites in December 1989 and March 1991 (Table 1). However, yolk stores were scored as absent or minimal in 96% (23 of 24) and 72% (13 of 18) of alevins with lesions from oiled sites in May–June 1990 and May 1991 samples, respectively. Yolk stores were not significantly different between oiled and reference sites in either 1990 or 1991 ($P < 0.40$ and $P < 0.60$, respectively).

Some infrequent lesions occurred only in alevins

TABLE 1.—Cytochrome P-4501A induction and histopathology in preemergent pink salmon alevins collected from within intertidal sediments of the streams at study sites in Prince William Sound, 1989–1991.

Location	Date	P-4501A score[a]	Tissue induced[b]	Lesions[c]	Median yolk score[d]
			1989–1990 brood year		
Reference					
Chenega Island R1	Dec 9	0	None	None	2.0
Herring Bay R2	May 30	0	None	None	2.0
Eshamy Bay R3	Jun 2	0	None	EA	1.0
Oiled					
Marsha Bay O1	Dec 8	2.5	K_t	None	2.0
Sleepy Bay O4	May 25	0	None	None	2.0
Sleepy Bay O4	May 25	1	VCC	None	2.0
Sleepy Bay O4	May 25	1	VCC	EA	2.0
Marsha Bay O1	May 30	2	VCC, $I_{m,c}$	EA, MDN, IHD	0.5
Herring Point O2	Jun 8	3	G_e, Sk, VCC	EA, MDN	0.5
Shelter Bay O3	Jun 8	4	G_e, $K_{s,t}$, H, B, PC, L_h, VCC, $I_{c,e,m}$	EA, MDN	1.0
			1990–1991 brood year		
Reference					
Shelter Bay R5	Mar 21	0	None	None	2.0
Kake Cove R4	Mar 22	0	None	VDGG	2.0
Shelter Bay R5	May 4	0	None	EA, VDGG	1.5
Kake Cove R4	May 5	0	None	None	2.0
Oiled					
Sleepy Bay O4	Mar 17	3	$G_{e,b}$, Y, $K_{c,s,v}$, H	None	2.0
Marsha Bay O1	Mar 21	3	L_s, P, G_e, I_e, $K_{s,v}$	None	2.0
Marsha Bay O1	Mar 21	2	I_e, K_c	None	2.0
Shelter Bay O3	Mar 21	2.5	Y, I_e, L_s, $K_{s,v}$	None	2.0
Shelter Bay O3	Mar 21	3	I_e, $K_{s,v}$, G_b, H, L_h, Y	None	2.0
Sleepy Bay O4	May 4	3	G_e, P, H, $K_{c,s,v}$, Y, $L_{h,c}$	EA	2.0
Shelter Bay O3	May 4	2	L_h, $K_{s,v}$	EA	2.0
Marsha Bay O1	May 14	0	None	EA, VDGG	1.0
Marsha Bay O1	May 14	0	None	EA, VDGG	1.0

[a] Sample scores of 0 (negative), 1 (very mild), 2 (mild), 3 (moderate), and 4 (strong) were based on the extent of occurrence and intensity of immunochemical staining of P-4501A in histological sections of six fish per sample.

[b] Tissue types that stained immunochemically for cytochrome P-4501A: $K_{c,s,t,v}$ = kidney collecting ducts, sinusoidal endothelium, tubular epithelium, and vascular endothelium; VCC = vertebral cord cartilage; $I_{e,m,c}$ = intestine enterocytes, midgut and cecal epithelium; $G_{b,e}$ = gill buds and epithelium; Sk = skin; H = atrial or ventricular endothelium; B = brain vessel endothelium; PC = peritoneal connective tissue; $L_{c,h,s}$ = liver central veins, hepatocytes, and sinusoidal endothelium; Y = yolk sac endothelium; P = pharyngeal epithelium.

[c] Indicates the presence of any of four lesion types in one or more of the fry in a sample: EA = epidermal atrophy; IHD = individual hepatocellular degeneration; MDN = myofiber degeneration and/or necrosis; VDGG = vacuolar degeneration of gastric glands.

[d] Yolk stores were qualitatively scored as 0 (none), 1 (minimal), or 2 (abundant).

from oiled sites. All six alevins from Marsha Bay (O1, May 1990) had mild to moderate vacuolar changes in the liver; spherical spaces, each about 25 μm in diameter and partly filled by basophilic wispy material, were scattered throughout the liver and were probably of hepatocellular or Ito cell origin (they were scored as IHD). Three of the six alevins from the same Marsha Bay sample (O1, May 1990) had mild to moderate amounts of individual necrotic cells (origin unknown) in the subepithelial connective tissue of the stomach. Two fish from Herring Point (O2, June 1990) had lesions: one had mild renal tubular necrosis; the other had mild thrombosis of a branchial vessel. One fish from Shelter Bay (O3, May 1991) had a 200-μm-diameter focus of bile duct hyperplasia in the liver.

Discussion

Immunohistochemical staining results from this study indicate that preemergent pink salmon from oiled sites in Prince William Sound responded to chronic PAH exposure by induction of cytochrome P-4501A up to 26 months after initial oiling of intertidal spawning sediments. Other studies have demonstrated that P-450 content drops to basal levels 3 weeks after removal of the inducing agent, although this rate may be temperature dependent (Kloepper-Sams and Stegeman 1989). By fall 1989, many argued that any *Exxon Valdez* oil residues remaining in and near salmon spawning sites were biologically inert. This study clearly shows that 2 years after the spill, oil residues contaminating in-

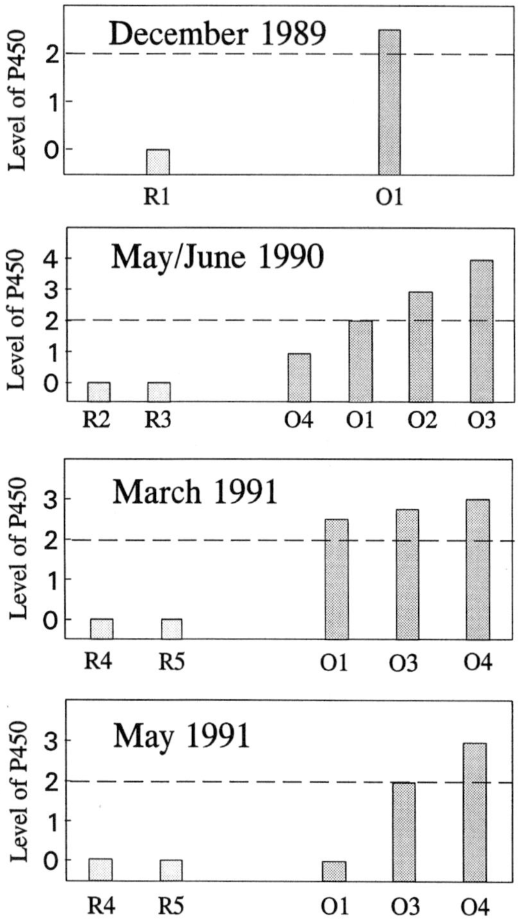

FIGURE 2.—Median cytochrome P-4501A induction scores for pink salmon alevin samples collected in December 1989, May–June 1990, March 1991, and May 1991 from reference (left) and oiled (right) sites. Induction scores are 0 = negative, 1 = very mild, 2 = mild, 3 = moderate, and 4 = severe. See Figure 1 for site locations.

tertidal substrates were not biologically inert and were still inducing measurable physiological responses in wild salmon.

Physiological systems of preemergent pink salmon alevins are competent to synthesize cytochrome P-4501A in response to oil contamination. Not only was P-4501A induced in hepatic tissues, but at least 11 tissues of nine organ systems were also inductively competent. Induction in the vascular endothelium of kidney, brain, heart, and ceca suggests a response to blood-borne xenobiotics. These results correspond well with reports of P-4501A localization in other fish species (Smolowitz et al. 1991). Induction in gill epithelium may indicate exposure to oil via water. However, P-4501A was also induced in gill epithelia when fish were exposed to xenobiotics via intraperitoneal injection (Smolowitz et al. 1991), indicating that the route of exposure may involve passage of PAHs via blood. Induction in cecal epithelium was observed by Smolowitz et al. (1991) in fish not exposed to food-borne xenobiotics but to intraperitoneal injection; inducers were thought to have entered the gut with the bile, followed by induction in mucosal tissue.

Elevated P-450 levels were not detected in eggs, not even in those collected at the same time and location as alevins with elevated P-450 levels. The absence of induction in eggs is consistent with other findings (Binder and Stegeman 1984; Goksøyr and Solberg 1987).

Studies have demonstrated that as a result of P-450-mediated metabolism, toxicity of PAHs greatly increases (Nebert and Gelboin 1968; Diamond and Clark 1970; Malaveille et al. 1975; Wood et al. 1976). The products of P-450-mediated metabolism bind to RNA (Blobstein et al. 1976), and DNA (Glover and Sims 1968; Sims et al. 1974) via arene oxide intermediates (Jerina and Daly 1974). The degree of metabolite binding is positively correlated to carcinogenic (Brookes and Lawley 1964) and mutagenic (Jerina and Daly 1974; Stegeman and Lech 1991) potential. These studies suggest that elevation of cytochrome P-4501A in preemergent pink salmon from oiled sites may create an increased potential for mutagenesis and carcinogenesis.

Fish exposed to xenobiotics have demonstrated an inverse correlation between MO activity and androgen, estrogen, and corticoid concentrations (Sivirajah et al. 1978), and testicular development (Truscott et al. 1983). Spies et al. (1988) found an inverse correlation between MO activity and parameters of female reproductive success (e.g., decreased proportion of viable eggs, decreased fertilization success, and decreased embryological success). Bue et al. (1996, this volume) found that pink salmon egg mortality was elevated in oil-contaminated streams during the falls of 1989, 1990, 1991, and 1992. These observations suggest that the reproductive potential of salmon developing in contaminated sediments may be diminished. As part of ongoing oil spill studies, continuing effort is being directed to evaluate the long-term reproductive impact to pink salmon resulting from exposure to oil-contaminated spawning substrates.

Unlike induction of P-4501A, the relationship of most histopathologic lesions was less clearly related to oil exposure. Lesions such as IHD, subepithelial gastric cell necrosis, renal tubular epithelial necrosis, and biliary epithelial hyperplasia occurred only in alevins from oiled sites. Because none of these lesions occurred at more than one site, the lesions cannot be unequivocally related to oil exposure. However, similar lesions have previously been related to oil exposure in fish: (1) hepatocellular microvesicles, a type of degenerative change, in Atlantic cod *Gadus morhua* exposed to Hibernian crude oil in the laboratory (Khan and Kiceniuk 1984) and (2) necrosis of renal tubular epithelial cells and interrenal tissue in the mummichog *Fundulus heteroclitus* exposed to naphthalene, a component of crude oil, in the laboratory (DiMichele and Taylor 1978).

Lesions occurring at more than one site were even less related to oil exposure. First, myodegeneration and necrosis were peracute lesions and probably were related to physical trauma of removing alevins from the gravel during collection. Second, VDGG occurred in alevins from both oiled and reference sites and was thought to represent a type of artifact or unrelated lesion rather than an oil-associated lesion. However, hydropic degeneration of gastric gland epithelial cells was described in plaice *Pleuronectes platessa* associated with the *Amoco Cadiz* oil spill (Haensly et al. 1982). And third, EA also occurred in alevins from both oiled and reference sites. Literature reports vary on the effects of oil exposure on epidermal thickness in fish. Studies with Venezuelan crude oil found decreased epidermal thickness (Burton et al. 1984, 1985), whereas studies with Hibernia crude oil (Khan 1991) and Louisiana crude oil (Solangi and Overstreet 1982) found gill lamellar hyperplasia. Decreased epidermal thickness and decreased mucous cell abundance, the primary components of EA, may be developmental changes associated with absorption of the yolk that are unrelated to oil exposure. This subject deserves further study.

Of interest was the absence of EA or other lesions in all alevins with abundant yolk stores. Lesion prevalence might have increased had all samples been collected after alevins had absorbed most of their yolk sacs. Korn and Rice (1981) found that fish with abundant yolk were most tolerant to aromatic hydrocarbons. The amount of yolk influenced sensitivity because the lipophilic, aromatic hydrocarbons were partitioned into the yolk, reducing the availability of the hydrocarbons to other embryonic tissues. Oil exposure may not, however, result in a pathognomonic lesion. As evidence, alevins from both oiled and reference streams had lesions when their yolk reserves were minimal.

In previous studies with the water-soluble fraction of Prudhoe Bay crude oil, sensitivity (mortality) to aromatic hydrocarbons in pink salmon increased from egg to fry (Rice et al. 1975; Moles et al. 1979; Korn and Rice 1981). Emergent fry were the most sensitive. However, no histopathology data is available on the effects of hydrocarbons on the early life stages of salmon. It should be reemphasized that all fish observed in the current study were collected early in their development and before emergence from spawning gravels. No attempt was made to collect older fish that were exposed as juveniles to contaminated substrates. Therefore, long-term effects of oil exposure on lesion development were not assessed. Laboratory studies are underway to determine more definitively the role of developmental age at oil exposure on the generation of histopathologic lesions in pink salmon alevins.

The current study demonstrates that pink salmon alevins in some heavily oiled sites continued to be exposed to physiologically relevant levels of hydrocarbons more than 2 years after the initial spill. Furthermore, this study demonstrates the utility of immunohistochemical detection of cytochrome P-4501A in fish as a marker of hydrocarbon exposure (Stegeman et al. 1991). Finally, these results should assist in developing appropriate standards for monitoring treatment of streams used by anadromous fish after major oil contamination.

Acknowledgments

We gratefully acknowledge project development and support from L. Trasky, M. Kuwada, J. Sullivan, D. Malins, and R. Spies. Assistance with fieldwork was provided by R. Talvi, R. Sinnott, R. Gustin, T. Crow, C. Rozen, A. Weseman, M. Erickson, E. Weiss, K. Sundet, G. Pavia, M. Abel, A. Butler, and J. Matthews. B. Bue provided statistical analysis. S. Peyer and N. Magnus provided clerical support.

References

Binder, R. L., and J. J. Stegeman. 1984. Microsomal electron transport and xenobiotic monooxygenase activities during the embryonic period of development in the killifish, *Fundulus heteroclitus*. Toxicology and Applied Pharmacology 73:432–443.

Blobstein, R., G. Harvey, C. Harris, H. Autrup, H. Kazai, and K. Nakanishi. 1976. Benzo[*a*]pyrene diol epoxides as intermediates in nucleic acid binding in vitro and in vivo. Science 193:592–595.

Brookes, P., and P. D. Lawley. 1964. Evidence for the binding of polynuclear aromatic hydrocarbons to the nucleic acids of mouse skins: relation between carcinogenic power of hydrocarbons and their binding to deoxyribonucleic acid. Nature 202:781–784.

Bue, B. G., S. Sharr, S. D. Moffitt, and A. K. Craig. 1996. Effects of the *Exxon Valdez* oil spill on pink salmon embryos and preemergent fry. American Fisheries Society Symposium 18:619–627.

Burns, K. A. 1976. Microsomal mixed function oxidases in an estuarine fish, *Fundulus heteroclitus*, and their induction as a result of environmental contamination. Comparative Biochemistry and Physiology 53(B): 443–446.

Burton, D., M. P. Burton, and D. R. Idler. 1984. Epidermal condition in post-spawned winter flounder, *Pseudopleuronectes americanus* (Walbaum), maintained in the laboratory and after exposure to crude petroleum. Journal of Fish Biology 25:593–606.

Burton, D., M. P. Burton, B. Truscott, and D. R. Idler. 1985. Epidermal cellular proliferation and differentiation in sexually mature male *Salmo salar* with androgen levels depressed by oil. Proceedings of the Royal Society of London B Biological Sciences 225: 121–128.

Cameron, J. A., and R. L. Smith. 1980. Ultrastructural effects of crude oil on early life stages of Pacific herring. Transactions of the American Fisheries Society 109:224–228.

Conover, W. J. 1980. Practical nonparametric statistics, 2nd edition. Wiley, New York.

Diamond, L., and H. F. Clark. 1970. Comparative studies of the interaction of benzo[*a*]pyrene with cells derived from the poikilothermic and homeothermic vertebrates. I. Metabolism of benzo[*a*]pyrene. Journal of the National Cancer Institute 45:945–946.

DiMichele, L., and M. H. Taylor. 1978. Histopathological and physiological responses of *Fundulus heteroclitus* to naphthalene exposure. Journal of the Fisheries Research Board of Canada 35:1060–1066.

Foureman, G. L., N. B. White Jr., and J. R. Bend. 1983. Biochemical evidence that winter flounder (*Pseudopleuronectes americanus*) have induced hepatic cytochrome P-450-dependent monooxygenase activities. Canadian Journal of Fisheries and Aquatic Sciences 40:854–865.

Glover, P. L., and P. Sims. 1968. Enzyme-catalyzed reactions of polycyclic hydrocarbons with deoxyribonucleic acid and protein in vitro. Biochemical Journal 110:159–160.

Goksøyr, A., and T. S. Solberg. 1987. Cytochrome P-450 in fish larvae: immunochemical detection of responses to oil pollution. Sarsia 72:405–407.

Haensly, W. E., J. M. Neff, J. R. Sharp, A. C. Morris, M. F. Bedgood, and P. D. Boem. 1982. Histopathology of *Pleuronectes platessa* L. from Aber Wrac'h and Aber Benoit, Brittany, France: long-term effects of the *Amoco Cadiz* crude oil spill. Journal of Fish Diseases 5:365–391.

Hawkes, J. W., and C. M. Stehr. 1982. Cytopathology of the brain and retina of embryonic surf smelt (*Hypomesus pretiosus*) exposed to crude oil. Environmental Research 27:164–178.

Hinton, D. E., and D. J. Lauren. 1990. Integrative histopathological approaches to detecting effects of environmental stressors on fishes. American Fisheries Society Symposium 8:51–66.

Jerina, D. M., and J. W. Daly. 1974. Arene oxides: a new aspect of drug metabolism. Science 185:573–582.

Jimenez, B. D., and J. J. Stegeman. 1990. Detoxication enzymes as indicators of environmental stress on fish. American Fisheries Society Symposium 8:67–79.

Khan, R. A. 1991. Effect of oil-contaminated sediment on the longhorn sculpin (*Myoxocephalus octodecemspinosus*) following chronic exposure. Bulletin of Environmental Contamination and Toxicology 47:63–69.

Khan, R. A., and J. Kiceniuk. 1984. Histopathological effects of crude oil on Atlantic cod following chronic exposure. Canadian Journal of Zoology 62:2038–2043.

Kloepper-Sams, P. J., and J. J. Stegeman. 1989. The temporal relationships between P-450 protein content, catalytic activity, and mRNA levels in the teleost *Fundulus heteroclitus* following treatment with B-naphthoflavone. Archives of Biochemistry and Biophysics 268:525–535.

Korn, S., and S. Rice. 1981. Sensitivity to, and accumulation and depuration of, aromatic petroleum components by early life stages of coho salmon *Oncorhynchus kisutch*. Rapports et Proces-Verbaux des Reunions Conseil International pour l'Exploration de la Mer 178:87–92.

Kurelec, B., S. Britvic, M. Rijavec, W. E. G. Muller, and R. K. Zahn. 1977. Benzo[*a*]pyrene monooxygenase induction in marine fish-molecular response to oil pollution. Marine Biology 44:211–216.

Malaveille, C., H. Bartsch, P. L. Glover, and P. Sims. 1975. Mutagenicity of non-K-region diols and diolepoxides of benz[*a*]anthracene and benz[*a*]pyrene. Biochemical and Biophysical Research Communications 66:693–700.

McKim, D. C. 1977. Evaluation of tests with early life stages of fish for predicting long-term toxicity. Journal of the Fisheries Research Board of Canada 43: 1148–1154.

Moles, A., S. D. Rice, and S. Korn. 1979. Sensitivity of Alaskan freshwater and anadromous fishes to Prudhoe Bay crude oil and benzene. Transactions of the American Fisheries Society 108:408–414.

Nebert, D. W., and H. V. Gelboin. 1968. Substrate inducible microsomal aryl hydroxylase in mammalian cell culture. II. Cellular response during enzyme induction. Journal of Biological Chemistry 243:6250–6261.

Park, S. S., H. Miller, A. V. Klotz, P. J. Kloepper-Sams, J. J. Stegeman, and H. V. Gelboin. 1986. Monoclonal antibodies to liver microsomal cytochrome P-450E of the marine fish *Stenotomus chrysops* (scup): cross-reactivity with 3-methylcholanthrene induced rat cytochrome P-450. Archives of Biochemistry and Biophysics 253:268–278.

Payne, J. F. 1976. Field evaluation of benzopyrene hydroxylase induction as a monitor for marine petroleum pollution. Science 191:945–946.

Rice, S. D., D. A. Moles, and J. W. Short. 1975. The effects of Prudhoe Bay crude oil on survival and growth of eggs, alevins, and fry of pink salmon, *Oncorhynchus gorbuscha*. Pages 503–507 *in* Proceedings, 1975 joint conference on prevention and control of oil pollution. American Petroleum Institute, Washington, DC.

Sims, P., P. L. Glover, A. Swaisland, K. Pal, and A. Hewer. 1974. Metabolic activation of benzo[*a*]pyrene proceeds by a diol-epoxide. Nature 252:326–328.

Sivirajah, K., C. S. Franklin, and W. P. Williams. 1978. The effects of polychlorinated biphenyls on plasma steroid levels and hepatic microsomal enzymes in fish. Journal of Fish Biology 13:401–409.

Smolowitz, R. M., M. E. Hahn, and J. J. Stegeman. 1991. Immunohistochemical localization of cytochrome P-450IA1 induced by 3,3′,4,4′-tetrachlorobiphenyl and by 2,3,7,8-tetrachlorodibenzo furan in liver and extrahepatic tissues of the teleost *Stenotomus chrysops* (scup). Drug Metabolism and Disposition 19: 113–123.

Solangi, M. A., and R. M. Overstreet. 1982. Histopathological changes in two estuarine fishes, *Menidia beryllina* (Cope) and *Trinectes maculatus* (Bloch and Schneider), exposed to crude oil and its water soluble fractions. Journal of Fish Diseases 5:13–35.

Spies, R. B., J. S. Felton, and L. Dillard. 1982. Hepatic mixed-function oxidases in California flatfishes are increased in contaminated environments and by oil and PCB ingestion. Marine Biology 70:117–127.

Spies, R. B., D. W. Rice Jr., and J. Felton. 1988. Effects of organic contaminants on reproduction of the starry flounder *Platichthys stellatus* in San Francisco Bay. II. Reproductive success of fish captured in San Francisco Bay and spawned in the laboratory. Marine Biology 98:191–200.

Stegeman, J. J., and J. J. Lech. 1991. Cytochrome P-450 monooxygenase systems in aquatic species: carcinogen metabolism and biomarkers for carcinogen and pollutant exposure. Environmental Health Perspectives 90:101–109.

Stegeman, J. J., R. M. Smolowitz, and M. E. Hahn. 1991. Immunohistochemical localization of cytochrome P-4501A1 induced by 3,3′4,4′-tetrachlorobiphenyl and by 2,3,7,8-tetrachlorodibenzofuran in extrahepatic tissues of a marine teleost (*Stenotomus chrysops*). Drug Metabolism and Disposition 19:113–123

Truscott, B., J. M. Walsh, M. P. Burton, J. F. Payne, and D. R. Idler. 1983. Effect of acute exposure to crude petroleum on some reproductive hormones in salmon and flounder. Comparative Biochemistry and Physiology 75(C):121–130.

Wood, A. W., and six coauthors. 1976. Metabolism of benzo[*a*]pyrene and benzo[*a*]pyrene derivatives to mutagenic products by highly purified hepatic microsomal enzymes. Journal of Biological Chemistry 251: 4882–4890.

American Fisheries Society Symposium 18:518–532, 1996

Abundance and Growth of Juvenile Pink Salmon in Oiled and Non-Oiled Locations of Western Prince William Sound after the *Exxon Valdez* Oil Spill

A. C. WERTHEIMER AND A. G. CELEWYCZ

National Marine Fisheries Service
Alaska Fisheries Science Center, Auke Bay Laboratory
11305 Glacier Highway, Juneau, Alaska 99801, USA

Abstract.—Juvenile pink salmon *Oncorhynchus gorbuscha* were sampled in oiled and non-oiled locations during their initial residency in the nearshore waters of Prince William Sound in 1989 and 1990. Pink salmon moved rapidly from sheltered bays to more exposed and steep beaches in migration corridors. Pink salmon were less abundant in oiled than in non-oiled locations in both 1989 and 1990; only 18% (1989) and 14% (1990) of pink salmon captured in systematic sampling were from oiled locations. Because the pattern of abundance did not change between years as oil contamination diminished, we attribute these differences to geographic variation in local production of pink salmon fry and in their migration pathways to the Gulf of Alaska, rather than to exposure to oil. Juvenile pink salmon were significantly smaller and had significantly lower apparent growth in oiled than in non-oiled locations in 1989, but not in 1990. In 1989, when growth was lower in oiled locations, hydrocarbon contamination of the environment was greatest, and direct contamination of juvenile pink salmon was observed. In 1990, when growth was similar between oiled and non-oiled locations, hydrocarbon contamination of the environment was greatly diminished, and no direct contamination of juvenile pink salmon was observed. For these reasons, we attribute the lower growth in oiled locations in 1989 to exposure to *Exxon Valdez* crude oil.

The salmon fishery is the largest and most valuable commercial fishery in Prince William Sound; in 1988, salmon had an ex-vessel value of $76 million dollars, over 80% of the total for all fisheries in the Sound (ADFG 1989). Most of the salmon landed are pink salmon *Oncorhynchus gorbuscha*. Besides their commercial value, salmon are harvested by subsistence and recreational fishers and are a critical element of the Prince William Sound ecosystem. After the *Exxon Valdez* oil spill, several Natural Resource Damage Assessment (NRDA) studies focused on pink salmon because of their life history, economic value, and numerical abundance.

Juvenile pink salmon in Prince William Sound were at risk of exposure and contamination by the *Exxon Valdez* oil spill because of their time of entry into the marine environment, their habitat use, and their migration routes to the Gulf of Alaska. The oil spill occurred just before juvenile pink salmon from freshwater habitats and hatcheries entered the marine environment. In Prince William Sound, wild pink salmon fry typically leave their natal streams in April and May (Kirkwood 1972). Hundreds of millions of pink salmon fry are released into the Sound from area hatcheries in late April and May (Olsen 1991). In 1989, juvenile salmon in the spill area encountered large quantities of oil in both the water and sediments of their nearshore habitats. The primary migration route of juvenile salmon from Prince William Sound to the Gulf of Alaska is through the Western Sound and the southwest passages, the same route the oil travelled into the Gulf of Alaska (Willette, 1996, this volume). Thus, large numbers of juvenile salmon from areas of the Sound not directly impacted by oil were also potentially exposed as they migrated seaward.

Oil can disrupt growth and survival of juvenile salmon. Oil can be directly toxic to juvenile salmon (Rice et al. 1975, 1984) and to their epibenthic and zooplanktonic prey (Samain et al. 1980; Bodin 1988). Exposure to sublethal levels of hydrocarbons, either directly in seawater or via ingestion, can affect metabolism and reduce growth of juvenile salmon (Rice et al. 1975; Schwartz 1985; Vignier et al. 1992).

Early marine residency is a critical phase in the life history of salmon that significantly affects year-class strength (Parker 1968; Walters et al. 1978; Bax 1983; Nichelson 1986). Juvenile pink salmon generally occupy nearshore estuarine waters when they first emigrate to seawater (Healey 1980). Growth of pink salmon during the early marine phase is extremely rapid (LeBrasseur and Parker 1964; Healey 1980), which is important if the fish are to escape size-selective predation (Parker 1971; Hargreaves and LeBrasseur 1985); early marine growth has been directly linked to survival within a year-class (Mortensen et al. 1991). If the oil spill reduced growth during this life history phase, affected pop-

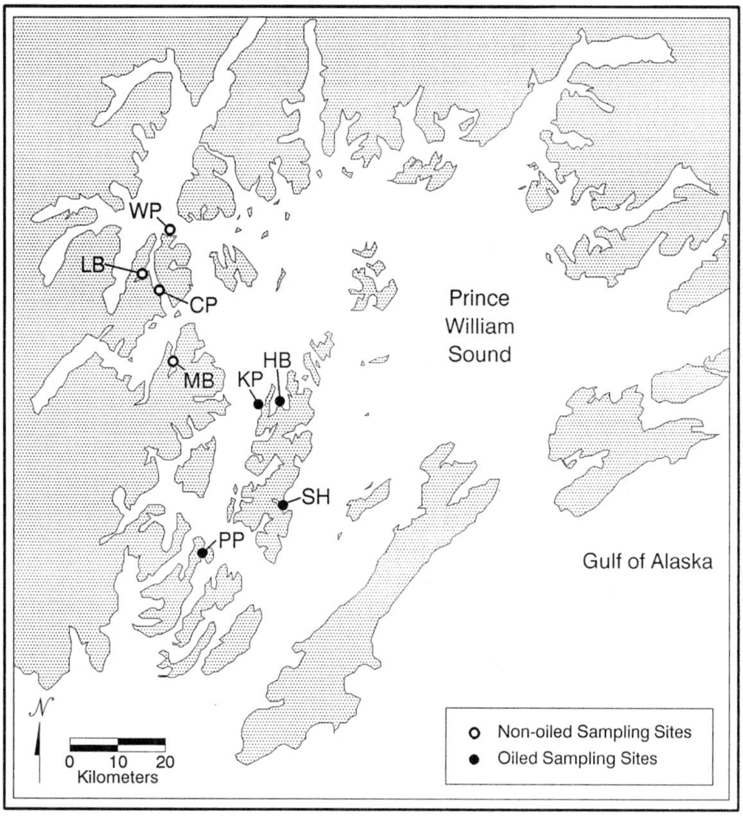

FIGURE 1.—Sampling locations in western Prince William Sound. Oiled locations—Herring Bay (HB), Snug Harbor (SH), Knight Island Passage (KP), and Prince of Wales Passage (PP)—are marked by filled circles; non-oiled sampling locations—McClure Bay (MB), Long Bay (LB), Wells Passage (WP), and Culross Passage (CP)—are marked by unfilled circles.

ulations of salmons would likely also have reduced survival from fry to returning adults.

The purpose of this study was to determine the impact of the oil spill on juvenile pink salmon. Specific objectives were to compare abundance of juvenile pink salmon between oiled and non-oiled locations; to compare distribution and habitat use by juvenile pink salmon between 1989 and 1990; and to compare the size and growth of juvenile pink salmon between oiled and non-oiled locations. A companion NRDA study focused on the impact of the *Exxon Valdez* oil spill on the migration and growth of juvenile pink salmon tagged and released from Prince William Sound hatcheries (Willette 1996).

Methods

Sample Collection and Processing

The sampling design incorporated eight locations: four oiled and four non-oiled (Figure 1). Within both oiled and non-oiled categories, two of the four locations were in bays and two in migration corridors (passages). At each of the eight locations, three habitat types (low-, medium-, and steep-gradient beaches) were sampled. Low-gradient beaches were less than 10% grade with granule–pebble substrate, medium-gradient beaches were 12–25% grade, with pebble–cobble substrate, and steep-gradient beaches were more than 50% grade with bedrock or large boulder substrate.

The study locations were paired a priori for pairwise comparisons between oiled and non-oiled locations. These pairings were (non-oiled first) McClure Bay–Herring Bay, Long Bay–Snug Harbor, Culross Passage–Prince of Wales Passage, and Wells Passage–Knight Island Passage. Sample sites for each habitat type within pairs of oiled and non-oiled locations were selected for similarity in wave exposure, macrophyte coverage, and substrate. In 1989, one beach of each habitat type was sampled at each location, for a total of 24 systematically sampled sites; in 1990, two beaches of each habitat type were sampled at each location, for a total of 48 systematically sampled sites. Beaches were sampled on each of five cruises between 10 April and 26 June 1989 and on each of four cruises between 16 April and 14 June 1990.

In addition to these systematically sampled sites, juvenile pink salmon were also sampled along 3–5 km of adjacent shoreline at each location. This supplemental sampling was intended to collect salmon tagged with coded wire tags as well as to supplement samples for hydrocarbon analysis when insufficient numbers of fish were collected at the systematically sampled beaches. During supplemental sampling, adjacent shoreline was visually surveyed to locate aggregations of juvenile salmon; then fish from these aggregations were captured with either beach seines or dip nets.

Beach seine hauls at the systematic sites were restricted to tide levels between −0.3 and +0.9 m to minimize tidal effects between sites. Fish were captured with 37-m beach seines at low- and medium-gradient sites. At the steep-gradient sites, we used a 37-m seine modified with a floor of 6-mm green mesh formed by a 9-m lead line connecting the bottom intersections of the wings with the bunt.

Catches were sorted by species and enumerated; all salmon were checked for the presence of coded wire tags. Samples of juvenile pink salmon (up to 60 fish from each haul) were preserved in buffered 10% formalin for later fork length (FL) and weight measurements. Fish collected for size analysis were retained in formalin for at least 45 d to assure uniform shrinkage. Coded-wire-tagged fish were stored frozen until processed by the Tag Processing Laboratory of the Alaska Department of Fish and Game in Juneau; data from tagged fish were incorporated in the analysis by Willette (1996).

Statistical Analysis

The univariate approach to analysis of variance (ANOVA) of a repeated measures design (Frane 1980) was used to analyze the systematic catch data (catch per unit effort, CPUE: number of fish per haul) for pink salmon. In the overall analysis for 1989 and 1990 combined, the factors considered were oil (dichotomous scale: oiled or non-oiled), year, time period (each sampling cruise was considered a time period), bay or corridor, location (nested within oil and bay–corridor), and habitat. This analysis included only the same time periods and sites that were sampled in both years.

Because the data for the overall ANOVA made up only a subset of all the systematic seine hauls from each year, ANOVAs were also run on the complete data sets of systematic catches from each year separately. The pink salmon catch data were subdivided further because of the results of the overall ANOVA, and separate analyses were conducted on bays in 1989, corridors in 1989, bays in 1990, and corridors in 1990. For each of these analyses, the factors were oil, time period, location (nested within oil), and habitat.

Statistical distributions of CPUE were highly skewed because of a high number of zero catches. Transformations were not effective at eliminating this skewness. Thus, ranks of the CPUE were used in each of the analyses mentioned above. An ANOVA of ranks is conditionally distribution free and usually has good efficiency; the approximate level of significance used in the test is usually close to the true level of significance, no matter what the underlying population distribution may be (Conover 1980).

To corroborate the results of the ANOVA on ranked CPUE, a nonparametric analysis was also used to test the hypothesis of no difference in abundance of juvenile pink salmon between oiled and non-oiled locations. Differences in CPUE between matched cells of the a priori pairs of oiled and non-oiled locations were compared with the nonparametric Wilcoxon matched-pairs, signed-ranks test. Cells of the paired locations were matched by time period, bay–corridor, and habitat. In 1989, 56 such comparisons were possible, and in 1990, 95 were possible. Differences in CPUE were also tested separately for bays and corridors.

Frequency of occurrence is a measure of the patchiness of schools of salmon. In both 1989 and 1990, frequency of occurrence of pink salmon was examined in terms of catches exceeding 0 fish, 100 fish, and 1,000 fish. Frequencies of occurrence were analyzed with a chi-square test.

Effects of the oil spill on size, growth, and condition of juvenile pink salmon were examined by comparing mean FL, apparent growth rate (the change in mean weight over time), and weight–length relationships between oiled and non-oiled locations. Mean FL was analyzed by two statistical methods: ANOVA (1990 only) and the nonparametric Wilcoxon matched-pairs, signed-ranks test (both years). An ANOVA could not be used for analysis of the 1989 FL data because of the large number of empty cells due to zero catches. The ANOVA used for the analysis of FL in 1990 included the factors oil, time period, bay–corridor, location (nested within oil and bay–corridor), and habitat. The nonparametric method tested only the null hypothesis of no difference in FL between oiled and non-oiled locations. It preserved possible location and habitat differences by comparing samples from the same time period and habitat type for the a priori matched pairs of oiled and non-oiled locations.

Apparent growth rate of pink salmon at each habitat type within a location was calculated as the slope of the regression of natural logarithm of weight as a function of time in days. Analysis of covariance was used to determine if fish could be pooled over habitats within a sampling location. Pooling was rejected. Analysis of variance was used to compare the apparent growth rates of pink salmon in corridors; factors considered were year, oil, location (nested within oil), and habitat. Bays were not considered in the ANOVA because there were too many empty cells to calculate valid growth rates in bays in 1989.

Fish condition was examined by comparing least-squares regression parameters between oiled and non-oiled locations as recommended by Cone (1989). The relationship between fish weight and length was described by:

$$\log_e(w) = a + b \cdot \log_e(\mathrm{FL}),$$

where w is weight, a is the intercept, FL is fork length, and b is the slope, the exponential rate of increase of weight with length. In this analysis, condition of the fish is a measure of the weight at a given length and is determined by the slope and the relative elevation of the regression line.

Analysis of covariance was used to test for homogeneity of slopes and equality of adjusted means between bays and corridors for non-oiled locations and for oiled locations. If the slopes and adjusted means were not significantly different ($P > 0.1$), data were pooled as to bay–corridor and tested between oiled and non-oiled locations. If slopes or adjusted means were significantly different between bays and corridors, data for oiled and non-oiled locations were tested separately for bays and corridors.

Results

Abundance and Habitat Utilization

More pink salmon were captured in non-oiled than in oiled locations in the systematic hauls in both 1989 and 1990. In 1989, 33,290 pink salmon were captured in 120 systematic hauls; 43% of hauls made zero catches, and the high catch was over 8,000 fish. More than four times as many pink salmon were captured in non-oiled (27,200 fish) than in oiled locations (6,090 fish). In 1990, 81,869 pink salmon were captured in 191 hauls; 28% of hauls made zero catches and the high catch was 22,977 fish. More than six times as many pink salmon were captured in non-oiled (70,496 fish) than in oiled locations (11,373 fish). Pink salmon CPUE was significantly greater in non-oiled locations for both years combined ($P = 0.03$, ANOVA). The nonsignificant year–oil interaction ($P > 0.1$) indicated that this trend was consistent in both years. Over both years, over five times more pink salmon were captured in non-oiled than in oiled locations (Figure 2). Additionally, in each year, the Wilcoxon test for matched pairs of hauls indicated higher CPUE in non-oiled locations ($P = 0.043$ in 1989 and 0.092 in 1990, Table 1).

Pink salmon CPUE was significantly greater in corridors than in bays for both years combined ($P = 0.004$, ANOVA). Over both years, 94% of the pink salmon were captured in corridors and only 6% were captured in bays (Figure 2). Because of the large difference in CPUE between bays and corridors, separate analyses were run on bays and corridors for each year.

When pink salmon CPUE in bays was examined separately by year, the data showed two consistent patterns. First, CPUE in both years was similar ($P > 0.1$) in oiled and non-oiled bays in both the ANOVA and the Wilcoxon test (Table 1). In both years, pink salmon occurred just as frequently in oiled as in non-oiled bays (Table 2). Second, medium-gradient bottom was the most important habitat for pink salmon in bays. In 1990, pink salmon CPUE differed significantly between habitats ($P = 0.05$); CPUE was greatest in the medium-gradient habitat (Figure 2). In 1989, habitat use did not differ significantly among gradients ($P > 0.1$), but CPUE in bays exhibited the same pattern in 1989 as in 1990: the sample mean CPUE was highest in the medium-gradient habitat (Figure 2; Table 3). In 1989, time period ($P = 0.023$) and the time period–habitat interaction ($P = 0.055$) were also significant in explaining differences in CPUE in bays. Catch rate peaked in early May on the medium-gradient habitat, in late May in the low-gradient habitat, and in early June in the steep-gradient habitat.

When pink salmon CPUE in corridors was examined separately by year, different factors were significant in explaining variations each year. In 1989, CPUE was higher in non-oiled than in oiled corridors in the ANOVA ($P = 0.035$) and the Wilcoxon test ($P = 0.013$; Table 1). Pink salmon occurred more frequently in non-oiled than oiled corridors in 1989 ($P < 0.050$; Table 2). In 1990, although more

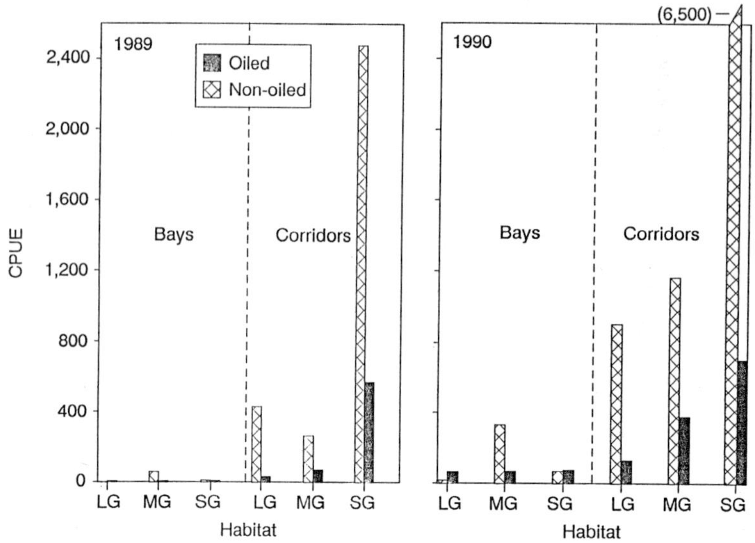

FIGURE 2.—Mean catches per unit effort (CPUEs) of juvenile pink salmon by habitat type in oiled and non-oiled bays and corridors of Prince William Sound, 1989 and 1990; LG = low-gradient, MG = medium-gradient, and SG = steep-gradient habitat.

than six times as many pink salmon were captured in non-oiled as in oiled corridors (Figure 2), this difference was not significant ($P > 0.1$; Table 2). Pink salmon occurred just as frequently in oiled as in non-oiled corridors in 1990 ($P > 0.1$). In corridors in 1990, oil was significant only in the context of the time period–oil interaction ($P = 0.026$). In non-oiled corridors, CPUE peaked in early May; in oiled corridors, CPUE peaked in late May.

In corridors, pink salmon schools were larger but occurred less frequently in the steep-gradient habitat than in other habitats. Although 74% of the pink salmon in corridors were captured in the steep-gradient habitat over both years, in neither year were the differences in CPUE between habitats significant ($P > 0.1$; Table 3). The highest CPUE and greatest variability in CPUE were in the steep-gradient habitat. Over both years, 25% of the hauls at steep-gradient sites in corridors captured more than 1,000 pink salmon, and 40% captured no pink salmon. In corridors, pink salmon were captured most frequently in the medium-gradient habitat and least frequently in the steep-gradient habitat in each year, although this trend was significant only in 1990 ($P < 0.05$; Table 3).

TABLE 1.—Summary of Wilcoxon matched-pairs, signed-ranks tests for catch per unit effort (CPUE) of juvenile pink salmon in Prince William Sound, 1989 and 1990. A negative value for the median difference indicates higher CPUE in non-oiled than in oiled locations. The CL values are 95% confidence limits on the median differences.

Year and location	Pairs	Wilcoxon statistic	P	Median difference	CL
1989					
All	56	289	0.043	−24	−107, 0
Bays	28	75	0.737	0	−2, 8
Corridors	28	77	0.013	−114	−655, −4
1990					
All	95	925	0.092	−4	−32, 0
Bays	48	328	0.273	−2	−14, 1
Corridors	7	260	0.112	−52	−290, 3

Timing of Catch

The pattern of the total catch (systematic and supplemental) of pink salmon over time differed between bays and corridors in both 1989 and 1990 (Figure 3). Catch in bays was generally lower than in corridors, except for the first sampling in early April. Catch in bays was small and generally stable in April and May. Catch in corridors was small in April, increased rapidly to peak in May, and declined in early June. In 1989, when sampling extended to late June, catch continued to decline in the non-oiled corridors but increased slightly in the oiled corridors.

TABLE 2.—Abundances of juvenile pink salmon in Prince William Sound, 1989 and 1990, as measured by five variables categorized by bays and corridors and by oiled and non-oiled locations. Significances (*P*) were based on analyses of variance (ANOVAs) for rank of catch per unit effort (CPUE) and on chi-square tests for frequencies.

Year and variable	Bays			Corridors		
	Non-oiled	Oiled	*P*	Non-oiled	Oiled	*P*
1989						
CPUE	23	8	[a]	1,057	225	[a]
Mean rank	24	25	NS	29	20	<0.05
Frequency of occurrence (%)	37	42	NS	83	54	<0.05
Frequency of >100 fish (%)	1	0	[b]	37	21	NS
Frequency of >1,000 fish (%)	0	0	[b]	25	8	[b]
1990						
CPUE	70	34	[a]	1,429	203	[a]
Mean rank	53	44	NS	55	41	NS
Frequency of occurrence (%)	75	62	NS	83	69	NS
Frequency of >100 fish (%)	10	8	[b]	43	25	<0.10
Frequency of >1,000 fish (%)	2	0	[b]	17	6	NS

[a]Not tested.
[b]Not enough samples for valid test.

Size, Growth, and Condition

Pink salmon were larger in non-oiled than oiled locations in 1989, but not in 1990. Based on comparisons of matched pairs, mean FL was significantly ($P = 0.066$) greater in non-oiled than in oiled locations in 1989 (Table 4). Mean FL was significantly ($P = 0.026$) greater in non-oiled than in oiled corridors, but mean FL was similar ($P > 0.1$) in oiled and non-oiled bays (Table 4). In 1990, pink salmon mean FL was similar ($P > 0.1$) in oiled and non-oiled locations overall, and also in bays and corridors considered separately (Table 4; the same result was obtained by ANOVA). Mean FL increased significantly ($P < 0.001$) with time in 1990.

In both years, mean FL of pink salmon was similar in oiled and non-oiled locations during April, but mean lengths were greater in non-oiled than in oiled locations during May and June (Figures 4 and 5). Likewise, fish in corridors had larger mean lengths than did fish in bays after early May. The interaction of time period with bay–corridor in the 1990 ANOVA for FL was significant ($P = 0.047$), indicating that the divergence in mean FL between bays and corridors in May (Figure 5) was significant.

Length frequency distributions of pink salmon differed between bays and corridors in both years (Figures 6 and 7). In bays, FL had a mode of 32–33 mm in April and May, which indicated that the fish were predominantly recent migrants from freshwater. The distributions of the few fish captured in

TABLE 3.—Abundances of juvenile pink salmon in Prince William Sound, 1989 and 1990, as measured by five variables categorized by habitats separately in bays and corridors; LG = low-gradient habitat, MG = medium-gradient habitat, and SG = steep-gradient habitat. Significances (*P*) were based on analyses of variance (ANOVAs) for rank of catch per unit effort (CPUE) and on chi-square tests for frequencies.

Year and variable	Bays				Corridors			
	LG	MG	SG	*P*	LG	MG	SG	*P*
1989								
CPUE	5	32	11	[a]	230	166	1,526	[a]
Mean rank	23	28	22	NS	24	25	25	NS
Frequency of occurrence (%)	31	56	31	NS	69	75	62	NS
Frequency of >100 fish (%)	0	6	0	[b]	31	31	31	NS
Frequency of >1,000 fish (%)	0	0	0	[b]	12	6	31	[b]
1990								
CPUE	20	100	35	[a]	259	449	1,793	[a]
Mean rank	80	104	75	<0.05	102	125	93	NS
Frequency of occurrence (%)	69	87	50	<0.01	81	87	59	<0.05
Frequency of >100 fish (%)	3	19	6	[b]	28	45	28	NS
Frequency of >1,000 fish (%)	0	3	0	[b]	3	9	22	[b]

[a]Not tested.
[b]Not enough samples for valid test.

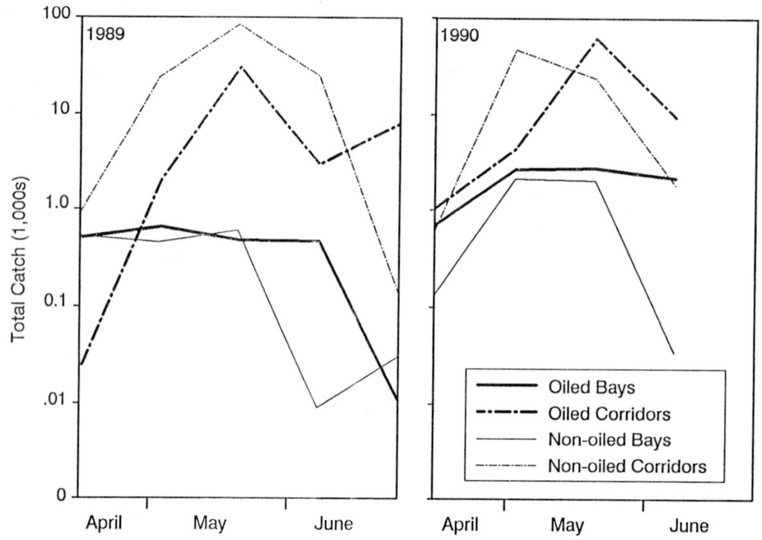

FIGURE 3.—Total catches (logarithmic scale) of juvenile pink salmon in Prince William Sound in April–June, 1989 and 1990. Fish captured in outer Snug Harbor were not included.

bays in June showed no distinct peak. In corridors, the mode of the FL distribution shifted from 31–32 mm in April to 40 mm in late May and to 45 mm in June. The corridor distributions also widened until June, when the tails began to truncate.

Apparent growth of pink salmon was lower in oiled than in non-oiled corridors in 1989 but not in 1990 (Table 5). The year–oil interaction was significant ($P = 0.054$), and growth was significantly ($P = 0.035$) higher in 1990 than in 1989. The year–oil interaction was significant because growth differed between oiled and non-oiled corridors in 1989 but not in 1990 (Figure 8). At each habitat type in 1989, mean growth of pink salmon was lower in oiled than in non-oiled corridors (Table 5). Over all habitat types, growth in 1989 was 1.5% of body weight per day in oiled corridors, significantly ($P < 0.05$) lower than the 2.9% of body weight per day in non-oiled corridors. In contrast, growth in both low- and medium-gradient habitats was higher in oiled than in non-oiled corridors in 1990 (Table 5). Growth over all habitat types in 1990 was slightly higher in oiled than in non-oiled corridors, 3.5% and 3.1% of body weight per day, respectively; these rates were not significantly ($P > 0.1$) different.

Apparent growth of pink salmon by habitat type in corridors also differed between years, as indicated by a significant year–habitat interaction ($P = 0.008$). This interaction was caused by higher growth in steep-gradient habitats of both oiled and non-oiled corridors in 1990 than in 1989. Growth in low- and medium-gradient habitats of both oiled and non-oiled corridors was similar in 1989 and 1990.

In both 1989 and 1990, condition (weight at a given length) of pink salmon was higher in oiled than in non-oiled locations. In 1989, the logarithmic weight–length regressions did not differ significantly ($P > 0.1$) between bays and corridors. Therefore, data from bays and corridors were pooled to compare condition between oiled and non-oiled locations. The resulting regression equations did not differ significantly in slope; however, the elevations of the lines differed significantly ($P < 0.001$; Table 6). The difference in condition, although significant,

TABLE 4.—Summary of Wilcoxon matched-pair, signed-rank tests for fork length (FL) of juvenile pink salmon in Prince William Sound, 1989 and 1990. A negative value for the median difference indicates larger FL in non-oiled than oiled locations; median values are in millimeters. The CL values are 95% confidence limits on the median differences.

Year and location	Pairs	Wilcoxon statistic	P	Median difference	CL
1989					
All	29	132	0.066	−1	−4, 0
Bays	10	30	0.838	1	−6, 3
Corridors	19	39	0.026	−2	−6, −1
1990					
All	39	346	0.544	0	−2, 1
Bays	18	80	0.828	0	−1, 2
Corridors	21	93	0.444	−1	−3, 2

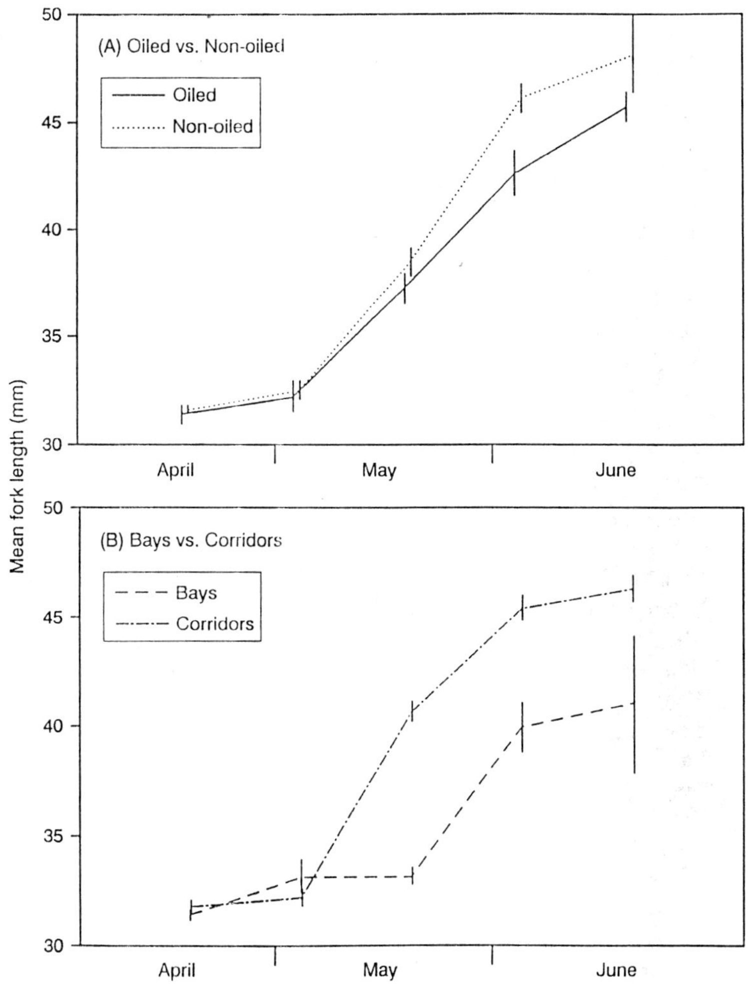

FIGURE 4.— Mean fork lengths of juvenile pink salmon captured in Prince William Sound, 1989, pooled (**A**) by oiled and non-oiled locations and (**B**) by bays and corridors. Vertical bars are 95% confidence intervals of the means.

was small; adjusted mean weights were 0.439 g in oiled locations and 0.431 g in non-oiled locations.

In 1990, bays and corridors were analyzed separately because of a significant ($P < 0.001$) difference in the weight–length regressions between bays and corridors in non-oiled locations. Regression slopes differed significantly ($P < 0.001$) between oiled and non-oiled bays, and between oiled and non-oiled corridors (Table 6). For both bays and corridors, the elevation of the regression line for oiled locations was above that for non-oiled locations over the size range of fish occurring in the habitats sampled. Thus, the interpretation is the same in 1990 as in 1989: pink salmon generally had higher condition in the oiled locations.

Discussion

Juvenile pink salmon were more abundant in non-oiled than in oiled locations in both 1989 and 1990. Avoidance of oiled habitats or direct mortality are possible explanations for the differences in abundance. However, no evidence of direct mortality was observed in oiled locations. In both years, large schools of juvenile pink salmon were observed and sampled in both oiled and non-oiled locations and did not appear to avoid oil. Schools were observed under large expanses of mousse accumulated along booms in outer Snug Harbor in 1989; the fish may actually have been using the mousse for cover.

Because the pattern of salmon abundance did not

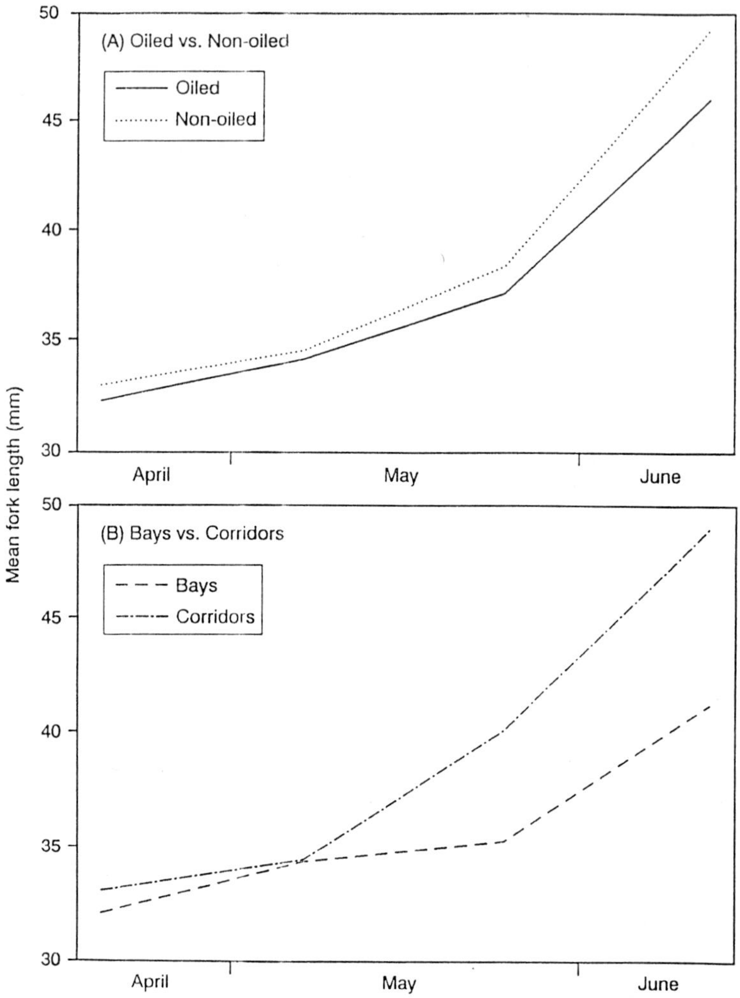

FIGURE 5.—Mean fork lengths of juvenile pink salmon captured in Prince William Sound, 1990, in (**A**) oiled and non-oiled areas and (**B**) bays and corridors; 95% confidence intervals were all within 0.5 mm of the indicated means.

change between years as the degree of oil contamination diminished (Carls et al. 1996b, this volume), we conclude that the differences in abundance between oiled and non-oiled locations were more likely due to geographic differences in the local production of salmon fry and in their migration pathways to the Gulf of Alaska, rather than to exposure to oil. The main criterion in selecting sampling locations in this study was the categorization as to "oiled" and "non-oiled." Because of the distribution of the spill, non-oiled study locations were clustered in the northwest region of Prince William Sound, on or close to the mainland, whereas oiled locations were generally more southerly and on islands (Figure 1). These geographic differences were reflected in physical differences between the locations. Although water temperature did not differ between oiled and non-oiled locations, salinity was lower in non-oiled locations, especially at the surface (Wertheimer et al. 1994). Spawning populations of pink salmon are substantially larger in the non-oiled portion of western Prince William Sound than in the oiled section (Pirtle 1977). Most of the hatchery production for these species is also away from the main spill area; only one (Armin Koernig Hatchery) of the four major pink salmon facilities in Prince William Sound was directly in the path of the spill (PWSRPT 1983).

After entering the nearshore marine waters of

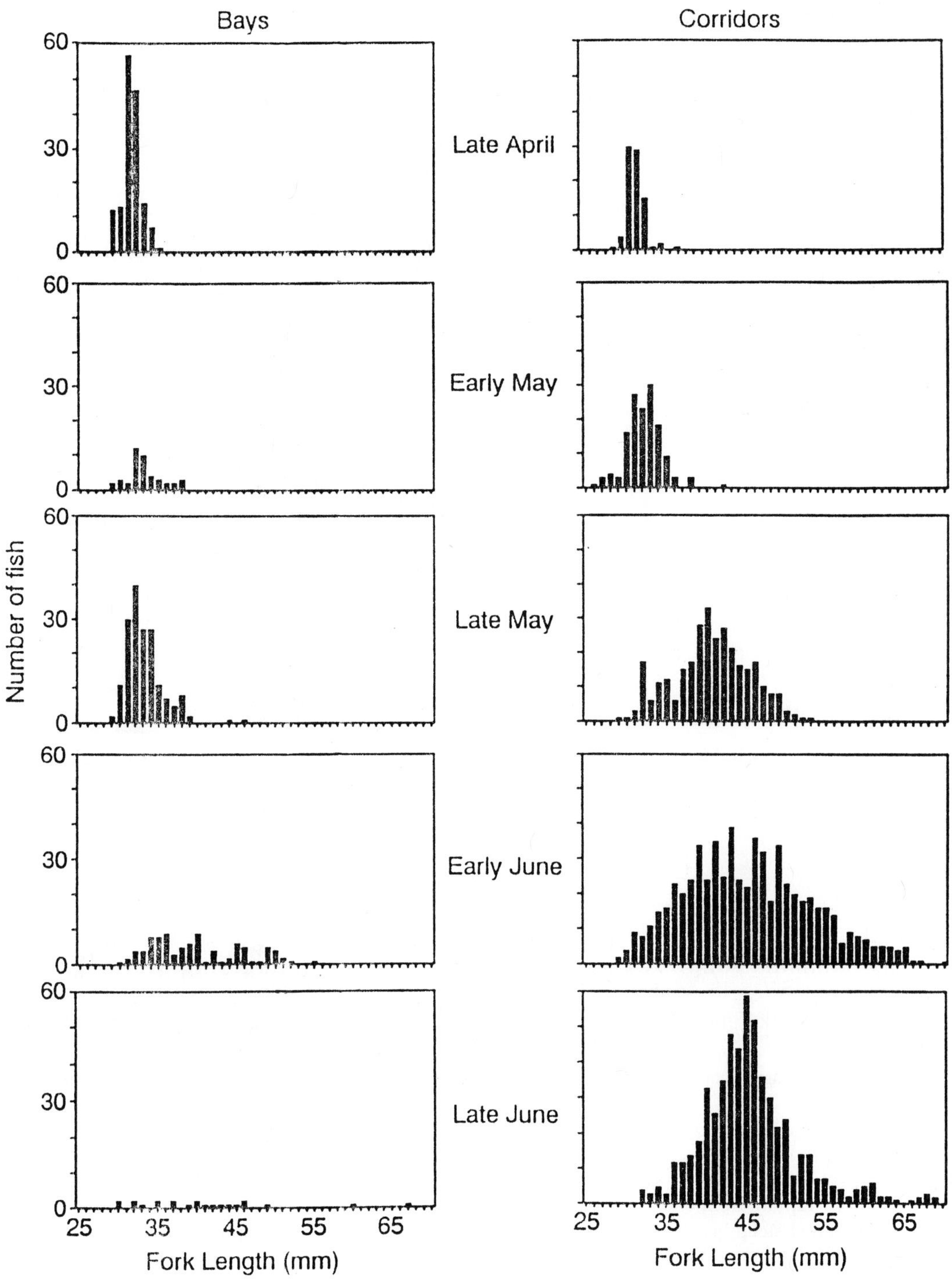

FIGURE 6.—Fork length frequencies of juvenile pink salmon captured in bays and corridors of Prince William Sound, 1989.

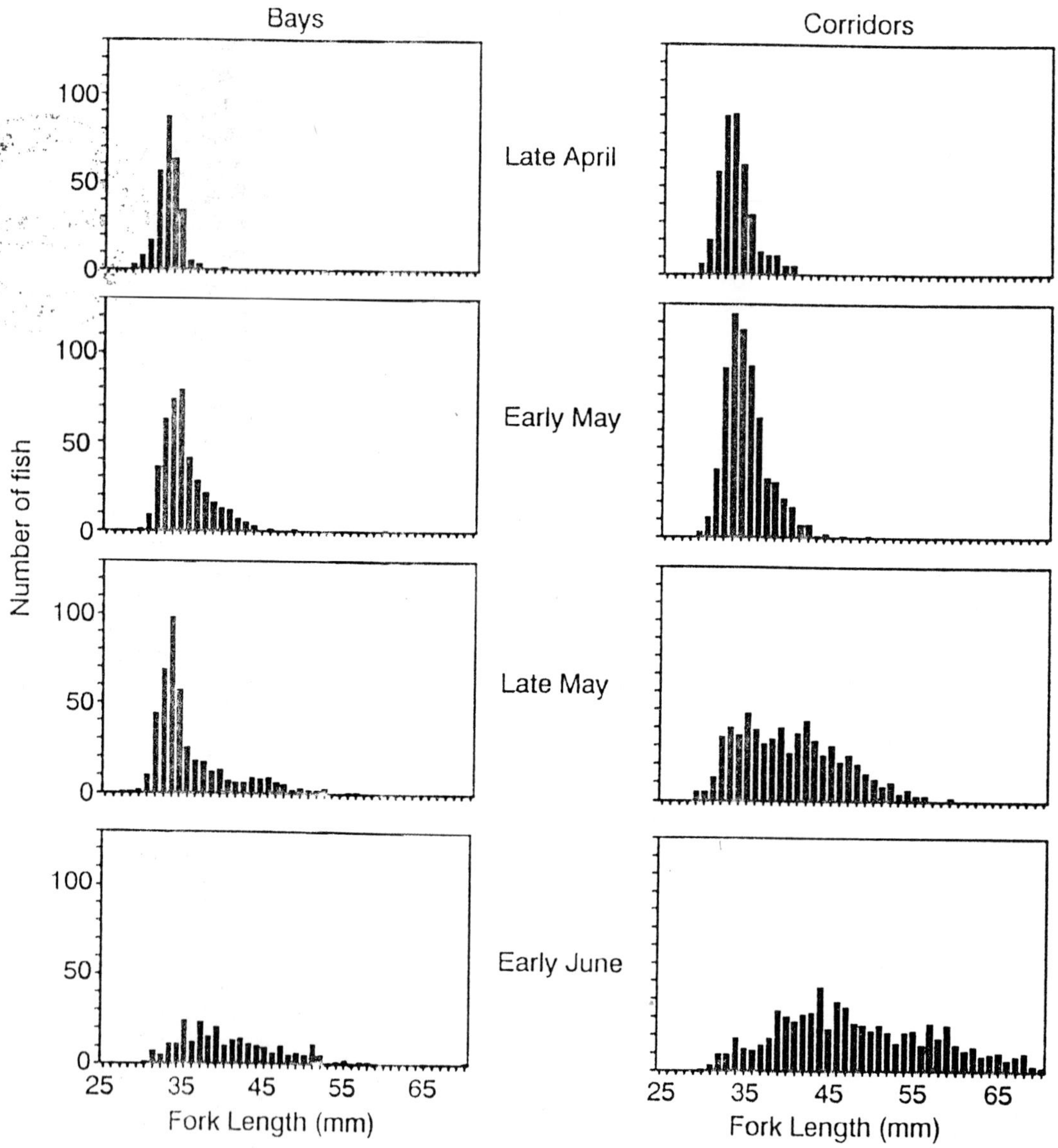

FIGURE 7.—Fork length frequencies of juvenile pink salmon captured in bays and corridors of Prince William Sound, 1990.

Prince William Sound in early spring, juvenile pink salmon moved rapidly from sheltered bays to the more exposed and steep beaches in migration corridors. Juvenile salmon in bays originated mostly from nearby streams, whereas fish in corridors represented several stocks of juvenile salmon that had left the bays and were migrating toward the Gulf of Alaska (Wertheimer et al. 1994). Previous work in Sawmill Bay in the southwestern Sound also showed that juvenile pink salmon moved rapidly from that bay to adjacent migration corridors (Cooney et al. 1981). Our observations of this behavior over a wider geographic range reinforce Cooney's (1990) conclusion that the presence of the oil-deflection booms in Port San Juan in 1989 disrupted the normal migration behavior of fish released from the Armin Koernig Hatchery into Sawmill Bay.

We did not detect any impact of the oil spill on either size or growth of juvenile pink salmon in bays. Pink salmon size was similar in oiled and

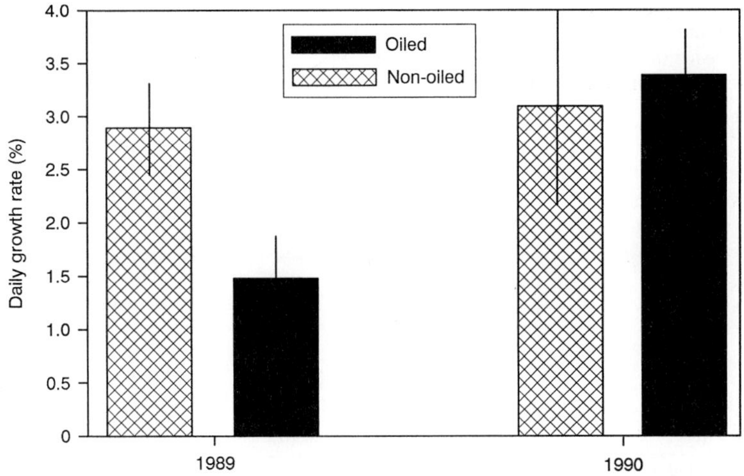

FIGURE 8.—Apparent daily growth (±1 SE) of juvenile pink salmon in corridors of Prince William Sound, 1989 and 1990.

non-oiled bays in both 1989 and 1990. The distribution and size analyses indicated short residence time and, thus, little opportunity for growth in either oiled or non-oiled bays before the fish moved into migration corridors. Apparent growth rates of pink salmon in bays could not be calculated because of insufficient data.

TABLE 5.—Apparent daily growth of juvenile pink salmon by habitat type in oiled and non-oiled corridors of Prince William Sound, 1989 and 1990. Growth was assumed to be exponential over time and was determined as the slope of the regression of the natural logarithm of weight on time in days. Numbers in table are percentages increase in body weight per day, with SDs in parentheses; LG = low gradient; MG = medium gradient; SG = steep gradient.

Year and location	Habitat		
	LG	MG	SG
1989, Oiled corridors			
Knight Island Passage	0.94 (0.17)	2.61 (0.17)	2.08 (0.34)
Prince of Wales Passage	2.18 (0.20)	0.68 (0.39)	0.45 (0.42)
Mean	1.56	1.64	1.26
1989, Non-oiled corridors			
Wells Passage	2.46 (0.26)	3.86 (0.35)	1.81 (0.40)
Culross Passage	2.05 (0.18)	2.61 (0.15)	4.53 (0.30)
Mean	2.25	3.23	3.17
1990, Oiled corridors			
Knight Island Passage	2.37 (0.20)	4.14 (0.32)	5.03 (0.24)
Prince of Wales Passage	2.36 (0.16)	2.87 (0.18)	3.86 (0.29)
Mean	2.36	3.50	4.44
1990, Non-oiled corridors			
Wells Passage	2.09 (0.11)	1.99 (0.11)	3.48 (0.17)
Culross Passage	1.68 (0.31)	1.74 (0.18)	7.69 (0.44)
Mean	1.88	1.86	5.58

Juvenile pink salmon were smaller and grew slower in oiled than non-oiled corridors of western Prince William Sound in 1989, but not in 1990. Because growth is based on change in size of unmarked fish over time, interpretation is complicated by recruitment of newly emerged fry into the marine environment and by size-specific movement from nearshore to offshore water (LeBrasseur and Parker 1964; Healey 1980) and between different nearshore habitats (Celewycz 1990). In this study, we did account for habitat differences in comparing apparent growth rate. Furthermore, our analysis of apparent growth rate of unmarked pink salmon is consistent with research on tagged fish, which showed that pink salmon from the same tag groups were significantly smaller when recovered in heavily-oiled than in lightly-oiled or non-oiled locations in 1989, but not in 1990 (Willette 1996).

We attribute the reduced growth of juvenile pink salmon in 1989 to hydrocarbon contamination in oiled locations. In contrast, Neff (1991) concluded that "It is extremely unlikely that hydrocarbon concentrations resulting from the spilled oil have had or will have any adverse effects on plants and animals living in the water column of Prince William Sound and the western Gulf of Alaska, including commercial fishery species." Hydrocarbon contamination was observed in tissues of juvenile pink salmon in 1989 (Carls et al. 1996b), in spite of the

TABLE 6.—Comparisons of weight–length relationships of juvenile pink salmon between oiled and non-oiled locations of Prince William Sound, 1989 and 1990; a = y-intercept, b = slope, $P(b)$ is the probability value for the tests of homogeneity of slopes for the pairs of regression lines shown, P(am) is the probability value for differences in adjusted means, and intersect is the length (mm) at which nonparallel regression lines intersect.

Year and location	N	a	b	R^2	$P(b)$	P (am)	Intersect
1989							
Non-oiled bays	236	−13.1693	3.3245	85.8			
Non-oiled corridors	899	−13.5611	3.4386	97.8	0.129	0.425	Parallel
Oiled bays	245	−13.7751	3.5010	93.1			
Oiled corridors	850	−13.5479	3.4408	97.1	0.276	0.239	Parallel
Non-oiled pooled bays and corridors	1,136	−13.5620	3.4385	97.7			
Oiled pooled bays and corridors	1,096	−13.6256	3.4610	97.1	0.343	0.000	Parallel
1990							
Non-oiled bays	791	−14.2775	3.6312	92.7			
Oiled bays	796	−13.5057	3.4208	95.3	0.000	[a]	40
Non-oiled corridors	1,464	−13.6495	3.4518	97.2			
Oiled corridors	1,067	−13.3544	3.3813	97.6	0.000	[a]	65

[a]No comparison of adjusted means is possible when slopes are not homogenous.

low concentrations of oil in the water column reported by Neff (1991). The low concentrations of oil in the water do not exclude other routes of contamination; ingestion of oil, either directly or via contaminated prey, was probably an important route of exposure for juvenile pink salmon (Carls et al. 1996b; Sturdevant et al. 1996, this volume). Recent laboratory studies have shown that ingestion of crude oil reduces growth rates of juvenile pink salmon (Carls et al. 1996a, this volume). Exposure and contamination of juvenile salmon diminished from 1989 to 1990 (Carls et al. 1996b). Because the differences in size and growth of juvenile pink salmon between oiled and non-oiled corridors also diminished in 1990 as exposure levels and contamination decreased, we conclude that the presence of oil inhibited the growth of pink salmon in 1989.

The relationship between weight and length of fish is frequently used as a measure of the effect of biotic or abiotic factors on the health or well-being of the fish. Juvenile pink salmon had consistently higher condition in oiled than in non-oiled locations in both 1989 and 1990. Other authors have argued that condition should be a sensitive indicator of effects of the oil spill on feeding and growth of juvenile pink salmon (Brannon et al. 1995). However, condition may not be a sensitive measure of growth reductions caused by the metabolic demands of depuration. Carls et al. (1996a) found that juvenile pink salmon ingesting low and moderate dosages of crude oil had similar or higher condition than control fish, even though growth was reduced at those dosages.

Although growth of juvenile pink salmon was reduced in the oiled locations in 1989, pink salmon there were still feeding and growing. Feeding was not reduced in oiled locations (Sturdevant et al. 1996). Growth was observed in heavily oiled locations by us in this study and by Willette (1996). The fish in oiled locations were not emaciated; as noted above, condition for pink salmon was actually higher in the oiled locations. Pink salmon fry can metabolize and depurate hydrocarbons (Thomas and Rice 1979). Depuration of hydrocarbons, however, has a metabolic cost, which can reduce food conversion efficiency in juvenile salmon (Moles and Rice 1983). For example, juvenile Atlantic salmon *Salmo salar* exposed to sublethal concentrations of crude oil had reduced growth that coincided with reduced food conversion efficiency rather than with reduced food intake (Vignier et al. 1992). The physiological load of oil contamination in Prince William Sound was apparently within the depuration capability of juvenile salmon, but the metabolic cost of depuration was sufficient to reduce growth of juvenile pink salmon in oiled locations.

Reduced growth during early marine residency can directly affect the survival of a cohort of pink salmon. Growth during this period is important if fish are to escape mortality from size-selective predation (Parker 1971; Hargreaves and LeBrasseur 1985). Within a year-class, early marine growth is positively correlated with survival (Mortensen et al. 1991). Geiger et al. (1996, this volume) estimated a loss of millions of pink salmon from populations in Prince William Sound due to oil-induced growth reduction during early marine residency.

Other species may have been affected by the oil spill in a similar manner as pink salmon. Juvenile chum salmon *Oncorhynchus keta* captured along oiled beaches in 1989 actually had higher levels of oil-induced enzymes than did pink salmon (Carls et al. 1996b). Thus, chum salmon probably suffered reduced growth and commensurate reductions in survival to adult return. A wide variety of other fishes use the nearshore environment of Prince William Sound and the adjacent Gulf of Alaska; many of these, especially pelagic schooling fishes and larval fishes, consume zooplankton as their principal prey (Rogers et al. 1986). We captured large numbers of Pacific herring *Clupea pallasi*, Pacific sand lance *Ammodytes hexapterus*, and juvenile Pacific cod *Gadus macrocephalus*, Pacific tomcod *Microgadus proximus*, and walleye pollock *Theragra chalcogramma* coincident with juvenile salmon. If ingestion of either whole oil or contaminated prey was the route of exposure for juvenile pink salmon, then other fishes that use the same habitats and that have overlapping feeding habits also may have been contaminated.

We conclude that the *Exxon Valdez* oil spill directly reduced growth of juvenile pink salmon in contaminated habitats, but that the differences in abundance that we observed between oiled and non-oiled locations were due to geographic variation in local production. The effect of hydrocarbon contamination on growth was limited to the year of the spill; growth was similar between oiled and non-oiled locations in 1990, when environmental contamination was greatly diminished. We also conclude that condition (the weight-at-length relationship) was not an effective measure of growth reduction caused by oil exposure.

References

ADFG (Alaska Department of Fish and Game). 1989. Wildlife resources in the Sound. Alaska Fish and Game 21(4):25–31.

Bax, N. J. 1983. Early marine mortality of marked juvenile chum salmon released into Hood Canal, Puget Sound, Washington, in 1980. Canadian Journal of Fisheries and Aquatic Sciences 40:426–435.

Bodin, P. 1988. Results of ecological monitoring of three beaches polluted by the *Amoco Cadiz* oil spill: development of meiofauna from 1978 to 1984. Marine Ecology Progress Series 42:105–123.

Brannon, E. J., L. L. Moulton, L. G. Gilbertson, A. W. Maki, and J. R. Skalaki. 1995. An assessment of oil spill effects on pink salmon populations following the *Exxon Valdez* oil spill—part 1: early life history. Pages 548–584 *in* P. G. Wells, J. N. Butler, and J. S. Hughes, editors. *Exxon Valdez* oil spill: fate and effects in Alaskan waters. American Society for Testing and Materials Special Technical Publication 1219, Philadelphia.

Carls, M. G., and six coauthors. 1996a. Growth, feeding, and survival of pink salmon fry exposed to food contaminated with crude oil. American Fisheries Society Symposium 18:608–618.

Carls, M. G., A. C. Wertheimer, J. W. Short, R. M. Smolowitz, and J. J. Stegeman. 1996b. Contamination of juvenile pink and chum salmon by hydrocarbons in Prince William Sound after the *Exxon Valdez* oil spill. American Fisheries Society Symposium 18: 593–607.

Celewycz, A. G. 1990. Distribution and sizes of juvenile pink and chum salmon in five habitats in Auke Bay, Alaska. Pages 54–57 *in* P. A. Knutson, editor. 14th northeast Pacific pink and chum salmon workshop. Washington Department of Fisheries, Olympia.

Cone, R. S. 1989. The need to reconsider the use of condition indices in fisheries science. Transactions of the American Fisheries Society 118:510–514.

Conover, W. J. 1980. Practical nonparametric statistics. Wiley, New York.

Cooney, R. T. 1990. Prey fields and the feeding behavior and growth of pink salmon fry released from the Armin F. Koening Hatchery at Sawmill Bay, Evans Island, Alaska, May and June 1989. Natural Resource Damage Assessment Status Report, Study 4, University of Alaska Fairbanks Component, Fairbanks.

Cooney, R. T., D. Uruquat, and D. Barnard. 1981. The behavior, feeding biology, and growth of hatchery released pink and chum salmon fry in Prince William Sound, Alaska. University of Alaska Fairbanks Sea Grant Report 81-5, Fairbanks.

Frane, J. 1980. The univariate approach to repeated measures—foundation, advantages, and caveats. Biomedical Data Processing Technical Report 69, University of California, Los Angeles.

Geiger, H. J., B. G. Bue, S. Sharr, A. C. Wertheimer, and T. M. Willette. 1996. A life history approach to estimating damage to Prince William Sound pink salmon caused by the *Exxon Valdez* oil spill. American Fisheries Society Symposium 18:487–498.

Hargreaves, N. B., and R. J. LeBrasseur. 1985. Species selective predation on juvenile pink (*Oncorhynchus gorbuscha*) and chum (*O. keta*) by coho salmon (*O. kisutch*). Canadian Journal of Fisheries and Aquatic Sciences 42:659–668.

Healey, M. C. 1980. The ecology of juvenile salmon in Georgia Strait, British Columbia. Pages 203–230 *in* W. J. McNeil and D. C. Himsworth, editors. Salmonid ecosystems of the north Pacific. Oregon State University Press, Corvallis.

Kirkwood, J. B. 1972. Inshore-marine and freshwater life history phases of the pink salmon. Doctoral dissertation. University of Louisville, Louisville, Kentucky.

LeBrasseur, R. J., and R. R. Parker. 1964. Growth rate of central British Columbia pink salmon (*Oncorhynchus gorbuscha*). Journal of the Fisheries Research Board of Canada 21:1101–1128.

Moles, A., and S. D. Rice. 1983. Effects of crude oil and naphthalene on growth, caloric content, and fat content of pink salmon juveniles in seawater. Transactions of the American Fisheries Society 112:205–211.

Mortensen, D. M., J. H. Landingham, A. C. Wertheimer, and S. G. Taylor. 1991. Relationship of early marine growth and survival of juvenile pink salmon to marine water temperature and secondary production in Auke Bay, Alaska. Pages 38–49 *in* I. Guthrie and B. Wright, editors. Proceedings, fifteenth northeast Pacific pink and chum salmon workshop. Pacific Salmon Commission, Parksville, British Columbia.

Murphy, M. L., J. F. Thedinga, and K. V. Koski. 1988. Size and diet of juvenile Pacific salmon during seaward migration through a small estuary in southeastern Alaska. U.S. National Marine Fisheries Service Fishery Bulletin 86:213–222.

Neff, J. M. 1991. Water quality in Prince William Sound and the Gulf of Alaska. Arthur D. Little, Inc., Cambridge, Massachusetts.

Nichelson, T. E. 1986. Influences of upwelling, ocean temperature, and smolt abundance on marine survival of coho salmon (*Oncorhynchus kisutch*) in the Oregon production area. Canadian Journal of Fisheries and Aquatic Sciences 43:527–535.

Olsen, J. 1991. Comparative performance of pink salmon from two hatcheries in Prince William Sound. Pages 17–29 *in* I. Guthrie and B. Wright, editors. Proceedings, fifteenth northeast Pacific pink and chum salmon workshop. Pacific Salmon Commission, Parksville, British Columbia.

Parker, R. R. 1968. Marine mortality schedules of pink salmon of the Bella Coola River, central British Columbia. Journal of the Fisheries Research Board of Canada 25:757–794.

Parker, R. R. 1971. Size selective predation among juvenile salmonid fishes in a British Columbia inlet. Journal of the Fisheries Research Board of Canada 28: 1503–1510.

Pirtle, R. B. 1977. Historical pink and chum salmon estimated spawning escapements from Prince William Sound, Alaska, streams, 1960–1975. Alaska Department of Fish Game Technical Report 35, Juneau.

PWSRPT (Prince William Sound Regional Plan Team). 1983. Prince William Sound–Copper River comprehensive salmon plan. Prince William Sound Regional Fisheries Plan Team, Cordova, Alaska.

Rice, S. D., and seven coauthors. 1984. Effects of petroleum hydrocarbons on Alaskan aquatic organisms. NOAA (National Oceanic and Atmospheric Administration) Technical Memorandum NMFS (National Marine Fisheries Service) F/NWC-67, Northwest Fisheries Science Center, Seattle.

Rice, S. D., D. A. Moles, and J. W. Short. 1975. The effect of Prudhoe Bay crude oil on survival and growth of eggs, alevins, and fry of pink salmon, *Oncorhynchus gorbuscha*. Pages 503–507 *in* Proceedings, 1975 joint conference on prevention and control of oil pollution. American Petroleum Institute, Washington, DC.

Rogers, D. E., B. J. Rogers, and R. J. Rosenthal. 1986. The nearshore fishes. Pages 399–415 *in* D. W. Hood and S. T. Zimmerman, editors. The Gulf of Alaska physical environment and biological resources. U.S. Department of the Interior Mineral Management Services (MMS) 86-0095, Washington, DC.

Samain, J. F., J. Moal, A. Coum, J. R. Le Coz, and J. Y. Daniel. 1980. Effects of the *Amoco Cadiz* oil spill on zooplankton. Helgolander Meeresuntersuchungen 33:225–235.

Schwartz, J. P. 1985. Effects of oil-contaminated prey on the feeding and growth rate of pink salmon fry *Oncorhynchus gorbuscha*. Pages 459–476 *in* F. J. Vernberg, F. Thurberg, A. Calabrese, and W. Vernberg, editors. Pollution and physiology of marine organisms. University of South Carolina Press, Columbia.

Sturdevant, M. V., A. C. Wertheimer, and J. L. Lum. 1996. Diets of juvenile pink and chum salmon in oiled and non-oiled nearshore habitats in Prince William Sound, 1989 and 1990. American Fisheries Society Symposium 18:578–592.

Thomas, R. E., and S. D. Rice. 1979. The effect of exposure temperature on oxygen consumption and opercular breathing rates of pink salmon fry exposed to toluene, napthalene, and water-soluble fractions of Cook Inlet crude oil and no. 2 fuel oil. Pages 39–52 *in* W. B. Vernberg, editor. Marine pollution: functional responses. Academic Press, New York.

Vignier, V., J. H. Vandermeulen, and A. J. Fraser. 1992. Growth and food conversion by Atlantic salmon parr during 40 days' exposure to crude oil. Transactions of the American Fisheries Society 121:322–332.

Walters, C. J., R. Hilborn, R. M. Peterman, and M. J. Staley. 1978. Model for examining early ocean limitations of Pacific salmon. Canadian Journal of Fisheries and Aquatic Sciences 41:126–133.

Wertheimer, A. C., A. G. Celewycz, M. G. Carls, and M. V. Sturdevant. 1994. Impact of the oil spill on juvenile pink and chum salmon and their prey in critical nearshore habitats. *Exxon Valdez* oil spill, Natural Resource Damage Assessment Final Report, Fish/ Shellfish Study 4, NMFS (National Marine Fisheries Service) Component, Juneau, Alaska.

Willette, M. 1996. Impacts of the *Exxon Valdez* oil spill on the migration, growth, and survival of juvenile pink salmon in Prince William Sound. American Fisheries Society Symposium 18:533–550.

American Fisheries Society Symposium 18:533–550, 1996

Impacts of the *Exxon Valdez* Oil Spill on the Migration, Growth, and Survival of Juvenile Pink Salmon in Prince William Sound

MARK WILLETTE

Alaska Department of Fish and Game, Commercial Fisheries Management and Development Division
P.O. Box 669, Cordova, Alaska 99574, USA

Abstract.—This study investigated the effects of the *Exxon Valdez* oil spill on the migration, growth, and survival of juvenile pink salmon *Oncorhynchus gorbuscha* during the first 2 months of their marine residence in Prince William Sound, Alaska. Coded-wire-tagged juvenile salmon released from hatcheries in the sound in 1989, 1990, and 1991 were the principal sources of data. The growth of juvenile pink salmon in Prince William Sound appeared to be reduced by oil contamination from the *Exxon Valdez* oil spill in 1989. Growth rates of juveniles released from Armin F. Koernig (AFK) Hatchery in 1989 were significantly lower ($P = 0.034$) in the moderately oiled area near the hatchery than along the lightly oiled southern coast of Knight Island. In 1990 and 1991, growth rates of juvenile pink salmon released from AFK Hatchery were again lower in the moderately oiled area compared with the lightly oiled southern coast of Knight Island. However, the growth difference was not significant in 1990 ($P = 0.103$), and was only marginally significant in 1991 ($P = 0.085$). The magnitude of the growth difference in 1990 and 1991 was also one half of the growth difference in 1989. Growth rates of juveniles released from Wally H. Noerenberg (WHN) Hatchery in 1989 were significantly lower ($P = 0.019$) in the moderately oiled area near Main Bay than in the non-oiled area near WHN Hatchery. Growth rates of juveniles released from WHN Hatchery were not significantly different between oiled and non-oiled areas in 1990 ($P = 0.767$) and 1991 ($P = 0.883$). Exposure to hydrocarbons from the *Exxon Valdez* oil spill appeared to reduce the growth rate of juvenile pink salmon by 0.76–0.94% body weight/day in 1989. The observed differences in growth rate did not appear to have been caused by measurement or sampling error or by differences in food consumption rate, prey composition, or water temperature. The observed reduction in growth rate was associated with a significantly greater ($P < 0.05$) frequency of cytochrome P-4501A enzyme induction in moderately oiled areas compared with non-oiled and lightly oiled areas in 1989. The greater frequency of P-4501A enzyme induction in moderately oiled areas indicates that fish in oil-contaminated habitats expended energy to metabolize and depurate hydrocarbons, leaving less energy available for somatic growth. Sufficient data were not available to determine if the level of hydrocarbon exposure was great enough to cause the estimated reduction in growth rate attributed to oil contamination. The growth of juvenile pink salmon in 1989 was significantly related ($P = 0.016$) to survival to the adult stage. The reduction in juvenile growth rate attributed to oil contamination in 1989 probably caused a 1.7–2.2% reduction in survival to the adult stage among fish that reared in oiled areas. The adult pink salmon return to Prince William Sound in 1990 was thus lower than if the *Exxon Valdez* oil spill had not occurred.

Approximately 1 month after the *Exxon Valdez* oil spill, 518 million juvenile pink salmon *Oncorhynchus gorbuscha* were released from four private nonprofit hatcheries in Prince William Sound, Alaska. At about the same time, a nearly equal number of wild pink salmon began to emigrate from numerous streams bordering the sound. This study focused on the effects of the *Exxon Valdez* oil spill on juvenile pink salmon during the first 2 months of their marine residence, because mortality at this life stage strongly affects recruitment to the adult stage (Parker 1968; Hartt 1980; Bax 1983). During this period, slow-growing individuals sustain a higher mortality, because they are vulnerable to predators for a longer time than are fast-growing individuals (Parker 1971; Healey 1982). Several studies have documented that hydrocarbon exposure reduces the growth of juvenile pink salmon in the laboratory (Rice et al. 1975; Moles and Rice 1983; Schwartz 1985). Thus, hydrocarbon exposure during the first 2 months of marine residence might reduce growth and subsequent survival to the adult stage.

The design of this study was based on a comparison of growth rates of coded-wire-tagged (CWT) juvenile pink salmon that were reared in oiled and non-oiled areas of Prince William Sound. A coded-wire tag is a nearly microscopic magnetic sliver of wire that is inserted in the fish's snout. A binary code etched onto the wire is used to identify fish from specific CWT groups. During each of the 3 years of this study (1989–1991), approximately 1 million CWT juvenile pink salmon were released from four hatcheries in Prince William Sound. Growth was estimated for CWT juvenile pink salmon recovered during the first 2 months of marine residence. Several CWT groups of juvenile salmon were released from hatcheries located in oiled and non-oiled areas of the sound, enabling comparison of juvenile growth among fish with different oil-exposure histories.

Determining the effect of hydrocarbon exposure on juvenile growth in nature also required an evaluation of natural environmental effects on growth. Stomach contents analysis, statistical analyses, and bioenergetic models were used to examine the effects of prey density, prey species composition, water temperature, and juvenile salmon density on growth. Feeding rate is strongly affected by the density of available prey (Holling 1966). Prey density also determines the net energetic gain from feeding, but the species and size compositions of available prey are also important because of differences in energy content (Raymont et al. 1969), visibility (Wright and O'Brien 1982), and handling times of prey (Werner 1974; Vinyard 1982). Water temperature affects activity level and thus feeding rate (Panadian and Vivekanandan 1985), but also largely determines both metabolic rate (Brett and Groves 1979) and gastric evacuation rate (Fange and Grove 1979). Fish density may also be a significant factor affecting growth and survival of juvenile salmon (Matthews 1980; Clark and McCarl 1983; McCarl and Rettig 1983; Nickelson 1986; Emlen et al. 1990; Walters and Juanes 1993).

The mortality of juvenile pink salmon during the first 2 months of marine residence was not directly measured in this study. Movements of fish through the study area and offshore precluded direct estimation of mortality using mark–recapture techniques (Ricker 1975). Mortality resulting from hydrocarbon exposure was assessed by a comparison of growth in oiled and non-oiled areas and by evaluation of relationships between growth and fry-to-adult survival.

The effect of the *Exxon Valdez* oil spill on migration patterns of juvenile pink salmon was examined because a disruption of migratory behavior during the critical juvenile lifestage may affect survival. Hydrocarbons may affect migratory behavior by damaging the olfactory lamellar surfaces (Babcock 1985) or by triggering an avoidance reaction (Rice 1973).

Our ability to estimate the level of exposure of juvenile pink salmon to hydrocarbons from the *Exxon Valdez* oil spill was critical to the other components of the study. Induction of cytochrome P-4501A enzymes was chosen as the best tool for this purpose. Detection of hydrocarbons by chemical analysis of tissue samples was not used, because fish, including pink salmon, metabolize hydrocarbons rapidly and accumulation in tissues is seldom detected. Metabolites in the bile indicate hydrocarbon exposure (Roubal et al. 1977; Thomas and Rice 1981), but juvenile pink salmon are too small to provide adequate bile samples. Cytochrome P-4501A enzymes are induced approximately 40 h after initial hydrocarbon exposure and can be detected in microscopic sections of tissues (Kloepper-Sams and Stegeman 1989). The presence of P-4501A enzymes indicates exposure to hydrocarbons, but the source of the hydrocarbons cannot be identified.

The methods and results of this paper are structured around the following study objectives:

(1) determine the effect of oil contamination from the *Exxon Valdez* oil spill on the migratory behavior of juvenile pink salmon in Prince William Sound.
(2) determine the effect of oil contamination from the *Exxon Valdez* oil spill on the growth of juvenile pink salmon that reared in oiled, nearshore nursery habitats in Prince William Sound:
   (a) test for differences in the growth rate of CWT juvenile pink salmon between oiled and non-oiled areas of the sound in 1989, 1990, and 1991;
   (b) test for differences in P-4501A enzyme induction in juvenile pink salmon tissues between oiled and non-oiled areas of the sound in 1989;
   (c) test for differences in prey composition and stomach fullness between oiled and non-oiled areas of the sound in 1989, 1990, and 1991; and
   (d) evaluate the relative effects of number of fry released, ocean temperature, prey density, and prey composition on the growth of juvenile pink salmon in the sound.
(3) determine the effect of oil contamination from the *Exxon Valdez* oil spill on the fry-to-adult survival of pink salmon that reared in oiled, nearshore nursery habitats in Prince William Sound.

Methods

This study focused on recovery of juvenile CWT pink salmon released from the Armin F. Koernig (AFK) and Wally H. Noerenberg (WHN) hatcheries in Prince William Sound. A half-length CWT was injected into each fish's head before the fish were released (Sharr 1993; Sharr and Peckham 1993). Juvenile CWT pink salmon were recovered using beach and purse seines deployed from a 6-m long aluminum skiff. Sampling was initiated each year during the third week of May and extended until about the beginning of July. Juvenile salmon were generally captured in areas where the fish were very abundant, because large numbers of fish needed to be recovered to obtain adequate samples of CWT fish. Laboratory methods used to extract CWTs and estimate growth are described in Willette (1994). Coded-wire tagged juvenile salmon were also recovered by National Marine Fisheries Service investigators in 1989 and 1990 (Wertheimer and Celewycz 1996, this volume).

Effect of Oil Contamination on Migratory Behavior

The numbers of CWT juvenile pink salmon recovered at each site were summed for each hatchery. Maps were prepared indicating the distribution of CWT recoveries in 1989, 1990, and 1991. A qualitative comparison of migration patterns in different years was made by simple visual evaluation of the distribution of CWT recoveries. An analysis of migration rate was not conducted, because sampling was not initiated until after most tag-code groups had been released from the hatcheries. Thus, the arrival of each tag-code group in each area could not be adequately documented. An analysis of catch per unit of effort (CPUE) was not conducted, because the catchability of the fish (Ricker 1975) varied considerably depending on weather and fish size. Also, CPUE varied considerably as the skill of the crew improved over time.

Effect of Oil Contamination on Growth

Test for differences in growth.—An exponential model was used to estimate growth rates (G_i) of individual CWT juvenile pink salmon:

$$G_i = \frac{\log_e(W_c) - \log_e(\bar{W}_r)}{t_c - t_r}, \qquad (1)$$

where W_c is the weight of individual fish at capture, $\bar{W}_r$ is the mean weight at release of the fish in a specific tag-code group, t_c is the date at capture, and t_r is the mean date of release for a specific tag-code group (Ricker 1975). In 1989, the exact date of release for individual fish was not known, because specific tag-code groups were released over a period of several days. In 1990 and 1991, specific tag-code groups were released on a single day, so the date of release for individual fish was known.

Analyses of growth were structured around three treatment groups receiving different feeding regimes at the hatcheries. An early-fed group was composed of individuals released during high zooplankton abundance after 1–2 weeks of feeding in net pens. An unfed group was released during high zooplankton abundance after only 2–5 d of feeding. A late-fed group was released during declining zooplankton abundance and increasing temperatures after 1–2 weeks of feeding. Analyses of variance indicated that, generally, growth was significantly different among treatment groups (Willette 1994). The number of tag recoveries in each treatment group and time period was examined to determine in which groups adequate sample sizes were available to test for differences between moderately oiled, and lightly and non-oiled areas. The early-fed groups released from AFK and WHN hatcheries consistently constituted the greatest proportion of the recoveries for any single group in each year (39 and 24% in 1989, 35 and 17% in 1990, and 42 and 32% in 1991, respectively). As a result, subsequent growth-rate analyses were restricted to the early-fed groups released from these two hatcheries. The majority of the early-fed CWT recoveries from the AFK Hatchery occurred in the moderately oiled area near the AFK Hatchery and in the lightly oiled area along the southern coast of Knight Island (Willette 1994). The majority of the early-fed CWT recoveries from the WHN Hatchery occurred in the non-oiled area near the WHN Hatchery and in the moderately oiled area near Main Bay. Subsequent tests for differences in growth were limited to these two areas for each hatchery (Figure 1). Recovery site was used as the sample unit in the analysis (Hurlbert 1984). A split-plot design was used to test for differences in growth of juvenile CWT salmon in the early-fed group released from the AFK Hatchery and recovered in moderately and lightly oiled areas in May and June, 1989. For this group, CWT juvenile pink salmon were recovered from a sufficient number of sites in each area during both May and June (Willette 1994). For other groups, a random design was used to test for differences in mean growth rate between moderately oiled, and lightly and non-oiled areas, because the number of sites repeatedly sampled in each area during each month was not adequate for a split-plot design. Probabilities for statistical tests were considered significant at $P < 0.05$, marginally significant at $0.05 < P < 0.10$, and not significant at $P > 0.10$.

Test for differences in P-4501A enzyme induction.—Induction of cytochrome P-4501A monooxygenases was used to estimate the degree of exposure of juvenile pink salmon to hydrocarbons (Stegeman 1992). Cytochrome P-4501A is an enzyme induced

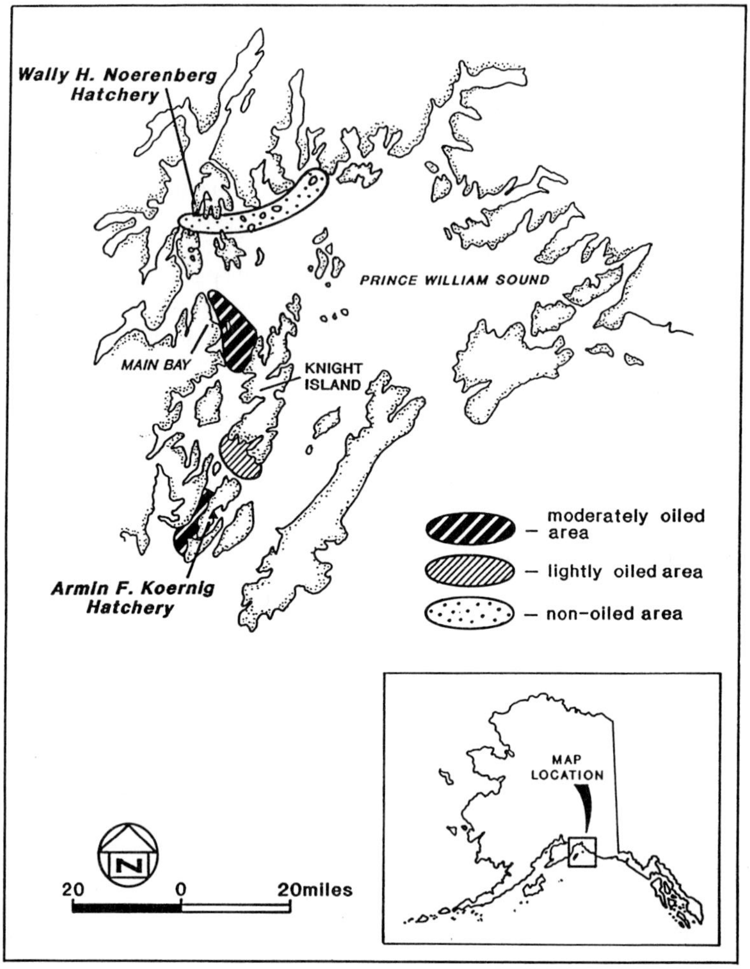

FIGURE 1.—Paired moderately oiled and lightly and non-oiled areas sampled for juvenile coded-wire tagged pink salmon released from the Wally H. Noerenberg and Armin F. Koernig hatcheries in Prince William Sound, 1989–1991.

when fish are exposed to hydrocarbons (Stegeman and Lech 1991). Untagged juvenile pink salmon corresponding in size to the dominant CWT group in the catch at each site were selected for analysis. Immunohistochemical methods were used to detect induction of P-4501A enzymes in various fish tissues (Smolowitz et al. 1989; Stegeman et al. 1991). Sample analysis involved microscopic examination of thin sections of each fish and documentation of the occurrence of a stain that indicates the presence of P-4501A enzymes in each tissue (Stegeman 1992). Analysis of samples for P-4501A induction was conducted by John Stegeman and Roxanna Smolowitz (Woods Hole Oceanographic Institute, Woods Hole, Massachusetts).

Fisher's exact test (Kendall and Stuart 1979) was used to test for differences in the frequency of occurrence of P-4501A staining in various tissues between moderately oiled, and lightly and non-oiled areas. Individual fish were used as the sample unit in the analysis. For each tissue type, a minimum sample size of 10 (minimum expected value of five) was needed from each sampling area to conduct the analysis (Zar 1984). Adequate sample sizes were obtained for 11 tissues. Fisher's exact test was also used to test for changes in the frequency of occurrence of P-4501A staining over time, because induction of P-4501A monooxygenases declines over time after initial exposure to hydrocarbons (Kloepper-Sams and Stegeman 1989). The test was only conducted for fish recovered along the southern coast of Knight Island before and after 10 June 1989, because adequate samples were obtained only in this area during these two time periods.

Test for differences in prey composition and stomach fullness.—Stomach contents analysis was used to determine whether prey composition or stomach fullness was different between moderately oiled, and lightly and non-oiled areas in 1989, 1990, and 1991. Samples of untagged juvenile pink salmon ($N = 10$) were preserved at various sites where CWT fish were captured. Prey organisms in each stomach were enumerated in the following categories: large calanoid copepods (>2.5 mm), small calanoid copepods (<2.5 mm), harpacticoid copepods, and "other prey." This approach was taken because the feeding rate of juvenile pink salmon is reduced when only small prey organisms are available (Parsons and LeBrasseur 1973). Prey biomass in each category was estimated by the product of prey abundance and average prey wet weight (Coyle et al. 1990). The weight of total stomach contents was estimated by the sum of prey biomass in all categories. Stomach fullness was estimated by the weight of total stomach contents as a percent of fish body weight.

Analysis of variance was used to test for differences in stomach fullness and prey composition after the data was rank transformed (Conover and Iman 1981). A completely random two-way factorial design was used with sampling area (oiled, non-oiled) and date (May, June) as independent variables. Site was used as the sample unit. Time of day was not included as an independent variable in the model because preliminary data analyses indicated no significant difference in stomach fullness or prey composition between morning and evening (before or after 1400 hours).

Evaluation of effects of fry density, ocean temperature, prey density, and prey composition.—Analysis of covariance was used to estimate the relative effects of number of fry released, zooplankton abundance, and ocean temperature on the growth of juvenile CWT pink salmon. The independent variables in the model were (1) number of fry released from each hatchery in each treatment group, (2) mean zooplankton settled volume for the period from release to recovery, and (3) mean ocean temperature for the period from release to recovery. Three classes were established for number of fry released, and least-squares mean growth was estimated for each class. The ocean temperature and zooplankton data used in these analyses were collected by the Prince William Sound Aquaculture Corporation. Ocean temperatures were measured and zooplankton samples were collected twice each week at stations located in passages adjacent to each hatchery. Zooplankton settled volume was estimated from replicate 20-m vertical tows taken with a 0.5-m diameter ring net (mesh size = 243 μm). Zooplankton species were not identified or enumerated.

Analysis of covariance was also used to test for differences in ocean temperature between the moderately oiled, and lightly and non-oiled areas where juvenile salmon from the AFK and WHN hatcheries were recovered in 1989, 1990, and 1991. The analysis was conducted to determine whether differences in temperature between moderately oiled, and lightly and non-oiled areas may have caused differences in growth.

A bioenergetics model was constructed to evaluate whether the growth of juvenile pink salmon may have been limited by low prey biomass in 1989. The model estimated the time required for a 1-g pink salmon to obtain a maximum daily ration composed of either large or small copepods at temperatures ranging between 4 and 14°C and prey biomass between 0.001 and 2.00 g/m^3. This was the approximate range of temperature and zooplankton biomass observed in Prince William Sound during May and June, 1989 (Wertheimer et al. 1993). The time required to obtain a ration composed of mixed prey was estimated from prey composition data and model estimates of feeding times required for large and small copepods, respectively. It was assumed that approximately 20 h of daylight are available for feeding in Prince William Sound during May and June. Feeding times in excess of 20 h indicate that the fish may not have acquired maximum daily ration. Willette (1994) describes the bioenergetics model in greater detail.

Effect of Oil Contamination on Fry-to-Adult Survival

Analysis of covariance was used to evaluate the relationship between mean juvenile salmon growth and the fry-to-adult survival of tag-code groups released from Prince William Sound hatcheries in 1989, 1990, and 1991 (Sharr 1993; Sharr and Peckham 1993). The effect of oil contamination on fry-to-adult survival was evaluated using the growth–survival relationship for juveniles released in 1989 and the mean growth of juveniles recovered in moderately, lightly, and non-oiled areas in 1989. Analysis of covariance was used to adjust mean growth in moderately, lightly, and non-oiled areas for differences in ocean temperature. Mean growth by area and date was the dependent variable in the analysis of covariance and mean temperature from fry re-

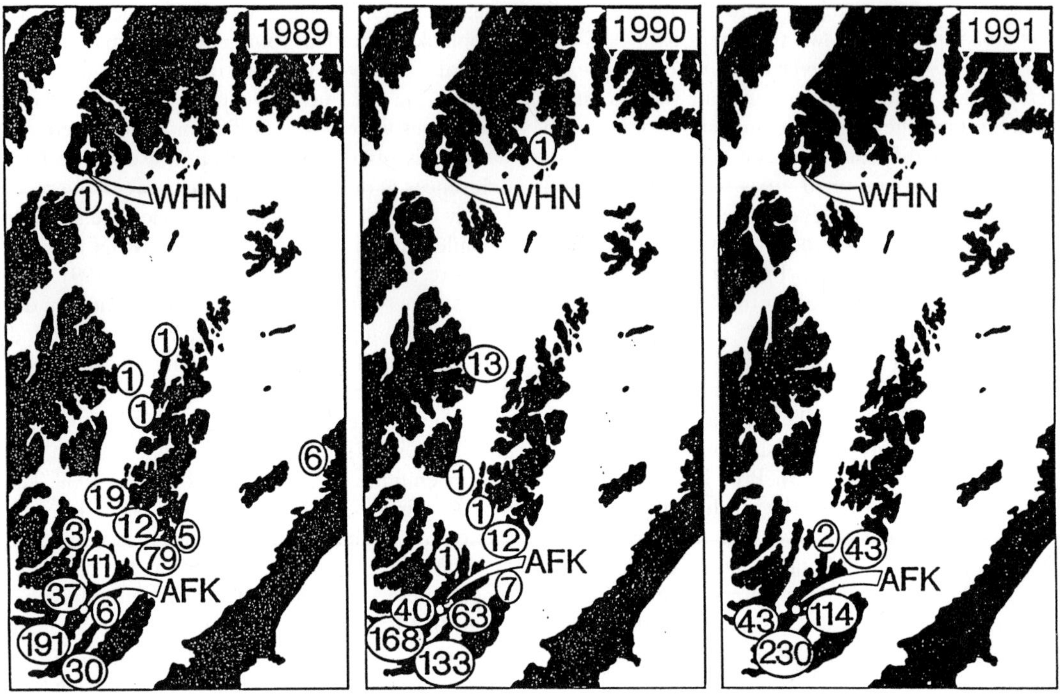

FIGURE 2.—Locations of recoveries of coded-wire tagged (CWT) juvenile pink salmon released from the Armin F. Koernig (AFK) Hatchery in 1989, 1990, and 1991. Enclosed values indicate the number of CWT juvenile pink salmon recovered at each site. The site of the Wally H. Noerenberg Hatchery is indicated by WHN.

lease to recapture was the covariate. Differences between the point estimates of fry-to-adult survival for fish being reared in moderately oiled, and lightly and non-oiled areas were used to assess the effect of oil contamination on fry-to-adult survival.

Results

Effect of Oil Contamination on Migratory Behavior

A relatively large number of juvenile pink salmon from AFK Hatchery migrated out of the moderately oiled area near the hatchery to the lightly oiled southern coast of Knight Island in 1989. One hundred and ten CWT juvenile pink salmon from AFK Hatchery were recovered along the southern coast of Knight Island in 1989 (Figure 2). Only 13 and 43 CWT fish from AFK Hatchery were recovered in this same area in 1990 and 1991, respectively (Figure 2). The relatively high catch of juvenile CWT pink salmon in this area in 1991 resulted from greater sampling effort on relatively few fish, which was done to obtain samples of juvenile CWT salmon for growth analyses. Visual observations also indicated a much greater abundance of juvenile salmon along the lightly oiled southern coast of Knight Island in 1989 compared with 1990 and 1991. There were no apparent interannual differences in the migration of juvenile CWT pink salmon from WHN Hatchery that appeared to be related to oil contamination.

Effect of Oil Contamination on Growth

Test for differences in growth.—The growth of juvenile CWT pink salmon released from AFK Hatchery in 1989 was significantly lower ($P = 0.034$) in the moderately oiled area near the hatchery than along the lightly oiled southern coast of Knight Island (Table 1). There was no difference in growth between these two areas in May 1989 ($P = 0.394$). However, growth in June 1989 was significantly lower ($P = 0.011$) in the moderately oiled area near AFK Hatchery compared with the lightly oiled southern coast of Knight Island. In 1990 and 1991, growth rates of juvenile CWT pink salmon released from AFK Hatchery were again lower in the moderately oiled area near the hatchery compared with

TABLE 1.—Mean growth rates (% body weight/d) of coded-wire-tagged (CWT) juvenile pink salmon in early-fed groups released from the Armin F. Koernig Hatchery and recovered during May and June, 1989. Significance levels (*P*) from an analysis of variance are indicated for the oil, time, and oil-by-time interaction terms in the model. Sample size (*N*) is expressed as number of sites. Number of individual CWT juvenile pink salmon included in the analysis is indicated in parentheses.

	May			Jun			Significance level		
Area	*N*	Mean	SE	*N*	Mean	SE	Oil	Time	Oil × time
Light oil	2 (66)	4.59	0.23	2 (23)	5.27	0.23	0.034	0.502	0.069
Oil	4 (33)	4.31	0.16	4 (48)	4.01	0.16			

the lightly oiled southern coast of Knight Island (Table 2). The growth difference was not significant in 1990 ($P = 0.103$), and marginally significant in 1991 ($P = 0.085$). The magnitude of the growth difference in 1990 and 1991 was one half of the difference in 1989.

Growth rates of juvenile CWT pink salmon released from WHN Hatchery in 1989 were significantly lower ($P = 0.019$) in the moderately oiled area near Main Bay than in the non-oiled area near WHN Hatchery (Table 2). Growth rates of juvenile pink salmon released from WHN Hatchery were not significantly different between moderately oiled and non-oiled areas in 1990 ($P = 0.767$) and 1991 ($P = 0.883$; Table 2).

TABLE 2.—Mean growth rates (% body weight/d) of coded-wire tagged (CWT) juvenile pink salmon in early-fed groups released from Armin F. Koernig (AFK) and Wally H. Noerenberg (WHN) hatcheries in Prince William Sound, 1989–1991. Sample size (*N*) is expressed as number of sites. Number of individual CWT juvenile pink salmon included in the analysis is indicated in parentheses.

Year	Area	*N*	Mean	SE	*P*
		AFK Hatchery			
1989	Light oil	4 (89)	4.93	0.20	0.034
	Oil	8 (81)	4.17	0.14	
1990	Light oil	3 (14)	4.80	0.28	0.103
	Oil	13 (225)	4.46	0.08	
1991	Light oil	1 (43)	3.38		0.085
	Oil	5 (272)	2.97	0.21	
		WHN Hatchery			
1989	Non-oil	4 (113)	5.16	0.33	0.019
	Oil	6 (29)	4.22	0.13	
1990	Non-oil	6 (129)	3.93	0.37	0.767
	Oil	3 (12)	3.81	1.46	
1991	Non-oil	4 (116)	3.59	0.19	0.883
	Oil	3 (118)	3.52	0.06	

Test for differences in P-4501A enzyme induction.—The frequency of occurrence of P-4501A enzyme induction closely coincided with the degree of oil contamination observed in each sampling area. The frequency of occurrence of P-4501A staining was significantly greater in gill pillar cells and cecal epithelium before 10 June 1989 than after this date along the southern coast of Knight Island (Table 3). Subsequent comparisons of the frequency of occurrence of staining between moderately oiled, and lightly and non-oiled areas were limited to the time period before 10 June 1989. The frequency of occurrence of P-4501A staining in gill pillar cells and gill epithelium was significantly greater ($P < 0.001$) in the moderately oiled area near Main Bay than in the non-oiled area near WHN Hatchery (Table 4). The frequency of occurrence of P-4501A staining in gill pillar cells was significantly greater ($P = 0.015$) in the moderately oiled area near AFK Hatchery than along the lightly oiled southern coast of Knight Island. The frequency of occurrence of staining was also greater in gill epithelium near AFK Hatchery, but the difference was only marginally significant ($P = 0.068$).

Test for differences in prey composition and stomach fullness.—Small copepods constituted a significantly greater ($P = 0.005$) proportion of the diet of juvenile pink salmon in the moderately oiled area near AFK Hatchery than along the lightly oiled southern coast of Knight Island in 1989 (Table 5). In 1990, large copepods and harpacticoid copepods constituted a significantly greater proportion ($P < 0.05$) of the diet in the moderately oiled area near AFK Hatchery than along the lightly oiled southern coast of Knight Island. Because of small sample sizes, no statistical comparison of stomach fullness or prey composition was possible between the two areas where fish from the AFK Hatchery were recovered in 1991. Prey composition and stomach fullness were not significantly different between the two areas where fish from the WHN Hatchery were

TABLE 3.—Percent frequency of occurrence of cytochrome P-4501A enzyme induction in tissues of juvenile pink salmon sampled along the southern coast of Knight Island during May and June, 1989. Fisher's exact test probabilities (*P*) of a type I error (two-tail) are shown for pairwise comparisons between months.

Organ	Tissue	May	Jun	*P*
Gill	Pillar cells	67	0	<0.001
	Epithelium	4	0	1.000
	Endothelium of gill arches	4	0	1.000
	Gill buds	0	0	
	Pharyngeal epithelium	4	0	1.000
Liver	Hepatocytes	6	9	1.000
	Sinusoidal endothelium	0	0	
	Central veins	0	0	
Kidney	Duct epithelium	0	0	
	Sinusoidal endothelium	0	0	
Intestine	Cecal epithelium	45	0	0.002

recovered in 1989 (Table 5). In 1990, stomach fullness was greater in the non-oiled area near the WHN Hatchery than in the moderately oiled area near Main Bay, but the difference was only marginally significant ($P = 0.061$). In 1991, there were no significant differences in stomach fullness or prey composition between areas where fish from the WHN Hatchery were recovered.

Evaluation of effects of fry density, ocean temperature, prey density, and prey composition.—Analysis of covariance indicated that the growth of juvenile CWT pink salmon was significantly related ($P = 0.050$) to mean ocean temperature measured in the passages adjacent to each hatchery (Table 6). Growth of juvenile CWT pink salmon was not significantly related to number of fry released ($P = 0.265$) or zooplankton volume ($P = 0.619$; Table 6). Further analyses of covariance indicated that ocean temperatures were generally warmer along the lightly oiled southern coast of Knight Island compared with the moderately oiled area near AFK Hatchery. The least-squares mean difference in temperature was 0.9°C ($P = 0.075$) in 1989, 2.4°C ($P < 0.001$) in 1990, and 0.7°C ($P = 0.290$) in 1991. There were no significant differences in ocean temperature between the non-oiled area near WHN Hatchery and the moderately oiled area near Main Bay in 1989, 1990, or 1991.

The bioenergetic model indicated that the growth of juvenile pink salmon probably was not affected by differences in prey density or prey composition between moderately oiled, and lightly and non-oiled areas of Prince William Sound during May 1989. Pink salmon diets at this time were composed of large copepods (40–52%), small copepods (21–27%), and various other invertebrates (Willette 1994). The biomass of large calanoid copepods ranged from 0.70 to 0.95 g/m^3 (Wertheimer et al. 1993), and water temperatures were generally less than 10°C. Under these conditions, juvenile pink salmon require a maximum of 10 to 12 h of feeding time to obtain maximum daily ration. Thus, juvenile salmon probably were able to obtain maximum daily ration at the lowest prey density measured, indicating that growth was not limited by feeding rate during May 1989.

The growth of juvenile pink salmon may have been affected by differences in prey density or prey composition between moderately oiled, and lightly and non-oiled areas of Prince William Sound during June 1989. Pink salmon diets at this time were composed of large copepods (1–24%), small cope-

TABLE 4.—Percent frequency of occurrence of cytochrome P-4501A enzyme induction in tissues of juvenile pink salmon sampled in non-oiled, lightly oiled, and moderately oiled nearshore habitats in Prince William Sound in 1989. Fisher exact test probabilities (*P*) of a type I error (two-tail) are shown for pairwise comparisons between areas.

		WHN Hatchery			AFK Hatchery		
Organ	Tissue	Moderately oiled	Non-oiled	*P*	Moderately oiled	Lightly oiled	*P*
Gill	Pillar cells	75	0	<0.001	100	67	0.015
	Epithelium	94	0	<0.001	27	4	0.068
	Endothelium of gill arches	6	0	1.000	0	4	1.000
	Gill buds	0	0		0	0	
	Pharyngeal epithelium	0	0		0	4	1.000
Liver	Hepatocytes	8	12	1.000	9	6	1.000
	Sinusoidal endothelium	0	0		9	0	0.379
	Central veins	0	0		9	0	0.423
Kidney	Duct epithelium	0	0		0	0	
	Sinusoidal endothelium	0	0		0	0	
Intestine	Cecal epithelium	14	9	1.000	21	45	0.175

TABLE 5.—Median prey composition (% stomach contents weight) and stomach fullness (% body weight) of juvenile pink salmon in moderately, lightly, and non-oiled areas of Prince William Sound, 1989–1991.

Prey type or statistic	WHN Hatchery: Moderately oiled	WHN Hatchery: Non-oiled	WHN Hatchery: P	AFK Hatchery: Moderately oiled	AFK Hatchery: Lightly oiled	AFK Hatchery: P
1989						
Large copepod	6.8	2.5	0.780	37.7	29.1	0.089
Small copepod	3.3	5.1	0.180	12.3	2.2	0.005
Harpacticoid copepod	13.1	1.4	0.397	1.1	16.3	0.245
Stomach fullness	4.0	3.5	0.916	2.4	2.0	0.337
Sample size (*N*)	10	20		20	15	
1990						
Large copepod	3.2	1.4	0.525	22.9	0.3	0.020
Small copepod	2.7	5.1	0.554	7.0	3.7	0.261
Harpacticoid copepod	0.6	0.0	0.234	2.0	0.0	0.017
Stomach fullness	0.8	4.2	0.061	2.4	3.1	0.179
Sample size (*N*)	9	7		34	13	
1991						
Large copepod	0.0	3.5	0.390	52.1	16.7	
Small copepod	1.6	7.6	0.211	30.0	47.8	
Harpacticoid copepod	6.3	6.1	0.590	5.9	32.0	
Stomach fullness	2.3	1.8	0.497	3.5	2.0	
Sample size (*N*)	6	7		10	1	

pods (3–10%), and various other invertebrates (Willette 1994). The biomass of large calanoid copepods decreased to less than 0.10 g/m^3 (Wertheimer et al. 1993), and water temperatures exceeded 10°C. Under these conditions, the growth of juvenile pink salmon may be limited by feeding rate. The proportion of the diet consisting of large copepods appears to be critical in determining whether growth is limited by feeding rate.

TABLE 6.—Analysis of covariance of mean growth of juvenile coded-wire tagged pink salmon in relation to number of fry released, ocean temperature, and zooplankton settled volume. Mean growth rates (% body weight/d) by size of release group are adjusted for covariates.

Variable	Mean growth	Parameter: Mean	Parameter: SE	Parameter: P
Intercept		1.83	1.08	0.096
Number of fry released:				
<50 million	4.70	0.82	0.73	0.265
>50 million and <150 million	4.15	0.27	0.59	0.653
>150 million	3.88			
Ocean temperature		0.30	0.15	0.050
Zooplankton volume		−0.06	0.13	0.619

Effect of Oil Contamination on Fry-to-Adult Survival

Analysis of covariance indicated that the mean growth of juvenile CWT pink salmon in each tag-code group was significantly related ($P < 0.001$) to fry-to-adult survival. However, the slopes of the growth–survival relationship were different among years ($P = 0.001$). Regression analysis was used to estimate the parameters of the growth–survival relationship for each year separately. The mean growth of juvenile CWT pink salmon in each tag-code group was significantly related ($P = 0.016$) to fry-to-adult survival in 1989 (Figure 3). The mean growth of juvenile CWT pink salmon in each tag-code group was not significantly related to fry-to-adult survival in 1990 ($P = 0.527$) and 1991 ($P = 0.774$; Figure 3).

An analysis of covariance was conducted to adjust the mean growth rates of juvenile CWT pink salmon in 1989 for the marginally significant difference ($P = 0.075$) in ocean temperature between the moderately oiled area near AFK Hatchery and the lightly oiled area along the southern coast of Knight Island. The results from the analysis indicated that growth was significantly different ($P = 0.010$) be-

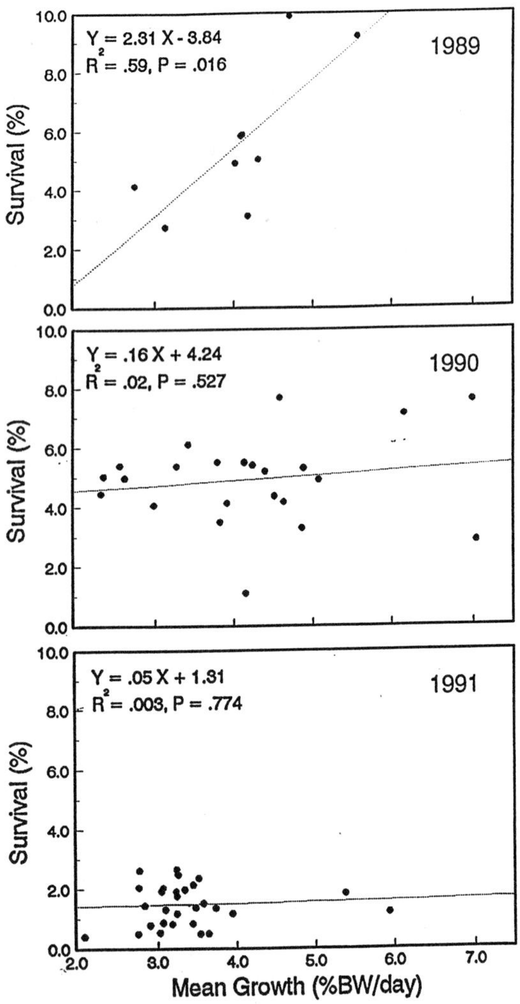

FIGURE 3.—Relationship between fry-to-adult survival and mean growth (BW = body weight) of coded-wire tagged juvenile pink salmon released from the Wally H. Noerenberg and Armin F. Koernig hatcheries in Prince William Sound in 1989, 1990, and 1991.

tween the two areas, but that ocean temperature was not significantly related ($P = 0.356$) to growth (Figure 4). The mean growth rates previously estimated from an analysis of variance (Table 2) were used to predict fry-to-adult survival for juveniles that were reared in moderately oiled, lightly oiled, and non-oiled areas using the growth–survival relationship for fish released in 1989 (Figure 3). In 1989, the differences in predicted fry-to-adult survival between moderately oiled, and lightly and non-oiled areas were 2.2% and 1.7% for juvenile CWT pink salmon released from the WHN and AFK hatcheries, respectively (Table 7).

Discussion

Effect of Oil Contamination on Migratory Behavior

Although a greater number of juveniles from AFK Hatchery migrated to the lightly oiled southern coast of Knight Island in 1989 compared with 1990 and 1991, no definitive conclusions can be drawn regarding the cause of these movements. Laboratory experiments have established that juvenile pink salmon will avoid sublethal concentrations of Prudhoe Bay crude oil in seawater. The avoidance thresholds of juvenile pink salmon ranged from 16.0 to 1.6 mg/L in June and August, respectively. Nonavoidance thresholds ranged from 8.8 to 0.75 mg/L in June and August, respectively (Rice 1973). These thresholds represent the levels for discrimination and not the concentrations actually avoided by motivated migratory juvenile pink salmon (Rice 1973). Hydrocarbon concentrations measured in Sawmill Bay within 2 weeks after the pink salmon release were substantially below the avoidance and nonavoidance thresholds (Short and Rounds 1993). However, a greater frequency of occurrence of P-4501A staining in juvenile pink salmon captured near the AFK Hatchery indicated that these fish were exposed to hydrocarbons (Table 4). Sufficient data are not available at present to determine if juvenile pink salmon migrating from the AFK Hatchery in 1989 encountered hydrocarbon concentrations that would cause an avoidance reaction.

The migratory behavior of juvenile pink salmon released from AFK Hatchery in 1989 was probably also affected by natural environmental conditions, however, the effect of natural environmental conditions on juvenile salmon migratory behavior is poorly understood. Simenstad et al. (1980) concluded that juvenile pink salmon and juvenile chum salmon *Oncorhynchus keta* emigrated rapidly from Hood Canal when food abundance was low. No data are available regarding relative food abundances at AFK Hatchery (moderately oiled area) compared with the southern coast of Knight Island (lightly oiled area) in 1989; however, stomach fullness in 1989 was not different between these two areas (Table 5). Hurley and Woodall (1968) concluded that juvenile pink salmon less than 60 mm in length prefer relatively warm temperatures (12.0–13.5°C). Cooler ocean temperatures near AFK Hatchery compared with the southern coast of

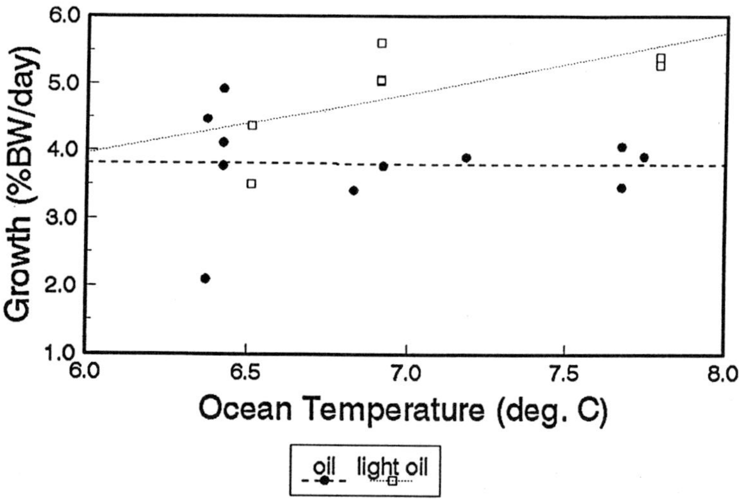

FIGURE 4.—Relationship between mean growth (BW = body weight) and ocean temperatures (°C for coded-wire tagged juvenile pink salmon in the early-fed group released from the Armin F. Koernig Hatchery and recovered in moderately and lightly oiled areas of Prince William Sound in 1989.

Knight Island may have caused juvenile pink salmon to migrate northward in 1989, but ocean temperatures were also cooler near AFK Hatchery in 1990 and 1991. Sufficient data are not available to determine the relative effects of oil contamination and natural environmental conditions on the migratory behavior of juvenile pink salmon in 1989.

Effect of Oil Contamination on Growth

Exposure to hydrocarbons from the *Exxon Valdez* oil spill appeared to reduce the growth of juvenile pink salmon by 0.76 to 0.94% body weight/day in 1989 (Table 7). The following alternative hypotheses must be evaluated to determine if these growth differences were caused by oil contamination from the spill:

TABLE 7.—Estimate of fry-to-adult survival (%) for juvenile coded-wire tagged pink salmon that reared in moderately, lightly, and non-oiled areas of Prince William Sound in 1989. Mean growth is % body weight/d. Mean difference is with respect to estimated survival. Hatcheries are Wally H. Noerenberg (WHN) and Armin F. Koernig (AFK).

Hatchery	Recovery area	Mean growth	Estimated survival	Mean difference
WHN	Non-oil	5.16	8.1	2.2
	Oil	4.22	5.9	
AFK	Light oil	4.93	7.5	1.7
	Oil	4.17	5.8	

(1) The differences in mean growth were caused by measurement or sampling error.
(2) The differences in mean growth were caused by differences in food consumption rate or prey composition.
(3) The differences in mean growth were caused by differences in ocean temperature and its effect on metabolic rate,
(4) The differences in mean growth were not associated with differential exposure to hydrocarbons from this oil spill.
(5) The estimated reduction in growth attributed to oil contamination is not associated with a level of hydrocarbon exposure sufficient to cause the estimated response.

Growth differences caused by measurement or sampling error.—It cannot be determined if the pattern of mean growth rates observed in 1989 is attributable to error in the estimation of growth rates of individual fish. The early-fed tag-code groups were released from the WHN and AFK hatcheries over a period of 8 and 5 d, respectively, in 1989. Because release date was not known for individual fish, mean release date for each tag-code group was used to estimate growth. Thus, this growth analysis is based on the assumption that fish released from the hatcheries on different days were distributed randomly between moderately oiled, and lightly and non-oiled areas. Also, body weight at release was not known for individual fish, so the mean body weight at release for each tag-code group was used to estimate growth. An analysis of the errors caused by this uncertainty in release date and size at release indicates that a nonrandom distribution of fish with different dates of release and body weights at release could have caused the observed differences in growth rate between moderately oiled, and lightly and non-oiled areas in 1989. However, juveniles released from the WHN and AFK hatcheries both exhibited a reduction in growth rate slightly less than 1% body weight/day in moderately oiled areas compared with non-oiled and lightly oiled areas in 1989. Untagged juvenile pink salmon also exhibited a reduction in growth of about 1.4% body weight/day (Wertheimer and Celewycz 1996). The similar apparent growth response of untagged juveniles as well as of fish from both WHN and AFK hatcheries indicates that a nonrandom distribution of fish with different release dates and sizes at release probably was not the cause of the observed difference in growth between areas.

The growth estimates obtained from equation (1) using mean body weight at release and mean date of release are not actual growth rates for individual fish. Consider the distributions of ($\log_e W_c - \log_e \bar{W}_r$) and ($\log_e W_c - \log_e W_r$), where W_c and $\bar{W}_r$ are defined as in equation (1), and W_r is the weight at release of individual fish. Any disparity in variance between them should be considered when interpreting a statistical test based on the former that is meant to make inference regarding the latter. Specifically, an analysis of variance based on a distribution with a smaller variance might indicate significant differences where in fact there are none. It can be shown that the difference in variance depends upon the relative magnitudes of the variance of $\log_e W_r$ and $\log_e W_c$ and the covariance of $\log_e W_r$ and $\log_e W_c$. Assuming equal variances of $\log_e W_r$ and $\log_e W_c$, the difference in the variances of ($\log_e W_c - \log_e W_r$) and ($\log_e W_c - \log_e \bar{W}_r$) depends on the magnitude of the covariance of $\log_e W_r$ and $\log_e W_c$. It is then true that for a correlation greater than 0.5, the variance based on $\log_e \bar{W}_r$ will actually be greater than that based on $\log_e W_r$, resulting in a conservative statistical test. In the present study, otolith analysis was used to back calculate the weight at release of individual juvenile CWT pink salmon. In three age groups, $\log_e W_r$ was significantly correlated with $\log_e W_c$, with correlations of 0.77 (N = 180), 0.66 (N = 231), and 0.80 (N = 21), all above 0.5. This suggests that our tests based on $\log_e \bar{W}_r$ are in fact conservative tests of the true growth differences.

The observed differences in growth do not appear to be caused by sampling error. Recovery site was used as the sample unit in this growth analysis. The analysis is based on the assumption that the various sites are a random sample of all possible sites within each area (Hurlbert 1984). The sites where large samples of CWT juvenile salmon were caught were generally places where fish were consistently abundant. This relationship between the sites and fish abundance was unavoidable, because only about one fish in a thousand was coded-wire tagged. As a result, a large number of fish needed to be scanned for CWTs. To our knowledge, there were no other sites within each of the areas where fish were consistently abundant. Thus, the group of sites included in the analysis probably represents a nearly complete set of all sites where fish were consistently abundant within each area.

Growth differences caused by food consumption rate or prey composition.—The observed growth differences do not appear to be caused by differences in food consumption rate between areas. Differences in food consumption rate between areas may be caused by differences in food abundance, fish density, water temperature, or the presence of currents that deliver food to the fish (Cooney et al. 1981). In the present study, stomach fullness and bioenergetics were used to evaluate whether food consumption rate was different between areas. Stomach samples were collected at various times during daylight hours, but a systematic study of diel feeding periodicity was not conducted. Diel feeding periodicity has been documented in juvenile pink salmon (Parker and Vanstone 1966; Parker 1969; Simenstad et al. 1980; Godin 1981). In general, stomach fullness is low at night, indicating a cessation of feeding, and then increases during the day (Bailey et al. 1975; Simenstad et al. 1980; Godin 1981). The magnitude of the change in stomach fullness with time of day has varied considerably among studies. In the present study, time of day was not a significant factor related to stomach fullness (Willette 1994). An evaluation of feeding bioenergetics indicated that differences in prey density between areas may not have affected food consumption rate during May 1989, because the fish probably were able to obtain maximum daily ration at the lowest prey density measured during this time period. Differences in prey density between areas may have affected food consumption rate in June 1989. However, stomach fullness was not significantly different between areas in either May or June, 1989 (Table 5). Sturdevant et al. (1996, this volume) also found no significant difference in stomach fullness of juvenile pink and chum salmon between oiled and non-oiled areas of Prince William Sound in 1989.

The observed growth differences do not appear to be caused by differences in prey composition between areas. The proportion of the diet consisting of large calanoid copepods may affect gastric evacuation rate and feeding rate. Gastric evacuation rate is generally greater when fish consume small versus large food particles (Fange and Grove 1979; Jobling 1980, 1981). However, the feeding rate of juvenile pink salmon was 10 and 30 mg/hr at maximum ration when the fish consumed small (*Pseudocalanus* spp.) and large (*Neocalanus plumchrus*) copepods, respectively (Parsons and LeBrasseur 1973). The handling time per calorie of food obtained is thus three times lower when large rather than small copepods are consumed (Parsons and LeBrasseur 1973). The greater net energetic gain associated with consumption of large copepods indicates that large copepods probably support a higher growth rate and are selected as prey (LeBrasseur 1969; Calow and Townsend 1981; Townsend and Winfield 1985). Results from laboratory studies indicate that juvenile pink salmon select large copepods (2–4 mm in length) over small copepods and over four other zooplankton species (Parsons and LeBrasseur 1973). In the present study, large copepods constituted a greater proportion of the diet in the moderately oiled area near the AFK Hatchery compared with the lightly oiled area along the southern coast of Knight Island in both 1989 and 1990 (Table 5). There were no differences in the proportion of the diet consisting of large copepods between areas where fish from the WHN Hatchery were recovered (Table 5).

Growth differences caused by ocean temperature.—Differences in ocean temperature cannot account entirely for growth differences between areas in 1989. Water temperature has a direct effect on the metabolic and gastric evacuation rates of fish (Brett and Groves 1979; Fange and Grove 1979). Under conditions of high food abundance, fish are able to acquire maximum daily ration, so temperature largely regulates growth (Shelbourn et al. 1973). It appears that juvenile pink salmon were able to acquire maximum daily ration in May 1989 when zooplankton abundance was very high in Prince William Sound. Ocean temperatures were lower ($P = 0.075$) in the moderately oiled area near the AFK Hatchery compared with the lightly oiled southern coast of Knight Island in 1989. However, analysis of covariance indicated that the mean growth of juvenile pink salmon adjusted for the effect of temperature was still significantly lower ($P = 0.010$) in the moderately oiled area near the AFK Hatchery compared with the lightly oiled southern coast of Knight Island in 1989. Continued lower growth in the moderately oiled area near AFK Hatchery in 1990 and 1991 was probably the result of lower temperatures in this area compared with the southern coast of Knight Island. There were no temperature differences between areas where fish from the WHN Hatchery were recovered in 1989, 1990, or 1991.

Growth differences associated with hydrocarbon exposure.—The significant differences in growth rate between moderately oiled, and lightly and non-oiled areas in 1989 are associated with significant differences in the frequency of P-4501A staining. The frequency of P-4501A staining in gill tissues closely coincided with observed levels of shoreline oil contamination in each sampling area (GIS Technical Group 1991). This association is evidence that the differences in P-4501A staining between areas were caused by oil contamination from the *Exxon Valdez* oil spill. Carls et al. (1996b, this volume) also found an association between P-4501A induction in juvenile pink salmon, shoreline oil contamination, and hydrocarbon tissue content. In the lightly oiled area along the southern coast of Knight Island, the frequency of P-4501A staining declined significantly between May and June (Table 3). This result is expected for fish that originated in the moderately oiled area near the AFK Hatchery and migrated into the lightly oiled area along the southern coast of Knight Island (Kloepper-Sams and Stegeman 1989). However, a decline in hydrocarbon concentrations between May and June may also explain the observed reduction in P-4501A staining among fish from the southern coast of Knight Island. Induction of P-4501A enzymes was not detected in juvenile pink salmon in 1990 (Carls et al. 1996b).

Level of hydrocarbon exposure as sufficient cause of reduced growth.—All necessary evidence is not available at present to determine if exposure to hydrocarbons from the *Exxon Valdez* oil spill was at a level sufficient to cause the estimated reduction in growth. The question of level of exposure of juvenile pink salmon to *Exxon Valdez* oil in 1989 is linked to questions regarding the route of contamination. Several routes of contamination probably caused the level of exposure to *Exxon Valdez* oil that triggered induction of P-4501A enzymes, leading to reduced growth in juvenile pink salmon.

It does not appear that the observed reduction in growth was caused solely by exposure to the water-soluble fraction (WSF) of crude oil. Exposure of juvenile pink salmon to 0.40 mg/L of the WSF of Cook Inlet crude oil caused a 0.8% body weight/day reduction in growth rate in the laboratory (Moles and Rice 1983). This concentration of the WSF is substantially greater than levels measured in the water column in Prince William Sound within a few weeks after juvenile pink salmon were released from hatcheries in the sound (Neff 1990; Short and Rounds 1993).

Ingestion of oil-contaminated prey or oil particles appears to be the most important route of contamination. The occurrence of oil particles in juvenile salmon stomachs (Sturdevant et al. 1996), the degree of hydrocarbon contamination of visceral tissues, the ratio of total aromatics to total hydrocarbons, and the frequency, intensity, and occurrence of tissue types staining for P-4501A activity led Carls et al. (1996b) to conclude that ingestion of oil particles or oil-contaminated prey was probably the primary route of contamination for both juvenile pink and chum salmon in Prince William Sound in 1989. Based on laboratory data, ingestion of prey contaminated with an estimated polyaromatic hydrocarbon (PAH) concentration of 0.29 milligrams per gram of food would cause a 1% body weight/day reduction in the growth of juvenile pink salmon (M. Carls, National Marine Fisheries Service, personal communication). Hydrocarbon concentrations were not measured in zooplankton or in epibenthic invertebrate samples collected in Prince William Sound in 1989. However, the estimated PAH concentration (393 mg/g) in carcasses of fish that exhibited a 1.4% body weight/day growth reduction in the laboratory was roughly equal to that measured in juvenile pink salmon carcasses (326 mg/g) from oiled areas of the sound in 1989 (M. Carls, personal communication). Ingestion of oil particles by zooplankton and epibenthic invertebrates has been documented in other areas. After the tanker *Arrow* spill, pelagic copepods ingested oil particles (Conover 1971) that were uniformly distributed to at least 80-m depth (Forrester 1971). Meiofauna and epibenthic invertebrates, such as harpacticoid copepods, are also known to ingest particulate oil (Landrum 1989).

In the present study, the frequency of P-4501A staining in gill tissues was significantly different between areas, but this was not the case in intestinal tissues. However, intensity of staining was not considered and this may influence how different tissues reflect hydrocarbon exposure. The route of contamination cannot be inferred from the frequency of occurrence of P-4501A staining in gill pillar and gill epithelial cells, because P-4501A enzymes are induced in these tissues by both intraperitoneal administration of PAH and exposure of fish to waterborne PAH (Miller et al. 1989; Stegeman et al. 1991).

A third possible route of contamination is linked to the behavior of juvenile pink salmon. Juvenile pink salmon commonly feed at the sea surface and leap from the water while feeding (Cooney et al. 1981; Heard 1991). The sea surface microlayer may concentrate nonpolar compounds such as petroleum hydrocarbons that are toxic to fish (Cross et al. 1987; Kocan et al. 1987). We frequently observed schools of juvenile pink salmon feeding and leaping through sheen in oiled areas of Prince William Sound in 1989. Further analysis of samples from an oil-ingestion experiment (Carls et al. 1996a, this volume) is expected to provide additional quantitative data to more firmly establish that exposure to hydrocarbons from the *Exxon Valdez* oil spill was at a level sufficient to cause the estimated reduction in growth.

The observed reduction in growth in oil-contaminated areas probably was caused primarily by the increased metabolic cost associated with hydrocarbon metabolism and depuration. The reduction in growth caused by oil exposure in the laboratory (Rice et al. 1975; Moles and Rice 1983; Schwartz 1985; Carls et al. 1996a) appears to be partly attributable to the increased metabolic demand associated with hydrocarbon depuration (Rice et al. 1977; Thomas and Rice 1979). Carls et al. (1996a) postulated that necrosis of the gastrointestinal tract may also reduce assimilation efficiency in juvenile pink salmon fed oil-contaminated prey. Veronique et al. (1992) documented a decline in food conversion efficiency but not feeding rate in the parr of Atlantic salmon *Salmo salar* exposed to a sublethal concentration of a Hibernia oil-water mixture. However, when high concentrations of hydrocarbons are introduced into the diet, feeding rate will decline (Schwartz 1985; Carls et al. 1996a). In the present study, stomach fullness was not different between areas in 1989 (Table 5). The greater frequency of P-4501A staining in moderately oiled areas compared with lightly and non-oiled areas (Table 4) is direct evidence that juvenile pink salmon in oiled areas expended greater amounts of energy to metabolize and depurate hydrocarbons, leaving less energy available for somatic growth.

Effect of Oil Contamination on Survival to the Adult Stage

Exposure of juvenile pink salmon to hydrocarbons from the *Exxon Valdez* oil spill appeared to reduce survival to the adult stage by 1.7 to 2.2% in

1989 (Table 7). Reduced growth in oiled nearshore habitats probably caused reduced survival, because slow-growing juvenile salmon are vulnerable to predators for a longer time (Parker 1971; Healey 1982; West and Larkin 1987). Mortenson et al. (1991) concluded that the growth of juvenile CWT pink salmon that were rearing in Auke Bay, Alaska, was significantly related to survival to the adult stage in 3 of 4 years. Mortenson et al. (1991) postulated that changes in predator abundance may have caused interannual changes in the growth–survival relationship. The observed interannual differences in the growth–survival relationship in the present study may have also been caused by interannual changes in predation rate. Predation rate may also differ among areas. The estimated reduction in survival attributed to oil contamination in the present study is based on the assumption that the growth–survival relationship was not different for fish being reared in different areas of Prince William Sound.

Conclusions

(1) A relatively large number of juvenile pink salmon from AFK Hatchery migrated out of the moderately oiled area near the hatchery to the lightly oiled southern coast of Knight Island in 1989. A similar northward migration of juveniles was not observed in 1990 and 1991. Although this behavior was associated with oil contamination, it cannot be attributed to it because of a lack of understanding regarding natural environmental effects on migratory behavior.

(2) The growth of juvenile pink salmon in Prince William Sound appeared to be reduced 0.76 to 0.94% body weight/d by oil contamination from the *Exxon Valdez* oil spill in 1989. Energy expenditures required to metabolize and depurate hydrocarbons probably resulted in less energy being available for somatic growth.

(3) The reduction in juvenile growth attributed to oil contamination in 1989 probably caused a 1.7 to 2.2% reduction in survival to the adult stage among fish that reared in oiled areas. The adult pink salmon return to Prince William Sound in 1990 was thus lower than if the *Exxon Valdez* oil spill had not occurred.

Acknowledgments

I thank Greg Carpenter, James A. Hasbrouck, Pat Shields, Stan Carlson, Edward Debevec, and other Alaska Department of Fish and Game staff who helped with collection of samples, sample processing, and data analysis. James A. Raymond was project leader in 1989 and 1990. A review provided by Phil Mundy improved the technical quality of the manuscript. This study would not have been possible without the help of Jeff Olsen and other staff at the Prince William Sound Aquaculture Corporation who kindly provided logistical support for our sampling crew in the field and data regarding their releases of juvenile salmon and environmental conditions in Prince William Sound. Alex Wertheimer and others from the National Marine Fisheries Service collected some of the CWT juvenile salmon used in this study in 1989 and 1990. John Stegeman and Roxanna Smolowitz conducted analyses of samples for P-4501A enzyme induction. This is contribution PP-084 of the Alaska Department of Fish and Game, Commercial Fisheries Management and Development Division, Juneau.

References

Babcock, M. M. 1985. Morphology of olfactory epithelium of pink salmon, *Oncorhynchus gorbuscha*, and changes following exposure to benzene: a scanning electron microscope study. Pages 259–267 *in* J. S. Gray and M. E. Christiansen, editors. Marine biology of polar regions and stress on marine organisms. Wiley, New York.

Bailey, J. E., B. L. Wing, and C. R. Mattson. 1975. Zooplankton abundance and feeding habits of fry of pink salmon, *Oncorhynchus gorbuscha*, and chum salmon, *Oncorhynchus keta*, in Traitors Cove, Alaska, with speculations on the carrying capacity of the area. U.S. National Marine Fisheries Service Fishery Bulletin 73:846–861.

Bax, N. J. 1983. Early marine mortality of marked juvenile chum salmon released into Hood Canal, Puget Sound, Washington, in 1980. Canadian Journal of Fisheries and Aquatic Sciences 40:426–435.

Brett, J. R., and T. D. D. Groves. 1979. Physiological energetics. Pages 280–344 *in* W. S. Hoar, D. J. Randall, and J. R. Brett, editors. Fish physiology, volume 8: bioenergetics and growth. Academic Press, New York.

Calow, P., and C. R. Townsend. 1981. Ecology, energetics and evolution. Pages 3–19 *in* C. R. Townsend and P. Calow, editors. Physiological ecology: an evolutionary approach to resource use. Blackwell Scientific Publications, Oxford.

Carls, M. G., and six coauthors. 1996a. Growth, feeding, and survival of pink salmon fry exposed to food contaminated with crude oil. American Fisheries Society Symposium 18:608–618.

Carls, M. G., A. C. Wertheimer, J. W. Short, R. M. Smolowitz, and J. J. Stegeman. 1996b. Contamination of juvenile pink and chum salmon by hydrocarbons in Prince William Sound after the *Exxon Valdez* oil spill. American Fisheries Society Symposium 18: 593–607.

Clark, J., and B. McCarl. 1983. An investigation of the relationship between Oregon coho salmon (*Oncorhynchus kisutch*) hatchery releases and adult production utilizing law of the minimum regression. Canadian Journal of Fisheries and Aquatic Sciences 40:516–523.

Conover, R. J. 1971. Some relations between zooplankton and bunker C oil in Chedabucto Bay following the wreck of the tanker *Arrow*. Journal of the Fisheries Research Board of Canada 28:1327–1330.

Conover, W. J., and R. L. Iman. 1981. Rank transformations as a bridge between parametric and nonparametric statistics. American Statistician 35(3):124–129.

Cooney, R. T., D. Urquhart, and D. Barnard. 1981. The behavior, feeding biology and growth of hatchery-released pink and chum salmon fry in Prince William Sound, Alaska. University of Alaska, Alaska Sea Grant Report 81-5, Fairbanks.

Coyle, K. O., A. J. Paul, and D. A. Ziemann. 1990. Copepod populations during the spring bloom in an Alaskan subarctic embayment. Journal of Plankton Research 12:759–797.

Cross, J. N., and six coauthors. 1987. Contaminant concentrations and toxicity of sea-surface microlayer near Los Angeles, California. Marine Environmental Research 23:307–323.

Emlen, J. M., R. R. Reisenbichler, A. M. McGie, and T. E. Nickelson. 1990. Density-dependence at sea for coho salmon (*Oncorhynchus kisutch*). Canadian Journal of Fisheries and Aquatic Sciences 47:1765–1772.

Fange, R., and D. Grove. 1979. Digestion. Pages 161–241 *in* W. S. Hoar, D. J. Randall, and J. R. Brett, editors. Fish physiology, volume 8: bioenergetics and growth. Academic Press, New York.

Forrester, W. D. 1971. Distribution of suspended oil particles following the grounding of the tanker *Arrow*. Journal of Marine Research 29:151–170.

GIS (Geographical Information Systems) Technical Group. 1991. GIS mapping and statistical analysis. Report to Natural Resource Damage Assessment, Technical Services Study 3, Anchorage, Alaska.

Godin, J. J. 1981. Daily patterns of feeding behavior, daily rations, and diets of juvenile pink salmon (*Oncorhynchus gorbuscha*) in two marine bays of British Columbia. Canadian Journal of Fisheries and Aquatic Sciences 38:10–15.

Hartt, A. C. 1980. Juvenile salmonids in the oceanic ecosystem—the critical first summer. Pages 25–57 *in* W. J. McNeil and D. C. Himsworth, editors. Salmonid ecosystems of the north Pacific. Oregon State University Press, Corvallis.

Healey, M. C. 1982. Timing and relative intensity of size-selective mortality of juvenile chum salmon during early sea life. Canadian Journal of Fisheries and Aquatic Sciences 39:952–957.

Heard, W. R. 1991. Life history of pink salmon (*Oncorhynchus gorbuscha*). Pages 119–231 *in* C. Groot and L. Margolis, editors. Pacific salmon life histories. University of British Columbia Press, Vancouver.

Holling, C. S. 1966. The functional response of invertebrate predators to prey density. Memoirs of the Entomological Society of Canada 48:1–86.

Hurlbert, S. 1984. Pseudo-replication and the design of ecological field experiments. Ecological Monographs 54(2):187–211.

Hurley, D. A., and W. L. Woodall. 1968. Responses of young pink salmon to vertical temperature and salinity gradients. International Pacific Salmon Fisheries Commission Progress Report 19.

Jobling, M. 1980. Gastric evacuation in plaice, *Pleuronectes platessa* L.: effects of dietary energy level and food consumption. Journal of Fish Biology 17:547–551.

Jobling, M. 1981. Mathematical models of gastric emptying and the estimation of daily rates of food consumption for fish. Journal of Fish Biology 19:245–257.

Kendall, M., and A. Stuart. 1979. The advanced theory of statistics, volume 2. Macmillan, New York.

Kloepper-Sams, P. J., and J. J. Stegeman. 1989. The temporal relationships between P450E protein content, catalytic activity and MRNA levels in the teleost *Fundulus heteroclitus* following treatment with B-naphthoflavone. Archives of Biochemistry and Biophysics 268:525–535.

Kocan, R. M., H. von Westernhagen, M. L. Landolt, and G. Furstenberg. 1987. Toxicity of sea-surface microlayer: effects of hexane extract on Baltic herring (*Clupea harengus*) and Atlantic cod (*Gadus morhua*) embryos. Marine Environmental Research 23:291–305.

Landrum, P. F. 1989. Bioavailability and toxikinetics of polycyclic aromatic hydrocarbons sorbed to sediments for the amphipod *Pontoporeia hoyi*. Environmental Science and Technology 23:588–595.

LeBrasseur, R. J. 1969. Growth of juvenile chum salmon (*Oncorhynchus keta*) under different feeding regimes. Journal of the Fisheries Research Board of Canada 26:1631–1645.

Matthews, S. B. 1980. Trends in Puget Sound and Columbia River coho salmon. Pages 133–146 *in* W. J. McNeil and D. C. Himsworth, editors. Salmonid ecosystems of the north Pacific. Oregon State University Press, Corvallis.

McCarl, B., and R. B. Rettig. 1983. Influence of hatchery smolt releases on adult salmon production and its variability. Canadian Journal of Fisheries and Aquatic Sciences 40:1880–1886.

Miller, M. R., D. E. Hinton, and J. J. Stegeman. 1989. Cytochrome P-450E induction and localization in gill pillar (endothelial) cells of scup and rainbow trout. Aquatic Toxicology 14:307–322.

Moles, A., and S. D. Rice. 1983. Effects of crude oil and naphthalene on growth, caloric content, and fat content of pink salmon juveniles in seawater. Transactions of the American Fisheries Society 112:205–211.

Mortenson, D., J. Landingham, A. Wertheimer, and S. Taylor. 1991. Relationship of early marine growth and survival of juvenile pink salmon to marine water temperature and secondary production in Auke Bay, Alaska. Pages 38–49 *in* B. White and I. Guthrie, editors. Proceedings of the 15th northeast Pacific pink and chum salmon workshop. Pacific Salmon Commission, Vancouver, British Columbia.

Neff, J. M. 1990. Water quality in Prince William Sound. Report of Battelle Ocean Sciences to Exxon Corp., Duxbury, Massachusetts.

Nickelson, T. E. 1986. Influence of upwelling, ocean temperature, and smolt abundance on marine survival of coho salmon (*Oncorhynchus kisutch*) in the Oregon production area. Canadian Journal of Fisheries and Aquatic Sciences 43:527–535.

Panadian, T. J., and E. Vivekanandan. 1985. Energetics of feeding and digestion. Pages 99–124 *in* P. Tytler and P. Calow, editors. Fish energetics: new perspectives. Johns Hopkins University Press, Baltimore, Maryland.

Parker, R. R. 1968. Marine mortality schedules of pink salmon of the Bella Coola River, central British Columbia. Journal of the Fisheries Research Board of Canada 25:757–794.

Parker, R. R. 1969. Foods and feeding of juvenile pink salmon in central British Columbia waters. I. 1966 diel series. Fisheries Research Board of Canada Manuscript Report 1017.

Parker, R. R. 1971. Size selective predation among juvenile salmonid fishes in a British Columbia inlet. Journal of the Fisheries Research Board of Canada 28: 1503–1510.

Parker, R. R., and W. E. Vanstone. 1966. Changes in chemical composition of central British Columbia pink salmon during early sea life. Journal of the Fisheries Research Board of Canada 23:1353–1384.

Parsons, T. R., and R. J. LeBrasseur. 1973. The availability of food to different trophic levels in the marine food chain. Pages 325–343 *in* J. H. Steele, editor. Marine food chains. Oliver and Boyd, Edinburgh, UK.

Raymont, J. E., R. T. Srinivasagam, and J. K. Raymont. 1969. Biochemical studies on marine zooplankton. VII. Observations on certain deep sea zooplankton. Internationale Revue der Gesamten Hydrobiologie 54:357–365.

Rice, S. D. 1973. Toxicity and avoidance tests with Prudhoe Bay oil and pink salmon fry. Pages 667–670 *in* Proceedings of the joint conference on prevention and control of oil spills. American Petroleum Institute, Washington, DC.

Rice, S. D., D. A. Moles, and J. W. Short. 1975. The effect of Prudhoe Bay crude oil on survival and growth of eggs, alevins, and fry of pink salmon, *Oncorhynchus gorbuscha*. Pages 503–507 *in* Proceedings, 1975 conference on prevention and control of oil pollution. American Petroleum Institute, Washington, DC.

Rice, S. D., R. E. Thomas, and J. W. Short. 1977. Effect of petroleum hydrocarbons on breathing and coughing rates and hydrocarbon uptake-depuration in pink salmon fry. Pages 259–277 *in* F. J. Vernberg, A. Calabrese, F. P. Thurberg, and W. B. Vernberg, editors. Physiological responses of marine biota to pollutants. Academic Press, New York.

Ricker, W. E. 1975. Computation and interpretation of biological statistics of fish populations. Fisheries Research Board of Canada Bulletin 191.

Roubal, W. T., T. K. Collier, and D. C. Malins. 1977. Accumulation and metabolism of carbon-14 labeled benzene, naphthalene, and anthracene by young coho salmon (*Oncorhynchus kisutch*). Archives of Environmental Contamination and Toxicology 5:513–529.

Schwartz, J. P. 1985. Effects of oil-contaminated prey on the feeding and growth rate of pink salmon fry *Oncorhynchus gorbuscha*. Pages 459–476 *in* J. F. Vernberg, F. Thurberg, A. Calabrese, and W. Vernberg, editors. Pollution and physiology of marine organisms. University of South Carolina Press, Columbia.

Sharr, S. 1993. Coded-wire tag studies on Prince William Sound salmon. Final Report. Natural Resource Damage Assessment, Fish/Shellfish Study 3. Alaska Department of Fish and Game, Cordova.

Sharr, S., and C. J. Peckham. 1993. Coded-wire tag studies on Prince William Sound salmon. Final Report. Natural Resource Damage Assessment, Fish/Shellfish Study R60A. Alaska Department of Fish and Game, Cordova.

Shelbourn, J. E., J. R. Brett, and S. Shirahata. 1973. Effect of temperature and feeding regime on the specific growth rate of sockeye salmon fry (*Oncorhynchus nerka*), with a consideration of size effect. Journal of the Fisheries Research Board of Canada 30:1191–1194.

Short, J. W., and P. Rounds. 1993. Petroleum hydrocarbons in near-surface seawater of Prince William Sound, Alaska following the *Exxon Valdez* oil spill I: chemical sampling and analysis. Natural Resources Damage Assessment Air/Water 3 Final Report. NOAA (National Oceanic and Atmospheric Administration) NMFS (National Marine Fisheries Service), Office of Oil Spill Damage Assessment and Restoration, Auke Bay, Alaska.

Simenstad, C. A., W. J. Kinney, S. S. Parker, E. O. Salo, J. R. Cordell, and H. Buechner. 1980. Prey community structure and trophic ecology of outmigrating juvenile chum and pink salmon in Hood Canal, Washington. University of Washington, College of Fisheries, Fisheries Research Institute Report FRI-UW-8026, Seattle.

Smolowitz, R. M., M. J. Moore, and J. J. Stegeman. 1989. Cellular distribution of cytochrome P450E in winter flounder liver with degenerative and neoplastic disease. Marine Environmental Research 28:441–446.

Stegeman, J. J. 1992. Cytochrome P450 (mixed function oxidases - MFO) in fish from Prince William Sound. Report of the Woods Hole Oceanographic Institute to the Alaska Department of Fish and Game, Anchorage.

Stegeman, J. J., and J. J. Lech. 1991. Cytochrome P450 systems in aquatic species: carcinogen metabolism and biomarkers for carcinogen and pollutant exposure. Environmental Health Perspectives 90:101–109.

Stegeman, J. J., R. M. Smolowitz, and M. E. Hahn. 1991. Immunohistochemical localization of environmentally induced cytochrome P-4501A in multiple organs of marine teleost, *Stenotomus chrysops* (scup). Toxicology and Applied Pharmacology 110:486–504.

Sturdevant, M. V., A. C. Wertheimer, and J. L. Lum. 1996. Diets of juvenile pink and chum salmon in oiled and non-oiled nearshore habitats in Prince William Sound, 1989 and 1990. American Fisheries Society Symposium 18:578–592.

Thomas, R. E., and S. D. Rice. 1979. The effect of exposure temperatures on oxygen consumption and opercular breathing rates of pink salmon fry exposed to toluene, naphthalene, and water-soluble fractions of Cook Inlet crude oil and no. 2 fuel oil. Pages 39–52 *in* W. B. Vernberg, A. Calabrese, F. P. Thurberg, and F. J. Vernberg, editors. Marine pollution: functional responses. Academic Press, New York.

Thomas, R. E., and S. D. Rice. 1981. Excretion of aromatic hydrocarbons and their metabolites by freshwater and seawater Dolly Varden char. Pages 425–448 *in* F. J. Vernberg, A. Calabrese, F. P. Thurberg, and W. B. Vernberg, editors. Biological monitoring of marine pollutants. Academic Press, New York.

Townsend, C. R., and I. J. Winfield. 1985. The application of optimal foraging theory to feeding behavior in fish. Pages 67–98 *in* P. Tytler and P. Calow, editors. Fish energetics: new perspectives. Johns Hopkins University Press, Baltimore, Maryland.

Veronique, V., J. H. Vandermeulen, and A. J. Fraser. 1992. Growth and food conversion by Atlantic salmon parr during 40 days exposure to crude oil. Transactions of the American Fisheries Society 121: 322–332.

Vinyard, G. L. 1982. Variable kinematics of Sacramento perch (*Archoplites interruptus*) capturing evasive and nonevasive prey. Canadian Journal of Fisheries and Aquatic Sciences 39:208–211.

Walters, C. J., and F. Juanes. 1993. Recruitment limitation as a consequence of natural selection for use of restricted feeding habitats and predation risk taking by juvenile fishes. Canadian Journal of Fisheries and Aquatic Sciences 50:2058–2070.

Werner, E. E. 1974. The fish size, prey size, handling time relation in several sunfishes and some implications. Journal of the Fisheries Research Board of Canada 31:1531–1536.

Wertheimer, A. C., and A. G. Celewycz. 1996. Abundance and growth of juvenile pink salmon in oiled and non-oiled locations of western Prince William Sound after the *Exxon Valdez* oil spill. American Fisheries Society Symposium 18:518–532.

Wertheimer, A. C., A. G. Celewycz, M. G. Carls, and M. V. Sturdevant. 1993. Impact of the oil spill on juvenile pink and chum salmon and their prey in critical nearshore habitats. Final Report. Natural Resource Damage Assessment, Fish/Shellfish Study 4. National Marine Fisheries Service, Auke Bay, Alaska.

West, C. J., and P. A. Larkin. 1987. Evidence of size-selective mortality of juvenile sockeye salmon (*Oncorhynchus nerka*) in Babine Lake, British Columbia. Canadian Journal of Fisheries and Aquatic Sciences 44:712–721.

Willette, T. M. 1994. Early marine salmon injury assessment in Prince William Sound. Final Report. Natural Resource Damage Assessment, Fish/Shellfish Study 4A, ADFG component. Alaska Department of Fish and Game, Cordova.

Wright, D. I., and W. J. O'Brien. 1982. Differential location of *Chaoborus* larvae and *Daphnia* by fish: the importance of motion and visible size. American Midland Naturalist 108:68.

Zar, J. H. 1984. Biostatistical analysis. Prentice-Hall, Englewood Cliffs, New Jersey.

American Fisheries Society Symposium 18:551–563, 1996

Harpacticoid Copepod Abundance and Population Structure in Prince William Sound, One Year after the *Exxon Valdez* Oil Spill

ALEX C. WERTHEIMER, NICHOLAS J. BAX,[1] ADRIAN G. CELEWYCZ, MARK G. CARLS, AND JOYCE H. LANDINGHAM

National Marine Fisheries Service, Alaska Fisheries Science Center
Auke Bay Laboratory, 11305 Glacier Highway, Juneau, Alaska 99801, USA

Abstract.—Density and population structure of epibenthic harpacticoid copepods were compared between lightly oiled and heavily oiled beaches in two bays, Herring Bay and Bay of Isles, in Prince William Sound in 1990, 1 year after the *Exxon Valdez* oil spill. Level of oiling was categorized according to shoreline surveys in fall 1989. Mean densities of total epibenthic harpacticoid copepods in both bays and of *Harpacticus* spp. in Herring Bay were higher on heavily oiled beaches than on lightly oiled beaches. Proportions of copepodites and egg-bearing female copepods were similar at lightly and heavily oiled beaches. Level of oiling in 1989 explained more of the variance in copepod densities in 1990 than did substrate composition, macrophyte coverage, total organic carbon in the sediments, or polycyclic aromatic hydrocarbons in the sediments. Amount of oil in sediments was not significantly correlated with the 1989 oiling level; however, interpretation of the hydrocarbon data was complicated by cleanup activity and possible secondary contamination at the sites. The results of this study suggest that epibenthic harpacticoids, which are common prey for salmon fry, maintained or increased their abundance in response to direct and indirect impacts of the *Exxon Valdez* oil spill on their intertidal habitats during the year after the spill.

One concern following the *Exxon Valdez* oil spill was that integral parts of the food web could be damaged by the crude oil. Meiofauna, small benthic organisms comprising a large number of taxa, are potentially susceptible to oil pollution, especially in intertidal and shallow subtidal habitats. Meiofauna are fed upon by a wide variety of fishes; harpacticoid copepods are typically the preferred prey taxa (Gee 1989). Juvenile salmon are among the fishes that use harpacticoid copepods. Harpacticoid copepods are important in the diet of juvenile salmon in Alaska waters (Cooney et al. 1978, 1981; Landingham 1982; Cordell 1986; Murphy et al. 1988) and elsewhere (Kaczynski et al. 1973; Sibert et al. 1977; Godin 1981; Simenstad and Salo 1982; Webb 1991). Harpacticoid copepods are a major component of the meiofauna in Prince William Sound (Feder et al. 1976).

Harpacticoid copepod populations can be impacted by oil pollution; their numbers have been shown to be severely reduced after shoreline oiling (Giere 1979; Elmgren et al. 1983; Bodin 1991). Conversely, population increases of opportunistic, phytal species were observed following the *Amoco Cadiz* oil spill (Bodin 1988). Thus, the effect of oil contamination in Prince William Sound to this important prey suite was not predictable.

Because of the great value of the salmon resource in Prince William Sound, particular attention in the damage assessment research was focused on the effects of oil contamination on food webs that contribute to salmon growth and survival. Following the *Exxon Valdez* oil spill in 1989, biomass of epibenthic harpacticoid copepods was significantly higher in oiled than in non-oiled locations sampled in western Prince William Sound, suggesting that the effects of the oil may have actually enhanced productivity of these animals (Celewycz and Wertheimer 1996, this volume). However, the observed differences in harpacticoid copepod abundance between oiled and non-oiled locations may also have been caused by geographic variation (Celewycz and Wertheimer 1996).

Although the general level of oil contamination in Prince William Sound had diminished greatly from 1989 to 1990, large amounts of oil remained in the sediments of some intertidal habitats where it could affect production of epibenthic harpacticoids. Our objective in this study was to assess the impact of the oil spill on epibenthic harpacticoid copepods 1 year after the spill. Our focus was on the genera that are most common in the diet of juvenile pink salmon in Prince William Sound: *Harpacticus*, *Tisbe*, and *Dactylopusia* (Cooney et al. 1981; Sturdevant et al. 1996, this volume). Our approach was

[1]Permanent address: Commonwealth Scientific and Industrial Research Organization, Division of Fisheries, GPO Box 1538, Hobart, Tasmania, 7001, Australia.

to compare density and population structure of these genera between lightly oiled and heavily oiled beaches within two contaminated bays, Bay of Isles and Herring Bay. We used lightly oiled beaches as a control because all beaches within the contaminated bays were assumed to have been exposed to oil to some degree. We did not want to compare heavily oiled beaches in a contaminated bay to non-oiled beaches in an uncontaminated bay because of the effects of geographic and hydrographic variability on the comparisons.

Methods

Study Design and Sampling

Harpacticoid copepods were sampled in Bay of Isles and Herring Bay in spring 1990. Eight beaches in each bay were selected for sampling based on degree of oiling and superficial physical and biological characteristics (Figure 1). Beaches selected were in protected reaches of the bays and were predominately pebble and cobble substrate, 5 to 15% gradient, with *Fucus* and filamentous algae covering most of the surface at the 0-m tide level. Four beaches were located in heavily oiled areas and four in lightly oiled areas, as categorized on Alaska Department of Environmental Conservation (ADEC) survey maps for fall 1989 (ADEC 1989). The ADEC maps identify beaches by assigning unique alphanumeric codes to beach segments; the codes for the beach segments on which sample sites were located are shown in Table 1. The number of lightly oiled beaches suitable for epibenthic sampling was limited, and random selection of beaches was not practical.

After selecting lightly oiled beaches, we selected heavily oiled beaches that were similar to the lightly oiled beaches in gradient, substrate, exposure, and macrophyte coverage. Some oil cleanup activity took place on all beach segments in 1989 and 1990, except for the lightly oiled beaches in Herring Bay (Table 1). We could not determine whether any cleanup activity occurred in 1989 at the specific portion of the beaches we sampled for copepods; we sampled in 1990 before 1990 cleanup activity at our sample sites.

Harpacticoid copepods were sampled along a 40-m transect placed along the mean-low-tide (0-m) contour of each selected beach. Each 40-m transect was divided into six sections, and four or five pump-sampling sites were placed randomly at least 1 m apart within each section. Harpacticoids were sampled using an epibenthic suction pump (Simenstad et al. 1989) with a 15-cm sampling cylinder. Water intake ports in the sampling cylinder were covered with 0.123-mm-mesh screen. The sampling cylinder was set on the substrate, enclosing 0.0177-m^2 surface area. A suction pump then drew water for 30 s into the cylinder through the screened ports. The pumped water was strained with a 0.123-mm-mesh net, thus capturing the entrained epibenthos from the area enclosed by the cylinder. Samples were then rinsed into a 500-ml bottle and preserved in buffered 5% formalin. A total of 25 pump samples was taken per transect, except at one Bay of Isles beach, where we were able to pump only 24 samples before we were flooded out by the incoming tide.

At each transect, surface substrate composition and macrophyte coverage were estimated, and samples of surface sediments were collected for total organic carbon (TOC) and hydrocarbon analysis. One sediment sampling site was located randomly within each of the six sections of the 40-m transects. A 0.25-m^2 quadrat was placed at each sediment sampling site. Surface substrate composition within the quadrat was visually classified by the proportion <65 mm (pebble or smaller). Percent macrophyte coverage within the quadrat was also estimated visually. Approximately equal amounts of fine sediment from the upper 2 cm of the substrate were aggregated into four hydrocarbon-free glass jars from each of the six quadrat locations along the transect, resulting in four sediment samples per transect.

In the laboratory, harpacticoid copepods in each pump sample were separated from organic matter or sediment by rinsing and rescreening. Samples with large amounts of algae and organic debris were first rinsed through a 1-mm mesh screen to remove the larger debris. All samples were then rinsed on a 0.125-mm screen, and the organisms were picked or decanted from retained sediments. The copepods were then identified to taxa that are common dietary components of juvenile salmon and were counted. Sex and life stage (copepodite, adult, egg-carrying female, clasping pair) were also determined. Densities per square meter of total harpacticoids and of specific genera in each sample were calculated by dividing the counts by the area sampled (0.0177 m^2). Pentec Environmental, Inc., processed the 199 pump samples from Bay of Isles; Kinnetic Laboratories, Inc., processed pump samples from Herring Bay. Because of budget constraints, only 120 of the 200 samples from Herring Bay were processed; therefore, 15 samples were randomly chosen from the 25 taken from each transect.

Three of the sediment samples collected from

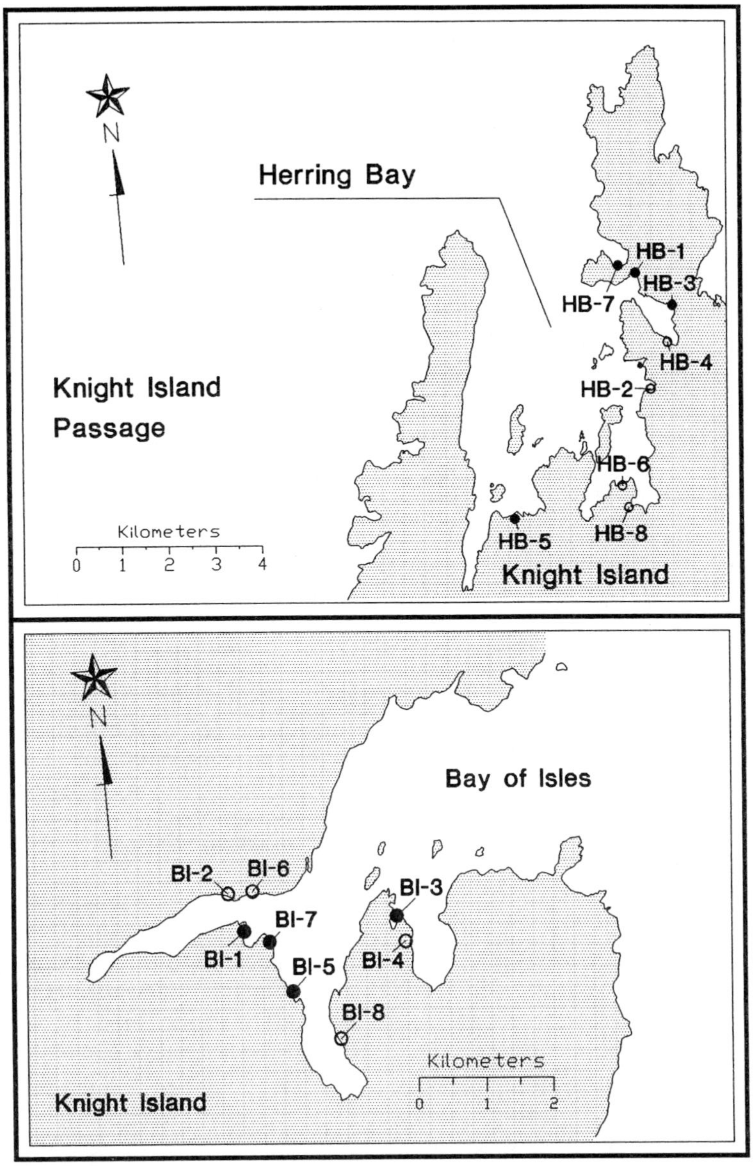

FIGURE 1.—Locations of beaches in Herring Bay (HB) and Bay of Isles (BI) sampled for epibenthic harpacticoid copepods in Prince William Sound in 1990. Open circles denote lightly oiled beaches; closed circles denote heavily oiled beaches.

each transect were submitted for hydrocarbon analysis; the fourth sediment sample was used for TOC analysis. Hydrocarbon content was measured by the Geochemical and Environmental Research Group at Texas A&M University as described by Short et al. (1996, this volume). Sediment TOC was determined by ignition by HUB Testing Laboratories.

Hydrocarbons compared between lightly oiled and heavily oiled beaches were phytane (an alkane) and five groups of polycyclic aromatics. The aromatic groups, in order from light to heavy, were naphthalenes, fluorenes, phenanthrenes, dibenzothiophenes, and chrysenes. These compounds do not occur in significant quantities in unpolluted sediments in Prince William Sound (Karinen et al. 1993).

TABLE 1.—Alaska Department of Environmental Conservation (ADEC) beach segment identification, degree of oiling, and cleanup activities associated with beaches sampled for harpacticoid copepods in spring 1990 in Prince William Sound. Degree of oiling was assigned in the ADEC surveys in fall 1989; HB beaches were in Herring Bay; BI beaches were in Bay of Isles.

Beach	ADEC segment	Degree of oiling	Cleanup history[a]	
			1989	1990[b]
HB-1	KN118	High	B	B
HB-2	KN5001	None	NTR	NTR
HB-3	KN119	High	B, M	B, M, ME, W
HB-4	KN120	VEry low	NTR	NTR
HB-5	KN126	High	B, W, T	B, M, ME, W
HB-6	KN5006	None	NTR	NTR
HB-7	KN115	High	B, M, TM	B, M, TM
HB-8	KN5004	None	NTR	NTR
BI-1	KN005	High	B, M, TM	B, M, TM
BI-2	KN006	Very low	TM	TM
BI-3	KN024	High	B, M	B, M, TM
BI-4	KN206	Very low	B, M	B, M, TM
BI-5	KN135	High	B, M, TM, W	B, M, TM
BI-6	KN006	Very low	TM	M, TM
BI-7	KN136	High	B, M, T, TM, W	B, M, TM, W
BI-8	KN205	None	B, M	M, TM

[a]B = bioremediation; M = manual removal; ME = mechanical; T = till; TM = tarmat removal; W = wash; NTR = no treatment.
[b]No cleanup activity occurred at the actual transect sites before sampling for harpacticoid copepods in 1990.

Statistical Analysis

Analysis of variance (ANOVA) was used to test whether lightly oiled and heavily oiled beaches within a bay differed in transect characteristics and harpacticoid abundance and population structure. Transect characteristics evaluated were (1) proportion of transect covered by small (<65 mm) substrate, (2) proportion of transect covered by macrophytes, (3) TOC, and (4) hydrocarbon analytes. Abundance and population structure variables evaluated were (1) density of combined harpacticoids; (2) separate densities of *Harpacticus*, *Tisbe*, and *Dactylopusia*; (3) proportion of adult females of these three genera with eggs; and (4) proportion of these three genera that were copepodites.

In each ANOVA, data were transformed to normalize the distribution and maximize variance homogeneity. Harpacticoid density was transformed to $\log_e(x + 1)$. Proportional data were transformed to arcsines. The factors in each ANOVA were oil (lightly or heavily oiled) and beach, with four beaches nested within each oil category. The significance of oil was tested by the F-ratio of the mean square of oil to the mean square of beach (nested within oil). The nested design removed the possibility of pseudoreplication (Hurlbert 1984). Because only one observation of TOC was made per beach, the ANOVA for TOC collapsed to a one-way test between levels of oiling.

If differences in total harpacticoids or specific taxa were identified by ANOVA, these variables were then examined with multiple linear regression to determine the proportion of the variance among transects explained by the degree of oiling relative to other variables. In each regression, the mean of the logarithmically transformed harpacticoid density by beach was the dependent variable; the independent variables were the arcsine-transformed mean proportion of substrate <65 mm, the arcsine-transformed mean proportion of macrophyte coverage, the TOC, the concentration of dibenzothiophenes and chrysenes in the sediments, and the categorical designation of heavily or lightly oiled. A stepwise regression analysis (Cruze and Weldon 1989) was used; the partial F-test had to be significant at $P < 0.1$ for the variable to remain in the regression model.

Results

Harpacticoid copepods were generally more abundant at the heavily oiled beaches than at the lightly oiled beaches at both Herring Bay and Bay of Isles (Table 2). At Herring Bay, densities of combined harpacticoids and of *Harpacticus* spp. were significantly higher ($P = 0.039$ and $P = 0.046$, respectively) at the heavily oiled beaches. Mean density over all heavily oiled beaches was 94,478/m^2 for all harpacticoids and 15,893/m^2 for *Harpacticus*, compared to 23,634/m^2 and 4,252/m^2, respectively, at lightly oiled beaches. Mean densities of *Dactylo-*

TABLE 2.—Mean densities (SEs in parentheses) of harpacticoid copepods sampled with an epibenthic pump in the Bay of Isles and Herring Bay in spring 1990. Number of samples at each transect were 15 for Herring Bay and 25 for Bay of Isles (except BI-6, with 24).

Condition or statistic	Beach	Density (number/m²) of: Harpacticus	Tisbe	Dactylopusia	Total harpacticoids
Bay of Isles					
Heavily oiled	BI-1	5,159 (1,197)	4,977 (1,263)	626 (275)	82,905 (12,801)
	BI-3	19,566 (2,944)	1,335 (216)	233 (110)	72,060 (7,881)
	BI-5	3,720 (628)	9,196 (2,451)	1,219 (331)	98,762 (18,848)
	BI-7	4,173 (769)	4,046 (614)	2,268 (815)	117,636 (16,921)
Mean		8,154 (3,816)	4,889 (1,630)	1,086 (443)	92,841 (9,918)
Lightly oiled	BI-2	7,336 (1,356)	2,290 (404)	1,431 (556)	70,905 (8,961)
	BI-4	2,600 (503)	4,096 (1,110)	152 (37)	35,398 (5,746)
	BI-6	2,586 (500)	2,652 (586)	881 (262)	39,763 (4,974)
	BI-8	5,579 (1,623)	22,655 (5,277)	3,846 (1,131)	127,142 (19,505)
Mean		4,525 (1,172)	7,923 (4,926)	1,578 (800)	68,302 (21,146)
ANOVA *P*		0.253	0.919	0.681	0.085
Herring Bay					
Heavily oiled	HB-1	24,744 (5,488)	15,275 (2,412)	1,966 (812)	227,575 (40,146)
	HB-3	6,979 (956)	7,318 (1,080)	7,262 (1,572)	53,168 (7,128)
	HB-5	13,944 (3,070)	18,320 (3,493)	4,664 (1,159)	60,218 (8,843)
	HB-7	17,906 (2,853)	2,068 (324)	2,121 (280)	36,953 (7,314)
Mean		15,893 (3,715)	10,112 (3,612)	3,622 (1,176)	94,478 (44,632)
Lightly oiled	HB-2	8,531 (1,318)	5,164 (617)	2,836 (447)	22,614 (2,276)
	HB-4	701 (119)	5,322 (744)	573 (76)	13,096 (1,935)
	HB-6	7,126 (1,756)	7,085 (1,088)	2,237 (527)	42,866 (7,903)
	HB-8	652 (113)	1,782 (517)	2,429 (424)	15,959 (2,791)
Mean		4,252 (2,084)	4,358 (993)	1,727 (481)	23,634 (6,714)
ANOVA *P*		0.046	0.278	0.859	0.039

pusia and *Tisbe* also tended to be higher at the heavily oiled beaches, although the differences were not significant. At Bay of Isles, mean density of combined harpacticoids was 92,841/m^2 at the heavily oiled beaches, compared to 68,302/m^2 at the lightly oiled beaches. This difference was marginally significant ($P = 0.085$). No significant difference between heavily and lightly oiled beaches was indicated for *Harpacticus*, *Tisbe*, or *Dactylopusia* at Bay of Isles.

Population structure was similar at heavily oiled and lightly oiled beaches. No significant difference ($P > 0.20$) was found between heavily oiled and lightly oiled beaches at Herring Bay or Bay of Isles in the proportion of adult females with eggs for *Harpacticus*, *Tisbe*, and *Dactylopusia* (Table 3). The proportion of copepodites for these three taxa also did not differ significantly ($P > 0.11$) between lightly and heavily oiled beaches (Table 4).

Physical characteristics were similar at lightly oiled and heavily oiled beaches. Substrate composition was not significantly different ($P = 0.522$) between the oiling categories in Bay of Isles, with substrate <65 mm constituting 66% on average (Table 5). In Herring Bay, the lightly oiled beaches tended to have coarser substrate, but the difference was not significant ($P = 0.106$). Heavily oiled beaches in Herring Bay had an overall mean of 77% substrate <65 mm, compared to 55% for lightly oiled beaches (Table 5). No significant difference in macrophyte coverage or TOC was indicated between heavily oiled and lightly oiled beaches in Herring Bay or Bay of Isles ($P > 0.40$ in all instances; Table 5). Macrophyte coverage averaged 87% at Bay of Isles and 71% at Herring Bay, where variability among beaches was higher. Sediment TOC varied considerably among beaches; levels tended to be higher at Herring Bay (513 mg/kg) than at Bay of Isles (350 mg/kg).

For the six groups of hydrocarbons evaluated, we found significantly higher concentrations at heavily oiled beaches for only dibenzothiophenes ($P = 0.029$) and chrysenes ($P = 0.053$) at Herring Bay (Figure 2). None of the other analytes at Herring

TABLE 3.—Mean proportions (SEs in parentheses) of adult females with eggs for three harpacticoid copepod taxa sampled with an epibenthic pump in the Bay of Isles and Herring Bay in spring 1990.

Condition or statistic	Beach	Proportion of adult:		
		Dactylopusia	Harpacticus	Tisbe
		Bay of Isles		
Heavily oiled	BI-1	0.34 (0.16)	0.00 (0.00)	0.17 (0.04)
	BI-3	0.49 (0.16)	0.00 (0.00)	0.13 (0.03)
	BI-5	0.50 (0.09)	0.00 (0.00)	0.20 (0.03)
	BI-7	0.16 (0.07)	0.00 (0.00)	0.11 (0.03)
Mean		0.37 (0.08)	0.00 (0.00)	0.15 (0.02)
Lightly oiled	BI-2	0.44 (0.14)	0.01 (0.01)	0.11 (0.02)
	BI-4	0.42 (0.12)	0.03 (0.01)	0.28 (0.04)
	BI-6	0.40 (0.11)	0.00 (0.00)	0.15 (0.03)
	BI-8	0.06 (0.04)	0.00 (0.00)	0.11 (0.02)
Mean		0.33 (0.09)	0.01 (0.01)	0.16 (0.04)
ANOVA *P*		0.616	0.234	0.797
		Herring Bay		
Heavily oiled	HB-1	0.24 (0.11)	0.01 (0.01)	0.32 (0.04)
	HB-3	0.53 (0.08)	0.02 (0.01)	0.36 (0.05)
	HB-5	0.46 (0.07)	0.03 (0.01)	0.17 (0.03)
	HB-7	0.34 (0.07)	0.12 (0.02)	0.25 (0.06)
Mean		0.39 (0.06)	0.04 (0.03)	0.27 (0.09)
Lightly oiled	HB-2	0.38 (0.06)	0.16 (0.04)	0.27 (0.04)
	HB-4	0.58 (0.09)	0.01 (0.01)	0.38 (0.02)
	HB-6	0.35 (0.10)	0.10 (0.05)	0.19 (0.03)
	HB-8	0.49 (0.06)	0.32 (0.10)	0.11 (0.05)
Mean		0.44 (0.05)	0.15 (0.07)	0.24 (0.06)
ANOVA *P*		0.910	0.205	0.621

Bay, and none of the analytes at Bay of Isles, differed significantly between heavily and lightly oiled beaches (Figures 2 and 3).

Naphthalene concentration was high in sediments from both Herring Bay and Bay of Isles. At Herring Bay, naphthalenes had the highest concentration of the hydrocarbon compounds measured at every beach except HB-1, ranging from 300 to 1,000 ng/g (Figure 2). Naphthalene concentration was highest at HB-6, on which no oil had been observed during 1989 shoreline surveys (Table 1), and lowest at HB-1, a heavily oiled beach. At Bay of Isles, the general profile of analytes by beach for all compounds except naphthalene showed a high spike at BI-1 (heavily oiled), with BI-6 (lightly oiled) having the second-highest level (Figure 3). For naphthalene, however, the lightly oiled BI-6 and BI-8 had the highest concentration, concentrations were generally elevated across all beaches, and BI-1 had the lowest concentration. Naphthalene concentration at Bay of Isles ranged from 600 to 2,500 ng/g (Figure 3).

The various beach characteristics (macrophyte coverage, proportion of substrate <65 mm, TOC, dibenzothiophene concentration, chrysene concentration, and level of oiling) generally had no significant association with each other. The exceptions involved the polynuclear aromatics and level of oiling. Chrysene and dibenzothiophene concentrations were significantly ($P < 0.05$) correlated in both Herring Bay ($r = 0.98$) and Bay of Isles ($r = 0.95$). These compounds were also significantly correlated with level of oiling at Herring Bay ($r = 0.74$ for chrysene, $r = 0.78$ for dibenzothiophene).

The three measures of harpacticoid copepod abundance that differed significantly between levels of oil (total harpacticoids in Herring Bay and Bay of Isles, and *Harpacticus* in Herring Bay) were tested against the other variables measured at each beach to determine if any variable other than level of oiling was correlated with the observed differences. The ADEC level of oiling was significantly correlated with these harpacticoid densities; the only other variable with a significant ($P < 0.10$) correlation with harpacticoid density was proportion of substrate <65 mm in Herring Bay (Table 6). When

TABLE 4.—Mean proportions (SEs in parentheses) of copepodites for three harpacticoid copepod taxa sampled with an epibenthic pump in the Bay of Isles and Herring Bay in spring of 1990.

Condition or statistic	Beach	Copepodite proportions of: Dactylopusia	Harpacticus	Tisbe
		Bay of Isles		
Heavily oiled	BI-1	0.05 (0.05)	0.09 (0.02)	0.44 (0.05)
	BI-3	0.07 (0.06)	0.10 (0.01)	0.40 (0.05)
	BI-5	0.09 (0.04)	0.04 (0.01)	0.24 (0.03)
	BI-7	0.00 (0.00)	0.02 (0.01)	0.24 (0.03)
Mean		0.05 (0.02)	0.06 (0.02)	0.33 (0.05)
Lightly oiled	BI-1	0.08 (0.06)	0.28 (0.03)	0.37 (0.03)
	BI-4	0.00 (0.00)	0.21 (0.03)	0.35 (0.03)
	BI-6	0.02 (0.02)	0.05 (0.02)	0.36 (0.03)
	BI-8	0.14 (0.05)	0.11 (0.02)	0.33 (0.03)
Mean		0.06 (0.03)	0.16 (0.05)	0.35 (0.01)
ANOVA *P*		0.950	0.110	0.950
		Herring Bay		
Heavily oiled	HB-1	0.16 (0.06)	0.73 (0.03)	0.64 (0.03)
	HB-3	0.60 (0.06)	0.65 (0.04)	0.33 (0.03)
	HB-5	0.49 (0.06)	0.57 (0.04)	0.31 (0.03)
	HB-7	0.61 (0.04)	0.71 (0.03)	0.30 (0.05)
Mean		0.46 (0.11)	0.66 (0.03)	0.40 (0.08)
Lighyly oiled	HB-2	0.57 (0.07)	0.82 (0.03)	0.46 (0.04)
	HB-4	0.20 (0.08)	0.62 (0.07)	0.23 (0.04)
	HB-6	0.41 (0.05)	0.61 (0.04)	0.44 (0.05)
	HB-8	0.36 (0.04)	0.50 (0.07)	0.64 (0.05)
Mean		0.38 (0.07)	0.63 (0.07)	0.45 (0.09)
ANOVA *P*		0.554	0.910	0.644

variance related to the level of oiling was removed from the data by entering level of oiling as the first variable in a stepwise multiple regression, the concentration of chrysenes in the sediments became a significant predictor of the density of total harpacticoids in Herring Bay ($P < 0.10$), increasing the variance explained by the regression from 53 to 78%. Chrysene concentration was not a significant predictor for the density of *Harpacticus* in Herring Bay or for the density of total harpacticoids in Bay of Isles. None of the other variables entered the stepwise regression models.

Discussion

The densities of total harpacticoids were significantly higher on heavily oiled than on lightly oiled beaches in both bays, suggesting that the *Exxon Valdez* oil spill impacted the abundance of epibenthic harpacticoid copepods on the heavily oiled beaches we studied in Prince William Sound. Not all genera were equally affected. Densities of *Harpacticus* at Herring Bay were higher on heavily oiled beaches, but the densities of *Tisbe* and *Dactylopusia* at both bays and of *Harpacticus* at Bay of Isles were similar between lightly and heavily oiled beaches.

We cannot separate the effects of the oil itself from those of oil spill cleanup activities. Such activities can have direct effects on the intertidal ecosystem. Houghton et al. (1991) found reduced species richness and infauna abundance, including the abundance of interstitial harpacticoids, the year after the *Exxon Valdez* oil spill on beaches treated with hot-water washes. We did not find epibenthic harpacticoid copepods to be reduced in association with increased cleanup activity. In Herring Bay, densities of total harpacticoids and *Harpacticus* were elevated on heavily oiled beaches in comparison to lightly oiled beaches, even though these heavily oiled beaches had been exposed to some cleanup activity, while lightly oiled beaches in the bay had no cleanup activity (Table 1). Records of beach treatment after the spill, however, are not sufficiently detailed to determine whether the specific transects that we sampled were treated. Because cleanup effects cannot be separated from oil

TABLE 5.—Mean physical characteristics (SEs in parentheses) of beaches sampled for harpacticoid copepods in the Bay of Isles and Herring Bay in spring 1990. Sample sizes per beach were six for substrate and macrophyte coverage and one for total organic carbon.

Condition or statistic	Beach	Substrate[a]	Macrophyte coverage[b]	Total organic carbon (mg/kg)
		Bay of Isles		
Heavily oiled	BI-1	0.77 (0.08)	0.83 (0.10)	110
	BI-3	0.80 (0.04)	0.88 (0.08)	148
	BI-5	0.70 (0.05)	0.82 (0.15)	560
	BI-7	0.50 (0.06)	0.87 (0.07)	596
Mean		0.69 (0.08)	0.85 (0.02)	354 (130)
Lightly oiled	BI-1	0.52 (0.07)	1.00 (0.0)	74
	BI-4	0.57 (0.11)	0.73 (0.11)	501
	BI-6	0.64 (0.06)	0.98 (0.02)	331
	BI-8	0.82 (0.04)	0.87 (0.08)	480
Mean		0.64 (0.08)	0.90 (0.07)	346 (98)
ANOVA *P*		0.522	0.417	0.967
		Herring Bay		
Heavily oiled	HB-1	0.90 (0.03)	1.00 (0.0)	494
	HB-3	0.82 (0.03)	0.48 (0.07)	328
	HB-5	0.65 (0.05)	0.85 (0.10)	1,009
	HB-7	0.70 (0.07)	0.58 (0.18)	223
Mean		0.77 (0.07)	0.73 (0.14)	514 (174)
Lightly oiled	HB-2	0.27 (0.10)	0.80 (0.06)	793
	HB-4	0.62 (0.08)	0.60 (0.08)	255
	HB-6	0.70 (0.07)	0.60 (0.10)	461
	HB-8	0.62 (0.10)	0.82 (0.13)	540
Mean		0.55 (0.12)	0.70 (0.07)	512 (111)
ANOVA *P*		0.106	0.613	0.995

[a]Proportion of substrate that was pebble or smaller (<65 mm).
[b]Proportion of substrate covered.

effects, we consider the observed differences in harpacticoid density to be related to the oil spill history of the heavily oiled sites, including both contamination and cleanup.

In general, the hydrocarbons measured in the sediments at the 0-m tide level did not correspond to the rating of amount of oiling by ADEC. Following the spill, oil was typically deposited in the upper intertidal zone (Owens 1991); visible oil pollution was the variable assessed by the ADEC surveys. We expected that oil would leach to the lower intertidal zone and contaminate sediments at the mean-low-tide transects to a greater degree at beaches rated as heavily oiled. We found this pattern only at Herring Bay for some of the more persistent aromatics (chrysene and dibenzothiophene). The lack of correspondence between the sediment hydrocarbon load and the ADEC rating of amount of oiling may result from the patchy distribution of hydrocarbons in the sediments (Owens 1991). Both the deposition pattern and differences in weathering associated with tidal flushing and winter storm exposure on individual beaches contributed to the high variability of residual hydrocarbons in the sediments.

Another explanation for the poor correspondence between the ADEC rating of level of oiling and the actual amounts of hydrocarbons measured in the sediments is contamination of the beaches subsequent to oiling from the *Exxon Valdez* spill. At both bays, we found high amounts of naphthalene at both lightly oiled and heavily oiled beaches. These concentrations were not consistent with the expected pattern from weathered Exxon Valdez crude oil (Short et al. 1996). The naphthalene may have been the result of recent pollution from other sources, such as boat traffic and cleanup activity. Such contamination was probably waterborne, because even beaches in Herring Bay with no cleanup activity had elevated naphthalene levels (e.g., beach 6; Table 1; Figure 2). Both bays were sites of significant vessel activity in 1990 for cleanup and dam-

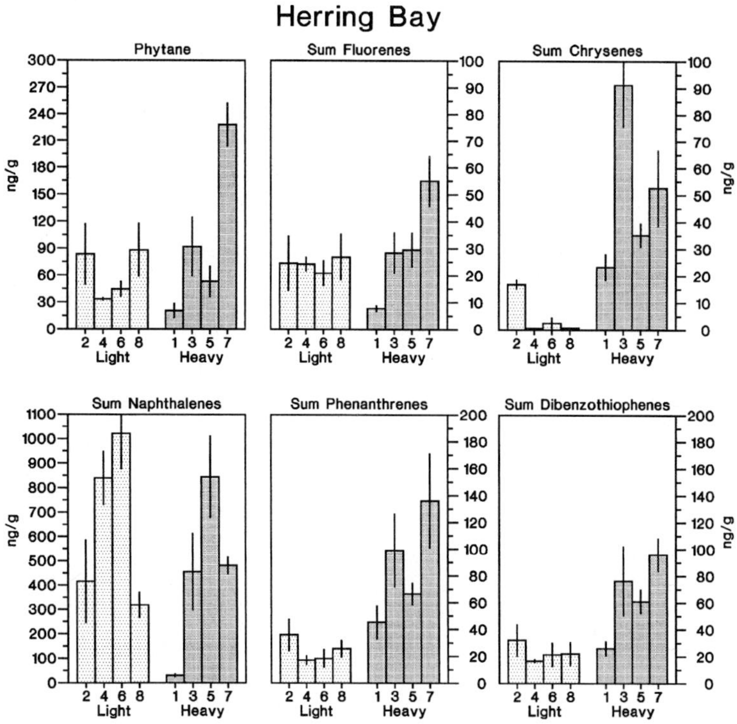

FIGURE 2.—Analyte concentrations in sediments from beaches in Herring Bay sampled for epibenthic harpacticoid copepods in 1990. The beaches had been classified as "light" or "heavy" according to the amount of oil visually observed during shoreline surveys in fall 1989. Numbers below bars refer to specific beaches listed in Table 1. Vertical lines at top of bars denote SE of the mean concentrations.

age assessment. At Bay of Isles, where the highest naphthalene concentrations were observed (Figure 2 versus Figure 3), a large cleanup fleet was present during our sampling in May 1990.

The level of oiling in the ADEC surveys was the measure that was most highly correlated with harpacticoid density in both Bay of Isles and Herring Bay. To conclusively assign causation from the spill for this correlation requires prespill data on the particular beaches sampled; these data do not exist. Making comparisons with controls from non-oiled areas does not eliminate the possibility of geographic variability (Celewycz and Wertheimer 1996). The differences in copepod densities can be attributed to the spill effects if we assume that similar beach types within the same bay would have similar densities of copepods in the absence of the oil spill. This assumption may not be valid; for example, the correlation could also be a coincidence of water circulation and hydrographic conditions in these bays that resulted in both deposition of drifting crude oil and conditions that favor epibenthic harpacticoids. However, no other beach characteristics that we measured (substrate, macrophyte coverage, and TOC) were significant in explaining the observed variance of epibenthic harpacticoid densities, nor were these other characteristics significantly correlated with the level of oiling. Thus we conclude that the expectation of similar densities in the absence of the oil spill is a reasonable assumption, indicating that the level of oiling did contribute to higher copepod densities observed on heavily oiled beaches.

The presence of petroleum hydrocarbons in intertidal areas may affect populations of harpacticoid copepods in different ways. After an oil spill, population abundances of harpacticoids may decline (Giere 1979; Bonsdorf 1981; Elmgren et al. 1983; Bodin 1988) or increase (Fricke et al. 1981; Bodin 1988). Increases have also been observed in experimental in situ applications of hydrocarbons (Fleeger and Chandler 1983; Gilfillan et al. 1983; Feder et al. 1990) and after the addition of oil to marine

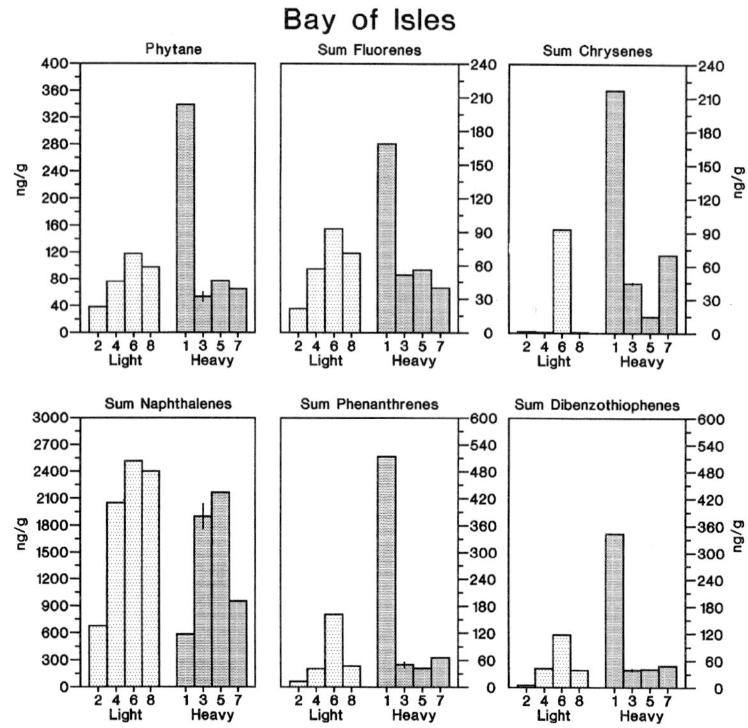

FIGURE 3.—Analyte concentrations in sediments from beaches in Bay of Isles sampled for epibenthic harpacticoid copepods in 1990. The beaches had been classified as "light" or "heavy" according to the amount of oil visually observed during shoreline surveys in fall 1989. Numbers below bars refer to specific beaches listed in Table 1. Vertical lines at top of bars for beach 3 denote SE of the mean concentrations; only one sediment sample from each of the other beaches was analyzed for hydrocarbons.

mesocosms (Frithsen et al. 1985; Stacey and Marcotte 1987).

Some of the differences between these studies may be explained by the different ecological groups of harpacticoid copepods sampled. After the *Amoco Cadiz* spill, populations of harpacticoid copepods in the interstitial zone declined at the same time as populations of epibenthic and epiphytal harpacticoids increased (Chasse 1978). Bodin (1988) categorized species of harpacticoid copepods by their rate of colonization after the *Amoco Cadiz* spill. First to appear after the initial toxic effect of the spill were opportunistic phytal species that exploit ephemeral algal blooms that often follow a spill and the tolerant generalist species regularly present in low densities that are stimulated by organic pollution. These groups peaked 2 years after the *Amoco Cadiz* spill. As the sediments

TABLE 6.—Correlation coefficients of harpacticoid copepod density with characteristics of beaches sampled in the Bay of Isles and Herring Bay in spring 1990; DBT = dibenzothiophene concentration; CHRY = chrysene concentration. Asterisks denote significant differences from zero at $P < 0.10$* or $P < 0.05$**.

Taxa	Level of oiling	Macrophyte coverage	Substrate <65 mm	Total organic carbon	DBT	CHRY
			Bay of Isles			
Total harpacticoids	0.656*	0.006	0.128	0.092	−0.087	0.118
			Herring Bay			
Total harpacticoids	0.713**	0.608	0.699*	0.154	0.274	0.212
Harpacticus	0.717**	0.284	0.205	0.226	0.558	0.515

became progressively cleaner, the opportunistic and generalist species were progressively replaced by a more sensitive group of harpacticoid copepods typical of similar sandy sites in unpolluted areas.

The higher abundance of epibenthic harpacticoids observed on heavily oiled beaches could have been a direct result of organic enrichment by oil or an indirect result of the disruption of the intertidal community by oil and the cleanup activity. In a simulated oil spill in Port Valdez, Alaska, Feder et al. (1990) observed an increase in abundance of harpacticoid copepods, including *Harpacticus uniremis*, and speculated that the availability of a new food resource (bacteria associated with the oil) may have led to improved survivorship of harpacticoids. After the *Exxon Valdez* oil spill, bacteria numbers in subtidal sediments were significantly greater in oiled locations than in control sites (Braddock et al. 1996, this volume). The abundance of macroinvertebrate grazers commonly decreases after an oil spill, leading to blooms of ephemeral algae (Foster et al. 1988), which may also have led to the higher abundance of the plant-associated harpacticoids that we observed. The fact that the concentrations of residual petroleum hydrocarbons were generally similar on heavily and lightly oiled beaches at the tide level sampled for copepods suggests that community disruption by the oil spill is a more likely causative mechanism of the differences observed than is organic enrichment by the oil itself.

Although other researchers have reported effects of oil pollution on the reproductive activity of various harpacticoid species (Ustach 1977; Frithsen et al. 1985; Bodin 1991), no significant differences between the lightly oiled and heavily oiled beaches were observed in this study with respect to either the proportion of females that were carrying eggs or the proportion of copepodites in the population. A single sampling date was probably not sufficient to track the reproductive dynamics of these animals in Prince William Sound. *H. uniremis* copepodites have a short and variable period of availability (Jewett and Feder 1977), which complicates comparison of their abundance at different sites. *H. uniremis* also produces only one brood per year in Prince William Sound; in Port Valdez ovigerous females were prevalent in the population only from November to May (Feder et al. 1976), and most ovigerous females occurred in February and March (Jewett and Feder 1977). This timing explains the higher density of ovigerous females in Herring Bay (sampled 24–27 April) compared with the Bay of Isles (sampled 24–27 May). It also suggests that the Herring Bay population was near the end of its reproductive period, which would cause the percentage of ovigerous females to be low and variable.

In conclusion, we found no indication that the epibenthic prey resources available to juvenile pink salmon *Oncorhynchus gorbuscha* in 1990 were reduced by continuing impacts of the *Exxon Valdez* oil spill. These animals are opportunistic harpacticoid copepods (sensu Bodin 1988) that are likely to increase in abundance after an oil spill. Indeed, we observed higher densities of epibenthic harpacticoids in general, and of *Harpacticus* in particular, at heavily oiled beaches. This conclusion does not exclude the possibility of reductions in abundance of these resources immediately after the spill in the spring of 1989. However, surveys of epibenthic prey in 1989 also indicated that epibenthic prey were as abundant or more abundant in oiled locations than in non-oiled locations (Celewycz and Wertheimer 1996). These results suggest that epibenthic harpacticoids common as prey to salmon fry maintained or increased their numbers in response to the direct and indirect impacts of the *Exxon Valdez* oil spill.

Acknowledgments

We thank Jeff Cordell for his help with both the sampling design and the field implementation of this project, and Skip Gish for his assistance in the laboratory and in the field.

References

ADEC (Alaska Department of Environmental Conservation). 1989. Impact maps and summary reports of shoreline surveys of the *Exxon Valdez* spill site. Report to the *Exxon Valdez* Oil Spill Trustee Council, Anchorage, Alaska.

Bodin, P. 1988. Results of ecological monitoring of three beaches polluted by the *Amoco Cadiz* oil spill: development of meiofauna from 1978 to 1984. Marine Ecology Progress Series 42:105–123.

Bodin, P. 1991. Perturbations in the reproduction cycle of some harpacticoid copepod species further to the *Amoco Cadiz* spill. Hydrobiologia 209:245–257.

Bonsdorf, E. 1981. The *Antonio Gramsci* oil spill. Impact on the littoral and benthic ecosystems. Marine Bulletin 12:301–305.

Braddock, J. F., J. E. Lindstrom, T. R. Yeager, B. T. Rasley, and E. J. Brown. 1996. Patterns of microbial activity in oiled and unoiled sediments in Prince William Sound. American Fisheries Society Symposium 18:94–108.

Celewycz, A. G., and A. C. Wertheimer. 1996. Prey availability to juvenile salmon after the *Exxon Valdez* oil spill. American Fisheries Society Symposium 18:564–577.

Chasse, C. 1978. The ecological impact on and near shores by the *Amoco Cadiz* oil spill. Marine Pollution Bulletin 9:298–301.

Cooney, R. T., D. Urquhart, and D. Barnard. 1981. The behavior, feeding biology, and growth of hatchery-released pink and chum salmon fry in Prince William Sound, Alaska. Alaska Sea Grant Report 81-5, University of Alaska, Fairbanks.

Cooney, R. T., D. Urquhart, R. Neve, J. Hilsinger, R. Clasby, and D. Barnard. 1978. Some aspects of the carrying capacity of Prince William Sound, Alaska, for hatchery released pink and chum salmon fry. Alaska Sea Grant Report 78-4, University of Alaska, Fairbanks.

Cordell, J. R. 1986. Structure and dynamics of an epibenthic harpacticoid assemblage and the role of predation by juvenile salmon. Master's thesis. University of Washington, Seattle.

Cruze, E., and J. Weldon. 1989. MINITAB reference manual. MINITAB, State College, Pennsylvania.

Elmgren, R., S. Hansson, U. Larsson, B. Sundelin, and P.D. Boehm. 1983. The *Tsesis* oil spill: acute and long-term impact on the benthos. Marine Biology 73:51–65.

Feder, H. M., and nine coauthors. 1976. The sediment environment of Port Valdez, Alaska: the effect of oil on this ecosystem. U.S. Environmental Protection Agency National Environmental Research Center Ecological Research Series, EPA-600/3-76-086.

Feder, H. M., A. S. Naidu, and A. J. Paul. 1990. Trace element and biotic changes following a simulated oil spill on a mudflat in Port Valdez, Alaska. Marine Pollution Bulletin 21:131–137.

Fleeger, J. W., and G. T. Chandler. 1983. Meiofauna responses to an experimental oil spill in a Louisiana salt marsh. Marine Ecology Progress Series 11:257–264.

Foster, M. S., A. P. De Vogelaere, C. Harrold, J. S. Pearse, and A. B. Thum. 1988. Causes of spatial and temporal patterns in rocky intertidal communities of central and northern California. California Academy of Sciences Memoir 9.

Fricke, A. H., H. F. O. Henning, and M. J. Orren. 1981. Relationship between oil pollution and psammolittoral meiofauna density of two South African beaches. Marine Environmental Research 5:59–77.

Frithsen, J. B., R. Elmgren, and D. T. Rudnick. 1985. Responses of benthic meiofauna to long-term low-level additions of no. 2 fuel oil. Marine Ecology Progress Series 23:1–14.

Gee, J. M. 1989. An ecological and economic review of meiofauna as food for fish. Zoological Journal of the Linnean Society 96:243–261.

Giere, O. 1979. The impact of oil pollution on intertidal meiofauna. Field studies after the *La Coruna* spill, May 1976. Cahiers de Biologie Marine 20:231–251.

Gilfillan, E. S., and six coauthors. 1983. Effects of spills of dispersed and non-dispersed oil on intertidal community structure. Pages 457–463 *in* Proceedings, 1983 oil spill conference. American Petroleum Institute Publication 4356, Washington, DC.

Godin, J. G. 1981. Daily patterns of feeding behavior, daily rations, and diets of juvenile pink salmon (*Oncorhynchus gorbuscha*) in two marine bays of British Columbia. Canadian Journal of Fisheries and Aquatic Sciences 38:10–15.

Houghton, J. P., W. B. Driskell, D. C. Lees, and A. J. Mearns. 1991. Impacts of the *Exxon Valdez* and subsequent cleanup on intertidal biota—1 year later. Pages 467–476 *in* Proceedings, 1991 international oil spill conference. American Petroleum Institute Publication 4529, Washington, DC.

Hurlbert, S. H. 1984. Pseudoreplication and the design of ecological field experiments. Ecological Monographs 54(2):187–211.

Jewett, S. C., and H. M. Feder. 1977. Biology of the harpacticoid copepod, *Harpacticus uniremis* Kröyer on Dayville Flats, Port Valdez, Alaska. Ophelia 16: 111–129.

Kaczynski, V. W., R. J. Feller, J. Clayton, and R. J. Gerke. 1973. Trophic analysis of juvenile pink and chum salmon (*Oncorhynchus gorbuscha* and *O. keta*) in Puget Sound. Journal of the Fisheries Research Board of Canada 30:1003–1008.

Karinen, J. F., M. M. Babcock, D. W. Brown, W. D. MacLeod, L. S. Ramos, and J. W. Short. 1993. Hydrocarbons in intertidal sediments and mussels from Prince William Sound, Alaska, prior to the *Exxon Valdez* oil spill: characterization and probable sources. NOAA (National Oceanic and Atmospheric Administration) Technical Memorandum NMFS (National Marine Fisheries Service)-AFSC-9, Seattle.

Landingham, J. H. 1982. Feeding ecology of pink and chum salmon fry in the nearshore habitat of Auke Bay, Alaska. Master's thesis. University of Alaska, Juneau.

Murphy, M. L., J. F. Thedinga, and K. V. Koski. 1988. Size and diet of juvenile Pacific salmon during seaward migration through a small estuary in southeastern Alaska. U.S. National Marine Fisheries Service Fishery Bulletin 86:213–222.

Owens, E. H. 1991. Changes in shoreline oiling conditions 1–1 1/2 years after the 1989 Prince William Sound spill. Woodward-Clyde, Seattle.

Short, J. W., T. J. Jackson, M. L. Larsen, and T. L. Wade. 1996. Analytical methods used for the analysis of hydrocarbons in crude oil, tissues, sediments, and seawater collected for the natural resources damage assessment of the *Exxon Valdez* oil spill. American Fisheries Society Symposium 18:140–148.

Sibert, J., T. J. Brown, M. C. Healey, B. A. Kask, and R. J. Naiman. 1977. Detritus-based food webs: exploitation by juvenile chum salmon (*Oncorhynchus keta*). Science 196:649–650.

Simenstad, C. A., and E. O. Salo. 1982. Foraging success as a determinant of estuarine and nearshore carrying capacity of juvenile chum salmon (*Oncorhynchus keta*) in Hood Canal, Washington. Pages 21–37 *in* B. R. Melteff and R. A. Neave, editors. Proceedings of the north Pacific aquaculture symposium. Alaska Sea Grant Report 82-2, University of Alaska, Fairbanks.

Simenstad, C. A., C. D. Tanner, and R. M. Thom. 1989. Estuarine wetland restoration monitoring protocol. University of Washington, College of Fisheries, Fisheries Research Institute, FRI-UW-8918, Seattle.

Stacey, B. M., and B. M. Marcotte. 1987. Chronic effect of no. 2 fuel oil on population dynamics of harpacticoid copepods in experimental marine mesocosms. Marine Ecology Progress Series 40:61–68.

Sturdevant, M. V., A. C. Wertheimer, and J. L. Lum. 1996. Diets of juvenile pink and chum salmon in oiled and non-oiled nearshore habitats in Prince William Sound, 1989 and 1990. American Fisheries Society Symposium 18:578–592.

Ustach, J. F. 1977. Effects of sub-lethal oil levels on the reproduction of a copepod, *Nitocra affinis*. North Carolina University Sea Grant Publication UNC-SG-76-10, Raleigh.

Webb, D. G. 1991. Effect of predation by juvenile Pacific salmon on marine harpacticoid copepods I. Comparisons of pattern of copepod mortality with patterns of salmon consumption. Marine Ecology Progress Series 72:25–36.

American Fisheries Society Symposium 18:564–577, 1996

Prey Availability to Juvenile Salmon after the *Exxon Valdez* Oil Spill

ADRIAN G. CELEWYCZ AND ALEX C. WERTHEIMER
National Marine Fisheries Service, Alaska Fisheries Science Center
Auke Bay Laboratory, 11305 Glacier Highway, Juneau, Alaska 99801, USA

Abstract.—The impact of the *Exxon Valdez* oil spill on potential prey resources of juvenile salmon was studied from April to June 1989 (zooplankton and epibenthic crustaceans) and 1990 (zooplankton only). Eight locations were sampled: two each in oiled and non-oiled bays, and two each in oiled and non-oiled corridors in western Prince William Sound. Zooplankton were sampled at each location with 20-m vertical hauls. Epibenthic crustaceans were sampled with 10-m hauls of an epibenthic sled at three nearshore habitat types: low-, medium-, and steep-gradient beaches. We did not detect any reduction in abundance of either zooplankton or epibenthic crustaceans important in the diet of juvenile salmon as a result of the *Exxon Valdez* oil spill. Density, biomass, and diversity (number of taxa) of zooplankton did not differ significantly between oiled and non-oiled locations in either 1989 or 1990. Biomass of harpacticoid copepods and all epibenthic crustaceans was significantly greater in oiled than in non-oiled locations. Diversity of epibenthic crustaceans, however, was similar in oiled and non-oiled locations. We conclude that the *Exxon Valdez* oil spill did not reduce the available prey resources of juvenile salmon in western Prince William Sound.

In spring 1989 much of the surface water and the littoral sediments of western Prince William Sound were contaminated by oil from the *Exxon Valdez*. Concern for catastrophic disruption of the food webs supporting salmon populations increased as the oil spread. Oil has been shown to have detrimental impacts on both zooplankton and epibenthic crustaceans (e.g., Samain et al. 1980; Bodin 1991). If the presence of oil in nearshore waters were to reduce availability of prey, growth of juvenile salmon could decline. Slower growth could result in a greater susceptibility of fry to predators because the fry would remain at a smaller, more vulnerable size for a longer period (Parker 1971).

During their first weeks of marine residency, pink salmon *Oncorhynchus gorbuscha* and chum salmon *O. keta* fry remain near shore, feeding on zooplankton and epibenthic organisms (Heard 1991; Salo 1991). Epibenthic prey, such as harpacticoid copepods, are the main food items in some areas (Kaczynski et al. 1973; Landingham 1982), whereas zooplankton, such as calanoid copepods and euphasiid eggs and larvae, are the predominant prey in other areas (Bailey et al. 1975; Healey 1980; Cooney et al. 1981).

The subarctic marine ecosystem has a highly seasonal production cycle, characterized by high levels of primary and secondary production in spring (Larrance 1977; Smetacek et al. 1984). The timing of pink and chum salmon emigration to seawater has presumably evolved to exploit this period of high productivity (Murphy et al. 1988; Holtby et al. 1989). The *Exxon Valdez* oil spill occurred just before the spring bloom of secondary production in Prince William Sound.

The purpose of this study was to assess the impact of the *Exxon Valdez* oil spill on prey resources of juvenile salmon in Prince William Sound. The specific objectives were to test the hypotheses that: 1) density, biomass, and diversity of zooplankton, in general, and small and large calanoid copepods, in particular, did not differ between oiled and non-oiled locations in western Prince William Sound in 1989 and 1990; and 2) density, biomass, and diversity of epibenthic crustaceans, in general, and harpacticoid copepods, in particular, did not differ between oiled and non-oiled littoral areas of western Prince William Sound in 1989.

Methods

We selected four sampling locations in the oiled area and four in the non-oiled area of western Prince William Sound (Figure 1). The oiled and non-oiled areas were identified by visual inspection of the actual distribution of oil in the first 2 weeks after the spill. Within each area, two bays and two migration corridors leading to the Gulf of Alaska were selected for a total of eight sampling locations. The non-oiled locations were in McClure Bay and Long Bay, and in Culross Passage and Wells Passage. The oiled locations were in Herring Bay and Snug Harbor, and in Prince of Wales Passage and Knight Island Passage. Prey were sampled in five trips between 10 April and 26 June 1989 and in four trips between 16 April and 14 June 1990. Confir-

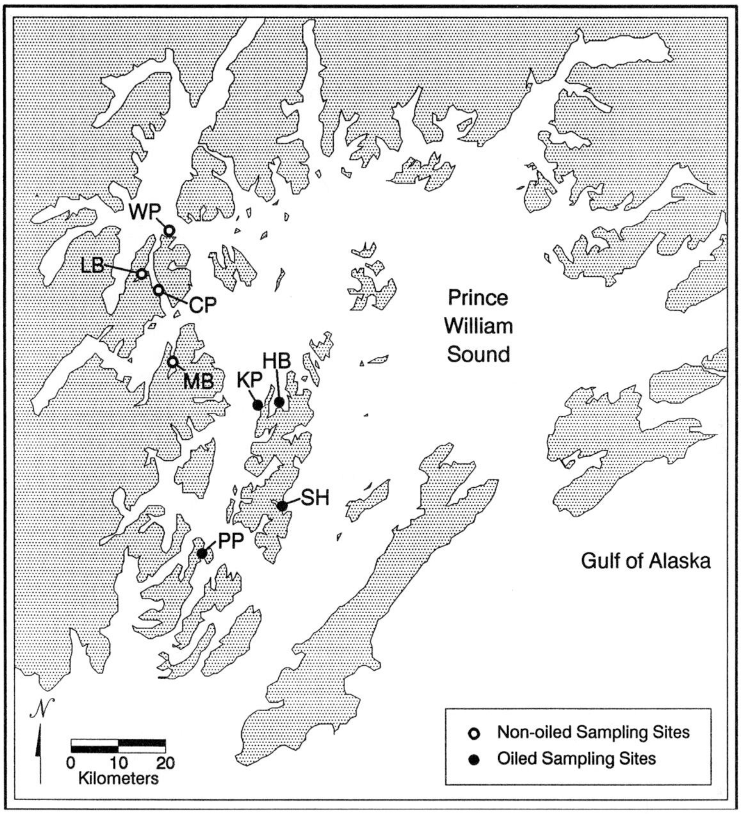

FIGURE 1.—Oiled and non-oiled sampling locations in western Prince William Sound. Oiled locations—Herring Bay (HB), Snug Harbor (SH), Knight Island Passage (KP), and Prince of Wales Passage (PP)—are marked by filled circles. Non-oiled sampling locations—McClure Bay (MB), Long Bay (LB), Wells Passage (WP), and Culross Passage (CP)—are marked by open circles.

mation of our designation of oiled versus non-oiled was provided by hydrocarbon analysis of mussel samples from each sampling location (Carls et al. 1996, this volume).

Zooplankton

Triplicate samples of zooplankton were taken at each location on each sampling trip with a 20-m vertical haul of a 0.5-m-diameter 243-micron-mesh net. Water of sufficient depth for the haul was located with a depth sounder or sounding line. After a haul, a seawater pump was used to wash down the outside of the sampling net. The contents were then rinsed with filtered seawater and preserved in 5% formalin solution.

Zooplankton samples were analyzed by the Alaska Department of Fish and Game (ADFG) Limnology Laboratory. Organisms in a subsample from each sample were identified and counted. Identification was generally to order level, except for calanoid and harpacticoid copepods, which were identified to genus or species level. Sex and life stage (copepodite, adult, egg-carrying females, clasping pairs) were also determined. A total number for the sample was computed by dividing the subsample count by the subsample fraction. Density (number/m^3) of organisms was computed by dividing the number of organisms in a sample by the volume of water sampled. In 1989, wet weights of organisms were determined for each sample by weighing up to 100 individuals from each taxon. Biomass (g/m^3) of an organism was computed by multiplying the mean weight of the organism by the number in the sample and then dividing by the volume of water sampled. In 1990, only counts were made. Mean weights of the taxa across all samples processed in 1989 were used to estimate biomass of taxa in 1990.

Three different sets of the zooplankton data were statistically analyzed. Because juvenile salmon prey on virtually all of the taxa we captured (Cooney et al. 1981; Brodeur and Pearcy 1990; J. H. Landingham, National Marine Fisheries Service, unpublished data), the first set consisted of all zooplankton. Further analysis focused on two size categories of calanoid copepods, small (<2.5-mm metasomal length) and large (≥2.5-mm metasomal length). These categories parallel the analysis of feeding habits of juvenile salmon (Sturdevant et al. 1996, this volume). Large calanoids included species such as *Neocalanus plumchrus*, *Neocalanus cristatus*, *Calanus* spp., and *Eucalanus bungii*; small calanoids included species such as *Pseudocalanus* spp., *Acartia longiremis*, and *Centropages abdominalis*.

The univariate approach to analysis of variance (ANOVA) of a repeated-measures design (Frane 1980) was used to analyze zooplankton data. Measures of abundance analyzed with ANOVA were biomass and density of total zooplankton, and biomass of small and large calanoid copepods. Factors in the ANOVA were time period, oil (dichotomous scale: oiled or non-oiled), bay–corridor, and location (nested in oil and bay–corridor). Each cell had three observations, except in one case where a replicate sample was lost because of improper preservation. Based on Box-Cox diagnostic plots (Dixon et al. 1988), density and biomass data were transformed to natural logarithms to normalize distribution and maximize variance homogeneity. The number of species or species groups was used as a measure of diversity (Pielou 1975) and was also analyzed by ANOVA.

Coefficients of variation were calculated for biomass and density of zooplankton at each location for each time period. Coefficient of variation (CV, corrected for bias) was calculated according to the formula

$$CV = \left(1 + \frac{1}{4N}\right)\left(100\frac{SD}{\bar{Y}}\right), \quad (1)$$

where N is the number of samples and $\bar{Y}$ is the mean density or biomass (Sokal and Rohlf 1981). The CV was used to compare variations in biomass and density between different locations and time periods independent of the magnitude of their means (Sokal and Rohlf 1981).

Epibenthic Crustaceans

Three habitat types (low-, medium-, and steep-gradient beaches) were sampled for epibenthic crustaceans at each location. Low-gradient beaches were <10% grade, with granule (2–4 mm)–pebble (4–64 mm) substrate; medium-gradient beaches were 12 to 25% grade, with pebble (4–64 mm)–cobble (64–256 mm) substrate; and steep-gradient beaches were >50% grade, with bedrock or boulder substrate (>256 mm). Sampling sites were selected for similarity in wave exposure, macrophyte coverage, and substrate.

Epibenthic crustaceans were sampled in 1989 with a 10-m horizontal haul of an epibenthic sled with an attached 0.3-m-diameter, 243-μm-mesh net. Runners on the sled were 11 cm high; thus, the bottom of the net was towed 11 cm above the substrate. Epibenthic sled samples were taken at 0.5-m water depth. After a haul, the outside of the sampling net was washed down with a seawater pump. The contents of the cup were then rinsed with filtered seawater and preserved in 5% formalin solution.

Two different series of epibenthic sled samples were collected in 1989. The first series, systematic samples, were taken adjacent to each systematic beach seining site immediately after the site was seined for juvenile salmon (Wertheimer and Celewycz 1996, this volume). Systematic sled samples were collected from −0.3-m to +0.9-m tide levels. The second series, tidal transect samples, were taken at 0.6-m intervals from −0.3-m to +2.7-m tide levels (actual water depth sampled was 0.5 m deeper than the nominal tide levels). Tidal transects were sampled at each bay on each sampling trip in 1989, except trip 1, when tidal transects were sampled only at McClure Bay and Herring Bay.

Epibenthic sled samples were analyzed by the Fisheries Research Institute, University of Washington. Crustaceans in a subsample of each sample were identified, counted, and weighed, as for zooplankton. A total number for the sample was computed by dividing the subsample count by the subsample fraction. Density (number/m^3) of organisms was computed by dividing the number of organisms in a sample by the volume of water sampled. Biomass (g/m^3) of an organism was computed by multiplying the mean weight of the organism by the number in the sample and then dividing by the volume of water sampled.

The epibenthic sled sampled organisms of both pelagic and epibenthic origin, but analysis was limited to those epibenthic crustaceans that are prey of juvenile salmon (Kaczynski et al. 1973; Bailey et al. 1975; Barnard 1981; Landingham and Mothershead 1988). Data were analyzed by repeated-measures ANOVA. Measures of abundance analyzed with ANOVA included total biomass of epibenthic crus-

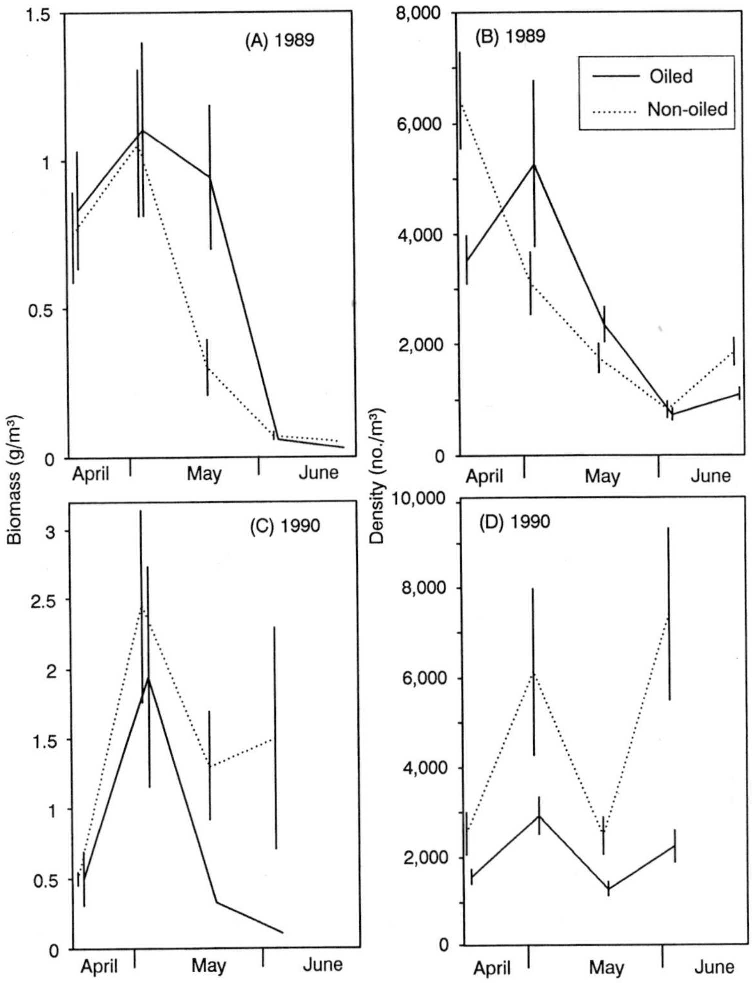

FIGURE 2.—Mean biomass (**A** and **C**) and mean density (**B** and **D**) of zooplankton in oiled and non-oiled locations of Prince William Sound in 1989 and 1990. Vertical bars indicate ±1 SE around the mean. Scales differ from graph to graph.

taceans and biomass of harpacticoid copepods. Based on Box-Cox diagnostic plots, biomass data were transformed to natural logarithms to normalize distribution and maximize variance homogeneity. For the systematic epibenthic samples, the factors were time period, oil, bay–corridor, habitat, and location (nested in oil and bay–corridor). Each cell had only one observation, except that six cells were empty because of samples destroyed in shipping. For the tidal transect samples, the factors were time period, oil, location, habitat, and tide level, with location nested in oil. Time period 1 was excluded from the ANOVA because all four bays were not sampled on the first sampling trip. The number of species or species groups of epibenthic crustaceans was used as a measure of diversity (Pielou 1975) and analyzed by ANOVA.

Results

Zooplankton

Biomass and density of zooplankton fluctuated widely between time periods (Figure 2) and locations in both 1989 and 1990. Variation between replicates at the same time and location was also large, with CV ranging from 5 to 85% for biomass and 5 to 66% for density.

Neither zooplankton density nor biomass differed significantly ($P > 0.1$) between oiled and non-oiled locations in 1989 or 1990. In 1989, mean biomass was similar in oiled and non-oiled locations in spring, peaking in early May, and declining to low levels by early June (Figure 2A). In 1989, density of zooplankton in oiled locations also peaked in early May, whereas in non-oiled locations, density peaked in mid-April (Figure 2B). Throughout May, zooplankton density declined in both oiled and non-oiled locations. Density increased slightly from early to late June in both oiled and non-oiled locations. In contrast to 1989, sample mean biomass of zooplankton in 1990 declined more rapidly in oiled than in non-oiled locations following the seasonal peak in early May (Figure 2C). Biomass in oiled locations continued to decline to seasonal lows through June, whereas in the non-oiled locations, sample mean biomass actually increased in June. Density of zooplankton in oiled locations also peaked in early May, but density in non-oiled locations peaked in early June (Figure 2D). In both oiled and non-oiled locations, mean zooplankton density increased from late May to early June. While peak density of zooplankton in non-oiled locations was similar between years, peak density in oiled locations was two times higher in 1989 than in 1990. Peak total biomass in both oiled and non-oiled locations was about two times higher in 1990 than in 1989.

In 1989, zooplankton biomass ($P = 0.042$) and density ($P = 0.021$) were significantly greater in corridors than in bays. Mean biomass was 3.4 times higher and mean density was 1.9 times higher overall in corridors than in bays. When examined over time (Figure 3), the biggest difference in zooplankton biomass between bays and corridors was in April and May. Overall, zooplankton biomass peaked in early May in both bays and corridors. By June, zooplankton biomass had declined sharply to similar levels in both bays and corridors.

In 1990, neither biomass nor density of zooplankton was significantly ($P > 0.1$) different overall between bays and corridors, although the sample mean biomass was 2.5 times higher and the sample mean density was 1.4 times higher in corridors than bays. During April and May, the sample mean biomass was higher in corridors than in bays (Figure 3); at the overall peak of zooplankton abundance in early May, the $\log_e$-transformed mean biomass differed between bays and corridors by more than 4 SE. In early June, however, while biomass declined sharply in corridors, it increased to the highest level observed in bays in 1990.

Calanoid copepods dominated the biomass and density of zooplankton in both 1989 and 1990. In 1989, 52 taxa of zooplankton were identified. In both bays and corridors, the dominant taxon by biomass was *Calanus* spp., and the dominant taxon by density was *Pseudocalanus* spp. In 1990, 43 taxa of zooplankton were identified. In overall biomass, the dominant taxa in corridors were *Neocalanus plumchrus* and *Calanus* spp., whereas the dominant taxon in bays was *Calanus* spp. In both bays and corridors, the dominant taxon by density again was *Pseudocalanus* spp.

When calanoid copepods were examined separately by size class, the biomass of large calanoid and of small calanoid copepods did not differ significantly ($P > 0.1$) between oiled and non-oiled locations in either 1989 or 1990. Only time period was significant in explaining variation in biomass in all four of these analyses. In general, both large and small calanoids closely tracked the pattern of total zooplankton biomass in both years (Figures 2, 4). The biomass of large calanoids peaked in May in oiled and non-oiled locations in both 1989 and 1990 (Figure 4A, 4C). The biomass of small calanoids peaked in early May in oiled locations in both 1989 and 1990, but in non-oiled locations, biomass of small calanoids peaked in mid-April in 1989 and early June in 1990 (Figure 4B, D).

Diversity of zooplankton, as measured by the number of identified taxa, did not differ significantly ($P > 0.1$) between oiled and non-oiled locations in 1989 or 1990. In 1989, the mean number of taxa was 15.9 in bays and 14.3 in corridors; this difference was marginally significant ($P = 0.082$). The sample mean number of taxa per haul was also higher in bays (17.7) than in corridors (17.2) in 1990, but this difference was not significant ($P > 0.1$).

Epibenthic Crustaceans

A total of 84 taxa of epibenthic origin and 42 taxa of pelagic origin were captured in the epibenthic sled in 1989; more than 70% of these organisms were considered to be prey of juvenile salmon. Harpacticoid copepods were the dominant epibenthic organisms at each location in terms of both density and biomass. They constituted 87% of the biomass of epibenthic crustaceans important in the diets of juvenile pink salmon. The most abundant harpacticoid copepods sampled by the sled were *Harpacticus* spp. and *Tisbe* spp., respectively. Because of the dominance in both the epibenthic samples and in the epibenthic crustaceans used by juvenile salmon as prey (Sturdevant et al. 1996), further analysis of epibenthic prey focused on harpacticoid copepods.

Biomass of harpacticoid copepods was greater in oiled than in non-oiled locations at each habitat

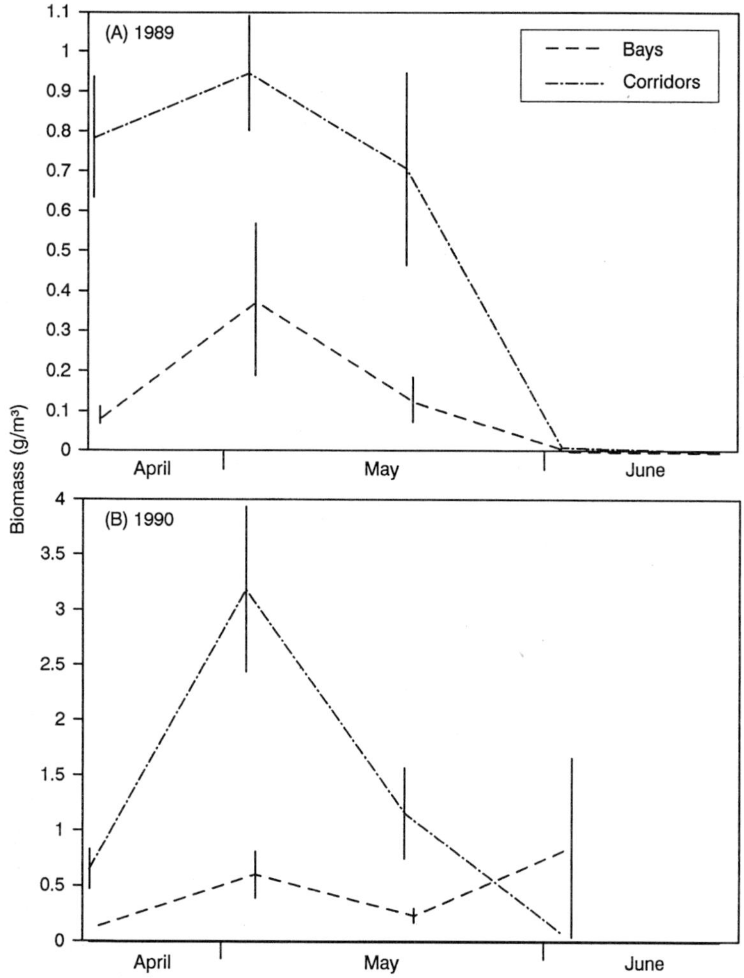

FIGURE 3.—Mean biomass (±1 SE) of zooplankton in bays and corridors of Prince William Sound in 1989 (**A**) and 1990 (**B**). Scales differ between the graphs.

type in 1989 (Figure 5). In the systematic samples, oil (P = 0.025) and habitat (P = 0.025) were each significant in explaining variation in biomass of harpacticoid copepods. Mean biomass of harpacticoids was 2.5 times higher overall in oiled than in non-oiled locations. Harpacticoid biomass was much lower in the steep-gradient than in either the low- or medium-gradient habitats.

Biomass of the two major genera of harpacticoids, *Harpacticus* and *Tisbe*, showed different patterns over time in oiled and non-oiled locations in 1989 (Figure 6). *Harpacticus* spp. in oiled locations peaked in mid-May, whereas the peak in non-oiled locations was in mid-April. *Tisbe* spp. in oiled locations peaked twice, once in mid-April and once in mid-May, whereas it peaked only in mid-May in non-oiled locations. For neither taxon was biomass significantly different between oiled and non-oiled locations ($P > 0.1$).

Time period (P = 0.066) and bay–corridor (P = 0.057) were marginally significant in explaining variation in biomass of harpacticoid copepods in the systematic samples. Biomass was three times higher overall in corridors than in bays, mainly because of the large peak in biomass in late May (Figure 7). By June, the biomass of harpacticoids was uniformly low in both bays and corridors.

In the tidal transect samples from bays in 1989, oil (P = 0.021), tide level ($P < 0.001$), and the oil–tide interaction (P = 0.001) were significant in explaining variation in biomass of harpacticoid copepods. Both density and biomass of harpacticoid

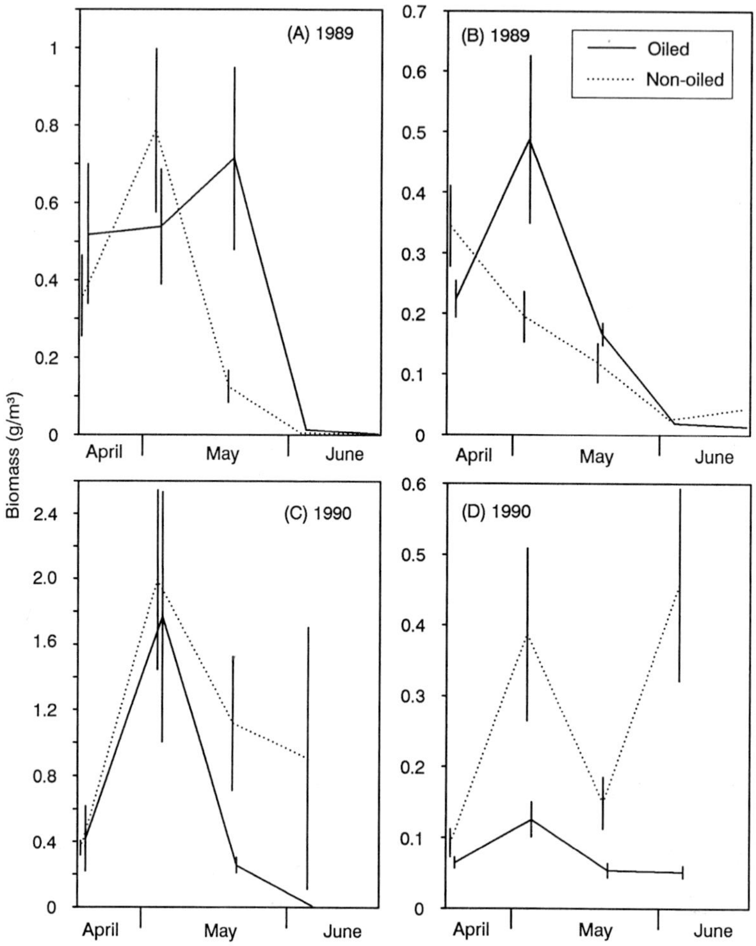

FIGURE 4.—Mean biomass (±1 SE) of large calanoid copepods (**A** and **C**) and small calanoid copepods (**B** and **D**) in oiled and non-oiled locations of Prince William Sound in 1989 and 1990. Scales differ between the graphs.

copepods were higher overall in oiled than in non-oiled locations, and both measures decreased with increasing tide level (Figure 8). The biomass of both major harpacticoid genera, *Harpacticus* and *Tisbe*, showed similar patterns with respect to the factors oil and tide level. The time period-habitat-tide level interaction was also significant ($P = 0.016$) in the analysis of biomass of harpacticoid copepods, but this third-order interaction was difficult to interpret.

Diversity of epibenthic crustaceans did not differ significantly ($P > 0.1$) between oiled and non-oiled locations in the systematic samples. Time period ($P < 0.001$), habitat ($P = 0.015$), and the time–oil interaction ($P = 0.047$) were significant in explaining the variation in number of taxa. Although the number of taxa exhibited a different pattern over time at oiled and non-oiled locations, the number of taxa generally increased over time (Figure 9). The number of taxa was highest in the low-gradient habitat, intermediate in the medium-gradient habitat, and lowest in the steep-gradient habitat.

Diversity of epibenthic crustaceans also did not differ significantly ($P > 0.1$) between oiled and non-oiled locations in the tidal transect samples. The number of taxa at both oiled and non-oiled locations decreased with elevation of tidal transects.

Discussion

Zooplankton

This study could demonstrate no detrimental effect of the *Exxon Valdez* oil spill on zooplankton in Prince William Sound. We observed no significant differences in either the density or biomass of zoo-

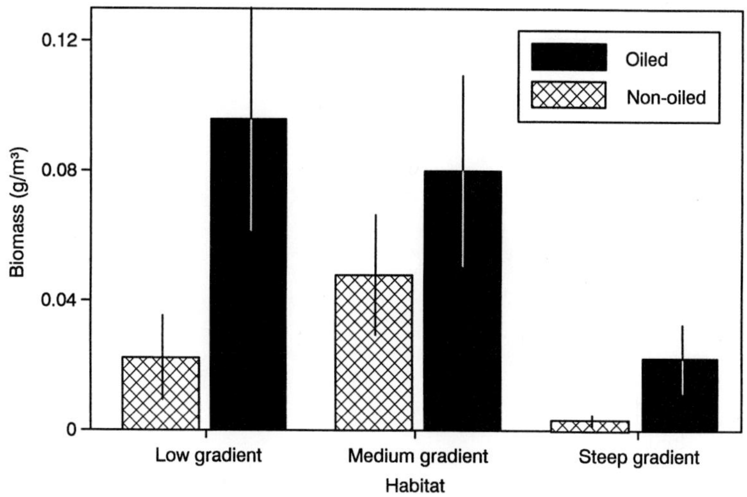

FIGURE 5.—Harpacticoid copepod mean biomass (±1 SE) in the epibenthic sled systematic samples in oiled and non-oiled habitats in Prince William Sound, April–June 1989.

plankton in general, or in large and small calanoid copepods in particular, between oiled and non-oiled locations in either 1989 or 1990. In addition, diversity of zooplankton did not differ between oiled and non-oiled locations.

Other studies have shown that crude oil and its derivatives can negatively impact zooplankton. Lee and Nicol (1977) found zooplankton to be susceptible to the water-soluble fraction of fuel oil, especially oceanic zooplankton compared to the more neritic forms. Samain et al. (1980) concluded that there were short-term (15–30 d) detrimental effects on zooplankton after the *Amoco Cadiz* spill off Brittany. However, Conover (1971) observed no apparent effect of bunker C oil on zooplanktors in Chedabucto Bay, even though these organisms were consuming large amounts of oil. Lee (1975) and Corner et al. (1976) demonstrated that certain species of calanoids will metabolize hydrocarbons with no apparent ill effect.

One possible reason why oil did not reduce zooplankton abundance is that the concentration of oil in the water column did not reach sufficient levels to cause direct mortality. Acutely lethal concentrations of crude oil to zooplankton have been reported to range from 100 to 43,000 μg/L (Corner 1978; Johnston 1984; Capuzzo 1987). Sublethal responses, such as decrease in growth rate, usually were observed at exposure concentrations that ranged from only slightly lower than the lethal concentration to about one-tenth the lethal concentration (Anderson et al. 1987; Neff 1987). In western Prince William Sound, however, the total concentration of polynuclear aromatics measured in the water column did not exceed 6.2 μg/L following the spill (Neff 1991; Short and Harris 1996, this volume).

A second possible explanation for no reduction of zooplankton is that advection of zooplankton from deeper waters of Prince William Sound and the Gulf of Alaska may have obscured any localized impacts. Physical transport of zooplankton can substantially modify populations in bays and inlets (Lindahl and Hernroth 1988; Aksnes et al. 1989; Coyle et al. 1990). The shelf and coastal food webs in the northern Gulf of Alaska are supplemented by an immense amount of biomass from the bordering ocean region (Cooney 1986), which was outside the area affected by the *Exxon Valdez* oil spill. Zooplankton abundance in nearshore waters of Prince William Sound is strongly influenced by water circulation both within the sound and from the Gulf of Alaska; large calanoids, such as *Neocalanus plumchrus* and *N. cristatus*, overwinter and reproduce at depths below 300 m, migrate to the surface as copepodids, and are advected into shallow (<20 m) nearshore waters of Prince William Sound in late winter and early spring (Cooney et al. 1981; Cooney 1986). The migration of large copepods into shallow nearshore waters occurs before the beginning of seasonal phytoplankton growth, rather than in response to it (Cooney 1986).

Was there a period of high initial mortality of zooplankton because of oil? Production of large

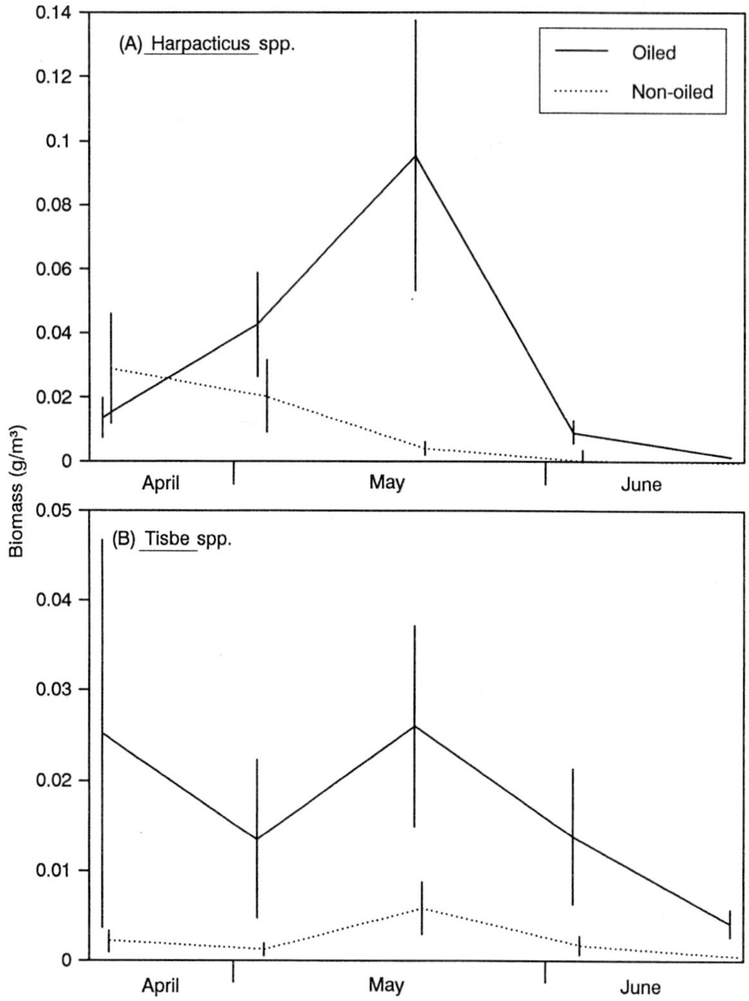

FIGURE 6.—Mean biomass (±1 SE) of *Harpacticus* spp. (**A**) and *Tisbe* spp. (**B**) in the epibenthic sled systematic samples in oiled and non-oiled locations of Prince William Sound in 1989. Scales differ between the graphs.

calanoids occurs in deep offshore waters that were outside the area of the oil spill. However, small neritic copepods, such as *Pseudocalanus* spp., which increase in abundance in response to localized production of the spring phytoplankton bloom (Cooney 1986), had slightly lower biomass in oiled than in non-oiled locations in mid-April 1989. If a period of initial mortality of small calanoids occurred, it was extremely short-lived; by early May, the biomass of small calanoids was higher in oiled than in non-oiled locations. Whether juvenile salmon were affected by this slightly lower biomass of small calanoids in oiled locations in April is doubtful; the total biomass of large and small calanoids together was no different between oiled and non-oiled locations in April, and few salmon fry were present in the nearshore marine environment at that time (Wertheimer and Celewycz 1996).

Epibenthic Crustaceans

Neither abundance nor diversity of epibenthic prey available to juvenile salmon was reduced in western Prince William Sound as a result of the *Exxon Valdez* oil spill. On the contrary, in both the systematic and tidal transect samples, the biomass of total prey epibenthic crustaceans in general and harpacticoid copepods in particular was significantly greater in oiled than in non-oiled locations.

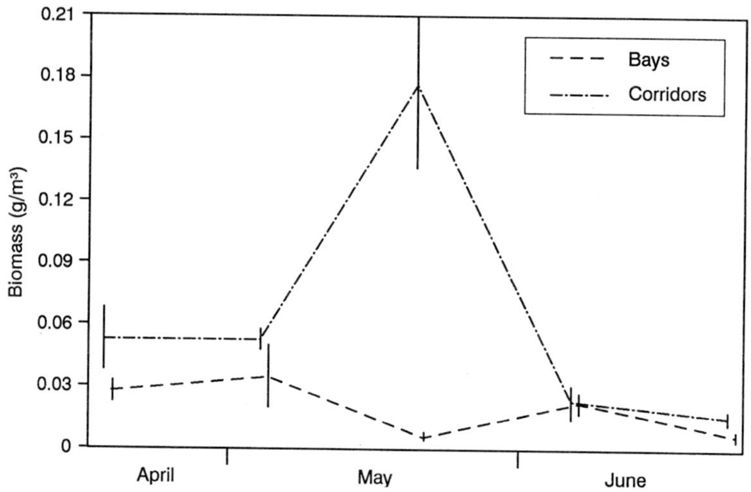

FIGURE 7.—Harpacticoid copepod mean biomass (±1 SE) in the epibenthic sled systematic samples in bays and corridors over time in Prince William Sound, April–June 1989.

Assemblages of epibenthic crustaceans were equally diverse at oiled and non-oiled locations.

The presence of petroleum hydrocarbons in intertidal areas has been shown to affect populations of harpacticoid copepods in different ways. The immediate response of harpacticoid copepod populations to an oil spill is frequently a decline in numbers (Wormald 1976; Giere 1979; Elmgren et al. 1983; Bodin 1991). However, increases in harpacticoid copepods and other meiofauna associated with oil contamination have also been observed (Naidu et al. 1978; Fricke et al. 1981; Fleeger and Chandler 1983; Feder et al. 1990). Stacey and Marcotte (1987) found that oil contamination had a detrimental effect on some species of harpacticoids, a beneficial effect on one species, and no effect on others.

We found no evidence of high initial mortality of harpacticoid copepods in response to the *Exxon Valdez* oil spill. During our first sampling trip in 1989 from 12 April to 20 April (19–27 days after the spill), the biomass of *Harpacticus* spp. was not lower in oiled than in non-oiled locations. The biomass of *Harpacticus* spp. did peak at a later time in the oiled area (mid-May) than in the non-oiled area (mid-April), a possible indication of suppressed production in oiled locations immediately after the spill. However, the biomass of *Tisbe* spp. was higher in oiled than in non-oiled locations in all time periods that we sampled throughout the spring of 1989.

Harpacticoid copepods are a diverse taxon; different species exploit different habitats and respond differently to oil spills. For example, 4 months after the *Amoco Cadiz* oil spill, numbers of free-living harpacticoids on the surface of algae were 20 times higher in polluted areas than in comparable unpolluted areas (Chasse 1978). At the same time, epipsammic harpacticoids in nearby sediments were virtually absent, perhaps because of high concentrations of hydrocarbons in interstitial water (Chasse 1978). Bodin (1988) also observed that after the *Amoco Cadiz* oil spill, interstitial taxa of harpacticoids declined, whereas plant-associated taxa of harpacticoids increased.

The epibenthic crustaceans sampled with the epibenthic sled were dominated by the free-living harpacticoid copepods *Harpacticus* spp. and *Tisbe* spp. These taxa are often associated with plants above the sediment interface (Hicks and Coull 1983). We did not sample the sediments for the interstitial taxa of harpacticoids because the harpacticoid copepods most important in the diets of juvenile salmon in Prince William Sound are the free-living forms that are available in the water column. At our study locations, *Harpacticus* spp. and *Tisbe* spp. were the most important harpacticoid copepods in the diets of pink and chum salmon fry (Sturdevant et al. 1996), as well as the dominant component of epibenthic prey sampled by the sled.

Reasons for the increased abundance of epibenthic crustaceans in oiled locations are not clear. Increased abundance could have been directly or indirectly caused by oil, or could have been unre-

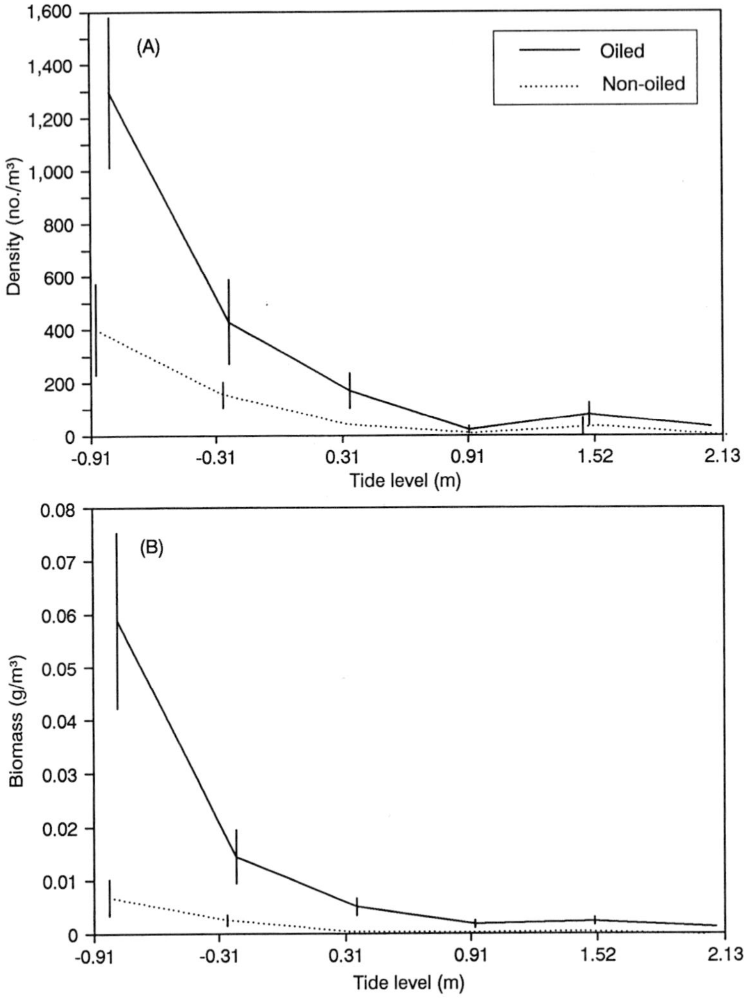

FIGURE 8.—Mean (±1 SE) density (**A**) and biomass (**B**) of harpacticoid copepods at different tide levels in the epibenthic sled tidal transect samples from oiled and non-oiled habitats in bays of Prince William Sound, April–June 1989.

lated to the spill. In a simulated oil spill in Port Valdez, Alaska, Feder et al. (1990) observed an increase in abundance of harpacticoid copepods, including *H. uniremis*, and speculated that the availability of a new food resource (bacteria associated with the oil) may have led to improved survival of harpacticoids. After the *Exxon Valdez* oil spill, bacteria numbers in subtidal sediments were significantly greater in oiled locations than in control sites (Braddock et al. 1996, this volume). The abundance of macroinvertebrate grazers commonly decreases after an oil spill, leading to blooms of ephemeral algae (Foster et al. 1988), which also may have led to the increased abundance of the plant-associated harpacticoids that we observed.

We cannot, however, exclude geographic variation as an explanation of the higher abundance of epibenthic crustaceans in the oiled locations. Abundance of epibenthic crustaceans may have been affected by the higher nearshore salinity at oiled as compared with non-oiled locations (Wertheimer and Celewycz 1996). Abundance of salmon was also significantly lower in oiled than in non-oiled locations (Wertheimer and Celewycz 1996); thus, reduced cropping by juvenile salmon could have increased the abundance of epibenthic crustaceans.

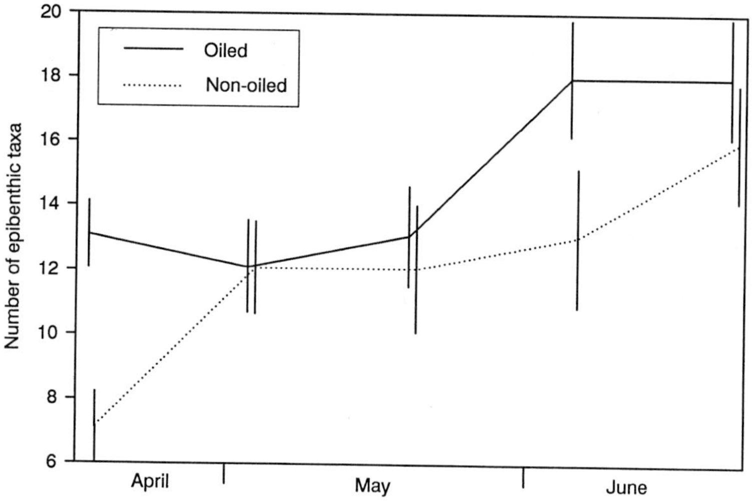

FIGURE 9.—Number of epibenthic taxa (±1 SE) captured in the epibenthic sled systematic samples in oiled and non-oiled locations over time in Prince William Sound, April–June 1989.

However, even when juvenile salmon are using epibenthic crustaceans as their primary prey, densities of the prey populations have not been affected significantly (Webb 1991).

Although we found no short-term reduction of epibenthic prey in oiled locations in 1989, longer-term effects could potentially be caused by contamination of sediments that persisted in western Prince William Sound in 1990. However, previous research has shown that even when the presence of oil has had a detrimental effect on harpacticoid copepod populations, recovery has often been rapid (Wormald 1976; Giere 1979; Elmgren et al. 1983; Bodin 1991). Azoic sediments treated with *Exxon Valdez* crude oil and buried in intertidal beaches in Prince William Sound in 1990 showed rapid colonization by harpacticoid copepods from the surrounding areas; even in the most heavily oiled sediments, colonization approached ambient levels within 28 d (Fleeger et al. 1996, this volume). The abundance of epibenthic harpacticoid copepods was higher at heavily oiled than at lightly oiled beaches in Prince William Sound in 1990, indicating that the epibenthic prey resources available to juvenile salmon in 1990 were not directly reduced by any continuing impact of the *Exxon Valdez* oil spill (Wertheimer et al. 1996, this volume).

Acknowledgments

We thank John Bortniak, Mark Carls, Jeff Cordell, Tom Dunatov, Robert Gish, Charles Guthrie III, James Hasbrook, Arleen Jones, Ben Kirkpatrick, Joyce Landingham, Michael Murphy, Joseph Orsi, Mark Pritchett, Molly Sturdevant, and Bob Weise for their assistance.

References

Aksnes, D. L., J. Aure, S. Kaartvedt, T. Magnesen, and J. Richard. 1989. Significance of advection for the carrying capacities of fjord populations. Marine Ecology Progress Series 50:263–274.

Anderson, J. W., J. M. Neff, and P. D. Boehm. 1987. Sources, fates and effects of aromatic hydrocarbons in the Alaskan marine environment with recommendations for monitoring strategies. U.S. Environmental Protection Agency, Report EPA/600/3–86/018 (NTIS PB 86-168291), Washington, DC.

Bailey, J. E., B. L. Wing, and C. R. Mattson. 1975. Zooplankton abundance and feeding habits of fry of pink salmon, *Oncorhynchus gorbuscha*, and chum salmon, *O. keta*, in Traitor's Cove, Alaska, with speculations on the carrying capacity of the area. U.S. National Marine Fisheries Service Fishery Bulletin 73:946–961.

Barnard, D. R. 1981. Prey relationships between juvenile pink (*Oncorhynchus gorbuscha*) and chum (*O. keta*) salmon in Prince William Sound, Alaska. Master's thesis. University of Alaska, Fairbanks.

Bodin, P. 1988. Results of ecological monitoring of three beaches polluted by the *Amoco Cadiz* oil spill: development of meiofauna from 1978 to 1984. Marine Ecology Progress Series 42:105–123.

Bodin, P. 1991. Perturbations in the reproduction cycle of some harpacticoid copepod species further to the *Amoco Cadiz* spill. Hydrobiologia 209:245–257.

Braddock, J. F., J. E. Lindstrom, T. R. Yeager, B. T.

Rasley, and E. J. Brown. 1996. Patterns of microbial activity in oiled and unoiled sediments in Prince William Sound. American Fisheries Society Symposium 18:94–108.

Brodeur, R. D., and W. G. Pearcy. 1990. Trophic relations of juvenile Pacific salmon off the Oregon and Washington coast. U.S. National Marine Fisheries Service Fishery Bulletin 88:617–636.

Capuzzo, J. M. 1987. Biological effects of petroleum hydrocarbons: assessments from experimental results. Pages 343–410 *in* D. F. Boesch and N. N. Rabalais, editors. Long-term environmental effects of offshore oil and gas development. Elsevier, New York.

Carls, M. G., A. C. Wertheimer, J. W. Short, R. M. Smolowitz, and J. J. Stegeman. 1996. Contamination of juvenile pink and chum salmon by hydrocarbons in Prince William Sound after the *Exxon Valdez* oil spill. American Fisheries Society Symposium 18:593–607.

Chasse, C. 1978. The ecological impact on and near shores by the *Amoco Cadiz* oil spill. Marine Pollution Bulletin 9:298–301.

Conover, R. J. 1971. Some relations between zooplankton and bunker C oil in Chedabucto Bay following the wreck of the tanker *Arrow*. Journal of the Fisheries Research Board of Canada 28:1327–1330.

Cooney, R. T. 1986. Zooplankton. Pages 285–304 *in* D. W. Hood and S. T. Zimmerman, editors. The Gulf of Alaska physical environment and biological resources. U.S. Department of the Interior, Mining and Managing Service. OCS (Outer Continental Shelf) Study, MMS 86-0095, Washington, DC.

Cooney, R. T., D. Urquhart, and D. Barnard. 1981. The behavior, feeding biology, and growth of hatchery released pink and chum salmon fry in Prince William Sound, Alaska. University of Alaska, Sea Grant Report 81-5, Fairbanks.

Corner, E. D. S. 1978. Pollution studies with marine plankton. Part I. Petroleum hydrocarbons and related compounds. Advances in Marine Biology 15:289–380.

Corner, E. D. S., R. P. Harris, C. C. Kilvington, and S. C. M. O'Hara. 1976. Petroleum compounds in the marine food web: short-term experiments on the fate of naphthalene in *Calanus*. Journal of the Marine Biological Association of the United Kingdom 56: 121–133.

Coyle, K. O., A. J. Paul, and D. A. Ziemann. 1990. Copepod populations during the spring bloom in an Alaskan subarctic embayment. Journal of Plankton Research 12:759–797.

Dixon, W. J., P. Sampson, and P. Mundle. 1988. One- and two-way analysis of variance with data screening. Pages 187–208 *in* W. J. Dixon, editor. BMDP statistical software manual. University of California Press, Berkeley.

Elmgren, R., S. Hansson, U. Larsson, B. Sundelin, and P. D. Boehm. 1983. The *Tsesis* oil spill: acute and long-term impact on the benthos. Marine Biology 73:51–65.

Feder, H. M., A. S. Naidu, and A. J. Paul. 1990. Trace element and biotic changes following a simulated oil spill on a mudflat in Port Valdez, Alaska. Marine Pollution Bulletin 21:131–137.

Fleeger, J. W., and G. T. Chandler. 1983. Meiofauna responses to an experimental oil spill in a Louisiana salt marsh. Marine Ecology Progress Series 11:257–264.

Fleeger, J. W., T. C. Shirley, M. G. Carls, and M. A. Todaro. 1996. Meiofaunal recolonization experiment with oiled sediments. American Fisheries Society Symposium 18:271–285.

Foster, M. S., A. P. De Vogelaere, C. Harrold, J. S. Pearse, and A. B. Thum. 1988. Causes of spatial and temporal patterns in rocky intertidal communities of central and northern California. California Academy of Sciences Memoir Number 9.

Frane, J. 1980. The univariate approach to repeated measures—foundation, advantages, and caveats. BMDP Technical Report 69, University of California Press, Berkeley.

Fricke, A. H., H. F. O. Henning, and M. J. Orren. 1981. Relationship between oil pollution and psammolittoral meiofauna density of two South African beaches. Marine Environmental Research 5:59–77.

Giere, O. 1979. The impact of oil pollution on intertidal meiofauna. Field studies after the *La Coruna* spill, May 1976. Cahiers de Biologie Marine 20:231–251.

Healey, M. C. 1980. The ecology of juvenile salmon in Georgia Strait, British Columbia. Pages 203–230 *in* W. J. McNeil and D. C. Himsworth, editors. Salmonid ecosystems of the North Pacific. Oregon State University Press, Corvallis.

Heard, W. R. 1991. Life history of pink salmon (*Oncorhynchus gorbuscha*). Pages 119–230 *in* C. Groot and L. Margolis, editors. Pacific salmon life histories. University of British Columbia Press, Vancouver.

Hicks, G. R. F., and B. C. Coull. 1983. The ecology of marine meiobenthic harpacticoid copepods. Oceanography and Marine Biology Annual Review 21:67–175.

Holtby, L. B., T. E. McMahon, and J. C. Scrivener. 1989. Stream temperatures and inter-annual variability in the emigration timing of coho salmon (*Oncorhynchus kisutch*) smolts and fry and chum salmon (*O. keta*) fry from Carnation Creek, British Columbia. Canadian Journal of Fisheries and Aquatic Sciences 46:1396–1405.

Johnston, R. 1984. Oil pollution and its management. Pages 1497–1530 *in* O. Kinne, editor. Marine ecology, volume 5. Wiley, New York.

Kaczynski, V. W., R. J. Feller, and C. Clayton. 1973. Trophic analysis of juvenile pink and chum salmon (*Oncorhynchus gorbuscha* and *O. keta*) in Puget Sound. Journal of the Fisheries Research Board of Canada 30:1003–1008.

Landingham, J. H. 1982. Feeding ecology of pink and chum salmon fry in the nearshore habitat of Auke Bay, Alaska. Master's thesis. University of Alaska, Juneau.

Landingham, J. H., and P. D. Mothershead. 1988. Feeding habits of juvenile pink salmon in nearshore and offshore sites of Auke Bay. Pages 449–469 *in* APPRISE annual report. University of Alaska, School of Fisheries and Ocean Sciences (SFOS APP87-100), Fairbanks.

Larrance, J. D. 1977. Primary production in the mid-subarctic Pacific region 1966–68. U.S. National Marine Fisheries Service Fishery Bulletin 69:595–613.

Lee, R. F. 1975. Fate of petroleum hydrocarbons in marine zooplankton. Pages 549–553 *in* Proceedings, 1975 conference on prevention and control of oil pollution. American Petroleum Institute, Washington, DC.

Lee, W. Y., and J. A. C. Nicol. 1977. The effects of the water soluble fractions of no. 2 fuel oil on the survival and behavior of coastal and oceanic zooplankton. Environmental Pollution 12:279–292.

Lindahl, O., and L. Hernroth. 1988. Large-scale and long-term variations in the zooplankton community of the Gullmar fjord, Sweden, in relation to advective processes. Marine Ecology Progress Series 43:161–171.

Murphy, M. L., J. F. Thedinga, and K. V. Koski. 1988. Size and diet of juvenile Pacific salmon during seaward migration through a small estuary in southeastern Alaska. U.S. National Marine Fisheries Service Fishery Bulletin 86:213–222.

Naidu, A. S., H. M. Feder, and S. A. Norrell. 1978. The effect of Prudhoe Bay crude oil on a tidal-flat ecosystem in Port Valdez, Alaska. Pages 97–104 *in* Tenth Offshore Technology Conference, Richardson, Texas.

Neff, J. M. 1987. Biological effects of oil in the marine environment. Chemical Engineering Progress 83:27–33.

Neff, J. M. 1991. Water quality in Prince William Sound and the Gulf of Alaska. Arthur D. Little, Inc., Cambridge, Massachusetts.

Parker, R. R. 1971. Size selective predation among juvenile salmonid fishes in a British Columbia inlet. Journal of the Fisheries Research Board of Canada 28: 1503–1510.

Pielou, E. C. 1975. Ecological diversity. Wiley, New York.

Salo, E. O. 1991. Life history of chum salmon (*Oncorhynchus keta*). Pages 231–310 *in* C. Groot and L. Margolis, editors. Pacific salmon life histories. University of British Columbia Press, Vancouver.

Samain, J. F., J. Moal, A. Coum, J. R. Le Coz, and J. Y. Daniel. 1980. Effects of the *Amoco Cadiz* oil spill on zooplankton. Helgolaender Meeresuntersuchumgen 33:225–235.

Short, J. W., and P. M. Harris. 1996. Chemical sampling and analysis of petroleum hydrocarbons in near-surface seawater of Prince William Sound after the *Exxon Valdez* oil spill. American Fisheries Society Symposium 18:17–28.

Smetacek, V., and six coauthors. 1984. Seasonal stages characterizing the annual cycle of an inshore pelagic system. Rapports et Proces-Verbaux des Reunions Conseil International pour l'Exploration de la Mer 83:126–135.

Sokal, R. R., and F. J. Rohlf. 1981. Biometry, the principles and practice of statistics in biological research, 2nd edition. Freeman, New York.

Stacey, B. M., and B. M. Marcotte. 1987. Chronic effect of no. 2 fuel oil on population dynamics of harpacticoid copepods in experimental marine mesocosms. Marine Ecology Progress Series 40:61–68.

Sturdevant, M. V., A. C. Wertheimer, and J. L. Lum. 1996. Diets of juvenile pink and chum salmon in oiled and non-oiled nearshore habitats in Prince William Sound, 1989 and 1990. American Fisheries Society Symposium 18:578–592.

Webb, D. G. 1991. Effect of predation by juvenile Pacific salmon on marine harpacticoid copepods. II. Predator density manipulation experiments. Marine Ecology Progress Series 72:37–47.

Wertheimer, A. C., N. J. Bax, A. G. Celewycz, M. G. Carls, and J. H. Landingham. 1996. Harpacticoid copepod abundance and population structure in Prince William Sound, one year after the *Exxon Valdez* oil spill. American Fisheries Society Symposium 18:551–563.

Wertheimer, A. C., and A. G. Celewycz. 1996. Abundance and growth of juvenile pink salmon in oiled and non-oiled locations of western Prince William Sound after the *Exxon Valdez* oil spill. American Fisheries Society Symposium 18:518–532.

Wormald, A. P. 1976. Effects of a spill of marine diesel oil on the meiofauna of a sandy beach at Picnic Bay, Hong Kong. Environmental Pollution 11:117–130.

American Fisheries Society Symposium 18:578–592, 1996

Diets of Juvenile Pink and Chum Salmon in Oiled and Non-Oiled Nearshore Habitats in Prince William Sound, 1989 and 1990

M. V. STURDEVANT, A. C. WERTHEIMER, AND J. L. LUM

National Marine Fisheries Service, Alaska Fisheries Science Center, Auke Bay Laboratory
11305 Glacier Highway, Juneau, Alaska 99801, USA

Abstract.—The diets of juvenile pink salmon *Oncorhynchus gorbuscha* and chum salmon *O. keta* were studied in nearshore habitats in oiled and non-oiled areas of Prince William Sound after the March 1989 *Exxon Valdez* oil spill. Food consumption by pink and chum salmon did not decrease in the presence of oil. Stomach fullness and food quantity were similar in oiled and non-oiled areas in both 1989 and 1990. Oil globules or sheen were observed, however, in the stomachs of pink and chum salmon from oiled sites in 1989. Because no oil was observed in fish stomachs from non-oiled sites in 1989 or from any site in 1990, these observations suggest that ingested oil could be a route of hydrocarbon contamination. Zooplankton averaged 69% of pink and 56% of chum salmon diet biomass in the 2 years. Small and large calanoid copepods were the primary zooplankters consumed by pink and chum salmon, respectively. Diets in oiled and non-oiled areas changed from 1989 to 1990. Pink salmon fed more on zooplankton in oiled areas than in non-oiled areas in 1989. Both pink and chum salmon fed less on zooplankton in oiled areas than in non-oiled areas in 1990. Conversely, both species ate epibenthic prey less in oiled areas in 1989, and more in oiled areas in 1990. These interannual changes in diet between oiled and non-oiled areas could have been caused by differences in distribution of fish, by interannual variation in prey availability, or by effects of oil.

Abundant food resources are critical to the rapid growth of salmon fry during the nearshore life history stage (Bailey et al. 1975; Cooney et al. 1981). Rapid growth may allow escape from predation and enhance survival (Parker 1968; Mortensen et al. 1991). Contaminated marine environments have the potential to affect this critical growth by interfering with food and feeding processes. For example, oil may affect the diet and physiology of juvenile salmon by altering abundance and composition of prey (Conover 1971; Bodin 1988; Feder et al. 1990), by interfering with the fish's ability to feed in some habitats, or by reducing assimilation efficiency (Vignier et al. 1992). Although laboratory studies have documented physiological impacts of oil on growth (eg., Rice et al. 1977; Schwartz 1985), no studies exist that report on the feeding biology of juvenile salmon in the field following a spill.

The variety of nearshore areas used by juvenile salmon is reflected in their diets. Diets differ according to the composition and abundance of food organisms available in different foraging habitats, and foraging habitats may vary in different areas (Kaczynski et al. 1973; Heard 1991). In areas where sheltered, low-gradient beaches support substantial epibenthic production, the initial food of salmon often consists of epibenthic prey (Feller and Kaczynski 1975; Sibert et al. 1977; Cordell 1986; Webb 1991). Fish may consume more planktonic prey with growth or as they move away from shore (Cooney et al. 1981; Landingham and Mothershead 1988; Cooney 1990). In Alaska waters, where nearshore habitats are often dominated by steep, rocky beaches with relatively little epibenthic production, planktonic prey may be more common than epibenthic prey in diets throughout the fish's nearshore residency (Bailey et al. 1975; Kron and Yuen 1978; Murphy et al. 1988). The susceptibility of juvenile salmon to hydrocarbon contamination by ingestion may vary with where they forage, because oil is removed more rapidly and prey resources recover more quickly from the effects of oiling in exposed compared to sheltered habitats (Bodin 1988).

This study compared the feeding habits of juvenile pink salmon *Oncorhynchus gorbuscha* and chum salmon *O. keta* between oiled and non-oiled nearshore habitats in western Prince William Sound in 1989 and 1990 as part of the Natural Resource Damage Assessment (NRDA) response to the *Exxon Valdez* oil spill. Its purpose was to determine if oil affected feeding and diet of these two species in nearshore areas of the sound in the year of and the year after the spill. The objectives of the study were to (1) compare food consumption of juvenile pink and chum salmon between oiled and non-oiled areas; (2) compare their diet compositions between oiled and non-oiled areas, between bays and corridors, and between habitats; and (3) examine their stomach contents for evidence of oil ingestion.

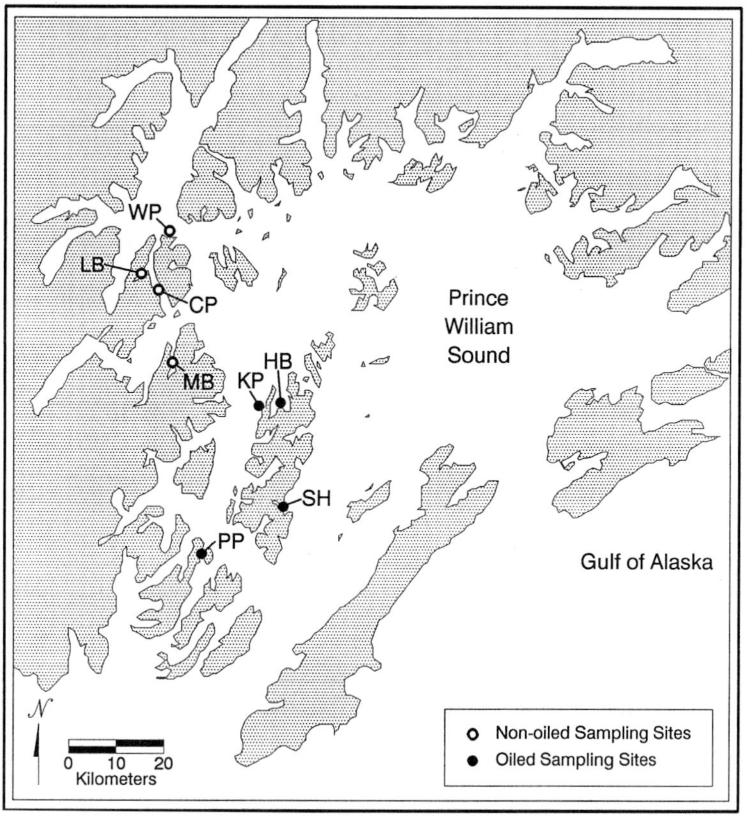

FIGURE 1.—Sampling locations in oiled (solid circles) and non-oiled (open circles) areas of western Prince William Sound, Alaska, 1989 and 1990. Paired bays (oiled first) were Herring Bay (HB)–McClure Bay (MB) and Snug Harbor (SH)–Long Bay (LB); paired corridors were Prince of Wales Passage (PP)–Culross Passage (CP) and Knight Island Passage (KP)–Wells Passage (WP).

Methods

Sample Collection

We selected four sampling locations in the oiled area and four in the non-oiled area of western Prince William Sound (Figure 1). Oiled and non-oiled areas were initially selected based on the projected path of oil dispersal and then were verified by visual inspection of the actual distribution of oil within 4 weeks after the spill. Confirmation of our designation of oiled versus non-oiled was provided by hydrocarbon analysis of mussel and sediment samples from each sampling location (Carls et al. 1996a, this volume). Within each area, two bays and two corridors leading to the Gulf of Alaska were selected. These eight locations were paired a priori for pairwise comparisons between oiled and non-oiled areas. The paired bays were Herring Bay–McClure Bay and Snug Harbor–Long Bay; the paired corridors were Prince of Wales Passage–Culross Passage and Knight Island Passage–Wells Passage (oiled and non-oiled areas, respectively).

In each location, we selected sites representing three different habitat types. Low-gradient habitats were <10% grade with granule–pebble substrate; medium-gradient habitats were 11 to 25% grade with pebble–cobble substrate; and steep-gradient habitats exceeded 45% grade with boulder or bedrock substrates. Other physical characteristics, such as macrophyte coverage and wave exposure, were also matched as closely as possible when sample sites were paired.

In 1989, we collected fish on five sampling trips from 10 April to 26 June; in 1990, we collected fish on four sampling trips from 16 April to 14 June. In 1989 and 1990, respectively, 24 and 48 sites were sampled at approximately 2-week intervals. Beach seines (37 m long, 6-mm mesh) were used to collect fish. Dip nets were also used to collect supplementary samples along 3 to 5 km of adjacent shoreline to supplement catches at specific habitat types. Sampling was restricted to tidal levels from −1 to +3 m to minimize tidal effects between sites; thus, sampling time-of-day varied.

Sample Processing

Up to 10 each of pink and chum salmon were randomly subsampled from each habitat type at each location and fixed in buffered 10% formalin for a standard period of ≥45 d to minimize variance in shrinkage (Gosho 1977). Fish were removed from the formalin, weighed to the nearest milligram wet weight, and measured to the nearest millimeter fork length (FL). Wet weight was converted to dry weight by a standard conversion factor of 0.21 (Volk et al. 1984). Specimens were then placed in 40% isopropyl alcohol until they could be processed for stomach contents.

During initial stomach processing, two measures of food consumption were obtained for each fish. First, when the stomach was excised (esophagus to pylorus), fullness was visually ranked on a scale from 0 to 6 as follows: empty, containing trace amount, 25%, 50%, 75%, or 100% full, and distended. Second, the intact stomach was weighed to the nearest milligram, stomach contents were removed, and the empty stomach was weighed again to estimate wet weight of stomach contents by subtraction.

Prey organisms were identified to the lowest practical taxon (generally to the level of order) and counted. Prey items were categorized by production system: planktonic zooplankton, epibenthos, or drift insects. Planktonic zooplankton were also categorized as small calanoid copepods (<2.5 mm metasomal length), large calanoid copepods (≥2.5 mm metasomal length), and other zooplankters. Calanoids were classified in this way because these sizes generally represent a mixed assemblage of copepods from shelf–coastal and oceanic domains: small calanoids, primarily *Pseudocalanus*, reproduce in neritic waters of Prince William Sound following the spring phytoplankton bloom, whereas large calanoids, primarily *Calanus* and *Neocalanus*, reproduce in oceanic waters during winter and are advected into the sound from the Gulf of Alaska (Cooney 1986). Epibenthos was divided into harpacticoid copepods and other epibenthic organisms (including intertidal chironomids). These subcategories were specified because harpacticoids are common in the diets of juvenile salmon in other regions (e.g., Kaczynski et al. 1973; Healey 1979; Landingham 1982; Webb 1991). Harpacticoids from 61 pink and 109 chum salmon stomachs from systematic and supplemental samples in 1989 were identified to family or genus to identify important taxa. Total dry weight of each prey taxon in a stomach was estimated by multiplying number in the stomach by its average dry weight obtained from the literature (Cooney et al. 1981) or from laboratory determinations.

After stomach analysis, individual fish diets were described by seven measures of food consumption (quantity) and five measures of diet composition (prey taxa). Measures of food consumption were (1) stomach fullness, (2) wet weight of contents, (3) content wet weight as a percent of fish wet weight, (4) total prey dry weight as a percent of fish dry weight, (5) percent of empty stomachs in a haul of fish, (6) total number of prey, and (7) total dry weight of prey. Measures of diet composition were (1) percent occurrence, (2) number, (3) percent total number, (4) dry weight, and (5) percent total dry weight represented by each of the three production systems and four prey subcategories.

We characterized salmon diets primarily in terms of prey biomass (dry weight) because this parameter best represents the energetic importance of prey to predators (Bowen 1983); percent number and percent frequency of occurrence were used to supplement description of the diets (Hyslop 1980). Relative importance diagrams with axes for each of the three components (percents biomass, number, and frequency of occurrence; Pinkas et al. 1971; Caillet 1976) were constructed to compare prey composition between oiled and non-oiled areas.

Statistical Analysis

The Wilcoxon matched-pairs, signed-ranks test (Conover 1980) was used to compare medians of food consumption, diet composition, and size of fish used in diet analyses between oiled and non-oiled locations. Medians were a more accurate measure of central tendency for diet parameters that were not normally distributed, and a parametric technique such as nested analysis of variance was not appropriate because stomach samples were frequently not available. The Wilcoxon test was used to compute the "estimated median difference" and its confidence interval for the ranked set of differences between matched pairs of diet parameters. For each set of comparisons, we tested the null hypothesis that the estimated median difference in the parameter was not different from zero. A sig-

TABLE 1.—Numbers of pink and chum salmon stomach samples from paired sites in oiled and non-oiled (in parentheses) areas of Prince William Sound. Collections were made in five sample periods in 1989 and four sample periods in 1990, at low-, medium-, and steep-gradient habitats in bays and corridors.

	1989			1990		
Habitat	Low	Medium	Steep	Low	Medium	Steep
Pink salmon—Bays						
Late April	0 (0)	10 (10)	0 (0)	10 (10)	20 (19)	10 (10)
Early May	0 (0)	6 (10)	0 (0)	10 (10)	20 (20)	10 (10)
Late May	17 (15)	20 (17)	10 (10)	5 (10)	17 (20)	7 (10)
Early June	0 (0)	0 (0)	0 (0)	10 (6)	10 (6)	10 (10)
Late June	0 (0)	8 (6)	0 (0)			
Pink salmon—Corridors						
Late April	10 (10)	0 (0)	0 (0)	0 (0)	10 (10)	0 (0)
Early May	0 (0)	10 (8)	0 (0)	20 (20)	20 (20)	20 (15)
Late May	10 (10)	10 (10)	20 (20)	10 (10)	20 (20)	0 (0)
Early June	10 (10)	20 (20)	10 (10)	20 (20)	20 (20)	20 (20)
Late June	10 (10)	10 (10)	10 (10)			
Chum salmon—Bays						
Late April	0 (0)	0 (0)		0 (0)	0 (0)	
Early May	0 (0)	6 (8)		6 (10)	10 (10)	
Late May	10 (10)	0 (0)		10 (10)	0 (0)	
Early June	10 (10)	0 (0)		10 (10)	0 (0)	
Late June	0 (0)	10 (10)				
Chum salmon—Corridors						
Late April	8 (10)			0 (0)		
Early May	0 (0)			10 (10)		
Late May	10 (10)			10 (10)		
Early June	0 (0)			10 (10)		
Late June	0 (0)					

nificant ($P < 0.1$) negative value for the difference indicated that the non-oiled parameter was greater than the oiled parameter, and vice versa.

The number of possible Wilcoxon comparisons varied between years for both pink and chum salmon because all habitats were not represented in all periods (Table 1). To ensure adequate representation of prey attributes, only hauls with five or more fish were used. Of the 608 pink salmon and 493 chum salmon examined in 1989, 397 pink and 112 chum salmon could be used in comparisons of fish from a priori paired locations matched by time period and habitat; the remainder were not matched by samples from companion sites–times. In 1990, all 595 pink and 136 chum salmon examined could be included in paired comparisons. For pink salmon, 21 paired comparisons were possible in 1989, and 31 were possible in 1990. Comparisons of chum salmon diet were limited to six pairs of hauls in 1989 and seven pairs in 1990, from low-gradient corridors and low- and medium-gradient bays (Table 1).

Fish FL was regressed on dry weight proportions of zooplankton and epibenthic prey in the diets to examine relationships between fish size and prey composition in 1989 and 1990.

Results

Food Consumption

Food consumption of pink salmon was not lower in the oiled area compared to the non-oiled area in either year. In 1989, a mean of 16% of pink salmon stomachs were empty in both oiled and non-oiled areas. Two measures of food consumption were significantly higher ($P = 0.046$, Wilcoxon test) in the oiled area; mean total prey biomass per fish was 3.2 mg versus 2.2 mg, representing 4.6% of dry body weight in the oiled area versus 3.4% in the non-oiled area. In 1990, empty stomachs occurred at rates of 9% in the oiled area and 10% in the non-oiled area, and no measures of food consumption differed significantly between the areas. Mean total prey biomass was 2.1 to 2.6 mg, representing approximately 3% of dry body weight in both oiled and non-oiled areas.

Food consumption of chum salmon also was not lower in oiled compared to non-oiled areas. In 1989, 3% of chum salmon stomachs were empty in the oiled area, compared to 8% in the non-oiled area; in 1990, 1% of the stomachs were empty in each area. In 1989, no significant differences were detected in any of the seven measures of food consumption; mean total prey biomass was 4.8 and 2.9 mg, constituting 2.7 to 3.2% dry body weight. In 1990, chum salmon stomachs contained an average of 2.7 to 5.9 mg of prey; percent dry weight of prey was significantly ($P = 0.052$) lower for fish from oiled areas than from non-oiled areas (1.5% versus 6.3%, respectively).

Diet Composition

Overview.—Zooplankton was the largest component of overall salmon diets in both years. For pink salmon, zooplankton was 66% of prey dry weight in 1989 and 71% in 1990; for chum salmon, it was 69% and 43%, for 1989 and 1990, respectively. Calanoid copepods made up most of the zooplankton biomass consumed by both species (Figures 2–4). Both large and small calanoids contributed large biomass components, but small calanoids were consumed by pink salmon more frequently and in greater numbers than large calanoids. In both years, patchy occurrence of large calanoids in pink salmon diets was indicated by median numbers and weights of zero coincident with large mean values. For chum

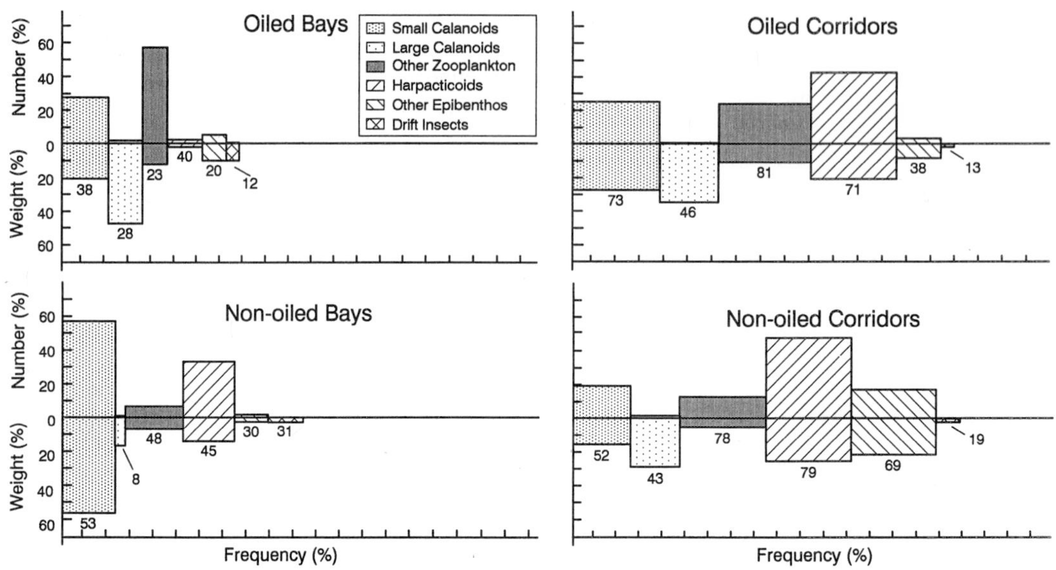

FIGURE 2.—Prey composition of 397 pink salmon from oiled and non-oiled bays and corridors in Prince William Sound, 1989. Height of bar above line indicates percent number; height of bar below line indicates percent dry weight; and width of bar and value indicate percent frequency of occurrence (magnitude indicated below each bar). Each division on the *x*-axis represents 20%.

salmon, however, large calanoids were consumed more frequently than small calanoids and contributed most of the calanoid biomass. Medians were often zero for small calanoid diet parameters among chum salmon. Large calanoids tended to be present in all stomachs or absent in all stomachs of chum salmon from different hauls, suggesting they were not always available, but were preferred when present. Neritic and oceanic calanoids occurred together in fish diets in corridors more often than in bays. Other zooplankters occurred in the diets frequently, but were rarely a large component by weight or number. Cladocerans, polychaetes (family Syllidae), and fish larvae were occasionally large components for both species.

Epibenthos was the second-largest category in salmon diets (Figures 2–4). Although harpacticoid copepods occurred frequently in diets of both species, usually they formed the majority of the epibenthic biomass in pink salmon and the minority of the epibenthic biomass in chum salmon. Both species consumed only epibenthic, not interstitial, harpacticoids. *Harpacticus*, *Tisbe*, and *Dactylopodia* were the most common genera, constituting approximately 85% of identifiable harpacticoid biomass in both pink and chum salmon diets. *Harpacticus* was the largest component in pink salmon (48%), whereas *Tisbe* was the largest component in chum salmon (47%; Figure 5). Juvenile mollusks, gammarid amphipods, and cumaceans contributed large proportions of epibenthic prey in both species. Epibenthic larvae of intertidal chironomids were also large components of chum salmon diets.

Drift insects (primarily unidentified adult Diptera) were generally a small proportion of the diets by dry weight. However, frequency of occurrence differed for pink and chum salmon; the frequency of drift insects was 21 to 35% in chum salmon diets, but was usually less than 20% in pink salmon diets (Figures 2–4).

Diet composition of pink salmon differed between bays and corridors and between low-, medium-, and steep-gradient habitats (Tables 2 and 3). Zooplankton was a larger proportion of prey dry weight among pink salmon in bays than in corridors, averaging 81% of the total in bays and 66% in corridors. Zooplankton was always the predominant prey consumed by pink salmon in steep-gradient habitats. The prey biomass consumed in low- and medium-gradient habitats in bays was mostly composed of zooplankton, but in these habitats in corridors, epibenthic prey contributed relatively more biomass.

Comparison of chum salmon diets between bays and corridors and between habitats was complicated by the paucity of samples from corridors and

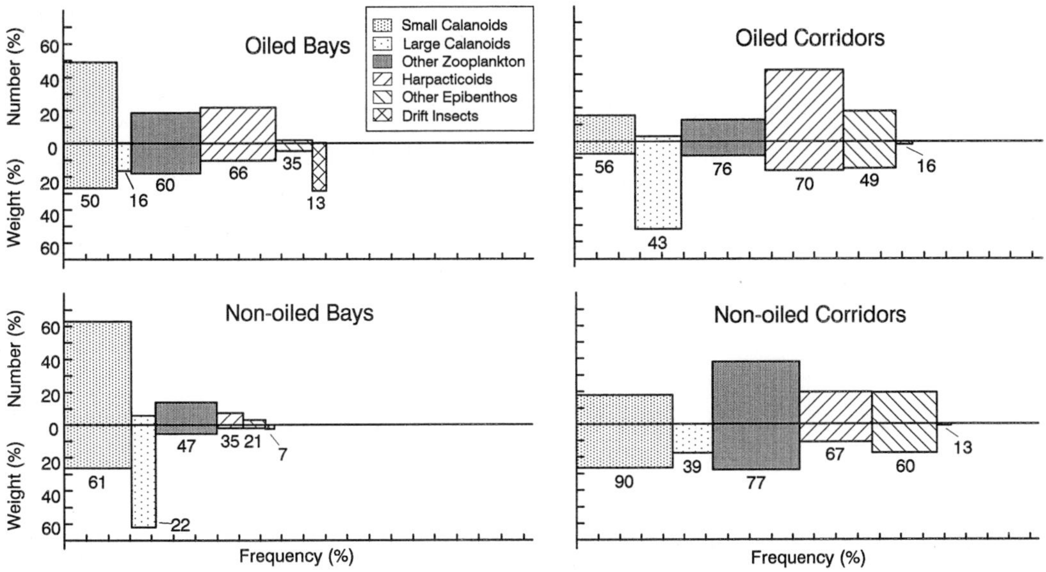

FIGURE 3.—Prey composition of 595 pink salmon from oiled and non-oiled bays and corridors in Prince William Sound, 1990. Height of bar above line indicates percent number; height of bar below line indicates percent dry weight; and width of bar and value indicate percent frequency of occurrence (magnitude indicated below each bar). Each division on the *x*-axis represents 20%.

the lack of samples from steep-gradient habitats. Zooplankton was a larger proportion of chum salmon diets than epibenthos at most low- and medium-gradient habitats (Table 4). However, epibenthos constituted similar or larger proportions of chum diets than zooplankton in one low- and two medium-gradient bay habitats and one low-gradient corridor habitat over the 2 years.

Oil effects: pink salmon.—Use of prey production systems by pink salmon differed between oiled and non-oiled areas in both years; however, trends were opposite in 1989 and 1990 (Figure 6). In 1989, zooplankton was higher in all five diet composition measures for the oiled area; percent number (P = 0.014), dry weight (P = 0.020), and percent dry weight (P = 0.042) of zooplankton were significantly higher in oiled areas than in non-oiled areas. Zooplankton averaged 71% and 53% of prey biomass in oiled and non-oiled areas, respectively. In 1990, most zooplankton composition measures were lower for the oiled area; percent number (P = 0.002), dry weight (P = 0.016), and percent dry weight (P = 0.004) of zooplankton were significantly lower. Zooplankton averaged 58% and 77% of prey dry weight in oiled and non-oiled areas, respectively.

The interannual shift in diets in oiled and non-oiled areas was also evident when prey subcategories were considered. Calanoids were consumed more in the oiled area in 1989 and less in the oiled area in 1990. In 1989, two measures of total calanoids were higher for the oiled area: percent number (P = 0.053) and dry weight (P = 0.037). In 1990, however, measures of total calanoids (except number of organisms) were lower for the oiled area ($P < 0.015$).

Interannual differences in zooplankton eaten in oiled versus non-oiled areas were most influenced by large calanoids in 1989 and by small calanoids in 1990. In 1989, all measures of large calanoids except percent frequency of occurrence were significantly higher in the oiled area ($P < 0.1$); small calanoid diet parameters tended to be higher in the oiled area, but no differences were significant. In 1990, all measures of small calanoids except number of organisms were significantly lower in the oiled area ($P < 0.064$). Diet parameters for large calanoids were not different.

The interannual shift in use of zooplankton by pink salmon in oiled and non-oiled areas was not uniform across bays and corridors. In 1989, the overall percent weights of zooplankton consumed in oiled bays and corridors were at least as great as in non-oiled bays and corridors (Figure 2). Zooplankton constituted 79% of the diet by weight in both oiled and non-oiled bays; it constituted 71% in oiled

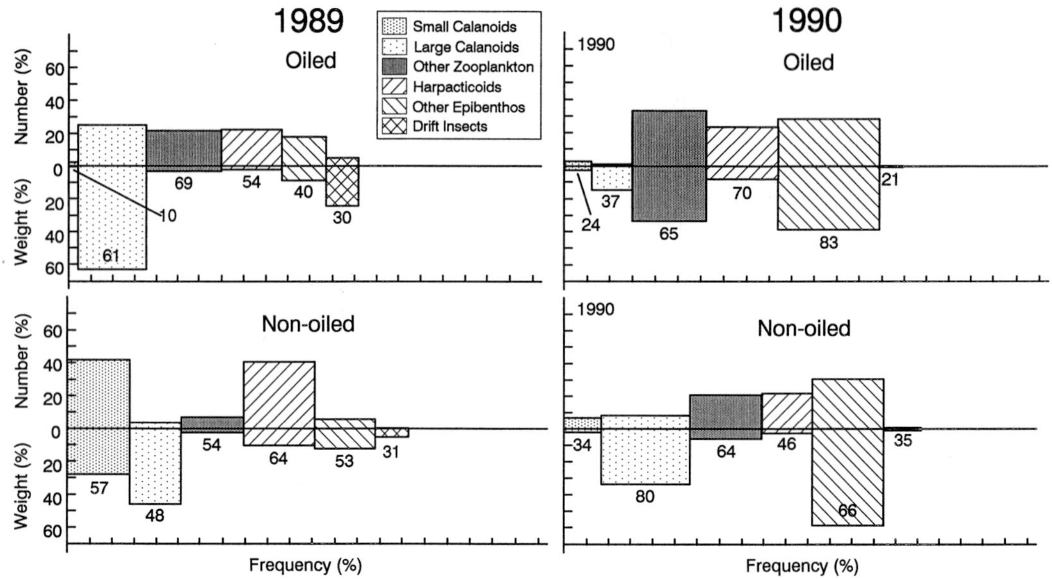

FIGURE 4.—Prey composition of 112 and 136 chum salmon from oiled and non-oiled areas in Prince William Sound, 1989 and 1990. Height of bar above line indicates percent number; height of bar below line indicates percent dry weight; and width of bar and value indicate percent frequency of occurrence (magnitude indicated below each bar). Each division on the *x*-axis represents 20%.

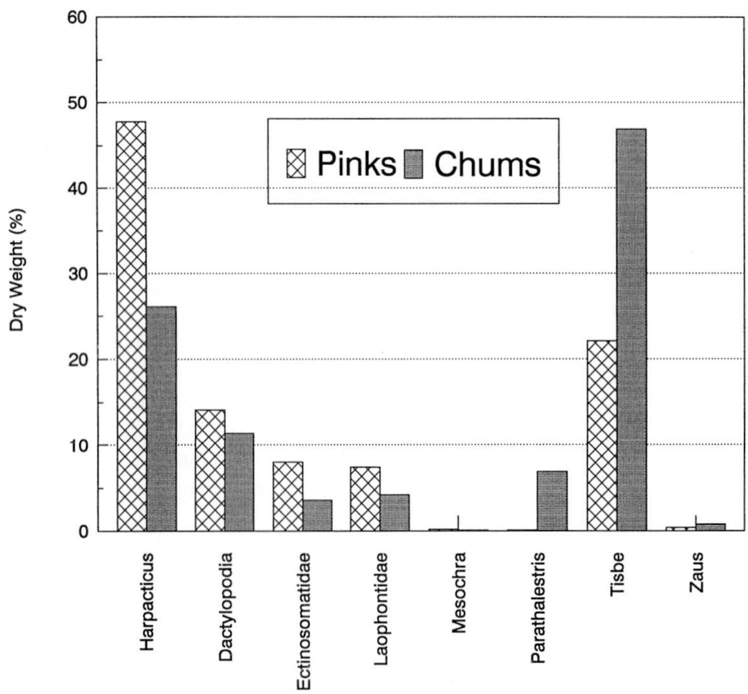

FIGURE 5.—Harpacticoid copepod composition (percent biomass of family or genus) in diets of pink and chum salmon in eight locations in Prince William Sound, 1989.

TABLE 2.—Percent dry weight of prey in the diet of 397 pink salmon fry collected in oiled and non-oiled areas of Prince William Sound, Alaska, April to June 1989, by low-, medium-, and steep-gradient habitats in bays and corridors; cal = calanoids; Zoop = pelagic zooplankton; epi = epibenthos.

	Oiled			Non-oiled		
Prey category	Low	Medium	Steep	Low	Medium	Steep
Bays						
Large cal	91.6	26.6	51.4	41.0	8.7	69.3
Small cal	5.8	8.7	37.0	46.5	62.1	10.5
Other zoop	1.6	28.8	0.0	0.2	7.2	4.5
Total zoop	99.0	64.2	88.3	87.8	78.0	84.3
Harpacticoids	0.9	1.8	0.8	4.7	17.4	0.2
Other epi	0.0	19.6	1.0	1.4	2.2	1.5
Total epi	0.9	21.5	1.8	6.2	19.6	1.7
Drift insects	0.0	14.3	9.9	6.0	2.3	14.0
Corridors						
Large cal	31.6	24.5	50.9	22.4	28.0	41.7
Small cal	13.8	43.8	23.3	13.5	3.5	30.2
Other zoop	7.9	12.1	10.5	4.1	8.2	6.7
Total zoop	53.4	80.4	84.7	40.1	39.8	78.6
Harpacticoids	35.4	14.3	6.6	26.5	44.2	6.1
Other epi	10.0	3.9	7.1	31.7	11.8	10.6
Total epi	45.4	18.2	13.8	58.2	56.0	16.7
Drift insects	1.1	1.5	1.5	1.6	4.1	4.6

TABLE 3.—Percent dry weight of prey categories in the diet of 595 pink salmon fry collected in oiled and non-oiled areas of Prince William Sound, Alaska, April to June 1990, by low-, medium-, and steep-gradient habitats in bays and corridors; cal = calanoids; zoop = pelagic zooplankton; epi = epibenthos.

	Oiled			Non-oiled		
Prey cateogry	Low	Medium	Steep	Low	Medium	Steep
Bays						
Large cal	0.0	18.1	17.8	57.6	70.2	21.5
Small cal	18.0	11.4	65.1	30.2	20.9	39.5
Other zoop	44.6	13.5	9.1	6.5	3.8	21.8
Total zoop	62.6	42.9	92.0	94.3	94.9	82.8
Harpacticoids	14.0	10.5	5.5	4.5	0.9	1.5
Other epi	4.4	5.0	1.6	1.0	0.9	5.9
Total epi	18.4	15.5	7.1	5.5	1.8	7.5
Drift insects	18.5	41.5	0.8	0.0	3.3	9.3
Corridors						
Large cal	49.8	57.8	45.8	7.8	27.9	12.6
Small cal	4.4	11.6	7.5	24.8	37.5	11.9
Other zoop	5.6	9.2	12.7	17.1	17.2	51.9
Total zoop	59.7	78.5	66.0	49.7	82.5	76.4
Harpacticoids	19.2	13.9	5.3	27.3	6.6	1.0
Other epi	18.5	2.3	23.6	21.5	10.1	21.8
Total epi	37.7	16.2	28.9	48.8	16.7	22.8
Drift insects	2.3	5.1	4.5	1.3	0.7	0.6

corridors and 50% in non-oiled corridors. Higher zooplankton consumption in the oiled area in 1989, therefore, was weighted by effects in corridors, rather than bays. In 1990, however, lower zooplankton consumption in the oiled area was weighted by lower consumption in bays, rather than corridors. The proportion of zooplankton was 58% in oiled bays and 94% in non-oiled bays, whereas it was 68% in oiled corridors and 72% in non-oiled corridors (Figure 3).

Differences in zooplankton consumption between oiled and non-oiled areas were generally consistent in the six different habitats. Compared to non-oiled habitats, percent biomass of total zooplankton consumed was higher in 1989 in all oiled habitats except medium-gradient habitats in bays (Table 2). In 1990, zooplankton was lower in all oiled habitats except steep-gradient habitats in bays and low-gradient habitats in corridors (Table 3).

Trends in consumption of epibenthos also were reversed from 1989 to 1990 (Figure 6). In 1989, all five measures of epibenthos tended to be lower in oiled areas than in non-oiled areas. In 1990, however, they tended to be higher in the oiled area; percent number and percent dry weight were significantly higher ($P = 0.005$ and 0.011, respectively). Interannual differences in use of the epibenthos in oiled versus non-oiled areas were most influenced by consumption of harpacticoid copepods. In 1989, diet parameters for harpacticoid copepods tended to be lower for the oiled area, but no comparisons were significant. In 1990, the pattern was reversed; all diet parameters for harpacticoid copepods were significantly higher for the oiled areas ($P < 0.085$).

The interannual shift in consumption of epibenthos between oiled and non-oiled areas also differed for bays and corridors. In 1989, epibenthos formed a smaller proportion of dietary weight in the oiled area in both bays and corridors. Epibenthos constituted 10% of the diet in oiled bays and 17% in non-oiled bays; it constituted 27% in oiled corridors and 47% in non-oiled corridors (Figure 2). Lower consumption of epibenthos in the oiled area in 1989, therefore, was weighted by differences in corridors. In 1990, however, greater differences in epibenthic consumption between oiled and non-oiled areas existed in bays than in corridors. The proportion of epibenthos in diets was 14% in oiled bays and 3% in non-oiled bays; it was 28% in oiled corridors and 27% in non-oiled corridors (Figure 3). In 1990, therefore, higher consumption of epibenthos in the oiled area resulted from differences in the diets of salmon from bays,

TABLE 4.—Percent dry weight of prey in the diets of 112 and 136 juvenile chum salmon in oiled and non-oiled areas of Prince William Sound, Alaska, April to June 1989 and 1990, by low- and medium-gradient habitats in bays and corridors; cal = calanoids; zoop = pelagic zooplankton; epi = epibenthos.

	Oiled			Non-oiled		
	Bay		Corridor	Bay		Corridor
Prey category	Low	Medium	Low	Low	Medium	Low
1989						
Large cal	74.3	36.4	63.6	0.0	64.7	65.0
Small cal	0.0	0.8	0.3	78.1	13.4	0.5
Other zoop	1.6	6.2	1.4	0.0	2.8	1.9
Total zoop	75.9	43.4	65.3	78.1	80.9	67.4
Harpacticoids	0.7	6.1	0.6	3.3	1.7	16.9
Other epi	6.5	47.8	0.1	13.6	1.4	14.1
Total epi	7.2	53.9	0.6	16.9	3.1	30.9
Drift insects	16.9	2.7	34.1	5.0	16.0	1.7
1990						
Large cal	3.0	0.0	36.9	69.4	92.8	19.3
Small cal	0.6	0.0	5.4	4.9	4.5	0.5
Other zoop	48.4	0.0	24.8	12.3	0.0	3.1
Total zoop	52.0	0.0	67.1	86.6	97.4	22.9
Harpacticoids	1.1	1.1	19.1	1.6	0.2	2.0
Other epi	46.0	98.8	10.2	9.8	0.3	74.2
Total epi	47.2	99.9	29.3	11.3	0.5	76.2
Drift insects	0.4	0.0	3.4	1.9	2.2	0.9

rather than differences in the diets of salmon from corridors.

Differences in consumption of epibenthos between oiled and non-oiled areas were usually consistent across habitats. In 1989, epibenthic prey were consumed less in all oiled habitats except medium- and steep-gradient habitats in bays (Table 2). In 1990, consumption of epibenthos in oiled habitats was higher or similar to that in non-oiled habitats for all except low-gradient habitats in corridors (Table 3).

In both years, drift insects were uncommon in diets and no significant difference existed between oiled and non-oiled areas. Drift insect consumption was similar among oiled and non-oiled habitats, except that insects constituted a higher percentage in medium-gradient habitats in oiled bays than in non-oiled bays in 1989 and the same was true for low- and medium-gradient habitats in 1990 (Tables 2 and 3).

Size of pink salmon did not affect comparisons of diets between oiled and non-oiled areas. Neither FL nor weight of fish used in diet studies differed between oiled and non-oiled areas in either 1989 or 1990 ($P > 0.1$). Mean FL of pink salmon was 38 mm in oiled areas and 39 mm in non-oiled areas in 1989; it was 38 mm in oiled areas and 37 mm in non-oiled areas in 1990. Fish wet weight was also similar between oiled and non-oiled areas. All sizes of pink salmon consumed both zooplankton and epibenthos. The proportion of zooplankton consumed tended to decrease as FL increased, whereas the proportion of epibenthos tended to increase as FL increased. Correlations between FL and percent dry weight of prey consumed were low but significant ($P < 0.05$) for zooplankton in 1990 ($r = -0.19$) and epibenthos in 1989 and 1990 ($r = 0.14$ and 0.18, respectively).

Oil effects: chum salmon.—Chum salmon also consumed prey resources differently in oiled and non-oiled areas, but fewer differences were significant than for pink salmon in both years. In 1989, total zooplankton consumption in the oiled area was not significantly different from the non-oiled area (Figure 7). Zooplankton was 62% of overall prey biomass in the oiled area and 59% in the non-oiled area. In 1990, zooplankton composition measures (except number of organisms) tended to be lower in the oiled area; percent number and percent dry weight of zooplankton were significantly ($P = 0.052$ and 0.076) lower in the oiled area (Figure 7). Zooplankton averaged 41% and 69% of prey biomass in oiled and non-oiled areas, respectively.

Trends in use of prey resources between oiled and non-oiled areas were also mirrored by prey subcategories. Although no difference between overall zooplankton consumption was observed in 1989, calanoids tended to be eaten less in the oiled area than in the non-oiled area in both 1989 and 1990. These differences were not significant in 1989, but in 1990 total calanoid measures (except number of organisms) were significantly lower in the oiled area ($P < 0.076$).

Differences between diets of chum salmon from oiled and non-oiled areas were caused by small calanoids in 1989 and large calanoids in 1990, the opposite of the pattern for pink salmon. Small calanoid composition measures tended to be lower in the oiled area, but only percent frequency of occurrence was significantly lower ($P = 0.059$). No significant difference existed for consumption of large calanoids in 1989. Conversely, no difference in small calanoid measures was observed in 1990; however, all large calanoid measures were significantly lower in the oiled area ($P < 0.059$) in 1990.

Differences in consumption of zooplankton between oiled and non-oiled areas were not consistent across habitats. In 1989, chum salmon consumed similar proportions of zooplankton in oiled and

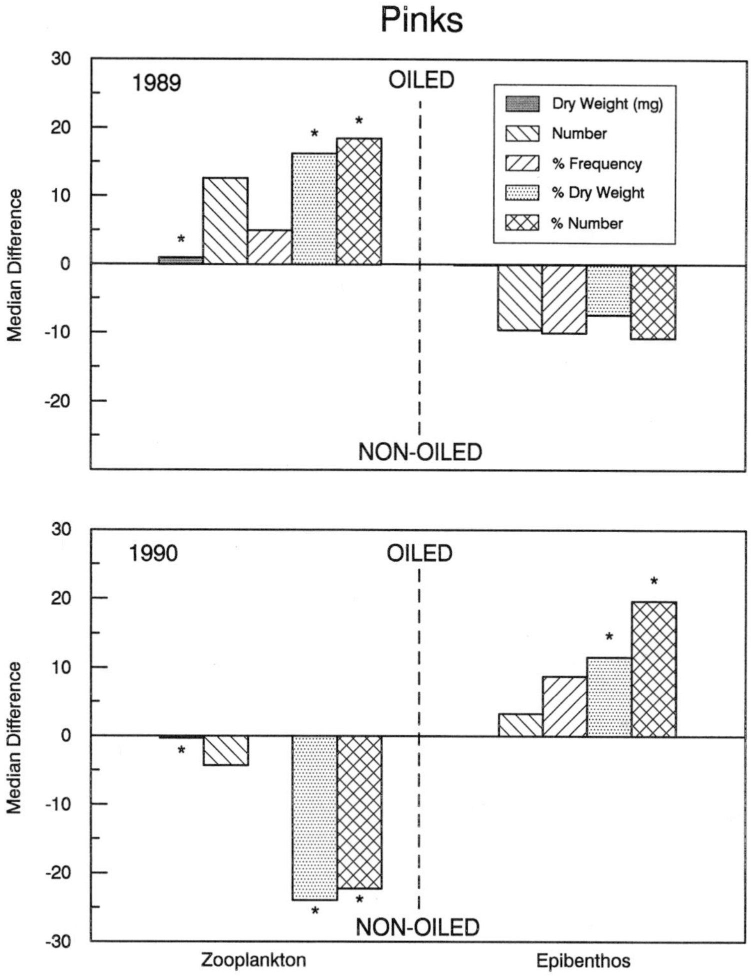

FIGURE 6.—Wilcoxon median differences in prey composition for pink salmon between oiled and non-oiled areas of Prince William Sound, 1989 and 1990. A negative median difference indicates non-oiled value > oiled value. An asterisk indicates a significant difference ($P < 0.1$).

non-oiled low-gradient habitats in both bays and corridors. In contrast, they consumed less zooplankton in oiled than in non-oiled medium-gradient habitats in bays (Table 4). In 1990, zooplankton consumption was substantially lower in the oiled area for both low-and medium-gradient habitats in bays, but was substantially higher in the oiled area for low-gradient habitats in corridors; the difference in the two bay habitats, therefore, accounted for significantly lower zooplankton use in the oiled area.

The pattern of consumption of epibenthos between oiled and non-oiled areas also reversed from 1989 to 1990 (Figure 7). In 1989, a weak trend for differential use of epibenthic prey was observed. Four of five epibenthic composition measures tended to be lower in the oiled area, but the differences were not significant. Epibenthos averaged 21% and 26% of the diet weight in the oiled and non-oiled areas, respectively. In 1990, epibenthic composition measures tended to be higher in the oiled area, but these differences were not significant. Epibenthos averaged 55% and 27% of diet weight in oiled and non-oiled areas, respectively. Interannual changes in consumption of epibenthos were the result of changes in consumption of harpacticoid copepods. Harpacticoids tended to be eaten less in the oiled area in 1989, but more in the oiled area in 1990. In 1990, number and dry weight of harpacticoids in the diet were significantly higher in the oiled area ($P < 0.063$).

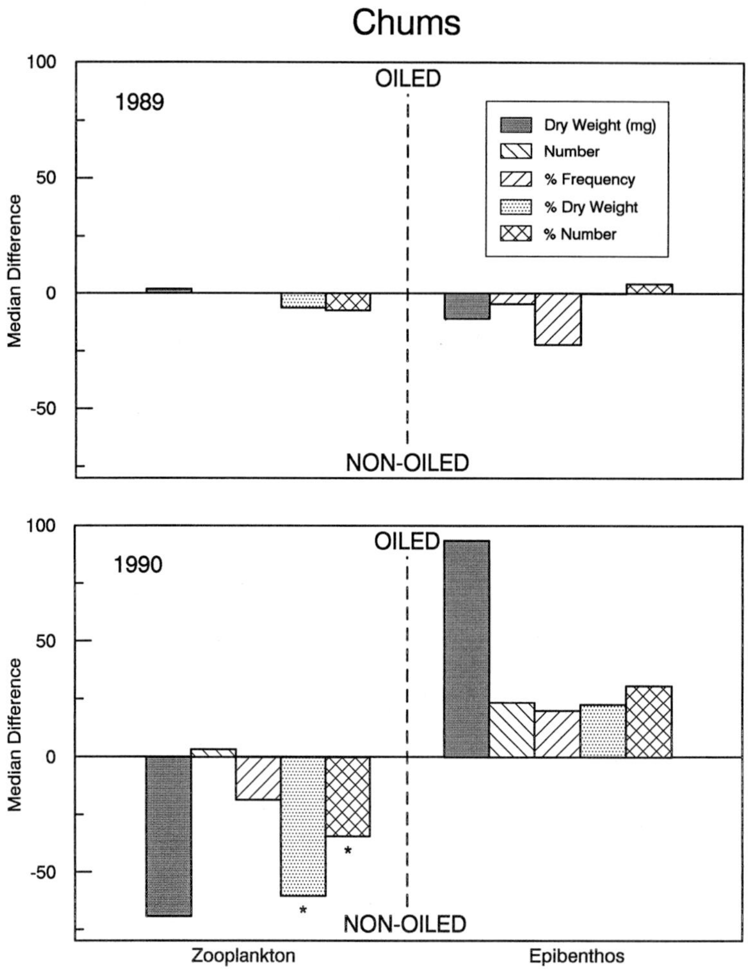

FIGURE 7.—Wilcoxon median differences in prey composition for chum salmon between oiled and non-oiled areas of Prince William Sound, 1989 and 1990. A negative median difference indicates non-oiled value > oiled value. An asterisk indicates a significant difference ($P < 0.1$).

Different proportions of epibenthos were consumed by chum salmon in different habitats (Table 4). Compared to non-oiled areas, smaller proportions of epibenthos were consumed in 1989 in low-gradient habitats in both oiled bays and oiled corridors, but larger proportions were consumed in medium-gradient habitats in oiled bays. The 1989 trend toward lower epibenthic consumption in the oiled compared to the non-oiled area, therefore, resulted from differences in low-gradient habitats. In 1990, however, epibenthos was higher in the oiled area for low- and medium-gradient habitats in bays, but was lower in low-gradient habitats in oiled corridors. Differences in bay habitats in 1990, therefore, accounted for the trend toward greater use of epibenthos in the oiled area.

No significant difference in use of drift insects was observed in 1989 or 1990. In 1989, the proportion of drift insects tended to be higher in oiled low-gradient habitats in both bays and corridors, but tended to be lower in oiled medium-gradient habitats in bays. Proportions were similar in all oiled habitats compared to non-oiled habitats in 1990 (Table 4).

Size of chum salmon differed between oiled and non-oiled areas, but the size difference probably did not affect diet comparisons because size and diet were not strongly correlated. Weight and FL of chum salmon were greater in the oiled area in both 1989 ($P = 0.036$) and 1990 ($P = 0.076$). Mean FL in 1989 was 48 mm and 39 mm in oiled and non-oiled areas, respectively. In 1990, mean FL was 52 mm in the oiled area and 42 mm in the non-oiled area. Chum salmon of all sizes ate both planktonic and epibenthic prey. Correlations between FL and prey composition were significant only for zooplankton in 1989. Percent dry weight of zooplankton decreased with increasing chum FL ($r = -0.25$; $P < 0.05$).

Oil Ingestion

Oil sheen and globules of tarry material were observed in the stomachs of salmon from the oiled area in 1989. One pink salmon stomach contained sheen and one had tar globules (0.7% of total). Three chum salmon stomachs (4.5%) contained sheen. No sheen or tar globules were observed in pink or chum salmon stomachs from non-oiled sites. The proportion of chum salmon stomachs containing oil residue was significantly greater than the proportion of pink salmon stomachs containing oil residue ($P = 0.001$; Fisher's exact test). In 1990, no evidence of oil was observed in any salmon stomachs. Fish that had ingested oil in 1989 were from both bays and corridors from late April to mid-June. Pink salmon were from steep-gradient habitats, and chum salmon were from low-and medium-gradient habitats. Only one of the five fish had an empty stomach; the other four had consumed pelagic, epibenthic, and drift insect prey.

Discussion

We found no indication that food consumption was lower in the oiled area compared to the non-oiled area for either pink or chum salmon. Fish fed similarly with or without oil present in both planktonic and epibenthic foraging areas. Stomachs were as full, as much total prey was consumed, and empty stomachs occurred as infrequently in the oiled area as in the non-oiled area in both 1989 and 1990. Indeed, some measures of food consumption were higher for pink salmon in oiled areas than in non-oiled areas in the year of the spill. However, lack of pre-spill information prevents a direct assessment of the potential change in food consumption within either area after the spill.

Although the quantity of food eaten was not lower in the presence of oil, we did find evidence that oil was ingested at all three habitat types in both bays and corridors in the oiled area in 1989. We attributed this oil to the *Exxon Valdez* spill because no oil was observed in stomachs from non-oiled areas in 1989 or from any 1990 collections. Although the incidence of oil observed in the stomachs was low, the occurrence of any observable oil supports the conclusion (Carls et al. 1996a) that ingestion was a route of contamination of these fish in Prince William Sound in 1989. The rate of contamination cannot be inferred by the incidence of observation because oil may be consumed intermittently or may not be recognizable as it is digested or if it is contained in prey.

Ingestion of oil can directly affect both feeding and growth rates of juvenile salmon. In laboratory studies, food contaminated with the water-soluble fraction and whole crude oil reduced pink salmon growth and feeding at high doses (Schwartz 1985; Carls et al. 1996b, this volume). At lower doses, however, consumption of oil-contaminated food reduced growth without lowering food consumption in both pink salmon and Atlantic salmon *Salmo salar* by reducing food conversion efficiency (Vignier et al. 1992; Carls et al. 1996b). Rice et al. (1977) speculated that energy demands remain elevated throughout the process of hydrocarbon depuration. Thus, the ingestion of oil in contaminated habitats in Prince William Sound, even at levels sufficient to cause physiological effects (e.g., reduced growth), is compatible with our observation that food consumption was not lowered in the oiled area.

We observed a shift in juvenile salmon diet composition between oiled and non-oiled areas from 1989 to 1990. A possible explanation for this change between years is that fish avoided epibenthic prey in contaminated nearshore habitats in 1989, and instead, shifted their feeding toward zooplankton prey. This argument is supported by two observations. First, the proportion of zooplankton consumed by both pink and chum salmon decreased from 1989 to 1990 in the oiled area, even though zooplankton was more abundant in the second year (Celewycz and Wertheimer 1996, this volume). Second, the epibenthic prey category was a smaller proportion of salmon diets in the oiled area in 1989, even though abundance of epibenthos was higher in the oiled area and abundance of zooplankton did not differ between oiled and non-oiled areas in 1989 (Celewycz and Wertheimer 1996).

Avoidance of epibenthic prey in contaminated habitats in 1989, however, is not supported by other observations in Prince William Sound or by laboratory feeding studies. Distribution of juvenile salmon in 1989 and 1990 did not indicate avoidance of oiled habitats (Wertheimer and Celewycz 1996, this volume). Furthermore, although consumption of epibenthic prey in the oiled area increased overall from 1989 to 1990, consumption of epibenthic prey in some oiled habitats in 1989 equaled or exceeded the proportion consumed in non-oiled habitats. The same hydrographic conditions that caused oiling may also have caused greater deposition of detritus, resulting in higher epibenthic production in these habitats. We were unable to assess potential changes in production or food consumption at specific habitats as a result of the spill. Nevertheless, in laboratory studies, juvenile salmon consumed oiled food at equal or greater rates than non-oiled food even at contamination doses that significantly reduced growth (Schwartz 1985; Vignier et al. 1992; Carls et al. 1996b).

An alternate explanation for the shifts in diet composition between 1989 and 1990 is that enhanced abundance of epibenthic prey influenced diet composition in 1990. In that year, both species ate proportionately more epibenthos in the oiled area than in the non-oiled area. Densities of harpacticoid copepods were higher on heavily oiled beaches in western Prince William Sound in 1990 (Wertheimer et al. 1996, this volume). We have no direct comparison with 1989 estimates of the epibenthic prey biomass available in foraging areas, however, because epibenthic prey were not sampled at the habitats where salmon diet studies were conducted in 1990.

The diets of juvenile pink and chum salmon were dominated by planktonic zooplankton, primarily calanoid copepods, in both oiled and non-oiled nearshore areas of western Prince William Sound in 1989 and 1990. Both neritic and oceanic calanoid resources were common. Although calanoids and other zooplankters are common in diets of juvenile salmon throughout their range (Heard 1991), epibenthic organisms, especially harpacticoid copepods, have been reported as the primary initial prey of salmon in many estuaries and nearshore marine habitats (Healey 1979; Godin 1981; Landingham 1982; Volk et al. 1984; Webb 1991). In Alaska waters, however, planktonic prey often predominate in the early diet of both pink and chum salmon. Epibenthic prey were not common in fish stomachs throughout the period of outmigration in an inlet in southeast Alaska (Bailey et al. 1975). High consumption of zooplankton was also reported from southwestern Prince William Sound, but salmon fed more on harpacticoids during their initial bay residency than in the corridor outside it (Cooney et al. 1981). In our study, zooplankton predominated in the diet of salmon in both bays and corridors.

Salmon diets differ by beach gradient. In studies reporting initial dependence of pink and chum salmon fry on harpacticoids in the Pacific Northwest, fish were sampled at beaches with substantial epibenthic production, habitats similar to our low- or medium-gradient habitats (e.g., Kaczynski et al. 1973; Godin 1981; Webb 1991). We also found that higher proportions of epibenthos were consumed by both species in low- and medium-gradient habitats than in steep-gradient habitats, reflecting the higher diversity of epibenthic taxa and the higher abundance of harpacticoids in these lower-gradient habitats (Celewycz and Wertheimer 1996). Information on the relative availability of zooplankton in different habitats was not available. In other Alaska studies, also, different prey were consumed in different types of habitat. Zooplankton (primarily calanoid copepods) predominated in the diets of pink salmon in Tutka Bay at steep-gradient beaches, while epibenthos (primarily harpacticoid copepods) predominated in the diets at low- and medium-gradient beaches (Kron and Yuen 1978). Low-gradient habitats were not available to salmon foraging in Traitor's Cove, Alaska, and diets predominately consisted of zooplankton (Bailey et al. 1975). In our study, zooplankton predominated in the diets of both pink and chum salmon in all habitat types, although some habitat specificity existed.

Because of their propensity to forage more heavily on the abundant epibenthos in lower-gradient habitats, chum salmon were more susceptible than pink salmon to hydrocarbon exposure in the oiled area. Oil deposits from the *Exxon Valdez* remained longer and natural cleaning processes occurred more slowly in protected low-energy beaches (Owens 1991), which would include our lower-gradient habitats. Chum salmon in oiled habitats generally had higher levels of mixed-function oxidase induction than did pink salmon (Carls et. al 1996a), which also supports the idea that they were more susceptible to hydrocarbon exposure. However, because oil was mixed in stomachs with prey from all three production systems, we could not determine which prey component fish were feeding on when the oil was ingested.

We conclude that oil did not interfere with the quantity of food consumed by pink and chum salmon in Prince William Sound, even though some fish ingested oil either directly or via contaminated prey. Changes in diet composition between oiled and non-oiled areas in 1989 and 1990 may have been caused by the oil spill or may have been caused by interannual or geographic variations in prey availability.

Acknowledgments

We are grateful for the assistance of S. Gish, M. Pritchett, J. Orsi, D. Mortensen, H. Jaenicke, M. Carls, B. Kirkpatrick, J. Cordell, S. McGregor, J. Landingham, R. Young, M. Murphy, and anonymous reviewers.

References

Bailey, J. E., B. L. Wing, and C. R. Mattson. 1975. Zooplankton abundance and feeding habits of fry of pink salmon and chum salmon in Traitor's Cove, Alaska, with speculations on the carrying capacity of the area. U.S. National Marine Fisheries Service Fishery Bulletin 73:946–961.

Bodin, P. 1988. Results of ecological monitoring of three beaches polluted by the *Amoco Cadiz* oil spill: development of meiofauna from 1978 to 1984. Marine Ecology Progress Series 42:105–123.

Bowen, S. H. 1983. Quantitative description of the diet. Pages 325–336 *in* L. A. Nielsen and D. L. Johnson, editors. Fisheries techniques. American Fisheries Society, Bethesda, Maryland.

Caillet, G. M. 1976. Several approaches to the feeding ecology of fishes. Pages 1–13 *in* C. A. Simenstad and S. J. Lipovsky, editors. Proceedings, fish food habits studies: first Pacific Northwest technical workshop, WSG-WO 77-2. Washington Sea Grant, University of Washington, Seattle.

Carls, M. G., and six coauthors. 1996b. Growth, feeding, and survival of pink salmon fry exposed to food contaminated with crude oil. American Fisheries Society Symposium 18:608–618.

Carls, M. G., A. C. Wertheimer, J. W. Short, R. M. Smolowitz, and J. J. Stegeman. 1996a. Contamination of juvenile pink and chum salmon by hydrocarbons in Prince William Sound after the *Exxon Valdez* oil spill. American Fisheries Society Symposium 18: 593–607.

Celewycz, A. G., and A. C. Wertheimer. 1996. Prey availability to juvenile salmon after the *Exxon Valdez* oil spill. American Fisheries Society Symposium 18:564–577.

Conover, R. J. 1971. Some relations between zooplankton and bunker C oil in Chedabucto Bay following the wreck of the tanker *Arrow*. Journal of the Fisheries Research Board of Canada 28:1327–1330.

Conover, W. J. 1980. Practical nonparametric statistics. Wiley, New York.

Cooney, R. T. 1986. Zooplankton. Pages 285–304 *in* D. W. Hood and S. T. Zimmerman, editors. The Gulf of Alaska physical environment and biological resources. U.S. Department of the Interior Minerals Management Service OCS (Outer Continental Shelf) Study, MMS 86-0095, U.S. Government Printing Office, Washington, DC.

Cooney, R. T. 1990. Prey fields and the feeding behavior and growth of pink salmon fry released from the Armin F. Koening Hatchery at Sawmill Bay, Evans Island, Alaska: May and June 1989. Natural Resources Damage Assessment Status Report, Study 4, University of Alaska Fairbanks Component, Fairbanks.

Cooney, R. T., D. Urquhart, and D. Barnard. 1981. The behavior, feeding biology, and growth of hatchery released pink and chum salmon fry in Prince William Sound, Alaska. University of Alaska, Sea Grant Report 81-5, Fairbanks.

Cordell, J. R. 1986. Structure and dynamics of an epibenthic harpacticoid assemblage and the role of predation by juvenile salmon. Master's thesis. University of Washington, Seattle.

Feder, H. M., A. S. Naidu, and A. J. Paul. 1990. Trace element and biotic changes following a simulated oil spill on a mudflat in Port Valdez, Alaska. Marine Pollution Bulletin 21:131–137.

Feller, R. J., and V. W. Kaczynski. 1975. Size selective predation by juvenile chum salmon (*Oncorhynchus keta*) on epibenthic prey in Puget Sound. Journal of the Fisheries Research Board of Canada 32:1419–1429.

Godin, J. J. 1981. Daily patterns of feeding behavior, daily rations, and diets of juvenile pink salmon (*Oncorhynchus gorbuscha*) in two marine bays of British Columbia. Canadian Journal Fisheries and Aquatic Sciences 38:10–15.

Gosho, M. E. 1977. The food and feeding habits of juvenile pink salmon in the estuaries of Kodiak Island, Alaska. Master's thesis. University of Washington, Seattle.

Healey, M. C. 1979. Detritus and juvenile salmon production in the Nanaimo estuary: 1. Production and feeding rates of juvenile chum salmon (*Oncorhynchus keta*). Journal of the Fisheries Research Board Canada 36:488–496.

Heard, W. H. 1991. Life history of pink salmon. Pages 119–231 *in* C. Groot and L. Margolis, editors. Pacific salmon life histories. University of British Columbia Press, Vancouver.

Hyslop, E. J. 1980. Stomach contents analysis—a review of methods and their application. Journal of Fish Biology 17:411–429.

Kaczynski, V. W., R. J. Feller, and C. Clayton. 1973. Trophic analysis of juvenile pink and chum salmon in Puget Sound. Journal of the Fisheries Research Board of Canada 30:1003–1008.

Kron, T., and H. Yuen. 1978. Emigration and feeding of juvenile pink salmon *Oncorhynchus gorbuscha*, in relation to zooplankton abundance in Tutka Bay, Alaska, 1975. Alaska Department of Fish and Game, Lower Cook Inlet Data Report 85-5, Anchorage.

Landingham, J. H. 1982. Feeding ecology of pink and chum salmon fry in the nearshore habitat of Auke Bay, Alaska. Master's thesis. University of Alaska, Juneau.

Landingham, J. H., and P. D. Mothershead. 1988. Feeding habits of juvenile pink salmon in nearshore and offshore areas of Auke Bay. University of Alaska Fairbanks, School of Fisheries and Ocean Sciences, APPRISE (Association of Primary Production and Recruitment in a Subarctic Ecosystem) Annual Report, SFOS APP87-100, Fairbanks.

Mortensen, D. M., J. H. Landingham, A. C. Wertheimer,

and S. G. Taylor. 1991. Relationship of early marine growth and survival of juvenile pink salmon to marine water temperature and secondary production in Auke Bay, Alaska. *In* I. Guthrie and B. Wright, editors. Proceedings, fifteenth northeast Pacific pink and chum salmon workshop. Pacific Salmon Commission, Parksville, British Columbia.

Murphy, M. L., J. F. Thedinga, and K. K. Koski. 1988. Size and diet of juvenile Pacific salmon during seaward migration through a small estuary in southeastern Alaska. U.S. National Marine Fisheries Service Fishery Bulletin 86:213–222.

Owens, E. H. 1991. Changes in shoreline oiling conditions 1–1/2 years after the 1989 Prince William Sound spill. Woodward-Clyde, Seattle.

Parker, R. R. 1968. Marine mortality schedules of pink salmon of the Bella Coola River, central British Columbia. Journal of the Fisheries Research Board of Canada 25:757–794.

Pinkas L., M. S. Oliphant, and I. L. K. Iverson. 1971. Food habits of albacore, bluefin tuna, and bonito in California waters. California Fish and Game Fish Bulletin 152:1–83.

Rice, S. D., R. E. Thomas, and J. W. Short. 1977. Effect of petroleum hydrocarbons on breathing and coughing rates and hydrocarbon uptake-depuration in pink salmon fry. Pages 259–277 *in* F. J. Vernberg, A. Calabrese, P. Thurberg, and W. B. Vernberg, editors. Physiological responses of marine biota to pollutants. Academic Press, New York.

Schwartz, J. P. 1985. Effects of oil-contaminated prey on the feeding and growth rate of pink salmon fry *Oncorhynchus gorbuscha*. Pages 459–476 *in* F. Vernberg, R. Thurberg, A. Calabrese, and W. Vernberg, editors. Pollution and physiology of marine organisms. University of South Carolina Press, Columbia.

Sibert, J. R., T. J. Brown, M. C. Healey, and B. A. Kask. 1977. Detritus-based food webs: exploitation by juvenile chum salmon (*Oncorhynchus keta*). Science 106: 649–650.

Vignier, V., J. H. Vandermeulen, and A. J. Fraser. 1992. Growth and food conversion by Atlantic salmon parr during 40 days' exposure to crude oil. Transactions of the American Fisheries Society 121:322–332.

Volk, E. C., R. C. Wissmar, C. A. Simenstad, and D. M. Eggers. 1984. Relationship between otolith microstructure and the growth of juvenile chum salmon (*Oncorhynchus keta*) under different prey rations. Canadian Journal of Fisheries and Aquatic Sciences 41:126–133.

Webb, D. G. 1991. Effect of predation by juvenile Pacific salmon on marine harpacticoid copepods. I. Comparisons of patterns of copepod mortality with patterns of salmon consumption. Marine Ecology Progress Series 72:25–36.

Wertheimer, A. C., N. J. Bax, A. G. Celewycz, M. G. Carls, and J. H. Landingham. 1996. Harpacticoid copepod abundance and population structure in Prince William Sound, one year after the *Exxon Valdez* oil spill. American Fisheries Society Symposium 18:551–563.

Wertheimer, A. C., and A. G. Celewycz. 1996. Abundance and growth of juvenile pink salmon in oiled and non-oiled locations of western Prince William Sound after the *Exxon Valdez* oil spill. American Fisheries Society Symposium 18:518–532.

American Fisheries Society Symposium 18:593–607, 1996

Contamination of Juvenile Pink and Chum Salmon by Hydrocarbons in Prince William Sound after the *Exxon Valdez* Oil Spill

MARK G. CARLS, ALEX C. WERTHEIMER, JEFFREY W. SHORT, ROXANNA M. SMOLOWITZ, AND JOHN J. STEGEMAN

National Oceanic and Atmospheric Administration, National Marine Fisheries Service
Auke Bay Laboratory, Alaska Fisheries Science Center
11305 Glacier Highway, Juneau, Alaska 99801, USA

Abstract.—*Juvenile pink salmon Oncorhynchus gorbuscha* and chum salmon *O. keta* were collected from oiled nearshore marine habitats in Prince William Sound in 1989 and 1990 to determine if they were contaminated by exposure to *Exxon Valdez* crude oil. For comparison, fish were also collected from non-oiled, control areas located as close as possible to oiled areas. Mussels were routinely sampled to obtain site-specific measures of biological availability of hydrocarbons. Mussels were contaminated with polynuclear aromatic hydrocarbons (PAHs) in oiled areas in 1989 (6,942 ± 1,427 ng/g); much lower concentrations were observed in control locations (25 ± 13 ng/g). Contamination of mussel tissues persisted into 1990, but at greatly reduced concentrations. In 1989, PAH concentrations were significantly greater in pink salmon tissues from oiled areas (181 ± 28 ng/g) than from non-oiled areas (54 ± 8 ng/g). In 1990, PAH concentrations (51 ± 7 ng/g) were not significantly different in oiled and non-oiled locations. Most of the PAHs detected in pink salmon and mussel tissues were also found in *Exxon Valdez* crude oil. As may be expected after a single perturbation, PAH concentrations in mussels and pink salmon increased to peak levels after the spill and then declined. Cytochrome P4501A was significantly induced in juvenile pink and chum salmon from oiled locations in 1989 but not in 1990. Based on these observations, we conclude that *Exxon Valdez* crude oil caused these differential changes in PAH concentrations and induced cytochrome P4501A enzyme activity. The degree of contamination of visceral tissues and the frequency of tissue types staining for P4501A indicated that ingestion of whole oil was an important route of contamination for juvenile pink and chum salmon. We believe contamination of juvenile salmon was physiologically significant and contributed to the reduced growth observed in juvenile pink salmon from oiled areas of Prince William Sound in 1989.

When the *Exxon Valdez* grounded on Bligh Reef on 24 March 1989, the living marine resources of a large ecosystem were placed in jeopardy from direct and indirect impacts of the resulting oil spill. The effect of the oil on the salmon resources of the region was of particular concern because of their enormous biological, economic, recreational, and cultural value. Because early marine residency is a critical phase in the life history of salmon and significantly affects year-class strength (Parker 1968; Walters et al. 1978; Bax 1983; Nichelson 1986), we focused on juvenile salmon, hereafter referred to as fry.

Oil in the marine environment can affect fry in a variety of ways. It can be directly toxic to salmon fry, which are especially susceptible when first in seawater (Rice et al. 1975, 1984). Sublethal levels of hydrocarbons can affect metabolism, reduce growth of fry (Rice et al. 1975), and damage olfactory lamellar surfaces, conceivably impacting migratory behavior and feeding patterns (Babcock 1985). Reduction in growth of salmon fry can be caused by direct exposure to water-soluble fractions (WSF) of oil (Moles and Rice 1983), contamination of prey by WSF (Schwartz 1985), and direct ingestion of crude oil (Carls et al. 1996, this volume).

In 1989, the fry in the spill area encountered an environment with large quantities of oil in both the water and the sediments of the nearshore habitats that these fish utilize. The oil spill occurred just before entry of pink salmon fry from freshwater habitats and hatcheries into the marine environment of Prince William Sound. Because the primary migration route of salmon fry from the sound to the Gulf of Alaska (ADFG 1989) probably coincided with the route that the oil traveled as it dispersed into the Gulf of Alaska, large numbers of salmon fry from areas not directly affected by oil were also potentially exposed.

The purpose of this study was to determine if salmon fry inhabiting the nearshore marine environment of Prince William Sound were contaminated by exposure to *Exxon Valdez* crude oil in 1989 and whether such contamination continued in 1990 because of the persistence of oil in the nearshore environment. We focused on pink salmon *Oncorhynchus gorbuscha*, the most abundant salmonid in the region, but also studied chum salmon *O. keta*. Fry tissues were analyzed for the presence of hydrocarbons as a direct measure of contamination and for the induction of the enzyme cytochrome P4501A as an indirect measure of exposure and contamination by polynuclear aromatic hydrocarbons (PAHs) (Stegeman 1990). Classification of collection sites as reference or oiled was judged by analysis of hydrocarbons from tissues of the mussel *Mytilus trossulus*.

Methods

Site Selection

Eight locations in western Prince William Sound were chosen for study based on the distribution of *Exxon Valdez* crude oil in early April 1989. Four of these locations were oiled (Herring Bay, Snug Harbor, Knight Island Passage, and Prince of Wales Passage) and four were not oiled (McClure Bay, Long Bay, Culross Passage, and Wells Passage) (Figure 1). We chose non-oiled reference locations as close as possible to oiled locations and attempted to match site characteristics. Non-oiled locations were clustered in the northwest region of Prince William Sound, on or close to the mainland, whereas oiled locations were generally more southerly and on islands (Figure 1). Pink salmon, mussel, and sediment samples were collected for hydrocarbon analysis during five trips between 10 April and 26 June 1989 and during four trips between 16 April and 14 June 1990. Field blanks (jars that were opened and closed empty at sample collection sites) were regularly collected and frozen for quality control of jars, methods, storage, and processing.

Animal Collection and Handling

To evaluate the degree of hydrocarbon contamination at each location, mussel samples were collected throughout the sampling period. Sufficient quantities of whole mussels required to yield at least 10 g of tissue were collected by hand, placed in hydrocarbon-free glass jars, and frozen until analysis for hydrocarbon content. Mussel shell lengths ranged from 17.2 to 49.9 mm.

Pink salmon fry were collected in nearshore habitats by beach seine or dip net. Because sheen and tar balls were often trapped in the seine at the oiled locations in 1989, care was taken to herd the fry away from such sources of external contamination. Contaminated nets were not used to collect fish in non-oiled areas. Our intent was to collect sufficient numbers for triplicate samples of approximately 10 g of fry at each location on each sampling trip. When numbers of fry collected by seine at routinely sampled beaches were insufficient, additional fry were collected by seine or dip net from other littoral sites within the sampling location, as available. Live fry were maintained in buckets of seawater, transported back to the ship, placed in hydrocarbon-free glass jars, and frozen until analysis for hydrocarbons. Chum salmon fry were not collected for hydrocarbon analysis. Because a freezer was not available on the first sampling trip (April 1989), samples were packed in ice until they could be frozen. Samples of pink and chum salmon fry were also preserved in buffered 10% formalin for analysis of cytochrome P4501A enzyme induction. Pink salmon lengths ranged from a minimum of 28 mm in April to a maximum of 94 mm in June; mean lengths increased from 32.4 ± 0.1 mm in April to 47.1 ± 0.2 mm in June. Chum salmon lengths ranged from a minimum of 30 mm in April to a maximum of 85 mm in June; mean lengths increased from 36.1 ± 0.1 mm in April to 45.0 ± 0.2 mm in June.

To ensure that hydrocarbons detected in pink salmon tissues did not come from external contamination, a subset of pink salmon samples collected from oiled locations in 1989 (N = 17, involving approximately 797 fry) were dissected with hydrocarbon-free equipment and separated into visceral and carcass subsamples; the carcass samples contained integument and muscle tissues sans heads. We hypothesized that if observed hydrocarbon contamination were merely an external artifact of sampling in polluted water where sheen and mousse were often present, viscera would show little or no contamination relative to carcasses. Separated samples spanned the period 15 April through 21 May 1989 (22–58 days after the spill).

Hydrocarbon Analysis

Analysis of samples for hydrocarbon content was determined by gas chromatography followed by mass spectrometry for aromatics or flame ionization detection for aliphatics at either of two independent laboratories: the Geochemical and Environmental Research Group (GERG) at Texas A&M University, and the National Marine Fisheries Service (NMFS), Auke Bay Laboratory (Short et al. 1996, this volume). The majority of samples were analyzed by GERG. Data for an unweathered *Exxon Valdez* crude oil sample were supplied by NMFS. In each sample, hydrocarbon concentrations were ac-

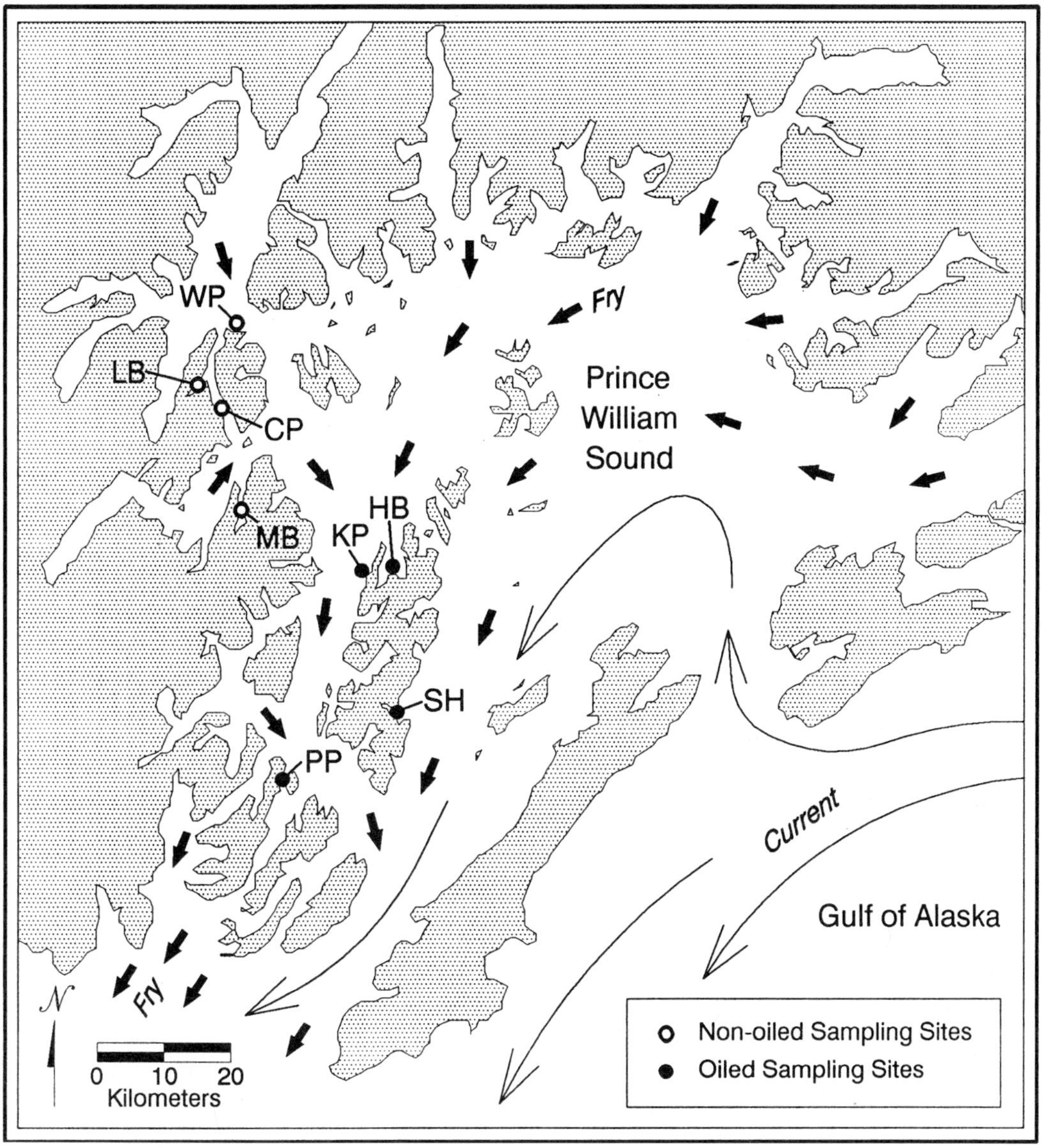

FIGURE 1.— Map of Prince William Sound, depicting sample locations and hypothesized migratory routes of salmon fry (short, solid arrows) from the sound to the Gulf of Alaska. Non-oiled locations (open circles) were Culross Passage (CP), Long Bay (LB), McClure Bay (MB), and Wells Passage (WP); oiled locations (solid circles) were Herring Bay (HB), Knight Island Passage (KP), Prince of Wales Passage (PP), and Snug Harbor (SH).

cepted only if corresponding deuterated internal standards ranged from 30 to 150%; records with concentrations outside these limits were deleted (7% of the data). Minimum detection limits (MDL) for each hydrocarbon were reported by the analytical laboratories; concentrations below the MDL were set to zero.

Polynuclear aromatic hydrocarbons (PAHs) were chosen as the primary hydrocarbon measurement in this study because (1) PAHs are potentially toxic to fish and other marine life, (2) PAHs are not normally present in Prince William Sound, and (3) natural sources of alkanes are present in the sound (Karinen et al. 1993). All aromatic hydrocarbons reported by the analytical laboratories were included in the filtered PAH concentration; of these, only 1.5% ± 0.8 (mussels, $N = 53$) or 3.8% ± 0.6 (pink salmon, $N = 95$) did not occur in *Exxon Valdez* crude oil. Reported conclusions were not changed when unfiltered PAH or filtered concentrations restricted to only those aromatics present in *Exxon Valdez* crude oil were used as the hydrocarbon measurement. Concentration changes in various subcomponents of PAHs (naphthalenes, fluorenes, phenanthrenes, dibenzothiophenes, and chrysenes) were similar to changes in PAHs, thus the more generalized PAH measurement was used to summarize the data. The mean PAH concentration in field blanks analyzed for hydrocarbon content was 5 ± 1 ng/g ($N = 11$).

Concentrations of hydrocarbons in mussel and pink salmon samples were analyzed with single-factor (oil code) analysis of variance (ANOVA). Because variance was proportional to the means, data were log transformed before analysis ($\log_e$ [concentration + 1]). Outcomes of all statistical tests were judged significant where $P \leq 0.05$. Numerical results are reported as means ± SE.

Hydrocarbons detected in the carcass and visceral tissues of pink salmon were analyzed with two-factor (oil code and tissue type) ANOVA. Data were log transformed before analysis. Significance of differences between treatment means was determined with a least-significant-difference option (SAS 1989). In an additional test designed to verify that observed differences in PAH concentration between carcass and visceral tissues at oiled locations were not caused by natural differences in tissue concentration, data from oiled locations were reanalyzed with single-factor (tissue type) ANOVA after subtracting mean concentrations at non-oiled locations from the concentrations observed at oiled locations (by tissue type as appropriate). One of the non-oiled visceral subsamples was apparently contaminated, possibly as a result of procedural errors, and was not included in the analysis. Concentrations of individual PAH compounds in fry tissues were compared to those in *Exxon Valdez* crude oil by normalizing all concentrations in each respective sample to the total PAH concentration in that sample.

Immunohistochemistry

Samples of pink and chum salmon fry preserved in formalin were analyzed blind for cytochrome P4501A induction. Approximately six fish per sample group were processed. Fry were sectioned with standard histologic techniques and stained with monoclonal antibody 1–12-3 and a peroxidase-labeled second antibody according to the procedures of Smolowitz et al. (1991). Slides were stained in duplicate for both specific antibody and control antibody. Staining intensity and occurrence were determined for tissues from each fry. Staining intensity is the density of the peroxide-stained cells within a tissue, ranked from 0 (negative) to 5 (heavy). Occurrence is the relative frequency of stained cells in a tissue, ranked from 0 (no staining) to 3 (diffuse). Each sample was assigned an overall induction rank, based on the frequency of induction of fish in the sample, staining intensity, and staining occurrence. Induction ranks were 0 = negative, 1 = very mild, 2 = mild, 3 = mild to moderate, 4 = moderate, 5 = moderate to strong, and 6 = strong.

Additional comparisons among tissues processed for cytochrome P4501A were necessary because of procedural differences in the preservation of fry samples. In 1989, all fry remained in formalin until analysis, but in 1990, fry were transferred to isopropanol after 6–8 weeks of preservation in formalin. Four comparisons were possible where subsamples of pink salmon from the same seine set in 1989 had been preserved in both formalin and isopropanol. On average, preservation in isopropanol reduced the sensitivity of the immunohistochemical assay by one overall rank, with a range of −2 levels to equal.

Results

Site Contamination

Mussels from locations that were not visibly oiled (Culross Passage, Long Bay, McClure Bay, and Wells Passage) contained relatively few hydrocarbons compared to mussels from visibly oiled locations (Herring Bay, Knight Island Passage, Snug Harbor, and Prince of Wales Passage) (Figure 2). An exception occurred in Culross Passage, where slightly elevated concentrations were observed in tissues collected on 4 May 1989. Small amounts of mousse were observed on beaches in the vicinity of the 4 May mussel collection, but because the contamination was low and did not persist, we continued to consider Culross Passage as a reference location.

In 1989, alkane and PAH concentrations in mussel tissues from oiled locations were significantly greater than concentrations in tissues from non-oiled locations (Figure 2). Results were not altered by including or excluding the contaminated Culross Passage sample. In non-oiled locations, only 6 ± 2% of the PAHs were above minimum detection limits (MDL), but 69 ± 3% of the PAHs were above MDL in oiled locations.

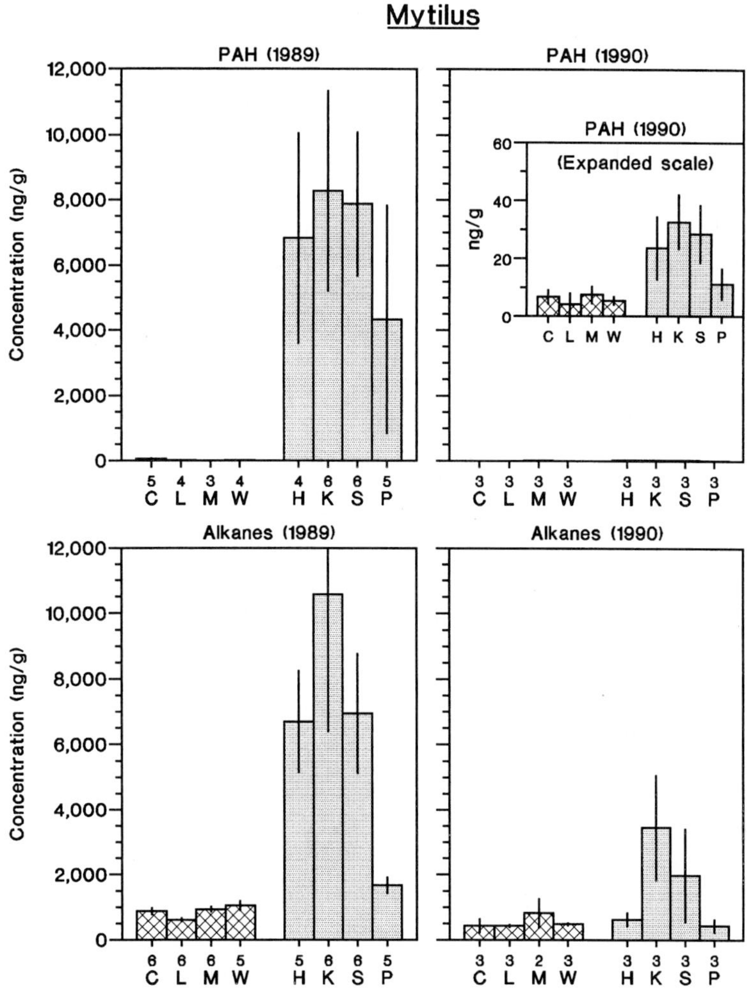

FIGURE 2.—Polynuclear aromatic hydrocarbon (PAH) and alkane concentrations in tissues of the mussel *Mytilus trossulus* from non-oiled and oiled locations in Prince William Sound in 1989 and 1990. Non-oiled locations were Culross Passage (C), Long Bay (L), McClure Bay (M), and Wells Passage (W); oiled locations were Herring Bay (H), Knight Island Passage (K), Snug Harbor (S), and Prince of Wales Passage (P). Numbers along the *x*-axis indicate sample size. Error bars are ±1 SE.

Concentrations of PAHs in mussels increased over time at all oiled sites except Snug Harbor, peaked above 20,000 ng/g 22–73 d after the spill, then declined to less than 2,001 ng/g by day 131. Peak hydrocarbon concentrations may have occurred before our first collection in Snug Harbor (22 d). At control sites, PAH concentrations were ≤52 ng/g and did not exhibit any trends, except (as noted previously) there was a minor peak (205 ng/g) in Culross Passage 41 d after the spill. Contamination of mussel tissues at oiled locations persisted, at much reduced levels, into 1990, while non-oiled locations remained clean. Differences in PAH concentrations in mussels were significant between non-oiled and oiled locations in 1990 ($P = 0.026$; Figure 2).

Salmon Fry

In 1989, PAH concentrations in pink salmon fry from oiled locations were significantly greater than in fry from non-oiled locations (Figure 3). Alkane concentrations did not differ significantly between oiled and non-oiled locations (Figure 3).

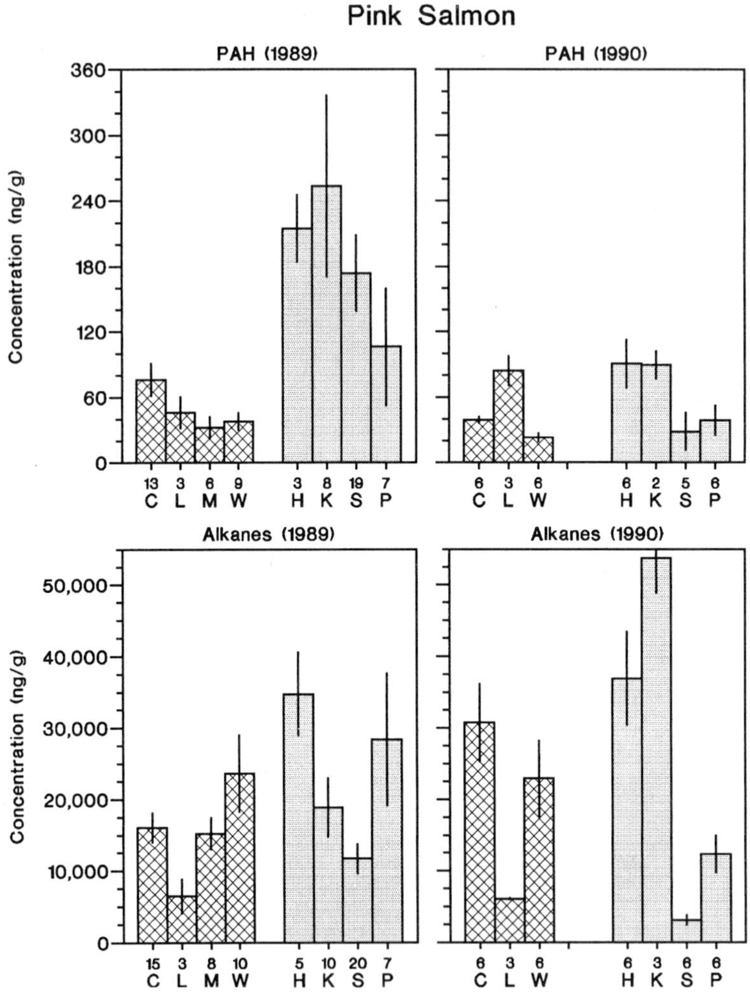

FIGURE 3.—Polynuclear aromatic hydrocarbon (PAH) and alkane concentrations in pink salmon tissues from non-oiled and oiled locations in Prince William Sound in 1989 and 1990. Non-oiled locations were Culross Passage (C), Long Bay (L), McClure Bay (M), and Wells Passage (W); oiled locations were Herring Bay (H), Knight Island Passage (K), Snug Harbor (S) and Prince of Wales Passage (P). Numbers along the *x*-axis indicate sample size. Error bars are ±1 SE.

Concentrations of PAHs in pink salmon fry at oiled sites either increased to a peak and then declined, or generally declined after the first sampling. Mean PAH concentrations in pink salmon fry increased to a maximum (510 and 263 ng/g) 46 and 58 d after the spill in Snug Harbor and Herring Bay, respectively, and then declined. In Prince of Wales Passage and Knight Island Passage, maximum concentrations (314 ± 5 and 675 ng/g) were observed in the first samples collected (39 and 40 d after the spill), and generally declined thereafter. Mean PAH concentrations at reference sites remained below 74 ng/g, except concentrations rose to a minor peak (154 ± 2 ng/g) in Culross Passage 57 d after the spill. In 1990, there were no significant differences between oiled and non-oiled locations in alkane and PAH concentrations in pink salmon (Figure 3).

In 1989, PAH concentrations in pink salmon carcass and visceral tissues were significantly greater in tissues from oiled locations than in corresponding tissues from non-oiled locations (Figure 4). In fry collected from oiled locations, concentrations of PAHs were significantly greater in visceral than carcass tissues; PAH concentrations in viscera re-

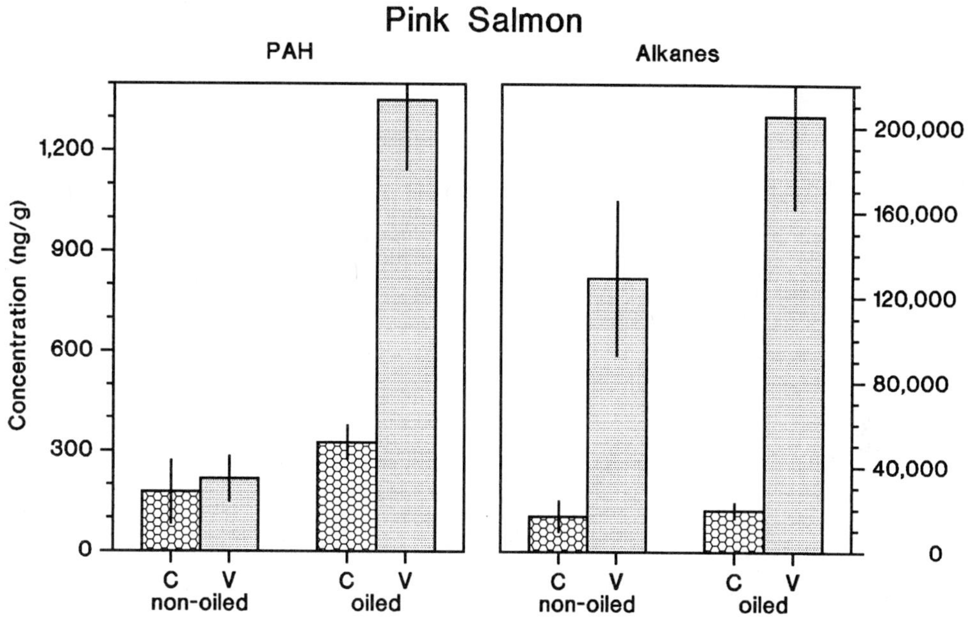

FIGURE 4.—Polynuclear aromatic hydrocarbon (PAH) and alkane concentrations in carcass (C) and visceral (V) tissues of pink salmon fry from non-oiled and oiled locations in Prince William Sound in 1989. Error bars are ±1 SE.

mained significantly greater than in carcasses when concentrations were corrected by mean non-oiled concentrations. Alkane concentrations in corresponding tissues did not differ significantly between oiled and non-oiled locations, but were significantly greater in visceral tissues than in carcass tissues (Figure 4).

Relative PAH concentrations in fry carcass and visceral tissues were consistent with the relative PAH concentrations in *Exxon Valdez* crude oil (Figure 5). Naphthalenes, biphenyl, fluorenes, dibenzothiophenes, and chrysenes were important components in *Exxon Valdez* crude oil. These compounds were also present in carcass and visceral tissue. Concentrations of fluorenes through C-4 phenanthrenes were proportionately higher in tissue samples than in the parent oil, especially in the viscera. Chrysenes were present in both types of tissue. Similarly, relative PAH concentrations in whole pink salmon fry and in mussels were also consistent with relative PAH concentrations in *Exxon Valdez* crude oil.

Cytochrome P4501A Induction

Pink salmon fry from oiled locations in 1989 generally had higher staining ranks and showed staining in more cell types than did fry from non-oiled locations (Table 1). In 1989, the median rank for samples from oiled locations was 4 (range was 2–6); the median rank from non-oiled locations was 1 (range was 1–2) (Figure 6). The median rank was significantly greater in oiled locations than in non-oiled locations ($P < 0.01$, Mann–Whitney *U*-test). We infer, therefore, that the minimum hydrocarbon induction rank was 3.

Chum salmon were rare in the oiled locations sampled; samples were collected only in Herring Bay. In 1989, these samples had higher ranks and more tissue types stained than did samples of chum salmon from non-oiled locations (Table 2). In 1989, the median rank for samples from oiled locations was 6 (range was 5–6); the median rank from non-oiled locations was 1 (range was 0–1) (Figure 6). The median rank was significantly greater in oiled locations than in non-oiled locations ($P < 0.05$, Mann–Whitney *U*-test).

Among pink or chum salmon in each sample with an overall staining rank of 3 or higher, frequency, intensity, and diffuseness (occurrence) of staining were compared for gill epithelial tissue, gill endothelial tissue (pillar cells), and cecal epithelial tissue (Table 3). In pink salmon, the median frequency of P4501A staining was 0.8 for gill pillar cells and cecal epithelial cells, and 0.3 for gill epithelial cells. Median intensity and occurrence was highest in gill pillar cells and lower, but similar to each other, for

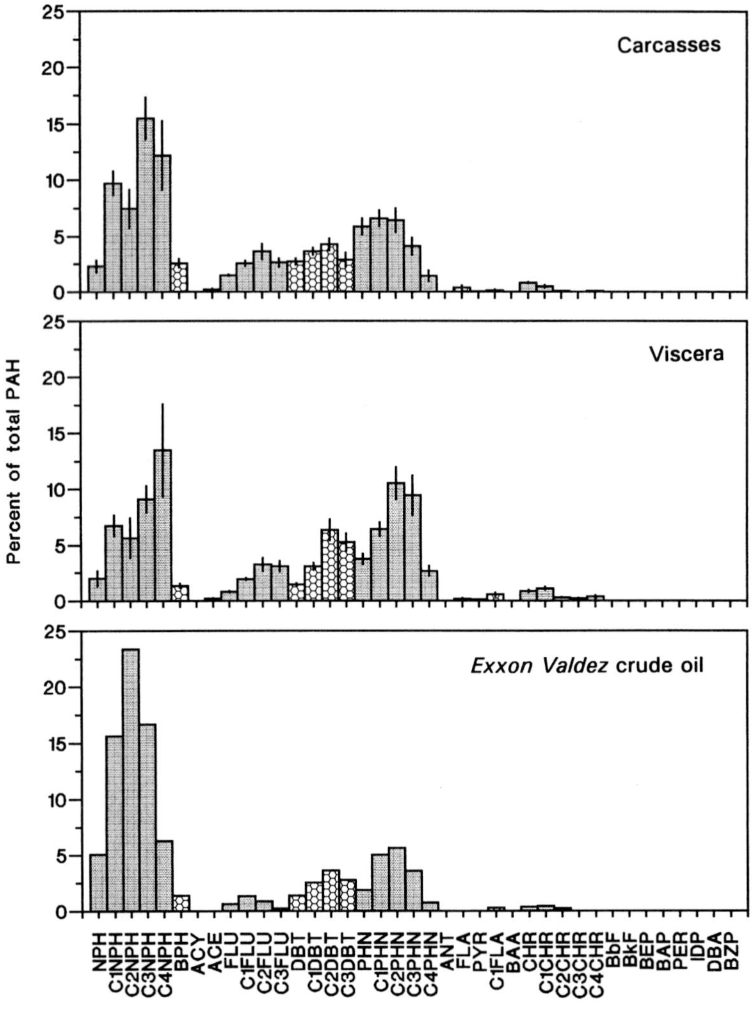

FIGURE 5.—Composition of polynuclear aromatic hydrocarbons (PAH) in pink salmon carcasses and viscera from oiled locations in Prince William Sound, and in *Exxon Valdez* crude oil. Error bars are ±1 SE.

gill and cecal epithelial cells. In chum salmon, staining was observed in gill pillar cells and cecal epithelial cells in all fish analyzed from the oiled location (median frequency = 1.0), whereas median frequency of staining for gill epithelial cells was 0.5. Median intensity and occurrence of staining for these tissues in chum salmon was highest in cecal epithelial cells, intermediate in gill pillar cells, and lowest in gill epithelial cells.

Differential staining of P4501A was not detected in pink and chum salmon tissues in nearshore habitats in 1990. All samples ranked either 1 or 0, and the number of tissue types staining was greatly reduced in the oiled locations compared to 1989 (Tables 1 and 2). Although the median rank for pink salmon in the oiled locations was 1 in 1990 and 0 in the non-oiled locations, ranks were not significantly different ($P > 0.3$, Mann–Whitney U-test) (Figure 6). In 1990, the median rank for chum salmon was 0 for both oiled and non-oiled locations (Figure 6). Assuming preservation in isopropanol reduced staining by one rank, no samples would be promoted to the previously defined hydrocarbon induction range (3–6) if activity level were increased by one.

Discussion

Bivalve molluscs generally are considered an excellent measure of the biological availability of pe-

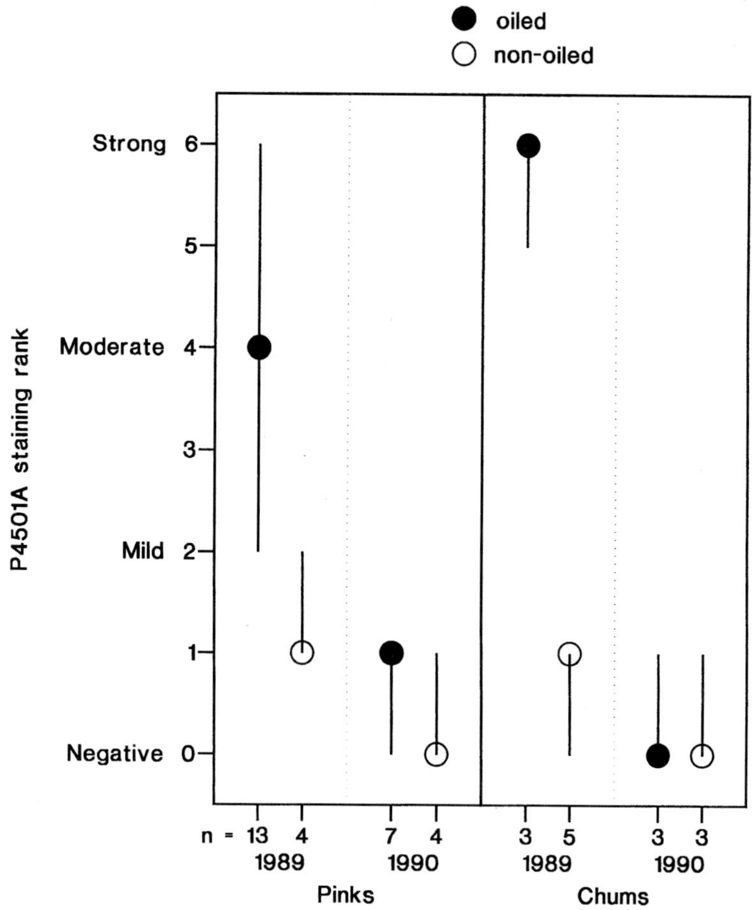

FIGURE 6.—Medians and ranges of ranks of P4501A staining in pink and chum salmon fry sampled in nearshore waters in oiled and non-oiled areas of Prince William Sound in 1989 and 1990. Sample sizes are indicated along the *x*-axis.

troleum hydrocarbons at specific sites because of their sessile nature and low capacity to metabolize hydrocarbons (Vandermeulen and Penrose 1978; Stegeman 1985; Livingstone et al. 1989). Mussels can bioaccumulate petroleum hydrocarbons by more than 1,000 times the exposure level (Fossato and Canzonier 1976). Despite the extremely low concentrations of hydrocarbons in the water column in the spill area (Short and Harris 1996a, this volume), hydrocarbon concentrations in mussel tissues were dramatically higher in oiled locations than in reference locations in 1989. As may be expected after a single spill perturbation, concentrations of PAHs increased in mussels from oiled locations after the *Exxon Valdez* spill, peaked, and then declined. Composition of PAHs in mussels from oiled locations were characteristic of *Exxon Valdez* crude oil (98.5 ± 0.8 percent *Exxon Valdez* PAHs) and included chrysenes and substituted chrysenes. Differences observed in hydrocarbon concentrations in mussels, therefore, can generally be attributed to the differential contamination of the nearshore marine environment by *Exxon Valdez* crude oil.

Concentrations of PAHs were also significantly elevated in pink salmon from oiled locations, peaked after the *Exxon Valdez* spill and then declined. Composition of PAHs in juvenile salmon tissues from oiled locations exhibited characteristics of *Exxon Valdez* crude oil (96.2 ± 0.6 percent *Exxon Valdez* PAH), including chrysenes, and was similar to the more clearly defined PAH composition in mussel tissues collected at the same locations. Because pink salmon fry are pelagic and are capable of actively metabolizing hydrocarbons (Rice et al.

TABLE 1.—Ranking for cytochrome P4501A staining in tissues of pink salmon fry from oiled and non-oiled nearshore marine habitats in Prince William Sound in 1989 and 1990. Sample ranks ranging from 0 (negative) to 6 (strong) were based on extent of occurrence and intensity of immunochemical staining for P4501A in histological sections of three to six fish per sample.

			Tissues stained	
Sampling location	Date	P4501A rank	*N*	Type[a]
		1989, non-oiled		
McClure Bay	15 May	1	1	L_h
Wells Passage	21 May	1	3	L_h; G_p; I_c
Culross Passage	20 May	2	3	L_h; G_p; I_c
Culross Passage	31 May	1	3	L_h; K_c; I_c
		1989, oiled		
Herring Bay	14 May	2	4	L_h; G_p; K_c; I_c
Herring Bay	30 May	6	11	L_h; $G_{p,a,e}$; $K_{t,s,g,c}$; $I_{p,c}$; H
Snug Harbor–Inner Bay	17 May	3	9	$L_{b,h}$; $G_{p,e}$; $K_{t,s,c}$; I_c; P
Snug Harbor–Inner Bay	2 Jun	4	11	L_h; $G_{p,e}$; $K_{t,s,c}$; $I_{a,p,c}$; H; P
Snug Harbor–Outer Bay	16 May	4	10	L_h; $G_{p,a,e}$; $K_{v,s,c}$; $I_{a,p,c}$
Snug Harbor–Outer Bay	8 Jun	5	6	L_h; $G_{a,p}$; K_c; I_c; H
Snug Harbor–Outer Bay	18 Jun	2	3	L_h; G_p; H
Knight Island Passage	19 May	4	11	L_h; $G_{p,a,e}$; $K_{s,g,c}$; $I_{a,p,c}$; H
Knight Island Passage	19 May	5	11	L_h; $G_{p,a,e}$; $K_{s,g,c}$; I_c; H; P
Knight Island Passage	18 Jun	4	6	L_h; $G_{p,e}$; K_c; I_c; H
Prince Wales Passage	17 May	6	5	L_h; $G_{p,e}$; I_c; H
Prince Wales Passage	9 Jun	4	7	L_h; G_p; K_c; $I_{a,p,c}$; H
Prince Wales Passage	25 Jun	2	2	L_h; K_c
		1990, non-oiled		
McClure Bay	22 Apr	0	0	
Long Bay	5 May	0	0	
Wells Passage	21 Apr	0	0	
Culross Passage	8 May	1	1	I_c
		1990, oiled		
Herring Bay	3 May	1	2	L_h; I_c
Herring Bay	2 Jun	0	0	
Herring Bay	11 Jun	0	0	
Snug Harbor–Inner Bay	19 Apr	1	3	L_h; G_p; K_c
Snug Harbor–Inner Bay	19 May	0	0	
Knight Island Passage	7 May	1	5	G_p; $K_{s,c}$; $I_{a,c}$
Prince Wales Passage	20 Apr	1	1	I_c

[a]Tissue types that immunochemically stained for cytochrome P4501A: L_h = liver hepatocytes; $G_{p,a,e}$ = gill pillar cells (endothelium), gill buds or arch endothelium, gill epithelium; $K_{c,g,s,t,v}$ = kidney collecting ducts, glomerular endothelium, sinusoidal endothelium, tubular epithelium, vascular endothelium; $I_{a,p,c}$ = anterior intestinal epithelium, posterior intestinal epithelium, cecal epithelium; H = heart endothelium; P = pharyngeal epithelium.

1977), we expected greater variability in salmon tissues than in mussel tissues. Induction of P4501A in salmon was another indicator of oiling. Although other compounds, including polychlorinated biphenyls (PCBs) and other halogenated hydrocarbons, are inducers of P4501A in fish (Stegeman 1990), the presence of a strong effect only in the year of the spill, and the absence of strong induction in nearby, but non-oiled, reference locations, indicated that *Exxon Valdez* crude oil was the inducer. Differences between oiled and reference locations in PAH concentrations and in P4501A induction in salmon fry in 1989, therefore, were also the result of differential contamination by *Exxon Valdez* crude oil.

Not all pink salmon fry collected at contaminated locations had hydrocarbons in their tissues, and not all fry collected in control locations were hydrocarbon-free. Uncontaminated samples from oiled locations may indicate that fry had not been exposed long enough to become contaminated, or fry had been exposed for sufficient time that metabolism and depuration had reduced hydrocarbon body burden below detectable limits, or that hydrocarbon concentrations in the area had declined significantly. In one instance, salmon tissues from a non-oiled reference site (Wells Passage, 5 May 1989) showed possible contamination. The migratory nature of pink salmon fry may also explain this observation. Alternatively, the Wells Passage locations may have been exposed to a small amount of oil for a short time; small amounts of oil were observed by us nearby (Culross Passage on 4 May), and slick was

TABLE 2.—Ranking for cytochrome P4501A staining in tissues of chum salmon fry from oiled and non-oiled nearshore marine habitats in Prince William Sound in 1989 and 1990. Sample ranks ranging from 0 (negative) to 6 (strong) were based on extent of occurrence and intensity of immunochemical staining for P4501A in histological sections of three to six fish per sample. Abbreviations for tissue types are explained in footnote to Table 1.

Sampling location	Date	P4501A rank	Tissues stained	
			N	Type
		1989, non-oiled		
McClure Bay	15 May	0	0	
Long Bay	16 May	1	1	L_h
Culross Passage	20 May	1	1	L_h
Wells Passage	21 May	1	1	L_h
Culross Passage	23 Jun	1	1	L_h
		1989, oiled		
Herring Bay	14 May	6	10	L_h; $G_{e,a,p}$; $K_{g,c}$; $I_{a,p,c}$; H
Herring Bay	30 May	6	11	L_h; $G_{e,p}$; $K_{g,c,s}$; $I_{a,p,c}$; H; P
Herring Bay	15 Jun	5	8	L_h; $G_{e,a,p}$; $I_{a,c}$; H; P
		1990, non-oiled		
McClure Bay	9 May	0	0	
McClure Bay	17 May	1	2	L_h; I_c
Long Bay	5 May	0	0	
		1990, oiled		
Herring Bay	3 May	0	0	
Herring Bay	16 May	1	1	L_h
Herring Bay	2 Jun	0	0	

TABLE 3.—Frequency, intensity, and occurrence of cytochrome P4501A staining in gill pillar cells (GP), gill epithelial cells (GE), and cecal epithelial cells (CE) in pink and chum salmon fry in which P4501A activity had been induced by crude oil contamination in Prince William Sound in 1989. Frequency is the proportion of fish in a sample showing induction in that tissue. Intensity is the median degree of staining in fish in a sample in which induction was observed, ranked from 1 (very mild) to 5 (strong). Occurrence is the median of the relative frequency of stained cells within a tissue, ranked from 1 (rare) to 3 (diffuse).

Location or statistic	Date	Frequency			Intensity			Occurrence		
		GP	GE	CE	GP	GE	CE	GP	GE	CE
					Chum salmon					
Herring Bay	15 Jun	1.0	0.2	1.0	2.5	2.0	2.0	2.5	1.0	2.0
Herring Bay	30 May	1.0	0.5	1.0	2.0	2.0	3.5	2.0	1.0	3.0
Herring Bay	14 May	1.0	0.5	1.0	3.5	2.5	3.0	2.5	1.0	3.0
Median		1.0	0.5	1.0	2.5	2.0	3.0	2.5	1.0	3.0
					Pink salmon					
Herring Bay	30 May	1.0	0.5	1.0	3.0	2.5	3.0	3.0	1.5	2.5
Snug Harbor	16 May	0.8	0.8	0.8	4.0	2.0	4.0	3.0	2.0	3.0
Snug Harbor	17 May	0.4	0.4	0.7	2.0	2.5	2.5	2.0	2.5	2.0[a]
Snug Harbor	2 Jun	1.0	0.2	1.0	2.5	2.0	2.0	2.5	1.0	2.5
Snug Harbor	8 Jun	0.8	0.0	0.3	3.0		2.0	2.0		1.5
Knight Island Passage	19 May	1.0	1.0	1.0	4.0	2.0	3.0	3.0	2.0	2.0
Knight Island Passage	19 May	1.0	1.0	0.3[a]	4.0	3.0	1.0[a]	3.0	1.0	2.0
Knight Island Passage	18 Jun	0.3	0.2	0.5	1.5	2.0	1.0[a]	1.5	1.0	2.0
Prince Wales Passage	17 May	0.5	0.2	0.8	3.0	3.0	2.0[a]	3.0	2.0	2.0
Prince Wales Passage	9 Jun	0.3	0.0	1.0	1.0		1.5	2.5		2.0
Median		0.8	0.3	0.8	3.0	2.2	2.0	2.8	1.8	2.0

[a]Cases in which frequency, median intensity, or median occurrence of P4501A staining within a sample was greater for gill epithelial cells than for cecal epithelial cells.

observed along the eastern shore of Culross Island (Pavia et al. 1991) within 5 km of our sampling location.

The observed contamination of juvenile salmon was not superficial contamination caused by collection techniques because the viscera contained significantly higher hydrocarbon concentrations than the carcasses. Likewise, the induction of cytochrome P4501A could not have been caused by superficial exposure at the time of sampling; it is clear evidence that the tissues of pink and chum salmon fry were exposed to hydrocarbons in the oiled locations.

Other investigators also found evidence of hydrocarbons from the *Exxon Valdez* oil spill in fish, particularly in the spill path. Bile of salmonids and bottomfishes contained petroleum-related fluorescent aromatic compounds (FACs) (Haynes and Urban 1993; Collier et al. 1996; Hom et al. 1996; both this volume). Consistent with concentration declines in pink salmon tissue, FAC concentrations in Dolly Vardens *Salvelinus malma* declined from 1989 to 1990 (Collier et al. 1996). In contrast, FAC concentrations in 1990 remained elevated in yellowfin soles *Pleuronectes asper*, rock soles *P. bilineatus*, and flathead soles *Hippoglossoides elassodon* (Collier et al. 1996), and contamination of bottomfishes was more widespread in 1990 than 1989 (Haynes and Urban 1993). Consistent with induction of P4501A by petroleum hydrocarbons in pink salmon, induction of hepatic aryl hydrocarbon hydroxylase (AHH) activity (associated with induction of P4501A) was observed in Dolly Vardens, and in yellowfin and rock soles; AHH activity in Dolly Vardens dropped substantially from 1989 to 1990 (Collier et al. 1996); however, induction of P4501A was observed in preemergent pink salmon from oiled sites more than 2 years after the spill (Wiedmer and Fink 1996, this volume).

There were several possible avenues of exposure of juvenile salmon to *Exxon Valdez* crude oil. Fry may have been exposed directly to whole oil as they fed at the water surface: we frequently observed dense schools of juvenile salmon in water covered by sheen and mousse from *Exxon Valdez* crude oil, with individuals jumping through the contaminated surface. Prey-sized oil particles could have been mistaken for prey and ingested directly: oil sheen or droplets were observed in stomachs of several fry from oiled locations, but not in fish from non-oiled locations (Sturdevant et al. 1996, this volume). In Prince William Sound, particulate oil characteristic of *Exxon Valdez* crude oil was detected in mussels deployed at depths of 25 m (Short and Harris 1996b, this volume). Fry may also have ingested oil-contaminated prey. Uptake of hydrocarbons by planktonic crustaceans, and their ingestion of particulate oil has been well documented (e.g., Conover 1971; Corner et al. 1976; Harris et al. 1977; Macek et al. 1979; Schwartz 1985; Carls 1987). In Puget Sound, Washington, aromatic hydrocarbon concentrations were significantly higher in stomach contents of juvenile chinook salmon Oncorhynchus tshawytscha from a contaminated estuary than from a more pristine estuary (McCain et al. 1990). Significant exposure to hydrocarbons dissolved in water (WSF) was less likely because total concentrations of aromatic hydrocarbons dissolved in the water column were less than 7 μg/L (Maki 1991; Short and Rounds 1993a; Neff and Stubblefield 1995).

The pattern of P4501A staining in the tissues of pink and chum salmon from oiled locations suggest that ingestion was a route of oil contamination. Endothelial cells, such as gill pillar cells, are strongly induced in teleosts by both intraperitoneal administration of PAHs (mimicking the ingestion route; Miller et al. 1989) and by exposure of fish to waterborne PAHs (Smolowitz et al. 1992). Induction of gill and gut epithelial cells, however, may differ because initial cell exposure depends on the route of contamination (Miller et al. 1989; Stegeman et al. 1991). Cecal epithelial cells were stained in both species. In chum salmon, the frequency, intensity, and occurrence of staining was greater in cecal epithelial cells than in gill epithelial cells, and similar to staining in gill endothelial (pillar) cells, thus indicating ingestion was an important route of contamination. However, in pink salmon fry, median intensity and occurrence of staining was similar for gill and cecal epithelial cells, and in one sample the frequency of induction was highest in gill epithelial cells, indicating that direct exposure to waterborne oil was also a contamination route in some circumstances. Experiments in which salmon fry are exposed to petroleum hydrocarbons directly via water or diet are necessary to confirm that the relative induction of gill and gut epithelia distinguish between these routes of contamination. Such studies are in progress.

Differences in feeding behavior between pink and chum salmon in the nearshore marine environment may be important in causing differences in the degree of exposure to different routes of contamination. Pink and chum salmon fry co-occur in shallow littoral habitats, but epibenthic prey are more important as food for chum salmon than for pink salmon (Kaczynski et al. 1973; Bailey et al. 1975;

Cooney et al. 1981). Thus, chum salmon may be more likely to feed near the bottom in littoral habitats and ingest prey contaminated by oiled sediments. Conversely, pelagic zooplankton are more important as food for pink salmon, and therefore, pink salmon may be more likely to encounter surface contamination.

The physiological consequences of the contamination of juvenile salmon in Prince William Sound in 1989 can be inferred both from laboratory studies and from empirical field evidence. Direct exposure to WSF of crude oil and the consumption of prey contaminated with WSF have been shown to reduce growth of pink salmon fry (Rice et al. 1975; Schwartz 1985). Ingestion of food contaminated with whole crude oil has also been shown to reduce growth and survival of pink salmon fry (Wang et al. 1993; Carls et al. 1996). In 1989, the apparent growth rate of the pink salmon fry we sampled was lower in oiled locations in Prince William Sound than in non-oiled locations (Wertheimer and Celewycz 1996, this volume); growth rates of specific tagged groups of pink salmon fry were also reduced in heavily oiled locations in 1989 (Willette 1996, this volume). Consistent with observations of reduced growth in pink salmon, growth of two other salmonids that forage along the shoreline, Dolly Varden and cutthroat trout *Oncorhynchus clarki*, was also reduced (Hepler et al. 1996, this volume). We conclude that contamination of pink and chum salmon in Prince William Sound with *Exxon Valdez* crude oil was physiologically significant and affected their capacity for growth and survival.

Acknowledgments

We thank Adrian Celewycz, Skip Gish, Herbert Jaenicke, Ben Kirkpatrick, Mark Pritchett, and Joe Orsi for their assistance. In addition, we acknowledge the contributions of Bob Wiese, skipper of the FV *Kanak*; John Bortniak, master of the National Oceanic and Atmospheric Administration (NOAA) RV *Murre II*; and the master and crew of the NOAA RV *John N. Cobb*.

References

ADFG (Alaska Department of Fish and Game). 1989. *Exxon Valdez* oil spill. Alaska Fish and Game 21(4): 4–37.

Babcock, M. M. 1985. Morphology of olfactory epithelium of pink salmon, *Oncorhynchus gorbuscha*, and changes following exposure to benzene: a scanning electron microscope study. Pages 259–267 *in* J. S. Gray and M. E. Christiansen, editors. Marine biology of polar regions and stress on marine organisms. Wiley, London.

Bailey, J. E., B. L. Wing, and C. R. Mattson. 1975. Zooplankton abundance and feeding habits of fry of pink salmon, *Oncorhynchus gorbuscha*, and chum salmon, *Oncorhynchus keta*, in Traitor's Cove, Alaska, with speculations on the carrying capacity of the area. U.S. National Marine Fisheries Service Fishery Bulletin 73:946–961.

Bax, N. J. 1983. Early marine mortality of marked juvenile chum salmon (*Oncorhynchus keta*) released into Hood Canal, Puget Sound, Washington, in 1980. Canadian Journal of Fisheries and Aquatic Sciences 40:426–435.

Carls, M. G. 1987. Effects of direct and indirect oil exposure on larval Pacific herring (*Clupea harengus pallasi*). Marine Environmental Research 22:253–270.

Carls, M. G., and six coauthors. 1996. Growth, feeding, and survival of pink salmon fry exposed to food contaminated with crude oil. American Fisheries Society Symposium 18:608–618.

Collier, T. K., C. A. Krone, M. M. Krahn, J. E. Stein, S.-L. Chan, and U. Varanasi. 1996. Petroleum exposure and associated biochemical effects in subtidal fish after the *Exxon Valdez* oil spill. American Fisheries Society Symposium 18:671–683.

Conover, R. J. 1971. Some relations between zooplankton and bunker C oil in Chedabucto Bay following the wreck of the tanker *Arrow*. Journal of the Fisheries Research Board of Canada 28:1327–1330.

Cooney, R. T., D. Uruquat, and D. Barnard. 1981. The behavior, feeding biology, and growth of hatchery released pink and chum salmon fry in Prince William Sound, Alaska. University of Alaska, Sea Grant Report 81-5.

Corner, E. D. S., R. P. Harris, C. C. Kilvington, and S. C. M. O'Hara. 1976. Petroleum compounds in the marine food web: short-term experiments on the fate of naphthalene in *Calanus*. Journal of the Marine Biological Association of the United Kingdom 56:121–133.

Fossato, V. U., and W. J. Canzonier. 1976. Hydrocarbon uptake and loss by the mussel *Mytilus edulis*. Marine Biology 36:243–250.

Harris, R. P., V. Berdugo, E. D. S. Corner, C. C. Kilvington, and S. C. M. O'Hara. 1976. Factors affecting the retention of a petroleum hydrocarbon by marine planktonic copepods. Pages 286–304 in D. A. Wolfe, editor. Fate and effects of petroleum hydrocarbons in marine organisms and ecosystems. Pergamon Press, New York.

Haynes, E., and D. Urban. 1993. Prince William Sound trawl assessment. Fish/shellfish study 18, state/federal natural resource damage assessment final report. NOAA (National Oceanic and Atmospheric Administration)/NMFS (National Marine Fisheries Service), Alaska Fisheries Science Center, Auke Bay Laboratory, Juneau.

Hepler, K. R., P. A. Hansen, and D. R. Bernard. 1996. Impact of oil spilled from the *Exxon Valdez* on survival and growth of Dolly Varden and cutthroat trout in Prince William Sound. American Fisheries Society Symposium 18:645–658.

Hom, T., U. Varanasi, J. E. Stein, C. A. Sloan, K. L.

Tilbury, and S.-L. Chan. 1996. Assessment of the exposure of subsistence fish to aromatic compounds after the *Exxon Valdez* oil spill. American Fisheries Society Symposium 18:856–866.

Kaczynski, V. W., R. J. Feller, and C. Clayton. 1973. Trophic analysis of juvenile pink and chum salmon (*Oncorhynchus gorbuscha* and *O. keta*) in Puget Sound. Journal of the Fisheries Research Board of Canada 30:1003–1008.

Karinen, J. F., M. M. Babcock, D. W. Brown, W. D. MacLeod, Jr., L. S. Ramos, and J. W. Short. 1993. Hydrocarbons in intertidal sediments and mussels from Prince William Sound, Alaska, 1977–1980: characterization and probable sources. NOAA (National Oceanic and Atmospheric Administration) Technical Memorandum NMFS (National Marine Fisheries Service)-AFSC-9, Seattle.

Livingstone, D. R., M. A. Kirchin, and A. Wiseman. 1989. Cytochrome P450 and oxidative metabolism in molluscs. Xenobiotica 19:1041–1062.

Macek, K. J., S. R. Petrocelli, and B. H. Sleight, III. 1979. Considerations in assessing the potential for, and significance of, biomagnification of chemical residues in aquatic food chains. Pages 251–268 *in* Marking and Kimerle, editors. Aquatic toxicology, 2nd symposium. American Society for Testing and Materials Special Technical Publication 667, Philadelphia.

Maki, A. W. 1991. The *Exxon Valdez* oil spill: initial environmental impact assessment. Environmental Science and Technology 25:24–29.

McCain, B. B., and six coauthors. 1990. Uptake of aromatic and chlorinated hydrocarbons by juvenile chinook salmon (*Oncorhynchus tshawytscha*) in an urban estuary. Archives of Environmental Contamination and Toxicology 19:10–16.

Miller, M. R., D. E. Hinton, and J. J. Stegeman. 1989. Cytochrome P-450E induction and localization in gill pillar (endothelial) cells of scup and rainbow trout. Aquatic Toxicology 14:307–322.

Moles, A., and S. D. Rice. 1983. Effects of crude oil and naphthalene on growth, caloric content, and fat content of pink salmon juveniles in seawater. Transactions of the American Fisheries Society 112:205–211.

Neff, J. M., and W. A. Stubblefield. 1995. Chemical and toxicological evaluation of water quality following the *Exxon Valdez* oil spill. Pages 141–177 *in* P. G. Wells, J. N. Butler, and J. S. Hughes, editors. *Exxon Valdez* oil spill: fate and effects in Alaskan waters. American Society for Testing and Materials, Special Technical Publication 1219, Philadelphia.

Nichelson, T. E. 1986. Influences of upwelling, ocean temperature, and smolt abundance on marine survival of coho salmon (*Oncorhynchus kisutch*) in the Oregon production area. Canadian Journal of Fisheries and Aquatic Sciences 43:527–535.

Parker, R. R. 1968. Marine mortality schedules of pink salmon of the Bella Coola River, central British Columbia. Journal of the Fisheries Research Board of Canada 25:757–794.

Pavia, E., C. Crosby, P. Endres, W. Lane, and M. Siddiqui. 1991. The Prince William Sound and Homer regions evaluation group final report. Alaska Department of Environmental Conservation, Anchorage.

Rice, S. D., D. A. Moles, and J. W. Short. 1975. The effect of Prudhoe Bay crude oil on survival and growth of eggs, alevins, and fry of pink salmon, *Oncorhynchus gorbuscha*. Pages 503–507 *in* Proceedings of the 1975 conference on prevention and control of oil pollution. American Petroleum Institute, Washington, DC.

Rice, S. D., R. E. Thomas, and J. W. Short. 1977. Effect of petroleum hydrocarbons on breathing and coughing rates and hydrocarbon uptake-depuration in pink salmon fry. Pages 259–277 *in* F. J. Vernberg, A. Calabrese, F. P. Thurberg, and W. B. Vernberg, editors. Physiological responses of marine biota to pollutants. Academic Press, New York.

Rice, S. D., and seven coauthors. 1984. Effects of petroleum hydrocarbons on Alaskan aquatic organisms. NOAA (National Oceanic and Atmospheric Administration) Technical Memorandum NMFS (National Marine Fisheries Service) F/NWC-67, Seattle.

SAS. 1989. SAS/STAT user's guide, version 6, 4th edition, volume 2. SAS Institute Inc., Cary, North Carolina.

Schwartz, J. P. 1985. Effect of oil-contaminated prey on the feeding and growth rate of pink salmon fry (*Oncorhynchus gorbuscha*). Pages 459–476 *in* F. Vernberg, R. Thurberg, A. Calabrese, and W. Vernberg, editors. Pollution and physiology of marine organisms. University of South Carolina Press, Columbia.

Short, J. W., and P. M. Harris. 1996a. Chemical sampling and analysis of petroleum hydrocarbons in near-surface seawater of Prince William Sound after the *Exxon Valdez* oil spill. American Fisheries Society Symposium 18:17–28.

Short, J. W., and P. M. Harris. 1996b. Petroleum hydrocarbons in caged mussels deployed in Prince William Sound after the *Exxon Valdez* oil spill. American Fisheries Society Symposium 18:29–39.

Short, J. W., T. J. Jackson, M. L. Larsen, and T. L. Wade. 1996. Analytical methods used for the analysis of hydrocarbons in crude oil, tissues, sediments, and seawater collected for the natural resources damage assessment of the *Exxon Valdez* oil spill. American Fisheries Society Symposium 18:140–148.

Smolowitz, R. M., M. E. Hahn, and J. J. Stegeman. 1991. Immunohistochemical localization of cytochrome P-450IA1 induced by 3,3',4,4'-tetrachlorobiphenyl and by 2,3,7,8-tetrachlorodibenzo-furan in liver and extrahepatic tissues of the teleost *Stenotomus chrysops* (Scup). Drug Metabolism and Disposition 19: 113–123.

Smolowitz, R. M., M. Schultz, and J. J. Stegeman. 1992. Cytochrome P-450IA induction in tissues, including olfactory epithelium, of top minnows (*Poeciliopsis* spp.) by waterborne benzo[*a*]pyrene. Carcinogenesis 13:2395–2402.

Stegeman, J. J. 1985. Benzo[*a*]pyrene oxidation and microsomal enzyme activity in the mussel (*Mytilus edulis*) and other bivalve mollusc species from the western north Atlantic. Marine Biology 89:21–30.

Stegeman, J. J. 1990. Detecting the biological effects of deep sea waste disposal. Oceanus 33(2):54–60.
Stegeman, J. J., R. M. Smolowitz, and M. E. Hahn. 1991. Immunohistochemical localization of environmentally induced cytochrome P450IA1 in multiple organs of the marine teleost *Stenotomus chrysops* (Scup). Toxicology and Applied Pharmacology 110:486–504.
Sturdevant, M. V., A. C. Wertheimer, and J. L. Lum. 1996. Diets of juvenile pink and chum salmon in oiled and non-oiled nearshore habitats in Prince William Sound, 1989 and 1990. American Fisheries Society Symposium 18:578–592.
Vandermeulen, J., and W. R. Penrose. 1978. Absence of aryl hydrocarbon hydroxylase (AHH) in three marine bivalves. Journal of the Fisheries Research Board of Canada 35:643–647.
Walters, C. J., R. Hilborn, R. M. Peterman, and M. J. Staley. 1978. Model for examining early ocean limitations of Pacific salmon. Canadian Journal of Fisheries and Aquatic Sciences 41:126–133.
Wang, S. Y., J. L. Lum, M. G. Carls, and S. D. Rice. 1993. The relationship between growth and total nucleic acids in juvenile pink salmon *Oncorhynchus gorbuscha* fed crude oil-contaminated food. Canadian Journal of Fisheries and Aquatic Sciences 50:996–1001.
Wertheimer, A. C., and A. G. Celewycz. 1996. Abundance and growth of juvenile pink salmon in oiled and non-oiled locations of western Prince William Sound after the *Exxon Valdez* oil spill. American Fisheries Society Symposium 18:518–532.
Wiedmer, M., M. J. Fink, J. J. Stegeman, R. Smolowitz, G. D. Marty, and D. E. Hinton. 1996. Cytochrome P-450 induction and histopathology in preemergent pink salmon from oiled spawning sites in Prince William Sound. American Fisheries Society Symposium 18:509–517.
Willette, M. 1996. Impacts of the *Exxon Valdez* oil spill on the migration, growth, and survival of juvenile pink salmon in Prince William Sound. American Fisheries Society Symposium 18:533–550.

American Fisheries Society Symposium 18:608–618, 1996

Growth, Feeding, and Survival of Pink Salmon Fry Exposed to Food Contaminated with Crude Oil

M. G. Carls, L. Holland, M. Larsen, J. L. Lum, D. G. Mortensen, S. Y. Wang, and A. C. Wertheimer

National Marine Fisheries Service, Auke Bay Laboratory
Alaska Fisheries Science Center, 11305 Glacier Highway, Juneau, Alaska 99801, USA

Abstract.—Growth of pink salmon fry *Oncorhynchus gorbuscha* was reduced by oil-contaminated food in an 8-week laboratory experiment, and feeding and survival were reduced at high oil concentrations (34.8 mg total oil per gram of food). Total concentrations of Alaska North Slope crude oil in the food ranged from 0.00 (control) to 34.8 mg/g. In the highest oil treatment, mortality became significant after 2 weeks and increased rapidly until fry began feeding on clean food. Growth reductions in exposed fry were significant after 1 week in mid- (2.8 mg/g) and high-oil treatments and became more pronounced over time. The estimated feeding rate was inhibited only in the highest oil treatment, but increased in the lowest oil concentration (0.37 mg/g). Although feeding rate did not decline in the low- and mid-oil treatments, growth was significantly reduced in these groups. This suggests that changes in growth caused by oil were not caused simply by starvation, but rather were the result of increased metabolic demand or reduced assimilation efficiency. Results of this experiment support the hypothesis that juvenile pink salmon in Prince William Sound were contaminated with hydrocarbons from the 1989 *Exxon Valdez* oil spill by ingesting oiled prey and that this contamination was responsible for reduced growth of fry in oiled habitats.

Early marine residency is a critical phase in the life history of salmon that significantly affects year-class strength (Parker 1968; Walters et al. 1978; Bax 1983; Nichelson 1986). Growth during the early marine phase of pink salmon *Oncorhynchus gorbuscha* is extremely rapid (LeBrasseur and Parker 1964; Healey 1980) and is important to escape mortality from size-selective predation (Parker 1971; Hargreaves and LeBrasseur 1985; Mortensen et al. 1991). Slower-growing groups of pink salmon fry within a year-class exhibit lower marine survival than their faster-growing cohorts (Mortensen et al. 1991; Willette 1996, this volume). Among the many factors affecting growth, exposure of pink salmon fry to petroleum hydrocarbons can reduce growth (Rice et al. 1975; Moles and Rice 1983; Schwartz 1985).

After the *Exxon Valdez* oil spill in 1989, tissues of pink salmon fry from contaminated nearshore marine habitats in Prince William Sound contained petroleum hydrocarbons, and cytochrome P4501A enzymes were elevated (Carls et al. 1996, this volume). Because concentrations of petroleum hydrocarbons dissolved in the water column of Prince William Sound were very low (Maki 1991; Neff 1991; Short and Harris 1996a, this volume), the water-soluble fraction (WSF) of oil was probably not the primary route of contamination. Alternative routes of hydrocarbon contamination included direct contact and ingestion. In fry collected from oiled habitats in Prince William Sound, there was no visual evidence of coating, and hydrocarbon concentrations were higher in viscera than in carcass tissues, suggesting that ingestion of oil was a more likely mechanism of contamination than was coating (Carls et al. 1996). Induction of P4501A also signaled the presence of internal petroleum hydrocarbons (Carls et al. 1996). Oil was visible in stomachs of several fry from oiled locations, but not in fry from non-oiled locations (Sturdevant et al. 1996, this volume). In contaminated intertidal habitats of Prince William Sound, oiled substrate may have served as an indirect source of oil contamination, either by providing particulate oil for direct fry consumption or by contaminating planktonic prey (Carls et al. 1996). Composition of polycyclic aromatic hydrocarbons (PAH) in contaminated intertidal sediments generally matched that of *Exxon Valdez* crude oil and contained a broad spectrum of aromatic hydrocarbons, including naphthalene through chrysenes (Carls et al. 1994). The PAH composition in contaminated fry also matched that of *Exxon Valdez* crude oil (Carls et al. 1996).

The organisms that pink salmon fry prey upon may have also been contaminated by the *Exxon Valdez* oil spill. In nearshore marine habitats, pink salmon fry feed primarily on epibenthic meiofauna (e.g., harpacticoid copepods) and pelagic zooplankton, such as calanoid copepods and euphausiid eggs and larvae (Kaczynski et al. 1973; Bailey et al. 1975; Healey 1980; Landingham 1982; Volk et al. 1984). After an oil spill, pelagic zooplankton and meio-

TABLE 1.—Mean concentrations (mg/g) of hydrocarbon components in the food fed to pink salmon fry. The major components were polycyclic aromatic hydrocarbons (PAH), alkanes, and unresolved complex mixture (UCM). Raw total = PAH + alkanes + UCM. Adjusted total = total hydrocarbon content attributable to Alaska North Slope crude oil, calculated by subtracting the raw mean control concentration from all other concentrations. Treated control = TC; SE = standard error; N = number of samples analyzed.

		PAH		Alkanes		UCM		Raw total		Adjusted total	
Treatment	N	Mean	SE	Mean	SE	Mean	SE	Sum	SE	Sum	SE
Control	3	0.001	0.000	0.069	0.004	0.132	0.016	0.201	0.014	0.000	0.025
TC	3	0.001	0.000	0.050	0.025	0.174	0.026	0.225	0.026	0.024	0.037
Low oil	2	0.013	0.001	0.154	0.004	0.403	0.019	0.571	0.021	0.370	0.027
Mid oil	6	0.132	0.002	0.792	0.026	2.052	0.083	2.976	0.111	2.775	0.229
High oil	3	1.357	0.137	8.451	0.474	25.220	2.587	35.028	3.114	34.827	3.814

fauna may be contaminated by direct ingestion of particulate crude oil (Conover 1971) or by association with contaminated sediments (Landrum 1989). In Prince William Sound, oil particles were observed in mussels deployed at depths of 25 m (Short and Harris 1996b, this volume), thus oil may have been widely available as a source of contamination for pelagic zooplankton.

Although previous research (Schwartz 1985) has shown that prey contaminated with the WSF of crude oil can reduce the feeding and growth of pink salmon fry, the effects of WSF-contaminated prey are probably not representative of prey contaminated by whole oil because many of the larger, more toxic PAH are not present in WSF-contaminated prey. The purpose of this study, therefore, was to determine the effects of ingestion of whole crude oil on growth, feeding, and survival of juvenile pink salmon. To our knowledge, this is the first study of ingestion of whole oil by pink salmon.

Methods

Pink salmon fry were obtained from the Auke Creek Hatchery (Juneau, Alaska) on 16 April 1991. Before the experiment, fry were reared in 800-L cylindrical tanks receiving approximately 20 L/min single-pass seawater for 3 weeks at 4.3 ± 0.1°C. Fry were fed Biodiet[1] pellets 0.6 to 0.8 mm in diameter. Four days before treatment, 16,145 fish were randomly allocated into five treatment groups, with three replicate 65-L tanks per group. Groups were untreated control, treated control, and low-, mid-, and high-oil treatments (0.00 ± 0.02; 0.02 ± 0.04; 0.37 ± 0.03; 2.8 ± 0.2; and 35 ± 4 mg of crude oil per gram of food, respectively). The purpose of the treated control was to determine if dichloromethane, used as a carrier solvent to facilitate the mixing of crude oil and food pellets, affected the nutritional quality of the feed. Each tank received 1.4 L/min of single-pass seawater filtered through a 25-μm filter to ensure that potential prey were not introduced into the tanks. Mean water temperature was maintained at 7.8 ± 0.02°C; temperature ranged from 6 to 10°C. After 6 weeks of treatment (13 May–26 June), all groups received untreated food for 2 weeks to investigate possible recovery.

Food pellets were contaminated by dissolving Alaska North Slope crude oil obtained from the Kenai Refinery of Chevron USA (Kenai, Alaska) in dichloromethane carrier solvent and mixing it with 200-g batches of food pellets. The solvent was then removed with a rotary evaporator, followed by 1 h of air drying. Contaminated food was prepared weekly as needed and frozen in glass jars sealed with Teflon lids until use. Food for the treated controls was processed as above, except no oil was added. Control food was not modified in any way. The high-oil treatment contained the maximum amount of oil that could be absorbed by the food; the mid- and low-oil treatments were approximately one and two orders of magnitude less than the high treatment, respectively.

The amount of crude oil in the food was determined by gas chromatography, followed by mass spectrometry for aromatics or flame ionization detection for aliphatics (Short et al. 1996, this volume). Raw total hydrocarbon concentrations in all samples, including controls, were greater than zero because the food contained alkanes from natural sources. Because controls did not contain any crude oil, the mean control concentration was subtracted from all other concentrations to determine total hydrocarbon content (aromatic and aliphatic hydrocarbons, including the unresolved complex mixture [UCM]) in the food attributable to Alaska North Slope crude oil (adjusted total in Table 1). Hydro-

[1]Use of trade names does not imply endorsement by the National Marine Fisheries Service.

carbon concentrations in treated controls were not significantly different from control concentrations. The relationship among aromatic and aliphatic concentrations in food treatments is summarized in Table 1. The PAH fraction in oiled food comprised 57 ± 1% naphthalenes, 19.1 ± 0.4% phenanthrenes, 10.3 ± 0.5% dibenzothiophenes, 9.0 ± 0.5% fluorenes, 2.0 ± 0.1% chrysenes, and 2.2% other PAH.

Each day, food was provided in excess at the rate of 10% of the estimated total biomass, with a minimum of 10 g per tank per day. Biomass estimates were updated weekly to compensate for fry growth, mortality, and sampling rates. Food was provided continuously with belt feeders for approximately 16 h per day. Excess food and feces were removed twice each day, and algae was occasionally removed from the sides of the tank.

Growth, condition, and mortality were monitored for each treatment. In this article, we define growth as the change in size (length and weight) from day 0 to the date of observation. Initial fry size was determined on day 0 from three random samples of approximately 50 fry pooled from all tanks. Each week, 50 fry were randomly sampled from each tank, narcotized with MS-222 (tricaine), measured (fork length), blotted dry, and weighed. Allometric condition factor (CF) was calculated from length (mm) and weight (mg):

$$CF = \frac{\text{weight}}{\text{length}^{3.576}} \times 1{,}000. \qquad (1)$$

The exponent was determined empirically from our data. Because fry were regularly removed for analysis, cumulative percent mortality (Σp_i) was calculated as

$$\Sigma p_i = \frac{m_i}{(n_0 - \Sigma s_i - \Sigma d_i)} \times 100 + \Sigma p_{i-1}, \qquad (2)$$

where m_i = number of dead fry removed on day i, n_0 = initial number of fry, Σs_i = cumulative number sampled for experimentation up to day i, and Σd_i = cumulative number of fry removed for disease control up to day i.

Feeding rate was measured by percent stomach weight, fecal production, and stomach fullness. Each week, stomachs were excised from 15 fry, blotted, and weighed (wet) to determine percentage of total body weight. Stomach fullness was judged qualitatively, and the percentage of fry with full stomachs was calculated by tank each week. Feces were collected from groups of approximately 50 fry isolated for 1.8 days in 4-L beakers on weeks 2, 3, 6, and 7. Feces were filtered through a 200-μm nylon net, washed onto tared glass-fiber filters, dried for 24 h at 60°C, and weighed. One observation was made per replicate tank, except for week 2, when only one observation was made per treatment.

Growth coefficients were computed for the treatment period (0–6 weeks) with an exponential model:

$$\text{weight} = a \times e^{\kappa t}, \qquad (3)$$

where a = fitted intercept, t = time in days, and κ is the growth coefficient. To estimate κ, $\log_e$ (weight) was regressed on time independently by treatment and replicate to yield three estimates of κ per treatment.

Data were analyzed by week with nested analysis of variance (ANOVA) (replicate tanks were nested in treatment) or single-factor ANOVA where insufficient data (stomach fullness, fecal mass, and growth coefficients) existed for the nested model (SAS 1989). Percentage data were arcsine transformed before analysis (Kleinbaum and Kupper 1983). Relationships between control, treated control, and oil-treatment groups were examined with a priori contrasts. Results were considered statistically significant where $P \leq 0.05$. Results as presented in this paper compare treatment groups to the untreated control groups. Rarely were control and treated control responses significantly different; conclusions reached were not altered by the choice of control used for analysis. To represent nested ANOVA data graphically, tank means were computed, then averaged; N = 3 for each error bar. Median lethal concentrations were estimated with logit analysis (Berkson 1957).

Results

Mortality

Mortality in the high-oil group was significantly greater than in controls after 2 weeks and increased rapidly until fry began feeding on clean food (day 45). Mortality remained low (about 4%) in all other treatments and controls (Figure 1). Mortality in the high-oil treatment did not change much after the oiled food was discontinued until the end of the experiment (day 58), but mortality in all other groups increased gradually (Figure 1). The difference in survival between mid- and high-oil treatments was abrupt; the estimated median lethal concentration (LC50) was 29 ± 1 mg total

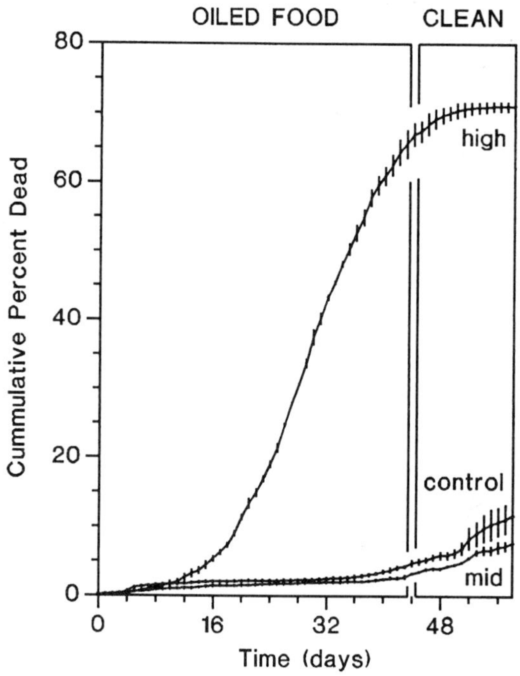

FIGURE 1.—Cumulative deaths of pink salmon fry as a function of time. The shaded region indicates the period when fry received oiled food. For clarity, only control, mid-, and high-oil treatments are displayed. The low-oil treatment mortality was intermediate between the control and mid-oil treatments after day 32. Error bars are ±1 SE.

hydrocarbons per gram of food at the end of the treatment period (1.11 ± 0.04 mg PAH/g), and 25 ± 1 mg/g at the end of the experiment (0.97 ± 0.05 mg PAH/g).

Feeding

Feeding rate, as estimated by percent stomach weight, tended to increase at low doses of oil but was depressed at high doses (Figure 2). Percent stomach weight declined immediately in the high-oil treatment and remained significantly depressed until clean food was available. Fry in the high-oil treatment eventually became lethargic and struck food less frequently than in controls and other treatments (qualitative observation), but did continue to consume food through the 6-week treatment period. Percent stomach weight was also significantly depressed during the first week (only) in the mid-oil treatment. In the low-oil treatment, percent stomach weight was significantly greater than in controls on weeks 4 and 5. No other groups differed from the controls. After oiled food was discontinued, feeding in the high-oil treatment increased rapidly and merged with controls in weeks 7 to 8 (Figure 2). Feeding rates, averaged over weeks 1–6, were significantly depressed in the high-oil treatment, but not in other treatments (Figure 3).

Feeding response, as measured by percentage of fry with full stomachs, was similar to the response measured by stomach weight (Figure 2). Overall (weeks 0–8), the mean percentage of control fry with full stomachs was 52 ± 4 (SE). The mean percentage of fry with full stomachs in the high-oil treatment was significantly reduced after 1 week and remained below 10% until oiled food was discontinued. The mean percentage of fry with full stomachs was also reduced in the low- and mid-oil treatments on week 1, but was not significantly different from controls on other weeks.

Fecal Production

Fecal production was also affected by consumption of oiled food; this response was generally similar to changes in percent stomach weight (Figure 3). On week 3, and for weeks 3 and 6 combined, fecal production in the high-oil group was significantly depressed. On week 2, where only one sample was collected per treatment, production declined uniformly with increasing oil concentration. Statistical discrimination power was poor because fewer fecal data were collected than for any other response variable.

Growth

Growth of fry was inhibited by oiled food (Figure 4). Fry in the mid- and high-oil treatments grew significantly less than controls after 1 week of exposure and throughout the experiment. By week 2, growth was significantly depressed in all oil treatment groups and remained significantly depressed through the end of the experiment, except for weeks 4 and 5 in the low-oil treatment. Significant reductions in growth occurred at oil concentrations as low as 1.3% of the LC50.

Exponential growth coefficients, estimated during the exposure period (weeks 0–6), were reduced by exposure to oiled food (Figure 5). Coefficients in the mid- and high-oil treatments were significantly less than control and treated control coefficients. Estimated growth (percent body weight per day) was 2.2 ± 0.1% for control fry, and 0.4 ± 0.1% for high-oil-treatment fry. Estimated linearly, control fry grew 0.25 ± 0.01 mm/d, and fry in the high-oil treatment grew 0.07 ± 0.01 mm/d.

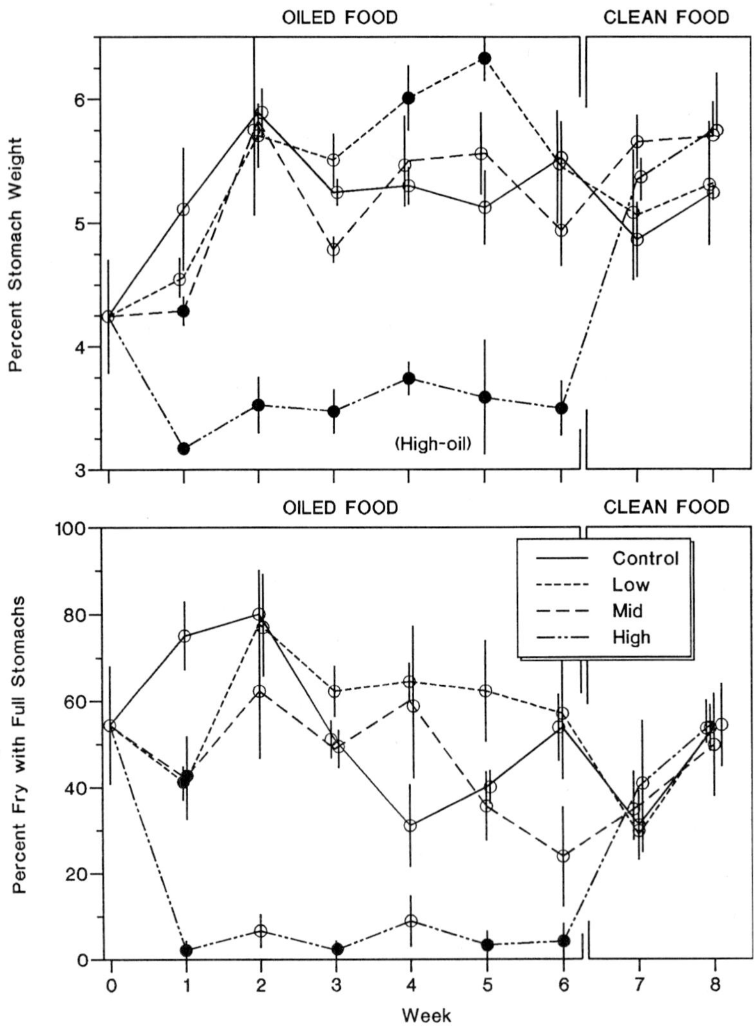

FIGURE 2.—Time series showing changes in percent stomach weight and percent of juvenile pink salmon with full stomachs. The shaded region indicates the period when fry received oiled food. Solid symbols indicate significant differences from control. For clarity, treated controls were not plotted. Error bars are ±1 SE.

Pink salmon length–weight relationships (where $y = \log_e$ [weight] and $x = \log_e$ [length]), for weeks 0 to 8, analyzed with analysis of covariance, were not significantly different ($P_{slope} = 0.402$, $P_{intercept} = 0.398$) among treatments. For all treatments (weeks 0–8), weight $= 0.001 \cdot \text{length}^{3.576}$. Except for the high-oil treatment, each treatment fit the model well; $r^2 = 0.73$ for the high-oil treatment, but ranged from 0.94 to 0.96 for all other treatments, including control.

Because the length–weight relationship in the high-oil treatment was more variable than in other treatments, we analyzed condition factor weekly, reasoning that condition might change with time. Changes in *CF* and percent stomach weight were similar (compare Figures 6 and 2). In the high-oil treatment, *CF* was significantly less than in the control by week 1 and remained depressed until oiled food was discontinued (Figure 6). On week 7, *CF* in the high-oil treatment was still significantly depressed, but had recovered substantially. On week 8, *CF* in the high-oil treatment was not significantly different from the control, but was more similar to the *CF* in the mid-oil treatment. In the mid-oil treatment, *CF* became significantly elevated (weeks 5 through 7); this trend began after 4 weeks and continued through the remainder of the experiment.

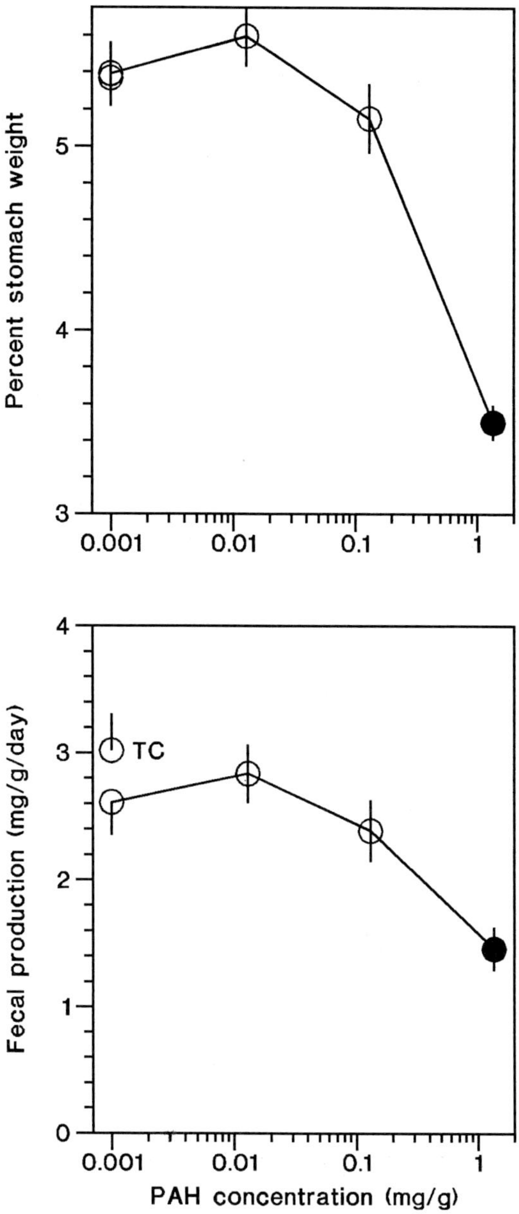

FIGURE 3.—Percent stomach weight (weeks 1–6; means by treatment) and fecal dry weight (weeks 3 and 6; means by treatment) as a function of PAH concentration in oiled food. Solid symbols indicate significant differences from control; TC = treated control. Error bars are ±1 SE.

Discussion

Results of this study demonstrate that direct ingestion of oil-contaminated food can negatively impact the growth and survival of salmon fry. Feeding rate and survival of pink salmon fry given food contaminated with high oil concentrations (34.8 mg oil/g food) was reduced. Moreover, at lower concentrations (0.37 and 2.8 mg oil/g food), growth of fry was reduced even though feeding rates did not decline. Although direct exposure of pink salmon fry to the WSF of crude oil can lead to death (Rice 1973; Rice et al. 1975; Rice et al. 1979; Moles and Rice 1983), we are unaware of previous reports documenting that oiled food is toxic to salmon.

Previous studies documenting ingestion of petroleum hydrocarbons by salmon have shown more moderate effects than those in this study, probably because (1) the composition of the aromatic hydrocarbons in the food was different, (2) total aromatic hydrocarbon concentrations were lower, and (3) exposure durations were more limited. In 50-d tests, Schwartz (1985) found that prey contaminated with the WSF of Cook Inlet crude oil was not acutely toxic to pink salmon fry. However, in the Schwartz (1985) study, monoaromatic hydrocarbons constituted 35% of the aromatics in the prey, and compounds larger than two rings probably were negligible. In our study the food contained primarily two- to four-ring PAH compounds; monoaromatic compounds probably were removed by food preparation procedures. Because the toxicity of aromatic hydrocarbons increases with molecular size (Rice et al. 1977a), the food in our study was more inherently toxic than the WSF-contaminated prey in the Schwartz (1985) study. Additionally, the maximum hydrocarbon concentration tested by Schwartz (1985) was similar to the low-oil treatment in this study. In another study, no mortality was found in juvenile coho salmon *Oncorhynchus kisutch* force-fed 1 g Prudhoe Bay crude oil per 100 g body weight (Morrow 1974), but they were observed for only 4 d. Mortality in our experiments became significant only after 2 weeks of exposure.

Mortality in the high-oil treatment probably was caused by a combination of starvation and oil toxicity. Because separate starvation tests were not conducted, we were unable to clearly separate starvation and toxicity effects. Starvation is a probable contributing factor because feeding was significantly depressed for 5 to 6 weeks and the fish showed symptoms of starvation, such as depressed condition factor and watery flesh. Nucleic acid content in the fry indicated that metabolic catabolism occurred at the cellular level as the result of nutritional stress or starvation (Wang et al. 1993), particularly in the high-oil treatment. In addition, the aromatic compounds in the oiled food were probably toxic. In tests involving alkanes and monoaro-

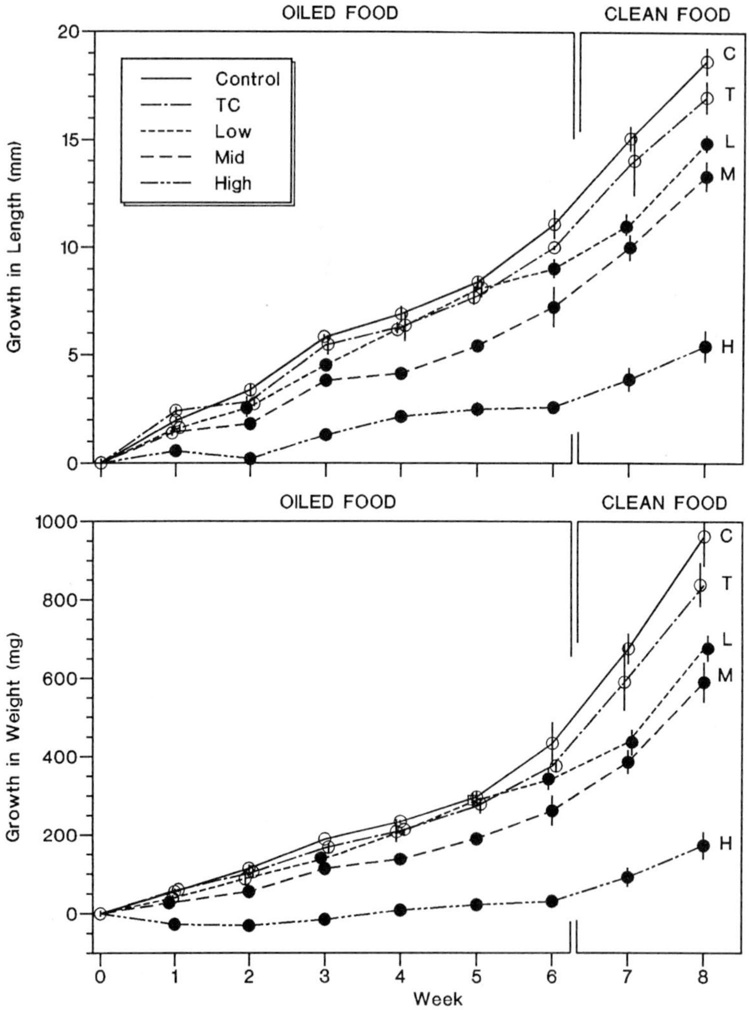

FIGURE 4.—Time series showing growth of juvenile pink salmon in length and wet weight. The shaded region indicates the period when fry received oiled food. Solid symbols indicate significant differences from control. Error bars are ±1 SE.

matic hydrocarbons, Morrow et al. (1975) concluded that unsaturated cyclic hydrocarbons were the most toxic. In WSF and oil-in-water dispersion tests, Anderson et al. (1974) concluded that toxicity was primarily a function of di- and triaromatic content. In our study, two- to four-ring aromatic compounds (PAH) were present in the oiled food and accounted for 3.9 ± 0.3% of all hydrocarbons.

Reductions in salmon feeding as a result of exposure to crude oil has also been reported by other researchers. For example, when offered prey contaminated with WSF, the feeding rate of pink salmon fry declined (Schwartz 1985). Schwartz (1985) found that feeding rate was reduced at concentrations below those in our high-oil treatment, but the technique used to estimate feeding rate was more sensitive than our method. Declines in feeding have also been observed in adult salmonids exposed to WSF of crude oil. For example, most oil-exposed coho adults showed little or no interest in prey when exposed to 343 ± 93 μg WSF/L of Cook Inlet crude oil (Folmar et al. 1981).

Fry in our high-oil treatment showed remarkable ability to recover when offered clean food; mortality ceased abruptly, swimming behavior became more normal, feeding and growth rates increased, and condition factor increased rapidly. This recovery was facilitated by rapid elimination of petroleum

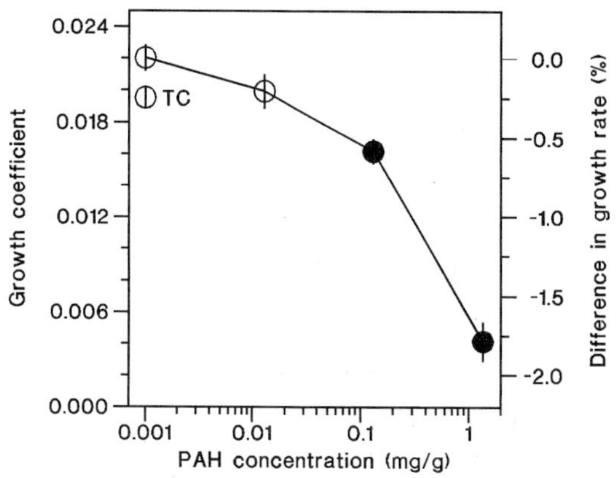

FIGURE 5.—Growth coefficient (κ) as a function of exposure (during weeks 0–6) to PAH concentrations in oiled food. Solid symbols indicate significant differences from control; TC = treated control. Percent change in growth rate with respect to mean control growth rate is indicated. Error bars are ±1 SE.

hydrocarbons from body tissues and probably was enhanced by metabolism of aromatic hydrocarbons by mixed function oxidase enzymes. For example, pink salmon alevins and fry exposed to WSF begin metabolizing petroleum hydrocarbons at rates high enough to reduce hydrocarbon concentrations in tissues while still exposed to WSF or oil-contaminated prey (Rice et al. 1977b; Schwartz 1985; Moles et al. 1987).

In this study, feeding rate did not decline in the low- and mid-oil treatments, but growth was significantly depressed. This suggests that reduced somatic growth was not simply the result of starvation, but rather to increased metabolic demand or changes in digestive efficiency. In a study where juvenile pink salmon consumed prey contaminated with the WSF of crude oil, somatic growth was reduced (Schwartz 1985); direct exposure of pink salmon fry to WSF also reduced growth (Rice et al. 1975; Moles and Rice 1983). Energy available for growth could have been reduced by metabolic demands, reduced feeding, or reduced digestive assimilation efficiency. Synthesis of enzymes necessary to metabolize petroleum hydrocarbons may require energy otherwise available for somatic growth. Pink salmon are capable of actively metabolizing hydrocarbons (Rice et al. 1977b; Schwartz 1985). One known enzyme that is induced by PAH in salmonids and other fish is cytochrome P450 (Payne and Penrose 1975; Miller et al. 1989). For example, P450 activity increased more than threefold in livers of coho salmon smolts exposed to WSF for 30 d (Collodi et al. 1984). Exposure of pink salmon to the WSF of crude oil may also cause elevated breathing rates and oxygen consumption, further evidence of increased metabolic demand (Thomas and Rice

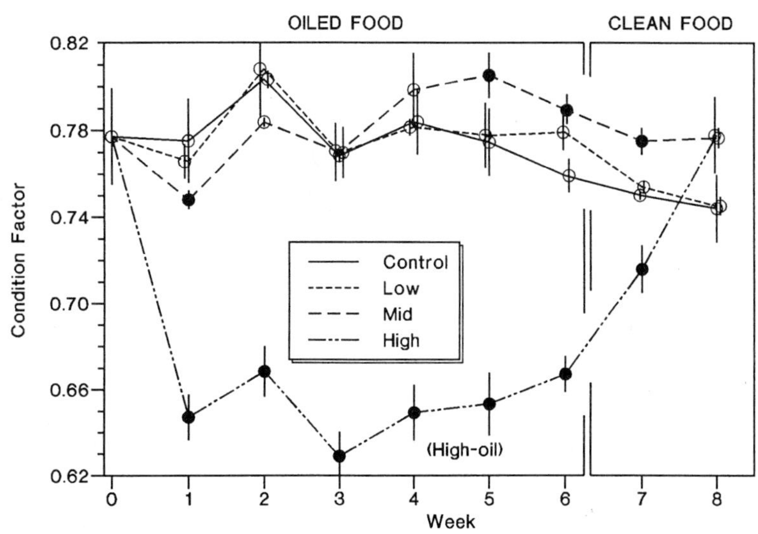

FIGURE 6.—Time series showing changes in allometric condition factor of juvenile pink salmon. The shaded region indicates the period when fry received oiled food. Solid symbols indicate significant differences from control. For clarity, treated controls were not plotted. Error bars are ±1 SE.

1979). Exposure to sublethal hydrocarbon concentrations elevates metabolism and energy demands and requires increased food intake (Rice et al. 1977b). Reduction in digestive efficiency may have also been a factor in growth reduction, as found in Atlantic salmon parr (Vignier et al. 1992).

Change in fry growth in this experiment was consistent with that observed in fry from oiled areas of Prince William Sound; in both cases, exposure to petroleum hydrocarbons reduced growth. In western Prince William Sound in 1989, the apparent daily growth rate of fry was lower in oiled areas (1.5% body weight per day) than in non-oiled areas (2.9% body weight per day), namely, a difference of 1.4% (Wertheimer and Celewycz 1996, this volume). The difference in the actual growth rate of specific tagged groups of pink salmon fry was about 1% lower for fish recovered in oiled areas than in non-oiled areas (Willette 1996). In our experiment, the 1% difference in growth rate occurred between the mid- and high-oil treatments at an estimated concentration of 7 mg total petroleum hydrocarbons per gram of food (0.3 mg PAH/g food). This implies that fry in oiled areas in Prince William Sound ingested or otherwise acquired substantial quantities of oil.

Exposure to petroleum hydrocarbons can affect juvenile pink salmon populations through predation effects in two ways. First, the lethargic behavior of fry in the high-oil treatment suggests that salmon exposed to high oil levels would be less able to avoid predation. Second, a reduction in the growth rate of juvenile salmon may prolong exposure to predation, and thus lead to increased mortality. Because year-class strength may be affected by early marine residency (Parker 1968; Walters et al. 1978; Bax 1983; Nichelson 1986), the reduced growth caused by exposure of pink salmon fry to *Exxon Valdez* crude oil in Prince William Sound after the 1989 spill (Wertheimer and Celewycz 1996; Willette 1996) may have reduced population survival (Geiger et al. 1996, this volume).

In conclusion, this study demonstrated that ingestion of oil-contaminated food can reduce growth and survival of juvenile pink salmon; growth was reduced at concentrations as low as 1.3% of the LC50. This research strongly supports the hypothesis that ingestion of petroleum hydrocarbons was a cause of reduced growth in pink salmon fry in Prince William Sound in the marine environment in 1989 (Carls et al. 1996).

Acknowledgments

We thank Joe Alesky, Kitty Bauer, Tony Chan, Gennifer Moreau, Marc Pritchett, Lori Quintrell, and Irene Strafford for their efforts, talents, and time dedicated to the successful completion of this study. For reviewing and improving the manuscript, we thank Mike Murphy. We also thank the *Exxon Valdez* Oil Spill Trustee Council for funding this research.

References

Anderson, J. W., J. M. Neff, B. A. Cox, H. E. Tatem, and G. M. Hightower. 1974. Characteristics of dispersions and water-soluble extracts of crude and refined oils and their toxicity to estuarine crustaceans and fish. Marine Biology 27:75–88.

Bailey, J. E., B. L. Wing, and C. R. Mattson. 1975. Zooplankton abundance and feeding habits of fry of pink salmon and chum salmon in Traitor's Cove, Alaska, with speculations on the carrying capacity of the area. U.S. National Marine Fisheries Service Fishery Bulletin 73:946–961.

Bax, N. J. 1983. Early marine mortality of marked juvenile chum salmon *Oncorhynchus keta* released into Hood Canal, Puget Sound, Washington, in 1980. Canadian Journal of Fisheries and Aquatic Sciences 40:426–435.

Berkson, J. 1957. Tables for the maximum likelihood estimate of the logistic function. Biometrics 13:28–34.

Carls, M. G., A. G. Celewycz, and A. C. Wertheimer. 1994. Contamination of surficial sediments by *Exxon Valdez* crude oil in nearshore habitats of juvenile salmon in Prince William Sound. Pages 3B1–3B17 *in* A. C. Wertheimer, A. G. Celewycz, M. G. Carls, and M. V. Sturdevant, editors. Impact of the oil spill on juvenile pink and chum salmon and their prey in critical nearshore habitats—fish/shellfish study 4. State–Federal Natural Resource Damage Assessment Final Report to *Exxon Valdez* Oil Spill Trustee Council, Anchorage.

Carls, M. G., A. C. Wertheimer, J. W. Short, R. M. Smolowitz, and J. J. Stegeman. 1996. Contamination of juvenile pink and chum salmon by hydrocarbons in Prince William Sound after the *Exxon Valdez* oil spill. American Fisheries Society Symposium 18:593–607.

Collodi, P., M. S. Stekoll, and S. D. Rice. 1984. Hepatic aryl hydrocarbon hydroxylase activities in coho salmon (*Oncorhynchus kisutch*) exposed to petroleum hydrocarbons. Comparative Biochemistry and Physiology 79C(2):337–341.

Conover, R. J. 1971. Some relations between zooplankton and bunker C oil in Chedabucto Bay following the wreck of the tanker *Arrow*. Journal of the Fisheries Research Board of Canada 28:1327–1330.

Folmar, L. C., D. R. Craddock, J. W. Blackwell, G. Joyce, and H. O. Hodgins. 1981. Effects of petroleum exposure on predatory behavior of coho salmon *Oncorhynchus kisutch*. Bulletin of Environmental Contamination and Toxicology 27:458–462.

Geiger, H. J., B. G. Bue, S. Sharr, A. C. Wertheimer, and

T. M. Willette. 1996. A life history approach to estimating damage to Prince William Sound pink salmon caused by the *Exxon Valdez* oil spill. American Fisheries Society Symposium 18:487–498.

Hargreaves, N. B., and R. J. LeBrasseur. 1985. Species selective predation on juvenile pink (*Oncorhynchus gorbuscha*) and chum (*O. keta*) by coho salmon (*O. kisutch*). Canadian Journal of Fisheries and Aquatic Sciences 42:659–668.

Healey, M. C. 1980. The ecology of juvenile salmon in Georgia Strait, British Columbia. Pages 203–230 *in* W. J. McNeil and D. C. Himsworth, editors. Salmonid ecosystems of the north Pacific. Oregon State University Press, Corvallis.

Kaczynski, V. W., R. J. Feller, and C. Clayton. 1973. Trophic analysis of juvenile pink and chum salmon in Puget Sound. Journal of the Fisheries Research Board of Canada 30:1003–1008.

Kleinbaum, D. G., and L. L. Kupper. 1983. Applied regression analysis and other multivariable methods. Duxbury Press, Boston.

Landingham, J. H. 1982. Feeding ecology of pink and chum salmon fry in the nearshore habitat of Auke Bay, Alaska. Master's thesis. University of Alaska, Juneau.

Landrum, P. F. 1989. Bioavailability and toxikinetics of polycyclic aromatic hydrocarbons sorbed to sediments for the amphipod *Pontoporeia hoyi*. Environmental Science and Technology 23:588–595.

LeBrasseur, R. J., and R. R. Parker. 1964. Growth rate of central British Columbia pink salmon (*Oncorhynchus gorbuscha*). Journal of the Fisheries Research Board of Canada 21:1101–1128.

Maki, A. W. 1991. The *Exxon Valdez* oil spill: initial environmental impact assessment. Environmental Science and Technology 25:24–29.

Miller, M. R., D. E. Hinton, and J. J. Stegeman. 1989. Cytochrome P-450E induction and localization in gill pillar (endothelial) cells of scup and rainbow trout. Aquatic Toxicology 14:307–322.

Moles, A., M. M. Babcock, and S. D. Rice. 1987. Effects of oil exposure on pink salmon *Oncorhynchus gorbuscha* alevins in a simulated intertidal environment. Marine Environmental Research 21:49–58.

Moles, A., and S. D. Rice. 1983. Effects of crude oil and naphthalene on growth, caloric content, and fat content of pink salmon juveniles in seawater. Transactions of the American Fisheries Society 112:205–211.

Morrow, J. E. 1974. Effects of crude oil and some of its components on young coho and sockeye salmon. U.S. Environmental Protection Agency Report (EPA 660/3-73-018), Washington, DC.

Morrow, J. E., R. L. Gritz, and M. P. Kirton. 1975. Effects of some components of crude oil on young coho salmon. Copeia 1975:326–331.

Mortensen, D. G., J. H. Landingham, A. C. Wertheimer, and S. G. Taylor. 1991. Relationship of early marine growth and survival of juvenile pink salmon to marine water temperature and secondary production in Auke Bay, Alaska. Pages 38–49 *in* I. Guthrie and B. Wright, editors. Proceedings, fifteenth northeast Pacific pink and chum salmon workshop. Pacific Salmon Commission, Parksville, British Columbia.

Neff, J. M. 1991. Water quality in Prince William Sound and the Gulf of Alaska (an update of the April, 1990 report). Arthur D. Little, Cambridge, Massachusetts.

Nichelson, T. E. 1986. Influences of upwelling, ocean temperature, and smolt abundance on marine survival of coho salmon (*Oncorhynchus kisutch*) in the Oregon production area. Canadian Journal of Fisheries and Aquatic Sciences 43:527–535.

Parker, R. R. 1968. Marine mortality schedules of pink salmon of the Bella Coola River, central British Columbia. Journal of the Fisheries Research Board of Canada 25:757–794.

Parker, R. R. 1971. Size selective predation among juvenile salmonid fishes in a British Columbia inlet. Journal of the Fisheries Research Board of Canada 28: 1503–1510.

Payne, J. F., and W. R. Penrose. 1975. Induction of aryl hydrocarbon (benzo[*a*]pyrene) hydroxylase in fish by petroleum. Bulletin of Environmental Contamination and Toxicology 14:112–116.

Rice, S. D. 1973. Toxicity and avoidance tests with Prudhoe Bay oil and pink salmon fry. Pages 667–670 *in* Proceedings, 1973 joint conference on prevention and control of oil spills. American Petroleum Institute, Washington, DC.

Rice, S. D., D. A. Moles, and J. W. Short. 1975. The effect of Prudhoe Bay crude oil on survival and growth of eggs, alevins, and fry of pink salmon, *Oncorhynchus gorbuscha*. Pages 503–507 *in* API (American Petroleum Institute), EPA (U.S. Environmental Protection Agency), and USCG (U.S. Coast Guard) 1975 conference on prevention and control of oil pollution, proceedings. API, Washington, DC.

Rice, S. D., J. W. Short, and J. F. Karinen. 1977a. Comparative oil toxicity and comparative animal sensitivity. Pages 78–94 *in* D. A. Wolfe, editor. Fate and effects of petroleum hydrocarbons in marine organisms and ecosystems. Pergamon Press, New York.

Rice, S. D., R. E. Thomas, and J. W. Short. 1977b. Effect of petroleum hydrocarbons on breathing and coughing rates and hydrocarbon uptake-depuration in pink salmon fry. Pages 259–277 *in* F. J. Vernberg, A. Calabrese, F. P. Thurberg, and W. B. Vernberg, editors. Physiological responses of marine biota to pollutants. Academic Press, New York.

SAS. 1989. SAS/STAT user's guide, version 6, 4th edition, volume 2. SAS Institute, Inc., Cary, North Carolina.

Schwartz, J. P. 1985. Effects of oil-contaminated prey on the feeding and growth rate of pink salmon fry *Oncorhynchus gorbuscha*. Pages 459–476 *in* F. J. Vernberg, F. Thurberg, A. Calabrese, and W. Vernberg, editors. Pollution and physiology of marine organisms. University of South Carolina Press, Columbia.

Short, J. W., T. J. Jackson, M. L. Larsen, and T. L. Wade. 1996. Analytical methods used for the analysis of hydrocarbons in crude oil, tissues, sediments, and seawater collected for the natural resources damage assessment of the *Exxon Valdez* oil spill. American Fisheries Society Symposium 18:140–148.

Short, J. W., and P. M. Harris. 1996a. Chemical sampling and analysis of petroleum hydrocarbons in near-surface seawater of Prince William Sound after the *Exxon Valdez* oil spill. American Fisheries Society Symposium 18:17–28.

Short, J. W., and P. M. Harris. 1996b. Petroleum hydrocarbons in caged mussels deployed in Prince William Sound after the *Exxon Valdez* oil spill. American Fisheries Society Symposium 18:29–39.

Sturdevant, M. V., A. C. Wertheimer, and J. L. Lum. 1996. Diets of juvenile pink and chum salmon in oiled and non-oiled nearshore habitats in Prince William Sound, 1989 and 1990. American Fisheries Society Symposium 18:578–592.

Thomas, R. E., and S. D. Rice. 1979. The effect of exposure temperatures on oxygen consumption and opercular breathing rates of pink salmon fry exposed to toluene, naphthalene, and water-soluble fractions of Cook Inlet crude oil and no. 2 fuel oil. Pages 39–52 *in* W. B. Vernberg, A. Calabrese, F. P. Thurberg, and F. J. Vernberg, editors. Marine pollution: functional responses. Academic Press, New York.

Vignier, V., J. H. Vandermeulen, and A. J. Fraser. 1992. Growth and food conversion by Atlantic salmon parr during 40 days' exposure to crude oil. Transactions of the American Fisheries Society 121:322–332.

Volk, E. C., R. C. Wissmar, C. A. Simenstad, and D. M. Eggers. 1984. Relationship between otolith microstructure and the growth of juvenile chum salmon (*Oncorhynchus keta*) under different prey rations. Canadian Journal of Fisheries and Aquatic Sciences 41:126–133.

Walters, C. J., R. Hilborn, R. M. Peterman, and M. J. Staley. 1978. Model for examining early ocean limitations of Pacific salmon. Canadian Journal of Fisheries and Aquatic Sciences 41:126–133.

Wang, S. Y., J. L. Lum, M. G. Carls, and S. D. Rice. 1993. The relationship between growth and total nucleic acids in juvenile pink salmon, *Oncorhynchus gorbuscha*, fed crude oil contaminated food. Canadian Journal of Fisheries and Aquatic Sciences 50:996–1001.

Wertheimer, A. C., and A. G. Celewycz. 1996. Abundance and growth of juvenile pink salmon in oiled and non-oiled locations of western Prince William Sound after the *Exxon Valdez* oil spill. American Fisheries Society Symposium 18:518–532.

Willette, M. 1996. Impacts of the *Exxon Valdez* oil spill on the migration, growth, and survival of juvenile pink salmon in Prince William Sound. American Fisheries Society Symposium 18:533–550.

American Fisheries Society Symposium 18:619–627, 1996

Effects of the *Exxon Valdez* Oil Spill on Pink Salmon Embryos and Preemergent Fry

BRIAN G. BUE
Alaska Department of Fish and Game, 333 Raspberry Road, Anchorage, Alaska 99518, USA

SAMUEL SHARR, STEVE D. MOFFITT, AND ANDREW K. CRAIG
Alaska Department of Fish and Game, P.O. Box 669, Cordova, Alaska 99574, USA

Abstract.—We examined pink salmon *Oncorhynchus gorbuscha* preemergent fry mortality, embryo mortality, and survival from embryo to preemergent fry in intertidal and upstream areas of Prince William Sound that were affected by the *Exxon Valdez* oil spill. No change in preemergent fry density or increase in numbers of dead fry was detected within 1 month of the spill. Pink salmon embryo mortality was elevated in oil-affected streams during fall of each year from 1989 through 1992 ($P <$ 0.023 for all years). In 1989 increased embryo mortality was detected in the lower intertidal zones, and in 1990 elevated mortalities were observed at the highest intertidal zone. In 1991 embryo mortality was significantly higher in all intertidal zones, as well as upstream of oil contamination. The same patterns of mortalities were observed in 1992, although they were not as extreme as that observed in 1991. However, no difference in survival of embryo to preemergent fry was detected for the 1989–1991 brood years.

Wild Pacific salmon play a major role in the Prince William Sound ecosystem while also contributing to the region's commercial fisheries. Migrating salmon fry are an important spring food source for many animals. Marine mammals, fish, and birds prey on the ocean stages of salmon; terrestrial species, such as bears, river otters, eagles, and gulls depend on salmon for their summer diet. Salmon also provide a pathway for transferring nutrients from marine ecosystems to nearshore and terrestrial ecosystems. Total returns of wild pink salmon *Oncorhynchus gorbuscha* have ranged from 1.8 to 21.0 million since 1978.

Up to 75% of spawning pink salmon in Prince William Sound use intertidal areas (Helle et al. 1964), which by nature are highly susceptible to contamination from marine oil spills. Moles et al. (1987) and Rice et al. (1975) found that pink salmon embryos and preemergent fry were adversely affected by exposure to crude oil and that the effect was most acute in intertidal environments. The 24 March 1989 spill from the *Exxon Valdez* occurred just before the spring migration of salmon fry and contaminated many intertidal spawning areas in central and southwestern Prince William Sound.

We evaluated (1) the short-term effect of oil-contaminated incubation substrate on preemergent pink salmon fry, (2) the effect of intertidal oiling on pink salmon embryo mortality, and (3) the effect of intertidal oiling on the survival of pink salmon embryos to preemergent fry. This study is part of the work done by Sharr et al. (1994a; 1994b).

Materials and Methods

Study Sites

Study streams were selected using the following criteria: (1) adult salmon returns were expected to be large enough to provide a high probability of success in embryo and fry sampling, (2) embryo and fry sampling had been done in past years, and (3) streams which showed no oil contamination (reference) were selected near oil-contaminated (oiled) streams. Because pink salmon embryos and preemergent fry remain in the area of the stream where they were deposited as eggs, we were able to select oiled and reference sites in close proximity to each other, thus reducing the geographical effect on the findings.

Stream oiling was assessed through visual observations of the stream and the adjacent area. The observations were supported by photographs, observation maps, and hydrocarbon analysis of mussels (*Mytilus* sp.) collected near the stream mouths. These data were collected as part of another Natural Resource Damage Assessment study (NRDA F/S study 1).

Sample Design

The methods used for both embryo and preemergent fry sampling were similar to those described by Pirtle and McCurdy (1977). Sample sites were stratified by tide zone to account for differences in embryo mortality or overwinter survival caused by salinity, temperature, predation, and oiling gradient. Four zones, three intertidal and one above the mean high tide line, were sampled whenever possible for each stream: 1.8–2.4 m, 2.4–3.0 m, 3.0–3.7 m above mean low water, and upstream of mean high tide (3.7 m). Zone boundaries were staked before sampling and established by elevation above mean low water using appropriate tide tables, a surveyor's level, and stadia rod. No sampling was done below the 1.8–2.4-m zone because survival was expected to be low (Helle et al. 1964). Upstream sample areas were often within the reach of extreme high tides (3.7–4.6 m) because ice and snow often limited the extent of upstream sampling.

Separate linear transects were established for each zone on the embryo and preemergent fry surveys. Although most transects were 30.5 m long, some were shorter because of steep stream gradients. Transects were placed in riffle areas where spawning was observed during escapement surveys conducted by NRDA F/S study 1. Transects ran diagonally across the river: preemergent fry survey transects started downstream against the right bank and moved upstream to the left bank, whereas embryo survey transects started downstream against the left bank and moved upstream to the right bank. This placement of embryo and fry transects reduced sampling overlap and the influence of fall embryo sampling on spring fry abundance. A map was drawn for each stream, indicating the tide zones and transect locations in relation to major landmarks. Each embryo transect was marked with surveyor's flagging and photographed to ensure the transects would be in the same immediate area.

Fourteen circular samples, each 0.186 m^2, were systematically collected at approximately equal intervals along each transect. The number of samples was a compromise between reducing variance and the practicality and expense of conducting the study. To avoid excessive sampling of the stream, fewer samples were completed on narrow stream channels. Streams that split into two or more channels within a zone were sampled either by allocating samples among channels based on spawner distribution observed during NRDA F/S study 1 or, where spawner distribution was unknown, by an equal allocation of samples among channels.

Sampling date, time, tide zone, the number of live and dead embryos, the number of live and dead preemergent fry, and an estimate of the overall yolk sac absorption percentage were collected for each transect during both embryo and fry sampling. Pink salmon embryos were separated from chum salmon *Oncorhynchus keta* and coho salmon *O. kisutch* salmon embryos by their smaller size. Pink salmon fry were differentiated from chum salmon fry by their smaller size and lack of parr marks. An embryo was considered dead if it was opaque or discolored. Sampling often killed preemergent fry (especially newly hatched fry), so fry were only considered dead if decomposition was evident.

1989 Preemergent Studies

The short-term effect of oil-contaminated incubation substrate on pink salmon preemergent fry was examined during spring 1989. The annual preemergent pink salmon fry density survey conducted by the Alaska Department of Fish and Game was underway at the time of the *Exxon Valdez* spill; consequently, 39 streams were sampled for preemergent fry density before or shortly after being oiled. An additional session of sampling was initiated approximately 2 weeks later. In this second session, 14 streams examined during the first session were reexamined to assess the immediate effect of oiling. Differences in pink salmon preemergent fry density and mortality rate between the first and second survey were tested for each stream using a fixed-effects, two-way analysis of variance. The two factors were sampling period (first and second sampling) and height in the intertidal zone (1.8–2.4 m, 2.4–3.0 m, and 3.0–3.7 m above mean low water and upstream).

Embryo Mortality

The effect of intertidal oiling on pink salmon embryo mortality was assessed by examining the number of dead and live embryos present in the stream gravel shortly before freeze-up in the fall in 10 oiled and 15 reference streams (Figure 1). Pink salmon embryo mortality rates ($\hat{M}^e_{ij}$) were estimated for each stream and zone using the following relationship:

$$\hat{M}^e_{ij} = \frac{\sum_k (\mathrm{DE}^e_{ijk} + \mathrm{DF}^e_{ijk})}{\sum_k (\mathrm{LE}^e_{ijk} + \mathrm{DE}^e_{ijk} + \mathrm{LF}^e_{ijk} + \mathrm{DF}^e_{ijk})}, \quad (1)$$

where DE^e_{ijk}, DF^e_{ijk}, LE^e_{ijk}, and LF^e_{ijk} are the number of dead embryos, dead fry, live embryos, and live fry from dig k, stream i, zone j, collected during the embryo survey, respectively.

Differences in embryo mortality rate were examined using a mixed-effects, two-factor experiment with repeated measures on one factor (Neter et al. 1990):

$$Y_{ijk} = \mu_{...} + O_i + Z_j + (OZ)_{ij} + S_{k(i)} + \epsilon_{(ijk)}. \quad (2)$$

The two treatments were extent of oiling (O_i; oiled and reference) and height in the intertidal zone (Z_j;

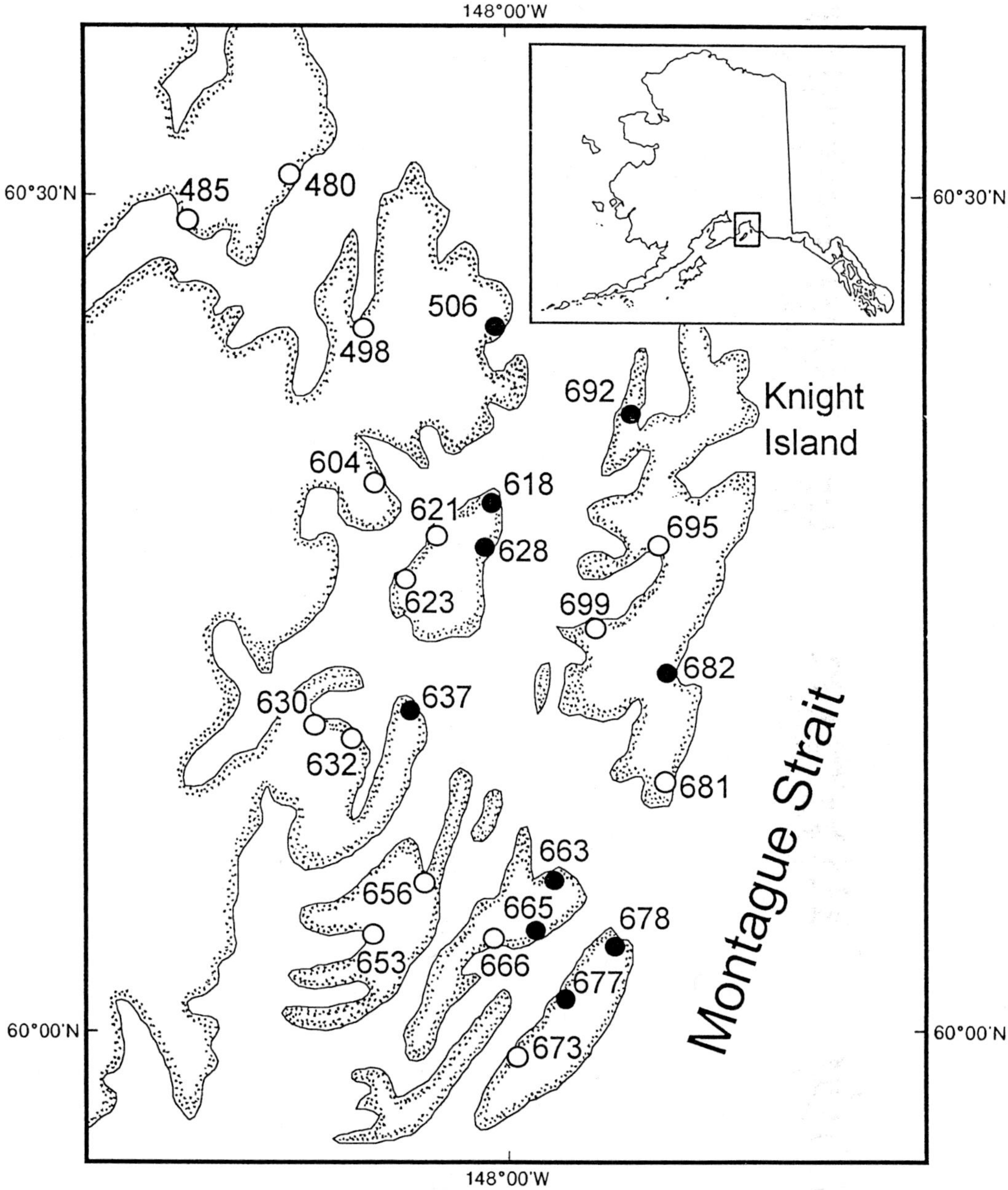

FIGURE 1.—Location of 10 oil-contaminated (solid circles) and 15 reference streams (open circles) used for assessment of pink salmon embryo mortality and survival from embryo to preemergent fry in Prince William Sound, Alaska. The number next to the stream locator is the stream identifier used by the Alaska Department of Fish and Game.

1.8–2.4 m, 2.4–3.0 m, and 3.0–3.7 m above mean low water and upstream). The data were blocked by stream ($S_{k(i)}$), a random effect that was nested within extent of oiling. The interaction of extent of oiling and height in the intertidal zone was also examined. The assumption of constant variance for error terms was tested using the F_{max}-test (Sokal and Rohlf 1969), and the normality of error term was visually assessed using scatter plots, box plots, and normal probability plots. The logit transform was used when analyzing the embryo mortality data. Assumptions relating to a valid split-plot analysis of

TABLE 1.—Results of the repeated pink salmon preemergent fry sampling during spring 1989 in Prince William Sound. Oiling (N = no, Y = yes) is based on visual observations from Natural Resource Damage Assessment F/S study 1.

Stream			Date of sample		Probability of a more extreme result (P value)					
					Live fry density			Fry mortality		
Number	Name	Oiling	First	Second	Sample[a]	Zone[b]	Interaction[c]	Sample[d]	Zone[b]	Interaction[c]
35	Koppen	N	8 Apr	24 Apr	0.33	0.00[e]	0.98	0.70	0.26	0.98
455	Paulson	N	30 Mar	14 Apr	0.73	0.01[e]	0.78	[f]	[f]	[f]
480	Mink	N	30 Mar	15 Apr	0.00[eg]	0.00[e]	0.00[e]	0.39	0.01[e]	0.32
604	Erb	N	31 Mar	16 Apr	0.66	0.00[e]	0.33	0.30	0.13	0.11
621	Totemoff	N	31 Mar	16 Apr	0.97	0.00[e]	0.51	0.14	0.41	0.61
632	Claw	N	1 Apr	17 Apr	0.36	0.00[e]	0.34	0.14	0.41	0.39
673	Falls	N	1 Apr	18 Apr	0.79	0.00[e]	0.78	0.28	0.41	0.38
677	Hayden	Y	1 Apr	18 Apr	0.32	0.30	0.81	0.12	0.13	0.13
681	Hogan	N	4 Apr	21 Apr	0.36	0.16	0.81	0.31	0.29	0.29
682	Snug	Y	3 Apr	21 Apr	0.87	0.00[e]	0.31	0.78	0.01[e]	1.00
692	Herring	Y	3 Apr	19 Apr	0.73	0.46	0.33	0.80	0.62	0.92
749	Shad	N	5 Apr	22 Apr	0.48	0.26	0.06	[f]	[f]	[f]
828	Cook	N	6 Apr	23 Apr	0.84	0.05[e]	0.20	0.21	0.32	0.32
861	Bernard	N	9 Apr	23 Apr	0.15	0.00[e]	0.46	0.55	0.99	0.60

[a] H_0: No change in fry density between the first and second sample.
[b] H_0: No difference between stream zones.
[c] H_0: No sample-by-stream zone interaction.
[d] H_0: No change in the percent dead fry between the first and second sample.
[e] Statistically significant at $\alpha = 0.05$.
[f] Low sample size because of missing cells.
[g] Statistically different, but more fry present in second sample than in first.

the repeated measures factor, zone, were also examined. Tests of homogeneity of between-treatment covariance matrices and the degree of sphericity of the pooled covariance matrix (Mauchly 1940) were effected. Four contrasts (oil versus reference for the four stream zones) and corresponding Bonferroni family confidence intervals (α = 0.10 overall) were estimated if a significant difference attributable to oiling was detected. The SAS (SAS Institute Inc. 1988) General Linear Models Procedure was used to analyze the data. The statistical power of the test for oil effects was examined if no statistical difference attributable to oil was observed. This analysis helped to evaluate whether sufficient sampling had been performed to detect a biologically meaningful difference if one existed.

Overwinter Survival

Differences in survival of pink salmon embryos to preemergent fry between the 10 oiled and 15 reference streams (Figure 1) were assessed by comparing preemergent fry densities in the spring with embryo densities observed the previous fall. Pink salmon survival rates from embryo to preemergent fry ($\hat{S}_{ij}$) were estimated by

$$\hat{S}_{ij} = \frac{\dfrac{\sum_k \mathrm{LF}^f_{ijk}}{n^f_{ij}}}{\dfrac{\sum_k (\mathrm{LE}^e_{ijk} + \mathrm{DE}^e_{ijk} + \mathrm{LF}^e_{ijk} + \mathrm{DF}^e_{ijk})}{n^e_{ij}}}, \quad (3)$$

where LF^f_{ijk} is the number of live fry from dig k, stream i, zone j, collected during the fry survey; n^e_{ij} and n^f_{ij} are the number of digs for stream i, zone j for the embryo and fry surveys, respectively. The same mixed-effects, two-factor experiment (equation 2) used for the examination of embryo mortality was employed to test for differences in embryo-to-preemergent fry survival.

Results

1989 Preemergent Studies

Few dead pink salmon fry were found during the 1989 preemergent fry surveys, although only 3 of the 14 resampled streams were observed to be oiled (Table 1). Only 9 of the 52 transects examined yielded more than five dead fry. No increase in fry mortality or decrease in fry density was detected between the first and second survey. The date of first sampling averaged 8 d after the spill (range of

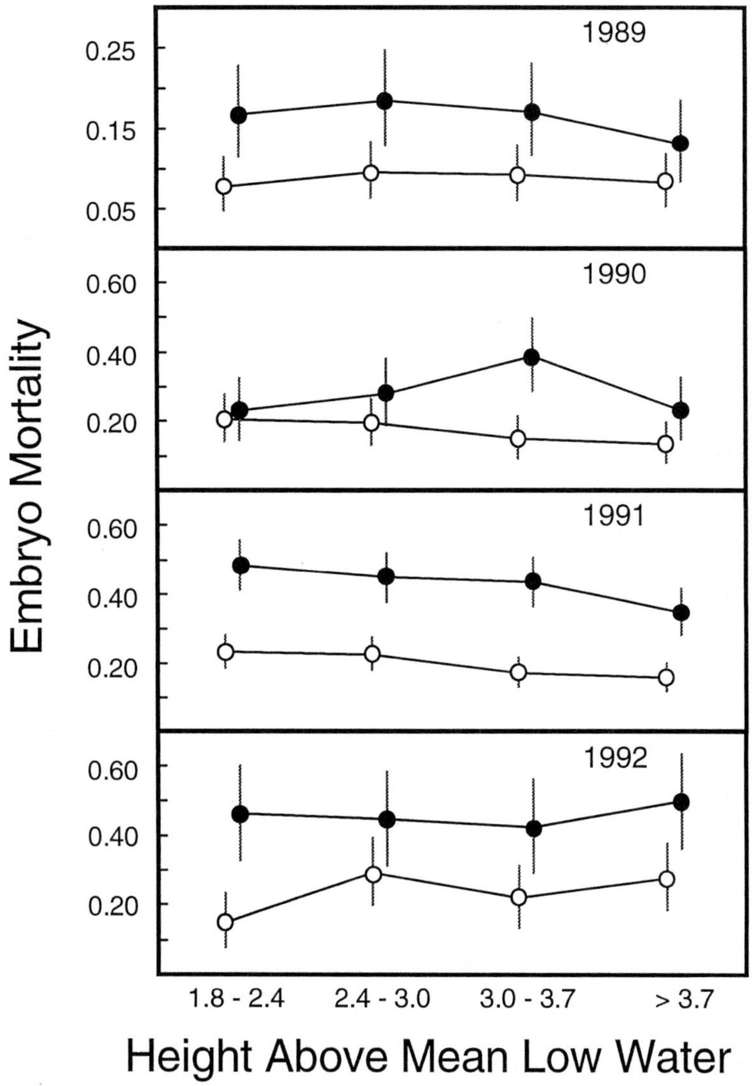

FIGURE 2.—Mean pink salmon embryo mortality (proportion) observed during field sampling in the falls of 1989–1992. Solid circles indicate data for the 10 oil-contaminated streams; open circles indicate data for the 15 reference streams. Standard error bars represent the 90% family confidence interval. Note that the scale of embryo mortality is not the same for all years. Height above mean low water is in meters.

6–16 d), and the second sampling occurred an average of 16 d after the first sampling (range of 14–18 d).

Embryo Mortality

The 1989 analysis revealed a statistically significant difference in embryo mortality (P = 0.004) between oiled and reference streams (Figure 2). No zone effect (P = 0.499) or oil-by-zone interaction (P = 0.863) was evident. Examination of estimated contrasts indicated the differences in mortality were in the intertidal zones. The overall mean embryo mortalities for the oiled and reference streams were 0.174 and 0.104, respectively.

The 1990 embryo mortality data also showed a statistically significant difference (P = 0.023) between oiled and reference streams (Figure 2). No zone effect (P = 0.595) or oil-by-zone interaction (P = 0.159) was observed. Estimated contrasts indicated that differences attributable to oiling were in the upper intertidal zone. The overall mean em-

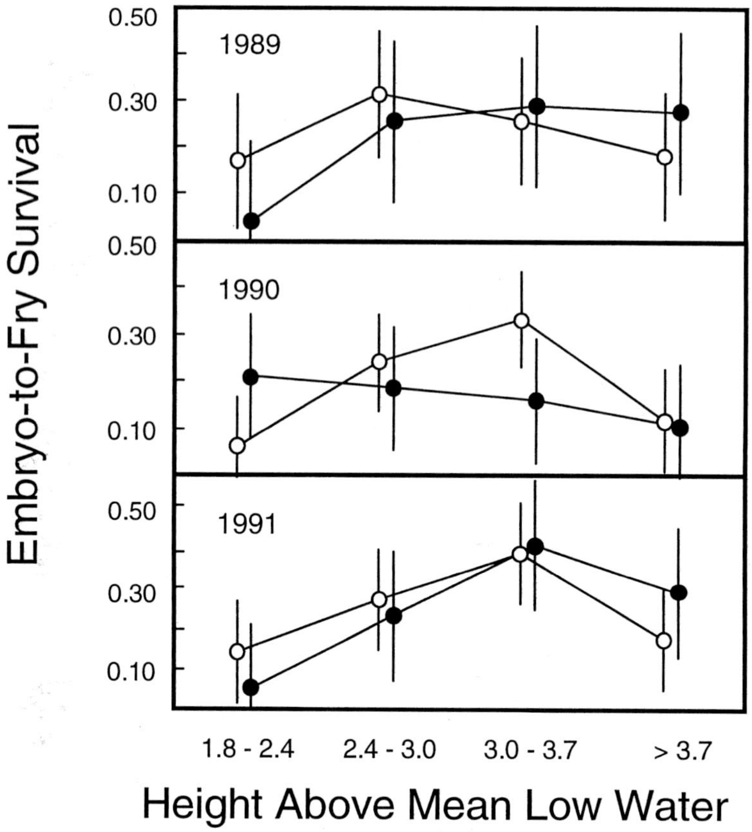

FIGURE 3.—Mean pink salmon survival from embryo to preemergent fry (proportion) observed during field sampling of the 1989–1991 brood years. Solid circles indicate data for the 10 oil-contaminated streams; open circles indicate data for the 15 reference streams. Standard error bars represent the 90% family confidence interval. Height above mean low water is in meters.

bryo mortalities for the oiled and reference streams were 0.295 and 0.195, respectively.

The 1991 embryo mortality data indicated very significant differences between the oiled and reference streams ($P = 0.003$; Figure 2). A significant zone effect ($P = 0.011$) was evident, although no oil-by-zone interaction was found ($P = 0.816$). Estimated contrasts indicated differences attributable to oiling in all three intertidal zones as well as in the upstream zone. The overall mean embryo mortalities for the oiled and reference streams were 0.433 and 0.221, respectively.

Significant differences in embryo mortality were again detected in 1992 between the oiled and reference streams ($P = 0.010$; Figure 2). No zone effect ($P = 0.167$) or oil-by-zone interaction ($P = 0.788$) was detected. Estimated contrasts indicated that the statistical differences were in two of the three intertidal zones as well as in the upstream zone. The overall mean embryo mortalities for the oiled and reference streams were 0.448 and 0.250, respectively.

Overwinter Survival

Embryo-to-preemergent fry survival data were edited before analysis to remove values greater than 2.0 (i.e., overwinter survivals greater than 200%). Only one data point in the 1990 brood year embryo-to-fry survival was removed (9,200% survival).

No significant effects attributable to oil ($P = 0.820$) were found for survival of pink salmon embryos to preemergent fry for the 1989 brood year (Figure 3). No significant zone or oil-by-zone interaction was detected ($P = 0.155$ and $P = 0.594$, respectively). Similar results were obtained for the

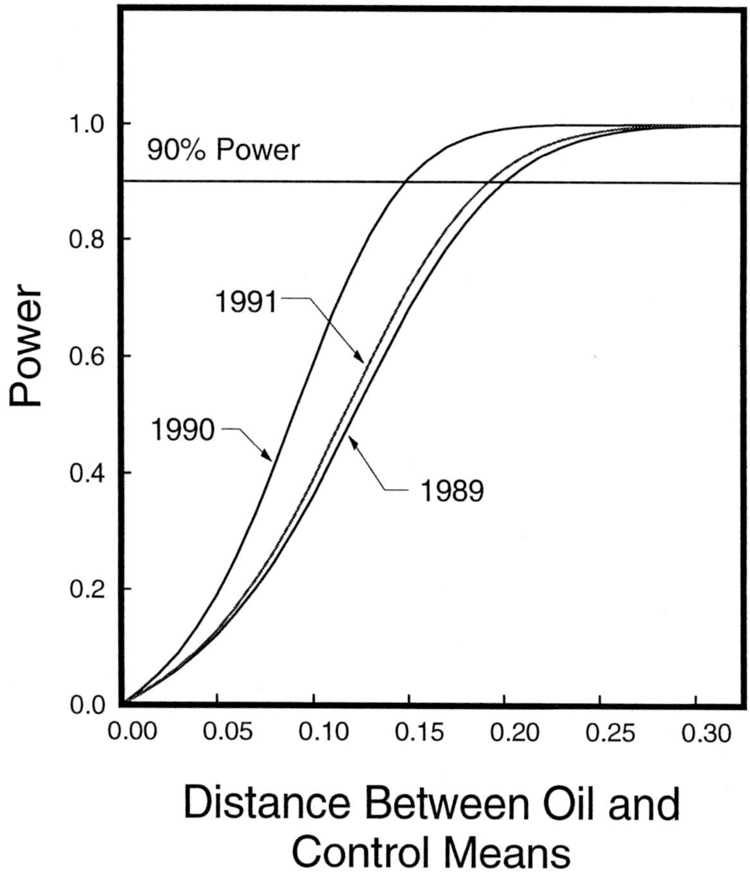

FIGURE 4.—Estimated statistical power for the 1989–1991 brood year comparisons of survival of pink salmon embryo to preemergent fry between oil-contaminated and reference streams.

survival of embryos to fry for the 1990 brood year (Figure 3). No significant oil, zone, or oil-by-zone effect was found (P = 0.713, P = 0.147, and P = 0.112, respectively). The 1991 brood year embryo-to-fry survival information again indicated no oil or oil-by-zone interaction (P = 0.983 and P = 0.596; Figure 3), although there was evidence for a zone effect (P = 0.004).

Statistical power was estimated for all 3 years of embryo-to-preemergent fry data. The test for differences in survival from embryo to preemergent fry between oiled and reference streams for the 1990 brood was more powerful than for the other 2 years (Figure 4). The test for 1990 brood year survival should have detected a difference of ±0.15 with a probability of 0.90, whereas the 1989 and 1991 brood year tests were able to detect a difference of ±0.20 with the same probability.

Discussion

Pink salmon embryos that incubated in oil-contaminated spawning areas in Prince William Sound in 1989–1992 appear to have been adversely affected by the *Exxon Valdez* oil spill. The elevated mortalities in 1989 and 1990 were confined to the intertidal zones whereas those detected in 1991 and 1992 were in all zones. We believe the elevated mortalities observed in 1989 and 1990 were caused by direct exposure to oil; those in 1991 and 1992 may have resulted from genetic damage sustained during early fish development. The pink salmon that spawned during fall 1991 were the fry that incubated in oil-contaminated streams during winter 1989–1990, the first winter after the spill. Likewise, the pink salmon that spawned during fall 1992 were the same fry that incubated in oiled stream gravel during fall 1990 and spring 1991. There is a possibility that the surviving embryos sustained sublethal genetic damage that was manifested in the form of functional sterility in 1991 and 1992. Moles et al. (1987) confirmed that pink salmon embryos take up polynuclear aromatic hydrocarbons, a major component of crude oil, and McBee and Bickham (1988) showed that these compounds are capable of inducing chromosomal lesions. Also, Hose et al. (in press) found chromosome damage in Pacific herring *Clupea pallasi* exposed to *Exxon Valdez* crude oil.

An alternative to the genetic hypothesis is that the observed differences are caused by environmental variation. This study is based on observational data; as such, we were unable to randomize stream oiling to account for environmental differences between streams. We addressed this concern in our original experimental design by selecting reference streams in close proximity to oil-contaminated streams and by including as many streams as possible in the analysis. However, there is a definite oiling pattern in southwestern Prince William Sound: streams on points that faced northeast were heavily oiled, and streams that faced west and southwest were most likely not oiled. The consequences of this difference in stream orientation were not accounted for and at present are unknown.

No difference in survival from embryo to preemergent fry was detected. Survival was expected to be reduced in oil-contaminated streams because an increase in embryo mortality had already been detected. This anomaly can potentially be explained by compensation in the intergravel environment or problems in the experimental design. Geiger et al. (1996, this volume) found no evidence to suggest that compensation in the intergravel life stages is playing a role in determining the number of emerging fry for the years in the study. The power analysis for the survival from embryo-to-preemergent fry test indicated statistical power was adequate to detect a biologically meaningful difference if present. However, unexpected changes in stream characteristics may have prevented sampling the same areas for embryos in the fall and fry in the spring. This would provide unrepresentative estimates of survival because survival was estimated as the density of live fry in the spring divided by the density of total embryos and fry in the fall. Runoff from fall rains increases stream depth and width, whereas spring water levels are usually low because the majority of the winter precipitation is tied up in ice and snow. Also, stream channels in Prince William Sound are not well defined in intertidal areas. Some intertidal stream segments were found to migrate along the beach, especially if the beach is usually exposed to winter storms. The magnitude of these changes was unexpected when this study was designed and initiated.

This project is ongoing: embryo mortality is monitored annually and experiments designed to evaluate the genetic hypothesis have been initiated. We believe this study illustrates the importance of long-term damage assessment and monitoring projects.

Acknowledgments

This article is contribution PP-102 of the Alaska Department of Fish and Game, Commercial Fisheries Management and Development Division, Juneau. This work was funded by the *Exxon Valdez* Oil Spill Trustee Council.

References

Geiger, H. J., B. G. Bue, S. Sharr, A. C. Wertheimer, and T. M. Willette. 1996. A life history approach to estimating damage to Prince William Sound pink salmon caused by the *Exxon Valdez* oil spill. American Fisheries Society Symposium 18:487–498.

Helle, J. H., R. S. Williamson, and J. E. Bailey. 1964. Intertidal ecology and life history of pink salmon at Olsen Creek, Prince William Sound, Alaska. U.S. Fish and Wildlife Service Special Scientific Report Fisheries 483.

Hose, J. E., M. D. McGurk, G. D. Marty, E. D. Biggs, and T. T. Baker. In press. Sublethal effects of the *Exxon Valdez* oil spill on herring larvae: morphologic, cytogenetic, and histopathological assessments, 1989–1991. Canadian Journal of Fisheries and Aquatic Sciences.

Mauchly, J. W. 1940. Significance test for sphericity of a normal n-variate distribution. Annals of Mathematical Statistics 11:204–209.

McBee, K., and J. W. Bickham. 1988. Petrochemical-related DNA damage in wild rodents detected by flow cytometry. Bulletin of Environmental Contamination and Toxicology 40:343–349.

Moles, A., M. M. Babcock, and S. D. Rice. 1987. Effects of oil exposure on pink salmon, *O. gorbuscha*, alevins in a simulated intertidal environment. Marine Environmental Research 21:49–58.

Neter, J., W. Wasserman, and M. H. Kutner. 1990. Applied linear statistical models—third edition. Irwin, Homewood, Illinois.

Pirtle, R. B., and M. L. McCurdy. 1977. Prince William Sound general districts 1976 pink and chum salmon aerial and ground escapement surveys and consequent brood year egg deposition and preemergent fry index programs. Alaska Department of Fish and Game, Division of Commercial Fisheries, Technical Data Report 9, Juneau.

Rice, S. D., D. A. Moles, and J. W. Short. 1975. The

effect of Prudhoe Bay crude oil on survival and growth of eggs, alevins, and fry of pink salmon, *Oncorhynchus gorbuscha*. Pages 503–507 *in* API (American Petroleum Institute), EPA (Environmental Protection Agency), and USCG (U.S. Coast Guard) 1975 conference on prevention and control of oil pollution, proceedings. API, Washington, DC.

SAS Institute, Inc. 1988. SAS/STAT™ user's guide, release 6.03 edition. SAS Institute Inc., Cary, North Carolina.

Sharr, S., B. G. Bue, S. D. Moffitt, A. K. Craig, and D. G. Evans. 1994a. Injury to salmon eggs and preemergent fry in Prince William Sound—fish/shellfish study 2. Alaska Department of Fish and Game, Division of Commercial Fisheries Management and Development, Regional Informational Report ZA94-52, Anchorage.

Sharr, S., J. E. Seeb, B. G. Bue, S. D. Moffitt, A. K. Craig, and G. D. Miller. 1994b. Injury to salmon eggs and preemergent fry in Prince William Sound—restoration study R60C. Alaska Department of Fish and Game, Division of Commercial Fisheries Management and Development, Regional Information Report ZA94-50, Anchorage.

Sokal, R. R., and F. J. Rohlf. 1969. Biometry. Freeman, San Francisco.

American Fisheries Society Symposium 18:628–638, 1996

Kenai River Sockeye Salmon: An Assessment of Overescapements as a Cause of the Decline

D. C. Schmidt, K. E. Tarbox, and B. E. King
Alaska Department of Fish and Game, 34828 Kalifornski Beach Road, Soldotna, Alaska 99669, USA

L. K. Brannian
Alaska Department of Fish and Game, 333 Raspberry Road, Anchorage, Alaska 99518, USA

G. B. Kyle and S. R. Carlson
Alaska Department of Fish and Game, 333 Raspberry Road, Anchorage, Alaska 99518, USA

Abstract.—In the Kenai River, the estimated number of smolts of sockeye salmon *Oncorhynchus nerka* decreased from 30 million in 1989 to less than 500,000 in 1992, while parent-year escapements were high. In 1989, an estimated 1.4 million sockeye salmon entered the Kenai River as a result of curtailed fishing because of oil on the fishing grounds from the *Exxon Valdez* oil spill. We examined the causes for declining smolt production and identified decreased winter survival of age-0 juvenile sockeye salmon rearing in Skilak and Kenai lakes as the major contributor to this decline. We compared smolt production in the Kenai system with that from Tustumena Lake, a nearby system that had an average escapement in 1989 that did not exhibit major changes in smolt production. Adult returns from years with extremely low smolt production in the Kenai system are needed to verify smolt estimation methods. Preliminary examination of limnological data (chemical, physical, and zooplankton forage base) from Skilak and Kenai lakes provided no obvious explanation for the major decline in winter survival of juvenile sockeye salmon. Because diel vertical migration of zooplankton differed among Kenai Peninsula glacial lakes with different densities of planktivores, we hypothesize that behavioral changes in the zooplankton induced by predation was a major contributor to the decreased production of juvenile sockeye salmon in the Kenai River lakes.

Most damages associated with the *Exxon Valdez* oil spill have been directly linked to the toxicity of oil on animals or on some component of their food chain. The potential impacts on the stocks of Kenai River and Kodiak Island sockeye salmon *Oncorhynchus nerka* are in contrast to the direct toxic effect of oil. The primary impact was not from oil toxicity, but rather from the disruption of commercial fisheries, thereby causing harvestable salmon to escape into their natal streams beyond the need for spawning. The closure of the normal commercial fishery resulted in a large number of spawners, which has the potential to overload lakes with juvenile salmon. Large numbers of juvenile salmon have the potential to overexploit food such as zooplankton and initiate multiple-year population cycles, typical of unregulated populations of northern terrestrial animals, such as snowshoe hares and lemmings. The commercial fishery was closed not because of fear of oil damage to the stocks, but because of possible contamination of gear and catch by floating oil collected in the fishing nets. Oiled fish potentially would have destroyed markets if any contaminated product was distributed to the public. Therefore, when oil was present, preventative closures were instituted to ensure that no contaminated product would enter the marketplace.

This investigation assesses the impacts of high escapements on rearing juveniles and the ecology of major sockeye salmon-rearing lakes on the Kenai Peninsula (Figure 1). During 1987–1992, Kenai River sockeye salmon escapements averaged more than 750,000, which were nearly twice that seen during 1981–1986 (Figure 2). An estimated 1.4 million sockeye salmon were available to spawn in the Kenai River system in 1989, which is about three times the desired management goal of 550,000 (range of 400,000–700,000) fish. These large escapements resulted from many factors, including larger than expected runs and closure of fishing areas affected by oil from the 1987 *Glacier Bay* and 1989 *Exxon Valdez* spills. Also, management was complicated in 1988 because multispecies management plans required managers to attempt to provide higher escapements of sockeye salmon into the Susitna River and chinook salmon *O. tshawytscha* into the Kenai River, while simultaneously harvesting surplus sockeye salmon destined for the Kenai River. Our studies were designed to characterize populations of sockeye salmon smolts resulting from escapements before and after the 1989 *Exxon Valdez* oil spill. In contrast to the escapements for the Kenai River in 1989, escapements did not greatly exceed targeted values in the nearby Kasilof

River–Tustumena Lake sockeye system during this same period (Figure 3) when oiled fishing grounds curtailed commercial fishing in 1989. In the last decade, sockeye salmon escapements into the Kasilof River have averaged about 230,000 sockeye salmon. Because of physical proximity, similar size, and glacial influence, Tustumena Lake was compared with Skilak and Kenai lakes. Limnological assessments were initiated to assess change in the rearing habitat resulting from large numbers of rearing fish. Because high densities of planktivorous fish can exert top-down control over lower trophic levels, measurable ecosystem changes within the affected lakes were expected to occur (Carpenter et al. 1985). For example, major forage items within the zooplankton community may be reduced or eliminated, prey body sizes can be reduced, and preferred prey can be replaced by forms resistant to predation. Large sockeye escapements into Frazer Lake on Kodiak Island resulted in subsequent reductions in smolt size that were correlated to a depressed zooplankton community (Kyle et al. 1988). Density-dependent mechanisms may decrease production whereby predation exhausts or alters the availability of prey. This reduction in food resources may subsequently lead to limited survival. Alternatively, climatic variation, such as extended winters, could be a cause of poor survival (Burgner 1987).

Although measuring the magnitude of losses to fish production is a high priority, measurement of nutrient status and zooplankton community attributes provides information to support potential rehabilitation programs required to restore lost productivity. Our investigation examined the availability of food resources to sockeye salmon juveniles in five glacial lake systems on the Kenai Peninsula. The studies attempted to link measurements of food supply with fall size and winter survival of juvenile sockeye salmon. These data also are essential to determine if production changes in fish are related to density-dependent rather than climatic or other density-independent sources.

Methods

Study Area

The five lakes in this study are located on the Kenai Peninsula in south-central Alaska (Figure 1). Copepods *Diaptomus* spp. and *Cyclops* spp. constitute almost all of the zooplankton in these lakes. Skilak and Kenai lakes are the major sockeye salmon nursery lakes in the Kenai River watershed, and both are large and turbid from glacial silt (nephelometric turbidity units [NTU = 5–15]). Skilak Lake supports the vast majority of sockeye salmon production (rearing fish) within the drainage (Tarbox et al. 1993).

Tustumena Lake, which is drained by the Kasilof River, is located approximately 28 km southeast of Skilak Lake and is the most turbid of the study lakes (NTU = 40–50). This lake has been stocked with sockeye salmon fry produced from Crooked Creek Hatchery since the early 1980s (Kyle 1992), but most of the fry are produced from natural spawning.

Ptarmigan and Grant lakes are located approximately 50 km east of Skilak Lake in the upper drainage of the Kenai River. These lakes are also glacially turbid, with Ptarmigan Lake being the least turbid of all the study lakes. These lakes lack significant planktivore populations because of passage barriers to anadromous species.

Adult Sockeye Salmon Assessment

Escapements into the Kenai and Kasilof rivers were estimated by sonar counters; species apportionment, sex ratios, and lengths were estimated with samples from fish wheels (King and Tarbox 1991). Sonar counts are used for in-season management of the sockeye salmon runs bound for the Kenai and Kasilof rivers. Spawners in the Kenai River were estimated from the sonar counts (apportioned to sockeye salmon) minus the estimated sport-fishing harvests upstream of the counting sites. Kasilof River sport-fishing harvests were considered negligible; therefore, sonar counts of sockeye salmon were considered equal to spawners. Kenai River spawner estimates were further reduced by removing escapement estimates based on weir counts for the Russian River and Hidden Lake systems.

Juvenile Sockeye Salmon Assessment

Annual fall hydroacoustic surveys have been conducted in Kenai and Skilak lakes since 1986 (Tarbox et al. 1993). Fish density was estimated for each transect of a stratified random sampling design by echo integration. Dual-beam acoustic data were used to estimate mean back-scattering cross section. Densities were expanded to population estimates for each stratum (lake basin) and summed for a total population estimate. A trawl was used to collect juveniles for size, age, and species composition. Hydroacoustics were also used to estimate numbers of sockeye salmon in Tustumena Lake (Kyle 1992). Freshwater growth and age of juvenile sockeye salmon from all study systems were determined

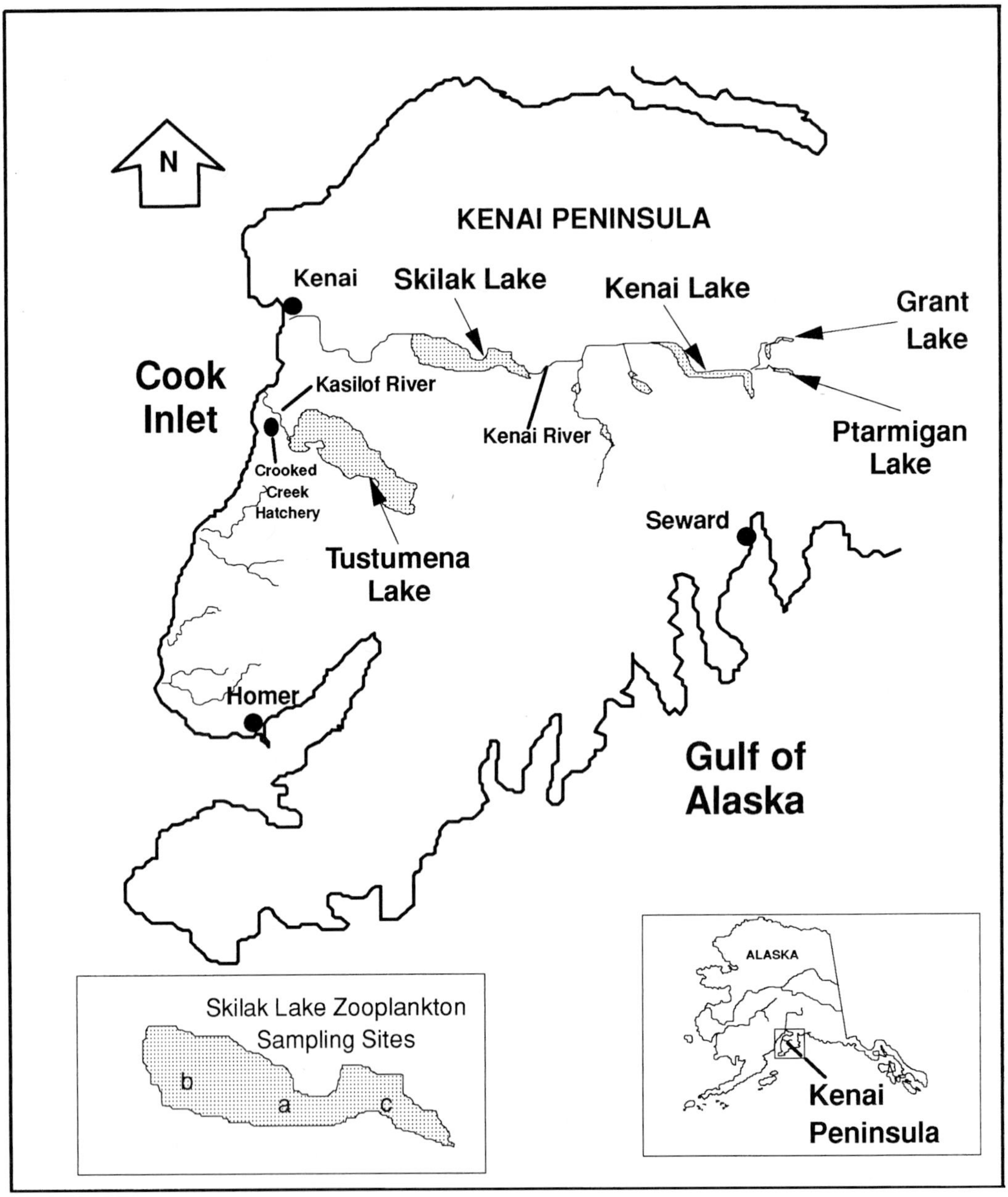

FIGURE 1.—Location of study lakes within the Kenai River drainage, Kenai Peninsula, Alaska.

from scale and otolith measurements made either by direct visual analysis of scales or with an optical pattern recognition system for otoliths.

Additional collection of hydroacoustic data and analyses were initiated in 1992 to determine the vertical distribution of sockeye salmon juveniles within Skilak, Kenai, and Tustumena lakes. One of the hydroacoustic transects used in fall 1991 for generating the juvenile estimate in Skilak Lake was surveyed again in May 1992. Multiple recordings of hydroacoustic data from this transect were obtained from twilight through darkness to evaluate diel changes in vertical distribution of juveniles. Density estimates were made by echo integration.

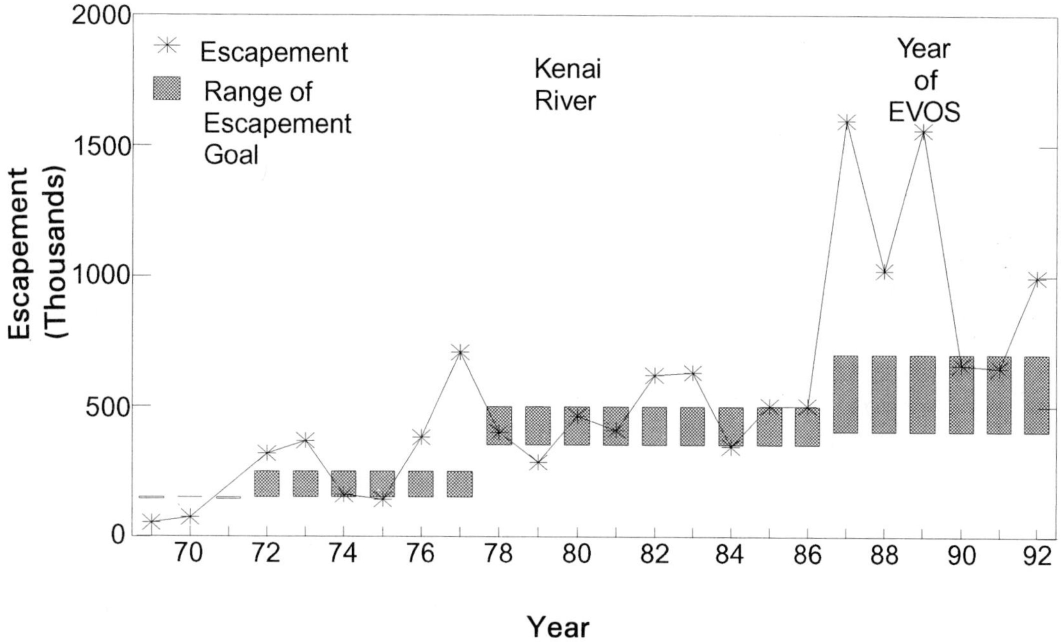

FIGURE 2.—Summary of sockeye salmon escapement estimates and historical escapement goal ranges for the Kenai River. Data represent total cumulative daily apportioned sonar counts at mile 19 of the Kenai River. Top of bars represents maximum escapement goals, and bottom of bars represents minimum escapement goal. Year of EVOS means year of the *Exxon Valdez* oil spill.

Sockeye Salmon Smolt Enumeration

The total number of sockeye salmon smolts migrating from each of the lake systems was estimated (with 95% confidence intervals) by a mark–recapture technique (Rawson 1984). Smolt numbers have been estimated since 1989 in the Kenai River (King et al. 1994) and since 1981 in the Kasilof River (Kyle 1992). Smolts were caught with inclined-plane traps. At weekly intervals, a sample of fish was marked by immersion in a water bath with Bismark Brown dye and released upstream. Recapture rates of dyed fish were used to estimate trap efficiency. This efficiency rate was used to expand the unmarked catch and estimate the total smolt outmigration. Estimated size and age of sockeye salmon smolts were compared by brood year.

Survival

We evaluated the relationship between fall fry abundance and potential egg deposition (PED) in the Skilak–Kenai lake system from 1985 to 1991 using linear regression analysis. Measurement errors in the estimate of fall fry abundance were assumed random, uncorrelated, and unbiased, and thus absorbed in the model error term (Neter et al. 1989). Fall fry counts (dependent variable) were based on combined estimates from the two lakes (1986–1992), and PED (independent variable) was approximated by multiplying the number of mainstem female spawners by a constant 3,500 eggs/female for the corresponding brood years (1985–1991).

Sockeye salmon juveniles were collected from Skilak Lake in 1992 and screened for general disease and infectious hematopoietic necrosis (IHN) virus by the Alaska Department of Fish and Game (ADFG) pathology laboratory in Anchorage using ADFG methods (Burke and Mulcahy 1980; Fried 1984; Meyers et al. 1990).

Limnology

Limnological sampling was conducted at three stations in Skilak and Kenai lakes since 1990 and at Tustumena Lake since 1981, while two stations were sampled in Skilak and Kenai lakes during 1986–1989. Zooplankton data were collected from Skilak Lake at 5 stations in 1990, 10 stations in 1991, and 3 stations in 1992. Zooplankton was sampled in Kenai Lake at three stations during 1990–1992. Wa-

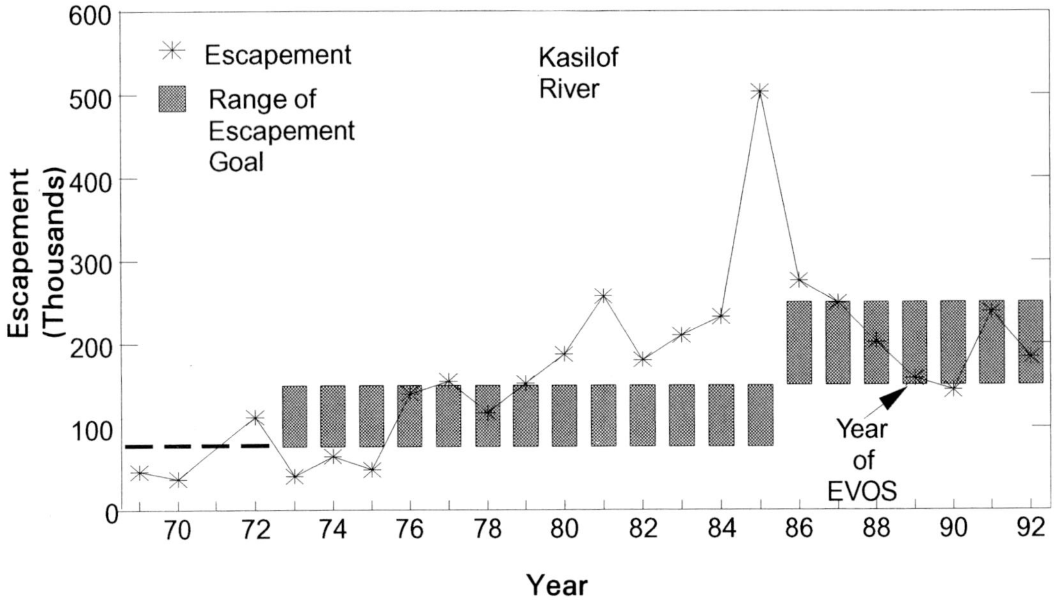

FIGURE 3.—Summary of sockeye salmon escapement estimates and historical escapement goal ranges for the Kasilof River. Data represent total cumulative daily apportioned sonar counts at mile 11 of the Kasilof River. Top of bars represents maximum escapement goal, and bottom of the bars represents minimum escapement goal. Year of EVOS means year of the *Exxon Valdez* oil spill.

ter nutrients, basic physical characteristics, chlorophyll *a*, and zooplankton samples were analyzed using standardized laboratory and quality-assurance methods (Koenings et al. 1987). The three lakes were sampled approximately once every 3 weeks from May through October (Schmidt and Tarbox 1993). When available, limnological data from pre-1989 were compared with post-1989 data.

The vertical distribution of zooplankton was measured in all lakes in 1992. Two locations were sampled in Skilak Lake during daylight on 19 May, 8 June, 1 July, 30 July, 26 August, 25 September, and 27 October. Zooplankton was sampled at night in Skilak Lake on 19 May, 11 June, 17 July, 6 August, 2 September, 8 October, 12 October, and 27 October. Tustumena Lake was sampled at a single location during daylight hours on 7 October and 26 October and during the night on 7 October only. Kenai Lake was sampled at one location during daylight on 21 August, 28 September, and 11 November, and during the night on 12 October and 11 November. Ptarmigan and Grant lakes were sampled during daylight at one location on 14 September and 3 September, respectively.

A closing zooplankton net was used to collect samples for vertical distribution information. The net was constructed of 153-μm Nitex mesh with a 0.5-m stainless steel ring at the mouth and a 200-mL collection bucket. The net was lowered vertically to the desired depth measured with a towline marked in 1-m increments. After vertically retrieving a 5- or 10-m tow, the line was jerked sharply, triggering the release of a closure mechanism at the opening of the net. Triggering this mechanism caused the net to fold over on itself, thus stopping further collection of zooplankton. After the collection bucket was thoroughly rinsed with distilled water, the release mechanism was reset, and the net was lowered to the depth at which collection had previously been halted and the procedure was repeated.

Results

Juvenile and Smolt Production

The relationship between PED and fall fry numbers in Skilak and Kenai lakes over the available time series (Figure 4) was assessed by applying two models: one with and one without the *y*-intercept (constant) term. The first model (constant included) gave the highest possible correlation within the range of PED estimates and provided a way to test the constant. The 0-intercept model adhered to the biological requirement of including the origin, assuming a linear model below the range of PED.

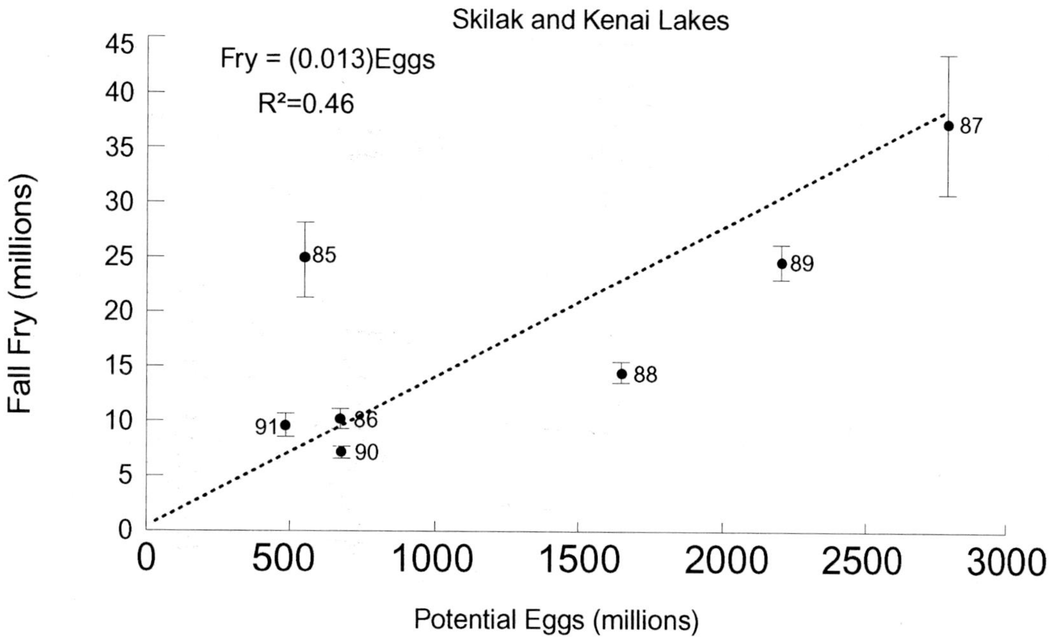

FIGURE 4.—Relationship between number of age-0 fall juveniles in Skilak and Kenai lakes and mainstem potential egg deposition. Years (next to data points) indicate brood years of eggs and fry. Vertical bars are standard errors of mean estimated fry abundances.

Both models had significant slopes ($P < 0.05$), although the model with the constant accounted for more variation in fall fry abundance ($R^2 = 0.591$) than the 0-intercept model ($R^2 = 0.461$). However, the constant term in the former was not significantly larger than zero ($P = 0.26$). The Durbin-Watson statistic indicated no autocorrelation ($P > 0.05$) among years. Age-0 fall juvenile abundance in 1986 (1985 brood year) had a studentized residual of 4.06 and was therefore a possible outlier. These data indicate that the age-0 fall juvenile abundance in the lakes was a linear function of escapement. To test the hypothesis of compensatory juvenile production, we applied Ricker's nonlinear stock recruitment model (Ricker 1975), which resulted in a nonsignificant slope ($P = 0.27$). Consequently, we conclude that during this time period, no measurable compensation was observed in fall juvenile production related to variation in escapement. Thus, the egg-to-fall fry survival during this limited time period is probably the result of density-independent factors.

Skilak Lake fry numbers per unit surface area and zooplankton biomass per fall fry were higher than those in Tustumena Lake except in 1993 (Table 1). Fall juveniles were generally smaller in Skilak and Kenai lakes than in Tustumena Lake (Table 2). The 1993 age-0 fall fry from Tustumena Lake were more than twice the weight of those from the Kenai systems, which paralleled a near doubling of zooplankton density in Tustumena Lake (Table 1).

Estimates of sockeye salmon smolt migrants in the Kenai River declined from 30.2 million for brood year 1987 to 3.1 million for brood year 1989 (Tables 3, 4 and Figure 5). In contrast, smolt numbers in the Kasilof River peaked at 15 million for the 1982–1983 brood years and ranged from 4.4 to 6.6 million for the 1987–1989 brood years (Table 5 and Figure 6). Kenai River smolts were predominantly age-1, ranging from 81% for brood year 1987 to 92% for 1988 (Table 3). There was a significant yearly difference ($P < 0.05$) in the proportion of age-2 smolts, primarily because of the 1987 brood year. The variation in age-class was minor when compared with the Kasilof River smolts. The Kasilof River ranged from 70 to 84% age-1 smolt during 1983–1986 and ranged from 23 to 56% for smolt years 1987–1992 (Table 6). The 1989 Kenai River escapement had minimal impact on 1990 smolts, whereas 1991 and 1992 smolts were primarily affected by the 1988, 1989, and 1990 escapements.

TABLE 1.—Comparison of copepod biomass and fall density of juvenile sockeye salmon in Skilak and Tustumena lakes, 1986–1993. Standard error of fall fry density estimates are < 25% of the mean. The number of stations is indicated by n.

	Copepod biomass (mg/m^2)						Fall sockeye density (number/m^2)		Copepod biomass/fall fry (mg/fry)	
Year	Skilak	SE	n	Tustumena	SE	n	Skilak	Tustumena	Skilak	Tustumena
1986	514	46	2	115	13	3	0.22	0.056	2,370	2,062
1987	586	28	2	100	23	3	0.09	0.045	6,426	2,242
1988	565	6	2	75	10	3	0.31	0.051	1,805	1,472
1989	783	257	2	90	18	3	0.22	0.056	3,554	1,599
1990	417	55	5	74	5	5	0.23	0.062	1,833	1,186
1991	571	40	10	165	15	5	0.07	0.052	8,418	3,181
1992	637	191	3	110	9	5	0.09	0.051	7,428	2,136
1993[a]	710	134	3	204	29	5	0.34	0.049	2,082	4,198

[a]Preliminary data.

The above data indicate major decreases in smolt abundance in 1991 and 1992 in the Kenai River, but not in the Kasilof River. A shift of Kenai River smolt age composition in 1992 to predominantly age-2 (Table 4), probably was the result of an increased proportion of age-2 smolts from the Russian River and other minor systems. The low winter survival of juvenile sockeye salmon from Kenai and Skilak lakes apparently resulted in the Russian River lakes and other minor systems producing most of the sockeye salmon smolt migrating out of the Kenai River. Because the lake of origin of

TABLE 2.—Mean fork length and weight of fall juvenile sockeye salmon in Kenai Peninsula lakes. Missing values indicate no data are available.

	Age-0						Age-1					
	Length			Weight			Length			Weight		
Year	*N*	mm	SD	*N*	g	SD	*N*	mm	SD	*N*	g	SD
						Skilak						
1986	15	57					8	74				
1988	109	50	5.3	109	0.9	0.4						
1989	136	50	3.3	136	1.2	0.3	126	64	6	126	2.8	0.7
1990	928	49	4.3	290	1.3	0.3	34	72.8	3.3	20	4.0	0.4
1991	863	51	4.9	286	1.5	0.5	55	73.8	3.8	14	4.7	0.5
1992	883	54	6	883	1.8	0.6	10	89	3	10	7.0	0.8
1993	3,652	49	5	3,652	1.2	0.4	55	75	5	55	4.5	0.9
						Kenai						
1986	227	52					2	77				
1989	38	48	4.5	38	1.0	0.2	56	64	4.6	56	2.5	0.6
1990	1,484	52	4.6	1,484	1.5	0.4	62	69.4	4.2	22	3.6	0.6
1991	1,364	53.5	6.5	1,364	2.0	0.6	40	75.9	4.8	15	5.5	1.0
1992	1,492	56	7.3	1,492	2.0	0.8	12	78	10	12	5.6	1.7
1993	2,969	45	4	2,969	1.0	0.2	4	68	1	4	3.3	0.5
						Tustumena						
1980	222	59	6.1	222	2.3	0.7	20	80	3.5	20	5.7	0.7
1981	197	55	5.1	197	1.6	0.4	21	73	4.6	21	3.8	0.7
1982	194	54	5.1	194	1.8	0.5	17	74	3.9	17	4.0	0.9
1983	562	60	6.1	562	2.5	0.7	55	80	5.0	55	5.8	1.1
1984	388	61	4.6	388	2.5	0.6	186	79	3.7	186	5.3	0.8
1985	173	56	5.6	173	2.1	0.6	52	78	5.0	52	5.6	1.2
1986	156	50	6.4	156	1.3	0.5	92	73	4.5	92	4.1	0.7
1987	143	53	5.9	143	1.8	0.6	50	71	3.8	50	4.2	0.6
1988	303	55	5.3	303	1.8	0.5	89	75	3.6	89	4.5	0.6
1989	47	52	5.7	47	1.9	0.6	18	74	4.6	18	5.1	0.9
1990	200	57	5.5	200	1.5	0.4	50	75	2.9	50	3.4	0.5
1991	202	57	5.4	202	2.0	0.5	47	78	6.5	47	5.1	1.2
1992	323	59	4.4	323	2.0	0.4	21	79	4.1	21	4.52	0.7
1993	417	63	6.7	417	2.9	0.8	46	81	3.0	46	6.18	0.7

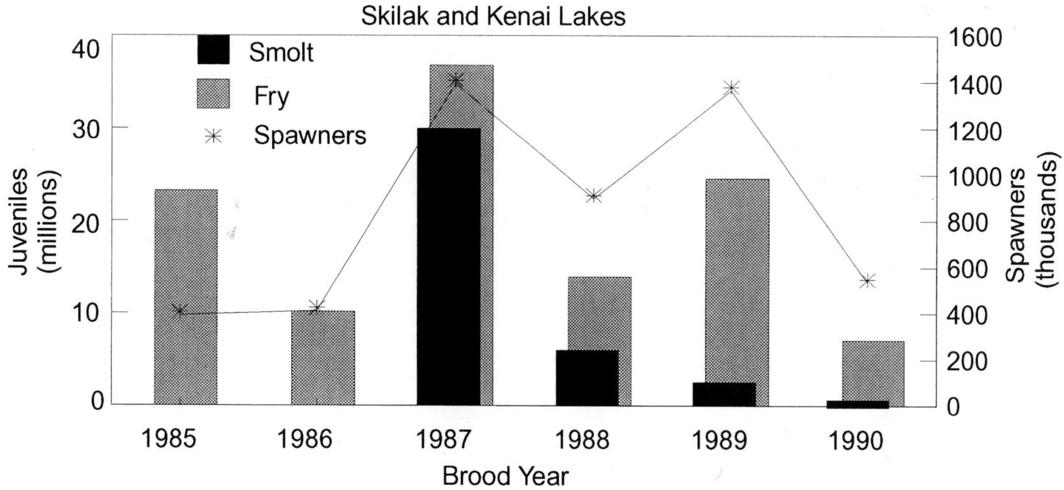

FIGURE 5.—Annual variation of mainstem Kenai River sockeye spawners, smolt production, and combined fall juvenile estimates from Skilak and Kenai lakes. Mainstem spawner abundance was estimated annually from apportioned sonar counts with sport fish harvests and estimated escapements into the Russian River and Hidden Lake subtracted. Smolt production is total smolt number estimated at Kenai River mile 19. Smolts were not counted before 1989 (brood year 1987).

smolts from the Kenai River is not known, the age, size, and condition of the smolts collected were not informative of the rearing conditions in the natal glacial lakes receiving high escapements.

The fall juvenile abundance generally reflects PED or the escapement (Figure 4), although the 1985 data suggest that before the high escapements egg-to-fall-fry survival may have been higher. Most of the decline in smolt numbers in the Kenai River system appears to be related to winter mortality (Figure 5). Smolt production, for the same time period from the Kasilof River system (Figure 6), which did not receive excess escapement in 1987 or 1989, did not decline appreciably. A hydroacoustic survey conducted during the day in May 1992, based on a single transect, indicated that juvenile sockeye salmon density in Skilak Lake was less than 10% of that observed during fall 1991 at the same location.

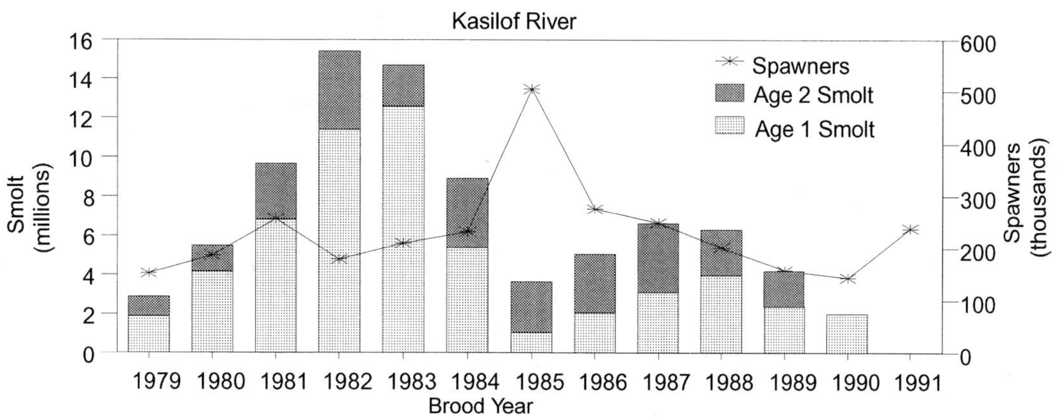

FIGURE 6.—Annual variation of the Kasilof River sockeye salmon escapements and brood year production of smolts from Tustumena Lake. Spawner abundance was estimated annually from apportioned sonar counts and sport fishing removals were considered negligible. Smolt production is total smolt number estimated at Kasilof River mile 10. The 1990 age-2 smolt data are unavailable.

TABLE 3.—Sockeye salmon adult escapement and smolt production in the Kenai River for brood years 1986–1993.

Brood year	Spawning escapement	Total number of smolts produced: Age-1	Age-2	Age-3	Brood year total
1986	422,000	[a]	115,000[b]	16,000	
1987	1,408,000	24,416,000[b]	5,807,000[b]	1,000	30,224,000
1988	910,000	5,249,000[b]	431,000[b]	0	5,680,000
1989	1,379,000	2,776,000[b]	312,000[c]	0	3,088,000
1990	519,000	253,000[c]	33,000[c]	[d]	286,000
1991	431,000	735,000[c]	[d]		
1992	798,000				
1993	814,000				

[a]In thousands.
[b]No data collected.
[b]Includes Hidden Lake migration not thought to be captured by the km 31 inclined-plane traps.
[c]Includes Hidden Lake and Moose River migration not thought to be captured by the km 31 inclined-plane traps.
[d]Migrate as smolt in 1994.

TABLE 5.—Sockeye salmon adult escapement and estimated smolt production in the Kasilof River for brood years 1981–1993.

Brood year	Spawning escapement	Numbers of smolt with percent hatchery contribution: Age-1[a]	% hatchery	Age-2[a]	% hatchery	Total[a]
1981	256,625	6,817	40.7	2,869	20.6	9,686
1982	180,239	11,390	36.3	4,001	7.7	15,391
1983	210,270	12,580	27.8	2,223	11.1	14,803
1984	231,685	5,268	30.2	3,540	31.1	8,808
1985	505,049	1,074	59.2	2,549	17.0	3,623
1986	275,963	2,056	52.0	3,009	10.2	5,065
1987	249,246[b]	3,109	30.3	3,521	3.0	6,630
1988	204,000	3,916	2.0	2,335	1.0	6,251
1989	158,206	2,400	7.0	2,012	1.3	4,412
1990	144,136	2,107	7.5	1,833	3.3	3,940
1991	238,269	7,189	6.3			
1992	184,178					
1993	149,939					

[a]In thousands.
[b]Because of sonar failure, escapement estimates were from stream surveys and spawning stream weir counts above Tustumena Lake. See Kyle (1992) for statistics of smolt abundance precision estimates.

This approximates the decline observed from the 1990 fall fry to 1991 smolt estimates, although the near-surface distribution of fry and its effects on abundance estimates have not been investigated.

A limited investigation of disease as a cause of the reduced spring smolt migration was undertaken. We screened 20 fish samples from juveniles collected within Skilak Lake in spring of 1992 for general disease and for the IHN virus and found that all of the specimens were IHN negative.

Limnology

The nutrient and physical data from Skilak and Tustumena lakes did not indicate trends among years or between the lakes that could explain the difference in production of sockeye salmon smolts (Schmidt et al. 1994). Primary production apparently was not affected, because reactive phosphorous and chlorophyll *a* indicate no trends among years. Turbidity and euphotic depth data indicate higher turbidity levels in Tustumena Lake than Skilak Lake but greater variability in Skilak Lake (see Figure 5 in Schmidt et al. 1994). In addition, surface temperatures in Skilak and Tustumena lakes were similar, indicating that heat budgets did not differentially affect productivity (see Figure 4 in Schmidt et al. 1994). Annual trends in ovigerous copepod densities in Skilak and Tustumena lakes demonstrate little variation since 1987 (Figure 7). However, during 1981–1992, winter survival of Tustumena Lake juvenile sockeye salmon was significantly correlated ($P = 0.019$; $R^2 = 0.52$) with ovigerous zooplankton densities (Schmidt and Tarbox 1993).

TABLE 4.—Smolt age composition in the Kenai River for smolt years 1989–1993.

Smolt year	Smolt age composition (%): Age-0	Age-1	Age-2	Age-3	*N*
1989	0	99.7	0.3	0	3,567
1990	0	46.7	53.1	0.2	3,422
1991	0	86.1	13.9	0	3,741
1992	0	17.3	82.7	0	981
1993	8.5	88.5	3.0	0	1,200

TABLE 6.—Sockeye salmon estimated smolt age composition and length in the Kasilof River for smolt years 1983–1993. Year indicates time of smolt outmigration. Mean length is in millimeters.

Smolt year	Age-1: *N*	%	Mean length	SD	Age-2: *N*	%	Mean length	SD
1983	712	84	70	3.8	451	16	83	5.4
1984	1,005	80	73	3.5	187	20	85	4.4
1985	981	76	70	2.8	282	24	84	5.4
1986	983	70	69	3.9	365	30	84	5.1
1987	412	23	64	6.1	1,223	77	77	4.8
1988	623	45	68	4.0	652	55	78	4.5
1989	609	51	66	4.4	516	49	81	4.5
1990	683	53	69	4.4	467	47	82	4.2
1991	529	51	68	3.5	489	49	80	3.2
1992	594	56	74	3.8	556	44	87	4.7
1993	755	80	71	3.5	187	20	82	4.0

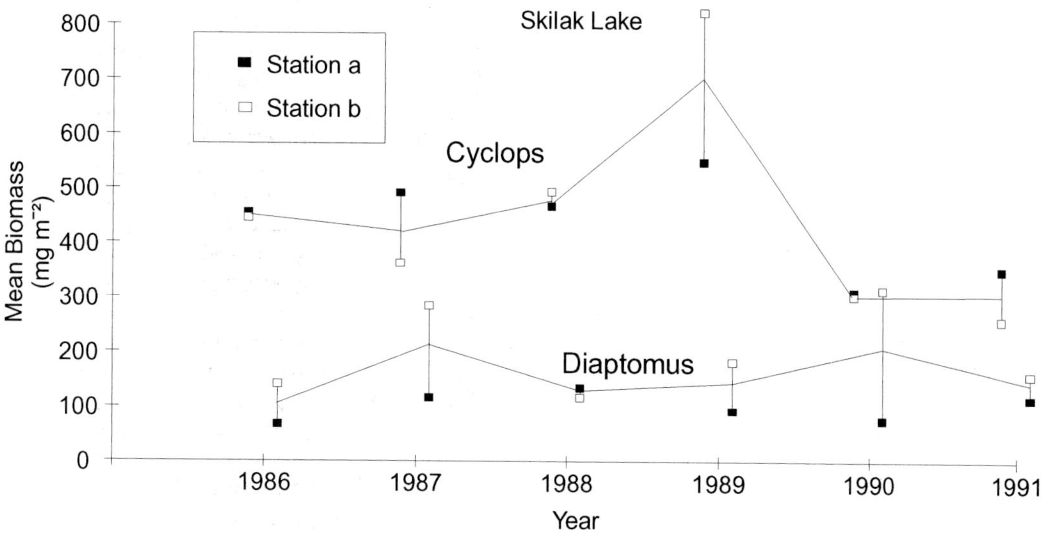

FIGURE 7.—Annual variation in seasonal mean biomass of copepods in Skilak Lake. Bars reflect ranges of the two stations.

Discussion

Because winter mortality is the leading hypothesis for the cause of poor smolt production from the Kenai River, we limited our investigations to those relationships that most likely explained high winter mortality. The limited disease screening of spring fry did not indicate the presence of disease. Low dissolved oxygen, commonly proposed as a cause for poor winter survival in lakes, was easily eliminated as a source of mortality because of the physical nature of the lakes, low loading of organic materials, and no history of low dissolved oxygen. We next examined correlative data with the most obvious physical factors affecting survival (Schmidt and Tarbox 1993). We found little or no correlation between winter survival and trends in physical factors, and there were no differences in trends between Tustumena and Skilak lakes.

In addition, we found that Skilak Lake had minimal annual variation in nutrients, chlorophyll *a*, and zooplankton biomass (Schmidt and Tarbox 1993), which indicated that changes in the availability of zooplankton may have been responsible for the decreased condition and subsequent reduced winter survival of juvenile sockeye salmon in this lake. Comparison among lakes of the daylight vertical distribution of copepod zooplankton suggests a hypothesis accounting for the differences in juvenile winter survival (Figure 11, Schmidt et al. 1994). This hypothesis suggests that increased predation reduced Skilak Lake food availability by inducing diel vertical migration (DVM) in the zooplankton (Schmidt et al. 1994). Reduced surface feeding of juvenile sockeye salmon may result in increased survival of zooplankton but probably at the expense of reduced reproduction because of nutritionally related decreases in fecundity. Increased feeding rates should cause higher fecundity. Because of the lack of pre-1989 DVM data, we can only test this hypothesis with experimental enclosure studies or by observing changes in the macro-zooplankton community when fish densities in the lake decrease.

If the hypothesized change in the plankton community proves to be a density-dependent mechanism responsible for the poor smolt production, we would expect recovery of the system, including changes in the availability of zooplankton, when escapements are reduced. These changes may be adaptive behavioral responses or may be caused by evolutionary changes in heritable characteristics of the populations. We would expect significantly longer recovery times if the sockeye salmon production is reduced because of changes in the heritable characteristics of the zooplankton community.

Conclusions

The investigation in progress has provided insight into the collapse of smolt production from the major nursery lakes of the Kenai River. Fall fry abundance compared with smolt data suggests that winter mortality of fry in the lake is primarily responsible for the collapse. This collapse was not paralleled in Tustumena Lake, a nearby glacial lake that had average escapements. The cause of the increased winter mortality has not been clearly defined, but we suggest that it relates directly to the size and condition of fall fry before winter. We further hypothesize that this decrease in condition is caused by density-dependent interactions of juvenile sockeye salmon and the zooplankton community during the first rearing year. The poor survival during the winters of 1991–1992 and 1992–1993 suggests that these changes are persistent. Data collected during future years when the density of sockeye salmon spawners and juveniles decrease in these lakes is essential to determine if subtle effects on the zooplankton community were responsible for decreased production or if other factors, unrelated to density, provide alternative explanations for the decline.

References

Burgner, R. L. 1987. Factors influencing age and growth of juvenile sockeye salmon (*Oncorhynchus nerka*) in lakes. Pages 129–142 *in* H. D. Smith, L. Margolis, and C. C. Wood, editors. Sockeye salmon (*Oncorhynchus nerka*) population biology and future management. Canadian Special Publication of Fisheries and Aquatic Sciences 96.

Burke, J. A., and D. Mulcahy. 1980. Plaquing procedure for infectious hematopoietic necrosis virus. Applied and Environmental Microbiology 39:872–876.

Carpenter, S. R., J. F. Kitchell, and J. R. Hodgson. 1985. Cascading trophic interactions and lake productivity. BioScience 35:634–639.

Fried, M. P. 1984. Laboratory techniques in virology. Alaska Department of Fish and Game, Fisheries Rehabilitation, Enhancement and Development Division Report 19, Juneau.

King, B. E., and K. E. Tarbox. 1991. Upper Cook Inlet salmon escapement studies, 1990. Alaska Department of Fish and Game, Division of Commercial Fisheries Technical Fisheries Report 91-21, Juneau.

King, B. E., L. K. Brannian, and K. E. Tarbox. 1994. Kenai River sockeye salmon smolt studies, 1992. Alaska Department of Fish and Game, Division of Commercial Fisheries, Management and Development, Regional Information Report 2A93-28, Anchorage.

Koenings, J. P., J. E. Edmundson, G. B. Kyle, and J. M. Edmundson. 1987. Limnology field and laboratory manual: methods for assessing aquatic production. Alaska Department of Fish and Game, Fisheries Rehabilitation, Enhancement and Development Division Report 71, Juneau.

Kyle, G. B. 1992. Summary of sockeye salmon (*Oncorhynchus nerka*) investigations in Tustumena Lake, 1981–1991. Alaska Department of Fish and Game, Fisheries Rehabilitation, Enhancement and Development Division Report 122, Juneau.

Kyle, G. B., J. P. Koenings, and B. M. Barrett. 1988. Density-dependent, trophic level responses to an introduced run of sockeye salmon (*Oncorhynchus nerka*) at Frazer lake, Kodiak Island, Alaska. Canadian Journal of Fisheries and Aquatic Sciences 45: 856–867.

Meyers, T. R., Thomas, J. B., Follett, J. E., and R. R. Saft. 1990. Infectious hematopoietic necrosis virus: trends in prevalence and the risk management approach in Alaskan sockeye salmon culture. Journal of Aquatic Animal Health 2:85–98.

Neter, J., W. Wasserman, and M. H. Kutner. 1989. Applied Linear Regression Models. Irwin, Homewood, Illinois.

Rawson, K. 1984. An estimate of the size of a migrating population of juvenile salmon using an index of trap efficiency obtained by dye marking. Alaska Department of Fish and Game, Fisheries Rehabilitation, Enhancement and Development Division Report 28, Juneau.

Ricker, W. E. 1975. Computation and interpretation of biological statistics of fish populations. Fisheries Research Board of Canada Bulletin 191.

Schmidt, D. C., J. P. Koenings, and G. B. Kyle. 1994. Predator-induced changes in copepod vertical migration: explanations for decreased overwinter survival of sockeye salmon. Pages 187–210 *in* D. J. Stouder, K. L. Fresh, and R. J. Feller, editors. Theory and application in fish feeding ecology. University of South Carolina Press, Columbia.

Schmidt D. C., and K. E. Tarbox. 1993. Sockeye salmon overescapement. State/federal natural resource damage assessment status report. Alaska Department of Fish and Game, Division of Commercial Fisheries Management and Development, FRED (Fisheries Rehabilitation Enhancement and Development) Division Report Series 136, Juneau.

Tarbox, K. E., B. E. King, and L. K. Brannian. 1993. An estimate of juvenile fish densities in Skilak and Kenai lakes, Alaska, through the use of dual-beam hydroacoustic techniques in 1991. Alaska Department of Fish and Game, Division of Commercial Fisheries, Technical Fisheries Report 93-01, Juneau.

American Fisheries Society Symposium 18:639–644, 1996

Detecting Population Impacts from Oil Spills: A Comparison of Methodologies

RAY HILBORN

University of Washington, School of Fisheries WH-10, Seattle, Washington 98195, USA

Abstract.—Five alternative methods for determining oil spill impacts at the population level are compared: (1) counts of dead animals, (2) pre- and postspill comparison of abundance, (3) oiled versus non-oiled comparison of abundance, (4) oiled versus non-oiled comparison of vital rates, and (5) direct experimental oiling. Counts of dead animals do not provide evidence of population-level impact. Pre- and postspill comparisons of abundance have very low statistical power and require substantial baseline data. Oiled versus non-oiled comparisons suffer from lack of randomization of oiling treatments. Experimental oiling must be shown to be comparable to the actual infield oil exposure. A strong case for oil impacts will usually require several of these types of data. It is suggested that traditional hypothesis tests are inappropriate for determination of oil spill impacts and that explicit calculation of the likelihood of different levels of impact is more appropriate.

The *Exxon Valdez* oil spill occurred on 24 March 1989. Tens of millions of dollars were spent to determine the extent of damage caused by the spill. Although the damage from the spill takes many forms, most of the scientific research has been aimed at detecting evidence of population-level impacts on the invertebrates, fishes, birds, and marine mammals in the spill area. There is a widespread feeling that the studies are not nearly as conclusive as many had hoped, and this has raised concern about the ability to detect oil spill impacts by existing methods.

The purpose of this article is to review methodological approaches used to explore population-level impacts resulting from the *Exxon Valdez* oil spill, which have a general application to the problems of detecting any sort of impact, positive or negative, on natural populations. First, I discuss alternative experimental designs for assessment of impacts; next, I consider the nature of statistical tests that should or could be employed in such analyses. Finally, I consider the implications of the different methods to the design of baseline studies before disturbances and of monitoring studies to follow recovery of disturbed systems.

Alternative Methods

There are two general ways that oil spills affect the abundance of a population: through direct mortality or through impacts on reproduction and survival. In each case the impacts might be followed by recovery to pre-impact levels or by a permanent change in abundance. Although most normal models of population dynamics predict recovery once the oil-induced changes disappear, permanent habitat change or a change in competitive or predation pressure could result in a long-term change in the abundance of the species. Holling (1973) provides a review of the types of system behavior that could result from perturbations such as an oil spill. In this section I consider the experimental design of detecting changes in abundance.

There are at least five ways to detect population-level impacts: (1) direct body counts of the number of animals killed, (2) pre- versus postspill comparison of population sizes, (3) oiled versus non-oiled spatial comparison of abundance after a spill, (4) direct measurement of vital rates in oiled versus non-oiled sites, and (5) experimental oil treatments. Each of these methods has advantages and disadvantages and may be the appropriate method for detecting impacts in some circumstances.

Body Counts

Sometimes the number of individuals killed by the spill and can be measured or estimated directly. Such counts are available for several bird and marine mammal species killed as a result of the *Exxon Valdez* oil spill. However, body counts, by themselves, do not provide any evidence of population-level impact. The body counts must be considered in relation to population abundance, natality, mortality rate, and behavior. For instance, Bowman et al. (1995) estimated that 902 bald eagles *Haliaeetus leucocephalus* (11% of the population) were killed directly by the spill, but they could detect no differences between pre- and postspill abundance. This finding could be a result of the high variance in the surveys, or it could be attributable to the fact that a one-time mortality of 11% is not detectable given the background variation in year-to-year recruitment and survival, or it could result from compensatory changes in births and deaths. In contrast, DeGange et al. (1994) estimated that 4,028 sea otters were directly killed by the spill, which, combined with pre- and postspill surveys, indicated that the sea otter *Enhydra lutris* population dropped by roughly 50% because of the spill.

Any assessment of population-level impacts using body counts needs to be supported by either direct comparisons of pre- and postspill abundance, oiled versus non-oiled comparisons of abundance, or a population dynamics model that accounts for recruitment and survival.

Pre- and Postspill Comparison

When prespill abundance surveys are available, comparison of pre- and postspill numbers can be used to assess the change in population. The statistical power of such comparison will depend upon the reliability of the census method, the natural variability of the population, and the magnitude of change induced by the spill. This method clearly can not be used when no prespill abundance data are available, as was the case with many fish species affected by the *Exxon Valdez* oil spill. However, pre- and postspill comparisons were effective in showing changes for many species, including sea otters (Garrott et al. 1993), and pigeon guillemots *Cepphus columba* (Oakley and Kuletz 1996, this volume).

Even when prespill data are available, the comparison may be of little value. Pink salmon *Oncorhynchus gorbuscha*, for instance, show very high year-to-year variability, and unless a spill is catastrophic, there is little chance of detecting its impact. In fact, pre- and postspill comparisons may be deceiving. Geiger et al. (1996, this volume) estimated a loss of several million pink salmon because of oiling in 1990, even though the run of pink salmon in 1990 was the largest in history.

A significant decline in abundance after a spill is not necessarily evidence of an oil impact. Populations often vary in abundance and without evidence of a mechanism for oil impact, such as direct body counts, or oiled and non-oiled comparisons, a decline in abundance is not strong proof of an impact of the spill. The disappearance of a number of killer whales from a single pod could be considered evidence of an oil impact, but without supporting mechanisms this evidence of a spill impact is subject to question.

Oiled versus Non-oiled Comparison of Abundance

Probably the most common technique used in assessing damage from the *Exxon Valdez* oil spill is postspill comparison of the abundance of a species in oiled sites to its abundance in non-oiled sites. This technique formed the basis of most intertidal and subtidal assessments (Collier et al. 1996, this volume; Highsmith et al. 1996, this volume; Jewett et al. 1996, this volume). As with pre- and postspill comparison, the power of this method depends on the reliability of the census method, the natural variability from site to site, and the magnitude of the change induced by the spill. A key problem posed by this approach is the fact that beaches were oiled as a result of physical processes, whereas in a designed experiment they would have been oiled by random assignment. Thus, postspill differences may reflect underlying habitat differences rather than the impacts of oiling. This nonrandomization of treatments is in itself not correctable by postspill analysis, but most investigators attempt to determine if there are other differences between sites and have generally tried to choose control sites (non-oiled sites) that are, to the human observer, as comparable as possible to the oiled sites.

The most convincing data from oiled versus non-oiled comparisons occur when the oiled site recovers to the same abundance as the non-oiled site during the course of the postspill evaluation, this strongly suggests that the observed differences immediately after the spill were caused by an oil impact.

Oiled versus Non-oiled Comparison of Vital Rates

An alternative approach is to measure life history parameters in oiled and non-oiled sites. The estimated parameters are used in a life history model to estimate population-level impacts. The differences in growth observed in oiled versus non-oiled sites (Hepler et al. 1996, this volume), and in egg survival for pink salmon (Bue et al. 1996, this volume) are examples of how this approach can provide evidence of damage even where population-level damages may be difficult to measure directly. The weakness of this approach is that it depends on the validity of the population dynamics models used, and in most cases the extent of damage depends on the level of compensatory mortality in the life history after the damage. If there is high density-dependent mortality, the population-level impacts will be much less than the mortality caused by oiling. The potential for compensatory mortality significantly decreases the power of this approach.

Comparison of vital rates between oiled versus non-oiled sites also suffers from the weaknesses of non-randomization of treatments discussed in the previous section. Again, if vital rates in oiled sites recover to the levels of vital rates in non-oiled sites, the argument that the differences were caused by oil effects rather than pre-existing is much stronger. The continued differences in pink salmon egg survival between oiled and non-oiled sites (Bue et al. 1996) need a more complex mechanism to explain the oil impact and actually weaken the argument that the differences are attributable to oil.

Experimental Oil Treatments

A final approach is the experimental application of oil to determine impacts. Such experimentation normally takes the form of laboratory studies where individuals are exposed to oil at concentrations comparable to that found in the field. This approach was used by Carls et al. (1996, this volume) on juvenile pink salmon. The advantage of this method is that direct experimental treatment and control allow clear determination of the oil impact. The disadvantage is that even when oil impacts are determined, it must be established that the individuals in the field received the same oil exposure and that the measured impacts would have a population-level effect. These two conditions are rarely met.

A more direct approach would be field application of oil to randomly selected sites, with similar randomly selected control sites. Although such ideas were frequently suggested by scientists during the damage assessment process, it was widely felt that infield oiling was unacceptable to the public.

Appropriate Statistics

A final issue to be considered is the statistical framework for analysis of damage studies. Almost all scientists are taught a very rigorous paradigm for testing hypotheses, accepting or rejecting the null hypothesis by comparing it to a working hypothesis. This method is found in every statistics textbook. An obvious question encountered in any study of oil spill impacts is what level of significance should be used. Scientific journals normally accept the 0.05 or 0.01 level of significance. If we were to apply this to an oil spill, the obvious conclusion would be that if you could not show an impact at the 0.05 level, you should conclude that there was no oil impact. This stands in contrast to the legal criterion that damage claims need to show that there was "more likely than not" an oil impact. The statistics taught in our

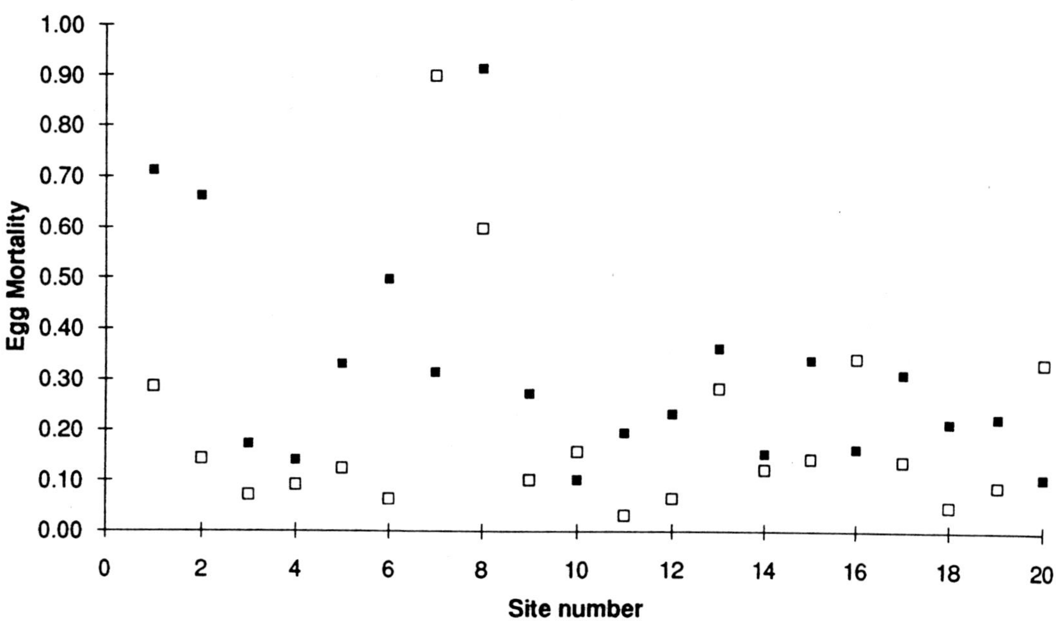

FIGURE 1.—Hypothetical data for pink salmon egg mortality in 20 paired oiled (black squares) and non-oiled (open squares) sites.

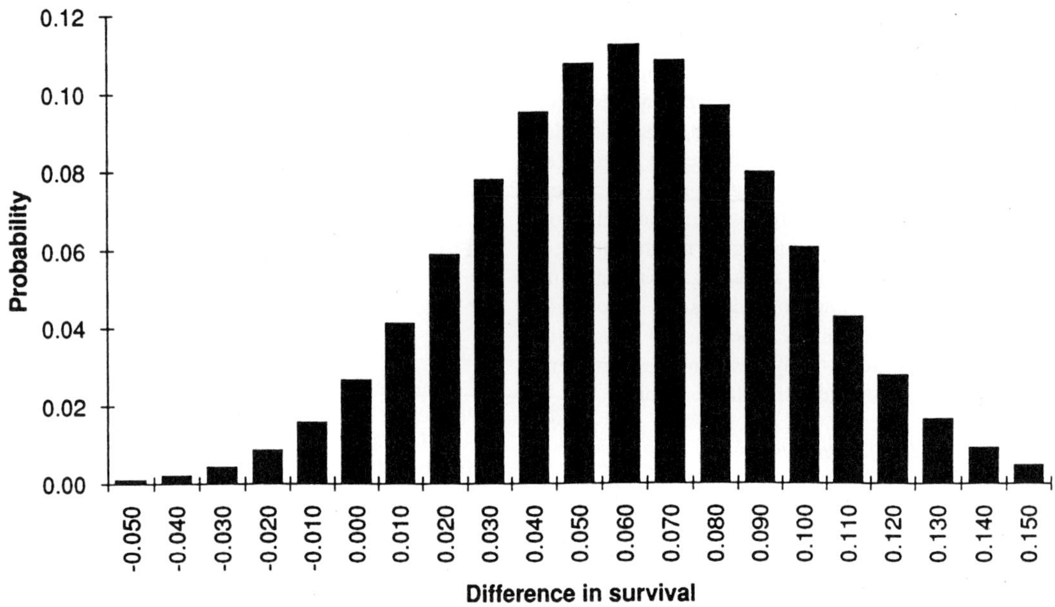

FIGURE 2.—The probability of differences in survival of pink salmon eggs between oiled and non-oiled sites from the data shown in Figure 1.

universities simply do not prepare the scientist to answer this type of question in a rigorous way.

Fortunately, almost all of the body of current statistics can be used to ask this question in a statistically rigorous fashion if we abandon the format of comparing a working hypothesis to a null hypothesis and simply look at the relative likelihood or Bayesian posterior distribution of different levels of oil impact. Methods of likelihood and Bayesian statistics are becoming more common in journal articles, but the statistics books in general use have not treated these methods in much detail. The most popular reference work on likelihood is Edwards (1972); Berger (1985) is a standard reference on Bayesian methods. The essence of likelihood is to look at the relative support provided for different values of a parameter. For instance, Figure 1 shows some hypothetical data on pink salmon egg mortality from 20 oiled (black squares) and 20 non-oiled sites (open squares). The "traditional" statistical approach would be to determine if the average mortality was different at the 0.05 or 0.01 level (see Bue et al. 1996). The relative likelihood of different levels of oil impact on mortality is shown in Figure 2. We can see that the data suggest the most likely impact of oil was a 6% increase in egg mortality, but the difference could be between −1% and 10%. Using Figure 2, a scientist could state that the most likely impact was a 6% difference in survival, but that there was a small chance (about 6%) that the average survival in oiled sites was as high or higher than in non-oiled sites. In contrast, a traditional statistical analysis would state that the null hypothesis that there was no difference between oiled and non-oiled sites could not be rejected at the 0.05 level.

There are two major advantages to be obtained by looking at likelihoods or Bayesian statistics rather than hypothesis tests. First, you avoid the common confusion between statistical and biological significance. Oiled versus non-oiled sites may be statistically significantly different, even if the difference is negligible. Any difference can be statistically significant if the sample size is large enough. When you look at the likelihood, you can see the magnitude of the difference (the units on the x-axis). A second advantage of using likelihoods is that if restoration actions are considered, there is a need to consider the consequences under different hypotheses about the intensity of damage. Likelihood lends itself to the machinery of statistical decision theory, whereas hypothesis testing does not.

Another statistical problem posed in damage assessment is pseudo-replication (Hurlbert 1984). As an example, consider a pre- and postspill comparison of abundance. The more data points that are available pre-and postspill, the more powerful the comparison would appear. Even if there are large differences between pre-and postspill, you really only have one observation at each site. Unless you can compare multiple sites pre- and postspill, the ability to say the differences result from oiling is weak.

Implications for Baseline Studies and Postspill Monitoring

Perhaps the biggest lesson learned from damage assessment of the *Exxon Valdez* oil spill is the usefulness of prespill baseline data. When only postspill data are available, the results are always suspect because of the nonrandomization of treatments. The most convincing case for oil damage is to show abundance in oiled versus non-oiled areas to be the same before a spill, a postspill decline in the oiled areas, and a gradual recovery to prespill levels. This is an ideal that I believe was not obtained with any study in Prince William Sound. However, prespill baseline studies on numerous birds and mammals in particular were important parts of the evidence for oil damage.

The limitations of baseline studies were illustrated in the case of salmon and herring. In both cases, prespill data existed as part of the normal commercial fisheries management. However, the high year-to-year variability in recruitment and survival meant that pre- and postspill comparisons could only detect highly significant impacts. Even if postspill abundance decreased by half, a number of other mechanisms, including climatic shift, fishery harvests, and hatcheries could be invoked to question the impact of oil. Without oiled versus non-oiled comparisons, changes in pre-and postspill abundance would be weak evidence for oil spill impacts.

In the absence of prespill baseline data, a long time series of postspill monitoring assumes great importance. If the oiled sites gradually converge toward the abundance in non-oiled sites, the case that the differences observed postspill were the result of oiling is strengthened. If the oiled sites do not converge toward the non-oiled sites, we are trapped in the dilemma that if the difference is from oil the damage is greater but our degree of belief that the damage was from oil will be weaker. In either case, we need to follow the oiled and non-oiled sites for many years after the spill to determine if convergence has occurred.

Discussion

When it is technically and politically possible, experimental oiling in the field would provide the most direct, incontestable evidence of oil impacts. Such experiments were not seriously considered in the case of the *Exxon Valdez* oil spill, but from a scientific perspective this would be the most satisfying method for providing evidence of damage.

Thus far, the discussion in this paper has concentrated on changes in population abundance. I have not considered ecosystem-level effects, that is, changes in the structure of the ecosystem. Indeed, most of the damage assessments were structured around single species; the intertidal and subtidal community studies are the major exception to this. In part this bias arose from the early legal demands for damage claims: It was desirable to assess damages for individual species where changes in abundance could be demonstrated. Resource managers have been managing species, and prespill information was on individual species, not ecosystems. The legal precedents for claiming ecosystem-level damages were not felt to be as strong, and the ecologists were less certain they knew how to measure such damage.

There is a danger, however, in ignoring ecosystem-level changes. It is not clear how to measure or value changes in the ecosystems, but it should be a subject for consideration. For instance, it would be possible to select oiled and non-oiled bays for an ecosystem-wide comparison. Such studies are fraught with difficulties, however, particularly because most of the species of public interest are quite mobile, and more individuals of these species will move in and out of any spatial area than can be readily studied.

To make a good case that oil caused changes in population abundance, no single source of evidence will be absolutely convincing. The strength of the case will depend upon the number of links that can be connected in a chain of evidence. Any single source of data, whether body counts, before and after comparisons, postspill oiled versus non-oiled comparisons, or experimentally induced oil effects will not provide a convincing case on its own. Ideally, data would show that oil reached individuals, that oil killed individuals, and that abundances in oiled sites were lower than in either non-oiled sites or in the same sites before the spill. Design of future damage assessments should be built upon consideration of these links in a chain of evidence.

References

Berger, J. O. 1985. Statistical decision theory and Bayesian analysis. Springer-Verlag, New York.

Bowman, T. D., P. F. Schempf, and J. A. Bernatowicz. 1995. Bald eagle survival and population dynamics in Alaska after the *Exxon Valdez* oil spill. Journal of Wildlife Management 59:317–324.

Bue, B. G., S. Sharr, S. D. Moffitt, and A. K. Craig. 1996. Effects of the *Exxon Valdez* oil spill on pink salmon embryos and preemergent fry. American Fisheries Society Symposium 18:619–627.

Carls, M. G., and six coauthors. 1996. Growth, feeding, and survival of pink salmon fry exposed to food contaminated with crude oil. American Fisheries Society Symposium 18:608–618.

Collier, T. K., C. A. Krone, M. M. Krahn, J. E. Stein, S.-L. Chan, and U. Varanasi. 1996. Petroleum exposure and associated biochemical effects in subtidal fish after the *Exxon Valdez* oil spill. American Fisheries Society Symposium 18:671–683.

DeGange, A. R., A. M. Doroff, and D. H. Monson. 1994. Experimental recovery of sea otter carcasses at Kodiak Island, Alaska, following the *Exxon Valdez* oil spill. Marine Mammal Science 10:492–496.

Edwards, A. W. F. 1972. Likelihood. Cambridge University Press, Cambridge.

Garrott, R. A., L. L. Eberhardt, and D. M. Burn. 1993. Mortality of sea otters in Prince William sound following the *Exxon Valdez* oil spill. Marine Mammal Science 9:343–359.

Geiger, H. J., B. G. Bue, S. Sharr, A. C. Wertheimer, and T. M. Willette. 1996. A life history approach to estimating damage to Prince William Sound pink salmon caused by the *Exxon Valdez* oil spill. American Fisheries Society Symposium 18:487–498.

Hepler, K. R., P. A. Hansen, and D. R. Bernard. 1996. Impact of oil spilled from the *Exxon Valdez* on survival and growth of Dolly Varden and cutthroat trout in Prince William Sound. American Fisheries Society Symposium 18:645–658.

Highsmith, R. C., and six coauthors. 1996. Impact of the *Exxon Valdez* oil spill on intertidal biota. American Fisheries Society Symposium 18:212–237.

Holling, C. S. 1973. Resilience and stability of ecological systems. Annual Review of Ecology and Systematics 4:1–23.

Hurlbert, S. H. 1984. Pseudoreplication and the design of ecological field experiments. Ecological Monographs 54:187–211.

Jewett, S. C., T. A. Dean, and D. R. Laur. 1996. Effects of the *Exxon Valdez* oil spill on benthic invertebrates in an oxygen-deficient embayment in Prince William Sound, Alaska. American Fisheries Society Symposium 18:440–447.

Oakley, K. L., and K. J. Kuletz. 1996. Population, reproduction, and foraging of pigeon guillemots at Naked Island, Alaska, before and after the *Exxon Valdez* oil spill. American Fisheries Society Symposium 18: 759–769.

American Fisheries Society Symposium 18:645–658, 1996

Impact of Oil Spilled from the *Exxon Valdez* on Survival and Growth of Dolly Varden and Cutthroat Trout in Prince William Sound

KELLY R. HEPLER AND PATRICIA A. HANSEN

Alaska Department of Fish and Game, Division of Oil Spill Impact Assessment and Restoration 333 Raspberry Road, Anchorage, Alaska 99518, USA

DAVID R. BERNARD

Alaska Department of Fish and Game, Division of Sport Fish, Anchorage, Alaska 99518, USA

Abstract.—When crude oil spilled from the grounded tanker *Exxon Valdez* into Prince William Sound, Alaska, in late March 1989, anadromous Dolly Varden *Salvelinus malma* and cutthroat trout *Oncorhynchus clarki* were residing in lakes. Five populations of each species were intercepted with five weirs in 1989, 1990, and 1991 during their annual seaward emigration in the spring. Two populations emigrated into the wake of the spill, while three emigrated into waters free of spilled crude oil. Because anadromous Dolly Varden and cutthroat trout feed during the summer near the shores of Prince William Sound, fish emigrating into the path of the spill spent their summers near stranded oil and possibly ingested contaminated food. Of the 11,995 Dolly Varden and cutthroat trout marked at weirs in 1989 and the 41,550 marked in 1990, 24% and 18% were recaptured in 1990 and 1991. Almost all (97%) recaptured emigrants were recaptured at the same weirs in 1990 and 1991 at which they were released 1 year earlier. However, on average an estimated 39% of survivors of both species evaded recapture in 1990 at fish-tight weirs that year. In populations associated with spilled oil, growth from 1989–1990 was 24% and 22% slower for recaptured subadult and adult Dolly Varden, respectively, and 36%–43% slower for subadult and adult cutthroat trout, respectively. These differences persisted through 1991 for cutthroat trout, but not for Dolly Varden; growth was slower during 1990–1991 for recaptured Dolly Varden in populations not associated with oil. Estimates of mean length by age-group for emigrating cutthroat trout in 1989 indicated that their growth had been uniform across Prince William Sound before the spill. Although averages of estimated survival rates from 1989 to 1990 are less in study populations associated with spilled oil (32% and 22% less for subadult and adult Dolly Varden, respectively, and 28% less for adult cutthroat trout), none of the differences are statistically significant. Chronic starvation and direct exposure to petrogenic hydrocarbons are hypothesized as the pathways by which spilled crude oil could have slowed the growth of Dolly Varden and cutthroat trout.

In 1989, many Dolly Varden *Salvelinus malma* and cutthroat trout *Oncorhynchus clarki* in Prince William Sound, Alaska, were exposed to crude oil spilled from the tanker *Exxon Valdez*. Forty thousand metric tons of unrefined oil composed of aliphatic hydrocarbons (such as the paraffins) and water-soluble aromatic hydrocarbons (such as benzene, toluene, xylene, and the napthalenes) spread across Prince William Sound and out into the Gulf of Alaska. As the oil spread to the southwest, anadromous Dolly Varden and cutthroat trout were still in the lakes and streams that surround Prince William Sound. By late April, Dolly Varden and cutthroat trout began leaving these sanctuaries and moving into estuaries and bays and along shorelines. During summer, members of both species use nearshore and estuarine waters for feeding (Scott and Crossman 1979; Armstrong and Morrow 1980; Morrow 1980; see Trotter 1989), areas in which the spillage of petrogenic hydrocarbons can be especially perturbing to biological communities (Teal and Howarth 1984; Highsmith et al. 1996, this volume).

Under controlled conditions, exposure to crude oil has proven toxic or lethal to Dolly Varden, cutthroat trout, and to members of closely related species. In static bioassays using the water-soluble fractions of crude oils from Prudhoe Bay and Cook Inlet in seawater, the 96-h median tolerance limit was 1.30–1.80 mg/L for smolts, juveniles, and adults of Dolly Varden from southeastern Alaska (Moles et al. 1979; Rice et al. 1979). In 90-d tests in flow-through bioassays with freshwater, 48% of the cutthroat trout fry held at concentrations of 0.52 mg/L of the water-soluble fraction of crude oil died (Woodward et al. 1981). Although there were few deaths of fry held at lower concentrations, growth was inhibited even at the lowest concentration in this study (0.10 mg/L). Growth of rainbow trout *O. mykiss* held in freshwater was significantly slowed when the fish were fed ad libitum a 1% diet of some of the aliphatic compounds found in crude oil (Luquet et al. 1983, 1984).

Toxicity under controlled conditions in the laboratory, however, is not a sufficient reason to expect impairment in the productivity of unconfined Dolly Varden and cutthroat trout in Prince William Sound. By June 1989, levels of aromatic hydrocarbons suspended in the water column in Prince William Sound (Neff 1990; Short and Harris 1996, this volume) were a few micrograms per liter, well below levels found toxic in bioassays for Dolly Varden. Many species of fish, including Dolly Varden and rainbow trout, can readily metabolize and excrete the aromatic and aliphatic compounds in crude oil if concentrations are sublethal (Pedersen et al. 1976; Varanasi and Malins 1977; Neff 1979; Thomas and Rice 1981). Because salmonids can detect concentrations of aromatic compounds down to a few milligrams per liter (Rice 1973; Maynard and Weber 1981; Weber et al. 1981), members of this family might have avoided any significant contact with crude oil in Prince William Sound. Yet these detectable levels of petrogenic hydrocarbons did not affect migrations of salmonids in experiments (Brannon et al. 1986; Nakatani and Nevissi 1991). Dolly Varden and cutthroat trout, especially immature individuals, do not feed as much in the water column as along the shoreline (Narver and Dahlberg 1965; see Trotter 1989), where crude oil is entrapped in sediments and macrophytes and transported through the food chain (see Varanasi and Malins 1977; Lee 1978; Teal and Howarth 1984). Relevance of these and similar studies is ambiguous because the studies are based on individual fish under circumstances dissimilar to those in Prince William Sound in 1989. Before the productivity of populations of Dolly Varden and cutthroat trout in Prince William Sound can be judged damaged, there must be a demonstrable diminution in survival or growth of fish in populations in the path of the spill and not elsewhere.

The objectives of our study were to test the hypotheses that growth and survival of Dolly Varden and cutthroat trout were significantly diminished in the path of the spill. A coordinated suite of studies to provide specific information on the impact of crude oil from the *Exxon Valdez* on the productivity of natural resources, including Dolly Varden and cutthroat trout in Prince William Sound, began immediately after the spill (Trustee Council 1989). Studies were planned in accordance with the U.S. Clean Water Act and the Comprehensive Environmental Response, Compensation, and Liability Act (CERCLA). Because baseline data on populations of Dolly Varden and cutthroat trout in Prince William Sound were limited, and because there was no control of which populations were exposed to crude oil, our investigation was observational, not experimental (Cochran 1983; Eberhardt and Thomas 1991). Our sampling was predicated on the unusual migratory behavior of both of these species.

Methods

Our sampling design was based upon the model of migratory behavior of anadromous Dolly Varden developed by Armstrong (1970, 1974, 1984) and Armstrong and Morrow (1980). Juvenile Dolly Varden remain in their natal stream for up to 4 years, then develop into smolts and migrate to sea during spring. In late summer or early fall, all surviving Dolly Varden return to freshwater to spend the winter. Dolly Varden spawned in watersheds with lakes (lacustrine watersheds) return to their natal watershed to overwinter in one of its lakes; fish that spawned in watersheds without lakes (fluvial watersheds) enter lacustrine watersheds in search of lakes suitable as their winter residence. Each spring, adult and immature Dolly Varden again migrate to the sea with adults preceding younger fish. Once a Dolly Varden has spent the winter in a lake, it returns to that lake each winter if it survives. Mature Dolly Varden return to their natal streams to spawn in the fall. The migratory habits of anadromous cutthroat trout are similar to those of Dolly Varden in that they both emigrate from lakes in the spring and return in the summer and fall, however, cutthroat trout spawn in the spring (Scott and Crossman 1979; Morrow 1980; see Trotter 1989). Timing of spring emigration of cutthroat trout corresponds to that of smolting Dolly Varden. Because Dolly Varden and cutthroat trout annually leave the ocean and return to the same lacustrine watershed to overwinter, all survivors of the emigration from such a watershed in one year would be expected to emigrate from that watershed in the next year.

Our study was designed to detect differences in survival rates and growth among several groups of marked fish (study populations). Dolly Varden and cutthroat trout emigrating from five lacustrine watersheds were marked in 1989 and 1990, and the survivors were recaptured during their spring emigrations in 1990 and 1991. Study populations were named according to where they were captured as emigrants: in Makaka Creek (Makaka), in unnamed streams flowing into Boswell and Rocky bays (Boswell and Rocky), in an unnamed stream on Green

Island (Green), and in Eshamy Creek (Eshamy). Of the 143 streams surveyed in Prince William Sound during 1989 in the search for spawning salmon (Sharr 1990), only 14 lacustrine watersheds contained anadromous Dolly Varden and 10 contained anadromous cutthroat trout (S. Sharr, Alaska Department of Fish and Game, unpublished data). Only two of these lacustrine watersheds with cutthroat trout emptied into the path of the spill (the watersheds on Green Island and Eshamy Creek).

Emigrating Dolly Varden and cutthroat trout were captured during the spring with weirs located just above tidewater. Weirs were constructed of inclined aluminum panels comprised of parallel vertical rods 1.25 cm in diameter set with 1.25-cm gaps. Each panel measured 2.0 m long and 1.0 m high and rested vertically against wooden tripods spaced approximately 2 m apart. Each weir completely blocked the stream and directed emigrating fish into a holding pen. Weirs were installed from 22 April through 15 May in 1989 and as soon as streams were ice-free in 1990 and 1991; weirs were dismantled at the end of June each year. Large numbers of Dolly Varden were captured as soon as weirs became fish-tight in 1989; few or no cutthroat trout were caught until several days had passed after installation of the weirs in all years. Weirs remained fish-tight through the emigrations in 1990 except for 2 d in mid-June on Makaka Creek and after 14 June for the Boswell population, when high water compromised these weirs. These breaches occurred after the emigration of Dolly Varden, but during the end of the emigration of cutthroat trout. In 1991, weirs lost integrity because of high water at Eshamy Creek (5 d in late May–early June) and at Green Island (4 d in early May); all other weirs remained fish-tight through the emigration that year.

Every captured fish was measured to the nearest millimeter of fork length (FL), and those 200 mm and larger were marked with an individually numbered Floy, T-bar tag (model 68-B) and by excision of their adipose fin. Tags were inserted between fin rays at the base of the dorsal fin approximately 15 mm forward of the posterior end of the fin on the left side. Fish were not anesthetized. Marked fish were retained for 5 min to cull any immediate mortalities. Study populations consisted of all marked fish released alive in 1989 and 1990 to continue their emigration. Eight percent of Dolly Varden recaptured from the Boswell population in 1990 had lost their tags, whereas <2% recaptured at other weirs in 1990 and 1991 had shed tags. All recaptured cutthroat trout retained their tags. According to observations from other studies on both species (Armstrong 1970, 1974; see Trotter 1989), almost all fish from 200 to 269 mm FL emigrating in the spring are immature (subadults); smaller fish are smolts and most larger fish are adults.

Ages of fish emigrating in 1989 and 1990 were determined from scales collected from samples of cutthroat trout and from sagittal otoliths from samples of Dolly Varden. Cutthroat trout were systematically sampled. The lowest sampling rate across populations and years was one of every three captured cutthroat trout; the highest rate was almost every emigrating fish. Scales were collected from the left side of cutthroat trout approximately two rows above their lateral line and on the diagonal row downward from the posterior insertion of the dorsal fin. Scales were impressed onto acetate under heat or mounted whole on glass slides. All impressions and whole mounts were scanned under 40× magnification in a microfiche reader by the same person within the last 3 months of 1992. Recognition of annuli followed criteria in Fuss (1982) and Jearld (1983). Ages were determined twice for each sampled cutthroat trout; all samples were scanned once before any sample was scanned twice. Only samples with matching estimates were used in subsequent analyses; all samples with regenerated scales or discordant estimates were excluded. The weekly frequency of samples retained for the analysis remained roughly proportional to the weekly frequency of cutthroat trout emigrating past weirs in 1989 and 1990. Although otoliths were collected from Dolly Varden in 1989, these samples proved to be unrepresentative of emigrating populations. Analysis of samples indicated that larger Dolly Varden emigrated first regardless of their age, and we probably did not sample the larger Dolly Varden in each age-group with our late, staggered start in 1989. Although weirs were installed in 1990 before the emigration of Dolly Varden began, sampling at Green Island and Eshamy Creek was delayed, and the largest Dolly Varden in these populations were not sampled. No otoliths and few scales were collected in 1991.

Analyses of variance (ANOVA) and covariance (ANCOVA) were used to test hypotheses that growth of fish had been the same regardless of their apparent association with spilled crude oil. Estimates of mean length by age-group of emigrant cutthroat trout sampled in 1989 and 1990 were compared across study populations in different analyses using the same design, an unbalanced two-way parametric ANOVA with study population (γ)

as the independent variable (a random effect) and age (ψ) as an exogenous variable; the model was

$$l_{ijk} = \mu + \gamma_i + \psi_j + \sigma_{ij} + \epsilon_{ijk}$$

with μ as the overall mean length and σ and ϵ as true error and sampling error, each distributed normally with mean 0. Because of their migratory behavior, information on mean length by age-group of cutthroat trout collected in 1989 is a priori information on populations never exposed to crude oil. Significance of differences in directly measured growth of recaptured Dolly Varden during each period of freedom (1989–1990 and 1990–1991) was tested with a two-way unbalanced parametric ANOVA with nested effects:

$$\Delta l_{ijtk} = \mu + \tau_i + \phi_j + \gamma(\tau_i)_t + \sigma_{ijt} + \epsilon_{ijtk}.$$

The difference between the length of an individual fish as measured in consecutive years (Δl) was the dependent variable, and association with spilled crude oil was the independent variable (a random effect). Study populations were nested within association with spilled oil (τ) as random effects. Starting length of Dolly Varden in 1989 or 1990 was the exogenous variable. Because starting length proved poorly correlated with growth for Dolly Varden, the Δl were segregated according to their starting lengths into two blocks (ϕ: adults and subadults) to maximize power in tests of significance (Cox 1957). A nested ANOVA could not be used to test the significance of differences in growth for cutthroat trout because few cutthroat trout in the Green population were recaptured in either 1990 or 1991. An unbalanced parametric ANCOVA with study population as the independent variable (a random effect) and Δl the dependent variable was used instead:

$$\Delta l_{ijk} = \mu + \gamma_i + \beta(l_{ijk}) + \sigma_{ij} + \epsilon_{ijk}.$$

Because starting length (l) of recaptured cutthroat trout was highly correlated with growth, starting length was treated as a covariate to maximize power of tests (Cox 1957). Before any analysis of growth for either species began, all Δl that fell outside 1.5 interquartiles away from the edge of box plots for each group (combination of population, species, and period of freedom) were considered as outliers because of measurement error (as per guidelines in Tukey 1977:30–44) and were not used in analyses. Approximately 6% of the data were excluded on this basis, with no more than 19% coming from a single study population (4 Dolly Varden removed from a sample of 25 from Green from 1989 to 1990).

Estimated survival rates were averaged across study populations with similar association to the spill to determine if there were similar expectations of mortality for the fish regardless of their association. Because immature and mature Dolly Varden and cutthroat trout typically have different survival rates (Sumner 1953; Armstrong 1974), study populations were divided into adults and subadults, and the analysis was repeated for each group. Because high water had compromised some of the weirs for a few days in 1990, survival rates were estimated with log-linear models following Cormack (1989). The M tagged and released fish that constituted each group in a study population in 1989 were tallied according to their capture history: (1) never recaptured (m_{100}); (2) recaptured in 1991, but not in 1990 (m_{101}); (3) recaptured in 1990, but not in 1991 (m_{110}); and (4) recaptured in 1990 and 1991 (m_{111}). The maximum likelihood estimates (MLE) of the probability of capturing a marked fish at the weir in 1990 (q) and the survival rate from 1989 to 1990 (S) for each group are

$$\text{MLE}\,[q] = \hat{q} = \frac{m_{111}}{m_{101} + m_{111}} \quad (1)$$

and

$$\text{MLE}\,[S] = \hat{S} = \frac{m_{111} + m_{110}}{M\,\hat{q}}. \quad (2)$$

Precision of average survival rates across populations was estimated through a two-stage Monte Carlo simulation based on empirical distributions for $\hat{S}$ derived through bootstrapping capture histories (Efron 1982; Sauermann 1989; Buckland and Garthwaite 1991). Each empirical distribution of estimated survival rates consisted of statistics calculated from 1,000 bootstrap samples for each study population. Each bootstrap sample consisted of M records randomly drawn with replacement from the original data to create a new set of capture histories $\{m^*\}$. Estimates of q and S were calculated for each bootstrap sample in which $m^*_{111} > 0$ ($m^*_{111} = 0$ in only 0.23% of the bootstrap samples drawn). The difference between the average of the survival rates from the bootstrap samples and the original estimate for the same population is a measure of the mathematical bias in the latter statistic (Efron 1982). These empirical distributions of survival rates based on bootstrapping were used in eight Monte Carlo simulations, one for each combination of species, maturity, and association with spilled oil.

TABLE 1.—Numbers of Dolly Varden and cutthroat trout in study populations in Prince William Sound that were tagged and released during their spring emigrations in 1989 and 1990 and were inspected for tags in 1990 and 1991. Adults of both species were 270 mm fork length (FL) and larger upon release; subadults were 200–269 mm FL.

	Dolly Varden				Cutthroat trout			
	Tagged and released		Inspected		Tagged and released		Inspected	
Population and group	1989	1990	1990	1991	1989	1990	1990	1991
Not associated with spilled oil								
Boswell								
Subadults	68	2,678	2,911	927	59	468	498	465
Adults	7	4,051	4,373	2,260	53	853	868	1,220
Makaka								
Subadults	4,114	3,188	3,290	6,717	491	263	264	1,487
Adults	1,178	9,311	9,660	7,862	254	570	571	1,667
Rocky								
Subadults	3,721	6,745	6,977	3,080	7	16	16	0
Adults	692	8,960	9,220	6,016	3	8	9	0
Associated with spilled oil								
Green								
Subadults	180	1,006	1,040	301	4	6	6	10
Adults	83	106	111	427	5	2	3	2
Eshamy								
Subadults	684	2,148	2,290	817	75	48	48	36
Adults	182	967	1,042	250	135	156	158	175

Each simulation had 1,000 iterations, and each iteration had two stages. Choosing n study populations at random with replacement from the N available populations in the combination was the first stage in each iteration; the second consisted of drawing a single statistic at random from each of the empirical distributions of survival rates for the chosen populations, then averaging these statistics across populations to produce a simulated survival rate. Each simulation produced 1,000 such simulated rates for each combination of species, maturity, and association with oil.

The significance of differences in estimated survival rates across associations with spilled oil was judged relative to their precision. Each of the four combinations of species and maturity corresponded to two distributions of simulated rates: one for study populations that were associated with spilled oil and one for study populations that were not. An angular transformation was applied to all simulated rates to create normally distributed data, and a mean θ and variance $V[\theta]$ were calculated for each transformed distribution. The difference between means for populations associated with spilled oil (θ_t) and those not (θ_c) is a measure of the potential impact of spilled oil. The probability P of making a type I error in judging a difference in survival rates to be real was based on a one-tailed alternative hypothesis H_a: $\theta_c > \theta_t$ and the test statistic

$$Z = \frac{\theta_c - \theta_t}{\sqrt{V[\theta_c] + V[\theta_t]}}. \quad (3)$$

Results

Size and Migrations of Study Populations

Table 1 presents the numbers of Dolly Varden and cutthroat trout that were tagged and released (1989 and 1990) and inspected (1990 and 1991). Because of the late start of the project in 1989 and the proclivity of larger Dolly Varden to emigrate earlier, disproportionately more immature fish were captured in 1989 than in other years (Figure 1). The trimodal distributions of emigrating fish of both species corresponded to smolts (80–160 mm FL), immature fish (160–269 mm FL), and adults (≥270 mm FL). Except in the Green population, adults of both species made up a higher fraction of study populations than did subadults in 1990 (Figure 1) and 1991.

There was little evidence that Dolly Varden and cutthroat trout had strayed into or out of areas in Prince William Sound covered by the spill. Of those fish recaptured during emigration in 1990 and 1991 (10,242), 97% were recaptured at the same weir at which they had been released 1 year earlier (Figure 2). Of those fish that had strayed (298 of

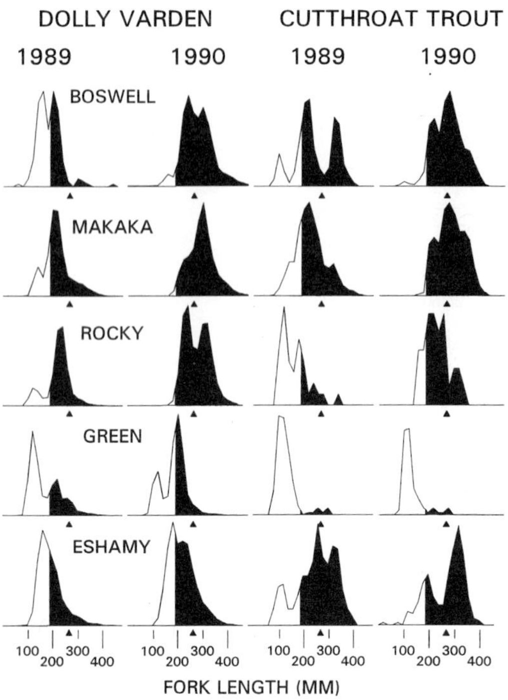

FIGURE 1.—Relative length frequency, by study population, of Dolly Varden and cutthroat trout caught at weirs while emigrating to Prince William Sound in spring 1989 and 1990. Individual frequencies are based on 20-mm cells. Black areas correspond to fish released with tags. Black triangles along the abscissa correspond to the length that separates adults from subadults (270 mm fork length).

10,242), few Dolly Varden and no cutthroat trout had been recaptured in watersheds across the boundary of the spill. Of the 336 fish that were recaptured and reported by persons working or recreating around Prince William Sound during the summers of 1989 and 1990, few Dolly Varden (18) and no cutthroat trout had been caught across the boundary of the spill. The median distances between release and recapture for these 336 fish were 26 km for Dolly Varden and 2 km for cutthroat trout; the maximum distances were 248 km and 55 km, respectively.

Differences in Growth

Estimates of mean length by age-group for 1989 indicate cutthroat trout had similar rates of growth across Prince William Sound before that year (Figure 3). Estimates of mean length at age 4 (years) are similar for emigrants from Eshamy and Makaka creeks and emigrants to Boswell Bay; for older ages, estimates from cutthroat trout sampled at Eshamy Creek are midway between statistics for emigrants at the other two sampling locations. Although the means are significantly different among these three locations (two-way ANOVA, $P < 0.001$), the only paired comparison of significance (least significant difference, $\alpha = 0.05$) is between statistics for emigrants to Boswell Bay and emigrants from Makaka Creek. Averages of statistics for emigrants at these two eastern locations followed the same pattern as did statistics for emigrants from Eshamy Creek on the western rim of Prince William Sound. Comparison of statistics was limited to emigrants past these three sampling locations because there were only a few emigrants leaving Green Island or entering Rocky Bay in 1989.

Growth of fish recaptured in 1990 and 1991 and estimates of mean length by age-group in 1990 showed (1) that growth of Dolly Varden and cut-

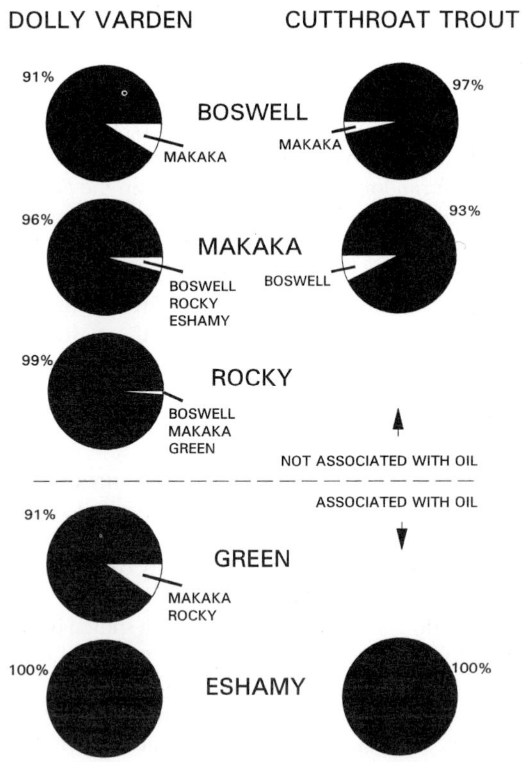

FIGURE 2.—Distribution, by study population, of Dolly Varden and cutthroat trout recaptured during emigration in 1990 and 1991 (combined) relative to where they were released 1 year earlier. Black areas correspond to fish recaptured at the same location 1 year later; white areas correspond to fish recaptured elsewhere at the specified locations.

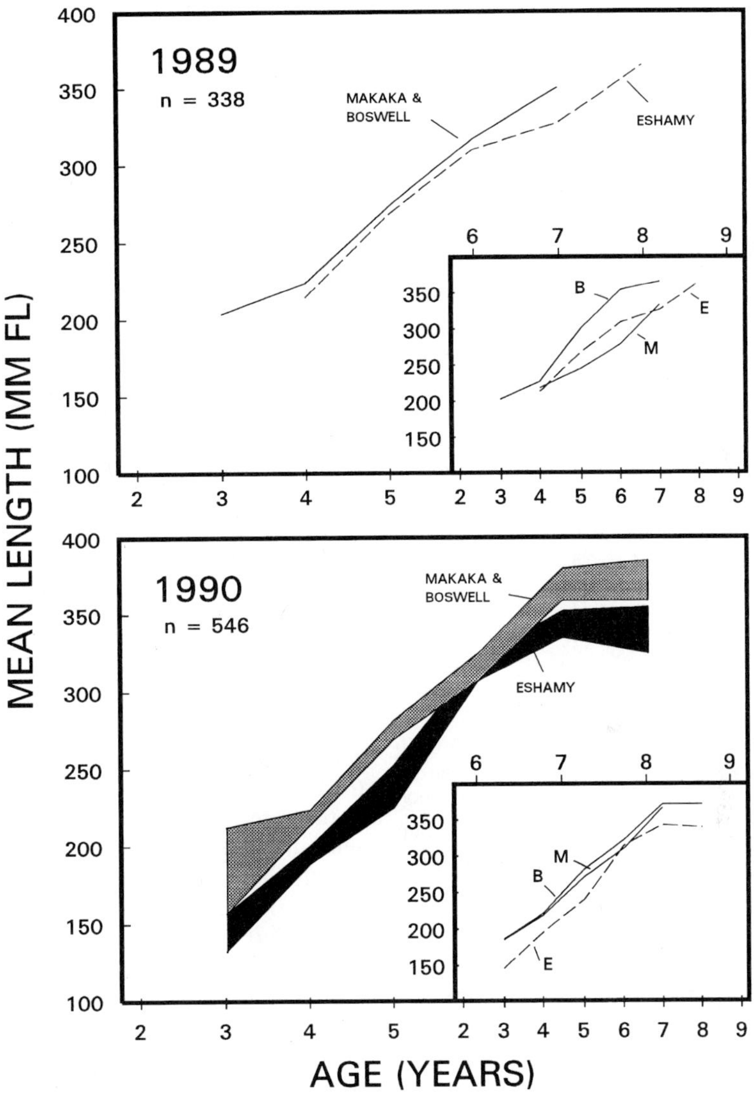

FIGURE 3.—Estimated mean fork length (FL) by age-group in 1989 and 1990 for cutthroat trout sampled at weirs on Eshamy (E) and Makaka (M) creeks and on a stream near Boswell Bay (B). Lines in insets correspond to statistics for the three individual sampling locations. The lines and the bands in the larger panels correspond to statistics for emigrants from Eshamy Creek and averages of statistics for emigrants at the other two locations. Bands in the lower panel correspond to approximate confidence intervals (±2 SE).

throat trout during 1989–1990 was appreciably slower in study populations associated with spilled oil and (2) that this depression in growth persisted into the next year for cutthroat trout, but not for Dolly Varden (Figure 3 and Table 2). Although estimates of mean length by age-group in 1990 are significantly different among cutthroat trout emigrating from Eshamy and Makaka creeks and to Boswell Bay ($P < 0.001$ in a two-way ANOVA with all possible paired comparisons significantly different: Scheffé's test, $\alpha = 0.05$), only statistics for emigrants from Eshamy Creek are meaningfully different (Figure 3). Statistics for 1990 are 21 and 28 mm lower and 5 mm higher for emigrants ages 4 through 6 from Eshamy Creek than are the corresponding statistics for 1989. Average statistics from 1989 and 1990 are similar for study populations not associated with spilled oil: 223 versus 219 mm FL

TABLE 2.—Mean annual growth (mm) of recaptured Dolly Varden and cutthroat trout in study populations associated with spilled oil (Green and Eshamy) relative to study populations that were not associated with oil (Boswell, Makaka, and Rocky), 1989–1990 and 1990–1991. Growth of individual fish were averaged within each study population, and these statistics were averaged across study populations of similar oil association. Adults of both species were ≥270 mm fork length (FL) upon release; while subadults were 200–269 mm FL.

	Growth in 1989–1990		Growth in 1990–1991	
Group and statistic	Dolly Varden	Cutthroat trout	Dolly Varden	Cutthroat trout
Subadults				
Not associated with oil (mm)	77	69	55	65
Associated with oil (mm)	59	44	51	45
Relative difference (%)	−24	−36	−7	−30
Adults				
Not associated with oil (mm)	57	44	44	42
Associated with oil (mm)	44	25	41	22
Relative difference (%)	−22	−43	−9	−47

for age 4, 272 versus 275 mm FL for age 5, and 316 versus 316 mm FL for age 6. Directly measured growth of recaptured fish during 1989–1990 (Figure 4) was significantly less for Dolly Varden ($P = 0.01$ in a nested ANOVA) and for cutthroat trout ($P < 0.001$ in an ANCOVA) in populations associated with oil than in populations encountering oil-free areas. Paired comparisons based on least-squares means in the ANCOVA showed similar growth by cutthroat trout in Makaka and Boswell populations and less growth in the Eshamy population ($\alpha = 0.05$). Growth was 24% less for subadult and 22% less for adult Dolly Varden associated with oil; growth was 36% less for subadult and 43% less for adult cutthroat trout. One year later (1990–1991), growth of recaptured cutthroat trout was still significantly less in the population associated with spilled oil ($P < 0.001$ in an ANCOVA), but not growth of recaptured Dolly Varden ($P = 0.88$ in a nested ANOVA) (Figure 4). Parity in growth of Dolly Varden across Prince William Sound in 1990–1991 came not through increased growth of Dolly Varden in the Eshamy and Green populations, but through decreased growth by Dolly Varden emigrating into areas not in the wake of the spill (Figure 4). Adult Dolly Varden marked in 1989 were similar in size to those marked in 1990 (324 versus 329 mm FL on average across all five study populations).

Differences in Survival Rates

Averages of estimated survival rates from 1989 to 1990 were 24–33% lower in those populations of Dolly Varden and adult cutthroat trout associated with spilled oil when compared with their respective oil-free populations (Table 3); however, these differences were not significant (all $P \geq 0.16$, Figure 5) when compared against standard thresholds of acceptable risk ($\alpha = 0.10$). Under the precision attained in this study, the chances of wrongly concluding lower survival rates for study populations associated with spilled oil are 2 in 7 for adult Dolly Varden, 1 in 6 for subadult Dolly Varden, and 1 in 3 for adult cutthroat trout (Figure 5). Because of small sample sizes, survival rates could not be estimated for some study populations, and these populations were excluded from the Monte Carlo simulation (Table 4). Differences between estimated survival rates and average rates from bootstrapping ranged from 0.01 to 0.02, indicating that bias in estimated rates was negligible for all populations except one. This exception concerns subadult cutthroat trout, for which the difference between the average rate from bootstrapping for the Eshamy population (0.60) is considerably lower than the corresponding estimated rate from the original data (0.72 in Table 3). Precision and bias in the latter statistic are such that the frequency distribution of simulated averages for the Eshamy completely swamps the corresponding distribution for the other populations (Figure 5). Even though weirs remained fish-tight through the spring emigration at most sites in 1990, relatively large numbers of fish recaptured in 1991 were not recaptured in 1990 (m_{101}) in many populations (Table 4). Average estimates of the probability of capture ($\hat{q}$) in 1990 at all weirs are 65 and 68% for subadult and adult Dolly Varden, respectively, and 49 and 58% for subadult and adult cutthroat trout, respectively.

Discussion

The reduced growth of Dolly Varden and cutthroat trout that we observed in the wake of the spill meet the necessary conditions to show that exposure to crude oil had damaged the resources that these fish represent. If growth of Dolly Varden had been similar across Prince William Sound after 1989, the hypothesis of damage would have been rejected; the same consequence holds for cutthroat trout. In fact, we found that growth was consistently better for those fish not associated with spilled oil in 1989. Estimates of mean length by age-group are consistent with populations of cutthroat trout hav-

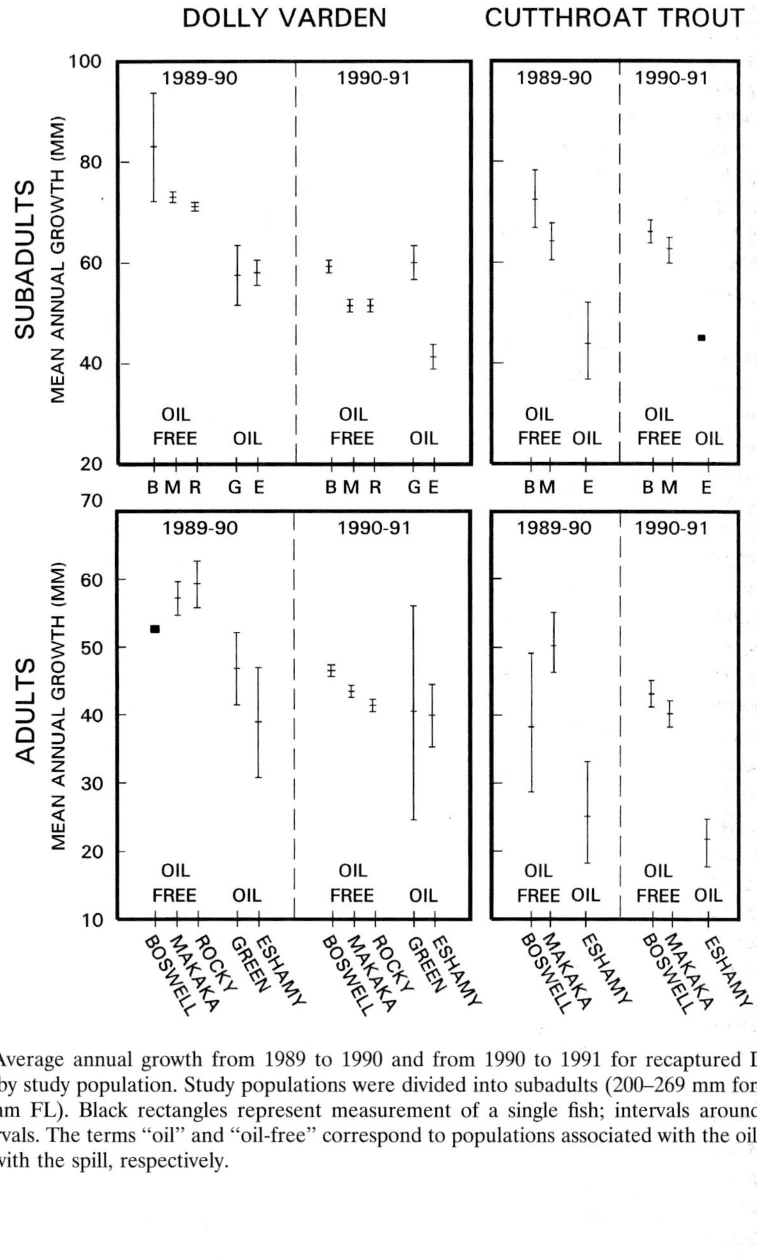

FIGURE 4.—Average annual growth from 1989 to 1990 and from 1990 to 1991 for recaptured Dolly Varden and cutthroat trout by study population. Study populations were divided into subadults (200–269 mm fork length, FL) and adults (≥270 mm FL). Black rectangles represent measurement of a single fish; intervals around means are 95% confidence intervals. The terms "oil" and "oil-free" correspond to populations associated with the oil spill and to those not associated with the spill, respectively.

TABLE 3.—Estimated survival rates ($\hat{S}$) for study populations of adults (≥270 mm fork length, FL) and subadult (200–269 mm FL) Dolly Varden and cutthroat trout from 1989 to 1990 averaged over populations associated with spilled crude oil (Green and Eshamy) and those not associated with oil (Boswell, Makaka, and Rocky).

	Dolly Varden		Cutthroat trout	
Group and statistic	Adults	Subadults	Adults	Subadults
Not associated with oil ($\hat{S}$)	0.21	0.39	0.31	0.39
Associated with oil ($\hat{S}$)	0.16	0.26	0.22	0.72
Relative difference (%)	−24	−33	−29	+85

ing similar growth rates across Prince William Sound before the spill, yet after the spill, growth was slower for those populations in its wake. Differences in growth according to association with crude oil persisted for cutthroat trout into 1990, but not for Dolly Varden. Surprisingly, the difference for Dolly Varden was negated, not through acceleration of growth for fish from Green and Eshamy populations, but through apparent deceleration of growth in populations not associated with spilled oil. Either Dolly Varden experienced annual varia-

TABLE 4.—Number of Dolly Varden and cutthroat trout released in 1989 (M), number of these fish recaptured in 1990 and 1991 (m_{111}), number recaptured in 1991 but not in 1990 (m_{101}), number recaptured in 1990 but not in 1991 (m_{110}), estimated probabilities of being recaptured in 1990 ($\hat{q}$), and estimated survival rates from 1989 to 1990 ($\hat{S}$) for adults (≥270 mm fork length, FL) and subadults (200–269 m FL) released in 1989. Those groups with no estimated survival rates $\hat{S}$ were excluded from the Monte Carlo simulation to measure precision of average statistics.

Population and group	M	m_{111}	m_{101}	m_{110}	$\hat{q}$	$\hat{S}$
Dolly Varden not associated with spilled oil						
Boswell						
Subadults	68	2	1	20	0.67	0.49
Adults	7	0	0	1		
Makaka						
Subadults	4,114	130	52	830	0.71	0.32
Adults	1,178	18	4	149	0.81	0.17
Rocky						
Subadults	3,721	153	40	921	0.79	0.36
Adults	692	23	5	111	0.82	0.24
Dolly Varden associated with spilled oil						
Green						
Subadults	180	2	3	15	0.40	0.24
Adults	83	3	4	1	0.43	0.11
Eshamy						
Subadults	684	4	2	128	0.67	0.29
Adults	182	2	1	24	0.67	0.21
Cutthroat trout not associated with spilled oil						
Boswell						
Subadults	59	10	3	10	0.77	0.44
Adults	53	3	3	6	0.50	0.34
Makaka						
Subadults	491	33	30	54	0.52	0.34
Adults	254	12	4	40	0.75	0.27
Rocky						
Subadults	7	0	0	1		
Adults	3	0	0	0		
Cutthroat trout associated with spilled oil						
Green						
Subadults	4	0	0	0		
Adults	5	0	1	0		
Eshamy						
Subadults	75	1	5	8	0.17	0.72
Adults	135	3	3	12	0.50	0.22

tion in growth rates that cutthroat trout did not, or spilled crude oil extended its influence in 1990.

There is evidence supporting both reduction of food and direct exposure as mechanisms by which spilled oil could have affected growth of Dolly Varden and cutthroat trout. Crustaceans and other small invertebrates dominate the diet of subadult Dolly Varden and cutthroat trout whereas adults eat mostly fish. During residence in Alaskan estuaries, smaller Dolly Varden mostly eat gammarid amphipods, euphausiids, megalopae, isopods, mysids, and polychaetes whereas larger Dolly Varden prey on capelin *Mallotus villosus*, Pacific sand lance *Ammodytes hexapterus*, Pacific herring *Clupea pallasi*, whitespotted greenling *Hexagrammos stelleri*, young salmon *Oncorhynchus* spp., rockfish larvae *Sebastes* spp., and silverspotted sculpin *Blepsias cirrhosus* (Lagler and Wright 1962; Narver and Dahlberg 1965). In Alaska, small cutthroat trout eat some of the same estuarine foods as small Dolly Varden eat: gammarid amphipods, euphausiids, capelin, Pacific sand lance, and young salmon (Armstrong 1971). Estuarine diets of cutthroat trout in Canada, Washington, and Oregon are similar to those of cutthroat trout and Dolly Varden in Alaska (see Trotter 1989). When exposed to petrogenic hydrocarbons, the subtidal invertebrate community shifts to smaller epifauna, usually with markedly fewer amphipods (Teal and Howarth 1984). Such a shift occurred in Prince William Sound in 1989 and 1990 in the path of the spill (Jewett et al. 1996, this volume; S. C. Jewett, University of Alaska, and T. A. Dean, Coastal Resources, Inc., unpublished data).

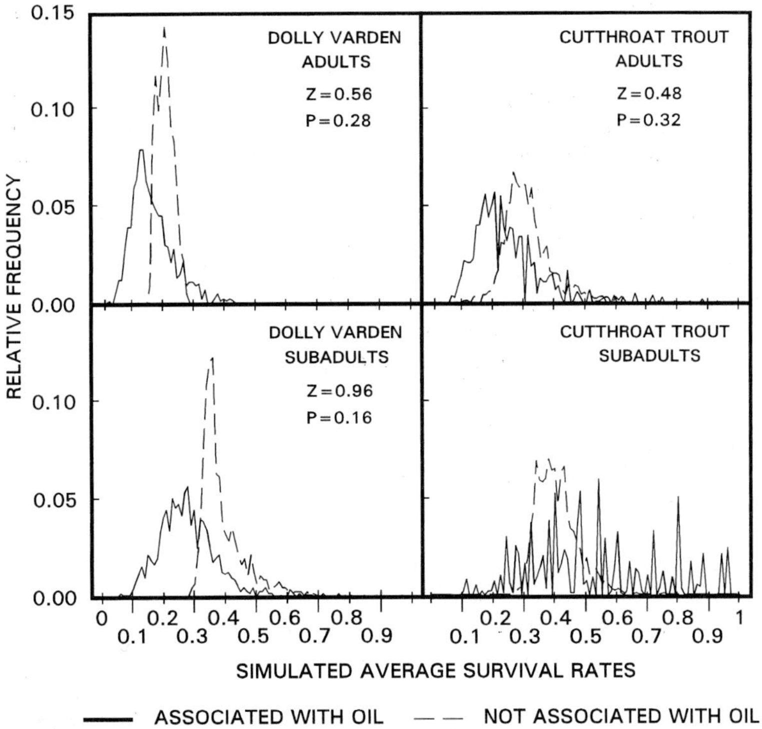

FIGURE 5.—Distributions of simulated average survival rates from 1989 to 1990 for study populations of Dolly Varden and cutthroat trout associated with spilled crude oil and those not associated with the oil. Distributions were built on 1,000 simulated averages each based on random sampling of empirical distributions of estimated survival rates built by bootstrapping the capture histories of individual study populations. Adults of both species were ≥270 mm fork length (FL) upon release; subadults were 200–269 mm FL. The *Z*-statistic could not be calculated for subadult cutthroat trout because of small sample sizes.

Dolly Varden collected along shorelines in the wake of the spill in another study had been exposed to crude oil (Collier et al. 1996, this volume). Large Dolly Varden were captured in shallow subtidal areas with gill nets and beach seines. Those fish captured in the wake of the spill in 1989 had exceptionally high levels of metabolites of petrogenic hydrocarbons, such as biliary fluorescent aromatic hydrocarbons (FACs). Enzymes for metabolizing aromatic hydrocarbons, such as hepatic cytochrome P-4501A, were also elevated. In contrast, contents of their stomachs were free of petrogenic hydrocarbons, indicating that absorption is more likely than digestion as the pathway of their exposure. Levels of FACs and hepatic P-4501 in Dolly Varden captured along shorelines in 1990 were considerably reduced from levels observed 1 year earlier. No cutthroat trout or small Dolly Varden were tested for FACs or hepatic P-4501A in this or any other study.

Although estimated differences in growth met the necessary conditions to show damage from oil spilled by the *Exxon Valdez*, estimated differences in survival rates did not. The late installation of weirs in 1989 and the breach of the weir on Eshamy Creek in 1991 kept sample sizes low and subsequently reduced the power of hypothesis tests for Dolly Varden. These factors were not as critical in studies of cutthroat trout; sample sizes for this species were small because their populations are small. Prince William Sound is the extreme northern edge of the observed range of coastal cutthroat trout (Scott and Crossman 1979). Unexpected migratory behavior of Dolly Varden may have also affected estimation of survival rates with log-linear models. Equations (1) and (2) are based on all tagged fish experiencing the same survival rates during the winter of 1989–1990 regardless of their location. Dolly Varden with capture history (101) may have resided in habitats different from other tagged fish. If survival rates had been less for wayward Dolly Varden than for their peers, m_{101} would be depressed, the estimated probability of capture $\hat{q}$ would be biased high, and the estimated survival rate $\hat{S}$ would be biased low as calculated with equations (1) and (2). If survival rates for wayward Dolly Varden had been better, m_{101} would be elevated, $\hat{q}$ biased high, and $\hat{S}$ biased low. In either instance, there would be an inverse relationship between $\hat{q}$ and $\hat{S}$. Casual inspection of Table 4 shows no such relationship, indicating that if there had been different survival rates for Dolly Varden with different capture histories during the winter of 1989–1990, all study populations were similarly affected. Because different survival rates only invalidates the log-linear model for capture history (101) and because only the value of m_{101} is affected, $\hat{q}$ and $\hat{S}$ are relatively insensitive to bias from all but the largest differences in rates.

Because our study was observational and not experimental, interpretation of its results is subject to two kinds of criticism: the first concerns experimental design and the second, natural variation. Whenever treatments (in our case, association with crude oil) are not randomly assigned to experimental units (populations), spatial or temporal trends in parameters (survival rates and growth) can be confused with effects of treatments (Green 1979; Hurlbert 1984). One way to avoid this problem is to estimate parameters for several years before the treatment is applied. Although we could not do this for survival rates, we could and did for growth. Our a priori information on mean length by age-group in 1989 showed a history of near uniform growth across Prince William Sound for cutthroat trout. Although we did not directly measure growth before 1989, mean length by age-group is an integration of growth over the extant year classes extending backward several years. This uniformity of growth is the consequence of the size of Prince William Sound; its uniformity of climate, limnology, and oceanography; and the mixture of reproductively distinct stocks in most of our study populations. There were no a priori estimates of survival rates. The second common criticism of observational studies is that natural variation in parameters can be confused with the effects of treatments when sample size is small (Eberhardt and Thomas 1991). Although populations in our study are few, they represented much of the suitable winter habitat, and by inference, many of the anadromous Dolly Varden and cutthroat trout in Prince William Sound. Of the 14 lacustrine watersheds with anadromous Dolly Varden and 10 with anadromous cutthroat trout, 5 are in this study, including the only 2 lacustrine watersheds that empty into the path of the spill (Green and Eshamy) and contain anadromous cutthroat trout.

One final note is warranted on those cutthroat trout and Dolly Varden with capture history (101). Fish with this history could only have escaped capture at fish-tight weirs in 1990 by remaining in the lake past spring, by spending the winter in a different watershed, or by spending the winter at sea. Observations of cutthroat trout at sea during the winter have been reported in the scientific literature (see Trotter 1989) but observations of Dolly Varden have not. All three of these possible explanations for not recapturing significant numbers of Dolly Varden in 1990 are at odds with the "migration hypothesis" for this species (see Armstrong and Morrow 1980), thereby making anomalous the low probabilities of capture that year.

Conclusions

Our findings are consistent with oil spilled from the *Exxon Valdez* having retarded the body growth of anadromous Dolly Varden and cutthroat trout in 1989. Our evidence is also consistent with lingering effects of this spilled oil having reduced body growth of cutthroat trout through at least 1990. Although estimated survival rates of Dolly Varden and cutthroat trout are lower for those fish emigrating into the path of the spill in 1989, differences are not significant given a reasonable statistical risk.

Acknowledgments

We thank Andy Hoffman, Suzie McCarron, Celia Rozen, Keith Webster, and Gail Heineman for their assistance with data and literature and Doug McBride for his continued support of this project. We also thank those persons who braved the inclement weather of Prince William Sound to collect our data. We are grateful to Phil Mundy, Ray Hilborn, Stanley Rice, and Robert Armstrong for their reviews of earlier drafts of the manuscript. We also acknowledge Robert Armstrong for his pioneering work in unraveling the mystery surrounding the migration of Dolly Varden.

References

Armstrong, R. H. 1970. Age, food, and migration of Dolly Varden smolts in southeastern Alaska. Journal of the Fisheries Research Board of Canada 27:991–1004.

Armstrong, R. H. 1971. Age, food, and migration of sea-run cutthroat trout, *Salmo clarki*, at Eva Lake, southeastern Alaska. Transactions of the American Fisheries Society 100:302–306.

Armstrong, R. H. 1974. Migration of anadromous Dolly

Varden (*Salvelinus malma*) in southeastern Alaska. Journal of the Fisheries Research Board of Canada 31:435–444.

Armstrong, R. H. 1984. Migration of anadromous Dolly Varden char in southeastern Alaska—a manager's nightmare. Pages 559–570 *in* L. Johnson and B. L. Burns, editors. Biology of the arctic char. University of Manitoba Press, Winnipeg.

Armstrong, R. H., and J. E. Morrow. 1980. The Dolly Varden charr. Pages 99–104 *in* E. K. Balon, editor. Charrs: salmonid fishes of the genus *Salvelinus*. Dr. W. Junk, The Hague, Netherlands.

Brannon, E. L., T. P. Quinn, R. P. Whitman, A. E. Nevissi, and R. E. Nakatani. 1986. Homing of adult chinook salmon after brief exposure to whole and dispersed crude oil. Transactions of the American Fisheries Society 115:823–827.

Buckland, S. T., and P. H. Garthwaite. 1991. Quantifying precision of mark–recapture estimates using the bootstrap and related methods. Biometrics 47:255–268.

Cochran, W. G. 1983. Planning and analysis of observational studies. Wiley, New York.

Collier, T. K., C. A. Krone, M. M. Krahn, J. E. Stein, S.-L. Chan, and U. Varanasi. 1996. Petroleum exposure and associated biochemical effects in subtidal fish after the *Exxon Valdez* oil spill. American Fisheries Society Symposium 18:671–683.

Cormack, R. M. 1989. Log-linear models for capture–recapture. Biometrics 45:395–413.

Cox, D. R. 1957. The use of a concomitant variable in selecting an experimental design. Biometrika 44:150–158.

Eberhardt, L. L., and J. M. Thomas. 1991. Designing environmental field studies. Ecological Monographs 61:53–73.

Efron, B. 1982. The jackknife, the bootstrap and other resampling plans. Society for Industrial and Applied Mathematics Publication 38, Philadelphia.

Fuss, H. J. 1982. Age, growth and instream movement of Olympic Peninsula coastal cutthroat trout (*Salmo clarki clarki*). Master's thesis. University of Washington, Seattle.

Green, R. H. 1979. Sampling design and statistical methods for environmental biologists. Wiley, New York.

Highsmith, R. C., and six coauthors. 1996. Impact of the *Exxon Valdez* oil spill on intertidal biota. American Fisheries Society Symposium 18:212–237.

Hurlbert, S. H. 1984. Pseudoreplication and the design of ecological field experiments. Ecological Monographs 54:187–211.

Jearld, A., Jr. 1983. Age determination. Pages 301–324 *in* L. A. Nielsen and D. L. Johnson, editors. Fisheries techniques. American Fisheries Society, Bethesda, Maryland.

Jewett, S. C., T. A. Dean, and D. R. Laur. 1996. Effects of the *Exxon Valdez* oil spill on benthic invertebrates in an oxygen-deficient embayment in Prince William Sound, Alaska. American Fisheries Society Symposium 18:440–447.

Lagler, K. F., and A. T. Wright. 1962. Predation of the Dolly Varden, *Salvelinus malma*, on young salmons, *Oncorhynchus* spp., in an estuary of southeastern Alaska. Transactions of the American Fisheries Society 91:90–43.

Lee, R. F. 1978. Accumulation and turnover of petroleum hydrocarbons in marine organisms. Pages 60–70 *in* D. Wolfe, editor. Fate and effects of petroleum hydrocarbons in marine organisms and ecosystems. Pergamon Press, New York.

Luquet, P., J. P. Cravedi, G. Choubert, J. Tulliez, and G. Bories. 1983. Long-term ingestion by rainbow trout of saturated hydrocarbons: effects of n-paraffins, pristane, and dodecyclcyclohexane on growth, feed intake, lipid digestibility and canthaxanthin deposition. Aquaculture 34:15–25.

Luquet, P., J. P. Cravedi, J. Tulliez, and G. Bories. 1984. Growth reduction in trout induced by naphthenic and isoprenoid hydrocarbons (dodecyclychohexan and pristane). Ecotoxicology and Environmental Safety 8:219–226.

Maynard, D. J., and D. D. Weber. 1981. Avoidance reaction of juvenile coho salmon (*Oncorhynchus kisutch*) to monocyclic aromatics. Canadian Journal of Fisheries and Aquatic Sciences 38:772–778.

Moles, A., S. D. Rice, and S. Korn. 1979. Sensitivity of Alaskan freshwater and anadromous fishes to Prudhoe Bay crude oil and benzene. Transactions of the American Fisheries Society 108:408–414.

Morrow, J. E. 1980. The freshwater fishes of Alaska. Alaska Northwest Publishing Company, Anchorage.

Nakatani, R. E., and A. E. Nevissi. 1991. Effects of Prudhoe Bay crude oil on the homing of coho salmon in marine waters. North American Journal of Fisheries Management 11:160–166.

Narver, D. W., and M. L. Dahlberg. 1965. Estuarine food of Dolly Varden at Chignik, Alaska. Transactions of the American Fisheries Society 94:405–408.

Neff, J. M. 1979. Polycyclic aromatic hydrocarbons in the aquatic environment: sources, fates, and biological effects. Applied Science Publishers, Ltd. London.

Neff, J. M. 1990. Water quality in Prince William Sound. Battelle Ocean Sciences, Duxbury, Massachusetts.

Pedersen, M. G., W. K. Hershberger, P. K. Zachariah, and M. R. Juchau. 1976. Hepatic biotransformation of environmental xenobiotics in six strains of rainbow trout (*Salmo gairdneri*). Journal of the Fisheries Research Board of Canada 33:666–675.

Rice, S. D. 1973. Toxicity and avoidance tests with Prudhoe Bay crude oil and pink salmon fry. Pages 667–770 *in* Proceedings, 1973 joint conference on prevention and control of oil spills. American Petroleum Institute, Washington, DC.

Rice, S. D., A. Moles, T. L. Taylor, and J. F. Karinen. 1979. Sensitivity of 39 Alaskan marine species to Cook Inlet crude oil and no. 2 fuel oil. Pages 549–554 *in* Proceedings, 1979 oil spill conference: prevention, behavior, control, cleanup. American Petroleum Institute Publication 4308, Washington, DC.

Sauermann, W. 1989. Bootstrapping the maximum likelihood estimator in high-dimensional log-linear models. The Annals of Statistics 17:1198–1216.

Scott, W. B., and E. J. Crossman. 1979. Freshwater fishes

of Canada. Fisheries Research Board of Canada Bulletin 184.

Sharr, S. 1990. Injury to salmon spawning areas in Prince William Sound. State/federal Natural Resource Damage Assessment, Fish/Shellfish Study 1. Progress Report to *Exxon Valdez* Oil Spill Trustee Council, Anchorage.

Short, J. W., and P. M. Harris. 1996. Chemical sampling and analysis of petroleum hydrocarbons in near-surface seawater of Prince William Sound after the *Exxon Valdez* oil spill. American Fisheries Society Symposium 18:17–28.

Sumner, F. H. 1953. Migration of salmonids in Sand Creek, Oregon. Transactions of the American Fisheries Society 82:139–150.

Teal, J. M., and R. W. Howarth. 1984. Oil spill studies: a review of ecological effects. Environmental Management 8:27–44.

Thomas, R. E., and S. D. Rice. 1981. Excretion of aromatic hydrocarbons and their metabolites by freshwater and seawater Dolly Varden char. Pages 425–448 *in* J. Vernberg, A. Calabrese, F. P. Thurbert, and W. B. Vernberg, editors. Biological monitoring of marine pollutants. Academic Press, New York.

Trotter, P. C. 1989. Coastal cutthroat trout: a life history compendium. Transactions of the American Fisheries Society 118:463–473.

Trustee Council. 1989. State/federal natural resources damage assessment plan for the *Exxon Valdez* oil spill, public review draft, August 1989. Trustee Council for the *Exxon Valdez* Oil Spill, Anchorage, Alaska.

Tukey, J. W. 1977. Exploratory data analysis. Addison-Wesley, Reading, Massachusetts.

Varanasi, U., and D. C. Malins. 1977. Metabolism of petroleum hydrocarbons: accumulation and biotransformation in marine organisms. Pages 175–270 *in* D. C. Malins, editor. Effects of petroleum on arctic and subarctic marine environments and organisms, volume II. Biological effects. Academic Press, New York.

Weber, D. D., and D. J. Maynard, W. D. Gronlund, and V. Konchin. 1981. Avoidance reactions of migrating adult salmon to petroleum hydrocarbons. Canadian Journal of Fisheries and Aquatic Sciences 38:779–781.

Woodward, D. F., P. Mehrle, Jr., and W. L. Mauck. 1981. Accumulation and sublethal effects of a Wyoming crude oil in cutthroat trout. Transactions of the American Fisheries Society 110:437–445.

American Fisheries Society Symposium 18:659–670, 1996

Coastal Habitat Studies: The Effect of the *Exxon Valdez* Oil Spill on Shallow Subtidal Fishes in Prince William Sound

David Laur

University of California, Marine Science Institute, Santa Barbara, California 93106, USA

Lewis Haldorson

University of Alaska, School of Fisheries and Ocean Sciences, Juneau, Alaska 99801, USA

Abstract.—Fish communities were monitored in shallow (<20 m) depths of paired oiled and non-oiled (control) study sites within Prince William Sound in 1990 using diving transect techniques. Habitats studied were defined by dominant macrophytes, including eelgrass *Zostera marina* and *Laminaria–Agarum*. In eelgrass, young-of-the-year Pacific cod *Gadus macrocephalus* were the most abundant fish in both oiled and control sites; they occurred at significantly higher densities at oiled sites. In *Laminaria–Agarum* habitats, the Arctic shanny *Stichaeus punctatus* and small sculpins were the most common fishes, and young-of-the-year Arctic shannies were more abundant in shallow depths at oiled sites. In both eelgrass and *Laminaria–Agarum* habitats, oiled sites had overall higher abundances of fishes than did control sites mainly because of the higher numbers of young-of-the-year fishes. Young-of-the-year Pacific cod from oiled eelgrass sites tended to have higher volumes of stomach contents and had eaten proportionally more mollusk larvae than had fish from control sites, where crustaceans were more common in diets. Differences in diet appear related to the relative availability of prey in oiled versus control eelgrass study sites, suggesting that increased abundance of young-of-the-year Pacific cod at oiled study sites is related to the increased diversity and abundance of epifaunal prey taxa. Increased numbers of young-of-the-year fishes at oiled sites could also be the result of enhanced supply of planktonic larvae or decreases in predation at oiled sites.

The Gulf of Alaska is a large, subarctic marine ecosystem with benthic and pelagic fish communities that support large populations of marine mammals and birds and are the basis for very valuable commercial fisheries. Many of the fish species that form the forage base for higher trophic levels or are targets of commercial fisheries have life history phases that depend on shallow nearshore habitats. For example, Pacific herring *Clupea pallasi*, a major forage fish as well as a commercially valuable species, spawns in shallow subtidal and intertidal areas of protected bays and sounds. Other species, such as rockfishes *Sebastes* spp. and cods (family Gadidae) use shallow nearshore waters as young juveniles. In addition, there is a diverse group of fishes that live in nearshore areas as juveniles and adults and have specific habitat requirements that determine their distributions along the extensive shoreline of the Gulf of Alaska. Large embayments, such as Prince William Sound, are particularly valuable areas because they provide extensive nearshore habitats that are protected from the extreme turbulence found on exposed outer coastlines of the gulf.

Prince William Sound has an extensive crenulated shoreline with numerous islands, embayments and fjords. Shallow subtidal waters of the sound typically have the rugged and rocky substrates that characterize the surrounding mountainous terrain, although there are occasional soft-bottom areas that are usually associated with mouths of streams entering the sound. A dominant feature of shallow subtidal habitats are marine macrophytes, including eelgrass *Zostera marina* growing in sand and mud substrates, canopy-forming bull kelp *Nereocystis leutkeana* on rocky substrates in exposed high-current areas, and algae with broad blades (*Laminaria saccharina* and *Agarum cribrosum* are most prevalent) that form a dense, low-growing cover over less-exposed rocky substrates (Rosenthal et al. 1977). The macrophytes contribute much of the spatial heterogeneity to shallow habitats as well as high levels of productivity (Mann 1973). These habitats are occupied by a diverse assemblage of fishes, including resident benthic species and pelagic transients (Rosenthal 1983), and are seasonally important habitats for young-of-the-year fishes because they provide rich foraging areas and refuge from predators (Carr 1983; Jones 1984; Haldorson and Richards 1987).

The *Exxon Valdez* oil spill in March 1989 resulted in oil contamination of nearshore environments in the western part of Prince William Sound. Cleanup activities flushed considerable quantities of oil from intertidal areas back into nearshore waters during summer months of the first two years (1989 and 1990) following the spill. As part of the Coastal

Habitat Studies program, we conducted surveys in 1990 to assess the effects of the oil spill on shallow-water fish communities. Our specific objectives were to determine if abundances of fishes differed in oiled and non-oiled habitats and if feeding and condition were affected by the presence of oil.

Methods

Sampling Strata

We stratified subtidal habitats by dominant macrophyte types, by location in the sound, by shoreline physiography, and by depth. The effects of oil were examined by pairwise comparison of oiled and control sites within macrophyte–shoreline strata. Our sampling design included random selection of oiled sites with selection of a paired control site that matched the oiled site as closely as possible with respect to nonbiological factors other than oiling.

The first level of stratification was based on macrophyte type. Shallow subtidal areas in Prince William Sound are dominated by three macrophyte communities: (1) eelgrass—found in soft-bottom bays near the mouths of streams; (2) *Laminaria–Agarum*—the most widespread macrophyte community in Prince William Sound, occurring on rocky substrates in all but the most exposed areas; and (3) bull kelp—forming a floating canopy with an understory of algae typical of areas with strong water motion.

The bull kelp community is rare in Prince William Sound, and its fishes were dominated by occasional large schools of pelagic fishes (Pacific herring, Pacific sand lance *Ammodytes hexapterus*, young walleye pollock *Theragra chalcogramma*), resulting in extremely high variances in abundance estimates. Other fishes were quite rare. Consequently, we concentrated our sampling and analyses on the other two habitat types and do not include results from the bull kelp habitat in the following sections.

For the *Laminaria–Agarum* habitat we further stratified into islands, mainland, and outer sound areas, and within each area into shoreline type: bays, points, and straight shoreline runs. We selected the island point and island bay strata for study. Therefore, the study was based on three habitat types: (1) eelgrass, (2) *Laminaria–Agarum* island points (henceforth L-A points), and (3) *Laminaria–Agarum* bays (L-A bays). We also stratified the L-A point and L-A bay habitats by depth. Eelgrass was not depth stratified because the plants occur in a relatively narrow depth range. The two L-A habitats were subdivided into shallow (2–11 m) and deep (12–20 m) strata. Consequently, there were five major sampling strata with three oil–control site pairs in each, except for the eelgrass habitat, where we examined four site pairs (Table 1).

TABLE 1.—Study design for assessment of the *Exxon Valdez* oil spill on shallow subtidal fishes in Prince William Sound in 1990.

Habitat	Depths (m)	Site pairs	Transects per site	Total transects
Eelgrass		4	3	12
Laminaria–Agarum				
Bays	2–11	3	3	9
	12–20	3	3	9
Points	2–11	3	3	9
	12–20	3	3	9
Total		16		48

Site Selection

We identified sites in each strata based on an overlay of oil information and habitat types. Oil distribution was identified based on 1989 oil maps and a walking survey of beach areas in September 1989. Areas that were identified as moderately to heavily oiled in both surveys were included in the pool of potential study sites. Habitat types were assigned based on navigational charts, field work conducted in fall 1989, and information gathered from individuals familiar with the sound. Oiled sites were randomly selected from the group of available locations.

Control sites were selected that matched oiled sites as closely as possible with regard to aspect, proximity to sources of freshwater input, slope, wave exposure, and circulation. We randomly selected a matched control site if more than one was identified. All of the oil–control site pairs are located on the western side of the sound (Table 2 and Figure 1).

Sampling Methods

Each site consisted of a 200-m section of shoreline. Transects at each site were located randomly in the 200-m section. Field work began in May 1990. Site pairs were sampled as synoptically as possible to reduce variability caused by seasonal changes in fish communities. Also, within a habitat we sampled all pairs consecutively to reduce temporal variation among site pairs within strata (Table 2).

Eelgrass.—At each site we established three 30-m transects parallel to the shoreline at randomly selected locations along the 200-m shoreline section

TABLE 2.—Study sites in shallow subtidal zone of Prince William Sound with oil status, date of survey (in 1990), and site name. Locations are indicated in Figure 1 by site number.

Site pair	Oil status	Site number	Survey dates	Site name
		Eelgrass		
1	Oiled	13	4 Jul	Bay of Isles
	Control	14	6 Jul	Drier Bay
2	Oiled	16	15 Jul	Herring Bay
	Control	15	13 Jul	Lower Herring Bay
3	Oiled	17	8 Jul	Sleepy Bay
	Control	18	11 Jul	Moose Lips Bay
4	Oiled	25	9, 10 Aug	Clammy Bay
	Control	26	11 Aug	Puffin Bay
		***Laminaria–Agarum* bays**		
1	Oiled	2	30, 31 May	Northwest Bay
	Control	1	23, 24 May	Cabin Bay
2	Oiled	3	30, 31 May	Herring Bay
	Control	4	3, 4 Jun	Lower Herring Bay
3	Oiled	6	11, 12 Jun	Bay of Isles
	Control	5	7, 8 Jun	Mummy Bay
		***Laminaria–Agarum* points**		
1	Oiled	19	22, 23 Jul	Disk Lagoon
	Control	20	18, 19 Jul	Lucky Bay
2	Oiled	22	4–6 Aug	Outer Herring Bay
	Control	21	7, 8 Aug	Outer Lower Herring Bay
3	Oiled	23	2, 3 Aug	Ingot Point
	Control	24	27, 29, 30 Jul	Peak Point

and approximately in the middle of the depth range of the eelgrass bed. Two divers swam each transect and counted fishes by species within 1 m on either side of the transect line and within 3 m of the bottom.

Laminaria–Agarum points and bays.—At each site three transects were randomly placed parallel to the shoreline within shallow and deep depths strata. Fish were counted as in the eelgrass habitat.

Sediment Hydrocarbon Analyses

Sediments were collected from substrates at eelgrass study sites for measurement of hydrocarbon content. Divers collected 10 to 100 g of sediment in small jars approximately in the middle of each sampling site. Samples were numbered, sealed with evidence tape, and frozen in the field. Frozen samples were sent to the Technical Services Task Force, Analytical Chemistry Group, National Marine Fisheries Service (NMFS), Auke Bay, Alaska, for processing.

Data Analyses

The null hypothesis of no difference between oiled and control sites was tested with blocked analysis of variance (ANOVA), with oil category as the main effect and pair as a blocking factor. Homogeneity of variance was tested with Levene's test (Milliken and Johnson 1984). If data failed to meet the assumption of homogeneity, and if the significance level for the ANOVA was less than 0.10, we tested the significance of the ANOVA F-statistic with a randomization procedure (Manly 1991).

Analyses of Young-of-the-Year Pacific Cod

We examined the feeding habits and length–weight condition index of young-of-the-year Pacific cod *Gadus macrocephalus* from three pairs of oiled–control eelgrass sites. Fish were collected by divers with small spears. Fish were frozen until examined. In the laboratory, each fish was weighed (wet weight) and measured. The stomach contents were removed, and the fish minus the contents was dried for 24 to 30 h in a drying oven and weighed (dry weight). The stomach contents were identified (generally to order), counted, and the volume of each taxon estimated by area covered to uniform depth on a grid of 1-mm squares. Condition was quantified with Fulton-type condition index using dry weight (Anderson and Gutreuter 1983).

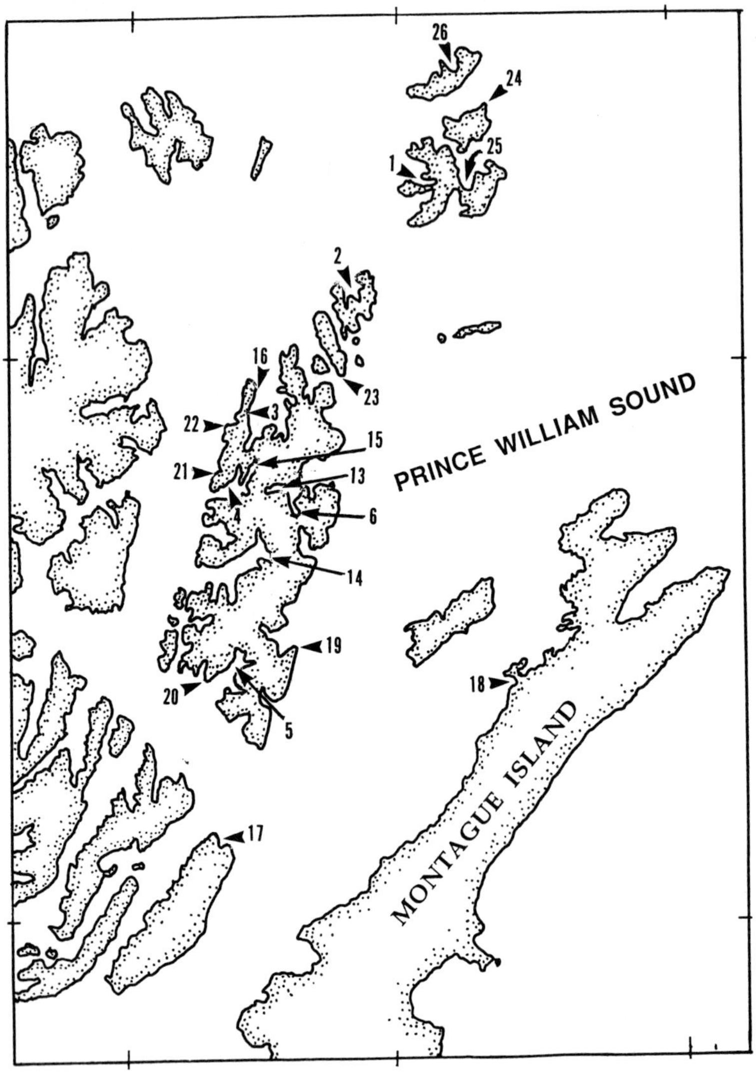

FIGURE 1.—The western side of Prince William Sound, with locations of the study sites. The numbers on the map are site numbers, identified in Table 2.

Results

Eelgrass

Over 15 species of fishes were observed in eelgrass habitats. For analyses, the less common species were combined into family-level taxa (Table 3). Young-of-the-year Pacific cod was the most abundant taxon, with the remainder of the community consisting of a diverse group of generally benthic fishes (Table 4). The overall abundance of fishes was significantly ($P < 0.01$) higher at oiled sites; however, when Pacific cod were removed from the analyses the abundances were similar (Figure 2A). There were significantly ($P < 0.01$) more young-of-the-year Pacific cod at oiled sites (Figure 2B, $P < 0.01$). Pacific cod adults were also significantly ($P < 0.01$) more abundant at oiled sites (data not shown). Polycyclic aromatic hydrocarbon (PAH) levels tended to be higher at the oiled sites (Figure 2C).

Laminaria–Agarum Habitats

Over 30 species of fishes occurred in L-A habitats, with generally similar species found at points

TABLE 3.—Composition of fish groups for analyses of eelgrass habitats in Prince William Sound in 1990.

Analysis group	Species
Young-of-the-year Pacific cod	Pacific cod *Gadus macrocephalus*
Gunnels	Crescent gunnel *Pholis laeta* Penpoint gunnel *Apodichthys flavidus* Unidentified gunnels
Sculpins	Red Irish lord *Hemilepidotus hemilepidotus* Great sculpin *Myoxocephalus polyacanthocepthalus* Silverspotted sculpin *Blepsias cirrhosus* Unidentified sculpins
Arctic shanny	Arctic shanny *Stichaeus punctatus*
Greenlings	Masked greenling *Hexagrammos octogrammus* Kelp greenling *Hexagrammos decagrammus* Whitespotted greenling *Hexagrammos stelleri*
Lingcod	Lingcod *Ophiodon elongatus*
Other	Snake prickleback *Lumpenus sagitta* Slender eelblenny *Lumpenus fabricii* Pacific sand lance *Ammodytes hexapterus* Tube-snout *Aulorhynchus flavidus* Rockfishes *Sebastes* spp. Flatfishes

and bays. The less common species were grouped into family-level taxa for analyses (Table 5). Fish communities in *Laminaria–Agarum* habitats were dominated by Arctic shannies and a mixed group of sculpins. Other fishes commonly observed were ronquils and various greenlings. The point habitats tended to have greater abundances and a higher diversity of fishes than the bays; however, the fish assemblages in the two habitats were otherwise quite similar.

Laminaria–Agarum Points

Shallow stratum.—Arctic shannies, both young-of-the-year and older ages, dominated the fish community in both oiled and control study sites (Table 6). In the shallow stratum, when young-of-the-year Arctic shannies were not included in the analyses, the abundances of fishes were similar at oiled and control sites (Figure 3A); however, when young-of-the-year Arctic shannies were included, the differences were significant. Young-of-the-year Arctic shannies were in higher abundances ($P < 0.10$) at oiled sites (Figure 3B).

Deep stratum.—In the deep stratum, the fish communities in oiled and control sites were generally similar to those at shallow depths, except for a sharp increase in numbers of sculpins (Table 6). As in the shallow depths at L-A points, there were more fishes ($P = 0.03$) observed in the oiled sites, but the numbers were similar when young-of-the-year Arctic shannies were not included (Figure 3C). The difference again was attributable largely to higher numbers of young-of-the-year Arctic shannies ($P < 0.10$, Figure 3D).

TABLE 4.—Relative abundances of species groups in eelgrass habitats, including all fish (with young-of-the-year cod) and all other species excluding young-of-the-year Pacific cod (no young-of-the-year cod), in Prince William Sound in 1990, with sample size (N) for each.

	With young-of-the-year cod		No young-of-the-year cod	
Group	Oil (N = 983)	Control (N = 759)	Oil (N = 120)	Control (N = 120)
Young-of-the-year cod	0.84	0.46		
Gunnels	0.05	0.12	0.30	0.22
Greenlings	0.03	0.14	0.17	0.27
Older cod	0.05	0.03	0.31	0.06
Arctic shanny	0.02	0.09	0.12	0.17
Sculpins	0.01	0.04	0.04	0.07
Lingcod	0.01	0.02	0.04	0.04
Other	0.004	0.10	0.02	0.18

Laminaria–Agarum Bays

Shallow stratum.—The Arctic shanny and a mixed group of sculpins were the most frequently observed taxa in the shallow stratum (Table 7). When Arctic shannies were not included, the numbers of fishes again were similar in oiled and control site pairs (Figure 4A). However, there was a significantly ($P < 0.01$) higher abundance of Arctic shannies at oiled sites (Figure 4B). Greenlings were also more abundant ($P = 0.07$) at oiled sites.

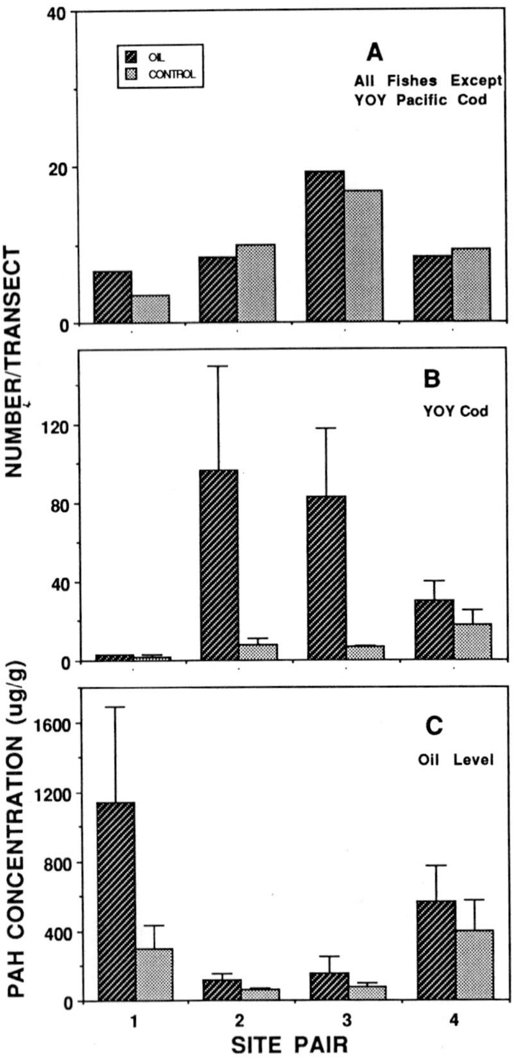

FIGURE 2.—The mean numbers of all fishes except young-of-the-year Pacific cod (**A**) and only young-of-the-year Pacific cod (**B**) at four pairs of oiled and control study sites in eelgrass habitats of Prince William Sound in 1990. Error bars indicate 1 SE. Panel **C** displays the concentration of polycyclic aromatic hydrocarbons (PAH, μg/g) from the same sites.

Deep stratum.—The fish community observed in the deep depth stratum of L-A bays was dominated by the Arctic shanny and a mixed group of sculpins (Table 7). When small sculpins were not included, the numbers of fish were similar (Figure 4C). Small sculpins were more abundant at oiled sites in all three site pairs, but not significantly (Figure 4D).

Because of the similarity of the fish assemblages in point and bay habitats, we also tested for differences between control and oiled sites within combined habitats and depths. In that overall analysis, greenlings and small sculpins were found to be more abundant at oiled sites ($P = 0.03$ and $P = 0.07$, respectively).

Among all habitats, the most consistent trend in abundance of fishes was the higher numbers, especially young-of-the-year fishes, at oiled sites. In combined eelgrass and L-A sites, oiled sites had higher abundances ($P = 0.02$, sign test) than control sites in 13 of 16 site pairs (Table 8).

Pacific Cod Young-of-the-Year Feeding and Condition

In eelgrass habitats, young-of-the-year Pacific cod fed on a variety of crustaceans, molluscs, and polychaetes; prey consumed in the greatest numbers and volumes were calanoid and harpacticoid copepods, gammarid amphipods, and bivalves (Table 9). Diets of fish at oiled and control sites differed significantly: fish from control sites ate more calanoid copepods, while at oiled sites they ate more bivalves and polychaetes. At all three site pairs, the average volume of stomach contents was higher at the oiled site (Figure 5A); however, there

TABLE 5.—Groupings of fishes found in *Laminaria–Agarum* habitats in Prince William Sound in 1990.

Analysis group	Species
Pacific cod	Pacific cod *Gadus macrocephalus*
Arctic shanny	Arctic shanny *Stichaeus punctatus*
Greenlings	Kelp greenling *Hexagrammos decagrammus* Rock greenling *Hexagrammos lagocephalus* Whitespotted greenling *Hexagrammos stelleri* Masked greenling *Hexagrammos octogrammus*
Gunnels	Crescent gunnel *Pholis laeta* Saddleback gunnel *Pholis ornata* Unidentified gunnels
Ronquils	Searcher *Bathymaster signatus* Northern ronquil *Ronquilus jordani*
Small sculpins	Grunt sculpin *Rhamphocottus richardsoni* Pit-head sculpin *Icelinus cavrifrons* *Artedius* sp. Unidentified small sculpins
Large sculpins	Buffalo sculpin *Enophrys bison* Red Irish lord *Hemilepidotus hemilepidotus* Yellow Irish lord *Hemilepidotus jordani* Unidentified large sculpins
Other	Slender eelblenny *Lumpenus fabricii* Rock prickleback *Xiphister mucosus* Yelloweye rockfish *Sebastes ruberrimus* Dusky rockfish *Sebastes ciliatus* Copper rockfish *Sebastes caurinus* Flatfishes, unidentified Poachers, unidentified Eelpouts, unidentified

TABLE 6.—Relative abundances (proportions observed on transects) of species groups in *Laminaria–Agarum* point habitats in Prince William Sound in 1990, with sample size (*N*).

Group	Shallow (2–11 m) Oil (*N* = 635)	Shallow (2–11 m) Control (*N* = 598)	Deep (12–20 m) Oil (*N* = 550)	Deep (12–20 m) Control (*N* = 448)
Arctic shanny	0.20	0.42	0.19	0.35
Arctic shanny young-of-the-year	0.34	0.24	0.23	0.14
Sculpins	0.09	0.10	0.34	0.38
Gunnels	0.03	0.04	<0.01	0.03
Dusky rockfish	0.02	0.12	0.12	0.02
Greenlings	0.03	0.03	0.01	0.01
Ronquils	0.04	0	0.06	0.04
Others	0.23	0.05	0.05	0.04

were only minor differences in condition factor between oiled and control sites at the three site pairs (Figure 5B).

Discussion

With a few exceptions, the fish communities we observed in vegetated habitats of Prince William Sound were similar in abundance, species richness, and species composition to those surveyed in a diving study conducted in the late 1970s (Rosenthal 1983). In an eelgrass habitat in Constantine Harbor on Hinchinbrook Island, Rosenthal (1983) observed 15 fish species with about 10 individuals per 60 m^2 on transects counted in the summer. That

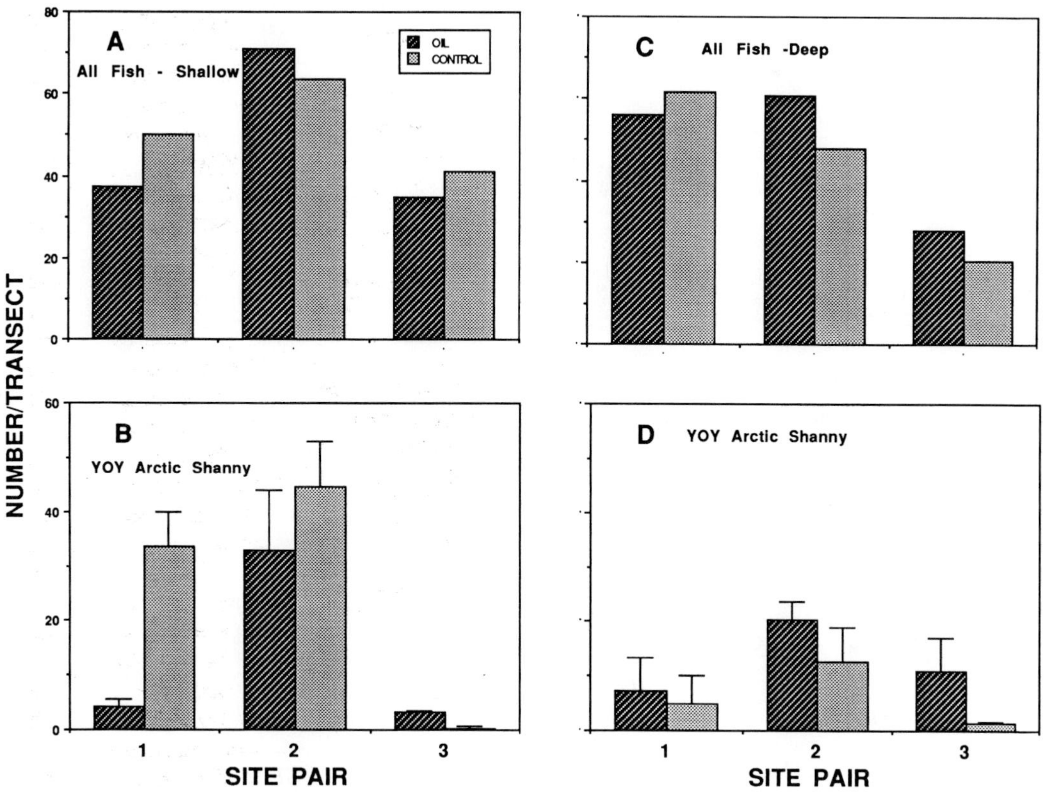

FIGURE 3.—Mean numbers of all fishes except young-of-the-year (YOY) Arctic shannies (**A**) and only young-of-the-year Arctic shannies (**B**) at three pairs of oiled and control study sites in the shallow (2–11 m) zone; and mean numbers of all fishes except young-of-the-year Arctic shannies (**C**) and only young-of-the-year Arctic shannies (**D**) in the deep zone (12–20 m) in *Laminaria–Agarum* point habitats of Prince William Sound in 1990. Error bars indicate 1 SE.

TABLE 7.—Relative abundances (proportions observed on transects) of species groups in *Laminaria–Agarum* bay habitats in Prince William Sound in 1990, with sample size (*N*).

Group	Shallow (2–11 m) Oil (N = 294)	Shallow (2–11 m) Control (N = 231)	Deep (12–20 m) Oil (N = 181)	Deep (12–20 m) Control (N = 88)
Arctic shanny	0.63	0.31	0.27	0.56
Arctic shanny young-of-the-year	0.01	0.29	0	0
Small sculpins	0.20	0.10	0.57	0.27
Large sculpins	0.03	0.02	0.02	0.03
Gunnels	0.04	0.22	0.04	0.04
Ronquils	0.05	<0.01	0.02	0
Greenlings	0.02	0.01	0.03	0.01
Others	0.02	0.04	0.04	0.08

species richness was very similar to that found in our study (Table 3), while our numbers per transect (60 m^2) were higher (about 10–100). However, the communities we sampled were dominated by young-of-the-year Pacific cod, a species group that did not occur in the earlier surveys. If young-of-the-year Pacific cod are removed from our analyses, the numbers per transect are very similar to those observed by Rosenthal (1983) for the same area (60 m^2) (Figure 2A).

Over rocky substrates, Rosenthal (1983) observed fish communities that also were similar to

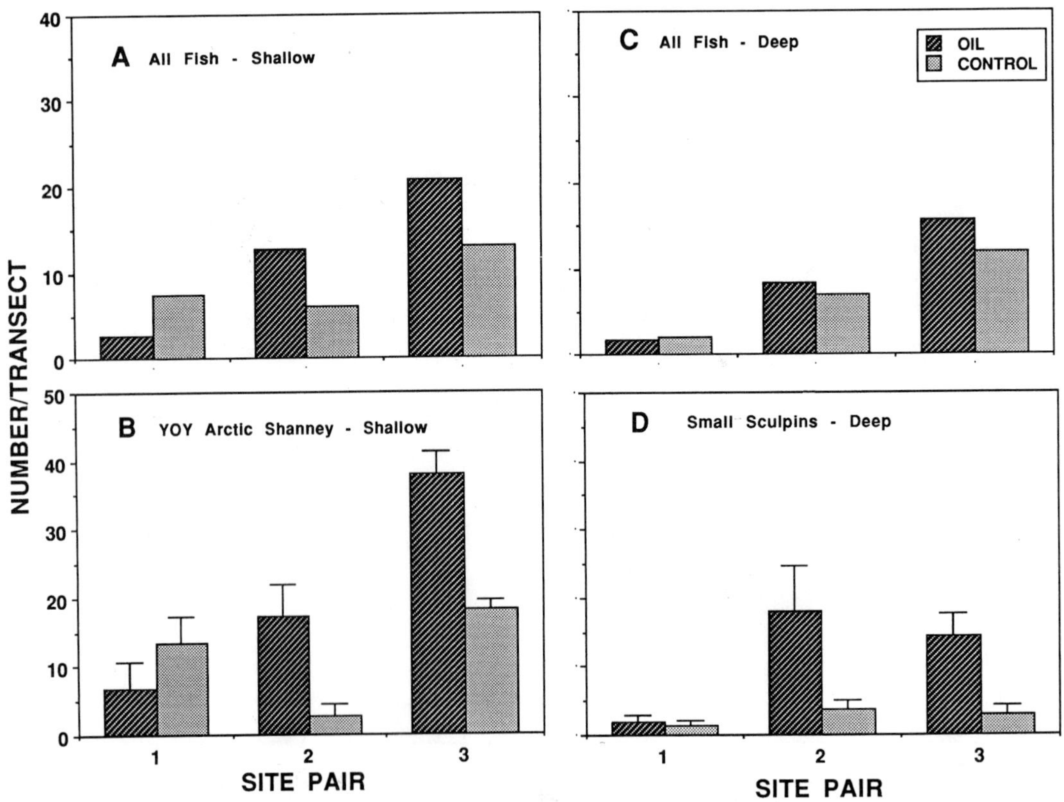

FIGURE 4.—Mean numbers of all fishes except young-of-the-year (YOY) Arctic shannies (**A**) and only young-of-the-year Arctic shannies (**B**) at three pairs of oiled and control study sites in the shallow (2–11 m) zone; and mean numbers of all fishes except small sculpins (**C**) and only small sculpins (**D**) in the deep zone (12–20 m) in *Laminaria–Agarum* bay habitats of Prince William Sound in 1990. Error bars indicate 1 SE.

TABLE 8.—Abundance trends of fishes in oiled and control study site pairs in Prince William Sound in 1990, indicating the number of site pairs where abundance of fishes was higher at oiled sites and with an indication of the taxa mainly responsible for the difference.

Habitat	Site pairs where oil > control	Species involved
Eelgrass	4 of 4	Young-of-the-year Pacific cod
Laminaria–Agarum bays		
Shallow	2 of 3	Arctic shanny
Deep	3 of 3	Small sculpins
Laminaria–Agarum points		
Shallow	2 of 3	Young-of-the-year Arctic shanny
Deep	2 of 3	Young-of-the-year Arctic shanny
All sites	13 of 16	

those we sampled after the *Exxon Valdez* oil spill. His Schooner Rock site on the northeast side of Montague Island is perhaps most comparable to our L-A point habitats. At Schooner Rock, 29 species of fishes were observed, with densities of about 25 to 50 individuals/60 m^2 in the summer (Rosenthal 1983). In our study the number of species at L-A Points (Table 6) was very similar to Rosenthal's (1983) counts, although our numbers per transect were somewhat higher, ranging from 20 to 80 in shallow depths and 40 to 100 in deeper depths (Figure 3A). Because of the general similarities in fish communities, we conclude that there were no major changes in nearshore fish communities in the 15 years between Rosenthal's work and ours.

Our results indicate that there were differences in oiled and non-oiled nearshore fish communities in Prince William Sound, mainly because of elevated abundances of young-of-the-year fishes at oiled

TABLE 9.—Stomach contents for young-of-the-year Pacific cod from three oiled–control site pairs in Prince William Sound in 1990, in mean (SE) percent volumes and mean (SE) percent numbers. Gammarids and caprellids are amphipods; calanoids and harpacticoids are copepods.

Prey and measure	Site pair 2		Site pair 3		Site pair 4	
	Site 16, oiled (*N* = 12)	Site 15, control (*N* = 13)	Site 17, oiled (*N* = 17)	Site 18, control (*N* = 18)	Site 25, oiled (*N* = 16)	Site 26, control (*N* = 12)
Gammarids						
Volume	0	4 (4)	0	7 (4)	21 (8)	14 (9)
Number	0	<1	0	1 (1)	15 (4)	11 (8)
Caprellids						
Volume	0	0	0	0	7 (5)	4 (2)
Number	0	0	0	0	7 (5)	2 (1)
Calanoids						
Volume	28 (10)	8 (3)	41 (11)	60 (11)	6 (4)	34 (9)
Number	31 (11)	12 (4)	50 (12)	67 (11)	6 (3)	54 (9)
Harpacticoids						
Volume	2 (1)	8 (2)	12 (5)	2 (1)	6 (3)	8 (3)
Number	2 (1)	16 (5)	14 (6)	2 (1)	29 (6)	13 (5)
Cladacerans						
Volume	3 (1)	9 (3)	0	2 (1)	5 (2)	1 (1)
Number	7 (4)	19 (4)	0	2 (1)	18 (6)	1 (1)
Bivalves						
Volume	28 (8)	17 (6)	17 (8)	3 (2)	<1	0
Number	43 (10)	36 (8)	19 (8)	11 (6)	1 (<1)	0
Gastropods						
Volume	7 (2)	0	<1	<1	6 (3)	7 (7)
Number	8 (2)	0	1 (1)	1 (<1)	5 (2)	2 (2)
Polychaetes						
Volume	6 (6)	2 (1)	5 (4)	0	22 (8)	4 (4)
Number	<1	1 (<1)	1 (1)	0	6 (2)	1 (<1)
Other identified						
Volume	12	24	12	4	15	5
Unidentified						
Volume	14	23	13	22	12	23

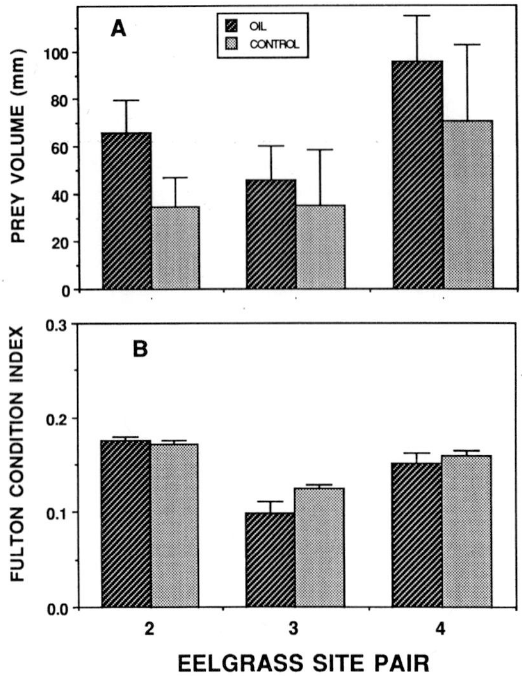

FIGURE 5.—Mean volumes of prey in stomach contents (**A**) and mean values of Fulton's condition index (**B**) of young-of-the-year Pacific cod at three pairs of oiled and control study sites in Prince William Sound in 1990.

sites. There are a number of possible explanations for this pattern, including (1) seasonal differences in time of sampling oiled and control sites, (2) differences in supply of planktonic fish larvae to oiled and control sites, (3) differential mortality of young-of-the-year fishes, and (4) differential rates of emigration–immigration by young-of-the-year fishes.

Abundance of fishes in macrophyte habitats varies seasonally in Prince William Sound (Rosenthal 1983) and elsewhere (Adams 1976). Consequently, consistent differences between oiled and control site pairs would be expected if all sites of one type (i.e., oiled or control sites) were surveyed before or after all sites of the other type. To reduce temporal variability between oiled and control sites within a pair and among site pairs, we sampled all site pairs as closely together as possible and attempted to sample each habitat type (eelgrass, L-A bays, L-A points) within relatively short spans of time (Table 2). Therefore, we conclude that temporal variation was not a cause of the increased abundance of fishes observed in oiled habitats.

Locations may differ in abundance of young-of-the-year juvenile fishes if the distribution of planktonic larvae is not uniform (Bell and Westoby 1986). In Prince William Sound this effect could have been responsible for the increase in young-of-the-year fishes we observed at oiled sites. The prevailing currents in the sound flow from east to west, exiting the sound through Montague Strait and other passageways on the southwest side. The trajectory of the oil is consistent with this circulation pattern. Because both oil and planktonic larvae are passively transported by currents, it is reasonable to suppose that areas contaminated with oil were also the sites where more larval fishes would occur, with subsequent higher abundances of juveniles after larvae metamorphose into young-of-the-year fishes.

Consistently lower abundance of young-of-the-year fishes at non-oiled sites could result from higher mortality. Because predation is a major cause of mortality at this life stage, the difference may reflect higher predator numbers or increased vulnerability to predators in non-oiled sites. We have no evidence that predators of young-of-the-year fishes were reduced at oiled sites, although this remains a possibility. Predators are larger fishes, marine mammals (including river otters *Lutra canadensis*), and birds. Fish communities at oiled and control sites were generally very similar when young-of-the-year fishes were not included in analyses; consequently, it seems unlikely that fish predation differed markedly in oiled and non-oiled sites. It has been shown, however, that the *Exxon Valdez* oil spill was very detrimental to both birds and mammals, therefore it is possible that in oiled habitats young-of-the-year fishes experienced reduced predation from those predators.

Higher abundances of young-of-the-year fishes could also occur where they are less vulnerable to predators. Structural complexity has been associated with increased abundances of young fish (Ebeling and Laur 1985), and increased vulnerability may be caused by a reduction in cover provided by the dominant macrophytes. Also, young fish are likely to be more vulnerable to predators if they are in poorer condition. In eelgrass, control sites had about one-third higher density of eelgrass blades than did the oiled sites (Dean et al. 1996, this volume), and the condition of young-of-the-year Pacific cod was similar in all site pairs (Figure 5B). Consequently, we conclude that differential vulnerability to predation was probably not a cause of the increased abundance of young-of-the-year fishes at oiled study sites.

Young-of-the-year fishes may move between habitats (Carr 1983; Haldorson and Richards 1987), probably in response to changes in value of prey

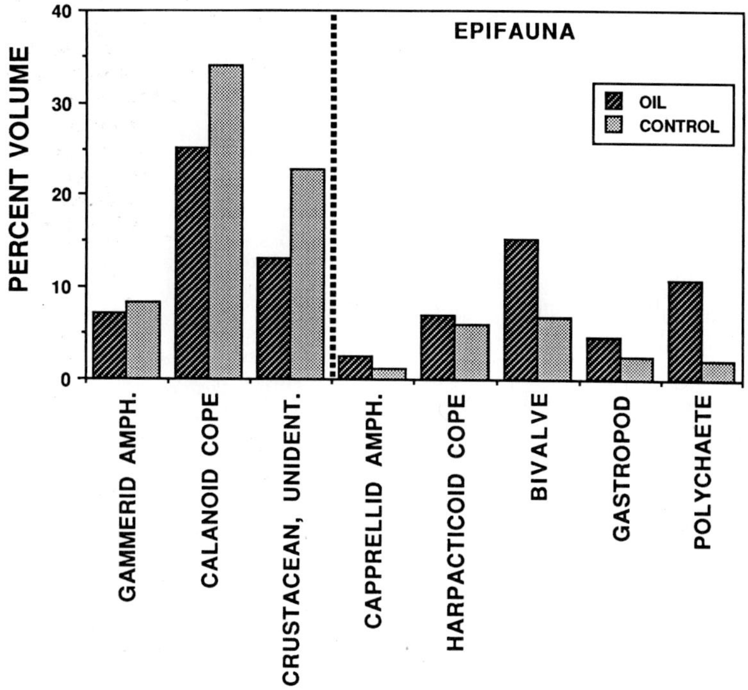

FIGURE 6.—Percent volumes of prey taxa in stomach contents of young-of-the-year Pacific cod from oiled and control study sites in Prince William Sound in 1990. Those taxa considered to be epifaunal are grouped to the right.

resources or differences in predation risk. Increased abundances of young-of-the-year fishes in oiled habitats could result from differences in emigration–immigration in those habitats relative to non-oiled areas. In that case, oiled sites would be expected to have higher abundances of prey species and/or reduced numbers of predators. At oiled eelgrass sites, young-of-the-year Pacific cod had higher volumes of stomach contents and fed on more epifaunal prey than did young-of-the-year cod at control sites (Figure 6). The sites did differ in prey availability, as oiled sites a had higher abundances of epifaunal organisms such as Mytilid bivalves, than did control sites (Jewett et al. 1996, this volume). Our results suggest that foraging conditions may have been substantially better for young-of-the-year fishes in oiled habitats; therefore, the higher abundances of young-of-the-year fishes may reflect differences in movement among oiled and non-oiled areas, with accumulation of individuals in oiled areas because of enhanced foraging conditions for epifaunal prey.

In general, little *Exxon Valdez* oil was found subtidally (O'Clair et. al. 1996, this volume), and toxic levels were present intertidally but not subtidally. The levels of PAH we observed at eelgrass study sites were not all from *Exxon Valdez* oil. Only the sediments from Herring Bay (site 16 in site pair 2) and Sleepy Bay (site 17 in site pair 3) had PAH signatures consistent with weathered *Exxon Valdez* crude oil—the PAH levels at the oiled study sites in site pairs 1 and 4 were consistent with gasoline or diesel oil (S. Rice, Auke Bay Laboratory, Juneau, Alaska, personal communication).

Among the four eelgrass study site pairs, the oiled sites at pairs 2 and 3 (the ones with PAH signatures similar to *Exxon Valdez* oil) were the ones with the highest abundances of young-of-the-year Pacific cod. This strengthens the argument that increased abundance of young-of-the-year fishes in oiled habitats did occur in the year following the oil spill. Higher abundance of fishes in oiled subtidal habitats suggests that oil did not directly affect fishes through toxicity, as does the lack of oil-related trends in condition factor of young-of-the-year Pacific cod. The apparent effects of oil may be spurious if the increased abundance of fishes was caused by delivery of higher numbers of planktonic larvae through the same current systems that delivered the oil. Alternatively, enhanced numbers of fishes in oiled areas may result from differences in populations of prey organisms or predators. Increased availability of epifaunal prey at oiled sites has been implicated by our data; however, decreased predation in oiled areas may also have contributed to higher abundances of fishes at oiled study sites.

Acknowledgments

This study was part of an integrated program to assess the effects of the *Exxon Valdez* oil spill on shallow subtidal biota in Prince William Sound. We thank Lyman McDonald for his help in study design and statistical analyses, Tom Dean and Stephen Jewett for help in all aspects of the work, and divers Alfred Calip-DuBois, Kieran Donahue, Chris Gotschalk, Daniel Martin, and Langdon Quetin for their help in the field.

References

Adams, S. M. 1976. The ecology of eelgrass, *Zostera marina* (l.), fish communities. I. Structural analysis. Journal of Experimental Marine Biology and Ecology 22:269–291.

Anderson, R. O., and S. J. Gutreuter. 1983. Length, weight, and associated structural indices. Pages 283–300 *in* L. A. Nielsen and D. L. Johnson, editors. Fisheries techniques. American Fisheries Society, Bethesda, Maryland.

Bell, J. D., and M. Westoby. 1986. Variation in seagrass height and density over a wide spatial scale: effects on common fish and decapods. Journal of Experimental Marine Biology and Ecology 104:275–295.

Carr, M. H. 1983. Spatial and temporal patterns of recruitment of young-of-the-year rockfishes (genus Sebastes) into a central California kelp forest. Master's thesis. San Francisco State University, California.

Dean, T. A., M. S. Stekoll, and R. O. Smith. 1996. Kelps and oil: the effects of the *Exxon Valdez* oil spill on subtidal algae. American Fisheries Society Symposium 18:412–423.

Ebeling, A. W., and D. L. Laur. 1985. The influence of plant cover on surfperch abundance at an offshore temperate reef. Environmental Biology of Fishes 12(3):169–179.

Haldorson, L., and L. J. Richards. 1987. Post-larval copper rockfish in the Strait of Georgia: habitat use, feeding, and growth in the first year. Pages 129–142 *in* Proceedings of the International rockfish symposium. University of Alaska, Alaska Sea Grant Report 87-2, Fairbanks.

Jewett, S. C., T. A. Dean, and D. R. Laur. 1996. Effects of the *Exxon Valdez* oil spill on benthic invertebrates in an oxygen-deficient embayment in Prince William Sound, Alaska. American Fisheries Society Symposium 18:440–447.

Jones, G. P. 1984. Population ecology of the temperate reef fish *Pseudolabrus celidotus* Bloch & Schneider (Pisces: Labridae). I. Factors influencing recruitment. Journal of Experimental Marine Biology and Ecology 75:257–276.

Manly, B. 1991. Randomization and Monte Carlo methods in biology. Chapman and Hall, London.

Mann, K. H. 1973. Seaweeds: their productivity and strategy for growth. Science 182:975–981.

Milliken, G. A., and D. E. Johnson. 1984. Analysis of messy data, volume II. Designed experiments. Life Time Learning Publications, Belmont, California.

O'Clair, C. E., J. W. Short, and S. D. Rice. 1996. Contamination of intertidal and subtidal sediments by oil from the *Exxon Valdez* in Prince William Sound. American Fisheries Society Symposium 18:61–93.

Rosenthal, R. J. 1983. Shallow water fish assemblages in the northeastern Gulf of Alaska: habitat evaluation, species composition, abundance, spatial distribution and trophic interaction. Pages 451–540 *in* Environmental assessment of the Alaskan continental shelf. Final reports of principal investigators, volume 17, biological studies. NOAA (National Oceanic and Atmospheric Administration)/NOS (National Ocean Service), Office of Oceanography and Marine Services, Anchorage, Alaska.

Rosenthal, R. J., D. C. Lees, and T. M. Rosenthal. 1977. Ecological assessment of sublittoral plant communities in the northern Gulf of Alaska. Final report to NMFS (National Marine Fisheries Service), Auke Bay Laboratory, Juneau, Alaska.

American Fisheries Society Symposium 18:671–683, 1996

Petroleum Exposure and Associated Biochemical Effects in Subtidal Fish after the *Exxon Valdez* Oil Spill

TRACY K. COLLIER, CHERYL A. KRONE, MARGARET M. KRAHN, JOHN E. STEIN, SIN-LAM CHAN, AND USHA VARANASI

National Oceanic and Atmospheric Administration, National Marine Fisheries Service, Northwest Fisheries Science Center, Environmental Conservation Division, Seattle, Washington 98112, USA

Abstract.—After the *Exxon Valdez* oil spill, petroleum exposure in subtidal fish species was assessed by determining levels of fluorescent aromatic compounds (FACs) in fish bile. The biochemical effects of this exposure were assessed by measuring induction of a specific cytochrome P-450, cytochrome P-4501A (CYP1A), and levels of deoxyribonucleic acid (DNA) adducts by ^{32}P-postlabeling. These studies were conducted at sites inside Prince William Sound and along the Kenai and Alaska peninsulas. The species examined were Dolly Varden *Salvelinus malma*, yellowfin sole *Pleuronectes asper*, rock sole *P. bilineatus*, flathead sole *Hippoglossoides elassodon*, and walleye pollock *Theragra chalcogramma*. For the Dolly Varden, a littoral-zone species, exposure was greatest in the 1989 sampling and dropped considerably by 1990; however, for nearshore benthic species (i.e., yellowfin, rock, and flathead soles), exposure continued into 1990 and 1991. Walleye pollock showed evidence of exposure when they were first sampled in 1990, at sites more than 640 km from where the *Exxon Valdez* went aground. Generally, CYP1A induction showed temporal trends that were similar to trends in FAC data, and was shown to be a useful measure of oil exposure; however, in two of the most heavily exposed species, there was no evidence of increased formation of petroleum-derived DNA adducts. Overall, the results show a continuing exposure of several subtidal fish species, suggesting that petroleum contamination of subtidal areas needs to be considered in assessing the impacts of major oil spills. Still to be determined are the potential impacts on fishery resources of long-term exposure to petroleum, albeit at moderate to low levels.

Petroleum and its components have the potential to damage fishery resources, as numerous laboratory studies have demonstrated that exposure of fish to petroleum hydrocarbons can result in a variety of adverse effects (National Research Council 1985). Because of the high rate of metabolism and depuration of hydrocarbons by many species of fish (Varanasi et al. 1989a), direct measurement of tissue concentrations of parent compounds generally does not provide a useful indicator of exposure of fish to petroleum hydrocarbons. Therefore, it has been difficult to document exposure of fish to petroleum after oil spills. Assessing exposure of fish to polycyclic aromatic hydrocarbons (PAHs) is of special interest because of the general recognition that this class of petroleum hydrocarbons has high potential for exerting toxicity. Methods have been developed for determining such exposure, based on our increased understanding of the mechanisms and pathways of PAH metabolism in fish.

One such method, the measurement of fluorescent aromatic compounds (FACs) in bile, relies on the fact that many fluorescent metabolites of PAHs are primarily excreted via the hepatobiliary system in fish. This method has the added advantage of being able to measure different classes of PAH, such as 2–3-ring, 3–4-ring, or 4–5-ring structures (Krahn et al. 1992, 1984). Another method relies on the finding that hepatic cytochrome P-4501A (CYP1A), together with the associated monooxygenase activity aryl hydrocarbon hydroxylase (AHH), can be rapidly induced in many fish species after exposure to a wide variety of organic contaminants, including PAH (Payne and Penrose 1975; Payne et al. 1987; Collier and Varanasi 1991). Thus the induction of CYP1A represents an early biological effect that is also useful in estimating PAH exposure of fish. An additional concern is based on our knowledge that electrophilic intermediates of high-molecular-weight PAHs can form covalent adducts with nucleophilic centers within biological macromolecules, including hepatic deoxyribonucleic acid (DNA). A method using ^{32}P-postlabeling of deoxynucleotides and chromatographic separation of adducted nucleotides from normal nucleotides has been developed that is highly sensitive for detecting bulky hydrophobic DNA adducts (Gupta and Randerath 1988; Varanasi et al. 1989b). Laboratory studies with fish have shown that hepatic AHH activities and levels of biliary FACs and hepatic DNA adducts are highly dose-responsive to PAH exposure (Varanasi et al. 1989c; Collier and Varanasi 1991; Stein et al. 1994). The formation of DNA adducts additionally provides an indication of exposure to genotoxic contaminants (Stein et al. 1994).

In this paper we present the results of our assessments of petroleum exposure and associated biochemical effects in several species of fish for 3 years after the *Exxon Valdez* oil spill, using measurements of biliary FACs, hepatic AHH activity, and levels of xenobiotic DNA adducts. Most of these data were collected under the aegis of the Natural Resources Damage Assessment (NRDA) process. Data are presented for Dolly Varden *Salvelinus malma*, yellowfin sole *Pleuronectes asper*, rock sole *P. bilineatus*, flathead sole *Hippoglossoides elassodon*, and walleye pollock *Theragra chalcogramma*. Our studies of exposure of adult Pacific salmon species are being presented in another paper in this volume detailing extensive samplings of subsistence fisheries (Hom et al. 1996, this volume).

Methods

Fish were captured by otter trawl (yellowfin, flathead, and rock soles, sampled from 15- to 60-m depth), midwater trawls (walleye pollock), or gillnet–beach seine (Dolly Varden, sampled in the littoral zone). Primary sampling times were May–June for each year, although some limited collections also were done in July and September of 1989; February, July, and August of 1990; and February 1991. February samplings were for walleye pollock only. Coordinates of sampling sites are listed in Table 1. Fish were kept alive in running seawater from the time of capture until necropsy, generally less than 2 h. After fish were killed by severing the spine, samples of bile and liver were immediately taken and stored appropriately (i.e., liver samples for AHH and adduct analyses were kept at −80°C or colder, bile samples for FAC analyses were kept at −20°C or colder). Generally, all samples were collected and frozen within 5–10 min after fish were killed. In 1989, there was a substantial loss of samples stored at −80°C (e.g., AHH and adduct samples) because of failure of a cryogenic cooler onboard the ship. This limited our ability to compare results from the different methods in the first year of sampling, when exposure levels were generally highest. In addition, there were also instances where the gall bladder was empty, thus bile could not be collected from some fish.

Analyses for biliary FACs were carried out at fluorescence wavelengths appropriate for phenanthrene (PHN) and naphthalene. These two wavelengths have been shown to be most useful for determining petroleum exposure in fish (Krahn et al. 1992). Only PHN values are reported here, although data for analyses at naphthalene wavelengths can be found in the final report submitted to the Natural Resource Trustee Council for projects Fish/Shellfish 24b, Fish/Shellfish 24, and Subtidal 7. Methods used are essentially as described by Krahn et al. (1986a). Biliary FAC values were normalized against protein content of bile in order to partially account for variable water content of bile (Collier and Varanasi 1991). Hepatic CYP1A was measured

TABLE 1.—Names, latitudes, and longitudes for sites where fish samples were collected in 1989–1991. Asterisk indicates site generally acknowledged to have been heavily oiled after the *Exxon Valdez* oil spill.

Site	Latitude	Longitude
Afognak	58°20′N	152°23′W
Balboa Bay	55°34′N	160°36′W
Bay of Isles*	60°24′N	147°30′W
Black Bay	59°32′N	150°13′W
Bogoslof	54°05′N	168°28′W
Cape Ikolik	57°07′N	155°17′W
Cape Kekurnoi	57°42′N	155°02′W
Cape Ugat	57°57′N	153°48′W
Cape Uyak	57°43′N	154°34′W
Cape Paramanof	58°21′N	153°08′W
Chignik Bay	56°19′N	158°23′W
Chirikof Island	55°07′N	156°42′W
Discoverer Bay	58°23′N	152°23′W
Eastend Transect	55°54′N	165°59′W
Fox Farm Bay	59°58′N	148°10′W
Goose Island	60°46′N	146°55′W
Hallo Bay	58°27′N	154°02′W
Hogan Bay	60°10′N	147°43′W
Kachemak Bay	59°34′N	151°16′W
Kamishak Bay	59°15′N	153°40′W
Katmai Bay	57°56′N	155°04′W
Kinak Bay	58°01′N	154°15′W
Kodiak Island	57°44′N	152°25′W
Kukak Bay	58°17′N	154°18′W
Kuliak Bay	58°11′N	153°53′W
MacLeod Harbor	59°54′N	147°45′W
Malina Point	58°04′N	153°36′W
Moose Lips Bay	60°13′N	147°18′W
Mummy Bay*–Island*	60°18′N	147°58′W
Naked Island*	60°38′N	147°26′W
Olsen Bay–Port Gravina	60°44′N	146°14′W
Point Bazil*	59°59′N	147°47′W
Portage Bay	56°55′N	155°47′W
Port Fidalgo	60°44′N	146°48′W
Resurrection Bay	59°55′N	149°20′W
Rocky Bay	60°21′N	147°06′W
Sanak Island	54°05′N	162°33′W
Sleepy Bay*	60°04′N	147°50′W
Snug Harbor*	60°14′N	147°43′W
Squirrel Bay	60°00′N	148°04′W
Sturgeon Head	57°39′N	155°00′W
Sunny Cove	59°55′N	149°21′W
Sutwick Island	56°30′N	156°12′W
Tonsina Bay*	59°14′N	150°58′W
Trinity Islands	55°59′N	154°24′W
Tugidak Island	56°17′N	156°25′W
Uganik Island	58°07′N	153°37′W
Valdez	61°07′N	146°21′W
Windy Bay	59°14′N	151°32′W

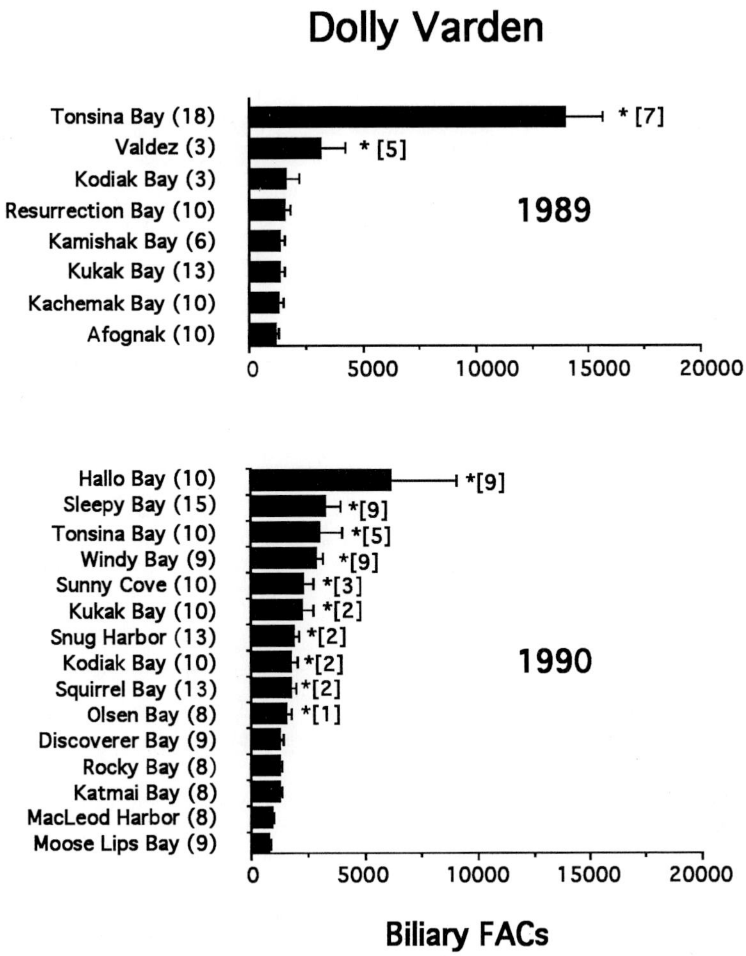

FIGURE 1.—Biliary fluorescent aromatic compounds (FACs) in Dolly Varden *Salvelinus malma* captured at sites in Alaska in 1989 and 1990. Units are nanograms phenanthrene (PHN) equivalents per milligram total biliary protein; values are means ± SEs. The number of analyses per site is shown in parentheses after the site name. The asterisk followed by a number in brackets indicates statistically significant site differences ($P \leq 0.05$). The bracketed values denote the number of sites with statistically significant, lower levels of FACs beginning at the site with the lowest concentration.

as AHH activity because this reaction is shown to be catalyzed almost solely by CYP1A in fish (Varanasi et al. 1986). This measurement was done by the methods reported in Collier et al. (1995), using ^{14}C-benzo[*a*]pyrene as the primary substrate. Levels of DNA adducts in liver samples of rock soles and Dolly Varden were determined by a ^{32}P-postlabeling assay with adduct enrichment by *n*-butanol extraction, as described by Stein et al. (1992).

All data were log-transformed before statistical analysis to reduce heteroscedasticity. Statistical analyses of data were done by one-way analysis of variance followed by Fisher's protected least-significant-difference test to determine site differences. Significance level for acceptance of all statistical tests was $P < 0.05$.

Results

Dolly Varden

Dolly Vardens showed some of the highest levels of exposure to oil (as measured by levels of FACs in bile) of any fish sampled in 1989 (Figure 1). Fish from Tonsina Bay, a heavily oiled site, had levels of biliary FACs more than twice as high as Dolly Vardens from any other site in 1989. Levels of biliary

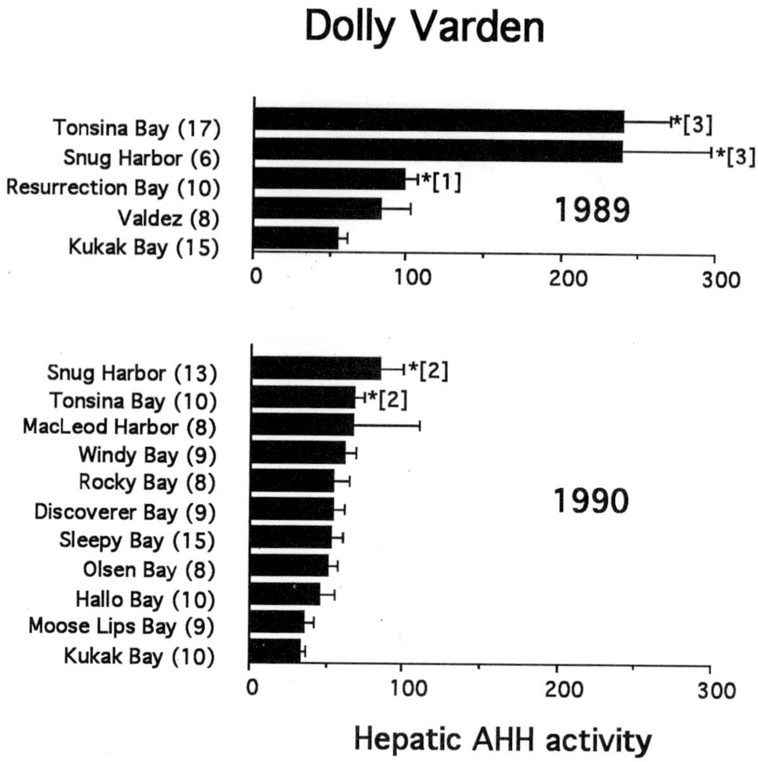

FIGURE 2.—Hepatic aryl hydrocarbon hydroxylase (AHH) activities in Dolly Varden *Salvelinus malma* captured at sites in Alaska in 1989 and 1990. Units are picomoles benzo[*a*]pyrene metabolized per milligram microsomal protein; values are means ± SEs. The number of analyses per site is shown in parentheses after the site name. The asterisk followed by a number in brackets indicates statistically significant site differences ($P \leq 0.05$). The bracketed values denote the number of sites with statistically significant, lower AHH activities beginning at the site with the lowest concentration.

FACs are not presented for Dolly Vardens from Snug Harbor in 1989 because biliary protein levels were not measured; however, levels on an uncorrected basis were approximately two-thirds the value for Dolly Vardens from Tonsina Bay (Varanasi et al. 1992). Moreover, hepatic AHH activities (Figure 2) were substantially induced in Dolly Vardens from Tonsina Bay and Snug Harbor compared to other sites. By 1990, the levels of biliary FACs and activities of hepatic AHH had dropped markedly at oiled sites, although there were still some statistically higher measurements at oiled sites. Dolly Vardens were not sampled in 1991.

These results for FACs and AHH suggest that the heavy oiling of the intertidal area seen in 1989 affected fish in the very nearshore subtidal area, such as Dolly Varden, and by 1990 the levels of petroleum hydrocarbon contamination of these areas were substantially reduced. As shown in Table 2, however, there was no increase in levels of DNA adducts in the livers of Dolly Vardens sampled in 1989 from either Snug Harbor or Tonsina Bay compared with levels in Dolly Vardens captured at Resurrection Bay. The levels reported are very near the background levels of DNA adducts, and there were no adduct patterns indicative of exposure to PAHs.

TABLE 2.—Levels of DNA adducts (nanomol DNA adducts per mol DNA nucleotides) in liver tissue of fishes sampled at four sites in 1989 after the *Exxon Valdez* oil spill. For data shown, $N = 4$. Values are means ± SEs. No significant differences between sites or species were found.

Site	Dolly Varden *Salvelinus malma*	Rock sole *Pleuronectes bilineatus*
Snug Harbor[a]	6.3 ± 0.7	
Tonsina Bay[a]	5.3 ± 0.7	6.7 ± 0.7
Resurrection Bay	6.6 ± 0.9	
Balboa Bay		7.4 ± 0.4

[a]Site generally acknowledged to have been heavily oiled after the oil spill.

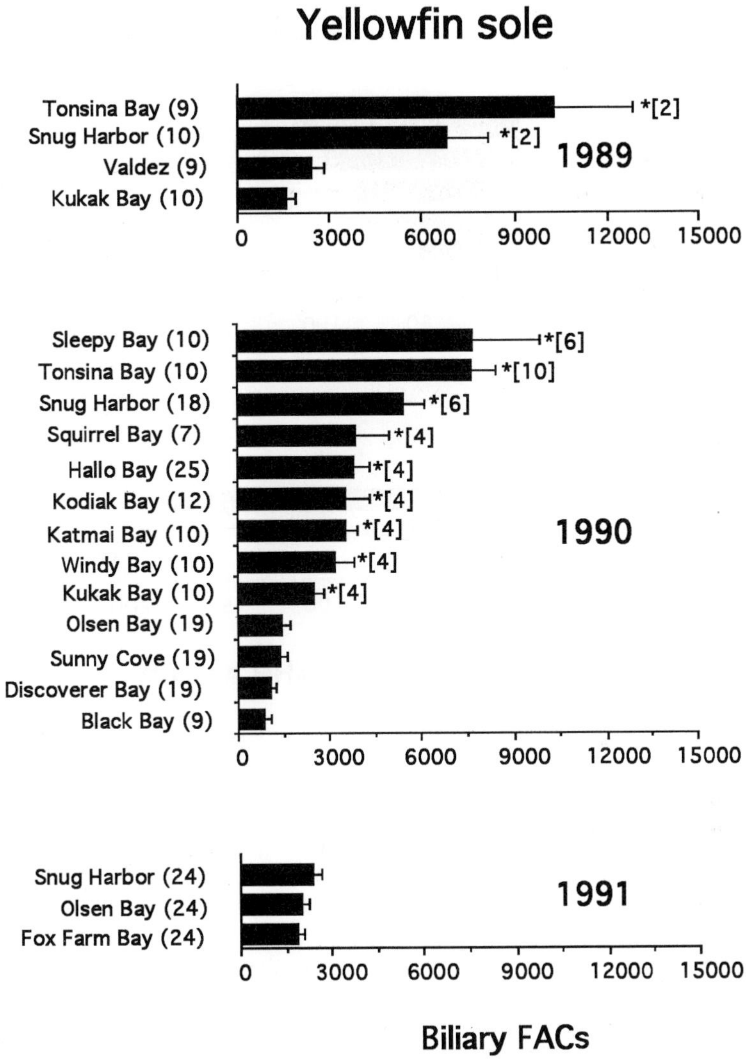

FIGURE 3.—Biliary fluorescent aromatic compounds (FACs) in yellowfin soles *Pleuronectes asper* captured at sites in Alaska in 1989–1991. Other details as in Figure 1.

Yellowfin Sole

Hepatic AHH activities were not measured in yellowfin soles in 1989, but measurements of biliary FACs showed that exposure to petroleum was readily discernible in this benthic flatfish species (Figure 3). The levels of biliary FACs in yellowfin soles were lower than those in Dolly Vardens, suggesting less exposure to oil. In contrast to levels in Dolly Vardens, levels in yellowfin soles did not drop markedly between 1989 and 1990 at oiled sites. Hepatic AHH activities in yellowfin soles sampled in 1990 (Figure 4) were significantly increased at Sleepy Bay, Tonsina Bay, and Snug Harbor, and fish from these three sites also showed the highest levels of biliary FACs in 1990. In 1991, yellowfin soles from Snug Harbor had induced hepatic AHH activities, but no increases in levels of biliary FACs, compared with yellowfin soles from other sites.

Rock Sole

As seen in yellowfin soles, there was evidence of exposure to petroleum in rock soles sampled near oiled sites in 1989 and 1990 (Figures 5 and 6). Decreased exposure was observed at oiled sites

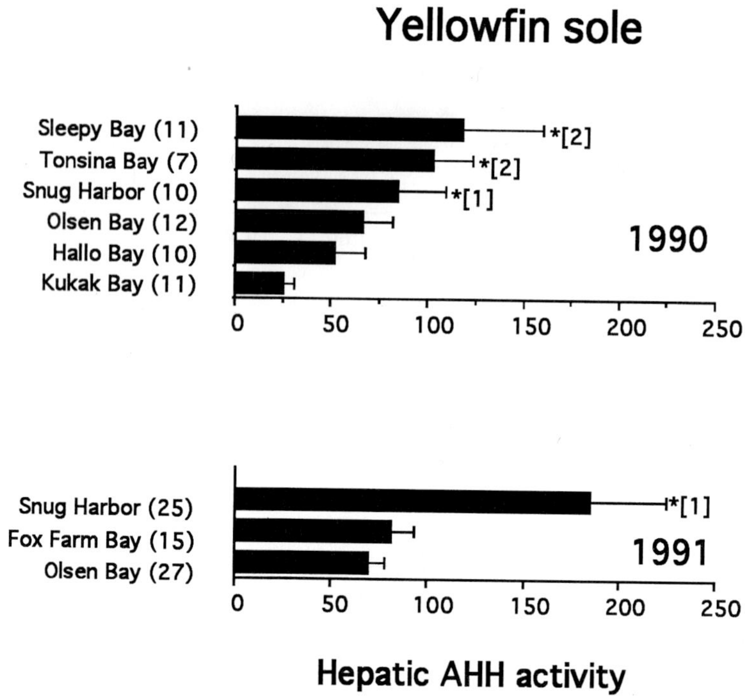

FIGURE 4.—Hepatic aryl hydrocarbon hydroxylase (AHH) activities in yellowfin soles *Pleuronectes asper* captured at sites in Alaska in 1990 and 1991. Other details as in Figure 2.

(e.g., Snug Harbor and Sleepy Bay) in the limited sampling done in 1991 as measured by levels of FACs in bile (Figure 5), but hepatic AHH activities still showed significant induction in rock soles sampled from Snug Harbor (Figure 6). The highest mean levels of biliary FACs in rock soles were measured in fish sampled from Squirrel Bay in 1991, although these levels were not significantly higher than those at other sites, because of large variance in the data. Levels of hepatic DNA adducts were measured in rock soles in 1989 (Table 2), and there were no increased levels of DNA adducts in liver tissue of rock soles sampled from Tonsina Bay when compared to rock soles from Balboa Bay.

Flathead Sole

Significantly increased levels of biliary FACs were measured in flathead soles from heavily oiled sites in 1989 and 1990 (Figure 7). These levels were similar in magnitude to those in other flatfish species (i.e., maximum mean values of almost 10,000 ng PHN equivalents/mg biliary protein. By 1991 levels of biliary FACs had dropped substantially in flathead soles from Snug Harbor but were still significantly higher than values for flathead soles from Olsen Bay. Hepatic AHH activities were not measured in flathead soles in 1989. In 1990 there were no significant differences in hepatic AHH activities for flathead soles captured at different sites, although the highest mean levels were seen at oiled sites (Figure 8). In 1991 hepatic AHH activities were significantly elevated in flathead soles from Snug Harbor compared with values in flathead soles from a less impacted site, Olsen Bay. Overall, these results are similar to the findings for yellowfin and rock soles, which is consistent with all three of these flatfish species being captured from similar habitats.

Walleye Pollock

Walleye pollock were not sampled until late winter 1990, and then they were only sampled for levels of biliary FACs. At that time increased levels of FACs were evident, especially in pollock from inside Prince William Sound, such as from Mummy Island, Naked Island, Point Bazil, and Goose Island (Figure 9). Outside Prince William Sound, statistically increased levels of biliary FACs were found in fish from as far away as Tugidak Island, a site 640 km west of Bligh Reef, where the *Exxon Valdez* went aground. By 1991 levels of biliary FACs in

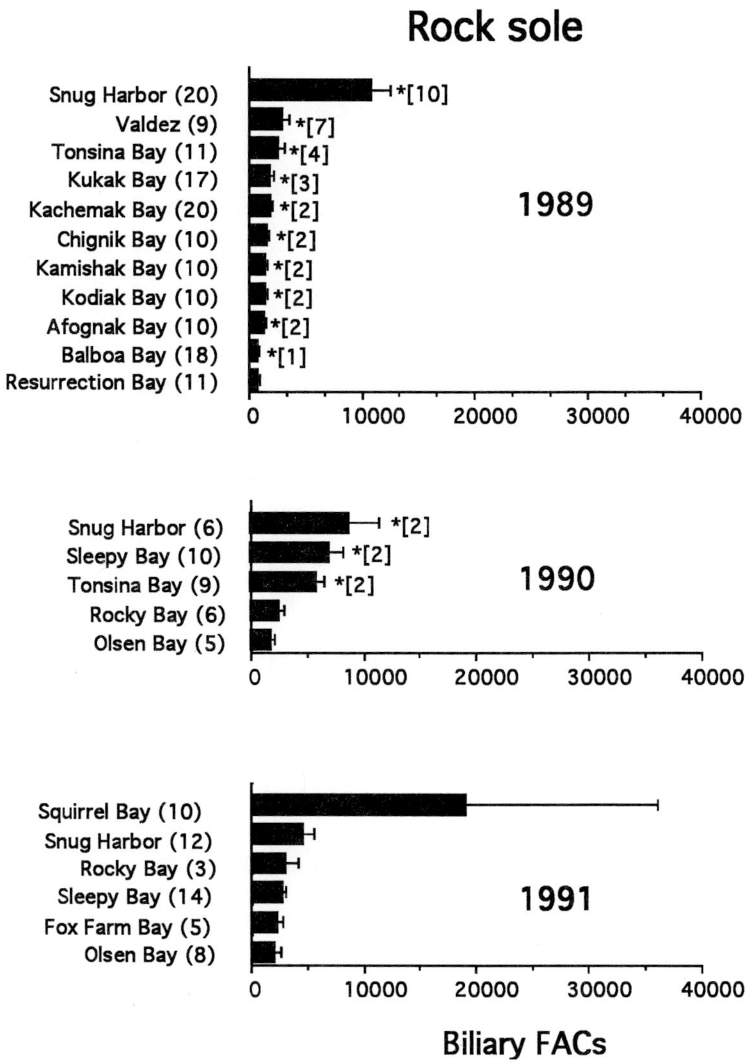

FIGURE 5.—Biliary fluorescent aromatic compounds (FACs) in rock soles *Pleuronectes bilineatus* captured at sites in Alaska in 1989–1991. Other details as in Figure 1.

walleye pollock had dropped substantially, although it was still possible to detect significantly increased levels at some sites inside Prince William Sound (e.g., Port Gravina, Bay of Isles, Hogan Bay, and Port Fidalgo), compared with levels in animals collected from unimpacted or distant sites. Mass spectrometric analysis has confirmed the presence of metabolites of petroleum hydrocarbons in the bile of four of the walleye pollock sampled in 1989 (Krahn et al. 1992).

Discussion

The results of analyses of oil exposure and associated biochemical effects in several species of subtidal fish after the *Exxon Valdez* oil spill definitively point to the necessity of monitoring the subtidal environment after major oil spills. The littoral zone appeared to be strongly affected in the months immediately after the spill, but 1 year later, after a winter storm cycle, exposure of fish inhabiting this zone (i.e., Dolly Vardens) was substantially decreased. Similar findings were reported for pink salmon *Oncorhynchus gorbuscha*, which were sampled from habitats similar to those of Dolly Vardens (Hom et al. 1996). In contrast, nearshore benthic fish species (up to approximately 30-m depth, species sampled were rock, yellowfin, and flathead

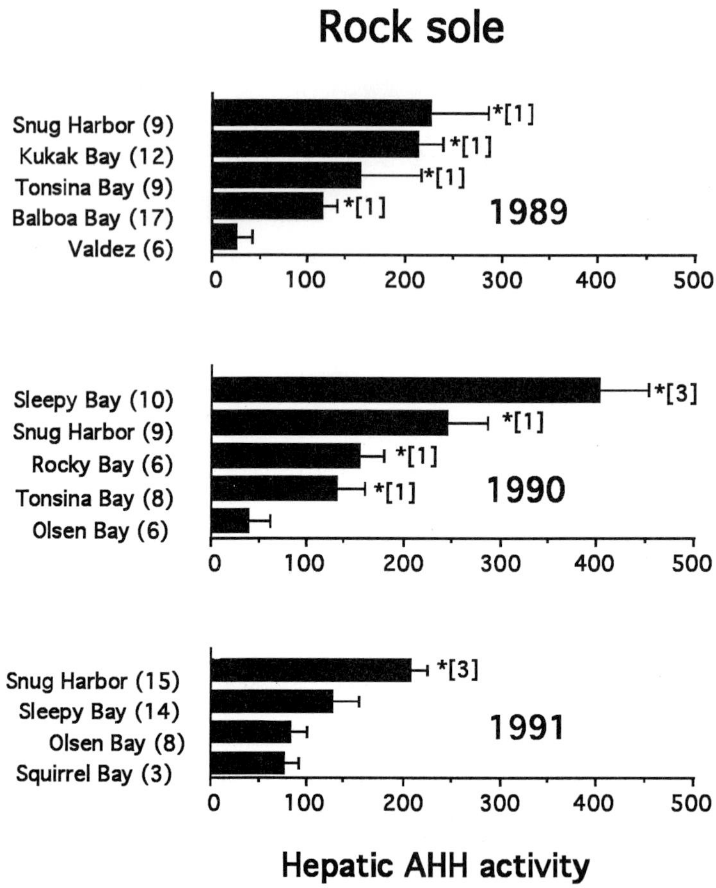

FIGURE 6.—Hepatic aryl hydrocarbon hydroxylase (AHH) activities in rock soles *Pleuronectes bilineatus* captured at sites in Alaska in 1989–1991. Other details as in Figure 2.

soles) showed continuing exposure through the first two field seasons after the *Exxon Valdez* oil spill, and even after more than 2 years there was some evidence of increased exposure of fish from these habitats. Chemical analyses of subtidal sediments (3- to 20-m depth) in Prince William Sound similarly showed the presence of oil contamination in 1989, 1990 and 1991, although levels of PAHs were steadily decreasing and the distribution of contamination became more patchy (O'Clair et al. 1996, this volume). To what degree the presence of oil at depth resulted from onshore activities, including hot-water washing of the intertidal zone, has not been determined.

A surprising finding was the evidence of exposure of walleye pollock to petroleum approximately 1 year after the spill at a site (e.g., Tugidak Island) more than 640 km from the grounding site. Walleye pollock, a major fisheries resource in Alaskan waters, are bathypelagic fish that feed in the water column. Thus these results suggest that the spilled oil affected either the water column or food supply of these fish at great distances from the spill and for some time after the spill. We have previously used the bile metabolite method for tracing petroleum exposure in fish after an oil spill in the Columbia River. The results of that study showed that white sturgeons *Acipenser transmontanus* had increased levels of FACs in their bile (levels more than 20-fold higher than in reference fish) 92 km downstream from the spill site, 5 d after the spill occurred (Krahn et al. 1986b). We are not aware of any other occasions, before the *Exxon Valdez* oil spill, where measurements of biliary FACs were used for determining oil exposure in feral fish.

The determination of CYP1A induction, whether measured as AHH activity or by other catalytic or immunologic assays (such as ethoxyresorufin-*O*-de-

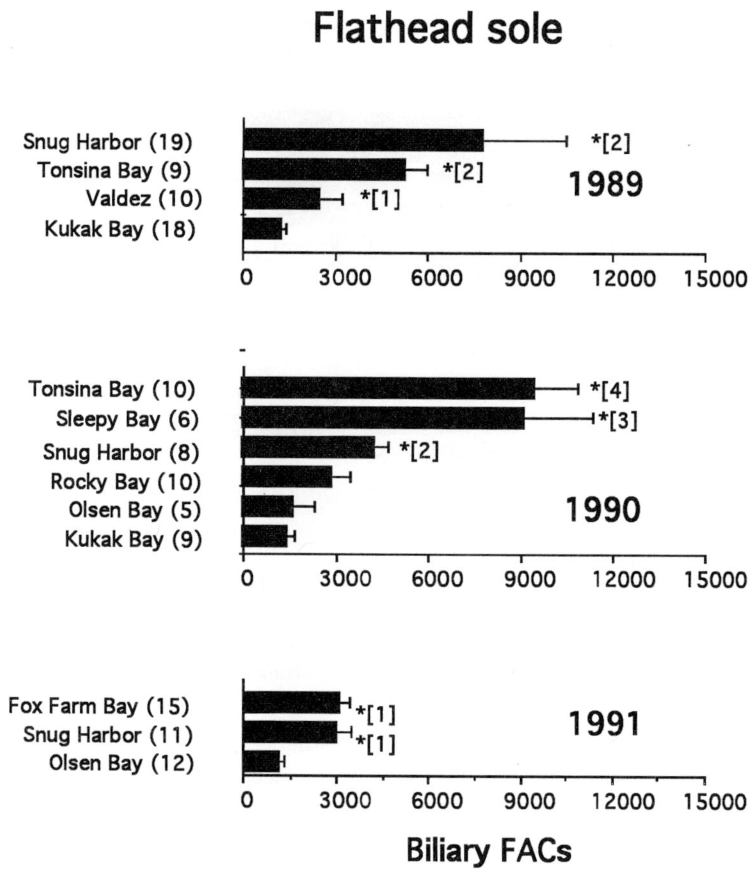

FIGURE 7.—Biliary fluorescent aromatic compounds (FACs) in flathead soles *Hippoglossoides elassodon* captured at sites in Alaska in 1989–1991. Other details as in Figure 1.

ethylase assay or enzyme-linked immunosorbent assay [ELISA]), has been used several times for assessing oil exposure of fish after oil releases, with the earliest such use more than 20 years ago (Payne and Penrose 1975). Recent dose–response studies have indicated that CYP1A induction may be more sensitive than increases in biliary levels of FACs for assessing exposure to a complex mixture of organic contaminants extracted from an urban sediment (Collier and Varanasi 1991). The current results showed significant site-specific increases in hepatic CYP1A activity but not biliary FACs in yellowfin and rock soles sampled in 1991, thus suggesting that similar dose–response curves may apply for petroleum exposure of fish. Alternatively, it is possible that by 1991 there was a preferential retention of compounds in the nearshore benthic environment that were capable of inducing CYP1A (e.g., higher-molecular-weight PAHs) and that the bile fluorescence method as we applied it, using wavelengths appropriate for phenanthrene, did not detect any exposure of fish to these compounds.

There was no evidence of oil-induced DNA adducts in the fish sampled in 1989, which was not unexpected given the comparatively low amount of high-molecular-weight PAHs in Prudhoe Bay crude oil (Clark and Brown 1977). To date there have been no published results showing increased levels of DNA adducts in feral teleost fish after petroleum releases, although there has been a preliminary report of increased DNA adducts resulting from petroleum exposure in a laboratory study (R. C. Gupta and associates, University of Kentucky, unpublished data). Thus, although DNA damage has been speculated to underlie deleterious effects in fishery resources in Prince William Sound after the *Exxon Valdez* oil spill, there is currently little direct evidence to support such a hypothesis; however, it

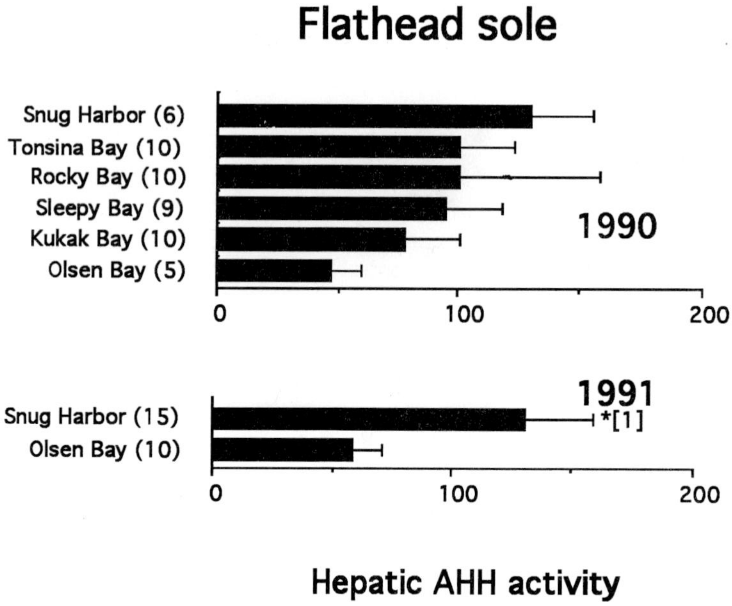

FIGURE 8.—Hepatic aryl hydrocarbon hydroxylase (AHH) activities in flathead soles *Hippoglossoides elassodon* captured at sites in Alaska in 1990 and 1991. Other details as in Figure 2.

should be remembered that the method used for assessing DNA damage in our study is directed toward detection of bulky hydrophobic adducts, and other types of adducts or damage may not be detected by this method.

Although the overall results show species- and site-specific patterns of oil exposure after the *Exxon Valdez* oil spill, substantially different conclusions might have been reached if just one marker of biological effects had been used. These results point to the importance of not relying on any individual marker of contaminant exposure and effects in conducting field surveys in the aquatic environment, which is in accord with our earlier studies (Stein et al. 1992; Collier et al. 1993). Evaluation of results from a suite of biological markers can often provide a clearer picture of the degree and timing of contaminant exposure, which can lead to a better understanding of any subsequent biological effects. Such was the case in recent studies concerning contaminant exposure and effects in juvenile salmon outmigrating through contaminated estuaries in Puget Sound, Washington (Stein et al. 1995).

The current findings of widespread and continuing exposure of subtidal fish after the *Exxon Valdez* oil spill raise several concerns. Perhaps foremost is the question of whether oil-exposed fish are safe for human consumption. Although this issue was not the subject of the Natural Resources Damage Assessment (NRDA) process after this oil spill, there was a substantial effort put forth to answer this question in association with the subsistence fisheries, after this spill. The findings from these analyses of subsistence foods (Hom et al. 1996) suggested that there was little risk involved in consumption of the flesh from oil-exposed fish species because of the extensive metabolism and excretion of petroleum compounds by fish; however, there was accumulation of hydrocarbons in various invertebrate species that do not possess substantial ability to metabolize these compounds (Brown et al. 1996, this volume). The finding of minimal risk from consumption of oil-exposed fish was supported by an advisory position of the Food and Drug Administration (Varanasi et al. 1990).

What remains to be determined are the potential impacts on fishery resources of long-term exposure to petroleum, albeit at moderate to low levels. To date, our studies have not shown any profound reproductive or histopathological effects in the species studied after the *Exxon Valdez* oil spill (Varanasi et al. 1992), but this finding is tempered by the delayed start of assessment of serious effects such as reproductive function and the relatively short time over which such analyses have been conducted. Although it will always be difficult to sample subtidal

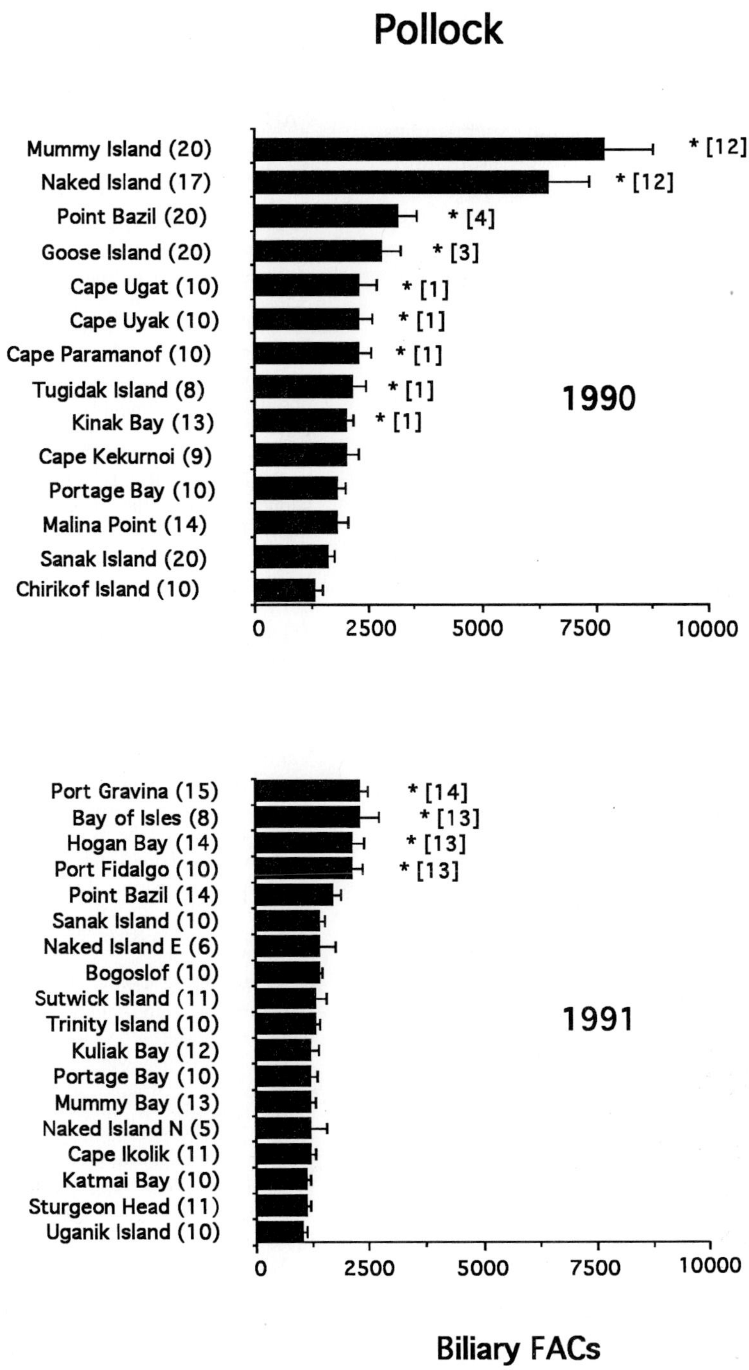

FIGURE 9.—Biliary fluorescent aromatic compounds (FACs) in walleye pollock *Theragra chalcogramma* captured at sites in Alaska in 1990 and 1991. Other details as in Figure 1.

fishery resources comprehensively and rapidly after an oil spill, a better understanding of the potential impacts of petroleum exposure on fishery resources can be obtained through careful and realistic laboratory exposure of fish to petroleum. Such studies will need to go beyond the relatively straightforward, short-term exposure studies that have been done in the past (National Research Council 1985). Advances in methodology for assessing oil exposure in fish, together with current knowledge of the biological processes involved in reproduction, immune function, and growth and survival of fish species (Arkoosh et al. 1991; Casillas et al. 1991; Collier et al. 1992; Johnson et al. 1994), make this an appropriate course of action.

Acknowledgments

We are grateful to the many members of the Environmental Conservation Division (ECD) who performed sample collections and analyses. This was truly a cooperative effort of many ECD scientists from different disciplines, but special thanks are due to Robert Clark, Paul Plesha, and Bill Gronlund for their efforts in organizing and overseeing many aspects of this work in the first two field seasons. Also, the officers and crew of the National Oceanic and Atmospheric Administration ships *Fairweather* and *Davidson* and the fishing vessel *Big Valley* contributed greatly to the success of this work. Funding approval for much of this work came from the Trustee Council, via the *Exxon Valdez* oil spill damage assessment process.

References

Arkoosh, M. R., E. Casillas, E. Clemons, B. B. McCain, and U. Varanasi. 1991. Suppression of immunological memory in juvenile chinook salmon (*Oncorhynchus kisutch*) from an urban estuary. Fish and Shellfish Immunology 4:261–278.

Brown, D. W., and nine coauthors. 1996. Survey of Alaskan subsistence invertebrate seafoods collected in 1989–1991 to determine exposure to oil spilled from the *Exxon Valdez*. American Fisheries Society Symposium 18:844–855.

Casillas, E., and seven coauthors. 1991. Inducibility of spawning and reproductive success of female English sole (*Parophrys vetulus*) from urban and nonurban areas of Puget Sound, Washington. Marine Environmental Research 31:99–122.

Clark, R. C., Jr., and D. W. Brown. 1977. Petroleum: properties and analyses in biotic and abiotic systems. Pages 1–89 *in* D. C. Malins, editor. Effects of petroleum on arctic and subarctic marine environments. Academic Press, New York.

Collier, T. K., B. F. Anulacion, J. E. Stein, A. Goksøyr, and U. Varanasi. 1995. A field evaluation of cytochrome P-4501A as a biomarker of contaminant exposure in three species of flatfish. Environmental Toxicology and Chemistry 14:143–152.

Collier, T. K., and six coauthors. 1993. Biomarkers of PAH exposure in oyster toadfish (*Opsanis tau*) from the Elizabeth River, Virginia. Environmental Science 2:161–177.

Collier, T. K., J. E. Stein, H. R. Sanborn, T. Hom, M. S. Myers, and U. Varanasi. 1992. Field studies of reproductive success and bioindicators of maternal contaminant exposure in English sole (*Parophrys vetulus*). Science of the Total Environment 116:169–185.

Collier, T. K., and U. Varanasi. 1991. Hepatic activities of xenobiotic metabolizing enzymes and biliary levels of xenobiotics in English sole (*Parophrys vetulus*) exposed to environmental contaminants. Archives of Environmental Contamination and Toxicology 20:462–473.

Gupta, R. C., and K. Randerath. 1988. Analysis of DNA adducts by ^{32}P-postlabeling and thin layer chromatography. Pages 399–418 *in* E. C. Friedberg and P. H. Hanawalt, editors. DNA repair. Marcel Dekker, New York.

Hom, T., U. Varanasi, J. E. Stein, C. A. Sloan, K. L. Tilbury, and S.-L. Chan. 1996. Assessment of the exposure of subsistence fish to aromatic compounds after the *Exxon Valdez* oil spill. American Fisheries Society Symposium 18:856–866.

Johnson, L. J., J. E. Stein, T. K. Collier, E. Casillas, and U. Varanasi. 1994. Indicators of reproductive development in prespawning female winter flounder (*Pleuronectes americanus*) from urban and non-urban estuaries on the Northeast coast of the United States. Science of the Total Environment 141:241–260.

Krahn, M. M., L. J. Kittle, Jr., and W. D. MacLeod, Jr. 1986b. Evidence for exposure of fish to oil spilled into the Columbia River. Marine Environmental Research 20:291–298.

Krahn, M. M., M. S. Myers, D. G. Burrows, and D. C. Malins. 1984. Determination of metabolites of xenobiotics in the bile of fish from polluted waterways. Xenobiotica 14:633–646.

Krahn, M. M., L. D. Rhodes, M. S. Myers, L. K. Moore, W. D. MacLeod, Jr., and D. C. Malins. 1986a. Associations between metabolites of aromatic compounds in bile and the occurrence of hepatic lesions in English sole (*Parophrys vetulus*) from Puget Sound, Washington. Archives of Environmental Contamination and Toxicology 15:61–67.

Krahn, M. M., and seven coauthors. 1992. Mass spectrometric determination of metabolites of aromatic compounds in bile of fish captured from Prince William Sound, Alaska, after the *Exxon Valdez* oil spill. Environmental Science and Technology 26:116–126.

National Research Council. 1985. Oil in the sea: inputs, fates, and effects. National Academy Press, Washington, DC.

O'Clair, C. E., J. W. Short, and S. D. Rice. 1996. Contamination of intertidal and subtidal sediments by oil from the *Exxon Valdez* in Prince William Sound. American Fisheries Society Symposium 18:61–93.

Payne, J. F., L. L. Fancey, A. D. Rahimtula, and E. L.

Porter. 1987. Review and perspective on the use of mixed-function oxygenase enzymes in biological monitoring. Comparative Biochemistry and Physiology 86(C):233–245.

Payne, J. F., and W. R. Penrose. 1975. Induction of aryl hydrocarbon (benzo[a]pyrene) hydroxylase in fish by petroleum. Bulletin of Environmental Contamination and Toxicology 14:112–116.

Stein, J. E., T. K. Collier, W. L. Reichert, E. Casillas, T. Hom, and U. Varanasi. 1992. Bioindicators of contaminant exposure and sublethal effects: studies with benthic fish in Puget Sound, Washington. Environmental Toxicology and Chemistry 11:701–714.

Stein, J. E., T. Hom, T. K. Collier, D. W. Brown, and U. Varanasi. 1995. Contaminant exposure and biochemical effects in outmigrant juvenile chinook salmon from urban and non-urban estuaries of Puget Sound, WA. Environmental Toxicology and Chemistry 14:1019–1029.

Stein, J. E., W. L. Reichert, and U. Varanasi. 1994. Molecular epizootiology: assessment of exposure to genotoxic compounds in teleosts. Environmental Health Perspectives 102(Supplement 12):19–23.

Varanasi, U., T. K. Collier, D. E. Williams, and D. R. Buhler. 1986. Hepatic cytochrome P-450 isozymes and aryl hydrocarbon hydroxylase in English sole (*Parophrys vetulus*). Biochemical Pharmacology 35: 2967–2971.

Varanasi, U., W. L. Reichert, B.-T. E. Eberhart, and J. E. Stein. 1989c. Formation and persistence of benzo[*a*]pyrene-diolepoxide-DNA adducts in liver of English sole (*Parophrys vetulus*). Chemico-Biological Interactions 69:203–216.

Varanasi, U., W. L. Reichert, and J. E. Stein. 1989b. ^{32}P-postlabeling analysis of DNA adducts in liver of wild English sole (*Parophrys vetulus*) and winter flounder (*Pseudopleuronectes americanus*). Cancer Research 49:1171–1177.

Varanasi, U., and six coauthors. 1992. Final report for NRDA subtidal 7 project—assessment of oil spill impacts on fishery resources: measurement of hydrocarbons and their metabolites, and their effects, in important species. NOAA (National Oceanic and Atmospheric Administration) NMFS (National Marine Fisheries Service) NWFSC (Northwest Fisheries Science Center), Environmental Conservation Division, Seattle.

Varanasi, U., J. E. Stein, and M. Nishimoto. 1989a. Biotransformation and disposition of polycyclic aromatic hydrocarbons (PAH) in fish. Pages 93–150 *in* U. Varanasi, editor. Metabolism of polycyclic aromatic hydrocarbons in the aquatic environment. CRC Press, Boca Raton, Florida.

Varanasi, U., and ten coauthors. 1990. Survey of subsistence fish and shellfish for exposure to oil spilled from the *Exxon Valdez*. First year: 1989. NOAA (National Oceanic and Atmospheric Administration) Technical Memorandum NMFS (National Marine Fisheries Service) F/NWC-191:1–151, Seattle.

American Fisheries Society Symposium 18:684–711, 1996

Total Direct Mortality of Seabirds from the *Exxon Valdez* Oil Spill

R. Glenn Ford, Michael L. Bonnell, and Daniel H. Varoujean
Ecological Consulting, Inc., 2735 NE Weidler Street, Portland, Oregon 97232, USA

Gary W. Page and Harry R. Carter
Point Reyes Bird Observatory, 4990 Shoreline Highway, Stinson Beach, California 94970, USA

Brian E. Sharp, Dennis Heinemann, and Janet L. Casey
Ecological Consulting, Inc.

Abstract.—An estimate of total mortality of seabirds from the *Exxon Valdez* oil spill was derived from numbers of carcasses collected on shore. The processes leading to carcass collection were modeled to account for sinking of oiled birds at sea, removal from the beach by scavengers, and incomplete search effort. To determine rates of loss of oiled birds at sea, radio-tagged carcasses were tracked by aircraft and boat. Eight groups totaling 184 carcasses were tracked in spring and summer 1990; four groups were released in Prince William Sound and four in the Gulf of Alaska off Kodiak Island. Values of median persistence time for groups released in Prince William Sound ranged from 15 to 20 d, whereas those of releases in the Gulf of Alaska ranged from 7 to 18 d. Assuming that birds drifted passively with the oil, the rate function for loss of birds at sea was applied to probable point-of-oiling and time to landfall at a particular beach. Concentrations of birds subject to oiling were located with survey data gathered in the Outer Continental Shelf Environmental Assessment Program and by the U.S. Fish and Wildlife Service. Point of oiling and path of drift to each beach were determined with the National Oceanic and Atmospheric Administration (NOAA) HAZMAT oil trajectory model and NOAA overflights. Scavenging rates were estimated from field studies in Prince William Sound and on the outer coast of the Olympic Peninsula, Washington, in which oiled carcasses were monitored daily. In each study, it was found that about 50% of carcasses were removed by scavengers within 2 d and most of the remainder were taken within 5 d. To model beach search effort and its relation to scavenging rate, it was assumed that (1) carcasses were deposited with equal frequency on any nonreflective substrate, (2) that all carcasses on nonreflective but unsearched beaches were missed, and (3) that a fraction of those on searched beaches were lost as a result of scavenging. Estimates of proportion of carcasses lost because of scavenging were made considering the interval between deposition of oiled birds and initial search and the interval between successive searches of a given beach. Based on best estimates of model parameters, 375,000 birds were directly killed by the *Exxon Valdez* oil spill. Uncertainty in model parameters was examined by randomly selecting from the probability distribution of possible values. Among 1,250 model runs, the 5% lower bound and 95% upper bound were 300,000 and 645,000 birds, respectively. Uncertainty in parameters tended to skew model results toward higher estimates of total mortality.

Prince William Sound and the western Gulf of Alaska include some of the most densely occupied marine bird habitat in the world (e.g., Gould et al. 1982). The number of seabirds present at the time of the *Exxon Valdez* oil spill is unknown; however, Piatt et al. (1990) estimated that the spring population in Prince William Sound alone ranged from 300,000 to 400,000, consisting principally of sea ducks, gulls, cormorants, grebes, loons, and alcids. Waters of the western Gulf of Alaska support colonies totaling more than 2 million birds (Sowls et al. 1978; U.S. Fish and Wildlife Service [USFWS] Colony Database).

Oil from the 1989 *Exxon Valdez* spill contacted islands and parts of the mainland shore in Prince William Sound in April and then drifted south, contaminating shoreline of the Kenai Peninsula, Barren Islands, Kodiak Archipelago, and Alaska Peninsula through August. Throughout the spill episode, portions of the affected shoreline were searched for oiled birds. Although incomplete in coverage, this effort resulted in collection of nearly 35,000 bird carcasses from more than 600 locations. In some areas, oil and oiled birds were deposited over a prolonged period of time; however, few beaches were searched daily. In the intervals between searches, carcasses were subject to loss from scavenging by eagles, bears, and foxes and from burial or rewashing from the beach by waves. Many instances of scavenging or other sources of undercounting of beached birds were reported by workers on the scene (Department of the Interior archives).

Oil continued to drift at sea in the western Gulf of Alaska through the summer. Thus, some oiled birds presumably never reached shore to be counted. Birds oiled in April may have drifted with oil for 6 weeks or longer before the oil reached shore. During this time, many carcasses probably sank (Hope-Jones et al. 1970; Page et al. 1990; Ford et al. 1991a). The proportion of total mortality represented by carcasses that sank before reaching land was unknown in the months after the spill.

The only data on direct mortality of birds from the spill were numbers of carcasses collected on shore and the distribution of search effort. To derive an estimate of total mortality, it was necessary to account for incomplete coverage and inadequate search of beaches and to estimate rates of carcass loss through scavenging on land and sinking at sea. Field studies were necessary to address questions of carcass loss and these were conducted in late spring and in summer 1990. The findings were used to place values on input parameters of an injury assessment model. The model was similar to that used for the 1984 *Puerto Rican*, 1986 *Apex Houston*, and 1988–1989 *Nestucca* oil spills (e.g., Ford et al. 1987, 1991a; Page et al. 1990). Steps in the model were sequential and the inverse of processes leading to collection of oiled birds on the beach. Beginning with the number of carcasses collected on a particular beach on a particular day, an estimate was made of the number of carcasses that may have been deposited before the toll taken by scavengers. Next, the estimate of numbers of carcasses originally present on a beach was extrapolated to nearby unsearched beaches. Lastly, the path of oil reaching each beach was determined along with the time course of oiling of birds at sea. A rate function was applied to correct numbers collected at each beach for loss of birds at sea attributable to sinking as they drifted with oil from the point of initial contact.

Studies Conducted to Provide Values of Model Parameters

Portion of Beached Birds Killed by Direct Oiling

Methods.—Nearly 35,000 bird carcasses were collected after the spill and logged into four morgues—Valdez, Seward, Homer, and Kodiak. Some of the birds collected on shore may have died from other causes and been oiled secondarily. In an earlier analysis, Piatt et al. (1990) assumed that birds collected after 31 July died of natural causes rather than from direct oiling from the spill. This approximation was based on the lack of appreciable amounts of oil on the plumage of many birds collected after this date, changes in species composition of the fauna through the season, and reports of a natural die-off (G. A. Sanger, U.S. Fish and Wildlife Service, unpublished data). Because the number of beached birds killed by direct oiling was fundamental to an estimate of total mortality, it was necessary to reexamine some of the frozen carcasses. Data from this sample were used to determine if a cutoff date should be imposed for direct mortality from the spill. To accomplish this, 3,243 birds were thawed and scored for degree of oiling. Selection of bags of frozen birds from the morgues was made to ensure representation of most geographic sites and the entire time period of carcass collection. For each bird in the sample, the following information was recorded: collection site and date, completeness of the carcass, degree of decomposition, and extent and location of oiled plumage. Carcasses consisting only of bones were not used in the analysis; these represented about 10% of the total. The approximate area of plumage visibly oiled was scored for each relatively complete carcass. Categories included no oil or only a few spots, oiled only where in contact with other oiled carcasses, lightly oiled (<33% of plumage area), moderately oiled (33–66% of area), and heavily oiled (>66% of area). Numbers and percent of carcasses classified as not oiled, oiled, and possibly oiled are presented in Table 1. The majority of birds (65%) were moderately to heavily oiled, leaving little doubt that they were contacted directly by the spill. Other feathered carcasses had only small amounts of oil where they were in contact with other oiled carcasses. These may have become oiled secondarily and were excluded from our calculations. The proportion of carcasses in this category was relatively low, ranging from 1% for murres, murrelets, and auklets to 7% for gulls.

Results.—The proportion of carcasses that were recovered in an oiled state over time was plotted for each of several species groups (Figure 1). These included murres, puffins, other alcids (auklets, murrelets, and pigeon guillemots *Cepphus columba*), tubenoses (fulmars, shearwaters, and storm-petrels), gulls, other swimming aquatic birds (cormorants, grebes, ducks, and loons), and all other species (principally shorebirds, passerines, raptors, and galliformes). For species groups represented by large numbers of oiled birds, data were combined for 10-d collection periods, referred to as early, mid, or late month; for species groups represented by fewer birds, data were combined by month. As noted by Piatt et al. (1990), the proportion of oiled

TABLE 1.—Number (*N*) and percent (%) of seabird carcasses with no visibly oiled plumage (not oiled), with visibly oiled plumage (oiled), with visible oil that was possibly from other carcasses in bags (possibly oiled), and for which only skeletons remained (unknown oil condition). Carcasses had been collected on beaches around Prince William Sound and the Gulf of Alaska during spring and summer 1989 and frozen. The "other alcids" category excludes murres, puffins, or any carcasses that could have been murres or puffins. The "other swimmers" category includes loons, grebes, cormorants, and ducks. The "other species" category includes mainly passerines, shorebirds, raptors, and galliformes.

		Not oiled		Oiled		Possibly oiled		Unknown oil condition	
Birds	Total	*N*	(%)	*N*	(%)	*N*	(%)	*N*	(%)
Tubenoses	641	486	(76)	127	(20)	18	(3)	10	(2)
Gulls	328	230	(70)	55	(17)	24	(7)	19	(6)
Murres	1,788	41	(2)	1,489	(83)	10	(1)	248	(14)
Puffins	119	97	(82)	13	(11)	7	(6)	2	(2)
Other alcids	192	47	(24)	131	(68)	2	(1)	12	(6)
Other swimmers	87	24	(28)	53	(61)	2	(2)	8	(9)
Other species	88	41	(47)	28	(32)	5	(6)	14	(16)

birds in the sample changed over time. For all species groups except murres, contact with oil appeared to occur predominantly in May and June. On this basis, taxa other than murres were combined and a cutoff date for direct mortality of 1 July imposed; birds collected on or after this date were presumed to have died from other causes. Murres, in contrast, appeared to constitute a distinct category. Because the proportion of carcasses that were recovered in an oiled state remained high throughout the entire carcass collection period, no cutoff date for murres was imposed. Numbers of murres and combined numbers of other species used in the model are summarized in Table 2.

Beach Search Effort

Methods.—To describe the spatial and temporal distribution of search effort, interviews were conducted with individuals responsible for carcass collection. Data from interviews were used to supplement information on beach search effort compiled

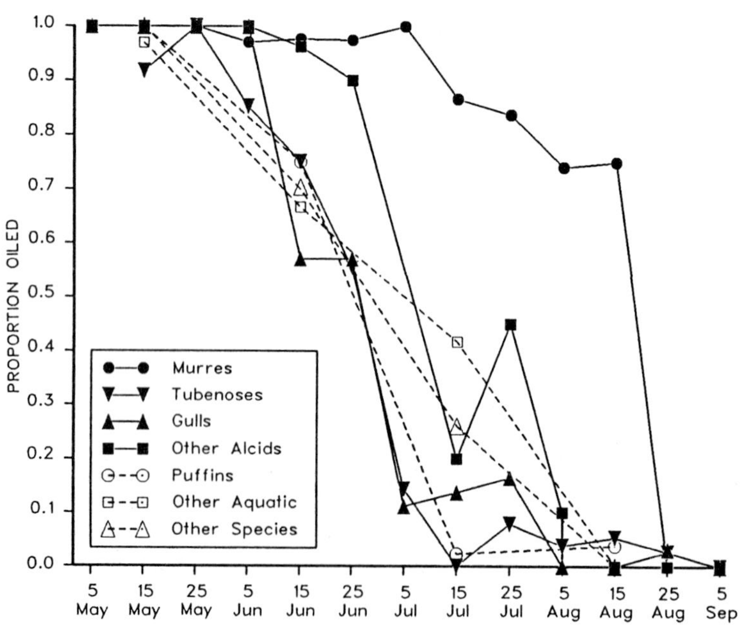

FIGURE 1.—Proportions of seabird carcasses that were recovered from Alaskan beaches in an oiled state, 1989. These data are based on the reexamination of bird carcasses in the freezer vans. Total sample sizes are as follows: murres, 1,530; tubenoses, 790; gulls, 285; other alcids, 178; puffins, 109; other aquatic birds, 77; other species, 69.

TABLE 2.—Numbers and proportions (%) of murres and non-murres collected from beaches in 1989 and used as a best-estimate base for model calculations. Murres include carcasses coded as common murre (*Uria aalge*), thick-billed murre (*Uria lomvia*), or unidentified murre, as well as an allocated proportion of carcasses coded as unidentified alcid or unidentified bird. Non-murres include all other taxa.

	Murres		Non-murres		Total birds	
Morgue	Number	(%)	Number	(%)	Number	(%)
Valdez	437	(13.07%)	2,907	(86.93%)	3,344	(100.0%)
Seward	1,869	(54.75%)	1,545	(45.25%)	3,414	(100.0%)
Homer	3,546	(67.08%)	1,740	(32.92%)	5,286	(100.0%)
Kodiak	15,752	(88.47%)	2,052	(11.53%)	17,804	(100.0%)
Total	21,604	(72.38%)	8,244	(27.62%)	29,848	(100.0%)

by the U.S. Department of the Interior (DOI). Interviews focused on the percent of shoreline searched and intervals between the probable arrival of oiled birds and searches of a given beach. Findings were used to characterize beach search effort within nine geographical sectors (Figure 2; east and west sides of Cook Inlet were combined and the Semidi Islands were combined with the Alaska Peninsula). Each sector was divided into blocks of 15 min of latitude and longitude and, within each block, the shoreline was classified according to likelihood of deposition of oiled birds and accessibility for beach search. Length and type of shoreline was determined using the Environmental Sensitivity Index (ESI) classification (RPI 1983, 1985; M. O. Hayes and C. H. Ruby, Research Planning Institute, unpublished data). A digital database of shoreline types was obtained from the Alaska Department of Natural Resources. These data were projected to latitude and longitude for calculations of total length of shoreline types in each geographical sector (Table 3). The potential for deposition of oiled birds was assumed to be negligible for reflective rocky shoreline (exposed or sheltered rock faces) and ice, and uniform for all other shoreline types. Shoreline types accessible to beach search included sand and gravel beaches, wave-cut platforms, and exposed tidal flats; other classifications were assumed to be inaccessible to search.

Results.—The results of interviews are summarized in Table 4. Estimates of the proportion of total shoreline consisting of walkable, relatively easy-to-search beaches ranged from 85 to 95% in Prince William Sound to 10 to 20% along the Kenai Fjords coast. Most of these beaches were searched at least once for oiled birds. For Prince William Sound, estimates of the total length of coastline searched ranged from 70 to 100% for walkable sandy or cobble beaches and from 50 to 95% for rocky shore. Along the Kenai Fjords, all sandy or cobble beaches were searched, but only about 10% of the rocky shore was searched. At the Barren Islands, all sandy beaches were searched repeatedly; however, rocky shore, which accounts for 87% of the total shoreline of the Barren Islands, was not searched at all.

In most areas, the time of arrival of oiled birds on particular shores was not known precisely. The interval between the arrival of oiled birds and the first bird collections was estimated by many observers as about 6 d. The shortest lag time was 2 d (at the Barren Islands) and the longest was 20 d (for parts of the Kenai Fjords). The beach search interval in Prince William Sound was about 2–4 d, whereas the Kenai Fjords and Katmai National Park were searched about once per week. Most walkable beaches along the rest of the Alaska Peninsula and Kodiak Island were searched less than three times during the entire summer. One exception was Puale Bay, which was searched every 2 d by USFWS personnel.

Effects of Scavenging

Methods.—After the oil spill, many instances of scavenging of beachcast birds were noted by workers on the scene. Observations and comments were recorded in DOI archives. More than 11,000 microfilm frames were reviewed and follow-up interviews were conducted with individuals to whom the records referred. Although scavenging was recognized as a common and widespread occurrence, studies were necessary to quantitate the proportion of beached birds that might have been scavenged beyond recognition or removed from the beach entirely.

Field studies of scavenging were made in Prince William Sound in summer 1990 and along the coast of the Olympic Peninsula, Washington, in winter 1991. Nine radio-tagged murre carcasses released in Prince William Sound were tracked by aircraft or boat to the shoreline. Once beached, they were

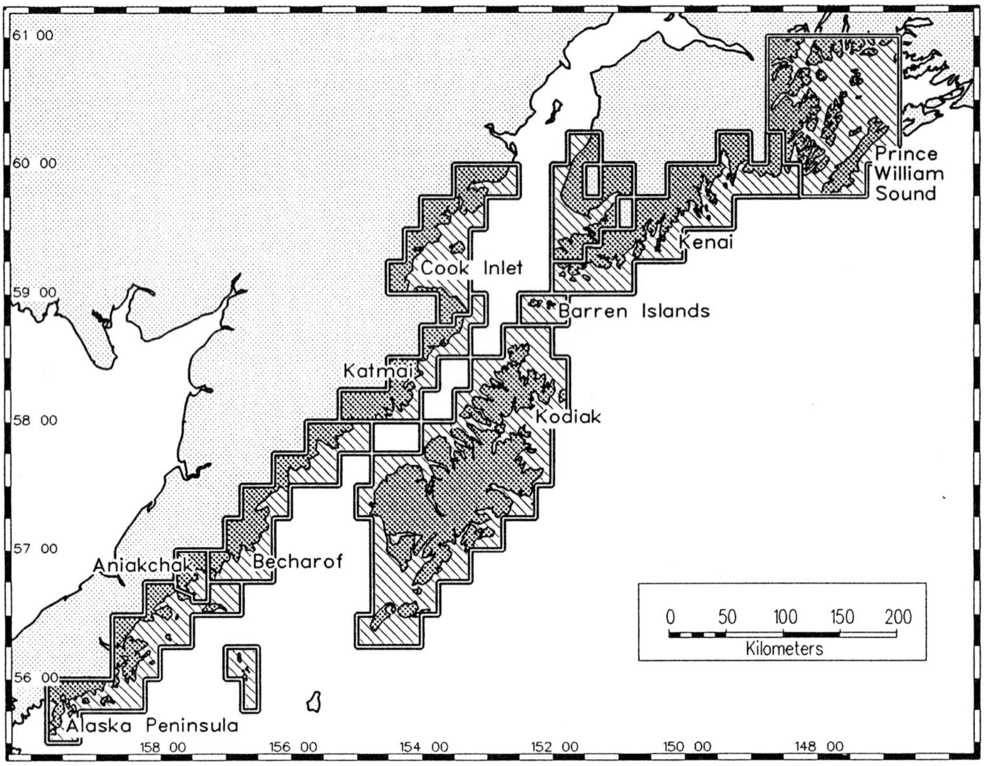

FIGURE 2.—Location of sectors used for segmentation of shoreline-type data.

checked visually every 24 h until they were scavenged or refloated. Carcasses that refloated and were cast up again on another beach were treated as separate trials resulting in 14 instances of beaching for the nine carcasses. Remains of scavenged carcasses were located by radio tracking and inspected. In another study, 23 crested auklets *Aethia cristatella* were placed above the high-tide line at six sites on the shore of Naked Island and monitored each day.

The study conducted along the outer coast of the Olympic Peninsula used three lines of 27 carcasses each (lines A, B, and C; Ford et al. 1991a). Lines A and B were on a gently sloping sand beach protected from surf by offshore rocks and terraces; line C was on a steeply sloping cobble beach subject to greater wave action. Carcasses were tagged with a leg band and placed on top of a cork float. If a carcass was removed by scavengers, only the cork float would remain. If a carcass was rewashed by the surf, both the carcass and the float would be missing. Each line was monitored in the morning and afternoon over a 5-d period. Carcasses were scored as present, removed by scavengers, scavenged in place, or removed by waves.

Results.—In Prince William Sound, radio-tagged murre carcasses disappeared from the beach at a rate of about 50%/d, with no carcass lasting more than 4 d on any beach (Figure 3). Most carcasses were removed from the beach and consumed in the forest margin. In two instances, the scavenger carried away the body but left the legs and radio tags behind on the beach. Of the crested auklet carcasses placed along the shore of Naked Island, 21 of the 23 carcasses were carried off the beach within the first 2 d (Figure 3).

In the Olympic Peninsula study, scavenging proceeded rapidly, with 50% of all carcasses removed after 1.5 d, 1.2 d, and 3.7 d for lines A, B, and C, respectively (Figure 4). Carcasses remained untouched on the beach for a mean of 2.55 d (1.66 SD), 1.08 d (0.65 SD), and 3.00 d (0.80 SD) for lines A, B, and C, respectively. The distribution of first scavenging events over time varied significantly from line to line ($G = 62.63$; df = 8; $P < 0.05$). Line A was untouched for 0.5 d, line B was scavenged

TABLE 3.—Summary of lineal coastal distances (km of shoreline) by sector in various areas of southern Alaska from environmental sensitivity index (ESI) classification (proportions in parentheses). Because of rounding, proportions may not total to 1.

ESI classification	Prince William Sound	Kenai	Cook Inlet	Barren Islands	Katmai	Kodiak	Becharof	Aniakchak	Alaska Peninsula
Fine–medium sand beach	26.45 (0.01)	0.00 (0.00)	24.18 (0.01)	0.00 (0.00)	0.00 (0.00)	5.51 (0.00)	22.18 (0.02)	6.48 (0.04)	17.18 (0.01)
Coarse sand beach	5.54 (0.00)	9.61 (0.00)	17.55 (0.01)	0.00 (0.00)	17.48 (0.02)	2.93 (0.00)	21.79 (0.02)	7.87 (0.05)	49.94 (0.03)
Mixed sand–gravel beach	932.85 (0.25)	288.7 (0.14)	430.5 (0.25)	19.52 (0.18)	237.7 (0.30)	1,298.4 (0.26)	183.6 (0.20)	20.50 (0.14)	203.5 (0.13)
Gravel beach	619.60 (0.17)	46.38 (0.02)	35.53 (0.02)	2.06 (0.02)	7.04 (0.01)	593.6 (0.12)	67.82 (0.07)	15.03 (0.10)	153.7 (0.10)
Exposed rocky shore	409.30 (0.11)	675.64 (0.34)	44.2 (0.03)	64.38 (0.59)	67.24 (0.09)	388.5 (0.08)	82.99 (0.09)	0.14 (0.00)	230.3 (0.14)
Exposed wave-cut platform	276.5 (0.08)	63.59 (0.03)	285.6 (0.17)	7.87 (0.07)	102.4 (0.13)	777.6 (0.15)	252.1 (0.27)	39.63 (0.27)	212.5 (0.13)
Sheltered rocky shore	1,076.3 (0.29)	560.4 (0.28)	288.7 (0.17)	0.89 (0.01)	99.1 (0.13)	1,115.2 (0.22)	23.39 (0.03)	3.07 (0.02)	147.2 (0.09)
Exposed tidal flat	2.64 (0.00)	37.21 (0.02)	188.9 (0.11)	0.00 (0.00)	72.72 (0.09)	65.81 (0.01)	55.96 (0.06)	32.27 (0.22)	325.1 (0.20)
Sheltered tidal flat	123.00 (0.03)	30.01 (0.02)	130.9 (0.08)	0.00 (0.00)	109.1 (0.14)	207.3 (0.04)	75.85 (0.08)	0.28 (0.00)	22.78 (0.01)
Marsh	84.46 (0.02)	39.1 (0.02)	206.0 (0.12)	0.00 (0.00)	33.35 (0.04)	52.29 (0.01)	31.85 (0.03)	7.2 (0.05)	70.61 (0.04)
Ice	22.27 (0.01)	6.68 (0.00)	0.00 (0.00)	0.00 (0.00)	0.00 (0.00)	0.00 (0.00)	0.00 (0.00)	0.00 (0.00)	0.00 (0.00)
Unclassified	89.38 (0.02)	234.4 (0.12)	75.21 (0.04)	14.23 (0.13)	37.99 (0.05)	526.34 (0.10)	109.7 (0.12)	14.88 (0.10)	183.9 (0.11)
Total	3,668.29	1,991.72	1,727.27	108.95	784.12	5,033.48	927.23	147.35	1,616.71

immediately, and line C was untouched for 1.5 d. For line A on a sandy beach, more than 60% of the carcasses were removed or almost completely consumed within 2 d and after 5 d about 77% were gone. For line B, on the same beach, 77% were removed or almost completely consumed after only 1.5 d and more than 95% were eaten or gone after 4.5 d. For line C on the steep cobbled beach, no scavenging occurred in the first 1.5 d; however, once the carcasses were discovered, 95% were removed or eaten within 4 d.

Adding to the generality of our findings, another study conducted during the same time period by A. E. Burger, University of Victoria, at Pachena Beach on Vancouver Island, British Columbia, produced similar results. Four carcasses (33%) were removed or eaten in place after 1 d, eight carcasses were taken within 3 d, and all carcasses were gone within 5 d (Burger 1993).

Carcass Loss at Sea

The length of time spent drifting at sea and the rate of disappearance at sea can be used to generate estimates of the proportion of carcasses enroute to a particular stretch of coastline that are lost before landfall (Ford et al. 1987, 1991a; Page et al. 1990). The greater the distance that a carcass must drift, the greater the chance that it will sink. Carcass recovery rates varying from 10 to 100% have been found in experiments using gull carcasses where trajectories seemed to be directed toward shore and where potential beaching areas were carefully searched (Hope-Jones et al. 1970; Bibby 1981).

Methods.—The rate of loss of oiled birds at sea was investigated by tracking radio-tagged carcasses from an aircraft or boat until they either beached or were lost through sinking. Prior experiments in Oregon waters showed that carcasses that had been frozen and thawed floated lower in the water and sank more rapidly than fresh carcasses (Ford et al. 1991a). Therefore, birds were collected so that fresh carcasses were available for radio-tracking and to ensure that each carcass had a known age and history. Collection was done only in waters well outside the area affected by the oil spill. Most birds were collected near large colonies in the Middleton

TABLE 4.—Summary of beach search effort data derived from microfilm and other documentary materials, unpublished reports, and telephone interviews pertaining to collection of oiled bird carcasses after the *Exxon Valdez* oil spill. Data were unavailable for Cook Inlet (not listed) and for a few other elements (empty data cells).

Section of coast	Time lag of initial search (days)	Sandy or walkable shorelines			Rocky or difficult-to-walk shorelines		
		Percent of shoreline	Percent searched	Mean search interval (days)	Percent of shoreline	Percent searched	Mean search interval (days)
Prince William Sound[a]	6–16[b]	85–95	70–100[c]	3.5[d]	5–15	50–95[c]	3.3[d]
Kenai Fjords area[e]	6–20[f]	10–20	100	7–11[g]	80–90	10[h]	
Barren Islands[i]	2[j]	13	100[j]	7.5[j]	87	0	0
Katmai National Park	8–15[k]	25[l]	70–90[m]	7–9[n]	75[k]	10–20[k]	0–30[k]
Becharof National Wildlife Refuge	7–15[o]	30	90	30–45	70	0–15	Once per summer[o]
Puale Bay	6	75	100	2	25	100	2
Aniakchak National Monument		50	100	6–10[q]	50	100	
Kodiak Island area[r]	6 (Min)		75 (Max)	Variable		25–50[s]	Variable
Alaska Peninsula National Wildlife Refuge area[t]	Midsummer	45	10	Once	55	0	0

[a]All data for Prince William Sound are from a telephone interview with K. Weaverling (VECO bird rescue coordinator in Prince William Sound), January 1991, unless otherwise noted.
[b]Average is closer to 6 than to 16 (E. Snyder-Conn and P. Scannell, U.S. Fish and Wildlife Service [USFWS], U.S. Department of the Interior archived microfilm 469-0583).
[c]Lower values are from a telephone interview with D. Nysewander (USFWS), January, 1991.
[d]Our estimate—1 to 3 times per week through 15 June.
[e]Most data for Kenai Fjords are from a telephone interview with B. Rice (National Park Service [NPS]), January 1991, unless otherwise noted.
[f]Lower values from microfilm 005-1216. Our best estimate is 8–10.
[g]Microfilms 005-1058 to 005-1216 and K. Jettmar (The Wilderness Society) trip report.
[h]Microfilm 003-0004 and 003-0017.
[i]Data for Barren Islands from E. Bailey (USFWS), trip reports.
[j]Actual interval, not estimate.
[k]Telephone interview with J. Wells (NPS), January 1991, and microfilms 001-0892, 001-0898, 001-0905, 001-0912, 001-1065, 001-1448, 003-0547, and 003-1201.
[l]Telephone interview with J. Meldrum (NPS), January 1991.
[m]Telephone interview with J. Wells (NPS), January 1991.
[n]Telephone interview with J. Meldrum (NPS), January 1991, and microfilm 001-1231.
[o]Data for Becharof National Wildlife Refuge, including Puale Bay, are from a telephone interview with R. Hood (USFWS), January 1991.
[p]Most data for Aniakchak National Monument are from a telephone interview with J. Bronson (NPS), January 1991.
[q]Telephone interview with J. Wells (NPS), January 1991, and microfilm 003-0998.
[r]Data for Kodiak Island area are from a telephone interview with J. Bellinger (USFWS), January 1991.
[s]Microfilms 003-0004, 003-0017, and 003-0998. Estimate is based on weather conditions.
[t]Most data for Alaska Peninsula National Wildlife Refuge are from a telephone interview with D. Dewhurst and J. Bellinger (USFWS), January 1991.

Island–Shumagin Island group (91%); the remainder were collected in the Cape Pierce–Bristol Bay–Togiak area. Under authorization of scientific collecting permits PRT-749140 (USFWS) and 90-82 (Alaska), 176 birds were collected. An additional 41 birds were taken under other permits and supplied to the project by the USFWS. All birds used in this study were intact (i.e., with integument and air sacs uncompromised). Any damaged specimens were used for scavenging studies. Carcasses were placed in plastic bags and kept on ice or refrigerated for a maximum of 10 d before being used for radiotracking.

For each carcass, a radio transmitter was embedded in a small polystyrene foam float and attached to the leg by an 8-cm length of stiff brass wire with swivels at both ends. Floats were hemispherical in shape, measuring about 6 cm at the base and 3.5 cm in height and weighted on the flat side with a lead weight to be self-righting. The transmitter (Telonics 1C; Telonics, Inc., Mesa, Arizona) had a mass of about 15 g and a 37-cm antenna oriented vertically. Each float was constructed to have slight positive buoyancy with the transmitter in place. To reduce any "sail" effect, floats extended only about 1 cm out of the water. Each transmitter had a unique frequency between 148 and 150 MHz and a unique pulse rate. Preliminary tests were conducted in Coos Bay, Oregon, to develop a size and shape such that the drift characteristics of carcasses with and without transmitters were indistinguishable (R. G. Ford, unpublished data; see also Ford et al. 1991a). Radio-tagged "dummies" made from styrofoam served as controls. These were designed to mimic the drift characteristics of radio-tagged carcasses. The dummies could be lost only as a result of transmitter failure or movement beyond the range of the tracking aircraft or boat. Thus, any difference

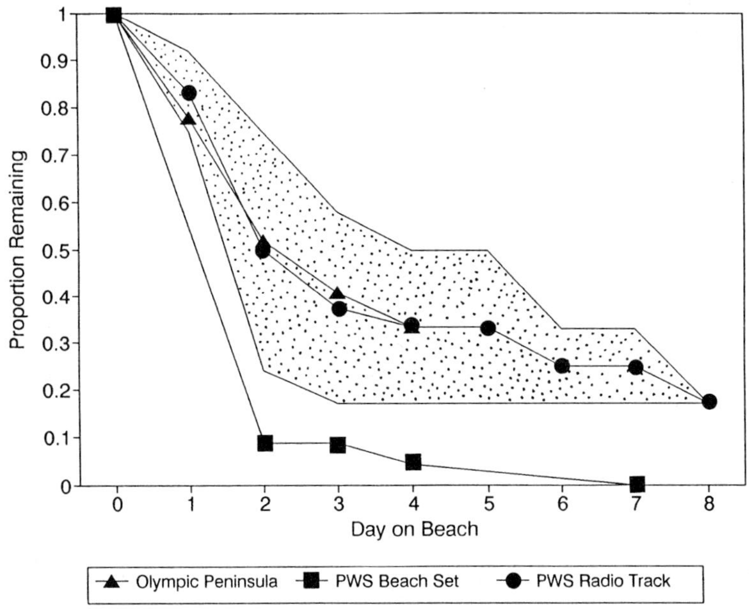

FIGURE 3.—The persistence of beached carcasses based on three separate studies. The Olympic Peninsula (Washington state) and the Prince William Sound (PWS) beach set studies were based on carcasses placed above the high-tide line and reexamined at regular intervals. The PWS radio-tracking study is based on carcasses that were tracked to the beach and reexamined at regular intervals. The stippled region indicates the interval of uncertainty about the exact time of arrival of the carcass on the beach for the radio-tracking study. The Olympic Peninsula study occurred in winter 1991 and the PWS studies occurred in winter 1991.

in rates of loss between carcasses and dummies could be attributed to natural processes rather than to factors associated with our methodology. Two sizes were constructed and weighted to approximate the size and buoyancy of the carcasses they were designed to simulate. Floats measuring 14 cm in length with a 115–170-g lead weight were used to simulate murres, ducks, and cormorants; smaller floats, 12.5 cm in length with a 115-g weight, were used to simulate auklets.

Carcasses and controls were released in eight groups—four in Prince William Sound and four in the Gulf of Alaska off Kodiak Island (Figure 5 and Table 5). Carcasses in each group were oiled to varying degrees with Prudhoe Bay crude oil. Fresh crude oil was used for the first two releases; however, observations of these carcasses from a boat showed that much of this oil washed off within a few days of drifting. For subsequent releases, we used oil that had been treated to simulate the aging of crude oil on open water. The treatment consisted of exposing oil to air and sunlight in galvanized tubs of seawater. For work in Prince William Sound, 1–3 L of oil were used in each of six tubs containing about 30 L of water. Oil used in the third release was weathered in this way for 5 d, whereas oil used in the fourth release was weathered for 10 d. For work in the Gulf of Alaska, the same amounts of oil were used, but they were placed in larger tubs and gently stirred four times each day over a period of 1 week. The larger surface area and stirring was used to accelerate the weathering process. Treated oil increased in viscosity and adhered more readily to the plumage of immersed carcasses. Two degrees of oiling were used, described on the basis of appearance as moderate or heavy. Moderate oiling of crested auklets consisted of immersion until 40–50 mL of oil coated the plumage, whereas heavily oiled crested auklets were repeatedly immersed until approximately 150 mL of oil adhered to the plumage. For common murres, moderate oiling consisted of approximately 150 mL, whereas heavy oiling consisted of 300–400 mL. Heavy oiling for ducks and cormorants consisted of about 400 mL; no ducks or cormorants received moderate oiling.

Radio-tagged carcasses and dummies were tracked in a turbine-powered Grumman Goose amphibious aircraft. The aircraft was equipped with

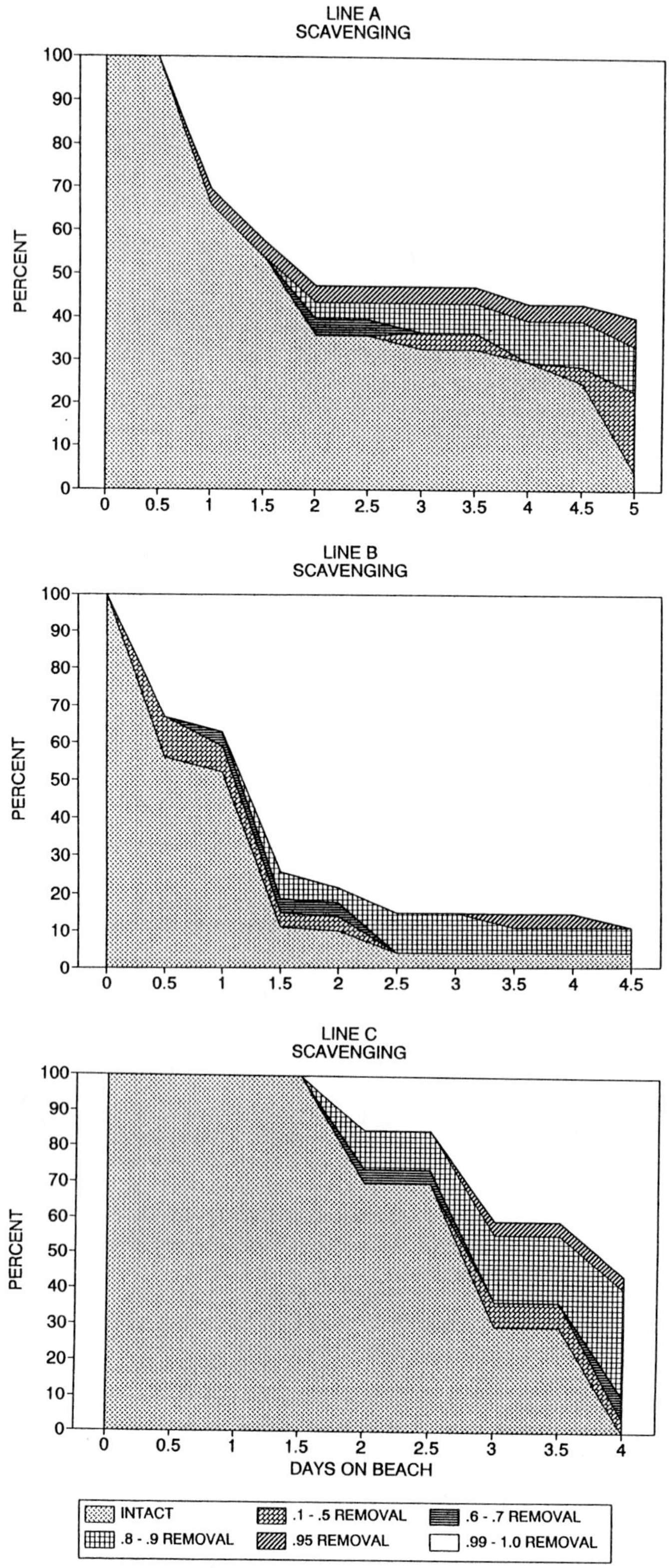
LINE A
SCAVENGING
PERCENT
100
90
80
70
60
50
40
30
20
10
0
0
0.5
1
1.5
2
2.5
3
3.5
4
4.5
5
LINE B
SCAVENGING
PERCENT
0
0.5
1
1.5
2
2.5
3
3.5
4
4.5
LINE C
SCAVENGING
PERCENT
DAYS ON BEACH
INTACT
.1 - .5 REMOVAL
.6 - .7 REMOVAL
.8 - .9 REMOVAL
.95 REMOVAL
.99 - 1.0 REMOVAL

FIGURE 4.—Results of a scavenging study carried out in Olympic National Park in January 1991. Lines A, B, and C represent three groups of 27 bird carcasses placed on three different beaches. The width of each stippled band represents the percentage of carcasses in each state as time progressed. A removal level of 0.99 indicates that only feathers or a fraction of a leg remained. A removal level of 1.0 indicates that no trace of the carcass remained on the beach.

high-sensitivity three-element Yagi antennae mounted near the distal end of the wings and low-sensitivity omnidirectional antennae 15 cm in length mounted beneath the wings near the fuselage. High-sensitivity antennae had a useful listening range of about 37 km at 150-m altitude and were used to provide a rapid inventory of transmitters and to determine their approximate locations. Low-sensitivity antennae had a range of about 13 km and were used to determine the location more precisely. The aircraft fuselage effectively shielded one low-sensitivity antenna such that only the antenna nearest the transmitter received a strong signal. Input from the antenna receiving the strongest signal was selected using a Telonics TAC-2-RLB switch box. Amplitude increased as the aircraft approached a transmitter and decreased as the aircraft passed the transmitter. Peak amplitude was reached when the bearing to the transmitter was at a right angle to the aircraft's heading. The bearing to the transmitter could therefore be calculated from the aircraft's

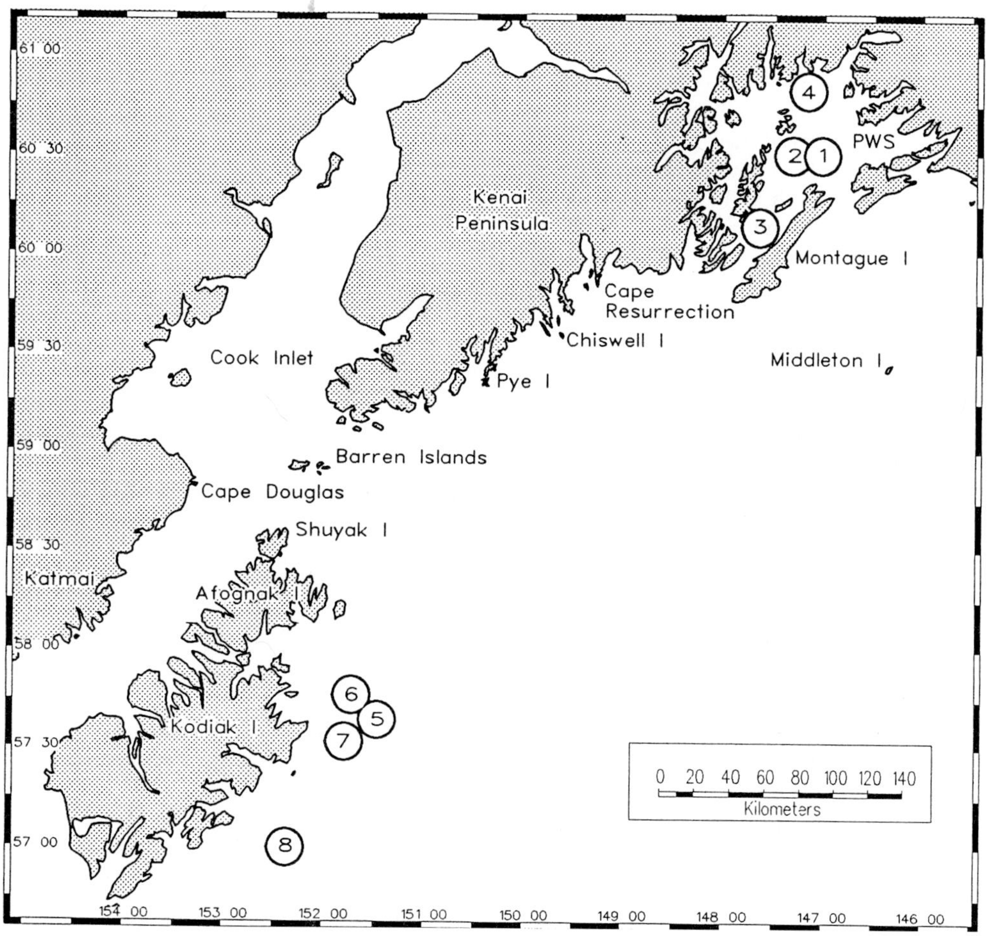

FIGURE 5.—Release sites for the carcass drift studies in Prince William Sound (PWS) and near Kodiak Island. Releases 1–4 were made between 25 May and 17 June 1990; Releases 5–8 were made between 18 July and 12 August 1990.

TABLE 5.—Releases of bird carcasses and dummies for at-sea loss rate studies, spring–summer 1990. Release sites 1–4 are in Prince William Sound and 5–8 are in the Gulf of Alaska off Kodiak Island.

Release number	Date	Number of carcasses	Bird species[a]	Number of dummies	Location of release
1	25 May	20	CRAU	5	60°30.29′N, 147°14.60′W
2	28 May	16	CRAU	4	60°30.04′N, 147°19.30′W
3	12 Jun	20	COMU	5	60°08.30′N, 147°39.20′W
4	17 Jun	9	WWSC	4	60°09.50′N, 147°10.50′W
		1	SUSC		
		5	HARL		
		3	PECO		
		2	DCCO		
5	18 Jul	18	COMU	9	57°39.80′N, 151°28.00′W
		9	CRAU		
6	24 Jul	18	COMU	9	57°46.20′N, 151°43.10′W
		9	CRAU		
7	6 Aug	18	COMU	9	57°32.00′N, 151°47.50′W
		9	CRAU		
8	12 Aug	18	COMU	9	57°00.00′N, 151°22.00′W
		9	CRAU		
Total		184		54	

[a]CRAU = creasted auklet *Aethia cristatella*; COMU = common murre *Uria aalge*; WWSC = white-winged scoter *Melanitta fusca*; SUSC = surf scoter *M. perspicillata*; HARL = harlequin duck *Histrionicus histrionicus*; PECO = pelagic cormorant *Phalacrocorax pelagicus*; DCCO = double-creasted cormorant *P. auritus*.

heading at the moment of peak amplitude. Receiving equipment consisted of a Telonics TR-2 receiver with a band width of 148–150 MHz, a TS-1 scanner–programmer, and a TDP-2 digital data processor. Output from the signal processor and from the aircraft's LORAN was provided directly to a laptop computer through an analog–digital interface. Records were logged automatically every 10 s that included the aircraft position (latitude and longitude) and heading and the signal amplitude. Signal amplitude was also displayed graphically on the computer screen for up to five transmitters at once, allowing the computer operator to log the aircraft's position and heading at the moment of peak amplitude.

The approximate distance from the aircraft to the transmitter at closest approach was estimated from a regression of distance against signal amplitude. Because the aircraft landed on the water to release a radio-tagged float, the initial position was known from LORAN-C and VLF-Omega navigation computers. The data for the regression analysis were obtained by flying past a floating transmitter along parallel track lines of increasing right-angle distance. Gain of the receiver and position of antennae were set and never changed during initial tests and all subsequent tracking work. From the regression analysis, tables were made of distance as a function of amplitude for two aircraft altitudes, 150 and 460 m above sea level (ASL). Close correspondence was found between the actual distances (known from LORAN fixes) and estimates obtained from the regression relationship (Table 6).

Reacquisition of each transmitter was attempted on each flight day and its position recorded. The last-recorded location was used to define the approximate area in which to begin searching for a carcass or dummy on the next flight day; however, the primary purpose in locating transmitters was to determine whether they were still afloat or had come ashore. If the transmitter was near land, the precise position was determined by boxing-in the signal at 150 m ASL using the low-gain antennae until the aircraft was directly over the transmitter. To determine if a transmitter had been beached, the aircraft was flown along the shoreline at 150 m ASL using the low-sensitivity antennae. If the signal was received only by the low-sensitivity antenna on the land side of the aircraft, the carcass or dummy

TABLE 6.—Agreement of the estimated distance derived from signal amplitude with the actual distance from calibration trials run in May 1990. The measure of the fit of regression equation to data is expressed as a coefficient of determination (r^2; N = number of trials).

Altitude above sea level	Aircraft antennae: Low sensitivity		Aircraft antennae: High sensitivity	
	r^2	N	r^2	N
150 m	0.91	42	0.93	56
460 m	0.86	56	0.91	68

was assumed to have been beached. In Prince William Sound, a team working from a boat was advised by radio-telephone of the probable location of each transmitter on or near land. Using a receiver (TRX1000S; Wildlife Materials, Inc., Carbondale, Illinois) and a four-element Yagi or two-element "H" antenna, the transmitter was located and, if beached, recovered.

Results.—From 25 May through 31 August 1990, 40 radio-tracking flights were conducted requiring about 175 h in the air. Twenty flights were conducted over Prince William Sound from mid-May to early July and 20 flights were conducted over the Gulf of Alaska from mid-July through the end of August. For 94 carcasses and controls released in Prince William Sound, more than 1,000 position fixes were recorded. Figure 6 shows the distribution of fixes of transmitters from one release off Glacier Island in northern Prince William Sound. In the Gulf of Alaska, more than 900 position fixes were recorded for 108 carcasses and controls released east of Kodiak Island. No radio-tagged controls, either in Prince William Sound or the Gulf of Alaska, were lost during the study. In contrast, radio-tagged carcasses were lost steadily in the days after release.

Median persistence time for each release was calculated from a data set that, for a given day, included only those carcasses that were still floating or had been beached after the last carcass sank. Median persistence time for releases in Prince William Sound ranged from 15 to 20 d, whereas that of releases in the Gulf of Alaska ranged from 7 to 18 d (Table 7). The generally shorter median persistence time of releases in the Gulf of Alaska probably resulted from higher winds and heavier sea conditions. Data collected from the Prince William Sound releases are shown in Figures 7 and 8. The rules used to assign a fate to carcasses between fixes are presented in Table 8. Data collected in the Gulf of Alaska are shown in Figures 9 and 10.

The statistical treatment of data collected in Prince William Sound presented difficulties because the beaching rate was so great—it was caused by the presence of many islands, coves, and points of land in Prince William Sound that snared drifting carcasses and controls. Of the 76 carcasses released in the sound, only 14 were known to have sunk. Most of the remainder were beached, but a few from later releases were still floating at the end of this phase of the fieldwork. For analysis of persistence times, a nonparametric technique was used that takes censored observations (carcasses that were beached or still floating on the last sampling date) into account (Gehan's test; Gross and Clark 1975). Because of the small number of carcasses that sank, the third and fourth releases were combined for this analysis. Three pairwise tests were conducted (the alpha was adjusted to preserve an experiment-wise type I rejection rate of 0.05; Sokal and Rohlf 1981, pp. 241–242). No significant differences were found in the persistence time in any comparison of releases in Prince William Sound (all $P > 0.40$). The same test was used to examine persistence times for releases in Prince William Sound relative to releases in the Gulf of Alaska. Carcasses released in the calm waters of the sound persisted significantly longer relative to those released in the Gulf of Alaska for all releases combined ($P < 0.0001$) and for each of the first three gulf releases ($P < 0.0001$). The persistence rates were not significantly different for Prince William Sound carcasses and those in the last release in the Gulf of Alaska ($P < 0.09$), probably because of the relatively mild sea conditions in the gulf at the time of the last release.

The effect of oiling on carcass persistence time was examined using only data collected in the Gulf of Alaska. Each of the four releases consisted of 18 common murres and 9 crested auklets; six murres and three auklets were non-oiled, six murres and three auklets were moderately oiled, and six murres and three auklets were heavily oiled. Examination of data showed a substantial degree of non-normality ($P < 0.001$, Kolmogorov–Smirnov test); also, data were heteroscedastic ($P = 0.004$, Bartlett's test). In a two-way analysis of variance (ANOVA), ranks of persistence time were used, a technique known to give robust results when the assumptions of ANOVA are not met (Conover 1980). Some of these releases had significantly longer persistence time than others (Tukey, multiple comparison test with alpha = 0.05), but degree of oiling had no significant effect ($P > 0.5$). Similarly, a two-way ANOVA showed no significant species difference in persistence time for the common murres and the crested auklets ($P > 0.9$).

Time Course of Oiling of Birds at Sea

Methods.—To describe the time course of oiling of birds at sea, a description of the path of movement and areal extent of the oil slick was needed. Overlaid with the distribution of birds, the path of the slick to a particular beach could then be used to identify locations where concentrations of birds at sea were most likely contacted by oil. Assuming that birds drifted passively with oil from the moment of initial contact, the duration of drift before landfall

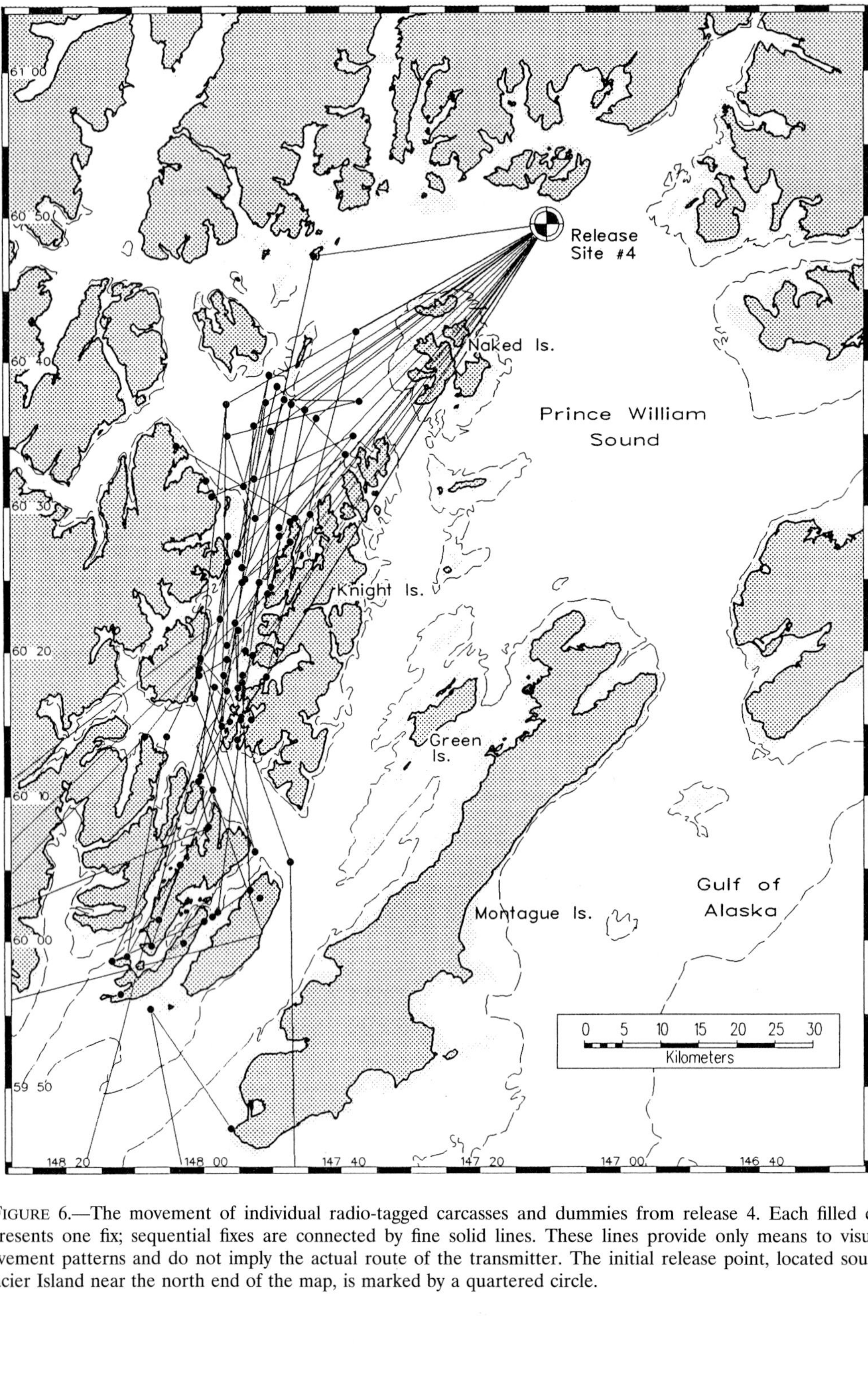

FIGURE 6.—The movement of individual radio-tagged carcasses and dummies from release 4. Each filled circle represents one fix; sequential fixes are connected by fine solid lines. These lines provide only means to visualize movement patterns and do not imply the actual route of the transmitter. The initial release point, located south of Glacier Island near the north end of the map, is marked by a quartered circle.

TABLE 7.—Median persistence time (rounded to nearest day) of radio-tagged carcasses in Prince William Sound and the Gulf of Alaska (includes carcasses that refloated after beachings of fewer than 5 d).

Release number	Duration of tracking period (days)	Sample size[a]	Median persistence time (days)[b]
	Prince William Sound		
1	32	12 (20)	20
2	38	6 (16)	18
3	21	3 (20)	15–18
4	16	4 (20)	15–18
	Gulf of Alaska		
5	20	27	11
6	14	27	7
7	17	27	9
8	19	27	18

[a]Numbers in parentheses indicate number of carcasses actually released. In Prince William Sound, many carcasses were beached.
[b]For calculation of median persistence time, releases 3 and 4 were combined because of the small sample sizes. Sample size is defined as the number of carcasses that were floating or had sunk as of the day of the last carcass sinking.

at a particular beach could be estimated and thus the proportion of the birds remaining afloat. The assumption that birds drift passively from the point of initial contact does not allow for the actual response of individual birds as they encountered oil. Depending on the degree of oiling, some birds may swim away from land for a period of time before dying. In the absence of data, the lag in time between contact with oil and death could not be estimated; however, given that these birds eventually died from oiling, this uncertainty would not be expected to have a pronounced effect on model results. The assumption that oiled birds drift passively with oil is generally supported by the data of Hope-Jones et al. (1970), as well as by observations during past spills that oil and birds tended to come ashore together (e.g., Burger 1993).

TABLE 8.—Assignment of the fates of carcasses for Figures 7 and 8, releases 1–4 in Prince William Sound. Letters in the body of this fate-transition matrix give the fates of carcasses on day between sampling dates (t and $t + n$) or after the last sampling date for that carcass (last column). Fates were floating (F), sunk (S), ashore (A), removed (R; retrieved or scavenged), and indeterminate (I); NP is not permitted. Categories A and R were combined in Figures 7 and 8.

Fate on day t	Fate on day $t + n$: F	S	A	R	Fate unknown[a]
F	F	I	I	I	I
S	NP	S	NP	NP	S
A	I	NP	A, I[b]	A	A, I[b]
R	NP	NP	NP	R	R

[a]This column gives the fates when t was the last day on which data were obtained for the given carcass.
[b]Carcasses that had been ashore more than 3 d before day t were classified as ashore, and indeterminate otherwise.

Output of the National Oceanic and Atmospheric Administration (NOAA) HAZMAT oil spill trajectory model was used to describe the movement of oil (Torgrimson 1981). This model tracks the position of many moving points (Lagrangian elements) that represent the distribution of quantities of oil at successive time intervals. The NOAA hindcast of the spill extended only through the end of April as oil was moving through the Gulf of Alaska into Shelikof Strait and along the eastern side of Kodiak Island. After April, data from NOAA overflights were used to describe the time course of oil movement in waters of the western Gulf of Alaska.

The distributions of birds in waters affected by the spill were mapped from data collected on Outer Continental Shelf Environmental Assessment Program (OCSEAP) and USFWS surveys. Data were obtained from the National Oceanographic Data Center, corrected for effort, and mapped as density distributions for three groups of birds: (1) murres, (2) small alcids (auklets and murrelets), and (3) sea ducks, cormorants, loons, and grebes (combined). These groups are believed to be especially vulnerable to oiling because they spend much time on the water (King and Sanger 1979; Speich et al. 1991). In past spills, they were the most common birds found on the beach (e.g., Page et al. 1990). Density of each group was estimated for blocks of 30 min of latitude and longitude for the periods of April–May and June–July (example distributions of murres are shown in Figure 11).

Calculation of relative numbers of birds contacted by oil in April was made from the intersection of the path of each Lagrangian element from the NOAA HAZMAT model with the OCSEAP distribution of birds. This analysis was extended into the Gulf of Alaska as far as the Shelikof Strait and Kodiak Island. In the absence of NOAA HAZMAT model results for spill movement from 1 May onward, relative numbers of birds contacted were calculated assuming an even distribution of Lagrangian elements in waters where NOAA overflights indicated that oil persisted. An even distribution of birds was also assumed, but at different mean densities, for the waters north of Kodiak Island, in Shelikof Strait, off the eastern side of Kodiak Island, and off the Alaska Peninsula. In computations, these simplifying assumptions had the

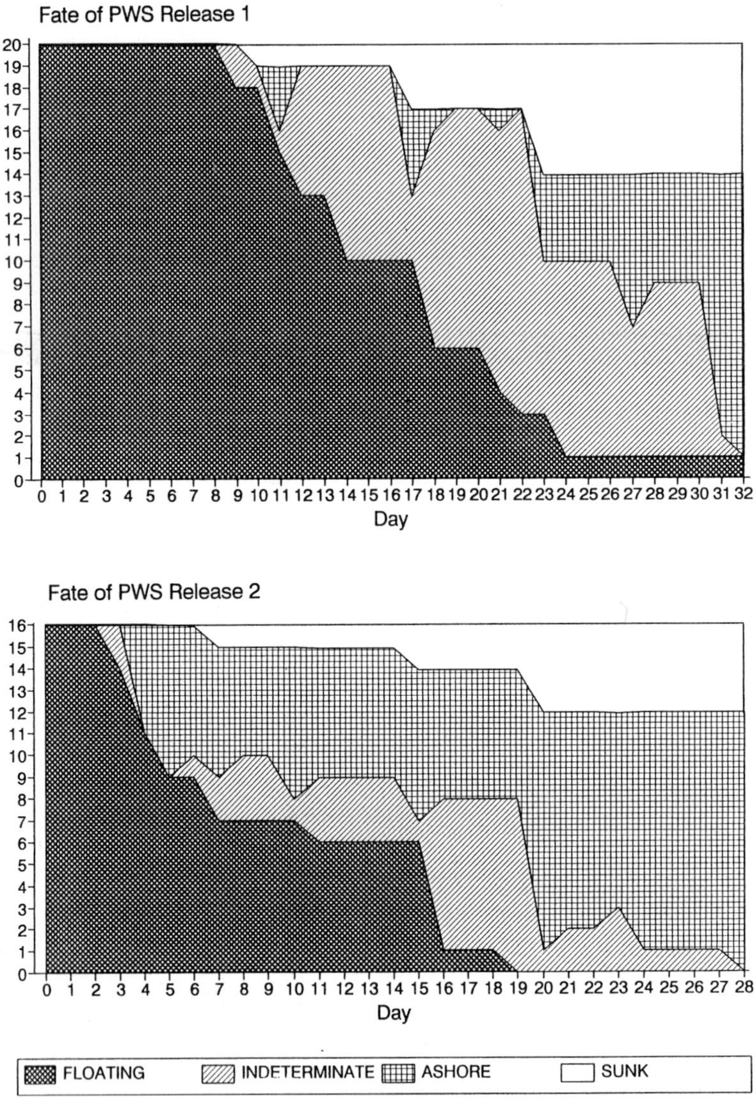

FIGURE 7.—Fate of carcasses in Prince William Sound (PWS), releases 1 and 2. The vertical height of each band indicates the number of carcasses in each state. The indeterminate state applies to carcasses between dates on which their status was known to have changed (Table 8). Fates of dummies were similar to those of carcasses, except that no dummies ever sank. Flights were conducted for data collection from mid-May to early July 1990.

same result as application of a constant rate function of persistence at sea for each beach where carcasses were collected.

Results.—From results of the trajectory model, oil in Prince William Sound was carried by an anticlockwise current toward the Knight Island Passage and Montague Strait, contacting the shoreline principally on the western side of the sound. By 29 March, oil entered the Gulf of Alaska and began moving southwest past Resurrection Bay and along the Kenai Peninsula. By 13 April, oil had moved past the Kenai Peninsula to the vicinity of the Barren Islands, and as far south as the northern end of Shuyak Island. The main mass of the slick remained near the Barren Islands for several days before moving into Cook Inlet and the Shelikof Strait (Figure 12). A portion of the oil moved southward during late April, passing along the eastern side of Kodiak Island. Relative numbers of birds contacted in these waters was calculated from the intersection

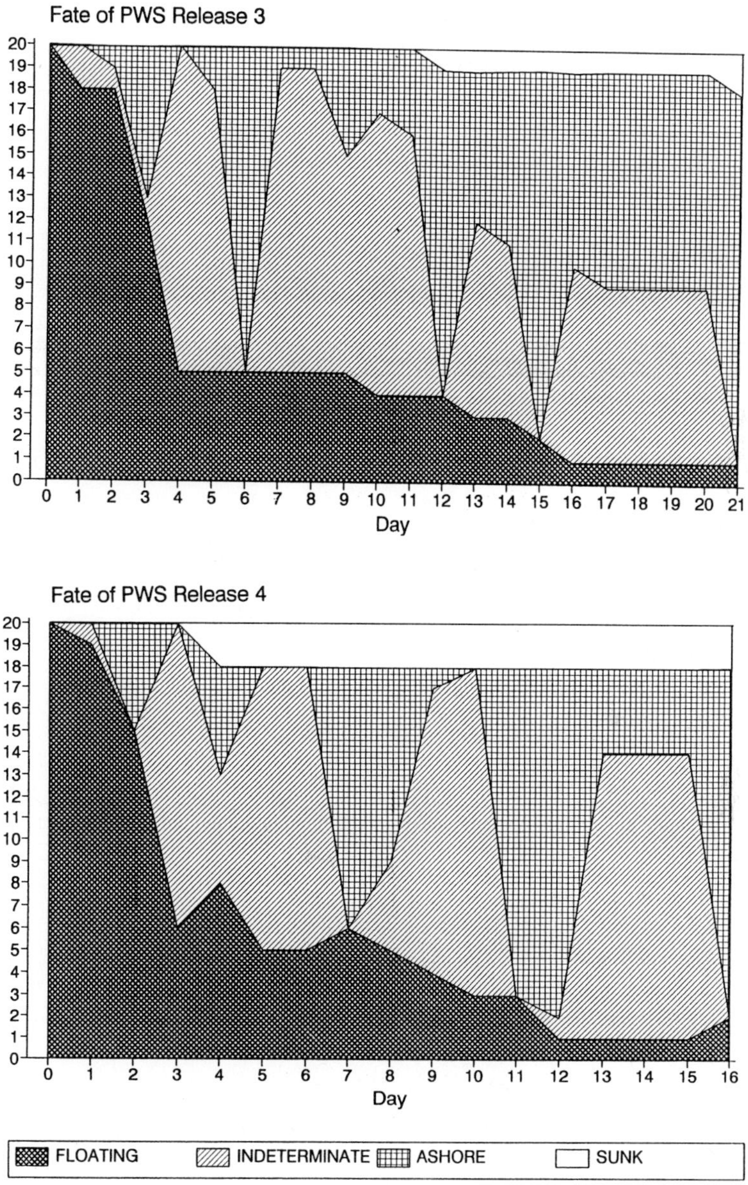

FIGURE 8.—Fate of carcasses in Prince William Sound (PWS), releases 3 and 4. The vertical height of each band indicates the number of carcasses in each state. The indeterminate state applies to carcasses between dates on which their status was known to have changed (Table 8). Fates of dummies were similar to those of carcasses, except that no dummies ever sank. Flights were conducted for data collection from mid-May to early July 1990.

of Lagrangian elements and the OCSEAP distribution of birds.

Observations made during NOAA overflights indicate that oil moved southwest through Shelikof Strait at a rate of about 10 km/d, reaching Sutwik Island on the Alaska Peninsula on 6 May. Oil drifted more slowly along the eastern side of Afognak and Kodiak islands. Oil movement became increasingly difficult to describe from overflights as it drifted southward; however, oiled birds were beached on Kodiak Island and on the Alaska Peninsula to as far as Perryville, about 200 km south of Kodiak Island, into September. Results of the carcass persistence study suggest that the chance of a

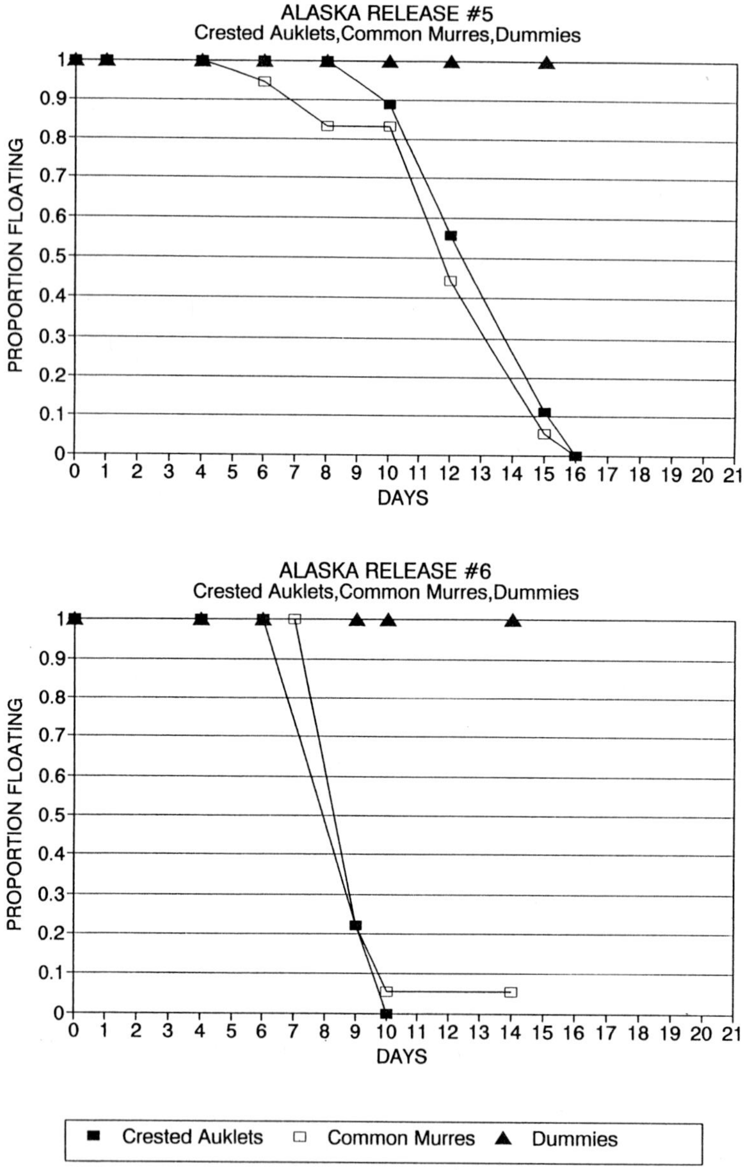

FIGURE 9.—The proportion of carcasses and dummies that were still floating as a function of the length of time since their release for Alaska releases 5 and 6. Data points indicate days on which flights were made to check for transmitter signals. Flights were conducted for data collection from mid-July to late-August 1990.

carcass drifting more than 3 weeks in these waters was very low. Thus, it is likely that slicks were still killing birds off Kodiak Island and the Alaska Peninsula at least into late July or early August.

In an event of such long duration as the *Exxon Valdez* spill, the distribution and mean density of birds in the affected waters may be assumed to change with time (see Ford et al. 1987; Carter 1989; Page et al. 1990). Quantitative data on the appropriate scale were not available to define the changes in distribution in such a way that they could be modeled as a time-varying function; however, the overall effect of assuming a static distribution was examined by comparing results of the model using the patchy distribution from OCSEAP data with that obtained using an even distribution. If high-density patches were to occur predominantly close to land, then use of an even-distributional model

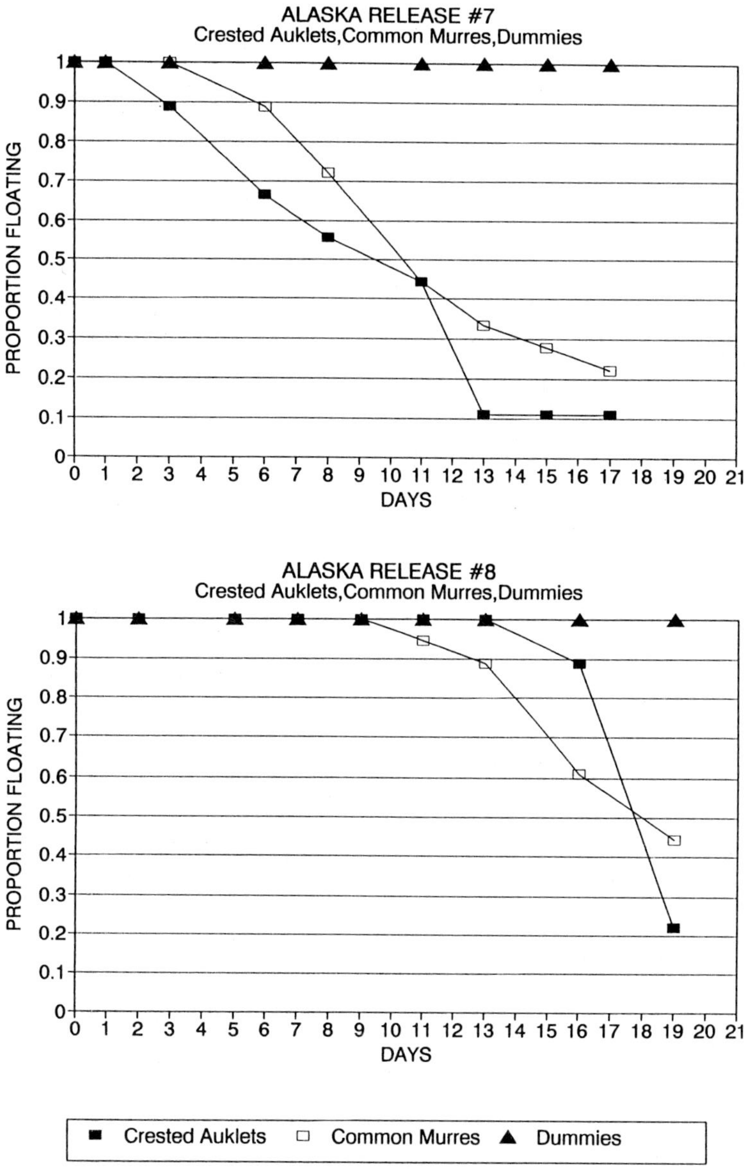

FIGURE 10.—The proportion of carcasses and dummies that were still floating as a function of the length of time since their release for Alaska releases 7 and 8. Data points indicate days on which flights were made to check for transmitter signals. Flights were conducted for data collection from mid-July to late-August 1990.

would overestimate mortality because birds recovered on the beach would have been contacted only a short time before the oil came ashore. On the other hand, if concentrations of birds were to occur farther from shore and were contacted earlier, fewer would reach shore with the oil and use of an even-distributional model would underestimate mortality. To construct an even-distributional model, it was assumed that bird density in the Shelikof Strait area was constant and equal to the densities recorded by J. F. Piatt (U.S. Fish and Wildlife Service, unpublished data) and that bird densities east of Kodiak Island were constant based on the OCSEAP data. The effect of using an even-distributional model rather than a patchy one did not result in a clear directional bias in the final

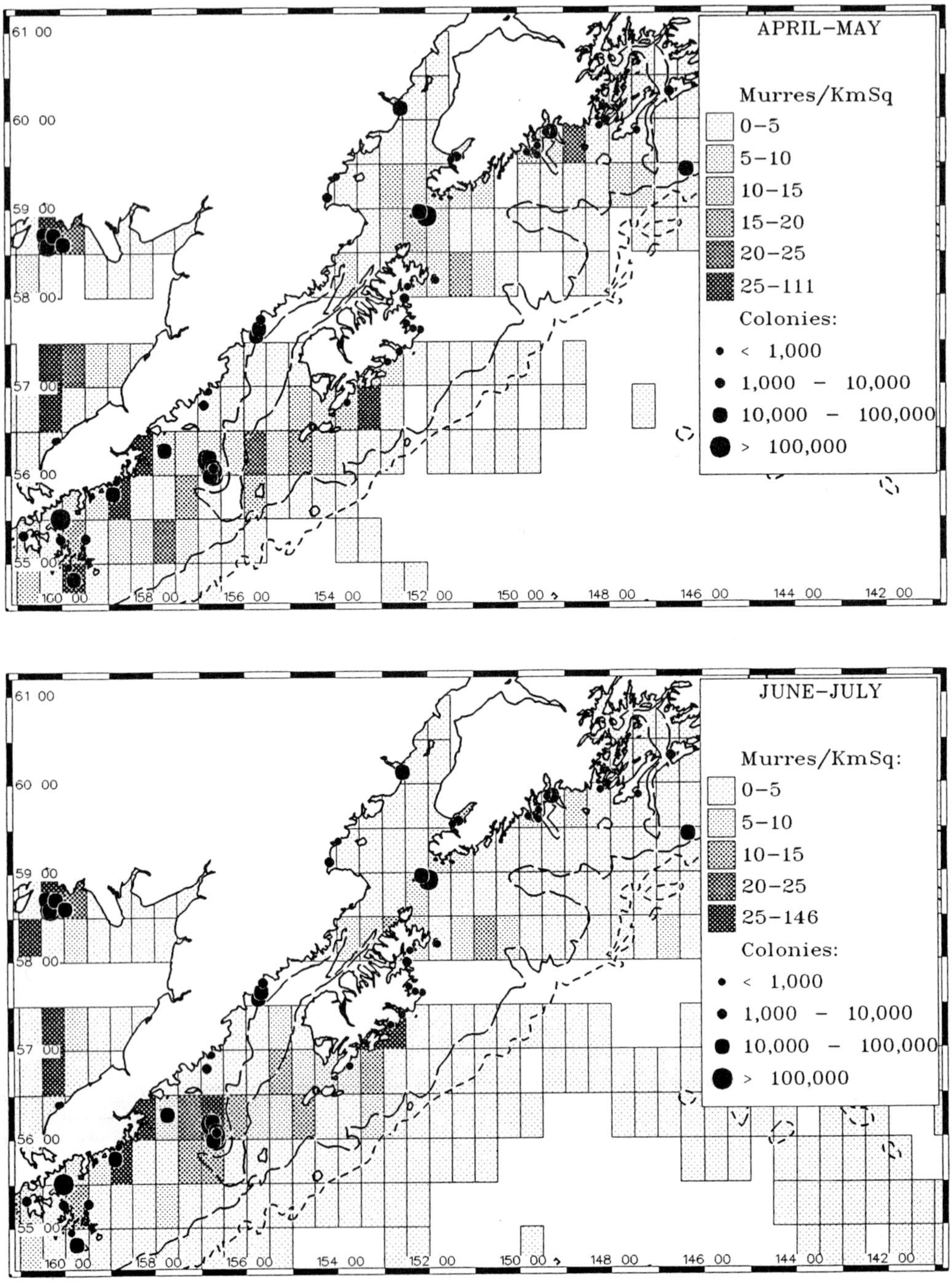

FIGURE 11.—Distribution of all murres during April–May (top) and June–July (bottom) 1989 based on the Outer Continental Shelf Environmental Assessment Program database. Data are summarized into 30-min blocks for display and analysis. Colonies are displayed as filled circles; the size of the circle indicates the size of a colony. Colony data are from Sowls et al. (1978), supplemented by more recent counts where available.

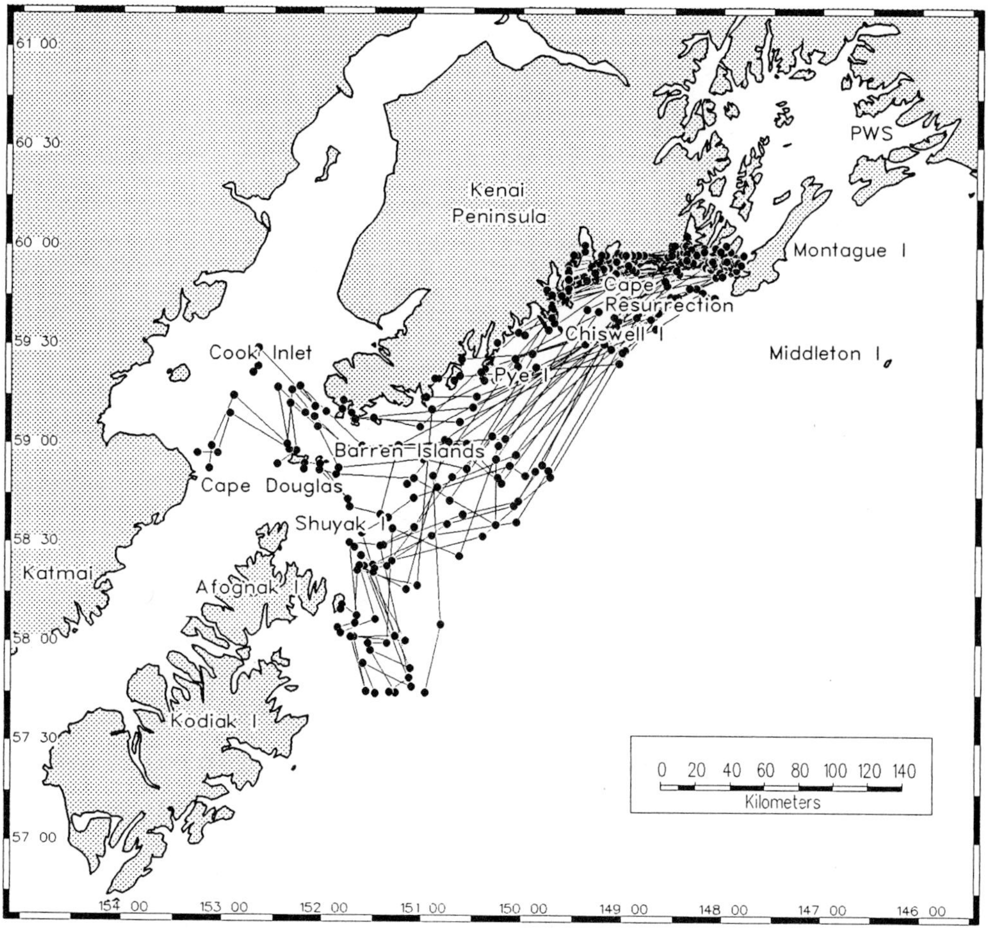

FIGURE 12.—The movement of Lagrangian elements simulated by the National Oceanic and Atmospheric Administration HAZMAT on-scene spill model for April 1989. The movement of these elements provides a way of simulating where and when seabirds would have encountered the oil slick and how long carcasses would have drifted. For clarity, this map contains a randomly selected subset of the Lagrangian elements simulated by the model. Lines connect the positions of each element at 3-d intervals; the model actually resolves down to 3-h intervals. Dots indicate the position at each 3-d interval. The abbreviation is PWS = Prince William Sound.

estimate of mortality. The number of murres lost at sea estimated using a patchy distributional model generated from OCSEAP data for April and May was about 6% lower than the estimate produced using an even-distributional model. Numbers of auklets and murrelets lost at sea was about 2% less. The at-sea loss estimates can therefore be assumed to be robust to the density-distributional model used, and the assumption of an even-distributional pattern of birds in the Shelikof Strait and waters off Kodiak Island was an acceptable alternative to the OCSEAP-based distributional model.

Structure of the Model and Use of Data

Estimation of Model Parameters

Beach search effort.—It was assumed that (1) carcasses were deposited with equal likelihood on any nonreflective substrate, (2) all carcasses on depositable but unsearchable beaches were missed, and (3) a fraction of those on searchable beaches were missed because of the time lag between carcass arrival on a beach and search of that beach. The following method was used to estimate the number of birds not enumerated because of incomplete search effort: Let k be 1 of 12 ESI shoreline types, L_k be the length of coastline within a 15-min latitude and longitude cell, R_k be a variable that assumes the value of 0 or 1 depending on whether the substrate was reflective or depositable (i.e., reflective beaches did not retain carcasses and were not searched; see Carter and Page 1989), Pw_k be the proportion of shoreline type k that was accessible, and Ps_k be the proportion of accessible shoreline type k that actually was searched. Then the proportion of the depositable fraction of the entire cell that was searched, P^*, is

$$P^* = \frac{\sum_{k=1}^{12} (R_k Pw_k Ps_k L_k)}{\sum_{k=1}^{12} (R_k L_k)} . \quad (1)$$

If Cb^* is the number of carcasses deposited on searched beaches between searches, those carcasses represent a subset of the larger number of carcasses that came ashore on all depositable beaches, Cs. It was assumed that some of these carcasses were not found because they were beached in areas that were never searched and that the likelihood of being beached within a sector was independent of the substrate except for reflective shoreline types. Then,

$$Cb^* = P^* Cs, \quad (2)$$

and solving for Cs;

$$Cs = \frac{Cb^*}{P^*} . \quad (3)$$

Scavenging of carcasses.—The rate of carcass loss attributable to scavenging was estimated as follows: let Cb_i represent the number of carcasses arriving on a given beach on day i, m be the number of days between searches of that beach, Cr be the number of carcasses recovered on day m, and Ss_j be the likelihood that a carcass will survive j days on the beach. Then, the number of carcasses recovered on day m will be

$$Cr = \sum_{i=1}^{m} (Cb_i \, Ss_{(m-i)}). \quad (4)$$

It was not possible to estimate precisely how many carcasses arrived on a given day on a given beach because many beaches were visited only sporadically. Depending on wind and tide, the carcasses might have been deposited in a single clump at any time during the m days between searches, or they might have been deposited in approximately equal numbers each day; however, if it is assumed that the deposition rate is constant, the expected value of Cr, when computed over many beaches, can be shown to be the same as if the number arriving on a given day was a random variable (Ford et al. 1991b, Appendix D).

Let Cb^* be the sum of the Cb_i (i.e., all carcasses beached between day 1 and day m), and assume that the rate of deposition is constant. Then,

$$Cr = \sum_{i=1}^{m} \left(\frac{Cb^*}{m}\right) (Ss_{(m-i)}), \quad (5)$$

and, solving for Cb^*;

$$Cb^* = \frac{mCr}{\sum_{i=1}^{m} (Ss_{(m-i)})} . \quad (6)$$

This holds as long as the number of carcasses deposited on day i, Cb_i, is independent of Ss_i, in which case the expectation of Cb^* is equal to the expectation of the sum of the Cb_i.

Persistence of drifting carcasses.—To estimate the persistence of floating carcasses (i.e., the time to sinking), a two-parameter survivorship curve based on the Weibull distribution was used:

$$S(t) = e^{-\lambda t^{\gamma}} . \quad (7)$$

In this equation, λ is a scale parameter, γ is a shape parameter, and t is the time of sinking. This function was chosen because it has a reputation of fitting survival data well (Gross and Clark 1975), is easily linearized, and provided a good fit to the carcass survival data.

Data were fit to the linearized form of the survivorship function, which is obtained by taking the natural log twice:

$$\log_e \{\log_e [1/S(t)]\} = \log_e (\lambda) + \gamma \log_e (t). \quad (8)$$

The fit was achieved using standard least-squares fitting techniques (i.e., linear regression), which gave the estimates λ and γ. This was achieved by regressing $\log_e \{\log_e [1/S(t)]\}$ on $\log_e(t)$, where the $S(t)$s were estimated from the proportions of carcasses surviving at time t, and the t values were the estimated times of sinking. Estimated times of sinking were obtained by linear interpolation between the last time the carcass was known to be floating and the first sampling time when it was determined to have sunk. The regression coefficients are related

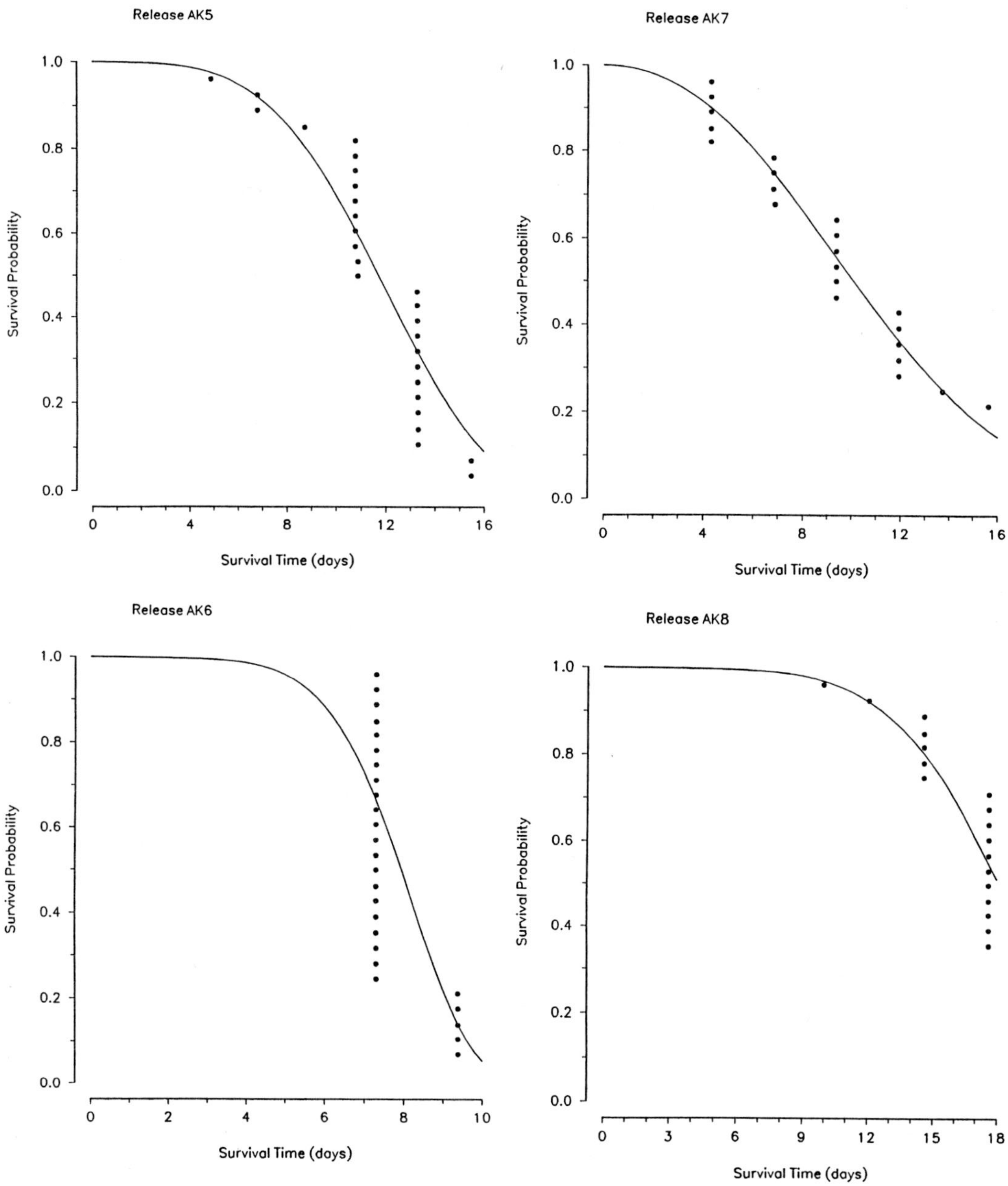

FIGURE 13.—Fit of the Weibull distribution to the carcass loss data from releases 5–8 in the Gulf of Alaska.

to the parameters as follows: $B_0 = \log_e(\lambda)$ and $b_1 = \gamma$. The fit of the Weibull distribution to the data for releases AK5–AK8 is shown in Figure 13.

At-sea loss of drifting birds.—Using output from the NOAA HAZMAT oil spill trajectory model and estimates of the spatial distribution of birds, the location was determined where birds might have been contacted by oil if they were beached along a particular portion of the coast. The at-sea loss model was constructed as follows: Let B_i be the density of birds at the position of a Lagrangian element on day i. Let r be a proportionality constant relating the density of birds, B_i, to the number of birds killed by that oil on day i, D_i. Then,

$$D_i = r\,B_i\,. \tag{9}$$

If a Lagrangian element is released on day i and reaches land on day n, the total mortality, T, attributed to that element is

$$T = r \sum_{i-1}^{n} B_i \,. \tag{10}$$

If some of the carcasses have been lost at sea, the number of carcasses beached will be less than T. To compute how many carcasses were lost by day j, let S_j be the probability that the carcass of a bird contacted j days ago will still be afloat (see equation 7), and let Cb_n be the number of carcasses beached on day n. Then,

$$Cb_n = \sum_{i=1}^{n} r\, B_i\, S_{(n-i)} \,. \tag{11}$$

The essence of the injury assessment problem relating to beachcast birds is to compute T, given Cb, B, S, and n. The ratio of recovered carcasses on day n, Cb_n, to total mortality, T, can be represented as

$$\frac{Cb_n}{T} = \frac{r \sum_{i=1}^{n} (B_i\, S_{(n-i)})}{r \sum_{i=1}^{n} B_i} \,, \tag{12}$$

and, solving for T,

$$T = Cb_n \frac{\sum_{i=1}^{n} B_i}{\sum_{i=1}^{n} (B_i\, S_{(n-i)})} \,. \tag{13}$$

If multiple Lagrangian elements arrive in the same beaching area on the same day, the two summation terms were computed over all Lagrangian elements. This is equivalent to the assumption that r does not vary over the Lagrangian elements (i.e., that each element represents the same quantity of oil).

Model Processes

Estimation of the total mortality of seabirds from numbers of carcasses was done using a computer program to implement the steps described above. The program proceeds by reading one record from the morgue database and sequentially estimating the number of birds not counted because of scavenging, the number of birds not counted because of incomplete search coverage, and the number of birds lost at sea. These computations were dependent on knowledge of the location where birds were collected; however, a portion of the carcasses in each morgue had no information regarding the collection location. For these, numbers were summed for each morgue and later apportioned after the entire database had been processed. Steps in the computer analysis are summarized in Table 9.

Results of Model Runs

For a carcass to have been logged into a morgue, three events must have occurred: (1) the carcass must have drifted ashore or been collected in adjacent nearshore waters before it sank; (2) the carcass must not have been consumed or carried away by scavengers on the beach where it was cast; and (3) searchers must have come to the beach, found the carcass, and collected it. The computations were therefore essentially multiplicative. The probability of recovering a carcass would be the probability that all three events occurred, that is, the product of probabilities of the three events. Using best estimates of input parameters, this probability was 8%. Thus, the total direct mortality represented by the nearly 30,000 oiled carcasses collected after the *Exxon Valdez* spill is estimated to be 375,000 birds.

Model calculations required a large number of input variables, some of which could be estimated only with considerable uncertainty. To evaluate the variability in model results occurring as a consequence of this uncertainty, a Monte Carlo procedure was used to simulate the effects of uncertainty on model outcomes. The model was run 1,250 times. For each run, a value was randomly selected from an appropriate probability distribution for each of the parameters considered to be a source of significant uncertainty. Repeated measurements of these variables were not available for most input parameters. In these cases, the probability density function of parameter values was modeled as a triangular distribution where the vertices of the triangle were defined by the minimum estimate of the parameter value, the best estimate, and the maximum estimate (Ford et al. 1982). This methodology was not used to model the probability density of the parameters defining the Weibull distribution, b_0 and b_1, which are known to be normally distributed. In the case of the scavenging rate function, there are minimum, best, and maximum esti-

TABLE 9.—Summary of model steps for generating an estimate of total mortality of seabirds from numbers of carcasses collected on the beach.

Step	Description
1	Compute the proportion of depositable coastline searched, P^*, for each 15 min of latitude and longitude cell using equation (1). These calculations were made on a cell-by-cell basis and remained fixed throughout the model run.
2	Read a record from the morgue database. If the collection location is not identified, add the number of birds in the record to a running total of such carcasses for the current morgue and repeat step 2.
3	If the species was not a murre, and the date of processing minus the estimated lag time between recovery and processing was later than the cutoff date for oil-induced mortality (1 July), return to step 2.
4	For the Barren Islands, or for carcasses recovered in the water, ignore this step. Use equation (6) to estimate the number of carcasses that would have been recovered were it not for loss attributable to scavenging: *Cr* is the number of carcasses recovered from the database record read in step 2, *m* is the estimated intersearch interval for a given sector, and *Ss* is the empirically estimated likelihood that a carcass will persist a given number of days on the beach in spite of scavenging. The result of this calculation is the number of carcasses deposited on the beach before scavenging, *Cb*.
5	For carcasses recovered in the water, ignore this step. Use equation (3) to estimate the number of carcasses that were beached, taking into account incomplete search effort. The value of P^* specific to the current 15 min of latitude by 15 min of longitude cell computed in step 1 and the value of Cb^*, the number of carcasses deposited on the beach before scavenging computed in step 4, are used as inputs. The result of this computation is *Cs*, the number of carcasses deposited on the beach after accounting for the effects of both scavenging and incomplete search effort.
6	For species with inshore distributions, or if the carcasses were recovered inside Prince William Sound, ignore this step. Use equation (13) to estimate the number of carcasses that would have come ashore were it not for loss at sea. Use *Cs* from step 5 as the estimate of Cb^*; *S* is an empirically derived function based on the Kodiak releases fitted using the methodology of the at-sea loss model. The variable B_i (density of birds at a particular location at sea on day *i*) is estimated as follows: (A) If the day of beaching (i.e., estimated as the date of processing minus the lag time between collection and the date of processing) is during April and the species is a murre, *B* is calculated using the Outer Continental Shelf Environmental Assessment Program (OCSEAP) mean abundance of murres for April and May for locations determined by intersection with the paths of Lagrangian elements from the HAZMAT model that later were beached in the current 30 min of latitude by 30 min of longitude cell on that date. If the beaching date is after April, then it is assumed that all Lagrangian elements that were beached in the current 30-min cell intersected a constant density of murres on each day from 1 May onward. The constant density encountered is assumed to be different depending on whether the beaching site from step 2 was located on the eastern or western side of Kodiak Island or in the area north of Kodiak Island. (B) If the species is an auklet, the method of step 6(A) is used with the OCSEAP distribution of auklets. (C) If the species is not an auklet or a murre, a constant density is assumed to have been encountered from 30 March to the beaching date.
7	Collect sums of numbers of birds missing because of scavenging, search effort, and loss at sea. Sums are collected separately for each geographic sector and morgue.
8	If all records in the morgue database have not been read, return to step 2 and repeat steps 2–8.
9	For each of the morgues, divide the number of carcasses missing (the sum from step 7) by the number of carcasses recovered with known locations. Multiply this result by the number of carcasses in the current morgue that were not associated with a location.
10	Output results.

mates of Ss_j for all *j*, but the Ss_j cannot be considered independent. In this case, the probability density function for each *j* was sampled using the same uniform random deviate so that all the Ss_j will tend to be high or low simultaneously. Table 10 presents a summary of model input values, including best estimates and ranges. Figure 14 shows the distribution of results of model runs.

Results of the Monte Carlo model showed that taking uncertainty in parameter values into account tended to skew model results toward higher estimates of mortality, which occurred partly because the interactions of the model processes were multiplicative and partly because uncertainty in some parameters was skewed in a direction that increases the estimate of mortality. The 5th percent lower-bound of mortality from model runs was 300,000 birds and the 95th percent upper-bound was 645,000. These are not true confidence limits, but rather are based on random sampling from the probability distribution of possible values of model input parameters. Given the uncertainty in these parameters, the best estimate of 375,000 birds directly killed by oil from the *Exxon Valdez* may be conservative.

An additional method was used to examine model sensitivity to input variables that was based on extreme values of each parameter or set of pa-

TABLE 10.—Summary of values input to the seabird mortality model. Symbols for model variables are given where appropriate.

Input category (source)	Best estimate (range) or expectation (variance)
Morgue data (U.S. Fish and Wildlife Service records)	
Collection date cutoff for non-murres (this study)	1 Jul 1989
Mean interval (days) between collection and processing (Piatt et al. 1990)	14 (7–14)
Persistence of carcasses on beach faces, Ss_j (this study; Burger 1993); proportion remaining on day:[a]	
0	1.000 (1.000–1.000)
1	0.759 (0.544–0.833)
2	0.381 (0.087–0.519)
3	0.301 (0.087–0.417)
4	0.198 (0.044–0.333)
5	0.178 (0.000–0.333)
6	0.083 (0.000–0.250)
33	0.000 (0.000–0.163)
99	0.000 (0.000–0.000)
Proportion of shoreline walkable, PW[b] (environmental sensitivity index; interviews), for shoreline type (reflective? R_k):	
Exposed rock shore (yes)	0.00
Sheltered rocky shore (yes)	0.00
Exposed wavecut platform (no)	0.50 (0.20–0.80)
Fine-medium-grain sand (no)	1.00
Coarse-grain sand (no)	1.00
Mixed sand-gravel (no)	1.00
Gravel (no)	0.50 (0.20–0.80)
Exposed tidal flat (no)	1.00
Sheltered tidal flat (no)	0.00
Marsh (no)	0.50 (0.00–1.00)
Ice (yes)	0.50 (0.00–1.00)
Unclassified (yes)	0.00
Proportion of walkable beach searched, Ps[c] (interviews), along:	
Prince William sound	0.85 (0.70–1.00)
Kenai Peninsula	1.00
Cook Inlet	1.00
Barren Islands	1.00
Katmai National Park	0.80 (0.70–0.90)
Becharof National Wildlife Refuge	0.95 (0.90–1.00)
Aniakchak National Monument	1.00
Kodiak Island	0.75 (0.50–1.00)
Alaska Peninsula National Wildlife Refuge	0.52 (0.05–1.00)
Frequency of search (days between searches), m[d] (interviews), along:	
Prince William Sound	3.0 (2–4)
Kenai Peninsula	9.0 (7–11)
Cook Inlet	3.0 (2–4)
Barren Islands	7.5
Katmai National Park	8.0 (7–9)
Becharof National Wildlife Refuge	37.5 (30–45)
Aniakchak National Monument	8.0 (6–10)
Kodiak Island	9.0 (7–9)
Alaska Peninsula National Wildlife Refuge	37.5 (30–45)
Outer Continental Shelf distributional data, for murres and auklets only (National Oceanographic Data Center; U.S. Fish and Wildlife Service)	
Loss-at-sea function (this study); expectation (variance) for:	
Scale parameter λ of Weibull distribution, b_0	−7.67 (0.0409)
Shape parameter γ of Weibull distribution, b_1	3.09 (0.0075)

[a]Ranges are for the four studies discussed. The best estimate is the mean over all studies for a given number of days on the beach.

[b]Estimates were derived from interviews with agency personnel and others familiar with the various areas and knowledgeable with regard to the search effort. For Prince William Sound, gravel beaches and sheltered rocky shores were considered walkable (this assumption is based on interview responses). Very little information was available on walkability of ice or marsh.

[c]Estimates were derived from interviews with agency personnel and others familiar with the various areas and knowledgeable with regard to the search effort. The values for the Kodiak and Alaska Peninsula sectors are the least well known. Cook Inlet was considered to be similar to Prince William Sound with respect to the search effort.

Estimates were derived from interviews with agency personnel and others familiar with the various areas and knowledgeable with regard to the search effort. The values for the Kodiak and Alaska Peninsula sectors are the least well known. Cook Inlet was considered to be similar to Prince William Sound, Kodiak was considered to be similar to Katmai, and the Alaska Peninsula area was considered to be similar to Becharof with respect to search frequency.

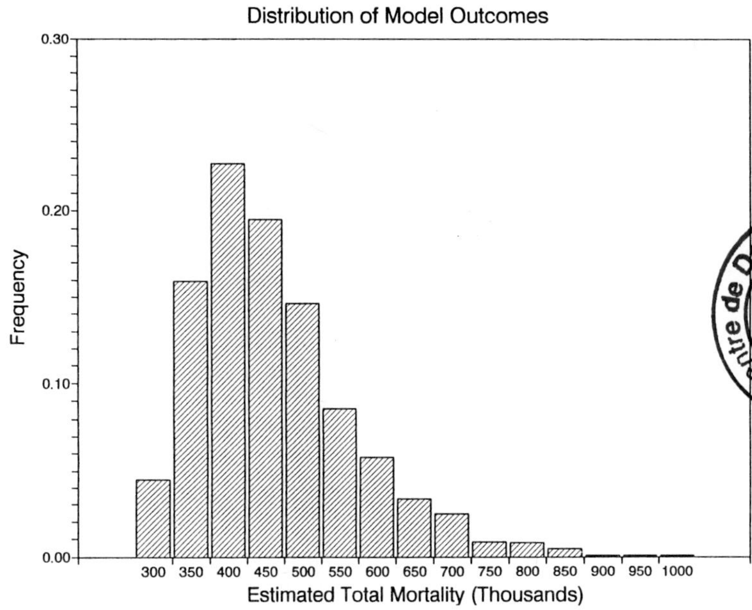

FIGURE 14.—Distribution of Monte Carlo model outcomes based on 1,250 model iterations. Input parameters were randomly selected for each iteration based on a range of possible values.

rameters. For this analysis, all model inputs were held at their best estimates except for one that was set at either its maximum or minimum value. The parameter ranges were used from Table 10 except for the parameters of the fitted Weibull distribution. In that case, the relatively rapid sinking rate observed for release AK-6 in the Gulf of Alaska was used as one extreme and the relatively slow sinking rate for release AK-8 as the other extreme. Results of this analysis are presented in Table 11. Because these extreme parameter values are somewhat unlikely, the results are best used for describing the relative contributions of various model inputs to overall model uncertainty. The factor most strongly influencing the estimate of direct mortality was the rate of scavenging of beached carcasses that determined persistence on the beach face. Use of the lower or upper bounds on this parameter resulted in a decrease of 38.1% or an increase of 105.4% in the estimate of direct mortality. In terms of increasing the mortality estimate, the next most important source of uncertainty was the proportion of the depositable beaches searched (36.9%). In terms of decreasing the mortality estimate, the next most important source of uncertainty was the rate of loss of birds at sea (24.3%).

TABLE 11.—Effects on model results (number of seabirds killed) when one input parameter at a time was varied while the other parameters were held at their best-estimate values.

Parameter varied	Low estimate of mortality	High estimate of mortality
Collection and processing interval	343,963	395,016
Persistence on the beach face	233,441	773,857
Proportion walkable	326,191	472,167
Walkable proportion searched	318,192	515,877
Frequency of beach searches	337,840	413,444
Loss-at-sea function	285,073	432,491

Discussion

Piatt et al. (1990) estimated that the total direct mortality of seabirds resulting from the *Exxon Valdez* oil spill was in the range of 100,000–300,000. This estimate was derived in part from the proportion of total estimated mortality that recovered carcasses have represented in several other oil spills; however, the recovery of carcasses after the *Exxon Valdez* spill was probably unusually low relative to other spills for several reasons. To collect dead birds, workers had to cover extensive stretches of coastline, most accessible only by boat. Using best estimates of beach search effort and deposition of carcasses, the highest carcass recovery rates would have occurred in Prince William Sound (35%) and at the Barren Islands (49%). These relatively high recovery rates resulted because most of the birds in these areas were beached relatively early, thus there was little time for loss at sea. In addition, the absence of mammalian scavengers on the Barren Islands would have resulted in a reduced loss attributable to scavenging. Cook Inlet and Kenai Fjords National Park had intermediate recovery rates of 21% and 14%, respectively, and Kodiak Island with 6% and the Alaska Peninsula with 2% had the lowest recovery rates. The low recovery rates on Kodiak Island and the Alaska Peninsula resulted, in part, from the long period of time that carcasses would have been at sea before being beached, and the relatively low level of search effort.

Another factor affecting recovery of carcasses was the density of scavengers along parts of the shore of coastal Alaska. In this study, the best estimate of daily persistence of carcasses was 31% once they were deposited on a beach. Similar studies after other spills in the eastern north Pacific have found daily persistence rates ranging from 45 to 84% (Burger 1992). The low estimate of persistence after the *Exxon Valdez* spill suggests that carcasses were rapidly removed by scavengers. Presence of oil did not appear to deter scavenging. After the 1989 *Nestucca* oil spill, Burger (1993) noted that even encasement of carcasses in moussed oil and extensive decomposition did little or nothing to inhibit scavenging.

Several other factors probably resulted in undercounts of carcasses but were not included in the model because quantitative data on the values of the relevant parameters were completely lacking. These included (1) rewash of carcasses by waves, (2) burial of carcasses, and (3) carcasses missed by searchers. These sources of undercounting may have been a common occurrence in the *Exxon Valdez* spill. During this study, extensive rewash of radio-tagged carcasses was observed in Prince William Sound, and burial of carcasses was noted in the U.S. Department of the Interior records. Page et al. (1990) also noted substantial rates of burial of carcasses after a spill in central California. In the model, it was assumed that all carcasses on searched beaches were counted; however, studies after other spills show that experienced workers still miss carcasses even on the most well-searched beaches (J. Y. Monnat and Y. Geurmeur, Institut d'Etudes Marines and Societe pour l'Etude et la Protection de la Nature en Bretagne, Brest, unpublished data).

Finally, the *Exxon Valdez* spill affected a large open-water area and persisted for months. Bird carcasses drifted longer than after other spills and were more likely to sink before reaching land. In the model, it was assumed that the proportionality between the density of birds and the mortality resulting from a given quantity of oil remained constant over time. Although this assumption may be valid over relatively short time-spans, over periods of months it is likely that oil becomes less lethal as it weathers and decreases in volume. Given weathering of the oil, birds collected on some beaches, especially along the shores of Kodiak Island and the Alaska Peninsula, might have been oiled earlier and drifted longer than the model predicted. Because the proportion of carcasses lost at sea increases with the length of time before being beached, carcasses recovered on those beaches may underrepresent numbers contacted at sea.

Acknowledgments

This study was conducted under contract with Walcoff and Associates, Alexandria, Virginia, for the U.S. Department of Justice; funding and administration of the study was provided by the USFWS, Migratory Birds Management, Anchorage, Alaska. We thank Lynn Denlinger of the USFWS for coordinating the logistics of the field studies and, with Tina Odenberg, for facilitating access to the morgue storage vans. Mary Cody assisted with morgue carcass re-examinations. Kent Wohl and Paul Gertler provided support and assistance, as did many other USFWS personnel. National Park Service and USFWS managers and researchers were of great help and were valuable sources of information on the extent and time course of beach search effort in their respective jurisdictions. We also thank the many volunteers for their tireless work collecting bird carcasses, which provided the baseline data for this study. Digital versions of the morgue data were supplied to us by John Piatt, USFWS, who also provided useful advice during all phases of the study. Aircraft time was provided by the Office of Aircraft Services. Our chief pilot, Dave Henley, was indispensable to the field operations both in Prince William Sound and the Gulf of Alaska and was a pleasure to work with. Programming for real-time data acquisition was provided by Ken Winnick, who helped greatly in making the data-logging system aboard the aircraft a reality. The field crew for radio-tracking studies, Chad Hewitt, Dave Roseneau, Jeff Tribble, Doug Warrick, and Wendy Williams, performed exceptional work under difficult and dangerous conditions. We also benefited from careful reviews of earlier drafts of this article by Alan Burger and Steve Speich.

References

Bibby, C. J. 1981. An experiment on the recovery of dead birds from the North Sea. Ornis Scandinavica 12:261–265.

Burger, A. E. 1992. The effects of oil pollution on seabirds off the west coast of Vancouver Island. Canadian Wildlife Service Occasional Paper 75:120–128.

Burger, A. E. 1993. Effects of the *Nestucca* oil spill on seabirds along the coast of Vancouver Island in 1989. Canadian Wildlife Service Technical Report Series 179, Pacific and Yukon Region, British Columbia.

Carter, H. R. 1989. Book review of "Bird communities at sea off california: 1975 to 1983." Auk 106:170–171.

Carter, H. R., and G. W. Page. 1989. Central California oil spill contingency plan: assessment of numbers and species composition of dead beached birds. NOAA (National Oceanic and Atmospheric Administration) Technical Memorandum 25, Washington, DC.

Conover, W. J. 1980. Practical nonparametric statistics. Wiley, New York.

Ford, R. G, G. W. Page, and H. R. Carter. 1987. Estimating mortality of seabirds from oil spills. Pages 848–851 *in* Proceedings, 1987 oil spill conference. American Petroleum Institute Publication 4452, Washington, DC.

Ford, R. G., and six coauthors. 1991a. Seabird mortality resulting from the *Nestucca* oil spill incident, winter 1988–89. Final report by Ecological Consulting, Inc. for the Washington Department of Wildlife, Olympia.

Ford, R. G., and six coauthors. 1991b. Assessment of direct seabird mortality in Prince William Sound and the western Gulf of Alaska resulting from the *Exxon Valdez* oil spill. Final report by Ecological Consulting, Inc. for U.S. Fish and Wildlife Service and the Department of Justice, Washington, DC.

Ford, R. G., J. A. Wiens, D. Heinemann, and G. L. Hunt, Jr. 1982. Modeling the sensitivity of colonially breeding marine birds to oil perturbation. Journal of Applied Ecology 19:1–31.

Gould, P. J., D. J. Forsell, and C. J. Lensink. 1982. Pelagic distribution and abundance of seabirds in the Gulf of Alaska and eastern Bering Sea. U.S. Fish and Wildlife Service, Office of Biological Services (FWS/OBS-82/48), Washington, DC.

Gross, A. J., and V. A. Clark. 1975. Survival distributions: reliability applications in the biomedical sciences. Wiley, New York.

Hope-Jones, P., G. Howells, E. I. S. Rees, and J. Wilson. 1970. Effect of the *Hamilton Trader* oil on birds in the Irish Sea in May 1969. British Birds 63:97–110.

King, J. G., and G. H. Sanger. 1979. An oil vulnerability index for marine oriented birds. U.S. Fish and Wildlife Service Wildlife Research Report 11:227–239.

Page, G. W., H. R. Carter, and R. G. Ford. 1990. Numbers of seabirds killed or debilitated in the 1986 *Apex Houston* oil spill in central California. Studies in Avian Biology 14:164–174.

Piatt, J. F., C. J. Lensink, W. Butler, M. Kendziorek, and D. R. Nysewander. 1990. Immediate impact of the *Exxon Valdez* oil spill on marine birds. Auk 107:387–397.

RPI (Research Planning Institute, Inc.). 1983. Sensitivity of coastal environments and wildlife to spilled oil, Prince William Sound, Alaska: an atlas of coastal resources. Prepared for NOAA (National Oceanic and Atmospheric Administration), Office of Oceanography and Marine Services, Seattle.

RPI (Research Planning Institute, Inc.). 1985. Sensitivity of coastal environments and wildlife to spilled oil, Cook Inlet/Kenai Peninsula, Alaska: an atlas of coastal resources. Prepared for NOAA (National Oceanic and Atmospheric Administration), Office of Oceanography and Marine Services (RPI/ESI/8570), Seattle.

Sokal, R. R., and F. J. Rohlf. 1981. Biometry. Freeman, San Francisco.

Sowls, A. L., S. A. Hatch, and C. J. Lensink. 1978. Catalog of Alaskan seabird colonies. U.S. Fish and Wildlife Service, Office of Biological Services (FWS/OBS-78/78), Washington, DC.

Speich, S. M., D. A. Manuwal, and T. R. Wahl. 1991. The bird/habitat oil index—a habitat vulnerability index based on avian utilization. Wildlife Society Bulletin 19:216–221.

Torgrimson, G. M. 1981. A comprehensive model for oil spill simulation. Pages 423–428 *in* Proceedings, 1981 oil spill conference. American Petroleum Institute Publication 4334, Washington, DC.

American Fisheries Society Symposium 18:712–719, 1996

How Many Seabirds Were Killed by the *Exxon Valdez* Oil Spill?

JOHN F. PIATT
National Biological Service, Alaska Science Center
1011 East Tudor Road, Anchorage, Alaska 99503, USA

R. GLENN FORD
Ecological Consulting Inc.
2735 N.E. Weidler Street Portland, Oregon 97232, USA

Abstract.—After the *Exxon Valdez* oil spill of 24 March 1989, 36,115 dead seabirds were recovered from beaches and processed at morgues. Most or all of 1,888 live oiled seabirds brought to rehabilitation centers also died and about 3,260 oiled carcasses were never delivered to morgues. Of these 41,263 carcasses accounted for, we estimated conservatively that only 30,000 were killed by oil pollution. Carcass drift and recovery experiments conducted in the affected area during the spill and 1 year later, along with historical experiments conducted in other oceanographic regions, suggest that only a fraction (range = 4–30%) of birds killed were likely to have been recovered from beaches. Regression analysis of these drift–recovery data predicts a 15% recovery rate ($r^2 = 0.38$, $P = 0.015$). We recognize uncertainties in the assumptions and parameter values used to extrapolate total mortality from drift–recovery data, but we have confidence that mortality fell within the extreme range of estimates (100,000–690,000 birds killed) because these reflect a very wide range of observed and experimentally determined parameter values. Total mortality can also be estimated by comparing pre- and postspill colony population estimates. Uncertainties about these census data are greater than those associated with drift–recovery data, but nonetheless provide an independent mortality estimate of similar magnitude. Taken together, all evidence suggests that about 250,000 seabirds were killed by *Exxon Valdez* oil.

It is difficult to measure the total mortality of seabirds from oil pollution. The number of oiled-bird carcasses found on beaches typically represents only a fraction of the total killed, because some sink at sea or drift offshore, and those that drift ashore may be removed rapidly by scavengers or be buried in sand and debris. Furthermore, the timing, frequency, and thoroughness of beach surveys determine the number of oiled carcasses likely to be counted. The ratio of birds recovered to those killed is a function of all these factors and is usually quite low (Ford et al. 1987).

There are three general approaches to estimating total seabird mortality resulting from an oil spill: (1) measuring the difference between pre- and postspill colony populations, (2) estimating carcass loss and recovery rates from experimental drift studies conducted during the actual spill, or (3) extrapolating from carcass drift and recovery experiments conducted at other times or places.

Estimating mortality from changes in colony population size is intuitively appealing but problematical. The effects of a spill (especially in winter) may be spread over many distantly spaced colonies, and postspill censuses may be difficult to interpret because of attendance by previously nonbreeding birds and immigrants from other colonies (Stowe 1982). Furthermore, colony censuses do not include the component of the population located at sea during the census, which can be larger than the component attending nest sites (Piatt and Ford 1993). Finally, prespill baseline data on colonies and associated populations at sea may be inadequate for statistical evaluation of population changes.

Estimating total mortality from carcass drift experiments does not require baseline data; however, processes governing the loss and recovery of carcasses vary considerably over time within and between areas, and it is often difficult to retroactively assess conditions that prevailed at the time of a spill. Conducting drift experiments during a spill (e.g., Hope-Jones et al. 1978) may be the best way to estimate total mortality because the recovery rates of marked carcasses will reflect actual oceanographic conditions and search effort at the time of the spill; however, this requires some degree of prespill readiness and knowledge of the distribution of spilled oil and birds at the time of the spill.

After the *Exxon Valdez* oil spill of 24 March 1989, 36,115 dead seabirds were recovered from beaches and processed at morgues. This was the highest recovery of carcasses ever documented after an acute oil pollution event (Piatt and Lensink 1989). In addition, most or all of 1,888 live oiled seabirds brought to rehabilitation centers also died (Piatt et al. 1990) and about 3,260 oiled carcasses were never delivered to morgues because they were discarded, burned, or buried (Piatt et al. 1990; Ecological Consulting, Inc. [ECI] 1991). Anecdotal reports suggested that thousands more carcasses were not accounted for (e.g., sucked into oil skimmers, otherwise recovered but discarded).

Based on the timing of recoveries, degree of oiling, and species composition, it was estimated conservatively (Piatt et al. 1990; ECI 1991) that about 30,000 of the total 41,263 carcasses accounted for died from oil pollution. Most of these were murres *Uria* spp. (74%), other alcids (7%), and seaducks (5.3%). After 1 August, most (72%) carcasses recovered were shearwaters and gulls that apparently died from natural causes (starvation). This die-off in late summer added uncertainty to the cause of death of other birds retrieved at that time. Thus, all specimens processed at morgues after 1 August were excluded from the tally of oil-killed birds even though some (e.g., including 702 seaducks and 440 murres) had probably been killed by oil and died earlier in summer (Piatt et al. 1990).

Based on this conservative estimate of birds killed by oil, and several drift–recovery experiments, initial extrapolations of total mortality ranged in the low hundreds of thousands of seabirds (100,000–300,000 by Piatt et al. 1990; 300,000–645,000 by ECI 1991). Despite extensive documentation in these original reports of the underlying assumptions and sources of uncertainty in the extrapolations, the high-end ranges of these estimates have been selectively quoted or criticized (e.g., Parrish and Boersma 1995a) in the popular press and used by some biologists as a basis for assessing impacts on populations. At the other extreme, even the low-end mortality estimates have been challenged because of uncertainties associated with them (Wells et al. 1995; Wiens 1995; Parrish and Boersma 1995a). Indeed, these biologists argue that only those carcasses recovered from beaches should be used to assess population impacts of the *Exxon Valdez* oil spill.

Thus, more than 6 years after the *Exxon Valdez* oil spill, we again ask the questions: What is a reasonable estimate of the number of seabirds killed, and how much confidence can we have in that estimate? In this article, we reevaluate our estimates of total mortality with respect to their underlying assumptions and conclusions and compare these with alternative estimates of mortality from other drift experiments and colony census data. Finally, we make some recommendations for estimating seabird mortality in future oil spills.

Estimating Total Seabird Mortality

Real-Time Drift Experiments

Most seabird mortality from the *Exxon Valdez* oil spill occurred during the 2 months after the spill on 24 March, although dead and dying birds were recovered from beaches for more than 5 months afterward (Piatt et al. 1990). A carcass drift experiment was initiated in Prince William Sound in April, but this experiment was aborted by administrators in the U.S. Fish and Wildlife Service because of negative public response (ECI 1991). Forty-seven carcasses were color-banded and released within 200 m of shore in western Prince William Sound. No banded birds were ever recovered; however, morgue attendants were never asked to look for bands, and extensive oiling of carcasses may have precluded their discovery.

Another drift experiment was conducted in the Gulf of Alaska about 6 weeks after the initial spill to estimate recovery rates (Piatt et al. 1990). Old (approximately 2–4 weeks), mousse-covered carcasses ($N = 100$; 89% murres) were recovered off a beach in Puale Bay along the Alaska Peninsula, cleaned of several inches of caked-on mousse, marked with wired "please report" tags that extended away from the body, and released at the Barren Islands in early May. Only three murres (3%) were later recovered, all from the very same beach in Puale Bay. This strongly suggested that prevailing currents had originally (in late April) carried oiled birds to this area from lower Cook Inlet.

Old carcasses sink faster than fresh ones (Burger and Fry 1993), and search effort on the remote beaches along the Alaska Peninsula was incomplete outside of Puale Bay (ECI 1991). Furthermore, most birds recovered west of the Kenai Peninsula (approximately 75% of total) were encased in thick mats of oil mousse, an emulsion that consists largely of trapped air and conceivably added buoyancy to the original bird carcasses.

For these reasons, Piatt et al. (1990) considered the 3% experimental recovery rate to be minimal. Also considering drift experiments conducted elsewhere in the world (Table 1), the estimated size of populations at risk, and the fact that a minimum 3% recovery was observed in the most distant recovery site (and was therefore likely to be much higher in areas closer to the spill source), the spillwide recovery rate was estimated conservatively at 10–30%, suggesting that 100,000–300,000 birds were killed by the oil spill (Table 2).

The problem with this estimate is that factors influencing recovery rates (such as survey effort,

TABLE 1.—Observed and estimated rates of recovery of seabirds on beaches from 17 carcass drift experiments. Predominant species used in experiments included alcids (a), gulls (g), and cormorants (c).

Region	Species	*N*	Recovery (%)	Reference
Irish Sea	a	410	20.0	Hope-Jones et al. (1970)
Irish Sea	a	319	7.5	Lloyd et al. (1974)
Bristol Channel	g	Unknown	10.0	Bibby and Lloyd (1977)
Irish Sea	g	300	11.0	Bibby and Lloyd (1977)
Irish Sea	g	305	44.0	Bibby and Lloyd (1977)
Irish Sea	g	347	59.0	Bibby and Lloyd (1977)
English Channel	g	144	20.0	Hope-Jones et al. (1978)
North Sea	g	600	9.8	Bibby (1981)
North Sea	g	40	40.8	Stowe (1982)
North Sea	g	150	11.2	Stowe (1982)
North Sea	c	130	14.6	Harris and Wanless (in press)
California	g	186	29.9	Page et al. (1982)
California	a	63	0.0	Page et al. (1982)
N. Grand Banks	a	115	0.0	Threlfall and Piatt (1982)
S. Grand Banks	a	129	0.0	Threlfall and Piatt (1982)
Gulf of Alaska	a	100	3.0	Piatt et al. (1990)
Gulf of Alaska	a	184	8.0	ECI (1991)
Total and mean		3,522	17.0	

sinking rates, scavenging rates, etc.) varied geographically, and the results of the drift experiment were directly applicable to only a portion of the affected region. As stated in Piatt et al. (1990), this conservative estimate of total mortality was put forward "tentatively" until such time that specific parameters needed to model total mortality could be measured. This was undertaken in summer 1990 by ECI (1991).

Postspill Drift and Carcass Experiments

In 1990, a series of carcass drift experiments were conducted in Prince William Sound and southeast of Kodiak Island in order to estimate some of the parameters needed to back-calculate total mortality (ECI 1991; Ford et al. 1996, this volume). In this study, fresh seabird carcasses ($N = 184$; 89% alcids) were coated with oil, radio-tagged, and released at sea. Subsequent drift and sinking rates were determined, and separate experiments to assess scavenging and disappearance rates on beaches were also conducted. Beach survey effort during the original spill was estimated from historical records.

Using the best estimates of input parameters, the joint probability of recovery was estimated at 8%, suggesting that 375,000 birds were killed. Using a range of possible values for each parameter estimate, the mean probability that a carcass would have been recovered was calculated by Monte Carlo analysis to be about 6.9%, suggesting that 435,000 (range, 300,000–645,000) seabirds died as a result of the spill (ECI 1991; Ford et al. 1996). The model was most sensitive to "sinking rate" and "persistence on beach" parameters.

There remain uncertainties about some parameter values, and there are gaps in our understanding of processes controlling the recovery of dead seabirds. If we could refine some parameter values, estimates of total mortality would probably decrease or increase, as the following examples suggest: (1) The results of drift experiments were highly variable (Table 1) and were determined largely by prevailing conditions at the time of a spill; were results obtained from experiments in 1990 applicable to the spill in 1989? Weather conditions were more severe in 1989 than during the study in 1990, and therefore the 1990 at-sea loss experiments may have underestimated actual losses in 1989. (2) Was the use of persistence rates obtained from experimental studies using dozens of carcasses applicable to the situation after the *Exxon Valdez* oil spill? Persistence rates obtained from recent study of a large (tens of thousands) murre die-off in Resurrection Bay in March 1993 (Piatt and van Pelt 1993; van Pelt and Piatt 1995) suggest a higher

TABLE 2.—Estimates of total seabird mortality from the *Exxon Valdez* oil spill.

Source	Estimate		
	Minimum	Mean	Maximum
Piatt et al. (1990)	100,000	200,000	300,000
ECI (1991)[a]	300,000	435,000	645,000
Drift experiments[b] ($N = 15$)	116,000	200,000	690,000
Murre colony data	51,000	146,000	241,000

[a]Minimum and maximum estimates include 95% of possible outcomes from 1,250 Monte Carlo simulations.

[b]Minimum and maximum estimates are 95% confidence limits.

persistence rate (89–96%/d) than the range of values used in the ECI model (54–83%/d). Use of these data would increase the estimated probability of carcass recovery and decrease the estimate of total mortality. (3) How efficiently were beaches checked for carcasses? Because data were not available, Ford et al. (1996) assumed 100% efficiency in locating carcasses, but unpublished studies conducted during the Amoco Cadiz oil spill suggested that beach surveys were only about 80% efficient under conditions that were more benign than those encountered by searchers after the *Exxon Valdez* oil spill. Furthermore, anecdotal reports from many areas of the Gulf of Alaska indicated that burial of carcasses in sand and gravel was frequent and may have "removed" 10–20% of carcasses from some beaches before searchers could recover them (ECI 1991). Incomplete recovery of carcasses from beaches would lead to underestimates of total mortality. (4) Did mousse-caked birds in 1989 (see above) have a lower sinking rate than oil-soaked birds in 1990? If so, then estimates of the at-sea loss of carcasses in some regions would be decreased, with a corresponding increase in estimates of total mortality. Experimental evaluation of these factors during future oil spill incidents would improve the accuracy of postspill mortality estimates.

Comparison of Real-Time and Postspill Experiments

There is at least one direct comparison to be made between the drift experiment conducted by Piatt et al. (1990) during the spill and that of ECI (1991) conducted in the following year. Based on the estimated parameter values in different areas, the ECI model (ECI 1991: 152) determined that the highest recovery rates would have occurred in Prince William Sound (35%) and at the Barren Islands (49%). Cook Inlet (21%) and Kenai Fjords National Park (14%) had intermediate recovery rates. Kodiak Island (6%) and the Alaska Peninsula (2%) had the lowest calculated recovery rates—largely because of the long periods of time carcasses would have been at sea and because of a relatively low level of search effort. Piatt et al. (1990) actually observed a 3% recovery rate along the Alaska Peninsula. Thus, despite uncertainties in assumptions of both studies, they were in remarkably good agreement about recovery rates along the Alaska Peninsula, providing some confidence that independent drift studies can yield repeatable results.

Nevertheless, the difference of 1% is noteworthy because 30% (N = 8,881) of all birds were recovered along the Alaska Peninsula and the extrapolation factor is very large. Thus, using the observed 3% recovery rate rather that the calculated 2% recovery rate would decrease the ECI best estimate of total mortality by 40% to about 227,000 birds. This is a simplistic analysis, however, because the ECI model calculates site-specific recovery rates along the peninsula, and 2% is an average value used for geographic comparisons. Nonetheless, it illustrates the need for caution when using low recovery rates to estimate total mortality.

Other Drift Experiments

Carcass drift and recovery experiments have been conducted in a variety of locations around the world (Table 1). Most of these experiments involved dropping marked carcasses at various distances offshore in regions where systematic beach surveys were being conducted (e.g., Bibby and Lloyd 1977; Page et al. 1982; Threlfall and Piatt 1982). Some were conducted during actual oil spills (e.g., Hope-Jones et al. 1978). In one case, a "natural" experiment involved the deaths and subsequent beach-recoveries of color-ringed cormorants after a large die-off of cormorants and murres at a well-studied colony (Harris and Wanless, in press). Finally, the recovery rate provided by ECI (1991) is actually a composite best estimate calculated from a number of independent experiments on drift, sinking, persistence, and so forth of oiled carcasses (see Ford et al. 1996).

The average (mean, 17%; median, 11%) recovery rate from all 17 studies is quite low (Table 1). Results indicate that recovery rates are highly dependent on local wind, sea, and current conditions at the time of the experiment (see especially Bibby and Lloyd 1977) and should be applied to particular spills with caution. Highest recovery rates have been observed where carcasses were released relatively close to shore during periods of onshore winds (Bibby and Lloyd 1977) or within confined oceanographic areas (e.g., the Irish Sea). Furthermore, gull carcasses typically have a higher recovery rate (10–59%) compared to those of alcids (0–20%), possibly because they are more buoyant and present more profile for wind-drift (Bibby and Lloyd 1977; Page et al. 1982).

Can drift experiments conducted at various times and locations be used to predict recovery rates in the case of the *Exxon Valdez* spill? Despite the many sources of variability, linear regression (Figure 1) of experimental drift data reveals that 31% of the variance in carcass recoveries can be accounted for

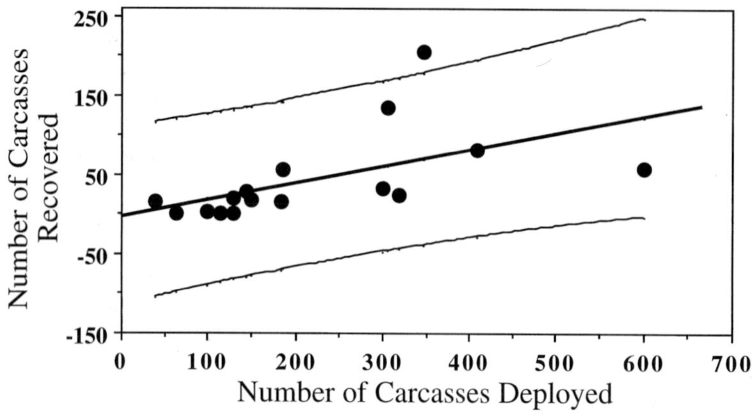

FIGURE 1.—Linear regression of seabird carcass recovery data from 16 drift experiments (Table 1; Bristol Channel data are not included because of unknown number of carcasses released). The best-fit regression line is bounded above and below by 95% confidence limits.

by the number of carcasses released (N = 16, r^2 = 0.31, P = 0.025). Note that the Bristol Channel data (Table 1) could not be included in this analysis because the number of carcasses released was unknown. The slope (0.209) of the regression line predicts a 21% recovery rate. If the lone outlier (above the upper 95% confidence limit in Figure 1) is removed from the analysis, the predicted recovery rate declines to 15% and the explained variance increases to 38% (slope ± SE = 0.151 ± 0.054, N = 15, r^2 = 0.38, P = 0.015). Removal of this outlier is justified because it was derived from an experiment in which gull carcasses were released relatively close to shore during a period when sustained winds were blowing directly into a large embayment (eastern Irish Sea; Bibby and Lloyd 1977). The next highest recovery rate (Figure 1, Table 1) was obtained under similar experimental conditions, but as it fell within the 95% confidence limits in the original analysis, we retain it in the final analysis. (Further removal of this data point would increase explained variance to 57% and decrease the predicted recovery rate to 12% at P = 0.002.)

Using a predicted recovery rate of 15%, we can therefore estimate that about 200,000 seabirds were killed by *Exxon Valdez* oil (Table 2). The range of predicted mortality (116,000–690,000; Table 2) is similar to that reported by ECI (1991). In any case, using these combined experimental data to predict mortality for the *Exxon Valdez* spill probably underestimates the true mortality because (1) most mortality occurred in open waters of the Gulf of Alaska, (2) search effort was relatively poor in this region (ECI 1991), (3) alcids were the predominant species affected, and (4) all these factors are known to decrease recovery rates.

Murre Colony Population Counts

It is also possible to make an estimate of the total mortality of seabirds from the spill by comparing pre- and postspill colony counts of murres (see Tables 1 and 2 of Piatt and Anderson 1996, this volume). Uncertainties associated with colony count data (see discussion in Piatt and Anderson 1996) are probably much greater than those associated with drift experiments, but it is worth examining these data for comparative purposes because they are an independent source of information about pre- and postspill seabird population changes.

Colony count data indicate that murre populations declined by 7–48% at individual colonies in the path of the spill (Nysewander et al. 1993; Piatt and Anderson 1996). Overall, "affected" murre populations declined by approximately 40% between pre- and postspill censuses; however, it appears that populations were already declining at some major colonies before the spill occurred, and not all of this decline should be attributed to the spill (Piatt and Anderson 1996). For example, murre populations at Puale Bay (in the spill path) declined by 5.8% per annum and at Chisik Island (in the spill region but outside the direct path) by about 7.5% per annum in years before the spill.

It is possible to calculate the maximum impact of the spill on Gulf of Alaska murre populations if we assume that: (1) declines at specific colonies occurred throughout the pelagic areas in which they were located; (2) all declines resulted from the oil spill (probably not true, see above); (3) colony counts include only birds found at the colony at the time of censusing, and about 1.4 times as many colony-associated birds (nonbreeders, subadults, etc.) are found at sea (Piatt and Ford 1993); (4) nonattending birds at sea were killed in the same proportion as those at colonies; and (5) no birds killed by the spill were in transit from areas unaffected by the spill.

The data suggest that the total population (at colony and at sea) of murres in affected areas declined by 178,490 birds between years before the spill (1975–1988) and years after the spill (1989–1991) (see Table 2 in Piatt and Anderson 1996). This represents the maximum possible number of murres that could have been killed by *Exxon Valdez* oil. Because 74% of all birds killed were murres (Piatt et al. 1990), this suggests that a maximum of about 241,200 (178,490/0.74) marine birds of all species were killed by oil. Historical data suggest, however, that murres were declining at some colonies before the spill, so the real number killed by the spill must be some fraction of this maximum estimate. For example, if the 5.8% per annum decline observed at Puale Bay before the spill persisted through and after the spill, then only 10% of the observed 48% decline there is unaccounted for by this trend. By extension, a minimum of about 37,500 (0.21×178,490) murres were killed by oil, and by further extrapolation, 50,650 birds of all species. Thus, the total bird kill estimated from colony censuses ranges from about 51,000 to 241,000 birds, depending on the assumptions used. The average of these estimates is 146,000 birds killed (Table 2).

The error associated with colony counts cannot be assessed with existing data (Piatt and Anderson 1996). Aside from that issue, the estimates of mortality from census data are probably underestimates because one assumption (number 5 above) is probably invalid. During the time (April–May) that oil was widely dispersed in the Gulf of Alaska, it was highly likely that many seabirds were in transit through the spill zone en route to colonies outside the spill zone (Piatt et al. 1990). Losses of these birds would not have been reflected at colonies censused within the spill zone.

Conclusions

Mean estimates of total seabird mortality (200,000–435,000) based on various drift experiments are all in general agreement as to the order of magnitude of the kill (Table 2). Although the individual estimates each depend on some uncertain assumptions, we have confidence that total mortality fell within the range of extreme estimates (i.e., 100,000–690,000) because these reflect a very wide range of observed or experimentally determined parameter estimates.

The estimated mortality derived from murre colony counts includes many potential sources of error that cannot be evaluated. Nonetheless, census data suggest a similar level of mortality as do drift experiments. Census data further suggest that if the highest estimate of seabird mortality (690,000) is correct, then the kill must have included many individuals in transit through the spill area because otherwise the entire prespill population of murres in the spill zone (approximately 446,000) would have been eliminated, which was not the case (Table 2; Piatt and Anderson 1996).

Taken together (Table 2), the evidence suggests that about one-quarter of a million seabirds died in the aftermath of the *Exxon Valdez* oil spill, which is by far the highest estimated level of seabird mortality ever reported from an acute oil spill (Piatt and Lensink 1989).

In a selective review of the evidence, Parrish and Boersma (1995a) assert that our estimates of total mortality were exaggerated and unscientific. We find it ironic that these statements are made in a non-peer-reviewed popular science magazine. Considering our conservative estimates of the proportion of recovered birds actually killed by oil and our clearly documented, conservative (and peer-reviewed) projections of total mortality (Piatt et al. 1990; ECI 1991; Ford et al. 1996), we are at a loss to understand the basis of Parrish and Boersma's conclusions. In contrast, their own "assessment" of mortality estimates (Parrish and Boersma 1995a) did not even reference Ecological Consulting, Inc.'s critical 1990 study nor mention even one of the many drift experiments conducted worldwide that clearly support the conclusion that only a fraction of seabirds were likely to have been recovered (Piatt 1995). Faced with these criticisms (Piatt 1995), Parrish and Boersma asserted (1995b) that drift experiments are much too variable to use for interpreting impacts of the *Exxon Valdez* oil spill. Again, the scientific facts (Figure 1 and above) are at odds with their opinions. We submit that Parrish and Boersma's assessment is seriously flawed and subjective.

In conclusion, a preponderance of evidence indicates that only a fraction of the seabirds killed by

Exxon Valdez oil were ever recovered from beaches. The exact proportion recovered will never be known, but to dismiss the range of possible mortality estimates as mere claims (Parrish and Boersma 1995; Wells et al. 1995) and insist on using only the number of birds retrieved from beaches as the basis for evaluating population impacts (Table 1 in Wiens 1995) is simply not credible: It ignores a large body of quantitative information and accepts an estimate that is severely biased in a known direction. On the other hand, stressing the high-end estimates of mortality ignores recognized uncertainties and accepts a known bias in the opposing direction.

Assessing the impacts of future oil spills on seabird populations is likely to include all the difficulties observed in the case of the *Exxon Valdez* spill. The kind of detailed baseline data needed to assess population changes at seabird colonies is rarely available to assess damages from acute or chronic oil pollution (Piatt et al. 1991); however, our understanding of processes affecting the recovery rates of seabirds from beaches has advanced considerably in recent years (Ford et al. 1987, 1996; Burger and Fry 1993; van Pelt and Piatt 1995), thereby providing a quantitative basis for estimating the population impacts of oil pollution.

We need to conduct more research on these processes so that we can better estimate those parameters that determine recovery of carcasses. More importantly, we should be prepared to conduct real-time drift experiments in the event of future oil spills. It is a relatively simple procedure to mark oiled birds and monitor their recovery on beach surveys. Information should be collected simultaneously on the at-sea distribution of birds and the trajectory of the oil. If properly executed, such experiments could provide relatively inexpensive and accurate estimates of total seabird mortality from individual oil spills and help in assessing mortality from spills where such experiments are not possible.

Acknowledgments

We thank Dee Boersma, Vernon Byrd, Scott Hatch, David Irons, David Roseneau, and Gerald Sanger for their constructive comments on a previous version of this article.

References

Bibby, C. J. 1981. An experiment on the recovery of dead birds from the North Sea. Ornis Scandinavica 12:261–265.

Bibby, C. J., and C. S. Lloyd. 1977. Experiments to determine the fate of dead birds at sea. Biological Conservation 12:295–309.

Burger, A. E., and M. Fry. 1993. Effects of oil pollution on seabirds in the northeast Pacific. Pages 254–263 *in* K. Vermeer, K. T. Briggs, K. H. Morgan, and D. Siegel-Causey, editors. The status, ecology, and conservation of marine birds of the north Pacific. Canadian Wildlife Service Special Publication, Ottawa.

Ecological Consulting, Inc. 1991. Assessment of direct seabird mortality in Prince William Sound and the western Gulf of Alaska resulting from the *Exxon Valdez* oil spill. Final report to the U.S. Department of Justice, Ecological Consulting Inc., Portland, Oregon.

Ford, R. G., G. W. Page, and H. R. Carter. 1987. Estimating mortality of seabirds from oil spills. Pages 747–757 *in* Proceedings, 1987 oil spill conference. American Petroleum Institute Publication, Washington, DC.

Ford, R. G., and seven coauthors. 1996. Total direct mortality of seabirds from the *Exxon Valdez* oil spill. American Fisheries Society Symposium 18: 684–711.

Harris, M. P., and S. Wanless. In press. Differential responses of guillemot *Uria* aalge and shag phalacrocorax aristotelis to a late winter wreck. Bird Study.

Hope-Jones, P., G. Howells, E. I. S. Rees, and J. Wilson. 1970. Effect of the *Hamilton Trader* oil on birds in the Irish Sea in May 1969. British Birds 63:97–110.

Hope-Jones, P., J. Y. Monnat, C. J. Cadbury, and T. J. Stowe. 1978. Birds oiled during the *Amoco Cadiz* incident—an interim report. Marine Pollution Bulletin 9:109–113.

Lloyd, C. S., J. A. Bogan, W. R. P. Bourne, P. Dawson, J. L. F. Parslow, and A. G. Stewart. 1974. Seabird mortality in the North Irish Sea and Firth of Clyde early 1974. Marine Pollution Bulletin 5:136–140.

Nysewander, D. R., C. Dippel, G. V. Byrd, and E. P. Knudtson. 1993. Effects of the T/V *Exxon Valdez* oil spill on murres: a perspective from observations at breeding colonies. Final report, Natural Resources Damage Assessment Bird Study 3. U.S. Fish and Wildlife Service, Homer, Alaska.

Page, G. W., L. E. Stenzel, and D. G. Ainley. 1982. Beached bird carcasses as a means of evaluating natural and human-caused seabird mortality. Final report to the Department of Energy, Point Reyes Bird Observatory, (Contract DE-ACO3–79EV10254), Stinson Beach, California.

Parrish, J. K., and P. D. Boersma. 1995a. Muddy waters. American Scientist 83:112–115.

Parrish, J. K., and P. D. Boersma. 1995b. Letters to the editors. American Scientist 83:396–400.

Piatt, J. 1995. Letters to the editors. American Scientist 83:396–398.

Piatt, J. F., and P. Anderson. 1996. Response of common murres to the *Exxon Valdez* oil spill and long-term changes in the Gulf of Alaska marine ecosystem. American Fisheries Society Symposium 18:720–737.

Piatt, J. F., H. R. Carter, and D. N. Nettleship. 1991. Effects of oil pollution on marine bird populations. Pages 125–141 *in* J. White, editor. The effects of oil on wildlife: research, rehabilitation and general concerns. Sheridan Press, Hanover, Pennsylvania.

Piatt, J. F., and R. G. Ford. 1993. Distribution and abundance of marbled murrelets in Alaska. Condor 95: 662–669.

Piatt, J. F., and C. J. Lensink. 1989. *Exxon Valdez* bird toll. Nature 342:865–866.

Piatt, J. F., C. J. Lensink, W. Butler, M. Kendziorek, and D. Nysewander. 1990. Immediate impact of the *Exxon Valdez* oil spill on marine birds. Auk 107:387–397.

Piatt, J. F., and T. van Pelt. 1993. Common murre die-off in Alaska. Pacific Seabird Group Bulletin 20:61.

Stowe, T. J. 1982. Beached bird surveys and surveillance of cliff-nesting seabirds. Royal Society for the Protection of Birds, Sandy, Bedfordshire, UK.

Threlfall, W., and J. F. Piatt. 1982. Assessment of offshore seabird oil mortality and corpse drift experiments. Final report for Mobile Oil Canada Ltd., Memorial University of Newfoundland, Department of Biology, St. John's, Newfoundland.

van Pelt, T., and J. F. Piatt. 1995. Deposition and persistence of beachcast seabird carcasses. Marine Pollution Bulletin 30:794–802.

Wells, P. G., J. N. Butler, and J. S. Hughes, editors. 1995. *Exxon Valdez* oil spill: fate and effects in Alaskan waters. American Society for Testing and Materials Special Technical Publication 1219, Philadelphia.

Wiens, J. A. 1993. Recovery of seabirds following the *Exxon Valdez* oil spill: an overview. Pages 854–893 *in* P. G. Wells, J. N. Butler, and J. S. Hughes, editors. *Exxon Valdez* oil spill: fate and effects in Alaskan waters. American Society for Testing and Materials Special Technical Publication 1219, Philadelphia.

American Fisheries Society Symposium 18:720–737, 1996

Response of Common Murres to the *Exxon Valdez* Oil Spill and Long-Term Changes in the Gulf of Alaska Marine Ecosystem

JOHN F. PIATT

National Biological Service, Alaska Science Center, 1011 East Tudor Road, Anchorage, Alaska 99503, USA

PAUL ANDERSON

National Marine Fisheries Service, Alaska Fisheries Science Center P.O. Box 1638, Kodiak, Alaska 99615, USA

Abstract.—Short-term effects of the 1989 TV *Exxon Valdez* oil spill on seabirds were dramatic and well documented. Seabird populations at sea in the spill zone were immediately depressed, and more than 30,000 dead, oiled seabirds were recovered from beaches within months of the spill. It is estimated that 250,000 seabirds were killed by oil, of which 74% were murres *Uria* spp. Based on comparisons of prespill (1970s) and postspill (1989–1994) data, long-term effects on murres attributed to oil pollution included population declines, reduced breeding success, and delayed breeding phenology. Populations remained depressed, but breeding success and phenology gradually returned to normal levels by 1993. An alternative hypothesis to explain these long-term effects is that murres were responding to natural events in their marine environment. Reduced flow of the Alaska Coastal Current (ACC) in 1989 may have reduced and delayed biological productivity in the ACC. On a broader time scale, marked changes in marine fish communities have occurred during the past 20 years. Coincident with cyclical fluctuations in seawater temperatures, the abundance of small forage species (e.g., northern pink shrimp *Pandalus borealis*, capelin *Mallotus villosus*, and Pacific sandfish *Trichodon trichodon*) declined precipitously in the late 1970s while populations of large predatory fish (e.g., walleye pollock *Theragra chalcogramma*, Pacific cod *Gadus macrocephalus*, and flatfish) increased dramatically. Correspondingly, seabird diets shifted from mostly capelin in the 1970s to mostly Pacific sand lance *Ammodytes hexapterus* and juvenile pollock in the late 1980s. Furthermore, a variety of seabirds and marine mammals both inside and outside of the oil spill zone exhibited signs of food stress (population declines, reduced productivity, die-offs) throughout the 1980s and early 1990s. We conclude that available data are inadequate to distinguish between long-term effects of the *Exxon Valdez* oil spill on murres and a natural response of murres to long-term changes in their marine environment.

On 24 March 1989, the TV *Exxon Valdez* grounded in Prince William Sound and spilled 43.9 million liters of crude oil into the marine environment. During the weeks that followed, winds and currents pushed oil out of Prince William Sound and into the Gulf of Alaska, where it eventually drifted 750 km to the southwest (Piatt et al. 1990; Galt et al. 1991). Pre- and postspill aerial surveys of Prince William Sound conducted in 1989 revealed that at-sea populations of loons, grebes, cormorants, and sea ducks declined by 44–84% in the immediate aftermath and trajectory of the spill (Piatt et al. 1990). In the gulf, censuses of murre colonies indicated that numbers were substantially lower (30–65%) at several colonies in 1989 than in previous years (Nysewander et al. 1993).

Between March and August, about 30,000 dead, oiled seabirds were retrieved from beaches (Piatt et al. 1990). Carcasses comprised mostly common murres *Uria aalge* (70%), thick-billed murres *U. lomvia* (4%), other alcids (7%), sea ducks (5.3%), and cormorants (3%). Extrapolating from numbers retrieved and the results of drift experiments, it was estimated that hundreds of thousands of seabirds were actually killed by oil (Piatt et al. 1990; ECI 1991; Piatt and Ford 1996, this volume). This was the highest toll of seabirds from a single oil spill ever documented (Piatt and Lensink 1989; Burger and Fry 1993).

Given these massive losses, Piatt et al. (1990) suggested that effects of the spill on murre populations might extend over many subsequent years because of the changes in demography, production, and recruitment that were predicted by population models (e.g., Ford et al. 1982; Samuels and Lanfear 1982). Furthermore, ingestion of oil by murres could have had some long-term effects on their breeding biology (e.g., Fry et al. 1986). Based on experiences elsewhere (e.g., Stowe 1982; Dunnet 1982), however, it seemed likely that population effects would be difficult to detect beyond the first year of impact (Piatt et al. 1990, 1991).

Various studies designed to assess long-term impacts on seabirds were conducted in the years

(1989–1994) after the *Exxon Valdez* oil spill, some under the auspices of the government-appointed *Exxon Valdez* Trustees Council (this volume), some under the auspices of the Exxon Corp. (Wells et al. 1995). Results of some of these studies suggested that the oil spill may have caused significant population declines, reduced breeding success, and delayed breeding phenology of murres in some colonies. In this overview, we consider results of these studies in light of information about the marine environment in the Gulf of Alaska and provide some alternate hypotheses to explain the observed changes in murre population biology.

Effects of the Spill on Murre Populations

Estimating the Size of Murre Populations

Murre census data must be interpreted with caution. We have no measure of the accuracy or precision of most prespill census data because they were usually gathered during one-time visits to colonies, and methods for censusing murres were not standardized. At least five counts per season at a murre colony are required for reliable ($P < 0.05$) detection of annual population changes on the order of ±25% (Wanless et al. 1982; Hatch and Hatch 1989). Large murre colonies are difficult to census completely in any case. At present, the preferred method is to assess population trends from small subsamples of total colony populations (e.g., Wanless et al. 1982; Piatt and McLagan 1987).

Furthermore, colony counts include only birds found at the colony at the time of censusing. In the case of murres, one bird of each breeding pair attends the nest site while the other forages. The population is also comprised of birds that do not attend the colony in any given year—including failed breeders, nonbreeders, and subadults. During the breeding season, about 1.4 times more murres than the number in attendance at colonies may be found at sea (Piatt and Ford 1993).

The choice of which pre- or postspill data to use in estimating murre population changes can be subjective. Widely varying annual population estimates may reflect real changes in murre populations over time, but they may also result from varying survey methods or conditions at the time of censusing. In the following analyses, we have selected the best census data available to assess population changes that occurred at major murre colonies (Figure 1) both before and after the oil spill (Table 1).

Determining which Colonies were "Affected"

There are no murre colonies within Prince William Sound (Figure 1). Porpoise Rocks and Middleton Island are located to the east and outside the path of the spill (see Piatt et al. 1990 and Galt et al. 1991 for discussion of the spill trajectory). About 25% of the oil escaped Prince William Sound (D. A. Wolfe and coworkers, unpublished data). In early April murres along the Kenai Peninsula (Resurrection Bay, Chiswell Islands) and in lower Cook Inlet (Barren Islands, Gull Island) were exposed to oil while it was still quite fluid, and this is when and where most murres were killed (Piatt et al. 1990). Small amounts of oil drifted south to the eastern side of Kodiak Island, possibly affecting murre colonies there (e.g., Triplet Island).

Effects of oil on murre colonies to the west of Cook Inlet (e.g., Puale Bay, Ugaiushak Island) were probably minimal. In mid- to late-April a series of storms emulsified the oil into "mousse." Between 29 April and 2 May, this mousse (then only 2% of the oil spilled) was swept up onto beaches of the Alaska Peninsula from Cape Douglas to Wide Bay (Piatt et al. 1990; Galt et al. 1991). Evidence suggests that both mousse and carcasses were carried over time by prevailing currents to Puale Bay from a point of origin in lower Cook Inlet (Piatt et al. 1990; Piatt and Ford 1996). Only small traces of oil reached the vicinity of Ugaiushak Island. Oil did not reach the Semidi Islands or coastal colonies farther west.

Murre Population Trends

Murres were the species most affected by the spill. If the total bird toll was 250,000 (Piatt and Ford 1996), then about 185,000 murres were killed. What would be the impact of such a kill on murre populations in the northern Gulf of Alaska? In a preliminary estimate, Piatt et al. (1990) suggested that the gulf murre population was reduced by about 10%. This estimate did not account for the at-sea component of populations associated with colonies (Table 2) and it used a few erroneous colony population estimates reported in Sowls et al. (1978). Including the at-sea component, and using revised historical colony estimates (Table 2), estimated losses (185,000 murres) comprised 7.0% of total prespill gulf populations (2,654,640 murres) and 41% of prespill populations in the oiled zone (446,200 murres). Actual censuses conducted after the spill suggest that murre populations in oiled areas declined by 40% and total gulf populations declined by 27% (Table 2).

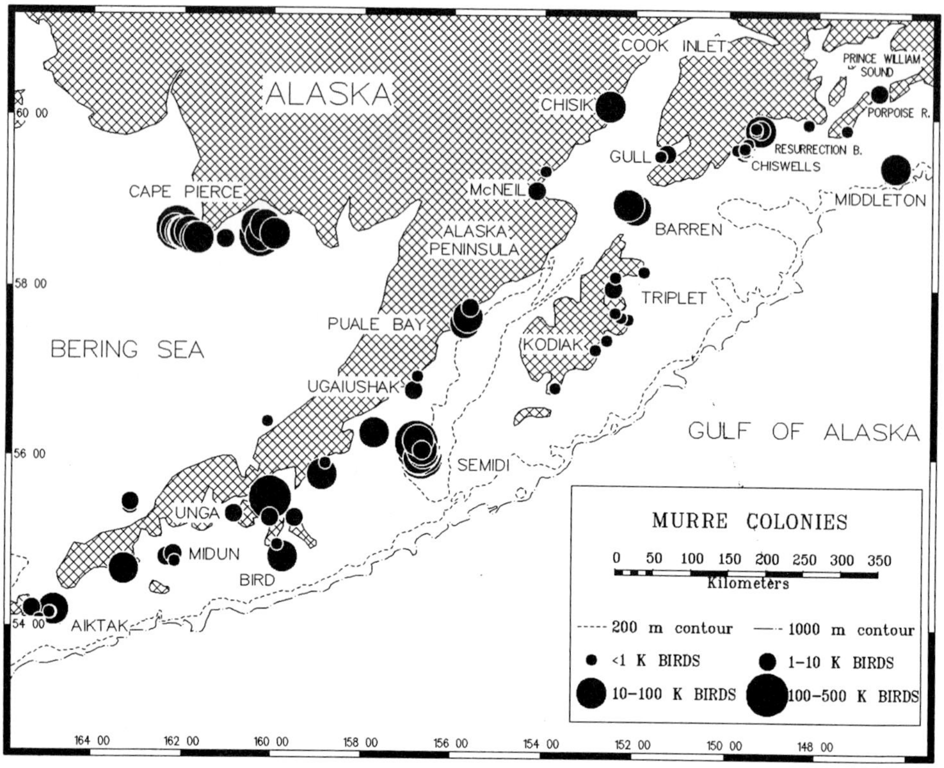

FIGURE 1.—Locations of common murre colonies in the Gulf of Alaska. The indicated colony sizes (K = 1,000) are based on census data obtained before the *Exxon Valdez* oil spill.

Nysewander et al. (1993) suggested that "since populations of murres at colonies [Middleton and Semidis] just outside the trajectory of the oil did not decline following the spill . . . it seems likely that oil mortality caused the population declines at affected colonies." It is tempting to conclude that the loss (−41%) estimated from recoveries of dead murres accounts fully for the observed decline (−40%) at "affected" colonies. However, this is probably coincidental because lower Cook Inlet is a regional "hotspot" that attracts large numbers of seabirds year-round (Gould et al. 1982; Piatt 1994), and it is likely that murres (particularly nonbreeders and subadults) from "unaffected" colonies outside the spill zone were among those killed. Furthermore, population declines occurred at "affected" colonies before the spill and at "unaffected" colonies well outside the range of the spill (Table 1).

More specifically, for colonies located within the path of the spill, major declines occurred (Figure 2) at Resurrection Bay, the Chiswell, Barren, and Triplet islands, and Puale Bay, but not at Gull Island. The average (unweighted by colony size) decline at these oiled colonies was −30%. However, at 10 colonies outside the spill path (Table 1; Figure 2), the average decline was even greater (−49%). The decline at Middleton Island represents only the most recent situation, because populations increased by about 14% between 1978 and 1988 (Hatch and Piatt 1995). Overall, however, the population decreased by 20% between 1978 and 1991.

Murre populations at the Semidi Islands appeared to be relatively stable over the time period (1975–1991) for which data are available. The Semidis may not be typical of other gulf colonies for several reasons. They are situated along the deep Shelikof trench (Figure 1), and wide-ranging murres forage over a large expanse of shelf and oceanic habitats (Gould et al. 1982; J. Piatt, unpublished data). The high diversity and abundance of breeding bird populations (Hatch and Hatch 1983, 1990a) and diversity of foraging habitats and prey types (Hatch and Hatch 1990b; Hatch and Sanger 1992; Piatt, unpublished data) all indicate that the Semidis are exceptional and stand in contrast to most coastal colonies that are relatively depauper-

TABLE 1.—Population size (individuals in attendance) of large (thousands) murre colonies in the northern Gulf of Alaska before and after the *Exxon Valdez* oil spill in 1989. Prespill changes were calculated from at least two census counts up to and including 1988. Postspill changes were calculated from the most recent census counts conducted before 1989 and counts conducted in 1989–1994. Values in parentheses are means of data from two or more consecutive years. "Years" indicate earliest and latest years of census data used in this analysis.[a]

Area	Years	Prespill change[b]		Postspill change	
		Number of birds	Percent	Number of birds	Percent
Middleton Island	1975–1991	(5,793) to (6,578)	+12%	(7,736) to (5,226)	−32%
Porpoise Rocks	1976–1991			(1,275) to 1,241	−3%
Resurrection Bay[c]	1976–1991			11,170 to 8,384	−25%
Chiswell Islands[c]	1976–1991			3,756 to (2,516)	−33%
Gull Island[c]	1984–1990	(2,323) to 5,500	+136%	5,500 to (5,125)	−7%
Chisik and Duck Islands	1978–1994	10,000 to 4,101	−59%	4,101 to (2,968)	−28%
McNeil Island	1976–1991	2,500 to 2,000	−20%	2,000 to 200	−90%
Barren Islands[c]	1975–1993			70,000 to (46,143)	−34%
Triplet Island[c]	1975–1989	(1,248) to 1,300	+4%	1,300 to 843	−35%
Puale Bay[c]	1976–1991	100,500 to 74,500	−26%	74,500 to (38,784)	−48%
Ugaiushak Island	1976–1991			8,340 to (5,073)	−39%
Semidi Islands	1977–1991	2,816 to 2,856	+1%	(2,613) to (2,973)	+12%
Bird Island	1973–1993			24,000 to 3,620	−85%
Unga Island	1973–1994			6,000 to 232	−96%
Midun Island	1978–1994			3,800 to 1,277	−57%
Aiktak Island	1981–1990			15,000 to 4,000	−73%

[a]All counts are whole-island censuses except for the Semidi Islands, which are counts of murres on plots. Census data are from the following sources and references therein: Sowls et al. (1978); Nysewander et al. (1993); Erickson (1995); Roseneau et al. (1994); USFWS, unpublished data files; S. A. Hatch, unpublished data (Middleton and Semidi Islands); J. Piatt, unpublished data (Bird, Unga, and Midun Islands), L. Slater, USFWS, unpublished data (Chisik and Gull Islands); and G. V. Byrd, USFWS, unpublished data (Aiktak Island).
[b]Absence of data in this column means data not available or not adequate for estimating prespill changes in population size.
[c]Colonies within the path of the spill (Piatt et al. 1990; Nysewander et al. 1993).

ate in seabird and prey species and are influenced primarily by a single water mass, namely, the Alaska Coastal Current (ACC) (Reed and Schumacher 1986). On the other hand, the breeding success of kittiwakes on the Semidis is often concordant with kittiwakes at coastal colonies (Hatch et al. 1993).

Owing to an influx of oceanic water from deep waters to the east, and tidal mixing at the mouth of Cook Inlet, the Barren Islands also support a diverse and abundant mix of seabird species (Piatt 1994). Similarly, Middleton Island provides access to productive shelf–edge habitat and oceanic prey (Hatch 1984; S. Hatch, National Biological Service, unpublished data). Like the Semidis, these colonies may not always reflect processes occurring in the ACC, but are often concordant with other gulf colonies with respect to kittiwake population biology (Hatch et al. 1993).

There is indirect evidence for oil spill effects on populations at some colonies. Boersma et al. (1995) noted a 21% increase in murre numbers on attendance plots at the Barren Islands (East Amatuli) between 1990 and 1991. Such a large "recovery" would be expected if postspill attendance on newly vacant ledges was supplemented by immigration from outside populations or higher recruitment of first-time breeders (Stowe 1982; Boersma et al.

TABLE 2.—Estimates of total murre populations in the northern Gulf of Alaska before (1975–1988) and after (1989–1994) the *Exxon Valdez* oil spill. Percentage changes in populations were calculated from the weighted (by colony size) average changes observed at individual colonies within each area (see Table 1).

Area	Prespill murre population	Change (%)	Postspill murre population
Middleton Island	7,740	−32	5,230
Prince William Sound and Kenai Peninsula[a]	16,410	−25	12,310
Cook Inlet[a]	81,600	−33	54,440
Kodiak Island[a]	5,070	−35	1,770
Puale Bay[a]	74,500	−48	38,780
Ugaiushak Island[a]	8,340	−39	5,070
Semidi Islands	567,090	+12	635,140
Alaska Peninsula	328,850	−85	49,330
East Aleutians	16,500	−73	4,460
Colony totals			
Oiled areas[a]	185,920	−40	112,370
Gulf of Alaska	1,106,100	−27	806,520
At-sea totals[b]			
Oiled areas[a]	260,290	−40	156,170
Gulf of Alaska	1,548,540	−27	1,130,430
Combined totals			
Oiled areas[a]	446,210	−40	268,540
Gulf of Alaska	2,654,640	−27	1,936,950

[a]Areas assumed to be affected by oil.
[b]That part of the population (nonattending breeders, failed or nonbreeders, and subadults) found at sea and not counted during colony censuses (see text).

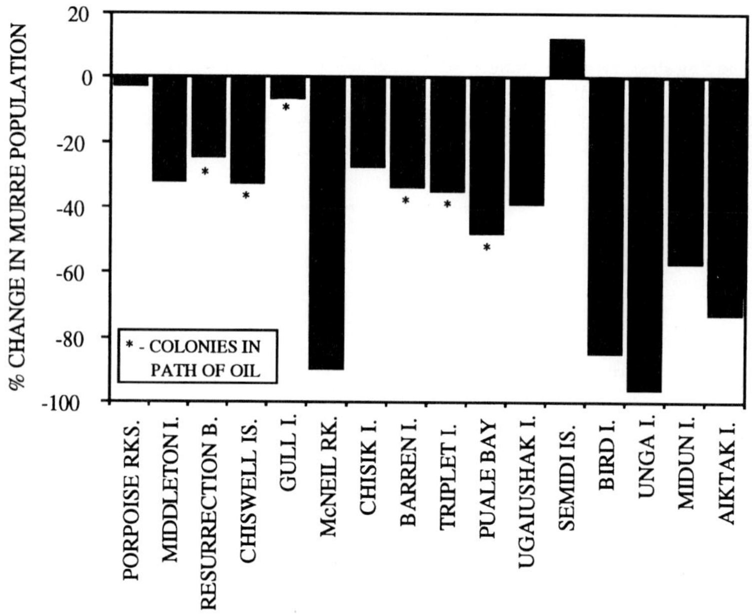

FIGURE 2.—Estimated changes in population size of common murre colonies in the Gulf of Alaska based on post- and prespill comparisons of colony census data (see also Figure 1, Table 1, and text).

1995). Similar postspill "recoveries" were noted at the Chiswell Islands (+28%), Barren Islands (Nord, East Amatuli, and Light Rock plots; +35%), and Puale Bay (+66%) by Nysewander et al. (1993) between 1989 and 1991. Changes of this magnitude (+21 to +66%) are larger than expected from normal annual variation at murre colonies (<15% per annum; Wanless et al. 1982; Harris et al. 1983; Piatt and McLagan 1987; Hatch and Hatch 1989). Postspill increases at the Semidi Islands (+10%, 1989–1991) were within the range of normal variation.

In summary, evidence for effects of the spill on murre populations is not conclusive. The accuracy of most historical census data and estimates of actual mortality from the spill are unknown. We cannot be certain of the degree to which different colonies were "affected" because the quality and toxicity of the oil changed over time, and large numbers of murres from "unaffected" colonies could have been killed while foraging in the spill zone. For this and other reasons, the Semidi and Middleton islands are probably not valid control colonies. Postspill "recoveries" at some affected colonies provide indirect evidence for oil spill effects. Murre populations were declining at some colonies before the spill, and major declines have occurred during the past 20 years at 13 of 16 major colonies in the northern Gulf of Alaska—8 of which were outside the spill zone, which suggests that other factors have contributed to declines of murre populations in the gulf.

Common Murre Breeding Success and Phenology

Prespill data from the Barren Islands and data collected there in 1989 and 1990 were not adequate to accurately estimate breeding parameters in murres (e.g., see Gaston et al. 1983; Hatch and Hatch 1989), thus, results must be interpreted with caution (Roseneau et al. 1994; Boersma et al. 1995). More extensive studies at the Barrens in 1993–1994 (Roseneau et al. 1994; D. G. Roseneau, U.S. Fish and Wildlife Service, personal communication), Puale Bay in 1989–1992 (J. H. McCarthy and D. A. Dewhurst, U.S. Fish and Wildlife Service, unpublished data), and the Semidi Islands before (Hatch and Hatch 1990a, 1990b) and after (Nysewander et al. 1993) the spill provide an accurate assessment of breeding success and phenology at these colonies.

Breeding success (chicks per site) on the Semidi Islands in both prespill (0.48–0.64) and postspill (0.52–0.60) years was close to average for common murres in Alaska (0.64 ± 0.14 [SD]; Byrd et al. 1993), as was prespill breeding success on the Bar-

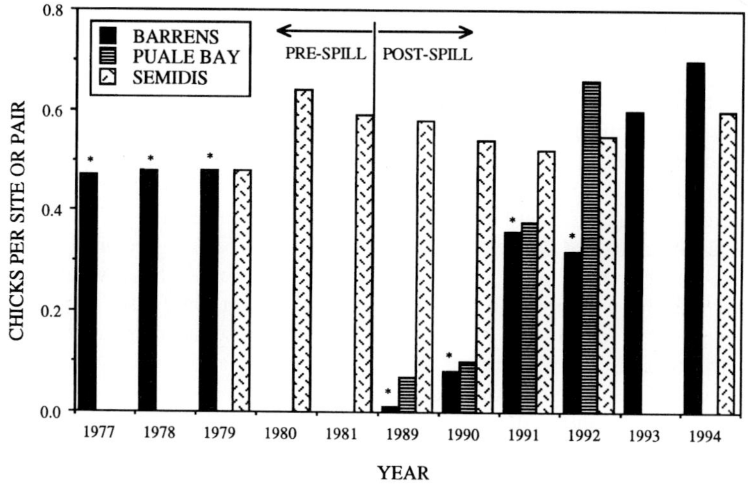

FIGURE 3.—Productivity of common murres at colonies in the Gulf of Alaska, 1977–1994. Barren Islands data in 1977–1980 is from D. A. Manuwal (U.S. Fish and Wildlife Service, unpublished data); in 1989–1992 from Nysewander et al. (1993) and Boersma et al. (1995); and in 1993–1994 from Roseneau et al. (1994). Estimates (Nysewander et al. 1993; Boersma et al. 1995) from Barren Islands in 1990 and 1991 were averaged; in 1992 only Nysewander et al. (1993) data was used. Puale Bay data is from J. H. McCarthy and D. A. Dewhurst (U.S. Fish and Wildlife Service, unpublished data). Semidis data is from Hatch and Hatch (1990b), Nyeswander et al. (1993) and J. Piatt (unpublished data). Asterisks indicate approximate or minimum values because of limited data collection (see text).

ren Islands (approximate minimum estimates; Figure 3). Phenology of first egg-laying on the Semidi and Middleton islands was likewise similar in years before (5–14 June) and after (9–24 June) the spill. First egg-laying at coastal murre colonies (Barren, Chisik, Hinchinbrook, and Ugaiushak islands) in the Gulf of Alaska before (1976–1979) the spill occurred between 17 and 30 June (Nysewander et al. 1993). These data indicate that hatching in prespill years usually occurred in late July through early August (Hatch and Hatch 1990a).

In years immediately after the spill, murres at "affected" colonies exhibited lower breeding success (Figure 3) and later breeding phenology (about 2–4 weeks) than murres in prespill years (Nysewander et al. 1993; Boersma et al. 1995; McCarthy and Dewhurst, unpublished data). Breeding success at sites on the Barren Islands (minimal data; Nysewander et al. 1993) and at Puale Bay was very low in both 1989 (<0.07 chicks/site) and 1990 (<0.10 chicks/site). In 1990, Boersma et al. (1995) also observed low levels of breeding success (about 0.15 chicks/site) on a 5 × 5-m plot at the Barrens. Breeding success recovered markedly in 1991 at the Barrens (0.47–0.72 chicks/site; Boersma et al. 1995) and at Puale Bay (0.48 chicks/site). In 1992, breeding success was higher still at Puale Bay (Figure 3). By 1993 and 1994, breeding success at the Barrens had returned to normal levels (about 0.6–0.7 chicks/site; Roseneau et al. 1994; Roseneau, personal communication).

After the spill, the onset of egg-laying at the Barren Islands and Puale Bay was later than in prespill years, occurring in early to mid-July (Nysewander et al. 1993; Boersma et al. 1995). In 1989–1991, median hatching dates for common murres were 20–23 d later at Puale Bay (4–13 September; McCarthy and Dewhurst, unpublished data) than at the Semidis (23–26 July) and at least 1 month later than normal for murres in the Gulf of Alaska (above). Median hatching dates at the Barrens were much earlier in 1993 (15–19 August) and 1994 (early to mid-August)—again suggesting a return to normal breeding conditions at the Barrens (Roseneau et al. 1994; Roseneau, personal communication).

Nysewander et al. (1993) advanced two hypotheses to explain the reduction of murre breeding success and delayed phenology at "affected" colonies in 1989–1991. First, the density of murres on breeding ledges may have been reduced drastically by the spill, leading to a reduction in breeding synchrony and success, which are often correlated with density (Birkhead 1977; Harris and Wanless 1988). Second, oil spill losses occurred early in the breeding season and therefore may have affected mostly experienced, adult breeders that return earlier to colonies than subadults (Birkhead 1977; Birkhead and Hudson 1977); this could have led to an increase in the proportion of inexperienced first-time breeders, which typically breed later and have lower breeding success compared to experienced murres (Birkhead and Hudson 1977; Gaston 1991; DeForest 1993). Unfortunately, there is no way to test either of these hypotheses because no data were collected to determine if age-classes were disproportionately affected by oil or whether pre- and postspill densities of murres on ledges were different.

An alternative hypothesis is that oil toxicity might have reduced breeding success in 1989 (Ainley et al. 1981; Trivelpiece et al. 1984) and perhaps the following year (Fry et al. 1986). It seems unlikely that toxic effects would have continued for three to four summers after the spill (Butler et al. 1988).

Another alternative hypothesis is that reduced breeding success and delayed phenology were caused by abnormal environmental conditions at the time of the spill. Breeding parameters at the Semidi Islands after the spill were within the range of prespill values there, but the Semidis may not be typical (see above) of most coastal murre colonies under the influence of the Alaska Coastal Current (ACC). It is more instructive to contrast observations at "affected" colonies with data on murres at Cape Pierce in Bristol Bay (Figure 1) and Bluff in Norton Sound. Although distant from the oil spill zone, these colonies are also bathed by the ACC, which enters the Bering Sea at Unimak Pass and flows north along the coast as far as the Chukchi Sea (Springer et al. 1984). Both colonies have similar seabird and forage fish communities to those found in ACC waters of the Gulf of Alaska (Springer et al. 1984, 1987; Murphy et al. 1986, 1991; L. M. Haggblom, U.S. Fish and Wildlife Service, unpublished data).

Murres at both Cape Pierce and Bluff also exhibited delayed phenology and reduced breeding success in 1989. At Cape Pierce, mean hatching was 27 d later in 1989 than it was in 1993, and, similarly to "affected" gulf colonies, hatching occurred progressively earlier from 1989 to 1993 (Haggblom, unpublished data). In 6 years (1989–1994) of study, breeding success was lowest in 1989 (0.47 chicks/site versus 0.55–0.64 chicks/site). During 5 years (1987–1991) of study at Bluff (Murphy and Schauer 1996), median fledging of murre chicks occurred later in 1989 (28 August) than in any other year (12–17 August), and breeding success was lowest in 1989 (0.51 chicks/site versus 0.66–0.78 chicks/site).

What natural factors might explain these observations at murre colonies in the ACC? Seawater temperatures in the Gulf of Alaska were not unusually warm or cold in 1989–1991 compared to surrounding years (Figure 4; Royer 1993). Owing to a major reduction of freshwater input in the Gulf of Alaska, however, the estimated transport of the ACC was lower in 1989 than in any year since 1932 (Niebauer et al. 1994; T. Royer, University of Alaska Fairbanks, personal communication). Effects of this reduced flow would include a reduction and delay in the onset of primary productivity (Royer, personal communication), which might have translated into reduced and delayed production by higher-trophic-level consumers such as murres. Such effects may have been attenuated with distance to colonies at Cape Pierce and Bluff. Estimated flow of the ACC returned to long-term average levels by 1993 (Royer, personal communication).

On a broader time scale, there is much evidence (see below) to suggest that the availability of forage fish species was much reduced in the late 1980s and early 1990s, and this may have had a detrimental effect on the reproductive biology of many piscivorous seabirds and marine mammals in the Gulf of Alaska during this time period.

In summary, a number of hypotheses can be advanced to explain the reduction in murre breeding success and delayed breeding phenology observed at "affected" colonies in years immediately after the oil spill. None are mutually exclusive, and it is impossible to segregate possible effects of oil pollution (reduced breeding densities, differential age-class effects, oil toxicity) from natural effects (food stress, oceanographic anomalies). However, the concordance between murres breeding in widely separated colonies within the ACC and evidence for broad-scale changes in marine food webs, suggest that natural factors were already having a strong negative influence on murres at "affected" colonies at the time of the spill and for some years later.

Changes in the Marine Environment

Oceanography and Fish Populations

Long-term oscillations in seawater temperatures in the northern North Pacific are well described (Niebauer 1983; Royer 1993). Water temperatures in the upper 250 m of the water column changed from being colder than average through the 1970s to warmer than average through the 1980s (Figure 4). The transition occurred around 1980. Niebauer (1983) linked these changes with El Niño Southern Oscillation (ENSO) events and shifts in

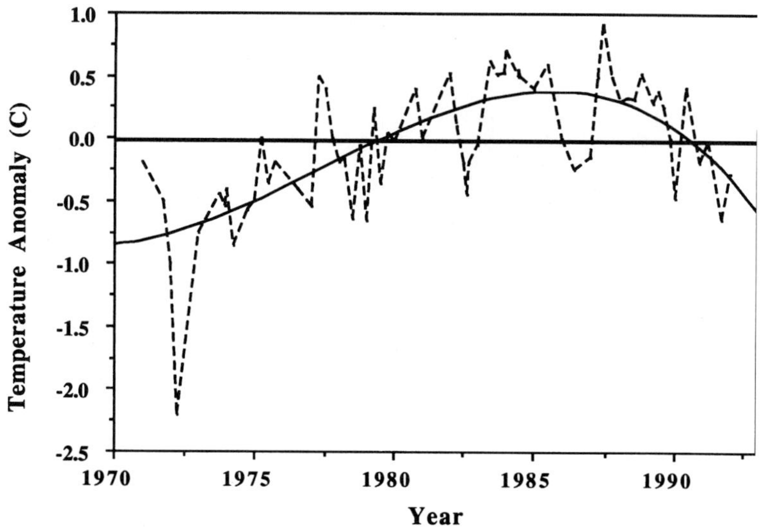

FIGURE 4.—Seawater temperature anomalies at 250-m depth in coastal waters of the northern Gulf of Alaska. Solid curved line shows third-order polynomial fit. (Redrawn from Royer 1993.)

the location and intensity of the Aleutian low pressure cell. He suggested that the shift in ocean conditions that occurred in the late 1970s was triggered by a strong 1976 ENSO event. Royer (1993) showed that this sea temperature cycle is correlated with an 18.6-year luni-solar tidal cycle. He further suggested that it could have important consequences for regional biological populations that are thermally sensitive.

Owing to an effect on metabolic rates, sea temperature anomalies of less than 2°C are sufficient to alter predator–prey relationships among fishes and cause major fluctuations in fish stocks (Laevastu 1984). Perhaps because of this (Albers and Anderson 1985), because many fishes have fairly narrow temperature preferences (e.g., capelins; Methven and Piatt 1991), and for other reasons yet to be determined, the composition and abundance of fish populations along the Alaska Peninsula changed abruptly in the late 1970s as seawater temperatures increased.

A 21-year time series of scientific trawl catches (Figures 5 and 6) in Pavlov Bay indicates that northern shrimp *Pandalus borealis* and capelins virtually disappeared between 1978 and 1979, while walleye pollock *Theragra chalcogramma*, Pacific cod *Gadus macrocephalus*, and flatfish populations increased greatly and persisted throughout the 1980s (Anderson 1991; P. J. Anderson, S. A. Payne, and B. A. Johnson, National Marine Fisheries Service, unpublished data). Mature pollock and cod stock fluctuations as determined by commercial fishery statistics show the same trends (Alverson 1992). The community transition was marked by abrupt changes in abundance of many other species as well (e.g., humpy shrimp *Pandalus goniurus*, longsnout prickleback *Lumpenella longirostris*, Pacific sandfish *Trichodon trichodon*, flathead sole *Hippoglossoides elassodon*, yellowfin sole *Pleuronectes asper*, and eulachon *Thaleichthys pacificus*; Anderson, Payne, and Johnson, unpublished data). This change in community composition was accompanied by about a 50% decrease in overall fish biomass.

The Pavlov Bay study is the longest continuous survey conducted at a single site in the Gulf of Alaska by the National Marine Fisheries Service (NMFS). But how applicable are these observations to other areas of the Gulf of Alaska? The National Marine Fisheries Service conducted trawls using the same gear in numerous bays, offshore gullies, and island passes from Unimak Pass to Castle Cape. Using trawl nets with the same design, the Alaska Department of Fish and Game (ADFG) and NMFS also sampled areas from Castle Cape to Cape Douglas (Cook Inlet), and more than 5,000 trawls were conducted in the bays and gullies around Kodiak and Afognak islands during 1969–1994. Analysis of these data are in progress. All surveys conducted in the western and central Gulf of Alaska reveal the same trend: Beginning around 1978, abrupt changes occurred in the species composition of trawl catches, and a variety of forage fishes such as cape-

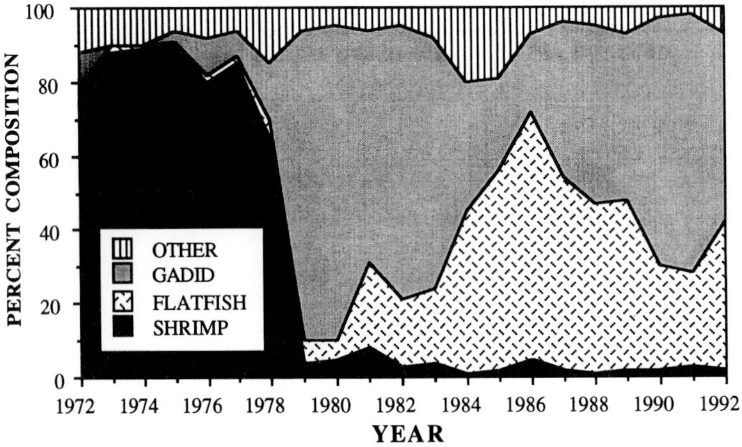

FIGURE 5.—Long-term changes in species composition (percent by weight) of taxa caught by midwater shrimp trawls in Pavlof Bay, Alaska, 1972–1992. (Taken from Anderson et al., unpublished.)

lin virtually disappeared from survey catches (Anderson et al., unpublished). Similar trends have been noted during 1969–1994 from ADFG shrimp trawls in Kachemak Bay (G. Kruse, ADFG personal communication).

In addition to natural factors influencing fish populations, human fisheries have probably had a major impact on trophic relationships in the Gulf of Alaska. During 1964–1994, ever-expanding domestic and foreign fisheries for salmon *Oncorhynchus* spp., Pacific herring *Clupea pallasi*, Atka mackerel *Pleurogrammus monopterygius*, walleye pollock, Pacific cod, Pacific halibut *Hippoglossus stenolepis*, flounders, shrimp, crab, and other species have extracted millions of tons of biomass from the Gulf of Alaska (Alverson 1992). The implications of this massive interference in natural food web relationships are unclear (Springer 1992). Large-scale removal of forage species such as herring and Atka mackerel probably has a direct impact on their predators (Alverson 1992). In some cases, removal of large predatory fish might actually benefit marine birds and mammals that compete for forage fish (Winters and Carscadden 1978; Alverson 1992;

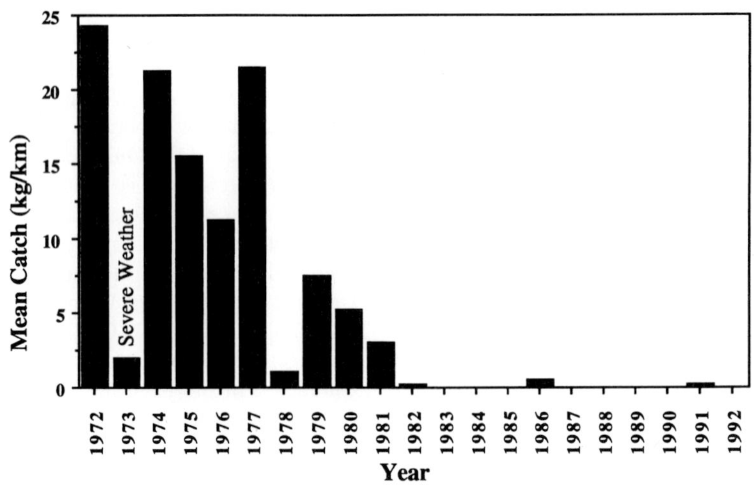

FIGURE 6.—Decline in mean catch-per-effort (kg/km) of capelins in midwater shrimp trawls in Pavlof Bay, Alaska, 1972–1992. Low catches in 1973 resulted from severe weather conditions during surveys. (Taken from Anderson et al., unpublished.)

Springer 1992). Conversely, the artificial introduction of large numbers of salmon from fish hatcheries may have a detrimental effect on competing predators. Because of environmental factors and management practices, salmon stocks were at record-high levels throughout the 1980s (Alverson 1992).

Marine Birds and Mammals

Abrupt changes in marine fish communities in the northern Gulf of Alaska are reflected in the diets and population biology of many marine birds and mammals. We consider here the diets of five abundant seabird species in the northwest Gulf of Alaska from samples (N = 1,231) collected during the Outer Continental Shelf Environmental Assessment Program (OCSEAP) (1975–1978; Sanger 1986) and in 1988–1991 (N = 560; Piatt, unpublished data). Some caution is required in comparing these samples because they were collected opportunistically at varying times of the year and in many different locations. Most OCSEAP samples were obtained in the Kodiak Island area, but locations sampled in both study periods included Prince William Sound, Kachemak Bay, Kodiak Island, the Semidi Islands, and the Shumagin Islands. In both study periods, samples were collected mostly during the summer breeding season.

Capelins were the dominant prey of seabirds in the late 1970s (Figure 7) but were absent or much reduced in seabird diets in the late 1980s and early 1990s. Capelins were replaced by Pacific sand lances *Ammodytes hexapterus* and pollock. Preliminary analysis of seabirds (N = 450) collected in 1992–1993 revealed a similar scarcity of capelins in diets (Piatt, unpublished data). Likewise, diets of northern fur seals *Callorhinus ursinus*, northern sea lions *Eumetopias jubatus*, and harbor seals *Phoca vitulina* collected in the 1970s were rich in fatty forage fish such as capelin and herring (Pitcher 1980, 1981; Castellini 1993). These prey were rare in stomachs of fur seals and Stellar sea lions collected in the 1980s (Alverson 1992; Castellini 1993; R. Merrick, NMFS, personal communication).

Coupled with these changes in diet, and probably because of them (Castellini 1993; Springer 1993), there have been marked changes in the population ecology of several marine bird and mammal species in the Gulf of Alaska. For example, breeding success of black-legged kittiwakes *Rissa tridactyla* at colonies in the Gulf of Alaska declined dramatically through the 1980s (Figure 8; Hatch et al. 1993; Hatch and Piatt 1995). Kittiwake populations have declined (about 50%) at Middleton Island since about 1980. Common murre populations have declined by up to 90% at many colonies outside the oil spill zone (Figure 2, Table 1) and were declining at some affected colonies before the spill occurred. Results of Prince William Sound boat surveys conducted in 1989–1991 suggest that there have been major declines (50–95%) in populations of 15 different coastal and marine bird species since surveys conducted in 1972–1973 (Klosiewski and Laing 1994). Species that have declined significantly include cormorants (−95%), *Larus* gulls (−69%), kittiwakes (−57%), pigeon guillemots *Cepphus columba* (−75%), marbled and Kittlitz's murrelets *Brachyramphus* spp. (−68%), and horned puffins *Fratercula corniculata* (−65%). Similarly, analysis of Christmas Bird Count data suggests that marbled murrelet populations in the Gulf of Alaska have declined by over 50% since the early 1970s (Piatt and Naslund 1995).

As another indication that food has been limiting in recent years, several large-scale die-offs of seabirds, mostly of surface-feeding species, have been observed in the Gulf of Alaska during 1983–1993, most notably in 1983, 1989, and 1993 (Hatch 1987; Piatt et al. 1990; D. R. Nysewander and J. L. Trapp, U.S. Fish and Wildlife Service, unpublished data). In March 1993, an estimated 100,000 common murres died throughout an area spanning southeast Alaska to the Alaska Peninsula (Piatt and van Pelt 1993). Starvation was determined to be the proximate cause of death. In contrast, there were no reported die-offs in the 1970s since Bailey and Davenport's (1972) report of a massive murre die-off along the Bering Sea side of the Alaska Peninsula in 1970.

Marine mammals have exhibited similar signs of food stress in recent years. Harbor seals at Tugidak Island in the Gulf of Alaska declined by about 85% between 1976 and 1988 (Figure 9; Pitcher 1990). Northern sea lion populations declined (Figure 9) by 36% in the Gulf of Alaska between 1977 and 1985 (Merrick et al. 1987) and by another 59% between 1985 and 1990 (Loughlin et al. 1992; Castellini 1993). Northern fur seals declined by about 35% by 1986 from their average numbers in the 1970s, although numbers had rebounded somewhat (20%) by 1990 (Castellini 1993). Associated with the declines in northern sea lions are declines in birth rate, fewer breeding females, fewer pups, decreased adult body condition, decreased juvenile survival, and a change in population age structure (Merrick et al. 1987; Loughlin et al. 1992). Declines in marine mammal populations may result in part

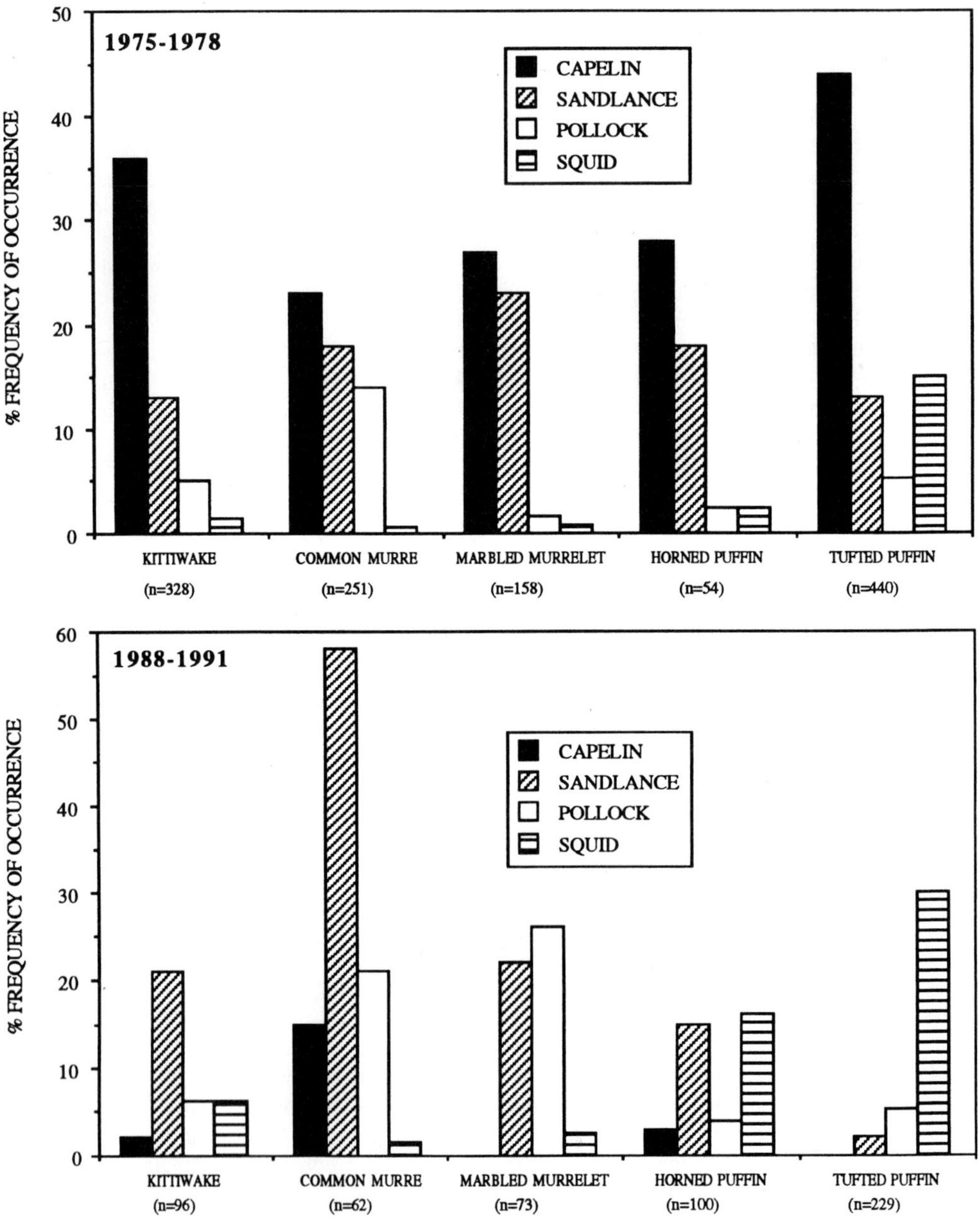

FIGURE 7.—Change in diet composition (% frequency of occurrence) of five seabird species collected in the Gulf of Alaska between 1975–1978 (upper panel; Outer Continental Shelf Environmental Assessment Program data; Sanger 1986) and 1988–1991 (lower panel; J. Piatt and A. Springer, unpublished data).

from harvest, incidental catch in fishing gear, and disease, but the concensus among researchers is that changing food availability (lower quality or biomass) is the most likely cause of recent declines (Castellini 1993).

Conclusions

A Changing Marine Environment

Various independent data indicate that there has been an abrupt shift in the marine ecosystem of the

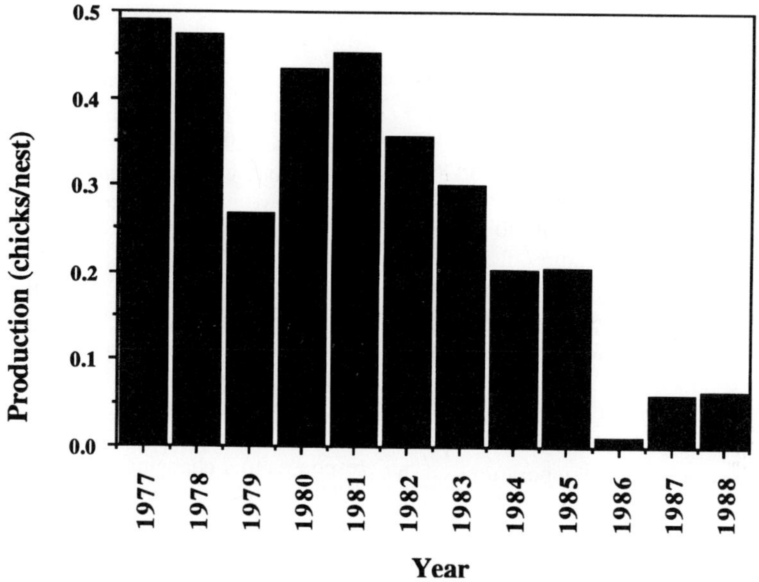

FIGURE 8.—Declining productivity of black-legged kittiwakes in the Gulf of Alaska, 1976–1989. Data from Middleton, Barren, and Semidi islands and Chiniak Bay (Hatch et al. 1993) were grouped for this analysis and plotted here as 3-year running-means from 1977 to 1988.

Gulf of Alaska between the mid-1970s and the mid-1990s. This shift has been manifested by marked changes in the composition of marine fish communities, reduced forage fish biomass, and dramatic changes in the population ecology of higher vertebrates that depend on those fish populations. Unlike short-term phenomena such as the El Niño, which may disrupt marine food webs and diminish seabird productivity for 1- to 2-year time periods (Ainley and Boekelheide 1990), this shift represents a more pervasive and persistent change in the ecosystem.

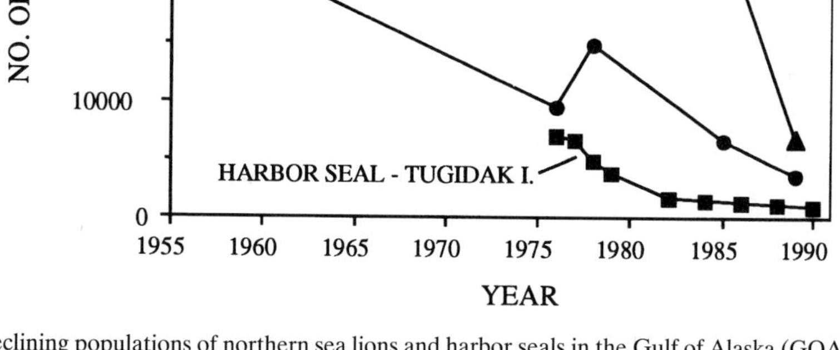

FIGURE 9.—Declining populations of northern sea lions and harbor seals in the Gulf of Alaska (GOA). (Derived from Pitcher 1990, Merrick et al. 1987, and Loughlin et al. 1992.)

It appears that a "change of state" was initiated in 1976, when atmospheric circulation in the north Pacific shifted and remained that way until the late 1980s (Trenberth 1990; Kerr 1992). Effects of this climate shift were evident in a number of environmental variables measured throughout the eastern North Pacific between 1968 and 1984 (Kerr 1992). It may have taken 3–5 years for this change of state to be manifested by changes in water temperatures, fish populations, and seabird demography in Alaska. Atmospheric and ocean climate conditions may now be returning to those observed before 1976 (Figure 1; Trenberth 1990; Royer 1993).

Although they do not appear to be synchronized with events in the Gulf of Alaska, similar temporal relationships between water temperatures, fish stocks, seabirds, and marine mammals have been observed in the Bering Sea (Alverson 1992; Springer 1992; Decker et al., in press). Productivity of seabirds nesting on the Pribilof Islands started declining in about 1978, as water temperatures rose above average and an unusually large year-class of pollock appeared. Seabird productivity remained low through the 1980s, while pollock stocks increased dramatically. Diet composition of murres and kittiwakes changed significantly, and indicator species such as capelins and hyperiid amphipods largely disappeared from diets in the 1980s (Decker et al., in press). By the late 1980s, seabird productivity was increasing again as pollock stocks declined. As in the Gulf of Alaska (above), a strong negative relationship between pollock biomass and kittiwake production may be an indication that predators compete for forage fish and that food webs are regulated by "top-down" interactions (Springer 1992). Alternatively, forage fish, pollock, kittiwakes, and other predators may be responding in their own fashion to changing oceanographic conditions and primary production ("bottom-up" control; Springer 1992).

Whatever the mechanism, natural long-term cycles in marine productivity may account for much of the variation observed in seabird population parameters. For example, Aebischer et al. (1990) demonstrated a remarkable similarity between long-term (33-year) trends in wind patterns, the abundance of phytoplankton, zooplankton, and Atlantic herring *Clupea harengus*, and kittiwake clutch size, phenology, and chick production in the North Sea. All these measures of biological production declined from about 1955 to the late 1970s, and have been increasing since that time.

Murre Population Declines

The magnitude of seabird mortality (about 250,000 killed) from the *Exxon Valdez* oil spill was high enough to expect that some seabird populations, especially those of murres, were damaged to a degree that was biologically significant. We conclude, however, that (1) coastal murre colonies under the influence of the ACC were already under stress and declining when the oil spill occurred; (2) a high toll of murres from the spill must have added to this decline, resulting in an overall decline of about 40% at affected colonies between the late 1970s and early 1990s; and (3) the quality and quantity of prespill colony data do not allow us to distinguish between effects of the spill and natural responses to a changing marine environment.

The magnitude of murre declines observed in the Gulf of Alaska are not unique. Massive declines (−70 to −90%) of common murre populations occurred quickly (1985–1987) in the Barents Sea, apparently in response to a collapse of capelin stocks (Vader et al. 1990). Murre populations at 28 colonies on the outer coast of Washington plummeted (about −90%) between 1982 and 1983, and populations have not recovered as of 1994 (Ainley et al. 1994). Similarly, common murre populations in central California declined 53% over a 6-year period (1980–1986), and populations are recovering very slowly (Takekawa et al. 1990; Ainley et al. 1994). The 1982–1983 ENSO event, gill-net mortality, oil spills, and competition from commercial fisheries are all implicated in the decline of murres in California and Washington.

In all these cases, it has not been possible to quantitatively apportion blame to any one cause of population declines. It is increasingly clear that large-scale fluctuations in prey abundance, usually mediated by oceanographic phenomena and commercial fisheries, have far greater influence on seabird population dynamics than local, one-time events such as oil spills. There is much circumstantial evidence to indicate that chronic oil pollution has long-term effects on seabird populations (Piatt et al. 1991), but it has been difficult to demonstrate long-term damages in acute cases (but see Stowe 1982).

Murre Breeding Success and Phenology

None of the postspill murre studies can show conclusively that the observed postspill reductions in murre breeding success and delayed phenology resulted directly from the oil spill. We can reasonably speculate about potential effects of changing demography and breeding densities, and toxic effects of oil, but a more plausible hypothesis is that changes in the marine environment of the Gulf of Alaska were primarily responsible for the observed changes in murre breeding biology.

Such effects have been documented repeatedly for murres and other seabirds in response to food stress or changes in oceanographic conditions (e.g., Boersma 1978; Hedgren 1979; Anderson and Gress 1984; Monaghan et al. 1989; Ainley and Boekelheide 1990; Murphy et al. 1991; Wilson 1991; Duffy 1993; Uttley et al. 1994). For example, median egg-laying of murres, Cassin's auklets *Ptychoramphus aleuticus*, and pigeon guillemots at the Farallon Islands were observed to be delayed by 2–6 weeks in association with ENSOs or anomalously warm water temperatures (Ainley and Boekelheide 1990). For common murres, mean dates of egg-laying varied by 1 month during 12 years of study (Boekelheide et al. 1990). Similarly, feeding rates and breeding success of several species at the Farallons declined markedly during ENSO events.

Reports of complete murre breeding failures are rare, however, and under a wide range of moderate to good conditions they exhibit low variance in breeding success—much lower than coexisting species (Piatt 1987, 1990; Boekelheide et al. 1990; Hatch and Hatch 1990b). For example, coefficients of variation in common murre breeding success at the Semidi (16%) and Farallon (29%) islands are far lower than those of coexisting kittiwakes (101%) and pelagic cormorants (85–97%). This situation probably occurs because murres can buffer against fluctuations in prey abundance by spending more or less time foraging (Piatt 1987, 1990; Burger and Piatt 1990; Uttley et al. 1994).

Despite their buffering capacity, murres at colonies in the ACC exhibited low levels of breeding success and markedly delayed phenology in 1989 and 1990. Kittiwakes also failed (0.0–0.06 chicks/nest) in 1989 (Hatch et al. 1993—no data for 1990) at most colonies in the Gulf of Alaska (Middleton, Barren, and Semidi islands) and Bering Sea (Cape Pierce, Pribilof Islands, and Bluff). Two different factors may have influenced murre foraging success in these years. First, major long-term changes in forage fish availability (above) may have already pushed murres to the limits of their capacity to deal with fluctuations in prey abundance in the late 1980s and early 1990s. Second, unusually low-volume transport of the ACC (above) may have reduced and delayed primary productivity in the coastal marine environment (Niebauer et al. 1994; Royer, personal communication), leading to a similar delay and reduction in productivity of murres.

Perspective on Oil Spill Impacts and Recovery

Our inability to demonstrate unequivocal long-term effects of the oil spill on murre populations should not lead us to a conclusion that there were none (e.g., Boersma et al. 1995; Erickson 1995). Wiens (1995) argued that biological systems are so variable that the effects of oil mortality were probably biologically insignificant and, in any case, statistically undetectable. We would argue that we lack the necessary data to draw strong conclusions one way or the other. We reach the heights of human arrogance when we assume that eliminating hundreds of thousands of healthy seabirds was unimportant to a species simply because we had few data on some narrowly defined parameters with which to "assess" damages. Furthermore, impacts of a single event like the *Exxon Valdez* oil spill may over the long term be additive to other sources of mortality (e.g., from gill-nets, starvation) and should not be evaluated in isolation (Piatt et al. 1991). On the other hand, the one-time loss of 7% of gulf murre populations is not a drastic occurrence for a species as resilient as the common murre. At least this percentage of murre populations dies annually from natural mortality (Birkhead and Hudson 1977; Sydeman 1993).

When will murre populations recover? Some early predictions (e.g., Piatt et al. 1990) about 20- to 70-year recovery times were based on demographic models (Ford et al. 1982) that assumed stability of the marine ecosystem. As a coauthor of these original models, Wiens (1995) more recently points out that this assumption is invalid. Nonetheless, evidence suggests that current conditions in the Gulf of Alaska are not conducive to a more rapid recovery of murre populations. Until we achieve a much better understanding of long-term cycles in the marine environment and factors influencing seabird demography, predictions about long-term impacts of oil pollution on seabird populations will remain largely speculative.

One positive consequence of the *Exxon Valdez* oil spill is that it forced us to gather more and better data on neglected seabird populations. Maybe we will be able to improve quantification of effects during the next oil spill. Meanwhile, the data collected on murres and other taxa provide insight into a phenomenon of greater impact and importance, namely, long-term changes in the marine ecosystem of the Gulf of Alaska. We need to study further the mechanisms that have led to chronic population declines of shrimp (>95%), capelins (>95%), cormorants (>90%), kittiwakes (>50%), murres (>50%), marbled murrelets (>60%), puffins (>60%), harbor seals (>85%), and sea lions (>90%) in some areas of the Gulf of Alaska from the mid-1970s to the mid-1990s.

Acknowledgments

For their thoughtful and constructive reviews of this paper, we thank Dee Boersma, Vernon Byrd, Scott Hatch, David Irons, David Roseneau, and Gerald Sanger. We are grateful to Bruce Wright for providing the opportunity to contribute this paper to the symposium proceedings.

References

Aebischer, N. J., J. C. Coulson, and J. M. Colebrook. 1990. Parallel long-term trends across four marine trophic levels and weather. Nature 347:753–755.

Ainley, D. G., and R. J. Boekelheide, editors. 1990. Seabirds of the Farallon Islands-ecology and dynamics of an upwelling-system community. Stanford University Press, Stanford, California.

Ainley, D. G., C. R. Grau, T. E. Roudybush, S. R. Morrell, and J. M. Utts. 1981. Petroleum ingestion reduces reproduction in Cassin's auklet. Marine Pollution Bulletin 12:314–317.

Ainley, D. G., W. J. Sydeman, S. A. Hatch, and U. W. Wilson. 1994. Seabird population trends along the west coast of North America: causes and the extent of regional concordance. Studies in Avian Biology 15: 119–133.

Albers, W. D., and P. J. Anderson. 1985. Diet of Pacific cod, *Gadus macrocephalus*, and predation on the northern pink shrimp, *Pandalus borealis*, in Pavlof Bay, Alaska. U.S. National Marine Fisheries Service Fishery Bulletin 83:601–610.

Alverson, D. L. 1992. A review of commercial fisheries and the Stellar sea lion (*Eumetopias jubatus*): the conflict arena. Reviews in Aquatic Sciences 6:203–256.

Anderson, P. J. 1991. Age, growth, and mortality of the northern shrimp *Pandalus borealis* Kröyer in Pavlof Bay, Alaska. U.S. National Marine Fisheries Service Fishery Bulletin 89:541–553.

Anderson, D. W., and F. Gress. 1984. Brown pelicans and the anchovy fishery off southern California. Pages 128–135 *in* D. N. Nettleship, G. A. Sanger, and P. F. Springer, editors. Marine birds: their feeding ecology and commercial fishery relationships. Canadian Wildlife Service Special Publication, Ottawa.

Bailey, E. P., and G. H. Davenport. 1972. Die-off of common murres on the Alaska Peninsula and Unimak Island. Condor 74:213–219.

Birkhead, T. R. 1977. The effect of habitat and density on breeding success in the common guillemot *Uria aalge*. Journal of Animal Ecology 46:751–764.

Birkhead, T. R., and P. J. Hudson. 1977. Population parameters for the common guillemot *Uria aalge*. Ornis Scandinavica 8:145–154.

Boekelheide, R. J., D. G. Ainley, S. H. Morrell, H. H. Huber, and T. J. Lewis. 1990. Common murre. Pages 245–275 *in* D. G. Ainley and R. J. Boekelheide, editors. Seabirds of the Farallon Islands-ecology and dynamics of an upwelling-system community. Stanford University Press, Stanford, California.

Boersma, P. D. 1978. Breeding patterns of Galapagos penguins as an indicator of oceanographic conditions. Science 200:1481–1483.

Boersma, P. D., J. K. Parrish, and A. B. Kettle. 1995. Common murre abundance, phenology, and productivity on the Barren Islands, Alaska: the *Exxon Valdez* oil spill and long-term environmental change. Pages 820–853 *in* P. G. Wells, J. N. Butler, and J. S. Hughes, editors. *Exxon Valdez* oil spill: fate and effects in Alaskan waters. American Society for Testing and Materials Special Technical Publication 1219, Philadelphia.

Burger, A. E., and M. Fry. 1993. Effects of oil pollution on seabirds in the northeast Pacific. Pages 254–263 *in* K. Vermeer, K. T. Briggs, K. H. Morgan, and D. Siegel-Causey, editors. The status, ecology, and conservation of marine birds of the north Pacific. Canadian Wildife Service Special Publication, Ottawa.

Burger, A. E., and J. F. Piatt. 1990. Flexible time budgets in breeding common murres: buffers against variable prey availability. Studies in Avian Biology 14:71–83.

Butler, R. G., A. Harfenist, F. A. Leighton, and D. B. Peakall. 1988. Impact of sublethal oil and emulsion exposure on the reproductive success of Leach's storm-petrels: short- and long-term effects. Journal of Applied Ecology 25:125–143.

Byrd, G. V., E. C. Murphy, G. W. Kaiser, A. J. Kondratyev, and Y. V. Shibaev. 1993. Status and ecology of offshore fish-feeding alcids (murres and puffins) in the north Pacific. Pages 176–186 *in* K. Vermeer, K. T. Briggs, K. H. Morgan, and D. Siegel-Causey, editors. The status, ecology, and conservation of marine birds of the north Pacific. Canadian Wildife Service Special Publication, Ottawa.

Castellini, M. 1993. Report of the marine mammal working group. Pages 4–13 *in* Is it food?: addressing marine mammal and seabird declines (workshop summary). University of Alaska, Alaska Sea Grant Report 93-01, Fairbanks.

Decker, M. B., G. L. Hunt, Jr., and G. V. Byrd. In press. The relationship between sea-surface temperature, the abundance of juvenile walleye pollock (*Theragra chalcogramma*), and the reproductive performance and diets of seabirds at the Pribilof Islands, in the southeastern Bering Sea. Canadian Journal of Fisheries and Aquatic Sciences.

DeForest, L. N. 1993. The effects of age, timing of breeding, and breeding-site characteristics on the reproductive success of the thick-billed murre (*Uria lomvia*). Master's thesis. University of Ottawa, Ottawa, Canada.

Duffy, D. C. 1993. Stalking the southern oscillation: environmental uncertainty, climate change, and north Pacific seabirds. Pages 61–67 *in* K. Vermeer, K. T.

Briggs, K. H. Morgan, and D. Siegel-Causey, editors. The status, ecology, and conservation of marine birds of the north Pacific. Canadian Wildife Service Special Publication, Ottawa.

Dunnet, G. M. 1982. Oil pollution and seabird populations. Philosophical Transactions of the Royal Society of London B Biological Sciences 297:413–427.

ECI (Ecological Consulting Inc.). 1991. Assessment of direct seabird mortality in Prince William Sound and the western Gulf of Alaska resulting from the *Exxon Valdez* oil spill. Final Report to the U.S. Department of Justice. Ecological Consulting, Inc., Portland, Oregon.

Erickson, D. E. 1995. Surveys of murre colony attendance in the northern Gulf of Alaska following the *Exxon Valdez* oil spill. Pages 780–819 *in* P. G. Wells, J. N. Butler, and J. S. Hughes, editors. *Exxon Valdez* oil spill: fate and effects in Alaskan waters. American Society for Testing and Materials Special Technical Publication 1219, Philadelphia.

Ford, R. G., J. A. Wiens, D. Heinemann, and G. L. Hunt. 1982. Modelling the sensitivity of colonially breeding marine birds to oil spills: guillemot and kittiwake populations on the Pribilof Islands, Bering Sea. Journal of Applied Ecology 19:1–31.

Fry, D. M., J. Swenson, G. A. Addiego, C. R. Grau, and A. Kang. 1986. Reduced reproduction of wedge-tailed shearwaters exposed to weathered Santa Barbara crude oil. Archives of Environmental Contamination and Toxicology 15:453–463.

Galt, J. A., W. J. Lehr, and D. L. Payton. 1991. Fate and transport of the *Exxon Valdez* oil spill. Environmental Science and Technology 25:202–209.

Gaston, A. J. 1991. Project on the population dynamics of the thick-billed murre, *Uria lomvia*: interim report on studies at Coats Island. Canadian Wildlife Service Technical Report Series 134.

Gaston, A. J., D. G. Noble, and M. A. Purdy. 1983. Monitoring breeding biology parameters for murres *Uria* spp.: levels of accuracy and sources of bias. Journal of Field Ornithology 54:275–282.

Gould, P. J., D. J. Forsell, and C. J. Lensink. 1982. Pelagic distribution and abundance of seabirds in the Gulf of Alaska and eastern Bering Sea. U.S. Fish and Wildlife Service, Biological Services Program, OBS 82/48.

Harris, M. P., and S. Wanless. 1988. The breeding biology of guillemots *Uria aalge* on the Isle of May over a six year period. Ibis 130:172–192.

Harris, M. P., S. Wanless, and P. Rothery. 1983. Assessing changes in numbers of guillemots *Uria aalge* at breeding colonies. Bird Study 30:57–66.

Hatch, S. A. 1984. Nestling diet and feeding rates of rhinoceros auklets in Alaska. Pages 106–115 *in* D. N. Nettleship, G. A. Sanger, and P. F. Springer, editors. Marine birds: their feeding ecology and commercial fishery relationships. Canadian Wildlife Service Special Publication, Ottawa.

Hatch, S. A. 1987. Did the 1982–1983 El Niño-Southern Oscillation affect seabirds in Alaska? Wilson Bulletin 99:468–474.

Hatch, S. A., G. V. Byrd, D. B. Irons, and G. L. Hunt. 1993. Status and ecology of kittiwakes (*Rissa tridactyla* and *R. brevirostris*) in the north Pacific. Pages 140–153 *in* K. Vermeer, K. T. Briggs, K. H. Morgan, and D. Siegel-Causey, editors. The status, ecology, and conservation of marine birds of the north Pacific. Canadian Wildife Service Special Publication, Ottawa.

Hatch, S. A., and M. A. Hatch. 1983. Populations and habitat use of marine birds in the Semidi Islands, Alaska. Murrelet 64:39–46.

Hatch, S. A., and M. A. Hatch. 1989. Attendance patterns of murres at breeding sites: implications for monitoring. Journal of Wildlife Management 53:483–493.

Hatch, S. A., and M. A. Hatch. 1990a. Breeding seasons of oceanic birds in a subarctic colony. Canadian Journal of Zoology 68:1664–1679.

Hatch, S. A., and M. A. Hatch. 1990b. Components of breeding productivity in a marine bird community—key factors and concordance. Canadian Journal of Zoology 68:1680–1690.

Hatch, S. A., and J. F. Piatt. 1995. Seabirds in Alaska. Pages 49–52 *in* E. T. La Roe, G. S. Farris, C. E. Puckett, P. D. Doran, and M. J. Mac, editors. Our living resources: a report to the nation on the distribution, abundance, and health of U.S. plants, animals, and ecosystems. U.S. Department of the Interior, National Biological Service, Washington, DC.

Hatch, S. A., and G. Sanger. 1992. Puffins as samplers of juvenile walleye pollock and other forage fish in the Gulf of Alaska. Marine Ecology Progress Series 80: 1–14.

Hedgren, S. 1979. Seasonal variation in fledging weights of guillemots, *Uria aalge*. Ibis 121:356–361.

Kerr, R. A. 1992. Unmasking a shifty climate system. Science 255:1508–1510.

Klosiewski, S. P., and K. K. Laing. 1994. Marine bird populations of Prince William Sound, Alaska, before and after the *Exxon Valdez* oil spill. Final report, Natural Resource Damage Assessment Bird Study 2. U.S. Fish and Wildlife Service, Migratory Bird Management, Anchorage, Alaska.

Laevastu, T. 1984. The effects of temperature anomalies on the fluctuation of fish stocks. Conseil International pour l'Exploration de la Mer 185:214–225.

Loughlin, T. R., A. S. Perlov, and V. A. Vladimirov. 1992. Range-wide survey and estimation of total number of Stellar sea lions in 1989. Marine Mammal Science 8:220–239.

Merrick, R. L., T. R. Loughlin, and D. G. Calkins. 1987. Decline in abundance of the northern sea lion, *Eumetopias jubatus*, in Alaska, 1956–86. U.S. National Marine Fisheries Service Fishery Bulletin 85:351–365.

Methven, D. A., and J. F. Piatt. 1991. Seasonal abundance and vertical distribution of capelin (*Mallotus villosus*) in relation to water temperature at a coastal site off eastern Newfoundland. ICES (International Council for the Exploration of the Sea) Journal of Marine Science 48:187–193.

Monaghan, P., J. D. Uttley, M. Burns, C. Thane, and J. Blackwood. 1989. The relationship between food supply, reproductive effort, and breeding success in

arctic terns *Sterna paradisea*. Journal of Animal Ecology 58:261–274.
Murphy, E. C., and J. H. Schauer. In press. The breeding biology of common murres at Bluff, Alaska in 1975–1991. Canadian Journal of Zoology.
Murphy, E. C., A. M. Springer, and D. G. Roseneau. 1986. Population status of common guillemots *Uria aalge* at a colony in western Alaska: results and simulations. Ibis 128:348–363.
Murphy, E. C., A. M. Springer, D. G. Roseneau, and B. A. Cooper. 1991. High annual variability in the reproductive success of kittiwakes (*Rissa tridactyla*) at a colony in western Alaska. Journal of Animal Ecology 60:515–534.
Niebauer, H. J. 1983. Multiyear sea ice variability in the eastern Bering Sea: an update. Journal of Geophysical Research 88:2733–2742.
Niebauer, H. J., T. C. Royer, and T. J. Weingartner. 1994. Circulation of Prince William Sound, Alaska. Journal of Geophysical Research 99:113–126.
Nysewander, D. R., C. Dippel, G. V. Byrd, and E. P. Knudtson. 1993. Effects of the T/V *Exxon Valdez* oil spill on murres: a perspective from observations at breeding colonies. Final report, Natural Resources Damage Assessment Bird Study 3. U.S. Fish and Wildlife Service, Migratory Bird Management, Homer, Alaska.
Piatt, J. F. 1987. Behavioral ecology of common murre and Atlantic puffin predation on capelin: implications for population biology. Doctoral dissertation. Memorial University of Newfoundland, St. John's.
Piatt, J. F. 1990. Aggregative response of common murres and Atlantic puffins to their prey. Studies in Avian Biology 14:36–51.
Piatt, J. F. 1994. Oceanic, shelf, and coastal seabird assemblages at the mouth of a tidally-mixed estuary (Cook Inlet, Alaska). OCS (Outer Continental Shelf) Study MMS-92, Final report for Minerals Management Service, Anchorage, Alaska.
Piatt, J. F., H. R. Carter, and D. N. Nettleship. 1991. Effects of oil pollution on marine bird populations. Pages 125–141 *in* J. White, editor. The effects of oil on wildlife: research, rehabilitation and general concerns. Sheridan Press, Hanover, Pennsylvania.
Piatt, J. F., and R. G. Ford. 1993. Distribution and abundance of marbled murrelets in Alaska. Condor 95: 662–669.
Piatt, J. F., and R. G. Ford. 1996. How many seabirds were killed by the *Exxon Valdez* oil spill? American Fisheries Society Symposium 18:712–719.
Piatt, J. F., and C. J. Lensink. 1989. *Exxon Valdez* bird toll. Nature 342:865–866.
Piatt, J. F., C. J. Lensink, W. Butler, M. Kendziorek, and D. Nysewander. 1990. Immediate impact of the *Exxon Valdez* oil spill on marine birds. Auk 107:387–397.
Piatt, J. F., and R. McLagan. 1987. Colony attendance patterns of common murres (*Uria aalge*) at Cape St. Mary's, Newfoundland. Canadian Journal of Zoology 65:1530–1534.
Piatt, J. F., and N. L. Naslund. 1995. Abundance, distribution and population status of marbled murrelets in Alaska. Pages 285–294 *in* C. J. Ralph, G. L. Hunt, M. Raphael, and J. F. Piatt, editors. Conservation assessment for the marbled murrelet. U.S. Forest Service Technical Report, PSW-152.
Piatt, J. F., and T. van Pelt. 1993. Common murre die-off in Alaska. Pacific Seabird Group Bulletin 20:61.
Pitcher, K. W. 1980. Food of the harbor seal, *Phoca vitulina richardsi*, in the Gulf of Alaska. U.S. National Marine Fisheries Service Fishery Bulletin 78:544–549.
Pitcher, K. W. 1981. Prey of the Stellar sea lion, *Eumetopias jubatus*, in the Gulf of Alaska. U.S. National Marine Fisheries Service Fishery Bulletin 79:467–472.
Pitcher, K. W. 1990. Major decline in number of harbor seals, *Phoca vitulina richardsi*, on Tugidak Island, Gulf of Alaska. Marine Mammal Science 6:121–134.
Reed, R. K., and J. D. Schumacher. 1986. Physical oceanography. Pages 57–76 *in* D. W. Hood and S. T. Zimmerman, editors. The Gulf of Alaska, physical environment and biological resources. U.S. Government Printing Office, Washington DC.
Roseneau, D. G., A. B. Kettle, and G. V. Byrd. 1994. Results of the common murre restoration monitoring project in the Barren Islands, Alaska, 1993. Final report to the *Exxon Valdez* Oil Spill Trustees Council (Project 93049), Homer, Alaska.
Royer, T. C. 1993. High-latitude oceanic variability associated with the 18.6-year nodal tide. Journal of Geophysical Research 98:4639–4644.
Samuels, W. B., and K. J. Lanfear. 1982. Simulations of seabird damage and recovery from oil spills in the northern Gulf of Alaska. Journal of Environmental Management 15:169–182.
Sanger, G. A. 1986. Diets and food-web relationships of seabirds in the Gulf of Alaska and adjacent marine regions. NOAA (National Oceanic and Atmospheric Adminsitration), OCSEAP (Outer Continental Shelf Environmental Assessment Program) Final Report 45:631–771.
Sowls, A. L., S. A. Hatch, and C. J. Lensink. 1978. Catalog of Alaskan seabird colonies. U.S. Fish and Wildlife Service catalog, FWS/OBS-78/78.
Springer, A .M. 1992. A review: walleye pollock in the north Pacific-how much difference do they really make? Fisheries Oceanography 1:80–96.
Springer, A. M. 1993. Report of the seabird working group. Pages 14–29 *in* Is it food?: addressing marine mammal and seabird declines (workshop summary). University of Alaska, Alaska Sea Grant Report 93-01, Fairbanks.
Springer, A. M., E. C. Murphy, D. G. Roseneau, C. P. McRoy, and B. A. Cooper. 1987. The paradox of pelagic food webs in the northern Bering Sea—I. Seabird food habits. Continental Shelf Research 7:895–911.
Springer, A. M., D. G. Roseneau, E. C. Murphy, and M. I. Springer. 1984. Environmental control of food webs: food habits of seabirds in the eastern Chukchi Sea. Canadian Journal of Fisheries and Aquatic Sciences 41:1202–1215.
Stowe, T. J. 1982. An oil spillage at a guillemot colony. Marine Pollution Bulletin 13:237–239.
Sydeman, W. J. 1993. Survivorship of common murres on

southeast Farallon Island, California. Ornis Scandinavica 24:135–141.

Takekawa, J. E., H. R. Carter, and T. E. Harvey. 1990. Decline of the common murre in central California, 1980–1986. Studies in Avian Biology 14:149–163.

Trenberth, K. E. 1990. Recent observed interdecadal climate changes in the northern hemisphere. Bulletin of the American Meteorological Society 71:988.

Trivelpiece, W. Z., R.G. Butler, D. S. Miller, and D. B. Peakall. 1984. Reduced survival of chicks of oil-dosed adult Leach's storm-petrels. Condor 86:81–82.

Uttley, J. D., P. Walton, P. Monaghan, and G. Austin. 1994. The effects of food abundance on breeding performance and adult time budgets of Guillemots *Uria aalge*. Ibis 136:205–213.

Vader, W., R. T. Barrett, K. E. Erikstad, and K.-B. Strann. 1990. Differential responses of common and thick-billed murres to a crash in the capelin stock in the southern Barents Sea. Studies in Avian Biology 14: 175–180.

Wanless, S., D. D. French, M. P. Harris, and D. R. Langslow. 1982. Detection of annual changes in the numbers of cliff-nesting seabirds in Orkney 1976–80. Journal of Animal Ecology 51:785–795.

Wells, P. G., J. N. Butler, and J. S. Hughes, editors. 1995. *Exxon Valdez* oil spill: fate and effects in Alaskan waters. American Society for Testing and Materials Special Technical Publication 1219, Philadelphia.

Wiens, J. A. 1995. Recovery of seabirds following the *Exxon Valdez* oil spill: an overview. Pages 854–893 *in* P. G. Wells, J. N. Butler, and J. S. Hughes, editors. *Exxon Valdez* oil spill: fate and effects in Alaskan waters. American Society for Testing and Materials Special Technical Publication 1219, Philadelphia.

Wilson, U. W. 1991. Responses of three seabird species to El Niño events and other warm episodes on the Washington coast, 1979–1990. Condor 93:853–858.

Winters, G. H., and J. E. Carscadden. 1978. Review of capelin ecology and estimation of surplus yield from predator dynamics. ICNAF (International Council for Northwest Atlantic Fisheries) Research Bulletin 13:21–30.

American Fisheries Society Symposium 18:738–747, 1996

Size and Productivity of Black-Legged Kittiwake Colonies in Prince William Sound before and after the *Exxon Valdez* Oil Spill

DAVID B. IRONS

U.S. Fish and Wildlife Service, Migratory Bird Management
1011 E. Tudor Road, Anchorage, Alaska 99503, USA

Abstract.—Black-legged kittiwakes *Rissa tridactyla* are the most abundant colonial seabirds nesting in Prince William Sound, Alaska. Approximately 20,000 pairs nest at 27 colonies in the sound. I counted the number of breeding pairs and determined productivity at 24 colonies between 1984 and 1994. Areas adjacent to 10 of these colonies were oiled by the *Exxon Valdez* oil spill and the areas adjacent to 14 colonies were not oiled. Data from the non-oiled colonies and from prespill years were used to determine if the oil spill had negative effects on the birds from oiled colonies. The number of breeding pairs did not decline at colonies in the oiled area after the *Exxon Valdez* oil spill compared to prespill years. Kittiwake productivity in 1989 in the oiled area was lower than expected, while productivity in the non-oiled area was the highest ever recorded. Low numbers of chicks in the oiled area appeared to be the result of low laying rate or low hatching success. Kittiwake productivity in the oiled area did not improve from 1989 to 1994 and productivity in the non-oiled area in Prince William Sound declined in postspill years. When compared to kittiwake colonies in other parts of the Gulf of Alaska, the dramatic difference in pre-and postspill productivity in Prince William Sound kittiwakes was unique. This indicates that factors regulating kittiwake productivity operated differently in Prince William Sound compared to other parts of the Gulf of Alaska.

The *Exxon Valdez* oil spill, which occurred on 24 March 1989, was the largest tanker oil spill in North America. Approximately 44 million liters of North Slope crude oil leaked from the tanker into the waters of Prince William Sound, Alaska. By the end of April much of the oil had flowed out of Prince William Sound or sunk; however, oil remained on the beaches throughout summer 1989. The beached oil often refloated during high tides and was redeposited on the same or different beaches. The currents and wind conditions at the time of the spill caused the oil to flow through the southwestern portion of Prince William Sound into the Gulf of Alaska, without oiling northern and eastern Prince William Sound. This oiled–non-oiled pattern allowed comparisons to be made between seabird colonies in the oiled and non-oiled areas within the sound.

Black-legged kittiwakes *Rissa tridactyla* are the most abundant colonial nesting seabirds in Prince William Sound. Approximately 20,000 pairs nest at 27 colonies distributed throughout the sound (Figure 1). Kittiwakes arrive in Prince William Sound in February or March (my personal observation). They attend the colonies shortly after they arrive, but nest building does not begin until mid to late May. Eggs are usually laid during the first 2 weeks of June. After chicks are fledged in August, colony attendance begins to decline (my unpublished data). I collected several years of data on the population size and productivity of 24 colonies before the *Exxon Valdez* oil spill, which flowed by 10 of these colonies (Figure 1).

The black-legged kittiwake is a cliff-nesting species of the gull family (Laridae), generally nesting in dense colonies on rocky cliffs adjacent to the ocean. They are surface feeders and generally spot prey from the air and then surface plunge or surface seize (Ashmole 1971) to catch their prey. Small forage fish are kittiwakes' primary prey (Sanger 1987), and in Prince William Sound diets consist chiefly of immature Pacific herring *Clupea pallasi*, Pacific sand lance *Ammodytes hexapterus*, capelin *Mallotus villosus*, and immature walleye pollock *Theragra chalcogramma* (Irons 1992). Kittiwakes that nest at colonies located in the sound's fjords fly farther to forage than do kittiwakes that nest at colonies located in central Prince William Sound (Irons 1992 and unpublished data).

Effects of oil spills on marine birds can be determined directly by counts of oiled birds or indirectly by documenting declines in the populations or reproduction of affected birds. Most studies on effects of oil spills on marine birds have quantified the damage by counts of oiled birds (Ford et al. 1987; Page et al. 1990; Piatt et al. 1991). To demonstrate that populations or reproductive output of birds affected by an oil spill declined, there must be sufficient prespill data (Wiens et al. 1984). Affected populations may remain stable if immigration has occurred and has

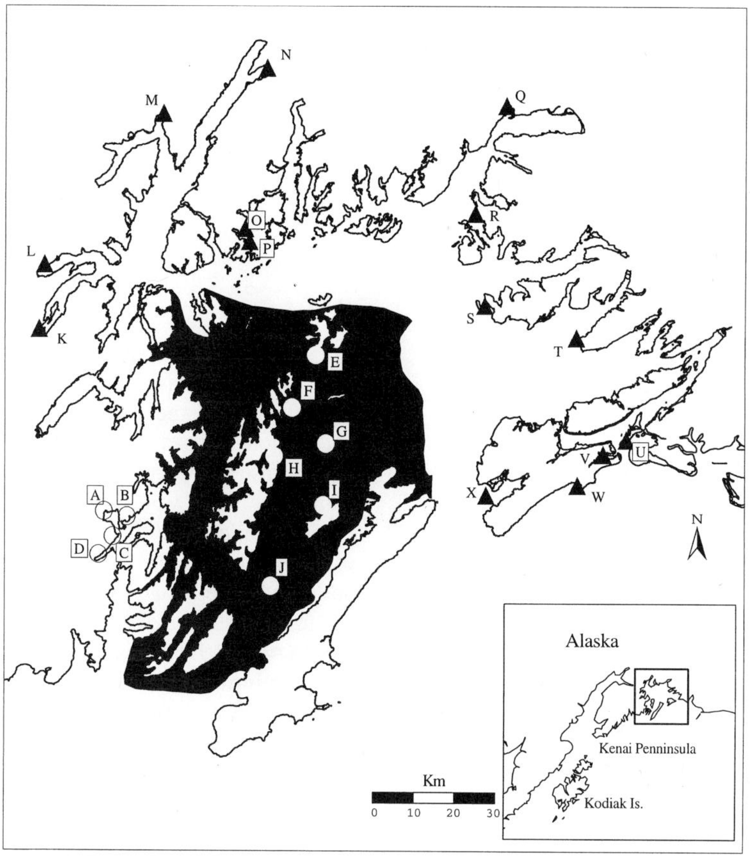

FIGURE 1.—Locations of surface oiling from the *Exxon Valdez* oil spill of March 1989 (ADEC 1989) in Prince William Sound and oiled (open circle) and non-oiled (black triangle) colonies of black-legged kittiwakes *Rissa tridactyla* in the sound. Oiled colonies: A = Chenega Glacier; B = North Icy Bay; C = Icy Bay; D = Tiger Glacier; E = Naked Island; F = Eleanor Island; G = Seal Island; H = Bay of Isles; I = Middle Green Island; J = The Needle. Non-oiled colonies: K = Blackstone Glacier; L = Passage Canal; M = Coxe Glacier; N = Yale Glacier; O = South Eaglek Bay; P = Eaglek Bay Entrance; Q = Shoup Bay; R = Ellamar; S = Gull Island; T = Gravina Rocks; U = Pinnacle Rocks; V = Boswell Rocks; W = Hook Point; X = Porpoise Rocks.

balanced spill-related mortality. Before the *Exxon Valdez* oil spill, few studies had demonstrated effects of oil spills on marine bird populations because prespill data usually did not exist for potentially affected populations (Piatt et al. 1991). The *Exxon Valdez* oil spill occurred in Prince William Sound, where there were prespill data on populations and reproduction of marine birds (Kuletz 1983; T. J. Dwyer, P. Isleib, D. A. Davenport, and J. L. Haddock, U.S. Fish and Wildlife Service [USFWS], unpublished data; K. L. Oakley and K. J. Kuletz, USFWS, unpublished data; D. B. Irons, D. R. Nysewander, and J. L. Trapp, USFWS, unpublished data). Therefore, there was an opportunity to determine whether bird populations or productivity declined after an oil spill.

The objectives of this study were to determine whether population size and productivity of kittiwakes at colonies in the area oiled by the *Exxon Valdez* spill declined in 1989 or in later years compared to those of birds that nested in the non-oiled area of Prince William Sound. I also determined if birds at oiled and non-oiled colonies were contaminated externally by oil in 1989. Because of an ongoing study on kittiwakes in Prince William Sound, there exist 5 years of prespill data and 5 years of postspill data available to help understand the effects of this oil spill.

Methods

Study Area

I studied black-legged kittiwakes breeding at 24 colonies located in Prince William Sound (61° 09′ N, 146° 35′ W) (Figure 1), which is a 10,000-km^2 body of protected water with irregular topography and 6-m tides, located along the northeastern coast of the Gulf of Alaska. Kittiwake colonies in the sound are located on cliffs on small islets and on mainland adjacent to glaciers in several fjords (Irons, Nysewander, and Trapp, unpublished data).

Data Collection

I counted the total number of nests annually at each colony in June or July to determine colony size. A nest was defined as any site with new nesting material. I counted the total number of chicks and recorded brood size annually at each colony in early to mid-August, just before the chicks fledged. Kittiwake productivity was defined as the average number of chicks per nest at a colony. All counts were made using binoculars from a boat. The same methods were used in all years of the study (Irons and coworkers, unpublished data). Total colony nest counts and chick counts were made at all 24 colonies in all years (1984–1994) with the following exception: In 1984 chicks were counted at only 13 colonies (5 oiled and 8 non-oiled), and brood size was not recorded. Productivity for the non-oiled area of Prince William Sound in 1993 and 1994 may have been elevated as a result of placing scarecrows at the Shoup Bay colony to try to keep predators away as part of another study; however, the effect of the scarecrows remains ambiguous.

As part of another study (Irons 1992), I collected data on several parameters of kittiwake reproductive biology for a sample of nests at two colonies, one that was oiled (Eleanor Island) and one that was not oiled (Shoup Bay). The number of nests with eggs, hatching success, and clutch size were determined by counting the eggs in a sample of accessible nests. Hatching success was defined as the percentage of eggs observed that hatched.

Kittiwake chick growth rates were also measured at the Eleanor Island and Shoup Bay colonies. Chicks were weighed to the nearest gram with 300- or 500-g Pesola scales every 3 d from hatching to just before fledging. Growth rates were calculated for the near-linear portion of the growth curve (i.e., 60–300 g) by dividing the weight gain by the number of days.

I demonstrated that brood reduction in kittiwakes is greater when food is scarce than when food is abundant, as a result of increased sibling aggression (Irons 1992). Therefore, when food is abundant, larger brood sizes at fledging would be expected than when food is scarce. I used data on brood size at fledging as an index of relative food availability.

Data Analysis

Colonies were considered to be in the oiled or non-oiled area based on the amount of oil that passed through the foraging area of the birds at each colony. I used foraging area to determine oiling because kittiwakes could have come into contact with oil on the water when they were foraging. Oiled colonies were those that had 50–100% of the foraging area used by birds covered by oil. All other colonies had less than 5% of the foraging area used by birds covered by oil and were considered non-oiled (ADEC 1989) (Figure 1). I defined foraging areas for birds in a colony based on results of a 1989 radio-telemetry foraging study and observations of birds traveling in and out of fjords. The radio-telemetry study showed that birds from Shoup Bay, a fjord colony, foraged an average of 40 km from their colony and that birds from Eleanor Island, a nonfjord colony, foraged an average of 5 km from their colony (Irons 1992). Based on observations in four fjords of large numbers of birds traveling out of the fjord to forage in central Prince William Sound and observations of birds foraging close to several nonfjord colonies, I assumed this to be a general pattern in the sound (my personal observation). Thus, areas within a radius of 40 km from fjord colonies and a radius of 5 km from nonfjord colonies were defined as foraging areas.

I used a paired t-test to determine whether productivity was different for kittiwakes in the oiled and non-oiled areas during prespill years (Sokal and Rohlf 1981). To determine whether kittiwake productivity in 1989 was lower than expected in the oiled area compared with productivity in the non-oiled area based on prespill data, I calculated the difference in productivity between oiled and non-oiled areas for all prespill years. I used a one-tailed t-test for a single observation to test whether the difference in kittiwake productivity between oiled and non-oiled areas in 1989 was greater than the prespill difference (Sokal and Rohlf 1981). I used a two-tailed t-test to determine if postspill (1990–1994) productivity was different from prespill

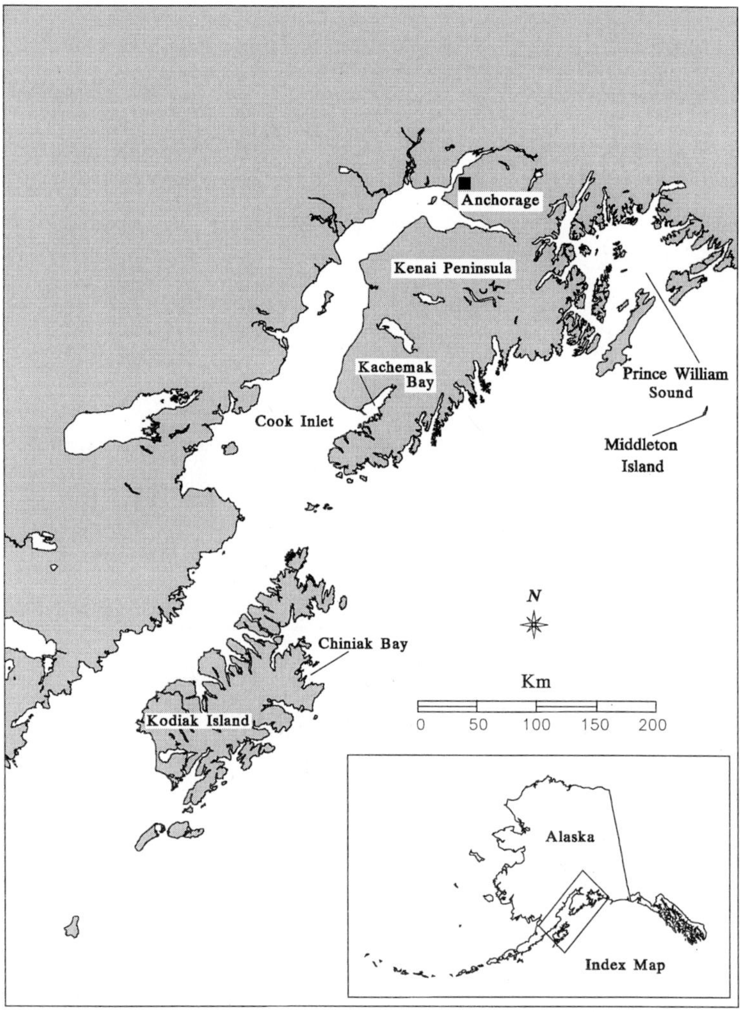

FIGURE 2.—Map showing four locations—Middleton Island, Kachemak Bay, Chiniak Bay, and Prince William Sound—where prespill and postspill productivity of black-legged kittiwake was compared.

(1984–1988) productivity for kittiwakes in the oiled area and to determine if postspill productivity was different from prespill and spill year (1984–1989) productivity in the non-oiled area of Prince William Sound and in other locations in the Gulf of Alaska (Figure 2).

Chi-square contingency tests were used to determine whether Eleanor Island, an oiled colony, and Shoup Bay, a non-oiled colony, had fewer nests with eggs in 1989 than before the spill in 1988. A two-tailed *t*-test was used to determine whether the proportion of successful nests with two chicks, an index of food availability (Irons 1992), changed in postspill years compared with the prespill years and the spill-year (Sokal and Rohlf 1981).

Results

Colony Size

The total number of kittiwake nests at colonies in the oiled area did not decline significantly after the oil spill in 1989 nor in the next 4 years (Figure 3). Colonies in the oiled and non-oiled areas increased by about a third from just before the oil spill in 1988 to 1991 and then declined in 1993 and increased in 1994.

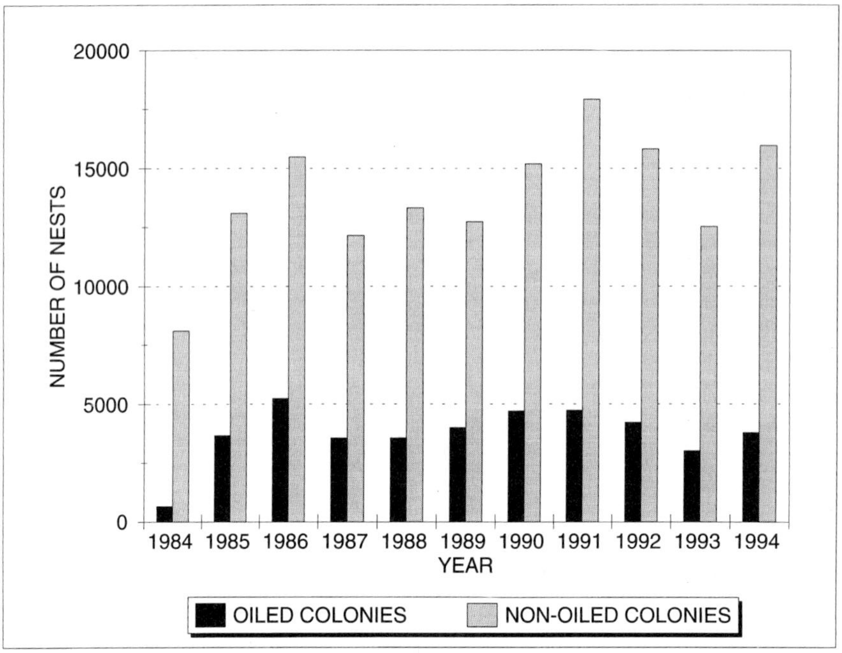

FIGURE 3.—Total numbers of black-legged kittiwake nests during 1984–1994 at 10 colonies in the area that was oiled by the *Exxon Valdez* spill of March 1989 and at 14 colonies in the area outside of the oiled area in Prince William Sound.

Productivity

Comparisons among 24 colonies in Prince William Sound.—For the 5 years before the spill, the productivity of kittiwakes in the oiled area was 0.16 ± 0.0372 chicks/nest lower than that of kittiwakes in the non-oiled area (paired-$t_{0.05(2),4}$ = 4.1903, P = 0.0138) (Figure 4).

In 1989 the productivity of birds at non-oiled colonies was the highest ever recorded at that time, but the productivity of birds at the oiled colonies was their second lowest (Figure 4). This difference between kittiwake productivity in the oiled and non-oiled areas in 1989 (0.33 chicks/nest) was greater than expected ($t_{0.05(1),4}$ = 1.8652, P = 0.0678), based on the 5 years of prespill data (Figure 4).

Productivity of birds in the oiled area of Prince William Sound during the 5 years after the oil spill was lower than prespill productivity (Table 1; Figure 4). Also, postspill productivity of birds in the non-oiled area was lower than prespill and spill-year productivity.

The proportion of successful nests at colonies in Prince William Sound that fledged two chicks was lower during postspill years than before and during the spill year ($t_{0.05(2)8}$ = 4.3894, P = 0.0023) (Figure 5).

Comparisons between one oiled and one non-oiled colony.—The oiled colony, Eleanor Island, had fewer nests with eggs in 1989 than expected when compared with 1988 (chi-square = 25.069, df = 1, $P < 0.0001$) (Table 2). Conversely, the non-oiled colony, Shoup Bay, had more nests with eggs in 1989 than expected when compared with 1988 (chi-square = 26.730, df = 1, $P < 0.0001$) (Table 2). Clutch sizes were similar in 1988 and 1989 at each colony (Table 2).

In 1989, hatching success for the nests in the study plots at the Eleanor Island colony was zero (N = 34 nests); however, eggs hatched in some nests that were not in the study plots and these chicks were used later to measure growth rates. Hatching success in the study plots at the Shoup Bay colony was 87% (N = 91 nests).

The total number of chicks fledged per nest at the Eleanor Island colony in 1989 was 0.24 chicks/nest lower than during the prespill years, and the total number of chicks fledged per nest at the Shoup Bay colony in 1989 was 0.13 chicks/nest higher than during prespill years; however, chicks at Eleanor Island grew at a normal rate (16.7 g/d, N = 38), which was almost identical to that of Shoup Bay chicks (16.6 g/d, N = 33) (Irons 1992). Also, of nests

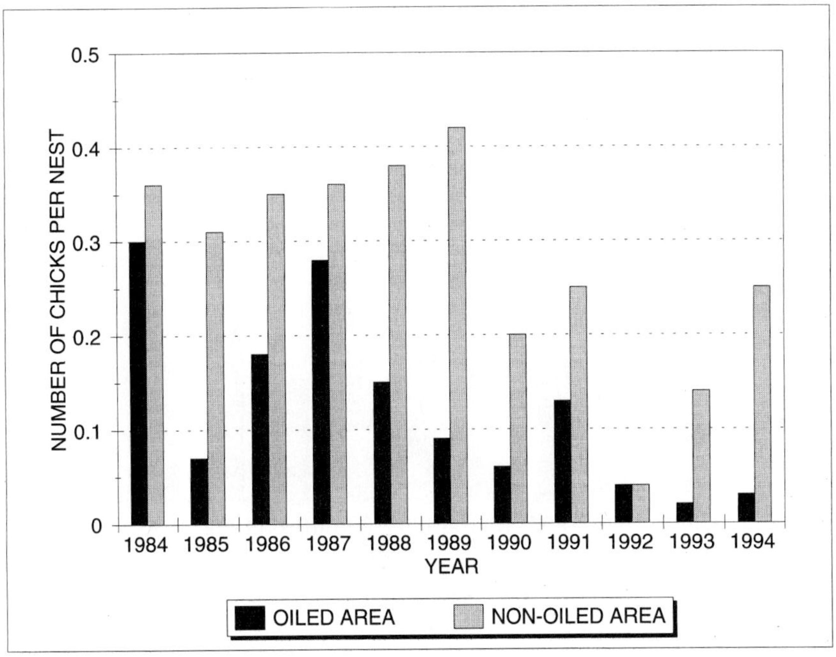

FIGURE 4.—Productivity (number of chicks per nest) of black-legged kittiwake colonies during 1984–1994 in the area oiled by the *Exxon Valdez* spill and in the non-oiled area.

that fledged chicks, two chicks fledged in almost one-half of the nests at both colonies; brood size at fledging averaged 1.41 chicks ($N = 17$) at Eleanor Island and 1.48 chicks ($N = 79$) at Shoup Bay (Irons 1992).

External Oil Contamination

Up to 37% of the birds observed at oiled colonies had oil on their breast feathers (Table 3). At one colony, Eleanor Island, repetitive observations were made from 23 June to 9 August. Fewer oiled birds were seen in July and August than in June. No oiled birds were observed at non-oiled colonies (Table 3).

Oil found on breast feathers was probably either picked up from the water or from oiled nest materials. Throughout summer 1989 oil sheens could be found on the coastal water in Prince William Sound

TABLE 1.—A comparison of 1984–1989 and 1990–1994 black-legged kittiwake productivity (chicks/nest) in the oiled and non-oiled areas in Prince William Sound, Middleton Island, Kachemak Bay in Cook Inlet, and Chiniak Bay on Kodiak Island.

	1984–1989			1990–1994					
	Productivity[a]			Productivity[b]					
Location	Mean	SE	N	Mean	SE	N	df	t^c	P
Prince William Sound oiled area	0.20	0.0425	5	0.05	0.0396	5	8	2.9903	0.0173
Prince William Sound non-oiled area	0.36	0.0144	6	0.18	0.039	5	9	4.7700	0.0010
Middleton Island[d]	0.16	0.1081	6	0.00	0.0	5	5	1.5110	0.1912
Kachemak Bay[e]	0.39	0.1147	5	0.25	0.0658	4	7	1.0401	0.3329
Chiniak Bay[f]	0.12	0.0762	3	0.34	0.0698	5	6	2.0234	0.0895

[a]1989 data were not used for the Prince William Sound oiled area because of effects from the oil spill. There were no data from Kachemak Bay for 1984 and there were no data from Chiniak Bay for 1984–1986.

[b]There were no data from Kachemak Bay for 1991.

[c]Satterthwaite's Approximation was used for Middleton Island data because variances were not equal (SAS 1988).

[d]Middleton Island data are from Hatch et al. (1993) and from S. Hatch (National Biological Service, unpublished data).

[e]Kachemak Bay data are from Slater et al. (1995).

[f]Chiniak Bay data are from D. Irons (U.S. Fish and Wildlife Service [USFWS], unpublished data), R. MacIntosh (National Marine Fisheries Service, unpublished data), D. Nysewander (USFWS, unpublished data), P. Seiser (USFWS, unpublished data), and D. Zwiefelhofer (USFWS, unpublished data).

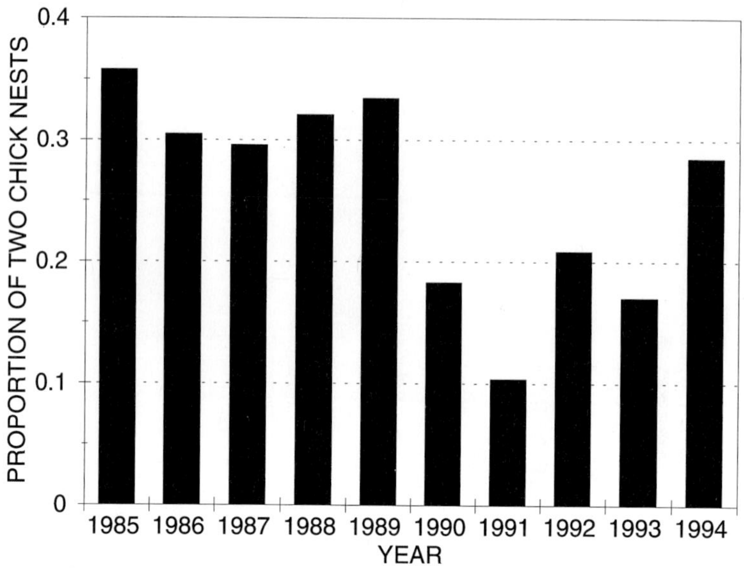

FIGURE 5.—Proportion of successful nests that fledged two chicks for 24 black-legged kittiwake colonies in Prince William Sound during 1985–1994.

(my personal observation). I noted several nests in which birds used pieces of oil booms as nesting material.

Discussion

The breeding population of black-legged kittiwakes in the oiled portion of Prince William Sound did not decline after the *Exxon Valdez* oil spill. It is possible that a substantial portion of the breeding population was killed by the oil spill and that their nest sites were filled with previously nonbreeding birds before I made my counts in 1989; however, two other sources of data suggest that this is unlikely. Data from boat-based marine bird surveys in Prince William Sound before and after the spill did not detect an oil spill effect on the kittiwake population (Klosiewski and Laing 1994). Of the approximately 100,000 kittiwakes that nested in the path of the oil spill (USFWS 1993), only 1.2% were retrieved after the spill, and some of these may have died from starvation rather than from oil (J. Piatt, USFWS, unpublished data). Results from these three studies provide evidence that the *Exxon Valdez* oil spill did not kill a large portion of the kittiwakes nesting in the oiled area in Prince William Sound. This relatively low mortality generally supports the hypothesis that surface-feeding seabirds are less susceptible to oil pollution than are seabirds that dive for food (King and Sanger 1979; Piatt et al. 1991).

The impact of the *Exxon Valdez* oil spill on kittiwake numbers might have been greater had the spill occurred during July and August, when chicks are present. Piatt et al. (1985) reported that thousands of juvenile kittiwakes were oiled from the 1979 *Kurdistan* oil spill shortly after they fledged. Adults may be more susceptible to oil spills while rearing chicks, because they must feed their chicks fre-

TABLE 2.—Proportion of nests with eggs and clutch sizes in 1988 and 1989 for one oiled and one non-oiled black-legged kittiwake colony in Prince William Sound.

Colony	Oiled in 1989	Year	Number of nests checked	Date checked	Proportion of nests with eggs	Clutch size
Shoup Bay	No	1988	415	18 June	0.68	1.65
		1989	120	25 June	0.92	1.75
Eleanor Island	Yes	1988	111	28 June	0.74	1.53
		1989	88	23 June	0.39	1.41

TABLE 3.—Number of black-legged kittiwakes at 18 colonies in Prince William Sound observed with oil on their breast feathers in summer 1989 after the *Exxon Valdez* oil spill.

Colony	Date	Number oiled	Total number observed	Proportion oiled
Oiled Colonies				
Bay of Isles	22 Jun	0	5	0.00
Bay of Isles	25 Jul	2	30	0.07
Eleanor Island	23 Jun	8	34	0.24
Eleanor Island	21 Jul	4	59	0.07
Eleanor Island	28 Jul	0	8	0.00
Eleanor island	29 Jul	0	39	0.00
Eleanor Island	4 Aug	1	62	0.02
Eleanor Island	9 Aug	1	57	0.02
Icy Bay	19 Jul	0	106	0.00
Middle Green Island	22 Jun	11	30	0.37
Naked Island	23 Jun	0	18	0.00
North Icy Bay	18 Jul	0	155	0.00
The Needle	22 Jun	5	30	0.17
Non-oiled Colonies				
Blackstone Glacier	18 Jul	0	174	0.00
Boswell Rocks	9 Jul	0	150	0.00
Coxe Glacier	12 Aug	0	81	0.00
Eaglek Bay Entrance	3 Aug	0	48	0.00
Ellamar	9 Jul	0	10	0.00
Gravina Rocks	9 Jul	0	37	0.00
Gull Island	9 Jul	0	24	0.00
Passage Canal	17 Jul	0	326	0.00
Shoup Bay	11 Aug	0	310	0.00
South Eaglek Bay	3 Aug	0	48	0.00
Yale Glacier	12 Aug	0	129	0.00

quently and individual birds may feed in a limited area (Irons 1992). During March, when adults can forage over a much larger area, perhaps kittiwakes more easily avoid being oiled.

To aid in assessing potential oil spill impacts on marine birds, it would be helpful to know if they avoid oil on the surface of the water. I did not collect data to determine if kittiwakes avoided oil, however, I suspect that kittiwakes may avoid foraging in water contaminated with crude oil based on two lines of reasoning. First, kittiwakes find their prey while flying and looking through the surface of the water (Irons, personal observation). If there is crude oil on the water's surface, it may inhibit birds from locating prey, therefore kittiwakes would be unlikely to forage in oiled water. Second, if kittiwakes did forage in oiled water, many of the birds in the path of the *Exxon Valdez* probably would have become completely coated with oil and died, because kittiwakes often forage by plunging into the water, completely submersing themselves (Townsend and Allen 1907; Belopol'skii 1957; Burtt 1974); however, there is no evidence that great numbers of kittiwakes became heavily oiled and died after the *Exxon Valdez* oil spill. Juvenile kittiwakes may behave differently from adults around oil spills. An oil spill in Newfoundland killed many recently fledged juveniles but few adults in August 1979 (Piatt et al. 1985).

Productivity of kittiwakes was lower than expected at oiled colonies in 1989 (Figure 4). Detailed information that might have revealed the mechanism of this reproductive failure was collected at only two colonies, Eleanor Island (oiled) and Shoup Bay (non-oiled). These data suggest that the low productivity at Eleanor Island in 1989 was a result of eggs not being laid, eggs not hatching, or predation during the egg stage, rather than food limitation during chick rearing. Fewer nests contained eggs than expected, but it is not known whether eggs were not laid or whether eggs were laid and then removed by predators. Nests that contained eggs had normal clutch sizes (Table 2), but few eggs survived to hatching. Eggs that survived to hatching produced chicks that did not appear to be food stressed, because Eleanor Island chicks grew at normal rates and several nests fledged two chicks.

Experimental studies have demonstrated reduced breeding success in seabirds because of oiling. Fry et al. (1986) showed that wedge-tailed shearwaters *Puffinus pacificus* exposed to oil exhibited a lower frequency of laying, lower hatching success, and reduced breeding success. Some birds abandoned their nests and others exhibited reduced attentiveness; however, chick growth rates were the same for exposed and control birds. Trivelpiece et al. (1984) demonstrated that Leach's storm-petrels *Oceanodroma leucorhoa* that were exposed to oil did not feed their young.

In this study, probable causes of egg mortality were oiling and predation. Many studies have documented that a small amount of toxic oil on eggs is lethal (Hartung 1965; Albers 1977; Szaro and Albers 1977; Albers 1978; White et al. 1979). If toxic North Slope crude oil contacted eggs from either the parents' plumage or from oiled nest material, it is likely that the eggs would die. Observations of birds at oiled colonies in late June and July showed that birds had oiled breast feathers (Table 3); however, these observations probably are underestimates of the proportion of birds that contacted oil, because seabirds are capable of preening oil off their feathers in several days (Fry 1987).

Predation by bald eagles *Haliaeetus leucocephalus* can be a locally and temporally important source of mortality for kittiwake eggs at colonies in Prince William Sound (Irons 1992 and unpublished data); however, colonies in the oiled area probably would not experience abnormally intense predation pressure compared to colonies outside the oiled area in 1989, unless adults were less attentive because of toxic effects of ingested oil.

Kittiwake productivity in the oiled area did not improve during the 5 years after the spill, and productivity in the non-oiled area in Prince William Sound was lower in postspill years than in prespill years (Table 1 and Figure 4). A lower percentage of two-chick broods throughout Prince William Sound (Figure 5) and lower chick growth rates at the Shoup Bay colony from 1990 to 1994 (Irons 1992 and unpublished data) suggest that kittiwakes were more food limited from 1990 to 1994 than in 1989 or during the prespill years. In 1992 and 1993, when pink salmon *Oncorhynchus gorbuscha* were late returning to Prince William Sound, there is anecdotal evidence that eagle predation on kittiwake eggs and chicks may have been higher than in previous years (Irons and G. Golet, USFWS, unpublished data). Food limitation coupled with predation may have been the reasons that oiled colonies did not recover and that non-oiled colonies had lowered productivity in postspill years.

Reasons for changes in food supply include both natural and anthropogenic factors. There is much natural variation in prey available to kittiwakes at some locations (Byrd 1989; Murphy et al. 1991). Some of the causes of this variation may be differences in year-class strength of prey species and changes in water temperature (Murphy et al. 1991). Potential anthropogenic changes in prey resources include negative effects on prey productivity and survival from the *Exxon Valdez* oil spill, from competition with high populations of hatchery salmon, or from overfishing of prey species by commercial fisheries.

To gain an understanding of the scope of the postspill decline in kittiwake productivity, I compared kittiwake productivity data from 1984–1989 with data from 1990–1994 for Prince William Sound and other areas in the northern Gulf of Alaska. In Prince William Sound there was a dramatic decrease in productivity from pre- to postspill (Table 1). There is weak evidence of lower productivity at colonies on Middleton Island and in Kachemak Bay, and production at Chiniak Bay colonies was higher after the spill (Table 1). These data indicate that for kittiwakes at these four locations in the northern Gulf of Alaska, only in Prince William Sound was kittiwake productivity consistently better prespill than postspill. Such results suggest that the factors that caused the reduced productivity in Prince William Sound did not have the same effect or were not present in the other areas.

Conclusions

The number of breeding pairs of black-legged kittiwakes in the oiled area of Prince William Sound did not decline after the *Exxon Valdez* oil spill. Kittiwake productivity was lower than expected in the oiled area after the spill in 1989, whereas productivity in the non-oiled portion of Prince William Sound was the highest ever recorded. Low productivity in the oiled area appeared to be the result of low laying rate or low hatching success. Kittiwake productivity in the oiled area did not improve from 1989 to 1994 and productivity in the non-oiled area in Prince William Sound declined in postspill years. When compared with kittiwake colonies in other parts of the Gulf of Alaska, the dramatic difference in pre- and postspill kittiwake productivity in Prince William Sound kittiwakes was unique. This indicates that factors regulating kittiwake productivity operated differently in Prince William Sound compared with other parts of the Gulf of Alaska.

Recommendations for Future Oil Spill Studies

If the effects of oil spills on seabirds are studied in the future, there is a need for prespill data on population size and reproductive parameters for birds in the oiled area and birds in a control area. Studies should begin on the day of the oil spill. To investigate direct mortality, data on population change, number of oiled carcasses, and observations of whether birds are repelled by, attracted to, or unconcerned about oil on the surface of the water are needed. To determine effects on productivity, data on behavior of birds, oiling of birds, eggs, and nests, and all productivity parameters such as phenology, clutch size, hatching success, fledging success, growth rates, and diets are needed. To assist in determining if birds and their prey are contaminated by oil, hydrocarbon data for adults, eggs, chicks, diets, and nests are needed.

Acknowledgments

I thank Dave Nysewander, Greg Golet, and many others who helped in the field during the several years of data collection. Dave Bowden and Steve Klosiewski provided technical expertise in statistical analysis. Scott Hatch, Rick MacIntosh, Dave Nysewander, Pam Seiser, and Denny Zwiefelhofer allowed me to use their unpublished data. John Chardine, John Coulson, Ada Fowler, Dennis Heinemann, Karen Oakley, John Piatt, Dan Roby, Bruce Wright, and two anonymous reviewers provided helpful comments on previous drafts of this article. The U.S. Fish and Wildlife Service and the *Exxon Valdez* Trustee Council provided funding for this study.

References

ADEC (Alaska Department of Environmental Conservation). 1989. *Exxon Valdez* oil spill: cumulative oiling on water through June 20, 1989. Arc info data files, Anchorage, Alaska.

Albers, P. H. 1977. Effects of external applications of fuel oil on hatchability of mallard eggs. Pages 158–163 *in* D. A. Wolfe, editor. Fate and effects of petroleum hydrocarbons in marine ecosystems and organisms. Pergamon Press, New York.

Albers, P. H. 1978. The effects of petroleum on different stages of incubation in bird eggs. Bulletin of Environmental Contamination and Toxicology 19:624–630.

Ashmole, N. P. 1971. Seabird ecology and the marine environment. Pages 224–286 *in* D. S. Farner and J. R. King, editors. Avian biology, volume I. Academic Press, New York.

Belopol'skii, L. C. 1957. Ecology of sea-colony birds of the Barents Sea. Izdatel'stvo Akademii Nauk S. S. R. Moscow-Lenningrad. Translated from Russian (1961), Israel Program for Scientific Translations, Jerusalem.

Burtt, E. H. 1974. Success of two feeding methods of the black-legged kittiwake. Auk 91:827–829.

Byrd, G. V. 1989. Seabirds in the Pribilof Islands, Alaska: trends and monitoring methods. Master's thesis. University of Idaho, Moscow.

Ford, R. G., G. W. Page, and H. R. Carter. 1987. Estimating mortality of seabirds from oil spills. Pages 547–555 *in* Proceedings, 1987 oil spill conference. American Petroleum Institute Publication 4452, Washington, DC.

Fry, D. M. 1987. Seabird oil toxicity study. Final report of Nero and Associates Inc. to U.S. Minerals Management Service, Washington DC.

Fry, D. M., J. Swenson, L. A. Addiego, C. R. Grau, and A. Kang. 1986. Reduced reproduction of wedge-tailed shearwaters exposed to weathered Santa Barbara crude oil. Archives of Environmental Contamination and Toxicology 15:453–463.

Hartung, R. 1965. Some effects of oiling on reproduction of ducks. Journal of Wildlife Management 29:872–874.

Hatch, S. A., G. V. Byrd, D. B. Irons, and G. L. Hunt. 1993. Status and ecology of kittiwakes in the North Pacific Ocean. Pages 140–153 *in* K. Vermeer, K. T. Briggs, K. H. Morgan, and D. Siegel-Causey, editors. The status, ecology, and conservation of marine birds of the North Pacific. Canadian Wildlife Service, Special Publication, Ottawa.

Irons, D. B. 1992. Aspects of foraging behavior and reproductive biology of the black-legged kittiwake. Doctoral dissertation. University of California, Irvine.

King, R. D., and G. A Sanger. 1979. Oil vulnerability index for marine birds. Pages 227–239 *in* J. C. Bartonek and D. N. Nettleship, editors. Conservation of marine birds of northern North America. U.S. Fish and Wildlife Service, Washington, DC.

Klosiewski, S. P., and K. K. Laing. 1994. Marine bird populations of Prince William Sound, Alaska, before and after the *Exxon Valdez* oil spill. Final report, Migratory Bird Management, Bird Study 2. U.S. Fish and Wildlife Service, Anchorage, Alaska.

Kuletz, K. J. 1983. Mechanisms and consequences of foraging behavior in a population of breeding pigeon guillemots. Master's thesis. University of California, Irvine.

Murphy, E. C., A. M. Springer, and D. G. Roseneau. 1991. High annual variability in reproductive success of kittiwakes (*Rissa tridactyla*) at a colony in western Alaska. Journal of Animal Ecology 60:515–534.

Page, G. W., H. R. Carter, and R. G. Ford. 1990. Numbers of seabirds killed or debilitated in the 1986 *Apex Houston* oil spill in central California. Studies in Avian Biology 14:164–174.

Piatt, J. F., H. R. Carter, and D. N. Nettleship. 1991. Effects of oil pollution on marine bird populations. Pages 125–141 *in* J. White, editor. The effects of oil on wildlife. The Sheridan Press, Hanover, Pennsylvania.

Piatt, J. F., R. D. Elliot, and A. MacCharles. 1985. Marine birds and oil pollution in Newfoundland, 1951–1984. Newfoundland Institute for Cold Ocean Science Report 105. Memorial University of Newfoundland, St. John's.

Sanger, G. A. 1987. Trophic levels and trophic relationships of seabirds in the Gulf of Alaska. Pages 229–257 *in* J. P. Croxall, editor. Seabirds: feeding ecology and role in marine ecosystems. Cambridge University Press, Cambridge, U.K.

SAS Institute. 1988. SAS user's guide: statistics, 6.03 edition. SAS Institute, Cary, North Carolina.

Sokal, R. R., and F. J. Rohlf. 1981. Biometry. Freeman, San Francisco.

Szaro, R. C., and P. H. Albers. 1977. Effects of external applications of no. 2 fuel oil on common eider eggs. Pages 164–167 *in* D. A. Wolfe, editor. Fate and effects of petroleum hydrocarbons in marine ecosystems and organisms. Pergamon Press, New York.

Townsend, C. W., and G. M. Allen. 1907. Birds of Labrador. Proceedings of the Boston Society of Natural History 33:312.

Trivelpiece, W., R. G. Butler, D. S. Miller, and D. B. Peakall. 1984. Reduced survival of chicks of oil-dosed adult Leach's storm-petrels. Condor 86:81–82.

USFWS (U.S. Fish and Wildlife Service). 1993. Catalog of Alaskan seabird colonies—computer archives. U.S. Fish and Wildlife Service, Migratory Bird Management, Anchorage, Alaska.

White, D. H., K. A. King, and N. C. Coon. 1979. Effects of no. 2 fuel oil on hatchability of marine and estuarine bird eggs. Bulletin of Environmental Contamination and Toxicology 21:7–10.

Wiens, J. A., R. G. Ford, and D. Heinemann. 1984. Information needs and priorities for assessing the sensitivity of marine birds to oil spills. Biological Conservation 28:21–49.

American Fisheries Society Symposium 18:748–758, 1996

Effects of the *Exxon Valdez* Oil Spill on the Black Oystercatcher

BRIAN E. SHARP

Ecological Perspectives, 2234 NE 9th Avenue, Portland, Oregon 97212, USA

MARY CODY

U.S. Fish and Wildlife Service, 1011 E. Tudor Road, Anchorage, Alaska 99503, USA

RICK TURNER[1]

Abstract.—Oil spilled from the *Exxon Valdez* in March 1989 affected an estimated 4,013 km of intertidal shoreline in Prince William Sound, the Kenai and Kodiak coasts, and the Alaska Peninsula. Intertidal habitats were still extensively oiled in 1990. The black oystercatcher *Haematopus bachmani*, a shorebird that forages in the intertidal zone, was studied in Prince William Sound at oiled sites (Green, Little Green, and Channel islands) in 1989 and 1990 and at a non-oiled control site (Montague Island) in 1989 to determine oil spill effects on reproduction and behavior. Oil cover on meter-square quadrats in oystercatcher-foraging areas on Green Island averaged 23 and 17% in 1989 and 1990, respectively. Relative to the control site, the proportion of nonbreeding oystercatcher pairs was higher on oiled sites in 1989, and eggs laid were smaller. Success of clutches to hatching was unexpectedly higher on oiled sites, and predators destroyed a smaller proportion of nests. In 1989, chick mortality was higher on oiled sites than on the non-oiled one and was positively correlated with amount of oil present in foraging territories. In 1990, chick production remained lower on oiled territories and failed completely on bioremediated shorelines.

The black oystercatcher *Haematopus bachmani* breeds throughout the spill-affected area (AOU 1957; Hatch and Hatch 1983; C. J. Martin, National Park Service, unpublished data) and probably winters there in at least low numbers, except perhaps in the Barren Islands (Bent 1927; Nysewander and Barbour 1979; Laing 1990; E. P. Bailey, U.S. Fish and Wildlife Service, unpublished data). Black oystercatchers feed exclusively in the intertidal zone, foraging on and affecting the density and distribution of mussels *Mytilus trossulus*, limpets, and other intertidal invertebrates (Webster 1941a; Hartwick 1976; Frank 1982; Marsh 1986a, 1986b). These birds nest on the ground on rocky, gravelly, or sandy substrates just above the high-tide line. Potential direct and indirect effects of shoreline oiling on black oystercatchers included (1) contact with oil, with lethal or sublethal effects; (2) transferral of oil to eggs; (3) ingestion of hydrocarbons from preening or consumption of contaminated prey; (4) disturbance to nesting pairs by shoreline cleanup crews, resulting in abandonment of nests or loss of eggs and chicks to predation; and (5) mortality of oystercatcher prey populations.

Study Area

Green Island, the primary oiled study area, is 13 km long and 4 km wide, with a maximum elevation of 150 m. Little Green Island and Channel Island, islets near Green Island, were included in the oiled study area (Figure 1). Shorelines predominantly consist of small and large cobble, boulder, exposed wave-cut platforms, with sandy beaches on Channel Island and the south shore of Green Island. Oil arrived on the shoreline of Green Island on 29 March 1989, 5 d after the spill (National Oceanic and Atmospheric Administration [NOAA], U.S. Coast Guard, and Exxon Corp. overflight maps). Much of the intertidal zone was coated with a 2- to 10-m wide band of oil. Channel Island, the south shore of Green Island, and Little Green Island were surrounded by floating oil and were oiled between 30 March and 5 April (Patten 1992; overflight maps; NOAA HAZMAT [hazardous material] model). Most floating oil passed Green Island by 5 April. Exxon Corp. attempted to clean oiled shorelines on Green Island during May 1989. Shoreline-cleaning crews, an armada of support vessels (barges, living accommodations, etc.), and aircraft were based at Green Island until 2 June.

Montague Island, 8 km southeast of Green Island, was the non-oiled control (Figure 1). Montague Island is 80 km long, 10 km wide, with a mountainous interior and elevations to 800 m. Shorelines in the vicinity of Port Chalmers, the center of the study area, consist of gravel, cobble, boulders, exposed rocky wave-cut platforms, and sheltered tidal flats. Most of the Montague Island

[1]Current address unknown.

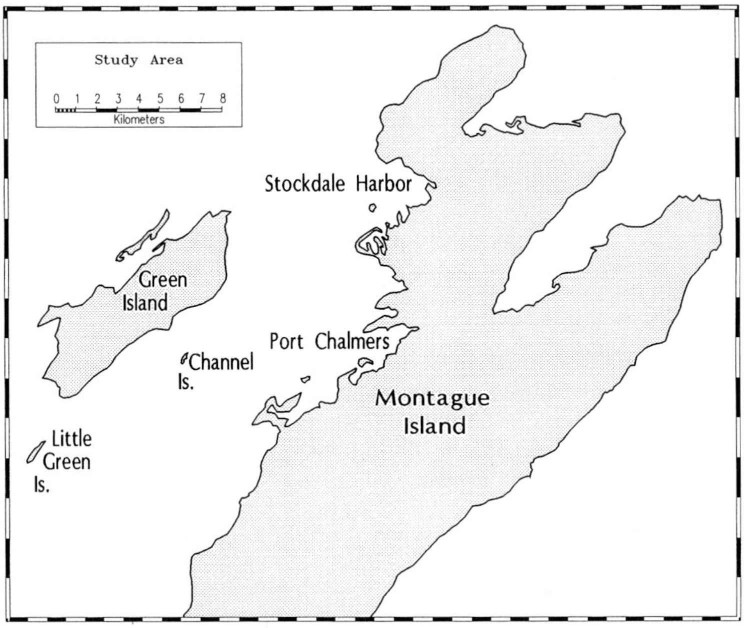

FIGURE 1.—Map of study areas in Prince William Sound. Oiled sites were Green, Little Green, and Channel islands; non-oiled site was west coast of Montague Island south of Stockdale Harbor.

shoreline was not oiled with the exception of Hinchinbrook Entrance and Stockdale Harbor, 7–14 km north of the study area.

Methods

Green and Montague Islands were staffed by two-person field teams in 1989. Except for locating a small number of nests in mid-May, fieldwork occurred from the beginning of June to mid-July. In 1990, the study was not funded, but M. Cody checked Green Island nesting sites for evidence of nesting activity and reproductive success. No data were obtained from Montague Island in 1990.

Effects of shoreline oiling were measured by comparing black oystercatcher population density, the proportion of nonbreeding pairs, clutch size, egg size, hatching success, behavior, chick survival, hydrocarbon levels, and prey population densities on oiled and non-oiled study areas of Prince William Sound. Null hypotheses were that estimates of parameters would not differ between oiled and non-oiled sites. Distribution and density of breeding and nonbreeding oystercatchers were determined from shoreline censuses. Nests were found by noting the location of the on-guard member of a pair and watching for the departure of the incubating bird. At nests, we recorded clutch size, examined egg surfaces for oil, and measured most eggs. Calipers were cleaned with acetone and hexane between egg measurements and sterile gloves were worn to avoid introducing oil onto eggs. Mean egg volumes for each clutch were calculated according to the formula $v = 0.48$ (length $\times$ breadth^2) $- 1.318$ (Nol 1989).

The fate of clutches was monitored at approximately 3-d intervals. Empty nests were examined for signs of predation. Unhatched eggs and shells from hatched and predated eggs were collected for hydrocarbon analysis. Data on chick survival were obtained by observing territorial pairs with broods during the first few weeks after hatching. Survival of clutches and chicks was estimated by the Mayfield method, with variances (var) calculated by var $=$ $(s)(m)$/exposure days (Mayfield 1960, 1975; Johnson 1979, where s = survival and m = mortality). Feeding rates were obtained by observing foraging birds at low tide.

In 1989 foraging–habitat quality was measured on 2–4 (mean = 3) transects from high-tide line to low-tide line in the feeding territories of each nesting pair, on meter-square quadrats located 1.7–5.25 m apart. Percent cover of invertebrates (mussels and barnacles *Balanus* sp.), oil, all algal species, *Fucus* sp., substrate (sand, gravel, cobble, rock, or water), dead *Fucus* sp, dead mussels, and dead barnacles was recorded in increments of 10%. Intertidal organisms present in trace amounts (cover $<1\%$) were noted. Ten of the larger mussels within the feeding territories of most breeding pairs were also measured in situ to obtain an approximation of the sizes of the largest mussels available in each territory. In 1990 oil cover within eight oystercatcher feeding-territories was measured using the same procedure as in 1989. Data were also obtained in 1990 relating nesting success to shoreline bioremediation treatments with Inipol.

Analysis of variance (ANOVA), *t*-tests, and chi-square tests were used to test for between-island differences in estimated parameters. Chick survival and brood reduction were tested for correlation with oil cover on those foraging areas accessible to chicks.

The extent of shoreline oiling in the Prince William Sound and Kenai Peninsula sectors of the *Exxon Valdez* oil spill-affected area was determined from maps and tables provided by the Alaska Department of Natural Resources based on data collected by Alaska Department of Environmental Conservation (ADEC). Shoreline oiling estimates were based on monthly aerial shoreline surveys in April–August 1989, on-ground surveys in September–October 1989, and aerial surveys in spring 1990. Data on shoreline oiling of the Kodiak–Alaska Peninsula coastline were obtained from on-ground surveys of Becharof and Kodiak National Wildlife Refuge shorelines by refuge personnel.

Morgue records of dead oystercatchers found by search-and-rescue workers and band recovery data were analyzed to estimate recovery rates for black oystercatchers. Data on black oystercatcher densities along the Kenai and Alaska peninsulas and evidence of population reduction and breeding failure were obtained from published and unpublished surveys and observations by U.S. Fish and Wildlife Service (USFWS), National Park Service, and other field personnel in 1989.

Results

Degree of Oiling

On Green Island, oil was present on 48 of 53 transects and on an average of 59% (range, 24–100%) of 703 plots within transects. Measurable oil ($>1\%$ cover) occurred on 43 of the 53 transects; on 264 plots that had measurable oil, the average oil cover was approximately 23%. Coverage by oil on some plots was 100%.

None of 1,123 quadrats along 17 transects in the intertidal zone of Montague Island contained visible surface or subsurface oil. Oil sheen was observed once on the surface of the intertidal zone within one feeding territory, and traces of oil were observed once in *Fucus* sp. drifting near another feeding area, after 2 d of strong, southerly winds.

On Green Island in 1990, on eight of the same territories examined in 1989, oil occurred on 41% of 241 meter-square quadrats within transects; on plots that had measurable oil ($>1\%$ cover), mean percent oil cover was 17% and the highest mean percent oil cover on plots within a transect was 38%. Shoreline oiling was reduced but pervasive in 1990; differences between years for various measures of oiling—percent of plots with any oil, mean percent oil cover, and highest percent oil cover—were not significant, and on two of eight territories mean percent oil cover was higher in 1990 than in 1989. According to ADEC data, oiling increased on 8 of 17 black oystercatcher territories on Green Island from fall 1989 to spring 1990. On the Kenai Peninsula, the known extent of oiling increased by 208% and on the Kodiak–Alaska Peninsula by 189% from 1989 to 1990 (Gundlach et al. 1991).

Direct Mortality

Nine oystercatcher carcasses representing a larger number of mortalities (see *Discussion*) were returned to the morgue from locations throughout the spill-affected area—two from Prince William Sound, two from the Kenai coast, and five from the Kodiak–Alaska Peninsula.

Nesting Density

We located 37 breeding pairs of black oystercatchers on oiled and non-oiled sites in Prince William Sound in 1989. Nesting densities on Green and Montague islands were approximately 0.88 and 0.89 pairs/km of occupied shoreline (17 and 20 pairs along 19.3 and 22.5 km of shoreline, respectively) and 0.37 and 0.49 pairs/km of total shoreline surveyed. Density of all oystercatchers, including nonbreeding birds, was 1.35 and 1.56 oystercatchers/km of total shoreline surveyed. Density at Channel Island was 6.33 pairs/km.

Reproductive Success

Eleven nonbreeding pairs were found on oiled areas (Green, Little Green, and Channel islands), and 2 nonbreeding pairs on the non-oiled site Montague Island (chi-square = 6.231, df = 1, $P < 0.025$) (Table 1). Mean hatching dates tended to be later on Green Island in 1989 than on Montague Island

TABLE 1.—Hatching success of black oystercatcher at nests of the oiled (Green, Little Green, and Channel islands) and non-oiled (Montague Island) sites in 1989.

Reproductive parameter	Oiled area	Non-oiled area
Hatched (number)	11	9
Failed (number)	4	8
Abandoned	0	1
Predated	2	6
Eggs cracked	0	1
Reason unknown	2	
Fate unknown (number)	4	5
Total breeding attempts (number)	21	22
Nonbreeding pairs (number)	11	2
Hatched[a]	73%	53%
Hatched including nonbreeding attempts[b]	42%	47%
Predated[c]	13%	35%

[a]Chi-square for number hatched = 0.707, $P < 0.50$.
[b]Chi-square for number hatched including nonbreeders = 0.057, $P < 0.75$.
[c]Chi-square for number predated = 1.02, $P < 0.25$.

(29 June versus 23 June) ($P = 0.10$), later than on Green Island in 1990 (most Green Island pairs in 1990 had chicks by 22 June), later than in Prince William Sound in 1991 (mean hatching date, 18 June) (Andres and DeZeeuw 1991), and later than on Hinchinbrook Island in 1976 (hatching, 12–17 June) (Nysewander and Knudtson 1977). A smaller proportion of nests found in June were located in heavily oiled areas than would be expected on the basis of availability (proportion of shoreline oiled) (chi-square = 4.74, df = 2, $P < 0.10$), and although the difference is not significant, some nests on oiled areas may have been lost before the study began.

Sizes of clutches found in June were not significantly different on Green and Montague islands (2.56 and 2.68 eggs/nest, respectively). Egg lengths and widths averaged 54.9 and 37.5 mm, respectively, on Green Island and 55.9 and 39.5 mm, respectively, on Montague Island. Egg volumes were smaller on Green Island when eggs were treated as independent units of comparison (ANOVA: $F = 6.37$, $P = 0.015$, df = 53), but mean egg volumes from 8 and 13 clutches from Green and Montague islands, respectively, were not significantly different when clutch was used as the unit of comparison ($P = 0.14$, df = 19). Egg volumes of two- and three-egg clutches were not significantly different ($P = 0.98$, df = 18). On Green Island, unusually small eggs from one pair (47 × 31 mm) hatched successfully.

Of nests with known outcomes, on Green Island 11 (73%) were successful in hatching at least one young, and 4 (27%) were unsuccessful ($N = 15$) (Table 1). On Montague Island, 9 nests (53%) were successful, and 8 (47%) failed. Hatching success was not significantly different between oiled and non-oiled sites (chi-square = 0.707, df = 1, $P < 0.500$). If the high proportion of nonbreeding pairs on oiled sites represented failed breeding attempts earlier in the season, hatching success was similar on oiled and non-oiled sites, 42 versus 47% (chi-square = 0.057, df = 1, $P < 0.900$). Daily survival of clutches calculated by the Mayfield method was not significantly different on Green and Montague islands (0.9823, 95% confidence interval = 0.9651–0.9994; and 0.9497, 95% confidence interval = 0.9177–0.9817, respectively).

Predation was the major reason for known nest failures on Montague Island, and investigator disturbance probably caused the losses of two nests. Predation was not significantly higher on Montague than on Green Island in 1989 at 35 versus 13%, respectively (chi-square = 1.02, df = 1, $P < 0.25$), but sample sizes were small (Table 1). Montague Island nests were visited more often, but the mean number of visits to hatched and failed nests within islands was not different (3.8 versus 3.5 on Montague Island, $P = 0.84$, df = 14; 1.4 versus 1.5 on Green Island, $P = 0.37$, df = 8).

We did not find oil on eggs in nests on Green Island. Two eggs did not hatch from two otherwise successful nests on Green Island. On Montague Island, no unhatched eggs were found in otherwise successful nests.

Chick Survival

Chicks of five pairs associated with heavily oiled foraging areas, of six pairs associated with moderately oiled foraging areas, and of four pairs associated with non-oiled foraging areas disappeared at rates directly proportional to the amount of oiling in the foraging area. From hatching to mid-July, brood size was reduced by 45% on the heavily oiled territories (mean percent oil cover = 42% with a range of 35–48% for plots with measurable oil), by 36% on moderately oiled territories (mean percent oil cover = 14% with a range of 7–28%), and there were no losses of chicks on Montague Island (non-oiled). The t-tests comparing chick mortality in non-oiled versus moderately oiled and non-oiled versus heavily oiled territories were not quite significant ($P < 0.08$, $P < 0.10$, respectively, df = 4); however, daily chick mortality rates calculated by the Mayfield method—6%/d in the heavily oiled feeding areas, 3%/d in moderately oiled areas, and no chick loss in non-oiled areas—were significantly different ($P < 0.05$). The Pearson correlation between degree of oiling and daily mortality rates of chicks was

0.985 (P = 0.109). Chick losses of 3%/d and 6%/d would result in losses of 66 and 89% at fledging age (35 d) on lightly and moderately–heavily oiled shorelines, respectively, if losses attributable to oiling are constant.

We correlated ADEC categorizations of shoreline oiling on Green Island with our quantified measurements of oil on quadrats. What ADEC defined as "heavy" or "moderate" oiling had severe brood reduction effects (6%/d), and "lightly" oiled shoreline had moderate brood reduction effects (3%/d). There was even some loss of chicks on shorelines where oil cover on quadrats was 7–10%, the upper limit of the ADEC "very light" category. We found significant oiling at 14 of 16 black oystercatcher territories, whereas ADEC aerial surveys in summer showed oil at only 5 of those sites. All ADEC surveys in 1989 combined (on-ground surveys in fall and monthly aerial surveys in summer) indicated that only 8 of 16 oystercatcher territories were oiled. Combined ADEC data for both 1989 and 1990 indicated oil on 13 of 16 territories, but eight shoreline segments classified by ADEC as very lightly oiled in 1990 were in fact moderately or heavily oiled. Surveys of oiled shoreline by ADEC thus underestimated both extent and degree of oiling at Green Island in 1989 and 1990.

The cause of chick mortalities may have been direct oiling. Oiled chicks were observed on two occasions: chicks crouched in oil in response to parental alarm calls, and we observed small pebbles attached to another chick by tar.

Chick mortality might also have been caused by ingestion of oiled prey. In 1989, 100% and 74% of mussel samples collected at Green Island in May and June, respectively, were oil-contaminated with a signature characteristic of *Exxon Valdez* oil (chi-square = 20.0, df = 1, $P < 0.005$, N = 20; and chi-square = 5.26, df = 1, $P < 0.05$, N = 23, respectively), whereas 70% of mussels collected at Montague Island (Stockdale Harbor) in May and 76% of those collected in June (Port Chalmers) were not oiled (chi-square = 4.17, df = 1, $P < 0.05$, N = 23; and chi-square = 6.76, df 1, $P < 0.01$, N = 25, respectively) (R. Britten, USFWS, personal communication 1992). At Green Island mussels collected in May were severely contaminated, containing as much as 28 mg oil/g, or 0.3% of body weight (J. Short, NOAA, personal communication, November 1992). Mean concentrations of total aromatics in mussels were lower at Stockdale Harbor, Montague Island, than at Green Island in May 1989 (249 versus 9,976 ng/g, P = 0.13, df = 6) (though Stockdale was lightly oiled) and lower at Port Chalmers, Montague Island, than at Green Island in June and early July (455 versus 1,754 ng/g, P = 0.0045, df = 37). Total aromatics declined from May to June–July 1989 at Green Island (mean = 9,976 versus 1,754 ng/g, P = 0.18, df = 6) and declined from June–July 1989 to August 1990 (mean = 1,754 versus 104.9 ng/g, P = 0.0002, df = 21). Hydrocarbons were at background levels in mussels at Green Island in 1990.

Mean concentrations of total dibenzothiophenes and of total aromatics in mussels from seven feeding territories (four on Green Island and three on Montague Island) where the fate of chicks was known were weakly correlated with brood reduction (Pearson r = 0.395 and 0.403, respectively). Total dibenzothiophenes and total aromatic concentrations were correlated with mean percent oil cover on quadrats (Pearson r = 0.890 and 0.899, respectively), which was correlated with chick mortality.

Reproductive Success in 1990

In 1990 brood size of oystercatchers on Green Island was 0.67 on nine heavily oiled territories (oil cover, 13–50%) and 1.0 on seven lightly to moderately oiled territories (oil cover, less than 13%) (t-test between brood sizes: P = 0.55, df = 14). An average of 0.3 fewer chicks fledged on heavily oiled than on moderately to lightly oiled territories in 1990, but regression of brood size on percent oil cover was not significant (r^2 = 14.6%, P = 0.310); however, no data were collected on non-oiled territories in 1990. There was no difference between mean brood sizes on Green Island in 1989 and 1990 (mean = 0.94 versus 0.81, P = 0.72, df = 29).

Mean brood sizes were smaller on territories that received bioremediation treatments (mean = 0.0) compared to those that were not treated (mean = 1.44) ($P < 0.01$). No broods fledged on any of the seven territories that were bioremediated. The only nest in a heavily oiled area that fledged any chicks was also the only heavily oiled nesting area that did not receive bioremediation treatment. The only moderately oiled nest to fail was also the only moderately oiled nest in a bioremediated area.

Behavior

Lack of parental attendance did not appear to contribute to chick loss on Green Island in 1989. Our observations indicated that Green Island breeding pairs with chicks spent more time flying (20 versus 13%) because some feeding grounds were separated from nesting sites, but distances to these feeding grounds were short and did not require extended absences from the brood. In addition, we did not observe simultaneous absences by both adults. Nevertheless, on Montague Island, of six nests predated during incubation, three (possibly five) involved "commuting" pairs.

Pairs with eggs spent similar amounts of time preening on oiled and non-oiled islands (9 versus 11%), flying (12 versus 14%), and socializing (5 versus 9%). Breeding pairs with chicks spent a relatively small proportion of time socializing on either island (3 and 0%). Observations of these categories of behavior did not indicate oiling effects, except perhaps for the observation that nonbreeding birds on Green Island (for which nesting was apparently disrupted early in the spill—see above) spent more time socializing than did breeding pairs with nests (17 versus 5%).

Average feeding rates were lower on Green Island than on Montague Island (1.49 versus 4.18 ingestions/min for pairs with eggs and 0.98 versus 4.52 ingestions/min for pairs with chicks and nonbreeding birds) ($P < 0.05$). The average duration of complete feeding bouts on Green Island was also longer than on Montague Island (21 min versus 11 min), although the number of complete samples was too small for statistical significance.

The quality of intertidal foraging habitat was not superior at non-oiled sites. Mean percent cover for marine algae other than *Fucus* sp. was somewhat greater on quadrats on Green Island, the oiled site, than on Montague Island (39 versus 28%, $P = 0.086$, df = 22), but mean percent cover was similar for *Fucus* sp. (34 versus 29%, $P = 0.068$, df = 22) and barnacles (*Balanus* sp.) (24 versus 21%, $P = 0.31$, df = 22), and mean cover for mussels, the main food for adult oystercatchers in the study areas, was greater on Green Island than on Montague Island (32 versus 21%, $P = 0.014$, df = 22). Lengths of the largest mussels tended to be less on Green Island (33.9 versus 38.2 mm, $P < 0.10$) but were within the range used by oystercatchers elsewhere (e.g., 41 mm, Sitka, Alaska, Webster 1941a; 39 mm, Cumberland, England, Norton-Griffiths 1967; >20 mm, Oregon, Marsh 1986b); however, mortality of intertidal organisms was higher on Green Island plots. Of mussels present in plots, 35% were dead on Green Island compared with 16% for Montague Island ($P = 0.0041$, df = 22), and 37% of barnacles were dead on Green Island versus 7% on Montague Island ($P = 0.00$, df = 22). On transects where *Fucus* mortality was recorded, the mean cover for dead *Fucus* was 78% on Green Island, whereas no *Fucus* mortality was observed on Montague Island ($P < 0.01$). Substrates differed: 75% ($N = 12$) on Green Island were cobble or rocky, whereas 83% ($N = 12$) on Montague Island were gravel. On Montague Island, we detected no relationship between success or failure of a clutch and the percent plots on which measurable numbers of mussels were present, the average percent cover of mussels in those plots, and the percent mortality of the mussels in meter-square plots on feeding-area transects of particular pairs.

Discussion

Our observed densities of 0.89 black oystercatcher pairs/km of occupied habitat and 0.49 black oystercatcher pairs/km of total shoreline on Montague Island probably were representative of prespill conditions. Boat surveys in Prince William Sound in July 1989 found 85 oystercatchers on 324.2 km of shoreline surveyed (mean density, 0.26 birds/km) (K. Laing, USFWS, personal communication, 1993). Those boat surveys detected 31% of the birds present in 1989 in our study areas (0.453 birds/km versus our 1.46 birds/km) ($P = 0.0085$, df = 6). Applying a correction of 0.31 for birds not detected, oystercatcher density in Prince William Sound derived from boat surveys was 0.84 birds/km in 1989, close to the densities in our study areas.

Similar densities were found from boat surveys of the Kenai coast in 1986 (0.10 birds/km; E. P. Bailey and B. Rice, USFWS, unpublished data), Katmai National Park in 1989 (0.32 birds/km; Martin, unpublished data), Kodiak Island in 1975–1978 (0.28 pairs/km; Nysewander and Barbour 1979), and the Semidi Islands between 1976–1981 (0.95 birds/km; Hatch and Hatch 1983).

The larger-than-expected number of nonbreeding pairs on Green Island, delayed hatching dates, smaller egg volumes, and an increase in the number of pairs breeding on Green Island in 1991 (Andres and DeZeeuw 1991) strongly suggest that many early nests failed on oiled sites in May, before our arrival. The high petroleum hydrocarbon concentrations found in mussels in May samples may be implicated in these early nesting failures. In addition, Green Island shorelines were cleaned in May, and Exxon Corp.'s cleanup effort, which treated a total of nearly 1,760 km of oil-impacted shoreline in 1989 (about 576 km in Prince William Sound and 1,168 km in the Gulf of Alaska), "involved more than 11,000 workers—of whom about 3,500 were working directly on the beach—as well as some 84 aircraft and more than 1,400 vessels of all types. . . ." Harrison (1991: 314). It is probable that this level of disturbance during the breeding season adversely affected breeding oystercatchers. Human disturbance has been cited as having an adverse effect on black oystercatchers elsewhere: on the Farallon Islands, California, and at Destruction Island, Washington, oystercatcher numbers increased greatly after cessation of disturbance (Ainley and Lewis 1974; Nysewander 1977); at Point Lobos, California, disturbance of an incubating bird caused eggs to be chilled (Legg 1954); in South Africa, numbers of African black oystercatchers *Haematopus moquini* decreased after disturbance (Jeffery 1987); and disturbance has been implicated in the extinction of the Canary Island black oystercatcher *H. meadewaldoi* (Hockey 1987).

Mean clutch size in Prince William Sound at Green and Montague islands combined, 2.62, was larger than that at Destruction Island, Washington, in two different years (2.0, 2.5; Nysewander 1977), in British Columbia (2.1; Hartwick 1974), and at Kodiak Island (2.24; Nysewander and Barbour 1979). Sizes of Montague Island oystercatcher eggs were similar to those reported elsewhere (56.2 × 39.5 mm, $N = 44$, Bent 1927; 56.2 × 39.1 mm, Destruction Island, Washington, Nysewander 1977). The smaller size of eggs on Green Island compared with that on Montague Island is probably attributable to a high number of second clutches resulting from the high incidence of early nesting failures, although it is also possible that oil ingestion by females physiologically resulted in smaller eggs (Harvey et al. 1982); however, hatching success at Green Island was not reduced and there was no reduction in egg viability.

Hatching success on both Green and Montague islands was comparable to or exceeded that reported by other workers: for example, 61% on Hinchinbrook Island in 1976 (Nysewander and Knudtson 1977); 52% on Kodiak Island in 1975–1978 (Nysewander and Barbour 1979); 62% and 77% on Destruction Island in 1974–1975 (Nysewander 1977); 25%, 34%, and 46% in 3 years in British Columbia (Hartwick 1974); and 23% in British Columbia (Groves 1984). Destruction of nests by tides, common elsewhere, was not found in this study.

Because small amounts (2–5 μL) of oil on eggs is sufficient to cause embryonic mortality (Albers 1978; Fry et al. 1986), we expected to find lower hatching success at oiled sites; however, hatching success at nests found in June was not lower at oiled sites on Green Island. Oystercatcher eggs on Green Island were, also surprisingly, not oiled. Oil was occasionally noticed on the feet of adult oystercatchers but evidently little or no oil was transferred to eggs. Oystercatchers on Green Island also did not spend an excessive amount of time preening, which would have been expected if plumage were oiled. Adult oystercatchers were not observed to feed in heavily oiled areas in 1989, and it appeared that adult oystercatchers on Green Island in June were staying relatively free of oil. They seem to have avoided, perhaps discriminated between, oiled and non-oiled areas. Both Montague and Green Island oystercatchers feed on gaping mussels using the "stabbing" technique, which severs the adductor muscle, rather than the "hammering" method (breaking the shell) used by some populations (Norton-Griffiths 1967; Hartwick 1978; this study). If mussels respond to severe environmental stress (e.g., oiling) by entering into an enforced state of inactivity, it is possible that the ostensible avoidance of oiled areas by Prince William Sound black oystercatchers was simply a consequence of feeding technique, because foraging on closed mussels in oiled beds would be unprofitable to oystercatchers that feed by stabbing.

Predation was the major source of egg losses on both study areas; although sample sizes are small, our data suggest that predation was lower on Green Island, the oiled site. Mustelids are mentioned as oystercatcher predators by Bianki (1977), Groves (1984), Heppleston (1972), and Vermeer et al. (1989). Both mink *Mustela vison* and river otters *Lutra canadensis* are intertidal foragers in Prince William Sound; mink patrol the high-tide line where oystercatcher nests are located, and mink prey remains in Prince William Sound included birds. Densities of river otters and mink in the sound were 0.3–0.85 and 1 animal/km of shoreline, respectively, in 1989 (Bowyer et al. 1993; M. Ben-David, Alaska Department of Fish and Game, personal communication, 1993), but Bowyer et al. (1993) demonstrated that river otters in the sound abandoned oiled intertidal areas, had home ranges that were twice as large on oiled areas, and avoided usually preferred shallowly sloping (oiled) intertidal habitat in 1989, 1990, and perhaps 1991. Observations of mink also indicated that populations were lower in oiled areas (Ben-David, personal communication). It seems likely that the avoidance of oiled shorelines by mustelids was responsible for the apparently lower predation rate and higher hatching success of oystercatcher nests at Green Island in 1989.

Adverse effects of shoreline oiling became evident after chicks were hatched. Heppleston (1972) and Groves (1984) found that the first 7 d of the posthatching period were critical to the ultimate reproductive output of oystercatchers in Scotland

and British Columbia. Our observed chick mortality rates of 66–89% on oiled shorelines in 1989 exceeded mortality reported in other studies, for example, 26% in Washington (Nysewander 1977) and 43% in Scotland (Heppleston 1972); however, chick mortality was high in British Columbia—72%, 63%, and 62% in three successive years (Hartwick 1974)—and 24 breeding pairs raised only 10 young near Sitka, Alaska, because of egg and chick loss in 1940 (Webster 1941b). Mean brood sizes on Green Island were not different in 1989 and 1990: the high chick mortality observed in 1989 continued into 1990.

Chicks disappeared at a rate proportional to the degree of intertidal-zone oiling. We are unsure whether the chick losses we observed were caused by direct oiling or by high levels of hydrocarbons in the mussels they were fed. Continued reduced reproductive success on oiled territories in 1990 and the fact that hydrocarbon concentrations were at background levels in 1990 might suggest that factors other than toxic hydrocarbon concentrations, such as direct oiling, may have caused the high rates of chick disappearance in 1989; however, Wolfe et al. (1996, this volume) asserted that toxicity of oiled intertidal sediments to intertidal organisms is caused by low-molecular-weight aromatic hydrocarbons (e.g., phenanthrene). Polynuclear aromatic hydrocarbon levels in mussels collected postspill at sites in Prince William Sound were significantly higher in 1989 than prespill (Karinen et al. 1993; J. Karinen, personal communication, National Marine Fisheries Service, May 1991).

Pressurized steam-cleaning of the intertidal zone removed oystercatcher prey organisms. Exxon Corp. stated that the "hot water treatment was used on about one fourth of the more than 350 miles . . . of coastline contaminated by the spill" (Anchorage Times, 10 April 1991). Reduced densities of limpets, mussels, and *Fucus* sp. were found on oiled sites in Prince William Sound, at lower Cook Inlet and Kenai Peninsula, and along Kodiak–Alaska Peninsula (Gibbons and Highsmith 1990). Oiled surfaces retarded the settlement of juvenile barnacles, and the reproductive cycle of mussels was delayed by several months. It is therefore possible that food limitation, in addition to direct oiling and/or hydrocarbon-induced mortality, affected oystercatchers adversely. Our observations of reduced feeding rates on Green Island (feeding rates on Montague Island were similar to those at an uncontaminated site in southeastern Alaska, where black oystercatchers took 3.3 mussels/min—Webster 1941a) suggest that (1) black oystercatchers were not able to forage as efficiently as usual, (2) they compensated by increasing their total feeding time, (3) they avoided foraging on oiled shorelines, (4) oiling may therefore have reduced the effective size of the foraging area, and (5) difficulty in obtaining food may thus have contributed to the observed high chick mortality. Groves (1984) demonstrated that delivery of less food to the smaller chick resulted in lighter chicks that survived less well to fledging. Our sample sizes suggested, but were not sufficient to demonstrate, that brood reduction in three-chick broods was greater than that for two-chick broods, and that three-chick broods lost a greater fraction of the brood on heavily oiled than on moderately oiled territories. Nol (1989) found that fledging success was correlated with size of foraging area but found that the match between food supply and reproductive performance was generally poor for American oystercatchers in Virginia.

There was a strong correlation between bioremediation and nest failure among the black oystercatcher nests examined on Green Island in 1990. Inipol EAP 22 was applied to 119 and 320 km of intertidal habitat in 1989 and 1990, respectively, including to some Green Island shorelines. The poor success of nesting attempts in bioremediated areas observed in 1990 indicated that bioremediation had adverse effects that were additive to oiling effects. Inipol is apparently toxic to wildlife, although the mechanism of its action is unclear. Mortality to mussel and oyster larvae has been demonstrated at Inipol concentrations in the 30–35 parts per million range (M. Fry, University of California at Davis, personal communication, June 1991), which would easily have been exceeded on Gulf of Alaska shorelines in 1989 and 1990.

A reduction in numbers and density of black oystercatchers in other parts of Prince William Sound and along the Kenai coast is indicated by prespill–postspill comparisons. Bailey and Rice (1989) found 27% fewer black oystercatchers on the Kenai Peninsula in 1989 than in 1986. S. Klosiewski and K. Laing (USFWS, unpublished data) found lower than expected numbers of black oystercatchers on oiled versus non-oiled areas in August 1989 and in July and August 1990 in the sound. They suggested population reductions of 50%, although we cannot corroborate losses of that magnitude at Green Island or from morgue data (see below). Andres and DeZeeuw (1991) found 24 breeding pairs on Green Island in 1991—7 more than bred in 1989—suggesting a population reduction in 1989; however, 4 or perhaps 5 of these pairs were present but not breeding in 1989, possibly having attempted and failed.

Our data on shoreline oiling of Green Island indicated that ADEC surveys underestimated both extent and intensity of oiling. Similarly, on the Alaska Peninsula, ADEC estimates of shoreline oiling for the Becharof National Wildlife Refuge (NWR) were 7.1% of USFWS estimates (12.7 versus 178.2 km), and ADEC surveys for the entire Kodiak–Alaska Peninsula sector provided shoreline oiling estimates that were 4% of USFWS refuge estimates for the Kodiak NWR alone (122.9 versus 3,146.3 km). Underestimation of oiled shoreline by ADEC resulted from a number of factors: (1) many oiled beaches surveyed by ADEC from the air appeared non-oiled; (2) on sandy shorelines, oil covered by sand by the next high tide was not visible from the air; (3) ADEC collected data from Kodiak–Alaska Peninsula in the fall, after oil had been buried in sediments, whereas USFWS surveys were conducted sooner after oil arrived in spring and early summer; and (4) a small proportion, 2.16% (247 of 11,457 km), of the Kodiak–Alaska Peninsula coastline was surveyed by ADEC. Similarly, more than 2,000 km of the Kenai Peninsula shoreline were not surveyed by ADEC, but independent estimates of oiling are not available for the Kenai coast.

That oystercatcher losses occurred throughout the Gulf of Alaska is predicated on the evidence of morgue data (adult mortalities from all sectors of the Gulf of Alaska coast, including Kodiak–Alaska Peninsula); boat surveys throughout Prince William Sound and along the Kenai coast, which demonstrated greater than expected reductions in numbers of black oystercatchers on oiled shorelines compared with non-oiled shorelines; anecdotal observations—of a reduction in the numbers of black oystercatchers on Perry Island (K. Weaverling, ex-Mayor, Cordova, Alaska, personal communication, January 1991) and the nesting failure of an externally oiled pair of black oystercatchers on the Barren Islands (E. Bailey, USFWS, unpublished data); and data on impaired reproductive success on Green Island, an oiled area in Prince William Sound (this study). An estimate of losses throughout the oil spill area is calculated from data on the densities of oystercatchers; from estimates of the extent of shorelines oiled from data provided by ADEC for the sound and for the Kenai coast and by Becharof and Kodiak NWRs for Kodiak–Alaska Peninsula; and from extrapolation of estimates of reproductive failure from this study in relation to the intensity of shoreline oiling using ADEC classifications of shoreline oiling. Data on the length of shoreline of the four sectors is summarized in Ford et al. (1991).

TABLE 2.—Estimated number of black oystercatcher chicks lost on oiled shorelines in Prince William Sound, Kenai Peninsula, and Kodiak–Alaska Peninsula in 1989.

	Alaska Department of Environmental Conservation oiling category			
Variable	Heavy	Moderate	Lights	Very light
Prince William Sound				
Kilometers oiled	144.8	113.1	192	129.5
Pairs exposed	71	55	94	63
Chicks lost	118	92	116	No data
Kenai Peninsula				
Kilometers oiled	28.9	75.2	107	76.4
Pairs exposed	11	29	41	29
Chicks lost	18	48	50	No data
Kodiak–Alaska Peninsula				
Kilometers oiled	27.4	93.3	397.5	2,628.1
Pairs exposed	8	26	111	736
Chicks lost	13	43	137	No data
Total chicks lost	149	183	303	No data

Black oystercatcher breeding pair densities for the four sectors (Prince William Sound, Kenai Peninsula, Kodiak Island and Alaska Peninsula) are conservatively estimated as 0.49, 0.38, 0.28, and 0.28 pairs/km, respectively; density on the Kenai is assumed to be intermediate between that of Prince William Sound and Kodiak–Alaska Peninsula and greater than the observed 0.1 pairs/km, which presumably is an underestimate. When one multiplies shoreline lengths and breeding pair densities, the black oystercatcher breeding population (in pairs) in the Gulf of Alaska is estimated to be as follows: Prince William Sound, 1,797; Kenai–Cook Inlet, 1,413; Kodiak–Alaska Peninsula, 2,382; total, 5,592.

Estimates by ADEC of the extent and degree of shoreline oiling were used for Prince William Sound and Kenai Peninsula, even though those data have been shown to be conservative. For the Kodiak and Alaska Peninsula coasts, the more extensive and detailed USFWS data on shoreline oiling provide estimates of 579, 288, and 3,146 km of oiled shoreline in Prince William Sound, Kenai Peninsula, and Kodiak–Alaska Peninsula sectors, respectively, for a total of 4,013 km. Of this, 201 km of shoreline were heavily, 282 km moderately, 697 km lightly, and 2,834 km were very lightly oiled according to ADEC oiling definitions (Table 2).

Combining available estimates of oiled shoreline and densities of oystercatchers, approximately 446 breeding pairs of black oystercatchers were exposed to heavy, moderate, and light shoreline oiling; if the very light oiling category is included, 1,274 breeding pairs were exposed. An estimated 8–23% of the black oystercatcher population in the Gulf of Alaska was therefore exposed to oil.

That some oystercatchers suffered some direct mortality is evidenced by the recovery of nine dead black oystercatchers throughout the Gulf of Alaska. The band recovery rate for black oystercatchers generally is 1.8% (of 495 banded, D. Bystrak, Bird Banding Laboratory, personal communication, April 1991); however, the recovery rate for oystercatchers on unpopulated Alaskan shorelines is probably lower than for oystercatchers in general but in 1989 was higher than usual because of the spill itself and to the search-and-rescue effort for oiled birds. Ford et al. (1991) calculated a recovery rate of 8% for common murres *Uria aalge* in the Gulf of Alaska in 1989, which was eight times the typical recovery rate of 1% (from 2,192 bandings, Bystrak, personal communication). Therefore, 62–500 black oystercatchers, or 0.5–4.5% of the gulf population, may have been directly killed by oil in 1989.

The lightly oiled and moderately to heavily oiled shoreline oiling categories of ADEC had moderate and severe brood reduction effects, respectively. Even though there was a somewhat higher proportion of nests hatching on oiled areas, most of those hatchlings did not fledge. Assuming that all chick mortality on oiled areas was attributable to oil in 1989, approximately 6% of the potential chick production in the Gulf of Alaska (an estimated 635 chicks) was lost on oiled shorelines in 1989 (Table 2). Chick production in 1989 and subsequent years obviated by deaths of adults is not included in this number.

Continued loss of chicks was documented at Green Island in 1990, but because no data were obtained from non-oiled shorelines, we cannot estimate losses attributable to oil in 1990. Bioremediation occurred on approximately 320 km of shoreline in Prince William Sound in 1990. If failed reproduction on our study area was representative of all bioremediated shorelines, approximately 157 breeding pairs would not have raised 293 chicks in 1990 in the sound because of side effects of bioremediation.

Acknowledgments

Rick Turner recruited volunteers from Oregon to respond to the oil spill in 1989, conducted field work in 1989–1990, and assisted with the bibliography. The U.S. Fish and Wildlife Service provided administrative support, equipment, and supplies. Philip Martin, Arctic NWR, collected May mussel samples and Marta McWhorter helped with fieldwork. Captains Ted Estrada and Chuck Osborne of MV *Curlew* supplied field camps, and cook Lesley Mulcahy's shipboard meals were greatly appreciated after camp fare. Local fishers helped us with emergencies. David Nysewander made available his library on oystercatchers, and Alaska Maritime NWR provided office space to analyze our data. Rich McMahon, Alaska Department of Natural Resources, provided ADEC shoreline oiling data, and Donna Dewhurst and Fred Zwiefelhofer provided shoreline oiling data for Becharof and Kodiak NWRs. Karen Laing provided raw data on numbers of black oystercatchers seen on boat surveys of Prince William Sound. Malin Babcock, NOAA, Auke Bay, and an anonymous reviewer commented on the manuscript.

References

Albers, P. H. 1978. The effects of petroleum of different stages of incubation in bird eggs. Bulletin of Environmental Contamination and Toxicology 19:624–630.

Ainley, D. G., and T. J. Lewis. 1974. The history of Farallon Island marine bird populations, 1854–1972. Condor 76:432–446.

Andres, B., and M. DeZeeuw. 1991. Feeding ecology and reproductive success of black oystercatchers in Prince William Sound. 1991 Restoration Project Report. U.S. Fish and Wildlife Service, Anchorage, Alaska.

AOU (American Ornithologists' Union). 1957. Check-list of North American birds, fifth edition. Port City Press, Baltimore, Maryland.

Bent, A. C. 1927. Life histories of North American shore birds. U.S. National Museum Bulletin 142.

Bianki, V. V. 1977. Gulls, shorebirds, and alcids of Kandalaksha Bay. Proc. Kandalaksha State Reservation 6. Translated from Russian, Israel Program for Scientific Translations, Jerusalem.

Bowyer, R. T., J. W. Testa, J. B. Faro, and L. K. Duffy. 1993. Effects of the *Exxon Valdez* oil spill on river otters in Prince William Sound. Pages 297–299 *in* R. Spies, L. J. Evans, B. Wright, M. Leonara, and C. Holba, editors. *Exxon Valdez* oil spill symposium. Anchorage, Alaska.

Ford, R. G., and six coauthors. 1991. Assessment of direct seabird mortality in Prince William Sound and the western Gulf of Alaska resulting from the *Exxon Valdez* oil spill. Final report to the Natural Resources Damage Assessment, Bird Damage Assessment Study 1. Final report U.S. Fish and Wildlife Service, Anchorage, Alaska.

Frank, P. W. 1982. Effects of winter feeding on limpets by black oystercatchers, *Haematopus bachmani*. Ecology 63:1352–1362.

Fry, D. M., J. Swenson, L. A. Addiego, C. R. Gray, and A. Kang. 1986. Reduced reproduction of wedge-tailed shearwaters exposed to weathered *Santa Barbara* crude oil. Archives of Environmental Contamination and Toxicology 15:453–463.

Gibbons, D., and R. C. Highsmith. 1990. Assessment of injury to coastal habitat. Natural Resource Damage Assessment Study 1, Anchorage, Alaska.

Groves, S. 1984. Chick growth, sibling rivalry, and chick production in American black oystercatchers. Auk 101:525–531.

Gundlach, E., E. A. Pavia, C. Robinson, and J. C. Gibeaut. 1991. Shoreline surveys at the *Exxon Valdez* oil spill: the state of Alaska response. Pages 519–529 *in* Proceedings, 1991 international oil spill conference. American Petroleum Institute Publication 4529, Washington, DC.

Harrison, O. R. 1991. An overview of the *Exxon Valdez* oil spill. Pages 313–319 *in* Proceedings, 1991 international oil spill conference. American Petroleum Institute Publication 4529, Washington, DC.

Hartwick, E. B. 1974. Breeding ecology of the black oystercatcher (*Haematopus bachmani* Audubon). Syesis 7:83–92.

Hartwick, E. B. 1976. Foraging strategy of the black oystercatcher (*Haematopus bachmani*). Canadian Journal of Zoology 54:142–155.

Hartwick, E. B. 1978. The use of feeding areas outside of the territory of breeding black oystercatchers. Wilson Bulletin 90:650–652.

Harvey, S., J. G. Phillips, and P. J. Sharp. 1982. Reproductive performance and endocrine responses to ingested North Sea oil. Pages 1–411 *in* C. G. Scanes et al., editors. Aspects of avian endocrinology: practical and theoretical implications. Texas Tech University, Lubbock.

Hatch, S. A., and M. A. Hatch. 1983. Populations and habitat use of marine birds in the Semidi Islands. Murrelet 64:39–46.

Heppleston, P. B. 1972. The comparative breeding ecology of oystercatchers (*Haematopus ostralegus* L.) in inland and coastal habitats. Journal of Animal Ecology 41:23–51.

Hockey, P. A. R. 1987. The influence of coastal utilization by man on the presumed extinction of the Canarian black oystercatcher *Haematopus meadewaldoi* Bannerman. Biological Conservation 39:49–62.

Jeffery, R. G. 1987. Influence of human disturbance on the nesting success of African black oystercatchers. South African Journal of Wildlife Research 17:71–72.

Johnson, D. H. 1979. Estimating nest success: the Mayfield method and an alternative. Auk 96:651–661.

Karinen, J. F., M. M. Babcock, D. W. Brown, W. D. MacLeod, Jr., L. S. Ramos, and J. W. Short. 1993. Hydrocarbons in intertidal sediments and mussels from Prince William Sound, Alaska, 1977–1980: characterization and probable sources. NOAA (National Oceanic and Atmospheric Administration) Technical Memorandum NMFS (National Marine Fisheries Service) AFCS-9, Seattle.

Laing, K. 1990. Assessment of injury to waterbirds from the *Exxon Valdez* oil spill: boat surveys to determine distribution and abundance of migratory birds in Prince William Sound and the northern Gulf of Alaska. U.S. Fish and Wildlife Service, Anchorage, Alaska.

Legg, K. 1954. Nesting and feeding of the black oystercatcher near Monterey, California. Condor 56:359–360.

Marsh, C. P. 1986a. Impact of avian predators on high intertidal limpet populations. Journal of Experimental Marine Biology and Ecology 104:185–202.

Marsh, C. P. 1986b. Rocky intertidal community organization: the impact of avian predators on mussel recruitment. Ecology 67:771–786.

Mayfield, H. F. 1960. Nesting success calculated from exposure. Wilson Bulletin 73:255–261.

Mayfield, H. F. 1975. Suggestions for calculating nest success. Wilson Bulletin 87:456–466.

Nol, E. 1989. Food supply and reproductive performance of the American oystercatcher in Virginia. Condor 91:429–435.

Norton-Griffiths, M. 1967. Some ecological aspects of the feeding behaviour of the oystercatcher (*Haematopus ostralegus*) on the edible mussel (*Mytilus edulis*). Ibis 109:412–424.

Nysewander, D. R. 1977. Reproductive success of the black oystercatcher in Washington state. Master's thesis. University of Washington, Seattle.

Nysewander, D. R., and D. B. Barbour. 1979. The breeding biology of marine birds associated with Chiniak Bay, Kodiak Island, 1975–1978. Annual report for NOAA (National Oceanic and Atmospheric Administration) OSCEAP (Outer Continental Shelf Environmental Assessment Program) Contract 01-5-022-2538. U.S. Fish and Wildlife Service, Anchorage, Alaska.

Nysewander, D., and P. Knudtson. 1977. The population ecology and migration of seabirds, shorebirds, and waterfowl associated with Constantine Harbor, Hinchinbrook Island, Prince William Sound, 1976. Pages *in* J. C. Bartonek et al., editors. Population dynamics and trophic relationships of marine birds in the Gulf of Alaska and southern Bering Sea. U.S. Fish and Wildlife Service, Anchorage, Alaska.

Patten, S. M. Jr. 1992. Injury assessment of hydrocarbon uptake by sea ducks in Prince William Sound and the Kodiak Archipelago, Alaska. Final report for Natural Resources Damage Assessment. Alaska Department of Fish and Game, Anchorage.

Vermeer, K., K. H. Morgan, and G. E. John Smith. 1989. Population and nesting habitat of American black oystercatchers in the Strait of Georgia. Pages 118–122 *in* K. Vermeer and R. W. Butler, editors. The ecology and status of marine and shoreline birds in the Strait of Georgia, British Columbia. Canadian Wildlife Service Special Publication, Ottawa.

Webster, J. D. 1941a. Feeding habits of the black oystercatcher. Condor 43:175–180.

Webster, J. D. 1941b. The breeding of the black oystercatcher. Wilson Bulletin 53:141–156.

Wolfe, D. A., and six coauthors. 1996. Toxicity of intertidal and subtidal sediments contaminated by the *Exxon Valdez* oil spill. American Fisheries Society Symposium 18:121–139.

American Fisheries Society Symposium 18:759–769, 1996

Population, Reproduction, and Foraging of Pigeon Guillemots at Naked Island, Alaska, before and after the *Exxon Valdez* Oil Spill

KAREN L. OAKLEY[1] AND KATHERINE J. KULETZ

U.S. Fish and Wildlife Service, Migratory Bird Management
1011 E. Tudor Road, Anchorage, Alaska 99503, USA

Abstract.—After the 1989 *Exxon Valdez* oil spill in Prince William Sound, Alaska, we studied pigeon guillemots *Cepphus columba* breeding just 30 km from the grounding site. The postspill population was 43% smaller than the prespill population, but we could not attribute the entire decline to the spill because a decline in this guillemot population may have predated the spill. However, relative declines in the population were greater along oiled shorelines, suggesting that the spill was responsible for some of the decline. Reproduction appeared largely unaffected, but the cryptic nature of guillemot nests undoubtedly reduced our ability to detect failed nests. Nesting success was lower, but the apparent cause—greater losses of chicks to predators—was not obviously related to the spill. Fledging weight and growth rates of chicks and the rate at which adults delivered food to their chicks were not lower after the spill. The most likely explanation for the few effects observed is that oil was present on the surface waters of the study area for a relatively short period before the guillemots returned to begin their annual reproductive activities.

In 1978, the U.S. Fish and Wildlife Service (USFWS) sent us to Naked Island, located in the center of Prince William Sound, Alaska, to study the local seabird population, and one of us (Kuletz) carried out studies there through 1981 (Oakley 1981; Kuletz 1983). Our studies centered on the pigeon guillemot *Cepphus columba*, a small diving seabird, which is abundant there. The impetus for these studies was the beginning of transport of Prudhoe Bay crude oil from the Trans-Alaska Pipeline System terminus in Valdez to West Coast and Gulf of Mexico refineries via supertankers. The tanker route lay 30 km to the east of Naked Island, and the USFWS wanted baseline data on the seabird population in the event of a major spill.

A major oil spill happened on 24 March 1989, when the TV *Exxon Valdez* ran aground on Bligh Reef, spilling 43.9 million liters of crude oil. Naked Island was the first land mass to be hit by the oil, and oil was observed on the waters surrounding Naked Island between 29 March and 19 April. Beach rocks on certain Naked Island guillemot colony sites remained oiled throughout the 1989 breeding season. We returned to Naked Island in 1989–1992 to duplicate our earlier work and learn to what extent the spill affected the pigeon guillemot population.

The pigeon guillemot, in the family Alcidae, breeds ubiquitously in small colonies along rocky shores throughout the eastern North Pacific (Ewins 1993). Guillemots feed in inshore waters, primarily on benthic fishes and invertebrates (Thoreson and Booth 1958; Drent 1965; Follett and Ainley 1976; Oakley 1981; Kuletz 1983; Ainley and Boekelheide 1990). Nest sites are found in preexisting cavities, generally rock crevices among talus boulders or on cliff faces (Drent 1965; Oakley 1981), but also in crevices in driftwood piles, sandy banks, docks, and bridge girders (Byrd et al. 1974; Campbell 1977; Vermeer et al. 1993b). Guillemots lay a clutch of one or two eggs, and the young are nidicolous, remaining in the nest for a little more than 1 month after hatching (Drent 1965). During this period, the adults feed their chicks by delivering single whole fishes to the nest throughout the day (Thoreson and Booth 1958; Drent 1965). The wintering habits of the pigeon guillemot are not well known, but these birds appear to favor sheltered inshore waters (Ewins 1993).

Guillemots, because of their diving habit and coastal distribution, are among the bird species considered most vulnerable to oil spills (King and Sanger 1979). The oil from the *Exxon Valdez* spill could have affected the guillemots breeding at Naked Island by killing a large percentage of the population. Hatchability of eggs contaminated by crude oil may have been reduced because of the embryotoxicity of aromatic compounds in the oil (Walters et al. 1987; Hoffman 1990).

The oil could also have had sublethal effects (Peakall et al. 1982, 1983; NRC 1985; Nero and Associates, Inc. 1987). Black guillemot *C. grylle*

[1]Present address: National Biological Service, Alaska Science Center, 1101 E. Tudor Road, Anchorage, Alaska 99503, USA.

chicks dosed with weathered South Carolina crude oil and raised by their parents in their own nests, grew significantly slower than controls and had impaired osmoregulatory function (Peakall et al. 1980). Cassin's auklets *Ptychoramphus aleuticus* that were dosed externally before egg laying abandoned nesting, and those that did not abandon nesting were delayed in egg laying by more than 20 d, presumably because egg formation was disrupted (Nero and Associates, Inc. 1987). Auklets that were dosed externally during incubation had a higher frequency of abandonment, lower hatching success, and lower net breeding success, than controls.

Oil from the *Exxon Valdez* was therefore most likely to have disrupted guillemot breeding at Naked Island by delaying egg laying, decreasing hatching success, increasing abandonment, and slowing the growth of chicks. Guillemot breeding could also have been disrupted if the oil spill caused changes in the distribution and abundance of prey.

We found that the population was reduced after the spill and that a portion of that decline was probably attributable to the spill. Reproduction was probably disrupted to some degree in 1989, but the nests we studied—primarily those that successfully hatched chicks—appeared largely unaffected. The rate at which adults carried food to their chicks was not lower after the spill, but some changes occurred in the prey species delivered. We concluded that the spill had few effects on Naked Island guillemots probably because the spill preceded the annual return of the guillemots to the area and because the study area was not heavily oiled.

Study Area

The study area comprised Naked, Peak, Storey, Smith, and Little Smith islands (Figure 1), located in the center of Prince William Sound, a fjord-type estuarine system adjoining the northern Gulf of Alaska. The Prince William Sound region is part of the Pacific coastal belt of spruce *Picea sitchensis* and hemlock *Tsuga heterophylla* forest (Cooper 1942). The islands are low (less than 400 m) and wooded to their summits. Shores in the area are rocky and consist of cliffs, broken cliffs, and escarpments interspersed with boulder beaches.

Naked Island was first hit with oil on 27 March 1989, 3 d after the initial spill, when strong winds moved the oil southwest (Galt et al. 1991). The entire island was surrounded with oil from 29 March through 1 April, and Naked Island waters were observed to be oiled on various days through 19 April. Oil exposure varied among the islands in the study area: Storey Island was largely spared, while the Smith islands were heavily oiled. A cold-water beach-washing technique was tested in McPherson Bay on Naked Island in April 1989, but, at the request of the USFWS, no Naked Island beaches were cleaned as part of the official cleanup until after seabirds had completed breeding. The most heavily oiled beaches on Naked Island were treated, apparently using the bioremediation technique, in September 1989.

Figure 1 shows the distribution of pigeon guillemot colonies in the study area and the locations of the colonies on the western side of Naked Island where our studies were focused. Beaches at the Parakeet Point, Row, Nomad, and South Cabin colonies had one or more patches of viscous oil that were several square meters in size, and we considered these colonies to be heavily oiled. The beach at the Hook colony had some oil, but not as much as at the other colonies, and we found negligible oil on beaches at the Thumb, North Cabin, and North Outside colonies.

Methods

One or both of us were present on Naked Island during 8 May–31 August 1978, 9 May–27 August 1979, 26 May–17 August 1980, 1 July–17 August 1981, 7–10 June and 13–17 August 1984, 8 June–14 August 1989, and 30 May–15 August 1990. In 1991 and 1992, we were only present during 3–7 June 1991 and 6–8 June 1992 to conduct population censuses.

Population

To determine if the pigeon guillemot population of the Naked Island area declined after the spill, we examined changes in population census counts for the entire study area and changes in the proportion of the population attending colonies located in oiled sections of the shoreline.

Population census.—The pigeon guillemot population of the study area was censused by circumnavigating each island in a small boat between 50 and 100 m from shore and counting all guillemots (Nettleship 1976; Cairns 1979; Ewins 1985; Ewins and Tasker 1985; Vermeer et al. 1993a). In all years except 1989 censuses were conducted during the first week of June (3–6 June), which, for Naked Island guillemots, is coincident with the median date of egg laying. All censuses were conducted between 0400 and 1000 hours (Alaska Daylight Time), and census dates were selected so that the morning high tide occurred roughly midway through the daily count window. All censuses were conducted when seas were calm.

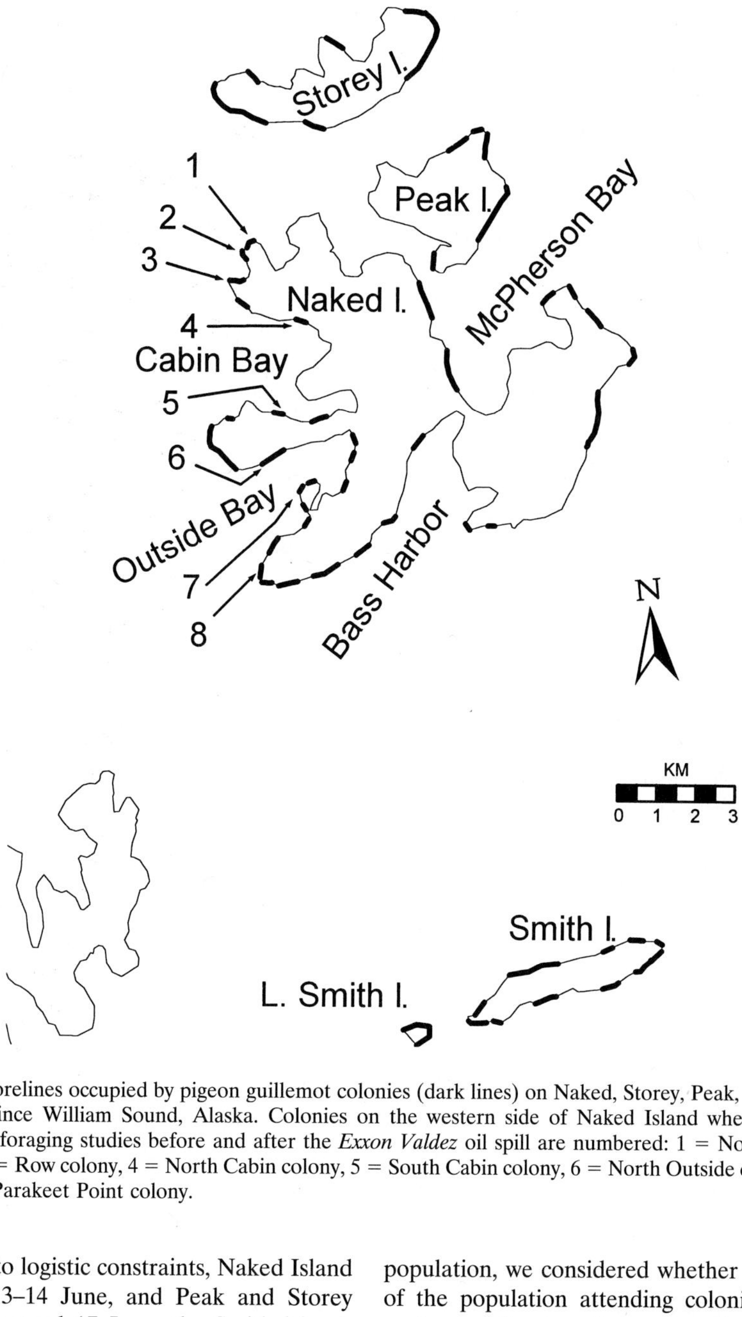

FIGURE 1.—Shorelines occupied by pigeon guillemot colonies (dark lines) on Naked, Storey, Peak, Smith, and Little Smith islands, Prince William Sound, Alaska. Colonies on the western side of Naked Island where we conducted reproductive and foraging studies before and after the *Exxon Valdez* oil spill are numbered: 1 = Nomad colony, 2 = Thumb colony, 3 = Row colony, 4 = North Cabin colony, 5 = South Cabin colony, 6 = North Outside colony, 7 = Hook colony, and 8 = Parakeet Point colony.

In 1989, due to logistic constraints, Naked Island was censused 13–14 June, and Peak and Storey islands were censused 17 June; the Smith islands were not censused. Because the 1989 counts were incomplete and were made later in the incubation period than the prespill counts, the 1989 counts were not used in comparing prespill to postspill counts.

Proportion of population that was attending colonies along oiled shorelines.—To determine whether shoreline oiling had local effects on the guillemot population, we considered whether the proportion of the population attending colonies along oiled sections of the shoreline was smaller than expected after the spill. We considered the shorelines of Cabin Bay, Outside Bay, Bass Harbor, McPherson Passage, Peak Island, Storey Island, and the southwestern side of Smith Island to be non-oiled, and the shorelines of McPherson Bay, the eastern and northwestern sides of Naked Island, Little Smith Island, and the northern and southeastern sides of Smith Island to be oiled (EVOSDAGG 1990).

Reproduction

At Naked Island, pigeon guillemots nest in talus crevices, cliff crevices, and cliff-edge burrows; the majority of nest sites are cryptic and inaccessible, or only partially accessible, to investigators (Oakley 1981). For this reason, estimates of reproductive parameters were based on observations at a relatively small number of accessible nests. In addition, these nests were spread out over 70 km of shoreline, making it difficult to visit them more than once every several days. Guillemot pairs typically reuse their nest sites (Drent 1965), and our sample of nests in each year included nests used previously. In 1979–1981 and in 1990, we probably studied some of the same pairs observed in the preceding year.

In the 1978–1981 studies, reproductive data were collected from 85 accessible nest sites on the western side of Naked Island. The majority of these nests were located at five colonies: Nomad, Thumb, Row, Hook, and Parakeet Point (Figure 1). In 1984, Kuletz made brief visits to the Row, Nomad, and Thumb colonies in June and August. In 1989, we were able to relocate only 40 of the known nest sites, and only 8 of these nests were active. We also found 64 new nest sites in 1989, of which 20 were accessible. In 1990, 71 of the nests known from all prior years were relocated, of which 62 were active. In addition, 34 nests were newly discovered, of which 18 were accessible. In all years, the status of each nest was checked every 3–5 d throughout the breeding season. Five aspects of reproduction were compared: the number of active nests, chronology, success, fledging weight, and chick weight growth rates.

Active nests.—We used the number of active nests, defined as any nest in which at least one egg was laid, as a measure of nesting effort in each year. We compared the number of active nests at Nomad, Thumb, Row, Hook, and Parakeet Point colonies in prior years with the number active in 1989 and 1990.

Chronology.—Because nests were only checked once every 3–5 d, exact laying, hatching, and fledging dates were rarely known. The chronology of nesting for each year was therefore constructed from estimated dates of laying, hatching, and fledging derived for all nests with sufficient nest-check data. For each egg or chick, we estimated the dates of laying, hatching, and fledging from whichever date was known with the least uncertainty. The mean length of the incubation period for Naked Island guillemots was unknown, so we used the mean values for incubation period of 32 d for first eggs and 30 d for second eggs found by Drent (1965) for British Columbia guillemots. From a small sample of chicks on Naked Island whose length of stay in the nest was known ($N = 8$), we estimated the mean length of the chick period for Naked Island guillemots to be 36 d, which was similar to the mean length of the chick period of 35 d for British Columbia guillemots (Drent 1965).

Success.—Mean clutch size and hatching, fledging, and nesting success rates were calculated for each year. To avoid biasing our estimates of hatching and nesting success toward successful nests, we defined hatching and nesting success as the average number of chicks hatched or fledged per nest for all nests found during the egg stage. We defined fledging success as the mean number of chicks fledged per nest with hatchlings for all nests, whether the nests were found in the chick or egg stage. Reasons for partial or complete nest failure were noted.

Fledging weight.—Chicks in accessible nests were weighed with Pesola hand-held scales, typically once every 3–5 d. Mean fledging weight for each year was calculated using the weights of all chicks weighed within 1 week of fledging.

Chick growth rates.—We calculated the rate of change in weight (g/d) for each accessible chick during the period of linear mass increase, reported by Koelink (1972) to occur between 8 and 18 d after hatching.

Foraging

We gathered information about guillemot foraging by observing adults feeding their chicks. We determined the rate of food delivery and the types of fishes fed to chicks during chick food watches. During a watch, an observer using a 15–45 power telescope and binoculars was stationed inside a blind for periods of 5–6 h. When an adult guillemot arrived with a fish, the time of arrival was recorded. The nest to which each fish was delivered and the times of delivery and departure were also recorded. Each fish was classified as either a schooling fish or a bottom fish and was identified to the lowest possible taxon. Schooling fish had a distinct silvery appearance and included Pacific sand lance *Ammodytes hexapterus*, Pacific herring *Clupea pallasi*, and smelts (Osmeridae); bottom fish had a variety of shapes, sizes, and colors and included blennies (Stichaeidae and Pholidae), sculpins (Cottidae), cods (Gadidae), ronquils (Bathymasteridae), and others (e.g., Zoarcidae, Agonidae, and Pleuronectidae). Watches were conducted at each colony about every 5 d throughout the nestling period (mid-July to late August).

TABLE 1.—Counts of pigeon guillemots during June censuses at Naked, Storey, Peak, Smith, and Little Smith islands, Prince William Sound, Alaska, before and after the *Exxon Valdez* oil spill. Blank spaces indicate no survey was conducted.

Year	Naked Island	Storey Island	Peak Island	Smith Island	Little Smith Island	Total
			Before spill			
1978	1,115	392	94	175	72	1,848
1979	1,226	495	150	301	58	2,230
1980	891					
			After spill			
1989[a]	615	193	73			
1990	729	293	102	124	31	1,279
1991	755	293	102	76	35	1,261
1992	586	266	87	100	23	1,062

[a]1989 counts were made later in the incubation period than counts in all other years and were not included in comparing prespill to postspill counts.

TABLE 2.—Numbers of pigeon guillemots counted during June censuses of oiled and non-oiled shoreline sections of Naked, Peak, Storey, Smith, and Little Smith island, Prince William Sound, Alaska, before and after the *Exxon Valdez* oil spill. Each section of shoreline was classified as oiled or non-oiled based on a map showing "Cumulative Oil Impact" as of 21 August 1989 (EVOSDAGG 1990). Blank spaces indicate no survey was conducted.

Year	Number in non-oiled	Number in oiled	Proportion in oiled
		Before spill	
1978	1,139	709	0.38
1979	1,339	891	0.40
		After spill	
1989			
1990	899	380	0.30
1991	895	366	0.29
1992	784	278	0.26

In 1979–1981, Kuletz (1983) conducted watches primarily at the Nomad, Thumb, and Row colonies. She observed 17 nests in 1979, 11 nests in 1980, and 9 nests in 1981. In 1989, we observed deliveries to 21 nests at five colonies. The colonies included two of the three colonies studied by Kuletz (Row and Nomad) as well as two colonies in Cabin Bay (North Cabin and South Cabin) and a single colony in Outside Bay (North Outside). We did not conduct watches at the Thumb colony because only one nest there had chicks in 1989. In 1990, we observed deliveries to 43 nests at the Nomad, Row, North Cabin, North Outside, and Parakeet Point colonies.

Data Analysis

To evaluate the effects of the spill on guillemots breeding at Naked Island, we compared prespill means to postspill means using a two-sample t-test. Because this study was carried out to support litigation, we were only interested in negative effects. Therefore, our alternative hypothesis was that the spill had a negative effect on guillemots, and all tests were one-tailed. We used $\alpha = 0.05$ as the level of significance.

Results

Population

Population census.—Guillemots breed ubiquitously in the Naked Island area (Figure 1), and their general breeding distribution in 1989–1992 was not different from that in 1978–1981. During postspill censuses, we found guillemots attending all colonies known from prior years. Although the distribution of breeding guillemots was not different, the number of guillemots attending colonies in the study area after the spill was 43% less than the number attending in prior years (Table 1; $t = -5.99$, df = 1.6, $P = 0.03$).

Proportion of population that was attending colonies along oiled shorelines.—The proportion of the population attending colonies along oiled shorelines declined after the spill (Table 2; $t = -8.10$, df = 3.0, $P = 0.002$).

Reproduction

Active nests.—After the spill, the total number of active nests at the five colonies we studied was less than the number present before the spill (Table 3; $t = -5.38$, df = 2.9, $P = 0.01$).

Chronology.—Nesting chronology of guillemots at Naked Island after the spill was not delayed in

TABLE 3.—Numbers of active pigeon guillemot nests at the Nomand, Thumb, Row, Hook, and Parakeet Point colonies on the western side of Naked Island, Prince William Sound, Alaska, before and after the *Exxon Valdez* oil spill. Blank spaces indicate no survey was conducted.

Year	Nomad	Thumb	Row	Hook	Parakeet Point	Total
			Before spill			
1978	7	11	9	12	19	58
1979	8	6	15	14	10	53
1980	5	4	10	8	6	33
1981	4	5	7			
			After spill			
1989	3	4	7	2	5	21
1990	4	3	13	2	12	34

TABLE 4.—Median laying, hatching, and fledging dates of pigeon guillemots at Naked Island, Prince William Sound, Alaska, before and after the *Exxon Valdez* oil spill.

Year	*N*	Laying date	*N*	Hatching date	*N*	Fledging date
			Before spill			
1978	48	5 Jun	40	3 Jul	38	10 Aug
1979	62	4 Jun	52	6 Jul	46	13 Aug
1980	40	5 Jun	19	7 Jul	15	11 Aug
1981	25	2 Jun	22	3 Jul	18	6 Aug
			After spill			
1989	22	3 Jun	20	5 Jul	16	8 Aug
1990	43	7 Jun	41	7 Jul	31	11 Aug

comparison to chronology in prespill years (Table 4; egg laying: $t = 0.47$, df = 1.3, $P = 0.36$; hatching: $t = 0.87$, df = 3.1, $P = 0.22$; fledging: $t = 0.24$, df = 2.9, $P = 0.41$).

Success.—As in prespill years, the majority of pairs laid two-egg clutches, and mean clutch size was not lower after the spill (Table 5; $t = -0.39$, df = 1.9, $P = 0.37$). The mean number of chicks hatched per nest was also not lower after the spill (Table 5; $t = 1.18$, df = 2.8, $P = 0.83$). The fledging success rate was lower after the spill (Table 5; $t = -2.93$, df = 4.9, $P = 0.02$), as was the overall nesting success rate (Table 5; $t = -2.38$, df = 4.3, $P = 0.04$). The lower rates of fledging success and overall nesting success after the spill appeared to be caused by a higher incidence of nest loss resulting from predation, particularly during the chick stage (Table 6).

Fledging weight.—Mean fledging weight of chicks after the spill was not less than the mean fledging weight of chicks raised in prior years (Table 7; $t = -0.18$, df = 1.7, $P = 0.44$).

Chick growth rate.—Mean growth rate after the spill was not less than the mean growth rate observed before the spill (Table 7; $t = -2.13$, df = 2.8, $P = 0.07$).

Foraging

Feeding rate.—In 1989 and 1990, guillemots delivered fish to their chicks at mean rates of 1.00 and 0.95 fish/nest/hour (f/n/h). Mean feeding rates in prior years were 1.06 f/n/h in 1979, 0.86 f/n/h in 1980, and 0.65 f/n/h in 1981. The mean rate at which guillemots delivered fish to their chicks after the spill was not less than the mean of mean feeding rates observed in prespill years ($t = 0.98$, df = 2.2, $P = 0.78$).

Prey species.—A relatively high percentage of all deliveries observed in 1990, 25%, were not identified, making rigorous comparisons of prey use in 1990 with that in other years difficult (Table 8). In general, the bottom fish species delivered to chicks were the same among years. Sculpins, gunnels, and pricklebacks were typically the most important bottom fish species in chick diets. Cods were the only bottom fish species that showed major changes in their relative importance in chick diets among years. Cods were delivered more frequently in 1989 and 1990 than in prespill years.

While the bottom fish species composing the chick diet were relatively stable among years, the schooling fish species delivered to chicks changed. Pacific sand lances, which were the most important schooling fish in 1978 (Oakley 1981) and in 1979–1981 (Kuletz 1983), represented only 14% of deliveries observed in 1989 and only 8% of deliveries observed in 1990. In 1989, Pacific herring were the most important schooling fish. Pacific herring were never observed as food for guillemot chicks in 1978 (Oakley 1981) or in 1979–1980. They were first observed in chick food deliveries in 1981.

TABLE 5.—Mean clutch sizes, and mean rates of hatching success, fledging success, and nesting success for pigeon guillemots at Naked Island, Alaska, before and after the *Exxon Valdez* oil spill.

Year	*N*	Clutch size	*N*	Hatching success	*N*	Fledging success	*N*	Nesting success
				Before spill				
1978	13	1.54	9	1.22	30	1.60	8	1.25
1979	33	1.85	32	1.34	23	1.48	30	1.13
1980	27	1.78	20	1.05	12	1.00	18	0.61
1981	22	1.59	21	1.14	17	1.06	19	0.74
1984	7	1.86	7	1.43	6	1.17	7	1.00
				After spill				
1989	7	1.57	7	1.43	14	0.93	6	0.50
1990	27	1.78	25	1.28	29	0.79	24	0.68

TABLE 6.—Numbers of pigeon guillemot nests in which at least one egg failed to hatch or chick failed to fledge because of failure to hatch, young chick death, depredation, starvation or exposure, and unknown reasons. "Unhatched egg" includes eggs that failed to hatch because of infertility, embryo death, or nest desertion. "Young chick death" refers to chicks, generally less than 1 week old, dying in the nest for no apparent reason. We rarely witnessed predators taking eggs or chicks—depredation was presumed based on disappearance of eggs or chicks.

Year	Number of nests	Unhatched egg	Young chick death	Depredation of: Egg	Depredation of: Chick	Starvation or exposure	Unknown reason
			Before spill				
1978	32	4	0	0	0	0	2
1979	30	6	1	2	1	1	4
1980	19	5	0	2	0	4	2
1981	22	6	1	2	0	6	0
			After spill				
1989	15	3	1	1	4	0	0
1990	38	2	1	3	4	1	5

Discussion

Population

Guillemot attendance at colonies varies in relation to the stage of the breeding season, time of day, weather conditions, and tidal stage (Ewins et al. 1993). The recommended census procedure involves counting guillemots in potential breeding habitat during early mornings in the prebreeding season (Ewins 1985; Ewins and Tasker 1985; Emms and Morgan 1989). However, Vermeer et al. (1993a) found that counts of British Columbia guillemots were less variable during incubation than during prebreeding. Our counts in the Naked Island area were made midway through incubation under similar conditions of time, tide, and weather; therefore, they provide a good basis for comparison of prespill and postspill populations.

It is tempting to conclude that the spill was the cause of the decline we observed in the Naked Island guillemot population. However, available evidence suggests that few guillemots were present at Naked Island when the oil went through and that few guillemots were killed there. In addition, the Prince William Sound population as a whole may have declined before the spill. These circumstances make it more difficult to reach a conclusion about the cause of the population decline we observed at Naked Island following the spill.

Oil from the *Exxon Valdez* was present on Naked Island waters from 29 March to 19 April 1989. At some locations, guillemots attend their colonies during the winter, and prebreeding activities may begin during late winter (Ewins 1985; Greenwood 1987; Nelson 1987). The earliest we ever visited Naked Island was in early May; thus, we have scant information about late winter activities of guillemots in the area. Isleib and Kessel (1973) suggested that the Prince William Sound guillemot population is smaller in winter than in summer and that the birds have a more evenly dispersed distribution.

The only information available to us suggests that few guillemots were in the Naked Island area at the time of the spill. In the days just before oil surrounded Naked Island (26–28 March 1989), fewer than 50 guillemots were counted in shoreline surveys around Naked, Peak, and Storey islands (S. P. Klosiewski and K. D. Wohl, USFWS, and K. Sundberg, Alaska Department of Fish and Game, personal communication). Similar surveys of this area in the summer, even in midday, would have probably yielded counts of several hundred guillemots.

Because so few guillemots were observed in the Naked Island area at the time of the spill, we suspect that only a small percentage of the Naked Island breeding guillemot population may have been killed directly by *Exxon Valdez* oil in that area. This conclusion is supported by the oiled bird re-

TABLE 7.—Mean fledging weights and mean growth rates of pigeon guillemot chicks raised at Naked Island, Prince William Sound, Alaska, before and after the *Exxon Valdez* oil spill.

Year	*N*	Fledging weight (g)	*N*	Growth rate (g/d)
		Before spill		
1978	29	467	15	19.6
1979	17	506	16	23.8
1980[a]	2	517	1	19.0
1981	13	428	11	19.2
		After spill		
1989	10	507	5	18.1
1990	13	438	12	16.6

[a]Few chicks were measured in 1980 because of loss of nests from netting and tagging of adults for foraging studies (Kuletz 1983).

TABLE 8.—Percentages of fish observed delivered by adult pigeon guillemots to their chicks at Naked Island, Prince William Sound, Alaska, that were schooling fish and bottom fish, before and after the *Exxon Valdez* oil spill. The numbers in parentheses are sample size.

Year and sample size	Schooling fish					Bottom fish					Other	
	SL[a]	HE[b]	SM[c]	UN[d]	TOT[e]	BL[f]	SC[g]	CO[h]	OT[i]	TOT	IN[j]	UN
Before spill												
1979 (579)	55	0	0	0	55	19	14	1	2	36	>1	9
1980 (711)	35	0	0	0	35	30	9	7	6	51	1	13
1981 (493)	23	8	8	0	38	19	11	1	17	49	>1	13
After spill												
1989 (565)	14	21	1	0	36	19	9	25	>1	54	0	10
1990 (906)	8	1	>1	3	13	28	9	14	10	61	>1	25

[a] SL = Pacific sand lance *Ammodytes hexapterus*.
[b] HE = Pacific herring *Clupea pallasi*.
[c] SM = smelts (*Mallotus villosus* and other Osmeridae).
[d] UN = unidentified.
[e] TOT = total.
[f] BL = blennies (*Lumpenus maculatus, L. sagitta, L. fabricii, Allolumpenus hypochromus, Pholis laeta, Apodichthys flavidus, Ronquilus jordani, Bathymaster signatus, Liparis* spp., and *Stichaeus punctatus*).
[g] SC = sculpins (*Oligocottus maculosus, Triglops pingeli*, and unidentified sculpins).
[h] CO = cods (*Microgadus proximus, Gadus macrocephalus*, and *Theragra chalcogramma*).
[i] OT = other bottom fish (*Ophiodon elongatus, Sebastes* spp., *Eopsetta exilis*, and unidentified bottom fish).
[j] IN = invertebrates (*Rossia pacifica, Pandalus* spp., and unidentified invertebrates).

covery data. Most of the oiled birds recovered from beaches in the Naked Island area were waterfowl; only 2 of the 69 dead birds recovered were identified as guillemots (USFWS, Anchorage, unpublished data). If Naked Island breeders were killed by the oil, they must have been killed elsewhere.

Possibly some Naked Island breeders were killed in the Naked Island area and their carcasses were carried south with the oil. Of the 136 guillemot carcasses recovered from Prince William Sound beaches, the majority were found in the Knight Island area just south of Naked Island (USFWS, Anchorage, unpublished data). Search effort in Prince William Sound was high, relative to other portions of the spill area, and Ecological Consulting, Inc. (1991) estimated that the recovery rate of carcasses in the sound may have been as high as 35%. Thus, the 136 guillemot carcasses recovered may represent a significant fraction of the total number killed by oil within Prince William Sound.

What portion of the reduced number of guillemots present at Naked Island in 1989–1992 was attributable to the oil spill is also difficult to assess because the guillemot population of Prince William Sound as a whole may have declined since the early 1970s. Dwyer et al. (ca. 1975) estimated the summer guillemot population in 1972–1973 to be 15,000 birds. Their surveys included shoreline and offshore strata and were designed to estimate populations of all marine bird species, including guillemots. David B. Irons, USFWS, Anchorage, (unpublished data) surveyed the entire Prince William Sound shoreline by boat in the summers of 1984 and 1985, counting 4,660 guillemots. In 1993, Sanger and Cody (1993) conducted the first, soundwide survey specifically for guillemots, and they found 3,028 guillemots. This estimate, based on appropriate guillemot census techniques, was similar to the postspill estimates of the guillemot population of Klosiewski and Laing (1994), which were based on stratified surveys similar to those of Dwyer et al. (ca. 1975). These data suggest that the Prince William Sound guillemot population declined, perhaps by as much as 50%, between the early 1970s and the 1990s. Irons's data are suggestive of the decline occurring before rather than after the spill.

The strongest evidence that the spill caused some of the population decline we found at Naked Island was the relatively greater decline in the number of guillemots attending colonies along shorelines that were oiled in 1989. How the spill would have caused this decline is unknown, given our assumption that few breeders were in the area when the oil went through.

Klosiewski and Laing (1994) also found evidence for a spill-related decline in the Prince William Sound guillemot population. The March guillemot population in the oiled zone of the sound declined more than expected, based on the prespill to postspill population change in the non-oiled zone. These observations suggest that the spill was responsible for some of the decline in the guillemot population at Naked Island and within Prince William Sound as a whole.

Oil spills and chronic oil pollution from shipping have killed large numbers of seabirds, but whether such pollution has directly caused declines in seabird populations is debatable (Evans and Nettleship 1985). The pigeon guillemot and its Atlantic counterpart, the black guillemot, are among the few seabird species for which researchers have attributed declines in local populations to oil pollution (Ainley and Lewis 1974; Asbirk 1978; Heubeck and Richardson 1980; Rov 1982; Ewins and Tasker 1985; Nettleship and Evans 1985). However, local guillemot populations that are known to have declined because of oil pollution appear to have recovered, in most cases, within a few years (Ainley and Lewis 1974; Evans and Nettleship 1985; Ewins and Tasker 1985).

The US$1.1 billion settlement reached in 1991 between the state and federal governments and Exxon Corp. required settlement funds to be spent to restore injured resources. Because the Prince William Sound guillemot population appears to have been declining, attempts as part of the restoration program to artificially restore this population to its prespill level may prove futile. At this point, the causes of the decline are unknown. The decline could be related to some of the disturbances, both natural and anthropogenic, that have affected the Prince William Sound ecosystem in recent years. Such disturbances include the Great Alaska Earthquake of 1964, a gradual increase in sea surface temperatures since 1975, enhancement of salmon stocks, and protection of sea otters *Enhydra lutris* (McRoy 1988). Further understanding of the causes of the long-term decline is required to properly design any guillemot restoration project for Prince William Sound.

Reproduction

We predicted that oil from the *Exxon Valdez* spill was most likely to have disrupted breeding by delaying egg laying, decreasing hatching success, increasing nest abandonment, and slowing the growth of chicks. We did not observe these effects at the nests we studied. Nesting chronology was identical to that observed in prespill years. Hatching success was not less, and chicks grew at similar rates and fledged at similar weights as in prespill years. We did not observe definitive cases of nest abandonment.

The primary difference in reproduction that we observed after the spill was a decreased nesting success rate, which was attributable to an increased rate of nest loss from predation during the chick stage. Nest predators at Naked Island could include common ravens *Corvus corax*, northwestern crows *C. caurinus*, black-billed magpies *Pica pica*, mink *Mustela vison*, and river otter *Lutra canadensis*. Losses resulting from predation might have been indirectly related to the spill if the spill had affected prey availability. Guillemots rely on cryptic and inaccessible nests to avoid predation. Hungry chicks will come to the nest entrance and thereby become more vulnerable to predators. However, based on our chick-feeding observations, food availability was not reduced. We found no obvious connection between the spill and the reduced nesting success that we observed after the spill.

Although reproduction appeared to be largely unaffected by the spill, some disruption of breeding probably occurred in 1989. We had difficulty finding nests in 1989, and some of this difficulty may have been because fewer pairs initiated nests. At the five colonies where we studied reproduction, the number of active nests increased 66% between 1989 and 1990. This increase suggests that in 1989 some birds did not breed at all. Failure to breed could have been the result of mate loss or physiological impairments related to ingestion of oil.

Alternatively, what appeared to us to be reduced nesting effort may have actually been an increased rate of nest loss during the egg stage. At Naked Island, guillemots nest in inaccessible and cryptic sites, and nests without hatchlings are difficult, if not impossible, to detect. Some of the attending birds that we thought failed to nest may have initiated nests but lost them early in the breeding season, possibly because of transfer of oil from the adults' legs and feet to the eggs.

Foraging

After the spill, guillemots fed chicks at an average rate that was not lower than that in prior years; the diet still comprised schooling fishes and bottom fishes. Chick growth rates and fledging weights, which would be expected to reflect any significant decreases in the amount or quality of foods delivered to chicks, were not lower after the spill. Starvation of chicks, which was a common cause of nest failure in 1981 when the weather was particularly bad (Kuletz 1983), was not observed as a cause of nest failure, except at one nest in 1990. These data support our conclusion that the spill did not cause any abrupt change in food availability.

Inferential information on relative prey abundance from chick-feeding observations suggested changes in the availability of cods, Pacific herring, and Pacific sand lance, prespill to postspill. Cods were used more frequently after the spill, and Pacific sand lance were used less frequently. Pacific herring were important only in 1989. Whether changes in the apparent availability of these species were caused by the oil spill is unknown.

Conclusions

This study examined the immediate effects of a major oil spill on the population size, reproduction, and foraging of pigeon guillemots breeding at one site at the upstream end of the spill trajectory. We found that the population declined by 43%, but because the regional guillemot population as a whole may have declined before the spill, we could not attribute the decline we observed in our study area entirely to the spill. The strongest evidence we found that the spill reduced the population was that the proportion of the population that was attending colonies along oiled shorelines was less than expected after the spill. This decline was still apparent in 1992—3 years after the spill.

The oil spill did not appear to significantly disrupt guillemot reproduction, but the cryptic nature of guillemot nests undoubtedly reduced our ability to detect one of the most likely effects of the spill—an increased incidence of unhatched eggs. Although there were some differences in prey species delivered to chicks after the spill, adults fed their chicks at the same rate.

The most likely explanation for the few effects we observed is that oil from the *Exxon Valdez* was on the surface waters around Naked Island for a relatively short period before most of the guillemots had returned to the area to begin their annual reproductive activities.

Acknowledgments

The U.S. Fish and Wildlife Service (USFWS), Office of Special Studies, funded our prespill studies at Naked Island, and we thank D. V. Derksen for initiating and continuing to support our studies there. The studies in 1980 and 1981 were also supported by grants (to K. J. K.) from the University of California Board of Regents and the Wilson Ornithological Society. The postspill work was part of the *Exxon Valdez* Oil Spill Natural Resources Damage Assessment and was funded by the *Exxon Valdez* Oil Spill Trustee Council. We thank the following people for their assistance in the field: A. Banks, A. Belleman, D. Brooks, D. Carr, M. Cody, W. Eldridge, G. Esslinger, K. Fine, L. Fuller, H. Good, N. Harrison, G. Kerndt, L. Ladenburg, I. Manley, D. Marks, N. Naslund, J. Pohl, C. Schneider, and L. Tyson. We also thank G. J. Divoky and an anonymous reviewer for their comments which greatly improved the manuscript.

References

Ainley, D. G., and R. J. Boekelheide. 1990. Seabirds of the Farallon Islands. Stanford University Press, Stanford, California.

Ainley, D. G., and T. J. Lewis. 1974. The history of Farallon Island marine bird populations, 1854–1972. Condor 76:432–446.

Asbirk, S. 1978. Tejsten *Cepphus grylle* som ynglefugl i Danmark. (English summary.) Dansk Ornithologisk Forenings Tidsskrift 72:161–178.

Byrd, G. V., D. D. Gibson, and D. L. Johnson. 1974. The birds of Adak Island, Alaska. Condor 76:288–300.

Cairns, D. C. 1979. Censusing hole-nesting auks by visual counts. Bird-Banding 50:358–364.

Campbell, R. W. 1977. Use of man-made structures as nest sites by pigeon guillemots. Canadian Field-Naturalist 91:193–194.

Cooper, W. S. 1942. Vegetation of the Prince William Sound region, Alaska; with a brief excursion into post-Pleistocene climatic history. Ecological Monographs 12:3–22.

Drent, R. H. 1965. Breeding biology of the pigeon guillemot, *Cepphus columba*. Ardea 53:99–160.

Dwyer, T. J., P. Isleib, D. A. Davenport, and J. L. Haddock. circa 1975. Marine bird populations in Prince William Sound, Alaska. U.S. Fish and Wildlife Service, Anchorage, Alaska.

Ecological Consulting, Inc. 1991. Assessment of direct seabird mortality in Prince William Sound and the western Gulf of Alaska resulting from the *Exxon Valdez* oil spill. Report to the U.S. Department of Justice and U.S. Fish and Wildlife Service, Ecological Consulting, Inc., Portland, Oregon.

Emms, S. K., and K. H. Morgan. 1989. The breeding biology and distribution of the pigeon guillemot *Cepphus columba* in the Strait of Georgia. Pages 100–106 *in* K. Vermeer and R. H. Butler, editors. The ecology and status of marine and shoreline birds in the Strait of Georgia, British Columbia. Canadian Wildlife Service Special Publication, Ottawa.

Evans, P. G. H., and D. N. Nettleship. 1985. Conservation of Atlantic alcidae. Pages 427–488 *in* D. N. Nettleship and T. R. Birkhead, editors. The Atlantic alcidae. The evolution, distribution and biology of auks inhabiting the Atlantic Ocean and adjacent water areas. Academic Press, London.

EVOSDAGG (*Exxon Valdez* Oil Spill Damage Assessment Geoprocessing Group). 1990. Cumulative oil impact as of August 21, 1989 (map). U.S. Fish and Wildlife Service and Alaska Department of Natural Resources, Anchorage.

Ewins, P. J. 1985. Colony attendance and censusing of black guillemots *Cepphus grylle* in England. Bird Study 32:176–185.

Ewins, P. J. 1993. Pigeon guillemot (*Cepphus columba*). Pages 1–23 *in* A. Poole and F. Gill, editors. The birds of North America, number 49. The Academy of Natural Sciences, Philadelphia, and The American Ornithologists' Union, Washington, DC.

Ewins, P. J., H. R. Carter, and Y. V. Shibaev. 1993. The status, distribution, and ecology of inshore fish-eating alcids (*Cepphus* guillemots and *Brachyramphus* murrelets) in the North Pacific. Pages 164–175 *in* K. Vermeer, K. T. Briggs, K. H. Morgan, and D. Siegel-Causey, editors. The status, ecology, and conservation of marine birds of the North Pacific. Canadian Wildlife Service Special Publication, Ottawa.

Ewins, P. J., and M. L. Tasker. 1985. The breeding distribution of black guillemots *Cepphus grylle* in Orkney and Shetland, 1982–84. Bird Study 32:186–193.

Follett, W. I., and D. G. Ainley. 1976. Fishes collected by pigeon guillemots, *Cepphus columba* (Pallas), nesting on southeast Farallon Island, California. California Fish and Game 62:28–31.

Galt, J. A., W. J. Lehr, and D. L. Payton. 1991. Fate and transport of the *Exxon Valdez* oil spill. Environmental Science and Technology 25:202–209.

Greenwood, J. G. 1987. Winter visits by black guillemots *Cepphus grylle* to an Irish breeding site. Bird Study 34:135–136.

Heubeck, M., and M. G. Richardson. 1980. Bird mortality following the *Esso Bernicia* oil spill, Shetland, December 1978. Scottish Birds 11:97–108.

Hoffman, D. J. 1990. Embryotoxicity and teratogenicity of environmental contaminants to bird eggs. Reviews of Environmental Contamination and Toxicology 115:39–89.

Isleib, M. E. P., and B. Kessel. 1973. Birds of the north gulf coast-Prince William Sound region, Alaska. Biological Papers of the University of Alaska 14:1–149.

King, J. G., and G. A. Sanger. 1979. Oil vulnerability index for marine oriented birds. Pages 227–239 *in* J. C. Bartonek and D. N. Nettleship, editors. Conservation of marine birds of northern North America. U.S. Fish and Wildlife Service Wildlife Research Report 11.

Klosiewski, S. P., and K. K. Laing. 1994. Marine bird populations of Prince William Sound, Alaska, before and after the *Exxon Valdez* oil spill. Final report. Natural Resources Damage Assessment Bird Study 2. U.S. Fish and Wildlife Service, Migratory Bird Management, Anchorage, Alaska.

Koelink, A. F. 1972. Bioenergetics of growth in the pigeon guillemot, *Cepphus columba*. Master's thesis. University of British Columbia, Vancouver.

Kuletz, K. J. 1983. Mechanisms and consequences of foraging behavior in a population of breeding pigeon guillemots. Master's thesis. University of California, Irvine.

McRoy, C. P. 1988. Natural and anthropogenic disturbances at the ecosystem level. Pages 329–344 *in* D. G. Shaw and M. J. Hameedi, editors. Environmental studies in Port Valdez, Alaska. A basis for management. Lecture notes on coastal and estuarine studies. Springer-Verlag, Berlin.

Nelson, D. A. 1987. Factors influencing colony attendance by pigeon guillemots on southeast Farallon Island, California. Condor 89:340–348.

Nero and Associates, Inc. 1987. Seabird oil toxicity study. Final Report. OCS (Outer Continental Shelf) Study MMS-87-0005. U.S. Department of the Interior, Minerals Management Service, Washington, DC.

Nettleship, D. N. 1976. Census techniques for seabirds of arctic and eastern Canada. Canadian Wildlife Service Occasional Paper 25.

Nettleship D. N., and P. G. H. Evans. 1985. Distribution and status of the Atlantic alcidae. Pages 53–154 *in* D. N. Nettleship and T. R. Birkhead, editors. The Atlantic alcidae. The evolution, distribution and biology of auks inhabiting the Atlantic Ocean and adjacent water areas. Academic Press, London.

NRC (National Research Council). 1985. Oil in the sea. Inputs, fates and effects. National Academy Press, Washington DC.

Oakley, K. L. 1981. Determinants of population size of pigeon guillemots *Cepphus columba* at Naked Island, Prince William Sound, Alaska. Master's thesis. University of Alaska, Fairbanks.

Peakall, D. B., D. Hallett, J. R. Bend, G. L. Foureman, and D. S. Miller. 1982. Toxicity of Prudhoe Bay crude oil and its aromatic fractions to nestling herring gulls. Environmental Research 27:206–215.

Peakall, D. B., D. Hallett, D. S. Miller, R. G. Butler, and W. B. Kinter. 1980. Effects of ingested crude oil on black guillemots: a combined field and laboratory study. Ambio 9:28–30.

Peakall D. B., D. S. Miller, and W. B. Kinter. 1983. Toxicity of crude oils and their fractions to nestling herring gulls—1. Physiological and biochemical effects. Marine Environmental Research 8:63–71.

Rov, N. 1982. Olje oy sjofugl pa Helgelandskysten 1981. Var Fuglefauna 5:91–95.

Sanger, G. A., and M. B. Cody. 1993. Survey of pigeon guillemot colonies in Prince William Sound, Alaska. *Exxon Valdez* oil spill restoration project (93034) final report. U.S. Fish and Wildlife Service, Migratory Bird Management, Anchorage, Alaska.

Thoreson, A. C., and E. S. Booth. 1958. Breeding activities of the pigeon guillemot, *Cepphus columba columba* (Pallas). Walla Walla College Publications 23:1–37.

Vermeer, K., K. H. Morgan, and G. E. J. Smith. 1993a. Colony attendance of pigeon guillemots as related to tide height and time of day. Colonial Waterbirds 16:1–8.

Vermeer, K., K. H. Morgan, and G. E. J. Smith. 1993b. Nesting biology and predation of pigeon guillemots in the Queen Charlotte islands, British Columbia. Colonial Waterbirds 16:119–129.

Walters, P., S. Khan, P. J. O'Brien, J. F. Payne, and A. D. Rahimtula. 1987. Effectiveness of a Prudhoe Bay crude oil and its aliphatic, aromatic and heterocyclic fractions in inducing mortality and aryl hydrocarbon hydroxylase in chick embryo in ovo. Archives of Toxicology 60:454–459.

American Fisheries Society Symposium 18:770–784, 1996

Marbled Murrelet Abundance and Breeding Activity at Naked Island, Prince William Sound, and Kachemak Bay, Alaska, before and after the *Exxon Valdez* Oil Spill

KATHERINE J. KULETZ

U.S. Fish and Wildlife Service, Migratory Bird Management
1011 E. Tudor Road, Anchorage, Alaska 99503, USA

Abstract.—The marbled murrelet *Brachyramphus marmoratus* is the most abundant seabird in Prince William Sound, Alaska, and breeds throughout the area affected by the 1989 *Exxon Valdez* oil spill. I compared pre- and postspill abundance and breeding activity of murrelets near the Naked Island group in central Prince William Sound, and in Kachemak Bay in lower Cook Inlet. Murrelet numbers at Naked Island were lower in 1989 than in 1978–1980 but were not lower in 1990–1992. At Kachemak Bay, where oiling was minimal, murrelet densities did not change between 1988 and 1989. I observed a lower ratio of juvenile to adult murrelets at Naked Island in postspill years, but data were insufficient to determine if reproduction was disrupted. There was no change in juvenile ratios in Kachemak Bay. My results suggest that the murrelet population at Kachemak Bay, further removed from the spill epicenter, was not as affected as the Naked Island population. The majority of murrelets breeding in Prince William Sound did not arrive until after most of the surface oil had dispersed. Murrelets near Kodiak Island and the southern Kenai Peninsula during April, including those migrating north, may have died in higher numbers than in the sound. A minimum of 8,400 *Brachyramphus* murrelets (including the Kittlitz's murrelet *B. brevirostris*) were killed by oil, perhaps 7% of the population in the spill zone. A 66% decline in the Prince William Sound murrelet population since the 1970s complicated the identification of spill-related effects on murrelets.

On 24 March 1989, the supertanker TV *Exxon Valdez* spilled 43.9 million liters of crude oil into Prince William Sound, Alaska. The oil eventually evaporated or traveled 750 km south to the Kenai Peninsula, Kodiak Island, and the Alaska Peninsula (Ecological Consulting, Inc. 1991). At the time of the spill, an estimated 1 million birds were at risk of being oiled (Piatt et al. 1990). Alcids, being diving seabirds, had the highest rate of mortality of the population at risk in the *Exxon Valdez* oil spill (Piatt et al. 1990; Ecological Consulting, Inc. 1991). I studied abundance of the marbled murrelet *Brachyramphus marmoratus*, a small nearshore alcid, in two areas in the *Exxon Valdez* spill zone for the purpose of assessing injury to this species from the spill. The Kittlitz's murrelet (*B. brevirostris*) also occurs in the spill zone, but constitutes less than 8% of the *Brachyramphus* murrelets (U.S. Fish and Wildlife Service, unpublished data). Because the two species were difficult to differentiate during surveys, they were often recorded as *Brachyramphus* murrelets. Although this study deals specifically with marbled murrelets, I refer to the genus name where applicable and the conclusions may be relevant to both species.

The marbled murrelet is considered to be one of the alcids most susceptible to oil spills because of their small size and nearshore foraging habits (King and Sanger 1979). Exposure of birds to surface oil can cause death by drowning or hypothermia. Secondary and long-term impacts to seabirds can occur from oil effects on reproduction and metabolism. Adult birds transfer oil to the surface of eggs during incubation, which can reduce embryo survival and hatching success (Szaro and Albers 1977; Nero and Associates, Inc. 1987). Birds can ingest oil by preening or by consuming contaminated prey; this process can lead to impairment of metabolic rates, potentially altering adult behavior (Cavanaugh et al. 1983; Nero and Associates, Inc. 1987; Eppley and Rubega 1990) and retarding chick growth (Miller et al. 1978; Peakall et al. 1980).

The marbled murrelet is the most abundant seabird in Prince William Sound in the summer, and it is common year-round throughout the spill zone (Isleib and Kessel 1973). The *Brachyramphus* murrelet population of the sound in the early 1990s ranges from 97,000 in July to 25,000 in March (Klosiewski and Laing 1994). An estimated 95% of marbled murrelets in U.S. waters occur in Alaska (Mendenhall 1992), and major centers of the population are in southeastern Alaska, Prince William Sound, and the Kodiak Archipelago (Piatt and Ford 1993). Thus, a large portion of the U.S. marbled murrelet population was at risk from the spill. The marbled murrelet is a Category II species (a species of concern for the U.S. Fish and Wildlife Service) in

Alaska and is listed as threatened in California, Oregon, and Washington under the Endangered Species Act.

Little is known about the life history of marbled murrelets, but, like other alcids, they are probably long-lived and rely on high adult survivorship to compensate for low reproductive potential (Nettleship and Birkhead 1985; Beissinger 1995). Marbled murrelets exhibit strong pair bonds, and the loss of adults could have long-term consequences by affecting breeding and social structures. The females each lay one egg and usually nest inland on the branches of conifers (Marshall 1988; Singer et al. 1991; Naslund et al. 1995). In south-central Alaska they also nest on the ground (Day et al. 1983; Johnston and Carter 1985).

Long-term impacts might result from reduced prey abundance or from consumption of contaminated prey. Murrelets typically forage <2 km from shore and feed primarily on midwater and surface schooling fish, such as Pacific sand lance *Ammodytes hexapterus*, capelin *Mallotus villosus*, juvenile Pacific herring *Clupea pallasi*, and cods *Gadidae* spp. (Sealy 1975; Sanger 1987). Occasionally they feed on salmon fry *Salmonidae* spp. (Carter and Sealy 1986). All of these fishes have life stages that occur in subtidal or nearshore waters.

Investigating injury to murrelets was difficult because few areas had adequate prespill counts for examining population trends, and little was known about the biology or migratory patterns of murrelets in Alaska. In Prince William Sound, the winter population is 20–30% of the summer population (Klosiewski and Laing 1994), but overwintering grounds are largely unknown. Unlike many seabirds, which nest on open cliff faces or in burrows, we could not count birds at their nests or measure reproductive success directly. I had monitored seabirds at the Naked Island group, in central Prince William Sound, from 1978 to 1980 as part of a U.S. Fish and Wildlife Service project to gather baseline data along the Prince William Sound oil tanker route. In 1988 I conducted at-sea counts of murrelets in Kachemak Bay for the Alaska Maritime National Wildlife Refuge. Thus, both the Naked Island and Kachemak Bay areas had prespill data. In this study I compared at-sea murrelet counts before and after the spill at these two areas. For the discussion, I also synthesized information on the abundance and distribution of murrelets to infer migration patterns and estimated direct mortality from carcasses recovered after the spill.

Study Area

The primary study area was the Naked Island group in central Prince William Sound, which includes Naked, Storey, and Peak islands, 30 km southwest of Bligh Reef (Figure 1). The three islands are forested and the surrounding waters relatively shallow (see Kuletz et al. 1994a). Marbled murrelets and pigeon guillemots *Cepphus columba* are the two most common seabirds (Oakley and Kuletz 1979). Approximately 3,000 murrelets occur within 5 km of the Naked Island group in summer (Kuletz et al. 1994b). Naked Island was hit by oil from the TV *Exxon Valdez* 3 d after the tanker's grounding. Between 27 March and 2 April, shorelines on the northern and eastern sides of Naked Island were lightly to heavily oiled (Figure 1). The oil moved south with prevailing winds, and small sections of the southern and western sides of Naked Island were lightly oiled about 7–9 April, when southerly winds brought oil back into the area. The oil largely missed Storey Island. Peak Island, although surrounded by oil sheen from 27 to 30 March, only had shoreline oil in one cove on the north side.

The second study area was Kachemak Bay, a 64-km embayment on the southern Kenai Peninsula in lower Cook Inlet (Figure 2). In the summer there are approximately 58,000 *Brachyramphus* murrelets in Cook Inlet (Agler et al. 1995) with an estimated 11,000 in Kachemak Bay (S. J. Kendall, U.S. Fish and Wildlife Service, unpublished data). The southern shoreline of the bay is deeper than the northern part and has sheltered passages and deep bays, bordered by steep mountains and glaciers (Trasky et al. 1977). There were two study sites in Kachemak Bay—the south inner bay and the south outer bay (Figure 2). Kachemak Bay received oil sheen from northward currents in mid-April, with tar balls reaching some sections of shoreline in the outer bay and on Grewingk Glacier Spit in the inner bay (Figure 2).

Methods

Abundance Indices at Naked Island

I used two types of surveys to determine the abundance of murrelets at the Naked Island group: (1) offshore counts of murrelets on five midbay transects and (2) shoreline surveys of all three islands. Prespill data were from 1978 to 1980 and postspill data were from 1989 to 1992. Few Kittlitz's murrelets were observed at Naked Island, and marbled murrelets were probably >99% of all murrelets.

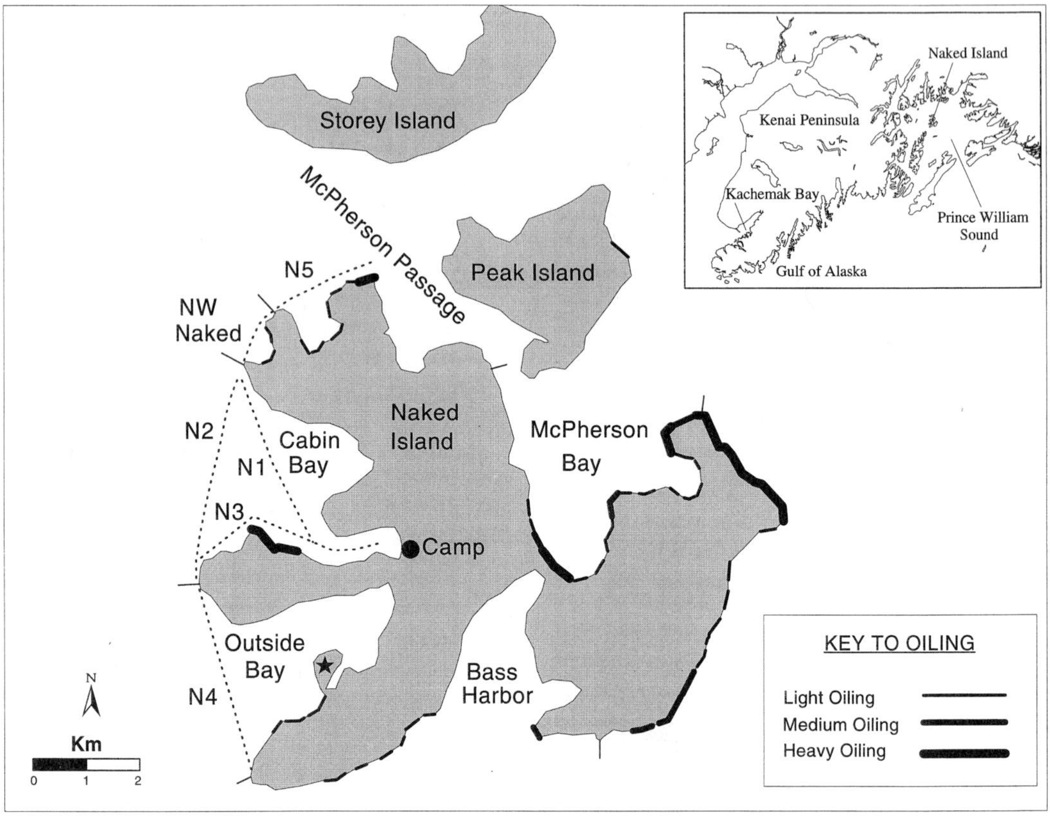

FIGURE 1.—The Naked Island study area and extent of shoreline oiling, cumulative to August 1989 (EVOSDAGP 1990). The five Naked Island transects (N1–N5), boundaries of seabird survey areas, and the land-based observation point (asterisk) are also shown.

Offshore counts.—Marbled murrelet abundance indices were determined for five transects (N1–N5) on the western side of Naked Island (Figure 1). These transects were derived from frequently used boat routes in 1978–1980 and were conducted from a 5-m inflatable boat. I defined transect routes by prominent landmarks; because they were within bays or across a bay mouth, routes were probably consistent.

Time of day, weather, and sea conditions were recorded before each survey. Most transects were surveyed before 1200 h from an inflatable boat in fair conditions. In postspill years, some surveys were done from a 7.7-m whaler. Preliminary analyses were performed to determine if seasonal, diel, weather, and platform conditions significantly affected murrelet counts (Appendices, Kuletz 1994). Based on these results, the final analyses included surveys done between 6 June and 6 August and between 0500 and 1700 h on days when seas were <0.6 m in height. Seventy-five prespill and 207 postspill counts satisfied these criteria.

Transect-viewing distance was not consistent among years. In 1978, birds were counted out to 100 m from the boat. In 1979–1980 and 1989–1990, birds were counted as far as 200 m from the boat. In 1990 we recorded the bird's distance as either ≤100 m from the boat or 101–200 m from the boat and determined the percentage of birds in each zone. Murrelet detectability declines dramatically beyond 100 m (C. J. Ralph, U.S. Forest Service, unpublished data). In 1990, I found that on average 84% of the murrelets we counted were <100 m from the boat, but the ratio varied with sea conditions. Therefore, rather than use densities or a correction factor, I used the total number of murrelets counted on each survey, with the assumption that the majority of murrelets detected were ≤100 m from the boat. For 1990 surveys, I used the number of birds ≤100 m from the boat.

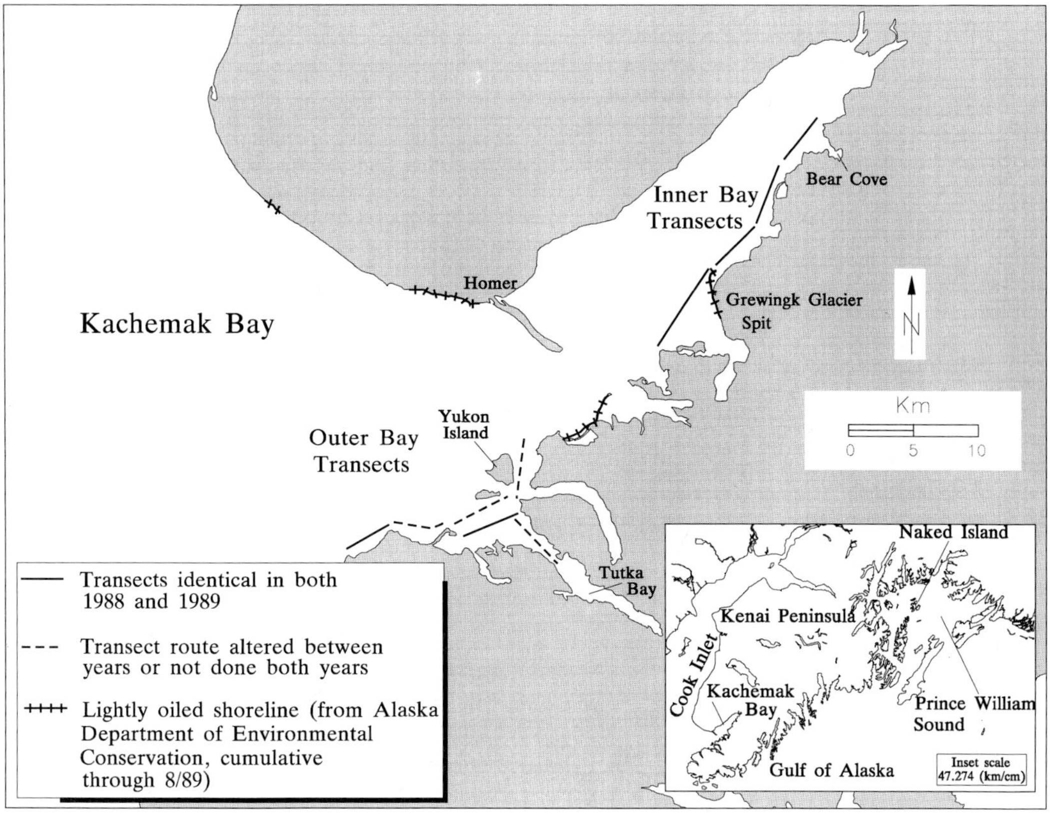

FIGURE 2.—Kachemak Bay transects surveyed in 1988 and 1989 and the cumulative amount of shoreline oiling.

Shoreline surveys.—The shoreline surveys were conducted by circumnavigating Naked, Storey, and Peak islands. Early June surveys were done in 1978–1980 and 1989–1992 at Naked Island and in 1978–1979 and 1989–1992 at Storey and Peak islands. Partial shoreline surveys were also done in late July or August in 1978, 1979, and 1989–1992 at Naked Island and at Storey and Peak islands in 1978, 1980, 1989, and 1990. The surveys were conducted between 0500 and 1000 h, during early morning high tides. Two observers in a 5-m or a 7.7-m boat traveled at 5–7 knots within 50–100 m of the shoreline and counted all seabirds. In 1978, all murrelets within view were counted, which could have included birds >200 m from shore. However, because the focus in 1978 was guillemot colonies, surveys tended to be near shoreline, and because of the low detectability of murrelets beyond 100 m I assumed that counts were comparable to those of later years, when we counted birds ≤200 m from the shoreline.

Data analysis.—Murrelet counts were transformed (natural log) before all tests to eliminate the likely correlation between sample means and variances, and because there was no guarantee of normality in the counts. Pre-and postspill means were tested with two-sample t-tests. I also used an outlier t-test to determine if 1989 counts alone were lower than prespill counts. One-tailed tests at $\alpha = 0.05$ were considered significant.

For each offshore transect, I used the yearly mean number of murrelets counted on the surveys. First, to test for a general decline in all five transects, the 1989 mean was subtracted from each transect's prespill mean, and the differences between 1989 and prespill means ($N = 5$) were tested for deviation from zero. Second, I used the outlier t-test on transects with at least 2 years prespill data (N1, N3, and N4); the 1989 mean number of murrelets per survey was tested for significant deviation from the annual prespill means. Finally, the mean number of murrelets counted for each of these three transects was compared between pre- and postspill years.

For the shoreline surveys, I tested for a decline in the numbers of murrelets after the spill for each of the three islands surveyed. For Naked Island, I also tested the seven geographic areas separately.

Oiled and non-oiled shorelines.—Since 1978 the shoreline of Naked Island has been divided into seven areas defined by landmarks (indicated by lines extending from shore in Figure 1). After the spill, these areas were designated as primarily oiled (>50% oiled shoreline) or non-oiled (<50% oiled shoreline), based on a map of "Cumulative Oil Impact" as of 21 August 1989 (EVOSDAGP 1990). The non-oiled areas were McPherson Passage, Bass Harbor, Outside Bay, and Cabin Bay of Naked Island, and Storey and Peak islands. The oiled areas were NW Naked, McPherson Bay, and East Naked. I tested for a decline in the proportion of murrelets counted in oiled areas before and after the spill with a two-sample *t*-test. Because data for all three islands were available only for prespill years 1978 and 1980, the test included two prespill and four postspill years.

Abundance Indices at Kachemak Bay

At Kachemak Bay murrelets were counted from a 7.7-m boat on four transects in the south inner bay and six transects in the south outer bay (Figure 2). Kittlitz's murrelets were up to 20% of the *Brachyramphus* murrelets identified to species, and these were primarily in the inner bay. Six of the 10 transects were surveyed in both years. To locate the transects we used landmarks and marine radar to stay a fixed distance from shore. Transects were surveyed between 0600 and 1000 h in calm seas. I conducted surveys on 22 d between 17 May and 5 August 1988 and on 16 days between 4 May and 18 August 1989.

The viewing distance was not the same for the two years. In 1988, I experimented with a distance used by Carter (1984), counting murrelets out to 500 m from the boat. In 1989, I returned to using a viewing distance of 200 m. In this case, to be conservative I compared murrelet densities (birds/km^2) rather than numbers. Sample size for each transect ranged from four to sixteen survey days. For the six transects that were the same in both years, the difference in annual mean murrelet density between 1988 and 1989 (N = 6) was tested for deviation from zero. Because the range of survey dates was greater at Kachemak Bay than at Naked Island and sampling was more clumped, I examined the mean daily densities of all transects in the inner bay and outer bay (see Figure 2) within four periods.

Effects of Boat Traffic

In 1989, vessels ≤200 m from the transect were counted to test for an effect on murrelet numbers. Transects with sufficient boat traffic for these tests were N4, across Outside Bay at Naked Island, and the outer bay transects in Kachemak Bay. Land-based counts of murrelets were used to determine if hourly boat and plane activity influenced murrelet numbers. Murrelets, boats, and low-flying aircraft were counted each hour within a portion of Outside Bay defined by landmarks. An observer would slowly scan the area once using a 20× spotting scope. Land-based counts were done on 9 d (N = 55 h, x = 6.1 h/d) between 0500 and 2000 h. I used Kendall's τ_b statistic to test for a correlation between the number of boats and aircraft present and the number of murrelets counted that hour, or in the case of the offshore surveys, on each transect.

Breeding Activity

Once fledged, juvenile murrelets are independent (Sealy 1975). I recorded the number of juvenile plumaged birds on the water during surveys. Because I did not know the percentage of nonbreeding murrelets on the water, I could not estimate the number of juveniles per breeding pair; instead, I used the percentage of juveniles of the total murrelet counts as an index of reproductive success.

Juveniles are difficult to distinguish from adults in winter plumage, but few adults are in winter plumage before mid-August (Carter and Stein 1995; Ralph and Long 1995). To minimize the possibility of encountering adults in winter plumage, I used Naked Island surveys conducted between 15 July, (when juveniles were first observed at Naked Island) and 15 August. This time period also avoided the decrease in adult murrelets after mid-August at Naked Island (Oakley and Kuletz 1979; Kuletz, unpublished data), which would have inflated the ratio of juveniles. All murrelets in juvenile or winter plumage were scored as juveniles. The late-summer shoreline surveys were conducted between 25 July and 15 August. Because few complete late surveys were done, I included partial-island surveys. At Naked Island, juveniles were recorded in 1978, 1979, and 1989–1992.

For offshore transects, I used those conducted between 15 July and 15 August, and I used the total number of murrelets counted on a day (all transects combined) to calculate the daily percentage of juveniles. These daily percentages were used to calculate the yearly average percentage of juveniles.

TABLE 1.—Mean number of marbled murrelets counted on five Naked Island offshore transects in Prince William Sound before and after the *Exxon Valdez* oil spill. Means before and after the oil spill for each transect were tested with a one-tailed *t*-test. Counts from 1989 were tested for deviation from prespill means with an outlier *t*-test. All numbers were log-transformed for analysis. Absence of data means no counts were made. In this table, N = number of individual surveys and $\overline{X}$ = mean number of murrelets counted per survey.

	Transect N1			Transect N2			Transect N3			Transect N4			Transect N5		
Year or statistic	N	$\overline{X}$	SD	N	$\overline{X}$	SD	N	$\overline{X}$	SD	N	$\overline{X}$	SD	N	$\overline{X}$	SD
Before spill															
1978	9	40.3	30.9	4	43.3	25.6	9	75.8	50.0	8	34.6	21.6	6	40.5	21.2
1979	13	44.6	18.3				8	54.8	25.6	2	127.5	106.7			
1980	14	51.3	21.6				2	66.0	29.7						
After spill															
1989	26	43.4	30.0	25	40.4	26.2	29	21.0	22.8	22	27.5	22.3	22	27.3	35.7
1990	18	113.0	57.8	17	95.9	63.8	17	106.8	92.6	16	23.1	18.7	15	69.0	43.1
Two-sample *t*-test															
t		−0.89			−0.66			0.71			1.77			0.17	
P		0.220			0.314			0.303			0.109			0.447	
df		3.0			1.0			3.0			2			1	
1989 outlier *t*-test															
t		−0.58			Insufficient data			−10.95			−1.07			Insufficient data	
P		0.309						0.004			0.239				
df		1						1			1				

Differences between pre- and postspill years were tested with a two-sample *t*-test.

In Kachemak Bay, I observed juveniles in the first week of July, and adult murrelets remained abundant and in breeding plumage into late August (personal observation). Therefore, I included data from 4 July to 18 August for comparing juvenile ratios. Because the species of murrelet could not be identified at a distance, I only used the number of murrelets identified as marbled murrelets, which also minimized the problem of identifying plumages when using the wide transect in 1988. Again, the total number of marbled murrelets counted on all transects combined was used to derive a daily percentage of juveniles. The percentage of juveniles observed before and after the spill was tested with a two-sample *t*-test of the daily means.

Results

Abundance Indices at Naked Island

Offshore counts.—Of the individual transects at Naked Island, only one (N3) of the five showed a significant decline in the number of murrelets between 1989 and prespill years (Table 1). However, all five transects had lower counts in 1989 than their prespill averages, and the pooled differences for the five transects showed a significant decline (t = 2.31, P = 0.04). Because of high numbers of murrelets in 1990 (or for N4, high interannual variance), the postspill mean number of murrelets on each transect was not significantly lower than the prespill mean (Table 1).

Shoreline surveys.—Compared to prespill years, the total number of murrelets counted during shoreline surveys of Naked Island was significantly lower in 1989 (Table 2; $t = -3.56$, df = 2, P = 0.04). Murrelet numbers increased in the following years, and were variable. Thus, although there was a 33% decline after the spill, the mean number of murrelets in postspill years was not significantly lower than in prespill years. When shoreline areas were analyzed separately, only McPherson Bay murrelet numbers remained significantly lower in postspill years (t = 2.61, df = 4.9, P = 0.05).

Both Peak and Storey islands had significantly fewer murrelets along their shoreline in June 1989 compared to prespill years (Table 3). However, murrelet numbers at both islands were back to prespill levels or higher in July 1989. For both Peak and Storey islands, the 1990 counts were the highest June counts on record.

Oiled and non-oiled shorelines.—The proportion of murrelets counted along oiled shorelines of the study area did not decline significantly after the spill (Table 4; $t = -0.23$, df = 4, P = 0.41).

Abundance Indices at Kachemak Bay

At Kachemak Bay, the sum of differences in murrelet density between 1988 and 1989 at six transects surveyed in both years was not significantly different

TABLE 2.—Number of marbled murrelets counted in June shoreline surveys of Naked Island Prince William Sound, before and after the *Exxon Valdez* oil spill. See Figure 1 for area locations.

Year	Naked Island study area							Total
	Cabin Bay	Outside Bay	Bass Harbor	East Side	McPherson Bay	McPherson Pass	NW Side	
Before spill								
1978	67	25	62	5	47	15	9	230
1979	77	17	7	20	71	6	1	199
1980	106	34	38	31	77	43	18	347
After spill								
1989	17	18	3	12	18	8	1	77
1990	42	9	25	6	27	122	14	245
1991	48	0	12	8	46	30	2	146
1992	88	25	13	18	44	28	14	230

from zero, indicating that there was no decline ($N = 6$, $t = 0.84$, df $= 5$, $P = 0.42$). For all of the surveys combined, average murrelet density after the spill in the inner bay was 37% lower, whereas murrelet density was 31% higher in the outer bay (Table 5). However, within a given survey period, murrelet densities were similar pre- and postspill for both inner and outer bay.

Effects of Boat Traffic

In 1989, the number of murrelets showed a negative relationship with boat traffic during offshore surveys (Figure 3). At Outside Bay, the number of boats on or near the transect ranged from 0 to 4 per day, and the number of murrelets was negatively correlated with the number of boats ($N = 27$, $\tau = -0.516$, $P = 0.0011$). At Kachemak Bay, the number of boats per transect ranged from 0 to 8 and was negatively correlated with the number of murrelets ($N = 70$, $\tau = -0.206$, $P = 0.030$). Hourly land-based counts of murrelets in Outside Bay were also negatively correlated with the number of boats and low-flying aircraft in the bay during each count (Figure 4; $N = 55$, $\tau_b = -0.214$, $P = 0.038$).

Breeding Activity

At Naked Island, a higher percentage of juveniles was recorded during shoreline surveys in prespill years than in postspill years (Table 6; $t = -2.46$, df $= 4$, $P = 0.035$). No juveniles were observed during shoreline surveys in 1989, although some

TABLE 3.—Number of marbled murrelets counted in shoreline surveys of Storey and Peak islands, Prince William Sound, in June and July. Mean counts before and after the spill were tested with a *t*-test, and 1989 data are tested separately against prespill counts with an outlier *t*-test. All numbers were log-transformed for analysis. Absence of data means no count was made.

Year or statistic	Storey Island		Peak Island	
	Jun	Jul	Jun	Jul
Before spill				
1978	37	11	49	69
1979	41		44	
1980		74		94
After spill				
1989	5	214	5	52
1990	132	361	73	59
1991	54		26	
1992	60		32	
t-test (2 means) (pre- versus postspill)				
t	0.026	−2.30	0.81	2.24
P	0.49	0.07	0.23	0.08
df	4	2	4	2
1989 outlier *t*-test				
t	−23.09	1.22	−23.91	−1.63
P	0.014	0.78	0.013	0.17
df	1	1	1	1

TABLE 4.—Number of marbled murrelets counted in June shoreline surveys of oiled and non-oiled areas of Naked, Storey, and Peak islands, Prince William Sound, before and after the *Exxon Valdez* oil spill. Each section of shoreline was classified as predominately oiled or non-oiled based on a map showing "Cumulative Oil Impact" as of 21 August 1989 (EVOSDAGP 1990).

Year	Number in non-oiled[a]	Number in oiled[b]	Proportion in oiled
Before spill			
1978	255	61	0.19
1979	192	91	0.32
After spill			
1989	56	30	0.35
1990	403	47	0.10
1991	170	58	0.25
1992	246	76	0.24

[a]Areas considered non-oiled were Cabin Bay, Outside Bay, Bass Harbor, and McPherson Passage on Naked, Peak, and Storey islands.

[b]Areas considered oiled were McPherson Bay, and East Naked and NW Naked Island.

TABLE 5.—Density of *Brachyramphus* murrelets (birds/km^2) by date and location (Inner and Outer Bay) in Kachemak Bay in 1988 and 1989. In this table, mean = average murrelet density for the specified dates and N = number of days the transects (four in Inner Bay, six in Outer Bay) were surveyed. See Figure 2 for the transect locations. Absence of data means no counts were made.

Year	Statistic	Period 1 (7 May–6 Jun)	Period 2 (7 Jun–12 Jul)	Period 3 (13 Jul–5 Aug)	Period 4 (6 Aug–5 Sep)	Murrelets/km^2 for all survey days
			Inner Bay			
1988	Mean		35.0	57.5		50.0
	SD			28.9		24.3
	N		1	2		3
1989	Mean	35.0	33.7		19.0	31.7
	SD	20.5	2.31			10.9
	N	2	3		1	6
			Outer Bay			
1988	Mean	18.2	14.7	45.3		20.6
	SD	7.4	4.8	12.5		12.9
	N	6	10	3		19
1989	Mean	18.0	16.7		60.3	29.9
	SD				24.3	24.3
	N	1	6		3	10

were observed on transects. On the five offshore transects at Naked Island, the percentage of juveniles was also lower after the spill (Table 7; $t = -10.03$, df = 2, $P = 0.005$). At Kachemak Bay, there was no change in the percentage of juveniles observed per day between 1988 and 1989 (Table 7; $t = -0.122$, df = 15, $P = 0.45$).

Discussion

Murrelet Abundance at Naked Island

Results of both survey methods—offshore counts and shoreline surveys—indicated that murrelet numbers were lower in 1989 compared to those in prespill years in the nearshore waters of the Naked

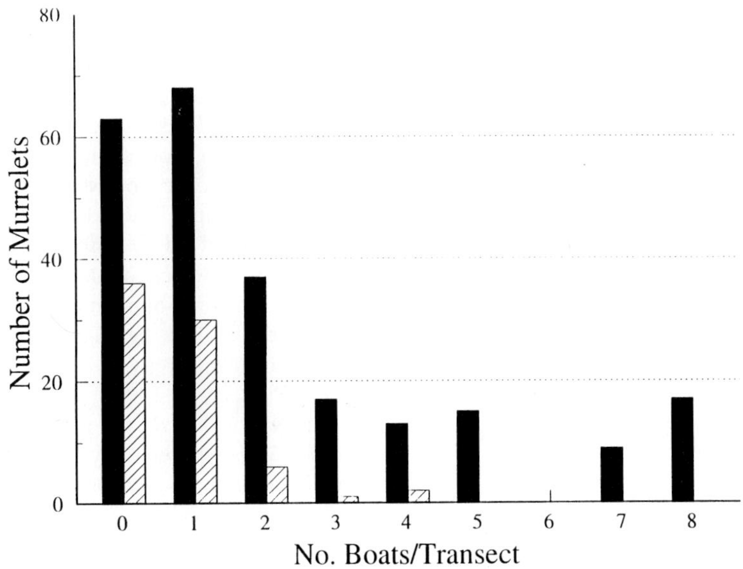

FIGURE 3.—Average number of *Brachyramphus* murrelets counted per transect relative to the number of boats counted on or near the transect at Naked Island, Prince William Sound (hatched bars; $N = 22$), and at Kachemak Bay (black bars; $N = 30$), Alaska, during summer 1989.

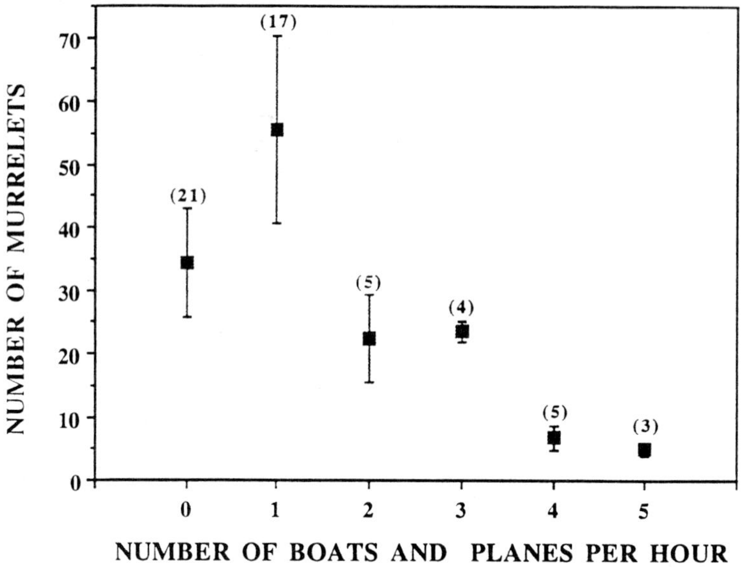

FIGURE 4.—Mean and standard error for the number of murrelets per hour (N = 55) counted from a land-based site, relative to the number of boats and low-flying planes in the bay during that hour. Counts were made in Outside Bay, Naked Island, Prince William Sound, in June and July 1989. Number in parentheses is the number of counts contributing to that data point.

Island group, at least through June. Both methods also showed a return to prespill numbers in the following years, suggesting that the local effect of the spill on the presence of murrelets was temporary. Murrelets were unlikely to have been killed at Naked Island, because few murrelets were there at the time of the spill. Only four murrelets were observed during a shoreline survey of the Naked Island group on 27–28 March 1989, before oil arrived (K. Wohl and S. Klosiewski, USFWS, Anchorage, unpublished data). Incidental observations suggest that few murrelets are in the area before late April (P. Isleib, personal communication), but they are abundant in early May (Oakley and Kuletz 1979). Surface oil was mostly absent from the area by 19 April, thus most of the summer population arrived after the main oil slick. Only 6 of the 229 marbled murrelet carcasses returned to Valdez, for which location of recovery was known, were found on Naked and Peak islands. Eighty percent of the Prince William Sound murrelet carcasses were found south of Eleanor Island (USFWS, Anchorage, Alaska, unpublished data).

TABLE 6.—Number of marbled murrelets and percentage of juveniles counted in shoreline surveys conducted between 25 July and 15 August at Naked, Storey, and Peak islands, Prince William Sound, before and after the *Exxon Valdez* oil spill.

Year	Survey area	Total number of murrelets	Percentage of juveniles
Before spill			
1978	Storey and Peak	80	6.25
1979	South Naked	14	14.29
After spill			
1989	Storey and Peak	266	0.0
1990	Naked, Storey, and Peak	592	0.67
1991	Naked	65	4.62
1992	Naked	222	4.95

Avoidance of oil by murrelets could have been one reason for lower counts in 1989. Aerial surveys in Prince William Sound in the first weeks after the spill revealed that the number of birds had decreased in oiled areas and increased in non-oiled areas (Piatt et al. 1990). Similarly, in the Kenai Fjords National Park, the number of murrelets increased 30% between early and late April 1989, but only in areas non-oiled between the two surveys; murrelet numbers decreased in oiled areas (Vequist and Nishimoto 1990). How much of the decrease resulted from mortality was unknown.

Another aspect of avoidance was the impact of cleanup and carcass recovery operations in oiled areas. Human disturbance probably reduced mur-

TABLE 7.—Mean percentage of juveniles per day, of the total number of marbled murrelets, counted on offshore transects at Naked Island and at Kachemak Bay before and after the *Exxon Valdez* oil spill. Surveys were conducted between 15 July and 15 August at Naked Island and between 4 July and 18 August at Kachemak Bay. Absence of data indicates no surveys were conducted.

	Naked Island transects			Kachemak Bay transects		
Year	Number of days	Total murrelets	Mean (±SD) percentage of juveniles/day	Number of days	Total murrelets	Mean (±SD) percentage of juveniles/day
Before spill						
1978	15	1,298	2.88 (3.29)			
1979	9	381	2.78 (4.12)			
1988				10	1,756	1.98 (1.80)
After spill						
1989	13	3,765	0.40 (0.78)	7	622	2.09 (1.82)
1990	10	2,279	0.83 (0.79)			

relet numbers near Naked Island. In 1989, there were cleanup activities in McPherson Bay and support activities in Outside Bay, where the TV *Exxon Valdez* was anchored until 22 June. The area that continued to have lower numbers of murrelets after 1989 was McPherson Bay, where an oil spill response barge has been anchored in summer months every year from 1989 to 1995. However, McPherson Bay was also the only large bay on the island with >50% of its shoreline oiled, so oil effects cannot be dismissed.

My surveys did not indicate an oil effect around Naked Island, suggesting that the amount of surface oil there in June did not affect murrelet distribution. However, the percentage of birds in non-oiled areas was unusually high in 1990, and some associated factor may have affected murrelet distribution that year. In 1990, the high number of murrelets on the offshore transects, which were in non-oiled bays, reflected this phenomenon.

Murrelet Abundance in Kachemak Bay

I found no changes in murrelet density between 1988 and 1989 in Kachemak Bay. Historical data on murrelet abundance in Kachemak Bay is anecdotal or fragmented, but there is no indication that the population has changed dramatically since the 1970s. The murrelet densities I observed in the south outer bay were comparable to those recorded for the bay mouth (12–20 murrelets/km^2) in 1976 (Erikson 1977).

Murrelets are present in Kachemak Bay in low numbers in the winter (Sanger 1987; M. Nishimoto, U.S. Fish and Wildlife Service, unpublished). In 1989, I made reconnaissance surveys in the bay by boat from mid-March through mid-May and recorded a steady increase in murrelets throughout April, with summer densities observed by May. Thus, when the oil entered the bay several weeks after the spill, most of the local murrelet population was in the bay. Few murrelet carcasses were recovered in Kachemak Bay, although carcasses found farther south could have originated there. Because of the small amount of weathered oil in Kachemak Bay, direct mortality of murrelets in the bay was probably low; however, birds that nested in Kachemak Bay could have been exposed to oil in the Gulf of Alaska in April.

Breeding Activity

My results, based on the offshore and shoreline surveys, suggest that reproduction was lower at Naked Island in postspill years. Studies at Naked Island of murrelet dawn activity support this deduction. Although the number of birds observed on the water in 1990 was high, there was no increase in juveniles or inland detection levels (dawn activity indicative of nesting). The amount of June (incubation phase) murrelet dawn activity in Cabin Bay increased in 1991 and 1992 compared to 1989 and 1990 (Kuletz 1994; Kuletz et al. 1994a). A similar association of low inland detections with high numbers at sea was observed in Oregon and California in 1992, an El Niño year (S. K. Nelson, Oregon Cooperative Wildlife Research Unit, personal communication). Nelson speculated that low inland activity, despite high numbers of birds at sea, reflected fewer breeding attempts. One explanation for the patterns I observed in the Naked Island area is that fewer murrelets attempted to breed or they failed early in 1989 and 1990.

At Kachemak Bay, the postspill ratio of juveniles to adults was higher than at Naked Island, and I found no difference between 1988 and 1989 at

Kachemak Bay. These results suggest that any disruption to murrelet breeding was limited to the heavily oiled spill zone. However, because the percentage of juvenile murrelets observed on the water is typically low, 2–5%, (Beissinger 1995), and the dispersal patterns of juveniles are unknown, no firm conclusions can be made about reproductive success in the spill zone.

Seabird breeding can be disrupted by a decrease in prey availability (review in Furness and Nettleship 1991), and prey abundance and distribution could have affected murrelets in the Naked Island area. Most of the prey species used by murrelets spawn intertidally and thus are highly susceptible to oil pollution (Trasky et al. 1977). At Naked Island, Pacific herring had sublethal damage and larval malformations (Brown et al. 1996, this volume). Fish used by murrelets, such as Pacific herring and Pacific sand lance, were infrequently delivered to pigeon guillemot chicks in 1989 and 1990 compared to their relative use in prespill years (Oakley and Kuletz 1996, this volume). Black-legged kittiwakes *Rissa tridactyla* in Prince William Sound also had reduced feeding rates of juvenile herring in 1990 (Irons 1992).

Breeding could also have been affected by direct mortality of breeding adults and disruption of breeding pairs. Marbled murrelets, like other alcids, appear to be monogamous (Sealy 1975). Marbled murrelets visit inland nesting areas year-round, suggesting strong pair-bonds or proprietorship of nesting sites (Naslund 1993). We do not know what proportion of the Prince William Sound winter population consisted of breeding adults, but in other alcids, breeding birds tend to winter near nesting areas while immatures move farther away (Nettleship and Birkhead 1985). If murrelets follow this pattern, those birds at risk in the sound in 1989 were probably resident breeding birds. Birds that lost mates may have delayed breeding, resulting in a late nestling phase relative to prey availability.

Sublethal oiling could have physiologically affected murrelets after the immediate oiling event. Oakley et al. (1996) identified compounds indicative of oil ingestion in murrelets collected in 1989 in heavily oiled areas but not in those from lightly oiled or non-oiled areas of Prince William Sound. Because these particular compounds are not lethal, their biological effects are uncertain. However, other studies found detrimental breeding behavior in seabirds that had ingested oil (Nero and Associates, Inc. 1987; Eppley and Rubega 1990).

Seasonal Migration and Direct Mortality

Of the six species of small alcids in the spill zone, marbled murrelets suffered the greatest mortality (Piatt et al. 1990; Ecological Consulting, Inc. 1991). However, U.S. Fish and Wildlife Service boat surveys found no oiling effect in murrelet population changes after the spill between oiled and non-oiled areas of Prince William Sound (Klosiewski and Laing 1994). I propose that the murrelets' foraging behavior and their dispersal to widely scattered breeding sites would mask any oiling effect but not necessarily preclude an effect on the population.

Fidelity to breeding sites is high among the Alcidae (review in Nettleship and Birkhead 1985) and murrelets returned to Prince William Sound breeding sites after most surface oil was gone; their July distribution would not necessarily be related to the degree of oiling at those locations. Species that had reduced colony numbers or breeding effects along oiled shorelines were the pigeon guillemot and the black oystercatcher *Haematopus bachmani*; both species rest on shoreline rocks and forage in inter- and subtidal kelp or rocks (Andres 1994; Oakley and Kuletz 1996). Klosiewski and Laing (1994) found oil effects on population abundance of 9 species or species groups, most of which are closely associated with intertidal habitat. Murrelets, in contrast, do not contact shoreline rocks; they forage widely and may make round trips between forage sites of 150 km (Burns et al. 1994). Such distances make it difficult to describe the home range of any murrelet as oiled or non-oiled.

An examination of morgue collection dates and locations, together with evidence for migration patterns, suggests that many murrelets encountered oil during spring migration in nearshore waters of south-central Alaska. Murrelets are present in nearshore waters of Alaska in winter at about 20–30% of the summer populations (Klosiewski and Laing 1994; Agler et al. 1995; Nishimoto, unpublished data), although high densities are reported at Kodiak Island in winter (Zwiefelhofer and Forsell 1989). During repeated transects in spring at Kachemak Bay (K. Kuletz, unpublished data) and the Kenai Peninsula (Vequist and Nishimoto 1990), murrelet numbers increased throughout April. Records kept by M. L. McAllister (personal communication) suggested that birds entering western Prince William Sound migrated along the southern Kenai Peninsula and through Montague Straight. Therefore, murrelets migrating north would have been affected throughout April, as the oil flowed south. This scenario is supported by the high pro-

TABLE 8.—Estimates of direct mortality (from carcass counts at receiving stations) of *Brachyramphus* murrelets from the *Exxon Valdez* oil spill in Alaska in 1989. Carcasses were identified as marbled murrelet (MAMU), Kittlitz's murrelet (KIMU), or unidentified murrelet (UNML). Totals are *Brachyramphus* murrelets.

Region	Estimated recovery rate[a]	MAMU carcasses	Estimated MAMU mortality[b]	KIMU carcasses	Estimated KIMU mortality[b]	UNML carcasses	Estimated UNML mortality[b,c]	Estimated total mortality[c]
Prince William Sound	0.35	289	826	23	66	21	60	952
Kenai Peninsula	0.14	113	807	23	164	73	521	1,492
Barren Islands	0.49	17	35	4	8	14	29	72
Kodiak Island	0.06	64	1,066	1	17	71	1,183	2,266
Alaska Peninsula	0.02	45	2,250	0	0	27	1,350	3,600
Total		528	4,984	51	255	206	3,143	8,382

[a]From Ecological Consulting, Inc., 1991.
[b]Based on recovery rate for that region and number of carcasses recovered.
[c]Does not include the proportion of ancient murrelets, based on carcasses identified to species.

portion of murrelet carcasses recovered in April (Piatt et al. 1990; Ecological Consulting, Inc. 1991) and the evidence of mortality outside of Prince William Sound.

Most of the marbled murrelet carcasses were retrieved in Prince William Sound, but recovery rates decreased with distance from the sound (Ecological Consulting, Inc. 1991). As a result, the estimated mortality for murrelets, based on carcass recoveries with known locations and recovery rates for each region, was actually higher south of the sound (Table 8). Using these figures, I estimated the *Brachyramphus* mortality at 8,382 birds, including 952 from Prince William Sound. This figure includes about 370 Kittlitz's murrelets, but not ancient murrelets *Synthliboramphus antiquus*. The estimated *Brachyramphus* mortality is probably low, because of the small size of murrelets and the low likelihood of their being recovered on cobble beaches (Ecological Consulting, Inc. 1991). Still, it is the highest estimated mortality for marbled murrelets from any single event (Carter and Kuletz 1995). Furthermore, a high proportion of unidentified small birds and alcids were probably *Brachyramphus* murrelets, based on a subsample reanalyzed to test identification accuracy (Ecological Consulting, Inc. 1991). When the percentage of unidentified small alcids and murrelets was added to that of identified *Brachyramphus* murrelets and applied to the mortality estimate for all birds developed by Ecological Consulting, Inc. (1991), the estimated mortality for all murrelets was 10,000–22,000, with a best approximation of 12,800–14,800.

The loss of 952 murrelets in Prince William Sound represents between 2 and 4% of the sound's March population, based on the average pre- and postspill March estimates of 44,000 and 24,000, respectively (Klosiewski and Laing 1994). It is probable that murrelets that breed in Prince William Sound were also killed in the southern portion of the spill zone. Assuming that the spill zone supported roughly half of the hypothesized 250,000 murrelets in Alaska (Mendenhall 1992; Piatt and Ford 1993), then 8,382 murrelets represents 7% of the murrelet population in the spill zone. The mortality estimate of 12,800–14,800 represents 10–12% of the murrelet population in the spill zone. Thus, I estimated that between 7 and 12% of the murrelet population in the spill zone was killed directly by oil.

The summer murrelet population has declined in Prince William Sound from 304,000 in the 1970s to 97,000 in the early 1990s, and there is evidence that the decline began before the spill (Klosiewski and Laing 1994). The murrelet population in the spill zone has been subject to other forms of environmental and anthropogenic stress. Murrelets are killed in gill nets in Prince William Sound (Wynne et al. 1991, 1992), and loss of nesting habitat by logging continues on private lands. Additionally, the simultaneous decline of other seabirds and marine mammals in the sound (Agler et al. 1994; Frost et al. 1994; Klosiewski and Laing 1994) suggest that certain forage fish have declined in the past decade, which could have reduced murrelet reproduction or increased adult mortality. Despite the lack of a clear effect of oil on the abundance of murrelets, the massive loss of adults from the spill contributed to the cumulative impact of perturbations on the murrelet population.

Conclusions

Adult murrelets at a nesting area in central Prince William Sound were temporarily displaced or killed in 1989 after the *Exxon Valdez* oil spill. There is evidence that cleanup and other spill-related activity disrupted nearshore murrelet distribution. Breeding may also have been disrupted at Naked Island, although the data are too limited to make firm conclusions. Numbers of adult murrelets in the Naked Island area returned to prespill levels between 1990 and 1992. At lightly oiled Kachemak Bay, there is no evidence that the spill significantly affected that population.

An estimated 7–12% of the murrelet population in the spill zone may have been killed directly by the oil spill, the largest single incident of murrelet mortality on record. It is possible that long-term effects would not be evident at the population level for many years, but our ability to separate spill-related causes from other perturbations is limited. The 66% decline in the Prince William Sound murrelet population since 1973, although not attributable to the spill, remains a concern for the recovery prospects of this species. The impact of the spill on murrelets was difficult to ascertain because of the ephemeral nature of this species along with its wide distribution, large foraging range, and secretive nesting habits. For these same reasons, oil-related effects on the murrelet population cannot be discounted. It is of primary importance to determine if there are long-term effects on the prey species on which murrelets depend.

Acknowledgments

For assistance in the field I am indebted to: A. Banks, D. Carr, M. Cody, L. Fuller, D. Marks, N. Naslund, and J. Pohl. Cooperation and logistical support were provided during the 1989 and 1990 field seasons by M. Nishimoto, D. Nysewander, and K. L. Laing. D. Bowden provided guidance in data analysis. I was assisted in the manuscript preparation by D. Marks, N. Naslund, M. Cody, D. Bothell, and L. Campbell. I thank H. Carter, M. Fry, G. L. Hunt, S. Klosiewski, K. Oakley, and B. Sharp for reviewing preliminary reports from which this article was derived and S. K. Nelson and S. Hatch for reviewing and improving an earlier draft of this manuscript.

References

Agler, B. A., P. E. Seiser, S. J. Kendall, and D. B. Irons. 1994. Marine bird and sea otter population abundance of Prince William Sound, Alaska: trends following the T/V *Exxon Valdez* oil spill, 1989–1993. U.S. Fish and Wildlife Service, *Exxon Valdez* Oil Spill Restoration Project Final Report, Project 93045, Anchorage, Alaska.

Agler, B. A., S. J. Kendall, P. E. Seiser, and D. B. Irons. 1995. Estimates of marine bird and sea otter abundance in Lower Cook Inlet, Alaska, during summer 1993 and winter 1994. U.S. Fish and Wildlife Service, Migratory Bird Management, OCS/MMS 94-0063, Anchorage, Alaska.

Andres, B. A. 1994. The effects of the *Exxon Valdez* oil spill on black oystercatchers breeding in Prince William Sound, Alaska. Final report. Bird Study 12 and Restoration Study 17. U.S. Fish and Wildlife Service, Migratory Bird Management, Anchorage, Alaska.

Beissinger, S. R. 1995. Population trends of the marbled murrelet projected from demographic analyses. Pages 385–394 *in* Ralph et al. (1995).

Brown, E. D., and seven coauthors. 1996. Injury to the early life history stages of Pacific herring in Prince William Sound after the *Exxon Valdez* oil spill. American Fisheries Society Symposium 18:448–462.

Burns, R. A., L. M. Prestash, and K. J. Kuletz. 1994. Pilot study on the capture and radio tagging of murrelets in Prince William Sound, Alaska, July and August, 1993. Project 93051B, Final report. U.S. Fish and Wildlife Service, Migratory Bird Management, Anchorage, Alaska.

Carter, H. R. 1984. At-sea biology of the marbled murrelet (*Brachyramphus marmoratus*) in Barkley Sound, British Columbia. Master's thesis. University of Manitoba, Winnipeg, Manitoba.

Carter, H. R., and K. Kuletz. 1995. Mortality of marbled murrelets due to oil pollution in North America. Pages 261–270 *in* Ralph et al. (1995).

Carter, H. R., and S. G. Sealy. 1986. Year-round use of coastal lakes by marbled murrelets. Condor 88:473–477.

Carter, H. R., and J. Stein. 1995. Molts and plumages in the annual cycle of the marbled murrelet. Pages 99–112 *in* Ralph et al. (1995).

Cavanaugh K. P., A. R. Goldsmith, W. N. Holmes, and B. K. Follett. 1983. Effects of ingested petroleum on the plasma prolactin levels during incubation and on the breeding success of paired mallard ducks. Archives of Environmental Contamination and Toxicology 12:335–341.

Day, R. H., K. L. Oakley, and D. R. Barnard. 1983. Nest sites and eggs of Kittlitz's and marbled murrelets. Condor 85:265–273.

Ecological Consulting, Inc. 1991. Assessment of direct seabird mortality in Prince William Sound and the western Gulf of Alaska resulting from the *Exxon Valdez* oil spill. Ecological Consulting, Inc., Portland, Oregon.

Eppley, Z. A., and M. A. Rubega. 1990. Indirect effects of an oil spill: reproductive failure in a population of south polar skuas following the *Bahia Paraiso* oil spill in Antarctica. Marine Ecology Progress Series 67:1–6.

Erikson, D. 1977. Distribution, abundance, migration and breeding locations of marine birds, Lower Cook Inlet, Alaska, 1976. Volume 3 *in* L. L. Trasky, L. B. Flagg, and D. C. Burbank, editors. Environmental studies of Kachemak Bay and lower Cook Inlet.

Alaska Department of Fish and Game, Marine/Coastal Habitat Management, Anchorage.

EVOSDAGP (*Exxon Valdez* Oil Spill Damage Assessment Geoprocessing Group). 1990. Cumulative oil impact as of August 21, 1989 (map). U.S. Fish and Wildlife Service and Alaska Department of Natural Resources, Anchorage.

Frost, K., L. F. Lowry, E. H. Sinclair, J. VerHoef, and D. C. McAllister. 1994. Impacts on distribution, abundance and productivity of harbor seals. Pages 97–118 *in* T. R. Loughlin, editor. Marine mammals and the *Exxon Valdez*. Academic Press, San Diego, California.

Furness, R. W., and D. N. Nettleship, editors. 1991. Seabirds as monitors of changing marine environments. Symposium 41 *in* Acta XX Congressus Internationalis Ornithologici. New Zealand Ornithological Congress Trust Board, Wellington, New Zealand.

Irons, D. B. 1992. Aspects of foraging behavior and reproductive biology of the black-legged kittiwake. Doctoral dissertation. University of California, Irvine.

Isleib, M. E., and B. Kessel. 1973. Birds of the north gulf coast-Prince William Sound region, Alaska. Biological Papers of University of Alaska 14.

Johnston, S., and H. R. Carter. 1985. Cavity-nesting marbled murrelets. Wilson Bulletin 97:1–3.

King, J. G., and G. A. Sanger. 1979. Oil vulnerability index for marine oriented birds. Pages 227–239 *in* J. C. Bartonek and D. N. Nettleship, editors. Conservation of marine birds of northern North America. U.S. Fish and Wildlife Service Wildlife Research Report 11:1–319.

Klosiewski, S. P., and K. K. Laing. 1994. Marine bird populations of Prince William Sound, Alaska, before and after the *Exxon Valdez* oil spill. Natural Resources Damage Assessment Bird Study 2. U.S. Fish and Wildlife Service, Migratory Bird Management, Anchorage, Alaska.

Kuletz, K. J. 1994. Marbled murrelet abundance and breeding activity at Naked Island, Prince William Sound, and Kachemak Bay, Alaska, before and after the *Exxon Valdez* oil spill. Final report. Natural Resources Damage Assessment Bird Study 6, U.S. Fish and Wildlife Service, Migratory Bird Management, Anchorage, Alaska.

Kuletz, K. J., D. K. Marks, and N. L. Naslund. 1994a. At-sea abundance and distribution of marbled murrelets in the Naked Island area in summer 1991 and 1992. U.S. Fish and Wildlife Service, *Exxon Valdez* Oil Spill Restoration Project, Final Report, Project R15, Anchorage, Alaska.

Kuletz, K. J., N. L. Naslund, and D. K. Marks. 1994b. Identification of marbled murrelet nesting habitat in the *Exxon Valdez* oil spill zone. U.S. Fish and Wildlife Service, *Exxon Valdez* Oil Spill Restoration Project, Final Report, Project R15, Anchorage, Alaska.

Marshall, D. B. 1988. Status of the marbled murrelet in North America with special emphasis on populations in Washington, Oregon and California. U.S. Fish and Wildlife Service Biological Report 88(30).

Mendenhall, V. M. 1992. Distribution, breeding records and conservation problems of the marbled murrelet in Alaska. Proceedings of the Western Foundation of Vertebrate Zoology 5:5–16.

Miller, D. S., D. B. Peakall, and W. B. Kinter. 1978. Ingestion of crude oil: sublethal effects in herring gull chicks. Science 199:315–317.

Naslund, N. L. 1993. Why do marbled murrelets attend old-growth forest nesting areas year-round? Auk 110: 594–602.

Naslund, N. L., K. J. Kuletz, D. K. Marks, and M. Cody. 1995. Tree and habitat characteristics and reproductive success of marbled murrelet tree nests in Alaska. Northwestern Naturalist 76(1):12–25.

Nero and Associates, Inc. 1987. Seabird oil toxicity study. Final report. Outer Continental Shelf Study MMS-87-0005. U.S. Department of the Interior, Minerals Management Service, Washington, DC.

Nettleship, D. N., and T. R. Birkhead, editors. 1985. The Atlantic alcidae. Academic Press, Harcourt Brace Jovanovich, London.

Oakley, K. L., and K. J. Kuletz. 1979. Summer distribution and abundance of marine birds and mammals near Naked Island, Alaska. U.S. Fish and Wildlife Service, Anchorage, Alaska.

Oakley, K. L., and K. J. Kuletz. 1996. Population, reproduction, and foraging of pigeon guillemots at Naked Island, Alaska, before and after the *Exxon Valdez* oil spill. American Fisheries Society Symposium 18:759–769.

Oakley, K. L., T. L. Wade, and K. J. Kuletz. 1996. Aliphatic and polycyclic aromatic hydrocarbons in livers of marbled murrelets in Prince William Sound, Alaska, after the *Exxon Valdez* oil spill. Final report. Natural Resources Damage Assessment Bird Study 6. U.S. Fish and Wildlife Service, Anchorage, Alaska.

Peakall, D. B., D. Hallett, D. S. Miller, R. G. Butler, and W. B. Kinter. 1980. Effects of ingested crude oil on black guillemots: a combined field and laboratory study. Ambio 9:28–30.

Piatt, J. F., and R. G. Ford. 1993. Distribution and abundance of marbled murrelets in Alaska. Condor 95: 662–669.

Piatt, J. F., C. J. Lensink, W. Butler, M. Kendziorek, and D. Nyswander. 1990. Immediate impact of the *Exxon Valdez* oil spill on marine birds. Auk 107(2):387–397.

Ralph, C. J., G. L. Hunt, Jr., M. G. Raphael, and J. F. Piatt, editors. 1995. Ecology and conservation of the marbled murrelet. U.S. Forest Service General Technical Report PSW-GTR-152.

Ralph, C. J., and L. L. Long. 1995. Productivity of marbled murrelets in California from observations of young at sea. Pages 371–380 *in* Ralph et al. (1995).

Sanger, G. 1987. Winter diets of common murres and marbled murrelets in Kachemak Bay, Alaska. Condor 89:426–430.

Sealy, S. G. 1975. Aspects of the breeding biology of the marbled murrelet in British Columbia. Bird-Banding 46:141–154.

Singer, S. W., N. L. Naslund, S. A. Singer, and C. J. Ralph. 1991. Discovery and observations of two tree nests of the marbled murrelet. Condor 93:330–339.

Szaro, R. C., and P. H. Albers. 1977. Effects of external applications of No. 2 fuel oil on Common Eider eggs. Pages 164–167 *in* D. A. Wolfe, editor. Fate and effects of petroleum hydrocarbons in marine ecosystems and organisms. Pergamon Press, New York.

Trasky, L. L., L. B. Flagg, and D. C. Burbank. 1977. Impact of oil on the Kachemak Bay environment. Volume 1 *in* L. L. Trasky, L. B. Flagg, and D. C. Burbank, editors. Environmental studies of Kachemak Bay and lower Cook Inlet. Alaska Department of Fish and Game, Marine/Coastal Habitat Management, Anchorage.

Vequist, G. W., and M. Nishimoto. 1990. Seabird surveys on the coast of Kenai Fiords during the *Exxon Valdez* oil spill. Natural resources final report AR-90/04. National Park Service, Anchorage, Alaska.

Wynne, K., D. Hicks, and N. Munro. 1991. Salmon gillnet-fisheries observer programs in Prince William Sound and South Unimak Alaska. Final report. Saltwater Inc., Anchorage, Alaska.

Wynne, K., D. Hicks, and N. Munro. 1992. Marine mammal observer program for the salmon driftnet fishery of Prince William Sound Alaska. Final report. Saltwater Inc., Anchorage, Alaska.

Zwiefelhofer, D. C., and D. J. Forsell. 1989. Marine birds and mammals wintering in selected bays of Kodiak Island, Alaska: a five-year study. U.S. Fish and Wildlife Service, Kodiak National Wildlife Refuge, Kodiak, Alaska.

American Fisheries Society Symposium 18:785–797, 1996

Bald Eagle Productivity in South-Central Alaska in 1989 and 1990 after the *Exxon Valdez* Oil Spill

JEFFREY A. BERNATOWICZ,[1] PHILIP F. SCHEMPF, AND TIMOTHY D. BOWMAN[2]
U.S. Fish and Wildlife Service
3000 Vintage Boulevard, Suite 240, Juneau, Alaska 99801, USA

Abstract.—Nesting surveys of bald eagles *Haliaeetus leucocephalus* were conducted in 1989 and 1990 in coastal south-central Alaska, including Prince William Sound, Kenai Peninsula, Kodiak Archipelago, Katmai National Monument, and Alaska Peninsula–Becharof National Wildlife Refuge, to assess the effects of the *Exxon Valdez* oil spill on reproductive success. Nesting success was appreciably lower only in 1989 within the spill area in western Prince William Sound, located at the upstream end of the spill trajectory. Productivity in western Prince William Sound in 1990 increased to levels that we believe are normal for the region. Reproductive success for regions outside of the sound appeared to be normal in both 1989 and 1990. Because of the lack of prespill data, survey biases, and the potential for wide natural fluctuations in reproductive success, only a substantial reduction in productivity would have been evident in this study. This study provides the first estimates of bald eagle productivity for many areas in south-central Alaska.

On 24 March 1989, the oil tanker *Exxon Valdez* ran aground spilling more than 40 million liters of crude oil into Prince William Sound, Alaska. The oil moved southwest, fouling approximately 2,500 km of shoreline from Prince William Sound to the Alaska Peninsula (Table 1 and Figure 1). In the cleanup effort during the following two springs and summers, thousands of workers and a variety of equipment scoured the oiled shoreline. An estimated 8,000 bald eagles *Haliaeetus leucocephalus* inhabit the spill area and their nesting season coincided with the arrival of oil and cleanup crews. Bald eagles in coastal areas are highly dependent on the nearshore ecosystem for nesting (Hodges and Robards 1982; J. I. Hodges, Jr., S. L. Cain, and P. F. Schempf, U.S. Fish and Wildlife Service [USFWS], unpublished data) and foraging (Imler and Kalmbach 1955; Garrett et al. 1993). Their opportunistic feeding strategy and willingness to enter the water make them susceptible to oiling (Todd et al. 1982).

There is no documentation on the effects of crude oil on bald eagle reproductive success; however, crude and refined oil have been found to cause significant embryo mortality in other avian species (Hartung 1965; Szaro 1977; Albers 1978; Albers and Szaro 1978; Szaro et al. 1978; Coon et al. 1979; King and Lefever 1979; Szaro 1979; Albers 1980; Szaro et al. 1980). Fresh and weathered crude oil in amounts greater than or equal to 5 μL applied to eggs significantly reduced hatching success of gulls and ducks, the only species tested. The long-term survival of avian nestlings ingesting oil is uncertain (Szaro 1977; Flemming et al. 1982; Pattee and Franson 1982; Leighton 1986), but oiled down and feathers lose insulation value and birds suffer from hypothermia (Szaro 1977).

Reproductive success of bald eagles is evaluated using two surveys. An occupancy survey is conducted to count the number of breeding pairs in the population (occupied nests) and identify the number of pairs that laid eggs (active nests). Occupancy surveys should be timed after all eggs have been laid and before any nesting failures have occurred (Postupalsky 1974; Fraser et al. 1983). In most cases, this will be a compromise, because failures probably begin before the last eggs are laid. A second survey, the productivity survey, is conducted to identify failed nests and to count older young that are likely to fledge. The productivity survey needs to be timed late enough so few additional nestling mortalities will occur but before any chicks leave the nest. Failure to time the surveys correctly will bias estimates of reproductive success (Steenhof and Kochert 1982; Fraser et al. 1983).

The standard measure of bald eagle reproductive success is young per occupied nest. Estimates should be based on all occupied nests, including those in which no eggs were laid (i.e., nonbreeding pairs) (Postupalsky 1974; Steenhof 1987). Determining the actual number of occupied nests is difficult because not all adults are visible (Bowman et al. 1993) or in the vicinity of the nest (Whitfield et al. 1974; Fraser et al. 1983). This determination also

[1]Present address: Washington Department of Fish and Wildlife, 16018 Mill Creek Boulevard, Mill Creek, Washington 98012, USA.
[2]Present address: U.S. Fish and Wildlife Service, 1011 East Tudor Road, Anchorage, Alaska 99503, USA.

TABLE 1.—Cumulative maximum oiled shorelines in south-central Alaska, by region, after the *Exxon Valdez* oil spill.

Region[a]	Heavy oiling[b]	Medium oiling[c]	Light oiling[d]	Very light oiling[e]	Total oiling	No oiling
PWS						
km	169.1	125.3	218.3	215.3	728.0	1,278.2
%[f]	8.4	6.2	10.9	10.7	36.3	63.7
Kenai						
km	36.1	51.6	111.6	136.7	336.0	1,514.0
%	1.9	2.8	6.0	7.4	18.2	81.8
Kodiak						
km	10.7	53.9	108.2	724.6	897.4	1,516.0
%	0.4	2.2	4.5	30.0	37.2	62.8
Katmai/APBNWR						
km	9.5	35.2	192.9	292.1	529.7	913.0
%	0.6	2.8	13.4	20.2	36.7	63.3

[a]PWS = Prince William Sound; Kenai = Kenai Peninsula; Kodiak = Kodiak Archipelago; Katmai/APBNWR = Katmai National Monument and Alaska Peninsula–Becharof National Wildlife Refuge.
[b]A band of surface and/or subsurface oil greater than 6 m wide or more than 50% coverage of the intertidal zone.
[c]An oil band of 3–6 m wide or 10–50% coverage of intertidal zone.
[d]An oil band less than 3 m wide or 10% coverage of intertidal zone.
[e]An oil band less than 1 m wide or beach having less than 1% coverage.
[f]Percent of all surveyed shoreline.

assumes the ability to delineate individual territories. There is potential to grossly underestimate the number of occupied territories, especially in areas with high bald eagle densities (Hansen and Hodges 1985).

Bowman et al. (1993) analyzed bald eagle productivity in relation to shoreline oiling in Prince William Sound, and D. Zwiefelhofer (USFWS, unpublished data) compared bald eagle productivity in 1989 to historical data in survey plots affected by oiling. The objectives of our study were to gather, collate, and reanalyze all available data on bald eagle productivity in the entire region affected by the *Exxon Valdez* oil spill during 1989 and 1990 to determine if the spill influenced reproductive success. This study was part of an assessment of potential damage caused by the spill to identify reparation and restoration needs.

Methods

Study Area

The study area encompassed all shorelines influenced by the oil spill which included: Prince William Sound (PWS), the southeastern coast of the Kenai Peninsula (Kenai), the Kodiak Archipelago (Kodiak), and the southern coast of the Alaska Peninsula within the Katmai National Monument (Katmai) and Alaska Peninsula–Becharof National Wildlife Refuge (APBNWR) (Figure 2). Bald eagles in Prince William Sound and Kenai nest primarily in western hemlock *Tsuga heterophylla* and Sitka spruce *Picea sitchensis*. In Kodiak, black cottonwood *Populus trichocarpa* is the predominant nest tree and approximately 18% of the active nests are on cliffs or sea stacks. The majority of nests within Katmai are ground nests, although alders *Alnus crispa* are occasionally used. All bald eagles within APBNWR nest on the ground. Ground nests are often built of driftwood, debris, and kelp but quickly become overgrown when they are inactive. The climate in the coastal areas is relatively mild, with an average range of 7°–25°C in summer and −5° to 1°C in winter. Only a few sheltered, brackish bays freeze in winter. Precipitation ranges from 120 cm on the Alaska Peninsula to more than 500 cm in some areas of Prince William Sound.

Survey Methods

Crews from the USFWS and National Park Service conducted the surveys. In 1989, nesting surveys were often conducted opportunistically while performing other duties related to the spill. Therefore, survey timing, area, and methods were not standardized. In 1990, at least two surveys were conducted to assess bald eagle productivity. An occupancy survey was conducted during the incubation period (May to early June), and a productivity survey was conducted after eggs had hatched but before young had fledged (July to early August). Survey methods and timing are given for each area in Table 2.

Prince William Sound.—Initial nest occupancy data were collected opportunistically during bald eagle population surveys in 1989. The population surveys followed methods presented by Hodges et al. (1984) and included all island shorelines and 23

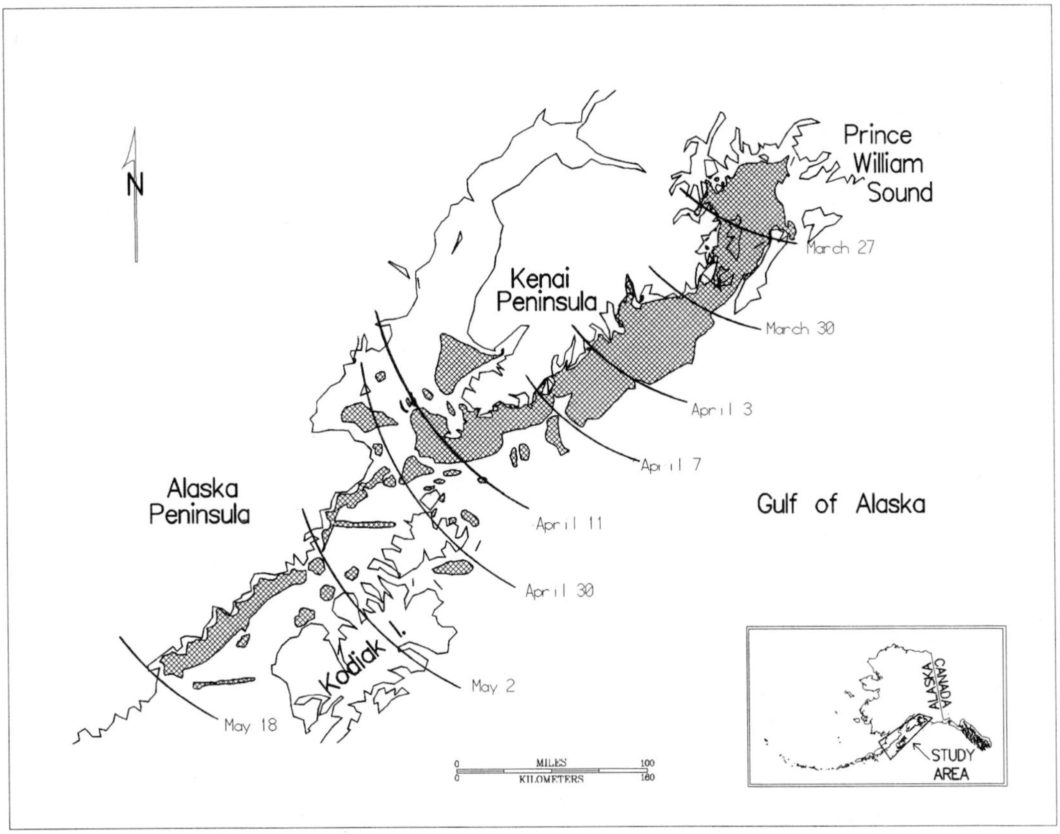

FIGURE 1.—Areas in south-central Alaska affected by the *Exxon Valdez* oil spill of 24 March 1989. Cross-hatched areas represent the range of the moving oil slick. Dated arcs represent the time of arrival of the oil slick at the indicated location.

random plots (16,828 ha) along the mainland shoreline. The locations of active nests were recorded as they were seen, but the primary effort was to count adults. The second occupancy survey was flown using mapped nest locations from the population survey and a study in the early 1980s of nest locations within Prince William Sound (USFWS, Juneau, Alaska, unpublished data). The second occupancy survey focused on finding occupied nests within the immediate spill area. Subsequent nest surveys were flown ad hoc throughout the remainder of the summer. The main purposes of the flights were to coordinate oil spill cleanup activities, collect addled eggs and prey remains, and attach a band and radio transmitter to nestlings. Some flights were made specifically to collect productivity information, but there was no systematic survey schedule or area because of the availability of personnel and aircraft. Surveys focused on the spill zone in western Prince William Sound.

In 1990, all shorelines directly affected by the spill were included in the survey area. Because of the low productivity observed in 1989, we also surveyed shorelines in eastern Prince William Sound that were not directly within the spill's path to determine the extent of damage. Approximately 80% of the nesting habitat in the sound was surveyed in 1990.

Kenai Peninsula.—No occupancy survey was flown in 1989. Limited nesting information was collected at random in June by personnel working on the oil spill cleanup. Productivity surveys in 1989 covered approximately 90% of the outer coast of the Kenai Peninsula (Figure 2) and concentrated on Kenai Fjords National Park. In 1990, single occupancy and productivity surveys were conducted along the same coast surveyed in 1989.

Kodiak Archipelago.—Approximately 60% of the Kodiak Archipelago shoreline (Figure 2) was surveyed once for occupancy and productivity in both

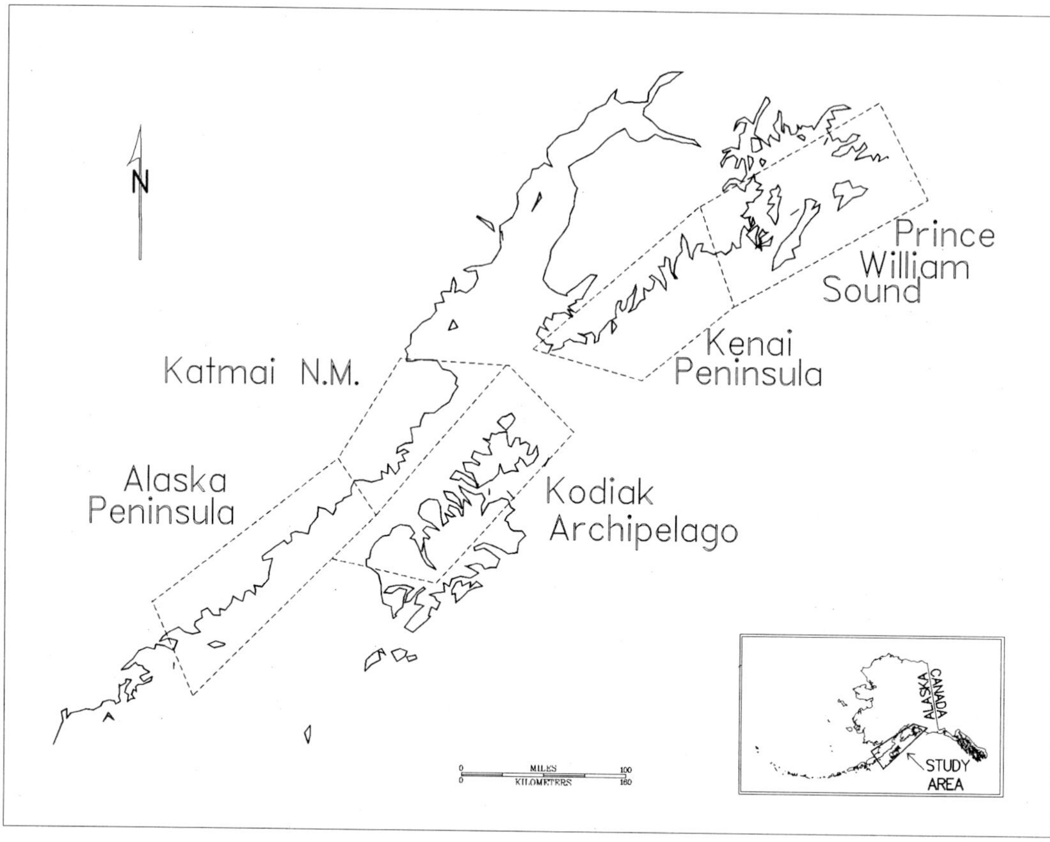

FIGURE 2.—Subregions surveyed for productivity of bald eagles *Haliaeetus leucocephalus* in south-central Alaska, 1989 and 1990, after the *Exxon Valdez* oil spill. (The abbreviation N. M. means National Monument.)

1989 and 1990. Active nests were not distinguished from those occupied by nonbreeding pairs in 1989. Surveys concentrated on the areas within the path of the oil spill (N, NW, and NE parts of the archipelago).

Katmai National Monument.—The survey area included the entire Katmai National Monument coastline on the south side of the Alaska Peninsula (Figure 2). In 1989, occupancy surveys were conducted by helicopter in June, but the data were lost. Productivity information in 1989 was collected by boat throughout the summer as time permitted, but data were recorded only in field notes, without using a data sheet, resulting in inconsistent records. In 1990, single occupancy and productivity surveys were conducted.

Alaska Peninsula–Becharof National Wildlife Refuge.—Surveys were conducted along the south side of the Alaska Peninsula within the APBNWR in 1989 and 1990. Only active nests were noted on the occupancy surveys in 1989.

Oiling Data

We examined reports issued by the Alaska Department of Environmental Conservation (ADEC), Alaska Department of Natural Resources, USFWS, U.S. Coast Guard, and Piatt et al. (1990) to determine the timing and extent of oil contamination. There were major differences in degree and extent of oiling between reports. We used a summary of extent and degree of oiling provided by ADEC (W. Lane, ADEC, unpublished data) because it was the most recent (and presumed to be the most accurate). Also, the data provided by ADEC pertained to areas that matched our nest survey areas to a large extent.

Data Analysis

We were responsible for collecting data in Prince William Sound. In other regions, all reports, data summaries, and available raw data came from other sources and were reviewed by us for this report

TABLE 2.—Dates and methods for nesting success and productivity surveys of bald eagles in south-central Alaska, 1989 and 1990, after the *Exxon Valdez* oil spill.

Region[a]	Occupancy survey: Dates	Occupancy survey: Method[b]	Occupied nests found after 1 Jun (%)	Productivity survey: Dates	Productivity survey: Method
PWS					
1989	18 Apr–5 May	Wing	0	15 Jul–8 Aug	Heli
	20–27 May	Heli[c]	0		
1990[d]	10–19 Apr	Heli	0	21–25 Jul	Heli
	13–19 May	Heli[c]	0	12–14 Aug	Heli[e]
	11–20 Jun	Heli[c]	0		
Kenai					
1989	Jun	Boat	ND[f]	27 Jul–8 Aug	Hewi
1990	19 May–7 Jun	Heli	30	4–5 Aug	Heli
Kodiak					
1989	10 May–7 Jun	Hewi	36	28 Jul–5 Aug	Heli
1990	7 May–3 Jun	Hewi	1	28 Jul–5 Aug	Heli
Katmai					
1989	11 Jun	Heli	ND[f]	26 Jun–15 Aug	Boat[f]
1990	26–27 May	Heli	0	24–25 Jul	Heli
APBNWR					
1989	10 May–3 Jun	Heli	78	4–5 Jul	Heli
	19–20 Jun	Heli	0	24–25 Jul	Wing[e]
1990	9–31 May	Heli	0	26–28 Jul	Heli
	19–21 Jun	Heli	0		

[a]PWS = Prince William Sound; Kenai = Kenai Peninsula; Kodiak = Kodiak Archipelago; Katmai = Katmai National Monument; APBNWR = Alaska Peninsula–Becharof National Wildlife Refuge.
[b]Wing = fixed-wing; Heli = helicopter; Boat = boat; Hewi = helicopter and fixed-wing.
[c]Sample contains nests that were originally found on previous survey.
[d]In addition to surveys listed, approximately 50% of the area was surveyed once every 10 d.
[e]Resurvey of nests with young that were not near fledging on previous survey.
[f]Surveys were conducted, but locations of individual nests were not recorded or data were lost.

(Table 4). Whenever possible, field investigators were interviewed. We mapped and assigned a number to each individual nest. Dates and nest status were assigned to each nest for individual observations. When discrepancies existed in the raw data, summaries, and reports for a nest, it was omitted from the data set. In 1989, only data summaries containing the final fate for each nest were available for the Kenai Peninsula and Katmai National Monument.

We used terminology and calculated measures of reproductive success based on definitions adapted from Postupalsky (1974) and Fraser et al. (1983) (Table 3 and Figure 3). We included hypothetical data (Figure 3) for clarification. Survey area, methodology, and timing were not consistent between areas and years. To reduce the potential biases, we used subsets of data for comparison between years. Within the spill zone (western Prince William Sound), we compared nests that were found on occupancy surveys before 1 June and on productivity surveys, in both 1989 and 1990. In eastern Prince William Sound, the 1989 sample was too small for meaningful comparison between years; thus, we included all nests with complete survey data from 1990. On Kenai, we were unable to determine which or how often areas were surveyed; therefore, we used only nests that were found in both 1989 and 1990 in the subset. Survey methods were similar between 1989 and 1990 in APBNWR, but the area was more extensive in 1990. There were also potential mapping errors in 1989. For comparison between years, we used only nests in the area surveyed in both years. Reported observations for individual nests were either not mapped or did not match available field notes for Katmai in 1989. Total young produced and number of occupied nests were the only data available for Katmai in 1989. No subsets were needed for Kodiak as methods and observer experience were similar between years. We do not think it is appropriate to compare survey results between years and among subregions using statistical techniques because of the biases involved in the way the data were collected.

Prespill estimates of bald eagle productivity and nesting success were lacking for most of the regions studied. We reviewed the literature and unpublished data of Alaska bald eagle nesting studies for comparative data. Most historical studies provide sampling data only for the actively nesting eagles. When nonbreeding birds were surveyed, there was little difference in productivity estimates based on occupied versus active nests. Therefore, we only

TABLE 3.—Definitions of terms used in calculations of nest success and productivity for bald eagles (after Postupalsky 1974 and Fraser et al. 1983) as used in this study.

Term	Definition
Occupied nest	An adult in incubating posture or two adults actively defending, near or at a nest. Includes nesting (active) and nonbreeding eagle pairs.
Active nest	A nest with an adult in incubation posture or with eggs or chicks.
Nonbreeding pair	A territorial pair of eagles that do not produce eggs during the survey year. Defined on a survey as two adult eagles obviously associated with a nest (generally within 300 m).
Total young produced[a]	Number of young produced to an advanced stage of development (fully feathered).
Young per occupied nest[b]	Total number of young produced divided by total number of occupied nests.
Young per active nest[b]	Total number of young produced divided by total number of active nests.
Percentage successful[b]	Number of nests in which at least one young was raised to an advanced stage of development divided by total number of active or occupied nests.

[a]Calculated using all nests found in survey area.
[b]Calculated using only nests found on both the occupancy survey and productivity survey.

included active nests in estimates of productivity and nest success in the comparative data.

Bald eagle nest locations were either not recorded, vague, or of questionable accuracy in many regions in 1989. The accuracy of the oiling data was also questionable. Furthermore, Bowman et al. (1993) found that even non-oiled beaches within the spill zone had high rates of nest failure. Therefore, we only considered bald eagle productivity in the context of oiling data on a regional basis rather than by nest site (Bowman et al. 1993) or sample plot (Zwiefelhofer, unpublished data).

Results

The oil remained in a smooth liquid form as it traveled NE to SW through Prince William Sound, reaching the Gulf of Alaska at the end of March (Figure 1). Approximately 36% of the Prince William Sound shoreline in the spill path was coated, with more than 14% receiving heavy or moderate oiling (Table 1). More than 7.6 million liters of oil (17% of original spill) escaped the sound and traveled the length of the Kenai Peninsula in 2 weeks, reaching the Barren Islands by 11 April (Figure 1). Less than 5% of the area surveyed along the Kenai Peninsula received heavy or moderate oil (Table 1). The oil became weathered, emulsified into "mousse," and broke up into patches before reaching Kodiak Island and the Alaska Peninsula by the end of April (Figure 1). Total percent of shorelines oiled on Kodiak Island and on the Alaska Peninsula were similar to the total percent in Prince William Sound (36–37%), but the majority in the former two areas were very lightly oiled (Table 1). Small patches of mousse, sheen, and tar balls were ob-

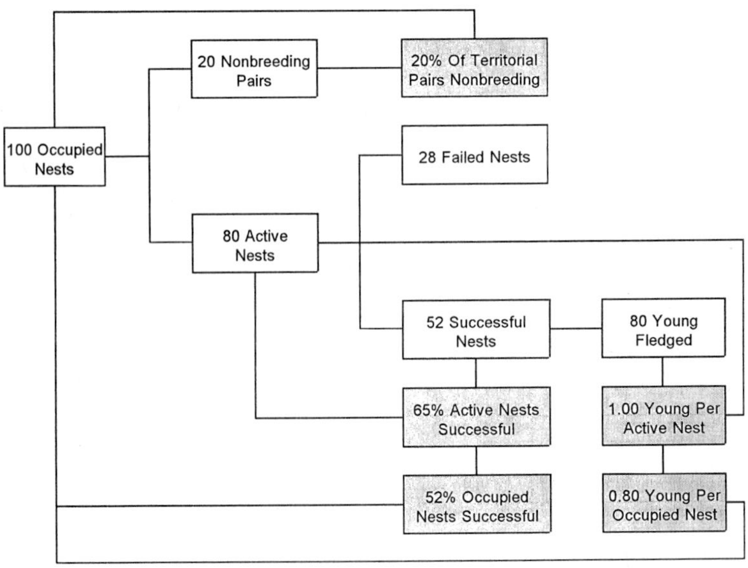

FIGURE 3.—Terminology used in reporting the reproductive patterns of nesting bald eagles *Haliaeetus leucocephalus*. Information gathered during surveys is in unshaded blocks, which allowed calculations of statistics (shaded blocks). The values presented are hypothetical.

TABLE 4.—Nesting success and productivity of bald eagles in south-central Alaska, 1989 and 1990, after the *Exxon Valdez* oil spill. Numbers in parentheses represent subsets of nests that minimized bias and were used for comparisons between years within a subregion; ND = not determined.

Region[a]	Occupied nests found on both surveys	Young per occupied nest[b]	Percent of occupied nests successful[b]	Young per active nest[b]	Percent of active nests successful[b]	Total occupied nests found[c]	Total young produced[c]
WPWS[d]							
1989	165	0.36	30	0.39	32	165	60
1990	165	1.04	64	1.19	72	165	172
EPWS							
1989	24	0.54	38	0.76	53	24	13
1990	348	0.74	53	0.89	64	372	276
Kenai[e]							
1989[f]	0	ND	ND	ND	ND	45 (29)	52 (36)
1990	45	0.92	51	1.00	58	63 (34)	58 (28)
Kodiak[g]							
1989	333	1.20	72	ND	ND	333	399
1990	412	1.05	64	1.07	66	412	431
Katmai							
1989[h]	0	ND	ND	ND	ND	47	54
1990	45	0.56	47	0.60	50	50	28
APBNWR[i]							
1989	62	ND	ND	1.00	55	69	62
1990	76	0.91	56	1.11	59	76	77

[a]WPWS = western Prince William Sound; EPWS = Eastern Prince William Sound; Kenai = Kenai Peninsula; Kodiak = Kodiac Archipelago; Katmai = Katmai National Monument; APBNWR = Alaska Peninsula–Becharof National Wildlife Refuge.
[b]Calculations based only on nests found on occupancy and productivity survey.
[c]Based on all nests found on the productivity survey (nests not found on occupancy survey included).
[d]Subset(s) includes only nests found on both occupancy and productivity surveys on both 1989 and 1990.
[e]Subset(s) includes only nests found on both the 1989 and 1990 productivity surveys.
[f]Kenai Fjords National Park, Seward, Alaska unpublished data.
[g]Kodiak National Wildlife Refuge and adjacent islands; Zwiefelhofer (unpublished).
[h]Yurick (unpublished).
[i]Dewhurst (unpublished). Subset(s) includes nests within the area surveyed in both 1989 and 1990.

served in June and July, but the majority of the oil either sank, evaporated, or had washed up on beaches by the end of May. Tides continued to move oil short distances, thereby coating new and previously cleaned shorelines throughout summer 1989. Winter storms dispersed the majority of surface oil that had not been scoured by cleanup crews. By June 1990 approximately 3% (390 km) of the shoreline in the spill area was considered oiled. Any remaining liquid oil was subsurface.

Logistical constraints incurred immediately after the oil spill in 1989 precluded an accurate measure of reproductive success for bald eagles in most areas. The majority of occupancy surveys were conducted 1–2 months after the spill, and oil-related mortalities and nest failures probably had occurred. Estimating the prespill breeding population was also hindered by the lack of information collected on pairs that were not actively nesting (nonbreeders and early failures). Young per occupied nest could only be calculated for Prince William Sound and Kodiak (Table 4) in 1989. Reproductive estimates for the remainder of the regions were based on actively nesting or successful pairs. Survey timing and techniques were improved in 1990.

The average for the eight comparison areas was 0.96 young per active nest and 54% nest success (Table 5). There was wide variation in both productivity (0.46–1.40 young per occupied nest) and active nest success (37–92%).

In general, reproductive success among regions was similar in 1989 and 1990 with one exception (Table 4) and similar to that of historical studies (Table 5). The only area obviously affected by the spill was western Prince William Sound in 1989. Only 30% of the occupied nests were successful and only 0.39 young per active nest were produced in 1989. All regions except Katmai produced more than 0.8 young per active nest. In 1990, nest success and production rebounded in Prince William Sound to a level that appeared to be normal. Katmai produced 0.6 young per occupied nest and had less than 50% occupied nest success, but we could not attribute the low productivity and nest success to the effects of oil.

Discussion

The extent of oiling given in Table 1 differs from that of other studies (Piatt et al. 1990; Bowman et al. 1993; Zwiefelhofer, unpublished). Early reports

TABLE 5.—Nesting success and productivity of actively nesting bald eagles in ecosystems not affected by organochloride chemicals. The abbreviation USFWS is U.S. Fish and Wildlife service.

Area	Years of study	Active nests	Young per active nest	Percent of active nests successful	Source
Kodiak National Wildlife Refuge[a]	25	24–413	1.00 (0.52–1.20)	65 (44–92)	Zwiefelhofer unpublished
Amchitka Island[a]	1	71	0.86	61	Sherrod et al. (1976)
Chatham Strait[a]	10	15–47	0.91 (0.46–1.23)	60 (37–82)	USFWS unpublished
Karluk Lake[b]	3	8–14	1.00 (0.60–1.40)		Hensel and Troyer (1964)
Kenai National Wildlife Refuge[b]	2	22–23	1.20 (1.00–1.40)	75 (72–78)	Bangs et al. (1982)
Tanana River[b,d]	3	13–15	1.07 (0.86–1.40)	64 (54–80)	Ritchie (1982)
Copper River Basin[b]	2	37–151	0.64 (0.59–0.84)	42 (40–51)	USFWS unpublished
Copper River Delta[b]	1	59	1.00	64	USFWS unpublished

[a]Marine ecosystem.
[b]Freshwater ecosystem.
[c]Occupancy surveys conducted after hatching, probably overestimated productivity and success.
[d]Productivity surveys conducted after some young fledged on year with lowest productivity.

issued by ADEC and the USFWS estimate 1,300–1,500 km of oiled shoreline in the spill path. We used recent updates by ADEC (Lane, unpublished data) and the U.S. Coast Guard (1993) that estimated 2,400–2,500 km of oiled shoreline. We assumed that the recent updates were the most accurate, but could not explain the differences between the reports.

Our bald eagle productivity estimates vary from results presented by others who participated in local bald eagle nesting studies after the spill (M. Amaral, USFWS, unpublished data; D. A. Dewhurst, USFWS, unpublished data; M. M. Yurick, National Park Service, unpublished data; Zwiefelhofer, unpublished data). We attribute these differences to interpretation of data. We deleted data that did not meet the criteria outlined in the *Methods*.

Our comparative sample of historical studies only sampled the actively nesting portion of the population, which probably resulted in an overestimated actual productivity of the breeding population because all pairs occupying territories were not considered. Hensel and Troyer (1964) suggested that 14–18 territorial pairs occupied their study area on Kodiak Island, but only 8–14 actively nested (Table 5). Assuming 18 pairs in the breeding population, the average productivity was 0.59 (range, 0.28–0.78) young per occupied nest. On Amchitka Island, the actual productivity may have been closer to 0.7 young per occupied nest because Sherrod et al. (1976) estimated that 19% of the breeding population never laid eggs. In Chatham Straight (USFWS, Juneau, Alaska, unpublished data), the number of active nests in the study area decreased from 35 in 1985 to 15 in 1987 and then increased to 47 in 1989. In 1987, productivity was reported at 0.8 young per active nest. Assuming there were 47 occupied territories in 1987, productivity was actually 0.25 young per occupied nest. Wide annual variations in the number of active nests in a study area have been documented in other parts of SE Alaska (Hodges 1982; Hansen and Hodges 1985). We do not know if the fluctuations in number of active nests were a result of survey error, birds using alternate sites, or territorial pairs not breeding; however, we believe the actual variation in productivity and nest success for occupied nests is greater than we report for active nests in Table 5.

The wide variations and lack of understanding of the breeding population make determining "normal" difficult. Recovery plans for bald eagles in the lower 48 specify 1.00 young per occupied nest and 65% occupied nest success for delisting (USFWS 1986). Anthony et al. (1994) suggested 1.00 young per occupied nest is a conservative goal. Populations in the other 48 continental states have been recovering from organochloride poisoning and their productivity rates are probably not comparable to those of Alaskan populations. Bowman et al. (1995) estimated that 0.86 young per occupied nest would lead to a population growth rate of 2% in Prince William Sound and that approximately 0.7 young per occupied nest would result in a stable population. Buehler et al. (1991) calculated that 0.74 young per occupied nest would result in a stable population in Chesapeake Bay. Both studies agreed with the findings of Grier (1980), who demonstrated that survival was more important than reproduction in bald eagle population dynamics. Short-term drops in the reproductive rate may be normal for populations that are near saturation and probably have little effect on the population as a whole. Thus, we believe that the variation in nest success and productivity in Table 5 is normal for the region.

Western Prince William Sound

In 1989, reproductive surveys focused on oiled areas in western Prince William Sound. Productivity (young per occupied or active nest) and nest success (Table 4) were lower than in any historical studies (Table 5) and comparable to those of populations affected by organochloride pesticides (Sprunt et al. 1973; Stalmaster 1987). The area surveyed also received the heaviest oiling of any region (Table 1). The poor reproductive success can be directly attributed to the spill. Piatt et al. (1990) found eagles scavenging on oiled carcasses in 1989. Bowman et al. (1993) found a higher failure rate and decreased production at nests exposed to oil compared to non-oiled nests in Prince William Sound in 1989. Nests with no oil, any oil, and heavy oil failed at rates of 65% (71 of 110), 76% (60 of 79), and 88% (36 of 41), respectively (Bowman et al. 1993). The high failure rates for nests on non-oiled beaches indicates either brief exposure may have been enough to cause nest failure or the oiling data were inaccurate (see oiling results above). Brief exposure to oil did cause significant embryo mortality in other avian species (Hartung 1965; Albers 1980), especially if exposure occurred early in incubation (Szaro 1977; Albers 1978).

We believe the majority of documented nest failures were the result of embryo mortality, and the toxic effects of oil were persistent throughout the egg-laying period. We compared nests first seen occupied in April ($N = 127$) to those first seen occupied in late May ($N = 64$). Nesting chronology data collected in Prince William Sound indicates egg laying starts in mid-April and continues through the end of May, with roughly 50% of the nests initiated after 1 May (USFWS, Juneau, Alaska, unpublished data). Productivity (0.34 versus 0.37 young per occupied nest) and nest success (27 versus 32%) was only slightly lower for the April subset. If the toxic effects of oil were short lived, we believe nests initiated later in the season would have had a higher success rate.

Human activity dramatically increased in western Prince William Sound in 1989 because of cleanup activity. Although disturbance early in the nesting season may cause nest failures (Steenhof and Kochert 1982), we do not believe it was a major factor in the poor nest success in 1989. We looked at the timing of nest failures by comparing a sample of nests that were documented as active between 18 April and 5 May (Table 2) and had failed by 1 June. In western Prince William Sound in 1989, 37% (10 of 27) of failures had occurred by 1 June. In eastern Prince William Sound in 1990 (where there was no cleanup activity), 33% (8 of 24) of nests initiated by 5 May had failed by 1 June. The mean difference between the April and May surveys was 30 (range, 15–43) d. Therefore, we believe the majority of birds continued to incubate dead embryos for the entire 34 to 35 d term. Albers (1980) documented that mallards *Anas platyrhynchos* also continue incubation for the full period, despite oil-related embryo mortality. In 1989 and 1990, we tried to collect eggs from failed nests and found that scavengers quickly consume eggs from abandoned nests. If failures in 1989 were attributable to disturbance, we believe clutches would have been destroyed and the percent failures observed in May would have been much higher.

Once the eggs hatched, survival was apparently high. Of 65 young seen in June, 57 (88%) survived to fledging. All 15 chicks (approximately 8 weeks old) that were radio-tagged in western Prince William Sound in 1989 survived to fledging (USFWS, Juneau, Alaska, unpublished data). Postfledging survival of radio-tagged chicks was also considered high (Bowman et al. 1995). Hatching weight (Szaro et al. 1980) and short-term survival (Albers 1980) were normal for chicks that survived oiling during incubation. Pattee and Franson (1982) and Flemming et al. (1982) found normal survival for young of other avian species that had ingested oil.

Oil was still present in 1990 but was mostly subsurface. We did not observe any significant sheens during nesting surveys and do not believe there was any impact on eagle nesting because productivity and nest success in western Prince William Sound (Table 4) were above average in 1990 compared with those of historical studies (Table 5).

Eastern Prince William Sound

We cannot attribute the relatively low productivity observed in eastern Prince William Sound in

1989 to the spill. Prevailing winds and currents pushed the oil and probably most oiled carcasses away from this area. Territorial eagles that were radio-tagged in Prince William Sound from 1989 to 1990 rarely moved more than 2 km during the nesting season, and no eagles ($N = 30$) from eastern Prince William Sound were relocated in the spill area during the breeding season (USFWS, Juneau, Alaska, unpublished data). Thus, eagles from eastern Prince William Sound probably never contacted oil. The low productivity was either attributable to the small sample size or to natural fluctuations in productivity.

In 1990, productivity was approximately 30% lower than in western Prince William Sound (Table 4) and was lower than average for other studies in the region (Table 5). Because productivity was so high in nearby oiled western Prince William Sound, the only probable explanation is natural variation.

Kenai Peninsula

In 1989, we could not calculate nest success or a productivity estimate because initial surveys did not commence until mid-June, and much of the original data were lost or erased from field maps; however, it appears that productivity was normal if we compare results to the 1990 data. The 1990 productivity (0.92 young per occupied nest) and nest success (58%) were near the average for all areas and years shown in Table 4. Only 6 fewer young (52 versus 58) were found in 1989. Assuming there were 63 (1990 survey) occupied nests in 1989, 0.83 young were produced per occupied nest in 1989; however, initial surveys were conducted sporadically by boat and only one aerial survey was flown. We believe some successful nests were overlooked and total young produced was underestimated because of the lack of an aerial occupancy survey and the inexperience of observers on the productivity survey. If we compare productivity based only on nests observed in both years (the data shown in parentheses), productivity was 51% (1.24 versus 0.82 young per occupied nest) higher in 1989 (Table 4).

Kodiak Archipelago

There was no obvious reduction in bald eagle productivity in Kodiak, despite sightings of oiled eagles (Zwiefelhofer, unpublished data). Zwiefelhofer (unpublished data) compared success of nests within oiled plots to historical data and found no significant difference. Estimates of reproductive success in the Kodiak area in 1989 and 1990 were higher than those for any other subregion and were above the historical average for the Kodiak National Wildlife Refuge (Table 5).

Katmai National Monument

In 1989, 47 occupied nests were seen between June and August and at least 54 young fledged (Table 4). This is a minimum estimate because field notes indicate the surveyors had a difficult time seeing nests from the water. We do not know the exact dates or locations of many observations because the original data were lost; however, if we assume there were 50 (1990 data) occupied nests, more than 1.00 young were produced per occupied nest in 1989. We believe total young produced were underestimated for the same reasons outline for Kenai.

The relatively poor nest success and productivity (Table 4) in 1990 is difficult to explain. In surrounding regions (Kodiak, Kenai, and APBNWR) and heavily oiled areas in western Prince William Sound reproductive estimates were above average for all survey areas and years. The most likely explanation for these findings is natural fluctuation.

Alaska Peninsula–Becharof National Wildlife Refuge

Productivity and nest success appeared to be normal in both 1989 and 1990 (Table 4) even though one oiled egg was collected in the area surveyed (Dewhurst, unpublished data). Six abandoned or failed nests were examined, and no sign of oiling was discovered; two of the nests had been destroyed by brown bears *Ursus arctos*. Nest success and total young produced were considered minimum estimates. Older young occasionally hide in grass surrounding the nest, giving observers the false impression that the nest had failed. The increase in occupied nests found in 1990 may have been caused by increased surveyor experience and consistency (Dewhurst, unpublished data).

Productivity Compared to Oiling

The apparent lack of reproductive failure outside of western Prince William Sound is difficult to interpret. The total shoreline oiled in Kodiak and along the Alaska Peninsula was similar to that in Prince William Sound. Although roughly five times (15 versus 3%) more shoreline in western Prince William Sound was moderately to heavily oiled in 1989, even nests along lightly oiled and non-oiled beaches had higher than expected failures (Bowman et al. 1993). Laboratory studies (Albers 1978; Albers and Szaro 1978; Coon et al. 1979; Szaro

1979; Szaro et al. 1980) have found that relatively small amounts of oil (5–20 μL) significantly increased embryo mortality. Hehnke (1973) found that eagles switch from a mostly fish to a bird diet on the Alaska Peninsula during a natural die-off of murres. Given the massive die-off (100,000–300,000) of seabirds and drift of carcasses (150 km) (Piatt et al. 1990), it seems likely that eagles would have contacted oil throughout the spill area. The majority (88%) of oiled carcasses were retrieved outside Prince William Sound.

One possible explanation for the apparently normal reproduction outside Prince William Sound is that the oil encountered by eagles was less toxic. Szaro et al. (1980) found Prudhoe Bay crude oil became increasingly less toxic after weathering 2–4 weeks in tap water. The review by Wolfe et al. (1994) suggests a fairly short time-frame for the degradation of *Exxon Valdez* oil after the spill. The oil that washed up in Prince William Sound had been in the water less than 1 week (Figure 1). The oil received some mixing with the tides, but oil and carcasses at the tide line were probably only exposed to water for a fraction of the day. Szaro et al. (1980) found that toxicity of crude oil is not greatly affected by exposure to sunlight and reported that oil at the tide line from a spill in Nova Scotia was largely unaltered after 5 years, in sharp contrast to the findings of Wolfe et al. (1994). Oil on Kenai shorelines had been at sea for 1–2 weeks (Figure 1). The Kenai is typified by steep rocky bluffs with few, small intertidal areas exposed to ocean swells. Oiled carrion was probably not as available to eagles, and shorelines received much heavier weathering than in Prince William Sound. Overall, the Kenai also received 50% less oiling (18 versus 36% total shoreline oiled) than other areas (Table 1). At the peak of nesting in early May (USFWS, Juneau, Alaska, unpublished data), the oil was lower in toxicity and more dispersed (Wolfe et al. 1994, 1996, this volume).

The oil turned to mousse before it reached Kodiak and the Alaska Peninsula 5–8 weeks after the spill (Figure 1). Szaro et al. (1980) found that, although less toxic, weathered oil still reduced hatching success if at least 20 μL were applied to eggs. Szaro et al. (1980) weathered the oil in small steel containers and observed reduced toxicity, apparently because of evaporation of volatile compounds and the dissolving of water-soluble compounds. In turning to mousse, the oil from Prince William Sound was not only diluted but was also exposed to more air and water per unit surface area. The added time and increased mixing greatly reduced the oil's toxicity (Wolfe et al. 1994).

Analysis of Survey Methodology

Surveys must be timed correctly to obtain an unbiased estimate of reproductive success (Postupalsky 1974; Steenhof and Kochert 1982; Fraser et al. 1983). In Prince William Sound and the Copper River Delta, egg laying starts in the second week of April and 90–95% of nests have been initiated by the third week in May (USFWS, Juneau, Alaska, unpublished data). Nests start failing in the third week of May, and approximately 35% of the failures have occurred by 1 June. The timing of nesting events in the other regions appears to be similar to that in Prince William Sound (Hehnke 1973; Zwiefelhofer, unpublished data), although some evidence collected by Dewhurst (unpublished data) suggests nesting may be slightly earlier on the Alaska Peninsula. In the sound, the majority of active nests were first observed in April. Occupancy surveys in Kodiak and the Alaska Peninsula did not conclude until after 1 June (Table 2). Therefore, nest success and productivity were probably overestimated in Kodiak and the Alaska Peninsula because of early failures being missed on the surveys. This bias may have been accentuated by the spill. Eagles nesting early in the season would have encountered oil on the open water, increasing the possibility of individual mortality and egg oiling. Failures of the early nesting eagles may not have been detected by surveys in late May or early June. By the end of May, the oil had passed through the area and few bird mortalities occurred after that time (Piatt et al. 1990). Thus, we suspect that active nests observed later in the 1989 breeding season had higher success.

Determining the size of the prespill breeding population is difficult because pairs without eggs are not easily recognized as failures (Whitfield et al. 1974; Fraser et al. 1983), and nests abandoned because of adult mortality were impossible to recognize. Failure to lay eggs accounted for 23% of all failures in Oregon (Anthony et al. 1994). In Alaska, where eagle populations may be near saturation, even higher rates of nonbreeding may occur (Hansen and Hodges 1985). The annual fluctuations in active nests seen in historical studies (Table 5; Hodges 1982) support the hypothesis that a significant number of Alaska eagles may not attempt to breed. We did not include an analysis based on occupied nests because few observers recorded these data and the abundance of eagles and nest sites makes the identification of individual occupied territories impossible. Ingestion of oil may inhibit egg laying (Hartung 1965), so a significant number of adults may never have attempted to lay eggs in 1989. Thus, we may have grossly underestimated reproductive failures attributable to the spill.

Kodiak serves as an example of the potential biases. The number of occupied nests increased 24% (333 to 412) from 1989 to 1990. Some increase was probably caused by increased surveyor experience, but there were obviously more active nests in 1990 (Zwiefelhofer, unpublished data). The breeding population on Kodiak is growing (Zwiefelhofer, unpublished data), but a 24% annual increase is not probable. We can also assume that there were more than 412 pairs in 1990 because of the difficulty in recognizing nonbreeding pairs (Whitfield et al. 1974; Fraser et al. 1983; Bowman et al. 1993). Thus, the actual productivity in 1989 was probably much lower than we report because of early failures and nonbreeding pairs being missed on the 1989 surveys. It is impossible to estimate the magnitude of the biases because the actual breeding population was not known in any region.

Population Implications

Bowman et al. (1995), Buehler et al. (1991), and Grier (1980) demonstrated that survival is more important than reproduction in population dynamics. One year of poor reproduction has little impact on the population as a whole. Bowman et al. (1993) predicted that the population in Prince William Sound would return to its prespill size in 4 years, despite oil-related mortality and poor reproductive success in 1989. Large-scale reproductive failure was not evident in any other region. Zwiefelhofer (unpublished data) reported that the nesting population in the Kodiak National Wildlife Refuge (approximately 50% was within the spill area) increased 47% from 1987 to 1992. Although we do not know survival rates of eagles outside of Prince William Sound, it is unlikely that the spill affected the populations as a whole, based on information from Prince William Sound and Kodiak National Wildlife Refuge.

Conclusions

Western Prince William Sound was the only region within the spill zone where a large-scale reproductive failure was evident. The poor reproduction was attributed to the oil spill. Reproduction rebounded in western Prince William Sound in 1990. All other regions appeared to have normal reproduction in both 1989 and 1990; however, because of the lack of prespill data, the wide range in nest success and productivity data from historical studies, and survey biases, only large-scale failures would have been evident. It is unlikely that there was any long-term effect on eagle populations in the region as a whole.

Acknowledgments

This study required the cooperation of many individuals from many agencies and private organizations. We thank the following people for participating in bald eagle surveys, or providing data or other essential information pertinent to specific regions in the oil spill area, and for providing assistance in many other ways: C. M. Adkins, M. Amaral, M. Bill, F. Bird, D. A. Boyce, B. A. Boyle, D. Dewhurst, K. Faber, J. B. Gray, J. I. Hodges, A. Hoover-Miller, J. Hughes, M. J. Jacobson, T. W. Jennings, R. King, K. Kozie, R. Mesta, M. Portner, T. V. Schumacher, M. Tetreau, S. L. Wilbor, D. Williamson, G. Wheeler, H. White, R. E. Yates, M. Yurick, and D. Zwiefelhofer. We thank J. D. Fraser and K. L. Oakley for helpful reviews of drafts of this article.

References

Albers, P. H. 1978. The effects of petroleum on different stages of incubation in bird eggs. Bulletin of Environmental Contamination and Toxicology 19:624–30.

Albers, P. H. 1980. Transfer of crude oil from contaminated water to birds eggs. Environmental Research 22:307–314.

Albers, P. H., and R. C. Szaro. 1978. Effects of no. 2 fuel oil on common eider eggs. Marine Pollution Bulletin 9:138–139.

Anthony, R. G., R. W. Frenzel, F. B. Isaacs, and M. G. Garrett. 1994. Probable causes of nesting failures in Oregon's bald eagle population. Wildlife Society Bulletin 22:576–582.

Bangs, E. E., T. N. Bailey, and V. D. Berns. 1982. Ecology of nesting bald eagles on the Kenai National Wildlife Refuge, Alaska. Pages 47–54 *in* W. N. Ladd and P. F. Schempf, editors. Raptor management and biology in Alaska and western Canada. U.S. Fish and Wildlife Service (FWS-AK-PROC-82), Anchorage, Alaska.

Bowman, T. D., P. F. Schempf, and J. A. Bernatowicz. 1993. Effects of the *Exxon Valdez* oil spill on bald eagles. Final report, Natural Resources Damage Assessment, Bird Study 4, U.S. Fish and Wildlife Service, Juneau, Alaska.

Bowman, T. D., P. F. Schempf, and J. A. Bernatowicz. 1995. Bald eagle survival and population dynamics in Alaska after the *Exxon Valdez* oil spill. Journal of Wildlife Management 59:317–324.

Buehler, D. A., J. D. Fraser, J. K. D. Seegar, G. D. Terres, and M. A. Byrd. 1991. Survival rates and population

dynamics of bald eagles on Chesapeake Bay. Journal of Wildlife Management 55:608–613.
Coon, N. C., P. H. Albers, and R. C. Szaro. 1979. No. 2 fuel oil decreases embryonic survival of great black-backed gulls. Bulletin of Environmental Contamination and Toxicology 21:152–156.
Flemming, W. J., L. Sileo, and J. C. Franson. 1982. Toxicity of Prudhoe Bay crude oil to sandhill cranes. Journal of Wildlife Management 46:474–478.
Fraser, J. D., L. D. Frenzel, J. E. Mathisen, F. Martin, and M. E. Shough. 1983. Scheduling bald eagle reproductive surveys. Wildlife Society Bulletin 11:13–16.
Garrett, M. G., J. W. Watson, and R. G. Anthony. 1993. Bald eagle home range and habitat use in the Columbia River estuary. Journal of Wildlife Management 57:19–26.
Grier, J. W. 1980. Modeling approaches to bald eagle population dynamics. Wildlife Society Bulletin 8:316–322.
Hansen, A. J., and J. I. Hodges. 1985. High rates of nonbreeding adult bald eagles in southeastern Alaska. Journal of Wildlife Management 49:454–458.
Hartung, R. 1965. Some effects of oiling on the reproduction of ducks. Journal of Wildlife Management 29: 872–874.
Hehnke, M. F. 1973. Nesting ecology and feeding behavior of bald eagles on the Alaska Peninsula. Master's thesis. Humboldt State University, Arcata, California.
Hensel, R. J., and W. A. Troyer. 1964. Nesting studies of the bald eagle in Alaska. Condor 66:282–286.
Hodges, J. I., Jr. 1982. Bald eagle nesting studies in Seymour Canal, southeast Alaska. Condor 84:125–127.
Hodges, J. I., Jr., J. G. King, and R. Davies. 1984. Bald eagle breeding population survey of coastal British Columbia. Journal of Wildlife Management 48:993–98.
Hodges, J. I., Jr., and F. C. Robards. 1982. Observations of 3,850 bald eagle nests in southeast Alaska. Pages 37–54 *in* W. N. Ladd and P. F. Schempf, editors. Raptor management and biology in Alaska and western Canada. U.S. Fish and Wildlife Service (FWS-AK-PROC-82), Anchorage, Alaska.
Imler, R. H., and E. R. Kalmbach. 1955. The bald eagle and its economic status. U.S. Fish and Wildlife Circular 30.
King, K. A., and C. A. Lefever. 1979. Effects of oil transferred from incubating gulls to their eggs. Marine Pollution Bulletin 10:319–321.
Leighton, F. A. 1986. Clinical, gross and histological findings in herring gulls and atlantic puffins that ingested Prudhoe Bay crude oil. Veterinary Pathology 23:254–263.
Pattee, O. H., and J. C. Franson. 1982. Short-term effects of oil ingestion on American kestrels (*Falco sparverius*). Journal of Wildlife Diseases 18:235–241.
Piatt, J. F., C. J. Lensink, W. Butler, M. Kendziorek, and D. R. Nysewander. 1990. Immediate impact of the *Exxon Valdez* oil spill on marine birds. Auk 107:387–397.
Postupalsky, S. 1974. Raptor reproductive success: some problems with methods, criteria, and terminology. Pages 21–31 *in* F. N. Hamerstrom, Jr., B. E. Harrell, and R. R. Olendorff, editors. Management of raptors, proceedings of the conference on raptor conservation techniques. Raptor Research Report 2, Raptor Research Foundation, Vermillion, South Dakota.
Ritchie, R. J. 1982. Investigations of bald eagles, Tanana River, Alaska, 1977–80. Pages 55–67 *in* W. N. Ladd and P. F. Schempf, editors. Raptor management and biology in Alaska and western Canada. U.S. Fish and Wildlife Service (FWS-AK-PROC-82), Anchorage, Alaska.
Sherrod, S. K., C. M. White, and F. S. L. Williamson. 1976. Biology of the bald eagle on Amchitka Island, Alaska. Living Bird 15:143–182.
Sprunt, A. I., IV, W. B. Robertson, Jr., S. Postupalsky, R. J. Hensel, C. E. Knoder, and F. J. Ligas. 1973. Comparative productivity of six bald eagle populations. Transactions of the North American Wildlife and Natural Resources Conference 38:96–105.
Stalmaster, M. V. 1987. The bald eagle. Universe Books, New York.
Steenhof, K. 1987. Assessing raptor reproductive success and productivity. Pages 157–170 *in* B. A. Giron Pendleton, B. A. Millsap, K. W. Cline, and D. M. Bird, editors. Raptor management techniques manual. National Wildlife Federation, Washington, DC.
Steenhof, K., and M. N. Kochert. 1982. An evaluation of methods used to estimate raptor nesting success. Journal of Wildlife Management 46:885–893.
Szaro, R. C. 1977. Effects of petroleum on birds. Transactions of the North American Wildlife and Natural Resources Conference 42:374–381.
Szaro, R. C. 1979. Bunker C fuel oil reduces mallard egg hatchability. Bulletin of Environmental Contamination and Toxicology 22:731–732.
Szaro, R. C., P. H. Albers, and N. C. Coon. 1978. Petroleum: effects on mallard egg hatchability. Journal of Wildlife Management 42:404–406.
Szaro, R. C., N. C. Coon, and W. Stout. 1980. Weathered petroleum: effects on mallard hatchability. Journal of Wildlife Management 44:709–713.
Todd, C. S., L. S. Young, R. B. Owen, Jr., and F. J. Gramlich. 1982. Food habits of bald eagles in Maine. Journal of Wildlife Management 46:636–645.
U.S. Coast Guard. 1993. T/V *Exxon Valdez* oil spill federal on scene coordinator's report. Vol. 1. U.S. Coast Guard, Department of Transportation, Washington, DC.
USFWS (U.S. Fish and Wildlife Service). 1986. Recovery plan for the Pacific bald eagle. U.S. Fish and Wildlife Service, Portland, Oregon.
Whitfield, D. W. A., J. M. Gerrard, W. J. Maher, and D. W. Davis. 1974. Bald eagle nesting habitat, density, and reproduction in central Saskatchewan and Manitoba. Canadian Field-Naturalist 88:399–407.
Wolfe, D. A., and eleven coauthors. 1994. The fate of the oil spilled from the *Exxon Valdez*. Environmental Science and Technology 28:561A–568A.
Wolfe, D. A., and six coauthors. 1996. Toxicity of intertidal and subtidal sediments contaminated by the *Exxon Valdez* oil spill. American Fisheries Society Symposium 18:121–139.

American Fisheries Society Symposium 18:798–808, 1996

Overview of Studies to Determine Injury Caused by the *Exxon Valdez* Oil Spill to Marine Mammals

THOMAS R. LOUGHLIN

National Marine Fisheries Service, National Marine Mammal Laboratory 7600 Sand Point Way, NE, Seattle, Washington, 98115, USA

BRENDA E. BALLACHEY

National Biological Service, Alaska Fish and Wildlife Research Center 1011 East Tudor Road, Anchorage, Alaska, 99503, USA

BRUCE A. WRIGHT

National Marine Fisheries Service, Office of Oil Spill Damage Assessment and Restoration 11305 Glacier Highway, Juneau, Alaska, 99801–8626, USA

Abstract.—Marine mammal damage assessment studies after the *Exxon Valdez* oil spill concentrated on sea otters *Enhydra lutris*, harbor seals *Phoca vitulina*, Steller sea lions *Eumetopias jubatus*, killer whales *Orcinus orca*, and humpback whales *Megaptera novaeangliae*. Sea otters and harbor seals were the most affected marine mammal; it was estimated that several thousand otters and several hundred harbor seals died within months of the spill. Steller sea lion, harbor seal, and sea otter numbers were monitored using aerial surveys. Studies of humpback whales and killer whales used photoidentification techniques to determine changes in abundance, distribution, mortality, and natality. Tissues from animals found dead in spill and control areas were analyzed for hydrocarbon levels. Sea otters, sea lions and harbor seals had elevated hydrocarbon levels, but only sea otters and harbor seals showed population declines associated with the spill. Humpback whales were not severely affected by the *Exxon Valdez* oil spill. Killer whale numbers in the resident AB pod declined after the spill. Coincidental evidence supports the oil spill as the causative agent.

More than 25 species of marine mammals are found in Prince William Sound, the Gulf of Alaska, and adjacent waters (Calkins 1986). The most abundant and conspicuous resident species are sea otters *Enhydra lutris*, Steller (northern) sea lions *Eumetopias jubatus*, harbor seals *Phoca vitulina*, Dall's porpoises *Phocoenoides dalli*, and killer whales *Orcinus orca*. A large proportion of these marine mammals occurs in areas affected by the *Exxon Valdez* oil spill, and the entire population of eastern Pacific grey whales *Eschrichtius robustus* migrates through the northern Gulf of Alaska. Killer whales, harbor porpoises *Phocoena phocoena*, sea otters, harbor seals, and Steller sea lions breed within or adjacent to Prince William Sound. Humpback whales *Megaptera novaeangliae*, an endangered species, use the sound as a summer feeding area, and several other cetacean species use the sound and adjacent waters intermittently. Several important pinniped rookeries occur within the sound and in adjacent waters. Both harbor seals and Steller sea lions are declining in numbers in the sound and in much of Alaska. Clearly, Prince William Sound and adjacent areas affected by the *Exxon Valdez* oil spill provide a variety of marine habitats seasonally critical for significant numbers of marine mammals that are of national and international importance.

This report summarizes results of studies conducted during 1989–1992 to determine spill-related injury to marine mammals. We summarize Natural Resource Damage Assessment (NRDA) study reports, published information, and chapters from Loughlin (1994a).

We summarize each study separately by species with the respective methods and results for each. The *Methods* section that follows here reports only those methods common to many studies. We report on sea otters, harbor seals, Steller sea lions, killer whales, humpback whales, and cetaceans generally.

Methods

Chemistry

Tissue samples from marine mammals found dead or collected were analyzed for levels of hydrocarbon contaminants to characterize the magnitude and extent of exposure to oil and to facilitate determination of short- and long-term effects resulting from exposure. Multiple tissue samples were collected according to a prescribed standard protocol (Loughlin and Sinclair 1994) and then maintained in frozen storage until analyzed. Measurements from dead animals were obtained according to accepted necropsy procedures (Committee on Marine Mammals 1961; Fay et al. 1979; Aguilar 1985).

Bile samples were analyzed for the fluorescent metabolites of aromatic hydrocarbons by the Environmental Conservation Division (ECD), Northwest Fisheries Center, National Oceanic and Atmospheric Administration, National Marine Fisheries Service, Seattle, Washington. Tissue samples were analyzed for a broad spectrum of aliphatic and aromatic hydrocarbons by Texas A&M University, College Station, and by ECD using capillary column gas chromatography with detection by flame ionization and mass spectrometry. Conclusions as to the presence or absence of petroleum in the sample were based on a preponderance of evidence, such as the presence of the aromatic compounds naphthalene, dibenzothiophene, phenanthrene, fluorenes, and chrysenes as both unsubstituted compounds and the alkylated homologues; the presence of an unresolved complex mixture; and a pristane: phytane ratio < 5 (Boehm et al. 1987; Wade et al. 1988; Varanasi et al. 1990; Wade et al. 1993; Krahn and associates, U.S. National Marine Fisheries Service, unpublished data). Because marine mammals extensively metabolize aromatic compounds in their livers and the metabolites are excreted, the metabolites were measured to establish exposure to Prudhoe Bay crude oil (Krahn and associates, unpublished data). Liver and blubber samples were usually analyzed as the most likely tissues to indicate the presence of petroleum hydrocarbons. In the case of harbor seals, brain and muscle were also analyzed.

Histopathology

Histopathology was studied in sea otters, harbor seals, and Steller sea lions following the methods in Lipscomb et al. (1994) and Spraker et al. (1994). Tissue samples were obtained from dead and collected animals as soon after death as possible; the postmortem interval ranged from 10 to 30 min for collected animals and an unknown amount of time for those found dead. Blood samples were taken from the extradural intervertebral vein or the jugular vein (sea otters) and serum, plasma, and whole blood were frozen for later analysis. Multiple tissues from organ systems (nervous, cardiovascular, respiratory, lymphohematopoietic, urogenital, integument, digestive, endocrine, musculoskeletal, and special senses) were taken for histopathology. Tissues were preserved in 10% neutral buffered formalin. Specimen slides for histological analysis are on file at the Armed Forces Institute of Pathology, Washington, DC.

Sea Otter

Sea otters occupy much of the coastal waters of south-central Alaska, from the Aleutian Islands to the Alaska panhandle (Kenyon 1969). The number of sea otters in Alaska is estimated at over 100,000 (Bodkin et al. 1994), with perhaps one-third of them inhabiting areas in the path of the spilled oil. They were the most abundant marine mammal affected by the spill and were particularly vulnerable because they rely on their fur for insulation, and oiling drastically reduces the insulative properties of their fur (Costa and Kooyman 1982). Nearly 1,000 sea otter carcasses were recovered within a few months of the spill. Efforts to capture and treat live, oiled sea otters began in late March 1989, within days of the spill (Bayha and Kormendy 1990). Soon thereafter, the U.S. Fish and Wildlife Service (USFWS) embarked on NRDA studies to assess the magnitude, extent, and duration of oil spill impacts on the otters (study 1) and to assess the fate of sea otters that had been oiled and then rehabilitated (study 2). Study 1 was separated into three major areas: (1) surveys of sea otter distribution and abundance throughout the spill area, (2) radiotelemetry studies to determine sea otter survival and reproduction rates and movements, and (3) studies of the pathological and toxicological effects of oil exposure. Studies relevant to item 2 and portions of item 3 were accomplished through a cooperative agreement with the Prince William Sound Science Center (PWSSC) and results of those studies will be presented by center scientists elsewhere. Injury to the population fell into two temporal components: acute injury, defined as deaths associated with oil exposure and occurring within 6 months of the spill; and chronic or long-term injury. The acute injury was partially reflected in carcass recovery. The information summarized here is principally from Ballachey et al. (1994), Burn (1994), Bodkin and Udevitz (1994), Doroff and Bodkin (1994), Lipscomb et al. (1994), and Mulcahy and Ballachey (1994).

Methods

Acute injury.—Two surveys were conducted to assess sea otter distribution and abundance. Helicopter surveys were conducted in April (before arrival of oil) and fall (after arrival of oil) 1989 to assess pre- and postspill abundance estimates. These surveys covered areas along the coasts of the

Kenai Peninsula, Kodiak Island, and Alaska Peninsula. Boat-based surveys of a portion of the coastline of Prince William Sound were conducted in fall 1989 (Burn 1994) and repeated in spring and summer 1990, 1991, and 1993 to estimate postspill abundance and distribution (Ballachey et al. 1994).

Twenty-five sea otter carcasses were marked and released in the water near Kodiak Island during May and June 1989 to assess the probability of recovering dead sea otters by survey groups or cleanup crews.

Chronic injury.—Sea otters in their first year of life, adult female otters, and otters released from rehabilitation centers were monitored with radio transmitters to assess survival and reproductive rates (adult females). Survival of juveniles was monitored during 1990–1993, and adult females and released otters were monitored from late 1989 to summer 1991. Results from these studies may be presented elsewhere by PWSSC scientists.

Prey selection and foraging success were studied at oiled and non-oiled sites during summer 1991 with concomitant analysis of bivalve prey for hydrocarbon concentrations.

Blood was sampled from male sea otters in Prince William Sound in 1990 and from females in 1991–1992 and submitted for hematologic and serum chemistry analysis. Analyses of blood from females and pups during 1989–1990 may be presented elsewhere by PWSSC scientists.

The methods used for chemical analysis and histopathology are described above.

Results

Acute injury.—The helicopter surveys did not detect statistically significant decreases in abundance, although decreases were noted for the Kenai Peninsula, Kodiak Island, and Alaska Peninsula (DeGange et al. 1994). The boat-based surveys indicated a 35% decline in oiled areas of Prince William Sound and a 13% increase in non-oiled areas relative to prespill data from 1984 to 1985 (Burn 1994). Using these data, Garrott et al. (1993) derived an estimate of 2,800 sea otters dying after the spill. However, Garshelis (unpublished) analyzed the same data set and concluded that the data do not support a significant decline in abundance.

Only 5 of the 25 relocated carcasses were found and returned to the collection centers, for a recovery rate of 20% (DeGange et al. 1994). The sample size was very small, limiting the validity of extrapolating from this study to sea otter carcasses throughout the spill zone.

Chronic injury.—Sea otter carcasses recovered in 1989–1991 included much higher proportions (1989, 48%; 1990–1991, 43%) of 2 to 8-year-old otters (prime age) than were collected before the spill (17%), suggesting unusual mortality patterns during and after the spill (Monson 1994).

Prey selection and foraging success were similar among oiled and non-oiled areas, and hydrocarbon levels in subtidal prey (clams) were generally low and similar among areas (Doroff and Bodkin 1994).

Differences were noted in blood chemistry between otters from eastern and western Prince William Sound, but the significance of these differences was undetermined (Rebar et al. 1994). Elevated serum transaminases occurred in blood from otters in western Prince William Sound in 1992 (USFWS, unpublished data); such elevations are consistent with liver damage, but corroborating histological evidence was not available.

Chemical analysis demonstrated that heavily oiled otters had elevated levels of both aliphatic and aromatic hydrocarbons in muscle, kidney, and liver compared with otters from southeastern Alaska (Mulcahy and Ballachey 1994). An interesting finding was that otters exhibiting similar degrees of external oiling can vary markedly in the hydrocarbon levels in internal tissues. Interstitial pulmonary emphysema was the most common lesion found during histological analysis. Gastric erosions were common in oiled otters dying at the rehabilitation centers. Damage to the liver and kidney (hepatic and renal lipidosis and centrilobular hepatic necrosis) was also observed in oiled otters found dead after the spill (Lipscomb et al. 1993, 1994). It seems reasonable to assume that similar pathological changes occurred in some otters that were exposed to sublethal levels of oil, compromising their survival and causing longer-term injury to the otter population.

Harbor Seal

In the mid-1970s harbor seals were abundant in Prince William Sound and the Gulf of Alaska; approximately 4,000 seals were counted in the sound in June 1973 and a minimum population of 13,000 was later estimated from harvest records based on bounty payments (see Frost et al. 1994a for sources). Trend-count surveys in 1984 and 1988–1992 indicated a 41% decline in mean counts from 1984 to 1988. By 1992, molt-period counts were 57% lower than in 1984 (Frost et al. 1994a). Because harbor seals are year-round residents of the sound, seals of all ages and both sexes were present and contacted spilled oil in the water and onshore. Pups were born in May and June, when some of the sites still had oil on them, and many pups became oiled and nursed on oiled mothers.

The NRDA studies were designed to quantify the effects of exposure to oil on the distribution, abundance, and health of harbor seals in Prince William Sound. Specific approaches included population assessment, chemical analysis of tissues, and histological examination of tissues to determine the nature and magnitude of tissue damage and to relate chemical findings to observed histopathology. The information summarized here is from Lowry et al. (1994), Frost and Lowry (1993), Frost et al. (1994a, 1994b), and Spraker et al. (1994).

Methods

Frost et al. (1994a) made repetitive aerial surveys during the annual molt (August–September 1989–1992) and pupping period (May–June 1989–1992) at 25 selected haulout sites established by the Alaska Department of Fish and Game in 1983–1984 (D. G. Calkins and K. W. Pitcher, Alaska Department of Fish and Game, unpublished data). The route included 7 oiled sites and 18 non-oiled sites. Fixed-wing aircraft (Cessna 180 or 185) were flown at an altitude of 150–300 m. Visual counts were made at each site and larger groups were photographed with a 35-mm camera with a 70–210 mm lens using ASA 400 film. Harbor seals were counted from the projected image. Mean counts from the repetitive surveys were analyzed using bootstrapping methods (to account for variability) and then analyzed using log-linear and LOGIT categorical models.

The methods used for chemical analysis and histopathology are described above.

Results

Population assessment.—Statistical analysis comparing counts obtained during the molt in 1983–1988 (prespill) to 1989 (just after the spill) showed a significant decline in seal numbers at oiled sites versus non-oiled sites in Prince William Sound. Frost et al. (1994a) estimated that a minimum of 302 harbor seals were missing in the sound after the oil spill.

Chemistry.—Frost et al. (1994b) conclude that tissues from seals found dead or collected in heavily oiled areas of Prince William Sound in 1989 contained elevated concentrations of petroleum-related hydrocarbons (i.e., phytane in brain samples and polycyclic aromatic hydrocarbons [PAHs] in blubber samples) as compared with tissues from seals collected in moderately oiled or non-oiled areas. They found that concentrations of three-ring fluorescent aromatic compounds (e.g., phenathrene [PHN]) measured in bile of seals collected in 1989 were significantly higher than PHN concentrations in the bile of seals collected in non-oiled areas (Gulf of Alaska and southeastern Alaska). One year after the spill, none of the tissues from seals collected in Prince William Sound showed elevated concentrations of petroleum-related hydrocarbons as related to seals from southeastern Alaska. However, the average PHN concentration in bile of harbor seals from the sound was still higher than that in animals in the Gulf of Alaska in 1989 or in control animals in 1990. Frost et al. (1994b) conclude that these chemical data support the hypothesis that Prince William Sound harbor seals were exposed to high concentrations of petroleum-related hydrocarbons in the spring and summer of 1989.

Histopathology.—Spraker et al. (1994) examined 19 seals that were found dead in Prince William Sound and the Gulf of Alaska or that had died in oil spill rehabilitation centers. Fifteen of the 19 were oiled, but because of carcass condition and the lengthy time between death and sampling, cause of death was equivocal. Thirteen of the 19 were pups, and stress or toxic effects of oil may have contributed to their deaths. Two older seals probably died from blunt trauma. Spraker et al. (1994) collected 27 additional seals in 1989, both oiled and non-oiled, and took tissue samples within minutes after death. Gross examination and detailed histopathology of these 27 seals resulted in documentation of several types of lesions that were probably caused by crude oil. Skin irritation, conjunctivitis, and liver lesions occurred more frequently in oiled seals. This damage was relatively mild, and probably reversible, in most cases. They found four types of lesions characteristic of hydrocarbon toxicity in the brains of oiled seals. These lesions occurred principally in the thalamus and probably account for the disorientation and lethargy observed in seals immediately after the spill (Lowry et al. 1994). The lesions were most acute in a seal collected 35 d after the spill and were milder in seals collected later.

Steller Sea Lion

The breeding range of the Steller sea lion extends around the north Pacific Ocean rim from the Kuril Islands and Okhotsk Sea, through the Aleutian Islands, along Alaska's southern coast, and south to central California (Loughlin et al. 1984). Results of a rangewide survey in 1989 resulted in an estimate of 116,000 Steller sea lions, down from 240,000–300,000 estimated 30 years earlier (Loughlin et al. 1992). In Alaska, declines have been reported for most of the state, except southeastern Alaska, including the areas in and adjacent to Prince William Sound (Merrick et al. 1992). The impact of the *Exxon Valdez* oil spill on sea lions in and adjacent to the sound was investigated by Calkins et al. (1994) by counting adult and juvenile sea lions at haulout sites in 1989, counting pups at rookeries in 1989 and 1990, and monitoring the rate of premature pupping at sites near and far from the spill area. The rationale for these studies was that exposure of pregnant females to spilled oil may affect development of the fetus and cause abortion and subsequent reduction in the number of counted pups. Chemical and histopathological analyses were conducted on tissues from animals found dead or that were collected for these studies.

Methods

Calkins et al. (1994) flew aerial surveys from 15 June to 15 July 1989 during the breeding season at rookeries and haulout sites from Cape St. Elias to the Semidi Islands in the Gulf of Alaska. Surveys were flown between 1000 and 1600 hours local time at approximately 90 knots and 150 to 300 m above the site. Replicate photographs were taken with a hand-held 35-mm camera and 70–210 lens using ASA 400 film. Sea lions were counted from the projected image in the laboratory. Pup counts were obtained in 1989 and 1990 at Seal Rocks (entrance to Prince William Sound) south to Chowiet Island (Marmot Island was excluded in 1990) by spook counts on land (Merrick et al. 1987). Premature pupping was measured by counting aborted fetuses and collection of tissue samples at Cape St. Elias (near the spill) and Chirikof Island (far from the spill) for a 4-week period between 5 April and 16 May 1989. The population data were analyzed using regression models.

The methods used for chemical analysis and histopathology are described above.

Results

Regression models constructed from historical data were used to predict numbers at rookeries and haulout sites. The predicted values were compared to actual counts and were not significantly different for adult and juvenile counts from 1989 and pup counts from 1989 and 1990, except at Sugarloaf Island at the tip of the Kenai Peninsula. Premature pup rates were significantly higher at Cape St. Elias than at Chirikof Island. None of these analyses, however, conclusively showed an impact from the *Exxon Valdez* oil spill and any affect of the spill was masked by the large Alaska-wide Steller sea lion decline.

Results from the chemical analysis were not consistently high enough to confirm contamination. No lesions related to hydrocarbon contamination were found during histological examinations. Chemical analysis of sea lions showed that some animals were exposed to oil and that they were metabolizing and excreting metabolites of aromatic hydrocarbons into the bile. However, no evidence indicated damage caused to sea lions from toxic effects of oil.

Killer Whale

In 1987, 221 individual killer whales were identified in Prince William Sound based on the shape and markings of the dorsal fin and saddle (Leatherwood et al. 1990). These whales occur in groups termed pods of which there are two types, resident and transient. Resident pods are composed of maternal groups that include a female and her offspring of both sexes. These pods have long-term stability and it is believed that whales are added only by birth and lost only by death (Bigg 1982). Matkin et al. (1994) studied the impact of the *Exxon Valdez* oil spill on Prince William Sound killer whales using photoidentification techniques to determine whether whales were missing from established resident pods after the spill, whether pod composition had been altered, and whether pod distribution had changed as a result of the spill.

Methods

During May and June to September 1989–1991 camps were established in southwestern Prince William Sound in areas where killer whales were known to occur. Small boats were then used to closely approach whales and obtain photographs of the dorsal fin and saddle for subsequent individual identification. These photographs were then compared to historical data to determine pod membership. Matkin et al. (1994) assumed that a whale not identified in a pod and missing for 2 or more years was dead. They photographed whales with 35-mm cameras, each equipped with a 300-mm lens, and high-speed black and white ASA 1600 film exposed at 1/1,000 s. They then calculated natality and mortality rates using the number of whales missing from or born to a pod. A simple binomial probability distribution was used to assess the likelihood of observed mortality rates, and a logistic model was used to model mortality rates.

Results

In 1989, killer whales were observed by Matkin et al. (1994) in all marine waters of Prince William Sound; the whales did not avoid using the oiled areas of the southwestern part of the sound and were observed swimming in oiled water. In September 1988, just before the spill, the AB resident pod was photographed; it contained 36 whales. When next photographed on 31 March 1989 (7 d after the spill), the pod contained 29 whales (7 missing), and an additional 6 were missing between September 1989 and June 1990. Matkin et al. (1994) report one additional missing whale in 1990–1991, and none missing in 1991–1992. They also report one calf born in AB pod in 1991 and two in 1992 when the pod totaled 25 whales. The other six frequently photographed resident pods totaled 86 whales in 1984 and 105 in 1992. The 14 missing whales from AB pod included four mature females, one mature male, and nine juveniles of which at least two were female and one male (Matkin et al. 1994). Mortality rates calculated for AB pod were 19.4% for 1988–1989 and 20.7% for 1989–1990. Matkin et al. (1994) concluded that these rates were significantly higher than expected when compared with mortality rates for killer whales in Puget Sound and British Columbia (Olesiuk et al. 1990).

Dahlheim and Matkin (1994) speculate on the possible causes for the whales to be missing from AB pod after the spill. Because carcasses were not found, no conclusive evidence exists to confirm that the animals had died or that the cause of death was a result of exposure to *Exxon Valdez* oil. However, the loss of 14 animals over a short time is unprecedented for killer whales. Dahlheim and Matkin (1994) contend that errors made during the identification process, bias in survey coverage, or movement of whales out of the area did not occur. They conclude that the missing whales died from a combination of natural causes, a result of fisheries interactions, or the oil spill. Dahlheim and Matkin (1994) suggest that the fact that the spill occurred at the same time as the loss of whales is strong coincidental evidence supporting the oil spill as the causative agent.

Humpback Whale

About 1,400 humpback whales occur in the north Pacific Ocean in a world population of about 10,000 (Baker et al. 1987). They have been documented throughout the north Pacific during summer, particularly in California, southeastern Alaska (Baker et al. 1992), and Prince William Sound (von Ziegesar and Matkin 1985). Approximately 60–100 humpback whales feed in the sound between June and September each year, and many of the same individuals return from year to year (von Ziegesar 1992). The impact of the *Exxon Valdez* oil spill was thought to affect humpback whales by causing them to leave Prince William Sound and feed elsewhere in Alaska (most likely southeastern Alaska) or by causing reduced natality and perhaps increased mortality. Population studies on humpback whales in the sound had been in progress since 1977 providing a baseline time series from which to compare studies after the oil spill (von Ziegesar and Matkin 1985). The objectives of the NRDA study were to count and individually identify humpback whales in Prince William Sound and southeastern Alaska, to determine if whale distribution within the sound had changed, and to determine if vital rates had changed (von Ziegesar et al. 1994).

Methods

Von Ziegesar et al. (1994) used photoidentification techniques similar to those used for killer whales, except that they used pigmentation patterns on the ventral side of the fluke and the shape of the trailing edge of the fluke to identify individuals (Katona et al. 1979; Hall 1981). Small boats were used to locate humpback whales by random surveys of southwestern Prince William Sound and northern southeastern Alaska at locations that were historically used by humpback whales (Hall 1979, 1981). When a whale was found, the researchers slowly approached from behind to a distance of 25–50 m and waited until the animal dove, at which time the ventral side of the animal's tail was photographed using the same methods as were used with killer whales. Humpback whale group size in the sound is about two to three animals per group, and calves stay close to their mothers, therefore a calf was assumed to be the offspring of the whale it was with. Individual whales were identified according to an alphanumeric system developed by von Ziegesar and Matkin (1985) and matched to all the whales photographed in Prince William Sound since 1977 (von Ziegesar 1992). Documented sightings and patterns of distribution were analyzed and compared with the subset of identified whales that were observed in the sound during 1988–1990. Sightings were analyzed using analysis of variance.

Results

Von Ziegesar et al. (1994) found that the number of individually identified humpback whales in Prince William Sound increased from 35 in 1988 to 59 in 1989 and 66 in 1990. Increased effort during 1989 and 1990 resulting from the *Exxon Valdez* oil spill accounts for some of this increase. Seventeen individual whales were seen each year from 1988 to 1990, suggesting that some whales return to the sound on an annual basis. These "core" whales averaged 49.1 d in the sound during the 3 years. Sightings of humpback whales in Prince William Sound were primarily in Knight Island Passage, but whale distribution was probably determined by prey distribution. Exodus from the sound to southeastern Alaska by humpback whales did not occur. Humpback whales were not observed swimming in oil and no mortalities were recorded. There was no apparent change in abundance or movement of humpback whales out of the sound resulting from the spill.

Other Cetaceans

Along the north coast of the Alaska Peninsula, carcasses of grey whales *Eschrichtius robustus* are routinely seen each spring, presumably a result of natural mortality. By mid-May 1989, four carcasses were found at Tugidak Island, south of Kodiak Island. These findings caused increased concern about the possible impact of the spilled oil on grey whales and other cetaceans and prompted researchers to locate and count stranded cetaceans from Kayak Island through Bristol Bay. If possible, carcasses were necropsied and tissue samples collected for chemical analysis to determine if the examined carcasses recently came in contact with oil and if stranded cetaceans died as a result of crude oil contamination (Loughlin 1994b).

Methods

Aerial surveys were used to search for stranded cetaceans on 13 June and from 23 to 30 June 1989. Surveys were flown from Kayak Island north to Prince William Sound, along all coastlines and island groups west to Cape Sarichef, and east to King Salmon in Bristol Bay. Unscheduled surveys occurred from March to October and were concentrated in Prince William Sound. Tissue samples were collected from specimens that were not decomposed and were analyzed for hydrocarbon concentrations following the methods described above.

Results

More than 6,000 miles of coastline were surveyed resulting in the finding of 37 dead cetaceans, of which 26 were grey whales, 5 were harbor porpoises, 3 were unidentified, 2 were minke whales *Balaenoptera acutorostrata*, and 1 was a fin whale *B. physalus*. Nearly all of the specimens found were moderately to severely decomposed, which roughly corresponds to several weeks to months postmortem. Only seven animals were in satisfactory condition (minimal tissue degradation) for tissue collection: three grey whales, one minke whale, and three harbor porpoises.

Tugidak and Sitkinak Islands had the greatest number of grey whale carcasses (10; 27%), probably because of the circulating current systems near the confluence of Shelikof Strait and the Alaska Stream (Reed and Schumacher 1986). During July 1990, 9 grey whales were stranded at Sitkinak Island and 14 were stranded at Tugidak Island; in August 3 more were seen at Tugidak Island. Thirty-six grey whale carcasses were counted at these two islands in 2 years.

Details of the necropsies and histological examination of tissues provide no indication of cause of death for the sampled cetaceans. Results from the chemical analyses of blubber from one grey whale showed a PAH level of 467 ng/g wet weight. These PAHs contained a suite of hydrocarbons broadly consistent with a petrogenic source, which may have been either a crude oil or diesel oil. Reported values for all other tissues were below detection limits or did not indicate exposure to petrogenic hydrocarbons.

Discussion

The *Exxon Valdez* oil spill caused a significant proportion of the exposed sea otter population, particularly in Prince William Sound, to die within several months of the spill. The total acute loss probably numbered several thousand otters, but a precise estimate cannot be made primarily because of the lack of prespill information. Longer-term studies indicated that the sea otter population in the sound suffered chronic effects from oil exposure, at least through 1991. Data collected in 1992 and preliminary results from 1993 suggest the occurrence of a decline in chronic effects, which may lead to recovery. Although there are many uncertainties in describing the effects of this spill on sea otter populations, there is incontrovertible evidence that the population suffered a major perturbation. The full impact on otter populations may never be known.

Continued monitoring is expected to provide additional information on the status of sea otter recovery in Prince William Sound.

In retrospect, many of the NRDA studies were driven largely by the impending litigation against Exxon Corp. Political as well as scientific considerations influenced the direction of the research. Strong emphasis was placed on estimating acute mortality and documenting continuing damages. These tasks were rendered difficult to impossible by a lack of prespill data and a lack of contingency plans for research, along with a failure to implement appropriate studies in a timely manner.

St. Aubin (1990) summarized available information on the physiological and toxic effects of oil on pinnipeds (see his Table 4–1). The record indicates that pinnipeds, in contrast to cetaceans, have been affected by oiling. Harbor seals and Steller sea lions were seen swimming in the *Exxon Valdez* oil, and seals were often seen with oiled pelage (Lowry et al. 1994). Young seals also had oiled pelage and were seen nursing on their oiled mothers (Frost et al. 1994a). It is unlikely that pinnipeds would consume lethal quantities of oil either through grooming or feeding, but animals may succumb to a combination of the effects of inhaling toxic oil vapors and of stress as a result of ingestion of oil, capture, preexisting disease, or reaction to oil cleanup activities.

A proportion of the hydrocarbons inhaled and ingested will deposit in lipid-rich tissues such as blubber (Geraci 1990). A delayed response to these stored hydrocarbons may affect pinnipeds when these fat reserves are later mobilized.

Geraci (1990) compiled available information on the physiological and toxic effects of oil on cetaceans, including a summary table (his Table 6–1) reporting cetaceans associated with oil. In spite of numerous observations of cetaceans in spills, no deleterious effects were recorded with certainty.

Cetaceans on the eastern Pacific Coast encounter natural oil seeps throughout their range, and ingestion of hydrocarbons through prey or in the water column is not uncommon (Loughlin 1994b). However, the most likely effects on cetaceans could result from inhalation of toxic vapors concentrated just above an oil slick (Geraci 1990).

Both mysticete and odontocete whales are able to detect oil, yet they seem not to avoid it (Geraci 1990). Geraci speculated that the stimulus from oil may not be noxious enough to elicit a negative (avoidance) response, or perhaps the cetacean disregards oil when engaged in more engrossing activities. Cetaceans were observed swimming and feeding in oiled areas during the *Exxon Valdez* oil spill-related studies (Harvey and Dahlheim 1994). The humpback whale was the only marine mammal species in Prince William Sound that was not seen swimming through or feeding in oiled waters, but these whales were observed in waters that had been oiled after the oil had passed. This lack of avoidance makes whales vulnerable to the toxic vapors immediately above an oil slick, if their timing places them in the vicinity of a recent spill.

Killer whales in AB pod appeared to succumb to the effects of the oil in two phases. Soon after the spill, seven members of AB pod were recorded as missing (they could have disappeared before the spill occurred). The loss of this first group of whales may have involved an immediate response to stress and the toxic vapors, resulting in sudden death. This response has been described in humans (Bass 1986). The second phase of deaths, where six additional whales were reported as missing, may have involved complications associated with mucous membrane damage, including damage to the airways (Carpenter et al. 1977). However, since none of the missing AB pod killer whales were found (they usually sink upon dying), the cause of death is unresolved.

Other cetaceans may have been effected by the *Exxon Valdez* oil spill, including harbor porpoises and Dall's porpoises. However, because of the lack of prespill data and the difficulty in studying these species, relatively little effort was made toward understanding the effects of the oil spill on them.

Summary

Studies on marine mammals after the *Exxon Valdez* oil spill indicate impacts on some species and no or uncertain impacts on others. Sea otters were the most abundant marine mammal in areas affected by the spill and the species most affected. Within 6 months of the spill, at least 994 otters had died (871 carcasses recovered and 123 died at rehabilitation centers). Several thousand sea otters probably died as a result of the spill, but a reliable estimate cannot be generated because of limitations on prespill data and failure to implement appropriate studies immediately after the spill.

Impacts on harbor seals and sea lions were more difficult to determine because these animals generally sink at death and because of declines in numbers in the affected areas unrelated to the spill (Pitcher 1990; Loughlin et al. 1992). Harbor seals were the only marine mammal species, with the exception of sea otters, for which adequate data existed to link the spill to deaths of individuals and to a reduction in population levels. Approximately 300 harbor seals were missing from trend sites in Prince William Sound after the spill. Data from chemical analysis showing high levels of hydrocarbons and tissue pathology in oiled seals suggests deleterious effects caused by the spill. Inadequate data exist to determine the impact of the *Exxon Valdez* oil spill on Steller sea lions because of the magnitude of the prespill decline in the Gulf of Alaska (Loughlin et al. 1992) and the difficulty in obtaining sufficient samples for chemical and histological examination (Calkins et al. 1994).

Killer whales missing from Prince William Sound pods after the *Exxon Valdez* oil spill are assumed to have died and oil may have been the cause, although alternative explanations exist. No evidence exists to demonstrate an adverse impact on humpback whales or other cetaceans (except killer whales) resulting from the oil spill.

References

Aguilar, A. 1985. Compartmentation and reliability of sampling procedures in organochloride pollution surveys of cetaceans. Residue Reviews 95:91–114.

Baker, S. C., A. Perry, and L. M. Herman. 1987. Reproductive histories of female humpback whales (*Megaptera novaeangliae*) in the north Pacific. Marine Ecology Progress Series 41:103–114.

Baker, S. C., J. M. Straley, and A. Perry. 1992. Population characteristics of individually identified humpback whales in southeast Alaska: summer and fall 1986. U.S. Marine Fisheries Service Fishery Bulletin 90:429–437.

Ballachey, B. E., J. L. Bodkin, and A. R. DeGange. 1994. An overview of sea otter studies in Prince William Sound. Pages 47–59 *in* Loughlin (1994a).

Bass, M. 1986. Sniffing gasoline. Journal of the American Medical Association 255:2604–2605.

Bayha, K., and J. Kormendy, editors. 1990. Sea otter symposium: proceedings of a symposium to evaluate the response effort on behalf of sea otters after the T/V *Exxon Valdez* oil spill into Prince William Sound, Anchorage, Alaska. U.S. Fish and Wildlife Service, Biological Report 90(12).

Bigg, M. A. 1982. An assessment of killer whales (*Orcinus orca*) stocks off Vancouver Island, British Columbia. Reports of the International Whaling Commission 32:655–666.

Bodkin, J. L., R. J. Jameson, and J. A. Estes. 1995. Sea otters in the North Pacific ocean. Pages 353–356 *in* E. T. LaRoe, editor. Our living resources. National Biological Service, Washington, DC.

Bodkin, J. L., and M. S. Udevitz. 1994. An intersection model for estimating sea otter mortality along the Kenai Peninsula. Pages 81–95 *in* Loughlin (1994a).

Boehm, P. D., and six coauthors. 1987. Comparative fate of chemically dispersed and beached crude oil in subtidal sediments of the Arctic nearshore. Arctic 40(Supplement 1):133–148.

Burn, D. M. 1994. Boat-based population surveys of sea otters in Prince William Sound. Pages 61–80 *in* Loughlin (1994a).

Calkins, D. G. 1986. Marine mammals. Pages 527–558 *in* D. W. Hood and S. T. Zimmerman, editors. The Gulf of Alaska, physical environment and biological resources. U.S. Government Printing Office, Washington, DC.

Calkins, D. G., E. Becker, and T. R. Loughlin. 1994. Impacts on the distribution and abundance of Steller sea lions. Pages 119–137 *in* Loughlin (1994a).

Carpenter, C. P., D. L. Geary, Jr., R. C. Meyers, D. J. Nachreiner, L. J. Sullivan, and J. M. King. 1977. Petroleum hydrocarbon toxicity studies. XIV. Animal and human response to vapor of high aromatic solvent. Toxicology and Applied Pharmacology 41(2): 235–249.

Committee on Marine Mammals. 1961. Standardized methods for measuring and recording data on the smaller cetaceans. Journal of Mammalogy 42:471–476.

Costa, D. P., and G. L. Kooyman. 1982. Oxygen consumption, thermoregulation, and the effect of fur oiling and washing on the sea otter, *Enhydra lutris*. Canadian Journal of Zoology 60:2761–2767.

Dahlheim, M. E., and C. O. Matkin. 1994. Assessment of injuries to Prince William Sound killer whales. Pages 163–171 *in* Loughlin (1994a).

DeGange, A. R., A. M. Doroff, and D. H. Monson. 1994. Experimental recovery of sea otter carcasses at Kodiak Island, Alaska, following the *Exxon Valdez* oil spill. Marine Mammal Science 10:496–501.

DeGange, A. R., D. C. Douglas, D. H. Monson, and C. Robbins. 1994. Surveys of sea otters in the Gulf of Alaska in response to the *Exxon Valdez* oil spill. Natural Resource Damage Assessment Report, Marine Mammal Study 6. U.S. Fish and Wildlife Service, Anchorage, Alaska.

Doroff, A., and J. L. Bodkin. 1994. Sea otter foraging behavior and hydrocarbon levels in prey. Pages 193–208 *in* Loughlin (1994a).

Fay, F. H., L. M. Shults, and R. A. Dieterich. 1979. A field manual of procedures for postmortem examination of Alaskan marine mammals. Institute of Marine Sciences Report 79–1. University of Alaska, Fairbanks.

Frost, K. J., and L. F. Lowry. 1993. Assessment of injury to harbor seals in Prince William Sound, Alaska, and adjacent areas following the *Exxon Valdez* oil spill. Natural Resource Damage Assessment Report, Marine Mammal Study 5, U.S. Fish and Wildlife Service, Anchorage, Alaska.

Frost, K. J., L. F. Lowry, E. Sinclair, J. Ver Hoef, and D. McAllister. 1994a. Impacts on distribution, abundance, and productivity of harbor seals. Pages 97–118 *in* Loughlin (1994a).

Frost, K. J., C.-A. Manen, and T. L. Wade. 1994b. Hydrocarbon contaminants in harbor seals in Prince William Sound and the Gulf of Alaska. Pages 331–358 *in* Loughlin (1994a).

Garrott, R. A., L. L. Eberhardt, and D. M. Burn. 1993. Mortality of sea otters in Prince William Sound following the *Exxon Valdez* oil spill. Marine Mammal Science 9:343–359.

Geraci, J. R. 1990. Physiologic and toxic effects on cetaceans. Pages 167–197 *in* J. R. Geraci and D. J. St. Aubin, editors. Sea mammals and oil, confronting the risks. Academic Press, San Diego, California.

Hall, J. D. 1979. A survey of cetaceans of Prince William Sound and adjacent vicinity, their numbers and seasonal movements. Final report of the principal investigators. Volume 6. Biological studies. Outer Continental Shelf Assessment Program, Boulder, Colorado.

Hall, J. D. 1981. Aspects of the natural history of cetaceans in Prince William Sound, Alaska. Doctoral dissertation. University of California, Santa Cruz.

Harvey, J. T., and M. E. Dahlheim. 1994. Cetaceans in oil. Pages 257–264 *in* Loughlin (1994a).

Katona, S., B. Baxter, O. Brazier, S. Kraus, J. Perkins, and H. Whitehead. 1979. Identification of humpback whales by fluke photographs. Pages 33–44 *in* H. E. Winn and B. L. Olla, editors. Behavior of marine animals-current perspectives in research, volume 3. Plenum Press, New York.

Kenyon, K. W. 1969. The sea otter in the eastern Pacific Ocean. North American Fauna 68.

Leatherwood, S., C. O. Matkin, J. D. Hall, and G. M. Ellis. 1990. Killer whales, *Orcinus orca*, photo-identified in Prince William Sound, Alaska, 1976 through 1987. Canadian Field-Naturalist 104:362–371.

Lipscomb, T. P., R. K. Harris, R. B. Moeller, J. M. Pletcher, R. J. Haebler, and B. E. Ballachey. 1993. Histopathologic lesions in sea otters exposed to crude oil. Veterinary Pathology 30:1–11.

Lipscomb, T. P., R. K. Harris, A. H. Rebar, B. E. Ballachey, and R. J. Haebler. 1994. Pathology of sea otters. Pages 265–279 *in* Loughlin (1994a).

Loughlin, T. R., editor. 1994a. Marine mammals and the *Exxon Valdez*. Academic Press, San Diego, California.

Loughlin, T. R. 1994b. Tissue hydrocarbon levels and the number of cetaceans found dead after the spill. Pages 359–369 in Loughlin (1994a).

Loughlin, T. R., A. S. Perlov, and V. A. Vladimirov. 1992. Range-wide survey and estimation of total number of Steller sea lions in 1989. Marine Mammal Science 8:220–239.

Loughlin, T. R., D. J. Rugh, and C. H. Fiscus. 1984. Northern sea lion distribution and abundance: 1956–80. Journal of Wildlife Management 48:729–740.

Loughlin, T. R., and E. H. Sinclair. 1994. Sample collection, storage, and documentation. Pages 377–382 *in* Loughlin (1994a).

Lowry, L. F., K. J. Frost, and K. W. Pitcher. 1994. Observations of oiling of harbor seals in Prince William Sound. Pages 209–225 *in* Loughlin (1994a).

Matkin, C. O., G. M. Ellis, M. E. Dahlheim, and J. Zeh. 1994. Status of killer whales in Prince William Sound 1985–92. Pages 141–162 *in* Loughlin (1994a).

Merrick, R. L., D. G. Calkins, and D. C. McAllister. 1992. Aerial and ship-based surveys of Steller sea lions (*Eumetopias jubatus*) in southeast Alaska, the Gulf of Alaska, and the Aleutian islands during June and July 1991. NOAA (National Oceanic and Atmospheric Administration) Technical Memorandum NMFS (National Marine Fisheries Service)-AFSC-1, Alaska Fisheries Science Center, Seattle.

Merrick, R. L., T. R. Loughlin, and D. G. Calkins. 1987. Decline in abundance of the northern sea lion, *Eumetopias jubatus*, in Alaska, 1956–86. U.S. National Marine Fisheries Service Fishery Bulletin 85:351–365.

Monson, D. H. 1994. Age distributions and sex ratios of sea otters found dead in Prince William Sound, Alaska, following the *Exxon Valdez* oil spill. Natural Resource Damage Assessment Report, Marine Mammal Study 6. U.S. Fish and Wildlife Service, Anchorage, Alaska.

Mulcahy, D. M., and B. E. Ballachey. 1994. Hydrocarbon residues in sea otter tissues. Pages 313–330 *in* Loughlin (1994a).

Olesiuk, P. F., M. A. Bigg, and G. M. Ellis. 1990. Life history and population dynamics of resident killer whales (*Orcinus orca*) in the coastal waters of British Columbia and Washington state. Reports of the International Whaling Commission Special Issue 12: 209–243.

Pitcher, K. W. 1990. Major decline in number of harbor seals, *Phoca vitulina richardsi*, on Tugidak Island, Gulf of Alaska. Marine Mammal Science 6:121–134.

Rebar, A. H., B. E. Ballachey, D. L. Bruden, and K. Kloecher. 1994. Hematology and clinical chemistry of sea otters captured in Prince William Sound, Alaska, following the *Exxon Valdez* oil spill. Natural Resource Damage Assessment Report, Marine Mammal Study 6. U.S. Fish and Wildlife Service, Anchorage, Alaska.

Reed, R. K., and J. D. Schumacher. 1986. Physical oceanography. Pages 57–75, *in* D. W. Hood and S. T. Zimmerman, editors. The Gulf of Alaska, physical environment and biological resources. U.S. Government Printing Office, Washington, DC.

Spraker, T. R., L. F. Lowry, and K. J. Frost. 1994. Gross necropsy and histopathological lesions found in harbor seals. Pages 281–311 *in* Loughlin (1994a).

St. Aubin, D. J. 1990. Physiologic and toxic effects on pinnipeds. Pages 103–127 *in* J. R. Geraci and D. J. St. Aubin, editors. Sea mammals and oil, confronting the risks. Academic Press, San Diego, California.

Varanasi, U., and ten coauthors. 1990. Survey of subsistence fish and shellfish for exposure to oil spilled from the *Exxon Valdez* first year: 1989. NOAA (National Oceanic and Atmospheric Administration) Technical Memorandum NMFS (National Marine Fisheries Service) F/NWC-191. Northwest Fisheries Science Center, Seattle.

von Ziegesar, O., editor. 1992. Humpback whale fluke catalogue for Prince William Sound, Alaska. North Gulf Oceanic Society, Homer, Alaska.

von Ziegesar, O., and C. O. Matkin, editors. 1985. Humpback whale fluke catalogue for Prince William Sound, Alaska identified by fluke photographs between the years 1977 and 1984. First edition. North Gulf Oceanic Society, Homer, Alaska.
von Ziegesar, O., E. Miller, and M. Dahlheim. 1994. Humpback whales in Prince William Sound, Alaska, 1988–90. Pages 173–191 *in* Loughlin (1994a).
Wade, T. L., and seven coauthors. 1988. NOAA Gulf of Mexico status and trends program: trace organic contaminant distribution in sediments and oysters. Estuaries 11:171–179.
Wade, T. L., T. J. Jackson, T. M. McDonald, D. L. Wilkenson, and J. M. Brooks. 1993. Oysters as biomonitors of the APEX Barge oil spill, Galveston Bay, Texas. Pages 313–317 *in* Proceedings, 1993 international oil spill conference. American Petroleum Institute Publication 4580, Washington, DC.

American Fisheries Society Symposium 18:809–813, 1996

Acute Phase Proteins and Cytokines in Alaskan Mammals as Markers of Chronic Exposure to Environmental Pollutants

LAWRENCE K. DUFFY AND R. TERRY BOWYER
University of Alaska Fairbanks, Institute of Arctic Biology Fairbanks, Alaska 99775-0180, USA

J. WARD TESTA AND JAMES B. FARO
Alaska Department of Fish and Game, Anchorage, Alaska 99503, USA

Abstract.—We examine the usefulness of developing acute phase proteins and cytokines as markers of biological effects for free-ranging Alaskan mammals. Blood assays for these markers are already in widespread use as indicators of chemical pollution exposure, disease, and stress in humans. The acute phase response includes fever and changes in the plasma concentration of several proteins that originate in the liver. The acute phase response is mediated by cytokines, such as interleukin-1 and interleukin-6, which are released by macrophages and other cells. The few data available indicate that these biochemical markers are valuable for documenting chemical exposure when used in conjunction with an overall biostatistical model. Additional baseline studies using this nonlethal, analytical method for mammal populations are needed to characterize age, sex, and season as sources of variation.

After stress or exposure to pollutants such as crude oil, animals may exhibit an immediate or acute response and a chronic response. Both of these responses affect various homeostatic mechanisms in vertebrates, the most significant being that on the immune system. Any secondary infection or tissue damage will lower the long-term survivorship of the organism and, if widespread, hold consequences for the population. Activity of the immune system can be monitored by measuring various blood components, including plasma proteins, enzymes, and cytokines. To develop mammalian plasma and cytokines as markers of biological effects, extensive laboratory and controlled field studies are needed to demonstrate that these markers show a reproducible response to oil exposure. Also, the variation in response attributable to differences in age, sex, and nutritional state must be characterized. Because these types of studies have only begun for Alaskan mammals, the major focus of this article is to examine changes observed in the pattern of acute phase proteins and cytokines after exposure to environmental hydrocarbons, as occurred after the *Exxon Valdez* oil spill in Prince William Sound.

The Acute Phase Response

The acute phase response (APR) is nonspecific and commences within hours of inflammatory stimuli and results in the increased synthesis of certain plasma proteins (Kushner 1982). During APR, a negative nitrogen balance and decreased synthesis of anabolic lipogenic enzymes occurs in association with fever (Atkins 1984; Lowry et al. 1992). Likewise, there is a change in the viscosity of blood as determined by the sedimentation rate of red blood cells (Bull 1981). This change in sedimentation rate has been well documented in man and more recently in sea otters *Enhydra lutris* exposed to crude oil spilled from the *Exxon Valdez*; Wilson et al. (1990) reported that this rate increased almost four fold in sea otters and observed that anemia and hemolysis, which could trigger APR, lasted for 90 d in oil-exposed sea otters.

Acute phase proteins are defined by the magnitude of change in their concentration after inflammation or infection (Atkins 1984; Kushner and MacKiewicz 1987). The change in concentration of some APR proteins can occur as early as 6 h and last up to 2 weeks under controlled laboratory conditions. Most APR proteins are glycoproteins, with the notable exceptions of albumin, C-reactive protein, and serum amyloid A protein (SAA). Although most APR proteins have been documented only in humans and laboratory animals (such as rats) SAA has been extensively studied in the mink *Mustela vison* (Marhaug et al. 1984), and in horses *Equus caballus* (Husebekk et al. 1986; Pepys et al. 1989). In the circumpolar north, the mink is a likely candidate for the monitoring of environmental changes caused by pollution (Edqvist et al. 1992). The ease of capture and recapture, widespread distribution, high reproductive rate, and abundance makes them effective biological monitors for long-term effects that may occur over several years and generations (Talmage and Walton 1991).

The increase of APR proteins in plasma after inflammatory stimulus results from increased rates of synthesis and secretion by liver parenchymal cells

TABLE 1.—Selected blood components for river otters inhabiting oiled (N = 11) and non-oiled (N = 11) areas of Prince William Sound in 1991. Part of this data has been published previously (Duffy et al. 1994). We examined relationships among blood values with simple correlations (Dixon 1985). These and other blood values were used to classify otters from oiled and non-oiled areas using logistic regression (Agresti 1990). Abbreviations are Hb = hemoglobin and IU = international unit.

Blood components	Oiled		Non-oiled	
	Mean	SE	Mean	SE
Interleukin (IL-6ir, pg/ml)	48.3	13.8	17.3	11.3
Interleukin (IL-1ir, pg/ml)	13.3	6.6	10.1	6.1
Haptoglobin (Hp, Hb binding/dL)	156.9	27.9	30.0	15.6
Alanine aminotransferase (ALT, IU)	152.7	8.8	138.5	14.6
Aspartate aminotransferase (AST, IU)	437.2	70.0	418.1	67.0
Lactate dehydrogenase (LDH, IU)	146.2	25.2	154.0	43.1
Creatine kinase (CK, IU)	3,038.6	820.8	1,885.8	516.4
Hemoglobin (Hb, g/dL)	16.3	0.6	15.7	0.6
Packed cell volume (PCV, ml/mm^3)	42.9	1.6	44.1	1.6
Sodium (Na, meq/L)	148.5	3.0	156.6	4.7
Potassium (K, meq/L)	4.1	0.1	4.5	0.1
Chloride (Cl, meq/L)	117.1	2.6	123.5	4.6
Calcium (Ca, mg/dL)	9.2	0.2	9.8	0.6
Phosphate (PO_4, mg/dL)	5.5	0.3	5.7	0.7
Blood urea nitrogen (BUN, mg/dL)	36.1	3.4	36.7	4.9
Creatine (CREA, mg/dL)	0.45	0.03	0.60	0.46
Cholesterol (Chol, mg/dL)	161.7	9.4	221.5	26.1
Bilirubin total (mg/dL)	0.27	0.05	0.32	0.07
Direct bilirubin (mg/dL)	0.16	0.05	0.20	0.08
Total protein (Prot., g/dL)	7.4	0.4	7.6	0.3
Albumin (Alb, g/dL)	3.1	0.2	3.1	0.1
Glutamyl transferase (GGT, IU)	20.7	2.8	20.1	2.7
Alkaline phosphatase (AP, IU)	226.2	25.3	262.9	55.3

(Pepys and Baltz 1983). Perlmutter et al. (1985) have shown that cells other than hepatocytes also synthesize these proteins, but that extrahepatic synthesis plays only a minor role in elevating blood levels of these proteins. Acute phase response proteins made in nonhepatocytes probably act locally in a particular tissue such as muscle or brain (Dinarello 1984; Tamm 1989).

Wide normal ranges and variability in the response of individual APR proteins has been observed between different species of animals (Sheldrick et al. 1982; Conner et al. 1986; Schreiber et al. 1989). In most animals studied, however, there always appears to be an enhanced synthesis of haptoglobin (Hp), alpha-1-acid glycoprotein (AG), and fibrinogen (Fb) and reduced albumin synthesis (Conner et al. 1986; Kushner and MacKiewicz 1987; Allen and Kold 1988; Schreiber et al. 1989). Other protein levels seem to vary between species; for example, fibronectin is an APR protein in rodents but not in man (Dyck and Rogers 1985). Approximately two-fold variability in response of individual APR proteins also has been related to the sex of the animal (Bosanquet et al. 1976), which may be in response to synergistic action of hormones and cytokines. There is a general lack of data on APR proteins and the level of concentration changes in marine species and aquatic wildlife such as mink, river otters *Lutra canadensis*, seals, and whales, which are very sensitive to environmental pollution (McIntyre and Pearce 1980; Dieter and DiGiulio 1989; Duffy et al. 1993). Limited comparative studies suggest that the physical structure of APR proteins may have been preserved through millions of years of evolution (Ledoux and Kenyon 1975; Kurosky et al. 1976; Atkins 1984; Wahl et al. 1992) and presumably is vital to survivorship of the organism.

In an *Exxon Valdez* study, river otters from oiled areas exhibited significantly higher levels of Hp, indicative of physiological damage, and were significantly lower in body mass than otters from non-oiled areas of Prince William Sound. River otters might be exposed to oil by grooming it from their fur, eating oil-covered prey, or ingesting food (e.g., mussels *Mytilus* sp.) that had concentrated hydrocarbons (Bowyer et al. 1994; Moore 1985). Elevated blood values (Table 1) and reduced body mass for otters from oiled sites might result from internal damage from oil, reduced digestibility of oiled food, a reduction in prey, or some combination of these (Duffy et al. 1993, 1994a). Haptoglobin values differed in animals from oiled and non-oiled regions of Prince William Sound for 2 years but may have recovered in the third year (Duffy et al. 1994b).

Haptoglobin values for river otters obtained in 1991 and 1992 are substantially lower than those from 1990. This may have occurred for two reasons: first, the Hp response elicited by exposure to crude oil may be waning over time; or second, samples were collected in late winter in 1990 but during summer in 1991 and 1992. Although the effects of exposure to oil probably would diminish over time, this does not explain why Hp levels declined markedly from 1990 levels in animals from both oiled and non-oiled areas. Thus, the most likely answer is that Hp levels are the result of complex interactions with other environmental factors that vary seasonally.

Cytokines and the Mediation of Acute Phase Response Protein Synthesis

Absorption and exposure of tissue to environmental pollutants such as crude oil can lead to local tissue injury. In Prince William Sound, tissue injury was observed in sea otters, as indicated by cell analysis and leakage of enzymes such as creatine kinase (Williams et al. 1990). Also, 1 year after the spill, increased level of liver enzymes, such as AST, leaking into the plasma was observed in river otters (Table 1; Duffy et al. 1994a). Injury of tissue is known to cause APR and is mediated by messengers called cytokines (Lowry et al. 1992).

Cytokines are small protein molecules (molecular weight <20 kilodaltons) that are synthesized and secreted by different cells, most notably those of the immune system such as macrophages. Interleukins are cytokines used by immune and inflammatory cells to communicate with each other and are produced at the site of infection or injury. They act locally on other cells (paracrine) or may circulate, causing systemic effects. There are now more than eight different cytokines with their receptor systems reported (Arai et al. 1990). Several of these cytokines induce APR, and it is becoming clear that the synthesis of individual APR proteins is a complex process of control. These molecules exert a wide range of biological effects on cells, allowing cells of immune, hemopoietic, and neuronal systems to communicate (Dinarello 1984; Machiewicz et al. 1991; Kishimoto et al. 1992). Not only do cytokines function pleiotropically, but they also operate in a redundant manner through a tyrosine kinase-linked receptor system (Mantovani et al. 1992; Taga and Kishimoto 1992).

Early interest focused on the role of interleukin-1 (IL-1) in mediating APR. Interleukin-1 leads to the production of IL-6, which has been shown to induce the synthesis of Hp. Detectable levels of IL-1-immunoreactive molecules (IL-1ir) have been observed in seals from the Prince William Sound area, which may have been exposed to crude oil (L. K. Duffy, Institute of Arctic Biology, and K. J. Frost, Alaska Department of Fish and Game, unpublished). Increased IL-1 levels may lead to local APR response in the brain and demyelination by macrophages (Said and Hontebbeyrie-Jaskowicz 1992). Furthermore, IL-1ir was detected in river otters but did not differ significantly between river otters living in oiled and non-oiled areas (Table 1; Duffy et al. 1994a). An IL-1-like substance also has been reported in fish and lizards (Kluger et al. 1975; Bernherm et al. 1978; S. Trask and Duffy, unpublished).

Our knowledge of the role of IL-6 in inflammation, viral infection, autoimmunity and cancer continues to increase. Abnormal IL-6 production has been associated with autoantibodies in human patients with benign heart tumors, and increased IL-6 production has been documented in rheumatoid arthritis. Viral infections such as that caused by the human immunodeficiency virus (HIV) lead to high serum concentrations of IL-6 (Breen et al. 1990). In wildlife populations, differences in IL-6ir have been observed between river otters living in oiled and those living in non-oiled areas of Prince William Sound in 1991 (Table 1; Duffy et al. 1994a). The importance of monitoring the immune system after chemical exposures has been repeatedly documented in humans and should be expanded to free-ranging mammals (Weeks et al. 1992).

Monitoring Acute Phase Response Proteins and Cytokines

To predict the environmental risk, the probability of deleterious effects must be estimated. Because the APR largely reflects plasma concentrations of Hp, fibrinogen, and other acute phase proteins, the direct measurement of these proteins to monitor chronic inflammation would be a useful diagnostic aid in following the course of exposure to pollutants and can be used to provide evidence of exposure of mammals to such pollutants (Duffy et al. 1994a). With respect to Prince William Sound, stepwise logistic regression using Hp, IL-6ir and a liver enzyme correctly classified 86.4% of 22 river otters as inhabiting oiled or non-oiled areas. Moreover, otters living in oiled areas abandoned use of existing latrine sites at a rate that was more than three times greater than in non-oiled areas, strongly suggesting that the predictive model based on blood values reflected changes in this otter population (Duffy et al. 1994a). More studies employing probability theory and different acute phase proteins will be required to identify how many acute phase proteins should be monitored. Haptoglobin and other acute phase proteins such as alpha-1-antichymotrysin inhibitor, whose function can be measured quantitatively, are useful in wildlife species where immunochemical probes are not yet available or validated. A probe for C-reactive protein levels, diagnostic of human patients with chronic inflammatory states and leukemia, would be especially useful in studying sea otters, river otters, or other large mammals exposed to environmental pollutants. Monitoring APR proteins would aid workers in oil spill recovery faculties when following the health of oiled animals before they are released into the wild. If long-term survivability is to be studied, measurement of cytokines such as IL-6 and tumor necrosis factor (TNF) would be useful to show abnormal, chronic stimulation of the immune system (Dinarello 1984; Breen et al. 1990; Lowry et al. 1992).

Understanding the responses of river otters to oil contamination obviously requires prespill data and long-term monitoring. Duffy et al. (1994a, 1994b) detected changes in blood values of river otters that are related to their health more than 2 years after the spill. This raises important questions about the long-term effects of oil on river otters and other mammals similarly exposed to crude oil. The validity of the approach is indicated by year-3 data, which show potential recovery in the oiled population of river otters. Because of their high position in the food chain, river otters and mink can serve as "indicator species" for monitoring the health and recovery of animals at lower trophic levels in Prince William Sound.

In the years immediately after the *Exxon Valdez* oil spill much attention has been given to the problem of restoring the damaged ecosystem in Prince William Sound. Nonlethal monitoring techniques using blood samples and recapture methods allow fish and wildlife managers to observe the health of the animals without removing them from the population (Fossi and Leonzio 1994). In the context of a nonlethal monitor of the environment, acute phase proteins and cytokines may be found to serve as an early warning signal of possibly more serious pathologies, especially if reexposure to chemical pollutants and environmental pathogens is prolonged.

Acknowledgments

We thank L. Morisky for her expert help with this manuscript. We thank A. Porchet for technical assistance and J. Blake, B. Ballachey and A. Rebar for helpful discussion. We are grateful to veterinarians W. Taylor and G. Grady for their help. This research was funded in part by the Institute of Arctic Biology, University of Alaska Fairbanks, and the Alaska Department of Fish and Game. Otters were trapped and blood was collected under permit 78-358, Alaska Department of Fish and Game; all procedures used in this study were approved by an Institutional Care and Use Committee at the University of Alaska Fairbanks (No. 89-022).

References

Agresti, A. 1990. Categorical data analysis. Wiley, New York

Allen, B. V., and S. E. Kold. 1988. Fibrinogen response to surgical tissue trauma in the horse. Equine Veterinary Journal 20:441–443.

Arai, K., F. Lee, A. Miyajima, S. Miyatake, N. Arai, and A. Yokota. 1990. Cytokines: coordinators of immune and inflammatory responses. Annual Review of Biochemistry 59:783–836.

Atkins, E. 1984. Fever: the old and the new. Journal of Infectious Diseases 149:339–348.

Bernherm, H. A., P. T. Bodel, P. W. Askevose, and E. Atkins. 1978. Effects of fever on host defence mechanisms after infection in the lizard (*D. dorsalis*). British Journal of Experimental Pathology 59:76–84.

Bosanquet, A. G., A. M. Chandler, and A. H. Gordon. 1976. Effects of injury on the concentration of alpha 1 macroglobulin and alpha 2 macroglobulin in the plasmas of male and female rats. Experientia 32: 1348–1349.

Bowyer, R. T., J. W. Testa, J. B. Faro, C. C. Schwartz, and J. B. Browning. 1994. Changes in diets of river otters in Prince William Sound, Alaska: effects of the *Exxon Valdez* oil spill. Canadian Journal of Zoology 72:970–976.

Breen, E. C., and seven coauthors. 1990. Infection with HIV is associated with elevated IL-6 levels and production. Journal of Immunology 144:480–484.

Bull, B. S. 1981. Clinical and laboratory implications of present ESR methodology. Clinical Laboratory Hematology 3:283–98.

Conner, J. G., P. D. Eckensall, and A. Wiseman. 1986. The acute phase response in the cow and dog. Protides of the Biological Fluids Proceedings of the Colloquium 34:509–512.

Dieter, M. P., and R. T. DiGiulo. 1989. A comparative biochemical approach for the evaluation of wildlife toxicity. Environmental Toxicology and Chemistry 8:1091–1092.

Dinarello, C. A. 1984. Interleukin-1 and the pathogenesis of the acute-phase response. New England Journal of Medicine 311:1413–1418.

Dixon, W. J. 1985. BMDP statistical software. University of California, Berkeley.

Duffy, L. K., R. T. Bowyer, J. Faro, and J. W. Testa. 1993. Differences in blood haptoglobin and length-mass relationships in river otters from oiled and nonoiled areas of Prince William Sound, Alaska. Journal of Wildlife Diseases 29:353–359.

Duffy, L. K., R. T. Bowyer, J. Faro, and J. W. Testa. 1994a. Chronic effects of *Exxon Valdez* oil spill on blood and enzyme chemistry of river otters. Environmental Toxicology and Chemistry 13:643–647.

Duffy, L. K., R. T. Bowyer, J. W. Testa, and J. B. Faro. 1994b. Evidence for recovery of body mass and haptoglobin values of river otters following the *Exxon Valdez* oil spill. Journal of Wildlife Diseases 30:421–425.

Dyck, R. F., and S. L. Rogers. 1985. Fibronection is an acute phase reactant in mice. Clinical and Investigative Medicine 8:148–151.

Edqvist, L. K., A. Modej, and M. Forsberg. 1992. Biochemical blood parameters in pregnant mink fed PCB and fractions of PCB. Ambio 21:577–581.

Fossi, M. C., and C. Leonzio. 1994. Nondestructive biomarkers in vertebrates. Lewis Publishers, Boca Raton, Florida.

Husebekk, A., G. Husby, K. Sletten, G. Marhaug, and K. Nordstoga. 1986. Characterization of amyloid protein AA and its serum precursor SAA in the horse. Scandinavian Journal of Immunology 23:703–709.

Kishimoto, T., S. Akire, and T. Toga. 1992. Interleukin-6 and its receptor: a paradigm for cytokines. Science 258:593–597.

Kluger, M. J., D. H. Ringler, and M. R. Anver. 1975. Fever and survival. Science 188:166–168.

Kurosky, A., H. H. Kim, and B. Touchstone. 1976. Comparative sequence analysis of N-terminal region of rat, rabbit and dog haptoglobin. Comparative Biochemistry and Physiology 55(B):453–459.

Kushner, I. 1982. The phenomenon of the acute phase response. Annals of the New York Academy of Sciences 389:39–48.

Kushner, I., and A. MacKiewicz. 1987. Acute phase proteins as disease markers. Disease Markers 5:1–11.

Ledoux, R. G., and A. J. Kenyon. 1975. Protides of the mustelidae II: immunologue relatedness. Comparative Biochemistry and Physiology 51(A):213–217.

Lowry, S. F., W. A. Thompson, D. Lazarus, and L. L. Moldawer. 1992. Interactions of cytokines and anabolic factors. Pages 71–76 *in* A. F. Roche, editor. Cytokines in critical illness. Report of the eleventh Ross conference on medical research. Ross Laboratories, Columbus, Ohio.

Machiewicz, A., T. Speroff, M. K. Ganapathi, and I. Kushner. 1991. Effects of cytokine combinations on acute phase protein production in two human hepatoma cell lines. Journal of Immunology 146:3032–3037.

Mantovani, A., F. Bussolino, and E. Dejana. 1992. Cytokine regulation of endothelial function. Federation of American Societies for Experimental Biology Journal 6:2591–2599.

Marhaug, G., A. Borresen, G. Husby, and K. Nordsioda. 1984. A comparative study of serum amyloid A protein from mink and man. Comparative Biochemistry and Physiology 78(B):401–406.

McIntyre, A. D., and J. B. Pearce. 1980. Biological effects of marine pollution and the problems of monitoring. Rapports et Proces-Verbaux des Reunions Conseil International pour l'Exploration de la Mer 179:1–346.

Moore, M. N. 1985. Cellular responses to pollutants. Marine Pollution Bulletin 16:134–139.

Pepys, M. B., and M. L. Baltz. 1983. Acute phase proteins with special reference to CRP, pentraxins and SAA. Advances in Immunology 34:141–212.

Pepys, M. B., M. L. Baltz, G. A. Tennent, J. Kent, J. Ousey, and P. D. Rossdale. 1989. Serum anyloid A protein (SAA) in horses: objective measurement of the acute phase response. Equine Veterinary Journal 21:106–109.

Perlmutter, D. P., F. S. Cole, P. Kilbridge, T. H. Rossing, and H. R. Colten. 1985. Expression of alpha 1 protease inhibitor gene in human monocytes and macrophages. Proceedings of the National Academy of Sciences of the United States of America 82:795–799.

Said, G. and A. Hontebbeyrie-Jaskowicz. 1992. Nerve lesions induced by macrophage activation. Research in Immunology 143:589–599.

Schreiber, G., and nine coauthors. 1989. The acute phase response in the rodent. Annals of the New York Academy of Sciences 557:61–83.

Sheldrick, R., J. E. Kent, and D. J. Blackmore. 1982. Hemoglobin binding capacity of serum as an indicator of infection in the horse. Veterinary Record 111: 128–129.

Taga, T. and T. Kishimoto. 1992. Cytokine receptors and signal transduction. Federation of American Societies for Experimental Biology Journal 6:3387–3396.

Talmage, S. S., and B. T. Walton. 1991. Small mammals as monitors of environmental contaminants. Reviews in Environmental Contamination and Toxicology 119:47–143.

Tamm, I. 1989. IL-6: current research and new questions. Annals of the New York Academy of Sciences 557: 478–489.

Wahl, S., J. Boger, V. Michael, and L. K. Duffy. 1992. Identification of a haptoglobin-hemoglobin complex in Alaskan Least Cisco (*C. sardinella*). Applied and Theoretical Electrophoresis 3:17–20.

Weeks, B. A., and six coauthors. 1992. Immunological biomarkers to assess environmental stress. Pages 5–86 *in* R. J. Huggett, R. A. Kimerle, P. M. Mehill, and H. S. Bergmen, editors. Biomarkers: biochemical, physiological and histological markers of anthropogenic stress. Lewis Publishers, Boca Raton, Florida.

Williams, T. M., R. Wilson, R. Tuomi, and L. Hunter. 1990. Critical care and toxicological evaluation of sea otters exposed to crude oil. Pages 236–258 *in* T. M. Williams and R. W. Davis, editors. Sea otter rehabilitation program: 1989 *Exxon Valdez* oil spill. International Wildlife Research.

Wilson, R. K., P. Tuomi, J. P. Schroeder, and T. Williams. 1990. Clinical treatment and rehabilitation of oiled sea otters. Pages 326–337 *in* T. M. Williams and R. W. Davis, editors. Sea otter rehabilitation program: 1989 *Exxon Valdez* oil spill. International Wildlife Research.

American Fisheries Society Symposium 18:814–818, 1996

Cultural Resources and the *Exxon Valdez* Oil Spill: An Overview

JUDITH E. BITTNER
Department of Natural Resources, Office of History and Archaeology
P.O. Box 107001, Anchorage, Alaska 99501-7001, USA

Abstract.—During 3 years of survey and monitoring after the 1989 *Exxon Valdez* oil spill, the number of known archaeological sites in the spill area along Prince William Sound and the Kenai Peninsula was more than doubled, reaching 609. Damage assessments revealed no contamination of the sites by oil, but considerable damage resulted from vandalism associated with cleanup activities and lesser amounts were caused by the cleanup process itself. Although the Exxon Corp. quickly addressed threats to cultural sites, timely field assessments were hindered by a shortage of funds for archaeological surveys and studies. Lessons learned from this Alaska experience can lead to better protection of cultural heritages elsewhere when future environmental emergencies occur.

The state of knowledge about human history in the northern Gulf of Alaska at the time of the *Exxon Valdez* oil spill was fragmentary at best. Except for excavations at a single site, no substantive archaeological work had been done in Prince William Sound since the 1930s. No serious large-scale inventory or testing had ever been conducted on the outer coast of the Kenai Peninsula. Kodiak Island and the Alaska Peninsula were the only known areas and data came from a small number of sites. Before the *Exxon Valdez* oil spill 283 sites were recorded in the tightly defined spill area on the inventory maintained by the state of Alaska Office of History and Archaeology. In the face of impending danger to sites, resource managers in the spill area were forced into the awkward position of protecting known archaeological sites while at the same time scrambling to locate undocumented sites. After three seasons of inventory and cleanup support activities, 326 new sites have been documented to raise the total of known sites in the spill area to 609.

Archaeologists were alarmed about potential damage to sites in the spill area because many sites had subsided into the intertidal zone during the 1964 earthquake. Those sites would be located in the path of the spreading oil slick. The primary concern was the effect of crude oil and cleanup procedures on the known sites; coupled with that was the concern that undocumented sites could unknowingly be destroyed during cleanup. The unsurveyed status of the area made the problem particularly disturbing.

The charge of protecting cultural resources during the spill and subsequent cleanup came from several authorities. Federal mandates included the Archaeological Resources Protection Act of 1979 (ARPA) and the National Historic Preservation Act of 1966 (NHPA). The Archaeological Resources Protection Act directs land-managing federal agencies to protect cultural sites on land that they manage. The National Historic Preservation Act directs federal agencies "undertaking actions," such as coordinating the spill cleanup, to consider the effects of their actions on cultural resources in consultation with the state historic preservation officer (SHPO). The U.S. Coast Guard (USCG) was the designated lead federal agency under the Section 106 project review process; the U.S. Forest Service, the primary federal land manager in Prince William Sound, advised the USCG on cultural resource matters. The Alaska Historic Preservation Act protects sites on state land, including the intertidal zone. The tidelands permit issued by the Department of Natural Resources directed state cultural resource consultation to the SHPO, who also serves as the chief of the state's history and archaeology programs. Thus, the cultural resource concerns of the state and federal agencies were combined into a single, coordinated program.

Close coordination between the Alaska SHPO, the USCG's federal on-scene coordinator, native representatives, federal representatives, and the Exxon Cultural Resource Program staff protected cultural resources during cleanup. Soon after the cleanup effort began, Exxon Corp. was informed of and recognized cultural resources as an important concern and included trained archaeologists in their beach assessment teams. A means for considering impacts on cultural resources was implemented early in Exxon Corp.'s response process (Mobley et al. 1990). During 1989 in Valdez, coordination occurred during cleanup planning between Exxon Corp. and a representative of the SHPO who reviewed Exxon Corp. archaeologists' reports for adequacy, evaluated proposed cleanup methods for each beach segment for possible impacts to sites, and, when appropriate, recommended alternative, less disruptive cleanup methods to avoid or reduce damage to known sites. To assure the federal on-scene coordinator that the work-plan review process was in compliance with the NHPA, each work order included a cultural resource comment section signed by an SHPO representative. The operating assumption was that beach segment cleanup was paramount and that protective efforts emphasized avoiding or minimizing damage to sites.

During Interagency Shoreline Cleanup Committee (ISCC) meetings in Valdez, federal, state, and native representatives reviewed and discussed actions, such as the status of shoreline cleanup assessment team (SCAT) surveys, cleanup methods, resource protection, and monitoring. Agency and native participants provided comments to committee representatives of the federal on-scene coordinator. As the cleanup operations spread to the outer Kenai Peninsula coast and the Alaska Peninsula, similar processes were followed at operation centers in Seward, Homer, and Kodiak but with slight differences and varied success. Logistics problems, lack of adequate staffing, and poor or incomplete information regarding field operations and cleanup plans hindered satisfactory coverage of field operations. In Kodiak, for example, work orders for many segments were prepared and reviewed after cleanup had taken place. Agency and native corporation archaeologists monitored cleanup activities on a sporadic basis, but not all culturally sensitive areas were covered.

In 1990 and 1991 the cleanup review process changed to involve a more active consultative role for federal archaeologists and native representatives. Also, Exxon cultural resource personnel became involved earlier in the summer cleanup planning process. Because there was no longer an emergency, there was more time to conduct surveys and review work orders in advance of field operations. Each successive cleanup season's review procedures benefited from the prior years' experience and the cumulative data collected (Betts et al. 1991; Haggerty et al. 1991). Monitoring by agency and native archaeologists also became more consistent during the second through fourth years of cleanup.

Determining injury and assessing damage to cultural resources as a result of the *Exxon Valdez* oil spill has been difficult. Unlike the clear legal authorities that protect cultural resources during oil spill response activities, the laws and regulations directing studies to determine injury and assess damage do not address cultural resources as clearly. Arguments and delays characterized exchanges between attorneys and resource managers on this issue. No cultural resource damage assessment studies were funded by the Trustees the first year of the spill, although at least one agency, the National Park Service, conducted prespill resource survey assessments in the outer Kenai Peninsula coast and the Alaska Peninsula before the oil moved into those areas.

Agency concerns about the effect of crude oil contamination on the radiocarbon dating process prompted a laboratory study initiated during 1990. The contract was funded by the state of Alaska and administered by the U.S. Forest Service's Alaska Region. Radiocarbon dating involves measuring the amount of disintegration by radioactive carbon to its nonradioactive isotopes. The dating method provides the most widely used, reliable means of estimating age. However, uncontrolled addition of carbon, such as from crude oil, to samples can seriously skew the age estimates and make the method unusable. The contractor for the radiocarbon study concluded that significant effect would occur, but that sample cleaning partially reduced the error (Mifflin and Associates 1991). One consequence of that finding was that the additional sample-cleaning steps are likely to increase costs of research on spill area sites. Testing the conclusion about radiocarbon dating on specific sites was the next step of injury assessment. Unfortunately, the radiocarbon study was not funded until 18 months after the spill, and follow-up field studies were delayed yet another year, although damage assessment study funds were requested during 1989 by the panel of archaeologists advising the Trustees' staff.

With funds provided by the state and the U.S. Forest Service, the contractor—State University of New York (SUNY) at Binghamton—began fieldwork near the end of the 1991 field season. The initial scope of the 1991 project was reduced when Trustee representatives decreased available funding. The project was aimed at testing a series of sites for injury, surveying to check the adequacy of the Exxon archaeological surveys, checking several soil chemicals for alteration by oiling, and testing a model developed to predict site locations. Testing radiocarbon dates was a basic part of the project.

The SUNY investigators found that the major direct impact of the *Exxon Valdez* spill on archaeological sites was confined to waterborne oil washing over intertidal archaeological deposits (RFSUNY 1993). This conclusion is partially at variance with Jesperson and Griffin's (1992) findings, which traced impacts to site disturbance by vandals. The SUNY team also concluded that presence of oil residue in site sediments resulted in the need for increased radiocarbon processing and thus higher research cost in the spill area. Human damage (vandalism) to sites investigated by the team could not conclusively be attributed to the oil spill or to related cleanup activities; this problem also was found in some instances during Jesperson and Griffin's (1992) study. Finally, the SUNY investigators concluded that sites in the spill area still retain good research potential and should be studied.

Delays and uncertainty in awarding the damage assessment contract during 1990 and 1991 prompted the state to begin a much smaller and more tightly focused study in May 1991 to assure that some damage assessment data were collected (Reger et al. 1992). The state study followed up on the conclusions of the earlier radiocarbon study by checking selected intertidal sites. Agency files were screened to select beach segments with sites that suffered oiling and had the best possibility of producing artifacts for comparison with existing dated collections. Alternative methods of dating archaeological remains were contrasted with radiocarbon estimates to check for injuries from oil contamination. The study was designed to complement the larger multiagency study, and Reger et al. (1992) concluded that no injury from oiling could be detected, particularly with regard to radiocarbon dating of the sites studied.

Another approach to assessing injuries to archaeological sites by spill-related activities was a compilation of information in agency and Exxon Corp. documentation. Field notes from various parties, monitoring reports, and shoreline cleanup assessment team reports were examined to determine kinds and degrees of impacts. This method of investigation was used partly because of the need for a more timely idea of impacts and partly to assess the adequacy of the documentation process. Although results of the document study provided only a rough idea of injuries, it was very useful in estimating injuries resulting from cleanup and attendant vandalism. Seventy-one percent of the 35 archaeological sites analyzed were injured because of oil spill response activities (such as cleanup, staging areas, and pedestrian traffic) and from vandalism (Jesperson and Griffin 1992). Valuable insights were gained into the types and detail of documentation needed from site identification surveys and monitoring. It is necessary for documentation of oiling on beaches associated with sites to specify the location of oil relative to the site so that cleanup activities and damage assessment studies can be more precisely designed. Also, subsurface testing of sites is necessary to determine whether and what protection is needed with beach-disturbing cleanup activities. For instance, if the exact distance of oiled debris from the limits of a site is known, specific constraints might be formulated that could easily be followed by the cleanup crew, which would be more effective than leaving decisions, at the time of cleanup, to an archaeological monitor with no real authority. The documents study provided most of the data used for assessing damage. Although hampered by documentation that was uneven in coverage and detail, the study allowed a rough estimate of damage to be made in time to be included in the oil spill restoration process, which started before the field studies were begun.

Assessment of monetary values for restitution followed the documents study. Findings of the state- and multiagency funded studies also provided data useful in determining levels of damage. A method of assessing damage based on procedures of the Archaeological Protection Act was used to provide a monetary estimate. The method of producing a valuation assessment based on an analysis of injury and determining the restoration required has been repeatedly used in other cases in the United States and provided very conservative cost estimates. The most clearly injured sites were assessed for damages, whereas the less adequately documented sites were deleted from the process. The 35 sites considered had sustained damage from vandalism or cleanup-related activities rather than from oiling.

Several useful observations about archaeology and the oil spill are possible based on damage assessment studies. The most important conclusion was that the sites generally were not directly affected by the spill. The most extensive damage resulted from vandalism, resulting from increased, widespread knowledge about sites. That knowledge increased because of expanded population and activity levels during cleanup. Another source of impact was directly from cleanup activities. Impact from cleanup was kept low because of the cleanup work-plan review process and the use of alternative cleanup techniques, but impacts did occur. Although the Exxon cultural resource program and agency cooperation was generally effective in keeping impacts small, lapses in control of cleanup procedures did happen. Most lapses occurred because of failure to communicate restrictions, such as at the Perevalnie Passage site, in which cleanup crews shoveled the oiled intertidal site, including artifacts, into bags for disposal. Occasional problems with control over local workers by Exxon Corp. or USCG also led to concern about unsupervised access to sites and unintentional damage to the remains.

Some aspects of the cultural resource program did not work well and must be improved. Lack of field assessment studies for the first 2 years after the spill squandered opportunities for collection of important injury data. Information collected by archaeologists and other monitors early in the cleanup phase could have included the data needed to protect the resource and answered important questions about potential injury. Also learned from the experience of the *Exxon Valdez* oil spill is the need for basic inventories of archaeological and historic sites, response plans tailored to protecting the cultural resources present, improved monitoring, and damage assessment studies begun in the first year. Inventory and planning are long-term projects that are typically difficult to fund. Improved monitoring requires analysis of existing studies and listing the information required during future emergencies. Consistent application of protective policies and procedures over the area of the spill needs to be addressed. Laws and regulations governing spill response, damage assessment, and restoration need to be revised to more clearly address cultural resources.

One positive aspect of the *Exxon Valdez* oil spill cleanup program is the demonstrated efficiency in preventing increased injury through active cooperation of the parties concerned with cultural resources. Even though serious questions existed regarding administrative authority over certain lands impacted by the spill and cleanup program, by adopting a reasonable and cooperative attitude toward each other, unnecessary delays and additional work were largely avoided. Selection of several acceptable cleanup methods in given situations allowed cleanup crews to complete their tasks more rapidly and minimize harm to sites. The cultural resource awareness training that was provided to cleanup workers and monitoring personnel also contributed to the understanding of agency concerns and made seemingly useless constraints understandable.

Finally, one of the important reasons the cultural resource program was successful was the early recognition by Exxon Corp. and the federal on-scene coordinator of their responsibilities and Exxon's recruitment of well qualified and committed archaeologists. Most archaeologists hired for survey or monitoring were selected for their experience in the spill area or in similar nearby environments. Their experience level enabled quick recognition of sites and artifacts not known in the area before the spill.

The outcome of future spills or similar emergencies will be improved if the lessons learned during the aftermath of the *Exxon Valdez* oil spill are applied. The importance of site location information in a project area and the need of a plan to deal with emergencies has been demonstrated. Inventory and planning will minimize injuries to cultural resources in future emergencies and will also reduce costs. Equally important is the need for revision of the laws and regulations governing response, damage assessment, and restoration processes so as to clearly address cultural resources as a class of resource needing protection, vulnerable to injury, and eligible for restoration consideration.

References

Betts, R., C. Wooley, C. Mobley, J. Haggerty, and A. Crowell. 1991. Site protection and oil spill treatment at SEL-188: an archaeological site in Kenai Fjords National Park, Alaska. Report submitted by Exxon Shipping Co. and Exxon Co., U.S.A. for Alaska Department of Natural Resources land use permit SCV 90-005, state of Alaska archaeology permit 90-2, National Park Service (NPS) special use permit ARO-9845-9500-008, and NPS Archaeological Resources Protection Act permit 89-Kenai Fjords-ARO-001.

Haggerty, J., C. Wooley, J. Erlandson, and A. Crowell. 1991. The 1990 Exxon cultural resource program: site protection and maritime cultural ecology in Prince William Sound and the Gulf of Alaska. Report submitted by Exxon Shipping Co. and Exxon Co., U.S.A. for state of Alaska archaeology permit 90-2, U.S. Forest Service special use permit 4118.01 (amendment 1), National Park Service (NPS) Archaeological Resources Protection Act (ARPA) permit 89-KATMAI-ARO-003, NPS special use permit ARO-KATMAI-9500-012, NPS ARPA permit 89-KENAI FJORDS-ARO-001, NPS special use permit ARO-9845-9500-008. NPS ARPA permit 90-KENAI FJORDS-ARO-001, U.S. Fish and Wildlife Service (USFWS) ARPA permit R7(MAR)90-1, USFWS special use permit 47921, and USFWS special use permit 50866.

Jesperson, M. M., and K. Griffin. 1992. An evaluation of archaeological injury documentation, *Exxon Valdez* oil spill. Manuscript report on file, Department of Natural Resources, Office of History and Archaeology, and National Park Service, Anchorage, Alaska.

Mifflin and Associates. 1991. *Exxon Valdez* oil spill damage assessment contamination of archeological materials, Chugach National Forest: radiocarbon experiments and related analyses. Final report. U.S. Forest Service Contract 53-0109-1-00305, Juneau.

Mobley, C., and nine coauthors. 1990. The 1989 *Exxon Valdez* cultural resource program. Report submitted by Exxon Shipping Co. and Exxon Co., U.S.A. for state of Alaska archaeology permit 89-5, Alaska Department of Natural Resources special use permit SCV-89-004, U.S. Forest Service special use permit 4118.01, National Park Service (NPS) Archaeological Resources Protection Act (ARPA) permit 89-KATMAI-ARO-003, NPS special use permit ARO-KATMAI-9500-019, NPS ARPA permit 89-KENAI FJORDS-ARO-001, NPS special use permit ARO-9845-9500-008. NPS ARPA Permit 89-KENAI FJORDS-ARO-001, U.S. Fish and Wildlife Service ARPA permit R7(MAR)89-4.

Reger, D.R., J. D. McMahan and C.E. Holmes. 1992. Effect of crude oil contamination on some archaeological sites in the Gulf of Alaska, 1991 investigations. Alaska Department of Natural Resources. Office of History and Archaeology Report 30, Anchorage.

RFSUNY (Research Foundation of the State University of New York). 1993. *Exxon Valdez* oil spill archaeological damage assessment. Final report. U.S. Forest Service Contract 53-0109-1-00325, Juneau, Alaska.

American Fisheries Society Symposium 18:819–836, 1996

Subsistence Uses of Fish and Wildlife before and after the *Exxon Valdez* Oil Spill

JAMES A. FALL

Division of Subsistence, Alaska Department of Fish and Game, 333 Raspberry Road, Anchorage, Alaska 99518, USA

L. JAY FIELD

National Oceanic and Atmospheric Administration, Hazardous Materials Response and Assessment Division, 7600 Sand Point Way, N.E., Seattle, Washington 98115, USA

Abstract.—The *Exxon Valdez* oil spill fouled waters and beaches used for subsistence hunting, fishing, and gathering by 2,200 residents of 15 Alaska Native communities. In the year after the spill, subsistence harvests declined by as much as 77% compared to pre-spill averages in 10 of these villages, primarily because of the villagers' uncertainty about the safety of using subsistence foods that might have been contaminated by the oil. An Oil Spill Health Task Force formed to respond to these concerns. The task force coordinated a subsistence foods testing program and communicated the program's findings to the affected users. During the 3 years of these studies, marine invertebrate, fish, and marine mammal tissue samples were analyzed by gas chromatography–mass spectrometry for aromatic contaminants (ACs). Most samples had AC concentrations less than 10 parts per billion (ppb). An expert toxicological committee convened to review the study findings. The committee concluded that finfishes from all areas were safe to eat, but that intertidal invertebrates from specific areas, which had reported concentrations of greater than 1,000 ppb, should not be consumed. The success of the risk communication program in the villages was mixed, in part because of the limited number of sampling locations and samples and the long intervals between the collection of samples and communication of the interpreted results. By the end of the third year after the spill, subsistence harvest levels had rebounded but remained below pre-spill norms in several communities, especially those of Prince William Sound. Although there was evidence of recovery of subsistence uses, some households returned to using subsistence foods despite continued misgivings because of economic and cultural reasons.

When the tanker *Exxon Valdez* ran aground off Bligh Reef on 24 March 1989, it dumped almost 44 million liters of crude oil into eastern Prince William Sound. Eventually, the currents and tides carried oil, mousse, sheen, and tar balls more than 900 km along Alaska's southern coast, fouling waters and beaches used for subsistence hunting, fishing, and gathering by 15 predominantly Alaska Native (Alutiiq or "Aleut") communities with about 2,200 people (Figure 1). For at least 7,000 years, Alaska's people have depended upon these lands and waters for survival (Clark 1984). Also critical to this survival has been the people's knowledge about the natural environment and their confidence in their abilities to understand and interpret this environment. In the 1980s, subsistence uses of natural resources remained culturally, socially, and economically important to the communities of the oil spill area and were particularly vital in the 15 Alaska Native communities.

This article reviews changes to subsistence uses of fish and wildlife resources in these communities in the 3 years after the oil spill. Concern about the safety of using subsistence foods that may have been contaminated by the oil was the primary reason for declines in subsistence harvests. The article also summarizes the findings of studies conducted to determine if subsistence resources had been contaminated and rendered unsafe for human consumption. We also describe the efforts to communicate these studies' findings to the affected communities.

Methods: Estimating Subsistence Harvests and Use

The 15 study communities were Tatitlek and Chenega Bay in Prince William Sound; Nanwalek (formerly English Bay) and Port Graham on lower Cook Inlet; Akhiok, Karluk, Larsen Bay, Old Harbor, Ouzinkie, and Port Lions in the Kodiak Island Borough; and Chignik Bay, Chignik Lagoon, Chignik Lake, Ivanof Bay, and Perryville along the Alaska Peninsula. Before the spill, the Division of Subsistence of the Alaska Department of Fish and Game (ADFG) had conducted baseline studies in all 15 study communities. Estimates of harvest quantities and levels of participation in subsistence uses were available for one pre-spill year for seven

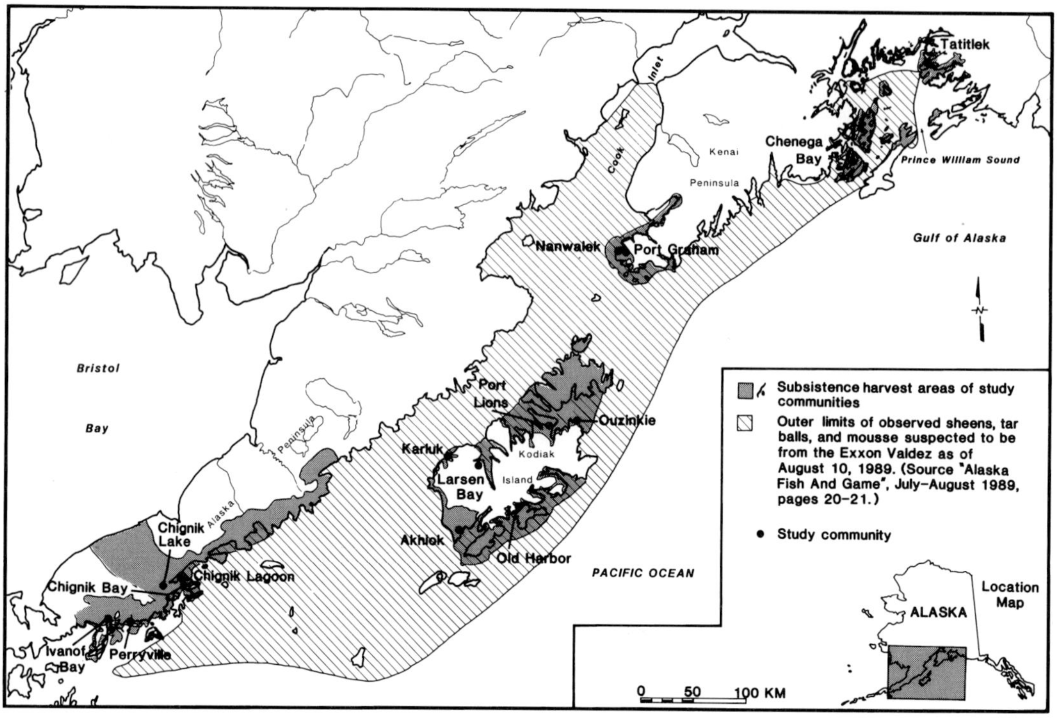

FIGURE 1.—Subsistence resource harvest areas of study communities and outer extent of *Exxon Valdez* oil spill.

of the study communities and for two pre-spill years for the other eight communities. These data are reported in the division's Community Profile Database (Scott et al. 1992) and in its reports in the Technical Paper Series (Stanek 1985; Stratton and Chisum 1986; Morris 1987; Schroeder et al. 1987; Stratton 1990).

Methods used for the postspill research were similar to those used in previous studies. Subsistence harvest and use data were collected during systematic household interviews using a standardized data-gathering instrument. Most interviews were conducted in the respondents' homes by Division of Subsistence personnel. The survey instrument lists each type of fish, marine invertebrate, mammal, and bird known to be used in the study communities. Respondents estimated their households' annual harvests of each resource, largely through retrospective recall, for a 12-month study period (usually a calendar year or from spring to spring). In addition to estimating harvest quantities, respondents reported whether their households used, attempted to harvest, received, or gave away each resource during the study year. Besides resource use data, demographic and employment information was also collected during the interviews (Fall 1990, 1991, 1992).

Through 1993, the Division of Subsistence had administered household harvest surveys in about 180 Alaska communities. It is the experience of the division, as well as of other researchers studying subsistence harvests by native communities of northern North America (Usher and Wenzel 1987: 156–157), that such surveys provide reliable estimates of subsistence harvests. For the oil spill area communities, there are a number of reasons for confidence in the harvest estimates. First, division researchers had worked in all of these communities before the spill and had generally developed a good level of trust with them. This is considered a key to the collection of accurate harvest information (Usher and Wenzel 1987:156). Second, levels of participation in the interviews were very high both before and after the spill, suggesting that households did not choose to participate in the postspill interviews for strategic purposes. Third, survey questions addressed subsistence harvests and uses directly without reference to the oil spill itself in order to avoid influencing responses. Because of the decentralized nature of subsistence harvesting in these communi-

TABLE 1.—Sample sizes (number of households) in Division of Subsistence harvest surveys in communities affected by the *Exxon Valdez* oil spill, 1989–1991. Random samples were selected in Old Harbor, Ouzinkie, and Port Lions for 1989, and in Old Harbor and Ouzinkie for 1991. In all other cases, an attempt was made to interview all households.

Community or variable	Study year 1989			Study year 1990			Study year 1991		
	Total	Surveyed	Percent	Total	Surveyed	Percent	Total	Surveyed	Percent
Chenega Bay	21	18	85.7%	21	18	85.7%	22	18	81.8%
Tatitlek	28	22	78.6%	28	17	60.7%	27	19	70.4%
Subtotal, Prince William Sound	49	40	81.6%	49	35	71.4%	49	37	75.5%
Nanwalek	41	33	80.5%	41	35	85.4%	41	29	70.7%
Port Graham	61	48	78.7%	55	46	83.6%	58	49	84.5%
Subtotal, Lower Cook Inlet	102	81	79.4%	96	81	84.4%	99	78	78.8%
Akhiok	13	10	76.9%						
Karluk	17	14	82.4%	19	17	89.5%	15	13	86.7%
Larsen Bay	39	34	87.2%	40	35	87.5%	43	38	88.4%
Old Harbor	93	48	51.6%				66	42	63.6%
Ouzinkie	69	35	50.7%	59	53	89.8%	55	32	58.2%
Port Lions	67	36	53.7%						
Subtotal, Kodiak Island	298	177	59.4%	118	105	89.0%	179	125	69.8%
Chignik Bay	39	35	89.7%				44	30	68.2%
Chignik Lagoon	15	15	100.0%						
Chignik Lake	28	21	75.0%				33	24	72.7%
Ivanof Bay	7	7	100.0%						
Perryville	31	27	87.1%						
Subtotal, Alaska Peninsula	120	105	87.5%				77	54	70.1%
Grand total	569	403	70.8%	263	221	84.0%	404	294	72.8%

ties (subsistence fishing and hunting are not community-wide activities, but are undertaken by smaller family groups), harvest estimates were obtained for each household and then aggregated, rather than coming from a single estimate from a community representative or key respondent, thereby making it extremely unlikely that communities could strategically manufacture the harvest estimates. In addition, division researchers assessed each household's harvest estimates to identify potentially anomalous or spurious data. In such cases, an attempt was made to obtain additional information to verify the response and, as necessary, the questionable data were not used. Finally, the relationships displayed by the data appear consistent with what one would expect intuitively, with communities more distant from the origin of the spill showing fewer effects, and communities through time showing reduced spill impacts. These findings of consistency with intuitive expectations further confirm the reliability of the information. In summary, the data on subsistence uses in the oil spill communities have been gathered using reliable methods that have been established through years of research in these villages as well as in other communities throughout the state.

Generally, researchers attempted to interview representatives of all permanent households (census sample) in the smaller communities and a 50% random sample in the larger communities. Random samples were chosen in Ouzinkie, Port Lions, and Old Harbor for 1989 and in Ouzinkie and Old Harbor for 1991. A census sample was attempted in Ouzinkie in 1991. Sample achievements for each round of interviews are summarized in Table 1.

In early 1990, household surveys were conducted in all 15 study communities (Table 1). Harvest and use data pertain to the 1989 calendar year with the exception of Chenega Bay and Tatitlek, for which an April 1989 through March 1990 study year was used. A total of 403 household interviews were conducted, for a sampling fraction of 70.8%. In early 1991, surveys pertaining to the second postspill year (April 1990 through March 1991) were conducted in seven communities. These were Chenega Bay, Tatitlek, Nanwalek, Port Graham, Karluk, Larsen Bay, and Ouzinkie. A total of 221 households were interviewed, for a sample of 84.0%. In 1992, interviews pertaining to the third postspill year (April 1991 through March 1992) took place in 10 of the communities. These were Chenega Bay, Tatitlek, Nanwalek, Port Graham, Karluk, Larsen Bay, Old Harbor, Ouzinkie, Chignik Bay, and Chignik Lake. The total number of household interviews completed in 1992 was 294, for a sample of 72.8%.

Data were coded for computer entry and analysis using the Statistical Package for the Social Sciences

TABLE 2.—Percentage of households participating in subsistence uses in the study communities before the *Exxon Valdez* oil spill. Sources for these data are Scott et al. (1992), Stratton (1990), Stratton and Chisum (1986), Schroeder et al. (1987), and Morris (1987).

Community	Study year	Using resources	Attempting a harvest	Harvesting resources	Receiving resources	Giving away resources
Chenega Bay	1985–1986	100%	100%	100%	94%	88%
Tatitlek	1988–1989	100%	100%	100%	100%	95%
Nanwalek	1987	97%	94%	94%	94%	94%
Port Graham	1987	100%	100%	100%	98%	82%
Akhiok	1982–1983	100%		100%	86%	76%
Karluk	1982–1983	100%		90%	100%	90%
Larsen Bay	1982–1983	100%		94%	97%	88%
Old Harbor	1982–1983	100%		100%	82%	78%
Ouzinkie	1982–1983	100%		97%	91%	84%
Port Lions	1982–1983	100%		95%	84%	76%
Chignik Bay	1984	100%	84%	84%	95%	79%
Chignik Lagoon	1984	100%	88%	88%	82%	71%
Chignik Lake	1984	100%	100%	100%	96%	83%
Ivanof Bay	1984	100%	100%	100%	100%	83%
Perryville	1984	100%	100%	100%	100%	100%

(SPSS) program. Harvest data reported in numbers of animals, gallons, or other units were converted to usable weights using standard conversion factors.

Subsistence Uses Before the Spill

Division of Subsistence research has demonstrated the continuing significance of subsistence hunting, fishing, and gathering for the communities of Prince William Sound, lower Cook Inlet, the Kodiak Island Borough, and the Alaska Peninsula. Tables 2, 3, and 4 summarize some information about subsistence uses in the 15 study communities in the 1980s. The studies found that participation in subsistence uses was almost universal in these villages in the 1980s. Almost every household used, harvested, and participated in the noncommercial exchange of wild foods (Table 2).

The same general categories of subsistence resources are available in all four regions. These categories are salmon (five species); other fish, such as

TABLE 3.—Average number of subsistence resources used, attempted to be harvested, and received per household in the study communities affected by the *Exxon Valdez* oil spill, from before the spill to 1991.[a]

	Used per household				Attempted to be harvested per household				Received per household			
Community	Pre-spill average	1989	1990	1991	Pre-spill average	1989	1990	1991	Pre-spill average	1989	1990	1991
Chenega Bay	18.8	7.2	9.8	13.2	14.1	5.1	6.2	9.6	9.8	3.5	5.4	8.0
Tatitlek	19.1	10.3	12.5	17.3	11.7	6.2	8.4	13.1	12.2	5.8	7.1	11.9
Nanwalek	25.0	13.7	22.4	21.2	18.0	9.9	15.4	14.9	15.0	6.9	13.1	12.8
Port Graham	21.5	11.2	17.4	22.0	14.9	7.6	12.1	14.7	10.6	6.4	9.3	13.4
Ouzinkie	18.3	7.6	13.9	15.2	14.0	5.7	9.5	11.1	7.9	3.2	6.7	7.4
Port Lions	14.0	9.4			10.5	6.8			5.6	3.8		
Larsen Bay	15.4	12.7	16.5	15.4	9.4	7.9	10.4	9.8	8.4	6.6	10.5	8.8
Karluk	17.0	10.9	13.2	13.5	11.6	8.3	9.2	10.5	6.6	5.4	8.4	8.7
Old Harbor	15.2	13.6		16.7	12.7	9.5		10.6	7.3	7.4		10.7
Akhiok	12.6	18.0			11.8	14.2			1.7	8.6		
Chignik Bay	12.5	14.6		14.7	8.4	10.9		9.4	7.1	7.1		8.7
Chignik Lagoon	10.7	13.6			6.9	7.8			5.2	8.3		
Chignik Lake	16.0	19.0		21.4	9.4	13.1		12.7	10.1	10.9		11.6
Perryville	21.6	19.9			14.1	13.4			13.4	12.5		
Ivanof Bay	18.8	26.9			15.2	22.3			9.0	16.7		

[a]In all regions except lower Cook Inlet (Nanwalek and Port Graham), the list of resources used to collect resource-use information was less detailed in some pre-spill years than that used to collect postspill data. Therefore, for comparative purposes, postspill estimates of household averages have been adjusted using the less detailed lists for the other 13 communities (total) in Prince William Sound, Kodiak Island Borough, and the Alaska Peninsula.

TABLE 4.—Subsistence harvests in kilograms per person in the study communities before and after the *Exxon Valdez* oil spill.[a]

Community	Pre-spill subsistence harvests			Postspill subsistence harvests		
	Year one	Year two	Average	1989	1990	1991
Chenega Bay	143.5	169.7	157.0	67.0	63.1	156.6
Tatitlek	159.5	291.9	219.0	97.4	69.3	156.9
Nanwalek	131.0			63.9	82.2	117.4
Port Graham	103.1			55.5	97.1	127.4
Akhiok	235.6	73.6	147.8	135.0		
Karluk	391.5	174.7	280.4	113.6	179.7	121.9
Larsen Bay	183.0	94.8	140.0	96.2	156.3	133.6
Old Harbor	222.8	192.0	206.9	123.6		177.3
Ouzinkie	167.4	182.7	174.4	40.3	93.1	95.0
Port Lions	126.9	151.1	139.2	66.5		
Chignik Bay	85.2			92.9		159.3
Chignik Lagoon	99.9			94.0		
Chignik Lake	126.6			202.1		197.7
Ivanof Bay	206.7			217.4		
Perryville	177.4			175.0		

[a]No pre-spill plant harvest estimates are available for the five Alaska Peninsula communities (Chignik Bay, Chignik Lagoon, Chignik Lake, Ivanof Bay, and Perryville). Therefore, for purposes of comparison, plant harvests for these communities have been removed from the 1989 and 1991 data.

halibut, rockfish, cod, and Dolly Vardens; marine invertebrates, such as clams, chitons, crabs, and octopuses; land mammals, such as deer in Prince William Sound and Kodiak Island, black bears and mountain goats on the lower Kenai Peninsula and Prince William Sound, and caribou on the Alaska Peninsula; marine mammals (harbor seals and sea lions); birds (ptarmigan, waterfowl, seabirds, gull and other eggs); and wild plants. Consequently, subsistence harvests in the 1980s were diverse in these villages. For example, in study years before the spill, households in Tatitlek used an average of 19.1 different kinds of wild foods, households in Perryville used 21.6 types, and families in Ouzinkie used 18.3 varieties (Table 3).

Also, in the 1980s, each community followed a patterned seasonal round of harvest activities that was largely conditioned by resource availability. Spring (March, April, May) is a period of renewed subsistence activities, during which harvests of herring, clams, birds, and other resources, occur. Summer is traditionally the busiest time of year, when people harvest and preserve large quantities of salmon for winter use. Fall is important for land-mammal hunting, waterfowl hunting, marine mammal hunting, and marine-invertebrate gathering.

Another important characteristic of subsistence activities in rural Alaska villages such as those in the oil spill area is that they are usually organized by kinship relationships. Harvesting and processing groups are generally composed of members of extended families, and subsistence foods are usually shared with relatives, elders, and others in need. In a case example from Nanwalek, one harbor seal was shared within an extended family of eight households with 25 people (Stanek 1985:170–171). Such extensive sharing is commonplace in all 15 villages.

Before the spill, subsistence harvests in the study communities ranged from about 70 kg per person per year to more than 275 kg per person annually (Table 4). These are substantial harvests, considering that the average family in the western United States purchased about 100 kg of meat, fish, and poultry per person each year in the late 1970s (USDA 1983). Subsistence harvests in the oil spill communities also far exceeded those of more populated, urbanized areas of Alaska (Wolfe and Walker 1987).

Subsistence Uses After the Spill: The First Year

This section will examine changes in four aspects of subsistence harvests and uses in the first year after the spill. These aspects are harvest quantities, diversity of resource uses, participation in hunting and fishing activities, and sharing of subsistence resources. It will also provide information on respondents' explanations of these changes.

Table 4 summarizes survey results on levels of subsistence harvests in kilograms usable weight per person. As shown in Figure 2, subsistence harvests declined in 1989 compared to pre-spill averages in all 10 of the study communities of Prince William

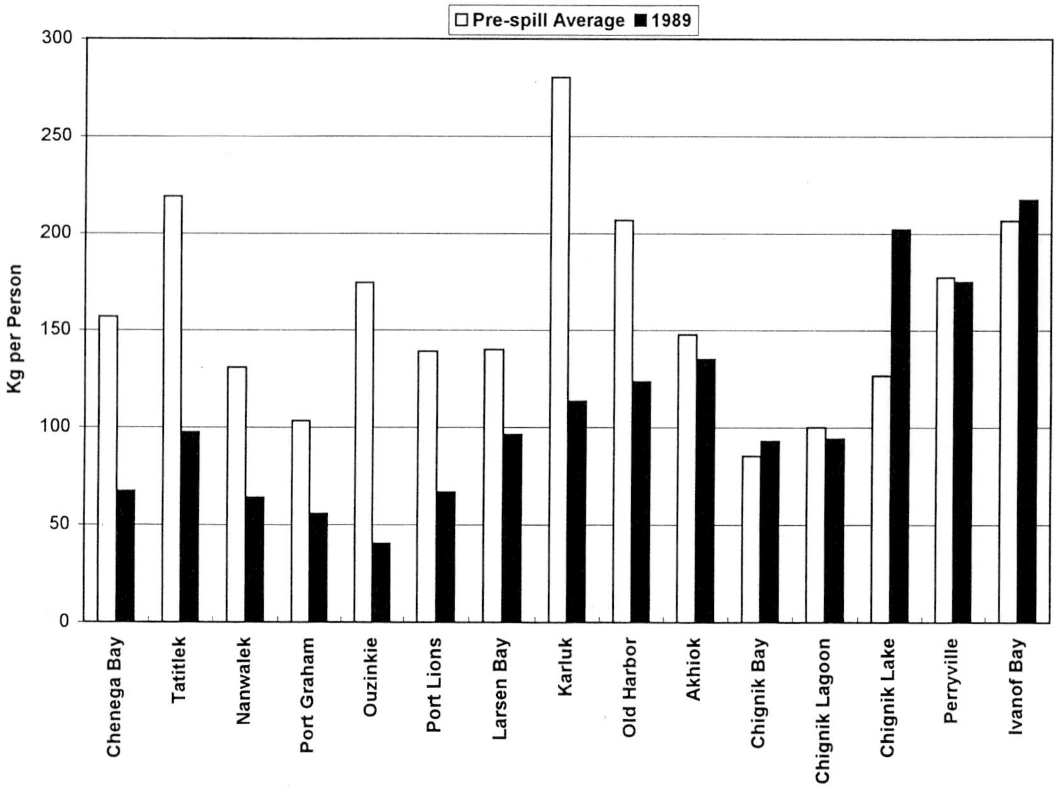

FIGURE 2.—Subsistence resource harvests before and after the 1989 *Exxon Valdez* oil spill.

Sound, lower Cook Inlet, and the Kodiak Island Borough. As illustrated in Figure 3, these declines were substantial (except at Akhiok), ranging from a drop of 31% at Larsen Bay to a reduction of 77% at Ouzinkie. Furthermore, these large declines occurred in almost all resource categories except wild plants in each of these communities (Scott et al. 1992).

These data may underestimate the extent of the decline in subsistence harvests after the spill in several communities. For example, the only pre-spill estimates for Chenega Bay pertain to the first 2 years after the village's resettlement on Evans Island in western Prince William Sound in 1984. The former village on Chenega Island had been destroyed by a tsunami after the great Alaska earthquake of 1964. As the residents of Chenega Bay regained familiarity with harvest areas after a 20-year absence, harvest levels probably increased (Stratton and Chisum 1986:113). Also, harvest data for 1989 (the postspill study year for all communities except Chenega Bay and Tatitlek) include resources that were taken before the spilled oil reached subsistence harvest areas. For example, when villagers from Nanwalek and Port Graham learned that the spilled oil would reach their communities, they made a strong effort to harvest marine invertebrates for fear that these resources would soon be unavailable.

A different pattern was found for the five Alaska Peninsula study communities, those farthest from the source of the spill. Subsistence harvest estimates for 1989 for four of these communities (Chignik Bay, Chignik Lagoon, Perryville, and Ivanof Bay) were virtually the same as the single (1984) pre-spill estimate (Table 4; Figures 2 and 3). In the fifth Alaska Peninsula community, Chignik Lake, the estimate for 1989 of 202.1 kg per person was 60% higher than the previous estimate of 126.6 kg, pertaining to 1984. Chignik Lake is the most inland of the 15 study communities. In addition to marine resources, this community is highly dependent upon caribou (a resource for which little concern about oil contamination was expressed in 1989; see below). The harvest estimate for caribou in 1989 for Chignik Lake was 78 kg per person, compared to

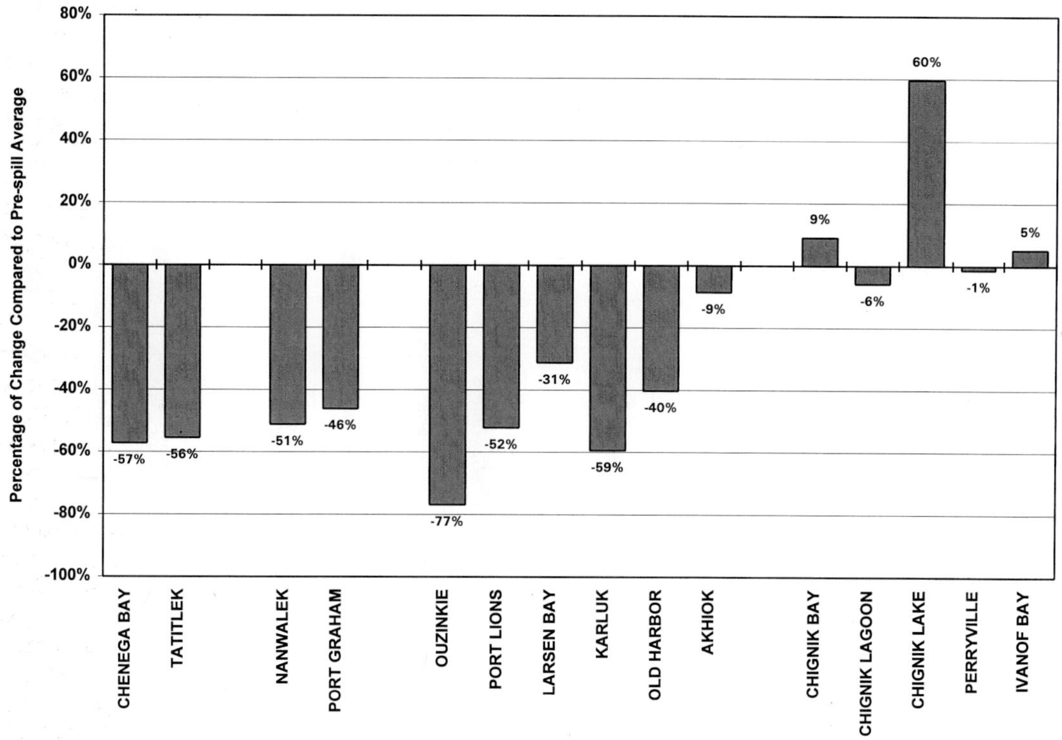

FIGURE 3.—Changes in subsistence resource harvest levels after the *Exxon Valdez* oil spill as compared to pre-spill averages (see Figure 2).

about 36 kg in 1984, accounting for much of the difference between the 2 years. Evidently, although the subsistence harvests of some resources by households of Chignik Lake and the other Alaska Peninsula communities were disrupted for at least a portion of 1989, they substituted other resources or made up for lost harvest opportunities later in the year. In contrast, households in the other regions had few alternative resources, and disruptions lasted for a full year or more.

The diversity of resources used for subsistence purposes also declined markedly in most Prince William Sound, lower Cook Inlet, and Kodiak Island Borough communities in the year after the spill. This is demonstrated by reductions in the average number of different kinds of subsistence resources used per household. For example, as shown in Table 3 and Figure 4, the range of wild foods used per household in Tatitlek dropped from an average of 19.1 kinds before the spill to 10.3 kinds in 1989–1990. Declines of a similar magnitude were also recorded for Chenega Bay, Nanwalek, Port Graham, and Ouzinkie. Lesser, but still notable, reductions in resource-use diversity also occurred in Port Lions, Larsen Bay, Karluk, and Old Harbor.

No direct estimates of subsistence harvest effort, such as days spent hunting and fishing, had been made in the study communities before the spill. However, indirect measures of levels of household and community involvement in hunting and fishing activities are available. One is the average number of kinds of resources that households attempted to harvest during the study year. As shown in Table 3 and Figure 5, this average dropped in 9 of the 10 study communities of Prince William Sound, lower Cook Inlet, and the Kodiak Island Borough in a pattern very similar to that of the average number of resources used per household. As with resource use diversity, the largest declines occurred in Tatitlek, Chenega Bay, Nanwalek, Port Graham, and Ouzinkie. These results strongly suggest that households in these communities spent less time engaged in subsistence activities during the oil spill year than before the spill.

Further evidence of reduced harvests and harvest

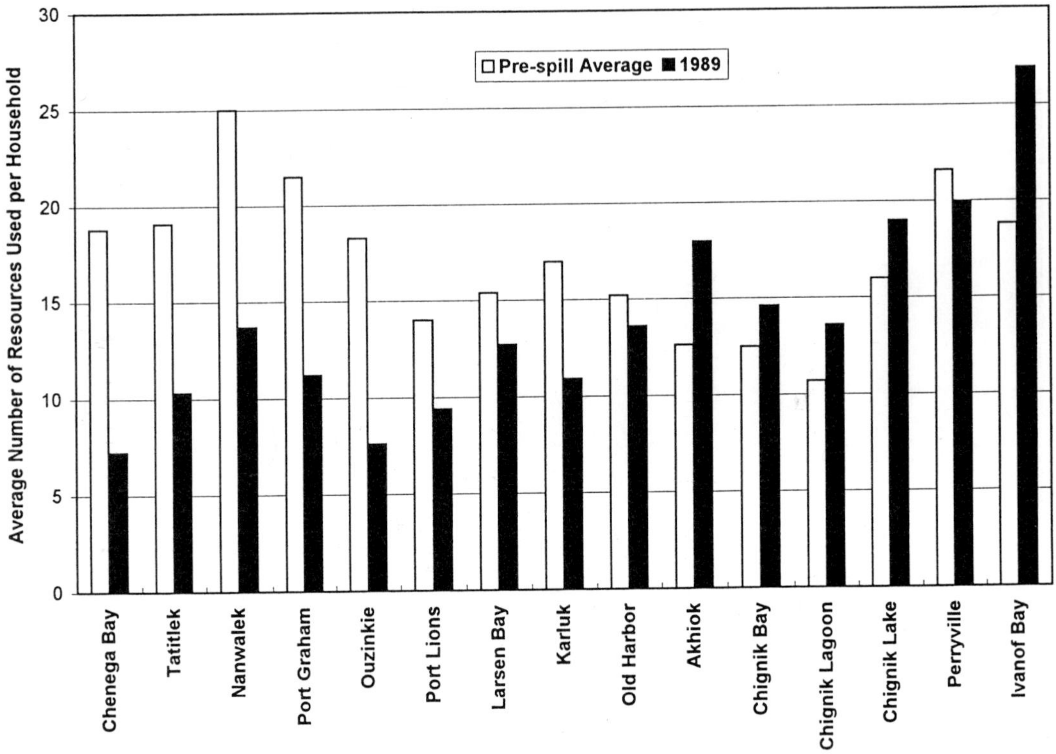

FIGURE 4.—Average number of subsistence resources used per household before (pre-spill average) and after (1989) the *Exxon Valdez* oil spill.

effort in the study communities in the year after the spill is provided in the declines in the average number of resources received per household (Table 3 and Figure 6). This average also declined in 1989 in a pattern similar to that of the average number of resources used per household and the average number attempted to harvest. On average, households received far fewer kinds of wild foods from other harvesters in 1989 than in earlier years in the 1980s. This reduction also indicates that resource-sharing patterns, central to community social organization and cultural values, were disrupted in the year after the spill.

During the interviews in early 1990, households were asked if, in their assessment, they had experienced a decline in their uses of particular subsistence resources. If so, they were asked to provide a reason for the decline in use. As shown in Figure 7, 92% of the households in the Prince William Sound communities and 78% in the lower Cook Inlet communities said that their harvests of at least one kind of subsistence food had gone down in 1989 as a result of their concerns about hydrocarbon contamination. In the Kodiak Island Borough communities, 30% of the households cited contamination concerns as a cause of reduced subsistence harvests, as did 23% of the interviewed households in the five Alaska Peninsula communities. Other reasons cited for reduced subsistence harvests included the time harvesters spent on the oil spill cleanup and the perception that fewer resources were available because of spill-induced mortality. Because of the latter, many households reduced their harvest effort as a voluntary conservation measure.

The study findings for the first postspill year demonstrate that the *Exxon Valdez* oil spill severely disrupted subsistence uses in the communities of the spill area, primarily because of oil contamination concerns. Members of most households in the study communities had direct contact with the effects of the spill during travel through their traditional harvest areas or while working on oil spill cleanup activities. They saw oil on the beaches, in the water, and on certain animals and birds. Others suspected oiling when they inspected resources they had harvested or had been given. Doubts arose

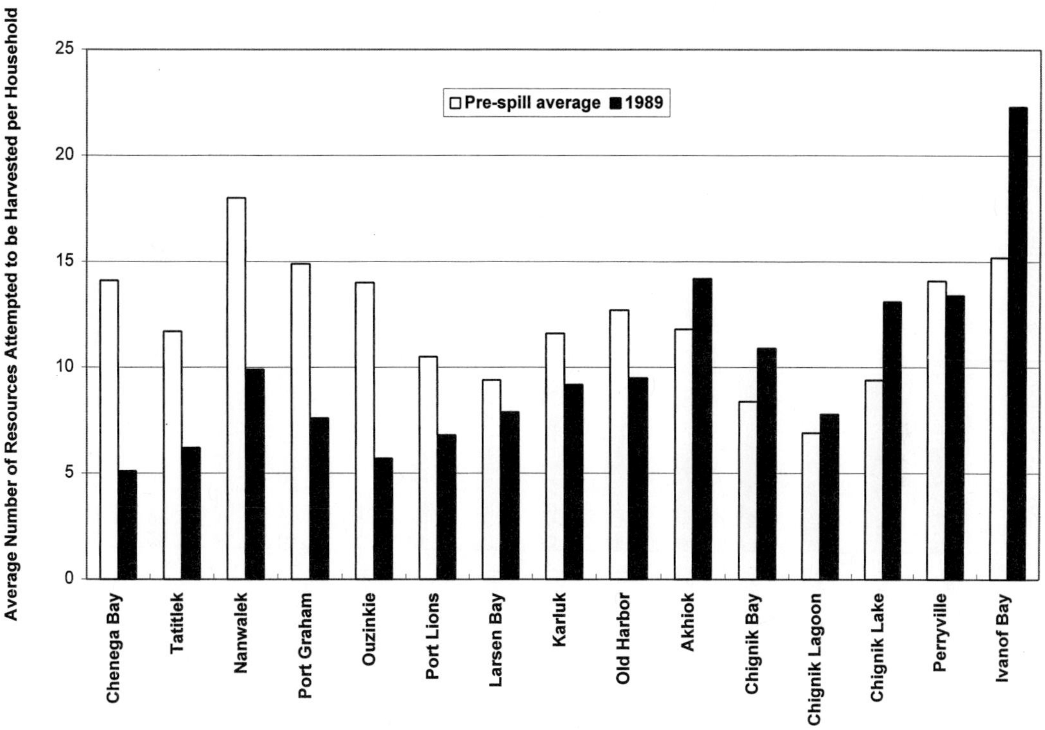

FIGURE 5.—Average number of subsistence resources attempted to be harvested per household before and after the *Exxon Valdez* oil spill.

about the safety of using any subsistence resource from the oil spill area, given the clear extent of the damage to some wildlife populations. In short, the oil spill had created conditions that were completely unfamiliar to the hunters and fishers of these villages. Their skills in understanding their environment and making informed decisions had been undermined. Consequently, subsistence harvesters acted in a culturally appropriate manner, with caution. In many cases, they refrained from harvesting or using traditional foods for fear that the resources had been poisoned. Because this prevailing concern about resource contamination included most kinds of resources (with the exception of wild plants and some large game such as caribou and moose), it also discouraged the substitution of one category of wild foods for another.

Response to Subsistence Food Safety Concerns

Oil Spill Health Task Force

One of the first responses to subsistence concerns was the formation of the Alaska Oil Spill Health Task Force (OSHTF), a group composed of representatives from the state of Alaska (departments of Fish and Game, Health and Social Services, and Environmental Conservation), federal agencies (Indian Health Service, National Oceanic and Atmospheric Administration [NOAA]), Alaska Native organizations (the North Pacific Rim and Kodiak Area Native Association), and Exxon Corp. The OSHTF was formed soon after the spill to address public health concerns resulting from the spill. The OSHTF served as a forum for the design of the subsistence food safety testing study and the communication of the results and conclusions derived from the study to the subsistence communities.

Resource Sampling and Testing Program

In order to put together the most comprehensive and credible study possible, NOAA and Exxon Corp. signed a Memorandum of Understanding that had three principal features: (1) a NOAA representative would accompany the Exxon Corp. collection team to the villages and participate in the sample collections; (2) tissue samples from fish and shellfish subsistence resources would be analyzed for aromatic contaminants by the NOAA National

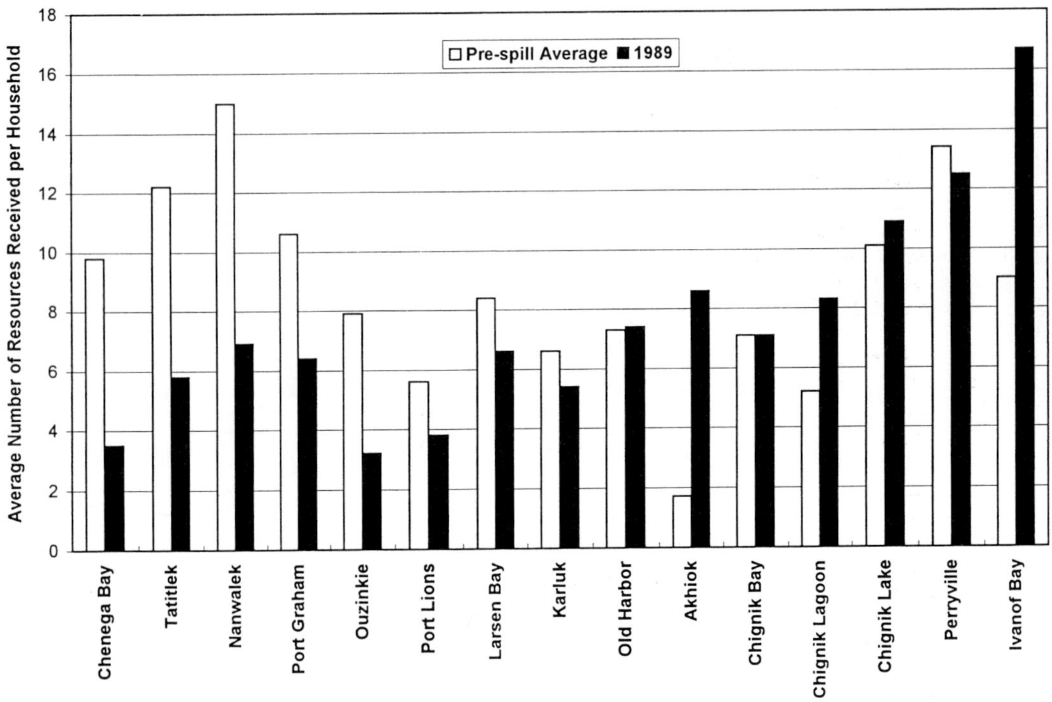

FIGURE 6.—Average number of subsistence resources received per household before and after the *Exxon Valdez* oil spill.

Marine Fisheries Service Environmental Conservation Division laboratory in Seattle, Washington; and, (3) all data from the study would be made public. These steps were taken to ensure that the validity of the analytical data would be unquestioned and that the results and their implications for food safety would be made available to public health officials and the communities in a timely manner. The OSHTF provided a forum for review of study plans and results and was instrumental in determining the best methods for communicating the goals and results of the study to the villages.

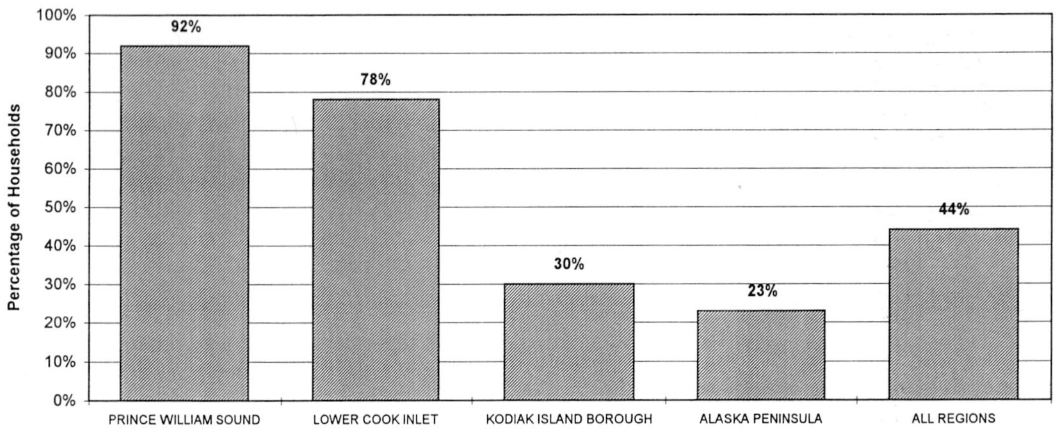

FIGURE 7.—Percentage of households reporting reductions in subsistence resource harvests after the *Exxon Valdez* oil spill because of contamination concerns.

The study area included subsistence seafood collection areas near villages in Prince William Sound (Tatitlek and Chenega Bay); lower Cook Inlet (Port Graham and Nanwalek); and Kodiak Island (Kodiak, Chiniak, Larsen Bay, Karluk, Akhiok, Old Harbor, Ouzinkie, and Port Lions). Four subsistence areas on the Alaska Peninsula (Chignik, Perryville, Ivanof Bay, and Kashvik Bay) were sampled by ADFG in 1990. Reference samples were collected outside the spill area in southeastern Alaska near the village of Angoon in 1989 and Yakutat in 1990.

The initial sampling plan called for approximately equal numbers of samples from each village area, although there were considerable differences in the potential degree of impact from *Exxon Valdez* oil. Sampling sites that represented important subsistence use areas were selected in consultation with village representatives. The degree of oiling was not a factor in site selection. Two subsistence beaches for the collection of intertidal shellfish were identified in each village area. Target species included several species of intertidal shellfish (mussels, two species of clams, and chitons), bottomfish (primarily halibut), and all five species of salmon. Other species were also collected but in much smaller numbers.

Between 1989 and 1991, a total of more than 1,000 shellfish tissue samples were collected from 79 stations (Table 5). Tissue samples were analyzed by gas chromatography–mass spectrometry (GC–MS) for major unsubstituted and alkylated aromatic compounds found in the Prudhoe Bay crude oil spilled by the *Exxon Valdez*. Fish were screened for exposure to petroleum hydrocarbons by analysis of bile samples from fish gall bladders for the presence of metabolites of aromatic contaminants. The results of the bile analyses were used primarily to prioritize samples of edible tissue for analysis by GC–MS. More than 300 samples of fish muscle tissue were analyzed for aromatic contaminants in 1989 and 1990. Liver, blubber, muscle, and kidney tissue samples from 33 harbor seals and 10 sea lions collected in 1989 and 1990 from Prince William Sound and the Gulf of Alaska were also analyzed for aromatic contaminants. Unlike the invertebrate and fish samples collected for the subsistence study, many of these animals were deliberately collected from areas with a high potential for exposure to *Exxon Valdez* oil as part of a Natural Resources Damage Assessment study. However, because of the importance of marine mammals as a subsistence resource, tissue samples were analyzed as part of the subsistence study in order to provide information on the potential contamination of marine mammal tissue to the subsistence communities in a timely manner. (For more details about the analytical methods and results from this study, see Varanasi et al. 1993; Brown et al. 1996 and Hom et al. 1996; both this volume).

TABLE 5.—The number of samples of invertebrates and fishes and the number of individual marine mammals analyzed in the subsistence study from 1989 to 1991.

Type of sample	1989	1990	1991	Total
Invertebrates	156	700	205	1,061
Fishes	210	102		312
Marine mammals	28	15		43

Results of Subsistence Food Safety Study

The aromatic contaminants (ACs) measured in the tissue samples were classified into two general groups: the low molecular weight ACs (LACs) and the high molecular weight ACs (HACs). The LACs are generally considered to be less persistent in the environment than the HACs. The analytical results mostly were presented as total ACs to assess the relative degree of contamination in tissue samples at different sampling locations. However, the HACs, which represented approximately 25% of the total AC concentration in mussels from oiled sites, were considered the most important for potential long-term health effects and were used in the quantitative risk assessment (see Bolger et al. 1996, this volume).

Invertebrates

During the 3 years of sample collections, a total of five stations had one or more invertebrate tissue samples exceeding 1,000 ppb (ng/g, wet weight). Three of these stations—two small islands in Windy Bay on the outer Kenai Peninsula and a beach at the southern end of Elrington Island—had clear evidence of *Exxon Valdez* oil on the beach. Total AC concentrations in invertebrates at these stations decreased considerably between 1989 and 1991. A few other stations had one or more samples that exceeded 100 ppb. Most of the samples analyzed had total AC concentrations of less than 10 ppb, including the samples from the reference areas.

Fish

Although the results of the bile analyses indicated that fish species from some locations were exposed to petroleum hydrocarbons, only one fish tissue sample (pink salmon *Oncorhynchus gorbuscha*) had a concentration of total ACs exceeding 100 ppb. Most of the fish tissue samples had total AC concentrations less than 10 ppb.

Marine Mammals

Several tissues were analyzed from each marine mammal collected. Concentrations of total ACs in muscle, kidney, and liver tissue from 33 harbor seals and 10 sea lions, including animals that were visibly oiled and/or collected from heavily oiled areas, were less than 20 ppb and presumed to be comparable to background levels. Blubber samples from four visibly oiled harbor seals collected in 1989 had total AC concentrations exceeding 100 ppb, which were composed of almost 100% LACs.

TABLE 6.—Members of the expert toxicological committee that provided an external review of this study.

Name	Affiliation
Dr. Mike Bolger	Division of Toxicological Review, Food and Drug Administration
Dr. Larry Curceo	Exxon Corp., USA
Dr. John French	University of Alaska
Dr. Richard Griesemer	National Institute of Environmental Health Sciences
Dr. Anthony Guarino	Fishery Research Branch, Food and Drug Administration
Dr. Steve Lewis	Exxon Biomedical Research Sciences
Dr. John Middaugh	Alaska Department of Health and Social Services
Dr. Tom Nighswander	Alaska Native Medical Center
Dr. Ken Orloff	Agency for Toxic Substances and Disease Registry
Dr. Bill Thar	Exxon Biomedical Research Sciences
Dr. Usha Varanasi	Northwest Fisheries Science Center, National Marine Fisheries Service
Dr. Gary Winston	Louisiana State University

Assessment of Subsistence Food Safety

A review of the literature on oil spills revealed that there was little information available on the effects of oil spills on the safety of oil-contaminated seafood (Winston 1990). The National Research Council (NRC 1985), in one of the most extensive summaries of scientific research available on oil spills, concluded that there was no evidence that long-term consumption of oil-contaminated seafood was related to an increased incidence of cancer or other diseases in humans. However, the National Research Council (1985:482) concluded that "exceptions . . . may arise in localized areas, as in the case of isolated fishing villages where seafood constitutes a major portion of the annual diet."

Expert Committee

In September 1989 and again in February 1990, NOAA convened a meeting in Seattle of nationally recognized experts in several disciplines related to assessing human health effects from fisheries consumption. This group evaluated the significance of the analytical data from the perspective of human health and provided an external review of the study. The group included representatives from several federal agencies, the University of Alaska, and Exxon Biomedical Research Sciences (Table 6).

The group concluded that finfishes from all areas were safe to consume, but that intertidal invertebrates from specific areas (e.g., Windy Bay on the outer Kenai Peninsula and Near Island adjacent to the Kodiak boat harbor) should not be consumed. Shellfishes tested from a few other beaches had higher concentrations of aromatic hydrocarbons than shellfishes from uncontaminated areas but at levels that were not considered to represent a serious health hazard.

Risk Assessment

No Food and Drug Administration (FDA) advisory levels or other guidelines for the safety of foods contaminated with polycyclic aromatic hydrocarbons (PAHs) were available at the time of the spill. Thus, there were no pre-existing standards to evaluate the analytical data on the concentrations of aromatic contaminants in edible tissue and determine whether the levels observed were safe to consume without restriction. Many villagers felt that any level of contamination was unsafe. The chairman of the OSHTF formally requested an advisory opinion from the U.S. FDA on the safety of aromatic contaminant residues in subsistence foods that were affected by the oil spill. The advisory opinion and accompanying report were received in late summer 1990 (USFDA 1990). The advisory was based on a risk analysis using a benzo[*a*]pyrene equivalency approach for the concentrations of aromatic contaminants observed in subsistence foods and the reported consumption patterns in the village of Chenega Bay. The report concluded that there was only a low risk associated with consumption of unsmoked salmon and other finfishes and shellfishes, including oil-contaminated bivalves (see Bolger et al. 1996).

Communications

Two health bulletins (ADHSS 1989a, 1989b) were released by the state of Alaska epidemiologist in May and July 1989 that addressed human health concerns related to consumption of seafood from the region affected by the spill. The primary message in these bulletins about food safety was based on organoleptic testing—if seafood did not appear to be oil-contaminated by visual observation, smell, or taste, then it was probably safe to eat. Following the expert committee's review of the results from the first-year study, additional bulletins containing the committee's recommendations were distributed to the villages. The OSHTF representatives, including members from ADFG, NOAA, North Pacific Rim, Kodiak Area Native Association, the Indian Health Service, and Exxon Corp., traveled to each village in October 1989 to present the results, provide background on the toxicology of petroleum hydrocarbons, interpret the results from the tissue analysis, and answer questions. Analytical results were displayed on individual maps for each village area so that villagers could see where specific samples were collected and the levels of aromatic contaminants associated with each sample. In June 1992, a follow-up trip was made to the individual Prince William Sound and lower Cook Inlet villages and the city of Kodiak to present the results of the 3 years of testing and the implications for food safety.

Beginning in early 1990, the ADFG Division of Subsistence published a monthly newsletter for the OSHTF that reported summaries and discussions of the results of the food-testing program; the newsletter was mailed directly to each village resident. A video was prepared by ADFG on subsistence food safety issues that was provided to each village and aired on the Rural Alaska Television Network.

Subsistence Uses After the Spill: The Second and Third Years

Table 4 provides estimates of subsistence harvest levels in the study communities for the second and third years after the spill. As shown in Figure 8, subsistence harvests in five of the seven communities that were surveyed for 1990 showed increased harvest levels over 1989. These were Nanwalek, Port Graham, Ouzinkie, Larsen Bay, and Karluk. In the case of Port Graham, Larsen Bay, and Karluk, subsistence harvests in 1990 matched at least one pre-spill year as well. Although harvests were up at Nanwalek and Ouzinkie, they remained well below pre-spill levels. The average number of resources used per household increased notably for all five of these communities as well (Figure 9), as did the average number of resources attempted to harvest and the average number received (Table 3), suggesting that more time was being spent in subsistence activities.

In contrast, very little evidence of recovery of subsistence harvests in the second postspill year was found for the Prince William Sound villages of Chenega Bay or Tatitlek. For the former, the harvest estimate of 63.1 kg/person for 1990–1991 was virtually identical to that of the first postspill year and remained about 60% below the average of pre-spill estimates (Table 4). For Tatitlek, a further decrease in subsistence harvests was recorded for 1990–1991, down to 69.3 kg/person, compared to 97.4 kg for 1989–1990 and a pre-spill average of 219.0 kg. Although the mean number of resources used per household increased in 1990–1991 for both Prince William Sound villages, the range of resources used remained below pre-spill estimates and was unusually lower than that for the other study communities (Table 3).

Interviewed households in Chenega Bay and Tatitlek continued to express concerns about oil contamination of subsistence foods in the second postspill year. In addition, respondents in these communities commented on the scarcity of certain resources, such as marine mammals, birds, and octopuses, which also affected their subsistence harvests. Even in the other five communities, which evidenced an increase in subsistence harvests in 1990, doubts remained in many households about the long-term health risks associated with using traditional foods from the spill area. Many respondents commented that prolonged periods of doing without subsistence foods had caused economic hardships, and, despite lingering concerns about food safety, they could no longer afford to refrain from subsistence harvests. Many also said they simply could no longer go without culturally important foods and returned to subsistence harvesting despite their misgivings.

Subsistence harvest estimates for the third postspill year are also presented in Table 4 and Figure 8. Subsistence harvests at Nanwalek and Port Graham continued their upward trend. The average number of resources used per household in each of these communities continued to rebound as well (Figure 9). With respect to subsistence harvest estimates (Figure 8), Old Harbor was restudied for 1991–1992 and showed a notable increase over the first postspill year, although this harvest was below the pre-spill average. In contrast, Ouzinkie's subsistence harvest in 1991–1992 was virtually identical to that of 1990–1991, being higher than the first postspill year but much lower than before the spill.

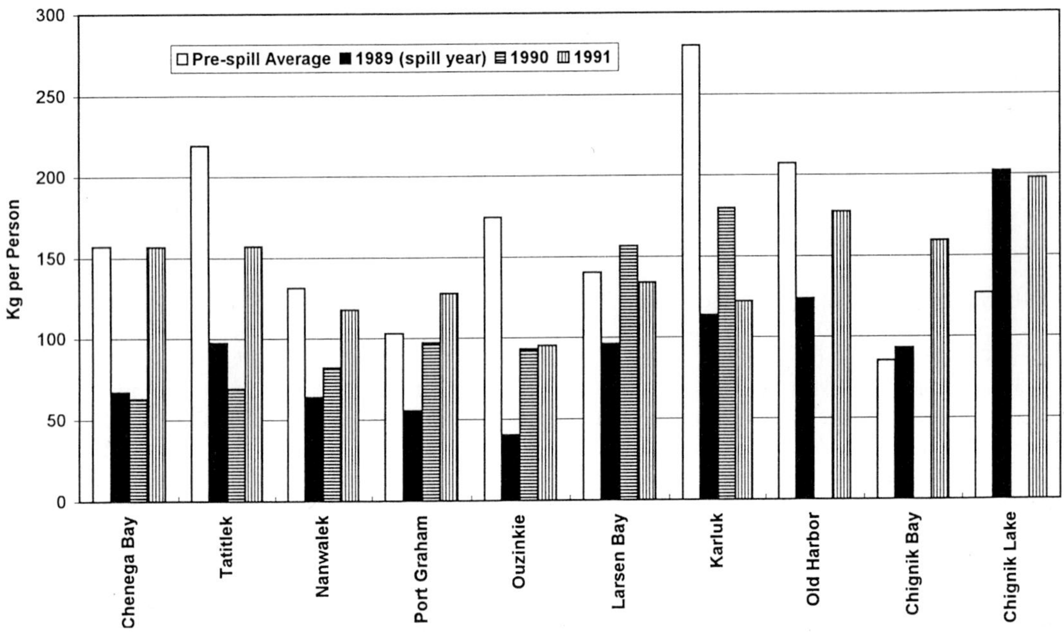

FIGURE 8.—Estimates of subsistence resource harvests before (pre-spill average) and after (1989–1991) the *Exxon Valdez* oil spill.

Estimates for Karluk and Larsen Bay for 1991–1992 decreased in comparison to the previous year, but remained above those of 1989. Chignik Lake and Chignik Bay were also resurveyed for 1991–1992. The subsistence harvest estimate for Chignik Lake of 197.7 kg per person was very similar to that of 1989, reinforcing a picture of relatively stable subsistence-use patterns in that community. There was a notable increase in Chignik Bay's subsistence harvest, however, from 92.9 kg/person in 1989 to 159.3 kg/person in 1991–1992. Possible reasons for this increase at Chignik Bay include reduced commercial fishing incomes and increasing costs of living in the community.

Most notable about the study findings for the third post-spill year was the large increase in harvest levels at Chenega Bay and Tatitlek (Figure 8). The average number of resources used, attempted to harvest, and received in the two Prince William Sound villages increased as well (Table 3 and Figure 9). In Tatitlek, the subsistence harvest estimate for 1991–1992 of 156.9 kg/person was well above that of the average of the first two postspill years but remained below the pre-spill average (Table 4 and Figure 8). At Chenega Bay, the estimate for 1991–1992 of 156.6 kg/person was virtually the same as that of Tatitlek and also indicates a large increase over the average of the previous 2 years. The 1991–1992 estimate was also about the same as the average of the two pre-spill estimates. As noted above, however, the two available pre-spill measures of subsistence harvests at Chenega Bay may underestimate levels of harvest in the years immediately preceding the spill.

Despite rebounding harvests and uses of subsistence foods in Chenega Bay and Tatitlek during the third postspill year, differences between subsistence uses before and after the spill are still evident. One difference is in harvest composition. As shown in Figure 10, marine mammals made up about 37% of the subsistence harvest at Chenega Bay in 1985–1986 compared to just 6% in 1991–1992. Conversely, the contribution of fishes increased from 38% of the harvest before the spill to 74% in 1991–1992. Marine mammal harvests at Tatitlek also remained below pre-spill averages, while the relative size of the harvest of fish increased there as well. Residents of both communities attribute reduced availability of seals and sea lions as the major cause of the decline. Thus, more effort was directed toward harvests of salmon and other fishes, which were also resources that the OSHTF had advised were generally free of oil contamination and were safe to eat.

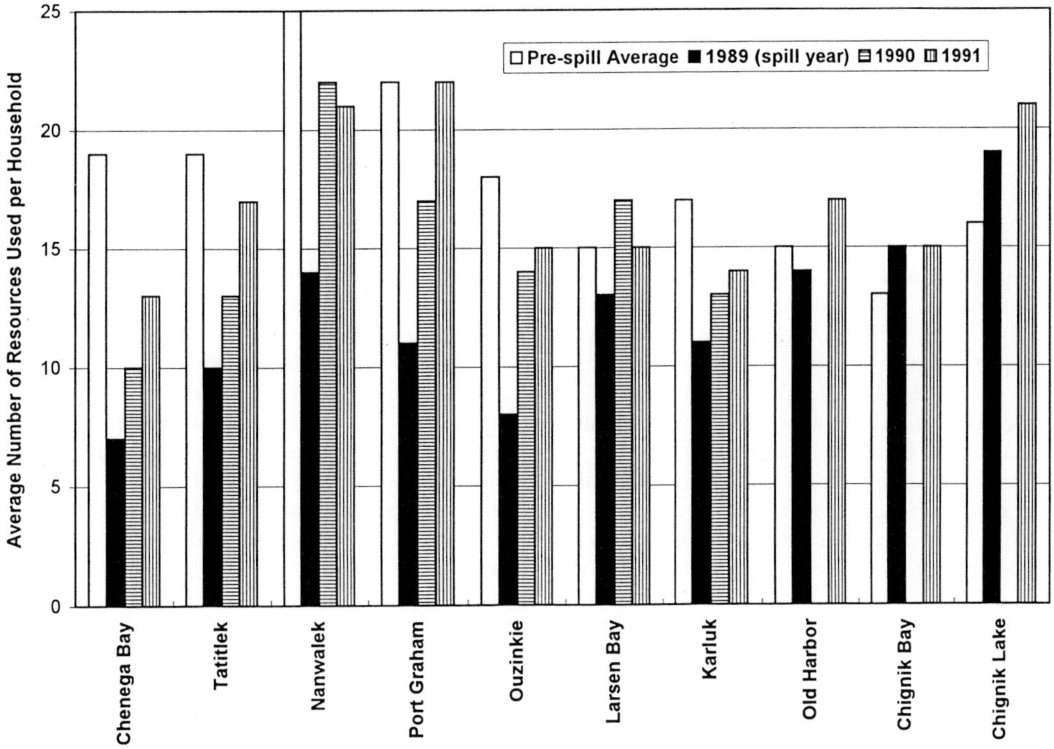

FIGURE 9.—Average number of subsistence resources used per household before (pre-spill average) and after (1989–1991) the *Exxon Valdez* oil spill.

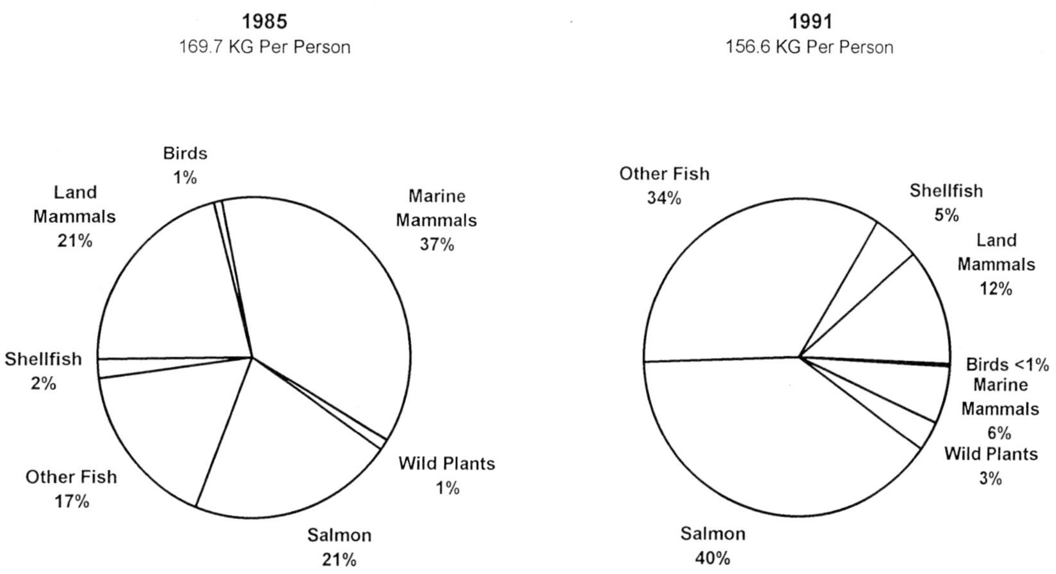

FIGURE 10.—Composition of subsistence resource harvests at Chenega Bay in 1985 and 1991.

Concerns about oil spill effects on subsistence resources remained strong in both Prince William Sound communities (Chenega Bay and Tatitlek) through the third postspill year. Respondents continued to report observations of oil in their traditional use areas. They expressed concern about long-term health effects of using subsistence foods, as well as perceptions of continued unusually low population levels of some resources, such as marine invertebrates and birds as well as marine mammals. Furthermore, these respondents described increased costs in time, effort, and money to harvest adequate supplies of subsistence foods. Some members of households traveled outside Prince William Sound to harvest subsistence foods, either to avoid oiled areas or because of scarce resources near their communities. Others reported that they were making more trips and traveling longer distances to harvest such important foods as harbor seals and deer. Overall, the assessment by virtually every household in Chenega Bay and Tatitlek was that, by 1992, subsistence harvests and uses had not yet recovered from the severe declines that followed the *Exxon Valdez* oil spill.

Conclusions

The *Exxon Valdez* oil spill provides a prime example of the consequences of a major environmental and industrial disaster for the traditional subsistence-based ways of life of Alaska's rural communities. As noted above, before the spill, subsistence uses were fundamental to the economic and cultural survival of the communities of Prince William Sound, lower Cook inlet, the Kodiak Island Borough, and the Alaska Peninsula. In the year after the spill, subsistence harvests in 10 communities in the spill area decreased by as much as 77% compared to pre-spill averages. The range of resources used for subsistence declined by half. In addition, the average number of resources that households attempted to harvest and that they received from other households dropped substantially, indicating reduced subsistence harvest efforts and effects on traditional sharing patterns. These results are supported by other research conducted in the oil spill communities, as summarized by Palinkas et al. (1993), where interviews with 432 residents of 11 communities found significant reductions in subsistence harvest effort and in sharing of subsistence foods after the oil spill. Furthermore, it must be noted that this article has focused primarily on the quantitative changes in subsistence uses after the spill. These changes had important cultural and social impacts that have been the subject of additional research in the communities of the oil spill region.[1]

The primary reason that interviewed households gave for reductions in subsistence harvests was concern that resources had been contaminated by the oil. This concern undermined people's confidence in their own abilities to discern if foods were safe to eat using traditional knowledge. During the first postspill year, little specific information was available to subsistence users to address these concerns.

The subsistence food safety study focused on sites selected by village representatives that were important subsistence collection areas. Edible tissue from more than 1,300 samples of invertebrates, fishes, and marine mammals were analyzed for aromatic contaminants. The results of this study were consistent with the prevailing scientific understanding at the time of the spill. Fish and marine mammals, because their ability to rapidly metabolize and excrete aromatic compounds, showed evidence of exposure to petroleum hydrocarbons but for the most part did not accumulate aromatic contaminants in their edible tissues. Invertebrates, which mostly lack the ability to metabolize ACs, accumulated high concentrations (greater than 1,000 ppb) at a few locations. Concentrations in shellfishes (mussels and chitons) from the most heavily oiled sites (two islands in Windy Bay) in the subsistence study decreased by about two orders of magnitude between 1989 and 1991 but were still well above background concentrations. Only a few of the sites sampled in the subsistence study area were moderately or heavily oiled by the *Exxon Valdez*.

No quick screening methods were available to provide a quantitative assessment and guidance for seafood safety. Initial guidance based on organolep-

[1]Through a cooperative agreement with the U.S. Department of the Interior, Minerals Management Service (MMS), in 1992 the ADF&G Division of Subsistence began a three-year study of "The Sociocultural Consequences of Outer Continental Shelf Development in Alaska," a major goal of which is to understand the sociocultural effects of the *Exxon Valdez* oil spill. A second major study of the sociocultural impacts of the oil spill, also funded by the MMS, is the Social Indicators study (Jorgensen 1994). A third major study, funded by the State of Alaska for the Alaska Conference of Mayors' "Oiled Mayors Subcommittee" investigated the social and psychological effects of the spill (Palinkas et al. 1993). As more complete findings from these projects become available, a fuller picture of the social and cultural effects of the disruption of subsistence activities on the communities of the oil spill area will emerge.

tic evaluation was greeted with skepticism. The approach followed by the OSHTF to address concerns about subsistence seafood safety—the collection and detailed chemical analysis of contaminant levels in subsistence resources from each village area, followed by the convening of an expert panel to provide interpretation of those results and their implications for food safety—did not provide a speedy resolution of the problem. By the time the results from the 1989 field sampling of finfish and shellfish resources and their implications for food safety were presented to the subsistence communities, the summer harvest season was past. For some subsistence users, their immediate concerns about subsistence food safety had shifted to other subsistence resources, such as marine mammals and deer.

The success of the risk-communication efforts to the villages was mixed. The analytical data and interpretation provided by the OSHTF were not entirely successful in mitigating the villagers' concerns about subsistence food safety. The high degree of alarm experienced by the village communities about subsistence food safety made them unreceptive to reassurances based on qualitative measures such as organoleptic testing. Although a large number of shellfish samples were analyzed for aromatic contaminants between 1989 and 1991, villagers pointed out that the numbers from each sampling location were limited, particularly in 1989, and that the results from one beach might not represent the conditions at other nearby areas. In addition, their apprehensions were reinforced by the long interval between the collection of tissue samples for analysis and the communication of interpreted results. A risk-communication workshop held in September 1990 with representatives from six villages revealed differences between individuals in the effectiveness of the different approaches to communication, as well as in the degree of acceptance of the conclusions about food safety.

Subsistence harvests and uses rebounded in the second and third postspill years. In all but the Prince William Sound villages of Chenega Bay and Tatitlek, increases in harvest levels, ranges of resources used, harvest effort, and resource sharing occurred in 1990, the second year after the spill. Substantial increases in these aspects of subsistence uses did not occur in Chenega Bay and Tatitlek until the third postspill year. Especially in these two villages, but among households in other communities as well, concerns about the long-term health risks associated with subsistence foods, as well as concerns about spill effects on the resource populations upon which subsistence uses depend, persisted. Although evidence was found of recovery of subsistence uses after the severe reductions of 1989, this evidence must be qualified in that many households returned to using subsistence foods, despite misgivings, because of economic and cultural reasons.

Acknowledgments

Funding to conduct the household surveys was provided by ADFG, the U.S. Fish and Wildlife Service, and the U.S. Minerals Management Service (MMS). We gratefully acknowledge the governments of each study community for granting approval of the research. Researchers from ADFG who conducted the household interviews and other fieldwork in the study communities included Jeff Barnhart, Janet Cohen, Philippa Coiley, Joe Dinnocenzo, James Fall, Rachel Mason, Susan McNeil, Rita Miraglia, Craig Mishler, Brad Palach, Deborah Robinson, Lisa Hutchinson-Scarbrough, Jody Seitz, Ronald Stanek, Lee Stratton, Lisa Tomrdle, and Vicki Vanek. Residents of the study communities who assisted with the research included James Kvasnikoff, Nick Tanape, Sr., Olga Fomin, Mary Malchoff, Violet Yeaton, Sheila Theriault, David Pestrikoff, John P. Moonin, Anna Meganack, Tamara Squartsoff, Angeline Anderson, Jennifer Brown, Elmer Anahonak, Fedora Hedrick, Dick Moonin, Vera Rose, and Nicholas Tanape. Other fieldworkers included Don Callaway (MMS) and Susan Savage (National Park Service). Division of Subsistence staff who were responsible for data management included Louis Brown, Gretchen Jennings, Sandra Skaggs, Robert Walker, Cheryl Scott, and Charles Utermohle. Research director Robert Wolfe made many important contributions to the research design and data analysis. Finally, we especially thank the many people who took the time to participate in the interviews.

References

ADHSS (Alaska Department of Health and Social Services). 1989a. Oil spill public health advice. Bulletin 6, Alaska Department of Health and Social Services, Division of Public Health, Section of Epidemiology, Anchorage.

ADHSS (Alaska Department of Health and Social Services). 1989b. Oil spill public health advice-report 2. Bulletin 8, Alaska Department of Health and Social Services, Division of Public Health, Section of Epidemiology, Anchorage.

Bolger, M., S. H. Henry, and C. D. Carrington. 1996. Hazard and risk assessment of crude oil contaminants in subsistence seafood samples from Prince William

Sound. American Fisheries Society Symposium 18: 837–843.
Brown, D. W., and nine coauthors. 1996. Survey of Alaskan subsistence invertebrate seafoods collected in 1989–1991 to determine exposure to oil spilled from the *Exxon Valdez*. American Fisheries Symposium 18:844–855.
Clark, D. W. 1984. Prehistory of the Pacific Eskimo region. Pages 136–148 *in* D. Damas, editor. Handbook of North American Indians. Volume V: Arctic. Smithsonian Institution, Washington, DC.
Fall, J. A. 1990. The Division of Subsistence of the Alaska Department of Fish and Game: an overview of its research program and findings: 1980–1990. Arctic Anthropology 27(2):68–92.
Fall, J. A. 1991. Subsistence uses of fish and wildlife and the *Exxon Valdez* oil spill. University of Alaska, Cooperative Extension Service. Arctic Issues Digest 1(1):12–25, Anchorage, Alaska.
Fall, J. A. 1992. Changes in subsistence uses of fish and wildlife resources in 15 Alaska Native villages following the *Exxon Valdez* oil spill. Pages 261–270 *in* K. L. Mitchell, editor. Fourth Information Transfer Meeting Conference Proceedings. MBC Applied Environmental Sciences Report (Contract 14-35-0001-30570) to U.S. Department of the Interior, Minerals Management Service, Alaska OCS (Outer Continental Shelf) Region, Anchorage.
Hom, T., U. Varanasi, J. E. Stein, C. A. Sloan, K. L. Tilbury, and S.-L. Chan. 1996. Assessment of the exposure of subsistence fish to aromatic compounds after the *Exxon Valdez* oil spill. American Fisheries Society Symposium 18:856–866.
Jorgensen, J. 1994. Social indicators study of Alaskan coastal villages, volume V: research methodology for the *Exxon Valdez* spill area, 1988–1992. U.S. Department of the Interior, Minerals Management Service, Outer Continental Shelf Study MMS 93-0071. Anchorage, Alaska.
Morris, J. M. 1987. Fish and wildlife uses in six Alaska Peninsula communities. Alaska Department of Fish and Game, Division of Subsistence Technical Paper 151, Juneau.
NRC (National Research Council). 1985. Oil in the sea: inputs, fates, and effects. National Research Council, Steering Committee for Petroleum in the Marine Environment Update. National Academy Press, Washington, DC.
Palinkas, L. A., M. A. Downs, J. S. Petterson, and J. Russell. 1993. Social, cultural, and psychological impacts of the *Exxon Valdez* oil spill. Human Organization 52:1–13.
Schroeder, R., D. Andersen, R. Bosworth, J. Morris, and J. Wright. 1987. Subsistence in Alaska: arctic, interior, southcentral, southwest, and western regional summaries. Alaska Department of Fish and Game, Division of Subsistence, Technical Paper 150, Juneau.
Scott, C. L., A. W. Paige, G. Jennings, and L. Brown. 1992. Community profile database catalog. Alaska Department of Fish and Game, Division of Subsistence, Juneau.
Stanek, R. T. 1985. Patterns of wild resource use in English Bay and Port Graham, Alaska. Alaska Department of Fish and Game, Division of Subsistence, Technical Paper 104, Juneau.
Stratton, L. 1990. Resource harvest and use in Tatitlek, Alaska. Alaska Department of Fish and Game, Division of Subsistence, Technical Paper 181, Juneau.
Stratton, L. and E. Chisum. 1986. Resource use patterns in Chenega, western Prince William Sound. Alaska Department of Fish and Game, Division of Subsistence, Technical Paper 139, Juneau.
USDA (U.S. Department of Agriculture). 1983. Food consumption: households in the west, seasons and year 1977–78. Nation wide food consumption survey 1977–78. Report H-10, Consumer Nutrition Division, Human Nutrition Information Service, Washington, DC.
USFDA (U.S. Food and Drug Administration). 1990. Advisory opinion on the safety of aromatic hydrocarbon residues found in subsistence foods that were affected by the *Exxon Valdez* oil spill. U.S. Food and Drug Administration, Center for Food Safety and Applied Nutrition, Washington, DC.
Usher, P., and G. Wenzel. 1987. Native harvest surveys and statistics: a critique of their construction and use. Arctic 40(2):145–160.
Varanasi, U., and nine coauthors. 1993. Survey of Alaskan subsistence fish, marine mammal, and invertebrate samples collected 1989–1991 for exposure to oil spilled from the *Exxon Valdez*. NOAA (National Oceanic and Atmospheric Administration) Technical Memorandum NMFS (National Marine Fisheries Service) NWFSC-13, Seattle.
Winston, G. 1990. Survey of health-related activities on previous petroleum spill: human, fish and fisheries, and ecosystems. Institute for Environmental Studies, Louisiana State University, Baton Rouge.
Wolfe, R. J., and R. J. Walker. 1987. Subsistence economies in Alaska: productivity, geography, and development impacts. Arctic Anthropology 24(2):56–81.

American Fisheries Society Symposium 18:837–843, 1996

Hazard and Risk Assessment of Crude Oil Contaminants in Subsistence Seafood Samples from Prince William Sound

MICHAEL BOLGER, SARA H. HENRY, AND CLARK D. CARRINGTON
U.S. Food and Drug Administration, Contaminants Standards Monitoring and Programs Branch Center for Food Safety and Applied Nutrition

Abstract.—One year after the *Exxon Valdez* oil spill, the commissioner of the Food and Drug Administration (FDA) received a request from the Alaska Oil Spill Health Task Force to assess the risk of oil contaminants in the subsistence seafood supply. The long-term health hazards of an oil spill reflect the processes of biodegradation and bioaccumulation. Of all the components of crude petroleum, the polynuclear aromatic hydrocarbons (PAHs) are toxic at the lowest doses and are the most likely to be passed through the marine food chain. The ability of fish to metabolize PAHs makes fish of considerably less concern as a dietary source of PAHs as compared to molluskan bivalves. A risk assessment using an equivalency approach was used for PAH concentrations in fish and shellfish for several reasons. The National Oceanic and Atmospheric Administration (NOAA) analyzed hundreds of subsistence fish and shellfish samples from 13 subsistence fishing grounds used by native Alaskans in or near Prince William Sound for aromatic contaminants, including aromatic hydrocarbons and dibenzothiophenes. Two surveys of fish and shellfish consumption in Chenega Bay on Evans Island in Prince William Sound and on Kodiak Island were used to estimate the risks of exposure to PAHs. Here we report the FDA's conclusion that the risk of contracting cancer from eating finfish collected from the spill area was, for practical purposes, equal to zero. Advisory levels for benzo[*a*]pyrene equivalencies (BaPEs) in fish and shellfish that correspond to a 10^{-6} risk were developed. As with all quantitative cancer risk assessments, this upper-bound risk estimate was produced with the knowledge that the true risk is somewhere between the upper-bound number presented and zero. Levels of BaPE in smoked salmon were shown to be higher than levels of BaPE in bivalve mollusks as a result of the oil spill. These conclusions were based on the higher levels of PAHs found in the 1989 samples. Recent samples have demonstrated a significant decrease in PAH levels since the oil spill.

In the summer of 1990, approximately 1 year after the *Exxon Valdez* oil spill of 24 March 1989, the commissioner of the Food and Drug Administration (FDA) received a request from the Alaska Oil Spill Health Task Force to assess the risk of oil contaminants in the subsistence seafood supply. The task force was charged with providing guidance for the sampling studies of subsistence foods that were affected by the oil spill and with interpreting the results of these studies for the public.

Contaminants in the Marine Food Chain

The long-term health hazards of an oil spill reflect the processes of biodegradation and bioaccumulation. Smaller aliphatic and aromatic hydrocarbons are readily degraded by bacteria. They are also more water soluble and therefore do not tend to bioaccumulate. Larger isoalkanes and polynuclear aromatic hydrocarbons (PAHs) are more hydrophobic and resistant to microbial degradation. Branched alkanes with quaternary carbon atoms at the end of an alkane chain and five-ring aromatic compounds may be totally resistant to degradation by bacteria (McKenna 1971; Gibson 1977; Gundlach et al. 1983; NRC 1985). These compounds can therefore accumulate and be passed through the food chain. For several reasons, concern for the contamination of seafood after past oil spills has been mainly restricted to the contribution of petroleum to levels of PAHs. First, PAHs are well characterized as carcinogens and potent metabolic enzyme inducers. They may therefore produce adverse health effects at very low levels. Second, as discussed above, PAHs are very hydrophobic and resistant to microbial degradation, which can result in a high degree of accumulation in biological tissues. Therefore, of all the components of crude petroleum, PAHs are toxic at the lowest doses and are the most likely to be passed through the marine food chain.

Metabolism and Bioaccumulation of Polynuclear Aromatic Hydrocarbons

The metabolism of PAHs by the cytochrome P450 systems of both fish and mammals is important for two reasons (Malins 1977; NRC 1982). First, metabolism of PAHs results in more polar metabolites, which may be conjugated and excreted. Second, epoxidation is required for production of the mutagenic metabolites of PAHs. Because the metabolites are unstable and are eliminated much more rapidly, they are hazardous in the organism in which they are produced, but they are not passed through the food chain. Because fish can metabolize PAHs, fish are of considerably less concern as a dietary source of PAHs than are molluskan bivalves (Gibson 1977; Gundlach et al. 1983).

Among marine organisms, bioaccumulation of PAHs occurs to the greatest extent in mollusks (Lee 1977), probably for the following reasons: First, because of the slow rate of conversion of PAHs to more polar metabolites in mollusks, PAHs are eliminated at a lower rate. Second, because of their immobility, mollusks are unable to move away from heavily polluted areas. Finally, shellfishes are in direct contact with sediment, where components of petroleum are known to accumulate.

Equivalency Approach in Estimating Risk Associated with Polycyclic Aromatic Hydrocarbon Content

The equivalency approach is a technique for assessing the risk from a number of different compounds that are present in a single exposure source. Instead of calculating individual risks for each component, one component of known potency is used as a standard and the concentration of each of the other components is adjusted on the basis of its estimated, comparative potency to calculate an equivalent concentration for the standard. Summation of the equivalent concentrations produces a single number from which the risk may be estimated (Slooff et al. 1989; ICF-Clements 1990).

Two rationales for using the equivalency approach in the case of a risk assessment for PAHs are applicable. First, there may not be fully adequate data for assessing risk for all of the individual components of the mixture, and it may not be feasible to obtain such data. However, there may be data from in vitro systems or from short-term in vivo exposures that measure a parameter thought to be relevant to the carcinogenic mechanism of the compounds that allows the potency of a data-deficient component to be quantitatively related to a compound for which a long-term study is available. Second, the data available for estimating risk may be much better for some of the components than for others. In such cases the potencies of the components for which there are relatively little data are highly subject to revision, whereas the potency number for a well-studied compound may be expected to be stable. Calculation of a tolerable intake based on the concentration of the component with the best available data will be the least likely to require revision. As long as the potency of the primary compound remains unchanged, the equivalency intake will remain unchanged. The equation by which the primary equivalents are calculated may change, but the number used to evaluate the sum will not.

The risk calculated by this procedure is only an estimate. It is invariably more reasonable than the estimates obtained either by ignoring all but a few well-described contaminants or than by assuming that all the congeners have equivalent potencies. The available toxicology data for benzo[*a*]pyrene (BaP) are much better than those data available for any of the other compounds detected in fish and shellfish. Short-term studies for most of the compounds are available that allow potencies relative to BaP to be approximated. These studies include (1) comparison of the initiating potency of individual PAHs after a single dermal application, followed by promotion; (2) comparison of the carcinogenic potency of individual PAHs after intraperitoneal injection in DMSO in newborn CD-1 and Swiss Webster mice; (3) comparison of the carcinogenic effects after implantation in the lungs of female Osborne-Mendel rats; and (4) comparison of the carcinogenic potency of individual PAHs after long-term dermal application (Slooff et al. 1989; ICF-Clements Associates 1990). Therefore, the equivalency approach is desirable in that it will emphasize the data considered to be the most sound and facilitate the incorporation of anticipated future revisions in the relative potencies of the PAHs other than BaP.

Estimation of the Benzo[a]pyrene Equivalents of Other Polynuclear Aromatic Hydrocarbons

Table 1 lists the potency numbers relative to BaP judged to be the best available for the nine PAHs measured in the Alaskan marine samples. For comparison, the table also lists the International Agency for Research on Cancer (IARC 1989) carcinogenicity class designations. The BaP equivalencies (BaPEs) are positively correlated with carcinogenic potency.

For the alkylated PAHs, the addition of the side chains was assumed to have no effect on carcinogenic potency. Because the alkylated compounds were reported as either mixtures of fluoranthenes and pyrenes or chrysenes and benz[*a*]anthracenes, the assumption was that the relative proportions of the alkylated mixtures was the same as the unsubstituted compounds that were measured separately. The alkylated PAHs were not placed in any IARC category because there are no data.

The FDA chose to use the equivalency approach to estimate the potencies of mixtures of PAHs from

TABLE 1.—Benzo[*a*]pyrene equivalencies (BaPEs) for nine polynuclear aromatic hydrocarbons (PAHs).

PAH	BaPE	IARC class[a]
Dibenz[*a,h*]anthracene	4.05	2A
Benzo[*a*]pyrene	1.00	2A
Indeno[1,2,3-*cd*]pyrene	0.25	2B
Pyrene[b]	0.13	3
Benzo[*b*]fluoranthene	0.11	2B
Benzo[*k*]fluoranthene	0.07	2B
Benzo[*g,h,i*]perylene	0.03	3
Fluoranthene[b]	0.02	3
Benz[*a*]anthracene	0.014	2A
Chrysene	0.013	3

[a]International Agency for Research on Cancer classification: 2A = probable human carcinogen, 2B = possible human carcinogen, 3 = unclassifiable.

[b]All values are from Clements Associates except for pyrene and fluoranthene, which were calculated by Janet Springer of the Division of Mathematics, Center for Food Safety and Applied Nutrition, Food and Drug Administration.

TABLE 2.—Number of Alaskan fish and shellfish samples, collected in 1989, analyzed over a range of benzo[*a*]pyrene equivalencies (BaPEs).

Source	BaPEs (parts per billion)				
	<0.5	1	2–5	8–17	>49
Fishes (including salmon)	>100	0	0	0	0
Molluskan bivalves (except Windy Bay)	139	11	4	1	0
Windy Bay molluskan bivalves[a]	1	0	2	2	2
Smoked salmon[b]	0	0	0	0	4

[a]Individual values (parts per billion) for Windy Bay molluskan bivalves: 0, 3, 4, 11, 17, 71, 107; mean = 30. [b]Individual values (parts per billion) for smoked salmon: 49, 52, 133, 215; mean = 112.

the Alaskan oil spill because this procedure is widely recognized as appropriate methodology in the international scientific community (Slooff et al. 1989; ICF-Clements Associates 1990). To date, better methodology has not been demonstrated.

Results

Levels of Benzo[a]pyrene Equivalents in Alaskan Fish and Shellfish Samples

The spreading of oil spilled from the *Exxon Valdez* led to concern among native Alaskans that their seafood could be contaminated by petroleum hydrocarbons. In response, the National Oceanic and Atmospheric Administration (NOAA) entered into a memorandum of understanding with Exxon Corp. to analyze several hundred subsistence fish and shellfish samples from 13 subsistence fishing grounds used by native Alaskans in or near Prince William Sound for aromatic contaminants, which included aromatic hydrocarbons and dibenzothiophenes. Fish and shellfish samples were collected near the village of Angoon in southeast Alaska to serve as a reference (Varanasi et al. 1990).

The NOAA study indicated that the pattern of aromatic contaminants in the fish samples is consistent with the known extensive metabolism of aromatic hydrocarbons by fish. In mollusks, which have little ability to metabolize aromatic contaminants, both alkylated and unsubstituted aromatic contaminants were present and the pattern resembled that of Prudhoe Bay crude oil.

Using the equivalency values presented in Table 1, BaPEs in the Alaskan fish and shellfish samples are shown in Table 2.

It should be noted that the BaPEs are higher in the smoked salmon. Polynuclear aromatic hydrocarbons in smoked foods are known to be generated by pyrolysis of the smoke constituents during preparation; searing or smoking food commonly results in two- to sixfold increase in PAH levels.

Fish Consumption Estimate for Alaskan Subsistence Fishermen

Two surveys of fish consumption were submitted to the FDA. One report surveyed the fish consumption patterns of the village of Chenega Bay on Evans Island in Prince William Sound and was conducted by the Alaska Department of Fish and Game, Division of Subsistence in 1984 and 1986 (Stratton and Chisum 1987). The residents, 77% of whom are native Alaskans, use traditional hunting and fishing practices to obtain a major portion of their food. This survey was particularly appropriate for FDA's purposes because this is an area where fish and shellfish were exposed to oil after the *Exxon Valdez* incident. In a second report, food consumption on Kodiak Island, Alaska, was surveyed by the Kodiak Area Native Association and the state of Alaska in 1983 (KIANA 1983).

The Food and Drug Administration also recently estimated freshwater fish consumption among subsistence fishermen in the contiguous United States. Consumption estimates for these three groups are tabulated for comparison in Table 3.

Fish and shellfish consumption in the isolated Alaskan communities depends on resources available, prevailing customs and tradition, and access to alternate food sources from outside the community. Average fish and shellfish consumption values may vary somewhat from community to community, but all are of the same order of magnitude, as is illustrated by the Kodiak Island study (KIANA 1983)

TABLE 3.—Fish and shellfish consumption for subsistence fishermen in Alaska and the contiguous United States.

Resource or statistic	g/person/d
Chenega Bay[a]	
Salmon[b]	89
All other fishes[c]	52
Subtotal	141
Bivalve mollusks[d]	2
Crabs[e]	8
Shrimp	11
Subtotal	21
Kodiak Island: most isolated population[a]	
Fishes (excluding shellfishes)	369
Butter clams	27
Kodiak Island: less isolated population	
Fishes (excluding shellfishes)[f]	24
Butter clams	10
Contiguous United States	
(Estimate for subsistence fisherman and family—Food and Drug Administration)	
Freshwater fishes	
Mean	69
90th percentile	116

[a]Per capita data.
[b]Includes chinook salmon *Oncorhynchus tshawytscha*, sockeye salmon *O. nerka*, pink salmon *O. gorbuscha*, chum salmon *O. keta*, and coho salmon *O. kisutch*.
[c]Includes halibut, herring and herring roe, gray cod, sablefish *Anoplopoma fimbria*, yelloweye rockfish *Sebastes ruberrimus*, black rockfish *S. melanops*, smelt–eulachon *Thaleichthys pacificus*, shark, eel, Dolly Varden *Salvelinus malma*, and trout.
[d]Includes razor and other clam species, cockles, and mussels.
[e]Includes Dungeness crabs *Cancer magister*, king crabs, and Tanner crabs *Chionoecetes bairdi*.
[f]Mostly sockeye and pink salmon.

when compared to the Chenega Bay study (Table 3). Isolated Kodiak Islanders apparently make much greater use of the naturally occurring food resources.

Risk Assessment for Cancer

After evaluating the available data, the FDA concluded that the most suitable chronic cancer bioassay for BaP, although less than ideal, is the one reported by Brune et al. in 1981. In this study, there was one dose group of rats that received BaP alone for their lifetime, a feature that was not available in the small number of other studies that might be considered for this purpose. The fact that there was only one dosage level administered makes a dose-response evaluation somewhat limited, however, the existence of a low incidence of treatment-related tumors that were mostly benign indicates that the response is at the lower end of the observable dose-response curve. This response, therefore, is at the point where FDA, using our usual methodology, would initiate a linear extrapolation to a zero dose, even if there were more data points (Gaylor and Kodell 1980).

TABLE 4.—Risk for Alaskan subsistence fishermen consuming polynuclear aromatic hydrocarbon-contaminated fish and shellfish. The abbreviation BaPE is benzo-[*a*]pyrene equivalency.

Source	Intake[a] (g/person/d)	BaPE[b] (mg/person/d)	Upper-bound lifetime risk[c]
Salmon[d]	89	ND[e]	NQ[e]
All other fishes[f]	52	ND	NQ
Smoked salmon[g]	44.5	5.0×10^{-3}	1.5×10^{-4}
Bivalve mollusks[g]	2	6.0×10^{-5}	1.75×10^{-6}

[a]Taken from Table 2.
[b]From Table 2, the mean concentration of BaPE for smoked salmon is 112 parts per billion BaPE $\times$ 44.5 g = 5.0×10^{-3} mg/person/d.
[c]For smoked salmon:

$$\frac{5.0 \times 10^{-3}\ \text{BaPE}}{60\text{-kg person}} \times \text{unit risk of } 1.75 = 1.5 \times 10^{-4}.$$

[d]Mostly sockeye and pink salmon.
[e]ND = not detectable; NQ = not quantifiable.
[f]Includes halibut, herring and herring roe, gray cod, sablefish, yelloweye and black rockfish, smelt–eulachon, shark, eel, Dolly Varden, and trout.
[g]It is estimated that Alaskan subsistence fishermen consume about 50% of their salmon catch smoked (R. Wolfe, Alaskan Department of Fish and Game, personal communication).
[h]Using the mean value at Windy Bay of 30 parts per billion for molluskan bivalves from Table 3.

The time-averaged daily dose was (bw = body weight):

$$0.15\ \text{mg} \cdot (\text{kg} \cdot \text{bw})^{-1} \cdot \text{d}^{-1} \times 5/7 = 0.107\ \text{mg} \cdot \text{kg}^{-1} \cdot \text{d}^{-1}.$$

$$\text{Unit risk (potency)} = \frac{8/32 - 2/32}{0.107\ \text{mg} \cdot \text{kg}^{-1} \cdot \text{d}^{-1}} = 1.75\ (\text{mg} \cdot \text{kg}^{-1} \cdot \text{d}^{-1})^{-1}.$$

The Food and Drug Administration chose to use the Chenega Bay fish and shellfish data for the estimation of upper-bound lifetime risks presented in Table 4. It was decided that the Chenega Bay residents were among the most likely consumers of fish and shellfish that might have come in contact with spilled oil, making the Chenega Bay data most useful and relevant for risk assessment.

The dose corresponding to a 10^{-6} lifetime upper-bound risk was calculated to be

$$10^{-6} / 1.75\ (\text{mg} \cdot \text{kg}^{-1} \cdot \text{d}^{-1})^{-1} = 0.57\ \text{ng} \cdot \text{kg}^{-1} \cdot \text{d}^{-1}.$$

For a 60-kg person, the dose corresponding to a 10^{-6} lifetime upper-bound risk would be

TABLE 5.—Food and Drug Administration advisory levels of benzo[*a*]pyrene equivalency (BaPE) for Alaskan subsistence fishery resources. Abbreviations are BaPE = benzo[*a*]pyrene equivalency and ppb = parts per billion.

Resource	Intake (g/person/d)	BaPE levels (ppb) for 10^{-6} risk	
		Lifetime	10 years[a]
Salmon	89	0.4	3
Other finfishes	52	0.7	5
Bivalve mollusks	2	17.0	120
Crustaceans	21	1.6	11

Advisory level corresponding to a 10^{-6} risk and a period of exposure of 10 years (1/7 of a lifetime of 70 years):

$$\text{Advisory level (ppb of ng/g)} = \frac{34 \text{ ng BaPE/person/d}}{\text{g of fishery resource/person/d}} \times 7$$

where 34 ng BaPE = dose corresponding to a 10^{-6} lifetime upper-bound risk and 7 = correction factor for less than lifetime exposure of 10 years.

$$0.57 \text{ ng} \cdot \text{kg}^{-1} \cdot \text{d}^{-1} \times 60 \text{ kg} = 34 \text{ ng} \cdot \text{d}^{-1}.$$

The levels of concern or advisory levels of BaPE in fish and shellfish that correspond to a 10^{-6} risk are shown in Table 5. For the purpose of estimating these advisory levels for BaPE from the *Exxon Valdez* oil spill it was assumed that the impact of the spill would occur for no more than 10 years. This assumption was based on data accumulated from prior spills (e.g., *Amoco Cadiz*) where it has been estimated that within approximately 10 years the environmental status of an impacted area returns to that which existed before the spill. The evidence from these spills indicates that oiled environments do cleanse themselves and that there are environmental mechanisms of oil degradation and biological recovery (National Research Council 1985). Therefore, the duration of exposure would not be for a 70-year lifetime, but rather for 10 years.

These advisory levels can be interpreted as concentrations that if not exceeded for the given conditions of exposure (lifetime or 10-years consumption at sevenfold higher levels) will conservatively limit any carcinogenic risk from ingested BaPE to less than 10^{-6}. The actual risk is undoubtedly lower than these figures suggest. Although PAHs produced by smoking fish are unrelated to the oil spill, an advisory level of 0.8 parts per billion (ppb) could be established for smoked salmon (0.8 ppb = 34 ng BaPE/45 g salmon). Because the mean concentration of BaPE in smoked salmon was reported to be 112 ppb, this illustrates the importance of this method of food preparation as a source of PAHs.

In interpreting these risk estimates, several points should be emphasized. First, the equivalency approach to estimating the potencies of various PAHs is based on several assumptions pointed out earlier; this approach cannot be assumed to produce exact potencies for the PAHs. Second, there are limitations in the laboratory animal data used to estimate the potency of BaP and other PAHs. Third, as with all FDA quantitative risk assessments, an upper-bound risk estimate has been produced with the knowledge that the true risk is somewhere between the upper-bound number presented and zero.

With regard to fresh salmon, the 10^{-6} risk level is at or below the detection limit of the PAHs. The risk associated with the consumption of salmon that is not smoked is insignificant relative to that associated with consuming smoked salmon; smoking significantly increases the PAH level.

The dose of BaPE corresponding to a 10^{-6} lifetime upper-bound risk for a 60-kg person of 34 ng/day was estimated based on a lifetime upper-bound exposure. Exposure to BaPE would be expected to decrease drastically as shown by the observed data in Figure 1. Figure 1 depicts the observed versus predicted decrease in BaPE in parts per billion in Windy Bay shellfish. The predicted model is based on first-order decay and was fitted using a nonlinear regression analysis. On the basis of the observed data, BaPEs are shown to have decreased from nearly 100 ppb at about 200 d after the spill to nondetectable levels at somewhere between 600 and 800 d after the spill. Thus consumption of BaPE from shellfish would become negligible 2 to 3 years after the spill, and, as a result, any associated health risks would be expected to decline concurrently.

Table 6 summarizes the average and upper percentile daily exposures to BaPEs from shellfish consumption from two locations, namely Chenega and Windy Bays, averaged over 10 years using a Monte Carlo analysis based on the results of a 500-iteration simulation. Based on these values, the BaP half-life in shellfish at Chenega Bay is 315 d and the half-life at Windy Bay is 117 d. Two levels of shellfish consumption are shown at each site: 2 g/d shellfish for an average consumer of shellfish and 8 g/d for a 90th-percentile consumer of shellfish. A meal size of 50 g was assumed, so that the number of meals was equal to

$$\frac{\text{daily intake} \times 10 \text{ years} \times 365 \text{ d/year}}{50 \text{ g}}.$$

The PAH concentration in each meal was estimated by randomly selecting a value from the distribution

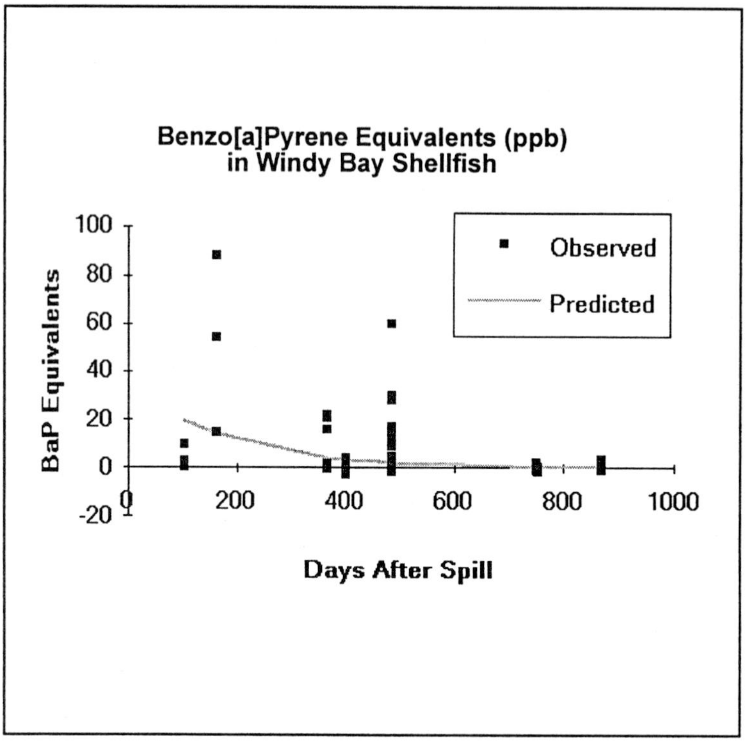

FIGURE 1.—Benzo[*a*]pyrene (BaP) equivalents (parts per billion) in Windy Bay shellfish.

of observed values and prorating the value based on the calculated half-life and the time after the spill.

Previously, the dose of BaPE corresponding to a 10^{-6} lifetime upper-bound risk for a 60-kg person was estimated to be 34 ng/day. Even at the higher shellfish consumption level in Table 6 of 8 g/day, shellfish consumers at Chenega Bay and Windy Bay would not be exposed to levels of BaPE significantly exceeding 34 ng/day. Furthermore, 34 ng/day was estimated for lifetime exposure, not 10 years.

TABLE 6.—Estimated daily exposure to benzo[*a*]pyrene equivalencies (BaPEs) (ng BaPE/d) from the consumption of contaminated shellfish for 10 years using a Monte Carlo analysis based on the results of a simulation of 500 individuals.

Statistic	Chenega Bay 2g/d[a]	Chenega Bay 8g/d	Windy Bay 2g/d	Windy Bay 8g/d
Mean	8.2	33.2	3.6	15.6
SD	1.3	2.8	2.0	4.8
Median	8.0	33.0	3.3	15.4
90th percentile	10.0	37.0	6.4	21.5
95th percentile	10.5	38.3	7.4	23.5
99th percentile	11.5	40.2	9.1	28.8
Maximum	12.5	41.4	10.2	32.4

[a]Shellfish intake.

Conclusions

At the request of the Oil Spill Health Task Force, the U.S. FDA assessed the long-term carcinogenic health risks of the consumption of shellfish and finfish by a subsistence population. The equivalency approach was used by the FDA in estimating risk associated with a mixture of PAHs that occurred because of a single source of exposure. The limitations and assumptions used in the risk assessment process were clearly identified.

In its final report, the FDA concluded that the risk of contracting cancer from eating finfish collected from the spill area was so low that it could not be calculated and was therefore, for practical purposes, equal to zero. It was also concluded that the risk of acquiring cancer from a lifetime of eating shellfish collected from even the most heavily impacted site (Windy Bay) was extremely low. Levels of BaPE in smoked salmon were shown to be higher than the levels of in bivalve mollusks as a result of the oil spill. These conclusions were based on the higher levels of PAHs found in the 1989 samples. Subsequent samples have demonstrated a significant decrease in PAH concentrations since the oil spill.

References

Brune, H., R. P. Deutsch-Wenzel, M. Habs, S. Ivankovic, and D. Schmahl. 1981. Investigation of the tumorigenic response to benzo[*a*]pyrene in aqueous caffeine solution applied orally to Sprague-Dawley rats. Journal of Cancer Research and Clinical Oncology 102: 153–157.

Gaylor, D. W., and R. L. Kodell. 1980. Linear interpolation algorithm for low dose assessment of toxic substances. Journal of Environmental Pathology and Toxicology 4:305.

Gibson, D. T. 1977. Biodegradation of aromatic hydrocarbon compounds. Pages 36–46 *in* D. A. Wolfe, editor. Fate and effects of petroleum hydrocarbons in marine organisms and ecosystems. Pergamon Press, New York.

Gundlach, E., P. Boehm, M. Marchand, R. Atlas, D. Ward, and D. Wolfe. 1983. The fate of *Amoco Cadiz* oil. Science 221:122–129.

IARC (International Agency for Research on Cancer). 1989. Crude oil. IARC Monographs on the Evaluation of Carcinogenic Risks to Humans 45:119–158.

ICF-Clements Associates. 1990. Comparative potency approach for estimating the cancer risk associated with exposure to mixtures of polycyclic aromatic hydrocarbons. Report to EPA (Environmental Protection Agency), Contract 68-02-4403, Fairfax, Virginia.

KIANA (Kodiak Island Area Native Association). 1983. Kodiak Island area local fish and game resource guide. KIANA, Kodiak Island, Alaska.

Lee, R. F. 1977. Accumulation and turnover of petroleum hydrocarbons in marine organisms. Pages 60–70 *in* D. A. Wolfe, editor. Fate and effects of petroleum hydrocarbons in marine organisms and ecosystems. Pergamon Press, New York.

Malins, D. C. 1977. Accumulation of petroleum hydrocarbons in marine organisms indigenous to the arctic and subarctic. Pages 47–59 *in* D. A. Wolfe, editor. Fate and effects of petroleum hydrocarbons in marine organisms and ecosystems. Pergamon Press, New York.

McKenna, E. J. 1971. Microbial degradation of normal and branched alkanes. Pages 73–97 *in* Degradation of synthetic organic molecules in the biosphere. National Academy Press, Washington, DC.

NRC (National Research Council). 1982. Drinking water and health, volume 4. National Academy Press, Washington, DC.

NRC (National Research Council). 1985. Oil in the sea. Inputs, fates and effects. National Academy Press, Washington, DC.

Slooff, W., J. A. Janus, A. J. C. M. Matthijen, G. K. Montizaan, and J. P. M. Ros, editors. 1989. Integrated criteria document PAHs. Report 758474011, National Institute of Public Health and Environmental Protection (Netherlands), Bilthoven, Netherlands.

Stratton, L., and E. B. Chisum. 1987. Resource and use patterns in Chenega, Western Prince William Sound: Chenega in the 1960s and Chenega Bay 1984–1986. Alaska Department of Fish and Game, Division of Subsistence, Technical Paper 139, Juneau.

Varanasi, U., and ten coauthors. 1990. Survey of subsistence fish and shellfish for exposure to oil spilled from the *Exxon Valdez*. Summary. Cycles I-III. NOAA (National Oceanic and Atmospheric Administration)/ NMFS (National Marine Fisheries Service) Environmental Conservation Division, Northwest Fisheries Science Center, Seattle.

American Fisheries Society Symposium 18:844–855, 1996

Survey of Alaskan Subsistence Invertebrate Seafoods Collected in 1989–1991 to Determine Exposure to Oil Spilled from the *Exxon Valdez*

DONALD W. BROWN, DOUGLAS G. BURROWS, CATHERINE A. SLOAN, RONALD W. PEARCE, SUSAN M. PIERCE, JENNIE L. BOLTON, KAREN L. TILBURY, KATHERINE L. DANA, SIN-LAM CHAN, AND USHA VARANASI

National Oceanic and Atmospheric Administration, National Marine Fisheries Service
Northwest Fisheries Science Center, Environmental Conservation Division
2725 Montlake Boulevard East, Seattle, Washington 98112, USA

Abstract.—In response to native Alaskans' concerns, subsistence seafoods were analyzed to determine levels of contamination and to assess the safety of consuming these foods after the *Exxon Valdez* oil spill. This article includes the results of analyses from approximately 1,000 samples of blue mussels *Mytilus edulis*, butter clams *Saxidomus giganteus*, Pacific littlenecks *Protothaca staminea*, and chitons (order Neoloricata). Between July 1989 and August 1991, samples of invertebrate species were collected from 80 stations near 22 native Alaskan subsistence seafood collection areas inside Prince William Sound and along the Kenai and Alaska peninsulas. Invertebrates were also collected from two sites far away from the spill to serve as reference samples. Samples were analyzed for aromatic contaminants consisting of aromatic hydrocarbons and dibenzothiophenes that are components of the Prudhoe Bay crude oil spilled by the *Exxon Valdez*.

Samples from some stations at Chenega, Windy Bay, Kodiak (Village), and Old Harbor consistently had more than 100 ng/g wet weight of total aromatic contaminants (moderately or highly contaminated). Generally, molluscs sampled from the majority of stations and from the reference areas near Angoon and Yakutat were not contaminated or were only minimally contaminated with aromatic contaminants (concentrations were less than 100 ng/g wet weight). In an advisory opinion, the Food and Drug Administration, in conjunction with the Alaska Oil Spill Health Task Force, indicated that little risk was involved in consumption of the subsistence seafoods studied, as long as visibly oiled organisms were avoided. Because invertebrates can serve as sentinel organisms for evaluating the decline in concentrations of aromatic contaminants over time, continued analyses of these organisms are also important for monitoring recovery from oil contamination.

The *Exxon Valdez* ran aground on Bligh Reef, Prince William Sound, Alaska, on 24 March 1989, spilling millions of liters of Prudhoe Bay crude oil (PBCO). Large amounts of floating oil spread to southwestern Prince William Sound, resulting in heavy oiling of several shorelines (Galt et al. 1991). The village of Tatitlek, which is only a few kilometers east of where the spill occurred, was not known to receive any direct shoreline oiling because the oil moved south and west (Fall and Field 1996, this volume). Several areas were targeted for sampling of subsistence seafood (Figure 1). Chenega, in southwestern Prince William Sound, was the first subsistence collection area in the path of the spilled oil. Oil was observed at three of the Chenega sampling stations (CHE1, CHE10, CHE24; Figure 2). As the spilled oil spread from Prince William Sound along the outer Kenai Peninsula, shoreline oiling was variable and some shorelines, such as Windy Bay, received relatively heavy oiling. By the time the oil reached Kodiak Island and the Alaska Peninsula, shoreline oiling was generally light because the oil was in the form of widely scattered patches of tar balls and mousse.

The spilled oil raised concerns among native Alaskans that their subsistence seafood (fishes, marine mammals, invertebrate organisms) was contaminated by the spilled petroleum. In response to the concerns of the villagers, the National Oceanic and Atmospheric Administration (NOAA) entered into memoranda of understanding with Exxon Corp. and the Alaska Department of Fish and Game (ADFG) for the Environmental Conservation Division to analyze subsistence seafoods sampled from sites near native Alaskan villages and from reference areas for evidence of contamination by the spilled oil. The study design and sampling plan for this study are described by Fall and Field (1996), and the risk assessment based on the data from this study is described by Bolger et al. (1996, this volume).

Petroleum, such as PBCO, contains many hundreds of organic compounds, generally classified into groups including aliphatic and polycyclic aro-

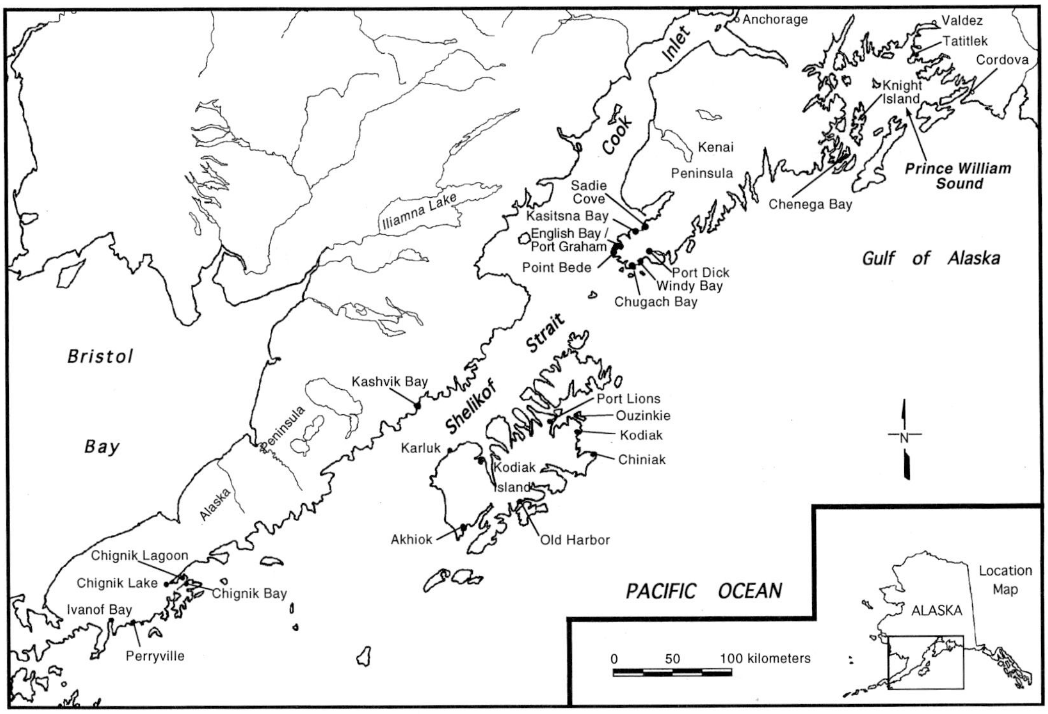

FIGURE 1.—Map of the sampling areas for subsistence seafoods after the *Exxon Valdez* oil spill.

matic hydrocarbons and compounds that contain sulfur, oxygen, and nitrogen (Clark and Brown 1977). Aromatic hydrocarbons and dibenzothiophenes constitute the group of aromatic contaminants (ACs) that are known for their toxicity and that can be monitored in biota as indicators of exposure to petroleum-related ACs. The concentrations of ACs were determined in molluscs (mussels, clams, chitons) and other species (e.g., crab, shrimp, herring roe, kelp) collected in 1989–1991 from 22 sampling areas and two reference areas. This report summarizes the results of these analyses (the complete data sets were reported by Varanasi et al. 1990, 1993a, 1993b). The data from this study were evaluated by the Alaska Oil Spill Health Task Force (Bolger et. al. 1996) for risks that may be involved in the consumption of these subsistence foods.

Methods

After the oil spill in March 1989, a program was organized and implemented to determine if subsistence foods were contaminated by the spilled oil. The philosophy for sampling design and the methods for sample collection are explained by Fall and Field (1996). Invertebrate samples were collected over a 3-year period (July 1989–August 1991) from a variety of sampling areas (Figures 1–3, Table 1). More than 900 samples were analyzed, with blue mussels *Mytilus edulis*, butter clams *Saxidomus giganteus*, Pacific littlenecks *Protothaca staminea*, and chitons (order Neoloricata) being the target invertebrate species. We sampled 80 stations in 22 areas, along with two reference areas. The largest number of mollusc samples was collected from the Chenega (187 samples) and Windy Bay (147 samples) sampling areas (Table 1) because these two subsistence food collection areas appeared to be most affected by the spill. At most sampling areas, more than one sampling station (beach area) was chosen (Fall and Field 1996).

Analyses of Edible Flesh

The edible flesh of mollusc samples was analyzed for the ACs listed in Table 2, using the procedures of Sloan et al. (1993). The analytical protocol consisted of four major steps: (1) extraction, (2) high performance liquid chromatography (HPLC) cleanup, (3) analyte determination by gas chromatography–mass spectrometry (GC–MS), and (4) quality assurance. In addition, samples of petroleum and

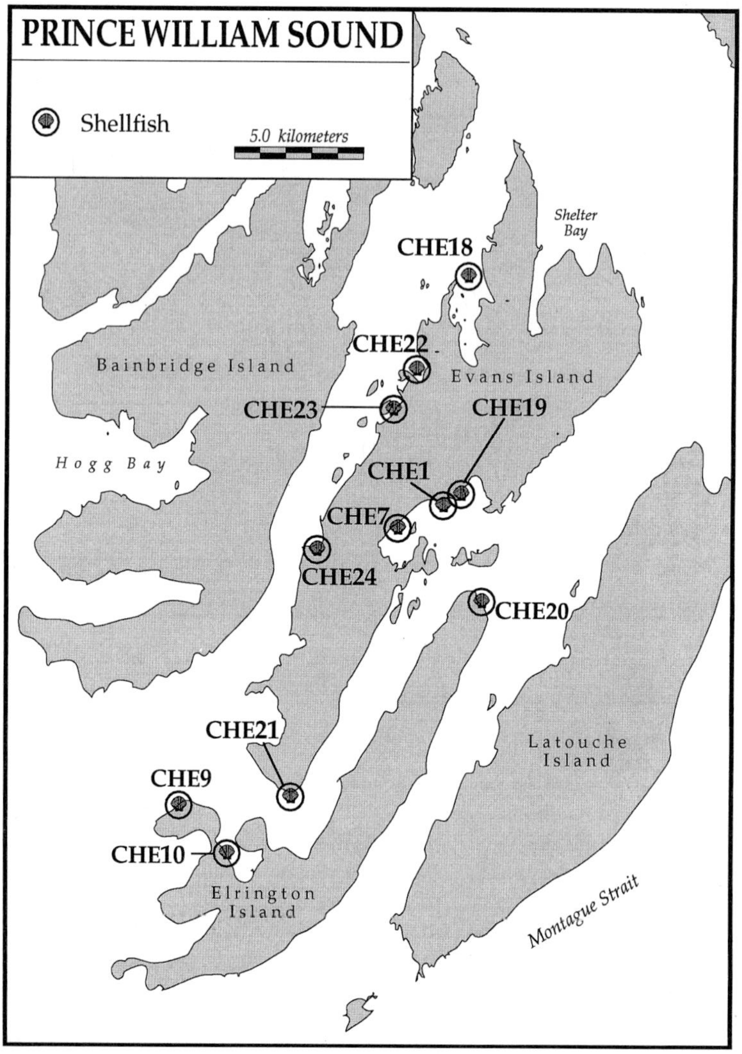

FIGURE 2.—Shellfish sampling stations in the vicinity of Chenega Bay.

related products were weighed, dissolved in methylene chloride, and analyzed, beginning with the HPLC step. Generally, 6–12 molluscs were collected for one representative sample. The edible tissues of about one-half of the animals in a sample were shucked, the composited tissue homogenized, and a 5-g portion of the homogenized tissue analyzed for ACs. The 5-g sample of tissue was added to a centrifuge tube containing sodium sulfate and methylene chloride. The method internal standards (surrogate standards) for ACs were added, and the mixture was macerated with a Tekmar Tissumizer. The extract was decanted into a centrifuge tube, and the extraction step was repeated one time. The resulting extract was filtered through a column of silica and alumina, and the extract was concentrated to 1 mL for cleanup by HPLC.

The extracts were chromatographed using an HPLC method based on size-exclusion chromatography with preparatory-size columns containing Phenogel 100-Å size-exclusion packing (Phenomenex, Rancho Palos Verdes, California) to obtain an extract fraction that contained the ACs separated from the lipids and other interferences (Krahn et al. 1988a, 1988b).

Sample extracts were analyzed using capillary-column GC–MS (Sloan et al. 1993). A sequence of selected ion groups that included 10 groups of

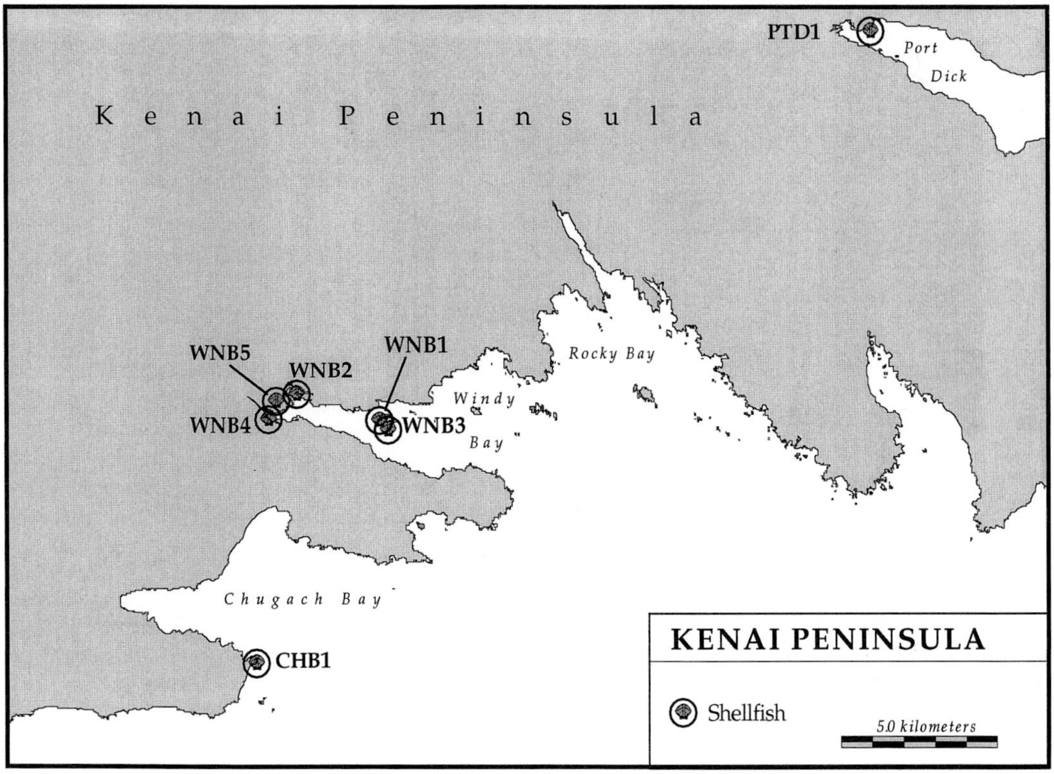

FIGURE 3.—Shellfish sampling stations in Windy Bay, Port Dick, and Chugach Bay.

about 20 ions each was used . The PBCO spilled from the *Exxon Valdez* was analyzed at the beginning of the study to determine the composition of the ACs that could be present in samples of subsistence foods. The availability of several substitution sites on parent ACs results in a variety of isomers of alkyl-substituted ACs denoted respectively as C1, C2, C3, and C4 (Table 2). For example, there are 12 isomers of C2-substituted naphthalenes (C2NPH). The concentration of C2NPH is reported as the sum of the concentrations of the individual C2NPH peaks in the sample chromatogram. The same process was followed for the other alkyl-substituted ACs. Accordingly, the 35 nonsubstituted and alkyl-substituted ACs consisted of some 270 individual components. In this article, the concentration of ACs in samples is the sum of the concentrations of all of the analytes listed in Table 2.

The concentrations of ACs were compared using Gabriel–Tukey 2 (GT2) comparison interval graphs. The GT2 comparison interval for each mean was calculated based on the number of samples for that mean, the variability about that mean, and the number of means being compared (Gabriel 1978; Sokal and Rohlf 1981). The advantage of using GT2 plots with comparison intervals is that they provide a graphical method of depicting significant differences between means (i.e., the comparison intervals do not overlap when the means are significantly different, $P \leq 0.05$).

Quality Assurance for Analyses for Aromatic Contaminants

An extensive quality assurance plan was followed that included analyses of method blanks, spiked blanks, matrix spikes, and National Institute of Standards and Technology Tissue Standard Reference Material 1974 (Sloan et al. 1993). The GC calibration standards generally included all surrogate standards and analytes of interest, except for some classes of alkylated aromatic hydrocarbons and dibenzothiophenes. In these cases, the response factor for the corresponding unsubstituted compound was used to calculate the analyte concentrations. Generally, method detection limits for indi-

TABLE 1.—Summary of sample collection after the *Exxon Valdez* oil spill of major invertebrate species by site, year, and number of samples. Name abbreviations used in other figures and tables are shown in parentheses.

		Number of samples collected			
Village sampling site	Year	Blue mussels	Butter clams	Pacific littlenecks	Chitons
Reference areas					
Angoon (AGN)	1989	2	1	0	2
Yakutat	1990	6	6	6	6
Prince William Sound					
Tatitlek (T or TAT)	1989	7	0	0	2
	1990	14	16	6	0
Chenega Bay (CHE)	1989	5	2	0	0
	1990	21	24	24	8
	1991	44	17	25	17
Lower Cook Inlet					
Port Graham–English Bay (PTG)	1989	10	0	1	3
	1990	24	0	2	33
Windy Bay (WNB)	1989	3	0	1	2
	1990	38	0	5	12
	1991	61	0	8	17
Port Dick (PTD)	1990	1	0	6	0
Chugach Bay (CHG)	1990	0	0	0	7
Port Chatham (PTC)	1990	2	0	5	9
Point Bede (BED)	1990	0	0	0	7
Kasitsna Bay (KAS)	1989	2	1	0	1
	1990	0	0	6	0
Sadie Cove (SAD)	1990	2	0	5	0
Kodiak Island					
Kodiak (K or KOD)	1989	1	1	2	2
	1990	7	16	8	8
Chiniak (CHI)	1989	2	7	3	5
	1990	16	8	4	4
Old Harbor (OHA)	1989	1	10	2	3
	1990	10	20	10	7
Akhiok (AKH)	1989	2	3	1	1
	1990	8	3	6	0
Karluk (KAR)	1989	4	1	0	2
	1990	8	4	4	0
Larsen Bay (LAB)	1989	1	7	0	3
	1990	4	14	7	1
Port Lions (PTL)	1989	0	7	2	5
	1990	14	11	14	4
Ouzinkie (OUZ)	1989	1	4	2	2
	1990	9	22	10	9
Alaska Peninsula					
Chignik (CHG)	1990	19	0	0	0
Ivanof Bay (IVB)	1990	9	3	0	0
Perryville (PER)	1990	9	0	0	1
Kashivak Bay (KAT)	1990	0	0	0	0

vidual analytes were lower than the target of ≤0.5 ng/g wet weight.

The percent recovery of the surrogate standards provides a measure of losses incurred for each sample during the extraction step. Ideally, recoveries would be near 100%, but, realistically, recoveries of about 90% are considered quite good because the procedures include many sample-handling steps. The mean percent recovery and SD for the surrogate standards (i.e., deuterated naphthalene, deuterated acenaphthene, deuterated benzo[*a*]pyrene) were 92 ± 15% (the SD was within our guidelines of ±25%).

Results and Discussion

Approximately 1,000 samples of molluscs were analyzed (e.g., blue mussels, butter clams, Pacific littlenecks, chitons). The concentrations of total ACs in molluscs from 66 of the 80 sample stations were rather low (<100 ng/g); however, the concentrations of ACs in some samples were elevated (>100 ng/g; the highest concentration of ACs in a mollusc sample was 18,000 ng/g). Because the concentrations of ACs covered such a wide range, for discussion purposes it was convenient to divide the concentrations into four categories of ranges:

Category	Description	Concentration of ACs (wet weight in ng/g)
1	Not contaminated by ACs	<10
2	Minimally contaminated by ACs	10–99
3	Moderately contaminated by ACs	100–1,000
4	Highly contaminated by ACs	>1,000

The focus of this article will be those stations where one or more samples were moderately or highly contaminated (category 3 or 4). All of the stations that had one or more invertebrate samples with concentrations of ACs in categories 3 or 4 are included in Table 3. Only 14 of the 80 sampling stations had at least one mollusc sample that contained 100 ng/g or higher concentration of ACs. Even at these 14 stations, in many cases the mean concentrations of ACs were less than 100 ng/g (Table 3), and the concentrations of ACs covered a wide range, as shown by the large SD (Varanasi et al. 1993a). Four stations had by far the most samples in categories 3 and 4 (CHE7 had 31, the combined WNB1 and WNB3 stations had 32, and KOD3 had 26). Of these four stations, only WNB1 and WNB3 were obviously oiled as a result of the spill.

Aromatic Contaminants in Mussels

Some blue mussel samples from Chenega station CHE10 (1990), Windy Bay stations WNB1 (1989), and WNB3 (1990), and Kodiak station KOD3 (1989) were highly contaminated with ACs (category 4, Table 3). Other blue mussel samples—such as from CHE1 (1989), CHE7 (1990), WNB3 (1991), KOD3 (1990), and OHA4 (1990)—were moder-

TABLE 2.—Aromatic contaminants (ACs: aromatic hydrocarbons and dibenzothiophenes) determined in edible tissue from the Exxon Corp. and Alaska Department of Fish and Game–National Oceanic and Atmospheric Administration Subsistence Studies. Abbreviations are as follows: LACs = lower molecular weight ACs; HACs = higher molecular weight ACs; C1-, C2-, etc. = methyl-, dimethyl or ethyl-, etc., respectively.

2- and 3-ring aromatic compounds (LACs)	4- to 6-ring aromatic compounds (HACs)
Naphthalene (NPH)	Fluoranthene (FLA)
C1-naphthalenes (C1NPH)	Pyrene (PYR)
C2-naphthalenes (C2NPH)	C1-fluoranthenes–pyrenes (C1FLA)
C3-naphthalenes (C3NPH)	Benz[*a*]anthracene (BAA)
C4-naphthalenes (C4NPH)	Chrysene (CHR)
Acenaphthylene (ACY)	C1-chrysenes–benz[*a*]anthracenes (C1CHR)
Acenaphthene (ACE)	C2-chrysenes–benz[*a*]anthracenes (C2CHR)
Fluorene (FLU)	C3-chrysenes–benz[*a*]anthracenes (C3CHR)
C1-fluorenes (C1FLU)	C4-chrysenes–benz[*a*]anthracenes (C4CHR)
C2-fluorenes (C2FLU)	Benzo[*b*]fluoranthene (BbF)
C3-fluorenes (C3FLU)	Benzo[*k*]fluoranthene (BkF)
Phenanthrene (PHN)	Benzo[*a*]pyrene (BAP)
C1-phenanthrenes–anthracenes (C1PHN)	Indeno[1,2,3-*cd*]pyrene (IDP)
C2-phenanthrenes–anthracenes (C2PHN)	Dibenz[*a,h*]anthracene (DBA)
C3-phenanthrenes–anthracenes (C3PHN)	Benzo[*ghi*]perylene (BZP)
C4-phenanthrenes–anthracenes (C4PHN)	
Dibenzothiophene (DBT)	
C1-dibenzothiophenes (C1DBT)	
C2-dibenzothiophenes (C2DBT)	
C3-dibenzothiophenes (C3DBT)	

ately contaminated with ACs (category 3). Blue mussels from the other samplings were not contaminated or were only minimally contaminated (category 1 or 2). As with molluscs in general, the majority of the blue mussel samples were not contaminated with ACs.

The two highest concentrations of ACs in individual blue mussel samples were from Windy Bay stations WNB1 (1989) and WNB3 (1990), with levels of 18,000 and 5,500 ng/g, respectively. Only 15 of the 367 blue mussel samples from this study had concentrations of ACs in the highly contaminated category, and they were from the Chenega Bay, Windy Bay, and Kodiak sampling areas. Forty-seven blue mussel samples were moderately contaminated (100–1,000 ng/g ACs) and were from these three areas plus the Old Harbor and Chignik sampling areas. It is notable that of these 47 mussel samples, many were in the low end of the moderately contaminated category (i.e., with concentrations of ACs <200 ng/g in 28 of the 47 samples). Even at Windy Bay stations WNB1 and WNB3, only 14 of the 31 moderately or highly contaminated mussel samples had concentrations greater than 200 ng/g. Blue mussel samples from all of the other sampling stations were not contaminated or were only minimally contaminated with ACs.

Aromatic Contaminants in Clams and Chitons

Overall, the mean concentrations of ACs in butter clams were mostly in the not-contaminated category or in the low end of the minimally contaminated category. The only butter clams that were highly contaminated based on mean concentrations of ACs (Table 3) were from Kodiak village (KOD3, 1990). Based on concentrations of ACs in individual samples, nine samples of butter clams (eight from KOD3 and one from CHE7, all collected in 1990) were highly contaminated with ACs. Only 18 samples of butter clams were moderately contaminated with ACs and they were from Tatitlek station T1 (two samples), Kodiak station KOD3 (five samples), Chenega station CHE7 (nine samples), and Old Harbor station OHA4 (two samples). Only two butter clam samples (both from T1) of 45 collected from the sampling stations in the Tatitlek sampling area had concentrations of ACs in category 3 (concentrations were 110 and 130 ng/g), and as explained in the patterns section (below), the ACs in these two samples probably were not related to the spilled oil.

None of the Pacific littleneck (clam) samples were highly contaminated with ACs. Only 21 samples were moderately contaminated with ACs, 13 of which were from CHE7 and 6 were from KOD3 (Table 3). Many of the Pacific littleneck samples were not contaminated with ACs, having concentrations similar to those from Yakutat, the reference area.

Only one sample of chitons was highly contaminated with ACs (WNB1, 1989), and only two samples were moderately contaminated (both from

TABLE 3.—Mean concentrations (ng/g wet weight ± SD) of aromatic contaminants (number of samples analyzed in parentheses) in the four target invertebrate species by area, station, and year collected after the *Exxon Valdez* oil spill.

Village sampling area and station	Year	Mean concentrations of aromatic contaminants			
		Blue mussels	Butter clams	Pacific littlenecks	Chitons
Prince William Sound					
Tatitlek					
T1	1990	23 ± 7 (3)	62 ± 46 (6)		
Chenega Bay					
CHE1	1989	190 ± 51 (5)	30 ± 16 (2)		
CHE1	1990	24 ± 35 (6)	33 ± 46 (7)	18 ± 18 (4)	4 ± 5 (2)
CHE1	1991	78 ± 18 (4)	20 ± 7 (4)	16 ± 4 (4)	
CHE7	1990	640 ± 830 (8)	330 ± 340 (9)	120 ± 44 (16)	
CHE7	1991	78 ± 33 (4)	110 ± 27 (4)	61 ± 5 (4)	
CHE10	1990	1,100 ± 620 (3)	27 ± 10 (8)	18 ± 11 (4)	2 ± 0.3 (2)
CHE10	1991	23 ± 11 (8)	5 ± 3 (8)	11 ± 18 (8)	
CHE24	1991	37 ± 100 (8)	0.4 (1)	0.4 (1)	
Lower Cook Inlet					
Windy Bay					
WNB1	1989	8,400 ± 8,800 (3)			2,900 ± 4,000 (2)
WNB1	1991	18 ± 12 (8)			0.6 ± 0.5 (4)
WNB2	1989			960 (1)	
WNB2	1990	51 ± 55 (6)			
WNB2	1991	14 ± 26 (5)			
WNB3	1990	1,600 ± 2,100 (24)			10 ± 25 (12)
WNB3	1991	110 ± 170 (38)			5 ± 8 (13)
WNB5	1991	20 ± 18 (5)		17 ± 2 (5)	
Kodiak Island					
Kodiak					
KOD3	1989	1,700 (1)	460 (1)	250 ± 150 (2)	190 ± 99 (2)
KOD3	1990	630 ± 200 (4)	1,200 ± 680 (12)	310 ± 83 (4)	43 ± 14 (4)
K7	1990	89 ± 31 (3)			
Old Harbor					
OHA4	1989		150 ± 150 (5)	220 (1)	
OHA4	1990	120 ± 190 (5)	17 ± 9 (9)	24 ± 22 (6)	
Alaska Peninsula					
Chignik					
CHG1	1990	42 ± 46 (9)			
CHG3	1990	40 ± 54 (10)			

KOD3, 1989). The mean concentrations of ACs in all the other chiton samples were in the not-contaminated category (Table 3).

The differences in concentrations of ACs in mussels, clams, and chitons may be related to differences in the level and mode of oil exposure and in the ability of organisms to take up and depurate ACs. Moreover, habitat requirements vary for different species, and individual stations sometimes encompassed a relatively large beach area over a range of tidal elevations, so that several species could be collected. For example, mussels were collected from higher tidal elevations than clams, whereas chitons were collected from rocky, lower intertidal areas.

Temporal Changes in the Concentrations of Aromatic Contaminants in Molluscs

The concentrations of ACs in mussels and clams from selected stations at Chenega Bay, Windy Bay, Kodiak village, and Old Harbor were statistically evaluated to determine if the concentrations of ACs declined during the period of this study. Both the number of samples of any one species analyzed from a station and the number of times samples were collected from a station were limited, which, in turn, limited the extent to which temporal trends could be evaluated. Given these limitations, the data were evaluated by GT2 statistics (Figures 4 and 5). Where the comparison interval bars in GT2 plots do not overlap, the concentrations of ACs being compared are significantly different.

The concentrations of ACs in blue mussels collected from Chenega station CHE1 in June 1991 were not significantly lower than AC concentrations for samples collected in August 1989 (Figure 4). Notably, the concentrations of ACs in blue mussel samples collected in March 1990 were significantly lower than for those collected in August 1989 and June 1991. Data from additional sample collections

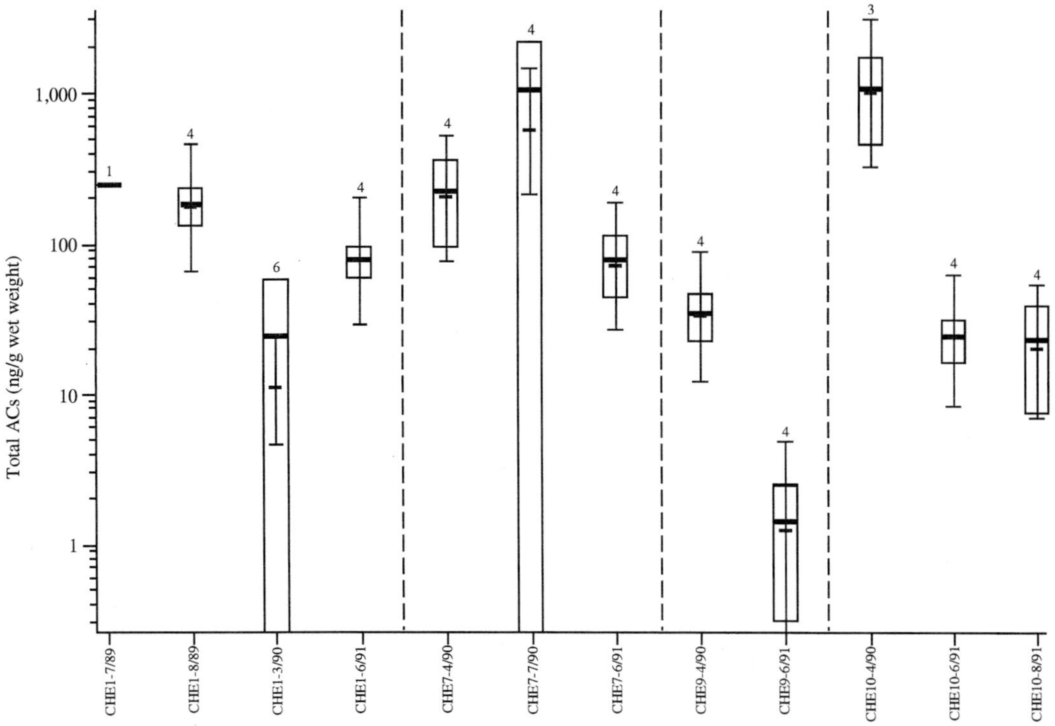

FIGURE 4.—Mean concentrations (ng/g wet weight) and GT2 comparison intervals for total aromatic contaminants (ACs) in Chenega Bay mussels by station and year. Arithmetic mean (black, longer bar) ±1 SD (rectangular box) and geometric mean (grey, shorter bar) ± comparison interval (grey). The number of samples is written above the SD box.

would be needed to better evaluate trends in concentrations of ACs. The concentrations of ACs in blue mussels from CHE7 in June 1991 were significantly lower than those in samples collected in July 1990, but not than those collected in April 1990. These results for CHE1 and CHE7 imply that the mussels from both of these stations were being exposed to a relatively constant source of ACs, because there was no consistent decline in the concentrations of ACs. Furthermore, there is evidence that PBCO may not have been the source of ACs in molluscs at CHE7 (see discussion below). The concentrations of ACs were significantly lower in mussel samples collected in 1991 from CHE9 and CHE10 compared to those collected in 1990.

The GT2 plots show that blue mussel samples from Windy Bay station WNB1 collected in April 1991 had significantly lower concentrations of ACs than those collected in 1989 (Figure 5). The concentrations of ACs were significantly lower in blue mussels from WNB3 collected in 1991 compared to those collected in July 1990; however, it is difficult to show a statistically significant declining trend in concentrations of ACs at WNB3, perhaps because there was a tar mat at this sampling station that resulted in continuing exposure. At WNB2, most of the concentrations of ACs in blue mussel samples were low (<100 ng/g), and there was no significant trend in the concentrations of ACs in the mussel samples. As at the Chenega sampling stations, evaluation of temporal trends in the concentrations of ACs was limited by the sampling scheme.

Most butter clam samples from CHE1 were minimally contaminated or not contaminated with ACs. There was no significant decline (not shown) in the concentrations of ACs in these samples over time (Varanasi et al. 1993a). However, this probably results from the fact that it becomes more difficult to show statistically significant differences when the concentrations of ACs are low. In contrast, the concentrations of ACs in butter clams from CHE7 were in the moderately or highly contaminated categories, and as seen for blue mussels at this station, there was no statistically significant continuing decline in the concentrations of ACs in butter clams from four sample collections in 1990 and 1991. A

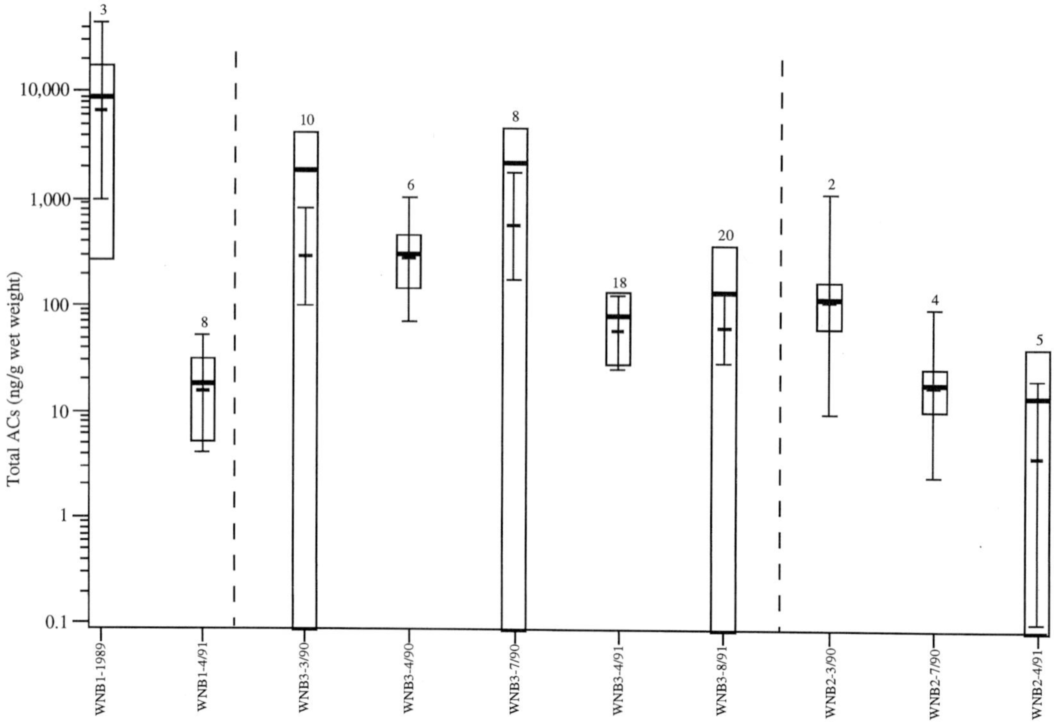

FIGURE 5.—Mean concentrations (ng/g wet weight) and GT2 comparison intervals for total aromatic contaminants (ACs) in Windy Bay mussels by station and year. Arithmetic mean (black, longer bar) ±1 SD (rectangular box) and geometric mean (grey, shorter bar) ± comparison interval (grey). The number of samples is written above the SD box.

local source of ACs may be responsible for the lack of a declining trend in the concentrations of ACs in mollusc samples at CHE7 (further discussion in the patterns section below). The concentrations of ACs were significantly lower in butter clams from CHE10 collected in June 1991 than in those collected in 1990, even though the concentrations were low in all of these samples. Although there was a tar mat on the beach at this station, the contamination of butter clams appeared to decline with time.

The concentrations of ACs in butter clams from Kodiak station KOD3 were mostly in the moderately or highly contaminated categories, and the concentrations did not decline significantly (not shown) over the four sampling events in 1990. Because the concentrations of ACs did not decline over time, a continuing source other than PBCO (i.e., a continuing local source) might be the reason for the continued high concentrations of ACs in butter clams from KOD3 (see patterns section below). Further investigation is warranted to determine the source of ACs in these clams.

Overall, most of the Pacific littleneck (clam) samples were not contaminated or were only minimally contaminated, and generally the concentrations of ACs did not decline significantly and will not be discussed here.

Patterns and Possible Sources of Aromatic Contaminants in Mollusc Samples

Comparing patterns of ACs from samples of biota and petroleums provides useful information about possible sources of contamination; however, certain limitations become apparent when comparing patterns of ACs in tissue samples to patterns in sources. Aromatic contaminants may come from a variety of sources, including petroleum, different fuel oils, and products of combustion, and there are distinct differences in the patterns of ACs among these materials. In addition, when released into the environment, all of these materials undergo a variety of changes (e.g., evaporation, dissolution in water, bacterial degradation, photochemical oxidation). Moreover, biogenic processes such as bioaccumulation, metabolism, and depuration of ACs

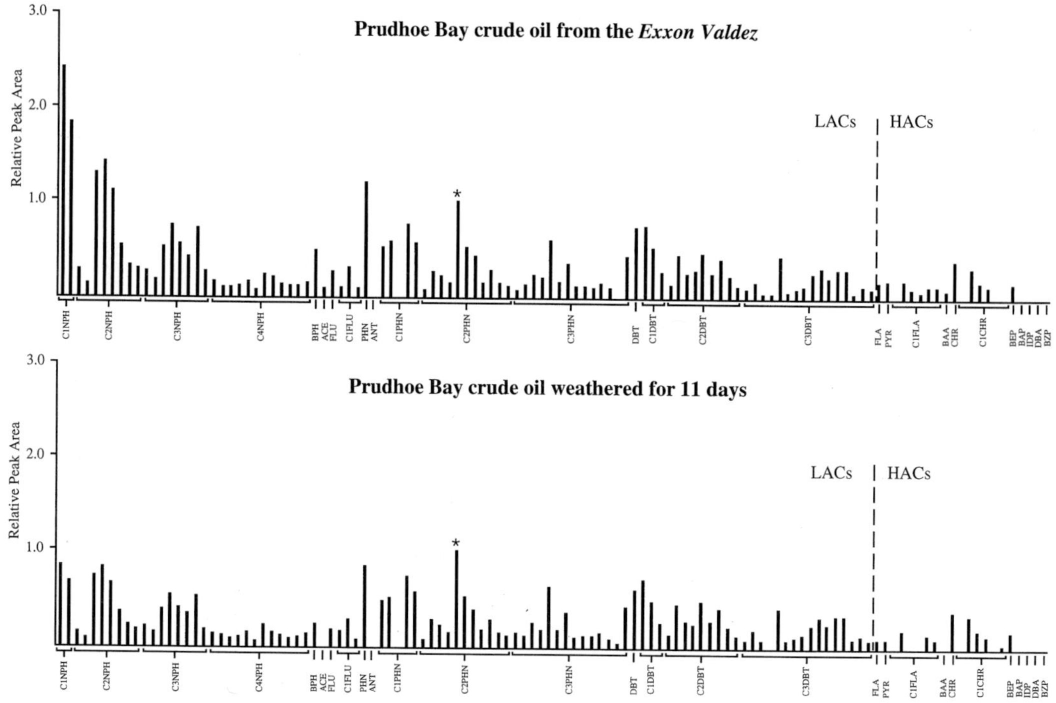

FIGURE 6.—Isomer patterns from gas chromatography–mass spectrometry, with peak areas normalized to the peak area of one of the C2-phenanthrene isomers (*), for weathered and unweathered Prudhoe Bay crude oil. Abbreviations are as in Table 2, except for biphenyl (BPH), anthracene (ANT), and benzo[*e*]pyrene (BEP).

can influence both the levels and patterns of ACs present in marine organisms. For these reasons, certain differences in patterns of ACs in different samples are expected. Thus, relating the presence of ACs in biota such as molluscs to sources is complicated. A thorough treatment of the data relative to patterns and possible sources is beyond the original scope of this study, but comparison of patterns of ACs provides valuable clues to the possible sources of ACs in molluscs.

In order to compare patterns of ACs in tissue samples from stations that had elevated concentrations of ACs, the GC–MS peak areas of selected ACs were graphed with the areas normalized to one specific alkyl-substituted phenanthrene peak area (C2PHN, Figure 6). Oil from the *Exxon Valdez* contained a higher proportion of the more volatile components, C1-naphthalenes (C1NPH) and C2-naphthalenes (C2NPH), than did oil collected from the sea surface 11 d after the spill (Figure 6). In a mussel sample from Windy Bay (Figure 7), the pattern of C2 and C3 alkyl-substituted phenanthrenes (C3PHN) and dibenzothiophenes (C2DBT, C3DBT) was very similar to the same components in the 11-d weathered oil; however, the relative amounts of the components to the left of the C2PHNs were much lower in the blue mussel sample than in the weathered oil sample (probably for reasons noted above). All factors considered, the source of the ACs in the Windy Bay mussel sample was probably the *Exxon Valdez* oil spill. In contrast, the pattern of ACs from butter clams from Tatitlek station T1 (Figure 7) contained minimal quantities of the alkyl-substituted components (C2 or C3NPH, PHN, or DBT) that were in the oil, but it did contain high relative concentrations of the nonsubstituted or parent ACs (PHN, FLA, PYR, BAA, CHR, BEP, BAP), which generally are indicative of combustion processes. Thus, although elevated concentrations of ACs were found in some molluscs from station T1, oil would not seem to be the source of the ACs in these samples. A somewhat similar situation was found at Chenega station CHE7. For example, the pattern of ACs in butter clams from CHE7 (Figure 7) had several similarities to the pattern for the T1 as discussed above. However, the

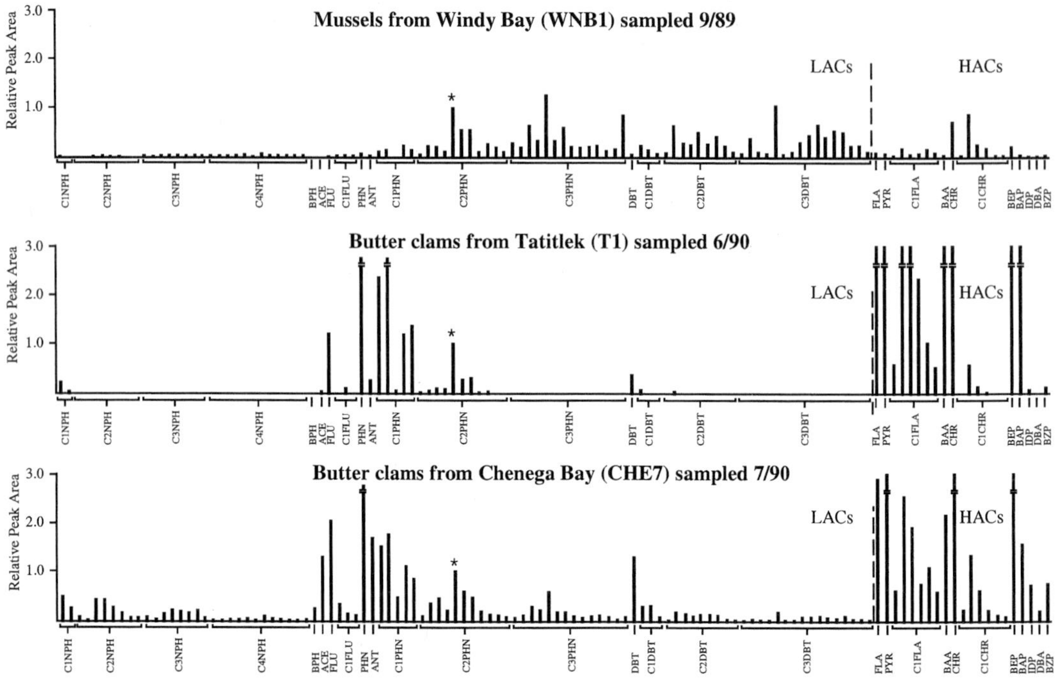

FIGURE 7.—Isomer patterns from gas chromatography–mass spectrometry, with peak areas normalized to the peak area of one of the C2-phenanthrene isomers (*), for mussels from WNB1 sampled in September 1989 (sample no. 60-243), butter clams from T1 sampled in June 1990 (sample no. 60–1596), and butter clams from CHE7 sampled in July 1990 (sample no. 60-1753). Abbreviations are as in Table 2, except for biphenyl (BPH), anthracene (ANT), and benzo[*e*]pyrene (BEP).

CHE7 sample also contained lower relative concentrations of the alkyl-substituted components that were present in the oil. One source of the ACs in the CHE7 clams may have been from nearby creosoted pilings. Another source of ACs in these clams could have been combustion products as well as oil, but the oil component appears minor.

The pattern of ACs in butter clams from Kodiak station KOD3 was similar to the pattern for blue mussels from Windy Bay (Varanasi et al. 1993a), except for the slightly higher relative proportion of alkyl-substituted naphthalenes, PHN, C1PHN, DBT, and C1DBT. This observation, combined with the fact that the concentration of ACs in butter clams from KOD3 did not decline over time and that the location of the sampling station was near a boat harbor, suggests that the ACs in butter clams were from a continuing local source.

Thus, evaluating patterns of ACs in samples can be very important in determining the sources of the ACs, but extreme caution is essential to prevent others from reaching false conclusions.

Conclusions

In summary, only 25 of the more than 900 mollusc samples analyzed were considered highly contaminated (>1,000 ng/g). Of these 25 samples, all were mussels or clams except for one sample of chitons, and all 25 samples were from three sampling areas (Chenega, Windy Bay, and Kodiak Island). Mollusc samples that were moderately or highly contaminated were primarily from Chenega stations CHE1, CHE7, CHE10, CHE24, Windy Bay stations WNB1, WNB3, Kodiak station KOD3, and Old Harbor station OHA4. Even so, the concentrations of ACs in many of the mollusc samples from the above stations were in the not-contaminated or minimally contaminated categories. Although ACs were detected in most mollusc samples, it is notable that mollusc samples from most of the 80 stations were not contaminated or were minimally contaminated with ACs (category 1 or 2) and that only a small number of samples was moderately or highly contaminated with ACs (category 3 or 4).

The concentrations of ACs in blue mussel samples declined over time at some of the stations (e.g., CHE9, CHE10, and WNB1). A statistically significant temporal trend in the concentrations of ACs in mollusc samples was not observed at other stations (e.g., CHE7 and KOD3), probably because there was a continuing presence of ACs from other sources.

Based on the patterns of ACs in some samples, the ACs in many mollusc samples (including those from CHE1, WNB1, and WNB3) are believed to have originated from the spilled oil, whereas the ACs in some mollusc samples (e.g., from CHE7, T1, and KOD3) may have been derived in part from other sources such as boat traffic or combustion products.

Acknowledgments

We appreciate the support of Charles Ehler, John Robinson, and Richard Berry for this study. We also thank Lew Consiglieri, Ann Hayward-Walker, L. Jay Field, Tom Nighswander, Tim Reilly and Dave Schmidt and his staff of field biologists from Dames and Moore. James Fall, Pippa Coiley, Rita Miraglia, Lisa Scarbrough, and Craig Mishler of the ADFG provided valuable assistance. Sharon Giese provided expert help in preparation of this manuscript. A number of Environmental Conservation Division scientists ably assisted this study in the sample analyses and the data management. In alphabetical order they are Nicolaus Adams, Stacie Aspey, Lisa Bogatzki, Richard Boyer, Kristin Bryant, Tho Dang, Tara Felix-Slinn, Jennifer Fisher, Barbara French, Rebecca Hastings, Deborah Holstad, Stephanie Johansen, Thomas Merculief, Hannah Morris, Andrew Nalin, David Rees, Eunice Schnell, John Shields, and Dana Whitney. We thank Bruce McCain and Larry Hufnagle for reviewing the manuscript.

References

Bolger, M., S. H. Henry, and C. D. Carrington. 1996. Hazard and risk assessment of crude oil contaminants in subsistence seafood samples from Prince William Sound. American Fisheries Society Symposium 18: 837–843.

Clark, R. C., Jr., and D. W. Brown. 1977. Petroleum: properties and analyses in biotic and abiotic systems. Pages 1–89 *in* D. C. Malins, editor. Effects of petroleum on arctic and subarctic marine environments and organisms, volume I. Academic Press, New York.

Fall, J. A., and L. J. Field. 1996. Subsistence uses of fish and wildlife before and after the *Exxon Valdez* oil spill. American Fisheries Society Symposium 18:819–836.

Gabriel, K. R. 1978. A simple method of multiple comparisons of means. Journal of the American Statistical Association 73:724–729.

Galt, J. A., W. J. Lehr, and D. L. Payton. 1991. Fate and transport of the *Exxon Valdez* oil spill. Environmental Science & Technology 25(2):202–209.

Krahn, M. M., L. K. Moore, R. G. Bogar, C. A. Wigren, S-L. Chan, and D. W. Brown. 1988b. High performance liquid chromatographic method for isolating organic contaminants from tissue and sediment extracts. Journal of Chromatography 437:161–175.

Krahn, M. M., and seven coauthors. 1988a. Standard analytical procedures of the NOAA National Analytical Facility. New HPLC cleanup and revised extraction procedures for organic contaminants. NOAA (National Oceanic and Atmospheric Administration) Technical Memorandum NMFS (National Marine Fisheries Service) F/NWC-153, Seattle.

Sloan, C. A., N. G. Adams, R. W. Pearce, D. W. Brown, and S-L. Chan. 1993. Northwest Fisheries Science Center organic analytical procedures. Sampling and analytical methods of the National Status and Trends Program National Benthic Surveillance and Mussel Watch Projects 1984–1992, volume IV comprehensive descriptions of trace organic analytical methods. NOAA (National Oceanic and Atmospheric Administration) Technical Memorandum NOS/ORCA/CMBAD 71, Silver Spring, Maryland.

Sokal, R., and F. Rohlf. 1981. Biometry (2nd edition). Freeman, San Francisco.

Varanasi, U., and nine coauthors. 1993a. Volume I: survey of Alaskan subsistence fish, marine mammal, and invertebrate samples collected 1989–91 for exposure to oil spilled from the *Exxon Valdez*. NOAA (National Oceanic and Atmospheric Administration) Technical Memorandum NMFS (National Marine Fisheries Service) F/NWC-12, Seattle.

Varanasi, U., and nine coauthors. 1993b. Volume II: supplemental information concerning a survey of Alaskan subsistence fish, marine mammal, and invertebrate samples collected 1989–91 for exposure to oil spilled from the *Exxon Valdez*. NOAA (National Oceanic and Atmospheric Administration) Technical Memorandum NMFS (National Marine Fisheries Service) F/NWC-13, Seattle.

Varanasi, U., and ten coauthors. 1990. Survey of subsistence fish and shellfish for exposure to oil spilled from the *Exxon Valdez*: first year: 1989. NOAA (National Oceanic and Atmospheric Administration) Technical Memorandum NMFS (National Marine Fisheries Service) F/NWC-191, Seattle.

American Fisheries Society Symposium 18:856–866, 1996

Assessment of the Exposure of Subsistence Fish to Aromatic Compounds after the *Exxon Valdez* Oil Spill

TOM HOM, USHA VARANASI, JOHN E. STEIN, CATHERINE A. SLOAN, KAREN L. TILBURY AND SIN-LAM CHAN

Environmental Conservation Division, Northwest Fisheries Science Center, National Marine Fisheries Service, National Oceanic and Atmospheric Administration, 2725 Montlake Boulevard East, Seattle, Washington 98112, USA

Abstract.—Previous research has shown that fish have the capacity to biotransform many aromatic compounds (ACs) to polar metabolites that are readily accumulated in the gall bladder for excretion and to greatly limit the deposition of ACs or their metabolites in edible muscle tissue. Accordingly, a rapid and sensitive method, developed in our laboratory for screening bile for fluorescent aromatic compounds (FACs) that are components of petroleum, was used to assess exposure of fish to ACs after the *Exxon Valdez* oil spill. The results of analysis of bile from nearly 500 fish provided evidence that many fish were exposed to ACs after the oil spill. Edible flesh samples from fishes showing a range of biliary FACs were analyzed for the presence of ACs by gas chromatography–mass spectrometry. Concentrations of FACs in bile of bottomfishes and salmon from areas affected by oil ranged from 10 to 17,000 ng phenanthrene equivalents/mg bile protein; however, no appreciable concentrations of ACs were detected in muscle of bottomfishes (<1 ng/g wet weight, parts per billion [ppb]), and, although concentrations of ACs in muscle of salmon were somewhat higher (most samples ranged from 0 to 20 ppb), they rarely exceeded 100 ppb. These findings were used by the Alaskan Oil Spill Health Task Force in arriving at an advisory opinion that consumption of the flesh of fish from areas affected by the oil spill posed minimal risk to native Alaskans. Furthermore, these results were used to verify findings of earlier laboratory studies on the pathways of metabolism and disposition of ACs in fish that were important in formulating a focused, scientific basis for the present investigation of oil-impacted subsistence fisheries.

Previous studies (Varanasi and Gmur 1981; Solbakken et al. 1984; Stein et al. 1984; Varanasi et al. 1989a) have shown that most marine organisms readily take up aromatic contaminants, including polycyclic aromatic hydrocarbons (PAHs) and structurally related heterocyclic compounds, from their environment. Marine organisms that are most susceptible to contaminant exposure are those that inhabit polluted waterways near urban centers or areas contaminated by an environmental catastrophe, such as the *Exxon Valdez* oil spill in Prince William Sound. Unlike many invertebrate species, such as molluscs, fish have the capacity to biotransform these compounds to polar metabolites that are released from the liver and concentrated into bile for excretion (Varanasi and Gmur 1981; Varanasi et al. 1985). Moreover, the extensive biotransformation of PAHs by fish greatly limits the accumulation of these compounds or their metabolites in extrahepatic tissues, including edible muscle (Stein et al. 1987). Thus, analyzing teleost tissues for parent PAHs is not an appropriate approach for assessing exposure to petroleum-related compounds.

The findings of laboratory studies showing high concentrations of PAH metabolites in bile led to the development of a method (Krahn et al. 1984, 1986) that measures fluorescent aromatic compounds (FACs) in bile by high-pressure liquid chromatography (HPLC). In addition, individual metabolites of several PAHs were identified by gas chromatography–mass spectrometry (GC–MS) in hydrolyzed bile of benthic fish sampled from contaminated urban waterways in Puget Sound, Washington, verifying that the measured fluorescence response reflected PAH compounds that were biotransformed (Krahn et al. 1984). The HPLC technique provides a rapid, semiquantitative means of assessing exposure of fish to PAHs and has been used extensively to monitor contaminant exposure in several fish species collected from sites along coastal areas of the United States as part of the National Oceanic and Atmospheric Administration's National Status and Trends Program (Varanasi et al. 1989b). Also, a recent laboratory study (Collier and Varanasi 1991) showed that biliary FACs are responsive to increasing doses of benzo[*a*]pyrene (BaP), a model PAH, and that substantial clearance of the BaP metabolites had occurred by 3 weeks postexposure. Thus, these results suggest that concentrations of biliary FACs could be used as an indicator of relatively recent exposure to PAHs in benthic fish.

In a study done in conjunction with the present one, the use of biliary FACs for estimating exposure of fish to aromatic compounds (ACs) was validated using fish collected from Prince William Sound shortly after the oil spill (Krahn et al. 1992). Several metabolites of ACs were identified by GC–MS in salmon and pollock captured in Prince William Sound several months after the oil spill. The identified AC metabolites were also present in bile of a Pacific halibut *Hippoglossus stenolepis* injected with weathered Prudhoe Bay crude oil (PBCO). In contrast, these metabolites were not present in bile of fish collected from reference sites not contaminated with crude oil. Thus, these findings further illustrate that estimating exposure by measuring biliary FACs is a valid measure of the metabolites resulting from the biotransformation by fish of petroleum-related ACs.

In the aftermath of the oil spill in Prince William Sound, there was considerable concern about deleterious ecological effects. The question of potential effects on human health from the contamination of fish used for subsistence by native Alaskans was also raised because of the possible uptake and accumulation of toxic compounds found in petroleum. Of particular concern was the accumulation of ACs in the edible tissues of fish. Some of the higher molecular weight ACs are suspected of being toxic or carcinogenic to humans (Dipple et al. 1984). The ACs include PAHs and sulfur-containing ACs (e.g., dibenzothiophenes), some of which are characteristic of certain crude oils, including PBCO (Krahn et al. 1993). Accordingly, the analyses of bile provided a cost-effective means of quickly estimating exposure of fish to ACs for prioritizing corresponding edible muscle samples for detailed analyses by GC–MS. The results of analyses of edible flesh of fish were then evaluated by the Alaska Oil Spill Health Task Force and the Food and Drug Administration to assess the potential risk to native Alaskans of consuming seafood from their traditional fishing sites after the spill.

Materials and Methods

Sample Collection

The area affected by the *Exxon Valdez* oil spill is shown in Figure 1. The oil spill spread in a southwest direction from Bligh Reef, through Prince William Sound, then into coastal areas along the Kenai Peninsula and Kodiak Island, affecting many native villages. Fish were caught by the consulting firm of Dames and Moore and the Alaska Department of Fish and Game in fishing grounds near these sites and sampled for bile and muscle. The gall bladder was excised and bile fluid was transferred to a 4-mL amber vial and frozen before being shipped to our laboratory in Seattle, Washington. At the Seattle facility, the bile samples were stored at −80°C until analysis by HPLC. Sections of muscle tissue were excised and wrapped in aluminum foil and frozen at −20°C before being shipped to Seattle. Muscle tissue samples were stored at −80°C until sample extraction and GC–MS analysis. Detailed locations of where subsistence fish were sampled have been presented in Varanasi et al. (1993).

Analysis of Bile and Edible Muscle

Bile of fish was analyzed for FACs by HPLC according to Krahn et al. (1986). Briefly, bile was injected directly into an HPLC equipped with a C_{18} reverse-phase column (Perkin-Elmer HCODS/PAH), and the analytes were monitored by fluorescence detectors connected in series. One detector was set at wavelength pairs for phenanthrenes (excitation/emission, 260/380 nm) and the other for naphthalenes (excitation/emission, 290/335 nm); these ACs and associated alkylated compounds are present in high proportions in PBCO (Varanasi et al. 1993). The concentrations of FACs in bile were determined using the PAHs phenanthrene and naphthalene as external standards and converting the fluorescence response of bile to phenanthrene or naphthalene equivalents. The bile results were reported on the basis of milligrams of biliary protein. A recent study (Collier and Varanasi 1991) has shown that such a normalization can, to a large extent, account for physiological changes in the composition of bile in fish. The concentrations of biliary protein were measured by the method of Lowry et al. (1951) using bovine serum albumin as the standard.

Samples of edible muscle in fish were analyzed for individual ACs (Table 1) according to Krahn et al. (1988) by GC–MS quantitation (D. G. Burrows, Northwest Fisheries Science Center, unpublished). The concentrations of ACs in muscle tissue are reported as the sum of low molecular weight ACs (2–3 rings) and high molecular weight ACs (4–6 rings), including several alkylated ACs (Table 1).

Statistical Analysis

The mean biliary FAC concentrations were compared using GT2 comparison interval plots (Gabriel 1978; Sokal and Rohlf 1981). The GT2 comparison intervals are similar to confidence intervals, but are more closely related to an analysis of variance. The

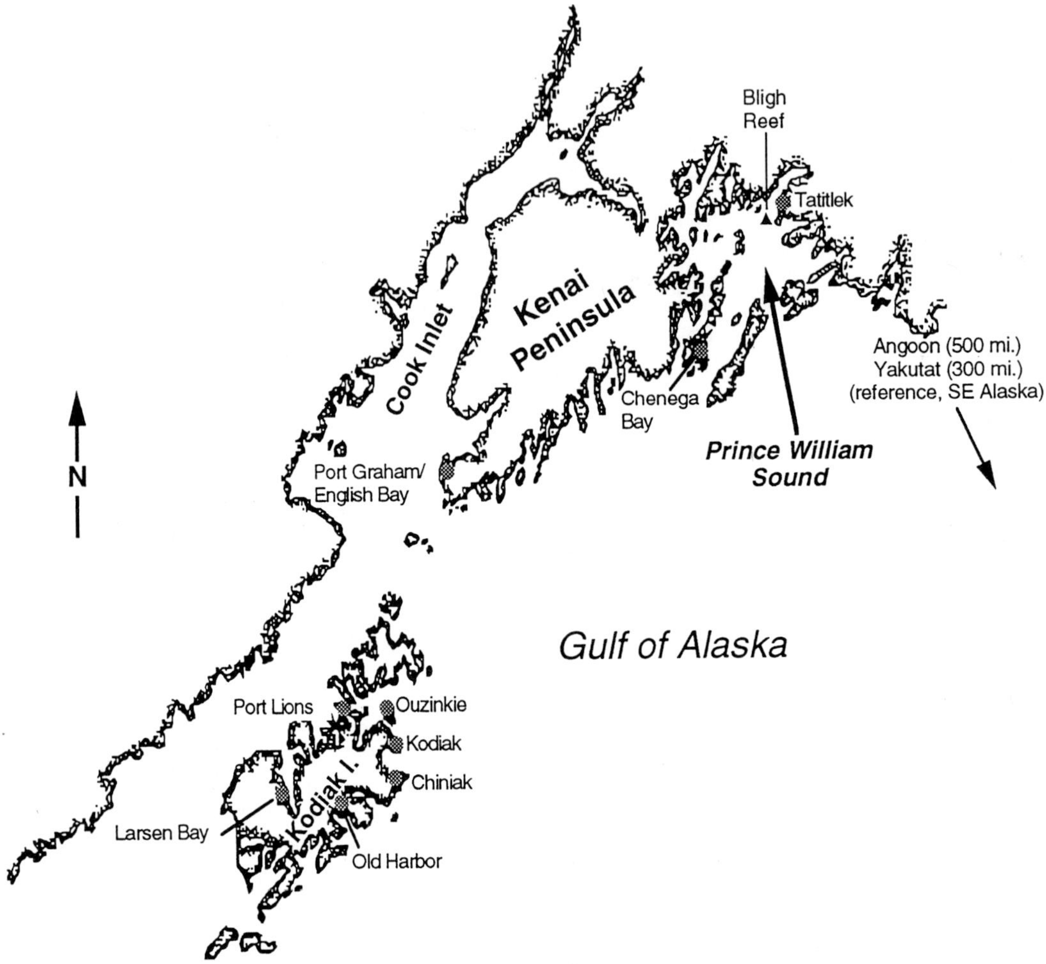

FIGURE 1.—Map of selected native Alaskan subsistence fishing sites sampled in 1989 and 1990. The oil spill spread in a southwestern direction from Bligh Reef, through Prince William Sound, then into coastal areas along the Kenai Peninsula and Kodiak Island.

comparison interval for each mean was calculated based on the number of samples for that mean, the variability about that mean, and the number of means being compared. Comparison intervals of means that do not overlap were considered significantly different. Statistical comparisons of other results were done using analysis of variance or Student's t-test; the significance level for these tests was set at $P \leq 0.05$.

Results

Fluorescent Aromatic Compounds in Bile of Fish

Bile was analyzed for FACs by HPLC-ultraviolet fluorescence at wavelength pairs for phenanthrenes and naphthalenes because these compounds are present in high proportions in PBCO (Varanasi et al. 1993). The concentrations of biliary FACs at phenanthrene and naphthalene wavelength pairs were highly correlated ($r = 0.91$, $P = 0.0001$), thus only the concentrations measured at the phenanthrene wavelength pair are reported. In addition, similar site differences were observed in biliary FAC results at both naphthalene (not shown) and phenanthrene wavelength pairs. Results from the analyses of bile for FACs from pink salmon *Oncorhynchus gorbuscha*, coho salmon *O. kisutch*, Pacific halibut, and Pacific cod *Gadus macrocephalus* collected in 1989 are summarized in Figure 2. The range of concentrations of FACs measured in the bile of individual fish sampled from subsistence fish-

TABLE 1.—Aromatic compounds (ACs) determined in tissues of salmon and bottomfishes sampled after the *Exxon Valdez* oil spill by gas chromatography–mass spectrometry. The C1, C2, C3, and C4 are abbreviations for alkylated groups such as methyl, dimethyl–ethyl, trimethyl–propyl, or tetramethyl–butyl that are substituted on the aromatic ring.

Low molecular weight ACs (LACs; 2–3 rings)	High molecular weight ACs (HACs; 4–6 rings)
Naphthalene	Fluoranthene
Acenaphthylene	Pyrene
Acenaphthene	Benz[*a*]anthracene
Fluorene	Chrysene
Phenanthrene	Benzo[*b*]fluoranthene
Dibenzothiophenes	Benzo[*k*]fluoranthene
	Benzo[*a*]pyrene
Alkylated LACs	Indeno[1,2,3-*cd*]pyrene
C1-C4 naphthalenes	Dibenz[*a,h*]anthracene
C1-C3 fluorenes	Benzo[*ghi*]perylene
C1-C4 phenanthrenes–anthracenes	
C1-C3 dibenzothiophenes	Alkylated HACs
	C1 fluoranthenes–pyrenes
	C1–C4 chrysenes–benz[*a*] anthracenes

ing grounds in 1989 and 1990 was 10–17,000 ng phenanthrene equivalents/mg bile protein.

Concentrations of FACs in the bile of other salmon and bottomfish species sampled in 1989 and 1990 were also determined but are not reported here because insufficient numbers of bile samples from these fishes were sampled in 1989 or 1990 to allow site comparisons. Results of analyses of these fish species were reported previously by Varanasi et al. (1990, 1993).

In 1989, the mean concentrations of biliary FACs in pink salmon from Chenega Bay, a site greatly affected by the spilled oil, were significantly higher than levels in fish from all other subsistence sites, except in fish from Kodiak (Figure 2). No significant differences in biliary FAC concentrations between pink salmon from Chenega Bay or Kodiak and Angoon, the reference site, were observed. This resulted from a single fish from Angoon having a biliary FAC level (12,000 ng phenanthrene equivalents/mg bile protein) that was 4- to 20-fold higher than levels in the other three pink salmon (1,800 ± 750 ng phenanthrene equivalents/mg bile protein) sampled from the reference site. However, if the reference sample showing the high biliary FAC value was removed from the data set, the concentrations of biliary FACs were significantly higher in pink salmon from Chenega Bay compared to Angoon. Furthermore, mean FAC concentrations in the bile of coho salmon (Figure 2) and chinook salmon *O. tshawytscha* (490 ± 170 ng phenanthrene equivalents/mg bile protein, $N = 3$) salmon from Angoon were considerably lower than FAC concentrations in pink salmon from Chenega Bay and Kodiak after the spill. Concentrations of biliary FACs in pink salmon from Chenega Bay and Kodiak were also significantly higher than levels in fish from Tatitlek, a site not in the path of the oil spill.

As with pink salmon, high concentrations of biliary FACs were also found in Pacific halibut and Pacific cod from Chenega Bay in 1989 (Figure 2). The concentrations of FACs in the bile of halibut from Chenega Bay were significantly higher than levels in halibut from Angoon, the reference site (Figure 2). Furthermore, significantly higher concentrations of biliary FACs were also observed in halibut from Ouzinkie and Port Lions compared to levels in halibut from Angoon. No statistical comparison could be made with halibut from Kodiak because only one fish was sampled; however, this fish also exhibited an elevated biliary FAC concentration (7,400 ng phenanthrene equivalents/mg bile protein) compared to levels in halibut from the reference site.

Although statistical comparisons could not be made with Pacific cod from Angoon because only one fish was collected, the concentration of FACs in the bile of this fish (480 ng phenanthrene equivalents/mg bile protein) was much lower than concentrations in cod from Chenega Bay (5,200 ng phenanthrene equivalents/mg bile protein) (Figure 2).

In 1989, concentrations of biliary FACs in individual coho salmon sampled from five sites ranged from 140 to 4,900 ng phenanthrene equivalents/mg bile protein (Figure 2). Unfortunately, no coho salmon were sampled from Chenega Bay or Kodiak, precluding comparisons of FAC levels with the other species sampled from these sites. Concentrations of biliary FACs were significantly higher in coho salmon from Ouzinkie and Tatitlek than levels in fish from Angoon, the reference site.

Pink salmon and Pacific cod were the only species collected in sufficient numbers in both sampling years to allow evaluation of temporal changes. Comparisons for other salmon and bottomfish species were not possible because of low sample size or lack of corresponding data from 1989.

Concentrations of biliary FACs in pink salmon and Pacific cod sampled in 1989 and 1990 are shown in Figure 3. Overall, statistical evaluation of the FAC results shows a significant ($P \leq 0.0001$) decline attributable to sampling year because the concentrations of FACs in the bile of pink salmon decreased in 1990 compared to 1989. Pink salmon

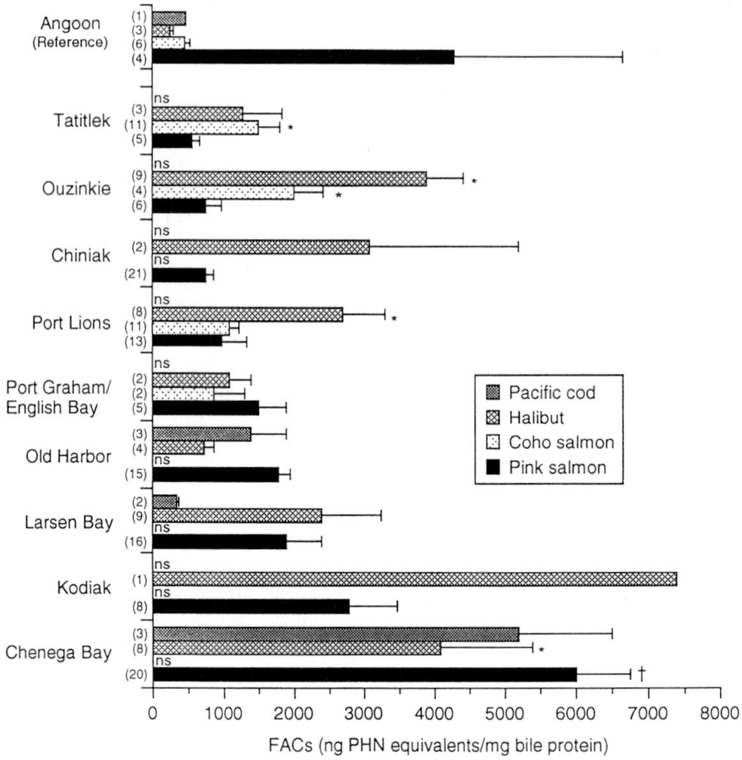

FIGURE 2.—Concentrations (mean ± SE) of fluorescent aromatic compounds (FACs; nanograms of phenanthrene [PHN] equivalents per milligram of bile protein) in salmon and bottomfishes sampled from Alaskan fishing sites in 1989. The numbers in parentheses indicate the number of samples that were analyzed. The abbreviation "ns" indicates a fish species that was not sampled. An asterisk (*) indicates a value that is significantly higher than the measured value in fish from Angoon, the designated reference site, by GT2 comparison intervals. The cross (†) denotes a value that is significantly higher than the measured value in fish from all other sites except Kodiak and Angoon by GT2 comparison intervals.

from Chenega Bay (Figure 3A) showed the greatest decline: a 12-fold decrease, from 6,000 ± 760 to 490 ± 100 ng phenanthrene equivalents/mg bile protein, in mean concentrations of biliary FACs was observed in fish collected in 1989 relative to those sampled in 1990. In addition, significant decreases in FACs from 1989 to 1990 were observed in pink salmon from Port Graham/English Bay, Old Harbor, and Larsen Bay (Figure 3A). Interestingly, in salmon sampled from sites near Kodiak in 1990 the concentration of FACs in salmon was comparable to the concentration in 1989, indicating that exposure to ACs was unchanged for salmon captured at this site.

Pacific cod were collected in both years at only a few subsistence sites. In contrast to pink salmon, no significant (P = 0.15) effect based on sampling year was observed. Pacific cod sampled in consecutive years from Chenega Bay and Old Harbor showed declines (> twofold) in mean concentrations of biliary FACs, but these were not statistically significant. An increase in biliary FACs from 1989 to 1990 was observed in Pacific cod from Larsen Bay; however, only two fish were sampled in each year (Figure 3B).

Aromatic Compounds in Edible Muscle Tissue of Fish

The summed concentrations of individual selected ACs (ΣACs) determined by GC–MS in muscle tissue of selected salmon and bottomfishes are summarized in Table 2. In muscle of halibut and Pacific cod, the mean concentrations of ΣACs were ≤3 ng/g wet weight in fish sampled from all subsistence sites in both years. Furthermore, mean concentrations of ΣACs in muscle of these bottomfishes were not significantly different from the

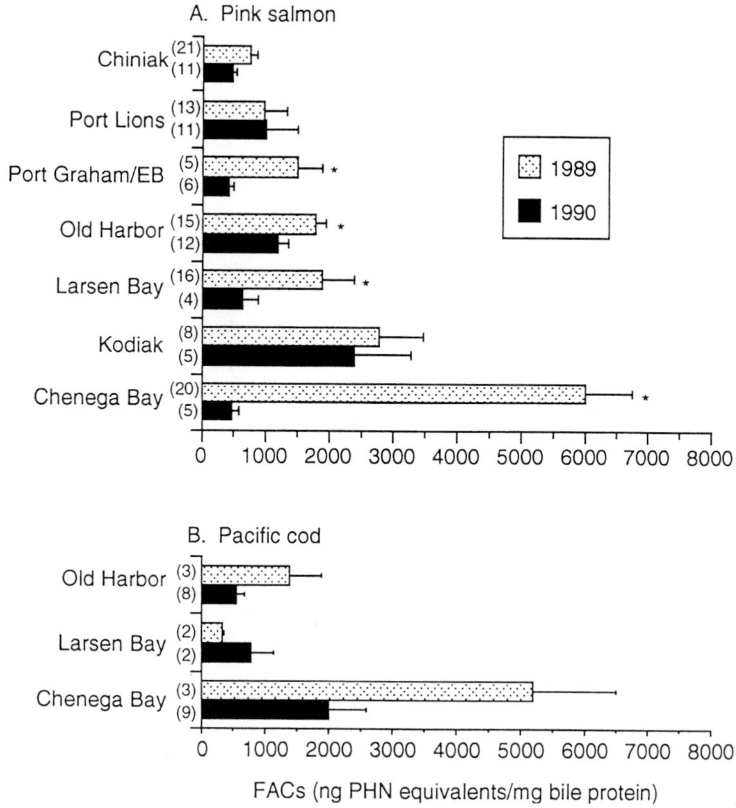

FIGURE 3.—Concentrations (mean ± SE) of fluorescent aromatic compounds (FACs; nanograms of phenanthrene [PAH] equivalents per milligram of bile protein) in pink salmon (**A**) and Pacific cod (**B**) sampled from Alaskan fishing sites in 1989 and 1990. The numbers in parentheses indicate the number of samples that were analyzed. An asterisk (*) indicates a value that is significantly different from the measured value in 1990 by Student's *t*-test.

concentrations of ΣACs found in halibut and Pacific cod from Angoon and Yakutat, the designated reference sites in southeastern Alaska. The range of concentrations of ΣACs in individual muscle samples of pink (not detected–100 ng/g wet weight) and coho (not detected–22 ng/g wet weight) salmon were somewhat higher than in bottomfishes; however, the concentrations of ΣACs in the majority of samples were less than 30 ng/g wet weight.

Of the several hundred muscle samples analyzed for individual ACs by GC–MS in 1989 and 1990, only 16 salmon (<6% of the total number of fish analyzed) had concentrations of ΣACs in muscle that exceeded 10 ng/g wet weight (Figure 4). Interestingly, the muscle samples that contained relatively higher concentrations of ΣACs were in pink salmon from Kodiak. Furthermore, GC–MS analyses of these fish showed that of the total ΣACs detected in muscle tissue, the amount consisting of high molecular weight ACs (listed in Table 1) ranged from 0 to less than 10% of the total concentrations of ACs.

Although sufficient samples of muscle from pink salmon and Pacific cod were collected in both sampling years to allow evaluation of temporal changes, comparisons in the concentrations of ΣACs in edible muscle were difficult because of the low amounts of ΣACs detected. For example, mean concentrations of ΣACs in muscle of pink salmon from Port Graham/English Bay, Old Harbor, Larsen Bay, and Chenega Bay were less than 3 ng/g wet weight in both sampling years (Table 2). However, in pink salmon from Kodiak the mean concentrations of ΣACs in muscle were slightly elevated (>30 ng/g wet weight) relative to levels in fishes from the other sites. Comparisons of mean concentrations of ΣACs in muscle of Kodiak fish between 1989 (31 ± 34 ng/g wet weight) and 1990 (34 ± 42 ng/g wet weight) showed no significant change in levels of ACs.

TABLE 2.—Concentrations (ng/g wet weight) of ΣACs in muscle of salmon and bottomfishes sampled after the *Exxon Valdez* oil spill. The concentration of ΣACs from each site is expressed as the mean ± SD. Of the total ΣACs detected in individual muscle, the amount attributable to high molecular weight ACs (listed in Table 1) ranged from 0 to less than 10%. Numbers in parentheses indicate the number of samples analyzed. The abbreviation "nd" indicates that the analyte was not detected above the limit of detection (ranging from 0.01 to 1.0 ng/g wet weight).

	Pink salmon		Coho salmon		Pacific halibut		Pacific cod	
Site	1989	1990	1989	1990	1989	1990	1989	1990
Angoon (reference)	2.0 ± 0.1 (2)		3.0 ± 1.0 (3)		0.4 (1)		0.6 (1)	
Yakutat (reference)					4.0 ± 3.9 (6)			
Tatitlek	0.01 ± 0.02 (6)		9.0 ± 7.7 (7)		nd (2)		nd (1)	nd (4)
Ouzinkie	2.0 ± 1.7 (10)	nd (1)	3.0 ± 2.3 (8)	0.5 ± 0.3 (3)	0.3 ± 0.5 (5)	1.0 ± 2.0 (2)		3.0 (1)
Chiniak	1.0 ± 1.4 (13)				1.0 ± 0.4 (2)			
Port Lions	0.7 ± 0.8 (10)	8.0 ± 8.3 (2)	1.0 ± 1.1 (6)		0.3 ± 0.3 (4)			1.0 ± 1.4 (2)
Port Graham/ English Bay	0.1 ± 0.3 (7)	0.4 ± 1.1 (5)	0.6 ± 0.8 (6)			0.1 ± 0.1 (2)		nd (1)
Old Harbor	0.6 ± 1.2 (18)	0.9 ± 1.1 (8)			0.7 ± 0.6 (3)		0.4 ± 0.5 (2)	1.0 (1)
Larsen Bay	2.0 ± 3.4 (15)	0.5 ± 0.5 (2)			0.4 ± 0.3 (5)		0.4 ± 0.1 (2)	0.7 (1)
Kodiak	31 ± 34 (15)	34 ± 42 (3)			0.4 ± 0.5 (3)			
Chenega Bay	2.0 ± 4.3 (21)	0.9 ± 0.7 (5)			0.1 ± 0.1 (4)	3.0 ± 3.7 (2)	0.1 ± 0.1 (4)	1.0 ± 1.0 (6)

Relationship between Biliary FACs and ACs in Muscle of Fish

Comparison of the concentrations of FACs in bile to the concentrations of ΣACs in corresponding muscle tissue of salmon and bottomfishes showed no significant correlation ($r = 0.06$, $P = 0.40$) (Figure 5). The concentrations of biliary FACs in fish from which samples of muscle tissue were analyzed for ACs ranged from 50 to 18,000 ng phenanthrene equivalents/mg bile protein. In contrast, the concentrations of ACs in corresponding muscle tissue of these fish ranged from not detected to only 100 ng/g wet weight. For example, pink salmon from Chenega Bay in 1989 exhibited high mean concentrations of biliary FACs of 6,000 ± 760 ng phenanthrene equivalents/mg bile protein, whereas low concentrations of parent ACs (1.6 ± 4.3 ng/g wet weight) were detected in corresponding muscle tissue.

Discussion

The results of the analyses of bile of fish species used for subsistence by native Alaskans showed substantial evidence of exposure to petroleum-related ACs after the *Exxon Valdez* oil spill. The high concentrations of biliary FACs detected in salmon and bottomfishes indicated that the fish were able to readily take up and metabolize ACs to polar com-

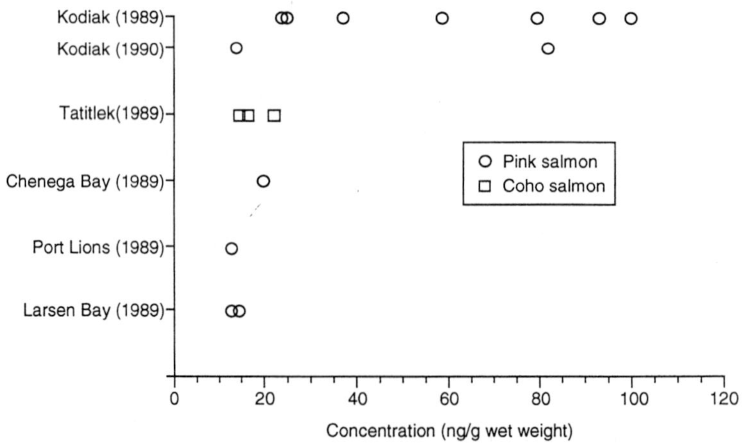

FIGURE 4.—Concentrations (nanograms per gram wet weight) of aromatic compounds in edible muscle tissues of salmon sampled from subsistence sites in 1989 and 1990 that exceeded 10 ng/g wet weight.

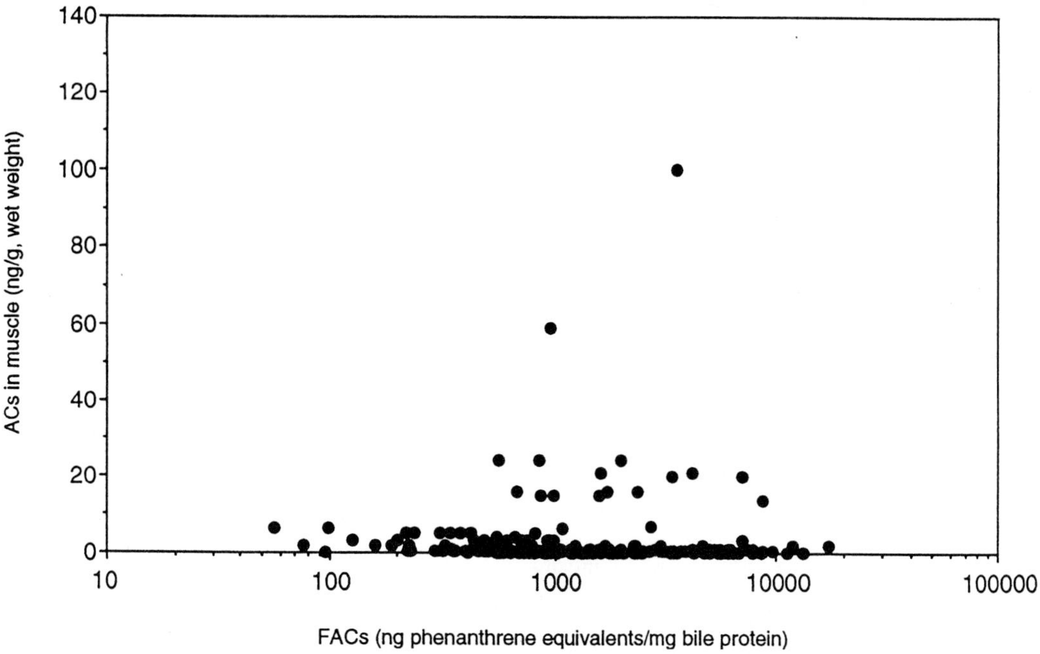

FIGURE 5.—Relationship between fluorescent aromatic compounds (FACs; nanograms of phenanthrene equivalents per milligram of bile protein) in bile and total aromatic compounds (ACs; nanograms per gram wet weight) in corresponding edible muscle determined in salmon and bottomfishes sampled from Alaskan fishing grounds in 1989 and 1990. Concentrations of FACs in bile are plotted on a semilogarithmic scale.

pounds for excretion. However, minimal contamination by ACs was detected in samples of corresponding edible muscle. These data showing low concentrations of ACs in muscle of fish exhibiting a substantial range in exposure to ACs corroborated results of earlier laboratory studies. For example, laboratory studies (Varanasi et al. 1981; Balk et al. 1984; Stein et al. 1984, 1987; Varanasi et al. 1989a) with fish exposed to benzo[*a*]pyrene (BaP) via water, food, or contaminated sediment showed that fish can readily take up and efficiently biotransform ACs to polar metabolites that are concentrated into bile for excretion, thus limiting the accumulation of parent ACs in extrahepatic tissues such as muscle. These findings were important in the development of techniques (Krahn et al. 1986) that would provide the necessary information for responding effectively to environmental emergencies and demonstrated the usefulness of rigorous laboratory studies for predicting the response of marine species to an environmental disaster such as an oil spill.

Salmon and bottomfishes sampled from selected subsistence sites exhibited a wide range of FACs in the bile. Those fish that exhibited high levels of biliary FACs were probably exposed to petroleum-related ACs. Fish captured in areas within Prince William Sound that were in the path of the spilled oil exhibited the highest levels of biliary FACs. For example, the highest mean concentrations of biliary FACs detected in pink salmon, Pacific halibut, and Pacific cod were collected from sites near Chenega Bay, which is located in southwestern Prince William Sound and was the first collection site in the path of the spilled oil. Fish captured at sites along Kodiak Island (i.e., Kodiak, Ouzinkie, Chiniak, Port Lions) also contained elevated concentrations of FACs in bile, but at concentrations below those found in fish from Chenega Bay. Surprisingly, coho salmon sampled at Tatitlek, which is located near Bligh Reef (the site where the *Exxon Valdez* ran aground) and not in the path of the spilled oil, contained relatively high concentrations of biliary FACs. Pelagic species such as salmon can migrate across large distances, thus the finding of elevated levels of FACs in the bile of coho salmon from Tatitlek may be the result of these fish passing through oil-contaminated waters before being captured. Generally, higher concentrations of biliary FACs were found in fish captured at sites in the path of the spilled oil, compared to concentrations in fish captured at Angoon, the designated reference site in southeastern Alaska not affected by the *Exxon Valdez* oil spill. In a separate but related study, Collier et al. (1996, this volume) reported that several subtidal fish species caught in sites heavily contaminated by spilled crude oil contained significantly higher concentrations of biliary FACs than those found in fishes sampled in areas that showed little or no signs of oiling.

The presence of FACs in the bile of fish is an indication of a relatively recent (within several weeks) exposure of fish to PAHs (Collier and Varanasi 1991). In the present study, significantly lower levels of biliary FACs were observed in pink salmon from Chenega Bay in 1990 compared to 1989. Furthermore, a trend of decreasing concentrations of biliary FACs in pink salmon and Pacific cod in 1990 compared to 1989 was also observed at most of the other subsistence sites that were sampled. Similar findings of markedly reduced biliary FACs in subtidal fish sampled in 1990 compared to 1989 from heavily oiled sites were also reported by Collier et al. (1996). Thus, the findings of lower concentrations of biliary FACs in these fish suggested a decrease in the availability of petroleum-related ACs at most sites 1 year after the oil spill. In contrast, concentrations of biliary FACs in pink salmon from the Kodiak site, a harbor with high commercial fishing-vessel traffic and fueling facilities, remained at a relatively high level in 1989 and 1990, indicating that the bioavailability of ACs remained unchanged in both sampling years.

Although levels of biliary FACs in several fish species sampled from oiled sites indicated significant exposure of fish to ACs, edible muscle tissues in fish were found to have concentrations of ACs that rarely exceeded the relatively low concentration of 30 ng/g wet weight. This finding was especially exemplified in fish from Chenega Bay, where fish showed the highest exposure to petroleum-related ACs in bile, but still contained minimal amounts of ΣACs in corresponding muscle tissue. Interestingly, a few samples of fish that contained the highest levels of ΣACs in muscle were captured from sites near Kodiak. These results, as well as those for FACs in bile of fish and data from analyses of shellfish from Kodiak (Brown et al. 1996, this volume), suggest exposure of marine species from Kodiak to a source of ACs other than spilled oil from the *Exxon Valdez*.

The findings of relatively minimal contamination of edible muscle of fresh fish sampled in the present study was further supported when these results were compared to the extremely high concentrations of ACs found in samples of smoked salmon from two Alaskan villages (Varanasi et al. 1990). Results of these analyses showed substantially higher concentrations of total ACs in edible tissues of smoked fish from Old Harbor (7,900 ng/g wet weight) and Tatitlek (23,000 ng/g wet weight) compared to levels found in fresh fish sampled from all subsistence fishing sites. Thus, the consistency of the results from this field study and those of previous laboratory studies provided evidence of minimal accumulation of ACs in edible flesh of fish captured at sites affected by the spilled oil.

The main use of bile analyses during environmental emergencies is to quickly estimate exposure of fish to ACs and then to use these results to prioritize corresponding samples (in this case, edible flesh) for more comprehensive analyses. Although analysis of bile for FACs provides a rapid, semiquantitative assessment of exposure to ACs that are present in oil, it does not allow identification of individual compounds found in bile. However, in a related study (Krahn et al. 1992) conducted with fish sampled following the *Exxon Valdez* oil spill, it was shown that the bile method is a sensitive indicator of exposure of fish to specific ACs present in oil. In the study of Krahn et al. (1992), metabolites of ACs were identified by GC–MS in the bile of fish sampled from sites in Prince William Sound shortly after the spill. Several of the AC metabolites were characteristic of ACs in PBCO as determined by analyses of both PBCO and bile from fish administered PBCO. Moreover, a significant correlation ($r = 0.93$, $P \leq 0.0001$) was found between the sum of selected individual biliary metabolites determined by GC–MS and the concentrations of FACs in bile determined by HPLC with fluorescence detection.

Overall, the present field study generated a large database on the exposure of fish to ACs from a relatively pristine area that was contaminated by spilled crude oil. Although the results of chemical analyses of the marine fish species showed evidence of exposure to ACs, the objective of these investigations was not to specifically determine the source of ACs in each sample. The results for these fishes, however, provided a large-scale field validation of findings of laboratory studies on the uptake and biotransformation of parent ACs by fish. This comparison further substantiated the important role of controlled laboratory studies in predicting the responses of marine species exposed to sources of chemical contaminants such as spilled crude oil, and also enhanced the interpretation of field data and assisted fishery resource managers during an environmental emergency.

Acknowledgments

We thank Anna Kagley and Sylvester Spencer for the analyses of fish bile by HPLC. We are also grateful to Don Brown for coordinating many aspects of this work; Doug Burrows, Jennie Bolten, and Ron Pearce for performing the GC–MS analyses; and Susan Pierce for calculating and compiling the data. We thank our colleagues in the National Marine Fisheries Service and the National Ocean Survey for their valuable assistance and support throughout these studies; and the biologists from Dames and Moore for the collection of fish samples. Funding for much of this work came from the Exxon Corp. and the Alaska Department of Fish and Game under a Memorandum of Understanding with the National Oceanic and Atmospheric Administration.

References

Balk, L., J. Meijer, J. W. DePierre, and L.-E. Appelgren. 1984. The uptake and distribution of [^{3}H]benzo-[*a*]pyrene in the northern pike (*Esox lucius*). Examination by whole-body autoradiography and scintillation counting. Toxicology and Applied Pharmacology 74:430–437.

Brown, D. W., and nine coauthors. 1996. Survey of Alaskan subsistence invertebrate seafoods collected in 1989–1991 to determine exposure to oil spilled from the *Exxon Valdez*. American Fisheries Society Symposium 18:844–855.

Collier, T. K., C. A. Krone, M. M. Krahn, J. E. Stein, S.-L. Chan, and U. Varanasi. 1996. Petroleum exposure and associated biochemical effects in subtidal fish after the *Exxon Valdez* oil spill. American Fisheries Society Symposium 18:671–683.

Collier, T. K., and U. Varanasi. 1991. Hepatic activities of xenobiotic metabolizing enzymes and biliary levels of xenobiotics in English sole (*Parophrys vetulus*) exposed to environmental contaminants. Archives of Environmental Contamination and Toxicology 20:462–473.

Dipple, A., R. C. Moschel, and C. A. H. Bigger. 1984. Polynuclear aromatic hydrocarbons. Pages 41–163 *in* C. E. Searle, editor. Chemical carcinogens. American Chemical Society, Washington, DC.

Gabriel, K. R. 1978. A simple method of multiple comparisons of means. Journal of the American Statistical Association 73:724–729.

Krahn, M. M., M. S. Myers, D. G. Burrows, and D. C. Malins. 1984. Determination of metabolites of xenobiotics in the bile of fish from polluted waterways. Xenobiotica 14:633–646.

Krahn, M. M., L. K. Moore, and W. D. MacLeod, Jr. 1986. Standard analytical procedures of the NOAA National Analytical Facility, 1986: metabolites of aromatic compounds in fish bile. NOAA (National Oceanic and Atmospheric Administration) Technical Memorandum NMFS (National Marine Fisheries Service) F/NWC-102, Seattle.

Krahn, M. M., and nine coauthors. 1993. Comparison of high-performance liquid chromatography/fluorescence screening and gas chromatography/mass spectrometry analysis for aromatic compounds in sediments sampled after the *Exxon Valdez* oil spill. Environmental Science and Technology 27:699–708.

Krahn, M. M., and seven coauthors. 1988. Standard analytical procedures of the NOAA National Analytical Facility, 1988: new HPLC cleanup and revised extraction procedures for organic contaminants. NOAA (National Oceanic and Atmospheric Administration) Technical Memorandum NMFS (National Marine Fisheries Service) F/NWC-153, Seattle.

Krahn, M. M., and seven coauthors. 1992. Mass spectrometric analysis for aromatic compounds in bile of fish sampled after the *Exxon Valdez* oil spill. Environmental Science and Technology 26:116–126.

Lowry, O. H., N. J. Rosebrough, A. L. Farr, and R. J. Randall. 1951. Protein measurement with the Folin phenol reagent. Journal of Biological Chemistry 193: 265–275.

Sokal, R. R., and F. J. Rohlf. 1981. Biometry. 2nd edition. Freeman, San Francisco.

Solbakken, J. E., M. Solberg, and K. H. Palmork. 1984. A comparative study on the disposition of three aromatic hydrocarbons in flounder (*Platichthys flesus*). Fiskeridirektoratets Skrifter Serie Havundersokelser 17:473–481.

Stein, J. E., T. Hom, and U. Varanasi. 1984. Simultaneous exposure of English sole (*Parophrys vetulus*) to sediment-associated xenobiotics: part 1-uptake and disposition of ^{14}C-polychlorinated biphenyls and ^{3}H-benzo[*a*]pyrene. Marine Environmental Research 13: 97–119.

Stein, J. E., T. Hom, E. Casillas, A.J. Friedman, and U. Varanasi. 1987. Simultaneous exposure of English sole (*Parophrys vetulus*) to sediment-associated xenobiotics: part 2-chronic exposure to an urban estuarine sediment with added ^{3}H-benzo[*a*]pyrene and ^{14}C-polychlorinated biphenyls. Marine Environmental Research 22:123–149.

Varanasi, U., and D. J. Gmur. 1981. Hydrocarbons and metabolites in English sole (*Parophrys vetulus*) exposed simultaneously to [^{3}H]benzo[a]pyrene and [^{14}C]naphthalene in oil-contaminated sediment. Aquatic Toxicology 1:49–67.

Varanasi, U., and nine coauthors. 1989b. National Benthic Surveillance Project: Pacific Coast, part II. Technical presentation of the results for cycles I to III (1984–86). NOAA (National Oceanic and Atmospheric Administration) Technical Memorandum NMFS (National Marine Fisheries Service) NWFSC-8, Seattle.

Varanasi, U., and nine coauthors. 1990. Survey of subsistence fish and shellfish for exposure to oil spilled from the *Exxon Valdez*, first year: 1989. NOAA (National

Oceanic and Atmospheric Administration) Technical Memorandum NMFS (National Marine Fisheries Service) F/NWC-191, Seattle.

Varanasi, U., and nine coauthors. 1993. Survey of Alaskan subsistence fish, marine mammal, and invertebrate samples collected 1989–91 for exposure to oil spilled from the *Exxon Valdez*. NOAA (National Oceanic and Atmospheric Administration) Technical Memorandum NMFS (National Marine Fisheries Service) F/NWC-12, Seattle.

Varanasi, U., W. L. Reichert, J. E. Stein, D. W. Brown, and H. R. Sanborn. 1985. Bioavailability and biotransformation of aromatic hydrocarbons in benthic organisms exposed to sediment from an urban estuary. Environmental Science and Technology 19:836–841.

Varanasi, U., J. E. Stein, and M. Nishimoto. 1989a. Biotransformation and disposition of polycyclic aromatic hydrocarbons (PAH) in fish. Pages 94–149 *in* U. Varanasi, editor. Metabolism of polycyclic aromatic hydrocarbons in the aquatic environment. CRC Press, Boca Raton, Florida.

American Fisheries Society Symposium 18:867–878, 1996

Psychological and Social Impacts of the *Exxon Valdez* Oil Spill and Cleanup

John C. Russell, Michael A. Downs, John S. Petterson, and Lawrence A. Palinkas

Impact Assessment, Inc., 2160 Avenida de la Playa, Suite A, La Jolla, California 92037, USA

Abstract.—A study of 22 communities around Prince William Sound and the Gulf of Alaska indicated that the *Exxon Valdez* oil spill and subsequent cleanup contributed to the occurrence of psychological and social impacts in the affected communities. Survey and ethnographic data collected for a study sponsored by the Oiled Mayors, a subcommittee of the Alaska Conference of Mayors, showed a significant relationship between measures of exposure to the spill and cleanup and measures of general anxiety disorder, post-traumatic stress disorder, symptoms of depression, substance abuse, and domestic violence. The analysis also indicated a significant relationship between exposure and social disruption following the oil spill and cleanup. The simultaneous occurrence of individual and social impacts impeded postspill recovery in the affected communities.

On the night of 23 March 1989, nothing indicated to the fishers of Prince William Sound and the Gulf of Alaska that their upcoming fishing season would not be even better than it had been in the summer of 1988, when the harvest was satisfactory and the price was excellent. Similarly, Alaska Natives in villages such as Tatitlek and Chenega Bay had no reason to think that the next day would not be like any other day of harvesting and sharing "Native foods" from the ocean and forests, activities that have become the embodiment of Native culture. Yet everyday life for Native and non-Native people of this region changed in the early hours of 24 March 1989, when the supertanker *Exxon Valdez* ran aground and spilled some 40 million liters of North Slope crude oil into the waters off Bligh Reef in Prince William Sound. One of the most unsettling changes for Natives and non-Natives alike was the disruption of familiar and traditional ways of life and uncertainty about the present and the future: Will there be fishing this year? Will there be long-term damage to fish resources? Are the clams safe to eat? How will boat payments and home mortgages be paid if there are no fish resources? Amidst uncertainties like these, the response effort directed by the Exxon Corp., which displaced the volunteer effort that had begun immediately after the spill, became troublesome (Davidson 1990).

Fishers and others began to look to "the cleanup" as one means to cope with the economic consequences of a lost fishing and harvesting season. Boat owners throughout the region sought contracts for cleanup work, leading to arguments and fights over why some owners received contracts and others did not. Some businesses lost money while others gained large profits, creating "spillionaires" out of former plumbers and refrigerator repairmen. Friends and neighbors argued over issues related to the spill, over inequities in opportunity or over money lost and gained in the cleanup. Long hours working on the cleanup and altered social roles brought arguments between parents and children. In some of the smaller, mostly Native, communities, almost all able-bodied adults worked on the cleanup, disrupting local government functions and the usual cycle of daily life.

Throughout the region, individuals reported feelings of anger, fear, and—in their word—"stress." Both Native and non-Native peoples were concerned about the present and future toxic effects of the spilled oil. Despite reports of social and psychological distress in human communities, the media, the spiller, and various governmental entities remained focused on natural resources damages, and the *Exxon Valdez* oil spill became known as an "environmental disaster." The purpose of this paper is to call attention to the impacts on human communities caused by this environmental disaster and to discuss some of the reasons for them.

The socioeconomic and psychological consequences of this oil spill and cleanup can be related to a growing body of scientific literature about disaster events in general (Drabek 1986) and "technological" disasters in particular (Baum 1987). Individuals or entire communities can experience psychological and social distress from a disaster itself and from altered social, economic, and cultural conditions induced by the event (Oliver-Smith 1986; Berren et al. 1989). Some work suggests that the effects of technological disasters on victims differ from those of natural disasters with respect to the phases of emotional response, the types of psychological distress experienced, and the psychosocial processes that occur during and after an event (Berren et al. 1989). Nonetheless, victims of both types of disasters report both psychological and social effects from these events. Psychological conditions such as depression (Baum et al. 1983), anxiety (Shore et al. 1986a, 1986b), and post-traumatic stress disorder (McFarlane 1987) have been reported as outcomes of disasters. Some research suggests such problems are of short duration (Poulshock and Cohen 1975), whereas other research suggests such effects can last several years after an event (Murphy 1986). Some research suggests that social bonds within communities (Edelstein 1988; Kroll-Smith and Couch 1990) or entire communities (Picou et al. 1992) are disrupted by disasters, especially technological disasters. Indeed, Erikson (1991) has argued that toxic chemical spills, oil spills, radiation leaks, and other such events that threaten neighborhoods or entire communities result in "collective trauma," defined as damage to the bonds that connect individuals within a social unit. Collective trauma was reported in Erikson's (1976) study of the Buffalo Creek flood, in Shkilnyk's (1985) study of the effects on an Ojibwa community of mercury poisoning of a river used for subsistence fishing, and in Kroll-Smith's and Couch's (1990) study of the social impacts of a Pennsylvania mine fire. Erikson (1991) argued that such collective trauma that coexists with individual psychological trauma results in a "new species of trouble."

In this paper our focus is to examine the co-occurrence of the psychological and social effects of the *Exxon Valdez* oil spill as an example of this "new species of trouble." We describe psychological and social effects of the oil spill and cleanup using data collected by a household survey within the affected communities (Impact Assessment, Inc. 1990d). The social and cultural processes that resulted from the spill and cleanup are then described with qualitative data collected by participant observation and ethnographic interviews with members of the affected communities. These qualitative data provide a means to interpret the quantitative findings that relate exposure to the oil spill to certain social, economic, and psychological impacts.

The data for this discussion were collected during a study of the social, economic, and psychological effects of the oil spill and cleanup on 22 communities of Prince William Sound, Kodiak Island, and the Alaska Peninsula. This study was funded by the City of Kodiak acting on behalf of the Alaska Conference of Mayors. The Conference formed a subcommittee of mayors from municipalities in the affected region. This subcommittee became known as the "Oiled Mayors," and although they were concerned about damages to natural resources, their focus was the various socioeconomic and psychosocial effects of the spill and cleanup.[1]

Study Sites

The Oiled Mayors' study was conducted in 22 communities of Prince William Sound, Kodiak Island, and the Alaska Peninsula. We refer to these as communities of the north gulf coast (NGC) region of the Gulf of Alaska. The NGC communities exist among bountiful natural resources, including some of the most productive salmon habitats in the world. The region is also especially valued for its pristine and scenic beauty that attracts tourists from all over the world. The population of the whole NGC region at the time of the oil spill was approximately 65,000 persons. The 22 communities included in the study contained about 32,000 persons before the spill (Table 1).

Both Native and non-Native communities occur throughout this region. The major non-Native communities are Valdez, Cordova, Homer, Soldotna, Whittier, Kenai, and Kodiak, although each of these communities has a Native population. The remaining communities are mostly composed of Aleuts, Koniags, and other Native Alaskan peoples. The non-Native communities of Cordova and Kodiak rely primarily upon fishing as their economic base. The economies in other non-Native communities are more diversified into tourism. The NGC Native communities also engage in commercial fishing, but subsistence hunting and fishing is the favored source of foodstuffs in these communities. But subsistence hunting and fishing are more than

[1]The Oiled Mayors appealed to the Alaska Department of Community and Regional Affairs for funding to respond to these impacts, including monies to fund a study of the socioeconomic and psychological effects of the spill and cleanup. In August 1989, the Oiled Mayors, through the City of Kodiak, issued a request for proposals to study the type and range of impacts occurring within the affected communities, including impacts on municipalities and organizations that provided services to Native communities of the region. The study was to be conducted over 1 year from December 1989 through November 1990. The study was intended to assess the social, economic, and psychological impacts on the affected communities as specified in a plan of work developed by the Oiled Mayors. A competitive bidding process resulted in the selection of Impact Assessment, Inc., a private applied social science research firm, to conduct the study.

TABLE 1.—Communities in the Oiled Mayors' study and their population sizes before the 1989 *Exxon Valdez* oil spill.

Kodiak Island		Prince William Sound and Kenai Peninsula		Alaska Peninsula	
Community	Population	Community	Population	Community	Population
Kodiak City	6,774	Cordova	2,048	Chignik Bay[a]	128
Akhiok[a]	93	Homer	4,338	Chignik Lagoon[a]	55
Karluk[a]	82	Kenai	6,543	Chignik Lake[a]	170
Larsen Bay[a]	149	Seward	2,400		
Old Harbor[a]	322	Soldotna	3,668		
Ouzinkie[a]	204	Valdez	3,313		
Port Lions[a]	300	Whittier	206		
		Chenega Bay[a]	84		
		English Bay[a]	200		
		Port Graham[a]	195		
		Seldovia[a]	565		
		Tatitlek[a]	120		

[a]Primarily Native Alaskan.

means of getting food; they have become the means of expressing and transmitting Native culture. Consequently, any actions that threaten subsistence practices or that endanger the natural resources of the region have implications for the economies and ways of life of both Native and non-Native communities.

Methods

The Oiled Mayors' study was designed to investigate the types and range of social, economic, and psychological impacts resulting from the oil spill and cleanup. The study was intended to provide practical and timely information to the affected communities regarding the full range of impacts they were experiencing so that mitigation actions could be planned and executed.[2] Three major tasks were specified for the study: assessment of fiscal and operational impacts on local governments, assessment of private sector economic impacts, and assessment of social and psychological impacts. The methods and procedures for each of these were specific to the task. Assessment of fiscal and operational impacts on local governments and assessment of private sector economic impacts are not the focus of this paper and their methodologies are not reviewed here. Social and psychological impacts were assessed by four methods. First, interviews were conducted with providers of social and psychological services in the affected communities to determine their perspectives on psychosocial impacts. Second, compiled statistics on mental health visits, social service usage, and police calls and arrests were assembled. Third, participant observation was conducted in most of the affected communities, including interviews with community leaders and other knowledgeable persons. Fourth, a face-to-face household survey concerning social and psychological impacts was administered to a randomly selected sample of approximately 600 persons from 11 affected and two "control" communities, one primarily Native (Angoon) and one primarily non-Native (Petersburg) (Table 2).

In presenting our findings about community and individual impacts, we focus on the results of the household survey. Consequently, we describe this methodology in more detail than the others.

The household survey was intended to provide data for an analysis of the relationship of "exposure" to the oil spill and various outcome measures. In addition to exposure questions and basic demographic information, data regarding the following outcome measures were collected: perceptions about the current and future types of oil spill impacts, social relationships and disruption, subsistence activities, family support, family and child relations, psychosocial conditions (depression symptoms, general anxiety disorder, post-traumatic stress disorder, health status, and perceptions of alcohol use, drug use, and domestic violence), perceptions of municipal service delivery, and income changes related to the oil spill.

Exposure Assessment

We constructed an oil spill exposure scale using the following six questions from the household survey. (1) Did you or anyone in your household use,

[2]The study was monitored by both a peer review committee and a working group of individuals from the affected communities. The peer review committee functioned to ensure the scientific quality of the study, whereas the working group provided input to ensure that the results were usable for the affected communities.

TABLE 2.—Communities and sample sizes in the Oiled Mayors' household survey.

Non-Native		Native		Control	
Community	Subjects	Community	Subjects	Community	Subjects
Cordova	66	Akhiok	11	Angoon	60
Kodiak City	119	Chenega Bay	11	Petersburg	102
Seward	60	Chignik Bay	29		
Valdez	65	English Bay	24		
		Karluk	11		
		Larsen Bay	22		
		Tatitlek	14		

before the spill, areas along the coast that were affected by the spill? (2) Did you work on any of the shoreline or water cleanup activities of the oil spill? (3) Are there any other ways that you came into contact with the oil spill or cleanup activities, such as during recreation, hunting, fishing, or gathering activities? (4) Did you have any property that was lost or damaged because of the oil spill or cleanup? (5) Did the oil spill cause any damage to the areas you or other household members fish commercially? (6) Has the oil spill directly affected the hunting, fishing or gathering activities of any members of this household? Each response was coded 0 for a "no" response and 1 for a "yes." Responses were then summed to provide a continuous measure of exposure with a range of 0 to 6. The exposure index had an internal consistency reliability (Cronbach's alpha) of 0.74 for this population. The mean exposure score for all study respondents was 1.97 (SD, 1.77). Subjects were classified into three groups on the basis of maximum level of exposure. Residents in the affected (noncontrol) communities first were grossly classified as being either exposed or unexposed, depending on whether their exposure scores fell above or below the group median (2.00). Then exposed residents were further dichotomized into a low-exposed group (scores of 2 or 3) and a high-exposed group (4 and above). Relationships between exposure and the various outcome measures were tested statistically.

Demographic Variables

Demographic data were obtained for age, gender, ethnicity, education, total household income in 1988, and marital status. Age had five categories: 18–24, 25–44, 45–64, and 65+ years. Ethnicity had 12 categories: white–Caucasian, Koniag–Aleut–Sugpiaq, Eyak, Athapaskan, Yup'ik–Inupiaq, other Alaskan Native, Filipino, Hispanic, Asian–Oriental, Black, American Indian, and other. These categories were grouped as Alaskan Natives or non-Natives. Education was an eight-item variable ranging from less than 6th grade to a graduate degree; respondents were secondarily grouped by whether they had or had not graduated from high school. Household income for 1988 had 13 categories from less than $5,000 to more than $150,000, which were divided at the median income of $40,000. The six marital status categories were married, never married, not married but living together, widowed, divorced, and separated.

Selected Outcome Measures

Below we briefly describe selected outcome measures that were used in the Oiled Mayors' household survey. The full range of measures and scales is given by Impact Assessment, Inc. (1990d).

Family and child relations.—The oil spill's impact on the family was examined with two measures. The first was a series of 11 questions relating to the behavior of children in the household. Responses to these questions ranged from 1 (strongly disagree) to 5 (strongly agree). These responses were coded and summed to comprise a "child dysfunctional behavior index." The inter-item reliability (Cronbach's alpha) of this scale was 0.65. A similar measure was developed to assess the degree of perceived family support. The 11-item family support scale had an internal reliability of 0.88 in this population.

Social relations and disruption.—The oil spill's impact on social relations was examined by asking respondents if relations with a spouse or partner, children living at home, other relatives, neighbors and friends, people from other communities, and coworkers had improved, stayed the same, or declined since the spill. Respondents were also asked whether they had experienced any problems with outsiders or with friends since the spill.

Subsistence activities.—The oil spill's impact on traditional subsistence production and distribution activities was assessed by asking respondents about time normally spent hunting, fishing, and gathering; time spent with people from other households on these activities; the amount of harvested resource foods shared with others and with elders; the amount of harvest resource foods received from other families; the number of household members participating in subsistence activities; and the opportunities for children to learn subsistence skills. Respondents were asked whether these activities increased, stayed the same, or decreased since the spill relative to the same time in 1988.

Depression.—Depressive symptoms were measured with the Center for Epidemiologic Studies Depression scale (CES-D; Radloff 1977). This scale has been used extensively in community studies to determine the prevalence of depressive symptoms. The scale has been shown to have high inter-item and test–retest reliability as well as sufficient convergent and discriminant validity for use as a measure of the level of depressive symptomatology in general populations (Radloff 1977; Weissman et al. 1977). In this population, the CES-D had an inter-item reliability (Cronbach's alpha) of 0.88; subpopulation reliabilities were 0.90 for Natives and 0.86 for non-Natives.[3]

In the CES-D, respondents describe their mood over the past week by rating each of 20 items on a scale from 0 (rarely or none, less than 1 d) to 3 (most or all, 5–7 d). A depression score was calculated for each respondent by summing the ratings (after ratings of four reverse-worded items were reversed). If a respondent completed 85% or more of the CES-D items but less than 100% ($N = 10$), responses to missing items were imputed by using the mean of that person's other answers.

Respondents with scores of 16 or above were classified as being depressed for the purpose of calculating postspill prevalence rates. Several studies have documented the validity of this cutpoint in distinguishing groups with high from those with low depressive symptomatology. Use of this cutpoint is not accurate enough to measure rates of clinical depression as defined by the *Diagnostic and Statistical Manual of Mental Disorders, Version III-R* (DSM III-R). Nevertheless, the extensive use of the CES-D scale in other community studies of depression gives us a basis of comparison with other segments of the general U.S. population.

The 16.6% prevalence of CES-D scores of 16 or greater in our study is much higher than the 6-month prevalences of DSM III-R depressive disorders (0.1–1.6% for major depressive episodes and 0.5–3.1% for dysthymia) found in the general U.S. population (Weissman et al. 1986). However, comparison between our results and those of other studies is limited by differences in study criteria (depressive symptoms versus DSM III-R diagnoses) and measurement (CES-D versus diagnostic interview schedule protocol). The CES-D was selected for use in this study because of its ability to describe that type of depressive symptomatology that would engage all aspects of the health care system and not merely mental health services.

Anxiety.—Unlike the diagnosis of depression, the diagnostic criteria for "generalized anxiety disorder" (GAD) were based on DSM III-R criteria and therefore reflected a clinical diagnosis. Respondents first had to acknowledge anxiety (Have you ever had a time when for a month or more most of the time you felt worried or anxious, perhaps afraid that something bad was going to happen either to you or someone that you cared about?). In addition, respondents had to acknowledge at least three additional symptoms of anxiety. These included one or more symptoms of motor tension (muscle tension, restlessness, easy fatigability), one or more symptoms of autonomic hyperactivity, and one or more symptoms of vigilance and scanning (feeling keyed up or on edge, difficulty concentrating, trouble falling or staying asleep, irritability).

Because disasters are known to lead to new cases of a psychiatric disorder as well as to exacerbate an existing disorder, we asked respondents when was the last time they experienced these symptoms. All GAD symptoms were used to calculate a lifetime prevalence rate and all symptoms reported within the past year were used to calculate a postspill prevalence rate. If respondents indicated that the first time they were worried or anxious or afraid most of the time for at least a month occurred within the past year, they were defined as new cases and used to calculate a postspill incidence rate. Lifetime prevalence refers to the total number of cases, both new and preexisting, within the lifetimes of respondents.

Post-traumatic stress disorder.—The diagnostic criteria for post-traumatic stress disorder (PTSD) were based on DSM III-R criteria. A respondent was diagnosed with PTSD if he or she experienced three or more of the following: persistent unpleasant memories, repeated bad dreams or nightmares, disturbing memories, feeling worse when reminded of a past event, flashback; three or more of the following: loss of interest in previously important activities, trying hard not to think of a past experience, loss of caring about previously important activities, avoidance of places or activities that remind

[3]Radloff (1977) obtained alphas of 0.84, 0.85, and 0.90 for her CES-D field trial data.

of a past event, avoidance of feelings about a past event, avoidance of other people, loss of feeling or reduction in emotion, change in future plans, inability to remember part of past; and two of the following: trouble concentrating, vigilance, insomnia, startled by noise, feeling panicky or fearful or anxious, irritability, autonomic hyperactivity (sweating, breathing heavily, heart pounding).

As with anxiety, we asked respondents when was the last time they experienced these symptoms. All PTSD symptoms were used to calculate a lifetime prevalence rate and all symptoms reported within the past year were used to calculate a postspill prevalence rate.

Substance abuse and domestic violence.—Because of the cultural sensitivity in rural Alaskan communities to the issues of alcohol and drug abuse and domestic violence, respondents were not asked about changes in their own patterns of drinking or drug use or their involvement in episodes of domestic violence. We felt that such questions would not lead to useful response rates or meaningful results. Instead, respondents were asked if they thought people in their community and their close friends and family were drinking, using drugs, and fighting more than, about the same as, or less than before the spill. They were also asked if the amount of drinking, drug use, and domestic violence was leading to problems that did not exist before the spill.

Analysis and Reporting

Household survey data, participant observations, and compiled municipal statistics were used to assess social and psychosocial impacts of the oil spill and cleanup (Impact Assessment, Inc. 1990c, 1990d; Palinkas, et al. 1993). Data from the household survey were subjected to statistical analysis to assess the relationship between exposure status and the various outcome measures. Prevalence rates of depressive symptoms, generalized anxiety disorder, and post-traumatic stress disorder were percentages of respondents or groups of respondents meeting the criteria for these conditions. Prevalence rates and proportional distributions across the three exposure categories were compared by chi-square tests for trends in proportions (Fleiss 1981). Prevalence rates were further adjusted to account for differences in the distribution of residents by ethnicity and marital status across the three exposure groups, for which the Mantel–Haeszel chi-square procedure was used. Comparisons of mean scores were performed by analysis of variance. Nonparametric tests (Wilcoxen) were used to compare changes in perceived health status before and after the spill in each exposure group.

This study resulted in many findings. Impacts on municipalities and private sector economic impacts were reported in detail by Impact Assessment, Inc. (1990a, 1990b, 1990c, 1990d). The full range of findings on psychosocial impacts was reported by Impact Assessment, Inc. (1990c, 1990d) and Palinkas et al. (1993). In this paper, we emphasize the co-occurrence of psychosocial and community-level impacts. In particular, we focus on findings from the household survey, because these data directly depict the association between exposure to the effects of the oil spill and the outcome measures of psychiatric condition, substance abuse and domestic violence, and social disruption. Although they receive less emphasis here, the ethnographic data and compiled statistics are rich sources of information about the experiences of individuals and communities with the effects of the oil spill and cleanup (Impact Assessment, Inc. 1990c, 1990d).

Results

Exposure

Human exposure to the effects of the oil spill and cleanup was not confined to the direct oiling of a community shoreline or beach. Indeed, one of the defining characteristics of this event was the absence of direct oiling as a condition of exposure. Individuals were exposed to the effects of the spill if their fishing grounds were exposed, if they worked on the cleanup, or if in some other fashion they came into direct contact with the oil spill or its effects. For example, residents of Cordova were considered exposed even though the community received no direct oiling, because the fishing grounds used by the Cordova fishing fleet were extensively oiled, and some Cordova residents worked with the "mosquito fleet" of small boats that collected dead or dying animals. The experiences of these individuals were relayed to their friends, family, and neighbors, thereby also exposing them to the effects of the spill. This notion of exposure was incorporated into the household survey.

Native communities had the highest overall exposure scores, although a majority of residents in the larger communities of Cordova, Kodiak, and Seward were exposed to one degree or another (Table 3). Approximately 48% of Valdez respondents were in the high- or low-exposed categories.

TABLE 3.—Mean oil spill and cleanup exposure index scores and percentages of community residents in high-, low-, and not-exposed groups.[a]

Community	*N*	Mean score	% high-exposed	% low-exposed	% not-exposed
English Bay[b]	24	3.92	62.5	37.5	0.0
Tatitlek[b]	14	3.79	71.4	21.4	7.1
Chenega Bay[b]	11	3.73	72.7	18.2	9.1
Larsen Bay[b]	22	3.59	54.5	36.4	9.1
Akhiok[b]	11	3.27	45.5	45.5	9.1
Karluk[b]	10	2.90	40.0	40.0	20.0
Chignik Bay[b]	30	2.53	23.3	50.0	26.7
Cordova	66	2.51	37.9	24.2	37.9
Kodiak	119	2.36	26.9	38.7	34.5
Seward	60	2.10	23.3	36.7	40.0
Valdez	65	1.77	16.9	30.8	53.3
Petersburg[c]	101	0.51	2.0	13.9	84.2
Angoon[c]	60	0.30	0.0	5.0	95.0
All	593	1.96	24.5	28.2	47.4

[a]Exposure scores: high-exposed, 4–6; low-exposed, 2–3; not-exposed, <2.
[b]Native community.
[c]Control community.

TABLE 4.—Demographic characteristics of Oiled Mayors' household survey respondents ($N = 593$)[a] by exposure status, 1990.

Demographic characteristic	Percent of respondents by exposure status: High-exposed ($N = 224$)	Low-exposed ($N = 209$)	Not-exposed ($N = 162$)
Sex and age			
Males	52.9	51.0	45.1
18–24	6.0	4.7	1.4
25–44	66.7	56.6	57.5
45–64	22.2	26.4	32.9
>65	5.1	12.3	8.2
Females	47.1	49.0	54.9
18–24	13.5	10.8	5.6
25–44	59.6	60.8	67.4
45–64	19.2	19.6	15.7
>65	7.7	8.8	11.2
Ethnicity			
Native	41.2	21.7	32.3
Non-Native	58.8	78.3	67.7
Education			
<12 years	49.5	51.5	49.4
High school graduate	50.5	48.5	50.6
Household income, 1988			
<$40,000	47.5	58.2	53.2
≥$40,000	52.5	41.8	46.8
Marital status			
Married or cohabiting	70.0	65.2	75.1
Single	14.5	9.2	11.2
Separated or divorced	10.0	19.3	6.8
Widowed	5.5	6.3	6.8

[a]Information on exposure status was missing for six respondents.

Demographic Characteristics

Of the 599 total interviews, 593 contained all of the exposure data necessary to perform the analysis of the relationship of exposure to outcome measures. Of the sample of 593 persons, 50.1% were males. The majority of the subjects (68.8%) were between the ages of 18 and 44 at the time of the interviews in spring 1990. The proportions of respondents with a high school education and incomes of $40,000 or more were greater in non-Native communities than in Native communities. Proportionally more men than women lived in the affected communities than in the control communities. Otherwise the sample showed no major differences between affected and control communities or between Natives and non-Natives. Table 4 shows the demographic characteristics of individuals interviewed for the household survey by exposure status.

Psychosocial Conditions

Psychiatric conditions.—No information exists about the prespill prevalence of psychiatric conditions within the communities included in this study. Postspill prevalence rates over all sampled communities and degrees of exposure were 20% for general anxiety disorder (GAD), 9.7% for post-traumatic stress disorder (PTSD), and 16.6% for depressive symptoms. Relative to control communities, subjects in the high-exposed group were twice as likely to have either GAD or PTSD and 1.8 times as likely to have depressive symptoms as measured by the CES-D. As indicated by the household survey, approximately 43% of individuals in the high-exposed group had one or more of these conditions as compared with 22.7% in the low-exposed and 22.7% in the not-exposed groups. Among exposure groups, GAD showed a statistically significant association with high exposure status for non-Natives and Natives, males and females, and ages below 45 (Table 5); PTSD was associated with high exposure for Natives and females, and depressive symptoms were associated with high exposure for females and younger respondents. In general, Natives had higher rates of psychiatric conditions than non-Natives.

Substance abuse and domestic violence.—Perceptions of drug and alcohol abuse and domestic violence, as well as perceptions of problems associated with these psychosocial conditions, were significantly associated with exposure to the oil spill and cleanup (Table 6). Compared with the control group, high-exposed residents were 11.4 times more

TABLE 5.—Postspill prevalence of psychiatric conditions by exposure status and sociocultural characteristics. Asterisks denote significant chi-square tests for trends at $P < 0.05$*, $P < 0.01$**, or $P < 0.001$***.

Outcome measure and sociocultural characteristic	Percent of respondents among: High-exposed	Low-exposed	Not-exposed
General anxiety disorder			
Native**	41.7	23.7	19.2
Non-Native**	24.6	14.8	10.1
Male***	25.4	10.4	5.5
Female**	36.9	19.6	19.1
18–44 years old***	35.2	15.2	13.0
≥45 years	18.3	14.3	13.0
Post-traumatic stress disorder			
Native**	25.0	5.3	5.8
Non-Native	10.0	6.8	8.3
Male	8.5	6.6	2.7
Female*	20.2	5.9	11.2
18–44 years old	15.4	6.5	9.3
≥45 years	10.0	5.7	3.7
Depression symptoms			
Native	28.8	21.6	17.6
Non-Native	17.3	12.7	10.2
Male	16.1	12.7	7.1
Female*	29.8	13.9	16.9
18–44 years old*	24.5	14.8	13.9
≥45 years	17.2	10.3	9.8

likely to report an increase in drinking in their community and 10.5 times more likely to report it among family and friends; 7.4 times more likely to report more drug use in their community and 2.5 times more likely to report more among family and friends; 11.6 times more likely to report an increase in domestic violence in their community and 1.9 times more likely to report it among family and friends; 6.6 times more likely to report an increase in problems caused by drinking in their community and 4.8 times more likely to report an increase among family and friends; 4.2 times more likely to report an increase in problems caused by drug use in their community and 12.9 times more likely to report an increase among family and friends; and 7.1 times more likely to report an increase in problems caused by domestic violence in their community and 22.1 times more likely to report an increase among family and friends.

Social Bonds

The effects of the oil spill and cleanup on social bonds within communities was measured in the household survey by questions regarding time spent visiting family and friends, participation in community activities, and social conflicts among family, friends, neighbors, and coworkers. Questions were also asked about subsistence activities that reinforce social bonds within Native communities (Jorgensen 1990). Generally, individuals in the exposed communities reported more difficulties in each of these areas than individuals in the control communities.[4]

TABLE 6.—Postspill problems with alcohol and drug abuse and domestic violence by exposure status. All chi-square tests for trends were significant at $P < 0.0001$*.

Social unit and problem	Percent of respondents reporting problem among: High-exposed	Low-exposed	Not-exposed
Community			
More drinking*	56.8	40.4	5.0
More drinking problems*	45.3	32.5	6.9
More drug use*	50.4	43.2	6.8
More drug problems*	39.5	30.8	9.4
More fighting*	40.5	32.3	3.5
More fighting problems*	33.9	27.7	4.8
Family and friends			
More drinking*	29.3	15.3	2.8
More drinking problems*	26.0	13.3	5.4
More drug use*	21.2	10.8	1.7
More drug problems*	19.4	9.1	1.5
More fighting*	19.7	3.8	0.9

Social interactions and conflicts.—Activities traditionally thought to promote social solidarity were disrupted by the oil spill and cleanup. Household survey and interview data indicated a significant reduction in the frequency of social visits between families and among friends and neighbors. For example, 20–40% of households reported spending less time visiting with friends and 15–30% of households reported decreases in family interaction time. In two Native villages, more than 70% of households reported decreased social visitations. Between 10 and 30% of households reported less time available for family vacations as a result of spill-related activities. Among respondents who actually worked on the cleanup, 42% reported fewer social visits with friends and 45% reported less time spent with other household members, compared with 19% and 16%, respectively, of those who did not work on the cleanup.

Because participation in community activities and celebrations reinforces social bonds, decreases in these activities may adversely affect community solidarity. Among respondents who worked on the cleanup, 38% reported reduced participation in

[4]In a report prepared for plaintiffs' counsel in *Exxon Valdez* litigation, three of us (J.C.R., L.A.P., and M.A.D.) constructed a "social disruption scale" using these measures (other than subsistence). A logistic regression analysis showed a statistically significant relationship between exposure status and the social disruption scale.

TABLE 7.—Postspill social conflicts by exposure status. Asterisks denote significant chi-square tests for trends at $P < 0.05^*$ or $P < 0.001^{**}$.

Problem since same period in 1988	Total responses	Number (%) expressing problem among:		
		High-exposed	Low-exposed	Not-exposed
Not getting along as well with:				
Spouse or partner**	444	25 (14.5)	7 (4.6)	0 (0.0)
Children living at home**	371	14 (10.1)	5 (4.2)	1 (0.9)
Other relatives living at home**	188	11 (17.2)	2 (3.7)	0 (0.0)
Relatives not living at home**	536	24 (11.6)	9 (4.9)	0 (0.0)
Neighbors and friends**	565	28 (13.1)	9 (4.5)	1 (0.7)
People from other communities**	447	28 (13.7)	14 (7.8)	2 (1.3)
Co-workers*	483	20 (10.6)	15 (8.9)	5 (4.0)
Increased conflicts with:				
Outsiders**	593	106 (47.5)	47 (22.6)	5 (3.1)
Friends**	591	89 (40.3)	30 (14.4)	4 (2.5)

community celebrations and 28% reported less participation in religious activities, compared with 12% and 6%, respectively, of those who did not work on cleanup.

These findings indicate an overall decrease in the social bonds that integrate communities. Furthermore, the household survey also indicated increased conflicts between friends, families, neighbors, and coworkers among those in the high-exposed communities (Table 7).

Subsistence activities.—An important cultural cycle within Native communities is subsistence harvesting, processing, and distribution. An argument can be sustained that disruption of this complex cultural system disrupts essential systems of meaning and social integration within Native communities. Household survey data and interviews with key informants in Native communities indicated that subsistence activities were significantly damaged by exposure to the oil spill and cleanup. For example, 35% of those interviewed said that the spill directly affected subsistence, 42% reported a decrease in time spent in hunting, fishing, and gathering activities, 35% indicated a decrease in time spent in joint subsistence activities with people from other households, 35% reported reductions in the amount of subsistence food shared with other households, 33% indicated a reduction in the amount of food available for sharing with elders, and 30% indicated that they received reduced amounts of subsistence foods from other families since the spill.

Disruption of subsistence activities causes more than the economic hardship that arises when food sharing declines. For many in the affected Native communities, the oil spill and cleanup threatened Native culture itself. A Native informant in one community observed,

> When we worry about losing our subsistence way of life, we worry about losing our identity.... It's that spirit that makes you who you are, makes you think the way you do and act the way you do and how you perceive the world and relate to the land. Ninety-five percent of our cultural tradition now is subsistence ... it's what we have left of our tradition.

Discussion

The picture of psychosocial and community-level impacts emerging from the data is that threads weaving the social fabric of communities into a whole were damaged at the same time that individuals were traumatized by the oil spill and cleanup. In disaster situations, undermining resources for collective action in favor of self protection and health promotion places communities at risk of adverse social and psychological outcomes. Erikson (1985), in his introduction to Shkilnyk's (1985) book about Ojibwa exposure to relocation stressors and then to mercury poisoning, observed that trauma to both individuals and the ties that integrate communities can be devastating:

> When survivors suffer from loss of community as well as from individual shock, it is not just a question of getting them back on their feet but of seeing to it that there is some kind of communal ground, as it were, for them to stand on once they are upright. We can dress their physical wounds, provide food and shelter and clothing, console them for their losses, ease their grief, find ways to calm their anxieties. But until we restore the communal surroundings that was so vital to their sense of health and security, they will remain like refugees in their own land, damaged in spirit long after they have been put together again in body, and feeling a long way from home. [Erikson 1985:xvii.]

Some communities and individuals affected by the *Exxon Valdez* oil spill and cleanup also remain "a long way from home." "Home" now seems changed, its "pristine" quality lost. The perception of home as a "safe haven" from the troubles and pollution of the "lower forty-eight" was changed by the events surrounding the oil spill and cleanup. In these new circumstances, individuals feel threatened by another oil spill and its consequences. This new sense of threat has many worried about what the future holds for them, their families, and those things they hold dear about rural Alaska.

Home has not only changed in people's minds, it has also been changed by the experiences of the oil spill and cleanup itself. The joint experiences of individual distress and an unraveling social fabric are problematic for at least two reasons. First, individuals rely upon community support systems for recovery from personal distress, and thus harm to these social support systems can impede or delay individual recovery. Second, damage to the ties that bind people together impedes cooperative action for the protection and safety of an entire town or village, and it may also impede the process of community recovery from the event itself. Each of these points is worth some elaboration.

Barton (1969) was among the first social scientists to elaborate the notion of a post-disaster "therapeutic community" that is characterized by altruism, cooperation, helping behavior, and social support. Not only does a therapeutic community provide social support and aid to those in need (Drabek and Key 1984), it also can act as a buffer that inhibits adverse psychological distress for individuals (Lindy and Grace 1985). The therapeutic community thus promotes the overall well-being of individuals and communities after a disaster. The household survey data presented in this paper and ethnographic data presented elsewhere (Impact Assessment, Inc. 1990c, 1990d) suggest that the social environment after the *Exxon Valdez* oil spill was not entirely therapeutic. Rather than cooperative pulling together, there was often conflict and divisiveness as well as charge and counter-charge among neighbors and friends about how some individuals were benefiting (it was said) at the expense of others.

The conflicts did not arise immediately. In the early days and weeks of the oil spill and cleanup, many residents and outsiders worked cooperatively in response to demands of the event. Cordova residents formed a "mosquito fleet" to rescue birds and other wildlife. Others took their boats and sometimes buckets out to collect oil. Local governments and private citizens worked cooperatively to do what they could to protect their communities and fishing grounds. But it soon became apparent that the vast quantity of spilled oil was more than any mosquito fleet or well-intentioned, bucket-bearing boat owner could cope with. Furthermore, individuals were threatened with the immediate loss of their incomes because of fishery closures.

Under these conditions the Exxon-sponsored cleanup undermined the spirit of cooperativeness that initially drew people together. It established a monetary basis for working on the cleanup rather than voluntary action. Furthermore, it generally undermined local ways of doing business, and it established hierarchies that residents perceived were more responsible to Exxon's needs than to community needs. The prevailing wage for cleanup workers was US$16.69 per hour, plus overtime. This was far above the earning potential of the clerks, laborers, waiters, and others that are the mainstay of business labor in the affected communities. This sometimes created conflicts among workers and business owners about "loyalty." Additionally, fishers who would otherwise have been fishing contracted with Exxon to work on the cleanup. In communities such as Cordova and Kodiak, there were intense arguments about the allocation of boat contracts, and charges of favoritism became rampant. There developed a sense of communities "unraveling" or at least becoming segmented. This was especially evident in Native communities. The following two quotations typify comments regarding the divisiveness and anomie that developed in these communities:

> It has drifted people apart. [It is] not the same as it was before. We used to help each other. . . .
>
> [Before] the village got to be a whole family. Before- . . . these people were one big family . . . but during the oil spill I noticed the village, that it's pulling away again, people started going into their own shells, and just pulling away. It was like people were mad at each other, they put a lot of stress on the workers . . .

This postspill environment also produced a moral discourse about participation in the cleanup that itself became divisive and impeded resolution of the event. This process is representative of other social conflicts and divisiveness that seem to keep an event alive. Disaster researchers have noted that when there is no apparent "low point" or where other social processes prolong an event, recovery is impeded or delayed (Couch and Kroll Smith 1985; Baum 1987). The moral discourse that developed around the *Exxon Valdez* spill had just this effect: it kept the event in the consciousness of people in affected communities and it contributed to an overall environment of conflict and divisiveness.

This moral discourse had at least two themes that generated conflict rather than cohesiveness in the affected communities. First, the "morality" of accepting what was sometimes termed "Exxon blood money" was widely debated. On one hand were "purists" who contended that accepting any money from Exxon was immoral because the intent of offering high-paying jobs was to "pay off" local residents and to keep them quiet. To the purists, those who took such money were "Exxon whores." On the other hand were the "realists," who perceived their livelihood and economic future as threatened, and who thought it necessary for survival to take Exxon's money, even if the cleanup effort did not appear sincere. The immediate need to survive and perhaps to prosper in these circumstances thus created a dialogue of conflict about the morality of participating in the cleanup. The rhetoric about this conflict was highly divisive and charged with moral negativity that loosened the bonds between individuals and their communities.

A second theme concerned the "morality" of participating in a perceived exploitation of Exxon. The perception of "gouging" Exxon had several forms: individuals or businesses making claims for spill-related damages that were not actually spill related; sending bills to Exxon for cleanup work far in excess of actual costs; receiving payment for labor or boat contracts but not actually doing the work, and diverting to personal use equipment, supplies, or other goods and services intended for the cleanup.

The morality of gouging Exxon had a significant place in the dialogue about the oil spill and its aftermath, and this issue has continued to be divisive. It has marked the social reputations of local residents and, in some instances, it has generated ongoing conflicts that have spilt communities into factions, further eroding social solidarity. In one community, for example, a resident commented that since the spill a "number" of people have been considered social pariahs because of their Exxon gouging. This resident said words to the effect that, "To this day there are people I no longer talk to because of how they acted, of what they did to take advantage of the situation. You just wonder where their morals went." It can be argued that the availability of large sums of money and how Exxon spent those monies created an environment of opportunism after the oil spill and that some individuals simply were overcome by this environment. However, because of the juxtaposition of the perceived threats to community well-being and survival, such individual gain was and is severely criticized. Indeed, the perceived gouging of Exxon by some residents has been evaluated more harshly than Exxon's actions. Such an evaluation is explained by the importance attached to maintaining the moral integrity of the group in extreme circumstances. Individuals whose actions threaten that integrity undermine the support functions of the group and thereby receive a strong negative evaluation.

The effect of this moral discourse has been mostly negative: it has divided rather than united communities and it has damaged social reputations. During the immediate aftermath of the spill, these kinds of conditions worked against the formation of a "therapeutic community" that could mitigate the social and psychological damages caused by the spill itself. This divisiveness itself spawned other arguments and alliances within communities, and the overall effect was one of social chaos. Residents perceived the chaos but they were not always aware of the reasons for it. Mostly, it seemed to them that unrelated events interacted and caused new sources of disorder, thereby adding to an already unsettled and tense social environment and prolonging resolution of the social trauma caused by the spill and cleanup.

Acknowledgements

We gratefully acknowledge the assistance of the Oiled Mayors and the people of the north gulf coast communities who participated in this study. Robert Bolin developed some of the measures used in the household survey and was instrumental in the construction and administration of that survey.

References

Barton, A. 1969. Communities in disaster. A sociological study of collective stress situations. Doubleday, Garden City, New York.

Baum, A., editor. 1987. Cataclysms, crises, and catastrophes. American Psychological Association, Washington, DC.

Baum, A., R. J. Gatchel, and J. E. Schaeffer. 1983. Emotional, behavioral and psychophysiological effects of chronic stress at Three Mile Island. Journal of Consulting and Clinical Psychology 51:565–572.

Berren, M. A., J. M. Santiago, A. Beigel, and S. A. Timmons. 1989. A classification scheme for disaster. Pages 40–58 *in* R. Gist and B. Lubin, editors. Psychosocial aspects of disasters. Wiley, New York.

Couch, S. R., and J. S. Kroll-Smith. 1985. The chronic technological disaster: toward a social scientific perspective. Social Science Quarterly 1:546–575.

Davidson, A. 1990. In the wake of the *Exxon Valdez*. Sierra Club Books, San Francisco.

Drabek, T. E. 1986. Human system responses to disaster. Springer-Verlag, London.

Drabek, T., and W. Key. 1984. Conquering disaster: family recovery and long-term consequence. Irvington Publishers, New York.

Edelstein, M. 1988. Contaminated communities: the social and psychological impact of residential toxic exposure. Westview Press, Boulder, Colorado.

Erikson, K. 1976. Loss of communality at Buffalo Creek. American Journal of Psychiatry 133(3):302–305.
Erikson, K. 1985. Forward. Pages xiii–xviii *in* A. M. Shkilnyk, editor. A poison stronger than love: the destruction of an Ojibwa community. Yale University Press, New Haven, Connecticut.
Erikson, K. 1991. A new species of trouble. Pages 11–29 *in* S. R. Couch and J. S. Kroll-Smith, editors. Communities at risk. Collective responses to technological disaster. Peter Lang, New York.
Fleiss, J. L. 1981. Statistical methods for rates and proportions. Wiley, New York.
Impact Assessment, Inc. 1990a. Analysis of fiscal impacts to local jurisdictions. First interim report. Prepared for the Oiled Mayors Subcommittee, Alaska Conference of Mayors. Impact Assessment, Inc., La Jolla, California.
Impact Assessment, Inc. 1990b. Public and private sector impacts of the *Exxon Valdez* oil spill. Second interim report. Prepared for the Oiled Mayors Subcommittee, Alaska Conference of Mayors. Impact Assessment, Inc., La Jolla, California.
Impact Assessment, Inc. 1990c. Social and psychological impacts of the *Exxon Valdez* oil spill. Third interim report. Prepared for the Oiled Mayors Subcommittee, Alaska Conference of Mayors. Impact Assessment, Inc., La Jolla, California.
Impact Assessment, Inc. 1990d. Economic, social, and psychological impact assessment of the *Exxon Valdez* oil spill. Final report. Prepared for Oiled Mayors Subcommittee, Alaska Conference of Mayors. Impact Assessment, Inc., La Jolla, California.
Jorgensen, J. 1990. Oil age Eskimos. University of California Press, Berkeley.
Kroll-Smith, J. S., and S. R. Couch. 1990. The real disaster is above ground. A mine fire and social conflict. University of Kentucky Press, Lexington.
Lindy, J. D., and M. Grace. 1985. The recovery environment: continuing stressor versus a healing psychosocial space. Pages 137–149 *in* B. Sowder, editor. Disaster and mental health: selected contemporary perspectives. Department of Health and Human Services Publication ADM 85-1421, U.S. Government Printing Office, Washington, DC.
McFarlane, A. C. 1987. Family functioning and overprotection following a natural disaster: the longitudinal effects of post-traumatic morbidity. Australian and New Zealand Journal of Psychiatry 21:210–218.
Murphy, S. 1986. Perceptions of stress, coping, and recovery one and three years after a natural disaster. Issues in Mental Health Nursing 8:63–77.
Oliver-Smith, A. 1986. The martyred city: death and rebirth in the Andes. University of New Mexico Press, Albuquerque.
Palinkas, L. A., M. A. Downs, J. S. Petterson, and J. Russell. 1993. Social, cultural, and psychological impacts of the *Exxon Valdez* oil spill. Human Organization 1:1–13.
Picou S. J., D. A. Gill, C. L. Dyer, and E. W. Curry. 1992. Disruption and stress in an Alaskan fishing community: initial and continuing impacts of the *Exxon Valdez* oil spill. Industrial Crisis Quarterly 6:1–22.
Poulshock, S. W., and E. S. Cohen. 1975. The elderly in the aftermath of disaster. Gerontologist 15:357–361.
Radloff, L. S. 1977. The CES-D scale: a new self-report depression scale for research in the general population. Applied Psychological Measurement 1:385–401.
Shkilnyk, A. M. 1985. A poison stronger than love: the destruction of an Ojibwa community. Yale University Press, New Haven, Connecticut.
Shore, J. H., E. L. Tatum, and W. M. Vollmer. 1986a. The Mount St. Helens stress response syndrome. Pages 77–98 *in* J. H. Shore, editor. Disaster stress studies: new methods and findings. American Psychiatric Press, Washington, DC.
Shore, J. H., E. L. Tatum, and W. M. Vollmer. 1986b. Psychiatric reactions to disaster: the Mount St. Helens experience. American Journal of Psychiatry 143:590–595.
Weissman, M. M., J. K. Myers, and C. Ross, editors. 1986. Community surveys of psychiatric disorder. Rutgers University Press, New Brunswick, New Jersey.
Weissman, M. M., D. Sholomskas, M. Pottenger, B. A. Prusoff, and B. Z. Locke. 1977. Assessing depressive symptoms in five psychiatric populations: a validation study. American Journal of Epidemiology 106:203–214.

American Fisheries Society Symposium 18:879–893, 1996

The *Exxon Valdez* Oil Spill and Chronic Psychological Stress

J. Steven Picou
University of South Alabama, Department of Sociology and Anthropology
Mobile, Alabama 36688-0002, USA

Duane A. Gill
Mississippi State University, Social Science Research Center and
Department of Sociology, Anthropology and Social Work
Mississippi State, Mississippi 39762, USA

Abstract.—We evaluated the long-term psychological impacts of the *Exxon Valdez* oil spill in terms of community structure and resource-based stress responses. Propositions and hypotheses that identify vulnerable communities and occupational groups were derived from an ecological–symbolic conceptualization of disaster impacts. Data collected from stratified random household samples in three Alaskan communities were used to empirically evaluate hypotheses. Chronic stress was documented for communities affected by the spill and for members of occupational groups most dependent on commercial fishing. These empirical findings support the need to recognize and evaluate stress in resource-dependent human communities when technological disasters, such as the *Exxon Valdez* oil spill, occur.

Over the last 25 years, the concept of disaster has been discussed, debated, and empirically studied in relation to an increasing variety of catastrophic events. As a result, the concept of "technological disaster" has emerged, bringing attention to the unique qualities and impacts of events that arise from failures of technology. The negative consequences of disasters stem from social responses to extreme stressors (Dynes 1993), yet some stress responses may vary according to the perceived source of the triggering event. For example, Tierney and Baisden (1979) identified greater long-term mental health impacts of the Buffalo Creek (West Virginia) dam collapse (a technological disaster) than of the Xenia (Ohio) tornado (a natural disaster). Researchers have extended this line of inquiry by investigating theoretical and applied issues related to the social and long-term psychological impacts of technological disasters (e.g., Erikson 1976, 1991, 1994; Titchener and Kapp 1976; Ahearn and Cohen 1984; Couch and Kroll-Smith 1985; Edelstein 1988; Bogard 1989; Kroll-Smith and Couch 1991a, 1993a, 1993b). In general, these studies consistently indicate that technological disasters produce chronic social and psychological impacts.

Research on well-publicized technological disasters at Three-Mile Island (Pennsylvania), Bhopal (India), Chernobyl (Russia), and Love Canal (New York) have documented a variety of initial and long-term psychosocial impacts (Brown and Harris 1979; Baum et al. 1982; 1983; Levine 1982; Shrivastava 1987; Houts et al. 1988; Bogard 1989; Davidson and Baum 1991; Weisaeth 1991; Baum and Fleming 1993). Furthermore, survey and case studies of less-publicized accidents involving toxic contamination and litigation in Livingston (Louisiana), Woburn (Massachusetts), and Legler (New Jersey) have shown prolonged community and neighborhood disruption as well as patterns of psychological stress and trauma (Picou 1984; Gill 1986; Edelstein 1988; Brown and Mikkelsen 1989; Gill and Picou 1991; Picou and Rosebrook 1993). Chronic stress resulting from technological disasters has been related to issues and perceptions of "uncertainty" regarding the extent and consequences of biophysical contamination (Vyner 1988; Davidson and Baum 1991; Erikson 1991) and protracted litigation (Brown and Mikkelsen 1989; Picou and Rosebrook 1993). Furthermore, technological disasters are characterized by a lack of finality which may result in the emergence of a "corrosive" or "abrasive" community context, and this, in turn, fosters the development of long-term stress and disruption (Freudenburg and Jones 1991; Kroll-Smith and Couch 1991b, 1993a).

This line of inquiry suggests that a paradigm shift is occurring in disaster research as a result of the persistent empirical anomaly presented by studies that consistently document long-term negative mental health outcomes for victims of technological disasters.[1] Such a pattern of chronic impacts is relatively rare for victims of "natural" disasters (Kreps 1980; Drabek 1986). This paradigm shift has led disaster researchers to propose alternative theoretical approaches (Kroll-Smith and Couch 1991a, 1993a) and to call for longitudinal research designs (Solomon 1989; Maser and Solomon 1990). Our study contributes to this line of inquiry by providing resource-based concepts of community stress response to evaluate hypotheses regarding long-term psychological stress resulting from the *Exxon Valdez* oil spill (EVOS).[2]

The *Exxon Valdez* Oil Spill

On 24 March 1989, the supertanker *Exxon Valdez* ran aground on Bligh Reef in Prince William Sound (PWS), Alaska, and released approximately 42 million liters of oil into several commercial fishing areas. Inadequate response to the spill exacerbated the situation and eventually over 1,900 km of rugged coastline were oiled (Skinner and Reilly 1989; Piper 1993). The EVOS occurred in the highly productive fishing grounds of PWS and spread to fishing grounds along the Kenai Peninsula, Cook Inlet, Kodiak Island, and the Alaska Peninsula. The spill occurred at the beginning of the most biologically active season and impacted ecosystem processes throughout the summer of 1989; havoc was wreaked through both contamination from oil and certain methods of cleanup (Hodgson 1989). Although the environmental disaster has been well documented and widely acknowledged, the human dimensions of the EVOS have been studied less and are perhaps most misunderstood.

The EVOS provides a case study of a technological disaster for evaluating emerging theoretical issues in disaster research. Social and organizational research on the EVOS has included studies of the political context and management of the accident, as well as cross-sectional analyses of initial social impacts (Impact Assessment, Inc. 1990; Browning and Shelter 1992; Clarke 1992; Gramling and Freudenburg 1992; Williams and Treadway 1992; Sellnow 1993). Studies conducted within the impacted region document that subsistence harvests by some groups of Alaskan Natives were severely disrupted (Fall 1990; Restoration Planning Work Group 1990) and that low harvest activities characterized "highly oiled" villages through 1991 (J. A. Fall, Alaska Department of Fish and Game, unpublished). Additional research suggests that social and cultural activities of Alaskan Natives were initially disrupted and threatened by the spill (Dyer et al. 1992; Dyer 1993). Furthermore, when compared to other ethnic groups, Alaskan Natives experienced relatively high levels of depression (Palinkas et al. 1992).

Studies of the economic impacts of the EVOS provide evidence that a dramatic increase in nonfishing wages characterized impacted regions immediately following the spill (Cohen 1993) and that losses incurred in 1989 and 1990 by commercial fishermen may have reached 155 million dollars (Cohen 1995). Furthermore, the influx of cleanup activity income in small Alaskan communities and Alaskan Native villages resulted in a variety of negative social impacts (Rodin et al. 1992). These impacts were associated with the destruction of the physical environment, the effects of the cleanup, and the organizational structure of Exxon and its subcontractor for the cleanup, VECO (Rodin, et al. 1992; Piper 1993).

Research findings also document the existence of a variety of initial negative mental health impacts resulting from the EVOS (Donald et al. 1990; Rodin et al. 1992; Palinkas et al. 1993). In terms of longitudinal data, one study found that patterns of disruption and stress persisted approximately 18 months following the spill (Picou et al. 1992). This research documented that high levels of intrusive stress and avoidance behavior characterized residents of an impacted fishing community (Cordova).[3] In addition, more personal, work, and community disruptions were found in Cordova than in a control community (Petersburg). These disruption trends were associated with stress levels in 1989 and 1990, suggesting the beginning of chronic psychological stress for the most threatened subpopulations (e.g., fishers, subsistence users) in impacted communities (Picou et al. 1992).

[1]The concept of paradigm shift identifies a reorientation of research questions in terms of alternative scientific explanations that incorporate recurring anomalies, or exceptions to traditional research (Kuhn 1962; McCann 1978). For examples of applications of this model of scientific development to sociology, see Wells and Picou (1981), Turk (1988), and Picou et al. (1990).

[2]We use EVOS to denote a range of activities associated with the *Exxon Valdez* oil spill, including the actual spill, the contamination, and cleanup activities.

[3]Intrusive stress and avoidance behavior are forms of psychological stress measured by the subscales of the impact of events scale (IES) (Horowitz 1974, 1976; Horowitz et al. 1979). The IES is listed as a component of the DSM-III criteria for post-traumatic stress disorder (Davidson et al. 1989; Shore et al. 1989).

Empirical documentation of the long-term impacts of technological disasters in general and the specific documentation of continuous social and psychological impacts resulting from the EVOS provide an opportunity to more fully examine long-term psychological impacts. In this paper, we report research that extends this line of inquiry. First, we develop the idea of a renewable resource community (RRC) to provide a theoretical basis for explaining chronic psychological stress. Next, we derive hypotheses regarding chronic stress from the conservation of resources (COR) stress model and empirically evaluate them from data collected 32 and 42 months following the EVOS. Finally, we discuss the results' theoretical and practical implications.

The Renewable Resource Community: An Ecological–Symbolic Approach

An RRC is a population of individuals who live within a bounded area and whose primary cultural, social, and economic existences are based on the harvest and use of renewable natural resources. The RRC perspective identifies direct links between seasonal ecosystem cycles and culturally-based community activities. Consistent with the ecological–symbolic approach to disasters, the RRC is a mediating social structure that organizes interpretive responses to the biophysical environment. The RRC contains a seasonal cycle for understanding community-ecosystem links through cultural and economic behaviors that occur at specific times during the year (Gill 1994).

A fundamental assumption of the RRC perspective is that, to varying degrees, all human communities are economically, socially, and culturally linked to biophysical resources. From the ecological–symbolic perspective, residents of an RRC exist in exchange relationships with their biophysical environment, and their collective interpretation of this relationship is essential for community equilibrium (Kroll-Smith and Couch 1991a; 1993a; 1993b).

The RRC is connected to the biophysical environment through resource harvests and uses. This connection can be economically structured on a cash or subsistence basis and socioculturally structured through behaviors, values, attitudes, and knowledge. Economic and sociocultural linkages are identifiable in terms of occupational and status roles associated with exchange networks of harvested resources. Further, sociocultural values found in subsistence activities provide a collective value set that links human conceptions of community culture to the biophysical environment. Consideration of the occupational nature of the RRC allows conceptualization of occupational behaviors that directly involve commercial harvests of renewable resources.

Cultural ecology provides a historical basis for understanding the early resource-dependent communities of the northwest coast of North America (Netting 1986). Direct links between culture, social organization, and resources have been identified by Jorgensen (1990:80–81) in his description of the "quintessential subsistence economy" once characteristic of Eskimo villages. Subsistence economics provide a cultural context for harvesting, consuming, exchanging, and maintaining renewable natural resources. With the absence of monetary capital, these communities are structured into relatively nonhierarchical systems in which the primary economic objectives are the maintenance of life, social organization, and culture (Jorgensen 1990).

The five communities in PWS all are small in population size and geographically isolated from each other. The Native Alaskan villages of Tatitlek and Chenega Bay are populated by fewer than 100 people and can be described as modern versions of traditional subsistence culture. Even though contemporary Native communities are far removed from traditional subsistence culture, subsistence still plays a vital role in terms of sociocultural structure and continuity with the past. The other three communities—Whittier, Valdez, and Cordova—have cash economies and predominantly non-Native (White) populations. To various degrees, these five communities are commercially and culturally linked to renewable natural resource harvests, particularly harvests of various fisheries in PWS.[4]

A 1993 social indicators study of Alaskan coastal villages (Anonymous 1993) classified PWS communities in terms of commercial fishing prevalence and transportation development. Prevalence of commercial fishing was based on the percentage of total community income derived from fishing related businesses. When cross-classified by the extent of transportation services, infrastructure, and social services, PWS communities can be delineated into

[4]Prince William Sound (PWS) is part of the "area E" commercial fishing region (Gill 1994). Of the three non-Native communities in PWS, Cordova best fits the renewable resource community concept. The character of Valdez is dominated by oil and the activities of the Aleyeska pipeline. Whittier's character is shaped by its historic role as a transportation and service center.

noncommercial fishing hubs, which are communities characterized by a fairly complex occupational division of labor and an "urban bias" afforded by transportation services, and commercial fishing periphery communities characterized by dependence on commercial fishing and a simpler division of labor.

The resource contamination that resulted from the EVOS simultaneously provided an immediate and a long-term threat to the commercial fishing industry in PWS. Cordova, the primary RRC in PWS, is an occupational community based on commercial fishing.[5] A wide variety of support occupations such as cannery packing, net mending, electronics repair, and boat repair are linked to and depend on commercial fishing. Limited cash revenues may be available to the RRC through tourism, harvests of nonrenewable resources, and government agencies. However, PWS communities dependent on renewable resources are geographically isolated, which severely restricts their economic diversification.

The Conservation of Resources Stress Model

Stress researchers have long noted the importance of social structure for understanding variations in stress responses (Kahn 1973; Pearlin 1989). The ecological–symbolic approach with the RRC perspective provides an understanding of community social structure in terms of threats posed by environmental contamination to renewable fishery resources. Environmental threats to human communities are stressors, and empirical studies clearly indicate that such stressors "cause measurable psychological problems" (Hallman and Wandersman 1992:114). The conservation of resources (COR) stress model (Hobfall 1988, 1989) takes an ecological perspective but recognizes that stress and coping occur within a social setting with characteristics that are as important as those of individuals. The COR stress model provides a more detailed specification of stress responses applicable to all types of disasters, and it is particularly appropriate for understanding the variations in stress response within an RRC.

The COR stress model integrates stress response approaches that have emphasized environmental and cognitive characteristics (Lazarus and Folkman 1984). A basic assumption of the COR explanation of stress "is that people strive to retain, protect and build resources and that what is threatening to them is the potential or actual loss of these valued resources" (Hobfall 1989:516). Stress occurs when people experience "(a) the threat of a net loss of resources, (b) the net loss of resources, or (c) a lack of resource gain following the investment of resources" (Hobfall 1989:516). Resources of direct concern for this model include "objects, personal characteristics, conditions, or energies" that have value and meanings for individuals (Hobfall 1989: 516).

In the context of the COR stress model, fisheries are "objective resources" that, within the EVOS impact area, are necessary for RRC economic, social, and cultural continuity. Residents of an RRC are directly linked to fishery resources through their social status (e.g., occupational role) and energy (time, money, skills, etc.). The threat to fishery resources posed by the EVOS is structurally linked to the social and energy resources of commercial fishers. Over time, this contamination threat to community-linked resources (i.e., objective resources) poses a threat to the economic infrastructure of RRCs in the impact region. For example, recent declines in PWS commercial harvests of pink salmon *Oncorhynchus gorbuscha* and the collapse of the fishery for Pacific herring *Clupea pallasi* provide an indicator of prolonged "objective" resource depletion (Ott 1992, 1994; Kizza 1993; Phillips 1993; Schneider 1993; Fried 1994; Steiner 1993). The actual and potential loss of fishery resources over time has, in part, been attributed to the EVOS. For this reason, the EVOS continues to be a source of actual and potential resource loss and continues to threaten the economic, social, and cultural structure of RRCs in the impact region.

Propositions and Hypothesis

Human impacts of the EVOS can be identified from a consideration of RRC dependency on renewable resources and in terms of the COR stress model. The EVOS threatened the essential economic, social, and cultural viability of RRCs in the impact region. No occupational group was more threatened by the EVOS than commercial fishers and those directly linked to commercial fishing. The oiled fishing grounds posed an immediate threat to the commercial fisheries and generated uncertainty regarding the long-term recovery of the resources. This threat and uncertainty were acute for individ-

[5]Oil spill damage to fisheries was not confined to Prince William Sound. In particular, Kodiak is a renewable resource community (RRC) that was impacted. However, the focus of our RRC analysis is Cordova.

uals in occupational roles related to commercial fishing.

Numerous types of losses are experienced as a result of any disaster (Dynes 1970). However, the loss or potential loss of renewable resources is a primary source of stress for victims of the EVOS. These renewable resources entail more than material values; they are important for establishing self-esteem and social standing as well as for transmitting cultural values and traditions. The uncertainty regarding the recovery and future use of renewable resources contributes to chronic stress. Furthermore, impacted communities and individuals have invested resources in litigation to recover damages and the lack of a timely final legal resolution further contributes to chronic stress (Gill 1994).

Occupational roles associated with commercial fishing link RRCs to the biophysical environment and establish direct socioeconomic consequences for deleterious effects of the EVOS. The theoretical conceptualization of the COR stress model and the RRC allows derivation of several propositions regarding environment-community linkages and impacts on humans in RRCs.

Proposition (1).—Renewable resource communities are directly linked to the biophysical environment by economic, social, and cultural structures.

Proposition (2).—The economic structure of RRCs consists of an occupational community built around commercial fishing.

Proposition (3).—The occupational role of commercial fishers structurally links the economic infrastructure of RRCs to resource harvest activities.

Proposition (4).—Contamination of fishing grounds is a direct threat to the economic viability of commercial fishing harvests in RRCs and is specifically a threat to occupational roles associated with fishing.

Proposition (5).—Real, potential, and perceived losses of resources resulting from contamination cause stress that becomes chronic as losses remain unresolved.

These propositions yield two general research hypotheses concerning the stress consequences of the EVOS. The first reflects that RRCs are more economically and culturally dependent than nonrenewable resource communities on fishing harvests.

Hypothesis (1).—Residents of RRCs will exhibit higher levels of oil-spill-related collective stress than residents of non-RRCs.

The second reflects that RRCs are structurally linked to the biophysical environment by the occupational role of commercial fishers.

Hypothesis (2).—People who work in commercial fishing occupations will exhibit higher levels of spill-related stress than people who work in nonfishing occupations.

These hypotheses provide a basis for assessing the validity of the RRC explanation of vulnerability to spill-related stress. As a group, occupations related to commercial fishing are seriously threatened by the spill and logically this occupational group becomes a focus for theoretically predicting empirical patterns of chronic stress impacts.

Methods

The strongest evidence concerning cause-and-effect relationships can be obtained when experimental designs are fully controlled and subjects are randomly assigned to treatment conditions manipulated directly by researchers (Campbell and Stanley 1963). However, environmental disasters cannot be manipulated directly by researchers, and subjects cannot be randomly assigned to treatment groups a priori. Instead, the community impacts of environmental disasters can only be assessed a posteriori with research designs and interview techniques adapted to the unique subject (Drabek 1970; Mileti 1987; Solomon 1989).

We relied on an ex post facto design (Spector 1981) in which the impacts of the EVOS were assessed by comparing communities and occupational groups that had experienced different potential adverse environmental impacts. Like most such field experiments, ours is subject to the criticism that preexisting group differences could account for the differences observed after the EVOS (Campbell and Stanley 1963; Spector 1981). However, the lack of prespill (baseline) data does not preclude evaluating theoretically based hypotheses within communities purposefully sampled in terms of conceptual and potential impact considerations. Using demographic and historical documents, we selected two communities within PWS and a third community from outside the EVOS impact region. Within PWS, Cordova was identified as an RRC and Valdez as a non-RRC. Petersburg was selected as an RRC outside the impact region.

Cordova: Impact Renewable Resource Community

Cordova was selected as an impact community because of its economic dependency on commercial fishing and its cultural heritage of subsistence. The community is at the southeastern edge of PWS, and it has remained geographically isolated since the

1964 earthquake (Janson 1975). Cordova has a cash economy derived primarily from commercial fishing (Anonymous 1993). Cordova fishers own 44% of all herring permits and 55% of all salmon fishery permits in the area E region (Stratton 1989). Approximately 50% of Cordova's labor force is employed in fisheries harvest and processing occupations (Fried 1994:1). The base population of 2,500 (18% Alaskan Natives) typically increases to over 4,500 during the summer commercial fishing season. Subsistence activities (harvesting and giving, or receiving fish, moose, deer, berries, etc.) characterize 90% of Cordova's households (Stratton 1989). Cordova can be classified as an RRC given the community's economic dependence on fishing grounds for commercial and subsistence harvests.

Although oil from the EVOS did not reach Cordova, many of the community's primary fishing areas were initially contaminated and closed, which severely disrupted the 1989 fishing season. Cordova was also a cleanup support community used by Exxon and its subcontractors. Some Cordova fishermen leased boats to assist in the cleanup and other residents were employed to clean beaches. During the first 3 years following the spill, residents experienced various disruptions in the fishing seasons and in their local economy as well as increased community conflict and segmentation (Gill 1994). Despite not being directly oiled, Cordova experienced significant shock, disruption, and uncertainty in the aftermath of the oil spill (Picou et al. 1992; Anonymous 1993).

Valdez: Impact Non-Renewable Resource Community

Valdez was selected as a second impact community. Prior to the 1964 earthquake, Valdez had little recorded history (Anonymous 1993). Following the total destruction and relocation of the community after the earthquake, Valdez experienced a modest economic boom from reconstruction activities. With the construction of the pipeline terminus in the 1970s, the Valdez population temporarily increased dramatically to over 9,000 residents. By the 1980s, however, the town's population had stabilized at approximately 3,500–3,900 residents (Anonymous 1993).

The Valdez economy is very different from the economy of Cordova. Valdez has three major employment sectors: transportation, primarily based on Alyeska's pipeline responsibilities; public sector employment; and seasonal tourism (Anonymous 1993). Valdez has a mixed economic infrastructure and is connected to other Alaskan communities by the Richardson highway. The economic significance of commercial fishing in Valdez is very small, accounting for less than 3% of employment and income (Anonymous 1993). Valdez was classified as a "hub, mixed" Alaskan community by Anonymous (1993). Thus Valdez is classified as a non-RRC in the impact region.

Petersburg: Control Renewable Resource Community

Petersburg was selected as the control community because it shares many economic, demographic, and geographical characteristics with Cordova. Petersburg is on an island in the southeastern part of the state. It is geographically isolated with no road connections outside of Mikoff Island. Petersburg has a population of 3,200, of which Alaskan Natives make up approximately 20%. Petersburg's economic base is commercial fishing, especially its $29.5 million salmon fishery. The community is home to a large commercial fishing fleet and various support businesses. Further, Petersburg residents engaged in subsistence activities at a rate similar to Cordova residents (Smythe 1988; Stratton 1989). These indicators provide a basis for classifying Petersburg as an RRC.

Although similar to Cordova in many ways, Petersburg has some distinctions. Petersburg has a distinct Norwegian heritage, whereas Cordova consists of an amalgamation of several European heritages. Petersburg is economically tied to Juneau (Alaska) and Seattle (Washington); Cordova's ties are to Anchorage (Alaska).

The most important distinction between these two communities, however, is that Petersburg experienced minimal direct impacts from the EVOS. The fishing grounds used by the vast majority of Petersburg's fishing fleet were not contaminated by the spill. Very few Petersburg fishers leased their boats during the cleanup and only a few own area E fishing permits. Thus, the use of Petersburg as a control community provided a very conservative standard for evaluating psychological consequences of the EVOS.

Data Collection

Data used in this analysis were collected in each of the three communities in 1991 and 1992. Data collection included face-to-face survey interviews, telephone interviews, and mail surveys. The research design in Cordova and Petersburg was based on an expansion of community research conducted in August, 1989 (Picou et al. 1992).[6] The 1989 sample comprised 118 respondents from 102 households in Cordova and 73 respondents from 73 households in Petersburg. In 1991, we attempted to survey all individuals who had participated in the 1989 surveys and to systematically expand the samples. The 1991 Cordova sample was expanded by interviewing panel members' spouses who had not participated in 1989 and by interviewing adults from additional randomly selected households within the same geographical areas, exclusive of households selected in 1989 and identified Alaskan Native households. This procedure resulted in 228 respondents from 184 households. Random-digit-dialing telephone sampling was used to expand the original Petersburg sample. This resulted in 101 respondents from 101 households.[7]

In Valdez (1991), city maps of residential units were used to identify five geographical areas containing approximately the same number of households. Households within these geographical areas were randomly selected and adult members were asked to be interviewed. One hundred nineteen respondents from 100 households completed the survey.

Surveys in Cordova and Valdez were administered by a field team consisting of the researchers and select volunteers from the Earthwatch Center for Field Research. Volunteers participated in an intensive training program that sensitized them to the local ecology and culture and provided them with information on the oil spill and the nature of the research project. Volunteers also participated in interviewer training sessions patterned after one developed at the University of Michigan Institute for Social Research (Guenzel et al. 1983). In addition to learning how to conduct a survey, volunteers were given detailed explanations of the intent and purpose of each section of the survey. Each volunteer was given a mock interview test and only a limited number were permitted to conduct household surveys. Surveys were screened by the researchers daily. Screening consisted of a review of the survey to determine completeness and a debriefing period with each interviewer to discuss open-ended responses.

In 1992, a mailed survey was sent to all previous participants (original and expansion) in Cordova, Petersburg, and Valdez. The mail survey had three scheduled mailings and was augmented by telephone and personal contacts by researchers in each community. As a result, 163 surveys from 131 households were completed in Cordova, 61 surveys from 42 households were completed in Valdez, and 60 surveys were completed in Petersburg.

Response rates for the 2 years are provided in Table 1. Respondents who participated in the original 1989 household survey (O) are distinguished from those initially selected in the 1991 expansion (E). The highest response rates were in the impact region (70% in Cordova and 62% in Valdez, compared to 57% in Petersburg) and among participants from the original study.

The estimated sample error for Table 1 was based on the random selection of households in Cordova and Valdez and the random selection of telephone respondents among Petersburg households.[8] This sample error provides an estimate of the chances that the sample represents the actual population of each community. In Cordova, the 1991 sample error with a 95% band of confidence was 7.2% for a variable that showed 50% of the sample having some specific characteristic. Thus, if 50% of the Cordova sample were male, one would be 95% confident that the population figure of males is between 42.8% and 57.2%. With random sampling procedures being followed, 1 of 20 samples taken in this manner would be outside this band of confidence.

Sociodemographic Characteristics

Sociodemographic characteristics of the total sample and each community for both years are presented in Table 2. In general, the sample characteristics are similar to known population characteristics although some categories may be over- or underrepresented. For example, the Valdez and Cordova samples underrepresent males and the Petersburg sample overrepresents males. Nonmarried individuals are also underrepresented in all samples. However, differences in representation among demographic categories were not deemed large enough to warrant weighting of the data for this analysis.

---

[6]We attempted to exclude any purposeful bias in the activities that resulted in household selection. Selection of all households in Cordova was based on random procedures. Selection of households in Petersburg was based on random-digit-dialing techniques.

[7]We assumed that random digit dialing resulted in only one household being contacted.

[8]Sample error was estimated at the 95% confidence interval for a 50:50 binomial distribution and expressed in percentage points (plus or minus).

TABLE 1.—Response rates within original (O, 1989) and expansion (E) samples for the 1991 and 1992 household surveys in Cordova, Petersburg, and Valdez.

Measure	1991 O	1991 E	1992 O	1992 E	1992 O and E
		Cordova			
Individual respondents					
Number (*N*)[a]	104	151	101	151	252
Completed	77	151	74	89	163
Not completed or returned	16	b	22	44	66
Refused	8	b	2	2	4
Deceased	1	b	1	0	1
Not found	2	b	2	16	18
Response rate	74.0%		73.3%	58.9%	64.7%
Adjusted response rate	76.3%		75.5%	65.9%	70.0%
Households					
Number (*N*)	95	116	92	119	208
Completed	72	116	67	72	139
Response rate	75.8%		72.8%	60.5%	66.8%
Estimated sample error[c]	7.2%				8.3%
		Valdez[d]			
Individual respondents					
Number (*N*)		119		119	119
Completed		119		63	63
Not completed or returned		b		37	37
Refused		b		1	1
Deceased				b	b
Not found				18	18
Response rate				52.9%	52.9%
Adjusted response rate				62.4%	62.4%
Households					
Number (*N*)		100		100	100
Completed		100		42	42
Response rate				42.0%	42.0%
Estimated sample error[c]	9.8%				15.1%
		Petersburg			
Individuals or households					
Number (*N*)[a]	67	68	66	68	134
Completed	34	68	43	16	59
Not completed or returned	20	b	21	22	43
Refused	12	b	2	0	2
Deceased	b		b	b	b
Not found	1		0	30	30
Response rate	50.7%		65.2%	23.5%	44.0%
Adjusted response rate	51.5%			42.1%	56.7%
Estimated sample error[c]	9.7%				12.8%

[a]Total number for original panel is adjusted for sample attrition. Those who died the previous year and those who were permanently lost (not found) were subtracted from the original number of respondents and households in Cordova (118,102) and Petersburg (73,73).
[b]Data are unavailable.
[c]Estimated sample error (alpha = 0.05) is reported for a 50:50 binomial distribution.
[d]Valdez was not part of the original 1989 study.

Indicators and Measures

Questionnaires containing similar indicators were used for data collection. Questionnaire items were selected because of their use in previous disaster research and their relevance to the EVOS. Our analysis focused on indicators of chronic stress, community type, and occupational role.

Psychological stress was measured by the intrusive stress subscale derived from the "impact of events scale" (IES). The IES consists of 15 items that elicit statements of potential responses to extraordinary events. The contents of questions making up the IES were originally derived from statements made by people who had experienced stressful life changes (Horowitz 1974, 1976; Horowitz et al. 1979). Responses to the instrument are anchored to a specific event as well as to time.[9]

[9]The exact wordings of the stem and the seven intrusive stress items were as follows. Stem: An accident like the *Exxon Valdez* oil spill can affect people in different ways. These effects may last long after the actual accident. Please read the following statements made by people after some other stressful life events. For each statement,

TABLE 2.—Sociodemographic characteristics of 1991 and 1992 total and community samples. Percentages, means, and medians have been adjusted for missing data, and actual sample size may vary among characteristics.

	1991				1992			
Characteristic	Total (N = 449)	Cordova (N = 228)	Petersburg (N = 102)	Valdez (N = 119)	Total (N = 285)	Cordova (N = 163)	Petersburg (N = 59)	Valdez (N = 63)
Percent male	48.1	42.9	62.7	45.4	46.7	46.6	50.8	42.9
Percent married[a]	76.6	78.0	74.0	76.3	81.9	83.4	85.0	76.2
Percent white[a]	90.6	91.2	89.7	89.8	90.9	90.7	93.8	90.5
Percent commercial fishing	30.1	39.5	37.3	5.9	31.2	38.1	37.9	6.6
Percent commercial fishing households	44.6	59.0	45.1	16.8	44.8	52.8	55.9	13.1
Percent with household income $100,000 or more	14.0	15.1	2.8	19.8	13.3	11.7	6.1	23.6
Percent residing in single-family dwelling[a]	63.1	62.5	70.3	58.0	65.7	67.3	73.2	57.1
Mean age	44.2	44.4	46.3	42.1	45.8	45.0	48.3	45.4
Mean years in community[a]	18.5	20.0	22.5	12.2	19.2	19.7	26.6	12.9
Mean years of education[a]	13.7	13.6	13.9	13.7	13.7	13.5	14.1	13.8
Mean household size[a]	3.2	3.2	2.9	3.5	3.2	3.1	3.1	3.4
Median income category ($1,000s)	50–60	60–70	40–50	70–80	50–60	50–60	40–50	70–80

[a]1992 values are based on 1991 data.

The rationale underlying the development of the IES is that the more stressful an event, the more likely it is to have two consequences: (a) a high incidence of recurring, unbidden, and distressing thoughts and feelings; (b) a high incidence of deliberate attempts to suppress these feelings and avoid reminders of the incident. Each of these potential consequences is tapped by a subset of IES items. Seven questions that operationalize the intrusive stress subscale asked about the frequency of such spill-related occurrences.

Psychological stress as measured by the IES is associated with patterns of community post-traumatic stress disorder produced by disasters (Davidson et al. 1989; Shore et al. 1989). Nonetheless, this indicator may actually underestimate the wide variety of stress that characterizes communities experiencing disasters (Shore et al. 1989). The IES provides a valid and reliable indicator of event-related stress that has been used in previous disaster studies (Shore et al. 1989; Solomon 1989).

The intrusive stress subscale is the dependent variable for the statistical analysis. The seven-item scale ranges in summed scores from 0 to 35. The reliability of this subscale was assessed by Cronbach's alpha (Cronbach 1951). The alpha coefficients for 1991 and 1992 were 0.8725 and 0.8956, respectively, indicating a high degree of scale reliability. Independent variables are community type (impacted RRC–Cordova; impacted non-RRC–Valdez; nonimpacted RRC–Petersburg), and fishing–nonfishing occupations. Occupation was determined by an analysis of open-ended responses. Respondents who reported occupations such as commercial fisher (gillnetter, seiner, crabber, etc.), deck hand, net mender, marine mechanic, or cannery worker, were coded as fishing occupations. Nonfishing occupations included government employee, business owner, service worker, professional among others, and those with no stated occupation (e.g., retired, housewife, student, unemployed).

Levels of psychological stress were calculated for an impacted RRC (Cordova) and an impacted non-RRC (Valdez), and both impact communities were contrasted with the control RRC (Petersburg) for the years 1991–1992. Levels of psychological stress also were compared among fishing and nonfishing occupational groups both within and across communities for 1991–1992. Psychological stress was calculated from the intrusive subscale, and the Mann–Whitney U-statistic was used to test whether the comparison groups were selected from the same population (Siegel 1956).

---

please indicate how often it was true for you *about the oil spill* during the past seven (7) days. It might not have happened at all during the past week, or during the past week, it might have occurred only *rarely*, *sometimes*, or *often*. Stress items: (1) I *thought about it* when I didn't mean to (the thought of the spill just popped into my mind). (2) Pictures about it popped into my mind. (3) Other things kept making me have *thoughts* about it (even when I didn't want to). (4) I had dreams about it. (5) I had *trouble falling asleep or staying asleep* because pictures or thoughts about it came into my mind. (6) I had waves of strong feelings about it (feelings about it just seemed to wash over me). (7) *Reminders* of it brought back feelings I first *felt* about it.

TABLE 3.—Mann–Whitney *U*-tests of differences in intrusive stress between communities, 1991 and 1992.

Comparison	Mean	Mean rank	*Z*-score	*P*[a]	*N*	Direction
			1991 intrusive stress			
Cordova (C)	9.50	195.61			224	
Petersburg (P)	2.43	90.47	−9.459	0.0000	101	C > P
Valdez (V)	7.13	134.17			116	
Petersburg (P)	2.43	80.09	−6.666	0.0000	101	V > P
Cordova (C)	9.50	181.94			224	
Valdez (V)	7.13	148.40	−2.992	0.0014	116	C > V
			1992 intrusive stress			
Cordova (C)	8.51	121.97			160	
Petersburg (P)	2.79	72.60	−5.204	0.0000	57	C > P
Valdez (V)	5.98	70.83			61	
Petersburg (P)	2.79	47.38	−3.856	0.0001	57	V > P
Cordova (C)	8.51	115.21			160	
Valdez (V)	5.98	99.95	−1.599	0.0549	61	NS

[a]One-tailed probability.

Results

Community Comparisons

We first establish the logic of the RRC framework by comparing stress levels by region of impact. Next, we compare stress levels within the impact region for the RRC (Cordova) and the non-RRC (Valdez). The following hypotheses reflect these contrasts.

Hypothesis (1).—Residents of communities in the region of the EVOS will have higher levels of stress than residents from communities outside of the spill region.

Hypothesis (2).—Within the impact region of the EVOS, residents of renewable resource communities will have higher levels of stress than residents of nonrenewable resource communities.

Worded more specifically, hypothesis (1) is that spill-related stress is higher in Cordova and Valdez (impact communities) than in Petersburg (control community). Hypothesis (2) states that psychological stress is greater in Cordova than in Valdez.

Results of this analysis support both hypotheses (Table 3). In both 1991 and 1992, levels of intrusive stress were significantly higher in Cordova (impact) and Valdez (impact), than in Petersburg (control). These results suggest that communities in the impact region experienced chronic stress for at least 3.5 years following the EVOS. In both 1991 and 1992, Cordova had higher mean stress scores than Valdez; the difference was statistically significant in 1991 but not in 1992 at an alpha level of 0.05. The absolute difference between the mean stress levels in Cordova and Valdez increased from 1991 (2.37) to 1992 (2.53), and the probability of no real difference (0.0549) was only slightly greater than threshold in 1992. We believe that these results provide evidence consistent with the RRC perspective.

Occupational Group Comparisons

We extended the logic of the RRC framework to include the COR stress model for predicting differences in psychological stress by occupational group. The relevant hypothesis is as follows.

Hypothesis (3).—People who work in commercial fishing occupations will have higher levels of spill-related stress than people who work in other occupations.

The RRC model for understanding spill-related social psychological impacts was consistently supported by the data (Table 4). Between occupational groups in the total sample and the impact RRC (Cordova), commercial fishers had significantly higher levels of spill-related intrusive stress than nonfishers. Differences were not significant in the Petersburg samples, although they were directionally consistent with hypothesis (3), and the low mean scores for both occupational sectors indicate that (as expected) the Petersburg control community suffered no substantive negative psychological impacts 2.5–3.5 years after the EVOS.

Demographic Comparisons

It is possible that the inferred causal relationships between stress and RRC characteristics are spurious. To examine this possibility, we compared intrusive stress within demographic categories traditionally subject to differential stress responses (Mirowsky and Ross 1989; Kessler et al. 1994) for

TABLE 4.—Mann–Whitney *U*-tests of differences in intrusive stress between commercial fishing and nonfishing occupations, 1991 and 1992.[a]

Community and comparison	Mean	Mean rank	*Z*-score	P^b	*N*	Direction
		1991 intrusive stress				
Total sample						
Fishing (F)	9.98	255.02			133	
Nonfishing (NF)	6.08	206.31	−3.730	0.0001	308	F > NF
Cordova						
Fishing (F)	12.39	131.64			88	
Nonfishing (NF)	7.63	100.11	−3.563	0.0002	136	F > NF
Petersburg						
Fishing (F)	3.16	55.71			38	
Nonfishing (NF)	1.98	48.16	−1.591	0.0558	63	NS
		1992 intrusive stress				
Total sample						
Fishing (F)	9.77	161.32			84	
Nonfishing (NF)	5.51	126.19	−3.451	0.0003	189	F > NF
Cordova						
Fishing (F)	11.92	95.32			59	
Nonfishing (NF)	6.59	69.17	−3.522	0.0002	98	F > NF
Petersburg						
Fishing (F)	3.68	30.98			22	
Nonfishing (NF)	2.29	26.90	−1.026	0.1526	34	NS

[a]The Valdez community sample was excluded because it included only seven commercial fishers.
[b]One-tailed probability.

each community and the total sample. Levels of intrusive stress did not vary significantly by age, education, ethnicity, and the presence or absence of dependent children (Table 5). The only significant community differences found (between genders and between marital groupings) were in Valdez, a non-RRC sample. These results provide evidence that the relationships observed in our analysis are not spurious.

Discussion

This research provides a conceptualization of social structure that, to varying degrees, characterizes communities located in the impact region of the EVOS. From an ecological–symbolic perspective, an RRC is defined in terms of economic, social, and cultural dependence on the biophysical environment. This dependence is reflected in role behaviors (e.g., commercial and subsistence harvesting) that were directly threatened by the EVOS. The COR stress model identified objective, personal, and social resources and corresponding losses of and threats to those resources as a context for understanding stress-related responses to the *Exxon Valdez* disaster. The resources and the corresponding human behaviors toward them are interpreted in terms of community social structure that is conditioned by its relationship with the biophysical environment.

The RRC perspective and the COR stress model provide a theoretical framework for deducing testable hypotheses regarding patterns of chronic psychological stress resulting from the EVOS. Postspill data collected in categorized communities in 1991 and 1992 provided the basis for an empirical eval-

TABLE 5.—Mann–Whitney two-tailed *U*-test probabilities of differences in demographic comparisons of intrusive stress for total and community samples, 1991. Directionality is shown parenthetically for significant differences (alpha = 0.05).

Demographic comparison	Total	Cordova	Petersburg	Valdez
Age (55 and over versus under 55)	0.1685	0.6719	0.7177	0.4480
Gender (female [F] versus male [M])	0.5980	0.7711	0.9040	0.0127 (F > M)
Marital status (nonmarried [NM] versus married [M])	0.0373 (NM > M)	0.1042	0.1626	0.0255 (NM > M)
Education (high school or less versus more than high school)	0.4978	0.3638	0.2813	0.6591
Ethnicity (White versus non-White)	0.4288	0.0865	0.3697	0.2728
Dependent children (no versus yes)	0.9048	0.3101	0.9648	0.3373

uation of the hypotheses. The data analysis indicated that the RRC model effectively differentiated the impacts of the EVOS on Alaskan communities. The COR stress model affirmed this differentiation and further showed that chronic patterns of psychological stress characterized groups occupationally tied to commercial fishing. These results provide evidence that the EVOS produced chronic patterns of psychological stress that differentially impacted communities and individuals. The COR stress model results suggest that the RRC classification captured, among other things, perceptions of vulnerability to and threats of future economic loss. The data analysis shows that, overall, mean levels of intrusive stress have been decreasing over time in the two PWS communities communities examined (Cordova and Valdez). However, for commercial fishing occupations in Cordova, high stress levels continued from 1991 to 1992, suggesting a chronic stress response based on resource depletion and continuing threats to fishery resources. Previous research on technological disasters and chronic stress has shown significant relationships between long-term psychological impairment and symptoms of post-traumatic stress disorder (Davidson and Baum 1986).

How should our findings of chronic psychological stress be interpreted? The intrusive stress subscale of the IES is an "... efficacious measure of post-traumatic adjustment in crime victims and other personal injury victims, combat veterans, and bereaved individuals" (Seidner et al. 1988:256). A longitudinal study of rape victims reported that initial intrusive stress scores of 20 and above signaled a need for clinical therapy and that intrusive stress scores averaged 11.40 two years after initial treatment (Seidner et al. 1988). In comparison, Cordova's commercial fishing occupations had a mean intrusive stress level of 12.39 in 1991 and 11.92 in 1992. The traumas of rape and oil spill are disparate, but the similarity in mean intrusive stress scores between these two cases is revealing: at least 2 years (rape) to 3.5 years (EVOS) after a destructive event, stress levels in the affected populations remain high.

The EVOS has had negative psychological impacts on communities affected by the oil spill and its aftermaths. Demographic comparisons convince us that our results are not spurious. As predicted from the RRC perspective, we observed consistent and statistically significant associations between intrusive stress, community type, and occupational groups. The attribution of "cause" is complex and, for many, an impossible task for social science (Cook and Campbell 1979). Nonetheless, causal inferences can be made when the criteria of rationale, temporal sequence, association, and nonspuriousness are met by social science (Labovitz and Hagedorn 1971). Our research design and analysis meets these criteria.

Our findings of elevated stress in RRC communities and within commercial fishing occupations are consistent with an ever-growing body of research on the long-term consequences of technological disasters (Baum and Fleming 1993; Kroll-Smith and Couch 1993a, 1993b). In addition, our results are consistent with other studies of the social and psychological impacts of the EVOS (Donald et al. 1990; Palinkas et al. 1992, 1993; Picou et al; 1992; Rodin et al. 1992). Although these studies have used various methodologies to collect data, all consistently identify social, psychological, and behavioral impacts resulting from the EVOS. These studies, combined with the present study, provide strong convergent evidence that the EVOS had negative social and psychological impacts on local communities. Although the empirical evidence in our study is limited to three communities, the theoretical basis for community selection allows our findings to be generalized in terms of renewable resource dependency and associated occupational roles.

The theoretical and methodological implications of this research are that the nature and significance of cultural and social links to the biophysical environment should be conceptually delineated and empirically evaluated in greater detail. We have considered only general measurement of RRC social structure and only one role linkage (commercial fishing) to the biophysical environment. The importance of social and cultural linkages for understanding negative impacts on traditional subsistence activities and behaviors, as well as on self-concept and belief systems, should be further developed and evaluated.

Longitudinal data collection is essential in extending this type of inquiry. However, the reality of litigation and the ability of adversarial parties to subpoena research documents virtually precludes this methodological strategy in the future (Marshall 1993). Alternatively, longitudinal studies of these events could be conducted under the auspices of a judicial court order (Picou and Rosebrook 1993) or if litigation were bypassed in mitigating negative impacts and resolving disputes.

Given the scientific findings on technological disasters in general and on the EVOS in particular, it is imperative to recognize legitimate long-term human impacts resulting from such events. It is important to understand the sources of distress and to put this knowledge to use in mitigating negative social and psychological impacts. A tremendous amount of money and effort has gone into restoring the environmental damage caused by the EVOS (*Exxon Valdez* Oil Spill Trustee Council 1994). However, as this study indicates, true restoration must also include the reestablishment of a social equilibrium between the biophysical environment and the human community. We have empirically documented that an imbalance exists in impacted RRCs and among individuals occupationally linked to commercial fishing. Community restoration would be directly facilitated by mitigation strategies designed to resolve community conflict and uncertainty and reduce chronic psychological stress. Without mitigated human restoration, the persistent threat, uncertainty, and lack of economic and ecological resolution resulting from the EVOS will continue to produce patterns of chronic stress. In summary, the restoration of renewable resources must be accompanied by the restoration of the quality of life in communities negatively affected by the EVOS.

Acknowledgments

This is a revised version of a presentation made at the *Exxon Valdez* Oil Spill Symposium, Anchorage, Alaska, February 1993. Funds for this research were provided by the National Science Foundation, Polar Social Science Division, grant DPP 9101093. Fieldwork in 1991 were supplemented by a grant from Earthwatch and the Center for Field Research. Onsite field support was provided by the Prince William Sound Science Center and the Copper River Delta Institute. We appreciate the invaluable support provided by the College of Arts and Science (Larry Allen, Dean), University of South Alabama, and by the Social Science Research Center (Arthur G. Cosby, Director), Mississippi State University. The review comments of Michael Downs and Steve Kroll-Smith improved the overall quality of the manuscript.

References

Ahearn, F. L., and R. E. Cohen. 1984. Disasters and mental health: an annotated bibliography. National Institute of Mental Health, Rockville, Maryland.

Anonymous. 1993. Post spill key informant studies, schedule C communities. U.S. Department of the Interior, Anchorage, Alaska.

Baum, A., and I. Fleming. 1993. Implications of psychological research on stress and technological accidents. American Psychologist 48:665–672.

Baum, A., R. Fleming, and J. E. Singer. 1982. Stress at Three Mile Island: applying psychological impact analysis. Pages 217–248 *in* L. Bickman, editor. Applied social psychology annual, volume 3. Sage, Beverly Hills, California.

Baum, A., R. Fleming, and J. E. Singer. 1983. Coping with victimization by technological disaster. Journal of Social Issues 39(2):117–138.

Bogard, W. 1989. The Bhopal tragedy. Westview, Boulder, Colorado.

Brown, G. W., and T. Harris. 1979. Laying waste: Love Canal and the poisoning of America. Random House, New York.

Brown, P., and E. J. Mikkelsen. 1989. No safe place: toxic waste leukemia and community action. University of California Press, Berkeley.

Browning, L. D., and J. C. Shelter. 1992. Communication in crisis, communication in recovery: a postmodern commentary on the *Exxon Valdez* oil spill. International Journal of Mass Emergencies and Disasters 10:477–498.

Campbell, D. T., and C. Stanley. 1963. Experimental and quasi-experimental designs for research. Rand McNally, Chicago.

Clarke, L. 1992. The wreck of the *Exxon Valdez*. Pages 80–96 *in* D. Nelkin, editor. Controversy: politics of technical decisions, 3rd edition. Sage, Newbury Park, California.

Cohen, M. J. 1993. Economic impact of an environmental accident: a time-series analysis of the *Exxon Valdez* oil spill in southcentral Alaska. Sociological Spectrum 13(1):35–64.

Cohen, M. J. 1995. Technological disaster and natural resource damage assessment: an evaluation of the *Exxon Valdez* oil spill. Land Economics 71(1):65–82.

Cook T. D., and D. T. Campbell. 1979. Quasi-experimentation: design and analysis issues for field settings. Rand McNally, Chicago.

Couch, S. R., and J. S. Kroll-Smith. 1985. The chronic technical disaster: toward a social scientific perspective. Social Science Quarterly 66:564–575.

Couch, S. R., and J. S. Kroll-Smith. 1991. Communities at risk. Peter Lang, New York.

Cronbach, L. J. 1951. Coefficient alpha and the internal structure of tests. Psychometrika 16:297–334.

Davidson, L. M., and A. Baum. 1986. Chronic stress and posttraumatic disorders. Journal of Consulting and Clinical Psychology 54:303–308.

Davidson, L. M., and A. Baum. 1991. Victimization and self-blame following a technological disaster. Pages 33–52 *in* Couch and Kroll-Smith (1991).

Davidson, J., R. Smith, and H. Kudler. 1989. Validity and reliability of the DSM-III criteria for posttraumatic stress disorder. The Journal of Nervous and Mental Disease 177:336–341.

Donald, R., R. Cook, R. F. Bixby, R. Benda, and A. Wolf. 1990. The stress related impact of the Valdez oil spill on the residents of Cordova and Valdez, Alaska: a comparative study conducted by the Valdez counseling center. Valdez Counseling Center, Valdez, Alaska.

Drabek, T. E. 1970. Methodology of studying disasters: past patterns and future possibilities. American Behavioral Scientist 13:331–343.

Drabek, T. E. 1986. Human system responses to disaster:

an inventory of sociological findings. Springer-Verlag, New York.

Dyer, C. L. 1993. Tradition loss as secondary disaster: long-term cultural impacts of the *Exxon Valdez* oil spill. Sociological Spectrum 13(1):65–88.

Dyer, C. L., D. A. Gill, and J. S. Picou. 1992. Social disruption and the Valdez oil spill: Alaskan natives in a natural resource community. Sociological Spectrum 12(2):105–126.

Dynes, R. R. 1970. Organized behavior in disaster. Heath Lexington, Lexington, Massachusetts.

Dynes, R. R. 1993. Disaster reduction: the importance of adequate assumptions about social organization. Sociological Spectrum 13(1):175–192.

Edelstein, M. R. 1988. Contaminated communities: the social and psychological impacts of residential toxic exposure. Westview, Boulder, Colorado.

Erikson, K. T. 1976. Everything in its path: destruction of community in the Buffalo Creek flood. Simon & Schuster, New York.

Erikson, K. T. 1991. A new species of trouble. Pages 11–29 *in* Couch and Kroll-Smith (1991).

Erikson, K. T. 1994. A new species of trouble. Norton, New York.

Exxon Valdez Oil Spill Trustee Council. 1994. Final environmental impact statement for the *Exxon Valdez* oil spill restoration plan. Anchorage, Alaska.

Fall, J. A. 1990. The division of subsistence of the Alaska department of fish and game: an overview of its research program and findings: 1980–1990. Arctic Anthropology 27(2):68–92.

Freudenburg, W. R., and T. R. Jones. 1991. Attitudes and stress in the presence of technological risk: a test of the supreme court hypothesis. Social Forces 69: 1143–1168.

Fried, N. 1994. A trends profile—the city of Cordova. Alaska Economic Trends (March):1–4.

Gill, D. A. 1986. A disaster impact assessment model: an empirical study of a technological disaster. Doctoral dissertation. Texas A&M University, College Station.

Gill, D. A. 1994. Environmental disaster and fishery co-management in a natural resource community: impacts of the *Exxon Valdez* oil spill. Pages 207–235 *in* C. L. Dyer and J. R. McGoodwin, editors. Folk management in the world fisheries: implications for fishery managers. University of Colorado Press, Boulder.

Gill, D. A., and J. S. Picou. 1991. The social psychological impacts of a technological accident: collective stress and perceived health risks. Journal of Hazardous Materials 27(1):77–89.

Gramling, R., and W. R. Freudenburg. 1992. The *Exxon Valdez* oil spill in the context of US petroleum politics. Industrial Crisis Quarterly 6(3):175–195.

Guenzel, P. J., T. R. Berckmans, and C. F. Cannell. 1983. General interviewing techniques. Survey Research Center, Institute for Social Research, University of Michigan, Ann Arbor.

Hallman, W. K., and A. Wandersman. 1992. Attribution of responsibility and individual and collective coping with environmental threats. Journal of Social Issues 48(4):101–118.

Hobfall, S. E. 1988. The ecology of stress. Hemisphere, Washington DC.

Hobfall, S. E. 1989. Conservation of resources: a new attempt at conceptualizing stress. American Psychologist 44:513–524.

Hodgson, B. 1989. Alaska's big spill: can the wilderness heal? National Geographic 177(1):5–43.

Horowitz, M. J. 1974. Stress response syndromes: character style and brief psychotherapy. Archives of General Psychiatry 31:768–781.

Horowitz, M. J. 1976. Stress response syndromes. Jason Aronson, New York.

Horowitz, M.J., N. Milner, and W. Alverez. 1979. Impact of event scale: a measure of subjective stress. Psychosomatic Medicine 41:209–218.

Houts, P. S., P. D. Cleary, and T. Hu. 1988. The Three Mile Island crisis. The Pennsylvania State University Press, Hershey.

Impact Assessment, Inc. 1990. Economic, social, and psychological impact assessment of the *Exxon Valdez* oil spill. Report to the Oiled Mayors Subcommittee of the Alaska Conference of Mayors, Anchorage.

Janson, L. E. 1975. The copper spike. Alaska Northwest, Anchorage.

Jorgensen, J. G. 1990. Oil age Eskimos. University of California Press, Berkeley.

Kahn, R. L. 1973. Conflict, ambiguity and overload: three elements in job related stress. Occupational Mental Health 3:2–9.

Kessler, R. C., and associates. 1994. Lifetime and 12-month prevalence of DSM-III-R psychiatric disorders in the United States. Archives of General Psychiatry 51:8–19.

Kizza, T. 1993. Measuring up the sound: biologists can't explain collapse of pink runs; fishermen blame spill. Anchorage Daily News (September 5):A1, A8.

Kreps, G. 1980. Research needs and policy issues on mass media. Pages 37–74 *in* Disaster and the mass media: proceedings of the Committee on Disasters and the Mass Media workshop. National Academy of Sciences, Washington, DC.

Kroll-Smith, J., S., and S. R. Couch. 1991a. What is a disaster? an ecological symbolic approach to resolving the definitional debate. International Journal of Mass Emergencies and Disasters 9:355–366.

Kroll-Smith, J., S., and S. R. Couch. 1991b. Technological hazards, adaptation and social change. Pages 293–316 *in* Couch and Kroll-Smith (1991).

Kroll-Smith, J., S., and S. R. Couch. 1993a. Symbols, ecology, and contamination: case studies in the ecological-symbolic approach to disaster. Research in Social Problems and Public Policy 5:47–73.

Kroll-Smith, J., S., and S. R. Couch. 1993b. Technological hazards: social responses as traumatic stressors. Pages 79–91 *in* J. P. Wilson and B. Raphael, editors. International handbook of traumatic stress syndromes. Plenum, New York.

Kuhn, T. S. 1962. The structure of scientific revolutions. University of Chicago Press, Chicago.

Labovitz S., and R. Hagedorn. 1971. Introduction to social research. McGraw-Hill, New York.

Lazarus, R. S., and S. Folkman. 1984. Stress, appraisal and coping. Springer-Verlag, New York.

Levine, A. 1982. Love Canal: power, politics and people. Lexington, Lexington, Massachusetts.

Marshall, E. 1993. Court orders 'sharing' of data. Science 261:284–286.

Maser, J. D., and S. D. Solomon. 1990. Psychosocial and psychobiological consequences of exposure to extreme stressors. Journal of Applied Social Psychology 20, 21:1725–1732.

McCann, H. G. 1978. Chemistry transformed: the paradigm shift from phlogeston to oxygen. Ablex, Norwood, New Jersey.

Mileti, D. 1987. Sociological methods and disaster research. Pages 57–69 *in* R. R. Dynes, B. De Marchi and C. Pelanda, editors. Sociology of disaster. Anseli, Milan, Italy.

Mirowsky, J., and C. Ross. 1989. Social causes of psychological distress. Aldine de Gruyter, New York.

Netting, R. M. 1986. Cultural ecology. Waveland, Prospect Heights, Illinois.

Ott, R. 1992. Could it be oil? 1992 Alaska pink salmon harvest shows possible impact of EVOS. Alaskan Fishermen's Journal (October):14–15,19.

Ott, R. 1994. Sound truth: Exxon's manipulation of science and the significance of the *Exxon Valdez* oil spill. Greenpeace, Alaska, Anchorage.

Palinkas, L. A., J. S. Petterson, J. Russell, and M. A. Downs. 1993. Community patterns of psychiatric disorders after the *Exxon Valdez* oil spill. American Journal of Psychiatry 150:1517–1523.

Palinkas, L. A., J. Russell, M. A. Downs, and J. S. Petterson. 1992. Ethnic differences on stress, coping and depressive symptoms after the *Exxon Valdez* oil spill. Journal of Nervous and Mental Disease 189:287–295.

Pearlin, L. T. 1989. The sociological study of stress. Journal of Health and Social Behavior 30:241–256.

Phillips, N. 1993. Five summers after the spill, scientists on each side still disagree. Anchorage Daily News (September 5):A1, A6.

Picou, J. S. 1984. Ecological, physical, economic, sociological and psychological assessment of the Illinois Central Gulf train derailment. Volume 5: sociological assessment. Report of Gulf South Research Institute to the Twenty-First District Court of the State of Louisiana, Baton Rouge.

Picou, J. S., E. W. Curry, and R. H. Wells. 1990. Partial paradigm shifts in the social sciences: twenty years of research in rural sociology. Rural Sociology 55(1): 101–111.

Picou, J. S., D. A. Gill, C. L. Dyer, and E. W. Curry. 1992. Stress and disruption in an Alaskan fishing community: initial and continuing impacts of the *Exxon Valdez* oil spill. Industrial Crisis Quarterly 6:235–257.

Picou, J. S., and D. R. Rosebrook. 1993. Technological accident, community class-action litigation, and scientific damage assessment: a case study of court-ordered research. Sociological Spectrum 13(1):117–138.

Piper, E. 1993. The *Exxon Valdez* oil spill: final report, state of Alaska response. Alaska Department of Environmental Conservation, Anchorage.

Restoration Planning Work Group. 1990. Restoration planning following the *Exxon Valdez* oil spill: August 1990 progress report. State of Alaska, Department of the Interior, Anchorage.

Rodin, M., M. Downs, J. Petterson, and J. Russell. 1992. Community impacts resulting from the *Exxon Valdez* oil spill. Industrial Crisis Quarterly 6:219–234.

Schneider, D. 1993. Alaska's unsound salmon fishery. Pacific Fishing (June):36–39,42.

Seidner, A. L., A. E. Amick, and D. G. Kilpatrick. 1988. Impact of event scale. Pages 255–257 *in* M. Hersen and A. S. Bellack, editors. Dictionary of behavioral assessment techniques. Pergamon, New York.

Sellnow, T. 1993. Scientific argument in organizational crisis contamination. Argumentation and Advocacy 30(1):28–42.

Shore, J. H., W. H. Vollmer, and E. L. Tatum. 1989. Community patterns of posttraumatic stress disorders. The Journal of Nervous and Mental Disease 177:681–685.

Shrivastava, P. 1987. Bhopal: an anatomy of a crisis. Harper and Row, Cambridge, Massachusetts.

Siegel, S. 1956. Nonparametric statistics for the behavioral sciences. McGraw-Hill, New York.

Skinner, S. K., and W. K. Reilly. 1989. The *Exxon Valdez* oil spill: a report to the president. National Response Team, Washington, DC.

Smythe, C. W. 1988. Harvest and use of fish and wildlife resources by residents of Petersburg, Alaska. Alaska Department of Fish and Game, Division of Subsistence, Technical Paper 164, Anchorage.

Solomon, S. D. 1989. Research issues in assessing disaster's effects. Pages 308–340 *in* R. Gist and B. Lubin, editors. Psychosocial aspects of disasters. Wiley, New York.

Spector, P. E. 1981. Research designs. Sage, Beverly Hills, California.

Steiner, R. 1993. Probing an oil-stained legacy. National Wildlife April/May:5–11.

Stratton, L. 1989. Resource use in Cordova: a coastal community of southcentral Alaska. Alaska Department of Fish and Game, Department of Subsistence Technical Paper No. 153, December, Anchorage.

Tierney, K. J., and B. Baisden. 1979. Crisis intervention program for disaster victims. U.S. Department of Health, Education and Welfare, Washington, DC.

Titchener, J. L., and F. T. Kapp. 1976. Family and character change at Buffalo Creek. American Journal of Psychiatry 133:295–299.

Turk, H. 1988. Sociological theory. Pages 18–41 *in* E. F. Borgotta and K. S. Cook, editors. The future of sociology. Sage, Newbury Park, California.

Vyner, H. M. 1988. Invisible trauma: the psychosocial effects of invisible environmental contaminants. Heath, Lexington, Massachusetts.

Weisaeth, L. 1991. Psychosocial reactions in Norway to nuclear fallout from the Chernobyl disaster. Pages 53–80 *in* Couch and Kroll-Smith (1991).

Wells, R. H., and J. S. Picou. 1981. American sociology: theoretical and methodological structure. University Press of America, Washington, DC.

Williams, D. E., and G. Treadway. 1992. Exxon and the Valdez accident: a failure in crisis communication. Communication Studies 43:55–64.

Author Index

Within this index, numbers are the first pages of respective articles.

Index